Third Edition

Algebra: Introductory and Intermediate

Instructor's Annotated Edition

Richard N. Aufmann
Palomar College, California

Vernon C. Barker
Palomar College, California

Joanne S. Lockwood
Plymouth State College, New Hampshire

HOUGHTON MIFFLIN COMPANY
Boston New York

Publisher: *Jack Shira*
Senior Sponsoring Editor: *Lynn Cox*
Senior Development Editor: *Dawn Nuttall*
Editorial Associate: *Melissa Parkin*
Senior Project Editor: *Nancy Blodget*
Editorial Assistant: *Celeste Ng*
Senior Production/Design Coordinator: *Carol Merrigan*
Manufacturing Manager: *Florence Cadran*
Senior Marketing Manager: *Ben Rivera*
Marketing Associate: *Alexandra Shaw*

Cover photographer: Harold Burch, Harold Burch Design/NYC.

Photo credits: p. 1: © Allan Tannenbaum/The Image Works; p. 36: © Atsushi Tsukada/AP-Wide World Photos; p. 53: © The Image Bank–Getty Images; p. 71: © Lester Lefkowitz/Getty Images; p. 96: Bettmann/Corbis Images; p. 109: © Pat Little/AP-Wide World Photos; p. 143: © Peter Vanderwalker/Stock Boston; p. 160: © 2002 Photodisc; p. 174: © Topham/The Image Works; p. 176: © Roger Ressmeyer/Corbis; p. 199: © Ed Horn/FPG–Getty Images; p. 242: © Frank Ward; p. 273: © Michael A. Keller Studios, Ltd./Corbis; p. 309: © Getty Images; p. 310: © Zefa Visual Media, Germany/Index Stock Imagery; p. 327: © AFP/Corbis; p. 340: © NASA/Reuters/Archive–Getty Images; p. 377: © Guy Sauvage/Vandystadt Agency/Photo Researchers; p. 429: © Jack Dempsey/AP-Wide World Photos; p. 466: © Cameraphoto/Art Resource, NY; p. 466: © Sean Clayton/The Image Works; p. 476: © AP/Wide World Photos; p. 483: © Amos Nachoum/Corbis; p. 495: © Hubertus Kanus/Photo Researchers; p. 532: © Tim Boyle/Newsmakers–Getty Images; p. 543: © David Young-Wolff/Photo Edit; p. 573: © Sean Clayton/The Image Works; p. 574: © Hugh Thomas/BWP Media–Getty Images; p. 589: © The Image Bank–Getty Images; p. 594: © AFP/Corbis; p. 603: © Oliver Benn/Getty Images; p. 649: © Francois Gohier/Photo Researchers; p. 680: © Layne Kennedy/Corbis; p. 684: © Craig Aurness/Corbis; p. 685: © Mike Johnson/Marine Natural History Photography, San Diego, CA; p. 685: © Leonard de Selva/Corbis.

Printed in the U.S.A.

Library of Congress Control Number: 2002109362

ISBN
Student Text: 0-618-293914
Instructor's Annotated Edition: 0-618-293922

123456789-VH-07 06 05 04 03

Contents

Preface

The third edition of *Algebra: Introductory and Intermediate* examines the fundamental ideas of algebra. Recognizing that the basic principles of geometry are a necessary part of mathematics, we have also included a separate chapter on geometry (Chapter 3) and have integrated geometry topics, where appropriate, throughout the text. The text has been designed not only to meet the needs of the traditional college student but also to serve the needs of returning students whose mathematical proficiency may have declined during years away from formal education.

In this new edition of *Algebra: Introductory and Intermediate*, we have continued to integrate the approaches suggested by AMATYC. Each chapter opens with a photo and a reference to a mathematical application within the chapter. At the end of each section, there are "Applying the Concepts" exercises that include writing, synthesis, critical thinking, and challenge problems. At the end of each chapter, there is a "Focus on Problem Solving" that introduces students to various problem-solving strategies. This is followed by "Projects and Group Activities" that can be used for cooperative-learning activities.

NEW! Changes to This Edition

New to this edition is the topic of graphing scatter diagrams. See Objective 4.1C in Chapter 4, *Linear Functions and Inequalities in Two Variables*. This objective reinforces the skill of graphing points in a rectangular coordinate system while providing interesting applications of the skill.

In response to user requests, we have expanded the material on functions to include the mapping of functions and illustrations of function machines. See, for example, Section 2 of Chapter 4, *Linear Functions and Inequalities in Two Variables*. This chapter also now includes graphing exercises that require critical thinking. See the exercises in Section 4.4.

In Chapter 5, *Systems of Linear Equations and Inequalities*, the discussion of solutions of systems of equations has been expanded and improved. This development should lead to greater student understanding of dependent, inconsistent, and independent systems of equations.

In Chapter 9, *Rational Exponents and Radicals*, nth roots are simplified by using perfect powers rather than by using prime factorization. Therefore, rather than rewriting the square root of 20 as the square root of 2^2 times 5, 20 is written as the product of 4 and 5; the cube root of 16 is rewritten as the cube root of 8 times 2, not 2^4 and then 2^3 times 2. This approach better mirrors what the student is thinking mentally when simplifying radical expressions.

A new chapter, Chapter R, *Review of Introductory Algebra Topics*, follows Chapter 12 and was written specifically for this edition of *Algebra: Introductory and Intermediate*. This chapter provides a review of the main topics of first-year algebra. It can be used at any point in the course. For students who need more in-depth review of a topic, there are references within the chapter that refer them back to the expanded coverage within the text.

Concept-based writing exercises have been added at the beginning of many of the objective-specific exercise sets. These exercises require a student to verbalize basic concepts presented in the lesson. These are concepts that the student must have an understanding of before attempting any exercises. For example, in Chapter 12, *Exponential and Logarithmic Functions*, the first exercise for Objective 12.2A asks, "What is a common logarithm? How is the common logarithm of $4z$ written?"

Many of the exercise sets now include developmental exercises, the intent of which is to reinforce the concepts underlying the skills presented in the lesson. For example, in Chapter 10, *Quadratic Equations*, the exercises for the first section require students to write quadratic equations in standard form and then to name the values of a, b, and c prior to having them solve any quadratic equations.

Throughout the text, data problems have been updated to reflect current data and trends. These application problems will demonstrate to students the variety of problems that require mathematical analysis. Instructors will find that many of these problems may lead to interesting class discussions. Besides the Index of Applications on the inside front cover, a chapter-by-chapter index of the variety of application problems can be found in the Table of Contents. This additional index highlights, in an easily accessible location, the importance and scope of the applications of mathematics.

Another new feature of this edition is *AIM for Success*, which explains what is required of a student to be successful and how this text has been designed to foster student success. *AIM for Success* can be used as a lesson on the first day of class or as a project for students to complete to strengthen their study skills. There are suggestions for teaching this lesson in the *Instructor's Resource Manual*.

Related to *AIM for Success* are *Prep Tests,* which occur at the beginning of each chapter. These tests focus on the particular prerequisite skills that will be used in the upcoming chapter. The answers to these questions can be found in the Answer Section along with a reference (except for Chapter 1) to the objective from which the question was taken. Students who miss a question are encouraged to review the objective from which the question was taken.

The *Go Figure* problem that follows the *Prep Test* is a puzzle problem designed to engage students in problem solving.

Chapter Opening Features

NEW! Chapter Opener

New, motivating chapter opener photos and captions have been added, illustrating and referencing specific applications from the chapter.

The WWW WEB at the bottom of the page lets students know of additional on-line resources at **math.college.hmco.com/students.**

Objective-Specific Approach

Each chapter begins with a list of learning objectives which form the framework for a complete learning system. The objectives are woven throughout the text (i.e., Exercises, Prep Tests, Chapter Reviews, Chapter Tests, and Cumulative Reviews) as well as through the print and multimedia ancillaries. This results in a seamless learning system delivered in one consistent voice.

Chapter 11 **Functions and Relations**

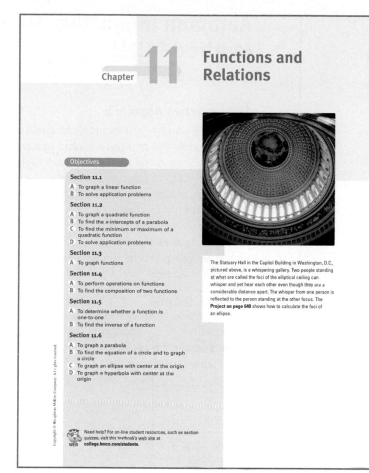

Objectives

Section 11.1
- A To graph a linear function
- B To solve application problems

Section 11.2
- A To graph a quadratic function
- B To find the *x*-intercepts of a parabola
- C To find the minimum or maximum of a quadratic function
- D To solve application problems

Section 11.3
- A To graph functions

Section 11.4
- A To perform operations on functions
- B To find the composition of two functions

Section 11.5
- A To determine whether a function is one-to-one
- B To find the inverse of a function

Section 11.6
- A To graph a parabola
- B To find the equation of a circle and to graph a circle
- C To graph an ellipse with center at the origin
- D To graph a hyperbola with center at the origin

The Statuary Hall in the Capitol Building in Washington, D.C., pictured above, is a whispering gallery. Two people standing at what are called the foci of the elliptical ceiling can whisper and yet hear each other even though they are a considerable distance apart. The whisper from one person is reflected to the person standing at the other focus. The **Project on page 640** shows how to calculate the foci of an ellipse.

WWW WEB Need help? For on-line student resources, such as section quizzes, visit this textbook's web site at **college.hmco.com/students.**

Page 589

Page 200

Prep Test

For Exercises 1 to 3, simplify.

1. $-4(x-3)$

2. $\sqrt{(-6)^2+(-8)^2}$

3. $\dfrac{3-(-5)}{2-6}$

4. Evaluate $-2x+5$ for $x=-3$.

5. Evaluate $\dfrac{2r}{r-1}$ for $r=5$.

6. Evaluate $2p^3-3p+4$ for $p=-1$.

7. Evaluate $\dfrac{x_1+x_2}{2}$ for $x_1=7$ and $x_2=-5$.

8. Given $3x-4y=12$, find the value of x when $y=0$.

Go Figure

If $\boxed{5}=4$ and ⑤ $=6$ and $y=x-1$, which of the following has the largest value?

$\boxed{x}$ ⓧ $\boxed{y}$ ⓨ

NEW! Prep Test and Go Figure

Prep Tests occur at the beginning of each chapter and test students on *previously covered* concepts that are required in the upcoming chapter. Answers are provided in the Answer Section. Objective references are also provided if a student needs to review specific concepts.

The **Go Figure** problem that follows the *Prep Test* is a playful puzzle problem designed to engage students in problem solving.

Aufmann Interactive Method (AIM)

An Interactive Approach

Algebra: Introductory and Intermediate uses an interactive style that provides a student with an opportunity to try a skill as it is presented. Each section is divided into objectives, and every objective contains one or more sets of matched-pair examples. The first example in each set is worked out; the second example, called "You Try It," is for the student to work. By solving this You Try It, the student actively practices concepts as they are presented in the text.

There are *complete, worked-out* solutions to these examples in an appendix. By comparing their solutions to the solution in the appendix, students obtain immediate feedback on, and reinforcement of, the concept.

Objective B **To solve application problems** VIDEO & DVD · CD TUTOR · WEB · SSM

A right triangle contains one 90° angle. The side opposite the 90° angle is called the **hypotenuse**. The other two sides are called **legs**.

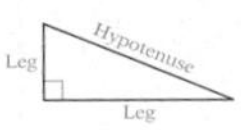

Point of Interest

The first known proof of this theorem occurs in a Chinese text, *Arithmetic Classic*, which was first written around 600 B.C.

Pythagoras, a Greek mathematician, discovered that the square of the hypotenuse of a right triangle is equal to the sum of the squares of the two legs. This is called the **Pythagorean Theorem**.

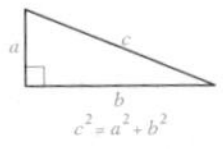

Example 3
A ladder 20 ft long is leaning against a building. How high on the building will the ladder reach when the bottom of the ladder is 8 ft from the building? Round to the nearest tenth.

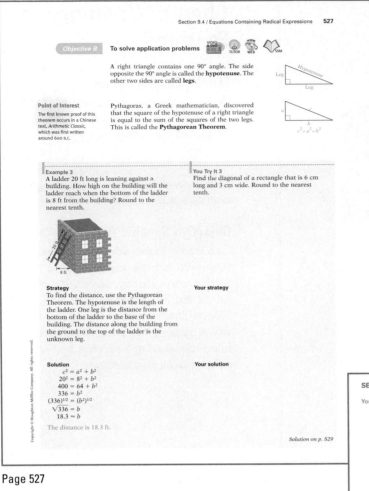

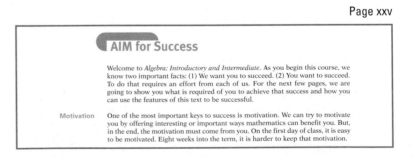

Strategy
To find the distance, use the Pythagorean Theorem. The hypotenuse is the length of the ladder. One leg is the distance from the bottom of the ladder to the base of the building. The distance along the building from the ground to the top of the ladder is the unknown leg.

Solution

$$c^2 = a^2 + b^2$$
$$20^2 = 8^2 + b^2$$
$$400 = 64 + b^2$$
$$336 = b^2$$
$$(336)^{1/2} = (b^2)^{1/2}$$
$$\sqrt{336} = b$$
$$18.3 \approx b$$

The distance is 18.3 ft.

You Try It 3
Find the diagonal of a rectangle that is 6 cm long and 3 cm wide. Round to the nearest tenth.

Your strategy

Your solution

Solution on p. S29

Page 527

SECTION 9.4

You Try It 3

Strategy To find the diagonal, use the Pythagorean Theorem. One leg is the length of the rectangle. The second leg is the width of the rectangle. The hypotenuse is the diagonal of the rectangle.

Solution

$$c^2 = a^2 + b^2$$
$$c^2 = (6)^2 + (3)^2$$
$$c^2 = 36 + 9$$
$$c^2 = 45$$
$$(c^2)^{1/2} = (45)^{1/2}$$
$$c = \sqrt{45}$$
$$c \approx 6.7$$

The diagonal is 6.7 cm.

Page S29

NEW! AIM for Success Student Preface

This new student "how to use this book" preface explains what is required of a student to be successful and how this text has been designed to foster student success, including the Aufmann Interactive Method (AIM). *AIM for Success* can be used as a lesson on the first day of class or as a project for students to complete to strengthen their study skills. There are suggestions for teaching this lesson in the *Instructor's Resource Manual.*

Page xxv

AIM for Success

Welcome to *Algebra: Introductory and Intermediate*. As you begin this course, we know two important facts: (1) We want you to succeed. (2) You want to succeed. To do that requires an effort from each of us. For the next few pages, we are going to show you what is required of you to achieve that success and how you can use the features of this text to be successful.

Motivation One of the most important keys to success is motivation. We can try to motivate you by offering interesting or important ways mathematics can benefit you. But, in the end, the motivation must come from you. On the first day of class, it is easy to be motivated. Eight weeks into the term, it is harder to keep that motivation.

Problem Solving

Focus on Problem Solving

At the end of each chapter is a **Focus on Problem Solving** feature, which introduces the student to various successful problem-solving strategies. Strategies such as drawing a diagram, applying solutions to other problems, working backwards, inductive reasoning, and trial and error are some of the techniques that are demonstrated.

Focus on Problem Solving

Solve an Easier Problem

Suppose you are in charge of your softball league, which consists of 15 teams. You must devise a schedule in which each team plays every other team once. How many games must be scheduled?

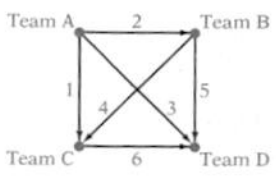

To solve this problem, we will attempt an easier problem first. Suppose that your league contains only a small number of teams. For instance, if there were only 1 team, you would schedule 0 games. If there were 2 teams, you would schedule 1 game. If there were 3 teams, you would schedule 3 games. The diagram at the left shows that 6 games must be scheduled when there are 4 teams in the league.

Here is a table of our results so far. (Remember that making a table is a strategy to be used in problem solving.)

Number of Teams	*Number of Games*	*Possible Pattern*
1	0	0
2	1	1
3	3	1 + 2
4	6	1 + 2 + 3

1. Draw a diagram with 5 dots to represent the teams. Draw lines from each dot to a second dot, and determine the number of games required.
2. What is the apparent pattern for the number of games required?
3. Assuming that the pattern continues, how many games must be scheduled for the 15 teams of the original problem?

After solving a problem, good problem solvers ask whether it is possible to solve the problem in a different manner. Here is a possible alternative method of solving the scheduling problem.

Begin with one of the 15 teams (say team A) and ask, "How many games must this team play?" Because there are 14 teams left to play, you must schedule 14 games. Now move to team B. It is already scheduled to play team A, and it does not play itself, so there are 13 teams left for it to play. Consequently, you must schedule 14 + 13 games.

4. Continue this reasoning for the remaining teams and determine the number of games that must be scheduled. Does this answer correspond to the answer you obtained using the first method?
5. Visit www.sports.com to get lists of professional sports teams, divisions, etc. Find the strategy used to create the schedule of your favorite team.

Projects and Group Activities

Using a Graphing Calculator to Solve a System of Equations

A graphing calculator can be used to solve a system of equations. For this procedure to work on most calculators, it is necessary that the point of intersection be on the screen. This means that you may have to experiment with Xmin, Xmax, Ymin, and Ymax values until the graphs intersect on the screen.

Page 317

Page 128

Problem-Solving Strategies

The text features a carefully developed approach to problem solving that emphasizes the importance of *strategy* when solving problems. Students are encouraged to develop their own strategies—to draw diagrams, to write out the solution steps in words—as part of their solution to a problem. In each case, model strategies are presented as guides for students to follow as they attempt the "You Try It" problem. Having students provide strategies is a natural way to incorporate writing into the math curriculum.

Objective C To solve application problems

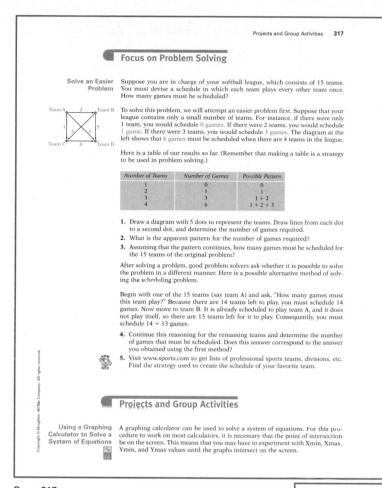

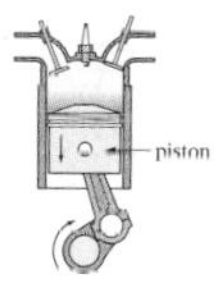

The **tolerance** of a component, or part, is the acceptable amount by which the component may vary from a given measurement. For example, the diameter of a piston may vary from the given measurement of 9 cm by 0.001 cm. This is written 9 cm ± 0.001 cm and is read "9 centimeters plus or minus 0.001 centimeter." The maximum diameter, or **upper limit**, of the piston is 9 cm + 0.001 cm = 9.001 cm. The minimum diameter, or **lower limit**, is 9 cm − 0.001 cm = 8.999 cm.

The lower and upper limits of the diameter of the piston could also be found by solving the absolute value inequality $|d - 9| \le 0.001$, where d is the diameter of the piston.

$$|d - 9| \le 0.001$$
$$-0.001 \le d - 9 \le 0.001$$
$$-0.001 + 9 \le d - 9 + 9 \le 0.001 + 9$$
$$8.999 \le d \le 9.001$$

The lower and upper limits of the diameter of the piston are 8.999 cm and 9.001 cm.

Example 5

A doctor has prescribed 2 cc of medication for a patient. The tolerance is 0.03 cc. (In the medical field, cubic centimeter is usually abbreviated cc.) Find the lower and upper limits of the amount of medication to be given.

Strategy

Let p represent the prescribed amount of medication, T the tolerance, and m the given amount of medication. Solve the absolute value inequality $|m - p| \le T$ for m.

Solution

$|m - p| \le T$
$|m - 2| \le 0.03$

$$-0.03 \le m - 2 \le 0.03$$
$$-0.03 + 2 \le m - 2 + 2 \le 0.03 + 2$$
$$1.97 \le m \le 2.03$$

The lower and upper limits of the amount of medication to be given are 1.97 cc and 2.03 cc.

You Try It 5

A machinist must make a bushing that has a diameter of 2.55 in. The tolerance of the bushing is 0.003 in. Find the lower and upper limits of the diameter of the bushing.

Your strategy

Your solution

Solution on p. S7

Real Data and Applications

Applications

One way to motivate an interest in mathematics is through applications. Wherever appropriate, the last objective of a section presents applications that require the student to use problem-solving strategies, along with the skills covered in that section, to solve practical problems. This carefully integrated applied approach generates student awareness of the value of algebra as a real-life tool.

Applications are taken from many disciplines, including agriculture, business, carpentry, chemistry, construction, Earth science, education, manufacturing, nutrition, real estate, and sociology.

Example 7

In a recent year, 238 U.S. airports collected $1,100 million in passenger taxes. Of this amount, $88 million was spent on noise reduction. What percent of the passenger taxes collected was spent on noise reduction?

Strategy

To find the percent, solve the basic percent equation using $B = 1{,}100$ million and $A = 88$ million. The percent is unknown.

Solution

$P \cdot B = A$

$P(1100) = 88$

$\frac{P(1100)}{1100} = \frac{88}{1100}$

$P = 0.08$

8% of the passenger taxes collected was spent on noise reduction.

You Try It 7

The total revenue for all football bowl games in 2000 was $158.3 million. (Source: *San Diego Union-Tribune*, January 18, 2001). Of that amount, the Big Ten conference received $22.45 million. What percent of the total revenue did the Big Ten conference receive? Round to the nearest tenth of a percent.

Your strategy

Your solution

Solutions on p. S4

Page 78

Real Data

Real data examples and exercises, identified by [icon], ask students to analyze and solve problems taken from actual situations. Students are often required to work with tables, graphs, and charts drawn from a variety of disciplines.

Page 32

190. At the close of the stock markets on May 9, 2001, the indexes were posted as shown below, along with the decreases for that day, shown as negative numbers (Source: **www.wsj.com**). At what level were the indexes at the close of the day on May 8, 2001?

Index	Points at Market Close	Decrease for the Day
Dow Jones Industrial Average	10866.98	−16.53
Standard and Poor 500	1255.54	−5.66
NASDAQ	2156.62	−42.15

191. A survey asked 600 small business owners about the impact of the Internet on their businesses. The results are shown in the graph at the right. (Source: Infoworld by Dun & Bradstreet. Copyright © 2000 by Infoworld Media Group, Inc. Reproduced with permission of Infoworld Media Group, Inc. via Copyright Clearance Center.) What percent of the owners responded that the Internet had helped business?

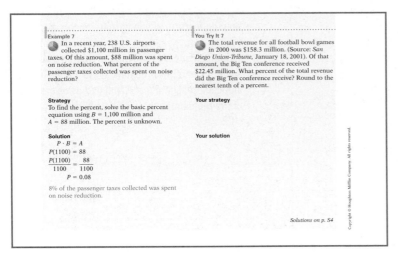

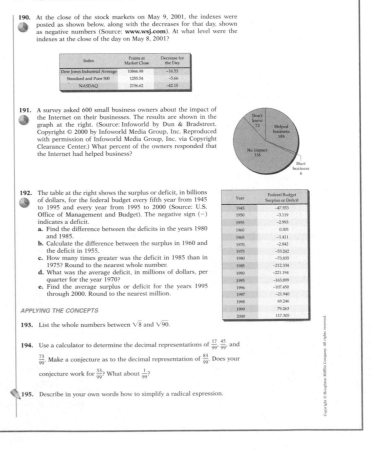

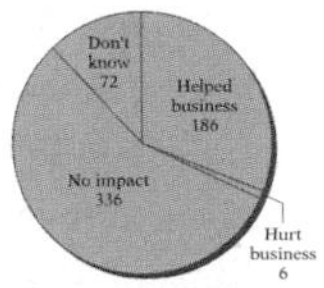

192. The table at the right shows the surplus or deficit, in billions of dollars, for the federal budget every fifth year from 1945 to 1995 and every year from 1995 to 2000 (Source: U.S. Office of Management and Budget). The negative sign (−) indicates a deficit.

a. Find the difference between the deficits in the years 1980 and 1985.

b. Calculate the difference between the surplus in 1960 and the deficit in 1955.

c. How many times greater was the deficit in 1985 than in 1975? Round to the nearest whole number.

d. What was the average deficit, in millions of dollars, per quarter for the year 1970?

e. Find the average surplus or deficit for the years 1995 through 2000. Round to the nearest million.

Year	Federal Budget Surplus or Deficit
1945	−47.553
1950	−3.119
1955	−2.993
1960	0.301
1965	−1.411
1970	−2.842
1975	−53.242
1980	−73.835
1985	−212.334
1990	−221.194
1995	−163.899
1996	−107.450
1997	−21.940
1998	69.246
1999	79.263
2000	117.305

APPLYING THE CONCEPTS

193. List the whole numbers between $\sqrt{8}$ and $\sqrt{90}$.

194. Use a calculator to determine the decimal representations of $\frac{17}{99}$, $\frac{45}{99}$, and $\frac{73}{99}$. Make a conjecture as to the decimal representation of $\frac{83}{99}$. Does your conjecture work for $\frac{33}{99}$? What about $\frac{1}{99}$?

195. Describe in your own words how to simplify a radical expression.

Student Pedagogy

Icons

The VIDEO & DVD, CD TUTOR, WEB, SSM at each objective head remind students of the many and varied additional resources available for each objective.

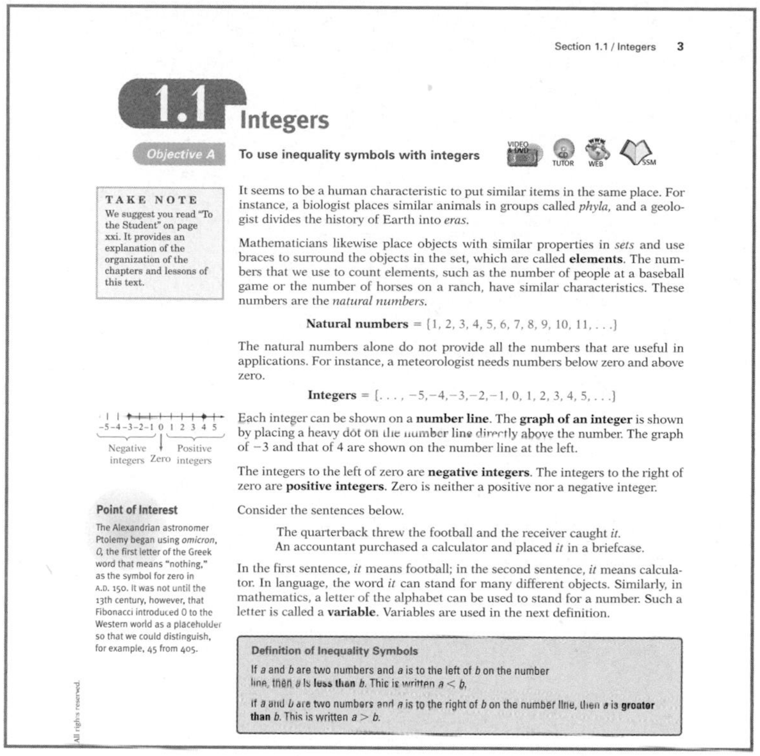

1.1 Integers

Objective A **To use inequality symbols with integers**

TAKE NOTE
We suggest you read "To the Student" on page xxi. It provides an explanation of the organization of the chapters and lessons of this text.

It seems to be a human characteristic to put similar items in the same place. For instance, a biologist places similar animals in groups called *phyla,* and a geologist divides the history of Earth into *eras.*

Mathematicians likewise place objects with similar properties in *sets* and use braces to surround the objects in the set, which are called **elements**. The numbers that we use to count elements, such as the number of people at a baseball game or the number of horses on a ranch, have similar characteristics. These numbers are the *natural numbers.*

Natural numbers = {1, 2, 3, 4, 5, 6, 7, 8, 9, 10, 11, . . .}

The natural numbers alone do not provide all the numbers that are useful in applications. For instance, a meteorologist needs numbers below zero and above zero.

Integers = {. . . , −5, −4, −3, −2, −1, 0, 1, 2, 3, 4, 5, . . .}

Negative integers Zero Positive integers

Each integer can be shown on a **number line**. The **graph of an integer** is shown by placing a heavy dot on the number line directly above the number. The graph of −3 and that of 4 are shown on the number line at the left.

The integers to the left of zero are **negative integers**. The integers to the right of zero are **positive integers**. Zero is neither a positive nor a negative integer.

Point of Interest
The Alexandrian astronomer Ptolemy began using *omicron,* O, the first letter of the Greek word that means "nothing," as the symbol for zero in A.D. 150. It was not until the 13th century, however, that Fibonacci introduced 0 to the Western world as a placeholder so that we could distinguish, for example, 45 from 405.

Consider the sentences below.

The quarterback threw the football and the receiver caught *it.*
An accountant purchased a calculator and placed *it* in a briefcase.

In the first sentence, *it* means football; in the second sentence, *it* means calculator. In language, the word *it* can stand for many different objects. Similarly, in mathematics, a letter of the alphabet can be used to stand for a number. Such a letter is called a **variable**. Variables are used in the next definition.

Definition of Inequality Symbols
If a and b are two numbers and a is to the left of b on the number line, then a is **less than** b. This is written $a < b$.
If a and b are two numbers and a is to the right of b on the number line, then a is **greater than** b. This is written $a > b$.

Page 3

Key Terms and Concepts

Key terms, in bold, emphasize important terms. The key terms are also provided in a **Glossary** at the back of the text.

Key concepts are presented in purple boxes in order to highlight these important concepts and to provide for easy reference.

Point of Interest

These margin notes contain interesting sidelights about mathematics, its history, or its application.

Take Note

These margin notes alert students to a point requiring special attention or are used to amplify the concepts under discussion.

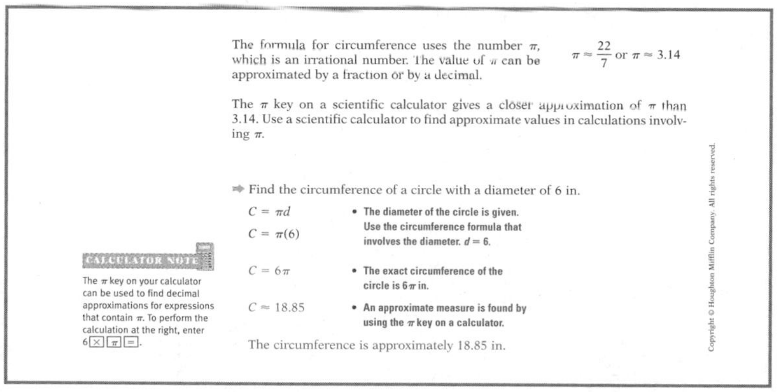
The formula for circumference uses the number π, which is an irrational number. The value of π can be approximated by a fraction or by a decimal.

$\pi \approx \frac{22}{7}$ or $\pi \approx 3.14$

The π key on a scientific calculator gives a closer approximation of π than 3.14. Use a scientific calculator to find approximate values in calculations involving π.

Find the circumference of a circle with a diameter of 6 in.

$C = \pi d$ • The diameter of the circle is given. Use the circumference formula that involves the diameter. $d = 6$.
$C = \pi(6)$

$C = 6\pi$ • The exact circumference of the circle is 6π in.

$C \approx 18.85$ • An approximate measure is found by using the π key on a calculator.

The circumference is approximately 18.85 in.

CALCULATOR NOTE
The π key on your calculator can be used to find decimal approximations for expressions that contain π. To perform the calculation at the right, enter 6 × π =.

Page 164

Annotated Examples

Examples indicated by ➡ use annotations in blue to explain what is happening in key steps of the complete, worked-out solutions.

Calculator Note

These margin notes provide suggestions for using a calculator in certain situations.

Exercises and Projects

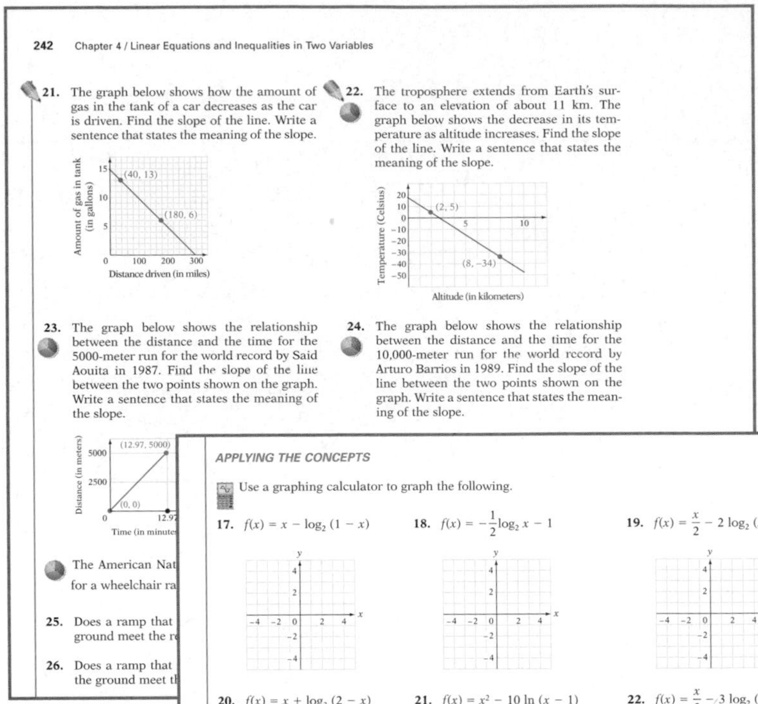
242 Chapter 4 / Linear Equations and Inequalities in Two Variables

21. The graph below shows how the amount of gas in the tank of a car decreases as the car is driven. Find the slope of the line. Write a sentence that states the meaning of the slope.

22. The troposphere extends from Earth's surface to an elevation of about 11 km. The graph below shows the decrease in its temperature as altitude increases. Find the slope of the line. Write a sentence that states the meaning of the slope.

23. The graph below shows the relationship between the distance and the time for the 5000-meter run for the world record by Said Aouita in 1987. Find the slope of the line between the two points shown on the graph. Write a sentence that states the meaning of the slope.

24. The graph below shows the relationship between the distance and the time for the 10,000-meter run for the world record by Arturo Barrios in 1989. Find the slope of the line between the two points shown on the graph. Write a sentence that states the meaning of the slope.

The American Nat
for a wheelchair ra

25. Does a ramp that
ground meet the r

26. Does a ramp that
the ground meet t

Page 242

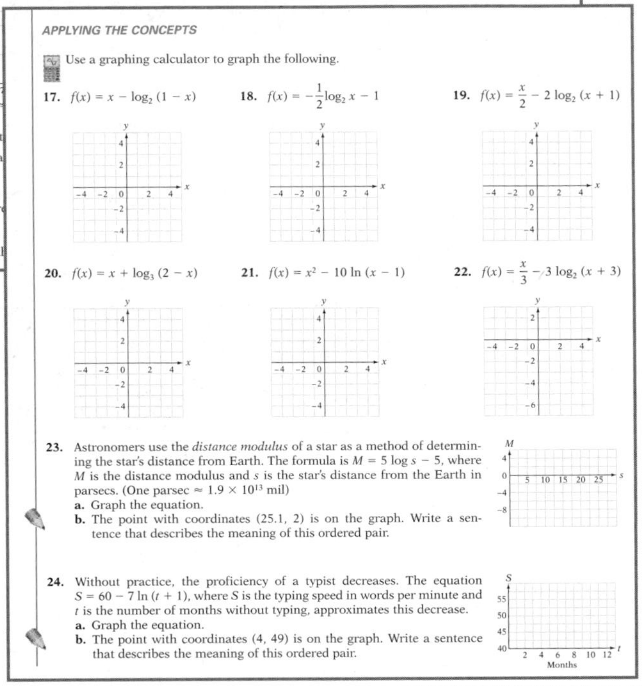
APPLYING THE CONCEPTS

Use a graphing calculator to graph the following.

17. $f(x) = x - \log_2(1 - x)$ **18.** $f(x) = -\frac{1}{2}\log_2 x - 1$ **19.** $f(x) = \frac{x}{2} - 2\log_2(x + 1)$

20. $f(x) = x + \log_3(2 - x)$ **21.** $f(x) = x^2 - 10\ln(x - 1)$ **22.** $f(x) = \frac{x}{3} - 3\log_2(x + 3)$

23. Astronomers use the *distance modulus* of a star as a method of determining the star's distance from Earth. The formula is $M = 5 \log s - 5$, where M is the distance modulus and s is the star's distance from the Earth in parsecs. (One parsec $\approx 1.9 \times 10^{13}$ mi)
a. Graph the equation.
b. The point with coordinates (25.1, 2) is on the graph. Write a sentence that describes the meaning of this ordered pair.

24. Without practice, the proficiency of a typist decreases. The equation $S = 60 - 7\ln(t + 1)$, where S is the typing speed in words per minute and t is the number of months without typing, approximates this decrease.
a. Graph the equation.
b. The point with coordinates (4, 49) is on the graph. Write a sentence that describes the meaning of this ordered pair.

Page 672

Exercises

The exercise sets of *Algebra: Introductory and Intermediate* emphasize skill building, skill maintenance, and applications.

Concept-based writing or developmental exercises have been integrated with the exercise sets.

Icons identify appropriate writing, data analysis, and calculator exercises.

Included in each exercise set is a section called **Applying the Concepts,** which presents extensions of topics, requires analysis, or offers challenge problems. The writing exercises ask students to explain answers, write about a topic in the section, or research and report on a related topic.

Page 421

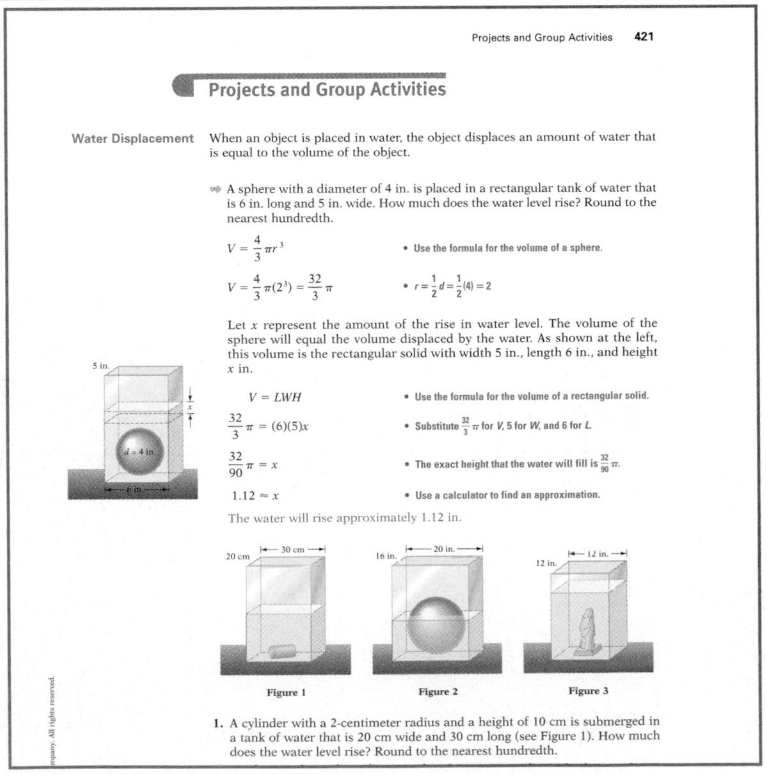
Projects and Group Activities 421

Projects and Group Activities

Water Displacement When an object is placed in water, the object displaces an amount of water that is equal to the volume of the object.

A sphere with a diameter of 4 in. is placed in a rectangular tank of water that is 6 in. long and 5 in. wide. How much does the water level rise? Round to the nearest hundredth.

$V = \frac{4}{3}\pi r^3$ • Use the formula for the volume of a sphere.

$V = \frac{4}{3}\pi(2^3) = \frac{32}{3}\pi$ • $r = \frac{1}{2}d = \frac{1}{2}(4) = 2$

Let x represent the amount of the rise in water level. The volume of the sphere will equal the volume displaced by the water. As shown at the left, this volume is the rectangular solid with width 5 in., length 6 in., and height x in.

$V = LWH$ • Use the formula for the volume of a rectangular solid.

$\frac{32}{3}\pi = (6)(5)x$ • Substitute $\frac{32}{3}\pi$ for V, 5 for W, and 6 for L.

$\frac{32}{90}\pi = x$ • The exact height that the water will fill is $\frac{32}{90}\pi$.

$1.12 \approx x$ • Use a calculator to find an approximation.

The water will rise approximately 1.12 in.

Figure 1 Figure 2 Figure 3

1. A cylinder with a 2-centimeter radius and a height of 10 cm is submerged in a tank of water that is 20 cm wide and 30 cm long (see Figure 1). How much does the water level rise? Round to the nearest hundredth.

Projects and Group Activities

The **Projects and Group Activities** feature at the end of each chapter can be used as extra credit or for cooperative learning activities. The projects cover various aspects of mathematics, including the use of calculators, collecting data from the Internet, data analysis, and extended applications.

End of Chapter

Chapter Summary

At the end of each chapter there is a **Chapter Summary** that includes *Key Words* and *Essential Rules* that were covered in the chapter. These chapter summaries provide a single point of reference as the student prepares for a test. Each concept references the page number from the lesson where the concept is introduced.

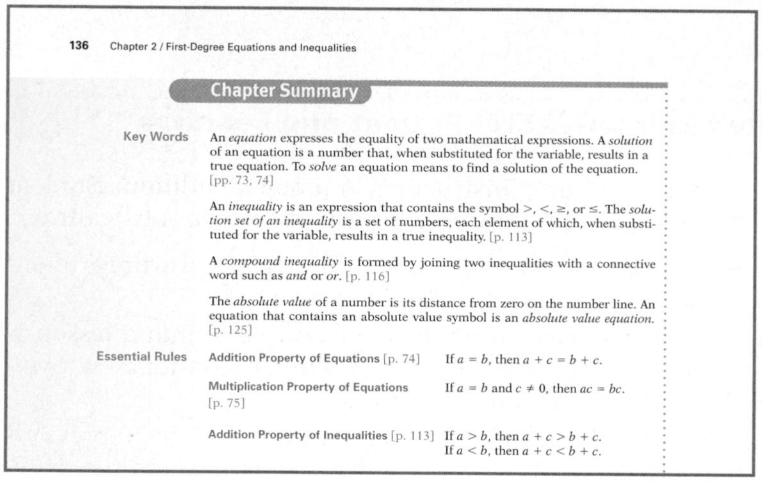
136 Chapter 2 / First-Degree Equations and Inequalities

Chapter Summary

Key Words	An *equation* expresses the equality of two mathematical expressions. A *solution* of an equation is a number that, when substituted for the variable, results in a true equation. To *solve* an equation means to find a solution of the equation. [pp. 73, 74]
	An *inequality* is an expression that contains the symbol $>$, $<$, $\geq$, or $\leq$. The *solution set of an inequality* is a set of numbers, each element of which, when substituted for the variable, results in a true inequality. [p. 113]
	A *compound inequality* is formed by joining two inequalities with a connective word such as *and* or *or*. [p. 116]
	The *absolute value* of a number is its distance from zero on the number line. An equation that contains an absolute value symbol is an *absolute value equation*. [p. 125]
Essential Rules	Addition Property of Equations [p. 74] — If $a = b$, then $a + c = b + c$.
	Multiplication Property of Equations [p. 75] — If $a = b$ and $c \neq 0$, then $ac = bc$.
	Addition Property of Inequalities [p. 113] — If $a > b$, then $a + c > b + c$. If $a < b$, then $a + c < b + c$.

Page 136

Chapter Review

Review exercises are found at the end of each chapter. These exercises are selected to help the student integrate all of the topics presented in the chapter.

Page 691

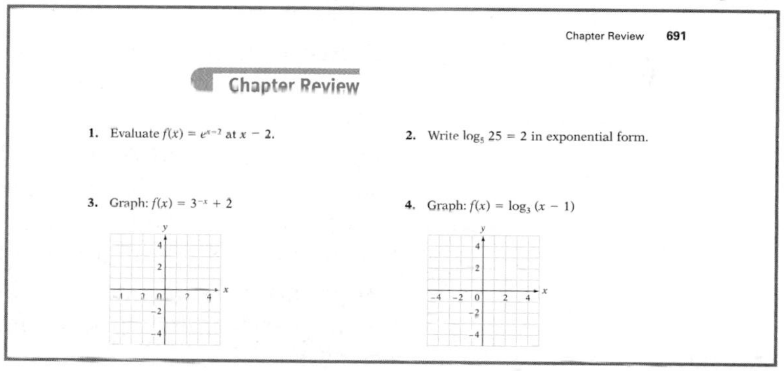
Chapter Review 691

Chapter Review

1. Evaluate $f(x) = e^{x-2}$ at $x = 2$.

2. Write $\log_5 25 = 2$ in exponential form.

3. Graph: $f(x) = 3^{-x} + 2$

4. Graph: $f(x) = \log_3 (x - 1)$

Chapter Test

The Chapter Test exercises are designed to simulate a possible test of the material in the chapter.

Page 693

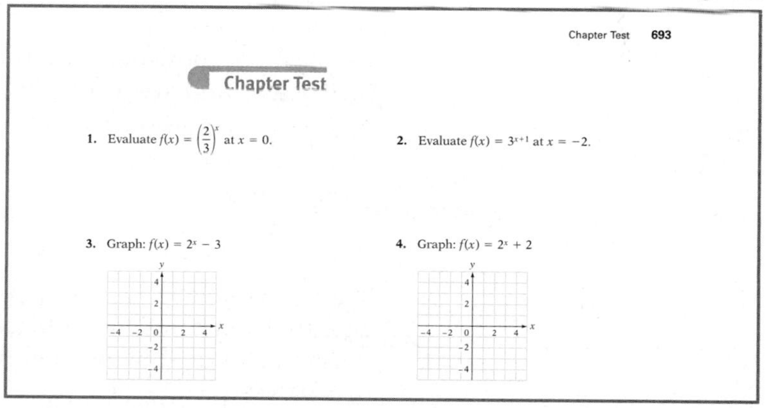
Chapter Test 693

Chapter Test

1. Evaluate $f(x) = \left(\frac{2}{3}\right)^x$ at $x = 0$.

2. Evaluate $f(x) = 3^{x+1}$ at $x = -2$.

3. Graph: $f(x) = 2^x - 3$

4. Graph: $f(x) = 2^x + 2$

Cumulative Review

Cumulative Review exercises, which appear at the end of each chapter (beginning with Chapter 2), help students maintain skills learned in previous chapters.

The answers to all Chapter Review exercises, all Chapter Test exercises, and all Cumulative Review exercises are given in the Answer Section. Along with the answer, there is a reference to the objective that pertains to each exercise.

Page 695

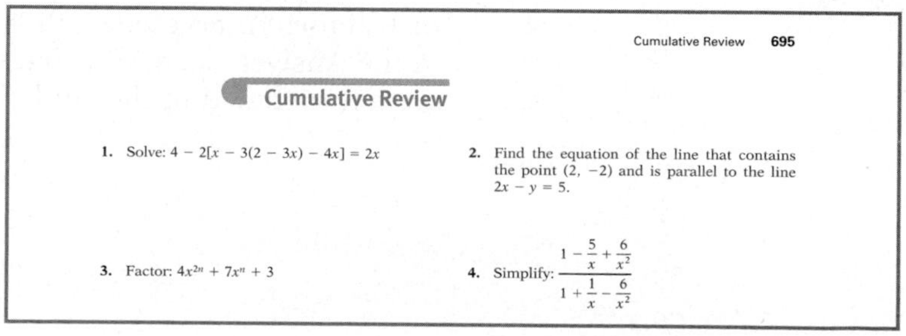
Cumulative Review 695

Cumulative Review

1. Solve: $4 - 2[x - 3(2 - 3x) - 4x] = 2x$

2. Find the equation of the line that contains the point $(2, -2)$ and is parallel to the line $2x - y = 5$.

3. Factor: $4x^{2n} + 7x^n + 3$

4. Simplify: $\dfrac{1 - \frac{5}{x} + \frac{6}{x^2}}{1 + \frac{1}{x} - \frac{6}{x^2}}$

Instructor's Annotated Edition—NEW! Format and Features

This edition offers a completely new Instructor's Annotated Edition. Student pages are reduced, creating a margin for the following instructor-only features.

- **Instructor Notes** Instructor notes include teaching ideas, warnings about common student errors, or historical notes.
- **New Vocabulary** A list of new vocabulary introduced within a lesson is provided for appropriate objectives. There are similar lists for *New Symbols, Formulas, Rules, Properties,* and *Equations.*
- **Vocabulary to Review** A list of vocabulary introduced in a previous objective that students will need to recall in order to understand the material in the present lesson is provided for appropriate objectives. There are similar lists for *Symbols, Formulas, Rules, Properties,* and *Equations.*
- **In-Class Examples** For *every* objective, extra examples are offered that can be used during the presentation of the lesson. These extra examples do not duplicate the examples presented in the student textbook.
- **Discuss the Concepts** These questions, or requests for an explanation, can be used for class discussion or for writing exercises. They require students to verbalize the basic concepts presented in the lesson.
- **Concept Check** These questions or exercises can be used after the presentation of a lesson to test student understanding of the concepts developed.
- **Optional Student Activity** These exercises can be assigned at the conclusion of the lesson. They can serve as a class activity, individual work, or as cooperative learning projects. In general, activities were written to be accomplished within about a five-minute period.
- **Suggested Assignment** At the beginning of *every* exercise set, there is a suggested homework assignment that covers the essential topics of the section.
- **Quick Quiz** For *every* objective, there is a short quiz that can be given to students. These quizzes are designed to check basic concepts. These quizzes can also be downloaded from our web site at math.college.hmco.com/instructors and are available on the Instructor *ClassPrep* CD.
- **Answers to Writing Exercises, Focus on Problem Solving, and Projects and Group Activities** Suggested answers to all the writing exercises are given in the Instructor's Annotated Edition. Also included, where appropriate, are answers to Focus on Problem Solving exercises and exercises in the Projects and Group Activities feature.
- **Transparencies** Next to many of the graphs or tables in the text, there is a T that indicates that a blackline master of that graph is available. These masters can be downloaded from our web site at math.college.hmco.com/instructors.
- **Answers to Exercises** Answers to all the exercises are given in the Instructor's Annotated Edition. Only answers to the odd exercises are given in the Answer Section in the student edition. Answers to writing exercises are not included in the student edition.

Instructor Resources

Algebra: Introductory and Intermediate has a complete set of teaching aids for the instructor.

Instructor's Annotated Edition This edition contains a replica of the student text and additional items just for the instructor. These include: *Instructor Notes, Transparency Master icons, In-Class Examples, Concept Checks, Discuss the Concepts, New Vocabulary/Symbols, etc., Vocabulary/Symbols, etc. to Review, Optional Student Activities, Quick Quizzes, Answers to Writing Exercises,* and *Suggested Assignments*. Answers to all exercises are also provided.

Instructor's Resource Manual with Solutions Manual The *Instructor's Resource Manual* includes a lesson plan for the *AIM for Success* as well as suggested course sequences. The *Instructor's Solutions Manual* contains worked-out solutions for all exercises in the text.

Printed Test Bank with Chapter Tests The *Printed Test Bank* provides a printout of one example of each of the algorithmic items in *HM Testing*. This resource also contains ready-to-use printed Chapter Tests as well as cumulative tests and final exams.

HM³ Tutorial (Instructor version) This tutorial CD software package for Microsoft Windows® was written and developed by the authors specifically for use with this text. For each objective, exercises and quizzes are algorithmically generated, solution steps are animated, lessons and exercises are presented in a colorful, lively manner, and an integrated classroom management system tracks student performance.

NEW! ***WebCT ePacks*** *WebCT ePacks* provide instructors with a flexible, Internet-based education platform providing multiple ways to present learning materials. The *WebCT ePacks* come with a full array of features to enrich the on-line learning experience.

NEW! ***Blackboard Cartridges*** The *Houghton Mifflin Blackboard Cartridge* allows flexible, efficient, and creative ways to present learning materials and opportunities. In addition to course management benefits, instructors may make use of an electronic grade book, receive papers from students enrolled in the course via the Internet, and track student use of the communication and collaboration functions.

NEW! ***HM ClassPrep with HM Testing CD-ROM*** *HM ClassPrep* contains a multitude of text-specific resources for instructors to use to enhance the classroom experience. These resources can be easily accessed by chapter or resource type and can also link you to the text's web site. *HM Testing* is our computerized test generator and contains a database of algorithmic test items as well as providing **on-line testing** and **gradebook** functions.

NEW! ***Instructor Text-Specific Web Site*** The resources available on the *ClassPrep CD* are also available on the instructor web site at math.college.hmco.com/instructors. Appropriate items are password protected. Instructors also have access to the student part of the text's web site.

Student Resources

Student Solutions Manual The *Student Solutions Manual* contains complete solutions to all odd-numbered exercises in the text.

Math Study Skills Workbook by Paul D. Nolting This workbook is designed to reinforce skills and minimize frustration for students in any math class, lab, or study skills course. It offers a wealth of study tips and sound advice on note taking, time management, and reducing math anxiety. In addition, numerous opportunities for self assessment enable students to track their own progress.

NEW! ***HM eduSpace*** *eduSpace* is a new content delivery system, combining an algorithmic tutorial program, on-line delivery of course materials, and classroom management functions. The interactive on-line content correlates directly to this text and can be accessed 24 hours a day.

HM³ Tutorial (Student version) This tutorial CD software package for Microsoft Windows® was written and developed by the authors specifically for use with this text. HM³ is an interactive tutorial containing lessons, exercises, and quizzes for every section of the text. Lessons are presented in a colorful, lively manner, and solution steps are animated. Exercises and quizzes are algorithmically generated. Next to every objective head, the serves as a reminder that there is an HM³ tutorial lesson corresponding to that objective.

NEW! ***SMARTHINKING™ Live, On-line Tutoring*** Houghton Mifflin has partnered with SMARTHINKING to provide an easy-to-use and effective on-line tutorial service. **Whiteboard Simulations** and **Practice Area** promote real-time visual interaction.

Three levels of service are offered.

- **Text-Specific Tutoring** provides real-time, one-on-one instruction with a specially qualified "e-structor."
- **Questions Any Time** allows students to submit questions to the tutor outside the scheduled hours and receive a reply within 24 hours.
- **Independent Study Resources** connect students with around-the-clock access to additional educational services, including interactive web sites, diagnostic tests, and Frequently Asked Questions posed to SMARTHINKING e-structors.

NEW! ***Videos and DVDs*** This edition offers brand new text-specific videos and DVDs, hosted by Dana Mosely, covering all sections of the text and providing a valuable resource for further instruction and review. Next to every objective head, the VIDEO & DVD serves as a reminder that the objective is covered in a video/DVD lesson.

NEW! ***Student Text-Specific Web Site*** On-line student resources, such as section quizzes, can be found at this text's web site at math.college.hmco.com/students.

Acknowledgments

The authors would like to thank the people who have reviewed this manuscript and provided many valuable suggestions.

Allen S. Aldridge, *College of the Albemarle, NC*
K. Castagna, *West Hills College, CA*
Oiyin Pauline Chow, *Harrisburg Area Community College, PA*
Sergio Loch, *Grand View College, IA*
Linda Marable, *Nashville State Technical Community College, TN*
Linda J. Murphy, *Northern Essex Community College, MA*
Nancy Nickerson, *Northern Essex Community College, MA*
Catherine H. Pirri, *Northern Essex Community College, MA*
Renea Reid, *Mississippi County Community College, AR*
Hanna Schott, *Kenai Peninsula College, AK*
Lauri Semarne
Jean M. Shutters, *Harrisburg Area Community College, PA*
Andrea Spratt, *University of Maryland-Baltimore County, MD*
David Verkonck, *Keuka College, NY*

Special thanks to Christi Verity for her diligent preparation of the solutions manuals and for her contribution to the accuracy of the textbook.

AIM for Success

Welcome to *Algebra: Introductory and Intermediate*. As you begin this course, we know two important facts: (1) We want you to succeed. (2) You want to succeed. To do that requires an effort from each of us. For the next few pages, we are going to show you what is required of you to achieve that success and how you can use the features of this text to be successful.

Motivation

One of the most important keys to success is motivation. We can try to motivate you by offering interesting or important ways mathematics can benefit you. But, in the end, the motivation must come from you. On the first day of class, it is easy to be motivated. Eight weeks into the term, it is harder to keep that motivation.

> **TAKE NOTE**
> Motivation alone will not lead to success. For instance, suppose a person who cannot swim is placed in a boat, taken out to the middle of a lake, and then thrown overboard. That person has a lot of motivation but there is a high likelihood the person will drown without some help. Motivation gives us the desire to learn but is not the same as learning.

To stay motivated, there must be outcomes from this course that are worth your time, money, and energy. List some reasons you are taking this course. Do not make a mental list—actually write them out.

____

____

____

Although we hope that one of the reasons you listed was an interest in mathematics, we know that many of you are taking this course because it is required to graduate, it is a prerequisite for a course you must take, or because it is required for your major. Although you may not agree that this course is necessary, it is! If you are motivated to graduate or complete the requirements for your major, then use that motivation to succeed in this course. Do not become distracted from your goal to complete your education!

Commitment

To be successful, you must make a commitment to succeed. This means devoting time to math so that you achieve a better understanding of the subject.

List some activities (sports, hobbies, talents such as dance, art, or music) that you enjoy and at which you would like to become better.

ACTIVITY	TIME SPENT	TIME WISHED SPENT

Thinking about these activities, put the number of hours that you spend each week practicing these activities next to the activity. Next to that number, indicate the number of hours per week you would like to spend on these activities.

Whether you listed surfing or sailing, aerobics or restoring cars, or any other activity you enjoy, note how many hours a week you spend doing it. To succeed in math, you must be willing to commit the same amount of time. Success requires some sacrifice.

The "I Can't Do Math" Syndrome

There may be things you cannot do, such as lift a two-ton boulder. You can, however, do math. It is much easier than lifting the two-ton boulder. When you first learned the activities you listed above, you probably could not do them well. With practice, you got better. With practice, you will be better at math. Stay focused, motivated, and committed to success.

It is difficult for us to emphasize how important it is to overcome the "I Can't Do Math Syndrome." If you listen to interviews of very successful atheletes after a particularly bad performance, you will note that they focus on the positive aspect of what they did, not the negative. Sports psychologists encourage athletes to always be positive—to have a "Can Do" attitude. Develop this attitude toward math.

Strategies for Success

Textbook Reconnaissance Right now, do a 15-minute "textbook reconnaissance" of this book. Here's how:

First, read the table of contents. Do it in three minutes or less. Next, look through the entire book, page by page. Move quickly. Scan titles, look at pictures, notice diagrams.

A textbook reconnaissance shows you where a course is going. It gives you the big picture. That's useful because brains work best when going from the general to the specific. Getting the big picture before you start makes details easier to recall and understand later on.

Your textbook reconnaissance will work even better if, as you scan, you look for ideas or topics that are interesting to you. List three facts, topics, or problems that you found interesting during your textbook reconnaissance.

____

____

____

The idea behind this technique is simple: It's easier to work at learning material if you know it's going to be useful to you.

Not all the topics in this book will be "interesting" to you. But that is true of any subject. Surfers find that on some days the waves are better than others, musicians find some music more appealing than other music, computer gamers find some computer games more interesting than others, car enthusiasts find some cars more exciting than others. Some car enthusiasts would rather have a completely restored 1957 Chevrolet than a new Ferrari.

Know the Course Requirements To do your best in this course, you must know exactly what your instructor requires. Course requirements may be stated in a *syllabus*, which is a printed outline of the main topics of the course, or they may be presented orally. When they are listed in a syllabus or on other printed pages, keep them in a safe place. When they are presented orally, make sure to take complete notes. In either case, it is important that you understand them completely and follow them exactly. Be sure you know the answer to each of the following questions.

1. What is your instructor's name?
2. Where is your instructor's office?
3. At what times does your instructor hold office hours?
4. Besides the textbook, what other materials does your instructor require?
5. What is your instructor's attendance policy?
6. If you must be absent from a class meeting, what should you do before returning to class? What should you do when you return to class?

7. What is the instructor's policy regarding collection or grading of homework assignments?
8. What options are available if you are having difficulty with an assignment? Is there a math tutoring center?
9. If there is a math lab at your school, where is it located? What hours is it open?
10. What is the instructor's policy if you miss a quiz?
11. What is the instructor's policy if you miss an exam?
12. Where can you get help when studying for an exam?

Remember: Your instructor wants to see you succeed. If you need help, ask! Do not fall behind. If you are running a race and fall behind by 100 yards, you may be able to catch up but it will require more effort than had you not fallen behind.

TAKE NOTE
Besides time management, there must be realistic ideas of how much time is available. There are very few people who can *successfully* work full-time and go to school full-time. If you work 40 hours a week, take 15 units, spend the recommended study time given at the right, and sleep 8 hours a day, you will use over 80% of the available hours in a week. That leaves less than 20% of the hours in a week for family, friends, eating, recreation, and other activities.

Time Management We know that there are demands on your time. Family, work, friends, and entertainment all compete for your time. We do not want to see you receive poor job evaluations because you are studying math. However, it is also true that we do not want to see you receive poor math test scores because you devoted too much time to work. When several competing and important tasks require your time and energy, the only way to manage the stress of being successful at both is to manage your time efficiently.

Instructors often advise students to spend twice the amount of time outside of class studying as they spend in the classroom. Time management is important if you are to accomplish this goal and succeed in school. The following activity is intended to help you structure your time more efficiently.

List the name of each course you are taking this term, the number of class hours each course meets, and the number of hours you should spend studying each subject outside of class. Then fill in a weekly schedule like the one printed below. Begin by writing in the hours spent in your classes, the hours spent at work (if you have a job), and any other commitments that are not flexible with respect to the time that you do them. Then begin to write down commitments that are more flexible, including hours spent studying. Remember to reserve time for activities such as meals and exercise. You should also schedule free time.

	Monday	Tuesday	Wednesday	Thursday	Friday	Saturday	Sunday
7–8 a.m.							
8–9 a.m.							
9–10 a.m.							
10–11 a.m.							
11–12 p.m.							
12–1 p.m.							
1–2 p.m.							
2–3 p.m.							
3–4 p.m.							
4–5 p.m.							
5–6 p.m.							
6–7 p.m.							
7–8 p.m.							
8–9 p.m.							
9–10 p.m.							
10–11 p.m.							
11–12 a.m.							

We know that many of you must work. If that is the case, realize that working 10 hours a week at a part-time job is equivalent to taking a three-unit class. If you must work, consider letting your education progress at a slower rate to allow you to be successful at both work and school. There is no rule that says you must finish school in a certain time frame.

Schedule Study Time As we encouraged you to do by filling out the time management form above, schedule a certain time to study. You should think of this time the way you would the time for work or class—that is, reasons for missing study time should be as compelling as reasons for missing work or class. "I just didn't feel like it" is not a good reason to miss your scheduled study time.

Although this may seem like an obvious exercise, list a few reasons you might want to study.

____

____

____

Of course we have no way of knowing the reasons you listed, but from our experience one reason given quite frequently is "To pass the course." There is nothing wrong with that reason. If that is the most important reason for you to study, then use it to stay focused.

One method of keeping to a study schedule is to form a ***study group***. Look for people who are committed to learning, who pay attention in class, and who are punctual. Ask them to join your group. Choose people with similar educational goals but different methods of learning. You can gain insight from seeing the material from a new perspective. Limit groups to four or five people; larger groups are unwieldy.

There are many ways to conduct a study group. Begin with the following suggestions and see what works best for your group.

1. Test each other by asking questions. Each group member might bring two or three sample test questions to each meeting.
2. Practice teaching each other. Many of us who are teachers learned a lot about our subject when we had to explain it to someone else.
3. Compare class notes. You might ask other students about material in your notes that is difficult for you to understand.
4. Brainstorm test questions.
5. Set an agenda for each meeting. Set approximate time limits for each agenda item and determine a quitting time.

And finally, probably the most important aspect of studying is that it should be done in relatively small chunks. If you can only study three hours a week for this course (probably not enough for most people), do it in blocks of one hour on three separate days, preferably after class. Three hours of studying on a Sunday is not as productive as three hours of paced study.

Text Features That Promote Success

There are 12 chapters in this text. Each chapter is divided into sections, and each section is subdivided into learning objectives. Each learning objective is labeled with a letter from A to G.

Preparing for a Chapter Before you begin a new chapter, you should take some time to review previously learned skills. There are two ways to do this. The first is to complete the ***Cumulative Review***, which occurs after every chapter (except Chapter 1). For instance, turn to page 325. The questions in this review are taken from the previous chapters. The answers for all these exercises can be found on page A16. Turn to that page now and locate the answers for the Chapter 5 Cumulative Review. After the answer to the first exercise, which is $-6\sqrt{10}$, you will see the objective reference [1.2F]. This means that this question was taken from Chapter 1, Section 2, Objective F. If you missed this question, you should return to that objective and restudy the material.

A second way of preparing for a new chapter is to complete the ***Prep Test***. This test focuses on the particular skills that will be required for the new chapter. Turn to page 274 to see a Prep Test. The answers for the Prep Test are the first set of answers in the answer section for a chapter. Turn to page A13 to see the answers for the Chapter 5 Prep Test. Note that an objective reference is given for each question. If you answer a question incorrectly, restudy the objective from which the question was taken.

Before the class meeting in which your professor begins a new section, you should read each objective statement for that section. Next, browse through the objective material, being sure to note each word in bold type. These words indicate important concepts that you must know in order to learn the material. Do not worry about trying to understand all the material. Your professor is there to assist you with that endeavor. The purpose of browsing through the material is so that your brain will be prepared to accept and organize the new information when it is presented to you.

Turn to page 3. Write down the title of the first objective in Section 1.1. Under the title of the objective, write down the words in the objective that are in bold print. It is not necessary for you to understand the meaning of these words. You are in this class to learn their meaning.

______	______	______	______
______	______	______	______
______	______	______	______
______	______	______	______
______	______	______	______

Math Is not a Spectator Sport To learn mathematics you must be an active participant. Listening and watching your professor do mathematics is not enough. Mathematics requires that you interact with the lesson you are studying. If you filled in the blanks above, you were being interactive. There are other ways this textbook has been designed to help you be an active learner.

Annotated Examples An orange arrow indicates an example with explanatory remarks to the right of the work. Using paper and pencil, you should work along as you go through the example.

$3x + 2 < -4$
$3x < -6$
$\frac{3x}{3} < \frac{-6}{3}$
$x < -2$
The solution set is $\{x|x < -2\}$.

➡ Solve: $3x + 2 < -4$

$3x + 2 < -4$

$3x < -6$ • **Subtract 2 from each side of the inequality.**

$\frac{3x}{3} < \frac{-6}{3}$ • **Divide each side of the inequality by the coefficient 3.**

$x < -2$

The solution set is $\{x|x < -2\}$.

Page 115

When you complete the example, get a clean sheet of paper. Write down the problem and then try to complete the solution without referring to your notes or the book. When you can do that, move on to the next part of the objective.

Leaf through the book now and write down the page numbers of two other occurrences of an arrowed example.

You Try Its One of the key instructional features of this text is the paired examples. Notice that in each example box, the example on the left is completely worked out and the "You Try It" example on the right is not. Study the worked-out example carefully by working through each step. Then work the You Try It. If you get stuck, refer to the page number at the end of the example, which directs you to the place where the You Try It is solved—a complete worked-out solution is provided. Try to use the given solution to get a hint for the step you are stuck on. Then try to complete your solution.

Example 2	Solve: $3x - 5 \le 3 - 2(3x + 1)$	You Try It 2	Solve: $5x - 2 \le 4 - 3(x - 2)$
Solution	$3x - 5 \le 3 - 2(3x + 1)$ $3x - 5 \le 3 - 6x - 2$ $3x - 5 \le 1 - 6x$ $9x - 5 \le 1$ $9x \le 6$ $\frac{9x}{9} \le \frac{6}{9}$ $x \le \frac{2}{3}$ $\left\{x\middle|x \le \frac{2}{3}\right\}$	**Your solution**	$5x - 2 \le 4 - 3x + 6$ $5x - 2 \le 10 - 3x$ $8x - 2 \le 10$ $8x \le 12$ $\frac{8x}{8} \le \frac{12}{8}$ $x \le \frac{3}{2}$ $\left\{x\middle|x \le \frac{3}{2}\right\}$

Solution on p. S6

Page 116

When you have completed your solution, check your work against the solution we provided. (Turn to page S6 to see the solution of You Try It 2.) Be aware that frequently there is more than one way to solve a problem. Your answer, however, should be the same as the given answer. If you have any question as to whether your method will "always work," check with your instructor or with someone in the math center.

Browse through the textbook and write down the page numbers where two other paired example features occur.

Remember: Be an active participant in your learning process. When you are sitting in class watching and listening to an explanation, you may think that you understand. However, until you actually try to do it, you will have no confirmation of the new knowledge or skill. Most of us have had the experience of sitting in class thinking we knew how to do something only to get home and realize that we didn't.

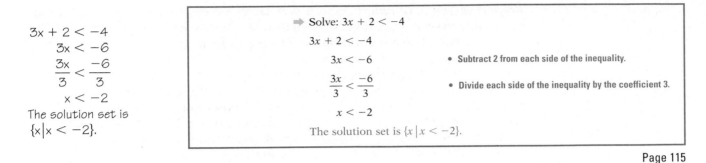

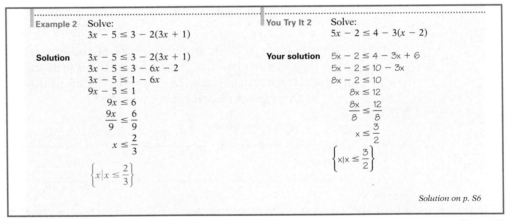

> **TAKE NOTE**
> There is a strong connection between reading and being a successful student in math or any other subject. If you have difficulty reading, consider taking a reading course. Reading is much like other skills. There are certain things you can learn that will make you a better reader.

Word Problems Word problems are difficult because we must read the problem, determine the quantity we must find, think of a method to do that, and then actually solve the problem. In short, we must formulate a *strategy* to solve the problem and then devise a *solution*.

Note in the paired example below that part of every word problem is a strategy and part is a solution. The strategy is a written description of how we will solve the problem. In the corresponding You Try It, you are asked to formulate a strategy. Do not skip this step, and be sure to write it out.

Example 7

In a recent year, 238 U.S. airports collected \$1,100 million in passenger taxes. Of this amount, \$88 million was spent on noise reduction. What percent of the passenger taxes collected was spent on noise reduction?

Strategy

To find the percent, solve the basic percent equation using $B = 1,100$ million and $A = 88$ million. The percent is unknown.

Solution

$$P \cdot B = A$$
$$P(1100) = 88$$
$$\frac{P(1100)}{1100} = \frac{88}{1100}$$
$$P = 0.08$$

8% of the passenger taxes collected was spent on noise reduction.

You Try It 7

The total revenue for all football bowl games in 2000 was \$158.3 million. (Source: *San Diego Union-Tribune*, January 18, 2001). Of that amount, the Big Ten conference received \$22.45 million. What percent of the total revenue did the Big Ten conference receive? Round to the nearest tenth of a percent.

Your strategy

Use the basic percent equation, $P \cdot B = A$, with $B = 158.3$ million and $A = 22.45$ million. Solve for P.

Your solution

$$P \cdot B = A$$
$$P(158.3) = 22.45$$
$$\frac{P(158.3)}{158.3} = \frac{22.56}{158.3}$$
$$P \approx 0.142$$

The Big Ten conference received approximately 14.2% of the total revenue.

Solutions on p. S4

Page 78

> **TAKE NOTE**
> If a rule has more than one part, be sure to make a notation to that effect.

Rule Boxes Pay special attention to rules placed in boxes. These rules give you the reasons certain types of problems are solved the way they are. When you see a rule, try to rewrite the rule in your own words.

Multiplying each side of an equation by the same number (NOT 0) will not change the solution.

> **Multiplication Property of Equations**
> Each side of an equation can be multiplied by the same *nonzero* number without changing the solution of the equation. In symbols, if $c \neq 0$, then the equation $a = b$ has the same solutions as the equation $ac = bc$.

Page 75

Chapter Exercises When you have completed studying an objective, do the exercises in the exercise set that correspond with that objective. The exercises are labeled with the same letter as the objective. Math is a subject that needs to be learned in small sections and practiced continually in order to be mastered. Doing all of the exercises in each exercise set will help you master the problem-solving techniques necessary for success. As you work through the exercises for an objective, check your answers to the odd-numbered exercises with those in the back of the book.

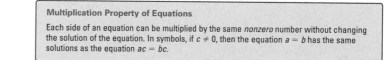

Preparing for a Test There are important features of this text that can be used to prepare for a test.

- Chapter Summary
- Chapter Review
- Chapter Test

After completing a chapter, read the Chapter Summary. (See page 320 for the Chapter 5 Summary.) This summary highlights the important topics covered in the chapter. The page number following each topic refers you to the page in the text on which you can find more information about the concept.

Following the Chapter Summary are a Chapter Review (see page 321) and a Chapter Test (see page 323). Doing the review exercises is an important way of testing your understanding of the chapter. The answer to each review exercise is given at the back of the book, along with its objective reference. After checking your answers, restudy any objective from which a question you missed was taken. It may be helpful to retry some of the exercises for that objective to reinforce your problem-solving techniques.

The Chapter Test should be used to prepare for an exam. We suggest that you try the Chapter Test a few days before your actual exam. Take the test in a quiet place and try to complete the test in the same amount of time you will be allowed for your exam. When taking the Chapter Test, practice the strategies of successful test takers: (1) scan the entire test to get a feel for the questions; (2) read the directions carefully; (3) work the problems that are easiest for you first; and perhaps most importantly, (4) try to stay calm.

When you have completed the Chapter Test, check your answers. If you missed a question, review the material in that objective and rework some of the exercises from that objective. This will strengthen your ability to perform the skills in that objective.

Your career goal goes here

Is it difficult to be successful? YES! Successful music groups, artists, professional athletes, chefs, and ________________ have to work very hard to achieve their goals. They focus on their goals and ignore distractions. The things we ask you to do to achieve success take time and commitment. We are confident that if you follow our suggestions, you will succeed.

Chapter 1

Real Numbers and Variable Expressions

Containers full of products to be shipped and sold in other countries are piled high onto huge cargo ships at ports around the United States. The depot at this port, near Newark, New Jersey, sees activity every day from trucks dropping off containers. The more goods a country exports, the better the likelihood of a favorable balance of trade. In order to avoid a trade deficit, a country needs to export more than it imports. The **Project on pages 63–64** shows how to calculate a country's balance of trade and determine whether it is favorable or unfavorable.

Objectives

Section 1.1

- **A** To use inequality symbols with integers
- **B** To find the additive inverse and absolute value of a number
- **C** To add or subtract integers
- **D** To multiply or divide integers
- **E** To solve application problems

Section 1.2

- **A** To write a rational number as a decimal
- **B** To convert among percents, fractions, and decimals
- **C** To add or subtract rational numbers
- **D** To multiply or divide rational numbers
- **E** To evaluate exponential expressions
- **F** To simplify numerical radical expressions
- **G** To solve application problems

Section 1.3

- **A** To use the Order of Operations Agreement to simplify expressions

Section 1.4

- **A** To evaluate a variable expression
- **B** To simplify a variable expression using the Properties of Addition
- **C** To simplify a variable expression using the Properties of Multiplication
- **D** To simplify a variable expression using the Distributive Property
- **E** To translate a verbal expression into a variable expression

Section 1.5

- **A** To write a set using the roster method
- **B** To write a set using set-builder notation
- **C** To graph an inequality on the number line

WWW WEB Need help? For on-line student resources, such as section quizzes, visit this textbook's web site at **math.college.hmco.com/students.**

Instructor Note

The *Prep Test* is a means to test your students' mastery of *prerequisite* material that is assumed in the coming chapter. All answers, along with the objective to review (if necessary), are provided in the Answers to Selected Exercises appendix.

Features of the IAE

Among the features included in the margins of this Instructor's Annotated Edition are the following:

Instructor Note

These notes offer suggestions for presenting the material in the objective or provide information that may be helpful.

In-Class Examples

These exercises can be used during your presentation of the lesson. They do not duplicate examples presented in the textbook.

Discuss the Concepts

These questions or requests for an explanation can be used for class discussion or for writing exercises. They require students to verbalize the basic concepts presented in the lesson.

Concept Check

These questions or exercises can be used after the presentation of a lesson to test student understanding of the concepts developed.

Optional Student Activity

This exercise can be assigned at the conclusion of the lesson. It can serve as a class activity, for individual work, or as a cooperative-learning project. In general, activities were written to be accomplished within about a 5-minute period.

Prep Test

1. What is 127.1649 rounded to the nearest hundredth?
127.16

2. Add: 49.147 + 5.96
55.107

3. Subtract: 5004 − 487
4517

4. Multiply: 407 × 28
11,396

5. Divide: 456 ÷ 19
24

6. What is the smallest number that both 8 and 12 divide evenly into?
24

7. What is the greatest number that divides evenly into both 16 and 20?
4

8. Without using 1, write 21 as a product of two whole numbers.
$3 \cdot 7$

9. Represent the shaded portion of the figure as a fraction.

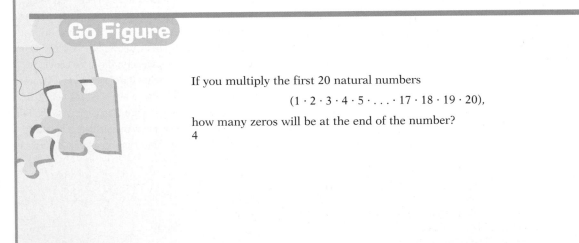

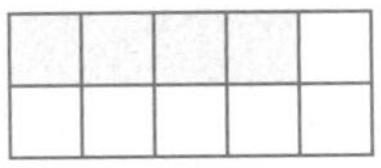

$\frac{2}{5}$

Go Figure

If you multiply the first 20 natural numbers

$$(1 \cdot 2 \cdot 3 \cdot 4 \cdot 5 \cdot \ldots \cdot 17 \cdot 18 \cdot 19 \cdot 20),$$

how many zeros will be at the end of the number?
4

Instructor Note

Margin notes entitled *Point of Interest, Take Note,* and *Calculator Note* are printed in the student text. The *Point of Interest* feature provides a historical note or mathematical fact of interest. The *Take Note* feature flags important information or provides assistance in understanding a concept. The *Calculator Note* feature describes some of the functions of a calculator.

1.1 Integers

Objective A **To use inequality symbols with integers**

VIDEO & DVD CD TUTOR WEB SSM

TAKE NOTE
We suggest you read "To the Student" on page xxi. It provides an explanation of the organization of the chapters and lessons of this text.

It seems to be a human characteristic to put similar items in the same place. For instance, a biologist places similar animals in groups called *phyla,* and a geologist divides the history of Earth into *eras.*

Mathematicians likewise place objects with similar properties in *sets* and use braces to surround the objects in the set, which are called **elements**. The numbers that we use to count elements, such as the number of people at a baseball game or the number of horses on a ranch, have similar characteristics. These numbers are the *natural numbers.*

Natural numbers $= \{1, 2, 3, 4, 5, 6, 7, 8, 9, 10, 11, \ldots\}$

The natural numbers alone do not provide all the numbers that are useful in applications. For instance, a meteorologist needs numbers below zero and above zero.

Integers $= \{\ldots, -5, -4, -3, -2, -1, 0, 1, 2, 3, 4, 5, \ldots\}$

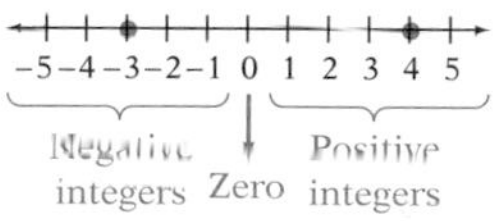

Each integer can be shown on a **number line**. The **graph of an integer** is shown by placing a heavy dot on the number line directly above the number. The graph of -3 and that of 4 are shown on the number line at the left.

The integers to the left of zero are **negative integers**. The integers to the right of zero are **positive integers**. Zero is neither a positive nor a negative integer.

Point of Interest
The Alexandrian astronomer Ptolemy began using *omicron,* *O,* the first letter of the Greek word that means "nothing," as the symbol for zero in A.D. 150. It was not until the 13th century, however, that Fibonacci introduced 0 to the Western world as a placeholder so that we could distinguish, for example, 45 from 405.

Consider the sentences below.

The quarterback threw the football and the receiver caught *it.*
An accountant purchased a calculator and placed *it* in a briefcase.

In the first sentence, *it* means football; in the second sentence, *it* means calculator. In language, the word *it* can stand for many different objects. Similarly, in mathematics, a letter of the alphabet can be used to stand for a number. Such a letter is called a **variable**. Variables are used in the next definition.

> **Definition of Inequality Symbols**
>
> If a and b are two numbers and a is to the left of b on the number line, then a is **less than** b. This is written $a < b$.
>
> If a and b are two numbers and a is to the right of b on the number line, then a is **greater than** b. This is written $a > b$.

There are also inequality symbols for **less than or equal to** ($\leq$) and **greater than or equal to** ($\geq$). For instance,

$6 \leq 6$ because $6 = 6$. $\quad 7 \leq 15$ because $7 < 15$.

It is convenient to use a variable to represent, or stand for, any one of the elements of a set. For instance, the statement "x is an element of the set $\{0, 2, 4, 6\}$" means that x can be replaced by 0, 2, 4, or 6. The symbol for "is an element of" is $\in$; the symbol for "is not an element of" is $\notin$. For example,

$2 \in \{0, 2, 4, 6\} \qquad 6 \in \{0, 2, 4, 6\} \qquad 7 \notin \{0, 2, 4, 6\}$

Objective 1.1A

New Vocabulary
set
elements
natural numbers
integers
number line
graph of an integer
negative integers
positive integers
variable

New Symbols
is less than ($<$)
is greater than ($>$)
is less than or equal to ($\leq$)
is greater than or equal to ($\geq$)
is an element of ($\in$)
is not an element of ($\notin$)

Discuss the Concepts
1. Which symbols, $>$, $<$, $\geq$, or $\leq$, can correctly be placed between the two numbers? Explain your answer.
   a. $-6 \quad -6$ $\geq$ or $\leq$
   b. $-3 \quad -9$ $>$ or $\geq$
2. True or false: If a number is an integer, then it is a natural number.

Concept Check
1. Write the set of the first four natural numbers. $\{1, 2, 3, 4\}$
2. Classify each number as a positive integer, a negative integer, or neither:
   a. -12 **b.** 18 **c.** -7
   d. 0 **e.** $\frac{3}{4}$ **f.** 365
   a. negative integer
   b. positive integer
   c. negative integer
   d. neither **e.** neither
   f. positive integer
3. Place the symbol $<$ or $>$ in the space provided.
   a. 0 ___ any positive number. $<$
   b. 0 ___ any negative number. $>$

In-Class Examples (Objective 1.1A)
1. Place the correct symbol, $<$ or $>$, between the two numbers.
   a. $-8 \quad 4$ $<$ **b.** $-9 \quad -3$ $<$
   c. $0 \quad -6$ $>$ **d.** $51 \quad -29$ $>$
2. Write the set of natural numbers less than 8. $\{1, 2, 3, 4, 5, 6, 7\}$
3. Given $A = \{-9, -6, -3, 0, 3\}$, which elements of set A are:
   a. greater than -2? 0, 3
   b. less than -5? -9, -6
   c. less than or equal to -6? -9, -6

Optional Student Activity

1. Graph the numbers −5 and 3 on a number line. Use the graph to explain why −5 is less than 3 and to explain why 3 is greater than −5.
   $-5 < 3$ because −5 is to the left of 3 on the number line.
   $3 > -5$ because 3 is to the right of −5 on the number line.
2. Graph the numbers 1 and −2 on a number line. Use the graph to explain why 1 is greater than −2 and to explain why −2 is less than 1.
   $1 > -2$ because 1 is to the right of −2 on the number line. $-2 < 1$ because −2 is to the left of 1 on the number line.

Objective 1.1B

New Vocabulary

opposite
additive inverse
absolute value

Symbols

absolute value ($| \, |$)

Discuss the Concepts

1. How can you use a number line to show that $|4| = 4$ and $|-4| = 4$?
2. When is the absolute value of a number equal to itself?
3. If the absolute value of a number is 5, what are the possible values of the number?

Example 1

Let $x \in \{-6, -2, 0\}$. For which values of x is the inequality $x \le -2$ a true statement?

Solution

Replace x by each element of the set and determine whether the inequality is true.

$x \le -2$

$-6 \le -2$ True. $-6 < -2$

$-2 \le -2$ True. $-2 = -2$

$0 \le -2$ False.

The inequality is true for −6 and −2.

You Try It 1

Let $y \in \{-5, -1, 5\}$. For which values of y is the inequality $y > -1$ a true statement?

Your solution

5

Solution on p. S1

Objective B **To find the additive inverse and absolute value of a number**

VIDEO & DVD CD TUTOR WEB SSM

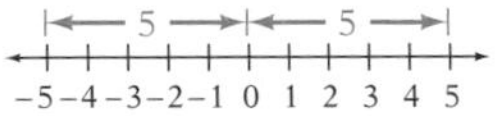

On the number line, the numbers 5 and −5 are the same distance from zero but on opposite sides of zero. The numbers 5 and −5 are called **opposites** or **additive inverses** of each other. (See the number line at the left.)

TAKE NOTE
The distance from 0 to 5 is 5; $|5| = 5$. The distance from 0 to −5 is 5; $|-5| = 5$.

The opposite (or additive inverse) of 5 is −5. The opposite of −5 is 5. The symbol for opposite is −.

−(5) means the opposite of *positive* 5. $-(5) = -5$

−(−5) means the opposite of *negative* 5. $-(-5) = 5$

The **absolute value** of a number is its distance from zero on the number line. The symbol for absolute value is two vertical bars, | |.

Point of Interest

The definition of absolute value that we have given in the box is written in what is called "rhetorical style." That is, it is written without the use of variables. This is how *all* mathematics was written prior to the Renaissance. During that period, from the 14th to the 16th century, the idea of expressing a variable symbolically was developed.

Absolute Value

The absolute value of a positive number is the number itself. The absolute value of zero is zero. The absolute value of a negative number is the opposite of the negative number.

$|9| = 9$
$|0| = 0$
$|-7| = 7$

⇒ Evaluate: $-|-12|$

$-|-12| = -12$

• The absolute value sign does not affect the negative sign in front of the absolute value sign.

Example 2

Let $a \in \{-12, 0, 4\}$. Find the additive inverse of a and the absolute value of a for each element of the set.

Solution

Replace a by each element of the set.

$-a$	$\lvert a \rvert$
$-(-12) = 12$	$\lvert -12 \rvert = 12$
$-(0) = 0$	$\lvert 0 \rvert = 0$
$-(4) = -4$	$\lvert 4 \rvert = 4$

You Try It 2

Let $z \in \{-11, 0, 8\}$. Find the additive inverse of z and the absolute value of z for each element of the set.

Your solution

11; 11
0; 0
−8; 8

Solution on p. S1

In-Class Examples (Objective 1.1B)

1. Find the additive inverse.
   a. −12 12 b. 34 −34
2. Evaluate.
   a. $|50|$ 50 b. $|-9|$ 9
   c. $-|6|$ −6 d. $-|-78|$ −78

Objective C To add or subtract integers

VIDEO & DVD CD TUTOR WEB SSM

A number can be represented anywhere along the number line by an arrow. A positive number is represented by an arrow pointing to the right, and a negative number is represented by an arrow pointing to the left. The size of the number is represented by the length of the arrow.

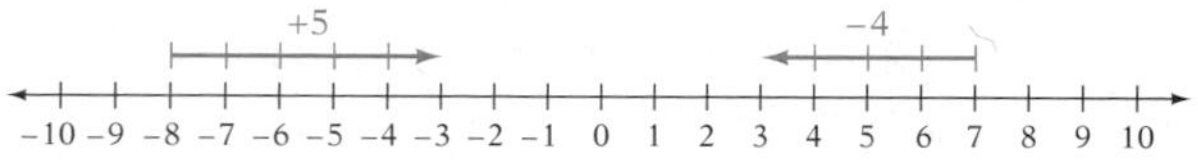

Addition of integers can be shown on the number line. To add integers, start at zero and draw an arrow representing the first number. At the tip of the first arrow, draw a second arrow representing the second number. The sum is below the tip of the second arrow.

$4 + 2 = 6$

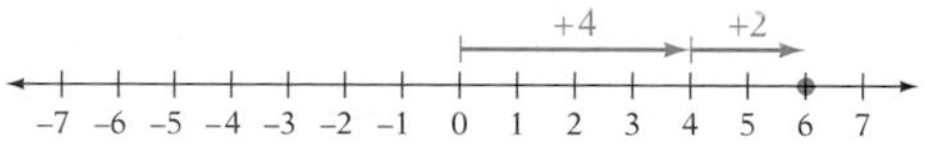

$-4 + (-2) = -6$

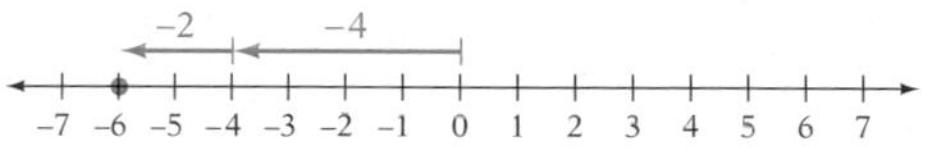

$-4 + 2 = -2$

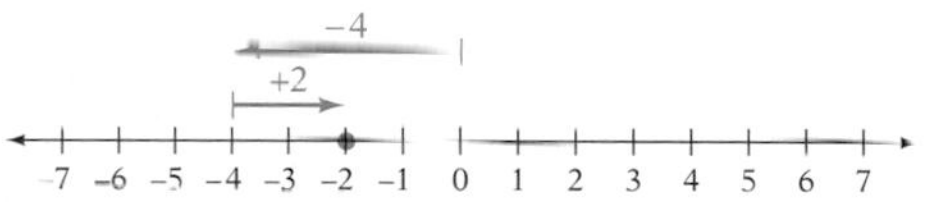

$4 + (-2) = 2$

The pattern for the addition of integers shown on the number line can be summarized in the following rule.

> **Addition of Integers**
>
> - *Numbers with the same sign*
>   To add two numbers with the same sign, add the absolute values of the numbers. Then attach the sign of the addends.
> - *Numbers with different signs*
>   To add two numbers with different signs, find the absolute value of each number. Then subtract the smaller of these numbers from the larger one. Attach the sign of the number with the larger absolute value.

⇒ Add: $(-9) + 8$

$|-9| = 9 \quad |8| = 8$ — • **The signs are different. Find the absolute value of each number.**

$9 - 8 = 1$ — • **Subtract the smaller number from the larger.**

$(-9) + 8 = -1$ — • **Attach the sign of the number with the larger absolute value. Because $|-9| > |8|$, use the sign of −9.**

In-Class Examples (Objective 1.1C)

Add.

1. $-8 + 6$ -2
2. $-43 + (-72)$ -115
3. $15 + (-26) + 9$ -2

Subtract.

4. $-3 - 18$ -21
5. $21 - (-17)$ 38
6. $-5 - 16 - (-14)$ -7

Concept Check

Determine whether the statement is true or false.

1. The additive inverse of a negative number is a positive number. True
2. The absolute value of a number is positive. False; $|0| = 0$
3. Given $|x| = 3$, then x is a positive integer. False

Optional Student Activity

Translate and evaluate.

1. The opposite of the additive inverse of 7 7
2. The absolute value of the opposite of −8 8
3. The opposite of the absolute value of 8 −8
4. The absolute value of the additive inverse of −6 6
5. The additive inverse of the absolute value of −6 −6

Objective 1.1C

New Rules

addition of integers
subtraction of integers

Instructor Note

There are a number of models of the addition of integers. Using arrows on the number line is just one of them. Another model is a checking account. If there is a balance of \$25 in a checking account, and a check is written for \$30, the account will be overdrawn by \$5 (−5).

Another model uses two colors of plastic chips, say blue for positive and red for negative, and the idea that a blue/red pair is equal to zero. To add $-8 + 3$, place 8 red chips and 3 blue chips in a circle. Pair as many red and blue chips as possible and remove the pairs from the region. There are 5 red chips remaining, or −5.

(Continued on next page)

(Continued)

To model $(-8) + (-3)$, place 8 red chips in the region and then 3 more red chips in the region. There are no pairs of red and blue chips, so there are 11 red chips. Therefore, the answer is -11.

A subtraction model based on blue and red chips similar to that of addition can be provided. Restrict the terms of the subtraction to, say, between -10 and 10, and start with 10 blue/red pairs in a circle. Because each blue/red pair is equal to zero, the circle contains 10 zeros. To model $-3 - (-7)$, place 3 more red chips in the circle and remove (subtract) any 7 red chips. Now pair as many blue and red chips as possible. There will be 4 blue chips left without a red chip. In other words, $-3 - (-7) = 4$.

Discuss the Concepts

1. Explain why the absolute value of -15 is greater than the absolute value of 3.
2. Explain why the difference between -18 and -4 is a positive number.
3. Explain why the difference between 9 and -12 is a positive number.

➡ Add: $(-23) + 47 + (-18) + 5$

To add more than two numbers, add the first two numbers. Then add the sum to the third number. Continue until all the numbers are added.

$$\begin{aligned}(-23) + 47 + (-18) + 5 &= 24 + (-18) + 5\\ &= 6 + 5\\ &= 11\end{aligned}$$

Look at the two expressions below and note that each expression equals the same number.

$8 - 3 = 5$ 8 minus 3 is 5.

$8 + (-3) = 5$ 8 plus the opposite of 3 is 5.

This example suggests that to subtract two numbers, we add the opposite of the second number to the first number.

first number	−	second number	=	first number	+	the opposite of the second number	
40	−	60	=	40	+	(−60)	= −20
−40	−	60	=	−40	+	(−60)	= −100
−40	−	(−60)	=	−40	+	60	= 20
40	−	(−60)	=	40	+	60	= 100

➡ Subtract: $-21 - (-40)$

Change this sign to plus.

$-21 - (-40) = -21 + 40 = 19$

Change -40 to the opposite of -40.

• Rewrite each subtraction as addition of the opposite. Then add.

➡ Subtract: $15 - 51$

Change this sign to plus.

$15 - 51 = 15 + (-51) = -36$

Change 51 to the opposite of 51.

• Rewrite each subtraction as addition of the opposite. Then add.

➡ Subtract: $-12 - (-21) - 15$

$$\begin{aligned}-12 - (-21) - 15 &= -12 + 21 + (-15)\\ &= 9 + (-15)\\ &= -6\end{aligned}$$

• Rewrite each subtraction as addition of the opposite. Then add.

Example 3 Add: $(-52) + (-39)$

Solution The signs are the same. Add the absolute values of the numbers: $52 + 39 = 91$

Attach the sign of the addends: $(-52) + (-39) = -91$

You Try It 3 Add: $100 + (-43)$

Your solution 57

Example 4 Add: $37 + (-52) + (-21) + (-7)$

Solution
$$\begin{aligned} &37 + (-52) + (-21) + (-7) \\ &= -15 + (-21) + (-7) \\ &= -36 + (-7) \\ &= -43 \end{aligned}$$

You Try It 4 Add: $(-51) + 42 + 17 + (-102)$

Your solution -94

Example 5 Subtract: $-11 - 15$

Solution
$$\begin{aligned} -11 - 15 &= -11 + (-15) \\ &= -26 \end{aligned}$$

You Try It 5 Subtract: $19 - (-32)$

Your solution 51

Example 6 Subtract: $-14 - 18 - (-21) - 4$

Solution
$$\begin{aligned} &-14 - 18 - (-21) - 4 \\ &= -14 + (-18) + 21 + (-4) \\ &= -32 + 21 + (-4) \\ &= -11 + (-4) \\ &= -15 \end{aligned}$$

You Try It 6 Subtract: $-9 - (-12) - 17 - 4$

Your solution -18

Solutions on p. S1

Objective D To multiply or divide integers

VIDEO & DVD CD TUTOR WEB SSM

Multiplication is the repeated addition of the same number. The product 3×5 is shown on the number line below.

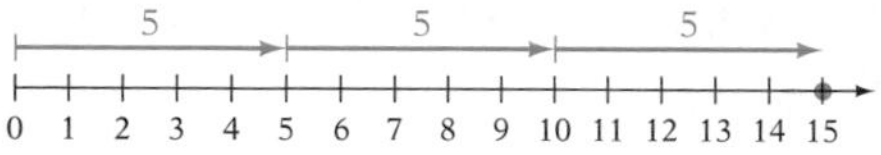

$$3 \times 5 = \underbrace{5 + 5 + 5}_{\text{5 is added 3 times.}} = 15$$

To indicate multiplication, several different symbols are used.

$3 \times 5 = 15$ $3 \cdot 5 = 15$ $(3)(5) = 15$ $3(5) = 15$ $(3)5 = 15$

Note that when parentheses are used and there is no arithmetic operation symbol, the operation is multiplication. Each number in a product is called a **factor**. For instance, 3 and 5 are factors of the product $3 \cdot 5 = 15$.

Point of Interest

The cross × was first used as a symbol for multiplication in 1631 in a book titled *The Key to Mathematics.* Also in that year, another book, *Practice of the Analytical Art,* advocated the use of a dot to indicate multiplication.

Concept Check

State whether or not the operation is addition. If the operation is addition, state whether the numbers being added have the same sign or different signs.

1. $3 + (-5)$ Yes; different signs
2. $-6 + 9$ Yes; different signs
3. $4 - 12$ No
4. $(-10) + (-8)$ Yes; same sign
5. $-14 + (-2)$ Yes; same sign
6. $-7(+1)$ No

State whether each "−" sign is a minus sign or a negative sign.

1. $8 - 5$ Minus
2. $2 - (-7)$ Minus; negative
3. $-6 - 1$ Negative; minus
4. $-4 - (-3)$ Negative; minus; negative

Optional Student Activity

On the number line, illustrate each of the following sums.

1. $-4 + 3$
2. $-5 + 8$
3. $2 + (-7)$
4. $1 + (-6)$
5. $-3 + (-4)$
6. $-2 + (-5)$

Objective 1.1D

New Vocabulary

multiplication
factors
product
quotient

New Rules

multiplication of integers
division of integers

In-Class Examples (Objective 1.1D)

Multiply.

1. $-5(8)$ -40
2. $-7(-12)$ 84
3. $-10(-6)(4)$ 240

Divide.

4. $18 \div (-3)$ -6
5. $-108 \div (-12)$ 9
6. $-64 \div 0$ undefined

New Properties

Properties of zero and one in division

New Symbols

is not equal to ($\neq$)

Discuss the Concepts

Have students show that a positive integer times a negative integer equals a negative integer by using repeated addition. For instance, have them describe $(-5)(3)$ as $(-5) + (-5) + (-5) = -15$ or $(-4)(2)$ as $(-4) + (-4) = -8$.

The idea that negative times negative equals positive seems arbitrary to students. You might relate it to using a double negative in English. For example, ask them the meaning of the sentence "It is not impossible to run a 4-minute mile" or "I never don't brush my teeth in the morning."

Discuss the Concepts

1. Describe the rules for dividing two integers.
2. Explain why the quotient of -18 and -6 is positive.
3. Is the quotient of two integers always less than either of the integers? Explain.
4. Every division problem has a related multiplication problem. Provide an example.

Concept Check

True or false:

1. The product of a nonzero number and its opposite is a negative number. True
2. The product of an odd number of negative numbers is a negative number. True
3. The product of an even number of negative numbers is a negative number. False
4. To find the opposite of a number, multiply the number by -1. True

(Continued on next page)

Now consider the product of a positive and a negative number.

TAKE NOTE
$3(-5)$ is 3 times -5.

$$3(-5) = \underbrace{(-5) + (-5) + (-5)}_{-5 \text{ is added 3 times.}} = -15$$

• **Multiplication is repeated addition.**

This suggests that the product of a positive number and a negative number is negative. Here are a few more examples.

$$4(-7) = -28 \qquad (-6)5 = -30 \qquad (-5) \cdot 7 = -35$$

To find the product of two negative numbers, look at the pattern at the right. As -5 multiplies a sequence of decreasing integers, the products increase by 5.

The pattern can be continued by requiring that the product of two negative numbers be positive.

These numbers decrease by 1. These numbers increase by 5.

$$-5(3) = -15$$
$$-5(2) = -10$$
$$-5(1) = -5$$
$$-5(0) = 0$$
$$-5(-1) = 5$$
$$-5(-2) = 10$$
$$-5(-3) = 15$$

Multiplication of Integers

- *Numbers with the same sign*
  To multiply two numbers with the same sign, multiply the absolute values of the numbers. The product is positive.
- *Numbers with different signs*
  To multiply two numbers with different signs, multiply the absolute values of the numbers. The product is negative.

⇒ Multiply: $-2(5)(-7)(-4)$

$$\begin{aligned} -2(5)(-7)(-4) &= -10(-7)(-4) \\ &= 70(-4) \\ &= -280 \end{aligned}$$

• **To multiply more than two numbers, multiply the first two. Then multiply the product by the third number. Continue until all the numbers are multiplied.**

For every division problem there is a related multiplication problem.

$$\frac{8}{2} = 4 \quad \text{because} \quad 4 \cdot 2 = 8$$

Division Related multiplication

This fact and the rules for multiplying integers can be used to illustrate the rules for dividing integers.

Note in the following examples that the quotient of two numbers with the same sign is positive.

$$\frac{12}{3} = 4 \text{ because } 4 \cdot 3 = 12 \qquad \frac{-12}{-3} = 4 \text{ because } 4 \cdot (-3) = -12$$

The next two examples illustrate that the quotient of two numbers with different signs is negative.

$$\frac{12}{-3} = -4 \text{ because } (-4)(-3) = 12 \qquad \frac{-12}{3} = -4 \text{ because } (-4) \cdot 3 = -12$$

Division of Integers

- *Numbers with the same sign*
  To divide two numbers with the same sign, divide the absolute values of the numbers. The quotient is positive.
- *Numbers with different signs*
  To divide two numbers with different signs, divide the absolute values of the numbers. The quotient is negative.

➡ Simplify: $-\frac{-56}{7}$

$$-\frac{-56}{7} = -\left(\frac{-56}{7}\right) = -(-8) = 8$$

Note that $\frac{-12}{3} = -4$, $\frac{12}{-3} = -4$, and $-\frac{12}{3} = -4$. This suggests the following rule.

TAKE NOTE
The symbol $\neq$ is read "is not equal to."

If a and b are integers, and $b \neq 0$, then $\frac{-a}{b} = \frac{a}{-b} = -\frac{a}{b}$.

PROPERTIES OF ZERO AND ONE IN DIVISION

- Zero divided by any number other than zero is zero.

  $\frac{0}{a} = 0$ because $0 \cdot a = 0$ For example, $\frac{0}{7} = 0$ because $0 \cdot 7 = 0$.

CALCULATOR NOTE
Enter 4 [÷] 0 [=] on your calculator. You will get an error message.

- Division by zero is not defined.

  To understand that division by zero is not permitted, suppose that $\frac{4}{0}$ were equal to n, where n is some number. Because each division problem has a related multiplication problem, $\frac{4}{0} = n$ means $n \cdot 0 = 4$. But $n \cdot 0 = 4$ is impossible because any number times 0 is 0. Therefore, division by 0 is not defined.

- Any number other than zero divided by itself is 1.

  $\frac{a}{a} = 1$, $a \neq 0$ For example, $\frac{-8}{-8} = 1$.

TAKE NOTE
"To the Student" on page xi explains how best to use the boxed examples in this text, such as the one below.

- Any number divided by one is the number.

  $\frac{a}{1} = a$ For example, $\frac{9}{1} = 9$.

Example 7 Multiply: $(-3)4(-5)$	**You Try It 7** Multiply: $8(-9)10$
Solution $(-3)4(-5) = (-12)(-5) = 60$	**Your solution** -720

Solution on p. S1

(Continued)
What property of division is illustrated by each of the following?

a. $\frac{0}{-12} = 0$
Zero divided by any number other than zero is zero.

b. $\frac{-18}{1} = -18$
Any number divided by 1 is the number.

c. $\frac{-5}{0}$ is undefined.
Division by zero is undefined.

d. $\frac{-22}{-22} = 1$
Any number other than zero divided by itself is 1.

Optional Student Activity
Illustrate each of the following on the number line. (Suggest to students that their illustrations of division should be similar to their illustrations of multiplication on the number line.)

1. $3 \times (-3) = -9$
2. $2 \times (-5) = -10$
3. $3 \times (-2) = -6$
4. $5 \times (-1) = -5$
5. $\frac{-12}{4} = -3$
6. $\frac{-10}{5} = -2$
7. $\frac{-15}{3} = -5$

Objective 1.1E

New Vocabulary
average

Concept Check
The list below provides temperatures either on the surface or in the atmospheres of bodies in our solar system.
Mercury's dark side: −346°F
Mercury's sunlit side: 950°F
Mars, daytime: −17°F
Saturn's moon Titan: −190°F
Neptune's atmosphere: −360°F
Pluto, atmospheric high: −163°F
Pluto, atmospheric low: −550°F

1. Find the difference between the temperature on Mercury's sunlit side and Mars' daytime temperature. 967°F
2. Find the difference between Pluto's atmospheric high and low temperatures. 387°F

Optional Student Activity

1. Determine whether the square is a magic square. In a magic square, the sum of each row, each column, and the two diagonals is the same.

−4	1	0
3	−1	−5
−2	−3	2

Yes; the sum is −3.

2. Complete the magic square.

	−4	1
−2	0	
−1		−3

Row 1:3; Row 2:2; Row 3:4

Example 8 Multiply: $12(-4)(-3)(-5)$

Solution
$12(-4)(-3)(-5)$
$= (-48)(-3)(-5)$
$= 144(-5) = -720$

You Try It 8 Multiply: $(-2)3(-8)7$

Your solution 336

Example 9 Divide: $(-120) \div (-8)$

Solution $(-120) \div (-8) = 15$

You Try It 9 Divide: $(-135) \div (-9)$

Your solution 15

Example 10
Divide: $\frac{95}{-5}$

Solution $\frac{95}{-5} = -19$

You Try It 10
Divide: $\frac{-72}{4}$

Your solution −18

Example 11
Divide: $-\frac{-81}{3}$

Solution $-\frac{-81}{3} = -(-27) = 27$

You Try It 11
Divide: $-\frac{36}{-12}$

Your solution 3

Solutions on p. S1

Objective E To solve application problems

VIDEO & DVD CD TUTOR WEB SSM

To solve an application problem, first read the problem carefully. The Strategy involves identifying the quantity to be found and planning the steps that are necessary to find that quantity. The Solution involves performing each operation stated in the Strategy and writing the answer.

Example 12
The average temperature on Mercury's sunlit side is 950°F. The average temperature on Mercury's dark side is −346°F. Find the difference between these two average temperatures.

Strategy
To find the difference, subtract the average temperature on the dark side (−346) from the average temperature on the sunlit side (950).

Solution
$950 - (-346) = 950 + 346$
$= 1296$
The difference between these average temperatures is 1296°F.

You Try It 12
The daily low temperatures (in degrees Celsius) during one week were recorded as follows: −6°, −7°, 0°, −5°, −8°, −1°, −1°. Find the average daily low temperature.

Your strategy

Your solution
−4°C

Solution on p. S1

In-Class Examples (Objective 1.1E)

1. When a certain amount of salt is added to water, the freezing point of the salt-water mixture is −3°C. The boiling point of the mixture is 103°C. What is the difference between the boiling point and the melting point of the salt-water mixture? 106°C
2. The combined scores, in relation to par, of the top nine golfers in a golf tournament equaled −54. What was the average score of the nine golfers? −6
3. In a card game of Hearts, you had a score of −12 points before you "shot the moon," entitling you to add 26 points to your score. What was your score after you "shot the moon"? 14 points

Section 1.1

Suggested Assignment
Exercises 1–55, odds
Exercises 59–111, odds
Exercises 115–187, odds
More challenging problems:
Exercises 189, 190

1.1 Exercises

TAKE NOTE
"To the Student" on page xxi discusses the exercise sets in this textbook.

Objective A

Place the correct symbol, $<$ or $>$, between the two numbers.

1. $8 > -6$ **2.** $-14 < 16$ **3.** $-12 < 1$ **4.** $35 > 28$ **5.** $42 > 19$

6. $-42 < 27$ **7.** $0 > -31$ **8.** $-17 < 0$ **9.** $53 > -46$ **10.** $-27 > -39$

Answer true or false.

11. $-13 > 0$ false **12.** $-20 > 3$ false **13.** $12 > -31$ true **14.** $9 > 7$ true **15.** $-5 > -2$ false

16. $-44 > -21$ false **17.** $-4 > -120$ true **18.** $0 > -8$ true **19.** $-1 > 0$ false **20.** $-10 > -88$ true

21. Let $x \in \{-23, -18, -8, 0\}$. For which values of x is the inequality $x < -8$ a true statement?
−23, −18

22. Let $w \in \{-33, -24, -10, 0\}$. For which values of w is the inequality $w < -10$ a true statement?
−33, −24

23. Let $a \in \{-33, -15, 21, 37\}$. For which values of a is the inequality $a > -10$ a true statement?
21, 37

24. Let $v \in \{-27, -14, 14, 27\}$. For which values of v is the inequality $v > -15$ a true statement?
−14, 14, 27

25. Let $n \in \{-23, -1, 0, 4, 29\}$. For which values of n is the inequality $-6 > n$ a true statement?
−23

26. Let $m \in \{-33, -11, 0, 12, 45\}$. For which values of m is the inequality $-15 > m$ a true statement?
−33

Objective B

Find the additive inverse.

27. 4 −4 **28.** 8 −8 **29.** −9 9 **30.** −12 12 **31.** −28 28 **32.** −36 36

Evaluate.

33. $-(-14)$ 14 **34.** $-(-40)$ 40 **35.** $-(77)$ −77 **36.** $-(39)$ −39 **37.** $-(0)$ 0 **38.** $-(-13)$ 13

39. $|-74|$ 74 **40.** $|-96|$ 96 **41.** $-|-82|$ −82 **42.** $-|-53|$ −53 **43.** $-|81|$ −81 **44.** $-|38|$ −38

Quick Quiz (Objective 1.1A)

1. Place the correct symbol, $<$ or $>$, between the two numbers.
   a. 5 −8 $>$
   b. −6 −2 $<$
2. Write the set of negative integers greater than or equal to −4. $\{-4, -3, -2, -1\}$
3. Given $B = \{-10, -5, 0, 5\}$, which elements of set B are less than −6? −10

Answers to Writing Exercises

57a. Students should rephrase the rule for adding two integers with the same sign: To add two integers with the same sign, add the absolute values of the numbers. Then attach the sign of the addends.

b. Students should rephrase the rule for adding two integers with different signs: To add two integers with different signs, find the absolute value of each number. Then subtract the smaller of these absolute values from the larger. Attach the sign of the integer with the larger absolute value.

58. To rewrite $8 - (-6)$ as addition of the opposite, change the subtraction to addition and change -6 to the opposite of -6: $8 - (-6) = 8 + 6$.

Place the correct symbol, $<$ or $>$, between the values of the two numbers.

45. $|-83| > |58|$ **46.** $|22| > |-19|$ **47.** $|43| < |-52|$ **48.** $|-71| < |92|$

49. $|-68| > |-42|$ **50.** $|12| < |-31|$ **51.** $|-45| < |-61|$ **52.** $|-28| < |43|$

53. Let $p \in \{-19, 0, 28\}$. Evaluate $-p$ for each element of the set.
19, 0, −28

54. Let $q \in \{-34, 0, 31\}$. Evaluate $-q$ for each element of the set.
34, 0, −31

55. Let $x \in \{-45, 0, 17\}$. Evaluate $-|x|$ for each element of the set.
−45, 0, −17

56. Let $y \in \{-91, 0, 48\}$. Evaluate $-|y|$ for each element of the set.
−91, 0, −48

Objective C

57. **a.** Explain the rule for adding two integers with the same sign.
b. Explain the rule for adding two integers with different signs.

58. Explain how to rewrite the subtraction $8 - (-6)$ as addition of the opposite.

Add or subtract.

59. $-3 + (-8)$
−11

60. $-6 + (-9)$
−15

61. $-8 + 3$
−5

62. $-9 + 2$
−7

63. $-3 + (-80)$
−83

64. $-12 + (-1)$
−13

65. $-23 + (-23)$
−46

66. $-12 + (-12)$
−24

67. $16 + (-16)$
0

68. $-17 + 17$
0

69. $48 + (-53)$
−5

70. $19 + (-41)$
−22

71. $-17 + (-3) + 29$
9

72. $13 + 62 + (-38)$
37

73. $-3 + (-8) + 12$
1

74. $-27 + (-42) + (-18)$
−87

75. $16 - 8$
8

76. $12 - 3$
9

77. $7 - 14$
−7

78. $6 - 9$
−3

79. $-7 - 2$
−9

80. $-9 - 4$
−13

81. $7 - (-2)$
9

82. $3 - (-4)$
7

83. $-6 - (-3)$
−3

84. $-4 - (-2)$
−2

85. $6 - (-12)$
18

86. $-12 - 16$
−28

Quick Quiz (Objective 1.1B)

1. Find the additive inverse.
a. 45 −45
b. −27 27

2. Evaluate.
a. $|-16|$ 16
b. $-|8|$ −8
c. $-|-30|$ −30

87. $13 + (-22) + 4 + (-5)$
-10

88. $-14 + (-3) + 7 + (-21)$
-31

89. $-16 + (-17) + (-18) + 10$
-41

90. $-25 + (-31) + 24 + 19$
-13

91. $26 + (-15) + (-11) + (-12)$
-12

92. $-32 + 40 + (-8) + (-19)$
-19

93. $-14 + (-15) + (-11) + 40$
0

94. $28 + (-19) + (-8) + (-1)$
0

95. $-4 - 3 - 2$
-9

96. $4 - 5 - 12$
-13

97. $12 - (-7) - 8$
11

98. $-12 - (-3) - (-15)$
6

99. $-19 - (-19) - 18$
-18

100. $-8 - (-8) - 14$
-14

101. $-17 - (-8) - (-9)$
0

102. $7 - 8 - (-1)$
0

103. $-30 - (-65) - 29 - 4$
2

104. $42 - (-82) - 65 - 7$
52

105. $-16 - 47 - 63 - 12$
-138

106. $42 - (-30) - 65 - (-11)$
18

107. $-47 - (-67) - 13 - 15$
-8

108. $-18 - 49 - (-84) - 27$
-10

109. $-19 - 17 - (-36) - 12$
-12

110. $48 - 19 - 29 - 51$
-51

111. $21 - (-14) - 43 - 12$
-20

112. $17 - (-17) - 14 - 21$
-1

Objective D

113. Describe the rules for multiplying two integers.

114. Name the operation in each expression. Justify your answer.
a. $8(-7)$ **b.** $8 - 7$ **c.** $8 - (-7)$ **d.** $-xy$ **e.** $x(-y)$ **f.** $-x - y$

Answers to Writing Exercises

113. Students should rephrase the rule for multiplying two integers with the same sign: To multiply two integers with the same sign, multiply the absolute values of the factors; the product is positive.

114. a. The operation in the expression $8(-7)$ is multiplication. There is no operation symbol between the 8 and the left parenthesis.

b. The operation in the expression $8 - 7$ is subtraction. There is a space before and after the minus sign. 7 is subtracted from 8.

c. The operation in the expression $8 - (-7)$ is subtraction. -7 is subtracted from 8.

d. The operation in the expression $-xy$ is multiplication. The x and y are right next to each other with no sign in between.

e. The operation in the expression $x(-y)$ is multiplication. There is no operation symbol between the x and the left parenthesis.

f. The operation in the expression $-x - y$ is subtraction. There is a space before and after the minus sign. y is subtracted from $-x$.

Quick Quiz (Objective 1.1C)

Add.

1. $7 + (-12)$ -5

2. $-14 + (-26)$ -40

3. $35 + (-10) + (-8)$ 17

Subtract.

4. $4 - 17$ -13

5. $-62 - (-53)$ -9

6. $11 - 28 - (-9)$ -8

Multiply or divide.

115. (14)3 42 **116.** (17)6 102 **117.** $-7 \cdot 4$ -28 **118.** $-8 \cdot 7$ -56 **119.** $(-12)(-5)$ 60 **120.** $(-13)(-9)$ 117

121. $-11(23)$ -253 **122.** $-8(21)$ -168 **123.** $(-17)14$ -238 **124.** $(-15)12$ -180 **125.** $6(-19)$ -114 **126.** $17(-13)$ -221

127. $12 \div (-6)$ -2 **128.** $18 \div (-3)$ -6 **129.** $(-72) \div (-9)$ 8 **130.** $(-64) \div (-8)$ 8 **131.** $-42 \div 6$ -7

132. $(-56) \div 8$ -7 **133.** $(-144) \div 12$ -12 **134.** $(-93) \div (-3)$ 31 **135.** $48 \div (-8)$ -6 **136.** $57 \div (-3)$ -19

137. $\frac{-49}{7}$ -7 **138.** $\frac{-45}{5}$ -9 **139.** $\frac{-44}{-4}$ 11 **140.** $\frac{-36}{-9}$ 4 **141.** $\frac{98}{-7}$ -14

142. $\frac{85}{-5}$ -17 **143.** $-\frac{-120}{8}$ 15 **144.** $-\frac{-72}{4}$ 18 **145.** $-\frac{-80}{-5}$ -16 **146.** $-\frac{-114}{-6}$ -19

147. $0 \div (-9)$ 0 **148.** $0 \div (-14)$ 0 **149.** $\frac{-261}{9}$ -29 **150.** $\frac{-128}{4}$ -32 **151.** $9 \div 0$ undefined

152. $(-21) \div 0$ undefined **153.** $\frac{132}{-12}$ -11 **154.** $\frac{250}{-25}$ -10 **155.** $\frac{0}{0}$ undefined **156.** $\frac{-58}{0}$ undefined

157. $7(5)(-3)$ -105 **158.** $(-3)(-2)8$ 48 **159.** $9(-7)(-4)$ 252 **160.** $(-2)(6)(-4)$ 48

161. $16(-3)5$ -240 **162.** $20(-4)3$ -240 **163.** $-4(-3)8$ 96 **164.** $-5(-9)6$ 270

165. $-3(-8)(-9)$ -216 **166.** $-7(-6)(-5)$ -210 **167.** $(-9)7(5)$ -315 **168.** $(-8)7(10)$ -560

169. $7(-2)(5)(-6)$ 420 **170.** $(-3)7(-2)8$ 336 **171.** $-9(-4)(-8)(-10)$ 2880 **172.** $-11(-3)(-5)(-2)$ 330

173. $7(9)(-11)4$ -2772 **174.** $-12(-4)7(-2)$ -672 **175.** $(-14)9(-11)0$ 0 **176.** $(-13)(15)(-19)0$ 0

Quick Quiz (Objective 1.1D)

Multiply.

1. $6(-9)$ -54

2. $-5(-6)$ 30

3. $15(-3)(10)$ -450

Divide.

4. $-48 \div 8$ -6

5. $-27 \div (-9)$ 3

6. $0 \div (-5)$ 0

15

Objective E Application Problems

The elevation, or height, of places on Earth is measured in relation to sea level, or the average level of the ocean's surface. The table below shows height above sea level as a positive number and depth below sea level as a negative number. Use the table for Exercises 177 to 179.

Continent	Highest Elevation (in meters)		Lowest Elevation (in meters)	
Africa	Mt. Kilimanjaro	5895	Qattara Depression	−133
Asia	Mt. Everest	8848	Dead Sea	−400
Europe	Mt. Elbrus	5634	Caspian Sea	−28
America	Mt. Aconcagua	6960	Death Valley	−86

177. Find the difference in elevation between Mt. Aconcagua and Death Valley.
7046 m

178. What is the difference in elevation between Mt. Kilimanjaro and the Qattara Depression?
6028 m

179. For which continent shown is the difference between the highest and lowest elevations greatest?
Asia

The table at the right shows the boiling point and the melting point in degrees Celsius of three chemical elements. Use this table for Exercises 180 and 181.

Chemical Element	Boiling Point	Melting Point
Mercury	357	−39
Radon	−62	−71
Xenon	−107	−112

180. Find the difference between the boiling point and the melting point of mercury.
396°C

181. Find the difference between the boiling point and the melting point of xenon.
5°C

182. To discourage random guessing on a multiple-choice exam, a professor assigns 5 points for a correct answer, −2 points for an incorrect answer, and 0 points for leaving the question blank. What is the score for a student who had 20 correct answers, had 13 incorrect answers, and left 7 questions blank?
74

183. To discourage random guessing on a multiple-choice exam, a professor assigns 7 points for a correct answer, −3 points for an incorrect answer, and −1 point for leaving the question blank. What is the score for a student who had 17 correct answers, had 8 incorrect answers, and left 2 questions blank?
93

Quick Quiz (Objective 1.1E)

1. The temperature at which ethyl alcohol boils is 78°C. Ethyl alcohol freezes at −117°C. Determine the difference between the temperature at which ethyl alcohol boils and the temperature at which it freezes. 195°C

2. The daily low temperatures, in degrees Fahrenheit, during one week were recorded as follows: 4°, −6°, 8°, −2°, −9°, −11°, −5°. Find the average daily low temperature for the week. −3°F

Answers to Writing Exercises

189. The product of two negative integers is a positive number. Because -4 is a negative number, it must be multiplied by a negative integer in order for the product to be positive. Therefore, if $-4x$ equals a positive integer, then x must be a negative integer.

The table at the right shows the average temperatures at different cruising altitudes for airplanes. Use the table for Exercises 184 to 186.

Cruising Altitude	Average Temperature
12,000 ft	16°F
20,000 ft	−12°F
30,000 ft	−48°F
40,000 ft	−70°F
50,000 ft	−70°F

184. What is the difference between the average temperatures at 12,000 ft and at 40,000 ft?
86°F

185. What is the difference between the average temperatures at 40,000 ft and at 50,000 ft?
0°F

186. How much colder is the average temperature at 30,000 ft than at 20,000 ft?
36°F

A meteorologist may report a wind-chill temperature. This is the equivalent temperature, including the effects of wind and temperature, that a person would feel in calm air conditions. The table below gives the wind-chill temperature for various wind speeds and temperatures. For instance, when the temperature is 5°F and the wind is blowing at 15 mph, the wind-chill temperature is −25°F. Use this table for Exercises 187 and 188.

Wind-Chill Factors

Wind Speed (mph)	Thermometer Reading (degrees Fahrenheit)																
	35	30	25	20	15	10	5	0	−5	−10	−15	−20	−25	−30	−35	−40	−45
5	33	27	21	19	12	7	0	−5	−10	−15	−21	−26	−31	−36	−42	−47	−52
10	22	16	10	3	−3	−9	−15	−22	−27	−34	−40	−46	−52	−58	−64	−71	−77
15	16	9	2	−5	−11	−18	−25	−31	−38	−45	−51	−58	−65	−72	−78	−85	−92
20	12	4	−3	−10	−17	−24	−31	−39	−46	−53	−60	−67	−74	−81	−88	−95	−103
25	8	1	−7	−15	−22	−29	−36	−44	−51	−59	−66	−74	−81	−88	−96	−103	−110
30	6	−2	−10	−18	−25	−33	−41	−49	−56	−64	−71	−79	−86	−93	−101	−109	−116
35	4	−4	−12	−20	−27	−35	−43	−52	−58	−67	−74	−82	−89	−97	−105	−113	−120
40	3	−5	−13	−21	−29	−37	−45	−53	−60	−69	−76	−84	−92	−100	−107	−115	−123
45	2	−6	−14	−22	−30	−38	−46	−54	−62	−70	−78	−85	−93	−102	−109	−117	125

187. When the thermometer reading is −5°F, what is the difference between the wind-chill factor when the wind is blowing 10 mph and when the wind is blowing 30 mph?
29°F

188. When the thermometer reading is −20°F, what is the difference between the wind-chill factor when the wind is blowing 15 mph and when the wind is blowing 25 mph?
16°F

APPLYING THE CONCEPTS

189. If $-4x$ equals a positive integer, is x a positive or a negative integer? Explain your answer.
x is negative.

190. Is the difference between two integers always smaller than either of the integers? If not, give an example for which the difference between two integers is greater than either integer.
No. For example, $10 - (-8) = 18$.

1.2 Rational and Irrational Numbers

Objective A
To write a rational number as a decimal

Point of Interest
As early as A.D. 630, the Hindu mathematician Brahmagupta wrote a fraction as one number over another separated by a space. The Arab mathematician al Hassar (around A.D. 1050) was the first to show a fraction with the horizontal bar separating the numerator and denominator.

A *rational number* is the quotient of two integers. A rational number written in this way is commonly called a fraction. Here are some examples of rational numbers.

$$\frac{3}{4}, \quad \frac{-4}{9}, \quad \frac{15}{-4}, \quad \frac{8}{1}, \quad -\frac{5}{6}$$

Rational Numbers

A **rational number** is a number that can be written in the form $\frac{a}{b}$, where a and b are integers and $b \neq 0$.

Because an integer can be written as the quotient of the integer and 1, every integer is a rational number. For instance,

$$\frac{6}{1} = 6 \qquad \frac{-8}{1} = -8$$

Point of Interest
Simon Stevin (1548–1620) was the first to name decimal numbers. He wrote the number 2.345 as 2 0 3 1 4 2 5 3. He called the whole number part the *commencement*, the tenths digit was *primo*, the hundredths digit was *second*, the thousandths digit was *third*, and so on.

A number written in **decimal notation** is also a rational number.

three-tenths $\quad 0.3 = \frac{3}{10}$ $\qquad$ forty-three thousandths $\quad 0.043 = \frac{43}{1000}$

A rational number written as a fraction can be written in decimal notation.

➡ Write $\frac{5}{8}$ as a decimal.

The fraction bar can be read "÷".

$\frac{5}{8} = 5 \div 8$

```
   0.625  ← This is called a terminating decimal.
8)5.000
 -4 8
    20
   -16
     40
    -40
      0  ← The remainder is zero.
```

$\frac{5}{8} = 0.625$

Write $\frac{4}{11}$ as a decimal.

```
    0.3636 . . .  ← This is called a repeating decimal.
11)4.0000
  -3 3
     70
    -66
      40
     -33
       70
      -66
        4  ← The remainder is never zero.
```

$\frac{4}{11} = 0.\overline{36}$ ← The bar over the digits 3 and 6 is used to show that these digits repeat.

Objective 1.2A

New Vocabulary
rational number
decimal notation
terminating decimal
repeating decimal

New Symbols
bar over repeating digits of a decimal, e.g., $0.\overline{6}$

Discuss the Concepts

1. Do all proper fractions with a denominator of 6 have a decimal equivalent that contains repeating digits?
   No. $\frac{3}{6} = 0.5$
2. Do all proper fractions with a denominator of 11 have a decimal equivalent that contains repeating digits? Yes

Optional Student Activity
Show that the number is a rational number by writing it as the quotient of two integers. (*Note:* Answers may vary. Possible answers are given.)

1. $-\frac{5}{6}$ $\frac{-5}{6}$
2. $-\frac{7}{25}$ $\frac{-7}{25}$
3. $3\frac{2}{9}$ $\frac{29}{9}$
4. -0.16 $\frac{-4}{25}$
5. -0.75 $\frac{-3}{4}$
6. 5.2 $\frac{26}{5}$
7. -2.4 $\frac{-12}{5}$
8. -6.25 $\frac{-25}{4}$

In-Class Examples (Objective 1.2A)

1. Write $\frac{3}{8}$ as a decimal. 0.375
2. Write $\frac{7}{15}$ as a decimal. Place a bar over the repeating digits. $0.4\overline{6}$

Objective 1.2B

New Vocabulary
percent

Discuss the Concepts

1. Why do we multiply a percent by $\frac{1}{100}$ in order to rewrite it as a fraction and by 0.01 in order to rewrite it as a decimal?
2. Is 9.4% the same as $9\frac{2}{5}\%$? If not, what is the difference between the decimal equivalent of 9.4% and $9\frac{2}{5}\%$? Yes
3. Is $\frac{1}{2}\%$ the same as 0.5? If not, what is the difference between 0.5 and the decimal equivalent of $\frac{1}{2}\%$? No. 0.495

Concept Check
Have students complete a table of equivalent fractions, decimals, and percents. For example:

Fraction	*Decimal*	*Percent*
$\frac{1}{2}$	0.5	50%
$\frac{3}{4}$	0.75	75%
$\frac{2}{5}$	0.4	40%
$\frac{3}{8}$	0.375	37.5%
$\frac{13}{25}$	0.52	52%
$\frac{9}{16}$	0.5625	56.25%
$\frac{11}{20}$	0.55	55%
$\frac{3}{5}$	0.6	60%
$\frac{1}{3}$	$0.33\frac{1}{3}$	$33\frac{1}{3}\%$

Example 1
Write $\frac{8}{11}$ as a decimal. Place a bar over the repeating digits of the decimal.

Solution $\frac{8}{11} = 8 \div 11 = 0.7272\ldots = 0.\overline{72}$

You Try It 1
Write $\frac{4}{9}$ as a decimal. Place a bar over the repeating digits of the decimal.

Your solution $0.\overline{4}$

Solution on p. S1

Objective B **To convert among percents, fractions, and decimals**

VIDEO & DVD CD TUTOR WEB SSM

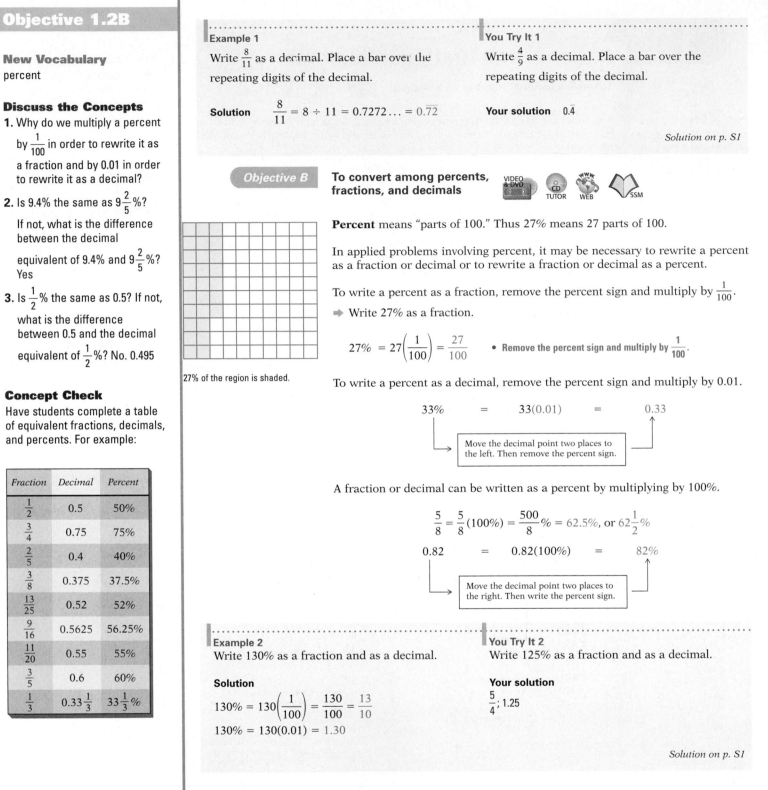

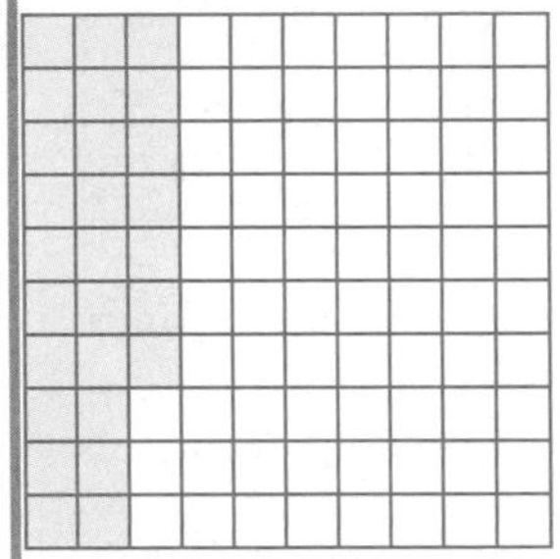

27% of the region is shaded.

Percent means "parts of 100." Thus 27% means 27 parts of 100.

In applied problems involving percent, it may be necessary to rewrite a percent as a fraction or decimal or to rewrite a fraction or decimal as a percent.

To write a percent as a fraction, remove the percent sign and multiply by $\frac{1}{100}$.

➡ Write 27% as a fraction.

$$27\% = 27\left(\frac{1}{100}\right) = \frac{27}{100}$$ • **Remove the percent sign and multiply by $\frac{1}{100}$.**

To write a percent as a decimal, remove the percent sign and multiply by 0.01.

$$33\% = 33(0.01) = 0.33$$

Move the decimal point two places to the left. Then remove the percent sign.

A fraction or decimal can be written as a percent by multiplying by 100%.

$$\frac{5}{8} = \frac{5}{8}(100\%) = \frac{500}{8}\% = 62.5\%, \text{ or } 62\frac{1}{2}\%$$

$$0.82 = 0.82(100\%) = 82\%$$

Move the decimal point two places to the right. Then write the percent sign.

Example 2
Write 130% as a fraction and as a decimal.

Solution
$$130\% = 130\left(\frac{1}{100}\right) = \frac{130}{100} = \frac{13}{10}$$
$$130\% = 130(0.01) = 1.30$$

You Try It 2
Write 125% as a fraction and as a decimal.

Your solution
$\frac{5}{4}$; 1.25

Solution on p. S1

In-Class Examples (Objective 1.2B)

1. Write $8\frac{1}{3}\%$ as a fraction. $\frac{1}{12}$
2. Write $\frac{7}{8}$ as a percent. 87.5% or $87\frac{1}{2}\%$
3. Write 1.5 as a percent. 150%
4. Write 40% as a decimal. 0.4

Example 3

Write $\frac{5}{6}$ as a percent.

Solution $\frac{5}{6} = \frac{5}{6}(100\%) = \frac{500}{6}\% = 83\frac{1}{3}\%$

You Try It 3

Write $\frac{1}{3}$ as a percent.

Your solution $33\frac{1}{3}\%$

Example 4 Write 0.092 as a percent.

Solution $0.092 = 0.092(100\%) = 9.2\%$

You Try It 4 Write 0.043 as a percent.

Your solution 4.3%

Solutions on p. S1

Objective C To add or subtract rational numbers

VIDEO & DVD CD TUTOR WEB SSM

Fractions with the same denominator are added by adding the numerators and placing the sum over the common denominator.

Addition of Fractions

To add two fractions with the same denominator, add the numerators and place the sum over the common denominator.

$$\frac{a}{c} + \frac{b}{c} = \frac{a+b}{c}$$

To add fractions with different denominators, first rewrite the fractions as equivalent fractions with a common denominator. Then add the fractions.

The least common denominator is the **least common multiple** (LCM) of the denominators. This is the smallest number that is a multiple of each of the denominators.

TAKE NOTE

You can find the LCM by multiplying the denominators and then dividing by the *common factor* of the two denominators. In the case of 6 and 10, $6 \cdot 10 = 60$. Now divide by 2, the common factor of 6 and 10.

$60 \div 2 = 30.$

➡ Add: $-\frac{5}{6} + \frac{3}{10}$

The LCM of 6 and 10 is 30. Rewrite the fractions as equivalent fractions with the denominator 30. Then add the fractions.

$$-\frac{5}{6} + \frac{3}{10} = -\frac{5}{6}\cdot\frac{5}{5} + \frac{3}{10}\cdot\frac{3}{3} = -\frac{25}{30} + \frac{9}{30} = \frac{-25+9}{30} = \frac{-16}{30} = -\frac{8}{15}$$

To subtract fractions with the same denominator, subtract the numerators and place the difference over the common denominator.

TAKE NOTE

The least common multiple of the denominators is frequently called the **least common denominator** (LCD).

➡ Subtract: $-\frac{4}{9} - \left(-\frac{7}{12}\right)$

The LCM of 9 and 12 is 36. Rewrite the fractions as equivalent fractions with the denominator 36. Then subtract the fractions.

$$-\frac{4}{9} - \left(-\frac{7}{12}\right) = -\frac{16}{36} - \left(-\frac{21}{36}\right) = \frac{-16-(-21)}{36} = \frac{-16+21}{36} = \frac{5}{36}$$

Optional Student Activity

Create grids and ask the students to shade a given percent of the region. For example: Shade 40% of the region below. 8 squares should be shaded.

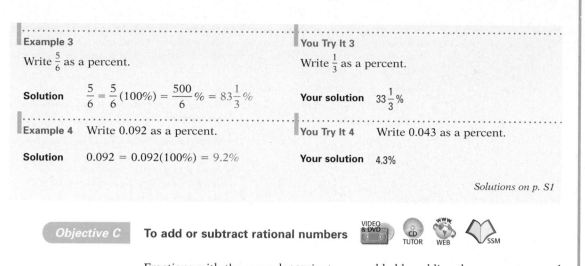

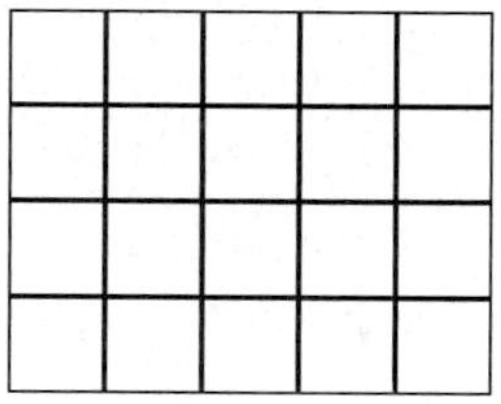

Objective 1.2C

Vocabulary to Review

common denominator
least common multiple
equivalent fraction

New Rules

addition of fractions with the same denominator

Discuss the Concepts

1. Estimate $\frac{5}{6} + \frac{8}{9}$. Is the sum closest to 1, 2, 13, or 15? Why? 2
2. Estimate $\frac{4}{9} - \frac{5}{11}$. Is the difference closest to 0, 1, −1, or −2? Why? 0
3. Estimate $-\frac{1}{7} - \left(-\frac{1}{8}\right)$. Is the difference closest to 0, 1, −1, −2, or −15? Why? 0

In-Class Examples (Objective 1.2C)

Simplify.

1. $-\frac{3}{4} + \left(-\frac{1}{6}\right)$ $-\frac{11}{12}$
2. 3.8 − 7.4 −3.6
3. 6.2 − (−4.61) 10.81

Concept Check

When two rational numbers are added, it is possible for the sum to be less than either addend, greater than either addend, or a number between the two addends. Give examples of each of these occurrences. Answers will vary. For example:

$-\frac{1}{2}+\left(-\frac{1}{4}\right)=-\frac{3}{4}$;

$\frac{1}{2}+\frac{1}{4}=\frac{3}{4}$;

$\frac{3}{4}+\left(-\frac{1}{4}\right)=\frac{1}{2}$

Optional Student Activity

1. The least common multiple of two numbers is 60 and the greatest common divisor is 6. What are the two numbers? 60 and 6
2. Find two fractions with different denominators, one positive and one negative, whose sum is $\frac{1}{2}$. Answers will vary. One example is $\frac{4}{5}+\left(-\frac{3}{10}\right)=\frac{1}{2}$.

Objective 1.2D

New Vocabulary

reciprocal

New Rules

$\frac{a}{b}\cdot\frac{c}{d}=\frac{ac}{bd}$

New Symbols

is approximately equal to ($\approx$)

Discuss the Concepts

1. Given a nonzero number, what is the reciprocal of the reciprocal of the number? The original number

(Continued on next page)

To add or subtract decimals, write the numbers so that the decimal points are in a vertical line. Then proceed as in the addition or subtraction of integers. Write the decimal point in the answer directly below the decimal points in the problem.

➡ Add: $-114.039 + 84.76$

$|-114.039| = 114.039$
$|84.76| = 84.76$

• The signs are different. Find the absolute value of each number.

$$\begin{array}{r} 114.039 \\ -\ \ 84.76 \\ \hline 29.279 \end{array}$$

• Subtract the smaller of these numbers from the larger.

• Attach the sign of the number with the larger absolute value. Because $|-114.039| > |84.76|$, use the sign of -114.039.

$-114.039 + 84.76 = -29.279$

Example 5

Simplify: $-\frac{3}{4}+\frac{1}{6}-\frac{5}{8}$

Solution The LCM of 4, 6, and 8 is 24.

$$-\frac{3}{4}+\frac{1}{6}-\frac{5}{8}=-\frac{18}{24}+\frac{4}{24}-\frac{15}{24}$$
$$=\frac{-18+4-15}{24}$$
$$=\frac{-29}{24}=-\frac{29}{24}$$

You Try It 5

Simplify: $-\frac{7}{8}-\frac{5}{6}+\frac{3}{4}$

Your solution $-\frac{23}{24}$

Example 6 Subtract: $42.987 - 98.61$

Solution
$42.987 - 98.61$
$= 42.987 + (-98.61)$
$= -55.623$

You Try It 6 Subtract: $16.127 - 67.91$

Your solution -51.783

Solutions on p. S2

Objective D To multiply or divide rational numbers

VIDEO & DVD CD TUTOR WEB SSM

The product of two fractions is the product of the numerators divided by the product of the denominators. $\frac{a}{b}\cdot\frac{c}{d}=\frac{ac}{bd}$

➡ Multiply: $\frac{3}{8}\cdot\frac{12}{17}$

$\frac{3}{8}\cdot\frac{12}{17}=\frac{3\cdot 12}{8\cdot 17}$

• Multiply the numerators. Multiply the denominators.

$=\frac{3\cdot \overset{1}{\cancel{2}}\cdot \overset{1}{\cancel{2}}\cdot 3}{\underset{1}{\cancel{2}}\cdot \underset{1}{\cancel{2}}\cdot 2\cdot 17}$

• Write the prime factorization of each factor. Divide by the common factors.

$=\frac{9}{34}$

• Multiply the factors in the numerator and in the denominator.

In-Class Examples (Objective 1.2D)

Simplify.

1. $-\frac{5}{6}\left(\frac{3}{10}\right)$ $-\frac{1}{4}$
2. $-6.8(-2.1)$ 14.28
3. $9.44 \div (-8)$ -1.18
4. $-\frac{5}{6}\div\frac{5}{9}$ $-\frac{3}{2}$

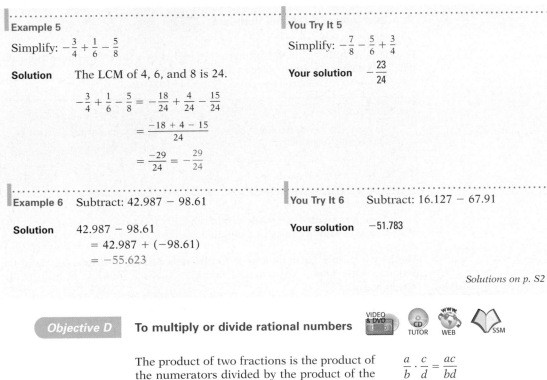

TAKE NOTE
To invert the divisor means to write its reciprocal. The reciprocal of $\frac{18}{25}$ is $\frac{25}{18}$.

To divide fractions, invert the divisor. Then multiply the fractions.

➡ Divide: $\frac{3}{10} \div \left(-\frac{18}{25}\right)$

The signs are different. The quotient is negative.

$$\frac{3}{10} \div \left(-\frac{18}{25}\right) = -\left(\frac{3}{10} \div \frac{18}{25}\right) = -\left(\frac{3}{10} \cdot \frac{25}{18}\right) = -\left(\frac{3 \cdot 25}{10 \cdot 18}\right)$$

$$= -\left(\frac{\overset{1}{\cancel{3}} \cdot \overset{1}{\cancel{5}} \cdot 5}{2 \cdot \underset{1}{\cancel{5}} \cdot 2 \cdot \underset{1}{\cancel{3}} \cdot 3}\right) = -\frac{5}{12}$$

To multiply decimals, multiply as with integers. Write the decimal point in the product so that the number of decimal places in the product equals the sum of the decimal places in the factors.

➡ Multiply: $-6.89(0.00035)$

6.89	2 decimal places	
× 0.00035	5 decimal places	• Multiply the absolute values.
3445		
2067		
0.0024115	7 decimal places	

$-6.89(0.00035) = -0.0024115$ • The signs are different. The product is negative.

To divide decimals, move the decimal point in the divisor to the right to make it a whole number. Move the decimal point in the dividend the same number of places to the right. Place the decimal point in the quotient directly over the decimal point in the dividend. Then divide as with whole numbers.

TAKE NOTE
The symbol ≈ is used to indicate that the quotient is an approximate value that has been rounded off.

➡ Divide: $1.32 \div 0.27$. Round to the nearest tenth.

```
         4.88 ≈ 4.9
0.27.)1.32.00
     -1 08
        240
       -216
        240
       -216
         24
```

• Move the decimal point 2 places to the right in the divisor and then in the dividend. Place the decimal point in the quotient.

Example 7

Divide: $-\frac{5}{8} \div \left(-\frac{5}{40}\right)$

Solution The quotient is positive.

$$-\frac{5}{8} \div \left(-\frac{5}{40}\right) = \frac{5}{8} \div \frac{5}{40} = \frac{5}{8} \cdot \frac{40}{5} = \frac{5 \cdot 40}{8 \cdot 5}$$

$$= \frac{\overset{1}{\cancel{5}} \cdot \overset{1}{\cancel{2}} \cdot \overset{1}{\cancel{2}} \cdot \overset{1}{\cancel{2}} \cdot 5}{\underset{1}{\cancel{2}} \cdot \underset{1}{\cancel{2}} \cdot \underset{1}{\cancel{2}} \cdot \underset{1}{\cancel{5}}} = \frac{5}{1} = 5$$

You Try It 7

Divide: $-\frac{3}{8} \div \left(-\frac{5}{12}\right)$

Your solution $\frac{9}{10}$

Solution on p. S2

(Continued)

2. Suppose the numerator of a fraction is a fixed number, for instance, 5. How does the value of the fraction change as the denominator increases? The value decreases.

3a. Find the product of 1.0035 and 1.00079 without using a calculator. Then find the product using a calculator and compare the two numbers. Some calculators truncate the product, which means that the digits that cannot be displayed are discarded. Other calculators round the answer to the leftmost place value in the calculator's display. Determine which method your calculator uses to handle approximate answers. 1.004292765; truncate: 1.0042927; round: 1.0042928

b. If the decimal places in a negative number are truncated, is the resulting number greater than, less than, or equal to the original number? Greater than

Concept Check

Replace the question mark with the number that makes the equation true.

1. $4.3 \times ? = -2.58$ −0.6

2. $32.4 = -9 \times ?$ −3.6

3. $? \times (-0.2) = -1.6$ 8

4. $\frac{?}{0.4} = -20$ 8

5. $\frac{?}{-8} = -3.1$ 24.8

6. $\frac{?}{0.48} = -12.5$ −6

Optional Student Activity

1. A package of meat is marked 1.385 lb. How many ounces over 1 lb does the package contain? 6.16 oz
2. Multiply.
   a. $2 \times 0.333\ldots$ $\frac{2}{3}$
   b. $3 \times 0.666\ldots$ 2
3. Given that $x = 0.999\ldots$, which of the following is true?
   a. $x = 1$
   b. $x < 1$
   c. x is the largest number less than 1 a

Objective 1.2E

New Vocabulary

exponent
base
factored form
exponential form

Instructor Note

You may want to introduce the alternative wording for exponential expressions, reading 2^4 as "two to the fourth power," for example, and b^5 as "b to the fifth power."

Discuss the Concepts

1. How are exponential expressions related to multiplication?
2. For the expression 5^4, which is the base and which is the exponent? What does the base represent? What does the exponent represent?

Example 8 Multiply: $-4.29(8.2)$

Solution The product is negative.

$$\begin{array}{r} 4.29 \\ \times\ 8.2 \\ \hline 858 \\ 3432 \\ \hline 35.178 \end{array}$$

$-4.29(8.2) = -35.178$

You Try It 8 Multiply: $-5.44(3.8)$

Your solution -20.672

Solution on p. S2

Objective E **To evaluate exponential expressions**

VIDEO & DVD CD TUTOR WEB SSM

Point of Interest

René Descartes (1596–1650) was the first mathematician extensively to use exponential notation as it is used today. However, for some unknown reason, he always used xx for x^2.

Repeated multiplication of the same factor can be written using an exponent.

$2 \cdot 2 \cdot 2 \cdot 2 \cdot 2 = 2^5$ ← exponent; 2 ← base

$a \cdot a \cdot a \cdot a = a^4$ ← exponent; a ← base

The **exponent** indicates how many times the factor, called the **base**, occurs in the multiplication. The multiplication $2 \cdot 2 \cdot 2 \cdot 2 \cdot 2$ is in **factored form**. The exponential expression 2^5 is in **exponential form**.

2^1 is read "the first power of 2" or just 2. Usually the exponent 1 is not written.

2^2 is read "the second power of 2" or "2 squared."

2^3 is read "the third power of 2" or "2 cubed."

2^4 is read "the fourth power of 2."

a^4 is read "the fourth power of a."

There is a geometric interpretation of the first three natural-number powers.

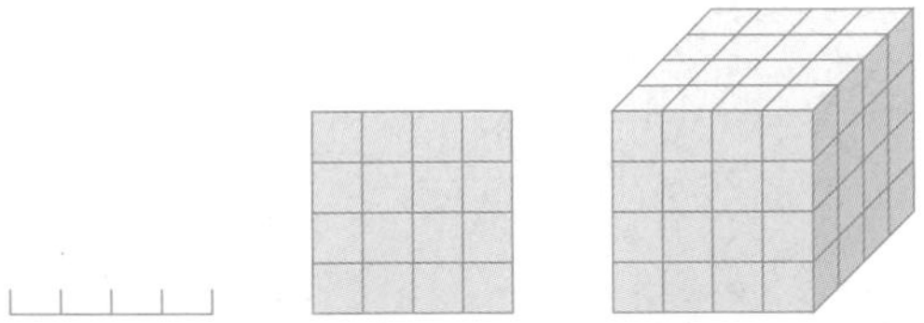

$4^1 = 4$ Length: 4 ft | $4^2 = 16$ Area: 16 ft^2 | $4^3 = 64$ Volume: 64 ft^3

In-Class Examples (Objective 1.2E)

1. Evaluate 4^3. 64
2. Evaluate $(-6)^2$. 36
3. Evaluate -6^2. -36
4. Evaluate $-3^2 \cdot 5^2$. -225

To evaluate an exponential expression, write each factor as many times as indicated by the exponent. Then multiply.

➡ Evaluate $(-2)^4$.

$(-2)^4 = (-2)(-2)(-2)(-2)$ • Write (-2) as a factor 4 times.
$= 16$ • Multiply.

➡ Evaluate -2^4.

$-2^4 = -(2 \cdot 2 \cdot 2 \cdot 2)$ • Write 2 as a factor 4 times.
$= -16$ • Multiply.

From these last two examples, note the difference between $(-2)^4$ and -2^4.

$$(-2)^4 = 16$$
$$-2^4 = -(2^4) = -16$$

Example 9 Evaluate -5^3.

Solution $-5^3 = -(5 \cdot 5 \cdot 5) = -125$

You Try It 9 Evaluate -6^3.

Your solution -216

Example 10 Evaluate $(-4)^4$.

Solution $(-4)^4 = (-4)(-4)(-4)(-4)$
$= 256$

You Try It 10 Evaluate $(-3)^4$.

Your solution 81

Example 11 Evaluate $(-3)^2 \cdot 2^3$.

Solution $(-3)^2 \cdot 2^3 = (-3)(-3) \cdot (2)(2)(2)$
$= 9 \cdot 8 = 72$

You Try It 11 Evaluate $(3^3)(-2)^3$.

Your solution -216

Example 12
Evaluate $\left(-\frac{2}{3}\right)^3$.

Solution $\left(-\frac{2}{3}\right)^3 = \left(-\frac{2}{3}\right)\left(-\frac{2}{3}\right)\left(-\frac{2}{3}\right)$
$= -\frac{2 \cdot 2 \cdot 2}{3 \cdot 3 \cdot 3} = -\frac{8}{27}$

You Try It 12
Evaluate $\left(-\frac{2}{5}\right)^2$.

Your solution $\frac{4}{25}$

Example 13 Evaluate $-4(0.7)^2$.

Solution $-4(0.7)^2 = -4(0.7)(0.7)$
$= -2.8(0.7) = -1.96$

You Try It 13 Evaluate $-3(0.3)^3$.

Your solution -0.081

Solutions on p. S2

Concept Check

Rewrite each expression as an exponential expression.

1. nine to the fifth power 9^5
2. y to the fourth power y^4
3. seven to the nth power 7^n
4. $b \cdot b \cdot b \cdot b \cdot b \cdot b \cdot b \cdot b$ b^8

Optional Student Activity

In which column is the number 1 million, column A, B, or C?

A	B	C
1	8	27
64	125	216
⋮	⋮	⋮

Column A; $1{,}000{,}000 = 100^3$

Objective 1.2F

New Vocabulary
square root
radical sign
principal square root
radicand
perfect square
irrational number
real number

Vocabulary to Review
rational number

New Symbols
$\sqrt{\ }$

New Properties
Product Property of Square Roots

Discuss the Concepts

1. Explain the difference between *square root* and *principal square root.*
2. Explain the difference between *simplifying* and *approximating* principal square roots. Then simplify $\sqrt{20}$ and approximate $\sqrt{20}$.
3. Is $\sqrt{-49}$ a real number? Explain.
4. You are to grade this solution to the problem "Write $\sqrt{48}$ in simplest form."
$\sqrt{48} = \sqrt{4 \cdot 12} = \sqrt{4}\sqrt{12} = 2\sqrt{12}$
Is the solution correct? If not, what error was made? What is the correct solution?

Objective F To simplify numerical radical expressions

VIDEO & DVD CD TUTOR WEB SSM

A **square root** of a positive number x is a number whose square is x.

A square root of 16 is 4 because $4^2 = 16$.
A square root of 16 is -4 because $(-4)^2 = 16$.

Square Roots of Perfect Squares

$\sqrt{1} = 1$
$\sqrt{4} = 2$
$\sqrt{9} = 3$
$\sqrt{16} = 4$
$\sqrt{25} = 5$
$\sqrt{36} = 6$
$\sqrt{49} = 7$
$\sqrt{64} = 8$
$\sqrt{81} = 9$
$\sqrt{100} = 10$
$\sqrt{121} = 11$
$\sqrt{144} = 12$

T

Every positive number has two square roots, one a positive and one a negative number. The symbol "$\sqrt{\ }$," called a **radical sign**, is used to indicate the positive or **principal square root** of a number. For example, $\sqrt{16} = 4$ and $\sqrt{25} = 5$. The number under the radical sign is called the **radicand**.

When the negative square root of a number is to be found, a negative sign is placed in front of the radical. For example, $-\sqrt{16} = -4$ and $-\sqrt{25} = -5$.

The square of an integer is a **perfect square**. 49, 81, and 144 are examples of perfect squares.

$7^2 = 49$
$9^2 = 81$
$12^2 = 144$

The principal square root of an integer that is a perfect square is a positive integer.

$\sqrt{49} = 7$
$\sqrt{81} = 9$
$\sqrt{144} = 12$

If a number is not a perfect square, its square root can only be approximated. For example, 2 and 7 are not perfect squares. The square roots of these numbers are **irrational numbers**. Their decimal representations never terminate or repeat.

$\sqrt{2} \approx 1.4142135\ldots \qquad \sqrt{7} \approx 2.6457513\ldots$

Recall that rational numbers are fractions such as $-\frac{6}{7}$ or $-\frac{10}{3}$, where the numerator and denominator are integers. Rational numbers are also represented by repeating decimals, such as $0.25767676\ldots$, and by terminating decimals, such as 1.73. An irrational number is neither a repeating nor a terminating decimal. For instance, $2.45445444544445\ldots$ is an irrational number.

> **Real Numbers**
> The rational numbers and the irrational numbers taken together are called the **real numbers**.

TAKE NOTE
Recall that a factor of a number divides the number evenly. For instance, 6 is a factor of 18. The perfect square 9 is also a factor of 18. 9 is a *perfect-square factor* of 18, whereas 6 is not a perfect-square factor of 18.

Radical expressions that contain radicands that are not perfect squares are frequently written in simplest form. A radical expression is in simplest form when the radicand contains no factor greater than 1 that is a perfect square. For instance, $\sqrt{50}$ is not in simplest form because 25 is a perfect square factor of 50. The radical expression $\sqrt{15}$ is in simplest form because there are no perfect-square factors of 15 that are greater than 1.

The Product Property of Square Roots and a knowledge of perfect squares are used to simplify radicands that are not perfect squares.

In-Class Examples (Objective 1.2F)
Simplify.

1. $\sqrt{675}$ $15\sqrt{3}$
2. $7\sqrt{48}$ $28\sqrt{3}$
3. $-3\sqrt{24}$ $-6\sqrt{6}$

The Product Property of Square Roots

If a and b are positive real numbers, then $\sqrt{ab} = \sqrt{a} \cdot \sqrt{b}$.

TAKE NOTE

From the example at the right, $\sqrt{72} = 6\sqrt{2}$. The two expressions are different representations of the same number. Using a calculator, we find that $\sqrt{72} \approx 8.485281$ and $6\sqrt{2} \approx 8.485281$.

➡ Simplify: $\sqrt{72}$

$\sqrt{72} = \sqrt{36 \cdot 2}$
- Write the radicand as the product of a perfect square and a factor that does not contain a perfect square.

$= \sqrt{36}\sqrt{2}$
- Use the Product Property of Square Roots to write the expression as a product.

$= 6\sqrt{2}$
- Simplify $\sqrt{36}$.

Note that 72 must be written as the product of a perfect square and *a factor that does not contain a perfect square*. Therefore, it would not be correct to rewrite $\sqrt{72}$ as $\sqrt{9 \cdot 8}$ and simplify the expression as shown at the right. Although 9 is a perfect-square factor of 72, 8 contains a perfect square ($8 = 4 \cdot 2$). Therefore, $\sqrt{8}$ is not in simplest form. Remember to find the *largest* perfect-square factor of the radicand.

$$\sqrt{72} = \sqrt{9 \cdot 8} = \sqrt{9}\sqrt{8} = 3\sqrt{8}$$

Not in simplest form

➡ Simplify: $\sqrt{-16}$

Because the square of any real number is positive, there is no real number whose square is -16. $\sqrt{-16}$ is not a real number.

Example 14 Simplify: $3\sqrt{90}$

Solution $3\sqrt{90} = 3\sqrt{9 \cdot 10}$
$= 3\sqrt{9}\sqrt{10}$
$= 3 \cdot 3\sqrt{10}$
$= 9\sqrt{10}$

You Try It 14 Simplify: $-5\sqrt{32}$

Your solution $-20\sqrt{2}$

Example 15 Simplify: $\sqrt{252}$

Solution $\sqrt{252} = \sqrt{36 \cdot 7}$
$= \sqrt{36}\sqrt{7}$
$= 6\sqrt{7}$

You Try It 15 Simplify: $\sqrt{216}$

Your solution $6\sqrt{6}$

Solutions on p. S2

Instructor Note

Have students who are having difficulty finding a perfect-square factor of a radicand write the prime factorization of the number. For example,

$\sqrt{288} = \sqrt{2^5 3^2} = \sqrt{2^4 3^2 \cdot 2} = \sqrt{2^4 3^2}\sqrt{2} = 2^2 \cdot 3\sqrt{2} = 12\sqrt{2}$

Optional Student Activity

1. The distance from home plate to second base on a major league baseball field is $\sqrt{16{,}200}$ ft. Find the distance from home plate to second base to the nearest hundredth of a foot. 127.28 ft
2. Find the two-digit perfect square that has exactly nine factors. 36
3. Find two whole numbers such that their difference is 10, the smaller number is a perfect square, and the larger number is two less than a perfect square. 4, 14
4. Simplify.
   a. $\sqrt{\sqrt{16}}$ 2
   b. $\sqrt{\sqrt{81}}$ 3
5. Approximate to the nearest ten-thousandth.
   a. $\sqrt{\sqrt{17}}$ 2.0305
   b. $\sqrt{\sqrt{77}}$ 2.9623

Objective 1.2G

Concept Check
The table below shows a person's investments. What percent of the person's investments are in certificates of deposit (CDs)? Round to the nearest tenth of a percent.

Account	*Investment*
Stocks	$53,096
Bonds	$10,425
CDs	$48,231

43.2%

Optional Student Activity
Consult a map of the United States.

1. What percent of the states share parts of their borders with Mexico? 8%
2. What percent of the states share parts of their borders with Canada? 22%
3. What percent of the states have names that begin with the letter A? 8%
4. What percent of the states have names that begin with the letter M? 16%

Objective G To solve application problems

VIDEO & DVD CD TUTOR WEB SSM

One of the applications of percent is to express a portion of a total as a percent. For instance, a recent survey of 450 mall shoppers found that 270 preferred the mall closest to their home even though it did not have as much store variety as a mall farther from home. The percent of shoppers who preferred the mall closest to home can be found by converting a fraction to a percent.

$$\frac{\text{Portion preferring mall closest to home}}{\text{Total number surveyed}} = \frac{270}{450}$$
$$= 0.60 = 60\%$$

The Congressional Budget Office projected that the total surpluses for 2001 through 2011 would be $5.6 trillion. The number 5.6 trillion means

$$5.6 \times \underbrace{1{,}000{,}000{,}000{,}000}_{\text{1 trillion}} = 5{,}600{,}000{,}000{,}000$$

Numbers such as 5.6 trillion are used in many instances because they are easy to read and offer an approximation of the actual number.

The table below shows the net incomes, in millions of dollars, for the first quarter of 2001 and the first quarter of 2000 for two U.S. companies. Profits are shown as positive numbers; losses are shown as negative numbers. One quarter of a year is three months. Use this table for Example 16 and You Try It 16.

Company	1st Quarter 2001 Net Income	1st Quarter 2000 Net Income
Cisco Systems, Inc.	−2,693	641
Friendly Ice Cream	−3.203	−18.510

Source: **www.wsj.com**

Example 16
If earnings continued throughout the year at the same level, what would be the 2001 annual net income for Cisco Systems?

Strategy
To find the annual net income, multiply the net income in the first quarter of 2001 (−2,693) by the number of quarters in one year (4).

Solution
$-2{,}693(4) = -10{,}772$

The annual net income for Cisco Systems in 2001 would be −$10,772 million.

You Try It 16
For the first quarter of 2000, what was the average monthly net income for Friendly Ice Cream?

Your strategy

Your solution
−$6.17 million

Solution on p. S2

In-Class Examples (Objective 1.2G)
The circle graph at the right, based on the 2000 census, shows the population of the United States, in millions of residents, by region. What percent of the population of the United States in the year 2000 were residents of the Midwest? Round to the nearest tenth of a percent. 22.9%

U.S. Population in 2000
(millions of residents) by Region

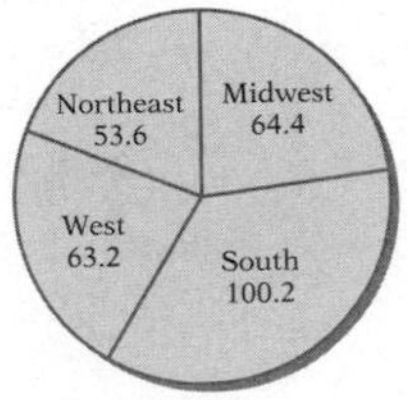

1.2 Exercises

Objective A

Write as a decimal. Place a bar over the repeating digits of a repeating decimal.

1. $\frac{1}{8}$ 0.125
2. $\frac{7}{8}$ 0.875
3. $\frac{2}{9}$ $0.\overline{2}$
4. $\frac{8}{9}$ $0.\overline{8}$
5. $\frac{1}{6}$ $0.1\overline{6}$
6. $\frac{5}{6}$ $0.8\overline{3}$
7. $\frac{9}{16}$ 0.5625
8. $\frac{15}{16}$ 0.9375
9. $\frac{7}{12}$ $0.58\overline{3}$
10. $\frac{11}{12}$ $0.91\overline{6}$
11. $\frac{6}{25}$ 0.24
12. $\frac{14}{25}$ 0.56
13. $\frac{9}{40}$ 0.225
14. $\frac{21}{40}$ 0.525
15. $\frac{5}{11}$ $0.\overline{45}$

Objective B

16. a. Explain how to convert a fraction to a percent.
    b. Explain how to convert a percent to a fraction.
    c. Explain how to convert a decimal to a percent.
    d. Explain how to convert a percent to a decimal.

17. Explain why multiplying a number by 100% does not change the value of the number.

Write as a fraction and as a decimal.

18. 75% $\frac{3}{4}$, 0.75
19. 40% $\frac{2}{5}$, 0.40
20. 64% $\frac{16}{25}$, 0.64
21. 88% $\frac{22}{25}$, 0.88
22. 125% $\frac{5}{4}$, 1.25
23. 160% $\frac{8}{5}$, 1.6
24. 19% $\frac{19}{100}$, 0.19
25. 87% $\frac{87}{100}$, 0.87
26. 5% $\frac{1}{20}$, 0.05
27. 450% $\frac{9}{2}$, 4.50

Write as a fraction.

28. $11\frac{1}{9}\%$ $\frac{1}{9}$
29. $4\frac{2}{7}\%$ $\frac{3}{70}$
30. $12\frac{1}{2}\%$ $\frac{1}{8}$
31. $37\frac{1}{2}\%$ $\frac{3}{8}$
32. $66\frac{2}{3}\%$ $\frac{2}{3}$
33. $\frac{1}{4}\%$ $\frac{1}{400}$
34. $\frac{1}{2}\%$ $\frac{1}{200}$
35. $6\frac{1}{4}\%$ $\frac{1}{16}$
36. $83\frac{1}{3}\%$ $\frac{5}{6}$
37. $5\frac{3}{4}\%$ $\frac{23}{400}$

Section 1.2

Suggested Assignment

Exercises 1–15, odds
Exercises 19–191, odds
More challenging problems:
Exercises 193, 194

Answers to Writing Exercises

16a. To convert a fraction to a percent, multiply the fraction by 100%.
b. To convert a percent to a fraction, remove the percent sign and multiply by $\frac{1}{100}$.
c. To convert a decimal to a percent, multiply the decimal by 100%.
d. To convert a percent to a decimal, remove the percent sign and multiply by 0.01.

17. Since 100% = 100 × 0.01 = 1, multiplying a number by 100% is the same as multiplying the number by 1. Multiplying a number by 1 does not change the value of the number.

Quick Quiz (Objective 1.2A)

1. Write $\frac{5}{16}$ as a decimal. 0.3125

2. Write $\frac{8}{11}$ as a decimal. $0.\overline{72}$

Quick Quiz (Objective 1.2B)

1. Write 80% as a fraction and as a decimal. $\frac{4}{5}$; 0.8

2. Write 0.9 and $\frac{7}{8}$ as percents. 90%; 87.5%

Write as a decimal.

38. 7.3%
0.073

39. 9.1%
0.091

40. 15.8%
0.158

41. 16.7%
0.167

42. 0.3%
0.003

43. 0.9%
0.009

44. 9.9%
0.099

45. 9.15%
0.0915

46. 121.2%
1.212

47. 18.23%
0.1823

Write as a percent.

48. 0.15
15%

49. 0.37
37%

50. 0.05
5%

51. 0.02
2%

52. 0.175
17.5%

53. 0.125
12.5%

54. 1.15
115%

55. 1.36
136%

56. 0.008
0.8%

57. 0.004
0.4%

58. $\frac{27}{50}$
54%

59. $\frac{83}{100}$
83%

60. $\frac{1}{3}$
$33\frac{1}{3}\%$

61. $\frac{3}{8}$
$37\frac{1}{2}\%$

62. $\frac{5}{11}$
$45\frac{5}{11}\%$

63. $\frac{4}{9}$
$44\frac{4}{9}\%$

64. $\frac{7}{8}$
$87\frac{1}{2}\%$

65. $\frac{9}{20}$
45%

66. $1\frac{2}{3}$
$166\frac{2}{3}\%$

67. $2\frac{1}{2}$
250%

Objective C

Add or subtract.

68. $-\frac{5}{6}-\frac{5}{9}$
$-\frac{25}{18}$

69. $-\frac{6}{13}+\frac{17}{26}$
$\frac{5}{26}$

70. $-\frac{7}{12}+\frac{5}{8}$
$\frac{1}{24}$

71. $\frac{5}{8}-\left(-\frac{3}{4}\right)$
$\frac{11}{8}$

72. $\frac{3}{5}-\frac{11}{12}$
$-\frac{19}{60}$

73. $\frac{11}{12}-\frac{5}{6}$
$\frac{1}{12}$

74. $-\frac{2}{3}-\left(-\frac{11}{18}\right)$
$-\frac{1}{18}$

75. $-\frac{5}{8}-\left(-\frac{11}{12}\right)$
$\frac{7}{24}$

76. $\frac{1}{3}+\frac{5}{6}-\frac{2}{9}$
$\frac{17}{18}$

77. $\frac{1}{2}-\frac{2}{3}+\frac{1}{6}$
0

78. $-\frac{5}{16}+\frac{3}{4}-\frac{7}{8}$
$-\frac{7}{16}$

79. $\frac{1}{2}-\frac{3}{8}-\left(-\frac{1}{4}\right)$
$\frac{3}{8}$

Quick Quiz (Objective 1.2C)

Simplify.

1. $-\frac{3}{8}-\frac{1}{2}-\frac{7}{8}$

2. $5.63-(-2.1)$ 7.73

3. $-\frac{5}{6}-\left(-\frac{2}{9}\right)$ $-\frac{11}{18}$

29

80. $\frac{3}{4}-\left(-\frac{7}{12}\right)-\frac{7}{8}$
$\frac{11}{24}$

81. $\frac{1}{3}-\frac{1}{4}-\frac{1}{5}$
$-\frac{7}{60}$

82. $\frac{2}{3}-\frac{1}{2}+\frac{5}{6}$
1

83. $\frac{5}{16}+\frac{1}{8}-\frac{1}{2}$
$-\frac{1}{16}$

84. $-13.092 + 6.9$
-6.192

85. $2.54 - 3.6$
-1.06

86. $5.43 + 7.925$
13.355

87. $-16.92 - 6.925$
-23.845

88. $-3.87 + 8.546$
4.676

89. $6.9027 - 17.692$
-10.7893

90. $2.09 - 6.72 - 5.4$
-10.03

91. $-18.39 + 4.9 - 23.7$
-37.19

92. $19 - (-3.72) - 82.75$
-60.03

93. $-3.07 - (-2.97) - 17.4$
-17.5

94. $16.4 - (-3.09) - 7.93$
11.56

95. $-3.09 - 4.6 - (-27.3)$
19.61

96. $2.66 - (-4.66) - 8.2$
-0.88

Objective D

Multiply or divide.

97. $\frac{1}{2}\left(-\frac{3}{4}\right)$
$-\frac{3}{8}$

98. $-\frac{2}{9}\left(-\frac{3}{14}\right)$
$\frac{1}{21}$

99. $\left(-\frac{3}{8}\right)\left(-\frac{4}{15}\right)$
$\frac{1}{10}$

100. $\left(-\frac{3}{4}\right)\left(-\frac{8}{27}\right)$
$\frac{2}{9}$

101. $-\frac{1}{2}\left(\frac{8}{9}\right)$
$-\frac{4}{9}$

102. $\frac{5}{12}\left(-\frac{8}{15}\right)$
$-\frac{2}{9}$

103. $\frac{5}{8}\left(-\frac{7}{12}\right)\frac{16}{25}$
$-\frac{7}{30}$

104. $\left(\frac{5}{12}\right)\left(-\frac{8}{15}\right)\left(-\frac{1}{3}\right)$
$\frac{2}{27}$

105. $\frac{1}{2}\left(-\frac{3}{4}\right)\left(-\frac{5}{8}\right)$
$\frac{15}{64}$

106. $\frac{3}{8} \div \frac{1}{4}$
$\frac{3}{2}$

107. $\frac{5}{6} \div \left(-\frac{3}{4}\right)$
$-\frac{10}{9}$

108. $-\frac{5}{12} \div \frac{15}{32}$
$-\frac{8}{9}$

109. $-\frac{7}{8} \div \frac{4}{21}$
$-\frac{147}{32}$

110. $\frac{7}{10} \div \frac{2}{5}$
$\frac{7}{4}$

111. $-\frac{15}{64} \div \left(-\frac{3}{40}\right)$
$\frac{25}{8}$

Quick Quiz (Objective 1.2D)

Simplify.

1. $-\frac{7}{12}\left(-\frac{4}{9}\right)\frac{7}{27}$

2. $-9.3(12.7) - 118.11$

3. $15.33 \div (-7) - 2.19$

4. $-\frac{2}{3} \div \left(-\frac{5}{6}\right)\frac{4}{5}$

30

112. $\frac{1}{8} \div \left(-\frac{5}{12}\right)$ $-\frac{3}{10}$

113. $-\frac{4}{9} \div \left(-\frac{2}{3}\right)$ $\frac{2}{3}$

114. $-\frac{6}{11} \div \frac{4}{9}$ $-\frac{27}{22}$

115. $1.2(3.47)$ 4.164

116. $(-0.8)6.2$ -4.96

117. $(-1.89)(-2.3)$ 4.347

118. $(6.9)(-4.2)$ -28.98

119. $1.06(-3.8)$ -4.028

120. $-2.7(-3.5)$ 9.45

121. $1.2(-0.5)(3.7)$ -2.22

122. $-2.4(6.1)(0.9)$ -13.176

123. $2.3(-0.6)(0.8)$ -1.104

Divide. Round to the nearest hundredth.

124. $-1.27 \div (-1.7)$ 0.75

125. $9.07 \div (-3.5)$ -2.59

126. $0.0976 \div 0.042$ 2.32

127. $-6.904 \div 1.35$ -5.11

128. $-7.894 \div (-2.06)$ 3.83

129. $-354.2086 \div 0.1719$ -2060.55

Objective E

Evaluate.

130. 6^2 36

131. 7^4 2401

132. -7^2 -49

133. -4^3 -64

134. $(-3)^2$ 9

135. $(-2)^3$ -8

136. $(-3)^4$ 81

137. $(-5)^3$ -125

138. $\left(\frac{1}{2}\right)^2$ $\frac{1}{4}$

139. $\left(-\frac{3}{4}\right)^3$ $-\frac{27}{64}$

140. $(0.3)^2$ 0.09

141. $(1.5)^3$ 3.375

142. $\left(\frac{2}{3}\right)^2 \cdot 3^3$ 12

143. $\left(-\frac{1}{2}\right)^3 \cdot 8$ -1

144. $(0.3)^3 \cdot 2^3$ 0.216

145. $(-2) \cdot (-2)^2$ -8

146. $2^3 \cdot 3^3 \cdot (-4)$ -864

147. $(-3)^3 \cdot 5^2 \cdot 10$ -6750

148. $(-7) \cdot 4^2 \cdot 3^2$ -1008

149. $(-2) \cdot 2^3 \cdot (-3)^2$ -144

150. $\left(\frac{2}{3}\right)^2 \cdot \frac{1}{4} \cdot 3^3$ 3

151. $\left(\frac{3}{4}\right)^2 \cdot (-4) \cdot 2^3$ -18

152. $8^2 \cdot (-3)^5 \cdot 5$ $-77{,}760$

Quick Quiz (Objective 1.2E)

1. Evaluate 2^7. 128
2. Evaluate $(-5)^2$. 25
3. Evaluate -5^2. -25
4. Evaluate $-4^2 \cdot 3^3$. -432

Objective F

Simplify.

153. $\sqrt{16}$ 4
154. $\sqrt{64}$ 8
155. $\sqrt{49}$ 7
156. $\sqrt{144}$ 12
157. $\sqrt{32}$ $4\sqrt{2}$
158. $\sqrt{50}$ $5\sqrt{2}$

159. $\sqrt{8}$ $2\sqrt{2}$
160. $\sqrt{12}$ $2\sqrt{3}$
161. $6\sqrt{18}$ $18\sqrt{2}$
162. $-3\sqrt{48}$ $-12\sqrt{3}$
163. $5\sqrt{40}$ $10\sqrt{10}$
164. $2\sqrt{28}$ $4\sqrt{7}$

165. $\sqrt{15}$ $\sqrt{15}$
166. $\sqrt{21}$ $\sqrt{21}$
167. $\sqrt{29}$ $\sqrt{29}$
168. $\sqrt{13}$ $\sqrt{13}$
169. $-9\sqrt{72}$ $-54\sqrt{2}$
170. $11\sqrt{80}$ $44\sqrt{5}$

171. $\sqrt{45}$ $3\sqrt{5}$
172. $\sqrt{225}$ 15
173. $\sqrt{0}$ 0
174. $\sqrt{210}$ $\sqrt{210}$
175. $6\sqrt{128}$ $48\sqrt{2}$
176. $9\sqrt{288}$ $108\sqrt{2}$

Find the decimal approximation rounded to the nearest thousandth.

177. $\sqrt{240}$ 15.492
178. $\sqrt{300}$ 17.321
179. $\sqrt{288}$ 16.971
180. $\sqrt{600}$ 24.495
181. $\sqrt{256}$ 16
182. $\sqrt{324}$ 18

183. $\sqrt{275}$ 16.583
184. $\sqrt{450}$ 21.213
185. $\sqrt{245}$ 15.652
186. $\sqrt{525}$ 22.913
187. $\sqrt{352}$ 18.762
188. $\sqrt{363}$ 19.053

Objective G *Application Problems*

189. The table below shows the net incomes for the first quarter of 2001 and the first quarter of 2000 for three companies in the entertainment industry (Source: **www.wsj.com**). Figures are in millions of dollars. Profits are shown as positive numbers; losses are shown as negative numbers.

a. If earnings were to continue throughout the year at the same level, what would the 2001 annual net income be for Midway Games? −\$103.408 million

b. For the first quarter of 2001, what was the average monthly net income for Six Flags? −\$43.584 million

c. Find the difference between Fox Entertainment's first-quarter net income for 2000 and Midway Games's first-quarter net income for 2000. \$30.481 million

Company	1st Quarter 2001 Net Income	1st Quarter 2000 Net Income
Fox Entertainment	−9.0	19.0
Midway Games, Inc.	−25.852	−11.481
Six Flags, Inc.	−130.752	−113.892

Quick Quiz (Objective 1.2F)

Simplify.

1. $\sqrt{36}$ 6
2. $\sqrt{18}$ $3\sqrt{2}$
3. $\sqrt{11}$ $\sqrt{11}$
4. $-6\sqrt{75}$ $-30\sqrt{3}$

Answers to Writing Exercises

195. Students should explain that to simplify a radical expression, we first write the radicand as a product of perfect-square factors and factors that do not contain a perfect square. Take the square root of the perfect-square factors; write the square root in front of the radical sign. The other factors remain under the radical sign.

190. At the close of the stock markets on May 9, 2001, the indexes were posted as shown below, along with the decreases for that day, shown as negative numbers (Source: **www.wsj.com**). At what level were the indexes at the close of the day on May 8, 2001?

Index	Points at Market Close	Decrease for the Day
Dow Jones Industrial Average	10866.98	−16.53
Standard and Poor 500	1255.54	−5.66
NASDAQ	2156.62	−42.15

DJIA: 10883.51; S&P: 1261.20; NASDAQ: 2198.77

191. A survey asked 600 small business owners about the impact of the Internet on their businesses. The results are shown in the graph at the right. (Source: Infoworld by Dun & Bradstreet. Copyright © 2000 by Infoworld Media Group, Inc. Reproduced with permission of Infoworld Media Group, Inc. via Copyright Clearance Center.) What percent of the owners responded that the Internet had helped business? 31%

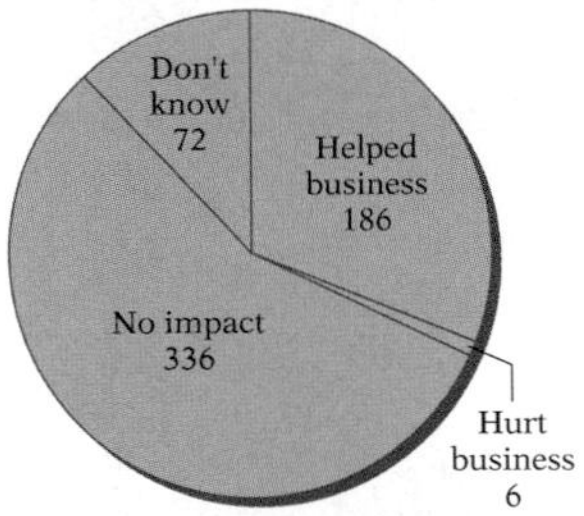

192. The table at the right shows the surplus or deficit, in billions of dollars, for the federal budget every fifth year from 1945 to 1995 and every year from 1995 to 2000 (Source: U.S. Office of Management and Budget). The negative sign (−) indicates a deficit.

a. Find the difference between the deficits in the years 1980 and 1985. $138.499 billion

b. Calculate the difference between the surplus in 1960 and the deficit in 1955. $3.294 billion

c. How many times greater was the deficit in 1985 than in 1975? Round to the nearest whole number. 4 times greater

d. What was the average deficit, in millions of dollars, per quarter for the year 1970? −$710.5 million

e. Find the average surplus or deficit for the years 1995 through 2000. Round to the nearest million. −$4,579 million

Year	Federal Budget Surplus or Deficit
1945	−47.553
1950	−3.119
1955	−2.993
1960	0.301
1965	−1.411
1970	−2.842
1975	−53.242
1980	−73.835
1985	−212.334
1990	−221.194
1995	−163.899
1996	−107.450
1997	−21.940
1998	69.246
1999	79.263
2000	117.305

APPLYING THE CONCEPTS

193. List the whole numbers between $\sqrt{8}$ and $\sqrt{90}$.
3, 4, 5, 6, 7, 8, 9

194. Use a calculator to determine the decimal representations of $\frac{17}{99}$, $\frac{45}{99}$, and $\frac{73}{99}$. Make a conjecture as to the decimal representation of $\frac{83}{99}$. Does your conjecture work for $\frac{33}{99}$? What about $\frac{1}{99}$?

195. Describe in your own words how to simplify a radical expression.

Quick Quiz (Objective 1.2G)

According to the U.S. Census Bureau, the resident population of the United States on April 1, 2000, was 281,421,906. The Census Bureau reported that California had the highest population of all the states, with 33,871,648 residents. What percent of the U.S. population were residents of California on April 1, 2000? Round to the nearest tenth of a percent. 12.0%

1.3 The Order of Operations Agreement

Objective A

To use the Order of Operations Agreement to simplify expressions

VIDEO & DVD CD TUTOR WEB SSM

Let's evaluate $2 + 3 \cdot 5$.

There are two arithmetic operations, addition and multiplication, in this expression. The operations could be performed in different orders.

Multiply first.	$2 + \underbrace{3 \cdot 5}$	Add first.	$\underbrace{2 + 3} \cdot 5$
Then add.	$\underbrace{2 + 15}$	Then multiply.	$\underbrace{5 \cdot 5}$
	17		25

In order to prevent there being more than one answer for a numerical expression, an Order of Operations Agreement has been established.

T

THE ORDER OF OPERATIONS AGREEMENT

Step 1 Perform operations inside grouping symbols. Grouping symbols include parentheses ( ), brackets [ ], braces { }, absolute value symbols | |, and the fraction bar.

Step 2 Simplify exponential expressions.

Step 3 Do multiplication and division as they occur from left to right.

Step 4 Do addition and subtraction as they occur from left to right.

➡ Evaluate $12 - 24(8 - 5) \div 2^2$.

$$12 - 24(8 - 5) \div 2^2 = 12 - 24(3) \div 2^2$$
• Perform operations inside grouping symbols.

$$= 12 - 24(3) \div 4$$
• Simplify exponential expressions.

$$= 12 - 72 \div 4$$
• Do multiplication and division as they occur from left to right.

$$= 12 - 18$$

$$= -6$$
• Do addition and subtraction as they occur from left to right.

One or more of the above steps may not be needed to evaluate an expression. In that case, proceed to the next step in the Order of Operations Agreement.

In-Class Examples (Objective 1.3A)

Simplify.

1. $(-4)^2 - 8 + 9(-3)$ -19
2. $7(3 - 6) - (-1)$ -20
3. $(-6)^2 \div (6 - 4)^2 - (-12)4$ 57

Objective 1.3A

New Vocabulary

Order of Operations Agreement
grouping symbols

Instructor Note

You might tell students that, when simplifying an expression inside grouping symbols, we use the Order of Operations Agreement.

Discuss the Concepts

Without using a calculator, determine which of the following expressions are equal. Explain how you arrived at your answer.

a. $587 - 98 \cdot 417$
b. $(587 - 98) \cdot 417$
c. $587 - (98 \cdot 417)$

a and c are equal.

Concept Check

Arrange the expressions in order from the least value to the greatest value.

a. $3[(12 \div 4) - (-2)] + 5$

b. $30 \div 5\left(\frac{16 + 4}{-3^2 + 4}\right) - 2$

c. $\frac{2}{3} - \frac{\frac{13}{24}}{3 - \frac{1}{3}} \div \frac{1}{8}$

d. $0.4(1.5 - 2.1)^2 + 6.3$

b (-26), c $\left(-\frac{23}{24}\right)$, d (6.444), a (20)

Optional Student Activity

1. What is the smallest multiple of 3 greater than $32 \div (1 - 9) + (-3)^2$? 6
2. What is the smallest integer greater than $-2^2 - (-3)^2 + 5(4) \div 10 - (-6)$? -4
3. What is the smallest natural number greater than 2 that divides evenly into the sum of 3^9 and 5^{11}? 4
4. Simplify: $2\left(1 - \frac{1}{2}\right) + 3\left(1 - \frac{1}{3}\right) + 4\left(1 - \frac{1}{4}\right) + \ldots + 10\left(1 - \frac{1}{10}\right)$ 45

When an expression has grouping symbols inside grouping symbols, perform the operations inside the inner grouping symbols first.

➡ Evaluate $6 \div [4 - (6 - 8)] + 2^2$.

$6 \div [4 - (6 - 8)] + 2^2 = 6 \div [4 - (-2)] + 2^2$ • Perform operations inside grouping symbols.

$= 6 \div 6 + 2^2$

$= 6 \div 6 + 4$ • Simplify exponential expressions.

$= 1 + 4$ • Do multiplication and division as they occur from left to right.

$= 5$ • Do addition and subtraction as they occur from left to right.

Example 1

Evaluate $4 - 3[4 - 2(6 - 3)] \div 2$.

Solution

$4 - 3[4 - 2(6 - 3)] \div 2$
$= 4 - 3[4 - 2 \cdot 3] \div 2$
$= 4 - 3[4 - 6] \div 2$
$= 4 - 3[-2] \div 2$
$= 4 + 6 \div 2$
$= 4 + 3$
$= 7$

You Try It 1

Evaluate $18 - 5[8 - 2(2 - 5)] \div 10$.

Your solution

11

Example 2

Evaluate $27 \div (5 - 2)^2 + (-3)^2 \cdot 4$.

Solution

$27 \div (5 - 2)^2 + (-3)^2 \cdot 4$
$= 27 \div 3^2 + (-3)^2 \cdot 4$
$= 27 \div 9 + 9 \cdot 4$
$= 3 + 9 \cdot 4$
$= 3 + 36$
$= 39$

You Try It 2

Evaluate $36 \div (8 - 5)^2 - (-3)^2 \cdot 2$.

Your solution

-14

Example 3

Evaluate $(1.75 - 1.3)^2 \div 0.025 + 6.1$.

Solution

$(1.75 - 1.3)^2 \div 0.025 + 6.1$
$= (0.45)^2 \div 0.025 + 6.1$
$= 0.2025 \div 0.025 + 6.1$
$= 8.1 + 6.1$
$= 14.2$

You Try It 3

Evaluate $(6.97 - 4.72)^2 \cdot 4.5 \div 0.05$.

Your solution

455.625

Solutions on p. S2

35

1.3 Exercises

Objective A

1. Why do we need an Order of Operations Agreement?

2. Describe each step in the Order of Operations Agreement.

Evaluate by using the Order of Operations Agreement.

3. $4 - 8 \div 2$
   0

4. $2^2 \cdot 3 - 3$
   9

5. $2(3 - 4) - (-3)^2$
   −11

6. $16 - 32 \div 2^3$
   12

7. $24 - 18 \div 3 + 2$
   20

8. $8 - (-3)^2 - (-2)$
   1

9. $8 - 2(3)^2$
   −10

10. $16 - 16 \cdot 2 \div 4$
    8

11. $12 + 16 \div 4 \cdot 2$
    20

12. $16 - 2 \cdot 4^2$
    −16

13. $27 - 18 \div (-3^2)$
    29

14. $4 + 12 \div 3 \cdot 2$
    12

15. $16 + 15 \div (-5) - 2$
    11

16. $14 - 2^2 - (4 - 7)$
    13

17. $14 - 2^2 - |4 - 7|$
    7

18. $10 - |5 - 8| + 2^3$
    15

19. $3 - 2[8 - (3 - 2)]$
    −11

20. $-2^2 + 4[16 \div (3 - 5)]$
    −36

21. $6 + \dfrac{16 - 4}{2^2 + 2} - 2$
    6

22. $24 \div \dfrac{3^2}{8 - 5} - (-5)$
    13

23. $18 \div |9 - 2^3| + (-3)$
    15

24. $96 \div 2[12 + (6 - 2)] - 3^2$
    759

25. $4[16 - (7 - 1)] \div 10$
    4

26. $18 \div 2 - 4^2 - (-3)^2$
    −16

27. $20 \div (10 - 2^3) + (-5)$
    5

28. $16 - 3(8 - 3)^2 \div 5$
    1

29. $4(-8) \div [2(7 - 3)^2]$
    −1

Section 1.3

Suggested Assignment

Exercises 3–35, odds
More challenging problems:
Exercises 36–40

Answers to Writing Exercises

1. We need an Order of Operations Agreement to ensure that there is only one way in which an expression can be correctly simplified.
2. Students should describe the steps in the Order of Operations Agreement: Step 1: Perform operations inside grouping symbols. Step 2: Simplify exponential expressions. Step 3: Do multiplication and division as they occur from left to right. Step 4: Do addition and subtraction as they occur from left to right.

Quick Quiz (Objective 1.3A)

Simplify.

1. $-5(1 - 4) \div (-15)$ −1
2. $3 - (-5)^2 - 6 + 2(-4)$ −36
3. $24 \div (5 - 7) + (-2)^2(-3)$ −24

30. $\dfrac{(-10)+(-2)}{6^2-30} \div |2-4|$
−1

31. $16 - 4 \cdot \dfrac{3^3-7}{2^3+2} - (-2)^2$
4

32. $(0.2)^2 \cdot (-0.5) + 1.72$
1.70

33. $0.3(1.7 - 4.8) + (1.2)^2$
0.51

34. $(1.8)^2 - 2.52 \div 1.8$
1.84

35. $(1.65 - 1.05)^2 \div 0.4 + 0.8$
1.7

APPLYING THE CONCEPTS

36. Find two fractions between $\frac{2}{3}$ and $\frac{3}{4}$. (There is more than one answer to this question.)
Answers will vary. For example, $\frac{17}{24}$ and $\frac{33}{48}$.

37. A **magic square** is one in which the numbers in every row, column, and diagonal sum to the same number. Complete the magic square at the right.

T

$\frac{2}{3}$	$-\frac{1}{6}$	0
$-\frac{1}{2}$	$\frac{1}{6}$	$\frac{5}{6}$
$\frac{1}{3}$	$\frac{1}{2}$	$-\frac{1}{3}$

38. For each part below, find a rational number r that satisfies the condition.
a. $r^2 < r$ **b.** $r^2 = r$ **c.** $r^2 > r$
Answers will vary. For example, a. $\frac{1}{2}$, b. 1, c. 2

39. In a survey of consumers, 9% said they would buy an electric vehicle. Approximately 43% said they would be willing to pay between \$1000 and \$2000 more for a new car if the car had an EPA rating of 80 mpg. If your car now gets 28 mpg and you drive approximately 10,000 mi per year, in how many months would your savings on gasoline pay for the increased cost of such a car? Assume the average cost for gasoline is \$1.60 per gallon.
32 to 65 months

40. Find three different natural numbers a, b, and c such that $\frac{1}{a} + \frac{1}{b} + \frac{1}{c}$ is a natural number.
$a = 2$, $b = 3$, $c = 6$

41. The following was offered as the simplification of $6 + 2(4 - 9)$.

$$\begin{aligned} 6 + 2(4-9) &= 6 + 2(-5) \\ &= 8(-5) \\ &= -40 \end{aligned}$$

Is this a correct simplification? Explain your answer.

42. The following was offered as the simplification of $2 \cdot 3^3$.

$$2 \cdot 3^3 = 6^3 = 216$$

Is this is a correct simplification? Explain your answer.

Answers to Writing Exercises

41. The Order of Operations Agreement was not followed in the given simplification because the addition 6 + 2 was performed before the multiplication 2(−5). The expression should be simplified as follows:
$6 + 2(4 - 9) =$
$6 + 2(-5) =$
$6 + (-10) = -4$

42. The Order of Operations Agreement was not followed in the given simplification because the multiplication $2 \cdot 3$ was performed before the exponential expression 3^3 was simplified. The expression should be simplified as follows:
$2 \cdot 3^3 = 2 \cdot 27 = 54$

1.4 Variable Expressions

Objective A To evaluate a variable expression

VIDEO & DVD CD TUTOR WEB SSM

Often we discuss a quantity without knowing its exact value—for example, the price of gold next month, the cost of a new automobile next year, or the tuition cost for next semester. Recall that a letter of the alphabet can be used to stand for a quantity that is unknown or that can change, or *vary*. Such a letter is called a variable. An expression that contains one or more variables is called a **variable expression**.

Point of Interest

Historical manuscripts indicate that mathematics is at least 4000 years old. Yet it was only 400 years ago that mathematicians started using variables to stand for numbers. The idea that a letter can stand for some number was a critical turning point in mathematics.

A variable expression is shown at the right. The expression can be rewritten by writing subtraction as the addition of the opposite.

$3x^2 - 5y + 2xy - x - 7$

$3x^2 + (-5y) + 2xy + (-x) + (-7)$

Note that the expression has 5 addends. The **terms** of a variable expression are the addends of the expression. The expression has 5 terms.

5 terms

$3x^2 \; - \; 5y \; + \; 2xy \; - \; x \; - \; 7$

variable terms: $3x^2 - 5y + 2xy - x$; constant term: -7

The terms $3x^2$, $-5y$, $2xy$, and $-x$ are **variable terms.**

The term -7 is a **constant term**, or simply a **constant.**

Each variable term is composed of a **numerical coefficient** and a **variable part** (the variable or variables and their exponents).

numerical coefficient

$3x^2 - 5y + 2xy - 1x - 7$

variable part

When the numerical coefficient is 1 or -1, the 1 is usually not written ($x = 1x$ and $-x = -1x$).

Replacing each variable by its value and then simplifying the resulting numerical expression is called **evaluating the variable expression**.

CALCULATOR NOTE

See the appendix "Guidelines for Using Graphing Calculators" for instructions on using a graphing calculator to evaluate variable expressions.

⇒ Evaluate $ab - b^2$ when $a = 2$ and $b = -3$.

$ab - b^2$

$2(-3) - (-3)^2$ • Replace each variable in the expression by its value.

$= 2(-3) - 9$ • Use the Order of Operations Agreement to simplify the resulting numerical expression.

$= -6 - 9$

$= -15$

Objective 1.4A

Vocabulary to Review

variable

New Vocabulary

variable expression
terms
variable terms
constant term
numerical coefficient
variable part
evaluate a variable expression

Instructor Note

The calculator logo indicates instruction on the use of a graphing calculator or computer graphing utility, or its use in solving problems or illustrating a point. Coverage of this material is optional.

Discuss the Concepts

1. What is the goal in evaluating a variable expression?
2. What are the steps involved in evaluating a variable expression?
   (1) Write the variable expression.
   (2) Substitute the given values for the variables.
   (3) Simplify the resulting numerical expression, using the Order of Operations Agreement, so that the final answer is one number.

Concept Check

1. Find the smallest possible value for the expression $x + \frac{3}{x}$ when x is a positive integer. $3\frac{1}{2}$

(Continued on next page)

In-Class Examples (Objective 1.4A)

1. Name the variable terms in the expression $3b^3 - 4b - 2$. $3b^3, -4b$
2. Evaluate $3a^2 - 4ab$ when $a = 5$ and $b = -4$. 155
3. Evaluate $\frac{x^3 + y^3}{x + y}$ when $x = 2$ and $y = -3$. 19
4. Evaluate $a^2 - 5(a - 2b) - c^2$ when $a = -3$, $b = 2$, and $c = -1$. 43

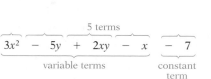

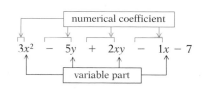

(Continued)

2. Find the largest possible value for the expression $x + \frac{3}{x}$ when x is a negative integer. $-3\frac{1}{2}$

Note: Students will probably need assistance in solving these exercises. For example, in Exercise 1, suggest that they start with the smallest positive integer (1) and then try the successive positive integers (2, 3, 4, . . .). Have them observe the pattern of the resulting values so that they can determine the smallest possible value.

Optional Student Activity

Have students complete the following exercises. After each exercise, have them evaluate the variable expression for some values of the variable that you provide.

1. A community college charges each student a \$25 student activity fee, which is added to the student's cost for tuition. Let T represent the tuition. Write a variable expression that represents the final bill after the student activity fee has been added. $T + 25$
2. The instructor of a sociology class is grading the first exam on a curve. The recorded grade on each exam is calculated by adding 8 points to the earned grade on the exam. Let G represent the earned grade on an exam. Write a variable expression that represents the recorded exam grade. $G + 8$
3. The varsity basketball coach at a high school must order jerseys for the team. The price of one jersey is \$34. Let N represent the number of players on the basketball team. Write a variable expression that represents the total cost for jerseys for the team. $34N$

(Continued on next page)

Example 1

Evaluate $\frac{a^2 - b^2}{a - b}$ when $a = 3$ and $b = -4$.

Solution

$$\frac{a^2 - b^2}{a - b}$$

$$\frac{3^2 - (-4)^2}{3 - (-4)} = \frac{9 - 16}{3 - (-4)}$$

$$= \frac{-7}{7} = -1$$

You Try It 1

Evaluate $\frac{a^2 + b^2}{a + b}$ when $a = 5$ and $b = -3$.

Your solution 17

Example 2 Evaluate $x^2 - 3(x - y) - z^2$ when $x = 2$, $y = -1$, and $z = 3$.

Solution

$$x^2 - 3(x - y) - z^2$$

$$2^2 - 3[2 - (-1)] - 3^2$$
$$= 2^2 - 3(3) - 3^2$$
$$= 4 - 3(3) - 9$$
$$= 4 - 9 - 9$$
$$= -5 - 9$$
$$= -14$$

You Try It 2 Evaluate $x^3 - 2(x + y) + z^2$ when $x = 2$, $y = -4$, and $z = -3$.

Your solution 21

Solutions on p. S2

Objective B **To simplify a variable expression using the Properties of Addition**

VIDEO & DVD CD TUTOR WEB SSM

Like terms of a variable expression are terms with the same variable part. (Because $x^2 = x \cdot x$, x^2 and x are not like terms.)

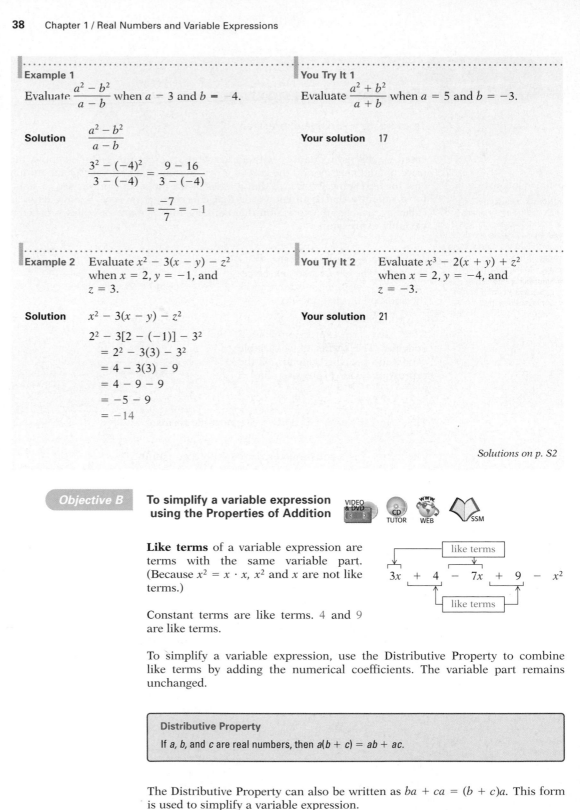

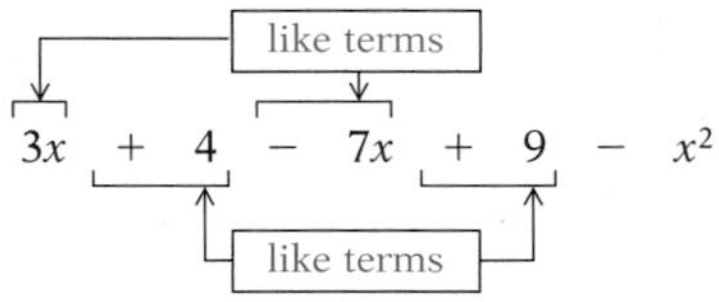

Constant terms are like terms. 4 and 9 are like terms.

To simplify a variable expression, use the Distributive Property to combine like terms by adding the numerical coefficients. The variable part remains unchanged.

Distributive Property

If a, b, and c are real numbers, then $a(b + c) = ab + ac$.

The Distributive Property can also be written as $ba + ca = (b + c)a$. This form is used to simplify a variable expression.

➡ Simplify: $2x + 3x$

Use the Distributive Property to add the numerical coefficients of the like variable terms. This is called **combining like terms**.

$2x + 3x = (2 + 3)x$ • Use the Distributive Property.
$= 5x$

➡ Simplify: $5y - 11y$

$5y - 11y = (5 - 11)y$ • Use the Distributive Property. This step is usually done mentally.
$= -6y$

TAKE NOTE
Simplifying an expression means combining like terms. A constant term (5) and a variable term ($7p$) are not like terms and therefore cannot be combined.

➡ Simplify: $5 + 7p$

The terms 5 and $7p$ are not like terms.

The expression $5 + 7p$ is in simplest form.

In simplifying variable expressions, the following Properties of Addition are used.

The Associative Property of Addition
If a, b, and c are real numbers, then $(a + b) + c = a + (b + c)$.

When three or more like terms are added, the terms can be grouped (with parentheses, for example) in any order. The sum is the same. For example,

$$(3x + 5x) + 9x = 3x + (5x + 9x)$$
$$8x + 9x = 3x + 14x$$
$$17x = 17x$$

The Commutative Property of Addition
If a and b are real numbers, then $a + b = b + a$.

When two like terms are added, the terms can be added in either order. The sum is the same. For example,

$$2x + (-4x) = -4x + 2x$$
$$-2x = -2x$$

The Addition Property of Zero
If a is a real number, then $a + 0 = 0 + a = a$.

The sum of a term and zero is the term. For example,

$$5x + 0 = 0 + 5x = 5x$$

(Continued)

4. You have a niece who is 16 years younger than you are. Let A represent your age. Write a variable expression that represents your niece's age. $A - 16$

Objective 1.4B

New Vocabulary
like terms
combining like terms

Vocabulary to Review
additive inverse

New Properties
Distributive Property
Associative Property of Addition
Commutative Property of Addition
Addition Property of Zero
Inverse Property of Addition

Instructor Note
Combining like terms can be related to many everyday experiences. For instance, 5 bricks plus 7 bricks is 12 bricks. But 5 bricks plus 7 nails is 5 bricks plus 7 nails. Students need to be constantly reminded that algebra is a reflection of our experiences, not some arbitrarily made-up system of rules.

Discuss the Concepts

1. Suppose you correctly simplify an expression and write the answer as $b + 9$, and another person writes the answer as $9 + b$. Are both answers correct?

(Continued on next page)

In-Class Examples (Objective 1.4B)

Simplify.

1. $4a - 5b - 3a + 2b$ $a - 3b$

2. $y^2 + 2 + 9y^2 - 14$ $10y^2 - 12$

3. $2z^2 - 3z - 5 - 3z^2 - 5$ $-z^2 - 3z - 10$

(Continued)

2. Why are $3x^2$ and $-4x^2$ like terms?
3. Why are b and b^2 not like terms?

Concept Check

1. Which of the following pairs of terms are like terms?
   a. $4a$ and $4b$
   b. $8z^3$ and $8z^2$
   c. $6ab$ and $3a$
   d. $-5c^2$ and $6c^2d$
2. Determine the additive inverse of each of the following.
   a. $4xy$ $-4xy$
   b. $-6b^2$ $6b^2$
   c. $-cd$ cd

Optional Student Activity

1. Make up three sentences similar to "5 bricks plus 7 bricks is 12 bricks" that illustrate that like terms can be combined.
2. Make up three sentences similar to "3 cows plus 4 boys is NOT 7 cowboys" that illustrate that unlike terms cannot be combined.

Objective 1.4C

New Vocabulary
multiplicative inverse

Vocabulary to Review
reciprocal

New Properties
Associative Property of Multiplication
Commutative Property of Multiplication
Multiplication Property of One
Inverse Property of Multiplication

> **The Inverse Property of Addition**
> If a is a real number, then $a + (-a) = (-a) + a = 0$.

The sum of a term and its opposite is zero. The opposite of a number is called its **additive inverse**.

$$7x + (-7x) = -7x + 7x = 0$$

➡ Simplify: $8x + 4y - 8x + y$

Use the Commutative and Associative Properties of Addition to rearrange and group like terms. Then combine like terms.

$$\begin{aligned} 8x + 4y - 8x + y &= (8x - 8x) + (4y + y) \\ &= 0 + 5y \\ &= 5y \end{aligned}$$

• This step is usually done mentally.

➡ Simplify: $4x^2 + 5x - 6x^2 - 2x + 1$

Use the Commutative and Associative Properties of Addition to rearrange and group like terms. Then combine like terms.

$$\begin{aligned} 4x^2 + 5x - 6x^2 - 2x + 1 &= (4x^2 - 6x^2) + (5x - 2x) + 1 \\ &= -2x^2 + 3x + 1 \end{aligned}$$

Example 3 Simplify: $3x + 4y - 10x + 7y$

Solution $3x + 4y - 10x + 7y = -7x + 11y$

You Try It 3 Simplify: $3a - 2b - 5a + 6b$

Your solution $-2a + 4b$

Example 4 Simplify: $x^2 - 7 + 4x^2 - 16$

Solution $x^2 - 7 + 4x^2 - 16 = 5x^2 - 23$

You Try It 4 Simplify: $-3y^2 + 7 + 8y^2 - 14$

Your solution $5y^2 - 7$

Solutions on p. S3

Objective C **To simplify a variable expression using the Properties of Multiplication**

In simplifying variable expressions, the following Properties of Multiplication are used.

> **The Associative Property of Multiplication**
> If a, b, and c are real numbers, then $(a \cdot b) \cdot c = a \cdot (b \cdot c)$.

When three or more factors are multiplied, the factors can be grouped in any order. The product is the same. For example,

$$2(3x) = (2 \cdot 3)x = 6x$$

In-Class Examples (Objective 1.4C)

Simplify.

1. $-6(-4y^2)$ $24y^2$
2. $-3(-12b)$ $36b$
3. $(5a)(-6)$ $-30a$
4. $-\frac{1}{9}(-9d)$ d

The Commutative Property of Multiplication

If a and b are real numbers, then $a \cdot b = b \cdot a$.

Two factors can be multiplied in either order. The product is the same. For example,

$$(2x) \cdot 3 = 3 \cdot (2x) = 6x$$

The Multiplication Property of One

If a is a real number, then $a \cdot 1 = 1 \cdot a = a$.

The product of a term and one is the term. For example,

$$(8x)(1) = (1)(8x) = 8x$$

The Inverse Property of Multiplication

If a is a real number, and a is not equal to zero, then
$a \cdot \frac{1}{a} = \frac{1}{a} \cdot a = 1$.

$\frac{1}{a}$ is called the **reciprocal** of a. $\frac{1}{a}$ is also called the **multiplicative inverse** of a.

The product of a number and its reciprocal is one. For example,

$$7 \cdot \frac{1}{7} = \frac{1}{7} \cdot 7 = 1$$

The multiplication properties just discussed are used to simplify variable expressions.

➡ Simplify: $2(-x)$

$$2(-x) = 2(-1 \cdot x) = [2(-1)]x = -2x$$

- Use the Associative Property of Multiplication to group factors.

➡ Simplify: $\frac{3}{2}\left(\frac{2x}{3}\right)$

Use the Associative Property of Multiplication to group factors.

$$\frac{3}{2}\left(\frac{2x}{3}\right) = \frac{3}{2}\left(\frac{2}{3}x\right) = \left(\frac{3}{2} \cdot \frac{2}{3}\right)x = 1 \cdot x = x$$

- Note that $\frac{2x}{3} = \frac{2}{3}x$.
- The steps in the dashed box are usually done mentally.

Discuss the Concepts

Classify each statement below as illustrating the Commutative Property of Addition, the Associative Property of Addition, the Commutative Property of Multiplication, or the Associative Property of Multiplication. Explain your answer.

a. $14 + 9 = 9 + 14$
b. $(7 + 3) + 5 = 7 + (3 + 5)$
c. $20(6) = 6(20)$
d. $(4 \cdot 8)7 = 4(8 \cdot 7)$
e. $x + 35 = 35 + x$
f. $y + (1 + 9) = (y + 1) + 9$
g. $rs = sr$
h. $x(yz) = (xy)z$

a and e are the Commutative Property of Addition. b and f are the Associative Property of Addition. c and g are the Commutative Property of Multiplication. d and h are the Associative Property of Multiplication.

Instructor Note

Simplifying expressions such as this one prepares the student for solving equations.

Concept Check

Determine whether each of the following mathematical statements is true. In each case, tell which property supports your answer.

1. $-4y \cdot 1 = -4y$ True. The Multiplication Property of One.
2. $(6x) \cdot 2 = 2 \cdot (6x)$ True. The Commutative Property of Multiplication.
3. $5 \cdot \frac{1}{5} = 5$ False.
4. $3(4x) = (3 \cdot 4)x$ True. The Associative Property of Multiplication.

Optional Student Activity

1. Simplify: $(2x) \div \frac{1}{2}$ $4x$
2. Simplify: $\left(\frac{1}{2} \div 2\right)x$ $\frac{1}{4}x$
3. Replace the ? to make a true statement.

   $? \cdot (2x) = x$ $\frac{1}{2}$

Objective 1.4D

Properties to Review

Distributive Property

➡ Simplify: $(16x)2$

Use the Commutative and Associative Properties of Multiplication to rearrange and group factors.

$$(16x)2 = 2(16x) = (2 \cdot 16)x = 32x$$

• The steps in the dashed box are usually done mentally.

Example 5 Simplify: $-2(3x^2)$

Solution $-2(3x^2) = -6x^2$

You Try It 5 Simplify: $-5(4y^2)$

Your solution $-20y^2$

Example 6 Simplify: $-5(-10x)$

Solution $-5(-10x) = 50x$

You Try It 6 Simplify: $-7(-2a)$

Your solution $14a$

Example 7 Simplify: $(6x)(-4)$

Solution $(6x)(-4) = -24x$

You Try It 7 Simplify: $(-5x)(-2)$

Your solution $10x$

Solutions on p. S3

Objective D To simplify a variable expression using the Distributive Property

VIDEO & DVD CD TUTOR WEB SSM

Recall that the Distributive Property states that if *a*, *b*, and *c* are real numbers, then

$$a(b + c) = ab + ac$$

The Distributive Property is used to remove parentheses from a variable expression.

➡ Simplify: $3(2x + 7)$

$$3(2x + 7) = 3(2x) + 3(7) = 6x + 21$$

• Use the Distributive Property. Do this step mentally.

➡ Simplify: $-5(4x + 6)$

$$-5(4x + 6) = -5(4x) + (-5) \cdot 6 = -20x - 30$$

• Use the Distributive Property. Do this step mentally.

In-Class Examples (Objective 1.4D)

Simplify.

1. $-6ac - 9ac$ $-15ac$
2. $-5a + 9b + 10a$ $5a + 9b$
3. $4x - 3(5x - 6)$ $-11x + 18$
4. $6[-3 - 5(a - 2)]$ $-30a + 42$
5. $-3(x - 2y) + 5(3x - 4y)$ $12x - 14y$

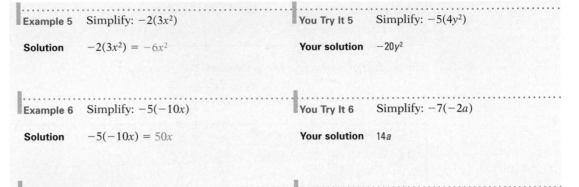

➡ Simplify: $-(2x - 4)$

$$-(2x - 4) = \boxed{\begin{array}{l}-1(2x - 4)\\ -1(2x) - (-1)(4)\end{array}} = -2x + 4$$

• Use the Distributive Property. Do these steps mentally.

Note: When a negative sign immediately precedes the parentheses, the sign of each term inside the parentheses is changed.

➡ Simplify: $-\frac{1}{2}(8x - 12y)$

$$-\frac{1}{2}(8x - 12y) = \boxed{-\frac{1}{2}(8x) - \left(-\frac{1}{2}\right)(12y)} = -4x + 6y$$

• Use the Distributive Property. Do this step mentally.

➡ Simplify: $4(x - y) - 2(-3x + 6y)$

$$4(x - y) - 2(-3x + 6y) = 4x - 4y + 6x - 12y$$

• Use the Distributive Property twice.

$$= 10x - 16y$$

• Combine like terms.

An extension of the Distributive Property is used when an expression inside parentheses contains more than two terms. See the example below.

➡ Simplify: $3(4x - 2y - z)$

$$3(4x - 2y - z) = \boxed{3(4x) - 3(2y) - 3(z)} = 12x - 6y - 3z$$

• Use the Distributive Property. Do this step mentally.

Example 8 Simplify: $-3(-5a + 7b)$

Solution $-3(-5a + 7b) = 15a - 21b$

You Try It 8 Simplify: $-8(-2a + 7b)$

Your solution $16a - 56b$

Example 9 Simplify: $(2x - 6)2$

Solution $(2x - 6)2 = 4x - 12$

You Try It 9 Simplify: $(3a - 1)5$

Your solution $15a - 5$

Solutions on p. S3

Discuss the Concepts

1. When do we use the Distributive Property? What is its purpose? (You might emphasize here that the Distributive Property is used to remove parentheses from a variable expression, and it enables us to write an expression equivalent to the original one. This is an important concept in simplifying expressions.)
2. Is it appropriate to use the Distributive Property to simplify $2(3 \cdot x)$? No. The expression inside the parentheses is a product, not a sum or difference. The Associative Property of Multiplication is used.
3. Does the expression $-(5x - 2)$ mean the opposite of $5x - 2$ or does it represent $-1(5x - 2)$? Both are correct.

Instructor Note

A common mistake for students to make is to combine x^2 terms with x terms. Citing a physical example, such as trying to combine feet and square feet, may help these students.

Concept Check

Each of the following has an error. Correct the right-hand side of the statement.

1. $-(3x - 5) = 3x + 5$
   $3x + 5$
2. $4(3 + 9y) = 12 + 9y$
   $12 + 36y$
3. $(8w - 3)7 = 56w + 21$
   $56w - 21$
4. $-6(3x - 2y + z) = -18x - 12y - 6z$
   $-18x + 12y - 6z$

Optional Student Activity

Which of the following expressions are equivalent?

1. $2x + 4(2x + 1)$
2. $x - (4 - 9x) + 8$
3. $7(x - 4) - 3(2x + 6)$
4. $3(2x + 8) + 4(x - 5)$
5. $6 - 2[x + (3x - 4)] + 2(9x - 5)$ 1, 2, 4, and 5 are equivalent; they are equal to $10x + 4$.

Objective 1.4E

New Vocabulary

See the list of verbal phrases that translate into mathematical operations.

Instructor Note

Translating the words *sum*, *difference*, *product*, and *quotient* is difficult for students. Suggest to students that these words imply the use of parentheses. As an example, 7 less than twice x is $2x - 7$, whereas twice the difference between x and 7 is $2(x - 7)$.

Example 10 Simplify: $3(x^2 - x - 5)$

Solution $3(x^2 - x - 5) = 3x^2 - 3x - 15$

You Try It 10 Simplify: $2(x^2 - x + 7)$

Your solution $2x^2 - 2x + 14$

Example 11 Simplify: $2x - 3(2x - 7y)$

Solution $2x - 3(2x - 7y) = 2x - 6x + 21y$
$= -4x + 21y$

You Try It 11 Simplify: $3y - 2(y - 7x)$

Your solution $14x + y$

Example 12 Simplify: $7(x - 2y) - (-x - 2y)$

Solution $7(x - 2y) - (-x - 2y)$
$= 7x - 14y + x + 2y$
$= 8x - 12y$

You Try It 12 Simplify: $-2(x - 2y) - (-x + 3y)$

Your solution $-x + y$

Example 13 Simplify: $2x - 3[2x - 3(x + 7)]$

Solution $2x - 3[2x - 3(x + 7)]$
$= 2x - 3[2x - 3x - 21]$
$= 2x - 3[-x - 21]$
$= 2x + 3x + 63$
$= 5x + 63$

You Try It 13 Simplify: $3y - 2[x - 4(2 - 3y)]$

Your solution $-2x - 21y + 16$

Solutions on p. S3

Objective E **To translate a verbal expression into a variable expression**

VIDEO & DVD CD TUTOR WEB SSM

One of the major skills required in applied mathematics is the ability to translate a verbal expression into a variable expression. This requires recognizing the verbal phrases that translate into mathematical operations. A partial list of the verbal phrases used to indicate the different mathematical operations follows.

Addition	added to	6 added to y	$y + 6$
	more than	8 more than x	$x + 8$
	the sum of	the sum of x and z	$x + z$
	increased by	t increased by 9	$t + 9$
	the total of	the total of 5 and y	$5 + y$

In-Class Examples (Objective 1.4E)

1. A mixture of nuts contains 5 lb more peanuts than cashews. Express the amount of peanuts in the mixture in terms of the amount of cashews in the mixture. Pounds of cashews in the mixture: c; pounds of peanuts: $c + 5$
2. A wire 8 ft long is cut into two pieces, one shorter than the other. Express the length of the shorter piece in terms of the length of the longer piece. Length of the longer piece: L; length of the shorter piece: $8 - L$

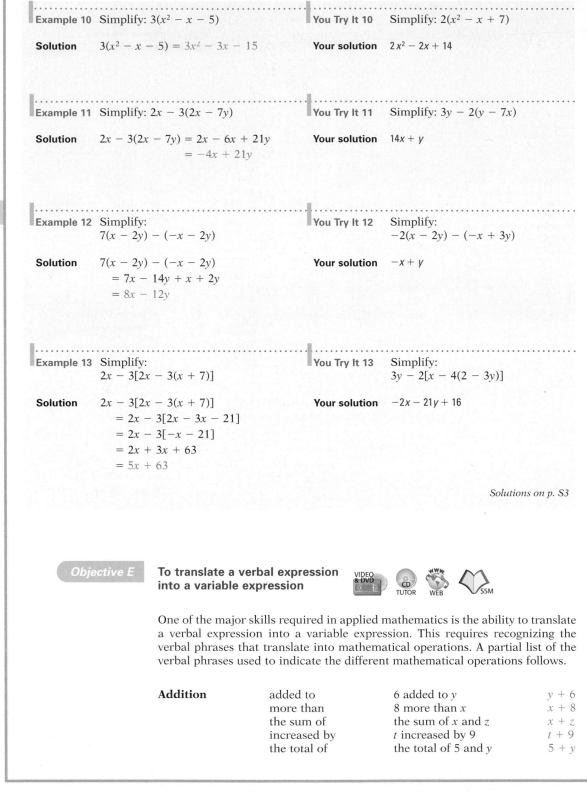

Point of Interest

The way in which expressions are symbolized has changed over time. Here is how some of the expressions shown at the right may have appeared in the early 16th century.

R p. 8 for $x + 8$. The symbol R was used for a variable to the first power. The symbol p. was used for plus.

R m. 2 for $x - 2$. The symbol R is still used for the variable. The symbol m. was used for minus.

The square of a variable was designated by Q and the cube was designated by C. The expression $x^3 + x^2$ was written **C p. Q.**

Subtraction	minus	x minus 2	$x - 2$
	less than	7 less than t	$t - 7$
	decreased by	m decreased by 3	$m - 3$
	the difference between	the difference between y and 4	$y - 4$
Multiplication	times	10 times t	$10t$
	of	one-half of x	$\frac{1}{2}x$
	the product of	the product of y and z	yz
	multiplied by	y multiplied by 11	$11y$
	twice	twice d	$2d$
Division	divided by	x divided by 12	$\frac{x}{12}$
	the quotient of	the quotient of y and z	$\frac{y}{z}$
	the ratio of	the ratio of t to 9	$\frac{t}{9}$
Power	the square of	the square of x	x^2
	the cube of	the cube of a	a^3

➡ Translate "14 less than the cube of x" into a variable expression.

14 *less than* the *cube* of x • **Identify the words that indicate the mathematical operations.**

$x^3 - 14$ • **Use the identified operations to write the variable expression.**

In most applications that involve translating phrases into variable expressions, the variable to be used is not given. To translate these phrases, a variable must be assigned to an unknown quantity before the variable expression can be written.

➡ Translate "the sum of two consecutive integers" into a variable expression. Then simplify.

the first integer: n • **Assign a variable to one of the unknown quantities.**

the next consecutive integer: $n + 1$ • **Use the assigned variable to write an expression for any other unknown quantity.**

$n + (n + 1)$ • **Use the assigned variable to write the variable expression.**

$(n + n) + 1$ • **Simplify the variable expression.**

$2n + 1$

Discuss the Concepts

In each of the following, two verbal expressions are given. Do they represent the same algebraic expression? Explain your answer.

1. the sum of y and eight; eight more than y Same
2. three times the sum of b and five; the sum of three times b and five Different
3. the product of six and x subtracted from nine; nine less than six times x Different
4. the ratio of twice c to d; the quotient of two times c and d Same
5. ten less than m; ten less m Different

Concept Check

1. Write five phrases that would translate into the expression $y + 6$. For example: the sum of y and 6; the total of y and 6; 6 more than y; 6 added to y; y increased by 6
2. Write four phrases that would translate into the expression $c - 18$. For example: 18 less than c; the difference between c and 18; c minus 18; c decreased by 18
3. Write three phrases that would translate into the expression $5b$. For example: 5 times b; the product of 5 and b; 5 multiplied by b
4. Write three phrases that would translate into the expression $\frac{x}{4}$. For example: x divided by 4; the quotient of x and 4; the ratio of x to 4

Optional Student Activity

1. Given that x is equal to −3, what is the value of x raised to the −x power? −27
2. Given that y is equal to 3x and that z is equal to 2y, find the sum of x, y, and z in terms of x. 10x
3. At a small college, there are three times as many commuters as boarders and nine times as many commuters as faculty members. Using C to represent the number of commuters, write an expression in simplest form for the total number of commuters, boarders, and faculty at the college. $\frac{13}{9}C$

Many of the applications of mathematics require that you identify an unknown quantity, assign a variable to that quantity, and then attempt to express another unknown quantity in terms of the variable.

⇒ A confectioner makes a mixture of candy that contains 3 lb more of milk chocolate than of caramel. Express the amount of milk chocolate in the mixture in terms of the amount of caramel in the mixture.

Amount of caramel in the mixture: c • Assign a variable to the amount of caramel in the mixture.

Amount of milk chocolate in the mixture: $c + 3$ • Express the amount of milk chocolate in the mixture in terms of c.

Example 14
Translate "four times the sum of half of a number and fourteen" into a variable expression. Then simplify.

Solution
the unknown number: n
half of the number: $\frac{1}{2}n$
the sum of half of the number and fourteen: $\frac{1}{2}n + 14$
$4\left(\frac{1}{2}n + 14\right)$
$2n + 56$

You Try It 14
Translate "five times the difference between a number and sixty" into a variable expression. Then simplify.

Your solution
$5(x - 60)$; $5x - 300$

Example 15
The length of a swimming pool is 4 ft less than two times the width. Express the length of the pool in terms of the width.

Solution
the width of the pool: w
the length is 4 ft less than two times the width: $2w - 4$

You Try It 15
The speed of a new printer is twice the speed of an older model. Express the speed of the new model in terms of the speed of the older model.

Your solution
the speed of the older model: s
the speed of the new model: $2s$

Example 16
A banker divided $5000 between two accounts, one paying 10% annual interest and the second paying 8% annual interest. Express the amount invested in the 10% account in terms of the amount invested in the 8% account.

Solution
the amount invested at 8%: x
the amount invested at 10%: $5000 - x$

You Try It 16
A guitar string 6 ft long was cut into two pieces. Express the length of the shorter piece in terms of the length of the longer piece.

Your solution
the length of the longer piece: L
the length of the shorter piece: $6 - L$

Solutions on p. S3

1.4 Exercises

Objective A

Evaluate the variable expression when $a = 2$, $b = 3$, and $c = -4$.

1. $6b \div (-a)$
-9

2. $bc \div (2a)$
-3

3. $b^2 - 4ac$
41

4. $a^2 - b^2$
-5

5. $b^2 - c^2$
-7

6. $(a + b)^2$
25

7. $a^2 + b^2$
13

8. $2a - (c + a)^2$
0

9. $(b - a)^2 + 4c$
-15

10. $b^2 - \frac{ac}{8}$
10

11. $\frac{5ab}{6} - 3cb$
41

12. $(b - 2a)^2 + bc$
-11

Evaluate the variable expression when $a = -2$, $b = 4$, $c = -1$, and $d = 3$.

13. $\frac{b + c}{d}$
1

14. $\frac{d - b}{c}$
1

15. $\frac{2d + b}{-a}$
5

16. $\frac{b + 2d}{b}$
$\frac{5}{2}$

17. $\frac{b - d}{c - a}$
1

18. $\frac{2c - d}{-ad}$
$-\frac{5}{6}$

19. $(b + d)^2 - 4a$
57

20. $(d - a)^2 - 3c$
28

21. $(d - a)^2 \div 5$
5

22. $3(b - a) - bc$
22

23. $\frac{b - 2a}{bc^2 - d}$
8

24. $\frac{b^2 - a}{ad + 3c}$
-2

25. $\frac{1}{3}d^2 - \frac{3}{8}b^2$
-3

26. $\frac{5}{8}a^4 - c^2$
9

27. $\frac{-4bc}{2a - b}$
-2

28. $-\frac{3}{4}b + \frac{1}{2}(ac + bd)$
4

29. $-\frac{2}{3}d - \frac{1}{5}(bd - ac)$
-4

30. $(b - a)^2 - (d - c)^2$
20

31. $(b + c)^2 + (a + d)^2$
10

32. $4ac + (2a)^2$
24

33. $3dc - (4c)^2$
-25

Section 1.4

Suggested Assignment

Exercises 1–169, every other odd
Exercises 171–209, odds
More challenging problems:
Exercises 211–214, 216

Quick Quiz (Objective 1.4A)

1. Name the terms of the variable expression $3x^2 - 4x - 5$. Then underline the constant term. $3x^2, -4x, \underline{-5}$
2. Name the coefficients of the variable terms in the expression $5x^3 - x^2 + 3x - 4$. $5, -1, 3$
3. Evaluate $a^2 - ab$ when $a = -2$ and $b = -7$. -10
4. Evaluate $2x - (y + z)^2$ when $x = 5$, $y = -3$, and $z = 6$. 1

Objective B

Simplify.

34. $6x + 8x$
$14x$

35. $12x + 13x$
$25x$

36. $9a - 4a$
$5a$

37. $12a - 3a$
$9a$

38. $4y + (-10y)$
$-6y$

39. $8y + (-6y)$
$2y$

40. $-3b - 7$
$-3b - 7$

41. $-12y - 3$
$-12y - 3$

42. $-12a + 17a$
$5a$

43. $-3a + 12a$
$9a$

44. $5ab - 7ab$
$-2ab$

45. $9ab - 3ab$
$6ab$

46. $-12xy + 17xy$
$5xy$

47. $-15xy + 3xy$
$-12xy$

48. $-3ab + 3ab$
0

49. $-7ab + 7ab$
0

50. $-\frac{1}{2}x - \frac{1}{3}x$
$-\frac{5}{6}x$

51. $-\frac{2}{5}y + \frac{3}{10}y$
$-\frac{1}{10}y$

52. $\frac{3}{8}x^2 - \frac{5}{12}x^2$
$-\frac{1}{24}x^2$

53. $\frac{2}{3}y^2 - \frac{4}{9}y^2$
$\frac{2}{9}y^2$

54. $3x + 5x + 3x$
$11x$

55. $8x + 5x + 7x$
$20x$

56. $5a - 3a + 5a$
$7a$

57. $10a - 17a + 3a$
$-4a$

58. $-5x^2 - 12x^2 + 3x^2$
$-14x^2$

59. $-y^2 - 8y^2 + 7y^2$
$-2y^2$

60. $7x + (-8x) + 3y$
$-x + 3y$

61. $8y + (-10x) + 8x$
$-2x + 8y$

62. $7x - 3y + 10x$
$17x - 3y$

63. $8y + 8x - 8y$
$8x$

64. $3a + (-7b) - 5a + b$
$-2a - 6b$

65. $-5b + 7a - 7b + 12a$
$19a - 12b$

66. $3x + (-8y) - 10x + 4x$
$-3x - 8y$

67. $3y + (-12x) - 7y + 2y$
$-12x - 2y$

68. $x^2 - 7x + (-5x^2) + 5x$
$-4x^2 - 2x$

69. $3x^2 + 5x - 10x^2 - 10x$
$-7x^2 - 5x$

Quick Quiz (Objective 1.4B)

Simplify.

1. $9b - 5b$ $4b$

2. $8x^2 - x^2 + 2x^2$ $9x^2$

3. $4y^2 + 3y - 6y^2 + 2y$ $-2y^2 + 5y$

49

Objective C

Simplify.

70. $4(3x)$ $12x$
71. $12(5x)$ $60x$
72. $-3(7a)$ $-21a$
73. $-2(5a)$ $-10a$
74. $-2(-3y)$ $6y$
75. $-5(-6y)$ $30y$
76. $(4x)2$ $8x$
77. $(6x)12$ $72x$
78. $(3a)(-2)$ $-6a$
79. $(7a)(-4)$ $-28a$
80. $(-3b)(-4)$ $12b$
81. $(-12b)(-9)$ $108b$
82. $-5(3x^2)$ $-15x^2$
83. $-8(7x^2)$ $-56x^2$
84. $\frac{1}{3}(3x^2)$ x^2
85. $\frac{1}{6}(6x^2)$ x^2
86. $\frac{1}{5}(5a)$ a
87. $\frac{1}{8}(8x)$ x
88. $-\frac{1}{2}(-2x)$ x
89. $-\frac{1}{4}(-4a)$ a
90. $-\frac{1}{7}(-7n)$ n
91. $-\frac{1}{9}(-9b)$ b
92. $(3x)\left(\frac{1}{3}\right)$ x
93. $(12x)\left(\frac{1}{12}\right)$ x
94. $(-6y)\left(-\frac{1}{6}\right)$ y
95. $(-10n)\left(-\frac{1}{10}\right)$ n
96. $\frac{1}{3}(9x)$ $3x$
97. $\frac{1}{7}(14x)$ $2x$
98. $-\frac{1}{5}(10x)$ $-2x$
99. $-\frac{1}{8}(16x)$ $-2x$
100. $-\frac{2}{3}(12a^2)$ $-8a^2$
101. $-\frac{5}{8}(24a^2)$ $-15a^2$
102. $-\frac{1}{2}(-16y)$ $8y$
103. $-\frac{3}{4}(-8y)$ $6y$
104. $(16y)\left(\frac{1}{4}\right)$ $4y$
105. $(33y)\left(\frac{1}{11}\right)$ $3y$
106. $(-6x)\left(\frac{1}{3}\right)$ $-2x$
107. $(-10x)\left(\frac{1}{5}\right)$ $-2x$
108. $(-8a)\left(-\frac{3}{4}\right)$ $6a$
109. $(21y)\left(-\frac{3}{7}\right)$ $-9y$

Objective D

Simplify.

110. $-(x + 2)$ $-x - 2$
111. $-(x + 7)$ $-x - 7$
112. $2(4x - 3)$ $8x - 6$
113. $5(2x - 7)$ $10x - 35$
114. $-2(a + 7)$ $-2a - 14$
115. $-5(a + 16)$ $-5a - 80$
116. $-3(2y - 8)$ $-6y + 24$
117. $-5(3y - 7)$ $-15y + 35$

Quick Quiz (Objective 1.4C)

Simplify.

1. $9(6a)$ $54a$
2. $-\frac{1}{8}(-8b)$ b
3. $(-12c)\frac{2}{3}$ $-8c$

118. $(5 - 3b)7$
$35 - 21b$

119. $(10 - 7b)2$
$20 - 14b$

120. $\frac{1}{3}(6 - 15y)$
$2 - 5y$

121. $\frac{1}{2}(-8x + 4y)$
$-4x + 2y$

122. $3(5x^2 + 2x)$
$15x^2 + 6x$

123. $6(3x^2 + 2x)$
$18x^2 + 12x$

124. $-2(-y + 9)$
$2y - 18$

125. $-5(-2x + 7)$
$10x - 35$

126. $(-3x - 6)5$
$-15x - 30$

127. $(-2x + 7)7$
$-14x + 49$

128. $2(-3x^2 - 14)$
$-6x^2 - 28$

129. $5(-6x^2 - 3)$
$-30x^2 - 15$

130. $-3(2y^2 - 7)$
$-6y^2 + 21$

131. $-8(3y^2 - 12)$
$-24y^2 + 96$

132. $3(x^2 - y^2)$
$3x^2 - 3y^2$

133. $5(x^2 + y^2)$
$5x^2 + 5y^2$

134. $-\frac{2}{3}(6x - 18y)$
$-4x + 12y$

135. $-\frac{1}{2}(x - 4y)$
$-\frac{1}{2}x + 2y$

136. $-(6a^2 - 7b^2)$
$-6a^2 + 7b^2$

137. $3(x^2 + 2x - 6)$
$3x^2 + 6x - 18$

138. $4(x^2 - 3x + 5)$
$4x^2 - 12x + 20$

139. $-2(y^2 - 2y + 4)$
$-2y^2 + 4y - 8$

140. $\frac{1}{2}(2x - 6y + 8)$
$x - 3y + 4$

141. $-\frac{1}{3}(6x - 9y + 1)$
$-2x + 3y - \frac{1}{3}$

142. $4(-3a^2 - 5a + 7)$
$-12a^2 - 20a + 28$

143. $-5(-2x^2 - 3x + 7)$
$10x^2 + 15x - 35$

144. $-3(-4x^2 + 3x - 4)$
$12x^2 - 9x + 12$

145. $3(2x^2 + xy - 3y^2)$
$6x^2 + 3xy - 9y^2$

146. $5(2x^2 - 4xy - y^2)$
$10x^2 - 20xy - 5y^2$

147. $-(3a^2 + 5a - 4)$
$-3a^2 - 5a + 4$

148. $-(8b^2 - 6b + 9)$
$-8b^2 + 6b - 9$

149. $4x - 2(3x + 8)$
$-2x - 16$

150. $6a - (5a + 7)$
$a - 7$

151. $9 - 3(4y + 6)$
$-12y - 9$

152. $10 - (11x - 3)$
$-11x + 13$

153. $5n - (7 - 2n)$
$7n - 7$

154. $8 - (12 + 4y)$
$-4y - 4$

Quick Quiz (Objective 1.4D)

Simplify.

1. $-3xy + 7xy$ $4xy$

2. $3b - 7a - 10b$ $-7a - 7b$

3. $3x - 2(x - 4y)$ $x + 8y$

4. $7[y - 4(2y - x)]$ $-49y + 28x$

5. $6(-a - 3b) - 4(2a - 4b)$ $-14a - 2b$

51

155. $3(x + 2) - 5(x - 7)$
$-2x + 41$

156. $2(x - 4) - 4(x + 2)$
$-2x - 16$

157. $12(y - 2) + 3(7 - 3y)$
$3y - 3$

158. $6(2y - 7) - (3 - 2y)$
$14y - 45$

159. $3(a - b) - (a + b)$
$2a - 4b$

160. $2(a + 2b) - (a - 3b)$
$a + 7b$

161. $4[x - 2(x - 3)]$
$-4x + 24$

162. $2[x + 2(x + 7)]$
$6x + 28$

163. $-2[3x + 2(4 - x)]$
$-2x - 16$

164. $-5[2x + 3(5 - x)]$
$5x - 75$

165. $-3[2x - (x + 7)]$
$-3x + 21$

166. $-2[3x - (5x - 2)]$
$4x - 4$

167. $2x - 3[x - (4 - x)]$
$-4x + 12$

168. $-7x + 3[x - (3 - 2x)]$
$2x - 9$

169. $-5x - 2[2x - 4(x + 7)] - 6$
$-x + 50$

Objective E

Translate into a variable expression. Then simplify.

170. twelve minus a number
$12 - x$

171. a number divided by eighteen
$\frac{x}{18}$

172. two-thirds of a number
$\frac{2}{3}x$

173. twenty more than a number
$x + 20$

174. the quotient of twice a number and nine
$\frac{2x}{9}$

175. ten times the difference between a number and fifty
$10(x - 50)$; $10x - 500$

176. eight less than the product of eleven and a number
$11x - 8$

177. the sum of five-eighths of a number and six
$\frac{5}{8}x + 6$

178. nine less than the total of a number and two
$(x + 2) - 9$; $x - 7$

179. the difference between a number and three more than the number
$x - (x + 3)$; -3

Quick Quiz (Objective 1.4E)

1. The measure of angle A is 40 degrees more than the measure of angle B. Express the measure of angle A in terms of the measure of angle B.
Measure of angle B: B; measure of angle A: $B + 40$

2. A rope 9 ft long is cut into two pieces, one shorter than the other. Express the length of the shorter piece in terms of the length of the longer piece.
Length of the longer piece: L; length of the shorter piece: $9 - L$

180. the quotient of seven and the total of five and a number
$\frac{7}{5 + x}$

181. four times the sum of a number and nineteen
$4(x + 19); 4x + 76$

182. five increased by one-half of the sum of a number and three
$5 + \frac{1}{2}(x + 3); \frac{1}{2}x + \frac{13}{2}$

183. the quotient of fifteen and the sum of a number and twelve
$\frac{15}{x + 12}$

184. a number added to the difference between twice the number and four
$(2x - 4) + x; 3x - 4$

185. the product of two-thirds and the sum of a number and seven
$\frac{2}{3}(x + 7); \frac{2}{3}x + \frac{14}{3}$

186. the product of five less than a number and seven
$(x - 5)7; 7x - 35$

187. the difference between forty and the quotient of a number and twenty
$40 - \frac{x}{20}$

188. the quotient of five more than twice a number and the number
$\frac{2x + 5}{x}$

189. the sum of the square of a number and twice the number
$x^2 + 2x$

190. a number decreased by the difference between three times the number and eight
$x - (3x - 8); -2x + 8$

191. the sum of eight more than a number and one-third of the number
$(x + 8) + \frac{1}{3}x; \frac{4}{3}x + 8$

192. a number added to the product of three and the number
$3x + x; 4x$

193. a number increased by the total of the number and nine
$x + (x + 9); 2x + 9$

194. five more than the sum of a number and six
$(x + 6) + 5; x + 11$

195. a number decreased by the difference between eight and the number
$x - (8 - x); 2x - 8$

196. a number minus the sum of the number and ten
$x - (x + 10); -10$

197. the difference between one-third of a number and five-eighths of the number
$\frac{1}{3}x - \frac{5}{8}x; -\frac{7}{24}x$

198. the sum of one-sixth of a number and four-ninths of the number
$\frac{1}{6}x + \frac{4}{9}x; \frac{11}{18}x$

199. two more than the total of a number and five
$(x + 5) + 2; x + 7$

200. the sum of a number divided by three and the number
$\frac{x}{3} + x; \frac{4x}{3}$

201. twice the sum of six times a number and seven
$2(6x + 7); 12x + 14$

53

Instructor Note
Jupiter has 39 moons. There were 898 tornadoes in 2000.

202. The planet Saturn has 9 more moons than Jupiter (Source: NASA). Express the number of moons Saturn has in terms of the number of moons Jupiter has.
Let m be the number of moons Jupiter has; $m + 9$

203. According to the National Oceanic and Atmospheric Administration, the number of tornadoes in 1990 was 235 more than the number of tornadoes in 2000. Express the number of tornadoes in 1990 in terms of the number of tornadoes in 2000.
Let t be the number of tornadoes in 2000; $t + 235$

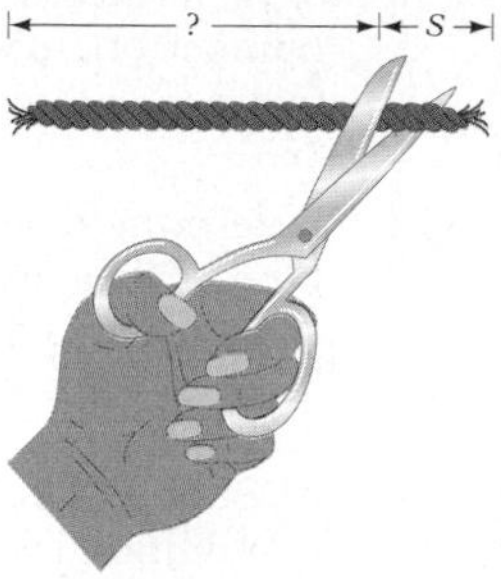

204. A rope 12 ft long was cut into two pieces of different lengths. Use one variable to express the lengths of the two pieces.
Let S be the length of one piece; S, $12 - S$

205. Twenty gallons of crude oil were poured into two containers of different sizes. Use one variable to express the amount of oil poured into each container.
Let g be the amount of oil in one container: g, $20 - g$

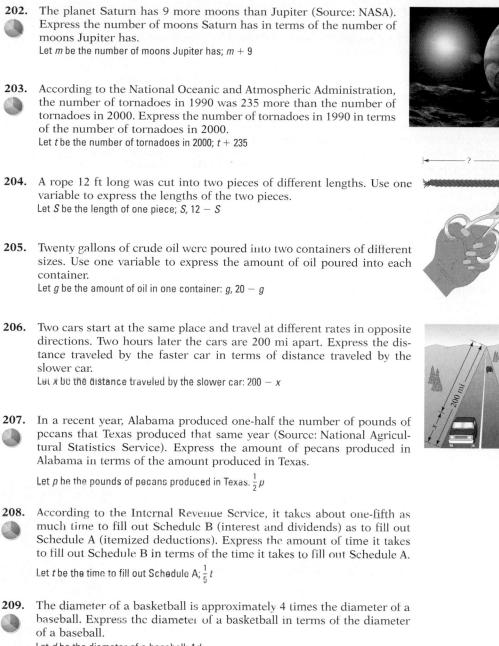

206. Two cars start at the same place and travel at different rates in opposite directions. Two hours later the cars are 200 mi apart. Express the distance traveled by the faster car in terms of distance traveled by the slower car.
Let x be the distance traveled by the slower car: $200 - x$

207. In a recent year, Alabama produced one-half the number of pounds of pecans that Texas produced that same year (Source: National Agricultural Statistics Service). Express the amount of pecans produced in Alabama in terms of the amount produced in Texas.
Let p be the pounds of pecans produced in Texas. $\frac{1}{2}p$

208. According to the Internal Revenue Service, it takes about one-fifth as much time to fill out Schedule B (interest and dividends) as to fill out Schedule A (itemized deductions). Express the amount of time it takes to fill out Schedule B in terms of the time it takes to fill out Schedule A.
Let t be the time to fill out Schedule A; $\frac{1}{5}t$

209. The diameter of a basketball is approximately 4 times the diameter of a baseball. Express the diameter of a basketball in terms of the diameter of a baseball.
Let d be the diameter of a baseball: $4d$

210. According to the U.S. Bureau of the Census, the world population in the year 2050 is expected to be twice the world population in 1980. Express the world population in 2050 in terms of the world population in 1980.
Let p be the world population in 1980; $2p$

Answers to Writing Exercises

218. Examples of two operations that occur in everyday experience and are not commutative are (1) unlocking the car door and starting the car and (2) taking a shower and drying oneself off.

219. Students should provide examples of evaluating each of the two expressions using the same value for *a*. Students should note that for each number chosen, the values of the two expressions are the same. They should reach the conclusion that the two variable expressions are equivalent.

APPLYING THE CONCEPTS

211. Does every number have an additive inverse? If not, which real numbers do not have an additive inverse?
Yes

212. Does every number have a multiplicative inverse? If not, which real numbers do not have a multiplicative inverse?
No. 0 does not have a multiplicative inverse.

213. The chemical formula for glucose (sugar) is $C_6H_{12}O_6$. This formula means that there are twelve hydrogen atoms, six carbon atoms, and six oxygen atoms in each molecule of glucose. If x represents the number of atoms of oxygen in a pound of sugar, express the number of hydrogen atoms in the pound of sugar.
$2x$

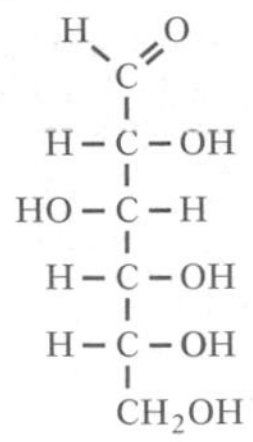

214. Determine whether the statement is true or false. If the statement is false, give an example that illustrates that it is false.
a. Division is a commutative operation. false
b. Division is an associative operation. false
c. Subtraction is an associative operation. false
d. Subtraction is a commutative operation. false
e. Addition is a commutative operation. true

215. A wire whose length is given as x inches is bent into a square. Express the length of a side of the square in terms of x.
$\frac{1}{4}x$

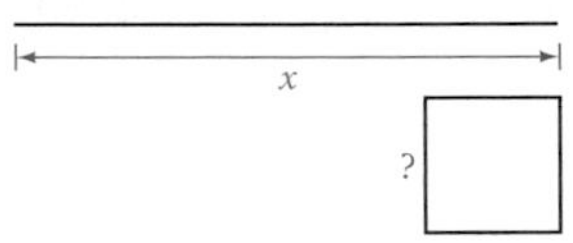

216. For each of the following, determine the first natural number x, greater than 2, for which the second expression is larger than the first.

a. x^3, 3^x 4 **b.** x^4, 4^x 5 **c.** x^5, 5^x 6 **d.** x^6, 6^x 7

On the basis of your answers, make a conjecture that appears to be true about the expressions x^n and n^x, where $n = 3, 4, 5, 6, 7, \ldots$ and x is a natural number greater than 2.
$n^x > x^n$ if $x \geq n + 1$

217. A block-and-tackle system is designed so that pulling five feet on one end of a rope will move a weight on the other end a distance of three feet. If x represents the distance the rope is pulled, express the distance the weight moves in terms of x. $\frac{3}{5}x$

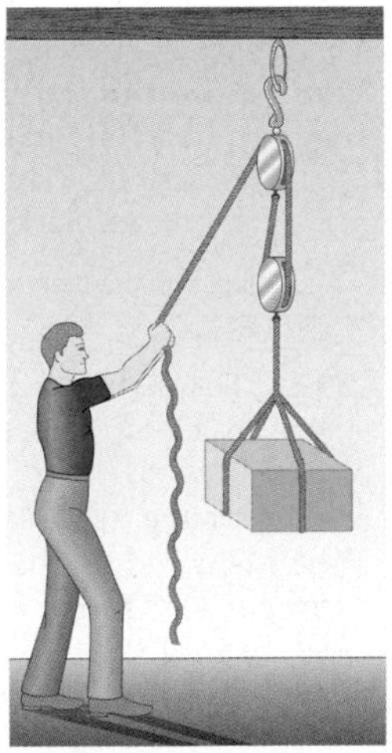

218. Give examples of two operations that occur in everyday experience that are not commutative (for example, putting on socks and then shoes).

219. Choose any number a. Evaluate the expressions $6a^2 + 2a - 10$ and $2a(3a - 4) + 10(a - 1)$. Now choose a different number and evaluate the expressions again. Repeat this two more times with different numbers. What conclusions might you draw from your evaluations?

1.5 Sets

Objective A To write a set using the roster method

VIDEO & DVD CD TUTOR WEB SSM

A **set** is a collection of objects, which are called the **elements** of the set. The **roster method** of writing a set encloses a list of the elements in braces.

The set of the last three letters of the alphabet is written {x, y, z}.

The set of the positive integers less than 5 is written {1, 2, 3, 4}.

➡ Use the roster method to write the set of integers between 0 and 10.

$A = \{1, 2, 3, 4, 5, 6, 7, 8, 9\}$

A set can be designated by a capital letter. Note that 0 and 10 are not elements of set A.

➡ Use the roster method to write the set of natural numbers.

$A = \{1, 2, 3, 4, \ldots\}$ • The three dots mean that the pattern of numbers continues without end.

The **empty set**, or **null set**, is the set that contains no elements. The symbol $\varnothing$ or { } is used to represent the empty set.

The set of people who have run a two-minute mile is the empty set.

The **union** of two sets, written $A \cup B$, is the set that contains the elements of A and the elements of B.

➡ Find $A \cup B$, given $A = \{1, 2, 3, 4\}$ and $B = \{3, 4, 5, 6\}$.

$A \cup B = \{1, 2, 3, 4, 5, 6\}$ • The union of A and B contains all the elements of A and all the elements of B. Any elements that are in both A and B are listed only once.

The **intersection** of two sets, written $A \cap B$, is the set that contains the elements that are common to both A and B.

➡ Find $A \cap B$, given $A = \{1, 2, 3, 4\}$ and $B = \{3, 4, 5, 6\}$.

$A \cap B = \{3, 4\}$ • The intersection of A and B contains the elements common to A and B.

Example 1
Use the roster method to write the set of the odd positive integers less than 12.

Solution
$A = \{1, 3, 5, 7, 9, 11\}$

You Try It 1
Use the roster method to write the set of the odd negative integers greater than −10.

Your solution
$A = \{-9, -7, -5, -3, -1\}$

Solution on p. S3

Objective 1.5A

Vocabulary to Review
set
element

New Vocabulary
roster method
empty set
null set
union
intersection

Instructor Note
Students want to write the empty set as $\{\varnothing\}$. It is very difficult to convince them that this is incorrect. Students, in general, have a difficult time with the idea that a set can be an element of another set.

Concept Check
Let N be the set of all positive even integers and let O be the set of all positive odd integers.
1. What is $N \cup O$? The positive integers
2. What is $N \cap O$? $\varnothing$

Discuss the Concepts
Decide whether each statement is true or false. Explain your reasoning.
1. For any set A, $A \cup \varnothing = A$. True
2. For any set A, $A \cap \varnothing = A$. False

In-Class Examples (Objective 1.5A)
1. Use the roster method to write the set of even integers between 20 and 30. $A = \{22, 24, 26, 28\}$
2. Find $A \cup B$ and $A \cap B$, given that $A = \{a, e, i, o, u\}$ and $B = \{a, b, c, d, e\}$. $\{a, b, c, d, e, i, o, u\}$; $\{a, e\}$
3. Find $C \cup D$ and $C \cap D$, given that $C = \{10, 20, 30, 40\}$ and $D = \{5, 15, 25, 35\}$. $\{5, 10, 15, 20, 25, 30, 35, 40\}$; $\varnothing$

Optional Student Activity

1. Find two sets A and B such that $A \cup B = \{-4, -2, 0, 1, 2, 3, 4, 5, 6\}$ and $A \cap B = \{2, 4\}$. For example, $A = \{1, 2, 3, 4, 5, 6\}$ and $B = \{-4, -2, 0, 2, 4\}$
2. Find two sets C and D such that $C \cup D = \{10, 20, 30, 40, 50, 60, 70, 80\}$ and $C \cap D = \varnothing$. For example, $A = \{10, 20, 30, 40\}$ and $B = \{50, 60, 70, 80\}$

Optional Student Activity

Use these sets:
$A = \{1, 2, 3, 4, 5, 6\}$
$B = \{2, 4, 6, 8, 10\}$
$C = \{1, 3, 5, 7, 9\}$

1. Find $(A \cup B) \cap C$. $\{1, 3, 5\}$
2. Find $A \cup (B \cap C)$. A
3. Find $(A \cap C) \cup B$. $\{1, 2, 3, 4, 5, 6, 8, 10\}$
4. Find $A \cap (C \cup B)$. A

Objective 1.5B

New Vocabulary

set-builder notation

Concept Check

Use these two sets, which are written in set-builder notation.
$A = \{x \mid x > -2, x \in \text{integers}\}$
$B = \{x \mid x < 4, x \in \text{integers}\}$

1. Read each set out loud.
2. Find $A \cup B$. Write your answer using set-builder notation. $\{x \mid x \in \text{integers}\}$
3. Find $A \cap B$. Write your answer using the roster method. $\{-1, 0, 1, 2, 3\}$

Example 2
Use the roster method to write the set of the even positive integers.

Solution
$A = \{2, 4, 6, \ldots\}$

You Try It 2
Use the roster method to write the set of the odd positive integers.

Your solution
$A = \{1, 3, 5, \ldots\}$

Example 3
Find $D \cup E$, given $D = \{6, 8, 10, 12\}$ and $E = \{-8, -6, 10, 12\}$.

Solution
$D \cup E = \{-8, -6, 6, 8, 10, 12\}$

You Try It 3
Find $A \cup B$, given $A = \{-2, -1, 0, 1, 2\}$ and $B = \{0, 1, 2, 3, 4\}$.

Your solution
$A \cup B = \{-2, -1, 0, 1, 2, 3, 4\}$

Example 4
Find $A \cap B$, given $A = \{5, 6, 9, 11\}$ and $B = \{5, 9, 13, 15\}$.

Solution
$A \cap B = \{5, 9\}$

You Try It 4
Find $C \cap D$, given $C = \{10, 12, 14, 16\}$ and $D = \{10, 16, 20, 26\}$.

Your solution
$C \cap D = \{10, 16\}$

Example 5
Find $A \cap B$ given $A = \{1, 2, 3, 4\}$ and $B = \{8, 9, 10, 11\}$.

Solution
$A \cap B = \varnothing$

You Try It 5
Find $A \cap B$, given $A = \{-5, -4, -3, -2\}$ and $B = \{2, 3, 4, 5\}$.

Your solution
$A \cap B = \varnothing$

Solutions on p. S3

Objective B — To write a set using set-builder notation

VIDEO & DVD · CD TUTOR · WEB · SSM

Point of Interest

The symbol $\in$ was first used in the book *Arithmeticae Principia*, published in 1889. It was the first letter of the Greek word εστι, which means "is." The symbols for union and intersection were also introduced at that time.

Another method of representing sets is called **set-builder notation.** Using set-builder notation, the set of all positive integers less than 10 is as follows:

$\{x \mid x < 10, x \in \text{positive integers}\}$, which is read "the set of all x such that x is less than 10 and x is an element of the positive integers."

➡ Use set-builder notation to write the set of real numbers greater than 4.

$\{x \mid x > 4, x \in \text{real numbers}\}$ • "$x \in$ real numbers" is read "x is an element of the real numbers."

In-Class Examples (Objective 1.5B)

Use set-builder notation to write each set.

1. The even integers greater than 10
   $\{x \mid x > 10, x \in \text{even integers}\}$
2. The positive real numbers less than 25
   $\{x \mid x < 25, x \in \text{positive real numbers}\}$
3. The real numbers greater than -12
   $\{x \mid x > -12, x \in \text{real numbers}\}$
4. The integers less than -100
   $\{x \mid x < -100, x \in \text{integers}\}$

Example 6
Use set-builder notation to write the set of negative integers greater than -100.

Solution
$\{x|x > -100, x \in \text{negative integers}\}$

You Try It 6
Use set-builder notation to write the set of positive even integers less than 59.

Your solution
$\{x|x < 59, x \in \text{positive even integers}\}$

Example 7
Use set-builder notation to write the set of real numbers less than 60.

Solution
$\{x|x < 60, x \in \text{real numbers}\}$

You Try It 7
Use set-builder notation to write the set of real numbers greater than -3.

Your solution
$\{x|x > -3, x \in \text{real numbers}\}$

Solutions on p. S3

Objective C — To graph an inequality on the number line

VIDEO & DVD — CD TUTOR — WEB — SSM

Point of Interest
The symbols for "is less than" and "is greater than" were introduced by Thomas Harriot around 1630. Before that, ⊏ and ⊐ were used for > and <, respectively.

An expression that contains the symbol >, <, ≥, or ≤ is called an **inequality.** An inequality expresses the relative order of two mathematical expressions. The expressions can be either numerical or variable.

$$\left.\begin{array}{l} 4 > 2 \\ 3x \le 7 \\ x^2 - 2x > y + 4 \end{array}\right\} \text{Inequalities}$$

An inequality can be graphed on the number line.

TAKE NOTE
In many cases, we assume that the real numbers are being used and omit "$x \in$ real numbers" from set builder notation. Using this convention, $\{x|x > 1, x \in \text{real numbers}\}$ is written $\{x|x > 1\}$.

➡ Graph: $\{x|x > 1\}$

The graph is the real numbers greater than 1. The parenthesis at 1 indicates that 1 is not included in the graph.

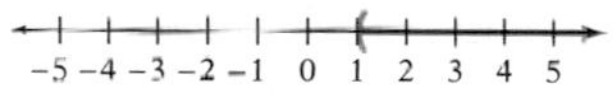

➡ Graph: $\{x|x \ge 1\}$

The bracket at 1 indicates that 1 is included in the graph.

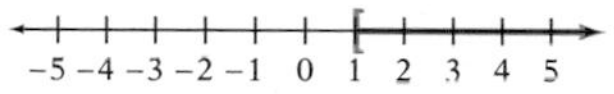

➡ Graph: $\{x|x < -1\}$

The numbers less than -1 are to the left of -1 on the number line.

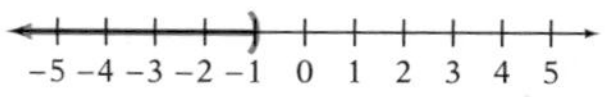

The union of two sets is the set that contains all the elements of each set.

➡ Graph: $\{x|x > 4\} \cup \{x|x < 1\}$

The graph is the numbers greater than 4 and the numbers less than 1.

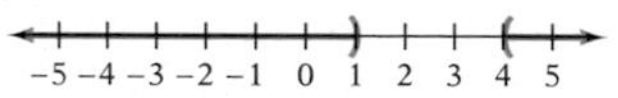

Objective 1.5C

New Vocabulary
inequality

Discuss the Concepts
Discuss why $-1 > x$ is equivalent to $x < -1$.

Concept Check
Rewrite each of the following inequalities so that the variable appears on the left.

1. $2 < x$ $x > 2$
2. $-3 \ge a$ $a \le -3$
3. $6 \le n$ $n \ge 6$
4. State a rule for switching the sides of an inequality. When the two sides of an inequality are switched, the inequality symbol must be reversed; for example, if $a > b$, then $b < a$.

Instructor Note
Students need to distinguish between > and ≥ and between < and ≤. Use the Concept Check and Discuss the Concepts questions on the next page to help students understand the important differences between these pairs of symbols.

In-Class Examples (Objective 1.5C)
Graph.

1. $\{x|x \ge -3\}$

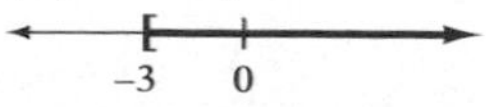

2. $\{x|x \ge 0\} \cap \{x|x \le 5\}$

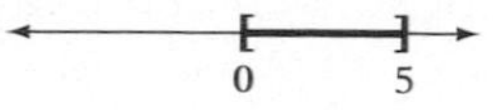

3. $\{x|x < -1\} \cup \{x|x > 3\}$

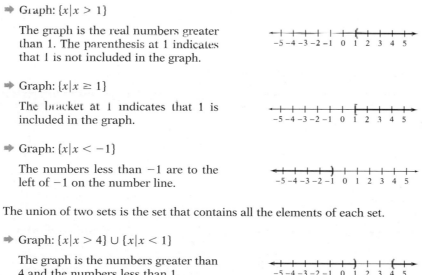

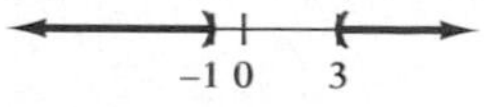

Concept Check

Which of the numbers 3, 2.5, and 2 satisty the given inequality?

1. $x > 2$ 3, 2.5
2. $x \le 3$ 3, 2.5, 2

Discuss the Concepts

Have students describe when to use a parenthesis and when to use a bracket when graphing an inequality.

Optional Student Activity

1. Write an inequality that describes the graph.

a. 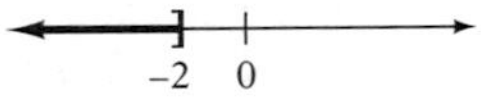

$\{x \mid x \le -2\}$

b. 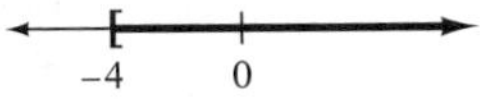

$\{x \mid x \ge -4\}$

2. Use set-builder notation to write a union or intersection of two inequalities that describes the graph.

a. 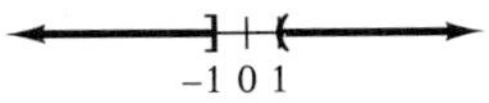

$\{x \mid x \le -1\} \cup \{x \mid x > 1\}$

b. 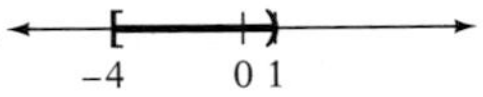

$\{x \mid x \ge -4\} \cap \{x \mid x < 1\}$

Optional Student Activity

1. Write an inequality that describes the given situation.
   a. In order to avoid monthly fees, one must maintain a minimum balance, b, of \$1000. $b \ge 1000$
   b. The temperature, t, never got above freezing (32°F). $t \le 32$
2. Use set-builder notation to write a union or intersection of two inequalities that describes this situation: The daily high temperatures in March ranged from 18°F to 57°F. $\{t \mid t \ge 18\} \cap \{t \mid t \le 57\}$
3. Make up your own examples like the ones in parts 1 and 2.

The intersection of two sets is the set that contains the elements common to both sets.

➡ Graph: $\{x \mid x > -1\} \cap \{x \mid x < 2\}$

The graphs of $\{x \mid x > -1\}$ and $\{x \mid x < 2\}$ are shown at the right.

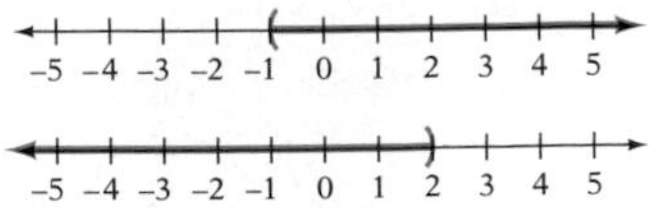

The graph of $\{x \mid x > -1\} \cap \{x \mid x < 2\}$ is the numbers between -1 and 2.

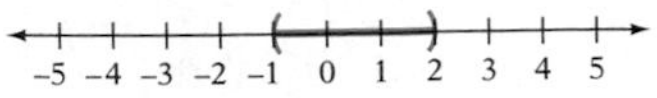

Example 8
Graph: $\{x \mid x < 5\}$

Solution
The graph is the numbers less than 5.

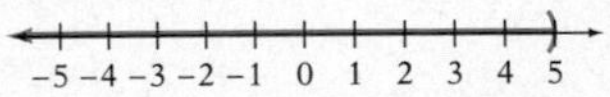

You Try It 8
Graph: $\{x \mid x > -2\}$

Your solution

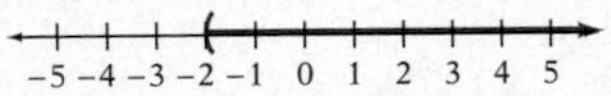

Example 9
Graph: $\{x \mid x > 3\} \cup \{x \mid x < 1\}$

Solution
The graph is the numbers greater than 3 and the numbers less than 1.

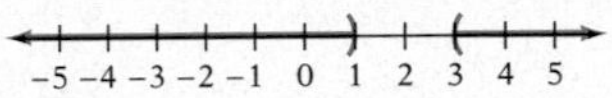

You Try It 9
Graph: $\{x \mid x > -1\} \cup \{x \mid x < -3\}$

Your solution

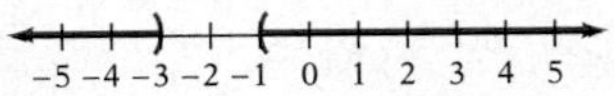

Example 10
Graph: $\{x \mid x > -2\} \cap \{x \mid x < 1\}$

Solution
The graph is the numbers between -2 and 1.

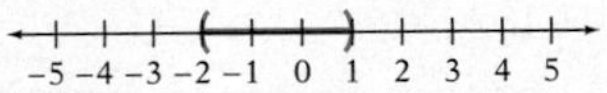

You Try It 10
Graph: $\{x \mid x \le 4\} \cap \{x \mid x \ge -4\}$

Your solution

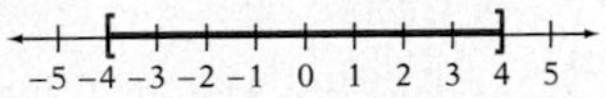

Example 11
Graph: $\{x \mid x \le 5\} \cup \{x \mid x \ge -3\}$

Solution
The graph is the real numbers.

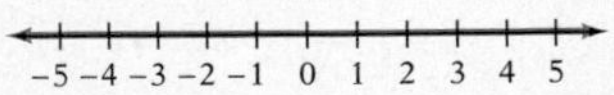

You Try It 11
Graph: $\{x \mid x < 2\} \cup \{x \mid x \ge -2\}$

Your solution

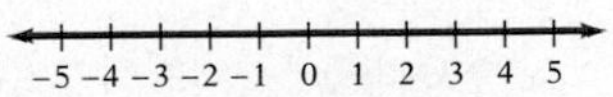

Solutions on p. S3

1.5 Exercises

Objective A

Use the roster method to write the set.

1. the integers between 15 and 22
$A = \{16, 17, 18, 19, 20, 21\}$

2. the integers between -10 and -4
$A = \{-9, -8, -7, -6, -5\}$

3. the odd integers between 8 and 18
$A = \{9, 11, 13, 15, 17\}$

4. the even integers between -11 and -1
$A = \{-10, -8, -6, -4, -2\}$

5. the letters of the alphabet between a and d
$A = \{b, c\}$

6. the letters of the alphabet between p and v
$A = \{q, r, s, t, u\}$

7. Explain how to find the union of two sets.

8. Explain how to find the intersection of two sets.

Find $A \cup B$.

9. $A = \{3, 4, 5\}$ $B = \{4, 5, 6\}$
$A \cup B = \{3, 4, 5, 6\}$

10. $A = \{-3, -2, -1\}$ $B = \{-2, -1, 0\}$
$A \cup B = \{-3, -2, -1, 0\}$

11. $A = \{-10, -9, -8\}$ $B = \{8, 9, 10\}$
$A \cup B = \{-10, -9, -8, 8, 9, 10\}$

12. $A = \{a, b, c\}$ $B = \{x, y, z\}$
$A \cup B = \{a, b, c, x, y, z\}$

13. $A = \{a, b, d, e\}$ $B = \{c, d, e, f\}$
$A \cup B = \{a, b, c, d, e, f\}$

14. $A = \{m, n, p, q\}$ $B = \{m, n, o\}$
$A \cup B \{m, n, o, p, q\}$

15. $A = \{1, 3, 7, 9\}$ $B = \{7, 9, 11, 13\}$
$A \cup B = \{1, 3, 7, 9, 11, 13\}$

16. $A = \{-3, -2, -1\}$ $B = \{-1, 1, 2\}$
$A \cup B = \{-3, -2, -1, 1, 2\}$

Find $A \cap B$.

17. $A = \{3, 4, 5\}$ $B = \{4, 5, 6\}$
$A \cap B = \{4, 5\}$

18. $A = \{-4, -3, -2\}$ $B = \{-6, -5, -4\}$
$A \cap B = \{-4\}$

19. $A = \{-4, -3, -2\}$ $B = \{2, 3, 4\}$
$A \cap B = \varnothing$

20. $A = \{1, 2, 3, 4\}$ $B = \{1, 2, 3, 4\}$
$A \cap B = \{1, 2, 3, 4\}$

21. $A = \{a, b, c, d, e\}$ $B = \{c, d, e, f, g\}$
$A \cap B = \{c, d, e\}$

22. $A = \{m, n, o, p\}$ $B = \{k, l, m, n\}$
$A \cap B = \{m, n\}$

Objective B

Use set-builder notation to write the set.

23. the negative integers greater than -5
$\{x \mid x > -5, x \in \text{negative integers}\}$

24. the positive integers less than 5
$\{x \mid x < 5, x \in \text{positive integers}\}$

Section 1.5

Suggested Assignment
Exercises 1–39, odds
More challenging problems:
Exercises 41–43

Answers to Writing Exercises

7. Student explanations should include the idea that to find the union of two sets, we list all the elements of the first set and then list all the elements of the second set that are not elements of the first set.

8. Student explanations should include the idea that to find the intersection of two sets, we list only those elements that are elements of both of the sets.

Quick Quiz (Objective 1.5A)

1. Use the roster method to write the set of integers between -6 and 0. $A = \{-5, -4, -3, -2, -1\}$
2. Find $A \cup B$ and $A \cap B$, given that $A = \{p, q, r\}$ and $B = \{m, n, o\}$. $\{m, n, o, p, q, r\}$; $\varnothing$
3. Find $C \cup D$ and $C \cap D$, given that $C = \{-3, -2, -1, 0\}$ and $D = \{0, 1, 2, 3\}$. $\{-3, -2, -1, 0, 1, 2, 3\}$; $\{0\}$

Quick Quiz (Objective 1.5B)

Use set-builder notation to write each set.

1. The odd integers less than 20
$\{x \mid x < 20, x \in \text{odd integers}\}$
2. The real numbers greater than -15
$\{x \mid x > -15, x \in \text{real numbers}\}$
3. The positive integers less than 32
$\{x \mid x < 32, x \in \text{positive integers}\}$
4. The even integers greater than 50
$\{x \mid x > 50, x \in \text{even integers}\}$

25. the integers greater than 30
$\{x|x > 30, x \in \text{integers}\}$

26. the integers less than -70
$\{x|x < -70, x \in \text{integers}\}$

27. the even integers greater than 5
$\{x|x > 5, x \in \text{even integers}\}$

28. the odd integers less than -2
$\{x|x < -2, x \in \text{odd integers}\}$

29. the real numbers greater than 8
$\{x|x > 8, x \in \text{real numbers}\}$

30. the real numbers less than 57
$\{x|x < 57, x \in \text{real numbers}\}$

Objective C

Graph.

31. $\{x|x > 2\}$

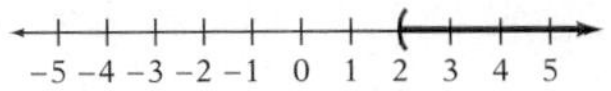

32. $\{x|x \geq -1\}$

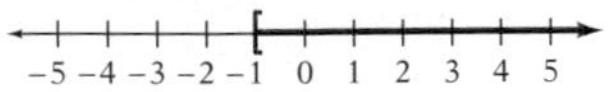

33. $\{x|x \leq 0\}$

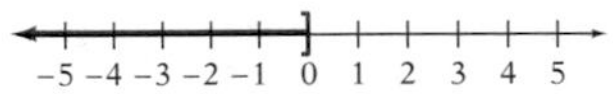

34. $\{x|x < 4\}$

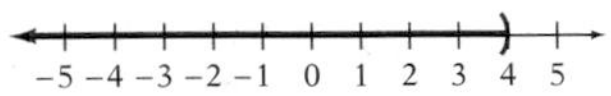

35. $\{x|x > -2\} \cup \{x|x < -4\}$

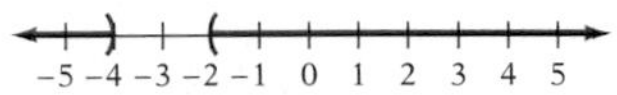

36. $\{x|x > 4\} \cup \{x|x < -2\}$

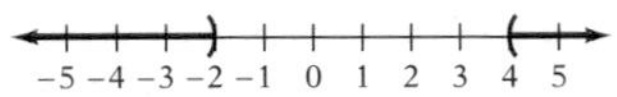

37. $\{x|x > -2\} \cap \{x < 4\}$

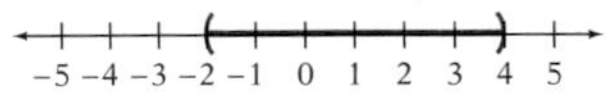

38. $\{x|x > -3\} \cap \{x|x < 3\}$

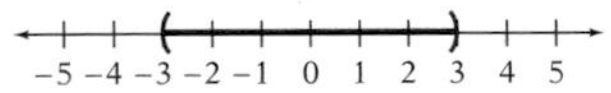

39. $\{x|x \geq -2\} \cup \{x|x < 4\}$

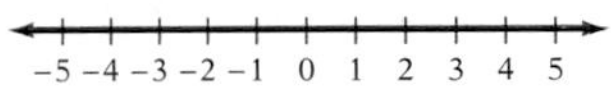

40. $\{x|x > 0\} \cup \{x|x \leq 4\}$

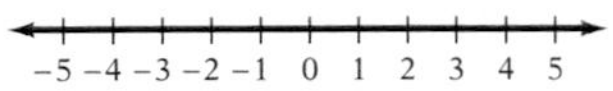

APPLYING THE CONCEPTS

41. Determine whether the statement is always true, sometimes true, or never true.

a. Given that $a > 0$ and $b < 0$, then $ab > 0$. never true

b. Given that $a < 0$, then $a^2 > 0$. always true

c. Given that $a > 0$ and $b < 0$, then $a^2 > b$. always true

42. By trying various sets, make a conjecture as to whether the union of two sets is

a. a commutative operation yes

b. an associative operation yes

43. By trying various sets, make a conjecture as to whether the intersection of two sets is

a. a commutative operation yes

b. an associative operation yes

Quick Quiz (Objective 1.5C)

Graph.

1. $\{x|x < 4\}$

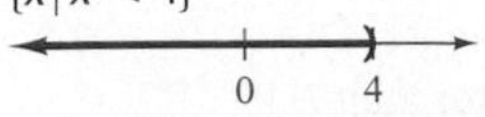

2. $\{x|x \geq 0\} \cup \{x|x \leq -3\}$

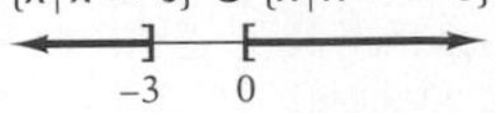

3. $\{x|x < 2\} \cap \{x|x > -4\}$

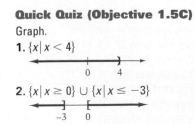

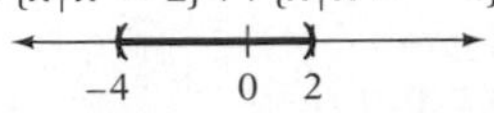

Focus on Problem Solving

Inductive Reasoning Suppose you take 9 credit hours each semester. The total number of credit hours you have taken at the end of each semester can be described in a list of numbers.

$$9, 18, 27, 36, 45, 54, 63, \ldots$$

The list of numbers that indicates the total credit hours is an ordered list of numbers, called a **sequence.** Each number in a sequence is called a **term** of the sequence. The list is ordered because the position of a number in the list indicates the semester in which that number of credit hours has been taken. For example, the 7th term of the sequence is 63, and a total of 63 credit hours have been taken after the 7th semester.

Assuming the pattern is continued, find the next three numbers in the pattern

$$-6, -10, -14, -18, \ldots$$

This list of numbers is a sequence. The first step in solving this problem is to observe the pattern in the list of numbers. In this case, each number in the list is 4 less than the previous number. The next three numbers are -22, -26, -30.

This process of discovering the pattern in a list of numbers is inductive reasoning. **Inductive reasoning** involves making generalizations from specific examples; in other words, we reach a conclusion by making observations about particular facts or cases.

Try the following exercises. Each exercise requires inductive reasoning.

Name the next two terms in the sequence.

1. 1, 3, 5, 7, 1, 3, 5, 7, 1, . . .

2. 1, 4, 2, 5, 3, 6, 4, . . .

3. 1, 2, 4, 7, 11, 16, . . .

4. A, B, C, G, H, I, M, . . .

Draw the next shape in the sequence.

5. 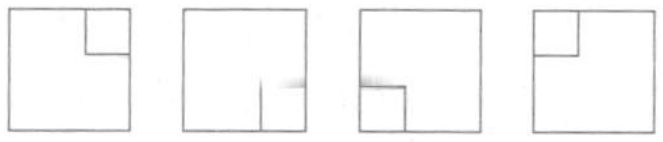

6. |• ||• ||•• |||•• |||•••

Solve.

7. Convert $\frac{1}{11}, \frac{2}{11}, \frac{3}{11}, \frac{4}{11}$, and $\frac{5}{11}$ to decimals. Then use the pattern you observe to convert $\frac{6}{11}, \frac{7}{11}$, and $\frac{9}{11}$ to decimals.

8. Convert $\frac{1}{33}, \frac{2}{33}, \frac{4}{33}, \frac{5}{33}$, and $\frac{7}{33}$ to decimals. Then use the pattern you observe to convert $\frac{8}{33}, \frac{13}{33}$, and $\frac{19}{33}$ to decimals.

Instructor Note

The feature entitled "Focus on Problem Solving" appears at the end of every chapter of the text. It provides optional material that can be used to enhance your students' problem-solving skills.

Focus on Problem Solving: Inductive Reasoning

1. 3, 5

2. 7, 5

3. 22, 29

4. N, O

5. 

6. ||||•••

7. The first digit is 1 less than the numerator. The second digit is 9 minus the first digit. The two digits then repeat.

$\frac{6}{11} = 0.\overline{54}$, $\frac{7}{11} = 0.\overline{63}$, $\frac{9}{11} = 0.\overline{81}$

8. There are two digits that repeat. The repeating digits are three times the numerator, where a 0 is inserted if necessary.

$\frac{8}{33} = 0.\overline{24}$, $\frac{13}{33} = 0.\overline{39}$, $\frac{19}{33} = 0.\overline{57}$

Instructor Note
Project and Group Activities appear at the end of each chapter in this text. The feature can be used for individual assignments, such as extra credit, or for cooperative learning exercises, such as small-group projects, or for class discussions.

Answers to Projects and Group Activities: Calculators

1. 15
2. −5
3. 21
4. 18
5. −8
6. −24
7. 56
8. −36
9. −5
10. 21

Projects and Group Activities

Calculators

Does your calculator use the Order of Operations Agreement? To find out, try this problem:

$$2 + 4 \cdot 7$$

If your answer is 30, then the calculator uses the Order of Operations Agreement. If your answer is 42, it does not use that agreement.

Even if your calculator does not use the Order of Operations Agreement, you can still correctly evaluate numerical expressions. The parentheses keys, [(] and [)], are used for this purpose.

Remember that $2 + 4 \cdot 7$ means $2 + (4 \cdot 7)$ because the multiplication must be completed before the addition. To evaluate this expression, enter the following:

Enter:	2	[+]	[(]	4	[×]	7	[)]	[=]
Display:	2	2	[(]	4	4	7	28	30

When using your calculator to evaluate numerical expressions, insert parentheses around multiplications or divisions. This has the effect of forcing the calculator to do the operations in the order you want rather than in the order the calculator wants.

Evaluate.

1. $3 \cdot (15 - 2 \cdot 3) - 36 \div 3$

2. $4 \cdot 2^2 - (12 + 24 \div 6) - 5$

3. $16 \div 4 \cdot 3 + (3 \cdot 4 - 5) + 2$

4. $15 \cdot 3 \div 9 + (2 \cdot 6 - 3) + 4$

Using your calculator to simplify numerical expressions sometimes requires use of the [+/−] key or, on some calculators, the negative key, which is frequently shown as [(−)]. These keys change the sign of the number currently in the display. To enter −4:

- For those calculators with [+/−], press 4 and then [+/−].
- For those calculators with [(−)], press [(−)] and then 4.

Here are the keystrokes for evaluating the expression $3(-4) - (-5)$.

Calculators with [+/−] key: 3 [×] 4 [+/−] − 5 [+/−] [=]

Calculators with [(−)] key: 3 [×] [(−)] 4 [−] [(−)] 5 [=]

This example illustrates that calculators make a distinction between negative and minus. To perform the operation $3 - (-3)$, you cannot enter 3 [−] [−] 3. This would result in 0, which is not the correct answer. You must enter

3 [−] 3 [+/−] [=] or 3 [−] [(−)] 3 [=]

Use a calculator to evaluate each of the following exercises.

5. $-16 \div 2$

6. $3(-8)$

7. $47 - (-9)$

8. $-50 - (-14)$

9. $4 - (-3)^2$

10. $-8 + (-6)^2 - 7$

Addition and Multiplication Properties

The chart below is an addition table. Use it to answer Exercises 1 to 7.

+	Δ	‡	◇
Δ	‡	◇	Δ
‡	◇	Δ	‡
◇	Δ	‡	◇

1. Find the sum of Δ and ‡.

2. What is ◇ plus ◇?

3. In our number system, 0 can be added to any number without changing that number; 0 is called the **additive identity.** What is the additive identity for the system in the chart above? Explain your answer.

4. Does the Commutative Property of Addition apply to this system? Explain your answer.

5. What is −Δ (the opposite of Δ) equal to? Explain your answer.

6. What is −‡ (the opposite of ‡) equal to? Explain your answer.

7. Simplify −Δ + ‡ − ◇. Explain how you arrived at your answer.

The chart below is a multiplication table. Use it to answer Exercises 8 to 14.

×	£	¿	&
£	£	¿	&
¿	¿	¿	¿
&	&	¿	*

8. Find the product of £ and &.

9. What is ¿ times £?

10. Find the square of &.

11. Does the Commutative Property of Multiplication apply to this system? Explain your answer.

12. In our number system, the product of a number and 0 is 0. Is there an element in this system that corresponds to 0 in our system? Explain your answer.

13. In our number system, 1 can be multiplied by any number without changing that number; 1 is called the **multiplicative identity.** What is the multiplicative identity for the system in the chart above? Explain your answer.

14. Simplify & ÷ £ × ¿. Explain how you arrived at your answer.

Balance of Payments

An **export** is a good or service produced in one's own country and sold for consumption in another country. An **import** is a good or service that is consumed in one's own country but was bought from another country. A nation's **balance of trade** is the difference between the value of its exports and the value of its imports during a particular period of time.

Instructor Note

Before having students work on this activity, you might show them traditional addition and multiplication tables. Discuss the Commutative Properties of Addition and Multiplication, the Addition Property of Zero, the Multiplication Property of Zero, and the Multiplication Property of One.

Answers to Projects and Group Activities: Addition and Multiplication Properties

1. ◇
2. ◇
3. ◇
4. Yes. For example, ‡ + Δ = Δ + ‡. This is true for all elements in the table.
5. −Δ = ‡ because Δ + ‡ = 0.
6. −‡ = Δ because ‡ + Δ = 0.
7. −Δ + ‡ − ◇ = ‡ + ‡ − ◇ − Δ − ◇ = Δ − 0 = Δ
8. &
9. ¿
10. *
11. Yes. For example, ¿ × £ = £ × ¿. This is true for all elements in the table.
12. Yes, ¿. The product of any symbol and ¿ is equal to ¿.
13. £. The product of £ and any symbol is equal to the symbol.
14. & ÷ £ × ¿ = & × ¿ = ¿

Answers to Projects and Group Activities: Balance of Payments

Answers will vary depending on the year.

A **favorable balance of trade** exists when the value of the exports is greater than the value of the imports. In this case, the balance of trade is a positive number. An **unfavorable balance of trade** exists when the value of the imports is greater than the value of the exports. In this case, the balance of trade is a negative number. An unfavorable balance of trade is referred to as a **trade deficit.** A trade deficit is considered unfavorable because more money is going out of the country to pay for goods imported than is coming into the country to pay for goods exported.

The U.S. government provides data on international trade. On the Internet, go to **www.fedstats.gov** and do a search for "balance of payments." Find tables that provide data on the value of U.S. imports and exports. All figures in the tables are in millions of dollars. The first two columns are annual figures. Subsequent columns are for quarters (3-month periods).

We located the following data for the first quarter of the year 2000.

Exports: Total, all countries	183,659
Imports: Total, all countries	289,699

We then calculated the balance of trade as follows:

$$\begin{aligned}\text{Balance of trade} &= \text{value of exports} - \text{value of imports}\\ &= 183{,}659 - 289{,}699\\ &= 183{,}659 + (-289{,}699)\\ &= -106{,}040\end{aligned}$$

The balance of trade for the first quarter of 2000 was $-\$106{,}040$ million. This figure is provided in the table under "Balance: Total, all countries."

a. Show the calculation of the balance of trade for each of the four quarters of last year. Use the above calculation as a model.

b. Show the calculation of the annual balance of trade for last year.

c. Show that the sum of the four quarterly figures is equal to the annual figure.

Chapter Summary

Key Words

A *set* is a collection of objects. The objects in the set are called the *elements* of the set. The *roster method* of writing sets encloses a list of the elements in braces. [p. 55]

The set of *natural numbers* is {1, 2, 3, 4, 5, 6, 7, . . .}. The set of *integers* is {. . . , −4, −3, −2, −1, 0, 1, 2, 3, 4, . . .}. [p. 3]

The *empty set* or *null set,* written ∅ or { }, is the set that contains no elements. [p. 55]

A number *a is less than* another number *b,* written $a < b$, if a is to the left of b on the number line. A number *a is greater than* another number *b,* written $a > b$, if a is to the right of b on the number line. The symbol ≤ means *is less than or equal to.* The symbol ≥ means *is greater than or equal to.* An expression that contains the symbol >, <, ≥, or ≤ is an *inequality.* [pp. 3, 57]

Two numbers that are the same distance from zero on the number line but on opposite sides of zero are *opposite numbers,* or *opposites.* The *additive inverse* of a number is the opposite of the number. [p. 4]

TAKE NOTE

"To the Student" on page xxi provides suggestions on how best to make use of the "Chapter Summary," the "Chapter Review," and the "Chapter Test."

65

The *multiplicative inverse* of a number is the reciprocal of the number. [p. 41]

The *absolute value* of a number is its distance from zero on the number line. [p. 4]

A *rational number* is a number that can be written in the form $\frac{a}{b}$, where a and b are integers and $b \neq 0$. An *irrational number* is a number that has a decimal representation that never terminates or repeats. The rational numbers and the irrational numbers taken together are called the *real numbers*. [pp. 17, 24]

Percent means "parts of 100." [p. 18]

An expression of the form a^n is in *exponential form*, where a is the base and n is the exponent. [p. 22]

A *square root* of a positive number x is a number whose square is x. The *principal square root* of a number is the positive square root. The symbol $\sqrt{\ }$ is called a *radical sign* and is used to indicate the principal square root of a number. The *radicand* is the number under the radical sign. [p. 24]

The square of an integer is a *perfect square*. If a number is not a perfect square, its square root can only be approximated. [p. 24]

A *variable* is a letter that is used to stand for a quantity that is unknown or that varies. A *variable expression* is an expression that contains one or more variables. [p. 37]

The *terms* of a variable expression are the addends of the expression. A *variable term* is composed of a numerical coefficient and a variable part. *Like terms* of a variable expression are terms with the same variable part. [pp. 37, 38]

The *union* of two sets, written $A \cup B$, is the set that contains all the elements of A and all the elements of B. (Any elements that are in both set A and set B are listed only once.) The *intersection* of two sets, written $A \cap B$, is the set that contains the elements that are common to both A and B. [p. 55]

Essential Rules

To add two numbers with the same sign, add the absolute values of the numbers. Then attach the sign of the addends. [p. 5]

To add two numbers with different signs, find the absolute value of each number. Subtract the smaller of these from the larger. Attach the sign of the addend with the larger absolute value. [p. 5]

To subtract two numbers, add the opposite of the second number to the first number. [p. 6]

To multiply two numbers with the same sign, multiply the absolute values of the factors. The product is positive. [p. 8]

To multiply two numbers with different signs, multiply the absolute values of the factors. The product is negative. [p. 8]

To divide two numbers with the same sign, divide the absolute values of the numbers. The quotient is positive. [p. 9]

To divide two numbers with different signs, divide the absolute values of the numbers. The quotient is negative. [p. 9]

To convert a percent to a decimal, remove the percent sign and multiply by 0.01. [p. 18]

To convert a percent to a fraction, remove the percent sign and multiply by $\frac{1}{100}$. [p. 18]

To convert a decimal or a fraction to a percent, multiply by 100%. [p. 18]

Product Property of Square Roots [p. 25]	$\sqrt{ab} = \sqrt{a} \cdot \sqrt{b}$
Addition Property of Zero [p. 39]	$a + 0 = a$ or $0 + a = a$
Commutative Property of Addition [p. 39]	$a + b = b + a$
Associative Property of Addition [p. 39]	$(a + b) + c = a + (b + c)$
Inverse Property of Addition [p. 40]	$a + (-a) = 0$ or $-a + a = 0$
Multiplication Property of One [p. 41]	$a \cdot 1 = a$ or $1 \cdot a = a$
Commutative Property of Multiplication [p. 41]	$a \cdot b = b \cdot a$
Associative Property of Multiplication [p. 40]	$(a \cdot b) \cdot c = a \cdot (b \cdot c)$
Inverse Property of Multiplication [p. 41]	$a \cdot \frac{1}{a} = \frac{1}{a} \cdot a = 1, a \neq 0$
Division Properties of Zero and One [p. 9]	If $a \neq 0$, $0 \div a = 0$. If $a \neq 0$, $a \div a = 1$. $a \div 1 = a$ $a \div 0$ is undefined.
Distributive Property [p. 38]	$a(b + c) = ab + ac$

The Order of Operations Agreement [p. 33]

Step 1 Perform operations inside grouping symbols. The grouping symbols include parentheses, brackets, braces, absolute value symbols, and the fraction bar.

Step 2 Simplify exponential expressions.

Step 3 Do multiplication and division as they occur from left to right.

Step 4 Do addition and subtraction as they occur from left to right.

67

Chapter Review

1. Let $x \in \{-4, 0, 11\}$. For what values of x is the inequality $x < 1$ a true statement?
   $-4, 0$ [1.1A]

2. Find the additive inverse of -4.
   4 [1.1B]

3. Evaluate $-|-5|$.
   -5 [1.1B]

4. Add: $-3 + (-12) + 6 + (-4)$
   -13 [1.1C]

5. Subtract: $16 - (-3) - 18$
   1 [1.1C]

6. Multiply: $(-6)(7)$
   -42 [1.1D]

7. Divide: $-100 \div 5$
   -20 [1.1D]

8. Write $\frac{7}{25}$ as a decimal.
   0.28 [1.2A]

9. Write 6.2% as a decimal.
   0.062 [1.2B]

10. Write $\frac{5}{8}$ as a percent.
    62.5% [1.2B]

11. Simplify: $\frac{1}{3} - \frac{1}{6} + \frac{5}{12}$
    $\frac{7}{12}$ [1.2C]

12. Subtract: $5.17 - 6.238$
    -1.068 [1.2C]

13. Divide: $-\frac{18}{35} \div \frac{17}{28}$
    $-\frac{72}{85}$ [1.2D]

14. Multiply: $4.32(-1.07)$
    -4.6224 [1.2D]

15. Evaluate $\left(-\frac{2}{3}\right)^4$.
    $\frac{16}{81}$ [1.2E]

16. Simplify: $2\sqrt{36}$
    12 [1.2F]

17. Simplify: $-3\sqrt{120}$
    $-6\sqrt{30}$ [1.2F]

18. Evaluate $-3^2 + 4[18 + (12 - 20)]$.
    31 [1.3A]

Instructor Note

The notation [1.1A] following the answer to Exercise 1 indicates the objective that the student should review if that question is answered incorrectly. The notation [1.1A] means Chapter 1, Section 1, Objective A. This notation is used following every answer in all of the Prep Tests, Chapter Reviews, Chapter Tests, and Cumulative Reviews throughout the text.

19. Evaluate $(b - a)^2 + c$ when $a = -2$, $b = 3$, and $c = 4$.
29 [1.4A]

20. Simplify: $6a - 4b + 2a$
$8a - 4b$ [1.4B]

21. Simplify: $-3(-12y)$
$36y$ [1.4C]

22. Simplify: $5(2x - 7)$
$10x - 35$ [1.4D]

23. Simplify: $-4(2x - 9) + 5(3x + 2)$
$7x + 46$ [1.4D]

24. Simplify: $5[2 - 3(6x - 1)]$
$-90x + 25$ [1.4D]

25. Use the roster method to write the set of odd positive integers less than 8.
$\{1, 3, 5, 7\}$ [1.5A]

26. Find $A \cap B$, given $A = \{1, 5, 9, 13\}$ and $B = \{1, 3, 5, 7, 9\}$.
$A \cap B = \{1, 5, 9\}$ [1.5A]

27. Graph: $\{x|x > 3\}$

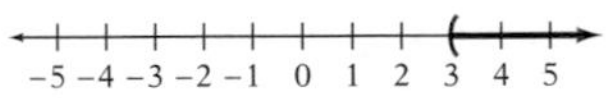

[1.5C]

28. Graph $\{x|x \leq 3\} \cup \{x|x < -2\}$.

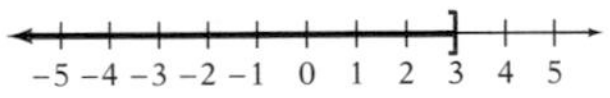

[1.5C]

29. To discourage random guessing on a multiple-choice exam, a professor assigns 6 points for a correct answer, −4 points for an incorrect answer, and −2 points for leaving a question blank. What is the score for a student who had 21 correct answers, had 5 incorrect answers, and left 4 questions blank?
98 [1.1E]

30. The circle graph shows the amount of candy consumed by Americans during a recent year (Source: Candy USA). What percent of the candy consumed was chocolate? Round to the nearest tenth of a percent.
50.8% [1.2G]

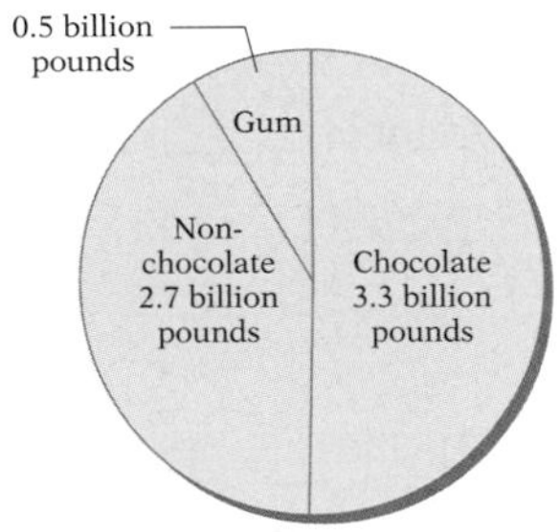

31. Translate "the difference between twice a number and one-half of the number" into a variable expression. Then simplify.
$2x - \frac{1}{2}x$; $\frac{3}{2}x$ [1.4E]

32. A baseball card collection contains five times as many National League players' cards as American League players' cards. Express the number of National League players' cards in terms of the number of American League players' cards.
Let A be the number of American League playing cards; $5A$ [1.4E]

33. A club treasurer has some five-dollar bills and some ten-dollar bills. The treasurer has a total of 35 bills. Express the number of five-dollar bills in terms of the number of ten-dollar bills.
Let T be the number of ten-dollar bills; $35 - T$ [1.4E]

69

Chapter Test

1. Place the correct symbol, < or >, between the two numbers.
$-2 > -40$ [1.1A]

2. Find the opposite of -4.
4 [1.1B]

3. Evaluate $-|-4|$.
-4 [1.1B]

4. Subtract: $16 - 30$
-14 [1.1C]

5. Add: $-22 + 14 + (-8)$
-16 [1.1C]

6. Subtract: $16 - (-30) - 42$
4 [1.1C]

7. Divide: $-561 \div (-33)$
17 [1.1D]

8. Write $\frac{7}{9}$ as a decimal. Place a bar over the repeating digit of the decimal.
$0.\overline{7}$ [1.2A]

9. Write 45% as a fraction and as a decimal.
$\frac{9}{20}$, 0.45 [1.2B]

10. Add: $-\frac{2}{5} + \frac{7}{15}$
$\frac{1}{15}$ [1.2C]

11. Multiply: $6.02(-0.89)$
-5.3578 [1.2D]

12. Divide: $\frac{5}{12} \div \left(-\frac{5}{6}\right)$
$-\frac{1}{2}$ [1.2D]

13. Evaluate $\frac{3}{4} \cdot (4)^2$
12 [1.2E]

14. Simplify: $-2\sqrt{45}$
$-6\sqrt{5}$ [1.2F]

15. Evaluate $16 \div 2[8 - 3(4 - 2)] + 1$.
17 [1.3A]

16. Evaluate $b^2 - 3ab$ when $a = 3$ and $b = -2$.
22 [1.4A]

17. Simplify: $3x - 5x + 7x$
$5x$ [1.4B]

18. Simplify: $\frac{1}{5}(10x)$
$2x$ [1.4C]

70 Chapter 1 / Real Numbers and Variable Expressions

19. Simplify: $-3(2x^2 - 7y^2)$
$-6x^2 + 21y^2$ [1.4D]

20. Simplify: $2x - 3(x - 2)$
$-x + 6$ [1.4D]

21. Simplify: $2x + 3[4 - (3x - 7)]$

$-7x + 33$ [1.4D]

22. Use the roster method to write the set of integers between -3 and 4.
$\{-2, -1, 0, 1, 2, 3\}$ [1.5A]

23. Use set-builder notation to write the set of real numbers less than -3.
$\{x \mid x < -3, x \in \text{real numbers}\}$ [1.5B]

24. Find $A \cup B$ given $A = \{1, 3, 5, 7\}$ and $B = \{2, 4, 6, 8\}$.
$A \cup B = \{1, 2, 3, 4, 5, 6, 7, 8\}$ [1.5A]

25. Graph $\{x \mid x < 1\}$.

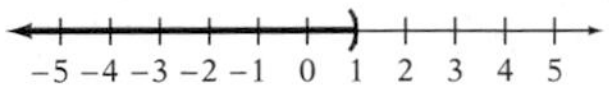

[1.5C]

26. Graph $\{x \mid x \le -3\} \cup \{x \mid x > 0\}$.

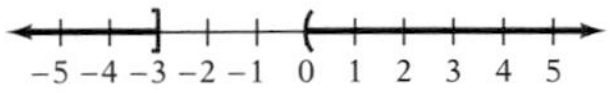

[1.5C]

27. Translate "ten times the difference between a number and 3" into a variable expression. Then simplify.
$10(x - 3)$; $10x - 30$ [1.4E]

28. The speed of a pitcher's fastball is twice the speed of the catcher's return throw. Express the speed of the fastball in terms of the speed of the return throw.
Let s be the speed of the catcher's return throw; $2s$ [1.4E]

29. The table at the right shows the U.S. balance of trade, in billions of dollars, for the years 1980 to 2000 (Source: U.S. Dept. of Commerce). See the Projects and Group Activities for a discussion of balance of trade.

a. In which years did the trade balance increase from the previous year? '81, '88, '89, '90, '91, '95 [1.2G]

b. Calculate the difference between the trade balance in 1990 and that in 2000. $288.6 billion [1.2G]

c. During which two consecutive years was the difference in the trade balance greatest? 1999 and 2000 [1.2G]

d. How many times greater was the trade balance in 1990 than in 1980? Round to the nearest whole number.
4 times greater [1.2G]

e. Calculate the average trade balance per quarter for the year 2000. −$92.425 billion [1.2G]

Year	Trade Balance
1980	−19.4
1981	−16.2
1982	−24.2
1983	−57.8
1984	−109.2
1985	−122.1
1986	−140.6
1987	−153.3
1988	−115.9
1989	−92.2
1990	−81.1
1991	−30.7
1992	−35.7
1993	−68.9
1994	−97.0
1995	−95.9
1996	−102.1
1997	−104.7
1998	−166.9
1999	−265.0
2000	−369.7

30. The boiling point of mercury is 356.58°C. The melting point of mercury is −38.87°C. Find the difference between the boiling point and the melting point of mercury. 395.45°C [1.2G]

Chapter 2

First-Degree Equations and Inequalities

To ensure that heavy equipment, such as a truck transmission, functions properly, each component must be built precisely to the manufacturer's specifications. For example, the thickness and length of the teeth in a gear is specified. A vernier dial caliper, as shown above, is used to measure the components to the nearest thousandth of an inch. A slight variance from the given measurement is allowed. This margin of error is known as tolerance. Given the tolerance of a component, you can calculate the upper and lower limits, or the maximum and minimum allowable measurements. This is illustrated in **Exercises 91–96 on pages 131 and 132.**

Objectives

Section 2.1

- **A** To determine whether a given number is a solution of an equation
- **B** To solve an equation of the form $x + a = b$
- **C** To solve an equation of the form $ax = b$
- **D** To solve application problems using the basic percent equation

Section 2.2

- **A** To solve an equation of the form $ax + b = c$
- **B** To solve an equation of the form $ax + b = cx + d$
- **C** To solve an equation containing parentheses
- **D** To translate a sentence into an equation and solve

Section 2.3

- **A** To solve value mixture problems
- **B** To solve percent mixture problems
- **C** To solve investment problems
- **D** To solve uniform motion problems

Section 2.4

- **A** To solve an inequality in one variable
- **B** To solve a compound inequality
- **C** To solve application problems

Section 2.5

- **A** To solve an absolute value equation
- **B** To solve an absolute value inequality
- **C** To solve application problems

WWW WEB Need help? For on-line student resources, such as section quizzes, visit this textbook's web site at **college.hmco.com/students.**

Features of the IAE

For a description of the features included in the margins of this Instructor's Annotated Edition, see page 2.

Instructor Note

The *Prep Test* is a means to test your students' mastery of *prerequisite* material that is assumed in the coming chapter. All answers, along with the objective to review (if necessary), are provided in the Answers to Selected Exercises appendix.

Prep Test

For Exercises 1 to 5, add, subtract, multiply, or divide.

1. $8 - 12$
-4 [1.1C]

2. $-9 + 3$
-6 [1.1C]

3. $\dfrac{-18}{-6}$
3 [1.1D]

4. $-\dfrac{3}{4}\left(-\dfrac{4}{3}\right)$
1 [1.2D]

5. $-\dfrac{5}{8}\left(\dfrac{4}{5}\right)$
$-\dfrac{1}{2}$ [1.2D]

For Exercises 6 to 9, simplify.

6. $3x - 5 + 7x$
$10x - 5$ [1.4B]

7. $6(x - 2) + 3$
$6x - 9$ [1.4D]

8. $n + (n + 2) + (n + 4)$
$3n + 6$ [1.4B]

9. $0.08x + 0.05(400 - x)$
$0.03x + 20$ [1.4D]

10. A 20-ounce package of a snack mixture contains nuts and pretzels. Let n represent the number of ounces of nuts in the mixture. Express the number of ounces of pretzels in the mixture in terms of n.
$20 - n$ [1.4E]

Go Figure

Two fractions are inserted between $\frac{1}{4}$ and $\frac{1}{2}$ so that the difference between any two successive fractions is the same. Find the sum of the four fractions.
$\dfrac{3}{2}$

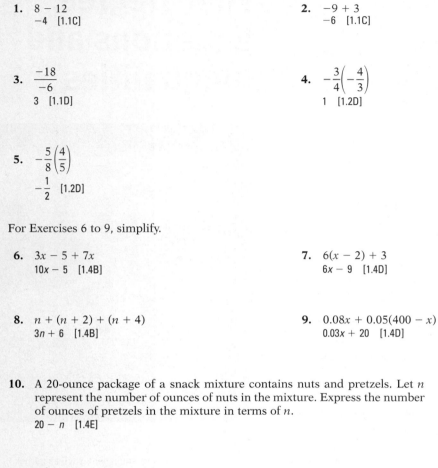

2.1 Introduction to Equations

Objective A To determine whether a given number is a solution of an equation

VIDEO & DVD CD TUTOR WEB SSM

Point of Interest

One of the most famous equations ever stated is $E = mc^2$. This equation, stated by Albert Einstein, shows that there is a relationship between mass m and energy E. As a side note, the chemical element einsteinium was named in honor of Einstein.

An **equation** expresses the equality of two mathematical expressions. The expressions can be either numerical or variable expressions.

$$\left.\begin{array}{l} 9 + 3 = 12 \\ 3x - 2 = 10 \\ y^2 + 4 = 2y - 1 \\ z = 2 \end{array}\right\} \text{Equations}$$

The equation at the right is true if the variable is replaced by 5.

$x + 8 = 13$

$5 + 8 = 13$ A true equation

The equation is false if the variable is replaced by 7.

$7 + 8 = 13$ A false equation

A **solution** of an equation is a number that, when substituted for the variable, results in a true equation. 5 is a solution of the equation $x + 8 = 13$. 7 is not a solution of the equation $x + 8 = 13$.

➡ Is -2 a solution of $2x + 5 = x^2 - 3$?

TAKE NOTE

The Order of Operations Agreement applies to evaluating $2(-2) + 5$ and $(-2)^2 - 3$.

$$\begin{array}{r|l} \multicolumn{2}{c}{2x + 5 = x^2 - 3} \\ \hline 2(-2) + 5 & (-2)^2 - 3 \\ -4 + 5 & 4 - 3 \\ \multicolumn{2}{c}{1 = 1} \end{array}$$

- Replace x by -2.
- Evaluate the numerical expressions.
- If the results are equal, -2 is a solution of the equation. If the results are not equal, -2 is not a solution of the equation.

Yes, -2 is a solution of the equation.

Example 1 Is -4 a solution of $5x - 2 = 6x + 2$?

Solution

$$\begin{array}{r|l} \multicolumn{2}{c}{5x - 2 = 6x + 2} \\ \hline 5(-4) - 2 & 6(-4) + 2 \\ -20 - 2 & -24 + 2 \\ \multicolumn{2}{c}{-22 = -22} \end{array}$$

Yes, -4 is a solution.

You Try It 1 Is $\frac{1}{4}$ a solution of $5 - 4x = 8x + 2$?

Your solution yes

Example 2 Is -4 a solution of $4 + 5x = x^2 - 2x$?

Solution

$$\begin{array}{r|l} \multicolumn{2}{c}{4 + 5x = x^2 - 2x} \\ \hline 4 + 5(-4) & (-4)^2 - 2(-4) \\ 4 + (-20) & 16 - (-8) \\ \multicolumn{2}{c}{-16 \neq 24} \end{array}$$

($\neq$ means "is not equal to")

No, -4 is not a solution.

You Try It 2 Is 5 a solution of $10x - x^2 = 3x - 10$?

Your solution no

Solutions on p. S3

Objective 2.1A

New Vocabulary

equation
solution

Instructor Note

Explain that determining whether a given number is a solution of an equation is a way to check an answer after an equation has been solved.

Discuss the Concepts

1. What is a solution of an equation?
2. How can we determine whether a number is a solution of an equation?

Concept Check

1. Label each of the following as either an expression or an equation.
   a. $3x + 7 = 9$
   b. $3x + 7$
   c. $4 - 6(y + 5)$
   d. $a + b = 8$
   e. $a + b - 8$
   a and d are equations. b, c, and e are expressions.
2. Which of the numbers in parentheses are solutions of the equation?
   a. $x^2 - 7x = 8$
   $(-1, 2, 1, 8)$ $-1, 8$
   b. $y^2 = 2y + 15$
   $(5, 3, 5, 3)$ $3, 5$
   c. $2x(x + 3) = x + 12$
   $\left(-4, -\frac{3}{2}, \frac{3}{2}, 4\right)$ $-4, \frac{3}{2}$
   d. $2x^2 - 6x = 5x - 15$
   $\left(-3, -\frac{5}{2}, \frac{5}{2}, 3\right)$ $\frac{5}{2}, 3$

In-Class Examples (Objective 2.1A)

1. Is -6 a solution of the equation $4x + 3 = 2x - 9$? Yes
2. Is $-\frac{2}{3}$ a solution of the equation $4 - 6x = 9x + 1$? No
3. Is -2 a solution of $3(x - 2) = x^2 - 16$? Yes

Objective 2.1B

New Vocabulary
solve an equation
equivalent equations

New Properties
Addition Property of Equations

Discuss the Concepts

1. What is the solution of the equation $x = 8$? Use the answer to explain why the goal in solving the equations is to get the variable alone on one side of the equation.
2. Which of the following are equations of the form $x + a = b$? Explain your answer. If it is an equation of the form $x + a = b$, what would you do to solve the equation?
   a. $d + 7.8 = -9.2$
   b. $0.3 = a + 1.4$
   c. $-9 = 3y$
   d. $-8 + c = -5.6$
   a, b, and d are equations of the form $x + a = b$.

Optional Student Activity

1. If $y - 3 = 2$ and $x + 7 = 13$, then is $x > y$ or $y > x$? $x > y$
2. Two numbers form a "two-pair" if the sum of their reciprocals equals 2. For example, $\frac{8}{15}$ and 8 are a two-pair because $\frac{15}{8} + \frac{1}{8} = 2$. If two numbers, a and b, form a two-pair and $a = \frac{7}{3}$, what is the value of b? $\frac{7}{11}$
3. Make equations using the numbers 5, 10, and 15 to fill the boxes in $x + \square = \square - \square$.
   a. What is the largest number solution possible? 0
   b. What is the smallest number solution possible? −20

Objective B To solve an equation of the form $x + a = b$

Point of Interest
Finding solutions of equations has been a principal aim of mathematics for thousands of years. However, the equals sign did not occur in any text until 1557.

To **solve an equation** means to find a solution of the equation. The simplest equation to solve is an equation of the form *variable* = *constant*, because the constant is the solution.

The solution of the equation $x = 5$ is 5 because $5 = 5$ is a true equation.

The solution of the equation at the right is 7 because $7 + 2 = 9$ is a true equation. $\quad x + 2 = 9 \qquad 7 + 2 = 9$

Note that if 4 is added to each side of the equation $x + 2 = 9$, the solution is still 7.

$$x + 2 = 9$$
$$x + 2 + 4 = 9 + 4$$
$$x + 6 = 13 \qquad 7 + 6 = 13$$

If −5 is added to each side of the equation $x + 2 = 9$, the solution is still 7.

$$x + 2 = 9$$
$$x + 2 + (-5) = 9 + (-5)$$
$$x - 3 = 4 \qquad 7 - 3 = 4$$

Equations that have the same solution are **equivalent equations**. The equations $x + 2 = 9$, $x + 6 = 13$, and $x - 3 = 4$ are equivalent equations; each equation has 7 as its solution. These examples suggest that adding the same number to each side of an equation produces an equivalent equation. This is called the *Addition Property of Equations.*

> **Addition Property of Equations**
> The same number can be added to each side of an equation without changing its solution. In symbols, the equation $a = b$ has the same solution as the equation $a + c = b + c$.

In solving an equation, the goal is to rewrite the given equation in the form *variable* = *constant*. The Addition Property of Equations is used to remove a *term* from one side of the equation by adding the opposite of that term to each side of the equation.

TAKE NOTE
Think of an equation as a balance scale. If the weights added to each side of an equation are not the same, the "pans" no longer balance.

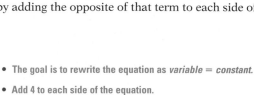

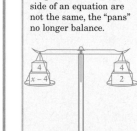

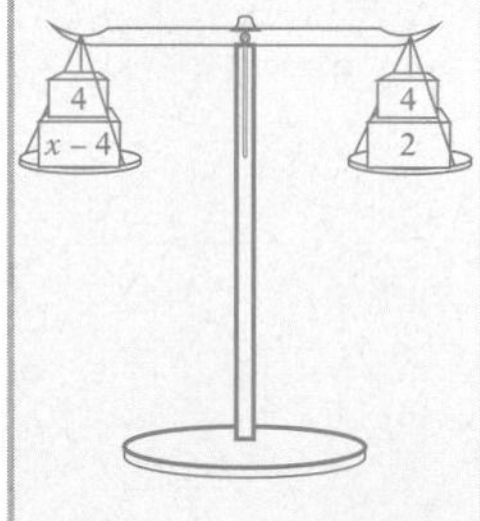

➡ Solve: $x - 4 = 2$

$x - 4 = 2$ • **The goal is to rewrite the equation as *variable* = *constant*.**
$x - 4 + 4 = 2 + 4$ • **Add 4 to each side of the equation.**
$x + 0 = 6$ • **Simplify.**
$x = 6$ • **The equation is in the form *variable* = *constant*.**

Check: $x - 4 = 2$
$6 - 4 \mid 2$
$2 = 2$ A true equation

The solution is 6.

Because subtraction is defined in terms of addition, the Addition Property of Equations also makes it possible to subtract the same number from each side of an equation without changing the solution of the equation.

In-Class Examples (Objective 2.1B)
Solve.
1. $3 + x = 9$ 6
2. $y - \frac{1}{4} = \frac{5}{6}$ $\frac{13}{12}$
3. $5 = b + 5$ 0
4. $d + 4 = -8$ −12

➡ Solve: $y + \frac{3}{4} = \frac{1}{2}$

$$y + \frac{3}{4} = \frac{1}{2}$$

• The goal is to rewrite the equation in the form *variable = constant.*

$$y + \frac{3}{4} - \frac{3}{4} = \frac{1}{2} - \frac{3}{4}$$

• Subtract $\frac{3}{4}$ from each side of the equation.

$$y + 0 = \frac{2}{4} - \frac{3}{4}$$

• Simplify.

$$y = -\frac{1}{4}$$

• The equation is in the form *variable = constant.*

The solution is $-\frac{1}{4}$. You should check this solution.

Example 3

Solve: $x + \frac{3}{4} = \frac{1}{3}$

Solution

$$x + \frac{3}{4} = \frac{1}{3}$$
$$x + \frac{3}{4} - \frac{3}{4} = \frac{1}{3} - \frac{3}{4}$$
$$x + 0 = \frac{4}{12} - \frac{9}{12}$$
$$x = -\frac{5}{12}$$

The solution is $-\frac{5}{12}$.

You Try It 3

Solve: $\frac{5}{6} = y - \frac{3}{8}$

Your solution $\frac{29}{24}$

Solution on p. S4

Objective C To solve an equation of the form $ax = b$

VIDEO & DVD · CD TUTOR · WWW WEB · SSM

The solution of the equation at the right is 3 because $2 \cdot 3 = 6$ is a true equation.

$$2x = 6 \qquad 2 \cdot 3 = 6$$

Note that if each side of $2x = 6$ is multiplied by 5, the solution is still 3.

$$2x = 6$$
$$5(2x) = 5 \cdot 6$$
$$10x = 30 \qquad 10 \cdot 3 = 30$$

If each side of $2x = 6$ is multiplied by -4, the solution is still 3.

$$2x = 6$$
$$(-4)(2x) = (-4) \cdot 6$$
$$-8x = -24 \qquad -8 \cdot 3 = -24$$

The equations $2x = 6$, $10x = 30$, and $-8x = -24$ are equivalent equations; each equation has 3 as its solution. These examples suggest that multiplying each side of an equation by the same number produces an equivalent equation.

Multiplication Property of Equations

Each side of an equation can be multiplied by the same *nonzero* number without changing the solution of the equation. In symbols, if $c \neq 0$, then the equation $a = b$ has the same solutions as the equation $ac = bc$.

Objective 2.1C

New Properties

Multiplication Property of Equations

Instructor Note

The requirement that each side be multiplied by a *nonzero* number is important. Later in the text, students will solve some equations by multiplying each side by a variable expression whose value may be zero. The result may be extraneous solutions.

Discuss the Concepts

Which of the following are equations of the form $ax = b$? Explain your answer. If it is an equation of the form $ax = b$, what would you do to solve the equation?

a. $3y = -12$
b. $2.4 = 0.6a$
c. $-5 = z - 10$
d. $-8c = -56$

a, b, and d are equations of the form $ax = b$.

Concept Check

Equations that have the same solution are equivalent equations. Which of the following are equivalent equations?

1. $5x = -20$
2. $-2x = 8$
3. $24 = 6x$
4. $\frac{3}{4}x = -3$
5. $-1 = -\frac{1}{4}x$
6. $x = 4$
7. $x = -4$

1, 2, 4, and 7 are equivalent equations. 3, 5, and 6 are equivalent equations.

In-Class Examples (Objective 2.1C)

Solve.

1. $-\frac{5}{8}x = 25$ −40
2. $4y - 10y = -42$ 7
3. $9 = \frac{3x}{4}$ 12
4. $2z = 0$ 0

Optional Student Activity

Match each numbered equation with a lettered question that can be used to solve it.

1. $x + 3 = 8$ d
2. $x - 5 = 20$ b
3. $4x = 16$ a
4. $\frac{x}{7} = 1$ e
5. $99 = -9x$ c

a. 4 times what number is equal to 16?

b. What number minus 5 is equal to 20?

c. 99 is equal to -9 times what number?

d. What number plus 3 is equal to 8?

e. What number divided by 7 is equal to 1?

The Multiplication Property of Equations is used to remove a coefficient by multiplying each side of the equation by the reciprocal of the coefficient.

➡ Solve: $\frac{3}{4}z = 9$

$$\frac{3}{4}z = 9$$
• The goal is to rewrite the equation in the form *variable = constant.*

$$\frac{4}{3} \cdot \frac{3}{4}z = \frac{4}{3} \cdot 9$$
• Multiply each side of the equation by $\frac{4}{3}$.

$$1 \cdot z = 12$$
• Simplify.

$$z = 12$$
• The equation is in the form *variable = constant.*

The solution is 12. You should check this solution.

Because division is defined in terms of multiplication, each side of an equation can be divided by the same nonzero number without changing the solution of the equation.

➡ Solve: $6x = 14$

$$6x = 14$$
• The goal is to rewrite the equation in the form *variable = constant.*

$$\frac{6x}{6} = \frac{14}{6}$$
• Divide each side of the equation by 6.

$$x = \frac{7}{3}$$
• Simplify. The equation is in the form *variable = constant.*

The solution is $\frac{7}{3}$.

TAKE NOTE
Remember to check the solution.

Check: $6x = 14$

$6\left(\frac{7}{3}\right) \mid 14$

$14 = 14$

When using the Multiplication Property of Equations, multiply each side of the equation by the reciprocal of the coefficient when the coefficient is a fraction. Divide each side of the equation by the coefficient when the coefficient is an integer or decimal.

Example 4

Solve: $\frac{3x}{4} = -9$

Solution

$$\frac{3x}{4} = -9$$

$$\frac{4}{3} \cdot \frac{3}{4}x = \frac{4}{3}(-9) \quad \bullet \left[\frac{3x}{4} = \frac{3}{4}x\right]$$

$$x = -12$$

The solution is -12.

You Try It 4

Solve: $-\frac{2x}{5} = 6$

Your solution -15

Example 5 Solve: $5x - 9x = 12$

Solution

$$5x - 9x = 12$$

$$-4x = 12 \quad \bullet \text{ Combine like terms.}$$

$$\frac{-4x}{-4} = \frac{12}{-4}$$

$$x = -3$$

The solution is -3.

You Try It 5 Solve: $4x - 8x = 16$

Your solution -4

Solutions on p. S4

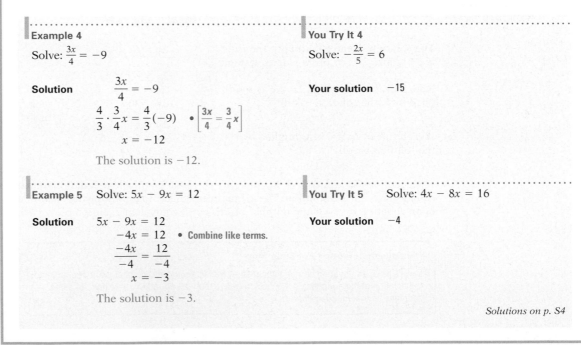

2.2 General Equations

Objective A — To solve an equation of the form $ax + b = c$

VIDEO & DVD · CD TUTOR · WEB · SSM

Point of Interest

Evariste Galois, despite being killed in a duel at the age of 21, made significant contributions to solving equations. In fact, there is a branch of mathematics called Galois Theory showing what kinds of equations can and what kinds cannot be solved.

In solving an equation of the form $ax + b = c$, the goal is to rewrite the equation in the form *variable* = *constant*. This requires the application of both the Addition and the Multiplication Properties of Equations.

⇒ Solve: $\frac{3}{4}x - 2 = -11$

The goal is to write the equation in the form *variable* = *constant*.

$$\frac{3}{4}x - 2 = -11$$

$$\frac{3}{4}x - 2 + 2 = -11 + 2$$ • Add 2 to each side of the equation.

$$\frac{3}{4}x = -9$$ • Simplify.

$$\frac{4}{3} \cdot \frac{3}{4}x = \frac{4}{3}(-9)$$ • Multiply each side of the equation by $\frac{4}{3}$.

$$x = -12$$ • The equation is of the form *variable* = *constant*.

Check: • Check the solution.

$\frac{3}{4}x - 2 = -11$	
$\frac{3}{4}(-12) - 2$	-11
$-9 - 2$	-11
$-11 = -11$	• A true equation

The solution is −12.

Example 1 Solve: $3x - 7 = -5$

Solution

$$3x - 7 = -5$$
$$3x - 7 + 7 = -5 + 7$$
$$3x = 2$$
$$\frac{3x}{3} = \frac{2}{3}$$
$$x = \frac{2}{3}$$

The solution is $\frac{2}{3}$.

You Try It 1 Solve: $5x + 7 = 10$

Your solution $\frac{3}{5}$

Solution on p. S4

Objective 2.2A

Instructor Note

The objective for this lesson is to "solve equations of the form $ax + b = c$." If you ask students to name the variables in this equation, they may answer *a*, *b*, *c*, and *x*, which is not an unreasonable answer. However, as writing symbolic math evolved, it became customary to think of letters at the beginning of the alphabet as constants and coefficients and of those at the end of the alphabet as variables. This kind of implicit understanding is often lost on students.

For this objective, the goal is to solve for *x* given the coefficient *a* and the constants *b* and *c*. In the next objective, students will solve equations of the form $ax + b = cx + d$, again with the implicit understanding that *a* and *c* are coefficients and *b* and *d* are constants.

Later in the text, students will be introduced to the equation $y = mx + b$, which also makes implicit assumptions about variables, coefficients, and constants.

Discuss the Concepts

Explain the steps you would take to solve each equation.

a. $5y + 1 = 11$
b. $2z - 9 = 11$
c. $12 = 2 + 5a$
d. $-7v + 6 = -8$
e. $8 - 5x = -12$
f. $9 = 15 - 2y$

Concept Check

Match the equation with the first step in solving that equation.

a. $3x - 7 = 5$
b. $4x + 7 = -5$
c. $7x - 5 = 2$
d. $-7x + 5 = -2$
1. Add 7 to each side.
2. Add 5 to each side.
3. Subtract 7 from each side.
4. Subtract 5 from each side.

a and 1, b and 3, c and 2, d and 4

In-Class Examples (Objective 2.2A)

Solve.

1. $8a + 3 = 10$ $\frac{7}{8}$

2. $7 = 12 + 5h$ -1

3. $3 - 5y = 2$ $\frac{1}{5}$

4. $\frac{2}{5}x - \frac{1}{4} = \frac{3}{2}$ $\frac{35}{8}$

Optional Student Activity

1. Create a three-question quiz on the material in this objective. Provide solutions for all three questions.
2. Make up an equation of the form:
   a. $x + a = b$ that has -4 as a solution.
   b. $ax = b$ that has -5 as a solution.
   c. $ax + b = c$ that has -3 as a solution.

Objective 2.2B

Discuss the Concepts

1. When solving an equation of the form $ax + b = cx + d$, would it be correct always to start by adding or subtracting the variable term
   a. on the right side of the equation? Yes
   b. on the left side of the equation? Yes
   c. with the smaller coefficient? Yes
2. Can the equation at the right be solved by subtracting 3 from each side before subtracting $5x$ from each side? If so, complete the solution using this approach.

Example 2
Solve: $5 = 9 - 2x$

Solution

$$5 = 9 - 2x$$
$$5 - 9 = 9 - 9 - 2x$$
$$-4 = -2x$$
$$\frac{-4}{-2} = \frac{-2x}{-2}$$
$$2 = x$$

The solution is 2.

You Try It 2
Solve: $2 = 11 + 3x$

Your solution
-3

Example 3
Solve: $2x + 4 - 5x = 10$

Solution

$$2x + 4 - 5x = 10$$
$$-3x + 4 = 10 \quad \text{• Combine like terms.}$$
$$-3x + 4 - 4 = 10 - 4$$
$$-3x = 6$$
$$\frac{-3x}{-3} = \frac{6}{-3}$$
$$x = -2$$

The solution is -2.

You Try It 3
Solve: $x - 5 + 4x = 25$

Your solution
6

Solutions on p. S4

Objective B To solve an equation of the form $ax + b = cx + d$

VIDEO & DVD CD TUTOR WEB SSM

In solving an equation of the form $ax + b = cx + d$, the goal is to rewrite the equation in the form *variable* = *constant*. Begin by rewriting the equation so that there is only one variable term in the equation. Then rewrite the equation so that there is only one constant term.

➡ Solve: $2x + 3 = 5x - 9$

$$2x + 3 = 5x - 9$$
$$2x - 5x + 3 = 5x - 5x - 9 \quad \text{• Subtract } 5x \text{ from each side of the equation.}$$
$$-3x + 3 = -9 \quad \text{• Simplify.}$$
$$-3x + 3 - 3 = -9 - 3 \quad \text{• Subtract 3 from each side of the equation.}$$
$$-3x = -12 \quad \text{• Simplify.}$$
$$\frac{-3x}{-3} = \frac{-12}{-3} \quad \text{• Divide each side of the equation by } -3.$$
$$x = 4 \quad \text{• The equation is in the form } variable = constant.$$

The solution is 4. You should verify this by checking this solution.

In-Class Examples (Objective 2.2B)

Solve.

1. $5x - 4 = 3x - 10$ -3
2. $8x + 3 - 4x = 5 + x$ $\frac{2}{3}$
3. $3x - 7 = 5x - 7$ 0

Example 4
Solve: $4x - 5 = 8x - 7$

Solution

$$4x - 5 = 8x - 7$$
$$4x - 8x - 5 = 8x - 8x - 7$$
$$-4x - 5 = -7$$
$$-4x - 5 + 5 = -7 + 5$$
$$-4x = -2$$
$$\frac{-4x}{-4} = \frac{-2}{-4}$$
$$x = \frac{1}{2}$$

The solution is $\frac{1}{2}$.

You Try It 4
Solve: $5x + 4 = 6 + 10x$

Your solution

$-\frac{2}{5}$

Example 5
Solve: $3x + 4 - 5x = 2 - 4x$

Solution

$$3x + 4 - 5x = 2 - 4x$$
$$-2x + 4 = 2 - 4x$$
$$-2x + 4x + 4 = 2 - 4x + 4x$$
$$2x + 4 = 2$$
$$2x + 4 - 4 = 2 - 4$$
$$2x = -2$$
$$\frac{2x}{2} = \frac{-2}{2}$$
$$x = -1$$

The solution is -1.

You Try It 5
Solve: $5x - 10 - 3x = 6 - 4x$

Your solution

$\frac{8}{3}$

Solutions on p. S4

Objective C To solve an equation containing parentheses

VIDEO & DVD CD TUTOR WEB SSM

When an equation contains parentheses, one of the steps in solving the equation requires the use of the Distributive Property. The Distributive Property is used to remove parentheses from a variable expression.

Concept Check

1. A friend needs help solving the equation $3x - 7 = 5x + 1$. Write an explanation that your friend could use in solving the equation.
2. Make up an equation of the form $ax + b = cx + d$ that has -3 as a solution.

Optional Student Activity

1. Given $2x + 1 = 3a + 5 = 4x - 9$, find the value of a. 2
2. At the local gym, either you can pay \$75 for a year's membership and \$5 per visit, or you can pay \$150 for a year's membership plus \$2 per visit. How many times during the year must you go to the gym for the two plans to be equal in price? 25 visits
3. The gauge on a water tank shows that the tank is $\frac{5}{8}$ full. After 18 more gallons are drained from the tank, the gauge shows that it is $\frac{1}{4}$ full. How many gallons of water were in the tank when it was $\frac{1}{2}$ full? 24 gallons

Objective 2.2C

Properties to Review
Distributive Property

Instructor Note
Remind students that the goal is still *variable = constant.*

In-Class Examples (Objective 2.2C)

Solve.

1. $5 - 4(x + 6) = 21$ -10
2. $9x - 3(2x + 5) = 4(5x + 2) - 6$ -1
3. $5[6 - 2(5x + 1)] = 8x - 9$ $\frac{1}{2}$

Discuss the Concepts
What are the steps involved in solving an equation containing parentheses?

Concept Check
Which of the following equations have a solution of 2?
1. $3 + 2(4x - 1) = 5(2x - 1) + 2$
2. $x - (3x + 2) = 4(1 - x) - 2$
3. $2[5 - 3(x + 1)] = 4(x - 4)$

Equations 1, 2, and 3 have a solution of 2.

Optional Student Activity
1. a. Solve: $3(2x - 1) - (6x - 4) = -9$ No solution
   b. Solve: $2(5x - 6) - 3(x - 4) = 7x + 14$ No solution
2. I am thinking of a number. When I subtract 4 from the number and then take 300% of the result, it is equal to the original number. What is the original number? 6
3. If $s = 5x - 3$ and $t = x + 4$, find the value x for which $s = 3t - 1$. 7
4. The population of the town of Hampton increased by 10,000 people during the 1990s. In the first decade of the new millennium, the population of Hampton decreased by 10%, at which time the town had 6000 more people than at the beginning of the 1990s. Find Hampton's population at the beginning of the 1990s. 30,000 people

➡ Solve: $4 + 5(2x - 3) = 3(4x - 1)$

$4 + 5(2x - 3) = 3(4x - 1)$	
$4 + 10x - 15 = 12x - 3$	• Use the Distributive Property. Then simplify.
$10x - 11 = 12x - 3$	
$10x - 12x - 11 = 12x - 12x - 3$	• Subtract $12x$ from each side of the equation.
$-2x - 11 = -3$	• Simplify.
$-2x - 11 + 11 = -3 + 11$	• Add 11 to each side of the equation.
$-2x = 8$	• Simplify.
$\frac{-2x}{-2} = \frac{8}{-2}$	• Divide each side of the equation by -2.
$x = -4$	• The equation is in the form *variable = constant.*

The solution is -4. You should verify this by checking this solution.

Example 6
Solve: $3x - 4(2 - x) = 3(x - 2) - 4$

Solution

$$3x - 4(2 - x) = 3(x - 2) - 4$$
$$3x - 8 + 4x = 3x - 6 - 4$$
$$7x - 8 = 3x - 10$$
$$7x - 3x - 8 = 3x - 3x - 10$$
$$4x - 8 = -10$$
$$4x - 8 + 8 = -10 + 8$$
$$4x = -2$$
$$\frac{4x}{4} = \frac{-2}{4}$$
$$x = -\frac{1}{2}$$

The solution is $-\frac{1}{2}$.

You Try It 6
Solve: $5x - 4(3 - 2x) = 2(3x - 2) + 6$

Your solution
2

Solution on p. S5

Example 7
Solve: $3[2 - 4(2x - 1)] = 4x - 10$

Solution

$$3[2 - 4(2x - 1)] = 4x - 10$$
$$3[2 - 8x + 4] = 4x - 10$$
$$3[6 - 8x] = 4x - 10$$
$$18 - 24x = 4x - 10$$
$$18 - 24x - 4x = 4x - 4x - 10$$
$$18 - 28x = -10$$
$$18 - 18 - 28x = -10 - 18$$
$$-28x = -28$$
$$\frac{-28x}{-28} = \frac{-28}{-28}$$
$$x = 1$$

The solution is 1.

You Try It 7
Solve: $-2[3x - 5(2x - 3)] = 3x - 8$

Your solution
2

Solution on p. S5

Objective D To translate a sentence into an equation and solve

VIDEO & DVD CD TUTOR WEB SSM

An equation states that two mathematical expressions are equal. Therefore, to translate a sentence into an equation requires recognition of the words or phrases that mean "equals." Some of these words and phrases are listed below.

equals *is* *totals* *is the same as*

Point of Interest

Number puzzle problems similar to the one on this page have appeared in textbooks for hundreds of years. Here is one from a 1st-century Chinese textbook: "When a number is divided by 3, the remainder is 2; when it is divided by 5, the remainder is 3; when it is divided by 7, the remainder is 2. Find the number." There are actually an infinite number of solutions to this problem. See if you can find one of them.

⇒ Translate "five less than four times a number is four more than the number" into an equation and solve.

the unknown number: n

- **Assign a variable to the unknown number.**

five less than four times a number	is	four more than the number

- **Find two verbal expressions for the same value.**

$$4n - 5 = n + 4$$
$$4n - n - 5 = n - n + 4$$
$$3n - 5 = 4$$
$$3n - 5 + 5 = 4 + 5$$
$$3n = 9$$
$$\frac{3n}{3} = \frac{9}{3}$$
$$n = 3$$

- **Write an equation.**
- **Solve the equation.**
- ***Check:***

5 less than 4 times 3	4 more than 3
$4 \cdot 3 - 5$	$3 + 4$
$12 - 5$	7

$7 = 7$

The number is 3.

In-Class Examples (Objective 2.2D)

Translate into an equation and solve.

1. The sum of a number and six is fifteen. Find the number. 9
2. The product of 4 and a number is negative twenty-four. Find the number. -6
3. The difference between nine and four times a number is three. Find the number. $1\frac{1}{2}$
4. The sum of two-fifths of a number and three is negative one. Find the number. -10

Solve.

5. The relationship between Celsius temperature and Fahrenheit temperature is given by the formula $C = \frac{5}{9}(F - 32)$, where C is the Celsius temperature and F is the Fahrenheit temperature. Find the Fahrenheit temperature when the Celsius temperature is $-22°$. $-7.6°F$

Objective 2.2D

New Vocabulary

See the list of words and phrases that mean "equals."

Concept Check

In building highways, engineers provide for expansion joints so the highway won't break when it expands on very hot days. For a particular 1-mile portion of a two-lane highway, an engineer used the equation $I = 0.06336T - 0.00084L$, where I is the expansion in feet, T is the temperature in degrees Fahrenheit, and L is the length of the highway in feet. If 1 mi of the highway expanded 2.5 ft, what was the temperature? Round to the nearest tenth. (*Note:* 1 mi = 5280 ft) 109.5°F

Optional Student Activity

1. Two people decide to open a business to recondition toner cartridges for copy machines. They rent a building for \$7000 per year and estimate that the building maintenance, taxes, and insurance will cost \$6500 per year. Each person wants to make \$12 per hour in the first year and will work 10 hours per day for 260 days of the year. Assume that it costs \$20 to restore a cartridge and that they can sell the restored catridge for \$45.
   a. How many cartridges must they restore and sell annually to break even, not including the hourly wage they wish to earn? Approximately 794 cartridges
   b. How many cartridges must they restore and sell annually to earn the hourly wage they desire? Approximately 3671 cartridges
   c. Suppose the entrepreneurs are successful in their business and are restoring and selling 25 cartridges each day of the 260 days they are open. What will be their hourly wage for the year? Approximately \$18.65 per hour

Many of the applications of mathematics require that you identify an unknown quantity, assign a variable to that quantity, and then attempt to express another unknown quantity in terms of that variable.

TAKE NOTE

The sum of two numbers is 10.

One number	*Other number*
4	$10 - 4 = 6$
7	$10 - 7 = 3$
2	$10 - 2 = 8$
x	$10 - x$

Suppose we know that the sum of two numbers is 10 and that one of the two numbers is 4. We can find the other number by subtracting 4 from 10.

one number: 4
other number: $10 - 4 = 6$
The two numbers are 4 and 6.

Now suppose we know that the sum of two numbers is 10, we don't know either number, and we want to express *both* numbers in terms of the *same* variable. Let one number be x. Again, we can find the other number by subtracting x from 10.

one number: x
other number: $10 - x$
The two numbers are x and $10 - x$.

Note that the sum of x and $10 - x$ is 10.

$x + (10 - x) = x + 10 - x = 10$

Example 8

The sum of two numbers is 9. Eight times the smaller number is five less than three times the larger number. Find the numbers.

Solution

the smaller number: p
the larger number: $9 - p$

eight times the smaller number	is	five less than three times the larger number

$$8p = 3(9 - p) - 5$$
$$8p = 27 - 3p - 5$$
$$8p = 22 - 3p$$
$$8p + 3p = 22 - 3p + 3p$$
$$11p = 22$$
$$\frac{11p}{11} = \frac{22}{11}$$
$$p = 2$$
$$9 - p = 9 - 2 = 7$$

These numbers check as solutions.

The smaller number is 2.
The larger number is 7.

You Try It 8

The sum of two numbers is 14. One more than three times the smaller number equals the sum of the larger number and three. Find the two numbers.

Your solution

4 and 10

Solution on p. S5

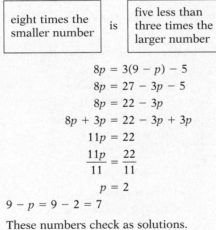

Section 2.2

Suggested Assignment

Exercises 1–129, every other odd
Exercises 133–155, odds
More challenging problems:
Exercises 157–163, odds

2.2 Exercises

Objective A

Solve and check.

1. $3x + 1 = 10$
   3
2. $4y + 3 = 11$
   2
3. $2a - 5 = 7$
   6
4. $5m - 6 = 9$
   3
5. $5 = 4x + 9$
   -1
6. $2 = 5b + 12$
   -2
7. $2x - 5 = -11$
   -3
8. $3n - 7 = -19$
   -4
9. $4 - 3w = -2$
   2
10. $5 - 6x = -13$
    3
11. $8 - 3t = 2$
    2
12. $12 - 5x = 7$
    1
13. $4a - 20 = 0$
    5
14. $3y - 9 = 0$
    3
15. $6 + 2b = 0$
    -3
16. $10 + 5m = 0$
    -2
17. $-2x + 5 = -7$
    6
18. $-5d + 3 = -12$
    3
19. $-12x + 30 = -6$
    3
20. $-13 = -11y + 9$
    2
21. $2 = 7 - 5a$
    1
22. $3 = 11 - 4n$
    2
23. $-35 = -6b + 1$
    6
24. $-8x + 3 = -29$
    4
25. $-3m - 21 = 0$
    -7
26. $-5x - 30 = 0$
    -6
27. $-4y + 15 = 15$
    0
28. $-3x + 19 = 19$
    0
29. $9 - 4x = 6$
    $\frac{3}{4}$
30. $3t - 2 = 0$
    $\frac{2}{3}$
31. $9x - 4 = 0$
    $\frac{4}{9}$
32. $7 - 8z = 0$
    $\frac{7}{8}$
33. $1 - 3x = 0$
    $\frac{1}{3}$
34. $9d + 10 = 7$
    $-\frac{1}{3}$
35. $12w + 11 = 5$
    $-\frac{1}{2}$
36. $6y - 5 = -7$
    $-\frac{1}{3}$
37. $8b - 3 = -9$
    $-\frac{3}{4}$
38. $5 - 6m = 2$
    $\frac{1}{2}$
39. $7 - 9a = 4$
    $\frac{1}{3}$
40. $9 = -12c + 5$
    $-\frac{1}{3}$

Quick Quiz (Objective 2.2A)

Solve.

1. $7b + 5 = 61$ 8
2. $12 - 4c = 15$ $-\frac{3}{4}$
3. $\frac{4}{5}x + 3 = 7$ 5
4. $-3 = 6m + 4 + m$ -1

41. $10 = -18x + 7$
$-\frac{1}{6}$

42. $2y + \frac{1}{3} = \frac{7}{3}$
1

43. $4a + \frac{3}{4} = \frac{19}{4}$
1

44. $2n - \frac{3}{4} = \frac{13}{4}$
2

45. $3x - \frac{5}{6} = \frac{13}{6}$
1

46. $5y + \frac{3}{7} = \frac{3}{7}$
0

47. $9x + \frac{4}{5} = \frac{4}{5}$
0

48. $8 = 7d - 1$
$\frac{9}{7}$

49. $8 = 10x - 5$
$\frac{13}{10}$

50. $4 = 7 - 2w$
$\frac{3}{2}$

51. $7 = 9 - 5a$
$\frac{2}{5}$

52. $8t + 13 = 3$
$-\frac{5}{4}$

53. $12x + 19 = 3$
$-\frac{4}{3}$

54. $-6y + 5 = 13$
$-\frac{4}{3}$

55. $-4x + 3 = 9$
$-\frac{3}{2}$

56. $\frac{1}{2}a - 3 = 1$
8

57. $\frac{1}{3}m - 1 = 5$
18

58. $\frac{2}{5}y + 4 = 6$
5

59. $\frac{3}{4}n + 7 = 13$
8

60. $-\frac{2}{3}x + 1 = 7$
-9

61. $-\frac{3}{8}b + 4 = 10$
-16

62. $\frac{x}{4} - 6 = 1$
28

63. $\frac{y}{5} - 2 = 3$
25

64. $\frac{2x}{3} - 1 = 5$
9

65. $\frac{3c}{7} - 1 = 8$
21

66. $4 - \frac{3}{4}z = -2$
8

67. $3 - \frac{4}{5}w = -9$
15

68. $5 + \frac{2}{3}y = 3$
-3

69. $17 + \frac{5}{8}x = 7$
-16

70. $17 = 7 - \frac{5}{6}t$
-12

71. $9 = 3 - \frac{2x}{7}$
-21

72. $3 = \frac{3a}{4} + 1$
$\frac{8}{3}$

73. $7 = \frac{2x}{5} + 4$
$\frac{15}{2}$

74. $5 - \frac{4c}{7} = 8$
$-\frac{21}{4}$

75. $7 - \frac{5}{9}y = 9$
$-\frac{18}{5}$

76. $6a + 3 + 2a = 11$
1

77. $5y + 9 + 2y = 23$
2

78. $7x - 4 - 2x = 6$
2

79. $11z - 3 - 7z = 9$
3

80. $2x - 6x + 1 = 9$
-2

81. $b - 8b + 1 = -6$
1

82. $3 = 7x + 9 - 4x$
-2

83. $-1 = 5m + 7 - m$
-2

84. $8 = 4n - 6 + 3n$
2

Quick Quiz (Objective 2.2B)

Solve.

1. $7x + 4 = 3x - 20$ -6

2. $4x + 5 = 23 - 2x$ 3

3. $9 - 4x = -7 - 2x$ 8

93

Objective B

Solve and check.

85. $8x + 5 = 4x + 13$
2

86. $6y + 2 = y + 17$
3

87. $5x - 4 = 2x + 5$
3

88. $13b - 1 = 4b - 19$
−2

89. $15x - 2 = 4x - 13$
−1

90. $7a - 5 = 2a - 20$
−3

91. $3x + 1 = 11 - 2x$
2

92. $n - 2 = 6 - 3n$
2

93. $2x - 3 = -11 - 2x$
−2

94. $4y - 2 = -16 - 3y$
−2

95. $2b + 3 = 5b + 12$
−3

96. $m + 4 = 3m + 8$
−2

97. $4y - 8 = y - 8$
0

98. $5a + 7 = 2a + 7$
0

99. $6 - 5x = 8 - 3x$
−1

100. $10 - 4n = 16 - n$
−2

101. $5 + 7x = 11 + 9x$
−3

102. $3 - 2y = 15 + 4y$
−2

103. $2x - 4 = 6x$
−1

104. $2b - 10 = 7b$
−2

105. $8m = 3m + 20$
4

106. $9y = 5y + 16$
4

107. $8b + 5 = 5b + 7$
$\frac{2}{3}$

108. $6y - 1 = 2y + 2$
$\frac{3}{4}$

109. $7x - 8 = x - 3$
$\frac{5}{6}$

110. $2y - 7 = -1 - 2y$
$\frac{3}{2}$

111. $2m - 1 = -6m + 5$
$\frac{3}{4}$

Objective C

Solve and check.

112. $5x + 2(x + 1) = 23$
3

113. $6y + 2(2y + 3) = 16$
1

114. $9n - 3(2n - 1) = 15$
4

115. $12x - 2(4x - 6) = 28$
4

116. $7a - (3a - 4) = 12$
2

117. $9m - 4(2m - 3) = 11$
−1

Quick Quiz (Objective 2.2C)

Solve.

1. $2x + 5(x + 1) = -9$ −2

2. $4x + 2(x + 3) = 3(x - 1)$ −3

3. $3 + 2[4x - 3(5 - x)] = 3(x - 20)$ −3

118. $5(3 - 2y) + 4y = 3$
2

119. $4(1 - 3x) + 7x = 9$
-1

120. $5y - 3 = 7 + 4(y - 2)$
2

121. $5 + 2(3b + 1) = 3b + 5$
$-\frac{2}{3}$

122. $6 - 4(3a - 2) = 2(a + 5)$
$\frac{2}{7}$

123. $7 - 3(2a - 5) = 3a + 10$
$\frac{4}{3}$

124. $2a - 5 = 4(3a + 1) - 2$
$-\frac{7}{10}$

125. $5 - (9 - 6x) = 2x - 2$
$\frac{1}{2}$

126. $7 - (5 - 8x) = 4x + 3$
$\frac{1}{4}$

127. $3[2 - 4(y - 1)] = 3(2y + 8)$
$-\frac{1}{3}$

128. $5[2 - (2x - 4)] = 2(5 - 3x)$
5

129. $3a + 2[2 + 3(a - 1)] = 2(3a + 4)$
$\frac{10}{3}$

130. $5 + 3[1 + 2(2x - 3)] = 6(x + 5)$
$\frac{20}{3}$

131. $-2[4 - (3b + 2)] = 5 - 2(3b + 6)$
$-\frac{1}{4}$

132. $-4[x - 2(2x - 3)] + 1 = 2x - 3$
2

Objective D

Translate into an equation and solve.

133. The sum of a number and twelve is twenty. Find the number.
$x + 12 = 20; 8$

134. The difference between nine and a number is seven. Find the number.
$9 - x = 7; 2$

135. Three-fifths of a number is negative thirty. Find the number.
$\frac{3}{5}x = -30; -50$

136. The quotient of a number and six is twelve. Find the number.
$\frac{x}{6} = 12; 72$

137. Four more than three times a number is thirteen. Find the number.
$3x + 4 = 13; 3$

138. The sum of twice a number and five is fifteen. Find the number.
$2x + 5 = 15; 5$

139. The difference between nine times a number and six is twelve. Find the number.
$9x - 6 = 12; 2$

Quick Quiz (Objective 2.2D)

Translate into an equation and solve.

1. Three-fifths of a number is fifteen. Find the number. 25
2. Three more than the product of six and a number is twenty-one. Find the number. 3
3. Nine less than three-fourths of a number is six. Find the number. 20

Solve.

4. The pressure at a certain depth in the ocean can be approximated by the equation $P = \frac{1}{2}D + 15$, where P is the pressure in pounds per square inch, and D is the depth in feet. Find the depth of a diver when the pressure on the diver is 40 lb/in^2. 50 ft

95

140. Six less than four times a number is twenty-two. Find the number.
$4x - 6 = 22$; 7

141. The sum of a number and twice the number is nine. Find the number.
$x + 2x = 9$; 3

142. Eleven more than negative four times a number is three. Find the number.
$-4x + 11 = 3$; 2

143. Seventeen less than the product of five and a number is two. Find the number.
$5x - 17 = 2$; $\frac{19}{5}$

144. Eight less than the product of eleven and a number is negative nineteen. Find the number.
$11x - 8 = -19$; -1

145. Seven more than the product of six and a number is eight less than the product of three and the number. Find the number.
$6x + 7 = 3x - 8$; -5

146. Fifteen less than the product of four and a number is the difference between six times the number and eleven. Find the number.
$4x - 15 = 6x - 11$; -2

147. Thirty equals nine less than the product of seven and a number. Find the number.
$30 = 7x - 9$; $\frac{39}{7}$

148. Twenty-three equals the difference between eight and the product of five and a number. Find the number.
$23 = 8 - 5x$; -3

149. The sum of two numbers is twenty-one. Twice the smaller number is three more than the larger number. Find the two numbers.
$2x = (21 - x) + 3$; 8, 13

150. The sum of two numbers is thirty. Three times the smaller number is twice the larger number. Find the two numbers.
$3x = 2(30 - x)$; 12, 18

151. The sum of two numbers is twenty-three. The larger number is five more than twice the smaller number. Find the two numbers.
$23 - x = 2x + 5$; 6, 17

152. The sum of two numbers is twenty-five. The larger number is ten less than four times the smaller number. Find the two numbers.
$25 - x = 4x - 10$; 7, 18

Instructor Note
Here is an extension for Exercise 155. Will the car slide farther as the temperature increases or as it decreases? As it increases

Answers to Writing Exercises

165. The sentence "Solve $2x - 3(4x + 1)$" does not make sense because $2x - 3(4x + 1)$ is not an equation; it is an expression. Equations can be solved. Expressions cannot be solved.

166. Many beginning algebra students do not differentiate between an equation that has no solution and an equation whose solution is zero. Students should explain that zero is a (real) number and that the solution of the equation $2x + 3 = 3$ is the (real) number zero. However, there is no solution to the equation $x = x + 1$ because there is no (real) number that is equal to the number plus 1.

96

153. Anthropologists can approximate the height of a primate from the size of its humerus (the bone extending from the shoulder to the elbow) by using the equation $H = 1.2L + 27.8$, where L is the length of the humerus in inches and H is the height of the primate in inches. If the height of a primate is estimated to be 62 in., what is the approximate length of the humerus of this primate?
28.5 in.

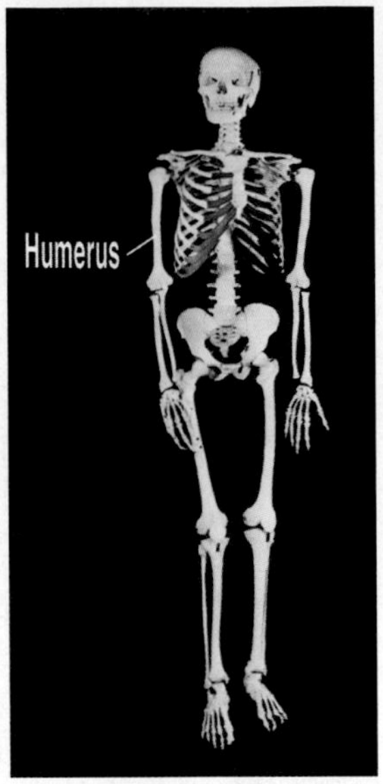

154. The fare, F, to be charged a customer by a taxi company is calculated using the formula $F = 1.50 + 0.95(m - 1)$, where m is the number of miles traveled. If a passenger is charged \$9.10, how many miles was the passenger driven?
9 mi

155. Black ice is an ice covering on roads that is especially difficult to see and therefore extremely dangerous for motorists. The distance that a car traveling 30 mph will slide after its brakes are applied is related to the outside temperature by the formula $C = \frac{1}{4}D - 45$, where C is the Celsius temperature and D is the distance in feet that the car will slide. Determine the distance a car will slide on black ice when the outside temperature is $-11°C$.
136 ft

APPLYING THE CONCEPTS

156. If $2x - 3 = 7$, evaluate $3x + 4$.
19

157. If $5x = 3x - 8$, evaluate $4x + 2$.
-14

158. Solve: $x \div 28 = 1481$ remainder 25
41,493

159. If $3 + 2(4a - 3) = 5$ and $4 - 3(2 - 3b) = 11$, which is larger, a or b?
b

Solve. If the equation has no solution, write "no solution."

160. $3(2x - 1) - (6x - 4) = -9$
no solution

161. $7(3x + 6) - 4(3 + 5x) = 13 + x$
no solution

162. $\frac{1}{5}(25 - 10a) + 4 = \frac{1}{3}(12a - 15) + 14$
0

163. $5[m + 2(3 - m)] = 3[2(4 - m) - 5]$
-21

164. Solve the equation $x + a = b$ for x. Is the solution you have written valid for all real numbers a and b?
$x = b - a$. Yes.

165. Does the sentence "Solve $2x - 3(4x + 1)$" make sense? Why or why not?

166. The equation $x = x + 1$ has no solution, whereas the solution of the equation $2x + 3 = 3$ is zero. Is there a difference between no solution and a solution of zero? Explain your answer.

2.3 Mixture, Investment, and Motion Problems

Objective A To solve value mixture problems

VIDEO & DVD CD TUTOR WEB SSM

TAKE NOTE
The equation $AC = V$ is used to find the value of an ingredient. For example, the value of 4 lb of cashews costing \$6 per pound is

$$AC = V$$
$$4 \cdot 6 = V$$
$$\$24 = V$$

A value mixture problem involves combining two ingredients that have different prices into a single blend. For example, a coffee merchant may blend two types of coffee into a single blend, or a candy manufacturer may combine two types of candy to sell as a "variety pack."

The solution of a value mixture problem is based on the equation $V = AC$, where V is the value of an ingredient, A is the amount of the ingredient, and C is the cost per unit of the ingredient.

➡ A coffee merchant wants to make 6 lb of a blend of coffee costing \$5 per pound. The blend is made using a \$6-per-pound grade of coffee and a \$3-per-pound grade. How many pounds of each of these grades should be used?

Strategy for Solving a Value Mixture Problem

1. For each ingredient in the mixture, write a numerical or variable expression for the amount of the ingredient used, the unit cost of the ingredient, and the value of the amount used. For the blend, write a numerical or variable expression for the amount, the unit cost of the blend, and the value of the amount. The results can be recorded in a table.

Amount of \$6 coffee: x
Amount of \$3 coffee: $6 - x$

• The sum of the amounts is 6 lb.

TAKE NOTE
Use the information given in the problem to fill in the "Amount" and "Unit Cost" columns of the table. Fill in the "Value" column by multiplying the two expressions you wrote in each row. Use the expressions in the last column to write the equation.

	Amount, A	·	*Unit Cost, C*	=	*Value, V*
\$6 grade	x	·	6	=	$6x$
\$3 grade	$6 - x$	·	3	=	$3(6 - x)$
\$5 blend	6	·	5	=	$5(6)$

2. Determine how the values of the individual ingredients are related. Use the fact that the sum of the values of these ingredients is equal to the value of the blend.

$$6x + 3(6 - x) = 5(6)$$
$$6x + 18 - 3x = 30$$
$$3x + 18 = 30$$
$$3x = 12$$
$$x = 4$$

• The sum of the values of the \$6 grade and the \$3 grade is equal to the value of the \$5 blend.

$$6 - x = 6 - 4 = 2$$

• Find the amount of \$3 coffee.

Objective 2.3A

New Vocabulary
value mixture problem

Formulas
$AC = V$
(Amount of ingredient · unit cost = value of ingredient)

Instructor Note
For value mixture problems, the sum of the values of the ingredients equals the value of the mixture. If students organize these problems in a table such as the one at the left, the last column can be used to write the equation. The sum of the values of the ingredients, $6x + 3(6 - x)$, equals the value of the mixture, $5(6)$.

In-Class Examples (Objective 2.3A)

1. A coffee merchant wants to make 30 lb of a blend of coffee costing \$4.80 per pound. The blend is made using a \$6 grade and a \$4 grade of coffee. How many pounds of each grade of coffee should be used? 12 lb of the \$6 grade; 18 lb of the \$4 grade
2. How many ounces of pure silver that costs \$11 per ounce must be mixed with 60 oz of a silver alloy costing \$7.20 per ounce to make an alloy that costs \$9.50 per ounce? 92 oz
3. A butcher combined 90 lb of meat that cost \$1.50 per pound with 50 lb of meat that cost \$3.60 per pound. Find the cost of the meat mixture. \$2.25 per pound

Discuss the Concepts

1. If a grocer blended peanuts costing \$3.50 per pound with almonds costing \$6.00 per pound and then sold the mixture for \$3.50 per pound, would the grocer make money or lose money? Lose money
2. Suppose a coffee merchant blends Ethiopian Mocha Java coffee beans costing \$7 per pound with Hawaiian Kona coffee beans costing \$12 per pound. Will the coffee merchant always make a profit if the price of the blend is \$13 per pound? Will the coffee merchant make a profit if the price of the blend is \$10 per pound? Will the coffee merchant make a profit if the price of the blend is \$6 per pound? Yes; sometimes; no

Concept Check

Have your students solve the problem on page 97 using x to represent the number of pounds of \$3.00 coffee. Have them discuss the differences and similarities between the two solutions.

Example 1

How many ounces of a silver alloy that costs \$4 an ounce must be mixed with 10 oz of an alloy that costs \$6 an ounce to make a mixture that costs \$4.32 an ounce?

Strategy

- Ounces of \$4 alloy: x

	Amount	*Cost*	*Value*
\$4 alloy	x	4	$4x$
\$6 alloy	10	6	$6(10)$
\$4.32 mixture	$10 + x$	4.32	$4.32(10 + x)$

- The sum of the values before mixing equals the value after mixing.

Solution

$$4x + 6(10) = 4.32(10 + x)$$
$$4x + 60 = 43.2 + 4.32x$$
$$-0.32x + 60 = 43.2$$
$$-0.32x = -16.8$$
$$x = 52.5$$

52.5 oz of the \$4 silver alloy must be used.

You Try It 1

A gardener has 20 lb of a lawn fertilizer that costs \$.80 per pound. How many pounds of a fertilizer that costs \$.55 per pound should be mixed with this 20 lb of lawn fertilizer to produce a mixture that costs \$.75 per pound?

Your strategy

Your solution

5 lb

Solution on p. S5

Objective B To solve percent mixture problems

VIDEO & DVD CD TUTOR WEB SSM

The amount of a substance in a solution can be given as a percent of the total solution. For example, a 5% salt-water solution means that 5% of the total solution is salt. The remaining 95% is water.

TAKE NOTE
The equation $Ar = Q$ is used to find the amount of a substance in a mixture. For example, the number of grams of silver in 50 g of a 40% alloy is

$$Ar = Q$$
$$(50\text{ g})(0.40) = Q$$
$$20\text{ g} = Q$$

The solution of a percent mixture problem is based on the equation $Q = Ar$, where Q is the quantity of a substance in the solution, r is the percent of concentration, and A is the amount of solution.

➡ A 500-milliliter bottle contains a 4% solution of hydrogen peroxide. Find the amount of hydrogen peroxide in the solution.

$Q = Ar$ • Given: $A = 500$; $r = 4\% = 0.04$

$Q = 500(0.04)$

$Q = 20$

The bottle contains 20 ml of hydrogen peroxide.

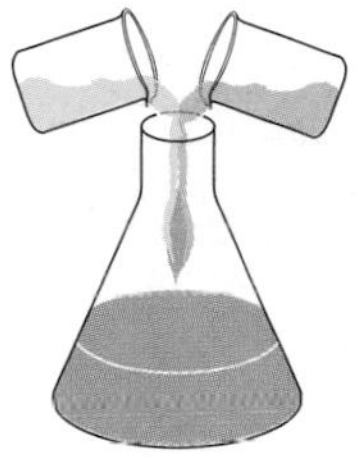

➡ How many gallons of a 20% salt solution must be mixed with 6 gal of a 30% salt solution to make a 22% salt solution?

Strategy for Solving a Percent Mixture Problem

1. For each solution, use the equation $Ar = Q$. Write a numerical or variable expression for the amount of solution, the percent of concentration, and the quantity of the substance in the solution. The results can be recorded in a table.

The unknown quantity of 20% solution: x

TAKE NOTE
Use the information given in the problem to fill in the "Amount" and "Percent" columns of the table. Fill in the "Quantity" column by multiplying the two expressions you wrote in each row. Use the expressions in the last column to write the equation.

	Amount of Solution, A	·	Percent of Concentration, r	=	Quantity of Substance, Q
20% solution	x	·	0.20	=	$0.20x$
30% solution	6	·	0.30	=	$0.30(6)$
22% solution	$x + 6$	·	0.22	=	$0.22(x + 6)$

2. Determine how the quantities of the substance in the individual solutions are related. Use the fact that the sum of the quantities of the substances being mixed is equal to the quantity of the substance after mixing.

$$0.20x + 0.30(6) = 0.22(x + 6)$$
• The sum of the quantities of the substance in the 20% solution and the 30% solution is equal to the quantity of the substance in the 22% solution.
$$0.20x + 1.80 = 0.22x + 1.32$$
$$-0.02x + 1.80 = 1.32$$
$$-0.02x = -0.48$$
$$x = 24$$

24 gal of the 20% solution must be used.

Objective 2.3B

New Vocabulary
percent mixture problem

New Formulas
$Ar = Q$ (Amount of solution · percent of concentration = quantity of a substance in the solution)

Discuss the Concepts

1. Suppose orange juice is added to a solution of sugar and water. Does the percent concentration of the juice increase or decrease? Increase
2. Pure gold is added to a mixture of silver and copper. Does the percent concentration of silver increase or decrease? Decrease

Instructor Note
These problems are similar to the value mixture problems. The last column of the table can be used to write the equation.

Optional Student Activity
When 1 oz of water is added to an acid solution, the new mixture is 20% acid. When 1 oz of pure acid is added to this new mixture, the result is $33\frac{1}{3}\%$ acid. What was the percent concentration of acid in the original mixture? 25%

In-Class Examples (Objective 2.3B)

1. A silversmith mixed 40 g of a 70% silver alloy with 60 g of a 30% silver alloy. What is the percent concentration of the resulting alloy? 46%
2. A butcher has some hamburger that is 24% fat and some that is 16% fat. How many pounds of each should be mixed to make 80 lb of hamburger that is 18% fat? 20 lb of 24%; 60 lb of 16%
3. How many grams of a 4% salt solution must be mixed with 50 g of a 9% salt solution to make a 6% salt solution? 75 g

Optional Student Activity

To form alloys of gold for jewelry, some other metals are added. These metals are typically silver, copper, zinc, nickel, and palladium. The amount of each of these metals that is added to the pure gold depends on the desired final karat weight of the gold. In its pure form, gold is defined as 24-karat. The percent concentrations of the metals in an 18-karat yellow gold necklace are gold, 75%; silver, 16%; and copper, 9%. (White gold differs in makeup from yellow gold. An 18-karat white gold necklace contains 75% gold, 4% silver, 4% copper, and 17% palladium.) Note that 18-karat gold is $\frac{18\text{ karat}}{24\text{ karat}} = 0.75 = 75\%$ gold.

1. Suppose a jeweler wants to make a 15-gram alloy that is 18-karat gold. How many grams of gold, silver, and copper are required? 11.25 g of gold, 2.4 g of silver, 1.35 g of copper
2. What is the percent concentration, to the nearest tenth of a percent, of gold in a 14-karat gold alloy? 58.3%
3. A 14-karat gold alloy is 4% silver, 31.2% copper, and 6.3% nickel. How many grams of each are there in 12 g of a 14-karat gold alloy? 0.48 g of silver, 3.744 g of copper, 0.756 g of nickel

Example 2

A chemist wishes to make 2 L of an 8% acid solution by mixing a 10% acid solution and a 5% acid solution. How many liters of each solution should the chemist use?

Strategy

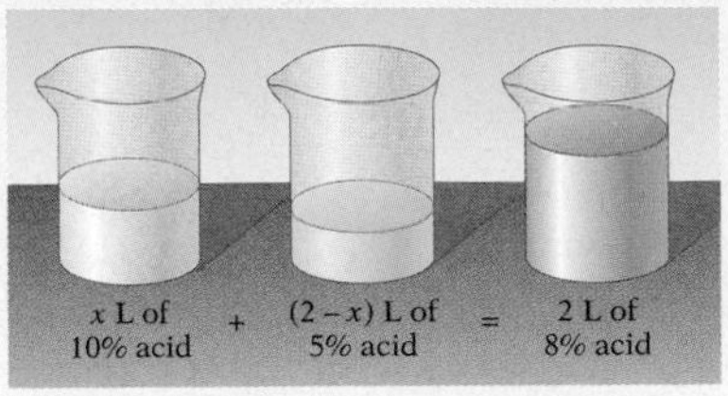

- Liters of 10% solution: x
  Liters of 5% solution: $2 - x$

	Amount	*Percent*	*Quantity*
10% solution	x	0.10	$0.10x$
5% solution	$2 - x$	0.05	$0.05(2 - x)$
8% solution	2	0.08	0.08(2)

- The sum of the quantities before mixing is equal to the quantity after mixing.

Solution

$$0.10x + 0.05(2 - x) = 0.08(2)$$
$$0.10x + 0.10 - 0.05x = 0.16$$
$$0.05x + 0.10 = 0.16$$
$$0.05x = 0.06$$
$$x = 1.2$$

$$2 - x = 2 - 1.2 = 0.8$$

The chemist needs 1.2 L of the 10% solution and 0.8 L of the 5% solution.

You Try It 2

A pharmacist dilutes 5 L of a 12% solution with a 6% solution. How many liters of the 6% solution are added to make an 8% solution?

Your strategy

Your solution

10 L

Solution on p. S5

Objective 2.3C

Objective C To solve investment problems

VIDEO & DVD CD TUTOR WEB SSM

Point of Interest

You may be familiar with the simple interest formula $I = Prt$. If so, you know that t represents time. In the problems in this section, time is always 1 (one year), so the formula $I = Prt$ simplifies to

$I = Pr(1)$

$I = Pr$

The annual simple interest that an investment earns is given by the equation $I = Pr$, where I is the simple interest, P is the principal, or the amount invested, and r is the simple interest rate.

➡ The annual interest rate on a \$2500 investment is 8%. Find the annual simple interest earned on the investment.

$I = Pr$ • Given: $P = \$2500$; $r = 8\% = 0.08$

$I = 2500(0.08)$

$I = 200$

The annual simple interest is \$200.

➡ An investor has a total of \$10,000 deposited in two simple interest accounts. On one account, the annual simple interest rate is 6%. On the second account, the annual simple interest rate is 10%. How much is invested in the 6% account if the total annual interest earned is \$900?

Strategy for Solving a Problem Involving Money Deposited in Two Simple Interest Accounts

1. For each amount invested, use the equation $Pr = I$. Write a numerical or variable expression for the principal, the interest rate, and the interest earned. The results can be recorded in a table.

Amount invested at 6%: x • The sum of the amounts invested is \$10,000.

Amount invested at 10%: $\$10{,}000 - x$

TAKE NOTE

Use the information given in the problem to fill in the "Principal" and "Interest Rate" columns of the table. Fill in the "Interest Earned" column by multiplying the two expressions you wrote in each row.

	Principal, P	·	*Interest Rate, r*	=	*Interest Earned, I*
Amount at 6%	x	·	0.06	=	$0.06x$
Amount at 10%	$10{,}000 - x$	·	0.10	=	$0.10(10{,}000 - x)$

2. Determine how the amounts of interest earned on the individual amounts are related. For example, the total interest earned by both accounts may be known, or it may be known that the interest earned on one account is equal to the interest earned on the other account.

Point of Interest

One way to invest money is to buy U.S. Treasury Bills, often called T-Bills. T-Bills are issued by the U.S. Treasury, and the purchaser of a T-Bill is actually lending money to the federal government. The simple interest formula is used to compute the interest on a T-Bill.

$$
\begin{aligned}
0.06x + 0.10(10{,}000 - x) &= 900 \\
0.06x + 1000 - 0.10x &= 900 \\
-0.04x + 1000 &= 900 \\
-0.04x &= -100 \\
x &= 2500
\end{aligned}
$$

• The sum of the interest earned on the two accounts is \$900.

The amount invested at 6% is \$2500.

Instructor Note

Some students may recall the simple interest equation as $I = Prt$. Because we are discussing *annual* interest, $t = 1$. Thus we just write $I = Pr$.

Instructor Note

Students may not realize that investors may not always choose to put all their money into the account with the greatest interest rate, because that account usually has the most risk. Placing money in different accounts allows the investor to diversify.

Optional Student Activity

Have students use the simple interest equation to check the solution of this problem. They should show that the interest earned on the \$2500 plus the interest earned on the \$7500 is \$900.

In-Class Examples (Objective 2.3C)

A web page designer invested a total of \$10,000 in two accounts: a money market account and a high-yield corporate bond fund. The annual interest rate on the money market account was 3.5%, and the annual interest rate on the high-yield corporate bond fund was 9.25%. If the designer received annual income of \$723.75, how much was invested in each account? \$3500 at 3.5%; \$6500 at 9.25%

Example 3

An investment counselor invested 75% of a client's money in a 9% annual simple interest money market fund. The remainder was invested in 7% annual simple interest government securities. Find the amount invested in each if the total annual interest earned is $3825.

Strategy

- Amount invested: x
  Amount invested at 7%: $0.25x$
  Amount invested at 9%: $0.75x$

	Principal	*Rate*	*Interest*
Amount at 7%	$0.25x$	0.07	$0.0175x$
Amount at 9%	$0.75x$	0.09	$0.0675x$

- The sum of the interest earned by the two investments equals the total annual interest earned ($3825).

Solution

$$0.0175x + 0.0675x = 3825$$
$$0.085x = 3825$$
$$x = 45{,}000$$

$$0.25x = 0.25(45{,}000) = 11{,}250$$

$$0.75x = 0.75(45{,}000) = 33{,}750$$

The amount invested at 7% is $11,250.
The amount invested at 9% is $33,750.

You Try It 3

An investment of $5000 is made at an annual simple interest rate of 8%. How much additional money must be invested at 11% so that the total interest earned will be 9% of the total investment?

Your strategy

Your solution

$2500

Solution on pp. S5–S6

Objective D **To solve uniform motion problems** VIDEO & DVD CD TUTOR WEB SSM

A train that travels constantly in a straight line at 50 mph is in *uniform motion.* **Uniform motion** means that the speed or direction of an object does not change.

The solution of a uniform motion problem is based on the equation $d = rt$, where d is the distance traveled, r is the rate of travel, and t is the time traveled.

A train traveled at a speed of 55 mph for 3 h. The distance traveled by the train can be found by using the equation $d = rt$, as shown at the right.

$$d = rt$$
$$d = 55(3)$$
$$d = 165$$

The distance traveled by the train was 165 mi.

➡ A car leaves a town traveling at 40 mph. Two hours later, a second car leaves the same town, on the same road, traveling at 60 mph. In how many hours will the second car pass the first car?

Strategy for Solving a Uniform Motion Problem

1. For each object, use the equation $d = rt$. Write a numerical or variable expression for the distance, rate, and time. The results can be recorded in a table.

The first car traveled 2 h longer than the second car.

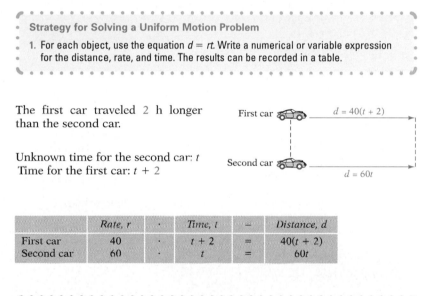

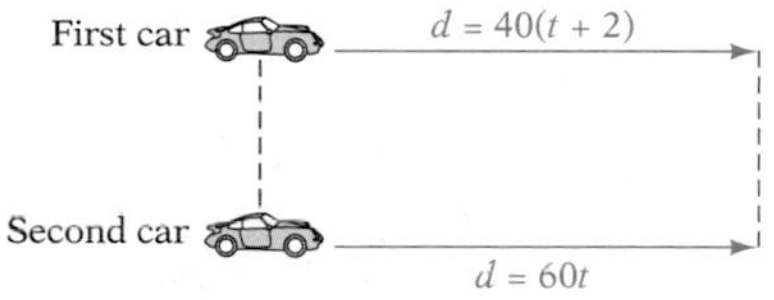

Unknown time for the second car: t
Time for the first car: $t + 2$

TAKE NOTE
Use the information given in the problem to fill in the "Rate" and "Time" columns of the table. Fill in the "Distance" column by multiplying the two expressions you wrote in each row.

	Rate, r	·	Time, t	=	Distance, d
First car	40	·	$t + 2$	=	$40(t + 2)$
Second car	60	·	t	=	$60t$

2. Determine how the distances traveled by the individual objects are related. For example, the total distance traveled by both objects may be known, or it may be known that the two objects traveled the same distance.

The two cars travel the same distance.

$$40(t + 2) = 60t$$
$$40t + 80 = 60t$$
$$80 = 20t$$
$$4 = t$$

The second car will pass the first car in 4 h.

Objective 2.3D

New Vocabulary
uniform motion problem

New Formulas
$rt = d$ (Rate · time = distance)

Instructor Note
One of the complications of distance–rate problems is that the variable may not directly represent the unknown. Example 5 on page 104 illustrates this. The unknown is distance, but the variable is time.

Discuss the Concepts
You might find the problem below to be an interesting problem to discuss with the class. Having different students present alternative solutions can help students expand their problem-solving skills.

"Jason and Ryan have only one bicycle. Jason rides for an agreed-upon distance and then locks up the bike for Ryan, who has been walking. Meanwhile, Jason walks on ahead. They alternate walking and riding. If they both walk at a rate of 4 mph and ride at a rate of 12 mph, what part of the time has the bike been locked up when Jason and Ryan meet again?" Half of the time

In-Class Examples (Objective 2.3D)

1. Two buses start from the same station and drive in opposite directions. The express bus is traveling 10 mph faster than the local bus. In 3 h the buses are 300 mi apart. Find the rate of each bus. Express bus: 55 mph; local bus: 45 mph
2. A speeding car going 75 mph has a 1-hour head start on a helicopter trying to overtake the car. The helicopter is traveling at 150 mph. How far from the starting point does the helicopter overtake the car? 150 mi

Concept Check

Traveling at a constant speed, it took you 5 h to drive from your home to Westport. If you had driven 2 mph faster, you would have arrived in Westport 12 min earlier. How many miles is it from your home to Westport? 240 mi

Optional Student Activity

1. Three mechanical rabbits run a race. Rabbit A runs 2 m/s slower than Rabbit B. Rabbit A runs three times as fast as Rabbit C. If Rabbit C runs at x m/s, how many meters ahead of Rabbit C will Rabbit B be after 10 s? Express the answer in terms of x. $(20x + 20)$ m
2. Amanda can walk at a rate of 100 yd/min and ride her scooter at a rate of 300 yd/min. Amanda must deliver a package to a destination 6000 yd away. If she walks for 12 min on her delivery route, for how many minutes must she ride her scooter to reach her destination? 16 min

Example 4

Two cars, one traveling 10 mph faster than the other, start at the same time from the same point and travel in opposite directions. In 3 h they are 300 mi apart. Find the rate of each car.

Strategy

- Rate of 1st car: r
- Rate of 2nd car: $r + 10$

	Rate	*Time*	*Distance*
1st car	r	3	$3r$
2nd car	$r + 10$	3	$3(r + 10)$

- The total distance traveled by the two cars is 300 mi.

Solution

$$3r + 3(r + 10) = 300$$
$$3r + 3r + 30 = 300$$
$$6r + 30 = 300$$
$$6r = 270$$
$$r = 45$$

$r + 10 = 45 + 10 = 55$

The first car is traveling 45 mph.
The second car is traveling 55 mph.

You Try It 4

Two trains, one traveling at twice the speed of the other, start at the same time on parallel tracks from stations that are 288 mi apart and travel toward each other. In 3 h, the trains pass each other. Find the rate of each train.

Your strategy

Your solution

32 mph; 64 mph

Example 5

How far can the members of a bicycling club ride out into the country at a speed of 12 mph and return over the same road at 8 mph if they travel a total of 10 h?

Strategy

- Time spent riding out: t
  Time spent riding back: $10 - t$

	Rate	*Time*	*Distance*
Out	12	t	$12t$
Back	8	$10 - t$	$8(10 - t)$

- The distance out equals the distance back.

Solution

$$12t = 8(10 - t)$$
$$12t = 80 - 8t$$
$$20t = 80$$
$$t = 4 \quad \text{(The time is 4 h.)}$$

The distance out $= 12t = 12(4) = 48$ mi.
The club can ride 48 mi into the country.

You Try It 5

A pilot flew out to a parcel of land and back in 5 h. The rate out was 150 mph, and the rate returning was 100 mph. How far away was the parcel of land?

Your strategy

Your solution

300 mi

Solutions on p. S6

105

Section 2.3

Suggested Assignment
Exercises 1–69, odds
More challenging problems:
Exercises 71–74

2.3 Exercises

Objective A *Application Problems*

1. An herbalist has 30 oz of herbs costing $2 per ounce. How many ounces of herbs costing $1 per ounce should be mixed with the 30 oz to produce a mixture costing $1.60 per ounce?
20 oz

2. The manager of a farmer's market has 500 lb of grain that costs $1.20 per pound. How many pounds of meal costing $.80 per pound should be mixed with the 500 lb of grain to produce a mixture that costs $1.05 per pound?
300 lb

3. Find the cost per pound of a meatloaf mixture made from 3 lb of ground beef costing $3.99 per pound and 1 lb of ground turkey costing $1.79 per pound.
$3.44

4. Find the cost per ounce of a sunscreen made from 100 oz of a lotion that costs $2.50 per ounce and 50 oz of a lotion that costs $4.00 per ounce.
$3

5. A snack food is made by mixing 5 lb of popcorn that costs $.80 per pound with caramel that costs $2.40 per pound. How much caramel is needed to make a mixture that costs $1.40 per pound?
3 lb

6. A wild birdseed mix is made by combining 100 lb of millet seed costing $.60 per pound with sunflower seeds costing $1.10 per pound. How many pounds of sunflower seeds are needed to make a mixture that costs $.70 per pound?
25 lb

7. Ten cups of a restaurant's house Italian dressing is made by blending olive oil costing $1.50 per cup with vinegar that costs $.25 per cup. How many cups of each are used if the cost of the blend is $.50 per cup?
olive oil: 2 c; vinegar: 8 c

8. A high-protein diet supplement that costs $6.75 per pound is mixed with a vitamin supplement that costs $3.25 per pound. How many pounds of each should be used to make 5 lb of a mixture that costs $4.65 per pound?
diet supplement: 2 lb; vitamin supplement: 3 lb

9. Find the cost per ounce of a mixture of 200 oz of a cologne that costs $5.50 per ounce and 500 oz of a cologne that costs $2.00 per ounce.
$3.00

10. Find the cost per pound of a trail mix made from 40 lb of raisins that cost $4.40 per pound and 100 lb of granola that costs $2.30 per pound.
$2.90

Quick Quiz (Objective 2.3A)

1. To make a flour mix, a miller combined soybeans that cost $7.50 per bushel with wheat that cost $4.50 per bushel. How many bushels of each were used to make a mixture of 800 bushels that costs $5.25 per bushel? 200 bushels of soybeans; 600 bushels of wheat
2. Forty liters of pure maple syrup that cost $18 per liter were mixed with imitation maple syrup that cost $8.50 per liter. How much imitation maple syrup is needed to make a mixture costing $11 per liter? 112 L

106

11. Twenty ounces of a platinum alloy that costs \$220 per ounce are mixed with an alloy that costs \$400 per ounce. How many ounces of the \$400 alloy should be used to make an alloy that costs \$300 per ounce?
16 oz

12. How many liters of a blue dye that costs \$1.60 per liter must be mixed with 18 L of anil that costs \$2.50 per liter to make a mixture that costs \$1.90 per liter?
36 L

13. The manager of a specialty food store combined almonds that cost \$4.50 per pound with walnuts that cost \$2.50 per pound. How many pounds of each were used to make a 100-pound mixture that costs \$3.24 per pound?
almonds: 37 lb; walnuts: 63 lb

14. A goldsmith combined an alloy that cost \$4.30 per ounce with an alloy that cost \$1.80 per ounce. How many ounces of each were used to make a mixture of 200 oz costing \$2.50 per ounce?
\$4.30 alloy: 56 oz; \$1.80 alloy: 144 oz

15. Adult tickets for a play cost \$6.00 and children's tickets cost \$2.50. For one performance, 370 tickets were sold. Receipts for the performance were \$1723. Find the number of adult tickets sold.
228 adult tickets

16. Tickets for a piano concert sold for \$4.50 for each adult. Student tickets sold for \$2.00 each. The total receipts for 1720 tickets were \$5980. Find the number of adult tickets sold.
1016 adult tickets

17. Find the cost per pound of sugar-coated breakfast cereal made from 40 lb of sugar that costs \$1.00 per pound and 120 lb of corn flakes that cost \$.60 per pound.
\$.70

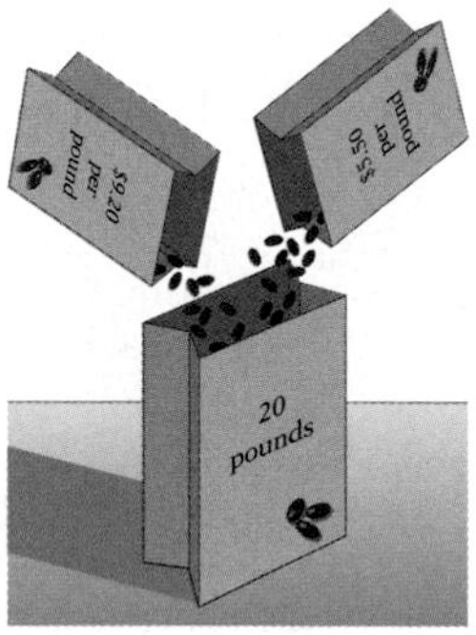

18. Find the cost per pound of a coffee mixture made from 8 lb of coffee that costs \$9.20 per pound and 12 lb of coffee that costs \$5.50 per pound.
\$6.98

Objective B *Application Problems*

19. Forty ounces of a 30% gold alloy are mixed with 60 oz of a 20% gold alloy. Find the percent concentration of the resulting gold alloy.
24%

20. One hundred ounces of juice that is 50% tomato juice are added to 200 oz of a vegetable juice that is 25% tomato juice. What is the percent concentration of tomato juice in the resulting mixture?
$33\frac{1}{3}\%$

Quick Quiz (Objective 2.3B)

1. A goldsmith mixed 20 g of a 60% gold alloy with 80 g of a 20% gold alloy. What is the percent concentration of the resulting alloy? 28%
2. A hospital staff mixed a 70% disinfectant solution with a 30% disinfectant solution. How many liters of each were used to make 30 L of a 45% disinfectant solution? 11.25 L of 70%; 18.75 L of 30%
3. How many pounds of a 12% aluminum alloy must be mixed with 400 lb of a 25% aluminum alloy to make a 17% aluminum alloy? 640 lb

107

21. How many gallons of a 15% acid solution must be mixed with 5 gal of a 20% acid solution to make a 16% acid solution?
20 gal

22. How many pounds of a chicken feed that is 50% corn must be mixed with 400 lb of a feed that is 80% corn to make a chicken feed that is 75% corn?
80 lb

23. A rug is made by weaving 20 lb of yarn that is 50% wool with a yarn that is 25% wool. How many pounds of the yarn that is 25% wool are used if the finished rug is 35% wool?
30 lb

24. Five gallons of a dark green latex paint that is 20% yellow paint are combined with a lighter green latex paint that is 40% yellow paint. How many gallons of the lighter green paint must be used to create a green paint that is 25% yellow paint?
$1\frac{2}{3}$ gal

25. How many gallons of a plant food that is 9% nitrogen must be combined with another plant food that is 25% nitrogen to make 10 gal of a solution that is 15% nitrogen?
6.25 gal

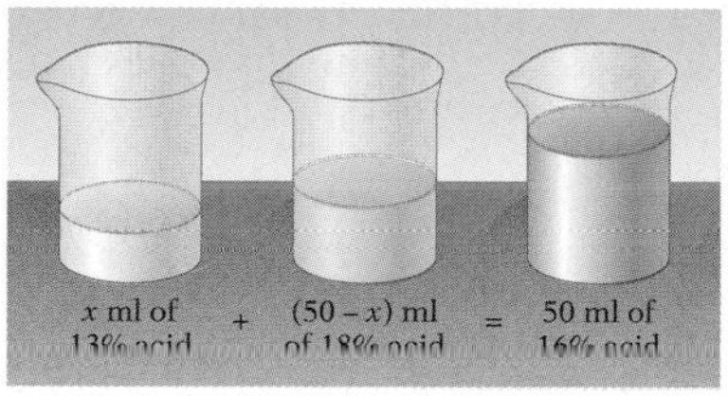

26. A chemist wants to make 50 ml of a 16% acid solution by mixing a 13% acid solution and an 18% acid solution. How many milliliters of each solution should the chemist use?
13% solution: 20 ml; 18% solution: 30 ml

27. Five grams of sugar are added to a 45-gram serving of a breakfast cereal that is 10% sugar. What is the percent concentration of sugar in the resulting mixture?
19%

28. A goldsmith mixes 8 oz of a 30% gold alloy with 12 oz of a 25% gold alloy. What is the percent concentration of the resulting alloy?
27%

29. How many pounds of coffee that is 40% java beans must be mixed with 80 lb of coffee that is 30% java beans to make a coffee blend that is 32% java beans?
20 lb

30. The manager of a garden shop mixes grass seed that is 60% rye grass with 70 lb of grass seed that is 80% rye grass to make a mixture that is 74% rye grass. How much of the 60% rye grass is used?
30 lb

31. A hair dye is made by blending a 7% hydrogen peroxide solution and a 4% hydrogen peroxide solution. How many milliliters of each are used to make a 300-milliliter solution that is 5% hydrogen peroxide?
7% solution: 100 ml; 4% solution: 200 ml

32. A tea that is 20% jasmine is blended with a tea that is 15% jasmine. How many pounds of each tea are used to make 5 lb of tea that is 18% jasmine?
20% jasmine: 3 lb; 15% jasmine: 2 lb

33. How many ounces of pure chocolate must be added to 150 oz of chocolate topping that is 50% chocolate to make a topping that is 75% chocolate?
150 oz

34. How many ounces of pure bran flakes must be added to 50 oz of cereal that is 40% bran flakes to produce a mixture that is 50% bran flakes?
10 oz

35. Thirty ounces of pure silver are added to 50 oz of a silver alloy that is 20% silver. What is the percent concentration of the resulting alloy?
50%

36. A clothing manufacturer has some pure silk thread and some thread that is 85% silk. How many kilograms of each must be woven together to make 75 kg of cloth that is 96% silk?
pure silk: 55 kg; 85% silk: 20 kg

Objective C *Application Problems*

37. An investment of $3000 is made at an annual simple interest rate of 5%. How much additional money must be invested at an annual simple interest rate of 9% so that the total annual interest earned is 7.5% of the total investment?
$5000

38. A total of $6000 is invested into two simple interest accounts. The annual simple interest rate on one account is 9%; on the second account, the annual simple interest rate is 6%. How much should be invested in each account so that both accounts earn the same amount of annual interest?
$2400 at 9%; $3600 at 6%

39. An engineer invested a portion of $15,000 in a 7% annual simple interest account and the remainder in a 6.5% annual simple interest government bond. The amount of interest earned for one year was $1020. How much was invested in each account?
$9000 at 7%; $6000 at 6.5%

40. An investment club invested part of $20,000 in preferred stock that pays 8% annual simple interest and the remainder in a municipal bond that pays 7% annual simple interest. The amount of interest earned each year is $1520. How much was invested in each account?
$12,000 at 8%; $8000 at 7%

Quick Quiz (Objective 2.3C)

A landscape architect invested a total of $20,000 in two accounts: a municipal bond fund and a real estate investment trust. The annual interest rate on the municipal bond fund was 4.25%, and the annual interest rate on the real estate investment trust was 9%. If the designer received annual income of $1230, how much was invested in each account? $12,000 at 4.25%; $8000 at 9%

109

41. A grocery checker deposited an amount of money into a high-yield mutual fund that returns a 9% annual simple interest rate. A second deposit, $2500 more than the first, was placed in a certificate of deposit that returns a 5% annual simple interest rate. The total interest earned on both investments for one year was $475. How much money was deposited in the mutual fund?
$2500

42. A deposit was made into a 7% annual simple interest account. Another deposit, $1500 less than the first deposit, was placed in a 9% annual simple interest certificate of deposit. The total interest earned on both accounts for one year was $505. How much money was deposited in the certificate of deposit?
$2500

43. A corporation gave a university $300,000 to support product safety research. The university deposited some of the money in a 10% simple interest account and the remainder in an 8.5% simple interest account. How much was deposited in each account if the annual interest earned is $28,500?
$200,000 at 10%; $100,000 at 8.5%

44. A financial consultant advises a client to invest part of $30,000 in municipal bonds that earn 6.5% annual simple interest and the remainder of the money in 8.5% corporate bonds. How much should be invested in each type of bond so that the total annual interest earned each year is $2190?
$18,000 at 6.5%; $12,000 at 8.5%

45. To provide for retirement income, an auto mechanic purchases a $5000 bond that earns 7.5% annual simple interest. How much money must be invested in additional bonds that have an interest rate of 8% so that the total annual interest earned from the two investments is $615?
$3000

46. The portfolio manager for an investment group invested $40,000 in a certificate of deposit that earns 7.25% annual simple interest. How much money must be invested in additional certificates that have an interest rate of 8.5% so that the total annual interest earned from the two investments is $5025?
$25,000

47. A charity deposited a total of $54,000 into two simple interest accounts. The annual simple interest rate on one account is 8%. The annual simple interest rate on the second account is 12%. How much was invested in each account if the total interest earned is 9% of the total investment?
$40,500 at 8%; $13,500 at 12%

48. A college sports foundation deposited a total of $24,000 into two simple interest accounts. The annual simple interest rate on one account is 7%. The annual simple interest rate on the second account is 11%. How much was invested in each account if the total annual interest earned is 10% of the total investment?
$6000 at 7%; $18,000 at 11%

49. An investment banker invested 55% of the bank's available cash in an account that earns 8.25% annual simple interest. The remainder of the cash was placed in an account that earns 10% annual simple interest. The interest earned in one year was $58,743.75. Find the total amount invested.
$650,000

50. A financial planner invested 40% of a client's cash account in preferred stock that earns 9% annual simple interest. The remainder of the client's cash was placed in treasury bonds that earn 7% annual interest. The total annual interest earned from the two investments was $2496. What was the total amount invested?
$32,000

51. The manager of a mutual fund placed 30% of the fund's available cash in a 6% simple interest account, 25% in 8% corporate bonds, and the remainder in a money market fund that earns 7.5% annual simple interest. The total annual interest from the investments was $35,875. What was the total amount invested?
$500,000

52. The manager of a trust decided to invest 30% of a client's cash in government bonds that earn 6.5% annual simple interest. Another 30% was placed in utility stocks that earn 7% annual simple interest. The remainder of the cash was placed in an account earning 8% annual simple interest. The total annual interest earned from the investments was $5437.50. What was the total amount invested?
$75,000

Objective D Application Problems

53. Two small planes start from the same point and fly in opposite directions. The first plane is flying 25 mph slower than the second plane. In 2 h, the planes are 470 mi apart. Find the rate of each plane.
105 mph, 130 mph

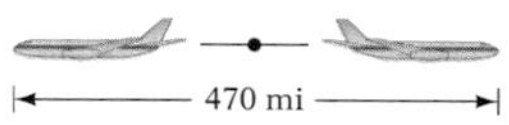

54. Two cyclists start from the same point and ride in opposite directions. One cyclist rides twice as fast as the other. In 3 h, they are 81 mi apart. Find the rate of each cyclist.
9 mph, 18 mph

55. Two planes leave an airport at 8 A.M., one flying north at 480 km/h and the other flying south at 520 km/h. At what time will they be 3000 km apart?
11 A.M.

56. A long-distance runner started on a course running at an average speed of 6 mph. One-half hour later, a second runner began the same course at an average speed of 7 mph. How long after the second runner started will the second runner overtake the first runner?
3 h

111

57. A motorboat leaves a harbor and travels at an average speed of 9 mph toward a small island. Two hours later a cabin cruiser leaves the same harbor and travels at an average speed of 18 mph toward the same island. In how many hours after the cabin cruiser leaves will the cabin cruiser be alongside the motorboat?
2 h

58. A 555-mile, 5-hour plane trip was flown at two speeds. For the first part of the trip, the average speed was 105 mph. For the remainder of the trip, the average speed was 115 mph. For how long did the plane fly at each speed?
105 mph: 2 h; 115 mph: 3 h

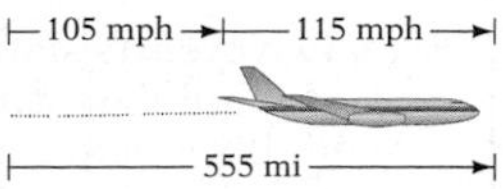

59. An executive drove from home at an average speed of 30 mph to an airport where a helicopter was waiting. The executive boarded the helicopter and flew to the corporate offices at an average speed of 60 mph. The entire distance was 150 mi. The entire trip took 3 h. Find the distance from the airport to the corporate offices.
120 mi

60. After a sailboat had been on the water for 3 h, a change in the wind direction reduced the average speed of the boat by 5 mph. The entire distance sailed was 57 mi. The total time spent sailing was 6 h. How far did the sailboat travel in the first 3 h?
36 mi

61. A car and a bus set out at 3 P.M. from the same point headed in the same direction. The average speed of the car is twice the average speed of the bus. In 2 h the car is 68 mi ahead of the bus. Find the rate of the car.
68 mph

62. A passenger train leaves a train depot 2 h after a freight train leaves the same depot. The freight train is traveling 20 mph slower than the passenger train. Find the rate of each train if the passenger train overtakes the freight train in 3 h.
passenger train: 50 mph; freight train: 30 mph

63. As part of flight training, a student pilot was required to fly to an airport and then return. The average speed on the way to the airport was 100 mph, and the average speed returning was 150 mph. Find the distance between the two airports if the total flying time was 5 h.
300 mi

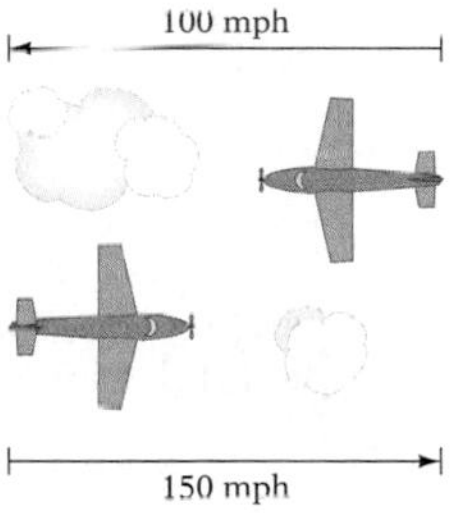

64. A ship traveling east at 25 mph is 10 mi from a harbor when another ship leaves the harbor traveling east at 35 mph. How long does it take the second ship to catch up to the first ship?
1 h

65. At 10 A.M. a plane leaves Boston, Massachusetts, for Seattle, Washington, a distance of 3000 mi. One hour later a plane leaves Seattle for Boston. Both planes are traveling at a speed of 500 mph. How many hours after the plane leaves Seattle will the planes pass each other?
2.5 h

Quick Quiz (Objective 2.3D)

1. Two planes start from the same point and fly in opposite directions. The first plane is flying 40 mph slower than the second plane. In 2 h the planes are 1040 mi apart. Find the rate of each plane. First plane: 240 mph; second plane: 280 mph
2. A cabin cruiser left a harbor and traveled to a small island at an average speed of 24 mph. On the return trip, the cabin cruiser traveled at an average speed of 16 mph. The total time for the trip was 5 h. How far was the island from the harbor? 48 mi

66. At noon a train leaves Washington, D.C., headed for Charleston, South Carolina, a distance of 500 mi. The train travels at a speed of 60 mph. At 1 P.M. a second train leaves Charleston headed for Washington, D.C., traveling at 50 mph. How long after the train leaves Charleston will the two trains pass each other?
4 h

67. Two cyclists start at the same time from opposite ends of a course that is 51 mi long. One cyclist is riding at a rate of 16 mph, and the second cyclist is riding at a rate of 18 mph. How long after they begin will they meet?
1.5 h

68. A bus traveled on a straight road for 2 h at an average speed that was 20 mph faster than its average speed on a winding road. The time spent on the winding road was 3 h. Find the average speed on the winding road if the total trip was 210 mi.
34 mph

69. A bus traveling at a rate of 60 mph overtakes a car traveling at a rate of 45 mph. If the car had a 1-hour head start, how far from the starting point does the bus overtake the car?
180 mi

70. A car traveling at 48 mph overtakes a cyclist who, riding at 12 mph, had a 3-hour head start. How far from the starting point does the car overtake the cyclist?
48 mi

APPLYING THE CONCEPTS

71. A radiator contains 15 gal of a 20% antifreeze solution. How many gallons must be drained from the radiator and replaced by pure antifreeze so that the radiator will contain 15 gal of a 40% antifreeze solution?
3.75 gal

72. A plant manager invested $3000 more in preferred stocks than in bonds. The stocks paid 8% annual simple interest, and the bonds paid 9.5% annual simple interest. Both investments yielded the same income. Find the total annual interest received on both investments.
$3040

73. A bicyclist rides for 2 h at a speed of 10 mph and then returns at a speed of 20 mph. Find the cyclist's average speed for the trip.
$13\frac{1}{3}$ mph

74. A car travels a 1-mile track at an average speed of 30 mph. At what average speed must the car travel the next mile so that the average speed for the 2 mi is 60 mph?
It is impossible to average 60 mph.

2.4 Inequalities in One Variable

Objective A **To solve an inequality in one variable**

The **solution set of an inequality** is a set of numbers, each element of which, when substituted for the variable, results in a true inequality.

The inequality at the right is true if the variable is replaced by (for instance) 3, -1.98, or $\frac{2}{3}$.

$$x - 1 < 4$$
$$3 - 1 < 4$$
$$-1.98 - 1 < 4$$
$$\frac{2}{3} - 1 < 4$$

There are many values of the variable x that will make the inequality $x - 1 < 4$ true. The solution set of the inequality is any number less than 5. The solution set can be written in set-builder notation as $\{x \mid x < 5\}$.

The graph of the solution set of $x - 1 < 4$ is shown at the right.

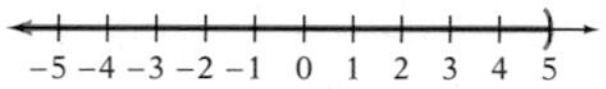

In solving an inequality, we use the Addition and Multiplication Properties of Inequalities to rewrite the inequality in the form *variable* < *constant* or in the form *variable* > *constant*.

The Addition Property of Inequalities

If $a > b$, then $a + c > b + c$.
If $a < b$, then $a + c < b + c$.

The Addition Property of Inequalities states that the same number can be added to each side of an inequality without changing the solution set of the inequality. This property is also true for an inequality that contains the symbol $\leq$ or $\geq$.

The Addition Property of Inequalities is used to remove a term from one side of an inequality by adding the additive inverse of that term to each side of the inequality. Because subtraction is defined in terms of addition, the same number can be subtracted from each side of an inequality without changing the solution set of the inequality.

⇒ Solve: $x + 2 \geq 4$

$$x + 2 \geq 4$$
$$x + 2 - 2 \geq 4 - 2 \quad \text{• Subtract 2 from each side of the inequality.}$$
$$x \geq 2 \quad \text{• Simplify.}$$

The solution set is $\{x \mid x \geq 2\}$.

Objective 2.4A

New Vocabulary
solution set of an inequality

New Properties
Addition Property of Inequalities
Multiplication Property of Inequalities

Discuss the Concepts

1. How does the solution set of $x \leq 5$ differ from the solution set of $x < 5$?
2. True or false: The solution set of $x \geq 5$ is the set $\{5, 6, 7, 8, 9, \ldots\}$. False

Optional Student Activity

Determine whether the following statements are always true, sometimes true, or never true. If a statement is sometimes true, find conditions that will make it always true.

1. If $ax < bx$, then $a < b$. Sometimes true. $x > 0$ will make the statement always true.
2. If $a < b$, then $a^2 < b^2$. Sometimes true. $0 < a < b$ will make the statement always true.
3. If $a < b$, then $ax^2 < bx^2$. Sometimes true. $x \neq 0$ will make the statement always true.
4. If $a < b$ and $a \neq 0$, $b \neq 0$, then $\frac{1}{a} > \frac{1}{b}$. Sometimes true. $0 < a < b$ will make the statement always true.
5. If $a > b > 0$, then $\frac{1}{a} < \frac{1}{b}$. Always true

In-Class Examples (Objective 2.4A)

Solve.

1. $x - 1 < 3$ $\{x \mid x < 4\}$
2. $-2x \leq 10$ $\{x \mid x \geq -5\}$
3. $6 - 4x \geq 30$ $\{x \mid x \leq -6\}$
4. $5x - 3 > 2x + 6$ $\{x \mid x > 3\}$

Instructor Note

Students are often confused by the need to change the inequality symbol when multiplying or dividing by a negative number. Emphasize that it is what you multiply or divide *by* that is important. For instance:

$$2x < -8$$
$$\frac{2x}{2} < \frac{-8}{2}$$
$$x < -4$$

Here, we divide *by* 2. It does not matter that we are dividing into a negative number.

Here is another example:

$$-2x < 8$$
$$\frac{-2x}{-2} > \frac{8}{-2}$$
$$x > -4$$

Now we have divided *by* a negative number; the inequality symbol must be reversed.

Another source of confusion is problems of the following type: $-4x > 0$ and $3x < 0$. You might solve a few inequalities of this type in class, emphasizing that the Multiplication Property of Inequalities still applies.

➡ Solve: $3x - 4 < 2x - 1$

$$3x - 4 < 2x - 1$$
$$3x - 2x - 4 < 2x - 2x - 1$$
• Subtract $2x$ from each side of the inequality.
$$x - 4 < -1$$
$$x - 4 + 4 < -1 + 4$$
• Add 4 to each side of the inequality.
$$x < 3$$

The solution set is $\{x | x < 3\}$.

The Multiplication Property of Inequalities is used to remove a coefficient from one side of an inequality by multiplying each side of the inequality by the reciprocal of the coefficient.

TAKE NOTE
$c > 0$ means c is a positive number.

$c < 0$ means c is a negative number.

The Multiplication Property of Inequalities

Rule 1 If $a > b$ and $c > 0$, then $ac > bc$.
If $a < b$ and $c > 0$, then $ac < bc$.

Rule 2 If $a > b$ and $c < 0$, then $ac < bc$.
If $a < b$ and $c < 0$, then $ac > bc$.

Here are some examples of this property.

Rule 1		Rule 2	
$3 > 2$	$2 < 5$	$3 > 2$	$2 < 5$
$3(4) > 2(4)$	$2(4) < 5(4)$	$3(-4) < 2(-4)$	$2(-4) > 5(-4)$
$12 > 8$	$8 < 20$	$-12 < -8$	$-8 > -20$

Rule 1 states that **when each side of an inequality is multiplied by a positive number, the inequality symbol remains the same.** However, Rule 2 states that **when each side of an inequality is multiplied by a negative number, the inequality symbol must be reversed.** Because division is defined in terms of multiplication, **when each side of an inequality is divided by a positive number, the inequality symbol remains the same. But when each side of an inequality is divided by a negative number, the inequality symbol must be reversed.**

The Multiplication Property of Inequalities is also true for the symbols $\le$ and $\ge$.

➡ Solve: $-3x > 9$

TAKE NOTE
Each side of the inequality is *divided by* a negative number; the inequality symbol must be reversed.

$$-3x > 9$$
$$\frac{-3x}{-3} < \frac{9}{-3}$$
• Divide each side of the inequality by the coefficient -3. Because -3 is a negative number, the inequality symbol must be reversed.
$$x < -3$$

The solution set is $\{x | x < -3\}$.

115

TAKE NOTE
Any time an inequality is multiplied or divided by a negative number, the inequality symbol must be reversed. Compare the next two examples.

$2x < -4$ Divide each side by *positive* 2.
$\frac{2x}{2} < \frac{-4}{2}$
$x < -2$ Inequality *is not* reversed.

$-2x < 4$ Divide each side by *negative* 2.
$\frac{-2x}{-2} > \frac{4}{-2}$
$x > -2$ Inequality *is* reversed.

➡ Solve: $3x + 2 < -4$

$3x + 2 < -4$

$3x < -6$ • Subtract 2 from each side of the inequality.

$\frac{3x}{3} < \frac{-6}{3}$ • Divide each side of the inequality by the coefficient 3.

$x < -2$

The solution set is $\{x|x < -2\}$.

➡ Solve: $2x - 9 > 4x + 5$

$2x - 9 > 4x + 5$

$-2x - 9 > 5$ • Subtract $4x$ from each side of the inequality.

$-2x > 14$ • Add 9 to each side of the inequality.

$\frac{-2x}{-2} < \frac{14}{-2}$ • Divide each side of the inequality by the coefficient −2. Reverse the inequality symbol.

$x < -7$

The solution set is $\{x|x < -7\}$.

TAKE NOTE
Solving these inequalities is similar to solving the equations solved in Section 2 *except* that when you multiply or divide the inequality by a negative number, you must reverse the inequality symbol.

➡ Solve: $5(x - 2) \geq 9x - 3(2x - 4)$

$5(x - 2) \geq 9x - 3(2x - 4)$

$5x - 10 \geq 9x - 6x + 12$ • Use the Distributive Property to remove parentheses.

$5x - 10 \geq 3x + 12$ • Combine like terms.

$2x - 10 \geq 12$ • Subtract $3x$ from each side of the inequality.

$2x \geq 22$ • Add 10 to each side of the inequality.

$\frac{2x}{2} \geq \frac{22}{2}$ • Divide each side of the inequality by the coefficient 2.

$x \geq 11$

The solution set is $\{x|x \geq 11\}$.

Example 1
Solve: $x + 3 > 4x + 6$

Solution

$x + 3 > 4x + 6$

$-3x + 3 > 6$ • Subtract $4x$ from each side.

$-3x > 3$ • Subtract 3 from each side.

$\frac{-3x}{-3} < \frac{3}{-3}$ • Divide each side by −3.

$x < -1$

The solution set is $\{x|x < -1\}$.

You Try It 1
Solve: $2x - 1 < 6x + 7$

Your solution
$\{x|x > -2\}$

Solution on p. S6

Objective 2.4B

New Vocabulary
compound inequality

Instructor Note
The compound inequalities in this section prepare the student to solve the absolute value inequalities they will study in the next section.

Concept Check
If n is an integer, determine the number of values of n that satisfy the inequality $0 \le n^2 \le 100$. 21

Optional Student Activity
Given the rules for rounding numbers, some possible values of the number 2.7 before it was rounded to the nearest tenth are 2.73, 2.68, 2.65, and 2.749. If V represents the exact value of 2.7 before it was rounded, then the inequality $2.65 \le V < 2.75$ represents all possible values of 2.7 before it was rounded.

Now suppose a rectangle is measured to be 3.4 m by 4.8 m, each measurement rounded to the nearest tenth of a meter. By using the smallest and largest possible values of each measurement, we can find the possible values of the area, A.

$$3.35(4.75) \le A < 3.45(4.85)$$
$$15.9125 \le A < 16.7325$$

The area is greater than or equal to 15.9125 square meters and less than 16.7325 square meters.
(Continued on next page)

Example 2 Solve: $3x - 5 \le 3 - 2(3x + 1)$

Solution

$$3x - 5 \le 3 - 2(3x + 1)$$
$$3x - 5 \le 3 - 6x - 2$$
$$3x - 5 \le 1 - 6x$$
$$9x - 5 \le 1$$
$$9x \le 6$$
$$\frac{9x}{9} \le \frac{6}{9}$$
$$x \le \frac{2}{3}$$
$$\left\{x \mid x \le \frac{2}{3}\right\}$$

You Try It 2 Solve: $5x - 2 \le 4 - 3(x - 2)$

Your solution

$$\left\{x \mid x \le \frac{3}{2}\right\}$$

Solution on p. S6

Objective B To solve a compound inequality

VIDEO & DVD CD TUTOR WEB SSM

A **compound inequality** is formed by joining two inequalities with a connective word such as *and* or *or.* The inequalities at the right are compound inequalities.

$2x < 4$ and $3x - 2 > -8$

$2x + 3 > 5$ or $x + 2 < 5$

The solution set of a compound inequality with the connective word *and* is the set of all elements that appear in the solution sets of both inequalities. Therefore, it is the intersection of the solution sets of the two inequalities.

TAKE NOTE
The intersection $\{x \mid x < 3\} \cap \{x \mid x > -2\}$ can be written $\{x \mid x > -2 \text{ and } x < 3\}$. However, it is more commonly written $\{x \mid -2 < x < 3\}$. This is read "the set of all x such that x is greater than -2 *and* less than 3."

➡ Solve: $2x < 6$ and $3x + 2 > -4$

$2x < 6$	and	$3x + 2 > -4$	
$x < 3$		$3x > -6$	• Solve each inequality.
$\{x \mid x < 3\}$		$x > -2$	
		$\{x \mid x > -2\}$	

The solution of a compound inequality with *and* is the intersection of the solution sets of the two inequalities.

$\{x \mid x < 3\} \cap \{x \mid x > -2\} = \{x \mid -2 < x < 3\}$

➡ Solve: $-3 < 2x + 1 < 5$

This inequality is equivalent to the compound inequality shown at the right.

$-3 < 2x + 1$ and $2x + 1 < 5$

$-3 < 2x + 1$	and	$2x + 1 < 5$	
$-4 < 2x$		$2x < 4$	• Solve each inequality.
$-2 < x$		$x < 2$	
$\{x \mid x > -2\}$		$\{x \mid x < 2\}$	

$\{x \mid x > -2\} \cap \{x \mid x < 2\} = \{x \mid -2 < x < 2\}$

In-Class Examples (Objective 2.4B)
Solve.
1. $2x < 8$ and $x + 3 > 1$ $\{x \mid -2 < x < 4\}$
2. $x + 1 \ge 6$ or $2x < 6$ $\{x \mid x \ge 5 \text{ or } x < 3\}$
3. $-1 < 2x + 5 < 9$ $\{x \mid -3 < x < 2\}$
4. $4x - 3 > 5$ or $4x - 3 \le -15$ $\{x \mid x > 2 \text{ or } x \le -3\}$

There is an alternative method for solving the inequality in the last example.

➡ Solve: $-3 < 2x + 1 < 5$

$$-3 < 2x + 1 < 5$$

$$-3 - 1 < 2x + 1 - 1 < 5 - 1$$ • Subtract 1 from each of the three parts of the inequality.

$$-4 < 2x < 4$$

$$\frac{-4}{2} < \frac{2x}{2} < \frac{4}{2}$$ • Divide each of the three parts of the inequality by the coefficient 2.

$$-2 < x < 2$$

The solution set is $\{x \mid -2 < x < 2\}$.

The solution set of a compound inequality with the connective word *or* is the union of the solution sets of the two inequalities.

➡ Solve: $2x + 3 > 7$ or $4x - 1 < 3$

$2x + 3 > 7$ or $4x - 1 < 3$

$2x > 4$ $\quad$ $4x < 4$ • Solve each inequality.

$x > 2$ $\quad$ $x < 1$

$\{x \mid x > 2\}$ $\quad$ $\{x \mid x < 1\}$

$\{x \mid x > 2\} \cup \{x \mid x < 1\} = \{x \mid x > 2 \text{ or } x < 1\}$ • Find the union of the solution sets.

Example 3
Solve: $1 < 3x - 5 < 4$

Solution

$$1 < 3x - 5 < 4$$
$$1 + 5 < 3x - 5 + 5 < 4 + 5$$
$$6 < 3x < 9$$
$$\frac{6}{3} < \frac{3x}{3} < \frac{9}{3}$$
$$2 < x < 3$$

$\{x \mid 2 < x < 3\}$

You Try It 3
Solve: $-2 \le 5x + 3 \le 13$

Your solution
$\{x \mid -1 \le x \le 2\}$

Example 4
Solve: $11 - 2x > -3$ and $7 - 3x < 4$

Solution

$11 - 2x > -3$ and $7 - 3x < 4$

$-2x > -14$ $\quad$ $-3x < -3$

$x < 7$ $\quad$ $x > 1$

$\{x \mid x < 7\}$ $\quad$ $\{x \mid x > 1\}$

$\{x \mid x < 7\} \cap \{x \mid x > 1\} = \{x \mid 1 < x < 7\}$

You Try It 4
Solve: $2 - 3x > 11$ or $5 + 2x > 7$

Your solution
$\{x \mid x < -3 \text{ or } x > 1\}$

Solutions on p. S6

(Continued)

1. The length of a line is 4.2 in., measured to the nearest tenth. Write an inequality that represents the possible lengths of the line.
$4.15 \le L < 4.25$
2. The length of a side of a square is 6.4 cm, measured to the nearest tenth. Write an inequality that represents the possible areas of the square.
$40.3225 \le A < 41.6025$
3. The base of a triangle is 5.43 m and the height is 2.47 m, measured to the nearest hundredth. Write an inequality that represents the possible areas of the triangle.
$6.6863125 \le A < 6.7258125$
4. Write an inequality that represents the possible areas of a rectangle that is 3.0 m by 4.0 m, measured to the nearest tenth.
$11.6525 \le A < 12.3525$

Instructor Note

Have students solve the compound inequality $2x - 1 > 5$ and $3x - 2 < 1$. The solution will require that they find the intersection of $\{x \mid x > 3\}$ and $\{x \mid x < 1\}$, which is the empty set. Suggest that another way to think about the solution is to ask, "What number is greater than 3 *and* less than 1?" Because there is no such number, the solution set is the empty set.

Next, ask students whether the solution set changes if the word *and* in the compound inequality is replaced by *or*. The answer is yes. The solution set is now the union of the two sets $\{x \mid x > 3\}$ and $\{x \mid x < 1\}$. Another way to think about this solution is to say, "We are looking for a number that is greater than 3 *or* less than 1."

Objective 2.4C

Discuss the Concepts

1. Explain why the inequality $x > x + 1$ has no solution.
2. Explain why the solution set of the inequality $x < x + 1$ is all real numbers.

Concept Check

Rewrite each inequality so that x is on the left side of the inequality.

1. $5 < x$ $x > 5$
2. $-3 \leq x$ $x \geq -3$
3. $-7 \leq x$ $x \geq -7$
4. $4 > x$ $x < 4$

Concept Check

Justin purchased some 23¢ stamps and some 37¢ stamps. He bought a total of 80 stamps for a total cost of less than $25. What is the maximum number of 37¢ stamps that Justin could have purchased? Forty-seven 37¢ stamps

Optional Student Activity

Have students translate the following sentences into inequalities.

1. x is at least 20. $x \geq 20$
2. The minimum value of y is 12. $y \geq 12$
3. b exceeds -3. $b > -3$
4. z is at most 7. $z \leq 7$
5. d is more than 15. $d > 15$
6. m is 17 or less. $m \leq 17$
7. The maximum value of n is 8. $n \leq 8$

Objective C To solve application problems

VIDEO & DVD CD TUTOR WEB SSM

Example 5

A rectangle is 10 ft wide and $(2x + 4)$ ft long. Express as an integer the maximum length of the rectangle when the area is less than 200 ft^2. (The area of a rectangle is equal to its length times its width.)

Strategy

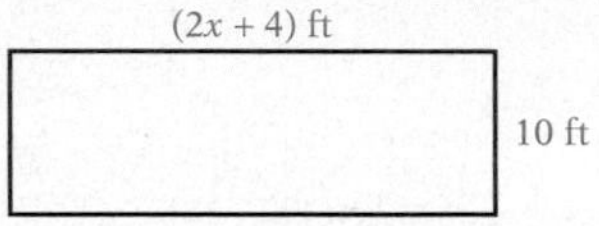

To find the maximum length:
- Replace the variables in the area formula by the given values and solve for x.
- Replace the variable in the expression $2x + 4$ with the value found for x.

Solution

Length times width	is less than	200 ft^2

$$(2x + 4)10 < 200$$
$$20x + 40 < 200$$
$$20x + 40 - 40 < 200 - 40$$
$$20x < 160$$
$$\frac{20x}{20} < \frac{160}{20}$$
$$x < 8$$

The length is $(2x + 4)$ ft. Because $x < 8$, $2x + 4 < 2(8) + 4 = 20$. Therefore, the length is less than 20 ft.

The maximum length is 19 ft.

You Try It 5

Company A rents cars for $24 a day and 10¢ for every mile driven. Company B rents cars for $30 a day and 8¢ per mile driven. You want to rent a car for one week. What is the maximum number of miles you can drive a Company A car if it is to cost you less than a Company B car?

Your strategy

Your solution

2099 mi

Solution on pp. S6–S7

In-Class Examples (Objective 2.4C)

1. Four times the difference between a number and six is less than or equal to six times the sum of the number and four. Find the smallest number that will satisfy the inequality. -24
2. Company A rents a car for $36 per day and 10¢ for every mile driven. Company B rents a car for $68.50 per day with unlimited mileage. How many miles per day can you drive a Company A car if it is to cost you less than a Company B car? Less than 325 mi

2.4 Exercises

Objective A

1. State the Addition Property of Inequalities and give numerical examples of its use.

2. State the Multiplication Property of Inequalities and give numerical examples of its use.

3. Which numbers are solutions of the inequality $x + 7 \le -3$?
a. -17 **b.** 8 **c.** -10 **d.** 0
a, c

4. Which numbers are solutions of the inequality $2x - 1 > 5$?
a. 6 **b.** -4 **c.** 3 **d.** 5
a, d

Solve.

5. $x - 3 < 2$
$\{x \mid x < 5\}$

6. $x + 4 \ge 2$
$\{x \mid x \ge -2\}$

7. $4x \le 8$
$\{x \mid x \le 2\}$

8. $6x > 12$
$\{x \mid x > 2\}$

9. $-2x > 8$
$\{x \mid x < -4\}$

10. $-3x \le -9$
$\{x \mid x \ge 3\}$

11. $3x - 1 > 2x + 2$
$\{x \mid x > 3\}$

12. $5x + 2 \ge 4x - 1$
$\{x \mid x \ge -3\}$

13. $2x - 1 > 7$
$\{x \mid x > 4\}$

14. $3x + 2 < 8$
$\{x \mid x < 2\}$

15. $5x - 2 \le 8$
$\{x \mid x \le 2\}$

16. $4x + 3 \le -1$
$\{x \mid x \le -1\}$

17. $6x + 3 > 4x - 1$
$\{x \mid x > -2\}$

18. $7x + 4 < 2x - 6$
$\{x \mid x < -2\}$

19. $8x + 1 \ge 2x + 13$
$\{x \mid x \ge 2\}$

20. $5x - 4 < 2x + 5$
$\{x \mid x < 3\}$

21. $4 - 3x < 10$
$\{x \mid x > -2\}$

22. $2 - 5x > 7$
$\{x \mid x < -1\}$

23. $7 - 2x \ge 1$
$\{x \mid x \le 3\}$

24. $3 - 5x \le 18$
$\{x \mid x \ge -3\}$

25. $-3 - 4x > -11$
$\{x \mid x < 2\}$

26. $-2 - x < 7$
$\{x \mid x > -9\}$

27. $4x - 2 < x - 11$
$\{x \mid x < -3\}$

28. $6x + 5 \le x - 10$
$\{x \mid x \le -3\}$

Section 2.4

Suggested Assignment

Exercises 3–4
Exercises 5–45, every other odd
Exercises 49–99, odds
More challenging problems:
Exercises 100, 102–104

Answers to Writing Exercises

1. The Addition Property of Inequalities states that the same number can be added to each side of an inequality without changing the solution set of the inequality.
Examples will vary. For instance:

$$8 > 6$$
$$8 + 4 > 6 + 4$$
$$12 > 10$$

and

$$-5 < -1$$
$$-5 + (-7) < -1 + (-7)$$
$$-12 < -8$$

2. Rule 1 of the Multiplication Property of Inequalities states that when each side of an inequality is multiplied by a positive number, the inequality symbol remains the same. Rule 2 states that when each side of an inequality is multiplied by a negative number, the inequality symbol must be reversed. Examples will vary. For instance:

$$8 > 6$$
$$3 \cdot 8 > 3 \cdot 6$$
$$24 > 18$$

and

$$8 > 6$$
$$-3 \cdot 8 < -3 \cdot 6$$
$$-24 < -18$$

Quick Quiz (Objective 2.4A)

Solve.

1. $x + 3 \ge 1$ $\{x \mid x \ge -2\}$
2. $-3x > 15$ $\{x \mid x < -5\}$
3. $2 - 7x \le 16$ $\{x \mid x \ge -2\}$
4. $6x + 1 \ge 4x - 3$ $\{x \mid x \ge -2\}$

Answers to Writing Exercises

46a. When a compound inequality is combined with *or,* the set operation union is used.

b. When a compound inequality is combined with *and,* the set operation intersection is used.

47. Writing $-3 > x > 4$ does not make sense because there is no number that is less than -3 *and* greater than 4.

29. $x + 7 \geq 4x - 8$
$\{x \mid x \leq 5\}$

30. $3x + 1 \leq 7x - 15$
$\{x \mid x \geq 4\}$

31. $3x + 2 \leq 7x + 4$
$\left\{x \mid x \geq -\frac{1}{2}\right\}$

32. $3x - 5 \geq -2x + 5$
$\{x \mid x \geq 2\}$

33. $\frac{3}{5}x - 2 < \frac{3}{10} - x$
$\left\{x \mid x < \frac{23}{16}\right\}$

34. $\frac{5}{6}x - \frac{1}{6} < x - 4$
$\{x \mid x > 23\}$

35. $\frac{2}{3}x - \frac{3}{2} < \frac{7}{6} - \frac{1}{3}x$
$\left\{x \mid x < \frac{8}{3}\right\}$

36. $\frac{7}{12}x - \frac{3}{2} < \frac{2}{3}x + \frac{5}{6}$
$\{x \mid x > -28\}$

37. $\frac{1}{2}x - \frac{3}{4} < \frac{7}{4}x - 2$
$\{x \mid x > 1\}$

38. $6 - 2(x - 4) \leq 2x + 10$
$\{x \mid x \geq 1\}$

39. $4(2x - 1) > 3x - 2(3x - 5)$
$\left\{x \mid x > \frac{14}{11}\right\}$

40. $2(1 - 3x) - 4 > 10 + 3(1 - x)$
$\{x \mid x < -5\}$

41. $2 - 5(x + 1) \geq 3(x - 1) - 8$
$\{x \mid x \leq 1\}$

42. $2 - 2(7 - 2x) < 3(3 - x)$
$\{x \mid x < 3\}$

43. $3 + 2(x + 5) \geq x + 5(x + 1) + 1$
$\left\{x \mid x \leq \frac{7}{4}\right\}$

44. $3x - 2(3x - 5) \leq 2 - 5(x - 4)$
$\{x \mid x \leq 6\}$

45. $12 - 2(3x - 2) \geq 5x - 2(5 - x)$
$\{x \mid x \leq 2\}$

Objective B

46. **a.** Which set operation is used when a compound inequality is combined with *or*?
b. Which set operation is used when a compound inequality is combined with *and*?

47. Explain why writing $-3 > x > 4$ does not make sense.

Solve.

48. $3x < 6$ and $x + 2 > 1$
$\{x \mid -1 < x < 2\}$

49. $x - 3 \leq 1$ and $2x \geq -4$
$\{x \mid -2 \leq x \leq 4\}$

50. $x + 2 \geq 5$ or $3x \leq 3$
$\{x \mid x \geq 3 \text{ or } x \leq 1\}$

51. $2x < 6$ or $x - 4 > 1$
$\{x \mid x < 3 \text{ or } x > 5\}$

Quick Quiz (Objective 2.4B)

Solve.

1. $x - 5 \leq 2$ and $3x > -12$ $\{x \mid -4 < x \leq 7\}$
2. $3x < 12$ or $x - 1 > 5$ $\{x \mid x < 4 \text{ or } x > 6\}$
3. $-2 < 3x + 1 < 10$ $\{x \mid -1 < x < 3\}$
4. $3x - 7 > 5$ or $3x - 7 < -10$
$\{x \mid x > 4 \text{ or } x < -1\}$

121

52. $-2x > -8$ and $-3x < 6$
$\{x \mid -2 < x < 4\}$

53. $\frac{1}{2}x > -2$ and $5x < 10$
$\{x \mid -4 < x < 2\}$

54. $\frac{1}{3}x < -1$ or $2x > 0$
$\{x \mid x < -3 \text{ or } x > 0\}$

55. $\frac{2}{3}x > 4$ or $2x < -8$
$\{x \mid x > 6 \text{ or } x < -4\}$

56. $x + 4 \geq 5$ and $2x \geq 6$
$\{x \mid x \geq 3\}$

57. $3x < -9$ and $x - 2 < 2$
$\{x \mid x < -3\}$

58. $-5x > 10$ and $x + 1 > 6$
$\varnothing$

59. $7x < 14$ and $1 - x < 4$
$\{x \mid -3 < x < 2\}$

60. $2x - 3 > 1$ and $3x - 1 < 2$
$\varnothing$

61. $4x + 1 < 5$ and $4x + 7 > -1$
$\{x \mid -2 < x < 1\}$

62. $3x + 7 < 10$ or $2x - 1 > 5$
$\{x \mid x < 1 \text{ or } x > 3\}$

63. $6x - 2 < -14$ or $5x + 1 > 11$
$\{x \mid x < -2 \text{ or } x > 2\}$

64. $-5 < 3x + 4 < 16$
$\{x \mid -3 < x < 4\}$

65. $5 < 4x - 3 < 21$
$\{x \mid 2 < x < 6\}$

66. $0 < 2x - 6 < 4$
$\{x \mid 3 < x < 5\}$

67. $-2 < 3x + 7 < 1$
$\{x \mid -3 < x < -2\}$

68. $4x - 1 > 11$ or $4x - 1 \leq -11$
$\left\{x \mid x > 3 \text{ or } x \leq -\frac{5}{2}\right\}$

69. $3x - 5 > 10$ or $3x - 5 < -10$
$\left\{x \mid x > 5 \text{ or } x < -\frac{5}{3}\right\}$

70. $2x - 3 \geq 5$ and $3x - 1 > 11$
$\{x \mid x > 4\}$

71. $6x - 2 < 5$ or $7x - 5 < 16$
$\{x \mid x < 3\}$

72. $9x - 2 < 7$ and $3x - 5 > 10$
$\varnothing$

73. $8x + 2 \leq -14$ and $4x - 2 > 10$
$\varnothing$

122

74. $3x - 11 < 4$ or $4x + 9 \geq 1$
$\{x | x \in \text{real numbers}\}$

75. $5x + 12 \geq 2$ or $7x - 1 \leq 13$
$\{x | x \in \text{real numbers}\}$

76. $-6 \leq 5x + 14 \leq 24$
$\{x | -4 \leq x \leq 2\}$

77. $3 \leq 7x - 14 \leq 31$
$\left\{x \middle| \frac{17}{7} \leq x \leq \frac{45}{7}\right\}$

78. $3 - 2x > 7$ and $5x + 2 > -18$
$\{x | -4 < x < -2\}$

79. $1 - 3x < 16$ and $1 - 3x > -16$
$\left\{x \middle| -5 < x < \frac{17}{3}\right\}$

80. $9 - x \geq 7$ and $9 - 2x < 3$
$\varnothing$

81. $6x + 5 < -1$ or $1 - 2x < 7$
$\{x | x \in \text{real numbers}\}$

Objective C *Application Problems*

82. Five times the difference between a number and two is greater than the quotient of two times the number and three. Find the smallest integer that will satisfy the inequality.
3

83. Two times the difference between a number and eight is less than or equal to five times the sum of the number and four. Find the smallest number that will satisfy the inequality.
−12

84. The length of a rectangle is 2 ft more than four times the width. Express as an integer the maximum width of the rectangle when the perimeter is less than 34 ft.
2 ft

85. The length of a rectangle is 5 cm less than twice the width. Express as an integer the maximum width of the rectangle when the perimeter is less than 60 cm.
11 cm

86. In 2002, the computer service America Online offered its customers the option of paying \$23.90 per month for unlimited use. Another option was a rate of \$4.95 per month with 3 free hours plus \$2.50 per hour thereafter. How many hours per month can you use this second option if it is to cost you less than the first option? Round to the nearest whole number.
10 h or less

Quick Quiz (Objective 2.4C)

1. The length of a rectangle is 3 m less than four times the width. Express as an integer the maximum width of the rectangle when the perimeter is less than 74 m. 7 m
2. A bank offers two types of checking accounts. One account has a charge of \$5 per month plus \$.04 per check. The second account has a charge of \$2 per month plus \$.10 per check. How many checks can a customer who has the second type of account write if it is to cost the customer less than the first type of account? Less than 50 checks

123

87. TopPage advertises local paging service for \$6.95 per month for up to 400 pages and \$.10 per page thereafter. A competitor advertises service for \$3.95 per month for up to 400 pages and \$.15 per page thereafter. For what number of pages per month is the TopPage plan less expensive?
more than 460 pages

88. Suppose PayRite Rental Cars rents compact cars for \$32 per day with unlimited mileage and Otto Rentals offers compact cars for \$19.99 per day but charges \$.19 for each mile beyond 100 mi driven per day. You want to rent a car for one week. How many miles can you drive during the week if Otto Rentals is to be less expensive than PayRite?
less than 1143 mi

89. During a weekday, to call a city 40 mi away from a certain pay phone costs \$.70 for the first 3 min and \$.15 for each additional minute. If you use a calling card, there is a \$.35 fee and then the rates are \$.196 for the first minute and \$.126 for each additional minute. How long must a call be for it to be cheaper to pay with coins rather than a calling card?
7 min or less

90. The temperature range for a week was between 14°F and 77°F. Find the temperature range in Celsius degrees. $F = \frac{9}{5}C + 32$
$-10° < C < 25°$

91. The temperature range for a week in a mountain town was between 0°C and 30°C. Find the temperature range in Fahrenheit degrees.
$C = \frac{5(F - 32)}{9}$
$32° < F < 86°$

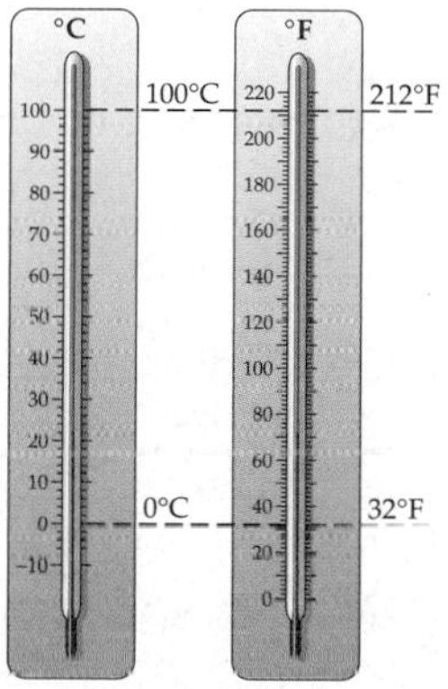

92. You are a sales account executive earning \$1200 per month plus 6% commission on the amount of sales. Your goal is to earn a minimum of \$6000 per month. What amount of sales will enable you to earn \$6000 or more per month?
\$80,000 or more

93. George Stoia earns \$1000 per month plus 5% commission on the amount of sales. George's goal is to earn a minimum of \$3200 per month. What amount of sales will enable George to earn \$3200 or more per month?
\$44,000 or more

94. Heritage National Bank offers two different checking accounts. The first charges \$3 per month and \$.50 per check after the first 10 checks. The second account charges \$8 per month with unlimited check writing. How many checks can be written per month if the first account is to be less expensive than the second account?
less than 20 checks

95. Glendale Federal Bank offers a checking account to small businesses. The charge is \$8 per month plus \$.12 per check after the first 100 checks. A competitor is offering an account for \$5 per month plus \$.15 per check after the first 100 checks. If a business chooses the first account, how many checks does the business write monthly if it is assumed that the first account will cost less than the competitor's account?
more than 200 checks

96. An average score of 90 or above in a history class receives an A grade. You have grades of 95, 89, and 81 on three exams. Find the range of scores on the fourth exam that will give you an A grade for the course.
$95 \le N \le 100$

97. An average of 70 to 79 in a mathematics class receives a C grade. A student has grades of 56, 91, 83, and 62 on four tests. Find the range of scores on the fifth test that will give the student a C for the course.
$58 \le N \le 100$

98. Grade A hamburger cannot contain more than 20% fat. How much fat can a butcher mix with 300 lb of lean meat to meet the 20% requirement?
75 lb or less

99. A shuttle service taking skiers to a ski area charges \$8 per person each way. Four skiers are debating whether to take the shuttle bus or rent a car for \$45 plus \$.25 per mile. The skiers will share the cost of the car, and they want the least expensive method of transportation. How far away is the ski area if they choose the shuttle service?
more than 38 mi away

APPLYING THE CONCEPTS

100. Determine whether the statement is always true, sometimes true, or never true, given that a, b, and c are real numbers.
a. If $a > b$, then $-a > -b$. never true
b. If $a < b$, then $ac < bc$. sometimes true
c. If $a > b$, then $a + c > b + c$. always true
d. If $a \neq 0$, $b \neq 0$, and $a > b$, then $\frac{1}{a} > \frac{1}{b}$. sometimes true

101. Use the roster method to list the set of positive integers that are solutions of the inequality $7 - 2b \le 15 - 5b$.
$\{1, 2\}$

102. Determine the solution set of $2 - 3(x + 4) < 5 - 3x$.
$\{x \mid x \in \text{real numbers}\}$

103. Determine the solution set of $3x + 2(x - 1) > 5(x + 1)$.
$\varnothing$

104. The average of two negative integers is less than or equal to -15. The smaller integer is 7 less than the larger integer. Find the greatest possible value for the smaller integer.
-19

2.5 Absolute Value Equations and Inequalities

Objective A **To solve an absolute value equation**

VIDEO & DVD · CD TUTOR · WEB · SSM

The **absolute value** of a number is its distance from zero on the number line. Distance is always a positive number or zero. Therefore, the absolute value of a number is always a positive number or zero.

The distance from 0 to 3 or from 0 to -3 is 3 units.

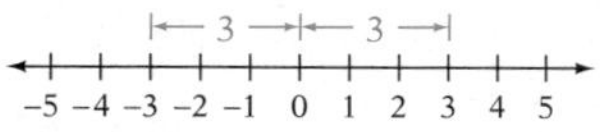

$|3| = 3 \qquad |-3| = 3$

An equation that contains an absolute value symbol is called an **absolute value equation**. The solution of an absolute value equation is based on the following property:

If $a \geq 0$ and $|x| = a$, then $x = a$ or $x = -a$.

For instance, given $|x| = 3$, then $x = 3$ or $x = -3$, because $|3| = 3$ and $|-3| = 3$.

TAKE NOTE
Because the absolute value of $x + 2$ is 8, the number $x + 2$ is 8 units from 0 on the number line. Therefore, $x + 2$ is equal to 8 or -8.

➡ Solve: $|x + 2| = 8$

$|x + 2| = 8$

$x + 2 = 8 \qquad x + 2 = -8$ • Remove the absolute value sign and rewrite as two equations.

$x = 6 \qquad x = -10$ • Solve each equation.

Check:

$\lvert x + 2\rvert = 8$		$\lvert x + 2\rvert = 8$	
$\lvert 6 + 2\rvert$	8	$\lvert -10 + 2\rvert$	8
$\lvert 8\rvert$	8	$\lvert -8\rvert$	8
$8 = 8$		$8 = 8$	

The solutions are 6 and -10.

➡ Solve: $|5 - 3x| - 8 = -4$

$|5 - 3x| - 8 = -4$

$|5 - 3x| = 4$ • Solve for the absolute value.

$5 - 3x = 4 \qquad 5 - 3x = -4$ • Remove the absolute value sign and rewrite as two equations.

$-3x = -1 \qquad -3x = -9$ • Solve each equation.

$x = \frac{1}{3} \qquad x = 3$

Check:

$\lvert 5 - 3x\rvert - 8 = -4$		$\lvert 5 - 3x\rvert - 8 = -4$	
$\left\lvert 5 - 3\left(\frac{1}{3}\right)\right\rvert - 8$	-4	$\lvert 5 - 3(3)\rvert - 8$	-4
$\lvert 5 - 1\rvert - 8$	-4	$\lvert 5 - 9\rvert - 8$	-4
$4 - 8$	-4	$4 - 8$	-4
$-4 = -4$		$-4 = -4$	

The solutions are $\frac{1}{3}$ and 3.

Objective 2.5A

New Vocabulary
absolute value equation

Vocabulary to Review
absolute value

Discuss the Concepts

1. In the definition of the solution of an absolute value equation given on this page, why must $a \geq 0$?
2. If the absolute value of a number must be positive or zero, why can the solution of an absolute value equation be a negative number?

Concept Check

1. Use absolute value to express the fact that both -7 and 3 are 5 units from -2. $|x + 2| = 5$
2. Use absolute value to express the fact that both -5 and 11 are 8 units from 3. $|x - 3| = 8$

Optional Student Activity

1. Solve: $|3 - |x|| = 1$ $\quad -4, -2, 2, 4$
2. Solve: $|x + 4| = 4x$ $\quad \frac{4}{3}$
3. Solve $x + |x| = 0$ and write the answer in set-builder notation. $\{x \mid x \leq 0\}$

In-Class Examples (Objective 2.5A)

Solve.

1. $|x + 3| = 1$ $\quad -2, -4$
2. $|2 - 3x| = 2$ $\quad 0, \frac{4}{3}$
3. $|7 - y| - 2 = 2$ $\quad 3, 11$

126

Example 1
Solve: $|2 - x| = 12$

Solution

$$|2 - x| = 12$$

$$2 - x = 12 \qquad\qquad 2 - x = -12$$
$$-x = 10 \qquad\qquad -x = -14$$
$$x = -10 \qquad\qquad x = 14$$

The solutions are −10 and 14.

You Try It 1
Solve: $|2x - 3| = 5$

Your solution
4, −1

Example 2
Solve: $3 - |2x - 4| = -5$

Solution

$$3 - |2x - 4| = -5$$
$$-|2x - 4| = -8$$
$$|2x - 4| = 8$$

$$2x - 4 = 8 \qquad\qquad 2x - 4 = -8$$
$$2x = 12 \qquad\qquad 2x = -4$$
$$x = 6 \qquad\qquad x = -2$$

The solutions are 6 and −2.

You Try It 2
Solve: $5 - |3x + 5| = 3$

Your solution
$-1, -\frac{7}{3}$

Solutions on p. S7

Objective B To solve an absolute value inequality

VIDEO & DVD CD TUTOR WEB SSM

Recall that absolute value represents the distance between two points. For example, the solutions of the absolute value equation $|x - 1| = 3$ are the numbers whose distance from 1 is 3. Therefore, the solutions are −2 and 4.

The solutions of the absolute value inequality $|x - 1| < 3$ are the numbers whose distance from 1 is less than 3. Therefore, the solutions are the numbers greater than −2 and less than 4. The solution set is $\{x \mid -2 < x < 4\}$.

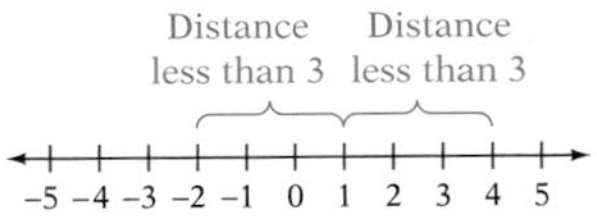

To solve an absolute value inequality of the form $|ax + b| < c$, solve the equivalent compound inequality $-c < ax + b < c$.

> **TAKE NOTE**
> Because the absolute value of $3x - 1$ is less than 5, the number $3x - 1$ is less than 5 units from 0 on the number line.
>
> 
>
>
> Therefore, $3x - 1$ is between −5 and 5.

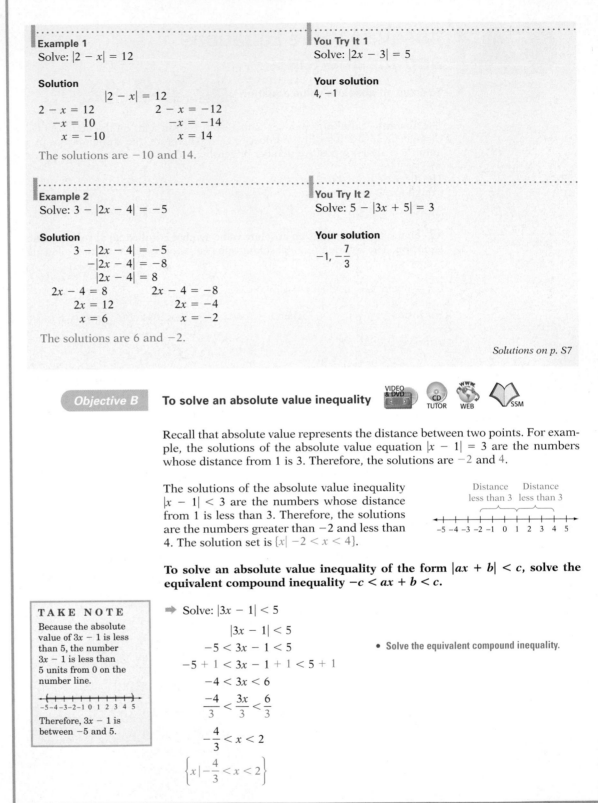

➡ Solve: $|3x - 1| < 5$

$$|3x - 1| < 5$$
$$-5 < 3x - 1 < 5$$ • Solve the equivalent compound inequality.
$$-5 + 1 < 3x - 1 + 1 < 5 + 1$$
$$-4 < 3x < 6$$
$$\frac{-4}{3} < \frac{3x}{3} < \frac{6}{3}$$
$$-\frac{4}{3} < x < 2$$
$$\left\{x \mid -\frac{4}{3} < x < 2\right\}$$

The solutions of the absolute value inequality $|x + 1| > 2$ are the numbers whose distance from -1 is greater than 2. Therefore, the solutions are the numbers that are less than -3 or greater than 1. The solution set of $|x + 1| > 2$ is $\{x|x < -3 \text{ or } x > 1\}$.

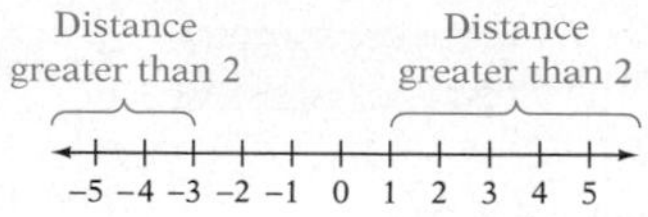

TAKE NOTE
Because the absolute value of $3 - 2x$ is greater than 1, the number $3 - 2x$ is more than 1 unit from 0 on the number line.

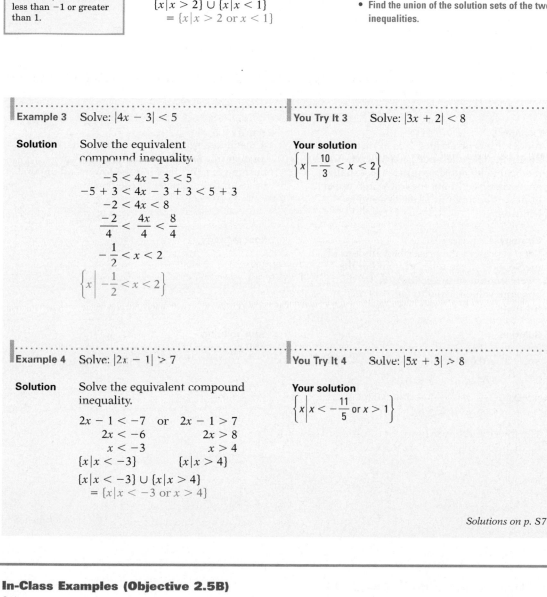

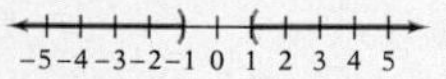

Therefore, $3 - 2x$ is less than -1 or greater than 1.

To solve an absolute value inequality of the form $|ax + b| > c$, solve the equivalent compound inequality $ax + b < -c$ or $ax + b > c$.

➡ Solve: $|3 - 2x| > 1$

$$3 - 2x < -1 \quad \text{or} \quad 3 - 2x > 1$$
$$-2x < -4 \qquad\qquad -2x > -2$$
$$x > 2 \qquad\qquad x < 1$$
$$\{x|x > 2\} \qquad\qquad \{x|x < 1\}$$

• Solve each inequality.

$$\{x|x > 2\} \cup \{x|x < 1\} = \{x|x > 2 \text{ or } x < 1\}$$

• Find the union of the solution sets of the two inequalities.

Example 3 Solve: $|4x - 3| < 5$

Solution Solve the equivalent compound inequality.

$$-5 < 4x - 3 < 5$$
$$-5 + 3 < 4x - 3 + 3 < 5 + 3$$
$$-2 < 4x < 8$$
$$\frac{-2}{4} < \frac{4x}{4} < \frac{8}{4}$$
$$-\frac{1}{2} < x < 2$$
$$\left\{x \middle| -\frac{1}{2} < x < 2\right\}$$

You Try It 3 Solve: $|3x + 2| < 8$

Your solution
$\left\{x \middle| -\frac{10}{3} < x < 2\right\}$

Example 4 Solve: $|2x - 1| > 7$

Solution Solve the equivalent compound inequality.

$$2x - 1 < -7 \quad \text{or} \quad 2x - 1 > 7$$
$$2x < -6 \qquad\qquad 2x > 8$$
$$x < -3 \qquad\qquad x > 4$$
$$\{x|x < -3\} \qquad\qquad \{x|x > 4\}$$
$$\{x|x < -3\} \cup \{x|x > 4\} = \{x|x < -3 \text{ or } x > 4\}$$

You Try It 4 Solve: $|5x + 3| > 8$

Your solution
$\left\{x \middle| x < -\frac{11}{5} \text{ or } x > 1\right\}$

Solutions on p. S7

Objective 2.5B

Discuss the Concepts

Explain how the solution set of $|x - 4| \le c$ changes for $c > 0$, $c = 0$, and $c < 0$.

Concept Check

1. For $c > 0$, how does the solution set of $|ax + b| > c$ differ from the solution set of $|ax + b| \ge c$? The second solution set contains the endpoints of the interval; the first one does not.
2. Which of the following inequalities have no solution? Which have all real numbers as the solution set?
   a. $|3x + 5| \ge -9$
   b. $|4x - 8| \le -7$
   c. $|2x + 1| > -6$
   d. $|-x - 10| < -3$
   e. $|7 - 5x| \ge 0$
   f. $|9 - x| \le -4$
   b, d, and f have no solution. For a, c, and e, the solution set is all real numbers.
3. Use absolute value to represent the inequality $-3 \le x \le 5$. $|x - 1| \le 4$

Optional Student Activity

Solve each inequality for x. In each inequality, $a > 0$, $b > 0$.

1. $|x + a| < b$ $-b - a \le x \le b - a$
2. $|x - a| > b$ $x < a - b$ or $x > a + b$
3. $|x + a| > a$ $x < -2a$ or $x > 0$
4. $|x - a| \le a$ $0 \le x \le 2a$

In-Class Examples (Objective 2.5B)

Solve.

1. $|x + 2| > 3$ $\{x|x > 1 \text{ or } x < -5\}$
2. $|x - 3| \le 5$ $\{x|-2 \le x \le 8\}$
3. $|10 - 2x| \le 0$ $\{x|x = 5\}$
4. $|9 - 3x| \ge 0$ $\{x|x \in \text{real numbers}\}$

Objective 2.5C

New Vocabulary
tolerance
upper limit
lower limit

Instructor Note
You might have students write absolute value inequalities from data you give them. Here are a couple of examples.

a. An adult's normal body temperature is within 1°F of 98.6°F. $|t - 98.6| \le 1$

b. The net weight of a cereal box labeled "20 oz" must be within 0.45 oz of 20 oz. $|w - 20| \le 0.45$

Optional Student Activity

1. A statistician is willing to state that a coin is fair if, when it is tossed 1000 times, the number of tails satisfies the inequality $\left|\frac{t - 500}{15.81}\right| < 2.33$. Determine the values of t that will allow the statistician to state that the coin is fair. 464 to 536 tails, inclusive
2. According to the registrar of a college, there is a 95% chance that a student applying for admission will have an SAT score, x, that satisfies the inequality $\left|\frac{x - 950}{98}\right| < 1.96$. Determine the values of x that the registrar expects from a student application. 758 to 1142, inclusive

Objective C To solve application problems

VIDEO & DVD CD TUTOR WEB SSM

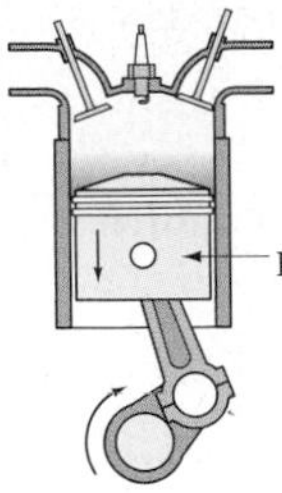

The **tolerance** of a component, or part, is the acceptable amount by which the component may vary from a given measurement. For example, the diameter of a piston may vary from the given measurement of 9 cm by 0.001 cm. This is written 9 cm ± 0.001 cm and is read "9 centimeters plus or minus 0.001 centimeter." The maximum diameter, or **upper limit**, of the piston is 9 cm + 0.001 cm = 9.001 cm. The minimum diameter, or **lower limit**, is 9 cm − 0.001 cm = 8.999 cm.

The lower and upper limits of the diameter of the piston could also be found by solving the absolute value inequality $|d - 9| \le 0.001$, where d is the diameter of the piston.

$$|d - 9| \le 0.001$$
$$-0.001 \le d - 9 \le 0.001$$
$$-0.001 + 9 \le d - 9 + 9 \le 0.001 + 9$$
$$8.999 \le d \le 9.001$$

The lower and upper limits of the diameter of the piston are 8.999 cm and 9.001 cm.

Example 5
A doctor has prescribed 2 cc of medication for a patient. The tolerance is 0.03 cc. (In the medical field, cubic centimeter is usually abbreviated cc.) Find the lower and upper limits of the amount of medication to be given.

Strategy
Let p represent the prescribed amount of medication, T the tolerance, and m the given amount of medication. Solve the absolute value inequality $|m - p| \le T$ for m.

Solution
$|m - p| \le T$
$|m - 2| \le 0.03$

$$-0.03 \le m - 2 \le 0.03$$
$$-0.03 + 2 \le m - 2 + 2 \le 0.03 + 2$$
$$1.97 \le m \le 2.03$$

The lower and upper limits of the amount of medication to be given are 1.97 cc and 2.03 cc.

You Try It 5
A machinist must make a bushing that has a diameter of 2.55 in. The tolerance of the bushing is 0.003 in. Find the lower and upper limits of the diameter of the bushing.

Your strategy

Your solution
2.547 in., 2.553 in.

Solution on p. S7

In-Class Examples (Objective 2.5C)

1. A machinist must make a bushing that has a tolerance of 0.004 in. The diameter of the bushing is 3.25 in. Find the lower and upper limits of the diameter of the bushing. Upper limit: 3.254 in.; lower limit: 3.246 in.

129

2.5 Exercises

Objective A

1. Is 2 a solution of $|x - 8| = 6$? yes
2. Is -2 a solution of $|2x - 5| = 9$? yes
3. Is -1 a solution of $|3x - 4| = 7$? yes
4. Is 1 a solution of $|6x - 1| = -5$? no

Solve.

5. $|x| = 7$ 7, −7
6. $|a| = 2$ 2, −2
7. $|b| = 4$ 4, −4
8. $|c| = 12$ 12, −12
9. $|-y| = 6$ 6, −6
10. $|-t| = 3$ 3, −3
11. $|-a| = 7$ 7, −7
12. $|-x| = 3$ 3, −3
13. $|x| = -4$ no solution
14. $|y| = -3$ no solution
15. $|x + 2| = 3$ 1, −5
16. $|x + 5| = 2$ −3, −7
17. $|y - 5| = 3$ 8, 2
18. $|y - 8| = 4$ 12, 4
19. $|r - 2| = -4$ no solution
20. $|x + 8| = -2$ no solution
21. $|3 - 4x| = 9$ $-\frac{3}{2}$, 3
22. $|2 - 5x| = 3$ $-\frac{1}{5}$, 1
23. $|2x - 3| = 0$ $\frac{3}{2}$
24. $|5x + 5| = 0$ −1
25. $|3x - 2| = -4$ no solution
26. $|2x + 5| = -2$ no solution
27. $|x - 2| - 2 = 3$ 7, −3
28. $|x - 9| - 3 = 2$ 14, 4
29. $|3a + 2| - 4 = 4$ 2, $-\frac{10}{3}$
30. $|2a + 9| + 4 = 5$ −4, −5
31. $|2 - y| + 3 = 4$ 1, 3
32. $|8 - y| - 3 = 1$ 4, 12
33. $|2x - 3| + 3 = 3$ $\frac{3}{2}$
34. $|4x - 7| - 5 = -5$ $\frac{7}{4}$
35. $|2x - 3| + 4 = -4$ no solution
36. $|3x - 2| + 1 = -1$ no solution

Section 2.5

Suggested Assignment

Exercises 1–99, odds

More challenging problems:
Exercises 101–103

Quick Quiz (Objective 2.5A)

Solve.

1. $|y - 4| = 7$ −3, 11
2. $|3 - 2x| = 3$ 0, 3
3. $|2 - b| + 5 = 6$ 1, 3

37. $|6x - 5| - 2 = 4$
$\frac{11}{6}, -\frac{1}{6}$

38. $|4b + 3| - 2 = 7$
$\frac{3}{2}, -3$

39. $|3t + 2| + 3 = 4$
$-\frac{1}{3}, -1$

40. $|5x - 2| + 5 = 7$
$\frac{4}{5}, 0$

41. $3 - |x - 4| = 5$
no solution

42. $2 - |x - 5| = 4$
no solution

43. $8 - |2x - 3| = 5$
3, 0

44. $8 - |3x + 2| = 3$
$1, -\frac{7}{3}$

45. $|2 - 3x| + 7 = 2$
no solution

46. $|1 - 5a| + 2 = 3$
$0, \frac{2}{5}$

47. $|8 - 3x| - 3 = 2$
$1, \frac{13}{3}$

48. $|6 - 5b| - 4 = 3$
$-\frac{1}{5}, \frac{13}{5}$

49. $|2x - 8| + 12 = 2$
no solution

50. $|3x - 4| + 8 = 3$
no solution

51. $2 + |3x - 4| = 5$
$\frac{7}{3}, \frac{1}{3}$

52. $5 + |2x + 1| = 8$
1, −2

53. $5 - |2x + 1| = 5$
$-\frac{1}{2}$

54. $3 - |5x + 3| = 3$
$-\frac{3}{5}$

55. $6 - |2x + 4| = 3$
$-\frac{1}{2}, -\frac{7}{2}$

56. $8 - |3x - 2| = 5$
$\frac{5}{3}, -\frac{1}{3}$

57. $8 - |1 - 3x| = -1$
$-\frac{8}{3}, \frac{10}{3}$

58. $3 - |3 - 5x| = -2$
$-\frac{2}{5}, \frac{8}{5}$

59. $5 + |2 - x| = 3$
no solution

60. $6 + |3 - 2x| = 2$
no solution

Objective B

Solve.

61. $|x| > 3$
$\{x \mid x > 3 \text{ or } x < -3\}$

62. $|x| < 5$
$\{x \mid -5 < x < 5\}$

63. $|x + 1| > 2$
$\{x \mid x > 1 \text{ or } x < -3\}$

64. $|x - 2| > 1$
$\{x \mid x > 3 \text{ or } x < 1\}$

65. $|x - 5| \le 1$
$\{x \mid 4 \le x \le 6\}$

66. $|x - 4| \le 3$
$\{x \mid 1 \le x \le 7\}$

67. $|2 - x| \ge 3$
$\{x \mid x \ge 5 \text{ or } x \le -1\}$

68. $|3 - x| \ge 2$
$\{x \mid x \le 1 \text{ or } x \ge 5\}$

69. $|2x + 1| < 5$
$\{x \mid -3 < x < 2\}$

Quick Quiz (Objective 2.5B)

Solve.

1. $|x - 1| > 2$ $\{x \mid x > 3 \text{ or } x < -1\}$
2. $|x - 5| \le 1$ $\{x \mid 4 \le x \le 6\}$
3. $|12 - 2x| \le 0$ $\{x \mid x = 6\}$

70. $|3x - 2| < 4$
$\left\{x \middle| -\frac{2}{3} < x < 2\right\}$

71. $|5x + 2| > 12$
$\left\{x \middle| x > 2 \text{ or } x < -\frac{14}{5}\right\}$

72. $|7x - 1| > 13$
$\left\{x \middle| x > 2 \text{ or } x < -\frac{12}{7}\right\}$

73. $|4x - 3| \le -2$
$\varnothing$

74. $|5x + 1| \le -4$
$\varnothing$

75. $|2x + 7| > -5$
the set of real numbers

76. $|3x - 1| > -4$
the set of real numbers

77. $|4 - 3x| \ge 5$
$\left\{x \middle| x \le -\frac{1}{3} \text{ or } x \ge 3\right\}$

78. $|7 - 2x| > 9$
$\{x | x < -1 \text{ or } x > 8\}$

79. $|5 - 4x| \le 13$
$\left\{x \middle| -2 \le x \le \frac{9}{2}\right\}$

80. $|3 - 7x| < 17$
$\left\{x \middle| -2 < x < \frac{20}{7}\right\}$

81. $|6 - 3x| \le 0$
$\{x | x = 2\}$

82. $|10 - 5x| \ge 0$
the set of real numbers

83. $|2 - 9x| > 20$
$\left\{x \middle| x < -2 \text{ or } x > \frac{22}{9}\right\}$

84. $|5x - 1| < 16$
$\left\{x \middle| -3 < x < \frac{17}{5}\right\}$

85. $|2x - 3| + 2 < 8$
$\left\{x \middle| -\frac{3}{2} < x < \frac{9}{2}\right\}$

86. $|3x - 5| + 1 < 7$
$\left\{x \middle| -\frac{1}{3} < x < \frac{11}{3}\right\}$

87. $|2 - 5x| - 4 > -2$
$\left\{x \middle| x < 0 \text{ or } x > \frac{4}{5}\right\}$

88. $|4 - 2x| - 9 > -3$
$\{x | x < -1 \text{ or } x > 5\}$

89. $8 - |2x - 5| < 3$
$\{x | x > 5 \text{ or } x < 0\}$

90. $12 - |3x - 4| > 7$
$\left\{x \middle| -\frac{1}{3} < x < 3\right\}$

Objective C Application Problems

91. The diameter of a bushing is 1.75 in. The bushing has a tolerance of 0.008 in. Find the lower and upper limits of the diameter of the bushing.
1.742 in.; 1.758 in.

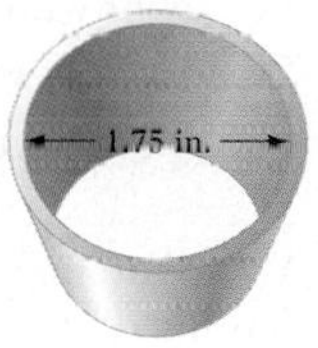

92. A machinist must make a bushing that has a tolerance of 0.004 in. The diameter of the bushing is 3.48 in. Find the lower and upper limits of the diameter of the bushing.
3.476 in.; 3.484 in.

93. An electric motor is designed to run on 220 volts plus or minus 25 volts. Find the lower and upper limits of voltage on which the motor will run.
195 volts; 245 volts

Quick Quiz (Objective 2.5C)

1. A doctor has prescribed 3 cc of medicine for a patient. The tolerance is 0.05 cc. Find the lower and upper limits of the amount of medication to be given. Upper limit: 3.05 cc; lower limit: 2.95 cc

132

94. A power strip is utilized on a computer to prevent the loss of programming by electrical surges. The power strip is designed to allow 110 volts plus or minus 16.5 volts. Find the lower and upper limits of voltage to the computer.
93.5 volts; 126.5 volts

95. A piston rod for an automobile is $9\frac{5}{8}$ in. long with a tolerance of $\frac{1}{32}$ in. Find the lower and upper limits of the length of the piston rod.
$9\frac{19}{32}$ in.; $9\frac{21}{32}$ in.

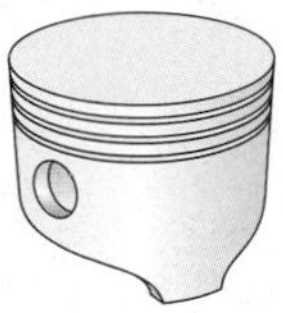

96. The diameter of a piston for an automobile is $3\frac{5}{16}$ in. with a tolerance of $\frac{1}{64}$ in. Find the lower and upper limits of the diameter of the piston.
$3\frac{19}{64}$ in.; $3\frac{21}{64}$ in.

The tolerance of the resistors used in electronics is given as a percent. Use your calculator for the following exercises.

97. Find the lower and upper limits of a 29,000-ohm resistor with a 2% tolerance.
28,420 ohms; 29,580 ohms

98. Find the lower and upper limits of a 15,000-ohm resistor with a 10% tolerance.
13,500 ohms; 16,500 ohms

99. Find the lower and upper limits of a 25,000-ohm resistor with a 5% tolerance.
23,750 ohms; 26,250 ohms

100. Find the lower and upper limits of a 56-ohm resistor with a 5% tolerance.
53.2 ohms; 58.8 ohms

APPLYING THE CONCEPTS

101. For what values of the variable is the equation true? Write the solution set in set-builder notation.
a. $|x + 3| = x + 3$ **b.** $|a - 4| = 4 - a$
a. $\{x \mid x \geq -3\}$ **b.** $\{a \mid a \leq 4\}$

102. Write an absolute value inequality to represent all real numbers within 5 units of 2.
$|x - 2| < 5$

103. Replace the question mark with $\leq$, $\geq$, or $=$.
a. $|x + y| \ ? \ |x| + |y|$ $\leq$
b. $|x - y| \ ? \ |x| - |y|$ $\geq$
c. $||x| - |y|| \ ? \ |x| - |y|$ $\geq$
d. $\left|\frac{x}{y}\right| \ ? \ \frac{|x|}{|y|}, y \neq 0$ $=$
e. $|xy| \ ? \ |x||y|$ $=$

133

Focus on Problem Solving

From Concrete to Abstract

In your study of algebra, you will find that the problems are less concrete than those you studied in arithmetic. Problems that are concrete provide information pertaining to a specific instance. Algebra is more abstract. Abstract problems are theoretical; they are stated without reference to a specific instance. Let's look at an example of an abstract problem.

How many minutes are in h hours?

A strategy that can be used to solve this problem is to solve the same problem after substituting a number for the variable.

How many minutes are in 5 hours?

You know that there are 60 minutes in 1 hour. To find the number of minutes in 5 hours, multiply 5 by 60.

$60 \cdot 5 = 300$ There are 300 minutes in 5 hours.

Use the same procedure to find the number of minutes in h hours: multiply h by 60.

$60 \cdot h = 60h$ There are $60h$ minutes in h hours.

This problem might be taken a step further:

If you walk 1 mile in x minutes, how far can you walk in h hours?

Consider the same problem using numbers in place of the variables.

If you walk 1 mile in 20 minutes, how far can you walk in 3 hours?

To solve this problem, you need to calculate the number of minutes in 3 hours (multiply 3 by 60), and divide the result by the number of minutes it takes to walk one mile (20 minutes).

$\frac{60 \cdot 3}{20} = \frac{180}{20} = 9$ If you walk 1 mile in 20 minutes, you can walk 9 miles in 3 hours.

Use the same procedure to solve the related abstract problem. Calculate the number of minutes in h hours (multiply h by 60), and divide the result by the number of minutes it takes to walk 1 mile (x minutes).

$\frac{60 \cdot h}{x} = \frac{60h}{x}$ If you walk 1 mile in x minutes, you can walk $\frac{60h}{x}$ miles in h hours.

At the heart of the study of algebra is the use of variables. It is the variables in the problems above that make them abstract. But it is variables that allow us to generalize situations and state rules about mathematics.

Try each of the following problems.

1. How many hours are in d days?
2. You earn d dollars an hour. What are your wages for working h hours?

Answers to Focus on Problem Solving: From Concrete to Abstract

1. $24d$
2. dh
3. $\frac{d}{p}$
4. $\frac{d}{s}$
5. $v - t$
6. $\frac{t}{g}$
7. $\frac{32q}{j}$
8. $\frac{60m}{s}$
9. $\frac{60hp}{m}$
10. $\frac{5q}{n}$

134

3. If p is the price of one share of stock, how many shares can you purchase with d dollars?
4. A company pays a television station d dollars to air a commercial lasting s seconds. What is the cost per second?
5. After every v videotape rentals, you are entitled to one free rental. You have rented t tapes, where $t < v$. How many more do you need to rent before you are entitled to a free rental?
6. Your car gets g miles per gallon. How many gallons of gasoline does your car consume traveling t miles?
7. If you drink j ounces of juice each day, how many days will q quarts of the juice last?
8. A TV station has m minutes of commercials each hour. How many ads lasting s seconds each can be sold for each hour of programming?
9. A factory worker can assemble p products in m minutes. How many products can the factory worker assemble in h hours?
10. If one candy bar costs n nickels, how many candy bars can be purchased with q quarters?

Projects and Group Activities

Prime and Composite Numbers

A **prime number** is a natural number greater than 1 whose only natural-number factors are itself and 1. The number 11 is a prime number because the only natural-number factors of 11 are 11 and 1.

Eratosthenes, a Greek philosopher and astronomer who lived from 270 to 190 B.C., devised a method of identifying prime numbers. It is called the **Sieve of Eratosthenes**. The procedure is illustrated below.

~~1~~	2	3	~~4~~	5	~~6~~	7	~~8~~	~~9~~	~~10~~
11	~~12~~	13	~~14~~	~~15~~	~~16~~	17	~~18~~	19	~~20~~
~~21~~	~~22~~	23	~~24~~	~~25~~	~~26~~	~~27~~	~~28~~	29	~~30~~
31	~~32~~	~~33~~	~~34~~	~~35~~	~~36~~	37	~~38~~	~~39~~	~~40~~
41	~~42~~	43	~~44~~	~~45~~	~~46~~	47	~~48~~	~~49~~	~~50~~
~~51~~	~~52~~	53	~~54~~	~~55~~	~~56~~	~~57~~	~~58~~	59	~~60~~
61	~~62~~	~~63~~	~~64~~	~~65~~	~~66~~	67	~~68~~	~~69~~	~~70~~
71	~~72~~	73	~~74~~	~~75~~	~~76~~	~~77~~	~~78~~	79	~~80~~
~~81~~	~~82~~	83	~~84~~	~~85~~	~~86~~	~~87~~	~~88~~	89	~~90~~
~~91~~	~~92~~	~~93~~	~~94~~	~~95~~	~~96~~	97	~~98~~	~~99~~	~~100~~

List all the natural numbers from 1 to 100. Cross out the number 1, because it is not a prime number. The number 2 is prime; circle it. Cross out all the other multiples of 2 (4, 6, 8, . . .), because they are not prime. The number 3 is prime; circle it. Cross out all the other multiples of 3 (6, 9, 12, . . .) that are not already crossed out. The number 4, the next consecutive number in the list, has already been crossed out. The number 5 is prime; circle it. Cross out all the other multiples of 5 that are not already crossed out. Continue in this manner until all the prime numbers less than 100 are circled.

A **composite number** is a natural number greater than 1 that has a natural-number factor other than itself and 1. The number 21 is a composite number because it has factors of 3 and 7. All the numbers crossed out in the preceding table, except the number 1, are composite numbers.

1. Use the Sieve of Eratosthenes to find the prime numbers between 100 and 200.
2. How many prime numbers are even numbers?
3. Find the "twin primes" between 1 and 200. Twin primes are two prime numbers whose difference is 2. For instance, 3 and 5 are twin primes; 5 and 7 are also twin primes.
4. **a.** List two prime numbers that are consecutive natural numbers.
   b. Can there be any other pairs of prime numbers that are consecutive natural numbers?
5. Some primes are the sum of a square and 1. For example, $5 = 2^2 + 1$. Find another prime p such that $p = n^2 + 1$, where n is a natural number.
6. Find a prime number p such that $p = n^2 - 1$, where n is a natural number.
7. **a.** 4! (which is read "4 factorial") is equal to $4 \cdot 3 \cdot 2 \cdot 1$. Show that $4! + 2$, $4! + 3$, and $4! + 4$ are all composite numbers.
   b. 5! (which is read "5 factorial") is equal to $5 \cdot 4 \cdot 3 \cdot 2 \cdot 1$. Will $5! + 2$, $5! + 3$, $5! + 4$, and $5! + 5$ generate four consecutive composite numbers?
   c. Use the notation 6! to represent a list of five consecutive composite numbers.

Investigation into Operations with Even and Odd Integers

Complete each statement with the word *even* or *odd*.

1. If k is an odd integer, then $k + 1$ is an _____ integer.
2. If k is an odd integer, then $k - 2$ is an _____ integer.
3. If n is an integer, then $2n$ is an _____ integer.
4. If m and n are even integers, then $m - n$ is an _____ integer.
5. If m and n are even integers, then mn is an _____ integer.
6. If m and n are odd integers, then $m + n$ is an _____ integer.
7. If m and n are odd integers, then $m - n$ is an _____ integer.
8. If m and n are odd integers, then mn is an _____ integer.
9. If m is an even integer and n is an odd integer, then $m - n$ is an _____ integer.
10. If m is an even integer and n is an odd integer, then $m + n$ is an _____ integer.

135

Answers to Projects and Group Activities: Prime and Composite Numbers

1. 101, 103, 107, 109, 113, 127, 131, 137, 139, 149, 151, 157, 163, 167, 173, 179, 181, 191, 193, 197, 199
2. One; the number 2
3. 3, 5; 5, 7; 11, 13; 17, 19; 29, 31; 41, 43; 59, 61; 71, 73; 101, 103; 107, 109; 137, 139; 149, 151; 179, 181; 191, 193; 197, 199
4. 2, 3; No, because any other two consecutive numbers include an even number.
5. Answers will vary. For instance, $17 = 4^2 + 1$
6. $3 = 2^2 - 1$

7a. 26, 27, and 28 are divisible by 2, 3, and 4, respectively.
b. Yes.
c. $6! + 2$, $6! + 3$, $6! + 4$, $6! + 5$, $6! + 6$

Answers to Projects and Group Activities: Investigation into Operations with Even and Odd Integers

1. even
2. odd
3. even
4. even
5. even
6. even
7. even
8. odd
9. odd
10. odd

136

Chapter Summary

Key Words

An *equation* expresses the equality of two mathematical expressions. A *solution* of an equation is a number that, when substituted for the variable, results in a true equation. To *solve* an equation means to find a solution of the equation. [pp. 73, 74]

An *inequality* is an expression that contains the symbol >, <, ≥, or ≤. The *solution set of an inequality* is a set of numbers, each element of which, when substituted for the variable, results in a true inequality. [p. 113]

A *compound inequality* is formed by joining two inequalities with a connective word such as *and* or *or*. [p. 116]

The *absolute value* of a number is its distance from zero on the number line. An equation that contains an absolute value symbol is an *absolute value equation*. [p. 125]

Essential Rules

Addition Property of Equations [p. 74]	If $a = b$, then $a + c = b + c$.
Multiplication Property of Equations [p. 75]	If $a = b$ and $c \neq 0$, then $ac = bc$.
Addition Property of Inequalities [p. 113]	If $a > b$, then $a + c > b + c$. If $a < b$, then $a + c < b + c$.
Multiplication Property of Inequalities [p. 114]	**Rule 1** If $a > b$ and $c > 0$, then $ac > bc$. If $a < b$ and $c > 0$, then $ac < bc$. **Rule 2** If $a > b$ and $c < 0$, then $ac < bc$. If $a < b$ and $c < 0$, then $ac > bc$.
Basic Percent Equation [p. 77]	Percent · base = amount $P \cdot B = A$
Value Mixture Equation [p. 97]	Amount · unit cost = value $A \cdot C = V$
Percent Mixture Equation [p. 99]	Amount of solution · percent of concentration = quantity of substance $A \cdot r = Q$
Annual Simple Interest Equation [p. 101]	Principal · interest rate = interest earned $P \cdot r = I$
Uniform Motion Equation [p. 103]	Rate · time = distance $r \cdot t = d$

To solve an absolute value inequality of the form $|ax + b| < c$, solve the equivalent compound inequality $-c < ax + b < c$. [p. 126]

To solve an absolute value inequality of the form $|ax + b| > c$, solve the equivalent compound inequality $ax + b < -c$ or $ax + b > c$. [p. 127]

137

Chapter Review

1. Is 3 a solution of $5x - 2 = 4x + 5$?
no [2.1A]

2. Solve: $x + 3 = 24$
21 [2.1B]

3. Solve: $\frac{3}{5}a = 12$
20 [2.1C]

4. Solve: $-4x - 2 = 10$
-3 [2.2A]

5. Solve: $14x + 7x + 8 = -10$
$-\frac{6}{7}$ [2.2A]

6. Solve: $12y - 1 = 3y + 2$
$\frac{1}{3}$ [2.2B]

7. Solve: $-6x + 16 = -2x$
4 [2.2B]

8. Solve: $6x + 3(2x - 1) = -27$
-2 [2.2C]

9. Solve: $x + 5(3x - 20) = 10(x - 4)$
10 [2.2C]

10. Solve: $3x - 7 > -2$
$\left\{x \mid x > \frac{5}{3}\right\}$ [2.4A]

11. Solve: $4 - 3(x + 2) < 2(2x + 3) - 1$
$\{x \mid x > -1\}$ [2.4A]

12. Solve: $3x < 4$ and $x + 2 > -1$
$\left\{x \mid -3 < x < \frac{4}{3}\right\}$ [2.4B]

13. Solve: $3x - 2 > x - 4$ or $7x - 5 < 3x + 3$
$\{x \mid x \in \text{real numbers}\}$ [2.4B]

14. Solve: $|3 - 5x| = 12$
$-\frac{9}{5}, 3$ [2.5A]

15. Solve: $|x - 4| - 8 = -3$
9, -1 [2.5A]

16. Solve: $|2x - 5| < 3$
$\{x \mid 1 < x < 4\}$ [2.5B]

17. Solve: $|4x - 5| \geq 3$
$\left\{x \mid x \geq 2 \text{ or } x \leq \frac{1}{2}\right\}$ [2.5B]

18. 30 is what percent of 12?
250% [2.1D]

19. $\frac{1}{2}$% of what is 8?
1600 [2.1D]

20. Translate "four less than the product of five and a number is sixteen" into an equation and solve.
$5x - 4 = 16$; $x = 4$ [2.2D]

21. The sum of two numbers is twenty-one. Three times the smaller number is two less than twice the larger number. Translate into an equation. Then find the two numbers.
$3x = 2(21 - x) - 2$; 8, 13 [2.2D]

138

22. An airline knowingly overbooks certain flights by selling 18% more tickets than there are available seats. How many tickets would this airline sell for an airplane that has 150 seats? 177 tickets [2.1D]

23. An auto manufacturer offers a rebate of $1000 on each car sold by a dealership. A customer bought a car from the dealership for $16,500. What percent of the cost is the $1000 rebate? Round to the nearest tenth of a percent. 6.1% [2.1D]

24. A health food store combined cranberry juice that costs $2.79 per quart with apple juice that costs $2.19 per quart. How many quarts of each were used to make 10 qt of a cranapple juice mixture that costs $2.31 per quart?
2 qt of cranberry juice; 8 qt of apple juice [2.3A]

25. Find the cost per ounce of a sun screen made from 100 oz of a lotion that costs $2.50 per ounce and 50 oz of a lotion that costs $4.00 per ounce. $3 [2.3A]

26. A dairy mixed 5 gal of cream that is 30% butterfat with 8 gal of milk that is 4% butterfat. What is the percent concentration of butterfat in the resulting mixture? 14% [2.3B]

27. An alloy containing 30% tin is mixed with an alloy containing 70% tin. How many pounds of each were used to make 500 lb of an alloy containing 40% tin? 375 lb of the 30% tin alloy; 125 lb of the 70% tin alloy [2.3B]

28. An investment banker invested 45% of the bank's available cash in an account earning 8.5% annual simple interest. The remainder of the cash was placed in an account earning 10% annual simple interest. The interest earned in one year was $41,962.50. What was the total amount invested?
$450,000 [2.3C]

29. A club treasurer deposited $2400 into two simple interest accounts. On one account the annual simple interest rate was 6.75%. The annual simple interest rate on the other account was 9.45%. How much was deposited in each account if both accounts earned the same amount of interest?
$1400 @ 6.75%; $1000 @ 9.45% [2.3C]

30. A jet plane traveling at 600 mph overtakes a propeller-driven plane that had a 2-hour headstart. The propeller-driven plane is traveling at 200 mph. How far from the starting point does the jet overtake the propeller-driven plane?
600 mi [2.3D]

31. A bus traveled on a straight road for 2 h at an average speed that was 20 mph faster than it traveled on a winding road. The time spent on the winding road was 3 h. Find the average speed on the winding road if the total trip was 200 mi. 32 mph [2.3D]

32. An average score of 80 to 90 in a psychology class receives a B grade. A student has grades of 92, 66, 72, and 88 on four tests. Find the range of scores on the fifth test that will earn the student a B for the course.
$82 \le x \le 100$ [2.4C]

33. A doctor has prescribed 2 cc of medication for a patient. The tolerance is 0.25 cc. Find the lower and upper limits of the amount of medication to be given. 1.75 cc and 2.25 cc [2.5C]

139

Chapter Test

1. Solve: $3x - 2 = 5x + 8$
-5 [2.2B]

2. Solve: $x - 3 = -8$
-5 [2.1B]

3. Solve: $3x - 5 = -14$
-3 [2.2A]

4. Solve: $4 - 2(3 - 2x) = 2(5 - x)$
2 [2.2C]

5. Is -2 a solution of $x^2 - 3x = 2x - 6$?
no [2.1A]

6. Solve: $7 - 4x = -13$
5 [2.2A]

7. What is 0.5% of 8?
0.04 [2.1D]

8. Solve: $5x - 2(4x - 3) = 6x + 9$
$-\frac{1}{3}$ [2.2C]

9. Solve: $5x + 3 - 7x = 2x - 5$
2 [2.2B]

10. Solve: $\frac{3}{4}x = -9$
-12 [2.1C]

11. Solve: $3x - 2 \geq 6x + 7$
$\{x \mid x \leq -3\}$ [2.4A]

12. Solve: $4 - 3(x + 2) < 2(2x + 3) - 1$
$\{x \mid x > -1\}$ [2.4A]

13. Solve: $4x - 1 > 5$ or $2 - 3x < 8$
$\{x \mid x > -2\}$ [2.4D]

14. Solve: $4 - 3x \geq 7$ and $2x + 3 \geq 7$
$\emptyset$ [2.4D]

15. Solve: $|3 - 5x| = 12$
$3, -\frac{9}{5}$ [2.5A]

16. Solve: $2 - |2x - 5| = -7$
$7, -2$ [2.5A]

17. Solve: $|3x - 5| \leq 4$
$\left\{x \,\middle|\, \frac{1}{3} \leq x \leq 3\right\}$ [2.5B]

18. Solve: $|4x - 3| > 5$
$\left\{x \,\middle|\, x > 2 \text{ or } x < -\frac{1}{2}\right\}$ [2.5B]

140 Chapter 2 / First-Degree Equations and Inequalities

19. The sum of two integers is fifteen. Eight times the smaller integer is one less than three times the larger integer. Find the integers.
4, 11 [2.2D]

20. A butcher combines 100 lb of hamburger that costs \$2.40 per pound with 60 lb of hamburger that costs \$3.20 per pound. Find the cost of the hamburger mixture.
\$2.70 per pound [2.3A]

21. How many gallons of water must be mixed with 5 gal of a 20% salt solution to make a 16% salt solution?
1.25 gal [2.3B]

22. A trust administrator divided \$20,000 between two accounts. One account earns an annual simple interest rate of 3%, and the second account earns an annual simple interest rate of 7%. The total annual income from the two accounts is \$1200. How much is invested in each account?
\$5000 at 3%; \$15,000 at 7% [2.3C]

23. A cross-country skier leaves a camp to explore a wilderness area. Two hours later a friend leaves the camp in a snowmobile, traveling 4 mph faster than the skier, and meets the skier 1 h later. Find the rate of the snowmobile.
6 mph [2.3D]

24. Gambelli Agency rents cars for \$24 a day and 10¢ for every mile driven. McDougal Rental rents cars for \$48 a day with unlimited mileage. How many miles a day can you drive a Gambelli Agency car if it is to cost you less than a McDougal Rental car?
less than 120 mi [2.4C]

25. A machinist must make a shoulder bushing that has a tolerance of 0.002 in. The outer diameter of the bushing is 2.65 in. Find the lower and upper limits of the outer diameter of the bushing.
2.648 in.; 2.652 in. [2.5C]

141

Cumulative Review

1. Subtract: $-6 - (-20) - 8$
6 [1.1C]

2. Multiply: $(-2)(-6)(-4)$
-48 [1.1D]

3. Subtract: $-\frac{5}{6} - \left(-\frac{7}{16}\right)$
$-\frac{19}{48}$ [1.2C]

4. Simplify: $-4^2 \cdot \left(-\frac{3}{2}\right)^3$
54 [1.2E]

5. Simplify: $\frac{5}{8} - \left(\frac{1}{2}\right)^2 \div \left(\frac{1}{3} - \frac{3}{4}\right)$
$\frac{49}{40}$ [1.3A]

6. Evaluate $3(a - c) - 2ab$ when $a = 2$, $b = 3$, and $c = -4$.
6 [1.4A]

7. Simplify: $3x - 8x + (-12x)$
$-17x$ [1.4B]

8. Simplify: $2a - (-b) - 7a - 5b$
$-5a - 4b$ [1.4B]

9. Simplify: $(16x)\left(\frac{1}{8}\right)$
$2x$ [1.4C]

10. Simplify: $-4(-9y)$
$36y$ [1.4C]

11. Simplify: $-2(-x^2 - 3x + 2)$
$2x^2 + 6x - 4$ [1.4D]

12. Simplify: $-2(x - 3) + 2(4 - x)$
$-4x + 14$ [1.4D]

13. Simplify: $-3[2x - 4(x - 3)] + 2$
$6x - 34$ [1.4D]

14. Find $A \cap B$ given $A = \{-4, -2, 0, 2\}$ and $B = \{-4, 0, 4, 8\}$.
$A \cap B = \{-4, 0\}$ [1.5B]

15. Graph: $\{x \mid x < 3\} \cap \{x \mid x > -2\}$
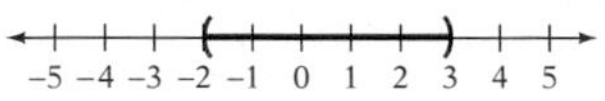

[1.5C]

16. Is -3 a solution of $x^2 + 6x + 9 = x + 3$?
yes [2.1A]

17. Solve: $\frac{3}{5}x = -15$
-25 [2.1C]

18. Solve: $7x - 8 = -29$
-3 [2.2A]

19. Solve: $13 - 9x = -14$
3 [2.2A]

20. Solve: $5x - 8 = 12x + 13$
-3 [2.2B]

21. Solve: $11 - 4x = 2x + 8$
$\frac{1}{2}$ [2.2B]

22. Solve: $8x - 3(4x - 5) = -2x - 11$
13 [2.2C]

23. Solve: $3 - 2(2x - 1) \geq 3(2x - 2) + 1$
$\{x \mid x \leq 1\}$ [2.4A]

24. Solve: $3x + 2 \leq 5$ and $x + 5 \geq 1$
$\{x \mid -4 \leq x \leq 1\}$ [2.4B]

25. Solve: $|3 - 2x| = 5$
$-1, 4$ [2.5A]

26. Solve: $|3x - 1| > 5$
$\left\{x \mid x > 2 \text{ or } x < -\frac{4}{3}\right\}$ [2.5B]

27. Write 55% as a fraction.
$\frac{11}{20}$ [1.2B]

28. Write 1.03 as a percent.
103% [1.2B]

29. 25% of what number is 30?
120 [2.1D]

30. Translate "the sum of six times a number and thirteen is five less than the product of three and the number" into an equation and solve.
$6x + 13 = 3x - 5$; -6 [2.2D]

31. How many pounds of an oat flour that costs \$.80 per pound must be mixed with 40 lb of a wheat flour that costs \$.50 per pound to make a blend that costs \$.60 per pound?
20 lb [2.3A]

32. How many grams of pure gold must be added to 100 g of a 20% gold alloy to make an alloy that is 36% gold?
25 g [2.3B]

33. A sprinter ran to the end of a track at an average rate of 8 m/s and then jogged back to the starting point at an average rate of 3 m/s. The sprinter took 55 s to run to the end of the track and jog back. Find the length of the track.
120 m [2.3D]

Chapter 3

Geometry

The best way to appreciate the different shapes and sizes of grassy areas in Boston's Public Garden is to view the garden from overhead. Each geometric shape, having its own set of dimensions, combines to form the entire park. **Exercise 98 on page 177** illustrates how to use a geometric formula to determine how large an area is and how much grass seed is needed for an area that size.

Objectives

Section 3.1

- **A** To solve problems involving lines and angles
- **B** To solve problems involving angles formed by intersecting lines
- **C** To solve problems involving the angles of a triangle

Section 3.2

- **A** To solve problems involving the perimeter of geometric figures
- **B** To solve problems involving the area of geometric figures

Section 3.3

- **A** To solve problems involving the volume of a solid
- **B** To solve problems involving the surface area of a solid

WWW WEB Need help? For online student resources, such as section quizzes, visit this textbook's website at **college.hmco.com/students.**

Prep Test

1. Solve: $x + 47 = 90$
   43 [2.1B]

2. Solve: $32 + (97 + x) = 180$
   51 [2.1B]

3. Simplify: $2(18) + 2(10)$
   56 [1.3A]

4. Evaluate abc when $a = 2$, $b = 3.14$, and $c = 9$.
   56.52 [1.4A]

5. Evaluate xyz^3 when $x = \frac{4}{3}$, $y = 3.14$, and $z = 3$.
   113.04 [1.4A]

6. Evaluate $\frac{1}{2}a(b + c)$ when $a = 6$, $b = 25$, and $c = 15$.
   120 [1.4A]

Go Figure

How can you cut a donut into 8 equal pieces with three cuts of the knife?
With the donut on a table, slice the donut parallel to the table. Now cut the halved donut into quarters.

3.1 Introduction to Geometry

Objective A To solve problems involving lines and angles

VIDEO & DVD CD TUTOR WEB SSM

Point of Interest

Geometry is one of the oldest branches of mathematics. Around 350 B.C., the Greek mathematician Euclid wrote the *Elements*, which contained all of the known concepts of geometry. Euclid's contribution was to unify various concepts into a single deductive system that was based on a set of axioms.

The word *geometry* comes from the Greek words for "earth" and "measure." The original purpose of geometry was to measure land. Today geometry is used in many fields, such as physics, medicine, and geology, and is applied in such areas as mechanical drawing and astronomy. Geometric forms are also used in art and design.

Three basic concepts of geometry are the point, line, and plane. A **point** is symbolized by drawing a dot. A **line** is determined by two distinct points and extends indefinitely in both directions, as the arrows on the line shown at the right indicate. This line contains points A and B and is represented by $\overleftrightarrow{AB}$. A line can also be represented by a single letter, such as ℓ.

A **ray** starts at a point and extends indefinitely in *one* direction. The point at which a ray starts is called the **endpoint** of the ray. The ray shown at the right is denoted by $\overrightarrow{AB}$. Point A is the endpoint of the ray.

A **line segment** is part of a line and has two endpoints. The line segment shown at the right is denoted by $\overline{AB}$.

The distance between the endpoints of $\overline{AC}$ is denoted by AC. If B is a point on $\overline{AC}$, then AC (the distance from A to C) is the sum of AB (the distance from A to B) and BC (the distance from B to C).

$AC = AB + BC$

➡ Given $AB = 22$ cm and $AC = 31$ cm, find BC.

$AC = AB + BC$ • Write an equation for the distances between points on the line segment.

$31 = 22 + BC$ • Substitute the given distances for AB and AC into the equation.

$9 = BC$ • Solve for BC.

$BC = 9$ cm

In this section we will be discussing figures that lie in a plane. A **plane** is a flat surface and can be pictured as a table top or blackboard that extends in all directions. Figures that lie in a plane are called **plane figures**.

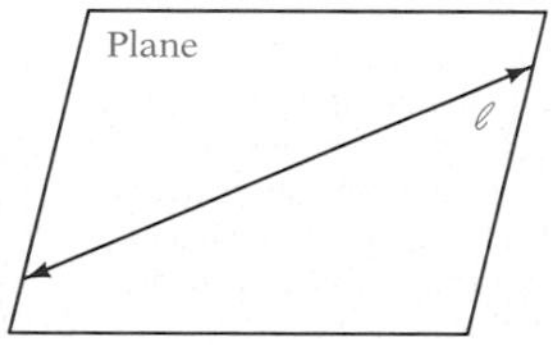

Objective 3.1A

New Vocabulary

point
line
ray
endpoint
line segment
plane
plane figures
space
intersecting lines
parallel lines
angle
vertex
side of an angle
degrees
protractor
right angle
perpendicular lines
complementary angles
straight angle
supplementary angles
acute angle
obtuse angle
adjacent angles

New Symbols

line, e.g. $\overleftrightarrow{AB}$
ray, e.g. $\overrightarrow{AB}$
line segment, e.g. $\overline{AB}$
$\parallel$ (is parallel to)
$\angle$ (angle)
° (degree)
∟ (right angle)
$\perp$ (is perpendicular to)

Instructor Note

Tell students that when no units (such as feet or meters) are given for lengths along a line segment, the distances are assumed to be in the same unit of length.

In-Class Examples (Objective 3.1A)

1. How many degrees are in two-thirds of a revolution? 240°
2. Find the complement of a 38° angle. 52°
3. Find the supplement of a 57° angle. 123°

Discuss the Concepts

1. Describe each of the following: ray, line, line segment.
2. Does the surface of Earth lie in a plane?
3. Using any objects in the classroom, provide examples of each of the following: a right angle, an acute angle, an obtuse angle, a plane, intersecting lines, parallel lines, perpendicular lines. (If students need some assistance, suggest they consider the windows.)

Concept Check

1. Provide three names for the angle below.

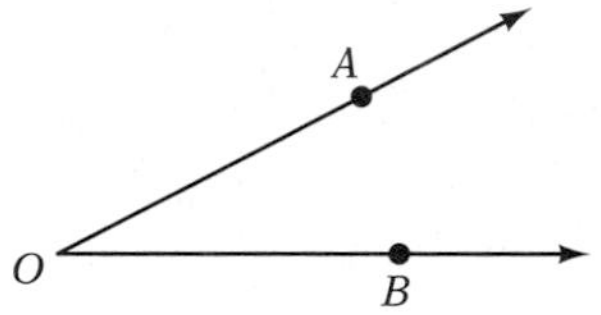

∠O, ∠AOB, ∠BOA

2. Name the number of degrees in a full circle, in a straight angle, and in a right angle. 360°, 180°, 90°
3. How many dimensions does a point have? a line? a line segment? a ray? an angle? 0; 1; 1; 1; 2
4. What is the name given to lines in a plane that do not intersect? parallel lines
5. What is the name given to two lines that intersect at right angles? perpendicular lines

Lines in a plane can be intersecting or parallel. **Intersecting lines** cross at a point in the plane. **Parallel lines** never meet. The distance between them is always the same.

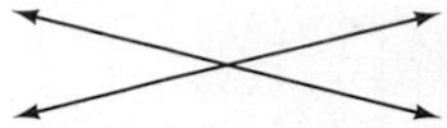

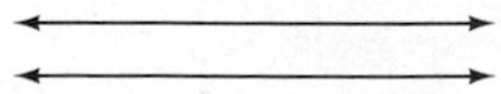

The symbol ∥ means "is parallel to." In the figure at the right, $j \parallel k$ and $\overline{AB} \parallel \overline{CD}$. Note that j contains $\overline{AB}$ and k contains $\overline{CD}$. Parallel lines contain parallel line segments.

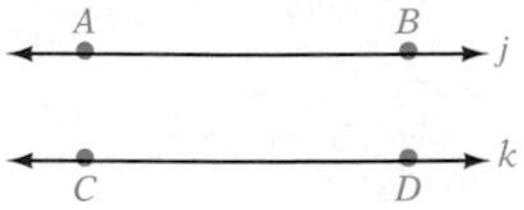

An **angle** is formed by two rays with the same endpoint. The **vertex** of the angle is the point at which the two rays meet. The rays are called the **sides** of the angle.

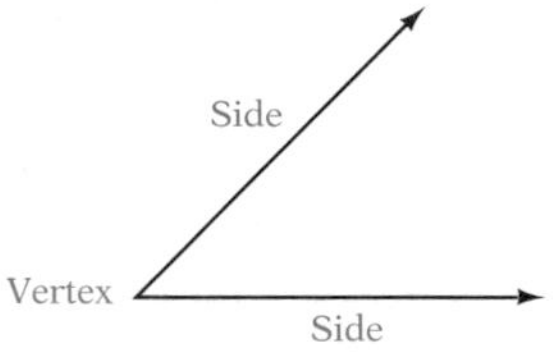

If A and C are points on rays r_1 and r_2, and B is the vertex, then the angle is called $\angle B$ or $\angle ABC$, where $\angle$ is the symbol for angle. Note that either the angle is named by the vertex, or the vertex is the second point listed when the angle is named by giving three points. $\angle ABC$ could also be called $\angle CBA$.

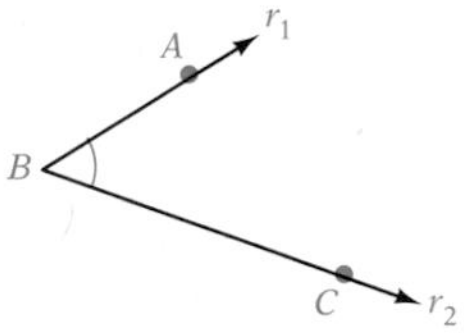

An angle can also be named by a variable written between the rays close to the vertex. In the figure at the right, $\angle x = \angle QRS$ and $\angle y = \angle SRT$. Note that in this figure, more than two rays meet at R. In this case, the vertex cannot be used to name an angle.

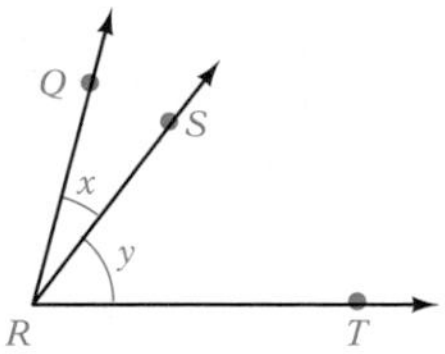

Point of Interest

The first woman mathematician for whom documented evidence exists is Hypatia (370–415). She lived in Alexandria, Egypt, and lectured at the Museum, the forerunner of our modern university. She made important contributions in mathematics, astronomy, and philosophy.

An angle is measured in **degrees**. The symbol for degrees is a small raised circle, °. Probably because early Babylonians believed that Earth revolves around the sun in approximately 360 days, the angle formed by a circle has a measure of 360° (360 degrees).

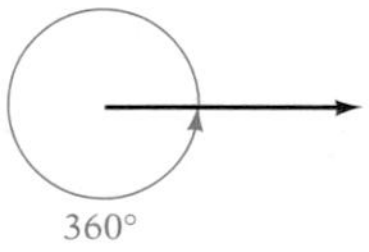

A **protractor** is used to measure an angle. Place the center of the protractor at the vertex of the angle with the edge of the protractor along a side of the angle. The angle shown in the figure below measures 58°.

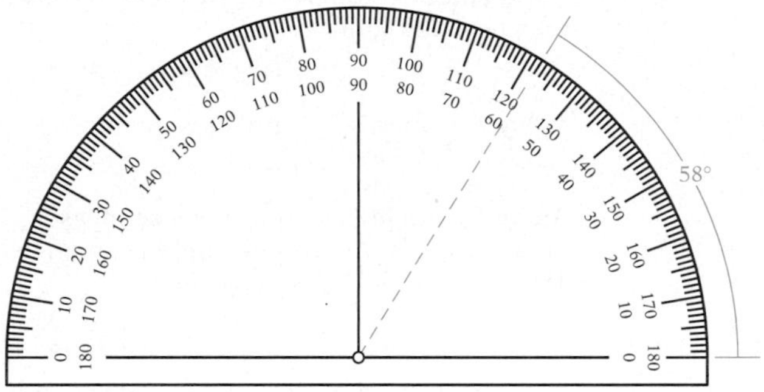

> **TAKE NOTE**
> The corner of a page of this book is a good example of a 90° angle.

A 90° angle is called a **right angle**. The symbol ∟ represents a right angle.

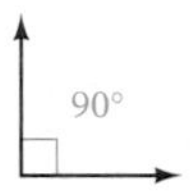

Perpendicular lines are intersecting lines that form right angles.

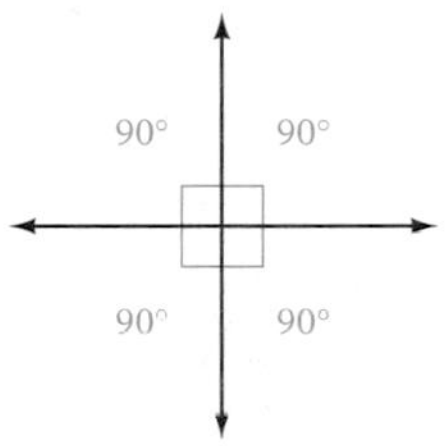

The symbol ⊥ means "is perpendicular to." In the figure at the right, $p \perp q$ and $\overline{AB} \perp \overline{CD}$. Note that line p contains $\overline{AB}$ and line q contains $\overline{CD}$. Perpendicular lines contain perpendicular line segments.

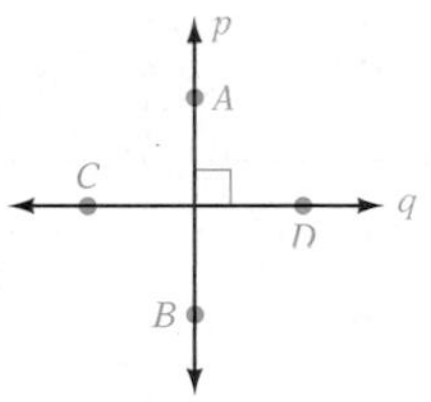

Complementary angles are two angles whose measures have the sum 90°.

$$\angle A + \angle B = 70° + 20° = 90°$$

$\angle A$ and $\angle B$ are complementary angles.

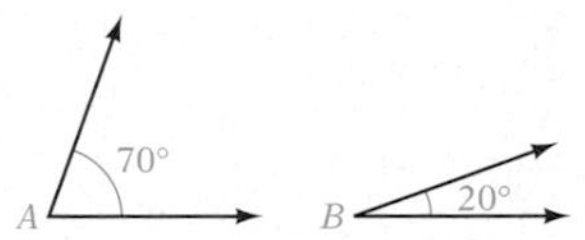

Optional Student Activity

1. Prepare a handout on which you have drawn four angles. You might include a 40° angle, a 60° angle, a 120° angle, and a 30° angle. Orient the angles in different directions. Make enough copies of the handout so that each student in the class will have one.

   During class provide students with protractors and instruction on how to measure angles. Give each student a copy of the handout. Have them measure each angle and then classify each angle as an acute angle or an obtuse angle.
2. On a number line, the points *A*, *B*, *C*, and *D* have coordinates −2.5, 2, 5, and 3.5, respectively. Which of these points is halfway between two others? *D* or 3.5
3. Find the measure of the smaller angle between the hands of a clock when the time is 5 o'clock. 150°
4. On a line, *Q* is between *P* and *S*. *R* is between *Q* and *S*. *S* is between *Q* and *T*. *PT* = 28, *QS* = 8, and *PQ* = *QR* = *RS*. Find *ST*. 16 units

A 180° angle is called a **straight angle**.

$\angle AOB$ is a straight angle.

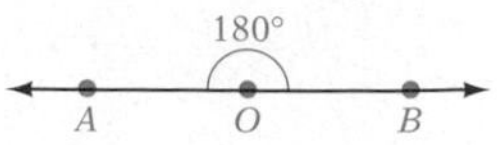

Supplementary angles are two angles whose measures have the sum 180°.

$$\angle A + \angle B = 130° + 50° = 180°$$

$\angle A$ and $\angle B$ are supplementary angles.

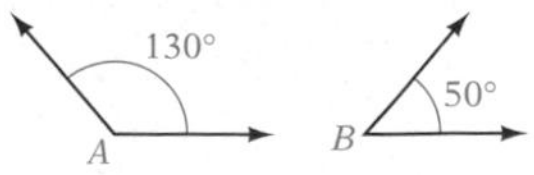

An **acute angle** is an angle whose measure is between 0° and 90°. $\angle B$ above is an acute angle. An **obtuse angle** is an angle whose measure is between 90° and 180°. $\angle A$ above is an obtuse angle.

Two angles that share a common side are **adjacent angles**. In the figure at the right, $\angle DAC$ and $\angle CAB$ are adjacent angles. $\angle DAC = 45°$ and $\angle CAB = 55°$.

$$\angle DAB = \angle DAC + \angle CAB$$
$$= 45° + 55° = 100°$$

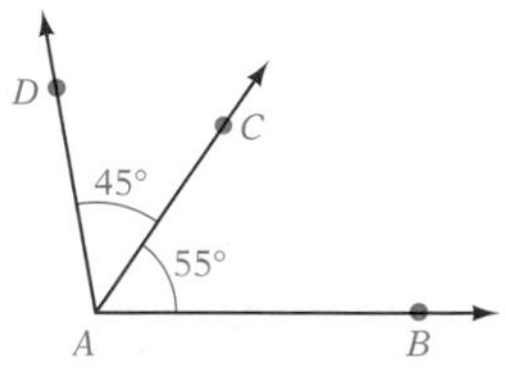

➡ In the figure at the right, $\angle EDG = 80°$. $\angle FDG$ is three times the measure of $\angle EDF$. Find the measure of $\angle EDF$.

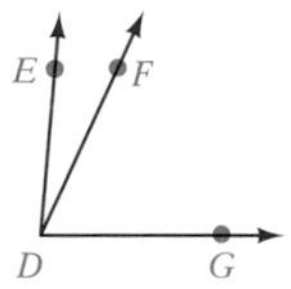

> **TAKE NOTE**
> Answers to application problems must have units, such as degrees, feet, dollars, or hours.

Let x = the measure of $\angle EDF$. Then $3x$ = the measure of $\angle FDG$. Write an equation and solve for x, the measure of $\angle EDF$.

$$\angle EDF + \angle FDG = \angle EDG$$
$$x + 3x = 80$$
$$4x = 80$$
$$x = 20$$

$\angle EDF = 20°$

Example 1

Given $MN = 15$ mm, $NO = 18$ mm, and $MP = 48$ mm, find OP.

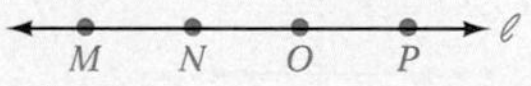

Solution

$$MN + NO + OP = MP$$
$$15 + 18 + OP = 48$$
$$33 + OP = 48$$
$$OP = 15$$

$OP = 15$ mm

You Try It 1

Given $QR = 24$ cm, $ST = 17$ cm, and $QT = 62$ cm, find RS.

Your solution

21 cm

Solution on p. S7

149

Example 2
Given $XY = 9$ m and YZ is twice XY, find XZ.

Solution
$XZ = XY + YZ$
$XZ = XY + 2(XY)$
$XZ = 9 + 2(9)$
$XZ = 9 + 18$
$XZ = 27$

$XZ = 27$ m

You Try It 2
Given $BC = 16$ ft and $AB = \frac{1}{4}(BC)$, find AC.

Your solution
20 ft

Example 3
Find the complement of a 38° angle.

Strategy
Complementary angles are two angles whose sum is 90°. To find the complement, let x represent the complement of a 38° angle. Write an equation and solve for x.

Solution
$x + 38° = 90°$
$x = 52°$

The complement of a 38° angle is a 52° angle.

You Try It 3
Find the supplement of a 129° angle.

Your strategy

Your solution
51°

Example 4
Find the measure of $\angle x$.

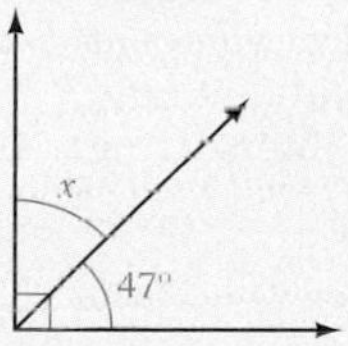

Strategy
To find the measure of $\angle x$, write an equation using the fact that the sum of the measure of $\angle x$ and 47° is 90°. Solve for $\angle x$.

Solution
$\angle x + 47° = 90°$
$\angle x = 43°$

The measure of $\angle x$ is 43°.

You Try It 4
Find the measure of $\angle a$.

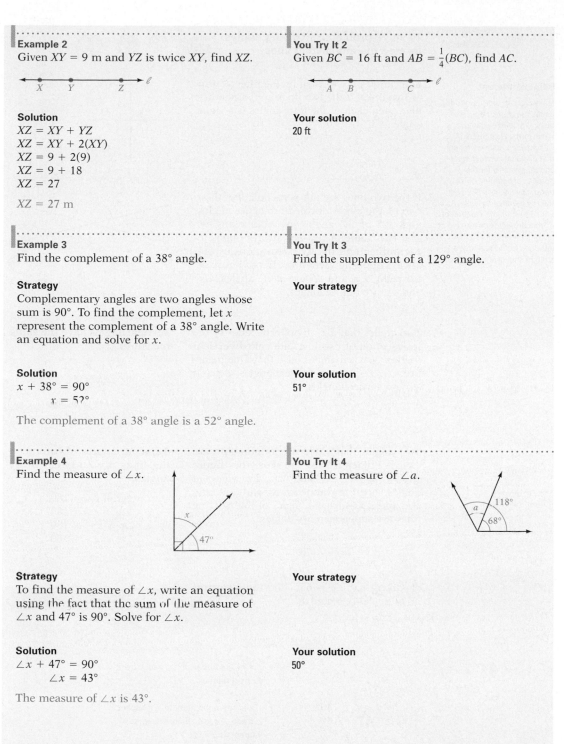

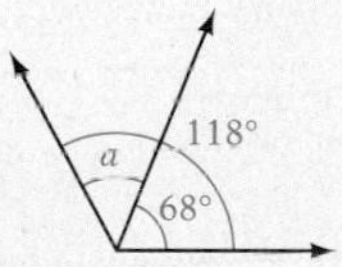

Your strategy

Your solution
50°

Solutions on pp. S7–S8

Objective 3.1B

New Vocabulary
vertical angles
adjacent angles
transversal
alternate interior angles
alternate exterior angles
corresponding angles

Vocabulary to Review
intersecting lines
parallel lines
supplementary

Discuss the Concepts
What is a transversal? Describe two different ways in which a transversal can intersect two other lines.

Concept Check
When a transversal intersects two parallel lines, which of the following are supplementary angles: vertical angles, adjacent angles, alternate interior angles, alternate exterior angles, corresponding angles? adjacent angles

Optional Student Activity
In the figure, $BD \parallel AC$ and $\angle BAC = \angle BCA$. Find the value of x. 65°

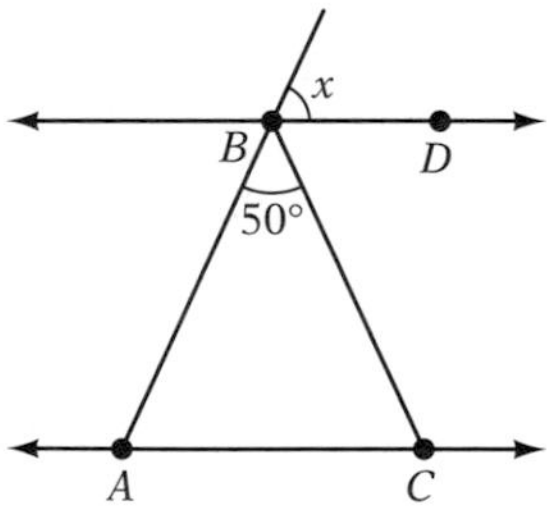

Objective B To solve problems involving angles formed by intersecting lines

VIDEO & DVD CD TUTOR WEB SSM

Point of Interest

Many cities in the New World, unlike those in Europe, were designed using rectangular street grids. Washington, D.C. was planned that way except that diagonal avenues were added, primarily for the purpose of enabling quick troop movement in the event the city required defense. As an added precaution, monuments of statuary were constructed at major intersections so that attackers would not have a straight shot down a boulevard.

Four angles are formed by the intersection of two lines. If the two lines are perpendicular, then each of the four angles is a right angle.

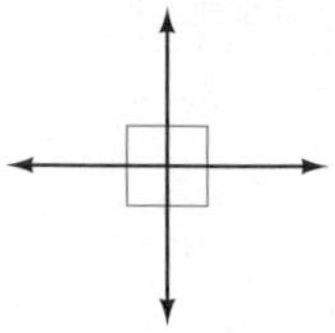

If the two lines are not perpendicular, then two of the angles formed are acute angles and two of the angles are obtuse angles. The two acute angles are always opposite each other, and the two obtuse angles are always opposite each other. In the figure at the right, $\angle w$ and $\angle y$ are acute angles. $\angle x$ and $\angle z$ are obtuse angles.

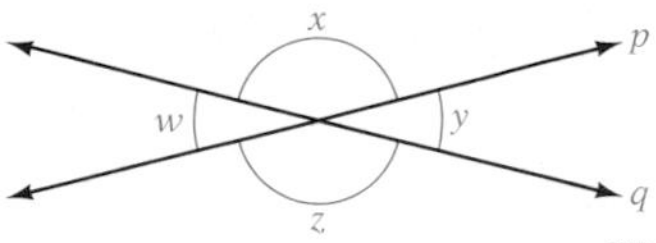

T

Two angles that are on opposite sides of the intersection of two lines are called **vertical angles**. Vertical angles have the same measure. $\angle w$ and $\angle y$ are vertical angles. $\angle x$ and $\angle z$ are vertical angles.

Vertical angles have the same measure.

$$\angle w = \angle y$$
$$\angle x = \angle z$$

Two angles that share a common side are called **adjacent angles**. For the figure shown above, $\angle x$ and $\angle y$ are adjacent angles, as are $\angle y$ and $\angle z$, $\angle z$ and $\angle w$, and $\angle w$ and $\angle x$. Adjacent angles of intersecting lines are supplementary angles.

Adjacent angles of intersecting lines are supplementary angles.

$$\angle x + \angle y = 180°$$
$$\angle y + \angle z = 180°$$
$$\angle z + \angle w = 180°$$
$$\angle w + \angle x = 180°$$

➡ Given that $\angle c = 65°$, find the measures of angles a, b, and d.

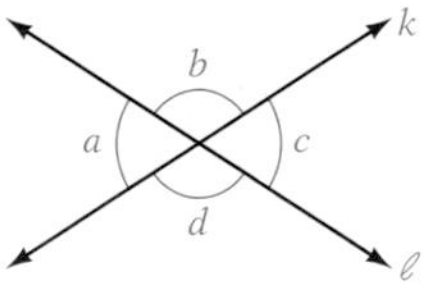

$\angle a = 65°$ • **$\angle a = \angle c$ because $\angle a$ and $\angle c$ are vertical angles.**

$\angle b + \angle c = 180°$
$\angle b + 65° = 180°$
$\angle b = 115°$ • **$\angle b$ is supplementary to $\angle c$ because $\angle b$ and $\angle c$ are adjacent angles of intersecting lines.**

$\angle d = 115°$ • **$\angle d = \angle b$ because $\angle d$ and $\angle b$ are vertical angles.**

In-Class Examples (Objective 3.1B)

1. In the figure, $\angle a = 62°$. Find $\angle b$. 118°

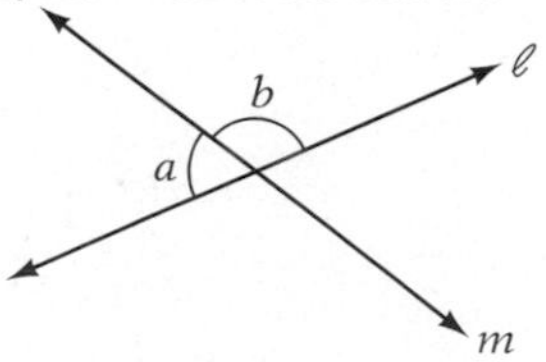

2. In the figure, $\ell_1 \parallel \ell_2$ and $\angle c = 103°$. Find $\angle b$. 77°

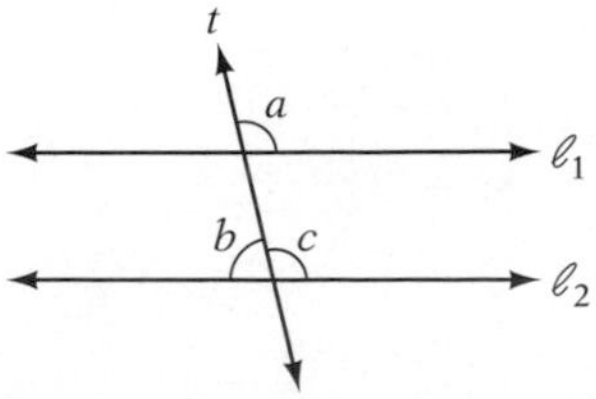

151

A line that intersects two other lines at different points is called a **transversal**.

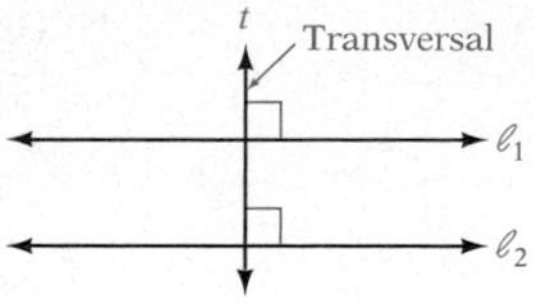

If the lines cut by a transversal t are parallel lines and the transversal is perpendicular to the parallel lines, all eight angles formed are right angles.

If the lines cut by a transversal t are parallel lines and the transversal is not perpendicular to the parallel lines, all four acute angles have the same measure and all four obtuse angles have the same measure. For the figure at the right,

$$\angle b = \angle d = \angle x = \angle z$$
$$\angle a = \angle c = \angle w = \angle y$$

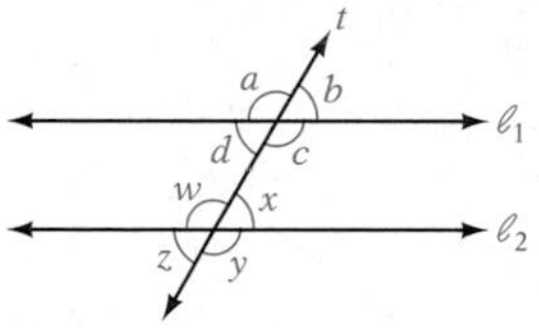

Alternate interior angles are two nonadjacent angles that are on opposite sides of the transversal and lie between the parallel lines. In the figure above, $\angle c$ and $\angle w$ are alternate interior angles; $\angle d$ and $\angle x$ are alternate interior angles. Alternate interior angles have the same measure.

Alternate interior angles have the same measure.

$$\angle c = \angle w$$
$$\angle d = \angle x$$

Alternate exterior angles are two nonadjacent angles that are on opposite sides of the transversal and lie outside the parallel lines. In the figure above, $\angle a$ and $\angle y$ are alternate exterior angles; $\angle b$ and $\angle z$ are alternate exterior angles. Alternate exterior angles have the same measure.

Alternate exterior angles have the same measure.

$$\angle a = \angle y$$
$$\angle b = \angle z$$

Corresponding angles are two angles that are on the same side of the transversal and are both acute angles or are both obtuse angles. For the figure above, the following pairs of angles are corresponding angles: $\angle a$ and $\angle w$, $\angle d$ and $\angle z$, $\angle b$ and $\angle x$, and $\angle c$ and $\angle y$. Corresponding angles have the same measure.

Corresponding angles have the same measure.

$$\angle a = \angle w$$
$$\angle d = \angle z$$
$$\angle b = \angle x$$
$$\angle c = \angle y$$

152 Chapter 3 / Geometry

➡ Given that $\ell_1 \parallel \ell_2$ and $\angle c = 58°$, find the measures of $\angle f$, $\angle h$, and $\angle g$.

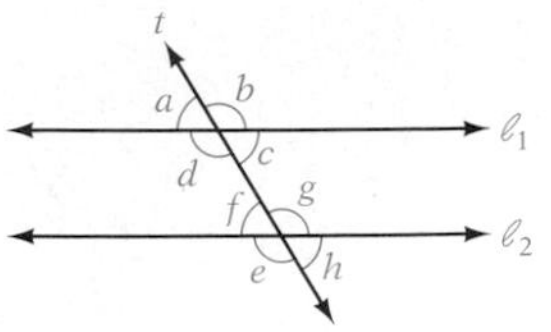

$\angle f = \angle c = 58°$ • **$\angle c$ and $\angle f$ are alternate interior angles.**

$\angle h = \angle c = 58°$ • **$\angle c$ and $\angle h$ are corresponding angles.**

$\angle g + \angle h = 180°$ • **$\angle g$ is supplementary to $\angle h$.**
$\angle g + 58° = 180°$
$\angle g = 122°$

Example 5
Find x.

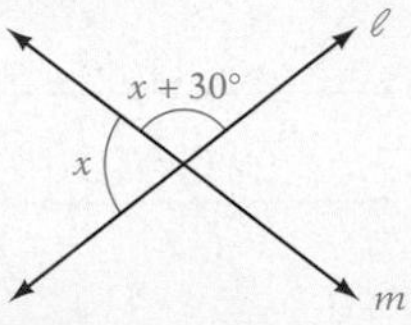

Strategy
The angles labeled are adjacent angles of intersecting lines and are, therefore, supplementary angles. To find x, write an equation and solve for x.

Solution

$$\begin{aligned} x + (x + 30°) &= 180° \\ 2x + 30° &= 180° \\ 2x &= 150° \\ x &= 75° \end{aligned}$$

You Try It 5
Find x.

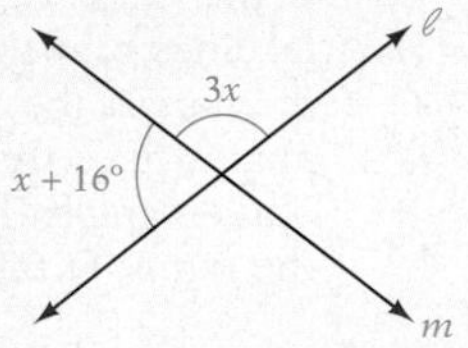

Your strategy

Your solution
41°

Example 6
Given $\ell_1 \parallel \ell_2$, find x.

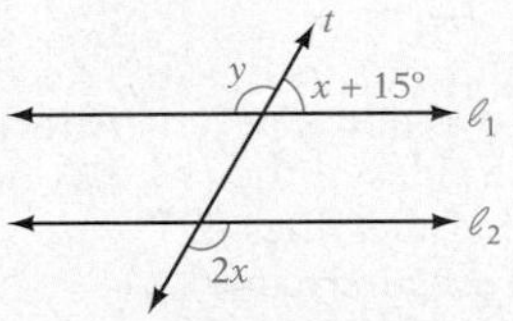

Strategy
$2x = y$ because alternate exterior angles have the same measure. $(x + 15°) + y = 180°$ because adjacent angles of intersecting lines are supplementary angles. Substitute $2x$ for y and solve for x.

Solution

$$\begin{aligned} (x + 15°) + 2x &= 180° \\ 3x + 15° &= 180° \\ 3x &= 165° \\ x &= 55° \end{aligned}$$

You Try It 6
Given $\ell_1 \parallel \ell_2$, find x.

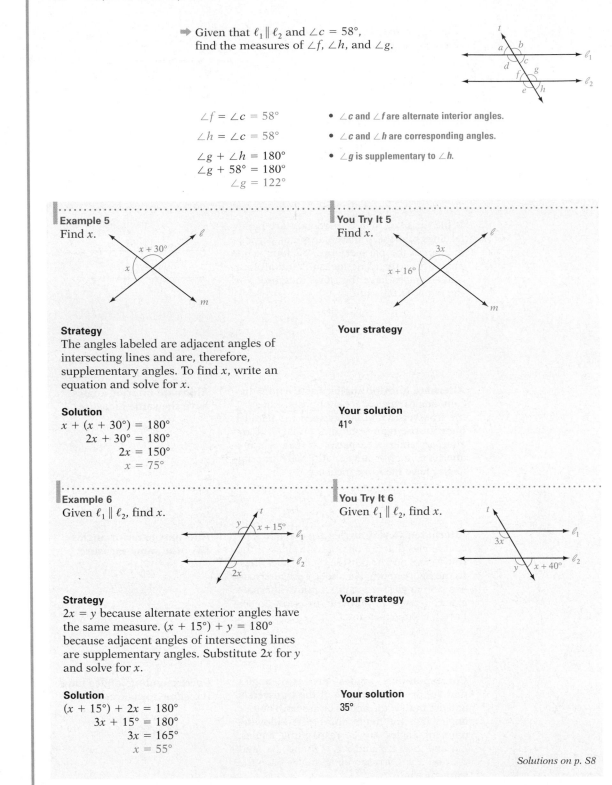

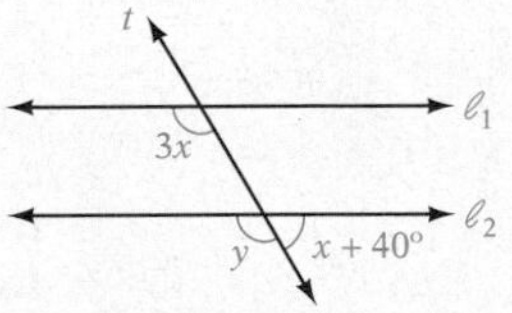

Your strategy

Your solution
35°

Solutions on p. S8

Objective C **To solve problems involving the angles of a triangle**

VIDEO & DVD CD TUTOR WEB SSM

If the lines cut by a transversal are not parallel lines, the three lines will intersect at three points. In the figure at the right, the transversal t intersects lines p and q. The three lines intersect at points A, B, and C. These three points define three line segments: $\overline{AB}$, $\overline{BC}$, and $\overline{AC}$. The plane figure formed by these three line segments is called a **triangle**.

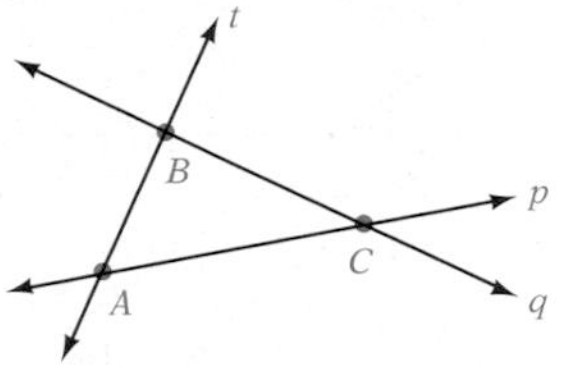

Each of the three points of intersection is the vertex of four angles. The angles within the region enclosed by the triangle are called **interior angles**. In the figure at the right, angles a, b, and c are interior angles. The sum of the measures of the interior angles of a triangle is 180°.

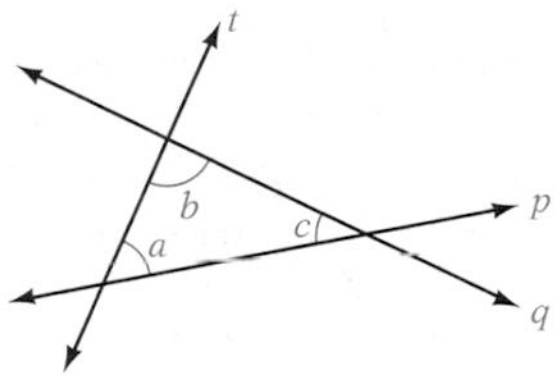

$\angle a + \angle b + \angle c = 180°$

The Sum of the Measures of the Interior Angles of a Triangle

The sum of the measures of the interior angles of a triangle is 180°.

An angle adjacent to an interior angle is an **exterior angle**. In the figure at the right, angles m and n are exterior angles for angle a. The sum of the measures of an interior and an exterior angle is 180°.

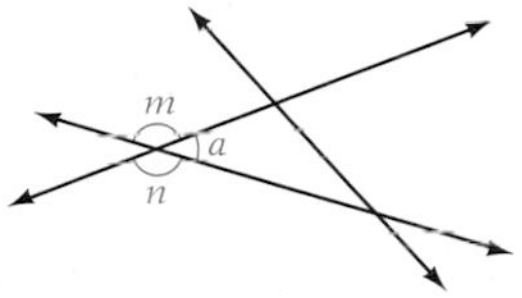

$\angle a + \angle m = 180°$
$\angle a + \angle n = 180°$

➡ Given that $\angle c = 40°$ and $\angle d = 100°$, find the measure of $\angle e$.

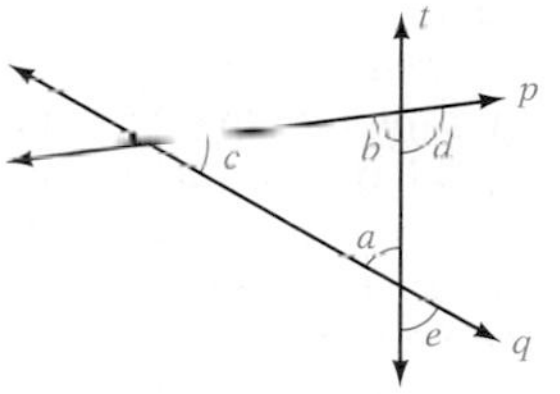

$\angle d$ and $\angle b$ are supplementary angles.

$$\angle d + \angle b = 180°$$
$$100° + \angle b = 180°$$
$$\angle b = 80°$$

The sum of the interior angles is 180°.

$$\angle c + \angle b + \angle a = 180°$$
$$40° + 80° + \angle a = 180°$$
$$120° + \angle a = 180°$$
$$\angle a = 60°$$

$\angle a$ and $\angle e$ are vertical angles.

$$\angle e = \angle a = 60°$$

Objective 3.1C

New Vocabulary

triangle
interior angles
exterior angle

New Properties

The sum of the measures of the interior angles of a triangle is 180°.

Instructor Note

Students will need to know that the sum of the measures of the interior angles of a triangle is 180° when working with similar triangles in the section "Rational Equations" in the chapter "Rational Expressions."

Discuss the Concepts

In a right triangle, why are the two acute angles complementary?

Concept Check

1. The angles of a triangle are in the ratio 2:3:7. Find the number of degrees in the largest angle. 105°
2. The measures of the angles of a triangle are consecutive integers. Find the measure of each angle. 59°, 60°, 61°
3. Determine the measures of the angles of an equilateral triangle. 60°, 60°, 60°

In-Class Examples (Objectives 3.1C)

1. A triangle has a 21° angle and a 64° angle. Find the measure of the third angle. 95°
2. A right triangle has a 52° angle. Find the measures of the other two angles. 90°, 38°
3. One angle of a right triangle is 2° less than three times the measure of the smallest angle. Find the measure of each angle. 23°, 67°, 90°

Discuss the Concepts

Use the figure in Example 7.

1. What does the square in the corner of the triangle mean?
2. If $\angle b = 30°$, what is the measure of $\angle d$? Why?
3. If $\angle b = 30°$, what is the measure of $\angle a$? Why?
4. If $\angle b = 30°$, what is the measure of $\angle y$? Why?

Optional Student Activity

Draw five triangles of different sizes. For each triangle, use a protractor to find the measure of each angle. (*Note:* You will want the triangles to be fairly large so that measuring the angles is not difficult.) Then find the sum of the angles of each triangle.

Example 7

Given that $\angle y = 55°$, find the measures of angles a, b, and d.

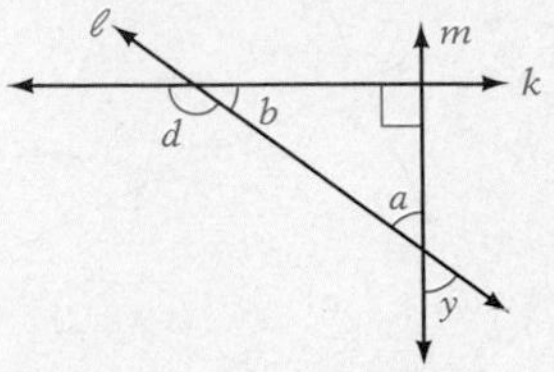

Strategy

- To find the measure of angle a, use the fact that $\angle a$ and $\angle y$ are vertical angles.
- To find the measure of angle b, use the fact that the sum of the measures of the interior angles of a triangle is 180°.
- To find the measure of angle d, use the fact that the sum of an interior and an exterior angle is 180°.

Solution

$\angle a = \angle y = 55°$

$$\begin{aligned} \angle a + \angle b + 90° &= 180° \\ 55° + \angle b + 90° &= 180° \\ \angle b + 145° &= 180° \\ \angle b &= 35° \end{aligned}$$

$$\begin{aligned} \angle d + \angle b &= 180° \\ \angle d + 35° &= 180° \\ \angle d &= 145° \end{aligned}$$

You Try It 7

Given that $\angle a = 45°$ and $\angle x = 100°$, find the measures of angles b, c, and y.

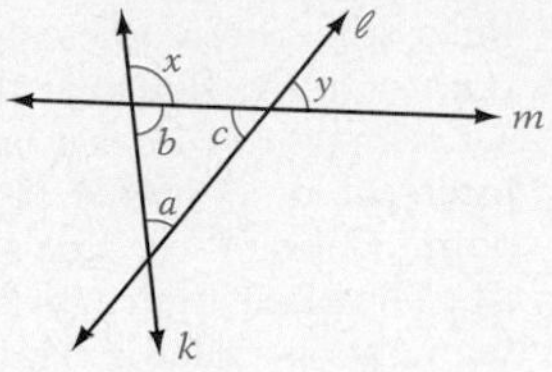

Your strategy

Your solution

$\angle b = 80°$,
$\angle y = \angle c = 55°$

Example 8

Two angles of a triangle measure 53° and 78°. Find the measure of the third angle.

Strategy

To find the measure of the third angle, use the fact that the sum of the measures of the interior angles of a triangle is 180°. Write an equation using x to represent the measure of the third angle. Solve the equation for x.

Solution

$$\begin{aligned} x + 53° + 78° &= 180° \\ x + 131° &= 180° \\ x &= 49° \end{aligned}$$

The measure of the third angle is 49°.

You Try It 8

One angle in a triangle is a right angle, and one angle measures 34°. Find the measure of the third angle.

Your strategy

Your solution

56°

Solutions on p. S8

3.1 Exercises

Objective A

Use a protractor to measure the angle. State whether the angle is acute, obtuse, or right.

1.

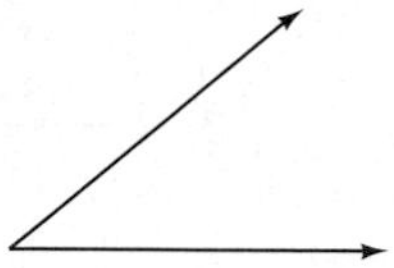

40°; acute

2.

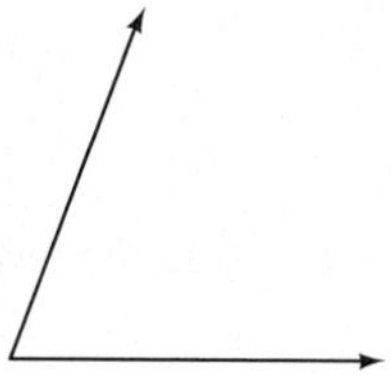

69°; acute

3.

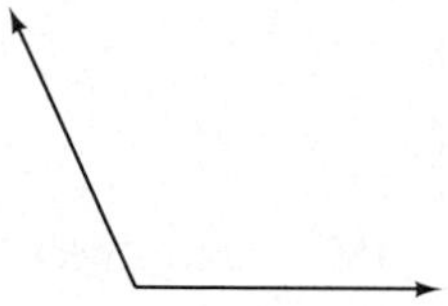

115°; obtuse

4.

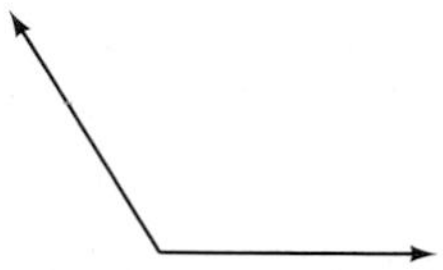

122°; obtuse

5.

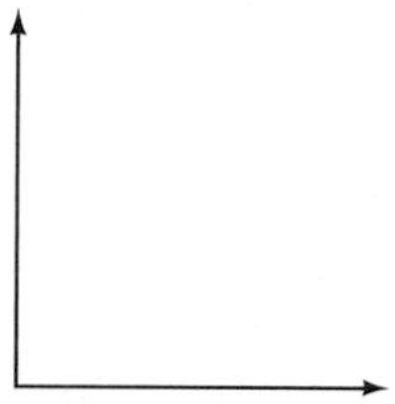

90°; right

6.

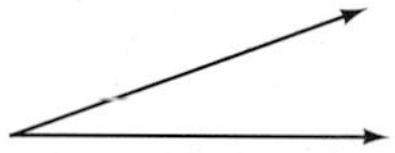

20°; acute

Solve.

7. Find the complement of a 62° angle. 28°

8. Find the complement of a 31° angle. 59°

9. Find the supplement of a 162° angle. 18°

10. Find the supplement of a 72° angle. 108°

11. Given $AB = 12$ cm, $CD = 9$ cm, and $AD = 35$ cm, find the length of BC. 14 cm

12. Given $AB = 21$ mm, $BC = 14$ mm, and $AD = 54$ mm, find the length of CD. 19 mm

13. Given $QR = 7$ ft and RS is three times the length of QR, find the length of QS. 28 ft

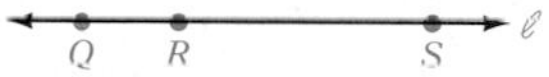

14. Given $QR = 15$ in. and RS is twice the length of QR, find the length of QS. 45 in.

15. Given $EF = 20$ m and FG is $\frac{1}{2}$ the length of EF, find the length of EG. 30 m

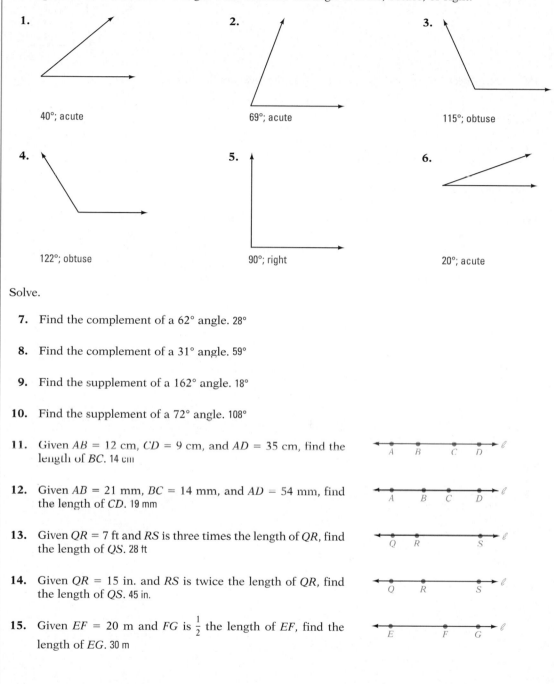

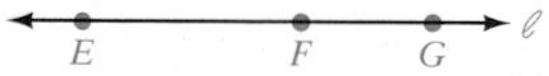

Section 3.1

Suggested Assignment

Exercises 1–59, odds

More challenging problems:
Exercises 61, 64, 65

Quick Quiz (Objective 3.1A)

1. How many degrees are in three-fifths of a revolution? 216°
2. Find the complement of a 54° angle. 36°
3. Find the supplement of a 22° angle. 158°

156 Chapter 3 / Geometry

16. Given $EF = 18$ cm and FG is $\frac{1}{3}$ the length of EF, find the length of EG.
24 cm

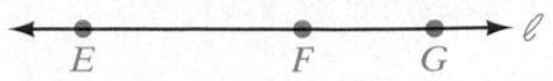

17. Given $\angle LOM = 53°$ and $\angle LON = 139°$, find the measure of $\angle MON$.
86°

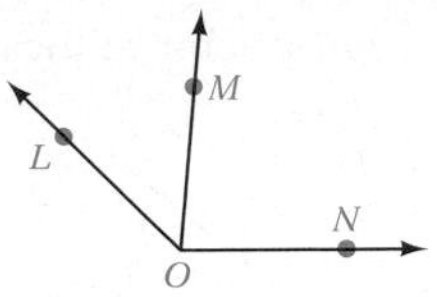

18. Given $\angle MON = 38°$ and $\angle LON = 85°$, find the measure of $\angle LOM$.
47°

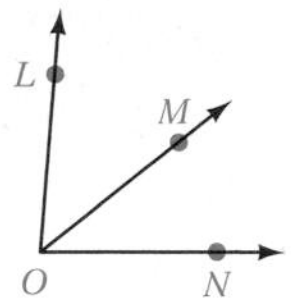

Find the measure of $\angle x$.

19.

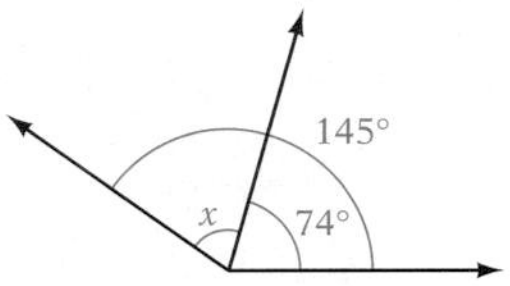

71°

20.

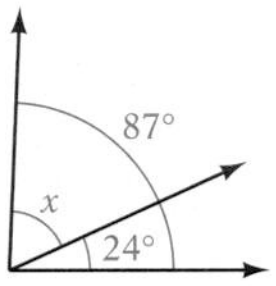

63°

Given that $\angle LON$ is a right angle, find the measure of $\angle x$.

21.

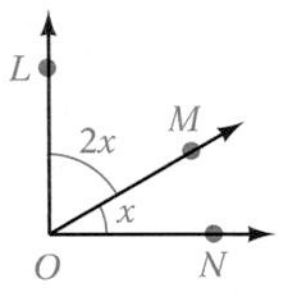

30°

22.

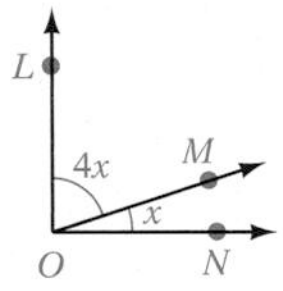

18°

23.

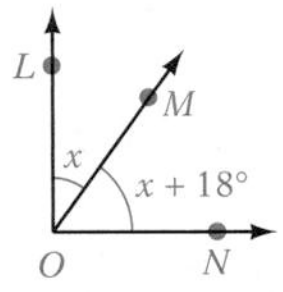

36°

24.

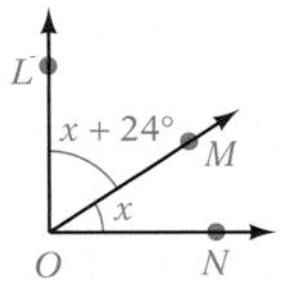

33°

Find the measure of $\angle a$.

25.

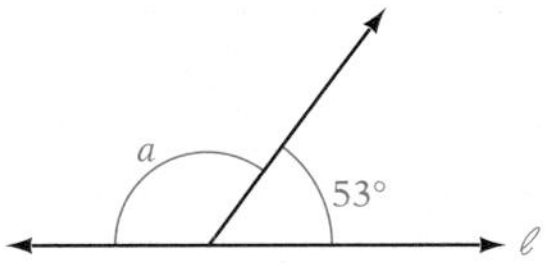

127°

26.

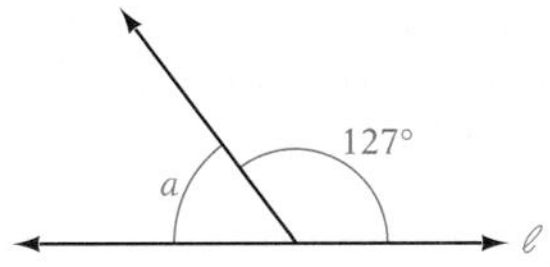

53°

27.

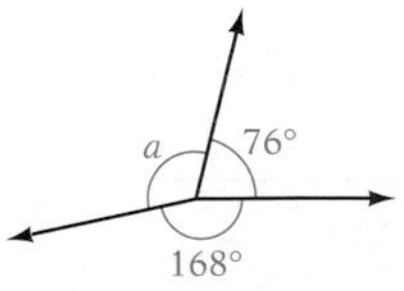

116°

28.

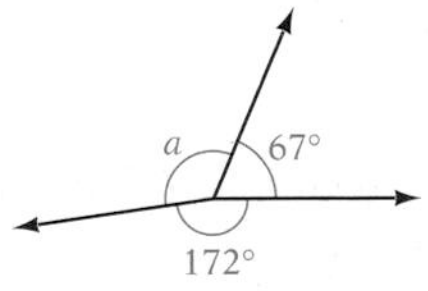

121°

157

Find x.

29.

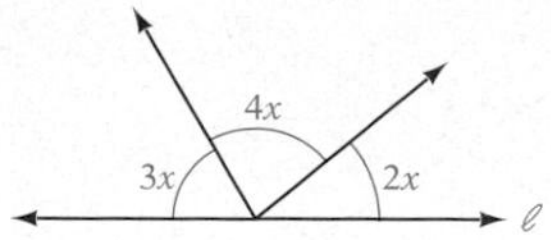

20°

30.

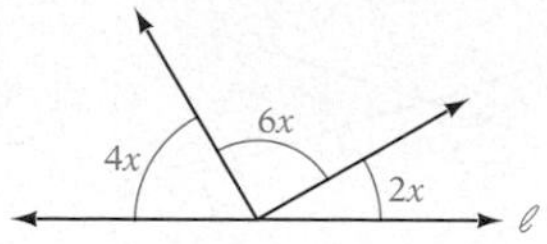

15°

31.

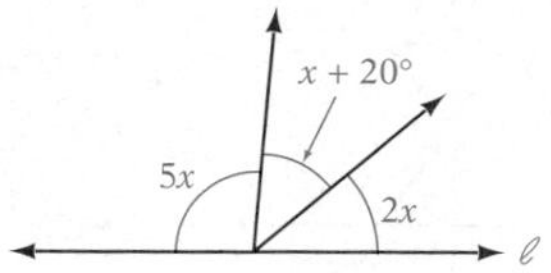

20°

32.

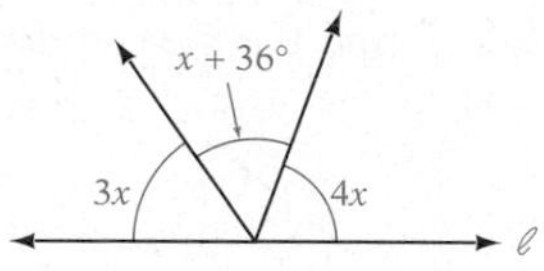

18°

33.

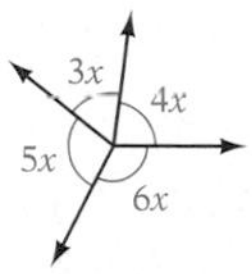

20°

34.

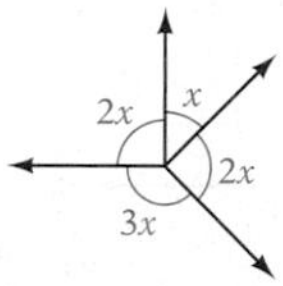

45°

Solve.

35. Given $\angle a = 51°$, find the measure of $\angle b$.

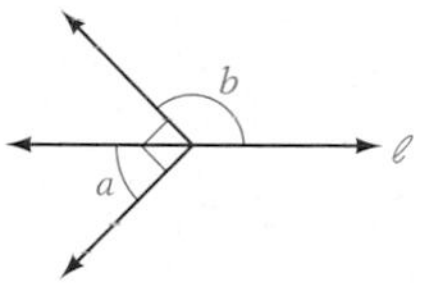

141°

36. Given $\angle a = 38°$, find the measure of $\angle b$.

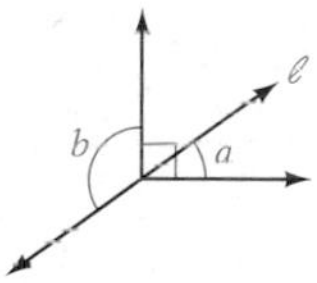

128°

Objective B

Find the measure of $\angle x$.

37.

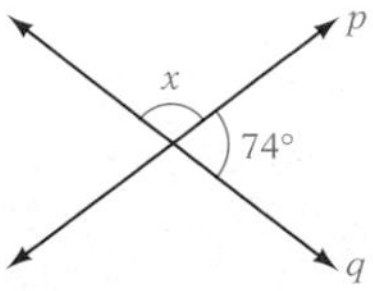

106°

38.

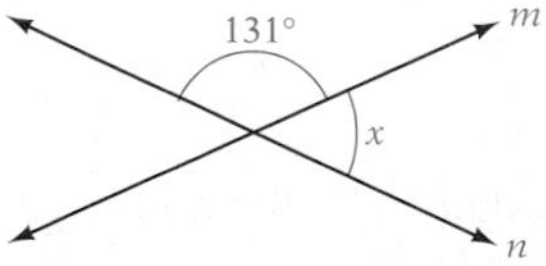

49°

Quick Quiz (Objective 3.1B)

1. In the figure, $\angle a = 118°$. Find $\angle b$. 62°

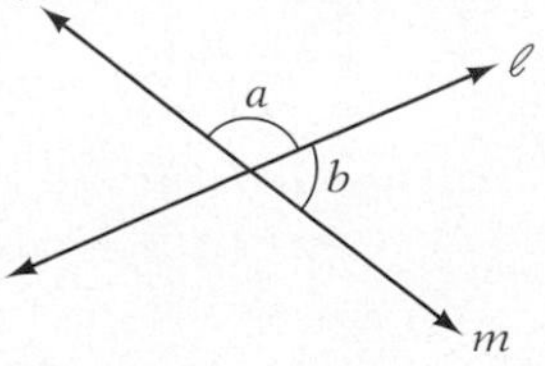

2. In the figure, $\ell_1 \parallel \ell_2$ and $\angle c = 67°$. Find $\angle b$. 113°

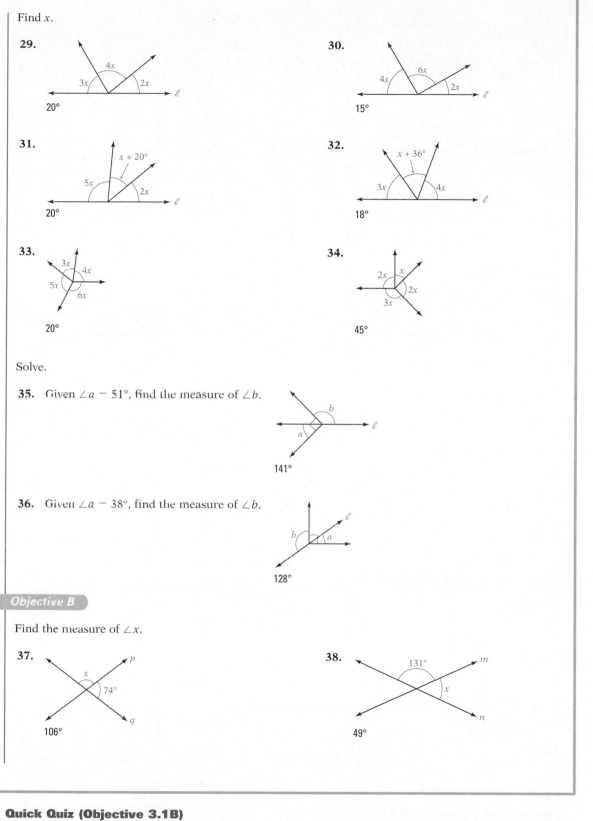

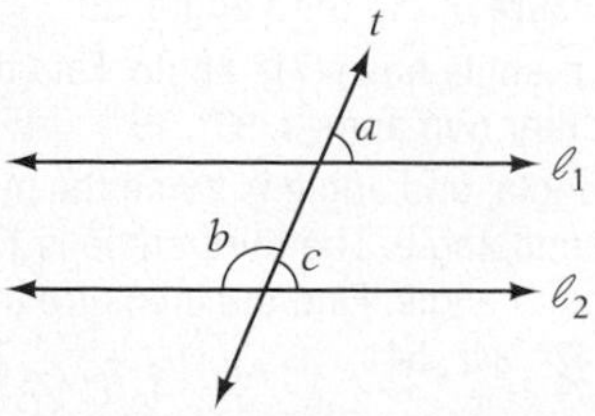

158

Find x.

39.

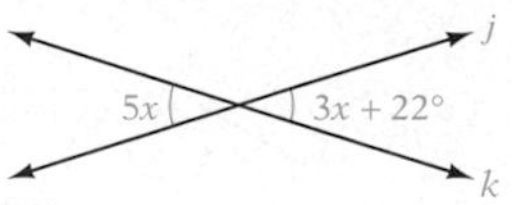

11°

40.

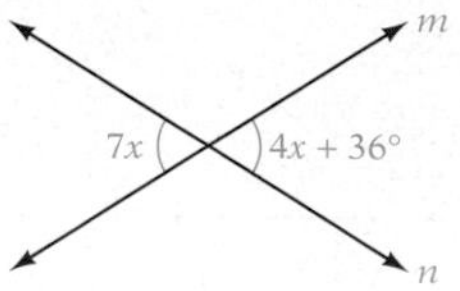

12°

Given that $\ell_1 \parallel \ell_2$, find the measures of angles a and b.

41.

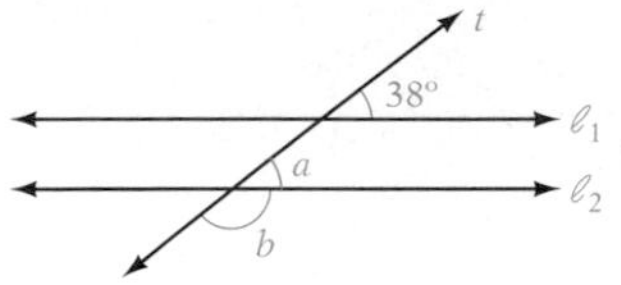

$\angle a = 38°$, $\angle b = 142°$

42.

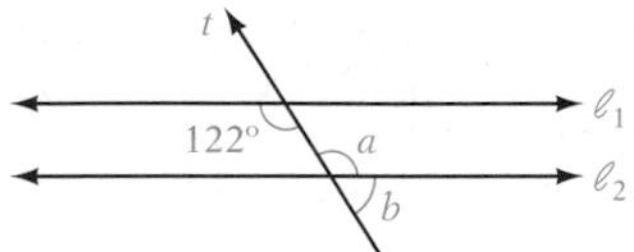

$\angle a = 122°$, $\angle b = 58°$

43.

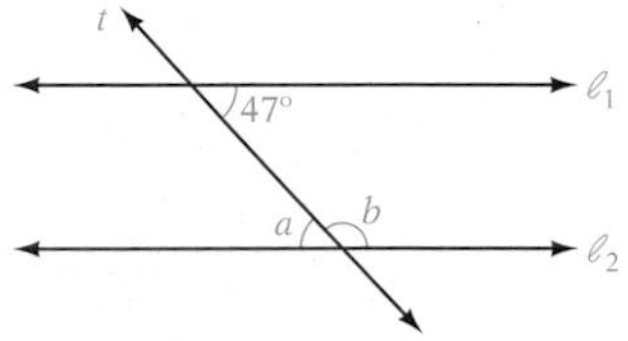

$\angle a = 47°$, $\angle b = 133°$

44.

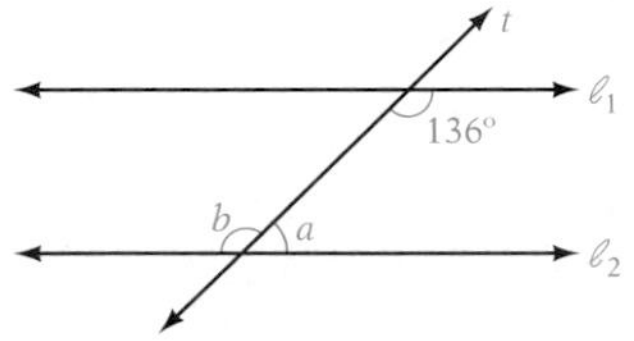

$\angle a = 44°$, $\angle b = 136°$

Given that $\ell_1 \parallel \ell_2$, find x.

45.

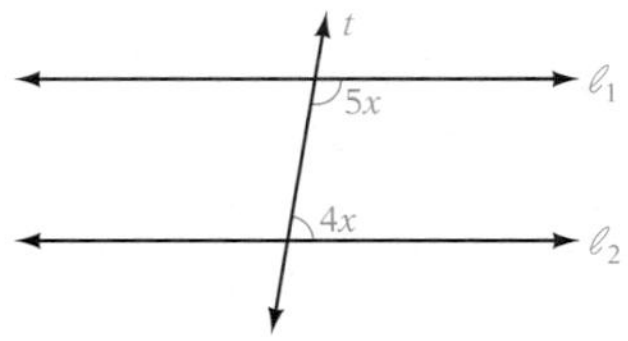

20°

46.

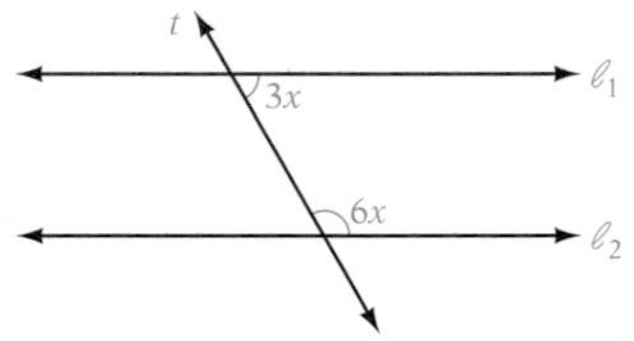

20°

47.

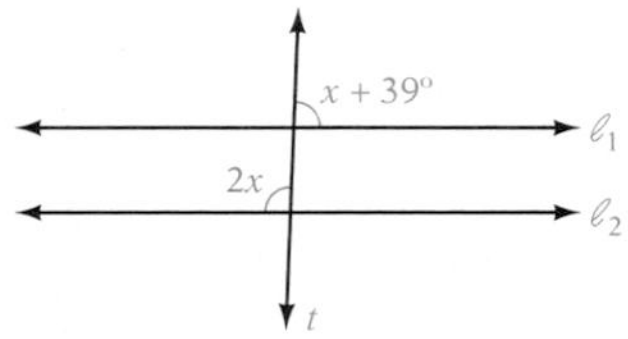

47°

48.

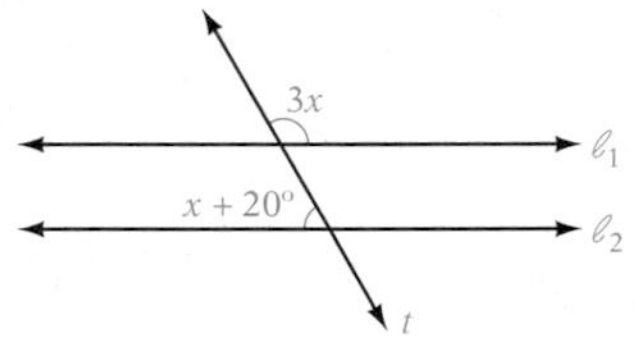

40°

Objective C

Solve.

49. Given that $\angle a = 95°$ and $\angle b = 70°$, find the measures of angles x and y.
$\angle x = 155°$, $\angle y = 70°$

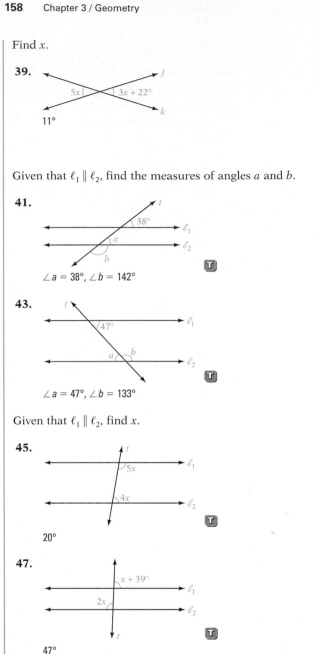

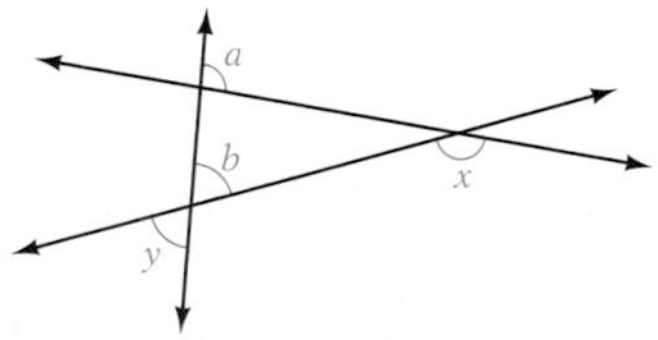

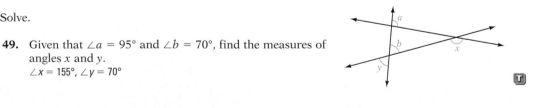

Quick Quiz (Objective 3.1C)

1. A triangle has a 110° angle and a 35° angle. Find the measure of the third angle. 35°
2. A right triangle has a 71° angle. Find the measure of the other two angles. 90°, 19°
3. In a triangle, one angle is twice the measure of the second angle. The third angle is 12° more than the second angle. Find the measure of each angle. 42°, 84°, 54°

159

50. Given that $\angle a = 35°$ and $\angle b = 55°$, find the measures of angles x and y.
$\angle x = 160°$,
$\angle y = 145°$

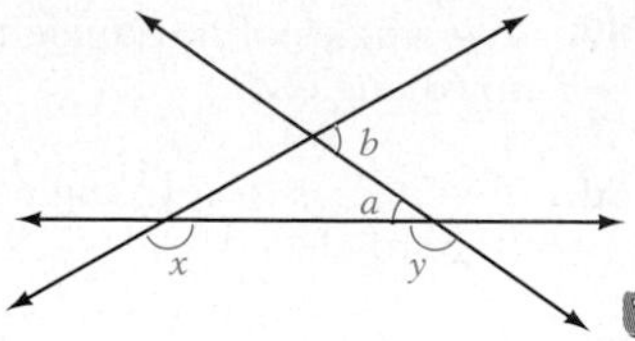

51. Given that $\angle y = 45°$, find the measures of angles a and b.
$\angle a = 45°$,
$\angle b = 135°$

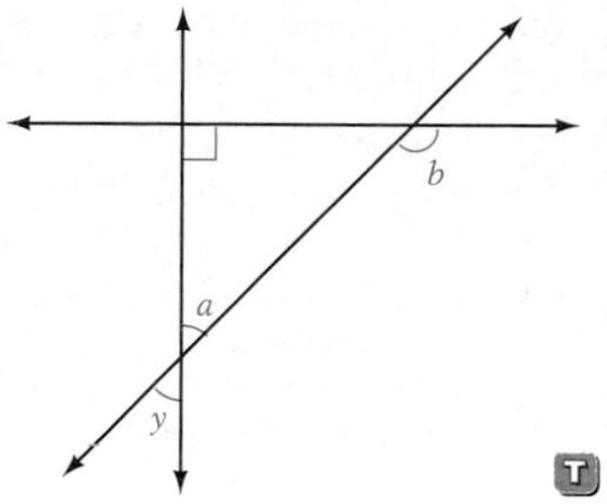

52. Given that $\angle y = 130°$, find the measures of angles a and b.
$\angle a = 40°$,
$\angle b = 140°$

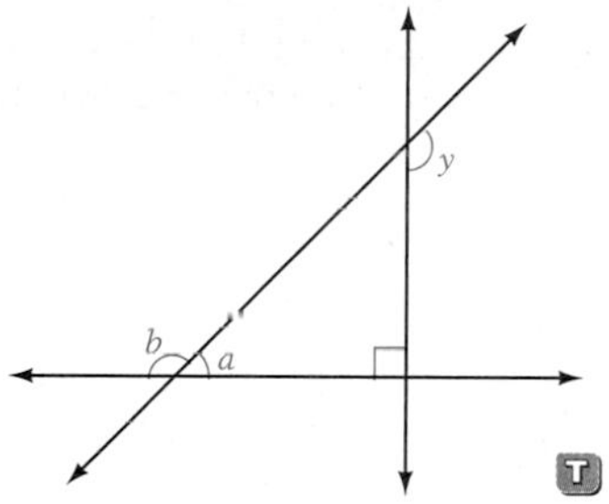

53. Given that $\overline{AO} \perp \overline{OB}$, express in terms of x the number of degrees in $\angle BOC$.
$90° - x$

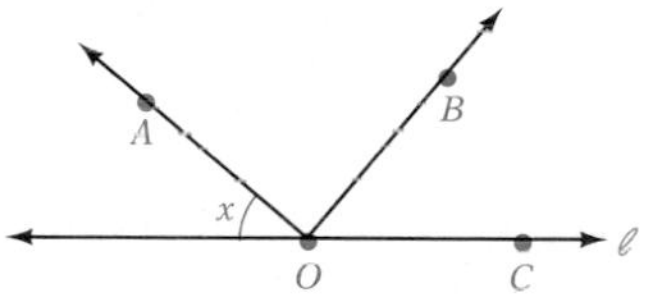

54. Given that $\overline{AO} \perp \overline{OB}$, express in terms of x the number of degrees in $\angle AOC$.
$75° - x$

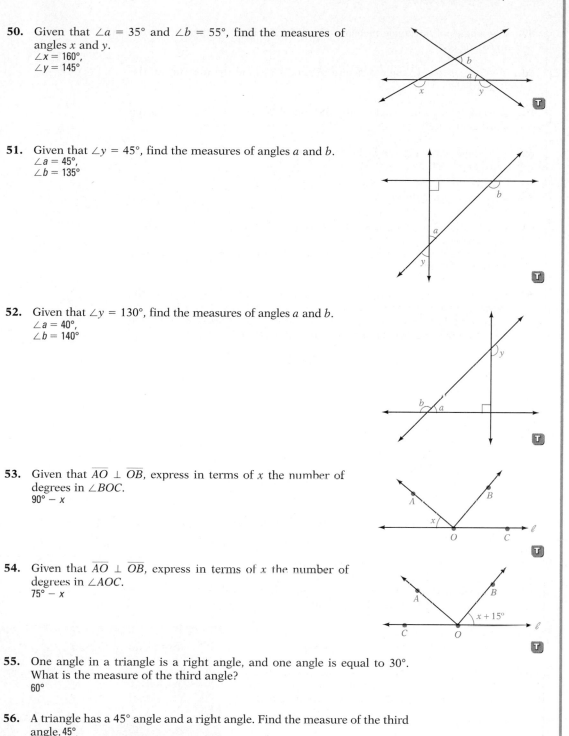

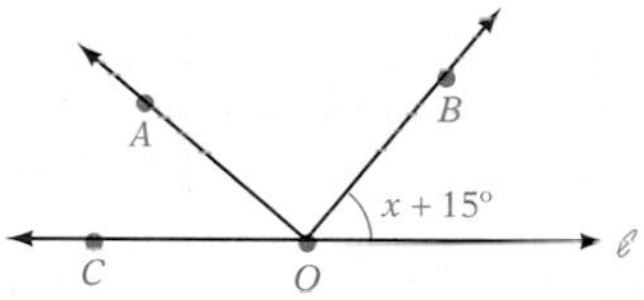

55. One angle in a triangle is a right angle, and one angle is equal to 30°. What is the measure of the third angle?
60°

56. A triangle has a 45° angle and a right angle. Find the measure of the third angle. 45°

57. Two angles of a triangle measure 42° and 103°. Find the measure of the third angle. 35°

58. Two angles of a triangle measure 62° and 45°. Find the measure of the third angle. 73°

59. A triangle has a 13° angle and a 65° angle. What is the measure of the third angle? 102°

60. A triangle has a 105° angle and a 32° angle. What is the measure of the third angle? 43°

APPLYING THE CONCEPTS

61. **a.** What is the smallest possible whole number of degrees in an angle of a triangle? 1°
    b. What is the largest possible whole number of degrees in an angle of a triangle? 179°

62. Cut out a triangle and then tear off two of the angles, as shown at the right. Position the pieces you tore off so that angle a is adjacent to angle b and angle c is adjacent to angle b (on the other side). Describe what you observe. What does this demonstrate? See the *Solutions Manual.*

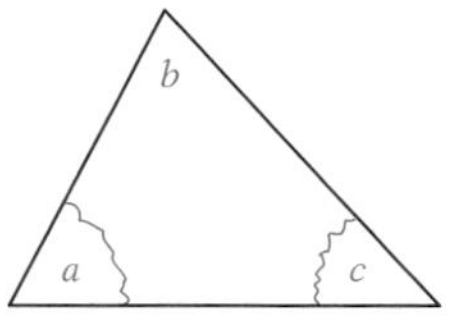

63. Construct a triangle with the given angle measures.
    a. 45°, 45°, and 90°
    b. 30°, 60°, and 90°
    c. 40°, 40°, and 100°

64. Determine whether the statement is always true, sometimes true, or never true.
    a. Two lines that are parallel to a third line are parallel to each other. always true
    b. A triangle contains two acute angles. always true
    c. Vertical angles are complementary angles. sometimes true

65. For the figure at the right, find the sum of the measures of angles x, y, and z. 360°

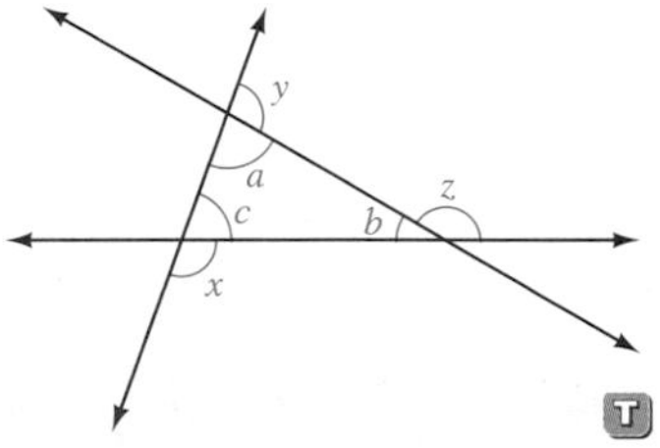

66. For the figure at the right, explain why $\angle a + \angle b = \angle x$. Write a rule that describes the relationship between an exterior angle of a triangle and the opposite interior angles. Use the rule to write an equation involving angles a, c, and z.

67. If $\overline{AB}$ and $\overline{CD}$ intersect at point O, and $\angle AOC = \angle BOC$, explain why $\overline{AB} \perp \overline{CD}$.

68. Do some research on the principle of reflection. Explain how this principle applies to the operation of a periscope and to the game of billiards.

Answers to Writing Exercises

66. The sum of the interior angles of a triangle is 180°; therefore, $\angle a + \angle b + \angle c = 180°$. The sum of an interior and an exterior angle is 180°; therefore, $\angle c + \angle x = 180°$. Solving this equation for $\angle c$, $\angle c = 180° - \angle x$. Substitute $180° - \angle x$ for $\angle c$ in the equation $\angle a + \angle b + \angle c = 180°$: $\angle a + \angle b + 180° - \angle x = 180°$. Add the measure of $\angle x$ to each side of the equation, and subtract 180° from each side of the equation: $\angle a + \angle b = \angle x$. The measure of an exterior angle of a triangle is equal to the sum of the measures of the two opposite interior angles: $\angle a + \angle c = \angle z$.

67. $\angle AOC$ and $\angle BOC$ are supplementary angles. Therefore, $\angle AOC + \angle BOC = 180°$. Because $\angle AOC = \angle BOC$, by substitution $\angle AOC + \angle AOC = 180°$. Therefore, $2(\angle AOC) = 180°$, and $\angle AOC = 90°$. Therefore, $\overline{AB}$ is perpendicular to $\overline{CD}$.

68. The student should make note that when a ray of light hits a surface, such as a mirror, the light is reflected at the same angle at which it hits the surface. In a periscope, light is reflected twice, with the result that light rays entering the periscope are parallel to the light rays at eye level. A ball bouncing off the side of a billiard table bounces off the side at the same angle at which it hit the side.

3.2 Plane Geometric Figures

Objective A — To solve problems involving the perimeter of geometric figures

VIDEO & DVD · CD TUTOR · WWW WEB · SSM

A **polygon** is a closed figure determined by three or more line segments that lie in a plane. The line segments that form the polygon are called its **sides**. The figures below are examples of polygons.

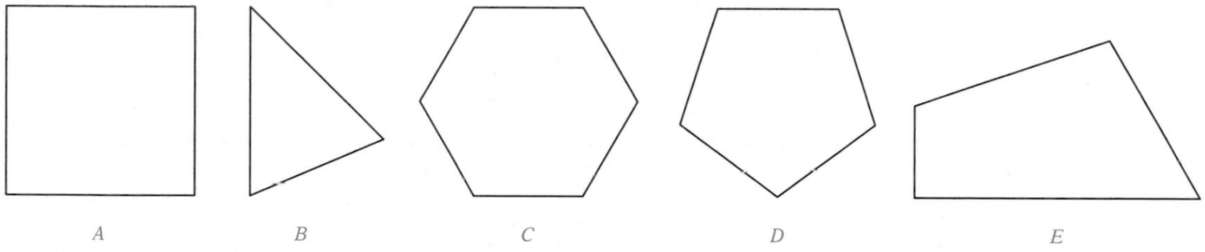

Point of Interest

Although a polygon is defined in terms of its *sides* (see the definition above), the word actually comes from the Latin word *polygonum*, which means "having many *angles*." This is certainly the case for a polygon.

A **regular polygon** is one in which each side has the same length and each angle has the same measure. The polygons in Figures *A*, *C*, and *D* above are regular polygons.

The name of a polygon is based on the number of its sides. The table below lists the names of polygons that have from 3 to 10 sides.

Number of Sides	*Name of the Polygon*
3	Triangle
4	Quadrilateral
5	Pentagon
6	Hexagon
7	Heptagon
8	Octagon
9	Nonagon
10	Decagon

Triangles and quadrilaterals are two of the most common types of polygons. Triangles are distinguished by the number of equal sides and also by the measures of their angles.

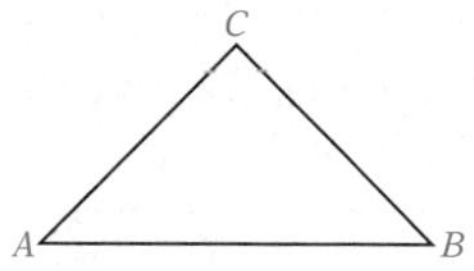

An **isosceles triangle** has two sides of equal length. The angles opposite the equal sides are of equal measure.
$AC = BC$
$\angle A = \angle B$

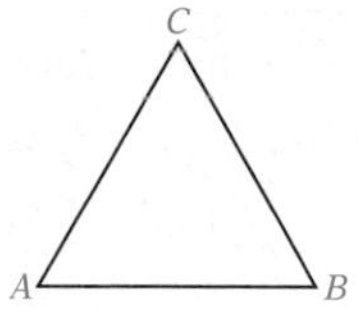

The three sides of an **equilateral triangle** are of equal length. The three angles are of equal measure.
$AB = BC = AC$
$\angle A = \angle B = \angle C$

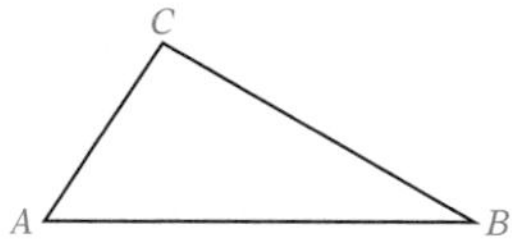

A **scalene triangle** has no two sides of equal length. No two angles are of equal measure.

T

161

Objective 3.2A

New Vocabulary
polygon
sides of a polygon
regular polygon
pentagon
hexagon
heptagon
octagon
nonagon
decagon
isosceles triangle
equilateral triangle
scalene triangle
acute triangle
obtuse triangle
quadrilateral
parallelogram
rectangle
square
rhombus
trapezoid
isosceles trapezoid
perimeter
circle
diameter
radius
circumference

Vocabulary to Review
plane geometric figure
triangle
right triangle

New Symbols
π (pi)

New Formulas
Perimeter of a triangle:
$P = a + b + c$
Perimeter of a rectangle:
$P = 2L + 2W$
Perimeter of a square: $P = 4s$
Diameter of a circle: $d = 2r$
Circumference of a circle:
$C = \pi d$ or $C = 2\pi r$

Discuss the Concepts

1. Is every square a rectangle? Is every rectangle a square?
2. In the definition of a polygon, what does the phrase *closed figure* mean? Draw a figure that is not closed.

In-Class Examples (Objective 3.2A)

1. Find the perimeter of a square in which the sides are equal to 15 m. 60 m
2. Find the perimeter of a rectangle with a length of 5 m and a width of 1.4 m. 12.8 m
3. Find the circumference of a circle with a radius of 11 cm. Round to the nearest hundredth. 69.12 cm

Instructor Note

The diagram at the right below shows the relationships among quadrilaterals. The description of each quadrilateral is within an example of that quadrilateral.

Concept Check

1. Figure A below is a rectangle. Label the length of the rectangle L and the width of the rectangle W.

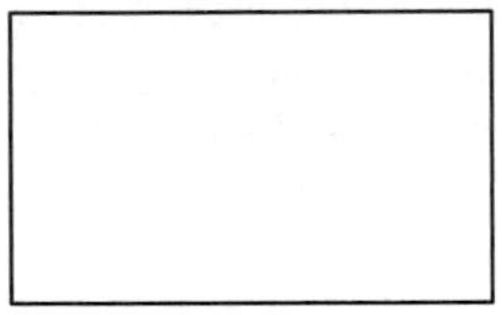

Figure A

2. What is the name of a regular polygon that has three sides? Equilateral triangle
3. What is the name of a parallelogram in which all angles are the same measure? Rectangle

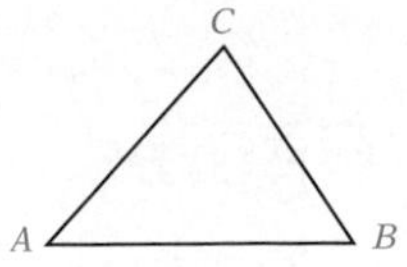

An **acute triangle** has three acute angles.

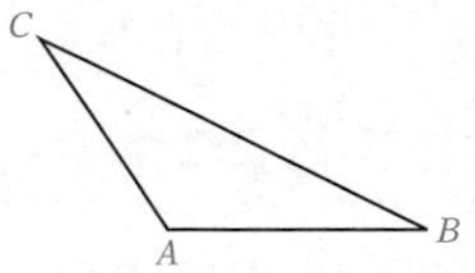

An **obtuse triangle** has one obtuse angle.

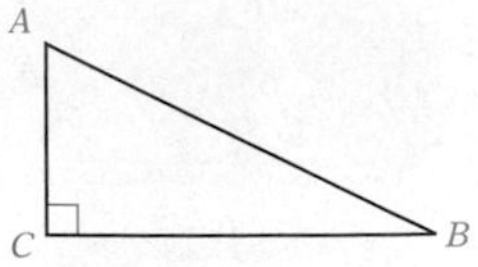

A **right triangle** has a right angle.

TAKE NOTE

The diagram below shows the relationships among all quadrilaterals. The description of each quadrilateral is within an example of that quadrilateral.

Quadrilaterals are also distinguished by their sides and angles, as shown below. Note that a rectangle, a square, and a rhombus are different forms of a parallelogram.

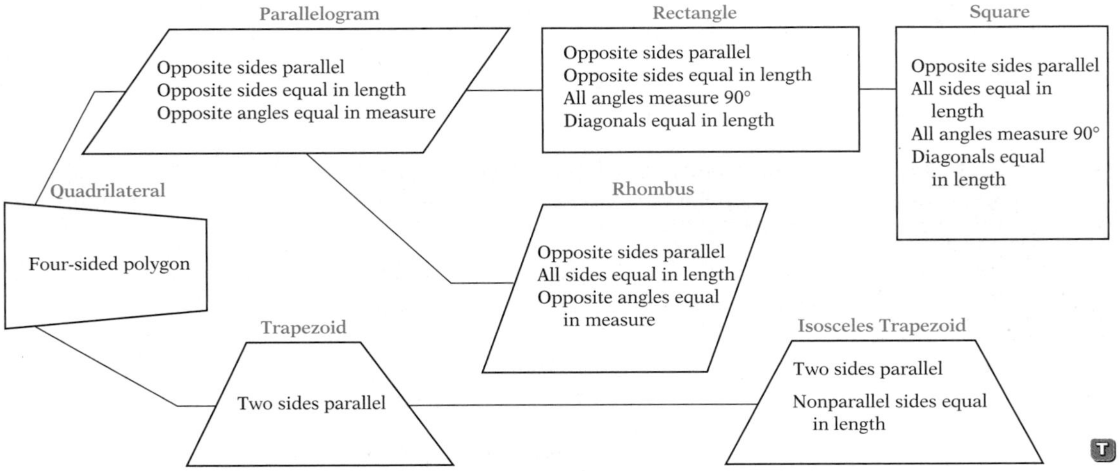

The **perimeter** of a plane geometric figure is a measure of the distance around the figure. Perimeter is used in buying fencing for a lawn or determining how much baseboard is needed for a room.

The perimeter of a triangle is the sum of the length of the three sides.

Perimeter of a Triangle

Let a, b, and c be the lengths of the sides of a triangle. The perimeter, P, of the triangle is given by $P = a + b + c$.

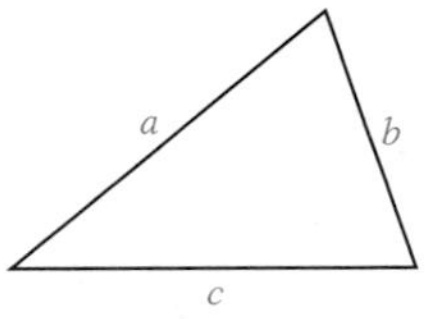

$P = a + b + c$

➡ Find the perimeter of the triangle shown at the right.

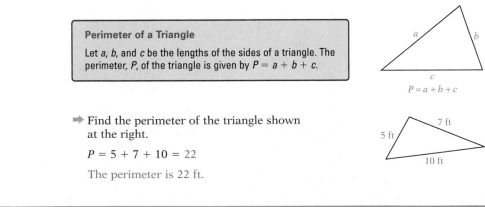

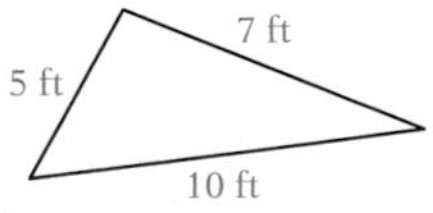

$P = 5 + 7 + 10 = 22$

The perimeter is 22 ft.

The perimeter of a quadrilateral is the sum of the lengths of its four sides.

Point of Interest

Leonardo DaVinci painted the *Mona Lisa* on a rectangular canvas whose height was approximately 1.6 times its width. Rectangles with these proportions, called golden rectangles, were used extensively in Renaissance art.

A rectangle is a quadrilateral with opposite sides of equal length. Usually the length, L, of a rectangle refers to the length of one of the longer sides of the rectangle, and the width, W, refers to the length of one of the shorter sides. The perimeter can then be represented $P = L + W + L + W$.

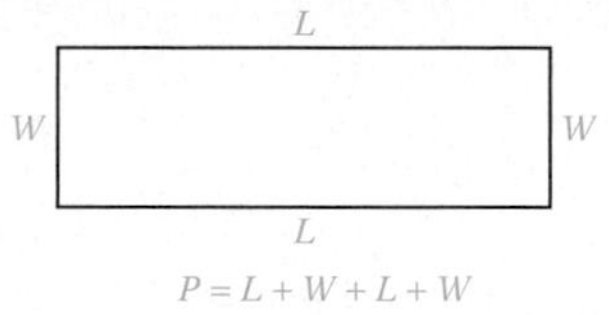

The formula for the perimeter of a rectangle is derived by combining like terms.

$P = 2L + 2W$

Perimeter of a Rectangle

Let L represent the length and W the width of a rectangle. The perimeter, P, of the rectangle is given by $P = 2L + 2W$.

➡ Find the perimeter of the rectangle shown at the right.

$P = 2L + 2W$

$P = 2(5) + 2(2)$ • The length is 5 m. Substitute 5 for L. The width is 2 m. Substitute 2 for W.

$P = 10 + 4$ • Solve for P.

$P = 14$

The perimeter is 14 m.

A square is a rectangle in which each side has the same length. Let s represent the length of each side of a square. Then the perimeter of a square can be represented $P = s + s + s + s$.

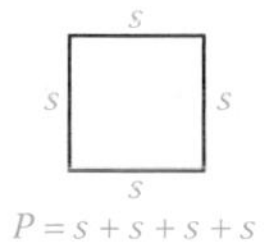

The formula for the perimeter of a square is derived by combining like terms.

$P = 4s$

Perimeter of a Square

Let s represent the length of a side of a square. The perimeter, P, of the square is given by $P = 4s$.

➡ Find the perimeter of the square shown at the right.

$P = 4s = 4(8) = 32$

The perimeter is 32 in.

Optional Student Activity

1. The base of isosceles triangle ABC (with $AB = BC$ and $\angle B = 42°$) and the base of equilateral triangle CDE lie on line segment AE. Find the measure of $\angle BCD$. 51°

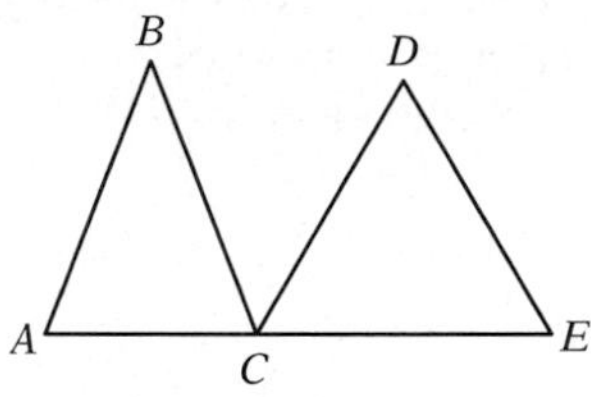

2. Three line segments are randomly chosen from line segments whose lengths are 1 cm, 2 cm, 3 cm, 4 cm, and 5 cm. What is the probability that a triangle can be formed from these line segments? $\frac{3}{10}$
3. Triangle FJH is an isosceles triangle in which $FJ = FH$, $FK = KJ$, and $FG = GH$. Find the perimeter of triangle FJH if $FK = 2x + 3$, $GH = 5x - 9$, and $JH = 4x$. 60 units

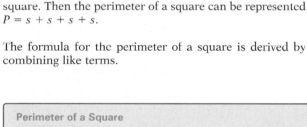

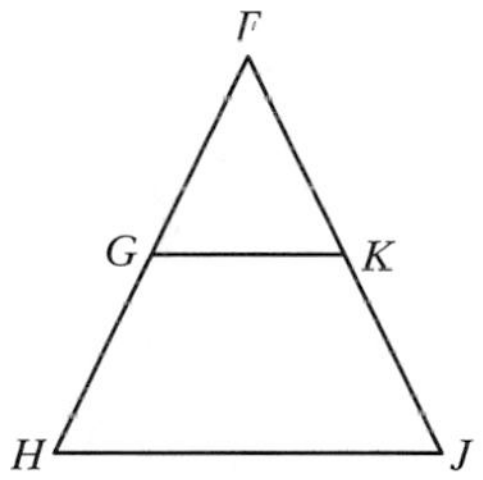

164 Chapter 3 / Geometry

A **circle** is a plane figure in which all points are the same distance from point O, which is called the **center** of the circle.

The **diameter** of a circle is a line segment across the circle through point O. AB is a diameter of the circle at the right. The variable d is used to designate the diameter of a circle.

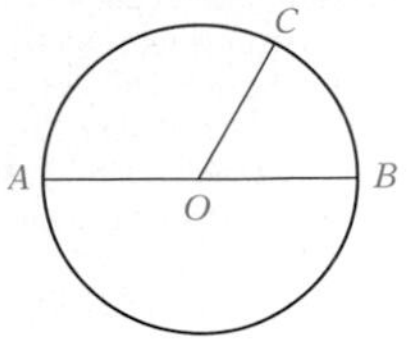

The **radius** of a circle is a line segment from the center of the circle to a point on the circle. OC is a radius of the circle at the right. The variable r is used to designate a radius of a circle.

The length of the diameter is twice the length of the radius.

$$d = 2r \text{ or } r = \frac{1}{2}d$$

Point of Interest

Archimedes (c. 287–212 B.C.) was the mathematician who gave us the approximate value of π as $\frac{22}{7} = 3\frac{1}{7}$. He actually showed that π was between $3\frac{10}{71}$ and $3\frac{1}{7}$. The approximation $3\frac{10}{71}$ is closer to the exact value of π, but it is more difficult to use.

The distance around a circle is called the **circumference**. The circumference, C, of a circle is equal to the product of π (pi) and the diameter.

$$C = \pi d$$

Because $d = 2r$, the formula for the circumference can be written in terms of r.

$$C = 2\pi r$$

The Circumference of a Circle

The circumference, C, of a circle with diameter d and radius r is given by $C = \pi d$ or $C = 2\pi r$.

The formula for circumference uses the number π, which is an irrational number. The value of π can be approximated by a fraction or by a decimal.

$$\pi \approx \frac{22}{7} \text{ or } \pi \approx 3.14$$

The π key on a scientific calculator gives a closer approximation of π than 3.14. Use a scientific calculator to find approximate values in calculations involving π.

CALCULATOR NOTE

The π key on your calculator can be used to find decimal approximations for expressions that contain π. To perform the calculation at the right, enter 6 [×] [π] [=].

➡ Find the circumference of a circle with a diameter of 6 in.

$C = \pi d$

$C = \pi(6)$

- **The diameter of the circle is given. Use the circumference formula that involves the diameter. $d = 6$.**

$C = 6\pi$

- **The exact circumference of the circle is 6π in.**

$C \approx 18.85$

- **An approximate measure is found by using the π key on a calculator.**

The circumference is approximately 18.85 in.

165

Example 1
A carpenter is designing a square patio with a perimeter of 44 ft. What is the length of each side?

Strategy
To find the length of each side, use the formula for the perimeter of a square. Substitute 44 for P and solve for s.

Solution
$P = 4s$
$44 = 4s$
$11 = s$

The length of each side of the patio is 11 ft.

You Try It 1
The infield for a softball field is a square with each side of length 60 ft. Find the perimeter of the infield.

Your strategy

Your solution
240 ft

Example 2
The dimensions of a triangular sail are 18 ft, 11 ft, and 15 ft. What is the perimeter of the sail?

Strategy
To find the perimeter, use the formula for the perimeter of a triangle. Substitute 18 for a, 11 for b, and 15 for c. Solve for P.

Solution
$P = a + b + c$
$P = 18 + 11 + 15$
$P = 44$

The perimeter of the sail is 44 ft.

You Try It 2
What is the perimeter of a standard piece of typing paper that measures $8\frac{1}{2}$ in. by 11 in.?

Your strategy

Your solution
39 in.

Example 3
Find the circumference of a circle with a radius of 15 cm. Round to the nearest hundredth.

Strategy
To find the circumference, use the circumference formula that involves the radius. An approximation is asked for; use the π key on a calculator. $r = 15$.

Solution
$C = 2\pi r = 2\pi(15) = 30\pi \approx 94.25$

The circumference is 94.25 cm.

You Try It 3
Find the circumference of a circle with a diameter of 9 in. Give the exact measure.

Your strategy

Your solution
9π in.

Solutions on pp. S8–S9

Objective 3.2B

New Vocabulary
area
base
height

New Symbols
in^2 (square inches)
cm^2 (square centimeters)
ft^2 (square feet)
m^2 (square meters)
mi^2 (square miles)

New Formulas
Area of a rectangle: $A = LW$
Area of a square: $A = s^2$
Area of a parallelogram: $A = bh$
Area of a triangle: $A = \frac{1}{2}bh$
Area of a trapezoid:
$A = \frac{1}{2}h(b_1 + b_2)$
Area of a circle: $A = \pi r^2$

Instructor Note
Ask students to find the next three square numbers. Have them form the square array, as well as represent the number in both standard form and exponential form.

You may want to introduce the students to triangular numbers. See the Focus on Problem Solving at the end of the chapter entitled "Rational Expressions."

Discuss the Concepts
What is wrong with the statement?
a. The perimeter is 40 m^2.
b. The area is 120 ft.

Objective B To solve problems involving the area of geometric figures

Area is the amount of surface in a region. Area can be used to describe the size of a rug, a parking lot, a farm, or a national park. Area is measured in square units.

Point of Interest

Polygonal numbers are whole numbers that can be represented as regular geometric figures. For example, a square number is one that can be represented as a square array.

The square numbers are 1, 4, 9, 16, 25, . . . They can be represented as 1^2, 2^2, 3^2, 4^2, 5^2, . . .

A square that measures 1 in. on each side has an area of 1 square inch, written 1 in^2.

A square that measures 1 cm on each side has an area of 1 square centimeter, written 1 cm^2.

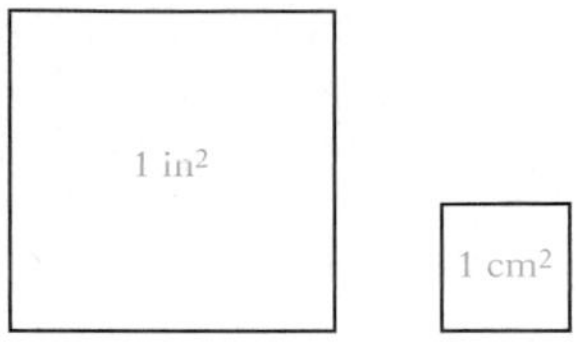

Larger areas can be measured in square feet (ft^2), square meters (m^2), square miles (mi^2), acres (43,560 ft^2), or any other square unit.

The area of a geometric figure is the number of squares that are necessary to cover the figure. In the figures below, two rectangles have been drawn and covered with squares. In the figure on the left, 12 squares, each of area 1 cm^2, were used to cover the rectangle. The area of the rectangle is 12 cm^2. In the figure on the right, 6 squares, each of area 1 in^2, were used to cover the rectangle. The area of the rectangle is 6 in^2.

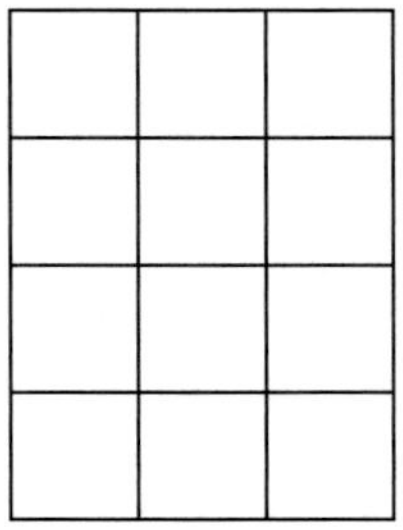

The area of the rectangle is 12 cm^2.

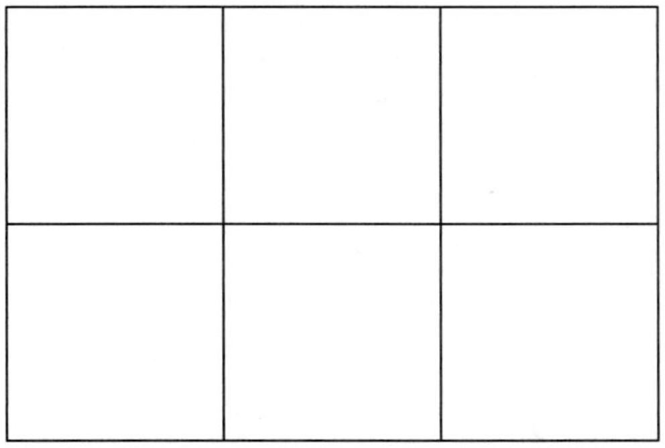

The area of the rectangle is 6 in^2.

Note from the above figures that the area of a rectangle can be found by multiplying the length of the rectangle by its width.

> **Area of a Rectangle**
>
> Let L represent the length and W the width of a rectangle. The area, A, of the rectangle is given by $A = LW$.

➡ Find the area of the rectangle shown at the right.

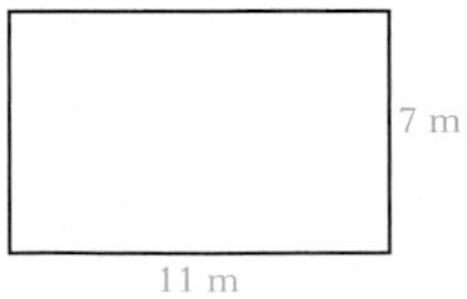

$A = LW = 11(7) = 77$

The area is 77 m^2.

In-Class Examples (Objective 3.2B)

1. Find the area of a triangle with a base of 5 cm and a height of 2.6 cm. 6.5 cm^2
2. Find the area of a square with a side of 8.5 ft. 72.25 ft^2
3. Find the area of a rectangle with a length of 37 in. and a width of 15 in. 555 in^2
4. Find the area of a circle with a diameter of 16 in. Round to the nearest hundredth. 201.06 in^2

167

A square is a rectangle in which all sides are the same length. Therefore, both the length and the width of a square can be represented by s, and $A = LW = s \cdot s = s^2$.

Area of a Square

Let s represent the length of a side of a square. The area, A, of the square is given by $A = s^2$.

$A = s \cdot s = s^2$

➡ Find the area of the square shown at the right.

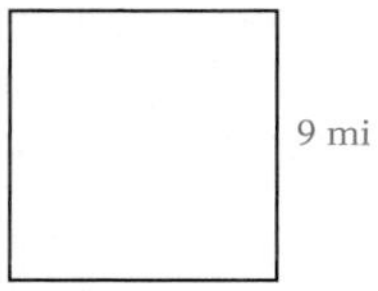

$A = s^2 = 9^2 = 81$

The area is 81 mi^2.

Figure $ABCD$ is a parallelogram. BC is the **base**, b, of the parallelogram. AE, perpendicular to the base, is the **height**, h, of the parallelogram.

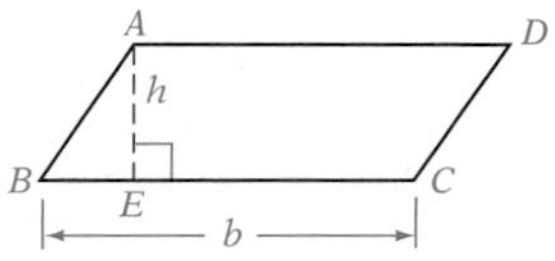

Any side of a parallelogram can be designated as the base. The corresponding height is found by drawing a line segment perpendicular to the base from the opposite side.

A rectangle can be formed from a parallelogram by cutting a right triangle from one end of the parallelogram and attaching it to the other end. The area of the resulting rectangle will equal the area of the original parallelogram.

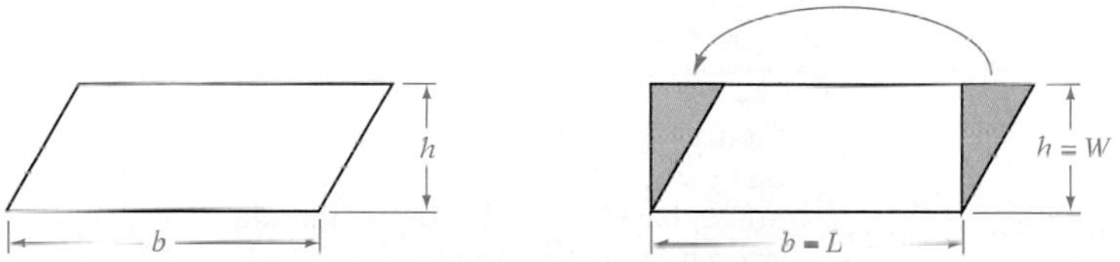

Area of a Parallelogram

Let b represent the length of the base and h the height of a parallelogram. The area, A, of the parallelogram is given by $A = bh$.

➡ Find the area of the parallelogram shown at the right.

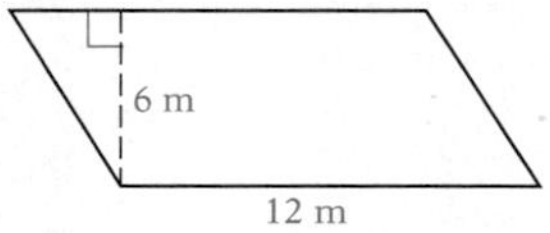

$A = bh = 12 \cdot 6 = 72$

The area is 72 m^2.

Concept Check

The concepts of square units and area are difficult for students. After introducing these ideas, ask students questions such as the following:

1. Would I measure the distance from Chicago to Boston in miles or square miles? miles
2. Is the amount of land cultivated by a gardener measured in feet or square feet? square feet
3. How is the size of a state park measured? acres or square miles
4. How is the length of a family room measured? feet or meters

168

Optional Student Activity

1. Use graph paper to draw different rectangles, each with a perimeter of 20 units. Investigate just whole-number dimensions. What dimensions will result in a rectangle with the greatest possible area? 5×5
2. Ancient Egyptians gave the formula for the area of a circle as $\left(\frac{8}{9}d\right)^2$, where d is the diameter. Does this formula give an area that is less than or greater than the area given by the correct formula? greater than

Figure ABC is a triangle. AB is the **base**, b, of the triangle. CD, perpendicular to the base, is the **height**, h, of the triangle.

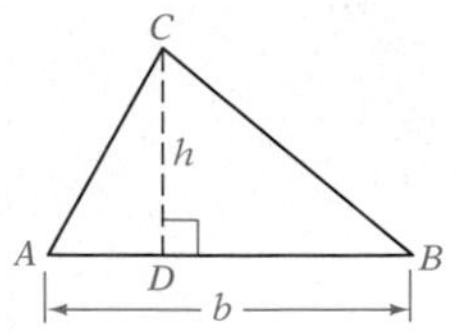

Any side of a triangle can be designated as the base. The corresponding height is found by drawing a line segment perpendicular to the base from the vertex opposite the base.

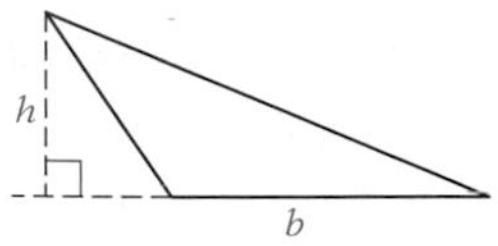

Consider the triangle with base b and height h shown at the right. By extending a line from C parallel to the base AB and equal in length to the base, and extending a line from B parallel to AC and equal in length to AC, a parallelogram is formed. The area of the parallelogram is bh and is twice the area of the triangle. Therefore, the area of the triangle is one-half the area of the parallelogram, or $\frac{1}{2}bh$.

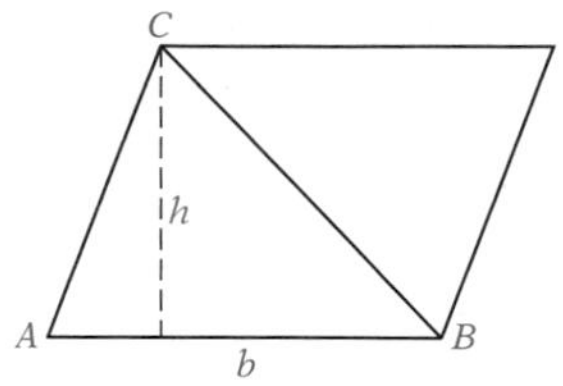

Area of a Triangle

Let b represent the length of the base and h the height of a triangle. The area, A, of the triangle is given by $A = \frac{1}{2}bh$.

➡ Find the area of a triangle with a base of 18 cm and a height of 6 cm.

$$A = \frac{1}{2}bh = \frac{1}{2} \cdot 18 \cdot 6 = 54$$

The area is 54 cm^2.

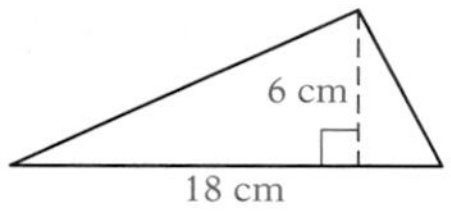

Figure $ABCD$ is a trapezoid. AB is one **base**, b_1, of the trapezoid, and CD is the other base, b_2. AE, perpendicular to the two bases, is the **height**, h.

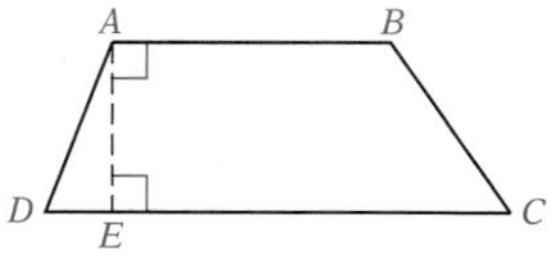

In the trapezoid at the right, the line segment BD divides the trapezoid into two triangles, ABD and BCD. In triangle ABD, b_1 is the base and h is the height. In triangle BCD, b_2 is the base and h is the height. The area of the trapezoid is the sum of the areas of the two triangles.

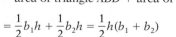

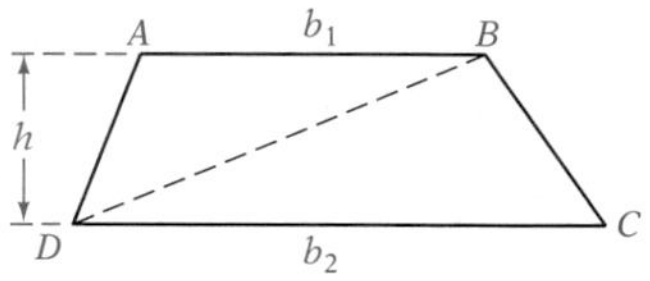

$$\text{Area of trapezoid } ABCD = \text{area of triangle } ABD + \text{area of triangle } BCD$$

$$= \frac{1}{2}b_1h + \frac{1}{2}b_2h = \frac{1}{2}h(b_1 + b_2)$$

169

Area of a Trapezoid

Let b_1 and b_2 represent the lengths of the bases and h the height of a trapezoid. The area, A, of the trapezoid is given by $A = \frac{1}{2}h(b_1 + b_2)$.

➡ Find the area of a trapezoid that has bases measuring 15 in. and 5 in. and a height of 8 in.

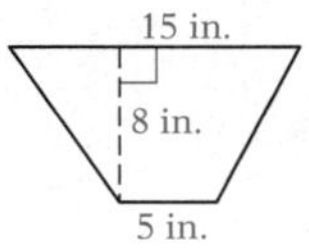

$$A = \frac{1}{2}h(b_1 + b_2)$$

$$= \frac{1}{2} \cdot 8(15 + 5) = 4(20) = 80$$

The area is 80 in².

The area of a circle is equal to the product of π and the square of the radius.

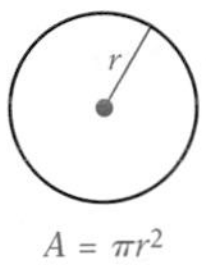

Area of a Circle

The area, A, of a circle with radius r is given by $A = \pi r^2$.

➡ Find the area of a circle that has a radius of 6 cm.

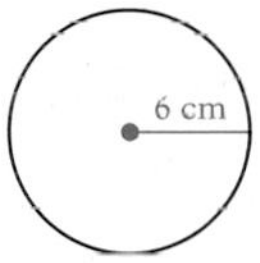

$A = \pi r^2$	• Use the formula for the area of a circle. $r = 6$
$A = \pi(6)^2$	
$A = \pi(36)$	
$A = 36\pi$	• The exact area of the circle is 36π cm².
$A \approx 113.10$	• An approximate measure is found by using the π key on a calculator.

The approximate area of the circle is 113.10 cm².

CALCULATOR NOTE

To approximate 36π on your calculator, enter 36 [×] [π] [=].

For your reference, all of the formulas for the perimeters and areas of the geometric figures presented in this section are listed in the Chapter Summary located at the end of this chapter.

Example 4
The Parks and Recreation Department of a city plans to plant grass seed in a playground that has the shape of a trapezoid, as shown below. Each bag of grass seed will seed 1500 ft^2. How many bags of grass seed should the department purchase?

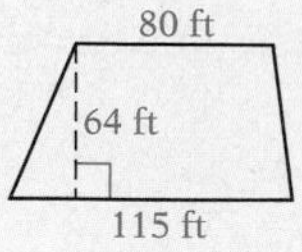

Strategy
To find the number of bags to be purchased:
- Use the formula for the area of a trapezoid to find the area of the playground.
- Divide the area of the playground by the area one bag will seed (1500).

Solution

$$A = \frac{1}{2}h(b_1 + b_2)$$

$$A = \frac{1}{2} \cdot 64(80 + 115)$$

$$A = 6240$$ • **The area of the playground is 6240 ft^2.**

$$6240 \div 1500 = 4.16$$

Because a portion of a fifth bag is needed, 5 bags of grass seed should be purchased.

You Try It 4
An interior designer decides to wallpaper two walls of a room. Each roll of wallpaper will cover 30 ft^2. Each wall measures 8 ft by 12 ft. How many rolls of wallpaper should be purchased?

Your strategy

Your solution
7 rolls

Example 5
Find the area of a circle with a diameter of 5 ft. Give the exact measure.

Strategy
To find the area:
- Find the radius of the circle.
- Use the formula for the area of a circle. Leave the answer in terms of π.

Solution

$$r = \frac{1}{2}d = \frac{1}{2}(5) = 2.5$$

$$A = \pi r^2 = \pi(2.5)^2 = \pi(6.25) = 6.25\pi$$

The area of the circle is 6.25π ft^2.

You Try It 5
Find the area of a circle with a radius of 11 cm. Round to the nearest hundredth.

Your strategy

Your solution
380.13 cm^2

Solutions on p. S9

Section 3.2

Suggested Assignment
Exercises 1–105, odds
More challenging problems:
Exercises 106, 110

3.2 Exercises

Objective A

Name each polygon.

1.
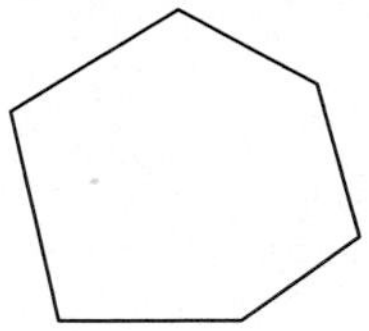
hexagon

2.
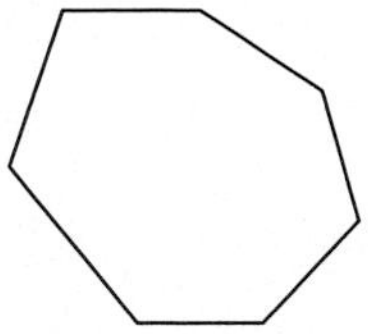
heptagon

3.
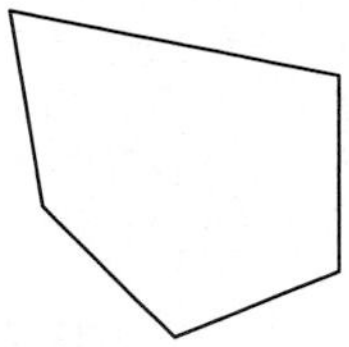
pentagon

4.
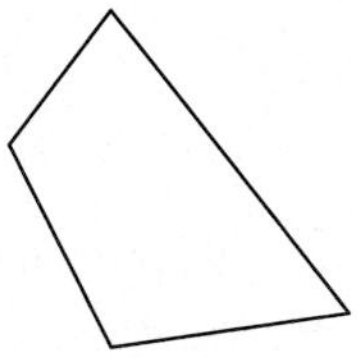
quadrilateral

Classify the triangle as isosceles, equilateral, or scalene.

5.
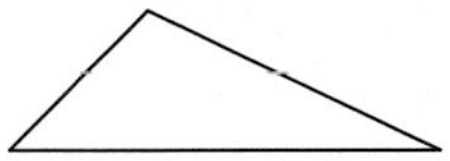
scalene

6.
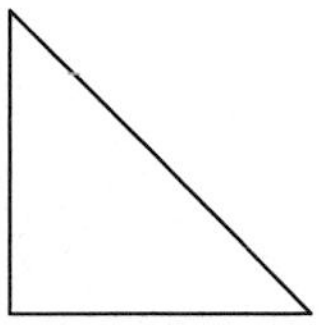
isosceles

7.
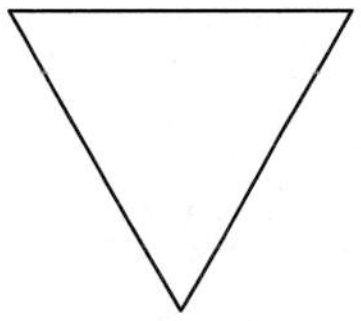
equilateral

8.
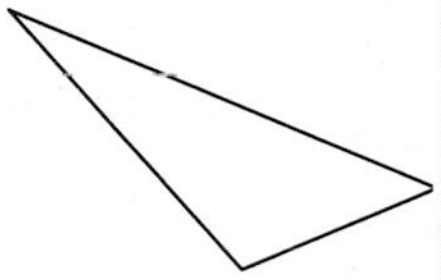
scalene

Classify the triangle as acute, obtuse, or right.

9.
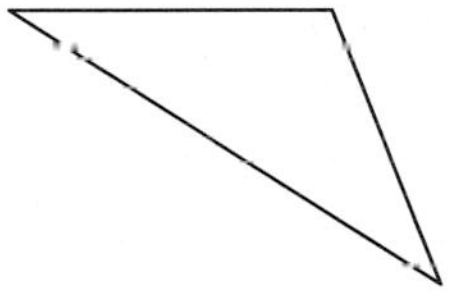
obtuse

10.
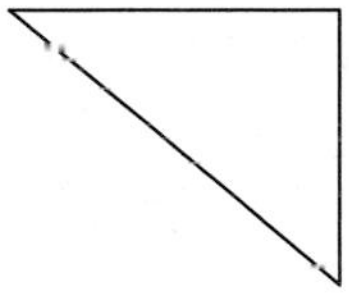
right

11.
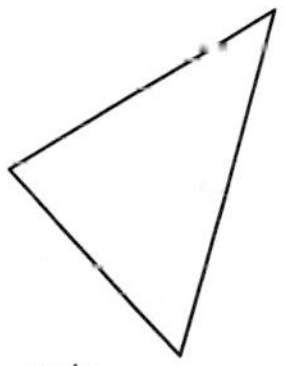
acute

12.
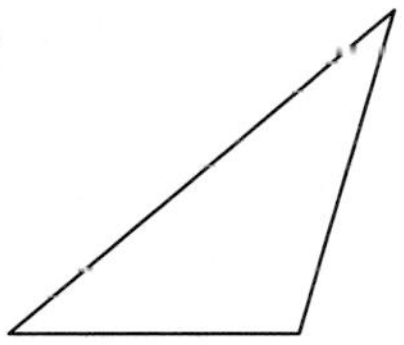
obtuse

Find the perimeter of the figure.

13.
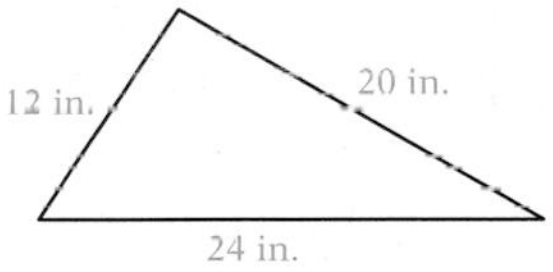

56 in.

14.
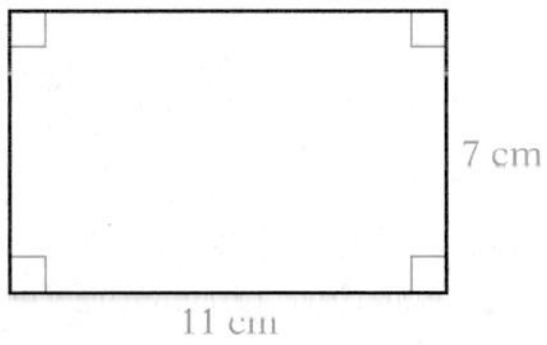

36 cm

15.
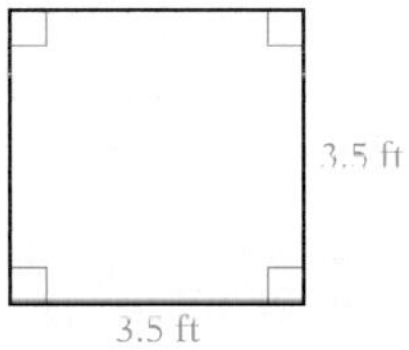

14 ft

16.
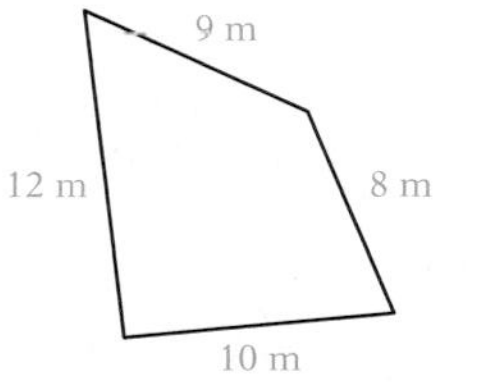

39 m

17.
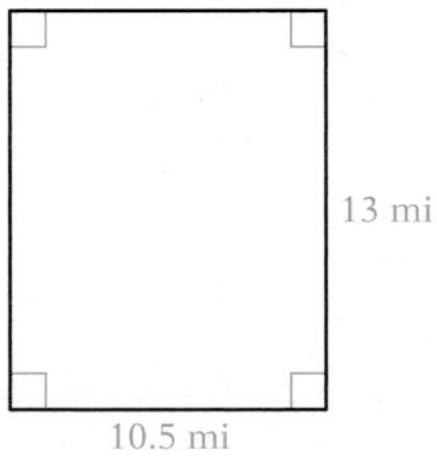

47 mi

18.
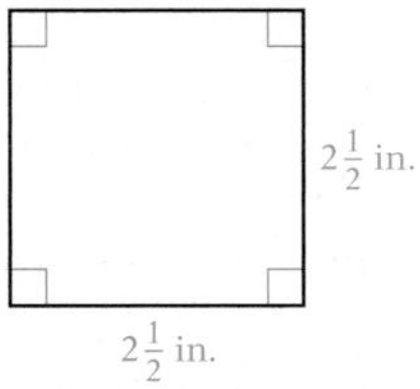

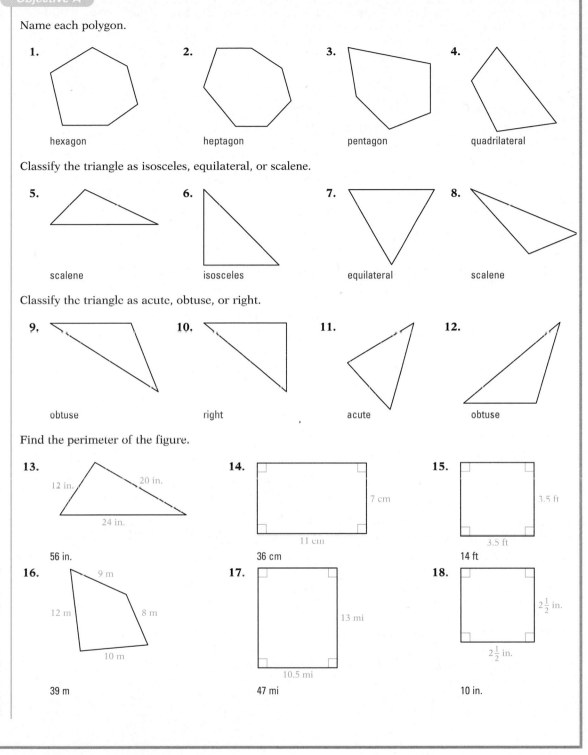

10 in.

Quick Quiz (Objective 3.2A)

1. Find the perimeter of a square in which each side is 13.5 cm long. 54 cm
2. Find the perimeter of a rectangle with a length of 3 m and a width of 0.75 m. 7.5 m
3. Find the circumference of a circle with a radius of 14 in. Round to the nearest hundredth. 87.96 in.

172 Chapter 3 / Geometry

Find the circumference of the figure. Give both the exact value and an approximation to the nearest hundredth.

19.

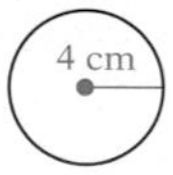

8π cm; 25.13 cm

20.

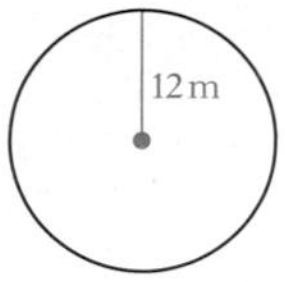

24π m; 75.40 m

21.

11π mi; 34.56 mi

22.

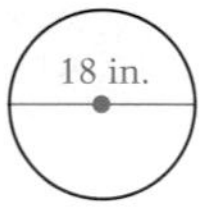

18π in.; 56.55 in.

23.

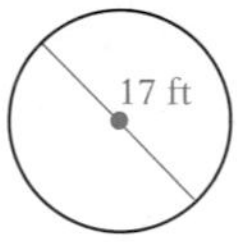

17π ft; 53.41 ft

24.

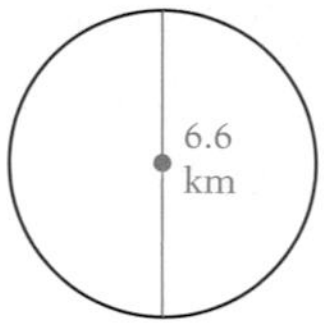

6.6π km; 20.73 km

Solve.

25. The lengths of the three sides of a triangle are 3.8 cm, 5.2 cm, and 8.4 cm. Find the perimeter of the triangle. 17.4 cm

26. The lengths of the three sides of a triangle are 7.5 m, 6.1 m, and 4.9 m. Find the perimeter of the triangle. 18.5 m

27. The length of each of two sides of an isosceles triangle is $2\frac{1}{2}$ cm. The third side measures 3 cm. Find the perimeter of the triangle. 8 cm

28. The length of each side of an equilateral triangle is $4\frac{1}{2}$ in. Find the perimeter of the triangle. $13\frac{1}{2}$ in.

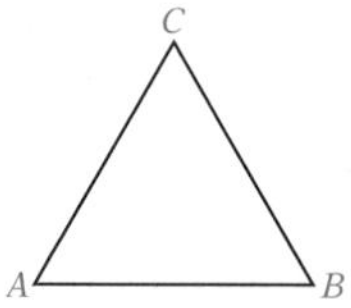

29. A rectangle has a length of 8.5 m and a width of 3.5 m. Find the perimeter of the rectangle. 24 m

30. Find the perimeter of a rectangle that has a length of $5\frac{1}{2}$ ft and a width of 4 ft. 19 ft

31. The length of each side of a square is 12.2 cm. Find the perimeter of the square. 48.8 cm

32. Find the perimeter of a square that measures 0.5 m on each side. 2 m

33. Find the perimeter of a regular pentagon that measures 3.5 in. on each side. 17.5 in.

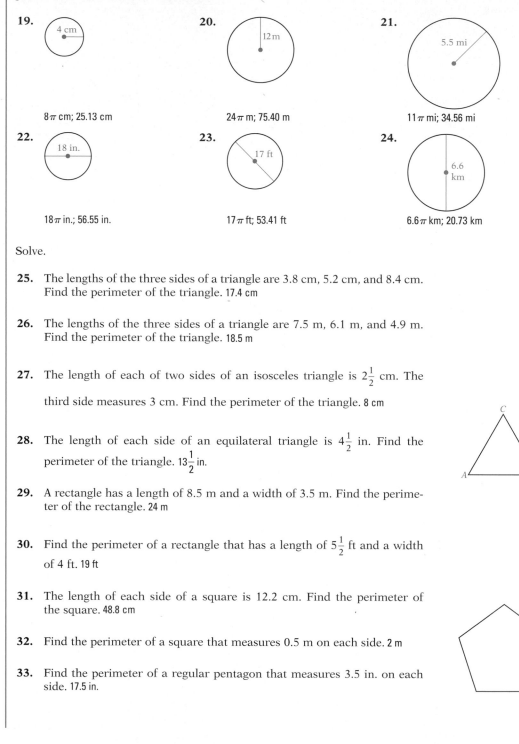

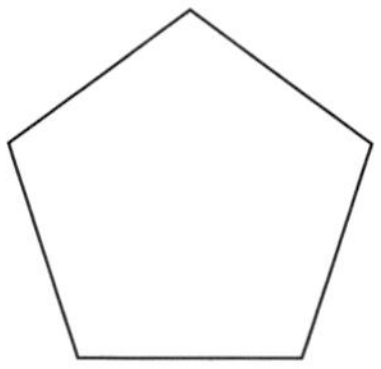

173

34. What is the perimeter of a regular hexagon that measures 8.5 cm on each side? 51 cm

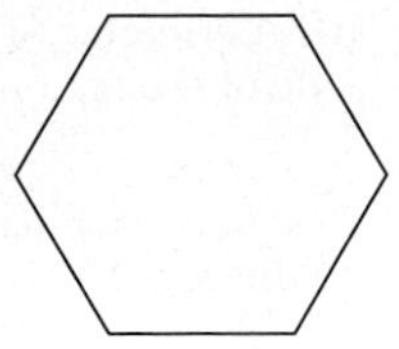

35. The radius of a circle is 4.2 cm. Find the length of a diameter of the circle. 8.4 cm

36. The diameter of a circle is 0.56 m. Find the length of a radius of the circle. 0.28 m

37. Find the circumference of a circle that has a diameter of 1.5 in. Give the exact value. 1.5π in.

38. The diameter of a circle is 4.2 ft. Find the circumference of the circle. Round to the nearest hundredth. 13.19 ft

39. The radius of a circle is 36 cm. Find the circumference of the circle. Round to the nearest hundredth. 226.19 cm

40. Find the circumference of a circle that has a radius of 2.5 m. Give the exact value. 5π m

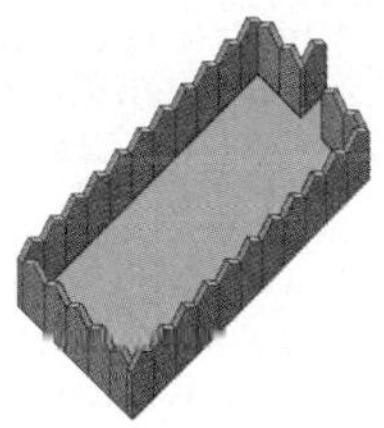

41. How many feet of fencing should be purchased for a rectangular garden that is 18 ft long and 12 ft wide? 60 ft

42. How many meters of binding are required to bind the edge of a rectangular quilt that measures 3.5 m by 8.5 m? 24 m

43. Wall-to-wall carpeting is installed in a room that is 12 ft long and 10 ft wide. The edges of the carpet are nailed to the floor. Along how many feet must the carpet be nailed down? 44 ft

44. The length of a rectangular park is 55 yd. The width is 47 yd. How many yards of fencing are needed to surround the park? 204 yd

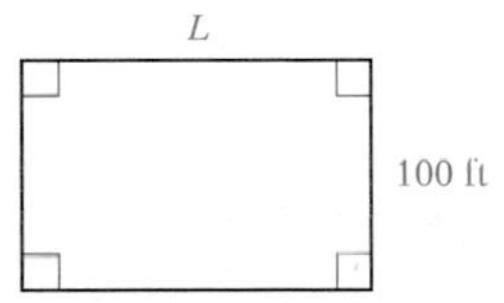

45. The perimeter of a rectangular playground is 440 ft. If the width is 100 ft, what is the length of the playground? 120 ft

46. A rectangular vegetable garden has a perimeter of 64 ft. The length of the garden is 20 ft. What is the width of the garden? 12 ft

47. Each of two sides of a triangular banner measures 18 in. If the perimeter of the banner is 46 in., what is the length of the third side of the banner? 10 in.

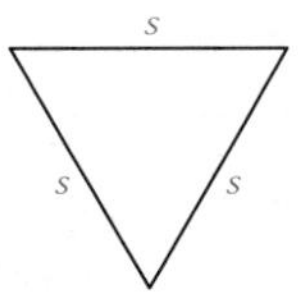

48. The perimeter of an equilateral triangle is 13.2 cm. What is the length of each side of the triangle? 4.4 cm

174 Chapter 3 / Geometry

49. The perimeter of a square picture frame is 48 in. Find the length of each side of the frame. 12 in.

50. A square rug has a perimeter of 32 ft. Find the length of each edge of the rug. 8 ft

51. The circumference of a circle is 8 cm. Find the length of a diameter of the circle. Round to the nearest hundredth. 2.55 cm

52. The circumference of a circle is 15 in. Find the length of a radius of the circle. Round to the nearest hundredth. 2.39 in.

53. Find the length of molding needed to put around a circular table that is 4.2 ft in diameter. Round to the nearest hundredth. 13.19 ft

54. How much binding is needed to bind the edge of a circular rug that is 3 m in diameter? Round to the nearest hundredth. 9.42 m

55. A bicycle tire has a diameter of 24 in. How many feet does the bicycle travel when the wheel makes eight revolutions? Round to the nearest hundredth. 50.27 ft

56. A tricycle tire has a diameter of 12 in. How many feet does the tricycle travel when the wheel makes twelve revolutions? Round to the nearest hundredth. 37.70 ft

57. The distance from the surface of Earth to its center is 6356 km. What is the circumference of Earth? Round to the nearest hundredth. 39,935.93 km

58. Bias binding is to be sewed around the edge of a rectangular tablecloth measuring 72 in. by 45 in. If the bias binding comes in packages containing 15 ft of binding, how many packages of bias binding are needed for the tablecloth? 2 packages

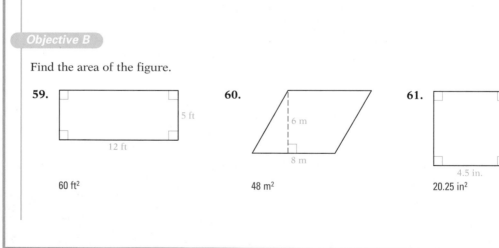

Objective B

Find the area of the figure.

59.

60 ft^2

60.

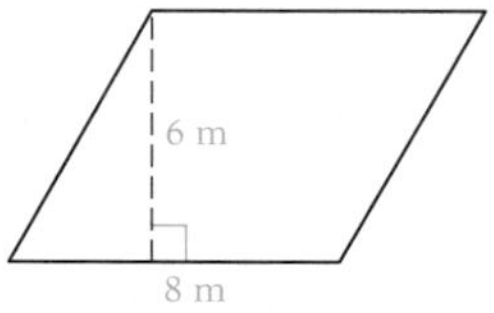

48 m^2

61.

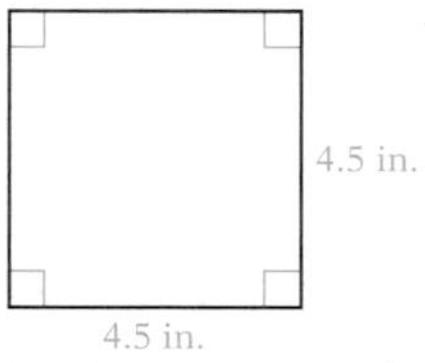

20.25 in^2

175

62.
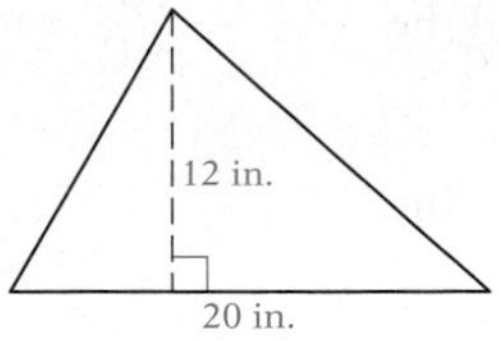

120 in²

63.
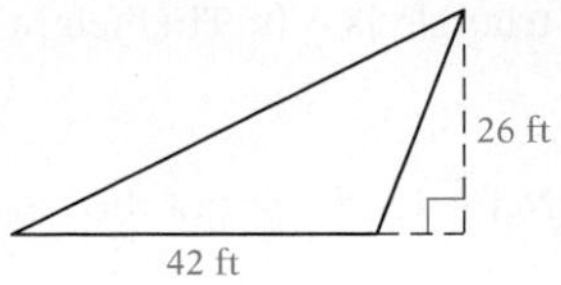

546 ft²

64.
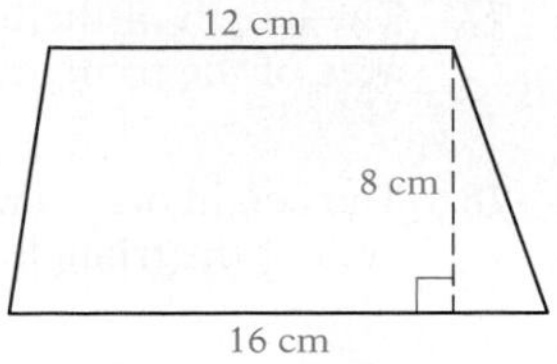

112 cm²

Find the area of the figure. Give both the exact value and an approximation to the nearest hundredth.

65.
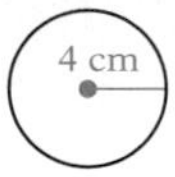

16π cm²; 50.27 cm²

66.
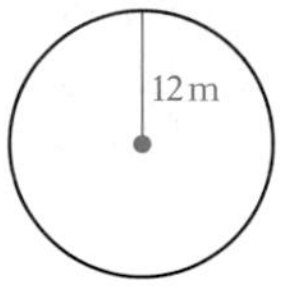

144π m²; 452.39 m²

67.

30.25π mi²; 95.03 mi²

68.
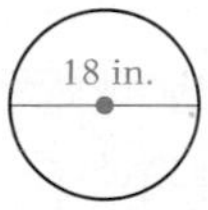

81π in²; 254.47 in²

69.
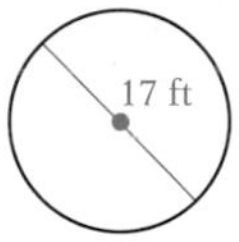

72.25π ft²; 226.98 ft²

70.
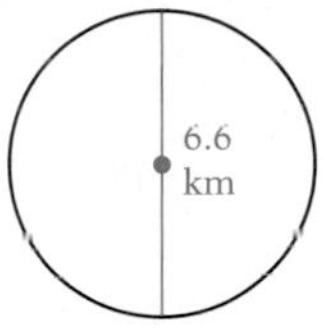

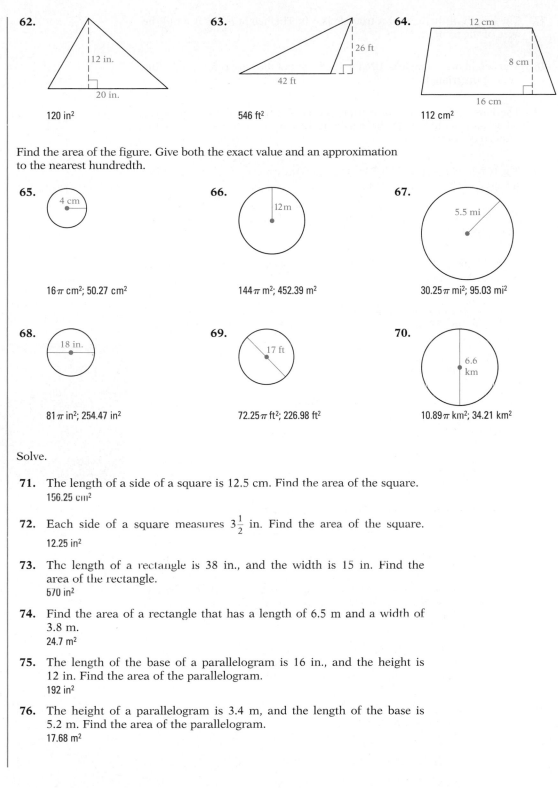

10.89π km²; 34.21 km²

Solve.

71. The length of a side of a square is 12.5 cm. Find the area of the square.
156.25 cm²

72. Each side of a square measures $3\frac{1}{2}$ in. Find the area of the square.
12.25 in²

73. The length of a rectangle is 38 in., and the width is 15 in. Find the area of the rectangle.
570 in²

74. Find the area of a rectangle that has a length of 6.5 m and a width of 3.8 m.
24.7 m²

75. The length of the base of a parallelogram is 16 in., and the height is 12 in. Find the area of the parallelogram.
192 in²

76. The height of a parallelogram is 3.4 m, and the length of the base is 5.2 m. Find the area of the parallelogram.
17.68 m²

Quick Quiz (Objective 3.2B)

1. Find the area of a triangle with a base of 10 ft and a height of 16 ft. 80 ft²
2. Find the area of a square with a side of 16 cm. 256 cm²
3. Find the area of a rectangle with a length of 64 cm and a width of 22 cm. 1408 cm²
4. Find the area of a circle with a diameter of 26 in. Round to the nearest hundredth. 530.93 in²

176 Chapter 3 / Geometry

77. The length of the base of a triangle is 6 ft. The height is 4.5 ft. Find the area of the triangle. 13.5 ft^2

78. The height of a triangle is 4.2 cm. The length of the base is 5 cm. Find the area of the triangle. 10.5 cm^2

79. The length of one base of a trapezoid is 35 cm, and the length of the other base is 20 cm. If the height is 12 cm, what is the area of the trapezoid? 330 cm^2

80. The height of a trapezoid is 5 in. The bases measure 16 in. and 18 in. Find the area of the trapezoid. 85 in^2

81. The radius of a circle is 5 in. Find the area of the circle. Give the exact value. 25π in^2

82. Find the area of a circle with a radius of 14 m. Round to the nearest hundredth. 615.75 m^2

83. Find the area of a circle that has a diameter of 3.4 ft. Round to the nearest hundredth. 9.08 ft^2

84. The diameter of a circle is 6.5 m. Find the area of the circle. Give the exact value. 10.5625π m^2

85. The dome of the Hale telescope at Mount Palomar, California, has a diameter of 200 in. Find the area across the dome. Give the exact value. $10{,}000\pi$ in^2

86. An irrigation system waters a circular field that has a 50-foot radius. Find the area watered by the irrigation system. Give the exact value. 2500π ft^2

87. Find the area of a rectangular flower garden that measures 14 ft by 9 ft. 126 ft^2

88. What is the area of a square patio that measures 8.5 m on each side? 72.25 m^2

89. Artificial turf is being used to cover a playing field. If the field is rectangular with a length of 100 yd and a width of 75 yd, how much artificial turf must be purchased to cover the field? 7500 yd^2

90. A fabric wall hanging is to fill a space that measures 5 m by 3.5 m. Allowing for 0.1 m of the fabric to be folded back along each edge, how much fabric must be purchased for the wall hanging? 19.24 m^2

91. The area of a rectangle is 300 in^2. If the length of the rectangle is 30 in., what is the width? 10 in.

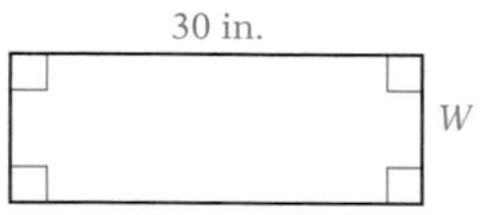

177

92. The width of a rectangle is 12 ft. If the area is 312 ft^2, what is the length of the rectangle? 26 ft

93. The height of a triangle is 5 m. The area of the triangle is 50 m^2. Find the length of the base of the triangle. 20 m

94. The area of a parallelogram is 42 m^2. If the height of the parallelogram is 7 m, what is the length of the base? 6 m

95. You plan to stain the wooden deck attached to your house. The deck measures 10 ft by 8 ft. If a quart of stain will cover 50 ft^2, how many quarts of stain should you buy? 2 qt

96. You want to tile your kitchen floor. The floor measures 12 ft by 9 ft. How many tiles, each a square with side $1\frac{1}{2}$ ft, should you purchase for the job? 48 tiles

97. You are wallpapering two walls of a child's room, one measuring 9 ft by 8 ft and the other measuring 11 ft by 8 ft. The wallpaper costs $24.50 per roll, and each roll of the wallpaper will cover 40 ft^2. What will it cost to wallpaper the two walls? $98

98. An urban renewal project involves reseeding a park that is in the shape of a square 60 ft on each side. Each bag of grass seed costs $9.75 and will seed 1200 ft^2. How much money should be budgeted for buying grass seed for the park? $29.25

99. A circle has a radius of 8 in. Find the increase in area when the radius is increased by 2 in. Round to the nearest hundredth. 113.10 in^2

100. A circle has a radius of 6 cm. Find the increase in area when the radius is doubled. Round to the nearest hundredth. 339.29 cm^2

101. You want to install wall-to-wall carpeting in your living room, which measures 15 ft by 24 ft. If the cost of the carpet you would like to purchase is $21.95 per square yard, what will be the cost of the carpeting for your living room? (*Hint:* 9 ft^2 = 1 yd^2) $878

102. You want to paint the walls of your bedroom. Two walls measure 15 ft by 9 ft, and the other two walls measure 12 ft by 9 ft. The paint you wish to purchase costs $12.98 per gallon, and each gallon will cover 400 ft^2 of wall. Find the total amount you will spend on paint. $25.96

103. A walkway 2 m wide surrounds a rectangular plot of grass. The plot is 30 m long and 20 m wide. What is the area of the walkway? 216 m^2

Answers to Writing Exercises

111. Students need to incorporate the following equations in their explanations: $b = \frac{1}{2}C$; $h = r$; $A = bh = \frac{1}{2}C \cdot r = \frac{1}{2}(\pi d)r = \frac{1}{2}\pi(2r) = \frac{1}{2}\pi r^2$.

112. Draw line segments from the center of the polygon to each of the vertices. The number of triangles formed is equal to the number of sides of the polygon. The area of the polygon is equal to the sum of the areas of the triangles. The area of each triangle is equal to one-half the product of the length of a side and the apothem. Multiply the area of one triangle by the number of triangles to find the area of the polygon. This can also be stated as "one-half the product of the apothem and the perimeter of the polygon."

104. Pleated draperies for a window must be twice as wide as the width of the window. Draperies are being made for four windows, each 2 ft wide and 4 ft high. Because the drapes will fall slightly below the window sill, and because extra fabric will be needed for hemming the drapes, 1 ft must be added to the height of the window. How much material must be purchased to make the drapes? 80 ft^2

105. Find the cost of plastering the walls of a room 22 ft long, 25 ft 6 in. wide, and 8 ft high. Subtract 120 ft^2 for windows and doors. The cost is \$2.50 per square foot. \$1600

APPLYING THE CONCEPTS

106. If both the length and the width of a rectangle are doubled, how many times larger is the area of the resulting rectangle? 4 times

107. A **hexagram** is a six-pointed star, formed by extending each of the sides of a regular hexagon into an equilateral triangle. A hexagram is shown at the right. Use a pencil, paper, protractor, and ruler to create a hexagram.

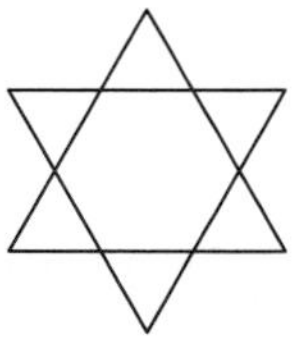

108. If the formula $C = \pi d$ is solved for π, the resulting equation is $\pi = \frac{C}{d}$. Therefore, π is the ratio of the circumference of a circle to the length of its diameter. Use several circular objects, such as coins, plates, tin cans, and wheels, to show that the ratio of the circumference of each object to its diameter is approximately 3.14.

109. Derive a formula for the area of a circle in terms of the diameter of the circle. $A = \frac{\pi d^2}{4}$

110. Determine whether the statement is always true, sometimes true, or never true.

a. If two triangles have the same perimeter, then they have the same area. sometimes true

b. If two rectangles have the same area, then they have the same perimeter. sometimes true

c. If two squares have the same area, then the sides of the squares have the same length. always true

d. An equilateral triangle is also an isosceles triangle. always true

e. All the radii (plural of radius) of a circle are equal. always true

f. All the diameters of a circle are equal. always true

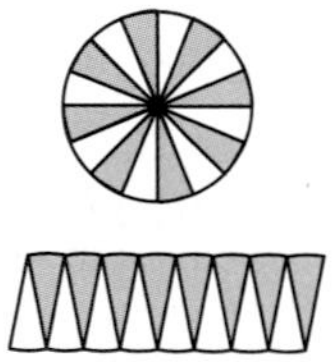

111. Suppose a circle is cut into 16 equal pieces, which are then arranged as shown at the right. The figure formed resembles a parallelogram. What variable expression could describe the base of the parallelogram? What variable could describe its height? Explain how the formula for the area of a circle is derived from this approach.

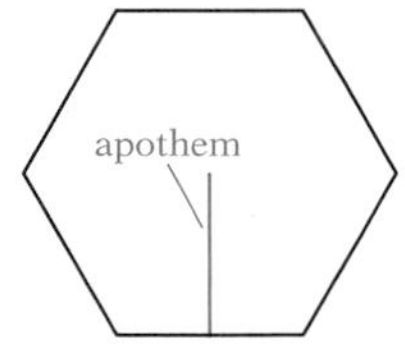

112. The **apothem** of a regular polygon is the distance from the center of the polygon to a side. Explain how to derive a formula for the area of a regular polygon using the apothem. (*Hint:* Use the formula for the area of a triangle.)

3.3 Solids

Objective A **To solve problems involving the volume of a solid**

VIDEO & DVD CD TUTOR WEB SSM

Geometric solids are figures in space. Five common geometric solids are the rectangular solid, the sphere, the cylinder, the cone, and the pyramid.

A **rectangular solid** is one in which all six sides, called **faces**, are rectangles. The variable L is used to represent the length of a rectangular solid, W its width, and H its height.

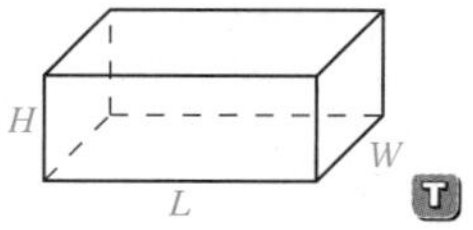

A **sphere** is a solid in which all points are the same distance from point O, which is called the **center** of the sphere. The **diameter**, d, of a sphere is a line across the sphere going through point O. The **radius**, r, is a line from the center to a point on the sphere. AB is a diameter and OC is a radius of the sphere shown at the right.

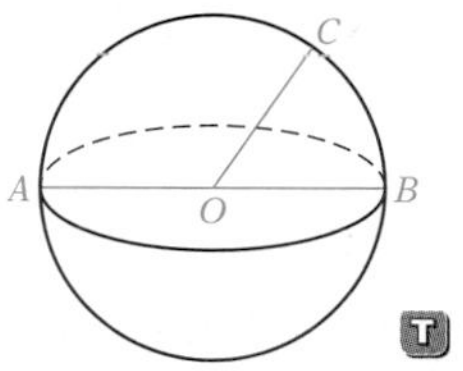

$$d = 2r \text{ or } r = \frac{1}{2}d$$

The most common cylinder, called a **right circular cylinder**, is one in which the bases are circles and are perpendicular to the height of the cylinder. The variable r is used to represent the radius of a base of a cylinder, and h represents the height. In this text, only right circular cylinders are discussed.

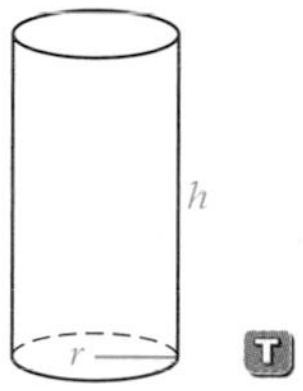

A **right circular cone** is obtained when one base of a right circular cylinder is shrunk to a point, called a **vertex**, V. The variable r is used to represent the radius of the base of the cone, and h represents the height. The variable ℓ is used to represent the **slant height**, which is the distance from a point on the circumference of the base to the vertex. In this text, only right circular cones are discussed.

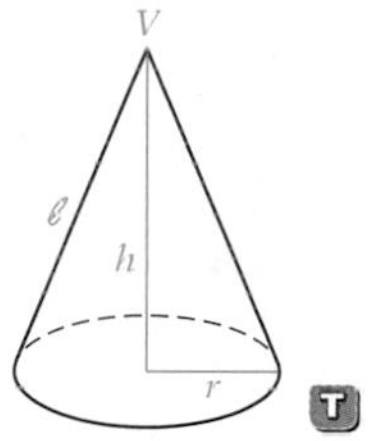

The base of a **regular pyramid** is a regular polygon, and the sides are isosceles triangles. The height, h, is the distance from the vertex, V, to the base and is perpendicular to the base. The variable ℓ is used to represent the **slant height**, which is the height of one of the isosceles triangles on the face of the pyramid. The regular square pyramid at the right has a square base. This is the only type of pyramid discussed in this text.

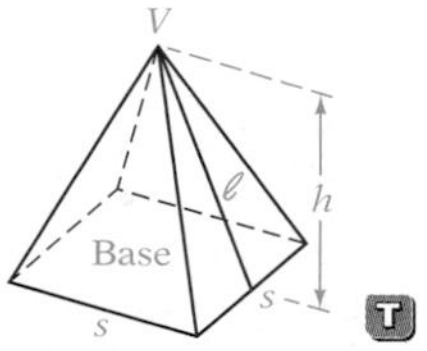

Objective 3.3A

New Vocabulary

geometric solids
rectangular solid
faces
sphere
center
diameter
radius
right circular cylinder
right circular cone
vertex
slant height
regular pyramid
cube
volume

New Symbols

in^3 (cubic inches)
cm^3 (cubic centimeters)
ft^3 (cubic feet)
m^3 (cubic meters)

New Formulas

Volume of a rectangular solid: $V = LWH$

Volume of a cube: $V = s^3$

Volume of a sphere: $V = \frac{4}{3}\pi r^3$

Volume of a right circular cylinder: $V = \pi r^2 h$

Volume of a right circular cone: $V = \frac{1}{3}\pi r^2 h$

Volume of a regular square pyramid: $V = \frac{1}{3}s^2 h$

Discuss the Concepts

The difficulty students have distinguishing linear measure from square measure is compounded with volume measure. Ask students to give examples of things that would be measured in, for instance, feet, square feet, and cubic feet—for example, the length of a room, the area of the floor, and the volume of air in the room. Here are some more examples.

a. The distance across a lake, the area of the surface of the lake, and the volume of water in the lake

b. The length of a driveway, the area of the driveway that needs to be plowed, and the volume of asphalt used to pave the driveway

In-Class Examples (Objective 3.3A)

Solve. Round to the nearest hundredth.

1. Find the volume of a rectangular solid with a length of 6 m, a width of 400 cm, and a height of 4.5 m. $108\ m^3$
2. Find the volume of a cube with a side of 5 ft 3 in. $144.70\ ft^3$
3. Find the volume of a sphere with a radius of 4 mm. $268.08\ mm^3$
4. Find the volume of a cylinder with a radius of 15 cm and a height of 14 cm. $9896.02\ cm^3$

Concept Check

1. Indicate which of the following are rectangular solids: a juice box, a milk carton, a can of soup, a compact disk, the plastic container a compact disk is packaged in. a juice box and the plastic container a compact disk is packaged in
2. Indicate which of the following units could not be used to measure the volume of a cylinder: ft^3, m^3, yd^2, cm^3, mi. yd^2, mi

A **cube** is a special type of rectangular solid. Each of the six faces of a cube is a square. The variable s is used to represent the length of one side of a cube.

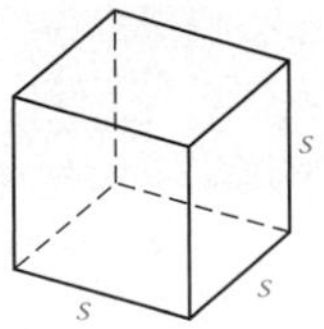

Volume is a measure of the amount of space inside a figure in space. Volume can be used to describe the amount of heating gas used for cooking, the amount of concrete delivered for the foundation of a house, or the amount of water in storage for a city's water supply.

A cube that is 1 ft on each side has a volume of 1 cubic foot, which is written 1 ft^3. A cube that measures 1 cm on each side has a volume of 1 cubic centimeter, which is written 1 cm^3.

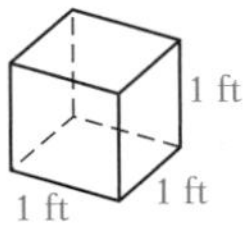

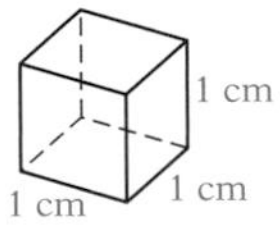

The volume of a solid is the number of cubes that are necessary to exactly fill the solid. The volume of the rectangular solid at the right is 24 cm^3 because it will hold exactly 24 cubes, each 1 cm on a side. Note that the volume can be found by multiplying the length times the width times the height.

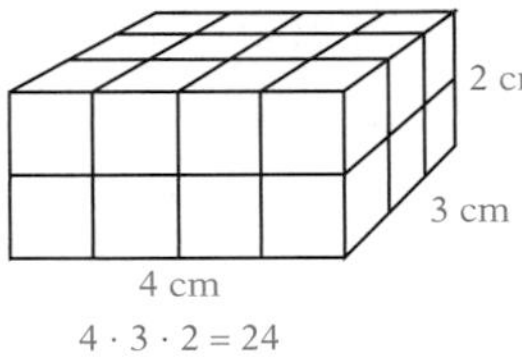

The formulas for the volumes of the geometric solids described above are given below.

Volumes of Geometric Solids

The volume, V, of a **rectangular solid** with length L, width W, and height H is given by $V = LWH$.

The volume, V, of a **cube** with side s is given by $V = s^3$.

The volume, V, of a **sphere** with radius r is given by $V = \frac{4}{3}\pi r^3$.

The volume, V, of a **right circular cylinder** is given by $V = \pi r^2 h$, where r is the radius of the base and h is the height.

The volume, V, of a **right circular cone** is given by $V = \frac{1}{3}\pi r^2 h$, where r is the radius of the circular base and h is the height.

The volume, V, of a **regular square pyramid** is given by $V = \frac{1}{3}s^2 h$, where s is the length of a side of the base and h is the height.

181

➡ Find the volume of a sphere with a diameter of 6 in.

$r = \frac{1}{2}d = \frac{1}{2}(6) = 3$ • **First find the radius of the sphere.**

$V = \frac{4}{3}\pi r^3$ • **Use the formula for the volume of a sphere.**

$V = \frac{4}{3}\pi(3)^3$

$V = \frac{4}{3}\pi(27)$

$V = 36\pi$ • **The exact volume of the sphere is 36π in^3.**

$V \approx 113.10$ • **An approximate measure can be found by using the π key on a calculator.**

The approximate volume is 113.10 in^3.

CALCULATOR NOTE

To approximate 36π on your calculator, enter 36 [×] [π] [=].

Example 1
The length of a rectangular solid is 5 m, the width is 3.2 m, and the height is 4 m. Find the volume of the solid.

Strategy
To find the volume, use the formula for the volume of a rectangular solid. $L = 5$, $W = 3.2$, $H = 4$.

Solution
$V = LWH = 5(3.2)(4) = 64$

The volume of the rectangular solid is 64 m^3.

You Try It 1
Find the volume of a cube that measures 2.5 m on a side.

Your strategy

Your solution
15.625 m^3

Example 2
The radius of the base of a cone is 8 cm. The height is 12 cm. Find the volume of the cone. Round to the nearest hundredth.

Strategy
To find the volume, use the formula for the volume of a cone. An approximation is asked for; use the π key on a calculator. $r = 8$, $h = 12$.

Solution

$V = \frac{1}{3}\pi r^2 h$

$V = \frac{1}{3}\pi(8)^2(12) = \frac{1}{3}\pi(64)(12) = 256\pi$

≈ 804.25

The volume is approximately 804.25 cm^3.

You Try It 2
The diameter of the base of a cylinder is 8 ft. The height of the cylinder is 22 ft. Find the exact volume of the cylinder.

Your strategy

Your solution
352π ft^3

Solutions on p. S9

Optional Student Activity

1. A foot is what fraction of a yard? $\frac{1}{3}$
2. A square foot is what fraction of a square yard? $\frac{1}{9}$
3. A cubic foot is what fraction of a cubic yard? $\frac{1}{27}$
4. An inch is what fraction of a foot? $\frac{1}{12}$
5. A square inch is what fraction of a square foot? $\frac{1}{144}$
6. A cubic inch is what fraction of a cubic foot? $\frac{1}{1728}$
7. A centimeter is what fraction of a meter? $\frac{1}{100}$
8. A square centimeter is what fraction of a square meter? $\frac{1}{10{,}000}$
9. A cubic centimeter is what fraction of a cubic meter? $\frac{1}{1{,}000{,}000}$

Objective 3.3B

New Vocabulary
surface area

New Formulas
Surface area of a rectangular solid:
$SA = 2LW + 2LH + 2WH$
Surface area of a cube:
$SA = 6s^2$
Surface area of a sphere:
$SA = 4\pi r^2$
Surface area of a right circular cylinder: $SA = 2\pi r^2 + 2\pi rh$
Surface area of a right circular cone: $SA = \pi r^2 + \pi rl$
Surface area of a regular pyramid: $SA = s^2 + 2sl$

Discuss the Concepts
1. Explain the difference between volume and surface area.
2. In what units is surface area measured? Why?

Concept Check
1. How much larger is the surface area of a sphere with a radius of 9 cm than that of a cube with side length 8 cm? Round to the nearest hundredth. 633.88 cm^2
2. How much larger is the surface area of a cone with a radius of 3 in. and a slant height of 4 in. than that of a pyramid in which the length of a side of the base is 3 in. and the slant height is 4 in.? Round to the nearest hundredth. 32.97 in^2

Objective B

To solve problems involving the surface area of a solid

VIDEO & DVD CD TUTOR WEB SSM

The **surface area** of a solid is the total area on the surface of the solid.

When a rectangular solid is cut open and flattened out, each face is a rectangle. The surface area, *SA*, of the rectangular solid is the sum of the areas of the six rectangles:

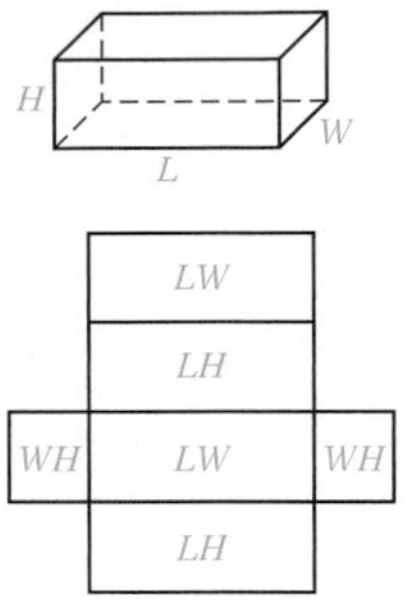

$$SA = LW + LH + WH + LW + WH + LH$$

which simplifies to

$$\mathbf{SA = 2LW + 2LH + 2WH}$$

The surface area of a cube is the sum of the areas of the six faces of the cube. The area of each face is s^2. Therefore, the surface area, *SA*, of a cube is given by the formula $\mathbf{SA = 6s^2}$.

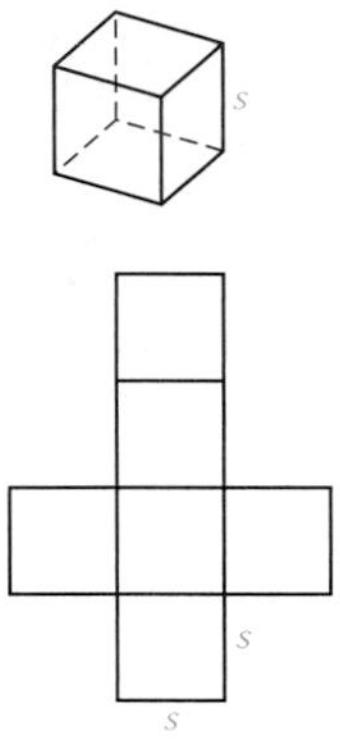

When a cylinder is cut open and flattened out, the top and bottom of the cylinder are circles. The side of the cylinder flattens out to a rectangle. The length of the rectangle is the circumference of the base, which is $2\pi r$; the width is h, the height of the cylinder. Therefore, the area of the rectangle is $2\pi rh$. The surface area, *SA*, of the cylinder is

$$SA = \pi r^2 + 2\pi rh + \pi r^2$$

which simplifies to

$$\mathbf{SA = 2\pi r^2 + 2\pi rh}$$

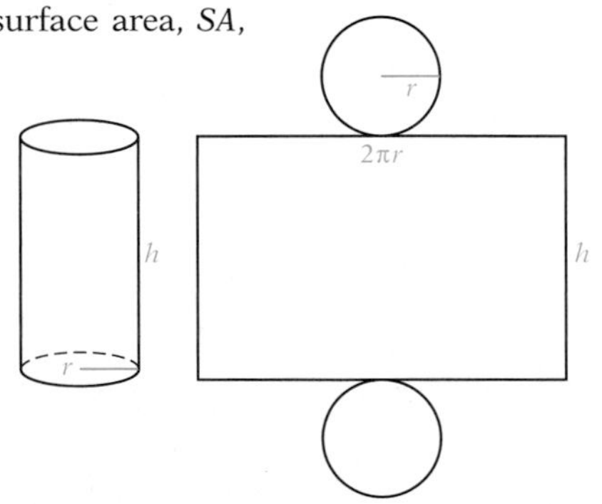

In-Class Examples (Objective 3.3B)
Solve. Round to the nearest hundredth.
1. Find the surface area of a cube that measures 2.5 in. on a side. 37.5 in^2
2. Find the surface area of a sphere that has a diameter of 8 cm. 201.06 cm^2
3. Find the surface area of a cone with a slant height of 3 in. The radius of the base is 2 in. 31.42 in^2

183

The surface area of a pyramid is the area of the base plus the area of the four isosceles triangles. A side of the square base is s; therefore, the area of the base is s^2. The slant height, ℓ, is the height of each triangle, and s is the base of each triangle. The surface area, SA, of a pyramid is

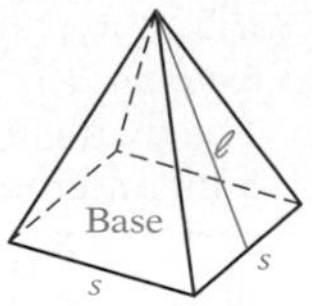

$$SA = s^2 + 4\left(\frac{1}{2}s\ell\right)$$

which simplifies to

$$\boldsymbol{SA = s^2 + 2s\ell}$$

Formulas for the surface areas of geometric solids are given below.

Surface Areas of Geometric Solids

The surface area, SA, of a **rectangular solid** with length L, width W, and height H is given by $SA = 2LW + 2LH + 2WH$.

The surface area, SA, of a **cube** with side s is given by $SA = 6s^2$.

The surface area, SA, of a **sphere** with radius r is given by $SA = 4\pi r^2$.

The surface area, SA, of a **right circular cylinder** is given by $SA = 2\pi r^2 + 2\pi rh$, where r is the radius of the base and h is the height.

The surface area, SA, of a **right circular cone** is given by $SA = \pi r^2 + \pi r\ell$, where r is the radius of the circular base and ℓ is the slant height.

The surface area, SA, of a **regular pyramid** is given by $SA = s^2 + 2s\ell$, where s is the length of a side of the base and ℓ is the slant height.

➡ Find the surface area of a sphere with a diameter of 18 cm.

$r = \frac{1}{2}d = \frac{1}{2}(18) = 9$ • **First find the radius of the sphere.**

$SA = 4\pi r^2$ • **Use the formula for the surface area of a sphere.**

$SA = 4\pi(9)^2$

$SA = 4\pi(81)$

$SA \approx 324\pi$ • **The exact surface area of the sphere is 324π cm².**

$SA \approx 1017.88$ • **An approximate measure can be found by using the π key on a calculator.**

The approximate surface area is 1017.88 cm².

Optional Student Activity

Find the surface area of the solid. Round to the nearest hundredth.

1.

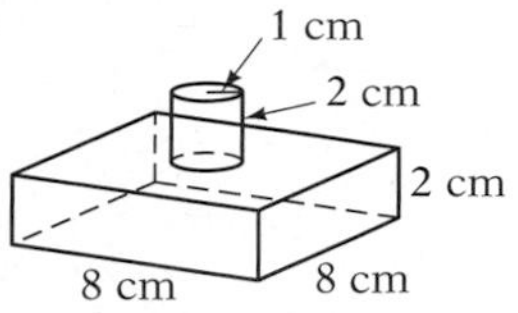

204.57 cm²

2.

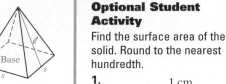

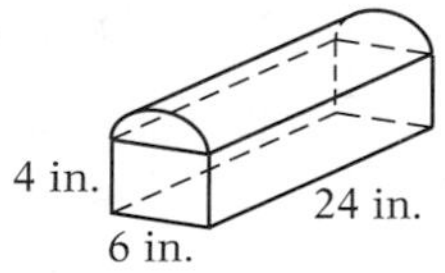

638.47 in²

Example 3
The diameter of the base of a cone is 5 m, and the slant height is 4 m. Find the surface area of the cone. Give the exact measure.

Strategy
To find the surface area of the cone:
- Find the radius of the base of the cone.
- Use the formula for the surface area of a cone. Leave the answer in terms of π.

Solution

$r = \frac{1}{2}d = \frac{1}{2}(5) = 2.5$

$SA = \pi r^2 + \pi r\ell$
$SA = \pi(2.5)^2 + \pi(2.5)(4)$
$SA = \pi(6.25) + \pi(2.5)(4)$
$SA = 6.25\pi + 10\pi$
$SA = 16.25\pi$

The surface area of the cone is 16.25π m^2.

You Try It 3
The diameter of the base of a cylinder is 6 ft, and the height is 8 ft. Find the surface area of the cylinder. Round to the nearest hundredth.

Your strategy

Your solution
207.35 ft^2

Example 4
Find the area of a label used to cover a soup can that has a radius of 4 cm and a height of 12 cm. Round to the nearest hundredth.

Strategy
To find the area of the label, use the fact that the surface area of the sides of a cylinder is given by $2\pi rh$. An approximation is asked for; use the π key on a calculator. $r = 4$, $h = 12$.

Solution
Area of the label $= 2\pi rh$
Area of the label $= 2\pi(4)(12) = 96\pi$
≈ 301.59

The area is approximately 301.59 cm^2.

You Try It 4
Which has a larger surface area, a cube with a side measuring 10 cm or a sphere with a diameter measuring 8 cm?

Your strategy

Your solution
the cube

Solutions on pp. S9–S10

185

Section 3.3

Suggested Assignment
Exercises 1–51, odds
More challenging problems:
Exercises 53, 54

3.3 Exercises

Objective A

Find the volume of the figure. For calculations involving π, give both the exact value and an approximation to the nearest hundredth.

1.
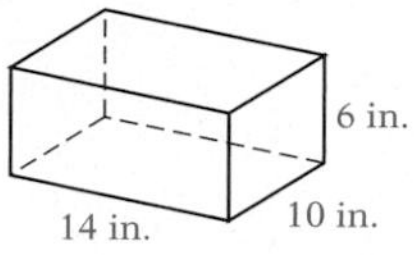

840 in^3

2.
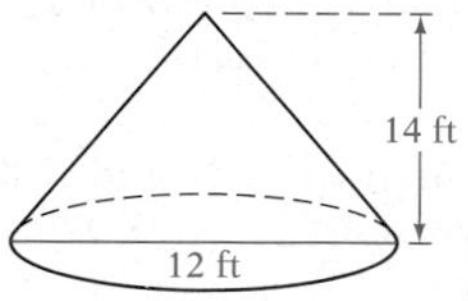

168π ft^3; 527.79 ft^3

3.
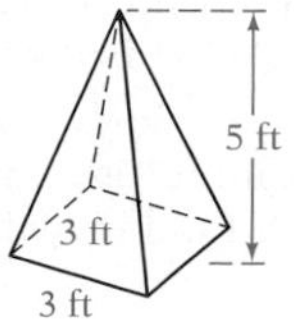

15 ft^3

4.
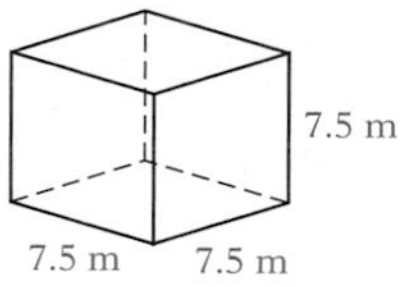

421.875 m^3

5.
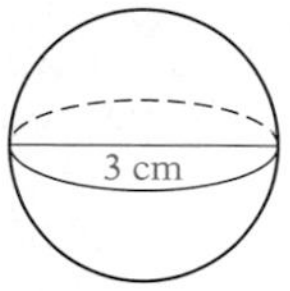

4.5π cm^3; 14.14 cm^3

6.
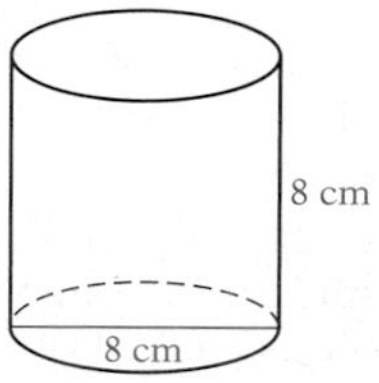

128π cm^3; 402.12 cm^3

Solve.

7. A rectangular solid has a length of 6.8 m, a width of 2.5 m, and a height of 2 m. Find the volume of the solid. 34 m^3

8. Find the volume of a rectangular solid that has a length of 4.5 ft, a width of 3 ft, and a height of 1.5 ft. 20.25 ft^3

9. Find the volume of a cube whose side measures 3.5 in. 42.875 in^3

10. The length of a side of a cube is 7 cm. Find the volume of the cube. 343 cm^3

11. The diameter of a sphere is 6 ft. Find the volume of the sphere. Give the exact measure. 36π ft^3

12. Find the volume of a sphere that has a radius of 1.2 m. Round to the nearest tenth. 7.2 m^3

13. The diameter of the base of a cylinder is 24 cm. The height of the cylinder is 18 cm. Find the volume of the cylinder. Round to the nearest hundredth. 8143.01 cm^3

14. The height of a cylinder is 7.2 m. The radius of the base is 4 m. Find the volume of the cylinder. Give the exact measure. 115.2π m^3

15. The radius of the base of a cone is 5 in. The height of the cone is 9 in. Find the volume of the cone. Give the exact measure. 75π in^3

Quick Quiz (Objective 3.3A)

Solve. Round to the nearest hundredth.

1. Find the volume of a rectangular solid with a length of 3 m, a width of 90 cm, and a height of 5 m. 13.5 m^3
2. Find the volume of a cube with a side of 8 ft. 512 ft^3
3. Find the volume of a sphere with a 5-foot diameter. 65.45 ft^3
4. Find the volume of a cylinder with a radius of 30 cm and a height of 42 cm. 118,752.20 cm^3

16. The height of a cone is 15 cm. The diameter of the cone is 10 cm. Find the volume of the cone. Round to the nearest hundredth. 392.70 cm^3

17. The length of a side of the base of a pyramid is 6 in., and the height is 10 in. Find the volume of the pyramid. 120 in^3

18. The height of a pyramid is 8 m, and the length of a side of the base is 9 m. What is the volume of the pyramid? 216 m^3

19. The volume of a freezer with a length of 7 ft and a height of 3 ft is 52.5 ft^3. Find the width of the freezer. 2.5 ft

20. The length of an aquarium is 18 in., and the width is 12 in. If the volume of the aquarium is 1836 in^3, what is the height of the aquarium? 8.5 in.

21. The volume of a cylinder with a height of 10 in. is 502.4 in^3. Find the radius of the base of the cylinder. Round to the nearest hundredth. 4.00 in.

22. The diameter of the base of a cylinder is 14 cm. If the volume of the cylinder is 2310 cm^3, find the height of the cylinder. Round to the nearest hundredth. 15.01 cm

23. A rectangular solid has a square base and a height of 5 in. If the volume of the solid is 125 in^3, find the length and the width. length: 5 in.; width: 5 in.

24. The volume of a rectangular solid is 864 m^3. The rectangular solid has a square base and a height of 6 m. Find the dimensions of the solid. length: 12 m; width: 12 m

25. An oil storage tank, which is in the shape of a cylinder, is 4 m high and has a diameter of 6 m. The oil tank is two-thirds full. Find the number of cubic meters of oil in the tank. Round to the nearest hundredth. 75.40 m^3

26. A silo, which is in the shape of a cylinder, is 16 ft in diameter and has a height of 30 ft. The silo is three-fourths full. Find the volume of the portion of the silo that is not being used for storage. Round to the nearest hundredth. 1507.96 ft^3

Objective B

Find the surface area of the figure.

27.
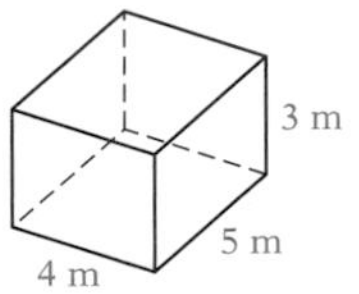

94 m^2

28.
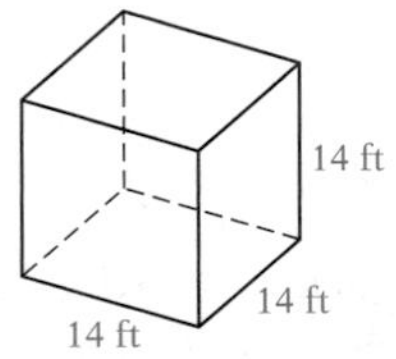

1176 ft^2

29.
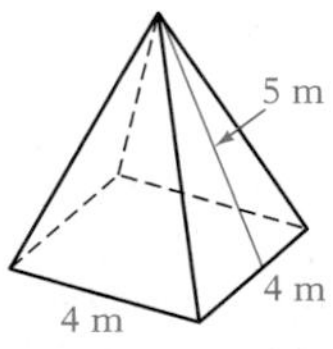

56 m^2

Quick Quiz (Objective 3.3B)

Solve. Round to the nearest hundredth.

1. Find the surface area of a rectangular solid that has a height of 4 ft, a length of 6 ft, and a width of 2 ft. 88 ft^2
2. Find the surface area of a sphere that has a diameter of 3 in. 28.27 in^2
3. Find the surface area of a cylinder with a height of 5 m. The radius of the base is 2 m. 87.96 m^2

187

Find the surface area of the figure. Give both the exact value and an approximation to the nearest hundredth.

30.

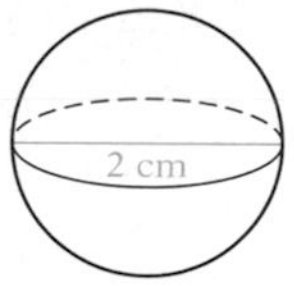

4π cm²; 12.57 cm²

31.

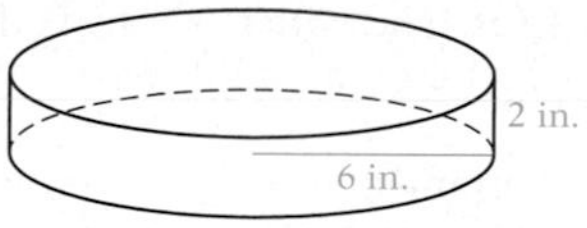

96π in²; 301.59 in²

32.

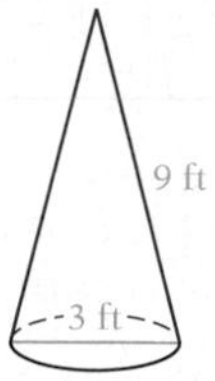

15.75π ft²; 49.48 ft²

Solve.

33. The height of a rectangular solid is 5 ft. The length is 8 ft, and the width is 4 ft. Find the surface area of the solid. 184 ft²

34. The width of a rectangular solid is 32 cm. The length is 60 cm, and the height is 14 cm. What is the surface area of the solid? 6416 cm²

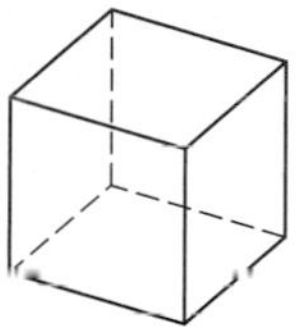

35. The side of a cube measures 3.4 m. Find the surface area of the cube. 69.36 m²

36. Find the surface area of a cube that has a side measuring 1.5 in. 13.5 in²

37. Find the surface area of a sphere with a diameter of 15 cm. Give the exact value. 225π cm²

38. The radius of a sphere is 2 in. Find the surface area of the sphere. Round to the nearest hundredth. 50.27 in²

39. The radius of the base of a cylinder is 4 in. The height of the cylinder is 12 in. Find the surface area of the cylinder. Round to the nearest hundredth. 402.12 in²

40. The diameter of the base of a cylinder is 1.8 m. The height of the cylinder is 0.7 m. Find the surface area of the cylinder. Give the exact value. 2.88π m²

41. The slant height of a cone is 2.5 ft. The radius of the base is 1.5 ft. Find the surface area of the cone. Give the exact value. 6π ft²

42. The diameter of the base of a cone is 21 in. The slant height is 16 in. What is the surface area of the cone? Round to the nearest hundredth. 874.15 in²

43. The length of a side of the base of a pyramid is 9 in., and the slant height is 12 in. Find the surface area of the pyramid. 297 in²

44. The slant height of a pyramid is 18 m, and the length of a side of the base is 16 m. What is the surface area of the pyramid? 832 m²

Answers to Writing Exercises

55. a. For example, cut perpendicular to the top and bottom faces and parallel to two of the sides.
b. For example, beginning at an edge that is perpendicular to the bottom face, cut at an angle through to the bottom face.
c. For example, beginning at the top face, at a distance *d* from a vertex, cut at an angle to the bottom face, ending at a distance greater than *d* from the opposite vertex.
d. For example, beginning on the top face, at a distance *d* from a vertex, cut across the cube to a point just below the opposite vertex, intersecting the bottom face.

45. The surface area of a rectangular solid is 108 cm^2. The height of the solid is 4 cm, and the length is 6 cm. Find the width of the rectangular solid. 3 cm

46. The length of a rectangular solid is 12 ft. The width is 3 ft. If the surface area is 162 ft^2, find the height of the rectangular solid. 3 ft

47. A can of paint will cover 300 ft^2. How many cans of paint should be purchased in order to paint a cylinder that has a height of 30 ft and a radius of 12 ft? 11 cans

48. A hot air balloon is in the shape of a sphere. Approximately how much fabric was used to construct the balloon if its diameter is 32 ft? Round to the nearest whole number. 3217 ft^2

49. How much glass is needed to make a fish tank that is 12 in. long, 8 in. wide, and 9 in. high? The fish tank is open at the top. 456 in^2

50. Find the area of a label used to cover a can of juice that has a diameter of 16.5 cm and a height of 17 cm. Round to the nearest hundredth. 881.22 cm^2

51. The length of a side of the base of a pyramid is 5 cm, and the slant height is 8 cm. How much larger is the surface area of this pyramid than the surface area of a cone with a diameter of 5 cm and a slant height of 8 cm? Round to the nearest hundredth. 22.53 cm^2

APPLYING THE CONCEPTS

52. Half of a sphere is called a **hemisphere**. Derive formulas for the volume and surface area of a hemisphere. See the *Solutions Manual.*

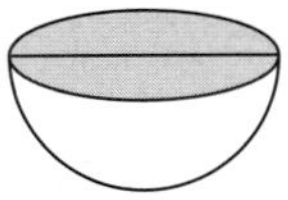

53. Determine whether the statement is always true, sometimes true, or never true.
a. The slant height of a regular pyramid is longer than the height. always true
b. The slant height of a cone is shorter than the height. never true
c. The four triangular faces of a regular pyramid are equilateral triangles. sometimes true

54. **a.** What is the effect on the surface area of a rectangular solid when the width and height are doubled? doubled plus 4*WH*
b. What is the effect on the volume of a rectangular solid when both the length and the width are doubled? quadrupled
c. What is the effect on the volume of a cube when the length of each side of the cube is doubled? 8 times as large
d. What is the effect on the surface area of a cylinder when the radius and height are doubled? 4 times as large

55. Explain how you could cut through a cube so that the face of the resulting solid is
a. a square
b. an equilateral triangle
c. a trapezoid
d. a hexagon

189

Focus on Problem Solving

Trial and Error Some problems in mathematics are solved by using **trial and error**. The trial-and-error method of arriving at a solution to a problem involves repeated tests or experiments until a satisfactory conclusion is reached.

Many of the Applying the Concepts exercises in this text require a trial and error method of solution. For example, an exercise in Section 3 of this chapter reads:

Explain how you could cut through a cube so that the face of the resulting solid is **(a)** a square, **(b)** an equilateral triangle, **(c)** a trapezoid, **(d)** a hexagon.

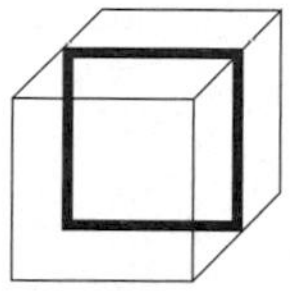

There is no formula to apply to this problem; there is no computation to perform. This problem requires picturing a cube and the results after cutting through it at different places on its surface and at different angles. For part (a), cutting perpendicular to the top and bottom of the cube and parallel to two of its sides will result in a square. The other shapes may prove more difficult.

When solving problems of this type, keep an open mind. Sometimes when using the trial-and-error method, we are hampered by narrowness of vision; we cannot expand our thinking to include other possibilities. Then when we see someone else's solution, it appears so obvious to us! For example, for the Applying the Concepts question above, it is necessary to conceive of cutting through the cube at places other than the top surface; we need to be open to the idea of beginning the cut at one of the corner points of the cube.

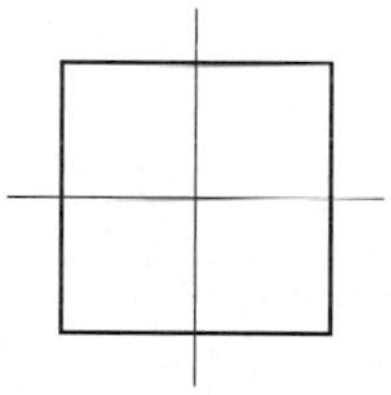

A topic of the Projects and Group Activities in this chapter is symmetry. Here again, trial and error is used to determine the lines of symmetry inherent in an object. For example, in determining lines of symmetry for a square, begin by drawing a square. The horizontal line of symmetry and the vertical line of symmetry may be immediately obvious to you.

But there are two others. Do you see that a line drawn through opposite corners of the square is also a line of symmetry?

Many of the questions in this text that require an answer of "always true, sometimes true, or never true" are best solved by the trial-and-error method. For example, consider the statement presented in Section 2 of this chapter.

If two rectangles have the same area, then they have the same perimeter.

Answers to Focus on Problem Solving: Trial and Error

No, it is not possible for two different pairs of factors of a number to have the same sum.

Try some numbers. Each of two rectangles, one measuring 6 units by 2 units and another measuring 4 units by 3 units, has an area of 12 square units, but the perimeter of the first is 16 units and the perimeter of the second is 14 units. So the answer "always true" has been eliminated. We still need to determine whether there is a case when it is true. After experimenting with a lot of numbers, you may come to realize that we are trying to determine if it is possible for two different pairs of factors of a number to have the same sum. Is it?

Don't be afraid to make many experiments, and remember that *errors*, or tests that "don't work," are a part of the trial-and-*error* process.

Projects and Group Activities

Investigating Perimeter

The perimeter of the square at the right is 4 units.

If two squares are joined along one of the sides, the perimeter is 6 units. Note that it does not matter which sides are joined; the perimeter is still 6 units.

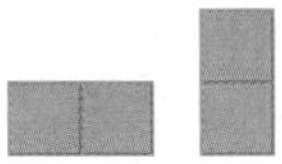

If three squares are joined, the perimeter of the resulting figure is 8 units for each possible placement of the squares.

Four squares can be joined in five different ways as shown. There are two possible perimeters, 10 units for A, B, C, and D, and 8 units for E.

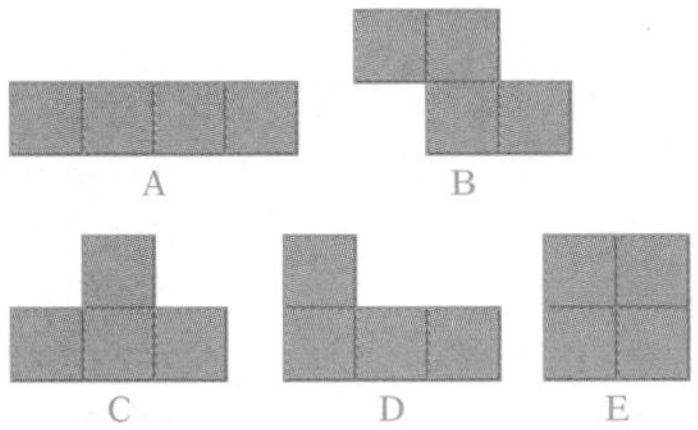

1. If five squares are joined, what is the maximum perimeter possible?

2. If five squares are joined, what is the minimum perimeter possible?

3. If six squares are joined, what is the maximum perimeter possible?

4. If six squares are joined, what is the minimum perimeter possible?

Answers to Projects and Group Activities: Investigating Perimeter

1. 12 units
2. 10 units
3. 14 units
4. 10 units

Symmetry Look at the letter A printed at the left. If the letter were folded along line ℓ, the two sides of the letter would match exactly. This letter has **symmetry** with respect to line ℓ. Line ℓ is called the **axis of symmetry**.

Now consider the letter H printed below at the left. Both lines ℓ_1 and ℓ_2 are axes of symmetry for this letter; the letter could be folded along either line and the two sides would match exactly.

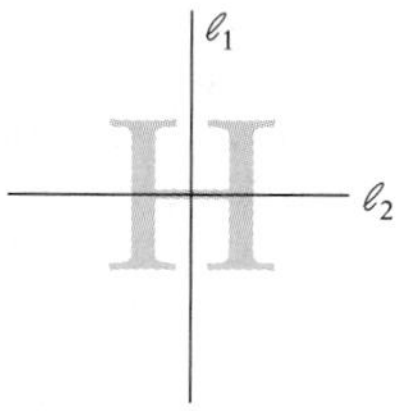

1. Does the letter A have more than one axis of symmetry?
2. Find axes of symmetry for other capital letters of the alphabet.
3. Which lowercase letters have one axis of symmetry?
4. Do any of the lowercase letters have more than one axis of symmetry?
5. Find the number of axes of symmetry for each of the plane geometric figures presented in this chapter.
6. There are other types of symmetry. Look up the meaning of point symmetry and rotational symmetry. Which plane geometric figures provide examples of these types of symmetry?
7. Find examples of symmetry in nature, art, and architecture.

Chapter Summary

Key Words A *line* is determined by two distinct points and extends indefinitely in both directions. A *line segment* is part of a line that has two endpoints. *Parallel lines* never meet; the distance between them is always the same. *Perpendicular lines* are intersecting lines that form right angles. [pp. 145–147]

A *ray* starts at a point and extends indefinitely in one direction. The point at which a ray starts is the *endpoint* of the ray. An *angle* is formed by two rays with the same endpoint. The *vertex* of an angle is the point at which the two rays meet. An angle is measured in *degrees*. A 90° angle is a *right angle*. A 180° angle is a *straight angle*. An *acute angle* is an angle whose measure is between 0° and 90°. An *obtuse angle* is an angle whose measure is between 90° and 180°. *Complementary angles* are two angles whose measures have the sum 90°. *Supplementary angles* are two angles whose measures have the sum 180°. [pp. 145–148]

Two angles that are on opposite sides of the intersection of two lines are *vertical angles;* vertical angles have the same measure. Two angles that share a common side are *adjacent angles;* adjacent angles of intersecting lines are supplementary angles. [p. 150]

A line that intersects two other lines at two different points is a *transversal.* If the lines cut by a transversal are parallel lines, equal angles are formed: *alternate interior angles, alternate exterior angles,* and *corresponding angles*. [p. 151]

A *polygon* is a closed figure determined by three or more line segments. The line segments that form the polygon are its *sides*. A *regular polygon* is one in which each side has the same length and each angle has the same measure. Polygons are classified by the number of sides. [p. 161]

Answers to Projects and Group Activities: Symmetry

1. No
2. Besides A and H, the capital letters B, C, D, E, I, M, O, T, V, W, X, and Y have axes of symmetry. Note that the letters I, O, and X have more than one axis of symmetry. (For some letters, it depends on how the letter is written; an example is the letter U.)
3. The lowercase letters c, i, l, t, v, and w have one axis of symmetry. (For some letters, it depends on how the letter is written; an example is the letter t.)
4. The lowercase letters o and x have more than one axis of symmetry.
5. An isosceles triangle has one axis of symmetry. An equilateral triangle has three axes of symmetry. A rectangle has two axes of symmetry. A square has four axes of symmetry. A circle has an infinite number of axes of symmetry. An isosceles trapezoid has one axis of symmetry. Other trapezoids have no axis of symmetry.
6. If a figure is unchanged after being rotated 180° about a point *O*, then the figure has point symmetry. A rectangle, a square, and a circle have point symmetry.
   If a figure is unchanged after being rotated more than 0° and less than 360°, then the figure has rotational symmetry. An equilateral triangle, a rectangle, a square, a circle, and a parallelogram have rotational symmetry. It may be helpful to note that a regular pentagon has rotational symmetry but not point symmetry.
7. Examples of symmetry in nature include people and animals, snowflakes, starfish, and many types of flowers. Examples of symmetry in art and architecture can be found in history texts and in art and architecture magazines.

A *triangle* is a plane figure formed by three line segments. An *isosceles triangle* has two sides of equal length. The three sides of an *equilateral triangle* are of equal length. A *scalene triangle* has no two sides of equal length. An *acute triangle* has three acute angles. An *obtuse triangle* has one obtuse angle. A *right triangle* has a right angle. [pp. 161–162]

A *quadrilateral* is a four-sided polygon. A parallelogram, a rectangle, a square, a rhombus, and a trapezoid are all quadrilaterals. [p. 162]

A *circle* is a plane figure in which all points are the same distance from the center of the circle. A *diameter* of a circle is a line segment across the circle through the center. A *radius* of a circle is a line segment from the center of the circle to a point on the circle. [p. 164]

The *perimeter* of a plane geometric figure is a measure of the distance around the figure. The distance around a circle is called the *circumference. Area* is the amount of surface in a region. *Volume* is a measure of the amount of space inside a figure in space. The *surface area* of a solid is the total area on the surface of the solid. [pp. 162, 164, 166, 179, 182]

Essential Rules

Triangles [p. 153]

Sum of the measures of the interior angles = 180°

Sum of an interior and corresponding exterior angle = 180°

Perimeter [pp. 162–164]	Triangle:	$P = a + b + c$
	Rectangle:	$P = 2L + 2W$
	Square:	$P = 4s$
	Circle:	$C = \pi d$ or $C = 2\pi r$
Area [pp. 166–169]	Triangle:	$A = \frac{1}{2}bh$
	Rectangle:	$A = LW$
	Square:	$A = s^2$
	Circle:	$A = \pi r^2$
	Parallelogram:	$A = bh$
	Trapezoid:	$A = \frac{1}{2}h(b_1 + b_2)$
Volume [p. 180]	Rectangular solid:	$V = LWH$
	Cube:	$V = s^3$
	Sphere:	$V = \frac{4}{3}\pi r^3$
	Right circular cylinder:	$V = \pi r^2 h$
	Right circular cone:	$V = \frac{1}{3}\pi r^2 h$
	Regular pyramid:	$V = \frac{1}{3}s^2 h$
Surface Area [p. 183]	Rectangular solid:	$SA = 2LW + 2LH + 2WH$
	Cube:	$SA = 6s^2$
	Sphere:	$SA = 4\pi r^2$
	Right circular cylinder:	$SA = 2\pi r^2 + 2\pi rh$
	Right circular cone:	$SA = \pi r^2 + \pi r\ell$
	Regular pyramid:	$SA = s^2 + 2s\ell$

193

Chapter Review

1. Given that $\angle a = 74°$ and $\angle b = 52°$, find the measures of angles x and y.

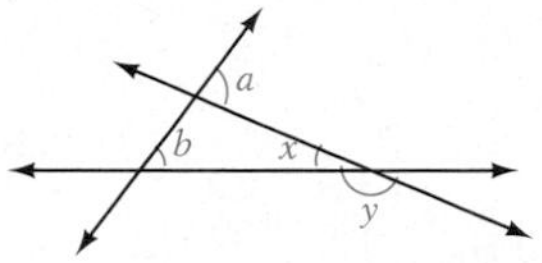

$\angle x = 22°$; $\angle y = 158°$ [3.1C]

2. Find the measure of $\angle x$.

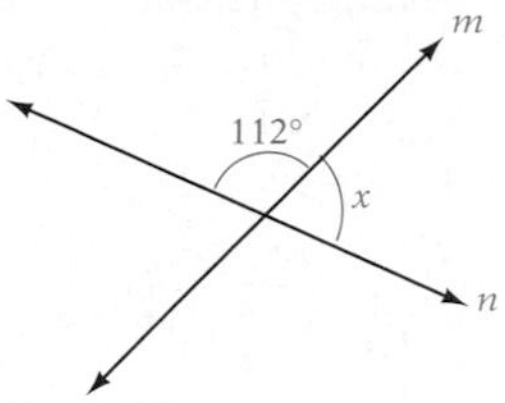

68° [3.1B]

3. Given that $BC = 11$ cm and AB is three times the length of BC, find the length of AC.

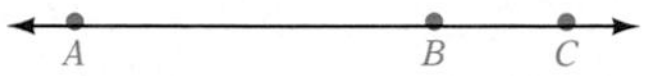

44 cm [3.1A]

4. Find x.

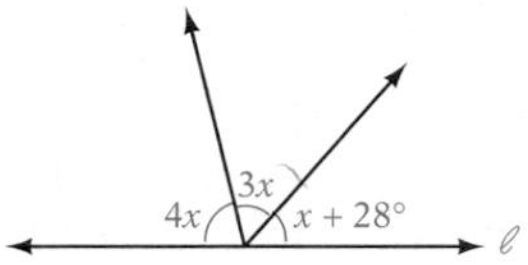

19° [3.1A]

5. Find the volume of the figure.

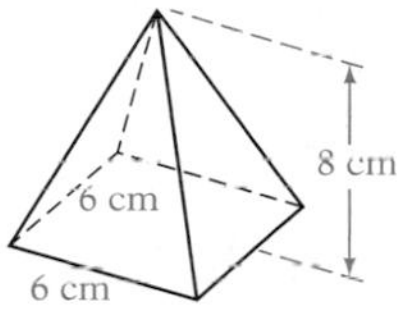

96 cm³ [3.3A]

6. Given that $\ell_1 \parallel \ell_2$, find the measures of angles a and b.

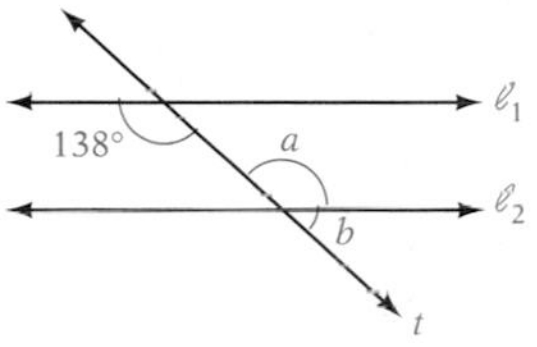

$\angle a = 138°$; $\angle b = 42°$ [3.1B]

7. Find the surface area of the figure.

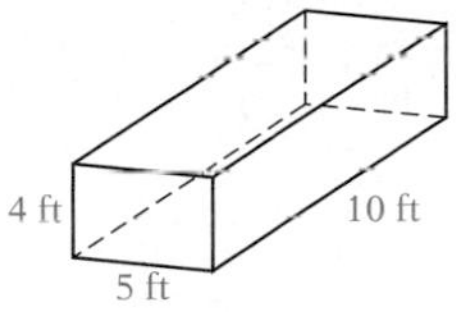

220 ft² [3.3B]

8. Find the supplement of a 32° angle.
148° [3.1A]

9. Determine the area of a rectangle with a length of 12 cm and a width of 6.5 cm.
78 cm² [3.2B]

10. Determine the area of a triangle whose base is 9 m and whose height is 14 m.
63 m² [3.2B]

11. Find the volume of a rectangular solid with a length of 6.5 ft, a width of 2 ft, and a height of 3 ft.
39 ft^3 [3.3A]

12. Two angles of a triangle measure 37° and 48°. Find the measure of the third angle.
95° [3.1C]

13. The height of a triangle is 7 cm. The area of the triangle is 28 cm^2. Find the length of the base of the triangle.
8 cm [3.2B]

14. Find the volume of a sphere that has a diameter of 12 mm. Find the exact value.
288π mm^3 [3.3A]

15. Determine the exact volume of a right circular cone whose radius is 7 cm and whose height is 16 cm.
$\frac{784\pi}{3}$ cm^3 [3.3A]

16. The perimeter of a square picture frame is 86 cm. Find the length of each side of the frame.
21.5 cm [3.2A]

17. A can of paint will cover 200 ft^2. How many cans of paint should be purchased in order to paint a cylinder that has a height of 15 ft and a radius of 6 ft?
4 cans [3.3B]

18. The length of a rectangular park is 56 yd. The width is 48 yd. How many yards of fencing are needed to surround the park?
208 yd [3.2A]

19. What is the area of a square patio that measures 9.5 m on each side?
90.25 m^2 [3.2B]

20. A walkway 2 m wide surrounds a rectangular plot of grass. The plot is 40 m long and 25 m wide. What is the area of the walkway?
276 m^2 [3.2B]

195

Chapter Test

1. The diameter of a sphere is 1.5 m. Find the radius of the sphere.
0.75 m [3.2A]

2. Find the circumference of a circle with a radius of 5 cm. Round to the nearest hundredth.
31.42 cm [3.2A]

3. Find the perimeter of the rectangle in the figure below.

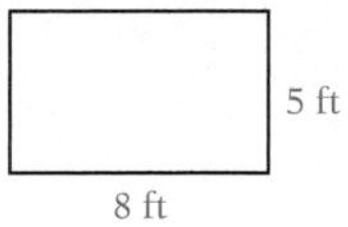

26 ft [3.2A]

4. Given $AB = 15$, $CD = 6$, and $AD = 24$, find the length of BC.

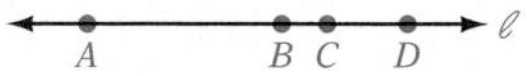

3 [3.1A]

5. Find the volume of a sphere with a diameter of 8 ft. Round to the nearest tenth.
268.08 ft^3 [3.3A]

6. Find the area of the circle shown below. Round to the nearest hundredth.

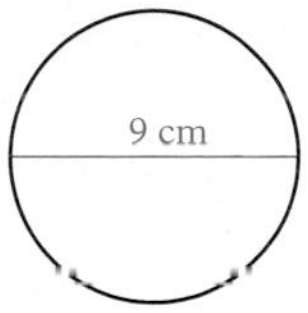

63.62 cm^2 [3.2B]

7. Given that $\ell_1 \parallel \ell_2$, find the measures of angles a and b.

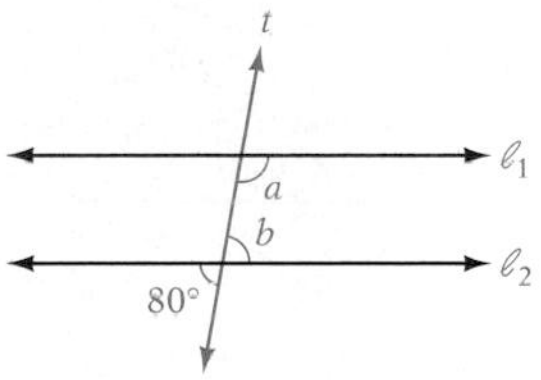

$a = 100°$, $b = 80°$ [3.1B]

8. Find the supplement of a 105° angle.
75° [3.1A]

9. Given that $\ell_1 \parallel \ell_2$, find the measures of angles a and b.

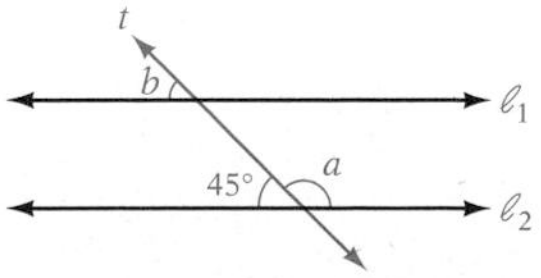

$a = 135°$, $b = 45°$ [3.1B]

10. Find the area of the rectangle shown below.

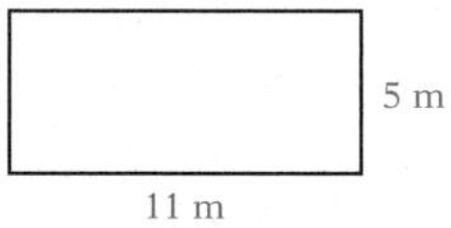

55 m^2 [3.2B]

11. Find the volume of a cylinder with a height of 6 m and a radius of 3 m. Round to the nearest hundredth.
169.65 m^3 [3.3A]

12. Find the perimeter of a rectangle that has a length of 2 m and a width of 1.4 m.
6.8 m [3.2A]

13. Find the complement of a 32° angle.
58° [3.1A]

14. Find the surface area of the figure. Round to the nearest hundredth.

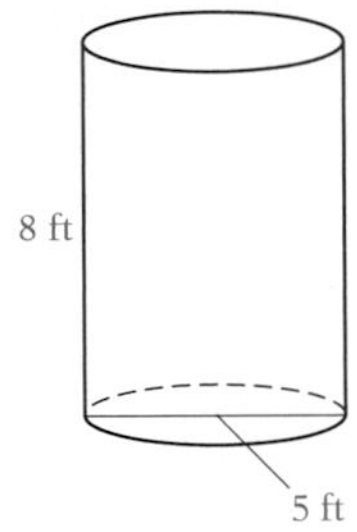

164.93 ft^2 [3.3B]

15. How much more pizza is contained in a pizza with radius 10 in. than in one with radius 8 in.? Round to the nearest hundredth.
113.10 in^2 [3.2B]

16. A right triangle has a 32° angle. Find the measures of the other two angles.
58° and 90° [3.1C]

17. A bicycle tire has a diameter of 28 in. How many feet does the bicycle travel if the wheel makes 10 revolutions? Round to the nearest tenth.
73.3 ft [3.2A]

18. New carpet is installed in a room measuring 18 ft by 14 ft. Find the area of the room in square yards. (9 ft^2 = 1 yd^2)
28 yd^2 [3.2B]

19. A silo, which is in the shape of a cylinder, is 9 ft in diameter and has a height of 18 ft. Find the volume of the silo. Round to the nearest hundredth.
1145.11 ft^3 [3.3A]

20. Find the area of a right triangle with a base of 8 m and a height of 2.75 m.
11 m^2 [3.2B]

197

Cumulative Review

1. Let $x \in \{-3, 0, 1\}$. For what values of x is the inequality $x \leq 1$ a true statement?
$-3, 0, 1$ [1.1A]

2. Write 8.9% as a decimal.
0.089 [1.2B]

3. Write $\frac{7}{20}$ as a percent.
35% [1.2B]

4. Divide: $-\frac{4}{9} \div \frac{2}{3}$
$-\frac{2}{3}$ [1.2D]

5. Multiply: $5.7\,(-4.3)$
-24.51 [1.2D]

6. Simplify: $-\sqrt{125}$
$-5\sqrt{5}$ [1.2F]

7. Evaluate $5 - 3[10 + (5 - 6)^2]$.
-28 [1.3A]

8. Evaluate $a(b - c)^3$ when $a = -1$, $b = -2$, and $c = -4$.
-8 [1.4A]

9. Simplify: $5m + 3n - 8m$
$-3m + 3n$ [1.4B]

10. Simplify: $-7(-3y)$
$21y$ [1.4C]

11. Simplify: $4(3x + 2) - (5x - 1)$
$7x + 9$ [1.4D]

12. Use the roster method to write the set of negative integers greater than or equal to -2.
$\{-2, -1\}$ [1.5A]

13. Find $C \cup D$, given $C = \{0, 10, 20, 30\}$ and $D = \{-10, 0, 10\}$.
$C \cup D = \{-10, 0, 10, 20, 30\}$ [1.5A]

14. Graph: $x \leq 1$

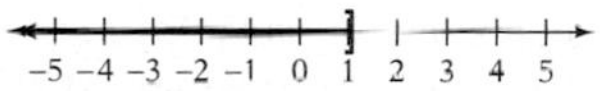

[1.5C]

15. Solve: $4x + 2 = 6x - 8$
5 [2.2B]

16. Solve: $3(2x + 5) = 18$
$\frac{1}{2}$ [2.2C]

17. Solve: $4y - 3 \geq 6y + 5$
$\{y \mid y \leq -4\}$ [2.4A]

18. Solve: $8 - 4(3x + 5) \leq 6(x - 8)$
$\{x \mid x \geq 2\}$ [2.4A]

19. Solve: $2x - 3 > 5$ or $x + 4 < 1$
$\{x \mid x < -3 \text{ or } x > 4\}$ [2.4B]

20. Solve: $-3 \le 2x - 7 \le 5$
$\{x \mid 2 \le x \le 6\}$ [2.4B]

21. Solve: $|3x - 1| = 2$
$1, -\frac{1}{3}$ [2.5A]

22. Solve: $|x - 8| \le 2$
$\{x \mid 6 \le x \le 10\}$ [2.5B]

23. Find the measure of $\angle x$.
131° [3.1B]

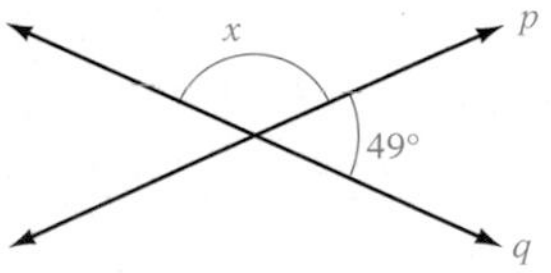

24. Translate "the difference between four times a number and ten is two" into an equation and solve.
$4x - 10 = 2; x = 3$ [2.2D]

25. Two angles of a triangle measure 37° and 21°. Find the measure of the third angle of the triangle.
122° [3.1C]

26. An engineering consultant invested $14,000 in a 5.5% annual simple interest account. How much should the consultant deposit in an account that pays 9.5% annual simple interest if the total interest earned on both accounts is to be $1245?
$5000 [2.3C]

27. Two sides of an isosceles triangle measure 7.5 m. The perimeter of the triangle is 19.5 m. Find the measure of the third side of the triangle.
4.5 m [3.2A]

28. According to the Census Bureau, the median annual earnings of a man with a bachelor's degree is $49,982, and the median earnings of a woman with a bachelor's degree is $35,408. What percent of the men's median annual earnings is the women's median annual earnings? Round to the nearest tenth of a percent. (Median is a type of average.)
70.8% [1.2G]

29. Find the exact area of a circle that has a diameter of 9 cm.
$20.25\pi \text{ cm}^2$ [3.2B]

30. The volume of a box is 144 ft^3. The length of the box is 12 ft, and the width is 4 ft. Find the height of the box.
3 ft [3.3A]

Chapter 4

Linear Equations and Inequalities in Two Variables

During portions of a flight, a passenger jet, like the one shown here, is in uniform motion, meaning that its speed and direction do not change. Uniform motion can be depicted in a graph to illustrate how fast or slow an object is moving and how much distance it has covered. By plotting the time and distance of two objects in uniform motion, it is clear which one is moving faster and has traveled the greater distance. The **Project on page 265** discusses objects in uniform motion and how to analyze speed and distance.

Objectives

Section 4.1

A To graph points in a rectangular coordinate system

B To determine ordered-pair solutions of an equation in two variables

C To graph a scatter diagram

Section 4.2

A To evaluate a function

Section 4.3

A To graph an equation of the form $y = mx + b$

B To graph an equation of the form $Ax + By = C$

C To solve application problems

Section 4.4

A To find the slope of a straight line

B To graph a line using the slope and the y-intercept

Section 4.5

A To find the equation of a line given a point and the slope

B To find the equation of a line given two points

C To solve application problems

Section 4.6

A To find equations of parallel and perpendicular lines

Section 4.7

A To graph an inequality in two variables

WWW WEB Need help? For online student resources, such as section quizzes, visit this textbook's website at **college.hmco.com/students.**

Prep Test

For Exercises 1 to 3, simplify.

1. $-4(x - 3)$
$-4x + 12$ [1.4D]

2. $\sqrt{(-6)^2 + (-8)^2}$
10 [1.2F]

3. $\dfrac{3 - (-5)}{2 - 6}$
-2 [1.3A]

4. Evaluate $-2x + 5$ for $x = -3$.
11 [1.4A]

5. Evaluate $\dfrac{2r}{r - 1}$ for $r = 5$.
2.5 [1.4A]

6. Evaluate $2p^3 - 3p + 4$ for $p = -1$.
5 [1.4A]

7. Evaluate $\dfrac{x_1 + x_2}{2}$ for $x_1 = 7$ and $x_2 = -5$.
1 [1.4A]

8. Given $3x - 4y = 12$, find the value of x when $y = 0$.
4 [2.2A]

Go Figure

If $\boxed{5} = 4$ and $\textcircled{5} = 6$ and $y = x - 1$, which of the following has the largest value?

$\boxed{x}$ $\textcircled{x}$ $\boxed{y}$ $\textcircled{y}$ $\textcircled{x}$

4.1 The Rectangular Coordinate System

Objective A

To graph points in a rectangular coordinate system

VIDEO & DVD CD TUTOR WEB SSM

Point of Interest

A rectangular coordinate system is also called a **Cartesian** coordinate system in honor of Descartes.

Before the 15th century, geometry and algebra were considered separate branches of mathematics. That all changed when René Descartes, a French mathematician who lived from 1596 to 1650, developed **analytic geometry**. In this geometry, a *coordinate system* is used to study relationships between variables.

A **rectangular coordinate system** is formed by two number lines, one horizontal and one vertical, that intersect at the zero point of each line. The point of intersection is called the **origin**. The two lines are called **coordinate axes**, or simply **axes**. Generally, the horizontal axis is labeled the *x*-axis and the vertical axis is labeled the *y*-axis.

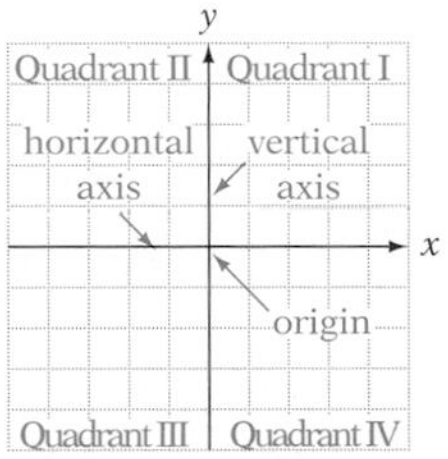

The axes determine a **plane**, which can be thought of as a large, flat sheet of paper. The two axes divide the plane into four regions called **quadrants**, which are numbered counterclockwise from I to IV starting from the upper right.

Point of Interest

Gottfried Leibnitz introduced the words *abscissa* and *ordinate*. *Abscissa* is from a Latin word meaning "to cut off." Originally, Leibnitz used *abscissa linea*, "cut off a line" (axis). The root of *ordinate* is also a Latin word used to suggest the sense of order.

Each point in the plane can be identified by a pair of numbers called an **ordered pair**. The first number of the ordered pair measures a horizontal distance and is called the **abscissa**, or ***x*-coordinate**. The second number of the pair measures a vertical distance and is called the **ordinate**, or ***y*-coordinate**. The ordered pair (*x*, *y*) associated with a point is also called the **coordinates** of the point.

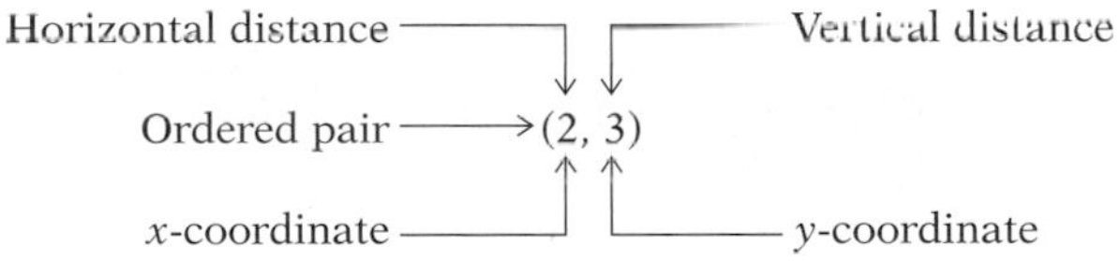

To **graph** or **plot** a point in the plane, place a dot at the location given by the ordered pair. The **graph of an ordered pair** is the dot drawn at the coordinates of the point in the plane. The points whose coordinates are (3, 4) and (−2.5, −3) are graphed in the figures below.

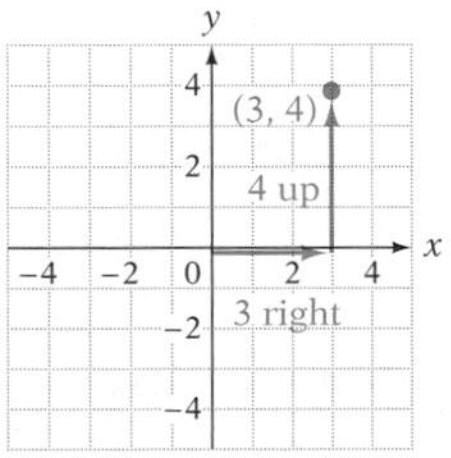

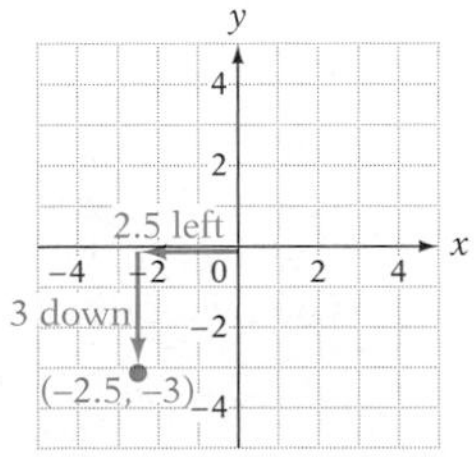

Objective 4.1A

New Vocabulary

rectangular coordinate system
origin
coordinate axes, or axes
plane
quadrant
ordered pair
abscissa
x-coordinate
ordinate
y-coordinate
coordinates
to graph, or plot, an ordered pair
graph of an ordered pair

Discuss the Concepts

1. Find the distance from the given point to the horizontal axis.
   - **a.** (−5, 1) 1 unit
   - **b.** (3, −4) 4 units
   - **c.** (−6, 0) 0 units
2. Find the distance from the given point to the vertical axis.
   - **a.** (−2, 4) 2 units
   - **b.** (1, −3) 1 unit
   - **c.** (5, 0) 5 units
3. What is the value of the *y*-coordinate of any point on the *x* axis? 0
4. What is the value of the *x*-coordinate of any point on the *y*-axis? 0
5. Name any two points on a horizontal line that is 2 units above the *x*-axis. Answers will vary. For example, (−5, 2) and (4, 2).
6. Name any two points on a vertical line that is 3 units to the left of the *y*-axis. Answers will vary. For example, (−3, 6) and (−3, −1).

In-Class Examples (Objective 4.1A)

1. Graph the ordered pairs (1, 4), (−1, −2), and (3, 0).

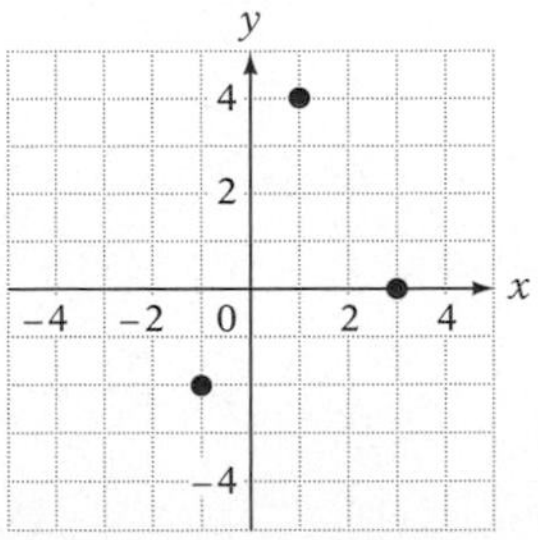

2. Find the coordinates of each point.
   A (3, −2)
   B (0, −3)
   C (−4, 3)
   D (−3, −4)

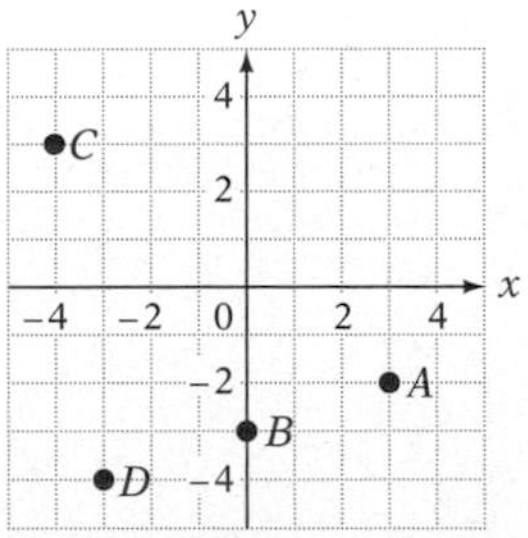

Optional Student Activity

1. If $(-2, 5)$ and $(4, -1)$ are the coordinates of two opposite vertices of a square, what are the coordinates of the other two vertices? $(4, 5)$ and $(-2, -1)$
2. If $(0, -3)$ and $(-6, 5)$ are the coordinates of two opposite vertices of a rectangle, what are the coordinates of the other two vertices? $(0, 5)$ and $(-6, -3)$
3. Consider the point (x, y) in Quadrant I. This point is reflected across the y-axis and then reflected across the x-axis. After these reflections, the coordinates of the point are (a, b). Find the value of $xy - ab$. 0
4. Consider the point (x, y) described in Exercise 3 above. If this point is reflected first across the x-axis and then across the y-axis, and the coordinates of the resulting point are (a, b), what is the value of $xy - ab$? 0

> **TAKE NOTE**
> This is very important. An **ordered pair** is a *pair* of coordinates, and the *order* in which the coordinates appear is important.

The points whose coordinates are $(3, -1)$ and $(-1, 3)$ are graphed at the right. Note that the graphed points are in different locations. *The order of the coordinates of an ordered pair is important.*

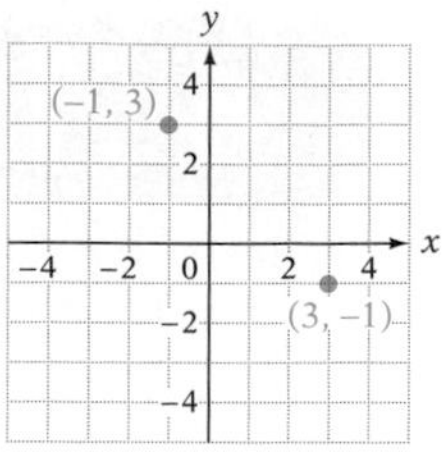

Each point in the plane is associated with an ordered pair, and each ordered pair is associated with a point in the plane. Although only the labels for integers are given on a coordinate grid, the graph of any ordered pair can be approximated. For example, the points whose coordinates are $(-2.3, 4.1)$ and $(\pi, 1)$ are shown on the graph at the right.

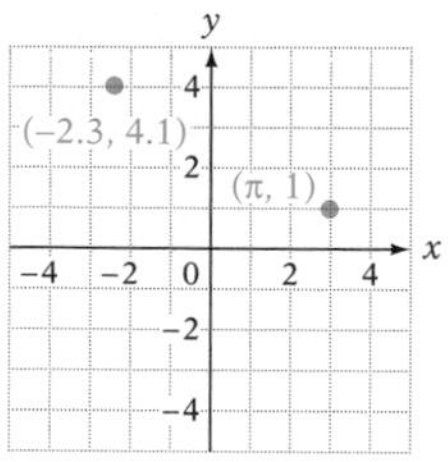

Example 1 Graph the ordered pairs $(-2, -3)$, $(3, -2)$, $(0, -2)$, and $(3, 0)$.

Solution

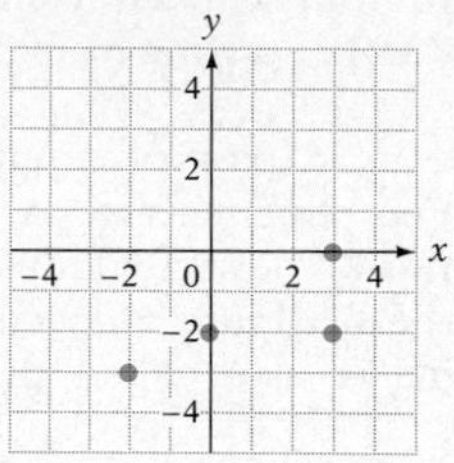

You Try It 1 Graph the ordered pairs $(-4, 1)$, $(3, -3)$, $(0, 4)$, and $(-3, 0)$.

Your solution

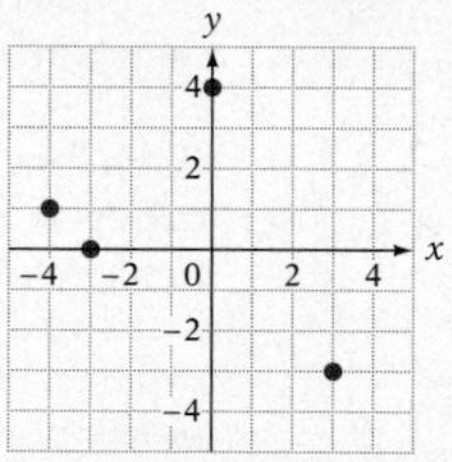

Example 2 Give the coordinates of the points labeled A and B. Give the abscissa of point C and the ordinate of point D.

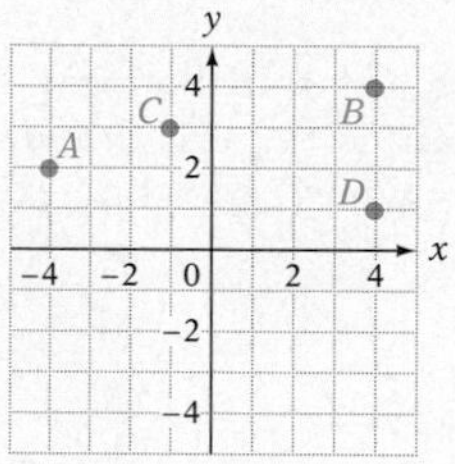

Solution The coordinates of A are $(-4, 2)$.
The coordinates of B are $(4, 4)$.
The abscissa of C is -1.
The ordinate of D is 1.

You Try It 2 Give the coordinates of the points labeled A and B. Give the abscissa of point D and the ordinate of point C.

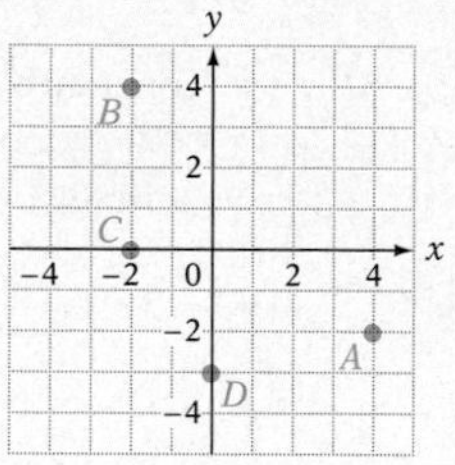

Your solution $A(4, -2)$, $B(-2, 4)$. The abscissa of D is 0. The ordinate of C is 0.

Solutions on p. S10

203

Objective B

To determine ordered-pair solutions of an equation in two variables

VIDEO & DVD CD TUTOR WEB SSM

When drawing a rectangular coordinate system, we often label the horizontal axis x and the vertical axis y. In this case, the coordinate system is called the ***xy*-coordinate system.** The coordinates of the points are given by ordered pairs (x, y), where the abscissa is called the x-coordinate and the ordinate is called the y-coordinate.

A coordinate system is used to study the relationship between two variables. Frequently this relationship is given by an equation. Examples of equations in two variables include

$$y = 2x - 3 \qquad 3x + 2y = 6 \qquad x^2 - y = 0$$

A **solution of an equation in two variables** is an ordered pair (x, y) whose coordinates make the equation a true statement.

TAKE NOTE
An ordered pair is of the form (x, y). For the ordered pair $(-3, 7)$, -3 is the x value and 7 is the y value. Substitute -3 for x and 7 for y.

➡ Is the ordered pair $(-3, 7)$ a solution of the equation $y = -2x + 1$?

$$\begin{array}{c|l} y & = -2x + 1 \\ \hline 7 & -2(-3) + 1 \\ 7 & 6 + 1 \\ \end{array}$$
$7 = 7$

- Replace x by -3 and y by 7.
- Simplify.
- Compare the results. If the resulting equation is a true statement, the ordered pair is a solution of the equation. If it is not a true statement, the ordered pair is not a solution of the equation.

Yes, the ordered pair $(-3, 7)$ is a solution of the equation.

Besides $(-3, 7)$, there are many other ordered-pair solutions of $y = -2x + 1$. For example, $(0, 1)$, $\left(-\frac{3}{2}, 4\right)$, and $(4, -7)$ are also solutions.

In general, an equation in two variables has an infinite number of solutions. By choosing any value of x and substituting that value into the equation, we can calculate a corresponding value of y.

➡ Find the ordered-pair solution of $y = \frac{2}{3}x - 3$ that corresponds to $x = 6$.

$$\begin{aligned} y &= \frac{2}{3}x - 3 \\ &= \frac{2}{3}(6) - 3 \\ &= 4 - 3 \\ &= 1 \end{aligned}$$

- Replace x by 6.
- Solve for y.

The ordered-pair solution is $(6, 1)$.

The solution of an equation in two variables can be graphed in an xy-coordinate system.

Objective 4.1B

New Vocabulary
xy-coordinate system
solution of an equation in two variables

Vocabulary to Review
x-coordinate
y-coordinate

Discuss the Concepts
Explain how to check whether the ordered pair $(5, 1)$ is a solution of the equation $y = 2x - 9$.

Concept Check
Find three ordered-pair solutions of the equation $y = x - 1$. For example, $(-1, -2)$, $(0, -1)$, $(2, 1)$.

Instructor Note
Problems such as the one at the bottom of page 203 are given to prepare the student to graph straight lines, which is the subject of the section "Linear Equations in Two Variables" of this chapter.

The example at the top of page 204 is preparing the student to see the relationship between a solution of an equation and the pictorial representation of that solution, its graph.

Optional Student Activity

The example at the top of page 204 and Example 5 both show points falling on a straight line. This is not always the case. Graph the ordered-pair solutions of $y = x^2$ when $x = -3, -2, -1, 0, 1, 2,$ and 3. How would you describe the shape on which these points seem to lie?

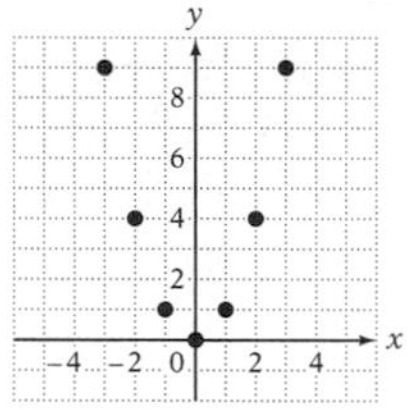

U-shaped

➡ Graph the ordered-pair solutions of $y = -2x + 1$ when $x = -2, -1, 0, 1,$ and 2.

Use the values of x to determine ordered-pair solutions of the equation. It is convenient to record these in a table.

x	$y = -2x + 1$	y	(x, y)
-2	$-2(-2) + 1$	5	$(-2, 5)$
-1	$-2(-1) + 1$	3	$(-1, 3)$
0	$-2(0) + 1$	1	$(0, 1)$
1	$-2(1) + 1$	-1	$(1, -1)$
2	$-2(2) + 1$	-3	$(2, -3)$

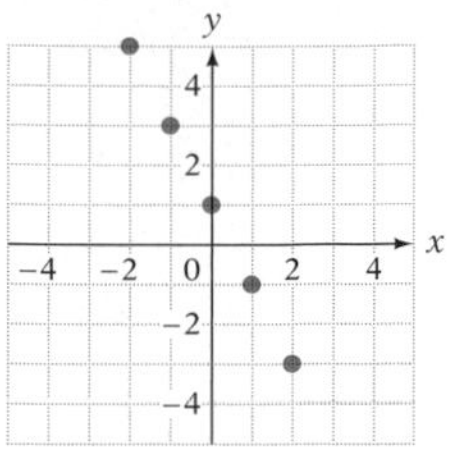

Example 3

Is $(3, -2)$ a solution of $3x - 4y = 15$?

Solution

$$\begin{array}{r|l} 3x - 4y & = 15 \\ \hline 3(3) - 4(-2) & 15 \\ 9 + 8 & 15 \\ 17 & \neq 15 \end{array}$$

• **Replace x by 3 and y by -2.**

No, $(3, -2)$ is not a solution of $3x - 4y = 15$.

You Try It 3

Is $(-2, 4)$ a solution of $x - 3y = -14$?

Your solution

yes

Example 4

Find the ordered-pair solution of $y = \frac{x}{x-2}$ corresponding to $x = 4$.

Solution

Replace x by 4 and solve for y.

$$y = \frac{x}{x-2} = \frac{4}{4-2} = \frac{4}{2} = 2$$

The ordered-pair solution is $(4, 2)$.

You Try It 4

Find the ordered-pair solution of $y = \frac{3x}{x+1}$ corresponding to $x = -2$.

Your solution

$(-2, 6)$

Solutions on p. S10

In-Class Examples (Objective 4.1B)

1. Is $(0, -1)$ a solution of $2x - y = 1$? Yes
2. Is $(2, -2)$ a solution of $y = 3x + 8$? No
3. Graph the ordered-pair solutions of $y = -3x$ when $x = -1, 0,$ and 1.

3.

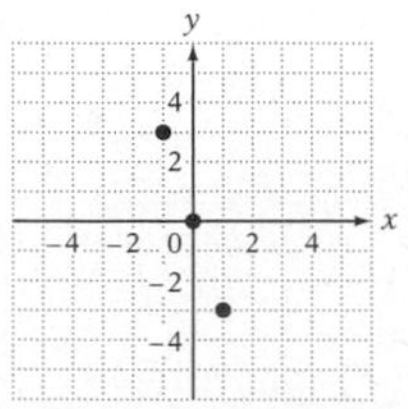

Example 5
Graph the ordered-pair solutions of $y = \frac{2}{3}x - 2$ when $x = -3, 0, 3, 6$.

Solution
Replace x in $y = \frac{2}{3}x - 2$ by $-3, 0, 3,$ and 6. For each value of x, determine the value of y.

x	$y = \frac{2}{3}x - 2$	y	(x, y)
-3	$\frac{2}{3}(-3) - 2$	-4	$(-3, -4)$
0	$\frac{2}{3}(0) - 2$	-2	$(0, -2)$
3	$\frac{2}{3}(3) - 2$	0	$(3, 0)$
6	$\frac{2}{3}(6) - 2$	2	$(6, 2)$

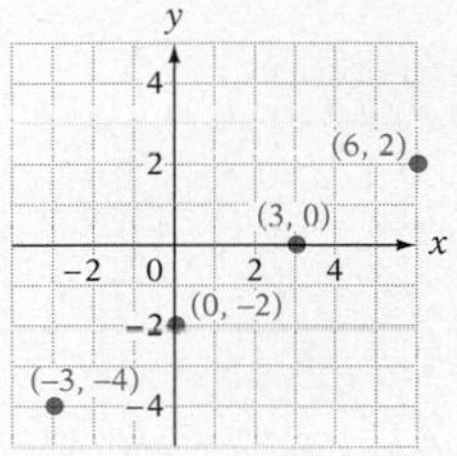

You Try It 5
Graph the ordered-pair solutions of $y = -\frac{1}{2}x + 2$ when $x = -4, -2, 0, 2$.

Your solution

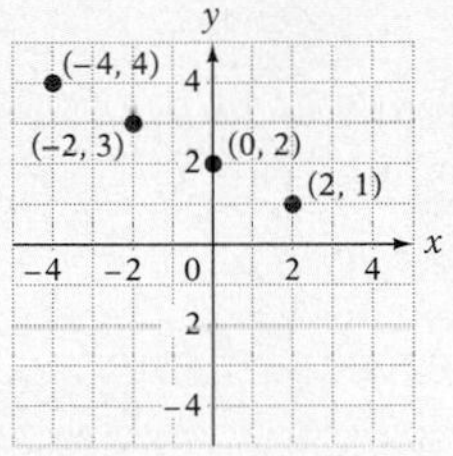

Solution on p. S10

Objective 4.1C

New Vocabulary
scatter diagram

Optional Student Activity
Ask students to bring data to class and then prepare a scatter diagram of the data. They can find data on the Internet or in the graphs printed in a newspaper such as *USA Today.* Be sure they label both the horizontal and the vertical axes. Also check that the distance between units is uniform.

You might ask students to determine whether there is a trend in the data they have graphed. For example, in You Try It 6 on page 206, students can see that the software rental market is increasing.

Objective C To graph a scatter diagram

VIDEO & DVD CD TUTOR WEB SSM

Discovering a relationship between two variables is an important task in the study of mathematics. These relationships occur in many forms and in a wide variety of applications. Here are some examples.

- A botanist wants to know the relationship between the number of bushels of wheat yielded per acre and the amount of watering per acre.
- An environmental scientist wants to know the relationship between the incidence of skin cancer and the amount of ozone in the atmosphere.
- A business analyst wants to know the relationship between the price of a product and the number of products that are sold at that price.

In-Class Examples (Objective 4.1C)

1. The number of days of school mandated by the government has changed throughout the years. These changes are recorded at the right. Use the grid at the right to graph the scatter diagram for the data.

Year	1870	1910	1930	1950
Number of School Days Mandated	130	155	170	180

Graph the points (1870, 130), (1910, 155), (1930, 170), (1950, 180).

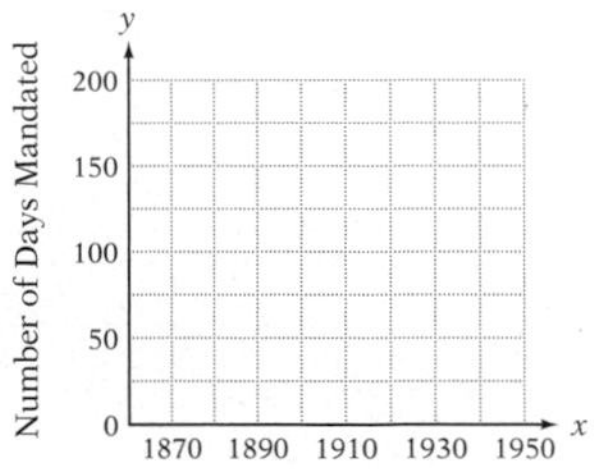

206

A researcher may investigate the relationship between two variables by means of *regression analysis,* which is a branch of statistics. The study of the relationship between the two variables may begin with a **scatter diagram,** which is a graph of the ordered pairs of the known data.

The following table shows randomly selected data from the participants 40 years old and older and their times (in minutes) for a recent Boston Marathon.

Age (x)	55	46	53	40	40	44	54	44	41	50
Time (y)	254	204	243	194	281	197	238	300	232	216

TAKE NOTE
The jagged portion of the horizontal axis in the figure at the right indicates that the numbers between 0 and 40 are missing.

The scatter diagram for these data is shown at the right. Each ordered pair represents the age and time for a participant. For instance, the ordered pair (53, 243) indicates that a 53-year-old participant ran the marathon in 243 min.

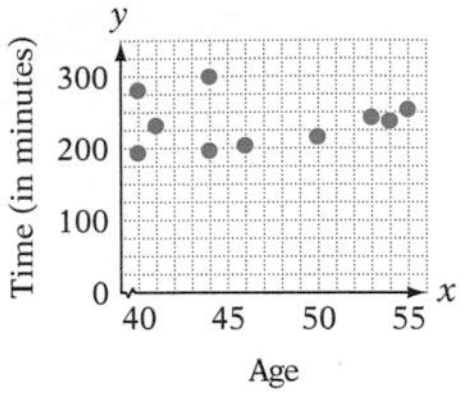

Example 6

The grams of sugar and the grams of fiber in a 1-ounce serving of six breakfast cereals are shown in the table below.

	Sugar (x)	*Fiber (y)*
Wheaties	4	3
Rice Krispies	3	0
Total	5	3
Life	6	2
Kix	3	1
Grape-Nuts	7	5

Draw a scatter diagram of these data.

Strategy

To draw a scatter diagram:

- Draw a coordinate grid with the horizontal axis representing the grams of sugar and the vertical axis the grams of fiber.
- Graph the ordered pairs (4, 3), (3, 0), (5, 3), (6, 2), (3, 1), and (7, 5).

Solution

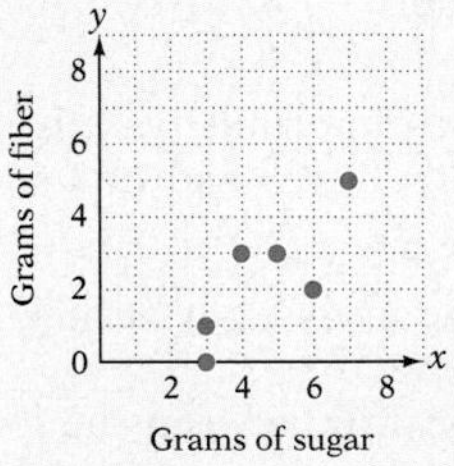

You Try It 6

According to Forester Research, the growth of the software rental market, in billions of dollars, is projected to be as shown in the table below.

Year	*Market (in billions of dollars)*
1999	1
2000	2
2001	4
2002	7
2003	11

Draw a scatter diagram of these data.

Your strategy

Your solution

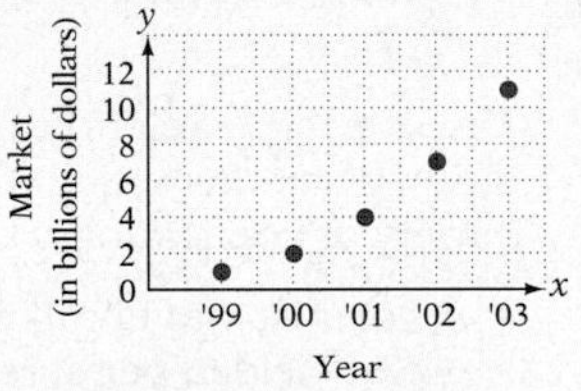

Solution on p. S10

207

4.1 Exercises

Objective A

1. Graph (−2, 1), (3, −5), (−2, 4), and (0, 3).

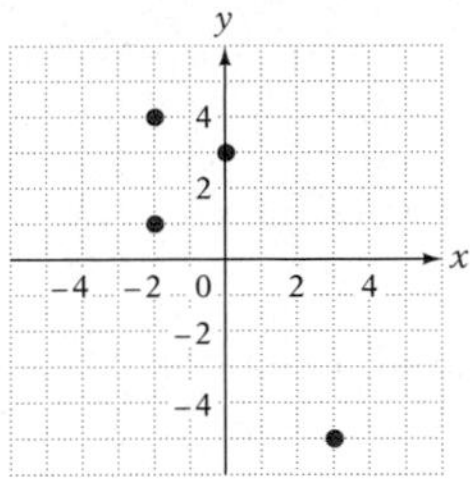

2. Graph (5, −1), (−3, −3), (−1, 0), and (1, −1).

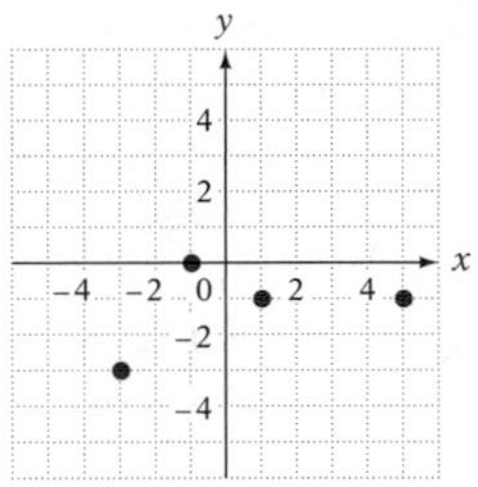

3. Graph (0, 0), (0, −5), (−3, 0), and (0, 2).

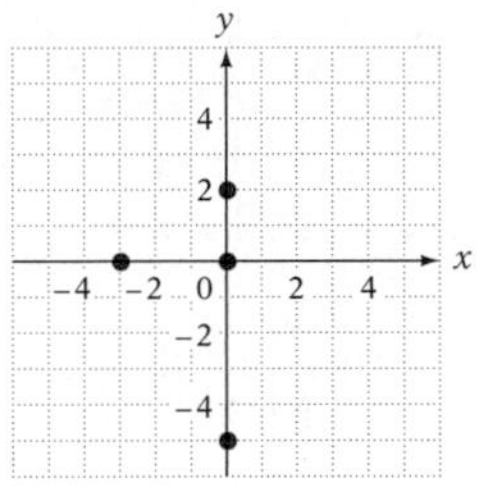

4. Graph (−4, 5), (−3, 1), (3, −4), and (5, 0).

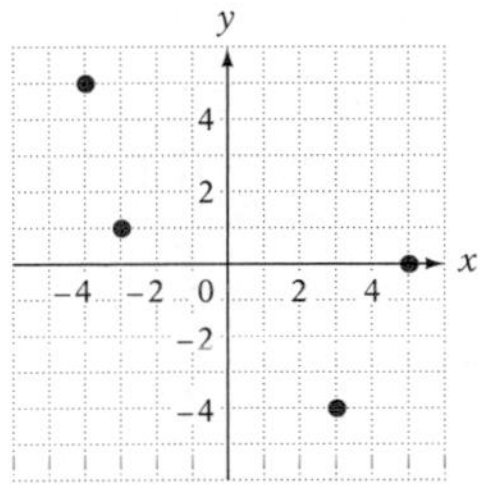

5. Graph (−1, 4), (−2, −3), (0, 2), and (4, 0).

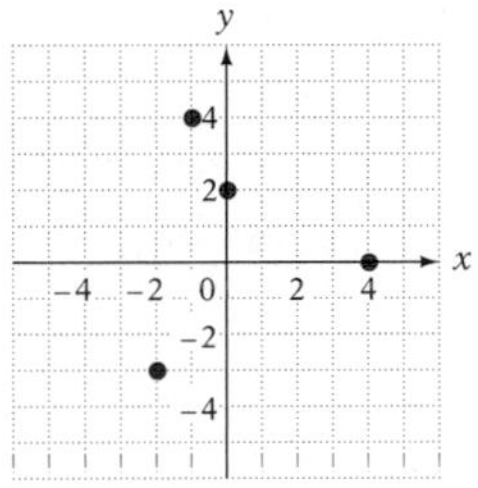

6. Graph (5, 2), (−4, −1), (0, 0), and (0, 3).

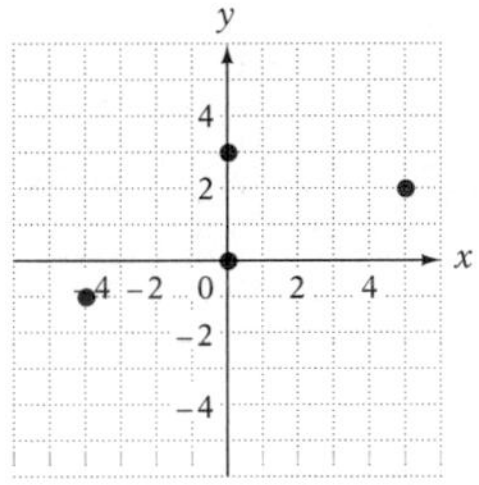

7. Find the coordinates of each of the points.

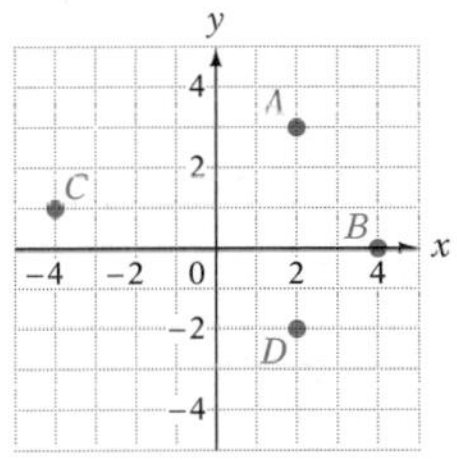

A(2, 3), *B*(4, 0), *C*(−4, 1), *D*(2, −2)

8. Find the coordinates of each of the points.

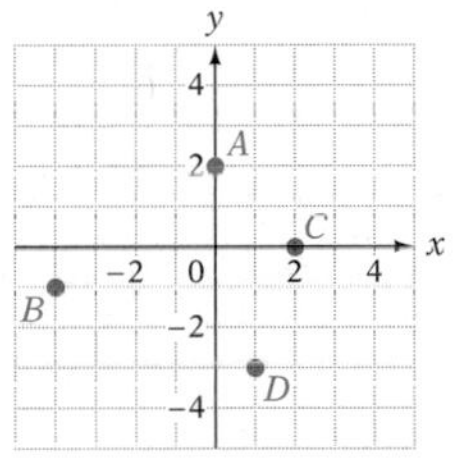

A(0, 2), *B*(−4, −1), *C*(2, 0), *D*(1, −3)

9. Find the coordinates of each of the points.

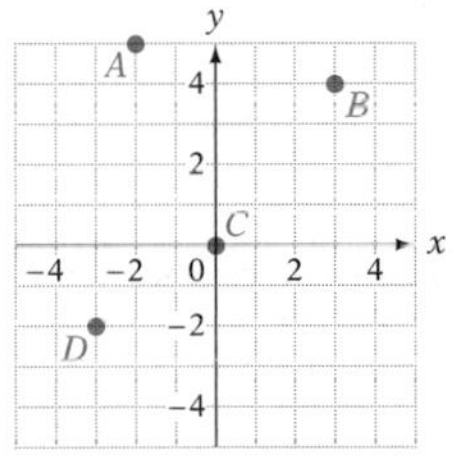

A(−2, 5), *B*(3, 4), *C*(0, 0), *D*(−3, −2)

10. Find the coordinates of each of the points.

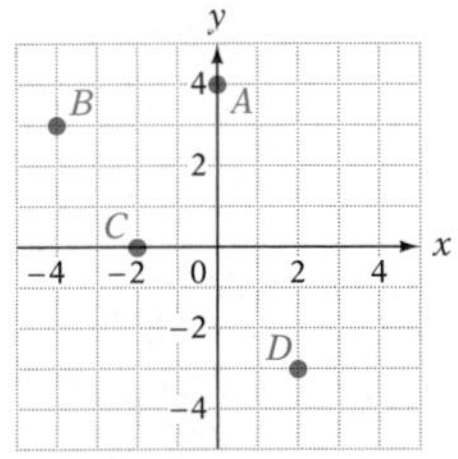

A(0, 4), *B*(−4, 3), *C*(−2, 0), *D*(2, −3)

11. Find the coordinates of each of the points.

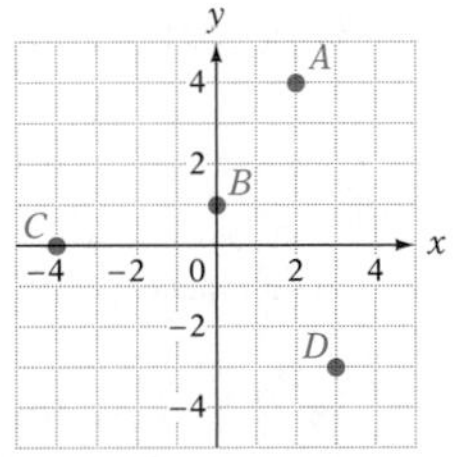

A(2, 4), *B*(0, 1), *C*(−4, 0), *D*(3, −3)

12. Find the coordinates of each of the points.

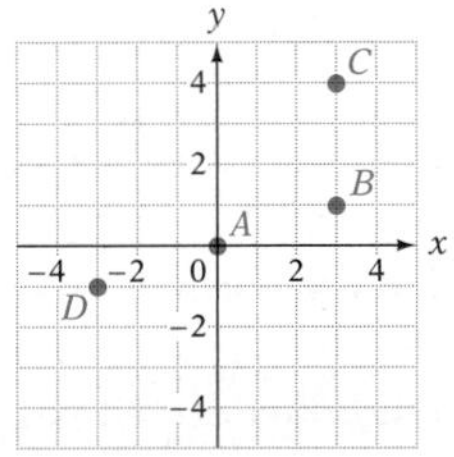

A(0, 0), *B*(3, 1), *C*(3, 4), *D*(−3, −1)

Section 4.1

Suggested Assignment

Exercises 1–43, odds

More challenging problems:
Exercises 45, 46

Quick Quiz (Objective 4.1A)

1. Graph the ordered pairs (4, −1), (1, 2), and (0, 3).

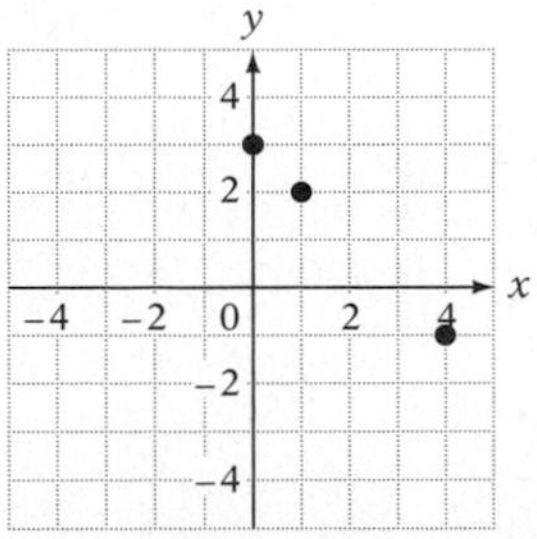

2. Find the coordinates of each of the points.
A (−2, 1)
B (1, 4)
C (−4, −2)
D (1, −4)

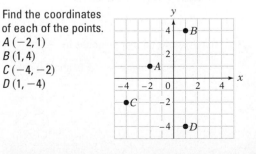

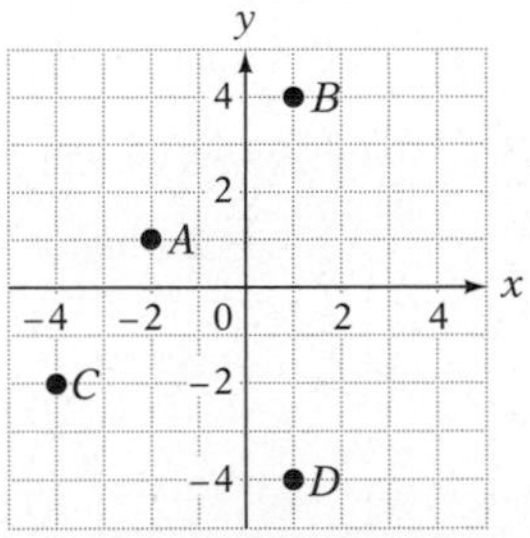

208

13. Draw a line through all points with an abscissa of 2.

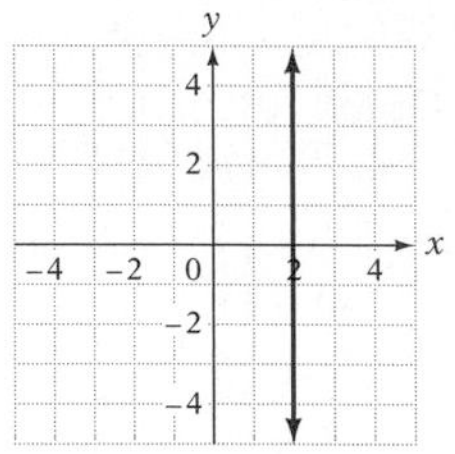

14. Draw a line through all points with an abscissa of -3.

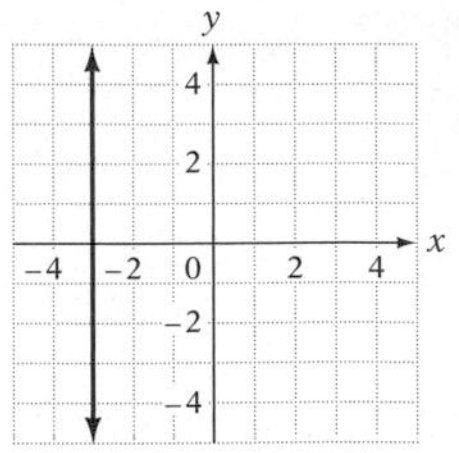

15. Draw a line through all points with an ordinate of -3.

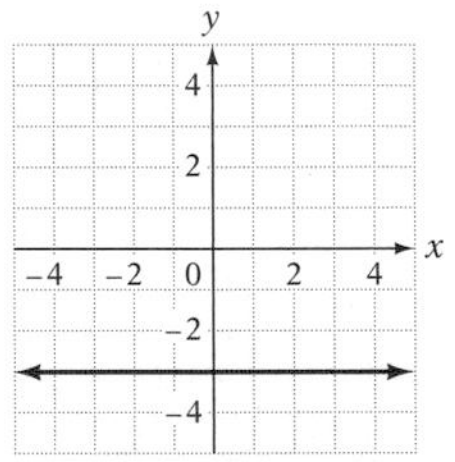

16. Draw a line through all points with an ordinate of 4.

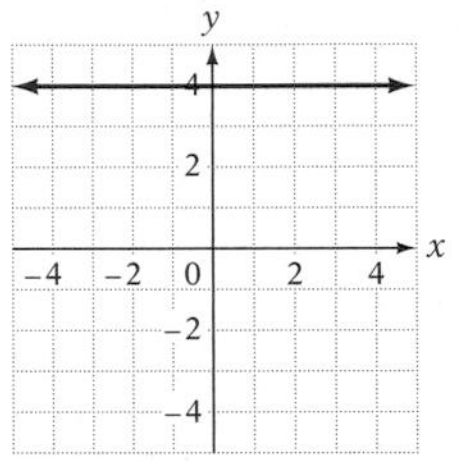

Objective B

17. Is (3, 4) a solution of $y = -x + 7$? yes

18. Is $(2, -3)$ a solution of $y = x + 5$? no

19. Is $(-1, 2)$ a solution of $y = \frac{1}{2}x - 1$? no

20. Is $(1, -3)$ a solution of $y = -2x - 1$? yes

21. Is (4, 1) a solution of $2x - 5y = 4$? no

22. Is $(-5, 3)$ a solution of $3x - 2y = 9$? no

23. Is (0, 4) a solution of $3x - 4y = -4$? no

24. Is $(-2, 0)$ a solution of $x + 2y = -1$? no

25. Find the ordered-pair solution of $y = 3x - 2$ corresponding to $x = 3$. (3, 7)

26. Find the ordered-pair solution of $y = 4x + 1$ corresponding to $x = -1$. $(-1, -3)$

27. Find the ordered-pair solution of $y = \frac{2}{3}x - 1$ corresponding to $x = 6$. (6, 3)

28. Find the ordered-pair solution of $y = \frac{3}{4}x - 2$ corresponding to $x = 4$. (4, 1)

29. Find the ordered-pair solution of $y = -3x + 1$ corresponding to $x = 0$. (0, 1)

30. Find the ordered-pair solution of $y = \frac{2}{5}x - 5$ corresponding to $x = 0$. $(0, -5)$

31. Find the ordered-pair solution of $y = \frac{2}{5}x + 2$ corresponding to $x = -5$. $(-5, 0)$

32. Find the ordered-pair solution of $y = -\frac{1}{6}x - 2$ corresponding to $x = 12$. $(12, -4)$

Quick Quiz (Objective 4.1B)

1. Is $(-3, 0)$ a solution of $y = \frac{1}{3}x - 1$? No

2. Graph the ordered-pair solutions of $y = -\frac{2}{3}x - 2$ when $x = -3$, 0, and 3.

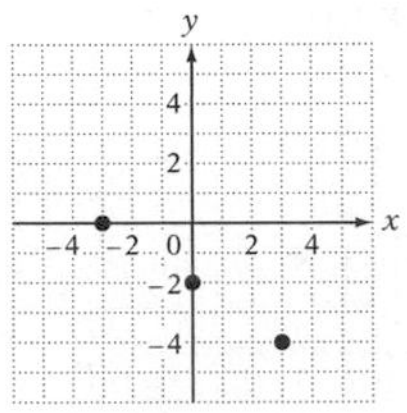

209

Graph the ordered-pair solutions for the given values of *x*.

33. $y = 2x;\ x = -2, -1, 0, 2$

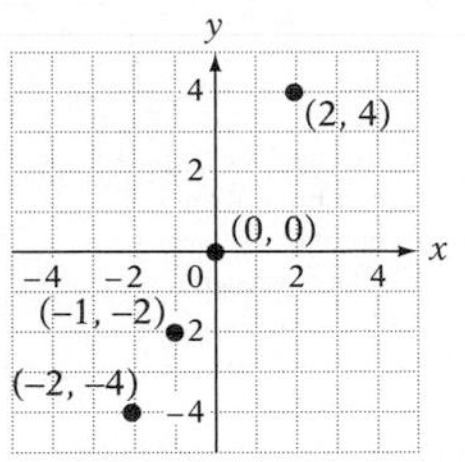

34. $y = -2x;\ x = -2, -1, 0, 2$

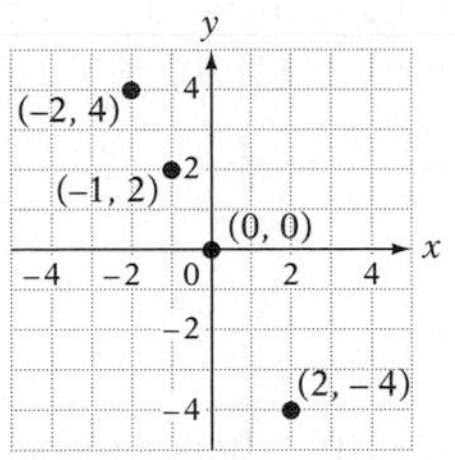

35. $y = x + 2;\ x = -4, -2, 0, 3$

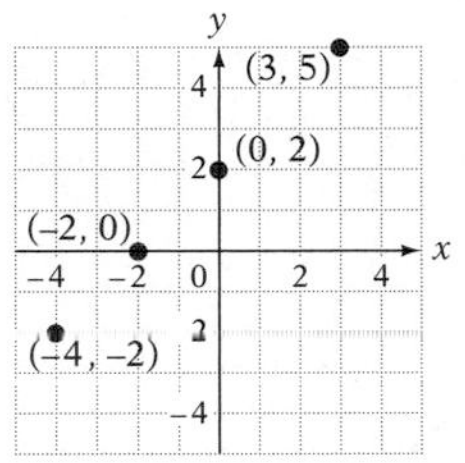

36. $y = \frac{1}{2}x - 1;\ x = -2, 0, 2, 4$

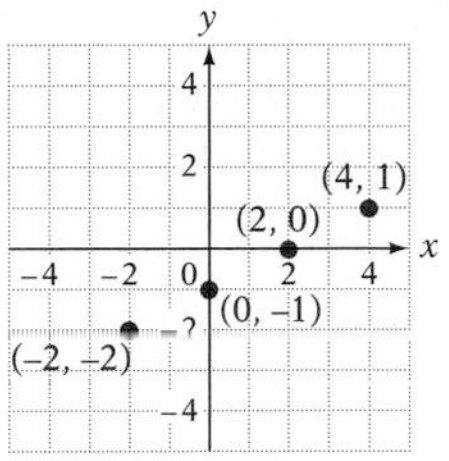

37. $y = \frac{2}{3}x + 1;\ x = -3, 0, 3$

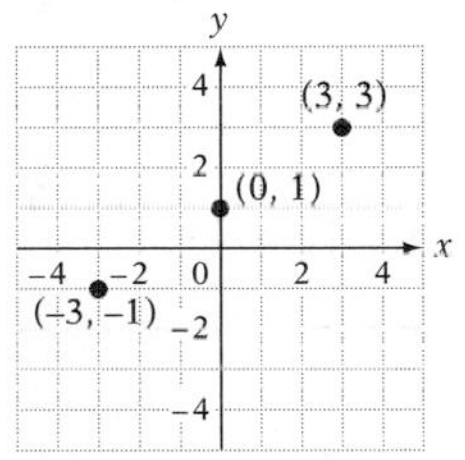

38. $y = -\frac{1}{3}x - 2;\ x = -3, 0, 3$

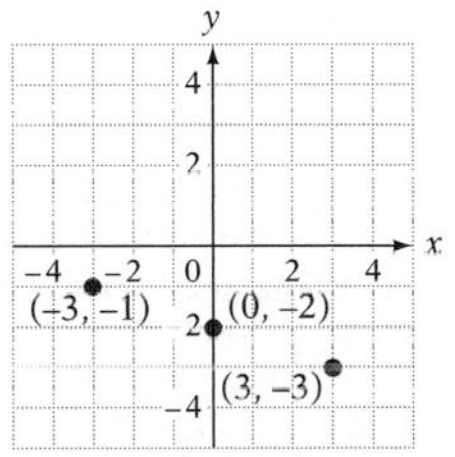

39. $y = x^2;\ x = -2, -1, 0, 1, 2$

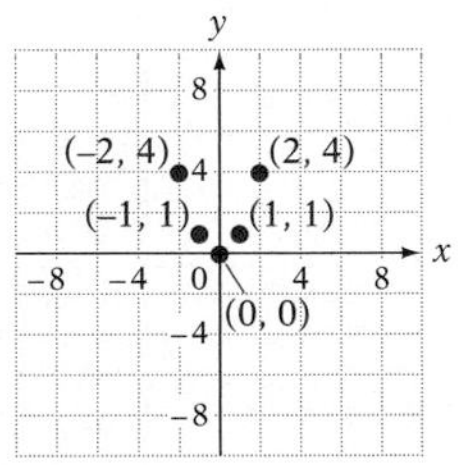

40. $y = |x + 1|;\ x = -5, -3, 0, 3, 5$

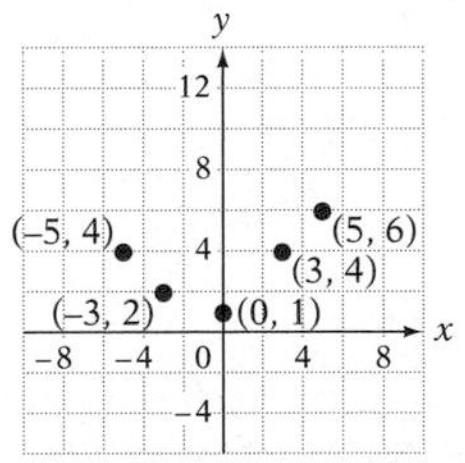

Answers to Writing Exercises

45. Students should explain that the ordered pairs are being plotted in reverse order. In an ordered pair, the first number indicates a movement to the left or right, and the second number indicates a movement up or down.

46. The graph of all ordered pairs (x, y) that are 5 units from the origin is a circle of radius 5 that has its center at $(0, 0)$.

Objective C

41. The temperature of a chemical reaction is measured at intervals of 10 min and recorded in the scatter diagram at the right.
a. Find the temperature of the reaction after 20 min.
b. After how many minutes is the temperature 160°F?
280°F; 50 min

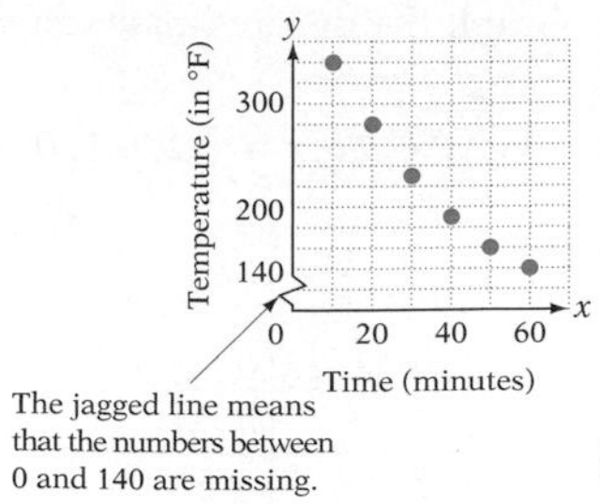

42. The amount of a substance that can be dissolved in a fixed amount of water usually increases as the temperature of the water increases. Cerium selenate, however, does not behave in this manner. The graph at the right shows the number of grams of cerium selenate that will dissolve in 100 mg of water for various temperatures, in degrees Celsius.
a. Determine the temperature at which 25 g of cerium selenate will dissolve. 50°C
b. Determine the number of grams of cerium selenate that will dissolve when the temperature is 80°C. 5 g

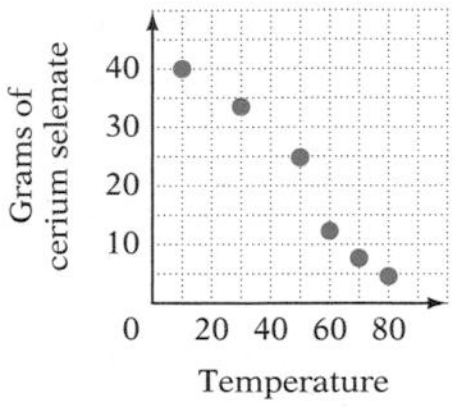

43. Past experience recorded by executives of a car company shows that the profit of a dealership will depend on the total income of all the residents of the town in which the dealership is located. The table below shows the profit of several dealerships and the total income of the towns. Draw the scatter diagram for these data.

Profit (thousands of $)	65	85	81	77	89	69
Total Income (billions of $)	2.2	2.6	2.5	2.4	2.7	2.3

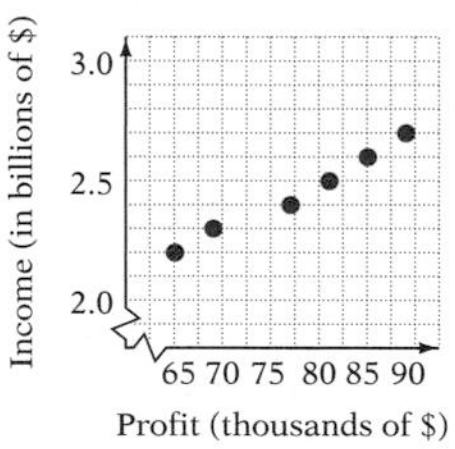

44. A power company suggests that a larger power plant can produce energy more efficiently and therefore at lower cost to consumers. The table below shows the output and average cost for power plants of various sizes. Draw the scatter diagram for these data.

Output (in millions of watts)	0.7	2.2	2.6	3.2	2.8	3.5
Average Cost (in dollars)	6.9	6.5	6.3	6.4	6.5	6.1

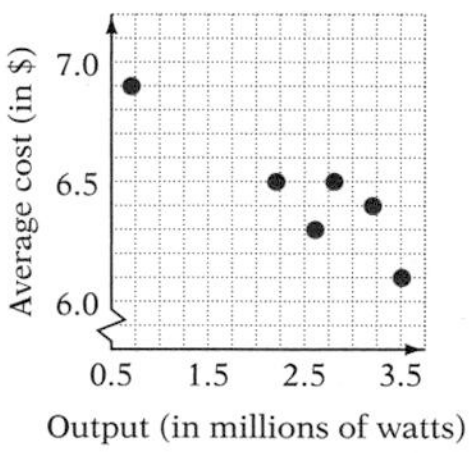

APPLYING THE CONCEPTS

45. Suppose you are helping a student who is having trouble graphing ordered pairs. The work of the student is at the right. What can you say to this student to correct the error that is being made?

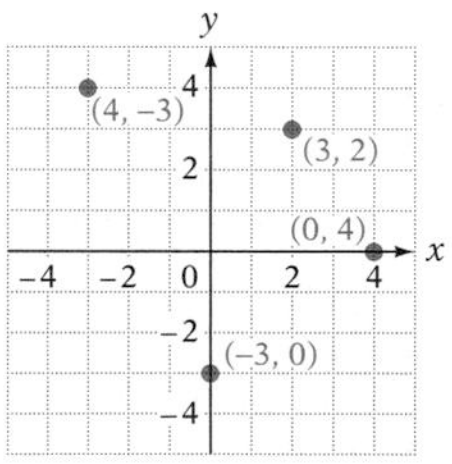

46. Describe the graph of all the ordered pairs (x, y) that are 5 units from the origin. A circle of radius 5, centered at the origin.

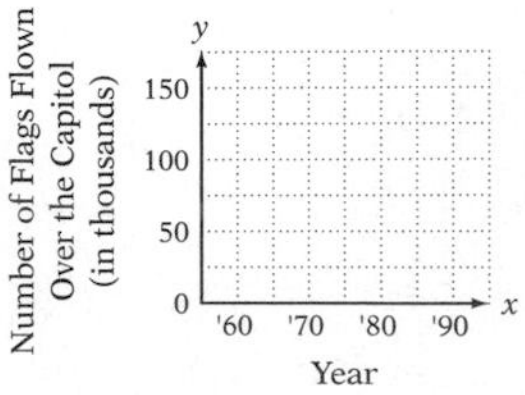

Quick Quiz (Objective 4.1C)

1. The table at the right shows the number of flags flown over the Capitol in Washington, DC, in 1960, 1970, 1980, and 1990. Use the grid at the left to graph a scatter diagram for the data.

Year	1960	1970	1980	1990
Number of Flags Flown Over the Capitol (in thousands)	16	32	70	142

Graph the points (60, 16), (70, 32), (80, 70), and (90, 142).

211

4.2 Introduction to Functions

Objective A To evaluate a function

VIDEO & DVD CD TUTOR WEB SSM

In mathematics and its applications, there are many times when it is necessary to investigate a relationship between two quantities. Here is a financial application: Consider a person who is planning to finance the purchase of a car. If the current interest rate for a five-year loan is 9%, the equation that describes the relationship between the amount that is borrowed B and the monthly payment P is $P = 0.020758B$.

For each amount the purchaser may borrow (B), there is a certain monthly payment (P). The relationship between the amount borrowed and the payment can be recorded as ordered pairs, where the first coordinate is the amount borrowed and the second coordinate is the monthly payment. Some of these ordered pairs are shown at the right.

$0.020758B = P$

(5000, 103.79)
(6000, 124.55)
(7000, 145.31)
(8000, 166.06)

A relationship between two quantities is not always given by an equation. The table at the right describes a grading scale that defines a relationship between a score on a test and a letter grade. For each score, the table assigns only one letter grade. The ordered pair (84, B) indicates that a score of 84 receives a letter grade of B.

Score	*Grade*
90–100	A
80–89	B
70–79	C
60–69	D
0–59	F

The graph at the right also shows a relationship between two quantities. It is a graph of the viscosity V of SAE 40 motor oil at various temperatures T. Ordered pairs can be approximated from the graph. The ordered pair (120, 250) indicates that the viscosity of the oil at 120°F is 250 units.

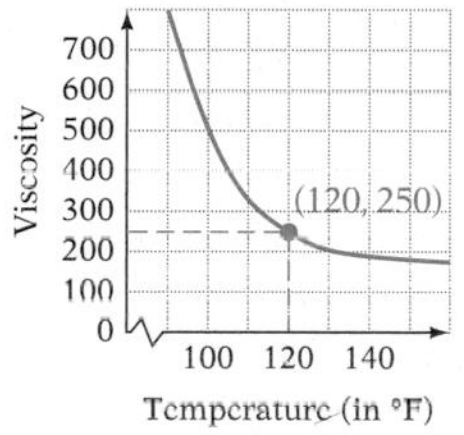

In each of these examples, there is a rule (an equation, a table, or a graph) that determines a certain set of ordered pairs.

Definition of Relation

A **relation** is a set of ordered pairs.

Here are some of the ordered pairs for the relations given above.

Relation	*Some of the Ordered Pairs of the Relation*
Car Payment	(2500, 51.90), (3750, 77.84), (4396, 91.25)
Grading Scale	(78, C), (98, A), (70, C), (81, B), (94, A)
Oil Viscosity	(100, 500), (120, 250), (130, 200), (150, 180)

Objective 4.2A

New Vocabulary

relation
function
domain
range
dependent variable
independent variable
functional notation
value of a function
evaluating a function

Discuss the Concepts

The examples on this page show various functions. Have students suggest other relationships between two quantities. Record these on the board. Then, when you have finished defining the term *function,* return to the suggestions and determine which are functions. Some possible relationships are

- The distance a rock falls in a given time interval
- The points scored by a football team and whether the team won or lost
- The age of a student and the score the student received on a test

In-Class Examples (Objective 4.2A)

1. Find the domain and range of the function $\{(3, 10), (4, 13), (5, 16)\}$. D: $\{3, 4, 5\}$; R: $\{10, 13, 16\}$
2. Evaluate $M(n) = 2n^2 - 3n - 5$ when $n = -1$. 0
3. Find the range of $f(x) = 4x - 1$ if the domain is $\{-2, -1, 0, 1, 2\}$. $\{-9, -5, -1, 3, 7\}$
4. What value is excluded from the domain of $\frac{3}{x-6}$? 6

Discuss the Concepts

1. What does it mean to evaluate a function? Explain how to evaluate $f(x) = 3x$ when $x = 2$.
2. What is the value of a function?
3. Are all functions relations?
4. Are all relations functions?
5. Is it possible for a function to have the same output value for two different input values?
6. Is it possible for a function to have two different output values for the same input value?

Instructor Note

Domain and range are difficult concepts for most students. We are introducing the topic at this point. We will continually revisit the concepts of domain and range as we progress through the text.

Concept Check

1. For $f(x) = \dfrac{x}{2x - 1}$, find $f(3) - f\left(\dfrac{1}{3}\right)$. $\dfrac{8}{5}$
2. A diagonal of a polygon is a line segment from one vertex to a nonadjacent vertex. The total number of diagonals for a polygon is given by $N(s) = \dfrac{s^2 - 3s}{2}$, where $N(s)$ is the total number of diagonals and s is the number of sides of the polygon. Find the total number of diagonals for a polygon with 12 sides. 54 diagonals
3. A function is defined by $\{(-4, -6), (-2, -2), (0, 2), (2, 6), (4, 10)\}$. Find $f(2)$. 6

212

These three relations are actually a special type of relation called a function. Functions play an important role in mathematics and its applications.

> **Definition of Function**
>
> A **function** is a relation in which no two ordered pairs have the same first coordinate and different second coordinates.

The **domain** of a function is the set of the first coordinates of all the ordered pairs of the function. The **range** is the set of the second coordinates of all the ordered pairs of the function.

For the function defined by the ordered pairs

$$\{(2, 3), (4, 5), (6, 7), (8, 9)\}$$

the domain is $\{2, 4, 6, 8\}$ and the range is $\{3, 5, 7, 9\}$.

➡ Find the domain and range of the function $\{(2, 3), (4, 6), (6, 8), (10, 6)\}$.

The domain is $\{2, 4, 6, 10\}$. • **The domain of the function is the set of the first components in the ordered pairs.**

The range is $\{3, 6, 8\}$. • **The range of the function is the set of the second components in the ordered pairs.**

For each element of the domain of a function there is a corresponding element in the range of the function. A possible diagram for the function in the example above is

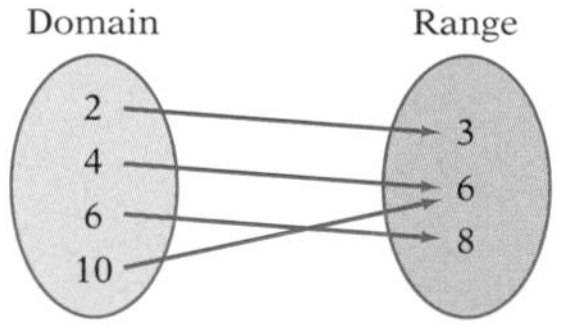

$\{(2, 3), (4, 6), (6, 8), (10, 6)\}$

Functions defined by tables or graphs, such as those described at the beginning of this section, have important applications. However, a major focus of this text is functions defined by equations in two variables.

The "square" function, which pairs each real number with its square, can be defined by the equation

$$y = x^2$$

This equation states that for a given value of x in the domain, the value of y in the range is the square of x. For instance, if $x = 6$, then $y = 36$, and if $x = -7$, then $y = 49$. Because the value of y *depends* on the value of x, y is called the **dependent variable** and x is called the **independent variable.**

4.2 Exercises

Objective A

1. In your own words, explain what a function is.

2. What is the domain of a function? What is the range of a function?

3. Does the diagram below represent a function? Explain your answer.

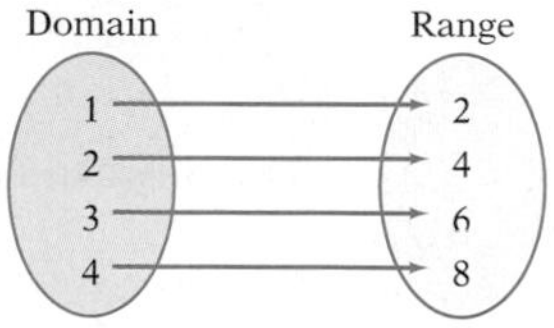

4. Does the diagram below represent a function? Explain your answer.

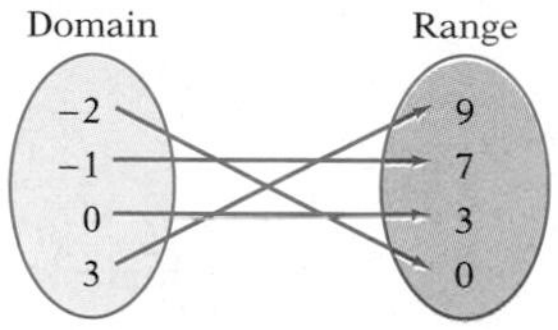

5. Does the diagram below represent a function? Explain your answer.

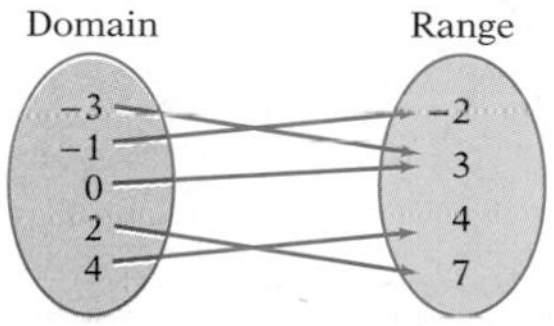

6. Does the diagram below represent a function? Explain your answer.

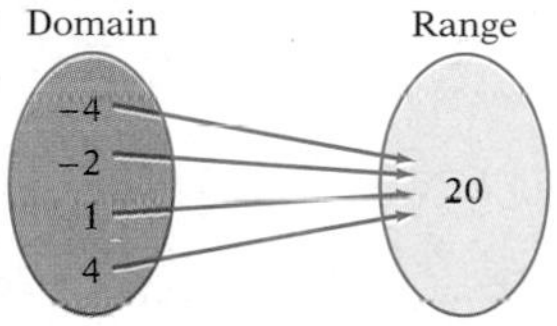

7. Does the diagram below represent a function? Explain your answer.

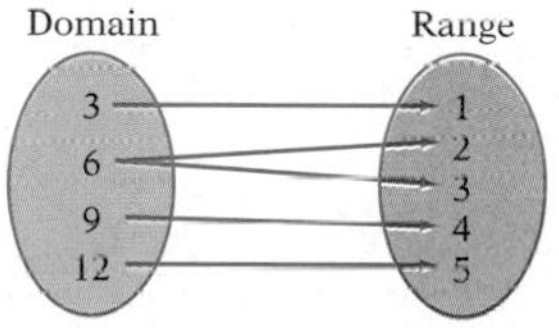

8. Does the diagram below represent a function? Explain your answer.

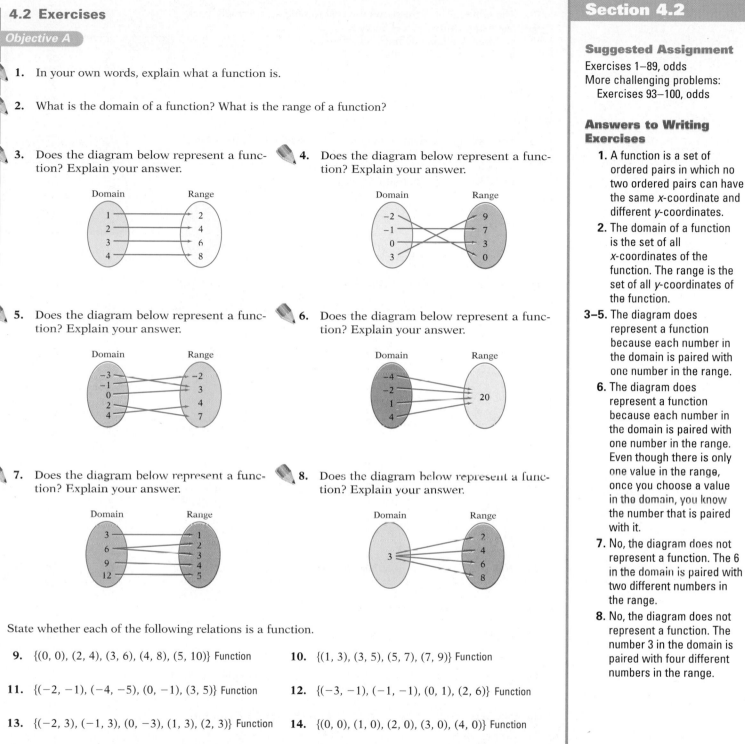

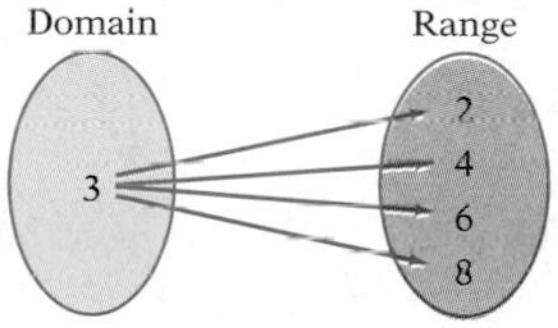

State whether each of the following relations is a function.

9. {(0, 0), (2, 4), (3, 6), (4, 8), (5, 10)} Function

10. {(1, 3), (3, 5), (5, 7), (7, 9)} Function

11. {(−2, −1), (−4, −5), (0, −1), (3, 5)} Function

12. {(−3, −1), (−1, −1), (0, 1), (2, 6)} Function

13. {(−2, 3), (−1, 3), (0, −3), (1, 3), (2, 3)} Function

14. {(0, 0), (1, 0), (2, 0), (3, 0), (4, 0)} Function

15. {(1, 1), (4, 2), (9, 3), (1, −1), (4, −2)} Not a function

16. {(3, 1), (3, 2), (3, 3), (3, 4)} Not a function

217

Section 4.2

Suggested Assignment

Exercises 1–89, odds
More challenging problems:
Exercises 93–100, odds

Answers to Writing Exercises

1. A function is a set of ordered pairs in which no two ordered pairs can have the same x-coordinate and different y-coordinates.
2. The domain of a function is the set of all x-coordinates of the function. The range is the set of all y-coordinates of the function.

3–5. The diagram does represent a function because each number in the domain is paired with one number in the range.

6. The diagram does represent a function because each number in the domain is paired with one number in the range. Even though there is only one value in the range, once you choose a value in the domain, you know the number that is paired with it.
7. No, the diagram does not represent a function. The 6 in the domain is paired with two different numbers in the range.
8. No, the diagram does not represent a function. The number 3 in the domain is paired with four different numbers in the range.

Quick Quiz (Objective 4.2A)

1. Find the domain and range of the function {(2, 9), (3, 13), (4, 17)}. D: {2, 3, 4}; R: {9, 13, 17}
2. Evaluate $f(z) = \frac{5z}{z+1}$ when $z = -2$. 10
3. Find the range of $h(x) = x^2 + x - 3$ if the domain is {−2, −1, 0, 1, 2}. {−3, −1, 3}
4. What values are excluded from the domain of $f(x) = 3x^2 - 4x + 1$? None

218

17. The table at the right shows the cost to send an overnight package using United Parcel Service.
a. Does this table define a function? Yes
b. Given $x = 2.75$ lb, find y. \$29.62

Weight in pounds (x)	Cost (y)
$0 < x \le 1$	\$24.05
$1 < x \le 2$	\$26.83
$2 < x \le 3$	\$29.62
$3 < x \le 4$	\$32.40
$4 < x \le 5$	\$35.18

18. The table at the right shows the cost to send an "Express Mail" package using the U.S. Postal Service.
a. Does this table define a function? Yes
b. Given $x = 0.5$ lb, find y. \$10.70

Weight in pounds (x)	Cost (y)
$0 < x \le 0.5$	\$10.70
$0.5 < x \le 2$	\$14.90
$2 < x \le 3$	\$14.90
$3 < x \le 4$	\$18.10
$4 < x \le 5$	\$21.25

Given $f(x) = 5x - 4$, evaluate:

19. $f(3)$
11

20. $f(-2)$
-14

21. $f(0)$
-4

22. $f(-1)$
-9

Given $G(t) = 4 - 3t$, evaluate:

23. $G(0)$
4

24. $G(-3)$
13

25. $G(-2)$
10

26. $G(4)$
-8

Given $q(r) = r^2 - 4$, evaluate:

27. $q(3)$
5

28. $q(4)$
12

29. $q(-2)$
0

30. $q(-5)$
21

Given $F(x) = x^2 + 3x - 4$, evaluate:

31. $F(4)$
24

32. $F(-4)$
0

33. $F(-3)$
-4

34. $F(-6)$
14

Given $H(p) = \frac{3p}{p+2}$, evaluate:

35. $H(1)$
1

36. $H(-3)$
9

37. $H(t)$
$\frac{3t}{t+2}$

38. $H(v)$
$\frac{3v}{v+2}$

Given $s(t) = t^3 - 3t + 4$, evaluate:

39. $s(-1)$
6

40. $s(2)$
6

41. $s(a)$
$a^3 - 3a + 4$

42. $s(w)$
$w^3 - 3w + 4$

43. Given $P(x) = 4x + 7$, write $P(-2 + h) - P(-2)$ in simplest form.
$4h$

44. Given $G(t) = 9 - 2t$, write $G(-3 + h) - G(-3)$ in simplest form.
$-2h$

219

45. Game Engineering has just completed the programming and testing for a new computer game. The cost to manufacture and package the game depends on the number of units Game Engineering plans to sell. The table at the right shows the cost per game for packaging various quantities. Evaluate this function when

a. $x = 7000$
$4.75 per game

b. $x = 20{,}000$
$4.00 per game

Number of Games Manufactured	Cost to Manufacture One Game
$0 < x \le 2500$	$6.00
$2500 < x \le 5000$	$5.50
$5000 < x \le 10{,}000$	$4.75
$10{,}000 < x \le 20{,}000$	$4.00
$20{,}000 < x \le 40{,}000$	$3.00

46. Airport administrators have a tendency to price airport parking at a rate that discourages people from using the parking lot for long periods of time. The rate structure for an airport is given in the table at the right. Evaluate this function when

a. $t = 2.5$ h
$6.50

b. $t = 7$ h
$10.00

Hours Parked	Cost
$0 < t \le 1$	$1.00
$1 < t \le 2$	$3.00
$2 < t \le 4$	$6.50
$4 < t \le 7$	$10.00
$7 < t \le 12$	$14.00

47. A real estate appraiser charges a fee that depends on the estimated value, V, of the property. A table giving the fees charged for various estimated values of the real estate appears at the right. Evaluate this function when

a. $V = \$5{,}000{,}000$
$3000

b. $V = \$767{,}000$
$950

Value of Property	Appraisal Fee
$V < 100{,}000$	$350
$100{,}000 \le V < 500{,}000$	$525
$500{,}000 \le V < 1{,}000{,}000$	$950
$1{,}000{,}000 \le V < 5{,}000{,}000$	$2500
$5{,}000{,}000 \le V < 10{,}000{,}000$	$3000

48. The cost to mail a priority overnight package by Federal Express depends on the weight, w, of the package. A table of the costs for selected weights is given at the right. Evaluate this function when

a. $w = 2$ lb 3 oz
$29.25

b. $w = 1.9$ lb
$26.75

Weight (lb)	Cost
$0 < w < 1$	$24.00
$1 \le w < 2$	$26.75
$2 \le w < 3$	$29.25
$3 \le w < 4$	$32.00
$4 \le w < 5$	$34.00

Find the domain and range of the function.

49. $\{(1, 1), (2, 4), (3, 7), (4, 10), (5, 13)\}$
D: $\{1, 2, 3, 4, 5\}$; R: $\{1, 4, 7, 10, 13\}$

50. $\{(2, 6), (4, 18), (6, 38), (8, 66), (10, 102)\}$
D: $\{2, 4, 6, 8, 10\}$; R: $\{6, 18, 38, 66, 102\}$

51. $\{(0, 1), (2, 2), (4, 3), (6, 4)\}$
D: $\{0, 2, 4, 6\}$; R: $\{1, 2, 3, 4\}$

52. $\{(0, 1), (1, 2), (4, 3), (9, 4)\}$
D: $\{0, 1, 4, 9\}$; R: $\{1, 2, 3, 4\}$

53. $\{(1, 0), (3, 0), (5, 0), (7, 0), (9, 0)\}$
D: $\{1, 3, 5, 7, 9\}$; R: $\{0\}$

54. $\{(-2, -4), (2, 4), (-1, 1), (1, 1), (-3, 9), (3, 9)\}$
D: $\{-3, -2, -1, 1, 2, 3\}$; R: $\{-4, 1, 4, 9\}$

55. {(0, 0), (1, 1), (−1, 1), (2, 2), (−2, 2)}
D: {−2, −1, 0, 1, 2}; R: {0, 1, 2}

56. {(0, −5), (5, 0), (10, 5), (15, 10)}
D: {0, 5, 10, 15}; R: {−5, 0, 5, 10}

57. {(−2, −3), (−1, 6), (0, 7), (2, 3), (1, 9)}
D: {−2, −1, 0, 1, 2}; R: {−3, 3, 6, 7, 9}

58. {(−8, 0), (−4, 2), (−2, 4), (0, −4), (4, 4)}
D: {−8, −4, −2, 0, 4}; R: {−4, 0, 2, 4}

What values are excluded from the domain of the function?

59. $f(x) = \frac{1}{x - 1}$
1

60. $g(x) = \frac{1}{x + 4}$
−4

61. $h(x) = \frac{x + 3}{x + 8}$
−8

62. $F(x) = \frac{2x - 5}{x - 4}$
4

63. $f(x) = 3x + 2$
None

64. $g(x) = 4 - 2x$
None

65. $G(x) = x^2 + 1$
None

66. $H(x) = \frac{1}{2}x^2$
None

67. $f(x) = \frac{x - 1}{x}$
0

68. $g(x) = \frac{2x + 5}{7}$
None

69. $H(x) = x^2 - x + 1$
None

70. $f(x) = 3x^2 + x + 4$
None

71. $f(x) = \frac{2x - 5}{3}$
None

72. $g(x) = \frac{3 - 5x}{5}$
None

73. $H(x) = \frac{x - 2}{x + 2}$
−2

74. $h(x) = \frac{3 - x}{6 - x}$
6

75. $f(x) = \frac{x - 2}{2}$
None

76. $G(x) = \frac{2}{x - 2}$
2

Find the range of the function defined by each equation and the given domain.

77. $f(x) = 4x - 3$; domain = {0, 1, 2, 3}
{−3, 1, 5, 9}

78. $G(x) = 3 - 5x$; domain = {−2, −1, 0, 1, 2}
{13, 8, 3, −2, −7}

79. $g(x) = 5x - 8$; domain = {−3, −1, 0, 1, 3}
{−23, −13, −8, −3, 7}

80. $h(x) = 3x - 7$; domain = {−4, −2, 0, 2, 4}
{−19, −13, −7, −1, 5}

81. $h(x) = x^2$; domain = {−2, −1, 0, 1, 2}
{0, 1, 4}

82. $H(x) = 1 - x^2$; domain = {−2, −1, 0, 1, 2}
{−3, 0, 1}

221

83. $f(x) = 2x^2 - 2x + 2$;
domain $= \{-4, -2, 0, 4\}$
$\{2, 14, 26, 42\}$

84. $G(x) = -2x^2 + 5x - 2$;
domain $= \{-3, -1, 0, 1, 3\}$
$\{-35, -9, -5, -2, 1\}$

85. $H(x) = \dfrac{5}{1 - x}$; domain $= \{-2, 0, 2\}$
$\left\{-5, \dfrac{5}{3}, 5\right\}$

86. $g(x) = \dfrac{4}{4 - x}$; domain $= \{-5, 0, 3\}$
$\left\{\dfrac{4}{9}, 1, 4\right\}$

87. $f(x) = \dfrac{2}{x - 4}$; domain $= \{-2, 0, 2, 6\}$
$\left\{-1, -\dfrac{1}{2}, -\dfrac{1}{3}, 1\right\}$

88. $g(x) = \dfrac{x}{3 - x}$; domain $= \{-2, -1, 0, 1, 2\}$
$\left\{-\dfrac{2}{5}, -\dfrac{1}{4}, 0, \dfrac{1}{2}, 2\right\}$

89. $H(x) = 2 - 3x - x^2$; domain $= \{-5, 0, 5\}$
$\{-38, -8, 2\}$

90. $G(x) = 4 - 3x - x^3$; domain $= \{-3, 0, 3\}$
$\{-32, 4, 40\}$

APPLYING THE CONCEPTS

91. Explain the words *relation* and *function*. Include in your explanation how the meanings of the two words differ.

92. Give a real-world example of a relation that is not a function. Is it possible to give an example of a function that is not a relation? If so, give one. If not, explain why it is not possible.

93. Find the set of ordered pairs (x, y) determined by the equation $y = x^3$, where $x \in \{-2, -1, 0, 1, 2\}$. Does the set of ordered pairs define a function? Why or why not?
$\{(-2, -8), (-1, -1), (0, 0), (1, 1), (2, 8)\}$. Yes, the set defines a function because each member of the domain is assigned to exactly one member of the range.

94. Find the set of ordered pairs (x, y) determined by the equation $|y| = x$, where $x \in \{0, 1, 2, 3\}$. Does the set of ordered pairs define a function? Why or why not?
$\{(0, 0), (1, -1), (1, 1), (2, -2), (2, 2), (3, -3), (3, 3)\}$. No, this is not a function because some points in the domain are assigned to two values in the range.

95. The power a windmill can generate is a function of the velocity of the wind. The function can be approximated by $P = f(v) = 0.015v^3$, where P is the power in watts and v is the velocity of the wind in meters per second. How much power will be produced by a windmill when the velocity of the wind is 15 m/s?
50.625 watts

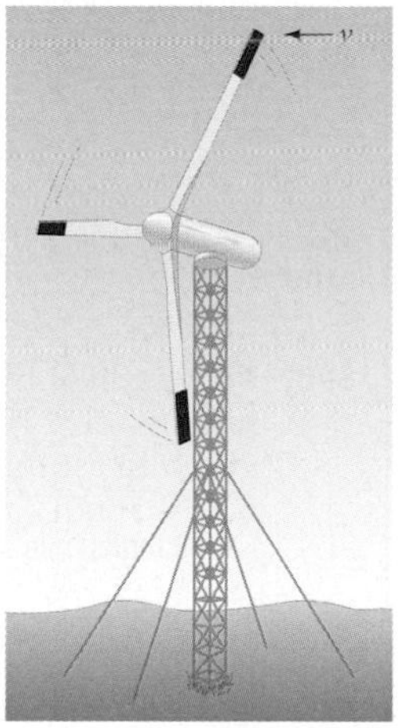

Answers to Writing Exercises

91. A relation and a function are similar in that both are sets of ordered pairs. A function is a specific type of relation. A function is a relation in which there are no two ordered pairs with the same first element. This can be alternatively stated as follows: A function is a relation in which no two ordered pairs with the same first component have different second components.

92. Students will provide different examples of relations that are not functions. Examples include a person's height and his or her shoe size, or a person's age and the educational level of that person. It is not possible to give an example of a function that is not a relation because every function is a relation.

96. The distance s, in feet, that a car will skid on a certain road surface after the brakes are applied is a function of the car's velocity, v (in miles per hour). The function can be approximated by $s = f(v) = 0.017v^2$. How far will a car skid after its brakes are applied if it is traveling 60 mph?
61.2 ft

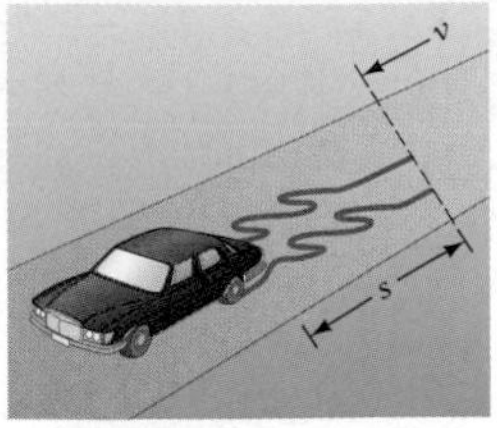

Each of the following graphs defines a function. Evaluate the function by estimating the ordinate (which is the value of the function) for the given value of t.

97. The graph at the right shows the speed v, in feet per second, that a parachutist is falling during the first 20 s after jumping out of a plane. Estimate the speed at which the parachutist is falling when

a. $t = 5$ s
22 ft/s

b. $t = 15$ s
30 ft/s

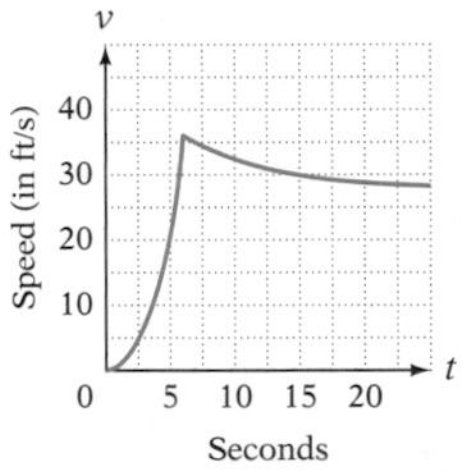

98. The graph at the right shows what an industrial psychologist has determined to be the average percent score, P, for an employee taking a performance test t weeks after training begins. Estimate the score an employee would receive on this test when

a. $t = 4$ weeks
90%

b. $t = 10$ weeks
100%

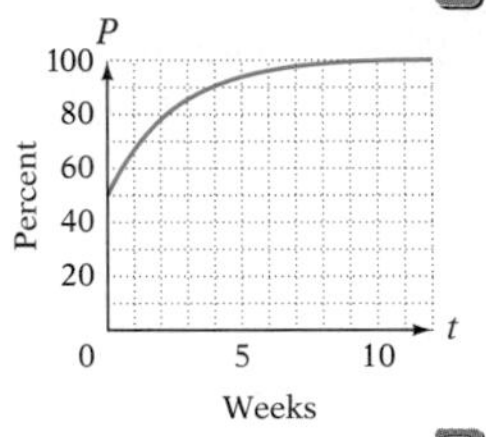

99. The graph at the right shows the temperature T, in degrees Fahrenheit, of a can of cola t hours after it is placed in a refrigerator. Use the graph to estimate the temperature of the cola when

a. $t = 5$ h
64°F

b. $t = 15$ h
52°F

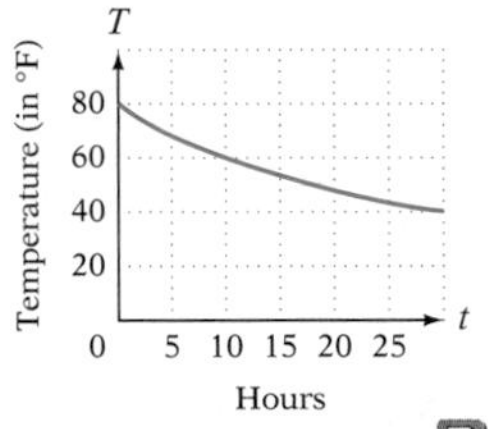

100. The graph at the right shows the decrease in the heart rate r, in beats per minute, of a runner t minutes after the completion of a race. Use the graph to estimate the heart rate of a runner when

a. $t = 5$ min
110 beats/min

b. $t = 20$ min
75 beats/min

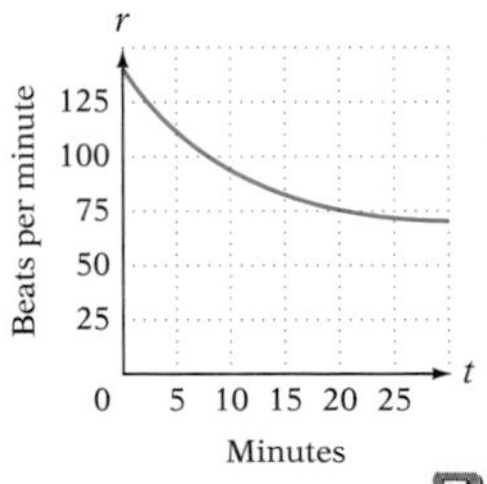

4.3 Linear Equations in Two Variables

Objective A To graph an equation of the form $y = mx + b$

VIDEO & DVD CD TUTOR WEB SSM

CALCULATOR NOTE

The Projects and Group Activities at the end of this chapter contain information on using calculators to graph an equation.

The **graph of an equation in two variables** is a graph of the ordered-pair solutions of the equation.

Consider $y = 2x + 1$. Choosing $x = -2$, -1, 0, 1, and 2 and determining the corresponding values of y produces some of the ordered pairs of the equation. These are recorded in the table at the right. See the graph of the ordered pairs in Figure 1.

x	$y = 2x + 1$	y	(x, y)
-2	$2(-2) + 1$	-3	$(-2, -3)$
-1	$2(-1) + 1$	-1	$(-1, -1)$
0	$2(0) + 1$	1	$(0, 1)$
1	$2(1) + 1$	3	$(1, 3)$
2	$2(2) + 1$	5	$(2, 5)$

Choosing values of x that are not integers produces more ordered pairs to graph, such as $\left(-\frac{5}{2}, -4\right)$ and $\left(\frac{3}{2}, 4\right)$, as shown in Figure 2. Choosing still other values of x would result in more and more ordered pairs being graphed. The result would be so many dots that the graph would appear as the straight line shown in Figure 3, which is the graph of $y = 2x + 1$.

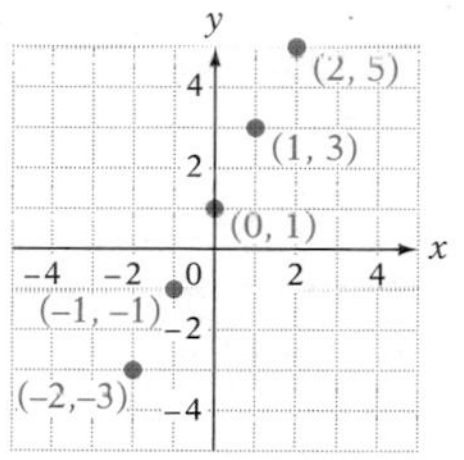

Figure 1

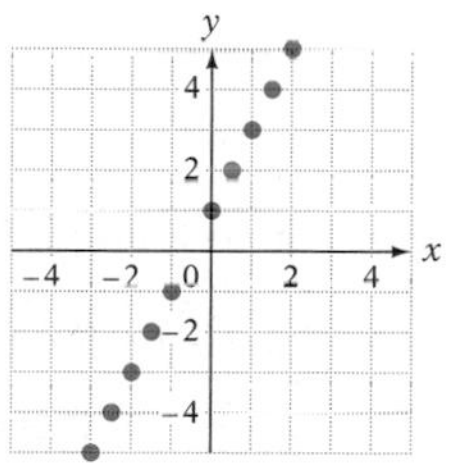

Figure 2

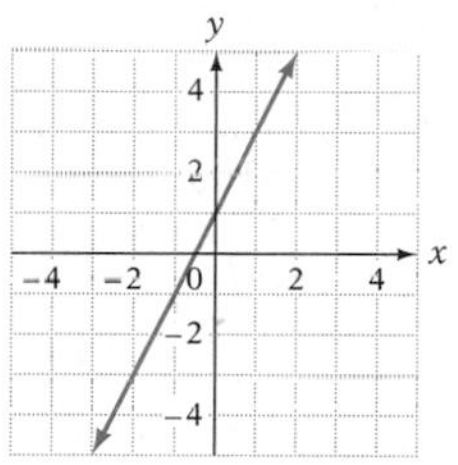

Figure 3

Equations in two variables have characteristic graphs. The equation $y = 2x + 1$ is an example of a *linear equation,* or *linear function,* because its graph is a straight line. It is also called a *first-degree equation* in two variables because the exponent on each variable is the first power.

> **Linear Equation in Two Variables**
>
> Any equation of the form $y = mx + b$, where m is the coefficient of x and b is a constant, is a **linear equation in two variables or a first-degree equation in two variables.** The graph of a linear equation in two variables is a straight line.

Examples of linear equations are shown at the right. These equations represent linear functions because there is only one possible y for each x. Note that for $y = 3 - 2x$, m is the coefficient of x and b is the constant.

$y = 2x + 1$ $(m = 2, b = 1)$

$y = x - 4$ $(m = 1, b = -4)$

$y = -\frac{3}{4}x$ $\left(m = -\frac{3}{4}, b = 0\right)$

$y = 3 - 2x$ $(m = -2, b = 3)$

The equation $y = x^2 + 4x + 3$ is not a linear equation in two variables because there is a term with a variable squared. The equation $y = \frac{3}{x - 4}$ is not a linear equation because a variable occurs in the denominator of a fraction.

Objective 4.3A

New Vocabulary

graph of an equation in two variables
linear equation in two variables
first-degree equation in two variables
linear function

Instructor Note

It is important for students to associate "graph is a straight line" with "$y = mx + b$." The Concept Check below gives exercises for students who may not see the connection.

Concept Check

1. Place an L to the left of each equation below that is a linear equation in two variables.

$y = x^2 + 1$

L $y = -x$ X

$y = \frac{1}{x}$

L $y = 2 - \frac{1}{2}x$ X

$y = \sqrt{x} - 1$

2. Place an X to the right of each equation above whose graph is a straight line.

224

To graph a linear equation, find ordered-pair solutions of the equation. Do this by choosing any value of x and finding the corresponding value of y. Repeat this procedure, choosing different values for x, until you have found the number of solutions desired.

Because the graph of a linear equation in two variables is a straight line, and a straight line is determined by two points, it is necessary to find only two solutions. However, it is recommended that at least three points be used to ensure accuracy.

Discuss the Concepts

In the example at the right, we have chosen three values of x to draw the graph, even though only two values are necessary. Have students explain why choosing three values of x and then plotting the resulting points helps ensure accuracy.

TAKE NOTE

If the three points you graph do not lie on a straight line, then either you have made an arithmetic error in calculating a point or you have plotted a point incorrectly.

➡ Graph $y = 2x + 1$.

Choose any values of x, and then find corresponding values of y. The numbers 0, 2, and -1 were chosen arbitrarily for x. It is convenient to record these solutions in a table.

x	$y = 2x + 1$	y
0	$2(0) + 1$	1
2	$2(2) + 1$	5
-1	$2(-1) + 1$	-1

Graph the ordered-pair solutions (0, 1), (2, 5), and $(-1, -1)$. Draw a line through the ordered-pair solutions.

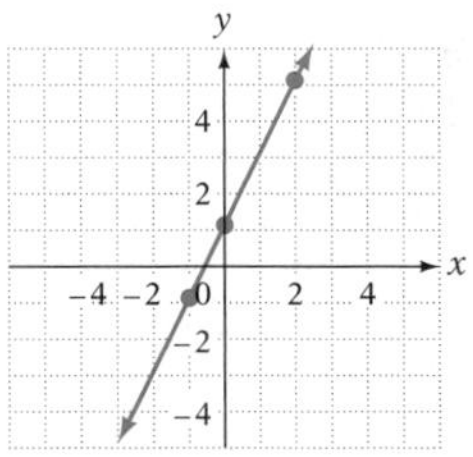

Concept Check

Is $(2, -3)$ a solution of the equation whose graph is shown below? No

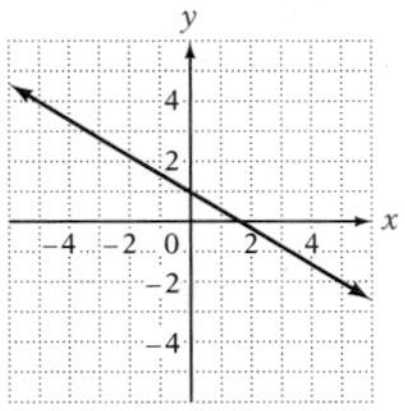

Remember that a graph is a drawing of the ordered-pair solutions of the equation. Therefore, every point on the graph is a solution of the equation, and every solution of the equation is a point on the graph.

The graph at the right is the graph of $y = x + 2$. Note that $(-4, -2)$ and $(1, 3)$ are points on the graph and that these points are solutions of $y = x + 2$. The point whose coordinates are $(4, 1)$ is not a point on the graph and is not a solution of the equation.

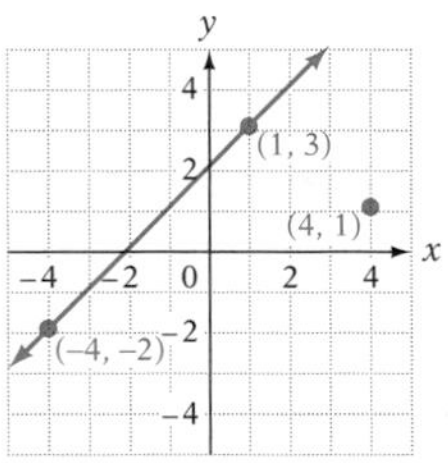

Discuss the Concepts

As noted in the example at the right, when m is a fraction, it is helpful to choose values of x that simplify the calculation by eliminating the fraction. Describe the values of x that you would choose in order to find integer solutions to each of the following equations.

1. $y = \frac{1}{3}x - 1$ Multiples of 3, such as -3, 0, 3, and 6
2. $y = -\frac{2}{5}x + 2$ Multiples of 5, such as -5, 0, 5, and 10

When m is a fraction in the equation $y = mx + b$, choose values of x that will simplify the evaluation.

➡ Graph $y = \frac{1}{3}x - 1$.

m is a fraction. $\left(m = \frac{1}{3}\right)$

Choose values of x that are multiples of the denominator. The numbers 0, 3, and -3 are used here.

x	y
0	-1
3	0
-3	-2

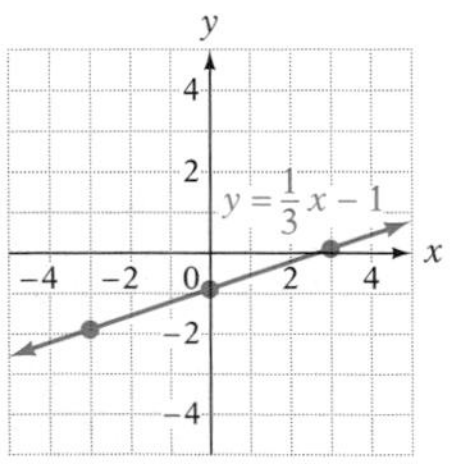

In-Class Examples (Objective 4.3A)

Graph.

1. $y = x - 2$
2. $y = -3x + 1$
3. $y = -\frac{4}{3}x$

1. 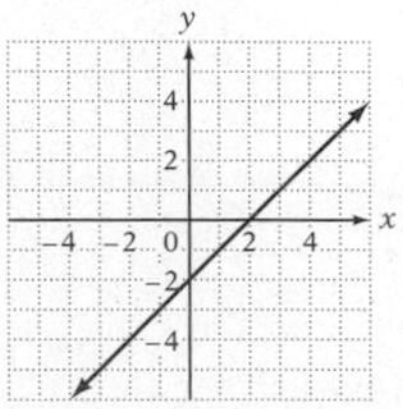

2. 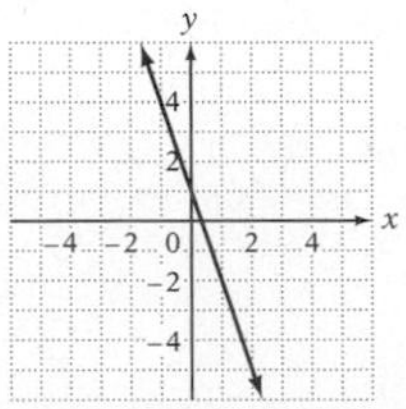

3. 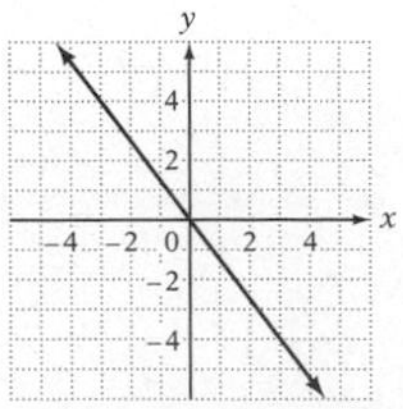

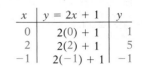

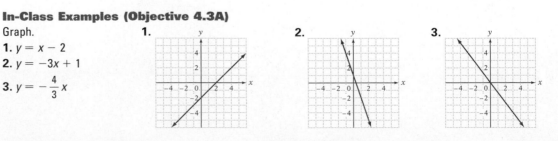

Example 1 Graph $y = 3x - 2$.

Solution

x	y
0	−2
−1	−5
2	4

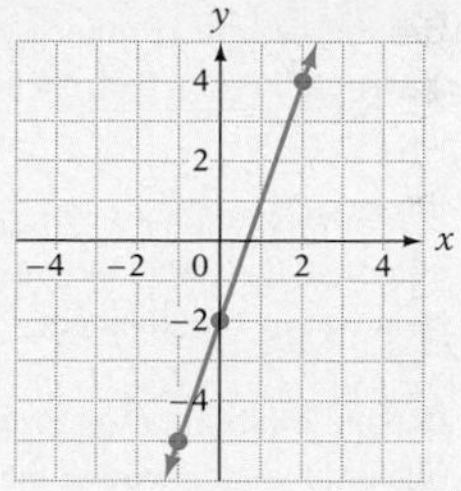

You Try It 1 Graph $y = 3x + 1$.

Your solution

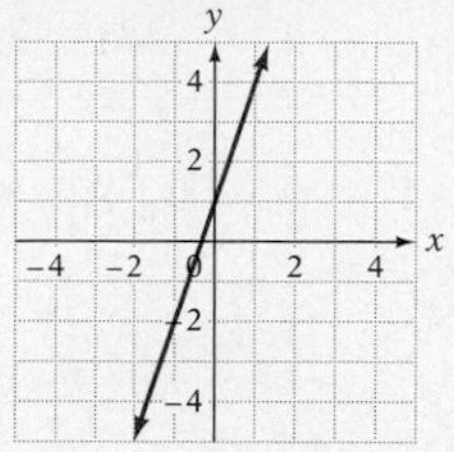

Example 2 Graph $y = 2x$.

Solution

x	y
0	0
2	4
−2	−4

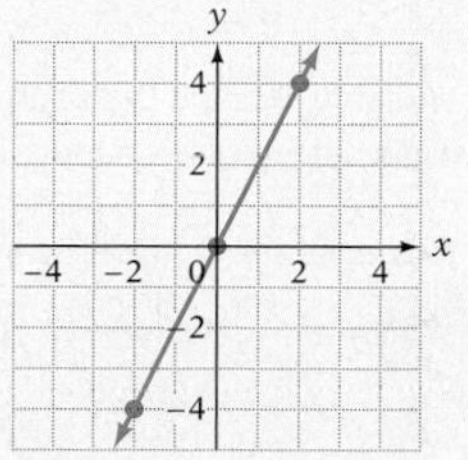

You Try It 2 Graph $y = -2x$.

Your solution

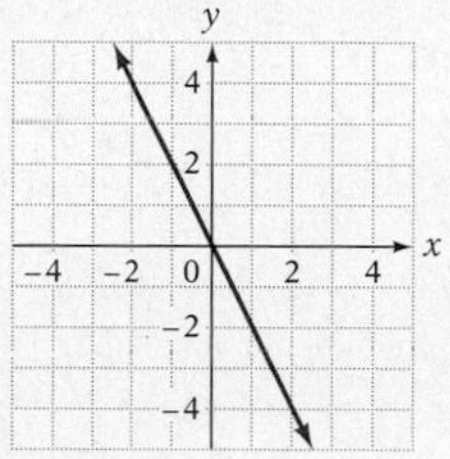

Example 3

Graph $y = \frac{1}{2}x - 1$.

Solution

x	y
0	−1
2	0
−2	−2

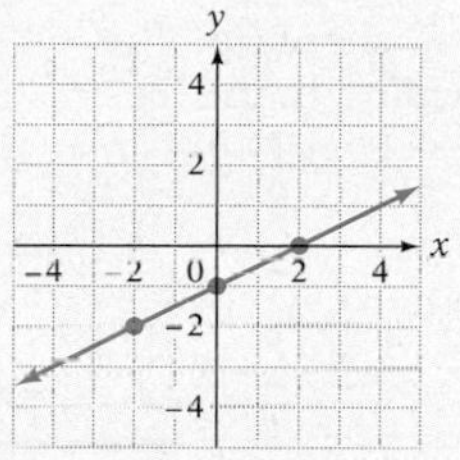

You Try It 3

Graph $y = \frac{1}{3}x - 3$.

Your solution

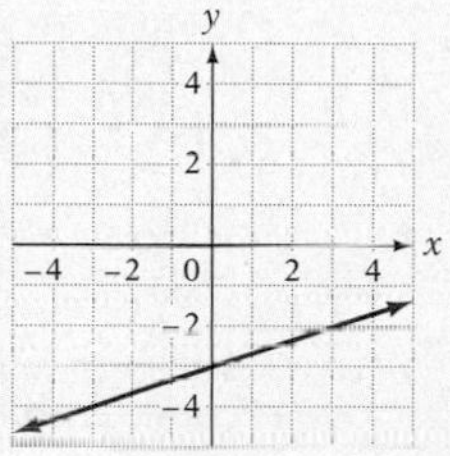

Solutions on p. S11

Objective B To graph an equation of the form $Ax + By = C$

VIDEO & DVD CD TUTOR WEB SSM

An equation in the form $Ax + By = C$ is also a linear equation. Examples of these equations are shown below.

$$\begin{array}{ll} 2x + 3y = 6 & (A = 2, B = 3, \ C = 6) \\ x - 2y = -4 & (A = 1, B = -2, C = -4) \\ 2x + y = 0 & (A = 2, B = 1, \ C = 0) \\ 4x - 5y = 2 & (A = 4, B = -5, C = 2) \end{array}$$

One method of graphing an equation of the form $Ax + By = C$ involves first solving the equation for y and then following the same procedure used for graphing an equation of the form $y = mx + b$. To solve the equation for y means to rewrite the equation so that y is alone on one side of the equation and the term containing x and the constant are on the other side of the equation. The Addition and Multiplication Properties of Equations are used to rewrite an equation of the form $Ax + By = C$ in the form $y = mx + b$.

225

Objective 4.3B

New Vocabulary

x-intercept
y-intercept
graph of $y = b$
graph of $x = a$

New Equations

$Ax + By = C$

Instructor Note

There are many times when an equation in two variables must be solved for y before it is graphed. For instance, most graphing calculators require that an equation be in this form. Rewriting linear equations in the form $y = mx + b$ is good practice for students.

Discuss the Concepts

1. Is the graph of $Ax + By = C$ always a straight line?
2. Describe the graph of $x = a$ and the graph of $y = b$.

Concept Check

1. Is the graph of a straight line always the graph of a function? If not, give an example of the graph of a line that is not a graph of a function. No. For example, the graph of $x = 3$.
2. For the function $f(x) = -3$, find $f(6)$ and $f(-5)$. −3; −3

In-Class Examples (Objective 4.3B)

1. Graph $2x - y = 4$.

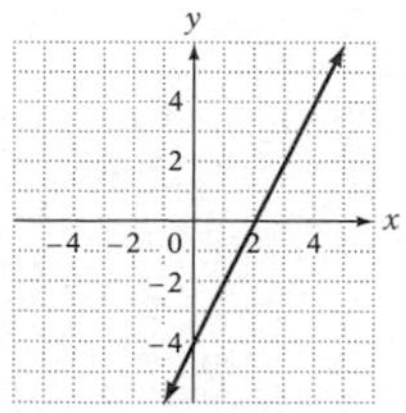

2. Graph $5x - 2y = 10$.

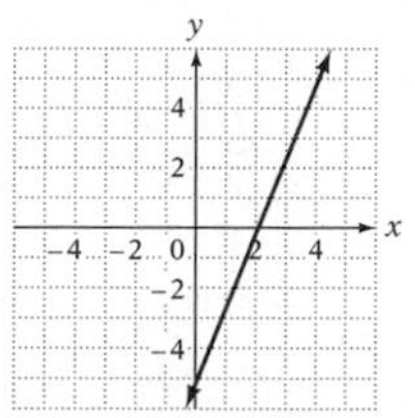

3. Graph $y = -3$.

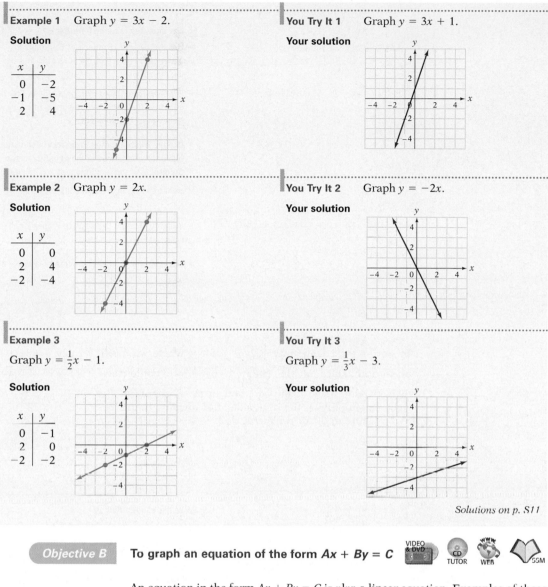

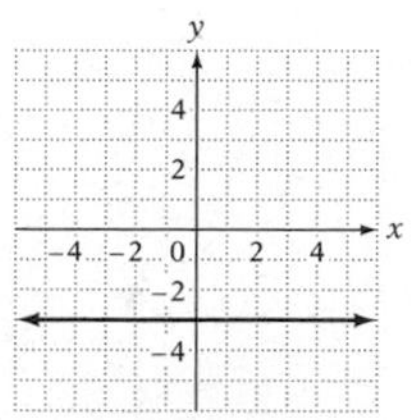

Optional Student Activity

1. Another form of the equation of a line is $\frac{x}{a} + \frac{y}{b} = 1$, where a and b are the x- and y-intercepts, respectively. Write this equation in the form $y = mx + b$. $y = -\frac{b}{a}x + b$
2. How many pairs of integers (m, n) satisfy the equation $m + n = mn$? Two: $(0, 0)$ and $(2, 2)$

➡ Solve the equation $3x + 2y = 4$ for y.

$3x + 2y = 4$ • The equation is in the form $Ax + By = C$.

$3x - 3x + 2y = -3x + 4$ • Use the Addition Property of Equations to subtract the term $3x$ from each side of the equation.

$2y = -3x + 4$ • Simplify. Note that on the right side of the equation, the term containing x is first, followed by the constant.

$\frac{1}{2} \cdot 2y = \frac{1}{2}(-3x + 4)$ • Use the Multiplication Property of Equations to multiply each side of the equation by the reciprocal of the coefficient of y. (The coefficient of y is 2; the reciprocal of 2 is $\frac{1}{2}$.)

$y = \frac{1}{2}(-3x) + \frac{1}{2}(4)$ • Simplify. Use the Distributive Property on the right side of the equation.

$y = -\frac{3}{2}x + 2$ • The equation is now in the form $y = mx + b$, with $m = -\frac{3}{2}$ and $b = 2$.

In solving the equation $3x + 2y = 4$ for y, where we multiplied both sides of the equation by $\frac{1}{2}$, we could have divided both sides of the equation by 2, as shown at the right. In simplifying the right side after dividing both sides by 2, be sure to divide *each term* by 2.

$$2y = -3x + 4$$
$$\frac{2y}{2} = \frac{-3x + 4}{2}$$
$$y = \frac{-3x}{2} + \frac{4}{2}$$
$$y = -\frac{3}{2}x + 2$$

➡ Graph $3x + 4y = 12$.

$3x + 4y = 12$ • Solve the equation for y.

$4y = -3x + 12$ • Subtract $3x$ from each side of the equation.

$y = -\frac{3}{4}x + 3$ • Divide each side of the equation by 4.

• Find three ordered-pair solutions of the equation.

x	y
0	3
4	0
−4	6

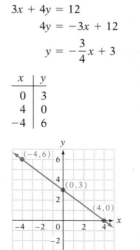

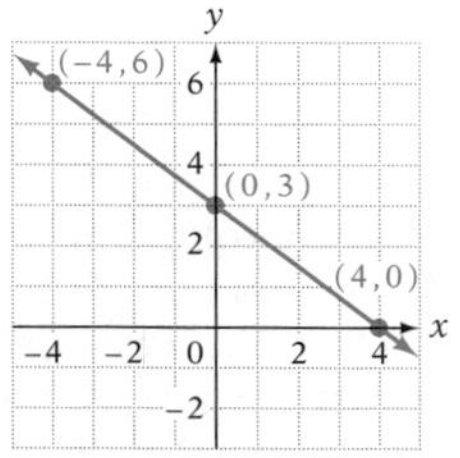

• Graph the ordered pairs and then draw a line through the points.

227

The graph of the equation $2x + 3y = 6$ is shown at the right. The graph crosses the x-axis at the point (3, 0). This point is called the **x-intercept**. The graph also crosses the y-axis at the point (0, 2). This point is called the **y-intercept**.

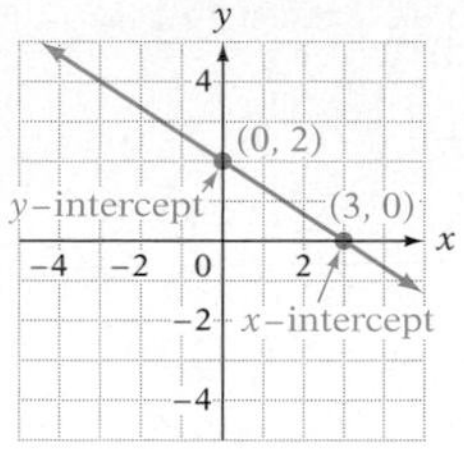

We can find the x-intercept and the y-intercept of the graph of the equation $2x + 3y = 6$ algebraically.

> **TAKE NOTE**
> Any point on the x-axis has y-coordinate 0. Any point on the y-axis has x-coordinate 0.

To find the x-intercept, let $y = 0$.
(Any point on the x-axis has y-coordinate 0.)

$$\begin{aligned} 2x + 3y &= 6 \\ 2x + 3(0) &= 6 \\ 2x &= 6 \\ x &= 3 \end{aligned}$$

The x-intercept is (3, 0).

To find the y-intercept, let $x = 0$.
(Any point on the y-axis has x-coordinate 0.)

$$\begin{aligned} 2x + 3y &- 6 \\ 2(0) + 3y &= 6 \\ 3y &= 6 \\ y &= 2 \end{aligned}$$

The y-intercept is (0, 2).

Another method of graphing an equation of the form $Ax + By = C$ is to find the x- and y-intercepts, plot both intercepts, and then draw a line through the two points. This method of graphing the equation $3x + 4y = 12$ is shown below. Note that this is the same equation graphed at the bottom of the previous page.

➡ Graph $3x + 4y = 12$ by using the x- and y-intercepts.

x-intercept: $3x + 4y = 12$
$3x + 4(0) = 12$ • To find the x-intercept, let $y = 0$.
$3x = 12$
$x = 4$ • The x-intercept is (4, 0).

y intercept: $3x \mid 4y = 12$
$3(0) + 4y = 12$ • To find the y-intercept, let $x = 0$.
$4y = 12$
$y = 3$ • The y-intercept is (0, 3).

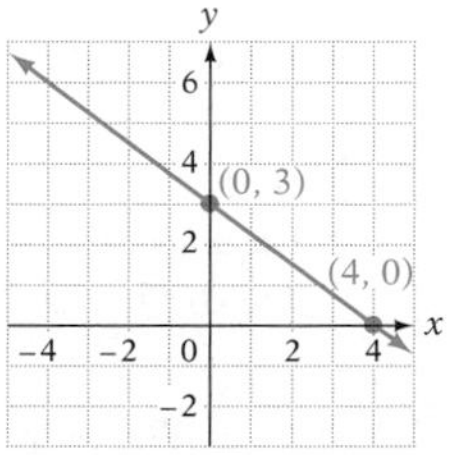

• Graph the ordered pairs (4, 0), and (0, 3). Draw a straight line through the points.

The graph of a linear equation with one of the variables missing is either a horizontal or a vertical line.

The equation $y = 2$ could be written $0x + y = 2$. Because $0x = 0$ for any value of x, the value of y is always 2, no matter what value of x is chosen. For instance, replace x by -4, -1, 0, or 3. In each case, $y = 2$.

$0x + y = 2$	
$0(-4) + y = 2$	$(-4, 2)$ is a solution.
$0(-1) + y = 2$	$(-1, 2)$ is a solution.
$0(0) + y = 2$	$(0, 2)$ is a solution.
$0(3) + y = 2$	$(3, 2)$ is a solution.

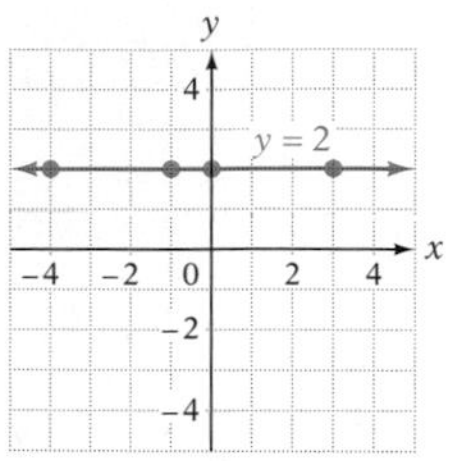

The solutions are plotted in the graph to the right, and a line is drawn through the plotted points. Note that the line is horizontal.

Graph of a Horizontal Line

The graph of $y = b$ is a horizontal line passing through $(0, b)$.

The equation $x = -2$ could be written $x + 0y = -2$. Because $0y = 0$ for any value of y, the value of x is always -2, no matter what value of y is chosen. For instance, replace y by -2, 0, 2, or 3. In each case, $x = -2$.

$x + 0y = -2$	
$x + 0(-2) = -2$	$(-2, -2)$ is a solution.
$x + 0(0) = -2$	$(-2, 0)$ is a solution.
$x + 0(2) = -2$	$(-2, 2)$ is a solution.
$x + 0(3) = -2$	$(-2, 3)$ is a solution.

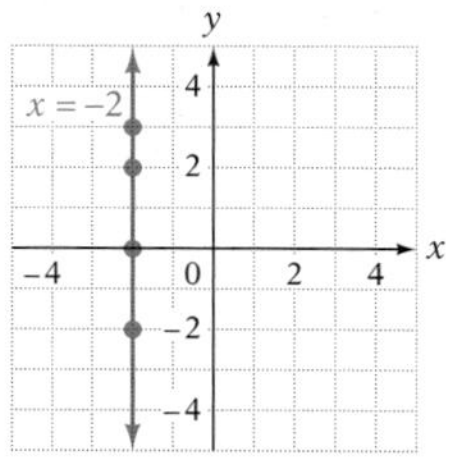

The solutions are plotted in the graph at the right, and a line is drawn through the plotted points. Note that the line is vertical.

Graph of a Vertical Line

The graph of $x = a$ is a vertical line passing through $(a, 0)$.

➡ Graph $x = -3$ and $y = -1$ on the same coordinate grid.

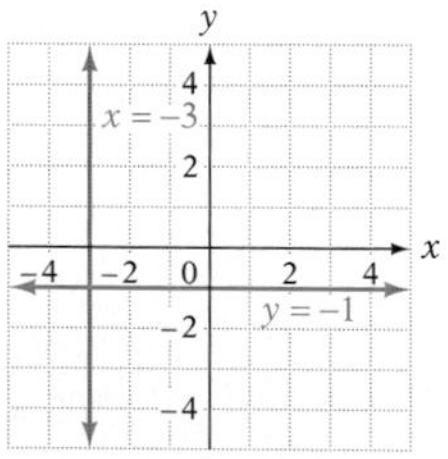

- **The graph of $x = -3$ is a vertical line passing through $(-3, 0)$.**
- **The graph of $y = -1$ is a horizontal line passing through $(0, -1)$.**

229

Example 4

Solve the equation $2x - 5y = 10$ for y. Then graph the equation.

Solution

$$2x - 5y = 10$$
$$-5y = -2x + 10$$
$$y = \frac{2}{5}x - 2$$

x	y
0	−2
5	0
−5	−4

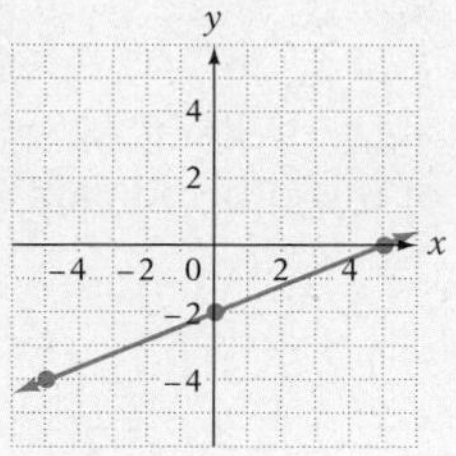

You Try It 4

Solve the equation $5x - 2y = 10$ for y. Then graph the equation.

Your solution

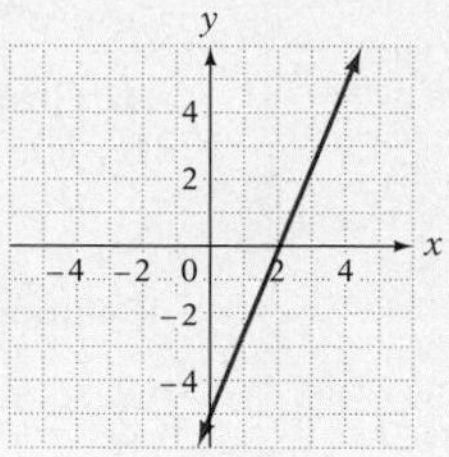

Example 5

Find the x- and y-intercepts of $x - 2y = 4$. Graph the line.

Solution

x-intercept:

$$x - 2y = 4$$
$$x - 2(0) = 4$$
$$x = 4$$

(4, 0)

y-intercept:

$$x - 2y = 4$$
$$0 - 2y = 4$$
$$-2y = 4$$
$$y = -2$$

(0, −2)

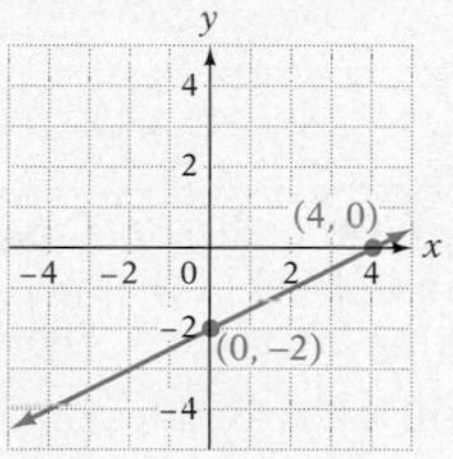

You Try It 5

Find the x- and y-intercepts of $x - 4y = -4$. Graph the line.

Your solution

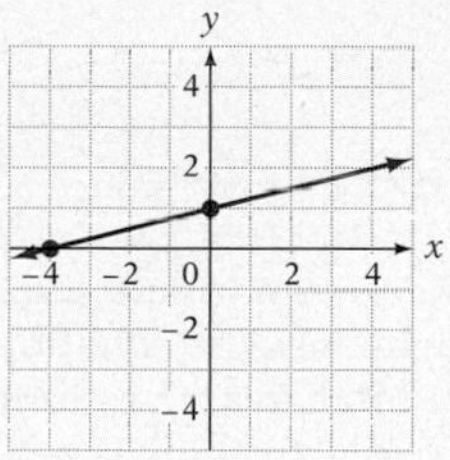

Example 6

Graph $y = -2$.

Solution

The graph of an equation of the form $y = b$ is a horizontal line passing through the point $(0, b)$.

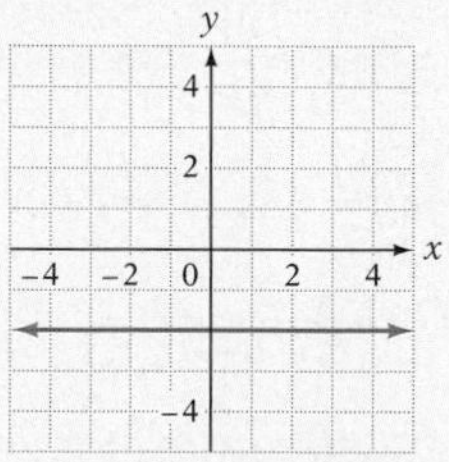

You Try It 6

Graph $x = -4$.

Your solution

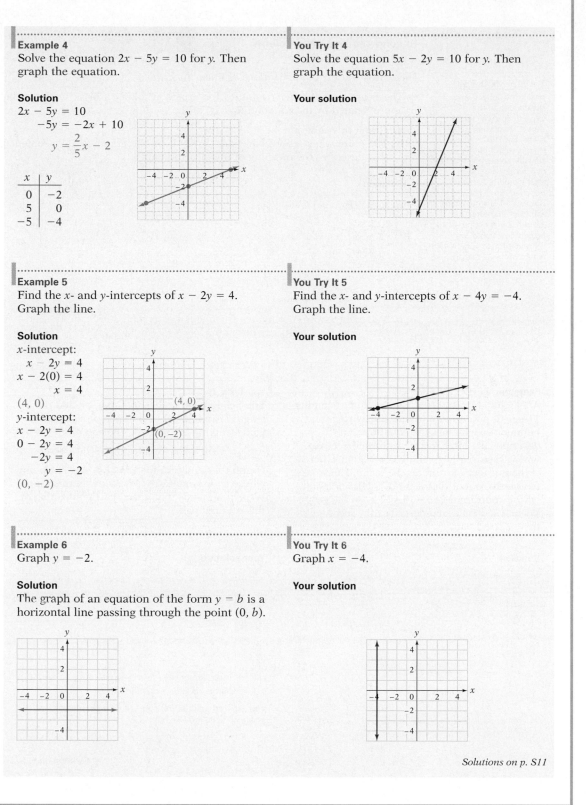

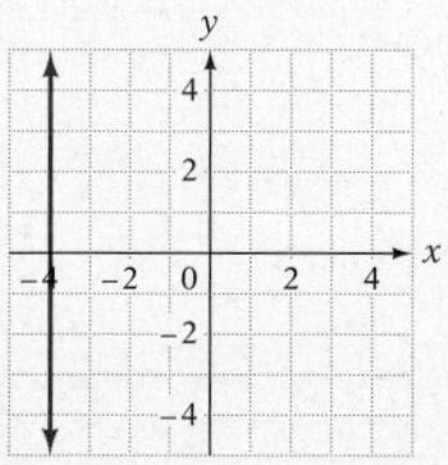

Solutions on p. S11

Objective 4.3C

Instructor Note

Part **b** of the example at the right asks the student to write a sentence that explains the meaning of an ordered pair. Questions like this one require the student to do more than just manipulate symbols. They require an understanding of the ordered pair in the context of the application.

Concept Check

Refer to the example at the right.

1. Find the person's heart rate after 8 min. 88 beats per minute
2. Verify that the point whose coordinates are (4, 80) is on the graph.

Discuss the Concepts

In Example 7, have students explain why the domain is chosen to be $0 \le t \le 60$. t represents the time it takes for the technician to repair the wiring. Generally the time required is one hour or less.

Optional Student Activity

A car dealer is advertising a 2-year lease for a new Honda Civic. Under one plan, the upfront cost is \$3000 with a monthly payment of \$150. Under a second plan, the upfront cost is \$2000 with a monthly payment of \$200.

a. Write two equations to represent the total cost C after x months of each lease. $C = 150x + 3000$; $C = 200x + 2000$

b. Graph the equations for $0 \le x \le 28$.

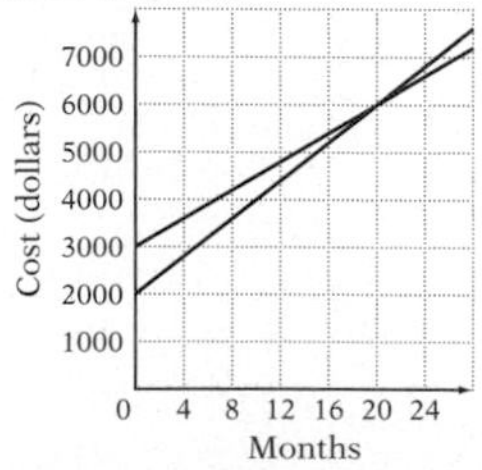

c. Which plan costs more after 12 months? after 24 months? The first plan; the second plan

Objective C To solve application problems

VIDEO & DVD CD TUTOR WEB SSM

> **TAKE NOTE**
> In many applications, the domain of the variable is given such that the equation makes sense. For this application, it would not be sensible to have values of t that are less than 0. This would indicate negative time! The number 12 is somewhat arbitrary, but after 12 min, most people's heart rates would level off and a linear function would no longer apply.

There are a variety of applications of linear functions.

➡ The heart rate, R, after t minutes for a person taking a brisk walk can be approximated by the equation $R = 2t + 72$.

a. Graph this equation for $0 \le t \le 12$.

b. The point whose coordinates are (5, 82) is on the graph. Write a sentence that describes the meaning of this ordered pair.

Solution

a.

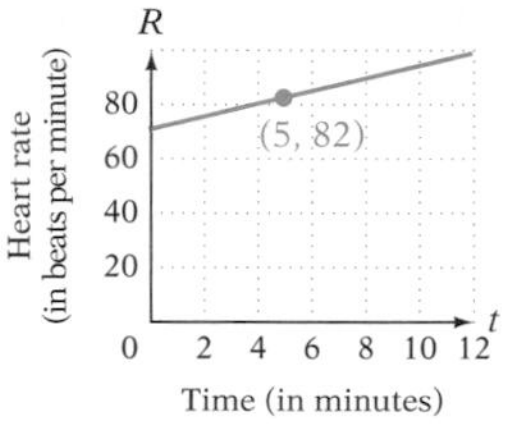

b. The point whose coordinates are (5, 82) means that after 5 min the person's heart rate is 82 beats per minute.

Example 7

A telephone repair technician charges \$45 plus \$1 per minute to repair defective wiring in a home or apartment. The equation that describes the total cost, C, to have defective wiring repaired is given by $C = t + 45$, where t is the number of minutes the technician works. Graph this equation for $0 \le t \le 60$. The point whose coordinates are (15, 60) is on the graph. Write a sentence that describes the meaning of this ordered pair.

Solution

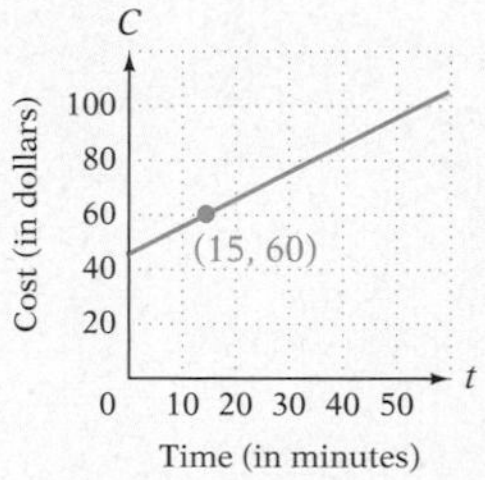

The ordered pair (15, 60) means that it costs \$60 for the technician to work 15 min.

You Try It 7

The height h (in inches) of a person and the length L (in inches) of that person's stride while walking are related. The equation $h = \frac{3}{4}L + 50$ approximates that relationship. Graph this equation for $15 \le L \le 40$. The point whose coordinates are (32, 74) is on this graph. Write a sentence that describes the meaning of this ordered pair.

Your solution

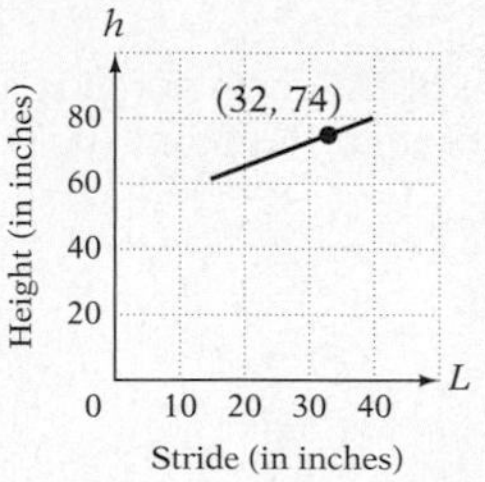

The ordered pair (32, 74) means that a person with a stride of 32 in. is 74 in. tall.

Solution on p. S11

In-Class Examples (Objective 4.3C)

A long-distance telephone service costs \$5.00 per month plus \$.10 per minute. The equation that describes the monthly cost C, in dollars, for m minutes of long-distance calls is $C = 0.10m + 5.00$. Graph this equation for $0 \le m \le 300$. The point whose coordinates are (120, 17.00) is on the graph. Write a sentence that describes the meaning of this ordered pair.

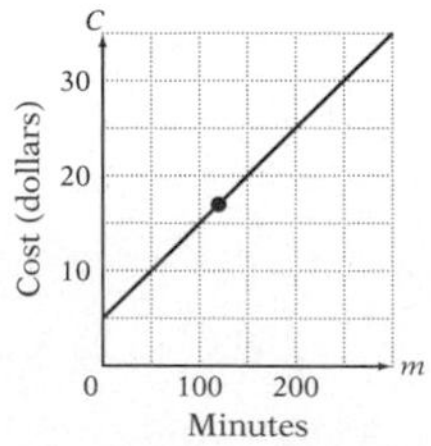

120 min of long-distance calls costs \$17.00.

231

4.3 Exercises

Objective A

Graph.

1. $y = 2x - 3$

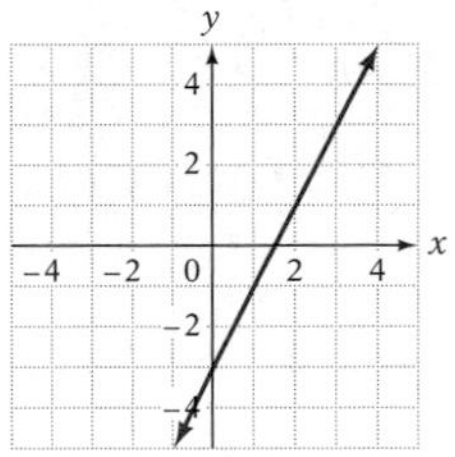

2. $y = -2x + 2$

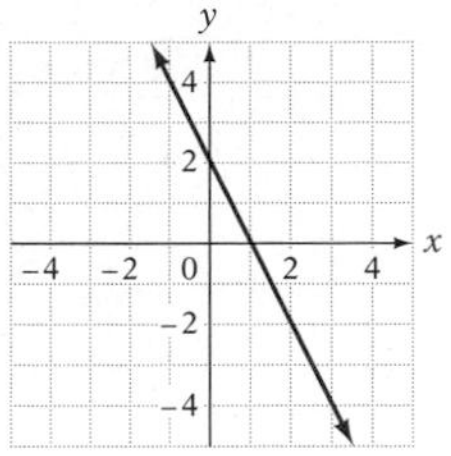

3. $y = \frac{1}{3}x$

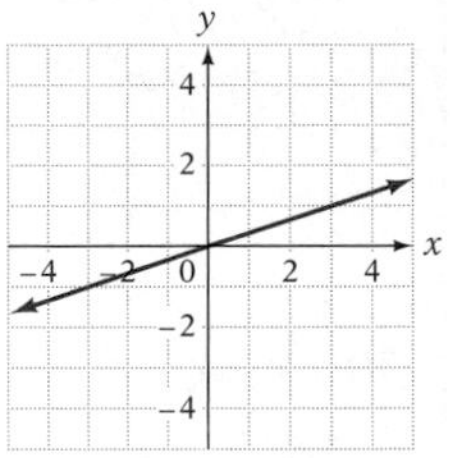

4. $y = -3x$

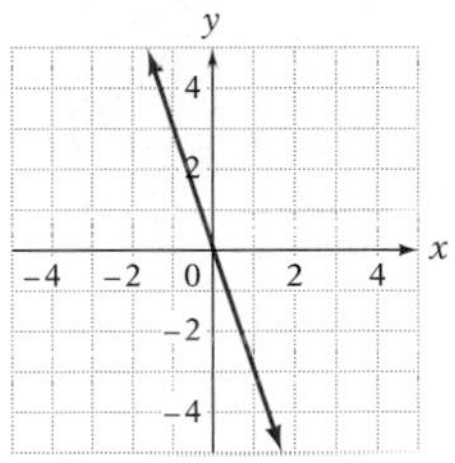

5. $y = \frac{2}{3}x - 1$

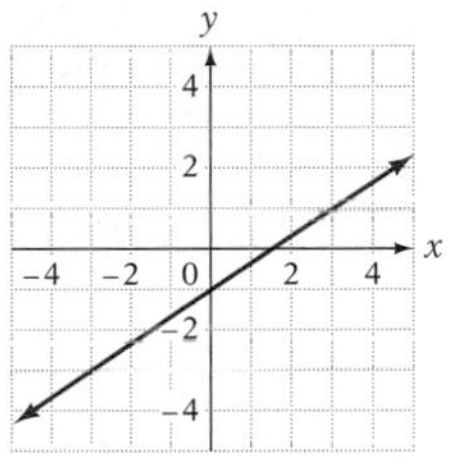

6. $y = \frac{3}{4}x + 2$

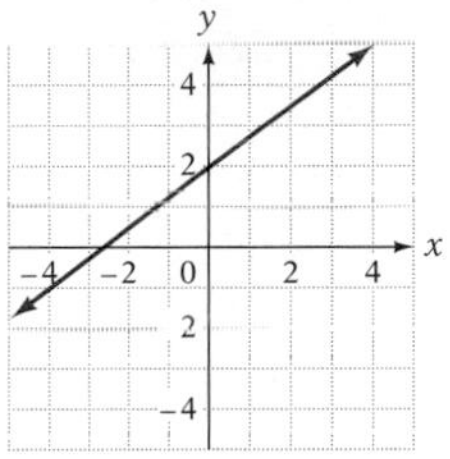

7. $y = -\frac{1}{4}x + 2$

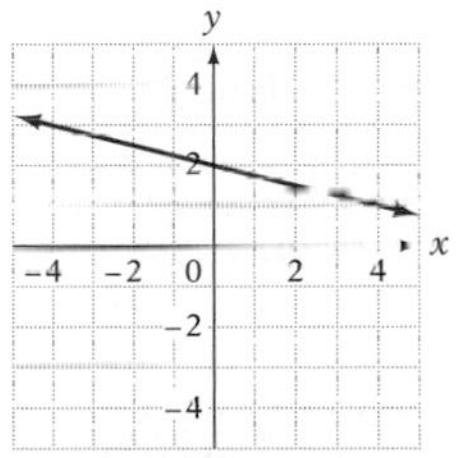

8. $y = -\frac{1}{3}x + 1$

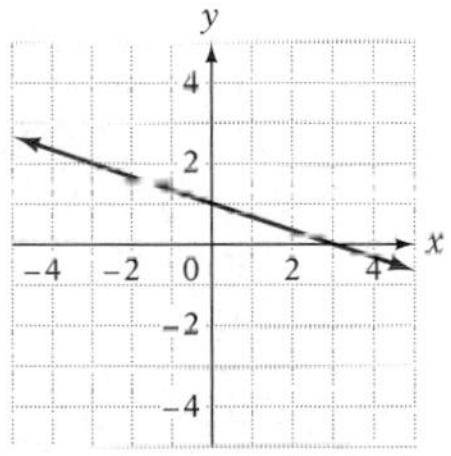

9. $y = -\frac{2}{5}x + 1$

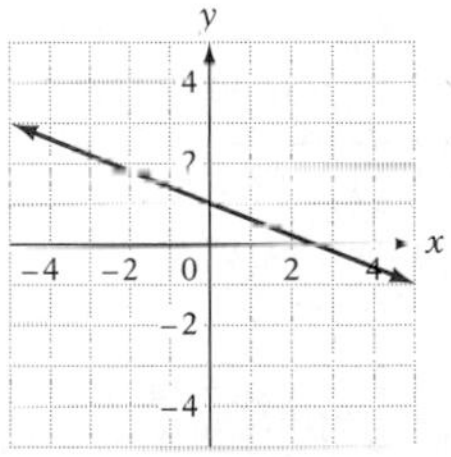

10. $y = -\frac{1}{2}x + 3$

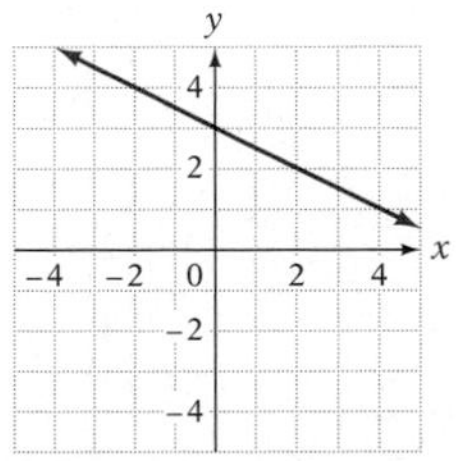

11. $y = 2x \quad 4$

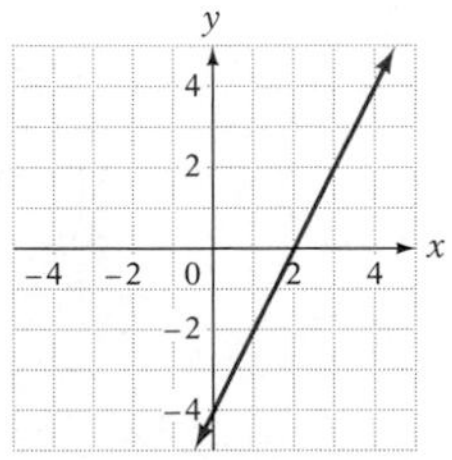

12. $y = 3x - 4$

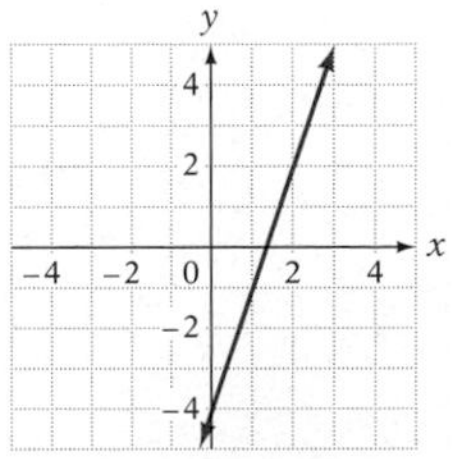

Section 4.3

Suggested Assignment

Exercises 1–43, odds

More challenging problems:
Exercise 46

Quick Quiz (Objective 4.3A)

Graph.

1. $y = -x + 3$
2. $y = 4x$
3. $y = \frac{3}{4}x + 1$

1.

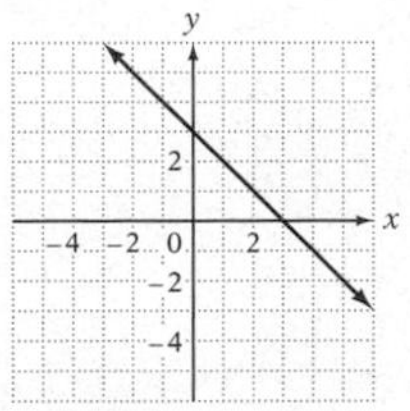

2.

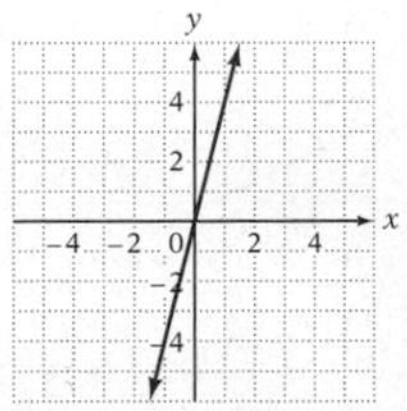

3.

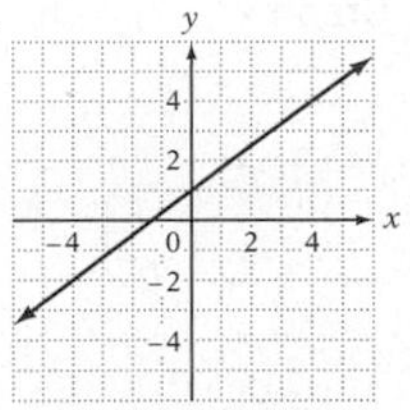

232 Chapter 4 / Linear Equations and Inequalities in Two Variables

13. $y = x - 3$

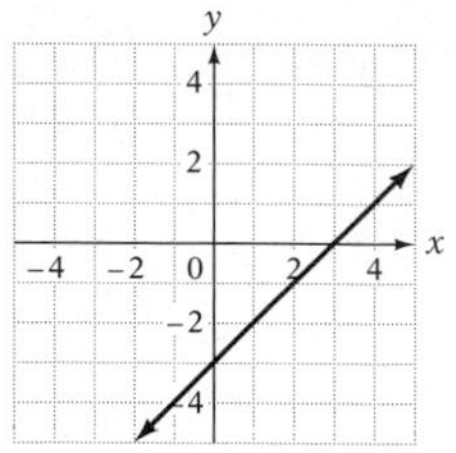

14. $y = x + 2$

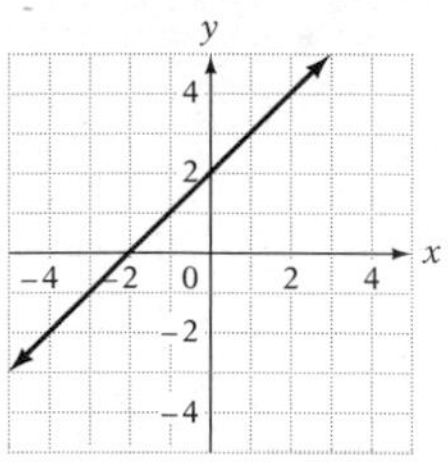

15. $y = -x + 2$

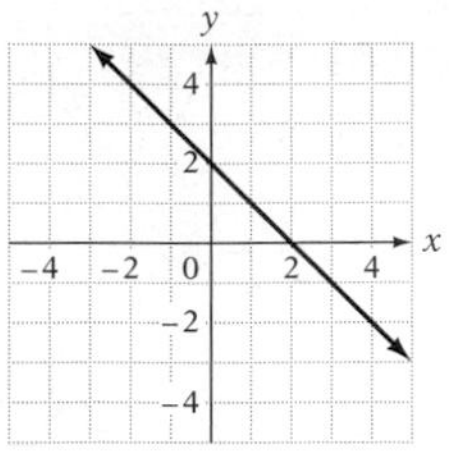

16. $y = -x - 1$

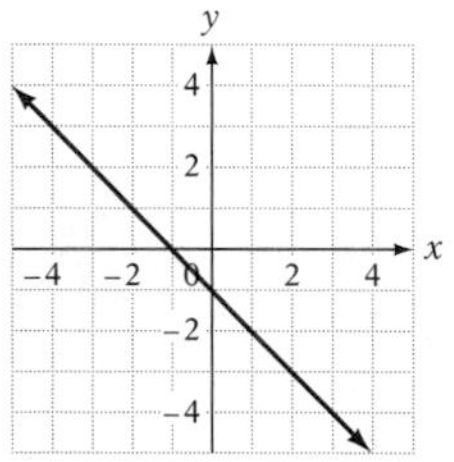

17. $y = -\frac{2}{3}x + 1$

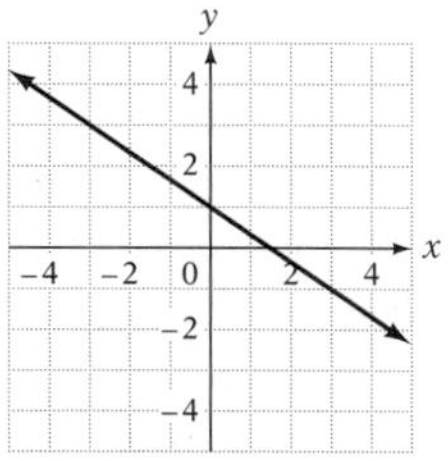

18. $y = 5x - 4$

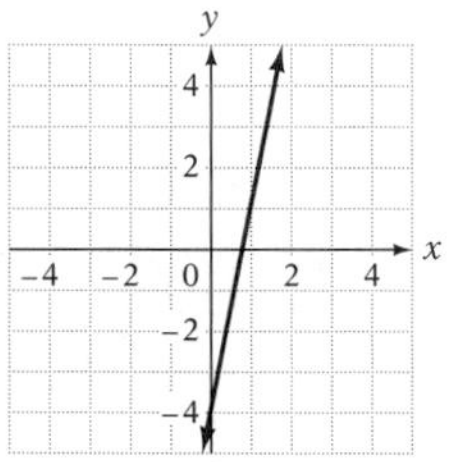

Objective B

Find the x- and y-intercepts.

19. $x - y = 3$
(3, 0), (0, −3)

20. $3x + 4y = 12$
(4, 0), (0, 3)

21. $x - 5y = 10$
(10, 0), (0, −2)

22. $3x + 2y = 12$
(4, 0), (0, 6)

23. $2x - 3y = 0$
(0, 0), (0, 0)

24. $3x + 4y = 0$
(0, 0), (0, 0)

Graph.

25. $3x + y = 3$

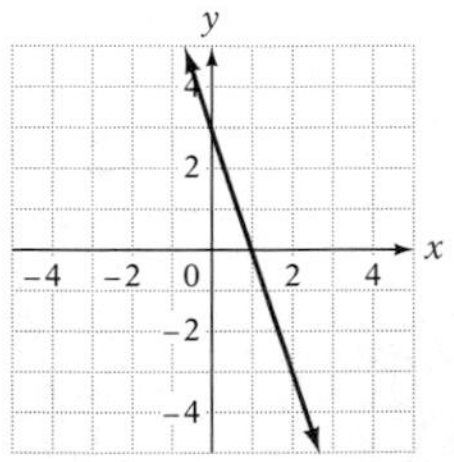

26. $2x + y = 4$

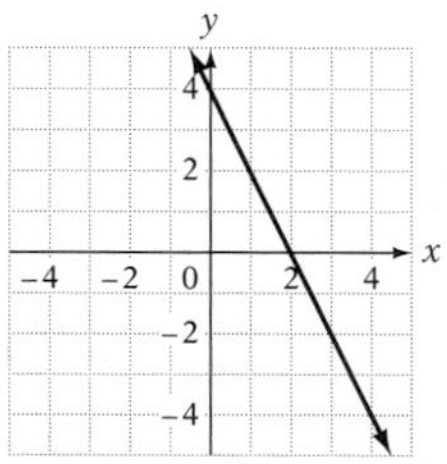

27. $2x + 3y = 6$

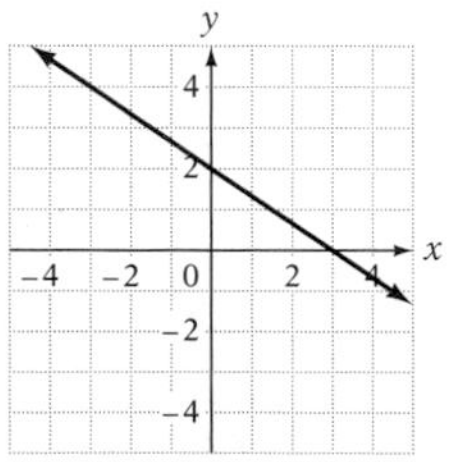

Quick Quiz (Objective 4.3B)

1. Graph $x + 2y = 6$.

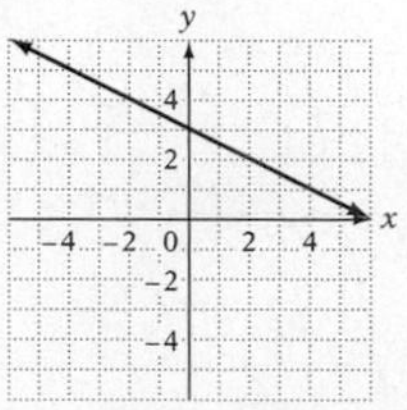

2. Graph $2x - 3y = 6$.

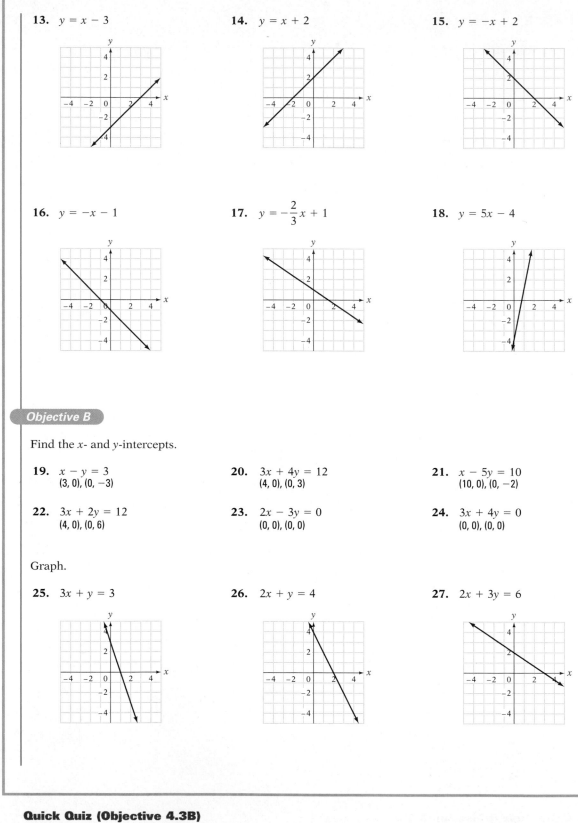

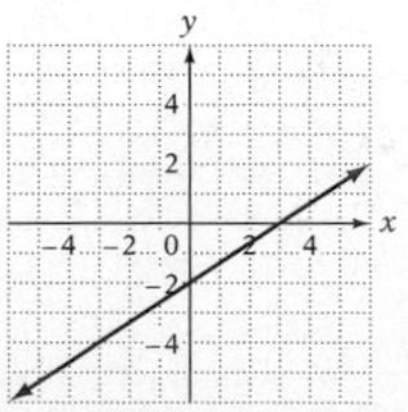

233

28. $x - 4y = 8$

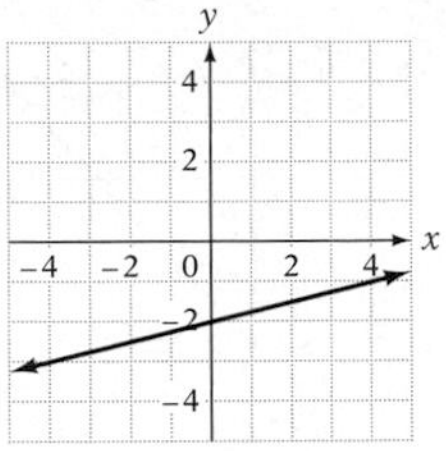

29. $2x - 3y = 12$

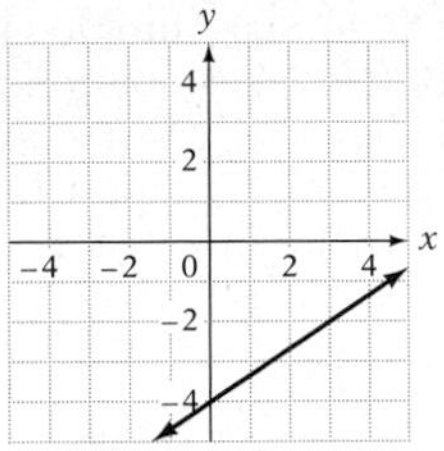

30. $x - 3y = 6$

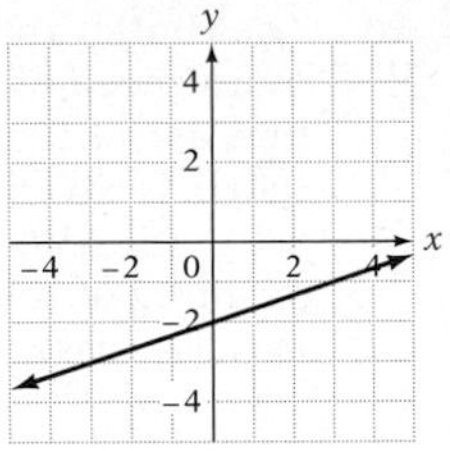

31. $2x - 3y = 6$

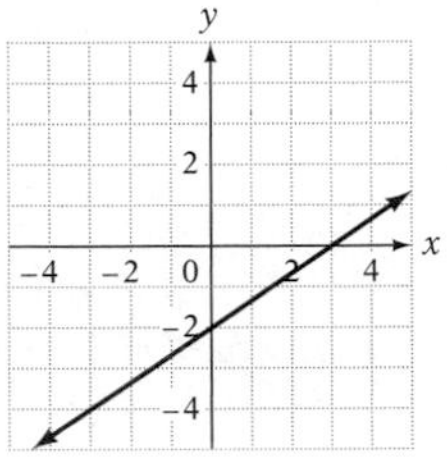

32. $x - 2y = -4$

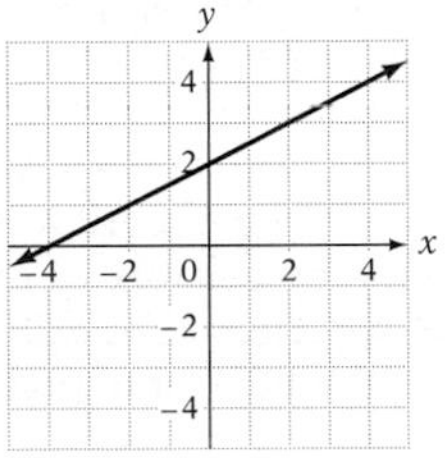

33. $2x + 5y = 10$

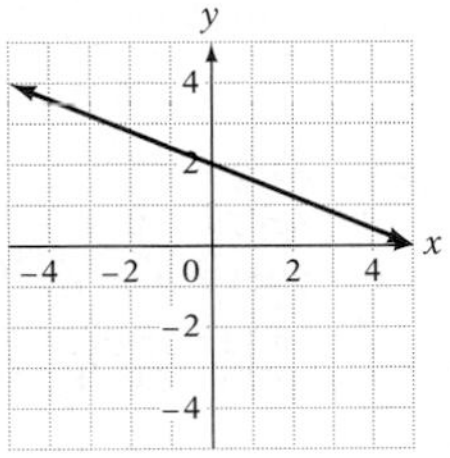

34. $2x - y = 4$

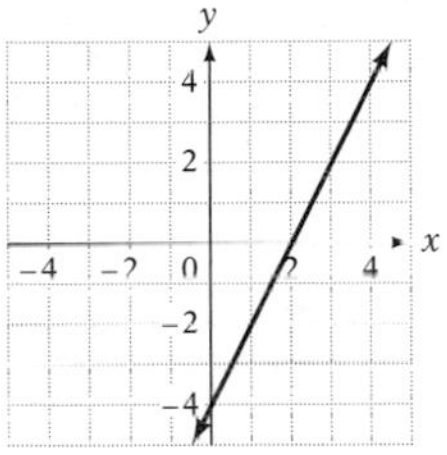

35. $x = 3$

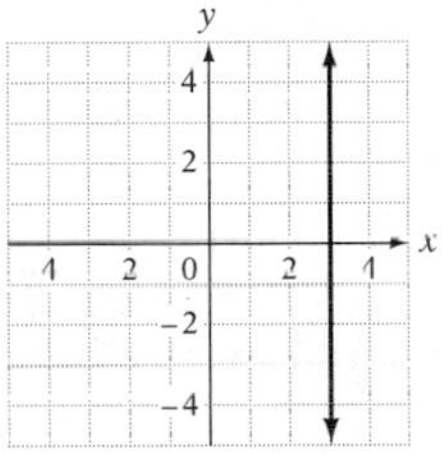

36. $y = -4$

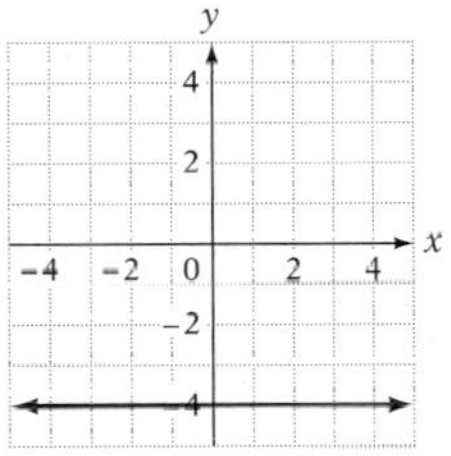

37. $x + 4y = 4$

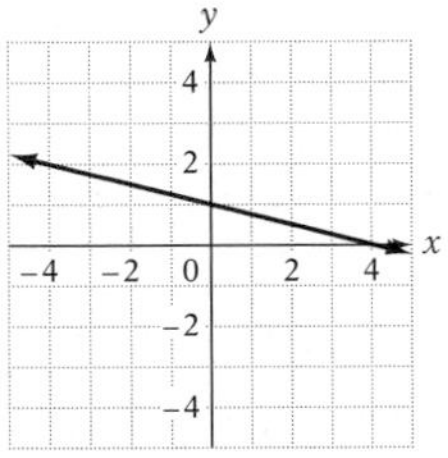

38. $4x - 3y = 12$

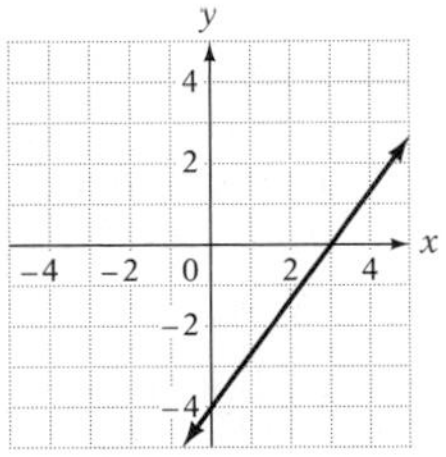

39. $y = 4$

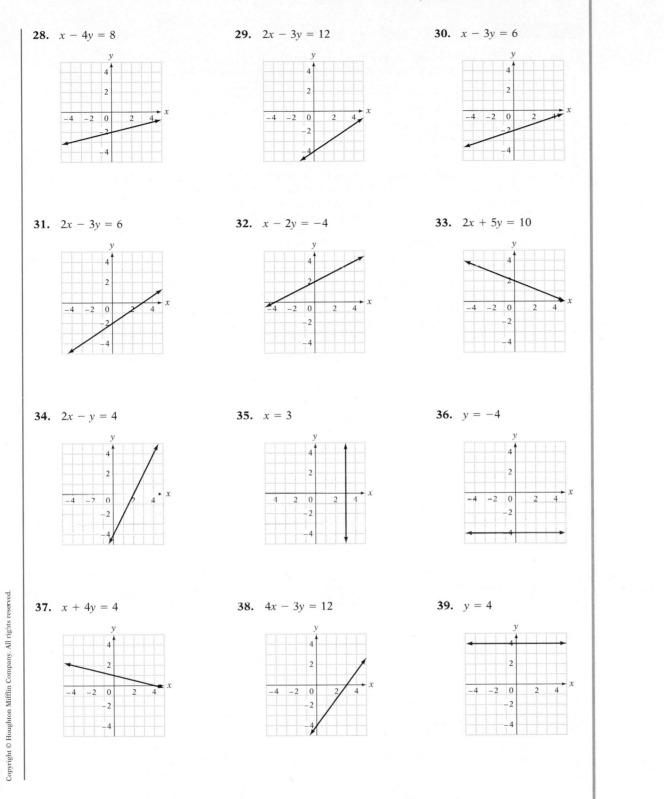

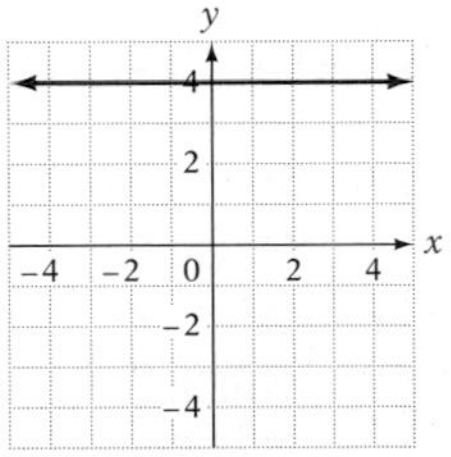

234

Objective C *Application Problems*

40. The Rattler roller coaster at Fiesta, in San Antonio, Texas, has a maximum speed of approximately 110 ft/s. The equation that describes the total number of feet traveled by the roller coaster in t seconds at this speed is given by $D = 110t$. Graph this equation for $0 \le t \le 10$. The point (5, 550) is on this graph. Write a sentence that describes the meaning of this ordered pair.
The roller coaster travels 550 ft in 5 s.

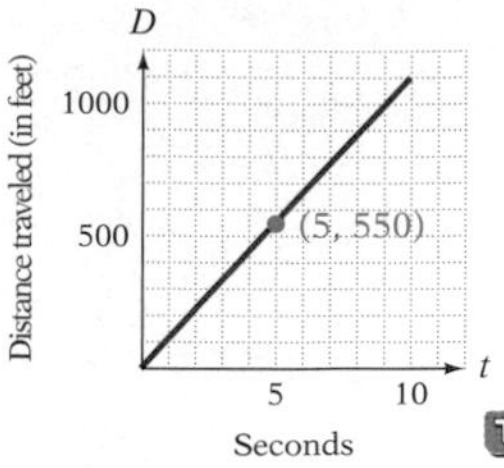

41. According to some veterinarians, the age, x, of a dog can be translated to "human years" by using the equation $H = 4x + 16$, where H is the human equivalent age for the dog. Graph this equation for $2 \le x \le 21$. The point whose coordinates are (6, 40) is on this graph. Write a sentence that explains the meaning of this ordered pair.
A dog 6 years old is equivalent in age to a human 40 years old.

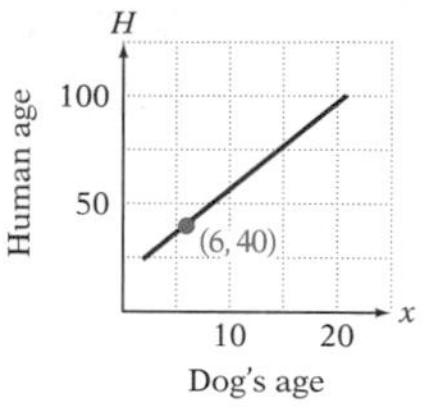

42. The monthly cost for receiving messages from a telephone answering service is \$8.00 plus \$.20 a message. The equation that describes the cost is $C = 0.20n + 8.00$, where n is the number of messages received. Graph this equation for $0 \le n \le 40$. The point (32, 14.40) is on the graph. Write a sentence that describes the meaning of this ordered pair.
The cost of receiving 32 messages is \$14.40.

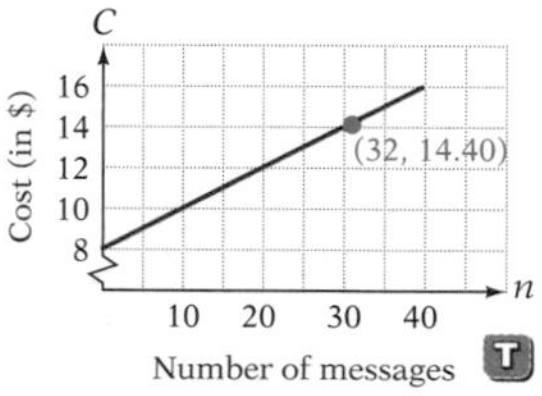

43. A custom-illustrated sign or banner can be commissioned for a cost of \$25 for the material and \$10.50 per square foot for the artwork. The equation that represents this cost is given by $y = 10.50x + 25$, where y is the cost and x is the number of square feet in the sign. Graph this equation for $0 \le x \le 20$. The point (15, 182.5) is on the graph. Write a sentence that describes the meaning of this ordered pair.
It costs \$182.50 for a custom sign 15 ft² in area.

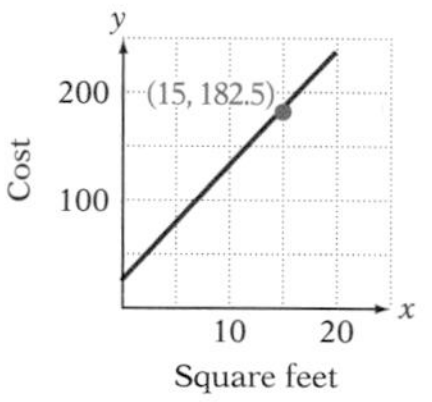

44. A rescue helicopter is rushing at a constant speed of 150 mph to reach several people stranded in the ocean 11 mi away after their boat sank. The rescuers can determine how far they are from the victims using the equation $D = 11 - 2.5t$, where D is the distance in miles and t is the time elapsed in minutes. Graph this equation for $0 \le t \le 4$. The point (3, 3.5) is on the graph. Write a sentence that describes the meaning of this ordered pair.
After flying for 3 min, the helicopter is 3.5 mi away from the victims.

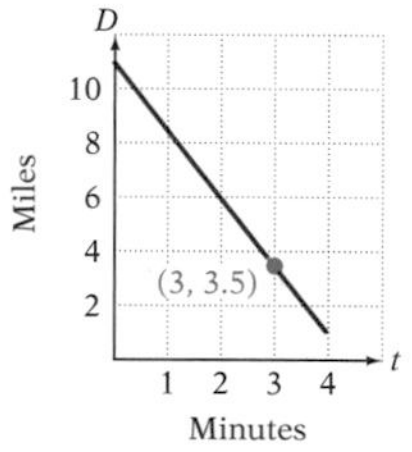

APPLYING THE CONCEPTS

45. Explain what the graph of an equation represents.

46. Explain why you cannot graph the equation $4x + 3y = 0$ by using just its intercepts.

Answers to Writing Exercises

45. Student answers will vary. However, you should look for the idea that the graph of an equation is a picture of all the ordered pairs (x, y) that are solutions of the equation.

46. The x-intercept of the graph of the equation $4x + 3y = 0$ is $(0, 0)$. The y-intercept of this graph is also $(0, 0)$. Therefore, the x-intercept and y-intercept are the same point. A straight line is determined by two points, so we must find another point on the line in order to graph the equation.

Quick Quiz (Objective 4.3C)

The temperature, F, in degrees Fahrenheit, that corresponds to a temperature, C, in degrees Celsius, is given by the equation $F = \frac{9}{5}C + 32$. Graph this equation for $0 \le x \le 100$. The point whose coordinates are (45, 113) is on the graph. Write a sentence that describes the meaning of this ordered pair.

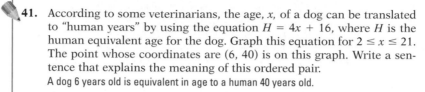

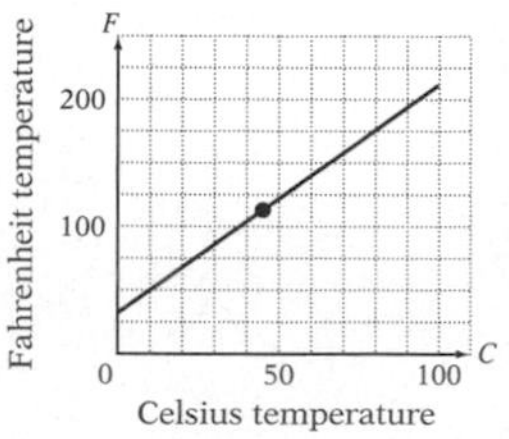

A temperature of 45 degrees Celsius is equivalent to a temperature of 113 degrees Fahrenheit.

4.4 Slopes of Straight Lines

Objective A To find the slope of a straight line

VIDEO & DVD CD TUTOR WEB SSM

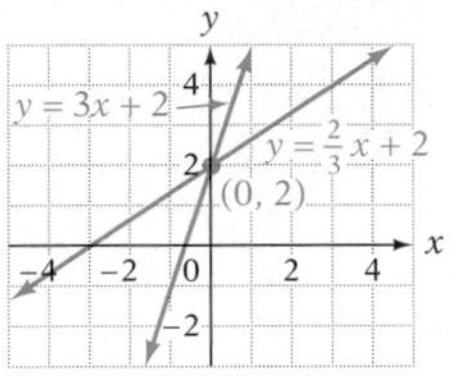

The graphs of $y = 3x + 2$ and $y = \frac{2}{3}x + 2$ are shown at the left. Each graph crosses the y-axis at the point (0, 2), but the graphs have different slants. The **slope** of a line is a measure of the slant of the line. The symbol for slope is m.

The slope of a line containing two points is the ratio of the change in the y values between the two points to the change in the x values. The line containing the points whose coordinates are $(-1, -3)$ and $(5, 2)$ is shown below.

The change in the y values is the difference between the y-coordinates of the two points.

Change in $y = 2 - (-3) = 5$

The change in the x values is the difference between the x-coordinates of the two points.

Change in $x = 5 - (-1) = 6$

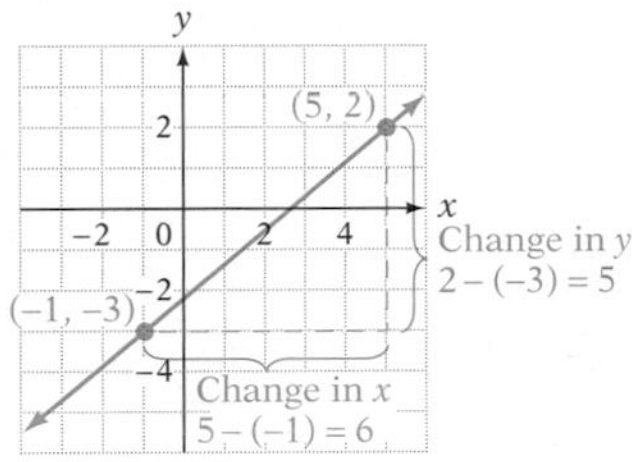

The slope of the line between the two points is the ratio of the change in y to the change in x.

$$\text{Slope} = m = \frac{\text{change in } y}{\text{change in } x} = \frac{5}{6} \qquad m = \frac{2 - (-3)}{5 - (-1)} = \frac{5}{6}$$

In general, if $P_1(x_1, y_1)$ and $P_2(x_2, y_2)$ are two points on a line, then

Change in $y = y_2 - y_1$ Change in $x = x_2 - x_1$

Using these ideas, we can state a formula for slope.

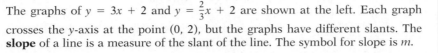

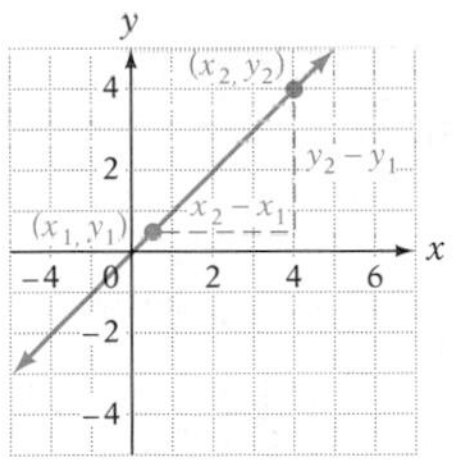

Slope Formula

The slope of the line containing the two points $P_1(x_1, y_1)$ and $P_2(x_2, y_2)$ is given by

$$m = \frac{y_2 - y_1}{x_2 - x_1}, \; x_1 \neq x_2$$

Frequently, the Greek letter Δ is used to designate the change in a variable. Using this notation, we can write the equations for the change in y and the change in x as follows:

Change in $y = \Delta y = y_2 - y_1$ Change in $x = \Delta x = x_2 - x_1$

With this notation, the slope formula is written $m = \frac{\Delta y}{\Delta x}$.

Objective 4.4A

New Vocabulary

slope
positive slope
negative slope
zero slope
undefined slope

New Symbols

m, for slope

New Formulas

slope formula

$$m = \frac{y_2 - y_1}{x_2 - x_1}$$

Discuss the Concepts

A warning sign for drivers on a mountain road might read, "Caution: 8% downgrade next 2 miles." Explain this statement.

Concept Check

When you have finished discussing positive and negative slope, draw some random lines and ask students whether the slope is positive or negative. Then draw a line and ask students how they could approximate the slope of the line. Do this for lines that have positive slope and lines that have negative slope.

In-Class Examples (Objective 4.4A)

Find the slope of the line containing the points.

1. $P_1(4, -2), P_2(-1, 3)$ -1
2. $P_1(-1, 2), P_2(5, 3)$ $\frac{1}{6}$
3. $P_1(2, 3), P_2(4, -2)$ $-\frac{5}{2}$
4. $P_1(1, 5), P_2(3, 5)$ 0
5. $P_1(-2, -1), P_2(-2, 6)$ Undefined

Instructor Note

When calculating the slope of a line, show students that the choice of P_1 and P_2 does not alter the value of the slope.

Instructor Note

Students have a tendency to think that zero slope means *no slope*. Emphasize that a slope of zero refers to a line that is parallel to the *x*-axis; a line whose slope is undefined is parallel to the *y*-axis.

Optional Student Activity

1. If (a, b) and (c, d) are the coordinates of different points on the graph of the equation $4x + 3y = 5$, what is the value of $\frac{b - d}{a - c}$? $-\frac{4}{3}$
2. Lois and Tanya start from the same place on a jogging course. Lois is jogging at 9 km/h, and Tanya is jogging at 6 km/h. The graph below shows the total distance traveled by each jogger and the total distance between Lois and Tanya. Which lines represent which distances? A: Lois; B: Tanya; C: Distance between

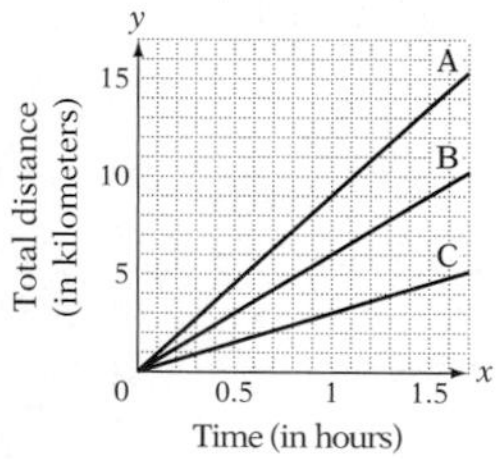

(Continued on next page)

⇒ Find the slope of the line containing the points whose coordinates are $(-2, 0)$ and $(4, 5)$.

Let $P_1 = (-2, 0)$ and $P_2 = (4, 5)$. (It does not matter which point is named P_1 or P_2; the slope will be the same.)

$$m = \frac{y_2 - y_1}{x_2 - x_1} = \frac{5 - 0}{4 - (-2)} = \frac{5}{6}$$

A line that slants upward to the right always has a **positive slope**.

TAKE NOTE
Positive slope means that the value of *y* increases as the value of *x* increases.

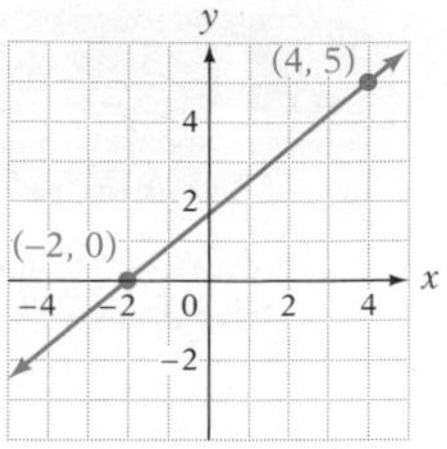

Positive slope

⇒ Find the slope of the line containing the points whose coordinates are $(-3, 4)$ and $(4, 2)$.

Let $P_1 = (-3, 4)$ and $P_2 = (4, 2)$.

$$m = \frac{y_2 - y_1}{x_2 - x_1} = \frac{2 - 4}{4 - (-3)} = \frac{-2}{7} = -\frac{2}{7}$$

A line that slants downward to the right always has a **negative slope**.

TAKE NOTE
Negative slope means that the value of *y* decreases as the value of *x* increases. Compare this to positive slope.

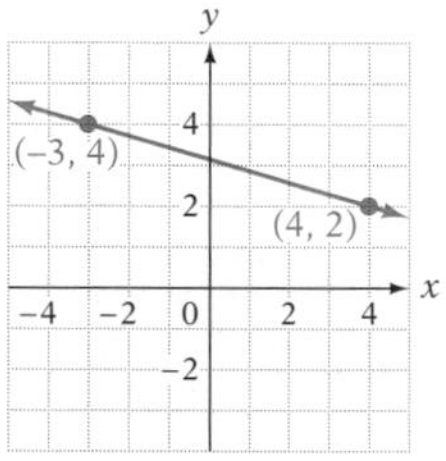

Negative slope

⇒ Find the slope of the line containing the points whose coordinates are $(-2, 2)$ and $(4, 2)$.

Let $P_1 = (-2, 2)$ and $P_2 = (4, 2)$.

$$m = \frac{y_2 - y_1}{x_2 - x_1} = \frac{2 - 2}{4 - (-2)} = \frac{0}{6} = 0$$

A horizontal line has **zero slope**.

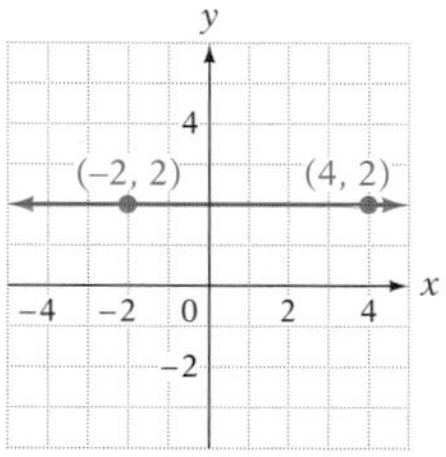

Zero slope

⇒ Find the slope of the line containing the points whose coordinates are $(1, -2)$ and $(1, 3)$.

Let $P_1 = (1, -2)$ and $P_2 = (1, 3)$.

$$m = \frac{y_2 - y_1}{x_2 - x_1} = \frac{3 - (-2)}{1 - 1} = \frac{5}{0} \quad \text{Not a real number}$$

The slope of a vertical line is **undefined**.

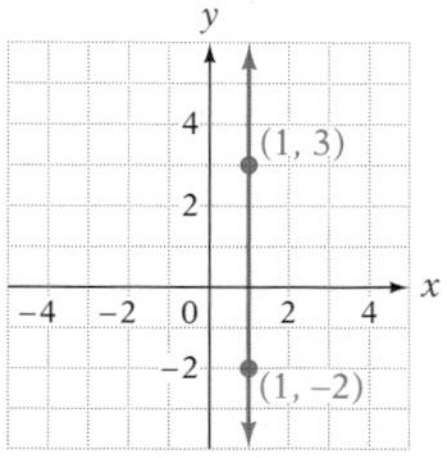

Undefined

Point of Interest

One of the motivations for the discovery of calculus was to solve a more complicated version of the distance-rate problem at the right.

You may be familiar with twirling a ball on the end of a string. If you release the string, the ball flies off in a path as shown below.

The question that mathematicians tried to answer was essentially "What is the slope of the line represented by the arrow?"

Answering questions similar to that led to the development of one aspect of calculus.

There are many applications of slope. Here are two possibilities.

The first record for the one-mile run was recorded in 1865 in England. Richard Webster ran the mile in 4 min 36.5 s. His average speed was approximately 19 feet per second.

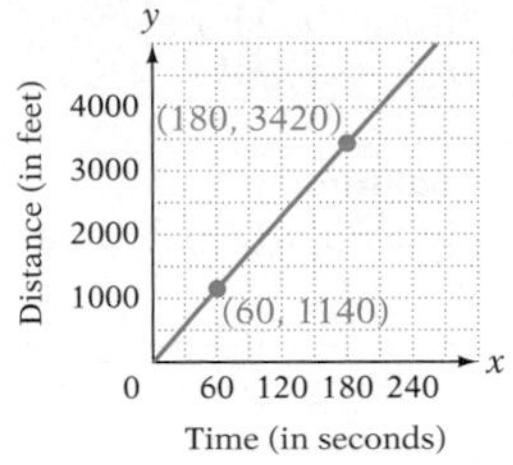

The graph at the right shows the distance Webster ran during that run. From the graph, note that after 60 s (1 min) he had traveled 1140 ft, and that after 180 s (3 min) he had traveled 3420 ft.

Let the point (60, 1140) be (x_1, y_1) and the point (180, 3420) be (x_2, y_2). The slope of the line between these two points is

$$m = \frac{y_2 - y_1}{x_2 - x_1}$$
$$= \frac{3420 - 1140}{180 - 60} = \frac{2280}{120} = 19$$

Note that the slope of the line is the same as Webster's average speed, 19 feet per second.

Average speed is related to slope.

Here is another example, this one related to economics.

As a result of cheaper digital networks and more competition, the cost of wireless phone use is decreasing. The average cost per minute in 1999 was 28¢. The projected average cost in 2003 is 20¢. (Source: The Strategic Group).

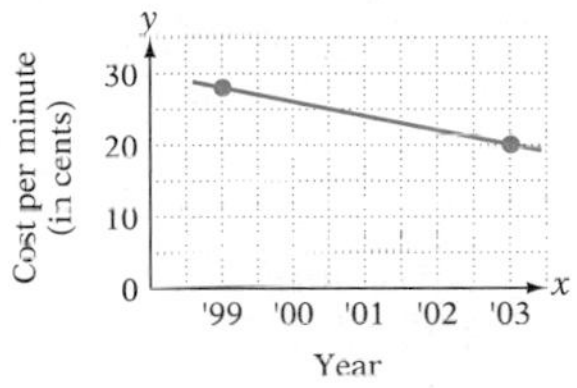

Let the point (1999, 28) be (x_1, y_1) and the point (2003, 20) be (x_2, y_2). The slope of the line between these two points is

$$m = \frac{y_2 - y_1}{x_2 - x_1}$$
$$= \frac{20 - 28}{2003 - 1999} = \frac{-8}{4} = -2$$

Note that if we interpret negative slope as decreasing, then the slope of the line is the same as the rate at which the per-minute cost of phone use is decreasing, 2¢ per year.

In general, any quantity that is expressed by using the word *per* is represented mathematically as slope. In the first example, the slope was 19 feet per second. In the second example, the slope was −2¢ per year.

(Continued)

3. A chemist is filling two cans from a faucet that releases water at a constant rate. Can 1 has a diameter of 20 mm, and can 2 has a diameter of 30 mm. The depth of the water in each can is measured at 5-second intervals. The graph of the results is shown below. On the graph, which line represents the depth of the water for which can? A: Can 1; B: Can 2

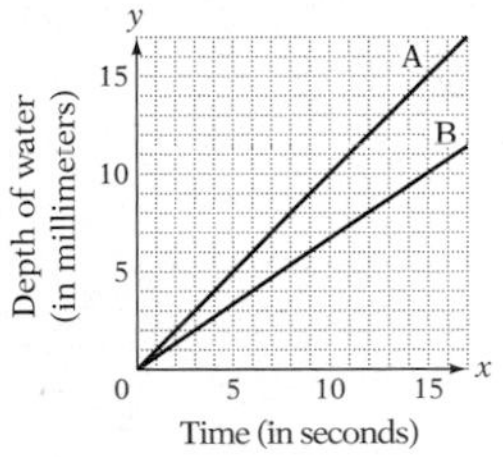

Discuss the Concepts

The two examples on this page illustrate that slope has real applications. You might work with the students to create a list of some quantities that we use every day that are treated mathematically as slope—for instance, miles per gallon, miles per hour, cost per unit. Explain that, just as we translate the phrase *the sum of* as addition, a phrase that contains *per* can be translated as slope.

Example 1
Find the slope of the line containing the points (2, −5) and (−4, 2).

Solution
Let $P_1 = (2, -5)$ and $P_2 = (-4, 2)$.

$$m = \frac{y_2 - y_1}{x_2 - x_1} = \frac{2 - (-5)}{-4 - 2} = \frac{7}{-6}$$

The slope is $-\frac{7}{6}$.

You Try It 1
Find the slope of the line containing the points (4, −3) and (2, 7).

Your solution
−5

Example 2
Find the slope of the line containing the points (−3, 4) and (5, 4).

Solution
Let $P_1 = (-3, 4)$ and $P_2 = (5, 4)$.

$$m = \frac{y_2 - y_1}{x_2 - x_1} = \frac{4 - 4}{5 - (-3)} = \frac{0}{8} = 0$$

The slope of the line is zero.

You Try It 2
Find the slope of the line containing the points (6, −1) and (6, 7).

Your solution
The slope of the line is undefined.

Example 3
The graph below shows the relationship between the cost of an item and the sales tax. Find the slope of the line between the two points shown on the graph. Write a sentence that states the meaning of the slope.

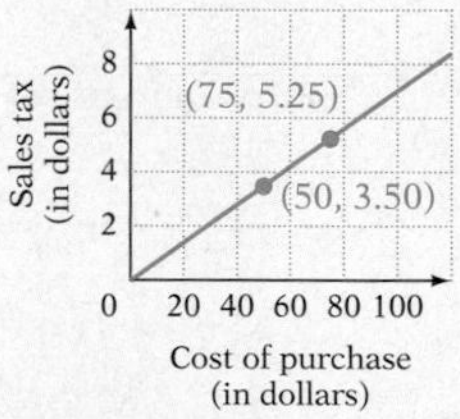

Solution

$$m = \frac{5.25 - 3.50}{75 - 50} = \frac{1.75}{25} = 0.07$$

A slope of 0.07 means that the sales tax is \$.07 per dollar.

You Try It 3
The graph below shows the decrease in the value of a printing press for a period of 6 years. Find the slope of the line between the two points shown on the graph. Write a sentence that states the meaning of the slope.

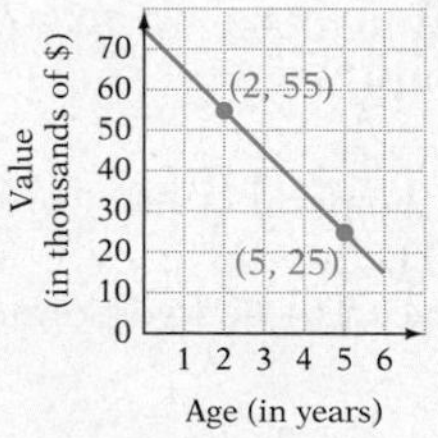

Your solution
$m = -10{,}000$
A slope of −10,000 means that the value of the printing press is decreasing by \$10,000 per year.

Solutions on p. S12

Objective B

To graph a line using the slope and the y-intercept

VIDEO & DVD CD TUTOR WEB SSM

Recall that we can find the y-intercept of a linear equation by letting $x = 0$.

➡ Find the y-intercept of $y = 3x + 4$.

$y = 3x + 4 = 3(0) + 4 = 4$ • Let $x = 0$.

The y-intercept is $(0, 4)$.

For any equation of the form $y = mx + b$, the y-intercept is $(0, b)$.

The graph of the equation $y = \frac{2}{3}x + 1$ is shown at the right. The points $(-3, -1)$ and $(3, 3)$ are on the graph. The slope of the line between the two points is

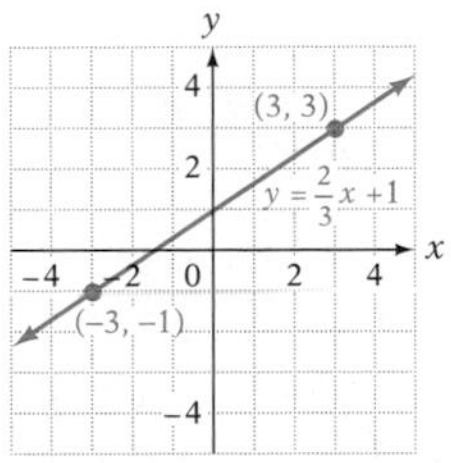

$$m = \frac{3 - (-1)}{3 - (-3)} = \frac{4}{6} = \frac{2}{3}$$

Observe that the slope of the line is the coefficient of x in the equation $y = \frac{2}{3}x + 1$.

Slope-Intercept Form of a Straight Line

An equation of the form $y = mx + b$ is called the **slope-intercept** form of a straight line. The slope of the line is m, the coefficient of x. The y-intercept is $(0, b)$, where b is the constant term of the equation.

The following equations are written in slope-intercept form.

$y = 2x - 3$	Slope $= 2$, y-intercept $= (0, -3)$
$y = -x + 2$	Slope $= -1$ $(-x = -1x)$, y-intercept $= (0, 2)$
$y = \frac{1}{2}x$	Slope $= \frac{1}{2}$, y-intercept $= (0, 0)$

When the equation of a straight line is in the form $y = mx + b$, the graph can be drawn by using the slope and the y-intercept. First locate the y-intercept. Use the slope to find a second point on the line. Then draw a line through the two points.

➡ Graph $y = \frac{5}{3}x - 4$ by using the slope and y-intercept.

The slope is the coefficient of x:

$$m = \frac{5}{3} = \frac{\text{change in } y}{\text{change in } x}$$

The y-intercept is $(0, -4)$.

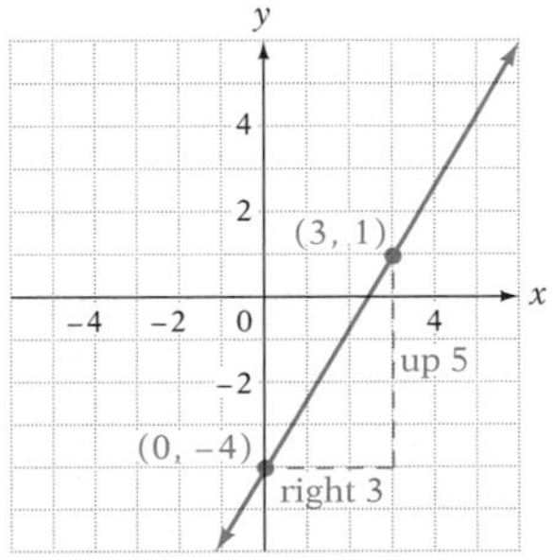

Beginning at the y-intercept $(0, -4)$, move right 3 units (change in x) and then up 5 units (change in y).

The point whose coordinates are $(3, 1)$ is a second point on the graph. Draw a line through the points $(0, -4)$ and $(3, 1)$.

TAKE NOTE

When graphing a line by using its slope and y-intercept, *always* start at the y-intercept.

Objective 4.4B

New Vocabulary

slope-intercept form

Discuss the Concepts

1. If two lines have the same slope and the same y-intercept, must the graphs of the two lines be the same?
2. Suppose two lines have a y-intercept of $(0, 4)$. Describe the difference between the appearance of these lines if one has a slope of 3 and the other has a slope of -3.

Instructor Note

For consistency, we always graph slope by moving to the right and then up or down, depending on whether the slope is positive or negative. An interesting class activity is to graph the slope in other ways.

For instance, a slope of $-\frac{1}{2}$ can be graphed by moving left 2 and up 1. It can also be graphed by moving right 1 and down $\frac{1}{2}$, right 4 and down 2, or in several other ways. This exercise will help reinforce the concept that slope measures the ratio of the change in y to the change in x.

In-Class Examples (Objective 4.4B)

Graph by using the slope and the y-intercept.

1. $y = \frac{2}{3}x - 2$

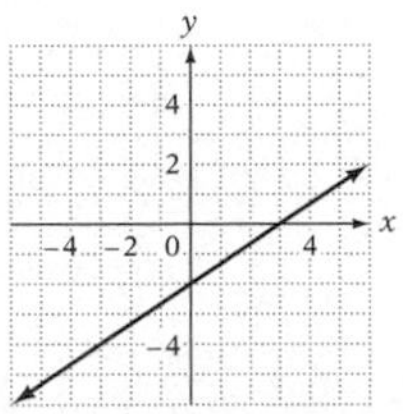

2. $y = 2x - 3$

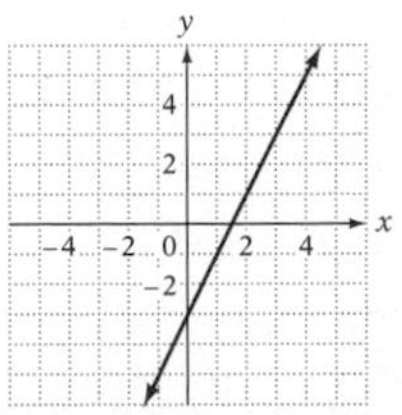

3. $y = -\frac{3}{4}x$

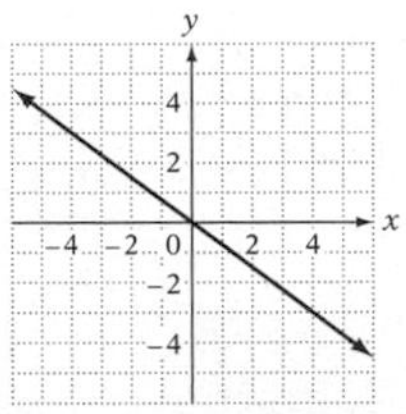

Concept Check

1. The line L in the xy-coordinate system has half the slope and twice the y-intercept of the line $y = \frac{2}{5}x + 6$. Find the equation of line L. Write the equation in slope-intercept form. $y = \frac{1}{5}x + 12$
2. Line A passes through the points $(-1, 7)$ and $(3, 9)$. Line B passes through the points $(1, -3)$ and $(6, -2)$. Which line is steeper, A or B? Line A

Optional Student Activity

1. If $(-1, 2)$ are the coordinates of a point on a line that has slope -3, what is the y-coordinate of the point on the line with x-coordinate $x = 1$? -4
2. If $(-2, -1)$ are the coordinates of a point on a line that has slope $\frac{3}{2}$, what is the y-coordinate of the point on the line with x-coordinate $x = -6$? -7

240

➡ Graph $y = 2x - 3$.

y-intercept $= (0, b) = (0, -3)$

$$m = 2 = \frac{2}{1} = \frac{\text{change in } y}{\text{change in } x}$$

Beginning at the y-intercept, move right 1 unit (change in x) and then up 2 units (change in y).

$(1, -1)$ is a second point on the graph.

Draw a line through the two points $(0, -3)$ and $(1, -1)$.

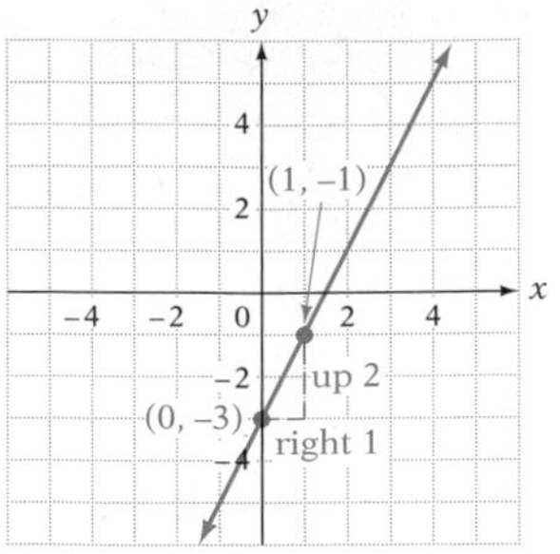

Example 4

Graph $y = -\frac{2}{3}x + 1$ by using the slope and y-intercept.

Solution y-intercept $= (0, b) = (0, 1)$

$$m = -\frac{2}{3} = \frac{-2}{3}$$

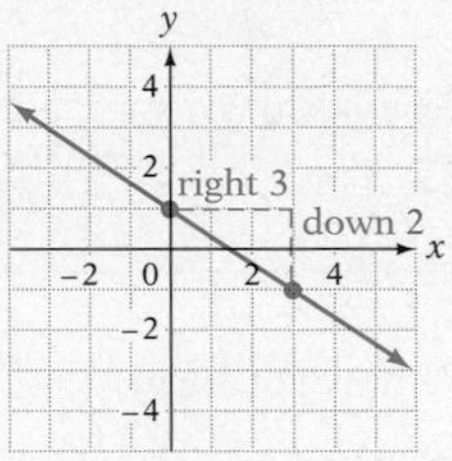

You Try It 4

Graph $y = -\frac{1}{4}x - 1$ by using the slope and y-intercept.

Your solution

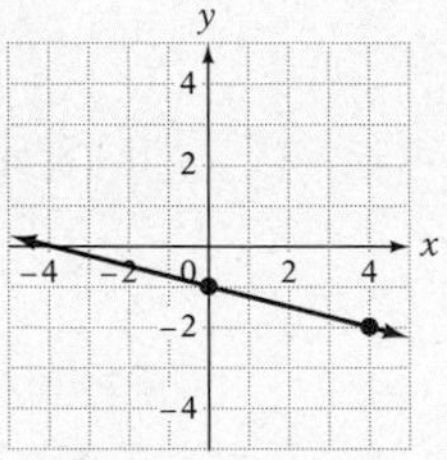

Example 5 Graph $2x - 3y = 6$ by using the slope and y-intercept.

Solution Solve the equation for y.

$$2x - 3y = 6$$
$$-3y = -2x + 6$$
$$y = \frac{2}{3}x - 2$$

y-intercept $= (0, -2)$; $m = \frac{2}{3}$

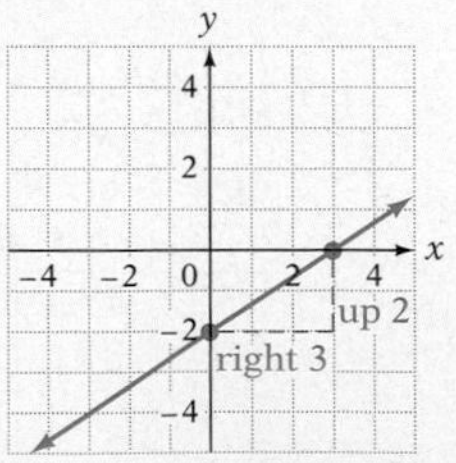

You Try It 5 Graph $x - 2y = 4$ by using the slope and y-intercept.

Your solution

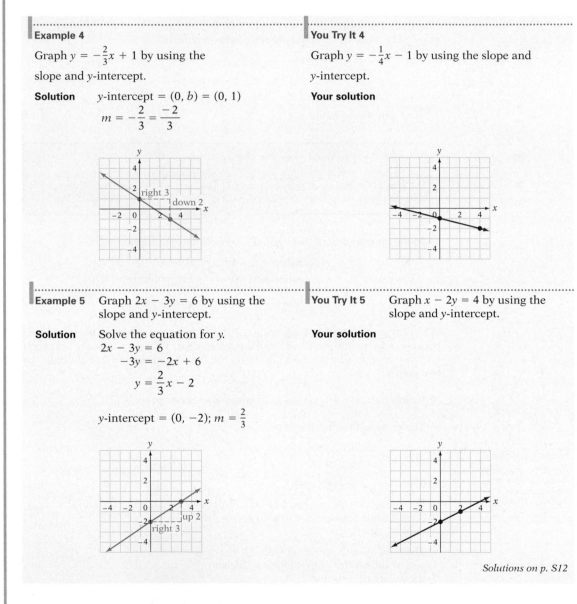

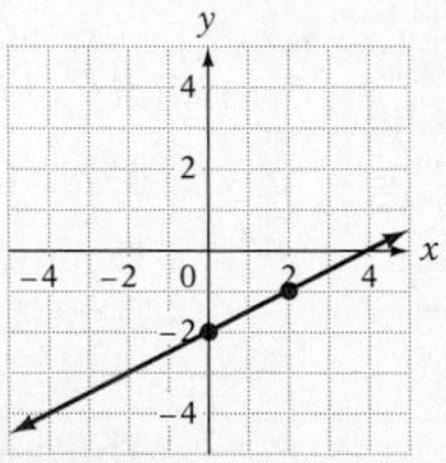

Solutions on p. S12

Section 4.4

Suggested Assignment
Exercises 1–43, odds
More challenging problems:
Exercises 45–50

4.4 Exercises

Objective A

Find the slope of the line containing the points.

1. $P_1(1, 3), P_2(3, 1)$
-1

2. $P_1(2, 3), P_2(5, 1)$
$-\frac{2}{3}$

3. $P_1(-1, 4), P_2(2, 5)$
$\frac{1}{3}$

4. $P_1(3, -2), P_2(1, 4)$
-3

5. $P_1(-1, 3), P_2(-4, 5)$
$-\frac{2}{3}$

6. $P_1(-1, -2), P_2(-3, 2)$
-2

7. $P_1(0, 3), P_2(4, 0)$
$-\frac{3}{4}$

8. $P_1(-2, 0), P_2(0, 3)$
$\frac{3}{2}$

9. $P_1(2, 4), P_2(2, -2)$
Undefined

10. $P_1(4, 1), P_2(4, -3)$
Undefined

11. $P_1(2, 5), P_2(-3, -2)$
$\frac{7}{5}$

12. $P_1(4, 1), P_2(-1, -2)$
$\frac{3}{5}$

13. $P_1(2, 3), P_2(-1, 3)$
0

14. $P_1(3, 4), P_2(0, 4)$
0

15. $P_1(0, 4), P_2(-2, 5)$
$-\frac{1}{2}$

16. $P_1(-2, 3), P_2(-2, 5)$
Undefined

17. $P_1(-3, -1), P_2(-3, 4)$
Undefined

18. $P_1(-2, -5), P_2(-4, -1)$
-2

19. The graph below shows the relationship between the distance traveled by a motorist and the time of travel. Find the slope of the line between the two points shown on the graph. Write a sentence that states the meaning of the slope.

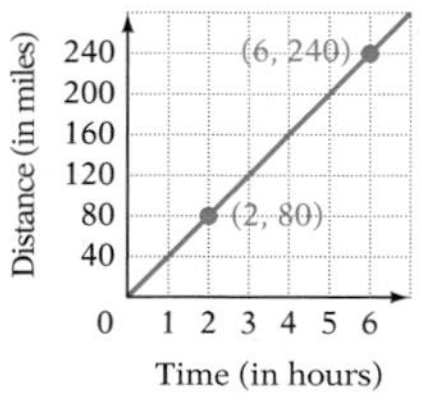

$m = 40$
The slope is the average speed of the motorist in miles per hour.

20. The graph below shows the number of people subscribing to a sports magazine whose popularity is increasing. Find the slope of the line between the two points shown on the graph. What is the meaning of the slope?

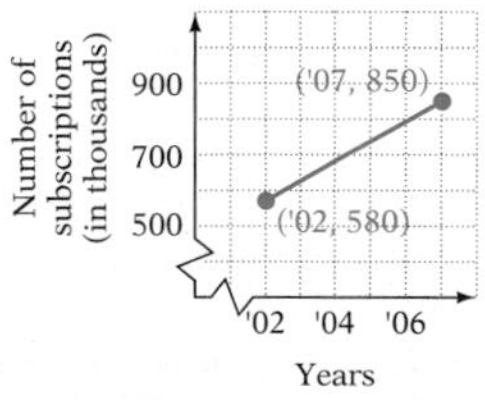

$m = 54{,}000$
The slope means that 54,000 subscribers are added each year.

Quick Quiz (Objective 4.4A)
Find the slope of the line containing the points.
1. $P_1(-1, 4), P_2(2, -2)$ -2
2. $P_1(1, -3), P_2(5, -13)$ $-\frac{5}{2}$
3. $P_1(-3, -1), P_2(-1, 4)$ $\frac{5}{2}$
4. $P_1(2, 3), P_2(2, -1)$ Undefined

Answers to Writing Exercises

27. The equation $y = mx + b$ is called the slope-intercept form of the equation of a straight line because m, the coefficient of x, is the slope of the line, and the y-intercept is $(0, b)$.

21. The graph below shows how the amount of gas in the tank of a car decreases as the car is driven. Find the slope of the line. Write a sentence that states the meaning of the slope.

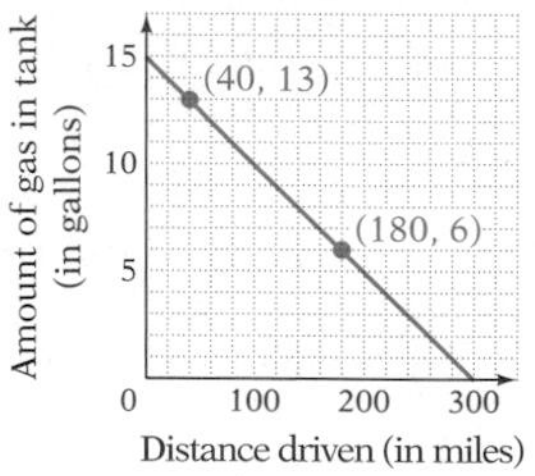

$m = -0.05$
For each mile the car is driven, approximately 0.05 gal of fuel is used.

22. The troposphere extends from Earth's surface to an elevation of about 11 km. The graph below shows the decrease in its temperature as altitude increases. Find the slope of the line. Write a sentence that states the meaning of the slope.

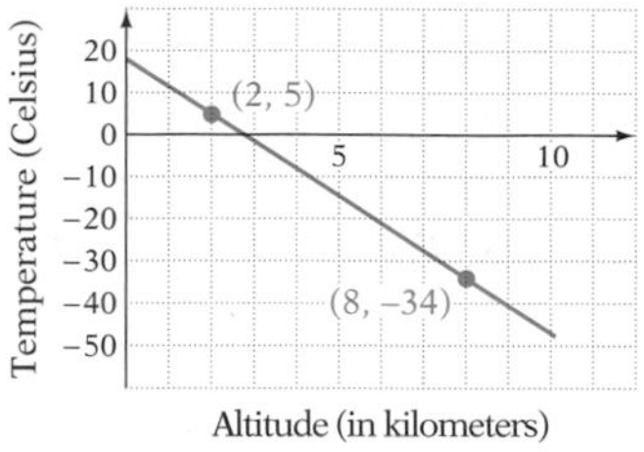

$m = -6.5$
The temperature of the troposphere decreases 6.5°C/km.

23. The graph below shows the relationship between the distance and the time for the 5000-meter run for the world record by Said Aouita in 1987. Find the slope of the line between the two points shown on the graph. Write a sentence that states the meaning of the slope.

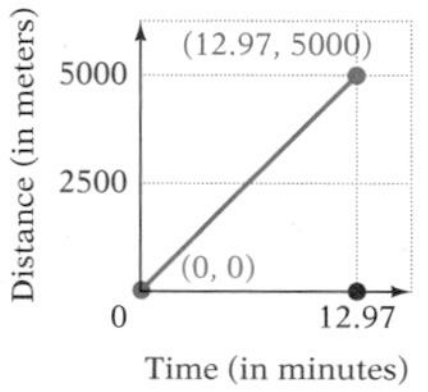

$m \approx 385.5$
The slope is the average speed of the runner in meters per minute.

24. The graph below shows the relationship between the distance and the time for the 10,000-meter run for the world record by Arturo Barrios in 1989. Find the slope of the line between the two points shown on the graph. Write a sentence that states the meaning of the slope.

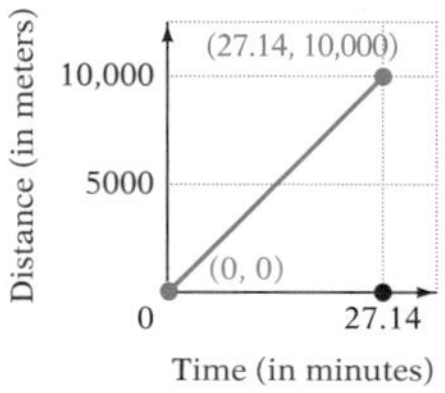

$m \approx 368.5$
The slope is the average speed of the runner in meters per minute.

The American National Standards Institute (ANSI) states that the slope for a wheelchair ramp must not exceed $\frac{1}{12}$.

25. Does a ramp that is 6 in. high and spans a distance of 5 ft along the ground meet the requirements of ANSI? no

26. Does a ramp that is 12 in. high and spans a distance of 170 in. along the ground meet the requirements of ANSI? yes

Objective B

27. Explain why $y = mx + b$ is called the slope-intercept form of the equation of a straight line.

28. Do all straight lines have a y-intercept? If not, give an example of one that does not. No. For instance, the graph of $x = 3$.

29. If two lines have the same slope and the same y-intercept, must the graphs of the lines be the same? If not, give an example. yes

Quick Quiz (Objective 4.4B)

Graph by using the slope and the y-intercept.

1. $y = \frac{1}{2}x - 1$

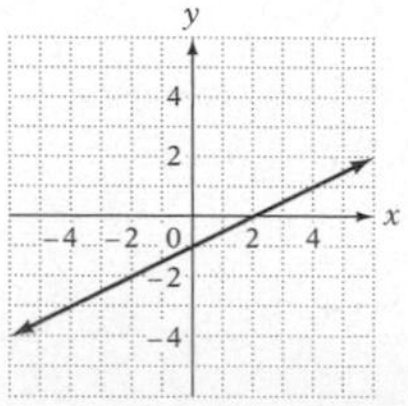

2. $y = 3x - 3$

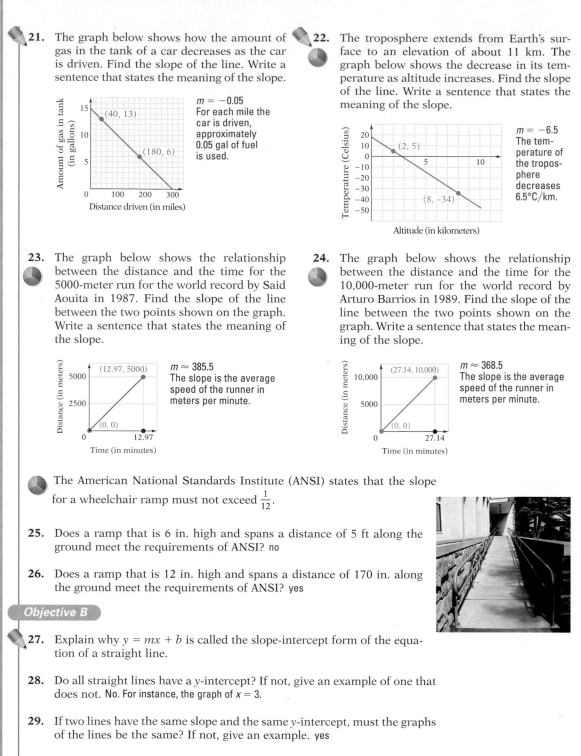

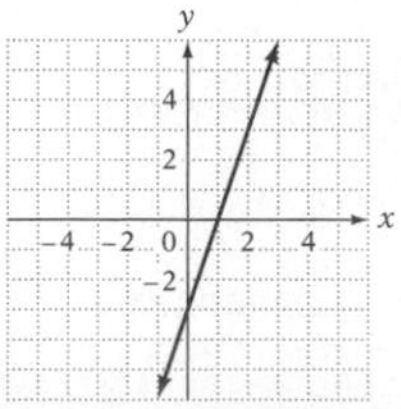

243

Graph by using the slope and y-intercept.

30. $y = 3x + 1$

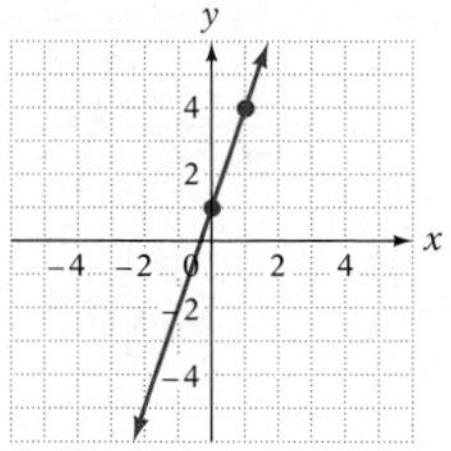

31. $y = -2x - 1$

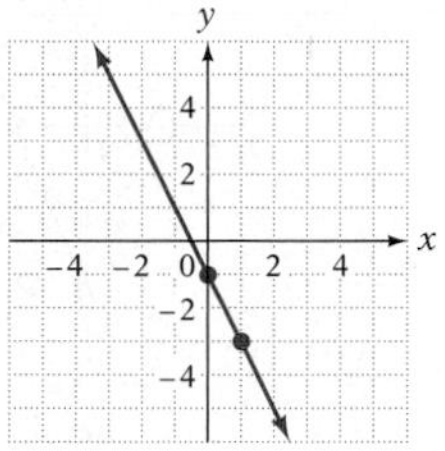

32. $y = \frac{2}{5}x - 2$

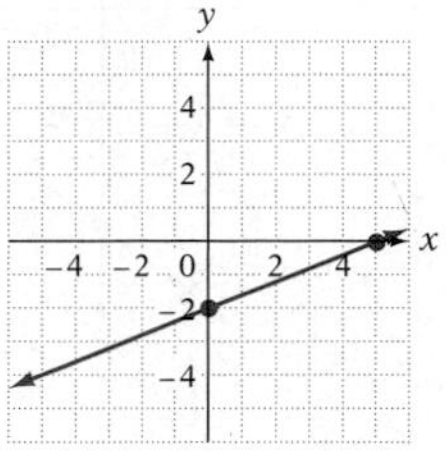

33. $y = \frac{3}{4}x + 1$

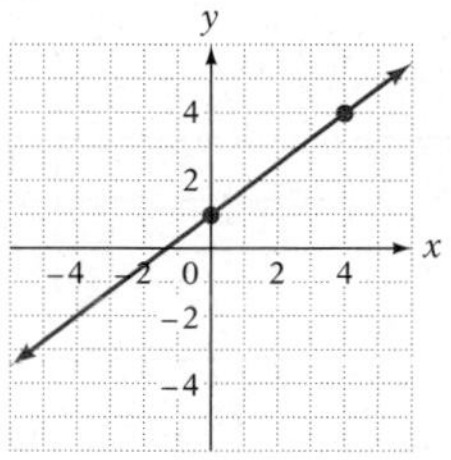

34. $2x + y = 3$

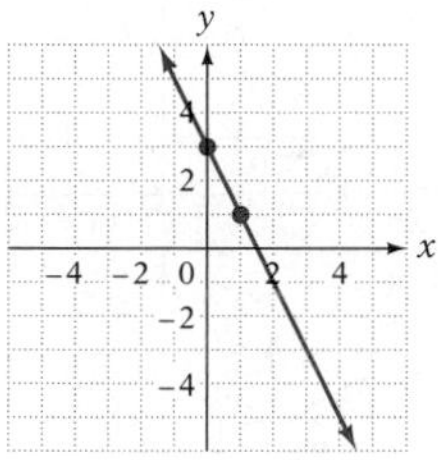

35. $3x - y = 1$

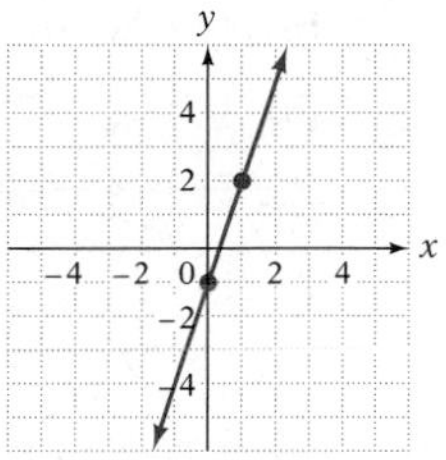

36. $x - 2y = 4$

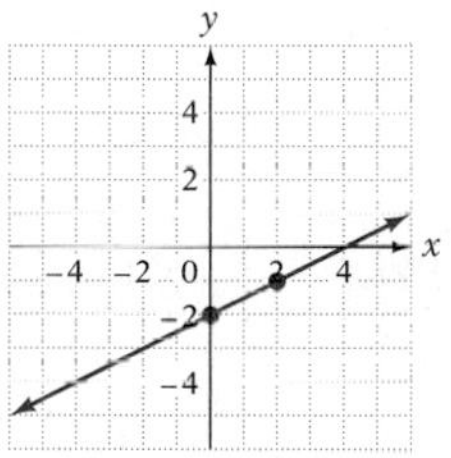

37. $x + 3y = 6$

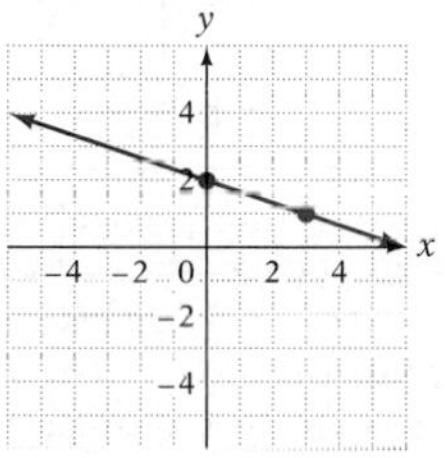

38. $y = \frac{2}{3}x$

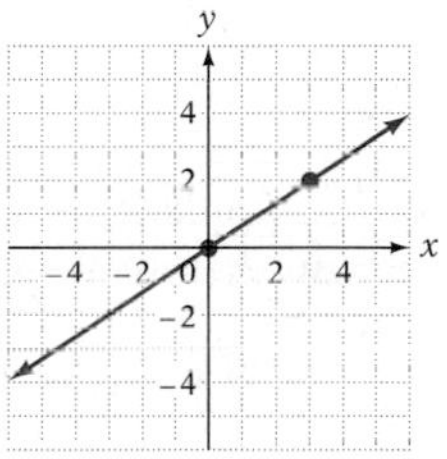

39. $y = \frac{1}{2}x$

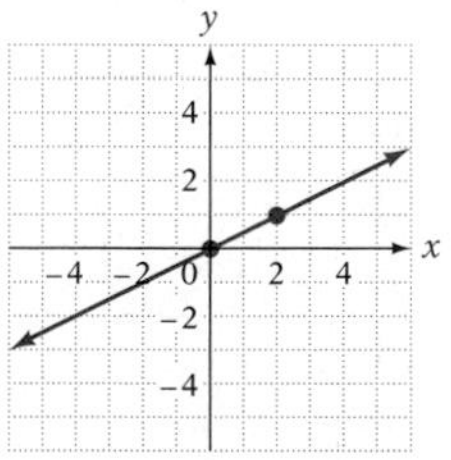

40. $y = \ \ x + 1$

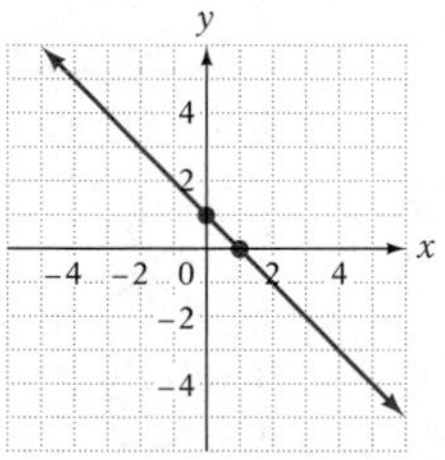

41. $y = -x - 3$

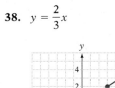

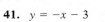

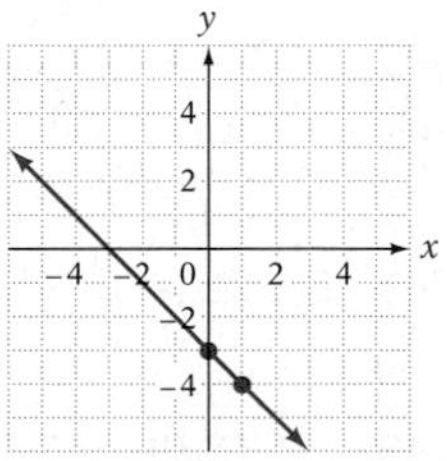

244 Chapter 4 / Linear Equations and Inequalities in Two Variables

42. $3x - 4y = 12$

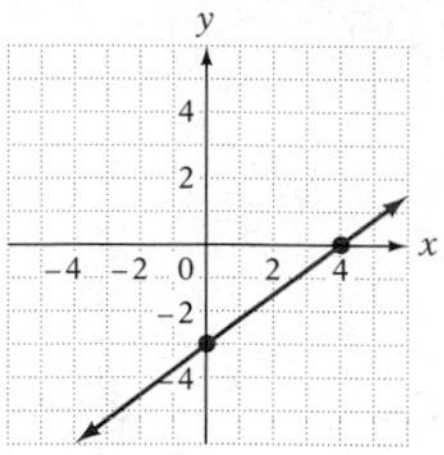

43. $5x - 2y = 10$

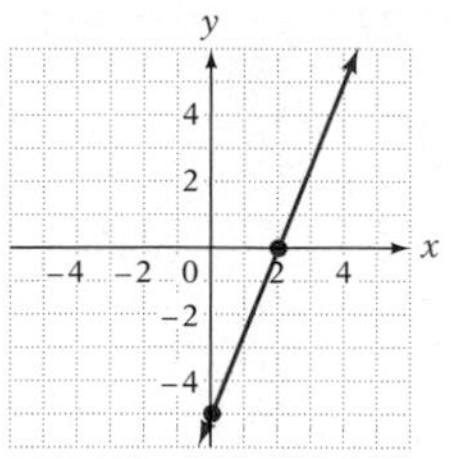

44. $y = -4x + 2$

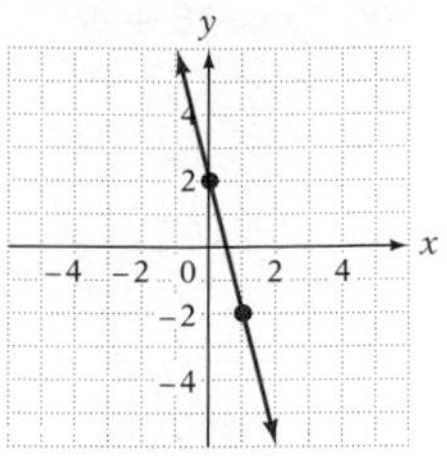

APPLYING THE CONCEPTS

45. Match each equation with its graph.

i. $y = -2x + 4$

ii. $y = 2x - 4$

iii. $y = 2$

iv. $2x + 4y = 0$

v. $y = \frac{1}{2}x + 4$

vi. $y = -\frac{1}{4}x - 2$

i and D; ii and C; iii and B; iv and F; v and E; vi and A

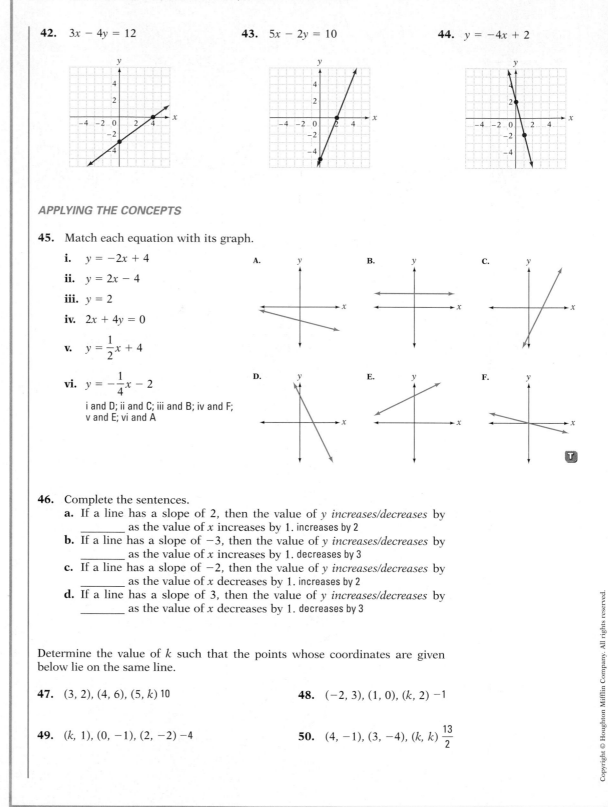

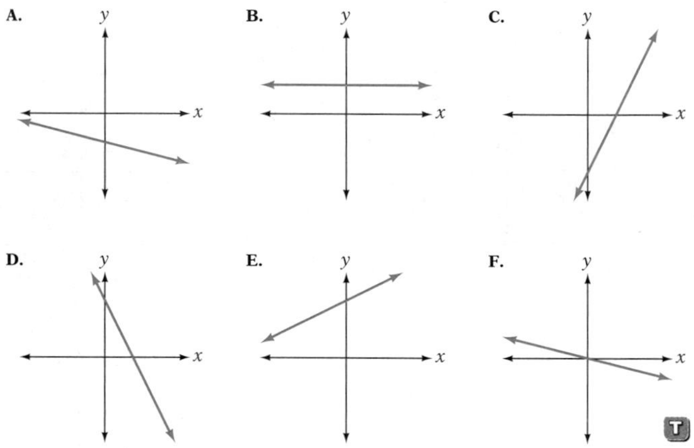

46. Complete the sentences.

a. If a line has a slope of 2, then the value of y *increases/decreases* by ________ as the value of x increases by 1. increases by 2

b. If a line has a slope of -3, then the value of y *increases/decreases* by ________ as the value of x increases by 1. decreases by 3

c. If a line has a slope of -2, then the value of y *increases/decreases* by ________ as the value of x decreases by 1. increases by 2

d. If a line has a slope of 3, then the value of y *increases/decreases* by ________ as the value of x decreases by 1. decreases by 3

Determine the value of k such that the points whose coordinates are given below lie on the same line.

47. $(3, 2), (4, 6), (5, k)$ 10

48. $(-2, 3), (1, 0), (k, 2)$ -1

49. $(k, 1), (0, -1), (2, -2)$ -4

50. $(4, -1), (3, -4), (k, k)$ $\frac{13}{2}$

4.5 Finding Equations of Lines

Objective A

To find the equation of a line given a point and the slope

VIDEO & DVD CD TUTOR WEB SSM

When the slope of a line and a point on the line are known, the equation of the line can be determined. If the particular point is the y-intercept, use the slope-intercept form, $y = mx + b$, to find the equation.

➡ Find the equation of the line that contains the point $(0, 3)$ and has slope $\frac{1}{2}$.

The known point is the y-intercept, (0, 3).

$y = mx + b$ • Use the slope-intercept form.

$y = \frac{1}{2}x + 3$ • Replace m with $\frac{1}{2}$, the given slope. Replace b with 3, the y-coordinate of the y-intercept.

The equation of the line is $y = \frac{1}{2}x + 3$.

One method of finding the equation of a line when the slope and *any* point on the line are known involves using the *point-slope* formula. This formula is derived from the formula for the slope of a line as follows.

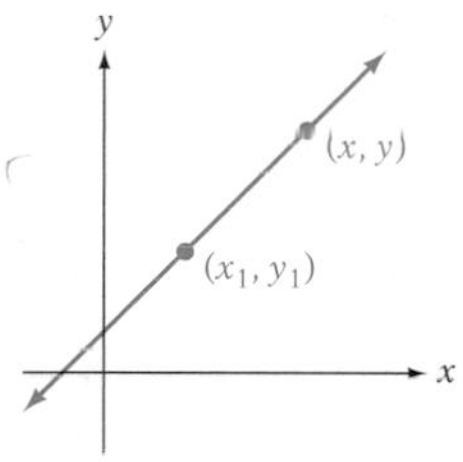

Let (x_1, y_1) be the given point on the line, and let (x, y) be any other point on the line. See the graph at the left.

$\frac{y - y_1}{x - x_1} = m$ • Use the formula for the slope of a line.

$\frac{y - y_1}{x - x_1}(x - x_1) = m(x - x_1)$ • Multiply each side by $(x - x_1)$.

$y - y_1 = m(x - x_1)$ • Simplify.

Point-Slope Formula

Let m be the slope of a line, and let (x_1, y_1) be the coordinates of a point on the line. The equation of the line can be found from the **point-slope formula:**

$$y - y_1 = m(x - x_1)$$

➡ Find the equation of the line that contains the point whose coordinates are $(4, -1)$ and has slope $-\frac{3}{4}$.

$y - y_1 = m(x - x_1)$ • Use the point-slope formula.

$y - (-1) = \left(-\frac{3}{4}\right)(x - 4)$ • $m = -\frac{3}{4}$, $(x_1, y_1) = (4, -1)$

$y + 1 = -\frac{3}{4}x + 3$ • Simplify.

$y = -\frac{3}{4}x + 2$ • Write the equation in the form $y = mx + b$.

The equation of the line is $y = -\frac{3}{4}x + 2$.

245

Objective 4.5A

Vocabulary to Review

slope-intercept form,
$y = mx + b$

New Formulas

point-slope formula
$y - y_1 = m(x - x_1)$

Discuss the Concepts

Why is having the equation of a line helpful?

Instructor Note

Write the point-slope formula as shown below. This will help some students make the correct substitutions, especially when a negative number is substituted.

$$y - y_1 = m(x - x_1)$$
$$\downarrow \quad \downarrow \quad \downarrow$$
$$y - (\ ) = (\ )[x - (\ )]$$

Concept Check

1. Given $F(x) = -20x + 15$,
   a. find $F(1)$. -5
   b. find $F(2)$. -25
   c. use parts **a** and **b** to verify that the slope of the graph of the function is -20.
2. Find the equation of the horizontal line that passes through the point $(4, -5)$. $y = -5$
3. Find the equation of the vertical line that passes through the point $(-3, 2)$. $x = -3$

In-Class Examples (Objective 4.5A)

Find the equation of the line that contains the given point and has the given slope.

1. Point $(0, 5)$, $m = -3$ $y = -3x + 5$
2. Point $(-2, 3)$, $m = -1$ $y = -x + 1$
3. Point $(4, -1)$, $m = \frac{3}{4}$ $y = \frac{3}{4}x - 4$

Optional Student Activity

1. A line with slope 2 goes through the point (4, 6). If the point $(x, -8)$ is on the line, find the value of x. -3
2. A line contains the point $(3, -2)$ and has slope -1. Write the equation of the line in $Ax + By = C$ form. $x + y = 1$

Objective 4.5B

Vocabulary to Review

formula for slope
point-slope formula

Discuss the Concepts

Given two points on a line, how can you tell if the points lie on a line whose equation is of the form $y = b$? How can you tell if the points lie on a line whose equation is of the form $x = a$?

Concept Check

A linear function includes the ordered pairs (1, 4) and (4, 10). Find the value of the function at $x = -1$. 0

246

➡ Find the equation of the line that passes through the point whose coordinates are (4, 3) and whose slope is undefined.

Because the slope is undefined, the point-slope formula cannot be used to find the equation. Instead, recall that when the slope is undefined, the line is vertical and that the equation of a vertical line is $x = a$, where a is the x-coordinate of the x-intercept. Because the line is vertical and passes through (4, 3), the x-intercept is (4, 0). The equation of the line is $x = 4$.

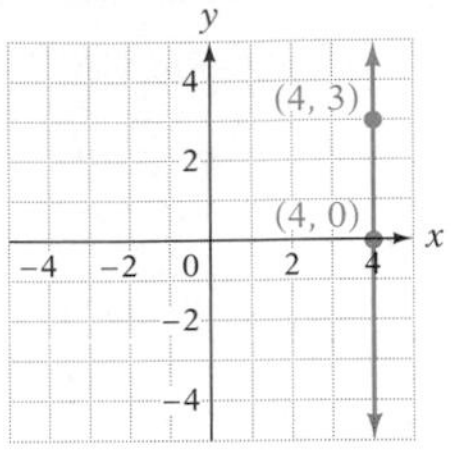

Example 1

Find the equation of the line that contains the point (3, 0) and has slope -4.

Solution

$m = -4 \qquad (x_1, y_1) = (3, 0)$

$$y - y_1 = m(x - x_1)$$
$$y - 0 = -4(x - 3)$$
$$y = -4x + 12$$

The equation of the line is $y = -4x + 12$.

You Try It 1

Find the equation of the line that contains the point $(-3, -2)$ and has slope $-\frac{1}{3}$.

Your solution

$y = -\frac{1}{3}x - 3$

Example 2

Find the equation of the line that contains the point $(-2, 4)$ and has slope 2.

Solution

$m = 2 \qquad (x_1, y_1) = (-2, 4)$

$$y - y_1 = m(x - x_1)$$
$$y - 4 = 2[x - (-2)]$$
$$y - 4 = 2(x + 2)$$
$$y - 4 = 2x + 4$$
$$y = 2x + 8$$

The equation of the line is $y = 2x + 8$.

You Try It 2

Find the equation of the line that contains the point $(4, -3)$ and has slope -3.

Your solution

$y = -3x + 9$

Solutions on p. S12

Objective B **To find the equation of a line given two points**

VIDEO & DVD CD TUTOR WEB SSM

The point-slope formula and the formula for slope are used to find the equation of a line when two points are known.

In-Class Examples (Objective 4.5B)

Find the equation of the line that contains the given points.

1. $P_1(1, 3)$, $P_2(2, 4)$ $y = x + 2$
2. $P_1(0, -2)$, $P_2(-3, 2)$ $y = -\frac{4}{3}x - 2$
3. $P_1(3, -3)$, $P_2(2, -3)$ $y = -3$

➡ Find the equation of the line containing the points (3, 2) and (−5, 6).

To use the point-slope formula, we must know the slope. Use the formula for slope to determine the slope of the line between the two given points. Let $(x_1, y_1) = (3, 2)$ and $(x_2, y_2) = (-5, 6)$.

$$m = \frac{y_2 - y_1}{x_2 - x_1} = \frac{6 - 2}{-5 - 3} = \frac{4}{-8} = -\frac{1}{2}$$

Now use the point-slope formula with $m = -\frac{1}{2}$ and $(x_1, y_1) = (3, 2)$.

$y - y_1 = m(x - x_1)$ • **Use the point-slope formula.**

$y - 2 = \left(-\frac{1}{2}\right)(x - 3)$ • **$m = -\frac{1}{2}$, $(x_1, y_1) = (3, 2)$**

$y - 2 = -\frac{1}{2}x + \frac{3}{2}$ • **Simplify.**

$y = -\frac{1}{2}x + \frac{7}{2}$

The equation of the line is $y = -\frac{1}{2}x + \frac{7}{2}$.

Example 3

Find the equation of the line containing the points (2, 3) and (4, 1).

Solution

Let $(x_1, y_1) = (2, 3)$ and $(x_2, y_2) = (4, 1)$.

$$m = \frac{y_2 - y_1}{x_2 - x_1} = \frac{1 - 3}{4 - 2} = \frac{-2}{2} = -1$$

$y - y_1 = m(x - x_1)$
$y - 3 = -1(x - 2)$
$y - 3 = -x + 2$
$y = -x + 5$

The equation of the line is $y = -x + 5$.

You Try It 3

Find the equation of the line containing the points (2, 0) and (5, 3).

Your solution

$y = x - 2$

Example 4

Find the equation of the line containing the points (2, −3) and (2, 5).

Solution

Let $(x_1, y_1) = (2, -3)$ and $(x_2, y_2) = (2, 5)$.

$$m = \frac{y_2 - y_1}{x_2 - x_1} = \frac{5 - (-3)}{2 - 2} = \frac{8}{0}$$

The slope is undefined, so the graph of the line is vertical.

The equation of the line is $x = 2$.

You Try It 4

Find the equation of the line containing the points (2, 3) and (−5, 3).

Your solution

$y = 3$

Solutions on p. S12

Instructor Note

Students need frequent reminders of the following fact: If the point-slope formula is used to determine the equation of a line, it is *always* necessary to know a point on the line and the slope of the line. For instance, to find the equation of a line given two points, the slope formula must first be used to determine m.

Instructor Note

You might have students use (x_2, y_2) in the point-slope formula at the left and verify that the same equation results.

Optional Student Activity

Assume that the maximum speed your car will go varies linearly with the steepness of the hill it is climbing or descending. If the hill is 5° up, your car can go 77 km per hour. If the hill is 2° down (−2°), your car can go 154 km per hour. When your top speed is 99 km per hour, how steep is the hill? State your answer in degrees and note whether it is degrees up or degrees down.
3° up

Objective 4.5C

Concept Check

In 1787 Jacques Charles, a French scientist, observed that when he plotted a graph of the volume of a fixed amount of air versus the temperature of the air, the points lay along a straight line. He found that at 27°C, a certain amount of air occupied 500 cm³. When he warmed the air to 90°C, its volume increased to 605 cm³. Write a linear equation expressing the volume, V, in terms of temperature, T, in slope-intercept form.

$V = \frac{5}{3}T + 455$

Optional Student Activity

1. A 20-gallon gas tank contains 2 gal of gas when a motorist decides to fill up the tank. The gas pump fills the tank at a rate of 0.08 gal/s. Write a linear function, in functional notation, that models the amount of fuel in the tank t seconds after refueling begins. $f(t) = 0.08t + 2$
2. The boiling point of water at sea level is 100°C. The boiling point decreases 3.5°C for each 1-kilometer increase in altitude. Write a linear function, in functional notation, that gives the boiling point of water as a function of altitude, A. $f(A) = -3.5A + 100$

Objective C To solve application problems

Example 5

On the basis of data from the Kelley Blue Book, the value of a certain car decreases approximately \$250 per month. The value of the car 2 years after it was purchased was \$14,000. Find a linear equation that models the value of the car after x months of ownership. Use the equation to find the value of the car after 3 years of ownership.

Strategy

- To find a linear equation that models the value of the car:
  Let y represent the value of the car after x months.
  Then $y = 14{,}000$ when $x = 24$. (*Note*: x represents *months* of ownership. Convert 2 years to 24 months.)
  The car is decreasing in value at a rate of \$250 per month. Therefore, the slope is -250.
  Use the point-slope formula to write the linear equation.
- To find the value of the car after 3 years, substitute 36 for x, and solve for y. (*Note*: 3 years is 36 months.)

Solution

$$y - y_1 = m(x - x_1)$$
$$y - 14{,}000 = -250(x - 24) \quad \bullet\ m = -250,\ (x_1, y_1) = (24, 14{,}000)$$
$$y - 14{,}000 = -250x + 6000$$
$$y = -250x + 20{,}000$$

A linear equation that models the value of the car after x months of ownership is $y = -250x + 20{,}000$.

$$y = -250x + 20{,}000$$
$$y = -250(36) + 20{,}000$$
$$y = -9000 + 20{,}000$$
$$y = 11{,}000$$

The value of the car after 3 years of ownership is \$11,000.

You Try It 5

A manufacturer of pickup trucks has determined that 40,000 trucks per month can be sold at a price of \$18,000. At a price of \$17,500, the number of trucks that can be sold per month increases to 44,000. Find a linear equation that models the number of trucks that can be sold at a given price. Use the equation to predict the number of trucks that can be sold at a price of \$16,500.

Your strategy

Your solution

$y = -8x + 184{,}000$

52,000 trucks

Solution on p. S13

In-Class Example (Objective 4.5C)

1. In 1950 there were 13 million adults 65 years old or older in the United States. Data from the Census Bureau show that the population of this group has been increasing at a constant rate of approximately 0.5 million per year. This rate of increase is expected to continue through the year 2010. Find the equation of the line that approximates the population of adults 65 years old or older in terms of the year. Use your equation to approximate the population of this group of adults in 2005. $y = 0.5x - 962$; 40.5 million people

4.5 Exercises

Objective A

1. Explain how to find the equation of a line given its slope and its *y*-intercept.

2. What is the point-slope formula and how is it used?

Find the equation of the line that contains the given point and has the given slope.

3. Point (0, 5), $m = 2$

   $y = 2x + 5$

4. Point (0, 3), $m = 1$

   $y = x + 3$

5. Point (2, 3), $m = \frac{1}{2}$

   $y = \frac{1}{2}x + 2$

6. Point (−1, 2), $m = -3$

   $y = -3x - 1$

7. Point (−2, 3), $m = \frac{1}{2}$

   $y = \frac{1}{2}x + 4$

8. Point (−2, 1), $m = \frac{3}{2}$

   $y = \frac{3}{2}x + 4$

9. Point (3, 0), $m = -\frac{5}{3}$

   $y = -\frac{5}{3}x + 5$

10. Point (−2, 0), $m = \frac{3}{2}$

    $y = \frac{3}{2}x + 3$

11. Point (2, 3), $m = -3$

    $y = -3x + 9$

12. Point (4, −2), $m = \frac{3}{4}$

    $y = \frac{3}{4}x - 5$

13. Point (−1, 7), $m = -3$

    $y = -3x + 4$

14. Point (−2, 4), $m = -4$

    $y = -4x - 4$

15. Point (3, 1), $m = \frac{1}{3}$

    $y = \frac{1}{3}x$

16. Point (5, −3), $m = -\frac{3}{5}$

    $y = -\frac{3}{5}x$

17. Point (0, 0), $m = \frac{1}{2}$

    $y = \frac{1}{2}x$

18. Point (0, 0), $m = \frac{3}{4}$

    $y = \frac{3}{4}x$

19. Point (2, −3), $m = 3$

    $y = 3x - 9$

20. Point (4, −5), $m = 2$

    $y = 2x - 13$

21. Point (3, 5), $m = -\frac{2}{3}$

    $y = -\frac{2}{3}x + 7$

22. Point (5, 1), $m = -\frac{4}{5}$

    $y = -\frac{4}{5}x + 5$

23. Point (0, −3), $m = -1$

    $y = -x - 3$

24. Point (2, 3), $m = -\frac{1}{2}$

    $y = -\frac{1}{2}x + 4$

25. Point (1, −4), $m = \frac{7}{5}$

    $y = \frac{7}{5}x - \frac{27}{5}$

26. Point (3, 5), $m = -\frac{3}{7}$

    $y = -\frac{3}{7}x + \frac{44}{7}$

249

Section 4.5

Suggested Assignment

Exercises 3–77, odds
More challenging problems:
Exercises 83–87, odds

Answers to Writing Exercises

1. When we know the slope and *y*-intercept, we can find the equation by using the slope-intercept form, $y = mx + b$. The slope can be substituted for *m*, and the *y*-coordinate of the *y*-intercept can be substituted for *b*.
2. The point-slope formula is $y - y_1 = m(x - x_1)$. It is used to find the equation of a line when one point that lies on the line and the slope of the line are known. The coordinates of the known point are substituted for (x_1, y_1) and the slope is substituted for *m*.

Quick Quiz (Objective 4.5A)

Find the equation of the line that contains the given point and has the given slope.

1. Point (0, 4), $m = -\frac{1}{2}$ $y = -\frac{1}{2}x + 4$
2. Point (2, 3), $m = -2$ $y = -2x + 7$
3. Point (4, 0), $m = -\frac{3}{2}$ $y = -\frac{3}{2}x + 6$

27. Point $(4, -1)$, $m = -\frac{2}{5}$
$y = -\frac{2}{5}x + \frac{3}{5}$

28. Point $(-3, 5)$, $m = -\frac{1}{4}$
$y = -\frac{1}{4}x + \frac{17}{4}$

29. Point $(3, -4)$, slope is undefined
$x = 3$

30. Point $(-2, 5)$, slope is undefined
$x = -2$

31. Point $(5, -1)$, $m = \frac{1}{5}$
$y = \frac{1}{5}x - 2$

32. Point $(-3, -2)$, $m = -\frac{2}{3}$
$y = -\frac{2}{3}x - 4$

33. Point $(-2, -3)$, $m = 0$
$y = -3$

34. Point $(-3, -2)$, $m = 0$
$y = -2$

35. Point $(4, -5)$, $m = -2$
$y = -2x + 3$

36. Point $(-3, 5)$, $m = 3$
$y = 3x + 14$

37. Point $(-5, -1)$, slope is undefined
$x = -5$

38. Point $(0, 4)$, slope is undefined
$x = 0$

Objective B

Find the equation of the line that contains the given points.

39. $(0, 2)$, $(3, 5)$
$y = x + 2$

40. $(0, 4)$, $(1, 5)$
$y = x + 4$

41. $(0, -3)$, $(-4, 5)$
$y = -2x - 3$

42. $(0, -2)$, $(-3, 4)$
$y = -2x - 2$

43. $(2, 3)$, $(-4, 0)$
$y = \frac{1}{2}x + 2$

44. $(4, 1)$, $(6, 3)$
$y = x - 3$

45. $(2, 3)$, $(3, 2)$
$y = -x + 5$

46. $(1, -1)$, $(-2, -7)$
$y = 2x - 3$

47. $(-2, 1)$, $(1, -5)$
$y = -2x - 3$

48. $(-3, -1)$, $(2, 4)$
$y = x + 2$

49. $(0, 3)$, $(2, 0)$
$y = -\frac{3}{2}x + 3$

50. $(0, 4)$, $(2, 0)$
$y = -2x + 4$

51. $(-3, -1)$, $(2, -1)$
$y = -1$

52. $(-3, -5)$, $(4, -5)$
$y = -5$

53. $(-2, -3)$, $(-1, -2)$
$y = x - 1$

Quick Quiz (Objective 4.5B)

Find the equation of the line that contains the given points.

1. $P_1(0, -1)$, $P_2(-2, 3)$ $y = -2x - 1$

2. $P_1(0, 3)$, $P_2(2, 4)$ $y = \frac{1}{2}x + 3$

3. $P_1(-1, -1)$, $P_2(3, 3)$ $y = x$

251

54. (4, 1), (3, −2)
$y = 3x - 11$

55. (−2, 3), (2, −1)
$y = -x + 1$

56. (−1, −3), (2, −12)
$y = -3x - 6$

57. (2, 3), (5, −5)
$y = -\frac{8}{3}x + \frac{25}{3}$

58. (7, 2), (4, 4)
$y = -\frac{2}{3}x + \frac{20}{3}$

59. (2, 0), (0, −1)
$y = \frac{1}{2}x - 1$

60. (0, 4), (−2, 0)
$y = 2x + 4$

61. (3, −4), (−2, −4)
$y = -4$

62. (−3, 3), (−2, 3)
$y = 3$

63. (0, 0), (4, 3)
$y = \frac{3}{4}x$

64. (2, −5), (0, 0)
$y = -\frac{5}{2}x$

65. (2, −1), (−1, 3)
$y = -\frac{4}{3}x + \frac{5}{3}$

66. (3, −5), (−2, 1)
$y = -\frac{6}{5}x - \frac{7}{5}$

67. (−2, 5), (−2, −5)
$x = -2$

68. (3, 2), (3, −4)
$x = 3$

69. (2, 1), (−2, −3)
$y = x - 1$

70. (−3, −2), (1, −4)
$y = -\frac{1}{2}x - \frac{7}{2}$

71. (−4, −3), (2, 5)
$y = \frac{4}{3}x + \frac{7}{3}$

72. (4, 5), (−4, 3)
$y = \frac{1}{4}x + 4$

73. (0, 3), (3, 0)
$y = -x + 3$

74. (−5, 0), (10, −3)
$y = -\frac{1}{5}x - 1$

Objective C *Application Problems*

75. The pilot of a Boeing 747 jet takes off from Boston's Logan Airport, which is at sea level, and climbs to a cruising altitude of 32,000 ft at a constant rate of 1200 ft/min. Write a linear equation for the height of the plane in terms of the time after takeoff. Use your equation to find the height of the plane 11 min after takeoff.
$y = 1200x$; 13,200 ft

76. The price of a new motorcycle this year is $18,000. According to a consumer group, the value of the motorcycle will be $7200 in 6 years. Write a linear equation that gives the value of the motorcycle in terms of its age. Use the equation to find the value of the motorcycle in 4 years.
$y = -1800x + 18{,}000$; $10,800

Quick Quiz (Objective 4.5C)

1. An Airbus 320 plane takes off from Denver International Airport, which is 5200 ft above sea level, and climbs to 30,000 ft at a constant rate of 1000 ft/min. Write a linear equation for the height of the plane in terms of the time after takeoff. Use your equation to find the height of the plane 8 min after takeoff. $y = 1000x + 5200, 0 \leq x \leq 24.8$; 13,200 ft

252 Chapter 4 / Linear Equations and Inequalities in Two Variables

Answers to Writing Exercises

79. The slope-intercept form of a straight line, $y = mx + b$, is the general form in which the equation of a straight line is written. Given the slope of a line and its y-intercept, the equation of the line can be written by substituting the slope for m and the y-coordinate of the y-intercept for b in the equation $y = mx + b$.
The point-slope formula, $y - y_1 = m(x - x_1)$, is used to find the equation of a line when the slope of the line and any point on the line other than the y-intercept are known. We substitute the slope and the coordinates of the known point into the point-slope formula and then rewrite the equation in the slope-intercept form of a straight line.

80. The slope of any line parallel to the y-axis is undefined. In order to use the point-slope formula, we must be able to substitute the slope of the line for m. In other words, in order to use the point-slope formula, the slope of the line must be defined.

81a. The slope of the line $y = -250x + 20{,}000$ is -250. The slope is the rate at which the the value of the car is decreasing per month: The value of the car is decreasing \$250 per month.

b. The y-intercept is $(0, 20{,}000)$. This means that the value of the car when new (after 0 months) was \$20,000.

c. The x-intercept is $(80, 0)$. This means that after 80 months, the value of the car is \$0.

77. The gas tank of a certain car contains 16 gal when the driver of the car begins a trip. Each mile driven by the driver decreases the amount of gas in the tank by 0.032 gal. Write a linear equation for the number of gallons of gas in the tank in terms of the number of miles driven. Use your equation to find the number of gallons in the tank after this driver has driven 150 mi.
$y = -0.032x + 16$; 11.2 gal

78. A jogger running at 9 mph burns approximately 14 calories per minute. Write a linear equation for the number of calories burned by the jogger in terms of the number of minutes run. Use your equation to find the number of calories one has burned after jogging for 32 min.
$y = 14x$; 448 calories

APPLYING THE CONCEPTS

79. Explain the similarities and differences between the point-slope formula and the slope-intercept form of a straight line.

80. Explain why the point-slope formula cannot be used to find the equation of a line that is parallel to the y-axis.

81. Refer to Example 5 in this section for each of the following.
a. Explain the meaning of the slope of the graph of the linear function.
b. Explain the meaning of the y-intercept.
c. Explain the meaning of the x-intercept.

82. For an equation of the form $y = mx + b$, how does the graph of this equation change if the value of b changes and the value of m remains constant?
Changing b results in the graph of the line moving up or down.

83. A line contains the points $(-3, 6)$ and $(6, 0)$. Find the coordinates of three other points that are on this line.
Answers will vary. Possible answers are $\{(0, 4), (3, 2), (9, -2)\}$.

84. A line contains the points $(4, -1)$ and $(2, 1)$. Find the coordinates of three other points that are on this line.
Answers will vary. Possible answers are $\{(0, 3), (1, 2), (3, 0)\}$.

85. Given that f is a linear function for which $f(1) = 3$ and $f(-1) = 5$, determine $f(4)$. 0

86. Given that f is a linear function for which $f(-3) = 2$ and $f(2) = 7$, determine $f(0)$. 5

87. Find the equation of the line that passes through the midpoint of the line segment between $P_1(2, 5)$ and $P_2(-4, 1)$ and has slope -2.
$y = -2x + 1$

4.6 Parallel and Perpendicular Lines

Objective A **To find parallel and perpendicular lines**

VIDEO & DVD CD TUTOR WEB SSM

Two lines that have the same slope do not intersect and are called **parallel lines**.

The slope of each of the lines at the right is $\frac{2}{3}$.

The lines are parallel.

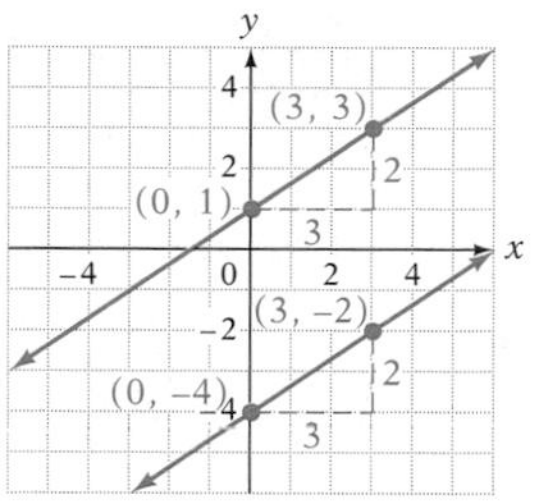

Slopes of Parallel Lines

Two nonvertical lines with slopes of m_1 and m_2 are parallel if and only if $m_1 = m_2$. Any two vertical lines are parallel.

➡ Is the line containing the points (−2, 1) and (−5, −1) parallel to the line that contains the points (1, 0) and (4, 2)?

$m_1 = \frac{-1-1}{-5-(-2)} = \frac{-2}{-3} = \frac{2}{3}$ • Find the slope of the line through (−2, 1) and (−5, −1).

$m_2 = \frac{2-0}{4-1} = \frac{2}{3}$ • Find the slope of the line through (1, 0) and (4, 2).

Because $m_1 = m_2$, the lines are parallel.

➡ Find the equation of the line that contains the point (2, 3) and is parallel to the line $y = \frac{1}{2}x - 4$.

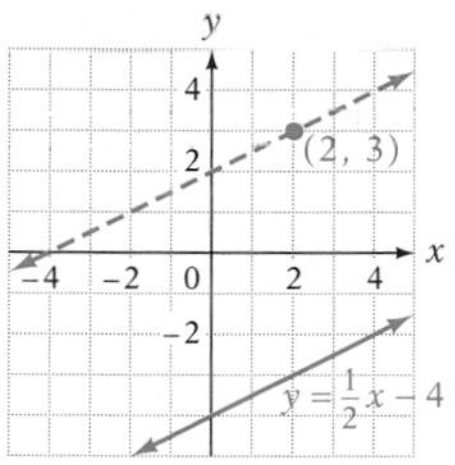

The slope of the given line is $\frac{1}{2}$. Because parallel lines have the same slope, the slope of the unknown line is also $\frac{1}{2}$.

$y - y_1 = m(x - x_1)$ • Use the point-slope formula.

$y - 3 = \frac{1}{2}(x - 2)$ • $m = \frac{1}{2}, (x_1, y_1) = (2, 3)$

$y - 3 = \frac{1}{2}x - 1$ • Simplify.

$y = \frac{1}{2}x + 2$ • Write the equation in the form $y = mx + b$.

The equation of the line is $y = \frac{1}{2}x + 2$.

Objective 4.6A

New Vocabulary

parallel lines
perpendicular lines
negative reciprocal

Instructor Note

When students are attempting to find the equations of parallel or perpendicular lines, it may help to use the model given in the last section.

$$y - y_1 = m(x - x_1)$$
$$\downarrow \quad \downarrow \quad \downarrow$$
$$y - (\ ) = (\ )[x - (\ )]$$

If the lines are parallel, m for the given line is the same as the slope of the line to which it is parallel. If the lines are perpendicular, m is the negative reciprocal of the slope of the line to which the given line is perpendicular.

Concept Check

1. Given the slope of one line, how can you find the slope of any line parallel to that line? The slopes are the same.
2. Given the slope of one line, how can you find the slope of any line perpendicular to that line? Find the reciprocal of the slope and change its sign.
3. If the slope of one line is positive, what do you know about the slope of any line perpendicular to that line? The slope is negative.
4. If the product of the slopes of two perpendicular lines does not equal −1, what is the slope of one of the lines? 0
5. What is the slope of any line parallel to the x-axis? 0

In-Class Examples (Objective 4.6A)

1. Is the line that contains the points (−2, 3) and (5, −2) perpendicular to the line that contains the points (2, 4) and (−3, −3)? Yes
2. Find the equation of the line containing the point (2, 1) and parallel to the line $3x + y = -1$. $y = -3x + 7$
3. Find the equation of the line containing the point (1, −4) and perpendicular to the line $y = \frac{3}{2}x - 2$. $y = -\frac{2}{3}x - \frac{10}{3}$

Discuss the Concepts

After presenting the concept of perpendicular lines, we suggest that you draw on a coordinate grid two perpendicular lines that are not parallel to the coordinate axes. Construct right triangles using the intersection of the lines as a vertex and with the hypotenuse on the line. Mark on one triangle the rise and run of the line. Ask the students what happens when this line is rotated 90° onto the other line. You want the students to see that the rotation results in the rise and run of the first line becoming the run and rise, respectively, of the second line and that the run changes direction.

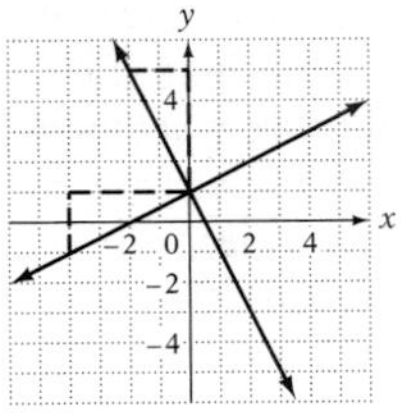

254

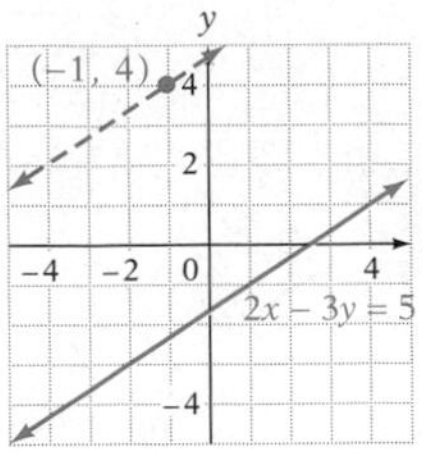

➡ Find the equation of the line that contains the point $(-1, 4)$ and is parallel to the line $2x - 3y = 5$.

Because the lines are parallel, the slope of the unknown line is the same as the slope of the given line. Solve $2x - 3y = 5$ for y and determine its slope.

$$2x - 3y = 5$$
$$-3y = -2x + 5$$
$$y = \frac{2}{3}x - \frac{5}{3}$$

The slope of the given line is $\frac{2}{3}$. Because the lines are parallel, this is the slope of the unknown line. Use the point-slope formula to determine the equation.

$y - y_1 = m(x - x_1)$ • **Use the point-slope formula.**

$y - 4 = \frac{2}{3}[x - (-1)]$ • **$m = \frac{2}{3}$, $(x_1, y_1) = (-1, 4)$**

$y - 4 = \frac{2}{3}x + \frac{2}{3}$ • **Simplify.**

$y = \frac{2}{3}x + \frac{14}{3}$ • **Write the equation in the form $y = mx + b$.**

The equation of the line is $y = \frac{2}{3}x + \frac{14}{3}$.

Two lines that intersect at right angles are **perpendicular lines.**

Any horizontal line is perpendicular to any vertical line. For example, $x = 3$ is perpendicular to $y = -2$.

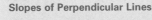

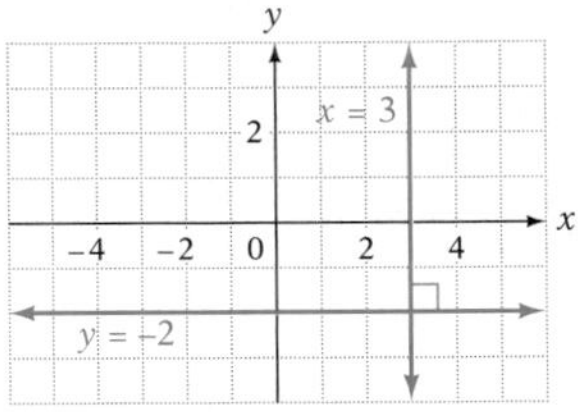

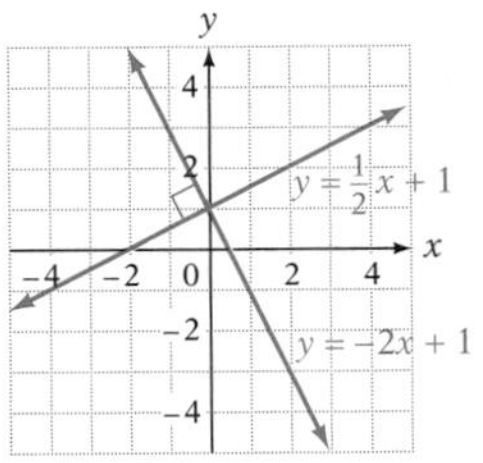

> **Slopes of Perpendicular Lines**
>
> If m_1 and m_2 are the slopes of two lines, neither of which is vertical, then the lines are perpendicular if and only if $m_1 \cdot m_2 = -1$.
>
> A vertical line is perpendicular to a horizontal line.

Solving $m_1 \cdot m_2 = -1$ for m_1 gives $m_1 = -\frac{1}{m_2}$. This last equation states that the slopes of perpendicular lines are *negative reciprocals* of each other.

➡ Is the line that contains the points $(4, 2)$ and $(-2, 5)$ perpendicular to the line that contains the points $(-4, 3)$ and $(-3, 5)$?

$m_1 = \frac{5 - 2}{-2 - 4} = \frac{3}{-6} = -\frac{1}{2}$ • **Find the slope of the line through (4, 2) and (−2, 5).**

$m_2 = \frac{5 - 3}{-3 - (-4)} = \frac{2}{1} = 2$ • **Find the slope of the line through (−4, 3) and (−3, 5).**

$m_1 \cdot m_2 = -\frac{1}{2}(2) = -1$ • **Find the product of the two slopes.**

Because $m_1 \cdot m_2 = -1$, the lines are perpendicular.

➡ Are the graphs of the equations $3x + 4y = 8$ and $8x + 6y = 5$ perpendicular?

To determine whether the lines are perpendicular, solve each equation for y and find the slope of each line. Then use the equation $m_1 \cdot m_2 = -1$.

$$\begin{aligned} 3x + 4y &= 8 \\ 4y &= -3x + 8 \\ y &= -\frac{3}{4}x + 2 \qquad m_1 = -\frac{3}{4} \end{aligned} \qquad \begin{aligned} 8x + 6y &= 5 \\ 6y &= -8x + 5 \\ y &= -\frac{4}{3}x + \frac{5}{6} \qquad m_2 = -\frac{4}{3} \end{aligned}$$

$$m_1 \cdot m_2 = \left(-\frac{3}{4}\right)\left(-\frac{4}{3}\right) = 1$$

Because $m_1 \cdot m_2 = 1 \neq -1$, the lines are not perpendicular.

➡ Find the equation of the line that contains the point $(-2, 1)$ and is perpendicular to the line $y = -\frac{2}{3}x + 2$.

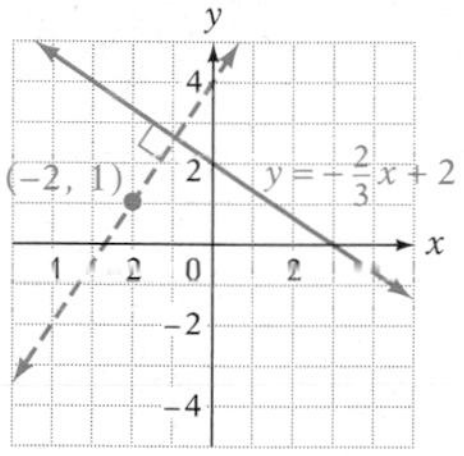

The slope of the given line is $-\frac{2}{3}$. The slope of the line perpendicular to the given line is the negative reciprocal of $-\frac{2}{3}$, which is $\frac{3}{2}$.

$y - y_1 = m(x - x_1)$ • Use the point-slope formula.

$y - 1 = \frac{3}{2}[x - (-2)]$ • $m = \frac{3}{2}$, $(x_1, y_1) = (-2, 1)$

$y - 1 = \frac{3}{2}x + 3$ • Write in the form $y = mx + b$.

$y = \frac{3}{2}x + 4$

The equation of the perpendicular line is $y = \frac{3}{2}x + 4$.

➡ Find the equation of the line that contains the point $(3, -4)$ and is perpendicular to the line $2x - y = -3$.

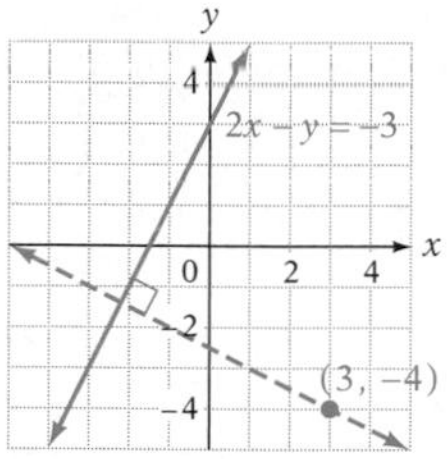

$2x - y = -3$ • Find the slope of the given line.

$y = -2x - 3$

$y = 2x + 3$ • The slope is 2. The slope of the line perpendicular to this line is $-\frac{1}{2}$.

$y - y_1 = m(x - x_1)$ • Use the point-slope formula.

$y - (-4) = -\frac{1}{2}(x - 3)$ • $m = -\frac{1}{2}$, $(x_1, y_1) = (3, -4)$

$y + 4 = -\frac{1}{2}x + \frac{3}{2}$ • Write in the form $y = mx + b$.

$y = -\frac{1}{2}x - \frac{5}{2}$

The equation of the perpendicular line is $y = -\frac{1}{2}x - \frac{5}{2}$.

Optional Student Activity

1. Find the coordinates of the y-intercept of the line parallel to $4x + 3y = 5$ that contains the point $(6, -1)$. $(0, 7)$
2. The graphs of $x + 2y = -3$ and $ax + 3y = -2$ are perpendicular lines. Find the value of a. -6

256 Chapter 4 / Linear Equations and Inequalities in Two Variables

Example 1

Is the line that contains the points (−4, 2) and (1, 6) parallel to the line that contains the points (2, −4) and (7, 0)?

Solution

$$m_1 = \frac{6-2}{1-(-4)} = \frac{4}{5}$$

$$m_2 = \frac{0-(-4)}{7-2} = \frac{4}{5}$$

$$m_1 = m_2 = \frac{4}{5}$$

The lines are parallel.

You Try It 1

Is the line that contains the points (−2, −3) and (7, 1) perpendicular to the line that contains the points (4, 1) and (6, −5)?

Your solution

no

Example 2

Are the lines $4x - y = -2$ and $x + 4y = -12$ perpendicular?

Solution

$$4x - y = -2$$
$$-y = -4x - 2$$
$$y = 4x + 2 \qquad \bullet\ m_1 = 4$$

$$x + 4y = -12$$
$$4y = -x - 12$$
$$y = -\frac{1}{4}x - 3 \qquad \bullet\ m_2 = -\frac{1}{4}$$

$$m_1 \cdot m_2 = 4\left(-\frac{1}{4}\right) = -1$$

The lines are perpendicular.

You Try It 2

Are the lines $5x + 2y = 2$ and $5x + 2y = -6$ parallel?

Your solution

yes

Example 3

Find the equation of the line that contains the point (3, −1) and is parallel to the line $y = \frac{3}{2}x - 2$.

Solution

$$y - y_1 = m(x - x_1)$$

$$y - (-1) = \frac{3}{2}(x - 3) \qquad \bullet\ m = \frac{3}{2}$$

$$y + 1 = \frac{3}{2}x - \frac{9}{2}$$

$$y = \frac{3}{2}x - \frac{11}{2}$$

The equation of the line is $y = \frac{3}{2}x - \frac{11}{2}$.

You Try It 3

Find the equation of the line that contains the point (−2, 2) and is perpendicular to the line $y = \frac{1}{4}x - 3$.

Your solution

$y = -4x - 6$

Solutions on p. S13

4.6 Exercises

Objective A

1. Explain how to determine whether the graphs of two lines are parallel.

2. Explain how to determine whether the graphs of two lines are perpendicular.

3. The slope of a line is -5. What is the slope of any line parallel to this line?
-5

4. The slope of a line is $\frac{3}{2}$. What is the slope of any line parallel to this line? $\frac{3}{2}$

5. The slope of a line is 4. What is the slope of any line perpendicular to this line?
$-\frac{1}{4}$

6. The slope of a line is $-\frac{4}{5}$. What is the slope of any line perpendicular to this line? $\frac{5}{4}$

7. Is the line $x = -2$ perpendicular to the line $y = 3$?
yes

8. Is the line $y = \frac{1}{2}$ perpendicular to the line $y = -4$?
no

9. Is the line $x = -3$ parallel to the line $y = \frac{1}{3}$?
no

10. Is the line $x = 4$ parallel to the line $x = -4$?
yes

11. Is the line $y = \frac{2}{3}x - 4$ parallel to the line $y = -\frac{3}{2}x - 4$?
no

12. Is the line $y = -2x + \frac{2}{3}$ parallel to the line $y = -2x + 3$?
yes

13. Is the line $y = \frac{4}{3}x - 2$ perpendicular to the line $y = -\frac{3}{4}x + 2$?
yes

14. Is the line $y = \frac{1}{2}x + \frac{3}{2}$ perpendicular to the line $y = -\frac{1}{2}x + \frac{3}{2}$?
no

15. Are the lines $2x + 3y = 2$ and $2x + 3y = -4$ parallel?
yes

16. Are the lines $2x - 4y = 3$ and $2x + 4y = -3$ parallel?
no

17. Are the lines $x - 4y = 2$ and $4x + y = 8$ perpendicular?
yes

18. Are the lines $4x - 3y = 2$ and $4x + 3y = -7$ perpendicular?
no

19. Is the line that contains the points (3, 2) and (1, 6) parallel to the line that contains the points $(-1, 3)$ and $(-1, -1)$?
no

20. Is the line that contains the points $(4, -3)$ and $(2, 5)$ parallel to the line that contains the points $(-2, -3)$ and $(-4, 1)$?
no

21. Is the line that contains the points $(-3, 2)$ and $(4, -1)$ perpendicular to the line that contains the points $(1, 3)$ and $(-2, -4)$?
yes

22. Is the line that contains the points $(-1, 2)$ and $(3, 4)$ perpendicular to the line that contains the points $(-1, 3)$ and $(-4, 1)$?
no

Section 4.6

Suggested Assignment

Exercises 3–27, odds
More challenging problems:
Exercises 29–34

Answers to Writing Exercises

1. The student should note that the slope of each graphed line must be determined. Look for the idea that two lines are parallel if they have the same slope and different y-intercepts.
2. The student should note that the slope of each graphed line must be determined. Look for the idea that two lines are perpendicular if their slopes are negative reciprocals of each other.

Quick Quiz (Objective 4.6A)

1. Is the line that contains the points (2, 1) and (0, 5) parallel to the line that contains the points $(-2, 4)$ and $(-2, -2)$? No
2. Find the equation of the line containing the point $(-1, -3)$ and parallel to the line $x - 2y = 3$. $y = \frac{1}{2}x - \frac{5}{2}$
3. Find the equation of the line containing the point (3, 1) and perpendicular to the line $y = -2x + 5$. $y = \frac{1}{2}x - \frac{1}{2}$

23. Find the equation of the line containing the point $(-2, -4)$ and parallel to the line $2x - 3y = 2$.
$y = \frac{2}{3}x - \frac{8}{3}$

24. Find the equation of the line containing the point $(3, 2)$ and parallel to the line $3x + y = -3$.
$y = -3x + 11$

25. Find the equation of the line containing the point $(4, 1)$ and perpendicular to the line $y = -3x + 4$.
$y = \frac{1}{3}x - \frac{1}{3}$

26. Find the equation of the line containing the point $(2, -5)$ and perpendicular to the line $y = \frac{5}{2}x - 4$.
$y = -\frac{2}{5}x - \frac{21}{5}$

27. Find the equation of the line containing the point $(-1, -3)$ and perpendicular to the line $3x - 5y = 2$.
$y = -\frac{5}{3}x - \frac{14}{3}$

28. Find the equation of the line containing the point $(-1, 3)$ and perpendicular to the line $2x + 4y = -1$.
$y = 2x + 5$

APPLYING THE CONCEPTS

For Exercises 29 and 30, suppose a ball is being twirled at the end of a string and the center of rotation is the origin of a coordinate system. If the string breaks, the initial path of the ball is on a line that is perpendicular to the radius of the circle.

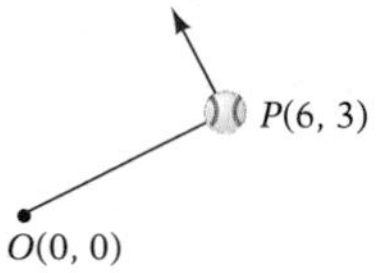

29. Suppose the string breaks when the ball is at the point whose coordinates are $P(6, 3)$. Find the equation of the line on which the initial path lies.
$y = -2x + 15$

30. Suppose the string breaks when the ball is at the point whose coordinates are $P(2, 8)$. Find the equation of the line on which the initial path lies.
$y = -\frac{1}{4}x + \frac{17}{2}$

31. If the graphs of $A_1x + B_1y = C_1$ and $A_2x + B_2y = C_2$ are perpendicular, express $\frac{A_1}{B_1}$ in terms of A_2 and B_2. $\frac{A_1}{B_1} = -\frac{B_2}{A_2}$

32. If the graphs of $A_1x + B_1y = C_1$ and $A_2x + B_2y = C_2$ are parallel, express $\frac{A_1}{B_1}$ in terms of A_2 and B_2. $\frac{A_1}{B_1} = \frac{A_2}{B_2}$

33. The graphs of $y = -\frac{1}{2}x + 2$ and $y = \frac{2}{3}x - 5$ intersect at the point whose coordinates are $(6, -1)$. Find the equation of a line whose graph intersects the graphs of the given lines to form a right triangle. (*Hint:* There is more than one answer to this question.)
Any equation of the form $y = 2x + b$ where $b \neq -13$ or of the form $y = -\frac{3}{2}x + c$ where $c \neq 8$.

34. A theorem from geometry states that a line passing through the center of a circle and through a point P on the circle is perpendicular to the tangent line at P. (See the figure at the right.) If the coordinates of P are $(5, 4)$ and the coordinates of C are $(3, 2)$, what is the equation of the tangent line?
$y = -x + 9$

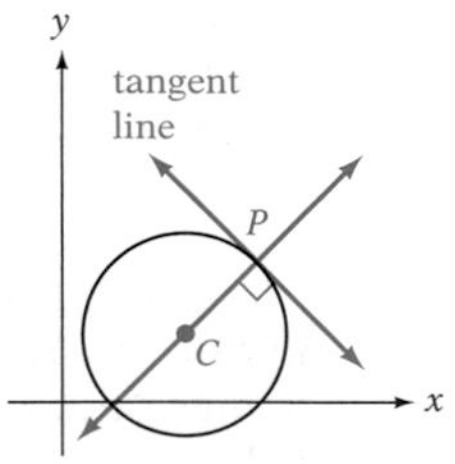

4.7 Graphing Linear Inequalities

Objective A To graph an inequality in two variables

VIDEO & DVD CD TUTOR WEB SSM

Point of Interest

Linear inequalities play an important role in applied mathematics. They are used in a branch of mathematics called *linear programming*, which was developed during World War II to solve problems in supplying the Air Force with the machine parts necessary to keep planes flying. Today, its applications have been broadened to many other disciplines.

The graph of the linear equation $y = x - 2$ separates a plane into three sets:

the set of points on the line
the set of points above the line
the set of points below the line

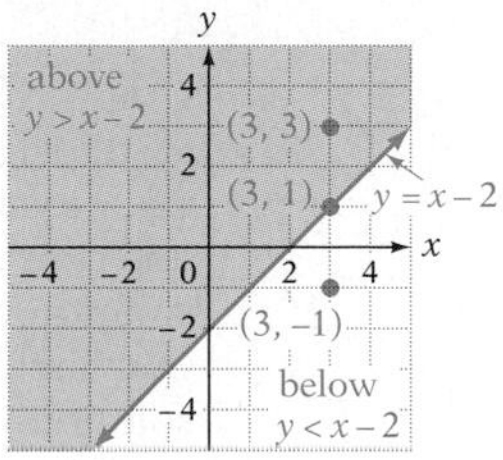

The point (3, 1) is a solution of $y = x - 2$.

$$\begin{array}{c|c} y & x - 2 \\ \hline 1 & 3 - 2 \\ 1 & 1 \end{array}$$

$y = x - 2$; $1 \;|\; 3 - 2$; $1 = 1$

The point (3, 3) is a solution of $y > x - 2$.

$y > x - 2$; $3 \;|\; 3 - 2$; $3 > 1$

Any point above the line is a solution of $y > x - 2$.

The point (3, −1) is a solution of $y < x - 2$.

$y < x - 2$; $-1 \;|\; 3 - 2$; $-1 < 1$

Any point below the line is a solution of $y < x - 2$.

The solution set of $y = x - 2$ is all points on the line. The solution set of $y > x - 2$ is all points above the line. The solution set of $y < x - 2$ is all points below the line. The solution set of an inequality in two variables is a **half-plane**.

The following illustrates the procedure for graphing a linear inequality.

➡ Graph the solution set of $2x + 3y \le 6$.

$$2x + 3y \le 6$$
$$2x - 2x + 3y \le -2x + 6 \quad \text{• Solve the inequality for } y. \text{ • Subtract } 2x \text{ from each side.}$$
$$3y \le -2x + 6 \quad \text{• Simplify.}$$
$$\frac{3y}{3} \le \frac{-2x + 6}{3} \quad \text{• Divide each side by 3.}$$
$$y \le -\frac{2}{3}x + 2 \quad \text{• Simplify.}$$

Change the inequality to an equality and graph $y = -\frac{2}{3}x + 2$. If the inequality is **≥ or ≤**, the line is in the solution set and is shown by a **solid line**. If the inequality is **> or <**, the line is not a part of the solution set and is shown by a **dotted line**.

If the inequality is **> or ≥**, shade the **upper half-plane**. If the inequality is **< or ≤**, shade the **lower half-plane**.

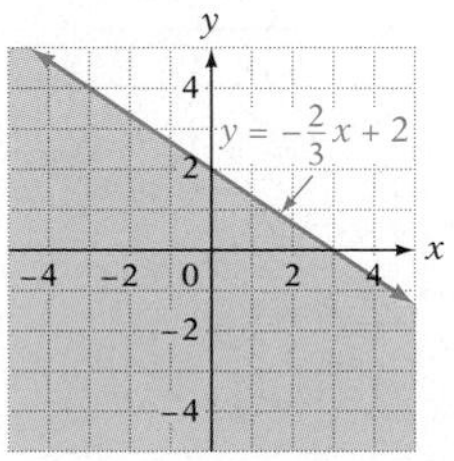

Objective 4.7A

New Vocabulary

half-plane
linear inequality in two variables

Concept Check

Determine whether the statement is true or false.

1. The graph of a linear inequality is a half-plane. True
2. The exponents on the variables in a linear inequality in two variables are 1. True
3. The solution of a linear inequality in two variables containing ≤ or ≥ includes the line separating the half-planes. True
4. The solution set of the inequality $y > x + 2$ is all the points above the line $y = x + 2$. True
5. It is possible to write a linear inequality in two variables that has no solution. False

Instructor Note

Show students that as long as the graph of the equation does not pass through the origin, the ordered pair (0, 0) affords an easy way to determine the correct region to shade. If (0, 0) satisfies the linear inequality, then the point whose coordinates are (0, 0) should be included in the shaded region. If (0, 0) does not satisfy the inequality, it should not be in the shaded region.

Optional Student Activity

Linear inequalities are used as constraints (conditions that must be satisfied) for some application problems. For these problems, the focus of attention is the first quadrant; a solution outside the first quadrant is not considered. For each of the following, shade the region of the first quadrant that satisfies the constraints.

(Continued on next page)

In-Class Examples (Objective 4.7A)

Graph the solution set.

1. $2x - 3y \ge 6$

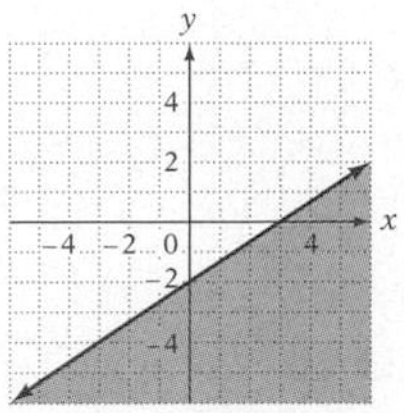

2. $x + 2y > 6$

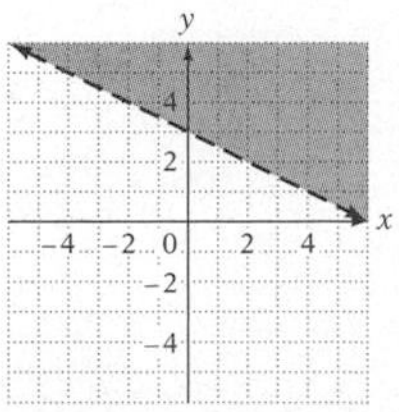

3. $y > 4$

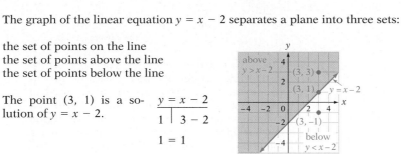

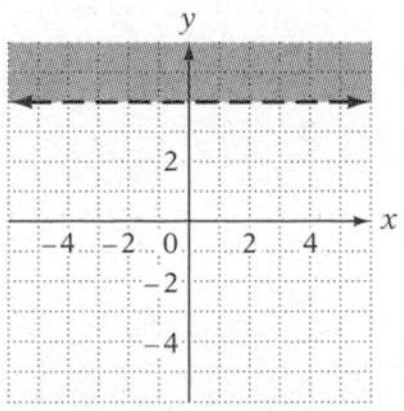

(Continued)

1. A manufacturer makes two types of computer monitors, 15-inch and 17-inch. Because of the production requirements for these monitors, the maximum number of monitors that can be produced in one day is 100. Shade the region of the first quadrant whose ordered pairs satisfy the constraint. For $x =$ the number of 15-inch monitors and $y =$ the number of 17-inch monitors, the solution set is $y \le 100 - x$.
2. A manufacturer makes two types of bicycle gears, standard and deluxe. It takes 4 hours of labor to produce a standard gear and 6 hours of labor to produce a deluxe gear. If there are a maximum of 480 hours of labor available, shade the region of the first quadrant whose ordered pairs satisfy the constraint. For $x =$ the number of standard gears and $y =$ the number of deluxe gears, the solution set is $y \le -\frac{2}{3}x + 80$.
3. Suppose a single tablet of the diet supplement SuperC contains 150 mg of calcium and one tablet of the diet supplement CalcPlus contains 200 mg of calcium. If a health care professional recommends that a patient take at least 500 mg of calcium per day but less than 1000 mg, shade the region of the first quadrant whose ordered pairs satisfy these constraints. For $x =$ the number of SuperC tablets and $y =$ the number of CalcPlus tablets, the solution set is $y \ge -0.75x + 2.5$ and $y < -0.75x + 5$.
4. In Exercise 3, what do the ordered pairs in the shaded region mean in the context of the problem? The ordered pairs in the shaded region represent the number of each type of calcium tablet needed to get at least 500 mg of calcium but less than 1000 mg.

The inequality $2x + 3y \le 6$ can also be graphed as shown below.

$2x + 3y = 6$ • Change the inequality to an equality.

$2x + 3(0) = 6$ • Find the x- and y-intercepts of the equation.
$2x = 6$
$x = 3$ • The x-intercept is (3, 0).

$2(0) + 3y = 6$
$3y = 6$
$y = 2$ • The y-intercept is (0, 2).

Graph the ordered pairs (3, 0) and (0, 2). Draw a solid line through the points because the inequality is $\le$. The point (0, 0) can be used to determine which region to shade. If (0, 0) is a solution of the inequality, then shade the region that includes the point (0, 0). If (0, 0) is not a solution of the inequality, then shade the region that does not include the point (0, 0). For this example, (0, 0) is a solution of the inequality. The region that contains the point (0, 0) is shaded.

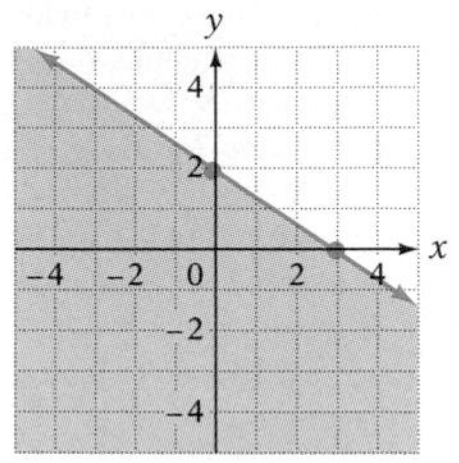

$2x + 3y \le 6$
$2(0) + 3(0) \le 6$
$0 \le 6$ True

TAKE NOTE
Any ordered pair is of the form (x, y). For the point (0, 0), substitute 0 for x and 0 for y in the inequality.

If the line passes through point (0, 0), another point must be used to determine which region to shade. For example, use the point (1, 0).

It is important to note that every point in the shaded region is a solution of the inequality and that every solution of the inequality is a point in the shaded region. No point outside the shaded region is a solution of the inequality.

Example 1
Graph the solution set of $3x + y > -2$.

Solution
$3x + y > -2$
$3x - 3x + y > -3x - 2$
$y > -3x - 2$
Graph $y = -3x - 2$ as a dashed line.
Shade the upper half-plane.

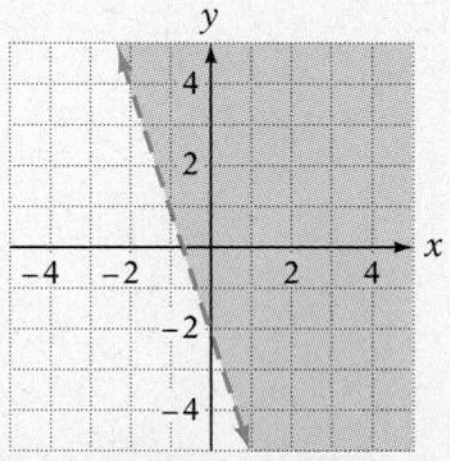

You Try It 1
Graph the solution set of $x - 3y < 2$.

Your solution

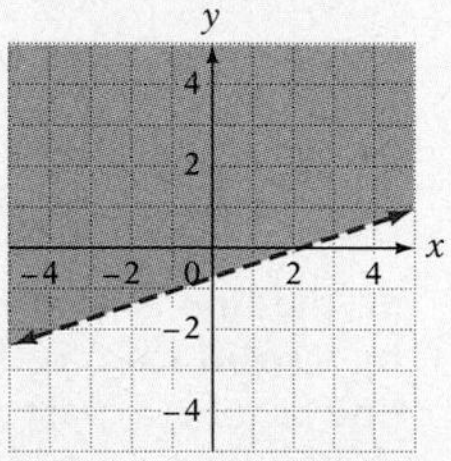

Solution on p. S13

261

4.7 Exercises

Objective A

1. What is a half-plane?

2. Explain a method you can use to check that the graph of a linear inequality in two variables has been shaded correctly.

3. Is (0, 0) a solution of $y > 2x - 7$? yes

4. Is (0, 0) a solution of $y < 5x + 3$? yes

5. Is (0, 0) a solution of $y \le -\frac{2}{3}x - 8$? no

6. Is (0, 0) a solution of $y \ge -\frac{3}{4}x + 9$? no

Graph the solution set.

7. $y > -x + 4$

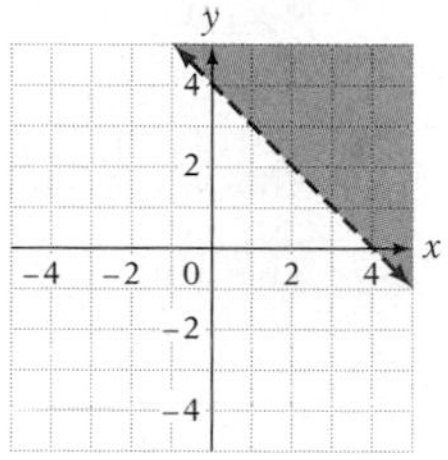

8. $y < x + 3$

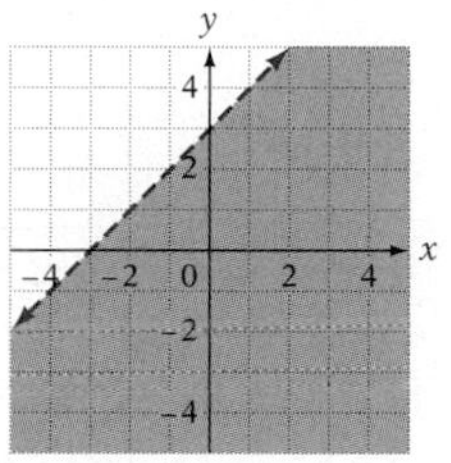

9. $y > 2x + 3$

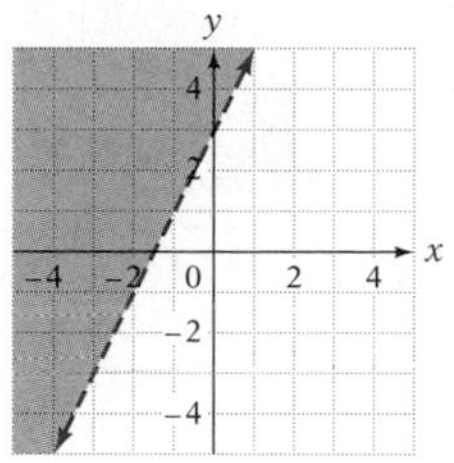

10. $y \ge -2x + 4$

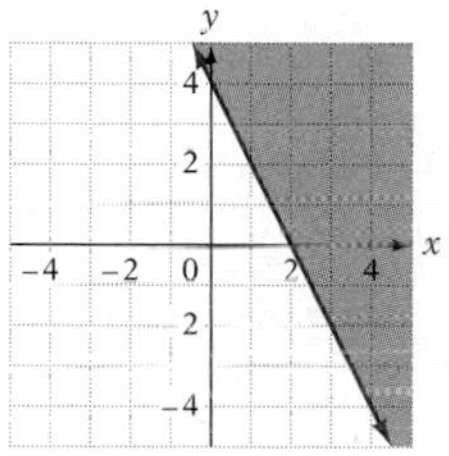

11. $y > \frac{3}{2}x - 4$

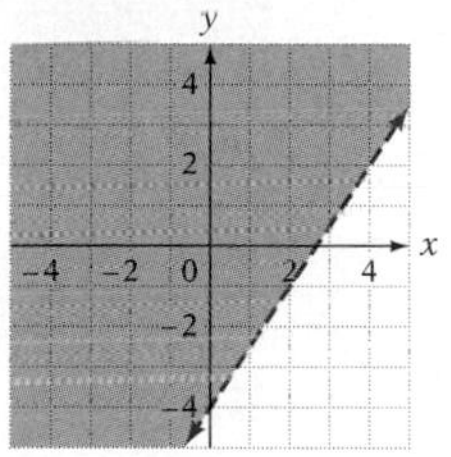

12. $y > -\frac{5}{4}x + 1$

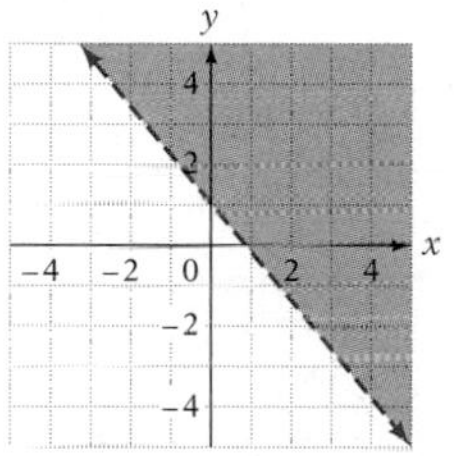

13. $y \le -\frac{3}{4}x - 1$

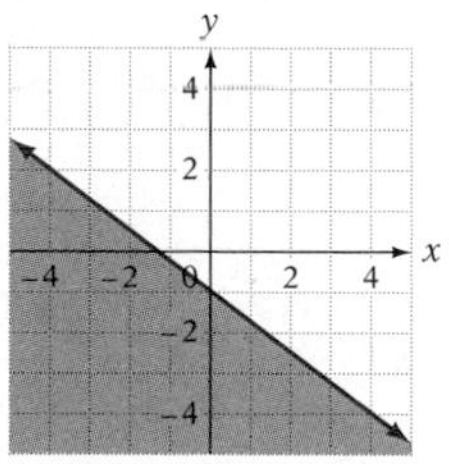

14. $y \le -\frac{5}{2}x - 4$

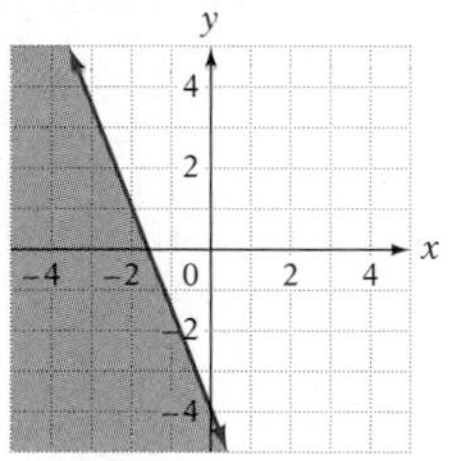

15. $y \le -2$

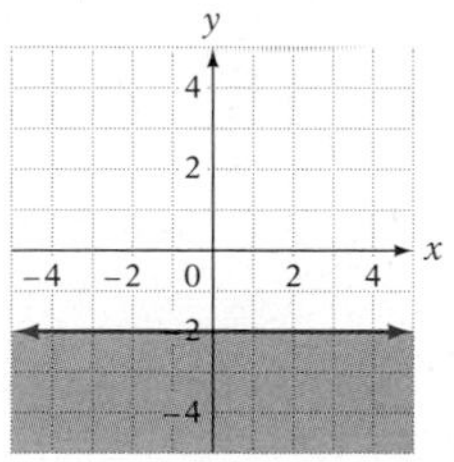

Section 4.7

Suggested Assignment

Exercises 3–23, odds

More challenging problems: Exercises 25–27

Answers to Writing Exercises

1. A half-plane is the set of points on one side of a line in the plane.
2. Choose a point that is not on the line corresponding to the inequality. If the point is a solution of the inequality, the half-plane containing the point should be shaded. If the point is not a solution of the inequality, the half-plane not containing the point should be shaded.

Quick Quiz (Objective 4.7A)

Graph the solution set.

1. $3x + 5y < 15$

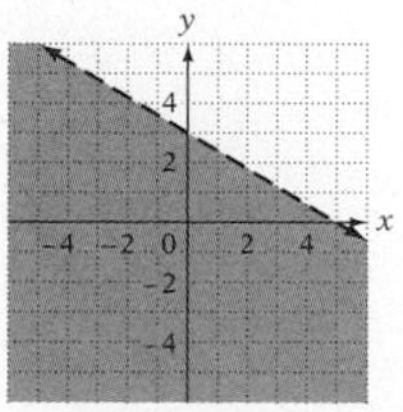

2. $2x - 5y \ge 10$

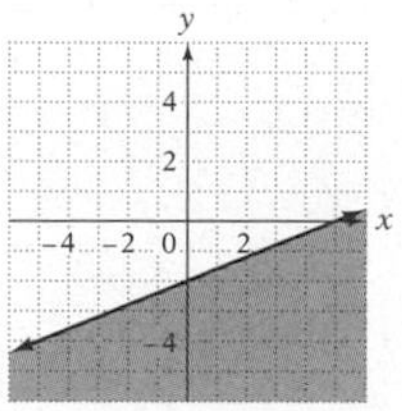

3. $x \ge 2$

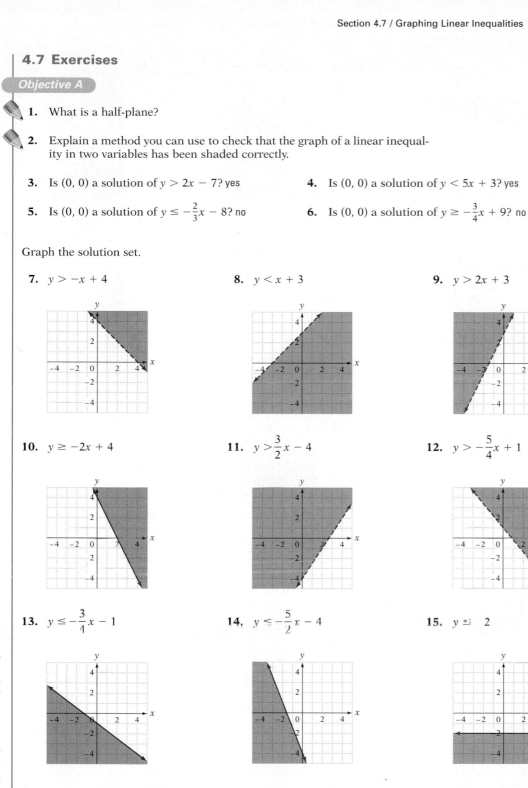

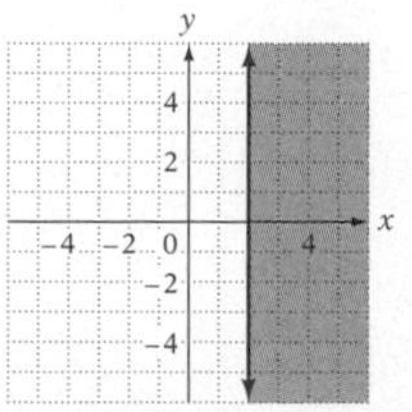

Answers to Writing Exercises

27. There are an infinite number of points whose coordinates satisfy both $y \le x + 3$ and $y \ge -\frac{1}{2}x + 1$. However, at this point, students should simply answer "yes" to the question of whether there are points whose coordinates satisfy both inequalities and cite three of them—for example, (2, 3), (6, 0), and (8, −1).

28. There are no points whose coordinates satisfy both $y \le x - 1$ and $y \ge x + 2$. The solution set of $y \le x - 1$ is all points on and below the line $y = x - 1$. The solution set of $y \ge x + 2$ is all points on and above the line $y = x + 2$. Since the lines $y = x - 1$ and $y = x + 2$ are parallel lines, and $y = x + 2$ is above the line $y = x - 1$, there are no points that lie both below $y = x - 1$ and above $y = x + 2$.

16. $y > 3$

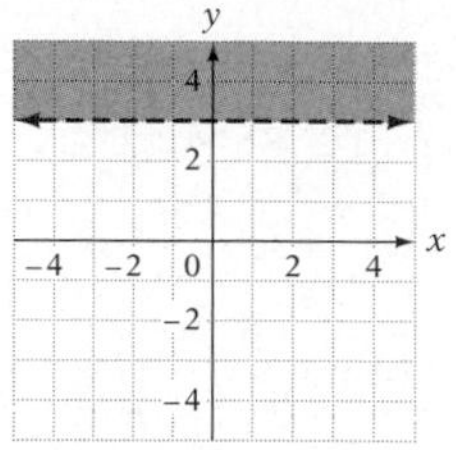

17. $y < \frac{4}{5}x + 3$

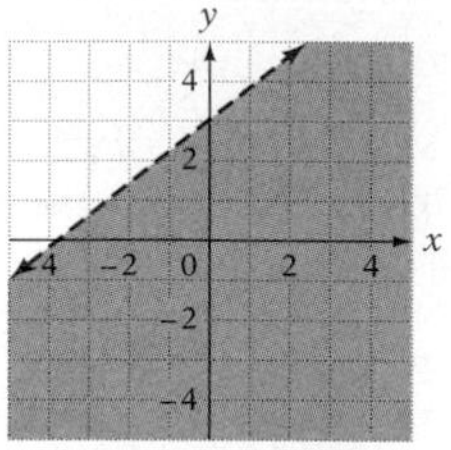

18. $y \le -\frac{6}{5}x - 2$

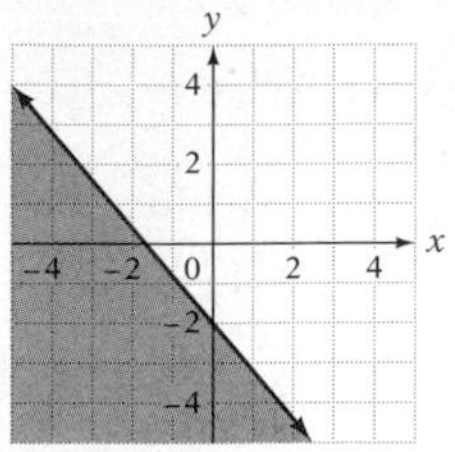

19. $3x - y < 9$

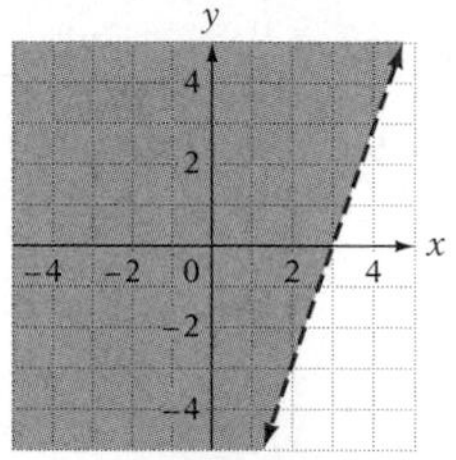

20. $3x + y \ge 6$

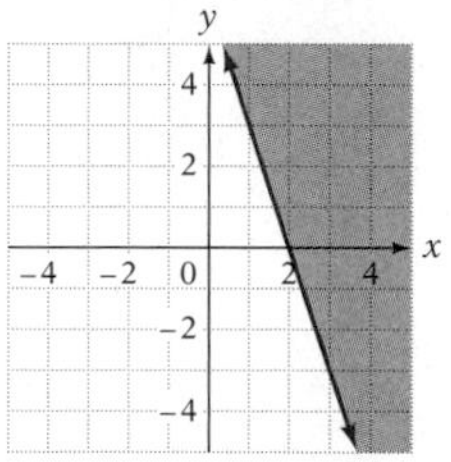

21. $2x + 2y \le -4$

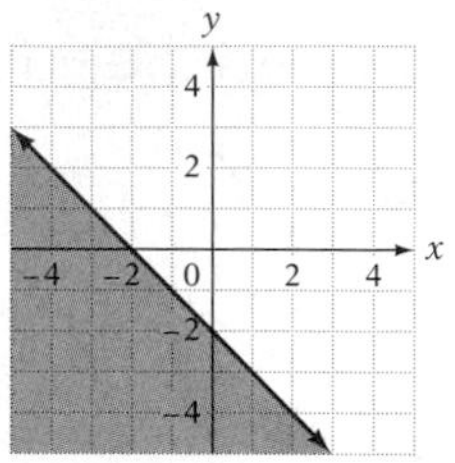

22. $-4x + 3y < -12$

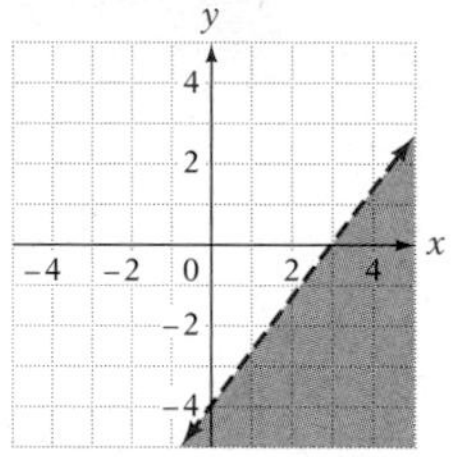

23. $-2x + 3y \le 6$

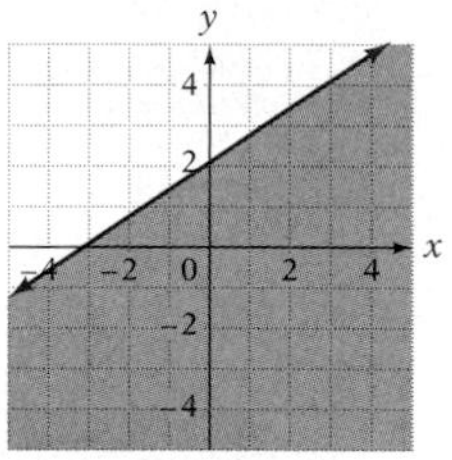

24. $3x - 4y > 12$

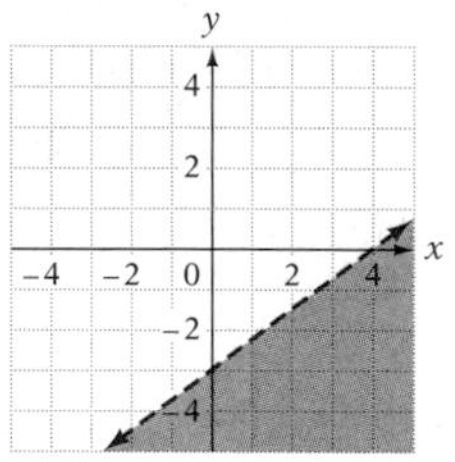

APPLYING THE CONCEPTS

Write the inequality given its graph.

25.

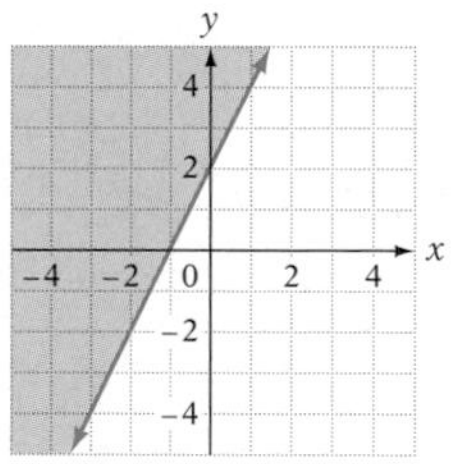

$y \ge 2x + 2$

26.

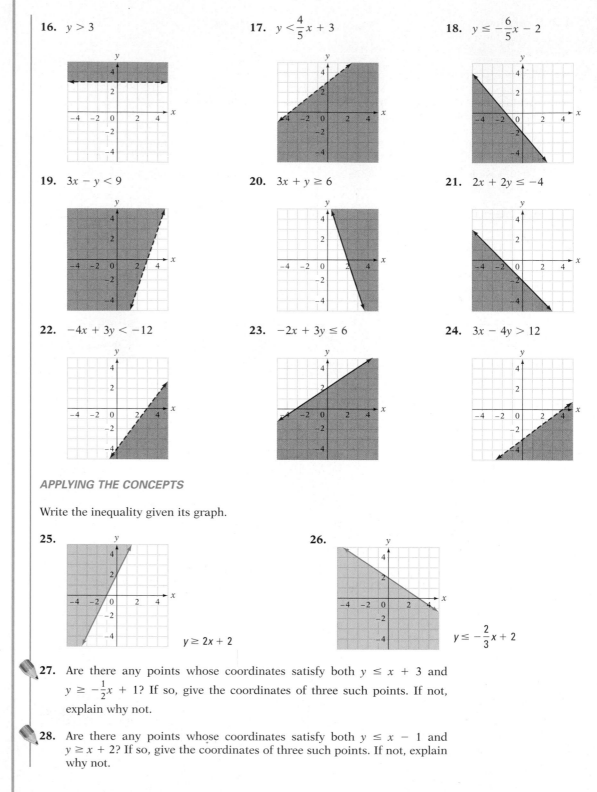

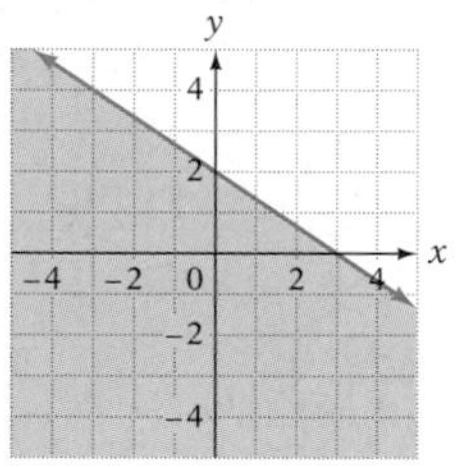

$y \le -\frac{2}{3}x + 2$

27. Are there any points whose coordinates satisfy both $y \le x + 3$ and $y \ge -\frac{1}{2}x + 1$? If so, give the coordinates of three such points. If not, explain why not.

28. Are there any points whose coordinates satisfy both $y \le x - 1$ and $y \ge x + 2$? If so, give the coordinates of three such points. If not, explain why not.

Focus on Problem Solving

Find a Counterexample

When you are faced with an assertion, it may be that the assertion is false. For instance, consider the statement "Every prime number is an odd number." This assertion is false because the prime number 2 is an even number.

Finding an example that illustrates that an assertion is false is called finding a *counterexample.* The number 2 is a counterexample to the assertion that every prime number is an odd number.

If you are given an unfamiliar problem, one strategy to consider as a means of solving the problem is to try to find a counterexample. For each of the following problems, answer *true* if the assertion is always true. If the assertion is not true, answer *false* and give a counterexample. If there are terms used that you do not understand, consult a reference to find the meaning of the term.

1. If x is a real number, then x^2 is always positive.
2. The product of an odd integer and an even integer is an even integer.
3. If m is a positive integer, then $2m + 1$ is always a positive odd integer.
4. If $x < y$, then $x^2 < y^2$.
5. Given any three positive numbers a, b, and c, it is possible to construct a triangle whose sides have lengths a, b, and c.
6. The product of two irrational numbers is an irrational number.
7. If n is a positive integer greater than 2, then $1 \cdot 2 \cdot 3 \cdot 4 \cdot \cdots \cdot n + 1$ is a prime number.
8. Draw a polygon with more than three sides. Select two different points inside the polygon and join the points with a line segment. The line segment always lies completely inside the polygon.
9. Let A, B, and C be three points in the plane that are not collinear. Let d_1 be the distance from A to B, and let d_2 be the distance from A to C. Then the distance between B and C is less than $d_1 + d_2$.

10. Consider the line segment AB shown at the left. Two points, C and D, are randomly selected on the line segment and three new segments are formed: AC, CD, and DB. The three new line segments can always be connected to form a triangle.

It may not be easy to establish that an assertion is true or to find a counterexample to the assertion. For instance, consider the assertion that every positive integer greater than 3 can be written as the sum of two primes. For example, $6 = 3 + 3$, $8 = 3 + 5$, $9 = 2 + 7$. Is this assertion always true? (*Note:* This assertion, called **Goldbach's conjecture,** has never been proved, nor has a counterexample been found!)

263

Answers to Focus on Problem Solving: Find a Counterexample

1. False. 0 is a real number, and $0^2 = 0$ is not positive.
2. True
3. True
4. False. Let $x = -4$ and $y = 2$. Then the expression $(-4)^2 < 2^2$ is not true.
5. False. It is impossible to construct a triangle with $a = 2$, $b = 3$, and $c = 10$. In a triangle, the sum of the lengths of two sides must be greater than the length of the third side.
6. False. The product $\sqrt{3} \cdot \sqrt{3} = 3$.
7. False. For $n = 4$, $1 \cdot 2 \cdot 3 \cdot 4 + 1 = 25$, which is not a prime number.
8. False. Part of line segment AB lies outside the polygon.

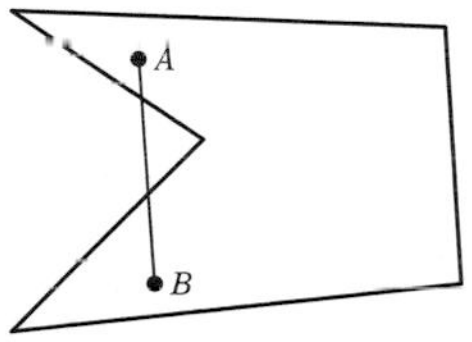

9. True
10. False. If the points are selected such that $AC + CD < DB$, a triangle cannot be formed.

Answers to Projects and Group Activities: Graphing Linear Equations with a Graphing Utility

1.
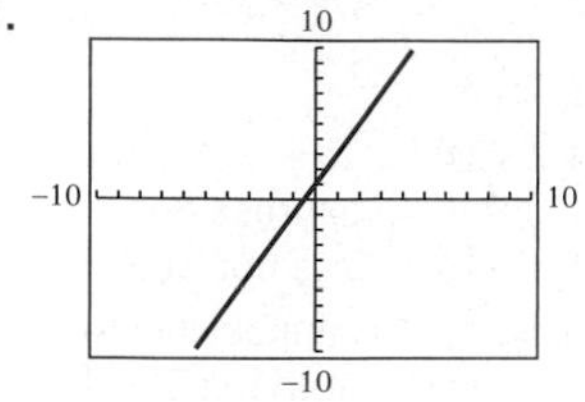

2.
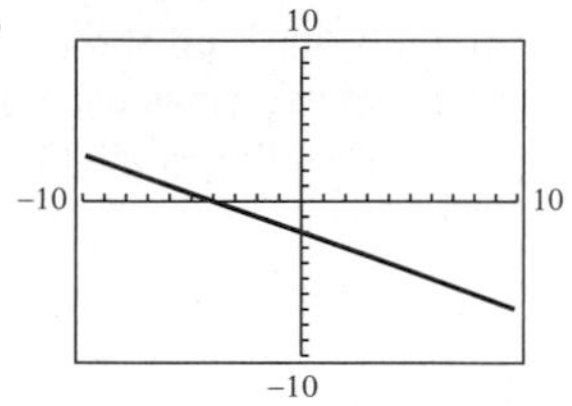

3.
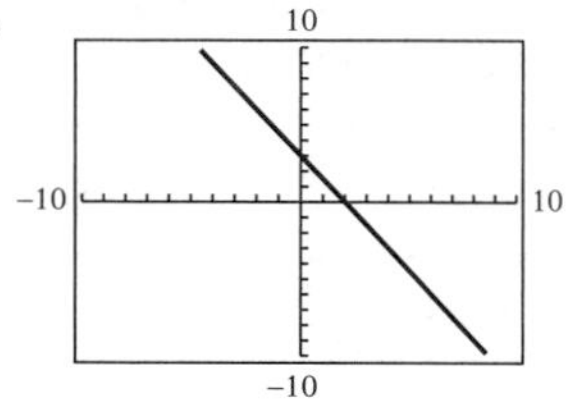

4.
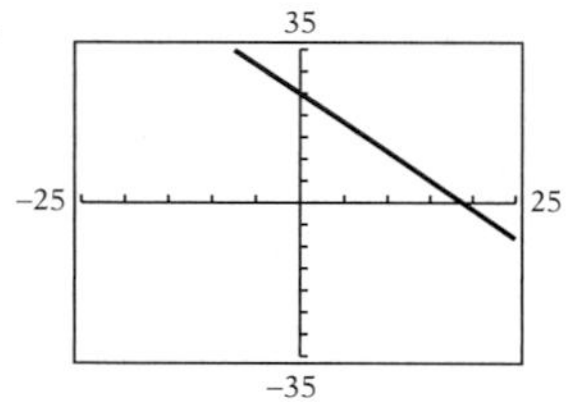

Projects and Group Activities

Graphing Linear Equations with a Graphing Utility

A computer or graphing calculator screen is divided into *pixels*. There are approximately 6000 to 790,000 pixels available on the screen (depending on the computer or calculator). The greater the number of pixels, the smoother a graph will appear. A portion of a screen is shown at the left. Each little rectangle represents one pixel.

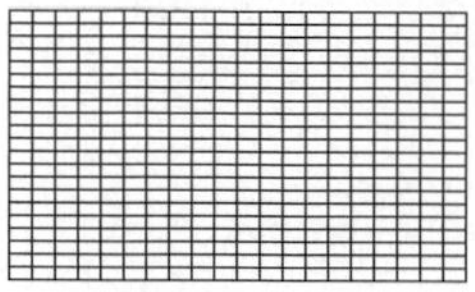

The graphing utilities that are used by computers or calculators to graph an equation do basically what we have shown in the text: They choose values of x and, for each, calculate the corresponding value of y. The pixel corresponding to the ordered pair is then turned on. The graph is jagged because pixels are much larger than the dots we draw on paper.

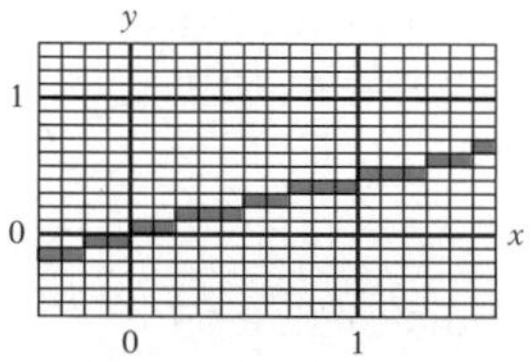

The graph of $y = 0.45x$ is shown at the left as the calculator drew it (jagged). The x- and y-axes have been chosen so that each pixel represents $\frac{1}{10}$ of a unit. Consider the region of the graph where $x = 1$, 1.1, and 1.2.

The corresponding values of y are 0.45, 0.495, and 0.54. Because the y-axis is in tenths, the numbers 0.45, 0.495, and 0.54 are rounded to the nearest tenth before plotting. Rounding 0.45, 0.495, and 0.54 to the nearest tenth results in 0.5 for each number. Thus the ordered pairs (1, 0.45), (1.1, 0.495), and (1.2, 0.54) are graphed as (1, 0.5), (1.1, 0.5), and (1.2, 0.5). These points appear as three illuminated horizontal pixels. The graph of the line appears horizontal. However, if you use the TRACE feature of the calculator (see the appendix), the actual y-coordinate for each value of x is displayed.

TAKE NOTE

Xmin and Xmax are the smallest and largest values of x that will be shown on the screen. Ymin and Ymax are the smallest and largest values of y that will be shown on the screen.

Here are the keystrokes to graph $y = \frac{2}{3}x + 1$ on a TI-83. First the equation is entered. Then the domain (Xmin to Xmax) and the range (Ymin to Ymax) are entered. This is called the **viewing window**. By changing the keystrokes 2 [X, T, θ, n] [÷] 3 [+] 1, you can graph different equations.

[Y =] [CLEAR] 2 [X, T, θ, n] [÷] 3
[+] 1 [WINDOW] [(−)] 10
[ENTER] 10 [ENTER] 1 [ENTER]
[(−)] 10 [ENTER] 10 [ENTER] 1
[ENTER] [GRAPH]

1. $y = 2x + 1$ — For $2x$, you may enter $2 \times x$ or just $2x$. The times sign $\times$ is not necessary on many graphing calculators.

2. $y = -\frac{1}{2}x - 2$ — Use the [(−)] key to enter a negative sign.

3. $3x + 2y = 6$ — Solve for y. Then enter the equation.

4. $4x + 3y = 75$ — You must adjust the viewing window. *Suggestion:* Xmin = −25, Xmax = 25, Xscl = 5, Ymin = −35, Ymax = 35, Yscl = 5. See the appendix for assistance.

Graphs of Motion

A graph can be useful in analyzing the motion of a body. For example, consider an airplane in uniform motion traveling at 100 m/s. The table at the right shows the distance, in meters, traveled by the plane at the end of each of five one-second intervals.

Time (in seconds)	*Distance (in meters)*
0	0
1	100
2	200
3	300
4	400
5	500

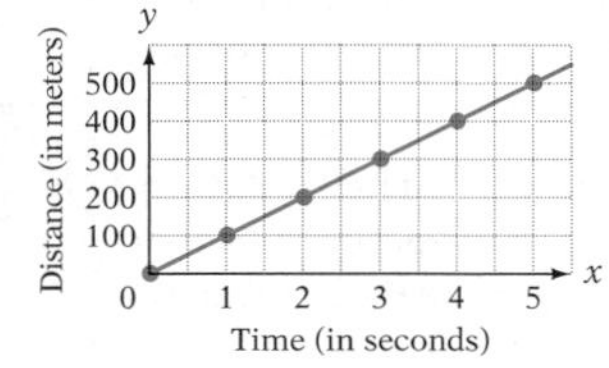

These data can be graphed on a rectangular coordinate system and a straight line drawn through the points plotted. The travel time is shown along the horizontal axis, and the distance traveled by the plane is shown along the vertical axis. (Note that the units along the two axes are not the same length.)

To write the equation for the line just graphed, use the coordinates of any two points on the line to find the slope. The y-intercept is (0, 0).

Let $(x_1, y_1) = (1, 100)$ and $(x_2, y_2) = (2, 200)$.

$$m = \frac{y_2 - y_1}{x_2 - x_1} = \frac{200 - 100}{2 - 1} = 100$$

$$y = mx + b$$
$$y = 100x + 0$$
$$y = 100x$$

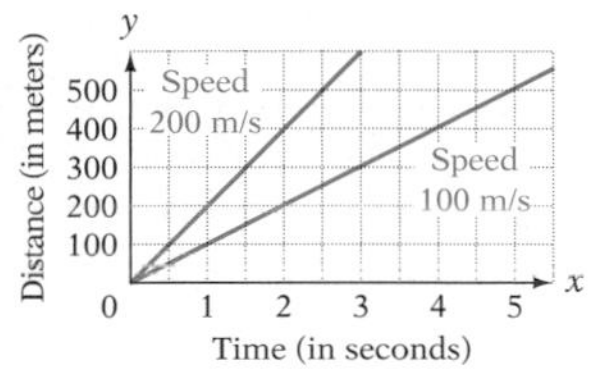

Note that the slope of the line, 100, is equal to the speed, 100 m/s. *The slope of a distance-time graph represents the speed of the object.*

The distance-time graphs for two planes are shown at the left. One plane is traveling at 100 m/s, and the other is traveling at 200 m/s. The slope of the line representing the faster plane is greater than the slope of the line representing the slower plane.

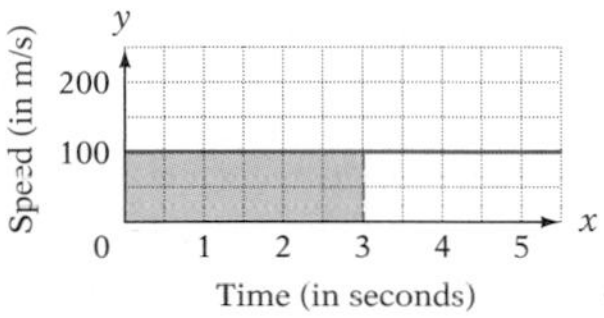

In the speed-time graph at the left, the time a plane has been flying at 100 m/s is shown along the horizontal axis and its speed is shown along the vertical axis. Because the speed is constant, the graph is a horizontal line.

The area between the horizontal line graphed and the horizontal axis is equal to the distance traveled by the plane up to that time. For example, the area of the shaded region on the graph is

$$\text{Length} \cdot \text{width} = (3\text{ s})(100\text{ m/s}) = 300\text{ m}$$

The distance traveled by the plane in 3 s is equal to 300 m.

1. A car in uniform motion is traveling at 20 m/s.
   a. Prepare a distance-time graph for the car for 0 s to 5 s.
   b. Find the slope of the line.
   c. Find the equation of the line.
   d. Prepare a speed-time graph for the car for 0 s to 5 s.
   e. Find the distance traveled by the car after 3 s.

2. One car in uniform motion is traveling at 10 m/s. A second car in uniform motion is traveling at 15 m/s.
   a. Prepare one distance-time graph for both cars for 0 s to 5 s.
   b. Find the slope of each line.
   c. Find the equation of each line graphed.
   d. Assuming that the cars started at the same point at 0 s, find the distance between the cars at the end of 5 s.

3. **a.** In a distance-time graph, is it possible for the graph to be a horizontal line?
   b. What does a horizontal line reveal about the motion of the object during that time period?

Answers to Projects and Group Activities: Graphs of Motion

1. a.

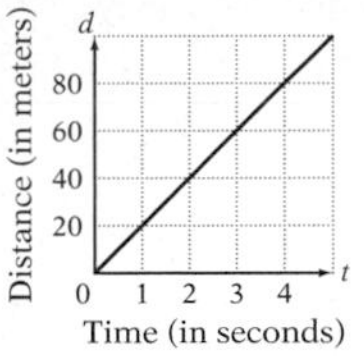

b. 20

c. $d = 20t$

d.

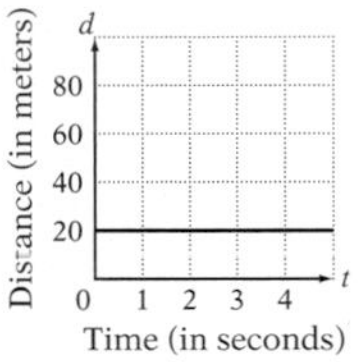

e. 60 mi

2. a.

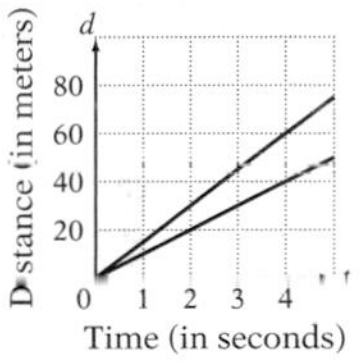

b. 10, 15

c. $d = 10t$, $d = 15t$

d. 25 m

3. a. Yes

b. The object is not moving

266 Chapter 4 / Linear Equations and Inequalities in Two Variables

Chapter Summary

Key Words

A *rectangular coordinate system* is formed by two number lines, one horizontal and one vertical, that intersect at the zero point of each line. A rectangular coordinate system divides the plane into four regions called *quadrants*. The number lines that make up a rectangular coordinate system are called the *coordinate axes*, or simply the *axes*. The *origin* is the point of intersection of the two coordinate axes. [p. 201]

Every point in the plane can be identified by an ordered pair (x, y). The first number in an ordered pair is called the *abscissa* or x-coordinate. The second number is called the *ordinate* or y-coordinate. The *coordinates* of a point are the numbers in the ordered pair associated with the point. [p. 201]

A *relation* is any set of ordered pairs. The *domain* of a relation is the set of first coordinates of the ordered pairs. The *range* is the set of second coordinates of the ordered pairs. A *function* is a relation in which no two ordered pairs have the same first coordinate and different second coordinates. A function designated by $f(x)$ is written in *functional notation*. The *value* of the function at x is $f(x)$. [pp. 211–214]

An equation of the form $y = mx + b$, where m is the coefficient of x and b is a constant, is a *linear equation in two variables*. m is the slope of the line, and $(0, b)$ is the y-intercept. An equation of the form $Ax + By = C$ is also a linear equation in two variables. A *solution* of a linear equation in two variables is an ordered pair (x, y) that makes the equation a true statement. The *graph of a linear equation* is a straight line. [pp. 203, 223]

The point at which a graph crosses the x-axis is called the *x-intercept*. The point at which a graph crosses the y-axis is called the *y-intercept*. [p. 227]

The *slope* of a line is a measure of the slant of the line. The symbol for slope is m. A line that slants upward to the right has a *positive slope*. A line that slants downward to the right has a *negative slope*. A horizontal line has *zero slope*. The slope of a vertical line is *undefined*. [pp. 235–236]

Two lines that have the same slope do not intersect and are *parallel lines*. Two lines that intersect at right angles are *perpendicular lines*. [pp. 253–254]

An inequality of the form $y > mx + b$ or $Ax + By > C$ is a *linear inequality in two variables*. The symbol $>$ here could be replaced by $\geq$, $<$, or $\leq$. The solution set of an inequality in two variables is a *half-plane*. [p. 259]

Essential Rules

To find the x-intercept, let $y = 0$
To find the y-intercept, let $x = 0$. [p. 227]

Slope of a Linear Equation [p. 235]	$\text{Slope} = m = \dfrac{y_2 - y_1}{x_2 - x_1}, x_1 \neq x_2$
Slope-Intercept Form of a Straight Line [p. 239]	$y = mx + b$
Point-Slope Formula [p. 245]	$y - y_1 = m(x - x_1)$
Slopes of Parallel Lines [p. 253]	$m_1 = m_2$
Slopes of Perpendicular Lines [p. 254]	$m_1 \cdot m_2 = -1$

267

Chapter Review

1. **a.** Graph the ordered pairs $(-2, 4)$ and $(3, -2)$.
b. Name the abscissa of point A. −2
c. Name the ordinate of point B. −4

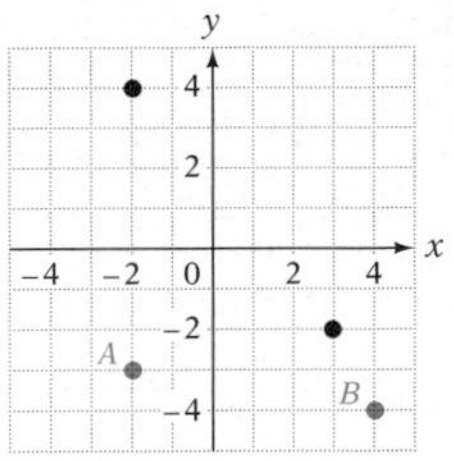

[4.1A]

2. Graph the ordered-pair solutions of $y = -\frac{1}{2}x - 2$ when $x = -4, -2, 0,$ and 2.

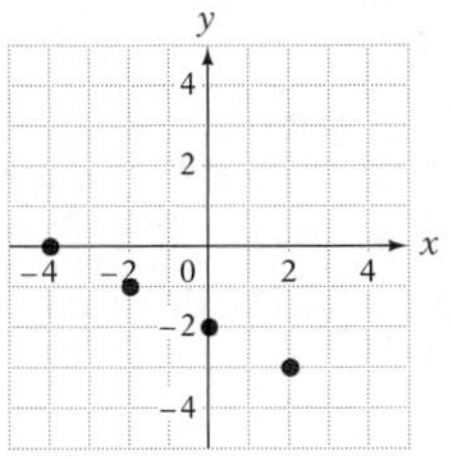

[4.1B]

3. Find the x- and y-intercepts of $3x - 2y = 24$.
(8, 0), (0, −12) [4.3B]

4. Given $f(x) = x^2 - 2$, find $f(-1)$.
−1 [4.2A]

5. Graph $x = -3$.

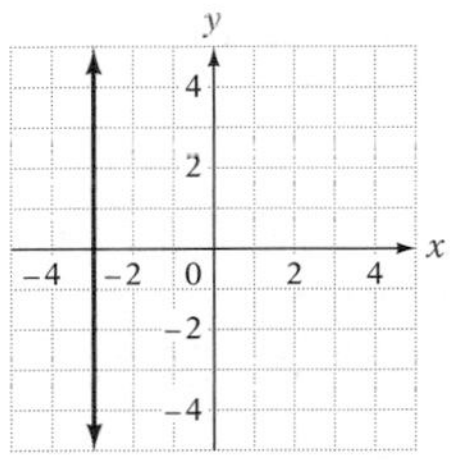

[4.3B]

6. Graph the line that has slope $-\frac{2}{3}$ and y-intercept $(0, 2)$.

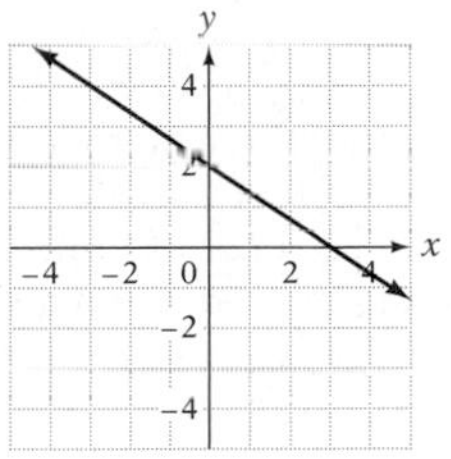

[4.4B]

7. Graph $y = -2x - 1$.

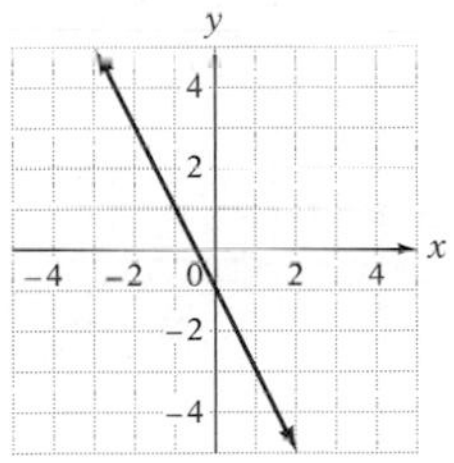

[4.3A]

8. Graph $y = \frac{1}{4}x + 3$.

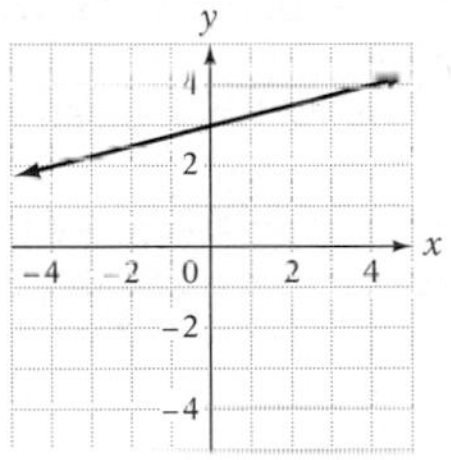

[4.3A]

9. Find the slope of the line containing the points $(-2, -3)$ and $(4, -3)$.
0 [4.4A]

10. Find the slope of the line containing the points $(9, 8)$ and $(-2, 1)$. $\frac{7}{11}$ [4.4A]

11. What value of x is excluded from the domain of $f(x) = \frac{2x + 1}{x}$?
0 [4.2A]

268

12. Graph $3x - 2y = -6$.

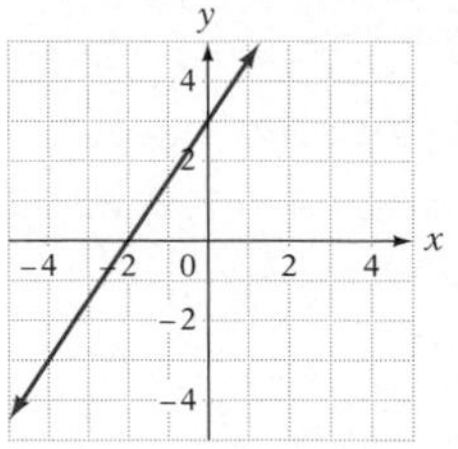

[4.3B]

13. Graph the solution set of $y < \frac{3}{4}x - 2$.

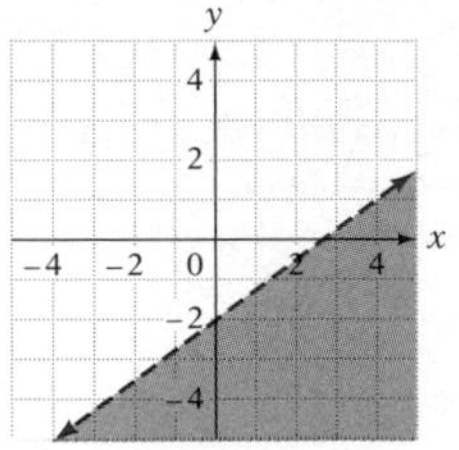

[4.7A]

14. Determine the equation of the line that passes through the points $(-1, 3)$ and $(2, -5)$.

$y = -\frac{8}{3}x + \frac{1}{3}$ [4.5B]

15. Determine the equation of the line that passes through the point $(6, 1)$ and has slope $-\frac{5}{2}$.

$y = -\frac{5}{2}x + 16$ [4.5A]

16. Find the equation of the line that contains the ordered pair $(-2, -4)$ and is parallel to the graph of $4x - 2y = 7$.

$y = 2x$ [4.6A]

17. Find the equation of the line that contains the point $(-2, -3)$ and is perpendicular to the line $y = -\frac{1}{2}x - 3$.

$y = 2x + 1$ [4.6A]

18. A building contractor estimates that the cost to build a new house is \$25,000 plus \$80 for each square foot of floor space. Determine a linear function that will give the cost to build a house that contains a given number of square feet. Use the model to determine the cost to build a house that contains 2000 ft^2.

$y = 80x + 25{,}000$; \$185,000 [4.5C]

19. An on-line research service charges a monthly access fee of \$75 plus \$.45 per minute to use the service. An equation that represents the monthly cost to use this service is $C = 0.45x + 75$, where C is the monthly cost and x is the number of minutes. Graph this equation for $0 \le x \le 100$. The point $(50, 97.5)$ is on the graph. Write a sentence that describes the meaning of this ordered pair.

The cost of 50 min of access time for one month is \$97.50. [4.3C]

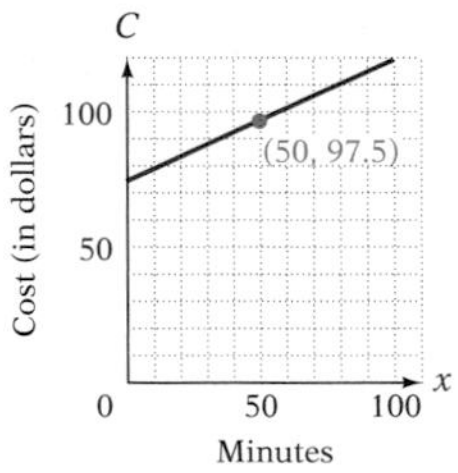

20. The data in the table below show the size of a house in square feet and the cost to build the house. Draw a scatter diagram of these data.

Square feet	1250	1400	1348	2675	2900
Cost	128,000	140,000	136,100	233,450	241,500

[4.1C]

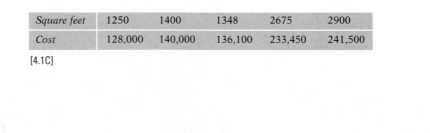

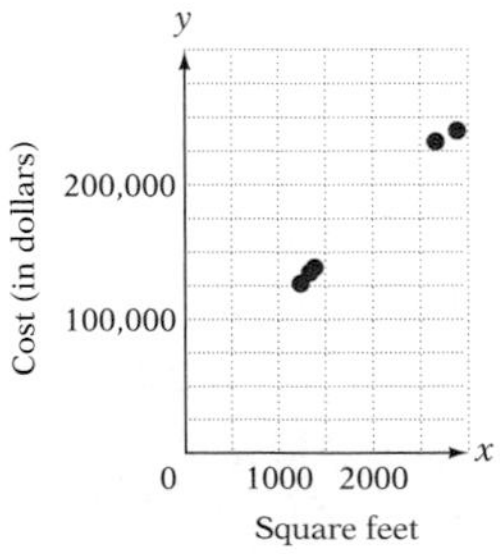

269

Chapter Test

1. Find the ordered-pair solution of $2x - 3y = 15$ corresponding to $x = 3$.
(3, −3) [4.1B]

2. Given $f(t) = t^2 + t$, find $f(2)$.
6 [4.2A]

3. The data in the table below show a reading test grade and the final exam grade in a history class. Draw a scatter diagram of these data.

Reading	8.5	9.4	10.0	11.4	12.0
History	64	68	76	87	92

[4.1C]

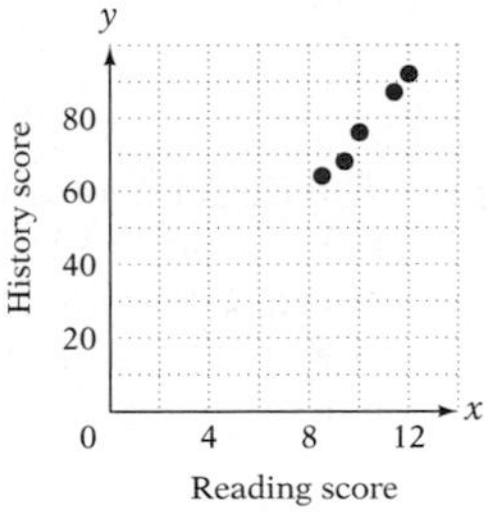

4. Graph the ordered-pair solutions of $y = -\frac{3}{2}x + 1$ when $x = -2$, 0, and 4.

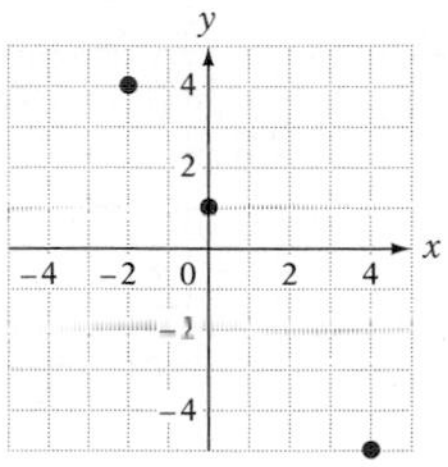

[4.1B]

5. Graph $y = -\frac{3}{4}x + 3$.

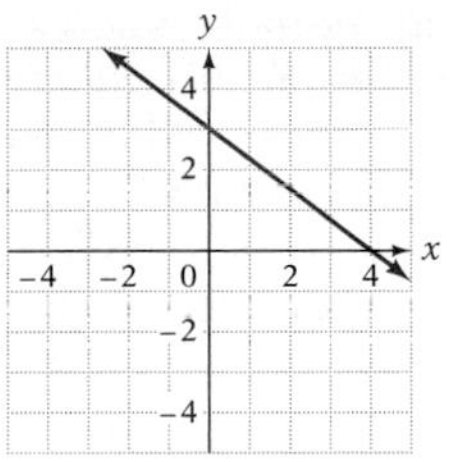

[4.3A]

6. Graph $3x - 2y = 6$.

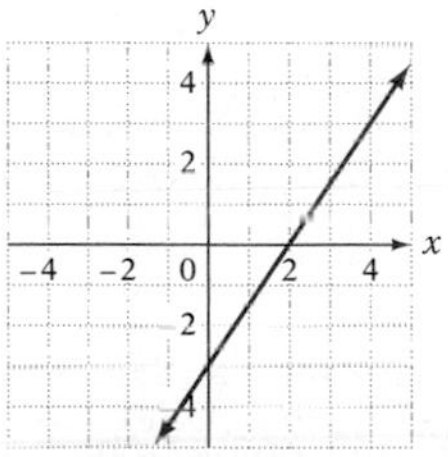

[4.3B]

7. Graph the line that has slope $-\frac{2}{3}$ and y-intercept (0, 4).

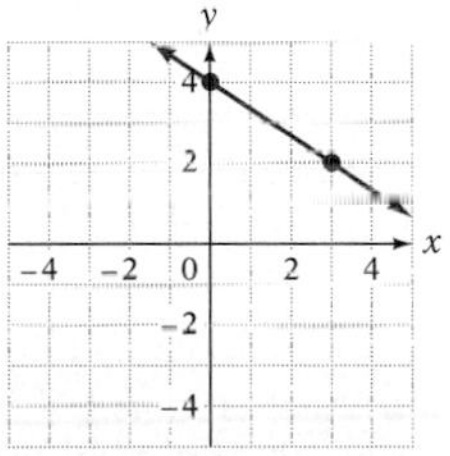

[4.4B]

8. Graph the solution set of $y \geq 2x - 3$.

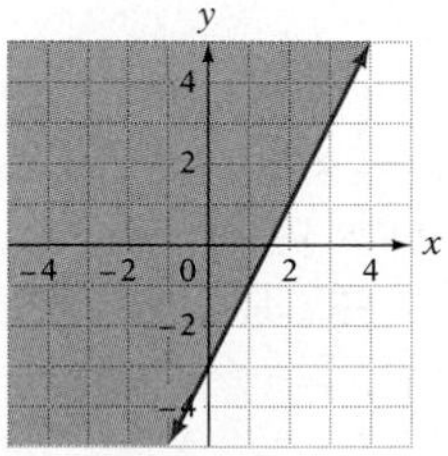

[4.7A]

9. Graph the solution set of $3x - 2y \leq 6$.

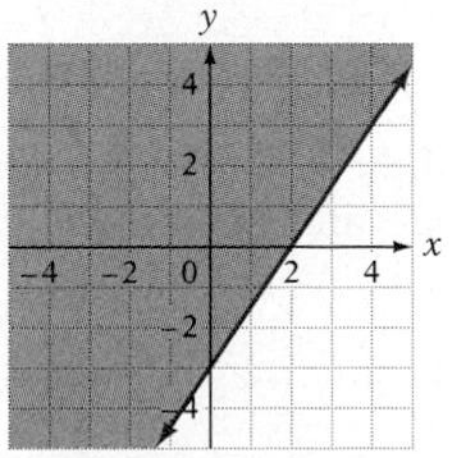

[4.7A]

10. The equation for the speed of a ball that is thrown straight up with an initial speed of 128 ft/s is $v = 128 - 32t$, where v is the speed of the ball after t seconds. Graph this equation for $0 \le t \le 4$. The point whose coordinates are (1, 96) is on the graph at the right. Write a sentence that describes this ordered pair.
After 1 s, the ball is traveling 96 ft/s. [4.3C]

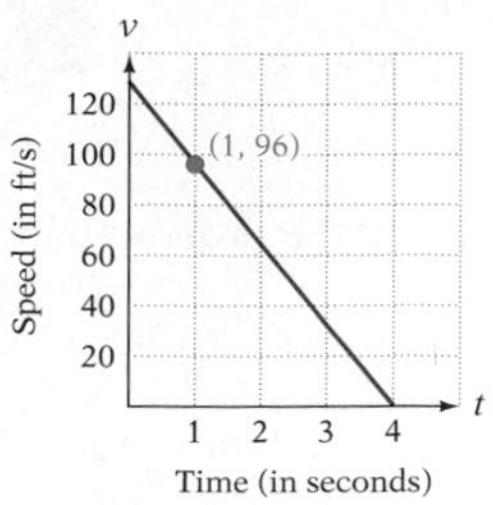

11. The director of a baseball school estimates that 100 students will enroll if the tuition is $250. For each $20 increase in tuition, 6 fewer students will enroll. Determine a linear function that will predict the number of students who will enroll at a given tuition. Use this model to predict enrollment when the tuition is $300.
$y = -\frac{3}{10}x + 175$; 85 students [4.5C]

12. A car is traveling at 55 mph. The equation that describes the distance traveled is $d = 55t$. Graph this equation for $0 \le t \le 6$. The point whose coordinates are (4, 220) is on the graph. Write a sentence that explains the meaning of this ordered pair.
After 4 h, the car has traveled 200 mi. [4.3C]

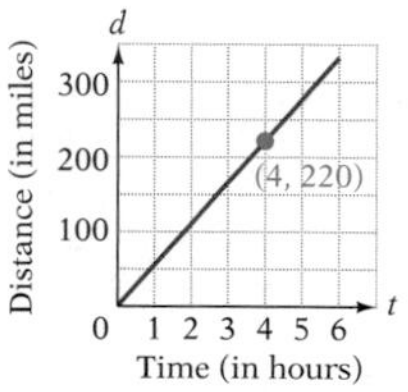

13. Find the x- and y-intercepts for $6x - 4y = 12$.
(2, 0), (0, −3) [4.3B]

14. Find the slope of the line containing the points (2, −3) and (4, 1).
2 [4.4A]

15. Find the slope of the line containing the points (−5, 2) and (−5, 7).
undefined [4.4A]

16. Find the slope of the line whose equation is $2x + 3y = 6$.
$-\frac{2}{3}$ [4.3B]

17. Find the equation of the line that contains the point (−3, 1) and has slope $\frac{2}{3}$.
$y = \frac{2}{3}x + 3$ [4.5A]

18. Find the equation of the line that passes through the points (−2, 0) and (5, −2).
$y = -\frac{2}{7}x - \frac{4}{7}$ [4.5B]

19. Find the equation of the line that contains the ordered pair (3, −2) and is parallel to the graph of $y = -3x + 4$.
$y = -3x + 7$ [4.6A]

20. Find the equation of the line that contains the ordered pair (2, 5) and is perpendicular to the graph of the line $y = -\frac{2}{3}x + 6$.
$y = \frac{3}{2}x + 2$ [4.6A]

271

Cumulative Review

1. Let $x \in \{-5, -3, -1\}$. For what values of x is the inequality $x \le -3$ a true statement?
$-5, -3$ [1.1A]

2. Write $\frac{17}{20}$ as a decimal.
0.85 [1.2B]

3. Simplify: $3\sqrt{45}$
$9\sqrt{5}$ [1.2F]

4. Simplify: $12 - 18 \div 3(-2)^2$
-12 [1.3A]

5. Evaluate $\frac{a - b}{a^2 - c}$ when $a = -2$, $b = 3$, and $c = -4$.
$-\frac{5}{8}$ [1.4A]

6. Simplify: $3d - 9 - 7d$
$-4d - 9$ [1.4B]

7. Simplify: $4(-8z)$
$-32z$ [1.4C]

8. Simplify: $2(x + y) - 5(3x - y)$
$-13x + 7y$ [1.4D]

9. Graph: $\{x \mid x < -2\} \cup \{x \mid x > 0\}$

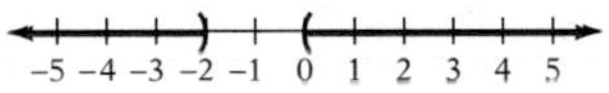

[1.5C]

10. Solve: $2x - \frac{2}{3} = \frac{7}{3}$
$\frac{3}{2}$ [2.2A]

11. Solve: $3x - 2(10x - 6) = x - 6$
1 [2.2C]

12. Solve: $4x - 3 < 9x + 2$
$\{x \mid x > -1\}$ [2.4A]

13. Solve: $3x - 1 < 4$ and $x - 2 > 2$
The solution set is $\varnothing$. [2.4B]

14. Solve: $|3x - 5| < 5$
$\left\{x \mid 0 < x < \frac{10}{3}\right\}$ [2.5B]

15. Given $f(t) = t^2 + t$, find $f(2)$.
6 [4.2A]

16. Find the slope of the line containing the points $(2, -3)$ and $(4, 1)$.
2 [4.4A]

272

17. Graph $y = 3x + 1$.

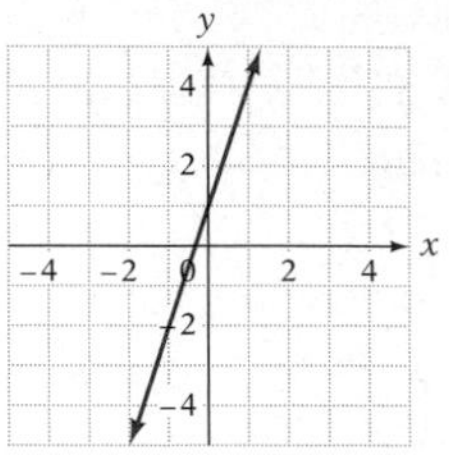

[4.3A]

18. Graph $x = -4$.

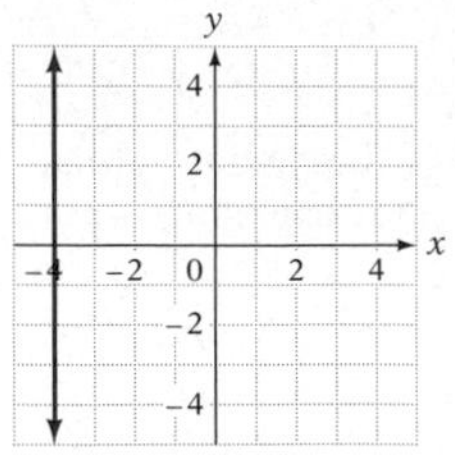

[4.3B]

19. Graph the line that has slope $\frac{1}{2}$ and y-intercept $(0, -1)$.

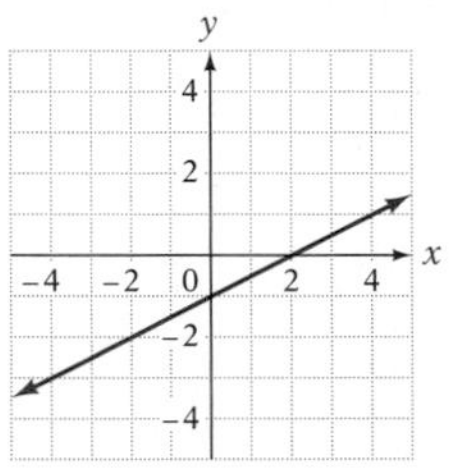

[4.4B]

20. Graph the solution set of $3x - 2y \geq 6$.

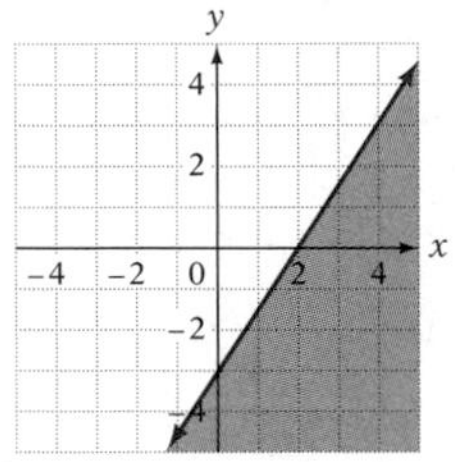

[4.7A]

21. Find the equation of the line that passes through the points $(6, -4)$ and $(-3, -1)$.

$y = -\frac{1}{3}x - 2$ [4.5B]

22. Find the equation of the line that contains the point $(2, 4)$ and is parallel to the line $y = -\frac{3}{2}x + 2$.

$y = -\frac{3}{2}x + 7$ [4.6A]

23. Two planes are 1800 mi apart and traveling toward each other. The first plane is traveling at twice the speed of the second plane. The planes meet in 3 h. Find the speed of each plane.

First plane: 400 mph; second plane: 200 mph [2.3D]

24. A grocer combines coffee that costs \$9 per pound with coffee that costs \$6 per pound. How many pounds of each should be used to make 60 lb of a blend that costs \$8 per pound?

20 lb of \$6 coffee; 40 lb of \$9 coffee [2.3A]

25. The graph at the right shows the relationship between the cost of a rental house and the depreciation allowed for income tax purposes. Find the slope of the line between the two points on the graph. Write a sentence that states the meaning of the slope.

$m = -\frac{10{,}000}{3}$. The value of the house decreases by \$3333.33 each year. [4.4A]

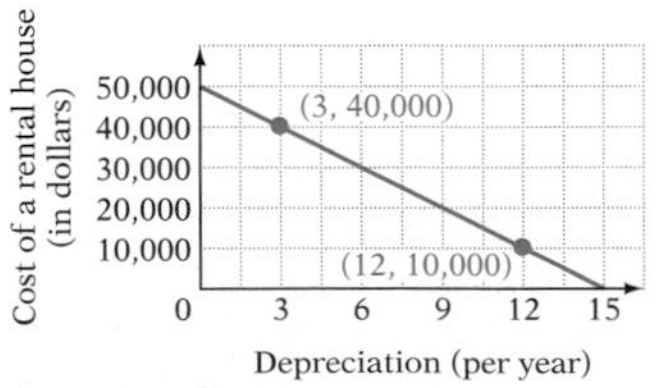

Chapter 5

Systems of Linear Equations and Inequalities

This surveyor is working on a construction site, using the tools necessary to find the area of a plot of land. When a plot is in the shape of a polygon, its vertices can be represented by rectangular coordinates. These coordinates are used in the surveyor's area formula to determine the area of the plot. **Exercise 37 on page 304,** which involves use of the surveyor's area formula, is an application of determinants.

Objectives

Section 5.1

A To solve a system of linear equations by graphing

B To solve a system of linear equations by the substitution method

Section 5.2

A To solve a system of two linear equations in two variables by the addition method

B To solve a system of three linear equations in three variables by the addition method

Section 5.3

A To evaluate a determinant

B To solve a system of equations by using Cramer's Rule

Section 5.4

A To solve rate-of-wind or rate-of-current problems

B To solve application problems using two variables

Section 5.5

A To graph the solution set of a system of linear inequalities

WWW WEB Need help? For online student resources, such as section quizzes, visit this textbook's website at **college.hmco.com/students.**

Prep Test

1. Simplify: $10\left(\frac{3}{5}x + \frac{1}{2}y\right)$
$6x + 5y$ [1.4D]

2. Evaluate $3x + 2y - z$ for $x = -1$, $y = 4$, and $z = -2$.
7 [1.4A]

3. Given $3x - 2z = 4$, find the value of x when $z = -2$.
0 [2.2A]

4. Solve: $3x + 4(-2x - 5) = -5$
-3 [2.2C]

5. Solve: $0.45x + 0.06(-x + 4000) = 630$
1000 [2.2C]

6. Graph: $y = \frac{1}{2}x - 4$

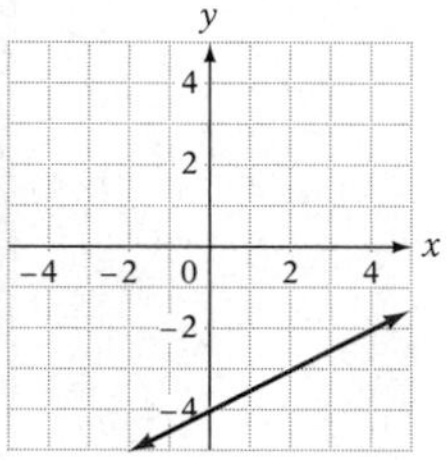

[4.3A]

7. Graph: $3x - 2y = 6$

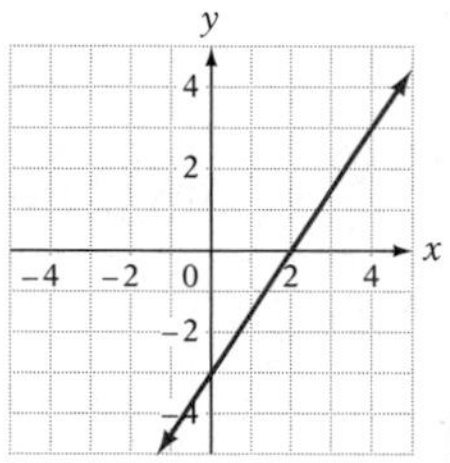

[4.3B]

8. Graph: $y > -\frac{3}{5}x + 1$

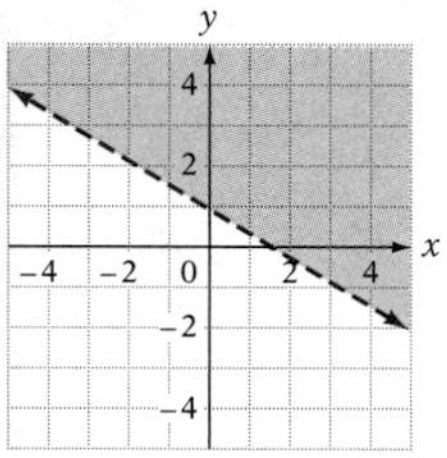

[4.7A]

Go Figure

Two children are on their way from school to home. Carla runs half the time and walks half the time. James runs half the distance and walks half the distance. If Carla and James walk at the same speed and run at the same speed, which child arrives home first?
Carla

275

5.1 Solving Systems of Linear Equations by Graphing and by the Substitution Method

Objective A

To solve a system of linear equations by graphing

VIDEO & DVD CD TUTOR WEB SSM

A **system of equations** is two or more equations considered together. The system at the right is a system of two linear equations in two variables. The graphs of the equations are straight lines.

$$3x + 4y = 7$$
$$2x - 3y = 6$$

A **solution of a system of equations in two variables** is an ordered pair that is a solution of each equation of the system.

➡ Is $(3, -2)$ a solution of the system $\begin{array}{l} 2x - 3y = 12 \\ 5x + 2y = 11? \end{array}$

$2x - 3y = 12$		$5x + 2y = 11$	
$2(3) - 3(-2)$	12	$5(3) + 2(-2)$	11
$6 - (-6)$		$15 + (-4)$	
$12 = 12$ True		$11 = 11$ True	

• Replace x by 3 and y by -2.

Yes, because $(3, -2)$ is a solution of each equation, it is a solution of the system of equations.

A solution of a system of linear equations can be found by graphing the lines of the system on the same coordinate axes. Three examples of linear equations in two variables are shown below, along with the graphs of the equations of the system.

System I

$x + 2y = 4$
$2x + y = -1$

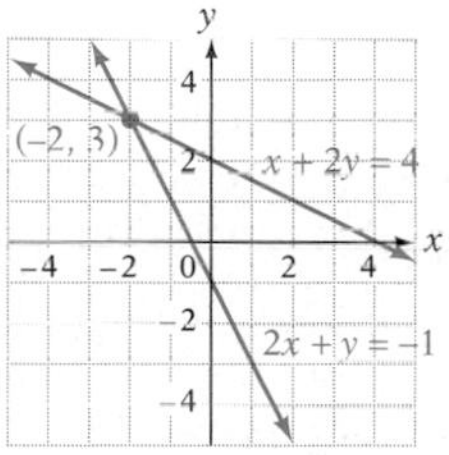

System II

$2x + 3y = 6$
$4x + 6y = -12$

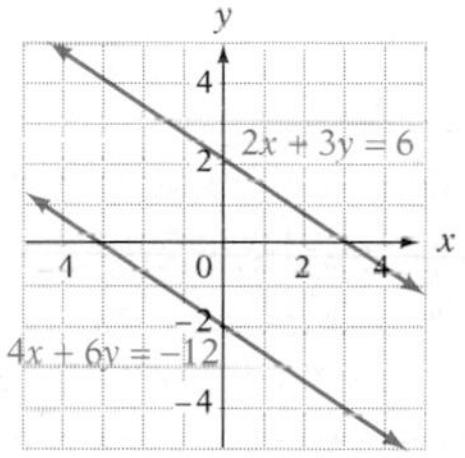

System III

$x - 2y = 4$
$2x - 4y = 8$

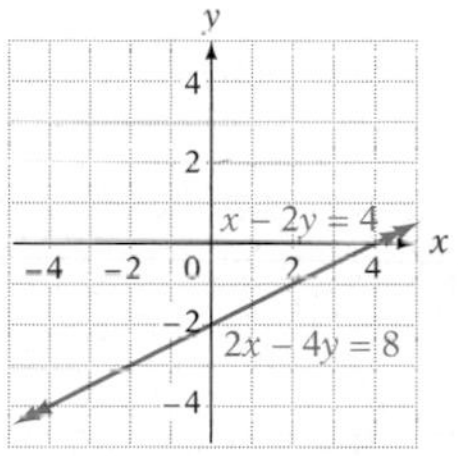

$x + 2y = 4$	
$-2 + 2(3)$	4
$-2 + 6$	4
$4 = 4$	

$2x + y = -1$	
$2(-2) + 3$	-1
$-4 + 3$	-1
$-1 = -1$	

For System I, the two lines intersect at a single point whose coordinates are $(-2, 3)$. Because this point lies on both lines, it is a solution of each equation of the system of equations. We can check this by replacing x by -2 and y by 3. The check is shown at the left. The ordered pair $(-2, 3)$ is a solution of System I.

When the graphs of a system of equations intersect at only one point, the system is called an **independent system of equations.** System I is an independent system of equations.

Objective 5.1A

New Vocabulary

system of equations
solution of a system of equations in two variables
independent system of equations
inconsistent system of equations
dependent system of equations

Vocabulary to Review

linear equation in two variables

Instructor Note

You might want to come back to the example at the left after solving a system of equations by graphing. Graph $2x - 3y = 12$ and $5x + 2y = 11$ and show that the point of intersection is $(3, -2)$.

Discuss the Concepts

For a system of two linear equations in two variables, explain, in geometric terms, each of the following: dependent system of equations, independent system of equations, and inconsistent system of equations.

Concept Check

1. How is the solution of a system of equations in two variables represented?
2. Can a system of two linear equations in two variables have exactly two solutions? Explain your answer.

In-Class Examples (Objective 5.1A)

Solve by graphing.

1. $x + y = 5$
   $2x - y = 1$ $(2, 3)$

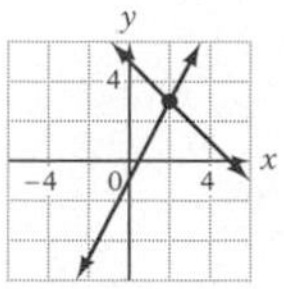

2. $x - y = 2$
   $x + 3y = -2$ $(1, -1)$

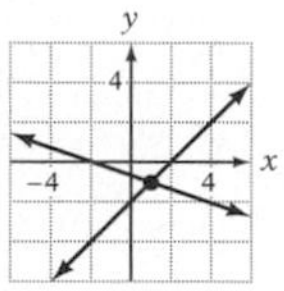

Instructor Note

The idea of a system of equations is fairly difficult for students. You might try to motivate the discussion by first asking, "Find two numbers whose sum is 20." It will not take students long to realize that there are infinitely many solutions to this question. By guiding your students a little further, you can show that the solution set is the set of points whose coordinates satisfy $x + y = 20$.

Now ask, "Find two numbers whose difference is 4." This will provide the equation of another line.

Finally ask, "Find two numbers whose sum is 20 and whose difference is 4." You might have students guess and check until they find the solution (8 and 12). You can then show students that the solution represents the coordinates of the point of intersection of the two lines.

You can extend this number problem to illustrate inconsistent and dependent systems of equations as well. For example:

"Find two numbers whose sum is 5 and whose sum is 8." There is no solution, and the graphs of the lines are parallel. Here is another example:

"Find two numbers whose sum is 10 such that the larger number is the difference between 10 and the smaller number." There are infinitely many solutions, and the graphs are the same line.

276

System II from the previous page and the graph of the equations of that system are shown again at the right. Note in this case that the graphs of the lines are parallel and do not intersect. Because the graphs do not intersect, there is no point that is on both lines. Therefore, the system of equations has no solution.

$$2x + 3y = 6$$
$$4x + 6y = -12$$

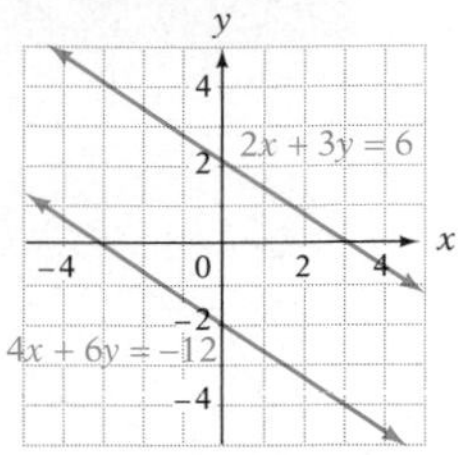

When a system of equations has no solution, it is called an **inconsistent system of equations.** System II is an inconsistent system of equations.

System III from the previous page and the graph of the equations of that system are shown again at the right. Note that the graph of $x - 2y = 4$ lies directly on top of the graph of $2x - 4y = 8$. Thus the two lines intersect at an infinite number of points, so there are an infinite number of solutions of this system of equations. Because each equation represents the same set of points, the solutions of the system of equations can be stated by using the ordered pairs of either one of the equations. Therefore, we can say, "The solutions are the ordered pairs that satisfy $x - 2y = 4$." Alternatively, we can solve the equation for y and say, "The solutions are the ordered pairs that satisfy $y = \frac{1}{2}x - 2$." We normally state this solution using ordered pairs. In this case, "The solutions are the ordered pairs $\left(x, \frac{1}{2}x - 2\right)$."

$$x - 2y = 4$$
$$2x - 4y = 8$$

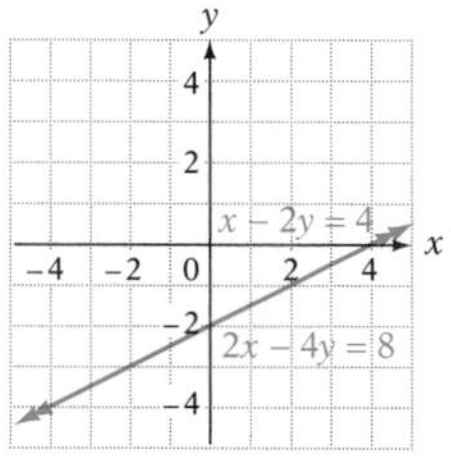

When the two equations in a system of equations represent the same line, the system is called a **dependent system of equations.** System III is a dependent system of equations.

The above systems illustrate the three possibilities for a system of linear equations in two variables:

1. The graphs intersect at one point.

The solution of the system of equations is the ordered pair (x, y) whose coordinates are the point of intersection.

The system of equations is independent.

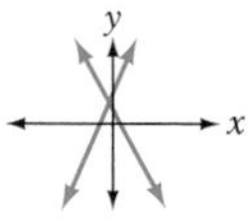

2. The lines are parallel and never intersect.

There is no solution of the system of equations.

The system of equations is inconsistent.

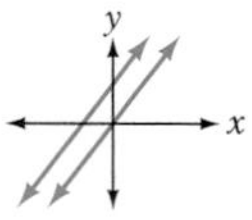

3. The graphs are the same line, and they intersect at infinitely many points.

There are an infinite number of solutions of the system of equations.

The system of equations is dependent.

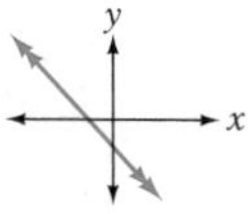

➡ Solve by the substitution method: (1) $6x + 2y = 8$
(2) $3x + y = 2$

$3x + y = 2$ • **We will solve Equation (2) for *y*.**

(3) $y = -3x + 2$ • **This is Equation (3).**

(1) $6x + 2y = 8$ • **This is Equation (1).**

$6x + 2(-3x + 2) = 8$ • **Equation (3) states that $y = -3x + 2$. Substitute $-3x + 2$ for y in Equation (1).**

$6x - 6x + 4 = 8$ • **Solve for *x*.**

$0x + 4 = 8$

$4 = 8$

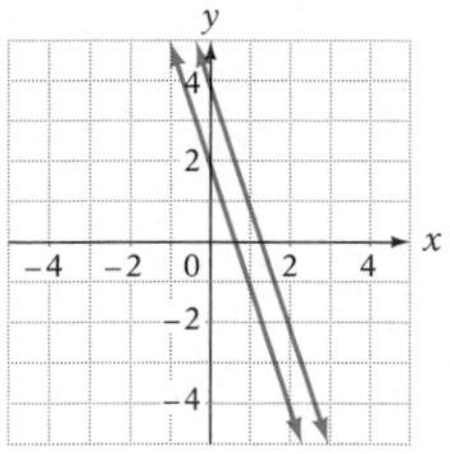

This is not a true equation.
The system of equations has no solution.
The system of equations is inconsistent.

The graph of the system of equations is shown at the left. Note that the lines are parallel.

Example 4 Solve by substitution:
(1) $3x - 2y = 4$
(2) $-x + 4y = -3$

Solution Solve Equation (2) for x.

$-x + 4y = -3$

$-x = -4y - 3$

$x = 4y + 3$ • **Equation (3)**

Substitute $4y + 3$ for x in Equation (1).

$3x - 2y = 4$ • **Equation (1)**

$3(4y + 3) - 2y = 4$ • **$x = 4y + 3$**

$12y + 9 - 2y = 4$

$10y + 9 = 4$

$10y = -5$

$y = -\frac{5}{10} = -\frac{1}{2}$

Substitute the value of y into Equation (3).

$x = 4y + 3$ • **Equation (3)**

$= 4\left(-\frac{1}{2}\right) + 3$ • **$y = -\frac{1}{2}$**

$= -2 + 3 = 1$

The solution is $\left(1, -\frac{1}{2}\right)$.

You Try It 4 Solve by substitution:
$3x - y = 3$
$6x + 3y = -4$

Your solution $\left(\frac{1}{3}, -2\right)$

Solution on p. S14

Concept Check

When you solve a system of equations by the substitution method, how do you determine whether the system of equations is inconsistent? How do you determine whether it is dependent?

Optional Student Activity

1. Write a system of equations for each case.
   a. The system has $(-3, 5)$ as its only solution.
   b. The system has no solution.
   c. The system is a dependent system of equations.
   Answers will vary. For example:
   a. $y = x + 8$
   $y = -x + 2$
   b. $y = 2x - 3$
   $y = 2x + 4$
   c. $y = 3x - 5$
   $3x - y = 5$
2. For what value of k is the system of equations inconsistent?
   a. $2x - 3y = 7$
   $kx - 3y = 4$ 2
   b. $8x - 4y = 1$
   $2x - ky = 3$ 1
   c. $x = 4y + 4$
   $kx - 8y = 4$ 2
3. Explain how to determine whether a system of linear equations in two variables is independent, inconsistent, or dependent without graphing or solving the system of equations. Answers will vary. For example, write each equation in the system in the form $y = mx + b$. If the equations have the same slope and the same y-intercept, the equations are dependent. If the equations have the same slope and different y-intercepts, the system of equations is inconsistent. If the equations in the system do not have the same slope, the system is independent.

280 Chapter 5 / Systems of Linear Equations and Inequalities

Instructor Note

After presenting Example 6, have students show that, for example, if $x = 1$, then $y = -3(1) + 4 = 1$. $(1, 1)$ is a solution of the system of equations. If $x = -2$, then $y = -3(-2) + 4 = 10$. $(-2, 10)$ is a solution of the system of equations.

Example 5 Solve by substitution and graph:

$$3x - 3y = 2$$
$$y = x + 2$$

Solution

$$3x - 3y = 2$$
$$3x - 3(x + 2) = 2$$
$$3x - 3x - 6 = 2$$
$$-6 = 2$$

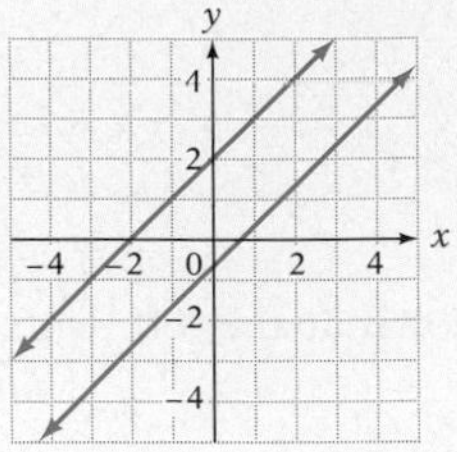

This is not a true equation. The lines are parallel, so the system is inconsistent. The system does not have a solution.

You Try It 5 Solve by substitution and graph:

$$y = 2x - 3$$
$$2x - 4y = 5$$

Your solution $\left(\frac{7}{6}, -\frac{2}{3}\right)$

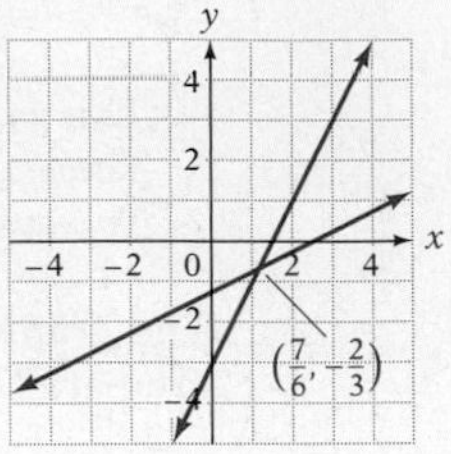

Example 6 Solve by substitution and graph:

$$9x + 3y = 12$$
$$y = -3x + 4$$

Solution

$$9x + 3y = 12$$
$$9x + 3(-3x + 4) = 12$$
$$9x - 9x + 12 = 12$$
$$12 = 12$$

This is a true equation. The system is dependent.

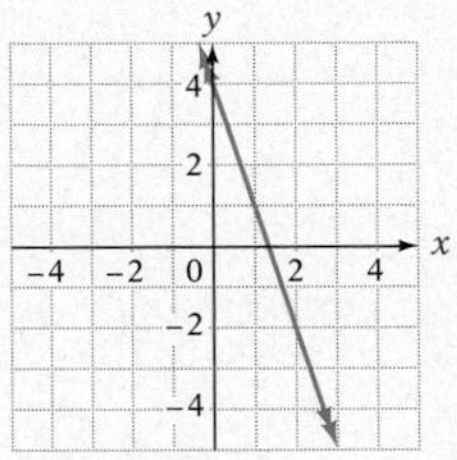

The solutions are the ordered pairs $(x, -3x + 4)$.

You Try It 6 Solve by substitution and graph:

$$6x - 3y = 6$$
$$2x - y = 2$$

Your solution Dependent system. The solutions are the ordered pairs $(x, 2x - 2)$.

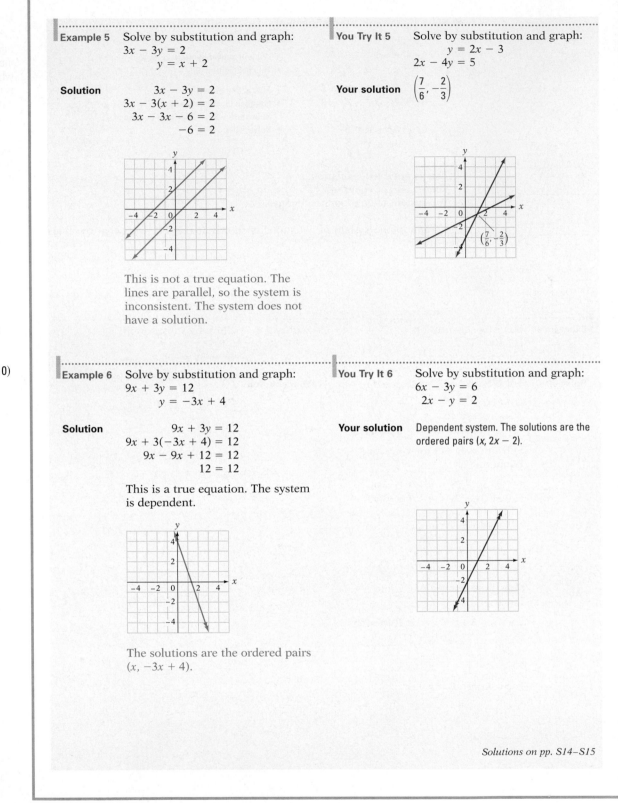

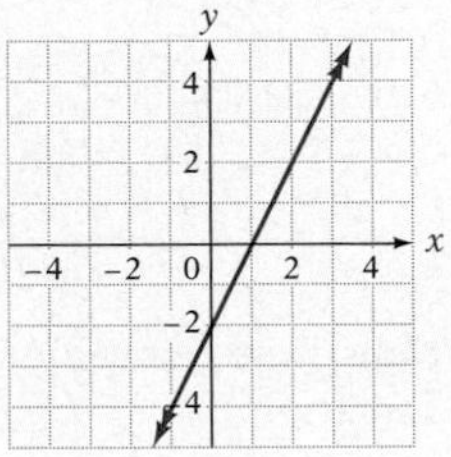

Solutions on pp. S14–S15

5.1 Exercises

Objective A

Is the ordered pair a solution of the system of equations?

1. $(0, -1)$; $3x - 2y = 2$, $x + 2y = 6$ — no

2. $(2, 1)$; $x + y = 3$, $2x - 3y = 1$ — yes

3. $(-3, -5)$; $x + y = -8$, $2x + 5y = -31$ — yes

4. $(1, -1)$; $3x - y = 4$, $7x + 2y = -5$ — no

5. $(4, 3)$; $5x - 2y = 14$, $x + y = 8$ — no

6. $(2, 5)$; $3x + 2y = 16$, $2x - 3y = 4$ — no

7. $(5, 2)$; $y = 2x - 8$, $y = 3x - 13$ — yes

8. $(-4, 3)$; $y = 2x + 11$, $y = x - 1$ — no

State whether the system of equations is independent, inconsistent, or dependent.

9.

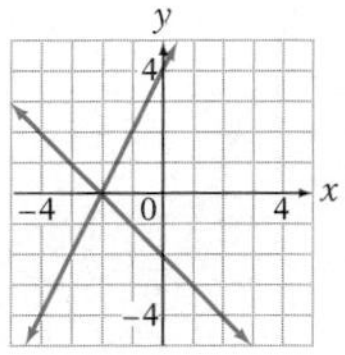

independent

10.

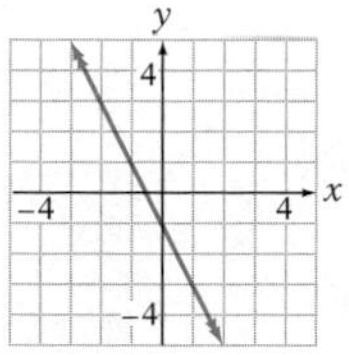

dependent

11.

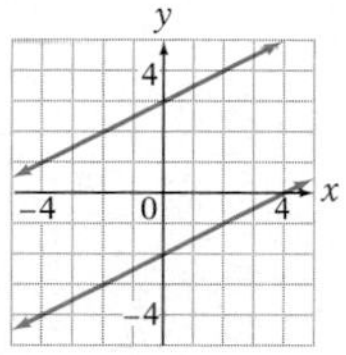

inconsistent

12.

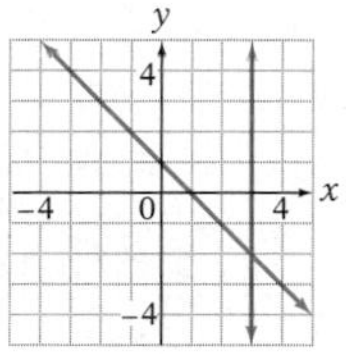

independent

Solve by graphing.

13. $x + y = 2$, $x - y = 4$

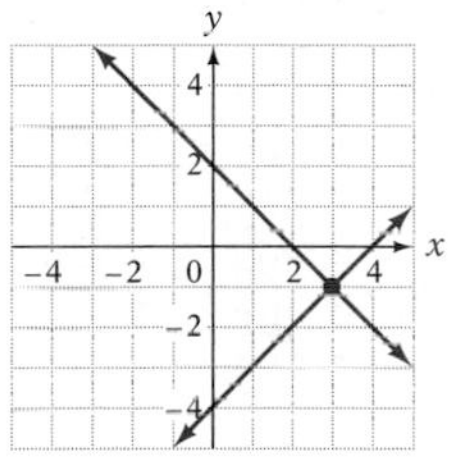

(3, −1)

14. $x + y = 1$, $3x - y = -5$

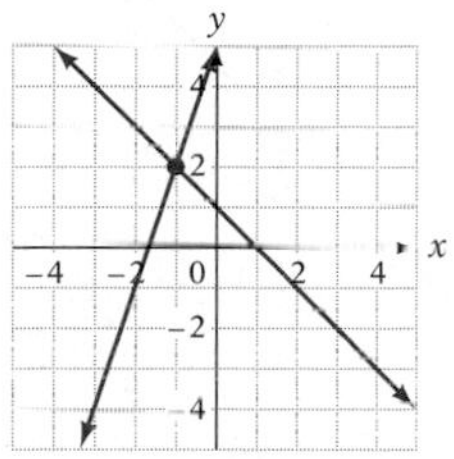

(−1, 2)

15. $x - y = -2$, $x + 2y = 10$

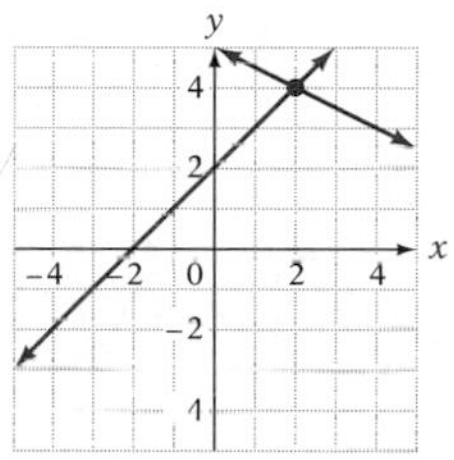

(2, 4)

16. $2x - y = 5$, $3x + y = 5$

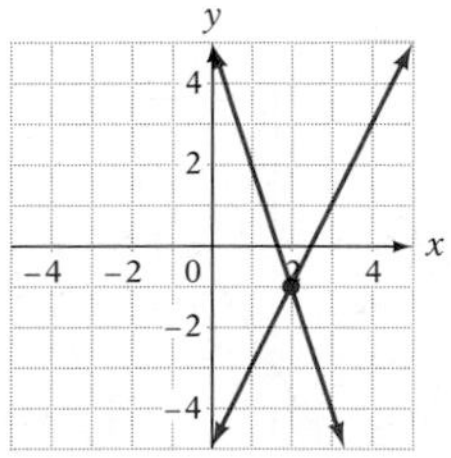

(2, −1)

17. $3x - 2y = 6$, $y = 3$

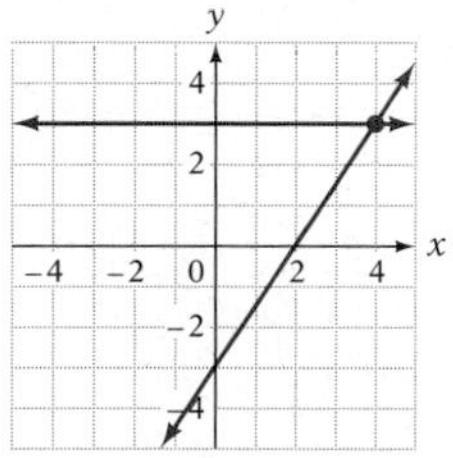

(4, 3)

18. $x = 4$, $3x - 2y = 4$

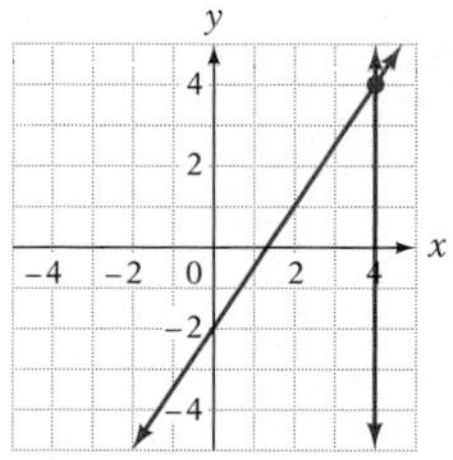

(4, 4)

Section 5.1

Suggested Assignment

Exercises 1–63, odds

More challenging problems:
Exercises 64, 66, 67, 73

Quick Quiz (Objective 5.1A)

Solve by graphing.

1. $x + y = 3$, $x - y = 1$ (2, 1)

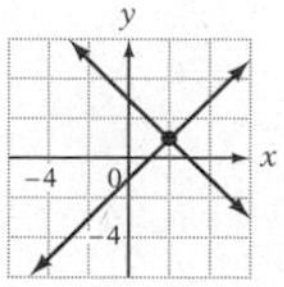

2. $x - 2y = -5$, $x + 3y = 5$ (−1, 2)

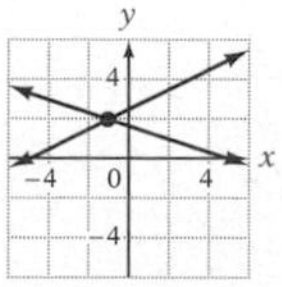

282

19. $x = 4$
$y = -1$

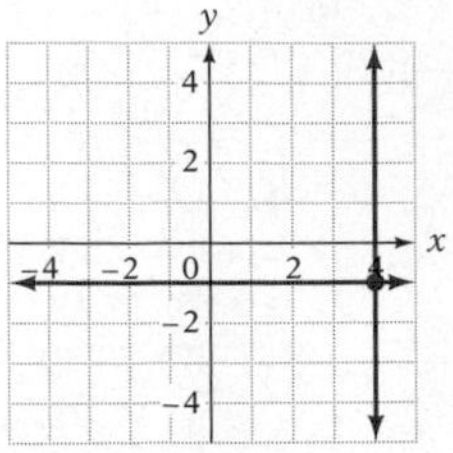

(4, −1)

20. $x + 2 = 0$
$y - 1 = 0$

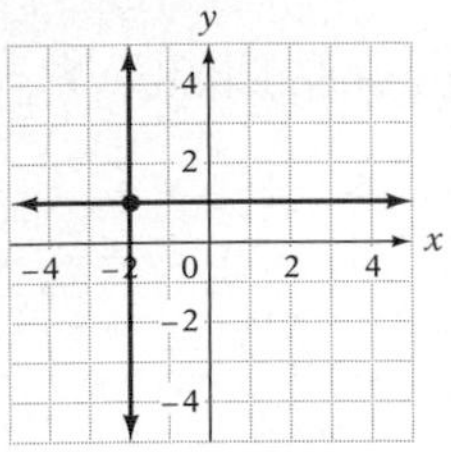

(−2, 1)

21. $2x + y = 3$
$x - 2 = 0$

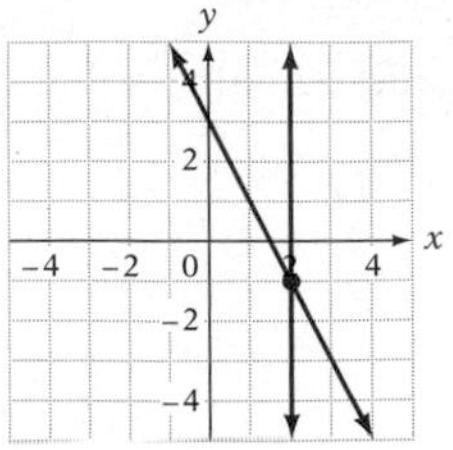

(2, −1)

22. $x - 3y = 6$
$y + 3 = 0$

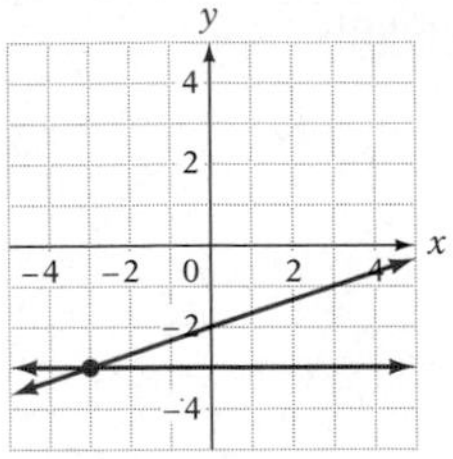

(−3, −3)

23. $x - y = 6$
$x + y = 2$

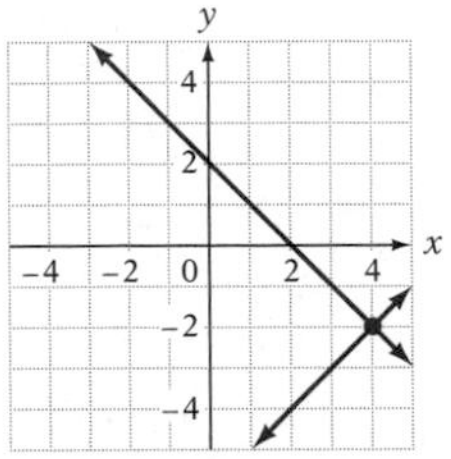

(4, −2)

24. $2x + y = 2$
$-x + y = 5$

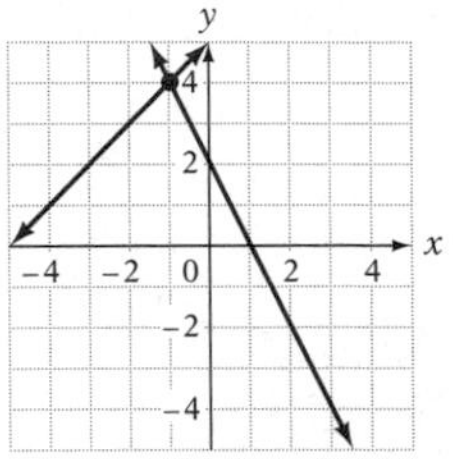

(−1, 4)

25. $y = x - 5$
$2x + y = 4$

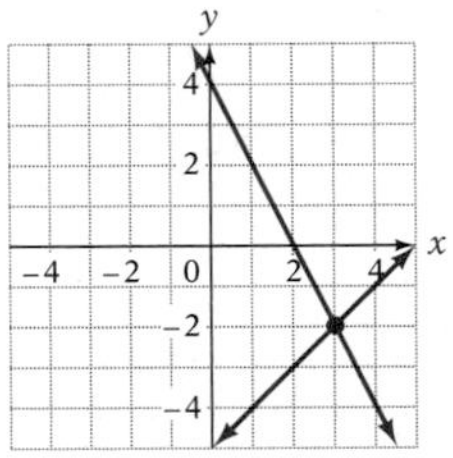

(3, −2)

26. $2x - 5y = 4$
$y = x + 1$

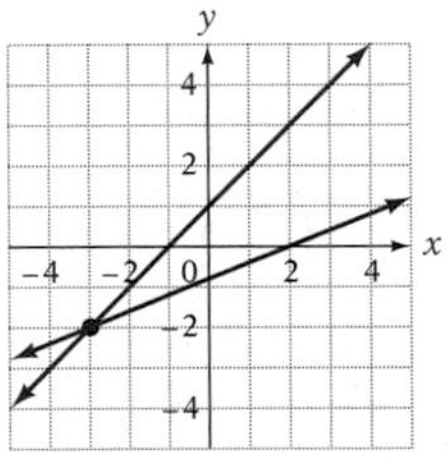

(−3, −2)

27. $y = \frac{1}{2}x - 2$
$x - 2y = 8$

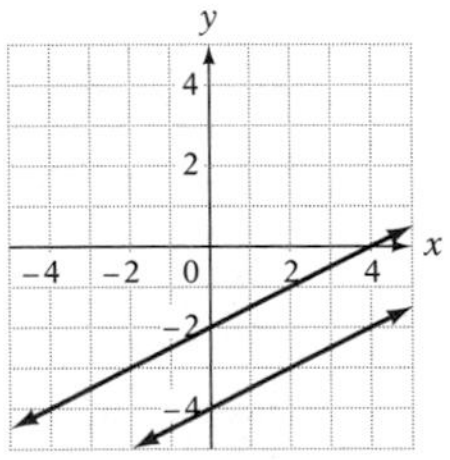

no solution

28. $2x + 3y = 6$
$y = -\frac{2}{3}x + 1$

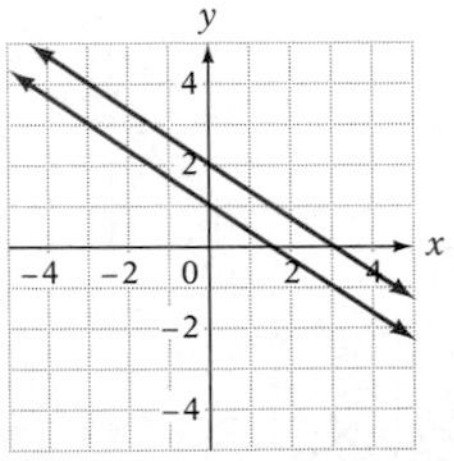

no solution

29. $2x - 5y = 10$
$y = \frac{2}{5}x - 2$

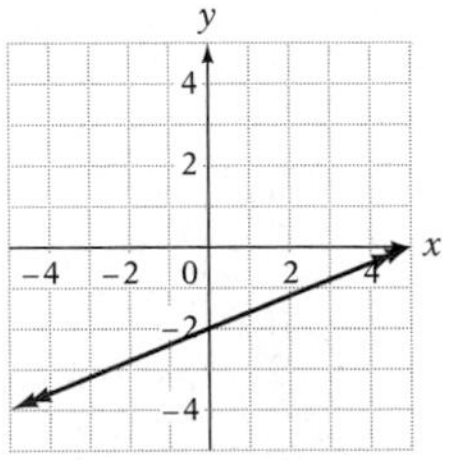

$\left(x, \frac{2}{5}x - 2\right)$

30. $3x - 2y = 6$
$y = \frac{3}{2}x - 3$

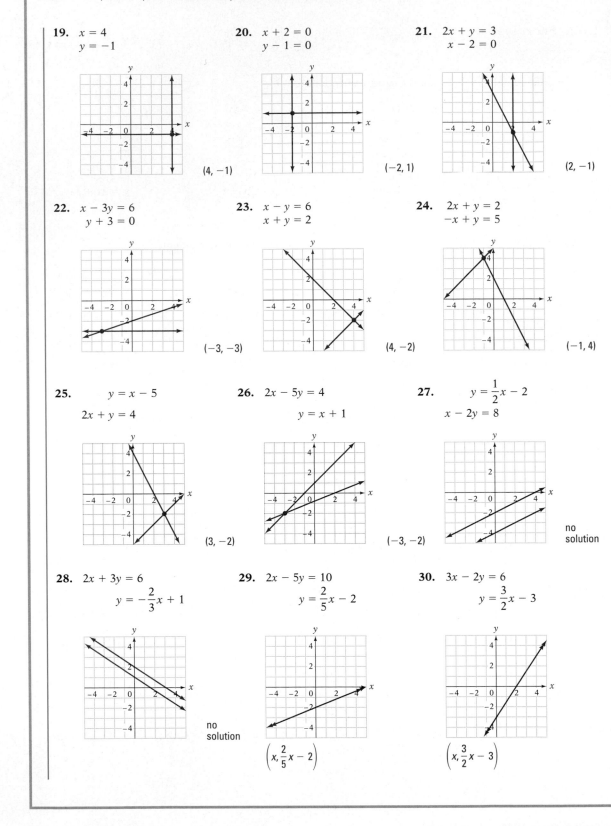

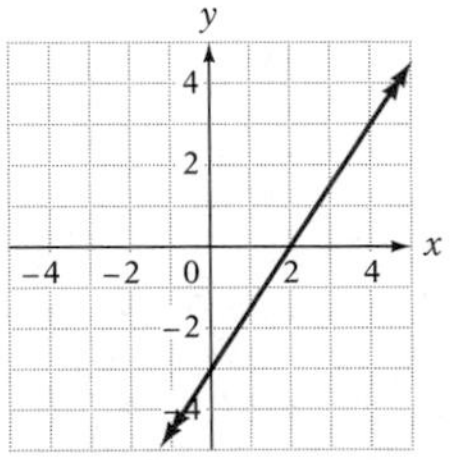

$\left(x, \frac{3}{2}x - 3\right)$

283

31. $3x - 4y = 12$
$5x + 4y = -12$

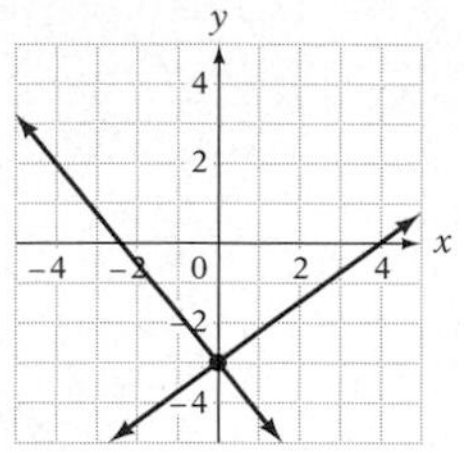

(0, −3)

32. $2x - 3y = 6$
$2x - 5y = 10$

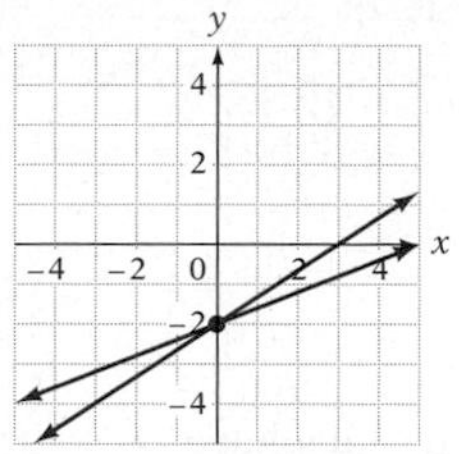

(0, −2)

33. $2x - 3y = 2$
$5x + 4y = 5$

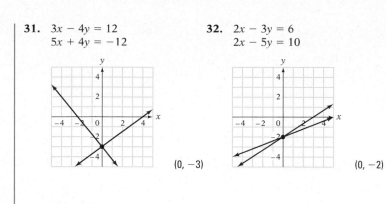

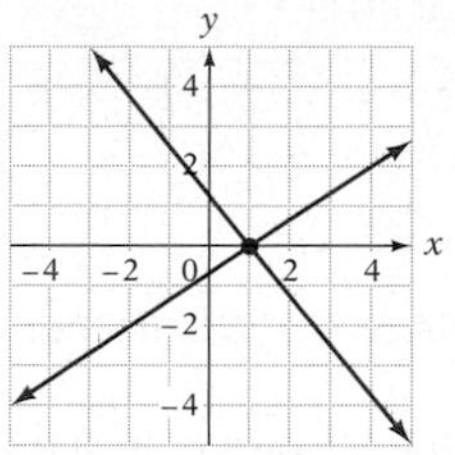

(1, 0)

Objective B

Solve by substitution.

34. $2x + 3y = 7$
$x = 2$
(2, 1)

35. $y = 3$
$3x - 2y = 6$
(4, 3)

36. $y = x - 3$
$x + y = 5$
(4, 1)

37. $y = x + 2$
$x + y = 6$
(2, 4)

38. $x = y - 2$
$x + 3y = 2$
(−1, 1)

39. $x = y + 1$
$x + 2y = 7$
(3, 2)

40. $y = 4 - 3x$
$3x + y = 5$
inconsistent

41. $y = 2 - 3x$
$6x + 2y = 7$
inconsistent

42. $x = 3y + 3$
$2x - 6y = 12$
inconsistent

43. $x = 2 - y$
$3x + 3y = 6$
$(x, 2-x)$

44. $3x + 5y = -6$
$x = 5y + 3$
$\left(-\frac{3}{4}, -\frac{3}{4}\right)$

45. $y = 2x + 3$
$4x - 3y = 1$
(−5, −7)

46. $3x + y = 4$
$4x - 3y = 1$
(1, 1)

47. $x - 4y = 9$
$2x - 3y = 11$
$\left(\frac{17}{5}, -\frac{7}{5}\right)$

48. $3x - y = 6$
$x + 3y = 2$
(2, 0)

49. $4x - y = -5$
$2x + 5y = 13$
$\left(-\frac{6}{11}, \frac{31}{11}\right)$

50. $3x - y = 5$
$2x + 5y = -8$
(1, −2)

51. $3x + 4y = 18$
$2x - y = 1$
(2, 3)

52. $4x + 3y = 0$
$2x - y = 0$
(0, 0)

53. $5x + 2y = 0$
$x - 3y = 0$
(0, 0)

54. $2x - y = 2$
$6x - 3y = 6$
$(x, 2x - 2)$

55. $3x + y = 4$
$9x + 3y = 12$
$(x, -3x + 4)$

56. $x = 3y + 2$
$y = 2x + 6$
(−4, −2)

57. $x = 4 - 2y$
$y = 2x - 13$
(6, −1)

Quick Quiz (Objective 5.1B)

Solve by substitution.

1. $y = -x + 3$
$2x - y = 6$ (3, 0)

2. $x + 2y = 7$
$2x - y = 4$ (3, 2)

3. $4x - y = 5$
$y = 4x - 1$ Inconsistent

Answers to Writing Exercises

73. The system of equations is dependent if the equation reduces to a true equation such as $4 = 4$. The system is inconsistent if the equation reduces to a false equation such as $0 = 2$. In both cases, the variable drops out and we are left with only constants.

58. $y = 2x + 11$
$y = 5x - 19$
(10, 31)

59. $y = 2x - 8$
$y = 3x - 13$
(5, 2)

60. $y = -4x + 2$
$y = -3x - 1$
(3, −10)

61. $x = 3y + 7$
$x = 2y - 1$
(−17, −8)

62. $x = 4y - 2$
$x = 6y + 8$
(−22, −5)

63. $x = 3 - 2y$
$x = 5y - 10$
$\left(-\frac{5}{7}, \frac{13}{7}\right)$

APPLYING THE CONCEPTS

For what value of k does the system of equations have no solution?

64. $2x - 3y = 7$
$kx - 3y = 4$
2

65. $8x - 4y = 1$
$2x - ky = 3$
1

66. $x = 4y + 4$
$kx - 8y = 4$
2

67. The following was offered as a solution to the system of equations

(1) $y = \frac{1}{2}x + 2$
(2) $2x + 5y = 10$

$2x + 5y = 10$ • **This is Equation (2).**
$2x + 5\left(\frac{1}{2}x + 2\right) = 10$ • **Substitute $\frac{1}{2}x + 2$ for y.**
$2x + \frac{5}{2}x + 10 = 10$ • **Solve for x.**
$\frac{9}{2}x = 0$
$x = 0$

At this point the student stated that because $x = 0$, the system of equations has no solution. If this assertion is correct, is the system of equations independent, dependent, or inconsistent? If the assertion is not correct, what is the correct solution?
The assertion is false. The solution is (0, 2).

Use a graphing calculator to solve each of the following systems of equations. Round answers to the nearest hundredth. See the Projects and Group Activities at the end of this chapter for assistance.

68. $y = -\frac{1}{2}x + 2$
$y = 2x - 1$
(1.20, 1.40)

69. $y = 1.2x + 2$
$y = -1.3x - 3$
(−2, −0.4)

70. $y = \sqrt{2}x - 1$
$y = -\sqrt{3}x + 1$
(0.64, −0.10)

71. $y = \pi x - \frac{2}{3}$
$y = -x + \frac{\pi}{2}$
(0.54, 1.03)

72. Write three different systems of equations: (a) one that has (−3, 5) as its only solution, (b) one for which there is no solution, and (c) one that is a dependent system of equations. Answers will vary.

73. When you solve a system of equations by the substitution method, how do you determine whether the system of equations is dependent? How do you determine whether the system of equations is inconsistent?

5.2 Solving Systems of Linear Equations by the Addition Method

Objective A

To solve a system of two linear equations in two variables by the addition method

VIDEO & DVD CD TUTOR WWW WEB SSM

The **addition method** is an alternative method for solving a system of equations. This method is based on the Addition Property of Equations. Use the addition method when it is not convenient to solve one equation for one variable in terms of the other variable.

TAKE NOTE

Equation (1) states that $5x - 3y$ equals 14, and Equation (2) states that $2x + 3y$ equals -7. Thus adding Equations (1) and (2) is like adding

$$\begin{array}{r} 14 = 14 \\ -7 = -7 \\ \hline 7 = 7 \end{array}$$ A true equation

The addition method of solving a system of equations is based on adding the same number to each side of the equation.

Note, for the system of equations at the right, the effect of adding Equation (2) to Equation (1). Because $-3y$ and $3y$ are additive inverses, adding the equations results in an equation with only one variable.

$$\begin{array}{lr} (1) & 5x - 3y = 14 \\ (2) & 2x + 3y = -7 \\ & 7x + 0y = 7 \\ & 7x = 7 \end{array}$$

The solution of the resulting equation is the first component of the ordered-pair solution of the system.

$$\begin{array}{r} 7x = 7 \\ x = 1 \end{array}$$

The second component is found by substituting the value of x into Equation (1) or (2) and then solving for y. Equation (1) is used here.

$$\begin{array}{lr} (1) & 5x - 3y = 14 \\ & 5(1) - 3y = 14 \\ & 5 - 3y = 14 \\ & -3y = 9 \\ & y = -3 \end{array}$$

The solution is $(1, -3)$.

Sometimes each equation of the system of equations must be multiplied by a constant so that the coefficients of one of the variables are opposites.

⇒ Solve by the addition method: (1) $3x + 4y = 2$
(2) $2x + 5y = -1$

To eliminate x, multiply Equation (1) by 2 and Equation (2) by -3. Note at the right how the constants are chosen.

$$2(3x + 4y) = 2 \cdot 2$$
$$-3(2x + 5y) = -3(-1)$$

• The negative is used so that the coefficients will be opposites.

$$\begin{array}{r} 6x + 8y = 4 \\ -6x - 15y = 3 \\ \hline -7y = 7 \\ y = -1 \end{array}$$

• 2 times Equation (1)
• -3 times Equation (2)
• Add the equations.
• Solve for y.

Substitute the value of y into Equation (1) or Equation (2) and solve for x. Equation (1) will be used here.

$$\begin{array}{lr} (1) & 3x + 4y = 2 \\ & 3x + 4(-1) = 2 \\ & 3x - 4 = 2 \\ & 3x = 6 \\ & x = 2 \end{array}$$

• Substitute -1 for y.
• Solve for x.

The solution is $(2, -1)$.

Objective 5.2A

New Vocabulary
addition method

Vocabulary to Review
system of two linear equations in two variables

Instructor Note
The addition method is not clear to many students. Try showing them that this method is really a variation of the substitution method.

Using the first system of equations at the left, add $2x + 3y$ to each side of $5x - 3y = 14$.

$$5x - 3y + (2x + 3y) = 14 + (2x + 3y)$$

Simplify the left side; substitute -7 for $2x + 3y$ on the right side. The result is $7x = 7$.

Discuss the Concepts
Ask students to explain how the following situation relates to an inconsistent system of equations.

The perimeter of a rectangle is 100 m. The sum of the length and width of the rectangle is 40 m. Find the dimensions of the rectangle.

In-Class Examples (Objective 5.2A)
Solve by the addition method.

1. $x - y = 6$
   $3x + y = 10$ (4, −2)
2. $2x + y = 3$
   $x + 2y = 0$ (2, −1)
3. $3x + 7y = 0$
   $4x + 5y = 0$ (0, 0)
4. $2x - 3y = 9$
   $4x - 6y = 4$ Inconsistent

Concept Check

Mark said to his brother John, "Give me eight of your Pokémon™ cards and then we'll each have the same number." John answered, "No, you give me eight of your Pokémon cards and then I will have twice as many as you." How many Pokémon cards did John have to start with? 56 cards

Instructor Note

Writing the solution set for the example at the bottom of this page is very difficult for students. Try approaching it by having students name an ordered-pair solution, then another, and another, and so on until they suspect that there are infinitely many solutions. To name all the solutions, we need to have variable coordinates in the ordered pair: $(x, 2x - 3)$.

Point of Interest

There are records of Babylonian mathematicians solving systems of equations 3600 years ago. Here is a system of equations from that time (in our modern notation):

$$\frac{2}{3}x = \frac{1}{2}y - 500$$
$$x - y = 1800$$

We say *modern notation* for many reasons. Foremost is the fact that using variables did not become widespread until the 17th century. There are many other reasons, however. The equals sign had not been invented, 2 and 3 did not look like they do today, and zero had not even been considered as a possible number.

➡ Solve by the addition method: (1) $\frac{2}{3}x + \frac{1}{2}y = 4$

(2) $\frac{1}{4}x - \frac{3}{8}y = -\frac{3}{4}$

Clear fractions. Multiply each equation by the LCM of the denominators.

$$6\left(\frac{2}{3}x + \frac{1}{2}y\right) = 6(4)$$
$$8\left(\frac{1}{4}x - \frac{3}{8}y\right) = 8\left(-\frac{3}{4}\right)$$

$$4x + 3y = 24$$
$$2x - 3y = -6$$

$6x = 18$ • Eliminate y. Add the equations.

$x = 3$ • Solve for x.

Substitute the value of x into Equation (1) and solve for y.

$\frac{2}{3}x + \frac{1}{2}y = 4$ • This is Equation (1).

$\frac{2}{3}(3) + \frac{1}{2}y = 4$ • $x = 3$

$2 + \frac{1}{2}y = 4$

$\frac{1}{2}y = 2$

$y = 4$

The solution is (3, 4).

TAKE NOTE

The result of adding Equations (3) and (2) is $0 = 0$. It is not $x = 0$, and it is not $y = 0$. There is no variable in the equation $0 = 0$. This does not indicate that the solution is (0, 0). Rather, it indicates a dependent system of equations.

➡ Solve by the addition method: (1) $2x - y = 3$

(2) $4x - 2y = 6$

Eliminate y. Multiply Equation (1) by -2.

(1) $-2(2x - y) = -2(3)$ • Multiply both sides of Equation (1) by -2.

(3) $-4x + 2y = -6$ • This is Equation (3).

Add Equation (3) to Equation (2).

(2) $4x - 2y = 6$

(3) $-4x + 2y = -6$

$0 = 0$ • This is a true equation.

The equation $0 = 0$ indicates that the system of equations is dependent. This means that the graphs of the two lines are the same. Therefore, the solutions of the system of equations are the ordered-pair solutions of the equation of the line. Solve Equation (1) for y.

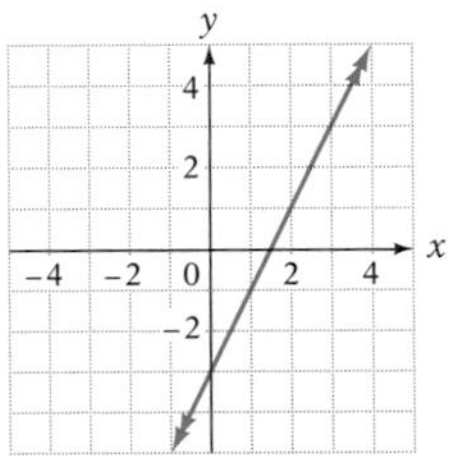

$$2x - y = 3$$
$$-y = -2x + 3$$
$$y = 2x - 3$$

The ordered-pair solutions are $(x, 2x - 3)$.

287

Example 1 Solve by the addition method:
(1) $3x - 2y = 2x + 5$
(2) $2x + 3y = -4$

Solution Write Equation (1) in the form $Ax + By = C$.

$$3x - 2y = 2x + 5$$
$$x - 2y = 5$$

Solve the system:

$$x - 2y = 5$$
$$2x + 3y = -4$$

Eliminate x.

$$-2(x - 2y) = -2(5)$$
$$2x + 3y = -4$$

$$-2x + 4y = -10$$
$$2x + 3y = -4$$

Add the equations.

$$7y = -14$$
$$y = -2$$

Replace y in Equation (2).

$$2x + 3y = -4$$
$$2x + 3(-2) = -4$$
$$2x - 6 = -4$$
$$2x = 2$$
$$x = 1$$

The solution is $(1, -2)$.

You Try It 1 Solve by the addition method:
$2x + 5y = 6$
$3x - 2y = 6x + 2$

Your solution $(-2, 2)$

Example 2 Solve by the addition method:
(1) $4x - 8y = 36$
(2) $3x - 6y = 27$

Solution Eliminate x.

$$3(4x - 8y) = 3(36)$$
$$-4(3x - 6y) = -4(27)$$

$$12x - 24y = 108$$
$$-12x + 24y = -108$$

Add the equations.

$$0 = 0$$

The system of equations is dependent. The solutions are the ordered pairs $\left(x, \frac{1}{2}x - \frac{9}{2}\right)$.

You Try It 2 Solve by the addition method:
$2x + y = 5$
$4x + 2y = 6$

Your solution The system of equations is inconsistent and therefore has no solution.

Solutions on p. S15

Optional Student Activity

1. The point of intersection of the graphs of the equations $Ax + 2y = 2$ and $2x + By = 10$ is $(2, -2)$. Find A and B. $A = 3$, $B = -3$
2. The point of intersection of the graphs of the equations $Ax - 4y = 9$ and $4x + By = -1$ is $(-1, -3)$. Find A and B. $A = 3$, $B = -1$
3. Given that the graphs of the equations $2x - y = 6$, $3x - 4y = 4$, and $Ax - 2y = 0$ all intersect at the same point, find A. $A = 1$
4. Given that the graphs of the equations $3x - 2y = -2$, $2x - y = 0$, and $Ax + y = 8$ all intersect at the same point, find A. $A = 2$
5. Find an equation such that the system of equations formed by your equation and $2x - 5y = 9$ will have $(2, -1)$ as a solution. Answers will vary. $3x + 2y = 4$ is a possibility.

Objective 5.2B

New Vocabulary

linear equation in three variables
xyz-coordinate system
ordered triple
solution of an equation in three variables
system of linear equations in three variables
solution of a system of equations in three variables

Vocabulary to Review

plane
independent system of equations
inconsistent system of equations
dependent system of equations

Discuss the Concepts

A good model of a three-dimensional coordinate system is the corner of the floor in a room. The xy-plane is the floor, the xz-plane is one wall, and the yz-plane is the other wall.

Give students various ordered triples with positive coordinates and have them indicate the location of the point in the room. Then ask where a point that has negative numbers as some or all of its coordinates would be located. Finally, ask students to identify the region for which (1) $x = 0$, (2) $y = 0$, (3) $z = 0$, (4) x and y are both 0, (5) x and z are both 0, (6) y and z are both 0, and (7) x, y, and z are 0.

Objective B

To solve a system of three linear equations in three variables by the addition method

VIDEO & DVD CD TUTOR WEB SSM

An equation of the form $Ax + By + Cz = D$, where A, B, and C are coefficients of the variables and D is a constant, is a **linear equation in three variables.** Examples of these equations are shown at the right.

$$2x + 4y - 3z = 7$$
$$x - 6y + z = -3$$

Graphing an equation in three variables requires a third coordinate axis perpendicular to the xy-plane. The third axis is commonly called the z-axis. The result is a three-dimensional coordinate system called the ***xyz*-coordinate system.** To help visualize a three-dimensional coordinate system, think of a corner of a room: the floor is the xy-plane, one wall is the yz-plane, and the other wall is the xz-plane. A three-dimensional coordinate system is shown at the right.

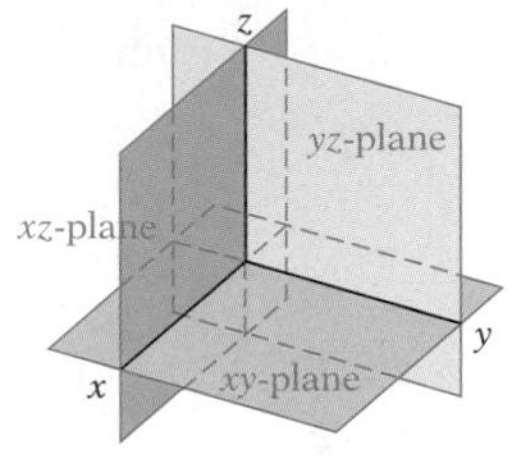

The graph of a point in an xyz-coordinate system is an **ordered triple** (x, y, z). Graphing an ordered triple requires three moves, the first along the x-axis, the second parallel to the y-axis, and the third parallel to the z-axis. The graphs of the points $(-4, 2, 3)$ and $(3, 4, -2)$ are shown at the right.

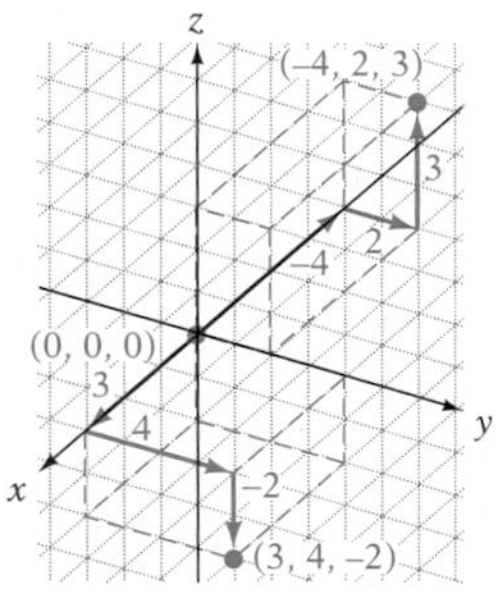

The graph of a linear equation in three variables is a plane. That is, if all the solutions of a linear equation in three variables were plotted in an xyz-coordinate system, the graph would look like a large piece of paper extending infinitely. The graph of $x + y + z = 3$ is shown at the right.

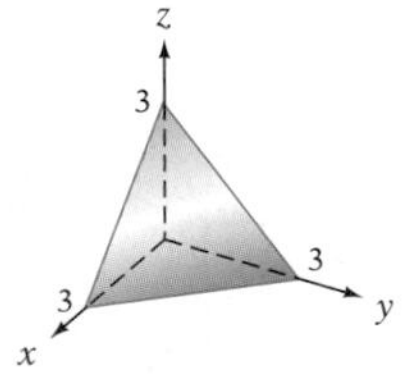

In-Class Examples (Objective 5.2B)

Solve by the addition method.

1. $x + 2y - z = 3$
   $2x - y + z = 3$
   $3x - 4y + 2z = -1$
   $(1, 3, 4)$

2. $2x - y + 4z = 7$
   $x + y + 2z = 2$
   $x + 2y - 2z = -1$
   $\left(2, -1, \frac{1}{2}\right)$

3. $x + 2y - z = 8$
   $2x + y + z = 1$
   $3x - y + 2z = -5$
   $(1, 2, -3)$

There are different ways in which three planes can be oriented in an *xyz*-coordinate system. The systems of equations represented by the planes below are inconsistent.

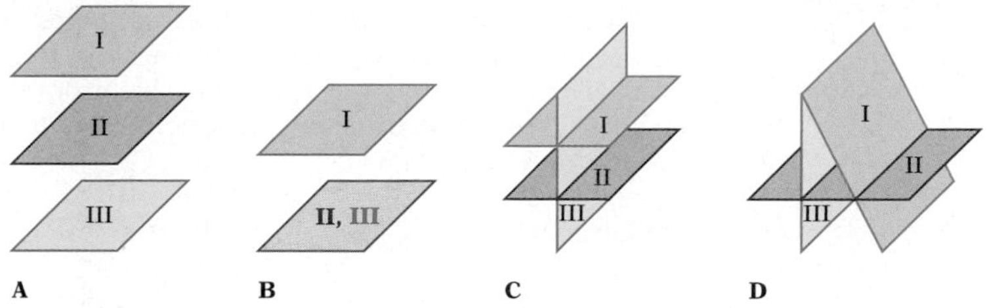

Graphs of Inconsistent Systems of Equations

For a system of three equations in three variables to have a solution, the graphs of the planes must intersect at a single point, they must intersect along a common line, or all equations must have a graph that is the same plane. These situations are shown in the figures below.

The three planes shown in Figure E intersect at a point. A system of equations represented by planes that intersect at a point is independent.

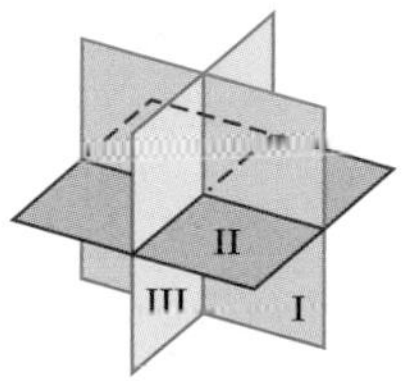

E

An Independent System of Equations

The planes shown in Figures F and G intersect along a common line. The system of equations represented by the planes in Figure H has a graph that is the same plane. The systems of equations represented by these three graphs are dependent.

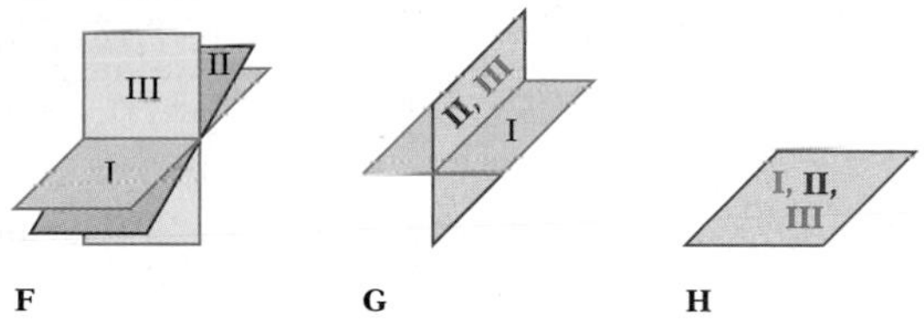

Dependent Systems of Equations

Just as a solution of an equation in two variables is an ordered pair (x, y), a **solution of an equation in three variables** is an ordered triple (x, y, z). For example, $(2, 1, -3)$ is a solution of the equation $2x - y - 2z = 9$. The ordered triple $(1, 3, 2)$ is not a solution.

Instructor Note

Your book and a desk or table top can be used to model various ways in which planes intersect.

Concept Check

1. The sum of three numbers is 62. The sum of the first and second numbers is equal to 10 less than the third. The first number minus the second number is equal to 34 less than the third number. Find the three numbers. 14, 12, 36
2. Jason and his older brother Aaron together have \$62. Jason and his sister Michelle together have \$67. Michelle and Aaron together have \$29. What is the least amount of money any of the three siblings has? \$12

Optional Student Activity

1. Find a three-digit number such that the sum of the digits is 7, the number is increased by 99 if the digits are reversed, and the hundreds digit is 3 less than the sum of the other two digits. 223
2. Let L be the line along which the planes $2x + y - z = 13$ and $x - 2y + z = 5$ intersect. If the point $(x, 3, z)$ lies on L, find the value of $(x - z)$. 3

Instructor Note

Show students that the goal of solving a system of three equations in three variables is to replace it with an equivalent system of two equations in two variables that can be solved by using the techniques of the preceding objective.

Point of Interest

In the early 1980s, Stephen Hoppe became interested in winning Monopoly strategies. Finding these strategies required solving a system that contained 123 equations in 123 variables!

A **system of linear equations in three variables** is shown at the right. A **solution of a system of equations in three variables** is an ordered triple that is a solution of each equation of the system.

$$\begin{aligned} x - 2y + z &= 6 \\ 3x + y - 2z &= 2 \\ 2x - 3y + 5z &= 1 \end{aligned}$$

A system of linear equations in three variables can be solved by using the addition method. First, eliminate one variable from any two of the given equations. Then eliminate the same variable from any other two equations. The result will be a system of two equations in two variables. Solve this system by the addition method.

➡ Solve:

$$\begin{aligned} (1)\quad x + 4y - z &= 10 \\ (2)\quad 3x + 2y + z &= 4 \\ (3)\quad 2x - 3y + 2z &= -7 \end{aligned}$$

Eliminate z from Equations (1) and (2) by adding the two equations.

$$\begin{aligned} x + 4y - z &= 10 \\ 3x + 2y + z &= 4 \\ (4)\quad 4x + 6y &= 14 \end{aligned}$$

• Add the equations. This is Equation (4).

Eliminate z from Equations (1) and (3). Multiply Equation (1) by 2 and add to Equation (3).

$$\begin{aligned} 2x + 8y - 2z &= 20 \\ 2x - 3y + 2z &= -7 \\ (5)\quad 4x + 5y &= 13 \end{aligned}$$

• 2 times Equation (1)
• This is Equation (3).
• Add the equations. This is Equation (5).

Using Equations (4) and (5), solve the system of two equations in two variables.

$$\begin{aligned} (4)\quad 4x + 6y &= 14 \\ (5)\quad 4x + 5y &= 13 \end{aligned}$$

Eliminate x. Multiply Equation (5) by -1 and add to Equation (4).

$$\begin{aligned} 4x + 6y &= 14 \\ -4x - 5y &= -13 \\ y &= 1 \end{aligned}$$

• This is Equation (4).
• -1 times Equation (5)
• Add the equations.

Substitute the value of y into Equation (4) or Equation (5) and solve for x. Equation (4) is used here.

$$\begin{aligned} 4x + 6y &= 14 \\ 4x + 6(1) &= 14 \\ 4x + 6 &= 14 \\ 4x &= 8 \\ x &= 2 \end{aligned}$$

• This is Equation (4).
• $y = 1$
• Solve for x.

Substitute the value of y and the value of x into one of the equations in the original system. Equation (2) is used here.

$$\begin{aligned} 3x + 2y + z &= 4 \\ 3(2) + 2(1) + z &= 4 \\ 6 + 2 + z &= 4 \\ 8 + z &= 4 \\ z &= -4 \end{aligned}$$

• $x = 2, y = 1$

The solution is $(2, 1, -4)$.

291

➡ Solve:
$$\begin{aligned} &(1) & 2x - 3y - z &= 1 \\ &(2) & x + 4y + 3z &= 2 \\ &(3) & 4x - 6y - 2z &= 5 \end{aligned}$$

Eliminate x from Equations (1) and (2).

$$\begin{aligned} 2x - 3y - z &= 1 && \bullet\ \text{This is Equation (1).} \\ -2x - 8y - 6z &= -4 && \bullet\ -2\text{ times Equation (2)} \\ -11y - 7z &= -3 && \bullet\ \text{Add the equations.} \end{aligned}$$

Eliminate x from Equations (1) and (3).

$$\begin{aligned} -4x + 6y + 2z &= -2 && \bullet\ -2\text{ times Equation (1)} \\ 4x - 6y - 2z &= 5 && \bullet\ \text{This is Equation (3).} \\ 0 &= 3 && \bullet\ \text{Add the equations.} \end{aligned}$$

The equation $0 = 3$ is not a true equation. The system of equations is inconsistent and therefore has no solution.

➡ Solve:
$$\begin{aligned} &(1) & 3x - z &= -1 \\ &(2) & 2y - 3z &= 10 \\ &(3) & x + 3y - z &= 7 \end{aligned}$$

Eliminate x from Equations (1) and (3). Multiply Equation (3) by -3 and add to Equation (1).

$$\begin{aligned} & & 3x - z &= -1 && \bullet\ \text{This is Equation (1).} \\ & & -3x - 9y + 3z &= -21 && \bullet\ -3\text{ times Equation (3)} \\ &(4) & -9y + 2z &= -22 && \bullet\ \text{Add the equations.} \end{aligned}$$

Use Equations (2) and (4) to form a system of equations in two variables.

$$\begin{aligned} &(2) & 2y - 3z &= 10 \\ &(4) & 9y + 2z &= -22 \end{aligned}$$

Eliminate z. Multiply Equation (2) by 2 and Equation (4) by 3.

$$\begin{aligned} 4y - 6z &= 20 && \bullet\ \text{2 times Equation (2)} \\ -27y + 6z &= -66 && \bullet\ \text{3 times Equation (4)} \\ -23y &= -46 && \bullet\ \text{Add the equations.} \\ y &= 2 && \bullet\ \text{Solve for } y. \end{aligned}$$

Substitute the value of y into Equation (2) or Equation (4) and solve for z. Equation (2) is used here.

$$\begin{aligned} &(2) & 2y - 3z &= 10 && \bullet\ \text{This is Equation (2).} \\ & & 2(2) - 3z &= 10 && \bullet\ y = 2 \\ & & 4 - 3z &= 10 && \bullet\ \text{Solve for } z. \\ & & -3z &= 6 \\ & & z &= -2 \end{aligned}$$

Substitute the value of z into Equation (1) and solve for x.

$$\begin{aligned} &(1) & 3x - z &= -1 && \bullet\ \text{This is Equation (1).} \\ & & 3x - (-2) &= -1 && \bullet\ z = -2 \\ & & 3x + 2 &= -1 && \bullet\ \text{Solve for } x. \\ & & 3x &= -3 \\ & & x &= -1 \end{aligned}$$

The solution is $(-1, 2, -2)$.

Example 3 Solve:

(1) $3x - y + 2z = 1$
(2) $2x + 3y + 3z = 4$
(3) $x + y - 4z = -9$

Solution Eliminate y. Add Equations (1) and (3).

$$\begin{aligned} 3x - y + 2z &= 1 \\ x + y - 4z &= -9 \\ 4x - 2z &= -8 \end{aligned}$$

Multiply each side of the equation by $\frac{1}{2}$.

(4) $2x - z = -4$

Multiply Equation (1) by 3 and add to Equation (2).

$$\begin{aligned} 9x - 3y + 6z &= 3 \\ 2x + 3y + 3z &= 4 \end{aligned}$$

(5) $11x + 9z = 7$

Solve the system of two equations.

(4) $2x - z = -4$
(5) $11x + 9z = 7$

Multiply Equation (4) by 9 and add to Equation (5).

$$\begin{aligned} 18x - 9z &= -36 \\ 11x + 9z &= 7 \\ 29x &= -29 \\ x &= -1 \end{aligned}$$

Replace x by -1 in Equation (4).

$$\begin{aligned} 2x - z &= -4 \\ 2(-1) - z &= -4 \\ -2 - z &= -4 \\ -z &= -2 \\ z &= 2 \end{aligned}$$

Replace x by -1 and z by 2 in Equation (3).

$$\begin{aligned} x + y - 4z &= -9 \\ -1 + y - 4(2) &= -9 \\ -9 + y &= -9 \\ y &= 0 \end{aligned}$$

The solution is $(-1, 0, 2)$.

You Try It 3 Solve:

$$\begin{aligned} x - y + z &= 6 \\ 2x + 3y - z &= 1 \\ x + 2y + 2z &= 5 \end{aligned}$$

Your solution $(3, -1, 2)$

Solution on p. S15

293

Section 5.2

Suggested Assignment
Exercises 1–65, odds
More challenging problems:
Exercises 69, 71

5.2 Exercises

Objective A

Solve by the addition method.

1. $x - y = 5$
$x + y = 7$
(6, 1)

2. $x + y = 1$
$2x - y = 5$
(2, −1)

3. $3x + y = 4$
$x + y = 2$
(1, 1)

4. $x - 3y = 4$
$x + 5y = -4$
(1, −1)

5. $3x + y = 7$
$x + 2y = 4$
(2, 1)

6. $x - 2y = 7$
$3x - 2y = 9$
(1, −3)

7. $2x + 3y = -1$
$x + 5y = 3$
(−2, 1)

8. $x + 5y = 7$
$2x + 7y = 8$
(−3, 2)

9. $3x - y = 4$
$6x - 2y = 8$
$(x, 3x - 4)$

10. $x - 2y = -3$
$-2x + 4y = 6$
$\left(x, \frac{1}{2}x + \frac{3}{2}\right)$

11. $2x + 5y = 9$
$4x - 7y = -16$
$\left(-\frac{1}{2}, 2\right)$

12. $8x - 3y = 21$
$4x + 5y = -9$
$\left(\frac{3}{2}, -3\right)$

13. $4x - 6y = 5$
$2x - 3y = 7$
inconsistent

14. $3x + 6y = 7$
$2x + 4y = 5$
inconsistent

15. $3x - 5y = 7$
$x - 2y = 3$
(−1, −2)

16. $3x + 4y = 25$
$2x + y = 10$
(3, 4)

17. $x + 3y = 7$
$-2x + 3y = 22$
(−5, 4)

18. $2x - 3y = 14$
$5x - 6y = 32$
(4, −2)

19. $3x + 2y = 16$
$2x - 3y = -11$
(2, 5)

20. $2x - 5y = 13$
$5x + 3y = 17$
(4, −1)

21. $4x + 4y = 5$
$2x - 8y = -5$
$\left(\frac{1}{2}, \frac{3}{4}\right)$

Quick Quiz (Objective 5.2A)

Solve by the addition method.

1. $x + y = 5$
$5x - y = 7$ (2, 3)

2. $x - 4y = -7$
$3x - 2y = -11$ (−3, 1)

3. $2x - y = 3$
$6x - 3y = 9$ $(x, 2x - 3)$

22. $3x + 7y = 16$
$4x - 3y = 9$
(3, 1)

23. $5x + 4y = 0$
$3x + 7y = 0$
(0, 0)

24. $3x - 4y = 0$
$4x - 7y = 0$
(0, 0)

25. $5x + 2y = 1$
$2x + 3y = 7$
(−1, 3)

26. $3x + 5y = 16$
$5x - 7y = -4$
(2, 2)

27. $3x - 6y = 6$
$9x - 3y = 8$
$\left(\frac{2}{3}, -\frac{2}{3}\right)$

28. $\frac{2}{3}x - \frac{1}{2}y = 3$
$\frac{1}{3}x - \frac{1}{4}y = \frac{3}{2}$
$\left(x, \frac{4}{3}x - 6\right)$

29. $\frac{3}{4}x + \frac{1}{3}y = -\frac{1}{2}$
$\frac{1}{2}x - \frac{5}{6}y = -\frac{7}{2}$
(−2, 3)

30. $\frac{2}{5}x - \frac{1}{3}y = 1$
$\frac{3}{5}x + \frac{2}{3}y = 5$
(5, 3)

31. $\frac{5x}{6} + \frac{y}{3} = \frac{4}{3}$
$\frac{2x}{3} - \frac{y}{2} = \frac{11}{6}$
(2, −1)

32. $\frac{3x}{4} + \frac{2y}{5} = -\frac{3}{20}$
$\frac{3x}{2} - \frac{y}{4} = \frac{3}{4}$
$\left(\frac{1}{3}, -1\right)$

33. $\frac{2x}{5} - \frac{y}{2} = \frac{13}{2}$
$\frac{3x}{4} - \frac{y}{5} = \frac{17}{2}$
(10, −5)

34. $\frac{x}{2} + \frac{y}{3} = \frac{5}{12}$
$\frac{x}{2} - \frac{y}{3} = \frac{1}{12}$
$\left(\frac{1}{2}, \frac{1}{2}\right)$

35. $\frac{3x}{2} - \frac{y}{4} = -\frac{11}{12}$
$\frac{x}{3} - y = -\frac{5}{6}$
$\left(-\frac{1}{2}, \frac{2}{3}\right)$

36. $\frac{3x}{4} - \frac{2y}{3} = 0$
$\frac{5x}{4} - \frac{y}{3} = \frac{7}{12}$
$\left(\frac{2}{3}, \frac{3}{4}\right)$

37. $4x - 5y = 3y + 4$
$2x + 3y = 2x + 1$
$\left(\frac{5}{3}, \frac{1}{3}\right)$

38. $5x - 2y = 8x - 1$
$2x + 7y = 4y + 9$
(−3, 5)

39. $2x + 5y = 5x + 1$
$3x - 2y = 3y + 3$
inconsistent

40. $4x - 8y = 5$
$8x + 2y = 1$
$\left(\frac{1}{4}, -\frac{1}{2}\right)$

41. $5x + 2y = 2x + 1$
$2x - 3y = 3x + 2$
(1, −1)

42. $3x + 3y = y + 1$
$x + 3y = 9 - x$
(−3, 5)

295

Objective B

Solve by the addition method.

43. $\begin{aligned} x + 2y - z &= 1 \\ 2x - y + z &= 6 \\ x + 3y - z &= 2 \end{aligned}$

(2, 1, 3)

44. $\begin{aligned} x + 3y + z &= 6 \\ 3x + y - z &= -2 \\ 2x + 2y - z &= 1 \end{aligned}$

(−1, 2, 1)

45. $\begin{aligned} 2x - y + 2z &= 7 \\ x + y + z &= 2 \\ 3x - y + z &= 6 \end{aligned}$

(1, −1, 2)

46. $\begin{aligned} x - 2y + z &= 6 \\ x + 3y + z &= 16 \\ 3x - y - z &= 12 \end{aligned}$

(6, 2, 4)

47. $\begin{aligned} 3x + y &= 5 \\ 3y - z &= 2 \\ x + z &= 5 \end{aligned}$

(1, 2, 4)

48. $\begin{aligned} 2y + z &= 7 \\ 2x - z &= 3 \\ x - y &= 3 \end{aligned}$

(4, 1, 5)

49. $\begin{aligned} x - y + z &= 1 \\ 2x + 3y - z &= 3 \\ -x + 2y - 4z &= 4 \end{aligned}$

(2, −1, −2)

50. $\begin{aligned} 2x + y - 3z &= 7 \\ x - 2y + 3z &= 1 \\ 3x + 4y - 3z &= 13 \end{aligned}$

(3, 1, 0)

51. $\begin{aligned} 2x + 3z &= 5 \\ 3y + 2z &= 3 \\ 3x + 4y &= -10 \end{aligned}$

(−2, −1, 3)

52. $\begin{aligned} 3x + 4z &= 5 \\ 2y + 3z &= 2 \\ 2x - 5y &= 8 \end{aligned}$

(−1, −2, 2)

53. $\begin{aligned} 2x + 4y - 2z &= 3 \\ x + 3y + 4z &= 1 \\ x + 2y - z &= 4 \end{aligned}$

inconsistent

54. $\begin{aligned} x - 3y + 2z &= 1 \\ x - 2y + 3z &= 5 \\ 2x - 6y + 4z &= 3 \end{aligned}$

inconsistent

55. $\begin{aligned} 2x + y - z &= 5 \\ x + 3y + z &= 14 \\ 3x - y + 2z &= 1 \end{aligned}$

(1, 4, 1)

56. $\begin{aligned} 3x - y - 2z &= 11 \\ 2x + y - 2z &= 11 \\ x + 3y - z &= 8 \end{aligned}$

(2, 1, −3)

57. $\begin{aligned} 3x + y - 2z &= 2 \\ x + 2y + 3z &= 13 \\ 2x - 2y + 5z &= 6 \end{aligned}$

(1, 3, 2)

58. $\begin{aligned} 4x + 5y + z &= 6 \\ 2x - y + 2z &= 11 \\ x + 2y + 2z &= 6 \end{aligned}$

(2, −1, 3)

59. $\begin{aligned} 2x - y + z &= 6 \\ 3x + 2y + z &= 4 \\ x - 2y + 3z &= 12 \end{aligned}$

(1, −1, 3)

60. $\begin{aligned} 3x + 2y - 3z &= 8 \\ 2x + 3y + 2z &= 10 \\ x + y - z &= 2 \end{aligned}$

(6, −2, 2)

Quick Quiz (Objective 5.2B)

Solve by the addition method.

1. $\begin{aligned} x - 2y - z &= -5 \\ 5x - 4y + z &= 5 \\ 2x + 3y + z &= 9 \end{aligned}$

(1, 1, 4)

2. $\begin{aligned} x - y + 2z &= 3 \\ 3x + 2y + 2z &= 2 \\ 2x + 3y - z &= -4 \end{aligned}$

(−2, 1, 3)

Answers to Writing Exercises

73. Student descriptions should include the following steps:
(1) If necessary, multiply one or both of the equations by a constant so that the coefficients of one variable will be opposites.
(2) Add the two equations and solve for the variable.
(3) Substitute the value of the variable into either equation in the system and solve for the second variable.
(4) Write the ordered-pair solution.
(5) Check the solution.

74a. The system of equations has no solution; it is inconsistent. See Figures A, B, C, and D on page 289.

b. The system of equations has exactly one solution; it is an independent system whose solution is a point in space. See Figure E on page 289.

c. The system of equations has infinitely many solutions; it is a dependent system. See Figures F, G, and H on page 289.

61. $3x - 2y + 3z = -4$
$2x + y - 3z = 2$
$3x + 4y + 5z = 8$
(0, 2, 0)

62. $3x - 3y + 4z = 6$
$4x - 5y + 2z = 10$
$x - 2y + 3z = 4$
(0, −2, 0)

63. $3x - y + 2z = 2$
$4x + 2y - 7z = 0$
$2x + 3y - 5z = 7$
(1, 5, 2)

64. $2x + 2y + 3z = 13$
$-3x + 4y - z = 5$
$5x - 3y + z = 2$
(2, 3, 1)

65. $2x - 3y + 7z = 0$
$x + 4y - 4z = -2$
$3x + 2y + 5z = 1$
(−2, 1, 1)

66. $5x + 3y - z = 5$
$3x - 2y + 4z = 13$
$4x + 3y + 5z = 22$
(1, 1, 3)

APPLYING THE CONCEPTS

67. The point of intersection of the graphs of the equations $Ax + 2y = 2$ and $2x + By = 10$ is $(2, -2)$. Find A and B.
$A = 3, B = -3$

68. The point of intersection of the graphs of the equations $Ax - 4y = 9$ and $4x + By = -1$ is $(-1, -3)$. Find A and B.
$A = 3, B = -1$

69. For what value of k is the system of equations dependent?

a. $2x + 3y = 7$
$4x + 6y = k$ 14

b. $y = \frac{2}{3}x - 3$
$y = kx - 3$ $\frac{2}{3}$

c. $x = ky - 1$
$y = 2x + 2$ $\frac{1}{2}$

70. For what values of k is the system of equations independent?

a. $x + y = 7$
$kx + y = 3$ $k \neq 1$

b. $x + 2y = 4$
$kx + 3y = 2$ $k \neq \frac{3}{2}$

c. $2x + ky = 1$
$x + 2y = 2$ $k \neq 4$

71. Given that the graphs of the equations $2x - y = 6$, $3x - 4y = 4$, and $Ax - 2y = 0$ all intersect at the same point, find A.
1

72. Given that the graphs of the equations $3x - 2y = -2$, $2x - y = 0$, and $Ax + y = 8$ all intersect at the same point, find A.
2

73. Describe in your own words the process of solving a system of two linear equations in two variables by the addition method.

74. Explain, graphically, the following situations when they are related to a system of three linear equations in three variables.
a. The system of equations has no solution.
b. The system of equations has exactly one solution.
c. The system of equations has infinitely many solutions.

5.3 Solving Systems of Equations by Using Determinants

Objective A **To evaluate a determinant**

VIDEO & DVD CD TUTOR WEB SSM

Point of Interest

The word *matrix* was first used in a mathematical context in 1850. The root of this word is the Latin *mater,* which means "mother." A matrix was thought of as an object from which something else originates. The idea was that a determinant, discussed below, originated (was born) from a matrix. Today, matrices are one of the most widely used tools of applied mathematics.

A **matrix** is a rectangular array of numbers. Each number in the matrix is called an **element** of the matrix. The matrix at the right, with three rows and four columns, is called a 3 × 4 (read "3 by 4") matrix.

$$A = \begin{pmatrix} 1 & -3 & 2 & 4 \\ 0 & 4 & -3 & 2 \\ 6 & -5 & 4 & -1 \end{pmatrix}$$

A matrix of m rows and n columns is said to be of **order** $m \times n$. The matrix above has order 3 × 4. The notation a_{ij} refers to the element of a matrix in the ith row and the jth column. For matrix A, $a_{23} = -3$, $a_{31} = 6$, and $a_{13} = 2$.

A **square matrix** is one that has the same number of rows as columns. A 2 × 2 matrix and a 3 × 3 matrix are shown at the right.

$$\begin{pmatrix} -1 & 3 \\ 5 & 2 \end{pmatrix} \qquad \begin{pmatrix} 4 & 0 & 1 \\ 5 & -3 & 7 \\ 2 & 1 & 4 \end{pmatrix}$$

Associated with every square matrix is a number called its **determinant**.

TAKE NOTE

Note that vertical bars are used to represent the determinant and that parentheses are used to represent the matrix.

Determinant of a 2 × 2 Matrix

The **determinant** of a 2 × 2 matrix $\begin{pmatrix} a_{11} & a_{12} \\ a_{21} & a_{22} \end{pmatrix}$ is written $\begin{vmatrix} a_{11} & a_{12} \\ a_{21} & a_{22} \end{vmatrix}$. The value of this determinant is given by the formula

$$\begin{vmatrix} a_{11} & a_{12} \\ a_{21} & a_{22} \end{vmatrix} = a_{11}a_{22} - a_{12}a_{21}$$

➡ Find the value of the determinant $\begin{vmatrix} 3 & 4 \\ -1 & 2 \end{vmatrix}$.

$$\begin{vmatrix} 3 & 4 \\ -1 & 2 \end{vmatrix} = 3 \cdot 2 - 4(-1) = 6 - (-4) = 10$$

The value of the determinant is 10.

For a square matrix whose order is 3 × 3 or greater, the value of the determinant of that matrix is found by using 2 × 2 determinants.

The **minor of an element** in a 3 × 3 determinant is the 2 × 2 determinant that is obtained by eliminating the row and column that contain that element.

➡ Find the minor of −3 for the determinant $\begin{vmatrix} 2 & -3 & 4 \\ 0 & 4 & 8 \\ -1 & 3 & 6 \end{vmatrix}$.

The minor of −3 is the 2 × 2 determinant created by eliminating the row and column that contain −3.

Eliminate the row and column as shown: $\begin{vmatrix} 2 & -3 & 4 \\ 0 & 4 & 8 \\ -1 & 3 & 6 \end{vmatrix}$

The minor of −3 is $\begin{vmatrix} 0 & 8 \\ -1 & 6 \end{vmatrix}$.

Objective 5.3A

New Vocabulary

matrix
element of a matrix
order $m \times n$
square matrix
determinant
minor of an element
cofactor of an element of a matrix
expanding by cofactors

Discuss the Concepts

1. What is a matrix?
2. What is a square matrix?
3. What is the difference between the notation for a matrix and the notation for a determinant?
4. How do you find the minor of an element in a 3 × 3 determinant?

Concept Check

1. Find the value of the expression.

$$\begin{vmatrix} 4 & 1 & 3 \\ 5 & -1 & 4 \\ 6 & 4 & -3 \end{vmatrix} - \begin{vmatrix} -2 & 3 & 7 \\ 3 & -8 & 4 \\ 1 & 5 & 6 \end{vmatrix}$$

−190

2. Find the value of the expression.

$$\begin{vmatrix} 3 & 1 & -2 \\ 0 & 4 & 5 \\ 1 & 2 & -3 \end{vmatrix} + \begin{vmatrix} 2 & -1 & 5 \\ -3 & 0 & 4 \\ 1 & 3 & 2 \end{vmatrix}$$

−146

In-Class Examples (Objective 5.3A)

Evaluate the determinant.

1. $\begin{vmatrix} 3 & -2 \\ 4 & 5 \end{vmatrix}$
23

2. $\begin{vmatrix} 3 & 2 & 4 \\ 1 & -1 & 2 \\ 2 & 2 & 1 \end{vmatrix}$
7

3. $\begin{vmatrix} 1 & -2 & 1 \\ 2 & 1 & 2 \\ -1 & 0 & 3 \end{vmatrix}$
20

Instructor Note

Emphasize to students that expanding by cofactors does not depend on the row or column that is selected. Students should choose the row or column with the most zeros; that will reduce the amount of computation.

Instructor Note

Another method for expanding a 3×3 determinant is to reproduce the first two columns, find the sum of the products along the downward arrows, and subtract the sum of the products along the upward arrows. This is shown below.

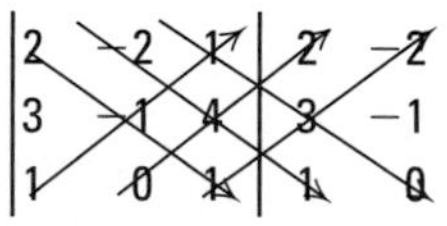

$[-2 + (-8) + 0] - [-1 + 0 + (-6)] = -3$

If you show students this method, be sure to mention that it works only for 3×3 determinants.

TAKE NOTE

The only difference between the cofactor and the minor is one of sign. The definition at the right can be stated symbolically as follows: If C_{ij} is the cofactor and M_{ij} is the minor of the matrix element a_{ij}, then $C_{ij} = (-1)^{i+j}M_{ij}$. If $i + j$ is an even number, then $(-1)^{i+j} = 1$ and $C_{ij} = M_{ij}$. If $i + j$ is an odd number, then $(-1)^{i+j} = -1$ and $C_{ij} = -M_{ij}$.

Cofactor of an Element of a Matrix

The **cofactor** of an element of a matrix is $(-1)^{i+j}$ times the minor of that element, where i is the row number of the element and j is the column number of the element.

➡ For the determinant $\begin{vmatrix} 3 & -2 & 1 \\ 2 & -5 & -4 \\ 0 & 3 & 1 \end{vmatrix}$, find the cofactor of -2 and of -5.

Because -2 is in the first row and the second column, $i = 1$ and $j = 2$. Thus $(-1)^{i+j} = (-1)^{1+2} = (-1)^3 = -1$. The cofactor of -2 is $(-1)\begin{vmatrix} 2 & -4 \\ 0 & 1 \end{vmatrix}$.

Because -5 is in the second row and the second column, $i = 2$ and $j = 2$. Thus $(-1)^{i+j} = (-1)^{2+2} = (-1)^4 = 1$. The cofactor of -5 is $1 \cdot \begin{vmatrix} 3 & 1 \\ 0 & 1 \end{vmatrix}$.

Note from this example that the cofactor of an element is -1 times the minor of that element or 1 times the minor of that element, depending on whether the sum $i + j$ is an odd or even integer.

The value of a 3×3 or larger determinant can be found by **expanding by cofactors** of *any* row or *any* column. The result of expanding by cofactors using the first row of a 3×3 matrix is shown below.

$$\begin{vmatrix} a_{11} & a_{12} & a_{13} \\ a_{21} & a_{22} & a_{23} \\ a_{31} & a_{32} & a_{33} \end{vmatrix} = a_{11}(-1)^{1+1}\begin{vmatrix} a_{22} & a_{23} \\ a_{32} & a_{33} \end{vmatrix} + a_{12}(-1)^{1+2}\begin{vmatrix} a_{21} & a_{23} \\ a_{31} & a_{33} \end{vmatrix} + a_{13}(-1)^{1+3}\begin{vmatrix} a_{21} & a_{22} \\ a_{31} & a_{32} \end{vmatrix}$$

$$= a_{11}\begin{vmatrix} a_{22} & a_{23} \\ a_{32} & a_{33} \end{vmatrix} - a_{12}\begin{vmatrix} a_{21} & a_{23} \\ a_{31} & a_{33} \end{vmatrix} + a_{13}\begin{vmatrix} a_{21} & a_{22} \\ a_{31} & a_{32} \end{vmatrix}$$

➡ Find the value of the determinant $\begin{vmatrix} 2 & -3 & 2 \\ 1 & 3 & -1 \\ 0 & -2 & 2 \end{vmatrix}$.

Expand by cofactors of the first row.

$$\begin{vmatrix} 2 & -3 & 2 \\ 1 & 3 & -1 \\ 0 & -2 & 2 \end{vmatrix} = 2\begin{vmatrix} 3 & -1 \\ -2 & 2 \end{vmatrix} - (-3)\begin{vmatrix} 1 & -1 \\ 0 & 2 \end{vmatrix} + 2\begin{vmatrix} 1 & 3 \\ 0 & -2 \end{vmatrix}$$

$$= 2(6 - 2) - (-3)(2 - 0) + 2(-2 - 0)$$

$$= 2(4) - (-3)(2) + 2(-2) = 8 - (-6) + (-4) = 10$$

To illustrate a statement made earlier, the value of this determinant will now be found by expanding by cofactors using the second column.

$$\begin{vmatrix} 2 & -3 & 2 \\ 1 & 3 & -1 \\ 0 & -2 & 2 \end{vmatrix} = -3(-1)^{1+2}\begin{vmatrix} 1 & -1 \\ 0 & 2 \end{vmatrix} + 3(-1)^{2+2}\begin{vmatrix} 2 & 2 \\ 0 & 2 \end{vmatrix} + (-2)(-1)^{3+2}\begin{vmatrix} 2 & 2 \\ 1 & -1 \end{vmatrix}$$

$$= -3(-1)\begin{vmatrix} 1 & -1 \\ 0 & 2 \end{vmatrix} + 3(1)\begin{vmatrix} 2 & 2 \\ 0 & 2 \end{vmatrix} + (-2)(-1)\begin{vmatrix} 2 & 2 \\ 1 & -1 \end{vmatrix}$$

$$= 3(2 - 0) + 3(4 - 0) + 2(-2 - 2)$$

$$= 3(2) + 3(4) + 2(-4) = 6 + 12 + (-8) = 10$$

Note that the value of the determinant is the same whether the first row or the second column is used to expand by cofactors. *Any row or column* can be used to evaluate a determinant by expanding by cofactors.

Example 1

Find the value of $\begin{vmatrix} 3 & -2 \\ 6 & -4 \end{vmatrix}$.

Solution

$\begin{vmatrix} 3 & -2 \\ 6 & -4 \end{vmatrix} = 3(-4) - (-2)(6) = -12 + 12 = 0$

The value of the determinant is 0.

You Try It 1

Find the value of $\begin{vmatrix} -1 & -4 \\ 3 & -5 \end{vmatrix}$.

Your solution

17

Example 2

Find the value of $\begin{vmatrix} -2 & 3 & 1 \\ 4 & -2 & 0 \\ 1 & -2 & 3 \end{vmatrix}$.

Solution

Expand by cofactors of the first row.

$$\begin{vmatrix} -2 & 3 & 1 \\ 4 & -2 & 0 \\ 1 & -2 & 3 \end{vmatrix}$$

$$= -2\begin{vmatrix} -2 & 0 \\ -2 & 3 \end{vmatrix} - 3\begin{vmatrix} 4 & 0 \\ 1 & 3 \end{vmatrix} + 1\begin{vmatrix} 4 & -2 \\ 1 & -2 \end{vmatrix}$$

$= -2(-6 - 0) - 3(12 - 0) + 1(-8 + 2)$
$= -2(-6) - 3(12) + 1(-6)$
$= 12 - 36 - 6$
$= -30$

The value of the determinant is -30.

You Try It 2

Find the value of $\begin{vmatrix} 1 & 4 & -2 \\ 3 & 1 & 1 \\ 0 & -2 & 2 \end{vmatrix}$.

Your solution

-8

Example 3

Find the value of $\begin{vmatrix} 0 & -2 & 1 \\ 1 & 4 & 1 \\ 2 & -3 & 4 \end{vmatrix}$.

Solution

$$\begin{vmatrix} 0 & -2 & 1 \\ 1 & 4 & 1 \\ 2 & -3 & 4 \end{vmatrix}$$

$$= 0\begin{vmatrix} 4 & 1 \\ -3 & 4 \end{vmatrix} - (-2)\begin{vmatrix} 1 & 1 \\ 2 & 4 \end{vmatrix} + 1\begin{vmatrix} 1 & 4 \\ 2 & -3 \end{vmatrix}$$

$= 0 - (-2)(4 - 2) + 1(-3 - 8)$
$= 2(2) + 1(-11)$
$= 4 - 11$
$= -7$

The value of the determinant is -7.

You Try It 3

Find the value of $\begin{vmatrix} 3 & -2 & 0 \\ 1 & 4 & 2 \\ -2 & 1 & 3 \end{vmatrix}$.

Your solution

44

Solutions on pp. S15–S16

Optional Student Activity

1. Solve for x.
   a. $\begin{vmatrix} 3 & 2 \\ 4 & x \end{vmatrix} = -11$
   b. $\begin{vmatrix} 1 & 0 & 2 \\ 4 & 3 & -1 \\ 0 & 2 & x \end{vmatrix} = -24$
   a. -1 b. -14

2a. Find the value of the determinant $\begin{vmatrix} x & x & a \\ y & y & b \\ z & z & c \end{vmatrix}$. 0

b. If two columns of a 3×3 matrix contain identical elements, what is the value of the determinant? 0

3. The area of a triangle with vertices (x_1, y_1), (x_2, y_2), and (x_3, y_3) can be given as one-half the absolute value of the determinant shown below. Find the area of the triangle whose vertices are $(-6, 3)$, $(1, 5)$, and $(4, -5)$.

$$\begin{vmatrix} x_1 & y_1 & 1 \\ x_2 & y_2 & 1 \\ x_3 & y_3 & 1 \end{vmatrix}$$

38 square units

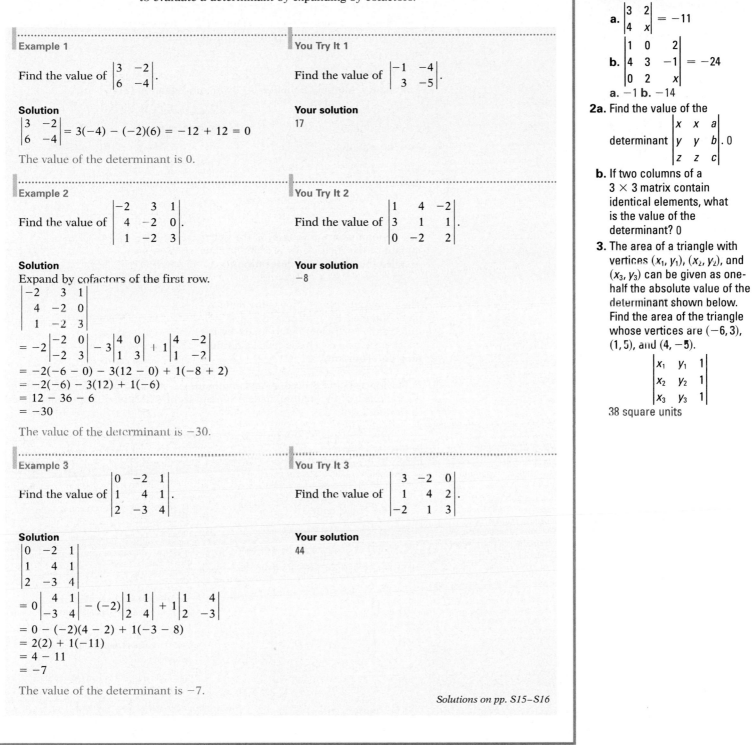

Objective 5.3B

New Vocabulary

coefficient determinant
numerator determinant

New Formulas

Cramer's Rule for a System of Two Equations in Two Variables
Cramer's Rule for a System of Three Equations in Three Variables

Discuss the Concepts

1. When can Cramer's Rule not be used to solve a system of equations?
2. If the determinant of the denominator is zero when using Cramer's Rule, the system of equations is either dependent or inconsistent. Explain how you can determine which it is.

Concept Check

Find the ordered pair of numbers (x, y) that satisfies the system

$$123x + 321y = 345$$
$$321x + 123y = 543$$

$\left(\frac{3}{2}, \frac{1}{2}\right)$

Objective B

To solve a system of equations by using Cramer's Rule

VIDEO & DVD CD TUTOR WEB SSM

The connection between determinants and systems of equations can be understood by solving a general system of linear equations.

Solve: (1) $a_1x + b_1y = c_1$
(2) $a_2x + b_2y = c_2$

Eliminate y. Multiply Equation (1) by b_2 and Equation (2) by $-b_1$.

$a_1b_2x + b_1b_2y = c_1b_2$ • b_2 times Equation (1).
$-a_2b_1x - b_1b_2y = -c_2b_1$ • $-b_1$ times Equation (2).

Add the equations.

$a_1b_2x - a_2b_1x = c_1b_2 - c_2b_1$
$(a_1b_2 - a_2b_1)x = c_1b_2 - c_2b_1$ • Solve for x, assuming $a_1b_2 - a_2b_1 \neq 0$.
$x = \dfrac{c_1b_2 - c_2b_1}{a_1b_2 - a_2b_1}$

The denominator $a_1b_2 - a_2b_1$ is the determinant of the coefficients of x and y. This is called the **coefficient determinant**.

$$a_1b_2 - a_2b_1 = \begin{vmatrix} a_1 & b_1 \\ a_2 & b_2 \end{vmatrix}$$

coefficients of x
coefficients of y

The numerator $c_1b_2 - c_2b_1$ is the determinant obtained by replacing the first column in the coefficient determinant by the constants c_1 and c_2. This is called a **numerator determinant**.

$$c_1b_2 - c_2b_1 = \begin{vmatrix} c_1 & b_1 \\ c_2 & b_2 \end{vmatrix}$$

constants of the equations

Following a similar procedure and eliminating x, it is also possible to express the y-component of the solution in determinant form. These results are summarized in Cramer's Rule.

Point of Interest

Cramer's Rule is named after Gabriel Cramer, who used it in a book he published in 1750. However, this rule was published in 1683 by the Japanese mathematician Seki Kown. That publication occurred seven years before Cramer's birth.

Cramer's Rule

The solution of the system of equations $\begin{matrix} a_1x + b_1y = c_1 \\ a_2x + b_2y = c_2 \end{matrix}$ is given by $x = \dfrac{D_x}{D}$ and $y = \dfrac{D_y}{D}$,

where $D = \begin{vmatrix} a_1 & b_1 \\ a_2 & b_2 \end{vmatrix}$, $D_x = \begin{vmatrix} c_1 & b_1 \\ c_2 & b_2 \end{vmatrix}$, $D_y = \begin{vmatrix} a_1 & c_1 \\ a_2 & c_2 \end{vmatrix}$, and $D \neq 0$.

➡ Solve by using Cramer's Rule: $3x - 2y = 1$
$2x + 5y = 3$

$D = \begin{vmatrix} 3 & -2 \\ 2 & 5 \end{vmatrix} = 19$ • Find the value of the coefficient determinant.

$D_x = \begin{vmatrix} 1 & -2 \\ 3 & 5 \end{vmatrix} = 11,\ D_y = \begin{vmatrix} 3 & 1 \\ 2 & 3 \end{vmatrix} = 7$ • Find the value of each of the numerator determinants.

$x = \dfrac{D_x}{D} = \dfrac{11}{19},\ y = \dfrac{D_y}{D} = \dfrac{7}{19}$ • Use Cramer's Rule to write the solution.

The solution is $\left(\frac{11}{19}, \frac{7}{19}\right)$.

In-Class Examples (Objective 5.3B)

Solve by using Cramer's Rule.

1. $3x + 5y = 8$
$5x - 7y = -2$
$(1, 1)$

2. $3x + 4y - 3z = 2$
$x - 2y + 3z = 6$
$2x + y - 3z = -5$
$(1, 2, 3)$

3. $x + 4y - 4z = 6$
$3x + 2y + 5z = -6$
$2x - 3y + 7z = -7$
$(2, -1, -2)$

301

A procedure similar to that followed for a system of two equations in two variables can be used to extend Cramer's Rule to a system of three equations in three variables.

Cramer's Rule for a System of Three Equations in Three Variables

The solution of the system of equations $\begin{aligned} a_1x + b_1y + c_1z &= d_1 \\ a_2x + b_2y + c_2z &= d_2 \\ a_3x + b_3y + c_3z &= d_3 \end{aligned}$ is given by

$$x = \frac{D_x}{D},\ y = \frac{D_y}{D},\ \text{and } z = \frac{D_z}{D},\ \text{where } D = \begin{vmatrix} a_1 & b_1 & c_1 \\ a_2 & b_2 & c_2 \\ a_3 & b_3 & c_3 \end{vmatrix},\ D_x = \begin{vmatrix} d_1 & b_1 & c_1 \\ d_2 & b_2 & c_2 \\ d_3 & b_3 & c_3 \end{vmatrix},$$

$$D_y = \begin{vmatrix} a_1 & d_1 & c_1 \\ a_2 & d_2 & c_2 \\ a_3 & d_3 & c_3 \end{vmatrix},\ D_z = \begin{vmatrix} a_1 & b_1 & d_1 \\ a_2 & b_2 & d_2 \\ a_3 & b_3 & d_3 \end{vmatrix},\ \text{and } D \neq 0.$$

➡ Solve by using Cramer's Rule: $\begin{aligned} 2x - y + z &= 1 \\ x + 3y - 2z &= -2 \\ 3x + y + 3z &= 4 \end{aligned}$

Find the value of the coefficient determinant.

$$D = \begin{vmatrix} 2 & -1 & 1 \\ 1 & 3 & -2 \\ 3 & 1 & 3 \end{vmatrix} = 2\begin{vmatrix} 3 & -2 \\ 1 & 3 \end{vmatrix} - (-1)\begin{vmatrix} 1 & -2 \\ 3 & 3 \end{vmatrix} + 1\begin{vmatrix} 1 & 3 \\ 3 & 1 \end{vmatrix}$$
$$= 2(11) + 1(9) + 1(-8)$$
$$= 23$$

Find the value of each of the numerator determinants.

$$D_x = \begin{vmatrix} 1 & -1 & 1 \\ -2 & 3 & -2 \\ 4 & 1 & 3 \end{vmatrix} = 1\begin{vmatrix} 3 & -2 \\ 1 & 3 \end{vmatrix} - (-1)\begin{vmatrix} -2 & -2 \\ 4 & 3 \end{vmatrix} + 1\begin{vmatrix} -2 & 3 \\ 4 & 1 \end{vmatrix}$$
$$= 1(11) + 1(2) + 1(-14)$$
$$= -1$$

$$D_y = \begin{vmatrix} 2 & 1 & 1 \\ 1 & -2 & -2 \\ 3 & 4 & 3 \end{vmatrix} = 2\begin{vmatrix} -2 & -2 \\ 4 & 3 \end{vmatrix} - 1\begin{vmatrix} 1 & -2 \\ 3 & 3 \end{vmatrix} + 1\begin{vmatrix} 1 & -2 \\ 3 & 4 \end{vmatrix}$$
$$= 2(2) - 1(9) + 1(10)$$
$$= 5$$

$$D_z = \begin{vmatrix} 2 & -1 & 1 \\ 1 & 3 & -2 \\ 3 & 1 & 4 \end{vmatrix} = 2\begin{vmatrix} 3 & -2 \\ 1 & 4 \end{vmatrix} - (-1)\begin{vmatrix} 1 & -2 \\ 3 & 4 \end{vmatrix} + 1\begin{vmatrix} 1 & 3 \\ 3 & 1 \end{vmatrix}$$
$$= 2(14) + 1(10) + 1(-8)$$
$$= 30$$

Use Cramer's Rule to write the solution.

$$x = \frac{D_x}{D} = \frac{-1}{23},\qquad y = \frac{D_y}{D} = \frac{5}{23},\qquad z = \frac{D_z}{D} = \frac{30}{23}$$

The solution is $\left(-\frac{1}{23}, \frac{5}{23}, \frac{30}{23}\right)$.

Optional Student Activity

Find the equation of a plane that contains the given points.

1. (2, 1, 1), (−1, 2, 12), (3, 2, 0)
   $3x - 2y + z = 5$
2. (1, −1, 5), (2, −2, 9), (−3, −1, −1)
   $\frac{3}{2}x - \frac{5}{2}y - z = -1$

Example 4

Solve by using Cramer's Rule:

$6x - 9y = 5$
$4x - 6y = 4$

Solution

$$D = \begin{vmatrix} 6 & -9 \\ 4 & -6 \end{vmatrix} = -36 + 36 = 0$$

Because $D = 0$, $\dfrac{D_x}{D}$ is undefined.

Therefore, the system is dependent or inconsistent.

It is not possible to solve this system by using Cramer's Rule.

You Try It 4

Solve by using Cramer's Rule:

$3x - y = 4$
$6x - 2y = 5$

Your solution

It is not possible to solve this system by using Cramer's Rule.

Example 5

Solve by using Cramer's Rule:

$3x - y + z = 5$
$x + 2y - 2z = -3$
$2x + 3y + z = 4$

Solution

$$D = \begin{vmatrix} 3 & -1 & 1 \\ 1 & 2 & -2 \\ 2 & 3 & 1 \end{vmatrix} = 28$$

$$D_x = \begin{vmatrix} 5 & -1 & 1 \\ -3 & 2 & -2 \\ 4 & 3 & 1 \end{vmatrix} = 28$$

$$D_y = \begin{vmatrix} 3 & 5 & 1 \\ 1 & -3 & -2 \\ 2 & 4 & 1 \end{vmatrix} = 0$$

$$D_z = \begin{vmatrix} 3 & -1 & 5 \\ 1 & 2 & -3 \\ 2 & 3 & 4 \end{vmatrix} = 56$$

$$x = \frac{D_x}{D} = \frac{28}{28} = 1, \quad y = \frac{D_y}{D} = \frac{0}{28} = 0,$$

$$z = \frac{D_z}{D} = \frac{56}{28} = 2$$

The solution is (1, 0, 2).

You Try It 5

Solve by using Cramer's Rule:

$2x - y + z = -1$
$3x + 2y - z = 3$
$x + 3y + z = -2$

Your solution

$\left(\frac{3}{7}, -\frac{1}{7}, -2\right)$

Solutions on p. S16

Section 5.3

Suggested Assignment

Exercises 3–33, odds

More challenging problems:
Exercises 36–39

Answers to Writing Exercises

1. The determinant associated with the 2 × 2 matrix $\begin{bmatrix} a & b \\ c & d \end{bmatrix}$ is $\begin{vmatrix} a & b \\ c & d \end{vmatrix}$. Its value is $ad - bc$.
2. The cofactor of a given element in a matrix is $(-1)^{i+j}$ times the minor of that element, where i is the row number of the element and j is the column number of the element.

5.3 Exercises

Objective A

1. How do you find the value of the determinant associated with a 2 × 2 matrix?

2. What is the cofactor of a given element in a matrix?

Evaluate the determinant.

3. $\begin{bmatrix} 2 & -1 \\ 3 & 4 \end{bmatrix}$ 11

4. $\begin{bmatrix} 5 & 1 \\ -1 & 2 \end{bmatrix}$ 11

5. $\begin{bmatrix} 6 & -2 \\ -3 & 4 \end{bmatrix}$ 18

6. $\begin{bmatrix} -3 & 5 \\ 1 & 7 \end{bmatrix}$ −26

7. $\begin{bmatrix} 3 & 6 \\ 2 & 4 \end{bmatrix}$ 0

8. $\begin{bmatrix} 5 & -10 \\ 1 & 2 \end{bmatrix}$ 0

9. $\begin{bmatrix} 1 & -1 & 2 \\ 3 & 2 & 1 \\ 1 & 0 & 4 \end{bmatrix}$ 15

10. $\begin{bmatrix} 4 & 1 & 3 \\ 2 & -2 & 1 \\ 3 & 1 & 2 \end{bmatrix}$ 3

11. $\begin{bmatrix} 3 & -1 & 2 \\ 0 & 1 & 2 \\ 3 & 2 & -2 \end{bmatrix}$ −30

12. $\begin{bmatrix} 4 & 5 & -2 \\ 3 & -1 & 5 \\ 2 & 1 & 4 \end{bmatrix}$ −56

13. $\begin{bmatrix} 4 & 2 & 6 \\ -2 & 1 & 1 \\ 2 & 1 & 3 \end{bmatrix}$ 0

14. $\begin{bmatrix} 3 & 6 & -3 \\ 4 & -1 & 6 \\ -1 & -2 & 3 \end{bmatrix}$ −54

Objective B

Solve by using Cramer's Rule.

15. $2x - 5y = 26$
$5x + 3y = 3$
(3, −4)

16. $3x + 7y = 15$
$2x + 5y = 11$
(−2, 3)

17. $x - 4y = 8$
$3x + 7y = 5$
(4, −1)

18. $5x + 2y = -5$
$3x + 4y = 11$
(−3, 5)

19. $2x + 3y = 4$
$6x - 12y = -5$
$\left(\frac{11}{14}, \frac{17}{21}\right)$

20. $5x + 4y = 3$
$15x - 8y = -21$
$\left(-\frac{3}{5}, \frac{3}{2}\right)$

21. $2x + 5y = 6$
$6x - 2y = 1$
$\left(\frac{1}{2}, 1\right)$

22. $7x + 3y = 4$
$5x - 4y = 9$
(1, −1)

23. $-2x + 3y = 7$
$4x - 6y = 9$
not independent

24. $9x + 6y = 7$
$3x + 2y = 4$
not independent

25. $2x - 5y = -2$
$3x - 7y = -3$
(−1, 0)

26. $8x + 7y = -3$
$2x + 2y = 5$
$\left(-\frac{41}{2}, 23\right)$

Quick Quiz (Objective 5.3A)

Evaluate the determinant.

1. $\begin{vmatrix} 5 & -1 \\ -2 & 3 \end{vmatrix}$
13

2. $\begin{vmatrix} 4 & -1 & 3 \\ 1 & 2 & 3 \\ 2 & 1 & -1 \end{vmatrix}$
−36

27. $2x - y + 3z = 9$
$x + 4y + 4z = 5$
$3x + 2y + 2z = 5$
(1, −1, 2)

28. $3x - 2y + z = 2$
$2x + 3y + 2z = -6$
$3x - y + z = 0$
(−1, −2, 1)

29. $3x - y + z = 11$
$x + 4y - 2z = -12$
$2x + 2y - z = -3$
(2, −2, 3)

30. $x + 2y + 3z = 8$
$2x - 3y + z = 5$
$3x - 4y + 2z = 9$
(−3, −2, 5)

31. $4x - 2y + 6z = 1$
$3x + 4y + 2z = 1$
$2x - y + 3z = 2$
not independent

32. $x - 3y + 2z = 1$
$2x + y - 2z = 3$
$3x - 9y + 6z = -3$
not independent

33. $5x - 4y + 2z = 4$
$3x - 5y + 3z = -4$
$3x + y - 5z = 12$
$\left(\frac{68}{25}, \frac{56}{25}, -\frac{8}{25}\right)$

34. $2x + 4y + z = 7$
$x + 3y - z = 1$
$3x + 2y - 2z = 5$
$\left(\frac{53}{19}, -\frac{1}{19}, \frac{31}{19}\right)$

35. $3x - 2y + 2z = 5$
$6x + 3y - 4z = -1$
$3x - y + 2z = 4$
$\left(\frac{2}{3}, -1, \frac{1}{2}\right)$

APPLYING THE CONCEPTS

36. Determine whether the following statements are always true, sometimes true, or never true.
a. The determinant of a matrix is a positive number. sometimes true
b. A determinant can be evaluated by expanding about any row or column of the matrix. always true
c. Cramer's Rule can be used to solve a system of linear equations in three variables. sometimes true

37. Surveyors use a formula to find the area of a plot of land. The **surveyor's area formula** states that if the vertices $(x_1, y_1), (x_2, y_2), \ldots, (x_n, y_n)$ of a simple polygon are listed counterclockwise around the perimeter, then the area of the polygon is

$$A = \frac{1}{2}\left\{\begin{bmatrix} x_1 & x_2 \\ y_1 & y_2 \end{bmatrix} + \begin{bmatrix} x_2 & x_3 \\ y_2 & y_3 \end{bmatrix} + \begin{bmatrix} x_3 & x_4 \\ y_3 & y_4 \end{bmatrix} + \ldots + \begin{bmatrix} x_n & x_1 \\ y_n & y_1 \end{bmatrix}\right\}$$

Use the surveyor's area formula to find the area of the polygon with vertices (9, −3), (26, 6), (18, 21), (16, 10), and (1, 11). Measurements are given in feet. 239 ft^2

Complete.

38. **a.** If all the elements in one row or one column of a 2×2 matrix are zeros, the value of the determinant of the matrix is ________. 0
b. If all the elements in one row or one column of a 3×3 matrix are zeros, the value of the determinant of the matrix is ________. 0

39. **a.** The value of the determinant $\begin{vmatrix} x & x & a \\ y & y & b \\ z & z & c \end{vmatrix}$ is ________. 0
b. If two columns of a 3×3 matrix contain identical elements, the value of the determinant is ________. 0

Quick Quiz (Objective 5.3B)
Solve by using Cramer's Rule.

1. $4x - 3y = 7$
$3x - 5y = 8$
(1, −1)

2. $x + 2y + 2z = 5$
$2x - y + 2z = 11$
$4x + 5y + z = 9$
(3, −1, 2)

5.4 Application Problems in Two Variables

Objective A **To solve rate-of-wind or rate-of-current problems**

VIDEO & DVD CD TUTOR WEB SSM

Motion problems that involve an object moving with or against a wind or current normally require two variables to solve.

➡ A motorboat traveling with the current can go 24 mi in 2 h. Against the current, it takes 3 h to go the same distance. Find the rate of the motorboat in calm water and the rate of the current.

Strategy for Solving Rate-of-Wind or Rate-of-Current Problems

1. Choose one variable to represent the rate of the object in calm conditions and a second variable to represent the rate of the wind or current. Using these variables, express the rate of the object with and against the wind or current. Use the equation $rt = d$ to write expressions for the distance traveled by the object. The results can be recorded in a table.

TAKE NOTE

The boat travels faster when it travels with the current than when it travels against the current. Note that the boat's rate is *increased* by the rate of the current when it is traveling with the current, and the boat's rate is *decreased* by the rate of the current when it is traveling against the current.

Rate of the boat in calm water: x
Rate of the current: y

	Rate	·	Time	=	Distance
With the current	$x + y$	·	2	=	$2(x + y)$
Against the current	$x - y$	·	3	=	$3(x - y)$

2. Determine how the expressions for distance are related.

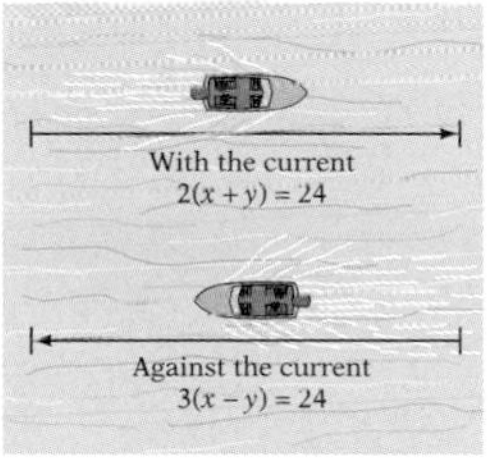

The distance traveled with the current is 24 mi: $2(x + y) = 24$
The distance traveled against the current is 24 mi: $3(x - y) = 24$

$2(x + y) = 24$
$3(x - y) = 24$ • Solve this system of equations.

$x + y = 12$ • Divide each side of the first equation by 2.
$x - y = 8$ • Divide each side of the second equation by 3.

$2x = 20$ • Add the equations.
$x = 10$ • Solve the equation for x.

$x + y = 12$
$10 + y = 12$ • Replace x by 10 in the equation $x + y = 12$ and solve for y.
$y = 2$

The rate of the boat in calm water is 10 mph.
The rate of the current is 2 mph.

Objective 5.4A

Formulas to Review

Rate · time = distance

Discuss the Concepts

In rate-of-wind and rate-of-current problems, how does the wind or current affect the rate of the object? (You want students to understand that when traveling with the wind or current, the speed of an object increases; traveling against the wind or current decreases the speed of the object.)

Concept Check

1. The speed of a plane is 500 mph. There is a headwind of 50 mph. What is the speed of the plane relative to an observer on the ground? 450 mph
2. The rate of a current in a river is x mph, and the rate of a boat in still water is y mph. How can you represent the rate of the boat going down the river? $y + x$ How can you represent the rate of the boat going up the river? $y - x$

In-Class Examples (Objective 5.4A)

1. Flying with the wind, a small plane flew 540 mi in 3 h. Against the wind, the plane could fly only 420 mi in the same amount of time. Find the rate of the plane in calm air and the rate of the wind. Plane: 160 mph; wind: 20 mph
2. A cabin cruiser traveling with the current went 60 mi in 3 h. Against the current, it took 5 h to travel the same distance. Find the rate of the cabin cruiser in calm water and the rate of the current. Cruiser: 16 mph; current: 4 mph

Optional Student Activity

A plane is flying 3500 mi from New York City to London. The speed of the plane in calm air is 375 mph, and there is a 50-mph tailwind. The *point of no return* is the point at which the flight time required to return to New York City is the same as the flight time required to travel on to London. For this flight, how far from New York is the point of no return? Round to the nearest whole number. 1517 mi

Example 1

Flying with the wind, a plane flew 1000 mi in 5 h. Flying against the wind, the plane could fly only 500 mi in the same amount of time. Find the rate of the plane in calm air and the rate of the wind.

Strategy

- Rate of the plane in calm air: p
  Rate of the wind: w

	Rate	*Time*	*Distance*
With wind	$p + w$	5	$5(p + w)$
Against wind	$p - w$	5	$5(p - w)$

- The distance traveled with the wind is 1000 mi.
  The distance traveled against the wind is 500 mi.

Solution

$$5(p + w) = 1000$$
$$5(p - w) = 500$$

$$p + w = 200$$
$$p - w = 100$$
• Divide each side of each equation by 5.

$$2p = 300$$
$$p = 150$$

$$p + w = 200$$
$$150 + w = 200$$
$$w = 50$$

The rate of the plane in calm air is 150 mph. The rate of the wind is 50 mph.

You Try It 1

A rowing team rowing with the current traveled 18 mi in 2 h. Against the current, the team rowed 10 mi in 2 h. Find the rate of the rowing team in calm water and the rate of the current.

Your strategy

Your solution

rate of the rowing team in calm water: 7 mph;
rate of the current: 2 mph

Solution on p. S16

Objective B **To solve application problems using two variables**

VIDEO & DVD CD TUTOR WEB SSM

The application problems in this section are varieties of problems solved earlier in the text. Each of the strategies for the problems in this section will result in a system of equations.

➡ A store owner purchased twenty 60-watt light bulbs and 30 fluorescent bulbs for a total cost of $80. A second purchase, at the same prices, included thirty 60-watt light bulbs and 10 fluorescent bulbs for a total cost of $50. Find the cost of a 60-watt bulb and that of a fluorescent bulb.

Objective 5.4B

Formulas to Review

Amount · unit cost = value
Principal · interest rate = interest earned
Number · value = total value

Instructor Note

The weights of the blocks described in the Point of Interest at the left are $\frac{35}{8}$ shequels and $\frac{33}{8}$ shequels.

Concept Check

1. A contractor bought 50 yd of nylon carpet for x dollars per yard and 100 yd of wool carpet for y dollars per yard. How can you represent the total cost of the carpet? $50x + 100y$
2. A coin bank contains only nickels and dimes. Using n for the number of nickels and d for the number of dimes, represent the total value of the coins in the bank. $5n + 10d$

Point of Interest

The Babylonians had a method for solving a system of equations. Here is an adaptation of a problem from an ancient (around 1500 B.C.) Babylonian text. "There are two silver blocks. The sum of $\frac{1}{7}$ of the first block and $\frac{1}{11}$ of the second block is one sheqel (a weight). The first block diminished by $\frac{1}{7}$ of its weight equals the second diminished by $\frac{1}{11}$ of its weight. What are the weights of the two blocks?"

Strategy for Solving an Application Problem in Two Variables

1. Choose one variable to represent one of the unknown quantities and a second variable to represent the other unknown quantity. Write numerical or variable expressions for all the remaining quantities. These results can be recorded in tables.

Cost of a 60-watt bulb: b
Cost of a fluorescent bulb: f

First purchase

	Amount	·	Unit Cost	=	Value
60-watt	20	·	b	=	$20b$
Fluorescent	30	·	f	=	$30f$

Second purchase

	Amount	·	Unit Cost	=	Value
60-watt	30	·	b	=	$30b$
Fluorescent	10	·	f	=	$10f$

2. Determine a system of equations. The strategies presented in previous chapters can be used to determine the relationships between the expressions in the tables.

The total of the first purchase was \$80: $20b + 30f = 80$
The total of the second purchase was \$50: $30b + 10f = 50$

Solve the system of equations: (1) $20b + 30f = 80$
(2) $30b + 10f = 50$

$$\begin{aligned} 60b + 90f &= 240 \quad &\text{• 3 times Equation (1)}\\ -60b - 20f &= -100 \quad &\text{• −2 times Equation (2)}\\ 70f &= 140 \\ f &= 2 \end{aligned}$$

Replace f by 2 in Equation (1). Solve for b.

$$\begin{aligned} 20b + 30f &= 80 \\ 20b + 30(2) &= 80 \\ 20b + 60 &= 80 \\ 20b &= 20 \\ b &= 1 \end{aligned}$$

The cost of a 60-watt bulb was \$1.00.
The cost of a fluorescent bulb was \$2.00.

In-Class Examples (Objective 5.4B)

1. A coin bank contains only quarters and dimes. The total value of the coins in the bank is \$10.40. If the dimes were quarters and the quarters were dimes, the total value of the coins would be \$9.20. Find the number of dimes in the bank. 24 dimes
2. A sheet metal shop ordered 60 lb of tin and 30 lb of a zinc alloy for a total cost of \$960. A second purchase, at the same prices, included 40 lb of tin and 70 lb of zinc alloy. The total cost was \$1540. Find the total cost per pound of the tin and the zinc alloy. Tin: \$7/lb; zinc: \$18/lb

Optional Student Activity

An artist is creating a mobile from which three objects will be suspended from a light rod that is 18 in. long, as shown below. The weight, in ounces, of each object is shown in the diagram. For the mobile to balance, the objects must be positioned such that $w_1d_1 + w_2d_2 = w_3d_3$. The artist wants d_1 to be 1.5 times d_2. Find the distances d_1, d_2, and d_3 such that the mobile will balance. $d_1 = 9$ in., $d_2 = 6$ in., $d_3 = 9$ in.

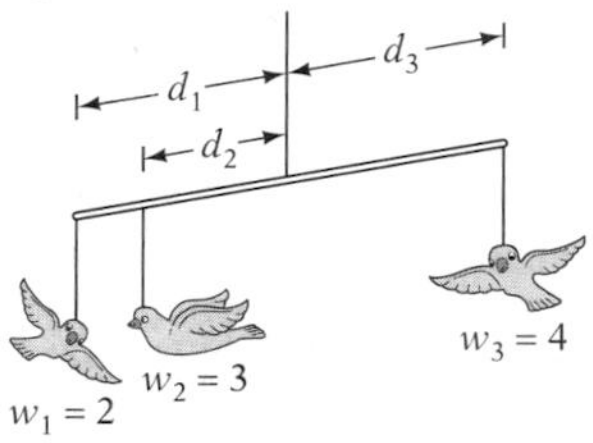

Example 2

A total of 260 tickets were sold for a softball game. Adult tickets sold for $6 each, and children's tickets sold for $2 each. If the total receipts were $1220, how many adult tickets and how many children's tickets were sold?

Strategy

- Number of adult tickets sold: *A*
  Number of children's tickets sold: *C*

Ticket	*Price*	*Number Sold*	*Total Receipts*
Adult	6	*A*	6*A*
Children	2	*C*	2*C*

- The total number of tickets sold was 260.
  The total receipts were $1220.

Solution

$$A + C = 260$$
$$6A + 2C = 1220$$

$$-2A - 2C = -520$$
$$6A + 2C = 1220$$
$$4A = 700$$
$$A = 175$$

$$A + C = 260$$
$$175 + C = 260$$
$$C = 85$$

There were 175 adult tickets sold.
There were 85 children's tickets sold.

You Try It 2

An investment adviser invested $15,000 in two accounts. One investment earned 8% annual simple interest. The other investment earned 7% annual simple interest. The total interest in one year was $1170. How much was invested in each account?

Your strategy

Your solution

$12,000 at 8%;
$3000 at 7%

Solution on pp. S16–S17

Section 5.4

Suggested Assignment
Exercises 1–27, odds
More challenging problems:
Exercises 29, 30

5.4 Exercises

Objective A *Application Problems*

1. A motorboat traveling with the current went 36 mi in 2 h. Against the current, it took 3 h to travel the same distance. Find the rate of the boat in calm water and the rate of the current.
rate of the motorboat in calm water: 15 mph;
rate of the current: 3 mph

2. A cabin cruiser traveling with the current went 45 mi in 3 h. Against the current, it took 5 h to travel the same distance. Find the rate of the cabin cruiser in calm water and the rate of the current.
rate of the cabin cruiser in calm water: 12 mph;
rate of the current: 3 mph

3. A jet plane flying with the wind went 2200 mi in 4 h. Against the wind, the plane could fly only 1820 mi in the same amount of time. Find the rate of the plane in calm air and the rate of the wind.
rate of the plane in calm air: 502.5 mph;
rate of the wind: 47.5 mph

4. Flying with the wind, a small plane flew 300 mi in 2 h. Against the wind, the plane could fly only 270 mi in the same amount of time. Find the rate of the plane in calm air and the rate of the wind.
rate of the plane in calm air: 142.5 mph;
rate of the wind: 7.5 mph

5. A rowing team rowing with the current traveled 20 km in 2 h. Rowing against the current, the team rowed 12 km in the same amount of time. Find the rate of the rowing team in calm water and the rate of the current.
rate of the team in calm water: 8 km/h;
rate of the current: 2 km/h

6. A motorboat traveling with the current went 72 km in 3 h. Against the current, the boat could go only 48 km in the same amount of time. Find the rate of the boat in calm water and the rate of the current.
rate of the boat in calm water: 20 km/h;
rate of the current: 4 km/h

7. A turbo-prop plane flying with the wind flew 800 mi in 4 h. Flying against the wind, the plane required 5 h to travel the same distance. Find the rate of the wind and the rate of the plane in calm air.
rate of the plane in calm air: 180 mph;
rate of the wind: 20 mph

8. Flying with the wind, a pilot flew 600 mi between two cities in 4 h. The return trip against the wind took 5 h. Find the rate of the plane in calm air and the rate of the wind.
rate of the plane in calm air: 135 mph;
rate of the wind: 15 mph

Quick Quiz (Objective 5.4A)

1. A jet plane flying with the wind went 2600 mi in 5 h. Against the wind, the plane could fly only 2200 mi in the same amount of time. Find the rate of the plane in calm air and the rate of the wind.
Plane: 480 mph; wind: 40 mph
2. A motorboat traveling with the current went 80 mi in 4 h. Against the current, it took 5 h to travel the same distance. Find the rate of the motorboat in calm water and the rate of the current.
Motorboat: 18 mph; current: 2 mph

310 Chapter 5 / Systems of Linear Equations and Inequalities

9. A plane flying with a tailwind flew 600 mi in 5 h. Against the wind, the plane required 6 h to fly the same distance. Find the rate of the plane in calm air and the rate of the wind.
rate of the plane in calm air: 110 mph;
rate of the wind: 10 mph

10. Flying with the wind, a plane flew 720 mi in 3 h. Against the wind, the plane required 4 h to fly the same distance. Find the rate of the plane in calm air and the rate of the wind.
rate of the plane in calm air: 210 mph;
rate of the wind: 30 mph

11. A motorboat traveling with the current went 48 mi in 3 h. Against the current, it took 4.8 h to travel the same distance. Find the rate of the boat in calm water and the rate of the current.
rate of the boat in calm water: 13 mph;
rate of the current: 3 mph

12. A plane traveling with the wind flew 3625 mi in 6.25 h. Against the wind, the plane required 7.25 h to fly the same distance. Find the rate of the plane in calm air and the rate of the wind.
rate of the plane in calm air: 540 mph;
rate of the wind: 40 mph

13. A cabin cruiser traveling with the current went 45 mi in 2.5 h. Against the current, the boat could go only 30 mi in the same amount of time. Find the rate of the cabin cruiser and the rate of the current.
rate of the cruiser in calm water: 15 mph;
rate of the current: 3 mph

14. Flying with the wind, a plane flew 450 mi in 3 h. Against the wind, the plane could fly only 270 mi in the same amount of time. Find the rate of the plane in calm air and the rate of the wind.
rate of the plane in calm air: 120 mph;
rate of the wind: 30 mph

Objective B *Application Problems*

15. A merchant mixed 10 lb of a cinnamon tea with 5 lb of spice tea. The 15-pound mixture cost \$40. A second mixture included 12 lb of the cinnamon tea and 8 lb of the spice tea. The 20-pound mixture cost \$54. Find the cost per pound of the cinnamon tea and the spice tea.
cinnamon tea: \$2.50/lb;
spice tea: \$3.00/lb

16. A carpenter purchased 60 ft of redwood and 80 ft of pine for a total cost of \$286. A second purchase, at the same prices, included 100 ft of redwood and 60 ft of pine for a total cost of \$396. Find the cost per foot of redwood and the cost per foot of pine.
pine: \$1.10 per foot;
redwood: \$3.30 per foot

Quick Quiz (Objective 5.4B)

1. The total value of the quarters and nickels in a coin bank is \$7. If the quarters were nickels and the nickels were quarters, the total value of the coins would be \$11. Find the number of quarters in the bank. 20 quarters
2. A restaurant manager buys 120 lb of hamburger and 60 lb of steak for a total cost of \$720. A second purchase, at the same prices, includes 200 lb of hamburger and 80 lb of steak. The total cost is \$1080. Find the cost of 1 lb of steak. \$6/lb

311

17. A contractor buys 16 yd of nylon carpet and 20 yd of wool carpet for \$1840. A second purchase, at the same prices, includes 18 yd of nylon carpet and 25 yd of wool carpet for \$2200. Find the cost per yard of the wool carpet.
\$52/yd

18. During one month, a homeowner used 500 units of electricity and 100 units of gas, for a total cost of \$352. The next month, 400 units of electricity and 150 units of gas were used, for a total cost of \$304. Find the cost per unit of gas.
\$.32

19. A company manufactures both mountain bikes and trail bikes. The cost of materials for a mountain bike is \$70, and the cost of materials for a trail bike is \$50. The cost of labor to manufacture a mountain bike is \$80, and the cost of labor to manufacture a trail bike is \$40. During a week in which the company has budgeted \$2500 for materials and \$2600 for labor, how many mountain bikes does the company plan to manufacture?
25 mountain bikes

20. A company manufactures both color and black-and-white monitors. The cost of materials for a black-and-white monitor is \$40, and the cost of materials for a color monitor is \$160. The cost of labor to manufacture a black-and-white monitor is \$60, and the cost of labor to manufacture a color monitor is \$100. During a week in which the company has budgeted \$8400 for materials and \$5600 for labor, how many color monitors does the company plan to manufacture?
50 color monitors

21. You have a total of \$5000 invested in two simple interest accounts. On one account, the annual simple interest rate is 6%. On the second account, the annual simple interest rate is 8%. The total annual interest earned in one year is \$370. How much is invested in each account?
\$1500 at 6%; \$3500 at 8%

22. Two investments earn total annual interest of \$970. One investment is in a 7.5% annual simple interest account. The other investment is in a 9.5% annual simple interest account. The total in the two investments is \$12,000. How much is invested in each account?
\$8500 at 7.5%; \$3500 at 9.5%

23. Two angles are complementary. The larger angle is 9° more than eight times the measure of the smaller angle. Find the measure of the two angles. (Complementary angles are two angles whose sum is 90°.)
9°, 81°

24. Two angles are supplementary. The larger angle is 40° more than three times the measure of the smaller angle. Find the measure of the two angles. (Supplementary angles are two angles whose sum is 180°.)
35°, 145°

25. On Monday, a computer-manufacturing company sent out three shipments. The first order, which contained a bill for $114,000, was for 4 Model II, 6 Model VI, and 10 Model IX computers. The second shipment, which contained a bill for $72,000, was for 8 Model II, 3 Model VI, and 5 Model IX computers. The third shipment, which contained a bill for $81,000, was for 2 Model II, 9 Model VI, and 5 Model IX computers. What does the manufacturer charge for each Model VI computer?
$4000

26. A relief organization supplies blankets, cots, and lanterns to victims of fires, floods, and other natural disasters. One week the organization purchased 15 blankets, 5 cots, and 10 lanterns for a total cost of $1250. The next week, at the same prices, the organization purchased 20 blankets, 10 cots, and 15 lanterns for a total cost of $2000. The next week, at the same prices, the organization purchased 10 blankets, 15 cots, and 5 lanterns for a total cost of $1625. Find the cost of one blanket, the cost of one cot, and the cost of one lantern.
$25 per blanket; $75 per cot; $50 per lantern

27. A science museum charges $10 for a regular admission ticket, but members receive a discount of $3 and students are admitted for $5. Last Saturday, 750 tickets were sold for a total of $5400. If 20 more student tickets than regular tickets were sold, how many of each type of ticket were sold?
190 regular; 210 student; 350 member-discount

28. An investor has a total of $25,000 deposited in three different accounts, which earn annual interest rates of 8%, 6%, and 4%. The amount deposited in the 8% account is twice the amount in the 6% account. If the three accounts earn total annual interest of $1520, how much money is deposited in each account?
$10,400 at 8%; 5200 at 6%; 9400 at 4%

APPLYING THE CONCEPTS

29. The sum of the digits of a two-digit number equals $\frac{1}{7}$ of the number. If the digits of the number are reversed, the new number is equal to 36 less than the original number. Find the original number.
84

30. The sum of the digits of a two-digit number equals $\frac{1}{5}$ of the number. If the digits of the number are reversed, the new number is equal to 9 more than the original number. Find the original number.
45

313

5.5 Solving Systems of Linear Inequalities

Objective A

To graph the solution set of a system of linear inequalities

VIDEO & DVD CD TUTOR WEB SSM

Two or more inequalities considered together are called a **system of inequalities.** The **solution set of a system of inequalities** is the intersection of the solution sets of the individual inequalities. To graph the solution set of a system of inequalities, first graph the solution set of each inequality. The solution set of the system of inequalities is the region of the plane represented by the intersection of the shaded areas.

TAKE NOTE

You can use a test point to check that the correct region has been denoted as the solution set. We can see from the graph that the point (2, 4) is in the solution set, and as shown below, it is a solution of each inequality in the system. This indicates that the solution set as graphed is correct.

$$2x - y \le 3$$
$$2(2) - (4) \le 3$$
$$0 \le 3 \quad \text{True}$$

$$3x + 2y > 8$$
$$3(2) + 2(4) > 8$$
$$14 > 8 \quad \text{True}$$

➡ Graph the solution set: $2x - y \le 3$
$3x + 2y > 8$

Solve each inequality for y.

$$2x - y \le 3 \qquad\qquad 3x + 2y > 8$$
$$-y \le -2x + 3 \qquad\qquad 2y > -3x + 8$$
$$y \ge 2x - 3 \qquad\qquad y > -\frac{3}{2}x + 4$$

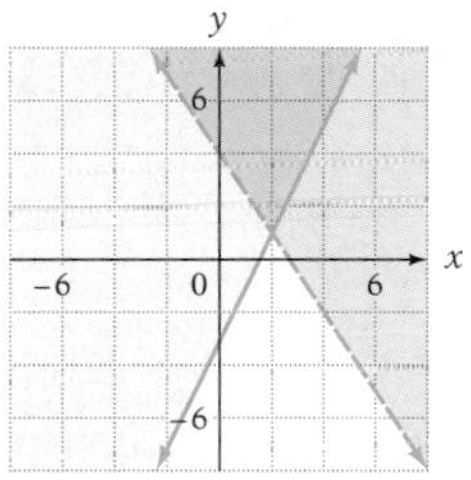

Graph $y = 2x - 3$ as a solid line. Because the inequality is $\ge$, shade above the line.

Graph $y = -\frac{3}{2}x + 4$ as a dashed line. Because the inequality is $>$, shade above the line.

T

The solution set of the system is the region of the plane represented by the intersection of the solution sets of the individual inequalities.

➡ Graph the solution set: $-x + 2y \ge 4$
$x - 2y \ge 6$

Solve each inequality for y.

$$-x + 2y \ge 4 \qquad\qquad x - 2y \ge 6$$
$$2y \ge x + 4 \qquad\qquad -2y \ge -x + 6$$
$$y \ge \frac{1}{2}x + 2 \qquad\qquad y \le \frac{1}{2}x - 3$$

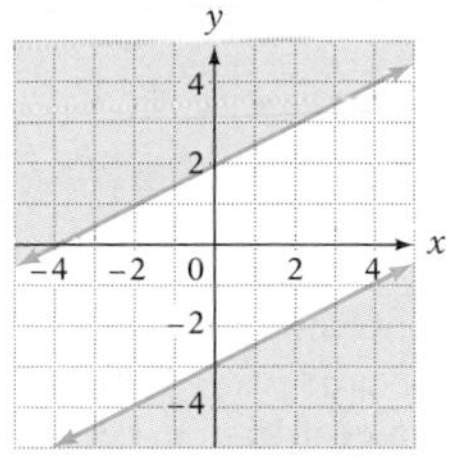

Shade above the solid line $y = \frac{1}{2}x + 2$.

Shade below the solid line $y = \frac{1}{2}x - 3$.

Because the solution sets of the two inequalities do not intersect, the solution of the system is the empty set.

Point of Interest

Large systems of linear inequalities containing over 100 inequalities have been used to solve application problems in such diverse areas as providing health care and hardening a nuclear missile silo.

Objective 5.5A

New Vocabulary

system of inequalities
solution set of a system of inequalities

Vocabulary to Review

inequality in two variables
half-plane

Instructor Note

Have students sketch the solution set of one inequality with parallel lines like //// and the solution set of the second inequality with parallel lines like \\\\. The area where the lines intersect (if they do intersect) is the solution set of the system of equations.

Discuss the Concepts

Determine whether the following statements are always true, sometimes true, or never true.

1. The solution set of a system of linear inequalities is an infinite set. Sometimes true
2. The solution set of a system of linear inequalities is the empty set. Sometimes true
3. The solution set of a system of linear inequalities is the intersection of the solution sets of the individual inequalities. Always true

In-Class Examples (Objective 5.5A)

Graph the solution set.

1. $3x - 2y < 12$
$x + 3y < 6$

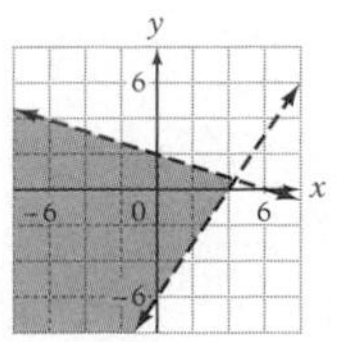

2. $x + 2y \le 4$
$x - y \le 3$

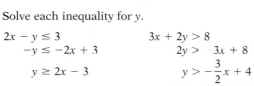

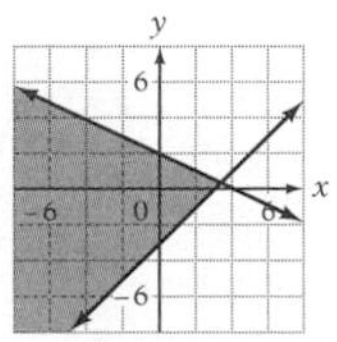

314

Concept Check

1. What must be true if an ordered pair is a solution of a system of linear inequalities? It must be a solution of every inequality in the system.
2. The set of points that satisfy the system

$$2x - y < 0$$
$$x + y > 3$$

is contained entirely in which quadrants? I and II

Optional Student Activity

Write a system of inequalities to represent the shaded region.

1. 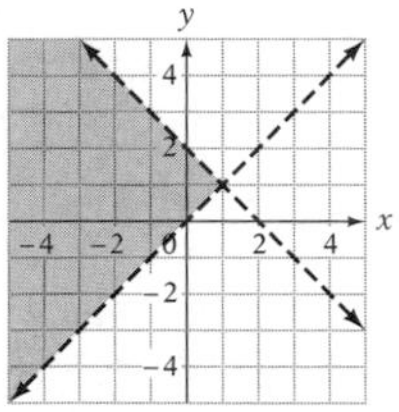

$y > x$
$y < -x + 2$

2. 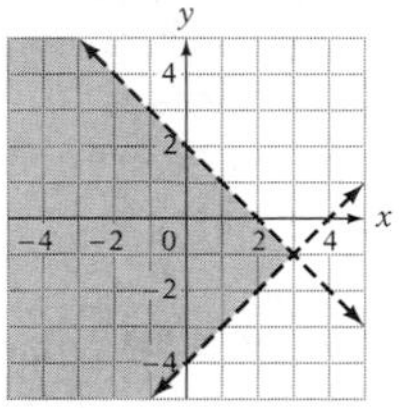

$x + y < 2$
$x - y < 4$

3. 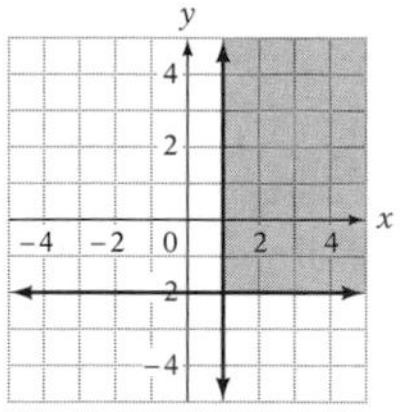

$y \geq -2$
$x \geq 1$

4. 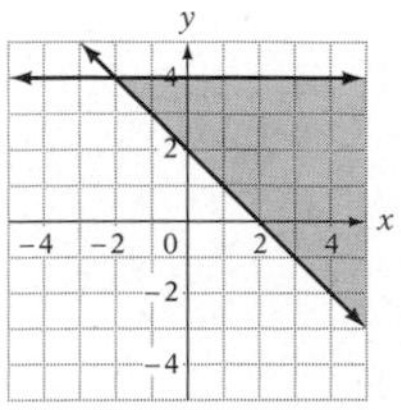

$y \leq 4$
$y \geq -x + 2$

Example 1

Graph the solution set: $y \geq x - 1$
$y < -2x$

Solution

Shade above the solid line $y = x - 1$.
Shade below the dashed line $y = -2x$.

The solution of the system is the intersection of the solution sets of the individual inequalities.

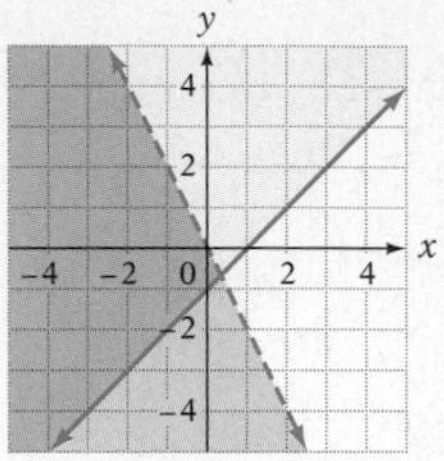

You Try It 1

Graph the solution set: $y \geq 2x - 3$
$y > -3x$

Your solution

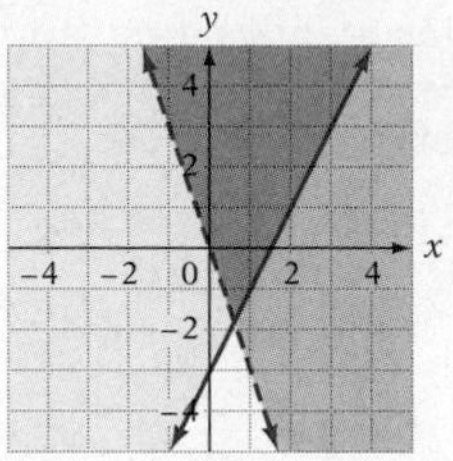

Example 2

Graph the solution set: $2x + 3y > 9$
$y < -\frac{2}{3}x + 1$

Solution

$$2x + 3y > 9$$
$$3y > -2x + 9$$
$$y > -\frac{2}{3}x + 3$$

Graph above the dashed line $y = -\frac{2}{3}x + 3$.

Graph below the dashed line $y = -\frac{2}{3}x + 1$.

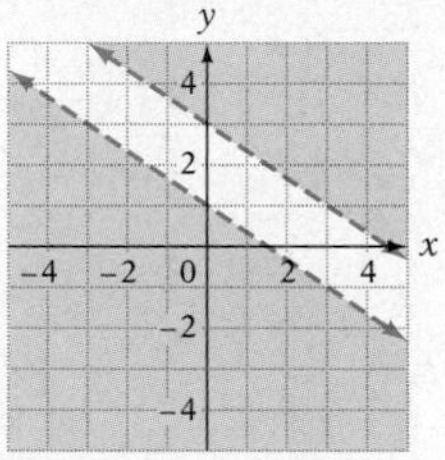

The intersection of the system is the empty set, because the solution sets of the two inequalities do not intersect.

You Try It 2

Graph the solution set: $3x + 4y > 12$
$y < \frac{3}{4}x - 1$

Your solution

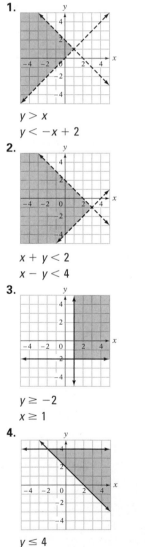

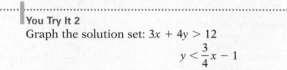

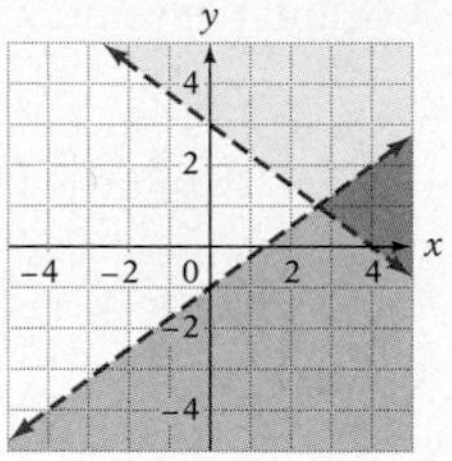

Solutions on p. S17

315

5.5 Exercises

Objective A

Graph the solution set.

1. $x - y \geq 3$
$x + y \leq 5$

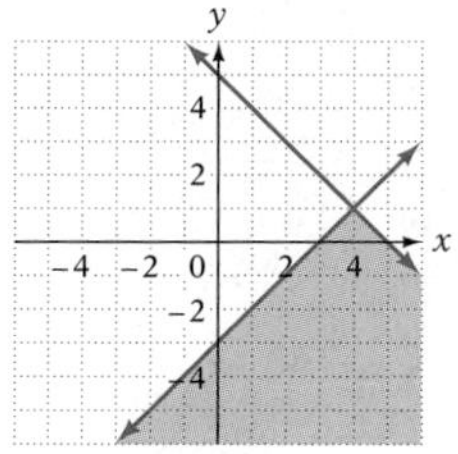

2. $2x - y < 4$
$x + y < 5$

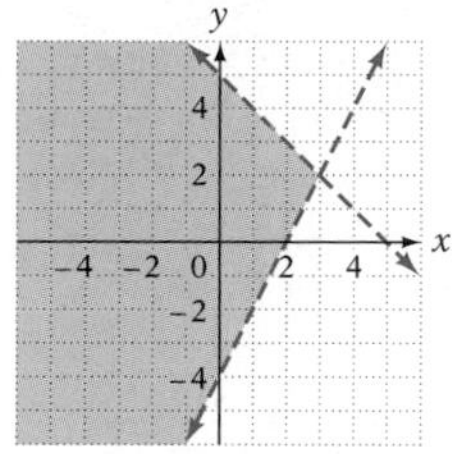

3. $3x - y < 3$
$2x + y \geq 2$

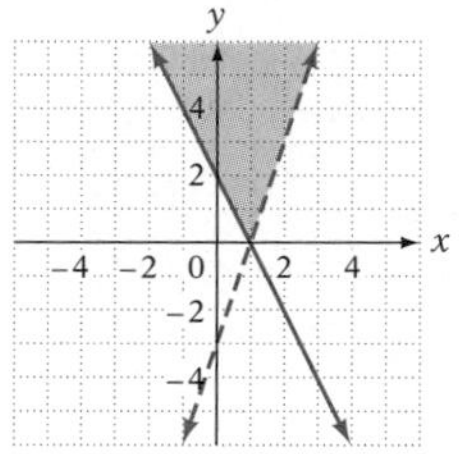

4. $x + 2y \leq 6$
$x - y \leq 3$

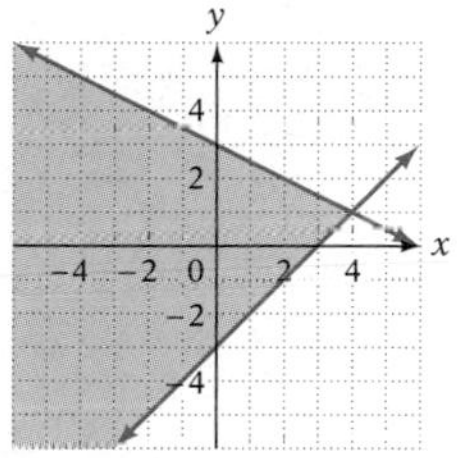

5. $2x + y \geq -2$
$6x + 3y \leq 6$

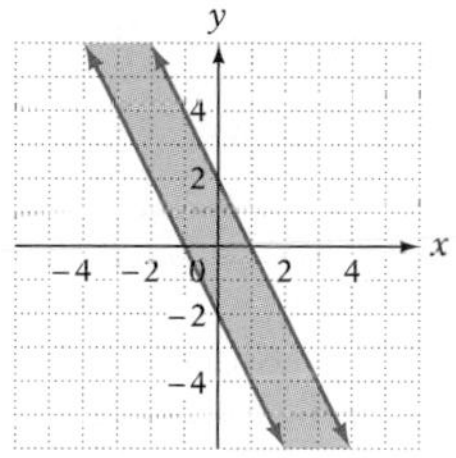

6. $x + y \geq 5$
$3x + 3y \leq 6$

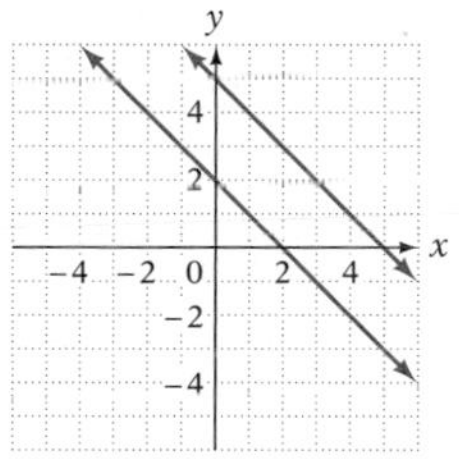

7. $3x - 2y < 6$
$y \leq 3$

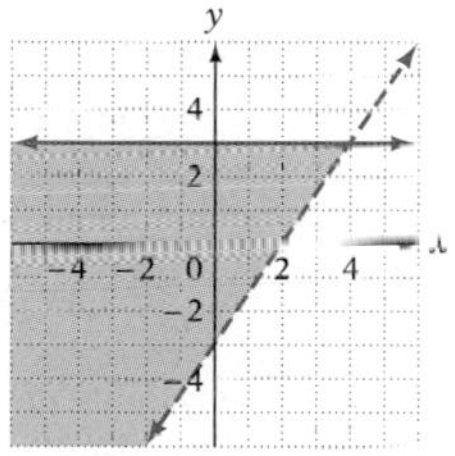

8. $x \leq 2$
$3x + 2y > 4$

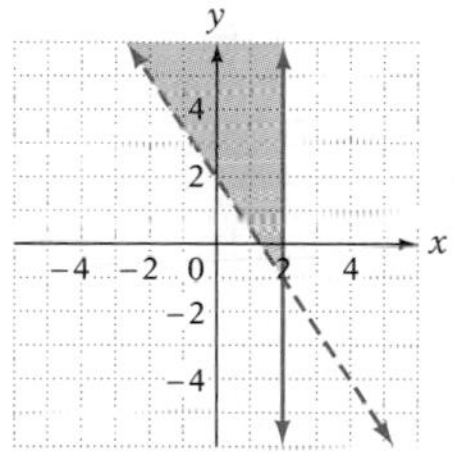

9. $y > 2x - 6$
$x + y < 0$

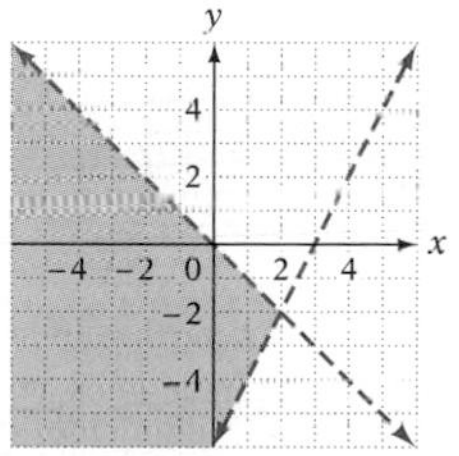

10. $x < 3$
$y < -2$

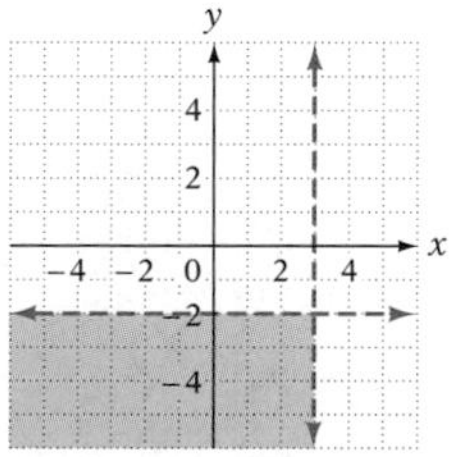

11. $x + 1 \geq 0$
$y - 3 \leq 0$

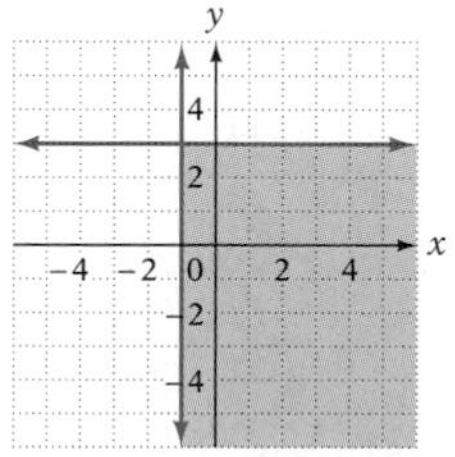

12. $5x - 2y \geq 10$
$3x + 2y \geq 6$

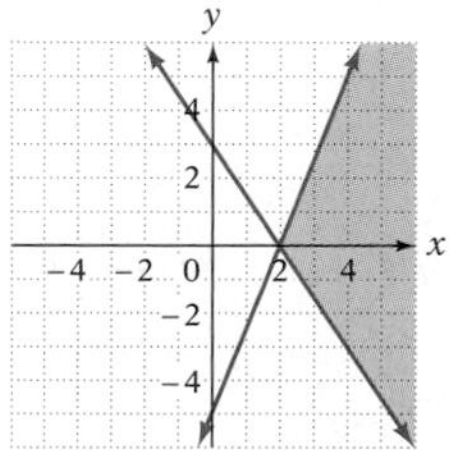

Section 5.5

Suggested Assignment

Exercises 1–17, odds

More challenging problems:
Exercises 19–23, odds

Quick Quiz (Objective 5.5A)

Graph the solution set.

1. $x + 2y \geq 4$
$4x + 3y < 12$

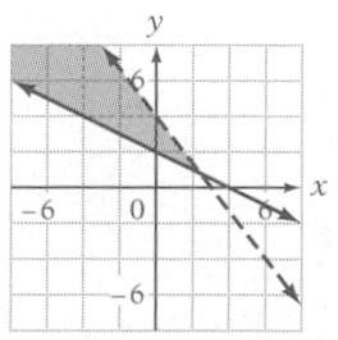

2. $x - 3y \leq 6$
$2x + y \leq 4$

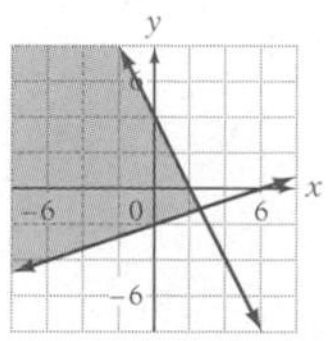

316

13. $2x + y \geq 4$
$3x - 2y < 6$

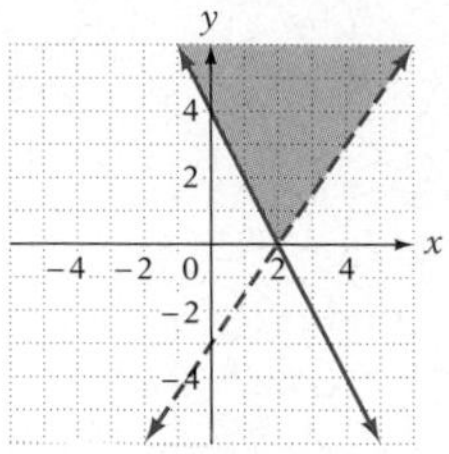

14. $3x - 4y < 12$
$x + 2y < 6$

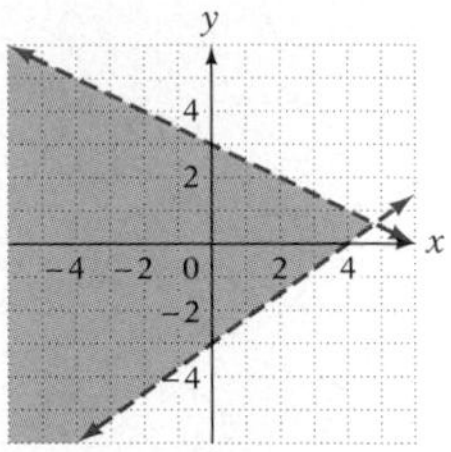

15. $x - 2y \leq 6$
$2x + 3y \leq 6$

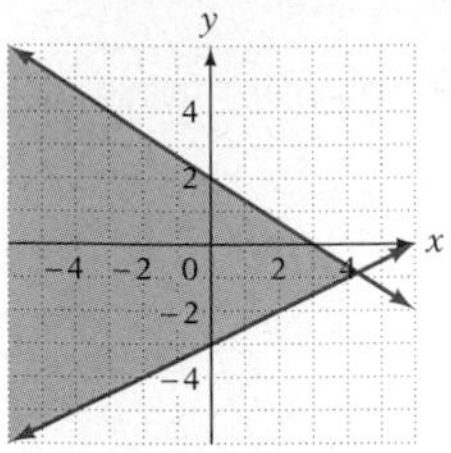

16. $x - 3y > 6$
$2x + y > 5$

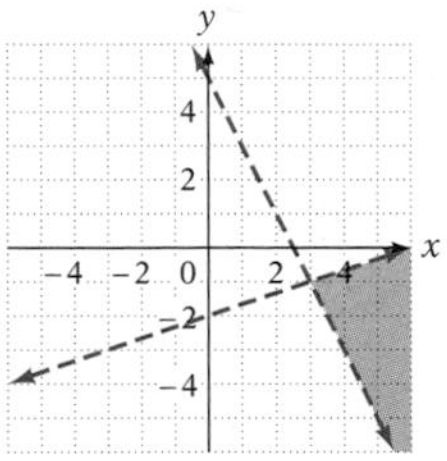

17. $x - 2y \leq 4$
$3x + 2y \leq 8$

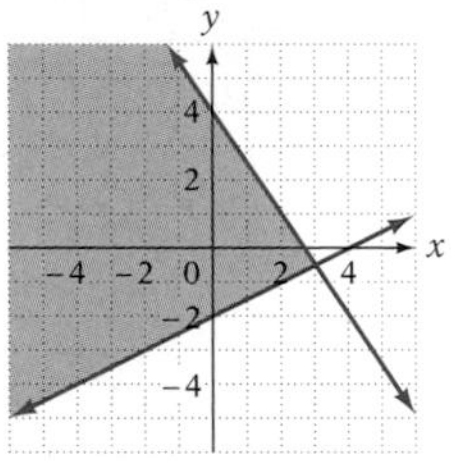

18. $3x - 2y < 0$
$5x + 3y > 9$

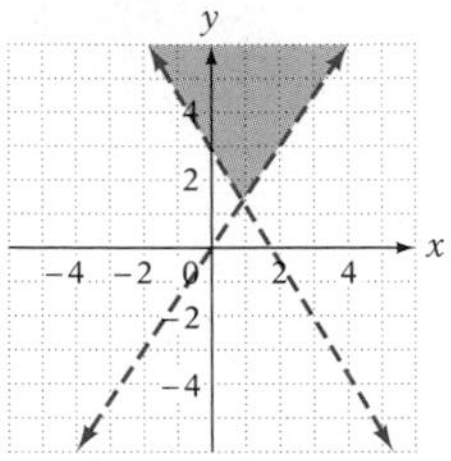

APPLYING THE CONCEPTS

Graph the solution set.

19. $2x + 3y \leq 15$
$3x - y \leq 6$
$y \geq 0$

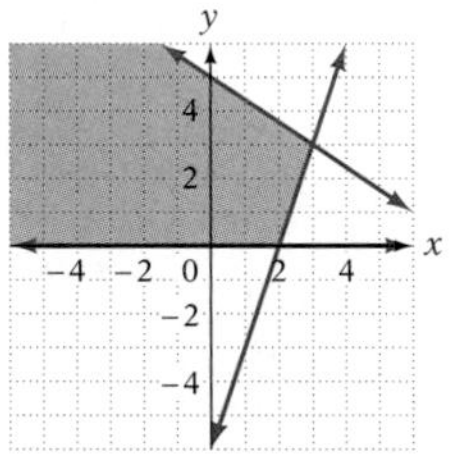

20. $x + y \leq 6$
$x - y \leq 2$
$x \geq 0$

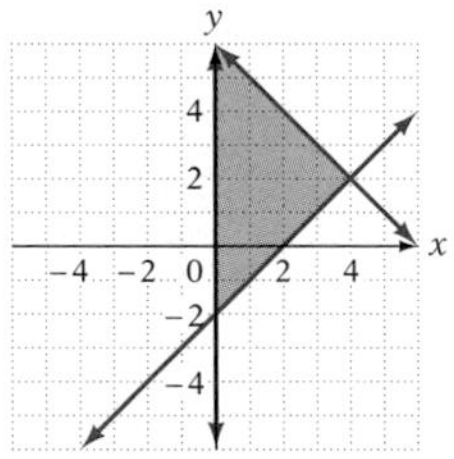

21. $x - y \leq 5$
$2x - y \geq 6$
$y \geq 0$

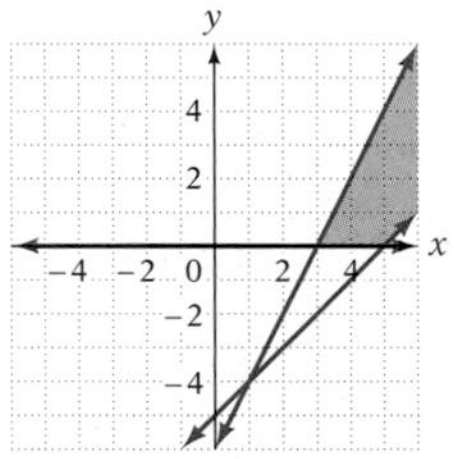

22. $x - 3y \leq 6$
$5x - 2y \geq 4$
$y \geq 0$

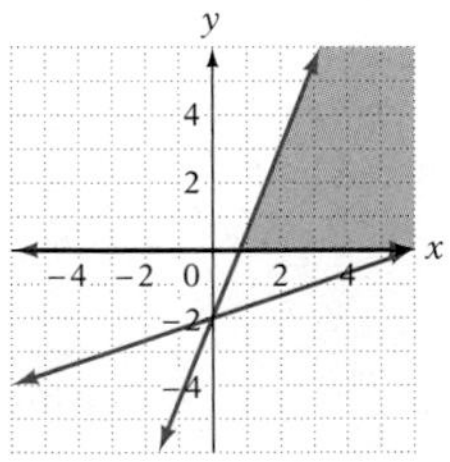

23. $2x - y \leq 4$
$3x + y < 1$
$y \leq 0$

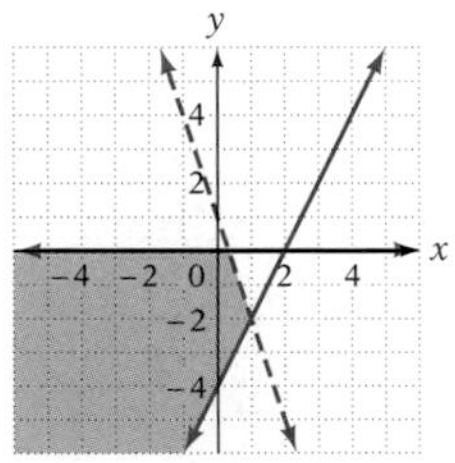

24. $x - y \leq 4$
$2x + 3y > 6$
$x \geq 0$

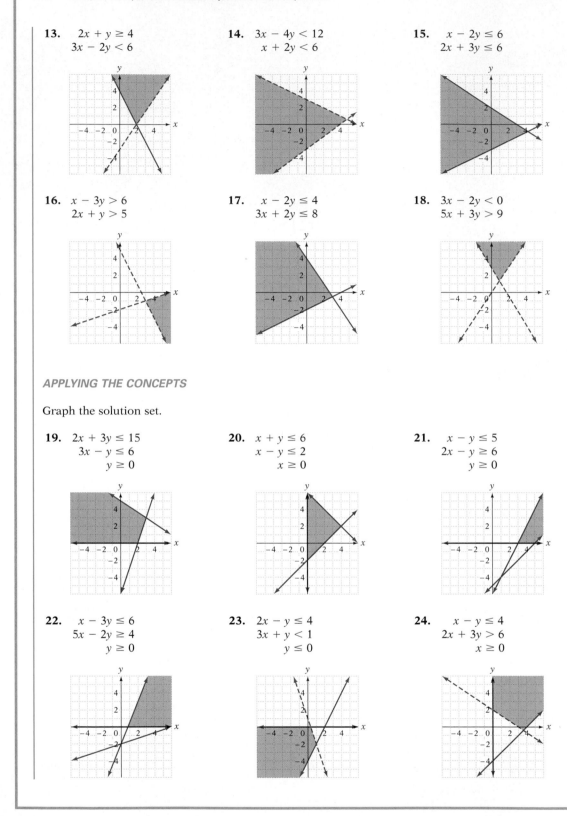

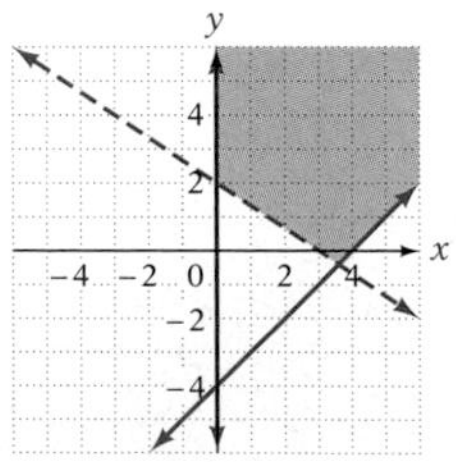

317

Focus on Problem Solving

Solve an Easier Problem

Suppose you are in charge of your softball league, which consists of 15 teams. You must devise a schedule in which each team plays every other team once. How many games must be scheduled?

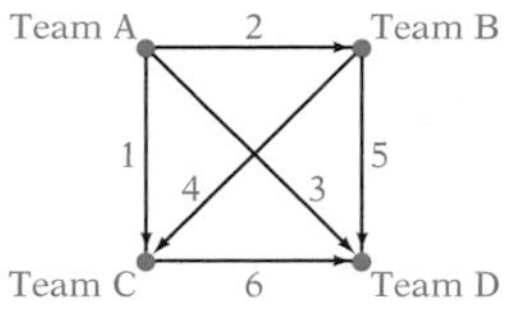

To solve this problem, we will attempt an easier problem first. Suppose that your league contains only a small number of teams. For instance, if there were only 1 team, you would schedule 0 games. If there were 2 teams, you would schedule 1 game. If there were 3 teams, you would schedule 3 games. The diagram at the left shows that 6 games must be scheduled when there are 4 teams in the league.

Here is a table of our results so far. (Remember that making a table is a strategy to be used in problem solving.)

Number of Teams	*Number of Games*	*Possible Pattern*
1	0	0
2	1	1
3	3	1 + 2
4	6	1 + 2 + 3

1. Draw a diagram with 5 dots to represent the teams. Draw lines from each dot to a second dot, and determine the number of games required.
2. What is the apparent pattern for the number of games required?
3. Assuming that the pattern continues, how many games must be scheduled for the 15 teams of the original problem?

After solving a problem, good problem solvers ask whether it is possible to solve the problem in a different manner. Here is a possible alternative method of solving the scheduling problem.

Begin with one of the 15 teams (say team A) and ask, "How many games must this team play?" Because there are 14 teams left to play, you must schedule 14 games. Now move to team B. It is already scheduled to play team A, and it does not play itself, so there are 13 teams left for it to play. Consequently, you must schedule 14 + 13 games.

4. Continue this reasoning for the remaining teams and determine the number of games that must be scheduled. Does this answer correspond to the answer you obtained using the first method?
5. Visit www.sports.com to get lists of professional sports teams, divisions, etc. Find the strategy used to create the schedule of your favorite team.

WEB

Projects and Group Activities

Using a Graphing Calculator to Solve a System of Equations

A graphing calculator can be used to solve a system of equations. For this procedure to work on most calculators, it is necessary that the point of intersection be on the screen. This means that you may have to experiment with Xmin, Xmax, Ymin, and Ymax values until the graphs intersect on the screen.

Answers to Focus on Problem Solving: Solve an Easier Problem

1. 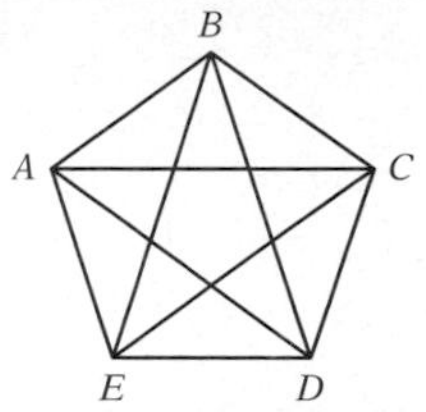

From the diagram, the number of games required if each of 5 teams plays every other team once is 10.
2. For n teams, the number of games is the sum of the first $n - 1$ whole numbers.
3. 105 games
4. The total number of games that must be scheduled is 105 games. This is the same result obtained using the first method.
5. Answers will vary.

318

To solve a system of equations graphically, solve each equation for y. Then graph the equations of the system. Their point of intersection is the solution.

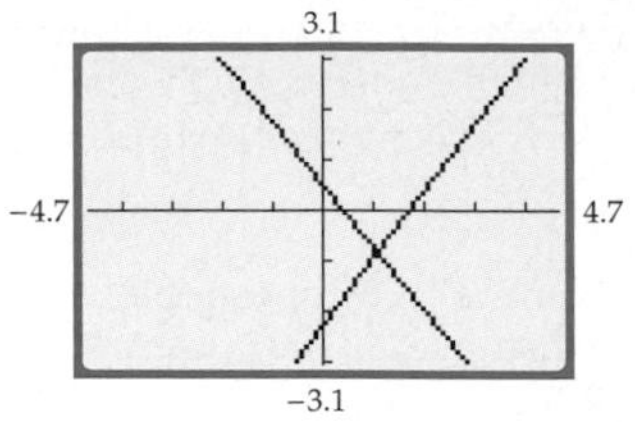

For instance, to solve the system of equations

$$4x - 3y = 7$$
$$5x + 4y = 2$$

first solve each equation for y.

$$4x - 3y = 7 \Rightarrow y = \frac{4}{3}x - \frac{7}{3}$$

$$5x + 4y = 2 \Rightarrow y = -\frac{5}{4}x + \frac{1}{2}$$

The keystrokes needed to solve this system using a TI-83 are given below. We are using the viewing window [−4.7, 4.7] by [−3.1, 3.1]. The approximate solution is (1.096774, −0.870968).

[Y=] [CLEAR] 4 [X, T, θ, n] [÷] 3 [−]
7 [÷] 3 [ENTER] [CLEAR]
[(−)] 5 [X, T, θ, n] [÷] 4 [+] 1 [÷]
2 [GRAPH]

Once the calculator has drawn the graphs, use the TRACE feature and move the cursor to the approximate point of intersection. This will give you an approximate solution of the system of equations. A more accurate solution can be found by using the following keystrokes.

[2nd] [CALC] 5 [ENTER] [ENTER] [ENTER]

Some of the exercises in the first section of this chapter asked you to solve a system of equations by graphing. Try those exercises again, this time using your graphing calculator.

Current models of calculators do not allow you to solve graphically a system of equations in three variables. However, these calculators do have matrix and determinant operations that can be used to solve these systems.

Solving a First-Degree Equation with a Graphing Calculator

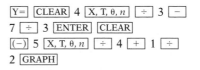

A first-degree equation in one variable can be solved graphically by using a graphing calculator. The idea is to rewrite the equation as a system of equations and then solve the system of equations as illustrated above. For instance, to solve the equation

$$2x + 1 = 5x - 5$$

write the equation as the following system of equations.

$$y = 2x + 1$$
$$y = 5x - 5$$

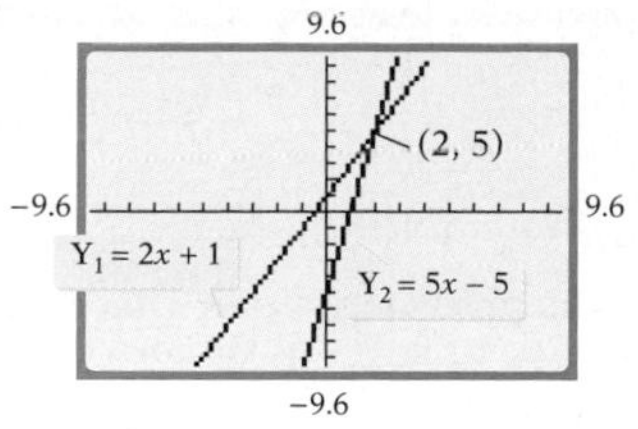

Note that we have used the left and right sides of the original equation to form the system of equations.

Now graph the equations and find the point of intersection. This is shown at the left using a viewing window of [−9.6, 9.6] by [−9.6, 9.6]. The solution of the original equation, $2x + 1 = 5x - 5$, is the x-coordinate of the point of intersection. Thus the solution of the equation is 2.

Recall that not all equations have a solution. Consider the equation

$$x - 3(x + 2) = 2x - 2(2x - 2)$$

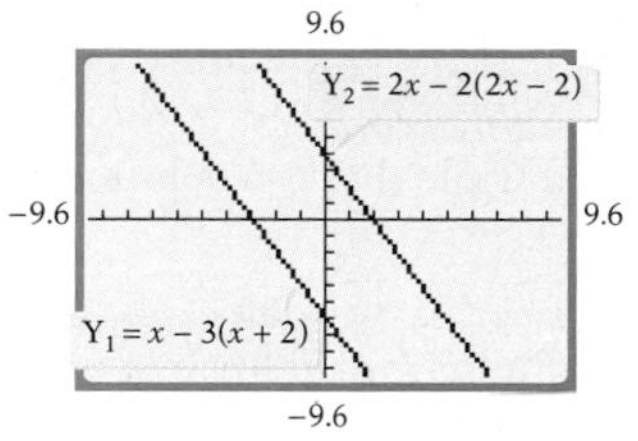

The graphs of the left and right sides of the equation are shown at the left. Note that the lines appear to be parallel and therefore do not intersect. Since the lines do not intersect, there is no solution of the system of equations and thus no solution of the original equation. We algebraically verify this result below.

$$x - 3(x + 2) = 2x - 2(2x - 2)$$
$$x - 3x - 6 = 2x - 4x + 4 \quad \text{• Use the Distributive Property.}$$
$$-2x - 6 = -2x + 4 \quad \text{• Simplify.}$$
$$-6 = 4 \quad \text{• Add } 2x \text{ to each side of the equation.}$$

Because $-6 = 4$ is not a true equation, there is no solution.

Note that in the third line of the solution we obtained the expressions $-2x - 6$ and $-2x + 4$. Consider the graphs of the equations $y = -2x - 6$ and $y = -2x + 4$; the slopes of the lines are equal (both are −2), and therefore the graphs of the lines are parallel.

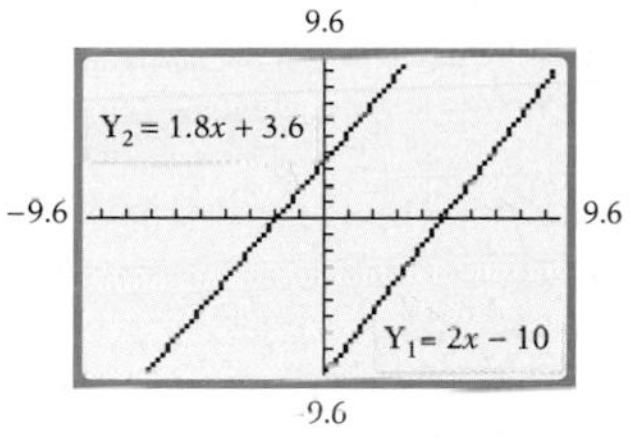

Before we leave graphical solutions, a few words of caution. Graphs can be deceiving and appear not to intersect when they do, and appear to intersect when they do not. For instance, an attempt to graphically solve

$$2x - 10 = 1.8x + 3.6$$

is shown at the left.

If you use the viewing window [−9.6, 9.6] by [−9.6, 9.6], the graphs will appear to be parallel. However, the graphs of $y = 2x - 10$ and $y = 1.8x + 3.6$ do not have the same slope and therefore must intersect. For this equation, you need a larger viewing window to see the point of intersection.

Solve the following equations by using a graphing calculator.

1. $3x - 1 = 5x + 1$
2. $3x + 2 = 4$
3. $3 + 2(2x - 4) = 5(x - 3)$
4. $2x - 4 = 5x - 3(x + 2) + 2$
5. Explain how problem 4 relates to an identity as explained in Section 2.1.
6. Find an appropriate viewing window so that you can determine the solution of the equation $2x - 10 = 1.8x + 3.6$ given above. What is the solution?

Answers to Projects and Group Activities: Solving a First-Degree Equation with a Graphing Calculator

1. −1
2. $\frac{2}{3}$
3. 10
4. All real numbers
5. Explanations will vary. For example, any replacement for x will result in a true equation.
6. [67.5, 68.5] by [123, 127]
   The solution is 68.

Chapter Summary

Key Words

Equations considered together are called a *system of equations*. [p. 275]

A *solution of a system of equations* in two variables is an ordered pair that is a solution of each equation of the system. [p. 275]

When the graphs of a system of equations intersect at only one point, the system is called an *independent system of equations*. When the graphs of a system of equations coincide, the system is called a *dependent system of equations*. When a system of equations has no solution, it is called an *inconsistent system of equations*. [pp. 275–276]

An equation of the form $Ax + By + Cz = D$, where A, B, and C are coefficients of the variables and D is a constant, is a *linear equation in three variables*. A *solution of a system of equations in three variables* is an ordered triple that is a solution of each equation of the system. [pp. 288–290]

A *matrix* is a rectangular array of numbers. Each number in the matrix is called an *element* of the matrix. A matrix of m rows and n columns is said to be of *order* $m \times n$. A *square matrix* has the same number of rows as columns. [p. 297]

A *determinant* is a number associated with a square matrix. [p. 297]

The *minor of an element* in a 3×3 determinant is the 2×2 determinant obtained by eliminating the row and column that contain that element. The *cofactor* of an element of a matrix is $(-1)^{i+j}$ times the minor of that element, where i is the row number of the element and j is its column number. [pp. 297–298]

The evaluation of the determinant of a 3×3 or larger matrix is accomplished by *expanding by cofactors*. [p. 298]

Inequalities considered together are called a *system of inequalities*. The *solution set of a system of inequalities* is the intersection of the solution sets of the individual inequalities. [p. 313]

Essential Rules

A system of equations can be solved by a graphing method, by the substitution method, or by the addition method. [pp. 275, 278, 285]

Cramer's Rule [pp. 300–301]

The solution of the system of equations $\begin{aligned} a_1x + b_1y &= c_1 \\ a_2x + b_2y &= c_2 \end{aligned}$ is

given by $x = \frac{D_x}{D}$ and $y = \frac{D_y}{D}$, where $D = \begin{vmatrix} a_1 & b_1 \\ a_2 & b_2 \end{vmatrix}$,

$D_x = \begin{vmatrix} c_1 & b_1 \\ c_2 & b_2 \end{vmatrix}$, $D_y = \begin{vmatrix} a_1 & c_1 \\ a_2 & c_2 \end{vmatrix}$, and $D \neq 0$.

The solution of the system of equations

$$\begin{aligned} a_1x + b_1y + c_1z &= d_1 \\ a_2x + b_2y + c_2z &= d_2 \\ a_3x + b_3y + c_3z &= d_3 \end{aligned}$$

is given by $x = \frac{D_x}{D}$, $y = \frac{D_y}{D}$, and

$z = \frac{D_z}{D}$, where $D = \begin{vmatrix} a_1 & b_1 & c_1 \\ a_2 & b_2 & c_2 \\ a_3 & b_3 & c_3 \end{vmatrix}$, $D_x = \begin{vmatrix} d_1 & b_1 & c_1 \\ d_2 & b_2 & c_2 \\ d_3 & b_3 & c_3 \end{vmatrix}$,

$D_y = \begin{vmatrix} a_1 & d_1 & c_1 \\ a_2 & d_2 & c_2 \\ a_3 & d_3 & c_3 \end{vmatrix}$, $D_z = \begin{vmatrix} a_1 & b_1 & d_1 \\ a_2 & b_2 & d_2 \\ a_3 & b_3 & d_3 \end{vmatrix}$, and $D \neq 0$.

321

Chapter Review

1. Solve by graphing: $2x - 3y = -6$
$2x - y = 2$

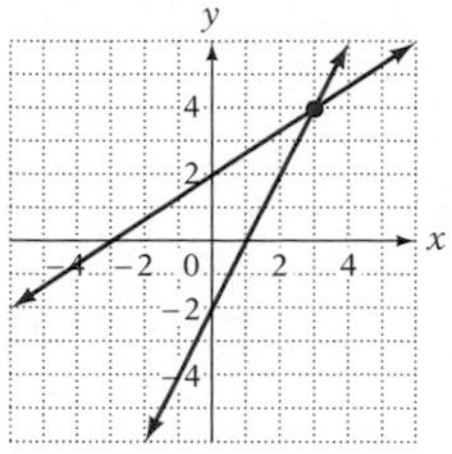

(3, 4) [5.1A]

2. Solve by graphing: $x - 2y = -5$
$3x + 4y = -15$

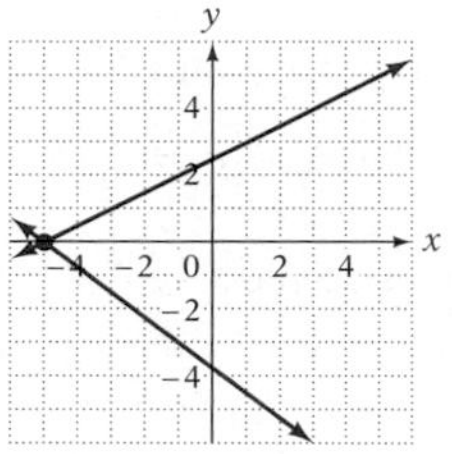

(−5, 0) [5.1A]

3. Graph the solution set: $2x - y < 3$
$4x + 3y < 11$

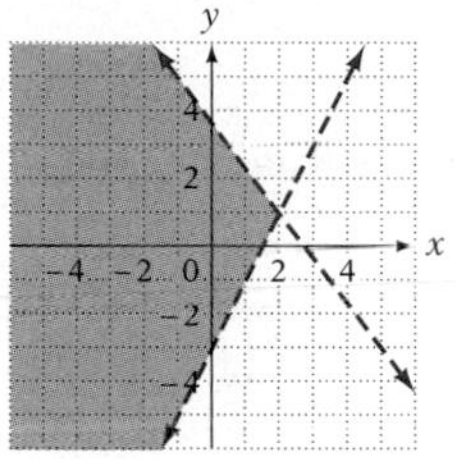

[5.5A]

4. Graph the solution set: $x + y > 2$
$2x - y < -1$

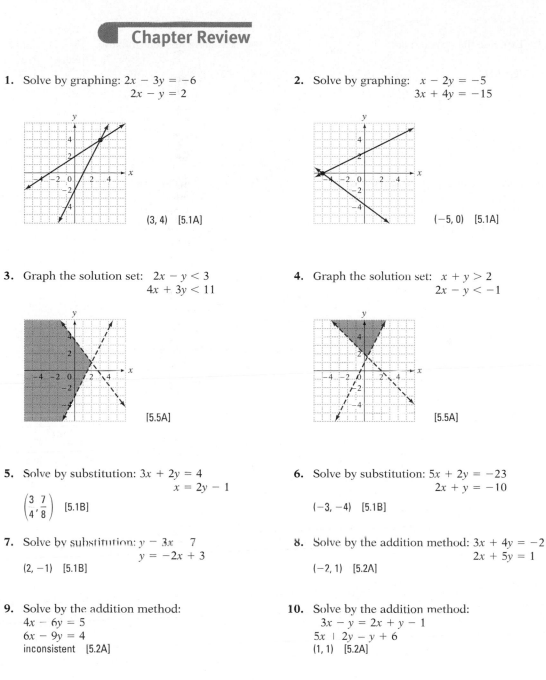

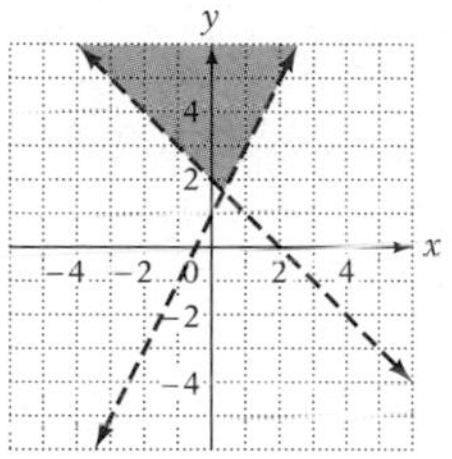

[5.5A]

5. Solve by substitution: $3x + 2y = 4$
$x = 2y - 1$
$\left(\frac{3}{4}, \frac{7}{8}\right)$ [5.1B]

6. Solve by substitution: $5x + 2y = -23$
$2x + y = -10$
(−3, −4) [5.1B]

7. Solve by substitution: $y = 3x - 7$
$y = -2x + 3$
(2, −1) [5.1B]

8. Solve by the addition method: $3x + 4y = -2$
$2x + 5y = 1$
(−2, 1) [5.2A]

9. Solve by the addition method:
$4x - 6y = 5$
$6x - 9y = 4$
inconsistent [5.2A]

10. Solve by the addition method:
$3x - y = 2x + y - 1$
$5x + 2y = y + 6$
(1, 1) [5.2A]

11. Solve by the addition method:
$2x + 4y - z = 3$
$x + 2y + z = 5$
$4x + 8y - 2z = 7$
inconsistent [5.2B]

12. Solve by the addition method:
$x - y - z = 5$
$2x + z = 2$
$3y - 2z = 1$
(2, −1, −2) [5.2B]

13. Evaluate the determinant: $\begin{vmatrix} 3 & -1 \\ -2 & 4 \end{vmatrix}$

10 [5.3A]

14. Evaluate the determinant: $\begin{vmatrix} 1 & -2 & 3 \\ 3 & 1 & 1 \\ 2 & -1 & -2 \end{vmatrix}$

−32 [5.3A]

15. Solve by using Cramer's Rule:

$x - y = 3$
$2x + y = -4$

$\left(-\frac{1}{3}, -\frac{10}{3}\right)$ [5.3B]

16. Solve by using Cramer's Rule:

$5x + 2y = 9$
$3x + 5y = -7$

$\left(\frac{59}{19}, -\frac{62}{19}\right)$ [5.3B]

17. Solve by using Cramer's Rule:

$x - y + z = 2$
$2x - y - z = 1$
$x + 2y - 3z = -4$

$\left(\frac{1}{5}, -\frac{6}{5}, \frac{3}{5}\right)$ [5.3B]

18. Solve by using Cramer's Rule:

$3x + 2y + 2z = 2$
$x - 2y - z = 1$
$2x - 3y - 3z = -3$

(0, −2, 3) [5.3B]

19. A plane flying with the wind went 350 mi in 2 h. The return trip, flying against the wind, took 2.8 h. Find the rate of the plane in calm air and the rate of the wind.

rate of the plane: 150 mph;
rate of the wind: 25 mph [5.4A]

20. A clothing manufacturer purchased 60 yd of cotton and 90 yd of wool for a total cost of $3600. Another purchase, at the same prices, included 80 yd of cotton and 20 yd of wool for a total cost of $2000. Find the cost per yard of the cotton and of the wool.

cost per yard of cotton: $18;
cost per yard of wool: $28 [5.4B]

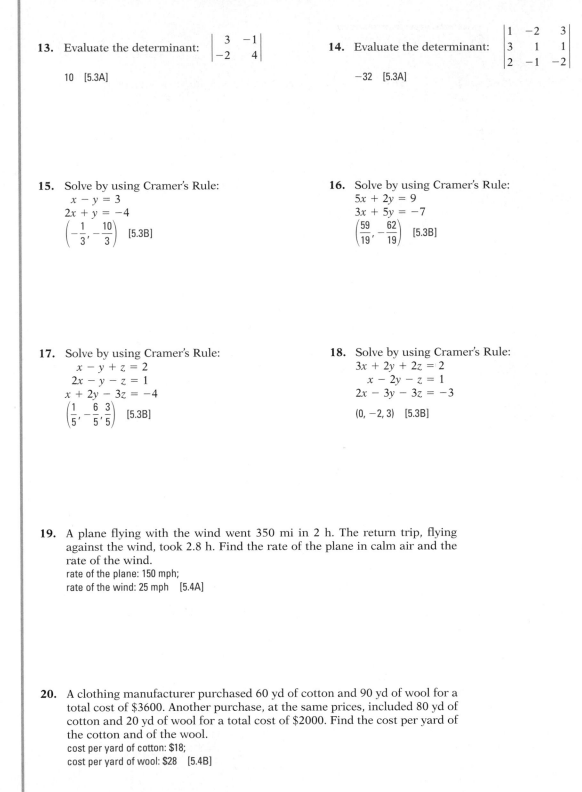

323

Chapter Test

1. Solve by substitution: $2x - 6y = 15$
$x = 4y + 8$

$\left(6, -\frac{1}{2}\right)$ [5.1B]

2. Solve by the addition method: $3x + 2y = 2$
$x + y = 3$

$(-4, 7)$ [5.2A]

3. Solve by graphing: $x + y = 3$
$3x - 2y = -6$

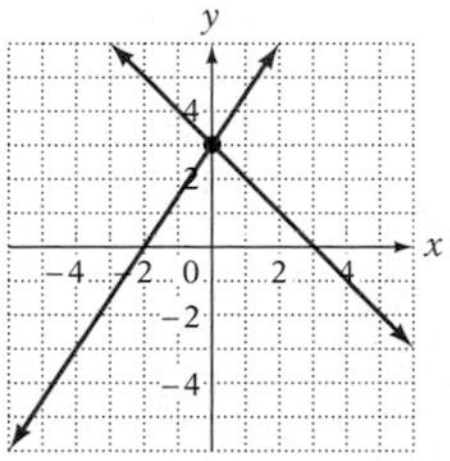

$(0, 3)$ [5.1A]

4. Solve by graphing: $2x - y = 4$
$y = 2x - 4$

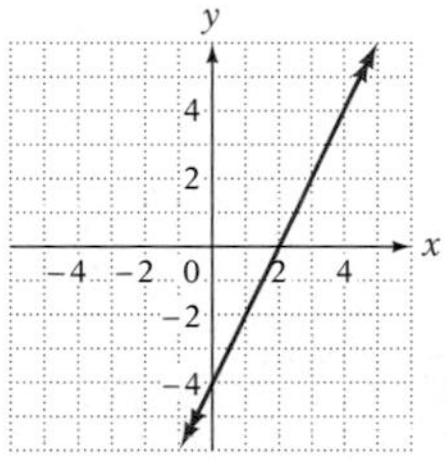

$(x, 2x - 4)$ [5.1A]

5. Solve by substitution: $3x + 12y = 18$
$x + 4y = 6$

$\left(x, -\frac{1}{4}x + \frac{3}{2}\right)$ [5.1B]

6. Solve by the addition method: $5x - 15y = 30$
$x - 3y = 6$

$\left(x, \frac{1}{3}x - 2\right)$ [5.2A]

7. Solve by the addition method:
$3x - 4y - 2z = 17$
$4x - 3y + 5z = 5$
$5x - 5y + 3z = 14$
$(3, -1, -2)$ [5.2B]

8. Solve by the addition method:
$3x + y = 13$
$2y + 3z = 5$
$x + 2z = 11$
$(5, -2, 3)$ [5.2B]

9. Evaluate the determinant:

$$\begin{vmatrix} 6 & 1 \\ 2 & 5 \end{vmatrix}$$

28 [5.3A]

10. Evaluate the determinant:

$$\begin{vmatrix} 1 & 5 & -2 \\ -2 & 1 & 4 \\ 4 & 3 & -8 \end{vmatrix}$$

0 [5.3A]

324

11. Solve by using Cramer's Rule:
$2x - y = 7$
$3x + 2y = 7$
(3, −1) [5.3B]

12. Solve by using Cramer's Rule:
$3x - 4y = 10$
$2x + 5y = 15$
$\left(\frac{110}{23}, \frac{25}{23}\right)$ [5.3B]

13. Solve by using Cramer's Rule:
$x + y + z = 0$
$x + 2y + 3z = 5$
$2x + y + 2z = 3$
(−1, −3, 4) [5.3B]

14. Solve by using Cramer's Rule:
$x + 3y + z = 6$
$2x + y - z = 12$
$x + 2y - z = 13$
(2, 3, −5) [5.3B]

15. Graph the solution set:
$x + 3y \leq 6$
$2x - y \geq 4$

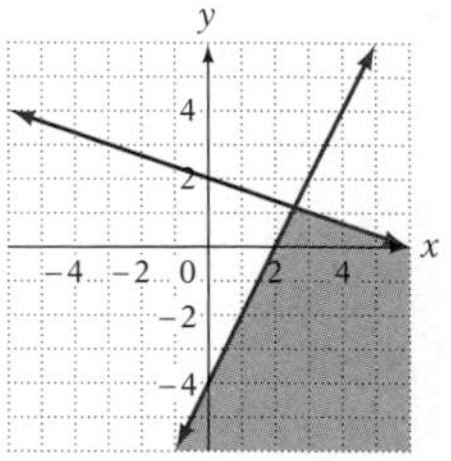

[5.5A]

16. Graph the solution set:
$2x + 4y \geq 8$
$x + y \leq 3$

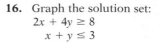

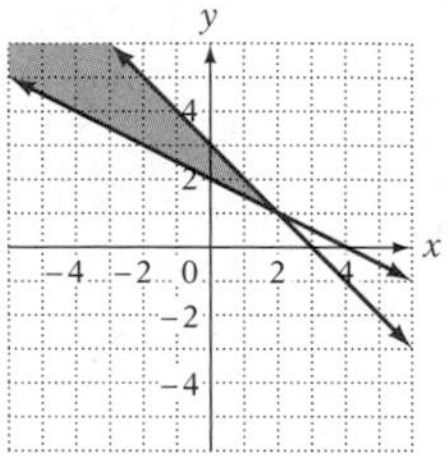

[5.5A]

17. Solve by substitution:
$3x - 2y = 4$
$y = 5x - 9$
(2, 1) [5.1B]

18. A cabin cruiser traveling with the current went 60 mi in 3 h. Against the current, it took 5 h to travel the same distance. Find the rate of the cabin cruiser in calm water and the rate of the current.
cabin cruiser: 16 mph; current: 4 mph [5.4A]

19. A pilot flying with the wind flew 600 mi in 3 h. Flying against the wind, the pilot required 4 h to travel the same distance. Find the rate of the plane in calm air and the rate of the wind.
plane: 175 mph; wind: 25 mph [5.4A]

20. At a movie theater, admission tickets are \$10 for children and \$16 for adults. The receipts for one Friday evening were \$5000. The next day there were three times as many children as the preceding evening and only half the number of adults as the night before, yet the receipts were still \$5000. Find the number of children who attended on Friday evening.
100 children [5.4B]

325

Cumulative Review

1. Simplify: $-2\sqrt{90}$
$-6\sqrt{10}$ [1.2F]

2. Solve: $3(x - 5) = 2x + 7$
22 [2.2C]

3. Simplify: $3[x - 2(5 - 2x) - 4x] + 6$
$3x - 24$ [1.4D]

4. Evaluate $a + bc \div 2$ when $a = 4$, $b = 8$, and $c = -2$.
-4 [1.4A]

5. Solve: $2x - 3 < 9$ or $5x - 1 < 4$
$\{x \mid x < 6\}$ [2.4B]

6. Solve: $|x - 2| - 4 < 2$
$\{x \mid -4 < x < 8\}$ [2.5B]

7. Solve: $|2x - 3| > 5$
$\{x \mid x > 4 \text{ or } x < -1\}$ [2.5B]

8. Given $F(x) = x^2 - 3$, find $F(2)$.
1 [4.2A]

9. Graph the solution set of $\{x \mid x \le 2\} \cap \{x \mid x > -3\}$.

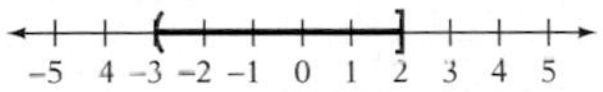

[1.5C]

10. Find the equation of the line that contains the point $(-2, 3)$ and has slope $-\frac{2}{3}$.
$y = -\frac{2}{3}x + \frac{5}{3}$ [4.5A]

11. Find the equation of the line that contains the points $(2, -1)$ and $(3, 4)$.
$y = 5x - 11$ [4.5B]

12. Find the equation of the line that contains the point $(-2, 2)$ and is perpendicular to the line $2x - 3y = 6$.
$y = -\frac{3}{2}x - 1$ [4.6A]

13. Graph $2x - 5y = 10$.

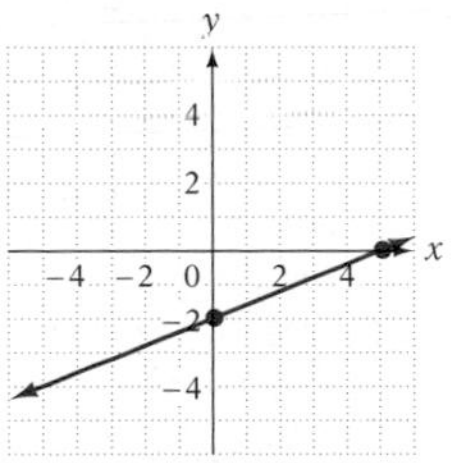

[4.4B]

14. Graph the solution set of $3x - 4y \ge 8$.

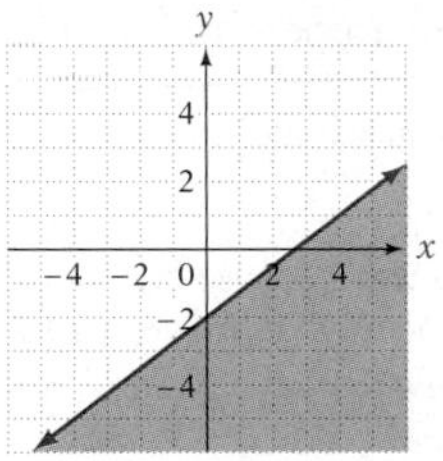

[4.7A]

15. Solve by graphing:
$5x - 2y = 10$
$3x + 2y = 6$

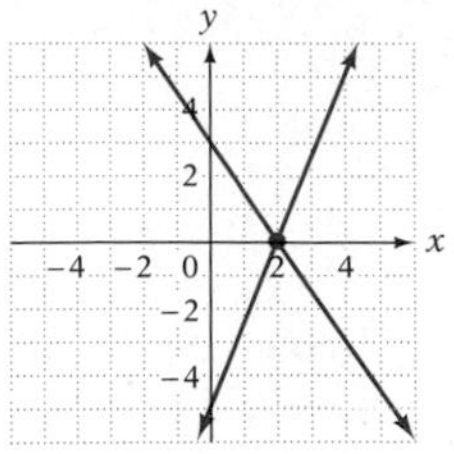

(2, 0) [5.1A]

16. Graph the solution set:
$3x - 2y \geq 4$
$x + y < 3$

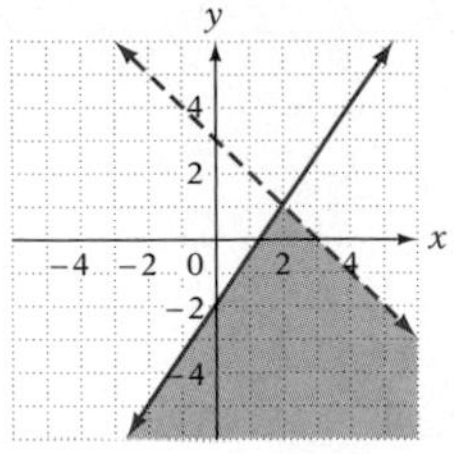

[5.5A]

17. Solve by the addition method:
$3x + 2z = 1$
$2y - z = 1$
$x + 2y = 1$

(1, 0, −1) [5.2B]

18. Evaluate the determinant:

$$\begin{vmatrix} 2 & -5 & 1 \\ 3 & 1 & 2 \\ 6 & -1 & 4 \end{vmatrix}$$

3 [5.3A]

19. Solve by using Cramer's Rule:
$4x - 3y = 17$
$3x - 2y = 12$
(2, −3) [5.3B]

20. Solve by substitution:
$3x - 2y = 7$
$y = 2x - 1$
(−5, −11) [5.1B]

21. How many milliliters of pure water must be added to 100 ml of a 4% salt solution to make a 2.5% salt solution?
60 ml [2.3B]

22. Flying with the wind, a small plane required 2 h to fly 150 mi. Against the wind, it took 3 h to fly the same distance. Find the rate of the wind.
12.5 mph [5.4A]

23. A restaurant manager buys 100 lb of hamburger and 50 lb of steak for a total cost of \$980. A second purchase, at the same prices, includes 150 lb of hamburger and 100 lb of steak. The total cost is \$1720. Find the price of 1 lb of steak.
\$10 [5.4B]

24. Find the lower and upper limits of a 12,000-ohm resistor with a 15% tolerance.
10,200 ohms and 13,800 ohms [2.5C]

25. The graph shows the relationship between the monthly income and the sales of an account executive. Find the slope of the line between the two points shown on the graph. Write a sentence that states the meaning of the slope.
$m = 40$. The slope represents the monthly income (in dollars) per thousand dollars in sales. [4.4A]

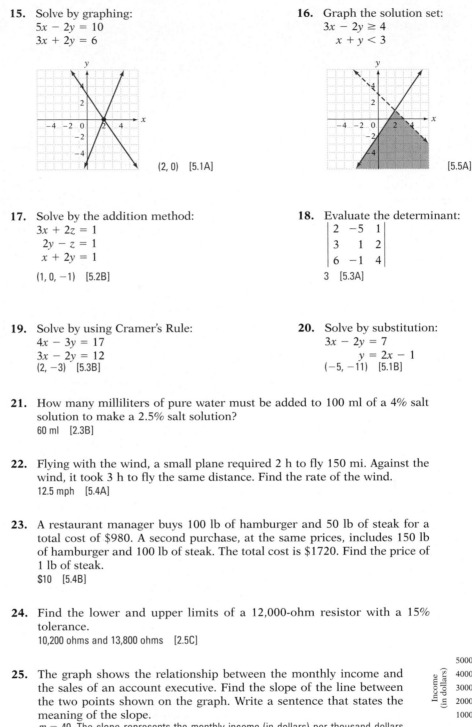

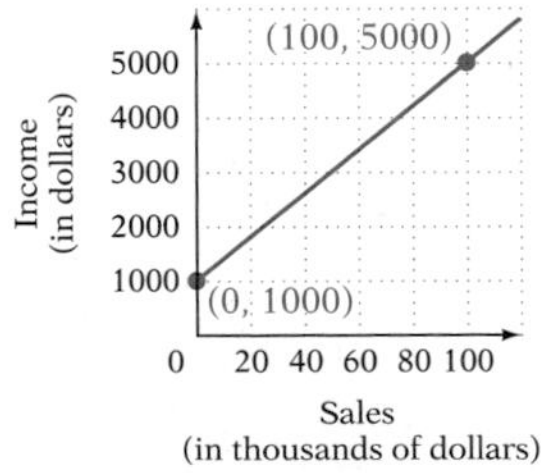

Chapter 6

Polynomials

The Pathfinder mission produced the closest look to date at the surface of Mars. This photo of the Pathfinder Sojourner rover was taken at Mermaid Dune on Mars on August 8, 1997. Images such as this, transmitted from Mars to Earth, traveled a distance of over 100 million miles. Distances this great, as well as very small measurements, are generally expressed in scientific notation, as illustrated in **Exercise 119 on page 340.**

Objectives

Section 6.1

- **A** To multiply monomials
- **B** To divide monomials and simplify expressions with negative exponents
- **C** To write a number using scientific notation
- **D** To solve application problems

Section 6.2

- **A** To evaluate polynomial functions
- **B** To add or subtract polynomials

Section 6.3

- **A** To multiply a polynomial by a monomial
- **B** To multiply two polynomials
- **C** To multiply two binomials
- **D** To multiply binomials that have special products
- **E** To solve application problems

Section 6.4

- **A** To divide polynomials
- **B** To divide polynomials using synthetic division
- **C** To evaluate a polynomial using synthetic division

WEB Need help? For on-line student resources, such as section quizzes, visit this textbook's web site at **college.hmco.com/students.**

Prep Test

1. Subtract: $-2 - (-3)$
1 [1.1C]

2. Multiply: $-3(6)$
-18 [1.1D]

For Exercises 3 to 8, simplify.

3. $-\dfrac{24}{-36}$
$\dfrac{2}{3}$ [1.1D]

4. $-4(3y)$
$-12y$ [1.4C]

5. $(-2)^3$
-8 [1.2E]

6. $-4a - 8b + 7a$
$3a - 8b$ [1.4B]

7. $3x - 2[y - 4(x + 1) + 5]$
$11x - 2y - 2$ [1.4D]

8. $-4y + 4y$
0 [1.4B]

9. Are $2x^2$ and $2x$ like terms?
no [1.4B]

Go Figure

A pair of perpendicular lines are drawn through the interior of a rectangle, dividing it into four smaller rectangles. The area of the smaller rectangles are x, 2, 3, and 6. Find the possible values of x.
1, 4, 9

329

6.1 Exponential Expressions

Objective A To multiply monomials

VIDEO & DVD CD TUTOR WEB SSM

A monomial is a number, a variable, or a product of a number and variables.

Point of Interest

Around A.D. 250, the monomial $3x^2$ shown at the right would have been written $\Delta^{Y}3$ or at least approximately like that. In A.D. 250, the symbol for 3 was not the one we use today.

The examples at the right are monomials. The **degree of a monomial** is the sum of the exponents of the variables.

x	degree 1 $(x = x^1)$
$3x^2$	degree 2
$4x^2y$	degree 3
$6x^3y^4z^2$	degree 9

In this chapter, the variable n is considered a positive integer when used as an exponent. x^n degree n

The degree of a nonzero constant term is zero. 6 degree 0

The expression $5\sqrt{x}$ is not a monomial because $\sqrt{x}$ cannot be written as a product of variables. The expression $\frac{x}{y}$ is not a monomial because it is a quotient of variables.

The expression x^4 is an exponential expression. The exponent, 4, indicates the number of times the base, x, occurs as a factor.

The product of exponential expressions with the *same* base can be simplified by writing each expression in factored form and writing the result with an exponent.

$$x^3 \cdot x^4 = \underbrace{\overbrace{(x \cdot x \cdot x)}^{3\text{ factors}} \cdot \overbrace{(x \cdot x \cdot x \cdot x)}^{4\text{ factors}}}_{7\text{ factors}}$$

$$= x^7$$

Note that adding the exponents results in the same product.

$$x^3 \cdot x^4 = x^{3+4} = x^7$$

> **Rule for Multiplying Exponential Expressions**
>
> If m and n are positive integers, then $x^m \cdot x^n = x^{m+n}$.

➡ Multiply: $(-4x^5y^3)(3xy^2)$

$(-4x^5y^3)(3xy^2) = (-4 \cdot 3)(x^5 \cdot x)(y^3 \cdot y^2)$
- Use the Commutative and Associative Properties of Multiplication to rearrange and group factors.

$= -12(x^{5+1})(y^{3+2})$
- Multiply variables with the same base by adding their exponents. Do these steps mentally.

$= -12x^6y^5$
- Simplify.

TAKE NOTE

The Rule for Multiplying Exponential Expressions requires the bases to be the same. The expression x^6y^5 cannot be simplified.

In-Class Examples (Objective 6.1A)

Simplify.

1. $(-2a^2b^3)(-4ab^2)$ $8a^3b^5$
2. $(3ab^2)^3$ $27a^3b^6$
3. $(-2a^2b)^2(-3ab^2)^2$ $36a^6b^6$
4. $x^{2n} \cdot x^{2n}$ x^{4n}

Objective 6.1A

New Vocabulary

monomial
degree of a monomial

New Rules

Rule for Multiplying Exponential Expressions
Rule for Simplifying Powers of Exponential Expressions
Rule for Simplifying Powers of Products

Instructor Note

You might prefer to have students develop the rule for multiplying exponential expressions themselves. Provide them with several expressions to simplify, for example,

x^4x^3
y^2y^6
a^5a^7

Ask them to simplify the expressions by writing each expression in factored form and then writing the result with an exponent. After they have completed several examples, ask them to write a rule for x^mx^n. This same approach can be used to derive the Rule for Simplifying Powers of Exponential Expressions and the Rule for Dividing Exponential Expressions.

Discuss the Concepts

1. Explain how to multiply two exponential expressions with the same base.
2. Why can't the exponential expression x^5y^3 be simplified?
3. Which expression is the product of two exponential expressions and which is the power of an exponential expression?
   a. $q^4 \cdot q^{10}$
   b. $(q^4)^{10}$
4. In the expression $(a^8b^6)^5$, what is the product and what is the power?

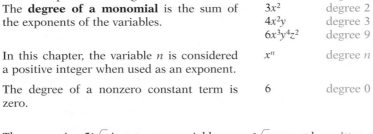

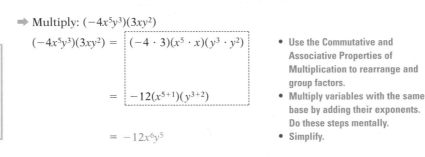

Concept Check

1. Find the area of the rectangle. The dimensions given are in kilometers. $54m^2n^4$ km^2

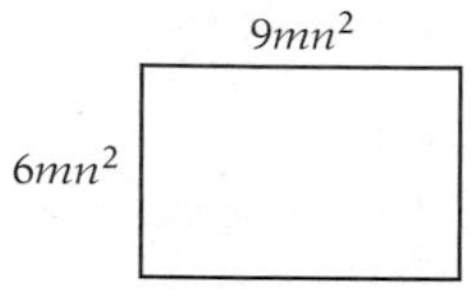

2. Find the area of the square. The dimension given is in meters. $64x^4y^2$ m^2

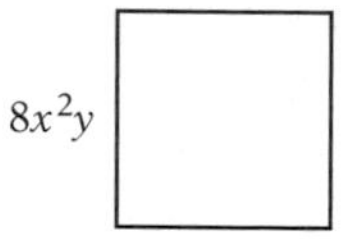

Optional Student Activity

1. Simplify: $(5xy^3)(3x^4y^2) - (2x^3y)(x^2y^4)$ $13x^5y^5$
2. Simplify: $4a^2(2ab)^3 - 5b^2(a^5b)$ $27a^5b^3$
3. If $3^{33} + 3^{33} + 3^{33} = 3^x$, find x. 34

Point of Interest

One of the first symbolic representations of powers was given by Diophantus (c. A.D. 250) in his book *Arithmetica*. He used Δ^Y for x^2 and κ^Y for x^3. The symbol Δ^Y was the first two letters of the Greek word *dunamis,* meaning "power"; κ^Y was from the Greek word *kubos,* meaning "cube." He also combined these symbols to denote higher powers. For instance, $\Delta\kappa^Y$ was the symbol for x^5.

The power of a monomial can be simplified by writing the power in factored form and then using the Rule for Multiplying Exponential Expressions.

$$(x^4)^3 = x^4 \cdot x^4 \cdot x^4 = x^{4+4+4} = x^{12}$$

$$(a^2b^3)^2 = (a^2b^3)(a^2b^3) = a^{2+2}b^{3+3} = a^4b^6$$

- Write in factored form.
- Use the Rule for Multiplying Exponential Expressions.

Note that multiplying each exponent inside the parentheses by the exponent outside the parentheses results in the same product.

$$(x^4)^3 = x^{4\cdot 3} = x^{12}$$

$$(a^2b^3)^2 = a^{2\cdot 2}b^{3\cdot 2} = a^4b^6$$

- Multiply each exponent inside the parentheses by the exponent outside the parentheses.

Rule for Simplifying Powers of Exponential Expressions

If m and n are positive integers, then $(x^m)^n = x^{mn}$.

Rule for Simplifying Powers of Products

If m, n, and p are positive integers, then $(x^my^n)^p = x^{mp}y^{np}$.

➡ Simplify: $(x^4)^5$

$$(x^4)^5 = x^{4\cdot 5} = x^{20}$$

- Use the Rule for Simplifying Powers of Exponential Expressions to multiply the exponents.

➡ Simplify: $(2a^3b^4)^3$

$$(2a^3b^4)^3 = 2^{1\cdot 3}a^{3\cdot 3}b^{4\cdot 3} = 2^3a^9b^{12} = 8a^9b^{12}$$

- Use the Rule for Simplifying Powers of Products to multiply each exponent inside the parentheses by the exponent outside the parentheses. Note that $2 = 2^1$.

Example 1 Simplify: $(2xy^2)(-3xy^4)^3$

Solution $(2xy^2)(-3xy^4)^3 = (2xy^2)[(-3)^3x^3y^{12}]$
$= (2xy^2)(-27x^3y^{12})$
$= -54x^4y^{14}$

You Try It 1 Simplify: $(-3a^2b^4)(-2ab^3)^4$

Your solution $-48a^6b^{16}$

Example 2 Simplify: $(x^{n+2})^5$

Solution $(x^{n+2})^5 = x^{5n+10}$

You Try It 2 Simplify: $(y^{n-3})^2$

Your solution y^{2n-6}

Solutions on p. S17

Example 3 Simplify: $[(2xy^2)^2]^3$

Solution $[(2xy^2)^2]^3 = [2^2x^2y^4]^3$
$= 2^6x^6y^{12} = 64x^6y^{12}$

You Try It 3 Simplify: $[(ab^3)^3]^4$

Your solution $a^{12}b^{36}$

Solution on p. S17

Objective B To divide monomials and simplify expressions with negative exponents

VIDEO & DVD CD TUTOR WEB SSM

The quotient of two exponential expressions with the same base can be simplified by writing each expression in factored form, dividing by the common factors, and then writing the result with an exponent.

$$\frac{x^5}{x^2} = \frac{\overset{1}{\cancel{x}} \cdot \overset{1}{\cancel{x}} \cdot x \cdot x \cdot x}{\underset{1}{\cancel{x}} \cdot \underset{1}{\cancel{x}}} = x^3$$

Note that subtracting the exponents gives the same result.

$$\frac{x^5}{x^2} = x^{5-2} = x^3$$

To divide two monomials with the same base, subtract the exponents of the like bases.

⇒ Simplify: $\frac{z^8}{z^2}$

$\frac{z^8}{z^2} = \boxed{z^{8-2}}$
$= z^6$

• **The bases are the same. Subtract the exponents. This step is often done mentally.**

⇒ Simplify: $\frac{a^5b^9}{a^4b}$

$\frac{a^5b^9}{a^4b} = \boxed{a^{5-4}b^{9-1}}$
$= ab^8$

• **Subtract the exponents of the like bases. This step is often done mentally.**

> **TAKE NOTE**
> A number divided by itself is 1. Therefore, $\frac{8}{8} = 1$, $\frac{3^4}{3^4} = 1$, and for $x \neq 0$, $\frac{x^4}{x^4} = 1$.

Consider the expression $\frac{x^4}{x^4}$, $x \neq 0$. This expression can be simplified, as shown below, by (1) subtracting exponents and (2) dividing by common factors.

$$\frac{x^4}{x^4} = x^{4-4} = x^0 \qquad \frac{x^4}{x^4} = \frac{\overset{1}{\cancel{x}} \cdot \overset{1}{\cancel{x}} \cdot \overset{1}{\cancel{x}} \cdot \overset{1}{\cancel{x}}}{\underset{1}{\cancel{x}} \cdot \underset{1}{\cancel{x}} \cdot \underset{1}{\cancel{x}} \cdot \underset{1}{\cancel{x}}} = 1$$

The equations $\frac{x^4}{x^4} = x^0$ and $\frac{x^4}{x^4} = 1$ suggest the following definition of x^0.

> **TAKE NOTE**
> For the example below, we indicated that $z \neq 0$. If we try to evaluate $(16z^5)^0$ when $z = 0$, we have $[16(0)^5]^0 = [16(0)]^0 = 0^0$. However, 0^0 is not defined. Therefore, we must assume that $z \neq 0$. To avoid stating this for every example or exercise, we will assume that variables do not have values that result in the expression 0^0.

> **Definition of Zero as an Exponent**
> If $x \neq 0$, then $x^0 = 1$. The expression 0^0 is not defined.

⇒ Simplify: $(16z^5)^0$, $z \neq 0$

$(16z^5)^0 = 1$

• **Any nonzero expression to the zero power is 1.**

Objective 6.1B

New Vocabulary

Definition of Zero as an Exponent
Definition of a Negative Exponent

New Rules

Rule for Simplifying Powers of Quotients
Rule for Dividing Exponential Expressions

Instructor Note

Here we are just verbalizing the rule for division of monomials. The theorem comes after we define negative exponents.

Have students copy this rule onto a piece of paper and then practice a few exercises such as $\frac{a^9}{a^2}$ and $\frac{y^8}{y}$. It may also help to give $\frac{a^9}{b^5}$ as an exercise to emphasize that the bases must be the same.

Discuss the Concepts

1. Explain how to divide two exponential expressions with the same base.
2. Why can't the expression $\frac{x^8}{y^2}$ be simplified?
3. Rewrite the following expressions with positive exponents.
   a. b^{-8}
   b. $\frac{1}{w^{-5}}$

In-Class Examples (Objective 6.1B)

Simplify.

1. $\frac{-24a^3b^9}{16a^6b^3} - \frac{3b^6}{2a^3}$
2. $\frac{a^2}{3b^{-1}} \; \frac{a^2b}{3}$
3. $(2x^{-3}y^3)^2 \; \frac{4y^6}{x^6}$
4. $\frac{x^{2n-1}}{x^{n-2}} \; x^{n+1}$

Concept Check

1. The area of the rectangle below is $24a^3b^5$ square yards. Find the length of the rectangle. $6ab^4$ yd

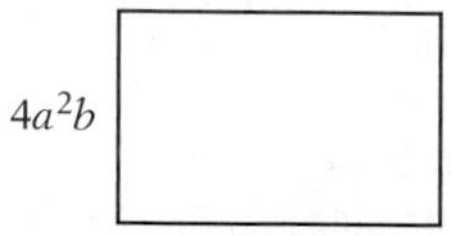

2. The area of the parallelogram below is $56w^4z^6$ square meters. Find the height of the parallelogram. $4w^2z$ m

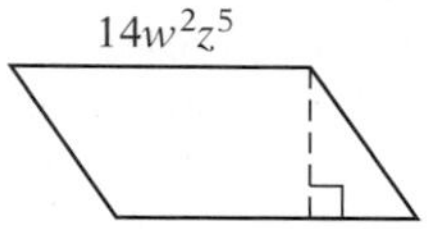

Optional Student Activity

1. The product of a monomial and $4b$ is $12a^2b$. Find the monomial. $3a^2$
2. The product of a monomial and $8y^2$ is $32x^2y^3$. Find the monomial. $4x^2y$
3. Simplify: $\frac{2^{40}}{4^{20}}$ 1
4. If $\frac{(a)(a)(a)}{a+a+a} = 3$, find the value of a^2. 9

Point of Interest

In the 15th century, the expression $12^{2\bar{m}}$ was used to mean $12x^{-2}$. The use of $\bar{m}$ reflects an Italian influence, where *m* was used for minus and *p* was used for plus. It was understood that $2\bar{m}$ referred to an unnamed variable. Isaac Newton, in the 17th century, advocated the use of a negative exponent, the symbol we use today.

➡ Simplify: $-(7x^4y^3)^0, x \neq 0, y \neq 0$

$-(7x^4y^3)^0 = -(1) = -1$ • **The negative outside the parentheses is not affected by the exponent.**

Consider the expression $\frac{x^4}{x^6}, x \neq 0$. This expression can be simplified, as shown below, by (1) subtracting exponents and (2) dividing by common factors.

$$\frac{x^4}{x^6} = x^{4-6} = x^{-2} \qquad \frac{x^4}{x^6} = \frac{\overset{1}{\cancel{x}} \cdot \overset{1}{\cancel{x}} \cdot \overset{1}{\cancel{x}} \cdot \overset{1}{\cancel{x}}}{\underset{1}{\cancel{x}} \cdot \underset{1}{\cancel{x}} \cdot \underset{1}{\cancel{x}} \cdot \underset{1}{\cancel{x}} \cdot x \cdot x} = \frac{1}{x^2}$$

The equations $\frac{x^4}{x^6} = x^{-2}$ and $\frac{x^4}{x^6} = \frac{1}{x^2}$ suggest that $x^{-2} = \frac{1}{x^2}$.

Definition of a Negative Exponent

If $x \neq 0$ and n is a positive integer, then

$$x^{-n} = \frac{1}{x^n} \quad \text{and} \quad \frac{1}{x^{-n}} = x^n$$

TAKE NOTE

Note from the example at the right that 2^{-4} is a *positive* number. A negative exponent does not indicate a negative number.

➡ Evaluate: 2^{-4}

$2^{-4} = \frac{1}{2^4}$ • **Use the Definition of a Negative Exponent.**

$= \frac{1}{16}$ • **Evaluate the expression.**

A power of the quotient of two exponential expressions can be simplified by multiplying each exponent in the quotient by the exponent outside the parentheses.

Rule for Simplifying Powers of Quotients

If m, n, and p are integers and $y \neq 0$, then $\left(\frac{x^m}{y^n}\right)^p = \frac{x^{mp}}{y^{np}}$.

➡ Simplify: $\left(\frac{a^2}{b^3}\right)^{-2}$

$\left(\frac{a^2}{b^3}\right)^{-2} = \frac{a^{2(-2)}}{b^{3(-2)}}$ • **Use the Rule for Simplifying Powers of Quotients.**

$= \frac{a^{-4}}{b^{-6}} = \frac{b^6}{a^4}$ • **Use the Definition of a Negative Exponent.**

Simplest Form of an Exponential Expression

An exponential expression is in simplest form when it is written with only positive exponents.

333

TAKE NOTE
The exponent on n is -5 (*negative* 5). The n^{-5} is written in the denominator as n^5. The exponent on 3 is 1 (*positive* 1). The 3 remains in the numerator.
Also, we indicated that $n \neq 0$. This is necessary because division by zero is not defined. In this textbook, we will assume that values of the variables are chosen so that division by zero does not occur.

➡ Simplify: $3n^{-5}, n \neq 0$

$$3n^{-5} = 3 \cdot \frac{1}{n^5} = \frac{3}{n^5}$$

• Use the Definition of a Negative Exponent to rewrite the expression with a positive exponent.

➡ Simplify: $\frac{2}{5a^{-4}}$

$$\frac{2}{5a^{-4}} = \frac{2}{5} \cdot \frac{1}{a^{-4}} = \frac{2}{5} \cdot a^4 = \frac{2a^4}{5}$$

• Use the Definition of a Negative Exponent to rewrite the expression with a positive exponent.

Now that zero and negative exponents have been defined, a rule for dividing exponential expressions can be stated.

Rule for Dividing Exponential Expressions

If m and n are integers and $x \neq 0$, then $\frac{x^m}{x^n} = x^{m-n}$.

➡ Simplify: $\frac{x^4}{x^9}$

$$\frac{x^4}{x^9} = \boxed{x^{4-9}}$$

• Use the Rule for Dividing Exponential Expressions. This step is often done mentally.

$$= x^{-5}$$

• Subtract the exponents.

$$= \frac{1}{x^5}$$

• Use the Definition of a Negative Exponent to rewrite the expression with a positive exponent.

The rules for simplifying exponential expressions and powers of exponential expressions are true for all integer exponents. These rules are restated here for your convenience.

Rules of Exponents

If m, n, and p are integers, then

$x^m \cdot x^n = x^{m+n}$ $(x^m)^n = x^{mn}$ $(x^m y^n)^p = x^{mp}y^{np}$

$\frac{x^m}{x^n} = x^{m-n}, x \neq 0$ $\left(\frac{x^m}{y^n}\right)^p = \frac{x^{mp}}{y^{np}}, y \neq 0$ $x^{-n} = \frac{1}{x^n}, x \neq 0$

$x^0 = 1, x \neq 0$

➡ Simplify: $(3ab^{-4})(-2a^{-3}b^7)$

$$(3ab^{-4})(-2a^{-3}b^7) = \boxed{[3 \cdot (-2)](a^{1+(-3)}b^{-4+7})}$$

• When multiplying expressions, add the exponents on like bases. Do this step mentally.

$$= -6a^{-2}b^3$$

$$= -\frac{6b^3}{a^2}$$

Optional Student Activity

Have your students complete the following equations and explain the pattern.

$10^5 =$
$10^4 =$
$10^3 =$
$10^2 =$
$10^1 =$

100,000; 10,000; 1000; 100; 10. The exponent on 10 decreases by 1 while each successive number is one-tenth of the number above it: $100{,}000 \div 10 = 10{,}000$; $10{,}000 \div 10 = 1000$; and so on. Now have them continue the established pattern. The next exponent on 10 is $1 - 1 = 0$. 10^0 is equal to $10 \div 10 = 1$. The next exponent is $0 - 1 = -1$, and 10^{-1} is equal to $1 \div 10 = 0.1$, and so on. The next eight equations are:

$10^0 = 1$
$10^{-1} = 0.1$
$10^{-2} = 0.01$
$10^{-3} = 0.001$
$10^{-4} = 0.0001$
$10^{-5} = 0.00001$
$10^{-6} = 0.000001$
$10^{-7} = 0.0000001$

This exercise will help students' understanding of x^0 and expressions with negative exponents.

Instructor Note

There are a few ways to simplify the expression at the right. Students could simplify it by starting as follows:

$$\left[\frac{6m^2n^3}{8m^7n^2}\right]^{-3} = \left[\frac{8m^7n^2}{6m^2n^3}\right]^3$$

➡ Simplify: $\dfrac{4a^{-2}b^5}{6a^5b^2}$

$$\frac{4a^{-2}b^5}{6a^5b^2} = \frac{2 \cdot 2a^{-2}b^5}{2 \cdot 3a^5b^2} = \frac{2a^{-2}b^5}{3a^5b^2}$$

- Divide the coefficients by their common factor.

$$= \frac{2a^{-2-5}b^{5-2}}{3}$$

- Use the Rule for Dividing Exponential Expressions.

$$= \frac{2a^{-7}b^3}{3} = \frac{2b^3}{3a^7}$$

- Use the Definition of a Negative Exponent to rewrite the expression with a positive exponent.

➡ Simplify: $\left(\dfrac{6m^2n^3}{8m^7n^2}\right)^{-3}$

$$\left(\frac{6m^2n^3}{8m^7n^2}\right)^{-3} = \left(\frac{3m^{2-7}n^{3-2}}{4}\right)^{-3}$$

- Simplify inside the brackets.

$$= \left(\frac{3m^{-5}n}{4}\right)^{-3}$$

- Subtract the exponents.

$$= \frac{3^{-3}m^{15}n^{-3}}{4^{-3}}$$

- Use the Rule for Simplifying Powers of Quotients.

$$= \frac{4^3m^{15}}{3^3n^3} = \frac{64m^{15}}{27n^3}$$

- Use the Definition of a Negative Exponent to rewrite the expression with positive exponents. Then simplify.

Example 4

Simplify: $\dfrac{-28x^6z^{-3}}{42x^{-1}z^4}$

Solution

$$\frac{-28x^6z^{-3}}{42x^{-1}z^4} = -\frac{14 \cdot 2x^{6-(-1)}z^{-3-4}}{14 \cdot 3} = -\frac{2x^7z^{-7}}{3} = -\frac{2x^7}{3z^7}$$

You Try It 4

Simplify: $\dfrac{20r^{-2}t^{-5}}{-16r^{-3}s^{-2}}$

Your solution $-\dfrac{5rs^2}{4t^5}$

Example 5

Simplify: $\dfrac{(3a^{-1}b^4)^{-3}}{(6^{-1}a^{-3}b^{-4})^3}$

Solution

$$\frac{(3a^{-1}b^4)^{-3}}{(6^{-1}a^{-3}b^{-4})^3} = \frac{3^{-3}a^3b^{-12}}{6^{-3}a^{-9}b^{-12}}$$
$$= 3^{-3} \cdot 6^3a^{12}b^0$$
$$= \frac{6^3a^{12}}{3^3} = \frac{216a^{12}}{27} = 8a^{12}$$

You Try It 5

Simplify: $\dfrac{(9u^{-6}v^4)^{-1}}{(6u^{-3}v^{-2})^{-2}}$

Your solution $\dfrac{4}{v^8}$

Example 6

Simplify: $\dfrac{x^{4n-2}}{x^{2n-5}}$

Solution

$$\frac{x^{4n-2}}{x^{2n-5}} = x^{4n-2-(2n-5)}$$
$$= x^{4n-2-2n+5} = x^{2n+3}$$

You Try It 6

Simplify: $\dfrac{a^{2n+1}}{a^{n+3}}$

Your solution a^{n-2}

Solutions on pp. S17–S18

Objective C To write a number using scientific notation

VIDEO & DVD CD TUTOR WEB SSM

Point of Interest

Astronomers measure the distance of some stars by using the parsec. One parsec is approximately 1.91×10^{13} mi.

Integer exponents are used to represent the very large and very small numbers encountered in the fields of science and engineering. For example, the mass of the electron is 0.000000000000000000000000000009 g. Numbers such as this are difficult to read and write, so a more convenient system for writing such numbers has been developed. It is called **scientific notation**.

To express a number in scientific notation, write the number as the product of a number between 1 and 10 and a power of 10. The form for scientific notation is $\boldsymbol{a \times 10^n}$, where $1 \le a < 10$.

TAKE NOTE

The numbers 9.65, 3.6, 9.2, 2, 9.74, and 8.6 are all numbers between 1 and 10. Each is multiplied by a power of 10. Note that there are two steps in writing a number in scientific notation: (1) Determine the number between 1 and 10, and (2) determine the exponent on 10.

For numbers greater than 10, move the decimal point to the right of the first digit. The exponent n is positive and equal to the number of places the decimal point has been moved.

$965{,}000 = 9.65 \times 10^5$

$3{,}600{,}000 = 3.6 \times 10^6$

$92{,}000{,}000{,}000 = 9.2 \times 10^{10}$

For numbers less than 1, move the decimal point to the right of the first nonzero digit. The exponent n is negative. The absolute value of the exponent is equal to the number of places the decimal point has been moved.

$0.0002 = 2 \times 10^{-4}$

$0.0000000974 = 9.74 \times 10^{-8}$

$0.000000000086 = 8.6 \times 10^{-11}$

Converting a number written in scientific notation to decimal notation requires moving the decimal point.

When the exponent is positive, move the decimal point to the right the same number of places as the exponent.

$1.32 \times 10^4 = 13{,}200$

$1.4 \times 10^8 = 140{,}000{,}000$

When the exponent is negative, move the decimal point to the left the same number of places as the absolute value of the exponent.

$1.32 \times 10^{-2} = 0.0132$

$1.4 \times 10^{-4} = 0.00014$

Numerical calculations involving numbers that have more digits than a hand-held calculator is able to handle can be performed using scientific notation.

➡ Simplify: $\dfrac{220{,}000 \times 0.000000092}{0.0000011}$

$$\frac{220{,}000 \times 0.000000092}{0.0000011} = \frac{2.2 \times 10^5 \times 9.2 \times 10^{-8}}{1.1 \times 10^{-6}}$$

• **Write the numbers in scientific notation.**

$$= \frac{(2.2)(9.2) \times 10^{5+(-8)-(-6)}}{1.1}$$

• **Simplify.**

$$= 18.4 \times 10^3 = 18{,}400$$

Example 7 Write 0.000041 in scientific notation.

Solution $0.000041 = 4.1 \times 10^{-5}$

You Try It 7 Write 942,000,000 in scientific notation.

Your solution 9.42×10^8

Solution on p. S18

Objective 6.1C

New Vocabulary

scientific notation

Discuss the Concepts

1. Name some situations in which scientific notation is used. Examples include molecular quantities and distances in the universe.
2. Determine whether the expression is written in scientific notation.
   a. 2.84×10^{-4} Yes
   b. 36.5×10 No. 36.5 is not a number between 1 and 10.
   c. 0.91×10^{-1} No. 0.91 is not a number between 1 and 10.

Instructor Note

For the example at the left, show students that $10^{-4} = \dfrac{1}{10^4} = \dfrac{1}{10{,}000} = 0.0001$. Stress that we are not changing the number, only writing the number in a different form.

Concept Check

Place the correct symbol, < or >, between the two numbers.

1. 5.23×10^{18} ? 5.23×10^{17}

2. 3.12×10^{11} ? 3.12×10^{12}

3. 3.45×10^{-14} ? 3.45×10^{-15}

1. > **2.** < **3.** >

Note: You might extend this exercise by asking students to write a rule for ordering two numbers written in scientific notation: the number with the larger power of ten is the larger number.

Instructor Note

You might ask students to explain why 18.4×10^3, shown at the left, is not in scientific notation.

In-Class Examples (Objective 6.1C)

Write in scientific notation.

1. 2,000,000,000 2×10^9

2. 0.000000016 1.6×10^{-8}

Write in decimal notation.

3. 7.4×10^8 740,000,000

4. 3.54×10^{-6} 0.00000354

Simplify. Write the answer in decimal notation.

5. (0.0000076) (35,000,000) 266

Objective 6.1D

Optional Student Activity

The rules for multiplying and dividing numbers written in scientific notation are the same as those for operating on algebraic expressions. The power of 10 corresponds to the variable expression and the number between 1 and 10 corresponds to the coefficient of the variable. For example:

$$(4x^{-3})(2x^5) = 8x^2$$

and

$$(4 \times 10^{-3})(2 \times 10^5) = 8 \times 10^2$$

or

$$\frac{6x^5}{3x^{-2}} = 2x^7$$

and

$$\frac{6 \times 10^5}{3 \times 10^{-2}} = 2 \times 10^7$$

Simplify the following.

1. $(1.9 \times 10^{12})(3.5 \times 10^7)$ 6.65×10^{19}
2. $(4.2 \times 10^7)(1.8 \times 10^{-5})$ 7.56×10^2
3. $(2.3 \times 10^{-8})(1.4 \times 10^{-6})$ 3.22×10^{-14}
4. $\dfrac{6.12 \times 10^{14}}{1.7 \times 10^9}$ 3.6×10^5
5. $\dfrac{6 \times 10^{-8}}{2.5 \times 10^{-2}}$ 2.4×10^{-6}
6. $\dfrac{5.58 \times 10^{-7}}{3.1 \times 10^{11}}$ 1.8×10^{-18}
7. $\dfrac{9.03 \times 10^6}{4.3 \times 10^{-5}}$ 2.1×10^{11}

Example 8 Write 3.3×10^7 in decimal notation.

Solution $3.3 \times 10^7 = 33{,}000{,}000$

You Try It 8 Write 2.7×10^{-5} in decimal notation.

Your solution 0.000027

Example 9

Simplify: $\dfrac{2{,}400{,}000{,}000 \times 0.0000063}{0.00009 \times 480}$

Solution

$$\frac{2{,}400{,}000{,}000 \times 0.0000063}{0.00009 \times 480}$$
$$= \frac{2.4 \times 10^9 \times 6.3 \times 10^{-6}}{9 \times 10^{-5} \times 4.8 \times 10^2}$$
$$= \frac{(2.4)(6.3) \times 10^{9+(-6)-(-5)-2}}{(9)(4.8)}$$
$$= 0.35 \times 10^6 = 350{,}000$$

You Try It 9

Simplify: $\dfrac{5{,}600{,}000 \times 0.000000081}{900 \times 0.000000028}$

Your solution $1.8 \times 10^4 = 18{,}000$

Solutions on p. S18

Objective D To solve application problems

VIDEO & DVD CD TUTOR WEB SSM

Example 10

How many miles does light travel in one day? The speed of light is 186,000 mi/s. Write the answer in scientific notation.

Strategy

To find the distance traveled:

- Write the speed of light in scientific notation.
- Write the number of seconds in one day in scientific notation.
- Use the equation $d = rt$, where r is the speed of light and t is the number of seconds in one day.

Solution

$r = 186{,}000 = 1.86 \times 10^5$

$t = 24 \cdot 60 \cdot 60 = 86{,}400 = 8.64 \times 10^4$

$$d = rt$$
$$d = (1.86 \times 10^5)(8.64 \times 10^4)$$
$$= 1.86 \times 8.64 \times 10^9$$
$$= 16.0704 \times 10^9$$
$$= 1.60704 \times 10^{10}$$

Light travels 1.60704×10^{10} mi in one day.

You Try It 10

A computer can do an arithmetic operation in 1×10^{-7} s. In scientific notation, how many arithmetic operations can the computer perform in 1 min?

Your strategy

Your solution

6×10^8 operations

Solution on p. S18

In-Class Examples (Objective 6.1D)

Solve. Write the answer in scientific notation.

1. How many kilometers does light travel in 6 h? The speed of light is 300,000 km/s. 6.48×10^9 km
2. A space vehicle travels 9.3×10^7 mi from Earth to Venus at an average velocity of 2.2×10^4 mph. How long does it take the space vehicle to reach Venus? 4.2×10^3 h

6.1 Exercises

Objective A

1. Which expression is the multiplication of two exponential expressions and which is the power of an exponential expression?
   a. $q^4 \cdot q^{10}$ **b.** $(q^4)^{10}$
   a is the multiplication of two exponential expressions. **b** is the power of an exponential expression.

2. In the expression $(a^8b^6)^5$, what is the product and what is the power?
   The product is a^8b^6. The power is 5.

Simplify.

3. $(ab^3)(a^3b)$ a^4b^4
4. $(-2ab^4)(-3a^2b^4)$ $6a^3b^8$
5. $(9xy^2)(-2x^2y^2)$ $-18x^3y^4$
6. $(x^2y)^2$ x^4y^2
7. $(x^2y^4)^4$ x^8y^{16}
8. $(-2ab^2)^3$ $-8a^3b^6$
9. $(-3x^2y^3)^4$ $81x^8y^{12}$
10. $(2^2a^2b^3)^3$ $64a^6b^9$
11. $(3^3a^5b^3)^2$ $729a^{10}b^6$
12. $(xy)(x^2y)^4$ x^9y^5
13. $(x^2y^2)(xy^3)^3$ x^5y^{11}
14. $[(2x)^4]^2$ $256x^8$
15. $[(3x)^3]^2$ $729x^6$
16. $[(x^2y)^4]^5$ $x^{40}y^{20}$
17. $[(ab)^3]^6$ $a^{18}b^{18}$
18. $[(2ab)^3]^2$ $64a^6b^6$
19. $[(2xy)^3]^4$ $4096x^{12}y^{12}$
20. $[(3x^2y^3)^2]^2$ $81x^8y^{12}$
21. $[(2a^4b^3)^3]^2$ $64a^{24}b^{18}$
22. $y^n \cdot y^{2n}$ y^{3n}
23. $x^n \cdot x^{n+1}$ x^{2n+1}
24. $y^{2n} \cdot y^{4n+1}$ y^{6n+1}
25. $y^{3n} \cdot y^{3n-2}$ y^{6n-2}
26. $(a^n)^{2n}$ a^{2n^2}
27. $(a^{n-3})^{2n}$ a^{2n^2-6n}
28. $(y^{2n-1})^3$ y^{6n-3}
29. $(x^{3n+2})^5$ x^{15n+10}
30. $(b^{2n-1})^n$ b^{2n^2-n}
31. $(2xy)(-3x^2yz)(x^2y^3z^3)$ $-6x^5y^5z^4$
32. $(x^2z^4)(2xyz^4)(-3x^3y^2)$ $-6x^6y^3z^8$
33. $(3b^5)(2ab^2)(-2ab^2c^2)$ $-12a^2b^9c^2$
34. $(-c^3)(-2a^2bc)(3a^2b)$ $6a^4b^2c^4$
35. $(-2x^2y^3z)(3x^2yz^4)$ $-6x^4y^4z^5$
36. $(2a^2b)^3(-3ab^4)^2$ $72a^8b^{11}$
37. $(-3ab^3)^3(-2^2a^2b)^2$ $-432a^7b^{11}$
38. $(4ab)^2(-2ab^2c^3)^3$ $-128a^5b^8c^9$
39. $(-2ab^2)(-3a^4b^5)^3$ $54a^{13}b^{17}$

Objective B

Simplify.

40. 2^{-3} $\frac{1}{8}$
41. $\frac{1}{3^{-5}}$ 243
42. $\frac{1}{x^{-4}}$ x^4
43. $\frac{1}{y^{-3}}$ y^3

Section 6.1

Suggested Assignment

Exercises 1–27, odds

More challenging problems:
Exercises 128, 129

Quick Quiz (Objective 6.1A)

Simplify.

1. $(8x^2y^2)(-3xy)$ $-24x^3y^3$
2. $(4a^4b^5)^3$ $64a^{12}b^{15}$
3. $(3ab^2)^2(-4a^2b)^2$ $144a^6b^6$
4. $x^{3n} \cdot x^{2n+3}$ x^{5n+3}

338

44. $\dfrac{2x^{-2}}{y^4}$ $\dfrac{2}{x^2y^4}$

45. $\dfrac{a^3}{4b^{-2}}$ $\dfrac{a^3b^2}{4}$

46. $x^{-3}y$ $\dfrac{y}{x^3}$

47. xy^{-4} $\dfrac{x}{y^4}$

48. $-5x^0$ -5

49. $\dfrac{1}{2x^0}$ $\dfrac{1}{2}$

50. $\dfrac{(2x)^0}{-2^3}$ $-\dfrac{1}{8}$

51. $\dfrac{-3^{-2}}{(2y)^0}$ $-\dfrac{1}{9}$

52. $\dfrac{y^{-7}}{y^{-8}}$ y

53. $\dfrac{y^{-2}}{y^6}$ $\dfrac{1}{y^8}$

54. $(x^2y^{-4})^2$ $\dfrac{x^4}{y^8}$

55. $(x^3y^5)^{-2}$ $\dfrac{1}{x^6y^{10}}$

56. $\dfrac{x^{-2}y^{-11}}{xy^{-2}}$ $\dfrac{1}{x^3y^9}$

57. $\dfrac{x^4y^3}{x^{-1}y^{-2}}$ x^5y^5

58. $\dfrac{a^{-1}b^{-3}}{a^4b^{-5}}$ $\dfrac{b^2}{a^5}$

59. $\dfrac{a^6b^{-4}}{a^{-2}b^5}$ $\dfrac{a^8}{b^9}$

60. $(2a^{-1})^{-2}(2a^{-1})^4$ $\dfrac{4}{a^2}$

61. $(3a)^{-3}(9a^{-1})^{-2}$ $\dfrac{1}{2187a}$

62. $(x^{-2}y)^2(xy)^{-2}$ $\dfrac{1}{x^6}$

63. $(x^{-1}y^2)^{-3}(x^2y^{-4})^{-3}$ $\dfrac{y^6}{x^3}$

64. $\dfrac{50b^{10}}{70b^5}$ $\dfrac{5b^5}{7}$

65. $\dfrac{x^3y^6}{x^6y^2}$ $\dfrac{y^4}{x^3}$

66. $\dfrac{x^{17}y^5}{-x^7y^{10}}$ $-\dfrac{x^{10}}{y^5}$

67. $\dfrac{-6x^2y}{12x^4y}$ $-\dfrac{1}{2x^2}$

68. $\dfrac{2x^2y^4}{(3xy^2)^3}$ $\dfrac{2}{27xy^2}$

69. $\dfrac{-3ab^2}{(9a^2b^4)^3}$ $-\dfrac{1}{243a^5b^{10}}$

70. $\left(\dfrac{-12a^2b^3}{9a^5b^9}\right)^3$ $-\dfrac{64}{27a^9b^{18}}$

71. $\left(\dfrac{12x^3y^2z}{18xy^3z^4}\right)^4$ $\dfrac{16x^8}{81y^4z^{12}}$

72. $\dfrac{(4x^2y)^2}{(2xy^3)^3}$ $\dfrac{2x}{y^7}$

73. $\dfrac{(3a^2b)^3}{(-6ab^3)^2}$ $\dfrac{3a^4}{4b^3}$

74. $\dfrac{(-4x^2y^3)^2}{(2xy^2)^3}$ $2x$

75. $\dfrac{(-3a^2b^3)^2}{(-2ab^4)^3}$ $-\dfrac{9a}{8b^6}$

76. $\dfrac{(-4xy^3)^3}{(-2x^7y)^4}$ $-\dfrac{4y^5}{x^{25}}$

77. $\dfrac{(-8x^2y^2)^4}{(16x^3y^7)^2}$ $\dfrac{16x^2}{y^6}$

78. $\dfrac{a^{5n}}{a^{3n}}$ a^{2n}

79. $\dfrac{b^{6n}}{b^{10n}}$ $\dfrac{1}{b^{4n}}$

80. $\dfrac{-x^{5n}}{x^{2n}}$ $-x^{3n}$

81. $\dfrac{y^{2n}}{-y^{8n}}$ $-\dfrac{1}{y^{6n}}$

82. $\dfrac{x^{2n-1}}{x^{n-3}}$ x^{n+2}

83. $\dfrac{y^{3n+2}}{y^{2n+4}}$ y^{n-2}

84. $\dfrac{a^{3n}b^n}{a^nb^{2n}}$ $\dfrac{a^{2n}}{b^n}$

85. $\dfrac{x^ny^{3n}}{x^ny^{5n}}$ $\dfrac{1}{y^{2n}}$

86. $\dfrac{a^{3n-2}b^{n+1}}{a^{2n+1}b^{2n+2}}$ $\dfrac{a^{n-3}}{b^{n+1}}$

87. $\dfrac{x^{2n-1}y^{n-3}}{x^{n+4}y^{n+3}}$ $\dfrac{x^{n-5}}{y^6}$

88. $\left(\dfrac{4^{-2}xy^{-3}}{x^{-3}y}\right)^3\left(\dfrac{8^{-1}x^{-2}y}{x^4y^{-1}}\right)^{-2}$ $\dfrac{x^{24}}{64y^{16}}$

89. $\left(\dfrac{9ab^{-2}}{8a^{-2}b}\right)^{-2}\left(\dfrac{3a^{-2}b}{2a^2b^{-2}}\right)^3$ $\dfrac{8b^{15}}{3a^{18}}$

90. $\left(\dfrac{2ab^{-1}}{ab}\right)^{-1}\left(\dfrac{3a^{-2}b}{a^2b^2}\right)^{-2}$ $\dfrac{a^8b^4}{18}$

Quick Quiz (Objective 6.1B)

Simplify.

1. $\dfrac{42a^3b^{10}}{-35a^5b^9}$ $-\dfrac{6b}{5a^2}$

2. $\dfrac{3x^{-1}}{x^6}$ $\dfrac{3}{x^7}$

3. $(3xy^{-2})^2$ $\dfrac{9x^2}{y^4}$

4. $\dfrac{x^{2n+3}}{x^{n+4}}$ x^{n-1}

339

Objective C

Write in scientific notation.

91. 0.00000467
4.67×10^{-6}

92. 0.00000005
5×10^{-8}

93. 0.00000000017
1.7×10^{-10}

94. 4,300,000
4.3×10^{6}

95. 200,000,000,000
2×10^{11}

96. 9,800,000,000
9.8×10^{9}

Write in decimal notation.

97. 1.23×10^{-7}
0.000000123

98. 6.2×10^{-12}
0.0000000000062

99. 8.2×10^{15}
8,200,000,000,000,000

100. 6.34×10^{5}
634,000

101. 3.9×10^{-2}
0.039

102. 4.35×10^{9}
4,350,000,000

Simplify. Write the answer in decimal notation.

103. $(3 \times 10^{-12})(5 \times 10^{16})$
150,000

104. $(8.9 \times 10^{-5})(3.2 \times 10^{-6})$
0.0000000002848

105. (0.0000065)(3,200,000,000,000)
20,800,000

106. (480,000)(0.0000000096)
0.004608

107. $\dfrac{9 \times 10^{-3}}{6 \times 10^{5}}$
0.000000015

108. $\dfrac{2.7 \times 10^{4}}{3 \times 10^{-6}}$
9,000,000,000

109. $\dfrac{0.0089}{500{,}000{,}000}$
0.0000000000178

110. $\dfrac{4800}{0.00000024}$
20,000,000,000

111. $\dfrac{0.00056}{0.000000000004}$
140,000,000

112. $\dfrac{0.000000346}{0.0000005}$
0.692

113. $\dfrac{(3.3 \times 10^{-11})(2.7 \times 10^{15})}{8.1 \times 10^{-3}}$
11,000,000

114. $\dfrac{(6.9 \times 10^{27})(8.2 \times 10^{-13})}{4.1 \times 10^{15}}$
1.38

115. $\dfrac{(0.00000004)(84{,}000)}{(0.0003)(1{,}400{,}000)}$
0.000008

116. $\dfrac{(720)(0.0000000039)}{(26{,}000{,}000{,}000)(0.18)}$
0.0000000000000006

Objective D *Application Problems*

Solve. Write the answer in scientific notation.

117. How many kilometers does light travel in one day? The speed of light is 300,000 km/s.
2.592×10^{10} km

118. How many meters does light travel in 8 h? The speed of light is 300,000,000 m/s.
8.64×10^{12} m

Quick Quiz (Objective 6.1C)

Write in scientific notation.

1. 87,000,000,000 8.7×10^{10}

2. 0.00000029 2.9×10^{-7}

Write in decimal notation.

3. 5.34×10^{6} 5,340,000

4. 6×10^{-9} 0.000000006

Simplify. Write the answer in decimal notation.

5. (36,000) (0.00000085) 0.0306

Answers to Writing Exercises

129a. The Distributive Property was used incorrectly. $2x + 3x = (2 + 3)x = 5x$

b. The Distributive Property was used incorrectly. $a - (b - c) = a - b + c$

c. The Definition of Zero as an Exponent was used incorrectly. $x^0 = 1$

d. The Rule for Simplifying Powers of Exponential Expressions was used incorrectly. $(x^4)^5 = x^{4(5)} = x^{20}$

e. The Rule for Multiplying Exponential Expressions was used incorrectly. $x^2 \cdot x^3 = x^{2+3} = x^5$

f. The Rule for Multiplying Exponential Expressions was used incorrectly. $b^{m+n} = b^m \cdot b^n$

119. The rover Sojourner landed on Mars in July of 1997. It took 11 min for the commands from a computer on Earth to travel to the Sojourner, a distance of 119 million miles. How fast did the signals from Earth to Mars travel? Write the answer in scientific notation.
$1.0\overline{81} \times 10^7$ mi/min

120. In the summer of 2001, the federal debt was 3×10^{12} dollars. How much would each American citizen have to contribute in order to pay off the debt? Use 300,000,000 as the number of citizens.
1×10^4 dollars

121. A high-speed centrifuge makes 9×10^7 revolutions each minute. Find the time in seconds for the centrifuge to make one revolution.
$6.\overline{6} \times 10^{-7}$ s

122. How long does it take light to travel to Earth from the sun? The sun is 9.3×10^7 mi from Earth, and light travels 1.86×10^5 mi/s.
5×10^2 s

123. The mass of Earth is 5.9×10^{27} g. The mass of the sun is 2×10^{33} g. How many times heavier is the sun than Earth?
3.38983×10^5 times heavier

124. One astronomical unit (A.U.) is 9.3×10^7 mi. The star Pollux in the constellation Gemini is 1.8228×10^{12} mi from Earth. Find the distance from Pollux to Earth in astronomical units.
1.96×10^4 A.U.

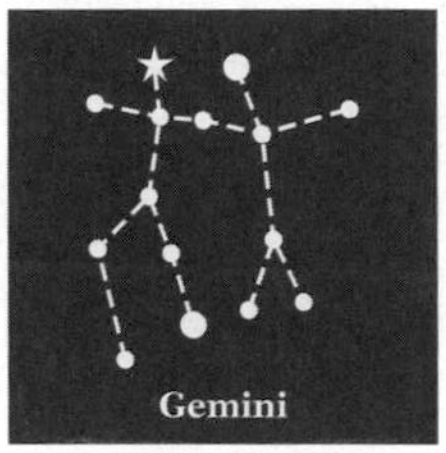

125. The weight of 31 million orchid seeds is one ounce. Find the weight of one orchid seed.
3.225806×10^{-8} oz

126. One light year, an astronomical unit of distance, is the distance that light will travel in one year. Light travels 1.86×10^5 mi/s. Find the measure of one light year in miles. Use a 365-day year.
5.865696×10^{12} mi

127. The Coma cluster of galaxies is approximately 2.8×10^8 light years from Earth. Find the distance, in miles, from the Coma cluster to Earth. (See Exercise 126 for the definition of one light year. Use a 365-day year.)
1.6423949×10^{21} mi

APPLYING THE CONCEPTS

128. Simplify.

a. $1 + (1 + (1 + 2^{-1})^{-1})^{-1}$
$\frac{8}{5}$

b. $2 - (2 - (2 - 2^{-1})^{-1})^{-1}$
$\frac{5}{4}$

129. Correct the error in each of the following expressions. Explain which rule or property was used incorrectly.

a. $2x + 3x = 5x^2$ **b.** $a - (b - c) = a - b - c$

c. $x^0 = 0$ **d.** $(x^4)^5 = x^9$

e. $x^2 \cdot x^3 = x^6$ **f.** $b^{m+n} = b^m + b^n$

Quick Quiz (Objective 6.1D)

Solve. Write the answer in scientific notation.

1. How many meters does light travel in 4 h? The speed of light is 300,000,000 m/s. 4.32×10^{12} m
2. A space vehicle travels at an average velocity of 2.5×10^4 mph. How far does it travel each day? 6.0×10^5 mi

6.2 Introduction to Polynomials

Objective A To evaluate polynomial functions

VIDEO & DVD CD TUTOR WEB SSM

TAKE NOTE
An analogy may help you understand these terms. Polynomial is similar to the word *car*; Chevrolet and Ford are types of cars, just as monomials and binomials are types of polynomials.

A **polynomial** is a variable expression in which the terms are monomials.

A polynomial of one term is a **monomial**. $5x$

A polynomial of two terms is a **binomial**. $5x^2y + 6x$

A polynomial of three terms is a **trinomial**. $3x^2 + 9xy - 5y$

Polynomials with more than three terms do not have special names.

The **degree of a polynomial** is the greatest of the degrees of any of its terms.

$3x + 2$	degree 1
$3x^2 + 2x - 4$	degree 2
$4x^3y^2 + 6x^4$	degree 5
$3x^{2n} - 5x^n - 2$	degree $2n$

The terms of a polynomial in one variable are usually arranged so that the exponents of the variable decrease from left to right. This is called **descending order**.

$2x^2 - x + 8$

$3y^3 - 3y^2 + y - 12$

For a polynomial in more than one variable, descending order may refer to any one of the variables.

The polynomial at the right is shown first in descending order of the x variable and then in descending order of the y variable.

$2x^2 + 3xy + 5y^2$

$5y^2 + 3xy + 2x^2$

Polynomial functions have many applications in mathematics. In general, a **polynomial function** is an expression whose terms are monomials. The **linear function** given by $f(x) = mx + b$ is an example of a polynomial function. It is a polynomial function of degree one. A second-degree polynomial function, called a **quadratic function**, is given by the equation $f(x) = ax^2 + bx + c$, $a \neq 0$. A third-degree polynomial function is called a **cubic function**.

To evaluate a polynomial function, replace the variable by its value and simplify.

➡ Given $P(x) = x^3 - 3x^2 + 4$, evaluate $P(-3)$.

$$\begin{aligned} P(x) &= x^3 - 3x^2 + 4 \\ P(-3) &= (-3)^3 - 3(-3)^2 + 4 \\ &= -27 - 3(9) + 4 \\ &= -27 - 27 + 4 \\ &= -50 \end{aligned}$$

• Substitute −3 for x and simplify.

The **leading coefficient** of a polynomial function is the coefficient of the variable with the largest exponent. The constant term is the term without a variable.

Objective 6.2A

New Vocabulary
polynomial
binomial
trinomial
degree of a polynomial
descending order
polynomial function
quadratic function
cubic function
leading coefficient

Vocabulary to Review
term
monomial
linear function
evaluate a function
constant term

Discuss the Concepts

1. State whether the polynomial is a monomial, a binomial, or a trinomial. Explain your answer.
   a. $8x^4 - 6x^2$
   b. $4a^2b^2 + 9ab + 10$
   c. $7x^3y^4$
2. State whether or not the expression is a polynomial. Explain your answer.
   a. $\frac{1}{5}x^3 + \frac{1}{2}x$
   b. $\frac{1}{5x^2} + \frac{1}{2x}$
   c. $x + \sqrt{5}$

In-Class Examples (Objective 6.2A)

1. Given $P(x) = 2x^2 - 4x + 5$, evaluate $P(3)$. 11
2. Given $f(x) = x^3 - 3x^2 + 6x - 7$, evaluate $f(-1)$. −17

Concept Check

Determine whether the statement is always true, sometimes true, or never true.

1. The terms of a polynomial are monomials. Always true
2. The leading coefficient of $4 - 2x - 3x^2$ is 4. Never true
3. $f(x) = 3x^2 - 2x + x^{-1}$ is a polynomial function. Never true
4. $f(x) = x^2 + 3x + 5x^3 - 2$ is a quadratic function. Never true
5. A binomial is a polynomial of degree 2. Sometimes true
6. A cubic polynomial is a polynomial that contains three terms. Sometimes true

➡ Find the leading coefficient, the constant term, and the degree of the polynomial function $P(x) = 7x^4 - 3x^2 + 2x - 4$.

The leading coefficient is 7, the constant term is −4, and the degree is 4.

The three equations below do not represent polynomial functions.

$f(x) = 3x^2 + 2x^{-1}$	A polynomial function does not have a variable raised to a negative power.
$g(x) = 2\sqrt{x} - 3$	A polynomial function does not have a variable expression within a radical.
$h(x) = \dfrac{x}{x-1}$	A polynomial function does not have a variable in the denominator of a fraction.

The graph of a linear function is a straight line and can be found by plotting just two points. The graph of a polynomial function of degree greater than one is a curve. Consequently, many points may have to be found before an accurate graph can be drawn.

Evaluating the quadratic function given by the equation $f(x) = x^2 - x - 6$ when $x = -3, -2, -1, 0, 1, 2, 3$, and 4 gives the points shown in Figure 1 below. For instance, $f(-3) = 6$, so $(-3, 6)$ is graphed; $f(2) = -4$, so $(2, -4)$ is graphed; and $f(4) = 6$, so $(4, 6)$ is graphed. Evaluating the function when x is not an integer, such as $x = -\frac{3}{2}$ and $x = \frac{5}{2}$, produces more points to graph, as shown in Figure 2. Connecting the points with a smooth curve results in Figure 3, which is the graph of f.

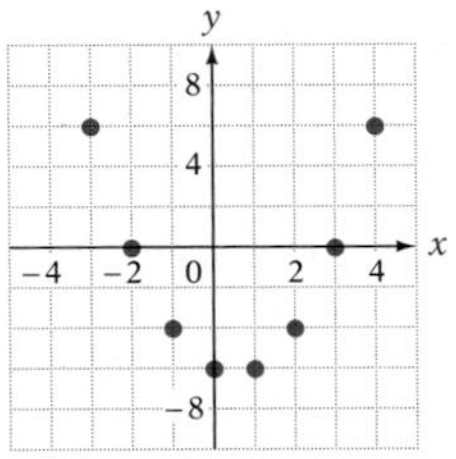

Figure 1

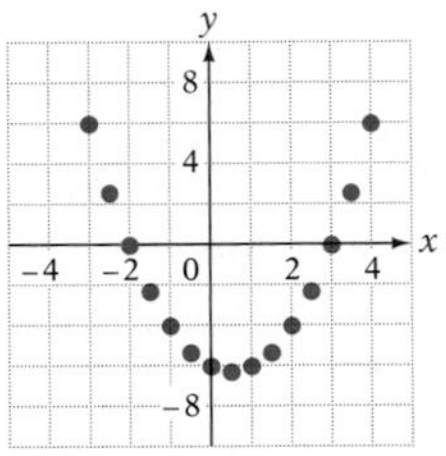

Figure 2

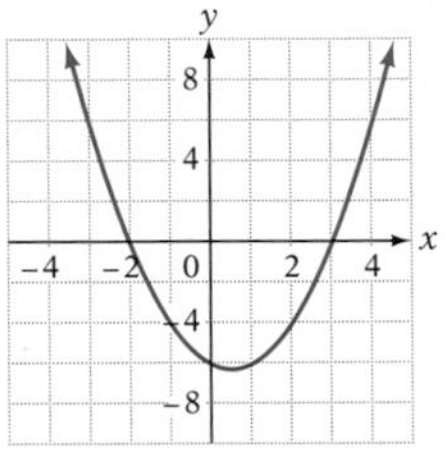

Figure 3

CALCULATOR NOTE

You can verify the graphs of these polynomial functions by using a graphing calculator. See the Appendix: Guidelines for Using a Graphing Calculator for instructions on graphing functions.

Here is an example of graphing a cubic function, $P(x) = x^3 - 2x^2 - 5x + 6$. Evaluating the function when $x = -2, -1, 0, 1, 2, 3$, and 4 gives the graph in Figure 4 below. Evaluating at some noninteger values gives the graph in Figure 5. Finally, connecting the dots with a smooth curve gives the graph in Figure 6.

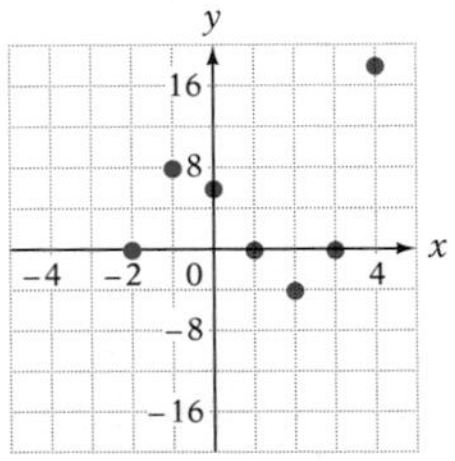

Figure 4

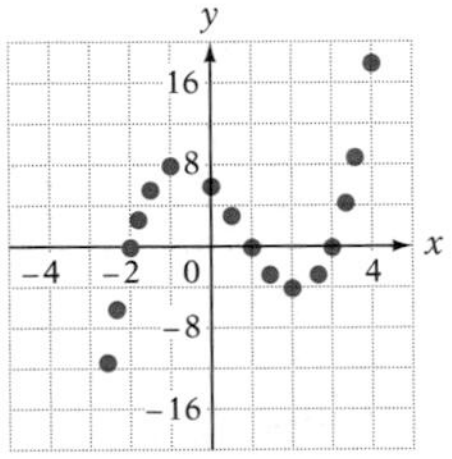

Figure 5

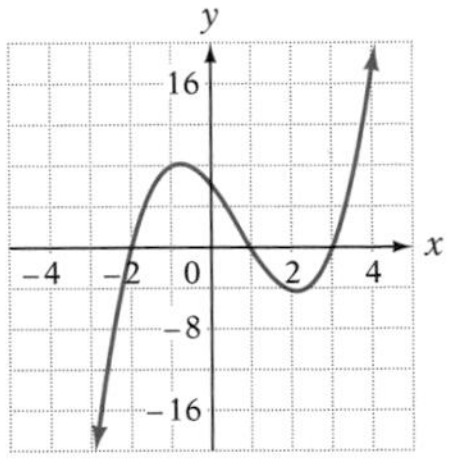

Figure 6

Example 1
Given $P(x) = x^3 + 3x^2 - 2x + 8$, evaluate $P(-2)$.

Solution

$$\begin{aligned} P(x) &= x^3 + 3x^2 - 2x + 8 \\ P(-2) &= (-2)^3 + 3(-2)^2 - 2(-2) + 8 \\ &= (-8) + 3(4) + 4 + 8 \\ &= -8 + 12 + 4 + 8 \\ &= 16 \end{aligned}$$

You Try It 1
Given $R(x) = -2x^4 - 5x^3 + 2x - 8$, evaluate $R(2)$.

Your solution
-76

Example 2
Find the leading coefficient, the constant term, and the degree of the polynomial.
$P(x) = 5x^6 - 4x^5 - 3x^2 + 7$

Solution
The leading coefficient is 5, the constant term is 7, and the degree is 6.

You Try It 2
Find the leading coefficient, the constant term, and the degree of the polynomial.
$r(x) = -3x^4 + 3x^3 + 3x^2 - 2x - 12$

Your solution
The leading coefficient is -3, the constant term is -12, and the degree is 4.

Example 3
Which of the following is a polynomial function?
a. $P(x) = 3x^{\frac{1}{2}} + 21x^2 - 3$
b. $T(x) = 3\sqrt{x} - 2x^2 - 3x + 2$
c. $R(x) = 14x^3 - \pi x^2 + 3x + 2$

Solution
a. This is not a polynomial function. A polynomial function does not have a variable raised to a fractional power.
b. This is not a polynomial function. A polynomial function does not have a variable expression within a radical.
c. This is a polynomial function.

You Try It 3
Which of the following is a polynomial function?
a. $R(x) = 5x^{14} - 5$
b. $V(x) = -x^{-1} + 2x - 7$
c. $P(x) = 2x^4 - 3\sqrt{x} - 3$

Your solution
a. yes
b. no
c. no

Example 4
Graph $f(x) = x^2 - 2$.

Solution

x	$y = f(x)$
-3	7
-2	2
-1	-1
0	-2
1	-1
2	2
3	7

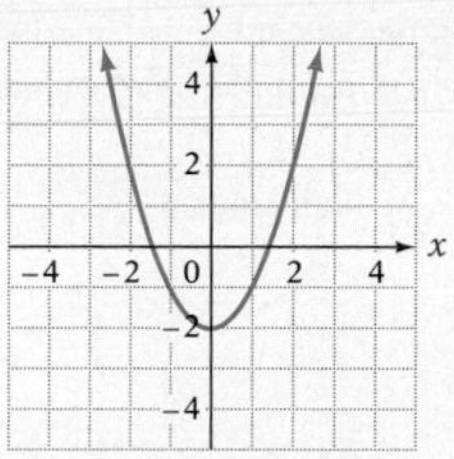

You Try It 4
Graph $f(x) = x^2 + 2x - 3$.

Your solution

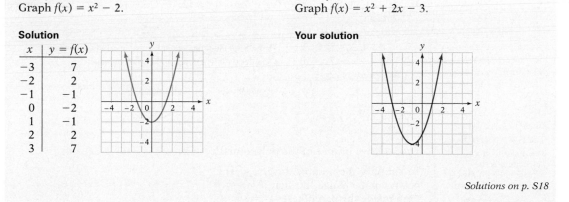

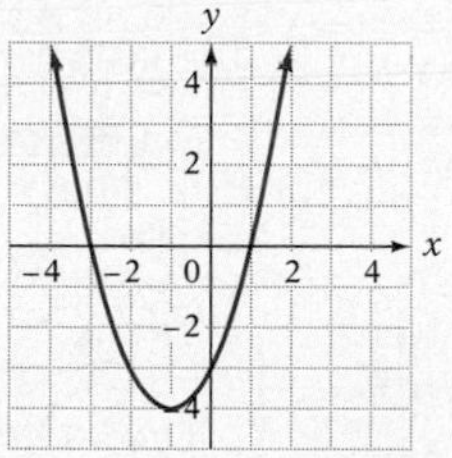

Solutions on p. S18

Optional Student Activity

1. The height h, in feet, of a golf ball t seconds after it has been struck is given by $h(t) = -16t^2 + 60t$. Determine the height of the ball 3 s after it is hit. 36 ft
2. Some forecasters predict that the revenue generated by business on the Internet from 1997–2002 can be approximated by the function $R(t) = 15.8t^2 - 17.2t + 10.2$, where R is the annual revenue in billions of dollars and t is the time in years, with $t = 0$ corresponding to the year 1997. Use this function to approximate the annual revenue in the year 2000. \$100.8 billion
3. If \$2000 is deposited into an Individual Retirement Account (IRA), then the value, V, of that investment three years later is given by the cubic polynomial function $V(r) = 2000r^3 + 6000r + 2000$, where r is the interest rate (as a decimal) earned on the investment. Determine the value after 3 years of \$2000 deposited in an IRA that earns an interest rate of 7%. \$2420.69

Objective 6.2B

New Vocabulary
additive inverse of a polynomial

Vocabulary to Review
additive inverse
definition of subtraction

Instructor Note
It may help to show students that adding polynomials is related to adding whole numbers. Adding the coefficients of like terms using a vertical format is similar to arranging whole numbers to be added in columns with the units digits aligned.

Concept Check

1. Find the length of $\overline{AC}$ given that the length of $\overline{AB}$ is $3x^2 - 4x + 5$ and the length of $\overline{BC}$ is $8x^2 + 6x - 1$. $11x^2 + 2x + 4$

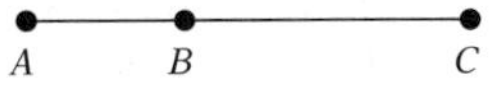

2. The length of $\overline{LN}$ is $7a^2 + 4a - 3$. Given that the length of $\overline{LM}$ is $2a^2 + a + 6$, find the length of $\overline{MN}$. $5a^2 + 3a - 9$

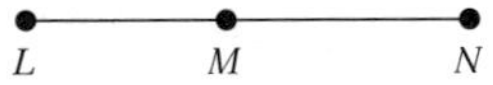

3. Find the perimeter of the rectangle. The dimensions given are in kilometers. $(8d^2 + 12d + 4)$ km

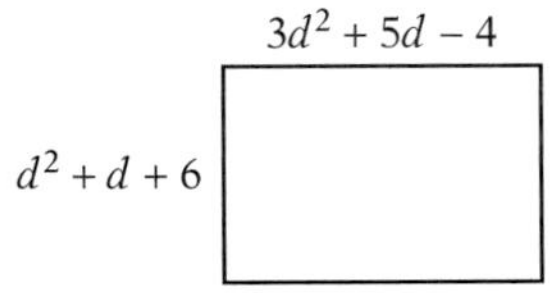

Example 5
Graph $f(x) = x^3 - 1$.

Solution

x	$y = f(x)$
-2	-9
-1	-2
0	-1
1	0
2	7

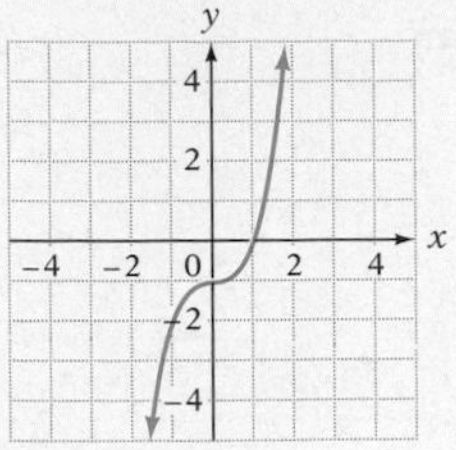

You Try It 5
Graph $f(x) = -x^3 + 1$.

Your solution

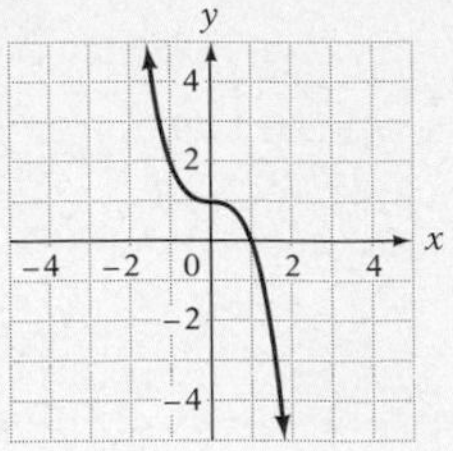

Solution on p. S18

Objective B To add or subtract polynomials

VIDEO & DVD · CD TUTOR · WEB · SSM

Polynomials can be added by combining like terms. Either a vertical or a horizontal format can be used.

➡ Add $(3x^3 - 7x + 2) + (7x^2 + 2x - 7)$. Use a horizontal format.

$(3x^3 - 7x + 2) + (7x^2 + 2x - 7)$
$= 3x^3 + 7x^2 + (-7x + 2x) + (2 - 7)$

- Use the Commutative and Associative Properties of Addition to rearrange and group like terms.

$= 3x^3 + 7x^2 - 5x - 5$

- Combine like terms.

➡ Add $(4x^2 + 5x - 3) + (7x^3 - 7x + 1) + (2x - 3x^2 + 4x^3 + 1)$. Use a vertical format.

$$\begin{array}{rrrr} & 4x^2 & + 5x & - 3 \\ 7x^3 & & - 7x & + 1 \\ 4x^3 & - 3x^2 & + 2x & + 1 \\ \hline 11x^3 & + x^2 & & - 1 \end{array}$$

- Arrange the terms of each polynomial in descending order with like terms in the same column.
- Add the terms in each column.

TAKE NOTE
The opposite of a polynomial is the polynomial with the sign of every term changed.

The additive inverse of the polynomial $x^2 + 5x - 4$ is $-(x^2 + 5x - 4)$.

To simplify the additive inverse of a polynomial, change the sign of every term inside the parentheses.

$-(x^2 + 5x - 4) = -x^2 - 5x + 4$

In-Class Examples (Objective 6.2B)
Simplify.

1. $(4x^2 + 3x - 5) + (x^2 - 7x + 10)$ $5x^2 - 4x + 5$
2. $(5x^2 - x + 6) + (-2x^2 + 3x - 11)$ $3x^2 + 2x - 5$
3. $(x^2 - 2x + 7) - (3x^2 - 4x + 7)$ $-2x^2 + 2x$
4. $(4a^2 - 7a) - (-6a^2 + 5a - 7)$ $10a^2 - 12a + 7$

TAKE NOTE
This is the same definition used for subtraction of integers: Subtraction is addition of the opposite.

To subtract two polynomials, add the additive inverse of the second polynomial to the first.

⇒ Subtract $(3x^2 - 7xy + y^2) - (-4x^2 + 7xy - 3y^2)$. Use a horizontal format.

$$\begin{aligned}&(3x^2 - 7xy + y^2) - (-4x^2 + 7xy - 3y^2)\\ &\quad = (3x^2 - 7xy + y^2) + (4x^2 - 7xy + 3y^2)\\ &\quad = 7x^2 - 14xy + 4y^2\end{aligned}$$

- Rewrite the subtraction as the addition of the additive inverse.
- Combine like terms.

⇒ Subtract $(6x^3 - 3x + 7) - (3x^2 - 5x + 12)$. Use a vertical format.

$$\begin{aligned}&(6x^3 - 3x + 7) - (3x^2 - 5x + 12)\\ &\quad = (6x^3 - 3x + 7) + (-3x^2 + 5x - 12)\end{aligned}$$

- Rewrite subtraction as the addition of the additive inverse.

$$\begin{array}{rrrr} 6x^3 & & -\,3x & +\ 7\\ & -3x^2 & +\,5x & -\ 12\\ \hline 6x^3 & -\,3x^2 & +\,2x & -\ 5\end{array}$$

- Arrange the terms of each polynomial in descending order with like terms in the same column.
- Combine the terms in each column.

Functional notation can be used when adding or subtracting polynomials.

⇒ Given $P(x) = 3x^2 - 2x + 4$ and $R(x) = -5x^3 + 4x + 7$, find $P(x) + R(x)$.

$$\begin{aligned}P(x) + R(x) &= (3x^2 - 2x + 4) + (-5x^3 + 4x + 7)\\ &= -5x^3 + 3x^2 + 2x + 11\end{aligned}$$

⇒ Given $P(x) = -5x^2 + 8x - 4$ and $R(x) = -3x^2 - 5x + 9$, find $P(x) - R(x)$.

$$\begin{aligned}P(x) - R(x) &= (-5x^2 + 8x - 4) - (-3x^2 - 5x + 9)\\ &= (-5x^2 + 8x - 4) + (3x^2 + 5x - 9)\\ &= -2x^2 + 13x - 13\end{aligned}$$

⇒ Given $P(x) = 3x^2 - 5x + 6$ and $R(x) = 2x^2 - 5x - 7$, find $S(x)$, the sum of the two polynomials.

$$\begin{aligned}S(x) = P(x) + R(x) &= (3x^2 - 5x + 6) + (2x^2 - 5x - 7)\\ &= 5x^2 - 10x - 1\end{aligned}$$

Note that evaluating $P(x) = 3x^2 - 5x + 6$ and $R(x) = 2x^2 - 5x - 7$ at, for example, $x = 3$, and then adding the values, is the same as evaluating $S(x) = 5x^2 - 10x - 1$ at 3.

$P(3) = 3(3)^2 - 5(3) + 6 = 27 - 15 + 6 = 18$

$R(3) = 2(3)^2 - 5(3) - 7 = 18 - 15 - 7 = -4$

$P(3) + R(3) = 18 + (-4) = 14$

$S(3) = 5(3)^2 - 10(3) - 1 = 45 - 30 - 1 = 14$

Optional Student Activity

A company's **revenue** is the money the company earns by selling its products. A company's **cost** is the money it spends to manufacture and sell its products. A company's **profit** is the difference between its revenue and its cost. This relationship is expressed by the formula $P = R - C$, where P is the profit, R is the revenue, and C is the cost.

1. A company manufactures and sells wood stoves. The total monthly cost, in dollars, to produce n wood stoves is $30n + 2000$. The company's revenue, in dollars, obtained from selling all n wood stoves is $-0.4n^2 + 150n$. Express the company's monthly profit in terms of n.
   $(-0.4n^2 + 120n - 2000)$ dollars
2. A company's total monthly cost, in dollars, for manufacturing and selling n videotapes is $35n + 2000$. The company's revenue, in dollars, from selling all n videotapes is $-0.2n^2 + 175n$. Express the company's monthly profit in terms of n.
   $(-0.2n^2 + 140n - 2000)$ dollars

Example 6
Add
$(4x^2 - 3xy + 7y^2) + (-3x^2 + 7xy + y^2)$.
Use a vertical format.

Solution

$$\begin{array}{r} 4x^2 - 3xy + 7y^2 \\ \underline{-3x^2 + 7xy + y^2} \\ x^2 + 4xy + 8y^2 \end{array}$$

You Try It 6
Add
$(-3x^2 - 4x + 9) + (-5x^2 - 7x + 1)$.
Use a vertical format.

Your solution
$-8x^2 - 11x + 10$

Example 7
Subtract
$(3x^2 - 2x + 4) - (7x^2 + 3x - 12)$.
Use a vertical format.

Solution
Add the additive inverse of $7x^2 + 3x - 12$ to $3x^2 - 2x + 4$.

$$\begin{array}{r} 3x^2 - 2x + 4 \\ \underline{-7x^2 - 3x + 12} \\ -4x^2 - 5x + 16 \end{array}$$

You Try It 7
Subtract
$(-5x^2 + 2x - 3) - (6x^2 + 3x - 7)$.
Use a vertical format.

Your solution
$-11x^2 - x + 4$

Example 8
Given $P(x) = -3x^2 + 2x - 6$ and $R(x) = 4x^3 - 3x + 4$, find $S(x) = P(x) + R(x)$. Evaluate $S(-2)$.

Solution

$$\begin{aligned} S(x) &= P(x) + R(x) \\ &= (-3x^2 + 2x - 6) + (4x^3 - 3x + 4) \\ &= 4x^3 - 3x^2 - x - 2 \end{aligned}$$

$$\begin{aligned} S(-2) &= 4(-2)^3 - 3(-2)^2 - (-2) - 2 \\ &= 4(-8) - 3(4) - (-2) - 2 \\ &= -32 - 12 + 2 - 2 \\ &= -44 \end{aligned}$$

You Try It 8
Given $P(x) = 4x^3 - 3x^2 + 2$ and $R(x) = -2x^2 + 2x - 3$, find $S(x) = P(x) + R(x)$. Evaluate $S(-1)$.

Your solution
$S(x) = 4x^3 - 5x^2 + 2x - 1$
$S(-1) = -12$

Example 9
Given $P(x) = 2x^{2n} - 3x^n + 7$ and $R(x) = 3x^{2n} + 3x^n + 5$, find $D(x) = P(x) - R(x)$.

Solution

$$\begin{aligned} D(x) &= P(x) - R(x) \\ &= (2x^{2n} - 3x^n + 7) - (3x^{2n} + 3x^n + 5) \\ &= (2x^{2n} - 3x^n + 7) + (-3x^{2n} - 3x^n - 5) \\ &= -x^{2n} - 6x^n + 2 \end{aligned}$$

You Try It 9
Given $P(x) = 5x^{2n} - 3x^n - 7$ and $R(x) = -2x^{2n} - 5x^n + 8$, find $D(x) = P(x) - R(x)$.

Your solution
$7x^{2n} + 2x^n - 15$

Solutions on p. S18

347

6.2 Exercises

Objective A

1. Given $P(x) = 3x^2 - 2x - 8$, evaluate $P(3)$.
13

2. Given $P(x) = -3x^2 - 5x + 8$, evaluate $P(-5)$.
-42

3. Given $R(x) = 2x^3 - 3x^2 + 4x - 2$, evaluate $R(2)$.
10

4. Given $R(x) = -x^3 + 2x^2 - 3x + 4$, evaluate $R(-1)$.
10

5. Given $f(x) = x^4 - 2x^2 - 10$, evaluate $f(-1)$.
-11

6. Given $f(x) = x^5 - 2x^3 + 4x$, evaluate $f(2)$.
24

Which of the following define a polynomial function? For those that are polynomial functions, identify **a.** the leading coefficient, **b.** the constant term, and **c.** the degree.

7. $P(x) = -x^2 + 3x + 8$
$-1, 8, 2$

8. $P(x) = 3x^4 - 3x - 7$
$3, -7, 4$

9. $R(x) = \dfrac{x}{x+1}$
not a polynomial

10. $R(x) = \dfrac{3x^2 - 2x + 1}{x}$
not a polynomial

11. $f(x) = \sqrt{x} - x^2 + 2$
not a polynomial

12. $f(x) = x^2 - \sqrt{x+2} - 8$
not a polynomial

13. $g(x) = 3x^5 - 2x^2 + \pi$
$3, \pi, 5$

14. $g(x) = -4x^5 - 3x^2 + x - \sqrt{7}$
$-4, -\sqrt{7}, 5$

15. $P(x) = 3x^2 - 5x^3 + 2$
$-5, 2, 3$

16. $P(x) = x^2 - 5x^4 - x^6$
$-1, 0, 6$

17. $R(x) = 14$
$14, 14, 0$

18. $R(x) = \dfrac{1}{x} + 2$
not a polynomial

Graph.

19. $P(x) = x^2 - 1$

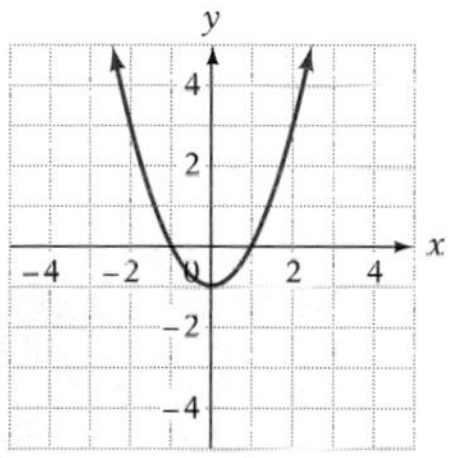

20. $P(x) = 2x^2 + 3$

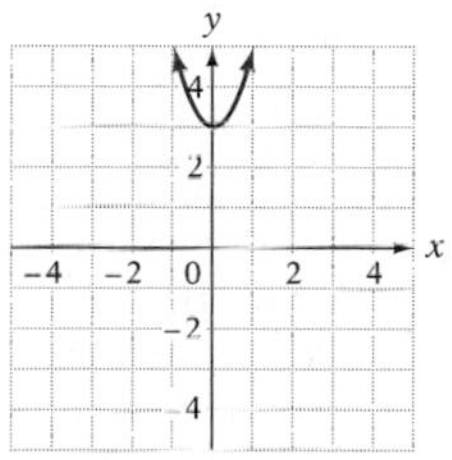

21. $R(x) = x^3 + 2$

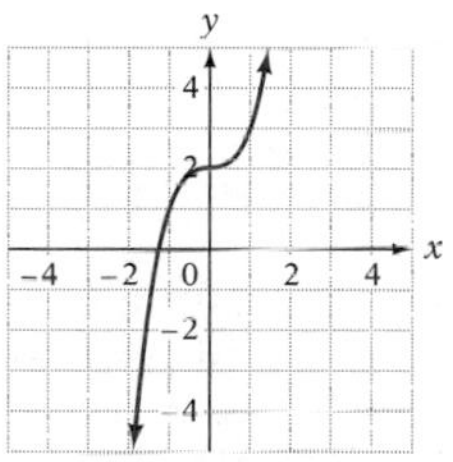

22. $R(x) = x^4 + 1$

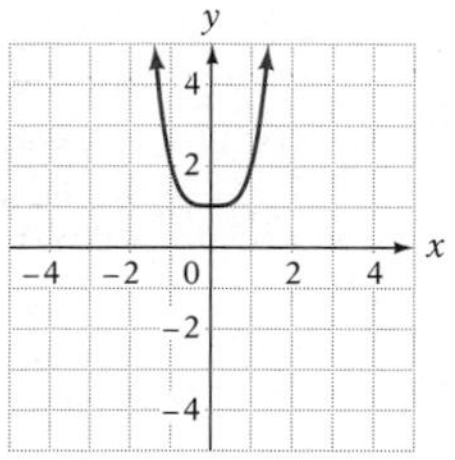

23. $f(x) = x^3 - 2x$

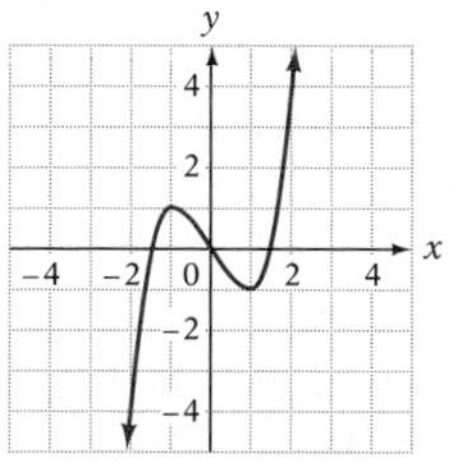

24. $f(x) = x^2 - x - 2$

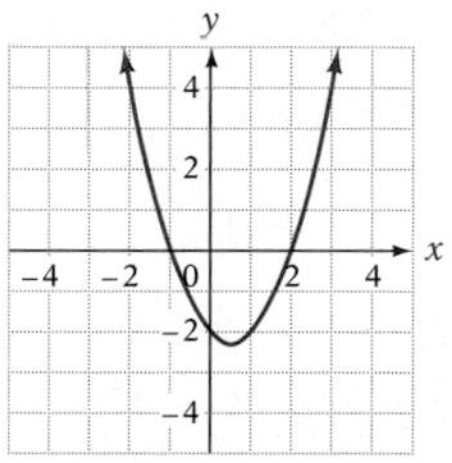

Section 6.2

Suggested Assignment
Exercises 1–37, odds
More challenging problems:
Exercises 39–42

Quick Quiz (Objective 6.2A)

1. Given $P(x) = 4x^2 - 3x + 6$, evaluate $P(2)$. 16
2. Given $f(x) = 2x^3 + 4x^2 - 5x + 8$, evaluate $f(-2)$. 18

Answers to Writing Exercises

41. If $P(x)$ is a third-degree polynomial and $Q(x)$ is a fourth-degree polynomial, then $P(x) + Q(x)$ is a fourth-degree polynomial. For example, let $P(x) = 2x^3 + 4x^2 - 3x + 6$ and let $Q(x) = x^4 - 5x + 1$. Then $P(x) + Q(x) = x^4 + 2x^3 + 4x^2 - 8x + 7$, a fourth-degree polynomial.

42. If $P(x)$ is a fifth-degree polynomial and $Q(x)$ is a fourth-degree polynomial, then $P(x) - Q(x)$ is a fifth-degree polynomial. For example, let $P(x) = 8x^5 - x^4 + 3x^2 - 1$ and let $Q(x) = -x^4 + x^3 + x^2 + 2x + 5$. Then $P(x) - Q(x) = 8x^5 - x^3 + 2x^2 - 2x - 6$, a fifth-degree polynomial.

Objective B

Add or subtract. Use a vertical format.

25. $(5x^2 + 2x - 7) + (x^2 - 8x + 12)$
$6x^2 - 6x + 5$

26. $(3x^2 - 2x + 7) + (-3x^2 + 2x - 12)$
-5

27. $(x^2 - 3x + 8) - (2x^2 - 3x + 7)$
$-x^2 + 1$

28. $(2x^2 + 3x - 7) - (5x^2 - 8x - 1)$
$-3x^2 + 11x - 6$

Add or subtract. Use a horizontal format for Exercises 29–32.

29. $(3y^2 - 7y) + (2y^2 - 8y + 2)$
$5y^2 - 15y + 2$

30. $(-2y^2 - 4y - 12) + (5y^2 - 5y)$
$3y^2 - 9y - 12$

31. $(2a^2 - 3a - 7) - (-5a^2 - 2a - 9)$
$7a^2 - a + 2$

32. $(3a^2 - 9a) - (-5a^2 + 7a - 6)$
$8a^2 - 16a + 6$

33. Given $P(x) = x^2 - 3x + 1$ and $R(x) = 2x^2 - 3$, find $P(x) + R(x)$.
$3x^2 - 3x - 2$

34. Given $P(x) = x^{2n} + 7x^n - 3$ and $R(x) = -x^{2n} + 2x^n + 8$, find $P(x) + R(x)$.
$9x^n + 5$

35. Given $P(x) = 3x^2 + 2$ and $R(x) = -5x^2 + 2x - 3$, find $P(x) - R(x)$.
$8x^2 - 2x + 5$

36. Given $P(x) = 2x^{2n} - x^n - 1$ and $R(x) = 5x^{2n} + 7x^n + 1$, find $P(x) - R(x)$.
$-3x^{2n} - 8x^n - 2$

37. Given $P(x) = 3x^4 - 3x^3 - x^2$ and $R(x) = 3x^3 - 7x^2 + 2x$, find $S(x) = P(x) + R(x)$. Evaluate $S(2)$.
$S(x) = 3x^4 - 8x^2 + 2x$;
$S(2) = 20$

38. Given $P(x) = 3x^4 - 2x + 1$ and $R(x) = 3x^5 - 5x - 8$, find $S(x) = P(x) + R(x)$. Evaluate $S(-1)$.
$S(x) = 3x^5 + 3x^4 - 7x - 7$;
$S(-1) = 0$

APPLYING THE CONCEPTS

39. For what value of k is the given equation an identity?
a. $(2x^3 + 3x^2 + kx + 5) - (x^3 + 2x^2 + 3x + 7) = x^3 + x^2 + 5x - 2$
b. $(6x^3 + kx^2 - 2x - 1) - (4x^3 - 3x^2 + 1) = 2x^3 - x^2 - 2x - 2$
a. $k = 8$ **b.** $k = -4$

40. The deflection D (in inches) of a beam that is uniformly loaded is given by the polynomial function $D(x) = 0.005x^4 - 0.1x^3 + 0.5x^2$, where x is the distance in feet from one end of the beam. See the figure at the right. The maximum deflection occurs when x is the midpoint of the beam. Determine the maximum deflection for the beam in the diagram.
3.125 in.

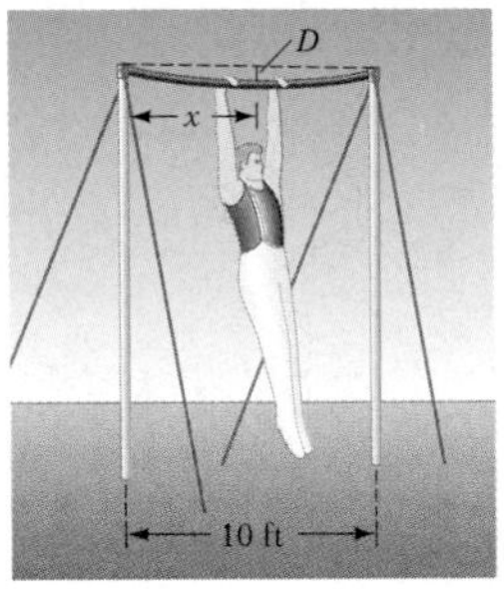

41. If $P(x)$ is a third-degree polynomial and $Q(x)$ is a fourth-degree polynomial, what can be said about the degree of $P(x) + Q(x)$? Give some examples of polynomials that support your answer.

42. If $P(x)$ is a fifth-degree polynomial and $Q(x)$ is a fourth-degree polynomial, what can be said about the degree of $P(x) - Q(x)$? Give some examples of polynomials that support your answer.

Quick Quiz (Objective 6.2B)

Simplify.

1. $(6x^2 + 4x - 3) + (2x^2 - 5x + 7)$ $8x^2 - x + 4$
2. $(7x^2 - 5x + 6) + (-3x^2 + 4x - 10)$ $4x^2 - x - 4$
3. $(3x^2 + 2x - 6) - (4x^2 - 7x - 3)$ $-x^2 + 9x - 3$
4. $(2a^4 - 3a + 2) - (-4a - 7)$ $2a^4 + a + 9$

349

6.3 Multiplication of Polynomials

Objective A To multiply a polynomial by a monomial

VIDEO & DVD CD TUTOR WEB SSM

To multiply a polynomial by a monomial, use the Distributive Property and the Rule for Multiplying Exponential Expressions.

➡ Multiply: $-3a(4a^2 - 5a + 6)$

$$-3a(4a^2 - 5a + 6) = -3a(4a^2) - (-3a)(5a) + (-3a)(6)$$
$$= -12a^3 + 15a^2 - 18a$$

- Use the Distributive Property. This step is frequently done mentally.

Example 1
Multiply: $(5x + 4)(-2x)$

Solution
$(5x + 4)(-2x) = -10x^2 - 8x$

You Try It 1
Multiply: $(-2y + 3)(-4y)$

Your solution
$8y^2 - 12y$

Example 2
Multiply: $2a^2b(4a^2 - 2ab + b^2)$

Solution
$2a^2b(4a^2 - 2ab + b^2)$
$= 8a^4b - 4a^3b^2 + 2a^2b^3$

You Try It 2
Multiply: $-a^2(3a^2 + 2a - 7)$

Your solution
$-3a^4 - 2a^3 + 7a^2$

Solutions on p. S18

Objective B To multiply two polynomials

VIDEO & DVD CD TUTOR WEB SSM

Multiplication of two polynomials requires the repeated application of the Distributive Property.

$$(y - 2)(y^2 + 3y + 1) = (y - 2)(y^2) + (y - 2)(3y) + (y - 2)(1)$$
$$= y^3 - 2y^2 + 3y^2 - 6y + y - 2$$
$$= y^3 + y^2 - 5y - 2$$

A convenient method of multiplying two polynomials is to use a vertical format similar to that used for multiplication of whole numbers.

$$\begin{array}{r} y^2 + 3y + 1 \\ y - 2 \\ \hline -2y^2 - 6y - 2 \\ y^3 + 3y^2 + y \\ \hline y^3 + y^2 - 5y - 2 \end{array}$$

- Multiply $-2(y^2 + 3y + 1)$.
- Multiply $y(y^2 + 3y + 1)$.
- Add the terms in each column.

Objective 6.3A

Vocabulary to Review
monomial
polynomial

Properties to Review
Distributive Property

Concept Check
1. Find the area of a rectangle that has a length of $5x$ mi and a width of $(2x - 7)$ mi.
   $(10x^2 - 35x)$ mi^2
2. The base of a triangle is $4x$ m and the height is $(2x + 5)$ m. Find the area of the triangle in terms of the variable x.
   $(4x^2 + 10x)$ m^2
3. An athletic field has dimensions of 25 yd by 50 yd. An end zone that is w yards wide borders each end of the field. Express the total area of the field and the end zones in terms of the variable w.
   $(50w + 1250)$ yd^2

Optional Student Activity
Have students explain why the following diagram represents $(a + b)^2 = a^2 + 2ab + b^2$.

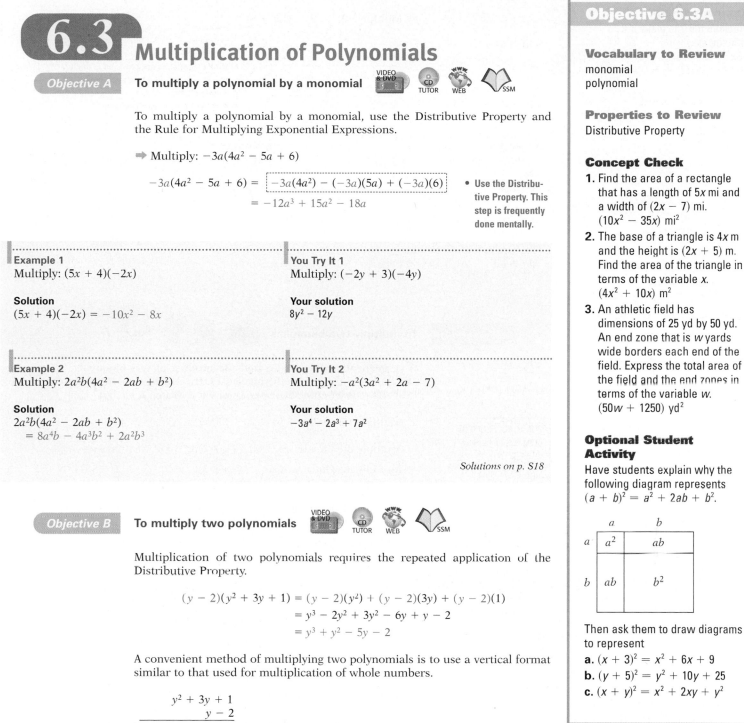

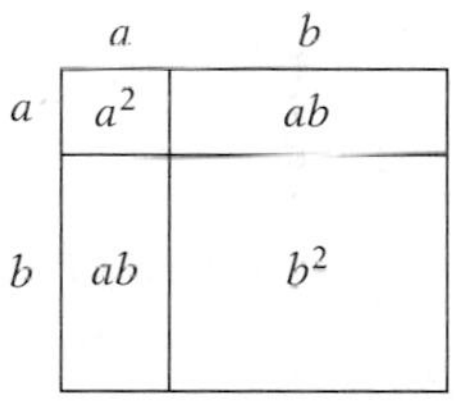

Then ask them to draw diagrams to represent
a. $(x + 3)^2 = x^2 + 6x + 9$
b. $(y + 5)^2 = y^2 + 10y + 25$
c. $(x + y)^2 = x^2 + 2xy + y^2$

Objective 6.3B

Instructor Note
Before doing an example similar to the one at the left, show students that the procedure for multiplication can be related to multiplying whole numbers such as 473×28.

In-Class Examples (Objective 6.3A)
Simplify.
1. $-5y^2(3y - 4y^2)$ $-15y^3 + 20y^4$
2. $3a + 2a(3 - a)$ $9a - 2a^2$
3. $2a^2b(4a^2 - 3ab + 2b^2)$ $8a^4b - 6a^3b^2 + 4a^2b^3$
4. $x^{3n}(x^{n-1} + x^n + x)$ $x^{4n-1} + x^{4n} + x^{3n+1}$
5. $2x^2 - x[x - 3(2x - 1)]$ $7x^2 - 3x$

In-Class Examples (Objective 6.3B)
Simplify.
1. $(2z^2 - 4z + 5)(4z - 2)$ $8z^3 - 20z^2 + 28z - 10$
2. $(6a^3 + 4a^2 - 3a)(3a - 2)$ $18a^4 - 17a^2 + 6a$
3. $(3x^3 - 2x + 5)(7x + 4)$
   $21x^4 + 12x^3 - 14x^2 + 27x + 20$

Concept Check
Replace the ? to make a true statement.
$(3x^2 + 2x + 5)(x + 3) =$
$3x^2(?) + 2x(?) + 5(?)$ $x + 3$

Optional Student Activity
Simplify: $(x^2 - xy + y^2)(x + y)$
$x^3 + y^3$

Objective 6.3C

New Vocabulary
FOIL

Concept Check
Simplify.
1. $(x - 3)(x - 6)$ $x^2 - 9x + 18$
2. $(x + 4)(x + 5)$ $x^2 + 9x + 20$
3. $(x - 6)(x + 2)$ $x^2 - 4x - 12$
4. $(x + 4)(x - 7)$ $x^2 - 3x - 28$

Optional Student Activity
Make up a problem involving the product of two binomials. Write an explanation using the FOIL method that demonstrates how to simplify the product.

➡ Multiply: $(2a^3 + a - 3)(a + 5)$

$$\begin{array}{r} 2a^3 + a - 3 \\ a + 5 \\ \hline 10a^3 \quad + 5a - 15 \\ 2a^4 \quad + a^2 - 3a \quad \\ \hline 2a^4 + 10a^3 + a^2 + 2a - 15 \end{array}$$

- Note that spaces are provided in each product so that like terms are in the same column.
- Add the terms in each column.

Example 3
Multiply: $(2b^3 - b + 1)(2b + 3)$

Solution

$$\begin{array}{r} 2b^3 - b + 1 \\ 2b + 3 \\ \hline 6b^3 \quad - 3b + 3 \\ 4b^4 + \quad - 2b^2 + 2b \quad \\ \hline 4b^4 + 6b^3 - 2b^2 - b + 3 \end{array}$$

You Try It 3
Multiply: $(2y^3 + 2y^2 - 3)(3y - 1)$

Your solution
$6y^4 + 4y^3 - 2y^2 - 9y + 3$

Solution on p. S19

Objective C **To multiply two binomials**

VIDEO & DVD CD TUTOR WWW WEB SSM

It is frequently necessary to find the product of two binomials. The product can be found using a method called **FOIL**, which is based on the Distributive Property. The letters of FOIL stand for **F**irst, **O**uter, **I**nner, and **L**ast.

> **TAKE NOTE**
> FOIL is not really a different way of multiplying. It is based on the Distributive Property.
> $(2x + 3)(x + 5)$
> $= 2x(x + 5) + 3(x + 5)$
> F O I L
> $= 2x^2 + 10x + 3x + 15$
> $= 2x^2 + 13x + 15$
>
> FOIL is an efficient way to remember how to do binomial multiplication.

➡ Multiply: $(2x + 3)(x + 5)$

Multiply the First terms.	$(2x + 3)(x + 5)$	$2x \cdot x = 2x^2$
Multiply the Outer terms.	$(2x + 3)(x + 5)$	$2x \cdot 5 = 10x$
Multiply the Inner terms.	$(2x + 3)(x + 5)$	$3 \cdot x = 3x$
Multiply the Last terms.	$(2x + 3)(x + 5)$	$3 \cdot 5 = 15$
		F O I L
Add the products.	$(2x + 3)(x + 5)$	$= 2x^2 + 10x + 3x + 15$
Combine like terms.		$= 2x^2 + 13x + 15$

➡ Multiply: $(4x - 3)(3x - 2)$

$$(4x - 3)(3x - 2) = \boxed{4x(3x) + 4x(-2) + (-3)(3x) + (-3)(-2)}$$
$$= 12x^2 - 8x - 9x + 6$$
$$= 12x^2 - 17x + 6$$

- Do this step mentally.

➡ Multiply: $(3x - 2y)(x + 4y)$

$$(3x - 2y)(x + 4y) = \boxed{3x(x) + 3x(4y) + (-2y)(x) + (-2y)(4y)}$$
$$= 3x^2 + 12xy - 2xy - 8y^2$$
$$= 3x^2 + 10xy - 8y^2$$

- Do this step mentally.

In-Class Examples (Objective 6.3C)
Simplify.
1. $(7x - 3)(5x - 6)$ $35x^2 - 57x + 18$
2. $(8y + 7)(3y - 4)$ $24y^2 - 11y - 28$
3. $(6w + 4)(5w - 2)$ $30w^2 + 8w - 8$

Example 4
Multiply: $(2a - 1)(3a - 2)$

Solution
$(2a - 1)(3a - 2) = 6a^2 - 4a - 3a + 2$
$= 6a^2 - 7a + 2$

You Try It 4
Multiply: $(4y - 5)(2y - 3)$

Your solution
$8y^2 - 22y + 15$

Example 5
Multiply: $(3x - 2)(4x + 3)$

Solution
$(3x - 2)(4x + 3) = 12x^2 + 9x - 8x - 6$
$= 12x^2 + x - 6$

You Try It 5
Multiply: $(3b + 2)(3b - 5)$

Your solution
$9b^2 - 9b - 10$

Solutions on p. S19

Objective D To multiply binomials that have special products

VIDEO & DVD CD TUTOR WEB SSM

Using FOIL, it is possible to find a pattern for the product of the sum and difference of two terms and for the square of a binomial.

The Sum and Difference of Two Terms

$$(a + b)(a - b) = a^2 - ab + ab - b^2$$
$$= a^2 - b^2$$

Square of the first term → a^2
Square of the second term → b^2

The Square of a Binomial

$$(a + b)^2 = (a + b)(a + b) = a^2 + ab + ab + b^2$$
$$= a^2 + 2ab + b^2$$

Square of the first term → a^2
Twice the product of the two terms → $2ab$
Square of the last term → b^2

➡ Multiply: $(2x + 3)(2x - 3)$

$(2x + 3)(2x - 3)$ is the sum and difference of two terms.

$(2x + 3)(2x - 3) = \boxed{(2x)^2 - 3^2}$ • **Do this step mentally.**
$= 4x^2 - 9$

➡ Multiply: $(3x - 2)^2$

$(3x - 2)^2$ is the square of a binomial.

$(3x - 2)^2 = \boxed{(3x)^2 + 2(3x)(-2) + (-2)^2}$ • **Do this step mentally.**
$= 9x^2 - 12x + 4$

351

Concept Check
Can FOIL be used to simplify $(x^2 - 5)(2x + 3)$? If so, what is the product? Yes.
$2x^3 + 3x^2 - 10x - 15$

Objective 6.3D

Instructor Note
Remind students that the rule that applies to $(ab)^2$ is different from the rule that applies to $(a + b)^2$.

Optional Student Activity
1. Demonstrate that $(a + b)^2 \neq a^2 + b^2$ by substituting numbers for a and b and then evaluating the expressions on each side of the equals sign.
2. Demonstrate that $(c - d)^2 \neq c^2 - d^2$ by substituting numbers for c and d and then evaluating the expressions on each side of the equals sign.

Concept Check
Simplify.
1. $(x + 3)(x - 3)$ $x^2 - 9$
2. $(x + 4)^2$ $x^2 + 8x + 16$

In-Class Examples (Objective 6.3D)
Simplify.
1. $(10x - 3)(10x + 3)$ $100x^2 - 9$
2. $(4x + 3y)^2$ $16x^2 + 24xy + 9y^2$

Concept Check

Simplify.

1. $(5b - 7c)(5b + 7c)$
   $25b^2 - 49c^2$
2. $(4y - 5z)^2$
   $16y^2 - 40yz + 25z^2$

Instructor Note

Binomials that are otherwise identical except that one is a sum and one is a difference are called *conjugates* of each other.

Optional Student Activity

Write an adaptation of the FOIL method that can be used to simplify the product of the sum and difference of two terms. Demonstrate your method on the product $(3x + 8)(3x - 8)$.
Students may refer to the method as the "FL" method, or the "$F^2 - L^2$" method.
$(3x + 8)(3x - 8) =$
$(3x)(3x) + (8)(-8) = 9x^2 - 64$

Objective 6.3E

Concept Check

1. The length of a rectangle is 6 in. The width is 4 in. Find the area of the rectangle. 24 in^2
2. The length of a rectangle is $(x + 3)$ in. The width is $(x - 2)$ in. Find the area of the rectangle in terms of the variable x. $(x^2 + x - 6)$ in^2

Example 6
Multiply: $(4z - 2w)(4z + 2w)$

Solution
$(4z - 2w)(4z + 2w) = 16z^2 - 4w^2$

You Try It 6
Multiply: $(2a + 5c)(2a - 5c)$

Your solution
$4a^2 - 25c^2$

Example 7
Simplify: $(2r - 3s)^2$

Solution
$(2r - 3s)^2 = 4r^2 - 12rs + 9s^2$

You Try It 7
Simplify: $(3x + 2y)^2$

Your solution
$9x^2 + 12xy + 4y^2$

Solutions on p. S19

Objective E To solve application problems

VIDEO & DVD CD TUTOR WEB SSM

Example 8
The length of a rectangle is $(2x + 3)$ ft. The width is $(x - 5)$ ft. Find the area of the rectangle in terms of the variable x.

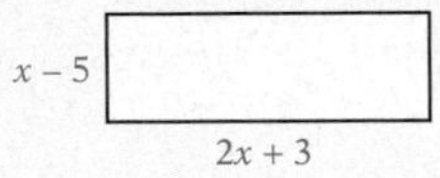

Strategy
To find the area, replace the variables L and W in the equation $A = L \cdot W$ by the given values and solve for A.

Solution
$A = L \cdot W$
$A = (2x + 3)(x - 5)$
$= 2x^2 - 10x + 3x - 15$
$= 2x^2 - 7x - 15$

The area is $(2x^2 - 7x - 15)$ ft^2.

You Try It 8
The base of a triangle is $(2x + 6)$ ft. The height is $(x - 4)$ ft. Find the area of the triangle in terms of the variable x.

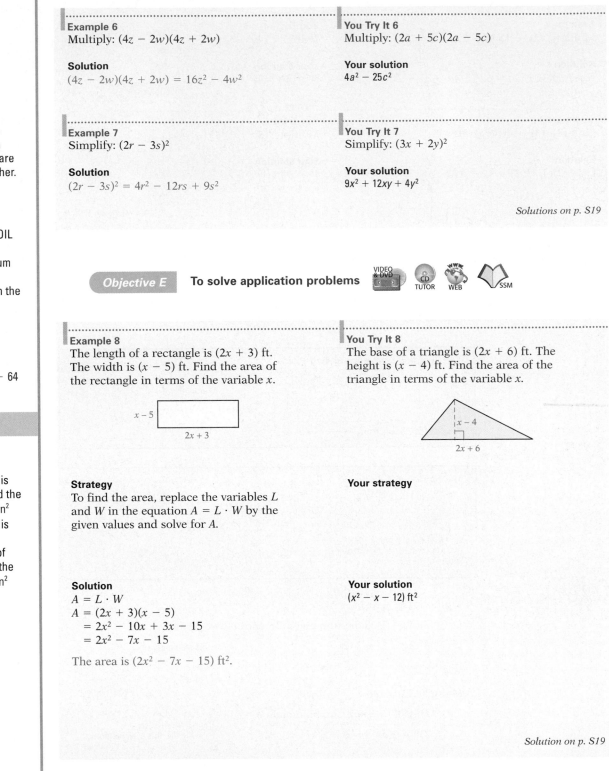

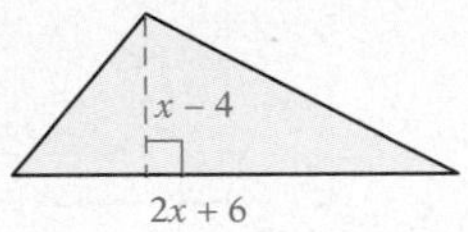

Your strategy

Your solution
$(x^2 - x - 12)$ ft^2

Solution on p. S19

In-Class Examples (Objective 6.3E)

The radius of a circle is $(x + 5)$ ft. Use the equation $A = \pi r^2$, where r is the radius, to find the area of the circle in terms of x. Leave the answer in terms of π. $(\pi x^2 + 10\pi x + 25\pi)$ ft^2

353

6.3 Exercises

Objective A

Multiply.

1. $x(x-2)$ x^2-2x
2. $y(3-y)$ $-y^2+3y$
3. $-x(x+7)$ $-x^2-7x$
4. $-y(7-y)$ y^2-7y
5. $3a^2(a-2)$ $3a^3-6a^2$
6. $4b^2(b+8)$ $4b^3+32b^2$
7. $-5x^2(x^2-x)$ $-5x^4+5x^3$
8. $-6y^2(y+2y^2)$ $-12y^4-6y^3$
9. $-x^3(3x^2-7)$ $-3x^5+7x^3$
10. $-y^4(2y^2-y^6)$ $y^{10}-2y^6$
11. $2x(6x^2-3x)$ $12x^3-6x^2$
12. $3y(4y-y^2)$ $-3y^3+12y^2$
13. $(2x-4)3x$ $6x^2-12x$
14. $(3y-2)y$ $3y^2-2y$
15. $(3x+4)x$ $3x^2+4x$
16. $(2x+1)2x$ $4x^2+2x$
17. $-xy(x^2-y^2)$ $-x^3y+xy^3$
18. $-x^2y(2xy-y^2)$ $-2x^3y^2+x^2y^3$
19. $x(2x^3-3x+2)$ $2x^4-3x^2+2x$
20. $y(-3y^2-2y+6)$ $-3y^3-2y^2+6y$
21. $-a(-2a^2-3a-2)$ $2a^3+3a^2+2a$
22. $-b(5b^2+7b-35)$ $-5b^3-7b^2+35b$
23. $x^2(3x^4-3x^2-2)$ $3x^6-3x^4-2x^2$
24. $y^3(-4y^3-6y+7)$ $-4y^6-6y^4+7y^3$
25. $2y^2(-3y^2-6y+7)$ $-6y^4-12y^3+14y^2$
26. $4x^2(3x^2-2x+6)$ $12x^4-8x^3+24x^2$
27. $(a^2+3a-4)(-2a)$ $-2a^3-6a^2+8a$
28. $(b^3-2b+2)(-5b)$ $-5b^4+10b^2-10b$
29. $-3y^2(-2y^2+y-2)$ $6y^4-3y^3+6y^2$
30. $-5x^2(3x^2-3x-7)$ $-15x^4+15x^3+35x^2$
31. $xy(x^2-3xy+y^2)$ $x^3y-3x^2y^2+xy^3$
32. $ab(2a^2-4ab-6b^2)$ $2a^3b-4a^2b^2-6ab^3$

Objective B

Multiply.

33. $(x^2+3x+2)(x+1)$ x^3+4x^2+5x+2
34. $(x^2-2x+7)(x-2)$ $x^3-4x^2+11x-14$
35. $(a^2-3a+4)(a-3)$ $a^3-6a^2+13a-12$

Section 6.3

Suggested Assignment

Exercises 1–81, every other odd
Exercises 85–109, odds
More challenging problems:
Exercises 111–115

Quick Quiz (Objective 6.3A)

Simplify.

1. $y(y+5)$ y^2+5y
2. $(a-3)2a^2$ $2a^3-6a^2$
3. $-5b^3(2b^2+3b-6)$ $-10b^5-15b^4+30b^3$

Quick Quiz (Objective 6.3B)

Simplify.

1. $(x^2-3x+2)(x+5)$ $x^3+2x^2-13x+10$
2. $(y^2+4)(y-1)$ y^3-y^2+4y-4
3. $(3a^3-2a^2+a-6)(2a-5)$ $6a^4-19a^3+12a^2-17a+30$

36. $(x^2 - 3x + 5)(2x - 3)$
$2x^3 - 9x^2 + 19x - 15$

37. $(-2b^2 - 3b + 4)(b - 5)$
$-2b^3 + 7b^2 + 19b - 20$

38. $(-a^2 + 3a - 2)(2a - 1)$
$-2a^3 + 7a^2 - 7a + 2$

39. $(-2x^2 + 7x - 2)(3x - 5)$
$-6x^3 + 31x^2 - 41x + 10$

40. $(-a^2 - 2a + 3)(2a - 1)$
$-2a^3 - 3a^2 + 8a - 3$

41. $(x^2 + 5)(x - 3)$
$x^3 - 3x^2 + 5x - 15$

42. $(y^2 - 2y)(2y + 5)$
$2y^3 + y^2 - 10y$

43. $(x^3 - 3x + 2)(x - 4)$
$x^4 - 4x^3 - 3x^2 + 14x - 8$

44. $(y^3 + 4y^2 - 8)(2y - 1)$
$2y^4 + 7y^3 - 4y^2 - 16y + 8$

45. $(5y^2 + 8y - 2)(3y - 8)$
$15y^3 - 16y^2 - 70y + 16$

46. $(3y^2 + 3y - 5)(4y - 3)$
$12y^3 + 3y^2 - 29y + 15$

47. $(5a^3 - 5a + 2)(a - 4)$
$5a^4 - 20a^3 - 5a^2 + 22a - 8$

48. $(3b^3 - 5b^2 + 7)(6b - 1)$
$18b^4 - 33b^3 + 5b^2 + 42b - 7$

49. $(y^3 + 2y^2 - 3y + 1)(y + 2)$
$y^4 + 4y^3 + y^2 - 5y + 2$

50. $(2a^3 - 3a^2 + 2a - 1)(2a - 3)$
$4a^4 - 12a^3 + 13a^2 - 8a + 3$

Objective C

Multiply.

51. $(x + 1)(x + 3)$
$x^2 + 4x + 3$

52. $(y + 2)(y + 5)$
$y^2 + 7y + 10$

53. $(a - 3)(a + 4)$
$a^2 + a - 12$

54. $(b - 6)(b + 3)$
$b^2 - 3b - 18$

55. $(y + 3)(y - 8)$
$y^2 - 5y - 24$

56. $(x + 10)(x - 5)$
$x^2 + 5x - 50$

57. $(y - 7)(y - 3)$
$y^2 - 10y + 21$

58. $(a - 8)(a - 9)$
$a^2 - 17a + 72$

59. $(2x + 1)(x + 7)$
$2x^2 + 15x + 7$

60. $(y + 2)(5y + 1)$
$5y^2 + 11y + 2$

61. $(3x - 1)(x + 4)$
$3x^2 + 11x - 4$

62. $(7x - 2)(x + 4)$
$7x^2 + 26x - 8$

63. $(4x - 3)(x - 7)$
$4x^2 - 31x + 21$

64. $(2x - 3)(4x - 7)$
$8x^2 - 26x + 21$

65. $(3y - 8)(y + 2)$
$3y^2 - 2y - 16$

66. $(5y - 9)(y + 5)$
$5y^2 + 16y - 45$

67. $(3x + 7)(3x + 11)$
$9x^2 + 54x + 77$

68. $(5a + 6)(6a + 5)$
$30a^2 + 61a + 30$

69. $(7a - 16)(3a - 5)$
$21a^2 - 83a + 80$

70. $(5a - 12)(3a - 7)$
$15a^2 - 71a + 84$

71. $(3a - 2b)(2a - 7b)$
$6a^2 - 25ab + 14b^2$

72. $(5a - b)(7a - b)$
$35a^2 - 12ab + b^2$

73. $(a - 9b)(2a + 7b)$
$2a^2 - 11ab - 63b^2$

Quick Quiz (Objective 6.3C)

Simplify.

1. $(x + 1)(x + 5)$ $x^2 + 6x + 5$

2. $(2x - 3)(3x - 4)$ $6x^2 - 17x + 12$

3. $(6x - 7)(4x + 3)$ $24x^2 - 10x - 21$

355

74. $(2a + 5b)(7a - 2b)$
$14a^2 + 31ab - 10b^2$

75. $(10a - 3b)(10a - 7b)$
$100a^2 - 100ab + 21b^2$

76. $(12a - 5b)(3a - 4b)$
$36a^2 - 63ab + 20b^2$

77. $(5x + 12y)(3x + 4y)$
$15x^2 + 56xy + 48y^2$

78. $(11x + 2y)(3x + 7y)$
$33x^2 + 83xy + 14y^2$

79. $(2x - 15y)(7x + 4y)$
$14x^2 - 97xy - 60y^2$

80. $(5x + 2y)(2x - 5y)$
$10x^2 - 21xy - 10y^2$

81. $(8x - 3y)(7x - 5y)$
$56x^2 - 61xy + 15y^2$

82. $(2x - 9y)(8x - 3y)$
$16x^2 - 78xy + 27y^2$

Objective D

83. What does it mean to square a binomial?

84. Why is $(a + b)^2$ not equal to $a^2 + b^2$?

Multiply.

85. $(y - 5)(y + 5)$
$y^2 - 25$

86. $(y + 6)(y - 6)$
$y^2 - 36$

87. $(2x + 3)(2x - 3)$
$4x^2 - 9$

88. $(4x - 7)(4x + 7)$
$16x^2 - 49$

89. $(x + 1)^2$
$x^2 + 2x + 1$

90. $(y - 3)^2$
$y^2 - 6y + 9$

91. $(3a - 5)^2$
$9a^2 - 30a + 25$

92. $(6x - 5)^2$
$36x^2 - 60x + 25$

93. $(3x - 7)(3x + 7)$
$9x^2 - 49$

94. $(9x - 2)(9x + 2)$
$81x^2 - 4$

95. $(2a + b)^2$
$4a^2 + 4ab + b^2$

96. $(x + 3y)^2$
$x^2 + 6xy + 9y^2$

97. $(4 - 3y)(4 + 3y)$
$16 - 9y^2$

98. $(4x - 9y)(4x + 9y)$
$16x^2 - 81y^2$

99. $(5x + 2y)^2$
$25x^2 + 20xy + 4y^2$

100. $(2a - 9b)^2$
$4a^2 - 36ab + 81b^2$

Objective E Application Problems

101. The length of a rectangle is $5x$ ft. The width is $(2x - 7)$ ft. Find the area of the rectangle in terms of the variable x.
$(10x^2 - 35x)$ ft^2

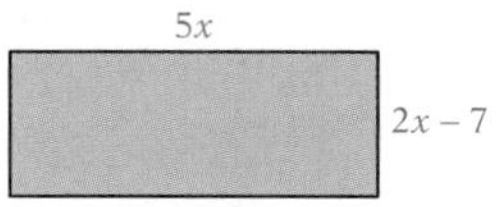

102. The width of a rectangle is $(x - 6)$ m. The length is $(2x + 3)$ m. Find the area of the rectangle in terms of the variable x.
$(2x^2 - 9x - 18)$ m^2

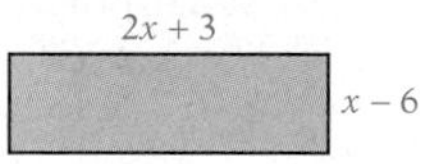

Answers to Writing Exercises

83. To square a binomial means to multiply the binomial times itself.

84. Students should explain that $(a + b)^2$ means to multiply $a + b$ times itself: $(a + b)(a + b)$. The product $(a + b)(a + b)$ is $a^2 + 2ab + b^2$, not $a^2 + b^2$.

Quick Quiz (Objective 6.3D)

Simplify.

1. $(5y + 3)(5y - 3)$ $25y^2 - 9$

2. $(x - 9)^2$ $x^2 - 18x + 81$

3. $(3x + 2y)^2$ $9x^2 + 12xy + 4y^2$

103. The length of a side of a square is $(2x + 1)$ km. Find the area of the square in terms of the variable x.
$(4x^2 + 4x + 1)$ km^2

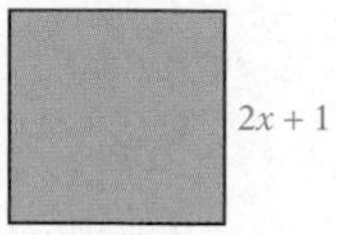

104. The length of a side of a square is $(2x - 3)$ yd. Find the area of the square in terms of the variable x.
$(4x^2 - 12x + 9)$ yd^2

105. The width of a rectangle is $(3x + 1)$ in. The length of the rectangle is twice the width. Find the area of the rectangle in terms of the variable x.
$(18x^2 + 12x + 2)$ in^2

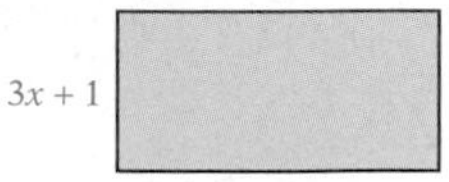

106. The width of a rectangle is $(4x - 3)$ cm. The length of the rectangle is twice the width. Find the area of the rectangle in terms of the variable x.
$(32x^2 - 48x + 18)$ cm^2

107. The base of a triangle is $4x$ m, and the height is $(2x + 5)$ m. Find the area of the triangle in terms of the variable x.
$(4x^2 + 10x)$ m^2

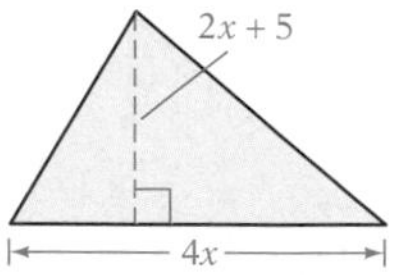

108. The base of a triangle is $(2x + 6)$ in., and the height is $(x - 8)$ in. Find the area of the triangle in terms of the variable x.
$(x^2 - 5x - 24)$ in^2

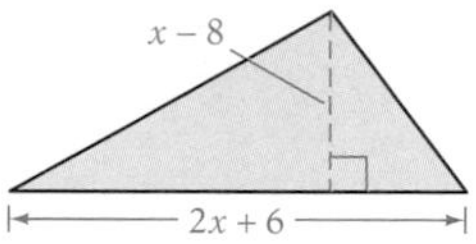

109. An athletic field has dimensions 30 yd by 100 yd. An end zone that is w yd wide borders each end of the field. Express the total area of the field and the end zones in terms of the variable w.
$(3000 + 60w)$ yd^2

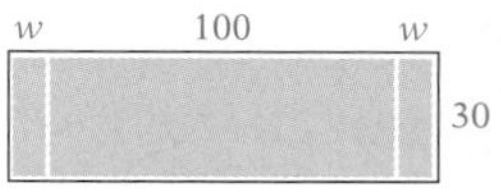

110. A softball diamond has dimensions 45 ft by 45 ft. A base path border x ft wide lies on both the first-base side and the third-base side of the diamond. Express the total area of the softball diamond and the base path in terms of the variable x.
$(2025 + 90x)$ ft^2

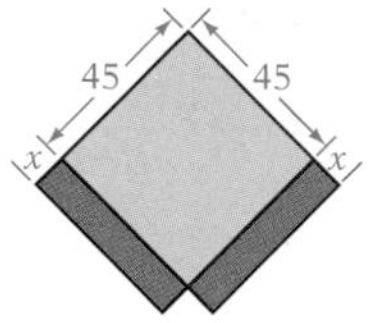

APPLYING THE CONCEPTS

Simplify.

111. $(a + b)^2 - (a - b)^2$
$4ab$

112. $(x^2 + x - 3)^2$
$x^4 + 2x^3 - 5x^2 - 6x + 9$

113. $(a + 3)^3$
$a^3 + 9a^2 + 27a + 27$

114. **a.** What polynomial has quotient $3x - 4$ when divided by $4x + 5$?
b. What polynomial has quotient $x^2 + 2x - 1$ when divided by $x + 3$?
a. $12x^2 - x - 20$ **b.** $x^3 + 5x^2 + 5x - 3$

115. **a.** Add $x^2 + 2x - 3$ to the product of $2x - 5$ and $3x + 1$.
b. Subtract $4x^2 - x - 5$ from the product of $x^2 + x + 3$ and $x - 4$.
a. $7x^2 - 11x - 8$ **b.** $x^3 - 7x^2 - 7$

Quick Quiz (Objective 6.3E)

1. The length of a rectangle is $(2x + 4)$ ft. The width is $(x - 5)$ ft. Find the area of the rectangle in terms of the variable x. $(2x^2 - 6x - 20)$ ft^2
2. The length of a side of a cube is $(x + 2)$ cm. Find the volume of the cube in terms of the variable x. $(x^3 + 6x^2 + 12x + 8)$ cm^3

6.4 Division of Polynomials

Objective A To divide polynomials

VIDEO & DVD CD TUTOR WEB SSM

To divide two polynomials, use a method similar to that used for long division of whole numbers. To check division of polynomials, use

Dividend = (quotient × divisor) + remainder

➡ Divide: $(x^2 + 5x - 7) \div (x + 3)$

Step 1

$$\begin{array}{r} x \\ x + 3\,\overline{)\,x^2 + 5x - 7} \\ \underline{x^2 + 3x} \downarrow \\ 2x - 7 \end{array}$$

Think: $x\overline{)x^2} = \frac{x^2}{x} = x$

Multiply: $x(x + 3) = x^2 + 3x$

Subtract: $(x^2 + 5x) - (x^2 + 3x) = 2x$

Bring down the -7.

Step 2

$$\begin{array}{r} x + 2 \\ x + 3\,\overline{)\,x^2 + 5x - 7} \\ \underline{x^2 + 3x} \\ 2x - 7 \\ \underline{2x + 6} \\ -13 \end{array}$$

Think: $x\overline{)2x} = \frac{2x}{x} = 2$

Multiply: $2(x + 3) = 2x + 6$

Subtract: $(2x - 7) - (2x + 6) = -13$

The remainder is -13.

Check: (Quotient)(divisor) + remainder

$$(x + 2)(x + 3) + (-13) = x^2 + 3x + 2x + 6 - 13 = x^2 + 5x - 7$$

$$(x^2 + 5x - 7) \div (x + 3) = x + 2 - \frac{13}{x + 3}$$

➡ Divide: $\dfrac{6 - 6x^2 + 4x^3}{2x + 3}$

Arrange the terms in descending order. Note that there is no term of x in $4x^3 - 6x^2 + 6$. Insert a $0x$ for the missing term so that like terms will be in the same columns.

$$\begin{array}{r} 2x^2 - 6x + 9 \\ 2x + 3\,\overline{)\,4x^3 - 6x^2 + 0x + 6} \\ \underline{4x^3 + 6x^2} \\ -12x^2 + 0x \\ \underline{-12x^2 - 18x} \\ 18x + 6 \\ \underline{18x + 27} \\ -21 \end{array}$$

$$\frac{4x^3 - 6x^2 + 6}{2x + 3} = 2x^2 - 6x + 9 - \frac{21}{2x + 3}$$

Objective 6.4A

New Formulas

Dividend = (quotient × divisor) + remainder

Instructor Note

Begin this section by showing an example of long division of whole numbers, for instance, $863 \div 57$.

Concept Check

1. Given that $\dfrac{x^3 + 1}{x + 1} = x^2 - x + 1$, name two factors of $x^3 + 1$. $x + 1$ and $x^2 - x + 1$
2. $3x + 1$ is a factor of $3x^3 - 8x^2 - 33x - 10$. Find a quadratic factor of $3x^3 - 8x^2 - 33x - 10$. $x^2 - 3x - 10$
3. $4x - 1$ is a factor of $8x^3 - 38x^2 + 49x - 10$. Find a quadratic factor of $8x^3 - 38x^2 + 49x - 10$. $2x^2 - 9x + 10$
4. Is $2x - 3$ a factor of $4x^3 + x - 12$? Explain your answer. No. There is a remainder when $4x^3 + x - 12$ is divided by $2x - 3$.

Instructor Note

After you have completed this objective, a possible extra-credit problem might be to divide $2x - 4\overline{)3x^3 - 5x^2 - 6x + 8}$. $\frac{3}{2}x^2 + \frac{1}{2}x - 2$

In-Class Examples (Objective 6.4A)

Divide using long division.

1. $(6x^2 + x - 5) \div (3x + 2)$ $2x - 1 - \dfrac{3}{3x + 2}$
2. $(x^3 - 2x^2 + 3) \div (x - 3)$ $x^2 + x + 3 + \dfrac{12}{x - 3}$
3. $\dfrac{3x^3 - 7x^2 - 22x + 8}{3x - 1}$ $x^2 - 2x - 8$

Optional Student Activity

1. Divide: $\dfrac{3x^2 - xy - 2y^2}{3x + 2y}$ $x - y$
2. Divide: $\dfrac{12x^2 + 11xy + 2y^2}{4x + y}$ $3x + 2y$
3. Divide: $\dfrac{a^4 + b^4}{a + b}$ $a^3 - a^2b + ab^2 - b^3 + \dfrac{2b^4}{a + b}$
4. When $x^2 + x + 2$ is divided by a polynomial, the quotient is $x + 4$ and the remainder is 14. Find the polynomial. $x - 3$

Optional Student Activity

Divide each polynomial given below by $x - y$.

a. $x^3 - y^3$
b. $x^5 - y^5$
c. $x^7 - y^7$
d. $x^9 - y^9$

a. $x^2 + xy + y^2$
b. $x^4 + x^3y + x^2y^2 + xy^3 + y^4$
c. $x^6 + x^5y + x^4y^2 + x^3y^3 + x^2y^4 + xy^5 + y^6$
d. $x^8 + x^7y + x^6y^2 + x^5y^3 + x^4y^4 + x^3y^5 + x^2y^6 + xy^7 + y^8$

Explain the pattern and use the pattern to write the quotient of $(x^{11} - y^{11}) \div (x - y)$.
$(x^{11} - y^{11}) \div (x - y) = x^{10} + x^9y + x^8y^2 + x^7y^3 + x^6y^4 + x^5y^5 + x^4y^6 + x^3y^7 + x^2y^8 + xy^9 + y^{10}$

Example 1

Divide: $\dfrac{12x^2 - 11x + 10}{4x - 5}$

Solution

$$\begin{array}{r} 3x + 1 \\ 4x - 5\,\overline{)\,12x^2 - 11x + 10} \\ \underline{12x^2 - 15x} \\ 4x + 10 \\ \underline{4x - 5} \\ 15 \end{array}$$

$$\frac{12x^2 - 11x + 10}{4x - 5} = 3x + 1 + \frac{15}{4x - 5}$$

You Try It 1

Divide: $\dfrac{15x^2 + 17x - 20}{3x + 4}$

Your solution

$5x - 1 - \dfrac{16}{3x + 4}$

Example 2

Divide: $\dfrac{x^3 + 1}{x + 1}$

Solution

$$\begin{array}{r} x^2 - x + 1 \\ x + 1\,\overline{)\,x^3 + 0x^2 + 0x + 1} \\ \underline{x^3 + x^2} \\ -x^2 + 0x \\ \underline{-x^2 - x} \\ x + 1 \\ \underline{x + 1} \\ 0 \end{array}$$

- Insert zeros for the coefficients of the missing terms.

$$\frac{x^3 + 1}{x + 1} = x^2 - x + 1$$

You Try It 2

Divide: $\dfrac{3x^3 + 8x^2 - 6x + 2}{3x - 1}$

Your solution

$x^2 + 3x - 1 + \dfrac{1}{3x - 1}$

Example 3

Divide:
$(2x^4 - 7x^3 + 3x^2 + 4x - 5) \div (x^2 - 2x - 2)$

Solution

$$\begin{array}{r} 2x^2 - 3x + 1 \\ x^2 - 2x - 2\,\overline{)\,2x^4 - 7x^3 + 3x^2 + 4x - 5} \\ \underline{2x^4 - 4x^3 - 4x^2} \\ -3x^3 + 7x^2 + 4x \\ \underline{-3x^3 + 6x^2 + 6x} \\ x^2 - 2x - 5 \\ \underline{x^2 - 2x - 2} \\ -3 \end{array}$$

$(2x^4 - 7x^3 + 3x^2 + 4x - 5) \div (x^2 - 2x - 2)$

$$= 2x^2 - 3x + 1 - \frac{3}{x^2 - 2x - 2}$$

You Try It 3

Divide:
$(3x^4 - 11x^3 + 16x^2 - 16x + 8) \div (x^2 - 3x + 2)$

Your solution

$3x^2 - 2x + 4$

Solutions on p. S19

359

Objective B To divide polynomials using synthetic division

Synthetic division is a shorter method of dividing a polynomial by a binomial of the form $x - a$.

➡ Divide $(3x^2 - 4x + 6) \div (x - 2)$ by using long division.

$$
\begin{array}{r}
3x + 2 \\
x - 2\overline{)\,3x^2 - 4x + 6} \\
\underline{3x^2 - 6x} \\
2x + 6 \\
\underline{2x - 4} \\
10
\end{array}
$$

$$(3x^2 - 4x + 6) \div (x - 2) = 3x + 2 + \frac{10}{x - 2}$$

The variables can be omitted because the position of a term indicates the power of the term.

$$
\begin{array}{rrrr}
 & & 3 & 2 \\ \hline
-2) & 3 & -4 & 6 \\
 & \underline{3} & \underline{-6} & \\
 & & 2 & 6 \\
 & & \underline{2} & \underline{-4} \\
 & & & 10
\end{array}
$$

Each number shown in color above is exactly the same as the number above it. Removing the colored numbers condenses the vertical spacing.

$$
\begin{array}{rrrr}
 & & 3 & 2 \\ \hline
-2) & 3 & -4 & 6 \\
 & & -6 & -4 \\ \hline
 & & 2 & 10
\end{array}
$$

The number in color on the top row is the same as the one in the bottom row. Writing the 3 from the top row in the bottom row allows the spacing to be condensed even further.

$$
\begin{array}{r|rrr}
-2 & 3 & -4 & 6 \\
 & & -6 & -4 \\ \hline
 & 3 & 2 & 10
\end{array}
$$

Terms of the quotient: 3, 2 Remainder: 10

Because the degree of the dividend $(3x^2 - 4x + 6)$ is 2 and the degree of the divisor $(x - 2)$ is 1, the degree of the quotient is $2 - 1 = 1$. This means that, using the terms of the quotient given above, the quotient is $3x + 2$. The remainder is 10.

In general, the degree of the quotient of two polynomials is the difference between the degree of the dividend and the degree of the divisor.

By replacing the constant term in the divisor by its additive inverse, we may add rather than subtract terms. This is illustrated in the following example.

Objective 6.4B

New Vocabulary

synthetic division

Vocabulary to Review

degree of a polynomial
additive inverse
binomial
coefficient

Discuss the Concepts

1. Suppose you are going to divide $2x^3 + 13x^2 + 15x - 5$ by $x + 5$ using synthetic division.
   a. What are the coefficients of the dividend?
   b. What is the value of a?
   c. What is the degree of the first term of the quotient?
2. Why, in synthetic division, is addition used rather than subtraction?
3. When synthetic division is used to divide a polynomial by a binomial of the form $x - a$, how is the degree of the quotient related to the degree of the dividend?
4. How can you check the answer to a synthetic division problem?

Concept Check

Which of the following divisions can be performed using synthetic division?

a. $(x^2 + 3x + 1) \div (x - 2)$
b. $(x^6 - 8x^4 + 3x^2 - 9) \div (x + 9)$
c. $(x^4 - 5x^3 - x^2 + 7x + 3) \div (x^2 + 1)$
d. $(x^8 - 2) \div (x^2 - 4)$
e. $(2x^2 + 6x + 7) \div (5 - x)$

a, b, and e

In-Class Examples (Objective 6.4B)

Divide using synthetic division.

1. $(3x^2 + 11x + 6) \div (x + 3)$ $3x + 2$
2. $(3x^2 - 6) \div (x - 2)$ $3x + 6 + \dfrac{6}{x - 2}$
3. $(3x^3 + 2x^2 - 8x + 4) \div (x - 1)$
   $3x^2 + 5x - 3 + \dfrac{1}{x - 1}$
4. $\dfrac{2x^4 - 5x^3 - 8x^2 - 17x + 4}{x - 4}$ $2x^3 + 3x^2 + 4x - 1$

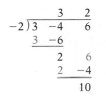

Instructor Note

Remind students that they can check an answer to a synthetic division problem in the same way that they check an answer to a long division problem.

Optional Student Activity

1. Two linear factors of $x^4 + x^3 - 7x^2 - x + 6$ are $x - 1$ and $x + 3$. Find the other two linear factors of $x^4 + x^3 - 7x^2 - x + 6$. $x - 2$ and $x + 1$
2. A rectangular box has a volume of $(x^3 + 11x^2 + 38x + 40)$ in^3. The height of the box is $(x + 2)$ in. Find the length and width of the box in terms of x. Length: $(x + 5)$ in.; width: $(x + 4)$ in.
3. The volume of a right circular cylinder is $\pi(x^3 + 7x^2 + 15x + 9)$ cm^3. The height of the cylinder is $(x + 1)$ cm. Find the radius of the cylinder in terms of x. $(x + 3)$ cm
4. When a polynomial $P(x)$ is divided by a polynomial $d(x)$, it produces a quotient $q(x)$ and a remainder $r(x)$. Either $r(x) = 0$ or the degree of $r(x)$ is less than the degree of the divisor $d(x)$. Why must the degree of $r(x)$ be less than the degree of $d(x)$? If the degree of $r(x)$ is not less than the degree of $d(x)$, then $r(x)$ is divisible by $d(x)$.

➡ Divide: $(3x^3 + 6x^2 - x - 2) \div (x + 3)$

The additive inverse of the binomial constant ↓ Coefficients of the polynomial

$$\begin{array}{r|rrrr} -3 & 3 & 6 & -1 & -2 \\ & \downarrow & & & \\ \hline & 3 & & & \end{array}$$

• Bring down the 3.

$$\begin{array}{r|rrrr} -3 & 3 & 6 & -1 & -2 \\ & & -9 & & \\ \hline & 3 & -3 & & \end{array}$$

• Multiply $-3(3)$ and add the product to 6.

$$\begin{array}{r|rrrr} -3 & 3 & 6 & -1 & -2 \\ & & -9 & 9 & \\ \hline & 3 & -3 & 8 & \end{array}$$

• Multiply $-3(-3)$ and add the product to -1.

$$\begin{array}{r|rrrr} -3 & 3 & 6 & -1 & -2 \\ & & -9 & 9 & -24 \\ \hline & 3 & -3 & 8 & -26 \end{array}$$

• Multiply $-3(8)$ and add the product to -2.

Terms of the quotient: 3, −3, 8 — Remainder: −26

TAKE NOTE
Remember that you can check the answer to a synthetic-division problem in the same way that you check the answer to a long-division problem.

The degree of the dividend is 3 and the degree of the divisor is 1. Therefore, the degree of the quotient is $3 - 1 = 2$.

$$(3x^3 + 6x^2 - x - 2) \div (x + 3) = 3x^2 - 3x + 8 - \frac{26}{x + 3}$$

➡ Divide: $(2x^3 - x + 2) \div (x - 2)$

The additive inverse of the binomial constant ↓ Coefficients of the polynomial

$$\begin{array}{r|rrrr} 2 & 2 & 0 & -1 & 2 \\ & \downarrow & & & \\ \hline & 2 & & & \end{array}$$

• Insert a 0 for the missing term and bring down the 2.

$$\begin{array}{r|rrrr} 2 & 2 & 0 & -1 & 2 \\ & & 4 & & \\ \hline & 2 & 4 & & \end{array}$$

• Multiply 2(2) and add the product to 0.

$$\begin{array}{r|rrrr} 2 & 2 & 0 & -1 & 2 \\ & & 4 & 8 & \\ \hline & 2 & 4 & 7 & \end{array}$$

• Multiply 2(4) and add the product to -1.

$$\begin{array}{r|rrrr} 2 & 2 & 0 & -1 & 2 \\ & & 4 & 8 & 14 \\ \hline & 2 & 4 & 7 & 16 \end{array}$$

• Multiply 2(7) and add the product to 2.

Terms of the quotient: 2, 4, 7 — Remainder: 16

$$(2x^3 - x + 2) \div (x - 2) = 2x^2 + 4x + 7 + \frac{16}{x - 2}$$

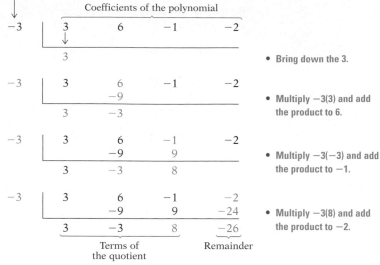

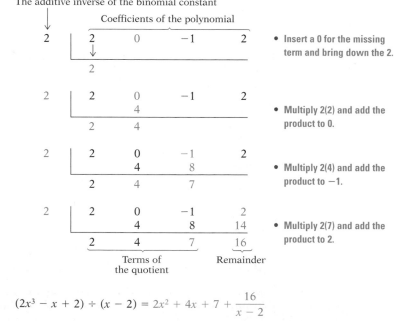

Example 4
Divide by using synthetic division:
$(7 - 3x + 5x^2) \div (x - 1)$

Solution
Arrange the coefficients in decreasing powers of x.

$$\begin{array}{r|rrr} 1 & 5 & -3 & 7 \\ & & 5 & 2 \\ \hline & 5 & 2 & 9 \end{array}$$

$$(5x^2 - 3x + 7) \div (x - 1) = 5x + 2 + \frac{9}{x - 1}$$

You Try It 4
Divide by using synthetic division:
$(8x + 6x^2 - 5) \div (x + 2)$

Your solution
$6x - 4 + \frac{3}{x + 2}$

Example 5
Divide by using synthetic division:
$(2x^3 + 4x^2 - 3x + 12) \div (x + 4)$

Solution

$$\begin{array}{r|rrrr} -4 & 2 & 4 & -3 & 12 \\ & & -8 & 16 & -52 \\ \hline & 2 & -4 & 13 & -40 \end{array}$$

$$(2x^3 + 4x^2 - 3x + 12) \div (x + 4)$$
$$= 2x^2 - 4x + 13 - \frac{40}{x + 4}$$

You Try It 5
Divide by using synthetic division:
$(5x^3 - 12x^2 - 8x + 16) \div (x - 2)$

Your solution
$5x^2 - 2x - 12 - \frac{8}{x - 2}$

Example 6
Divide by using synthetic division:
$(3x^4 - 8x^2 + 2x + 1) \div (x + 2)$

Solution
Insert a zero for the missing term.

$$\begin{array}{r|rrrrr} -2 & 3 & 0 & -8 & 2 & 1 \\ & & -6 & 12 & -8 & 12 \\ \hline & 3 & -6 & 4 & 6 & 13 \end{array}$$

$$(3x^4 - 8x^2 + 2x + 1) \div (x + 2)$$
$$= 3x^3 - 6x^2 + 4x - 6 + \frac{13}{x + 2}$$

You Try It 6
Divide by using synthetic division:
$(2x^4 - 3x^3 - 8x^2 - 2) \div (x - 3)$

Your solution
$2x^3 + 3x^2 + x + 3 + \frac{7}{x - 3}$

Solutions on p. S19

Objective 6.4C

Note: This objective begins on the next page.

New Vocabulary
Remainder Theorem

Vocabulary to Review
evaluate a polynomial
synthetic division

Discuss the Concepts

1. State the Remainder Theorem.
2. If the polynomial $3x^4 - 8x^2 + 2x + 1$ is divided by $x + 2$ and the remainder is 13, what do we know about $f(-2)$ for the function $f(x) = 3x^4 - 8x^2 + 2x + 1$?

Optional Student Activity
The Factor Theorem is a result of the Remainder Theorem. The Factor Theorem states that a polynomial $P(x)$ has a factor $(x - c)$ if and only if $P(c) = 0$. In other words, a remainder of zero means that the divisor is a factor of the dividend.

1. Determine whether $x + 5$ is a factor of $P(x) = x^4 + x^3 - 21x^2 - x + 20$.
2. Based on your answer to Exercise 1, is -5 a zero of $P(x)$? Explain your answer.
3. Explain why $P(x) = 4x^4 + 7x^2 + 12$ has no factor of the form $(x - c)$, where c is a real number.

(Continued on next page)

In-Class Examples (Objective 6.4C)
Use the Remainder Theorem to evaluate the polynomial.

1. $P(x) = 3x^2 - 4x + 5$; $P(3)$ 20
2. $f(x) = x^3 - 6x^2 + 3x - 8$; $f(2)$ -18

(Continued)

4. Determine whether the second polynomial is a factor of the first.

a. $x^3 + 8$; $x + 2$
b. $x^3 - 8$; $x + 2$
c. $x^3 + 8$; $x - 2$
d. $x^3 - 8$; $x - 2$
e. $x^4 + 16$; $x + 2$
f. $x^4 - 16$; $x + 2$
g. $x^4 + 16$; $x - 2$
h. $x^4 - 16$; $x - 2$

Use your answers to parts (a) through (h) to make a conjecture as to whether the statement is true or false.

i. For $n > 0$, $(x - y)$ is a factor of $(x^n - y^n)$.
j. For $n > 0$ and n an even integer, $(x + y)$ is a factor of $(x^n - y^n)$.
k. For $n > 0$ and n an odd integer, $(x + y)$ is a factor of $(x^n - y^n)$.
l. For $n > 0$ and n an even integer, $(x + y)$ is a factor of $(x^n + y^n)$.
m. For $n > 0$ and n an odd integer, $(x + y)$ is a factor of $(x^n + y^n)$.

1. Yes **2.** Yes **3.** The given polynomial has no factor of the form $(x - c)$ because the value of the polynomial is always greater than 0 and thus never equal to 0. **4a.** Yes **b.** No **c.** No **d.** Yes **e.** No **f.** Yes **g.** No **h.** Yes **i.** True **j.** True **k.** False **l.** False **m.** True

Objective C — To evaluate a polynomial using synthetic division

VIDEO & DVD · CD TUTOR · WEB · SSM

A polynomial can be evaluated by using synthetic division. Consider the polynomial $P(x) = 2x^4 - 3x^3 + 4x^2 - 5x + 1$. One way to evaluate the polynomial when $x = 2$ is to replace x by 2 and then simplify the numerical expression.

$$\begin{aligned} P(x) &= 2x^4 - 3x^3 + 4x^2 - 5x + 1 \\ P(2) &= 2(2)^4 - 3(2)^3 + 4(2)^2 - 5(2) + 1 \\ &= 2(16) - 3(8) + 4(4) - 5(2) + 1 \\ &= 32 - 24 + 16 - 10 + 1 \\ &= 15 \end{aligned}$$

Now use synthetic division to divide $(2x^4 - 3x^3 + 4x^2 - 5x + 1)$ by $(x - 2)$.

$$\begin{array}{r|rrrrr} 2 & 2 & -3 & 4 & -5 & 1 \\ & & 4 & 2 & 12 & 14 \\ \hline & 2 & 1 & 6 & 7 & 15 \end{array}$$

Terms of the quotient: 2, 1, 6, 7 — Remainder: 15

Note that the remainder is 15, which is the same value as $P(2)$. This is not a coincidence. The following theorem states that this situation is always true.

Remainder Theorem

If the polynomial $P(x)$ is divided by $x - a$, the remainder is $P(a)$.

➡ Use the Remainder Theorem to evaluate $P(x) = x^4 - 3x^2 + 4x - 5$ when $x = -2$.

The value at which the polynomial is evaluated: -2

$$\begin{array}{r|rrrrr} -2 & 1 & 0 & -3 & 4 & -5 \\ & & -2 & 4 & -2 & -4 \\ \hline & 1 & -2 & 1 & 2 & -9 \end{array}$$

• A 0 is inserted for the x^3 term.

-9 ← The remainder

$P(-2) = -9$

Example 7

Use the Remainder Theorem to evaluate $P(x) = -x^4 + 3x^3 + 2x^2 - x - 5$ when $x = -2$.

Solution

$$\begin{array}{r|rrrrr} -2 & -1 & 3 & 2 & -1 & -5 \\ & & 2 & -10 & 16 & -30 \\ \hline & -1 & 5 & -8 & 15 & -35 \end{array}$$

$P(-2) = -35$

You Try It 7

Use the Remainder Theorem to evaluate $P(x) = 2x^3 - 5x^2 + 7$ when $x = -3$.

Your solution

-92

Solution on p. S19

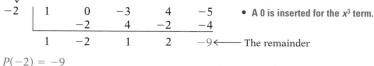

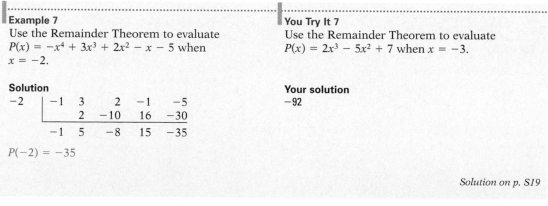

363

6.4 Exercises

Objective A

Divide by using long division.

1. $(x^2 + 3x - 40) \div (x - 5)$
$x + 8$

2. $(x^2 - 14x + 24) \div (x - 2)$
$x - 12$

3. $(x^3 - 3x^2 + 2) \div (x - 3)$
$x^2 + \dfrac{2}{x - 3}$

4. $(x^3 + 4x^2 - 8) \div (x + 4)$
$x^2 - \dfrac{8}{x + 4}$

5. $(6x^2 + 13x + 8) \div (2x + 1)$
$3x + 5 + \dfrac{3}{2x + 1}$

6. $(12x^2 + 13x - 14) \div (3x - 2)$
$4x + 7$

7. $(10x^2 + 9x - 5) \div (2x - 1)$
$5x + 7 + \dfrac{2}{2x - 1}$

8. $(18x^2 - 3x + 2) \div (3x + 2)$
$6x - 5 + \dfrac{12}{3x + 2}$

9. $(8x^3 - 9) \div (2x - 3)$
$4x^2 + 6x + 9 + \dfrac{18}{2x - 3}$

10. $(64x^3 + 4) \div (4x + 2)$
$16x^2 - 8x + 4 - \dfrac{4}{4x + 2}$

11. $(6x^4 - 13x^2 - 4) \div (2x^2 - 5)$
$3x^2 + 1 + \dfrac{1}{2x^2 - 5}$

12. $(12x^4 - 11x^2 + 10) \div (3x^2 + 1)$
$4x^2 - 5 + \dfrac{15}{3x^2 + 1}$

13. $\dfrac{-10 - 33x + 3x^3 - 8x^2}{3x + 1}$
$x^2 - 3x - 10$

14. $\dfrac{10 - 49x + 38x^2 - 8x^3}{1 - 4x}$
$2x^2 - 9x + 10$

15. $\dfrac{x^3 - 5x^2 + 7x - 4}{x - 3}$
$x^2 - 2x + 1 - \dfrac{1}{x - 3}$

16. $\dfrac{2x^3 - 3x^2 + 6x + 4}{2x + 1}$
$x^2 - 2x + 4$

17. $\dfrac{16x^2 - 13x^3 + 2x^4 + 20 - 9x}{x - 5}$
$2x^3 - 3x^2 + x - 4$

18. $\dfrac{x - x^2 + 5x^3 + 3x^4 - 2}{x + 2}$
$3x^3 - x^2 + x - 1$

Section 6.4

Suggested Assignment
Exercises 1–63, odds
More challenging problems:
Exercises 65–67

Quick Quiz (Objective 6.4A)
Divide using long division.

1. $(10x^2 + 14x - 15) \div (5x - 3)$ $2x + 4 - \dfrac{3}{5x - 3}$
2. $(x^3 + 4x^2 - 6) \div (x + 2)$ $x^2 + 2x - 4 + \dfrac{2}{x + 2}$
3. $\dfrac{8x^3 - 18x^2 + 27x + 8}{4x + 1}$ $2x^2 - 5x + 8$

19. $\dfrac{2x^3 + 4x^2 - x + 2}{x^2 + 2x - 1}$

$2x + \dfrac{x + 2}{x^2 + 2x - 1}$

20. $\dfrac{3x^3 - 2x^2 + 5x - 4}{x^2 - x + 3}$

$3x + 1 + \dfrac{-3x - 7}{x^2 - x + 3}$

21. $\dfrac{x^4 + 2x^3 - 3x^2 - 6x + 2}{x^2 - 2x - 1}$

$x^2 + 4x + 6 + \dfrac{10x + 8}{x^2 - 2x - 1}$

22. $\dfrac{x^4 - 3x^3 + 4x^2 - x + 1}{x^2 + x - 3}$

$x^2 - 4x + 11 + \dfrac{-24x + 34}{x^2 + x - 3}$

23. $\dfrac{x^4 + 3x^2 - 4x + 5}{x^2 + 2x + 3}$

$x^2 - 2x + 4 + \dfrac{-6x - 7}{x^2 + 2x + 3}$

24. $\dfrac{x^4 + 2x^3 - x + 2}{x^2 - x - 1}$

$x^2 + 3x + 4 + \dfrac{6x + 6}{x^2 - x - 1}$

Objective B

Divide by using synthetic division.

25. $(2x^2 - 6x - 8) \div (x + 1)$

$2x - 8$

26. $(3x^2 + 19x + 20) \div (x + 5)$

$3x + 4$

27. $(3x^2 - 14x + 16) \div (x - 2)$

$3x - 8$

28. $(4x^2 - 23x + 28) \div (x - 4)$

$4x - 7$

29. $(3x^2 - 4) \div (x - 1)$

$3x + 3 - \dfrac{1}{x - 1}$

30. $(4x^2 - 8) \div (x - 2)$

$4x + 8 + \dfrac{8}{x - 2}$

31. $(2x^3 - x^2 + 6x + 9) \div (x + 1)$

$2x^2 - 3x + 9$

32. $(3x^3 + 10x^2 + 6x - 4) \div (x + 2)$

$3x^2 + 4x - 2$

33. $(18 + x - 4x^3) \div (2 - x)$

$4x^2 + 8x + 15 + \dfrac{12}{x - 2}$

34. $(12 - 3x^2 + x^3) \div (x + 3)$

$x^2 - 6x + 18 - \dfrac{42}{x + 3}$

35. $(2x^3 + 5x^2 - 5x + 20) \div (x + 4)$

$2x^2 - 3x + 7 - \dfrac{8}{x + 4}$

36. $(5x^3 + 3x^2 - 17x + 6) \div (x + 2)$

$5x^2 - 7x - 3 + \dfrac{12}{x + 2}$

Quick Quiz (Objective 6.4B)

Divide using synthetic division.

1. $(3x^2 + 6x - 24) \div (x + 4)$ $3x - 6$

2. $(2x^3 + 3x^2 + 5x + 16) \div (x + 2)$

$2x^2 - x + 7 + \dfrac{2}{x + 2}$

3. $\dfrac{4x^4 + 7x^3 - 29x^2 + 22x - 24}{x + 4}$

$4x^3 - 9x^2 + 7x - 6$

365

37. $\dfrac{5 + 5x - 8x^2 + 4x^3 - 3x^4}{2 - x}$

$3x^3 + 2x^2 + 12x + 19 + \dfrac{33}{x - 2}$

38. $\dfrac{3 - 13x - 5x^2 + 9x^3 - 2x^4}{3 - x}$

$2x^3 - 3x^2 - 4x + 1$

39. $\dfrac{3x^4 + 3x^3 - x^2 + 3x + 2}{x + 1}$

$3x^3 - x + 4 - \dfrac{2}{x + 1}$

40. $\dfrac{4x^4 + 12x^3 - x^2 - x + 2}{x + 3}$

$4x^3 - x + 2 - \dfrac{4}{x + 3}$

41. $\dfrac{2x^4 - x^2 + 2}{x - 3}$

$2x^3 + 6x^2 + 17x + 51 + \dfrac{155}{x - 3}$

42. $\dfrac{x^4 - 3x^3 - 30}{x + 2}$

$x^3 - 5x^2 + 10x - 20 + \dfrac{10}{x + 2}$

Objective C

43. If the polynomial $2x^3 - 3x^2 - 4x - 15$ is divided by $x - 3$ and the remainder is 0, what do we know about $f(3)$ for the function $f(x) = 2x^3 - 3x^2 - 4x - 15$?
$f(3) = 0$

44. If the polynomial $3x^4 - 8x^2 + 2x + 1$ is divided by $x + 2$ and the remainder is 13, what do we know about $f(-2)$ for the function $f(x) = 3x^4 - 8x^2 + 2x + 1$?
$f(-2) = 13$

Use the Remainder Theorem to evaluate the polynomial.

45. $P(x) = 2x^2 - 3x - 1$; $P(3)$
8

46. $Q(x) = 3x^2 - 5x - 1$; $Q(2)$
1

47. $R(x) = x^3 - 2x^2 + 3x - 1$; $R(4)$
43

48. $F(x) = x^3 + 4x^2 - 3x + 2$; $F(3)$
56

49. $P(z) = 2z^3 - 4z^2 + 3z - 1$; $P(-2)$
-39

50. $R(t) = 3t^3 + t^2 - 4t + 2$; $R(-3)$
-58

51. $Z(p) = 2p^3 - p^2 + 3$; $Z(-3)$
-60

52. $P(y) = 3y^3 + 2y^2 - 5$; $P(-2)$
-21

Quick Quiz (Objective 6.4C)

Use the Remainder Theorem to evaluate the polynomial.

1. $R(x) = 4x^2 + 6x - 3$; $R(-2)$ 1

2. $F(x) = x^3 - 5x^2 + 2x - 10$; $F(3)$ -22

Answers to Writing Exercises

68. Synthetic division can be modified so that the divisor is of the form $ax + b$. Divide both the dividend and the divisor by a $\left(\text{or multiply both the dividend and the divisor by } \frac{1}{a}\right)$. The divisor is now in the form $x + \frac{b}{a}$ and the expression $\frac{b}{a}$ can be used for a in the $x - a$ of synthetic division.

53. $Q(x) = x^4 + 3x^3 - 2x^2 + 4x - 9$; $Q(2)$
31

54. $Y(z) = z^4 - 2z^3 - 3z^2 - z + 7$; $Y(3)$
4

55. $F(x) = 2x^4 - x^3 + 2x - 5$; $F(-3)$
178

56. $Q(x) = x^4 - 2x^3 + 4x - 2$; $Q(-2)$
22

57. $P(x) = x^3 - 3$; $P(5)$
122

58. $S(t) = 4t^3 + 5$; $S(-4)$
-251

59. $R(t) = 4t^4 - 3t^2 + 5$; $R(-3)$
302

60. $P(z) = 2z^4 + z^2 - 3$; $P(-4)$
525

61. $Q(x) = x^5 - 4x^3 - 2x^2 + 5x - 2$; $Q(2)$
0

62. $T(x) = 2x^5 + 4x^4 - x^2 + 4$; $T(3)$
805

63. $R(x) = 2x^5 - x^3 + 4x - 1$; $R(-2)$
-65

64. $P(x) = x^5 - x^3 + 4x + 1$; $P(-3)$
-227

APPLYING THE CONCEPTS

65. Divide by using long division.

a. $\dfrac{a^3 + b^3}{a + b}$
$a^2 - ab + b^2$

b. $\dfrac{x^5 + y^5}{x + y}$
$x^4 - x^3y + x^2y^2 - xy^3 + y^4$

c. $\dfrac{x^6 - y^6}{x + y}$
$x^5 - x^4y + x^3y^2 - x^2y^3 + xy^4 - y^5$

66. For what value of k will the remainder be zero?

a. $(x^3 - x^2 - 3x + k) \div (x + 3)$
$k = 27$

b. $(2x^3 - x + k) \div (x - 1)$
$k = -1$

67. Divide.

a. $(2x^3 + 7x^2 + 2x - 8) \div (4x + 8)$
$\frac{1}{2}x^2 + \frac{3}{4}x - 1$

b. $(4x^3 + 13x^2 - 22x + 24) \div (6x - 12)$
$\frac{2}{3}x^2 + \frac{7}{2}x + \frac{10}{3} + \frac{64}{6x - 12}$

c. $(2x^4 - 3x^3 + 4x^2 + x - 10) \div (x^2 - x + 1)$
$2x^2 - x + 1 + \dfrac{3x - 11}{x^2 - x + 1}$

d. $(x^4 + 4x^3 + 2x^2 - x + 5) \div (x^2 - 2x - 3)$
$x^2 + 6x + 17 + \dfrac{51x + 56}{x^2 - 2x - 3}$

68. Show how synthetic division can be modified so that the divisor can be of the form $ax + b$.

367

Focus on Problem Solving

Dimensional Analysis

In solving application problems, it may be useful to include the units in order to organize the problem so that the answer is in the proper units. Using units to organize and check the correctness of an application is called **dimensional analysis**. We use the operations of multiplying units and dividing units in applying dimensional analysis to application problems.

The Rule for Multiplying Exponential Expressions states that we multiply two expressions with the same base by adding the exponents.

$$x^4 \cdot x^6 = x^{4+6} = x^{10}$$

In calculations that involve quantities, the units are operated on algebraically.

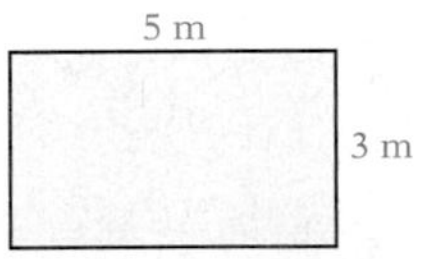

➡ A rectangle measures 3 m by 5 m. Find the area of the rectangle.

$A = LW = (3\text{ m})(5\text{ m}) = (3 \cdot 5)(\text{m} \cdot \text{m}) = 15\text{ m}^2$

The area of the rectangle is 15 m^2 (square meters).

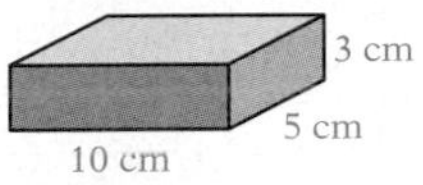

➡ A box measures 10 cm by 5 cm by 3 cm. Find the volume of the box.

$V = LWH = (10\text{ cm})(5\text{ cm})(3\text{ cm}) = (10 \cdot 5 \cdot 3)(\text{cm} \cdot \text{cm} \cdot \text{cm}) = 150\text{ cm}^3$

The volume of the box is 150 cm^3 (cubic centimeters).

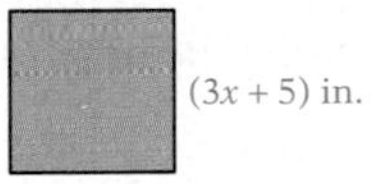

➡ Find the area of a square whose side measures $(3x + 5)$ in.

$A = s^2 = [(3x + 5)\text{ in.}]^2 = (3x + 5)^2\text{ in}^2 = (9x^2 + 30x + 25)\text{ in}^2$

The area of the square is $(9x^2 + 30x + 25)\text{ in}^2$ (square inches).

Dimensional analysis is used in the conversion of units.

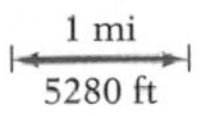

The following example converts the unit miles to feet. The equivalent measures 1 mi = 5280 ft are used to form the following rates, which are called conversion factors: $\frac{1\text{ mi}}{5280\text{ ft}}$ and $\frac{5280\text{ ft}}{1\text{ mi}}$. Because 1 mi = 5280 ft, both of the conversion factors $\frac{1\text{ mi}}{5280\text{ ft}}$ and $\frac{5280\text{ ft}}{1\text{ mi}}$ are equal to 1.

To convert 3 mi to feet, multiply 3 mi by the conversion factor $\frac{5280\text{ ft}}{1\text{ mi}}$.

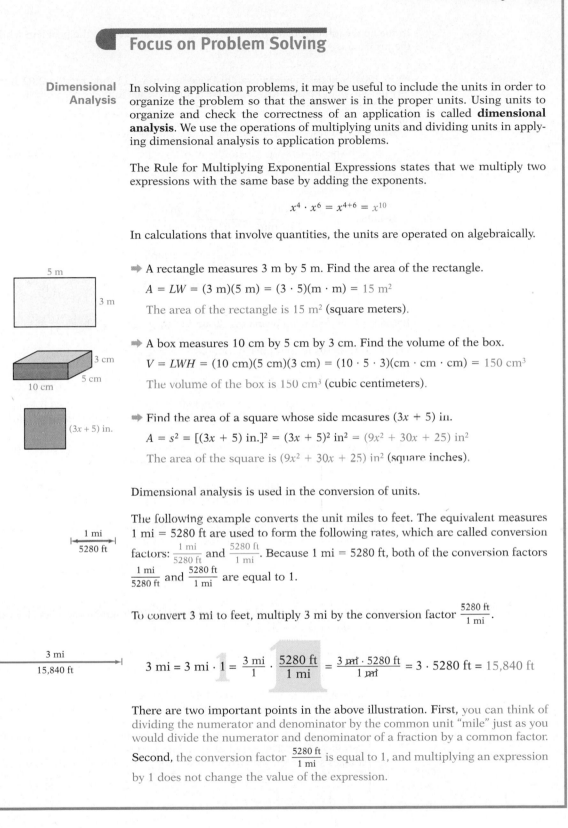

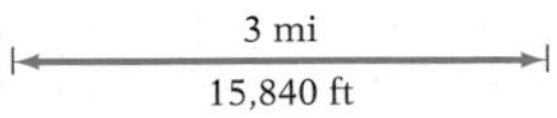

$$3\text{ mi} = 3\text{ mi} \cdot 1 = \frac{3\text{ mi}}{1} \cdot \frac{5280\text{ ft}}{1\text{ mi}} = \frac{3\text{ mi} \cdot 5280\text{ ft}}{1\text{ mi}} = 3 \cdot 5280\text{ ft} = 15{,}840\text{ ft}$$

There are two important points in the above illustration. First, you can think of dividing the numerator and denominator by the common unit "mile" just as you would divide the numerator and denominator of a fraction by a common factor. Second, the conversion factor $\frac{5280\text{ ft}}{1\text{ mi}}$ is equal to 1, and multiplying an expression by 1 does not change the value of the expression.

Answers to Focus on Problem Solving: Dimensional Analysis

1. 60 mph
2. 28.8 km/h
3. $9440
4. $2150
5. 59.8 gal
6. 22.4 gal
7. $23,413.50
8. $3/ft^2
9. 29.6 yd^2
10. 182.3 mph
11. 15.4 s

In the application problem that follows, the units are kept in the problem while the problem is worked.

➡ In 2000, a horse named Fusaichi Pegasus ran a 1.25-mile race in 2.02 min. Find Fusaichi Pegasus's average speed for that race in miles per hour. Round to the nearest tenth.

Strategy To find the average speed, use the formula $r = \frac{d}{t}$, where r is the speed, d is the distance, and t is the time. Use the conversion factor $\frac{60 \text{ min}}{1 \text{ h}}$.

Solution $$r = \frac{d}{t} = \frac{1.25 \text{ mi}}{2.02 \text{ min}} = \frac{1.25 \text{ mi}}{2.02 \text{ min}} \cdot \frac{60 \text{ min}}{1 \text{ h}}$$
$$= \frac{75 \text{ mi}}{2.02 \text{ h}} \approx 37.1 \text{ mph}$$

Fusaichi Pegasus's average speed was 37.1 mph.

Try each of the following problems. Round to the nearest tenth.

1. Convert 88 ft/s to miles per hour.
2. Convert 8 m/s to kilometers per hour (1 km = 1000 m).
3. A carpet is to be placed in a meeting hall that is 36 ft wide and 80 ft long. At $29.50 per square yard, how much will it cost to carpet the meeting hall?
4. A carpet is to be placed in a room that is 20 ft wide and 30 ft long. At $32.25 per square yard, how much will it cost to carpet the area?
5. Find the number of gallons of water in a fish tank that is 36 in. long and 24 in. wide and is filled to a depth of 16 in. (1 gal = 231 in^3).
6. Find the number of gallons of water in a fish tank that is 24 in. long and 18 in. wide and is filled to a depth of 12 in. (1 gal = 231 in^3).
7. A $\frac{1}{4}$-acre commercial lot is on sale for $2.15 per square foot. Find the sale price of the commercial lot (1 acre = 43,560 ft^2).
8. A 0.75-acre industrial parcel was sold for $98,010. Find the parcel's price per square foot (1 acre = 43,560 ft^2).
9. A new driveway will require 800 ft^3 of concrete. Concrete is ordered by the cubic yard. How much concrete should be ordered?
10. A piston-engined dragster traveled 440 yd in 4.936 s at Ennis, Texas, on October 9, 1988. Find the average speed of the dragster in miles per hour.
11. The Marianas Trench in the Pacific Ocean is the deepest part of the ocean. Its depth is 6.85 mi. The speed of sound under water is 4700 ft/s. Find the time it takes sound to travel from the surface to the bottom of the Marianas Trench and back.

369

Projects and Group Activities

Pascal's Triangle

Simplifying the power of a binomial is called *expanding the binomial.* The expansion of the first three powers of a binomial is shown below.

$(a + b)^1 = a + b$

$(a + b)^2 = (a + b)(a + b) = a^2 + 2ab + b^2$

$(a + b)^3 = (a + b)^2(a + b) = (a^2 + 2ab + b^2)(a + b) = a^3 + 3a^2b + 3ab^2 + b^3$

Find $(a + b)^4$. [*Hint*: $(a + b)^4 = (a + b)^3(a + b)$]

Find $(a + b)^5$. [*Hint*: $(a + b)^5 = (a + b)^4(a + b)$]

If we continue in this way, the results for $(a + b)^6$ are

$$(a + b)^6 = a^6 + 6a^5b + 15a^4b^2 + 20a^3b^3 + 15a^2b^4 + 6ab^5 + b^6$$

Point of Interest

Pascal did not invent the triangle of numbers known as Pascal's Triangle. It was known to mathematicians in China probably as early as A.D. 1050. But Pascal's *Traite du triangle arithmetique* (*Treatise Concerning the Arithmetical Triangle*) brought together all the different aspects of the numbers for the first time.

Now expand $(a + b)^8$. Before you begin, see whether you can find a pattern that will help you write the expansion of $(a + b)^8$ without having to multiply it out. Here are some hints.

1 1
1 2 1
1 3 3 1
1 4 6 4 1
1 5 10 10 5 1
1 6 15 20 15 6 1

1. Write out the variable terms of each binomial expansion from $(a + b)^1$ through $(a + b)^6$. Observe how the exponents on the variables change.
2. Write out the coefficients of all the terms without the variable parts. It will be helpful to make a triangular arrangement as shown at the left. Note that each row begins and ends with a 1. Also note (in the two shaded regions, for example) that any number in a row is the sum of the two closest numbers above it. For instance, $1 + 5 = 6$ and $6 + 4 = 10$.

The triangle of numbers shown at the left is called Pascal's Triangle. To find the expansion of $(a + b)^8$, you need to find the eighth row of Pascal's Triangle. First find row seven. Then find row eight and use the patterns you have observed to write the expansion $(a + b)^8$.

WWW WEB

Pascal's Triangle has been the subject of extensive analysis, and many patterns have been found. See whether you can find some of them. You might check the Internet, where you will find some web sites with information on Pascal's Triangle.

Answers to Projects and Group Activities: Pascal's Triangle

1. $(a + b)^1 = a + b$
$(a + b)^2 = a^2 + 2ab + b^2$
$(a + b)^3 = a^3 + 3a^2b + 3ab^2 + b^3$
$(a + b)^4 = a^4 + 4a^3b + 6a^2b^2 + 4ab^3 + b^4$
$(a + b)^5 = a^5 + 5a^4b + 10a^3b^2 + 10a^2b^3 + 5ab^4 + b^5$
$(a + b)^6 = a^6 + 6a^5b + 15a^4b^2 + 20a^3b^3 + 15a^2b^4 + 6ab^5 + b^6$
The exponents on a begin with the same exponent as that on the binomial and decrease by 1 each term.
The exponents on b increase by 1 each term, beginning with zero up to the exponent on the binomial. The sum of the exponents in each term is the same as the exponent on the binomial.
2. Row seven:
1 7 21 35 35 21 7 1
Row eight:
1 8 28 56 70 56 28 8 1
$(a + b)^8 = a^8 + 8a^7b + 28a^6b^2 + 56a^5b^3 + 70a^4b^4 + 56a^3b^5 + 28a^2b^6 + 8ab^7 + b^8$

Chapter Summary

Key Words

A *monomial* is a number, a variable, or a product of a number and variables. The *degree of a monomial* is the sum of the exponents of the variables. [p. 329]

An *exponential expression is in simplest form* when it is written with only positive exponents. [p. 332]

A number written in *scientific notation* is a number written in the form $a \times 10^n$, where $1 \le a < 10$. [p. 335]

A *polynomial* is a variable expression in which the terms are monomials. A polynomial of two terms is a *binomial.* A polynomial of three terms is a *trinomial.* The *degree of a polynomial* is the greatest of the degrees of any of its terms. A polynomial in one variable is usually written in *descending order,* which means that the terms are arranged so that the exponents of the variable decrease from left to right. [p. 341]

A *polynomial function* is an expression whose terms are monomials. A *linear function* is given by the equation $f(x) = mx + b$. A second-degree polynomial function, called a *quadratic function,* is given by the equation $f(x) = ax^2 + bx + c, a \neq 0$. A third-degree polynomial function is called a *cubic function*. The *leading coefficient* of a polynomial function is the coefficient of the variable with the largest exponent. [p. 341]

The product of two binomials can be found using the *FOIL* method, which is based on the Distributive Property. The letters of FOIL stand for First, Outer, Inner, and Last. [p. 350]

Synthetic division is a shorter method of dividing a polynomial by a binomial of the form $x - a$. This method uses only the coefficients of the variable terms. [p. 359]

Essential Rules

Rule for Multiplying Exponential Expressions [p. 329] $x^m \cdot x^n = x^{m+n}$

Rule for Simplifying Powers of Exponential Expressions [p. 330] $(x^m)^n = x^{mn}$

Rule for Simplifying Powers of Products [p. 330] $(x^m y^n)^p = x^{mp} y^{np}$

Definition of Zero as an Exponent [p. 331] If $x \neq 0$, then $x^0 = 1$.

Definition of a Negative Exponent [p. 332] $x^{-n} = \frac{1}{x^n}$ and $\frac{1}{x^{-n}} = x^n$, $x \neq 0$

Rule for Dividing Exponential Expressions [p. 333] $\frac{x^m}{x^n} = x^{m-n}, x \neq 0$

Rule for Simplifying Powers of Quotients [p. 332] $\left(\frac{x^m}{y^n}\right)^p = \frac{x^{mp}}{y^{np}}, y \neq 0$

The Sum and Difference of Two Terms [p. 351] $(a + b)(a - b) = a^2 - b^2$

The Square of a Binomial [p. 351] $(a + b)^2 = a^2 + 2ab + b^2$

Remainder Theorem [p. 362] If the polynomial $P(x)$ is divided by $x - a$, then the remainder is $P(a)$.

Dividend = (quotient × divisor) + remainder [p. 357]

371

Chapter Review

1. Multiply: $(-2a^2b^4)(3ab^2)$
$-6a^3b^6$ [6.1A]

2. Simplify: $(-3x^2y^3)^2$
$9x^4y^6$ [6.1A]

3. Simplify: $\dfrac{(2a^4b^{-3}c^2)^3}{(2a^3b^2c^{-1})^4}$

$\dfrac{c^{10}}{2b^{17}}$ [6.1B]

4. Write 2.54×10^{-3} in decimal notation.

0.00254 [6.1C]

5. Given $P(x) = 2x^3 - x + 7$, find $P(-2)$.
-7 [6.2A]

6. Subtract: $(5x^2 - 8xy + 2y^2) - (x^2 - 3y^2)$
$4x^2 - 8xy + 5y^2$ [6.2B]

7. Simplify: $(2x^{-1}y^2z^5)(-3x^3yz^{-3})^2$
$\dfrac{18x^5y^4}{z}$ [6.1B]

8. Multiply: $(2a^{12}b^3)(-9b^2c^6)(3ac)$
$-54a^{13}b^5c^7$ [6.1A]

9. Graph: $f(x) = x^2 + 1$

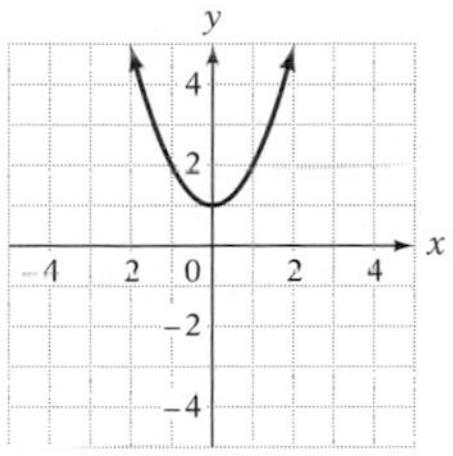

[6.2A]

10. Identify (a) the leading coefficient, (b) the constant term, and (c) the degree of the polynomial $P(x) = 3x^5 - 3x^2 + 7x + 8$.
a. 3
b. 8
c. 5 [6.2A]

11. Use the Remainder Theorem to evaluate $P(x) = x^3 - 2x^2 + 3x - 5$ when $x = 2$.
1 [6.4C]

12. Multiply: $-2x(4x^2 + 7x - 9)$
$-8x^3 - 14x^2 + 18x$ [6.3A]

13. Multiply: $2ab^3(4a^2 - 2ab + 3b^2)$
$8a^3b^3 - 4a^2b^4 + 6ab^5$ [6.3A]

14. Multiply: $(3y^2 + 4y - 7)(2y + 3)$
$6y^3 + 17y^2 - 2y - 21$ [6.3B]

15. Multiply: $(5a - 7)(2a + 9)$
$10a^2 + 31a - 63$ [6.3C]

16. Write 0.000000127 in scientific notation.
1.27×10^{-7} [6.1C]

372

17. Simplify: $(5y - 7)^2$

$25y^2 - 70y + 49$ [6.3D]

18. Divide: $\dfrac{15x^2 + 2x - 2}{3x - 2}$

$5x + 4 + \dfrac{6}{3x - 2}$ [6.4A]

19. Divide: $\dfrac{4x^3 + 27x^2 + 10x + 2}{x + 6}$

$4x^2 + 3x - 8 + \dfrac{50}{x + 6}$ [6.4B]

20. Divide: $\dfrac{x^4 - 4}{x - 4}$

$x^3 + 4x^2 + 16x + 64 + \dfrac{252}{x - 4}$ [6.4B]

21. Write 8.1×10^9 in decimal notation.

8,100,000,000 [6.1C]

22. Simplify: $\dfrac{-18a^6b}{27a^3b^4}$

$-\dfrac{2a^3}{3b^3}$ [6.1B]

23. Multiply: $(5a + 2b)(5a - 2b)$
$25a^2 - 4b^2$ [6.3D]

24. Use the Remainder Theorem to evaluate $P(x) = -2x^3 + 2x^2 - 4$ when $x = -3$.
68 [6.4C]

25. Add: $(12y^2 + 17y - 4) + (9y^2 - 13y + 3)$
$21y^2 + 4y - 1$ [6.2B]

26. Multiply: $(6b^3 - 2b^2 - 5)(2b^2 - 1)$
$12b^5 - 4b^4 - 6b^3 - 8b^2 + 5$ [6.3B]

27. Multiply: $(a + 7)(a - 7)$
$a^2 - 49$ [6.3D]

28. Write 765,000,000,000 in scientific notation.
7.65×10^{11} [6.1C]

29. The length of a Ping-Pong table is 1 ft less than twice the width of the table. Let w represent the width of the Ping-Pong table. Express the area of the table in terms of the variable w.
$(2w^2 - w)$ ft² [6.3E]

30. The most distant object visible from Earth without the aid of a telescope is the Great Galaxy of Andromeda. It takes light from this galaxy 2.2×10^6 years to travel to Earth. Light travels about 6.7×10^8 mph. How far from Earth is the Great Galaxy of Andromeda? Use a 365-day year.
1.291224×10^{19} mi [6.1D]

31. Write the number of seconds in one week in scientific notation.
6.048×10^5 s [6.1D]

32. The length of a side of a square checkerboard is $(3x - 2)$ in. Express the area of the checkerboard in terms of the variable x.
$(9x^2 - 12x + 4)$ in² [6.3E]

33. The length of a rectangle is $(5x + 3)$ cm. The width is $(2x - 7)$ cm. Find the area of the rectangle in terms of the variable x.
$(10x^2 - 29x - 21)$ cm² [6.3E]

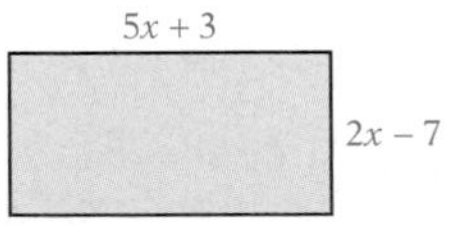

Objective B To factor by grouping

VIDEO & DVD CD TUTOR WEB SSM

In the examples at the right, the binomials in parentheses are called **binomial factors**.

$2a(a + b)^2$
$3xy(x - y)$

The Distributive Property is used to factor a common binomial factor from an expression.

The common binomial factor of the expression $6x(x - 3) + y^2(x - 3)$ is $(x - 3)$. To factor that expression, use the Distributive Property to write the expression as a product of factors.

$$6x(x - 3) + y^2(x - 3) = (x - 3)(6x + y^2)$$

Consider the following simplification of $-(a - b)$.

$$-(a - b) = -1(a - b) = -a + b = b - a$$

Thus, $$\mathbf{b - a = -(a - b)}$$

This equation is sometimes used to factor a common binomial from an expression.

➡ Factor: $2x(x - y) + 5(y - x)$

$$2x(x - y) + 5(y - x) = 2x(x - y) - 5(x - y)$$
$$= (x - y)(2x - 5)$$

- $5(y - x) = 5[(-1)(x - y)]$
  $= -5(x - y)$

Some polynomials can be factored by grouping terms in such a way that a common binomial factor is found.

➡ Factor: $ax + bx - ay - by$

$$ax + bx - ay - by = (ax + bx) - (ay + by)$$

- **Group the first two terms and the last two terms. Note that $-ay - by = -(ay + by)$.**

$$= x(a + b) - y(a + b)$$
$$= (a + b)(x - y)$$

- **Factor the GCF from each group.**

➡ Factor: $6x^2 - 9x - 4xy + 6y$

$$6x^2 - 9x - 4xy + 6y = (6x^2 - 9x) - (4xy - 6y)$$

- **Group the first two terms and the last two terms. Note that $-4xy + 6y = -(4xy - 6y)$.**

$$= 3x(2x - 3) - 2y(2x - 3)$$
$$= (2x - 3)(3x - 2y)$$

- **Factor the GCF from each group.**

Objective 7.1B

New Vocabulary
binomial factors

Instructor Note
It will help students if you give some examples of inserting parentheses into expressions before the examples on this page are attempted. Here are some suggestions.
$-a + 2b = -(a - 2b)$
$3x - 2y = -(-3x + 2y)$
$-4a - 3b = -(4a + 3b)$

Concept Check
Replace the ? to make a true statement.
1. $a - 3 = ?(3 - a)$ -1
2. $2 - (x - y) = 2 + (?)$ $y - x$
3. $4x + (3a - b) = 4x - (?)$
   $b - 3a$

Optional Student Activity
Factor:
$4x(a - b) - 5(b - a) + 4(a - b)$ $(a - b)(4x + 9)$

In-Class Examples (Objective 7.1B)
Factor.
1. $6x(4x + 3) - 5(4x + 3)$ $(4x + 3)(6x - 5)$
2. $8x^2 - 12x - 6xy + 9y$ $(2x - 3)(4x - 3y)$
3. $7xy^2 - 3y + 14xy - 6$ $(7xy - 3)(y + 2)$
4. $5xy - 9y - 18 + 10x$ $(5x - 9)(y + 2)$

Concept Check

Factor.

1. $5x(a + 7b) + 3(7b + a)$
   $(a + 7b)(5x + 3)$
2. $x^2 - 4x + 2xy - 8y$
   $(x - 4)(x + 2y)$
3. $3x^2y - 5x - 9xy + 15$
   $(3xy - 5)(x - 3)$
4. $10mn - 4 + 2n - 5mn^2$
   $(5mn - 2)(2 - n)$

Optional Student Activity

Factor:

$x^2 - 2y - x + 2xy$

$(x + 2y)(x - 1)$

Example 5

Factor: $4x(3x - 2) - 7(3x - 2)$

Solution

$4x(3x - 2) - 7(3x - 2)$

$= (3x - 2)(4x - 7)$

You Try It 5

Factor: $2y(5x - 2) - 3(2 - 5x)$

Your solution

$(5x - 2)(2y + 3)$

Example 6

Factor: $9x^2 - 15x - 6xy + 10y$

Solution

$9x^2 - 15x - 6xy + 10y$

$= (9x^2 - 15x) - (6xy - 10y)$

$= 3x(3x - 5) - 2y(3x - 5)$

$= (3x - 5)(3x - 2y)$

You Try It 6

Factor: $a^2 - 3a + 2ab - 6b$

Your solution

$(a - 3)(a + 2b)$

Example 7

Factor: $3x^2y - 4x - 15xy + 20$

Solution

$3x^2y - 4x - 15xy + 20$

$= (3x^2y - 4x) - (15xy - 20)$

$= x(3xy - 4) - 5(3xy - 4)$

$= (3xy - 4)(x - 5)$

You Try It 7

Factor: $2mn^2 - n + 8mn - 4$

Your solution

$(2mn - 1)(n + 4)$

Example 8

Factor: $4ab - 6 + 3b - 2ab^2$

Solution

$4ab - 6 + 3b - 2ab^2$

$= (4ab - 6) + (3b - 2ab^2)$

$= 2(2ab - 3) + b(3 - 2ab)$

$= 2(2ab - 3) - b(2ab - 3)$

$= (2ab - 3)(2 - b)$

You Try It 8

Factor: $3xy - 9y - 12 + 4x$

Your solution

$(x - 3)(3y + 4)$

Solutions on p. S20

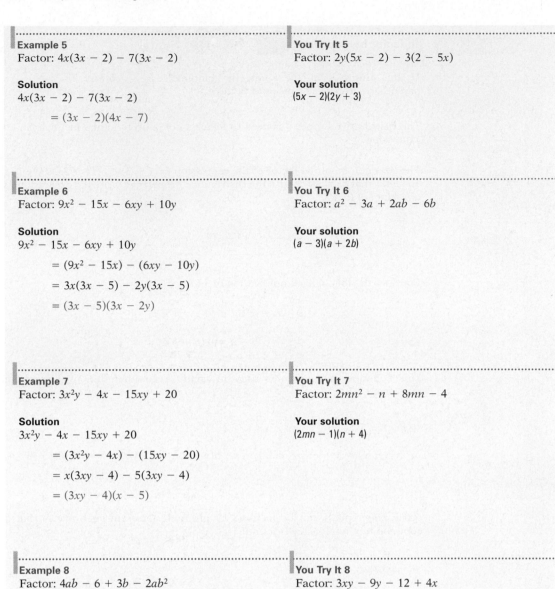

383

7.1 Exercises

Objective A

Factor.

1. $5a + 5$ — $5(a + 1)$
2. $7b - 7$ — $7(b - 1)$
3. $16 - 8a^2$ — $8(2 - a^2)$
4. $12 + 12y^2$ — $12(1 + y^2)$
5. $8x + 12$ — $4(2x + 3)$
6. $16a - 24$ — $8(2a - 3)$
7. $30a - 6$ — $6(5a - 1)$
8. $20b + 5$ — $5(4b + 1)$
9. $7x^2 - 3x$ — $x(7x - 3)$
10. $12y^2 - 5y$ — $y(12y - 5)$
11. $3a^2 + 5a^5$ — $a^2(3 + 5a^3)$
12. $9x - 5x^2$ — $x(9 - 5x)$
13. $14y^2 + 11y$ — $y(14y + 11)$
14. $6b^3 - 5b^2$ — $b^2(6b - 5)$
15. $2x^4 - 4x$ — $2x(x^3 - 2)$
16. $3y^4 - 9y$ — $3y(y^3 - 3)$
17. $10x^4 - 12x^2$ — $2x^2(5x^2 - 6)$
18. $12a^5 - 32a^2$ — $4a^2(3a^3 - 8)$
19. $8a^8 - 4a^5$ — $4a^5(2a^3 - 1)$
20. $16y^4 - 8y^7$ — $8y^4(2 - y^3)$
21. $x^2y^2 - xy$ — $xy(xy - 1)$
22. $a^2b^2 + ab$ — $ab(ab + 1)$
23. $3x^2y^4 - 6xy$ — $3xy(xy^3 - 2)$
24. $12a^2b^5 - 9ab$ — $3ab(4ab^4 - 3)$
25. $x^2y - xy^3$ — $xy(x - y^2)$
26. $3x^3 + 6x^2 + 9x$ — $3x(x^2 + 2x + 3)$
27. $5y^3 - 20y^2 + 10y$ — $5y(y^2 - 4y + 2)$
28. $2x^4 - 4x^3 + 6x^2$ — $2x^2(x^2 - 2x + 3)$
29. $3y^4 - 9y^3 - 6y^2$ — $3y^2(y^2 - 3y - 2)$
30. $2x^3 + 6x^2 - 14x$ — $2x(x^2 + 3x - 7)$
31. $3y^3 - 9y^2 + 24y$ — $3y(y^2 - 3y + 8)$
32. $2y^5 - 3y^4 + 7y^3$ — $y^3(2y^2 - 3y + 7)$
33. $6a^5 - 3a^3 - 2a^2$ — $a^2(6a^3 - 3a - 2)$
34. $x^3y - 3x^2y^2 + 7xy^3$ — $xy(x^2 - 3xy + 7y^2)$
35. $2a^2b - 5a^2b^2 + 7ab^2$ — $ab(2a - 5ab + 7b)$
36. $5y^3 + 10y^2 - 25y$ — $5y(y^2 + 2y - 5)$
37. $4b^5 + 6b^3 - 12b$ — $2b(2b^4 + 3b^2 - 6)$
38. $3a^2b^2 - 9ab^2 + 15b^2$ — $3b^2(a^2 - 3a + 5)$
39. $8x^2y^2 - 4x^2y + x^2$ — $x^2(8y^2 - 4y + 1)$
40. $x^{2n} - x^n$ — $x^n(x^n - 1)$
41. $2a^{5n} + a^{2n}$ — $a^{2n}(2a^{3n} + 1)$
42. $x^{3n} - x^{2n}$ — $x^{2n}(x^n - 1)$
43. $y^{4n} + y^{2n}$ — $y^{2n}(y^{2n} + 1)$
44. $a^{2n+2} + a^2$ — $a^2(a^{2n} + 1)$
45. $b^{n+5} - b^5$ — $b^5(b^n - 1)$

Section 7.1

Suggested Assignment

Exercises 1–71, odds

More challenging problems:

Exercises 73, 74

Quick Quiz (Objective 7.1A)

Factor.

1. $6ab + 9a$ $3a(2b + 3)$
2. $14x^2y^2 - 7xy$ $7xy(2xy - 1)$
3. $15r^2s + 20rs - 10rs^2$ $5rs(3r + 4 - 2s)$

384 Chapter 7 / Factoring

Objective B

Factor.

46. $x(b + 4) + 3(b + 4)$
$(b + 4)(x + 3)$

47. $y(a + z) + 7(a + z)$
$(a + z)(y + 7)$

48. $a(y - x) - b(y - x)$
$(y - x)(a - b)$

49. $3r(a - b) + s(a - b)$
$(a - b)(3r + s)$

50. $x(x - 2) + y(2 - x)$
$(x - 2)(x - y)$

51. $t(m - 7) + 7(7 - m)$
$(m - 7)(t - 7)$

52. $2x(7 + b) - y(b + 7)$
$(b + 7)(2x - y)$

53. $2y(4a - b) - (b - 4a)$
$(4a - b)(2y + 1)$

54. $8c(2m - 3n) + (3n - 2m)$
$(2m - 3n)(8c - 1)$

55. $x^2 + 2x + 2xy + 4y$
$(x + 2)(x + 2y)$

56. $x^2 - 3x + 4ax - 12a$
$(x - 3)(x + 4a)$

57. $p^2 - 2p - 3rp + 6r$
$(p - 2)(p - 3r)$

58. $t^2 + 4t - st - 4s$
$(t + 4)(t - s)$

59. $ab + 6b - 4a - 24$
$(a + 6)(b - 4)$

60. $xy - 5y - 2x + 10$
$(x - 5)(y - 2)$

61. $2z^2 - z + 2yz - y$
$(2z - 1)(z + y)$

62. $2y^2 - 10y + 7xy - 35x$
$(y - 5)(2y + 7x)$

63. $8v^2 - 12vy + 14v - 21y$
$(4v + 7)(2v - 3y)$

64. $21x^2 + 6xy - 49x - 14y$
$(7x + 2y)(3x - 7)$

65. $2x^2 - 5x - 6xy + 15y$
$(2x - 5)(x - 3y)$

66. $4a^2 + 5ab - 10b - 8a$
$(4a + 5b)(a - 2)$

67. $3y^2 - 6y - ay + 2a$
$(y - 2)(3y - a)$

68. $2ar + a^2 - 2r - a$
$(2r + a)(a - 1)$

69. $3xy - y^2 - y + 3x$
$(3x - y)(y + 1)$

70. $2ab - 3b^2 - 3b + 2a$
$(2a - 3b)(b + 1)$

71. $3st + t^2 - 2t - 6s$
$(3s + t)(t - 2)$

72. $4x^2 + 3xy - 12y - 16x$
$(4x + 3y)(x - 4)$

APPLYING THE CONCEPTS

73. A whole number is a perfect number if it is the sum of all of its factors less than itself. For example, 6 is a perfect number because all the factors of 6 that are less than 6 are 1, 2, and 3, and $1 + 2 + 3 = 6$.
a. Find the one perfect number between 20 and 30. 28
b. Find the one perfect number between 490 and 500. 496

74. Write the area of the shaded portion of each diagram in factored form.

a.

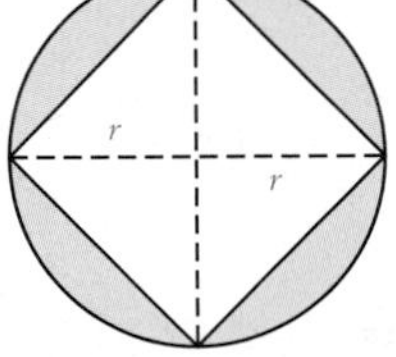

$r^2(\pi - 2)$

b.

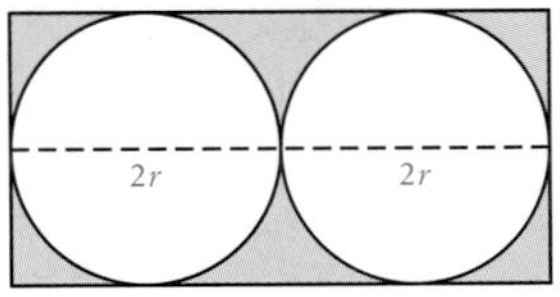

$2r^2(4 - \pi)$

c.

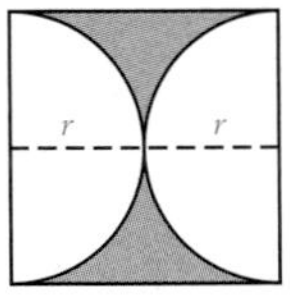

$r^2(4 - \pi)$

Quick Quiz (Objective 7.1B)
Factor.
1. $3xy - 9y + 2x - 6$ $(x - 3)(3y + 2)$
2. $2ay - a^2 - 3a + 6y$ $(2y - a)(a + 3)$

7.2 Factoring Polynomials of the Form $x^2 + bx + c$

Objective A **To factor a trinomial of the form $x^2 + bx + c$**

VIDEO & DVD CD TUTOR WEB SSM

Trinomials of the form $x^2 + bx + c$, where b and c are integers, are shown at the right.

$x^2 + 8x + 12; b = 8, c = 12$
$x^2 - 7x + 12; b = -7, c = 12$
$x^2 - 2x - 15; b = -2, c = -15$

To factor a trinomial of this form means to express the trinomial as the product of two binomials.

Trinomials expressed as the product of binomials are shown at the right.

$x^2 + 8x + 12 = (x + 6)(x + 2)$
$x^2 - 7x + 12 = (x - 3)(x - 4)$
$x^2 - 2x - 15 = (x + 3)(x - 5)$

The method by which factors of a trinomial are found is based on FOIL. Consider the following binomial products, noting the relationship between the constant terms of the binomials and the terms of the trinomials.

The signs in the binomials are the same.

$(x + 6)(x + 2) = x^2 + 2x + 6x + (6)(2) = x^2 + 8x + 12$
sum of 6 and 2
product of 6 and 2

$(x - 3)(x - 4) = x^2 - 4x - 3x + (-3)(-4) = x^2 - 7x + 12$
sum of −3 and −4
product of −3 and −4

The signs in the binomials are opposite.

$(x + 3)(x - 5) = x^2 - 5x + 3x + (3)(-5) = x^2 - 2x - 15$
sum of 3 and −5
product of 3 and −5

$(x - 4)(x + 6) = x^2 + 6x - 4x + (-4)(6) = x^2 + 2x - 24$
sum of −4 and 6
product of −4 and 6

IMPORTANT RELATIONSHIPS

1. When the constant term of the trinomial is positive, the constant terms of the binomials have the same sign. They are both positive when the coefficient of the x term in the trinomial is positive. They are both negative when the coefficient of the x term in the trinomial is negative.
2. When the constant term of the trinomial is negative, the constant terms of the binomials have opposite signs.
3. In the trinomial, the coefficient of x is the sum of the constant terms of the binomials.
4. In the trinomial, the constant term is the product of the constant terms of the binomials.

Objective 7.2A

New Vocabulary
nonfactorable over the integers
prime polynomial

Instructor Note
Students sometimes see factoring as unrelated to multiplication. Remind students that the relationships between the binomial factors and the terms of the polynomial are based on multiplying binomials.

Concept Check
Fill in the blanks.
1. To factor $x^2 - 3x - 18$, find two numbers whose product is ____ and whose sum is ____. −18, −3
2. To factor $x^2 - 9x + 18$, find two numbers whose product is ____ and whose sum is ____. 18, −9

Discuss the Concepts
How can you check the answer to a factoring problem? Answers will vary. Students should include the concept that the product of the factors should be equivalent to the original expression.

Discuss the Concepts
What clues in $x^2 - 4x + 5$ tell which signs should be used in the factored form? 5 is positive, so both binomial constants have the same sign. −4 is negative, so the signs are both negative.

In-Class Examples (Objective 7.2A)
Factor.
1. $x^2 - 8x + 12$ $(x - 6)(x - 2)$
2. $x^2 + 8x + 12$ $(x + 6)(x + 2)$
3. $x^2 + 7x + 12$ $(x + 4)(x + 3)$
4. $x^2 - 4x - 12$ $(x - 6)(x + 2)$
5. $x^2 - 11x - 12$ $(x - 12)(x + 1)$

Instructor Note

The phrase *nonfactorable over the integers* may require additional examples. Explain that it does not mean that the polynomial does not factor; it just does not factor if integers are used. An analogy to numbers may help. For instance, the only ways to write 7 as a product involving integers are $1 \cdot 7$ or $(-1)(-7)$. However, $\frac{21}{5} \cdot \frac{5}{3} = 7$. The ability to factor depends on the numbers that can be used.

Optional Student Activity

Factor $3x + 4$ by using 3 as one of the factors. $3\left(x + \frac{4}{3}\right)$

Discuss the Concepts

Explain why $x^2 + 3x + 1$ is a prime polynomial. There are no positive integer factors of 1 whose sum is 3.

TAKE NOTE
Always check your proposed factorization to ensure its accuracy.

➡ Factor: $x^2 - 7x + 10$

Because the constant term is positive and the coefficient of x is negative, the binomial constants will be negative. Find two negative factors of 10 whose sum is -7. The results can be recorded in a table.

Negative Factors of 10	*Sum*
$-1, -10$	-11
$-2, -5$	-7

• These are the correct factors.

$x^2 - 7x + 10 = (x - 2)(x - 5)$ • Write the trinomial as a product of its factors.

Check: $(x - 2)(x - 5) = x^2 - 5x - 2x + 10$
$= x^2 - 7x + 10$

• Check the proposed factorization by multiplying the two binomials.

➡ Factor: $x^2 - 9x - 36$

The constant term is negative. The binomial constants will have opposite signs. Find two factors of -36 whose sum is -9.

Factors of −36	*Sum*
$+1, -36$	-35
$-1, +36$	35
$+2, -18$	-16
$-2, +18$	16
$+3, -12$	-9

• Once the correct factors are found, it is not necessary to try the remaining factors.

$x^2 - 9x - 36 = (x + 3)(x - 12)$ • Write the trinomial as a product of its factors.

➡ Factor: $x^2 + 7x + 8$

Because the constant term is positive and the coefficient of x is positive, the binomial constants will be positive. Find two positive factors of 8 whose sum is 7.

Positive Factors of 8	*Sum*
1, 8	9
2, 4	6

There are no positive integer factors of 8 whose sum is 7. The trinomial $x^2 + 7x + 8$ is said to be **nonfactorable over the integers**. Just as 17 is a prime number, $x^2 + 7x + 8$ is a **prime polynomial**. Binomials of the form $x - a$ and $x + a$ are also prime polynomials.

Example 1
Factor: $x^2 - 8x + 15$

Solution

Factors	*Sum*
$-1, -15$	-16
$-3, -5$	-8

• Find two negative factors of 15 whose sum is -8.

$x^2 - 8x + 15 = (x - 3)(x - 5)$

You Try It 1
Factor: $x^2 + 9x + 20$

Your solution

$(x + 4)(x + 5)$

Solution on p. S20

Example 2
Factor: $x^2 + 6x - 27$

Solution

Factors	*Sum*
$+1, -27$	-26
$-1, +27$	26
$+3, -9$	-6

• Find two factors of -27 whose sum is 6.

$x^2 + 6x - 27 = (x - 3)(x + 9)$

You Try It 2
Factor: $x^2 + 7x - 18$

Your solution
$(x + 9)(x - 2)$

Solution on p. S20

Concept Check

Factor.

1. $x^2 - 5x - 6$ $(x - 6)(x + 1)$
2. $x^2 - 5x + 6$ $(x - 2)(x - 3)$

Objective 7.2B

Discuss the Concepts

Explain why $(4x + 2)(x - 5)$ is not factored completely. What is the correct factorization? The terms of $4x + 2$ have a common factor, namely 2. The complete factorization is $2(2x + 1)(x - 5)$.

Instructor Note

When the terms of a trinomial contain a common factor, some students will attempt to find the common factor and the two binomial factors in one step; they rarely obtain the correct final answer. Encourage the students to find only two factors at a time. Emphasize that the first step is to find the common factor and the resulting polynomial factor. After the first step is complete, the second step is to find the two binomial factors.

Objective B To factor completely

VIDEO & DVD CD TUTOR WEB SSM

A polynomial is factored completely when it is written as a product of factors that are nonfactorable over the integers.

> **TAKE NOTE**
> The first step in *any* factoring problem is to determine whether the terms of the polynomial have a **common factor.** If they do, factor it out first.

➡ Factor: $4y^3 - 4y^2 - 24y$

$4y^3 - 4y^2 - 24y = \boxed{4y(y^2) - 4y(y) - 4y(6)}$ • The GCF is $4y$. Do this step mentally.

$= 4y(y^2 - y - 6)$ • Use the Distributive Property to factor out the GCF.

$= 4y(y + 2)(y - 3)$ • Factor $y^2 - y - 6$. The two factors of -6 whose sum is -1 are 2 and -3.

It is always possible to check the proposed factorization by multiplying the polynomials. Here is the check for the last example.

Check: $4y(y + 2)(y - 3) = 4y(y^2 - 3y + 2y - 6)$
$= 4y(y^2 - y - 6)$
$= 4y^3 - 4y^2 - 24y$ • This is the original polynomial.

➡ Factor: $5x^2 + 60xy + 100y^2$

$5x^2 + 60xy + 100y^2 = \boxed{5(x^2) + 5(12xy) + 5(20y^2)}$ • The GCF is 5. Do this step mentally.

$= 5(x^2 + 12xy + 20y^2)$ • Use the Distributive Property to factor out the GCF.

$= 5(x + 2y)(x + 10y)$ • Factor $x^2 + 12xy + 20y^2$. The two factors of 20 whose sum is 12 are 2 and 10.

> **TAKE NOTE**
> $2y$ and $10y$ are placed in the binomials. This is necessary so that the middle term contains xy and the last term contains y^2.

Note that $2y$ and $10y$ were placed in the binomials. The following check shows that this was necessary.

Check: $5(x + 2y)(x + 10y) = 5(x^2 + 10xy + 2xy + 20y^2)$
$= 5(x^2 + 12xy + 20y^2)$
$= 5x^2 + 60xy + 100y^2$ • The original polynomial

In-Class Examples (Objective 7.2B)

Factor.

1. $3x^2 - 6x - 72$ $3(x - 6)(x + 4)$
2. $2x^2 + 6x - 20$ $2(x + 5)(x - 2)$
3. $-4y^3 + 28y^2 - 48y$ $-4y(y - 4)(y - 3)$
4. $5x^2 - 10xy - 15y^2$ $5(x + y)(x - 3y)$

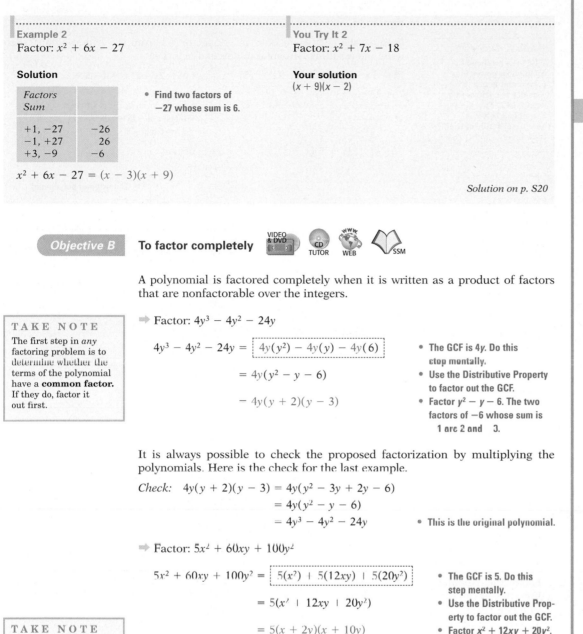

Concept Check

1. If $a(x + 3) = x^2 + 2x - 3$, find a. $x - 1$
2. If $-2x^3 - 6x^2 - 4x = a(x + 1)(x + 2)$, find a. $-2x$

Optional Student Activity

Two binomials and one monomial are multiplied together. The resulting product is $6x^4y - 18x^3y^2 - 60x^2y^3$. Write an explanation telling how to determine the original three factors. Answers will vary but should include determining the GCF first. The factored form is $6x^2y(x + 2y)(x - 5y)$.

TAKE NOTE

When the coefficient of the highest power in a polynomial is negative, consider factoring out a negative GCF. Example 3 below is another example of this technique.

➡ Factor: $15 - 2x - x^2$

Because the coefficient of x^2 is -1, factor -1 from the trinomial and then write the resulting trinomial in descending order.

$15 - 2x - x^2 = -(x^2 + 2x - 15)$ • $15 - 2x - x^2 = -1(-15 + 2x + x^2) = -(x^2 + 2x - 15)$

$= -(x + 5)(x - 3)$ • Factor $x^2 + 2x - 15$. The two factors of -15 whose sum is 2 are 5 and -3.

Check: $-(x + 5)(x - 3) = -(x^2 + 2x - 15)$
$= -x^2 - 2x + 15$
$= 15 - 2x - x^2$ • The original polynomial

Example 3

Factor: $-3x^3 + 9x^2 + 12x$

Solution

The GCF is $-3x$.

$-3x^3 + 9x^2 + 12x = -3x(x^2 - 3x - 4)$

Factor the trinomial $x^2 - 3x - 4$. Find two factors of -4 whose sum is -3.

Factors	*Sum*
$-2, +2$	0
$+1, -4$	-3

$-3x^3 + 9x^2 + 12x = -3x(x + 1)(x - 4)$

You Try It 3

Factor: $-2x^3 + 14x^2 - 12x$

Your solution

$-2x(x - 6)(x - 1)$

Example 4

Factor: $4x^2 - 40xy + 84y^2$

Solution

The GCF is 4.

$4x^2 - 40xy + 84y^2 = 4(x^2 - 10xy + 21y^2)$

Factor the trinomial $x^2 - 10xy + 21y^2$. Find two negative factors of 21 whose sum is -10.

Factors	*Sum*
$-1, -21$	-22
$-3, -7$	-10

$4x^2 - 40xy + 84y^2 = 4(x - 3y)(x - 7y)$

You Try It 4

Factor: $3x^2 - 9xy - 12y^2$

Your solution

$3(x + y)(x - 4y)$

Solutions on p. S20

389

7.2 Exercises

Objective A

Factor.

1. $x^2 + 3x + 2$
$(x + 1)(x + 2)$

2. $x^2 + 5x + 6$
$(x + 2)(x + 3)$

3. $x^2 - x - 2$
$(x + 1)(x - 2)$

4. $x^2 + x - 6$
$(x + 3)(x - 2)$

5. $a^2 + a - 12$
$(a + 4)(a - 3)$

6. $a^2 - 2a - 35$
$(a + 5)(a - 7)$

7. $a^2 - 3a + 2$
$(a - 1)(a - 2)$

8. $a^2 - 5a + 4$
$(a - 1)(a - 4)$

9. $a^2 + a - 2$
$(a + 2)(a - 1)$

10. $a^2 - 2a - 3$
$(a + 1)(a - 3)$

11. $b^2 - 6b + 9$
$(b - 3)(b - 3)$

12. $b^2 + 8b + 16$
$(b + 4)(b + 4)$

13. $b^2 + 7b - 8$
$(b + 8)(b - 1)$

14. $y^2 - y - 6$
$(y + 2)(y - 3)$

15. $y^2 + 6y - 55$
$(y + 11)(y - 5)$

16. $z^2 - 4z - 45$
$(z + 5)(z - 9)$

17. $y^2 - 5y + 6$
$(y - 2)(y - 3)$

18. $y^2 - 8y + 15$
$(y - 3)(y - 5)$

19. $z^2 - 14z + 45$
$(z - 5)(z - 9)$

20. $z^2 - 14z + 49$
$(z - 7)(z - 7)$

21. $z^2 - 12z - 160$
$(z + 8)(z - 20)$

22. $p^2 + 2p - 35$
$(p + 7)(p - 5)$

23. $p^2 + 12p + 27$
$(p + 3)(p + 9)$

24. $p^2 - 6p + 8$
$(p - 2)(p - 4)$

25. $x^2 + 20x + 100$
$(x + 10)(x + 10)$

26. $x^2 + 18x + 81$
$(x + 9)(x + 9)$

27. $b^2 + 9b + 20$
$(b + 4)(b + 5)$

28. $b^2 + 13b + 40$
$(b + 5)(b + 8)$

29. $x^2 - 11x - 42$
$(x + 3)(x - 14)$

30. $x^2 + 9x - 70$
$(x + 14)(x - 5)$

31. $b^2 - b - 20$
$(b + 4)(b - 5)$

32. $b^2 + 3b - 40$
$(b + 8)(b - 5)$

33. $y^2 - 14y - 51$
$(y + 3)(y - 17)$

34. $y^2 - y - 72$
$(y + 8)(y - 9)$

35. $p^2 - 4p - 21$
$(p + 3)(p - 7)$

36. $p^2 + 16p + 39$
$(p + 3)(p + 13)$

37. $y^2 - 8y + 32$
nonfactorable over the integers

38. $y^2 - 9y + 81$
nonfactorable over the integers

39. $x^2 - 20x + 75$
$(x - 5)(x - 15)$

40. $p^2 + 24p + 63$
$(p + 3)(p + 21)$

Section 7.2

Suggested Assignment

Exercises 1–133, every other odd

More challenging problems: Exercises 137–147, odds

Quick Quiz (Objective 7.2A)

Factor.

1. $x^2 + 7x + 10$ $(x + 5)(x + 2)$
2. $x^2 + 5x - 24$ $(x + 8)(x - 3)$
3. $x^2 - 13x + 36$ $(x - 9)(x - 4)$

41. $x^2 - 15x + 56$
$(x - 7)(x - 8)$

42. $x^2 + 21x + 38$
$(x + 2)(x + 19)$

43. $x^2 + x - 56$
$(x + 8)(x - 7)$

44. $x^2 + 5x - 36$
$(x + 9)(x - 4)$

45. $a^2 - 21a - 72$
$(a + 3)(a - 24)$

46. $a^2 - 7a - 44$
$(a + 4)(a - 11)$

47. $a^2 - 15a + 36$
$(a - 3)(a - 12)$

48. $a^2 - 21a + 54$
$(a - 3)(a - 18)$

49. $z^2 - 9z - 136$
$(z + 8)(z - 17)$

50. $z^2 + 14z - 147$
$(z + 21)(z - 7)$

51. $c^2 - c - 90$
$(c + 9)(c - 10)$

52. $c^2 - 3c - 180$
$(c + 12)(c - 15)$

53. $z^2 + 15z + 44$
$(z + 4)(z + 11)$

54. $p^2 + 24p + 135$
$(p + 9)(p + 15)$

55. $c^2 + 19c + 34$
$(c + 2)(c + 17)$

56. $c^2 + 11c + 18$
$(c + 2)(c + 9)$

57. $x^2 - 4x - 96$
$(x + 8)(x - 12)$

58. $x^2 + 10x - 75$
$(x + 15)(x - 5)$

59. $x^2 - 22x + 112$
$(x - 8)(x - 14)$

60. $x^2 + 21x - 100$
$(x + 25)(x - 4)$

61. $b^2 + 8b - 105$
$(b + 15)(b - 7)$

62. $b^2 - 22b + 72$
$(b - 4)(b - 18)$

63. $a^2 - 9a - 36$
$(a + 3)(a - 12)$

64. $a^2 + 42a - 135$
$(a + 45)(a - 3)$

65. $b^2 - 23b + 102$
$(b - 6)(b - 17)$

66. $b^2 - 25b + 126$
$(b - 7)(b - 18)$

67. $a^2 + 27a + 72$
$(a + 3)(a + 24)$

68. $z^2 + 24z + 144$
$(z + 12)(z + 12)$

69. $x^2 + 25x + 156$
$(x + 12)(x + 13)$

70. $x^2 - 29x + 100$
$(x - 4)(x - 25)$

71. $x^2 - 10x - 96$
$(x + 6)(x - 16)$

72. $x^2 + 9x - 112$
$(x + 16)(x - 7)$

Objective B

Factor.

73. $2x^2 + 6x + 4$
$2(x + 1)(x + 2)$

74. $3x^2 + 15x + 18$
$3(x + 2)(x + 3)$

75. $18 + 7x - x^2$
$-(x + 2)(x - 9)$

76. $12 - 4x - x^2$
$-(x - 2)(x + 6)$

77. $ab^2 + 2ab - 15a$
$a(b + 5)(b - 3)$

78. $ab^2 + 7ab - 8a$
$a(b + 8)(b - 1)$

79. $xy^2 - 5xy + 6x$
$x(y - 2)(y - 3)$

80. $xy^2 + 8xy + 15x$
$x(y + 3)(y + 5)$

81. $z^3 - 7z^2 + 12z$
$z(z - 3)(z - 4)$

391

82. $-2a^3 - 6a^2 - 4a$
$-2a(a + 1)(a + 2)$

83. $-3y^3 + 15y^2 - 18y$
$-3y(y - 2)(y - 3)$

84. $4y^3 + 12y^2 - 72y$
$4y(y + 6)(y - 3)$

85. $3x^2 + 3x - 36$
$3(x + 4)(x - 3)$

86. $2x^3 - 2x^2 + 4x$
$2x(x^2 - x + 2)$

87. $5z^2 - 15z - 140$
$5(z + 4)(z - 7)$

88. $6z^2 + 12z - 90$
$6(z + 5)(z - 3)$

89. $2a^3 + 8a^2 - 64a$
$2a(a + 8)(a - 4)$

90. $3a^3 - 9a^2 - 54a$
$3a(a + 3)(a - 6)$

91. $x^2 - 5xy + 6y^2$
$(x - 2y)(x - 3y)$

92. $x^2 + 4xy - 21y^2$
$(x + 7y)(x - 3y)$

93. $a^2 - 9ab + 20b^2$
$(a - 4b)(a - 5b)$

94. $a^2 - 15ab + 50b^2$
$(a - 5b)(a - 10b)$

95. $x^2 - 3xy - 28y^2$
$(x + 4y)(x - 7y)$

96. $s^2 + 2st - 48t^2$
$(s + 8t)(s - 6t)$

97. $y^2 - 15yz - 41z^2$
nonfactorable over the integers

98. $y^2 + 85yz + 36z^2$
nonfactorable over the integers

99. $z^4 - 12z^3 + 35z^2$
$z^2(z - 5)(z - 7)$

100. $z^4 + 2z^3 - 80z^2$
$z^2(z + 10)(z - 8)$

101. $b^4 - 22b^3 + 120b^2$
$b^2(b - 10)(b - 12)$

102. $b^4 - 3b^3 - 10b^2$
$b^2(b + 2)(b - 5)$

103. $2y^4 - 26y^3 - 96y^2$
$2y^2(y + 3)(y - 16)$

104. $3y^4 + 54y^3 + 135y^2$
$3y^2(y + 3)(y + 15)$

105. $-x^4 - 7x^3 + 8x^2$
$-x^2(x + 8)(x - 1)$

106. $-x^4 + 11x^3 + 12x^2$
$-x^2(x + 1)(x - 12)$

107. $4x^2y + 20xy - 56y$
$4y(x + 7)(x - 2)$

108. $3x^2y - 6xy - 45y$
$3y(x + 3)(x - 5)$

109. $c^3 + 18c^2 - 40c$
$c(c + 20)(c - 2)$

110. $-3x^3 + 36x^2 - 81x$
$-3x(x - 3)(x - 9)$

111. $-4x^3 - 4x^2 + 24x$
$-4x(x + 3)(x - 2)$

112. $x^2 - 8xy + 15y^2$
$(x - 3y)(x - 5y)$

113. $y^2 - 7xy - 8x^2$
$(y + x)(y - 8x)$

114. $a^2 - 13ab + 42b^2$
$(a - 6b)(a - 7b)$

115. $y^2 + 4yz - 21z^2$
$(y + 7z)(y - 3z)$

116. $y^2 + 8yz + 7z^2$
$(y + z)(y + 7z)$

117. $y^2 - 16yz + 15z^2$
$(y - z)(y - 15z)$

Quick Quiz (Objective 7.2B)

Factor.

1. $3a^3 + 15a^2 + 18a$ $3a(a + 3)(a + 2)$

2. $4 - 3x - x^2$ $-(x + 4)(x - 1)$

3. $5x^3 - 15x^2y + 10xy^2$ $5x(x - y)(x - 2y)$

118. $3x^2y + 60xy - 63y$
$3y(x + 21)(x - 1)$

119. $4x^2y - 68xy - 72y$
$4y(x + 1)(x - 18)$

120. $3x^3 + 3x^2 - 36x$
$3x(x + 4)(x - 3)$

121. $4x^3 + 12x^2 - 160x$
$4x(x + 8)(x - 5)$

122. $4z^3 + 32z^2 - 132z$
$4z(z + 11)(z - 3)$

123. $5z^3 - 50z^2 - 120z$
$5z(z + 2)(z - 12)$

124. $4x^3 + 8x^2 - 12x$
$4x(x + 3)(x - 1)$

125. $5x^3 + 30x^2 + 40x$
$5x(x + 2)(x + 4)$

126. $5p^2 + 25p - 420$
$5(p + 12)(p - 7)$

127. $4p^2 - 28p - 480$
$4(p + 8)(p - 15)$

128. $p^4 + 9p^3 - 36p^2$
$p^2(p + 12)(p - 3)$

129. $p^4 + p^3 - 56p^2$
$p^2(p + 8)(p - 7)$

130. $t^2 - 12ts + 35s^2$
$(t - 5s)(t - 7s)$

131. $a^2 - 10ab + 25b^2$
$(a - 5b)(a - 5b)$

132. $a^2 - 8ab - 33b^2$
$(a + 3b)(a - 11b)$

133. $x^2 + 4xy - 60y^2$
$(x + 10y)(x - 6y)$

134. $5x^4 - 30x^3 + 40x^2$
$5x^2(x - 2)(x - 4)$

135. $6x^3 - 6x^2 - 120x$
$6x(x - 5)(x + 4)$

APPLYING THE CONCEPTS

Factor.

136. $2 + c^2 + 9c$
nonfactorable over the integers

137. $x^2y - 54y - 3xy$
$y(x + 6)(x - 9)$

138. $45a^2 + a^2b^2 - 14a^2b$
$a^2(b - 5)(b - 9)$

Find all integers k such that the trinomial can be factored over the integers.

139. $x^2 + kx + 35$
$-36, 36, -12, 12$

140. $x^2 + kx + 18$
$-19, 19, -11, 11, -9, 9$

141. $x^2 + kx + 21$
$22, -22, 10, -10$

Determine the positive integer values of k for which the following polynomials are factorable over the integers.

142. $y^2 + 4y + k$
3, 4

143. $z^2 + 7z + k$
6, 10, 12

144. $a^2 - 6a + k$
5, 8, 9

145. $c^2 - 7c + k$
6, 10, 12

146. $x^2 - 3x + k$
2

147. $y^2 + 5y + k$
4, 6

148. In Exercises 142 to 147, there was the stated requirement that $k > 0$. If k is allowed to be any integer, how many different values of k are possible for each polynomial?
infinitely many

7.3 Factoring Polynomials of the Form $ax^2 + bx + c$

Objective 7.3A

Instructor Note

The first objective of this section is to factor by using trial factors. The second objective is to factor by grouping. You may skip one of these objectives or do both.

Discuss the Concepts

In multiplying two binomials, which portion of the FOIL method determines the middle term of the resulting trinomial? The OI part of FOIL (that is, the sum of the Outer and Inner products) determines the middle term.

Objective A

To factor a trinomial of the form $ax^2 + bx + c$ by using trial factors

VIDEO & DVD · CD TUTOR · WEB · SSM

Trinomials of the form $ax^2 + bx + c$, where a, b, and c are integers, are shown at the right.

$3x^2 - x + 4$; $a = 3, b = -1, c = 4$
$6x^2 + 2x - 3$; $a = 6, b = 2, c = -3$

These trinomials differ from those in the previous section in that the coefficient of x^2 is not 1. There are various methods of factoring these trinomials. The method described in this objective is factoring polynomials using trial factors.

To reduce the number of trial factors that must be considered, remember the following:

1. Use the signs of the constant term and the coefficient of x in the trinomial to determine the signs of the binomial factors. If the constant term is positive, the signs of the binomial factors will be the same as the sign of the coefficient of x in the trinomial. If the sign of the constant term is negative, the constant terms in the binomials have opposite signs.
2. If the terms of the trinomial do not have a common factor, then the terms of neither of the binomial factors will have a common factor.

⇒ Factor: $2x^2 - 7x + 3$

The terms have no common factor. The constant term is positive. The coefficient of x is negative. The binomial constants will be negative.

Positive Factors of 2 (coefficient of x^2)	*Negative Factors of 3* (constant term)
1, 2	−1, −3

Write trial factors. Use the **O**uter and **I**nner products of FOIL to determine the middle term, $-7x$, of the trinomial.

Trial Factors	*Middle Term*
$(x - 1)(2x - 3)$	$-3x - 2x = -5x$
$(x - 3)(2x - 1)$	$-x - 6x = -7x$

Write the factors of the trinomial. $2x^2 - 7x + 3 = (x - 3)(2x - 1)$

⇒ Factor: $3x^2 - 8x + 4$

The terms have no common factor. The constant term is positive. The coefficient of x is negative. The binomial constants will be negative.

Positive Factors of 3 (coefficient of x^2)	*Negative Factors of 4* (constant term)
1, 3	−1, −4 −2, −2

Write trial factors. Use the **O**uter and **I**nner products of FOIL to determine the middle term, $-8x$, of the trinomial.

Trial Factors	*Middle Term*
$(x - 1)(3x - 4)$	$-4x - 3x = -7x$
$(x - 4)(3x - 1)$	$-x - 12x = -13x$
$(x - 2)(3x - 2)$	$-2x - 6x = -8x$

Write the factors of the trinomial. $3x^2 - 8x + 4 = (x - 2)(3x - 2)$

In-Class Examples (Objective 7.3A)

Factor.

1. $5x^2 - 2x - 3$ $(5x + 3)(x - 1)$
2. $-12x^3 - 18x^2 + 30x$ $-6x(2x + 5)(x - 1)$

Instructor Note

Remind students that eliminating the trial factors that have a common factor requires that the terms of the polynomial do not have a common factor.

Concept Check

Factor.

1. $2x^2 - 5x + 2$
   $(2x - 1)(x - 2)$
2. $-4x^3 - 10x^2 - 6x$
   $-2x(2x + 3)(x + 1)$

Instructor Note

Some trial factors may be eliminated by realizing that if both first terms or both last terms of the binomials are even, the middle term's coefficient will never be odd. For example, in factoring $12x^2 - 17x + 6$, do not choose $6x$ and $2x$ as the factors of $12x^2$ because 6 and 2 are even but -17, the coefficient of the middle term, is odd.

Discuss the Concepts

Consider factoring $24x^2 - 29x - 4$.

1. Tell why you should *not* choose $12x$ and $2x$ as the factors of $24x^2$. Two evens never make an odd middle term.
2. Tell why you should *not* choose $6x$ and $4x$ as the factors of $24x^2$. Two evens never make an odd middle term.
3. Tell why you should *not* choose 2 and -2 as the factors of -4. Two evens never make an odd middle term.

⇒ Factor: $6x^3 + 14x^2 - 12x$

Factor the GCF, $2x$, from the terms. $6x^3 + 14x^2 - 12x = 2x(3x^2 + 7x - 6)$

Factor the trinomial $3x^2 + 7x - 6$. The constant term is negative. The binomial constants will have opposite signs.

Positive Factors of 3	*Factors of −6*
1, 3	−1, 6 1, −6 −2, 3 2, −3

TAKE NOTE
The binomial factor $3x + 6$ has a common factor of 3: $3x + 6 = 3(x + 2)$. Because $3x^2 + 7x - 6$ does not have a common factor, one of its binomial factors cannot have a common factor. For the examples and solutions in this text, all trial factors except those that have a common factor will be listed.

Write trial factors. Use the **O**uter and **I**nner products of FOIL to determine the middle term, $7x$, of the trinomial.

It is not necessary to test trial factors that have a common factor.

Trial Factors	*Middle Term*
$(x - 1)(3x + 6)$	Common factor
$(x + 6)(3x - 1)$	$-x + 18x = 17x$
$(x + 1)(3x - 6)$	Common factor
$(x - 6)(3x + 1)$	$x - 18x = -17x$
$(x - 2)(3x + 3)$	Common factor
$(x + 3)(3x - 2)$	$-2x + 9x = 7x$
$(x + 2)(3x - 3)$	Common factor
$(x - 3)(3x + 2)$	$2x - 9x = -7x$

Write the factors of the trinomial. $6x^3 + 14x^2 - 12x = 2x(x + 3)(3x - 2)$

For this example, all the trial factors were listed. Once the correct factors have been found, however, the remaining trial factors need not be checked.

Example 1

Factor: $3x^2 + x - 2$

Solution

Positive factors of 3: 1, 3 Factors of −2: 1, −2; −1, 2

Trial Factors	*Middle Term*
$(x + 1)(3x - 2)$	$-2x + 3x = x$
$(x - 2)(3x + 1)$	$x - 6x = -5x$
$(x - 1)(3x + 2)$	$2x - 3x = -x$
$(x + 2)(3x - 1)$	$-x + 6x = 5x$

$3x^2 + x - 2 = (x + 1)(3x - 2)$

You Try It 1

Factor: $2x^2 - x - 3$

Your solution

$(x + 1)(2x - 3)$

Example 2

Factor: $-12x^3 - 32x^2 + 12x$

Solution

The GCF is $-4x$.

$-12x^3 - 32x^2 + 12x = -4x(3x^2 + 8x - 3)$

Factor the trinomial.

Positive factors of 3: 1, 3 Factors of −3: 1, −3; −1, 3

Trial Factors	*Middle Term*
$(x - 3)(3x + 1)$	$x - 9x = -8x$
$(x + 3)(3x - 1)$	$-x + 9x = 8x$

$-12x^3 - 32x^2 + 12x = -4x(x + 3)(3x - 1)$

You Try It 2

Factor: $-45y^3 + 12y^2 + 12y$

Your solution

$-3y(3y - 2)(5y + 2)$

Solutions on pp. S20–S21

395

Objective 7.3B

Concept Check

Fill in the blanks.

1. To factor $2x^2 - 5x + 2$ by grouping, find two numbers whose product is ____ and whose sum is ____. 4, −5
2. To factor $6x^2 + 7x - 3$ by grouping, $7x$ must be written as ____ − ____. $9x$, $2x$

Objective B To factor a trinomial of the form $ax^2 + bx + c$ by grouping

VIDEO & DVD CD TUTOR WEB SSM

In the previous objective, trinomials of the form $ax^2 + bx + c$ were factored by using trial factors. In this objective, these trinomials will be factored by grouping.

To factor $ax^2 + bx + c$, first find two factors of $a \cdot c$ whose sum is b. Then use factoring by grouping to write the factorization of the trinomial.

⇒ Factor: $2x^2 + 13x + 15$

Find two positive factors of 30 $(2 \cdot 15)$ whose sum is 13.

Positive Factors of 30	*Sum*
1, 30	31
2, 15	17
3, 10	13

- When the required sum has been found, the remaining factors need not be checked.

$$2x^2 + 13x + 15 = 2x^2 + 3x + 10x + 15$$

- Use the factors of 30 whose sum is 13 to write $13x$ as $3x + 10x$.

$$= (2x^2 + 3x) + (10x + 15)$$
$$= x(2x + 3) + 5(2x + 3)$$
$$= (2x + 3)(x + 5)$$

- Factor by grouping.

Check: $(2x + 3)(x + 5) = 2x^2 + 10x + 3x + 15$
$$= 2x^2 + 13x + 15$$

⇒ Factor: $6x^2 - 11x - 10$

Find two factors of -60 $[6(-10)]$ whose sum is -11.

Factors of −60	*Sum*
1, −60	−59
−1, 60	59
2, −30	−28
2, 30	28
3, −20	−17
−3, 20	17
4, −15	−11

$$6x^2 - 11x - 10 = 6x^2 + 4x - 15x - 10$$

- Use the factors of −60 whose sum is −11 to write $-11x$ as $4x - 15x$.

$$= (6x^2 + 4x) - (15x + 10)$$
$$= 2x(3x + 2) - 5(3x + 2)$$
$$= (3x + 2)(2x - 5)$$

- Factor by grouping. Recall that $-15x - 10 = -(15x + 10)$.

Check: $(3x + 2)(2x - 5) = 6x^2 - 15x + 4x - 10$
$$= 6x^2 - 11x - 10$$

In-Class Examples (Objective 7.3B)

Factor.

1. $3x^2 + 7x + 4$ $(3x + 4)(x + 1)$
2. $72x^3 - 42x^2 - 72x$ $6x(4x + 3)(3x - 4)$

Optional Student Activity

The area of a rectangle is $(3x^2 + x - 2)$ ft^2. Find the dimensions of the rectangle in terms of the variable x. Given that $x > 0$, specify the dimension that is the length and the dimension that is the width. Can x be less than 0? Can x be equal to 0? The dimensions are $(3x - 2)$ ft by $(x + 1)$ ft. If $x = 1.5$, then the rectangle is a square. If $x < 1.5$, the length is $(x + 1)$ ft and the width is $(3x - 2)$ ft. If $x > 1.5$, the width is $(x + 1)$ ft and the length is $(3x - 2)$ ft. If $x < 0$, then $3x - 2$ is less than 0, which is not possible. Therefore, x cannot be less than 0. If $x = 0$, then the dimension $3x^2 + x - 2$ is negative, which is not possible. Therefore, x cannot be equal to 0.

➡ Factor: $3x^2 - 2x - 4$

Find two factors of -12 $[3(-4)]$ whose sum is -2.

Factors of -12	*Sum*
1, −12	−11
−1, 12	11
2, −6	−4
−2, 6	4
3, −4	−1
−3, 4	1

Because no integer factors of -12 have a sum of -2, $3x^2 - 2x - 4$ is nonfactorable over the integers. $3x^2 - 2x - 4$ is a prime polynomial.

Example 3

Factor: $2x^2 + 19x - 10$

Solution

Factors of -20 $[2(-10)]$	*Sum*
−1, 20	19

$$\begin{aligned}2x^2 + 19x - 10 &= 2x^2 - x + 20x - 10\\ &= (2x^2 - x) + (20x - 10)\\ &= x(2x - 1) + 10(2x - 1)\\ &= (2x - 1)(x + 10)\end{aligned}$$

You Try It 3

Factor: $2a^2 + 13a - 7$

Your solution

$(2a - 1)(a + 7)$

Example 4

Factor: $24x^2y - 76xy + 40y$

Solution

The GCF is $4y$.

$24x^2y - 76xy + 40y = 4y(6x^2 - 19x + 10)$

Negative Factors of 60 $[6(10)]$	*Sum*
−1, −60	−61
−2, −30	−32
−3, −20	−23
−4, −15	−19

$$\begin{aligned}6x^2 - 19x + 10 &= 6x^2 - 4x - 15x + 10\\ &= (6x^2 - 4x) - (15x - 10)\\ &= 2x(3x - 2) - 5(3x - 2)\\ &= (3x - 2)(2x - 5)\end{aligned}$$

$$\begin{aligned}24x^2y - 76xy + 40y &= 4y(6x^2 - 19x + 10)\\ &= 4y(3x - 2)(2x - 5)\end{aligned}$$

You Try It 4

Factor: $15x^3 + 40x^2 - 80x$

Your solution

$5x(3x - 4)(x + 4)$

Solutions on p. S21

397

Section 7.3

Suggested Assignment

Exercises 1–133, every other odd

More challenging problems: Exercises 134, 137, 140

7.3 Exercises

Objective A

Factor by using trial factors.

1. $2x^2 + 3x + 1$
$(x + 1)(2x + 1)$

2. $5x^2 + 6x + 1$
$(x + 1)(5x + 1)$

3. $2y^2 + 7y + 3$
$(y + 3)(2y + 1)$

4. $3y^2 + 7y + 2$
$(y + 2)(3y + 1)$

5. $2a^2 - 3a + 1$
$(a - 1)(2a - 1)$

6. $3a^2 - 4a + 1$
$(a - 1)(3a - 1)$

7. $2b^2 - 11b + 5$
$(b - 5)(2b - 1)$

8. $3b^2 - 13b + 4$
$(b - 4)(3b - 1)$

9. $2x^2 + x - 1$
$(x + 1)(2x - 1)$

10. $4x^2 - 3x - 1$
$(x - 1)(4x + 1)$

11. $2x^2 - 5x - 3$
$(x - 3)(2x + 1)$

12. $3x^2 + 5x - 2$
$(x + 2)(3x - 1)$

13. $2t^2 - t - 10$
$(t + 2)(2t - 5)$

14. $2t^2 + 5t - 12$
$(t + 4)(2t - 3)$

15. $3p^2 - 16p + 5$
$(p - 5)(3p - 1)$

16. $6p^2 + 5p + 1$
$(2p + 1)(3p + 1)$

17. $12y^2 - 7y + 1$
$(3y - 1)(4y - 1)$

18. $6y^2 - 5y + 1$
$(2y - 1)(3y - 1)$

19. $6z^2 - 7z + 3$
nonfactorable over the integers

20. $9z^2 + 3z + 2$
nonfactorable over the integers

21. $6t^2 - 11t + 4$
$(2t - 1)(3t - 4)$

22. $10t^2 + 11t + 3$
$(2t + 1)(5t + 3)$

23. $8x^2 + 33x + 4$
$(x + 4)(8x + 1)$

24. $7x^2 + 50x + 7$
$(x + 7)(7x + 1)$

25. $5x^2 - 62x - 7$
nonfactorable over the integers

26. $9x^2 - 13x - 4$
nonfactorable over the integers

27. $12y^2 + 19y + 5$
$(3y + 1)(4y + 5)$

28. $5y^2 - 22y + 8$
$(y - 4)(5y - 2)$

29. $7a^2 + 47a - 14$
$(a + 7)(7a - 2)$

30. $11a^2 - 54a - 5$
$(a - 5)(11a + 1)$

31. $3b^2 - 16b + 16$
$(b - 4)(3b - 4)$

32. $6b^2 - 19b + 15$
$(2b - 3)(3b - 5)$

33. $2z^2 - 27z - 14$
$(z - 14)(2z + 1)$

34. $4z^2 + 5z - 6$
$(z + 2)(4z - 3)$

35. $3p^2 + 22p - 16$
$(p + 8)(3p - 2)$

36. $7p^2 + 19p + 10$
$(p + 2)(7p + 5)$

Quick Quiz (Objective 7.3A)

Factor by using trial factors.

1. $2x^2 - 17x + 21$ $(2x - 3)(x - 7)$

2. $4x^2 + 11x - 20$ $(4x - 5)(x + 4)$

3. $12x^3y + x^2y - 6xy$ $xy(4x + 3)(3x - 2)$

37. $4x^2 + 6x + 2$
$2(x + 1)(2x + 1)$

38. $12x^2 + 33x - 9$
$3(x + 3)(4x - 1)$

39. $15y^2 - 50y + 35$
$5(y - 1)(3y - 7)$

40. $30y^2 + 10y - 20$
$10(y + 1)(3y - 2)$

41. $2x^3 - 11x^2 + 5x$
$x(x - 5)(2x - 1)$

42. $2x^3 - 3x^2 - 5x$
$x(x + 1)(2x - 5)$

43. $3a^2b - 16ab + 16b$
$b(a - 4)(3a - 4)$

44. $2a^2b - ab - 21b$
$b(a + 3)(2a - 7)$

45. $3z^2 + 95z + 10$
nonfactorable over the integers

46. $8z^2 - 36z + 1$
nonfactorable over the integers

47. $36x - 3x^2 - 3x^3$
$-3x(x - 3)(x + 4)$

48. $-2x^3 + 2x^2 + 4x$
$-2x(x + 1)(x - 2)$

49. $80y^2 - 36y + 4$
$4(4y - 1)(5y - 1)$

50. $24y^2 - 24y - 18$
$6(2y + 1)(2y - 3)$

51. $8z^3 + 14z^2 + 3z$
$z(2z + 3)(4z + 1)$

52. $6z^3 - 23z^2 + 20z$
$z(2z - 5)(3z - 4)$

53. $6x^2y - 11xy - 10y$
$y(2x - 5)(3x + 2)$

54. $8x^2y - 27xy + 9y$
$y(x - 3)(8x - 3)$

55. $10t^2 - 5t - 50$
$5(t + 2)(2t - 5)$

56. $16t^2 + 40t - 96$
$8(t + 4)(2t - 3)$

57. $3p^3 - 16p^2 + 5p$
$p(p - 5)(3p - 1)$

58. $6p^3 + 5p^2 + p$
$p(2p + 1)(3p + 1)$

59. $26z^2 + 98z - 24$
$2(z + 4)(13z - 3)$

60. $30z^2 - 87z + 30$
$3(2z - 5)(5z - 2)$

61. $10y^3 - 44y^2 + 16y$
$2y(y - 4)(5y - 2)$

62. $14y^3 + 94y^2 - 28y$
$2y(y + 7)(7y - 2)$

63. $4yz^3 + 5yz^2 - 6yz$
$yz(z + 2)(4z - 3)$

64. $12a^3 + 14a^2 - 48a$
$2a(2a - 3)(3a + 8)$

65. $42a^3 + 45a^2 - 27a$
$3a(2a + 3)(7a - 3)$

66. $36p^2 - 9p^3 - p^4$
$p^2(3 - p)(12 + p)$

67. $9x^2y - 30xy^2 + 25y^3$
$y(3x - 5y)(3x - 5y)$

68. $8x^2y - 38xy^2 + 35y^3$
$y(2x - 7y)(4x - 5y)$

69. $9x^3y - 24x^2y^2 + 16xy^3$
$xy(3x - 4y)(3x - 4y)$

70. $9x^3y + 12x^2y + 4xy$
$xy(3x + 2)(3x + 2)$

399

Objective B

Factor by grouping.

71. $6x^2 - 17x + 12$
$(2x - 3)(3x - 4)$

72. $15x^2 - 19x + 6$
$(3x - 2)(5x - 3)$

73. $5b^2 + 33b - 14$
$(b + 7)(5b - 2)$

74. $8x^2 - 30x + 25$
$(2x - 5)(4x - 5)$

75. $6a^2 + 7a - 24$
$(3a + 8)(2a - 3)$

76. $14a^2 + 15a - 9$
$(2a + 3)(7a - 3)$

77. $4z^2 + 11z + 6$
$(z + 2)(4z + 3)$

78. $6z^2 - 25z + 14$
$(2z - 7)(3z - 2)$

79. $22p^2 + 51p - 10$
$(2p + 5)(11p - 2)$

80. $14p^2 - 41p + 15$
$(2p - 5)(7p - 3)$

81. $8y^2 + 17y + 9$
$(y + 1)(8y + 9)$

82. $12y^2 - 145y + 12$
$(y - 12)(12y - 1)$

83. $18t^2 - 9t - 5$
$(6t - 5)(3t + 1)$

84. $12t^2 + 28t - 5$
$(2t + 5)(6t - 1)$

85. $6b^2 + 71b - 12$
$(b + 12)(6b - 1)$

86. $8b^2 + 65b + 8$
$(b + 8)(8b + 1)$

87. $9x^2 + 12x + 4$
$(3x + 2)(3x + 2)$

88. $25x^2 - 30x + 9$
$(5x - 3)(5x - 3)$

89. $6b^2 - 13b + 6$
$(2b - 3)(3b - 2)$

90. $20b^2 + 37b + 15$
$(4b + 5)(5b + 3)$

91. $33b^2 + 34b - 35$
$(3b + 5)(11b - 7)$

92. $15b^2 - 43b + 22$
$(3b - 2)(5b - 11)$

93. $18y^2 - 39y + 20$
$(3y - 4)(6y - 5)$

94. $24y^2 + 41y + 12$
$(3y + 4)(8y + 3)$

95. $15a^2 + 26a - 21$
$(3a + 7)(5a - 3)$

96. $6a^2 + 23a + 21$
$(2a + 3)(3a + 7)$

97. $8y^2 - 26y + 15$
$(2y - 5)(4y - 3)$

98. $18y^2 - 27y + 4$
$(3y - 4)(6y - 1)$

99. $8z^2 + 2z - 15$
$(2z + 3)(4z - 5)$

100. $10z^2 + 3z - 4$
$(5z + 4)(2z - 1)$

101. $15x^2 - 82x + 24$
nonfactorable over the integers

102. $13z^2 + 49z - 8$
nonfactorable over the integers

103. $10z^2 - 29z + 10$
$(2z - 5)(5z - 2)$

104. $15z^2 - 44z + 32$
$(3z - 4)(5z - 8)$

105. $36z^2 + 72z + 35$
$(6z + 5)(6z + 7)$

106. $16z^2 + 8z - 35$
$(4z + 7)(4z - 5)$

107. $3x^2 + xy - 2y^2$
$(x + y)(3x - 2y)$

108. $6x^2 + 10xy + 4y^2$
$2(3x + 2y)(x + y)$

109. $3a^2 + 5ab - 2b^2$
$(a + 2b)(3a - b)$

110. $2a^2 - 9ab + 9b^2$
$(a - 3b)(2a - 3b)$

Quick Quiz (Objective 7.3B)

Factor by grouping.

1. $10x^2 + x - 2$ $(5x - 2)(2x + 1)$
2. $12x^2 + 31x + 9$ $(3x + 1)(4x + 9)$
3. $12x^3y + 10x^2y - 8xy$ $2xy(3x + 4)(2x - 1)$

Answers to Writing Exercises

141. Students should explain that the sign of the product of the last terms of the two binomial factors must be the same as the sign of the last term of the trinomial. Thus if the last term of the trinomial is positive, the last terms of the two binomial factors are either both positive or both negative, depending on the middle term of the trinomial. If the last term of the trinomial is negative, the last terms of the two binomial factors will have different signs.

111. $4y^2 - 11yz + 6z^2$ $(y - 2z)(4y - 3z)$

112. $2y^2 + 7yz + 5z^2$ $(2y + 5z)(y + z)$

113. $28 + 3z - z^2$ $-(z - 7)(z + 4)$

114. $15 - 2z - z^2$ $-(z + 5)(z - 3)$

115. $8 - 7x - x^2$ $-(x - 1)(x + 8)$

116. $12 + 11x - x^2$ $-(x + 1)(x - 12)$

117. $9x^2 + 33x - 60$ $3(x + 5)(3x - 4)$

118. $16x^2 - 16x - 12$ $4(2x - 3)(2x + 1)$

119. $24x^2 - 52x + 24$ $4(2x - 3)(3x - 2)$

120. $60x^2 + 95x + 20$ $5(3x + 4)(4x + 1)$

121. $35a^4 + 9a^3 - 2a^2$ $a^2(5a + 2)(7a - 1)$

122. $15a^4 + 26a^3 + 7a^2$ $a^2(5a + 7)(3a + 1)$

123. $15b^2 - 115b + 70$ $5(b - 7)(3b - 2)$

124. $25b^2 + 35b - 30$ $5(b + 2)(5b - 3)$

125. $3x^2 - 26xy + 35y^2$ $(x - 7y)(3x - 5y)$

126. $4x^2 + 16xy + 15y^2$ $(2x + 5y)(2x + 3y)$

127. $216y^2 - 3y - 3$ $3(8y - 1)(9y + 1)$

128. $360y^2 + 4y - 4$ $4(9y + 1)(10y - 1)$

129. $21 - 20x - x^2$ $-(x - 1)(x + 21)$

130. $18 + 17x - x^2$ $-(x + 1)(x - 18)$

131. $15a^2 + 11ab - 14b^2$ $(5a + 7b)(3a - 2b)$

132. $15a^2 - 31ab + 10b^2$ $(3a - 5b)(5a - 2b)$

133. $33z - 8z^2 - z^3$ $-z(z + 11)(z - 3)$

APPLYING THE CONCEPTS

Factor.

134. $2(y + 2)^2 - (y + 2) - 3$ $(2y + 1)(y + 3)$

135. $3(a + 2)^2 - (a + 2) - 4$ $(3a + 2)(a + 3)$

136. $4(y - 1)^2 - 7(y - 1) - 2$ $(4y - 3)(y - 3)$

Find all integers k such that the trinomial can be factored over the integers.

137. $2x^2 + kx + 3$ 7, −7, 5, −5

138. $2x^2 + kx - 3$ 5, −5, 1, −1

139. $3x^2 + kx + 2$ 7, −7, 5, −5

140. Write the area of the shaded portion of each diagram in factored form.

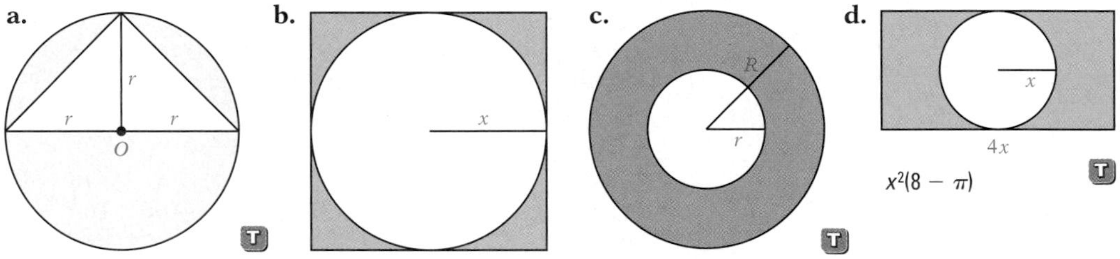

a. $r^2(\pi - 1)$ b. $x^2(4 - \pi)$ c. $\pi(R - r)(R + r)$ d. $x^2(8 - \pi)$

141. In your own words, explain how the signs of the last terms of the two binomial factors of a trinomial are determined.

7.4 Special Factoring

Objective A

To factor the difference of two perfect squares or a perfect-square trinomial

VIDEO & DVD · CD TUTOR · WEB · SSM

TAKE NOTE
Here are some more examples of perfect squares:
36, 49, x^8, y^{12}.

The product of a term and itself is called a **perfect square**. The exponents on variables of perfect squares are always even numbers.

Term		*Perfect Square*
5	$5 \cdot 5 =$	25
x	$x \cdot x =$	x^2
$3y^4$	$3y^4 \cdot 3y^4 =$	$9y^8$
x^n	$x^n \cdot x^n =$	x^{2n}

The **square root of a perfect square** is one of the two equal factors of the perfect square. "$\sqrt{\ }$" is the symbol for square root. To find the exponent of the square root of a variable term, divide the exponent by 2.

$\sqrt{25} = 5$
$\sqrt{x^2} = x$
$\sqrt{9y^8} = 3y^4$
$\sqrt{x^{2n}} = x^n$

The difference of two perfect squares is the product of the sum and difference of two terms. The factors of the difference of two perfect squares are the sum and difference of the square roots of the perfect squares.

TAKE NOTE
$(a + b)$ is the sum of the two terms a and b.
$(a - b)$ is the difference of the two terms a and b. a and b are the square roots of a^2 and b^2.

Factors of the Difference of Two Perfect Squares

$a^2 - b^2 = (a + b)(a - b)$

The sum of two perfect squares, $a^2 + b^2$, is nonfactorable over the integers.

⇒ Factor: $4x^2 - 81y^2$

$4x^2 - 81y^2 = (2x)^2 - (9y)^2$
- Write the binomial as the difference of two perfect squares.

$= (2x + 9y)(2x - 9y)$
- The factors are the sum and difference of the square roots of the perfect squares.

A perfect-square trinomial is the square of a binomial.

Factors of a Perfect-Square Trinomial

$a^2 + 2ab + b^2 = (a + b)^2$
$a^2 - 2ab + b^2 = (a - b)^2$

In factoring a perfect-square trinomial, remember that the terms of the binomial are the square roots of the perfect squares of the trinomial. The sign in the binomial is the sign of the middle term of the trinomial.

401

Objective 7.4A

New Vocabulary
perfect square
square root of a perfect square
difference of two perfect squares
sum of two perfect squares
product of the sum and difference of two terms
perfect-square trinomial

Vocabulary to Review
term
factor
square of a binomial

New Symbols
$\sqrt{\ }$

Discuss the Concepts
1. Is $x^2 + 9$ factorable? Why or why not?
2. Provide examples of the product of the sum and difference of two terms. For each example, state the two terms, the sum of the two terms, the difference of the two terms, and how the product is represented.
3. Is the product of the sum and difference of two terms always a binomial?
4. What is a perfect-square trinomial?
5. How can you determine the factors of a perfect-square trinomial?

In-Class Examples (Objective 7.4A)
Factor.
1. $25x^2 - 81$ $(5x + 9)(5x - 9)$
2. $x^2 + 16$ Nonfactorable
3. $36a^2 - 49b^2$ $(6a + 7b)(6a - 7b)$
4. $y^2 - 8y + 16$ $(y - 4)^2$
5. $25x^2 + 30x + 9$ $(5x + 3)^2$

Concept Check

1. The area of a square is $(4x^2 + 12x + 9)$ cm^2. Find the length of a side of the square in terms of the variable x. $(2x + 3)$ cm
2. The area of a square is $(9x^2 + 6x + 1)$ m^2. Find the length of a side of the square in terms of the variable x. $(3x + 1)$ m

Instructor Note

Another example of a trinomial that appears to be a perfect square trinomial, but does not factor, is $4x^2 - 17x + 25$.

Optional Student Activity

1. Find all integers k such that the trinomial is a perfect square.
   a. $x^2 - kx + 4$ $-4, 4$
   b. $9x^2 - kx + 1$ $6, -6$
   c. $16x^2 + kxy + y^2$ $8, -8$
   d. $x^2 + 8x + k$ 16
   e. $x^2 - 12x + k$ 36
   f. $x^2 + 4xy + ky^2$ 4
2. The area of a square is $(9x^2 + 24x + 16)$ ft^2. Find the dimensions of the square in terms of the variable x. Can $x = 0$? What are the possible values of x? $(3x + 4)$ ft by $(3x + 4)$ ft; Yes; $x > -\frac{4}{3}$

➡ Factor: $4x^2 + 12x + 9$

Because $4x^2$ is a perfect square $[4x^2 = (2x)^2]$ and 9 is a perfect square $(9 = 3^2)$, try factoring $4x^2 + 12x + 9$ as the square of a binomial.

$4x^2 + 12x + 9 \stackrel{?}{=} (2x + 3)^2$

Check: $(2x + 3)^2 = (2x + 3)(2x + 3) = 4x^2 + 6x + 6x + 9 = 4x^2 + 12x + 9$

The check verifies that $4x^2 + 12x + 9 = (2x + 3)^2$.

It is important to check a proposed factorization as we did above. The next example illustrates the importance of this check.

➡ Factor: $x^2 + 13x + 36$

Because x^2 is a perfect square and 36 is a perfect square, try factoring $x^2 + 13x + 36$ as the square of a binomial.

$x^2 + 13x + 36 \stackrel{?}{=} (x + 6)^2$

Check: $(x + 6)^2 = (x + 6)(x + 6) = x^2 + 6x + 6x + 36 = x^2 + 12x + 36$

In this case, the proposed factorization of $x^2 + 13x + 36$ does *not* check. Try another factorization. The numbers 4 and 9 are factors of 36 whose sum is 13.

$x^2 + 13x + 36 = (x + 4)(x + 9)$

Example 1
Factor: $25x^2 - 1$

Solution
$25x^2 - 1 = (5x)^2 - (1)^2$
$= (5x + 1)(5x - 1)$

You Try It 1
Factor: $x^2 - 36y^4$

Your solution
$(x + 6y^2)(x - 6y^2)$

Example 2
Factor: $4x^2 - 20x + 25$

Solution
$4x^2 - 20x + 25 = (2x - 5)^2$

You Try It 2
Factor: $9x^2 + 12x + 4$

Your solution
$(3x + 2)^2$

Example 3
Factor: $(x + y)^2 - 4$

Solution
$(x + y)^2 - 4 = (x + y)^2 - (2)^2$
$= (x + y + 2)(x + y - 2)$

You Try It 3
Factor: $(a + b)^2 - (a - b)^2$

Your solution
$4ab$

Solutions on p. S21

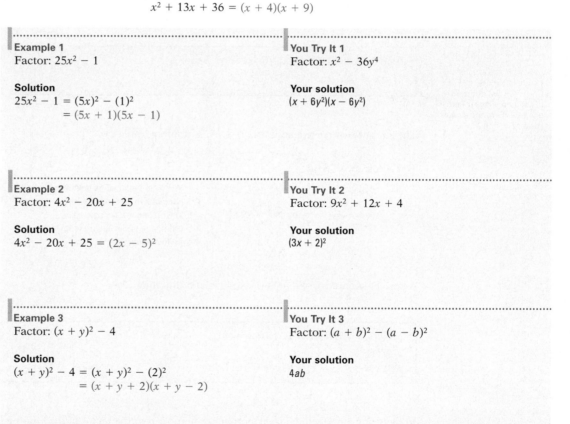

Objective B To factor the sum or the difference of two cubes

VIDEO & DVD CD TUTOR WEB SSM

TAKE NOTE
Here are some more examples of perfect cubes:
64, 1000, b^9, z^{15}

The product of the same three factors is called a **perfect cube**. The exponents on variables of perfect cubes are always divisible by 3.

Term		*Perfect Cube*
2	$2 \cdot 2 \cdot 2 = 2^3 =$	8
$3y$	$3y \cdot 3y \cdot 3y = (3y)^3 =$	$27y^3$
y^2	$y^2 \cdot y^2 \cdot y^2 = (y^2)^3 =$	y^6

The **cube root** of a perfect cube is one of the three equal factors of the perfect cube. "$\sqrt[3]{}$" is the symbol for cube root. To find the exponent of the cube root of a variable term, divide the exponent by 3.

$$\sqrt[3]{8} = 2$$
$$\sqrt[3]{27y^3} = 3y$$
$$\sqrt[3]{y^6} = y^2$$

The following rules are used to factor the sum or difference of two perfect cubes.

Factors of the Sum or Difference of Two Cubes

$a^3 + b^3 = (a + b)(a^2 - ab + b^2)$
$a^3 - b^3 = (a - b)(a^2 + ab + b^2)$

⇒ Factor: $8x^3 - 27$

Write the binomial as the difference of two perfect cubes.

$$8x^3 - 27 = (2x)^3 - 3^3$$

The terms of the binomial factor are the cube roots of the perfect cubes. The sign of the binomial factor is the same sign as in the given binomial. The trinomial factor is obtained from the binomial factor.

$$= (2x - 3)(4x^2 + 6x + 9)$$

Square of the first term ($4x^2$)
Opposite of the product of the two terms ($6x$)
Square of the last term (9)

TAKE NOTE
You can always check a proposed factorization by multiplying the factors.

Check:

$$\begin{array}{r} 4x^2 + 6x + 9 \\ 2x - 3 \\ \hline -12x^2 - 18x - 27 \\ 8x^3 + 12x^2 + 18x \\ \hline 8x^3 - 27 \end{array}$$

• The original polynomial

⇒ Factor: $a^3 + 64y^3$

$a^3 + 64y^3 = a^3 + (4y)^3$ • Write the binomial as the sum of two perfect cubes.

$= (a + 4y)(a^2 - 4ay + 16y^2)$ • Factor.

⇒ Factor: $64y^4 - 125y$

$64y^4 - 125y = y(64y^3 - 125)$ • Factor out y, the GCF.

$= y[(4y)^3 - 5^3]$ • Write the binomial as the difference of two cubes.

$= y(4y - 5)(16y^2 + 20y + 25)$ • Factor.

Objective 7.4B

New Vocabulary
perfect cube
cube root of a perfect cube
sum of two perfect cubes
difference of two perfect cubes

New Symbols
$\sqrt[3]{}$

Discuss the Concepts
1. Are both $x^2 + 16$ and $x^3 + 27$ factorable? Why or why not?
2. Which of the following are perfect cubes? Explain your answer.
   a. $125x^8$
   b. $1y^{12}$
   c. $8c^9$
   d. $9b^{27}$
3. How can you determine the factors of the sum of two perfect cubes?
4. How can you determine the factors of the difference of two perfect cubes?

Optional Student Activity
1. What is the smallest positive integer by which 252 should be multiplied to obtain a perfect cube? 294
2. Find the quotient when $x^6 - y^6$ is divided by $x - y$. $x^5 + x^4y + x^3y^2 + x^2y^3 + xy^4 + y^5$
3. Factor: $ax^3 - b + bx^3 - a$ $(x - 1)(x^2 + x + 1)(a + b)$

In-Class Examples (Objective 7.4B)
Factor.
1. $x^3 + 64y^3$ $(x + 4y)(x^2 - 4xy + 16y^2)$
2. $8a^3 - b^3$ $(2a - b)(4a^2 + 2ab + b^2)$
3. $x^{3n} - y^{6n}$ $(x^n - y^{2n})(x^{2n} + x^ny^{2n} + y^{4n})$

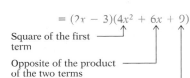

Objective 7.4C

New Vocabulary
quadratic in form

Vocabulary to Review
quadratic trinomial

Concept Check
Rewrite each of the following as the difference of two squares. Then factor.

1. $16x^2 - 1$ $(4x)^2 - 1^2$; $(4x + 1)(4x - 1)$
2. $9x^4 - 25$ $(3x^2)^2 - 5^2$; $(3x^2 + 5)(3x^2 - 5)$

Rewrite each of the following in quadratic form. Then factor.

3. $x^4 + 3x^2 + 2$ $u^2 + 3u + 2$; $(x^2 + 2)(x^2 + 1)$
4. $4x^4 - 9x^2 - 9$ $4u^2 - 9u - 9$; $(4x^2 + 3)(x^2 - 3)$

Optional Student Activity

1. Factor: $x^4 + 64$
   Suggestion: Add and subtract $16x^2$ so that the expression becomes $(x^4 + 16x^2 + 64) - 16x^2$. Now factor the difference of two squares.
   $(x^2 - 4x + 8)(x^2 + 4x + 8)$
2. Using the strategy in Exercise 1, factor $x^4 + x^2y^2 + y^4$.
   Suggestion: Add and subtract x^2y^2.
   $(x^2 + xy + y^2)(x^2 - xy + y^2)$

Example 4
Factor: $x^3y^3 - 1$

Solution
$x^3y^3 - 1 = (xy)^3 - 1^3$
$= (xy - 1)(x^2y^2 + xy + 1)$

You Try It 4
Factor: $8x^3 + y^3z^3$

Your solution
$(2x + yz)(4x^2 - 2xyz + y^2z^2)$

Example 5
Factor: $(x + y)^3 - x^3$

Solution
$(x + y)^3 - x^3$
$= [(x + y) - x][(x + y)^2 + x(x + y) + x^2]$
$= y(x^2 + 2xy + y^2 + x^2 + xy + x^2)$
$= y(3x^2 + 3xy + y^2)$

You Try It 5
Factor: $(x - y)^3 + (x + y)^3$

Your solution
$2x(x^2 + 3y^2)$

Solutions on p. S21

Objective C To factor a trinomial that is quadratic in form

VIDEO & DVD CD TUTOR WEB SSM

Certain trinomials that are not quadratic can be expressed in the form $ax^2 + bx + c$ by making suitable variable substitutions. A trinomial is **quadratic in form** if it can be written as $au^2 + bu + c$.

As shown below, the trinomials $x^4 + 5x^2 + 6$ and $2x^2y^2 + 3xy - 9$ are quadratic in form.

	$x^4 + 5x^2 + 6$		$2x^2y^2 + 3xy - 9$
	$(x^2)^2 + 5(x^2) + 6$		$2(xy)^2 + 3(xy) - 9$
Let $u = x^2$.	$u^2 + 5u + 6$	Let $u = xy$.	$2u^2 + 3u - 9$

When we use this method to factor a trinomial that is quadratic in form, the variable part of the first term in each binomial will be u.

For example, $x^4 + 5x^2 + 6$ is quadratic in form by letting $u = x^2$. Thus,
$x^4 + 5x^2 + 6 = (x^2)^2 + 5(x^2) + 6 = u^2 + 5u + 6$.

⇒ Factor: $x^4 + 5x^2 + 6$

$x^4 + 5x^2 + 6 = u^2 + 5u + 6$ • **Let $u = x^2$.**
$= (u + 3)(u + 2)$ • **Factor.**
$= (x^2 + 3)(x^2 + 2)$ • **Replace u by x^2.**

In-Class Examples (Objective 7.4C)
Factor.

1. $x^2y^2 - 9xy + 20$ $(xy - 5)(xy - 4)$
2. $y^4 - 8y^2 - 48$ $(y^2 - 12)(y^2 + 4)$
3. $a^{2n} + 5a^n - 36$ $(a^n + 9)(a^n - 4)$

Example 6
Factor: $6x^2y^2 - xy - 12$

Solution
Let $u = xy$.

$$6x^2y^2 - xy - 12 = 6u^2 - u - 12$$
$$= (3u + 4)(2u - 3)$$
$$= (3xy + 4)(2xy - 3)$$

You Try It 6
Factor: $6x^2y^2 - 19xy + 10$

Your solution
$(2xy - 5)(3xy - 2)$

Example 7
Factor: $2x^4 + 5x^2 - 12$

Solution
Let $u = x^2$.

$$2x^4 + 5x^2 - 12 = 2u^2 + 5u - 12$$
$$= (2u - 3)(u + 4)$$
$$= (2x^2 - 3)(x^2 + 4)$$

You Try It 7
Factor: $3x^4 + 4x^2 - 4$

Your solution
$(x^2 + 2)(3x^2 - 2)$

Example 8
Factor: $x^4y^4 + 3x^2y^2 - 10$

Solution
Let $u = x^2y^2$.

$$x^4y^4 + 3x^2y^2 - 10 = u^2 + 3u - 10$$
$$= (u + 5)(u - 2)$$
$$= (x^2y^2 + 5)(x^2y^2 - 2)$$

You Try It 8
Factor: $a^4b^4 + 6a^2b^2 - 7$

Your solution
$(a^2b^2 + 7)(a^2b^2 - 1)$

Solutions on p. S21

Objective D To factor completely

VIDEO & DVD CD TUTOR WWW SSM

TAKE NOTE
Remember that you may have to factor more than once in order to write the polynomial as a product of *prime* factors.

When factoring a polynomial completely, ask the following questions about the polynomial.

1. Is there a common factor? If so, factor out the GCF.
2. If the polynomial is a binomial, is it the difference of two perfect squares, the sum of two cubes, or the difference of two cubes? If so, factor.
3. If the polynomial is a trinomial, is it a perfect-square trinomial or the product of two binomials? If so, factor.
4. If the polynomial has four terms, can it be factored by grouping? If so, factor.
5. Is each factor nonfactorable over the integers? If not, factor.

Objective 7.4D

Vocabulary to Review
common factor
binomial
difference of two perfect squares
sum of two perfect cubes
difference of two perfect cubes
trinomial
perfect-square trinomial
factor by grouping
nonfactorable over the integers
prime factors

Discuss the Concepts
1. Provide an example of each of the following:
   the difference of two perfect squares
   the product of the sum and difference of two terms
   a perfect-square trinomial
   the square of a binomial
   the sum of two perfect squares
   the sum of two perfect cubes
   the difference of two perfect cubes
   a prime polynomial
2. Can a second-degree polynomial have factors $x + 2$, $x + 3$, and $x \quad 5$? Why or why not?

In-Class Examples (Objective 7.4D)
Factor.
1. $x^4 - 7x^3 + 10x^2$ $x^2(x - 2)(x - 5)$
2. $18x^2 + 24x + 8$ $2(3x + 2)^2$
3. $x^4 - 8x$ $x(x - 2)(x^2 + 2x + 4)$
4. $x^4 - 81$ $(x^2 + 9)(x + 3)(x - 3)$
5. $x^3 + x^2 - 4x - 4$ $(x - 2)(x + 2)(x + 1)$

Concept Check

1. The volume of a box is $(2xy^2 + 12xy + 10x)$ cubic inches. Find the dimensions of the box in terms of the variables x and y. $2x$ inches by $(y + 1)$ inches by $(y + 5)$ inches
2. The volume of a box is $(3x^2y + 21xy + 36y)$ cubic centimeters. Find the dimensions of the box in terms of the variables x and y. $3y$ centimeters by $(x + 4)$ centimeters by $(x + 3)$ centimeters

Optional Student Activity

1. Find the least common multiple of the polynomials $3x^2 + x - 2$, $3x^2 - 8x + 4$, and $x^3 - 2x^2 - x + 2$. $3x^4 - 8x^3 + x^2 + 8x - 4$
2. Factor: $x^2(x - 4) - 2x^2 + 5x + 12$ $(x - 4)(x - 3)(x + 1)$

Example 9
Factor: $6a^3 + 15a^2 - 36a$

Solution

$$6a^3 + 15a^2 - 36a = 3a(2a^2 + 5a - 12) = 3a(2a - 3)(a + 4)$$

You Try It 9
Factor: $18x^3 - 6x^2 - 60x$

Your solution
$6x(3x + 5)(x - 2)$

Example 10
Factor: $x^2y + 2x^2 - y - 2$

Solution

$$\begin{aligned} x^2y + 2x^2 - y - 2 &= (x^2y + 2x^2) - (y + 2) \\ &= x^2(y + 2) - 1(y + 2) \\ &= (y + 2)(x^2 - 1) \\ &= (y + 2)(x + 1)(x - 1) \end{aligned}$$

You Try It 10
Factor: $4x - 4y - x^3 + x^2y$

Your solution
$(x - y)(2 + x)(2 - x)$

Example 11
Factor: $x^{4n} - y^{4n}$

Solution

$$\begin{aligned} x^{4n} - y^{4n} &= (x^{2n})^2 - (y^{2n})^2 \\ &= (x^{2n} + y^{2n})(x^{2n} - y^{2n}) \\ &= (x^{2n} + y^{2n})[(x^n)^2 - (y^n)^2] \\ &= (x^{2n} + y^{2n})(x^n + y^n)(x^n - y^n) \end{aligned}$$

You Try It 11
Factor: $x^{4n} - x^{2n}y^{2n}$

Your solution
$x^{2n}(x^n + y^n)(x^n - y^n)$

Example 12
Factor: $x^{n+3} + x^ny^3$

Solution

$$\begin{aligned} x^{n+3} + x^ny^3 &= x^n(x^3 + y^3) \\ &= x^n(x + y)(x^2 - xy + y^2) \end{aligned}$$

You Try It 12
Factor: $ax^5 - ax^2y^6$

Your solution
$ax^2(x - y^2)(x^2 + xy^2 + y^4)$

Solutions on pp. S21–S22

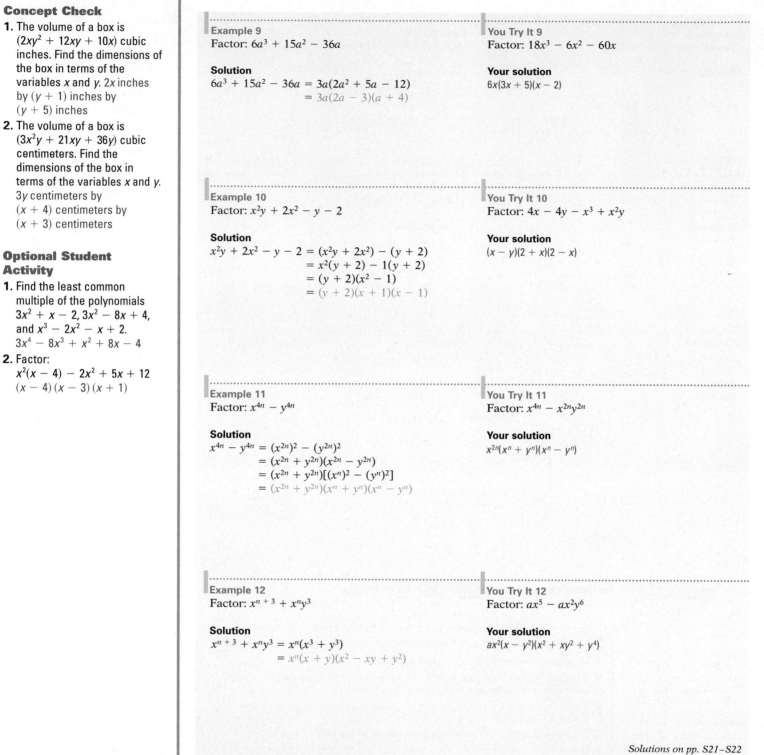

407

Section 7.4

Suggested Assignment
Exercises 1–127, odds
More challenging problems:
Exercises 128–131

7.4 Exercises

Objective A

Which of the expressions are perfect squares?

1. $4; 8; 25x^6; 12y^{10}; 100x^4y^4$
   $4; 25x^6; 100x^4y^4$
2. $9; 18; 15a^8; 49b^{12}; 64a^{16}b^2$
   $9; 49b^{12}; 64a^{16}b^2$

Name the square root of the expression.

3. $16z^8$
   $4z^4$
4. $36d^{10}$
   $6d^5$
5. $81a^4b^6$
   $9a^2b^3$
6. $25m^2n^{12}$
   $5mn^6$

Factor.

7. $x^2 - 16$
   $(x + 4)(x - 4)$
8. $y^2 - 49$
   $(y + 7)(y - 7)$
9. $4x^2 - 1$
   $(2x + 1)(2x - 1)$
10. $81x^2 - 4$
    $(9x + 2)(9x - 2)$
11. $16x^2 - 121$
    $(4x + 11)(4x - 11)$
12. $49y^2 - 36$
    $(7y + 6)(7y - 6)$
13. $1 - 9a^2$
    $(1 + 3a)(1 - 3a)$
14. $16 - 81y^2$
    $(4 + 9y)(4 - 9y)$
15. $x^2y^2 - 100$
    $(xy + 10)(xy - 10)$
16. $a^2b^2 - 25$
    $(ab + 5)(ab - 5)$
17. $x^2 + 4$
    nonfactorable over the integers
18. $a^2 + 16$
    nonfactorable over the integers
19. $25 - a^2b^2$
    $(5 + ab)(5 - ab)$
20. $64 - x^2y^2$
    $(8 + xy)(8 - xy)$
21. $a^{2n} - 1$
    $(a^n + 1)(a^n - 1)$
22. $b^{2n} - 16$
    $(b^n + 4)(b^n - 4)$
23. $x^2 - 12x + 36$
    $(x - 6)^2$
24. $y^2 - 6y + 9$
    $(y - 3)^2$
25. $b^2 - 2b + 1$
    $(b - 1)^2$
26. $a^2 + 14a + 49$
    $(a + 7)^2$
27. $16x^2 - 40x + 25$
    $(4x - 5)^2$
28. $49x^2 + 28x + 4$
    $(7x + 2)^2$
29. $4a^2 + 4a - 1$
    nonfactorable over the integers
30. $9x^2 + 12x - 4$
    nonfactorable over the integers
31. $b^2 + 7b + 14$
    nonfactorable over the integers
32. $y^2 - 5y + 25$
    nonfactorable over the integers
33. $x^2 + 6xy + 9y^2$
    $(x + 3y)^2$
34. $4x^2y^2 + 12xy + 9$
    $(2xy + 3)^2$
35. $25a^2 - 40ab + 16b^2$
    $(5a - 4b)^2$
36. $4a^2 - 36ab + 81b^2$
    $(2a - 9b)^2$

Quick Quiz (Objective 7.4A)
Factor.
1. $16x^2 - 49$ $(4x + 7)(4x - 7)$
2. $a^2 + 25$ Nonfactorable
3. $9x^2 - y^2$ $(3x + y)(3x - y)$
4. $b^2 - 16b + 64$ $(b - 8)^2$
5. $9x^2 - 24x + 16$ $(3x - 4)^2$

37. $x^{2n} + 6x^n + 9$
$(x^n + 3)^2$

38. $y^{2n} - 16y^n + 64$
$(y^n - 8)^2$

39. $(x - 4)^2 - 9$
$(x - 7)(x - 1)$

40. $16 - (a - 3)^2$
$(7 - a)(1 + a)$

41. $(x - y)^2 - (a + b)^2$
$(x - y + a + b)(x - y - a - b)$

42. $(x - 2y)^2 - (x + y)^2$
$(-3y)(2x - y)$

Objective B

Which of the expressions are perfect cubes?

43. $4;\ 8;\ x^9;\ a^8b^8;\ 27c^{15}d^{18}$
$8;\ x^9;\ 27c^{15}d^{18}$

44. $9;\ 27;\ y^{12};\ m^3n^6;\ 64mn^9$
$27;\ y^{12};\ m^3n^6$

Name the cube root of the expression.

45. $8x^9$
$2x^3$

46. $27y^{15}$
$3y^5$

47. $64a^6b^{18}$
$4a^2b^6$

48. $125c^{12}d^3$
$5c^4d$

Factor.

49. $x^3 - 27$
$(x - 3)(x^2 + 3x + 9)$

50. $y^3 + 125$
$(y + 5)(y^2 - 5y + 25)$

51. $8x^3 - 1$
$(2x - 1)(4x^2 + 2x + 1)$

52. $64a^3 + 27$
$(4a + 3)(16a^2 - 12a + 9)$

53. $x^3 - y^3$
$(x - y)(x^2 + xy + y^2)$

54. $x^3 - 8y^3$
$(x - 2y)(x^2 + 2xy + 4y^2)$

55. $m^3 + n^3$
$(m + n)(m^2 - mn + n^2)$

56. $27a^3 + b^3$
$(3a + b)(9a^2 - 3ab + b^2)$

57. $64x^3 + 1$
$(4x + 1)(16x^2 - 4x + 1)$

58. $1 - 125b^3$
$(1 - 5b)(1 + 5b + 25b^2)$

59. $27x^3 - 8y^3$
$(3x - 2y)(9x^2 + 6xy + 4y^2)$

60. $64x^3 + 27y^3$
$(4x + 3y)(16x^2 - 12xy + 9y^2)$

61. $x^3y^3 + 64$
$(xy + 4)(x^2y^2 - 4xy + 16)$

62. $8x^3y^3 + 27$
$(2xy + 3)(4x^2y^2 - 6xy + 9)$

63. $16x^3 - y^3$
nonfactorable over the integers

64. $27x^3 - 8y^2$
nonfactorable over the integers

65. $8x^3 - 9y^3$
nonfactorable over the integers

66. $27a^3 - 16$
nonfactorable over the integers

67. $(a - b)^3 - b^3$
$(a - 2b)(a^2 - ab + b^2)$

68. $a^3 + (a + b)^3$
$(2a + b)(a^2 + ab + b^2)$

69. $x^{6n} + y^{3n}$
$(x^{2n} + y^n)(x^{4n} - x^{2n}y^n + y^{2n})$

70. $x^{3n} + y^{3n}$
$(x^n + y^n)(x^{2n} - x^ny^n + y^{2n})$

71. $x^{3n} + 8$
$(x^n + 2)(x^{2n} - 2x^n + 4)$

72. $a^{3n} + 64$
$(a^n + 4)(a^{2n} - 4a^n + 16)$

Quick Quiz (Objective 7.4B)

Factor.

1. $a^3 + 8b^3$ $(a + 2b)(a^2 - 2ab + 4b^2)$

2. $125x^3 + 1$ $(5x + 1)(25x^2 - 5x + 1)$

3. $x^{3n} + 27$ $(x^n + 3)(x^{2n} - 3x^n + 9)$

409

Objective C

Factor.

73. $x^2y^2 - 8xy + 15$
$(xy - 3)(xy - 5)$

74. $x^2y^2 - 8xy - 33$
$(xy + 3)(xy - 11)$

75. $x^2y^2 - 17xy + 60$
$(xy - 5)(xy - 12)$

76. $a^2b^2 + 10ab + 24$
$(ab + 6)(ab + 4)$

77. $x^4 - 9x^2 + 18$
$(x^2 - 3)(x^2 - 6)$

78. $y^4 - 6y^2 - 16$
$(y^2 + 2)(y^2 - 8)$

79. $b^4 - 13b^2 - 90$
$(b^2 + 5)(b^2 - 18)$

80. $a^4 + 14a^2 + 45$
$(a^2 + 5)(a^2 + 9)$

81. $x^4y^4 - 8x^2y^2 + 12$
$(x^2y^2 - 2)(x^2y^2 - 6)$

82. $a^4b^4 + 11a^2b^2 - 26$
$(a^2b^2 + 13)(a^2b^2 - 2)$

83. $x^{2n} + 3x^n + 2$
$(x^n + 1)(x^n + 2)$

84. $a^{2n} - a^n - 12$
$(a^n + 3)(a^n - 4)$

85. $3x^2y^2 - 14xy + 15$
$(3xy - 5)(xy - 3)$

86. $5x^2y^2 - 59xy + 44$
$(5xy - 4)(xy - 11)$

87. $6a^2b^2 - 23ab + 21$
$(2ab - 3)(3ab - 7)$

88. $10a^2b^2 + 3ab - 7$
$(ab + 1)(10ab - 7)$

89. $2x^4 - 13x^2 - 15$
$(2x^2 - 15)(x^2 + 1)$

90. $3x^4 + 20x^2 + 32$
$(x^2 + 4)(3x^2 + 8)$

91. $2x^{2n} - 7x^n + 3$
$(2x^n - 1)(x^n - 3)$

92. $4x^{2n} + 8x^n - 5$
$(2x^n + 5)(2x^n - 1)$

93. $6a^{2n} + 19a^n + 10$
$(2a^n + 5)(3a^n + 2)$

Objective D

Factor.

94. $5x^2 + 10x + 5$
$5(x + 1)^2$

95. $12x^2 - 36x + 27$
$3(2x - 3)^2$

96. $3x^4 - 81x$
$3x(x - 3)(x^2 + 3x + 9)$

97. $27a^4 - a$
$a(3a - 1)(9a^2 + 3a + 1)$

98. $7x^2 - 28$
$7(x + 2)(x - 2)$

99. $20x^2 - 5$
$5(2x + 1)(2x - 1)$

100. $y^4 - 10y^3 + 21y^2$
$y^2(y - 7)(y - 3)$

101. $y^5 + 6y^4 - 55y^3$
$y^3(y + 11)(y - 5)$

102. $x^4 - 16$
$(x^2 + 4)(x + 2)(x - 2)$

103. $16x^4 - 81$
$(4x^2 + 9)(2x + 3)(2x - 3)$

104. $8x^5 - 98x^3$
$2x^3(2x + 7)(2x - 7)$

105. $16a - 2a^4$
$2a(2 - a)(4 + 2a + a^2)$

Quick Quiz (Objective 7.4C)

Factor.

1. $x^2y^2 - 7xy + 10$ $(xy - 5)(xy - 2)$

2. $x^4 - 14x^2 + 33$ $(x^2 - 11)(x^2 - 3)$

3. $x^{2n} + 7x^n + 12$ $(x^n + 4)(x^n + 3)$

Answers to Writing Exercises

130. No, a third-degree polynomial cannot have factors $(x - 1)$, $(x + 1)(x - 3)$, and $(x + 4)$, because the leading term in the product of these factors must be a fourth-degree term.

131. If $x - 3$ and $x + 4$ are factors of $x^3 + 6x^2 - 7x - 60$, then $x^3 + 6x^2 - 7x - 60$ is divisible by $x - 3$ and $x + 4$. Divide $x^3 + 6x^2 - 7x - 60$ by $x - 3$. The quotient is $x^2 + 9x + 20$. Divide this quotient by $x + 4$. The quotient is $x + 5$. $x + 5$ is a third first-degree factor of $x^3 + 6x^2 - 7x - 60$.

106. $x^3y^3 - x^3$
$x^3(y - 1)(y^2 + y + 1)$

107. $a^3b^6 - b^3$
$b^3(ab - 1)(a^2b^2 + ab + 1)$

108. $x^6y^6 - x^3y^3$
$x^3y^3(xy - 1)(x^2y^2 + xy + 1)$

109. $8x^4 - 40x^3 + 50x^2$
$2x^2(2x - 5)^2$

110. $6x^5 + 74x^4 + 24x^3$
$2x^3(3x + 1)(x + 12)$

111. $x^4 - y^4$
$(x^2 + y^2)(x + y)(x - y)$

112. $16a^4 - b^4$
$(4a^2 + b^2)(2a + b)(2a - b)$

113. $x^6 + y^6$
$(x^2 + y^2)(x^4 - x^2y^2 + y^4)$

114. $x^4 - 5x^2 - 4$
nonfactorable over the integers

115. $a^4 - 25a^2 - 144$
nonfactorable over the integers

116. $3b^5 - 24b^2$
$3b^2(b - 2)(b^2 + 2b + 4)$

117. $16a^4 - 2a$
$2a(2a - 1)(4a^2 + 2a + 1)$

118. $16x^3y + 4x^2y^2 - 42xy^3$
$2xy(4x + 7y)(2x - 3y)$

119. $24a^2b^2 - 14ab^3 - 90b^4$
$2b^2(3a + 5b)(4a - 9b)$

120. $x^3 - 2x^2 - x + 2$
$(x - 2)(x + 1)(x - 1)$

121. $x^3 - 2x^2 - 4x + 8$
$(x - 2)^2(x + 2)$

122. $4x^2y^2 - 4x^2 - 9y^2 + 9$
$(y + 1)(y - 1)(2x + 3)(2x - 3)$

123. $4x^4 - x^2 - 4x^2y^2 + y^2$
$(x + y)(x - y)(2x + 1)(2x - 1)$

124. $a^{2n+2} - 6a^{n+2} + 9a^2$
$a^2(a^n - 3)^2$

125. $x^{2n+1} + 2x^{n+1} + x$
$x(x^n + 1)^2$

126. $2x^{n+2} - 7x^{n+1} + 3x^n$
$x^n(2x - 1)(x - 3)$

127. $3b^{n+2} + 4b^{n+1} - 4b^n$
$b^n(3b - 2)(b + 2)$

APPLYING THE CONCEPTS

128. Find all integers k such that the trinomial is a perfect square.
a. $x^2 + kx + 36$
$-12, 12$
b. $4x^2 - kx + 25$
$-20, 20$
c. $49x^2 + kxy + 64y^2$
$-112, 112$

129. The area of a square is $(16x^2 + 24x + 9)$ m². Find the dimensions of the square in terms of the variable x. Can $x = 0$? What are the possible values of x?
$(4x + 3)$ m by $(4x + 3)$ m. Yes. $x > -\frac{3}{4}$

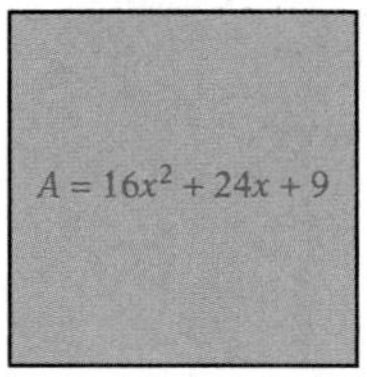

130. Can a third-degree polynomial have factors $(x - 1)$, $(x + 1)$, $(x - 3)$, and $(x + 4)$? Why or why not?

131. Given that $(x - 3)$ and $(x + 4)$ are factors of $x^3 + 6x^2 - 7x - 60$, explain how you can find a third *first-degree* factor of $x^3 + 6x^2 - 7x - 60$. Then find the factor.

Quick Quiz (Objective 7.4D)

Factor.

1. $x^5 + 6x^4 - 27x^3$ $x^3(x + 9)(x - 3)$

2. $x^2y^3 - 8x^2$ $x^2(y - 2)(y^2 + 2y + 4)$

3. $5x^5 - 5x$ $5x(x^2 + 1)(x + 1)(x - 1)$

4. $x^3 + 3x^2 - x - 3$ $(x + 1)(x - 1)(x + 3)$

7.5 Solving Equations

Objective A To solve equations by factoring

VIDEO & DVD CD TUTOR WEB SSM

The Multiplication Property of Zero states that the product of a number and zero is zero. This property is stated below.

$$\text{If } a \text{ is a real number, then } a \cdot 0 = 0 \cdot a = 0.$$

Now consider $x \cdot y = 0$. For this to be a true equation, then either $x = 0$ or $y = 0$.

> **Principle of Zero Products**
>
> If the product of two factors is zero, then at least one of the factors must be zero.
>
> If $a \cdot b - 0$, then $a - 0$ or $b - 0$.

The Principle of Zero Products is used to solve some equations.

⇒ Solve: $(x - 2)(x - 3) = 0$

By the Principle of Zero Products, if $(x - 2)(x - 3) = 0$, then $x - 2 = 0$ or $x - 3 = 0$.

$(x - 2)(x - 3) = 0$

$x - 2 = 0 \qquad x - 3 = 0$ • Let each factor equal zero (the Principle of Zero Products).

$x = 2 \qquad x = 3$ • Solve each equation for x.

Check:

$$\begin{array}{c|c} (x-2)(x-3) & 0 \\ \hline (2-2)(2-3) & 0 \\ 0(-1) & 0 \\ 0 = 0 & \text{True} \end{array} \qquad \begin{array}{c|c} (x-2)(x-3) & 0 \\ \hline (3-2)(3-3) & 0 \\ (1)(0) & 0 \\ 0 = 0 & \text{True} \end{array}$$

The solutions are 2 and 3.

An equation of the form $ax^2 + bx + c = 0$, $a \neq 0$, is a **quadratic equation**. A quadratic equation is in **standard form** when the polynomial is in descending order and equal to zero. The quadratic equations at the right are in standard form.

$3x^2 + 2x + 1 = 0$

$4x^2 - 3x + 2 = 0$

Objective 7.5A

New Vocabulary

Principle of Zero Products
quadratic equation
quadratic equation in standard form

Vocabulary to Review

descending order

Properties to Review

Multiplication Property of Zero

Instructor Note

As an application of factoring, quadratic equations are solved by using the Principle of Zero Products. A complete discussion of quadratic equations is presented later in the text.

Concept Check

1. Solve for the largest positive root of the equation $2x^3 + x^2 - 8x = 4$. 2
2. Show that the solutions of the equation $ax^2 + bx = 0$ are 0 and $-\frac{b}{a}$. Factor the left side: $x(ax + b) = 0$; then $x = 0$ or $ax + b = 0$. Solve $ax + b = 0$ for x: $x = -\frac{b}{a}$.

Optional Student Activity

Solve for x.

1. $x^2 - 9ax + 14a^2 = 0$ $2a, 7a$
2. $x^2 + 9xy - 36y^2 = 0$ $-12y, 3y$
3. $3x^2 - 4cx + c^2 = 0$ $\frac{c}{3}, c$
4. $2x^2 + 3bx + b^2 = 0$ $-\frac{b}{2}, -b$

In-Class Examples (Objective 7.5A)

Solve.

1. $x^2 + x - 12 = 0$ $-4, 3$
2. $y^2 - 3y = 0$ $0, 3$
3. $2x^2 - 5x = 12$ $-\frac{3}{2}, 4$
4. $t + 18 = t(t + 8)$ $-9, 2$

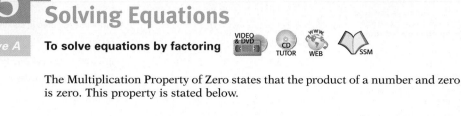

Concept Check

Which of the equations below can be solved by using the Principle of Zero Products without first rewriting the equation?

1. $4x(6x + 7) = 0$ Yes
2. $0 = (4x - 5)(3x + 8)$ Yes
3. $2x(x - 5) - 5 = 0$ No
4. $(x - 7)(y + 3) = 0$ Yes
5. $0 = (2x - 3)x + 3$ No
6. $0 = (2x - 3)(x + 3)$ Yes

Optional Student Activity

Examine the following solution of the equation $x^2 - 5x + 6 = 12$.

$(x - 3)(x - 2) = 3 \cdot 4$ Factor each side.

$x - 3 = 3$ or $x - 2 = 4$
$x = 6$ $\quad x = 6$

Check:

$6^2 - 5(6) + 6 = 12$
$12 = 12$

The solution is 6.
Do you agree with this solution? If not, show a correct solution.
No. The correct solutions are −1 and 6.

TAKE NOTE

Note the steps involved in solving a quadratic equation by factoring:

1. Write in standard form.
2. Factor.
3. Set each factor equal to 0.
4. Solve each equation.
5. Check the solutions.

➡ Solve: $2x^2 + x = 6$

$2x^2 + x = 6$
$2x^2 + x - 6 = 0$ • **Write the equation in standard form.**
$(2x - 3)(x + 2) = 0$ • **Factor.**
$2x - 3 = 0 \qquad x + 2 = 0$ • **Use the Principle of Zero Products.**
$2x = 3 \qquad x = -2$ • **Solve each equation for x.**
$x = \frac{3}{2}$

$\frac{3}{2}$ and -2 check as solutions. The solutions are $\frac{3}{2}$ and -2.

Example 1
Solve: $x(x - 3) = 0$

Solution
$x(x - 3) = 0$
$x = 0 \qquad x - 3 = 0$
$\qquad x = 3$

The solutions are 0 and 3.

You Try It 1
Solve: $2x(x + 7) = 0$

Your solution
0 and −7

Example 2
Solve: $2x^2 - 50 = 0$

Solution
$2x^2 - 50 = 0$
$2(x^2 - 25) = 0$
$2(x + 5)(x - 5) = 0$
$x + 5 = 0 \qquad x - 5 = 0$
$x = -5 \qquad x = 5$

The solutions are −5 and 5.

You Try It 2
Solve: $4x^2 - 9 = 0$

Your solution
$\frac{3}{2}$ and $-\frac{3}{2}$

Example 3
Solve: $(x - 3)(x - 10) = -10$

Solution
$(x - 3)(x - 10) = -10$
$x^2 - 13x + 30 = -10$ • **Multiply $(x - 3)(x - 10)$.**
$x^2 - 13x + 40 = 0$ • **Add 10 to each side of the equation. The equation is now in standard form.**
$(x - 8)(x - 5) = 0$
$x - 8 = 0 \qquad x - 5 = 0$
$x = 8 \qquad x = 5$

The solutions are 8 and 5.

You Try It 3
Solve: $(x + 2)(x - 7) = 52$

Your solution
−6 and 11

Solutions on p. S22

Objective 7.5B

Optional Student Activity

The sum of the squares of two consecutive odd integers is 202. Find all pairs of consecutive odd integers that make this statement true. −11, −9 and 9, 11

Objective B To solve application problems

VIDEO & DVD CD TUTOR WEB SSM

Recall that the integers are the numbers . . . , −3, −2, −1, 0, 1, 2, 3, ...

An **even integer** is an integer that is divisible by 2. Examples of even integers are −8, 0, and 22. An **odd integer** is an integer that is not divisible by 2. Examples of odd integers are −17, 1, and 39.

Consecutive integers are integers that follow one another in order. Examples of consecutive integers are shown at the right. (Assume that the variable n represents an integer.)

11, 12, 13
−8, −7, −6
$n, n + 1, n + 2$

Examples of **consecutive even integers** are shown at the right. (Assume that the variable n represents an even integer.)

24, 26, 28
−10, −8, −6
$n, n + 2, n + 4$

Examples of **consecutive odd integers** are shown at the right. (Assume that the variable n represents an odd integer.)

19, 21, 23
−1, 1, 3
$n, n + 2, n + 4$

Example 4

The sum of the squares of two consecutive positive even integers is equal to 100. Find the two integers.

Strategy

First positive even integer: n
Second positive even integer: $n + 2$

The sum of the square of the first positive even integer and the square of the second positive even integer is 100.

Solution

$$n^2 + (n + 2)^2 = 100$$
$$n^2 + n^2 + 4n + 4 = 100$$
$$2n^2 + 4n + 4 = 100$$
$$2n^2 + 4n - 96 = 0$$
$$2(n^2 + 2n - 48) = 0$$
$$n^2 + 2n - 48 = 0$$

• **Divide each side of the equation by 2.**

$$(n - 6)(n + 8) = 0$$

$n - 6 = 0 \qquad n + 8 = 0$
$n = 6 \qquad n = -8$

Because −8 is not a positive even integer, it is not a solution.

$n = 6$
$n + 2 = 6 + 2 = 8$

The two integers are 6 and 8.

You Try It 4

The sum of the squares of two positive consecutive integers is 61. Find the two integers.

Your strategy

Your solution

5 and 6

Solution on p. S22

In-Class Examples (Objective 7.5B)

1. The sum of the squares of two consecutive positive integers is 145. Find the two integers. 8, 9
2. The sum of the squares of two consecutive negative odd integers is 10. Find the two integers. −3, −1
3. A garden measures 12 ft by 16 ft. A uniform border around the garden increases the total area to 357 ft^2. What is the width of the border? 2.5 ft

Optional Student Activity

1. Find two consecutive integers whose cubes differ by 127. 6 and 7 or -7 and -6
2. The sum of the squares of three consecutive odd integers is 83. Find the three integers. 3, 5, and 7 or -3, -5, and -7
3. A model for the height above the ground of an arrow projected into the air with an initial velocity of 120 ft/s is $h = -16t^2 + 120t + 5$, where h is the height, in feet, of the arrow t seconds after it is released from the bow. Determine at what times the arrow is 181 ft above the ground. After 2 s and after 5.5 s
4. The base of a triangle is 2 in. less than four times the height. The area of the triangle is 45 in^2. Find the height and the length of the base of the triangle. Height: 5 in.; base: 18 in.
5. The height above Earth of a projectile fired upward is given by the formula $s = v_o t - 16t^2$, where s is the height in feet, v_0 is the initial velocity, and t is the time in seconds. Find the time for a projectile to return to Earth if it has an initial velocity of 200 ft/s. 12.5 s

Example 5

A stone is thrown into a well with an initial speed of 4 ft/s. The well is 420 ft deep. How many seconds later will the stone hit the bottom of the well? Use the equation $d = vt + 16t^2$, where d is the distance in feet, v is the initial speed, and t is the time in seconds.

Strategy

To find the time for the stone to drop to the bottom of the well, replace the variables d and v by their given values and solve for t.

Solution

$$d = vt + 16t^2$$
$$420 = 4t + 16t^2$$
$$0 = -420 + 4t + 16t^2$$
$$16t^2 + 4t - 420 = 0$$
$$4(4t^2 + t - 105) = 0$$
$$4t^2 + t - 105 = 0$$

• **Divide each side of the equation by 4.**

$$(4t + 21)(t - 5) = 0$$

$$4t + 21 = 0 \qquad t - 5 = 0$$
$$4t = -21 \qquad t = 5$$
$$t = -\frac{21}{4}$$

Because the time cannot be a negative number, $-\frac{21}{4}$ is not a solution.

The time is 5 s.

You Try It 5

The length of a rectangle is 4 in. longer than twice the width. The area of the rectangle is 96 in^2. Find the length and width of the rectangle.

Your strategy

Your solution

width: 6 in.;
length: 16 in.

Solution on p. S22

415

7.5 Exercises

Objective A

1. In your own words, explain why it is possible to solve a quadratic equation using the Principle of Zero Products.

Solve.

2. $(y + 3)(y + 2) = 0$ $-3, -2$
3. $(y - 3)(y - 5) = 0$ $3, 5$
4. $(z - 7)(z - 3) = 0$ $7, 3$
5. $(z + 8)(z - 9) = 0$ $-8, 9$
6. $x(x - 5) = 0$ $0, 5$
7. $x(x + 2) = 0$ $0, -2$
8. $a(a - 9) = 0$ $0, 9$
9. $a(a + 12) = 0$ $0, -12$
10. $y(2y + 3) = 0$ $0, -\frac{3}{2}$
11. $t(4t - 7) = 0$ $0, \frac{7}{4}$
12. $2a(3a - 2) = 0$ $0, \frac{2}{3}$
13. $4b(2b + 5) = 0$ $0, -\frac{5}{2}$
14. $(b + 2)(b - 5) = 0$ $-2, 5$
15. $(b - 8)(b + 3) = 0$ $8, -3$
16. $x^2 - 81 = 0$ $9, -9$
17. $x^2 - 121 = 0$ $11, -11$
18. $4x^2 - 49 = 0$ $\frac{7}{2}, -\frac{7}{2}$
19. $16x^2 - 1 = 0$ $\frac{1}{4}, -\frac{1}{4}$
20. $9x^2 - 1 = 0$ $\frac{1}{3}, -\frac{1}{3}$
21. $16x^2 - 49 = 0$ $\frac{7}{4}, -\frac{7}{4}$
22. $x^2 + 6x + 8 = 0$ $-4, -2$
23. $x^2 - 8x + 15 = 0$ $3, 5$
24. $z^2 + 5z - 14 = 0$ $2, -7$
25. $z^2 + z - 72 = 0$ $8, -9$
26. $2a^2 - 9a - 5 = 0$ $-\frac{1}{2}, 5$
27. $3a^2 + 14a + 8 = 0$ $-\frac{2}{3}, -4$
28. $6z^2 + 5z + 1 = 0$ $-\frac{1}{3}, -\frac{1}{2}$
29. $6y^2 - 19y + 15 = 0$ $\frac{5}{3}, \frac{3}{2}$
30. $x^2 - 3x = 0$ $0, 3$
31. $a^2 - 5a = 0$ $0, 5$
32. $x^2 - 7x = 0$ $0, 7$
33. $2a^2 - 8a = 0$ $0, 4$
34. $a^2 + 5a = -4$ $-1, -4$
35. $a^2 - 5a = 24$ $-3, 8$
36. $y^2 - 5y = -6$ $2, 3$
37. $y^2 - 7y = 8$ $-1, 8$
38. $2t^2 + 7t = 4$ $\frac{1}{2}, -4$
39. $3t^2 + t = 10$ $\frac{5}{3}, -2$
40. $3t^2 - 13t = -4$ $\frac{1}{3}, 4$
41. $5t^2 - 16t = -12$ $\frac{6}{5}, 2$

Section 7.5

Suggested Assignment
Exercises 3–81, odds
More challenging problems:
Exercises 82–86

Answers to Writing Exercises

1. Some students may explain that the Principle of Zero Products is based on the Multiplication Property of Zero, which states that the product of a number and zero is zero. The Principle of Zero Products begins with the conclusion of the Multiplication Property of Zero, stating that if the product of two or more factors is zero, then at least one of the factors must be equal to zero. Therefore, in order for us to use the Principle of Zero Products, a quadratic equation must be written as a product equal to zero. Then we can set each factor equal to zero and solve for the variable.

Quick Quiz (Objective 7.5A)
Solve.

1. $x^2 + 3x - 4 = 0$ $-4, 1$
2. $y^2 + 5y = 0$ $-5, 0$
3. $3x^2 + x = 2$ $-1, \frac{2}{3}$
4. $t + 12 = t(t - 3)$ $-2, 6$

416

42. $x(x - 12) = -27$
3, 9

43. $x(x - 11) = 12$
12, −1

44. $y(y - 7) = 18$
9, −2

45. $y(y + 8) = -15$
−3, −5

46. $p(p + 3) = -2$
−1, −2

47. $p(p - 1) = 20$
5, −4

48. $y(y + 4) = 45$
5, −9

49. $y(y - 8) = -15$
3, 5

50. $x(x + 3) = 28$
4, −7

51. $p(p - 14) = 15$
15, −1

52. $(x + 8)(x - 3) = -30$
−2, −3

53. $(x + 4)(x - 1) = 14$
−6, 3

54. $(z - 5)(z + 4) = 52$
−8, 9

55. $(z - 8)(z + 4) = -35$
1, 3

56. $(z - 6)(z + 1) = -10$
1, 4

57. $(a + 3)(a + 4) = 72$
−12, 5

58. $(a - 4)(a + 7) = -18$
−5, 2

59. $(2x + 5)(x + 1) = -1$
$-\frac{3}{2}, -2$

Objective B *Application Problems*

60. The square of a positive number is six more than five times the positive number. Find the number.
6

61. The square of a negative number is fifteen more than twice the negative number. Find the number.
−3

62. The sum of two numbers is six. The sum of the squares of the two numbers is twenty. Find the two numbers.
2, 4

63. The sum of two numbers is eight. The sum of the squares of the two numbers is thirty-four. Find the two numbers.
3, 5

64. The sum of the squares of two consecutive positive integers is forty-one. Find the two integers.
4, 5

65. The sum of the squares of two consecutive positive even integers is one hundred sixty-four. Find the two integers.
8, 10

66. The length of a rectangle is 2 ft more than twice the width. The area of the rectangle is 84 ft^2. Find the length and width of the rectangle.
length: 14 ft; width: 6 ft

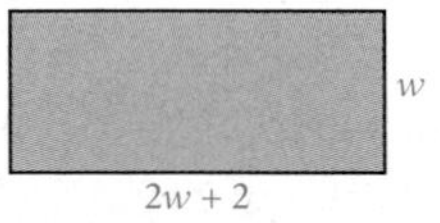

Quick Quiz (Objective 7.5B)

Solve.

1. Five times an integer plus three times the square of the integer is 22. Find the integer. 2
2. The length of a rectangle is 2 ft less than three times the width. The area of the rectangle is 40 ft^2. Find the length and width of the rectangle.
Length: 10 ft; width: 4 ft

417

67. The length of a rectangle is 8 cm more than three times the width. The area of the rectangle is 380 cm^2. Find the length and width of the rectangle.
length: 38 cm; width: 10 cm

The formula $S = \frac{n^2 + n}{2}$ gives the sum, S, of the first n natural numbers. Use this formula for Exercises 68 and 69.

68. How many consecutive natural numbers beginning with 1 will give a sum of 78?
12

69. How many consecutive natural numbers beginning with 1 will give a sum of 171?
18

The formula $N = \frac{t^2 - t}{2}$ gives the number, N, of football games that must be scheduled in a league with t teams if each team is to play every other team once. Use this formula for Exercises 70 and 71.

70. How many teams are in a league that schedules 15 games in such a way that each team plays every other team once?
6 teams

71. How many teams are in a league that schedules 45 games in such a way that each team plays every other team once?
10 teams

The distance, s, in feet, that an object will fall (neglecting air resistance) in t seconds is given by $s = vt + 16t^2$, where v is the initial velocity of the object in feet per second. Use this formula for Exercises 72 and 73.

72. An object is released from the top of a building 192 ft high. The initial velocity is 16 ft/s, and air resistance is neglected. How many seconds later will the object hit the ground?
3 s

73. An object is released from the top of a building 320 ft high. The initial velocity is 16 ft/s, and air resistance is neglected. How many seconds later will the object hit the ground?
4 s

The height, h, in feet, an object will attain (neglecting air resistance) in t seconds is given by $h = vt - 16t^2$, where v is the initial velocity of the object in feet per second. Use this formula for Exercises 74 and 75.

74. A golf ball is thrown onto a cement surface and rebounds straight up. The initial velocity of the rebound is 60 ft/s. How many seconds later will the golf ball return to the ground?
3.75 s

75. A foul ball leaves a bat and travels straight up with an initial velocity of 64 ft/s. How many seconds later will the ball be 64 ft above the ground?
2 s

Answers to Writing Exercises

87. The error in the solution
$(x + 2)(x - 3) = 6$
$x + 2 = 6 \quad x - 3 = 6$
$x = 4 \qquad x = 9$
occurs when the assumption is made that because $(x + 2)(x - 3) = 6$, $x + 2 = 6$ and/or $x - 3 = 6$. In other words, it is an error to say that if the product of two numbers is 6, then at least one of the numbers must be 6. The correct solution is
$(x + 2)(x - 3) = 6$
$x^2 - x - 6 = 6$
$x^2 - x - 12 = 0$
$(x + 3)(x - 4) = 0$
$x = -3 \quad x = 4$

76. The height of a triangle is 8 cm more than the length of the base. The area of the triangle is 64 cm². Find the base and height of the triangle.
base: 8 cm; height: 16 cm

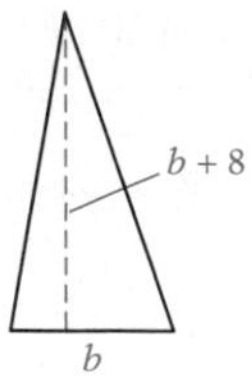

77. The height of a triangle is 4 m more than twice the length of the base. The area of the triangle is 35 m². Find the height of the triangle.
14 m

78. The length of each side of a square is extended 5 in. The area of the resulting square is 64 in². Find the length of a side of the original square.
3 in.

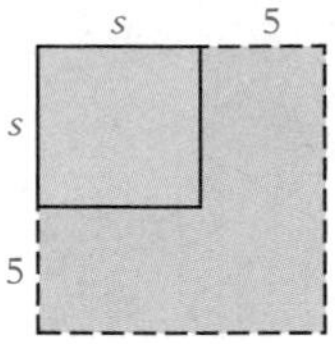

79. A small garden measures 8 ft by 10 ft. A uniform border around the garden increases the total area to 143 ft². What is the width of the border?
1.5 ft

80. The page of a book measures 6 in. by 9 in. A uniform border around the page leaves 28 in² for type. What are the dimensions of the type area?
4 in. by 7 in.

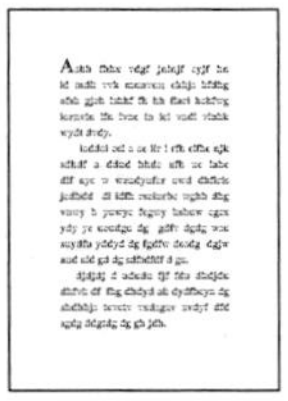

81. The radius of a circle is increased by 3 in.; this increases the area by 100 in². Find the radius of the original circle. Round to the nearest hundredth.
3.81 in.

APPLYING THE CONCEPTS

82. The length of a rectangle is 7 cm, and the width is 4 cm. If both the length and the width are increased by equal amounts, the area of the rectangle is increased by 42 cm². Find the length and width of the larger rectangle.
length: 10 cm; width: 7 cm

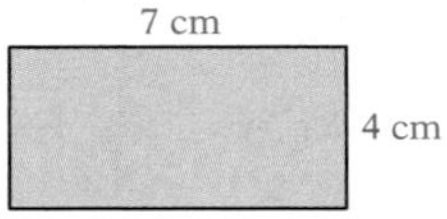

83. A rectangular piece of cardboard is 10 in. longer than it is wide. Squares 2 in. on a side are to be cut from each corner, and then the sides will be folded up to make an open box with a volume of 192 in³. Find the length and width of the piece of cardboard.
length: 20 in.; width: 10 in.

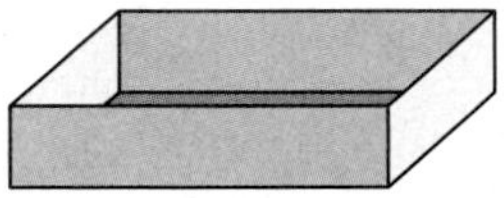

84. Solve: $p^3 = 9p^2$ 0, 9

85. Solve: $(x + 3)(2x - 1) = (3 - x)(5 - 3x)$ 1, 18

86. Find $3n^2$ if $n(n + 5) = -4$. 3, 48

87. Explain the error made in solving the equation at the right. Solve the equation correctly.

$(x + 2)(x - 3) = 6$
$x + 2 = 6 \quad x - 3 = 6$
$x = 4 \qquad x = 9$

Focus on Problem Solving

Polya's Four-Step Process

Your success in mathematics and your success in the workplace are heavily dependent on your ability to solve problems. One of the foremost mathematicians to study problem solving was George Polya (1887–1985). The basic structure that Polya advocated for problem solving has four steps, as outlined below.

WEB

(For more information on Polya and his work, conduct a search on the Internet for "Polya." You can also search the Math Forum at **www.forum.swarthmore.edu**.)

Point of Interest

George Polya was born in Hungary and moved to the United States in 1940. He lived in Providence, Rhode Island, where he taught at Brown University until 1942, when he moved to California. There he taught at Stanford University until his retirement. While at Stanford, he published 10 books and a number of articles for mathematics journals. Of the books Polya published, *How To Solve It* (1945) is one of the best known. In this book, Polya outlines a strategy for solving problems. This strategy, although frequently applied to mathematics, can be used to solve problems from virtually any discipline.

1. Understand the Problem

You must have a clear understanding of the problem. To help you focus on understanding the problem, here are some questions to think about.

- Can you restate the problem in your own words?
- Can you determine what is known about these types of problems?
- Is there missing information that you need in order to solve the problem?
- Is there information given that is not needed?
- What is the goal?

2. Devise a Plan

Successful problem solvers use a variety of techniques when they attempt to solve a problem. Here are some frequently used strategies.

- Make a list of the known information.
- Make a list of information that is needed to solve the problem.
- Make a table or draw a diagram.
- Work backwards.
- Try to solve a similar but simpler problem.
- Research the problem to determine whether there are known techniques for solving problems of its kind.
- Try to determine whether some pattern exists.
- Write an equation.

3. Carry Out the Plan

Once you have devised a plan, you must carry it out.

- Work carefully.
- Keep an accurate and neat record of all your attempts.
- Realize that some of your initial plans will not work and that you may have to return to Step 2 and devise another plan or modify your existing plan.

4. Review Your Solution

Once you have found a solution, check the solution against the known facts.

- Ensure that the solution is consistent with the facts of the problem.
- Interpret the solution in the context of the problem.
- Ask yourself whether there are generalizations of the solution that could apply to other problems.
- Determine the strengths and weaknesses of your solution. For instance, is your solution only an approximation to the actual solution?
- Consider the possibility of alternative solutions.

Answers to Focus on Problem Solving: Polya's Four-Step Process

1. (Exercise 1 is on page 420.)

Understand the problem.

We must determine the number of ounces of water the cup will hold. To do so, we need the volume of the cup. The dimensions of the cup are given in inches, so the volume will be in cubic inches. We will need to convert cubic inches to fluid ounces.

Devise a plan.

Consult a reference book to find the formula for the volume of a cone $\left(V = \frac{1}{3}\pi r^2 h\right)$. The conversion rate for cubic inches to fluid ounces is given on page 420 of the text: $1 \text{ in}^3 \approx 0.55 \text{ fl oz}$. The plan is to find the volume of the cup in cubic inches and then convert the volume to fluid ounces.

Carry out the plan.

Find the volume of the cone.
$r = 1.5$, $h = 4$

$$\begin{aligned} V &= \frac{1}{3}\pi r^2 h \\ &= \frac{1}{3}\pi(1.5)^2(4) \\ &\approx 9.424778 \text{ in}^3 \end{aligned}$$

Convert 9.424778 in^3 to fluid ounces.

$$\begin{aligned} V &= 9.424778(0.55) \\ &\approx 5.183628 \text{ fl oz} \end{aligned}$$

The cup will hold about 5.18 fl oz.

Review your solution.

It seems reasonable that a conical cup 4 in. tall would hold about 5 fl oz.

(Continued on next page)

(Continued)

2.

Understand the problem.

We are asked to determine the dimensions of a 12-ounce soft drink can. We need to approximate the length of a hand. We also need to know the formula for the volume of a right circular cylinder, and we need to convert 12 fl oz to cubic inches.

Devise a plan.

From a reference book, find the formula for the volume of a right circular cylinder ($V = \pi r^2 h$). We will approximate the length of a hand to be 8 in. We can use this approximation and the formula $C = 2\pi r$ to find the radius of the can. The conversion rate for cubic inches to fluid ounces is 0.55 fl oz $\approx$ 1 in^3. After finding the radius of the can and converting 12 fl oz to cubic inches, we will find the height of the can.

Carry out the plan.

The length of the hand is 75% of the circumference.

$$0.75C = 8$$
$$C \approx 10.67 \text{ in.}$$

Use the formula $C = 2\pi r$ to find the radius.

$$2\pi r = C$$
$$2\pi r = 10.67$$
$$r \approx 1.70 \text{ in.}$$

Convert 12 fl oz to cubic inches.

$$12 \div 0.55 \approx 21.81 \text{ in}^3$$

Use the formula for the volume of a right circular cylinder to find the height of the can.

$$\pi r^2 h = V$$
$$\pi(1.70)^2 h = 21.81$$
$$h \approx 2.40 \text{ in.}$$

The radius of the can is approximately 1.70 in. The height is approximately 2.40 in.

Review your solution.

The radius of a conventional soda can is smaller than 1.70 in. (only about 1.25 in.). The height is greater than 2.40 in. (almost 5 in.). But we can see that the dimensions are "in the right ball park."

We will use Polya's four-step process to solve the following problem.

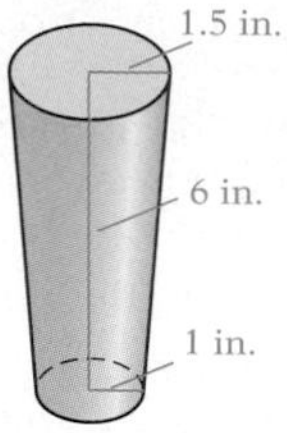

A large soft drink costs \$1.25 at a college cafeteria. The dimensions of the cup are shown at the left. Suppose you don't put any ice in the cup. Determine the cost per ounce for the soft drink.

1. *Understand the problem.* We must determine the cost per ounce for the soda. To do this, we need the dimensions of the cup (which are given), the cost of the drink (given), and a formula for the volume of the cup (unknown). Also, because the dimensions are given in inches, the volume will be in cubic inches; we need a conversion factor that will convert cubic inches to fluid ounces.
2. *Devise a plan.* Consult a resource book that gives the volume of the figure, which is called a **frustrum**. The formula for the volume is

$$V = \frac{\pi h}{3}(r^2 + rR + R^2)$$

where h is the height, r is the radius of the base, and R is the radius of the top. Also from a reference book, 1 $\text{in}^3 \approx 0.55$ fl oz. The general plan is to calculate the volume, convert the answer to fluid ounces, and then divide the cost by the number of fluid ounces.

3. *Carry out the plan.* Using the information from the drawing, evaluate the formula for the volume.

$$V = \frac{6\pi}{3}[1^2 + 1(1.5) + 1.5^2] = 9.5\pi \approx 29.8451 \text{ in}^3$$

$$V \approx 29.8451(0.55) \approx 16.4148 \text{ fl oz}$$ • Convert to fluid ounces.

$$\text{Cost per ounce} \approx \frac{1.25}{16.4148} \approx 0.07615$$ • Divide the cost by the volume.

The cost of the soft drink is approximately 7.62 cents per ounce.

4. *Review the solution.* The cost of a 12-ounce can of soda from a vending machine is generally about 75¢. Therefore, the cost of canned soda is 75¢ ÷ 12 = 6.25¢ per ounce. This is consistent with our solution. This does not mean our solution is correct, but it does indicate that it is at least reasonable. Why might soda from a cafeteria be more expensive per ounce than soda from a vending machine?

Is there an alternative way to obtain the solution? There are probably many, but one possibility is to get a measuring cup, pour the soft drink into it, and read the number of ounces. Name an advantage and a disadvantage of this method.

Use the four-step solution process for the following problems.

1. A cup dispenser next to a water cooler holds cups that have the shape of a right circular cone. The height of the cone is 4 in., and the radius of the circular top is 1.5 in. How many ounces of water can the cup hold?
2. Soft drink manufacturers do research into the preferences of consumers with regard to the look and feel and size of a soft drink can. Suppose that a manufacturer has determined that people want to have their hand reach around approximately 75% of the can. If this preference is to be achieved, how tall should the can be if it contains 12 oz of fluid? Assume the can is a right circular cylinder.

Projects and Group Activities

Water Displacement

When an object is placed in water, the object displaces an amount of water that is equal to the volume of the object.

➡ A sphere with a diameter of 4 in. is placed in a rectangular tank of water that is 6 in. long and 5 in. wide. How much does the water level rise? Round to the nearest hundredth.

$V = \frac{4}{3}\pi r^3$ • **Use the formula for the volume of a sphere.**

$V = \frac{4}{3}\pi(2^3) = \frac{32}{3}\pi$ • $r = \frac{1}{2}d = \frac{1}{2}(4) = 2$

Let x represent the amount of the rise in water level. The volume of the sphere will equal the volume displaced by the water. As shown at the left, this volume is the rectangular solid with width 5 in., length 6 in., and height x in.

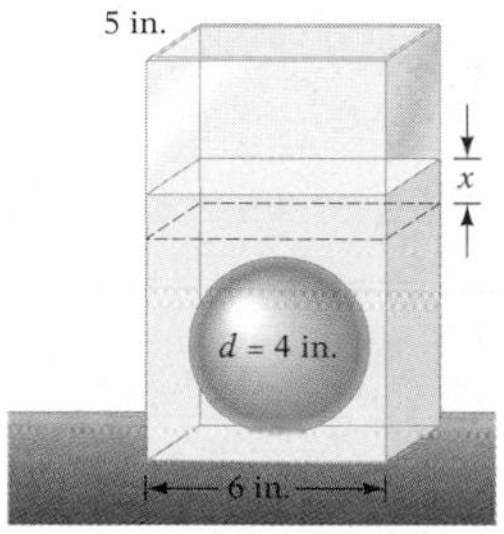

$V = LWH$ • **Use the formula for the volume of a rectangular solid.**

$\frac{32}{3}\pi = (6)(5)x$ • **Substitute $\frac{32}{3}\pi$ for V, 5 for W, and 6 for L.**

$\frac{32}{90}\pi = x$ • **The exact height that the water will fill is $\frac{32}{90}\pi$.**

$1.12 \approx x$ • **Use a calculator to find an approximation.**

The water will rise approximately 1.12 in.

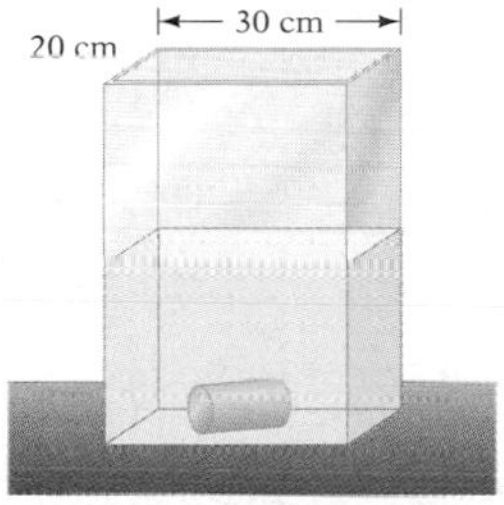

Figure 1

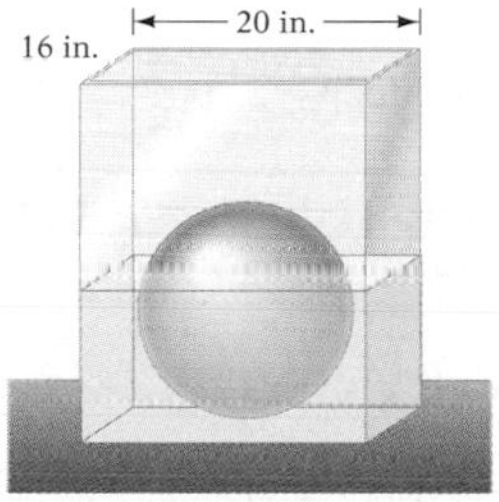

Figure 2

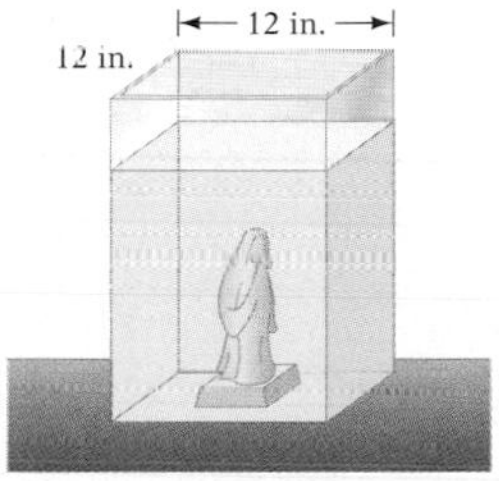

Figure 3

1. A cylinder with a 2-centimeter radius and a height of 10 cm is submerged in a tank of water that is 20 cm wide and 30 cm long (see Figure 1). How much does the water level rise? Round to the nearest hundredth.
2. A sphere with a radius of 6 in. is placed in a rectangular tank of water that is 16 in. wide and 20 in. long (see Figure 2). The sphere displaces water until two-thirds of the sphere is submerged. How much does the water level rise? Round to the nearest hundredth.
3. A chemist wants to know the density of a statue that weighs 15 lb. The statue is placed in a rectangular tank of water that is 12 in. long and 12 in. wide (see Figure 3). The water level rises 0.42 in. Find the density of the statue. Round to the nearest hundredth. (*Hint*: Density = weight ÷ volume)

Answers to Projects and Group Activities: Water Displacement

1. The volume of the cylinder is $V = \pi r^2 h$, where $r = 2$ and $h = 10$.

   $V = \pi r^2 h$
   $V = \pi(2)^2(10)$
   $V = 40\pi$

   The volume of water displaced is $V = LWH$, where $L = 30$, $W = 20$, and $H = x$.

   $V = LWH$
   $40\pi = (30)(20)x$
   $0.21 \approx x$

   The water will rise approximately 0.21 cm.

2. The volume of $\frac{2}{3}$ of the sphere is $V = \frac{2}{3}\left(\frac{4}{3}\pi r^3\right)$, where $r = 6$.

   $V = \frac{2}{3}\left(\frac{4}{3}\pi r^3\right)$
   $V = \frac{8}{9}\pi(6)^3 = 192\pi$

   The volume of the water displaced is $V = LWH$, where $L = 20$, $W = 16$, and $H = x$.

   $V = LWH$
   $192\pi = (20)(16)x$
   $1.88 \approx x$

   The water will rise approximately 1.88 in.

3. Find the volume of the statue by finding the volume of the water displaced by the statue.

   $V = LWH$, where $L = 12$, $W = 12$, and $H = 0.42$.
   $V = (12)(12)(0.42) = 60.48$

   The volume of the statue is 60.48 in^3.

   Density = weight ÷ volume
   Density = 15 ÷ 60.48 ≈ 0.25

   The density of the statue is approximately 0.25 lb/in^3.

Chapter Summary

Key Words

The *greatest common factor* (GCF) of two or more monomials is the product of the GCF of the coefficients and the common variable factors. [p. 379]

To *factor* a polynomial means to write the polynomial as a product of other polynomials. [p. 379]

To *factor* a trinomial of the form $ax^2 + bx + c$ means to express the trinomial as the product of two binomials. [p. 385]

A polynomial is *nonfactorable over the integers* if it does not factor using only integers. Such a polynomial is called a *prime polynomial.* [p. 386]

A product of a term and itself is a *perfect square.* [p. 401]

The product of the same three factors is a *perfect cube.* [p. 403]

An equation of the form $ax^2 + bx + c = 0$, $a \neq 0$, is a *quadratic equation.* [p. 411]

A quadratic equation is in *standard form* when the polynomial is in descending order and equal to zero. The quadratic equation $ax^2 + bx + c = 0$ is in standard form. [p. 411]

Essential Rules

Factors of the Difference of Two Perfect Squares [p. 401]

$$a^2 - b^2 = (a + b)(a - b)$$

Factors of a Perfect-Square Trinomial [p. 401]

$$a^2 + 2ab + b^2 = (a + b)^2$$
$$a^2 - 2ab + b^2 = (a - b)^2$$

Factors of the Sum or Difference of Two Cubes [p. 403]

$$a^3 + b^3 = (a + b)(a^2 - ab + b^2)$$
$$a^3 - b^3 = (a - b)(a^2 + ab + b^2)$$

Principle of Zero Products [p. 411]

If the product of two factors is zero, then at least one of the factors must be zero.

If $a \cdot b = 0$, then $a = 0$ or $b = 0$.

General Factoring Strategy [p. 405]

1. Is there a common factor? If so, factor out the GCF.
2. If the polynomial is a binomial, is it the difference of two perfect squares, the sum of two cubes, or the difference of two cubes? If so, factor.
3. If the polynomial is a trinomial, is it a perfect-square trinomial or the product of two binomials? If so, factor.
4. If the polynomial has four terms, can it be factored by grouping? If so, factor.
5. Is each factor nonfactorable over the integers? If not, factor.

423

Chapter Review

1. Factor: $5x^3 + 10x^2 + 35x$
$5x(x^2 + 2x + 7)$ [7.1A]

2. Factor: $12a^2b + 3ab^2$
$3ab(4a + b)$ [7.1A]

3. Factor: $14y^9 - 49y^6 + 7y^3$
$7y^3(2y^6 - 7y^3 + 1)$ [7.1A]

4. Factor: $4x(x - 3) - 5(3 - x)$
$(x - 3)(4x + 5)$ [7.1B]

5. Factor: $10x^2 + 25x + 4xy + 10y$
$(2x + 5)(5x + 2y)$ [7.1B]

6. Factor: $21ax - 35bx - 10by + 6ay$
$(3a - 5b)(7x + 2y)$ [7.1B]

7. Factor: $b^2 - 13b + 30$
$(b - 3)(b - 10)$ [7.2A]

8. Factor: $c^2 + 8c + 12$
$(c + 6)(c + 2)$ [7.2A]

9. Factor: $y^2 + 5y - 36$
$(y - 4)(y + 9)$ [7.2A]

10. Factor: $3a^2 - 15a - 42$
$3(a + 2)(a - 7)$ [7.2B]

11. Factor: $4x^3 - 20x^2 - 24x$
$4x(x - 6)(x + 1)$ [7.2B]

12. Factor: $n^4 - 2n^3 - 3n^2$
$n^2(n + 1)(n - 3)$ [7.2B]

13. Factor $6x^2 - 29x + 28$ by using trial factors.
$(2x - 7)(3x - 4)$ [7.3A]

14. Factor $12y^2 + 16y - 3$ by using trial factors.
$(6y - 1)(2y + 3)$ [7.3A]

15. Factor $2x^2 - 5x + 6$ by using trial factors.
nonfactorable over the integers [7.3A]

16. Factor $3x^2 - 17x + 10$ by grouping.
$(3x - 2)(x - 5)$ [7.3B]

424 Chapter 7 / Factoring

17. Factor $2a^2 - 19a - 60$ by grouping.
$(2a + 5)(a - 12)$ [7.3B]

18. Factor $18a^2 - 3a - 10$ by grouping.
$(6a - 5)(3a + 2)$ [7.3B]

19. Factor: $x^2y^2 - 9$
$(xy + 3)(xy - 3)$ [7.4A]

20. Factor: $4x^2 + 12xy + 9y^2$
$(2x + 3y)^2$ [7.4A]

21. Factor: $x^{2n} - 12x^n + 36$
$(x^n - 6)^2$ [7.4A]

22. Factor: $64a^3 - 27b^3$
$(4a - 3b)(16a^2 + 12ab + 9b^2)$ [7.4B]

23. Factor: $15x^4 + x^2 - 6$
$(3x^2 + 2)(5x^2 - 3)$ [7.4C]

24. Factor: $21x^4y^4 + 23x^2y^2 + 6$
$(7x^2y^2 + 3)(3x^2y^2 + 2)$ [7.4C]

25. Factor: $3a^6 - 15a^4 - 18a^2$

$3a^2(a^2 + 1)(a^2 - 6)$ [7.4D]

26. Solve: $4x^2 + 27x = 7$
$\frac{1}{4}, -7$ [7.5A]

27. Solve: $(x + 1)(x - 5) = 16$
$7, -3$ [7.5A]

28. The length of a hockey field is 20 yd less than twice the width of the hockey field. The area of the hockey field is 6000 yd^2. Find the length and width of the hockey field.
width: 60 yd; length: 100 yd [7.5B]

29. A rectangular photograph has dimensions 15 in. by 12 in. A picture frame around the photograph increases the total area to 270 in^2. What is the width of the frame?
1.5 in. [7.5B]

30. The lengths of two sides of a square garden plot are extended 4 ft, as shown in the figure at the right. The area of the resulting square is 576 ft^2. Find the length of a side of the original garden plot.
20 ft [7.5B]

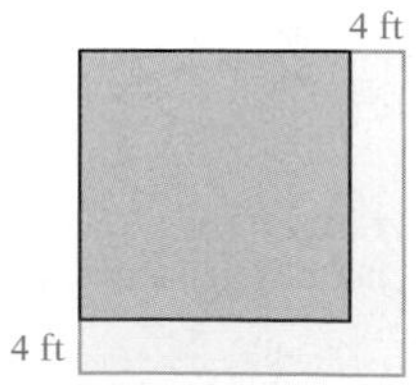

425

Chapter Test

1. Factor: $ab + 6a - 3b - 18$
$(b + 6)(a - 3)$ [7.1B]

2. Factor: $2y^4 - 14y^3 - 16y^2$
$2y^2(y + 1)(y - 8)$ [7.2B]

3. Factor $8x^2 + 20x - 48$ by grouping.
$4(x + 4)(2x - 3)$ [7.3B]

4. Factor $6x^2 + 19x + 8$ by using trial factors.
$(2x + 1)(3x + 8)$ [7.3A]

5. Factor: $a^2 - 19a + 48$
$(a - 3)(a - 16)$ [7.2A]

6. Factor: $6x^3 - 8x^2 + 10x$
$2x(3x^2 - 4x + 5)$ [7.1A]

7. Factor: $x^2 + 2x - 15$
$(x + 5)(x - 3)$ [7.2A]

8. Solve: $4x^2 - 1 = 0$
$\frac{1}{2}, -\frac{1}{2}$ [7.5A]

9. Factor: $5x^2 - 45x - 15$
$5(x^2 - 9x - 3)$ [7.1A]

10. Factor: $p^2 + 12p + 36$
$(p + 6)^2$ [7.4A]

11. Solve: $x(x - 8) = -15$
3, 5 [7.5A]

12. Factor: $3x^2 + 12xy + 12y^2$
$3(x + 2y)^2$ [7.4D]

13. Factor: $27x^3 - 8$
$(3x - 2)(9x^2 + 6x + 4)$ [7.4B]

14. Factor $6x^2y^2 + 9xy^2 + 3y^2$ by grouping.
$3y^2(2x + 1)(x + 1)$ [7.3B]

426 Chapter 7 / Factoring

15. Factor: $6a^4 - 13a^2 - 5$
$(2a^2 - 5)(3a^2 + 1)$ [7.4C]

16. Factor: $a(x - 2) + b(x - 2)$
$(x - 2)(a + b)$ [7.1B]

17. Factor: $x(p + 1) - (p + 1)$
$(p + 1)(x - 1)$ [7.1B]

18. Factor: $3a^2 - 75$
$3(a + 5)(a - 5)$ [7.4D]

19. Factor $2x^2 + 4x - 5$ by using trial factors.
nonfactorable over the integers [7.3A]

20. Factor: $x^2 - 9x - 36$
$(x + 3)(x - 12)$ [7.2A]

21. Factor: $4a^2 - 12ab + 9b^2$
$(2a - 3b)^2$ [7.4A]

22. Factor: $4x^2 - 49y^2$
$(2x + 7y)(2x - 7y)$ [7.4A]

23. Solve: $(2a - 3)(a + 7) = 0$
$\frac{3}{2}, -7$ [7.5A]

24. The length of a rectangle is 3 cm longer than twice its width. The area of the rectangle is 90 cm². Find the length and width of the rectangle.
length: 15 cm; width: 6 cm [7.5B]

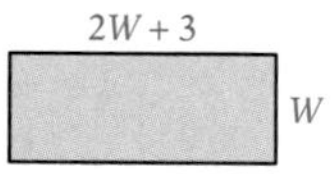

25. The sum of two numbers is ten. The sum of the squares of the two numbers is fifty-eight. Find the two numbers.
3, 7 [7.5B]

427

Cumulative Review

1. Subtract: $-2 - (-3) - 5 - (-11)$
   7 [1.1C]

2. Simplify: $(3 - 7)^2 \div (-2) - 3(-4)$
   4 [1.3A]

3. Evaluate $-2a^2 \div (2b) - c$ when $a = -4$, $b = 2$, and $c = -1$.
   -7 [1.4A]

4. Multiply: $-\frac{3}{4}(-20x^2)$
   $15x^2$ [1.4C]

5. Simplify: $-2[4x - 2(3 - 2x) - 8x]$
   12 [1.4D]

6. Solve: $-\frac{5}{7}x = -\frac{10}{21}$
   $\frac{2}{3}$ [2.1C]

7. Solve: $3x - 2 = 12 - 5x$
   $\frac{7}{4}$ [2.2B]

8. Solve: $-2 + 4[3x - 2(4 - x) - 3] = 4x + 2$
   3 [2.2C]

9. 120% of what number is 54?
   45 [2.1D]

10. Given $f(x) = -x^2 + 3x - 1$, find $f(2)$.
    1 [4.2A]

11. Graph $y = \frac{1}{4}x + 3$.

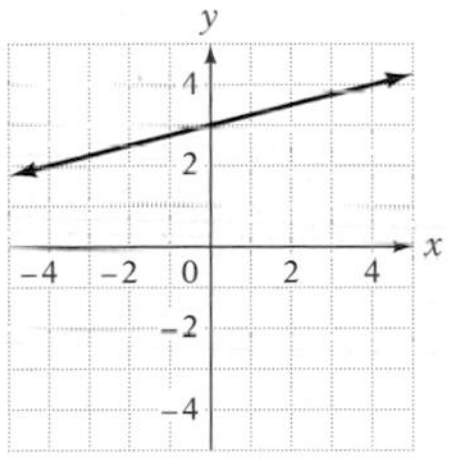

[4.3A]

12. Graph $5x + 3y = 15$.

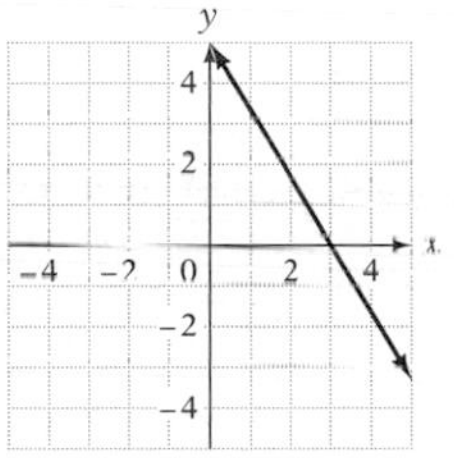

[4.3B]

13. Find the equation of the line that contains the point $(-3, 4)$ and has slope $\frac{2}{3}$.
    $y = \frac{2}{3}x + 6$ [4.5A]

14. Solve by substitution: $8x - y = 2$
    $y = 5x + 1$
    (1, 6) [5.1B]

428 Chapter 7 / Factoring

15. Solve by the addition method:
$5x + 2y = -9$
$12x - 7y = 2$
$(-1, -2)$ [5.2A]

16. Simplify: $(-3a^3b^2)^2$
$9a^6b^4$ [6.1A]

17. Multiply: $(x + 2)(x^2 - 5x + 4)$
$x^3 - 3x^2 - 6x + 8$ [6.3B]

18. Divide: $(8x^2 + 4x - 3) \div (2x - 3)$
$4x + 8 + \frac{21}{2x - 3}$ [6.4A]

19. Simplify: $(x^{-4}y^3)^2$
$\frac{y^6}{x^8}$ [6.1B]

20. Factor: $3a - 3b - ax + bx$
$(a - b)(3 - x)$ [7.1B]

21. Factor: $15xy^2 - 20xy^4$
$5xy^2(3 - 4y^2)$ [7.1A]

22. Factor: $x^2 - 5xy - 14y^2$
$(x - 7y)(x + 2y)$ [7.2A]

23. Solve: $3x^2 + 19x - 14 = 0$
$\frac{2}{3}, -7$ [7.5A]

24. Solve: $6x^2 + 60 = 39x$
$\frac{5}{2}, 4$ [7.5A]

25. A triangle has a 31° angle and a right angle. Find the measure of the third angle.
59° [3.1C]

26. A rectangular flower garden has a perimeter of 86 ft. The length of the garden is 28 ft. What is the width of the garden?
15 ft [3.2A]

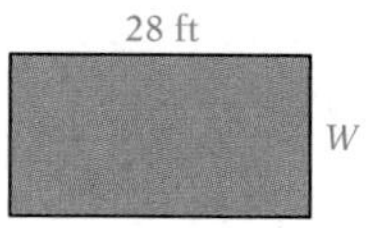

27. A board 10 ft long is cut into two pieces. Four times the length of the shorter piece is 2 ft less than three times the length of the longer piece. Find the length of each piece.
shorter piece: 4 ft; longer piece: 6 ft [2.2D]

28. An investment of \$4000 was made at an annual simple interest rate of 8%. How much more money was invested at an annual simple interest rate of 11% if the total interest earned in 1 year was \$1035?
\$6500 [2.3C]

29. A family drove to a resort at an average speed of 42 mph and later returned over the same road at an average speed of 56 mph. Find the distance to the resort if the total driving time was 7 h.
168 mi [2.3D]

30. The length of the base of a triangle is three times the height. The area of the triangle is 24 in². Find the length of the base of the triangle.
12 in. [7.5B]

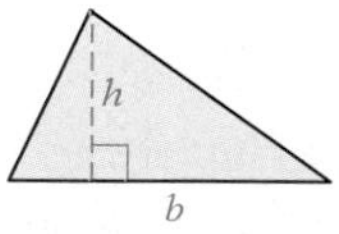

Chapter 8

Rational Expressions

In order to monitor species that are or are becoming endangered, scientists need to determine the present population of that species. Scientists catch and tag a certain number of the animals and then release them. Later, a group of the animals from that same habitat is caught and the number tagged is counted. A proportion is used to estimate the total population size in that region, as shown in **Exercise 70 on page 465.** Tracking the tagged animals also assists scientists in learning more about the habits of that species.

Objectives

Section 8.1

A To simplify a rational expression

B To multiply rational expressions

C To divide rational expressions

Section 8.2

A To find the least common multiple (LCM) of two or more polynomials

B To express two fractions in terms of the LCM of their denominators

C To add or subtract rational expressions with the same denominator

D To add or subtract rational expressions with different denominators

Section 8.3

A To simplify a complex fraction

Section 8.4

A To solve rational equations

B To solve proportions

C To solve problems involving similar triangles

D To solve application problems

Section 8.5

A To solve a literal equation for one of the variables

Section 8.6

A To solve work problems

B To solve uniform motion problems

Section 8.7

A To solve variation problems

WWW WEB Need help? For on-line student resources, such as section quizzes, visit this textbook's web site at **college.hmco.com/students.**

Prep Test

1. Find the LCM of 10 and 25.
50 [1.2C]

For Exercises 2 to 5, add, subtract, multiply, or divide.

2. $-\frac{3}{8} \cdot \frac{4}{9}$
$-\frac{1}{6}$ [1.2D]

3. $-\frac{4}{5} \div \frac{8}{15}$
$-\frac{3}{2}$ [1.2D]

4. $-\frac{5}{6} + \frac{7}{8}$
$\frac{1}{24}$ [1.2C]

5. $-\frac{3}{8} - \left(-\frac{7}{12}\right)$
$\frac{5}{24}$ [1.2C]

6. Evaluate $\frac{2x - 3}{x^2 - x + 1}$ for $x = 2$.
$\frac{1}{3}$ [1.4A]

7. Solve: $4(2x + 1) = 3(x - 2)$
-2 [2.2C]

8. Solve: $10\left(\frac{t}{2} + \frac{t}{5}\right) = 10(1)$
$\frac{10}{7}$ [2.2C]

9. Two planes start from the same point and fly in opposite directions. The first plane is flying 20 mph slower than the second plane. In 2 h, the planes are 480 mi apart. Find the rate of each plane.
110 mph, 130 mph [2.3D]

Go Figure

If 6 machines can fill 12 boxes of cereal in 7 min, how many boxes of cereal can 14 machines fill in 12 min?
48 boxes

8.1 Multiplication and Division of Rational Expressions

Objective A **To simplify a rational expression**

A fraction in which the numerator and denominator are polynomials is called a **rational expression**. Examples of rational expressions are shown at the right.

$$\frac{5}{z}, \quad \frac{x^2+1}{2x-1}, \quad \frac{y^2+y-1}{4y^2+1}$$

Care must be exercised with a rational expression to ensure that when the variables are replaced with numbers, the resulting denominator is not zero.

Consider the rational expression at the right. The value of x cannot be 3, because the denominator would then be zero.

$$\frac{4x^2-9}{2x-6}$$

$$\frac{4(3)^2-9}{2(3)-6} = \frac{27}{0} \quad \text{Not a real number}$$

A rational expression is in simplest form when the numerator and denominator have no common factors. The Multiplication Property of One is used to write a rational expression in simplest form.

➡ Simplify: $\frac{x^2-4}{x^2-2x-8}$

$$\frac{x^2-4}{x^2-2x-8} = \frac{(x-2)(x+2)}{(x-4)(x+2)}$$

• Factor the numerator and denominator.

$$= \frac{x-2}{x-4}\cdot\boxed{\frac{x+2}{x+2}} = \frac{x-2}{x-4}\cdot 1$$

$$= \frac{x-2}{x-4}, x \neq -2, 4$$

• The restrictions $x \neq -2, 4$ are necessary to prevent division by zero.

This simplification is usually shown with slashes through the common factors. The last simplification would be shown as follows:

$$\frac{x^2-4}{x^2-2x-8} = \frac{(x-2)\overset{1}{\cancel{(x+2)}}}{(x-4)\underset{1}{\cancel{(x+2)}}}$$

• Factor the numerator and denominator.

$$= \frac{x-2}{x-4}, x \neq -2, 4$$

• Divide by the common factors.

➡ Simplify: $\frac{10+3x-x^2}{x^2-4x-5}$

$$\frac{10+3x-x^2}{x^2-4x-5} = \frac{-(x^2-3x-10)}{x^2-4x-5}$$

• Factor −1 from the trinomial in the numerator.

$$= \frac{-(x+2)\overset{1}{\cancel{(x-5)}}}{(x+1)\underset{1}{\cancel{(x-5)}}}$$

• Factor the numerator and denominator. Divide by the common factors.

$$= -\frac{x+2}{x+1}, x \neq -1, 5$$

Objective 8.1A

New Vocabulary

rational expression
simplest form

Discuss the Concepts

Explain how to find the value of x that makes the denominator of the rational expression $\frac{5x}{3x+6}$ equal to 0.

Instructor Note

Simplifying a rational expression is closely related to simplifying a rational number: The common factors are removed. Making this connection will help some students.

Optional Student Activity

Have students evaluate $\frac{x^2-4}{x^2-2x-8}$ and $\frac{x-2}{x-4}$ for various values of x and determine that except for −2 and 4, the expressions are equal.

Concept Check

Explain why $x \neq 5$ in the example at the left, even though the rational expression $\frac{x+2}{x+1}$ is defined for $x = 5$.

In-Class Examples (Objective 8.1A)

Simplify.

1. $\frac{18x^5y^2}{12xy^3} \; \frac{3x^4}{2y}$
2. $\frac{x^2-1}{x^2+4x-5} \; \frac{x+1}{x+5}$
3. $\frac{a^2-2a}{4-2a} \; -\frac{a}{2}$

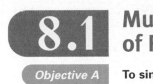

Concept Check

Simplify.

1. $\dfrac{(x+2)^3}{(x+2)^2}$ $x+2$
2. $\dfrac{4x-3}{3-4x}$ -1
3. $\dfrac{4a^2(a-b)}{6a(b-a)}$ $-\dfrac{2a}{3}$

Concept Check

Write two different rational expressions that equal $\dfrac{3a}{4}$ when simplified. Answers will vary. One example is $\dfrac{6a^2}{8a}$.

Instructor Note

It is important to emphasize that the numerator and denominator must be written in factored form before simplifying. This will help students avoid errors such as

$$\frac{x^2-x}{x^2}=\frac{\cancel{x^2}-x}{\cancel{x^2}}=1-x$$

Objective 8.1B

Discuss the Concepts

Have students describe the steps involved in simplifying the product of two rational expressions.

Concept Check

Simplify.

1. $\dfrac{8x^2}{y}\cdot\dfrac{y}{12x}$ $\dfrac{2x}{3}$
2. $\dfrac{a-b}{a}\cdot\dfrac{3a}{b-a}$ -3

For the remaining examples, we will omit the restrictions on the variables that prevent division by zero and assume that the values of the variables are such that division by zero is not possible.

Example 1

Simplify: $\dfrac{4x^3y^4}{6x^4y}$

Solution

$\dfrac{4x^3y^4}{6x^4y}=\dfrac{2y^3}{3x}$ • Use rules of exponents.

You Try It 1

Simplify: $\dfrac{6x^5y}{12x^2y^3}$

Your solution

$\dfrac{x^3}{2y^2}$

Example 2

Simplify: $\dfrac{9-x^2}{x^2+x-12}$

Solution

$$\frac{9-x^2}{x^2+x-12}=\frac{-(x^2-9)}{x^2+x-12}$$

$$=\frac{-(x+3)\overset{1}{\cancel{(x-3)}}}{\underset{1}{\cancel{(x-3)}}(x+4)}=-\frac{x+3}{x+4}$$

You Try It 2

Simplify: $\dfrac{x^2+2x-24}{16-x^2}$

Your solution

$-\dfrac{x+6}{x+4}$

Example 3

Simplify: $\dfrac{x^2+2x-15}{x^2-7x+12}$

Solution

$$\frac{x^2+2x-15}{x^2-7x+12}=\frac{(x+5)\overset{1}{\cancel{(x-3)}}}{\underset{1}{\cancel{(x-3)}}(x-4)}=\frac{x+5}{x-4}$$

You Try It 3

Simplify: $\dfrac{x^2+4x-12}{x^2-3x+2}$

Your solution

$\dfrac{x+6}{x-1}$

Solutions on pp. S22–S23

Objective B **To multiply rational expressions**

VIDEO & DVD · CD TUTOR · WEB · SSM

The product of two fractions is a fraction whose numerator is the product of the numerators of the two fractions and whose denominator is the product of the denominators of the two fractions.

Rule for Multiplying Fractions

If $\dfrac{a}{b}$ and $\dfrac{c}{d}$ are fractions and $b \neq 0$, $d \neq 0$, then $\dfrac{a}{b}\cdot\dfrac{c}{d}=\dfrac{ac}{bd}$.

$$\frac{2}{3}\cdot\frac{4}{5}=\frac{8}{15} \qquad \frac{3x}{y}\cdot\frac{2}{z}=\frac{6x}{yz} \qquad \frac{x+2}{x}\cdot\frac{3}{x-2}=\frac{3x+6}{x^2-2x}$$

➡ Multiply: $\frac{x^2+3x}{x^2-3x-4} \cdot \frac{x^2-5x+4}{x^2+2x-3}$

$$\frac{x^2+3x}{x^2-3x-4} \cdot \frac{x^2-5x+4}{x^2+2x-3}$$

$$= \frac{x(x+3)}{(x-4)(x+1)} \cdot \frac{(x-4)(x-1)}{(x+3)(x-1)}$$

• Factor the numerator and denominator of each fraction.

$$= \frac{x\overset{1}{\cancel{(x+3)}}\overset{1}{\cancel{(x-4)}}\overset{1}{\cancel{(x-1)}}}{\underset{1}{\cancel{(x-4)}}(x+1)\underset{1}{\cancel{(x+3)}}\underset{1}{\cancel{(x-1)}}}$$

• Multiply.

$$= \frac{x}{x+1}$$

• Write the answer in simplest form.

Example 4

Multiply: $\frac{10x^2-15x}{12x-8} \cdot \frac{3x-2}{20x-25}$

Solution

$$\frac{10x^2-15x}{12x-8} \cdot \frac{3x-2}{20x-25}$$

$$= \frac{5x(2x-3)}{4(3x-2)} \cdot \frac{(3x-2)}{5(4x-5)}$$

$$= \frac{\overset{1}{\cancel{5}}x(2x-3)\overset{1}{\cancel{(3x-2)}}}{4\underset{1}{\cancel{(3x-2)}}\underset{1}{\cancel{5}}(4x-5)} = \frac{x(2x-3)}{4(4x-5)}$$

You Try It 4

Multiply: $\frac{12x^2+3x}{10x-15} \cdot \frac{8x-12}{9x+18}$

Your solution

$$\frac{4x(4x+1)}{15(x+2)}$$

Example 5

Multiply: $\frac{x^2+x-6}{x^2+7x+12} \cdot \frac{x^2+3x-4}{4-x^2}$

Solution

$$\frac{x^2+x-6}{x^2+7x+12} \cdot \frac{x^2+3x-4}{4-x^2}$$

$$= \frac{(x+3)(x-2)}{(x+3)(x+4)} \cdot \frac{(x+4)(x-1)}{(2-x)(2+x)}$$

$$= \frac{\overset{1}{\cancel{(x+3)}}\overset{-1}{\cancel{(x-2)}}\overset{1}{\cancel{(x+4)}}(x-1)}{\underset{1}{\cancel{(x+3)}}\underset{1}{\cancel{(x+4)}}\underset{1}{\cancel{(2-x)}}(2+x)} = -\frac{x-1}{x+2}$$

You Try It 5

Multiply: $\frac{x^2+2x-15}{9-x^2} \cdot \frac{x^2-3x-18}{x^2-7x+6}$

Your solution

$$-\frac{x+5}{x-1}$$

Solutions on p. S23

Instructor Note

Remind students that when they carry out the multiplication step, writing the product as a single fraction, they should leave the numerator and denominator in factored form. The simplified answer, too, may be left in factored form, as in Example 4. (Note, however, that in Objective 8.2B, students will need to multiply out the numerators when they apply the skill of multiplying rational expressions to the skill of writing fractions in terms of the LCM of their denominators.)

Optional Student Activity

1. Find two rational expressions whose product is 1. Answers will vary; for example, $\frac{x+1}{5}$ and $\frac{5}{x+1}$.
2. Find two rational expressions whose product is $\frac{12a}{5b^3}$. Answers will vary; for example, $\frac{3a}{5b}$ and $\frac{4}{b^2}$.

In-Class Examples (Objective 8.1B)

Simplify.

1. $\frac{28a^5b^7}{5x^4} \cdot \frac{15x^2}{14ab} \quad \frac{6a^4b^6}{x^2}$
2. $\frac{6x^2-10x}{3-3x} \cdot \frac{x^2-1}{12x-20} \quad -\frac{x(x+1)}{6}$

Objective 8.1C

Vocabulary to Review
reciprocal

Discuss the Concepts
Express each division problem as a multiplication problem.

1. $\frac{3}{a} \div \frac{b}{7}$ $\quad \frac{3}{a} \cdot \frac{7}{b}$
2. $\frac{5x}{y} \div \frac{2x}{4y}$ $\quad \frac{5x}{y} \cdot \frac{4y}{2x}$
3. $\frac{x-1}{x} \div \frac{x+3}{x}$
   $\frac{x-1}{x} \cdot \frac{x}{x+3}$

Optional Student Activity

1. What should $\frac{5x^2}{3}$ be divided by to yield a quotient of 10? $\frac{x^2}{6}$
2. What should $\frac{2b}{a^3 - a^2}$ be divided by to yield a quotient of $\frac{2}{a-1}$? $\frac{b}{a^2}$

Objective C To divide rational expressions

VIDEO & DVD CD TUTOR WEB SSM

The reciprocal of a fraction is a fraction with the numerator and denominator interchanged.

Fraction	Reciprocal
$\frac{a}{b}$	$\frac{b}{a}$
$x^2 = \frac{x^2}{1}$	$\frac{1}{x^2}$
$\frac{x+2}{x}$	$\frac{x}{x+2}$

Rule for Dividing Fractions

Divide fractions by multiplying the dividend by the reciprocal of the divisor. $\quad \frac{a}{b} \div \frac{c}{d} = \frac{a}{b} \cdot \frac{d}{c} = \frac{ad}{bc}$

$$\frac{4}{x} \div \frac{y}{5} = \frac{4}{x} \cdot \frac{5}{y} = \frac{20}{xy} \qquad \frac{x+4}{x} \div \frac{x-2}{4} = \frac{x+4}{x} \cdot \frac{4}{x-2} = \frac{4(x+4)}{x(x-2)}$$

The basis for the division rule is shown at the right.

$$\frac{a}{b} \div \frac{c}{d} = \frac{\frac{a}{b}}{\frac{c}{d}} = \frac{\frac{a}{b}}{\frac{c}{d}} \cdot \frac{\frac{d}{c}}{\frac{d}{c}} = \frac{\frac{a}{b} \cdot \frac{d}{c}}{1} = \frac{a}{b} \cdot \frac{d}{c}$$

Example 6

Divide: $\frac{xy^2 - 3x^2y}{z^2} \div \frac{6x^2 - 2xy}{z^3}$

Solution

$$\frac{xy^2 - 3x^2y}{z^2} \div \frac{6x^2 - 2xy}{z^3}$$
$$= \frac{xy^2 - 3x^2y}{z^2} \cdot \frac{z^3}{6x^2 - 2xy}$$
$$= \frac{xy\overset{-1}{\cancel{(y - 3x)}} \cdot z^3}{z^2 \cdot 2x\underset{1}{\cancel{(3x - y)}}} = -\frac{yz}{2}$$

You Try It 6

Divide: $\frac{a^2}{4bc^2 - 2b^2c} \div \frac{a}{6bc - 3b^2}$

Your solution

$\frac{3a}{2c}$

Example 7

Divide: $\frac{2x^2 + 5x + 2}{2x^2 + 3x - 2} \div \frac{3x^2 + 13x + 4}{2x^2 + 7x - 4}$

Solution

$$\frac{2x^2 + 5x + 2}{2x^2 + 3x - 2} \div \frac{3x^2 + 13x + 4}{2x^2 + 7x - 4}$$
$$= \frac{2x^2 + 5x + 2}{2x^2 + 3x - 2} \cdot \frac{2x^2 + 7x - 4}{3x^2 + 13x + 4}$$
$$= \frac{(2x+1)\overset{1}{\cancel{(x+2)}}\overset{1}{\cancel{(2x-1)}}\overset{1}{\cancel{(x+4)}}}{\underset{1}{\cancel{(2x-1)}}\underset{1}{\cancel{(x+2)}}(3x+1)\underset{1}{\cancel{(x+4)}}} = \frac{2x+1}{3x+1}$$

You Try It 7

Divide: $\frac{3x^2 + 26x + 16}{3x^2 - 7x - 6} \div \frac{2x^2 + 9x - 5}{x^2 + 2x - 15}$

Your solution

$\frac{x+8}{2x-1}$

Solutions on p. S23

In-Class Examples (Objective 8.1C)
Simplify.

1. $\frac{12a^5b}{7xy^4} \div \frac{9ab^3}{35xy^2}$ $\quad \frac{20a^4}{3b^2y^2}$
2. $\frac{4x-8}{15+2x-x^2} \div \frac{x^2-4}{3x^2-15x}$ $\quad -\frac{12x}{(x+3)(x+2)}$

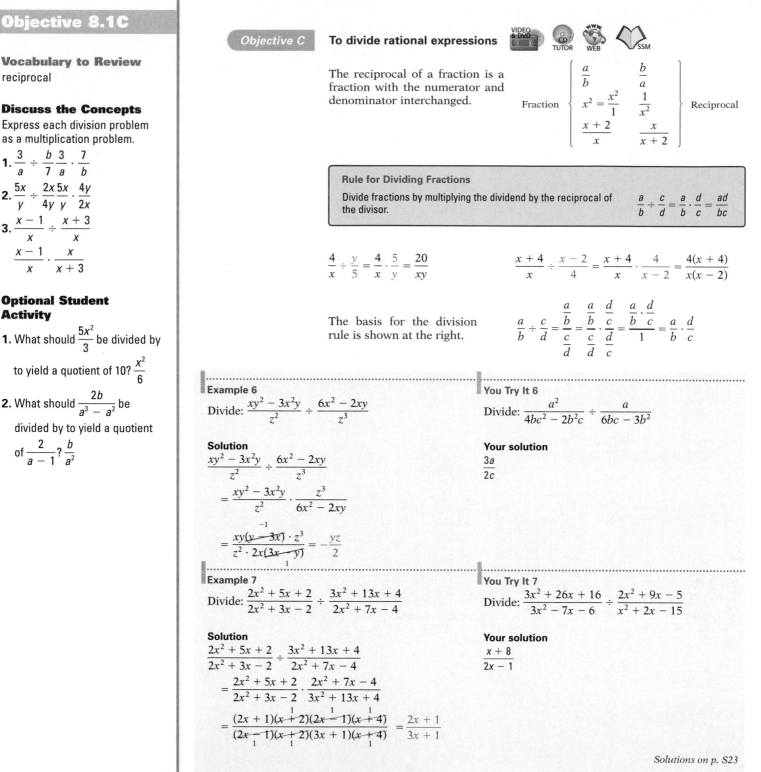

435

8.1 Exercises

Objective A

1. When is a rational expression in simplest form?

2. Are the rational expressions $\frac{x(x-2)}{2(x-2)}$ and $\frac{x}{2}$ equal for all values of x? Why or why not?

Simplify.

3. $\frac{9x^3}{12x^4}$ $\quad \frac{3}{4x}$

4. $\frac{16x^2y}{24xy^3}$ $\quad \frac{2x}{3y^2}$

5. $\frac{(x+3)^2}{(x+3)^3}$ $\quad \frac{1}{x+3}$

6. $\frac{(2x-1)^5}{(2x-1)^4}$ $\quad 2x-1$

7. $\frac{3n-4}{4-3n}$ $\quad -1$

8. $\frac{5-2x}{2x-5}$ $\quad -1$

9. $\frac{6y(y+2)}{9y^2(y+2)}$ $\quad \frac{2}{3y}$

10. $\frac{12x^2(3-x)}{18x(3-x)}$ $\quad \frac{2x}{3}$

11. $\frac{6x(x-5)}{8x^2(5-x)}$ $\quad -\frac{3}{4x}$

12. $\frac{14x^3(7-3x)}{21x(3x-7)}$ $\quad -\frac{2x^2}{3}$

13. $\frac{a^2+4a}{ab+4b}$ $\quad \frac{a}{b}$

14. $\frac{x^2-3x}{2x-6}$ $\quad \frac{x}{2}$

15. $\frac{4-6x}{3x^2-2x}$ $\quad -\frac{2}{x}$

16. $\frac{5xy-3y}{9-15x}$ $\quad -\frac{y}{3}$

17. $\frac{y^2-3y+2}{y^2-4y+3}$ $\quad \frac{y-2}{y-3}$

18. $\frac{x^2+5x+6}{x^2+8x+15}$ $\quad \frac{x+2}{x+5}$

19. $\frac{x^2+3x-10}{x^2+2x-8}$ $\quad \frac{x+5}{x+4}$

20. $\frac{a^2+7a-8}{a^2+6a-7}$ $\quad \frac{a+8}{a+7}$

21. $\frac{x^2+x-12}{x^2-6x+9}$ $\quad \frac{x+4}{x-3}$

22. $\frac{x^2+8x+16}{x^2-2x-24}$ $\quad \frac{x+4}{x-6}$

23. $\frac{x^2-3x-10}{25-x^2}$ $\quad -\frac{x+2}{x+5}$

24. $\frac{4-y^2}{y^2-3y-10}$ $\quad \frac{2-y}{y-5}$

25. $\frac{2x^3+2x^2-4x}{x^3+2x^2-3x}$ $\quad \frac{2(x+2)}{x+3}$

26. $\frac{3x^3-12x}{6x^3-24x^2+24x}$ $\quad \frac{x+2}{2(x-2)}$

27. $\frac{6x^2-7x+2}{6x^2+5x-6}$ $\quad \frac{2x-1}{2x+3}$

28. $\frac{2n^2-9n+4}{2n^2-5n-12}$ $\quad \frac{2n-1}{2n+3}$

29. $\frac{x^2+3x-28}{24-2x-x^2}$ $\quad -\frac{x+7}{x+6}$

Section 8.1

Suggested Assignment

Exercises 1–77, odd
More challenging problems:
Exercises 78–80, 83, 84

Answers to Writing Exercises

1. A rational expression is in simplest form when all common factors of the numerator and denominator are divided out.

2. The rational expression $\frac{x}{2}$ equals 1 when $x = 2$, whereas the expression $\frac{x(x-2)}{2(x-2)}$ is undefined when $x = 2$.

Quick Quiz (Objective 8.1A)

Simplify.

1. $\frac{8ab-2a}{6-24b}$ $\quad -\frac{a}{3}$

2. $\frac{x^2-x-12}{x^2+9x+18}$ $\quad \frac{x-4}{x+6}$

3. $\frac{25-n^2}{n^2+n-30}$ $\quad -\frac{n+5}{n+6}$

Objective B

Multiply.

30. $\dfrac{8x^2}{9y^3} \cdot \dfrac{3y^2}{4x^3}$

$\dfrac{2}{3xy}$

31. $\dfrac{14a^2b^3}{15x^5y^2} \cdot \dfrac{25x^3y}{16ab}$

$\dfrac{35ab^2}{24x^2y}$

32. $\dfrac{12x^3y^4}{7a^2b^3} \cdot \dfrac{14a^3b^4}{9x^2y^2}$

$\dfrac{8xy^2ab}{3}$

33. $\dfrac{18a^4b^2}{25x^2y^3} \cdot \dfrac{50x^5y^6}{27a^6b^2}$

$\dfrac{4x^3y^3}{3a^2}$

34. $\dfrac{3x-6}{5x-20} \cdot \dfrac{10x-40}{27x-54}$

$\dfrac{2}{9}$

35. $\dfrac{8x-12}{14x+7} \cdot \dfrac{42x+21}{32x-48}$

$\dfrac{3}{4}$

36. $\dfrac{3x^2+2x}{2xy-3y} \cdot \dfrac{2xy^3-3y^3}{3x^3+2x^2}$

$\dfrac{y^2}{x}$

37. $\dfrac{4a^2x-3a^2}{2by+5b} \cdot \dfrac{2b^3y+5b^3}{4ax-3a}$

ab^2

38. $\dfrac{x^2+5x+4}{x^3y^2} \cdot \dfrac{x^2y^3}{x^2+2x+1}$

$\dfrac{y(x+4)}{x(x+1)}$

39. $\dfrac{x^2+x-2}{xy^2} \cdot \dfrac{x^3y}{x^2+5x+6}$

$\dfrac{x^2(x-1)}{y(x+3)}$

40. $\dfrac{x^4y^2}{x^2+3x-28} \cdot \dfrac{x^2-49}{xy^4}$

$\dfrac{x^3(x-7)}{y^2(x-4)}$

41. $\dfrac{x^5y^3}{x^2+13x+30} \cdot \dfrac{x^2+2x-3}{x^7y^2}$

$\dfrac{y(x-1)}{x^2(x+10)}$

42. $\dfrac{2x^2-5x}{2xy+y} \cdot \dfrac{2xy^2+y^2}{5x^2-2x^3}$

$-\dfrac{y}{x}$

43. $\dfrac{3a^3+4a^2}{5ab-3b} \cdot \dfrac{3b^3-5ab^3}{3a^2+4a}$

$-ab^2$

44. $\dfrac{x^2-2x-24}{x^2-5x-6} \cdot \dfrac{x^2+5x+6}{x^2+6x+8}$

$\dfrac{x+3}{x+1}$

45. $\dfrac{x^2-8x+7}{x^2+3x-4} \cdot \dfrac{x^2+3x-10}{x^2-9x+14}$

$\dfrac{x+5}{x+4}$

46. $\dfrac{x^2+2x-35}{x^2+4x-21} \cdot \dfrac{x^2+3x-18}{x^2+9x+18}$

$\dfrac{x-5}{x+3}$

47. $\dfrac{y^2+y-20}{y^2+2y-15} \cdot \dfrac{y^2+4y-21}{y^2+3y-28}$

1

Quick Quiz (Objective 8.1B)

Simplify.

1. $\dfrac{x^2-5x+4}{2x^2+5x-3} \cdot \dfrac{x^2-9}{x^2-7x+12}$ $\dfrac{x-1}{2x-1}$

2. $\dfrac{10x^2-50x}{12x+24} \cdot \dfrac{2x+4}{5x-x^2}$ $-\dfrac{5}{3}$

437

48. $\frac{x^2-3x-4}{x^2+6x+5} \cdot \frac{x^2+5x+6}{8+2x-x^2}$
$-\frac{x+3}{x+5}$

49. $\frac{25-n^2}{n^2-2n-35} \cdot \frac{n^2-8n-20}{n^2-3n-10}$
$-\frac{n-10}{n-7}$

50. $\frac{12x^2-6x}{x^2+6x+5} \cdot \frac{2x^4+10x^3}{4x^2-1}$
$\frac{12x^4}{(x+1)(2x+1)}$

51. $\frac{8x^3+4x^2}{x^2-3x+2} \cdot \frac{x^2-4}{16x^2+8x}$
$\frac{x(x+2)}{2(x-1)}$

52. $\frac{16+6x-x^2}{x^2-10x-24} \cdot \frac{x^2-6x-27}{x^2-17x+72}$
$-\frac{x+3}{x-12}$

53. $\frac{x^2-11x+28}{x^2-13x+42} \cdot \frac{x^2+7x+10}{20-x-x^2}$
$-\frac{x+2}{x-6}$

54. $\frac{2x^2+5x+2}{2x^2+7x+3} \cdot \frac{x^2-7x-30}{x^2-6x-40}$
$\frac{x+2}{x+4}$

55. $\frac{x^2-4x-32}{x^2-8x-48} \cdot \frac{3x^2+17x+10}{3x^2-22x-16}$
$\frac{x+5}{x-12}$

Objective C

Divide.

56. $\frac{4x^2y^3}{15a^2b^3} \div \frac{6xy}{5a^3b^5}$
$\frac{2xy^2ab^2}{9}$

57. $\frac{9x^3y^4}{16a^4b^2} \div \frac{45x^4y^2}{14a^7b}$
$\frac{7a^3y^2}{40bx}$

58. $\frac{6x-12}{8x+32} \div \frac{18x-36}{10x+40}$
$\frac{5}{12}$

59. $\frac{28x+14}{45x-30} \div \frac{14x+7}{30x-20}$
$\frac{4}{3}$

60. $\frac{6x^3+7x^2}{12x-3} \div \frac{6x^2+7x}{36x-9}$
$3x$

61. $\frac{5a^2y+3a^2}{2x^3+5x^2} \div \frac{10ay+6a}{6x^3+15x^2}$
$\frac{3a}{2}$

62. $\frac{x^2+4x+3}{x^2y} \div \frac{x^2+2x+1}{xy^2}$
$\frac{y(x+3)}{x(x+1)}$

63. $\frac{x^3y^2}{x^2-3x-10} \div \frac{xy^4}{x^2-x-20}$
$\frac{x^2(x+4)}{y^2(x+2)}$

64. $\frac{x^2-49}{x^4y^3} \div \frac{x^2-14x+49}{x^4y^3}$
$\frac{x+7}{x-7}$

65. $\frac{x^2y^5}{x^2-11x+30} \div \frac{xy^6}{x^2-7x+10}$
$\frac{x(x-2)}{y(x-6)}$

Quick Quiz (Objective 8.1C)

Simplify.

1. $\frac{a^3b}{a^2-5a-14} \div \frac{ab^6}{a^2-3a-28}$ $\frac{a^2(a+4)}{b^5(a+2)}$

2. $\frac{10x+10}{3x-6} \div \frac{2x+2}{3xy-6y}$ $5y$

Answers to Writing Exercises

76. It is not possible to choose a value of x such that $\frac{9}{x^2+1}$ is greater than 10 because $x^2 + 1$ is greater than 1 for all nonzero values of x.

77. Choosing a value of y very close to 3 makes $y - 3$ very close to 0. Dividing 1 by a number close to 0 produces a very large number. For instance, when $y = 3.00000001$, $\frac{1}{y-3}$ is greater than 10,000,000.

66. $\frac{4ax - 8a}{c^2} \div \frac{2y - xy}{c^3}$
$-\frac{4ac}{y}$

67. $\frac{3x^2y - 9xy}{a^2b} \div \frac{3x^2 - x^3}{ab^2}$
$-\frac{3by}{ax}$

68. $\frac{x^2 - 5x + 6}{x^2 - 9x + 18} \div \frac{x^2 - 6x + 8}{x^2 - 9x + 20}$
$\frac{x-5}{x-6}$

69. $\frac{x^2 + 3x - 40}{x^2 + 2x - 35} \div \frac{x^2 + 2x - 48}{x^2 + 3x - 18}$
$\frac{(x-3)(x+6)}{(x+7)(x-6)}$

70. $\frac{x^2 + 2x - 15}{x^2 - 4x - 45} \div \frac{x^2 + x - 12}{x^2 - 5x - 36}$
1

71. $\frac{y^2 - y - 56}{y^2 + 8y + 7} \div \frac{y^2 - 13y + 40}{y^2 - 4y - 5}$
1

72. $\frac{8 + 2x - x^2}{x^2 + 7x + 10} \div \frac{x^2 - 11x + 28}{x^2 - x - 42}$
$-\frac{x+6}{x+5}$

73. $\frac{x^2 - x - 2}{x^2 - 7x + 10} \div \frac{x^2 - 3x - 4}{40 - 3x - x^2}$
$-\frac{x+8}{x-4}$

74. $\frac{2x^2 - 3x - 20}{2x^2 - 7x - 30} \div \frac{2x^2 - 5x - 12}{4x^2 + 12x + 9}$
$\frac{2x+3}{x-6}$

75. $\frac{6n^2 + 13n + 6}{4n^2 - 9} \div \frac{6n^2 + n - 2}{4n^2 - 1}$
$\frac{2n+1}{2n-3}$

APPLYING THE CONCEPTS

76. Given the expression $\frac{9}{x^2+1}$, choose some values of x and evaluate the expression for those values. Is it possible to choose a value of x for which the value of the expression is greater than 10? If so, give such a value. If not, explain why it is not possible.

77. Given the expression $\frac{1}{y-3}$, choose some values of y and evaluate the expression for those values. Is it possible to choose a value of y for which the value of the expression is greater than 10,000,000? If so, give such a value. If not, explain why it is not possible.

Complete each simplification.

78. $\frac{8x}{9y} \div \underline{\ ?\ } = \frac{10y}{9}$ $\quad \frac{4x}{5y^2}$

79. $\frac{n}{n+3} \div \underline{\ ?\ } = \frac{n}{n-2}$ $\quad \frac{n-2}{n+3}$

Simplify.

80. $\frac{xy}{3} \cdot \frac{x}{y^2} \div \frac{x}{4}$ $\quad \frac{4x}{3y}$

81. $\left(\frac{y}{3}\right) \div \left(\frac{y}{2} \cdot \frac{y}{4}\right)$ $\quad \frac{8}{3y}$

82. $\left(\frac{x-4}{y^2}\right)^3 \cdot \left(\frac{y}{4-x}\right)^3$ $\quad -\frac{1}{y^3}$

83. $\frac{x-2}{x+5} \div \frac{x-3}{x+5} \cdot \frac{x-3}{x-2}$ $\quad 1$

8.2 Addition and Subtraction of Rational Expressions

Objective A

To find the least common multiple (LCM) of two or more polynomials

VIDEO & DVD CD TUTOR WEB SSM

The **least common multiple (LCM)** of two or more numbers is the smallest number that contains the prime factorization of each number.

TAKE NOTE
The LCM of 12 and 18 is 36 because 36 is the smallest number that both 12 and 18 divide evenly into.

The LCM of 12 and 18 is 36 because 36 contains the prime factors of 12 and the prime factors of 18.

$$12 = 2 \cdot 2 \cdot 3$$
$$18 = 2 \cdot 3 \cdot 3$$
$$\text{LCM} = 36 = \overbrace{2 \cdot 2 \cdot 3}^{\text{Factors of 12}} \cdot 3 \quad (\underbrace{2 \cdot 3 \cdot 3}_{\text{Factors of 18}})$$

The least common multiple of two or more polynomials is the polynomial of least degree that contains the factors of each polynomial.

To find the LCM of two or more polynomials, first factor each polynomial completely. The LCM is the product of each factor the greatest number of times it occurs in any one factorization.

➡ Find the LCM of $4x^2 + 4x$ and $x^2 + 2x + 1$.

TAKE NOTE
The LCM must contain the factors of each polynomial. As shown with the braces at the right, the LCM contains the factors of $4x^2 + 4x$ and the factors of $x^2 + 2x + 1$.

The LCM of the polynomials is the product of the LCM of the numerical coefficients and each variable factor the greatest number of times it occurs in any one factorization.

$$4x^2 + 4x = 4x(x + 1) = 2 \cdot 2 \cdot x(x + 1)$$
$$x^2 + 2x + 1 = (x + 1)(x + 1)$$
$$\text{LCM} = \overbrace{2 \cdot 2 \cdot x(x + 1)}^{\text{Factors of } 4x^2 + 4x}(x + 1) = 4x(x + 1)(x + 1)$$
(Factors of $x^2 + 2x + 1$: $(x + 1)(x + 1)$)

Example 1
Find the LCM of $4x^2y$ and $6xy^2$.

Solution
$4x^2y = 2 \cdot 2 \cdot x \cdot x \cdot y$
$6xy^2 = 2 \cdot 3 \cdot x \cdot y \cdot y$
$\text{LCM} = 2 \cdot 2 \cdot 3 \cdot x \cdot x \cdot y \cdot y = 12x^2y^2$

You Try It 1
Find the LCM of $8uv^2$ and $12uw$.

Your solution
$24uv^2w$

Example 2
Find the LCM of $x^2 - x - 6$ and $9 - x^2$.

Solution
$x^2 - x - 6 = (x - 3)(x + 2)$
$9 - x^2 = -(x^2 - 9) = -(x + 3)(x - 3)$
$\text{LCM} = (x - 3)(x + 2)(x + 3)$

You Try It 2
Find the LCM of $m^2 - 6m + 9$ and $m^2 - 2m - 3$.

Your solution
$(m - 3)(m - 3)(m + 1)$

Solutions on p. S23

Objective 8.2A

New Vocabulary
least common multiple (LCM)

Discuss the Concepts
Have students describe the steps in the process of finding the LCM of two polynomials.

Concept Check
Find the LCM of the polynomials.
1. $3ab^3$ and $6a^2b$ $6a^2b^3$
2. $(x + 2)^2$ and $(x - 1)(x + 2)$ $(x + 2)^2(x - 1)$

Optional Student Activity
Find the LCM of $10x^5y$, $15y^3$, and $4x^2$. $60x^5y^3$

Instructor Note
In Example 2, point out that the factor -1 is not included in the LCM. The reason for this will be clearer to students in Objectives 8.2B and 8.2D.

In-Class Examples (Objective 8.2A)
Find the LCM of the polynomials.
1. $24a^3b^2$ and $9ab^5$ $72a^3b^5$
2. $2x^2 - 4x$ and $x^2 - 4$ $2x(x - 2)(x + 2)$
3. $(3x + 1)^2$ and $(x - 3)(3x + 1)$ $(3x + 1)^2(x - 3)$
4. $x^2 - 3x - 40$ and $8 + 7x - x^2$ $(x - 8)(x + 5)(x + 1)$

Objective 8.2B

Discuss the Concepts

Refer to the example at the right.

1. Explain why the LCM of the denominators is $12x^2(x-2)$.
2. Explain how to determine that the first fraction must be multiplied by $\frac{3(x-2)}{3(x-2)}$.
3. Explain how to determine that the second fraction must be multiplied by $\frac{2x}{2x}$.

Instructor Note

In Objective 8.1B, answers were left in factored form. Point out that now the numerators should be multiplied out (this is in preparation for actually adding fractions, as in Objective 8.3B).

Concept Check

Use the fractions $\frac{5}{a^2(a-4)}$ and $\frac{3}{a(a^2-16)}$.

1. Find the LCM of the denominators. $a^2(a-4)(a+4)$
2. What must the first fraction be multiplied by to change its denominator into the LCM? $\frac{a+4}{a+4}$
3. What must the second fraction be multiplied by to change its denominator into the LCM? $\frac{a}{a}$

Optional Student Activity

Write these three fractions in terms of the LCM of their denominators: $\frac{x-3}{x^2-1}$, $\frac{5x}{2-x-x^2}$, and $\frac{2x+1}{x^2+3x+2}$.

$\frac{(x-3)(x+2)}{(x+1)(x-1)(x+2)}$, $-\frac{5x(x+1)}{(x+1)(x-1)(x+2)}$, $\frac{(2x+1)(x-1)}{(x+1)(x-1)(x+2)}$

Objective B **To express two fractions in terms of the LCM of their denominators**

VIDEO & DVD CD TUTOR WEB SSM

When adding and subtracting fractions, it is frequently necessary to express two or more fractions in terms of a common denominator. This common denominator is the LCM of the denominators of the fractions.

➡ Write the fractions $\frac{x+1}{4x^2}$ and $\frac{x-3}{6x^2-12x}$ in terms of the LCM of the denominators.

> **TAKE NOTE**
> $\frac{3(x-2)}{3(x-2)} = 1$ and $\frac{2x}{2x} = 1$. We are multiplying each fraction by 1, so we are not changing the value of either fraction.

Find the LCM of the denominators. — The LCM is $12x^2(x-2)$.

For each fraction, multiply the numerator and denominator by the factors whose product with the denominator is the LCM.

$$\frac{x+1}{4x^2} = \frac{x+1}{4x^2}\cdot\frac{3(x-2)}{3(x-2)} = \frac{3x^2-3x-6}{12x^2(x-2)}$$

$$\frac{x-3}{6x^2-12x} = \frac{x-3}{6x(x-2)}\cdot\frac{2x}{2x} = \frac{2x^2-6x}{12x^2(x-2)}$$

LCM

Example 3

Write the fractions $\frac{x+2}{3x^2}$ and $\frac{x-1}{8xy}$ in terms of the LCM of the denominators.

Solution

The LCM is $24x^2y$.

$$\frac{x+2}{3x^2} = \frac{x+2}{3x^2}\cdot\frac{8y}{8y} = \frac{8xy+16y}{24x^2y}$$

$$\frac{x-1}{8xy} = \frac{x-1}{8xy}\cdot\frac{3x}{3x} = \frac{3x^2-3x}{24x^2y}$$

You Try It 3

Write the fractions $\frac{x-3}{4xy^2}$ and $\frac{2x+1}{9y^2z}$ in terms of the LCM of the denominators.

Your solution

$\frac{9xz-27z}{36xy^2z}$, $\frac{8x^2+4x}{36xy^2z}$

Example 4

Write the fractions $\frac{2x-1}{2x-x^2}$ and $\frac{x}{x^2+x-6}$ in terms of the LCM of the denominators.

Solution

$$\frac{2x-1}{2x-x^2} = \frac{2x-1}{-(x^2-2x)} = -\frac{2x-1}{x^2-2x}$$

The LCM is $x(x-2)(x+3)$.

$$\frac{2x-1}{2x-x^2} = -\frac{2x-1}{x(x-2)}\cdot\frac{x+3}{x+3} = -\frac{2x^2+5x-3}{x(x-2)(x+3)}$$

$$\frac{x}{x^2+x-6} = \frac{x}{(x-2)(x+3)}\cdot\frac{x}{x} = \frac{x^2}{x(x-2)(x+3)}$$

You Try It 4

Write the fractions $\frac{x+4}{x^2-3x-10}$ and $\frac{2x}{25-x^2}$ in terms of the LCM of the denominators.

Your solution

$\frac{x^2+9x+20}{(x+2)(x-5)(x+5)}$, $-\frac{2x^2+4x}{(x+2)(x-5)(x+5)}$

Solutions on p. S23

In-Class Examples (Objective 8.2B)

Write each fraction in terms of the LCM of the denominators.

1. $\frac{8}{3x^2y}$ and $\frac{x}{6xy^3}$ $\frac{16y^2}{6x^2y^3}$ and $\frac{x^2}{6x^2y^3}$
2. $\frac{a^3}{a(a+1)^2}$ and $\frac{2}{a^2(a+1)}$ $\frac{a^4}{a^2(a+1)^2}$ and $\frac{2a+2}{a^2(a+1)^2}$
3. $\frac{x+2}{x^2-3x-4}$ and $\frac{1}{16-x^2}$ $\frac{x^2+6x+8}{(x-4)(x+1)(x+4)}$ and $-\frac{x+1}{(x-4)(x+1)(x+4)}$

Objective C To add or subtract rational expressions with the same denominator

When adding rational expressions in which the denominators are the same, add the numerators. The denominator of the sum is the common denominator.

$$\frac{a}{b}+\frac{c}{b}=\frac{a+c}{b}$$

$$\frac{5x}{18}+\frac{7x}{18}=\frac{5x+7x}{18}=\frac{12x}{18}=\frac{2x}{3}$$

$$\frac{x}{x^2-1}+\frac{1}{x^2-1}=\frac{x+1}{x^2-1}=\frac{\overset{1}{\cancel{(x+1)}}}{(x-1)\underset{1}{\cancel{(x+1)}}}=\frac{1}{x-1}$$

Note that the sum is written in simplest form.

When subtracting rational expressions in which the denominators are the same, subtract the numerators. The denominator of the difference is the common denominator. Write the answer in simplest form.

> **TAKE NOTE**
> Be careful with signs when subtracting algebraic fractions. Note that we must subtract the *entire* numerator $2x + 3$.
> $(3x - 1) - (2x + 3) = 3x - 1 - 2x - 3.$

$$\frac{2x}{x-2}-\frac{4}{x-2}=\frac{2x-4}{x-2}=\frac{2\overset{1}{\cancel{(x-2)}}}{\underset{1}{\cancel{x-2}}}=2$$

$$\frac{3x-1}{x^2-5x+4}-\frac{2x+3}{x^2-5x+4}=\frac{(3x-1)-(2x+3)}{x^2-5x+4}=\frac{3x-1-2x-3}{x^2-5x+4}$$

$$=\frac{x-4}{x^2-5x+4}=\frac{\overset{1}{\cancel{(x-4)}}}{\underset{1}{\cancel{(x-4)}}(x-1)}=\frac{1}{x-1}$$

Example 5

Add: $\frac{7}{x^2}+\frac{9}{x^2}$

Solution

$$\frac{7}{x^2}+\frac{9}{x^2}=\frac{7+9}{x^2}=\frac{16}{x^2}$$

You Try It 5

Add: $\frac{3}{xy}+\frac{12}{xy}$

Your solution

$\frac{15}{xy}$

Example 6

Subtract: $\frac{3x^2}{x^2-1}-\frac{x+4}{x^2-1}$

Solution

$$\frac{3x^2}{x^2-1}-\frac{x+4}{x^2-1}=\frac{3x^2-(x+4)}{x^2-1}$$

$$=\frac{3x^2-x-4}{x^2-1}$$

$$=\frac{(3x-4)\overset{1}{\cancel{(x+1)}}}{(x-1)\underset{1}{\cancel{(x+1)}}}=\frac{3x-4}{x-1}$$

You Try It 6

Subtract: $\frac{2x^2}{x^2-x-12}-\frac{7x+4}{x^2-x-12}$

Your solution

$\frac{2x+1}{x+3}$

Solutions on p. S23

441

Objective 8.2C

Discuss the Concepts

Have students describe the steps in the process of adding or subtracting rational expressions with the same denominator.

Concept Check

Find the sum.

1. $\frac{8a}{15b}+\frac{2a}{15b}$ $\frac{2a}{3b}$
2. $\frac{x+4}{2x+1}+\frac{x-1}{2x+1}$ $\frac{2x+3}{2x+1}$

Optional Student Activity

In the first subtraction example at the left, some students are surprised that $\frac{2x}{x-2}-\frac{4}{x-2}=2$. Have students evaluate the expression $\frac{2x}{x-2}-\frac{4}{x-2}$ for several values of x to see that the calculation always yields 2 (except when $x = 2$).

Instructor Note

In the second subtraction example at the left, point out the importance of the parentheses around $2x + 3$. Emphasize that students must remember to subtract the whole numerator, which changes the sign of every term, not just the first term.

Concept Check

Find the difference.

1. $\frac{8a}{15b}-\frac{2a}{15b}$ $\frac{2a}{5b}$
2. $\frac{x+4}{2x+1}-\frac{x-1}{2x+1}$ $\frac{5}{2x+1}$

In-Class Examples (Objective 8.2C)

Simplify.

1. $\frac{2x^2}{x^2+7x-8}-\frac{x+1}{x^2+7x-8}$ $\frac{2x+1}{x+8}$
2. $\frac{3y^2-6}{y^2+y-20}+\frac{y-9}{y^2+y-20}-\frac{2y^2+y+1}{y^2+y-20}$ $\frac{y+4}{y+5}$

Objective 8.2D

Vocabulary to Review
least common multiple (LCM)

Concept Check

Use the fractions $\frac{3}{10x^2}$ and $\frac{y+1}{6x^2y}$.

1. Find the least common denominator. $30x^2y$
2. Rewrite each fraction with the common denominator. $\frac{9y}{30x^2y}$ and $\frac{5y+5}{30x^2y}$
3. Find the sum of the fractions. $\frac{14y+5}{30x^2y}$
4. Find the difference of the fractions. $\frac{4y-5}{30x^2y}$

Discuss the Concepts

Have students describe the steps involved in rewriting two or more rational expressions with a common denominator (this will review Objective 8.2B). The fractions in the example at the right can be used to illustrate the process.

Example 7
Simplify:

$$\frac{2x^2+5}{x^2+2x-3} - \frac{x^2-3x}{x^2+2x-3} + \frac{x-2}{x^2+2x-3}$$

Solution

$$\frac{2x^2+5}{x^2+2x-3} - \frac{x^2-3x}{x^2+2x-3} + \frac{x-2}{x^2+2x-3}$$

$$= \frac{(2x^2+5)-(x^2-3x)+(x-2)}{x^2+2x-3}$$

$$= \frac{2x^2+5-x^2+3x+x-2}{x^2+2x-3}$$

$$= \frac{x^2+4x+3}{x^2+2x-3} = \frac{\overset{1}{\cancel{(x+3)}}(x+1)}{\underset{1}{\cancel{(x+3)}}(x-1)} = \frac{x+1}{x-1}$$

You Try It 7
Simplify:

$$\frac{x^2-1}{x^2-8x+12} - \frac{2x+1}{x^2-8x+12} + \frac{x}{x^2-8x+12}$$

Your solution

$\frac{x+1}{x-6}$

Solution on p. S23

Objective D — To add or subtract rational expressions with different denominators

VIDEO & DVD · CD TUTOR · WEB · SSM

TAKE NOTE
This objective requires using the skills learned in Objective 8.2B (to express two fractions in terms of the LCM of their denominators) and 8.2C (to add or subtract algebraic fractions with the same denominator).
Note the steps involved in adding or subtracting rational expressions:

1. Find the LCM of the denominators.
2. Rewrite each fraction in terms of the common denominator.
3. Add or subtract the rational expressions.
4. Simplify the resulting sum or difference.

Before two fractions with unlike denominators can be added or subtracted, each fraction must be expressed in terms of a common denominator. This common denominator is the LCM of the denominators of the fractions.

➡ Add: $\frac{x-3}{x^2-2x} + \frac{6}{x^2-4}$

$x^2-2x = x(x-2);\ x^2-4 = (x+2)(x-2)$ The LCM is $x(x-2)(x+2)$.

$$\frac{x-3}{x^2-2x} + \frac{6}{x^2-4} = \frac{x-3}{x(x-2)}\cdot\frac{x+2}{x+2} + \frac{6}{(x-2)(x+2)}\cdot\frac{x}{x}$$

• **Rewrite each fraction with $x(x-2)(x+2)$ as the denominator.**

$$= \frac{x^2-x-6}{x(x-2)(x+2)} + \frac{6x}{x(x-2)(x+2)}$$

$$= \frac{(x^2-x-6)+6x}{x(x-2)(x+2)}$$

• **Add the fractions.**

$$= \frac{x^2+5x-6}{x(x-2)(x+2)}$$

$$= \frac{(x+6)(x-1)}{x(x-2)(x+2)}$$

The last step is to factor the numerator to determine whether there are common factors in the numerator and denominator. For this example there are no common factors, so the answer is in simplest form.

443

Example 8

Simplify: $\frac{y}{x} - \frac{4y}{3x} + \frac{3y}{4x}$

Solution

The LCM of the denominators is $12x$.

$$\frac{y}{x} - \frac{4y}{3x} + \frac{3y}{4x} = \frac{y}{x}\cdot\frac{12}{12} - \frac{4y}{3x}\cdot\frac{4}{4} + \frac{3y}{4x}\cdot\frac{3}{3}$$
$$= \frac{12y}{12x} - \frac{16y}{12x} + \frac{9y}{12x}$$
$$= \frac{12y - 16y + 9y}{12x} = \frac{5y}{12x}$$

You Try It 8

Simplify: $\frac{z}{8y} - \frac{4z}{3y} + \frac{5z}{4y}$

Your solution

$\frac{z}{24y}$

Example 9

Subtract: $\frac{2x}{x-3} - \frac{5}{3-x}$

Solution

Remember: $3 - x = -(x - 3)$.

Therefore, $\frac{5}{3-x} = \frac{5}{-(x-3)} = \frac{-5}{x-3}$.

$$\frac{2x}{x-3} - \frac{5}{3-x} = \frac{2x}{x-3} - \frac{-5}{x-3}$$
$$= \frac{2x-(-5)}{x-3} = \frac{2x+5}{x-3}$$

You Try It 9

Subtract: $\frac{5x}{x-2} - \frac{3}{2-x}$

Your solution

$\frac{5x+3}{x-2}$

Example 10

Subtract: $\frac{2x}{2x-3} - \frac{1}{x+1}$

Solution

The LCM is $(2x - 3)(x + 1)$.

$$\frac{2x}{2x-3} - \frac{1}{x+1}$$
$$= \frac{2x}{2x-3}\cdot\frac{x+1}{x+1} - \frac{1}{x+1}\cdot\frac{2x-3}{2x-3}$$
$$= \frac{2x^2+2x}{(2x-3)(x+1)} - \frac{2x-3}{(2x-3)(x+1)}$$
$$= \frac{(2x^2+2x)-(2x-3)}{(2x-3)(x+1)} = \frac{2x^2+3}{(2x-3)(x+1)}$$

You Try It 10

Subtract: $\frac{4x}{3x-1} - \frac{9}{x+4}$

Your solution

$\frac{4x^2 - 11x + 9}{(3x-1)(x+4)}$

Solutions on pp. S23–S24

Instructor Note

The process of adding or subtracting rational expressions is a complex one, involving many steps and the application of several skills. It may help students to keep a list of the basic steps in front of them as they work. Have them write out such a list now, or use Exercise 77 on page 447 as a way to develop a list of steps.

Optional Student Activity

Let x and y be positive integers. If $A = \frac{1}{x} + \frac{1}{y} + 1$ and $B = \frac{x+y}{xy}$, then which of the following is true? $A < B$, $A > B$, $A = B$ $A > B$

In-Class Examples (Objective 8.2D)

Simplify.

1. $\frac{1}{6a} + \frac{4}{9b} - \frac{5}{2a}$ $\frac{4a - 21b}{9ab}$
2. $\frac{x-2}{10x} - \frac{x-3}{15x}$ $\frac{1}{30}$
3. $\frac{7}{x+2} + 2$ $\frac{2x+11}{x+2}$
4. $\frac{a-1}{a^2b} - \frac{a-2}{ab}$ $\frac{-a^2+3a-1}{a^2b}$
5. $\frac{2x}{3x-2} + \frac{4}{x-3}$ $\frac{2(x+4)(x-1)}{(3x-2)(x-3)}$
6. $\frac{6y}{y^2-4} + \frac{3}{2-y}$ $\frac{3}{y+2}$

Example 11

Add: $1 + \frac{3}{x^2}$

Solution

The LCM is x^2.

$$1 + \frac{3}{x^2} = 1 \cdot \frac{x^2}{x^2} + \frac{3}{x^2}$$
$$= \frac{x^2}{x^2} + \frac{3}{x^2} = \frac{x^2 + 3}{x^2}$$

You Try It 11

Subtract: $2 - \frac{1}{x - 3}$

Your solution

$\frac{2x - 7}{x - 3}$

Example 12

Subtract: $\frac{x}{2x - 4} - \frac{4 - x}{x^2 - 2x}$

Solution

$2x - 4 = 2(x - 2)$; $x^2 - 2x = x(x - 2)$

The LCM is $2x(x - 2)$.

$$\frac{x}{2x - 4} - \frac{4 - x}{x^2 - 2x} = \frac{x}{2(x - 2)} \cdot \frac{x}{x} - \frac{4 - x}{x(x - 2)} \cdot \frac{2}{2}$$
$$= \frac{x^2 - (4 - x)2}{2x(x - 2)}$$
$$= \frac{x^2 - (8 - 2x)}{2x(x - 2)} = \frac{x^2 + 2x - 8}{2x(x - 2)}$$
$$= \frac{(x + 4)\overset{1}{\cancel{(x - 2)}}}{2x\underset{1}{\cancel{(x - 2)}}} = \frac{x + 4}{2x}$$

You Try It 12

Add: $\frac{a - 3}{a^2 - 5a} + \frac{a - 9}{a^2 - 25}$

Your solution

$\frac{2a + 3}{a(a + 5)}$

Example 13

Simplify: $\frac{3x + 2}{2x^2 - x - 1} - \frac{3}{2x + 1} + \frac{4}{x - 1}$

Solution

The LCM is $(2x + 1)(x - 1)$.

$$\frac{3x + 2}{2x^2 - x - 1} - \frac{3}{2x + 1} + \frac{4}{x - 1}$$
$$= \frac{3x + 2}{(2x + 1)(x - 1)} - \frac{3}{2x + 1} \cdot \frac{x - 1}{x - 1} + \frac{4}{x - 1} \cdot \frac{2x + 1}{2x + 1}$$
$$= \frac{3x + 2}{(2x + 1)(x - 1)} - \frac{3x - 3}{(2x + 1)(x - 1)} + \frac{8x + 4}{(2x + 1)(x - 1)}$$
$$= \frac{(3x + 2) - (3x - 3) + (8x + 4)}{(2x + 1)(x - 1)}$$
$$= \frac{3x + 2 - 3x + 3 + 8x + 4}{(2x + 1)(x - 1)} = \frac{8x + 9}{(2x + 1)(x - 1)}$$

You Try It 13

Simplify: $\frac{2x - 3}{3x^2 - x - 2} + \frac{5}{3x + 2} - \frac{1}{x - 1}$

Your solution

$\frac{2(2x - 5)}{(3x + 2)(x - 1)}$

Solutions on p. S24

445

8.2 Exercises

Objective A

Find the LCM of the expressions.

1. $8x^3y$, $12xy^2$ — $24x^3y^2$
2. $6ab^2$, $18ab^3$ — $18ab^3$
3. $10x^4y^2$, $15x^3y$ — $30x^4y^2$
4. $12a^2b$, $18ab^3$ — $36a^2b^3$
5. $8x^2$, $4x^2 + 8x$ — $8x^2(x + 2)$
6. $6y^2$, $4y + 12$ — $12y^2(y + 3)$
7. $2x^2y$, $3x^2 + 12x$ — $6x^2y(x + 4)$
8. $4xy^2$, $6xy^2 + 12y^2$ — $12xy^2(x + 2)$
9. $9x(x + 2)$, $12(x + 2)^2$ — $36x(x + 2)^2$
10. $8x^2(x - 1)^2$, $10x^3(x - 1)$ — $40x^3(x - 1)^2$
11. $3x + 3$, $2x^2 + 4x + 2$ — $6(x + 1)^2$
12. $4x - 12$, $2x^2 - 12x + 18$ — $4(x - 3)^2$
13. $(x - 1)(x + 2)$, $(x - 1)(x + 3)$ — $(x - 1)(x + 2)(x + 3)$
14. $(2x - 1)(x + 4)$, $(2x + 1)(x + 4)$ — $(2x - 1)(x + 4)(2x + 1)$
15. $(2x + 3)^2$, $(2x + 3)(x - 5)$ — $(2x + 3)^2(x - 5)$
16. $(x - 7)(x + 2)$, $(x - 7)^2$ — $(x - 7)^2(x + 2)$
17. $x - 1$, $x - 2$, $(x - 1)(x - 2)$ — $(x - 1)(x - 2)$
18. $(x + 4)(x - 3)$, $x + 4$, $x - 3$ — $(x + 4)(x - 3)$
19. $x^2 - x - 6$, $x^2 + x - 12$ — $(x - 3)(x + 2)(x + 4)$
20. $x^2 + 3x - 10$, $x^2 + 5x - 14$ — $(x + 5)(x + 7)(x - 2)$
21. $x^2 + 5x + 4$, $x^2 - 3x - 28$ — $(x + 4)(x + 1)(x - 7)$
22. $x^2 - 10x + 21$, $x^2 - 8x + 15$ — $(x - 7)(x - 3)(x - 5)$
23. $x^2 - 2x - 24$, $x^2 - 36$ — $(x - 6)(x + 6)(x + 4)$
24. $x^2 + 7x + 10$, $x^2 - 25$ — $(x + 5)(x - 5)(x + 2)$
25. $x^2 - 7x - 30$, $x^2 - 5x - 24$ — $(x - 10)(x - 8)(x + 3)$
26. $2x^2 - 7x + 3$, $2x^2 + x - 1$ — $(2x - 1)(x - 3)(x + 1)$
27. $3x^2 - 11x + 6$, $3x^2 + 4x - 4$ — $(3x - 2)(x - 3)(x + 2)$
28. $2x^2 - 9x + 10$, $2x^2 + x - 15$ — $(2x - 5)(x - 2)(x + 3)$
29. $6 + x - x^2$, $x + 2$, $x - 3$ — $(x + 2)(x - 3)$
30. $15 + 2x - x^2$, $x - 5$, $x + 3$ — $(x + 3)(x - 5)$
31. $5 + 4x - x^2$, $x - 5$, $x + 1$ — $(x - 5)(x + 1)$
32. $x^2 + 3x - 18$, $3 - x$, $x + 6$ — $(x + 6)(x - 3)$
33. $x^2 - 5x + 6$, $1 - x$, $x - 6$ — $(x - 3)(x - 2)(x - 1)(x - 6)$

Section 8.2

Suggested Assignment

Exercises 1–53, every other odd
Exercises 57–75, odds
Exercises 79–135, odds
More challenging problems:
Exercises 136–139

Quick Quiz (Objective 8.2A)

Find the LCM of the polynomials.

1. $14a^4b^2$ and $21ab^5$ $42a^4b^5$
2. $3x^2y$ and $2x^2 - 10x$ $6x^2y(x - 5)$
3. $x^2 - 4x - 12$ and $36 - x^2$ $(x - 6)(x + 6)(x + 2)$

Answers to Writing Exercises

56. Students should explain that when we subtract rational expressions with the same denominator, the numerators are subtracted and placed over the same denominator. The denominators are not changed.

Objective B

Write each fraction in terms of the LCM of the denominators.

34. $\frac{4}{x}, \frac{3}{x^2}$

$\frac{4x}{x^2}, \frac{3}{x^2}$

35. $\frac{5}{ab^2}, \frac{6}{ab}$

$\frac{5}{ab^2}, \frac{6b}{ab^2}$

36. $\frac{x}{3y^2}, \frac{z}{4y}$

$\frac{4x}{12y^2}, \frac{3yz}{12y^2}$

37. $\frac{5y}{6x^2}, \frac{7}{9xy}$

$\frac{15y^2}{18x^2y}, \frac{14x}{18x^2y}$

38. $\frac{y}{x(x-3)}, \frac{6}{x^2}$

$\frac{xy}{x^2(x-3)}, \frac{6x-18}{x^2(x-3)}$

39. $\frac{a}{y^2}, \frac{6}{y(y+5)}$

$\frac{ay+5a}{y^2(y+5)}, \frac{6y}{y^2(y+5)}$

40. $\frac{9}{(x-1)^2}, \frac{6}{x(x-1)}$

$\frac{9x}{x(x-1)^2}, \frac{6x-6}{x(x-1)^2}$

41. $\frac{a^2}{y(y+7)}, \frac{a}{(y+7)^2}$

$\frac{a^2y+7a^2}{y(y+7)^2}, \frac{ay}{y(y+7)^2}$

42. $\frac{3}{x-3}, \frac{5}{x(3-x)}$

$\frac{3x}{x(x-3)}, -\frac{5}{x(x-3)}$

43. $\frac{b}{y(y-4)}, \frac{b^2}{4-y}$

$\frac{b}{y(y-4)}, -\frac{b^2y}{y(y-4)}$

44. $\frac{3}{(x-5)^2}, \frac{2}{5-x}$

$\frac{3}{(x-5)^2}, -\frac{2x-10}{(x-5)^2}$

45. $\frac{3}{7-y}, \frac{2}{(y-7)^2}$

$-\frac{3y-21}{(y-7)^2}, \frac{2}{(y-7)^2}$

46. $\frac{3}{x^2+2x}, \frac{4}{x^2}$

$\frac{3x}{x^2(x+2)}, \frac{4x+8}{x^2(x+2)}$

47. $\frac{2}{y-3}, \frac{3}{y^3-3y^2}$

$\frac{2y^2}{y^2(y-3)}, \frac{3}{y^2(y-3)}$

48. $\frac{x-2}{x+3}, \frac{x}{x-4}$

$\frac{x^2-6x+8}{(x+3)(x-4)}, \frac{x^2+3x}{(x+3)(x-4)}$

49. $\frac{x^2}{2x-1}, \frac{x+1}{x+4}$

$\frac{x^3+4x^2}{(2x-1)(x+4)}, \frac{2x^2+x-1}{(2x-1)(x+4)}$

50. $\frac{3}{x^2+x-2}, \frac{x}{x+2}$

$\frac{3}{(x+2)(x-1)}, \frac{x^2-x}{(x+2)(x-1)}$

51. $\frac{3x}{x-5}, \frac{4}{x^2-25}$

$\frac{3x^2+15x}{(x-5)(x+5)}, \frac{4}{(x-5)(x+5)}$

52. $\frac{5}{2x^2-9x+10}, \frac{x-1}{2x-5}$

$\frac{5}{(2x-5)(x-2)}, \frac{x^2-3x+2}{(2x-5)(x-2)}$

53. $\frac{x-3}{3x^2+4x-4}, \frac{2}{x+2}$

$\frac{x-3}{(3x-2)(x+2)}, \frac{6x-4}{(3x-2)(x+2)}$

54. $\frac{x}{x^2+x-6}, \frac{2x}{x^2-9}$

$\frac{x^2-3x}{(x+3)(x-2)(x-3)}, \frac{2x^2-4x}{(x+3)(x-2)(x-3)}$

55. $\frac{x-1}{x^2+2x-15}, \frac{x}{x^2+6x+5}$

$\frac{x^2-1}{(x-3)(x+5)(x+1)}, \frac{x^2-3x}{(x-3)(x+5)(x+1)}$

Objective C

56. Explain the procedure for subtracting rational expressions with the same denominator.

Add or subtract.

57. $\frac{3}{y^2} + \frac{8}{y^2}$

$\frac{11}{y^2}$

58. $\frac{6}{ab} - \frac{2}{ab}$

$\frac{4}{ab}$

59. $\frac{3}{x+4} - \frac{10}{x+4}$

$-\frac{7}{x+4}$

Quick Quiz (Objective 8.2B)

Write each fraction in terms of the LCM of the denominators.

1. $\frac{y}{10x^2}$ and $\frac{x}{15y^3}$ $\frac{3y^4}{30x^2y^3}$ and $\frac{2x^2}{30x^2y^3}$
2. $\frac{4}{y(2-y)}$ and $\frac{1}{(y-2)^2}$ $-\frac{4y-8}{y(y-2)^2}$ and $\frac{y}{y(y-2)^2}$
3. $\frac{x+1}{x^2-x}$ and $\frac{x}{x^2+6x-7}$ $\frac{x^2+8x+7}{x(x-1)(x+7)}$ and $\frac{x^2}{x(x-1)(x+7)}$

60. $\dfrac{x}{x+6} - \dfrac{2}{x+6}$
$\dfrac{x-2}{x+6}$

61. $\dfrac{3x}{2x+3} + \dfrac{5x}{2x+3}$
$\dfrac{8x}{2x+3}$

62. $\dfrac{6y}{4y+1} - \dfrac{11y}{4y+1}$
$-\dfrac{5y}{4y+1}$

63. $\dfrac{2x+1}{x-3} + \dfrac{3x+6}{x-3}$
$\dfrac{5x+7}{x-3}$

64. $\dfrac{4x+3}{2x-7} + \dfrac{3x-8}{2x-7}$
$\dfrac{7x-5}{2x-7}$

65. $\dfrac{5x-1}{x+9} - \dfrac{3x+4}{x+9}$
$\dfrac{2x-5}{x+9}$

66. $\dfrac{6x-5}{x-10} - \dfrac{3x-4}{x-10}$
$\dfrac{3x-1}{x-10}$

67. $\dfrac{x-7}{2x+7} - \dfrac{4x-3}{2x+7}$
$\dfrac{-3x-4}{2x+7}$

68. $\dfrac{2n}{3n+4} - \dfrac{5n-3}{3n+4}$
$\dfrac{-3n+3}{3n+4}$

69. $\dfrac{x}{x^2+2x-15} - \dfrac{3}{x^2+2x-15}$
$\dfrac{1}{x+5}$

70. $\dfrac{3x}{x^2+3x-10} - \dfrac{6}{x^2+3x-10}$
$\dfrac{3}{x+5}$

71. $\dfrac{2x+3}{x^2-x-30} - \dfrac{x-2}{x^2-x-30}$
$\dfrac{1}{x-6}$

72. $\dfrac{3x-1}{x^2+5x-6} - \dfrac{2x-7}{x^2+5x-6}$
$\dfrac{1}{x-1}$

73. $\dfrac{4y+7}{2y^2+7y-4} - \dfrac{y-5}{2y^2+7y-4}$
$\dfrac{3}{2y-1}$

74. $\dfrac{x+1}{2x^2-5x-12} + \dfrac{x+2}{2x^2-5x-12}$
$\dfrac{1}{x-4}$

75. $\dfrac{2x^2+3x}{x^2-9x+20} + \dfrac{2x^2-3}{x^2-9x+20} - \dfrac{4x^2+2x+1}{x^2-9x+20}$
$\dfrac{1}{x-5}$

76. $\dfrac{2x^2+3x}{x^2-2x-63} - \dfrac{x^2-3x+21}{x^2-2x-63} - \dfrac{x-7}{x^2-2x-63}$
$\dfrac{x-2}{x-9}$

Objective D

77. Explain the procedure for adding rational expressions with different denominators.

Add or subtract.

78. $\dfrac{4}{x} + \dfrac{5}{y}$
$\dfrac{4y+5x}{xy}$

79. $\dfrac{7}{a} + \dfrac{5}{b}$
$\dfrac{7b+5a}{ab}$

80. $\dfrac{12}{x} - \dfrac{5}{2x}$
$\dfrac{19}{2x}$

Answers to Writing Exercises

77. Students should include each of the following steps:
(1) Find the LCM of the denominators.
(2) Write each rational expression as a fraction with the LCM as the denominator.
(3) Add the numerators and place the sum over the common denominator.
(4) Write the answer in factored form and express the sum in simplest form.

Quick Quiz (Objective 8.2C)

Simplify.

1. $\dfrac{8a}{3a-1} - \dfrac{10a}{3a-1}$ $-\dfrac{2a}{3a-1}$

2. $\dfrac{2x+3}{x-4} - \dfrac{x+5}{x-4}$ $\dfrac{x-2}{x-4}$

3. $\dfrac{2x-8}{3x^2-2x-1} + \dfrac{x+9}{3x^2-2x-1}$ $\dfrac{1}{x-1}$

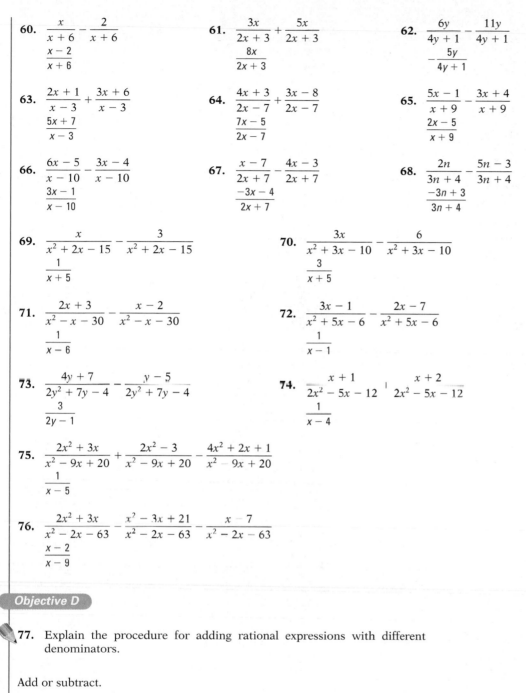

81. $\dfrac{5}{3a} - \dfrac{3}{4a}$ $\dfrac{11}{12a}$

82. $\dfrac{1}{2x} - \dfrac{5}{4x} + \dfrac{7}{6x}$ $\dfrac{5}{12x}$

83. $\dfrac{7}{4y} + \dfrac{11}{6y} - \dfrac{8}{3y}$ $\dfrac{11}{12y}$

84. $\dfrac{5}{3x} - \dfrac{2}{x^2} + \dfrac{3}{2x}$ $\dfrac{19x - 12}{6x^2}$

85. $\dfrac{6}{y^2} + \dfrac{3}{4y} - \dfrac{2}{5y}$ $\dfrac{120 + 7y}{20y^2}$

86. $\dfrac{2}{x} - \dfrac{3}{2y} + \dfrac{3}{5x} - \dfrac{1}{4y}$ $\dfrac{52y - 35x}{20xy}$

87. $\dfrac{5}{2a} + \dfrac{7}{3b} - \dfrac{2}{b} - \dfrac{3}{4a}$ $\dfrac{21b + 4a}{12ab}$

88. $\dfrac{2x + 1}{3x} + \dfrac{x - 1}{5x}$ $\dfrac{13x + 2}{15x}$

89. $\dfrac{4x - 3}{6x} + \dfrac{2x + 3}{4x}$ $\dfrac{14x + 3}{12x}$

90. $\dfrac{x - 3}{6x} + \dfrac{x + 4}{8x}$ $\dfrac{7}{24}$

91. $\dfrac{2x - 3}{2x} + \dfrac{x + 3}{3x}$ $\dfrac{8x - 3}{6x}$

92. $\dfrac{2x + 9}{9x} - \dfrac{x - 5}{5x}$ $\dfrac{x + 90}{45x}$

93. $\dfrac{3y - 2}{12y} - \dfrac{y - 3}{18y}$ $\dfrac{7}{36}$

94. $\dfrac{x + 4}{2x} - \dfrac{x - 1}{x^2}$ $\dfrac{x^2 + 2x + 2}{2x^2}$

95. $\dfrac{x - 2}{3x^2} - \dfrac{x + 4}{x}$ $\dfrac{-3x^2 - 11x - 2}{3x^2}$

96. $\dfrac{x - 10}{4x^2} + \dfrac{x + 1}{2x}$ $\dfrac{2x^2 + 3x - 10}{4x^2}$

97. $\dfrac{x + 5}{3x^2} + \dfrac{2x + 1}{2x}$ $\dfrac{6x^2 + 5x + 10}{6x^2}$

98. $\dfrac{4}{x + 4} - x$ $\dfrac{-x^2 - 4x + 4}{x + 4}$

99. $2x + \dfrac{1}{x}$ $\dfrac{2x^2 + 1}{x}$

100. $5 - \dfrac{x - 2}{x + 1}$ $\dfrac{4x + 7}{x + 1}$

101. $3 + \dfrac{x - 1}{x + 1}$ $\dfrac{4x + 2}{x + 1}$

102. $\dfrac{x + 3}{6x} - \dfrac{x - 3}{8x^2}$ $\dfrac{4x^2 + 9x + 9}{24x^2}$

103. $\dfrac{x + 2}{xy} - \dfrac{3x - 2}{x^2y}$ $\dfrac{x^2 - x + 2}{x^2y}$

104. $\dfrac{3x - 1}{xy^2} - \dfrac{2x + 3}{xy}$ $\dfrac{3x - 1 - 2xy - 3y}{xy^2}$

105. $\dfrac{4x - 3}{3x^2y} + \dfrac{2x + 1}{4xy^2}$ $\dfrac{16xy - 12y + 6x^2 + 3x}{12x^2y^2}$

106. $\dfrac{5x + 7}{6xy^2} - \dfrac{4x - 3}{8x^2y}$ $\dfrac{20x^2 + 28x - 12xy + 9y}{24x^2y^2}$

107. $\dfrac{x - 2}{8x^2} - \dfrac{x + 7}{12xy}$ $\dfrac{3xy - 6y - 2x^2 - 14x}{24x^2y}$

Quick Quiz (Objective 8.2D)

Simplify.

1. $\dfrac{y - 1}{5y^2} - \dfrac{y + 3}{10y} - \dfrac{y^2 + y + 2}{10y^2}$
2. $\dfrac{4x}{x^2 - 36} + \dfrac{3}{6 - x}\ \dfrac{x - 18}{(x - 6)(x + 6)}$
3. $\dfrac{a - 1}{a - 2} - \dfrac{3a + 1}{a^2 + 3a - 10}\ \dfrac{a + 3}{a + 5}$

108. $\dfrac{3x-1}{6y^2} - \dfrac{x+5}{9xy}$

$\dfrac{9x^2 - 3x - 2xy - 10y}{18xy^2}$

109. $\dfrac{4}{x-2} + \dfrac{5}{x+3}$

$\dfrac{9x+2}{(x-2)(x+3)}$

110. $\dfrac{2}{x-3} + \dfrac{5}{x-4}$

$\dfrac{7x-23}{(x-3)(x-4)}$

111. $\dfrac{6}{x-7} - \dfrac{4}{x+3}$

$\dfrac{2(x+23)}{(x-7)(x+3)}$

112. $\dfrac{3}{y+6} - \dfrac{4}{y-3}$

$\dfrac{-y-33}{(y+6)(y-3)}$

113. $\dfrac{2x}{x+1} + \dfrac{1}{x-3}$

$\dfrac{2x^2 - 5x + 1}{(x+1)(x-3)}$

114. $\dfrac{3x}{x-4} + \dfrac{2}{x+6}$

$\dfrac{3x^2 + 20x - 8}{(x-4)(x+6)}$

115. $\dfrac{4x}{2x-1} - \dfrac{5}{x-6}$

$\dfrac{4x^2 - 34x + 5}{(2x-1)(x-6)}$

116. $\dfrac{6x}{x+5} - \dfrac{3}{2x+3}$

$\dfrac{3(4x^2 + 5x - 5)}{(x+5)(2x+3)}$

117. $\dfrac{2a}{a-7} + \dfrac{5}{7-a}$

$\dfrac{2a-5}{a-7}$

118. $\dfrac{4x}{6-x} + \dfrac{5}{x-6}$

$\dfrac{-4x+5}{x-6}$

119. $\dfrac{x}{x^2-9} + \dfrac{3}{x-3}$

$\dfrac{4x+9}{(x+3)(x-3)}$

120. $\dfrac{y}{y^2-16} + \dfrac{1}{y-4}$

$\dfrac{2(y+2)}{(y+4)(y-4)}$

121. $\dfrac{2x}{x^2-x-6} - \dfrac{3}{x+2}$

$\dfrac{-x+9}{(x+2)(x-3)}$

122. $\dfrac{(x-1)^2}{(x+1)^2} - 1$

$\dfrac{4x}{(x+1)^2}$

123. $1 - \dfrac{(y-2)^2}{(y+2)^2}$

$\dfrac{8y}{(y+2)^2}$

124. $\dfrac{x}{1-x^2} - 1 + \dfrac{x}{1+x}$

$\dfrac{2x-1}{(1+x)(1\ \ x)}$

125. $\dfrac{y}{x-y} + 2 - \dfrac{x}{y-x}$

$\dfrac{3x-y}{x-y}$

126. $\dfrac{3x-1}{x^2-10x+25} - \dfrac{3}{x-5}$

$\dfrac{14}{(x-5)(x-5)}$

127. $\dfrac{2a+3}{a^2-7a+12} - \dfrac{2}{a-3}$

$\dfrac{11}{(a-3)(a-4)}$

128. $\dfrac{x+4}{x^2-x-42} + \dfrac{3}{7-x}$

$\dfrac{-2(x+7)}{(x+6)(x-7)}$

129. $\dfrac{x+3}{x^2-3x-10} + \dfrac{2}{5-x}$

$\dfrac{-x-1}{(x+2)(x-5)}$

Answers to Writing Exercises

140. The LCM of two expressions is equal to their product when the two expressions have no common factors.

130. $\dfrac{1}{x+1} + \dfrac{x}{x-6} - \dfrac{5x-2}{x^2-5x-6}$
$\dfrac{x-4}{x-6}$

131. $\dfrac{x}{x-4} + \dfrac{5}{x+5} - \dfrac{11x-8}{x^2+x-20}$
$\dfrac{x+3}{x+5}$

132. $\dfrac{3x+1}{x-1} - \dfrac{x-1}{x-3} + \dfrac{x+1}{x^2-4x+3}$
$\dfrac{2x+1}{x-1}$

133. $\dfrac{4x+1}{x-8} - \dfrac{3x+2}{x+4} - \dfrac{49x+4}{x^2-4x-32}$
$\dfrac{x-2}{x+4}$

134. $\dfrac{2x+9}{3-x} + \dfrac{x+5}{x+7} - \dfrac{2x^2+3x-3}{x^2+4x-21}$
$\dfrac{-3(x^2+8x+25)}{(x-3)(x+7)}$

135. $\dfrac{3x+5}{x+5} - \dfrac{x+1}{2-x} - \dfrac{4x^2-3x-1}{x^2+3x-10}$
$\dfrac{4(2x-1)}{(x+5)(x-2)}$

APPLYING THE CONCEPTS

136. Simplify.

a. $\left(\dfrac{b}{6} - \dfrac{6}{b}\right) \div \left(\dfrac{6}{b} - 4 + \dfrac{b}{2}\right)$
$\dfrac{b+6}{3(b-2)}$

b. $\dfrac{x^2+x-6}{x^2+2x-8} \cdot \dfrac{x^2+5x+4}{x^2+2x-3} - \dfrac{2}{x-1}$
1

137. Rewrite the fraction as the sum of two fractions in simplest form.

a. $\dfrac{3x+6y}{xy}$
$\dfrac{3}{y} + \dfrac{6}{x}$

b. $\dfrac{4a^2+3ab}{a^2b^2}$
$\dfrac{4}{b^2} + \dfrac{3}{ab}$

c. $\dfrac{3m^2n+2mn^2}{12m^3n^2}$
$\dfrac{1}{4mn} + \dfrac{1}{6m^2}$

138. Let $f(x) = \dfrac{x}{x+2}$, $g(x) = \dfrac{4}{x-3}$, and $S(x) = \dfrac{x^2+x+8}{x^2-x-6}$. Evaluate $f(4)$, $g(4)$, and $S(4)$. Does $f(4) + g(4) = S(4)$? Let a be a real number ($a \neq -2, a \neq 3$). Express $S(a)$ in terms of $f(a)$ and $g(a)$.

$f(4) = \dfrac{2}{3}$; $g(4) = 4$; $S(4) = 4\dfrac{2}{3}$; yes; $S(a) = f(a) + g(a)$

139. Suppose you drive about 12,000 mi per year and that the cost of gasoline averages $1.60 per gallon.

a. Let x represent the number of miles per gallon your car gets. Write a variable expression for the amount you spend on gasoline for 1 year.

b. Write and simplify a variable expression for the amount of money you will save each year if you can increase your gas mileage by 5 mi per gallon.

c. If you currently get 25 mi per gallon and you increase your gas mileage by 5 mi per gallon, how much will you save in 1 year?

a. $\dfrac{19{,}200}{x}$ dollars **b.** $\dfrac{96{,}000}{x(x+5)}$ dollars **c.** $128

140. When is the LCM of two expressions equal to their product?

8.3 Complex Fractions

Objective A To simplify a complex fraction

VIDEO & DVD CD TUTOR WEB SSM

Point of Interest

There are many instances of complex fractions in application problems. The fraction $\dfrac{1}{\dfrac{1}{R_1}+\dfrac{1}{R_2}}$ is used to determine the total resistance in certain electric circuits.

A **complex fraction** is a fraction whose numerator or denominator contains one or more fractions. Examples of complex fractions are shown at the right.

$$\frac{3}{2-\frac{1}{2}},\qquad \frac{4+\frac{1}{x}}{3+\frac{2}{x}},\qquad \frac{\frac{1}{x-1}+x+3}{x-3+\frac{1}{x+4}}$$

➡ Simplify: $\dfrac{1-\dfrac{4}{x^2}}{1+\dfrac{2}{x}}$

Find the LCM of the denominators of the fractions in the numerator and denominator. The LCM of x and x^2 is x^2.

$$\frac{1-\frac{4}{x^2}}{1+\frac{2}{x}} = \frac{1-\frac{4}{x^2}}{1+\frac{2}{x}}\cdot\frac{x^2}{x^2}$$

• Multiply the numerator and denominator by the LCM.

$$= \frac{1\cdot x^2-\frac{4}{x^2}\cdot x^2}{1\cdot x^2+\frac{2}{x}\cdot x^2}$$

• Simplify.

$$= \frac{x^2-4}{x^2+2x} = \frac{(x-2)\overset{1}{\cancel{(x+2)}}}{x\underset{1}{\cancel{(x+2)}}}$$

$$= \frac{x-2}{x}$$

TAKE NOTE

First of all, we are multiplying the complex fraction by $\frac{x^2}{x^2}$, which equals 1, so we are not changing the value of the fraction.

Second, we are using the Distributive Property to multiply $\left(1-\frac{4}{x^2}\right)x^2$ and $\left(1+\frac{2}{x}\right)x^2$.

The method shown above of simplifying a complex fraction by multiplying the numerator and denominator by the LCM of the denominators is used in the examples on the next page. However, a different approach is to rewrite the numerator and denominator of the complex fraction as single fractions and then divide the numerator by the denominator. The example shown above is simplified below using this alternative method.

$$\frac{1-\frac{4}{x^2}}{1+\frac{2}{x}} = \frac{1\cdot\frac{x^2}{x^2}-\frac{4}{x^2}}{1\cdot\frac{x}{x}+\frac{2}{x}} = \frac{\frac{x^2}{x^2}-\frac{4}{x^2}}{\frac{x}{x}+\frac{2}{x}} = \frac{\frac{x^2-4}{x^2}}{\frac{x+2}{x}}$$

• Rewrite the numerator and denominator of the complex fraction as single fractions.

$$= \frac{x^2-4}{x^2}\div\frac{x+2}{x} = \frac{x^2-4}{x^2}\cdot\frac{x}{x+2}$$

• Divide the numerator of the complex fraction by the denominator.

$$= \frac{(x^2-4)x}{x^2(x+2)} = \frac{(x+2)(x-2)x}{x^2(x+2)}$$

• Multiply the fractions. Factor the numerator.

$$= \frac{x-2}{x}$$

• Simplify.

TAKE NOTE

Recall that the fraction bar can be read "divided by."

Note that this is the same result as shown above.

Objective 8.3A

New Vocabulary

complex fraction

Discuss the Concepts

Solve Examples 1 and 2 of this objective by using the alternative method of rewriting the numerator and denominator of the complex fraction as a single fraction and then dividing the numerator by the denominator (see the bottom of page 451). Which method do you prefer? Why?

Concept Check

Have students find the reciprocal, in simplest form, of the complex fraction $\dfrac{\dfrac{x}{2x-5}}{\dfrac{x^2}{4x^2-25}}$ and show that the product of the complex fraction and its reciprocal is 1.

Note: The reciprocal is $\dfrac{x}{2x+5}$.

In-Class Examples (Objective 8.3A)

Simplify.

1. $\dfrac{2+\dfrac{1}{x}}{4-\dfrac{1}{x^2}}$ $\quad \dfrac{x}{2x-1}$

2. $\dfrac{1+\dfrac{2}{x}-\dfrac{8}{x^2}}{1+\dfrac{3}{x}-\dfrac{10}{x^2}}$ $\quad \dfrac{x+4}{x+5}$

3. $\dfrac{x+1-\dfrac{4}{x-2}}{x-4-\dfrac{24}{x-2}}$ $\quad \dfrac{x-3}{x-8}$

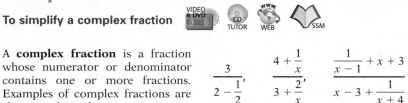

Optional Student Activity

According to the theory of relativity, the mass of a moving object is given by an equation that contains a complex fraction. The equation is $m = \dfrac{m_0}{\sqrt{1 - \dfrac{v^2}{c^2}}}$, where m is the mass in grams of the moving object, m_0 is the mass of the object at rest, v is the speed of the object, and c is the speed of light.

1. Evaluate the expression at speeds of $0.5c$, $0.75c$, $0.90c$, $0.95c$, and $0.99c$ when the mass of the object at rest is 10 g.
2. Explain how m changes as the speed of the object becomes closer to the speed of light.
3. Explain how this equation can be used to support the theory that an object cannot travel at the speed of light.

1. 11.547, 15.119, 22.942, 32.026, 70.888
2. As the velocity v of the object approaches the velocity of light c, the mass of the object increases.
3. According to the theory of relativity, the mass of an object at the speed of light would be infinite and would require an infinite force to accelerate it. Because this is impossible, the theory suggests that an object of any mass cannot attain the speed of light.

Example 1

Simplify: $\dfrac{\dfrac{1}{x} + \dfrac{1}{2}}{\dfrac{1}{x^2} - \dfrac{1}{4}}$

Solution

The LCM of x, 2, x^2, and 4 is $4x^2$.

$$\frac{\frac{1}{x} + \frac{1}{2}}{\frac{1}{x^2} - \frac{1}{4}} = \frac{\frac{1}{x} + \frac{1}{2}}{\frac{1}{x^2} - \frac{1}{4}} \cdot \frac{4x^2}{4x^2} = \frac{\frac{1}{x} \cdot 4x^2 + \frac{1}{2} \cdot 4x^2}{\frac{1}{x^2} \cdot 4x^2 - \frac{1}{4} \cdot 4x^2}$$

$$= \frac{4x + 2x^2}{4 - x^2} = \frac{2x\overset{1}{\cancel{(2 + x)}}}{(2 - x)\underset{1}{\cancel{(2 + x)}}} = \frac{2x}{2 - x}$$

You Try It 1

Simplify: $\dfrac{\dfrac{1}{3} - \dfrac{1}{x}}{\dfrac{1}{9} - \dfrac{1}{x^2}}$

Your solution

$\dfrac{3x}{x + 3}$

Example 2

Simplify: $\dfrac{x - 8 + \dfrac{20}{x + 4}}{x - 10 + \dfrac{24}{x + 4}}$

Solution

The LCM is $x + 4$.

$$\frac{x - 8 + \frac{20}{x + 4}}{x - 10 + \frac{24}{x + 4}}$$

$$= \frac{x - 8 + \frac{20}{x + 4}}{x - 10 + \frac{24}{x + 4}} \cdot \frac{x + 4}{x + 4}$$

$$= \frac{x(x + 4) - 8(x + 4) + \frac{20}{x + 4} \cdot (x + 4)}{x(x + 4) - 10(x + 4) + \frac{24}{x + 4} \cdot (x + 4)}$$

$$= \frac{x^2 + 4x - 8x - 32 + 20}{x^2 + 4x - 10x - 40 + 24} = \frac{x^2 - 4x - 12}{x^2 - 6x - 16}$$

$$= \frac{(x - 6)\overset{1}{\cancel{(x + 2)}}}{(x - 8)\underset{1}{\cancel{(x + 2)}}} = \frac{x - 6}{x - 8}$$

You Try It 2

Simplify: $\dfrac{x + 3 - \dfrac{20}{x - 5}}{x + 8 + \dfrac{30}{x - 5}}$

Your solution

$\dfrac{x - 7}{x - 2}$

Solutions on p. S24

8.3 Exercises

Objective A

Simplify.

1. $$\frac{1 + \frac{3}{x}}{1 - \frac{9}{x^2}}$$ $\frac{x}{x - 3}$

2. $$\frac{1 + \frac{4}{x}}{1 - \frac{16}{x^2}}$$ $\frac{x}{x - 4}$

3. $$\frac{2 - \frac{8}{x + 4}}{3 - \frac{12}{x + 4}}$$ $\frac{2}{3}$

4. $$\frac{5 - \frac{25}{x + 5}}{1 - \frac{3}{x + 5}}$$ $\frac{5x}{x + 2}$

5. $$\frac{1 + \frac{5}{y - 2}}{1 - \frac{2}{y - 2}}$$ $\frac{y + 3}{y - 4}$

6. $$\frac{2 - \frac{11}{2x - 1}}{3 - \frac{17}{2x - 1}}$$ $\frac{4x - 13}{2(3x - 10)}$

7. $$\frac{4 - \frac{2}{x + 7}}{5 + \frac{1}{x + 7}}$$ $\frac{2(2x + 13)}{5x + 36}$

8. $$\frac{5 + \frac{3}{x - 8}}{2 - \frac{1}{x - 8}}$$ $\frac{5x - 37}{2x - 17}$

9. $$\frac{1 - \frac{1}{x} - \frac{6}{x^2}}{1 - \frac{9}{x^2}}$$ $\frac{x + 2}{x + 3}$

10. $$\frac{1 + \frac{4}{x} + \frac{4}{x^2}}{1 - \frac{2}{x} - \frac{8}{x^2}}$$ $\frac{x + 2}{x - 4}$

11. $$\frac{1 - \frac{5}{x} - \frac{6}{x^2}}{1 + \frac{6}{x} + \frac{5}{x^2}}$$ $\frac{x - 6}{x + 5}$

12. $$\frac{1 - \frac{7}{a} + \frac{12}{a^2}}{1 + \frac{1}{a} - \frac{20}{a^2}}$$ $\frac{a - 3}{a + 5}$

13. $$\frac{1 - \frac{6}{x} + \frac{8}{x^2}}{\frac{4}{x^2} + \frac{3}{x} - 1}$$ $-\frac{x - 2}{x + 1}$

14. $$\frac{1 + \frac{3}{x} - \frac{18}{x^2}}{\frac{21}{x^2} - \frac{4}{x} - 1}$$ $-\frac{x + 6}{x + 7}$

15. $$\frac{x - \frac{4}{x + 3}}{1 + \frac{1}{x + 3}}$$ $x - 1$

16. $$\frac{y + \frac{1}{y - 2}}{1 + \frac{1}{y - 2}}$$ $y - 1$

17. $$\frac{1 - \frac{x}{2x + 1}}{x - \frac{1}{2x + 1}}$$ $\frac{1}{2x - 1}$

18. $$\frac{1 - \frac{2x - 2}{3x - 1}}{x - \frac{4}{3x - 1}}$$ $\frac{1}{3x - 4}$

Section 8.3

Suggested Assignment

Exercises 1–29, odds

More challenging problems:
Exercises 32–35

Quick Quiz (Objective 8.3A)

Simplify.

1. $$\frac{1 - \frac{2}{y} - \frac{8}{y^2}}{1 + \frac{5}{y} + \frac{6}{y^2}}$$ $\frac{y - 4}{y + 3}$

2. $$\frac{x - 2 + \frac{3}{x + 2}}{3x - 2 + \frac{5}{x + 2}}$$ $\frac{x - 1}{3x + 1}$

Answers to Writing Exercises

37. Students should explain that multiplying the numerator and denominator of a complex fraction by the LCM of the denominators of the fractions in the numerator and denominator eliminates fractions from the numerator and denominator. The result is a rational expression that is not a complex fraction.

19. $\dfrac{x - 5 + \dfrac{14}{x + 4}}{x + 3 - \dfrac{2}{x + 4}}$ $\quad \dfrac{x - 3}{x + 5}$

20. $\dfrac{a + 4 + \dfrac{5}{a - 2}}{a + 6 + \dfrac{15}{a - 2}}$ $\quad \dfrac{a - 1}{a + 1}$

21. $\dfrac{x + 3 - \dfrac{10}{x - 6}}{x + 2 - \dfrac{20}{x - 6}}$ $\quad \dfrac{x - 7}{x - 8}$

22. $\dfrac{x - 7 + \dfrac{5}{x - 1}}{x - 3 + \dfrac{1}{x - 1}}$ $\quad \dfrac{x - 6}{x - 2}$

23. $\dfrac{y - 6 + \dfrac{22}{2y + 3}}{y - 5 + \dfrac{11}{2y + 3}}$ $\quad \dfrac{2y - 1}{2y + 1}$

24. $\dfrac{x + 2 - \dfrac{12}{2x - 1}}{x + 1 - \dfrac{9}{2x - 1}}$ $\quad \dfrac{2x + 7}{2x + 5}$

25. $\dfrac{x - \dfrac{2}{2x - 3}}{2x - 1 - \dfrac{8}{2x - 3}}$ $\quad \dfrac{x - 2}{2x - 5}$

26. $\dfrac{x + 3 - \dfrac{18}{2x + 1}}{x - \dfrac{6}{2x + 1}}$ $\quad \dfrac{x + 5}{x + 2}$

27. $\dfrac{\dfrac{1}{x} - \dfrac{2}{x - 1}}{\dfrac{3}{x} + \dfrac{1}{x - 1}}$ $\quad -\dfrac{x + 1}{4x - 3}$

28. $\dfrac{\dfrac{3}{n + 1} + \dfrac{1}{n}}{\dfrac{2}{n + 1} + \dfrac{3}{n}}$ $\quad \dfrac{4n + 1}{5n + 3}$

29. $\dfrac{\dfrac{3}{2x - 1} - \dfrac{1}{x}}{\dfrac{4}{x} + \dfrac{2}{2x - 1}}$ $\quad \dfrac{x + 1}{2(5x - 2)}$

30. $\dfrac{\dfrac{4}{3x + 1} + \dfrac{3}{x}}{\dfrac{6}{x} - \dfrac{2}{3x + 1}}$ $\quad \dfrac{13x + 3}{2(8x + 3)}$

APPLYING THE CONCEPTS

Simplify.

31. $1 + \dfrac{1}{1 + \dfrac{1}{2}}$ $\quad \dfrac{5}{3}$

32. $1 + \dfrac{1}{1 + \dfrac{1}{1 + \dfrac{1}{2}}}$ $\quad \dfrac{8}{5}$

33. $1 - \dfrac{1}{1 - \dfrac{1}{x}}$ $\quad -\dfrac{1}{x - 1}$

34. $\dfrac{a^{-1} - b^{-1}}{a^{-2} - b^{-2}}$ $\quad \dfrac{ab}{a + b}$

35. $\left(\dfrac{y}{4} - \dfrac{4}{y}\right) \div \left(\dfrac{4}{y} - 3 + \dfrac{y}{2}\right)$ $\quad \dfrac{y + 4}{2(y - 2)}$

36. $\dfrac{1 + x^{-1}}{1 - x^{-1}}$ $\quad \dfrac{x + 1}{x - 1}$

37. How would you explain to a classmate why we multiply the numerator and denominator of a complex fraction by the LCM of the denominators of the fractions in the numerator and denominator?

8.4 Rational Equations

Objective A To solve rational equations

To solve an equation containing fractions, **clear denominators** by multiplying each side of the equation by the LCM of the denominators. Then solve for the variable.

➡ Solve: $\frac{3x-1}{4x} + \frac{2}{3x} = \frac{7}{6x}$

$$\frac{3x-1}{4x} + \frac{2}{3x} = \frac{7}{6x}$$

• The LCM of 4x, 3x, and 6x is 12x.

$$12x\left(\frac{3x-1}{4x} + \frac{2}{3x}\right) = 12x\left(\frac{7}{6x}\right)$$

• Multiply each side of the equation by the LCM of the denominators.

$$12x\left(\frac{3x-1}{4x}\right) + 12x\left(\frac{2}{3x}\right) = 12x\left(\frac{7}{6x}\right)$$

• Simplify using the Distributive Property.

$$\frac{12x}{1}\left(\frac{3x-1}{4x}\right) + \frac{12x}{1}\left(\frac{2}{3x}\right) = \frac{12x}{1}\left(\frac{7}{6x}\right)$$

$$3(3x-1) + 4(2) = 2(7)$$

• Solve for x.

$$9x - 3 + 8 = 14$$
$$9x + 5 = 14$$
$$9x = 9$$
$$x = 1$$

1 checks as a solution. The solution is 1.

TAKE NOTE
Note that we are now solving *equations*, not operating on *expressions*. We are not writing each fraction in terms of the LCM of the denominators; we are multiplying both sides of the equation by the LCM of the denominators.

Occasionally, a value of the variable that appears to be a solution of an equation will make one of the denominators zero. In this case, the equation has no solution for that value of the variable.

➡ Solve: $\frac{2x}{x-2} = 1 + \frac{4}{x-2}$

$$\frac{2x}{x-2} = 1 + \frac{4}{x-2}$$

$$(x-2)\frac{2x}{x-2} = (x-2)\left(1 + \frac{4}{x-2}\right)$$

• The LCM is $x - 2$. Multiply each side of the equation by the LCM.

$$(x-2)\frac{2x}{x-2} = (x-2)\cdot 1 + (x-2)\frac{4}{x-2}$$

• Simplify using the Distributive Property and the Properties of Fractions.

$$\frac{\overset{1}{\cancel{(x-2)}}}{1}\cdot\frac{2x}{\underset{1}{\cancel{x-2}}} = (x-2)\cdot 1 + \frac{\overset{1}{\cancel{(x-2)}}}{1}\cdot\frac{4}{\underset{1}{\cancel{x-2}}}$$

$$2x = x - 2 + 4$$

• Solve for x.

$$2x = x + 2$$
$$x = 2$$

When x is replaced by 2, the denominators of $\frac{2x}{x-2}$ and $\frac{4}{x-2}$ are zero. Therefore, the equation has no solution.

Objective 8.4A

New Vocabulary
clear denominators

Concept Check
Clear the denominators in each equation (rewrite each equation as an equation without fractions).

1. $\frac{x}{7} - \frac{2}{21} = \frac{x}{3}$ $3x - 2 = 7x$
2. $\frac{4}{y-3} - 1 = \frac{1}{y-3}$
   $4 - (y - 3) = 1$
3. $\frac{5}{a-1} = \frac{2}{a}$ $5a = 2(a - 1)$

Instructor Note
The example at the left illustrates the importance of checking a solution of a rational equation when each side is multiplied by a variable expression.

You might emphasize that the Multiplication Property of Equations states that each side of an equation can be multiplied by the same *nonzero* number. If x is allowed to be 2 in this example, each side of the equation is multiplied by zero.

Optional Student Activity
Find pairs of equivalent equations in the following list.

a. $3x = 2(x + 1) - 1$
b. $3 = 2 - \frac{x+1}{x}$
c. $\frac{3x}{x+1} = 2 - \frac{1}{x+1}$
d. $3x = 2x - (x + 1)$
a and c, b and d

In-Class Examples (Objective 8.4A)
Solve.

1. $\frac{8}{4x-3} = -4$ $\frac{1}{4}$
2. $\frac{3x}{2x+1} + \frac{1}{x+2} = \frac{4}{x+2}$ 1, −1

Instructor Note

After learning to clear denominators in order to solve an equation containing fractions or to simplify a complex fraction, some students may try using the same technique when simply *adding* fractions. The following Discuss the Concepts question should help students learn to avoid making this error.

Discuss the Concepts

Look at the following two problems involving the fractions $\frac{x}{2}, \frac{1}{3}$, and $\frac{5}{2}$. Explain why you can clear the denominators in Problem 1 but not in Problem 2.

1. Solve: $\frac{x}{2} + \frac{1}{3} = \frac{5}{2}$

2. Simplify: $\frac{x}{2} + \frac{1}{3} + \frac{5}{2}$

Example 1

Solve: $\frac{x}{x+4} = \frac{2}{x}$

Solution

The LCM is $x(x + 4)$.

$$\frac{x}{x+4} = \frac{2}{x}$$

$$x(x+4)\left(\frac{x}{x+4}\right) = x(x+4)\left(\frac{2}{x}\right)$$

$$\frac{x(x+4)}{1} \cdot \frac{x}{x+4} = \frac{x(x+4)}{1} \cdot \frac{2}{x}$$

$$x^2 = (x+4)2$$

$$x^2 = 2x + 8$$

Solve the quadratic equation by factoring.

$$x^2 - 2x - 8 = 0$$

$$(x-4)(x+2) = 0$$

$$x - 4 = 0 \qquad x + 2 = 0$$

$$x = 4 \qquad x = -2$$

Both 4 and −2 check as solutions.

The solutions are 4 and −2.

You Try It 1

Solve: $\frac{x}{x+6} = \frac{3}{x}$

Your solution

−3, 6

Example 2

Solve: $\frac{3x}{x-4} = 5 + \frac{12}{x-4}$

Solution

The LCM is $x - 4$.

$$\frac{3x}{x-4} = 5 + \frac{12}{x-4}$$

$$(x-4)\left(\frac{3x}{x-4}\right) = (x-4)\left(5 + \frac{12}{x-4}\right)$$

$$\frac{(x-4)}{1} \cdot \frac{3x}{x-4} = (x-4)5 + \frac{(x-4)}{1} \cdot \frac{12}{x-4}$$

$$3x = (x-4)5 + 12$$

$$3x = 5x - 20 + 12$$

$$3x = 5x - 8$$

$$-2x = -8$$

$$x = 4$$

4 does not check as a solution.

The equation has no solution.

You Try It 2

Solve: $\frac{5x}{x+2} = 3 - \frac{10}{x+2}$

Your solution

no solution

Solutions on p. S25

Objective B To solve proportions

Quantities such as 4 meters, 15 seconds, and 8 gallons are number quantities written with units. In these examples, the units are meters, seconds, and gallons.

A **ratio** is the quotient of two quantities that have the same unit.

The length of a living room is 16 ft and the width is 12 ft. The ratio of the length to the width is written

$\frac{16 \text{ ft}}{12 \text{ ft}} = \frac{16}{12} = \frac{4}{3}$ A ratio is in simplest form when the two numbers do not have a common factor. Note that the units are not written.

A **rate** is the quotient of two quantities that have different units.

There are 2 lb of salt in 8 gal of water. The salt-to-water rate is

$\frac{2 \text{ lb}}{8 \text{ gal}} = \frac{1 \text{ lb}}{4 \text{ gal}}$ A rate is in simplest form when the two numbers do not have a common factor. The units are written as part of the rate.

A **proportion** is an equation that states the equality of two ratios or rates. Examples of proportions are shown at the right.

$$\frac{30 \text{ mi}}{4 \text{ h}} = \frac{15 \text{ mi}}{2 \text{ h}}$$

$$\frac{4}{6} = \frac{8}{12}$$

$$\frac{3}{4} = \frac{x}{8}$$

➡ Solve: $\frac{4}{x} = \frac{2}{3}$

$$\frac{4}{x} = \frac{2}{3}$$

• The LCM of x and 3 is $3x$.

$$3x\left(\frac{4}{x}\right) = 3x\left(\frac{2}{3}\right)$$

• Multiply each side of the proportion by $3x$.

$$12 = 2x$$

• Solve the equation.

$$6 = x$$

The solution is 6.

Example 3

Solve: $\frac{8}{x+3} = \frac{4}{x}$

Solution

$$\frac{8}{x+3} = \frac{4}{x}$$

$$x(x+3)\frac{8}{x+3} = x(x+3)\frac{4}{x}$$

$$8x = 4(x+3)$$

$$8x = 4x + 12$$

$$4x = 12$$

$$x = 3$$

The solution is 3.

You Try It 3

Solve: $\frac{2}{x+3} = \frac{6}{5x+5}$

Your solution

2

Solution on p. S25

Objective 8.4B

New Vocabulary

ratio
rate
proportion

Discuss the Concepts

1. Explain the difference between a ratio and a rate.
2. Explain the difference between a ratio and a proportion.

Concept Check

Identify each of the following as a ratio or a rate and then write the expression in simplest form.

1. $\frac{50 \text{ ft}}{4 \text{ s}}$ Rate, $\frac{25 \text{ ft}}{2 \text{ s}}$
2. $\frac{28 \text{ in.}}{21 \text{ in.}}$ Ratio, $\frac{4}{3}$
3. $\frac{20 \text{ mi}}{2 \text{ h}}$ Rate, $\frac{10 \text{ mi}}{1 \text{ h}}$
4. $\frac{3 \text{ gal}}{18 \text{ gal}}$ Ratio, $\frac{1}{6}$

In-Class Examples (Objective 8.4B)

Solve.

1. $\frac{10}{x} = \frac{5}{7}$ 14
2. $\frac{3}{8} = \frac{12}{x-2}$ 34
3. $\frac{x-5}{4} = \frac{5x-1}{11}$ $-\frac{17}{3}$

Objective 8.4C

New Vocabulary
similar objects

Concept Check
Triangle *XYZ* and triangle *MNO* are similar.

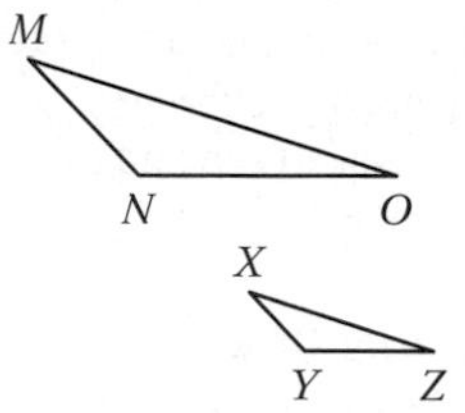

1. Name the corresponding part for side *XY*, for angle *O*, and for side *OM*. *MN*, angle *Z*, *ZX*
2. Write a proportion involving *YZ* and *NM*. $\frac{YZ}{NO} = \frac{YX}{NM}$

Discuss the Concepts
The figure at the top of the next page in which triangle *DEC* is similar to triangle *ABC* presents a situation that always produces similar triangles (a line drawn parallel to one side of a triangle cuts off a triangle similar to the original triangle). Have students try to express this idea in their own words.

Objective C **To solve problems involving similar triangles** VIDEO & DVD CD TUTOR WEB SSM

Similar objects have the same shape but not necessarily the same size. A tennis ball is similar to a basketball. A model ship is similar to an actual ship.

Similar objects have corresponding parts; for example, the rudder on the model ship corresponds to the rudder on the actual ship. The relationship between the sizes of each of the corresponding parts can be written as a ratio, and each ratio will be the same. If the rudder on the model ship is $\frac{1}{100}$ the size of the rudder on the actual ship, then the model wheelhouse is $\frac{1}{100}$ the size of the actual wheelhouse, the width of the model is $\frac{1}{100}$ the width of the actual ship, and so on.

The two triangles *ABC* and *DEF* shown at the right are similar. Side $\overline{AB}$ corresponds to $\overline{DE}$, side $\overline{BC}$ corresponds to $\overline{EF}$, and side $\overline{AC}$ corresponds to $\overline{DF}$. The height $\overline{CH}$ corresponds to the height $\overline{FK}$. The ratios of corresponding parts are equal.

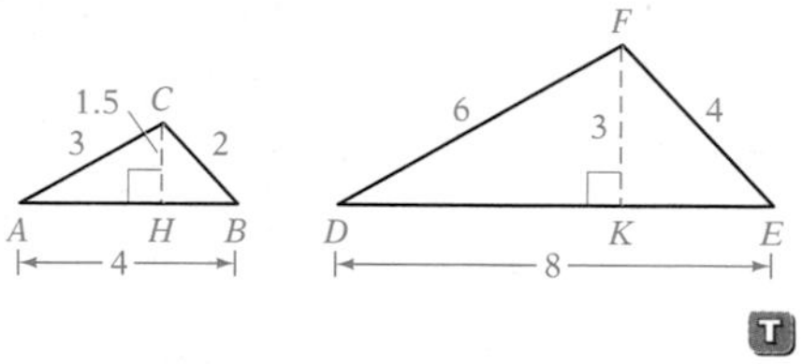

$$\frac{AB}{DE} = \frac{4}{8} = \frac{1}{2}, \quad \frac{AC}{DF} = \frac{3}{6} = \frac{1}{2}, \quad \frac{BC}{EF} = \frac{2}{4} = \frac{1}{2}, \quad \text{and} \quad \frac{CH}{FK} = \frac{1.5}{3} = \frac{1}{2}$$

Since the ratios of corresponding parts are equal, three proportions can be formed using the sides of the triangles.

$$\frac{AB}{DE} = \frac{AC}{DF}, \quad \frac{AB}{DE} = \frac{BC}{EF}, \quad \text{and} \quad \frac{AC}{DF} = \frac{BC}{EF}$$

Three proportions can also be formed by using the sides and height of the triangles.

$$\frac{AB}{DE} = \frac{CH}{FK}, \quad \frac{AC}{DF} = \frac{CH}{FK}, \quad \text{and} \quad \frac{BC}{EF} = \frac{CH}{FK}$$

➡ Triangles *ABC* and *DEF* at the right are similar. Find the height of triangle *ABC*.

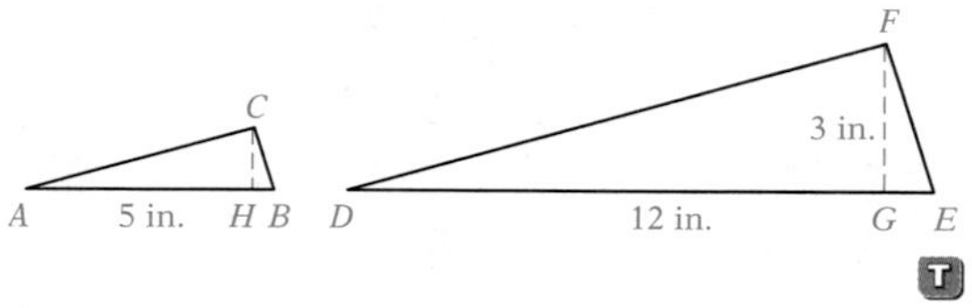

$$\frac{AB}{DE} = \frac{CH}{FG}$$

• Solve a proportion to find the height of triangle *ABC*.

$$\frac{5}{12} = \frac{CH}{3}$$

$$12 \cdot \frac{5}{12} = 12 \cdot \frac{CH}{3}$$

$$5 = 4(CH)$$

$$1.25 = CH$$

The height of triangle *ABC* is 1.25 in.

In-Class Examples (Objective 8.4C)
Triangles *ABC* and *DEF* are similar. Find the perimeter of triangle *DEF*. 33 cm

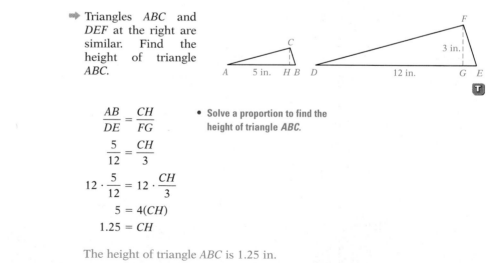

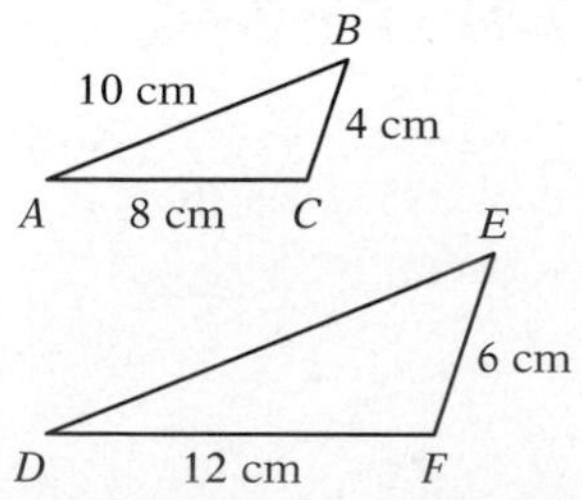

459

The corresponding angles in similar triangles are equal. Therefore, for the triangles in the last example,

$$\angle A = \angle D, \qquad \angle B = \angle E, \qquad \text{and} \qquad \angle C = \angle F$$

It is also true that if the three angles of one triangle are equal respectively to the three angles of another triangle, then the two triangles are similar.

TAKE NOTE
Vertical angles of intersecting lines, parallel lines, and angles of a triangle are discussed in the chapter titled "Geometry."

A line segment $\overline{DE}$ is drawn parallel to the base AB in the triangle at the right. $\angle x = \angle m$ and $\angle y = \angle n$ because corresponding angles are equal. $\angle C = \angle C$, and thus the three angles of triangle DEC are equal respectively to the three angles of triangle ABC. The triangle DEC is similar to the triangle ABC.

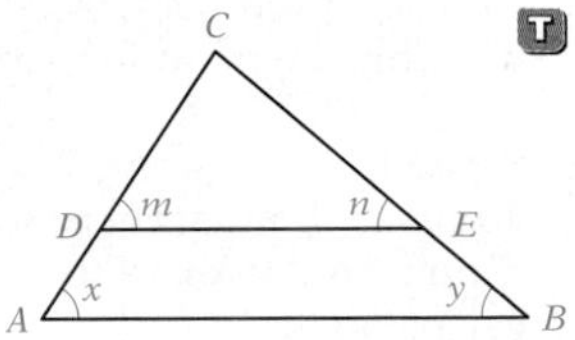

The sum of the three angles of a triangle is 180°. If two angles of one triangle are equal to two angles of another triangle, then the third angles must be equal. Thus we can say that if two angles of one triangle are equal to two angles of another triangle, then the two triangles are similar. This fact is used in Example 4.

Example 4
In the figure at the right, $\overline{AC}$ is parallel to $\overline{BD}$ and angles C and D are right angles. Find the length of DO.

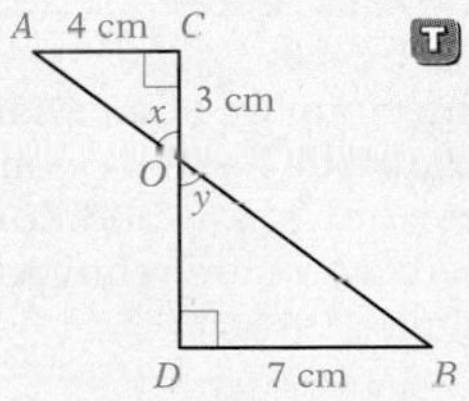

Strategy
$\angle C = \angle D$ because they are right angles.
$\angle x = \angle y$ because they are vertical angles.
Therefore, triangle AOC is similar to triangle BOD because two angles of one triangle are equal to two angles of the other. Use a proportion to find the length of DO.

Solution

$$\frac{AC}{DB} = \frac{CO}{DO}$$

$$\frac{4}{7} = \frac{3}{DO}$$

$$7(DO)\frac{4}{7} = 7(DO)\frac{3}{DO}$$

$$4(DO) = 7(3)$$

$$4(DO) = 21$$

$$DO = 5.25$$

The length of DO is 5.25 cm.

You Try It 4
In the figure at the right, $\overline{AB}$ is parallel to $\overline{DC}$ and angles A and D are right angles. Find the area of triangle AOB.

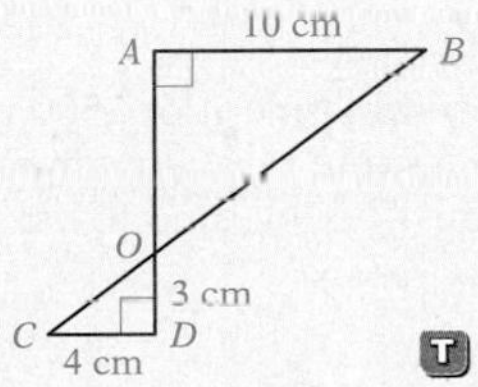

Your strategy

Your solution 37.5 cm²

Solution on p. S25

Optional Student Activity

1. For each of these pairs of similar triangles, write the ratio of corresponding sides in simplest form.

a. $\frac{1}{2}$

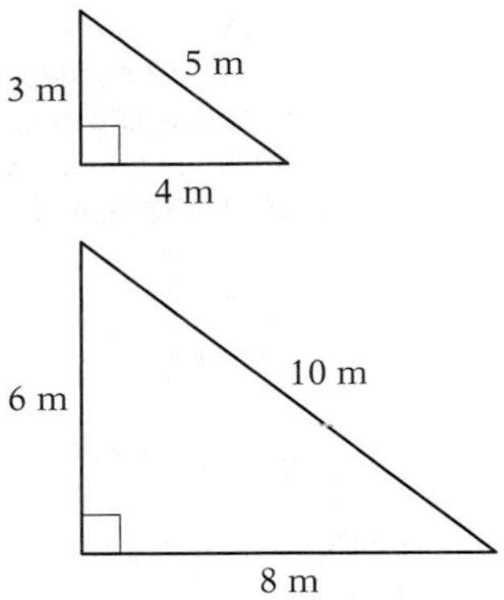

b. $\frac{1}{3}$

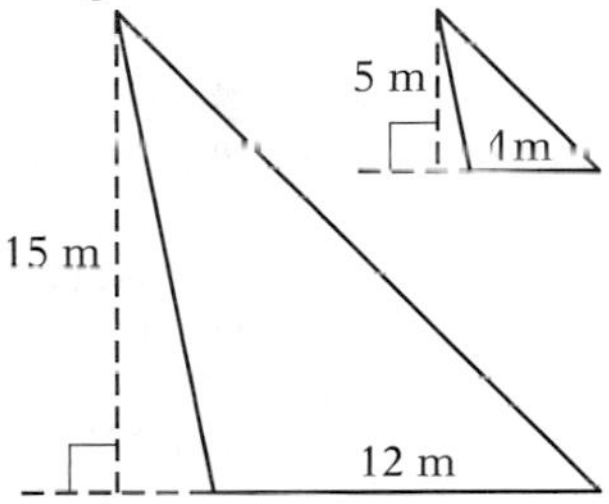

c. $\frac{2}{3}$

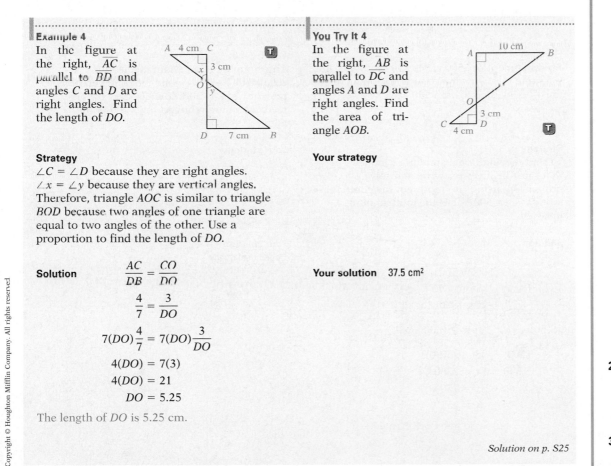

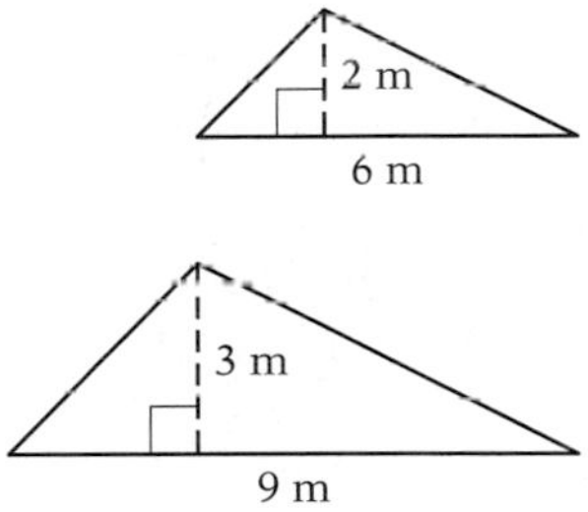

2. Find the area of each triangle in Activity 1.
a. 6 m² and 24 m²; **b.** 10 m² and 90 m²; **c.** 6 m² and $\frac{27}{2}$ m²

3. For each pair of similar triangles in Activity 1, write the ratio of the areas in simplest form.
a. $\frac{1}{4}$; **b.** $\frac{1}{9}$; **c.** $\frac{4}{9}$

4. Use your results from Activities 1 and 3 to make a conjecture about the ratio of the areas of two similar triangles.
The ratio of the areas is the square of the ratio of their corresponding sides.

Objective 8.4D

Concept Check

1. A 2-inch cube of silver weighs 4 lb and is worth \$280. How much is a 3-inch cube of silver worth? \$945
2. Assume that it takes a man 2 h to dig a hole that is 2 m wide, 2 m long, and 2 m deep. Digging at the same rate, how long would it take him to dig a hole 4 m wide, 4 m long, and 4 m deep? 16 h

Optional Student Activity

1. How many minutes does it take a clock's hour hand to move through one degree of revolution? 2 min
2. If $\frac{a \cdot a \cdot a}{a + a + a} = 3$, what is the value of a^2? 9
3. A team won 40 games out of 60 played. How many more games must the team win in succession to raise its record of wins to 80%? 40 games
4. A number h is the harmonic mean of the numbers a and b if the reciprocal of h is equal to the average of the reciprocals of a and b.
   a. Write an expression for the harmonic mean of a and b.
   b. Find the harmonic mean of 10 and 15.

   a. $\frac{2ab}{a + b}$ b. 12

Objective D To solve application problems

VIDEO & DVD CD TUTOR WEB SSM

Example 5
The monthly loan payment for a car is \$28.35 for each \$1000 borrowed. At this rate, find the monthly payment for a \$6000 car loan.

Strategy
To find the monthly payment, write and solve a proportion, using P to represent the monthly car payment.

Solution

$$\frac{28.35}{1000} = \frac{P}{6000}$$

$$6000\left(\frac{28.35}{1000}\right) = 6000\left(\frac{P}{6000}\right)$$

$$170.10 = P$$

The monthly payment is \$170.10.

You Try It 5
Sixteen ceramic tiles are needed to tile an area of 9 ft². At this rate, how many square feet can be tiled using 256 ceramic tiles?

Your strategy

Your solution
144 ft²

Example 6
An investment of \$500 earns \$60 each year. At the same rate, how much additional money must be invested to earn \$90 each year?

Strategy
To find the additional amount of money that must be invested, write and solve a proportion, using x to represent the additional money. Then $500 + x$ is the total amount invested.

Solution

$$\frac{60}{500} = \frac{90}{500 + x}$$

$$\frac{3}{25} = \frac{90}{500 + x}$$

$$25(500 + x)\left(\frac{3}{25}\right) = 25(500 + x)\left(\frac{90}{500 + x}\right)$$

$$(500 + x)3 = 25(90)$$

$$1500 + 3x = 2250$$

$$3x = 750$$

$$x = 250$$

An additional \$250 must be invested.

You Try It 6
Three ounces of a certain medication are required for a 150-pound adult. At the same rate, how many additional ounces of this medication are required for a 200-pound adult?

Your strategy

Your solution
1 additional ounce

Solutions on pp. S25–S26

In-Class Examples (Objective 8.4D)

1. A pre-election survey showed that 5 out of every 8 voters would vote in a special election. At this rate, how many people would be expected to vote in a city of 180,000? 112,500 people
2. In a wildlife preserve, 60 deer are captured, tagged, and then released. Later 200 deer are examined, and 8 of them are found to have tags. Estimate the number of deer in the preserve. 1500 deer

461

8.4 Exercises

Objective A

Solve.

1. $\frac{2x}{3} - \frac{5}{2} = -\frac{1}{2}$
3

2. $\frac{x}{3} - \frac{1}{4} = \frac{1}{12}$
1

3. $\frac{x}{3} - \frac{1}{4} = \frac{x}{4} - \frac{1}{6}$
1

4. $\frac{2y}{9} - \frac{1}{6} = \frac{y}{9} + \frac{1}{6}$
3

5. $\frac{2x-5}{8} + \frac{1}{4} = \frac{x}{8} + \frac{3}{4}$
9

6. $\frac{3x+4}{12} - \frac{1}{3} = \frac{5x+2}{12} - \frac{1}{2}$
2

7. $\frac{6}{2a+1} = 2$
1

8. $\frac{12}{3x-2} = 3$
2

9. $\frac{9}{2x-5} = -2$
$\frac{1}{4}$

10. $\frac{6}{4-3x} = 3$
$\frac{2}{3}$

11. $2 + \frac{5}{x} = 7$
1

12. $3 + \frac{8}{n} = 5$
4

13. $1 - \frac{9}{x} = 4$
-3

14. $3 - \frac{12}{x} = 7$
-3

15. $\frac{2}{y} + 5 = 9$
$\frac{1}{2}$

16. $\frac{6}{x} + 3 = 11$
$\frac{3}{4}$

17. $\frac{3}{x-2} = \frac{4}{x}$
8

18. $\frac{5}{x+3} - \frac{3}{x-1}$
7

19. $\frac{2}{3x-1} = \frac{3}{4x+1}$
5

20. $\frac{5}{3x-4} = \frac{-3}{1-2x}$
-7

21. $\frac{-3}{2x+5} = \frac{2}{x-1}$
-1

22. $\frac{4}{5y-1} = \frac{2}{2y-1}$
-1

23. $\frac{4x}{x-4} + 5 = \frac{5x}{x-4}$
5

24. $\frac{2x}{x+2} - 5 = \frac{7x}{x+2}$
-1

Section 8.4

Suggested Assignment

Exercises 1–79, odds

More challenging problems:
Exercises 81–83

Quick Quiz (Objective 8.4A)

Solve.

1. $5 - \frac{8}{x} = 1$ 2

2. $3 - \frac{4}{x+2} = \frac{2x}{x+2}$ No solution

Answers to Writing Exercises

34. A ratio is the quotient of two quantities that have the same units. A rate is the quotient of two quantities that have different units.

35. A proportion is an equation that states that two ratios or rates are equal.

25. $2 + \frac{3}{a-3} = \frac{a}{a-3}$
no solution

26. $\frac{x}{x+4} = 3 - \frac{4}{x+4}$
no solution

27. $\frac{x}{x-1} = \frac{8}{x+2}$
2, 4

28. $\frac{x}{x+12} = \frac{1}{x+5}$
−6, 2

29. $\frac{2x}{x+4} = \frac{3}{x-1}$
$-\frac{3}{2}, 4$

30. $\frac{5}{3n-8} = \frac{n}{n+2}$
$-\frac{2}{3}, 5$

31. $x + \frac{6}{x-2} = \frac{3x}{x-2}$
3

32. $x - \frac{6}{x-3} = \frac{2x}{x-3}$
6, −1

33. $\frac{8}{y} = \frac{2}{y-2} + 1$
4

Objective B

34. How does a ratio differ from a rate?

35. What is a proportion?

Solve.

36. $\frac{x}{12} = \frac{3}{4}$
9

37. $\frac{6}{x} = \frac{2}{3}$
9

38. $\frac{4}{9} = \frac{x}{27}$
12

39. $\frac{16}{9} = \frac{64}{x}$
36

40. $\frac{x+3}{12} = \frac{5}{6}$
7

41. $\frac{3}{5} = \frac{x-4}{10}$
10

42. $\frac{18}{x+4} = \frac{9}{5}$
6

43. $\frac{2}{11} = \frac{20}{x-3}$
113

44. $\frac{2}{x} = \frac{4}{x+1}$
1

45. $\frac{16}{x-2} = \frac{8}{x}$
−2

46. $\frac{x+3}{4} = \frac{x}{8}$
−6

47. $\frac{x-6}{3} = \frac{x}{5}$
15

48. $\frac{2}{x-1} = \frac{6}{2x+1}$
4

49. $\frac{9}{x+2} = \frac{3}{x-2}$
4

50. $\frac{2x}{7} = \frac{x-2}{14}$
$-\frac{2}{3}$

Quick Quiz (Objective 8.4B)

Solve the proportion.

1. $\frac{3}{15} = \frac{x}{45}$ 9

2. $\frac{x+4}{9} = \frac{2}{3}$ 2

3. $\frac{x}{3} = \frac{x-3}{6}$ −3

4. $\frac{12}{4-x} = \frac{5-x}{x}$ 1, 20

463

Objective C

Triangles *ABC* and *DEF* in Exercises 51 to 58 are similar. Round answers to the nearest tenth.

51. Find side *AC*.

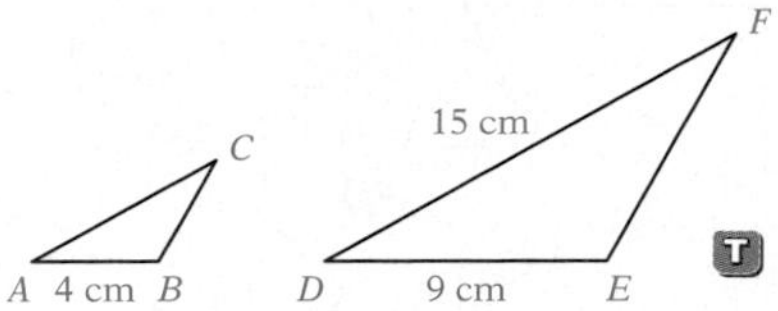

6.7 cm

52. Find side *DE*.

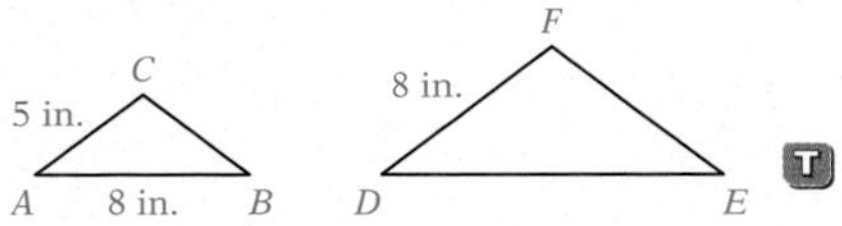

12.8 in.

53. Find the height of triangle *ABC*.

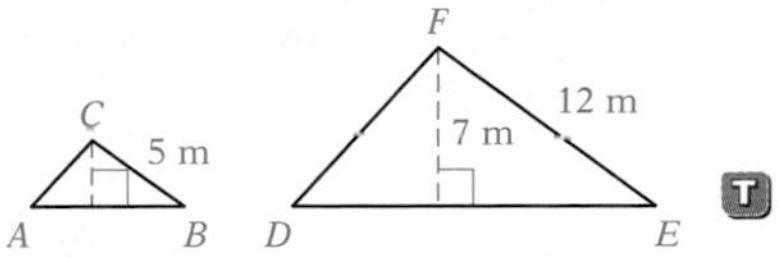

2.9 m

54. Find the height of triangle *DEF*.

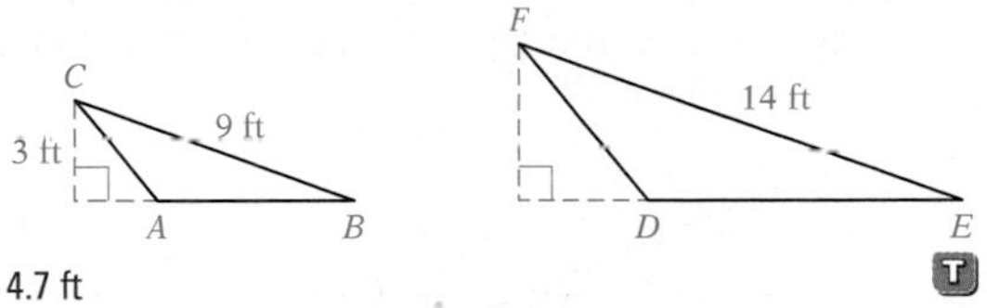

4.7 ft

55. Find the perimeter of triangle *DEF*.

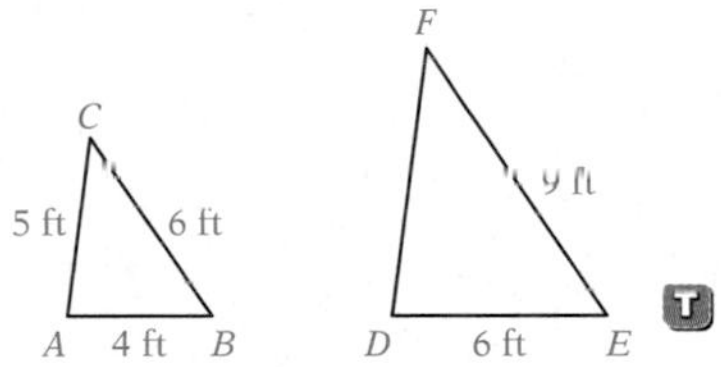

22.5 ft

56. Find the perimeter of triangle *ABC*.

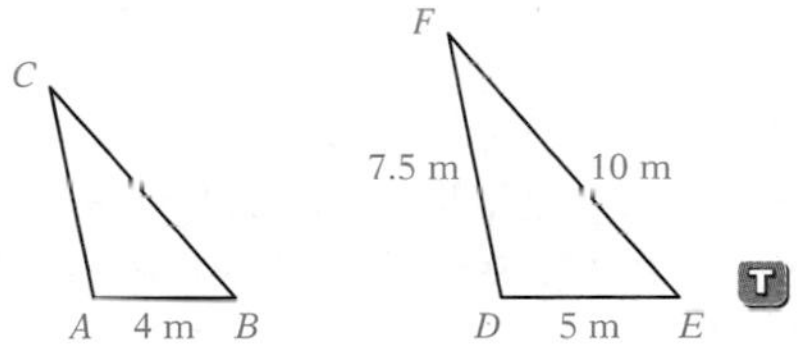

18 m

57. Find the area of triangle *ABC*.

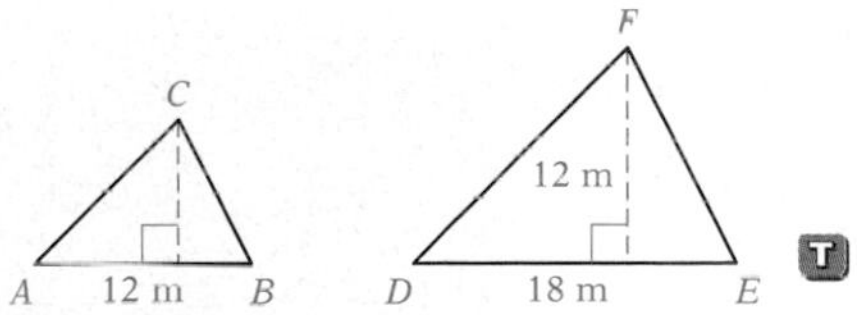

48 m²

58. Find the area of triangle *ABC*.

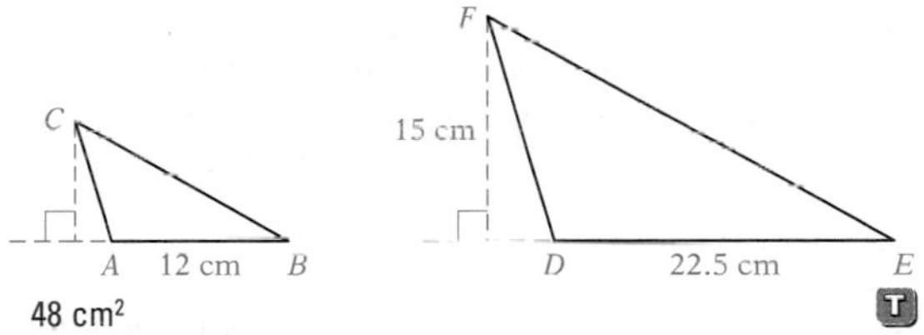

48 cm²

59. Given $\overline{BD} \parallel \overline{AE}$, *BD* measures 5 cm, *AE* measures 8 cm, and *AC* measures 10 cm, find the length of *BC*.

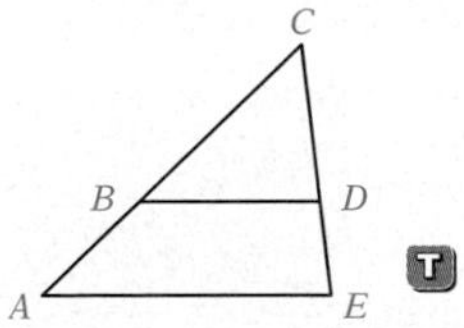

6.25 cm

60. Given $\overline{AC} \parallel \overline{DE}$, *BD* measures 8 m, *AD* measures 12 m, and *BE* measures 6 m, find the length of *BC*.

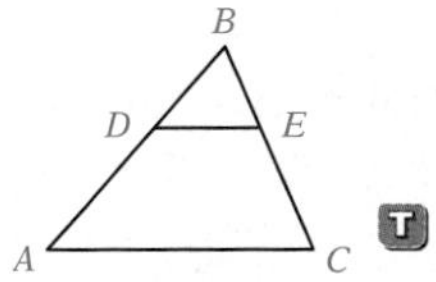

15 m

Quick Quiz (Objective 8.4C)
Given that $\overline{DE} \parallel \overline{AC}$, $BD = 2$ cm, $DA = 5$ cm, and $BE = 3$ cm, find the length of *BC*. 10.5 cm

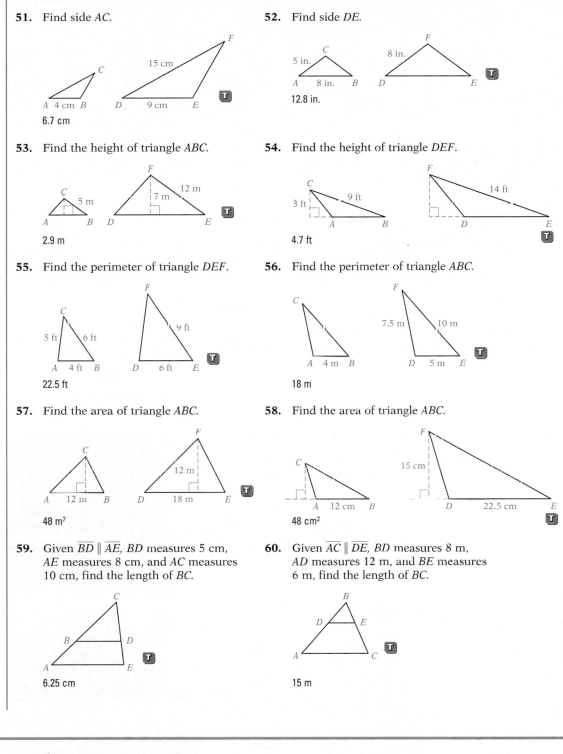

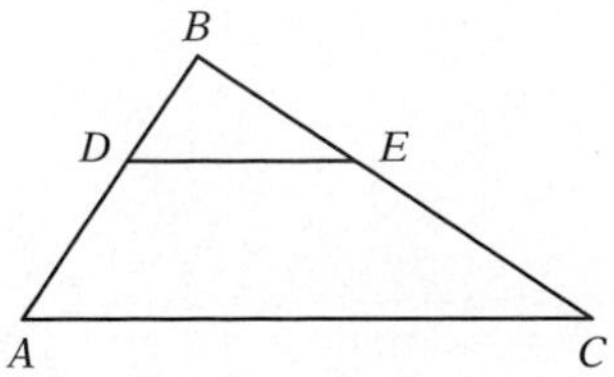

61. Given $\overline{DE} \parallel \overline{AC}$, DE measures 6 in., AC measures 10 in., and AB measures 15 in., find the length of DA.

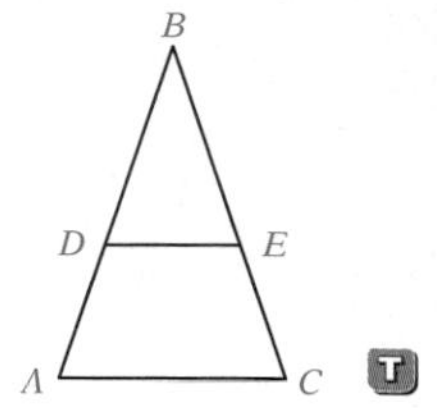

6 in.

62. Given $\overline{MP}$ and $\overline{NQ}$ intersect at O, NO measures 25 ft, MO measures 20 ft, and PO measures 8 ft, find the length of QO.

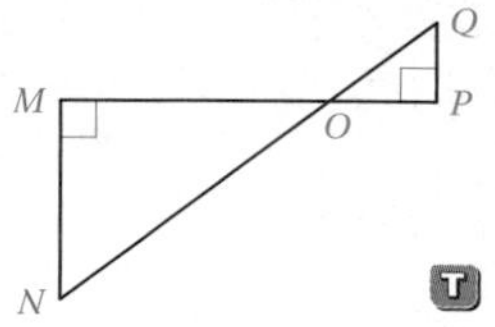

10 ft

63. Given $\overline{MP}$ and $\overline{NQ}$ intersect at O, NO measures 24 cm, MN measures 10 cm, MP measures 39 cm, and QO measures 12 cm, find the length of OP.

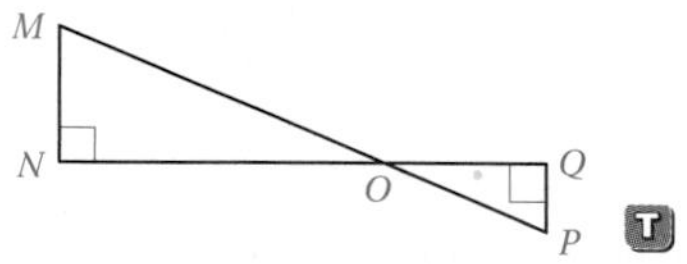

13 cm

64. Given $\overline{MQ}$ and $\overline{NP}$ intersect at O, NO measures 12 m, MN measures 9 m, PQ measures 3 m, and MQ measures 20 m, find the perimeter of triangle OPQ.

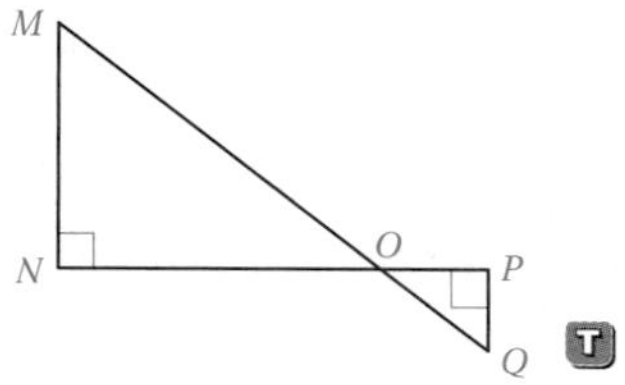

12 m

65. The sun's rays cast a shadow as shown in the diagram at the right. Find the height of the flagpole. Write the answer in terms of feet.
14.375 ft

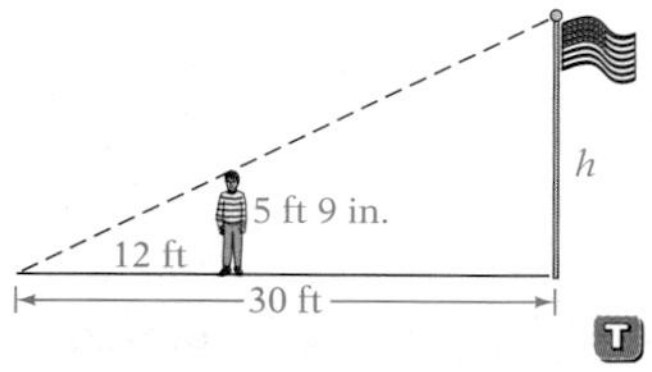

66. Similar triangles can be used as an indirect way to measure inaccessible distances. The diagram at the right represents a river of width DC. The triangles AOB and DOC are similar. The distances AB, BO, and OC can be measured. Find the width of the river.
35 m

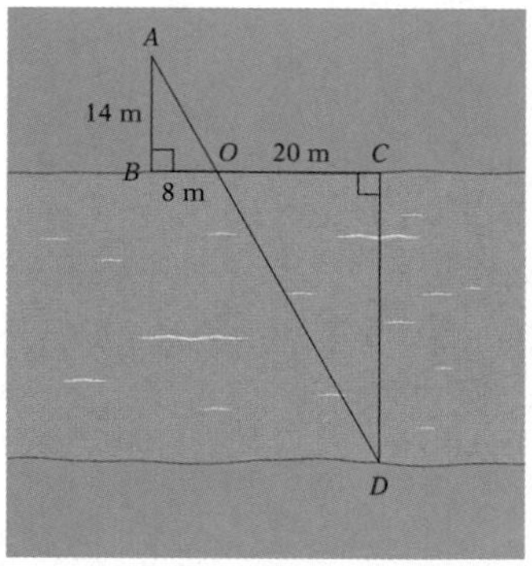

67. The diagram at the right shows how surveyors laid out similar triangles along a ravine. Find the width, w, of the ravine.
82.5 ft

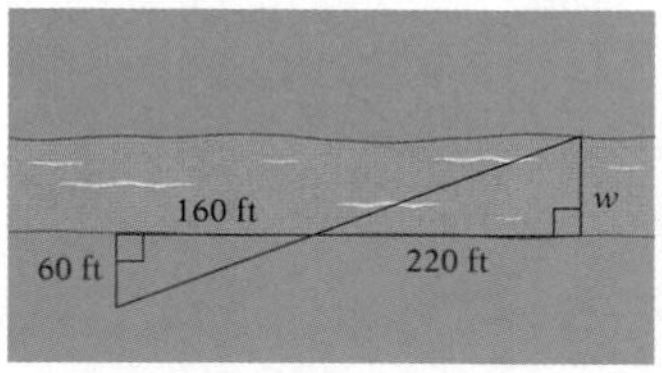

465

Objective D *Application Problems*

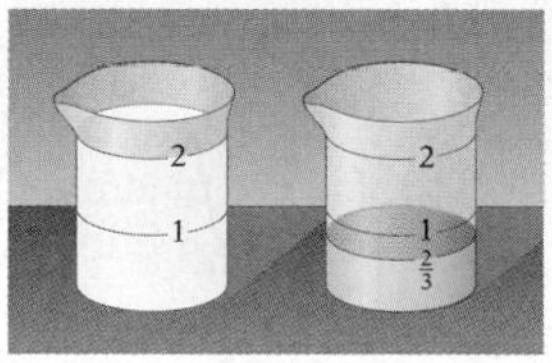

68. Simple syrup used in making some desserts requires 2 cups (c) of sugar for every $\frac{2}{3}$ c of boiling water. At this rate, how many cups of sugar are required for 2 c boiling water?
6 c

69. An exit poll survey showed that 4 out of every 7 voters cast a ballot in favor of an amendment to a city charter. At this rate, how many voters voted in favor of the amendment if 35,000 people voted?
20,000 voters

70. As part of a conservation effort for a lake, 40 fish are caught, tagged, and then released. Later 80 fish are caught. Four of the 80 fish are found to have tags. Estimate the number of fish in the lake.
800 fish

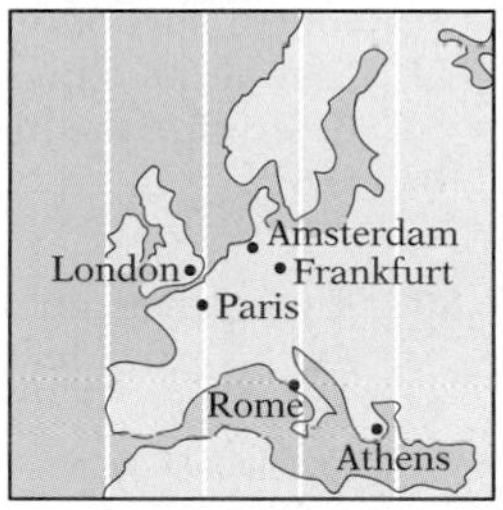

71. On a map, two cities are $2\frac{5}{8}$ in. apart. If $\frac{3}{8}$ in. on the map represents 25 mi, find the number of miles between the two cities.
175 mi

72. The lighting for some billboards is provided by using solar energy. If 3 small solar energy panels can generate 10 W of power, how many panels are necessary to provide 600 W of power?
180 panels

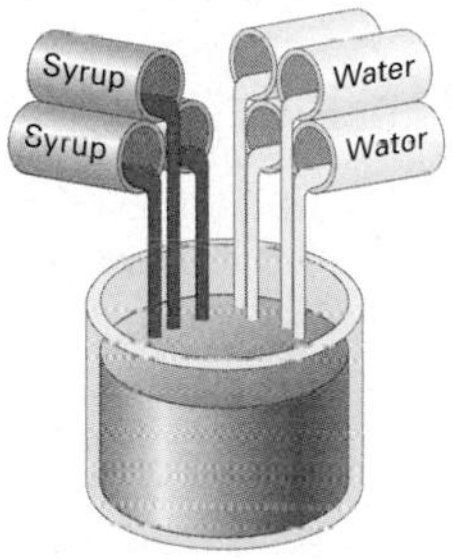

73. A soft drink is made by mixing 4 parts carbonated water with every 3 parts syrup. How many milliliters of water are in 280 ml of soft drink?
160 ml

74. An air conditioning specialist recommends 2 air vents for each 300 ft^2 of floor space. At this rate, how many air vents are required for a 21,000-square-foot office building?
140 air vents

75. A laser printer is rated by the number of pages per minute it can print. An inexpensive laser printer can print 5 pages every 2 min. At this rate, how long would it take to print a document 45 pages long?
18 min

76. A company decides to accept a large shipment of 10,000 computer chips if there are 2 or fewer defects in a sample of 100 randomly chosen chips. Assuming that there are 300 defective chips in the shipment and that the rate of defective chips in the sample is the same as the rate in the shipment, will the shipment be accepted?
no

Optional Student Activity

Place students in groups of three or four, and have them try the measurements in Exercises 70 and 71.

Another ratio that da Vinci recorded was that of the breadth of the shoulders at the widest part to the height of the person. According to da Vinci, this ratio is approximately $\frac{1}{4}$.

Quick Quiz (Objective 8.4D)

John begins a trip with a full tank of gas. When he stops at a service station, it takes 12 gal of gas to refill the 16-gallon tank. John checks his odometer and sees that he drove 276 mi on the 12 gal of gas. At this rate, how far can he drive on the full 16-gallon tank of gas? 368 mi

Answers to Writing Exercises

84. Students' explanations should include each of the following steps:
(1) Find the LCM of the denominators of the fractions in the equation.
(2) Multiply each side of the equation by the LCM of the denominators.
(3) Simplify each side of the equation.
(4) Solve for the variable.

77. To conserve energy and still allow for as much natural lighting as possible, an architect suggests that the ratio of the area of a window to the area of the total wall surface be 5 to 12. Using this ratio, determine the recommended area of a window to be installed in a wall that measures 8 ft by 12 ft.
40 ft^2

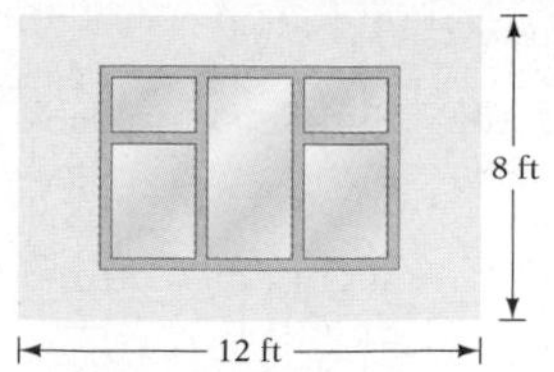

78. Leonardo da Vinci measured various distances on the human body in order to make accurate drawings. He determined that generally the ratio of the kneeling height of a person to the standing height of that person is $\frac{3}{4}$. Using this ratio, determine the height of a person who has a kneeling height of 48 in.
64 in.

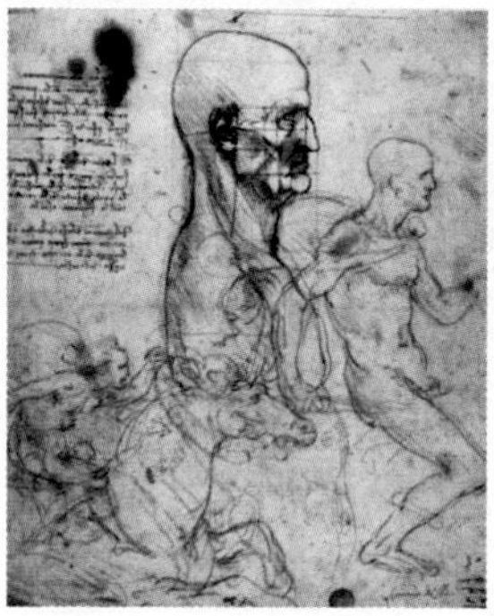

79. In one of Leonardo da Vinci's notebooks he wrote that "... from the top to the bottom of the chin is the sixth part of a face, and it is the fifty-fourth part of the man." Suppose the distance from the top to the bottom of the chin of a person is 1.25 in. Using da Vinci's measurements, determine the height of this person.
67.5 in.

80. A painter estimates that 5 gal of paint will cover 1200 ft^2 of wall space. At this rate, how many additional gallons will be necessary to cover 1680 ft^2?
2 additional gallons

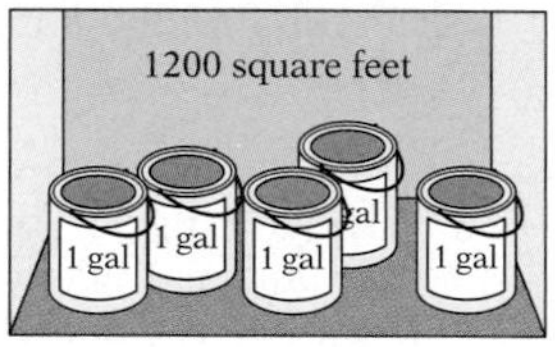

APPLYING THE CONCEPTS

81. Three people put their money together to buy lottery tickets. The first person put in \$25, the second person put in \$30, and the third person put in \$35. One of their tickets was a winning ticket. If they won \$4.5 million, what was the first person's share of the winnings?
\$1.25 million

82. No one belongs to both the Math Club and the Photography Club, but the two clubs join to hold a car wash. Ten members of the Math Club and 6 members of the Photography Club participate. The profits from the car wash are \$120. If each club's profits are proportional to the number of members participating, what share of the profits does the Math Club receive?
\$75

83. A basketball player has made 5 out of every 6 foul shots attempted in one year of play. If 42 foul shots were missed that year, how many shots did the basketball player make?
210 shots

84. Explain the procedure for solving an equation that contains fractions. Include in your discussion how the LCM of the denominators is used to eliminate fractions in the equation.

8.5 Literal Equations

Objective A

To solve a literal equation for one of the variables

VIDEO & DVD CD TUTOR WEB SSM

A **literal equation** is an equation that contains more than one variable. Examples of literal equations are shown at the right.

$2x + 3y = 6$
$4w - 2x + z = 0$

Formulas are used to express a relationship among physical quantities. A **formula** is a literal equation that states rules about measurements. Examples of formulas are shown at the right.

$\frac{1}{R_1} + \frac{1}{R_2} = \frac{1}{R}$ (Physics)
$s = a + (n - 1)d$ (Mathematics)
$A = P + Prt$ (Business)

The Addition and Multiplication Properties can be used to solve a literal equation for one of the variables. The goal is to rewrite the equation so that the variable being solved for is alone on one side of the equation and all the other numbers and variables are on the other side.

➡ Solve $A = P(1 + i)$ for i.

The goal is to rewrite the equation so that i is on one side of the equation and all other variables are on the other side.

$A = P(1 + i)$
$A = P + Pi$ • Use the Distributive Property to remove parentheses.
$A - P = P - P + Pi$ • Subtract P from each side of the equation.
$A - P = Pi$
$\frac{A - P}{P} = \frac{Pi}{P}$ • Divide each side of the equation by P.
$\frac{A - P}{P} = i$

Example 1
Solve $s = a + (n - 1)d$ for d.

Solution
$s = a + (n - 1)d$
$s - a = a - a + (n - 1)d$
$s - a = (n - 1)d$
$\frac{s - a}{n - 1} = \frac{(n - 1)d}{n - 1}$
$\frac{s - a}{n - 1} = d$

You Try It 1
Solve $A = P + Prt$ for r.

Your solution
$r = \frac{A - P}{Pt}$

Solution on p. S26

Objective 8.5A

New Vocabulary
literal equation
formula

Discuss the Concepts
Use the formula

$A = P + Prt$

which gives the value, A, of a principle, P, invested at a rate, r, for a given period of time, t.

1. In what circumstances would it be helpful to have the formula solved for t? for P?
2. Describe the steps you would use to solve the equation for t. What would you do to get the term involving t alone? What would you do then to solve for t?
3. If the formula is to be solved for P, P can appear only once in the formula. How can you rewrite the right side so that it has only one P? What would you do then to solve for P?

Instructor Note
Part 3 in the above Discuss the Concepts is difficult for students. This technique is also discussed in the Instructor Note on the next page.

Concept Check
Use the equation $5a = 7 - 2b$.

1. Solve for a. $a = \frac{7 - 2b}{5}$
2. Solve for b. $b = \frac{7 - 5a}{2}$

In-Class Examples (Objective 8.5A)

1. Solve for b: $A = \frac{1}{2}bh$ $b = \frac{2A}{h}$
2. Solve for x: $3x + 8y = 9$ $x = -\frac{8}{3}y + 3$
3. Solve for y: $7x - y = 12$ $y = 7x - 12$

Instructor Note

Example 4 will be difficult for students. Before doing that example or one similar to it, remind students that when solving $2x + 3x = 10$, they are using the Distributive Property to combine $2x$ and $3x$:

$$2x + 3x = (2 + 3)x = 5x$$

Now each side of the equation can be divided by 5.

For $ax + bx = 10$, the procedure is exactly the same, except that $a + b$ does not simplify further:

$$ax + bx = (a + b)x$$

Now each side of the equation can be divided by $(a + b)$.

Part 1 of the following Optional Student Activity can be used as another example illustrating this technique.

Optional Student Activity

1. Solve the equation $cx - y = bx + 5$ for x.
$$x = \frac{y + 5}{c - b}$$
2. Solve this physics formula for R_1:
$$\frac{1}{R_1} + \frac{1}{R_2} = \frac{1}{R}$$
$$R_1 = \frac{RR_2}{R_2 - R}$$

Example 2

Solve $I = \frac{E}{R + r}$ for R.

Solution

$$I = \frac{E}{R + r}$$
$$(R + r)I = (R + r)\frac{E}{R + r}$$
$$RI + rI = E$$
$$RI + rI - rI = E - rI$$
$$RI = E - rI$$
$$\frac{RI}{I} = \frac{E - rI}{I}$$
$$R = \frac{E - rI}{I}$$

You Try It 2

Solve $s = \frac{A + L}{2}$ for L.

Your solution

$L = 2s - A$

Example 3

Solve $L = a(1 + ct)$ for c.

Solution

$$L = a(1 + ct)$$
$$L = a + act$$
$$L - a = a - a + act$$
$$L - a = act$$
$$\frac{L - a}{at} = \frac{act}{at}$$
$$\frac{L - a}{at} = c$$

You Try It 3

Solve $s = a + (n - 1)d$ for n.

Your solution

$$n = \frac{s - a + d}{d}$$

Example 4

Solve $M = C - rC$ for C.

Solution

$$M = C - rC$$
$$M = (1 - r)C$$
$$\frac{M}{1 - r} = \frac{(1 - r)C}{1 - r}$$
$$\frac{M}{1 - r} = C$$

You Try It 4

Solve $S = C + rC$ for C.

Your solution

$$C = \frac{S}{1 + r}$$

Solutions on p. S26

Section 8.5

Suggested Assignment
Exercises 1–17, odds
More challenging problems:
Exercises 19, 21, 22

8.5 Exercises

Objective A

Solve the formula for the given variable.

1. $A = \frac{1}{2}bh; h$ (Geometry)
$h = \frac{2A}{b}$

2. $P = a + b + c; b$ (Geometry)
$b = P - a - c$

3. $d = rt; t$ (Physics)
$t = \frac{d}{r}$

4. $E = IR; R$ (Physics)
$R = \frac{E}{I}$

5. $PV = nRT; T$ (Chemistry)
$T = \frac{PV}{nR}$

6. $A = bh; h$ (Geometry)
$h = \frac{A}{b}$

7. $P = 2L + 2W; L$ (Geometry)
$L = \frac{P - 2W}{2}$

8. $F = \frac{9}{5}C + 32; C$ (Temperature conversion)
$C = \frac{5F - 160}{9}$

9. $A = \frac{1}{2}h(b_1 + b_2); b_1$ (Geometry)
$b_1 = \frac{2A - hb_2}{h}$

10. $C = \frac{5}{9}(F - 32); F$ (Temperature conversion)
$F = \frac{9C + 160}{5}$

11. $V = \frac{1}{3}Ah; h$ (Geometry)
$h = \frac{3V}{A}$

12. $P = R - C; C$ (Business)
$C = R - P$

13. $R = \frac{C - S}{t}; S$ (Business)
$S = C - Rt$

14. $P = \frac{R - C}{n}; R$ (Business)
$R = Pn + C$

15. $A = P + Prt; P$ (Business)
$P = \frac{A}{1 + rt}$

16. $T = fm - gm; m$ (Engineering)
$m = \frac{T}{f - g}$

17. $A = Sw + w; w$ (Physics)
$w = \frac{A}{S + 1}$

18. $a = S - Sr; S$ (Mathematics)
$S = \frac{a}{1 - r}$

APPLYING THE CONCEPTS

The surface area of a right circular cylinder is given by the formula $S = 2\pi rh + 2\pi r^2$, where r is the radius of the base, and h is the height of the cylinder.

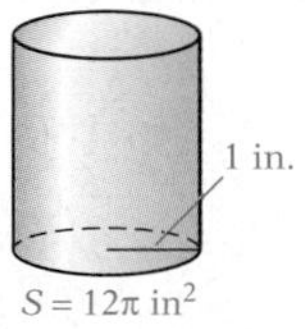

19. **a.** Solve the formula $S = 2\pi rh + 2\pi r^2$ for h. $h = \frac{S - 2\pi r^2}{2\pi r}$
b. Use your answer to part **a** to find the height of a right circular cylinder when the surface area is 12π in² and the radius is 1 in. 5 in.
c. Use your answer to part **a** to find the height of a right circular cylinder when the surface area is 24π in² and the radius is 2 in. 4 in.

Quick Quiz (Objective 8.5A)

1. Use the equation $4x + y = 8$.
a. Solve for y. $y = -4x + 8$
b. Solve for x. $x = -\frac{1}{4}y + 2$

2. Use the formula $P = C + Cr$.
a. Solve for r. $r = \frac{P - C}{C}$
b. Solve for C. $C = \frac{P}{1 + r}$

470 Chapter 8 / Rational Expressions

When markup is based on selling price, the selling price of a product is given by the formula $S = C + rC$, where C is the cost of the product, and r is the markup rate.

20. a. Solve the formula $S = C + rC$ for r. $r = \dfrac{S - C}{C}$
    b. Use your answer to part **a** to find the markup rate on a tennis racket when the cost is \$112 and the selling price is \$180. $66.\overline{6}\%$
    c. Use your answer to part **a** to find the markup rate on a radio when the cost is \$120 and the selling price is \$172. $43.\overline{3}\%$

Break-even analysis is a method used to determine the sales volume required for a company to break even, or experience neither a profit nor a loss on the sale of a product. The **break-even point** represents the number of units that must be made and sold for income from sales to equal the cost of the product. The break-even point can be calculated using the formula $B = \dfrac{F}{S - V}$, where F is the fixed costs, S is the selling price per unit, and V is the variable costs per unit.

21. a. Solve the formula $B = \dfrac{F}{S - V}$ for S. $S = \dfrac{F + BV}{B}$
    b. Use your answer to part **a** to find the required selling price per desk for a company to break even. The fixed costs are \$20,000, the variable costs per desk are \$80, and the company plans to make and sell 200 desks. \$180
    c. Use your answer to part **a** to find the required selling price per camera for a company to break even. The fixed costs are \$15,000, the variable costs per camera are \$50, and the company plans to make and sell 600 cameras. \$75

Resistors are used to control the flow of current. The total resistance of two resistors in a circuit can be given by the formula $R = \dfrac{1}{\dfrac{1}{R_1} + \dfrac{1}{R_2}}$, where R_1 and R_2 are the two resistors in the circuit. Resistance is measured in ohms.

22. a. Solve the formula $R = \dfrac{1}{\dfrac{1}{R_1} + \dfrac{1}{R_2}}$ for R_1. $R_1 = \dfrac{R_2 R}{R_2 - R}$
    b. Use your answer to part **a** to find the resistance in R_1 if the resistance in R_2 is 30 ohms and the total resistance is 12 ohms. 20 ohms
    c. Use your answer to part **a** to find the resistance in R_1 if the resistance in R_2 is 15 ohms and the total resistance is 6 ohms. 10 ohms

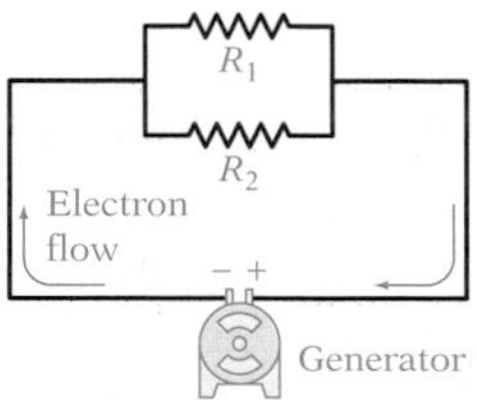

471

8.6 Work and Uniform Motion Problems

Objective A — To solve work problems

VIDEO & DVD CD TUTOR WEB SSM

Point of Interest

The following problem was recorded in the *Jiuzhang*, a Chinese text that dates to the Han dynasty (about 200 B.C. to A.D. 200). "A reservoir has 5 channels bringing water to it. The first can fill the reservoir in $\frac{1}{3}$ day, the second in 1 day, the third in $2\frac{1}{2}$ days, the fourth in 3 days, and the fifth in 5 days. If all channels are open, how long does it take to fill the reservoir?" This problem is the earliest known work problem.

If a painter can paint a room in 4 h, then in 1 h the painter can paint $\frac{1}{4}$ of the room. The painter's rate of work is $\frac{1}{4}$ of the room each hour. The **rate of work** is the part of a task that is completed in one unit of time.

A pipe can fill a tank in 30 min. This pipe can fill $\frac{1}{30}$ of the tank in 1 min. The rate of work is $\frac{1}{30}$ of the tank each minute. If a second pipe can fill the tank in x min, the rate of work for the second pipe is $\frac{1}{x}$ of the tank each minute.

In solving a work problem, the goal is to determine the time it takes to complete a task. The basic equation that is used to solve work problems is

Rate of work × time worked = part of task completed

For example, if a faucet can fill a sink in 6 min, then in 5 min the faucet can fill $\frac{1}{6} \times 5 = \frac{5}{6}$ of the sink. In 5 min the faucet completes $\frac{5}{6}$ of the task.

➡ A painter can paint a wall in 20 min. The painter's apprentice can paint the same wall in 30 min. How long will it take them to paint the wall when they work together?

Strategy for Solving a Work Problem

1. For each person or machine, write a numerical or variable expression for the rate of work, the time worked, and the part of the task completed. The results can be recorded in a table.

TAKE NOTE

Use the information given in the problem to fill in the "Rate" and "Time" columns of the table. Fill in the "Part Completed" column by multiplying the two expressions you wrote in each row.

Unknown time to paint the wall working together: t

	Rate of Work	·	Time Worked	=	Part of Task Completed
Painter	$\frac{1}{20}$	·	t	=	$\frac{t}{20}$
Apprentice	$\frac{1}{30}$	·	t	=	$\frac{t}{30}$

2. Determine how the parts of the task completed are related. Use the fact that the sum of the parts of the task completed must equal 1; the complete task.

$$\frac{t}{20} + \frac{t}{30} = 1$$

• The sum of the part of the task completed by the painter and the part of the task completed by the apprentice is 1.

$$60\left(\frac{t}{20} + \frac{t}{30}\right) = 60 \cdot 1$$

• Multiply by the LCM of 20 and 30.

$$3t + 2t = 60$$
$$5t = 60$$
$$t = 12$$

Working together, they will paint the wall in 12 min.

Objective 8.6A

New Vocabulary

rate of work

New Formulas

Rate of work × time worked = part of task completed

Instructor Note

The answer to the problem posed in the Point of Interest is $\frac{15}{74}$ day.

Discuss the Concepts

1. It takes a janitorial crew 5 h to clean a company's offices. What fraction of the job does the crew complete in x hours? $\frac{x}{5}$ of the job
2. Only two people worked on a job and together they completed it. One person completed $\frac{t}{30}$ of the job and the other person completed $\frac{t}{20}$ of the job. Write an equation to express the fact that together they completed the whole job. $\frac{t}{30} + \frac{t}{20} = 1$

Concept Check

A painter can paint a ceiling in 60 min. The painter's apprentice can paint the same ceiling in 90 min.

a. How long will it take to paint the ceiling if they work together? 36 min

(Continued on next page)

In-Class Examples (Objective 8.6A)

1. Pump A can fill a pool in 6 h, and Pump B can fill the same pool in 3 h. How long will it take to fill the pool if both pumps are used? 2 h
2. One member of a gardening team can mow and clean up a lawn in 9 h. With both members of the team working, the job can be done in 6 h. How long would it take the second member of the team, working alone, to do the job? 18 h

(Continued)

b. What fraction of the job does the painter complete? What fraction of the job does the apprentice complete? Are these fractions equal to the complete job? Painter: $\frac{3}{5}$; apprentice: $\frac{2}{5}$; Yes. $\frac{3}{5} + \frac{2}{5} = 1$, the complete job

c. Could the answer to part **a** ever be more than 60 min? Why or why not? No. The painter can complete the entire job in 60 min. With the apprentice helping, it must take less time to do the job than it does for the painter working alone.

Optional Student Activity

1. A bricklayer is installing a walkway in front of a house. She figures it will take her 6 days to complete the job. However, on the fifth day, the temperature rises and she works more slowly, completing only $\frac{1}{12}$ of the job. Her rate of work is back to normal the next day. How long does it take her to complete the entire job? $6\frac{1}{2}$ days
2. One press can print the weekly edition of a newspaper in 12 h, a second press can complete the job in 8 h, and a third press can complete the job in 6 h. How long would it take to print the newspaper with all three presses operating? $2\frac{2}{3}$ h

Example 1

A small water pipe takes three times longer to fill a tank than does a large water pipe. With both pipes open, it takes 4 h to fill the tank. Find the time it would take the small pipe, working alone, to fill the tank.

Strategy

- Time for large pipe to fill the tank: t
  Time for small pipe to fill the tank: $3t$

Fills tank in $3t$ hours — Fills tank in t hours

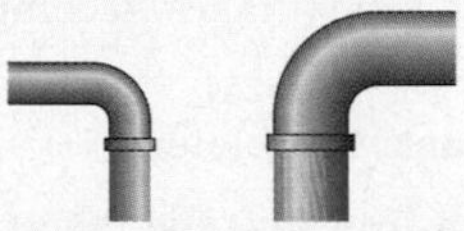

Fills $\frac{4}{3t}$ of the tank in 4 hours — Fills $\frac{4}{t}$ of the tank in 4 hours

	Rate	*Time*	*Part*
Small pipe	$\frac{1}{3t}$	4	$\frac{4}{3t}$
Large pipe	$\frac{1}{t}$	4	$\frac{4}{t}$

- The sum of the parts of the task completed by each pipe must equal 1.

Solution

$$\frac{4}{3t} + \frac{4}{t} = 1$$

$$3t\left(\frac{4}{3t} + \frac{4}{t}\right) = 3t \cdot 1$$

$$4 + 12 = 3t$$

$$16 = 3t$$

$$\frac{16}{3} = t$$

$$3t = 3\left(\frac{16}{3}\right) = 16$$

The small pipe working alone takes 16 h to fill the tank.

You Try It 1

Two computer printers that work at the same rate are working together to print the payroll checks for a large corporation. After they work together for 2 h, one of the printers quits. The second requires 3 h more to complete the payroll checks. Find the time it would take one printer, working alone, to print the payroll.

Your strategy

Your solution

7 h

Solution on p. S26

Objective B To solve uniform motion problems

VIDEO & DVD · CD TUTOR · WEB · SSM

A car that travels constantly in a straight line at 30 mph is in uniform motion. **Uniform motion** means that the speed or direction of an object does not change.

The basic equation used to solve uniform motion problems is

$$\textbf{Distance} = \textbf{rate} \times \textbf{time}$$

An alternative form of this equation can be written by solving the equation for time.

$$\frac{\textbf{Distance}}{\textbf{Rate}} = \textbf{time}$$

This form of the equation is useful when the total time of travel for two objects or the time of travel between two points is known.

➡ The speed of a boat in still water is 20 mph. The boat traveled 75 mi down a river in the same amount of time it took to travel 45 mi up the river. Find the rate of the river's current.

Strategy for Solving a Uniform Motion Problem

1. For each object, write a numerical or variable expression for the distance, rate, and time. The results can be recorded in a table.

TAKE NOTE
Use the information given in the problem to fill in the "Distance" and "Rate" columns of the table. Fill in the "Time" column by dividing the two expressions you wrote in each row.

The unknown rate of the river's current: r

	Distance	÷	*Rate*	=	*Time*
Down river	75	÷	$20 + r$	=	$\frac{75}{20 + r}$
Up river	45	÷	$20 - r$	=	$\frac{45}{20 - r}$

2. Determine how the times traveled by each object are related. For example, it may be known that the times are equal, or the total time may be known.

$$\frac{75}{20 + r} = \frac{45}{20 - r}$$

• The time down the river is equal to the time up the river.

$$(20 + r)(20 - r)\frac{75}{20 + r} = (20 + r)(20 - r)\frac{45}{20 - r}$$

• Multiply by the LCM of the denominators.

$$(20 - r)75 = (20 + r)45$$
$$1500 - 75r = 900 + 45r$$
$$-120r = -600$$
$$r = 5$$

The rate of the river's current is 5 mph.

Objective 8.6B

Vocabulary to Review
uniform motion

New Formulas
Distance = rate × time
or
Distance ÷ rate = time

Discuss the Concepts

1. A plane flies 300 mph in calm air and the rate of the wind is r mph.
   a. Write an expression to represent the rate of the plane flying with the wind. $300 + r$
   b. Write an expression to represent the rate of the plane flying against the wind. $300 - r$
2. The rate of a river's current is 2 mph. The rowing rate of a rowing crew in calm water is r.
   a. Write an expression to represent the rate of the rowing crew when traveling with the current. $r + 2$
   b. Write an expression to represent the rate of the rowing crew when traveling against the current. $r - 2$

Concept Check
Marlys can row a boat 3 mph faster than she can swim. She is able to row 10 mi in the same time it takes her to swim 4 mi. Find the rate at which Marlys swims. 2 mph

In-Class Examples (Objective 8.6B)

1. An express bus travels 300 mi in the same amount of time that a car travels 270 mi. The rate of the car is 6 mph less than the rate of the bus. Find the rate of the car. 54 mph
2. A plane can fly at a rate of 215 mph in calm air. Traveling with the wind, the plane flew 750 mi in the same amount of time that it flew 540 mi against the wind. Find the rate of the wind. 35 mph

Optional Student Activity

1. A plane flew from St. Louis to Boston, a distance of *d* miles, at an average rate of 400 mph. Due to prevailing winds, on the return trip the plane flew at an average rate of 500 mph.
   a. Write an expression for the total flying time. $\frac{9d}{2000}$
   b. Find the average rate for the entire round trip. 444.4 mph
2. For the first 5 mi of a 10-mile race, Ray's rate was 10 mph. For the last 5 mi, his rate slowed to 8 mph. How long did it take Ray to complete the race? What was his average rate for the race?
   $1\frac{1}{8}$ h; $8\frac{8}{9}$ mph
3. Friends who live 40 mi apart decide to bicycle toward each other and meet for a picnic lunch. The first friend bikes at a rate of 20 mph, and the second friend bikes at a rate of 15 mph. How far from the first friend's home will they meet for lunch?
   $22\frac{6}{7}$ mi
4. By increasing your speed by 5 mph, you can drive the 165-mile trip to your hometown in 15 min less time than it usually takes you to drive the trip. How fast do you usually drive? 55 mph

Example 2

A cyclist rode the first 20 mi of a trip at a constant rate. For the next 16 mi, the cyclist reduced the speed by 2 mph. The total time for the 36 mi was 4 h. Find the rate of the cyclist for each leg of the trip.

Strategy

- Rate for the first 20 mi: r
  Rate for the next 16 mi: $r - 2$

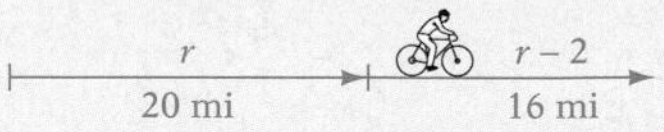

	Distance	*Rate*	*Time*
First 20 mi	20	r	$\frac{20}{r}$
Next 16 mi	16	$r - 2$	$\frac{16}{r-2}$

- The total time for the trip was 4 h.

Solution

$$\frac{20}{r} + \frac{16}{r-2} = 4$$
$$r(r-2)\left[\frac{20}{r} + \frac{16}{r-2}\right] = r(r-2) \cdot 4$$
$$(r-2)20 + 16r = (r^2 - 2r) \cdot 4$$
$$20r - 40 + 16r = 4r^2 - 8r$$
$$36r - 40 = 4r^2 - 8r$$

Solve the quadratic equation by factoring.

$$0 = 4r^2 - 44r + 40$$
$$0 = 4(r^2 - 11r + 10)$$
$$0 = 4(r - 10)(r - 1)$$

$r - 10 = 0 \qquad r - 1 = 0$

$r = 10 \qquad r = 1$

The solution $r = 1$ mph is not possible, because the rate on the last 16 mi would then be -1 mph.

10 mph was the rate for the first 20 mi.
8 mph was the rate for the next 16 mi.

You Try It 2

The total time it took for a sailboat to sail back and forth across a lake 6 km wide was 2 h. The rate sailing back was three times the rate sailing across. Find the rate sailing out across the lake.

Your strategy

Your solution

4 km/h

Solution on p. S26

475

8.6 Exercises

Objective A *Application Problems*

1. A park has two sprinklers that are used to fill a fountain. One sprinkler can fill the fountain in 3 h, whereas the second sprinkler can fill the fountain in 6 h. How long will it take to fill the fountain with both sprinklers operating?
2 h

	Rate	*Time*	*Part*
First	$\frac{1}{3}$	t	$\frac{t}{3}$
Second	$\frac{1}{6}$	t	$\frac{t}{6}$

2. One grocery clerk can stock a shelf in 20 min, whereas a second clerk requires 30 min to stock the same shelf. How long would it take to stock the shelf if the two clerks worked together?
12 min

3. One person with a skiploader requires 12 h to remove a large quantity of earth. A second, larger skiploader can remove the same amount of earth in 4 h. How long would it take to remove the earth with both skiploaders working together?
3 h

4. An experienced painter can paint a fence twice as fast as an inexperienced painter. Working together, the painters require 4 h to paint the fence. How long would it take the experienced painter working alone to paint the fence?
6 h

5. One computer can solve a complex prime factorization problem in 75 h. A second computer can solve the same problem in 50 h. How long would it take both computers, working together, to solve the problem?
30 h

6. A new machine can make 10,000 aluminum cans three times as fast as an older machine. With both machines working, 10,000 cans can be made in 9 h. How long would it take the new machine, working alone, to make the 10,000 cans?
12 h

7. A small air conditioner can cool a room 5° in 75 min. A larger air conditioner can cool the room 5° in 50 min. How long would it take to cool the room 5° with both air conditioners working?
30 min

8. One printing press can print the first edition of a book in 55 min, whereas a second printing press requires 66 min to print the same number of copies. How long would it take to print the first edition with both presses operating?
30 min

9. Two oil pipelines can fill a small tank in 30 min. Using one of the pipelines would require 45 min to fill the tank. How long would it take the second pipeline alone to fill the tank?
90 min

Section 8.6

Suggested Assignment
Exercises 1–33, odds
More challenging problems:
Exercises 34–37

Quick Quiz (Objective 8.6A)
One hose can fill a child's backyard pool in 24 min. A larger hose can fill the pool in 12 min. If the two hoses are used together, how long will it take to fill the pool? 8 min

10. Working together, two dock workers can load a crate in 6 min. One dock worker, working alone, can load the crate in 15 min. How long would it take the second dock worker, working alone, to load the crate?
10 min

11. A mason can construct a retaining wall in 10 h. With the mason's apprentice assisting, the task would take 6 h. How long would it take the apprentice working alone to construct the wall?
15 h

12. A mechanic requires 2 h to repair a transmission, whereas an apprentice requires 6 h to make the same repairs. The mechanic worked alone for 1 h and then stopped. How long will it take the apprentice, working alone, to complete the repairs?
3 h

13. One computer technician can wire a modem in 4 h, whereas it takes 6 h for a second technician to do the same job. After working alone for 2 h, the first technician quit. How long will it take the second technician to complete the wiring?
3 h

14. A wallpaper hanger requires 2 h to hang the wallpaper on one wall of a room. A second wallpaper hanger requires 4 h to hang the same amount of paper. The first wallpaper hanger worked alone for 1 h and then quit. How long will it take the second wallpaper hanger, working alone, to complete the wall?
2 h

15. Two welders who work at the same rate are welding the girders of a building. After they work together for 10 h, one of the welders quits. The second welder requires 20 more hours to complete the welds. Find the time it would have taken one of the welders, working alone, to complete the welds.
40 h

16. A large and a small heating unit are being used to heat the water of a pool. The larger unit, working alone, requires 8 h to heat the pool. After both units have been operating for 2 h, the larger unit is turned off. The small unit requires 9 h more to heat the pool. How long would it take the small unit, working alone, to heat the pool?
$14\frac{2}{3}$ h

17. Two machines that fill cereal boxes work at the same rate. After they work together for 7 h, one machine breaks down. The second machine requires 14 h more to finish filling the boxes. How long would it have taken one of the machines, working alone, to fill the boxes?
28 h

18. A large and a small drain are opened to drain a pool. The large drain can empty the pool in 6 h. After both drains have been open for 1 h, the large drain becomes clogged and is closed. The smaller drain remains open and requires 9 h more to empty the pool. How long would it have taken the small drain, working alone, to empty the pool?
12 h

477

Objective B Application Problems

19. Commuting from work to home, a lab technician traveled 10 mi at a constant rate through congested traffic. On reaching the expressway, the technician increased the speed by 20 mph. An additional 20 mi was traveled at the increased speed. The total time for the trip was 1 h. Find the rate of travel through the congested traffic.
20 mph

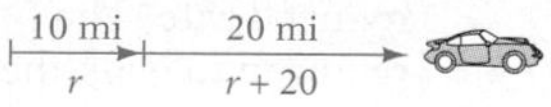

20. The president of a company traveled 1800 mi by jet and 300 mi on a prop plane. The rate of the jet was four times the rate of the prop plane. The entire trip took a total of 5 h. Find the rate of the jet plane.
600 mph

21. As part of a conditioning program, a jogger ran 8 mi in the same time a cyclist rode 20 mi. The rate of the cyclist was 12 mph faster than the rate of the jogger. Find the rate of the jogger and that of the cyclist.
jogger: 8 mph; cyclist: 20 mph

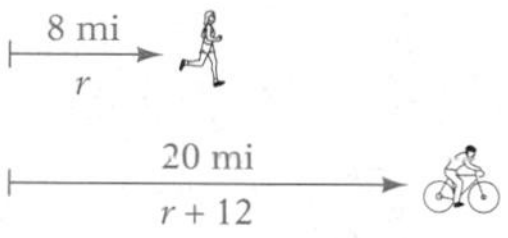

22. An express train travels 600 mi in the same amount of time it takes a freight train to travel 360 mi. The rate of the express train is 20 mph faster than that of the freight train. Find the rate of each train.
freight train: 30 mph; express train: 50 mph

23. To assess the damage done by a fire, a forest ranger traveled 1080 mi by jet and then an additional 180 mi by helicopter. The rate of the jet was 4 times the rate of the helicopter. The entire trip took a total of 5 h. Find the rate of the jet.
360 mph

24. A twin-engine plane can fly 800 mi in the same time that it takes a single-engine plane to fly 600 mi. The rate of the twin-engine plane is 50 mph faster than that of the single-engine plane. Find the rate of the twin-engine plane.
200 mph

25. The rate of a bicyclist is 7 mph greater than the rate of a long-distance runner. The bicyclist travels 30 mi in the same amount of time as it takes the runner to travel 16 mi. Find the rate of the runner.
8 mph

26. A car and a bus leave a town at 1 P.M. and head for a town 300 mi away. The rate of the car is twice the rate of the bus. The car arrives 5 h ahead of the bus. Find the rate of the car.
60 mph

27. A car is traveling at a rate that is 36 mph greater than the rate of a cyclist. The car travels 384 mi in the same time it takes the cyclist to travel 96 mi. Find the rate of the car.
48 mph

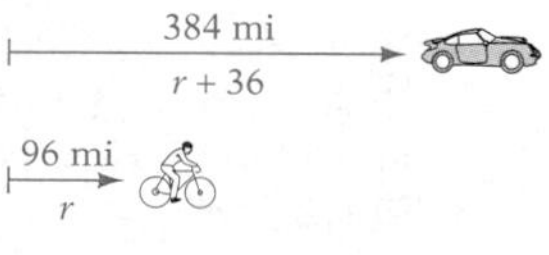

Quick Quiz (Objective 8.6B)

1. A plane flies 460 mph in calm air. Flying with the wind, the plane can travel 1560 mi in the same amount of time that it takes to travel 1200 mi flying against the wind. Find the rate of the wind. 60 mph

478 Chapter 8 / Rational Expressions

28. A backpacker hiking into a wilderness area walked 9 mi at a constant rate and then reduced this rate by 1 mph. Another 4 mi was hiked at this reduced rate. The time required to hike the 4 mi was 1 h less than the time required to walk the 9 mi. Find the rate at which the hiker walked the first 9 mi.
3 mph

29. A plane can fly 180 mph in calm air. Flying with the wind, the plane can fly 600 mi in the same amount of time it takes to fly 480 mi against the wind. Find the rate of the wind.
20 mph

30. A commercial jet can fly 550 mph in calm air. Traveling with the jet stream, the plane flew 2400 mi in the same amount of time it takes to fly 2000 mi against the jet stream. Find the rate of the jet stream.
50 mph

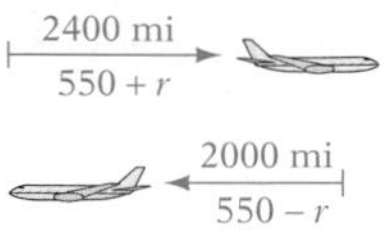

31. A cruise ship can sail at 28 mph in calm water. Sailing with the gulf current, the ship can sail 170 mi in the same amount of time that it can sail 110 mi against the gulf current. Find the rate of the gulf current.
6 mph

32. Rowing with the current of a river, a rowing team can row 25 mi in the same amount of time it takes to row 15 mi against the current. The rate of the rowing team in calm water is 20 mph. Find the rate of the current.
5 mph

33. On a recent trip, a trucker traveled 330 mi at a constant rate. Because of road construction, the trucker then had to reduce the speed by 25 mph. An additional 30 mi was traveled at the reduced rate. The total time for the entire trip was 7 h. Find the rate of the trucker for the first 330 mi.
55 mph

APPLYING THE CONCEPTS

34. One pipe can fill a tank in 2 h, a second pipe can fill the tank in 4 h, and a third pipe can fill the tank in 5 h. How long will it take to fill the tank with all three pipes working?
$1\frac{1}{19}$ h

35. A mason can construct a retaining wall in 10 h. The mason's more-experienced apprentice can do the same job in 15 h. How long would it take the mason's less-experienced apprentice to do the job if, working together, all three can complete the job in 5 h?
30 h

36. A surveyor traveled 32 mi by canoe and then hiked 4 mi. The rate by boat was four times the rate on foot. If the time spent walking was 1 h less than the time spent canoeing, find the amount of time spent traveling by canoe.
2 h

37. Because of bad weather, a bus driver reduced the usual speed along a 150-mile bus route by 10 mph. The bus arrived only 30 min later than its usual arrival time. How fast does the bus usually travel?
60 mph

8.7 Variation

Objective A To solve variation problems

Direct variation is a special function that can be expressed as the equation $y = kx$, where k is a constant. The equation $y = kx$ is read "y varies directly as x" or "y is directly proportional to x." The constant k is called the **constant of variation** or the **constant of proportionality**.

The circumference (C) of a circle varies directly as the diameter (d). The direct variation equation is written $C = \pi d$. The constant of variation is π.

A nurse makes \$25 per hour. The total wage (w) of the nurse is directly proportional to the number of hours (h) worked. The equation of variation is $w = 25h$. The constant of proportionality is 25.

A direct variation equation can be written in the form $y = kx^n$, where n is a positive number. For example, the equation $y = kx^2$ is read "y varies directly as the square of x."

The area (A) of a circle varies directly as the square of the radius (r) of the circle. The direct variation equation is $A = \pi r^2$. The constant of variation is π.

➡ Given that V varies directly as r and that $V = 20$ when $r = 4$, find the constant of variation and the equation of variation.

$V = kr$ • Write the basic direct variation equation.

$20 = k \cdot 4$ • Replace V and r by the given values. Then solve for k.

$5 = k$ • This is the constant of variation.

$V = 5r$ • Write the direct variation equation by substituting the value of k into the basic direct variation equation.

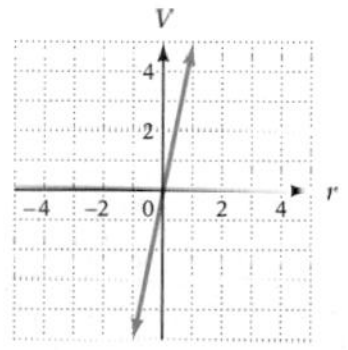

The Graph of $V = 5r$

➡ The tension (T) in a spring varies directly as the distance (x) it is stretched. If $T = 8$ lb when $x = 2$ in., find T when $x = 4$ in.

$T = kx$ • Write the basic direct variation equation.

$8 = k \cdot 2$ • Replace T and x by the given values.

$4 = k$ • Solve for the constant of variation.

$T = 4x$ • Write the direct variation equation.

$T = 4x = 4 \cdot 4 = 16$ • To find T when $x = 4$, substitute 4 for x in the equation and solve for T.

The tension is 16 lb.

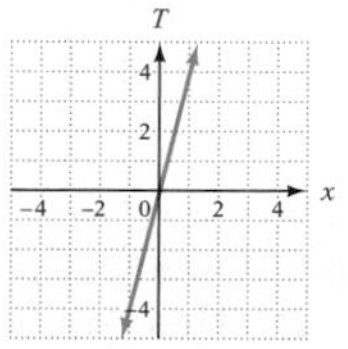

The Graph of $T = 4x$

Objective 8.7A

New Vocabulary
direct variation
constant of variation
constant of proportionality
y varies directly as x
y is directly proportional to x
inverse variation
y varies inversely as x
y is inversely proportional to x
joint variation
z varies jointly as x and y
combined variation

Vocabulary to Review
constant

Discuss the Concepts

1. What is the difference between direct variation and inverse variation?
2. Given that x and y vary directly, how can the constant of variation be defined?
$k = \frac{y}{x}$
3. Provide examples of two quantities that vary directly and examples of two quantities that vary inversely. Here are two examples that might interest students:
(1) The radius of a hailstone varies directly as the time that the hailstone spent in a high cloud.
(2) The wing flapping rate, in beats per second, of a bird is inversely proportional to the length of its wings.

In-Class Examples (Objective 8.7A)

1. The distance, d, a spring will stretch varies directly as the force, f, applied to the spring. If a force of 12 lb is required to stretch a spring 6 in., what force is required to stretch the spring 10 in.? 20 lb
2. At a constant temperature, the pressure, P, of a gas varies inversely as the volume, V. If the pressure of a gas is 75 lb/in^2 when the volume is 200 ft^3, find the pressure when the volume is 450 ft^3. $33.\overline{3}$ lb/in^2

Concept Check

Determine whether the following statements are true or false.

1. If x varies inversely as y, then when x is doubled, y is doubled. False
2. If a varies inversely as b, then ab is a constant. True
3. If a varies jointly as b and c, then $a = \frac{kb}{c}$. False
4. If the length of a rectangle is held constant, then the area of the rectangle varies directly as the width. True
5. If the area of a rectangle is held constant, then the length varies directly as the width. False
6. The circumference of a circle varies directly as the diameter. If the diameter of a circle is doubled, then the circumference of the circle will be doubled. True

Optional Student Activity

The number of times a wheel rotates in traveling one mile is inversely proportional to the diameter of the wheel. A bicycle tire with a 27-inch diameter makes 747 revolutions in traveling one mile. Find the diameter of a bicycle tire that would travel one mile in only 100 revolutions. Round to the nearest whole number. 202 in.

Inverse variation is a function that can be expressed as the equation $y = \frac{k}{x}$, where k is a constant. The equation $y = \frac{k}{x}$ is read "y varies inversely as x" or "y is inversely proportional to x."

In general, an inverse variation equation can be written $y = \frac{k}{x^n}$, where n is a positive number. For example, the equation $y = \frac{k}{x^2}$ is read "y varies inversely as the square of x."

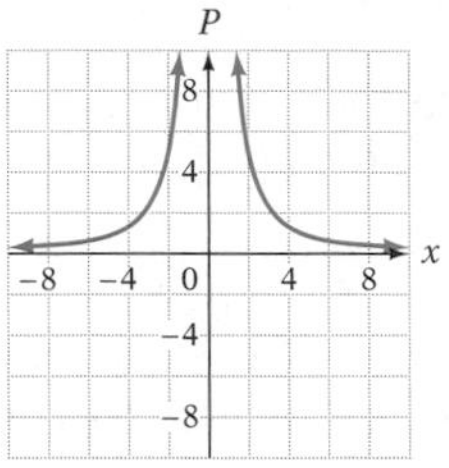

The Graph of $P = \frac{20}{x^2}$

➡ Given that P varies inversely as the square of x and that $P = 5$ when $x = 2$, find the constant of variation and the equation of variation.

$P = \frac{k}{x^2}$ • Write the basic inverse variation equation.

$5 = \frac{k}{2^2}$ • Replace P and x by the given values. Then solve for k.

$5 = \frac{k}{4}$

$20 = k$ • This is the constant of variation.

$P = \frac{20}{x^2}$ • Write the inverse variation equation by substituting the value of k into the basic inverse variation equation.

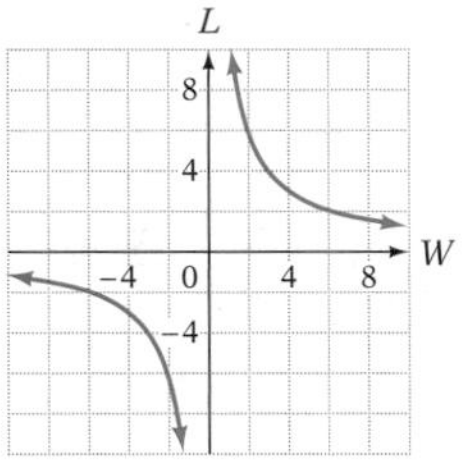

The Graph of $L = \frac{12}{W}$

➡ The length (L) of a rectangle with fixed area is inversely proportional to the width (W). If $L = 6$ ft when $W = 2$ ft, find L when $W = 3$ ft.

$L = \frac{k}{W}$ • Write the basic inverse variation equation.

$6 = \frac{k}{2}$ • Replace L and W by the given values.

$12 = k$ • Solve for the constant of variation.

$L = \frac{12}{W}$ • Write the inverse variation equation.

$L = \frac{12}{W} = \frac{12}{3} = 4$ • To find L when $W = 3$ ft, substitute 3 for W in the equation and solve for L.

The length is 4 ft.

Joint variation is a variation in which a variable varies directly as the product of two or more other variables. A joint variation can be expressed as the equation $z = kxy$, where k is a constant. The equation $z = kxy$ is read "z varies jointly as x and y."

The area (A) of a triangle varies jointly as the base (b) and the height (h). The joint variation equation is written $A = \frac{1}{2}bh$. The constant of variation is $\frac{1}{2}$.

A **combined variation** is a variation in which two or more types of variation occur at the same time. For example, in physics, the volume (V) of a gas varies directly as the temperature (T) and inversely as the pressure (P). This combined variation is written $V = \frac{kT}{P}$.

481

➡ A ball is being twirled on the end of a string. The tension (T) in the string is directly proportional to the square of the speed (v) of the ball and inversely proportional to the length (r) of the string. The tension is 96 lb when the length of the string is 0.5 ft and the speed is 4 ft/s. Find the tension when the length of the string is 1 ft and the speed is 5 ft/s.

$T = \frac{kv^2}{r}$ • Write the basic combined variation equation.

$96 = \frac{k \cdot 4^2}{0.5}$ • Replace T, v, and r by the given values.

$96 = \frac{k \cdot 16}{0.5}$

$96 = k \cdot 32$ • Solve for the constant of variation.

$3 = k$

$T = \frac{3v^2}{r}$ • Write the combined variation equation.

$T = \frac{3v^2}{r} = \frac{3 \cdot 5^2}{1} = 3 \cdot 25 = 75$ • To find T when $r = 1$ ft and $v = 5$ ft/s, substitute 1 for r and 5 for v and solve for T.

The tension is 75 lb.

Example 1

The amount (A) of medication prescribed for a person is directly related to the person's weight (W). For a 50-kilogram person, 2 ml of medication are prescribed. How many milliliters of medication are required for a person who weighs 75 kg?

Strategy

To find the required amount of medication:

- Write the basic direct variation equation, replace the variables by the given values, and solve for k.
- Write the direct variation equation, replacing k by its value. Substitute 75 for W and solve for A.

Solution

$A = kW$

$2 = k \cdot 50$

$\frac{1}{25} = k$

$A = \frac{1}{25}W = \frac{1}{25} \cdot 75 = 3$

The required amount of medication is 3 ml.

You Try It 1

The distance (s) a body falls from rest varies directly as the square of the time (t) of the fall. An object falls 64 ft in 2 s. How far will it fall in 5 s?

Your strategy

Your solution

400 ft

Solution on pp. S26–S27

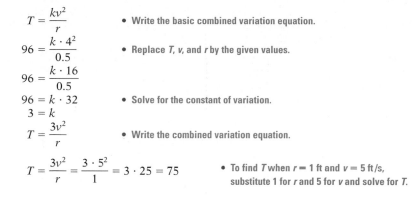

Example 2

A company that produces personal computers has determined that the number of computers it can sell (s) is inversely proportional to the price (P) of the computer. Two thousand computers can be sold when the price is \$2500. How many computers can be sold when the price of a computer is \$2000?

Strategy

To find the number of computers:

- Write the basic inverse variation equation, replace the variables by the given values, and solve for k.
- Write the inverse variation equation, replacing k by its value. Substitute 2000 for P and solve for s.

Solution

$$s = \frac{k}{P}$$
$$2000 = \frac{k}{2500}$$
$$5{,}000{,}000 = k$$
$$s = \frac{5{,}000{,}000}{P} = \frac{5{,}000{,}000}{2000} = 2500$$

At \$2000 each, 2500 computers can be sold.

You Try It 2

The resistance (R) to the flow of electric current in a wire of fixed length is inversely proportional to the square of the diameter (d) of a wire. If a wire of diameter 0.01 cm has a resistance of 0.5 ohm, what is the resistance in a wire that is 0.02 cm in diameter?

Your strategy

Your solution

0.125 ohm

Example 3

The pressure (P) of a gas varies directly as the temperature (T) and inversely as the volume (V). When $T = 50°$ and $V = 275$ in^3, $P = 20$ lb/in^2. Find the pressure of a gas when $T = 60°$ and $V = 250$ in^3.

Strategy

To find the pressure:

- Write the basic combined variation equation, replace the variables by the given values, and solve for k.
- Write the combined variation equation, replacing k by its value. Substitute 60 for T and 250 for V, and solve for P.

Solution

$$P = \frac{kT}{V}$$
$$20 = \frac{k \cdot 50}{275}$$
$$110 = k$$
$$P = \frac{110T}{V} = \frac{110 \cdot 60}{250} = 26.4$$

The pressure is 26.4 lb/in^2.

You Try It 3

The strength (s) of a rectangular beam varies jointly as its width (w) and the square of its depth (d) and inversely as its length (L). If the strength of a beam 2 in. wide, 12 in. deep, and 12 ft long is 1200 lb, find the strength of a beam that is 4 in. wide, 8 in. deep, and 16 ft long.

Your strategy

Your solution

800 lb

Solutions on p. S27

8.7 Exercises

Objective A

1. Which of the following represent(s) direct variation? Why?

   a. $y = kx$ b. $y = \dfrac{k}{x}$ c. $y = k + x$ d. $y = \dfrac{k}{x^2}$

2. Which of the following equations represent(s) "d varies directly as t"? Explain your answer.

   a. $d = 400t$ b. $d = \dfrac{16}{t}$ c. $d = 25t$ d. $d = t - 50$

3. The profit (P) realized by a company varies directly as the number of products it sells (s). If a company makes a profit of \$4000 on the sale of 250 products, what is the profit when the company sells 5000 products?
   \$80,000

4. The income (I) of a computer analyst varies directly as the number of hours (h) worked. If the analyst earns \$672 for working 8 h, how much will the analyst earn by working 36 h?
   \$3024

5. The pressure (p) on a diver in the water varies directly as the depth (d). If the pressure is 4.5 lb/in^2 when the depth is 10 ft, what is the pressure when the depth is 15 ft?
   6.75 lb/in^2

6. The distance (d) a spring will stretch varies directly as the force (f) applied to the spring. If a force of 6 lb is required to stretch a spring 3 in., what force is required to stretch the spring 4 in.?
   8 lb

7. The distance (d) an object will fall is directly proportional to the square of the time (t) of the fall. If an object falls 144 ft in 3 s, how far will the object fall in 10 s?
   1600 ft

8. The period (p) of a pendulum, or the time it takes the pendulum to make one complete swing, varies directly as the square root of the length (L) of the pendulum. If the period of a pendulum is 1.5 s when the length is 2 ft, find the period when the length is 5 ft. Round to the nearest hundredth.
   2.37 s

9. The distance (s) a ball will roll down an inclined plane is directly proportional to the square of the time (t). If the ball rolls 6 ft in 1 s, how far will it roll in 3 s?
   54 ft

Section 8.7

Suggested Assignment

Exercises 1–15, odds

More challenging problems:
Exercises 16–21

Answers to Writing Exercises

1. Only the equation in part **a** represents direct variation. Students may note that in this equation, the constant of variation k is multiplied times x, whereas in part **b**, k is divided by x; in part **c**, k is added to x; and in part **d**, k is divided by x^2.
2. The equations in parts **a** and **c** represent "d varies direct as t" because the constant of variation is multplied times x, whereas in part **b**, the constant of variation is divided by t, and in part **d**, the constant of variation is subtracted from t.

Quick Quiz (Objective 8.7A)

1. The pressure (P) on a diver in the water varies directly as the depth (d). If the pressure is 6 lb/in^2 when the depth is 12 ft, what is the pressure when the depth is 16 ft? 8 lb/in^2
2. The speed (v) of a gear varies inversely as the number of teeth (t). If a gear that has 36 teeth makes 30 revolutions per minute, how many revolutions will a gear that has 54 teeth make?
   20 rpm

10. The stopping distance (s) of a car varies directly as the square of its speed (v). If a car traveling 30 mph requires 63 ft to stop, find the stopping distance for a car traveling 55 mph.
211.75 ft

11. The time (t) for a car to travel between two cities is inversely proportional to the rate (r) of travel. If it takes 5 h to travel between the cities at a rate of 55 mph, find the time to travel between the two cities at a rate of 65 mph. Round to the nearest tenth.
4.2 h

12. The speed (v) of a gear varies inversely as the number of teeth (t). If a gear that has 45 teeth makes 24 revolutions per minute, how many revolutions per minute will a gear that has 36 teeth make?
30 revolutions/min

13. The pressure (p) of a liquid varies directly as the product of the depth (d) and the density (D) of the liquid. If the pressure is 150 lb/in^2 when the depth is 100 in. and the density is 1.2, find the pressure when the density remains the same and the depth is 75 in.
112.5 lb/in^2

14. The current (I) in a wire varies directly as the voltage (V) and inversely as the resistance (R). If the current is 10 amps when the voltage is 110 volts and the resistance is 11 ohms, find the current when the voltage is 180 volts and the resistance is 24 ohms.
7.5 amps

15. The repulsive force (f) between the north poles of two magnets is inversely proportional to the square of the distance (d) between them. If the repulsive force is 20 lb when the distance is 4 in., find the repulsive force when the distance is 2 in.
80 lb

APPLYING THE CONCEPTS

16. In the inverse variation equation $y = \frac{k}{x}$, what is the effect on x if y doubles?
x is halved.

17. In the direct variation equation $y = kx$, what is the effect on y when x doubles?
y is doubled.

Complete using the word *directly* or *inversely.*

18. If a varies directly as b and inversely as c, then c varies __directly__ as b and __inversely__ as a.

19. If a varies __inversely__ as b and c, then abc is constant.

20. If the length of a rectangle is held constant, the area of the rectangle varies __directly__ as the width.

21. If the area of a rectangle is held constant, the length of the rectangle varies __inversely__ as the width.

485

Focus on Problem Solving

Find a Pattern

Polya's four recommended problem-solving steps are stated below.

1. Understand the problem. **2.** Devise a plan.

3. Carry out the plan. **4.** Review your solution.

Point of Interest

Karl Friedrich Gauss (1777–1855) has been called the "Prince of Mathematicians" by some historians. He applied his genius to many areas of mathematics and science. A unit of magnetism, the *gauss*, is named in his honor. Some electronic equipment (the television, for instance) contains a *degausser* that controls magnetic fields.

One of the several ways of devising a plan is first to try to find a pattern. Karl Friedrich Gauss supposedly used this method to solve a problem that was given to his math class when he was in elementary school. As the story goes, his teacher wanted to grade some papers while the class worked on a math problem. The problem given to the class was to find the sum

$$1 + 2 + 3 + 4 + \cdots + 100$$

Gauss quickly solved the problem by seeing a pattern. Here is what he saw.

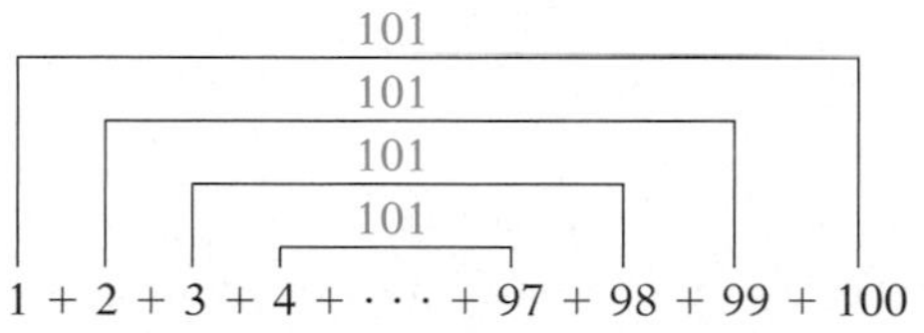

Note that

$$1 + 100 = 101$$
$$2 + 99 = 101$$
$$3 + 98 = 101$$
$$4 + 97 = 101$$

Gauss noted that there were 50 sums of 101. Therefore, the sum of the first 100 natural numbers is

$$1 + 2 + 3 + 4 + \cdots + 97 + 98 + 99 + 100 = 50(101) = 5050$$

Try to solve the following problems by finding a pattern.

1. Find the sum $2 + 4 + 6 + \cdots + 96 + 98 + 100$.

2. Find the sum $1 + 3 + 5 + \cdots + 97 + 99 + 101$.

3. Find another method of finding the sum $1 + 3 + 5 + \cdots + 97 + 99 + 101$ given in the previous exercise.

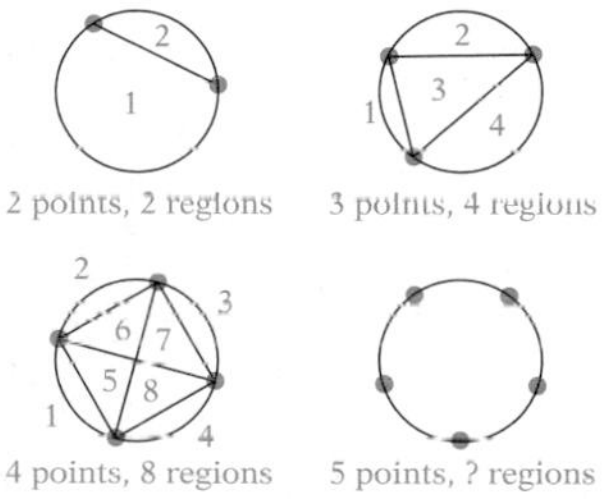

4. Find the sum $\frac{1}{1 \cdot 2} + \frac{1}{2 \cdot 3} + \frac{1}{3 \cdot 4} + \cdots + \frac{1}{49 \cdot 50}$.

Hint: $\frac{1}{1 \cdot 2} = \frac{1}{2}, \frac{1}{1 \cdot 2} + \frac{1}{2 \cdot 3} = \frac{2}{3}, \frac{1}{1 \cdot 2} + \frac{1}{2 \cdot 3} + \frac{1}{3 \cdot 4} = \frac{3}{4}$

5. The following problem shows that checking a few cases does not always result in a conjecture that is true for *all* cases. Select any two points on a circle and draw a *chord,* the line connecting the points (see the drawing in the left margin). The chord divides the circle into 2 regions. Now select 3 different points and draw chords connecting each of the three points with every other point. The chords divide the circle into 4 regions. Now select 4 points and connect each of the points with every other point. Make a conjecture about the relationship between the number of regions and the number of points on the circle. Does your conjecture work for 5 points? 6 points?

6. A *polygonal number* is a number that can be represented by arranging that number of dots in rows to form a geometric figure such as a triangle, square, pentagon, or hexagon. For instance, the first four *triangular numbers,* 3, 6, 10, and 15, are given (and shown) below. What are the next two triangular numbers?

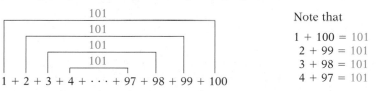

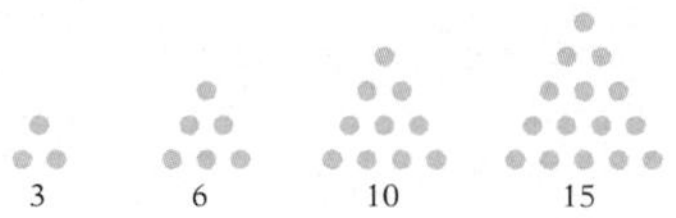

Answers to Focus on Problem Solving: Find a Pattern

1. Note that $2 + 100 = 102$, $4 + 98 = 102$, and $6 + 96 = 102$. There are 25 sums of 102. Thus the sum of the first 50 even integers is $25(102) = 2550$.

2. Note that $1 + 101 = 102$, $3 + 99 = 102$, and $5 + 97 = 102$. There is an odd number of integers, so all the numbers will not pair. The sum will be 25 pairs of 102 plus 51: $25(102) + 51 = 2601$.

3. Compare the sum listed in Exercise 1 and the sum listed in Exercise 2. For the first 50 numbers, each number in Exercise 3 is one less than the corresponding number in Exercise 1, so we can subtract 50 from the sum in Exercise 1. Then add 101, for the 51st number in the sum in Exercise 3. $2550 - 50 + 101 = 2601$.

4. From the pattern shown, $\frac{1}{1 \cdot 2} + \frac{1}{2 \cdot 3} + \frac{1}{3 \cdot 4} + \cdots + \frac{1}{49 \cdot 50} = \frac{49}{50}$.

5. The next two triangular numbers are 21 and 28.

6. With 2 points, the chord divides the circle into 2 regions. With 3 points, the chords divide the circle into $4 = 2^2$ regions. With 4 points, the chords divide the circle into $8 = 2^3$ regions. With 5 points, the chords divide the circle into $16 = 2^4$ regions. Conjecture: When n points are chosen, the chords will divide the circle into 2^{n-1} regions. However, with 6 points, the chords do not divide the circle into $32 = 2^5$ regions, but into 31 regions. The conjecture is not true.

Answers to Projects and Group Activities: Continued Fractions

1. $c_5 \approx 1.61538$
2. $c_5 \approx 3.13998$

486

Projects and Group Activities

Continued Fractions

The following complex fraction is called a **continued fraction**.

$$1 + \cfrac{1}{1 + \cfrac{1}{1 + \cfrac{1}{1 + \cfrac{1}{1 + \cdots}}}}$$

The dots indicate that the pattern continues to repeat forever.

A **convergent** of a continued fraction is an approximation of the repeated pattern. For instance,

$$c_2 = 1 + \cfrac{1}{1 + \cfrac{1}{1+1}} \qquad c_3 = 1 + \cfrac{1}{1 + \cfrac{1}{1 + \cfrac{1}{1+1}}} \qquad c_4 = 1 + \cfrac{1}{1 + \cfrac{1}{1 + \cfrac{1}{1 + \cfrac{1}{1+1}}}}$$

1. Calculate c_5 for the continued fraction above.

This particular continued fraction is related to the golden rectangle, which has been used in architectural designs as diverse as the Parthenon in Athens, built around 440 B.C., and the United Nations building. A golden rectangle is one for which

$$\frac{\text{length}}{\text{width}} = \frac{\text{length} + \text{width}}{\text{length}}$$

An example of a golden rectangle is shown at the right.

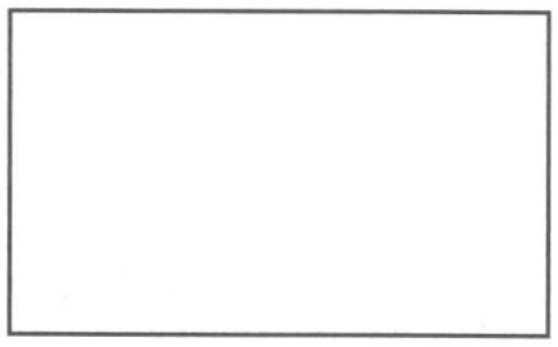

Here is another continued fraction that was discovered by Leonhard Euler (1707–1793). Calculating the convergents of this continued fraction yields approximations that are closer and closer to π.

$$\pi = 3 + \cfrac{1^2}{6 + \cfrac{3^2}{6 + \cfrac{5^2}{6 + \cfrac{7^2}{6 + \cdots}}}}$$

2. Calculate $c_5 = 3 + \cfrac{1^2}{6 + \cfrac{3^2}{6 + \cfrac{5^2}{6 + \cfrac{7^2}{6 + \cfrac{9^2}{6 + 11^2}}}}}$

Graphing Rational Functions

The domain of a function is the set of the first coordinates of all the ordered pairs of the function. When a function is given by an equation, the domain of the function is all real numbers for which the function evaluates to a real number. For rational functions, we must exclude from the domain all those values of the variable for which the denominator of the rational function is zero. The graphing calculator is a useful tool to show the graphs of functions with excluded values in the domain of the function.

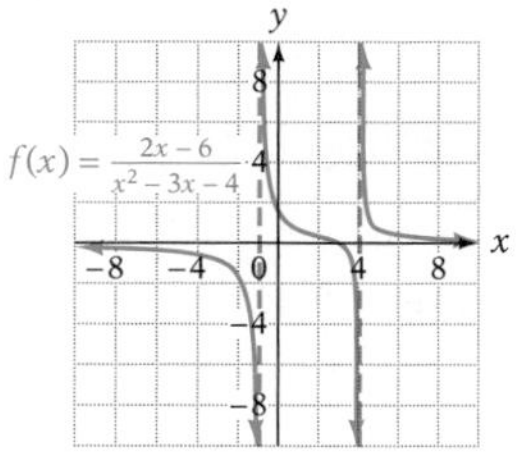

The domain is $\{x|x \neq -1, x \neq 4\}$.

The graph of the function $f(x) = \dfrac{2x-6}{x^2-3x-4}$ is shown at the left. Note that the graph never intersects the lines $x = -1$ and $x = 4$ (shown as dashed lines). These are the two values excluded from the domain of f. The graph of the function f gets closer to the dashed line $x = 4$ as x gets closer to 4. The graph of the function f also gets closer to $x = -1$ as x gets closer to -1. The lines that are "approached" by the function are called **asymptotes**.

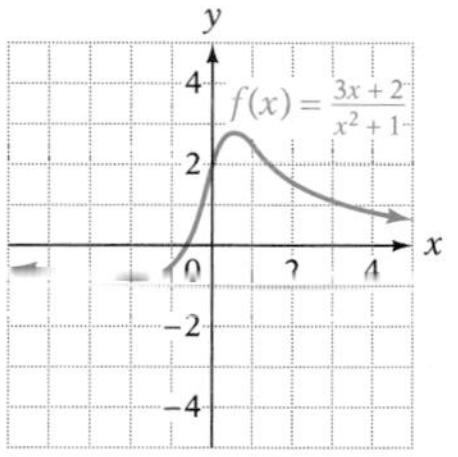

The domain is $\{x|x \in \text{real numbers}\}$.

The graph of the function $f(x) = \dfrac{3x+2}{x^2+1}$ is shown at the left. The domain must exclude values of x for which $x^2 + 1 = 0$. It is not possible that $x^2 + 1 = 0$, because $x^2 \geq 0$, and a positive number added to a number equal to or greater than zero cannot equal zero. Therefore, there are no real numbers that must be excluded from the domain of f.

Use a graphing calculator to find the domain of the rational function.

1. $h(x) = \dfrac{5x}{3x+9}$
2. $f(x) = \dfrac{-2x}{6-2x}$
3. $q(x) = \dfrac{2x-1}{x^2+x-6}$
4. $G(x) = \dfrac{3-4x}{x^2+4x-5}$
5. $H(x) = \dfrac{x+3}{x^2+2}$
6. $g(x) = \dfrac{x^2+1}{x^2}$

Graphing Variation Equations

1. Graph $y = kx$ when $k = 2$.
   What kind of function does the graph represent?

2. Graph $y = kx$ when $k = \dfrac{1}{2}$.
   What kind of function does the graph represent?

3. Graph $y = \dfrac{k}{x}$ when $k = 2$ and $x > 0$.
   Is this the graph of a function?

487

Answers to Projects and Group Activities: Graphing Rational Functions

1. $\{x|x \neq -3\}$
2. $\{x|x \neq 3\}$
3. $\{x|x \neq -3, 2\}$
4. $\{x|x \neq -5, 1\}$
5. $\{x|x \in \text{real numbers}\}$
6. $\{x|x \neq 0\}$

Answers to Projects and Group Activities: Graphing Variation Equations

1. 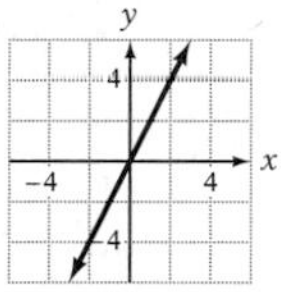

The graph represents a linear function.

2. 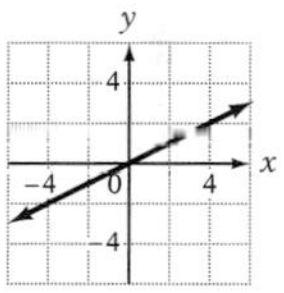

The graph represents a linear function.

3. 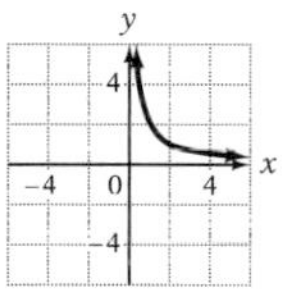

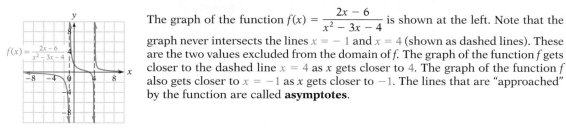

The graph is the graph of a function.

Chapter Summary

Key Words

A *rational expression* is a fraction in which the numerator and denominator are polynomials. A rational expression is in *simplest form* when the numerator and denominator have no common factors. [p. 431]

The *least common multiple* (LCM) of two or more polynomials is the simplest polynomial that contains the factors of each polynomial. [p. 439]

The *reciprocal* of a rational expression is the rational expression with the numerator and denominator interchanged. [p. 434]

A *complex fraction* is a fraction whose numerator or denominator contains one or more fractions. [p. 451]

A *ratio* is the quotient of two quantities that have the same unit. [p. 457]

A *rate* is the quotient of two quantities that have different units. [p. 457]

A *proportion* is an equation that states the equality of two ratios or rates. [p. 457]

A *literal equation* is an equation that contains more than one variable. [p. 467]

Direct variation is a special function that can be expressed as the equation $y = kx$, where k is a constant called the *constant of variation* or the *constant of proportionality.* [p. 479]

Inverse variation is a function that can be expressed as the equation $y = \frac{k}{x}$, where k is a constant. [p. 480]

Joint variation is a variation in which a variable varies directly as the product of two or more variables. A joint variation can be expressed as the equation $z = kxy$, where k is a constant. [p. 480]

Combined variation is a variation in which two or more types of variation occur at the same time. [p. 480]

Essential Rules

To Multiply Fractions [p. 432] $\frac{a}{b} \cdot \frac{c}{d} = \frac{ac}{bd}$

To Divide Fractions [p. 434] $\frac{a}{b} \div \frac{c}{d} = \frac{a}{b} \cdot \frac{d}{c} = \frac{ad}{bc}$

To Add Fractions [p. 441] $\frac{a}{c} + \frac{b}{c} = \frac{a+b}{c}$

To Subtract Fractions [p. 441] $\frac{a}{c} - \frac{b}{c} = \frac{a-b}{c}$

Equation for Work Problems [p. 471] $\text{Rate of work} \times \text{time worked} = \text{part of task completed}$

Uniform Motion Equation [p. 473] Distance = rate × time

489

Chapter Review

1. Divide: $\dfrac{6a^2b^7}{25x^3y} \div \dfrac{12a^3b^4}{5x^2y^2}$

$\dfrac{b^3y}{10ax}$ [8.1C]

2. Add: $\dfrac{x+7}{15x} + \dfrac{x-2}{20x}$

$\dfrac{7x+22}{60x}$ [8.2D]

3. Simplify: $\dfrac{x - \dfrac{16}{5x-2}}{3x - 4 - \dfrac{88}{5x-2}}$

$\dfrac{x-2}{3x-10}$ [8.3A]

4. Simplify: $\dfrac{x^2+x-30}{15+2x-x^2}$

$-\dfrac{x+6}{x+3}$ [8.1A]

5. Simplify: $\dfrac{16x^5y^3}{24xy^{10}}$

$\dfrac{2x^4}{3y^7}$ [8.1A]

6. Solve: $\dfrac{20}{x+2} = \dfrac{5}{16}$

62 [8.4B]

7. Divide: $\dfrac{10-23y+12y^2}{6y^2-y-5} \div \dfrac{4y^2-13y+10}{18y^2+3y-10}$

$\dfrac{(3y-2)^2}{(y-1)(y-2)}$ [0.1C]

8. Multiply: $\dfrac{8ab^2}{15x^3y} \cdot \dfrac{5xy^4}{16a^2b}$

$\dfrac{by^3}{6ax^2}$ [8.1B]

9. Simplify: $\dfrac{1-\dfrac{1}{x}}{1-\dfrac{8x-7}{x^2}}$

$\dfrac{x}{x-7}$ [8.3A]

10. Write each fraction in terms of the LCM of the denominators.

$\dfrac{x}{12x^2+16x-3}, \dfrac{4x^2}{6x^2+7x-3}$

$\dfrac{3x^2-x}{(2x+3)(6x-1)(3x-1)}, \dfrac{24x^3-4x^2}{(2x+3)(6x-1)(3x-1)}$ [8.2B]

11. Solve $T = 2(ab + bc + ca)$ for a.

$a = \dfrac{T-2bc}{2b+2c}$ [8.5A]

12. Solve: $\dfrac{5}{7} + \dfrac{x}{2} = 2 - \dfrac{x}{7}$

2 [8.4A]

13. Solve $i = \dfrac{100m}{c}$ for c.

$c = \dfrac{100m}{i}$ [8.5A]

14. Solve: $\dfrac{x+8}{x+4} = 1 + \dfrac{5}{x+4}$

no solution [8.4A]

15. Divide: $\dfrac{20x^2-45x}{6x^3+4x^2} \div \dfrac{40x^3-90x^2}{12x^2+8x}$

$\dfrac{1}{x^2}$ [8.1C]

16. Add: $\dfrac{2y}{5y-7} + \dfrac{3}{7-5y}$

$\dfrac{2y-3}{5y-7}$ [8.2D]

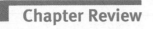

490

17. Subtract: $\dfrac{5x + 3}{2x^2 + 5x - 3} - \dfrac{3x + 4}{2x^2 + 5x - 3}$

$\dfrac{1}{x + 3}$ [8.2C]

18. Find the LCM of $10x^2 - 11x + 3$ and $20x^2 - 17x + 3$.

$(5x - 3)(2x - 1)(4x - 1)$ [8.2A]

19. Solve $4x + 9y = 18$ for y.

$y = -\dfrac{4}{9}x + 2$ [8.5A]

20. Multiply: $\dfrac{24x^2 - 94x + 15}{12x^2 - 49x + 15} \cdot \dfrac{24x^2 + 7x - 5}{4 - 27x + 18x^2}$

$\dfrac{8x + 5}{3x - 4}$ [8.1B]

21. Solve: $\dfrac{20}{2x + 3} = \dfrac{17x}{2x + 3} - 5$

5 [8.4A]

22. Add: $\dfrac{x - 1}{x + 2} + \dfrac{3x - 2}{5 - x} + \dfrac{5x^2 + 15x - 11}{x^2 - 3x - 10}$

$\dfrac{3x - 1}{x - 5}$ [8.2D]

23. Solve: $\dfrac{6}{x - 7} = \dfrac{8}{x - 6}$

10 [8.4B]

24. Solve: $\dfrac{3}{20} = \dfrac{x}{80}$

12 [8.4B]

25. Triangles ABC and DEF are similar. Find the perimeter of triangle ABC.

24 in. [8.4C]

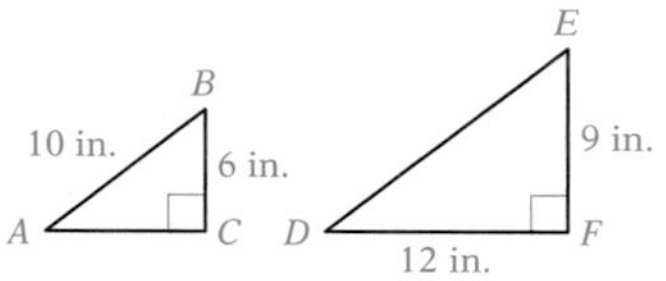

26. One hose can fill a pool in 15 h. The second hose can fill the pool in 10 h. How long would it take to fill the pool using both hoses?

6 h [8.6A]

27. A car travels 315 mi in the same amount of time that a bus travels 245 mi. The rate of the car is 10 mph greater than that of the bus. Find the rate of the car.

45 mph [8.6B]

28. The rate of a jet is 400 mph in calm air. Traveling with the wind, the jet can fly 2100 mi in the same amount of time it takes to fly 1900 mi against the wind. Find the rate of the wind.

20 mph [8.6B]

29. A pitcher's earned run average (ERA) is the average number of runs allowed in 9 innings of pitching. If a pitcher allows 15 runs in 100 innings, find the pitcher's ERA.

1.35 [8.4D]

30. The current (I) in an electric circuit varies inversely as the resistance (R). If the current in the circuit is 4 amps when the resistance is 50 ohms, find the current in the circuit when the resistance is 100 ohms.

2 amps [8.7A]

491

Chapter Test

1. Simplify: $\dfrac{16x^5y}{24x^2y^4}$

$\dfrac{2x^3}{3y^3}$ [8.1A]

2. Simplify: $\dfrac{x^2 + 4x - 5}{1 - x^2}$

$-\dfrac{x + 5}{x + 1}$ [8.1A]

3. Multiply: $\dfrac{x^3y^4}{x^2 - 4x + 4} \cdot \dfrac{x^2 - x - 2}{x^6y^4}$

$\dfrac{x + 1}{x^3(x - 2)}$ [8.1B]

4. Multiply: $\dfrac{x^2 + 2x - 3}{x^2 + 6x + 9} \cdot \dfrac{2x^2 - 11x + 5}{2x^2 + 3x - 5}$

$\dfrac{(x - 5)(2x - 1)}{(x + 3)(2x + 5)}$ [8.1B]

5. Divide: $\dfrac{x^2 + 3x + 2}{x^2 + 5x + 4} \div \dfrac{x^2 - x - 6}{x^2 + 2x - 15}$

$\dfrac{x + 5}{x + 4}$ [8.1C]

6. Find the LCM of $6x - 3$ and $2x^2 + x - 1$.

$3(2x - 1)(x + 1)$ [8.2A]

7. Write each fraction in terms of the LCM of the denominators.

$\dfrac{3}{x^2 - 2x}, \dfrac{x}{x^2 - 4}$

$\dfrac{3(x + 2)}{x(x + 2)(x - 2)}; \dfrac{x^2}{x(x + 2)(x - 2)}$ [8.2B]

8. Subtract: $\dfrac{2x}{x^2 + 3x - 10} - \dfrac{4}{x^2 + 3x - 10}$

$\dfrac{2}{x + 5}$ [8.2C]

9. Subtract: $\dfrac{2}{2x - 1} - \dfrac{3}{3x + 1}$

$\dfrac{5}{(2x - 1)(3x + 1)}$ [8.2D]

10. Subtract: $\dfrac{x}{x + 3} - \dfrac{2x - 5}{x^2 + x - 6}$

$\dfrac{x^2 - 4x + 5}{(x - 2)(x + 3)}$ [8.2D]

11. Simplify: $\dfrac{1 + \dfrac{1}{x} - \dfrac{12}{x^2}}{1 + \dfrac{2}{x} - \dfrac{8}{x^2}}$

$\dfrac{x - 3}{x - 2}$ [8.3A]

12. Solve: $\dfrac{6}{x} - 2 = 1$

2 [8.4A]

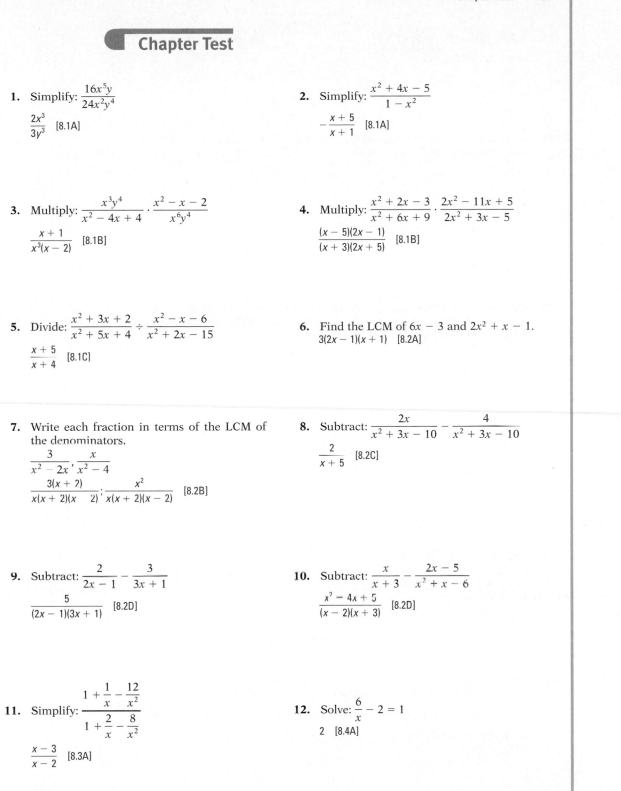

13. Solve: $\frac{2x}{x+1} - 3 = \frac{-2}{x+1}$

no solution [8.4A]

14. Solve: $\frac{3}{x+4} = \frac{5}{x+6}$

-1 [8.4B]

15. Triangles *ABC* and *DEF* are similar. Find the area of triangle *DEF*. 64.8 m^2 [8.4C]

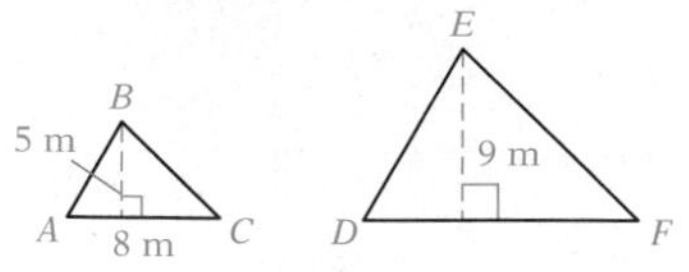

16. Solve $d = s + rt$ for t.

$t = \frac{d - s}{r}$ [8.5A]

17. An interior designer uses 2 rolls of wallpaper for every 45 ft^2 of wall space in an office. At this rate, how many rolls of wallpaper are needed for an office that has 315 ft^2 of wall space?

14 rolls [8.4D]

18. One landscaper can till the soil for a lawn in 30 min, whereas it takes a second landscaper 15 min to do the same job. How long would it take to till the soil for the lawn with both landscapers working together?

10 min [8.6A]

19. A cyclist travels 20 mi in the same amount of time as it takes a hiker to walk 6 mi. The rate of the cyclist is 7 mph faster than the rate of the hiker. Find the rate of the cyclist.

10 mph [8.6B]

20. The electrical resistance (r) of a cable varies directly as its length (l) and inversely as the square of its diameter (d). If a cable 16,000 ft long and $\frac{1}{4}$ in. in diameter has a resistance of 3.2 ohms, what is the resistance of a cable that is 8000 ft long and $\frac{1}{2}$ in. in diameter?

0.4 ohm [8.7A]

493

Cumulative Review

1. Simplify: $\left(\frac{2}{3}\right)^2 \div \left(\frac{3}{2} - \frac{2}{3}\right) + \frac{1}{2}$

$\frac{31}{30}$ [1.3A]

2. Evaluate $-a^2 + (a - b)^2$ when $a = -2$ and $b = 3$.

21 [1.4A]

3. Simplify: $-2x - (-3y) + 7x - 5y$

$5x - 2y$ [1.4B]

4. Simplify: $2[3x - 7(x - 3) - 8]$

$-8x + 26$ [1.4D]

5. Solve: $4 - \frac{2}{3}x = 7$

$-\frac{9}{2}$ [2.2A]

6. Solve: $3[x - 2(x - 3)] = 2(3 - 2x)$

-12 [2.2C]

7. Find $16\frac{2}{3}\%$ of 60.

10 [2.1D]

8. Solve: $x - 3(1 - 2x) \geq 1 - 4(3 - 2x)$

$\{x | x \leq 8\}$ [2.4A]

9. Find the volume of the rectangular solid shown in the figure.

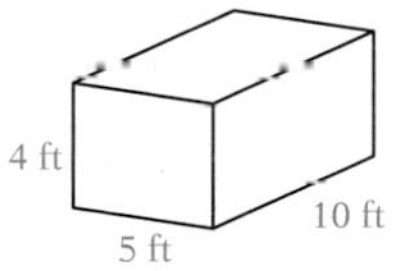

200 ft^3 [3.3A]

10. Graph: $x - 2y = 2$

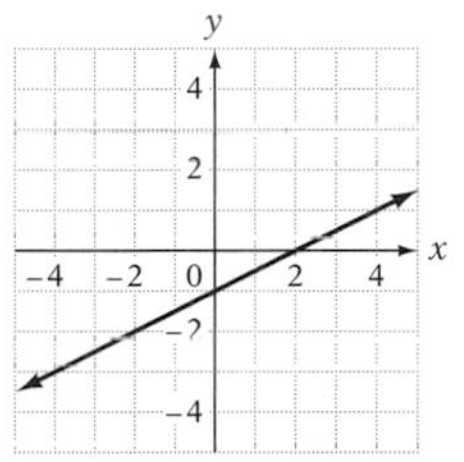

[4.3B]

11. Given $P(x) = \frac{x - 1}{2x - 3}$, find $P(-2)$.

$\frac{3}{7}$ [4.2A]

12. Find the equation of the line that contains the point $(-2, -1)$ and is parallel to the line $3x - 2y = 6$.

$y = \frac{3}{2}x + 2$ [4.6A]

13. Evaluate the determinant:

$$\begin{vmatrix} 6 & 5 \\ 2 & -3 \end{vmatrix}$$

-28 [5.3A]

14. Multiply: $(a^2b^5)(ab^2)$

a^3b^7 [6.1A]

15. Simplify: $\frac{(2a^{-2}b^3)^{-2}}{(4a)^{-1}}$

$\frac{a^5}{b^6}$ [6.1B]

16. Write 0.000000035 in scientific notation.

3.5×10^{-8} [6.1C]

17. Multiply: $(2a^2 - 3a + 1)(-2a^2)$

$-4a^4 + 6a^3 - 2a^2$ [6.3A]

18. Multiply: $(a - 3b)(a + 4b)$

$a^2 + ab - 12b^2$ [6.3C]

19. Divide: $(x^3 - 8) \div (x - 2)$
$x^2 + 2x + 4$ [6.4B]

20. Factor: $y^2 - 7y + 6$
$(y - 6)(y - 1)$ [7.2A]

21. Factor: $12x^2 - x - 1$
$(4x + 1)(3x - 1)$ [7.3A/7.3B]

22. Factor: $2a^3 + 7a^2 - 15a$
$a(2a - 3)(a + 5)$ [7.3A/7.3B]

23. Factor: $4b^2 - 100$
$4(b + 5)(b - 5)$ [7.4D]

24. Solve: $(x + 3)(2x - 5) = 0$
$-3, \frac{5}{2}$ [7.5A]

25. Simplify: $\frac{12x^4y^2}{18xy^7}$
$\frac{2x^3}{3y^5}$ [8.1A]

26. Simplify: $\frac{x^2 - 7x + 10}{25 - x^2}$
$-\frac{x - 2}{x + 5}$ [8.1A]

27. Divide: $\frac{x^2 - x - 56}{x^2 + 8x + 7} \div \frac{x^2 - 13x + 40}{x^2 - 4x - 5}$
1 [8.1C]

28. Subtract: $\frac{2}{2x - 1} - \frac{1}{x + 1}$
$\frac{3}{(2x - 1)(x + 1)}$ [8.2D]

29. Simplify: $\dfrac{1 - \frac{2}{x} - \frac{15}{x^2}}{1 - \frac{25}{x^2}}$
$\frac{x + 3}{x + 5}$ [8.3A]

30. Solve: $\frac{3x}{x - 3} - 2 = \frac{10}{x - 3}$
4 [8.4A]

31. A silversmith mixes 60 g of an alloy that is 40% silver with 120 g of another silver alloy. The resulting alloy is 60% silver. Find the percent of silver in the 120 g of alloy.
70% [2.3B]

32. A life insurance policy costs \$32 for every \$1000 of coverage. At this rate, how much money would a policy of \$5000 cost?
\$160 [8.4D]

33. One water pipe can fill a tank in 9 min, whereas a second pipe requires 18 min to fill the tank. How long would it take both pipes, working together, to fill the tank?
6 min [8.6A]

Example 1 Simplify: $64^{-2/3}$

Solution $64^{-2/3} = (2^6)^{-2/3} = 2^{-4}$
$= \frac{1}{2^4} = \frac{1}{16}$

You Try It 1 Simplify: $16^{-3/4}$

Your solution $\frac{1}{8}$

Example 2 Simplify: $(-49)^{3/2}$

Solution The base of the exponential expression is a negative number, while the denominator of the exponent is a positive even number.

Therefore, $(-49)^{3/2}$ is not a real number.

You Try It 2 Simplify: $(-81)^{3/4}$

Your solution not a real number

Example 3 Simplify: $(x^{1/2}y^{-3/2}z^{1/4})^{-3/2}$

Solution $(x^{1/2}y^{-3/2}z^{1/4})^{-3/2}$
$= x^{-3/4}y^{9/4}z^{-3/8}$
$= \frac{y^{9/4}}{x^{3/4}z^{3/8}}$

You Try It 3 Simplify: $(x^{3/4}y^{1/2}z^{-2/3})^{-4/3}$

Your solution $\frac{z^{8/9}}{xy^{2/3}}$

Example 4

Simplify: $\frac{x^{1/2}y^{-5/4}}{x^{-4/3}y^{1/3}}$

Solution $\frac{x^{1/2}y^{-5/4}}{x^{-4/3}y^{1/3}}$
$= x^{3/6 - (-8/6)}y^{-15/12 - 4/12}$
$= x^{11/6}y^{-19/12} = \frac{x^{11/6}}{y^{19/12}}$

You Try It 4

Simplify: $\left(\frac{16a^{-2}b^{4/3}}{9a^4b^{-2/3}}\right)^{-1/2}$

Your solution $\frac{3a^3}{4b}$

Solutions on p. S27

(Continued)

2. The Insurance Institute for Highway Safety has released data on the number of car accidents in which motorists of different ages are involved. The function that approximately models the data is $f(x) = 6434x^{-4/3}$, where x is the age of the driver in years and $f(x)$ is the number of crashes per 1000 licensed drivers.
   a. Use the model to approximate the accident rate per 1000 drivers for 18-year-olds. Round to the nearest tenth. 136.4
   b. Use the model to approximate the difference between the accident rate for 16-year-olds and the accident rate for 60-year-olds. Round to the nearest tenth. 132.2
   c. Provide an explanation for the decrease in the accident rate as age increases. Answers will vary. For example, older drivers have more driving experience.

Objective B

To write exponential expressions as radical expressions and to write radical expressions as exponential expressions

VIDEO & DVD CD TUTOR WEB SSM

Point of Interest

The radical sign was introduced in 1525 in a book by Christoff Rudolff called *Coss*. He modified the symbol to indicate square roots, cube roots, and fourth roots. The idea of using an index, as we do in our modern notation, did not occur until some years later.

Recall that $a^{1/n}$ is the nth root of a. The expression $\sqrt[n]{a}$ is another symbol for the nth root of a.

If a is a real number, then $a^{1/n} = \sqrt[n]{a}$.

In the expression $\sqrt[n]{a}$, the symbol $\sqrt{\ }$ is called a **radical sign**, n is the **index** of the radical, and a is the **radicand**. When $n = 2$, the radical expression represents a square root and the index 2 is usually not written.

An exponential expression with a rational exponent can be written as a radical expression.

Rule for Writing Exponential Expressions as Radical Expressions

If $a^{1/n}$ is a real number, then $a^{m/n} = a^{m \cdot 1/n} = (a^m)^{1/n} = \sqrt[n]{a^m}$.

The expression $a^{m/n}$ can also be written $a^{m/n} = (a^{1/n})^m = (\sqrt[n]{a})^m$.

Objective 9.1B

New Vocabulary

radical
index
radicand

Vocabulary to Review

nth root of a

New Symbols

$\sqrt[n]{a}$

New Rules

$a^{1/n} = \sqrt[n]{a}$
$a^{m/n} = \sqrt[n]{a^m}$

In-Class Examples (Objective 9.1B)

Rewrite the exponential expression as a radical expression.

1. $3^{1/2}$ $\sqrt{3}$
2. $a^{2/3}$ $\sqrt[3]{a^2}$
3. $(2x)^{5/3}$ $\sqrt[3]{(2x)^5}$
4. $-4a^{3/5}$ $-4\sqrt[5]{a^3}$
5. $x^{-2/5}$ $\frac{1}{\sqrt[5]{x^2}}$
6. $(a - b)^{1/2}$ $\sqrt{a - b}$

Rewrite the radical expression as an exponential expression.

7. $\sqrt[3]{x^2}$ $x^{2/3}$
8. $\sqrt[4]{5y^3}$ $(5y^3)^{1/4}$
9. $2x\sqrt[3]{y}$ $2xy^{1/3}$

Discuss the Concepts

1. Write two expressions that represent the nth root of a. For each expression, name the term that describes each part of the expression.
2. Write an exponential expression of the form $a^{m/n}$. Explain how to rewrite it as a radical expression.
3. Write a radical expression of the form $\sqrt[n]{a^m}$. Explain how to rewrite it as an exponential expression.

Concept Check

Write the expression in exponential form. Then simplify the resulting expression.

1. $\sqrt{16^{1/2}}$ $(16^{1/2})^{1/2}$, 2
2. $\sqrt[3]{4^{3/2}}$ $(4^{3/2})^{1/3}$, 2
3. $\sqrt[4]{32^{-4/5}}$ $(32^{-4/5})^{1/4}$, $\frac{1}{2}$
4. $\sqrt{243^{-4/5}}$ $(243^{-4/5})^{1/2}$, $\frac{1}{9}$

The exponential expression at the right has been written as a radical expression.

$$y^{2/3} = (y^2)^{1/3} = \sqrt[3]{y^2}$$

The radical expressions at the right have been written as exponential expressions.

$$\sqrt[5]{x^6} = (x^6)^{1/5} = x^{6/5}$$
$$\sqrt{17} = (17)^{1/2} = 17^{1/2}$$

➡ Write $(5x)^{2/5}$ as a radical expression.

$(5x)^{2/5} = \sqrt[5]{(5x)^2}$ • The denominator of the rational exponent is the index of the radical. The numerator is the power of the radicand.

$= \sqrt[5]{25x^2}$ • Simplify.

➡ Write $\sqrt[3]{x^4}$ as an exponential expression with a rational exponent.

$\sqrt[3]{x^4} = (x^4)^{1/3}$ • The index of the radical is the denominator of the rational exponent. The power of the radicand is the numerator of the rational exponent.

$= x^{4/3}$ • Simplify.

➡ Write $\sqrt[3]{a^3 + b^3}$ as an exponential expression with a rational exponent.

$\sqrt[3]{a^3 + b^3} = (a^3 + b^3)^{1/3}$

Note that $(a^3 + b^3)^{1/3} \neq a + b$.

Example 5 Write $(3x)^{5/4}$ as a radical expression.

Solution $(3x)^{5/4} = \sqrt[4]{(3x)^5} = \sqrt[4]{243x^5}$

You Try It 5 Write $(2x^3)^{3/4}$ as a radical expression.

Your solution $\sqrt[4]{8x^9}$

Example 6 Write $-2x^{2/3}$ as a radical expression.

Solution $-2x^{2/3} = -2(x^2)^{1/3} = -2\sqrt[3]{x^2}$

You Try It 6 Write $-5a^{5/6}$ as a radical expression.

Your solution $-5\sqrt[6]{a^5}$

Example 7 Write $\sqrt[4]{3a}$ as an exponential expression.

Solution $\sqrt[4]{3a} = (3a)^{1/4}$

You Try It 7 Write $\sqrt[3]{3ab}$ as an exponential expression.

Your solution $(3ab)^{1/3}$

Example 8 Write $\sqrt{a^2 - b^2}$ as an exponential expression.

Solution $\sqrt{a^2 - b^2} = (a^2 - b^2)^{1/2}$

You Try It 8 Write $\sqrt[4]{x^4 + y^4}$ as an exponential expression.

Your solution $(x^4 + y^4)^{1/4}$

Solutions on pp. S27–S28

Optional Student Activity

Simplify the product by first rewriting each radical expression as an exponential expression and then multiplying the resulting expressions.

$(\sqrt[4]{x} + \sqrt[4]{y})(\sqrt[4]{x^3} - \sqrt{x}\sqrt[4]{y} + \sqrt[4]{x}\sqrt{y} - \sqrt[4]{y^3})$

$x - y$

Objective C **To simplify radical expressions that are roots of perfect powers**

VIDEO & DVD CD TUTOR WEB SSM

Every positive number has two square roots, one a positive number and one a negative number. For example, because $(5)^2 = 25$ and $(-5)^2 = 25$, there are two square roots of 25: 5 and -5.

The symbol $\sqrt{}$ is used to indicate the positive or **principal square root**. To indicate the negative square root of a number, a negative sign is placed in front of the radical.

$\sqrt{25} = 5$

$-\sqrt{25} = -5$

The square root of zero is zero.

$\sqrt{0} = 0$

The square root of a negative number is not a real number, because the square of a real number must be positive.

$\sqrt{-25}$ is not a real number.

Note that

$\sqrt{(-5)^2} = \sqrt{25} = 5$ and $\sqrt{5^2} = \sqrt{25} = 5$

This is true for all real numbers and is stated as the following result.

For any real number a, $\sqrt{a^2} = |a|$ and $-\sqrt{a^2} = -|a|$. If a is a positive real number, then $\sqrt{a^2} = a$ and $(\sqrt{a})^2 = a$.

Besides square roots, we can also determine cube roots, fourth roots, and so on.

$\sqrt[3]{8} = 2$, because $2^3 = 8$. • The cube root of a positive number is positive.

$\sqrt[3]{-8} = -2$, because $(-2)^3 = -8$. • The cube root of a negative number is negative.

$\sqrt[4]{625} = 5$, because $5^4 = 625$.

$\sqrt[5]{243} = 3$, because $3^5 = 243$.

The following properties hold true for finding the nth root of a real number.

If n is an even integer, then $\sqrt[n]{a^n} = |a|$ and $-\sqrt[n]{a^n} = -|a|$. If n is an odd integer, then $\sqrt[n]{a^n} = a$.

For example,

$$\sqrt[6]{y^6} = |y| \qquad -\sqrt[12]{x^{12}} = -|x| \qquad \sqrt[5]{b^5} = b$$

For the remainder of this chapter, we will assume that variable expressions inside a radical represent positive numbers. Therefore, it is not necessary to use the absolute value signs.

TAKE NOTE

Note that when the index is an even natural number, the nth root requires absolute value symbols.

$\sqrt[6]{y^6} = |y|$ but $\sqrt[5]{y^5} = y$

Because we stated that variables within radicals represent *positive* numbers, we will omit the absolute value symbols when writing an answer.

➡ Simplify: $\sqrt[4]{x^4y^8}$

$\sqrt[4]{x^4y^8} = (x^4y^8)^{1/4}$ • The radicand is a perfect fourth power because the exponents on the variables are divisible by 4. Write the radical expression as an exponential expression.

$= xy^2$ • Use the Rule for Simplifying Powers of Products.

Objective 9.1C

New Vocabulary

principal square root
cube root
fourth root
fifth root
perfect cube
perfect fourth power
perfect fifth power

Vocabulary to Review

square root
perfect square

Instructor Note

All the expressions in this objective are perfect powers. Simplifying these expressions will prepare students for the next section.

Discuss the Concepts

1. Which of the following represent perfect squares? Why?
   a. d^8
   b. n^{11}
   c. a^{24}
2. Which of the following represent perfect cubes? Why?
   a. b^6
   b. c^{29}
   c. m^{24}
3. Explain how to determine whether a radical expression is the root of a perfect power.
4. Explain why $\sqrt[3]{x^3 + y^3} \neq x + y$. Because $(x + y)^3 \neq x^3 + y^3$

In-Class Examples (Objective 9.1C)

Simplify.

1. $\sqrt[3]{x^6y^9}$ x^2y^3
2. $\sqrt[3]{-8x^9y^{21}}$ $-2x^3y^7$
3. $\sqrt{-25x^2y^4}$ Not a real number
4. $-\sqrt[4]{a^{16}b^{24}}$ $-a^4b^6$

Concept Check

1. Evaluate the expression for the given values of the variables.
   a. $-4\sqrt{xy}$, where $x = 3$ and $y = 12$ −24
   b. $-3\sqrt{xy}$, where $x = 20$ and $y = 5$ −30
   c. $8\sqrt{x + y}$, where $x = 19$ and $y = 6$ 40
   d. $7\sqrt{x + y}$, where $x = 34$ and $y = 15$ 49
   e. $6\sqrt{ab} - 9$, where $a = 2$ and $b = 32$ 39
   f. $5 + 2\sqrt{ab}$, where $a = 27$ and $b = 3$ 23
   g. $\sqrt{a^2 + b^2}$, where $a = 3$ and $b = 4$ 5
   h. $\sqrt{c^2 - a^2}$, where $a = 6$ and $c = 10$ 8
2. If $\frac{\sqrt[3]{x}}{3}$ is an even integer, what is a possible value of x? Answers will vary. For example, 216, 1728, or 5832.
3. If $\sqrt[5]{x} = 4$, what is the value of $\sqrt{x}$? 32

Optional Student Activity

1. If $A \Delta B$ means A^B and $A \nabla B$ means $\sqrt[B]{A}$, what is the value of the expression $[(2 \Delta 6) \nabla 3] \Delta 2$? 16
2. For how many real numbers x is the expression $\sqrt{-(x+1)^2}$ a real number? One
3. By what factor must you multiply a number in order to double its square root? To triple its square root? To double its cube root? To triple its cube root? 4; 9; 8; 27

➡ Simplify: $\sqrt[3]{125c^9d^6}$

$\sqrt[3]{125c^9d^6} = (5^3c^9d^6)^{1/3}$

- **The radicand is a perfect cube because 125 is a perfect cube ($125 = 5^3$) and all the exponents on the variables are divisible by 3.**

$= 5c^3d^2$

- **Use the Rule for Simplifying Powers of Products.**

Note that **a variable expression is a perfect power if the exponents on the factors are evenly divisible by the index of the radical.**

The chart below shows roots of perfect powers. Knowledge of these roots is very helpful when simplifying radical expressions.

Square Roots		*Cube Roots*	*Fourth Roots*	*Fifth Roots*
$\sqrt{1} = 1$	$\sqrt{36} = 6$	$\sqrt[3]{1} = 1$	$\sqrt[4]{1} = 1$	$\sqrt[5]{1} = 1$
$\sqrt{4} = 2$	$\sqrt{49} = 7$	$\sqrt[3]{8} = 2$	$\sqrt[4]{16} = 2$	$\sqrt[5]{32} = 2$
$\sqrt{9} = 3$	$\sqrt{64} = 8$	$\sqrt[3]{27} = 3$	$\sqrt[4]{81} = 3$	$\sqrt[5]{243} = 3$
$\sqrt{16} = 4$	$\sqrt{81} = 9$	$\sqrt[3]{64} = 4$	$\sqrt[4]{256} = 4$	
$\sqrt{25} = 5$	$\sqrt{100} = 10$	$\sqrt[3]{125} = 5$	$\sqrt[4]{625} = 5$	

T

TAKE NOTE
From the chart, $\sqrt[5]{243} = 3$, which means that $3^5 = 243$. From this we know that $(-3)^5 = -243$, which means $\sqrt[5]{-243} = -3$.

➡ Simplify: $\sqrt[5]{-243x^5y^{15}}$

$\sqrt[5]{-243x^5y^{15}} = -3xy^3$

- **From the chart, 243 is a perfect fifth power, and each exponent is divisible by 5. Therefore, the radicand is a perfect fifth power.**

Example 9 Simplify: $\sqrt[3]{-125a^6b^9}$

Solution The radicand is a perfect cube.

$\sqrt[3]{-125a^6b^9} = -5a^2b^3$ • **Divide each exponent by 3.**

You Try It 9 Simplify: $\sqrt[3]{-8x^{12}y^3}$

Your solution $-2x^4y$

Example 10 Simplify: $-\sqrt[4]{16a^4b^8}$

Solution The radicand is a perfect fourth power.

$-\sqrt[4]{16a^4b^8} = -2ab^2$ • **Divide each exponent by 4.**

You Try It 10 Simplify: $-\sqrt[4]{81x^{12}y^8}$

Your solution $-3x^3y^2$

Solutions on p. S28

503

9.1 Exercises

Objective A

Simplify.

1. $8^{1/3}$ — 2

2. $16^{1/2}$ — 4

3. $9^{3/2}$ — 27

4. $25^{3/2}$ — 125

5. $27^{-2/3}$ — $\frac{1}{9}$

6. $64^{-1/3}$ — $\frac{1}{4}$

7. $32^{2/5}$ — 4

8. $16^{3/4}$ — 8

9. $(-25)^{5/2}$ — not a real number

10. $(-36)^{1/4}$ — not a real number

11. $\left(\frac{25}{49}\right)^{-3/2}$ — $\frac{343}{125}$

12. $\left(\frac{8}{27}\right)^{-2/3}$ — $\frac{9}{4}$

13. $x^{1/2}x^{1/2}$ — x

14. $a^{1/3}a^{5/3}$ — a^2

15. $y^{-1/4}y^{3/4}$ — $y^{1/2}$

16. $x^{2/5} \cdot x^{-4/5}$ — $\frac{1}{x^{2/5}}$

17. $x^{-2/3} \cdot x^{3/4}$ — $x^{1/12}$

18. $x \cdot x^{-1/2}$ — $x^{1/2}$

19. $a^{1/3} \cdot a^{3/4} \cdot a^{-1/2}$ — $a^{7/12}$

20. $y^{-1/6} \cdot y^{2/3} \cdot y^{1/2}$ — y

21. $\frac{a^{1/2}}{a^{3/2}}$ — $\frac{1}{a}$

22. $\frac{b^{1/3}}{b^{4/3}}$ — $\frac{1}{b}$

23. $\frac{y^{-3/4}}{y^{1/4}}$ — $\frac{1}{y}$

24. $\frac{x^{-3/5}}{x^{1/5}}$ — $\frac{1}{x^{4/5}}$

25. $\frac{y^{2/3}}{y^{-5/6}}$ — $y^{3/2}$

26. $\frac{b^{3/4}}{b^{-3/2}}$ — $b^{9/4}$

27. $(x^2)^{-1/2}$ — $\frac{1}{x}$

28. $(a^8)^{-3/4}$ — $\frac{1}{a^6}$

29. $(x^{-2/3})^6$ — $\frac{1}{x^4}$

30. $(y^{-5/6})^{12}$ — $\frac{1}{y^{10}}$

31. $(a^{-1/2})^{-2}$ — a

32. $(b^{-2/3})^{-6}$ — b^4

33. $(x^{-3/8})^{-4/5}$ — $x^{3/10}$

34. $(y^{-3/2})^{-2/9}$ — $y^{1/3}$

35. $(a^{1/2} \cdot a)^2$ — a^3

36. $(b^{2/3} \cdot b^{1/6})^6$ — b^5

37. $(x^{-1/2} \cdot x^{3/4})^{-2}$ — $\frac{1}{x^{1/2}}$

38. $(a^{1/2} \cdot a^{-2})^3$ — $\frac{1}{a^{9/2}}$

39. $(y^{-1/2} \cdot y^{2/3})^{2/3}$ — $y^{1/9}$

40. $(b^{-2/3} \cdot b^{1/4})^{-4/3}$ — $b^{5/9}$

41. $(x^8y^2)^{1/2}$ — x^4y

42. $(a^3b^9)^{2/3}$ — a^2b^6

Section 9.1

Suggested Assignment

Exercises 1–135, odds
More challenging problems:
Exercises 137–138

Quick Quiz (Objective 9.1A)

Simplify.

1. $16^{3/2}$ 64

2. $(a^4b^8)^{3/4}$ a^3b^6

3. $\left(\frac{x^{1/3}}{y^{-2}}\right)^6$ x^2y^{12}

4. $\frac{x^{3/4}}{x^{-1/4}}$ x

504 Chapter 9 / Rational Exponents and Radicals

43. $(x^4y^2z^6)^{3/2}$
$x^6y^3z^9$

44. $(a^8b^4c^4)^{3/4}$
$a^6b^3c^3$

45. $(x^{-3}y^6)^{-1/3}$
$\frac{x}{y^2}$

46. $(a^2b^{-6})^{-1/2}$
$\frac{b^3}{a}$

47. $(x^{-2}y^{1/3})^{-3/4}$
$\frac{x^{3/2}}{y^{1/4}}$

48. $(a^{-2/3}b^{2/3})^{3/2}$
$\frac{b}{a}$

49. $\left(\frac{x^{1/2}}{y^2}\right)^4$
$\frac{x^2}{y^8}$

50. $\left(\frac{b^{-3/4}}{a^{-1/2}}\right)^8$
$\frac{a^4}{b^6}$

51. $\frac{x^{1/4} \cdot x^{-1/2}}{x^{2/3}}$
$\frac{1}{x^{11/12}}$

52. $\frac{b^{1/2} \cdot b^{-3/4}}{b^{1/4}}$
$\frac{1}{b^{1/2}}$

53. $\left(\frac{y^{2/3} \cdot y^{-5/6}}{y^{1/9}}\right)^9$
$\frac{1}{y^{5/2}}$

54. $\left(\frac{a^{1/3} \cdot a^{-2/3}}{a^{1/2}}\right)^4$
$\frac{1}{a^{10/3}}$

55. $\left(\frac{b^2 \cdot b^{-3/4}}{b^{-1/2}}\right)^{-1/2}$
$\frac{1}{b^{7/8}}$

56. $\frac{(x^{-5/6} \cdot x^3)^{-2/3}}{x^{4/3}}$
$\frac{1}{x^{25/9}}$

57. $(a^{2/3}b^2)^6(a^3b^3)^{1/3}$
a^5b^{13}

58. $(x^3y^{-1/2})^{-2}(x^{-3}y^2)^{1/6}$
$\frac{y^{4/3}}{x^{13/2}}$

59. $(16m^{-2}n^4)^{-1/2}(mn^{1/2})$
$\frac{m^2}{4n^{3/2}}$

60. $(27m^3n^{-6})^{1/3}(m^{-1/3}n^{5/6})^6$
$\frac{3n^3}{m}$

61. $\left(\frac{x^{1/2}y^{-3/4}}{y^{2/3}}\right)^{-6}$
$\frac{y^{17/2}}{x^3}$

62. $\left(\frac{x^{1/2}y^{-5/4}}{y^{-3/4}}\right)^{-4}$
$\frac{y^2}{x^2}$

63. $\left(\frac{2^{-6}b^{-3}}{a^{-1/2}}\right)^{-2/3}$
$\frac{16b^2}{a^{1/3}}$

64. $\left(\frac{49c^{5/3}}{a^{-1/4}b^{5/6}}\right)^{-3/2}$
$\frac{b^{5/4}}{343a^{3/8}c^{5/2}}$

65. $y^{3/2}(y^{1/2} - y^{-1/2})$
$y^2 - y$

66. $y^{3/5}(y^{2/5} + y^{-3/5})$
$y + 1$

67. $a^{-1/4}(a^{5/4} - a^{9/4})$
$a - a^2$

68. $x^{4/3}(x^{2/3} + x^{-1/3})$
$x^2 + x$

69. $x^n \cdot x^{3n}$
x^{4n}

70. $a^{2n} \cdot a^{-5n}$
$\frac{1}{a^{3n}}$

71. $x^n \cdot x^{n/2}$
$x^{3n/2}$

72. $a^{n/2} \cdot a^{-n/3}$
$a^{n/6}$

73. $\frac{y^{n/2}}{y^{-n}}$
$y^{3n/2}$

74. $\frac{b^{m/3}}{b^m}$
$\frac{1}{b^{2m/3}}$

75. $(x^{2n})^n$
x^{2n^2}

76. $(x^{5n})^{2n}$
x^{10n^2}

77. $(x^{n/4}y^{n/8})^8$
$x^{2n}y^n$

78. $(x^{n/2}y^{n/3})^6$
$x^{3n}y^{2n}$

79. $(x^{n/5}y^{n/10})^{20}$
$x^{4n}y^{2n}$

80. $(x^{n/2}y^{n/5})^{10}$
$x^{5n}y^{2n}$

505

Objective B

Rewrite the exponential expression as a radical expression.

81. $3^{1/4}$ $\sqrt[4]{3}$

82. $5^{1/2}$ $\sqrt{5}$

83. $a^{3/2}$ $\sqrt{a^3}$

84. $b^{4/3}$ $\sqrt[3]{b^4}$

85. $(2t)^{5/2}$ $\sqrt{32t^5}$

86. $(3x)^{2/3}$ $\sqrt[3]{9x^2}$

87. $-2x^{2/3}$ $-2\sqrt[3]{x^2}$

88. $-3a^{2/5}$ $-3\sqrt[5]{a^2}$

89. $(a^2b)^{2/3}$ $\sqrt[3]{a^4b^2}$

90. $(x^2y^3)^{3/4}$ $\sqrt[4]{x^6y^9}$

91. $(a^2b^4)^{3/5}$ $\sqrt[5]{a^6b^{12}}$

92. $(a^3b^7)^{3/2}$ $\sqrt{a^9b^{21}}$

93. $(4x - 3)^{3/4}$ $\sqrt[4]{(4x-3)^3}$

94. $(3x - 2)^{1/3}$ $\sqrt[3]{3x-2}$

95. $x^{-2/3}$ $\dfrac{1}{\sqrt[3]{x^2}}$

96. $b^{-3/4}$ $\dfrac{1}{\sqrt[4]{b^3}}$

Rewrite the radical expression as an exponential expression.

97. $\sqrt{14}$ $14^{1/2}$

98. $\sqrt{7}$ $7^{1/2}$

99. $\sqrt[3]{x}$ $x^{1/3}$

100. $\sqrt[4]{x}$ $y^{1/4}$

101. $\sqrt[3]{x^4}$ $x^{4/3}$

102. $\sqrt[4]{a^3}$ $a^{3/4}$

103. $\sqrt[5]{b^3}$ $b^{3/5}$

104. $\sqrt[4]{b^5}$ $b^{5/4}$

105. $\sqrt[3]{2x^2}$ $(2x^2)^{1/3}$

106. $\sqrt[5]{4y^7}$ $(4y^7)^{1/5}$

107. $-\sqrt{3x^5}$ $-(3x^5)^{1/2}$

108. $-\sqrt[4]{4x^5}$ $-(4x^5)^{1/4}$

109. $3x\sqrt[3]{y^2}$ $3xy^{2/3}$

110. $2y\sqrt{x^3}$ $2x^{3/2}y$

111. $\sqrt{a^2 - 2}$ $(a^2 - 2)^{1/2}$

112. $\sqrt{3 - y^2}$ $(3 - y^2)^{1/2}$

Quick Quiz (Objective 9.1B)

Rewrite the exponential expression as a radical expression.

1. $x^{3/4}$ $\sqrt[4]{x^3}$
2. $-3x^{2/3}$ $-3\sqrt[3]{x^2}$
3. $a^{-1/2}$ $\dfrac{1}{\sqrt{a}}$

Rewrite the radical expression as an exponential expression.

4. $\sqrt[5]{a^4}$ $a^{4/5}$
5. $\sqrt[3]{3x^2}$ $(3x^2)^{1/3}$
6. $4x\sqrt[4]{y}$ $4xy^{1/4}$

Answers to Writing Exercises

139. No. If $x \geq 0$, the statement is true. However, if $x < 0$, then $\sqrt{x^2} = |x|$. For example, if $x = -2$, then $\sqrt{x^2} = \sqrt{(-2)^2} = \sqrt{4} = 2$, not -2.

Objective C

Simplify.

113. $\sqrt{x^{16}}$ x^8

114. $\sqrt{y^{14}}$ y^7

115. $-\sqrt{x^8}$ $-x^4$

116. $-\sqrt{a^6}$ $-a^3$

117. $\sqrt[3]{x^3y^9}$ xy^3

118. $\sqrt[3]{a^6b^{12}}$ a^2b^4

119. $-\sqrt[3]{x^{15}y^3}$ $-x^5y$

120. $-\sqrt[3]{a^9b^9}$ $-a^3b^3$

121. $\sqrt{16a^4b^{12}}$ $4a^2b^6$

122. $\sqrt{25x^8y^2}$ $5x^4y$

123. $\sqrt{-16x^4y^2}$ not a real number

124. $\sqrt{-9a^4b^8}$ not a real number

125. $\sqrt[3]{27x^9}$ $3x^3$

126. $\sqrt[3]{8a^{21}b^6}$ $2a^7b^2$

127. $\sqrt[3]{-64x^9y^{12}}$ $-4x^3y^4$

128. $\sqrt[3]{-27a^3b^{15}}$ $-3ab^5$

129. $-\sqrt[4]{x^8y^{12}}$ $-x^2y^3$

130. $-\sqrt[4]{a^{16}b^4}$ $-a^4b$

131. $\sqrt[5]{x^{20}y^{10}}$ x^4y^2

132. $\sqrt[5]{a^5b^{25}}$ ab^5

133. $\sqrt[4]{81x^4y^{20}}$ $3xy^5$

134. $\sqrt[4]{16a^8b^{20}}$ $2a^2b^5$

135. $\sqrt[5]{32a^5b^{10}}$ $2ab^2$

136. $\sqrt[5]{-32x^{15}y^{20}}$ $-2x^3y^4$

APPLYING THE CONCEPTS

137. Determine whether the following statements are true or false. If the statement is false, correct the right-hand side of the equation.

a. $\sqrt{(-2)^2} = -2$ false; 2

b. $\sqrt[3]{(-3)^3} = -3$ true

c. $\sqrt[n]{a} = a^{1/n}$ true

d. $\sqrt[n]{a^n + b^n} = a + b$ false; $(a^n + b^n)^{1/n}$

e. $(a^{1/2} + b^{1/2})^2 = a + b$ false; $a + 2a^{1/2}b^{1/2} + b$

f. $\sqrt[m]{a^n} = a^{mn}$ false: $a^{n/m}$

138. Simplify.

a. $\sqrt[3]{\sqrt{x^6}}$ x

b. $\sqrt[4]{\sqrt{a^8}}$ a

c. $\sqrt{\sqrt{81y^8}}$ $3y^2$

d. $\sqrt{\sqrt[n]{a^{4n}}}$ a^2

e. $\sqrt[n]{\sqrt{b^{6n}}}$ b^3

f. $\sqrt{\sqrt[3]{x^{12}y^{24}}}$ x^2y^4

139. If x is any real number, is $\sqrt{x^2} = x$ always true? Show why or why not.

Quick Quiz (Objective 9.1C)

Simplify.

1. $\sqrt[4]{a^8b^{12}}$ a^2b^3

2. $\sqrt[3]{-x^{12}y^{24}}$ $-x^4y^8$

3. $\sqrt{-81x^4y^6}$ Not a real number

4. $-\sqrt[5]{a^5b^{15}}$ $-ab^3$

9.2 Operations on Radical Expressions

Objective A To simplify radical expressions

VIDEO & DVD CD TUTOR WEB SSM

Point of Interest

The Latin expression for irrational numbers was *numerus surdus*, which literally means "inaudible number." A prominent 16th-century mathematician wrote of irrational numbers, "Just as an infinite number is not a number, so an irrational number is not a true number, but lies hidden in some sort of cloud of infinity." In 1872, Richard Dedekind wrote a paper that established the first logical treatment of irrational numbers.

If a number is not a perfect power, its root can only be approximated; examples include $\sqrt{5}$ and $\sqrt[3]{3}$. These numbers are **irrational numbers**. Their decimal representations never terminate or repeat.

$$\sqrt{5} = 2.2360679\ldots \qquad \sqrt[3]{3} = 1.4422495\ldots$$

A radical expression is not in simplest form when the radicand contains a factor greater than 1 that is a perfect power. The Product Property of Radicals is used to simplify radical expressions whose radicands are not perfect powers.

The Product Property of Radicals

If $\sqrt[n]{a}$ and $\sqrt[n]{b}$ are positive real numbers, then $\sqrt[n]{ab} = \sqrt[n]{a} \cdot \sqrt[n]{b}$ and $\sqrt[n]{a} \cdot \sqrt[n]{b} = \sqrt[n]{ab}$.

➡ Simplify: $\sqrt{48}$

$\sqrt{48} = \sqrt{16 \cdot 3}$ • **Write the radicand as the product of a perfect square and a factor that does not contain a perfect square.**

$= \sqrt{16}\sqrt{3}$ • **Use the Product Property of Radicals to write the expression as a product.**

$= 4\sqrt{3}$ • **Simplify $\sqrt{16}$.**

Note that 48 must be written as the product of a perfect square and *a factor that does not contain a perfect square.* Therefore, it would not be correct to rewrite $\sqrt{48}$ as $\sqrt{4 \cdot 12}$ and simplify the expression as shown at the right. Although 4 is a perfect-square factor of 48, 12 contains a perfect square ($12 = 4 \cdot 3$) and $\sqrt{12}$ can be simplified. Remember to find the largest perfect power that is a factor of the radicand.

$$\sqrt{48} = \sqrt{4 \cdot 12} = \sqrt{4}\sqrt{12} = 2\sqrt{12}$$

Not in simplest form

➡ Simplify: $\sqrt{18x^2y^3}$

$\sqrt{18x^2y^3} = \sqrt{9x^2y^2 \cdot 2y}$ • **Write the radicand as the product of a perfect square and factors that do not contain a perfect square.**

$= \sqrt{9x^2y^2}\sqrt{2y}$ • **Use the Product Property of Radicals to write the expression as a product.**

$= 3xy\sqrt{2y}$ • **Simplify.**

Objective 9.2A

Vocabulary to Review

irrational number
perfect square
perfect cube
perfect fourth power
perfect fifth power

New Properties

Product Property of Radicals

Instructor Note

Provide numerical examples of the Product Property of Radicals. For example:

$\sqrt{4 \cdot 9} = \sqrt{4} \cdot \sqrt{9}$ or
$\sqrt[3]{8 \cdot 64} = \sqrt[3]{8} \cdot \sqrt[3]{64}$

Discuss the Concepts

1. Which of the following represent irrational numbers?
   a. $\sqrt{18}$
   b. $\sqrt[3]{6}$
   c. $\sqrt[4]{81}$
   d. $\sqrt{24}$
   e. $\sqrt[3]{9}$
   f. $\sqrt[5]{32}$
2. Is the radical expression $\sqrt[3]{16}$ in simplest form? Why or why not?
3. Explain how to write $\sqrt{32a^5}$ in simplest form.
4. Explain how the Product Property of Radicals is used when we write a radical expression in simplest form.

In-Class Examples (Objective 9.2A)

Simplify.

1. $\sqrt{x^3y^4z^6}$ $xy^2z^3\sqrt{x}$
2. $\sqrt{27a^5b^4}$ $3a^2b^2\sqrt{3a}$
3. $\sqrt[3]{-250x^5y^7}$ $-5xy^2\sqrt[3]{2x^2y}$
4. $\sqrt[4]{32x^8y^{10}}$ $2x^2y^2\sqrt[4]{2y^2}$

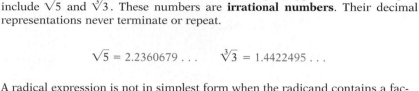

Concept Check

Determine whether the statement is always true, sometimes true, or never true.

1. The fourth root of a positive number is a positive number, and the fourth root of a negative number is a negative number. Never true
2. Every positive number has two cube roots, one of which is the opposite of the other. Never true
3. The square root of a number that is not a perfect square is an irrational number. Always true
4. If the radicand of a radical expression is evenly divisible by a perfect square greater than 1, then the radical expression is not in simplest form. Always true
5. If a and b are real numbers, then $\sqrt{ab} = \sqrt{a} \cdot \sqrt{b}$. Sometimes true

Objective 9.2B

Vocabulary to Review

radicand
index

Properties to Review

Distributive Property

Instructor Note

Mention to students that adding and subtracting radical expressions is similar to combining like terms.

➡ Simplify: $\sqrt[3]{x^7}$

$\sqrt[3]{x^7} = \sqrt[3]{x^6 \cdot x}$ • Write the radicand as the product of a perfect cube and a factor that does not contain a perfect cube.

$= \sqrt[3]{x^6}\sqrt[3]{x}$ • Use the Product Property of Radicals to write the expression as a product.

$= x^2\sqrt[3]{x}$ • Simplify.

➡ Simplify: $\sqrt[4]{32x^7}$

$\sqrt[4]{32x^7} = \sqrt[4]{2^5x^7}$ • Write the prime factorization of the coefficient of the radicand in exponential form.

$= \sqrt[4]{2^4x^4(2x^3)}$ • Write the radicand as the product of a perfect fourth power and factors that do not contain a perfect fourth power.

$= \sqrt[4]{2^4x^4}\sqrt[4]{2x^3}$ • Use the Product Property of Radicals to write the expression as a product.

$= 2x\sqrt[4]{2x^3}$ • Simplify.

Example 1 Simplify: $\sqrt[4]{x^9}$

Solution $\sqrt[4]{x^9} = \sqrt[4]{x^8 \cdot x} = \sqrt[4]{x^8}\sqrt[4]{x}$
$= x^2\sqrt[4]{x}$

You Try It 1 Simplify: $\sqrt[5]{x^7}$

Your solution $x\sqrt[5]{x^2}$

Example 2 Simplify: $\sqrt[3]{-27a^5b^{12}}$

Solution $\sqrt[3]{-27a^5b^{12}} = \sqrt[3]{(-3)^3a^5b^{12}}$
$= \sqrt[3]{(-3)^3a^3b^{12}(a^2)}$
$= \sqrt[3]{(-3)^3a^3b^{12}}\sqrt[3]{a^2}$
$= -3ab^4\sqrt[3]{a^2}$

You Try It 2 Simplify: $\sqrt[3]{-64x^8y^{18}}$

Your solution $-4x^2y^6\sqrt[3]{x^2}$

Solutions on p. S28

Objective B **To add or subtract radical expressions**

VIDEO & DVD CD TUTOR WEB SSM

The Distributive Property is used to simplify the sum or difference of radical expressions that have the same radicand and the same index. For example,

$$3\sqrt{5} + 8\sqrt{5} = (3 + 8)\sqrt{5} = 11\sqrt{5}$$
$$2\sqrt[3]{3x} - 9\sqrt[3]{3x} = (2 - 9)\sqrt[3]{3x} = -7\sqrt[3]{3x}$$

TAKE NOTE
Adding and subtracting radicals is similar to combining like terms.

Radical expressions that are in simplest form and have unlike radicands or different indices cannot be simplified by the Distributive Property. The expressions below cannot be simplified by the Distributive Property.

$$3\sqrt[4]{2} - 6\sqrt[4]{3} \qquad 2\sqrt[4]{4x} + 3\sqrt[3]{4x}$$

In-Class Examples (Objective 9.2B)

Simplify.

1. $\sqrt{32} - \sqrt{50}$ $-\sqrt{2}$
2. $2b\sqrt{12a^3b} - a\sqrt{192ab^3}$ $-4ab\sqrt{3ab}$
3. $6\sqrt{75} - 5\sqrt{48} + 2\sqrt{32}$ $10\sqrt{3} + 8\sqrt{2}$
4. $x\sqrt[3]{16y^4} + 4\sqrt[3]{54x^3y^4} - xy\sqrt[3]{250y}$ $9xy\sqrt[3]{2y}$

➡ Simplify: $3\sqrt{32x^2} - 2x\sqrt{2} + \sqrt{128x^2}$

$$\begin{aligned} 3\sqrt{32x^2} - 2x\sqrt{2} + \sqrt{128x^2} &= 3\sqrt{16x^2}\,\sqrt{2} - 2x\sqrt{2} + \sqrt{64x^2}\,\sqrt{2} \\ &= 3 \cdot 4x\sqrt{2} - 2x\sqrt{2} + 8x\sqrt{2} \\ &= 12x\sqrt{2} - 2x\sqrt{2} + 8x\sqrt{2} \\ &= 18x\sqrt{2} \end{aligned}$$

• First simplify each term. Then combine like terms by using the Distributive Property.

Example 3

Simplify: $5b\sqrt[4]{32a^7b^5} - 2a\sqrt[4]{162a^3b^9}$

Solution

$$\begin{aligned} &5b\sqrt[4]{32a^7b^5} - 2a\sqrt[4]{162a^3b^9} \\ &= 5b\sqrt[4]{16a^4b^4}\,\sqrt[4]{2a^3b} - 2a\sqrt[4]{81b^8}\sqrt[4]{2a^3b} \\ &= 5b \cdot 2ab\sqrt[4]{2a^3b} - 2a \cdot 3b^2\sqrt[4]{2a^3b} \\ &= 10ab^2\sqrt[4]{2a^3b} - 6ab^2\sqrt[4]{2a^3b} \\ &= 4ab^2\sqrt[4]{2a^3b} \end{aligned}$$

You Try It 3

Simplify: $3xy\sqrt[3]{81x^5y} - \sqrt[3]{192x^8y^4}$

Your solution

$5x^2y\sqrt[3]{3x^2y}$

Solution on p. S28

Discuss the Concepts

1. Why must radical expressions have the same index and the same radicand before they can be added or subtracted?
2. Which of the following expressions cannot be simplified?
   a. $\sqrt[4]{8y} + \sqrt[5]{8y}$
   b. $\sqrt[3]{a^2b} + \sqrt[3]{ab^2}$
   c. $\sqrt{x+y} + \sqrt{x+y}$

Concept Check

Given $f(x) = \sqrt{x+5} + \sqrt{5x+3}$, write $f(3)$ in simplest form. $5\sqrt{2}$

Optional Student Activity

1. Write a paragraph that compares adding two monomials to adding two radical expressions. For example, compare the addition of $7y + 9y$ to the addition of $7\sqrt{y} + 9\sqrt{y}$.
2. Write a paragraph that compares simplifying a variable expression such as $7x + 9y$ to simplifying a radical expression such as $7\sqrt{x} + 9\sqrt{y}$.

Objective C To multiply radical expressions

VIDEO & DVD CD TUTOR WEB SSM

The Product Property of Radicals is used to multiply radical expressions with the same index.

$$\sqrt{3x} \cdot \sqrt{5y} = \sqrt{3x \cdot 5y} = \sqrt{15xy}$$

➡ Simplify: $\sqrt[3]{2a^5b}\sqrt[3]{16a^2b^2}$

$$\begin{aligned} \sqrt[3]{2a^5b}\sqrt[3]{16a^2b^2} &= \sqrt[3]{32a^7b^3} \\ &= \sqrt[3]{8a^6b^3}\,\sqrt[3]{4a} \\ &= 2a^2b\sqrt[3]{4a} \end{aligned}$$

• Use the Product Property of Radicals to multiply the radicands.
• Simplify.

➡ Simplify: $\sqrt{2x}(\sqrt{8x} - \sqrt{3})$

$$\begin{aligned} \sqrt{2x}(\sqrt{8x} - \sqrt{3}) &= \sqrt{2x}(\sqrt{8x}) - \sqrt{2x}(\sqrt{3}) \\ &= \sqrt{16x^2} - \sqrt{6x} \\ &= 4x - \sqrt{6x} \end{aligned}$$

• Use the Distributive Property.
• Simplify.

➡ Simplify: $(2\sqrt{5} - 3)(3\sqrt{5} + 4)$

$$\begin{aligned} (2\sqrt{5} - 3)(3\sqrt{5} + 4) &= 6(\sqrt{5})^2 + 8\sqrt{5} - 9\sqrt{5} - 12 \\ &= 30 + 8\sqrt{5} - 9\sqrt{5} - 12 \\ &= 18 - \sqrt{5} \end{aligned}$$

• Use the FOIL method to multiply the numbers.
• Combine like terms.

Objective 9.2C

New Vocabulary

conjugate

Vocabulary to Review

The FOIL method

Properties to Review

Product Property of Radicals
Distributive Property

In-Class Examples (Objective 9.2C)

Simplify.

1. $\sqrt[3]{4}\,\sqrt[3]{16}$ 4
2. $\sqrt{3xy^2}\,\sqrt{27xy^2}$ $9xy^2$
3. $\sqrt[4]{27a^2b^3}\,\sqrt[4]{3a^6b^5}$ $3a^2b^2$
4. $\sqrt{5a}\left(\sqrt{20a} - \sqrt{a}\right)$ $10a - a\sqrt{5}$
5. $\left(\sqrt{x} - 4\right)^2$ $x - 8\sqrt{x} + 16$
6. $\left(5\sqrt{x} + 2\sqrt{y}\right)\left(5\sqrt{x} - 2\sqrt{y}\right)$ $25x - 4y$

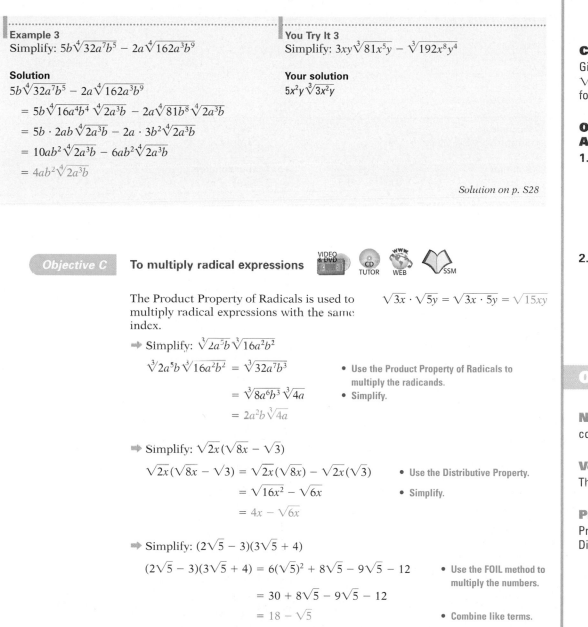

Discuss the Concepts

1. Why is it not necessary for two radical expressions that are to be multiplied to have the same radicand?
2. Must two radical expressions have the same index if they are to be multiplied? Why or why not?
3. How do you determine the conjugate of $\sqrt{5a} + \sqrt{b}$?
4. Why does the product of two conjugates involving square roots produce an expression without a radical?

Concept Check

1. Rectangle $ABCD$ in the rectangular coordinate system has a length of $\sqrt{50}$ units and a width of $\sqrt{32}$ units. Find the area of the rectangle. 40 square units
2. Verify that $\sqrt{5} - \sqrt{3}$ is a square root of $8 - 2\sqrt{15}$. $(\sqrt{5} - \sqrt{3})^2 = 8 - 2\sqrt{15}$

Optional Student Activity

1. Factor each expression over the set of real numbers. For example, since $\sqrt{8} = 2\sqrt{2}$, $x^2 - 8$ can be factored as $(x + 2\sqrt{2})(x - 2\sqrt{2})$.
   a. $y^2 - 27$
      $(y + 3\sqrt{3})(y - 3\sqrt{3})$
   b. $x^2 - 18$
      $(x + 3\sqrt{2})(x - 3\sqrt{2})$
   c. $y^2 + 4\sqrt{2}y + 8$
      $(y + 2\sqrt{2})^2$
   d. $x^2 + 6\sqrt{3}x + 27$
      $(x + 3\sqrt{3})^2$
2. Simplify: $(\sqrt{2} - 2)^3$
   $14\sqrt{2} - 20$

➡ Simplify: $(4\sqrt{a} - \sqrt{b})(2\sqrt{a} + 5\sqrt{b})$

$(4\sqrt{a} - \sqrt{b})(2\sqrt{a} + 5\sqrt{b})$

$= 8(\sqrt{a})^2 + 20\sqrt{ab} - 2\sqrt{ab} - 5(\sqrt{b})^2$ • Use the FOIL method.

$= 8a + 18\sqrt{ab} - 5b$

TAKE NOTE

The concept of conjugate is used in a number of different instances. Make sure you understand this idea.

The conjugate of $\sqrt{3} - 4$ is $\sqrt{3} + 4$.

The conjugate of $\sqrt{3} + 4$ is $\sqrt{3} - 4$.

The conjugate of $\sqrt{5a} + \sqrt{b}$ is $\sqrt{5a} - \sqrt{b}$.

The expressions $a + b$ and $a - b$ are **conjugates** of each other. Recall that $(a + b)(a - b) = a^2 - b^2$. This identity is used to simplify conjugate radical expressions.

➡ Simplify: $(\sqrt{11} - 3)(\sqrt{11} + 3)$

$(\sqrt{11} - 3)(\sqrt{11} + 3) = (\sqrt{11})^2 - 3^2 = 11 - 9$ • The radical expressions are conjugates.

$= 2$

Example 4

Simplify: $\sqrt{3x}(\sqrt{27x^2} - \sqrt{3x})$

Solution

$\sqrt{3x}(\sqrt{27x^2} - \sqrt{3x}) = \sqrt{81x^3} - \sqrt{9x^2}$

$= \sqrt{81x^2}\sqrt{x} - \sqrt{9x^2}$

$= 9x\sqrt{x} - 3x$

You Try It 4

Simplify: $\sqrt{5b}(\sqrt{3b} - \sqrt{10})$

Your solution

$b\sqrt{15} - 5\sqrt{2b}$

Example 5

Simplify: $(2\sqrt[3]{x} - 3)(3\sqrt[3]{x} - 4)$

Solution

$(2\sqrt[3]{x} - 3)(3\sqrt[3]{x} - 4)$

$= 6\sqrt[3]{x^2} - 8\sqrt[3]{x} - 9\sqrt[3]{x} + 12$

$= 6\sqrt[3]{x^2} - 17\sqrt[3]{x} + 12$

You Try It 5

Simplify: $(2\sqrt[3]{2x} - 3)(\sqrt[3]{2x} - 5)$

Your solution

$2\sqrt[3]{4x^2} - 13\sqrt[3]{2x} + 15$

Example 6

Simplify: $(2\sqrt{x} - \sqrt{2y})(2\sqrt{x} + \sqrt{2y})$

Solution

$(2\sqrt{x} - \sqrt{2y})(2\sqrt{x} + \sqrt{2y})$

$= (2\sqrt{x})^2 - (\sqrt{2y})^2$

$= 4x - 2y$

You Try It 6

Simplify: $(\sqrt{a} - 3\sqrt{y})(\sqrt{a} + 3\sqrt{y})$

Your solution

$a - 9y$

Solutions on p. S28

Objective D To divide radical expressions

VIDEO & DVD CD TUTOR WEB SSM

The Quotient Property of Radicals is used to divide radical expressions with the same index.

The Quotient Property of Radicals

If $\sqrt[n]{a}$ and $\sqrt[n]{b}$ are real numbers, and $b \neq 0$, then

$$\sqrt[n]{\frac{a}{b}} = \frac{\sqrt[n]{a}}{\sqrt[n]{b}} \quad \text{and} \quad \frac{\sqrt[n]{a}}{\sqrt[n]{b}} = \sqrt[n]{\frac{a}{b}}$$

➡ Simplify: $\sqrt[3]{\dfrac{81x^5}{y^6}}$

$$\sqrt[3]{\frac{81x^5}{y^6}} = \frac{\sqrt[3]{81x^5}}{\sqrt[3]{y^6}}$$

• Use the Quotient Property of Radicals.

$$= \frac{\sqrt[3]{27x^3}\sqrt[3]{3x^2}}{\sqrt[3]{y^6}} = \frac{3x\sqrt[3]{3x^2}}{y^2}$$

• Simplify each radical expression.

➡ Simplify: $\dfrac{\sqrt{5a^4b^7c^2}}{\sqrt{ab^3c}}$

$$\frac{\sqrt{5a^4b^7c^2}}{\sqrt{ab^3c}} = \sqrt{\frac{5a^4b^7c^2}{ab^3c}}$$

• Use the Quotient Property of Radicals.

$$= \sqrt{5a^3b^4c}$$

$$= \sqrt{a^2b^4}\sqrt{5ac} = ab^2\sqrt{5ac}$$

• Simplify the radicand.

A radical expression is in simplest form when no radical remains in the denominator of the radical expression. The procedure used to remove a radical from the denominator is called **rationalizing the denominator.**

➡ Simplify: $\dfrac{5}{\sqrt{2}}$

$$\frac{5}{\sqrt{2}} = \frac{5}{\sqrt{2}} \cdot 1 = \frac{5}{\sqrt{2}} \cdot \frac{\sqrt{2}}{\sqrt{2}}$$

• Multiply by $\frac{\sqrt{2}}{\sqrt{2}}$, which equals 1.

$$= \frac{5\sqrt{2}}{2}$$

• $\sqrt{2} \cdot \sqrt{2} = (\sqrt{2})^2 = 2$

TAKE NOTE

Multiplying by $\dfrac{\sqrt[3]{4x}}{\sqrt[3]{4x}}$ will not rationalize the denominator of $\dfrac{3x}{\sqrt[3]{4x}}$.

$$\frac{3}{\sqrt[3]{4x}} \cdot \frac{\sqrt[3]{4x}}{\sqrt[3]{4x}} = \frac{3\sqrt[3]{4x}}{\sqrt[3]{16x^2}}$$

Because $\sqrt[3]{16x^2}$ is not a perfect cube, the denominator still contains a radical expression.

➡ Simplify: $\dfrac{3}{\sqrt[3]{4x}}$

$$\frac{3}{\sqrt[3]{4x}} = \frac{3}{\sqrt[3]{4x}} \cdot \frac{\sqrt[3]{2x^2}}{\sqrt[3]{2x^2}}$$

• Because $\sqrt[3]{4x} \cdot \sqrt[3]{2x^2} = \sqrt[3]{8x^3}$, a perfect cube, multiply the expression by $\frac{\sqrt[3]{2x^2}}{\sqrt[3]{2x^2}}$, which equals 1.

$$= \frac{3\sqrt[3]{2x^2}}{\sqrt[3]{8x^3}} = \frac{3\sqrt[3]{2x^2}}{2x}$$

• Simplify.

Objective 9.2D

New Vocabulary

rationalizing the denominator

Vocabulary to Review

conjugate

New Properties

The Quotient Property of Radicals

Instructor Note

Students need assurance that rationalizing a denominator is nothing more than multiplying an expression by 1. Have students evaluate $\dfrac{5}{\sqrt{2}}$ and $\dfrac{5\sqrt{2}}{2}$ with their calculators to see that the decimal representations displayed are the same.

Discuss the Concepts

1. Must two radical expressions have the same index if they are to be divided? Why or why not?
2. Is the expression in simplest form? Why or why not?
   a. $\sqrt{\dfrac{2}{5}}$
   b. $\dfrac{3}{\sqrt[3]{x}}$
   c. $\dfrac{\sqrt{3}}{9}$
   d. $\dfrac{\sqrt[3]{x^4}}{13}$
3. Why can we multiply $\dfrac{2}{\sqrt[3]{3x}}$ by $\dfrac{\sqrt[3]{9x^2}}{\sqrt[3]{9x^2}}$ without changing the value of the expression?

In-Class Examples (Objective 9.2D)

Simplify.

1. $\dfrac{\sqrt{50x^3}}{\sqrt{2x}}$ $5x$
2. $\dfrac{8}{\sqrt{2a}}$ $\dfrac{4\sqrt{2a}}{a}$
3. $\sqrt{\dfrac{x}{3}}$ $\dfrac{\sqrt{3x}}{3}$
4. $\dfrac{4}{\sqrt[3]{2x^2}}$ $\dfrac{2\sqrt[3]{4x}}{x}$
5. $\dfrac{\sqrt{32a^5b}}{\sqrt{12ab^3}}$ $\dfrac{2a^2\sqrt{6}}{3b}$
6. $\dfrac{2}{\sqrt{3}+2}$ $4 - 2\sqrt{3}$
7. $\dfrac{\sqrt{3}-\sqrt{2}}{\sqrt{3}+\sqrt{2}}$ $5 - 2\sqrt{6}$
8. $\dfrac{8x^3 - \sqrt{32x}}{\sqrt{2x}}$ $4x^2\sqrt{2x} - 4$

Concept Check

Simplify each of the following expressions. Then write a rule for simplifying $\frac{a}{\sqrt{a}}$.

a. $\frac{3}{\sqrt{3}}$ $\sqrt{3}$

b. $\frac{5}{\sqrt{5}}$ $\sqrt{5}$

c. $\frac{7}{\sqrt{7}}$ $\sqrt{7}$

d. $\frac{11}{\sqrt{11}}$ $\sqrt{11}$

The rule is $\frac{a}{\sqrt{a}} = \sqrt{a}$.

Optional Student Activity

1. Simplify: $\frac{2}{\sqrt{x+4}+2}$

$\frac{2\sqrt{x+4}-4}{x}$

2. Simplify: $\frac{\sqrt{b+9}-3}{\sqrt{b+9}+3}$

$\frac{b+18-6\sqrt{b+9}}{b}$

To simplify a fraction that has a square-root expression with two terms in the denominator, multiply the numerator and denominator by the conjugate of the denominator. Then simplify.

➡ Simplify: $\frac{\sqrt{x}-\sqrt{y}}{\sqrt{x}+\sqrt{y}}$

TAKE NOTE
Here is an example of using a conjugate to simplify a radical expression.

$$\frac{\sqrt{x}-\sqrt{y}}{\sqrt{x}+\sqrt{y}} = \frac{\sqrt{x}-\sqrt{y}}{\sqrt{x}+\sqrt{y}} \cdot \frac{\sqrt{x}-\sqrt{y}}{\sqrt{x}-\sqrt{y}}$$
$$= \frac{(\sqrt{x})^2 - \sqrt{xy} - \sqrt{xy} + (\sqrt{y})^2}{(\sqrt{x})^2 - (\sqrt{y})^2} = \frac{x - 2\sqrt{xy} + y}{x - y}$$

Example 7

Simplify: $\frac{5}{\sqrt{5x}}$

Solution

$$\frac{5}{\sqrt{5x}} = \frac{5}{\sqrt{5x}} \cdot \frac{\sqrt{5x}}{\sqrt{5x}} = \frac{5\sqrt{5x}}{(\sqrt{5x})^2}$$
$$= \frac{5\sqrt{5x}}{5x} = \frac{\sqrt{5x}}{x}$$

You Try It 7

Simplify: $\frac{y}{\sqrt{3y}}$

Your solution

$\frac{\sqrt{3y}}{3}$

Example 8

Simplify: $\frac{3x}{\sqrt[4]{2x}}$

Solution

$$\frac{3x}{\sqrt[4]{2x}} = \frac{3x}{\sqrt[4]{2x}} \cdot \frac{\sqrt[4]{8x^3}}{\sqrt[4]{8x^3}}$$
$$= \frac{3x\sqrt[4]{8x^3}}{\sqrt[4]{16x^4}} = \frac{3x\sqrt[4]{8x^3}}{2x}$$
$$= \frac{3\sqrt[4]{8x^3}}{2}$$

You Try It 8

Simplify: $\frac{3x}{\sqrt[3]{3x^2}}$

Your solution

$\sqrt[3]{9x}$

Example 9

Simplify: $\frac{3}{5-2\sqrt{3}}$

Solution

$$\frac{3}{5-2\sqrt{3}} = \frac{3}{5-2\sqrt{3}} \cdot \frac{5+2\sqrt{3}}{5+2\sqrt{3}} = \frac{15+6\sqrt{3}}{5^2-(2\sqrt{3})^2}$$
$$= \frac{15+6\sqrt{3}}{25-12} = \frac{15+6\sqrt{3}}{13}$$

You Try It 9

Simplify: $\frac{3+\sqrt{6}}{2-\sqrt{6}}$

Your solution

$-\frac{12+5\sqrt{6}}{2}$

Solutions on p. S28

Section 9.2

Suggested Assignment
Exercises 1–73, 77–113, odds
More challenging problems:
Exercises 114, 115

9.2 Exercises

Objective A

Simplify.

1. $\sqrt{x^4y^3z^5}$ $x^2yz^2\sqrt{yz}$
2. $\sqrt{x^3y^6z^9}$ $xy^3z^4\sqrt{xz}$
3. $\sqrt{8a^3b^8}$ $2ab^4\sqrt{2a}$
4. $\sqrt{24a^9b^6}$ $2a^4b^3\sqrt{6a}$
5. $\sqrt{45x^2y^3z^5}$ $3xyz^2\sqrt{5yz}$
6. $\sqrt{60xy^7z^{12}}$ $2y^3z^6\sqrt{15xy}$
7. $\sqrt{-9x^3}$ not a real number
8. $\sqrt{-x^2y^5}$ not a real number
9. $\sqrt[3]{a^{16}b^8}$ $a^5b^2\sqrt[3]{ab^2}$
10. $\sqrt[3]{a^5b^8}$ $ab^2\sqrt[3]{a^2b^2}$
11. $\sqrt[3]{-125x^2y^4}$ $-5y\sqrt[3]{x^2y}$
12. $\sqrt[3]{-216x^5y^9}$ $-6xy^3\sqrt[3]{x^2}$
13. $\sqrt[3]{a^4b^5c^6}$ $abc^2\sqrt[3]{ab^2}$
14. $\sqrt[3]{a^8b^{11}c^{15}}$ $a^2b^3c^5\sqrt[3]{a^2b^2}$
15. $\sqrt[4]{16x^9y^5}$ $2x^2y\sqrt[4]{xy}$
16. $\sqrt[4]{64x^8y^{10}}$ $2x^2y^2\sqrt[4]{4y^2}$

Objective B

Add or subtract.

17. $2\sqrt{x} - 8\sqrt{x}$ $-6\sqrt{x}$
18. $3\sqrt{y} + 12\sqrt{y}$ $15\sqrt{y}$
19. $\sqrt{8} - \sqrt{32}$ $-2\sqrt{2}$
20. $\sqrt{27a} - \sqrt{8a}$ $3\sqrt{3a} - 2\sqrt{2a}$
21. $\sqrt{18b} + \sqrt{75b}$ $3\sqrt{2b} + 5\sqrt{3b}$
22. $2\sqrt{2x^3} + 4x\sqrt{8x}$ $10x\sqrt{2x}$
23. $3\sqrt{8x^2y^3} - 2x\sqrt{32y^3}$ $-2xy\sqrt{2y}$
24. $2\sqrt{32x^2y^3} - xy\sqrt{98y}$ $xy\sqrt{2y}$
25. $2a\sqrt{27ab^5} + 3b\sqrt{3a^3b}$ $6ab^2\sqrt{3ab} + 3ab\sqrt{3ab}$
26. $\sqrt[3]{128} + \sqrt[3]{250}$ $9\sqrt[3]{2}$
27. $\sqrt[3]{16} - \sqrt[3]{54}$ $-\sqrt[3]{2}$
28. $2\sqrt[3]{3a^4} - 3a\sqrt[3]{81a}$ $-7a\sqrt[3]{3a}$
29. $2b\sqrt[3]{16b^2} + \sqrt[3]{128b^5}$ $8b\sqrt[3]{2b^2}$
30. $3\sqrt[3]{x^5y^7} - 8xy\sqrt[3]{x^2y^4}$ $-5xy^2\sqrt[3]{x^2y}$

Quick Quiz (Objective 9.2A)

Simplify.

1. $\sqrt{x^5y^6z^7}$ $x^2y^3z^3\sqrt{xz}$
2. $\sqrt{48a^7b^3}$ $4a^3b\sqrt{3ab}$
3. $\sqrt[3]{-128x^7y^9}$ $-4x^2y^3\sqrt[3]{2x}$
4. $\sqrt[4]{162a^3b^5}$ $3b\sqrt[4]{2a^3b}$

31. $3\sqrt[4]{32a^5} - a\sqrt[4]{162a}$
$3a\sqrt[4]{2a}$

32. $2a\sqrt[4]{16ab^5} + 3b\sqrt[4]{256a^5b}$
$16ab\sqrt[4]{ab}$

33. $2\sqrt{50} - 3\sqrt{125} + \sqrt{98}$
$17\sqrt{2} - 15\sqrt{5}$

34. $3\sqrt{108} - 2\sqrt{18} - 3\sqrt{48}$
$6\sqrt{3} - 6\sqrt{2}$

35. $\sqrt{9b^3} - \sqrt{25b^3} + \sqrt{49b^3}$
$5b\sqrt{b}$

36. $\sqrt{4x^7y^5} + 9x^2\sqrt{x^3y^5} - 5xy\sqrt{x^5y^3}$
$6x^3y^2\sqrt{xy}$

37. $2x\sqrt{8xy^2} - 3y\sqrt{32x^3} + \sqrt{4x^3y^3}$
$-8xy\sqrt{2x} + 2xy\sqrt{xy}$

38. $5a\sqrt{3a^3b} + 2a^2\sqrt{27ab} - 4\sqrt{75a^5b}$
$-9a^2\sqrt{3ab}$

39. $\sqrt[3]{54xy^3} - 5\sqrt[3]{2xy^3} + y\sqrt[3]{128x}$
$2y\sqrt[3]{2x}$

40. $2\sqrt[3]{24x^3y^4} + 4x\sqrt[3]{81y^4} - 3y\sqrt[3]{24x^3y}$
$10xy\sqrt[3]{3y}$

41. $2a\sqrt[4]{32b^5} - 3b\sqrt[4]{162a^4b} + \sqrt[4]{2a^4b^5}$
$-4ab\sqrt[4]{2b}$

42. $6y\sqrt[4]{48x^5} - 2x\sqrt[4]{243xy^4} - 4\sqrt[4]{3x^5y^4}$
$2xy\sqrt[4]{3x}$

Objective C

Multiply.

43. $\sqrt{8}\sqrt{32}$
16

44. $\sqrt{14}\sqrt{35}$
$7\sqrt{10}$

45. $\sqrt[3]{4}\sqrt[3]{8}$
$2\sqrt[3]{4}$

46. $\sqrt[3]{6}\sqrt[3]{36}$
6

47. $\sqrt{x^2y^5}\sqrt{xy}$
$xy^3\sqrt{x}$

48. $\sqrt{a^3b}\sqrt{ab^4}$
$a^2b^2\sqrt{b}$

49. $\sqrt{2x^2y}\sqrt{32xy}$
$8xy\sqrt{x}$

50. $\sqrt{5x^3y}\sqrt{10x^3y^4}$
$5x^3y^2\sqrt{2y}$

51. $\sqrt[3]{x^2y}\sqrt[3]{16x^4y^2}$
$2x^2y\sqrt[3]{2}$

52. $\sqrt[3]{4a^2b^3}\sqrt[3]{8ab^5}$
$2ab^2\sqrt[3]{4b^2}$

53. $\sqrt[4]{12ab^3}\sqrt[4]{4a^5b^2}$
$2ab\sqrt[4]{3a^2b}$

54. $\sqrt[4]{36a^2b^4}\sqrt[4]{12a^5b^3}$
$2ab\sqrt[4]{27a^3b^3}$

55. $\sqrt{3}(\sqrt{27} - \sqrt{3})$
6

56. $\sqrt{10}(\sqrt{10} - \sqrt{5})$
$10 - 5\sqrt{2}$

57. $\sqrt{x}(\sqrt{x} - \sqrt{2})$
$x - \sqrt{2x}$

58. $\sqrt{y}(\sqrt{y} - \sqrt{5})$
$y - \sqrt{5y}$

59. $\sqrt{2x}(\sqrt{8x} - \sqrt{32})$
$4x - 8\sqrt{x}$

60. $\sqrt{3a}(\sqrt{27a^2} - \sqrt{a})$
$9a\sqrt{a} - a\sqrt{3}$

Quick Quiz (Objective 9.2B)

Simplify.

1. $\sqrt{27} - \sqrt{12}$ $\sqrt{3}$

2. $5x\sqrt{x^3y^5} - 3y\sqrt{x^5y^3}$ $2x^2y^2\sqrt{xy}$

3. $6\sqrt{28} - 3\sqrt{112} + 2\sqrt{63}$ $6\sqrt{7}$

4. $3\sqrt[3]{24x^4y} + x\sqrt[3]{192xy} - \sqrt[3]{375x^4y}$ $5x\sqrt[3]{3xy}$

515

61. $(\sqrt{x} - 3)^2$
$x - 6\sqrt{x} + 9$

62. $(\sqrt{2x} + 4)^2$
$2x + 8\sqrt{2x} + 16$

63. $(4\sqrt{5} + 2)^2$
$84 + 16\sqrt{5}$

64. $2\sqrt{3x^2} \cdot 3\sqrt{12xy^3} \cdot \sqrt{6x^3y}$
$36x^3y^2\sqrt{6}$

65. $2\sqrt{14xy} \cdot 4\sqrt{7x^2y} \cdot 3\sqrt{8xy^2}$
$672x^2y^2$

66. $\sqrt[3]{8ab}\sqrt[3]{4a^2b^3}\sqrt[3]{9ab^4}$
$2ab^2\sqrt[3]{36ab^2}$

67. $\sqrt[3]{2a^2b}\sqrt[3]{4a^3b^2}\sqrt[3]{8a^5b^6}$
$4a^3b^3\sqrt[3]{a}$

68. $(\sqrt{2} - 3)(\sqrt{2} + 4)$
$-10 + \sqrt{2}$

69. $(\sqrt{5} - 5)(2\sqrt{5} + 2)$
$-8\sqrt{5}$

70. $(\sqrt{y} - 2)(\sqrt{y} + 2)$
$y - 4$

71. $(\sqrt{x} - y)(\sqrt{x} + y)$
$x - y^2$

72. $(\sqrt{2x} - 3\sqrt{y})(\sqrt{2x} + 3\sqrt{y})$
$2x \quad 9y$

73. $(2\sqrt{3x} - \sqrt{y})(2\sqrt{3x} + \sqrt{y})$
$12x \quad y$

Objective D

74. When is a radical expression in simplest form?

75. Explain what it means to rationalize the denominator of a radical expression and how to do so.

Simplify.

76. $\dfrac{\sqrt{32x^2}}{\sqrt{2x}}$
$4\sqrt{x}$

77. $\dfrac{\sqrt{60y^4}}{\sqrt{12y}}$
$y\sqrt{5y}$

78. $\dfrac{\sqrt{42a^3b^5}}{\sqrt{14a^2b}}$
$b^2\sqrt{3a}$

79. $\dfrac{\sqrt{65ab^4}}{\sqrt{5ab}}$
$b\sqrt{13b}$

80. $\dfrac{1}{\sqrt{5}}$
$\dfrac{\sqrt{5}}{5}$

81. $\dfrac{1}{\sqrt{2}}$
$\dfrac{\sqrt{2}}{2}$

82. $\dfrac{1}{\sqrt{2x}}$
$\dfrac{\sqrt{2x}}{2x}$

83. $\dfrac{2}{\sqrt{3y}}$
$\dfrac{2\sqrt{3y}}{3y}$

84. $\dfrac{5}{\sqrt{5x}}$
$\dfrac{\sqrt{5x}}{x}$

85. $\dfrac{9}{\sqrt{3a}}$
$\dfrac{3\sqrt{3a}}{a}$

86. $\sqrt{\dfrac{x}{5}}$
$\dfrac{\sqrt{5x}}{5}$

87. $\sqrt{\dfrac{y}{2}}$
$\dfrac{\sqrt{2y}}{2}$

88. $\dfrac{3}{\sqrt[3]{2}}$
$\dfrac{3\sqrt[3]{4}}{2}$

89. $\dfrac{5}{\sqrt[3]{9}}$
$\dfrac{5\sqrt[3]{3}}{3}$

90. $\dfrac{3}{\sqrt[3]{4x^2}}$
$\dfrac{3\sqrt[3]{2x}}{2x}$

91. $\dfrac{5}{\sqrt[3]{3y}}$
$\dfrac{5\sqrt[3]{9y^2}}{3y}$

Answers to Writing Exercises

74. A radical expression is in simplest form when:
(1) The radicand contains no factor greater than 1 that is a perfect power of the index.
(2) There is no fraction under the radical sign.
(3) There is no radical in the denominator of a fraction.

75. To rationalize the denominator of a radical expression means to rewrite the expression with no radicals in the denominator. It is accomplished by multiplying both the numerator and denominator by the same expression, one that removes the radical(s) from the denominator of the original expression.

Quick Quiz (Objective 9.2C)

Simplify.

1. $\sqrt{6x^5y^3}\ \sqrt{8xy}$ $4x^3y^2\sqrt{3}$

2. $\sqrt[4]{18a^3b^5}\ \sqrt[4]{6a^5b^3}$ $a^2b^2\sqrt[4]{108}$

3. $\sqrt{3x}(\sqrt{27x} - \sqrt{12})$ $9x - 6\sqrt{x}$

4. $(3\sqrt{2x} - \sqrt{y})(3\sqrt{2x} + \sqrt{y})$ $18x - y$

92. $\dfrac{\sqrt{40x^3y^2}}{\sqrt{80x^2y^3}}$ $\dfrac{\sqrt{2xy}}{2y}$

93. $\dfrac{\sqrt{15a^2b^5}}{\sqrt{30a^5b^3}}$ $\dfrac{b\sqrt{2a}}{2a^2}$

94. $\dfrac{\sqrt{24a^2b}}{\sqrt{18ab^4}}$ $\dfrac{2\sqrt{3ab}}{3b^2}$

95. $\dfrac{\sqrt{12x^3y}}{\sqrt{20x^4y}}$ $\dfrac{\sqrt{15x}}{5x}$

96. $\dfrac{5}{\sqrt{3}-2}$ $-5\sqrt{3}-10$

97. $\dfrac{-2}{1-\sqrt{2}}$ $2+2\sqrt{2}$

98. $\dfrac{-3}{2-\sqrt{3}}$ $-6-3\sqrt{3}$

99. $\dfrac{-4}{3-\sqrt{2}}$ $\dfrac{-12-4\sqrt{2}}{7}$

100. $\dfrac{2}{\sqrt{5}+2}$ $2\sqrt{5}-4$

101. $\dfrac{5}{2-\sqrt{7}}$ $-\dfrac{10+5\sqrt{7}}{3}$

102. $\dfrac{3}{\sqrt{y}-2}$ $\dfrac{3\sqrt{y}+6}{y-4}$

103. $\dfrac{-7}{\sqrt{x}-3}$ $-\dfrac{7\sqrt{x}+21}{x-9}$

104. $\dfrac{\sqrt{2}-\sqrt{3}}{\sqrt{2}+\sqrt{3}}$ $-5+2\sqrt{6}$

105. $\dfrac{\sqrt{3}+\sqrt{4}}{\sqrt{2}+\sqrt{3}}$ $-\sqrt{6}+3-2\sqrt{2}+2\sqrt{3}$

106. $\dfrac{2+3\sqrt{7}}{5-2\sqrt{7}}$ $-\dfrac{52+19\sqrt{7}}{3}$

107. $\dfrac{2+3\sqrt{5}}{1-\sqrt{5}}$ $-\dfrac{17+5\sqrt{5}}{4}$

108. $\dfrac{2\sqrt{3}-1}{3\sqrt{3}+2}$ $\dfrac{20-7\sqrt{3}}{23}$

109. $\dfrac{2\sqrt{a}-\sqrt{b}}{4\sqrt{a}+3\sqrt{b}}$ $\dfrac{8a-10\sqrt{ab}+3b}{16a-9b}$

110. $\dfrac{2\sqrt{x}-4}{\sqrt{x}+2}$ $\dfrac{2x-8\sqrt{x}+8}{x-4}$

111. $\dfrac{3\sqrt{y}-y}{\sqrt{y}+2y}$ $\dfrac{3-7\sqrt{y}+2y}{1-4y}$

112. $\dfrac{3\sqrt{x}-4\sqrt{y}}{3\sqrt{x}-2\sqrt{y}}$ $\dfrac{9x-6\sqrt{xy}-8y}{9x-4y}$

APPLYING THE CONCEPTS

113. Determine whether the following statements are true or false. If the statement is false, correct the right side of the equation.

a. $\sqrt[2]{3}\cdot\sqrt[3]{4}=\sqrt[5]{12}$
false; $\sqrt[6]{432}$

b. $\sqrt{3}\cdot\sqrt{3}=3$
true

c. $\sqrt[3]{x}\cdot\sqrt[3]{x}=x$
false; $\sqrt[3]{x^2}$

d. $\sqrt{x}+\sqrt{y}=\sqrt{x+y}$
false; $\sqrt{x}+\sqrt{y}$

e. $\sqrt[2]{2}\cdot\sqrt[3]{3}=\sqrt[5]{2+3}$
false; $\sqrt[2]{2}+\sqrt[3]{3}$

f. $8\sqrt[5]{a}-2\sqrt[5]{a}=6\sqrt[5]{a}$
true

114. Multiply: $(\sqrt[3]{a}+\sqrt[3]{b})(\sqrt[3]{a^2}-\sqrt[3]{ab}+\sqrt[3]{b^2})$
$a+b$

115. Rewrite $\dfrac{\sqrt[4]{(a+b)^3}}{\sqrt{a+b}}$ as an expression with a single radical.
$\sqrt[4]{a+b}$

Quick Quiz (Objective 9.2D)

Simplify.

1. $\dfrac{\sqrt{48x^5}}{\sqrt{3x^2}}$ $4x\sqrt{x}$

2. $\dfrac{7}{\sqrt{7x}}$ $\dfrac{\sqrt{7x}}{x}$

3. $\dfrac{6}{\sqrt[3]{4y}}$ $\dfrac{3\sqrt[3]{2y^2}}{y}$

4. $\dfrac{\sqrt{25x^7y}}{\sqrt{40xy}}$ $\dfrac{x^3\sqrt{10}}{4}$

5. $\dfrac{\sqrt{5}+\sqrt{3}}{\sqrt{5}-\sqrt{3}}$ $4+\sqrt{15}$

6. $\dfrac{\sqrt{x^5}-\sqrt{x^7}}{\sqrt{x}}$ x^2-x^3

517

9.3 Complex Numbers

Objective A **To simplify a complex number**

VIDEO & DVD CD TUTOR WEB SSM

The radical expression $\sqrt{-4}$ is not a real number, because there is no real number whose square is -4. However, the solution of an algebraic equation is sometimes the square root of a negative number.

For example, the equation $x^2 + 1 = 0$ does not have a real number solution, because there is no real number whose square is a negative number.

$$x^2 + 1 = 0$$
$$x^2 = -1$$

Around the 17th century, a new number, called an **imaginary number**, was defined so that a negative number would have a square root. The letter i was chosen to represent the number whose square is -1.

$$i^2 = -1$$

An imaginary number is defined in terms of i.

Point of Interest

The first written occurrence of an imaginary number was in a book published in 1545 by Girolamo Cardan, where he wrote (in our modern notation) $5 + \sqrt{-15}$. He went on to say that the number "is as refined as it is useless." It was not until the 20th century that applications of complex numbers were found.

Definition of $\sqrt{-a}$

If a is a positive real number, then the principal square root of negative a is the imaginary number $i\sqrt{a}$.

$$\sqrt{-a} = i\sqrt{a}$$

Here are some examples.

$$\sqrt{-16} = i\sqrt{16} = 4i$$
$$\sqrt{-12} = i\sqrt{12} = 2i\sqrt{3}$$
$$\sqrt{-21} = i\sqrt{21}$$
$$\sqrt{-1} = i\sqrt{1} = i$$

It is customary to write i in front of a radical to avoid confusing $\sqrt{a}\, i$ with $\sqrt{ai}$.

The real numbers and imaginary numbers make up the complex numbers.

Complex Number

A **complex number** is a number of the form $\boldsymbol{a + bi}$, where a and b are real numbers and $i = \sqrt{-1}$. The number a is the **real part** of $a + bi$, and b is the **imaginary part**.

TAKE NOTE

The *imaginary part* of a complex number is a real number. As another example, the imaginary part of $6 - 8i$ is -8.

Examples of complex numbers are shown at the right.

Real Part	*Imaginary Part*
a	$+ bi$
3	$+ 2i$
8	$- 10i$

Objective 9.3A

New Vocabulary

imaginary number
complex number
real part of a complex number
imaginary part of a complex number

Vocabulary to Review

real number

New Symbols

i

Instructor Note

Complex numbers were not accepted by most mathematicians until late in the 19th century. Part of the difficulty was that there were no physical examples of these numbers.

Discuss the Concepts

1. What does the variable i represent?
2. What is a complex number?
3. What is **a.** the real part and **b.** the imaginary part of $-1 + 10i$?

Optional Student Activity

Find a complex number z such that $3z = -10 + 4iz$. Express z in the form $a + bi$. $-\frac{6}{5} - \frac{8}{5}i$

In-Class Examples (Objective 9.3A)

Simplify.

1. $\sqrt{-16}$ $4i$
2. $\sqrt{-50}$ $5i\sqrt{2}$
3. $\sqrt{36} + \sqrt{-49}$ $6 + 7i$
4. $\sqrt{8} - \sqrt{-32}$ $2\sqrt{2} - 4i\sqrt{2}$

Instructor Note

You may want to refer to the chart of the real numbers presented in Section 1 in the chapter titled "Real Numbers and Variable Expressions" to remind students of the subsets of the real numbers.

Objective 9.3B

Vocabulary to Review

real part of a complex number
imaginary part of a complex number

Concept Check

One area in which complex numbers are applied is the field of electrical engineering. In an alternating current (AC) circuit, the **impedance** is the amount by which the circuit resists the flow of electricity. It is a measure of the opposition to the flow of electricity. It is measured in ohms and is described by a complex number.

The total impedance Z_T in a circuit is a function of the impedances Z_1 and Z_2 of the individual circuits. In a series circuit, $Z_T = Z_1 + Z_2$.

1. Find the total impedance Z_T in a series circuit when $Z_1 = (6 + i)$ ohms and $Z_2 = (6 - i)$ ohms. 12 ohms
2. Find the total impedance Z_T in a series circuit when $Z_1 = (8 - 4i)$ ohms and $Z_2 = (8 + 4i)$ ohms. 16 ohms

Complex numbers $a + bi$
- Real Numbers $a + 0i$ — A *real number* is a complex number in which $b = 0$.
- Imaginary Numbers $0 + bi$ — An *imaginary number* is a complex number in which $a = 0$.

⇒ Simplify: $\sqrt{20} - \sqrt{-50}$

$\sqrt{20} - \sqrt{-50} = \sqrt{20} - i\sqrt{50}$
- Write the complex number in the form $a + bi$.

$= \sqrt{4 \cdot 5} - i\sqrt{25 \cdot 2}$
- Use the Product Property of Radicals to simplify each radical.

$= 2\sqrt{5} - 5i\sqrt{2}$

Example 1
Simplify: $\sqrt{-80}$

Solution
$\sqrt{-80} = i\sqrt{80} = i\sqrt{16 \cdot 5} = 4i\sqrt{5}$

You Try It 1
Simplify: $\sqrt{-45}$

Your solution
$3i\sqrt{5}$

Example 2
Simplify: $\sqrt{25} + \sqrt{-40}$

Solution
$\sqrt{25} + \sqrt{-40} = \sqrt{25} + i\sqrt{40}$
$= \sqrt{25} + i\sqrt{4 \cdot 10}$
$= 5 + 2i\sqrt{10}$

You Try It 2
Simplify: $\sqrt{98} - \sqrt{-60}$

Your solution
$7\sqrt{2} - 2i\sqrt{15}$

Solutions on p. S28

Objective B To add or subtract complex numbers

VIDEO & DVD · CD TUTOR · WEB · SSM

Addition and Subtraction of Complex Numbers

To add two complex numbers, add the real parts and add the imaginary parts. To subtract two complex numbers, subtract the real parts and subtract the imaginary parts.

$$(a + bi) + (c + di) = (a + c) + (b + d)i$$
$$(a + bi) - (c + di) = (a - c) + (b - d)i$$

⇒ Subtract: $(3 - 7i) - (4 - 2i)$

$(3 - 7i) - (4 - 2i) = (3 - 4) + [-7 - (-2)]i$
- Subtract the real parts and subtract the imaginary parts of the complex numbers.

$= -1 - 5i$

In-Class Examples (Objective 9.3B)

Simplify.

1. $(3 + 5i) + (4 - 4i)$ $7 + i$
2. $(-6 - 5i) - (5 - 9i)$ $-11 + 4i$
3. $(7 - \sqrt{-9}) + (3 + \sqrt{-25})$ $10 + 2i$
4. $(9 - \sqrt{-36}) - (3 + \sqrt{-81})$ $6 - 15i$

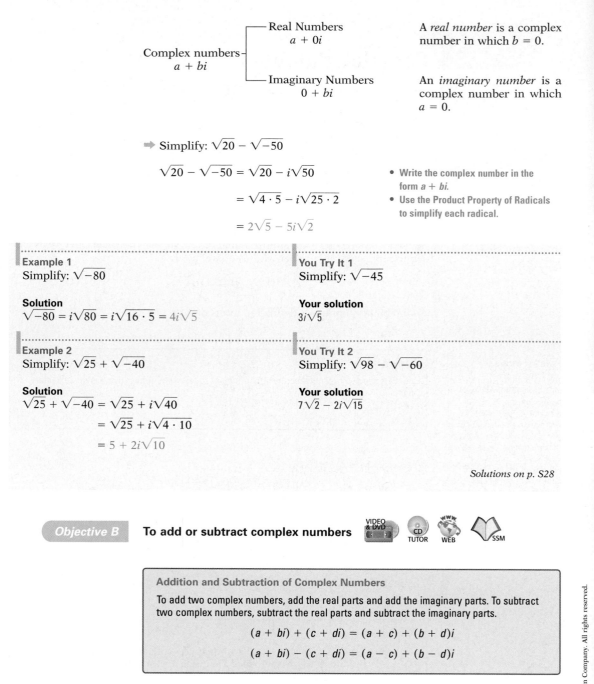

➡ Add: $(3 + \sqrt{-12}) + (7 - \sqrt{-27})$

$$
\begin{aligned}
(3 + \sqrt{-12}) &+ (7 - \sqrt{-27}) \\
&= (3 + i\sqrt{12}) + (7 - i\sqrt{27}) && \text{• Write each complex number in the form } a + bi. \\
&= (3 + i\sqrt{4 \cdot 3}) + (7 - i\sqrt{9 \cdot 3}) && \text{• Use the Product Property of Radicals to simplify each radical.} \\
&= (3 + 2i\sqrt{3}) + (7 - 3i\sqrt{3}) \\
&= 10 - i\sqrt{3} && \text{• Add the complex numbers.}
\end{aligned}
$$

Example 3

Add: $(3 + 2i) + (6 - 5i)$

Solution

$(3 + 2i) + (6 - 5i) = 9 - 3i$

You Try It 3

Subtract: $(-4 + 2i) - (6 - 8i)$

Your solution

$-10 + 10i$

Example 4

Subtract: $(9 - \sqrt{-8}) - (5 + \sqrt{-32})$

Solution

$$
\begin{aligned}
(9 - \sqrt{-8}) &- (5 + \sqrt{-32}) \\
&= (9 - i\sqrt{8}) - (5 + i\sqrt{32}) \\
&= (9 - i\sqrt{4 \cdot 2}) - (5 + i\sqrt{16 \cdot 2}) \\
&= (9 - 2i\sqrt{2}) - (5 + 4i\sqrt{2}) \\
&= 4 - 6i\sqrt{2}
\end{aligned}
$$

You Try It 4

Subtract: $(16 - \sqrt{-45}) - (3 + \sqrt{-20})$

Your solution

$13 - 5i\sqrt{5}$

Example 5

Add: $(6 + 4i) + (-6 - 4i)$

Solution

$(6 + 4i) + (-6 - 4i) = 0 + 0i = 0$

This illustrates that the additive inverse of $a + bi$ is $-a - bi$.

You Try It 5

Add: $(3 - 2i) + (-3 + 2i)$

Your solution

0

Solutions on p. S28

Objective C To multiply complex numbers

VIDEO & DVD CD TUTOR WEB SSM

When multiplying complex numbers, we often find that i^2 is a part of the product. Recall that $i^2 = -1$.

➡ Multiply: $2i \cdot 3i$

$$
\begin{aligned}
2i \cdot 3i &= 6i^2 && \text{• Multiply the imaginary numbers.} \\
&= 6(-1) && \text{• Replace } i^2 \text{ by } -1. \\
&= -6 && \text{• Simplify.}
\end{aligned}
$$

Optional Student Activity

Fractal geometry is the study of nonlinear dimensions. Fractal images are generated by substituting an initial value into a complex function, calculating the output, and then using the output as the next value to substitute into the function. The second output is then substituted into the function, and the process is repeated. This continual recycling of outputs is called **iteration,** and each output is called an **iterate.** Complex numbers are usually symbolized by the variable z, so z is used in the function below.

Let $f(z) = z + 4 + 3i$. Begin with the initial value $z = -2 + i$. Determine the first four iterates of the function. $2 + 4i$, $6 + 7i$, $10 + 10i$, $14 + 13i$

Objective 9.3C

Vocabulary to Review

conjugate

Symbols to Review

i

$i^2 = -1$

Properties to Review

Product Property of Radicals

New Rules

$(a + bi)(a - bi) = a^2 + b^2$

In-Class Examples (Objective 9.3C)

Simplify.

1. $(6i)(-7i)$ 42
2. $\sqrt{-1}\,\sqrt{-16}$ -4
3. $2i(4 + i)$ $-2 + 8i$
4. $\sqrt{-3}\left(\sqrt{27} + \sqrt{-3}\right)$ $-3 + 9i$
5. $(7 - i)(3 + i)$ $22 + 4i$
6. $(2 + i)^2$ $3 + 4i$

Discuss the Concepts

1. How do you determine the conjugate of $7 - 6i$?
2. What is the value of i^2? Give an example of the use of i^2 when multiplying complex numbers.
3. Explain how to find the product of $\sqrt{-2}$ and $\sqrt{-8}$.

Concept Check

Determine whether the following statements are always true, sometimes true, or never true.

1. The product of two imaginary numbers is a real number. Always true
2. The product of two complex numbers is a real number. Sometimes true
3. The product of a complex number and its conjugate is a real number. Always true

TAKE NOTE
This example illustrates an important point. When working with an expression that has a square root of a negative number, always rewrite the number as the product of a real number and i before continuing.

➡ Multiply: $\sqrt{-6} \cdot \sqrt{-24}$

$\sqrt{-6} \cdot \sqrt{-24} = i\sqrt{6} \cdot i\sqrt{24}$ • Write each radical as the product of a real number and i.

$= i^2\sqrt{144}$ • Multiply the imaginary numbers.

$= -\sqrt{144}$ • Replace i^2 by -1.

$= -12$ • Simplify the radical expression.

Note from the last example that it would have been incorrect to multiply the radicands of the two radical expressions. To illustrate,

$$\sqrt{-6} \cdot \sqrt{-24} = \sqrt{(-6)(-24)} = \sqrt{144} = 12, \textit{ not } -12$$

➡ Multiply: $4i(3 - 2i)$

$4i(3 - 2i) = 12i - 8i^2$ • Use the Distributive Property to remove parentheses.

$= 12i - 8(-1)$ • Replace i^2 by -1.

$= 8 + 12i$ • Write the answer in the form $a + bi$.

The product of two complex numbers is defined as follows.

The Product of Two Complex Numbers
$(a + bi)(c + di) = (ac - bd) + (ad + bc)i$

One way to remember this rule is to use the FOIL method.

➡ Multiply: $(2 + 4i)(3 - 5i)$

$(2 + 4i)(3 - 5i) = 6 - 10i + 12i - 20i^2$ • Use the FOIL method to find the product.

$= 6 + 2i - 20i^2$

$= 6 + 2i - 20(-1)$ • Replace i^2 by -1.

$= 26 + 2i$ • Write the answer in the form $a + bi$.

The conjugate of $a + bi$ is $a - bi$.
The product of conjugates, $(a + bi)(a - bi)$, is the real number $a^2 + b^2$.

$$(a + bi)(a - bi) = a^2 - b^2i^2$$
$$= a^2 - b^2(-1)$$
$$= a^2 + b^2$$

➡ Multiply: $(2 + 3i)(2 - 3i)$

$(2 + 3i)(2 - 3i) = 2^2 + 3^2$ • The product of conjugates is $a^2 + b^2$.

$= 4 + 9$

$= 13$

Note that the product of a complex number and its conjugate is a real number.

521

Example 6
Multiply: $(2i)(-5i)$

Solution
$(2i)(-5i) = -10i^2 = (-10)(-1) = 10$

You Try It 6
Multiply: $(-3i)(-10i)$

Your solution
-30

Example 7
Multiply: $\sqrt{-10} \cdot \sqrt{-5}$

Solution
$$\sqrt{-10} \cdot \sqrt{-5} = i\sqrt{10} \cdot i\sqrt{5}$$
$$= i^2\sqrt{50} = -\sqrt{25 \cdot 2} = -5\sqrt{2}$$

You Try It 7
Multiply: $-\sqrt{-8} \cdot \sqrt{-5}$

Your solution
$2\sqrt{10}$

Example 8
Multiply: $3i(2 - 4i)$

Solution:
$$3i(2 - 4i) = 6i - 12i^2 = 6i - 12(-1)$$
$$= 12 + 6i$$

You Try It 8
Multiply: $-6i(3 + 4i)$

Your solution
$24 - 18i$

Example 9
Multiply: $\sqrt{-8}(\sqrt{6} - \sqrt{-2})$

Solution
$$\sqrt{-8}(\sqrt{6} - \sqrt{-2}) = i\sqrt{8}(\sqrt{6} - i\sqrt{2})$$
$$= i\sqrt{48} - i^2\sqrt{16}$$
$$= i\sqrt{16 \cdot 3} - (-1)\sqrt{16}$$
$$= 4i\sqrt{3} + 4 = 4 + 4i\sqrt{3}$$

You Try It 9
Multiply: $\sqrt{-3}(\sqrt{27} - \sqrt{-6})$

Your solution
$3\sqrt{2} + 9i$

Example 10
Multiply: $(3 - 4i)(2 + 5i)$

Solution
$$(3 - 4i)(2 + 5i) = 6 + 15i - 8i - 20i^2$$
$$= 6 + 7i - 20i^2$$
$$= 6 + 7i - 20(-1) = 26 + 7i$$

You Try It 10
Multiply: $(4 - 3i)(2 - i)$

Your solution
$5 - 10i$

Example 11
Multiply: $(4 + 5i)(4 - 5i)$

Solution
$(4 + 5i)(4 - 5i) = 4^2 + 5^2 = 16 + 25 = 41$

You Try It 11
Multiply: $(3 + 6i)(3 - 6i)$

Your solution
45

Example 12
Multiply: $\left(\frac{9}{10} + \frac{3}{10}i\right)\left(1 - \frac{1}{3}i\right)$

Solution
$$\left(\frac{9}{10} + \frac{3}{10}i\right)\left(1 - \frac{1}{3}i\right) = \frac{9}{10} - \frac{3}{10}i + \frac{3}{10}i - \frac{1}{10}i^2$$
$$= \frac{9}{10} - \frac{1}{10}i^2 = \frac{9}{10} - \frac{1}{10}(-1)$$
$$= \frac{9}{10} + \frac{1}{10} = 1$$

You Try It 12
Multiply: $(3 - i)\left(\frac{3}{10} + \frac{1}{10}i\right)$

Your solution
1

Solutions on pp. S28–S29

Optional Student Activity

1. The property that $(a + bi)(a - bi) = a^2 + b^2$ can be used to factor the sum of two perfect squares over the set of complex numbers. For instance, $x^2 + 4 = (x + 2i)(x - 2i)$. Factor the following expressions over the set of complex numbers.
   a. $y^2 + 1$ $(y + i)(y - i)$
   b. $49x^2 + 16$
   $(7x + 4i)(7x - 4i)$
   c. $9a^2 + 64$
   $(3a + 8i)(3a - 8i)$
2. Two complex numbers have a sum of $1 + 3i$ and a difference of $7 - 5i$. What is the product of the two complex numbers? $-8 + 19i$

Objective 9.3D

Vocabulary to Review
conjugate

Instructor Note
Students think that division of complex numbers is somehow different from division of other types of numbers. Show them that just as $\frac{12}{4} = 3$ because $4 \cdot 3 = 12$, $\frac{3 + 2i}{1 + i} = \frac{5}{2} - \frac{1}{2}i$ because $(1 + i)\left(\frac{5}{2} - \frac{1}{2}i\right) = 3 + 2i$.

Concept Check
Find the reciprocal of $2 - 5i$. $\frac{2}{29} + \frac{5}{29}i$
Then show that the product of $2 - 5i$ and its reciprocal is 1.

Optional Student Activity

1. Show that $\sqrt{i} = \frac{\sqrt{2}}{2} + \frac{\sqrt{2}}{2}i$ by simplifying $\left(\frac{\sqrt{2}}{2} + \frac{\sqrt{2}}{2}i\right)^2$.
2. Given that $\sqrt{i} = \frac{\sqrt{2}}{2} + \frac{\sqrt{2}}{2}i$, find $\sqrt{-i}$. $-\frac{\sqrt{2}}{2} + \frac{\sqrt{2}}{2}i$
3. Use division of polynomials to find the remainder when
   a. $x^3 + 3x^2 + x - 3$ is divided by $x - i$.
   b. $x^3 - 3x^2 + x - 3$ is divided by $x - i$.
   a. -6 b. 0

Objective D To divide complex numbers

VIDEO & DVD CD TUTOR WEB SSM

A rational expression containing one or more complex numbers is in simplest form when no imaginary number remains in the denominator.

➡ Simplify: $\frac{2 - 3i}{2i}$

$$\frac{2 - 3i}{2i} = \frac{2 - 3i}{2i} \cdot \frac{i}{i}$$ • Multiply the expression by $\frac{i}{i}$.

$$= \frac{2i - 3i^2}{2i^2}$$

$$= \frac{2i - 3(-1)}{2(-1)}$$ • Replace i^2 by -1.

$$= \frac{3 + 2i}{-2}$$ • Simplify.

$$= -\frac{3}{2} - i$$ • Write the answer in the form $a + bi$.

➡ Simplify: $\frac{3 + 2i}{1 + i}$

$$\frac{3 + 2i}{1 + i} = \frac{3 + 2i}{1 + i} \cdot \frac{1 - i}{1 - i}$$ • Multiply the numerator and denominator by the conjugate of $1 + i$.

$$= \frac{3 - 3i + 2i - 2i^2}{1^2 + 1^2}$$ • In $1 + i$, $a = 1$ and $b = 1$. $a^2 + b^2 = 1^2 + 1^2$.

$$= \frac{3 - i - 2(-1)}{2}$$ • Replace i^2 by -1 and simplify.

$$= \frac{5 - i}{2} = \frac{5}{2} - \frac{1}{2}i$$ • Write the answer in the form $a + bi$.

Example 13
Simplify: $\frac{5 + 4i}{3i}$

Solution
$$\frac{5 + 4i}{3i} = \frac{5 + 4i}{3i} \cdot \frac{i}{i} = \frac{5i + 4i^2}{3i^2}$$
$$= \frac{5i + 4(-1)}{3(-1)} = \frac{-4 + 5i}{-3} = \frac{4}{3} - \frac{5}{3}i$$

You Try It 13
Simplify: $\frac{2 - 3i}{4i}$

Your solution
$-\frac{3}{4} - \frac{1}{2}i$

Example 14
Simplify: $\frac{5 - 3i}{4 + 2i}$

Solution
$$\frac{5 - 3i}{4 + 2i} = \frac{5 - 3i}{4 + 2i} \cdot \frac{4 - 2i}{4 - 2i}$$
$$= \frac{20 - 10i - 12i + 6i^2}{4^2 + 2^2}$$
$$= \frac{20 - 22i + 6(-1)}{20}$$
$$= \frac{14 - 22i}{20} = \frac{14}{20} - \frac{22}{20}i = \frac{7}{10} - \frac{11}{10}i$$

You Try It 14
Simplify: $\frac{2 + 5i}{3 - 2i}$

Your solution
$-\frac{4}{13} + \frac{19}{13}i$

Solutions on p. S29

In-Class Examples (Objective 9.3D)
Simplify.

1. $\frac{6}{i}$ $-6i$
2. $\frac{3 - 2i}{6i}$ $-\frac{1}{3} - \frac{1}{2}i$
3. $\frac{8}{5 + 4i}$ $\frac{40}{41} - \frac{32}{41}i$
4. $\frac{1 - 2i}{2 + i}$ $-i$
5. $\frac{\sqrt{-8}}{\sqrt{6} - \sqrt{-2}}$ $-\frac{1}{2} + \frac{\sqrt{3}}{2}i$

9.3 Exercises

Objective A

1. What is an imaginary number? What is a complex number?

2. Are all real numbers also complex numbers? Are all complex numbers also real numbers?

Simplify.

3. $\sqrt{-4}$
$2i$

4. $\sqrt{-64}$
$8i$

5. $\sqrt{-98}$
$7i\sqrt{2}$

6. $\sqrt{-72}$
$6i\sqrt{2}$

7. $\sqrt{-27}$
$3i\sqrt{3}$

8. $\sqrt{-75}$
$5i\sqrt{3}$

9. $\sqrt{16} + \sqrt{-4}$
$4 + 2i$

10. $\sqrt{25} + \sqrt{-9}$
$5 + 3i$

11. $\sqrt{12} - \sqrt{-18}$
$2\sqrt{3} - 3i\sqrt{2}$

12. $\sqrt{60} - \sqrt{-48}$
$2\sqrt{15} - 4i\sqrt{3}$

13. $\sqrt{160} - \sqrt{-147}$
$4\sqrt{10} - 7i\sqrt{3}$

14. $\sqrt{96} - \sqrt{-125}$
$4\sqrt{6} - 5i\sqrt{5}$

Objective B

Add or subtract.

15. $(2 + 4i) + (6 - 5i)$
$8 - i$

16. $(6 - 9i) + (4 + 2i)$
$10 - 7i$

17. $(-2 - 4i) - (6 - 8i)$
$-8 + 4i$

18. $(3 - 5i) + (8 - 2i)$
$11 - 7i$

19. $(8 - \sqrt{-4}) - (2 + \sqrt{-16})$
$6 - 6i$

20. $(5 - \sqrt{-25}) - (11 - \sqrt{-36})$
$0 + i$

21. $(12 - \sqrt{-50}) + (7 - \sqrt{-8})$
$19 - 7i\sqrt{2}$

22. $(5 - \sqrt{-12}) - (9 + \sqrt{-108})$
$-4 - 8i\sqrt{3}$

23. $(\sqrt{8} + \sqrt{-18}) + (\sqrt{32} - \sqrt{-72})$
$6\sqrt{2} - 3i\sqrt{2}$

24. $(\sqrt{40} - \sqrt{-98}) - (\sqrt{90} + \sqrt{-32})$
$-\sqrt{10} - 11i\sqrt{2}$

Objective C

Multiply.

25. $(7i)(-9i)$
63

26. $(-6i)(-4i)$
-24

27. $\sqrt{-2}\sqrt{-8}$
-4

523

Section 9.3

Suggested Assignment

Exercises 3–57, odds
More challenging problems:
Exercises 59, 60

Answers to Writing Exercises

1. An imaginary number is a number whose square is a negative number. Imaginary numbers are defined in terms of i, the number whose square is -1.
A complex number is a number of the form $a + bi$, where a and b are real numbers and $i = \sqrt{-1}$.
2. All real numbers are complex numbers; they are complex numbers of the form $a + bi$ where $b = 0$. Not all complex numbers are real numbers; any complex number of the form $a + bi$, $b \neq 0$, is not a real number.

Quick Quiz (Objective 9.3A)

Simplify.

1. $\sqrt{-9}$ $3i$
2. $\sqrt{-48}$ $4i\sqrt{3}$
3. $\sqrt{20} - \sqrt{-45}$ $2\sqrt{5} - 3i\sqrt{5}$

Quick Quiz (Objective 9.3B)

Simplify.

1. $(-3 - 3i) - (7 - 7i)$ $-10 + 4i$
2. $(10 - \sqrt{-32}) + (9 - \sqrt{-18})$ $19 - 7i\sqrt{2}$

28. $\sqrt{-5}\sqrt{-45}$
-15

29. $\sqrt{-3}\sqrt{-6}$
$-3\sqrt{2}$

30. $\sqrt{-5}\sqrt{-10}$
$-5\sqrt{2}$

31. $2i(6 + 2i)$
$-4 + 12i$

32. $-3i(4 - 5i)$
$-15 - 12i$

33. $\sqrt{-2}(\sqrt{8} + \sqrt{-2})$
$-2 + 4i$

34. $\sqrt{-3}(\sqrt{12} - \sqrt{-6})$
$3\sqrt{2} + 6i$

35. $(5 - 2i)(3 + i)$
$17 - i$

36. $(2 - 4i)(2 - i)$
$-10i$

37. $(6 + 5i)(3 + 2i)$
$8 + 27i$

38. $(4 - 7i)(2 + 3i)$
$29 - 2i$

39. $(1 - i)\left(\frac{1}{2} + \frac{1}{2}i\right)$
1

40. $\left(\frac{4}{5} - \frac{2}{5}i\right)\left(1 + \frac{1}{2}i\right)$
1

41. $\left(\frac{6}{5} + \frac{3}{5}i\right)\left(\frac{2}{3} - \frac{1}{3}i\right)$
1

42. $(2 - i)\left(\frac{2}{5} + \frac{1}{5}i\right)$
1

Objective D

Simplify.

43. $\frac{3}{i}$
$-3i$

44. $\frac{4}{5i}$
$-\frac{4}{5}i$

45. $\frac{2 - 3i}{-4i}$
$\frac{3}{4} + \frac{1}{2}i$

46. $\frac{16 + 5i}{-3i}$
$-\frac{5}{3} + \frac{16}{3}i$

47. $\frac{4}{5 + i}$
$\frac{10}{13} - \frac{2}{13}i$

48. $\frac{6}{5 + 2i}$
$\frac{30}{29} - \frac{12}{29}i$

49. $\frac{2}{2 - i}$
$\frac{4}{5} + \frac{2}{5}i$

50. $\frac{5}{4 - i}$
$\frac{20}{17} + \frac{5}{17}i$

51. $\frac{1 - 3i}{3 + i}$
$-i$

52. $\frac{2 + 12i}{5 + i}$
$\frac{11}{13} + \frac{29}{13}i$

53. $\frac{\sqrt{-10}}{\sqrt{8} - \sqrt{-2}}$
$-\frac{\sqrt{5}}{5} + \frac{2\sqrt{5}}{5}i$

54. $\frac{\sqrt{-2}}{\sqrt{12} - \sqrt{-8}}$
$-\frac{1}{5} + \frac{\sqrt{6}}{10}i$

55. $\frac{2 - 3i}{3 + i}$
$\frac{3}{10} - \frac{11}{10}i$

56. $\frac{3 + 5i}{1 - i}$
$-1 + 4i$

57. $\frac{5 + 3i}{3 - i}$
$\frac{6}{5} + \frac{7}{5}i$

58. $\frac{3 - 2i}{2i + 3}$
$\frac{5}{13} - \frac{12}{13}i$

APPLYING THE CONCEPTS

59. **a.** Is $3i$ a solution of $2x^2 + 18 = 0$? yes
b. Is $3 + i$ a solution of $x^2 - 6x + 10 = 0$? yes

60. Evaluate i^n for $n = 0, 1, 2, 3, 4, 5, 6$, and 7. Make a conjecture about the value of i^n for any natural number. Using your conjecture, evaluate i^{76}.
$i^{76} = 1$

Quick Quiz (Objective 9.3C)

Simplify.

1. $\sqrt{-2}\left(\sqrt{18} - \sqrt{-8}\right)$ $4 + 6i$
2. $(5 - 2i)(3 + i)$ $17 - i$
3. $(5 - i)^2$ $24 - 10i$

Quick Quiz (Objective 9.3D)

Simplify.

1. $\frac{4 - 6i}{-2i}$ $3 + 2i$
2. $\frac{5}{5 + 2i}$ $\frac{25}{29} - \frac{10}{29}i$
3. $\frac{1 - 5i}{5 + i}$ $-i$
4. $\frac{\sqrt{-2}}{\sqrt{8} - \sqrt{-2}}$ $-\frac{1}{5} + \frac{2}{5}i$

9.4 Equations Containing Radical Expressions

Objective A To solve a radical equation

VIDEO & DVD CD TUTOR WEB SSM

An equation that contains a variable expression in a radicand is a **radical equation**.

$\sqrt[3]{2x-5}+x=7$
$\sqrt{x+1}-\sqrt{x}=4$ } Radical Equations

The following property is used to solve a radical equation.

The Property of Raising Each Side of an Equation to a Power

If two numbers are equal, then the same powers of the numbers are equal.

If $a=b$, then $a^n=b^n$.

➡ Solve: $\sqrt{x-2}-6=0$

$\sqrt{x-2}-6=0$

$\sqrt{x-2}=6$ • Isolate the radical by adding 6 to each side of the equation.

$(\sqrt{x-2})^2=6^2$ • Square each side of the equation.

$x-2=36$ • Simplify and solve for x.

$x=38$

Check:

$\sqrt{x-2}-6$	$=0$
$\sqrt{38-2}-6$	0
$\sqrt{36}-6$	0
$6-6$	0
$0=0$	

38 checks as a solution. The solution is 38.

➡ Solve: $\sqrt[3]{x+2}=-3$

$\sqrt[3]{x+2}=-3$

$(\sqrt[3]{x+2})^3=(-3)^3$ • Cube each side of the equation.

$x+2=-27$ • Solve the resulting equation.

$x=-29$

Check:

$\sqrt[3]{x+2}$	$=-3$
$\sqrt[3]{-29+2}$	-3
$\sqrt[3]{-27}$	-3
$-3=-3$	

-29 checks as a solution. The solution is -29.

Objective 9.4A

New Vocabulary

radical equation
extraneous solution

New Properties

Property of Raising Each Side of an Equation to a Power: If $a=b$, then $a^n=b^n$.

Instructor Note

Some students may assume that the converse of this result is also true. That is, if $a^2=b^2$, then $a=b$. Use $a=-4$ and $b=4$ to show that $(-4)^2=4^2$ but $-4\neq 4$. The fact that the converse is not true is what creates the potential for extraneous solutions.

Discuss the Concepts

1. What is the first step in solving $\sqrt{x}+3=9$? Why?
2. Why is the first step in solving the equation $\sqrt{x+7}-\sqrt{x}=1$ not to square each side of the equation?
3. What does the Property of Raising Each Side of an Equation to a Power state?
4. When both sides of an equation are raised to an even power, why is it necessary to check the solutions?
5. Suppose you solve the equation $\sqrt{2x+1}+3=6$ and the result is $x=4$. Describe how to check the solution.

In-Class Examples (Objective 9.4A)

Solve.

1. $\sqrt{2y}=10$ 50
2. $\sqrt[3]{2x}=-2$ -4
3. $\sqrt{3x-5}=5$ 10
4. $\sqrt[3]{x-3}=2$ 11
5. $\sqrt[4]{3x+1}=4$ 85
6. $\sqrt{2x-1}-1=2$ 5

Concept Check

Determine whether the statement is always true, sometimes true, or never true.

1. We can square both sides of an equation without changing the solutions of the equation. Sometimes true
2. The Property of Raising Each Side of an Equation to a Power is used to eliminate a radical expression from an equation. Always true
3. If $a^n = b^n$, then $a = b$. Sometimes true
4. When you raise both sides of an equation to an even power, the resulting equation has a solution that is not a solution of the original equation. Sometimes true
5. The first step in solving a radical equation is to square both sides of the equation. Sometimes true

Optional Student Activity

1. Explain how solving the equation $3\sqrt{x-8} - 5 = 7$ is similar to solving the equation $3x - 5 = 7$.
2. The equation $s = 16.97\sqrt{n}$ can be used to predict the maximum speed, s (in feet per second), of n rowers on a scull.
   a. How many rowers are needed to travel at a speed of 35 ft/s? Round to the nearest whole number. 4 rowers
   b. Does doubling the number of rowers double the maximum speed of the scull? No

Raising each side of an equation to an even power may result in an equation that has a solution that is not a solution of the original equation. This is called an **extraneous solution**. Here is an example:

TAKE NOTE
Note that
$(2-\sqrt{x})^2 = (2-\sqrt{x})(2-\sqrt{x}) = 4 - 4\sqrt{x} + x$

TAKE NOTE
You must always check the proposed solutions to radical equations. The proposed solutions of the equation on the right were 1 and 25. However, 25 did not check as a solution. Here 25 is an extraneous solution.

Solve: $\sqrt{2x-1} + \sqrt{x} = 2$

$$\sqrt{2x-1} + \sqrt{x} = 2$$
$$\sqrt{2x-1} = 2 - \sqrt{x}$$
• Solve for one of the radical expressions.
$$(\sqrt{2x-1})^2 = (2-\sqrt{x})^2$$
• Square each side. Recall that $(a-b)^2 = a^2 - 2ab + b^2$.
$$2x - 1 = 4 - 4\sqrt{x} + x$$
$$x - 5 = -4\sqrt{x}$$
$$(x-5)^2 = (-4\sqrt{x})^2$$
• Square each side.
$$x^2 - 10x + 25 = 16x$$
$$x^2 - 26x + 25 = 0$$
$$(x-25)(x-1) = 0$$
• Solve the quadratic equation by factoring.
$$x = 25 \quad \text{or} \quad x = 1$$

Check:

$\sqrt{2x-1} + \sqrt{x}$	$= 2$
$\sqrt{2(25)-1} + \sqrt{25}$	2
$7 + 5$	2
$12 \neq 2$	

$\sqrt{2x-1} + \sqrt{x}$	$= 2$
$\sqrt{2(1)-1} + \sqrt{1}$	2
$1 + 1$	2
$2 = 2$	

25 does not check as a solution. 1 checks as a solution. The solution is 1.

Example 1

Solve: $\sqrt[3]{3x-1} = -4$

Solution

$$\sqrt[3]{3x-1} = -4$$
$$(\sqrt[3]{3x-1})^3 = (-4)^3$$
$$3x - 1 = -64$$
$$3x = -63$$
$$x = -21$$

Check:

$\sqrt[3]{3x-1}$	$= -4$
$\sqrt[3]{3(-21)-1}$	-4
$\sqrt[3]{-64}$	-4
$-4 = -4$	

The solution is -21.

You Try It 1

Solve: $\sqrt[4]{x-8} = 3$

Your solution

89

Example 2

Solve: $\sqrt{x-1} + \sqrt{x+4} = 5$

Solution

$$\sqrt{x-1} + \sqrt{x+4} = 5$$
$$\sqrt{x+4} = 5 - \sqrt{x-1}$$
$$(\sqrt{x+4})^2 = (5 - \sqrt{x-1})^2$$
$$x + 4 = 25 - 10\sqrt{x-1} + x - 1$$
$$2 = \sqrt{x-1}$$
$$2^2 = (\sqrt{x-1})^2$$
$$4 = x - 1$$
$$5 = x$$

5 checks as a solution. The solution is 5.

You Try It 2

Solve: $\sqrt{x} - \sqrt{x+5} = 1$

Your solution

no solution

Solutions on p. S29

Objective B To solve application problems

VIDEO & DVD CD TUTOR WEB SSM

A right triangle contains one 90° angle. The side opposite the 90° angle is called the **hypotenuse**. The other two sides are called **legs**.

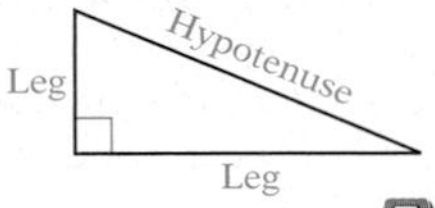

Point of Interest

The first known proof of this theorem occurs in a Chinese text, *Arithmetic Classic*, which was first written around 600 B.C.

Pythagoras, a Greek mathematician, discovered that the square of the hypotenuse of a right triangle is equal to the sum of the squares of the two legs. This is called the **Pythagorean Theorem**.

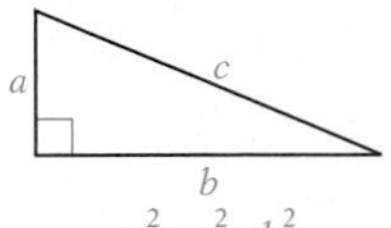

$c^2 = a^2 + b^2$

Example 3

A ladder 20 ft long is leaning against a building. How high on the building will the ladder reach when the bottom of the ladder is 8 ft from the building? Round to the nearest tenth.

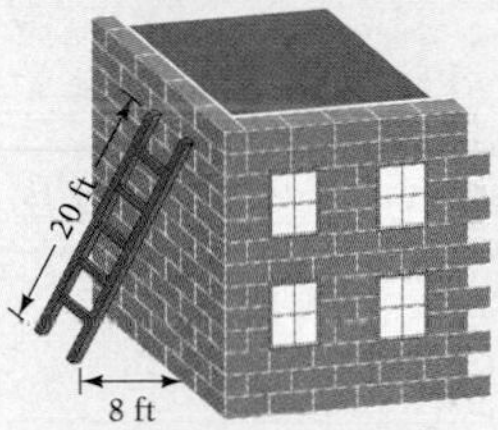

Strategy

To find the distance, use the Pythagorean Theorem. The hypotenuse is the length of the ladder. One leg is the distance from the bottom of the ladder to the base of the building. The distance along the building from the ground to the top of the ladder is the unknown leg.

Solution

$$\begin{aligned} c^2 &= a^2 + b^2 \\ 20^2 &= 8^2 + b^2 \\ 400 &= 64 + b^2 \\ 336 &= b^2 \\ (336)^{1/2} &= (b^2)^{1/2} \\ \sqrt{336} &= b \\ 18.3 &\approx b \end{aligned}$$

The distance is 18.3 ft.

You Try It 3

Find the diagonal of a rectangle that is 6 cm long and 3 cm wide. Round to the nearest tenth.

Your strategy

Your solution

6.7 cm

Solution on p. S29

Objective 9.4B

Vocabulary to Review

right triangle
hypotenuse
legs of a right triangle
Pythagorean Theorem

Concept Check

Hydroplaning occurs when, rather than gripping the road's surface, a tire slides on the surface of water that is on the pavement.

The equation $v = 8.6\sqrt{p}$ gives the relationship between v, the minimum hydroplaning speed in miles per hour, and p, the tire pressure in pounds per square inch.

1. As the tire pressure increases, does the minimum hydroplaning speed increase or decrease? How did you determine this? Increases
2. As the minimum hydroplaning speed increases, does the tire pressure increase or decrease? How did you determine this? Increases
3. Is there more danger of hydroplaning when the tire pressure is low or when the tire pressure is high? How did you determine this? When the tire pressure is low
4. What implications does this formula have for drivers with respect to checking the tires on their vehicles? Answers will vary.

In-Class Examples (Objective 9.4B)

1. How far would a submarine periscope have to be above the water to locate a ship 5.6 mi away? The equation for the distance in miles that the lookout can see is $d = 1.4\sqrt{h}$, where h is the height in feet above the surface of the water. 16 ft
2. Find the distance required for a car to reach a velocity of 48 m/s when the acceleration is 12 m/s^2. Use the equation $v = \sqrt{2as}$, where v is the velocity in meters per second, a is the acceleration, and s is the distance in meters. 96 m

Optional Student Activity

(*Note:* You will need a ruler and a compass for this activity.)
In this activity you will graph square roots on the number line. We will begin by explaining how to graph $\sqrt{2}$ on the number line.

Draw a number line from -2 to 2. Leave 1 inch between each number. Starting at 0, construct triangle *ABC*. Leg $\overline{AC}$, from 0 to 1 on the number line, is 1 unit long. Leg $\overline{BC}$ is perpendicular to $\overline{AC}$ and equal in length to $\overline{AC}$. Draw $\overline{AB}$ from point *A* to point *B*. Triangle *ABC* is a right triangle. Use the Pythagorean Theorem to find the length of the hypotenuse. (The length of the hypotenuse is $\sqrt{2}$ units.) Place the point of your compass at *A* (0 on the number line) and the compass pencil at point *B*. Draw a circle with radius *AB*. Label the point at which the circle intersects the number line as point *D*. Draw a dot at *D*. This is the graph of $\sqrt{2}$ on the number line.

Use the procedure outlined above to graph $\sqrt{5}$ and $\sqrt{8}$ on the number line.

Example 4

An object is dropped from a high building. Find the distance the object has fallen when its speed reaches 96 ft/s. Use the equation $v = \sqrt{64d}$, where v is the speed of the object in feet per second and d is the distance in feet.

Strategy

To find the distance the object has fallen, replace v in the equation with the given value and solve for d.

Solution

$$v = \sqrt{64d}$$
$$96 = \sqrt{64d}$$
$$(96)^2 = (\sqrt{64d})^2$$
$$9216 = 64d$$
$$144 = d$$

The object has fallen 144 ft.

You Try It 4

How far would a submarine periscope have to be above the water for the lookout to locate a ship 5.5 mi away? The equation for the distance in miles that the lookout can see is $d = \sqrt{1.5h}$, where h is the height in feet above the surface of the water. Round to the nearest hundredth.

Your strategy

Your solution

20.17 ft

Example 5

Find the length of a pendulum that makes one swing in 1.5 s. The equation for the time of one swing is given by $T = 2\pi\sqrt{\frac{L}{32}}$, where T is the time in seconds and L is the length in feet. Round to the nearest hundredth.

Strategy

To find the length of the pendulum, replace T in the equation with the given value and solve for L.

Solution

$$T = 2\pi\sqrt{\frac{L}{32}}$$
$$1.5 = 2\pi\sqrt{\frac{L}{32}}$$
$$\frac{1.5}{2\pi} = \sqrt{\frac{L}{32}}$$
- Divide each side by 2π.

$$\left(\frac{1.5}{2\pi}\right)^2 = \frac{L}{32}$$
- Square each side.

$$32\left(\frac{1.5}{2\pi}\right)^2 = L$$
- Multiply each side by 32.

$$1.82 \approx L$$
- Use the π key on your calculator.

The length of the pendulum is 1.82 ft.

You Try It 5

Find the distance required for a car to reach a velocity of 88 ft/s when the acceleration is 22 ft/s². Use the equation $v = \sqrt{2as}$, where v is the velocity in feet per second, a is the acceleration, and s is the distance in feet.

Your strategy

Your solution

176 ft

Solutions on p. S29

529

9.4 Exercises

Objective A

Solve.

1. $\sqrt{x} = 5$ 25
2. $\sqrt{y} = 2$ 4
3. $\sqrt[3]{a} = 3$ 27
4. $\sqrt[3]{y} = 5$ 125
5. $\sqrt{3x} = 12$ 48
6. $\sqrt{5x} = 10$ 20
7. $\sqrt[3]{4x} = -2$ -2
8. $\sqrt[3]{6x} = -3$ $-\frac{9}{2}$
9. $\sqrt{2x} = -4$ no solution
10. $\sqrt{5x} = -5$ no solution
11. $\sqrt{3x - 2} = 5$ 9
12. $\sqrt{5x - 4} = 9$ 17
13. $\sqrt{3 - 2x} = 7$ -23
14. $\sqrt{9 - 4x} = 4$ $-\frac{7}{4}$
15. $7 = \sqrt{1 - 3x}$ -16
16. $6 = \sqrt{8 - 7x}$ -4
17. $\sqrt[3]{4x - 1} = 2$ $\frac{9}{4}$
18. $\sqrt[3]{5x + 2} = 3$ 5
19. $\sqrt[3]{1 - 2x} = -3$ 14
20. $\sqrt[3]{3 - 2x} = -2$ $\frac{11}{2}$
21. $\sqrt[3]{9x + 1} = 4$ 7
22. $\sqrt{3x + 9} - 12 = 0$ 45
23. $\sqrt{4x - 3} - 5 = 0$ 7
24. $\sqrt{x - 2} = 4$ 18
25. $\sqrt[3]{x - 3} + 5 = 0$ -122
26. $\sqrt[3]{x - 2} = 3$ 29
27. $\sqrt[3]{2x - 6} = 4$ 35
28. $\sqrt{x^2 - 8x} = 3$ $-1, 9$
29. $\sqrt{x^2 + 7x + 11} = 1$ $-5, -2$
30. $\sqrt[4]{4x + 1} = 2$ $\frac{15}{4}$
31. $\sqrt[4]{2x - 9} = 3$ 45
32. $\sqrt{2x - 3} - 2 = 1$ 6
33. $\sqrt{3x - 5} - 5 = 3$ 23

Section 9.4

Suggested Assignment

Exercises 1–57, every other odd
Exercises 61–73, odds
More challenging problems:
Exercises 75–79

Quick Quiz (Objective 9.4A)

Solve.

1. $\sqrt{3y} = 9$ 27
2. $\sqrt[3]{3x} = -6$ -72
3. $\sqrt{5x + 1} = 6$ 7
4. $\sqrt[3]{2x - 1} = 3$ 14
5. $\sqrt[4]{2x - 1} = 3$ 41
6. $\sqrt{3x - 6} - 4 = 2$ 14

530

34. $\sqrt[3]{2x-3}+5=2$
-12

35. $\sqrt[3]{x-4}+7=5$
-4

36. $\sqrt{5x-16}+1=4$
5

37. $\sqrt{3x-5}-2=3$
10

38. $\sqrt{2x-1}-8=-5$
5

39. $\sqrt{7x+2}-10=-7$
1

40. $\sqrt[3]{4x-3}-2=3$
32

41. $\sqrt[3]{1-3x}+5=3$
3

42. $1-\sqrt{4x+3}=-5$
$\frac{33}{4}$

43. $7-\sqrt{3x+1}=-1$
21

44. $\sqrt{x+1}=2-\sqrt{x}$
$\frac{9}{16}$

45. $\sqrt{2x+4}=3-\sqrt{2x}$
$\frac{25}{72}$

46. $\sqrt{x^2+3x-2}-x=1$
3

47. $\sqrt{x^2-4x-1}+3=x$
5

48. $\sqrt{x^2-3x-1}=3$
$5, -2$

49. $\sqrt{x^2-2x+1}=3$
$4, -2$

50. $\sqrt{2x+5}-\sqrt{3x-2}=1$
2

51. $\sqrt{4x+1}-\sqrt{2x+4}=1$
6

52. $\sqrt{5x-1}-\sqrt{3x-2}=1$
2, 1

53. $\sqrt{5x+4}-\sqrt{3x+1}=1$
0, 1

54. $\sqrt[3]{x^2+2}-3=0$
$5, -5$

55. $\sqrt[3]{x^2+4}-2=0$
$-2, 2$

56. $\sqrt[4]{x^2+2x+8}-2=0$
$-4, 2$

57. $\sqrt[4]{x^2+x-1}-1=0$
$-2, 1$

58. $4\sqrt{x+1}-x=1$
$15, -1$

59. $3\sqrt{x-2}+2=x$
2, 11

60. $x+3\sqrt{x-2}=12$
6

Objective B *Application Problems*

61. Find the width of a rectangle that has a diagonal of 10 ft and a length of 8 ft.
6 ft

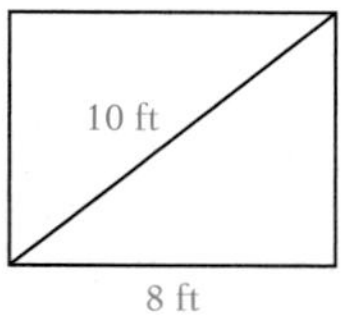

Quick Quiz (Objective 9.4B)

1. An 18-foot ladder is leaning against a building. How high on the building will the ladder reach when the bottom of the ladder is 6 ft from the building? Round to the nearest tenth. 17.0 ft
2. An object is dropped from a bridge. Find the distance the object has fallen when its speed reaches 120 ft/s. Use the equation $v=\sqrt{64d}$, where v is the speed of the object in feet per second and d is the distance in feet. 225 ft

531

62. Find the length of a rectangle that has a diagonal of 15 m and a width of 9 m.
12 m

63. A 26-foot ladder is leaning against a building. How far is the bottom of the ladder from the building when the ladder reaches a height of 24 ft on the building?
10 ft

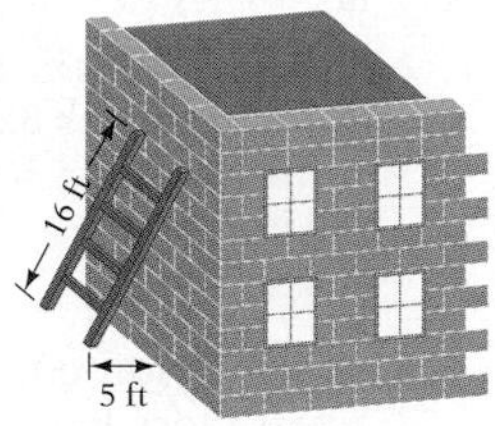

64. A 16-foot ladder is leaning against a building. How high on the building will the ladder reach when the bottom of the ladder is 5 ft from the building? Round to the nearest tenth.
15.2 ft

65. An object is dropped from an airplane. Find the distance the object has fallen when its speed reaches 400 ft/s. Use the equation $v = \sqrt{64d}$, where v is the speed of the object and d is the distance.
2500 ft

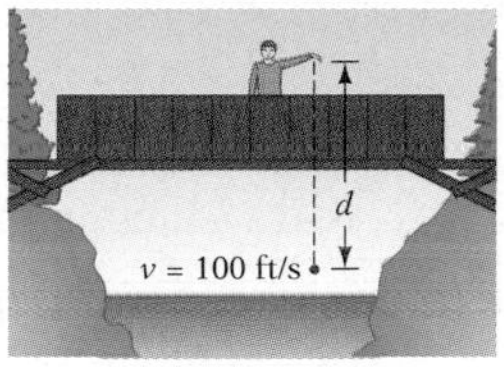

66. An object is dropped from a bridge. Find the distance the object has fallen when its speed reaches 100 ft/s. Use the equation $v = \sqrt{64d}$, where v is the speed of the object and d is the distance.
156.25 ft

67. How far would a submarine periscope have to be above the water for the lookout to locate a ship 3.6 mi away? The equation for the distance in miles that the lookout can see is $d = \sqrt{1.5h}$, where h is the height in feet above the surface of the water.
8.64 ft

68. How far would a submarine periscope have to be above the water for the lookout to locate a ship 4.2 mi away? The equation for the distance in miles that the lookout can see is $d = \sqrt{1.5h}$, where h is the height in feet above the surface of the water.
11.76 ft

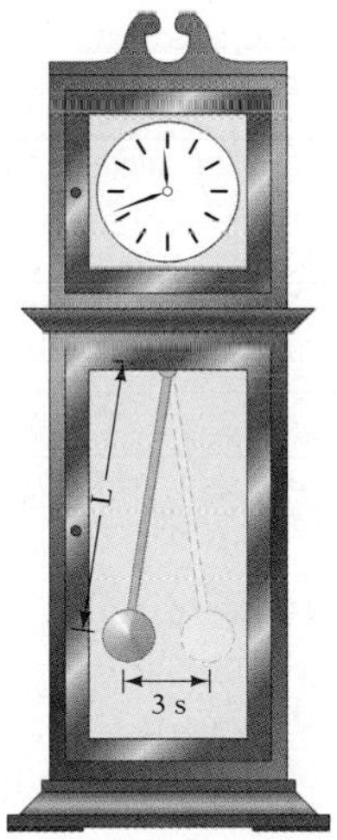

69. Find the length of a pendulum on a clock that makes one swing in 2.4 s. The equation for the time of one swing of a pendulum is given by $T = 2\pi\sqrt{\frac{L}{32}}$, where T is the time in seconds and L is the length in feet. Round to the nearest hundredth.
4.67 ft

70. Find the length of a pendulum that makes one swing in 3 s. The equation for the time of one swing of a pendulum is given by $T = 2\pi\sqrt{\frac{L}{32}}$, where T is the time in seconds and L is the length in feet. Round to the nearest hundredth.
7.30 ft

71. Find the distance required for a car to reach a velocity of 60 m/s when the acceleration is 10 m/s^2. Use the equation $v = \sqrt{2as}$, where v is the velocity, a is the acceleration, and s is the distance.
180 m

72. The time it takes for an object to fall a certain distance is given by the equation $t = \sqrt{\frac{2d}{g}}$, where t is the time in seconds, d is the distance in feet, and g is the acceleration due to gravity. The acceleration due to gravity on Earth is 32 feet per second. If an object is dropped from the top of a tall building, how far will it fall in 6 s?
576 ft

73. High definition television (HDTV) gives consumers a wider viewing area, more like a film in a theater. A regular television with a 27-inch diagonal measurement has a screen 16.2 in. tall. An HDTV screen with the same 16.2-inch height would have a diagonal measuring 33 in. How many inches wider is the HDTV screen? Round to the nearest hundredth.
7.15 in.

74. At what height above Earth's surface would a satellite be in orbit if it is traveling at a speed of 7500 m/s? Use the equation $v = \sqrt{\frac{4 \times 10^{14}}{h + 6.4 \times 10^6}}$, where v is the speed of the satellite in meters per second and h is the height above Earth's surface in meters. Round to the nearest thousand.
711,000 m

APPLYING THE CONCEPTS

75. Solve the following equations. Describe the solution by using the following terms: integer, rational number, irrational number, real number, and imaginary number. Note that more than one term may be used to describe the answer.

a. $x^2 + 3 = 7$
2, −2; integer, rational number, real number

b. $x^2 + 1 = 0$
$i, -i$; imaginary

c. $x^{3/4} = 8$
16; integer, rational number, real number

76. Solve: $\sqrt{3x - 2} = \sqrt{2x - 3} + \sqrt{x - 1}$
2

77. Solve $a^2 + b^2 = c^2$ for a.
$a = \sqrt{c^2 - b^2}$ or $a = -\sqrt{c^2 - b^2}$

78. Solve $V = \frac{4}{3}\pi r^3$ for r.
$r = \sqrt[3]{\frac{3V}{4\pi}}$

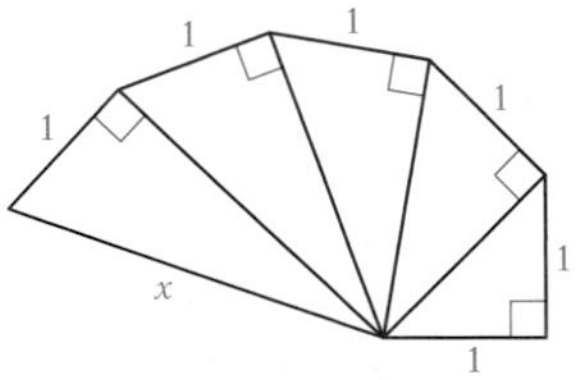

79. Find the length of the side labeled x in the diagram at the right.
$x = \sqrt{6}$

Focus on Problem Solving

Another Look at Polya's Four-Step Process

Polya's four general steps to follow when attempting to solve a problem are to understand the problem, devise a plan, carry out the plan, and review the solution. (See the Focus on Problem Solving in the chapter entitled "Factoring.") In the process of devising a plan (Step 2), it may be appropriate to write a mathematical expression or an equation. We will illustrate this with the following problem.

Number the letters of the alphabet in sequence from 1 to 26. (See the list at the left.) Find a word for which the product of the numerical values of the letters of the word equals 1,000,000. We will agree that a "word" is any sequence of letters that contains at least one vowel; it need not be in the dictionary.

A = 1
B = 2
C = 3
D = 4
E = 5
F = 6
G = 7
H = 8
I = 9
J = 10
K = 11
L = 12
M = 13
N = 14
O = 15
P = 16
Q = 17
R = 18
S = 19
T = 20
U = 21
V = 22
W = 23
X = 24
Y = 25
Z = 26

1. **Understand the Problem**
Consider REZB. The product of the values of the letters is $18 \cdot 5 \cdot 26 \cdot 2 = 4680$. This "word" is a sequence of letters with at least one vowel. However, the product of the numerical values of the letters is not 1,000,000. Thus this word does not solve our problem.

2. **Devise a Plan**
Actually, we should have known that the product of the values of the letters in REZB could not equal 1,000,000. The letter R has a factor of 9, and the letter Z has a factor of 13. Neither of these two numbers is a factor of 1,000,000. Consequently, R and Z cannot be letters in the word we are trying to find. This observation leads to an important observation: Each of the letters that make up our word must be a factor of 1,000,000. To find these letters, consider the prime factorization of 1,000,000.

$$1{,}000{,}000 = 2^6 \cdot 5^6$$

Looking at the prime factorization, we note that only letters that contain 2 or 5 as factors are possible candidates. These letters are B, D, E, H, J, P, T, and Y. One additional point: Because 1 times any number is the number, the letter A can be part of any word we construct.

Our task is now to construct a word from these letters such that the product is 1,000,000. From the prime factorization above, we must have 2 as a factor six times and 5 as a factor six times.

3. **Carry Out the Plan**
We must construct a word with the characteristics described in our plan. Here is a possibility:

THEBEYE

4. **Review Your Solution**
You should multiply the values of all the letters and verify that the product is 1,000,000. To ensure that you have an understanding of the problem, try to find other "words" that satisfy the conditions of the problem.

Answers to Focus on Problem Solving: Another Look at Polya's Four-Step Process

Here are two more examples of "words" for which the product of the numerical values of the letters equals 1,000,000:

PAYJAJY
DETTEY

Answers to Projects and Group Activities: Distance to the Horizon

1. Answers will vary. Students should conclude that $(h/5280)^2$ is a very small number and has very little influence on the answer.
2. The distance is approximately 3.9 mi.

Projects and Group Activities

Distance to the Horizon

In Section 9.4, we used the formula $d = \sqrt{1.5h}$ to calculate the approximate distance d (in miles) that a person using a periscope h feet above the water could see. That formula is derived by using the Pythagorean Theorem.

Consider the diagram (not to scale) at the right, which shows Earth as a sphere and the periscope extending h feet above the surface. From geometry, because AB is tangent to the circle and OA is a radius, triangle AOB is a right triangle. Therefore,

$$(OA)^2 + (AB)^2 = (OB)^2$$

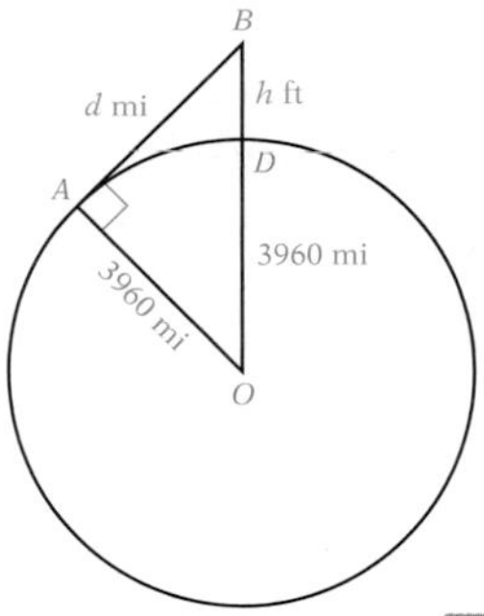

Substituting into this formula, we have

$$3960^2 + d^2 = \left(3960 + \frac{h}{5280}\right)^2$$

• Because h is in feet, $\frac{h}{5280}$ is in miles.

$$3960^2 + d^2 = 3960^2 + \frac{2 \cdot 3960}{5280}h + \left(\frac{h}{5280}\right)^2$$

$$d^2 = \frac{3}{2}h + \left(\frac{h}{5280}\right)^2$$

$$d = \sqrt{\frac{3}{2}h + \left(\frac{h}{5280}\right)^2}$$

At this point, an assumption is made that $\sqrt{\frac{3}{2}h + \left(\frac{h}{5280}\right)^2} \approx \sqrt{1.5h}$, where we have written $\frac{3}{2}$ as 1.5. Thus $d \approx \sqrt{1.5h}$ is used to approximate the distance that can be seen using a periscope h feet above the water.

1. Write a paragraph that justifies the assumption that

$$\sqrt{\frac{3}{2}h + \left(\frac{h}{5280}\right)^2} \approx \sqrt{1.5h}$$

(*Suggestion:* Evaluate each expression for various values of h. Because h is the height of a periscope above the water, it is unlikely that $h > 25$ ft.)

2. The distance d is the distance from the top of the periscope to A. The distance along the surface of the water is given by arc AD. This distance, L, can be approximated by the equation

$$L \approx \sqrt{1.5h} + 0.306186\left(\sqrt{\frac{h}{5280}}\right)^3$$

Using this formula, calculate L when $h = 10$.

Pythagorean Triples The Pythagorean Theorem states that if a and b are the legs of a right triangle and c is the length of the hypotenuse, then $c^2 = a^2 + b^2$.

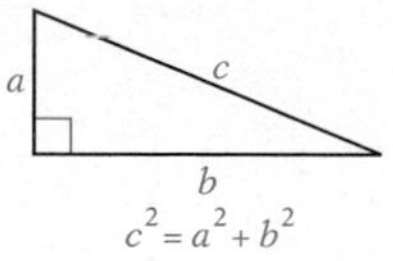

For instance, the triangle with legs 3 and 4 and hypotenuse 5 is a right triangle because $3^2 + 4^2 = 5^2$. The numbers 3, 4, and 5 are called a **Pythagorean triple** because they are natural numbers that satisfy the equation of the Pythagorean Theorem.

1. Determine whether the numbers are a Pythagorean triple.
a. 5, 7, and 9 **b.** 8, 15, and 17
c. 11, 60, and 61 **d.** 28, 45, and 53

Mathematicians have investigated Pythagorean triples and have found formulas that will generate these triples. One such formula is

$$a = m^2 - n^2 \qquad b = 2mn \qquad c = m^2 + n^2, \text{ where } m > n$$

For instance, let $m = 2$ and $n = 1$. Then $a = 2^2 - 1^2 = 3$, $b = 2(2)(1) = 4$, and $c = 2^2 + 1^2 = 5$. This is the Pythagorean triple given above.

2. Find the Pythagorean triple produced by each of the following.
a. $m = 3$ and $n = 1$ **b.** $m = 5$ and $n = 2$
c. $m = 4$ and $n = 2$ **d.** $m = 6$ and $n = 1$

3. Find values of m and n that yield the Pythagorean triple 11, 60, 61.

4. Verify that $a^2 + b^2 = c^2$ when $a = m^2 - n^2$, $b = 2mn$, and $c = m^2 + n^2$.

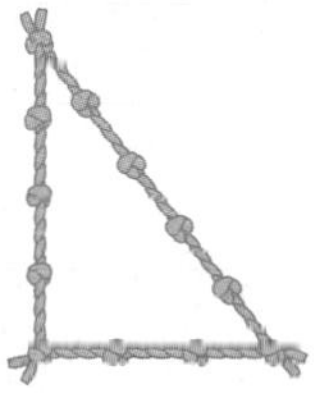

5. The early Greek builders used a rope with 12 equally spaced knots to make right-angle corners for buildings. Explain how they used the rope.

6. Find three odd integers, a, b, c, such that $a^2 + b^2 = c^2$.

Chapter Summary

Key Words The *nth root of a* is $a^{1/n}$. The expression $\sqrt[n]{a}$ is another symbol for the nth root of a. [pp. 497, 499]

In the expression $\sqrt[n]{a}$, the symbol $\sqrt{\ }$ is called a *radical sign*, n is the *index*, and a is the *radicand*. [p. 499]

Answers to Projects and Group Activities: Pythagorean Triples

1. a. No
b. Yes
c. Yes
d. Yes

2. a. 6, 8, 10
b. 20, 21, 29
c. 12, 16, 20
d. 12, 35, 37

3. $m = 6$, $n = 5$

4.
$$\begin{aligned} a^2 &= (m^2 - n^2)^2 \\ &= (m^2 - n^2)(m^2 - n^2) \\ &= m^4 - 2m^2n^2 + n^4 \\ b^2 &= (2mn)^2 = 4m^2n^2 \\ c^2 &= (m^2 + n^2)^2 \\ &= (m^2 + n^2)(m^2 + n^2) \\ &= m^4 + 2m^2n^2 + n^4 \\ a^2 + b^2 &= m^4 - 2m^2n^2 + n^4 \\ &\quad + 4m^2n^2 \\ &= m^4 + 2m^2n^2 + n^4 \\ &= c^2 \end{aligned}$$

5. Using the Pythagorean triple 3, 4, and 5, they could have held the rope in the shape of a right triangle with legs of lengths 3 and 4 and hypotenuse of length 5.

6. It is not possible. (The formula $b = 2mn$ ensures that one number will be even.)

The symbol $\sqrt{}$ is used to indicate the positive or *principal square root* of a number. [p. 501]

The expressions $a + b$ and $a - b$ are called *conjugates* of each other. The product of conjugates of the form $(a + b)(a - b)$ is $a^2 - b^2$. [p. 510]

The procedure used to remove a radical from the denominator of a radical expression is called *rationalizing the denominator.* [p. 511]

A *complex number* is a number of the form $a + bi$, where a and b are real numbers and $i = \sqrt{-1}$. For the complex number $a + bi$, a is the *real part* of the complex number, and b is the *imaginary part* of the complex number. [p. 517]

A *radical equation* is an equation that contains a variable expression in a radicand. [p. 525]

Essential Rules

Rule for Rational Exponents [p. 497]

If m and n are positive integers and $a^{1/n}$ is a real number, then $a^{m/n} = (a^{1/n})^m$.

Rule for Writing Exponential Expressions as Radical Expressions [p. 499]

If $a^{1/n}$ is a real number, then $a^{m/n} = \sqrt[n]{a^m}$.

The Product Property of Radicals [p. 507]

If a and b are positive real numbers, then $\sqrt[n]{ab} = \sqrt[n]{a}\sqrt[n]{b}$.

The Quotient Property of Radicals [p. 511]

If a and b are positive real numbers, then $\sqrt[n]{\frac{a}{b}} = \frac{\sqrt[n]{a}}{\sqrt[n]{b}}$.

Definition of $\sqrt{-a}$ [p. 517]

If a is a positive real number, then the principal square root of negative a is the imaginary number $i\sqrt{a}$: $\sqrt{-a} = i\sqrt{a}$.

Addition of Complex Numbers [p. 518]

If $a + bi$ and $c + di$ are complex numbers, then
$(a + bi) + (c + di) = (a + c) + (b + d)i$.

Subtraction of Complex Numbers [p. 518]

If $a + bi$ and $c + di$ are complex numbers, then
$(a + bi) - (c + di) = (a - c) + (b - d)i$.

The Property of Raising Each Side of an Equation to a Power [p. 525]

If a and b are real numbers and $a = b$, then $a^n = b^n$.

The Pythagorean Theorem [p. 527]

The square of the hypotenuse of a right triangle is equal to the sum of the squares of the two legs.
$c^2 = a^2 + b^2$

537

Chapter Review

1. Simplify: $(16x^{-4}y^{12})^{1/4}(100x^{6}y^{-2})^{1/2}$
$20x^2y^2$ [9.1A]

2. Solve: $\sqrt[4]{3x-5}=2$
7 [9.4A]

3. Multiply: $(6-5i)(4+3i)$
$39-2i$ [9.3C]

4. Rewrite $7y\sqrt[3]{x^2}$ as an exponential expression.
$7x^{2/3}y$ [9.1B]

5. Multiply: $(\sqrt{3}+8)(\sqrt{3}-2)$
$6\sqrt{3}-13$ [9.2C]

6. Solve: $\sqrt{4x+9}+10=11$
-2 [9.4A]

7. Simplify: $\dfrac{x^{-3/2}}{x^{7/2}}$
$\dfrac{1}{x^5}$ [9.1A]

8. Simplify: $\dfrac{8}{\sqrt{3y}}$
$\dfrac{8\sqrt{3y}}{3y}$ [9.2D]

9. Simplify: $\sqrt[3]{-8a^6b^{12}}$
$-2a^2b^4$ [9.1C]

10. Subtract: $\sqrt{50a^4b^3}-ab\sqrt{18a^2b}$
$2a^2b\sqrt{2b}$ [9.2B]

11. Simplify: $\dfrac{x+2}{\sqrt{x}+\sqrt{2}}$
$\dfrac{x\sqrt{x}-x\sqrt{2}+2\sqrt{x}-2\sqrt{2}}{x-2}$ [9.2D]

12. Simplify: $\dfrac{5+2i}{3i}$
$\dfrac{2}{3}-\dfrac{5}{3}i$ [9.3D]

13. Simplify: $\sqrt{18a^3b^6}$
$3ab^3\sqrt{2a}$ [9.2A]

14. Subtract: $(\sqrt{50}+\sqrt{-72})-(\sqrt{162}-\sqrt{-8})$
$-4\sqrt{2}+8i\sqrt{2}$ [9.3B]

15. Subtract: $3x\sqrt[3]{54x^8y^{10}}-2x^2y\sqrt[3]{16x^5y^7}$
$5x^3y^3\sqrt[3]{2x^2y}$ [9.2B]

16. Multiply: $\sqrt[3]{16x^4y}\,\sqrt[3]{4xy^5}$
$4xy^2\sqrt[3]{x^2}$ [9.2C]

17. Multiply: $i(3-7i)$
$7+3i$ [9.3C]

18. Rewrite $3x^{3/4}$ as a radical expression.
$3\sqrt[4]{x^3}$ [9.1B]

19. Simplify: $\sqrt[5]{-64a^8b^{12}}$
$-2ab^2\sqrt[5]{2a^3b^2}$ [9.2A]

20. Simplify: $\frac{5+9i}{1-i}$
$-2+7i$ [9.3D]

21. Multiply: $\sqrt{-12}\sqrt{-6}$
$-6\sqrt{2}$ [9.3C]

22. Solve: $\sqrt{x-5}+\sqrt{x+6}=11$
30 [9.4A]

23. Simplify: $\sqrt[4]{81a^8b^{12}}$
$3a^2b^3$ [9.1C]

24. Simplify: $\sqrt{-50}$
$5i\sqrt{2}$ [9.3A]

25. Subtract: $(-8+3i)-(4-7i)$
$-12+10i$ [9.3B]

26. Simplify: $(5-\sqrt{6})^2$
$31-10\sqrt{6}$ [9.2C]

27. Simplify: $4x\sqrt{12x^2y}+\sqrt{3x^4y}-x^2\sqrt{27y}$
$6x^2\sqrt{3y}$ [9.2B]

28. The velocity of the wind determines the amount of power generated by a windmill. A typical equation for this relationship is $v = 4.05\sqrt[3]{P}$, where v is the velocity in miles per hour and P is the power in watts. Find the amount of power generated by a 20-mph wind. Round to the nearest whole number.
120 watts [9.4B]

29. Find the distance required for a car to reach a velocity of 88 ft/s when the acceleration is 16 ft/s². Use the equation $v = \sqrt{2as}$, where v is the velocity in feet per second, a is the acceleration, and s is the distance in feet.
242 ft [9.4B]

30. A 12-foot ladder is leaning against a building. How far from the building is the bottom of the ladder when the top of the ladder touches the building 10 ft above the ground? Round to the nearest hundredth.
6.63 ft [9.4B]

539

Chapter Test

1. Write $\frac{1}{2}\sqrt[4]{x^3}$ as an exponential expression.

$\frac{1}{2}x^{3/4}$ [9.1B]

2. Subtract: $\sqrt[3]{54x^7y^3} - x\sqrt[3]{128x^4y^3} - x^2\sqrt[3]{2xy^3}$

$-2x^2y\sqrt[3]{2x}$ [9.2B]

3. Write $3y^{2/5}$ as a radical expression.

$3\sqrt[5]{y^2}$ [9.1B]

4. Multiply: $(2 + 5i)(4 - 2i)$

$18 + 16i$ [9.3C]

5. Simplify: $(2\sqrt{x} + \sqrt{y})^2$

$4x + 4\sqrt{xy} + y$ [9.2C]

6. Simplify: $\frac{r^{2/3}r^{-1}}{r^{-1/2}}$

$r^{1/6}$ [9.1A]

7. Solve: $\sqrt{x + 12} - \sqrt{x} = 2$

4 [9.4A]

8. Simplify: $\sqrt[3]{8x^3y^6}$

$2xy^2$ [9.1C]

9. Multiply: $\sqrt{3x}(\sqrt{x} - \sqrt{25x})$

$-4x\sqrt{3}$ [9.2C]

10. Subtract: $(5 - 2i) - (8 - 4i)$

$-3 + 2i$ [9.3B]

11. Simplify: $\sqrt{32x^4y^7}$

$4x^2y^3\sqrt{2y}$ [9.2A]

12. Multiply: $(2\sqrt{3} + 4)(3\sqrt{3} - 1)$

$14 + 10\sqrt{3}$ [9.2C]

13. Add: $(2 + i) + (2 - i)$

4 [9.3B]

14. Simplify: $\frac{4 - 2\sqrt{5}}{2 - \sqrt{5}}$

2 [9.2D]

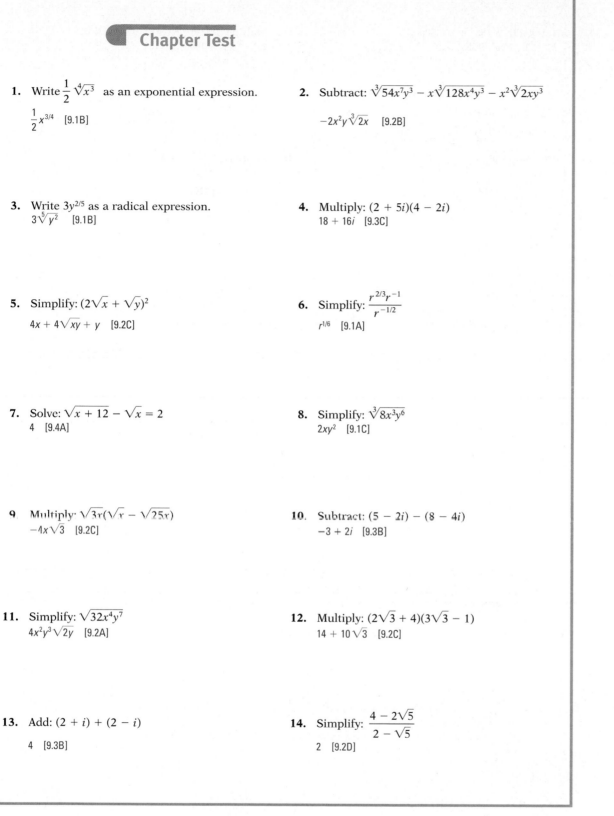

15. Add: $\sqrt{18a^3} + a\sqrt{50a}$
$8a\sqrt{2a}$ [9.2B]

16. Multiply: $(\sqrt{a} - 3\sqrt{b})(2\sqrt{a} + 5\sqrt{b})$
$2a - \sqrt{ab} - 15b$ [9.2C]

17. Simplify: $\dfrac{(2x^{1/3}y^{-2/3})^6}{(x^{-4}y^8)^{1/4}}$

$\dfrac{64x^3}{y^6}$ [9.1A]

18. Simplify: $\dfrac{\sqrt{x}}{\sqrt{x} - \sqrt{y}}$

$\dfrac{x + \sqrt{xy}}{x - y}$ [9.2D]

19. Simplify: $\dfrac{2 + 3i}{1 - 2i}$

$-\dfrac{4}{5} + \dfrac{7}{5}i$ [9.3D]

20. Solve: $\sqrt[3]{2x - 2} + 4 = 2$

-3 [9.4A]

21. Simplify: $\left(\dfrac{4a^4}{b^2}\right)^{-3/2}$

$\dfrac{b^3}{8a^6}$ [9.1A]

22. Simplify: $\sqrt[3]{27a^4b^3c^7}$

$3abc^2\sqrt[3]{ac}$ [9.2A]

23. Simplify: $\dfrac{\sqrt{32x^5y}}{\sqrt{2xy^3}}$

$\dfrac{4x^2}{y}$ [9.2D]

24. Multiply: $(\sqrt{-8})(\sqrt{-2})$

-4 [9.3C]

25. An object is dropped from a high building. Find the distance the object has fallen when its speed reaches 192 ft/s. Use the equation $v = \sqrt{64d}$, where v is the speed of the object in feet per second and d is the distance in feet.
576 ft [9.4B]

541

Cumulative Review

1. Simplify: $2^3 \cdot 3 - 4(3 - 4 \cdot 5)$
92 [1.3A]

2. Evaluate $4a^2b - a^3$ when $a = -2$ and $b = 3$.
56 [1.4A]

3. Simplify: $-3(4x - 1) - 2(1 - x)$

$-10x + 1$ [1.4D]

4. Solve: $5 - \frac{2}{3}x = 4$

$\frac{3}{2}$ [2.2A]

5. Solve: $2[4 - 2(3 - 2x)] = 4(1 - x)$
$\frac{2}{3}$ [2.2C]

6. Solve: $6x - 3(2x + 2) > 3 - 3(x + 2)$
$\{x | x > 1\}$ [2.4A]

7. Solve: $2 + |4 - 3x| = 5$
$\frac{1}{3}, \frac{7}{3}$ [2.5A]

8. Solve: $|2x + 3| \le 9$
$\{x | -6 \le x \le 3\}$ [2.5B]

9. Find the area of the triangle shown in the figure below.

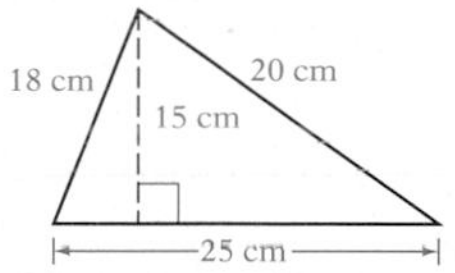

187.5 cm² [3.2 B]

10. Find the value of the determinant:

$$\begin{vmatrix} 1 & 2 & -3 \\ 0 & -1 & 2 \\ 3 & 1 & -2 \end{vmatrix}$$

3 [5.3A]

11. Graph $3x - 2y = -6$. State the slope and y-intercept.

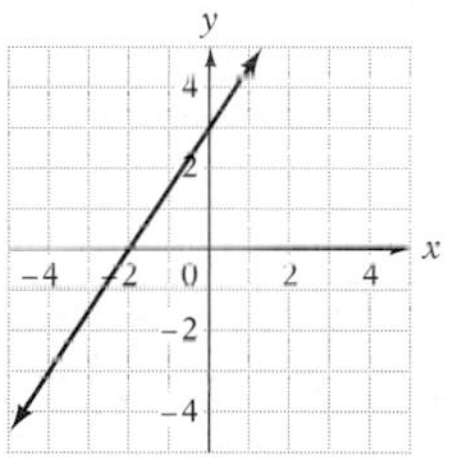

$m = \frac{3}{2}, b = 3$ [4.3B]

12. Graph the solution set of $3x + 2y \le 4$.

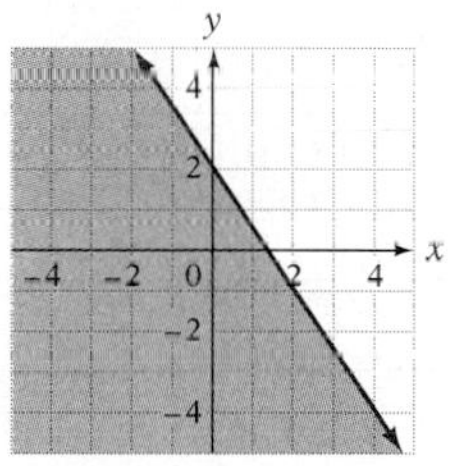

[4.7A]

13. Find the equation of the line that passes through the points (2, 3) and (−1, 2).

$y = \frac{1}{3}x + \frac{7}{3}$ [4.5B]

14. Solve by using Cramer's Rule:
$2x - y = 4$
$-2x + 3y = 5$
$\left(\frac{17}{4}, \frac{9}{2}\right)$ [5.3B]

15. Simplify: $(2^{-1}x^2y^{-6})(2^{-1}y^{-4})^{-2}$
$2x^2y^2$ [6.1B]

16. Factor: $81x^2 - y^2$
$(9x + y)(9x - y)$ [7.4A]

542 Chapter 9 / Rational Exponents and Radicals

17. Factor: $x^5 + 2x^3 - 3x$

$x(x^2 + 3)(x + 1)(x - 1)$ [7.4D]

18. Solve $P = \dfrac{R - C}{n}$ for C.

$C = R - nP$ [8.5A]

19. Simplify: $\left(\dfrac{x^{-2/3}y^{1/2}}{y^{-1/3}}\right)^6$

$\dfrac{y^5}{x^4}$ [9.1A]

20. Subtract: $\sqrt{40x^3} - x\sqrt{90x}$

$-x\sqrt{10x}$ [9.2B]

21. Multiply: $(\sqrt{3} - 2)(\sqrt{3} - 5)$

$13 - 7\sqrt{3}$ [9.2C]

22. Simplify: $\dfrac{4}{\sqrt{6} - \sqrt{2}}$

$\sqrt{6} + \sqrt{2}$ [9.2D]

23. Simplify: $\dfrac{2i}{3 - i}$

$-\dfrac{1}{5} + \dfrac{3}{5}i$ [9.3D]

24. Solve: $\sqrt[3]{3x - 4} + 5 = 1$

-20 [9.4A]

25. The two triangles are similar triangles. Find the length of side DE.
27 m [8.4C]

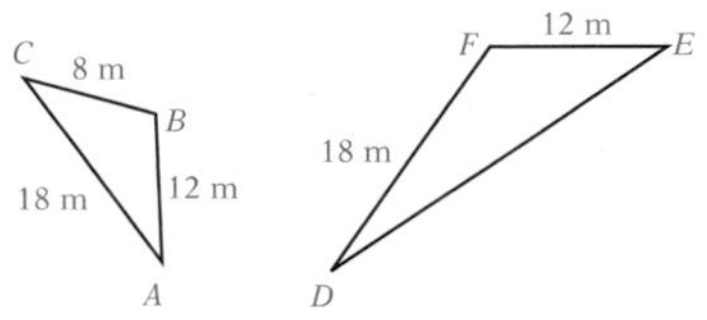

26. An investment of \$2500 is made at an annual simple interest rate of 7.2%. How much additional money must be invested at an annual simple interest rate of 8.4% so that the total interest earned is \$516?
\$4000 [2.3C]

27. A sales executive traveled 25 mi by car and then an additional 625 mi by plane. The rate of the plane was five times the rate of the car. The total time of the trip was 3 h. Find the rate of the plane.
250 mph [8.6B]

28. How long does it take light to travel to Earth from the moon when the moon is 232,500 mi from Earth? Light travels 1.86×10^5 mi/s.
1.25 s [6.1D]

29. The graph shows the amount invested and the annual interest earned on the investment. Find the slope of the line between the two points shown on the graph. Then write a sentence that states the meaning of the slope.
$m = 0.08$. The slope represents the simple interest rate on the investment. The interest rate is 8%. [4.4A]

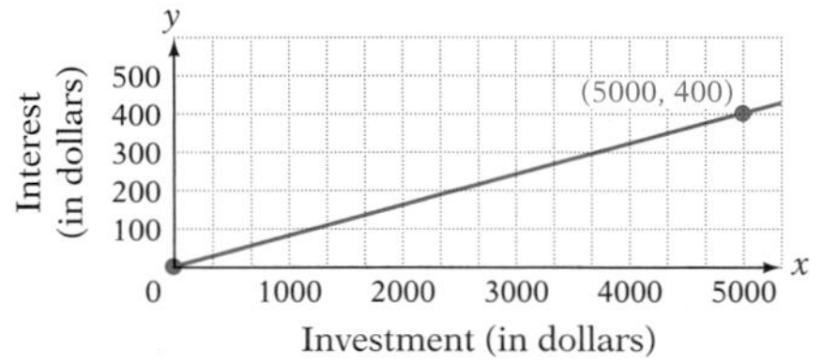

30. How far would a submarine periscope have to be above the water for the lookout to locate a ship 7 mi away? The equation for the distance in miles that the lookout can see is $d = \sqrt{1.5h}$, where h is the height in feet above the surface of the water. Round to the nearest tenth of a foot.
32.7 ft [9.4B]

Chapter 10

Quadratic Equations

A rainstorm like the one shown here directly affects highway driving conditions. How quickly a car is able to stop once the brakes are applied is determined by many factors, including how wet the road is, the type of pavement, and the speed of the car. Some factors, such as the tread on a car's tires, can be controlled, and good tread can minimize the adverse effects of braking on dangerously wet roads. The braking distance of a car can be modeled by quadratic equations such as the ones in **Exercises 5 and 9 on page 573.**

Objectives

Section 10.1

A To solve a quadratic equation by factoring

B To write a quadratic equation given its solutions

C To solve a quadratic equation by taking square roots

Section 10.2

A To solve a quadratic equation by completing the square

Section 10.3

A To solve a quadratic equation by using the quadratic formula

Section 10.4

A To solve an equation that is quadratic in form

B To solve a radical equation that is reducible to a quadratic equation

C To solve a fractional equation that is reducible to a quadratic equation

Section 10.5

A To solve application problems

Section 10.6

A To solve a nonlinear inequality

WEB Need help? For on-line student resources, such as section quizzes, visit this textbook's web site at **college.hmco.com/students.**

Prep Test

1. Simplify: $\sqrt{18}$
$3\sqrt{2}$ [1.2F]

2. Simplify: $\sqrt{-9}$
$3i$ [9.3A]

3. Simplify: $\dfrac{3x - 2}{x - 1} - 1$
$\dfrac{2x - 1}{x - 1}$ [8.2D]

4. Evaluate $b^2 - 4ac$ when $a = 2$, $b = -4$, and $c = 1$.
8 [1.4A]

5. Is $4x^2 + 28x + 49$ a perfect-square trinomial?
yes [7.4A]

6. Factor: $4x^2 - 4x + 1$
$(2x - 1)^2$ [7.4A]

7. Factor: $9x^2 - 4$
$(3x + 2)(3x - 2)$ [7.4A]

8. Graph: $\{x \mid x < -1\} \cap \{x \mid x < 4\}$

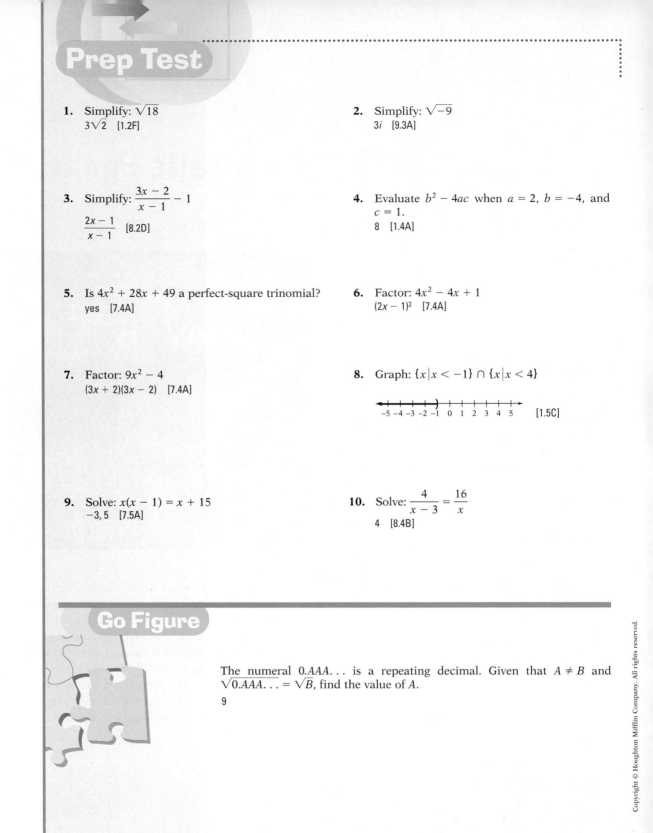

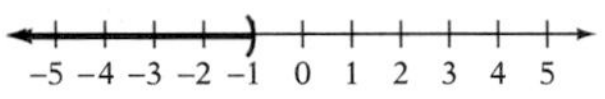

[1.5C]

9. Solve: $x(x - 1) = x + 15$
$-3, 5$ [7.5A]

10. Solve: $\dfrac{4}{x - 3} = \dfrac{16}{x}$
4 [8.4B]

Go Figure

The numeral $0.AAA\ldots$ is a repeating decimal. Given that $A \neq B$ and $\sqrt{0.AAA\ldots} = \sqrt{B}$, find the value of A.
9

10.1 Solving Quadratic Equations by Factoring or by Taking Square Roots

Objective A To solve a quadratic equation by factoring

VIDEO & DVD CD TUTOR WEB SSM

Recall that a **quadratic equation** is an equation of the form $ax^2 + bx + c = 0$, where a and b are coefficients, c is a constant, and $a \neq 0$.

Quadratic equations
$$\begin{cases} 3x^2 - x + 2 = 0, & a = 3, & b = -1, & c = 2 \\ -x^2 + 4 = 0, & a = -1, & b = 0, & c = 4 \\ 6x^2 - 5x = 0, & a = 6, & b = -5, & c = 0 \end{cases}$$

A quadratic equation is in **standard form** when the polynomial is in descending order and equal to zero. Because the degree of the polynomial $ax^2 + bx + c$ is 2, a quadratic equation is also called a **second-degree equation**.

As we discussed earlier, quadratic equations sometimes can be solved by using the Principle of Zero Products. This method is reviewed here.

The Principle of Zero Products

If a and b are real numbers and $ab = 0$, then $a = 0$ or $b = 0$.

The Principle of Zero Products states that if the product of two factors is zero, then at least one of the factors must be zero.

TAKE NOTE
Recall that the steps involved in solving a quadratic equation by factoring are
1. Write the equation in standard form.
2. Factor.
3. Use the Principle of Zero Products to set each factor equal to 0.
4. Solve each equation.
5. Check the solutions.

➡ Solve by factoring: $3x^2 = 2 - 5x$

$$3x^2 = 2 - 5x$$
$$3x^2 + 5x - 2 = 0 \qquad \text{• Write the equation in standard form.}$$
$$(3x - 1)(x + 2) = 0 \qquad \text{• Factor.}$$
$$3x - 1 = 0 \qquad x + 2 = 0 \qquad \text{• Use the Principle of Zero Products to write two equations.}$$
$$3x = 1 \qquad x = -2 \qquad \text{• Solve each equation.}$$
$$x = \frac{1}{3}$$

$\frac{1}{3}$ and -2 check as solutions. The solutions are $\frac{1}{3}$ and -2.

TAKE NOTE
When a quadratic equation has two solutions that are the same number, the solution is called a **double root** of the equation. 3 is a double root of $x^2 - 6x = -9$.

➡ Solve by factoring: $x^2 - 6x = -9$

$$x^2 - 6x = -9$$
$$x^2 - 6x + 9 = 0 \qquad \text{• Write the equation in standard form.}$$
$$(x - 3)(x - 3) = 0 \qquad \text{• Factor.}$$
$$x - 3 = 0 \qquad x - 3 = 0 \qquad \text{• Use the Principle of Zero Products.}$$
$$x = 3 \qquad x = 3 \qquad \text{• Solve each equation.}$$

3 checks as a solution. The solution is 3.

Objective 10.1A

New Vocabulary
second-degree equation

Vocabulary to Review
quadratic equation
standard form

Rules to Review
Principle of Zero Products

Instructor Note
The material on solving a quadratic equation by factoring was covered earlier in the text. It is reviewed here for completeness.

Discuss the Concepts
1. How does a quadratic equation differ from a linear equation?
2. Explain why the equation $x(x + 4) = 12$ cannot be solved by solving the two equations $x = 12$ and $x + 4 = 12$.

Concept Check
1. How many consecutive natural numbers, beginning with 1, will give a sum of 78? Use the formula $S = \frac{n^2 + n}{2}$, where S is the sum of the first n natural numbers. 12
2. A football league has 28 games scheduled. How many teams are in the league if each team plays every other team once? Use the formula $N = \frac{t^2 - t}{2}$, where N is the number of football games that must be scheduled in a league having t teams if each team is to play every other team once. 8 teams

In-Class Examples (Objective 10.1A)

Solve by factoring.
1. $x^2 - 2x = 15$ $-3, 5$
2. $x^2 + 12 = 8x$ $2, 6$
3. $4x^2 + 11x - 3 = 0$ $-3, \frac{1}{4}$
4. $x - 9 = x(x - 5)$ 3

Solve for x by factoring.
5. $x^2 + 12ax + 32a^2 = 0$ $-4a, -8a$
6. $2x^2 - 7cx + 3c^2 = 0$ $3c, \frac{1}{2}c$

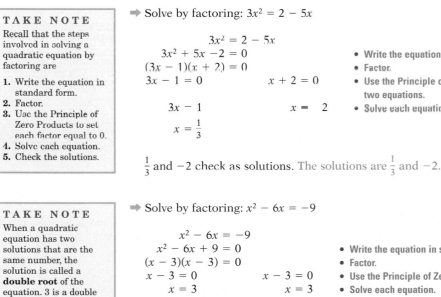

Instructor Note

If you ask your students to write an equation such as $4x = 5x^2 - 3$ in standard form, some students may ask which side to make zero. Show them that it does not matter.

Optional Student Activity

1. For how many integer values of b does the equation $x^2 - bx - 16 = 0$ have integer solutions? Five
2. One solution of the equation $2x^2 - 5x + c = 0$ is 4. What is the other solution? $-\frac{3}{2}$
3. Given $x = t^2 + 2t$, $y = t^3 + t^2$, and $x = 3$, find the possible values of y. $-18, 2$

Objective 10.1B

Concept Check

1. A quadratic equation has two roots, one of which is twice the other. If the sum of the roots is 30, find the quadratic equation and express it in the form $ax^2 + bx + c = 0$, where a, b, and c are integers and $c > 0$. $x^2 - 30x + 200 = 0$
2. Marc and Elena tried to solve the same equation, one of the form $ax^2 + bx + c = 0$. Unfortunately, each person made just one copying error. The solutions of Marc's equation were -2 and $-\frac{3}{2}$ but his value of a was incorrect. The solutions of Elena's equation were 2 and 3 but her value of b was incorrect. Find the solutions of the correct equation. $-6, -1$

Example 1

Solve by factoring: $2x(x - 3) = x + 4$

Solution

$$2x(x - 3) = x + 4$$
$$2x^2 - 6x = x + 4$$
$$2x^2 - 7x - 4 = 0$$
$$(2x + 1)(x - 4) = 0$$
$$2x + 1 = 0 \qquad x - 4 = 0$$
$$2x = -1 \qquad x = 4$$
$$x = -\frac{1}{2}$$

The solutions are $-\frac{1}{2}$ and 4.

You Try It 1

Solve by factoring: $2x^2 = 7x - 3$

Your solution

$\frac{1}{2}, 3$

Example 2

Solve for x by factoring: $x^2 - 4ax - 5a^2 = 0$

Solution

This is a literal equation. Solve for x in terms of a.

$$x^2 - 4ax - 5a^2 = 0$$
$$(x + a)(x - 5a) = 0$$
$$x + a = 0 \qquad x - 5a = 0$$
$$x = -a \qquad x = 5a$$

The solutions are $-a$ and $5a$.

You Try It 2

Solve for x by factoring: $x^2 - 3ax - 4a^2 = 0$

Your solution

$-a, 4a$

Solutions on p. S30

Objective B To write a quadratic equation given its solutions

VIDEO & DVD CD TUTOR WEB SSM

As shown below, the solutions of the equation $(x - r_1)(x - r_2) = 0$ are r_1 and r_2.

$$(x - r_1)(x - r_2) = 0$$
$$x - r_1 = 0 \qquad x - r_2 = 0$$
$$x = r_1 \qquad x = r_2$$

Check:

$(x - r_1)(x - r_2)$	$= 0$	$(x - r_1)(x - r_2)$	$= 0$
$(r_1 - r_1)(r_1 - r_2)$	0	$(r_2 - r_1)(r_2 - r_2)$	0
$0 \cdot (r_1 - r_2)$	0	$(r_2 - r_1) \cdot 0$	0
$0 = 0$		$0 = 0$	

Using the equation $(x - r_1)(x - r_2) = 0$ and the fact that r_1 and r_2 are solutions of this equation, it is possible to write a quadratic equation given its solutions.

➡ Write a quadratic equation that has solutions 4 and -5.

$$(x - r_1)(x - r_2) = 0$$
$$(x - 4)[x - (-5)] = 0 \qquad \text{• Replace } r_1 \text{ by 4 and } r_2 \text{ by } -5.$$
$$(x - 4)(x + 5) = 0 \qquad \text{• Simplify.}$$
$$x^2 + x - 20 = 0 \qquad \text{• Multiply.}$$

In-Class Examples (Objective 10.1B)

Write a quadratic equation that has integer coefficients and has as solutions the given pair of numbers.

1. -1 and 5 $x^2 - 4x - 5 = 0$
2. -3 and 0 $x^2 + 3x = 0$
3. $\frac{1}{3}$ and 2 $3x^2 - 7x + 2 = 0$
4. $-\frac{4}{5}$ and $\frac{2}{3}$ $15x^2 + 2x - 8 = 0$

547

➡ Write a quadratic equation with integer coefficients and solutions $\frac{2}{3}$ and $\frac{1}{2}$.

$$(x - r_1)(x - r_2) = 0$$

$$\left(x - \frac{2}{3}\right)\left(x - \frac{1}{2}\right) = 0$$ • Replace r_1 by $\frac{2}{3}$ and r_2 by $\frac{1}{2}$.

$$x^2 - \frac{7}{6}x + \frac{1}{3} = 0$$ • Multiply.

$$6\left(x^2 - \frac{7}{6}x + \frac{1}{3}\right) = 6 \cdot 0$$ • Multiply each side of the equation by the LCM of the denominators.

$$6x^2 - 7x + 2 = 0$$

Example 3

Write a quadratic equation with integer coefficients and solutions $\frac{1}{2}$ and -4.

Solution

$$(x - r_1)(x - r_2) = 0$$

$$\left(x - \frac{1}{2}\right)[x - (-4)] = 0$$

$$\left(x - \frac{1}{2}\right)(x + 4) = 0$$

$$x^2 + \frac{7}{2}x - 2 = 0$$

$$2\left(x^2 + \frac{7}{2}x - 2\right) = 2 \cdot 0$$

$$2x^2 + 7x - 4 = 0$$

You Try It 3

Write a quadratic equation with integer coefficients and solutions 3 and $-\frac{1}{2}$.

Your solution

$2x^2 - 5x - 3 = 0$

Solution on p. S30

Optional Student Activity

Here is an extension of the concepts developed in this objective.

$$(x - r_1)(x - r_2) = x^2 - (r_1 + r_2)x + r_1r_2$$

Thus the coefficient of x is the opposite of the sum of the roots, and the constant term is the product of the roots. This result can be used to check the solutions of a quadratic equation.

1. Check that $1 + \sqrt{2}$ and $1 - \sqrt{2}$ are solutions of the equation $x^2 - 2x - 1 = 0$.
   $-[(1 + \sqrt{2}) + (1 - \sqrt{2})] = -2$, the coefficient of x; $(1 + \sqrt{2})(1 - \sqrt{2}) = -1$, the constant term
2. Check that $2 + \sqrt{3}$ and $2 - \sqrt{3}$ are solutions of the equation $x^2 - 4x + 1 = 0$.
   $-[(2 + \sqrt{3}) + (2 - \sqrt{3})] = -4$, the coefficient of x; $(2 + \sqrt{3})(2 - \sqrt{3}) = 1$, the constant term

Objective C — To solve a quadratic equation by taking square roots

VIDEO & DVD · CD TUTOR · WEB · SSM

Objective 10.1C

New Symbols

$\pm a$

The solution of the quadratic equation $x^2 = 16$ is shown at the right.

$$x^2 = 16$$
$$x^2 - 16 = 0$$
$$(x - 4)(x + 4) = 0$$
$$x - 4 = 0 \qquad x + 4 = 0$$
$$x = 4 \qquad x = -4$$

The solutions can also be found by taking the square root of each side of the equation and writing the positive and negative square roots of the number. The notation $x = \pm 4$ means $x = 4$ or $x = -4$.

$$x^2 = 16$$
$$\sqrt{x^2} = \sqrt{16}$$
$$x = \pm 4$$

The solutions are 4 and -4.

➡ Solve by taking square roots: $3x^2 = 54$

$$3x^2 = 54$$

$$x^2 = 18$$ • Solve for x^2.

$$\sqrt{x^2} = \sqrt{18}$$ • Take the square root of each side of the equation.

$$x = \pm\sqrt{18} = \pm 3\sqrt{2}$$ • Simplify.

The solutions are $3\sqrt{2}$ and $-3\sqrt{2}$. • $3\sqrt{2}$ and $-3\sqrt{2}$ check as solutions.

In-Class Examples (Objective 10.1C)

Solve by taking square roots.

1. $4x^2 - 25 = 0$ $-\frac{5}{2}, \frac{5}{2}$
2. $y^2 + 4 = 0$ $-2i, 2i$
3. $v^2 - 12 = 0$ $2\sqrt{3}, -2\sqrt{3}$
4. $z^2 + 8 = 0$ $2i\sqrt{2}, -2i\sqrt{2}$
5. $(x + 1)^2 = 9$ $-4, 2$
6. $3(y - 3)^2 = 12$ $1, 5$
7. $(x + 3)^2 - 8 = 0$ $-3 + 2\sqrt{2}, -3 - 2\sqrt{2}$
8. $\left(t - \frac{1}{4}\right)^2 - 32 = 0$ $\frac{1 + 16\sqrt{2}}{4}, \frac{1 - 16\sqrt{2}}{4}$

Instructor Note

Another way to demonstrate that $x^2 = a$ implies $x = \pm\sqrt{a}$ is to rely on the fact that $\sqrt{x^2} = |x|$. Thus $x^2 = a$ implies $|x| = \sqrt{a}$, or $x = \pm\sqrt{a}$.

Discuss the Concepts

1. What are the solutions of $x^2 = 25$, and why?
2. If $x^2 = 16$, then $\sqrt{x^2} = \sqrt{16}$ and therefore $x = \pm 4$. Explain why the $\pm$ sign is necessary.
3. Given that $x^{400} = 400^{400}$ and $x \neq 400$, what is the value of x? -400

Concept Check

1. The equation $d = 0.071v^2$ can be used to approximate the distance d, in feet, a car traveling v miles per hour will skid before stopping after its brakes are applied. An officer investigating an auto accident noted that the vehicle involved required 144 ft to stop. At what speed was the vehicle traveling before its brakes were applied? Round to the nearest whole number. 45 mph
2. The value P of an initial investment of A dollars after 2 years is given by $P = A(1 + r)^2$, where r is the annual percentage rate earned by the investment. If an initial investment of \$5000 grew to a value of \$5724.50 in 2 years, what was the annual percentage rate? 7%

Solving a quadratic equation by taking the square root of each side of the equation can lead to solutions that are complex numbers.

➡ Solve by taking square roots: $2x^2 + 18 = 0$

$$\begin{aligned} 2x^2 + 18 &= 0 \\ 2x^2 &= -18 && \bullet \text{ Solve for } x^2. \\ x^2 &= -9 \\ \sqrt{x^2} &= \sqrt{-9} && \bullet \text{ Take the square root of each side of the equation.} \\ x &= \pm\sqrt{-9} && \bullet \text{ Simplify.} \\ x &= \pm 3i \end{aligned}$$

Check:

$2x^2 + 18 = 0$		$2x^2 + 18 = 0$	
$2(3i)^2 + 18$	0	$2(-3i)^2 + 18$	0
$2(-9) + 18$	0	$2(-9) + 18$	0
$-18 + 18$	0	$-18 + 18$	0
$0 = 0$		$0 = 0$	

The solutions are $3i$ and $-3i$.

An equation containing the square of a binomial can be solved by taking square roots.

➡ Solve by taking square roots: $(x + 2)^2 - 24 = 0$

$$\begin{aligned} (x + 2)^2 - 24 &= 0 \\ (x + 2)^2 &= 24 && \bullet \text{ Solve for } (x + 2)^2. \\ \sqrt{(x + 2)^2} &= \sqrt{24} && \bullet \text{ Take the square root of each side of the equation. Then simplify.} \\ x + 2 &= \pm\sqrt{24} \\ x + 2 &= \pm 2\sqrt{6} \end{aligned}$$

$$x + 2 = 2\sqrt{6} \qquad x + 2 = -2\sqrt{6} \qquad \bullet \text{ Solve for } x.$$
$$x = -2 + 2\sqrt{6} \qquad x = -2 - 2\sqrt{6}$$

The solutions are $-2 + 2\sqrt{6}$ and $-2 - 2\sqrt{6}$.

Example 4

Solve by taking square roots: $3(x - 2)^2 + 12 = 0$

Solution

$$\begin{aligned} 3(x - 2)^2 + 12 &= 0 \\ 3(x - 2)^2 &= -12 \\ (x - 2)^2 &= -4 \\ \sqrt{(x - 2)^2} &= \sqrt{-4} \\ x - 2 &= \pm\sqrt{-4} \\ x - 2 &= \pm 2i \end{aligned}$$

$$x - 2 = 2i \qquad x - 2 = -2i$$
$$x = 2 + 2i \qquad x = 2 - 2i$$

The solutions are $2 + 2i$ and $2 - 2i$.

You Try It 4

Solve by taking square roots: $2(x + 1)^2 - 24 = 0$

Your solution

$-1 + 2\sqrt{3}, -1 - 2\sqrt{3}$

Solution on p. S30

549

10.1 Exercises

Objective A

1. Explain why the restriction $a \neq 0$ is necessary in the definition of a quadratic equation.

2. What does the Principle of Zero Products state? How is it used to solve a quadratic equation?

Write the quadratic equation in standard form with the coefficient of x^2 positive. Give the values of a, b, and c.

3. $2x^2 - 4x = 5$
$2x^2 - 4x - 5 = 0$
$a = 2, b = -4, c = -5$

4. $x^2 = 3x + 1$
$x^2 - 3x - 1 = 0$
$a = 1, b = -3, c = -1$

5. $5x = 4x^2 + 6$
$4x^2 - 5x + 6 = 0$
$a = 4, b = -5, c = 6$

6. $3x^2 = 7$
$3x^2 - 7 = 0$
$a = 3, b = 0, c = -7$

Solve by factoring.

7. $x^2 - 4x = 0$
0, 4

8. $y^2 + 6y = 0$
0, −6

9. $t^2 - 25 = 0$
5, −5

10. $p^2 - 81 = 0$
9, −9

11. $s^2 - s - 6 = 0$
3, −2

12. $v^2 + 4v - 5 = 0$
1, −5

13. $y^2 - 6y + 9 = 0$
3

14. $x^2 + 10x + 25 = 0$
−5

15. $9z^2 - 18z = 0$
0, 2

16. $4y^2 + 20y = 0$
0, −5

17. $r^2 - 3r = 10$
5, −2

18. $p^2 + 5p = 6$
1, −6

19. $v^2 + 10 = 7v$
2, 5

20. $t^2 - 16 = 15t$
16, −1

21. $2x^2 - 9x - 18 = 0$
$6, -\frac{3}{2}$

22. $3y^2 - 4y - 4 = 0$
$-\frac{2}{3}, 2$

23. $4z^2 - 9z + 2 = 0$
$2, \frac{1}{4}$

24. $2s^2 - 9s + 9 = 0$
$\frac{3}{2}, 3$

25. $3w^2 + 11w = 4$
$\frac{1}{3}, -4$

26. $2r^2 + r = 6$
$\frac{3}{2}, -2$

27. $6x^2 = 23x + 18$
$\frac{9}{2}, -\frac{2}{3}$

28. $6x^2 = 7x - 2$
$\frac{2}{3}, \frac{1}{2}$

29. $4 - 15u - 4u^2 = 0$
$\frac{1}{4}, -4$

30. $3 - 2y - 8y^2 = 0$
$-\frac{3}{4}, \frac{1}{2}$

31. $x + 18 = x(x - 6)$
9, −2

32. $t + 24 = t(t + 6)$
3, −8

33. $4s(s + 3) = s - 6$
$-2, -\frac{3}{4}$

Section 10.1

Suggested Assignment

Exercises 1–37, every other odd
Exercises 39–51, odds
Exercises 53–81, every other odd
Exercises 83–111, odds
More challenging problems:
Exercises 113–121, odds

Answers to Writing Exercises

1. If $a = 0$ in the equation $ax^2 + bx + c = 0$, then there is no second-degree term in the equation, and it is therefore not a quadratic equation.
2. The Principle of Zero Products states that if the product of two factors is zero, then at least one of the factors must be zero. It is used to solve a quadratic equation of the form $ax^2 + bx + c = 0$ in which $ax^2 + bx + c$ can be written as a product of factors.

Quick Quiz (Objective 10.1A)

Solve by factoring.

1. $x^2 + 3x = 4$ −4, 1
2. $x^2 - 12 = 11x$ −1, 12
3. $3x^2 - 10x - 8 = 0$ $-\frac{2}{3}, 4$
4. $x + 3 = x(x + 3)$ −3, 1

Solve for x by factoring.

5. $x^2 - 5bx + 6b^2 = 0$ $2b, 3b$
6. $3x^2 + 8ax + 4a^2 = 0$ $-2a, -\frac{2}{3}a$

550 Chapter 10 / Quadratic Equations

34. $3v(v - 2) = 11v + 6$
$-\frac{1}{3}, 6$

35. $u^2 - 2u + 4 = (2u - 3)(u + 2)$
$2, -5$

36. $(3v - 2)(2v + 1) = 3v^2 - 11v - 10$
$-2, -\frac{4}{3}$

37. $(3x - 4)(x + 4) = x^2 - 3x - 28$
$-4, -\frac{3}{2}$

Solve for x by factoring.

38. $x^2 + 14ax + 48a^2 = 0$
$-6a, -8a$

39. $x^2 - 9bx + 14b^2 = 0$
$2b, 7b$

40. $x^2 + 9xy - 36y^2 = 0$
$-12y, 3y$

41. $x^2 - 6cx - 7c^2 = 0$
$7c, -c$

42. $x^2 - ax - 20a^2 = 0$
$-4a, 5a$

43. $2x^2 + 3bx + b^2 = 0$
$-\frac{b}{2}, -b$

44. $3x^2 - 4cx + c^2 = 0$
$\frac{c}{3}, c$

45. $3x^2 - 14ax + 8a^2 = 0$
$4a, \frac{2a}{3}$

46. $3x^2 - 11xy + 6y^2 = 0$
$\frac{2y}{3}, 3y$

47. $3x^2 - 8ax - 3a^2 = 0$
$-\frac{a}{3}, 3a$

48. $3x^2 - 4bx - 4b^2 = 0$
$-\frac{2b}{3}, 2b$

49. $4x^2 + 8xy + 3y^2 = 0$
$-\frac{3y}{2}, -\frac{y}{2}$

50. $6x^2 - 11cx + 3c^2 = 0$
$\frac{c}{3}, \frac{3c}{2}$

51. $6x^2 + 11ax + 4a^2 = 0$
$-\frac{a}{2}, -\frac{4a}{3}$

52. $12x^2 - 5xy - 2y^2 = 0$
$-\frac{y}{4}, \frac{2y}{3}$

Objective B

Write a quadratic equation that has integer coefficients and has as solutions the given pair of numbers.

53. 2 and 5
$x^2 - 7x + 10 = 0$

54. 3 and 1
$x^2 - 4x + 3 = 0$

55. −2 and −4
$x^2 + 6x + 8 = 0$

56. −1 and −3
$x^2 + 4x + 3 = 0$

57. 6 and −1
$x^2 - 5x - 6 = 0$

58. −2 and 5
$x^2 - 3x - 10 = 0$

59. 3 and −3
$x^2 - 9 = 0$

60. 5 and −5
$x^2 - 25 = 0$

61. 4 and 4
$x^2 - 8x + 16 = 0$

62. 2 and 2
$x^2 - 4x + 4 = 0$

63. 0 and 5
$x^2 - 5x = 0$

64. 0 and −2
$x^2 + 2x = 0$

Quick Quiz (Objective 10.1B)

Write a quadratic equation that has integer coefficients and has as solutions the given pair of numbers.

1. −3 and 1 $x^2 + 2x - 3 = 0$
2. −4 and 0 $x^2 + 4x = 0$
3. $-\frac{2}{3}$ and 2 $3x^2 - 4x - 4 = 0$
4. $-\frac{1}{8}$ and $\frac{2}{5}$ $40x^2 - 11x - 2 = 0$

551

65. 0 and 3
$x^2 - 3x = 0$

66. 0 and -1
$x^2 + x = 0$

67. 3 and $\frac{1}{2}$
$2x^2 - 7x + 3 = 0$

68. 2 and $\frac{2}{3}$
$3x^2 - 8x + 4 = 0$

69. $-\frac{3}{4}$ and 2
$4x^2 - 5x - 6 = 0$

70. $-\frac{1}{2}$ and 5
$2x^2 - 9x - 5 = 0$

71. $-\frac{5}{3}$ and -2
$3x^2 + 11x + 10 = 0$

72. $-\frac{3}{2}$ and -1
$2x^2 + 5x + 3 = 0$

73. $-\frac{2}{3}$ and $\frac{2}{3}$
$9x^2 - 4 = 0$

74. $-\frac{1}{2}$ and $\frac{1}{2}$
$4x^2 - 1 = 0$

75. $\frac{1}{2}$ and $\frac{1}{3}$
$6x^2 - 5x + 1 = 0$

76. $\frac{3}{4}$ and $\frac{2}{3}$
$12x^2 - 17x + 6 = 0$

77. $\frac{6}{5}$ and $-\frac{1}{2}$
$10x^2 - 7x - 6 = 0$

78. $\frac{3}{4}$ and $-\frac{3}{2}$
$8x^2 + 6x - 9 = 0$

79. $-\frac{1}{4}$ and $-\frac{1}{2}$
$8x^2 + 6x + 1 = 0$

80. $-\frac{5}{6}$ and $-\frac{2}{3}$
$18x^2 + 27x + 10 = 0$

81. $\frac{3}{5}$ and $-\frac{1}{10}$
$50x^2 - 25x - 3 = 0$

82. $\frac{7}{2}$ and $-\frac{1}{4}$
$8x^2 - 26x - 7 = 0$

Objective C

Solve by taking square roots.

83. $y^2 = 49$
$7, -7$

84. $x^2 = 64$
$8, \ 8$

85. $z^2 = -4$
$2i, -2i$

86. $v^2 = -16$
$4i, -4i$

87. $s^2 - 4 = 0$
$2, -2$

88. $r^2 - 36 = 0$
$6, -6$

89. $4x^2 - 81 = 0$
$\frac{9}{2}, -\frac{9}{2}$

90. $9x^2 - 16 = 0$
$\frac{4}{3}, -\frac{4}{3}$

91. $y^2 + 49 = 0$
$7i, -7i$

92. $z^2 + 16 = 0$
$4i, -4i$

93. $v^2 - 48 = 0$
$4\sqrt{3}, -4\sqrt{3}$

94. $s^2 - 32 = 0$
$4\sqrt{2}, -4\sqrt{2}$

95. $r^2 - 75 = 0$
$5\sqrt{3}, -5\sqrt{3}$

96. $u^2 - 54 = 0$
$3\sqrt{6}, -3\sqrt{6}$

97. $z^2 + 18 = 0$
$3i\sqrt{2}, -3i\sqrt{2}$

Quick Quiz (Objective 10.1C)
Solve by taking square roots.

1. $9x^2 - 25 = 0$ $-\frac{5}{3}, \frac{5}{3}$
2. $s^2 - 18 = 0$ $3\sqrt{2}, -3\sqrt{2}$
3. $s^2 + 50 = 0$ $5i\sqrt{2}, -5i\sqrt{2}$
4. $2(x + 2)^2 = 18$ $-5, 1$
5. $(t - 2)^2 - 20 = 0$ $2 + 2\sqrt{5}, 2 - 2\sqrt{5}$
6. $\left(y - \frac{3}{5}\right)^2 - 75 = 0$ $\frac{3 + 25\sqrt{3}}{5}, \frac{3 - 25\sqrt{3}}{5}$

Answers to Writing Exercises

120.

$ax^2 + bx = 0, a > 0, b > 0$

$x(ax + b) = 0$

$x = 0 \qquad ax + b = 0$

$ax = -b$

$x = -\frac{b}{a}$

The solutions are 0 and $-\frac{b}{a}$.

121.

$ax^2 + c = 0, a > 0, c > 0$

$ax^2 = -c$

$x^2 = -\frac{c}{a}$

$\sqrt{x^2} = \pm\sqrt{-\frac{c}{a}}$

$x = \pm i\sqrt{\frac{c}{a}}$

$x = \pm i\sqrt{\frac{c}{a}\cdot\frac{a}{a}}$

$x = \pm i\sqrt{\frac{ca}{a^2}}$

$x = \pm i\frac{\sqrt{ca}}{a}$

$x = \pm\frac{\sqrt{ca}}{a}i$

The solutions are $\frac{\sqrt{ca}}{a}i$ and $-\frac{\sqrt{ca}}{a}i$.

98. $t^2 + 27 = 0$
$3i\sqrt{3}, -3i\sqrt{3}$

99. $(x - 1)^2 = 36$
$7, -5$

100. $(x + 2)^2 = 25$
$3, -7$

101. $3(y + 3)^2 = 27$
$0, -6$

102. $4(s - 2)^2 = 36$
$5, -1$

103. $5(z + 2)^2 = 125$
$3, -7$

104. $2(y - 3)^2 = 18$
$6, 0$

105. $\left(v - \frac{1}{2}\right)^2 = \frac{1}{4}$
$1, 0$

106. $\left(r + \frac{2}{3}\right)^2 = \frac{1}{9}$
$-\frac{1}{3}, -1$

107. $(x + 5)^2 - 6 = 0$
$-5 + \sqrt{6}, -5 - \sqrt{6}$

108. $(t - 1)^2 - 15 = 0$
$1 + \sqrt{15}, 1 - \sqrt{15}$

109. $(v - 3)^2 + 45 = 0$
$3 + 3i\sqrt{5}, 3 - 3i\sqrt{5}$

110. $(x + 5)^2 + 32 = 0$
$-5 + 4i\sqrt{2}, -5 - 4i\sqrt{2}$

111. $\left(u - \frac{2}{3}\right)^2 - 18 = 0$
$\frac{2 - 9\sqrt{2}}{3}, \frac{2 + 9\sqrt{2}}{3}$

112. $\left(z - \frac{1}{2}\right)^2 - 20 = 0$
$\frac{1 + 4\sqrt{5}}{2}, \frac{1 - 4\sqrt{5}}{2}$

APPLYING THE CONCEPTS

Write a quadratic equation that has as solutions the given pair of numbers.

113. $\sqrt{2}$ and $-\sqrt{2}$
$x^2 - 2 = 0$

114. $2i$ and $-2i$
$x^2 + 4 = 0$

115. $3\sqrt{2}$ and $-3\sqrt{2}$
$x^2 - 18 = 0$

116. $2i\sqrt{3}$ and $-2i\sqrt{3}$
$x^2 + 12 = 0$

Solve for x.

117. $4a^2x^2 = 36b^2, a > 0, b > 0$
$\frac{3b}{a}, -\frac{3b}{a}$

118. $(x + a)^2 - 4 = 0$
$2 - a, -2 - a$

119. $(2x - 1)^2 = (2x + 3)^2$
$-\frac{1}{2}$

120. Show that the solutions of the equation $ax^2 + bx = 0, a > 0, b > 0$, are 0 and $-\frac{b}{a}$.

121. Show that the solutions of the equation $ax^2 + c = 0, a > 0, c > 0$, are $\frac{\sqrt{ca}}{a}i$ and $-\frac{\sqrt{ca}}{a}i$.

553

10.2 Solving Quadratic Equations by Completing the Square

Objective A

To solve a quadratic equation by completing the square

VIDEO & DVD CD TUTOR WEB SSM

Recall that a perfect-square trinomial is the square of a binomial.

Perfect-Square Trinomial		*Square of a Binomial*
$x^2 + 8x + 16$	=	$(x + 4)^2$
$x^2 - 10x + 25$	=	$(x - 5)^2$
$x^2 + 2ax + a^2$	=	$(x + a)^2$

For each perfect-square trinomial, the square of $\frac{1}{2}$ of the coefficient of x equals the constant term.

$$\left(\frac{1}{2} \text{ coefficient of } x\right)^2 = \text{constant term}$$

$$x^2 + 8x + 16, \quad \left(\frac{1}{2} \cdot 8\right)^2 = 16$$

$$x^2 - 10x + 25, \quad \left[\frac{1}{2}(-10)\right]^2 = 25$$

$$x^2 + 2ax + a^2, \quad \left(\frac{1}{2} \cdot 2a\right)^2 = a^2$$

This relationship can be used to write the constant term for a perfect-square trinomial. Adding to a binomial the constant term that makes it a perfect-square trinomial is called **completing the square.**

Point of Interest

Early attempts to solve quadratic equations were primarily geometric. The Persian mathematician al-Khowarizmi (c. A.D. 800) essentially completed a square of $x^2 + 12x$ as follows.

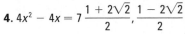

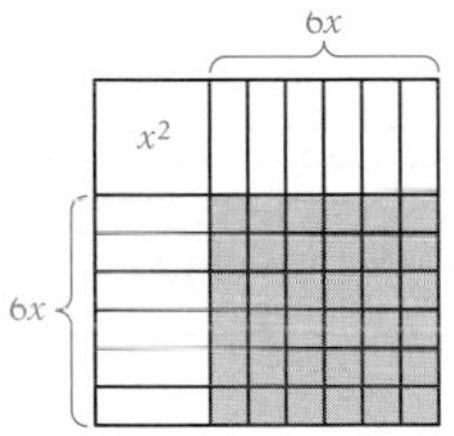

➡ Complete the square on $x^2 + 12x$. Write the resulting perfect-square trinomial as the square of a binomial.

$\left[\frac{1}{2}(12)\right]^2 = (6)^2 = 36$ • Find the constant term.

$x^2 + 12x + 36$ • Complete the square on $x^2 + 12x$ by adding the constant term.

$x^2 + 12x + 36 = (x + 6)^2$ • Write the resulting perfect-square trinomial as the square of a binomial.

➡ Complete the square on $z^2 - 3z$. Write the resulting perfect-square trinomial as the square of a binomial.

$\left[\frac{1}{2} \cdot (-3)\right]^2 = \left(-\frac{3}{2}\right)^2 = \frac{9}{4}$ • Find the constant term.

$z^2 - 3z + \frac{9}{4}$ • Complete the square on $z^2 - 3z$ by adding the constant term.

$z^2 - 3z + \frac{9}{4} = \left(z - \frac{3}{2}\right)^2$ • Write the resulting perfect-square trinomial as the square of a binomial.

Any quadratic equation can be solved by completing the square. Add to each side of the equation the term that completes the square. Rewrite the equation in the form $(x + a)^2 = b$. Then take the square root of each side of the equation.

Objective 10.2A

New Vocabulary

completing the square

New Equations

$\left(\frac{1}{2} \text{ coefficient of } x\right)^2 =$ constant term

Vocabulary to Review

perfect-square trinomial
square of a binomial

Instructor Note

To say that students will find completing the square a daunting task is an understatement. It may be necessary to give quite a few examples before students can do this effectively. The Point of Interest at the left may help some students.

Discuss the Concepts

1. To complete the square on $x^2 + 6x$, what number must be added to the expression? Explain how you arrived at the answer.
2. If you attempt to solve $x^2 - 6x + 9 = 0$ by completing the square, the result after simplifying is $(x - 3)^2 = 0$. Does this mean that the original equation has no solution? If not, what are the solutions of the equation? No. The equation has a double root, 3.
3. What is the next step when using the method of completing the square to solve $x^2 + 6x = 4$?
4. What is the meaning of the phrase *complete the square*?

In-Class Examples (Objective 10.2A)

Solve by completing the square.

1. $x^2 - 6x - 7 = 0$ $-1, 7$
2. $x^2 + 2x - 5 = 0$ $-1 + \sqrt{6}, -1 - \sqrt{6}$
3. $x^2 - 4x + 6 = 0$ $2 + i\sqrt{2}, 2 - i\sqrt{2}$
4. $4x^2 - 4x = 7$ $\frac{1 + 2\sqrt{2}}{2}, \frac{1 - 2\sqrt{2}}{2}$

Concept Check

1. What number is equal to one less than its square? $\frac{1+\sqrt{5}}{2}$ or $\frac{1-\sqrt{5}}{2}$
2. The equation $x^2 - 7x + 11 = 0$ has solutions r_1 and r_2. Find the value of $r_1^2 + r_2^2$. 27

Instructor Note

The Optional Student Activity on page 547 presented an alternative method of checking the solutions of a quadratic equation. You may want to use that method to check the solutions to this equation.

Optional Student Activity

1. If r_1 and r_2 are the two solutions of $x^2 - 5x + 3 = 0$, find the value of $(r_1 + r_2)(r_1r_2)$. 15
2. You have been hired by an observatory to track meteorites and determine whether they will strike Earth. The equation of Earth's path is $x^2 + y^2 = 40$. The first meteorite you observe is moving along a path whose equation is $18x - y^2 = -144$. Will the meteorite strike Earth? No. When $18x + 144$ is substituted for y^2 in the equation $x^2 + y^2 = 40$, the solutions of the resulting equation are complex numbers.

➡ Solve by completing the square: $x^2 - 6x - 15 = 0$

$x^2 - 6x - 15 = 0$

$x^2 - 6x = 15$ • **Add 15 to each side of the equation.**

$x^2 - 6x + 9 = 15 + 9$ • **Complete the square. Add $\left[\frac{1}{2}(-6)\right]^2 = (-3)^2 = 9$ to each side of the equation.**

$(x-3)^2 = 24$ • **Factor the perfect-square trinomial.**

$\sqrt{(x-3)^2} = \sqrt{24}$ • **Take the square root of each side of the equation.**

$x - 3 = \pm\sqrt{24}$ • **Solve for *x*.**

$x - 3 = \pm 2\sqrt{6}$

$x - 3 = 2\sqrt{6}$ $\quad$ $x - 3 = -2\sqrt{6}$

$x = 3 + 2\sqrt{6}$ $\quad$ $x = 3 - 2\sqrt{6}$

Check:

$x^2 - 6x - 15 = 0$	
$(3+2\sqrt{6})^2 - 6(3+2\sqrt{6}) - 15$	0
$9 + 12\sqrt{6} + 24 - 18 - 12\sqrt{6} - 15$	0
$0 = 0$	

$x^2 - 6x - 15 = 0$	
$(3-2\sqrt{6})^2 - 6(3-2\sqrt{6}) - 15$	0
$9 - 12\sqrt{6} + 24 - 18 + 12\sqrt{6} - 15$	0
$0 = 0$	

The solutions are $3 + 2\sqrt{6}$ and $3 - 2\sqrt{6}$.

TAKE NOTE
The exact solutions of the equation $x^2 - 6x - 15 = 0$ are $3 + 2\sqrt{6}$ and $3 - 2\sqrt{6}$. 7.899 and −1.899 are approximate solutions of the equation.

In the example above, the solutions of the equation $x^2 - 6x - 15 = 0$ are $3 + 2\sqrt{6}$ and $3 - 2\sqrt{6}$. These are the exact solutions. However, in some situations it may be preferable to have decimal approximations of the solutions to a quadratic equation. Approximate solutions can be found by using a calculator and then rounding to the desired degree of accuracy.

$$3 + 2\sqrt{6} \approx 7.899 \quad \text{and} \quad 3 - 2\sqrt{6} \approx -1.899$$

To the nearest thousandth, the solutions of the equation $x^2 - 6x - 15 = 0$ are 7.899 and −1.899.

➡ Solve $2x^2 - x - 2 = 0$ by completing the square. Find the exact solutions and approximate the solutions to the nearest thousandth.

In order for us to complete the square on an expression, the coefficient of the squared term must be 1. After adding the opposite of the constant term to each side of the equation, multiply each side of the equation by $\frac{1}{2}$.

$2x^2 - x - 2 = 0$

$2x^2 - x = 2$ • **Add 2 to each side of the equation.**

$\frac{1}{2}(2x^2 - x) = \frac{1}{2} \cdot 2$ • **Multiply each side of the equation by $\frac{1}{2}$.**

555

TAKE NOTE
This example illustrates all the steps required to solve a quadratic equation by completing the square.

1. Write the equation in the form $ax^2 + bx = -c$.
2. Multiply both sides of the equation by $\frac{1}{a}$.
3. Complete the square on $x^2 + \frac{b}{a}x$. Add the number that completes the square to both sides of the equation.
4. Factor the perfect-square trinomial.
5. Take the square root of each side of the equation.
6. Solve the resulting equation for x.
7. Check the solutions.

$$x^2 - \frac{1}{2}x = 1$$

- **The coefficient of x^2 is now 1.**

$$x^2 - \frac{1}{2}x + \frac{1}{16} = 1 + \frac{1}{16}$$

- **Complete the square. Add $\left[\frac{1}{2}\left(-\frac{1}{2}\right)\right]^2 = \left(-\frac{1}{4}\right)^2 = \frac{1}{16}$ to each side of the equation.**

$$\left(x - \frac{1}{4}\right)^2 = \frac{17}{16}$$

- **Factor the perfect-square trinomial.**

$$\sqrt{\left(x - \frac{1}{4}\right)^2} = \sqrt{\frac{17}{16}}$$

- **Take the square root of each side of the equation.**

$$x - \frac{1}{4} = \pm\frac{\sqrt{17}}{4}$$

- **Solve for x.**

$$x - \frac{1}{4} = \frac{\sqrt{17}}{4} \qquad x - \frac{1}{4} = -\frac{\sqrt{17}}{4}$$

$$x = \frac{1}{4} + \frac{\sqrt{17}}{4} \qquad x = \frac{1}{4} - \frac{\sqrt{17}}{4}$$

The exact solutions are $\frac{1 + \sqrt{17}}{4}$ and $\frac{1 - \sqrt{17}}{4}$.

$$\frac{1 + \sqrt{17}}{4} \approx 1.281 \qquad \frac{1 - \sqrt{17}}{4} \approx -0.781$$

To the nearest thousandth, the solutions are 1.281 and −0.781.

Point of Interest

Mathematicians have studied quadratic equations for centuries. Many of the initial equations were a result of trying to solve a geometry problem. One of the most famous, which dates from around 500 B.C., is "squaring the circle." The question was "Is it possible to construct a square whose area is that of a given circle?" For these early mathematicians, to *construct* meant to draw with only a straightedge and a compass. It was approximately 2300 years later that mathematicians were able to prove that such a construction is impossible.

➡ Solve by completing the square: $x^2 - 3x + 5 = 0$

$$x^2 - 3x + 5 = 0$$

$$x^2 - 3x = -5$$

- **Add the opposite of the constant term to each side of the equation.**

$$x^2 - 3x + \frac{9}{4} = -5 + \frac{9}{4}$$

- **Add to each side of the equation the term that completes the square on $x^2 - 3x$.**

$$\left(x - \frac{3}{2}\right)^2 = -\frac{11}{4}$$

- **Factor the perfect-square trinomial.**

$$\sqrt{\left(x - \frac{3}{2}\right)^2} = \sqrt{-\frac{11}{4}}$$

- **Take the square root of each side of the equation.**

$$x - \frac{3}{2} = \pm\frac{i\sqrt{11}}{2}$$

- **Simplify.**

$$x - \frac{3}{2} = \frac{i\sqrt{11}}{2} \qquad x - \frac{3}{2} = -\frac{i\sqrt{11}}{2}$$

- **Solve for x.**

$$x = \frac{3}{2} + \frac{i\sqrt{11}}{2} \qquad x = \frac{3}{2} - \frac{i\sqrt{11}}{2}$$

$\frac{3}{2} + \frac{\sqrt{11}}{2}i$ and $\frac{3}{2} - \frac{\sqrt{11}}{2}i$ check as solutions.

The solutions are $\frac{3}{2} + \frac{\sqrt{11}}{2}i$ and $\frac{3}{2} - \frac{\sqrt{11}}{2}i$.

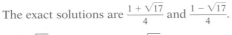

Example 1
Solve by completing the square:
$4x^2 - 8x + 1 = 0$

Solution

$$4x^2 - 8x + 1 = 0$$
$$4x^2 - 8x = -1$$
$$\frac{1}{4}(4x^2 - 8x) = \frac{1}{4}(-1)$$

• The coefficient of x^2 must be 1.

$$x^2 - 2x = -\frac{1}{4}$$
$$x^2 - 2x + 1 = -\frac{1}{4} + 1$$

• Complete the square.

$$(x - 1)^2 = \frac{3}{4}$$
$$\sqrt{(x - 1)^2} = \sqrt{\frac{3}{4}}$$
$$x - 1 = \pm\frac{\sqrt{3}}{2}$$

$$x - 1 = \frac{\sqrt{3}}{2} \qquad x - 1 = -\frac{\sqrt{3}}{2}$$
$$x = 1 + \frac{\sqrt{3}}{2} \qquad x = 1 - \frac{\sqrt{3}}{2}$$
$$= \frac{2 + \sqrt{3}}{2} \qquad = \frac{2 - \sqrt{3}}{2}$$

The solutions are $\frac{2 + \sqrt{3}}{2}$ and $\frac{2 - \sqrt{3}}{2}$.

You Try It 1
Solve by completing the square:
$4x^2 - 4x - 1 = 0$

Your solution
$\frac{1 + \sqrt{2}}{2}, \frac{1 - \sqrt{2}}{2}$

Example 2
Solve by completing the square:
$x^2 + 4x + 5 = 0$

Solution

$$x^2 + 4x + 5 = 0$$
$$x^2 + 4x = -5$$
$$x^2 + 4x + 4 = -5 + 4$$

• Complete the square.

$$(x + 2)^2 = -1$$
$$\sqrt{(x + 2)^2} = \sqrt{-1}$$
$$x + 2 = \pm i$$

$$x + 2 = i \qquad x + 2 = -i$$
$$x = -2 + i \qquad x = -2 - i$$

The solutions are $-2 + i$ and $-2 - i$.

You Try It 2
Solve by completing the square:
$2x^2 + x - 5 = 0$

Your solution
$\frac{-1 + \sqrt{41}}{4}, \frac{-1 - \sqrt{41}}{4}$

Solutions on p. S30

10.2 Exercises

Objective A

1. What is the meaning of the phrase *complete the square*?

2. Write out the steps for solving a quadratic equation by completing the square.

Solve by completing the square.

3. $x^2 - 4x - 5 = 0$
   $5, -1$

4. $y^2 + 6y + 5 = 0$
   $-1, -5$

5. $v^2 + 8v - 9 = 0$
   $-9, 1$

6. $w^2 - 2w - 24 = 0$
   $6, -4$

7. $z^2 - 6z + 9 = 0$
   3

8. $u^2 + 10u + 25 = 0$
   -5

9. $r^2 + 4r - 7 = 0$
   $-2 + \sqrt{11}, -2 - \sqrt{11}$

10. $s^2 + 6s - 1 = 0$
    $-3 + \sqrt{10}, -3 - \sqrt{10}$

11. $x^2 - 6x + 7 = 0$
    $3 + \sqrt{2}, 3 - \sqrt{2}$

12. $y^2 + 8y + 13 = 0$
    $-4 + \sqrt{3}, -4 - \sqrt{3}$

13. $z^2 - 2z + 2 = 0$
    $1 + i, 1 - i$

14. $t^2 - 4t + 8 = 0$
    $2 + 2i, 2 - 2i$

15. $s^2 - 5s - 24 = 0$
    $8, -3$

16. $v^2 + 7v - 44 = 0$
    $4, -11$

17. $x^2 + 5x - 36 = 0$
    $4, -9$

18. $y^2 - 9y + 20 = 0$
    $5, 4$

19. $p^2 - 3p + 1 = 0$
    $\frac{3 + \sqrt{5}}{2}, \frac{3 - \sqrt{5}}{2}$

20. $r^2 - 5r - 2 = 0$
    $\frac{5 + \sqrt{33}}{2}, \frac{5 - \sqrt{33}}{2}$

21. $t^2 - t - 1 = 0$
    $\frac{1 + \sqrt{5}}{2}, \frac{1 - \sqrt{5}}{2}$

22. $u^2 - u - 7 = 0$
    $\frac{1 + \sqrt{29}}{2}, \frac{1 - \sqrt{29}}{2}$

23. $y^2 - 6y = 4$
    $3 + \sqrt{13}, 3 - \sqrt{13}$

24. $w^2 + 4w = 2$
    $-2 + \sqrt{6}, -2 - \sqrt{6}$

25. $x^2 = 8x - 15$
    $5, 3$

26. $z^2 = 4z - 3$
    $3, 1$

27. $v^2 = 4v - 13$
    $2 + 3i, 2 - 3i$

28. $x^2 = 2x - 17$
    $1 + 4i, 1 - 4i$

29. $p^2 + 6p = -13$
    $-3 + 2i, -3 - 2i$

30. $x^2 + 4x = -20$
    $-2 + 4i, -2 - 4i$

31. $y^2 - 2y = 17$
    $1 + 3\sqrt{2}, 1 - 3\sqrt{2}$

32. $x^2 + 10x = 7$
    $-5 + 4\sqrt{2}, -5 - 4\sqrt{2}$

Section 10.2

Suggested Assignment

Exercises 3–55, odds
More challenging problems:
Exercises 57–60, odds

Answers to Writing Exercises

1. *Complete the square* means to add to a binomial the constant term that makes it a perfect-square trinomial.
2. (1) Write the equation in the form $x^2 + kx = c$.
   (2) Find the constant term that will complete the square on $x^2 + kx$; it is equal to $\left(\frac{1}{2} \cdot k\right)^2$.
   (3) Complete the square on the left side of the equation by adding the constant term found in Step 2 to both sides of the equation.
   (4) Factor the trinomial. Write it as the square of a binomial.
   (5) Take the square root of each side of the equation.
   (6) Solve the equation for x.

Quick Quiz (Objective 10.2A)

Solve by completing the square.

1. $x^2 + 4x + 3 = 0$ $-3, -1$
2. $x^2 + 4x - 1 = 0$ $-2 + \sqrt{5}, -2 - \sqrt{5}$
3. $x^2 + 6x + 10 = 0$ $-3 + i, -3 - i$
4. $2x^2 = 5 - x$ $\frac{-1 + \sqrt{41}}{4}, \frac{-1 - \sqrt{41}}{4}$

33. $z^2 = z + 4$
$\frac{1+\sqrt{17}}{2}, \frac{1-\sqrt{17}}{2}$

34. $r^2 = 3r - 1$
$\frac{3+\sqrt{5}}{2}, \frac{3-\sqrt{5}}{2}$

35. $x^2 + 13 = 2x$
$1 + 2i\sqrt{3}, 1 - 2i\sqrt{3}$

36. $6v^2 - 7v = 3$
$\frac{3}{2}, -\frac{1}{3}$

37. $4x^2 - 4x + 5 = 0$
$\frac{1}{2} + i, \frac{1}{2} - i$

38. $4t^2 - 4t + 17 = 0$
$\frac{1}{2} + 2i, \frac{1}{2} - 2i$

39. $9x^2 - 6x + 2 = 0$
$\frac{1}{3} + \frac{1}{3}i, \frac{1}{3} - \frac{1}{3}i$

40. $9y^2 - 12y + 13 = 0$
$\frac{2}{3} + i, \frac{2}{3} - i$

41. $2s^2 = 4s + 5$
$\frac{2+\sqrt{14}}{2}, \frac{2-\sqrt{14}}{2}$

42. $3u^2 = 6u + 1$
$\frac{3+2\sqrt{3}}{3}, \frac{3-2\sqrt{3}}{3}$

43. $2r^2 = 3 - r$
$1, -\frac{3}{2}$

44. $2x^2 = 12 - 5x$
$\frac{3}{2}, -4$

45. $y - 2 = (y - 3)(y + 2)$
$1 + \sqrt{5}, 1 - \sqrt{5}$

46. $8s - 11 = (s - 4)(s - 2)$
$7 + \sqrt{30}, 7 - \sqrt{30}$

47. $6t - 2 = (2t - 3)(t - 1)$
$\frac{1}{2}, 5$

48. $2z + 9 = (2z + 3)(z + 2)$
$\frac{1}{2}, -3$

49. $(x - 4)(x + 1) = x - 3$
$2 + \sqrt{5}, 2 - \sqrt{5}$

50. $(y - 3)^2 = 2y + 10$
$4 + \sqrt{17}, 4 - \sqrt{17}$

Solve by completing the square. Approximate the solutions to the nearest thousandth.

51. $z^2 + 2z = 4$
1.236, −3.236

52. $t^2 - 4t = 7$
5.317, −1.317

53. $2x^2 = 4x - 1$
1.707, 0.293

54. $3y^2 = 5y - 1$
1.434, 0.232

55. $4z^2 + 2z = 1$
0.309, −0.809

56. $4w^2 - 8w = 3$
2.323, −0.323

APPLYING THE CONCEPTS

Solve for x by completing the square.

57. $x^2 - ax - 2a^2 = 0$
$x = 2a, x = -a$

58. $x^2 + 3ax - 4a^2 = 0$
$x = a, x = -4a$

59. $x^2 + 3ax - 10a^2 = 0$
$x = 2a, x = -5a$

60. When a baseball is hit, the height h, in feet, of the ball above the ground t seconds after it is hit can be approximated by the equation $h = -16t^2 + 70t + 4$. The distance s, in feet, that the ball is from home plate t seconds after it is hit can be approximated by the equation $s = 44.5t$.

a. Determine when the ball will hit the ground. Round to the nearest hundredth. (*Hint*: The ball hits the ground when $h = 0$.) 4.43 s

b. Determine whether the ball will clear a 6-foot fence 325 ft from home plate. No. The ball will have gone only 197.2 ft when it hits the ground.

559

10.3 Solving Quadratic Equations by Using the Quadratic Formula

Objective A

To solve a quadratic equation by using the quadratic formula

VIDEO & DVD CD TUTOR WWW WEB SSM

A general formula known as the **quadratic formula** can be derived by applying the method of completing the square to the standard form of a quadratic equation. This formula can be used to solve any quadratic equation. The equation $ax^2 + bx + c = 0$ is solved by completing the square as follows.

$$ax^2 + bx + c = 0$$

Add the opposite of the constant term to each side of the equation.

$$ax^2 + bx + c + (-c) = 0 + (-c)$$

$$ax^2 + bx = -c$$

Multiply each side of the equation by the reciprocal of a, the coefficient of x^2.

$$\frac{1}{a}(ax^2 + bx) = \frac{1}{a}(-c)$$

$$x^2 + \frac{b}{a}x = -\frac{c}{a}$$

Complete the square by adding $\left(\frac{1}{2} \cdot \frac{b}{a}\right)^2 = \frac{b^2}{4a^2}$ to each side of the equation.

$$x^2 + \frac{b}{a}x + \frac{b^2}{4a^2} = \frac{b^2}{4a^2} - \frac{c}{a}$$

Simplify the right side of the equation.

$$x^2 + \frac{b}{a}x + \frac{b^2}{4a^2} = \frac{b^2}{4a^2} - \left(\frac{c}{a} \cdot \frac{4a}{4a}\right)$$

$$x^2 + \frac{b}{a}x + \frac{b^2}{4a^2} = \frac{b^2}{4a^2} - \frac{4ac}{4a^2}$$

$$x^2 + \frac{b}{a}x + \frac{b^2}{4a^2} = \frac{b^2 - 4ac}{4a^2}$$

Factor the perfect-square trinomial on the left side of the equation.

$$\left(x + \frac{b}{2a}\right)^2 = \frac{b^2 - 4ac}{4a^2}$$

Take the square root of each side of the equation.

$$\sqrt{\left(x + \frac{b}{2a}\right)^2} = \sqrt{\frac{b^2 - 4ac}{4a^2}}$$

$$x + \frac{b}{2a} = \pm\frac{\sqrt{b^2 - 4ac}}{2a}$$

Solve for x.

$$x + \frac{b}{2a} = \frac{\sqrt{b^2 - 4ac}}{2a} \qquad x + \frac{b}{2a} = -\frac{\sqrt{b^2 - 4ac}}{2a}$$

$$x = -\frac{b}{2a} + \frac{\sqrt{b^2 - 4ac}}{2a} \qquad x = -\frac{b}{2a} - \frac{\sqrt{b^2 - 4ac}}{2a}$$

$$= \frac{-b + \sqrt{b^2 - 4ac}}{2a} \qquad = \frac{-b - \sqrt{b^2 - 4ac}}{2a}$$

Point of Interest

Although mathematicians have studied quadratic equations since around 500 B.C., it was not until the 18th century that the formula was written as it is today. Of further note, the word *quadratic* has the same Latin root as does the word *square*.

The Quadratic Formula

The solutions of $ax^2 + bx + c = 0$, $a \neq 0$, are

$$\frac{-b + \sqrt{b^2 - 4ac}}{2a} \quad \text{and} \quad \frac{-b - \sqrt{b^2 - 4ac}}{2a}$$

Objective 10.3A

New Vocabulary

discriminant

New Formulas

quadratic formula

$$x = \frac{-b \pm \sqrt{b^2 - 4ac}}{2a}$$

Discuss the Concepts

1. If a quadratic equation is solved using the quadratic formula, and the result is $x = \frac{1 \pm \sqrt{23}}{3}$, what are the solutions of the equation?
2. If a quadratic equation is solved using the quadratic formula, and the result is $x = \frac{2 \pm 6}{4}$, what are the solutions of the equation?
3. Suppose you must solve the quadratic equation $x^2 = 3x + 5$. Does it matter whether you rewrite the equation as $x^2 - 3x - 5 = 0$ or as $0 = -x^2 + 3x + 5$ before you begin?
4. Which method of solving a quadratic equation do you prefer, completing the square or using the quadratic formula? Why?
5. Explain why the discriminant determines whether a quadratic equation has one real number solution, two real number solutions, or two complex number solutions.

In-Class Examples (Objective 10.3A)

Solve by using the quadratic formula.

1. $x^2 + 3x - 10 = 0$ $-5, 2$
2. $x^2 - 4x - 30 = 0$ $2 + \sqrt{34}, 2 - \sqrt{34}$
3. $x^2 = 2x - 15$ $1 + i\sqrt{14}, 1 - i\sqrt{14}$
4. $4x^2 + 8x = 3$ $\frac{-2 + \sqrt{7}}{2}, \frac{-2 - \sqrt{7}}{2}$

Use the discriminant to determine whether the equation has one real number solution, two real number solutions, or two complex number solutions.

5. $2x^2 + x + 7 = 0$ Two complex
6. $3x^2 - x - 2 = 0$ Two real

Instructor Note

You may want to review the methods of solving quadratic equations presented in this chapter: factoring, taking square roots, completing the square, and using the quadratic formula. Discuss when each method is appropriate. You might want to refer to the list provided in the Chapter Summary at the end of this chapter.

Instructor Note

One of the difficulties students have when using the quadratic formula is making correct substitutions. Have them first make sure that the equation is in standard form. Then circling the values of *a*, *b*, and *c* may help. They must remember that the sign that precedes a number is the sign of the number. For the equation at the right, we have

$$\overset{a}{(3)}x^2 \overset{b}{(-4)}x \overset{c}{(-6)} = 0$$

If you did not cover the section on solving quadratic equations by completing the square, you might want to discuss exact solutions versus decimal approximations at this point.

The quadratic formula is frequently written as $x = \dfrac{-b \pm \sqrt{b^2 - 4ac}}{2a}$.

TAKE NOTE
The solutions of this quadratic equation are rational numbers. When this happens, the equation could have been solved by factoring and using the Principle of Zero Products. This may be easier than applying the quadratic formula.

➡ Solve by using the quadratic formula: $2x^2 + 5x + 3 = 0$

$$x = \frac{-b \pm \sqrt{b^2 - 4ac}}{2a}$$

- **The equation $2x^2 + 5x + 3 = 0$ is in standard form. $a = 2$, $b = 5$, $c = 3$ Replace *a*, *b*, and *c* in the quadratic formula with these values.**

$$= \frac{-(5) \pm \sqrt{(5)^2 - 4(2)(3)}}{2(2)}$$

$$= \frac{-5 \pm \sqrt{25 - 24}}{4}$$

$$= \frac{-5 \pm \sqrt{1}}{4} = \frac{-5 \pm 1}{4}$$

$$x = \frac{-5 + 1}{4} = \frac{-4}{4} = -1 \qquad x = \frac{-5 - 1}{4} = \frac{-6}{4} = -\frac{3}{2}$$

The solutions are -1 and $-\frac{3}{2}$.

➡ Solve $3x^2 = 4x + 6$ by using the quadratic formula. Find the exact solutions and approximate the solutions to the nearest thousandth.

$$3x^2 = 4x + 6$$

$$3x^2 - 4x - 6 = 0$$

- **Write the equation in standard form. Subtract $4x$ and 6 from each side of the equation. $a = 3$, $b = -4$, $c = -6$ Replace *a*, *b*, and *c* in the quadratic formula with these values.**

$$x = \frac{-b \pm \sqrt{b^2 - 4ac}}{2a}$$

$$= \frac{-(-4) \pm \sqrt{(-4)^2 - 4(3)(-6)}}{2(3)}$$

$$= \frac{4 \pm \sqrt{16 - (-72)}}{6}$$

$$= \frac{4 \pm \sqrt{88}}{6} = \frac{4 \pm 2\sqrt{22}}{6}$$

$$= \frac{\cancel{2}(2 \pm \sqrt{22})}{\cancel{2} \cdot 3} = \frac{2 \pm \sqrt{22}}{3}$$

Check:

$3x^2$	$= 4x + 6$
$3\left(\frac{2 + \sqrt{22}}{3}\right)^2$	$4\left(\frac{2 + \sqrt{22}}{3}\right) + 6$
$3\left(\frac{4 + 4\sqrt{22} + 22}{9}\right)$	$\frac{8}{3} + \frac{4\sqrt{22}}{3} + \frac{18}{3}$
$3\left(\frac{26 + 4\sqrt{22}}{9}\right)$	$\frac{26}{3} + \frac{4\sqrt{22}}{3}$
$\frac{26 + 4\sqrt{22}}{3}$	$= \frac{26 + 4\sqrt{22}}{3}$

$3x^2$	$= 4x + 6$
$3\left(\frac{2 - \sqrt{22}}{3}\right)^2$	$4\left(\frac{2 - \sqrt{22}}{3}\right) + 6$
$3\left(\frac{4 - 4\sqrt{22} + 22}{9}\right)$	$\frac{8}{3} - \frac{4\sqrt{22}}{3} + \frac{18}{3}$
$3\left(\frac{26 - 4\sqrt{22}}{9}\right)$	$\frac{26}{3} - \frac{4\sqrt{22}}{3}$
$\frac{26 - 4\sqrt{22}}{3}$	$= \frac{26 - 4\sqrt{22}}{3}$

The solutions are $\frac{2 + \sqrt{22}}{3}$ and $\frac{2 - \sqrt{22}}{3}$.

$$\frac{2 + \sqrt{22}}{3} \approx 2.230 \qquad \frac{2 - \sqrt{22}}{3} \approx -0.897$$

To the nearest thousandth, the solutions are 2.230 and -0.897.

561

⇒ Solve by using the quadratic formula: $4x^2 = 8x - 13$

$$4x^2 = 8x - 13$$
$$4x^2 - 8x + 13 = 0$$ • **Write the equation in standard form.**

$$x = \frac{-b \pm \sqrt{b^2 - 4ac}}{2a}$$ • **Use the quadratic formula.**

$$= \frac{-(-8) \pm \sqrt{(-8)^2 - 4 \cdot 4 \cdot 13}}{2 \cdot 4}$$ • $a = 4, b = -8, c = 13$

$$= \frac{8 \pm \sqrt{64 - 208}}{8} = \frac{8 \pm \sqrt{-144}}{8}$$
$$= \frac{8 \pm 12i}{8} = \frac{2 + 3i}{2}$$

The solutions are $1 + \frac{3}{2}i$ and $1 - \frac{3}{2}i$.

Of the three preceding examples, the first two had real number solutions; the last one had complex number solutions.

In the quadratic formula, the quantity $b^2 - 4ac$ is called the **discriminant**. When a, b, and c are real numbers, the discriminant determines whether a quadratic equation will have a double root, two real number solutions that are not equal, or two complex number solutions.

THE EFFECT OF THE DISCRIMINANT ON THE SOLUTIONS OF A QUADRATIC EQUATION

1. If $b^2 - 4ac = 0$, the equation has one real number solution, a double root.
2. If $b^2 - 4ac > 0$, the equation has two unequal real number solutions.
3. If $b^2 - 4ac < 0$, the equation has two complex number solutions.

T

⇒ Use the discriminant to determine whether $x^2 - 4x - 5 = 0$ has one real number solution, two real number solutions, or two complex number solutions.

$b^2 - 4ac$ • **Evaluate the discriminant.** $a = 1, b = -4, c = -5$
$(-4)^2 - 4(1)(-5) = 16 + 20 = 36$
$36 > 0$

Because $b^2 - 4ac > 0$, the equation has two real number solutions.

Example 1 Solve by using the quadratic formula: $2x^2 - x + 5 = 0$

Solution $2x^2 - x + 5 = 0$
$a = 2, b = -1, c = 5$

$$x = \frac{-b \pm \sqrt{b^2 - 4ac}}{2a}$$
$$= \frac{-(-1) \pm \sqrt{(-1)^2 - 4(2)(5)}}{2 \cdot 2}$$
$$= \frac{1 \pm \sqrt{1 - 40}}{4} = \frac{1 \pm \sqrt{-39}}{4}$$
$$= \frac{1 \pm i\sqrt{39}}{4}$$

The solutions are $\frac{1}{4} + \frac{\sqrt{39}}{4}i$ and $\frac{1}{4} - \frac{\sqrt{39}}{4}i$.

You Try It 1 Solve by using the quadratic formula: $x^2 - 2x + 10 = 0$

Your solution $1 + 3i, 1 - 3i$

Solution on p. S31

Concept Check

1. For what value of k are the roots of $3x^2 - 4x + k = 0$ equal? $\frac{4}{3}$
2. Solve.
   a. $x^2 + x\sqrt{3} + 1 = 0$
   $-\frac{\sqrt{3}}{2} + \frac{1}{2}i$ and $-\frac{\sqrt{3}}{2} - \frac{1}{2}i$
   b. $x^2 + x\sqrt{7} + 2 = 0$
   $-\frac{\sqrt{7}}{2} + \frac{1}{2}i$ and $-\frac{\sqrt{7}}{2} - \frac{1}{2}i$

Optional Student Activity

1. For what values of k does the equation $2x^2 - kx + x + 8 = 0$ have a double root? $-7, 9$
2. Find the difference between the larger root and the smaller root of $x^2 - px + \frac{(p^2 - 1)}{4} = 0$. 1
3. What is the smallest integral value of K such that $2x(Kx - 4) - x^2 + 6 = 0$ has no real roots? 2
4. Show that the equation $x^2 + bx - 2 = 0$ always has real number solutions regardless of the value of b.

$$b^2 - 4ac = b^2 - 4(1)(-2)$$
$$= b^2 + 8$$

Because b^2 is always greater than zero, $b^2 + 8$ is always greater than zero. Therefore, $x^2 + bx - 2 = 0$ has real number solutions regardless of the value of b.

Example 2

Solve by using the quadratic formula:
$2x^2 = (x - 2)(x - 3)$

Solution

$2x^2 = (x - 2)(x - 3)$
$2x^2 = x^2 - 5x + 6$

$x^2 + 5x - 6 = 0$
$a = 1, b = 5, c = -6$

$$x = \frac{-b \pm \sqrt{b^2 - 4ac}}{2a}$$
$$= \frac{-5 \pm \sqrt{5^2 - 4(1)(-6)}}{2 \cdot 1}$$
$$= \frac{-5 \pm \sqrt{25 + 24}}{2} = \frac{-5 \pm \sqrt{49}}{2}$$
$$= \frac{-5 \pm 7}{2}$$
$$x = \frac{-5 + 7}{2} \qquad x = \frac{-5 - 7}{2}$$
$$= \frac{2}{2} = 1 \qquad = \frac{-12}{2} = -6$$

The solutions are 1 and −6.

You Try It 2

Solve by using the quadratic formula:
$4x^2 = 4x - 1$

Your solution

$\frac{1}{2}$

Example 3

Use the discriminant to determine whether $4x^2 - 2x + 5 = 0$ has one real number solution, two real number solutions, or two complex number solutions.

Solution

$a = 4, b = -2, c = 5$

$$b^2 - 4ac = (-2)^2 - 4(4)(5)$$
$$= 4 - 80$$
$$= -76$$

$-76 < 0$

Because the discriminant is less than zero, the equation has two complex number solutions.

You Try It 3

Use the discriminant to determine whether $3x^2 - x - 1 = 0$ has one real number solution, two real number solutions, or two complex number solutions.

Your solution

two real number solutions

Solutions on p. S31

563

10.3 Exercises

Objective A

1. Write the quadratic formula. What does each variable in the formula represent?

2. Write the expression that appears under the radical symbol in the quadratic formula. What is this quantity called? What can it be used to determine?

Solve by using the quadratic formula.

3. $x^2 - 3x - 10 = 0$
$5, -2$

4. $z^2 - 4z - 8 = 0$
$2 + 2\sqrt{3}, 2 - 2\sqrt{3}$

5. $y^2 + 5y - 36 = 0$
$4, -9$

6. $z^2 - 3z - 40 = 0$
$8, -5$

7. $w^2 = 8w + 72$
$4 + 2\sqrt{22}, 4 - 2\sqrt{22}$

8. $t^2 = 2t + 35$
$7, -5$

9. $v^2 = 24 - 5v$
$3, -8$

10. $x^2 = 18 - 7x$
$2, -9$

11. $2y^2 + 5y - 1 = 0$
$\frac{-5 + \sqrt{33}}{4}, \frac{-5 - \sqrt{33}}{4}$

12. $4p^2 - 7p + 1 = 0$
$\frac{7 + \sqrt{33}}{8}, \frac{7 - \sqrt{33}}{8}$

13. $8s^2 = 10s + 3$
$\frac{3}{2}, -\frac{1}{4}$

14. $12t^2 = 5t + 2$
$\frac{2}{3}, -\frac{1}{4}$

15. $x^2 = 14x - 4$
$7 + 3\sqrt{5}, 7 - 3\sqrt{5}$

16. $v^2 = 12v - 24$
$6 + 2\sqrt{3}, 6 - 2\sqrt{3}$

17. $2z^2 - 2z - 1 = 0$
$\frac{1 + \sqrt{3}}{2}, \frac{1 - \sqrt{3}}{2}$

18. $6w^2 = 9w - 1$
$\frac{9 + \sqrt{57}}{12}, \frac{9 - \sqrt{57}}{12}$

19. $z^2 + 2z + 2 = 0$
$-1 + i, -1 - i$

20. $p^2 - 4p + 5 = 0$
$2 + i, 2 - i$

21. $y^2 - 2y + 5 = 0$
$1 + 2i, 1 - 2i$

22. $x^2 + 6x + 13 = 0$
$-3 + 2i, -3 - 2i$

23. $s^2 - 4s + 13 = 0$
$2 + 3i, 2 - 3i$

24. $t^2 - 6t + 10 = 0$
$3 + i, 3 - i$

25. $2w^2 - 2w - 5 = 0$
$\frac{1 + \sqrt{11}}{2}, \frac{1 - \sqrt{11}}{2}$

26. $v^2 + 8v + 3 = 0$
$-4 + \sqrt{13}, -4 - \sqrt{13}$

27. $2x^2 + 6x + 5 = 0$
$-\frac{3}{2} + \frac{1}{2}i, -\frac{3}{2} - \frac{1}{2}i$

28. $2y^2 + 2y + 13 = 0$
$-\frac{1}{2} + \frac{5}{2}i, -\frac{1}{2} - \frac{5}{2}i$

29. $4t^2 - 6t + 9 = 0$
$\frac{3}{4} + \frac{3\sqrt{3}}{4}i, \frac{3}{4} - \frac{3\sqrt{3}}{4}i$

Section 10.3

Suggested Assignment
Exercises 1–41, odds
More challenging problems:
Exercises 43–49, odds

Answers to Writing Exercises

1. The quadratic formula is $x = \frac{-b \pm \sqrt{b^2 - 4ac}}{2a}$. In this formula a is the coefficient of x^2, b is the coefficient of x, and c is the constant term in the quadratic equation $ax^2 + bx + c = 0$, $a \neq 0$.
2. The expression $b^2 - 4ac$ appears under the radical symbol in the quadratic formula. It is called the discriminant. It can be used to determine whether a quadratic equation has one real number solution, two unequal real number solutions, or two complex number solutions.

Quick Quiz (Objective 10.3A)

Solve by using the quadratic formula.

1. $x^2 + 3x - 18 = 0$ $-6, 3$
2. $x^2 - 6x - 6 = 0$ $3 + \sqrt{15}, 3 - \sqrt{15}$
3. $x^2 = 4x - 6$ $2 + i\sqrt{2}, 2 - i\sqrt{2}$
4. $3x^2 + 6x = -1$ $\frac{-3 + \sqrt{6}}{3}, \frac{-3 - \sqrt{6}}{3}$

Use the discriminant to determine whether the equation has one real number solution, two real number solutions, or two complex number solutions.

5. $4x^2 - 12x + 9 = 0$ One real
6. $3x^2 - 4x + 5 = 0$ Two complex

Answers to Writing Exercises

42. Students might mention some of the following points.
Factoring
Advantage: It is the easiest of the three methods to use.
Disadvantage: Not every quadratic equation can be solved by factoring.
Completing the Square
Advantage: This method can be used to solve any quadratic equation.
Disadvantage: This method is long and sometimes difficult to use.
Quadratic Formula
Advantage: This method can be used to solve any quadratic equation.
Disadvantage: Simplification of the resulting expression is sometimes difficult, and the quadratic formula must be memorized.

49. The quadratic formula can be used to solve any quadratic equation. The only restriction on the expression $\frac{-b \pm \sqrt{b^2 - 4ac}}{2a}$ is that $a \neq 0$, because if $a = 0$ then the denominator is zero, and division by zero is undefined. But by definition a quadratic equation must have a term of degree 2. Therefore, $a \neq 0$ for any quadratic equation and so any quadratic equation can be solved by using the quadratic formula.

50. Student explanations should include the idea that the $\pm$ indicates both the positive and negative square root of the expression.

Solve by using the quadratic formula. Approximate the solutions to the nearest thousandth.

30. $x^2 - 6x - 6 = 0$
6.873, −0.873

31. $p^2 - 8p + 3 = 0$
7.606, 0.394

32. $r^2 - 2r = 4$
3.236, −1.236

33. $w^2 + 4w = 1$
0.236, −4.236

34. $3t^2 = 7t + 1$
2.468, −0.135

35. $2y^2 = y + 5$
1.851, −1.351

Use the discriminant to determine whether the quadratic equation has one real number solution, two real number solutions, or two complex number solutions.

36. $2z^2 - z + 5 = 0$
two complex

37. $3y^2 + y + 1 = 0$
two complex

38. $9x^2 - 12x + 4 = 0$
one real

39. $4x^2 + 20x + 25 = 0$
one real

40. $2v^2 - 3v - 1 = 0$
two real

41. $3w^2 + 3w - 2 = 0$
two real

APPLYING THE CONCEPTS

42. Name the three methods of solving a quadratic equation that have been discussed. What are the advantages and disadvantages of each?

For what values of p does the quadratic equation have two real number solutions that are not equal? Write the answer in set-builder notation.

43. $x^2 - 6x + p = 0$
$\{p \mid p < 9\}$

44. $x^2 + 10x + p = 0$
$\{p \mid p < 25\}$

For what values of p does the quadratic equation have two complex number solutions? Write the answer in set-builder notation.

45. $x^2 - 2x + p = 0$
$\{p \mid p > 1\}$

46. $x^2 + 4x + p = 0$
$\{p \mid p > 4\}$

47. Find all values of x that satisfy the equation $x^2 + ix + 2 = 0$.
$i, -2i$

48. Show that the equation $x^2 + bx - 1 = 0$ always has real number solutions regardless of the value of b.
The complete solution is in the *Solutions Manual.*

49. Can the quadratic formula be used to solve *any* quadratic equation? If so, explain why. If not, give an example of a quadratic equation that cannot be solved by using the quadratic formula.

50. One of the steps in the derivation of the quadratic formula is $x + \frac{b}{2a} = \pm\frac{\sqrt{b^2 - 4ac}}{2a}$ or $\pm\sqrt{\frac{b^2 - 4ac}{4a^2}}$. Carefully explain the occurrence of the $\pm$ sign. (*Hint:* Recall that $\sqrt{x^2} = |x|$.)

10.4 Solving Equations That Are Reducible to Quadratic Equations

Objective A **To solve an equation that is quadratic in form**

VIDEO & DVD · CD TUTOR · WEB · SSM

Certain equations that are not quadratic can be expressed in quadratic form by making suitable substitutions. An equation is **quadratic in form** if it can be written as $au^2 + bu + c = 0$.

The equation $x^4 - 4x^2 - 5 = 0$ is quadratic in form.

$$x^4 - 4x^2 - 5 = 0$$
$$(x^2)^2 - 4(x^2) - 5 = 0$$
$$u^2 - 4u - 5 = 0 \quad \text{• Let } x^2 = u.$$

The equation $y - y^{1/2} - 6 = 0$ is quadratic in form.

$$y - y^{1/2} - 6 = 0$$
$$(y^{1/2})^2 - (y^{1/2}) - 6 = 0$$
$$u^2 - u - 6 = 0 \quad \text{• Let } y^{1/2} = u.$$

Here is the key to recognizing equations that are quadratic in form: When the equation is written in standard form, the exponent on one variable term is $\frac{1}{2}$ the exponent on the other variable term.

➡ Solve: $z + 7z^{1/2} - 18 = 0$

$$z + 7z^{1/2} - 18 = 0 \quad \text{• The equation is quadratic in form.}$$
$$(z^{1/2})^2 + 7(z^{1/2}) - 18 = 0$$
$$u^2 + 7u - 18 = 0 \quad \text{• Let } z^{1/2} = u.$$
$$(u - 2)(u + 9) = 0 \quad \text{• Solve by factoring.}$$

$u - 2 = 0$ $\quad$ $u + 9 = 0$
$u = 2$ $\quad$ $u = -9$

$z^{1/2} = 2$ $\quad$ $z^{1/2} = -9$ $\quad$ • Replace u by $z^{1/2}$.
$\sqrt{z} = 2$ $\quad$ $\sqrt{z} = -9$

$(\sqrt{z})^2 = 2^2$ $\quad$ $(\sqrt{z})^2 = (-9)^2$ $\quad$ • Solve for z.
$z = 4$ $\quad$ $z = 81$

Check:

$z + 7z^{1/2} - 18 = 0$		$z + 7z^{1/2} - 18 = 0$	
$4 + 7(4)^{1/2} - 18$	0	$81 + 7(81)^{1/2} - 18$	0
$4 + 7 \cdot 2 - 18$		$81 + 7 \cdot 9 - 18$	
$4 + 14 - 18$		$81 + 63 - 18$	
$0 = 0$		$126 \neq 0$	

4 checks as a solution, but 81 does not check as a solution.
The solution is 4.

TAKE NOTE
When each side of an equation is squared, the resulting equation may have a solution that is not a solution of the original equation.

Objective 10.4A

Vocabulary to Review
quadratic in form

Discuss the Concepts
1. What does it mean for an equation to be quadratic in form?
2. Explain how to show that $x^4 - 3x^2 - 3 = 0$ is quadratic in form.
3. Is the equation $t^6 - 4t^3 + 4 = 0$ quadratic in form? Why or why not?
4. Why does the equation $x^2 + 5x - 6 = 0$ have two solutions while the equation $x + 5\sqrt{x} - 6 = 0$ has only one solution?

Concept Check
1. Write two equations that are not quadratic equations but that can be written in quadratic form. Then write them in quadratic form.
2. Solve: $x^{2/5} + 6 = 5x^{1/5}$ 32, 243
3. Solve: $\frac{x^4}{4} + 1 = \frac{5x^2}{4}$ $-1, 1, -2, 2$
4. Solve: $\frac{x^4}{3} - \frac{8x^2}{3} = 3$ $-3, 3, -i, i$

Optional Student Activity
1. Solve: $2^{2x} + 32 = 12(2^x)$ 2, 3
2. If $5^{2x} = 12 - 5^x$ and $5^{x+2} = N$, find the value of N. 75

In-Class Examples (Objective 10.4A)
Solve.
1. $x^4 - 10x^2 + 9 = 0$ $-3, 3, -1, 1$
2. $x - 3x^{1/2} - 4 = 0$ 16
3. $x^4 - 15x^2 - 16 = 0$ $-4, 4, -i, i$
4. $x^{2/3} - x^{1/3} - 2 = 0$ $-1, 8$

Objective 10.4B

Discuss the Concepts

1. What does the Property of Raising Each Side of an Equation to a Power state?
2. Is the equation $x + 3\sqrt{x} - 8 = 0$ quadratic in form? Explain why or why not.
3. Is the equation $\sqrt[4]{x} + 2\sqrt{3x} - 8 = 0$ quadratic in form? Explain why or why not.

Concept Check

Determine whether the statement is always true, sometimes true, or never true.

1. An equation that is quadratic in form can be solved by using the quadratic formula. Always true
2. Squaring both sides of a radical equation produces an extraneous root. Sometimes true
3. When the Property of Raising Both Sides of an Equation to a Power is used to solve an equation, the solutions must be checked. Always true

Example 1 Solve: $x^4 + x^2 - 12 = 0$

Solution

$$x^4 + x^2 - 12 = 0$$
$$(x^2)^2 + (x^2) - 12 = 0$$
$$u^2 + u - 12 = 0$$
$$(u - 3)(u + 4) = 0$$

$u - 3 = 0 \qquad u + 4 = 0$

$u = 3 \qquad u = -4$

Replace u by x^2.

$x^2 = 3 \qquad x^2 = -4$

$\sqrt{x^2} = \sqrt{3} \qquad \sqrt{x^2} = \sqrt{-4}$

$x = \pm\sqrt{3} \qquad x = \pm 2i$

The solutions are $\sqrt{3}$, $-\sqrt{3}$, $2i$, and $-2i$.

You Try It 1 Solve: $x - 5x^{1/2} + 6 = 0$

Your solution 4, 9

Solution on p. S31

Objective B To solve a radical equation that is reducible to a quadratic equation

VIDEO & DVD · CD TUTOR · WWW WEB · SSM

Certain equations that contain radicals can be expressed as quadratic equations.

➡ Solve: $\sqrt{x + 2} + 4 = x$

$\sqrt{x + 2} + 4 = x$

$\sqrt{x + 2} = x - 4$ • Solve for the radical expression.

$(\sqrt{x + 2})^2 = (x - 4)^2$ • Square each side of the equation.

$x + 2 = x^2 - 8x + 16$ • Simplify.

$0 = x^2 - 9x + 14$ • Write the equation in standard form.

$0 = (x - 7)(x - 2)$ • Solve for x.

$x - 7 = 0 \qquad x - 2 = 0$

$x = 7 \qquad x = 2$

Check:

$\sqrt{x + 2} + 4 = x$		$\sqrt{x + 2} + 4 = x$	
$\sqrt{7 + 2} + 4$	7	$\sqrt{2 + 2} + 4$	2
$\sqrt{9} + 4$		$\sqrt{4} + 4$	
$3 + 4$		$2 + 4$	
$7 = 7$		$6 \neq 2$	

7 checks as a solution, but 2 does not check as a solution.
The solution is 7.

TAKE NOTE
You should always check your solutions by substituting the proposed solutions back into the *original* equation.

In-Class Examples (Objective 10.4B)

Solve.

1. $\sqrt{x - 1} + x = 7$ 5
2. $\sqrt{3w + 7} = w - 1$ 6
3. $\sqrt{2x - 3} - 1 = \sqrt{x - 2}$ 2, 6
4. $\sqrt{2x + 1} - \sqrt{x} = 1$ 0, 4

Example 2
Solve: $\sqrt{7y - 3} + 3 = 2y$

Solution

$$\sqrt{7y - 3} + 3 = 2y$$
$$\sqrt{7y - 3} = 2y - 3$$
$$(\sqrt{7y - 3})^2 = (2y - 3)^2$$
$$7y - 3 = 4y^2 - 12y + 9$$
$$0 = 4y^2 - 19y + 12$$
$$0 = (4y - 3)(y - 4)$$

$$4y - 3 = 0 \qquad y - 4 = 0$$
$$4y = 3 \qquad y = 4$$
$$y = \frac{3}{4}$$

4 checks as a solution.

$\frac{3}{4}$ does not check as a solution.

The solution is 4.

You Try It 2
Solve: $\sqrt{2x + 1} + x = 7$

Your solution
4

Example 3
Solve: $\sqrt{2y + 1} - \sqrt{y} = 1$

Solution

$$\sqrt{2y + 1} - \sqrt{y} = 1$$

Solve for one of the radical expressions.

$$\sqrt{2y + 1} = \sqrt{y} + 1$$
$$(\sqrt{2y + 1})^2 = (\sqrt{y} + 1)^2$$
$$2y + 1 = y + 2\sqrt{y} + 1$$
$$y = 2\sqrt{y}$$

Square each side of the equation.

$$y^2 = (2\sqrt{y})^2$$
$$y^2 = 4y$$
$$y^2 - 4y = 0$$
$$y(y - 4) = 0$$
$$y = 0 \qquad y - 4 = 0$$
$$y = 4$$

0 and 4 check as solutions.

The solutions are 0 and 4.

You Try It 3
Solve: $\sqrt{2x - 1} + \sqrt{x} = 2$

Your solution
1

Solutions on p. S31

Optional Student Activity

[*Note:* Exercises 1 and 2 require use of the Pythagorean Theorem.]

1. The width of a rectangle is twice the square root of the length. A diagonal of the rectangle is 12 in. Find the length of the rectangle. Round to the nearest hundredth. 10.17 in.
2. The longer leg of a right triangle is four times the square root of the shorter leg. Find the lengths of the two legs if the hypotenuse is 6 ft. Round to the nearest hundredth. 2 ft and 5.66 ft
3. The total time from the instant you drop a rock into an abandoned mine shaft until the moment you hear its impact is 4.5 s. Find the depth of the mine shaft. Use the equation $1100T = 275\sqrt{s} + s$, where T is the time in seconds and s is the depth of the mine shaft in feet. Round to the nearest tenth. 287.5 ft

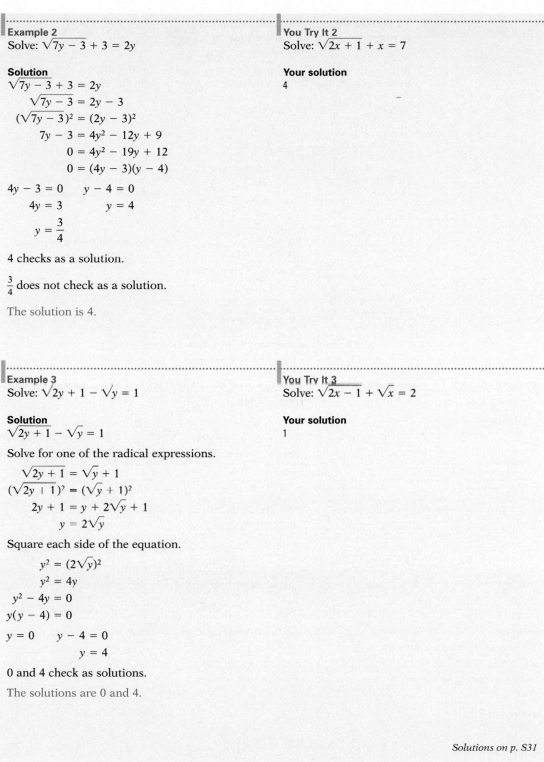

Objective 10.4C

Concept Check

1. For what values of x is the equation $x + \frac{1}{x} = 5 + \frac{1}{5}$ true? $5, \frac{1}{5}$
2. The reciprocal of $x + 1$ is equal to $x - 1$. Find the value of x. $\sqrt{2}, -\sqrt{2}$
3. Solve: $\frac{2}{x} + \frac{x}{2} = \frac{3}{x} + \frac{x}{3}$ $\sqrt{6}, -\sqrt{6}$

Optional Student Activity

1. Find the values of k such that the line containing the points $(-2, -4)$ and $(2 - k, k - 5)$ is parallel to the line containing the points $(1, -4)$ and $(k, k - 3)$. $-\frac{1}{2}, 3$
2. A chemical reaction between carbon monoxide and water vapor is used to increase the ratio of hydrogen gas in certain gas mixtures. In the process, carbon dioxide is also formed. For a certain reaction, the concentration of carbon dioxide, x, in moles per liter is given by the equation $0.58 = \frac{x^2}{(0.02 - x)^2}$. Solve this equation for x. Round to the nearest ten-thousandth. 0.0086 mole per liter

Objective C To solve a fractional equation that is reducible to a quadratic equation

VIDEO & DVD CD TUTOR WEB SSM

After each side of a fractional equation has been multiplied by the LCM of the denominators, the resulting equation may be a quadratic equation.

⇒ Solve: $\frac{1}{r} + \frac{1}{r+1} = \frac{3}{2}$

$$\frac{1}{r} + \frac{1}{r+1} = \frac{3}{2}$$

$$2r(r+1)\left(\frac{1}{r} + \frac{1}{r+1}\right) = 2r(r+1) \cdot \frac{3}{2}$$

• **Multiply each side of the equation by the LCM of the denominators.**

$$2(r+1) + 2r = r(r+1) \cdot 3$$
$$2r + 2 + 2r = 3r(r+1)$$
$$4r + 2 = 3r^2 + 3r$$
$$0 = 3r^2 - r - 2$$

• **Write the equation in standard form.**

$$0 = (3r+2)(r-1)$$

• **Solve for r by factoring.**

$$3r + 2 = 0 \qquad r - 1 = 0$$
$$3r = -2 \qquad r = 1$$
$$r = -\frac{2}{3}$$

$-\frac{2}{3}$ and 1 check as solutions. The solutions are $-\frac{2}{3}$ and 1.

Example 4

Solve: $\frac{9}{x-3} = 2x + 1$

Solution

$$\frac{9}{x-3} = 2x + 1$$
$$(x-3)\frac{9}{x-3} = (x-3)(2x+1)$$
$$9 = 2x^2 - 5x - 3$$
$$0 = 2x^2 - 5x - 12$$
$$0 = (2x+3)(x-4)$$

$$2x + 3 = 0 \qquad x - 4 = 0$$
$$2x = -3 \qquad x = 4$$
$$x = -\frac{3}{2}$$

$-\frac{3}{2}$ and 4 check as solutions.

The solutions are $-\frac{3}{2}$ and 4.

You Try It 4

Solve: $3y + \frac{25}{3y-2} = -8$

Your solution

-1

Solution on p. S31

In-Class Examples (Objective 10.4C)

Solve.

1. $x = \frac{9}{x-8}$ $-1, 9$
2. $\frac{x}{3x-4} = \frac{3}{x+2}$ $3, 4$
3. $\frac{x}{x+2} = \frac{-4}{x-2}$ $-1 + i\sqrt{7}, -1 - i\sqrt{7}$
4. $\frac{2x-3}{x-4} - x = 2$ $-1, 5$
5. $\frac{3x+4}{x+5} - 4x = 8$ $-\frac{9}{4}, -4$
6. $\frac{2}{x} - \frac{3}{2x+1} = 2$ $\frac{-1+\sqrt{33}}{8}, \frac{-1-\sqrt{33}}{8}$

569

10.4 Exercises

Objective A

Solve.

1. $x^4 - 13x^2 + 36 = 0$
   2, −2, 3, −3

2. $y^4 - 5y^2 + 4 = 0$
   1, −1, 2, −2

3. $z^4 - 6z^2 + 8 = 0$
   $2, -2, \sqrt{2}, -\sqrt{2}$

4. $t^4 - 12t^2 + 27 = 0$
   $3, -3, \sqrt{3}, -\sqrt{3}$

5. $p - 3p^{1/2} + 2 = 0$
   1, 4

6. $v - 7v^{1/2} + 12 = 0$
   9, 16

7. $x - x^{1/2} - 12 = 0$
   16

8. $w - 2w^{1/2} - 15 = 0$
   25

9. $z^4 + 3z^2 - 4 = 0$
   $2i, -2i, 1, -1$

10. $y^4 + 5y^2 - 36 = 0$
    $2, -2, 3i, -3i$

11. $x^4 + 12x^2 - 64 = 0$
    $4i, -4i, 2, -2$

12. $x^4 - 81 = 0$
    $3i, -3i, 3, -3$

13. $p + 2p^{1/2} - 24 = 0$
    16

14. $v + 3v^{1/2} - 4 = 0$
    1

15. $y^{2/3} - 9y^{1/3} + 8 = 0$
    1, 512

16. $z^{2/3} - z^{1/3} - 6 = 0$
    −8, 27

17. $9w^4 - 13w^2 + 4 = 0$
    $\frac{2}{3}, -\frac{2}{3}, 1, -1$

18. $4y^4 - 7y^2 - 36 = 0$
    $\frac{3}{2}i, -\frac{3}{2}i, 2, -2$

Objective B

Solve.

19. $\sqrt{x + 1} + x = 5$
    3

20. $\sqrt{x - 4} + x = 6$
    5

21. $x = \sqrt{x} + 6$
    9

22. $\sqrt{2y - 1} = y - 2$
    5

23. $\sqrt{3w + 3} = w + 1$
    2, −1

24. $\sqrt{2s + 1} = s - 1$
    4

25. $\sqrt{4y + 1} - y = 1$
    0, 2

26. $\sqrt{3s + 4} + 2s = 12$
    4

27. $\sqrt{10x + 5} - 2x = 1$
    $-\frac{1}{2}, 2$

Section 10.4

Suggested Assignment

Exercises 1–51, odds
Exercise 55
More challenging problems:
Exercises 52–54

Quick Quiz (Objective 10.4A)

Solve.

1. $y^4 - 13y^2 + 36 = 0$ $-2, 2, -3, 3$
2. $x - 2x^{1/2} - 8 = 0$ 16
3. $x^4 - 4x^2 - 5 = 0$ $i, -i, \sqrt{5}, -\sqrt{5}$
4. $x^{2/3} - 1 = 0$ $-1, 1$

Quick Quiz (Objective 10.4B)

Solve.

1. $\sqrt{x + 2} + x = 10$ 7
2. $\sqrt{x} + 2 = \sqrt{x + 12}$ 4
3. $\sqrt{3x + 4} - \sqrt{x} = 2$ 0, 4

28. $\sqrt{t+8} = 2t + 1$
1

29. $\sqrt{p+11} = 1 - p$
-2

30. $x - 7 = \sqrt{x-5}$
9

31. $\sqrt{x-1} - \sqrt{x} = -1$
1

32. $\sqrt{y} + 1 = \sqrt{y+5}$
4

33. $\sqrt{2x-1} = 1 - \sqrt{x-1}$
1

34. $\sqrt{x+6} + \sqrt{x+2} = 2$
-2

35. $\sqrt{t+3} + \sqrt{2t+7} = 1$
-3

36. $\sqrt{5-2x} = \sqrt{2-x} + 1$
$2, -2$

Objective C

Solve.

37. $x = \dfrac{10}{x-9}$
$10, -1$

38. $z = \dfrac{5}{z-4}$
$5, -1$

39. $\dfrac{t}{t+1} = \dfrac{-2}{t-1}$
$-\frac{1}{2} + \frac{\sqrt{7}}{2}i, -\frac{1}{2} - \frac{\sqrt{7}}{2}i$

40. $\dfrac{2v}{v-1} = \dfrac{5}{v+2}$
$\frac{1}{4} + \frac{\sqrt{39}}{4}i, \frac{1}{4} - \frac{\sqrt{39}}{4}i$

41. $\dfrac{y-1}{y+2} + y = 1$
$-3, 1$

42. $\dfrac{2p-1}{p-2} + p = 8$
$5, 3$

43. $\dfrac{3r+2}{r+2} - 2r = 1$
$0, -1$

44. $\dfrac{2v+3}{v+4} + 3v = 4$
$-\frac{13}{3}, 1$

45. $\dfrac{2}{2x+1} + \dfrac{1}{x} = 3$
$\frac{1}{2}, -\frac{1}{3}$

46. $\dfrac{3}{s} - \dfrac{2}{2s-1} = 1$
$\frac{3}{2}, 1$

47. $\dfrac{16}{z-2} + \dfrac{16}{z+2} = 6$
$-\frac{2}{3}, 6$

48. $\dfrac{2}{y+1} + \dfrac{1}{y-1} = 1$
$0, 3$

49. $\dfrac{t}{t-2} + \dfrac{2}{t-1} = 4$
$\frac{4}{3}, 3$

50. $\dfrac{4t+1}{t+4} + \dfrac{3t-1}{t+1} = 2$
$-\frac{11}{5}, 1$

51. $\dfrac{5}{2p-1} + \dfrac{4}{p+1} = 2$
$-\frac{1}{4}, 3$

APPLYING THE CONCEPTS

52. Solve: $(\sqrt{x} - 2)^2 - 5\sqrt{x} + 14 = 0$. (*Hint:* Let $u = \sqrt{x} - 2$.)
9, 36

53. Solve: $(\sqrt{x} + 3)^2 - 4\sqrt{x} - 17 = 0$. (*Hint:* Let $u = \sqrt{x} + 3$.)
4

54. The fourth power of a number is twenty-five less than ten times the square of the number. Find the number.
$\sqrt{5}$ or $-\sqrt{5}$

Quick Quiz (Objective 10.4C)

Solve.

1. $x = \dfrac{6}{x-5}$ $-1, 6$

2. $\dfrac{x}{x+2} = \dfrac{3}{x-2}$ $6, -1$

3. $\dfrac{x-2}{x+3} + x = 2$ $-4, 2$

4. $\dfrac{x}{x-3} + \dfrac{x}{x-1} = 5$ $\dfrac{8+\sqrt{19}}{3}, \dfrac{8-\sqrt{19}}{3}$

10.5 Applications of Quadratic Equations

Objective A To solve application problems

VIDEO & DVD CD TUTOR WEB SSM

The application problems in this section are similar to problems solved earlier in the text. Each of the strategies for the problems in this section will result in a quadratic equation.

➡ A small pipe takes 16 min longer to empty a tank than does a larger pipe. Working together, the pipes can empty the tank in 6 min. How long would it take the smaller pipe working alone to empty the tank?

Strategy for Solving an Application Problem

1. Determine the type of problem. Is it a uniform motion problem, a geometry problem, an integer problem, or a work problem?

The problem is a work problem.

2. Choose a variable to represent the unknown quantity. Write numerical or variable expressions for all the remaining quantities. These results can be recorded in a table.

The unknown time of the larger pipe: t
The unknown time of the smaller pipe: $t + 16$

	Rate of Work	·	*Time Worked*	=	*Part of Task Completed*
Larger pipe	$\frac{1}{t}$	·	6	=	$\frac{6}{t}$
Smaller pipe	$\frac{1}{t+16}$	·	6	=	$\frac{6}{t+16}$

3. Determine how the quantities are related.

$$\frac{6}{t} + \frac{6}{t+16} = 1$$

• **The sum of the parts of the task completed must equal 1.**

$$t(t+16)\left(\frac{6}{t} + \frac{6}{t+16}\right) = t(t+16) \cdot 1$$
$$(t+16)6 + 6t = t^2 + 16t$$
$$6t + 96 + 6t = t^2 + 16t$$
$$0 = t^2 + 4t - 96$$
$$0 = (t+12)(t-8)$$
$$t + 12 = 0 \qquad t - 8 = 0$$
$$t = -12 \qquad t = 8$$

Because time cannot be negative, the solution $t = -12$ is not possible.

The time for the smaller pipe is $t + 16$.

$t + 16 = 8 + 16 = 24$ • **Replace t by 8 and evaluate.**

The smaller pipe requires 24 min to empty the tank.

Objective 10.5A

Optional Student Activity

1. The sum of a number and twice its reciprocal is $\frac{33}{4}$. Find the number. $\frac{1}{4}$ or 8
2. Find two consecutive integers whose cubes differ by 127. 6 and 7 or −6 and −7
3. The *Tour de France* is a bicycle race that begins in Dublin, Ireland but is primarily a race through France. On each day, racers cover a certain distance depending on the steepness of the terrain. Suppose that on a particular day the racers must complete 210 mi. One cyclist, traveling 10 mph faster than a second cyclist, covers this distance in 2.4 h less time than the second cyclist. Find the rate of the first cyclist. 35 mph
4. A penalty kick in soccer is made from a penalty mark that is 36 ft from a goal that is 8 ft high. A possible equation for the flight of a penalty kick is $h = -0.002x^2 + 0.35x$, where h is the height, in feet, of the ball x feet from the penalty mark. Assuming that the flight of the kick is toward the goal and that the ball is not touched by the goalie, will the ball land in the net? No

In-Class Examples (Objective 10.5A)

1. An old pump requires 6 h longer to empty a pool than does a new pump. With both pumps working, the pool can be emptied in 4 h. Find the time required for the new pump, working alone, to empty the pool. 6 h
2. The rate of a jet in calm air is 225 mph. Flying with the wind, the jet can fly 1000 mi in 1 h less time than is required to make the return trip. Find the rate of the wind. 25 mph
3. The length of a rectangle is 2 ft less than three times the width. The area is 60 ft^2. Find the length of the rectangle. Round to the nearest hundredth. 12.45 ft

Example 1

In 8 h, two campers rowed 15 mi down a river and then rowed back to their campsite. The rate of the river's current was 1 mph. Find the rate at which the campers rowed.

Strategy

- This is a uniform motion problem.
- Unknown rowing rate of the campers: r

	Distance	*Rate*	*Time*
Down river	15	$r + 1$	$\frac{15}{r+1}$
Up river	15	$r - 1$	$\frac{15}{r-1}$

- The total time of the trip was 8 h.

Solution

$$\frac{15}{r+1} + \frac{15}{r-1} = 8$$

$$(r+1)(r-1)\left(\frac{15}{r+1} + \frac{15}{r-1}\right) = (r+1)(r-1)8$$

$$(r-1)15 + (r+1)15 = (r^2 - 1)8$$

$$15r - 15 + 15r + 15 = 8r^2 - 8$$

$$30r = 8r^2 - 8$$

$$0 = 8r^2 - 30r - 8$$

$$0 = 2(4r^2 - 15r - 4)$$

$$0 = 2(4r + 1)(r - 4)$$

$$4r + 1 = 0 \qquad r - 4 = 0$$

$$4r = -1 \qquad r = 4$$

$$r = -\frac{1}{4}$$

The solution $r = -\frac{1}{4}$ is not possible,

because the rate cannot be a negative number.

The rowing rate was 4 mph.

You Try It 1

The length of a rectangle is 3 m more than the width. The area is 54 m^2. Find the length of the rectangle.

Your strategy

Your solution

9 m

Solution on p. S32

10.5 Exercises

Objective A *Application Problems*

1. The base of a triangle is 1 cm less than five times the height of the triangle. The area of the triangle is 21 cm^2. Find the height and the length of the base of the triangle. height: 3 cm; base: 14 cm

2. The length of a rectangle is 2 ft less than three times the width of the rectangle. The area of the rectangle is 65 ft^2. Find the length and width of the rectangle. length: 13 ft; width: 5 ft

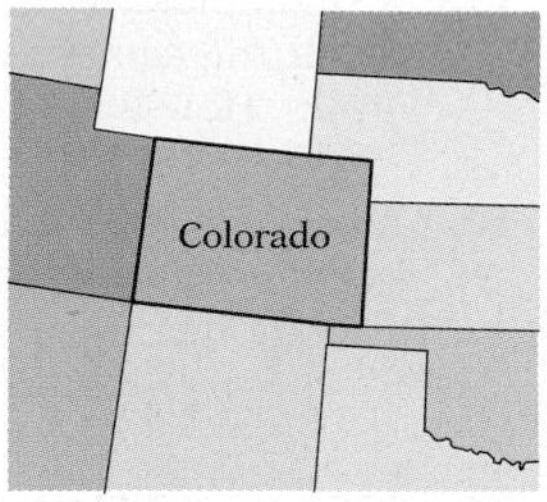

3. The state of Colorado is almost perfectly rectangular, with its north border 111 mi longer than its west border. If the state encompasses 104,000 mi^2, estimate the dimensions of Colorado. Round to the nearest mile. 272 mi by 383 mi

4. A square piece of cardboard is formed into a box by cutting 10-centimeter squares from each of the four corners and then folding up the sides, as shown in the figure. If the volume, V, of the box is to be 49,000 cm^3, what size square piece of cardboard is needed? Recall that $V = LWH$. 90 cm by 90 cm

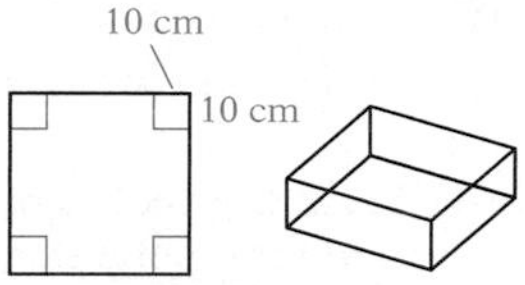

5. A car with good tire tread can stop in less distance than a car with poor tread. The formula for the stopping distance d, in feet, of a car with good tread on dry cement is approximated by $d = 0.04v^2 + 0.5v$, where v is the speed of the car. If the driver must be able to stop within 60 ft, what is the maximum safe speed, to the nearest mile per hour, of the car? 33 mph

6. A model rocket is launched with an initial velocity of 200 ft/s. The height h, in feet, of the rocket t seconds after the launch is given by $h(t) = -16t^2 + 200t$. How many seconds after the launch will the rocket be 300 ft above the ground? Round to the nearest hundredth of a second. 1.74 s, 10.76 s

7. The height of a projectile fired upward is given by the formula $s = v_0t - 16t^2$, where s is the height in feet, v_0 is the initial velocity, and t is the time in seconds. Find the time for a projectile to return to Earth if it has an initial velocity of 200 ft/s. 12.5 s

8. The height of a projectile fired upward is given by the formula $s = v_0t - 16t^2$, where s is the height in feet, v_0 is the initial velocity, and t is the time in seconds. Find the time for a projectile to reach a height of 64 ft if it has an initial velocity of 128 ft/s. Round to the nearest hundredth of a second.
0.54 s (on the way up) or 7.46 s (on the way down)

9. In Germany, there is no speed limit on some portions of the autobahn (highway). Other portions have a speed limit of 180 km/h (approximately 112 mph). The distance d, in meters, required to stop a car traveling at v kilometers per hour is $d(v) = 0.019v^2 + 0.69v$. Approximate, to the nearest tenth, the maximum speed a driver can be traveling and still be able to stop within 150 m. 72.5 km/h

German Autobahn System

Section 10.5

Suggested Assignment

Exercises 1–15, odds
More challenging problems:
Exercises 17, 18

Quick Quiz (Objective 10.5A)

1. A boat traveled 60 mi down a river and then returned. The total time for the round trip was 5 h, and the rate of the current was 5 mph. Find the rate of the boat in still water. 25 mph
2. The height of a triangle is 3 in. less than the length of the base of the triangle. The area of the triangle is 80 in^2. Find the height of the triangle. Round to the nearest hundredth. 11.24 in.

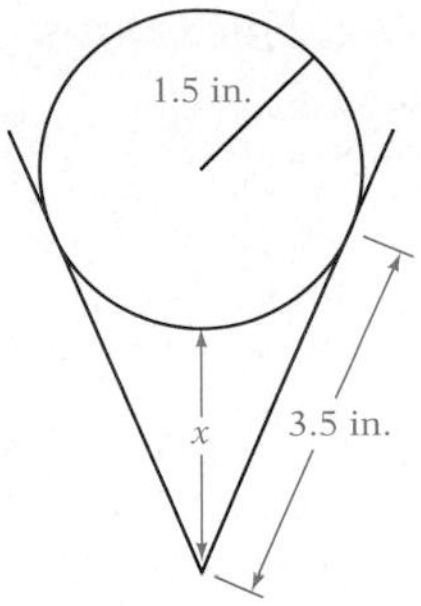

10. A perfectly spherical scoop of mint chocolate chip ice cream is placed in a cone, as shown in the figure. How far is the bottom of the scoop of ice cream from the bottom of the cone? Round to the nearest tenth. (*Hint:* A line segment from the center of the ice cream to the point at which the ice cream touches the cone is perpendicular to the edge of the cone.) 2.3 in.

11. A small pipe can fill a tank in 6 min more time than it takes a larger pipe to fill the same tank. Working together, both pipes can fill the tank in 4 min. How long would it take each pipe working alone to fill the tank? smaller pipe: 12 min; larger pipe: 6 min

12. A cruise ship made a trip of 100 mi in 8 h. The ship traveled the first 40 mi at a constant rate before increasing its speed by 5 mph. Then it traveled another 60 mi at the increased speed. Find the rate of the cruise ship for the first 40 mi. 10 mph

13. The Concorde's speed in calm air is 1320 mph. Flying with the wind, the Concorde can fly from New York to London, a distance of approximately 4000 mi, in 0.5 h less than the time required to make the return trip. Find the rate of the wind to the nearest mile per hour. 108 mph

14. A car travels 120 mi. A second car, traveling 10 mph faster than the first car, makes the same trip in 1 h less time. Find the speed of each car. first car: 30 mph; second car: 40 mph

15. For a portion of the Green River in Utah, the rate of the river's current is 4 mph. A tour guide can row 5 mi down this river and back in 3 h. Find the rowing rate of the guide in calm water. 6 mph

16. The height h, in feet, of an arch is given by the equation $h(x) = -\frac{3}{64}x^2 + 27$, where $|x|$ is the distance in feet from the center of the arch.

a. What is the maximum height of the arch? 27 ft

b. What is the height of the arch 8 ft to the right of the center? 24 ft

c. How far from the center is the arch 8 ft tall? 20.13 ft

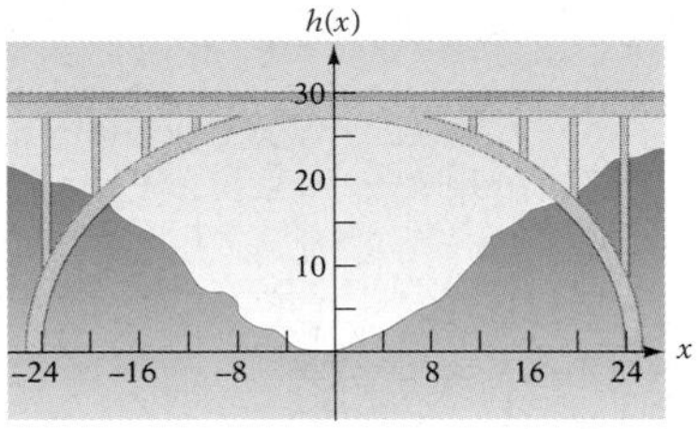

APPLYING THE CONCEPTS

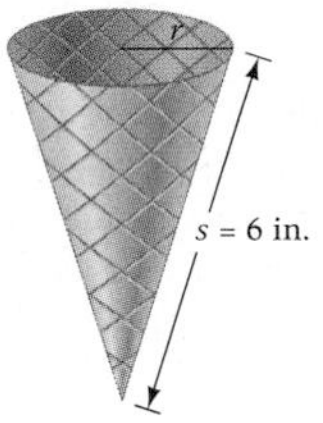

17. The surface area of the ice cream cone shown at the right is given by $A = \pi r^2 + \pi rs$, where r is the radius of the circular top of the cone and s is the slant height of the cone. If the surface area of the cone is 11.25π in^2 and its slant height is 6 in., find the radius of the cone. 1.5 in.

18. Using Torricelli's Principle, it can be shown that the depth, d, of a liquid in a bottle with a hole of area 0.5 cm^2 in its side can be approximated by $d = 0.0034t^2 - 0.52518t + 20$, where t is the time since a stopper was removed from the hole. When will the depth be 10 cm? Round to the nearest tenth of a second. 22.2 s

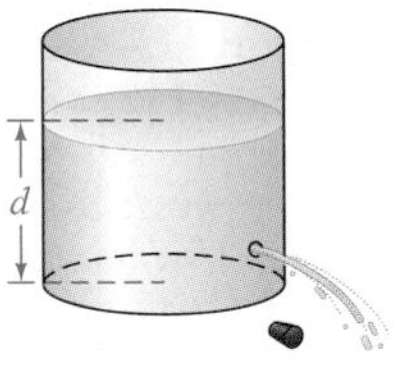

575

10.6 Quadratic Inequalities and Rational Inequalities

Objective A To solve a nonlinear inequality

VIDEO & DVD CD TUTOR WEB SSM

A **quadratic inequality** is one that can be written in the form $ax^2 + bx + c < 0$ or $ax^2 + bx + c > 0$, where $a \neq 0$. The symbols $\leq$ and $\geq$ can also be used. The solution set of a quadratic inequality can be found by solving a compound inequality.

To solve $x^2 - 3x - 10 > 0$, first factor the trinomial.

$$x^2 - 3x - 10 > 0$$
$$(x + 2)(x - 5) > 0$$

There are two cases for which the product of the factors will be positive: (1) both factors are positive, or (2) both factors are negative.

(1) $x + 2 > 0$ and $x - 5 > 0$
(2) $x + 2 < 0$ and $x - 5 < 0$

Solve each pair of compound inequalities.

(1) $x + 2 > 0$ and $x - 5 > 0$
$x > -2$ $\quad x > 5$
$\{x|x > -2\} \cap \{x|x > 5\} = \{x|x > 5\}$

(2) $x + 2 < 0$ and $x - 5 < 0$
$x < -2$ $\quad x < 5$
$\{x|x < -2\} \cap \{x|x < 5\} = \{x|x < -2\}$

Because the two cases for which the product will be positive are connected by *or*, the solution set is the union of the solution sets of the individual inequalities.

$$\{x|x > 5\} \cup \{x|x < -2\} = \{x|x > 5 \text{ or } x < -2\}$$

Although the solution set of any quadratic inequality can be found by using the method outlined above, a graphical method is often easier to use.

⇒ Solve and graph the solution set of $x^2 - x - 6 < 0$.

Factor the trinomial.

$$x^2 - x - 6 < 0$$
$$(x - 3)(x + 2) < 0$$

On a number line, draw lines indicating the numbers that make each factor equal to zero.

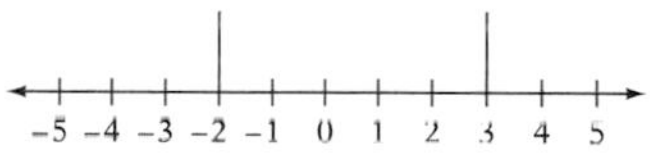

$x - 3 = 0$ $\quad x + 2 = 0$
$x = 3$ $\quad x = -2$

For each factor, place plus signs above the number line for those regions where the factor is positive and minus signs where the factor is negative.

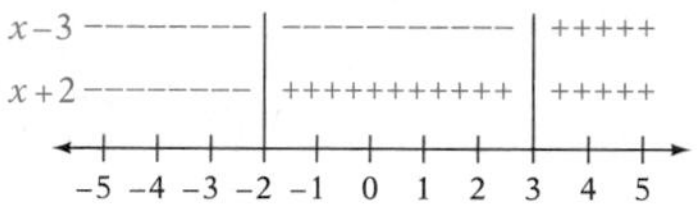

Because $x^2 - x - 6 < 0$, the solution set will be the regions where one factor is positive and the other factor is negative.

Write the solution set.

$$\{x|-2 < x < 3\}$$

The graph of the solution set of $x^2 - x - 6 < 0$ is shown at the right.

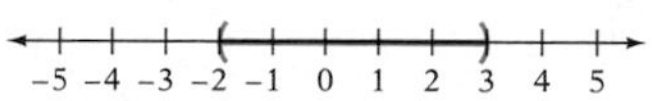

TAKE NOTE
For each factor, choose a number in each region. For example: when $x = -4$, $x - 3$ is negative; when $x = 1$, $x - 3$ is negative; and when $x = 4$, $x - 3$ is positive. When $x = -4$, $x + 2$ is negative; when $x = 1$, $x + 2$ is positive; and when $x = 4$, $x + 2$ is positive.

Objective 10.6A

New Vocabulary
quadratic inequality

Discuss the Concepts
1. For what values of x is $x - 5$ positive? For what values is it negative?
2. Is 4 an element in the solution set of $x^2 - 6x + 5 > 0$?
3. How does the solution set of $x^2 - 3x - 10 > 0$ differ from the solution set of $x^2 - 3x - 10 \geq 0$?

Concept Check
Determine whether the statement is always true, sometimes true, or never true.
1. The endpoints of the solution set of a quadratic inequality are not included in the solution set. Sometimes true
2. The solution set of $(x - 3)(x - 2)(x + 2) > 0$ is 3, 2, and −2. Never true
3. The solution set of a quadratic inequality is an infinite set. Sometimes true
4. The solution set of $\frac{1}{x - 1} > 0$ is $\{x|x > 1\}$. Always true

In-Class Examples (Objective 10.6A)
Solve.
1. $x^2 - 5x + 6 \geq 0$ $\{x|x \leq 2 \text{ or } x \geq 3\}$
2. $(x - 1)(x + 1)(x - 2) < 0$
$\{x|x < -1 \text{ or } 1 < x < 2\}$
3. $\frac{x - 3}{x + 1} > 0$ $\{x|x < -1 \text{ or } x > 3\}$
4. $\frac{(x - 2)(x + 1)}{x - 3} \leq 0$ $\{x|x \leq -1 \text{ or } 2 \leq x < 3\}$

Concept Check

1. How many integers satisfy the inequality $x^2 + 48 < 16x$? Seven
2. Find all values of x that satisfy both $x^2 - 8 \le 2x$ and $x^2 - 2x \ge 8$. $-2, 4$

Optional Student Activity

1. Solve: $x < x^2$ $\{x \mid x < 0 \text{ or } x > 1\}$
2. Solve: $x^3 > x$ $\{x \mid -1 < x < 0 \text{ or } x > 1\}$
3. Solve: $(x - 4)^2 > -2$ All real numbers
4. You shoot an arrow into the air with an initial velocity of 70 m/s. The distance up, in meters, is given by $d = rt - 5t^2$, where t is the number of seconds since the arrow was shot and r is the initial velocity of the arrow. Find the interval of time when the arrow will be more than 200 m high. $4 < t < 10$; between 4 s and 10 s

➡ Solve and graph the solution set of $(x - 2)(x + 1)(x - 4) > 0$.

On a number line, identify for each factor the regions where the factor is positive and those where the factor is negative.

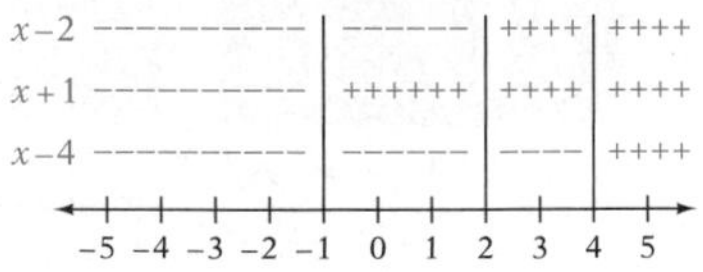

There are two regions where the product of the three factors is positive.

Write the solution set. $\{x \mid -1 < x < 2 \text{ or } x > 4\}$

The graph of the solution set of $(x - 2)(x + 1)(x - 4) > 0$ is shown at the right.

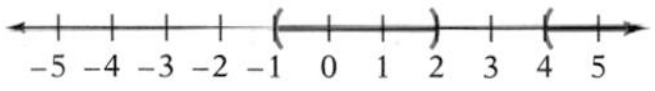

➡ Solve: $\dfrac{2x - 5}{x - 4} \le 1$

$$\frac{2x - 5}{x - 4} \le 1$$

Rewrite the inequality so that 0 appears on the right side of the inequality.

$$\frac{2x - 5}{x - 4} - 1 \le 0$$

Simplify.

$$\frac{2x - 5}{x - 4} - \frac{x - 4}{x - 4} \le 0$$

$$\frac{x - 1}{x - 4} \le 0$$

On a number line, identify for each factor of the numerator and each factor of the denominator the regions where the factor is positive and those where the factor is negative.

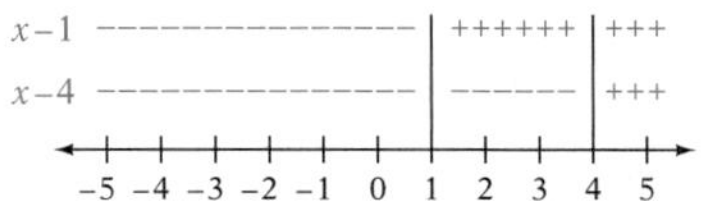

The region where the quotient of the two factors is negative is between 1 and 4.

Write the solution set. $\{x \mid 1 \le x < 4\}$

Note that 1 is part of the solution set but 4 is not because the denominator of the rational expression is zero when $x = 4$.

Example 1
Solve and graph the solution set of $2x^2 - x - 3 \ge 0$.

Solution

$$2x^2 - x - 3 \ge 0$$
$$(2x - 3)(x + 1) \ge 0$$

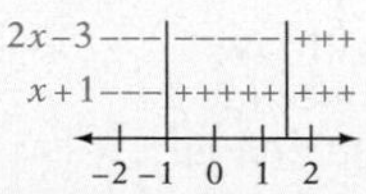

$$\left\{x \mid x \le -1 \text{ or } x \ge \frac{3}{2}\right\}$$

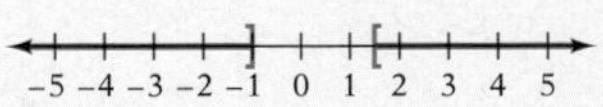

You Try It 1
Solve and graph the solution set of $2x^2 - x - 10 \le 0$.

Your solution

$$\left\{x \mid -2 \le x \le \frac{5}{2}\right\}$$

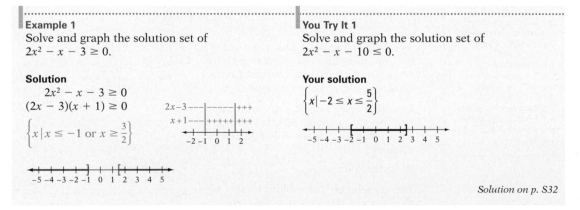

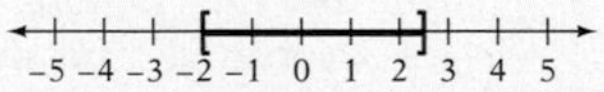

Solution on p. S32

10.6 Exercises

Objective A

1. If $(x - 3)(x - 5) > 0$, what must be true of the values of $x - 3$ and $x - 5$?

2. For the inequality $\frac{x-2}{x-3} \le 1$, which of the values 1, 2, and 3 is not a possible element of the solution set? Why?

Solve and graph the solution set.

3. $(x - 4)(x + 2) > 0$

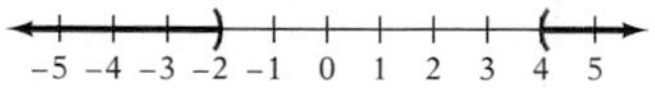

$\{x|x < -2 \text{ or } x > 4\}$

4. $(x + 1)(x - 3) > 0$

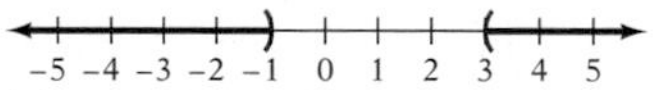

$\{x|x < -1 \text{ or } x > 3\}$

5. $x^2 - 3x + 2 \ge 0$

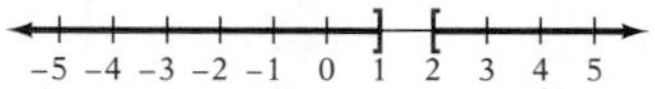

$\{x|x \le 1 \text{ or } x \ge 2\}$

6. $x^2 + 5x + 6 > 0$

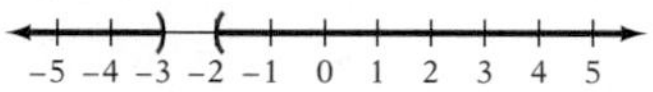

$\{x|x < -3 \text{ or } x > -2\}$

7. $x^2 - x - 12 < 0$

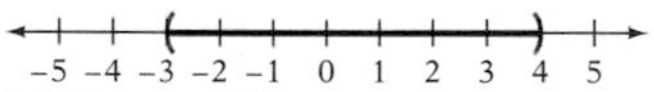

$\{x|-3 < x < 4\}$

8. $x^2 + x - 20 < 0$

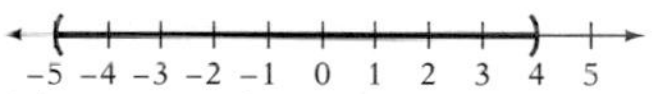

$\{x|-5 < x < 4\}$

9. $(x - 1)(x + 2)(x - 3) < 0$

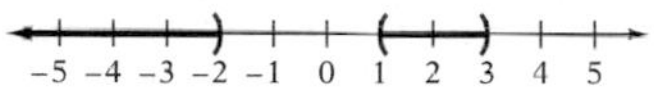

$\{x|x < -2 \text{ or } 1 < x < 3\}$

10. $(x + 4)(x - 2)(x + 1) > 0$

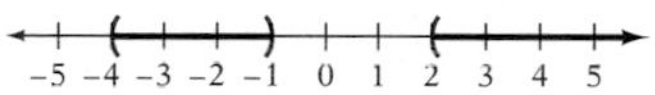

$\{x|-4 < x < -1 \text{ or } x > 2\}$

11. $(x + 4)(x - 2)(x - 1) \ge 0$

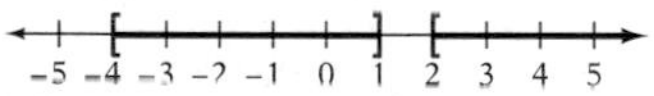

$\{x|-4 \le x \le 1 \text{ or } x \ge 2\}$

12. $(x - 1)(x + 5)(x - 2) \le 0$

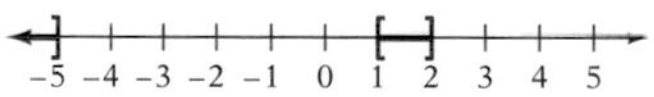

$\{x|x \le -5 \text{ or } 1 \le x \le 2\}$

13. $\frac{x-4}{x+2} > 0$

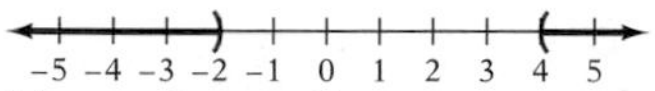

$\{x|x < -2 \text{ or } x > 4]$

14. $\frac{x+2}{x-3} > 0$

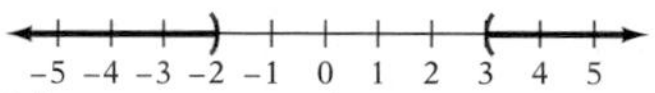

$\{x|x < -2 \text{ or } x > 3\}$

15. $\frac{x-3}{x+1} \le 0$

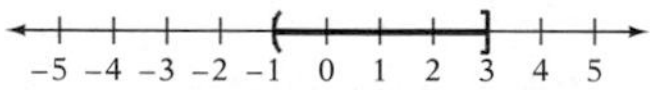

$\{x|-1 < x \le 3\}$

16. $\frac{x-1}{x} > 0$

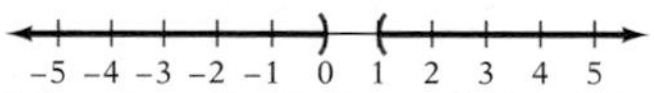

$\{x|x < 0 \text{ or } x > 1\}$

Quick Quiz (Objective 10.6A)

Solve and graph the solution set.

1. $x^2 + 2x - 8 < 0$
$\{x|-4 < x < 2\}$

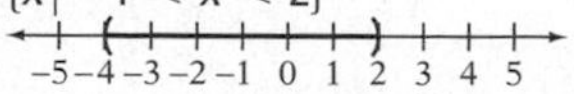

2. $(x + 3)(x - 1)(x + 2) > 0$
$\{x|x > 1 \text{ or } -3 < x < -2\}$

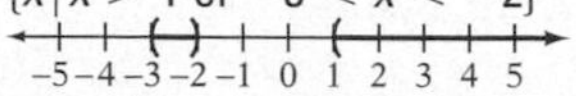

3. $\frac{x+1}{x-2} > 0$ $\{x|x < -1 \text{ or } x > 2\}$

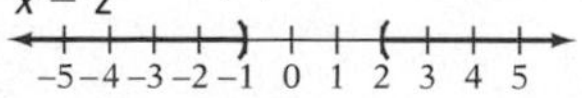

Section 10.6

Suggested Assignment

Exercises 1–33, odds
More challenging problems:
Exercises 35–39, odds

Answers to Writing Exercises

1. It must be true that $x - 3 > 0$ and $x - 5 > 0$ or that $x - 3 < 0$ and $x - 5 < 0$. In other words, either both factors are positive or both factors are negative.
2. 3 is not a possible element of the solution set because the value 3 makes the denominator equal to 0.

578 Chapter 10 / Quadratic Equations

17. $\dfrac{(x-1)(x+2)}{x-3} \le 0$

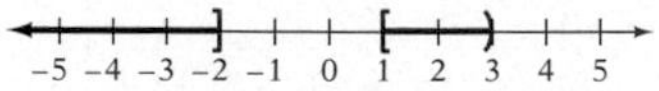

$\{x|x \le -2 \text{ or } 1 \le x < 3\}$

18. $\dfrac{(x+3)(x-1)}{x-2} \ge 0$

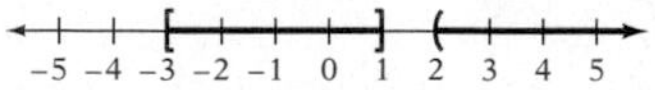

$\{x|-3 \le x \le 1 \text{ or } x > 2\}$

Solve.

19. $x^2 - 16 > 0$
$\{x|x > 4 \text{ or } x < -4\}$

20. $x^2 - 4 \ge 0$
$\{x|x \le -2 \text{ or } x \ge 2\}$

21. $x^2 - 9x \le 36$
$\{x|-3 \le x \le 12\}$

22. $x^2 + 4x > 21$
$\{x|x < -7 \text{ or } x > 3\}$

23. $4x^2 - 8x + 3 < 0$
$\left\{x\middle|\dfrac{1}{2} < x < \dfrac{3}{2}\right\}$

24. $2x^2 + 11x + 12 \ge 0$
$\left\{x\middle|x \le -4 \text{ or } x \ge -\dfrac{3}{2}\right\}$

25. $\dfrac{3}{x-1} < 2$
$\left\{x\middle|x < 1 \text{ or } x > \dfrac{5}{2}\right\}$

26. $\dfrac{x}{(x-1)(x+2)} \ge 0$
$\{x|x > 1 \text{ or } -2 < x \le 0\}$

27. $\dfrac{x-2}{(x+1)(x-1)} \le 0$
$\{x|x < -1 \text{ or } 1 < x \le 2\}$

28. $\dfrac{1}{x} < 2$
$\left\{x\middle|x > \dfrac{1}{2} \text{ or } x < 0\right\}$

29. $\dfrac{x}{2x-1} \ge 1$
$\left\{x\middle|\dfrac{1}{2} < x \le 1\right\}$

30. $\dfrac{x}{2x-3} \le 1$
$\left\{x\middle|x < \dfrac{3}{2} \text{ or } x \ge 3\right\}$

31. $\dfrac{x}{2-x} \le -3$
$\{x|2 < x \le 3\}$

32. $\dfrac{3}{x-2} > \dfrac{2}{x+2}$
$\{x|x > 2 \text{ or } -10 < x < -2\}$

33. $\dfrac{3}{x-5} > \dfrac{1}{x+1}$
$\{x|x > 5 \text{ or } -4 < x < -1\}$

APPLYING THE CONCEPTS

Graph the solution set.

34. $(x+2)(x-3)(x+1)(x+4) > 0$

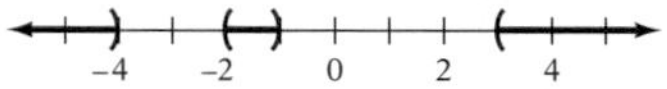

35. $(x-1)(x+3)(x-2)(x-4) \ge 0$

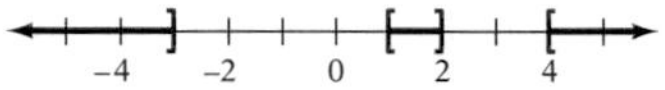

36. $(x^2 + 2x - 8)(x^2 - 2x - 3) < 0$

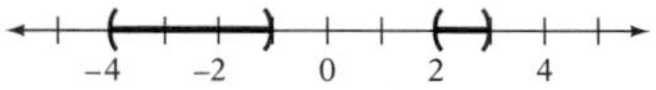

37. $(x^2 + 2x - 3)(x^2 + 3x + 2) \ge 0$

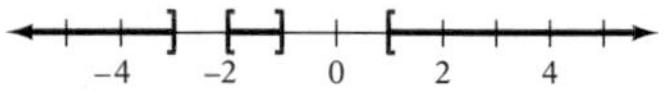

38. $(x^2 + 1)(x^2 - 3x + 2) > 0$

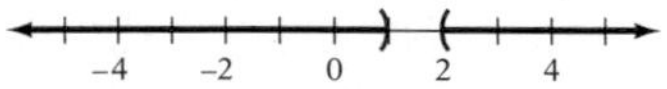

39. $\dfrac{x^2(3-x)(2x+1)}{(x+4)(x+2)} \ge 0$

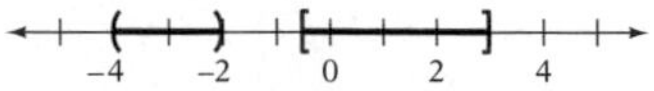

579

Focus on Problem Solving

Use a Variety of Problem-Solving Techniques

We have examined several problem-solving strategies throughout the text. See if you can apply those techniques to the following problems.

1. Eight coins look exactly alike, but one is lighter than the others. Explain how the different coin can be found in two weighings on a balance scale.

2. For the sequence of numbers 1, 1, 2, 3, 5, 8, 13, . . . , identify a possible pattern and then use that pattern to determine the next number in the sequence.

3. Arrange the numbers 1, 2, 3, 4, 5, 6, 7, 8, and 9 in the squares at the right so that the sum of any row, column, or diagonal is 15. (*Suggestion:* Note that 1, 5, 9; 2, 5, 8; 3, 5, 7; and 4, 5, 6 all add to 15. Because 5 is part of each sum, this suggests that 5 be placed in the center of the squares.)

	5	

4. A restaurant charges \$10.00 for a pizza that has a diameter of 9 in. Determine the selling price of a pizza with a diameter of 18 in. so that the selling price per square inch is the same as for the 9-inch pizza.

5. You have a balance scale and weights of 1 g, 4 g, 8 g, and 16 g. Using only these weights, can you weigh something that weighs 7 g? 9 g? 12 g? 19 g?

6. Can the checkerboard at the right be covered with dominos (which look like ) so that every square on the board is covered by a domino? Why or why not? (*Note:* The dominos cannot overlap.)

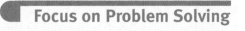

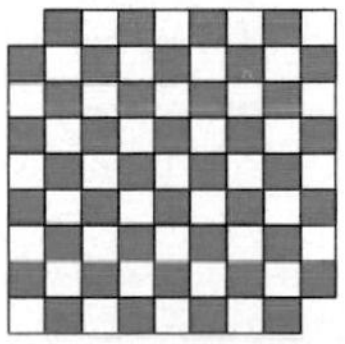

Projects and Group Activities

Completing the Square

Essentially all of the investigations into mathematics before the Renaissance were geometric. The solutions of quadratic equations were calculated from a construction of a certain area. Proofs of theorems, even theorems about numbers, were based entirely on geometry. In this project, we will examine the geometric solution of a quadratic equation.

Answers to Focus on Problem Solving: Use a Variety of Problem-Solving Techniques

1. Label the coins 1, 2, 3, 4, 5, 6, 7, and 8. Set coins 1 and 2 aside. Put coins 3, 4, and 5 on one side of the balance and coins 6, 7, and 8 on the other side of the balance. If the scale balances, then coin 1 or coin 2 is the lighter coin. You can determine which one by using the balance a second time.
   If the scale does not balance, choose the lighter set of coins. Assume the lighter set includes coins 6, 7, and 8. Choose two of these coins to put on the balance, say coins 6 and 7. If the coins do not balance, you have found the lighter coin.
   If coins 6 and 7 are equal in weight, then coin 8 is the lighter coin. Thus, the lighter coin can be found in two weighings.
2. Each term, after the first two, is the sum of the two preceding terms. Using that pattern, the next number in the sequence is 21.
3. For example:
   Row 1: 6, 7, 2
   Row 2: 1, 5, 9
   Row 3: 8, 3, 4
4. \$40.00
5. For 7 g, use 8 g on one side and 1 g plus the 7-gram weight on the other side of the balance. For 9 g, use 1-gram and 8-gram weights to balance the 9-gram weight.
   For 12 g, use 8 g and 4 g on one side and the 12-gram weight on the other side of the balance.
   For 19 g, use 4 g and 16 g on one side and 1 g plus the 19-gram weight on the other side of the balance.
6. No. There are an odd number of squares on the board.

Answers to Projects and Group Activities: Completing the Square

To solve $x^2 + 4x = 12$ by geometrically completing the square, begin with a line of unknown length, x, and form a square measuring x on a side. Draw a rectangle of width x and length 2 at the right of the square and another such rectangle below the square.

Complete the square with sides $x + 2$. The area of the large square is $(x + 2)^2$. From your diagram, note that

$$x^2 + 4x + 4 = (x + 2)^2$$

From the original equation, $x^2 + 4x = 12$. Thus

$$\begin{aligned} x^2 + 4x + 4 &= (x + 2)^2 \\ 12 + 4 &= (x + 2)^2 \\ 16 &= (x + 2)^2 \\ 4 &= x + 2 \\ 2 &= x \end{aligned}$$

The positive solution is 2.

➡ Solve: $x^2 + 6x = 7$

Begin with a line of unknown length, x, and one of length 6, the coefficient of x. Using these lines, construct a rectangle as shown.

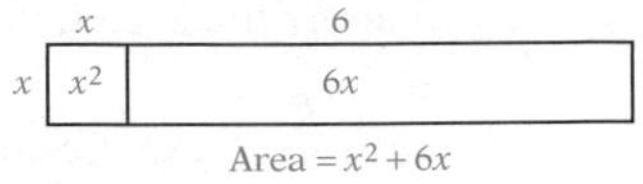

Figure 1

Now draw another area that has exactly the same area as Figure 1 by cutting off one-half of the rectangle of area $6x$ and placing it on the bottom of the square labeled x^2. See Figure 2.

The unshaded area in Figure 2 has exactly the same area as Figure 1. However, when the shaded area is added to Figure 2 to make a square, the total area is 9 square units larger than that of Figure 1. In equation form,

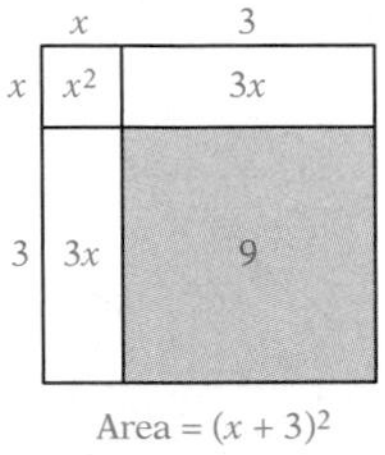

Figure 2

$$\text{(Area of Figure 1)} + 9 = \text{area of Figure 2}$$

or

$$x^2 + 6x + 9 = (x + 3)^2$$

From the original equation, $x^2 + 6x = 7$. Thus,

$$\begin{aligned} x^2 + 6x + 9 &= (x + 3)^2 \\ 7 + 9 &= (x + 3)^2 & \bullet\ x^2 + 6x = 7 \\ 16 &= (x + 3)^2 \\ 4 &= x + 3 & \bullet\ \text{See note below.} \\ 1 &= x \end{aligned}$$

Note: Although early mathematicians knew that a quadratic equation may have two solutions, both solutions were allowed only if they were positive. After all, a geometric construction could not have a negative length. Therefore, the solution of this equation was 1; the solution -7 would have been dismissed as *fictitious*, the actual word that was frequently used through the 15th century for negative-number solutions of an equation.

Try to solve the quadratic equation $x^2 + 4x = 12$ by geometrically completing the square.

Using a Graphing Calculator to Solve a Quadratic Equation

Recall that an x-intercept of the graph of an equation is a point at which the graph crosses the x-axis. For the graph in Figure 3, the x-intercepts are $(-2, 0)$ and $(3, 0)$.

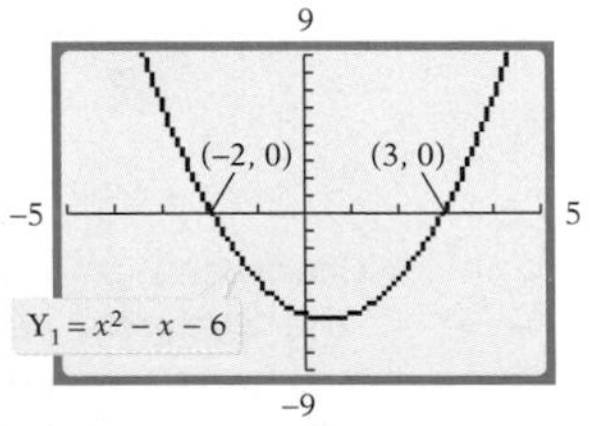

Figure 3

581

Recall also that to find the x-intercept of a graph, set $y = 0$ and then solve for x. For the equation in Figure 3, if we set $y = 0$, the resulting equation is $0 = x^2 - x - 6$, which is a quadratic equation. Solving this equation by factoring, we have

$$0 = x^2 - x - 6$$
$$0 = (x + 2)(x - 3)$$

$$x + 2 = 0 \qquad\qquad x - 3 = 0$$
$$x = -2 \qquad\qquad x = 3$$

Thus the solutions of the equation are the x-coordinates of the x-intercepts of the graph.

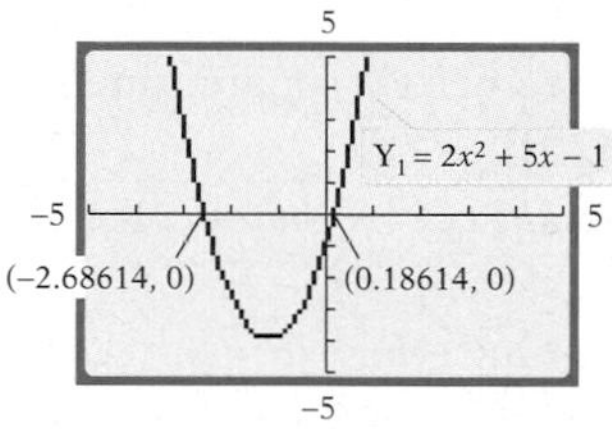

Figure 4

This connection between the solutions of an equation and the x-intercepts of its graph enables us to find approximations of the real number solutions of an equation graphically. For example, to use a TI-83 to approximate the solutions of $2x^2 + 5x - 1 = 0$ graphically, use the following keystrokes to graph $y = 2x^2 + 5x - 1$ and find the decimal approximations of the x-coordinates of the x-intercepts. Use the window shown in the graph in Figure 4.

[Y=] [CLEAR] 2 [X,T,θ,n] [x^2] [+] 5 [X,T,θ,n] [−] 1
[2nd] [CALC] 2

Use the arrow keys to move to the left of the leftmost x-intercept. Press ENTER. Use the arrow keys to move to a point just to the right of the leftmost x-intercept. Press ENTER twice. The x-coordinate at the bottom of the screen is the approximation of one solution of the equation. To find the other solution, use the same procedure, but move the cursor first to the left and then to the right of the rightmost x-intercept.

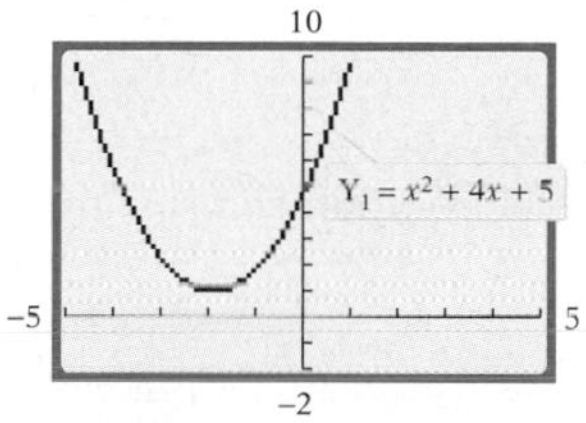

Figure 5

Attempting to find the solutions of an equation graphically will not necessarily find all the solutions. Because the x-coordinates of the x-intercepts of a graph are *real* numbers, only real number solutions can be found. For instance, consider the equation $x^2 + 4x + 5 = 0$. The graph of $y = x^2 + 4x + 5$ is shown in Figure 5. Note that the graph has no x-intercepts and, consequently, no real number solutions. However, $x^2 + 4x + 5 = 0$ does have complex number solutions that can be obtained by using the quadratic formula. They are $-2 + i$ and $-2 - i$.

Chapter Summary

Key Words

A *quadratic equation* is an equation of the form $ax^2 + bx + c = 0$, where a and b are coefficients, c is a constant, and $a \neq 0$. A quadratic equation is also called a *second-degree equation.* [p. 545]

A quadratic equation is in *standard form* when the polynomial is in descending order and equal to zero. [p. 545]

When a quadratic equation has two solutions that are the same number, the solution is called a *double root* of the equation. [p. 545]

Methods of Solving a Quadratic Equation:

1. **Factoring**
   This method is used only when the polynomial $ax^2 + bx + c$ is factorable.
2. **Taking square roots**
   This method is used only when the value of b in $ax^2 + bx + c$ is 0.
   It is also used when an equation can be written in the form

   Square of a binomial = constant
3. **Completing the square**
   This method can be used to solve any quadratic equation.
4. **Quadratic formula**
   This method can be used to solve any quadratic equation.
   [pp. 545, 547–548, 553–555, 559–561]

Adding to a binomial the constant term that makes it a perfect-square trinomial is called *completing the square.* [p. 553]

For an equation of the form $ax^2 + bx + c = 0$, the quantity $b^2 - 4ac$ is called the *discriminant.* [p. 561]

An equation is *quadratic in form* if it can be written as $au^2 + bu + c = 0$. [p. 565]

A *quadratic inequality* is one that can be written in the form $ax^2 + bx + c > 0$ or $ax^2 + bx + c < 0$, where $a \neq 0$. The symbols $\leq$ or $\geq$ can also be used. [p. 575]

Essential Rules

The Principle of Zero Products [p. 545] If $ab = 0$, then $a = 0$ or $b = 0$.

To Solve a Quadratic Equation by Factoring [p. 545]
1. Write the equation in standard form.
2. Factor.
3. Use the Principle of Zero Products to set each factor equal to 0.
4. Solve each equation.
5. Check the solutions.

To Write a Quadratic Equation Given Its Solutions [p. 547]
Use the equation $(x - r_1)(x - r_2) = 0$. Replace r_1 with one solution and r_2 with the other solution. Then multiply the two factors.

To Complete the Square [p. 553]
Add to a binomial of the form $x^2 + bx$ the square of $\frac{1}{2}$ of the coefficient of x.

To Solve a Quadratic Equation by Completing the Square [p. 555]
1. Write the equation in the form $ax^2 + bx = -c$.
2. Multiply both sides of the equation by $\frac{1}{a}$.
3. Complete the square on $x^2 + \frac{b}{a}x$. Add the number that completes the square to both sides of the equation.
4. Factor the perfect-square trinomial.
5. Take the square root of each side of the equation.
6. Solve the resulting equation for x.
7. Check the solutions.

The Quadratic Formula [p. 559]
The solutions of $ax^2 + bx + c = 0$, $a \neq 0$, are $x = \dfrac{-b \pm \sqrt{b^2 - 4ac}}{2a}$.

The Effect of the Discriminant on the Solutions of a Quadratic Equation [p. 561]
1. If $b^2 - 4ac = 0$, the equation has one real number solution, a double root.
2. If $b^2 - 4ac > 0$, the equation has two unequal real number solutions.
3. If $b^2 - 4ac < 0$, the equation has two complex number solutions.

583

Chapter Review

1. Solve by factoring: $2x^2 - 3x = 0$
$0, \frac{3}{2}$ [10.1A]

2. Solve by factoring: $6x^2 + 9cx = 6c^2$
$\frac{c}{2}, -2c$ [10.1A]

3. Solve by taking square roots:
$x^2 = 48$
$4\sqrt{3}, -4\sqrt{3}$ [10.1C]

4. Solve by taking square roots:
$\left(x + \frac{1}{2}\right)^2 + 4 = 0$
$-\frac{1}{2} + 2i, -\frac{1}{2} - 2i$ [10.1C]

5. Solve by completing the square:
$x^2 + 4x + 3 = 0$
$-3, -1$ [10.2A]

6. Solve by completing the square:
$7x^2 - 14x + 3 = 0$
$\frac{7 + 2\sqrt{7}}{7}, \frac{7 - 2\sqrt{7}}{7}$ [10.2A]

7. Solve by using the quadratic formula:
$12x^2 - 25x + 12 = 0$
$\frac{3}{4}, \frac{4}{3}$ [10.3A]

8. Solve by using the quadratic formula:
$x^2 - x + 8 = 0$
$\frac{1}{2} + \frac{\sqrt{31}}{2}i, \frac{1}{2} - \frac{\sqrt{31}}{2}i$ [10.3A]

9. Write a quadratic equation that has integer coefficients and has solutions 0 and -3.
$x^2 + 3x = 0$ [10.1B]

10. Write a quadratic equation that has integer coefficients and has solutions $\frac{3}{4}$ and $-\frac{2}{3}$
$12x^2 - x - 6 = 0$ [10.1B]

11. Solve by completing the square:
$x^2 - 2x + 8 = 0$
$1 + i\sqrt{7}, 1 - i\sqrt{7}$ [10.2A]

12. Solve by completing the square:
$(x - 2)(x + 3) = x - 10$
$2i, -2i$ [10.2A]

13. Solve by using the quadratic formula:
$3x(x - 3) = 2x - 4$
$\frac{11 + \sqrt{73}}{6}, \frac{11 - \sqrt{73}}{6}$ [10.3A]

14. Use the discriminant to determine whether $3x^2 - 5x + 1 = 0$ has one real number solution, two real number solutions, or two complex number solutions.
two real number solutions [10.3A]

15. Solve: $(x + 3)(2x - 5) < 0$
$\left\{x \middle| -3 < x < \frac{5}{2}\right\}$ [10.6A]

16. Solve: $(x - 2)(x + 4)(2x + 3) \le 0$
$\left\{x \middle| x \le -4 \text{ or } -\frac{3}{2} \le x \le 2\right\}$ [10.6A]

17. Solve: $x^{2/3} + x^{1/3} - 12 = 0$

27, −64 [10.4A]

18. Solve: $2(x - 1) + 3\sqrt{x - 1} - 2 = 0$

$\frac{5}{4}$ [10.4B]

19. Solve: $3x = \frac{9}{x - 2}$

−1, 3 [10.4C]

20. Solve: $\frac{3x + 7}{x + 2} + x = 3$

−1 [10.4C]

21. Solve and graph the solution set:

$\frac{x - 2}{2x - 3} \geq 0$

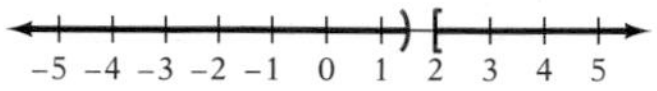

$\left\{x \mid x < \frac{3}{2} \text{ or } x \geq 2\right\}$ [10.6A]

22. Solve and graph the solution set:

$\frac{(2x - 1)(x + 3)}{x - 4} \leq 0$

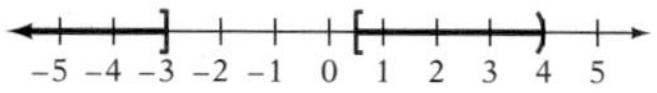

$\left\{x \mid x \leq -3 \text{ or } \frac{1}{2} \leq x < 4\right\}$ [10.6A]

23. Solve: $x = \sqrt{x} + 2$

4 [10.4B]

24. Solve: $2x = \sqrt{5x + 24} + 3$

5 [10.4B]

25. Solve: $\frac{x - 2}{2x + 3} - \frac{x - 4}{x} = 2$

$\frac{-3 + \sqrt{249}}{10}, \frac{-3 - \sqrt{249}}{10}$ [10.4C]

26. Solve: $1 - \frac{x + 4}{2 - x} = \frac{x - 3}{x + 2}$

$\frac{-11 + \sqrt{129}}{2}, \frac{-11 - \sqrt{129}}{2}$ [10.4C]

27. The length of a rectangle is two more centimeters than twice the width. The area of the rectangle is 60 cm^2. Find the length and width of the rectangle.

length: 12 cm; width: 5 cm [10.5A]

28. The sum of the squares of three consecutive even integers is fifty-six. Find the three integers.

2, 4, and 6 or −6, −4, and −2 [10.5A]

29. An older computer requires 12 min longer to print the payroll than does a newer computer. Together the computers can print the payroll in 8 min. Find the time for the new computer working alone to print the payroll.

12 min [10.5A]

30. A car travels 200 mi. A second car, making the same trip, travels 10 mph faster than the first car and makes the trip in 1 h less time. Find the speed of each car.

first car: 40 mph; second car: 50 mph [10.5A]

585

Chapter Test

1. Solve by factoring: $3x^2 + 10x = 8$
$\frac{2}{3}, -4$ [10.1A]

2. Solve by factoring: $6x^2 - 5x - 6 = 0$
$\frac{3}{2}, -\frac{2}{3}$ [10.1A]

3. Write a quadratic equation that has integer coefficients and has solutions 3 and −3.
$x^2 - 9 = 0$ [10.1B]

4. Write a quadratic equation that has integer coefficients and has solutions $\frac{1}{2}$ and −4.
$2x^2 + 7x - 4 = 0$ [10.1B]

5. Solve by taking square roots:
$3(x - 2)^2 - 24 = 0$
$2 + 2\sqrt{2}, 2 - 2\sqrt{2}$ [10.1C]

6. Solve by completing the square:
$x^2 - 6x - 2 = 0$
$3 + \sqrt{11}, 3 - \sqrt{11}$ [10.2A]

7. Solve by completing the square:
$3x^2 - 6x = 2$
$\frac{3 + \sqrt{15}}{3}, \frac{3 - \sqrt{15}}{3}$ [10.2A]

8. Solve by using the quadratic formula:
$2x^2 - 2x = 1$
$\frac{1 + \sqrt{3}}{2}, \frac{1 - \sqrt{3}}{2}$ [10.3A]

9. Solve by using the quadratic formula:
$x^2 + 4x + 12 = 0$
$-2 + 2i\sqrt{2}, -2 - 2i\sqrt{2}$ [10.3A]

10. Use the discriminant to determine whether $3x^2 - 4x = 1$ has one real number solution, two real number solutions, or two complex number solutions.
two real number solutions [10.3A]

11. Use the discriminant to determine whether $x^2 - 6x = -15$ has one real number solution, two real number solutions, or two complex number solutions.
two complex number solutions [10.3A]

12. Solve: $2x + 7x^{1/2} - 4 = 0$
$\frac{1}{4}$ [10.4A]

586

13. Solve: $x^4 - 4x^2 + 3 = 0$
$1, -1, \sqrt{3}, -\sqrt{3}$ [10.4A]

14. Solve: $\sqrt{2x + 1} + 5 = 2x$
4 [10.4B]

15. Solve: $\sqrt{x - 2} = \sqrt{x} - 2$

no solution [10.4B]

16. Solve: $\frac{2x}{x - 3} + \frac{5}{x - 1} = 1$

$2, -9$ [10.4C]

17. Solve and graph the solution set of $(x - 2)(x + 4)(x - 4) < 0$.

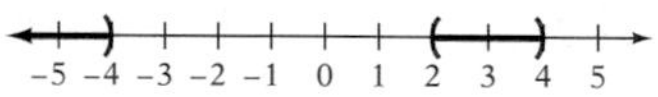

$\{x|x < -4 \text{ or } 2 < x < 4\}$ [10.6A]

18. Solve and graph the solution set of $\frac{2x - 3}{x + 4} \le 0$.

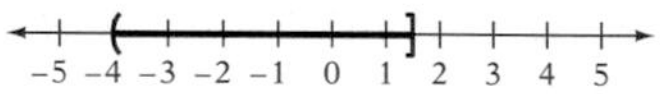

$\left\{x \middle| -4 < x \le \frac{3}{2}\right\}$ [10.6A]

19. The base of a triangle is 3 ft more than three times the height. The area of the triangle is 30 ft^2. Find the base and height of the triangle.
base: 15 ft; height: 4 ft [10.5A]

20. The rate of a river's current is 2 mph. A canoe was rowed 6 mi down the river and back in 4 h. Find the rowing rate in calm water.
4 mph [10.5A]

587

Cumulative Review

1. Evaluate $2a^2 - b^2 \div c^2$ when $a = 3$, $b = -4$, and $c = -2$.
14 [1.4A]

2. Solve: $|3x - 2| < 8$
$\left\{x \middle| -2 < x < \frac{10}{3}\right\}$ [2.5B]

3. Find the volume of a cylinder with a height of 6 m and a radius of 3 m. Give the exact measure.
54π m^3 [3.3A]

4. Given $f(x) = \frac{2x - 3}{x^2 - 1}$, find $f(-2)$.
$-\frac{7}{3}$ [4.2A]

5. Find the slope of the line containing the points $(3, -4)$ and $(-1, 2)$.
$-\frac{3}{2}$ [4.4A]

6. Find the x- and y-intercepts of the graph of $6x - 5y = 15$.
$\left(\frac{5}{2}, 0\right)$, $(0, -3)$ [4.3B]

7. Find the equation of the line that contains the point $(1, 2)$ and is parallel to the line $x - y = 1$.
$y = x + 1$ [4.6A]

8. Solve the system of equations.
$$x + y + z = 2$$
$$-x + 2y - 3z = -9$$
$$x - 2y - 2z = -1$$
$(1, -1, 2)$ [5.2B]

9. Graph the solution set:
$x + y \leq 3$
$2x - y < 4$

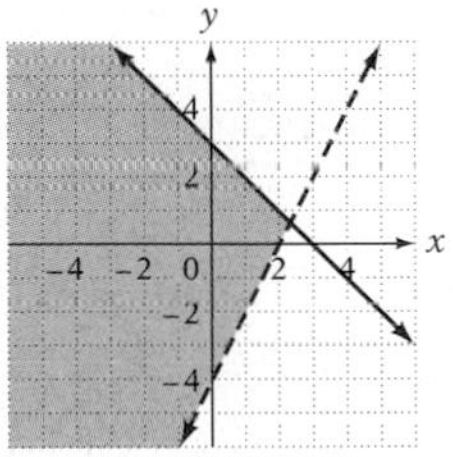

[5.5A]

10. Triangles ABC and DEF are similar. Find the height of triangle DEF.

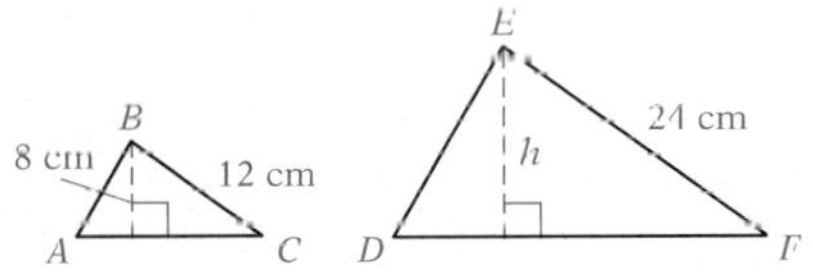

16 cm [8.4C]

11. Divide: $(3x^3 - 13x^2 + 10) \div (3x - 4)$
$x^2 - 3x - 4 - \frac{6}{3x - 4}$ [6.4A]

12. Factor: $-3x^3y + 6x^2y^2 - 9xy^3$
$-3xy(x^2 - 2xy + 3y^2)$ [7.1A]

13. Factor: $6x^2 - 7x - 20$
$(2x - 5)(3x + 4)$ [7.3A/7.3B]

14. Multiply: $\frac{x^2 + 2x + 1}{8x^2 + 8x} \cdot \frac{4x^3 - 4x^2}{x^2 - 1}$
$\frac{x}{2}$ [8.1B]

15. Solve: $\frac{x}{x+2} - \frac{4x}{x+3} = 1$

$-\frac{3}{2}, -1$ [8.4A]

16. Solve $S = \frac{n}{2}(a + b)$ for b.

$b = \frac{2S - an}{n}$ [8.5A]

17. Multiply: $a^{-1/2}(a^{1/2} - a^{3/2})$

$1 - a$ [9.1A]

18. Multiply: $-2i(7 - 4i)$

$-8 - 14i$ [9.3C]

19. Solve: $\sqrt{3x + 1} - 1 = x$

0, 1 [9.4A]

20. Solve: $x^4 - 6x^2 + 8 = 0$

$2, -2, \sqrt{2}, -\sqrt{2}$ [10.4A]

21. A piston rod for an automobile is $9\frac{3}{8}$ in. with a tolerance of $\frac{1}{64}$ in. Find the lower and upper limits of the length of the piston rod.

lower limit: $9\frac{23}{64}$ in.; upper limit: $9\frac{25}{64}$ in. [2.5C]

22. The base of a triangle is $(x + 8)$ ft. The height is $(2x - 4)$ ft. Find the area of the triangle in terms of the variable x.

$(x^2 + 6x - 16)$ ft^2 [6.3E]

23. The graph shows the relationship between the value of a building and the time, in years, since depreciation began. Find the slope of the line between the two points shown on the graph. Write a sentence that states the meaning of the slope.

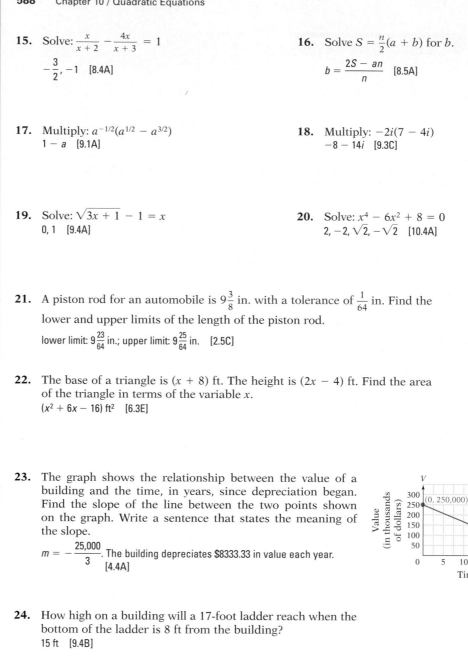

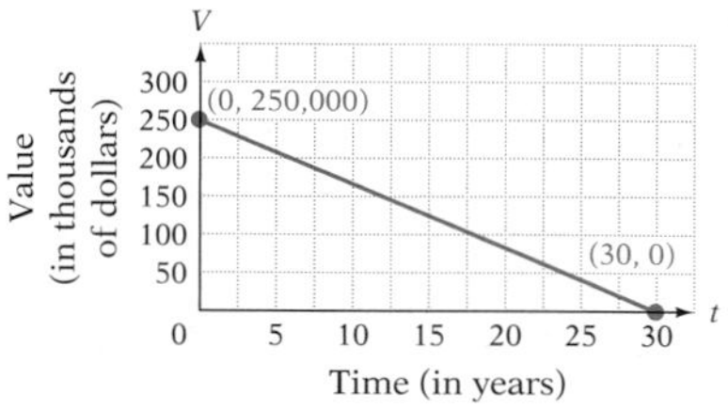

$m = -\frac{25{,}000}{3}$. The building depreciates $8333.33 in value each year. [4.4A]

24. How high on a building will a 17-foot ladder reach when the bottom of the ladder is 8 ft from the building?

15 ft [9.4B]

25. Use the discriminant to determine whether $2x^2 + 4x + 3 = 0$ has one real number solution, two real number solutions, or two complex number solutions.

two complex number solutions [10.3A]

Chapter 11

Functions and Relations

The Statuary Hall in the Capitol Building in Washington, D.C., pictured above, is a whispering gallery. Two people standing at what are called the foci of the elliptical ceiling can whisper and yet hear each other even though they are a considerable distance apart. The whisper from one person is reflected to the person standing at the other focus. The **Project on page 640** shows how to calculate the foci of an ellipse.

Objectives

Section 11.1

- **A** To graph a linear function
- **B** To solve application problems

Section 11.2

- **A** To graph a quadratic function
- **B** To find the *x*-intercepts of a parabola
- **C** To find the minimum or maximum of a quadratic function
- **D** To solve application problems

Section 11.3

- **A** To graph functions

Section 11.4

- **A** To perform operations on functions
- **B** To find the composition of two functions

Section 11.5

- **A** To determine whether a function is one-to-one
- **B** To find the inverse of a function

Section 11.6

- **A** To graph a parabola
- **B** To find the equation of a circle and to graph a circle
- **C** To graph an ellipse with center at the origin
- **D** To graph a hyperbola with center at the origin

WEB Need help? For on-line student resources, such as section quizzes, visit this textbook's web site at **college.hmco.com/students.**

Prep Test

1. Evaluate $-\frac{b}{2a}$ for $b = -4$ and $a = 2$.
1 [1.4A]

2. Given $y = -x^2 + 2x + 1$, find the value of y when $x = -2$.
-7 [1.4A]

3. Given $f(x) = x^2 - 3x + 2$, find $f(-4)$.
30 [4.2A]

4. Evaluate $p(r) = r^2 - 5$ when $r = 2 + h$.
$h^2 + 4h - 1$ [4.2A]

5. Solve: $0 = 3x^2 - 7x - 6$
$-\frac{2}{3}, 3$ [10.1A]

6. Solve: $0 = x^2 - 4x + 1$
$2 - \sqrt{3}, 2 + \sqrt{3}$ [10.2A, 10.3A]

7. Solve $x = 2y + 4$ for y.
$y = \frac{1}{2}x - 2$ [8.5A]

8. Find the domain and range of the relation $\{(-2, 4), (3, 5), (4, 6), (6, 5)\}$. Is the relation a function?
domain: $\{-2, 3, 4, 6\}$; range: $\{4, 5, 6\}$; yes [4.2A]

9. Graph: $x = -2$

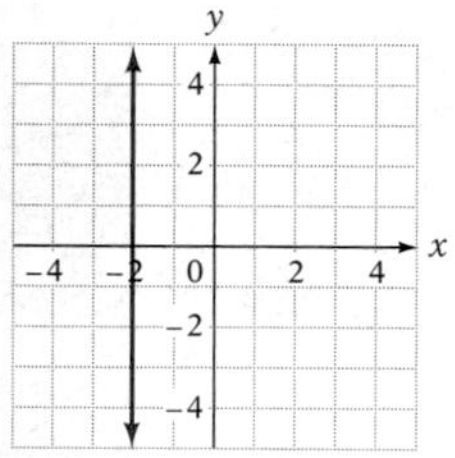

[4.3B]

Go Figure

Each time the two hands of a certain standard 12-hour clock form a 180° angle, a bell chimes once. From noon today until noon tomorrow, how many chimes will be heard?
22 chimes

11.1 Linear Functions

Objective A To graph a linear function

VIDEO & DVD · CD TUTOR · WEB · SSM

The graph of a function is the graph of the ordered pairs (x, y) that belong to the function. Because y and $f(x)$ are interchangeable, the ordered pairs of a function can be written as (x, y) or $(x, f(x))$. A function that can be written in the form $y = mx + b$ or $f(x) = mx + b$ is called a **linear function**. The graph of a linear function has certain characteristics. It is a straight line with slope m and y-intercept $(0, b)$.

⇒ Graph: $f(x) = 2x + 1$.

This is a linear function. You can think of the function as the equation $y = 2x + 1$. The y-intercept is $(0, 1)$. The slope is 2.

Beginning at the y-intercept, move right 1 and up 2. The point $(1, 3)$ is another point on the graph. Draw a straight line through the points $(0, 1)$ and $(1, 3)$.

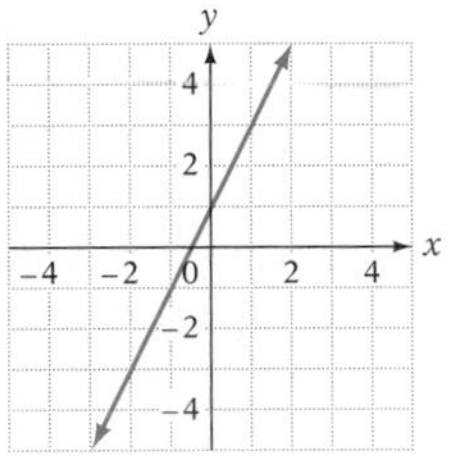

When a function is given by an equation, the domain of the function is all real numbers for which the function evaluates to a real number. For instance,

- The domain of $f(x) = 2x + 1$ is all real numbers because the value of $2x + 1$ is a real number for any value of x.
- The domain of $g(x) = \frac{1}{x - 2}$ is all real numbers except 2; when $x = 2$, $g(2) = \frac{1}{2 - 2} = \frac{1}{0}$, which is not a real number.

Example 1

Graph: $f(x) = \frac{2}{3}x$

Solution

$b = 0$

y-intercept: $(0, 0)$

$m = \frac{2}{3}$

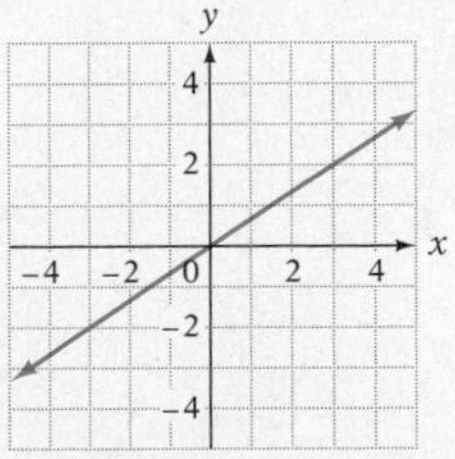

You Try It 1

Graph: $f(x) = \frac{3}{5}x - 4$

Your solution

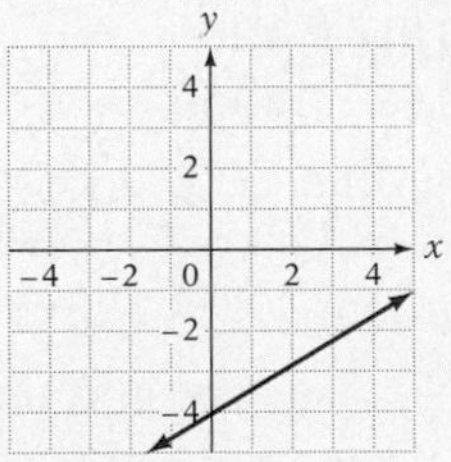

Solution on p. S32

591

Objective 11.1A

Vocabulary to Review

graph of a function
linear function

New Equations

$y = mx + b$ or $f(x) = mx + b$

Discuss the Concepts

1. What is a linear function?
2. Describe the shape of the graph of every linear function.

Concept Check

It is important for students to make the connection between a function and its graph. A linear function has a characteristic graph, which is different from the graph of a quadratic function, which is different from the graph of a cubic function, which is different from the graph of an absolute value function, and so on. Ask students whether

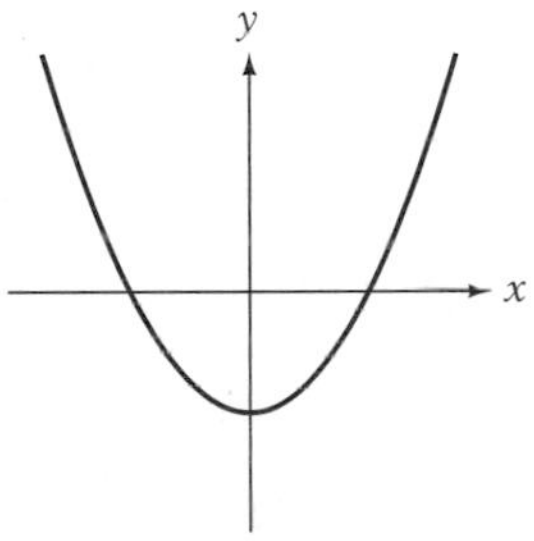

is the graph of a linear function. Intersperse examples of graphs of linear functions with graphs of other types of functions until you are confident that they understand the concept of a linear function.

Instructor Note

Students need to be reminded often that $f(x)$ and y have the same value for a given x.

In-Class Examples (Objective 11.1A)

1. Graph $y = 2x - 3$.

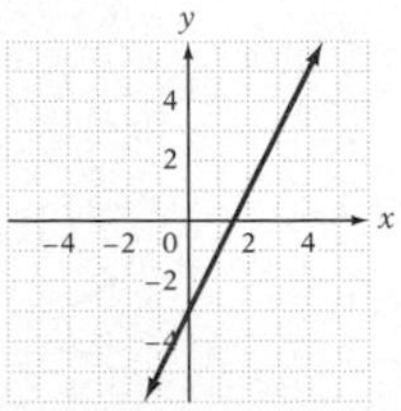

2. Graph $y = \frac{1}{2}x$.

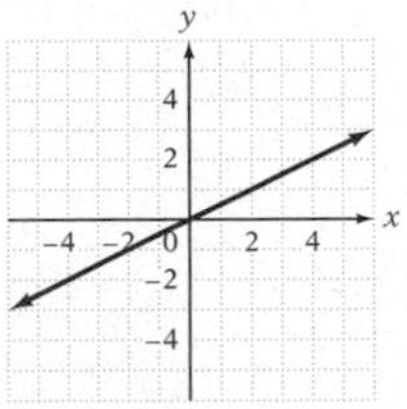

3. Graph $y = -\frac{2}{3}x + 2$.

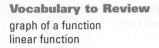

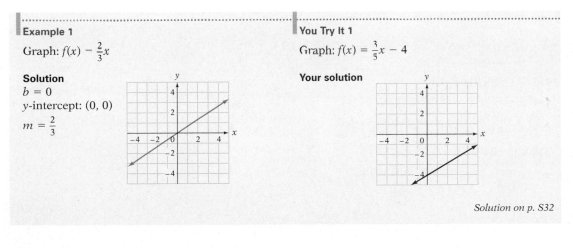

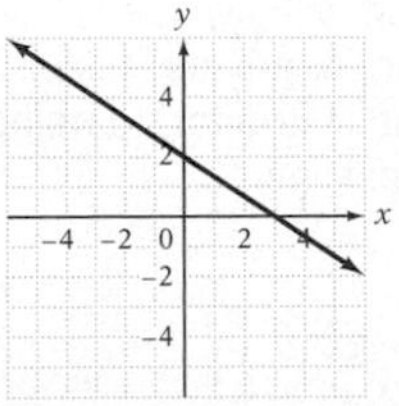

Objective 11.1B

Optional Student Activity

1. After a parachute is deployed, a function that models the height of the parachutist above the ground is $f(t) = -10t + 2800$, where $f(t)$ is the height, in feet, of the parachutist t seconds after the chute is deployed. Find the horizontal and vertical intercepts of the graph of this function and explain what they mean in the context of the problem. (0, 2800): The parachutist is 2800 ft above the ground when the parachute is deployed. (280, 0): The parachutist reached the ground 280 s after the parachute was deployed.
2. The temperature of an object taken from a freezer gradually rises and can be modeled by the function $T(x) = 20x - 100$, where $T(x)$ is the Fahrenheit temperature of the object x hours after being removed from the freezer. Find the intercepts of the graph of this function and explain what they mean in the context of the problem. (0, −100): When the object was taken from the freezer, its temperature was −100°F. (5, 0): Five hours after the object was removed from the freezer, its temperature was 0°F.

Objective B To solve application problems

VIDEO & DVD CD TUTOR WEB SSM

Linear functions can be used to model a variety of applications in science and business. For each application, data are collected and the independent and dependent variables are selected. Then a linear function is determined that models the data.

Example 2

Suppose a manufacturer has determined that at a price of \$115, consumers will purchase 1 million portable CD players and that at a price of \$90, consumers will purchase 1.25 million portable CD players. Describe this situation with a linear function. Use this function to predict how many portable CD players consumers will purchase if the price is \$80.

Strategy

- Select the independent and dependent variables. Because you are trying to determine the number of CD players, that quantity is the *dependent* variable, y. The price of CD players is the *independent* variable, x. From the given data, two ordered pairs are (115, 1) and (90, 1.25). (The ordinates are in millions of units.) Use these ordered pairs to determine the linear function.
- Evaluate the function for $x = 80$ to predict how many CD players consumers will purchase if the price is \$80.

Solution

Let $(x_1, y_1) = (115, 1)$ and $(x_2, y_2) = (90, 1.25)$.

$$m = \frac{y_2 - y_1}{x_2 - x_1} = \frac{1.25 - 1}{90 - 115} = -\frac{0.25}{25} = -0.01$$

$$y - y_1 = m(x - x_1)$$
$$y - 1 = -0.01(x - 115)$$
$$y - 1 = -0.01x + 1.15$$
$$y = -0.01x + 2.15$$

The linear function is $f(x) = -0.01x + 2.15$.

$$f(80) = -0.01(80) + 2.15 = 1.35$$

Consumers will purchase 1.35 million CD players at a price of \$80.

You Try It 2

Gabriel Daniel Fahrenheit invented the mercury thermometer in 1717. In terms of readings on this thermometer, water freezes at 32°F and boils at 212°F. In 1742 Anders Celsius invented the Celsius temperature scale. On this scale, water freezes at 0°C and boils at 100°C. Determine a linear function that can be used to predict the Celsius temperature when the Fahrenheit temperature is known.

Your strategy

Your solution

$f(F) = \frac{5}{9}(F - 32)$

Solution on p. S32

In-Class Examples (Objective 11.1B)

A tennis instructor is paid \$40 per hour for a private lesson by members of a condominium group. Determine a linear function that will give the instructor's earnings for h hours of private lessons. Use the model to determine the instructor's earnings for giving 25 hours of private lessons.
$f(h) = 40h$; \$1000

593

11.1 Exercises

Objective A

Graph.

1. $f(x) = 3x - 4$

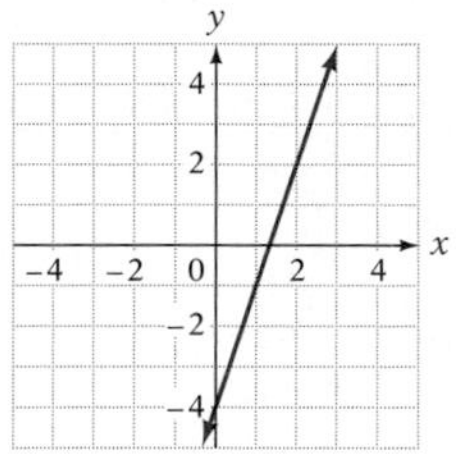

2. $f(x) = -2x + 3$

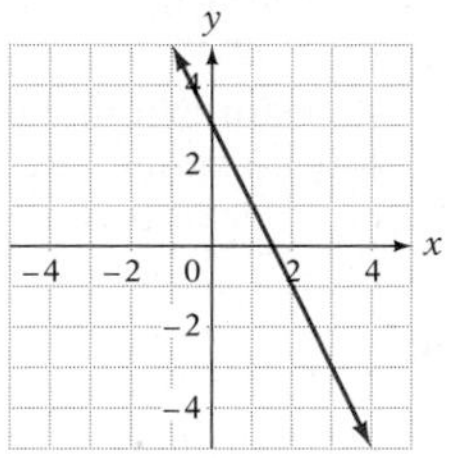

3. $f(x) = -\frac{2}{3}x$

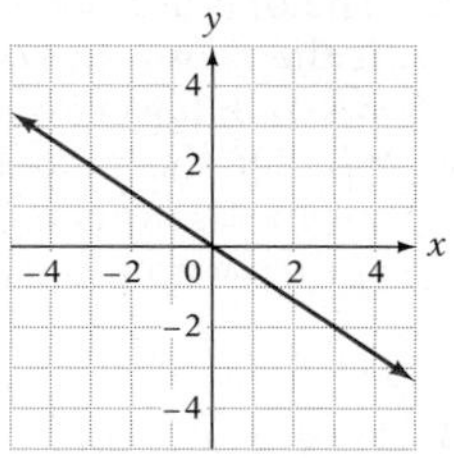

4. $f(x) = \frac{3}{2}x$

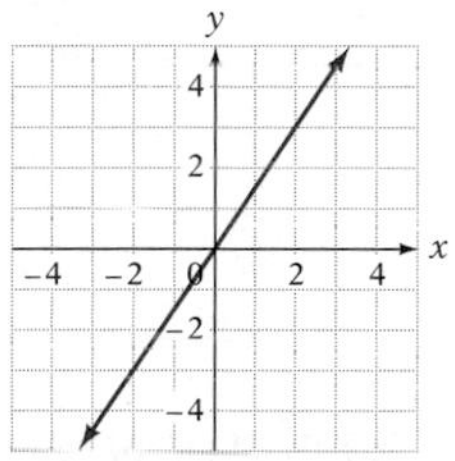

5. $f(x) = \frac{2}{3}x - 4$

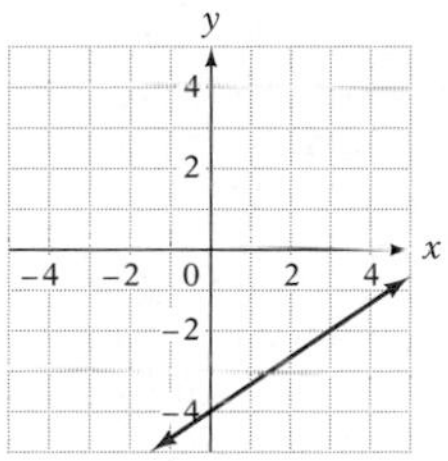

6. $f(x) = \frac{3}{4}x + 2$

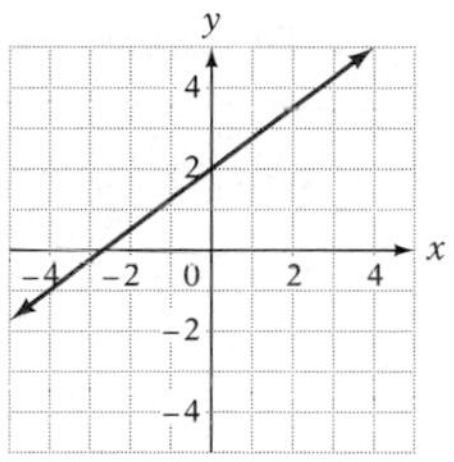

7. $f(x) = -\frac{1}{3}x + 2$

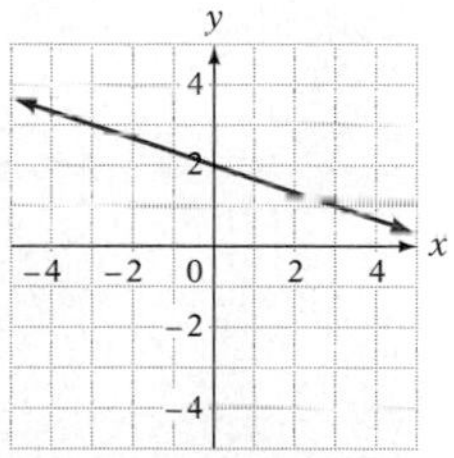

8. $f(x) = -\frac{3}{2}x - 3$

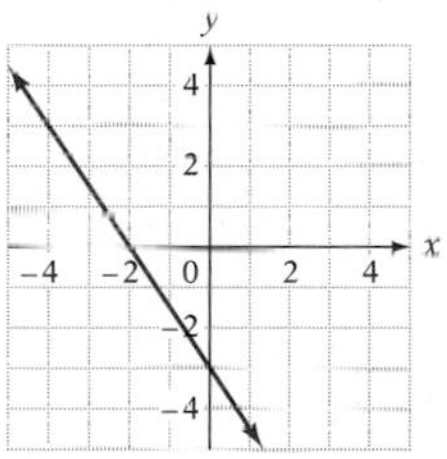

9. $f(x) = \frac{3}{5}x - 1$

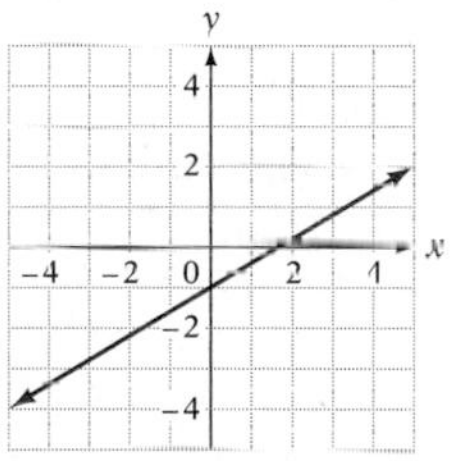

Objective B *Application Problems*

10. The sale price of an item is a function, s, of the original price, p, where $s(p) = 0.80p$. If an item's original price is \$200, what is the sale price of the item?
\$160

11. The markup on an item is a function, m, of its cost, c, where $m(c) = 0.25c$. If the cost of an item is \$150, what is the markup on the item?
\$37.50

Section 11.1

Suggested Assignment
Exercises 1–13, odds
More challenging problems:
Exercise 15

Quick Quiz (Objective 11.1A)

1. Graph $f(x) = 3x - 6$.

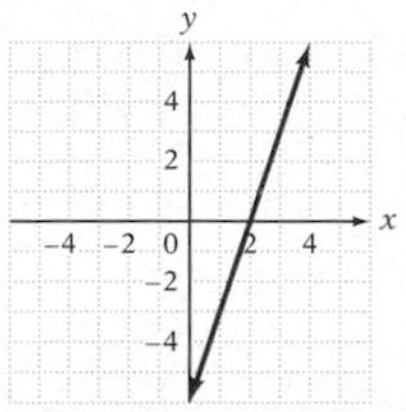

2. Graph $f(x) = -\frac{3}{4}x$.

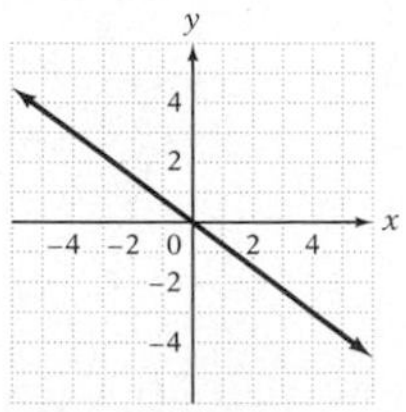

3. Graph $f(x) = \frac{1}{3}x - 2$.

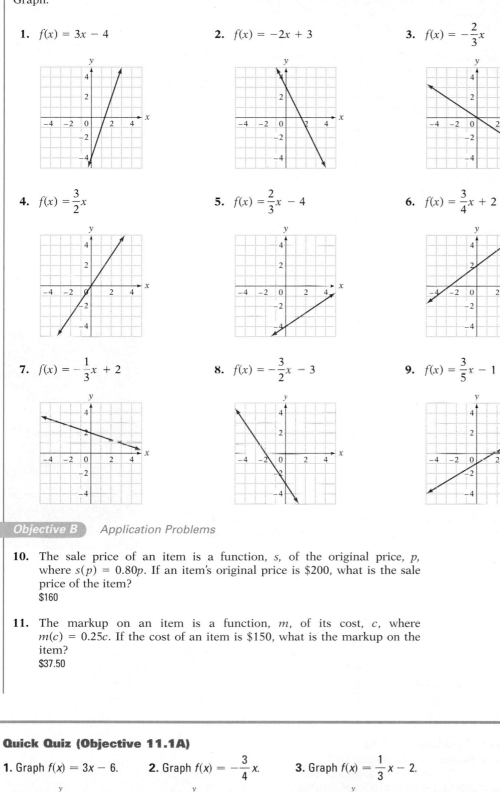

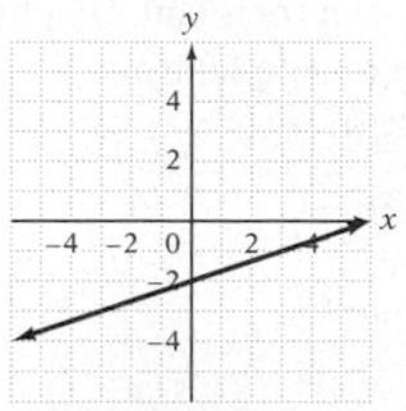

594 Chapter 11 / Functions and Relations

12. A manufacturer of graphing calculators has determined that 10,000 calculators per week will be sold at a price of \$95. At a price of \$90, it is estimated that 12,000 calculators would be sold. Determine a linear function that will predict the number of calculators that would be sold at a given price. Use this method to predict the number of calculators per week that would be sold at a price of \$75.
$y = -400x + 48{,}000$; 18,000 calculators

13. The operator of a hotel estimates that 500 rooms per night will be rented if the room rate per night is \$75. For each \$10 increase in the price of a room, 6 fewer rooms will be rented. Determine a linear function that will predict the number of rooms that will be rented for a given price per room. Use this model to predict the number of rooms that will be rented if the room rate is \$100.
$y = -\frac{3}{5}x + 545$; 485 rooms

14. A general building contractor estimates that the cost to build a new home is \$30,000 plus \$85 for each square foot of floor space in the house. Determine a linear function that will give the cost of building a house that contains a given number of square feet. Use this model to determine the cost to build a house that contains 1800 ft^2.
$y = 85x + 30{,}000$; \$183,000

APPLYING THE CONCEPTS

15. A child's height is a function of the child's age. The graph of this function is not linear, as children go through growth spurts as they develop. However, for the graph to be reasonable, the function must be an increasing function (that is, as the age increases, the height increases) because children do not get shorter as they grow older. Match each function described below with a reasonable graph of the function.
   a. The height of a plane above the ground during take-off depends on how long it has been since the plane left the gate.
   b. The height of a football above the ground is related to the number of seconds that have passed since it was punted.
   c. A basketball player is dribbling a basketball. The basketball's distance from the floor is related to the number of seconds that have passed since the player began dribbling the ball.
   d. Two children are seated together on a roller coaster. The height of the children above the ground depends on how long they have been on the ride.

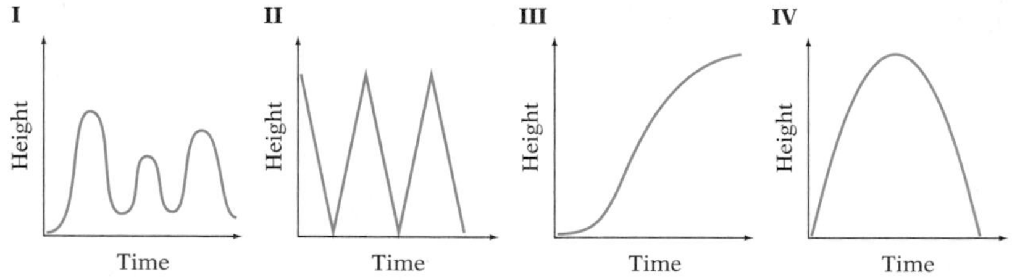

a. III; b. IV; c. II; d. I

Quick Quiz (Objective 11.1B)
A tree service charges a \$60 flat fee plus \$35 for each tree removed. Determine a linear function that will give the cost of having n trees removed. Use the model to determine the cost of having 50 trees removed. $f(n) = 35n + 60$; \$1810

11.2 Quadratic Functions

Objective A To graph a quadratic function

VIDEO & DVD CD TUTOR WEB SSM

Recall that a linear function is one that can be expressed by the equation $f(x) = mx + b$. The graph of a linear function has certain characteristics. It is a straight line with slope m and y-intercept $(0, b)$. A **quadratic function** is one that can be expressed by the equation $f(x) = ax^2 + bx + c$, $a \neq 0$. The graph of this function, called a **parabola**, also has certain characteristics. The graph of a quadratic function can be drawn by finding ordered pairs that belong to the function.

> **TAKE NOTE**
> Sometimes the value of the independent variable is called the **input** because it is *put in* place of the independent variable. The result of evaluating the function is called the **output**.
> An **input/output table** shows the results of evaluating a function for various values of the independent variable. An input/output table for $f(x) = x^2 - 2x - 3$ is shown at the right.

➡ Graph $f(x) = x^2 - 2x - 3$.

By evaluating the function for various values of x, find enough ordered pairs to determine the shape of the graph.

x	$f(x) = x^2 - 2x - 3$	$f(x)$	(x, y)
-2	$f(-2) = (-2)^2 - 2(-2) - 3$	5	$(-2, 5)$
-1	$f(-1) = (-1)^2 - 2(-1) - 3$	0	$(-1, 0)$
0	$f(0) = (0)^2 - 2(0) - 3$	-3	$(0, -3)$
1	$f(1) = (1)^2 - 2(1) - 3$	-4	$(1, -4)$
2	$f(2) = (2)^2 - 2(2) - 3$	-3	$(2, -3)$
3	$f(3) = (3)^2 - 2(3) - 3$	0	$(3, 0)$
4	$f(4) = (4)^2 - 2(4) - 3$	5	$(4, 5)$

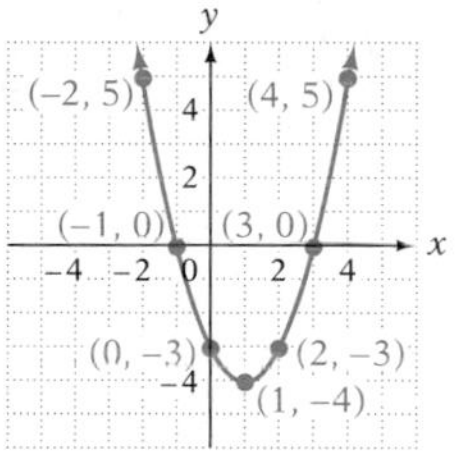

Because the value of $f(x) = x^2 - 2x - 3$ is a real number for all values of x, the domain of f is $\{x|x \in \text{real numbers}\}$. From the graph, it appears that no value of y is less than -4. Thus the range is $\{y|y \geq -4\}$. The range can also be determined algebraically, as shown below, by completing the square.

> **TAKE NOTE**
> In completing the square, 1 is both added and subtracted. Because $1 - 1 = 0$, the expression $x^2 - 2x - 3$ is not changed. Note that
> $(x - 1)^2 - 4$
> $= (x^2 - 2x + 1) - 4$
> $= x^2 - 2x - 3$
> which is the original expression.

$f(x) = x^2 - 2x - 3$

$= (x^2 - 2x) - 3$ • Group the variable terms.

$= (x^2 - 2x + 1) - 1 - 3$ • Complete the square of $x^2 - 2x$. Add and subtract $\left[\frac{1}{2}(-2)\right]^2 = 1$.

$= (x - 1)^2 - 4$ • Factor and combine like terms.

Because the square of a positive number is always positive, we have

$(x - 1)^2 \geq 0$

$(x - 1)^2 - 4 \geq -4$ • Subtract 4 from each side of the inequality.

$f(x) \geq -4$ • $f(x) = x^2 - 2x - 3 = (x - 1)^2 - 4$

$y \geq -4$

From the last inequality, the range is $\{y|y \geq -4\}$.

Objective 11.2A

New Vocabulary
parabola
input
output
input/output table
vertex
axis of symmetry

Vocabulary to Review
completing the square
domain
range
quadratic function

New Formulas
coordinates of the vertex of a parabola: $\left(-\frac{b}{2a}, f\left(-\frac{b}{2a}\right)\right)$

equation of the axis of symmetry of a parabola: $x = -\frac{b}{2a}$

Instructor Note
The graph of a quadratic function is discussed in this section, along with some of its properties and applications. A general discussion of parabolas as one of the conic sections appears in Section 6.

Discuss the Concepts
1. Describe how to find the vertex of a parabola.
2. Explain why the vertex of a parabola lies on the axis of symmetry.
3. Explain how, if the coordinates of the vertex of a parabola are known, the range of the quadratic function can be determined.

In-Class Examples (Objective 11.2A)
Find the vertex and axis of symmetry of the parabola given by the equation. Then graph the equation.

1. $y = x^2 - x - 2$
Vertex: $\left(\frac{1}{2}, -\frac{9}{4}\right)$;
Axis of symmetry:
$x = \frac{1}{2}$

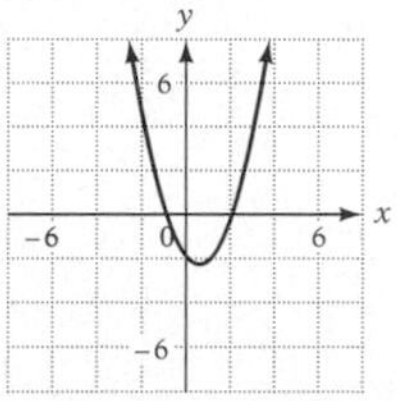

2. $f(x) = -x^2 + 2x + 4$
Vertex: $(1, 5)$;
Axis of symmetry:
$x = 1$

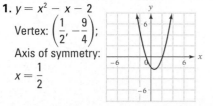

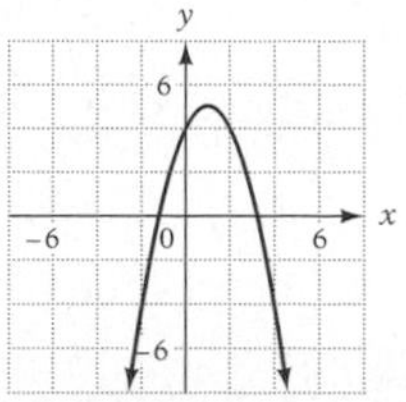

Concept Check

1. Draw the graphs of two different quadratic functions that have the line $x = 3$ as their line of symmetry. Name the coordinates of the vertex. State the domain and range of the function.
2. Draw the graphs of two different quadratic functions that have the line $x = -2$ as their line of symmetry. Name the coordinates of the vertex. State the domain and range of the function.

Instructor Note

The argument used to determine the vertex is difficult for students to understand. One approach is to have students evaluate $x^2 + 3$ and $x^2 - 1$ for various values of x and discover that the least value of the expression occurs when $x = 0$. Then repeat this process for $(x + 2)^2 + 3$ and $(x + 2)^2 - 1$.

Optional Student Activity

1. The axis of symmetry of a parabola is the line $x = 0$. The point $(-2, -3)$ lies on the parabola. Use the symmetry of a parabola to find a second point on the graph. $(2, -3)$
2. The axis of symmetry of a parabola is the line $x = 1$. The point $(3, 0)$ lies on the parabola. Use the symmetry of a parabola to find a second point on the graph. $(-1, 0)$
3. The axis of symmetry of a parabola is the line $x = 2$. The point $(4, -4)$ lies on the parabola. Use the symmetry of a parabola to find a second point on the graph. $(0, -4)$
4. The axis of symmetry of a parabola is the line $x = -1$. The point $(1, -1)$ lies on the parabola. Use the symmetry of a parabola to find a second point on the graph. $(-3, -1)$

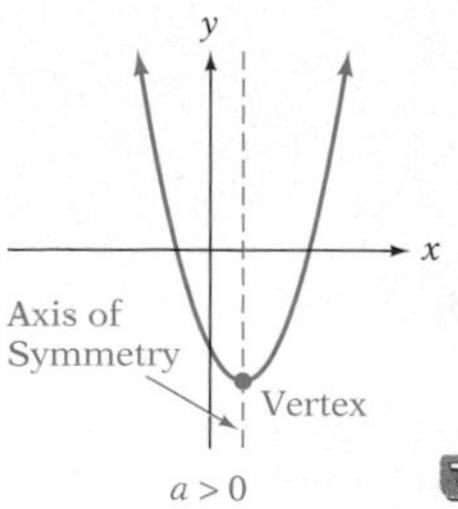

$a > 0$

In general, the graph of $f(x) = ax^2 + bx + c$, $a \neq 0$, resembles a "cup" shape as shown at the left. The parabola opens up when $a > 0$ (a is positive) and opens down when $a < 0$ (a is negative). When the parabola opens up, the **vertex** of the parabola is the point with the smallest y-coordinate. When the parabola opens down, the **vertex** is the point with the largest y-coordinate.

The **axis of symmetry** is a line that passes through the vertex of the parabola and is parallel to the y-axis. To understand the axis of symmetry, think of folding the graph along that line. The two portions of the graph will match up.

The following formulas can be used to find the axis of symmetry and the vertex of a parabola.

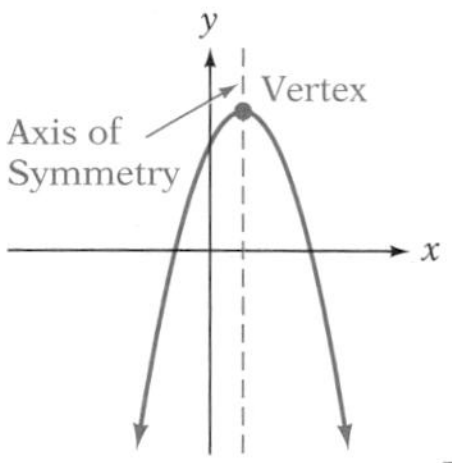

$a < 0$

The Axis of Symmetry and Vertex of a Parabola

Let $f(x) = ax^2 + bx + c$ be the equation of a parabola.

The equation of the axis of symmetry is $x = -\frac{b}{2a}$.

The coordinates of the vertex are $\left(-\frac{b}{2a}, f\left(-\frac{b}{2a}\right)\right)$.

➡ Find the axis of symmetry and the vertex of the parabola whose equation is $g(x) = -2x^2 + 3x + 1$. Then graph the equation.

From the equation $g(x) = -2x^2 + 3x + 1$, $a = -2$, $b = 3$, and $c = 1$.

Axis of symmetry: $x = -\dfrac{b}{2a} = -\dfrac{3}{2(-2)} = \dfrac{3}{4}$

The axis of symmetry is a vertical line passing through the point $\left(\frac{3}{4}, 0\right)$.

The x-coordinate of the vertex is $\frac{3}{4}$.

Find the y-coordinate of the vertex by replacing x with $\frac{3}{4}$ and evaluating.

$$y = -2x^2 + 3x + 1$$
$$= -2\left(\frac{3}{4}\right)^2 + 3\left(\frac{3}{4}\right) + 1 = \frac{17}{8}$$

The vertex is $\left(\frac{3}{4}, \frac{17}{8}\right)$.

Because a is negative ($a = -2$), the graph opens down. Find a few ordered pairs that belong to the function, and then sketch the graph.

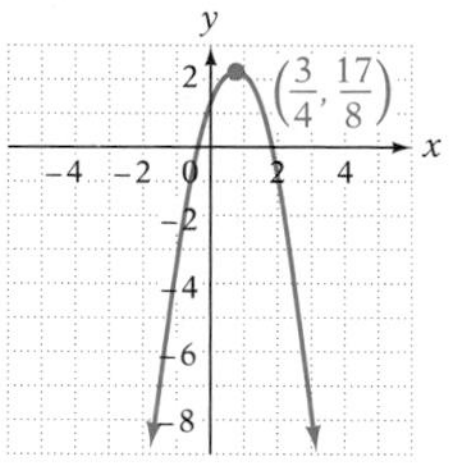

$g(x) = -2x^2 + 3x + 1$

TAKE NOTE

Once the coordinates of the vertex are found, the range of a quadratic function can be determined.

Once the y-coordinate of the vertex is known, the range of the function can be determined. Here, the graph of g opens down, so the y-coordinate of the vertex is the largest value of y. Therefore, the range of g is $\left\{y \mid y \leq \frac{17}{8}\right\}$. The value of $-2x^2 + 3x + 1$ is a real number for all values of x; the domain is $\{x \mid x \in \text{real numbers}\}$.

597

Example 1
Find the vertex and axis of symmetry of the parabola whose equation is $y = -x^2 + 4x + 1$. Then graph the equation.

Solution
x-coordinate of vertex:

$-\frac{b}{2a} = -\frac{4}{2(-1)} = 2$

y-coordinate of vertex:

$y = -x^2 + 4x + 1$
$= -(2)^2 + 4(2) + 1$
$= 5$

vertex: (2, 5)
axis of symmetry: $x = 2$

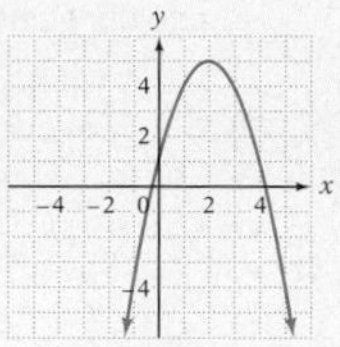

You Try It 1
Find the vertex and axis of symmetry of the parabola whose equation is $y = 4x^2 + 4x + 1$. Then graph the equation.

Your solution
vertex: $\left(-\frac{1}{2}, 0\right)$

axis of symmetry: $x = -\frac{1}{2}$

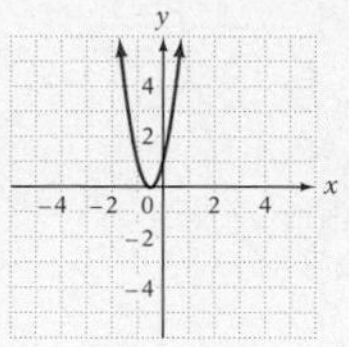

Solution on p. S32

Objective B To find the x-intercepts of a parabola

VIDEO & DVD CD TUTOR WEB SSM

Recall that a point at which a graph crosses the x- or y-axis is called an *intercept* of the graph. The x-intercepts of the graph of an equation occur when $y = 0$; the y-intercepts occur when $x = 0$.

The graph of $y = x^2 + 3x - 4$ is shown at the right. The points whose coordinates are $(-4, 0)$ and $(1, 0)$ are x-intercepts of the graph.

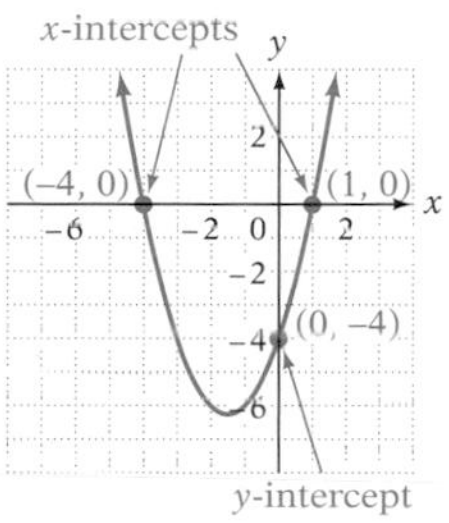

➡ Find the x-intercepts of the parabola whose equation is $y = 4x^2 - 4x + 1$.

To find the x-intercepts, let $y = 0$ and then solve for x.

$y = 4x^2 - 4x + 1$
$0 = 4x^2 - 4x + 1$ • **Let $y = 0$.**

$0 = (2x - 1)(2x - 1)$ • **Solve for x by factoring and using the Principle of Zero Products.**

$2x - 1 = 0 \qquad 2x - 1 = 0$
$2x = 1 \qquad 2x = 1$
$x = \frac{1}{2} \qquad x = \frac{1}{2}$

The x-intercept is $\left(\frac{1}{2}, 0\right)$.

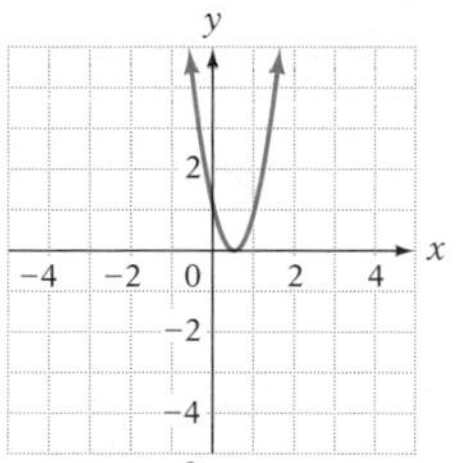
$y = 4x^2 - 4x + 1$

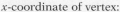

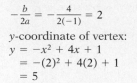

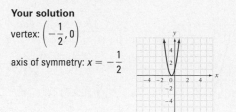

In this last example, the parabola has only one x-intercept. In this case, the parabola is said to be **tangent** to the x-axis at $x = \frac{1}{2}$.

Objective 11.2B

Vocabulary to Review
intercepts of a graph
discriminant

Instructor Note
Intercepts were discussed in the context of linear functions. Making a connection to those ideas may help students realize that this is not a new concept.

Discuss the Concepts
1. How are the x-intercepts of a graph and the solutions of the corresponding equation related?
2. When $f(x) = 0$, values of x for $f(x) = x^2 + 2x + 3$ are complex numbers. What does this say about the x-intercepts of the graph of f?

Concept Check
The concepts of the x-intercepts of the graph of a function and the solutions of an equation are different ways of discussing the same number. Use the term *intercepts* or *solutions* to complete the sentence.
1. If we are discussing graphing, then the ______ are our focus. intercepts
2. If we are discussing equations, then the ______ are our focus. solutions

Instructor Note
The discriminant was discussed in conjunction with the solution of a quadratic equation. It may be necessary to review that material at this time.

In-Class Examples (Objective 11.2B)
1. Find the x-intercepts of the parabola given by the equation $y = x^2 - 2x - 3$. (3, 0), (−1, 0)
2. Use the discriminant to determine the number of x-intercepts of the graph.
   a. $y = 2x^2 - x - 1$ Two x-intercepts
   b. $y = x^2 + x + 2$ No x-intercepts

Optional Student Activity

1. What effect does increasing the coefficient of x^2 have on the graph of $y = ax^2 + bx + c$, $a > 0$? The graph becomes thinner.
2. What effect does decreasing the coefficient of x^2 have on the graph of $y = ax^2 + bx + c$, $a > 0$? The graph becomes wider.
3. What effect does increasing the constant term have on the graph of $y = ax^2 + bx + c$, $a \neq 0$? The graph is higher on the rectangular coordinate system.
4. What effect does decreasing the constant term have on the graph of $y = ax^2 + bx + c$, $a \neq 0$? The graph is lower on the rectangular coordinate system.

➡ Find the x-intercepts of $y = x^2 - 2x - 1$.

To find the x-intercepts, let $y = 0$ and solve for x.

$y = x^2 - 2x - 1$

$0 = x^2 - 2x - 1$

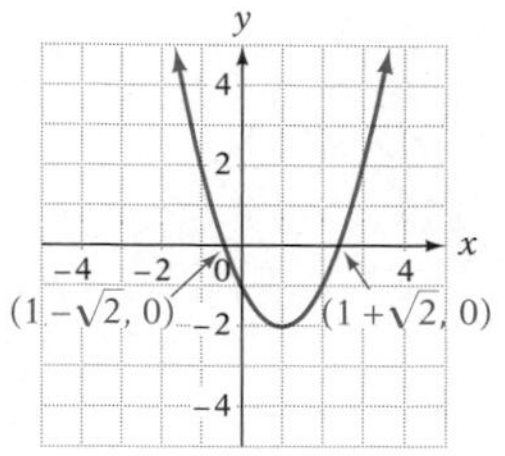

$$x = \frac{-b \pm \sqrt{b^2 - 4ac}}{2a}$$

• **Because $x^2 - 2x - 1$ does not easily factor, use the quadratic formula to solve for x.**

$$= \frac{-(-2) \pm \sqrt{(-2)^2 - 4(1)(-1)}}{2(1)}$$

• **$a = 1$, $b = -2$, $c = -1$**

$$= \frac{2 \pm \sqrt{4 + 4}}{2} = \frac{2 \pm \sqrt{8}}{2}$$

$$= \frac{2 \pm 2\sqrt{2}}{2} = 1 \pm \sqrt{2}$$

The x-intercepts are $(1 - \sqrt{2}, 0)$ and $(1 + \sqrt{2}, 0)$.

The graph of a parabola may not have x-intercepts. The graph of $y = -x^2 + 2x - 2$ is shown at the right. Note that the graph does not pass through the x-axis and thus there are no x-intercepts. This means there are no real number solutions of $-x^2 + 2x - 2 = 0$.

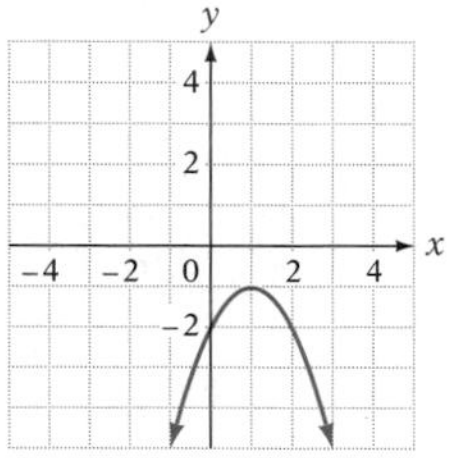

Using the quadratic formula, we find that the solutions of the equation $-x^2 + 2x - 2 = 0$ are the complex numbers $1 - i$ and $1 + i$.

Recall that the discriminant of the quadratic formula is the expression $b^2 - 4ac$ and that this expression can be used to determine whether $ax^2 + bx + c = 0$ has zero, one, or two real number solutions. Because there is a connection between the solutions of $ax^2 + bx + c = 0$ and the x-intercepts of the graph of $y = ax^2 + bx + c$, the discriminant can be used to determine the number of x-intercepts of a parabola.

The Effect of the Discriminant on the Number of x-Intercepts of a Parabola

1. If $b^2 - 4ac = 0$, the parabola has one x-intercept.
2. If $b^2 - 4ac > 0$, the parabola has two x-intercepts.
3. If $b^2 - 4ac < 0$, the parabola has no x-intercepts.

➡ Use the discriminant to determine the number of x-intercepts of the parabola whose equation is $y = 2x^2 - x + 2$.

$b^2 - 4ac$

$(-1)^2 - 4(2)(2) = 1 - 16 = -15$

$-15 < 0$

• **Evaluate the discriminant. $a = 2$, $b = -1$, $c = 2$**

The discriminant is less than zero, so the parabola has no x-intercepts.

Example 2
Find the x-intercepts of
$y = 2x^2 - 5x + 2$.

Solution
$y = 2x^2 - 5x + 2$
$0 = 2x^2 - 5x + 2$
$0 = (2x - 1)(x - 2)$

$2x - 1 = 0 \qquad x - 2 = 0$
$2x = 1 \qquad x = 2$
$x = \frac{1}{2}$

The x-intercepts are $\left(\frac{1}{2}, 0\right)$ and $(2, 0)$.

You Try It 2
Find the x-intercepts of
$y = x^2 + 3x + 4$.

Your solution
no x-intercepts

Example 3
Use the discriminant to determine the number of x-intercepts of $y = x^2 - 6x + 9$.

Solution
$a = 1, b = -6, c = 9$
$b^2 - 4ac = (-6)^2 - 4(1)(9) = 36 - 36 = 0$

Because the discriminant is equal to zero, the parabola has one x-intercept.

You Try It 3
Use the discriminant to determine the number of x-intercepts of $y = x^2 - x - 6$.

Your solution
two x-intercepts

Solutions on p. S33

Objective C — To find the minimum or maximum of a quadratic function

VIDEO & DVD CD TUTOR WEB SSM

Point of Interest

Calculus is a branch of mathematics that demonstrates, among other things, how to find the maximum or minimum of functions other than quadratic functions. These are very important problems in applied mathematics. For instance, an automotive engineer wants to design a car whose shape will *minimize* the effects of air flow. The same engineer tries to *maximize* the efficiency of a car's engine. Similarly, an economist may try to determine what business practices will *minimize* cost and *maximize* profit.

The graph of $f(x) = x^2 - 2x + 3$ is shown at the right. Because a is positive, the parabola opens up. The vertex of the parabola is the lowest point on the parabola. It is the point that has the minimum y-coordinate. Therefore, the value of the function at this point is a **minimum**.

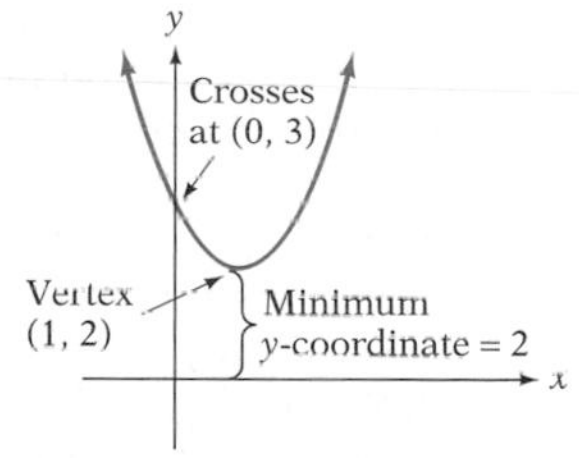

The graph of $f(x) = -x^2 + 2x + 1$ is shown at the right. Because a is negative, the parabola opens down. The vertex of the parabola is the highest point on the parabola. It is the point that has the maximum y-coordinate. Therefore, the value of the function at this point is a **maximum**.

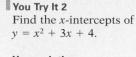

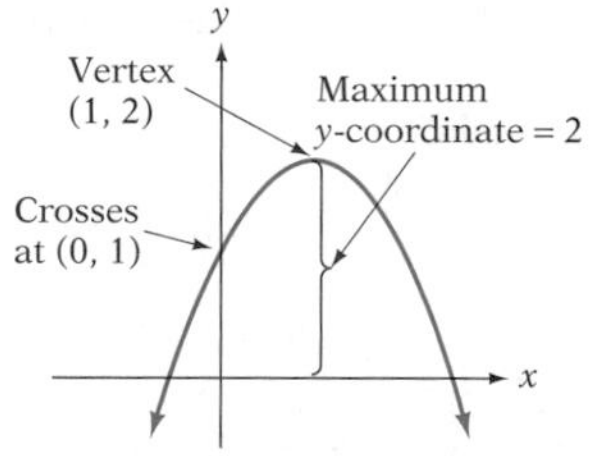

To find the minimum or maximum value of a quadratic function, first find the x-coordinate of the vertex. Then evaluate the function at that value.

Objective 11.2C

New Vocabulary
minimum value of a function
maximum value of a function

Vocabulary to Review
vertex

Instructor Note
One difficulty that students have with this concept is making the distinction between the maximum (or minimum) *value of the function* and the *value of x* that produces the maximum (or minimum).

Discuss the Concepts
Determine whether the following statements are always true, sometimes true, or never true.

1. The axis of symmetry of a parabola passes through the vertex. Always true
2. A parabola has two x-intercepts. Sometimes true
3. A quadratic function has a minimum value. Sometimes true

Concept Check

1. Which of the following functions has the highest minimum value?
   a. $f(x) = 3x^2 - 4$
   b. $f(x) = x^2 - 10x + 20$
   c. $f(x) = x^2 - 6x$
   a
2. What is the value of k if the vertex of the parabola given by the equation $y = x^2 + 4x + k$ is a point on the x-axis? 4

In-Class Examples (Objective 11.2C)
Find the minimum or maximum value of the quadratic function.

1. $f(x) = x^2 - 4x + 5$ Minimum value: 1
2. $f(x) = -2x^2 - 4x + 1$ Maximum value: 3
3. $f(x) = -3x^2 + 2x + 1$ Maximum value: $\frac{4}{3}$

Objective 11.2D

Instructor Note

This objective is frequently a very difficult one for students. You might want to use the problems provided in the Optional Student Activity below as extra problems for the students to work.

Optional Student Activity

1. The height s, in feet, of a ball thrown upward at an initial speed of 80 ft/s from a platform 50 ft high is given by the function $s(t) = -16t^2 + 80t + 50$, where t is the time in seconds. Find the maximum height above the ground that the ball will attain. 150 ft
2. Karen is throwing an orange to her brother Saul, who is standing on the balcony of their home. The height h, in feet, of the orange above the ground t seconds after it is thrown is given by $h(t) = -16t^2 + 32t + 4$. If Saul's outstretched arms are 18 ft above the ground, will the orange ever be high enough so that he can catch it? Yes

Example 4

Find the maximum or minimum value of $f(x) = 2x^2 - 3x + 1$.

Solution

$$x = -\frac{b}{2a} = -\frac{-3}{2(2)} = \frac{3}{4}$$

$$f(x) = 2x^2 - 3x + 1$$

$$f\left(\frac{3}{4}\right) = 2\left(\frac{3}{4}\right)^2 - 3\left(\frac{3}{4}\right) + 1$$

$$= \frac{9}{8} - \frac{9}{4} + 1 = -\frac{1}{8}$$

Because a is positive, the graph opens up. The function has a minimum value.

The minimum value of the function is $-\frac{1}{8}$.

You Try It 4

Find the maximum or minimum value of $f(x) = -3x^2 + 4x - 1$.

Your solution

maximum value at $\frac{1}{3}$

Solution on p. S33

Objective D To solve application problems

VIDEO & DVD CD TUTOR WEB SSM

Example 5

A mining company has determined that the cost C, in dollars per ton, of mining a mineral is given by the equation $C(x) = 0.2x^2 - 2x + 12$, where x is the number of tons of the mineral that are mined. Find the number of tons of the mineral that should be mined to minimize the cost. What is the minimum cost?

Strategy

- To find the number of tons that will minimize the cost, find the x-coordinate of the vertex.
- To find the minimum cost, evaluate the function at the x-coordinate of the vertex.

Solution

$$x = -\frac{b}{2a} = -\frac{-2}{2(0.2)} = 5$$

To minimize cost, 5 tons should be mined.

$$C(x) = 0.2x^2 - 2x + 12$$

$$C(5) = 0.2(5)^2 - 2(5) + 12 = 5 - 10 + 12 = 7$$

The minimum cost per ton is \$7.

You Try It 5

The height s, in feet, of a ball thrown straight up is given by the equation $s(t) = -16t^2 + 64t$, where t is the time in seconds. Find the time it takes the ball to reach its maximum height. What is the maximum height?

Your strategy

Your solution

2 s; 64 ft

Solution on p. S33

In-Class Examples (Objective 11.2D)

1. The height s, in feet, of a rocket after t seconds is given by the formula $s(t) = 192t - 16t^2$. Find the maximum height of the rocket. 576 ft
2. Find two numbers whose sum is 40 and whose product is a maximum. 20 and 20
3. A rectangle has a perimeter of 48 ft. What dimensions of the rectangle will produce a maximum area? What is the maximum area? 12 ft × 12 ft; 144 ft^2

11.2 Exercises

Objective A

1. Describe (a) the vertex and (b) the axis of symmetry of a parabola.

2. The axis of symmetry of a parabola is the line $x = -5$. What is the x-coordinate of the vertex of the parabola? -5

3. The axis of symmetry of a parabola is the line $x = 8$. What is the x-coordinate of the vertex of the parabola? 8

4. The vertex of a parabola is $(7, -9)$. What is the equation of the axis of symmetry of the parabola? $x = 7$

5. The vertex of a parabola is $(-4, 10)$. What is the equation of the axis of symmetry of the parabola? $x = -4$

Find the vertex and axis of symmetry of the parabola. Then sketch its graph.

6. $y = x^2 - 2x - 4$

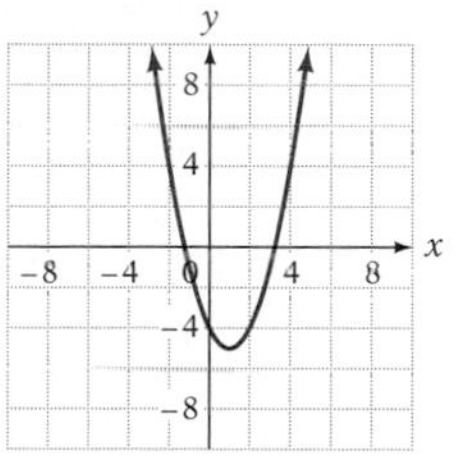

$(1, -5)$; $x = 1$

7. $y = x^2 + 4x - 4$

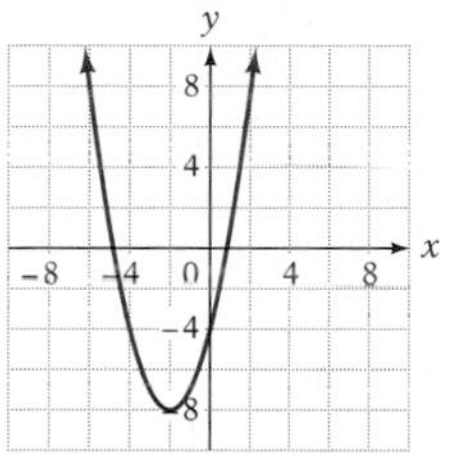

$(-2, -8)$; $x = -2$

8. $y = -x^2 + 2x - 3$

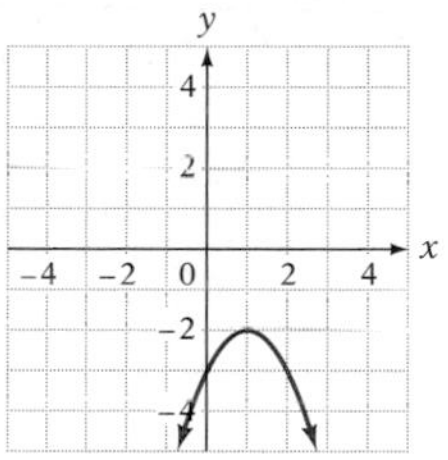

$(1, -2)$; $x = 1$

9. $y = -x^2 + 4x - 5$

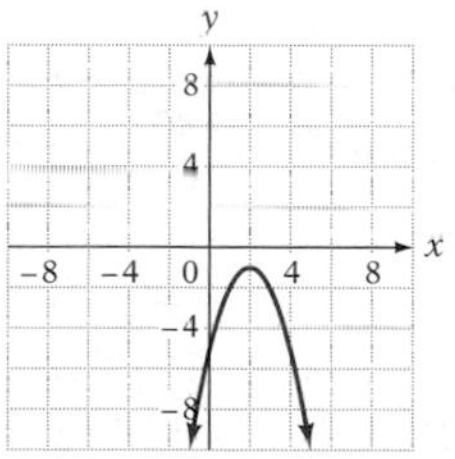

$(2, -1)$; $x = 2$

10. $f(x) = x^2 + 6x + 5$

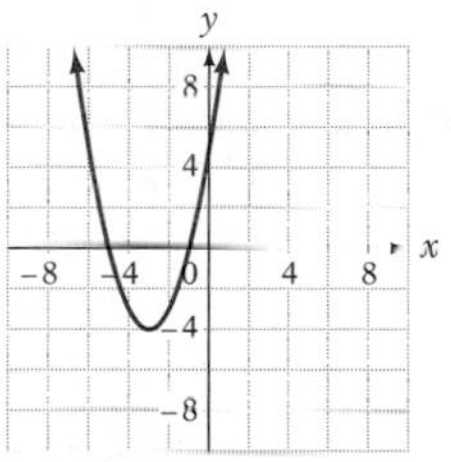

$(-3, -4)$; $x = -3$

11. $f(x) = x^2 - x - 6$

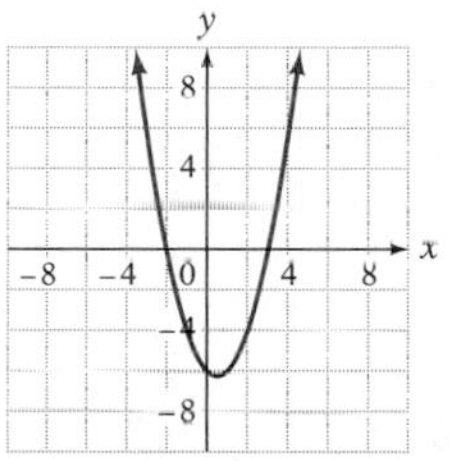

$\left(\frac{1}{2}, -\frac{25}{4}\right)$; $x = \frac{1}{2}$

12. $G(x) = x^2 - x - 2$

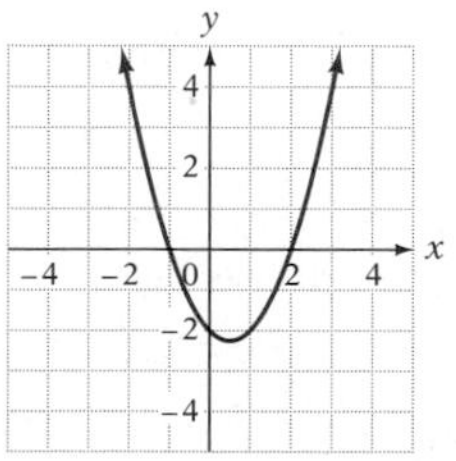

$\left(\frac{1}{2}, -\frac{9}{4}\right)$; $x = \frac{1}{2}$

13. $F(x) = x^2 - 3x + 2$

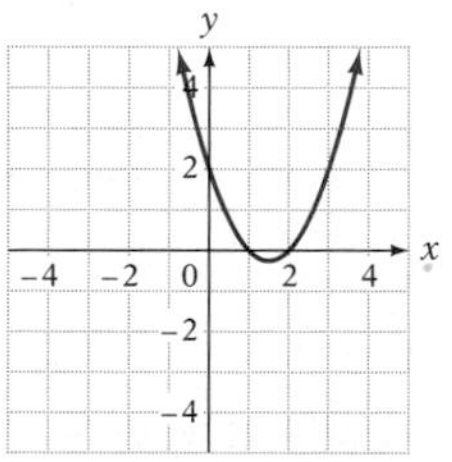

$\left(\frac{3}{2}, -\frac{1}{4}\right)$; $x = \frac{3}{2}$

14. $y = -2x^2 + 6x$

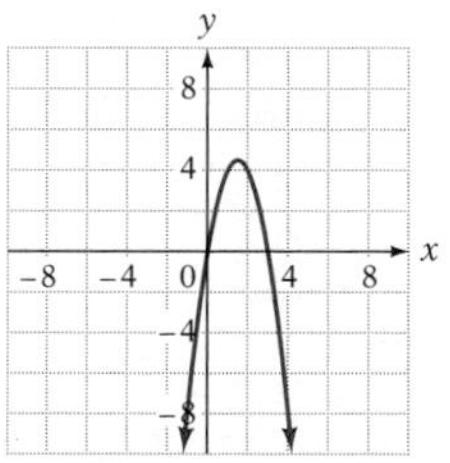

$\left(\frac{3}{2}, \frac{9}{2}\right)$; $x = \frac{3}{2}$

Section 11.2

Suggested Assignment

Exercises 1–65, odds

More challenging problems:
Exercises 67, 68

Answers to Writing Exercises

1a. The vertex of a parabola of the form $y = ax^2 + bx + c$ is the point with the smallest y-coordinate or the largest y-coordinate. When $a > 0$, the parabola opens up and the vertex of the parabola is the point with the smallest y-coordinate. When $a < 0$, the parabola opens down and the vertex is the point with the largest y-coordinate.

b. The axis of symmetry of the graph of a parabola of the form $y = ax^2 + bx + c$ is the vertical line that passes through the vertex of the parabola and is parallel to the y-axis.

Quick Quiz (Objective 11.2A)

Find the vertex and axis of symmetry of the parabola given by the equation. Then graph the equation.

1. $y = x^2 + 2x - 3$
Vertex: $(-1, -4)$;
Axis of symmetry:
$x = -1$

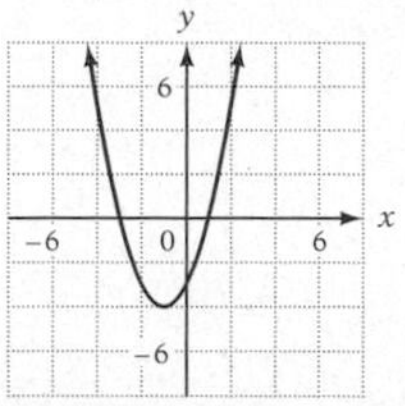

2. $f(x) = -x^2 + 4x - 3$
Vertex: $(2, 1)$;
Axis of symmetry:
$x = 2$

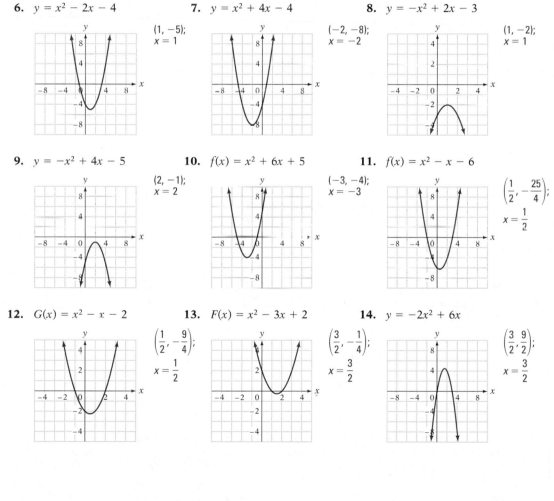

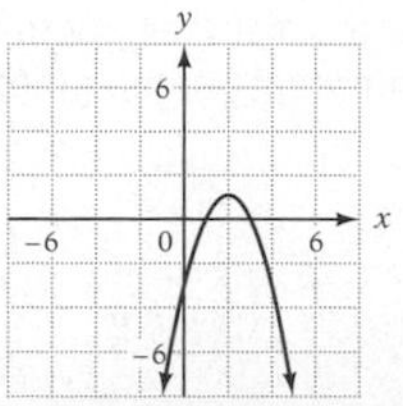

Answers to Writing Exercises

22. An x-intercept of the graph of a parabola is a point at which the graph crosses the x-axis. It is a point at which $y = 0$.

602

15. $y = -x^2 + 2x - 1$

(1, 0); $x = 1$

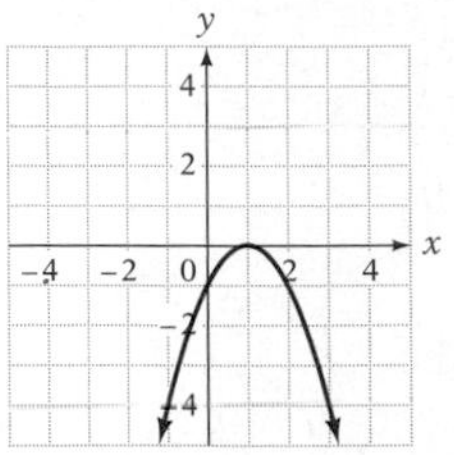

16. $h(x) = \frac{1}{2}x^2 - x + 1$

$\left(1, \frac{1}{2}\right)$; $x = 1$

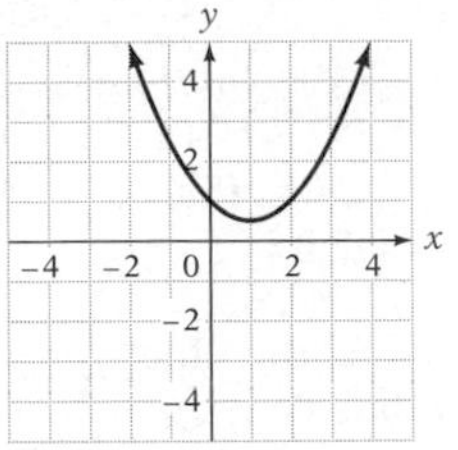

17. $P(x) = -\frac{1}{2}x^2 + 2x - 3$

(2, −1); $x = 2$

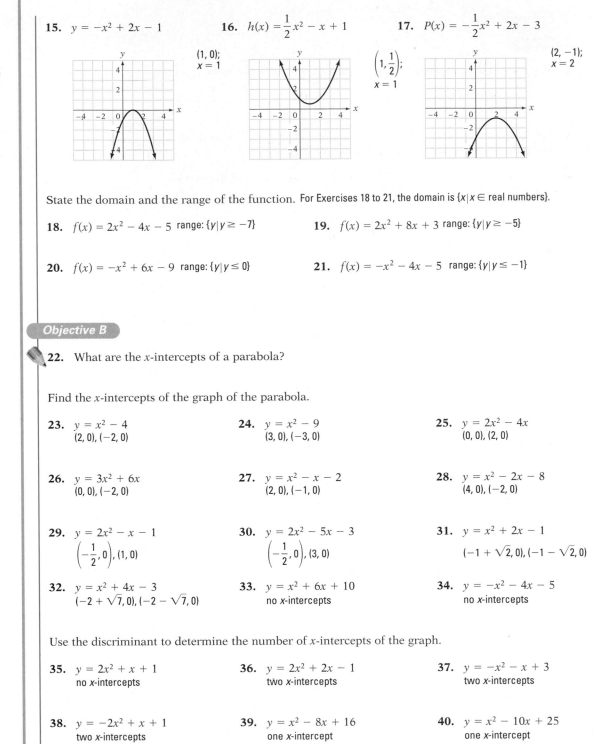

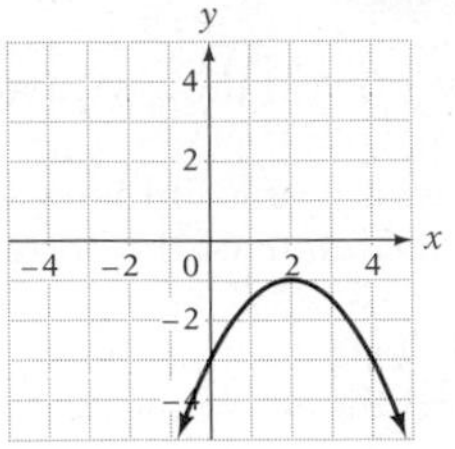

State the domain and the range of the function. For Exercises 18 to 21, the domain is $\{x \mid x \in \text{real numbers}\}$.

18. $f(x) = 2x^2 - 4x - 5$ range: $\{y \mid y \geq -7\}$

19. $f(x) = 2x^2 + 8x + 3$ range: $\{y \mid y \geq -5\}$

20. $f(x) = -x^2 + 6x - 9$ range: $\{y \mid y \leq 0\}$

21. $f(x) = -x^2 - 4x - 5$ range: $\{y \mid y \leq -1\}$

Objective B

22. What are the x-intercepts of a parabola?

Find the x-intercepts of the graph of the parabola.

23. $y = x^2 - 4$
(2, 0), (−2, 0)

24. $y = x^2 - 9$
(3, 0), (−3, 0)

25. $y = 2x^2 - 4x$
(0, 0), (2, 0)

26. $y = 3x^2 + 6x$
(0, 0), (−2, 0)

27. $y = x^2 - x - 2$
(2, 0), (−1, 0)

28. $y = x^2 - 2x - 8$
(4, 0), (−2, 0)

29. $y = 2x^2 - x - 1$
$\left(-\frac{1}{2}, 0\right)$, (1, 0)

30. $y = 2x^2 - 5x - 3$
$\left(-\frac{1}{2}, 0\right)$, (3, 0)

31. $y = x^2 + 2x - 1$
$(-1 + \sqrt{2}, 0), (-1 - \sqrt{2}, 0)$

32. $y = x^2 + 4x - 3$
$(-2 + \sqrt{7}, 0), (-2 - \sqrt{7}, 0)$

33. $y = x^2 + 6x + 10$
no x-intercepts

34. $y = -x^2 - 4x - 5$
no x-intercepts

Use the discriminant to determine the number of x-intercepts of the graph.

35. $y = 2x^2 + x + 1$
no x-intercepts

36. $y = 2x^2 + 2x - 1$
two x-intercepts

37. $y = -x^2 - x + 3$
two x-intercepts

38. $y = -2x^2 + x + 1$
two x-intercepts

39. $y = x^2 - 8x + 16$
one x-intercept

40. $y = x^2 - 10x + 25$
one x-intercept

Quick Quiz (Objective 11.2B)

1. Find the x-intercepts of the parabola given by the equation $y = x^2 - 3x - 18$. (6, 0), (−3, 0)
2. Use the discriminant to determine the number of x-intercepts of the graph of $y = x^2 - 2x + 1$. One

603

41. $y = -3x^2 - x - 2$
no x-intercepts

42. $y = -2x^2 + x - 1$
no x-intercepts

43. $y = 4x^2 - x - 2$
two x-intercepts

Objective C

44. When does a quadratic function have a minimum value and when does it have a maximum value?

45. Does the function have a minimum or a maximum value?
a. $f(x) = -x^2 + 6x - 1$ **b.** $f(x) = 2x^2 - 4$ **c.** $f(x) = -5x^2 + x$
a. maximum **b.** minimum **c.** maximum

46. Does the function have a minimum or a maximum value?
a. $f(x) = 3x^2 - 2x + 4$ **b.** $f(x) = -x^2 + 9$ **c.** $f(x) = 6x^2 - 3x$
a. minimum **b.** maximum **c.** minimum

Find the minimum or maximum value of the quadratic function.

47. $f(x) = x^2 - 2x + 3$
minimum: 2

48. $f(x) = x^2 + 3x - 4$
minimum: $-\frac{25}{4}$

49. $f(x) = -2x^2 + 4x - 3$
maximum: -1

50. $f(x) = -2x^2 - 3x + 4$
maximum: $\frac{41}{8}$

51. $f(x) = 2x^2 + 4x$
minimum: -2

52. $f(x) = -2x^2 - 3x$
maximum: $\frac{9}{8}$

53. $f(x) = -2x^2 + 4x - 5$
maximum: -3

54. $f(x) = -3x^2 + x - 6$
maximum: $-\frac{71}{12}$

55. $f(x) = 2x^2 + 3x - 8$
minimum: $-\frac{73}{8}$

56. Which of the following parabolas has the greatest minimum value?
a. $y = x^2 - 2x - 3$ **b.** $y = x^2 - 10x + 20$ **c.** $y = 3x^2 - 6$
a

57. Which of the following parabolas has the greatest maximum value?
a. $y = -2x^2 + 2x - 1$ **b.** $y = -x^2 + 8x - 2$ **c.** $y = -4x^2 + 3$
b

Objective D *Application Problems*

58. A tour operator believes that the profit, P, from selling x tickets is given by $P(x) = 40x - 0.25x^2$. Using this model, what is the maximum profit the tour operator can expect?
$1600

59. The suspension cable that supports a small footbridge hangs in the shape of a parabola. The height in feet of the cable above the bridge is given by the function $h(x) = 0.25x^2 - 0.8x + 25$, where x is the distance from one end of the bridge. What is the minimum height of the cable above the bridge?
24.36 ft

Answers to Writing Exercises

44. A quadratic function has a minimum value when $a > 0$ and a maximum value when $a < 0$ in the equation $f(x) = ax^2 + bx + c$.

Quick Quiz (Objective 11.2C)

Find the minimum or maximum value of the quadratic function.

1. $f(x) = -2x^2 + 4x - 1$ Maximum: 1

2. $f(x) = x^2 - 5x + 4$ Minimum: $-\frac{9}{4}$

60. The height s, in feet, of a rock thrown upward at an initial speed of 64 ft/s from a cliff 50 ft above an ocean beach is given by the function $s(t) = -16t^2 + 64t + 50$, where t is the time in seconds. Find the maximum height above the beach that the rock will attain. 114 ft

61. An event in the Summer Olympics is 10-meter springboard diving. In this event, the height s, in meters, of a diver above the water t seconds after jumping is given by $s(t) = -4.9t^2 + 7.8t + 10$. What is the maximum height that the diver will be above the water? Round to the nearest tenth. 13.1 m

62. A pool is treated with a chemical to reduce the amount of algae in the pool. The amount of algae in the pool t days after the treatment can be approximated by the function $A(t) = 40t^2 - 400t + 500$. How many days after treatment will the pool have the least amount of algae? 5 days

63. Some football fields are built in a parabolic mound shape so that water will drain off the field. A model for the shape of the field is given by $h(x) = -0.00023475x^2 + 0.0375x$, where h is the height of the field in feet at a distance of x feet from the sideline. What is the maximum height? Round to the nearest tenth. 1.5 ft

64. An equation that models the thickness h, in inches, of the mirror at the Palomar Mountain Observatory is given by

$$h(x) = 0.000379x^2 - 0.0758x + 24$$

where x is measured in inches from the edge of the mirror. Find the minimum thickness of the mirror. 20.21 in.

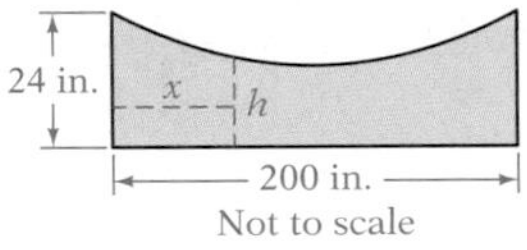

65. The fuel efficiency of an average car is given by the equation $E(v) = -0.018v^2 + 1.476v + 3.4$, where E is the fuel efficiency in miles per gallon and v is the speed of the car in miles per hour. What speed will yield the maximum fuel efficiency? What is the maximum fuel efficiency? 41 mph; 33.658 mi/gal

66. The height s, in feet, of water squirting from a fire hose nozzle is given by the equation $s(x) = -\frac{1}{30}x^2 + 2x + 5$, where x is the horizontal distance, in feet, from the nozzle. How high on a building 40 ft from the fire hose will the water land? $31\frac{2}{3}$ ft

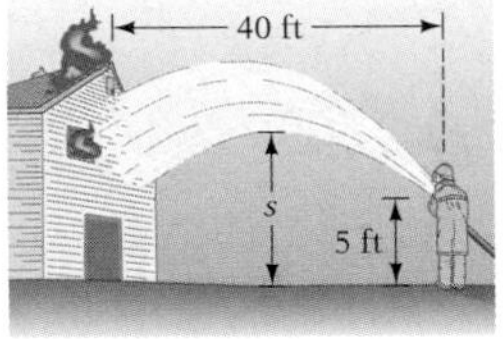

APPLYING THE CONCEPTS

67. One root of the quadratic equation $2x^2 - 5x + k = 0$ is 4. What is the other root? $-\frac{3}{2}$

68. What is the value of k if the vertex of the parabola $y = x^2 - 8x + k$ is a point on the x-axis? 16

Quick Quiz (Objective 11.2D)

1. A manufacturer of video cassette recorders believes that the revenue R, in dollars, the company receives is related to the price P of a recorder by the function $R(P) = 180P - \frac{1}{3}P^2$. What price will give the maximum revenue? $270
2. Find two numbers whose difference is 18 and whose product is a minimum. 9 and -9

605

11.3 Graphs of Functions

Objective A To graph functions

VIDEO & DVD · CD TUTOR · WEB · SSM

The graphs of the polynomial functions $f(x) = mx + b$ (a straight line) and $f(x) = ax^2 + bx + c, a \neq 0$ (a parabola) have been discussed. The graphs of other functions can be drawn by finding ordered pairs that belong to the function, plotting the points that correspond to the ordered pairs, and then drawing a curve through the points.

> **TAKE NOTE**
> The units along the x-axis are different from those along the y-axis. If the units along the x-axis were the same as those along the y-axis, the graph would appear narrower than the one shown.

⇒ Graph: $F(x) = x^3$

Select several values of x and evaluate the function.

x	$F(x) = x^3$	$F(x)$	(x, y)
-2	$(-2)^3$	-8	$(-2, -8)$
-1	$(-1)^3$	-1	$(-1, -1)$
0	0^3	0	$(0, 0)$
1	1^3	1	$(1, 1)$
2	2^3	8	$(2, 8)$

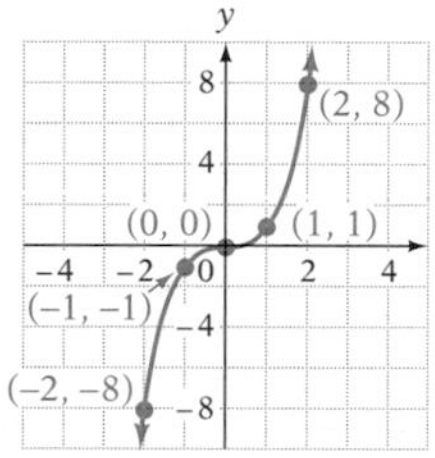

Plot the ordered pairs and draw a graph through the points.

⇒ Graph: $g(x) = x^3 - 4x + 5$

Select several values of x and evaluate the function.

x	$g(x) = x^3 - 4x + 5$	$g(x)$	(x, y)
-3	$(-3)^3 - 4(-3) + 5$	-10	$(-3, -10)$
-2	$(-2)^3 - 4(-2) + 5$	5	$(-2, 5)$
-1	$(-1)^3 - 4(-1) + 5$	8	$(-1, 8)$
0	$(0)^3 - 4(0) + 5$	5	$(0, 5)$
1	$(1)^3 - 4(1) + 5$	2	$(1, 2)$
2	$(2)^3 - 4(2) + 5$	5	$(2, 5)$

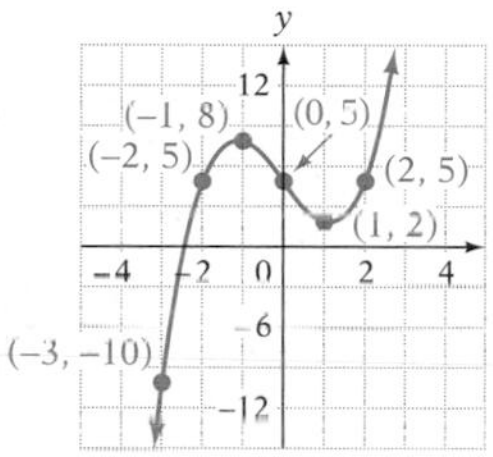

Plot the ordered pairs and draw a graph through the points.

Note from the graphs of the two cubic functions above that the shapes of the graphs can be different. The following graphs of typical cubic polynomial functions show their general shapes.

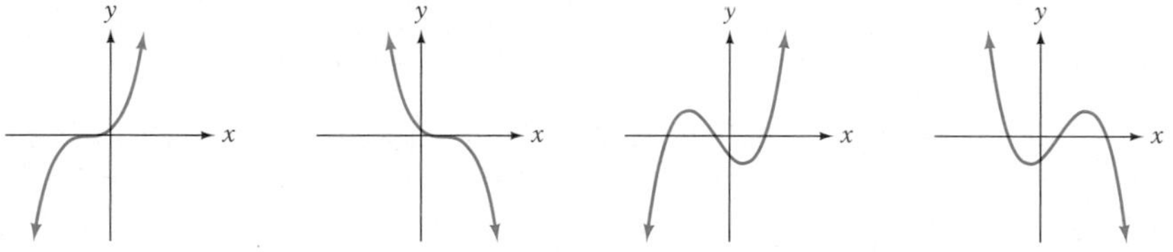

As the degree of a polynomial increases, the graph of the polynomial function can change significantly. In these cases, it may be necessary to plot many points before an accurate graph can be drawn. Only polynomials of degree 3 are considered here.

Objective 11.3A

New Vocabulary
absolute value function
radical function
vertical-line test

Vocabulary to Review
polynomial function
cubic function
domain
range
set-builder notation
interval notation

Instructor Note
The general idea presented here is that all functions have characteristic graphs, and knowing those characteristics can be useful when checking graphical solutions. For instance, if a student uses a graphing utility to graph a cubic polynomial and a graph such as the one below appears, the student will immediately know that there is some type of error.

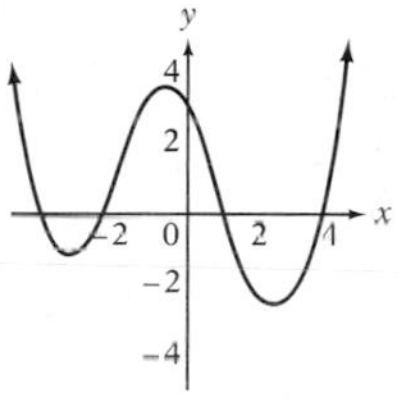

In-Class Examples (Objective 11.3A)

Graph the function and state its domain and range.

1. $f(x) = x^3 + x^2 - x + 1$
Domain: $\{x | x \in \text{real numbers}\}$;
Range: $\{y | y \in \text{real numbers}\}$

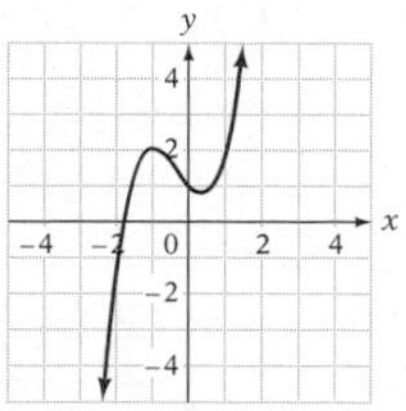

2. $f(x) = |x + 1|$
Domain: $\{x | x \in \text{real numbers}\}$;
Range: $\{y | y \geq 0\}$

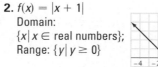

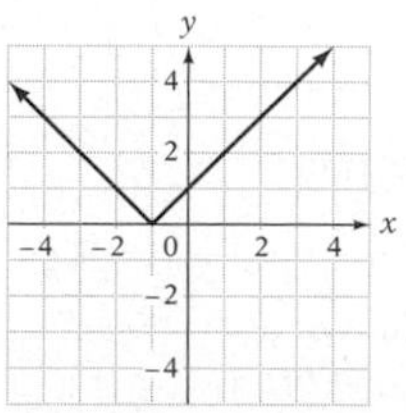

➡ Graph: $f(x) = |x + 2|$

This is an absolute value function.

x	$f(x) = \|x + 2\|$	$f(x)$	(x, y)
-3	$\|-3 + 2\|$	1	$(-3, 1)$
-2	$\|-2 + 2\|$	0	$(-2, 0)$
-1	$\|-1 + 2\|$	1	$(-1, 1)$
0	$\|0 + 2\|$	2	$(0, 2)$
1	$\|1 + 2\|$	3	$(1, 3)$
2	$\|2 + 2\|$	4	$(2, 4)$

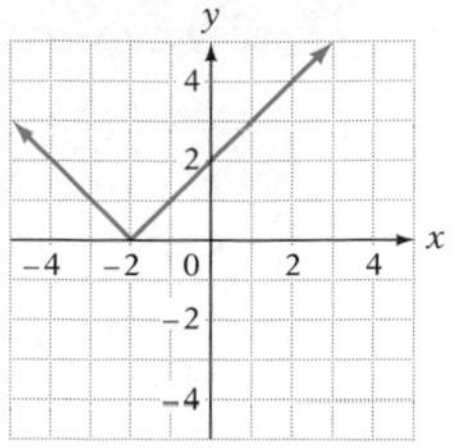

In general, the graph of the absolute value of a linear polynomial is V-shaped.

➡ Graph: $R(x) = \sqrt{2x - 4}$

This is a radical function. Because the square root of a negative number is not a real number, the domain of this function requires that $2x - 4 \geq 0$. Solve this inequality for x.

$$2x - 4 \geq 0$$
$$2x \geq 4$$
$$x \geq 2$$

The domain is $\{x|x \geq 2\}$. This means that only values of x that are greater than or equal to 2 can be chosen as values at which to evaluate the function. In this case, some of the y-coordinates must be approximated.

x	$R(x) = \sqrt{2x - 4}$	$R(x)$	(x, y)
2	$\sqrt{2(2) - 4}$	0	$(2, 0)$
3	$\sqrt{2(3) - 4}$	1.41	$(3, 1.41)$
4	$\sqrt{2(4) - 4}$	2	$(4, 2)$
5	$\sqrt{2(5) - 4}$	2.45	$(5, 2.45)$
6	$\sqrt{2(6) - 4}$	2.83	$(6, 2.83)$

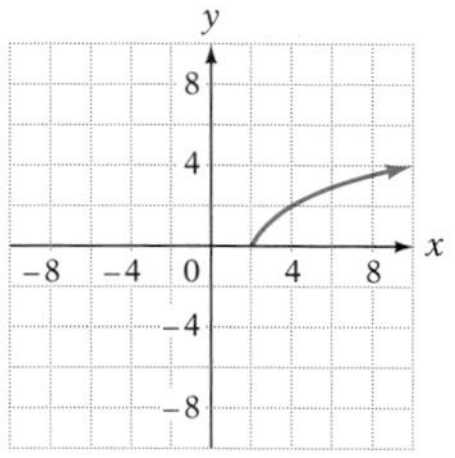

Recall that a function is a special type of relation, one for which no two ordered pairs have the same first coordinate. Graphically, this means that the graph of a function cannot pass through two points that have the same x-coordinate and different y-coordinates. For instance, the graph at the right is not the graph of a function because there are ordered pairs with the same x-coordinate and different y-coordinates.

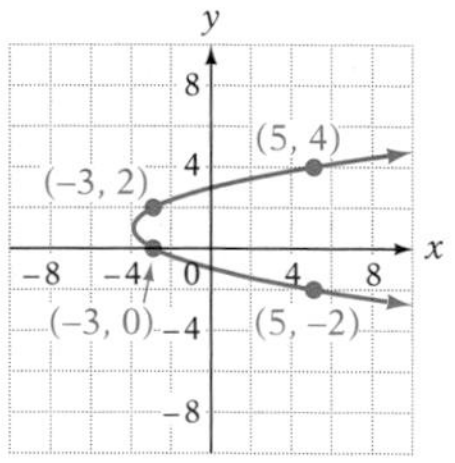

This last graph illustrates a general statement that can be made about whether a graph defines a function. It is called the *vertical-line test.*

Vertical-Line Test

A graph defines a function if any vertical line intersects the graph at no more than one point.

Optional Student Activity

Consider the graph of $y = x^2 - 2x - 3$ shown on page 595. As a point on the graph moves from left to right, the values of y are decreasing for $x < 1$. For $x > 1$, the values of y are increasing. The function is said to be *decreasing* for $x < 1$ and *increasing* for $x > 1$.

For each of the following, determine the interval on which the graph of the function is increasing and the interval on which the graph of the function is decreasing.

1. $f(x) = x^2 + 4x - 1$
   Decreasing: $(-\infty, -2)$;
   Increasing: $(-2, \infty)$
2. $f(x) = -x^2 + 4x - 3$
   Increasing: $(-\infty, 2)$;
   Decreasing: $(2, \infty)$
3. $f(x) = x^3 - 2$
   Increasing: $(-\infty, \infty)$
4. $f(x) = -x^3 + 3x^2 - 3x - 1$
   Decreasing: $(-\infty, \infty)$
5. $f(x) = |x - 4|$
   Decreasing: $(-\infty, 4)$;
   Increasing: $(4, \infty)$
6. $f(x) = -3|x| - 1$
   Increasing: $(-\infty, 0)$;
   Decreasing: $(0, \infty)$
7. $f(x) = \sqrt{x + 3}$
   Increasing: $(-3, \infty)$
8. $f(x) = -2\sqrt{x - 1}$
   Decreasing: $(1, \infty)$
9. $f(x) = 2x - 3$
   Increasing: $(-\infty, \infty)$
10. $f(x) = 1 - 3x$
    Decreasing: $(-\infty, \infty)$

For example, the graph of a nonvertical straight line is the graph of a function. Any vertical line intersects the graph no more than once. The graph of a circle, however, is not the graph of a function. There are vertical lines that intersect the graph at more than one point.

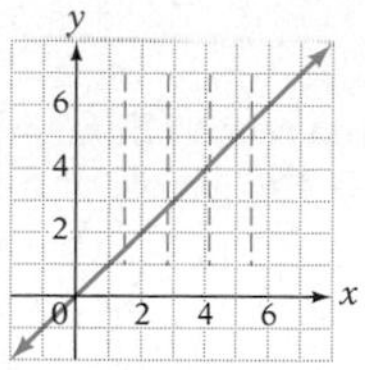

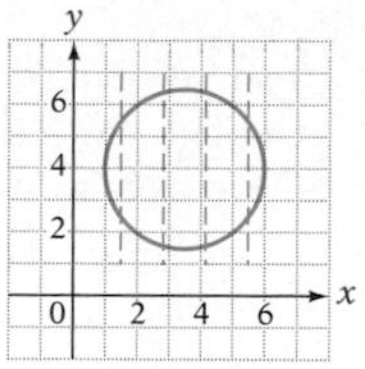

There are practical situations in which a graph is not the graph of a function. Below is an example of such an application.

One of the causes of smog is an inversion layer of air where temperatures at higher altitudes are warmer than those at lower altitudes. The graph at the right shows the altitudes at which various temperatures were recorded. As shown by the dashed lines in the graph, there are two altitudes at which the temperature was 25°C. This means that there are two ordered pairs (shown in the graph) with the same first coordinate but different second coordinates. The graph does not define a function.

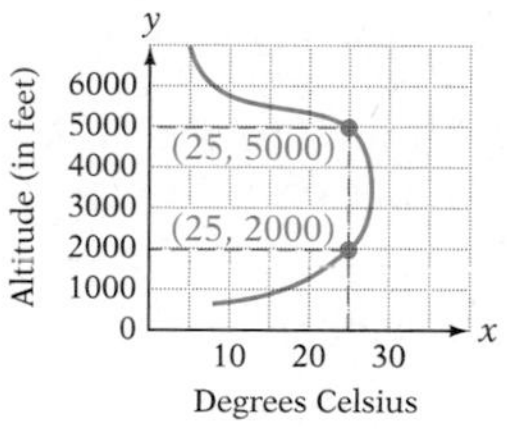

When a graph does define a function, the domain and range can be estimated from the graph.

TAKE NOTE

To determine the domain, think of collapsing the graph onto the x-axis and determining the interval on the x-axis that the graph covers. To determine the range, think of collapsing the graph onto the y-axis and determining the interval on the y-axis that the graph covers.

➡ Determine the domain and range of the function given by the graph at the right.

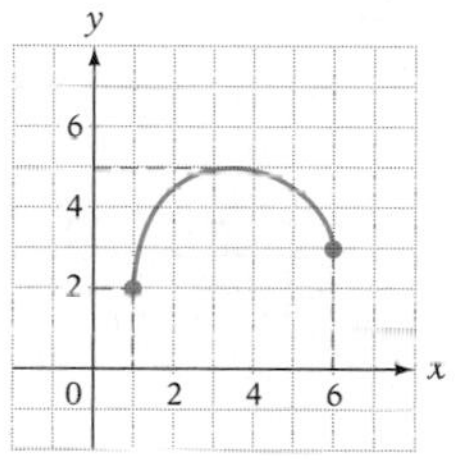

The solid dots on the graph indicate its beginning and ending points.

The domain is the set of x-coordinates.
Domain: $\{x | 1 \le x \le 6\}$

The range is the set of y-coordinates.
Range: $\{y | 2 \le y \le 5\}$

➡ Determine the domain and range of the function given by the graph at the right.

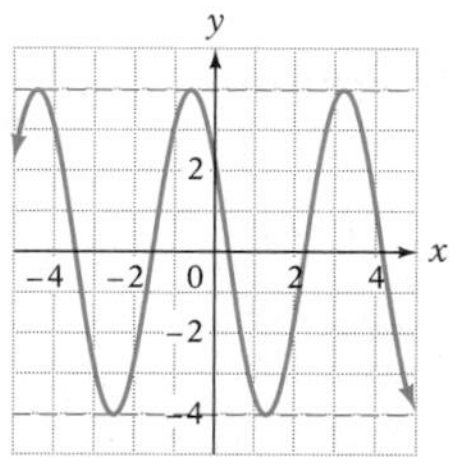

The arrows on the graph indicate that the graph continues in the same manner.

The domain is the set of x-coordinates.
Domain: $\{x | x \in \text{real numbers}\}$

The range is the set of y-coordinates.
Range: $\{y | -4 \le y \le 4\}$

Optional Student Activity

Graph $y = \sqrt{x+2}$ and $y^2 = x + 2$. Discuss the similarities and differences between the two graphs. The range of $y = \sqrt{x+2}$ is $\{y | y \ge 0\}$. The range of $y^2 = x + 2$ is $\{y | y \in \text{real numbers}\}$. The graphs are the same for $y \ge 0$, but the graph of $y^2 = x + 2$ includes the reflection of the graph of $y = \sqrt{x+2}$ over the x-axis.

Instructor Note

When discussing how to determine the domain and range of a function from its graph, you might tell students to think of "squishing" or "flattening" the graph onto the x-axis to determine the domain and "squishing" or "flattening" the graph onto the y-axis to determine the range.

608 Chapter 11 / Functions and Relations

Example 1
Use the vertical-line test to determine whether the graph shown is the graph of a function.

Solution

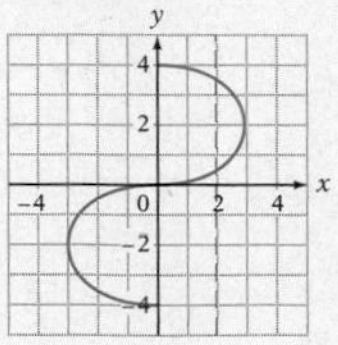

A vertical line intersects the graph more than once. The graph is not the graph of a function.

You Try It 1
Use the vertical-line test to determine whether the graph shown is the graph of a function.

Your solution

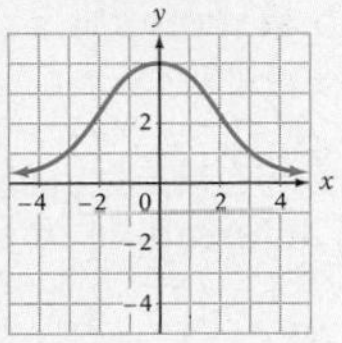

yes

Example 2
Graph $f(x) = x^3 - 3x$. State the domain and range of the function.

Solution

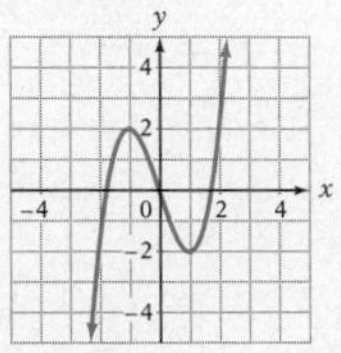

domain: $\{x \mid x \in \text{real numbers}\}$
range: $\{y \mid y \in \text{real numbers}\}$

You Try It 2
Graph $f(x) = -\frac{1}{2}x^3 + 2x$. State the domain and range of the function.

Your solution

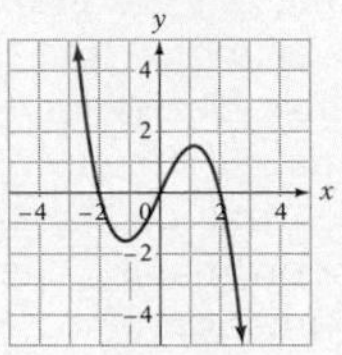

domain: $\{x \mid x \in \text{real numbers}\}$
range: $\{y \mid y \in \text{real numbers}\}$

Example 3
Graph $f(x) = |x| + 2$. State the domain and range of the function.

Solution

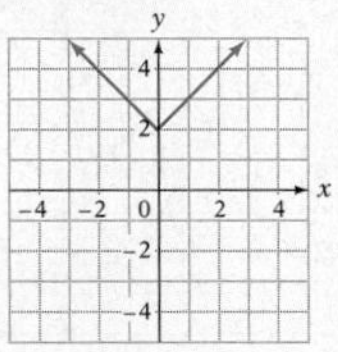

domain: $\{x \mid x \in \text{real numbers}\}$
range: $\{y \mid y \geq 2\}$

You Try It 3
Graph $f(x) = |x + 2|$. State the domain and range of the function.

Your solution

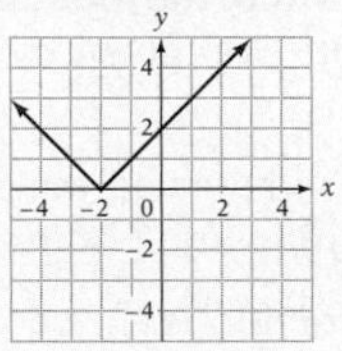

domain: $\{x \mid x \in \text{real numbers}\}$
range: $\{y \mid y \geq 0\}$

Example 4
Graph $f(x) = \sqrt{2 - x}$. State the domain and range of the function.

Solution

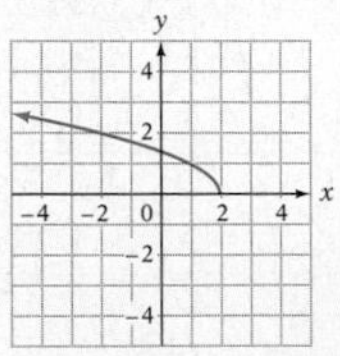

domain: $\{x \mid x \leq 2\}$
range: $\{y \mid y \geq 0\}$

You Try It 4
Graph $f(x) = -\sqrt{x - 1}$. State the domain and range of the function.

Your solution

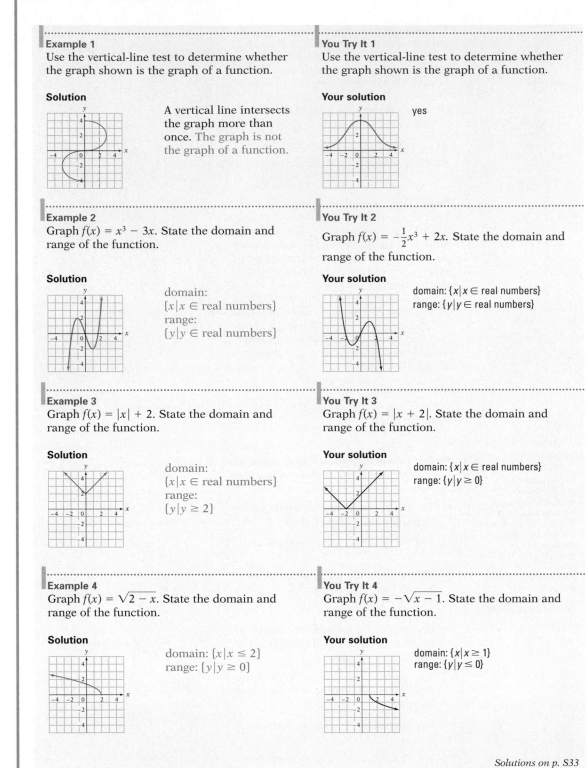

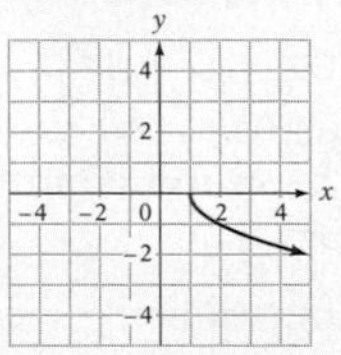

domain: $\{x \mid x \geq 1\}$
range: $\{y \mid y \leq 0\}$

Solutions on p. S33

11.3 Exercises

Objective A

Use the vertical-line test to determine whether the graph is the graph of a function.

1.

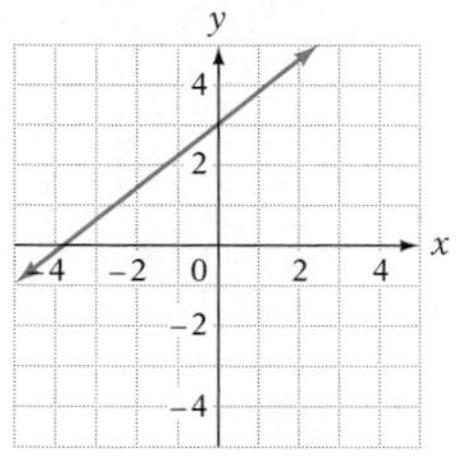

yes

2.

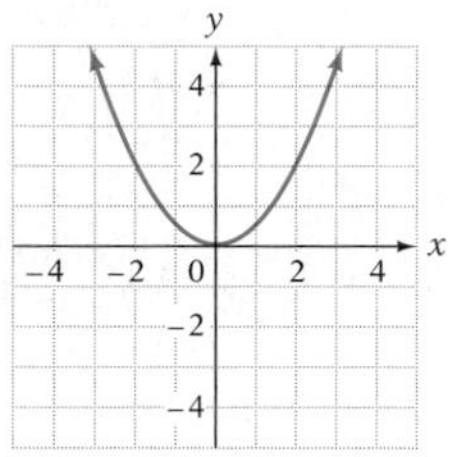

yes

3.

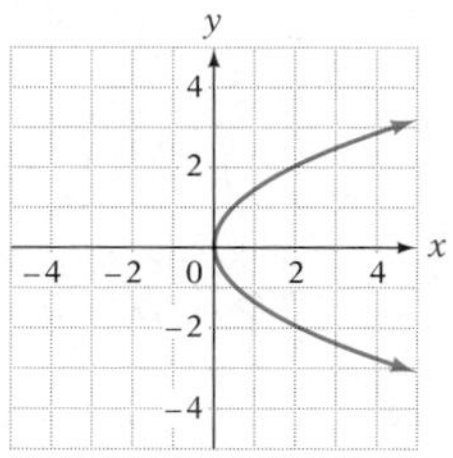

no

4.

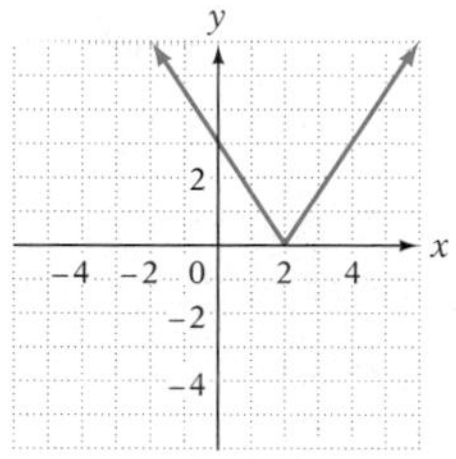

yes

5.

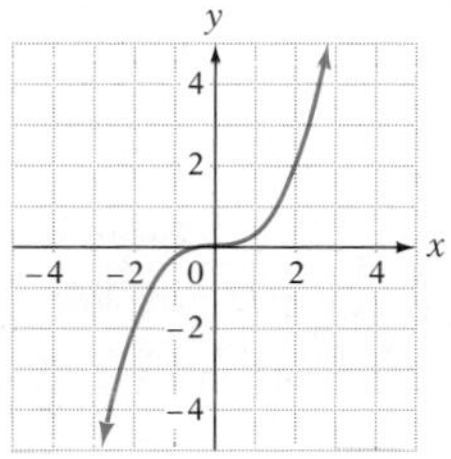

yes

6.

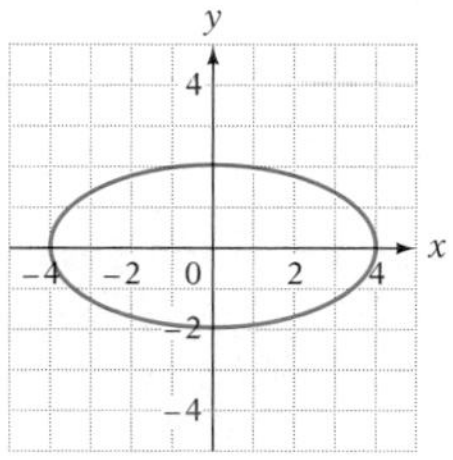

no

Graph the function and state its domain and range.

7. $f(x) = 3|2 - x|$

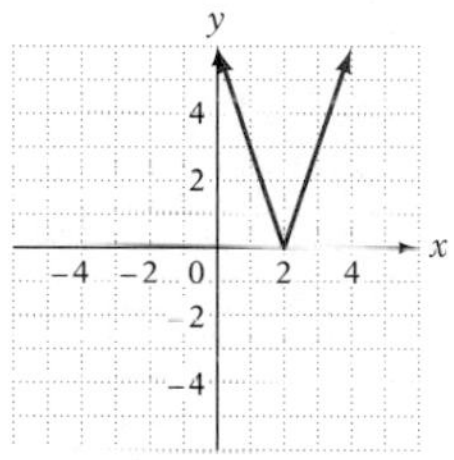

domain: $\{x|x \in \text{real numbers}\}$
range: $\{y|y \geq 0\}$

8. $f(x) = x^3 - 1$

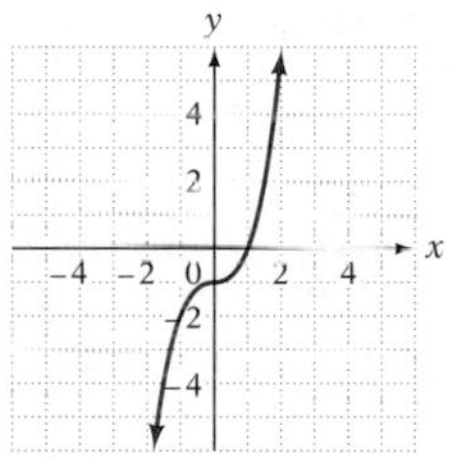

domain: $\{x|x \in \text{real numbers}\}$
range: $\{y|y \in \text{real numbers}\}$

9. $f(x) = 1 - x^3$

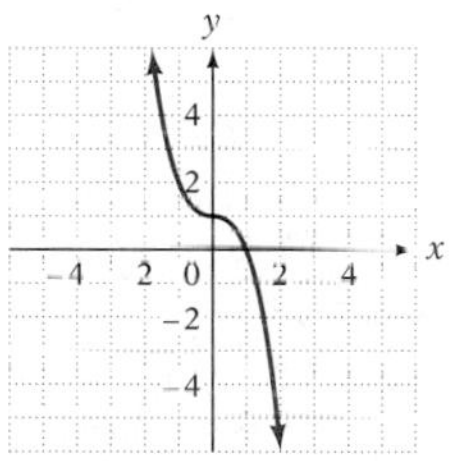

domain: $\{x|x \in \text{real numbers}\}$
range: $\{y|y \in \text{real numbers}\}$

10. $f(x) = \sqrt{1 + x}$

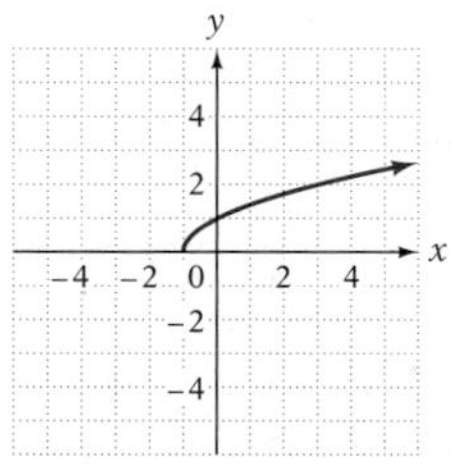

domain: $\{x|x \geq -1\}$
range: $\{y|y \geq 0\}$

11. $f(x) = \sqrt{4 - x}$

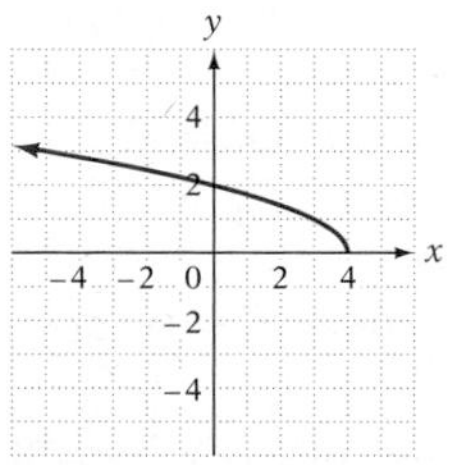

domain: $\{x|x \leq 4\}$
range: $\{y|y \geq 0\}$

12. $f(x) = |2x - 1|$

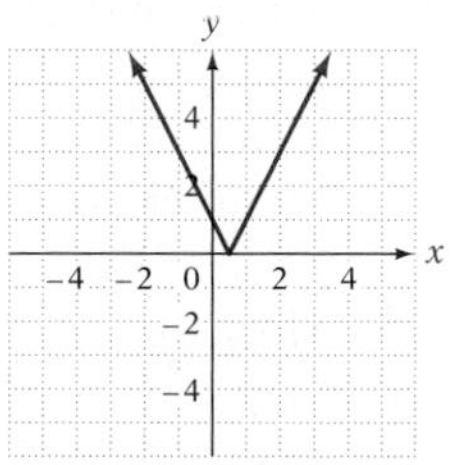

domain: $\{x|x \in \text{real numbers}\}$
range: $\{y|y \geq 0\}$

Section 11.3

Suggested Assignment

Exercises 1–17, odds
More challenging problems:
Exercises 19–25, odds;
Exercise 22

Quick Quiz (Objective 11.3A)

Graph the function and state its domain and range.

1. $f(x) = \sqrt{x + 2}$
Domain: $\{x|x \geq -2\}$;
Range: $\{y|y \geq 0\}$

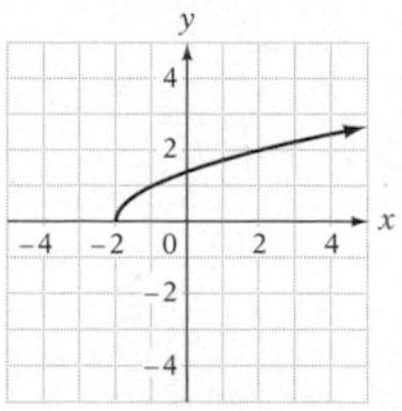

2. $f(x) = |x| + 2$
Domain:
$\{x|x \in \text{real numbers}\}$;
Range: $\{y|y \geq 2\}$

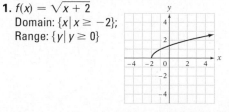

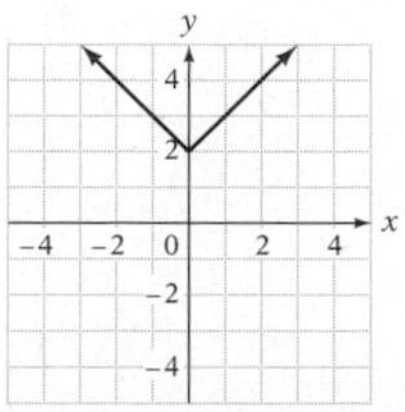

610

13. $f(x) = x^3 + 4x^2 + 4x$

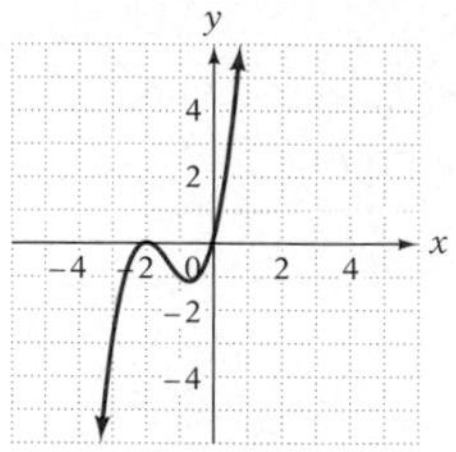

domain: $\{x \mid x \in \text{real numbers}\}$
range: $\{y \mid y \in \text{real numbers}\}$

14. $f(x) = x^3 - x^2 - x + 1$

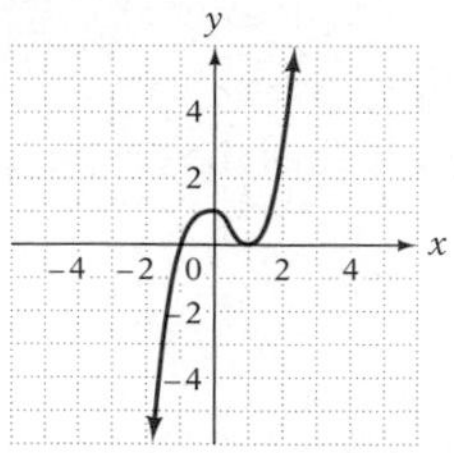

domain: $\{x \mid x \in \text{real numbers}\}$
range: $\{y \mid y \in \text{real numbers}\}$

15. $f(x) = -\sqrt{x + 2}$

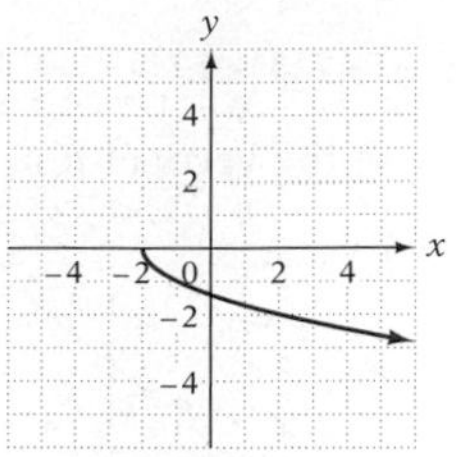

domain: $\{x \mid x \geq -2\}$
range: $\{y \mid y \leq 0\}$

16. $f(x) = -\sqrt{x - 3}$

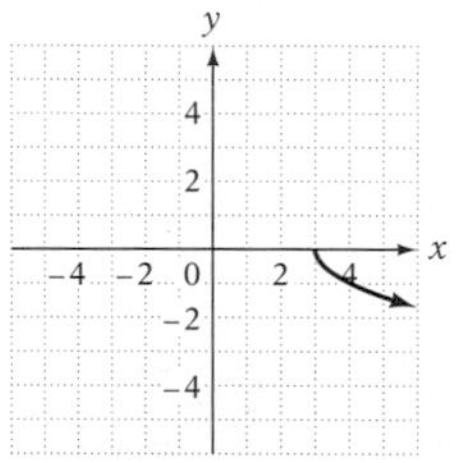

domain: $\{x \mid x \geq 3\}$
range: $\{y \mid y \leq 0\}$

17. $f(x) = |2x + 2|$

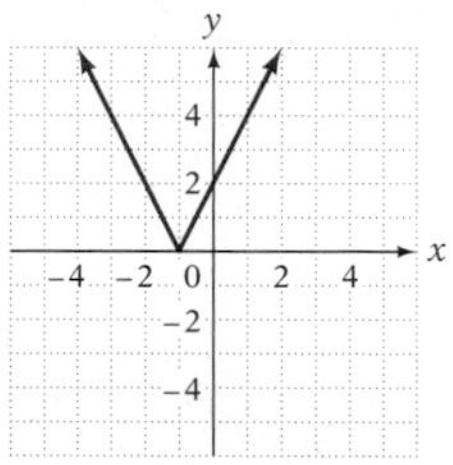

domain: $\{x \mid x \in \text{real numbers}\}$
range: $\{y \mid y \geq 0\}$

18. $f(x) = 2|x + 1|$

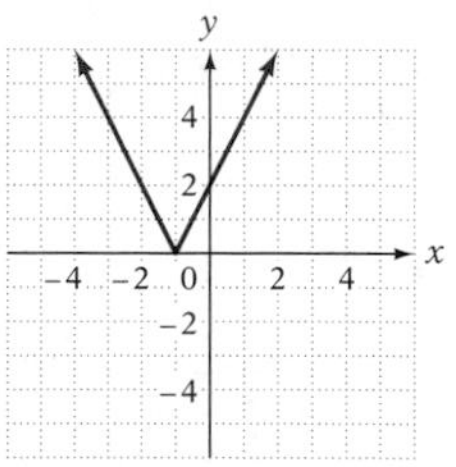

domain: $\{x \mid x \in \text{real numbers}\}$
range: $\{y \mid y \geq 0\}$

APPLYING THE CONCEPTS

19. If $f(x) = \sqrt{x - 2}$ and $f(a) = 4$, find a. $a = 18$

20. If $f(x) = \sqrt{x + 5}$ and $f(a) = 3$, find a. $a = 4$

21. $f(a, b)$ = the sum of a and b
$g(a, b)$ = the product of a and b
Find $f(2, 5) + g(2, 5)$. 17

22. The graph of the function f is shown at the right. For this function, which of the following are true?
a. $f(4) = 1$ **b.** $f(0) = 3$ **c.** $f(-3) = 2$ **a** and **c** are true.

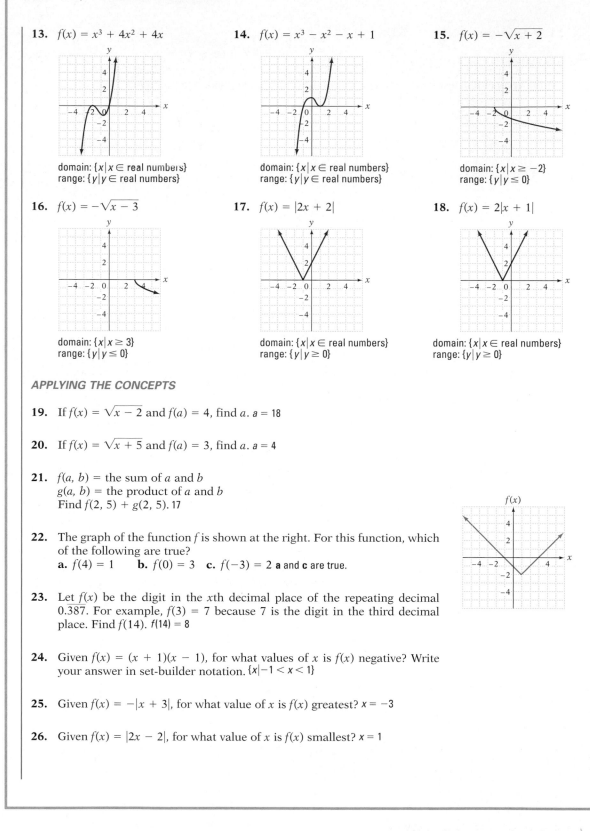

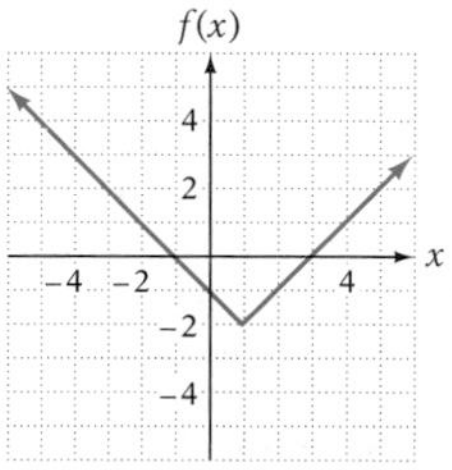

23. Let $f(x)$ be the digit in the xth decimal place of the repeating decimal $0.\overline{387}$. For example, $f(3) = 7$ because 7 is the digit in the third decimal place. Find $f(14)$. $f(14) = 8$

24. Given $f(x) = (x + 1)(x - 1)$, for what values of x is $f(x)$ negative? Write your answer in set-builder notation. $\{x \mid -1 < x < 1\}$

25. Given $f(x) = -|x + 3|$, for what value of x is $f(x)$ greatest? $x = -3$

26. Given $f(x) = |2x - 2|$, for what value of x is $f(x)$ smallest? $x = 1$

11.4 Algebra of Functions

Objective A **To perform operations on functions**

VIDEO & DVD · CD TUTOR · WEB · SSM

The operations of addition, subtraction, multiplication, and division of functions are defined as follows.

Operations on Functions

If f and g are functions and x is an element of the domain of each function, then

$(f+g)(x) = f(x) + g(x)$ $\quad (f \cdot g)(x) = f(x) \cdot g(x)$

$(f-g)(x) = f(x) - g(x)$ $\quad \left(\frac{f}{g}\right)(x) = \frac{f(x)}{g(x)}, \; g(x) \neq 0$

➡ Given $f(x) = x^2 + 1$ and $g(x) = 3x - 2$, find $(f+g)(3)$ and $(f \cdot g)(-1)$.

$$\begin{aligned}(f+g)(3) &= f(3) + g(3)\\ &= [(3)^2 + 1] + [3(3) - 2]\\ &= 10 + 7 = 17\end{aligned}$$

$$\begin{aligned}(f \cdot g)(-1) &= f(-1) \cdot g(-1)\\ &= [(-1)^2 + 1] \cdot [3(-1) - 2]\\ &= 2 \cdot (-5) = -10\end{aligned}$$

Consider the functions f and g from the last example. Let $S(x)$ be the sum of the two functions. Then

$$\begin{aligned}S(x) &= (f+g)(x) = f(x) + g(x) && \text{• The definition of addition of functions}\\ &= (x^2 + 1) + (3x - 2) && \text{• } f(x) = x^2 + 1,\ g(x) = 3x - 2\\ S(x) &= x^2 + 3x - 1\end{aligned}$$

Now evaluate $S(3)$.

$$\begin{aligned}S(3) &= (3)^2 + 3(3) - 1\\ &= 9 + 9 - 1\\ &= 17 = (f+g)(3)\end{aligned}$$

Note that $S(3) = 17$ and $(f+g)(3) = 17$. This shows that adding $f(x) + g(x)$ and then evaluating is the same as evaluating $f(x)$ and $g(x)$ and then adding. The same is true for the other operations on functions. For instance, let $P(x)$ be the product of the functions f and g. Then

$$\begin{aligned}P(x) &= (f \cdot g)(x) = f(x) \cdot g(x)\\ &= (x^2 + 1)(3x - 2)\\ &= 3x^3 - 2x^2 + 3x - 2\\ P(-1) &= 3(-1)^3 - 2(-1)^2 + 3(-1) - 2\\ &= -3 - 2 - 3 - 2\\ &= -10 && \text{• Note that } P(-1) = -10 \text{ and } (f \cdot g)(-1) = -10.\end{aligned}$$

611

Objective 11.4A

New Formulas

operations on functions:

$(f+g)(x) = f(x) + g(x)$

$(f-g)(x) = f(x) - g(x)$

$(f \cdot g)(x) = f(x) \cdot g(x)$

$\left(\frac{f}{g}\right)(x) = \frac{f(x)}{g(x)}, \; g(x) \neq 0$

Instructor Note

Definitions are usually difficult for students. After you have given some examples of operations on functions, you might point out the requirement in the definition that x must be an element of the domain of each function. For example, $(f+g)(1)$ is not a real number if $f(x) = x^2$ and $g(x) = \sqrt{x-2}$.

Concept Check

1. Given $f(x) = 4x^{1/2}$ and $g(x) = -2x^{1/2}$, find **a.** the sum of the functions, **b.** the difference of the functions, and **c.** the domains of the sum and difference. **a.** $2x^{1/2}$ **b.** $6x^{1/2}$ **c.** The domain of the sum is $\{x \mid x \geq 0\}$; the domain of the difference is $\{x \mid x \geq 0\}$.
2. Given $f(x) = 5x$ and $g(x) = x^{1/4}$, find **a.** the product of the functions, **b.** the quotient of the functions, and **c.** the domains of the product and quotient. **a.** $5x^{5/4}$ **b.** $5x^{3/4}$ **c.** The domain of the product is $\{x \mid x \geq 0\}$; the domain of the quotient is $\{x \mid x > 0\}$.

In-Class Examples (Objective 11.4A)

For $f(x) = x^2 - 4$ and $g(x) = -3x + 1$, find:

1. $f(4) + g(4)$ 1
2. $f(-1) - g(-1)$ -7
3. $(f \cdot g)(3)$ -40
4. $\left(\frac{g}{f}\right)(2)$ Cannot be evaluated

Optional Student Activity

The values of some functions may depend on several variables. For instance, the perimeter of a rectangle depends on the length L and the width W. We can write this as $P(L, W) = 2L + 2W$. To evaluate this function, we need to be given the values of L and W. For example, to find the perimeter of a rectangle whose length is 5 ft and whose width is 3 ft, evaluate $P(L, W)$ when $L = 5$ and $W = 3$.

$$\begin{aligned} P(L, W) &= 2L + 2W \\ P(5, 3) &= 2(5) + 2(3) \\ &= 10 + 6 = 16 \end{aligned}$$

The perimeter is 16 ft.

1. Evaluate $f(a, b) = 2a + 3b$ when $a = 3$ and $b = 4$. 18
2. Evaluate $R(s, t) = 2st - t^2$ when $s = -1$ and $t = 2$. −8
3. Although we normally do not think of addition as a function, it is a function of two variables. If we define the function *Add* as $Add(a, b) = a + b$, find $Add(3, 7)$. 10
4. Write the area of a triangle as a function of two variables. $A(b, h) = \frac{1}{2}bh$
5. Write the length of the hypotenuse of a right triangle as a function of two variables. $H(a, b) = \sqrt{a^2 + b^2}$
6. Give an example of a function whose value depends on more than one variable. Answers will vary. One example is the calculation of simple interest as a function of principal, interest rate, and time: $I(p, r, t) = prt$.

➡ Given $f(x) = 2x^2 - 5x + 3$ and $g(x) = x^2 - 1$, find $\left(\frac{f}{g}\right)(1)$.

$$\begin{aligned} \left(\frac{f}{g}\right)(1) &= \frac{f(1)}{g(1)} \\ &= \frac{2(1)^2 - 5(1) + 3}{(1)^2 - 1} = \frac{0}{0} \end{aligned}$$

• **Not a real number**

Because $\frac{0}{0}$ is not defined, the expression $\left(\frac{f}{g}\right)(1)$ cannot be evaluated.

Example 1

Given $f(x) = x^2 - x + 1$ and $g(x) = x^3 - 4$, find $(f - g)(3)$.

Solution

$$\begin{aligned} (f - g)(3) &= f(3) - g(3) \\ &= (3^2 - 3 + 1) - (3^3 - 4) \\ &= 7 - 23 \\ &= -16 \end{aligned}$$

$(f - g)(3) = -16$

You Try It 1

Given $f(x) = x^2 + 2x$ and $g(x) = 5x - 2$, find $(f + g)(-2)$.

Your solution

−12

Example 2

Given $f(x) = x^2 + 2$ and $g(x) = 2x + 3$, find $(f \cdot g)(-2)$.

Solution

$$\begin{aligned} (f \cdot g)(-2) &= f(-2) \cdot g(-2) \\ &= [(-2)^2 + 2] \cdot [2(-2) + 3] \\ &= 6 \cdot (-1) \\ &= -6 \end{aligned}$$

$(f \cdot g)(-2) = -6$

You Try It 2

Given $f(x) = 4 - x^2$ and $g(x) = 3x - 4$, find $(f \cdot g)(3)$.

Your solution

−25

Example 3

Given $f(x) = x^2 + 4x + 4$ and $g(x) = x^3 - 2$, find $\left(\frac{f}{g}\right)(3)$.

Solution

$$\begin{aligned} \left(\frac{f}{g}\right)(3) &= \frac{f(3)}{g(3)} \\ &= \frac{3^2 + 4(3) + 4}{3^3 - 2} \\ &= \frac{25}{25} \\ &= 1 \end{aligned}$$

$\left(\frac{f}{g}\right)(3) = 1$

You Try It 3

Given $f(x) = x^2 - 4$ and $g(x) = x^2 + 2x + 1$, find $\left(\frac{f}{g}\right)(4)$.

Your solution

$\frac{12}{25}$

Solutions on pp. S33–S34

Objective B To find the composition of two functions

VIDEO & DVD CD TUTOR WEB SSM

A function can be evaluated at the value of another function. Consider

$$f(x) = 2x + 7 \quad \text{and} \quad g(x) = x^2 + 1$$

The expression $f[g(-2)]$ means to evaluate the function f at $g(-2)$.

$$g(-2) = (-2)^2 + 1 = 4 + 1 = 5$$
$$f[g(-2)] = f(5) = 2(5) + 7 = 10 + 7 = 17$$

Definition of the Composition of Two Functions

Let f and g be two functions such that $g(x)$ is in the domain of f for all x in the domain of g. Then the **composition** of the two functions, denoted by $f \circ g$, is the function whose value at x is given by $(f \circ g)(x) = f[g(x)]$.

The function defined by $f[g(x)]$ is called the **composite** of f and g.

The function machine at the right illustrates the composition of two functions, $g(x) = x^2$ and $f(x) = 2x$. Note that a composite function combines two functions. First one function pairs an input with an output. Then that output is used as the input for a second function, which in turn produces a final output.

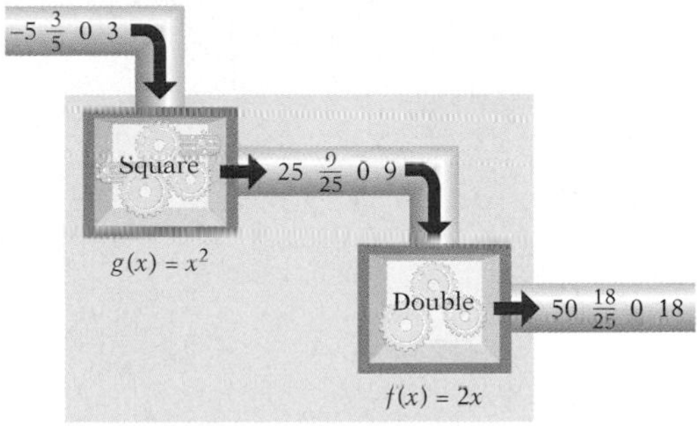

$(f \circ g)(x) = f[g(x)]$

The requirement in the definition of the composition of two functions that $g(x)$ be in the domain of f for all x in the domain of g is important. For instance, let

$$f(x) = \frac{1}{x-1} \quad \text{and} \quad g(x) = 3x - 5$$

When $x = 2$,

$$g(2) = 3(2) - 5 = 1$$
$$f[g(2)] = f(1) = \frac{1}{1-1} = \frac{1}{0}$$ • **This is not a real number.**

In this case, $g(2)$ is not in the domain of f. Thus the composition is not defined at 2.

➡ Given $f(x) = x^3 - x + 1$ and $g(x) = 2x^2 - 10$, evaluate $(g \circ f)(2)$.

$$f(2) = (2)^3 - (2) + 1 = 7$$
$$(g \circ f)(2) = g[f(2)]$$
$$= g(7)$$
$$= 2(7)^2 - 10 = 88$$
$$(g \circ f)(2) = 88$$

Objective 11.4B

New Vocabulary

composition of functions
composite function

New Symbols

$(f \circ g)(x) = f[g(x)]$

Instructor Note

This is a very difficult topic for many students. Using open brackets may help these students.

$$f(x) = 2x + 7$$
$$f[\quad] = 2[\quad] + 7$$
$$\downarrow \qquad \downarrow$$
$$f[g(-2)] = 2[g(-2)] + 7$$
$$f[g(-2)] = 2(5) + 7$$
$$f[g(-2)] = 17$$

Concept Check

Sets R, S, and T are shown in the figure at the bottom of this page, along with functions f and g.

1. Evaluate $f(4)$. 16
2. Evaluate $g(9)$. 5
3. Evaluate $(g \circ f)(2)$. 0
4. Evaluate $(g \circ f)(1)$. −3
5. Write a rule for f. The function f squares an element in set R.
6. Write a rule for g. The function g subtracts 4 from an element in set S.
7. Write a rule for $g \circ f$. The composite of functions f and g first squares an element in set R and then subtracts 4 from the result.
8. Create your own two functions f and g, diagram them as shown below, and find the composition of at least two elements of $g \circ f$.

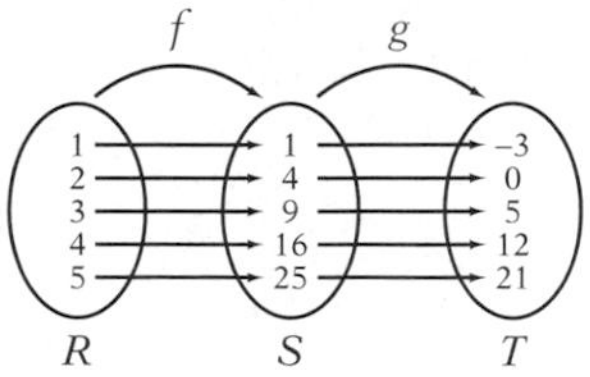

In-Class Examples (Objective 11.4B)

Given $f(x) = 3x - 2$ and $g(x) = \frac{1}{2}x + 4$, evaluate the composite function.

1. $f[g(2)]$ 13
2. $g[f(0)]$ 3
3. $f[g(-4)]$ 4

Given $f(x) = x^2 + 3$ and $h(x) = -x + 5$, evaluate the composite function.

4. $f[h(-3)]$ 67
5. $h[f(-3)]$ −7
6. $f[h(x)]$ $x^2 - 10x + 28$

Optional Student Activity

It is sometimes possible to "decompose" a composite function $f \circ g$ into two functions f and g. In the example below, we use the fact that $(f \circ g)(x) = f[g(x)]$.

Example:

Determine two functions f and g such that $(f \circ g)(x) = 8\sqrt{x} - 3$.

Solution:

Let $g(x) = \sqrt{x}$ and $f(x) = 8x - 3$.

Check:

$(f \circ g)(x) = f[g(x)] = f(\sqrt{x}) = 8\sqrt{x} - 3$

1. Determine two functions f and g such that $(f \circ g)(x) = 5x^{1/4} + 6$. Let $g(x) = x^{1/4}$ and $f(x) = 5x + 6$.
2. Determine two functions f and g such that $(f \circ g)(x) = -2x^3 - 7$. Let $g(x) = x^3$ and $f(x) = -2x - 7$.
3. Determine two functions f and g such that $(f \circ g)(x) = x^4 + x^2 + 9$. Let $g(x) = x^2$ and $f(x) = x^2 + x + 9$.

Instructor Note

You might present the functions $I(x) = 12x$, which converts x feet into I inches, and $F(y) = 3y$, which converts y yards into F feet. Ask students the following questions.

1. Explain the meaning of $I[F(y)]$. It converts y yards to inches.
2. Does $F[I(x)]$ make sense in the context of this exercise? No
3. Exercise 2 provides a real-world application of a property that composition of functions does not have. What is that property? Commutativity

➡ Given $f(x) = 3x - 2$ and $g(x) = x^2 - 2x$, find $(f \circ g)(x)$.

$$(f \circ g)(x) = f[g(x)] = 3(x^2 - 2x) - 2$$
$$= 3x^2 - 6x - 2$$

When we are evaluating compositions of functions, the order in which the functions are applied is important. In the two diagrams below, the order of the *square function*, $g(x) = x^2$, and the *double function*, $f(x) = 2x$, is interchanged. Note that the final outputs are different. Therefore, $(g \circ f)(x) \neq (f \circ g)(x)$.

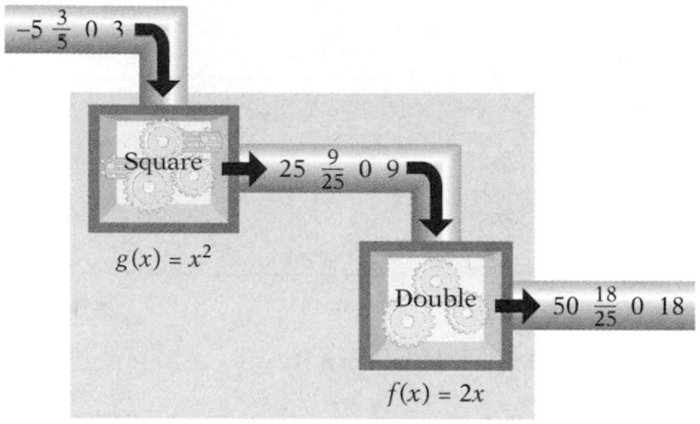

$(f \circ g)(x) = f[g(x)]$

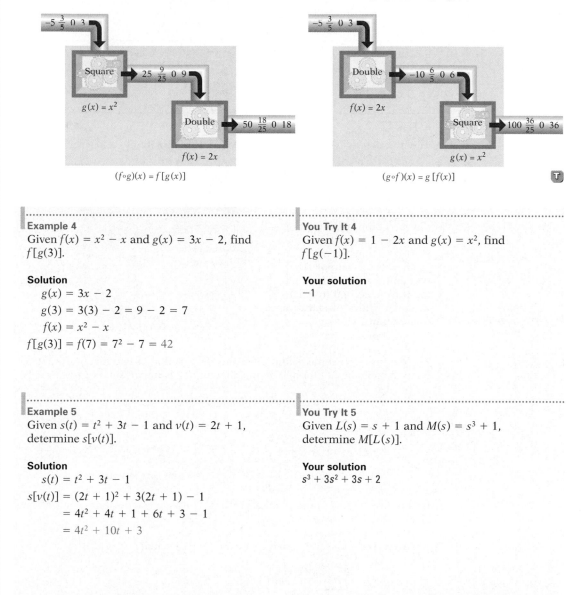

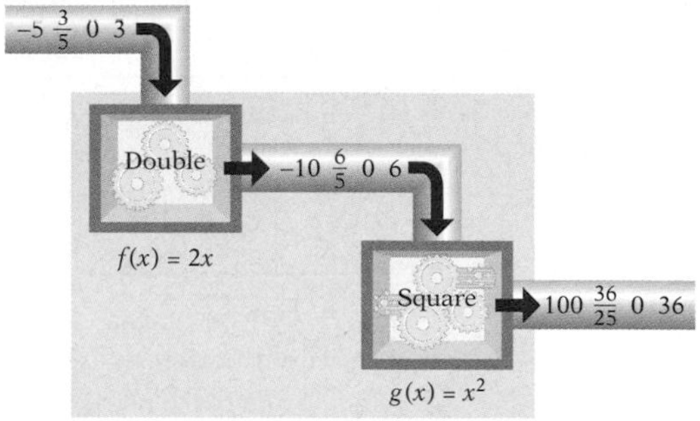

$(g \circ f)(x) = g[f(x)]$

Example 4

Given $f(x) = x^2 - x$ and $g(x) = 3x - 2$, find $f[g(3)]$.

Solution

$g(x) = 3x - 2$

$g(3) = 3(3) - 2 = 9 - 2 = 7$

$f(x) = x^2 - x$

$f[g(3)] = f(7) = 7^2 - 7 = 42$

You Try It 4

Given $f(x) = 1 - 2x$ and $g(x) = x^2$, find $f[g(-1)]$.

Your solution

-1

Example 5

Given $s(t) = t^2 + 3t - 1$ and $v(t) = 2t + 1$, determine $s[v(t)]$.

Solution

$s(t) = t^2 + 3t - 1$

$s[v(t)] = (2t + 1)^2 + 3(2t + 1) - 1$
$= 4t^2 + 4t + 1 + 6t + 3 - 1$
$= 4t^2 + 10t + 3$

You Try It 5

Given $L(s) = s + 1$ and $M(s) = s^3 + 1$, determine $M[L(s)]$.

Your solution

$s^3 + 3s^2 + 3s + 2$

Solutions on p. S34

Section 11.4

Suggested Assignment
Exercises 1–45, odds
More challenging problems:
Exercises 47–57, odds

11.4 Exercises

Objective A

For $f(x) = 2x^2 - 3$ and $g(x) = -2x + 4$, find:

1. $(f - g)(2)$
5

2. $(f - g)(3)$
17

3. $(f + g)(0)$
1

4. $(f + g)(1)$
1

5. $(f \cdot g)(2)$
0

6. $(f \cdot g)(-1)$
-6

7. $\left(\frac{f}{g}\right)(4)$
$-\frac{29}{4}$

8. $\left(\frac{f}{g}\right)(-1)$
$-\frac{1}{6}$

9. $\left(\frac{g}{f}\right)(-3)$
$\frac{2}{3}$

For $f(x) = 2x^2 + 3x - 1$ and $g(x) = 2x - 4$, find:

10. $(f + g)(-3)$
-2

11. $(f + g)(1)$
2

12. $(f - g)(-2)$
9

13. $(f - g)(4)$
39

14. $(f \cdot g)(-2)$
-8

15. $(f \cdot g)(1)$
-8

16. $\left(\frac{f}{g}\right)(2)$
undefined

17. $\left(\frac{f}{g}\right)(-3)$
$-\frac{4}{5}$

18. $(f \cdot g)\left(\frac{1}{2}\right)$
$-\frac{1}{3}$

For $f(x) = x^2 + 3x - 5$ and $g(x) = x^3 - 2x + 3$, find:

19. $(f - g)(2)$
-2

20. $(f \cdot g)(-3)$
90

21. $\left(\frac{f}{g}\right)(-2)$
7

Objective B

Given $f(x) = 2x - 3$ and $g(x) = 4x - 1$, evaluate the composite function.

22. $f[g(0)]$
-5

23. $g[f(0)]$
-13

24. $f[g(2)]$
11

25. $g[f(-2)]$
-29

26. $f[g(x)]$
$8x - 5$

27. $g[f(x)]$
$8x - 13$

Quick Quiz (Objective 11.4A)
For $f(x) = 2x^2 + x - 4$ and $g(x) = -x + 3$, find:
1. $f(2) + g(2)$ 7
2. $f(3) - g(3)$ 17
3. $(f \cdot g)(-1)$ -12
4. $\left(\frac{f}{g}\right)(-2)$ $\frac{2}{5}$

Given $h(x) = 2x + 4$ and $f(x) = \frac{1}{2}x + 2$, evaluate the composite function.

28. $(h \circ f)(0)$
8

29. $(f \circ h)(0)$
4

30. $(h \circ f)(2)$
10

31. $(f \circ h)(-1)$
3

32. $(h \circ f)(x)$
$x + 8$

33. $(f \circ h)(x)$
$x + 4$

Given $f(x) = x^2 + x + 1$ and $h(x) = 3x + 2$, evaluate the composite function.

34. $(f \circ h)(0)$
7

35. $(h \circ f)(0)$
5

36. $(f \circ h)(-1)$
1

37. $(h \circ f)(-2)$
11

38. $(f \circ h)(x)$
$9x^2 + 15x + 7$

39. $(h \circ f)(x)$
$3x^2 + 3x + 5$

Given $f(x) = x - 2$ and $g(x) = x^3$, evaluate the composite function.

40. $(f \circ g)(2)$
6

41. $(f \circ g)(-1)$
-3

42. $(g \circ f)(2)$
0

43. $(g \circ f)(-1)$
-27

44. $(f \circ g)(x)$
$x^3 - 2$

45. $(g \circ f)(x)$
$x^3 - 6x^2 + 12x - 8$

APPLYING THE CONCEPTS

For the function $g(x) = x^2 - 1$, find:

46. $g(2 + h)$
$h^2 + 4h + 3$

47. $g(3 + h) - g(3)$
$h^2 + 6h$

48. $g(-1 + h) - g(-1)$
$h^2 - 2h$

49. $\dfrac{g(1 + h) - g(1)}{h}$
$2 + h$

50. $\dfrac{g(-2 + h) - g(-2)}{h}$
$-4 + h$

51. $\dfrac{g(a + h) - g(a)}{h}$
$2a + h$

Given $f(x) = 2x$, $g(x) = 3x - 1$, and $h(x) = x - 2$, find:

52. $f(g[h(2)])$
-2

53. $g(h[f(1)])$
-1

54. $h(g[f(-1)])$
-9

55. $f(h[g(0)])$
-6

56. $f(g[h(x)])$
$6x - 14$

57. $g(f[h(x)])$
$6x - 13$

Quick Quiz (Objective 11.4B)
Given $g(x) = 2x^2 - 1$ and $h(x) = 3x + 5$, evaluate the composite function.
1. $g[h(-1)]$ 7
2. $h[g(0)]$ 2
3. $g[h(0)]$ 49
4. Given $h(x) = 3x + 5$ and $g(x) = 2x^2 - 1$, determine $h[g(x)]$. $6x^2 + 2$

11.5 One-to-One and Inverse Functions

Objective A — To determine whether a function is one-to-one

VIDEO & DVD CD TUTOR WEB SSM

Recall that a function is a set of ordered pairs in which no two ordered pairs that have the same first coordinate have different second coordinates. This means that given any x, there is only one y that can be paired with that x. A **one-to-one function** satisfies the additional condition that given any y, there is only one x that can be paired with the given y. One-to-one functions are commonly written as 1–1.

> **TAKE NOTE**
> Recall that a function is a set of ordered pairs in which no two ordered pairs have the same first coordinate. A 1–1 function is a function in which no two ordered pairs have the same second coordinate.

> **One-to-One Function**
> A function f is a 1–1 function if, for any a and b in the domain of f, $f(a) = f(b)$ implies that $a = b$.

This definition states that if the y-coordinates of an ordered pair are equal, $f(a) = f(b)$, then the x-coordinates must be equal, $a = b$.

The function defined by $f(x) = 2x + 1$ is a 1–1 function. To show this, determine $f(a)$ and $f(b)$. Then form the equation $f(a) = f(b)$.

$f(a) = 2a + 1 \qquad f(b) = 2b + 1$

$$f(a) = f(b)$$
$$2a + 1 = 2b + 1$$
$$2a = 2b \qquad \text{• Subtract 1 from each side of the equation.}$$
$$a = b \qquad \text{• Divide each side of the equation by 2.}$$

Because $f(a) = f(b)$ implies that $a = b$, the function is a 1–1 function.

Consider the function defined by $g(x) = x^2 - x$. Evaluate the function at -2 and 3.

$$g(-2) = (-2)^2 - (-2) = 6 \qquad g(3) = 3^2 - 3 = 6$$

From this evaluation, $g(-2) = 6$ and $g(3) = 6$, but $-2 \neq 3$. Thus g is not a 1–1 function.

The graphs of $f(x) = 2x + 1$ and $g(x) = x^2 - x$ are shown below. Note that a horizontal line intersects the graph of f at no more than one point. However, a horizontal line intersects the graph of g at more than one point.

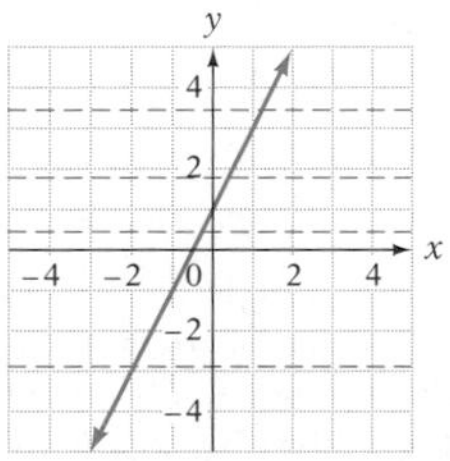

$f(x) = 2x + 1$

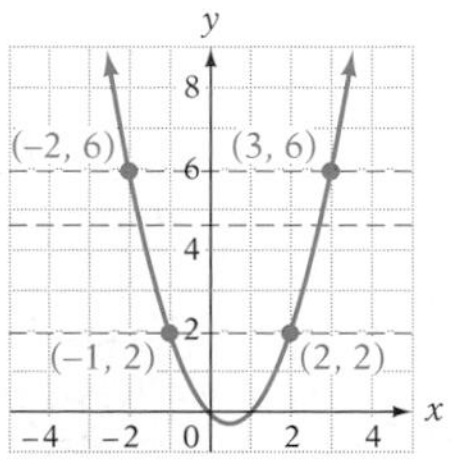

$g(x) = x^2 - x$

Objective 11.5A

New Vocabulary
one-to-one function (1–1 function)
horizontal-line test

Vocabulary to Review
function

Instructor Note
It may help students to recall that a function is a set of ordered pairs in which no two ordered pairs have the same first coordinate. A 1–1 function is one in which no two ordered pairs have the same second coordinate, either.

Discuss the Concepts
1. How can you determine whether a graph is the graph of a 1–1 function?
2. Why is a linear function always a 1–1 function?
3. Why is a quadratic function never a 1–1 function?

Concept Check
Which of the following functions are 1–1 functions?
a. $f(x) = \sqrt{0.5x}$
b. $f(x) = 3x + 1$
c. $f(x) = x^2 + 3x - 4$
d. $f(x) = |x| + 2$
e. $f(x) = 2x^3$
f. $f(x) = -x^3 + 4x^2 - 4x + 1$
a, b, and e

In-Class Examples (Objective 11.5A)
Determine whether the graph represents the graph of a 1–1 function.

1. 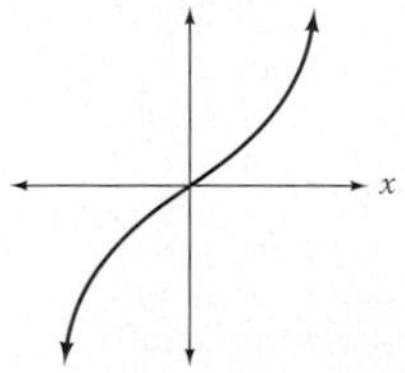

Yes

2. 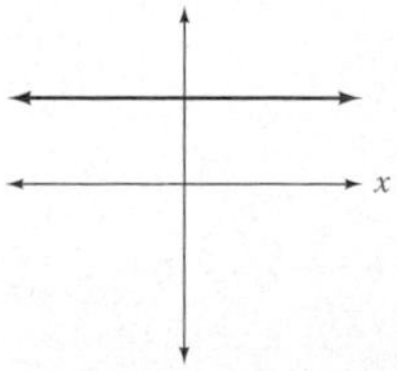

No

Optional Student Activity

The greatest-integer function is defined by the equation $f(x) = \lfloor x \rfloor$, where the symbol $\lfloor x \rfloor$ means the greatest integer less than or equal to x. Here are some examples.

$$\lfloor 4.2 \rfloor = 4$$
$$\lfloor 5.9999 \rfloor = 5$$
$$\lfloor \pi \rfloor = 3$$
$$\lfloor -2.3 \rfloor = -3$$
$$\lfloor 2 \rfloor = 2$$

In each of these examples, the value of the function is an integer that is less than or equal to the number in the symbol $\lfloor\ \rfloor$.

Graph the greatest-integer function. Explain why the greatest-integer function is sometimes called a step function. Is the greatest-integer function a 1–1 function? Why or why not?

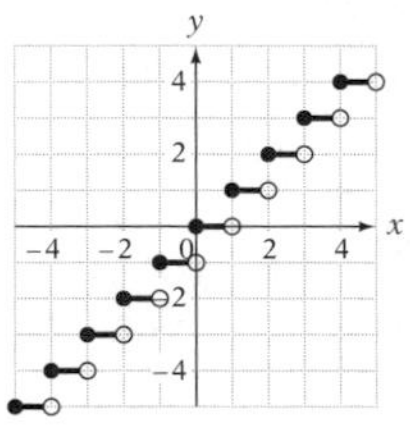

The greatest-integer function is sometimes called a step function because its graph resembles steps. The greatest-integer function is not a 1–1 function because it does not pass the horizontal-line test.

Looking at the graph of f on the previous page, note that for each y-coordinate there is only one x-coordinate. Thus f is a 1–1 function. From the graph of g, however, there are *two* x-coordinates for a given y-coordinate. For instance, $(-2, 6)$ and $(3, 6)$ are the coordinates of two points on the graph for which the y-coordinates are the same and the x-coordinates are different. Therefore, g is not a 1–1 function.

Horizontal-Line Test

The graph of a function represents the graph of a 1–1 function if any horizontal line intersects the graph at no more than one point.

Example 1

Determine whether the graph is the graph of a 1–1 function.

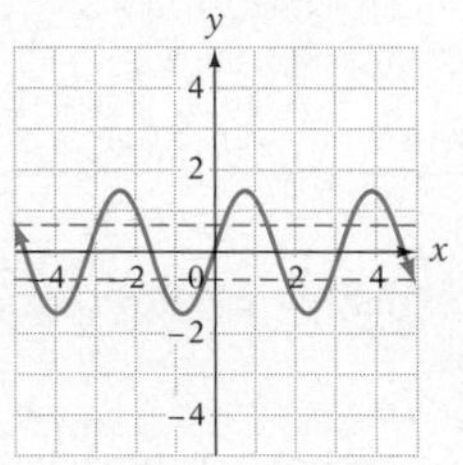

Solution

Because a horizontal line intersects the graph more than once, the graph is not the graph of a 1–1 function.

You Try It 1

Determine whether the graph is the graph of a 1–1 function.

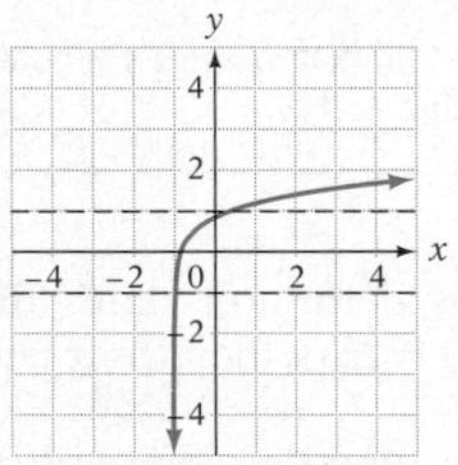

Your solution

Because any horizontal line intersects the graph at most once, the graph is the graph of a 1–1 function.

Solution on p. S34

Objective B To find the inverse of a function

VIDEO & DVD CD TUTOR WEB SSM

The **inverse of a function** is the set of ordered pairs formed by reversing the coordinates of each ordered pair of the function.

For example, the set of ordered pairs of the function defined by $f(x) = 2x$ with domain $\{-2, -1, 0, 1, 2\}$ is $\{(-2, -4), (-1, -2), (0, 0), (1, 2), (2, 4)\}$. The set of ordered pairs of the inverse function is $\{(-4, -2), (-2, -1), (0, 0), (2, 1), (4, 2)\}$.

From the ordered pairs of f, we have

$$\text{domain} = \{-2, -1, 0, 1, 2\} \quad \text{and} \quad \text{range} = \{-4, -2, 0, 2, 4\}$$

From the ordered pairs of the inverse function, we have

$$\text{domain} = \{-4, -2, 0, 2, 4\} \quad \text{and} \quad \text{range} = \{-2, -1, 0, 1, 2\}$$

Note that the domain of the inverse function is the range of the original function, and the range of the inverse function is the domain of the original function.

Now consider the function defined by $g(x) = x^2$ with domain $\{-2, -1, 0, 1, 2\}$. The set of ordered pairs of this function is $\{(-2, 4), (-1, 1), (0, 0), (1, 1), (2, 4)\}$. Reversing the ordered pairs gives $\{(4, -2), (1, -1), (0, 0), (1, 1), (4, 2)\}$. These ordered pairs do not satisfy the condition of a function, because there are ordered pairs with the same first coordinate and different second coordinates. This example illustrates that not all functions have an inverse function.

The graphs of $f(x) = 2x$ and $g(x) = x^2$ with the set of real numbers as the domain are shown at the right.

By the horizontal-line test, f is a 1–1 function but g is not.

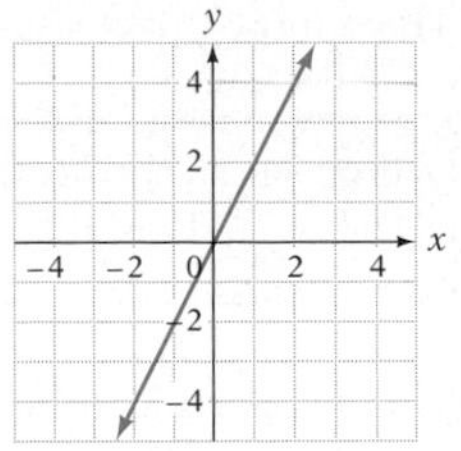

$f(x) = 2x$

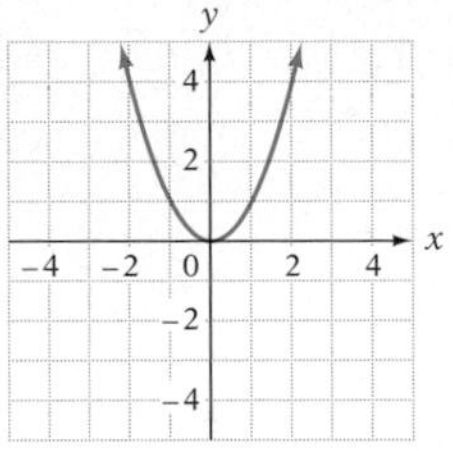

$g(x) = x^2$

Condition for an Inverse Function

A function f has an inverse function if and only if f is a 1–1 function.

TAKE NOTE

It is important to remember that f^{-1} is the symbol for the *inverse* function and does not denote reciprocal.

$$f^{-1}(x) \neq \frac{1}{f(x)}$$

The symbol f^{-1} is used to denote the inverse of the function f. The symbol $f^{-1}(x)$ is read "f inverse of x."

$f^{-1}(x)$ is *not* the reciprocal of $f(x)$, but rather is the notation for the inverse of a 1–1 function.

To find the inverse of a function, interchange x and y. Then solve for y.

⇒ Find the inverse of the function defined by $f(x) = 3x + 6$.

$f(x) = 3x + 6$

$y = 3x + 6$ • Replace $f(x)$ by y.

$x = 3y + 6$ • Interchange x and y.

$x - 6 = 3y$ • Solve for y.

$\frac{1}{3}x - 2 = y$

$f^{-1}(x) = \frac{1}{3}x - 2$ • Replace y by $f^{-1}(x)$.

The inverse of the function $f(x) = 3x + 6$ is given by $f^{-1}(x) = \frac{1}{3}x - 2$.

TAKE NOTE

If the ordered pairs of f are given by (x, y), then the ordered pairs of f^{-1} are given by (y, x). That is, x and y are interchanged. This is the reason for Step 2 at the right.

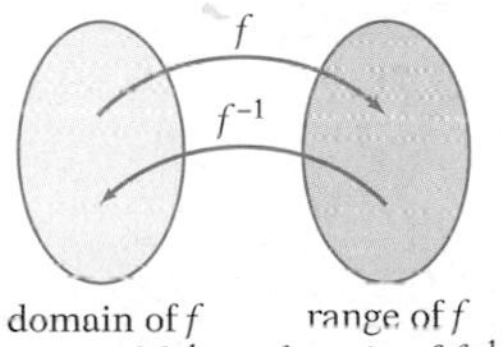

domain of f range of f^{-1} — range of f domain of f^{-1}

The fact that the ordered pairs of the inverse of a function are the reverse of those of the original function has a graphical interpretation. In the graph in the middle, the points with the coordinates reversed from the first graph are plotted. The inverse function is graphed by drawing a smooth curve through those points, as shown in the figure on the right.

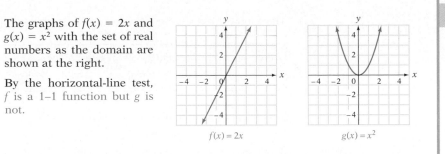

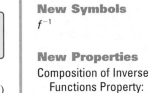

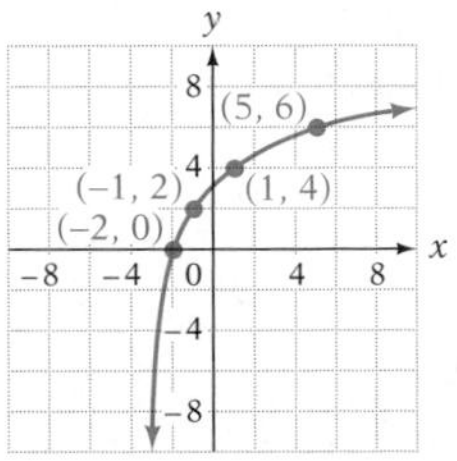

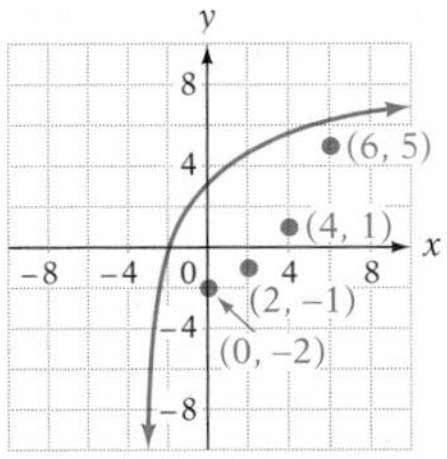

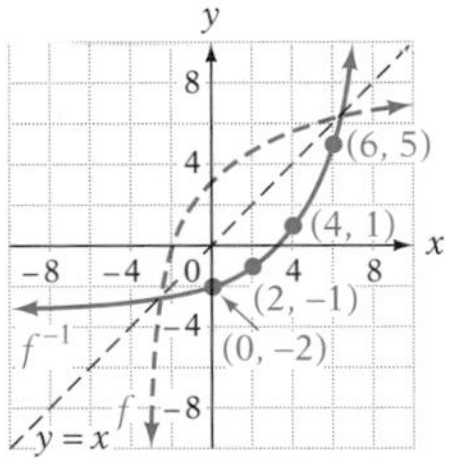

Note the dashed graph of $y = x$ that is shown in the figure on the right. If two functions are inverses of each other, their graphs are mirror images with respect to the graph of the line $y = x$.

Objective 11.5B

New Vocabulary

inverse of a function

Vocabulary to Review

domain
range

New Symbols

f^{-1}

New Properties

Composition of Inverse Functions Property:
$f^{-1}[f(x)] = x$ and
$f[f^{-1}(x)] = x$

Instructor Note

After discussing the example on this page (or a similar example), it may help some students to be reminded that the inverse of f is $f^{-1}(x) = \frac{x}{3} - 2$ but the reciprocal of f is $\frac{1}{3x + 6}$.

Discuss the Concepts

Determine whether the statement is always true, sometimes true, or never true.

1. A function has an inverse if and only if it is a 1–1 function. Always true
2. The inverse of the function {(2, 3), (4, 5), (6, 3)} is the function {(3, 2), (5, 4), (3, 6)}. Never true
3. The inverse of a function is a relation. Always true
4. The inverse of a function is a function. Sometimes true

In-Class Examples (Objective 11.5B)

Find $f^{-1}(x)$.

1. $f(x) = 2x - 6$ $f^{-1}(x) = \frac{1}{2}x + 3$
2. $f(x) = \frac{1}{3}x - 2$ $f^{-1}(x) = 3x + 6$
3. $f(x) = -3x + 3$ $f^{-1}(x) = -\frac{1}{3}x + 1$

Instructor Note

One way of helping students with the concept of mirror image is to have them think of one graph as being drawn with wet ink. Then folding the paper along the line $y = x$ produces a mirror image.

Concept Check

One length of an Olympic swimming pool is 50 m; therefore, swimming one lap is equal to swimming 100 m. We can refer to the distance swum as a function of the number of laps completed. Some of the ordered pairs of this function are $\{(0, 0), (1, 100), (2, 200), (3, 300), (4, 400)\}$. Provide some ordered pairs of the inverse function and state what the inverse function represents.

The equation that describes the distance swum as a function of the number of laps completed is $f(x) = 100x$. Write the inverse function. Some ordered pairs of the inverse function are $\{(0, 0), (100, 1), (200, 2), (300, 3), (400, 4)\}$. This function gives the number of laps completed as a function of the distance swum. The inverse function is $f^{-1}(x) = \frac{x}{100}$.

Optional Student Activity

1. To convert square feet to acres, we can use the equation $A = \frac{s}{43{,}560}$, where A is the number of acres and s is the number of square feet.
   a. Write an equation for the inverse function.
      $s = 43{,}560A$
   b. What does the inverse function represent?
      It represents the equation that converts acres to square feet.

(Continued on next page)

The composition of a function and its inverse have a special property.

Property of the Composition of Inverse Functions

$f^{-1}[f(x)] = x$ and $f[f^{-1}(x)] = x$

This property can be used to determine whether two functions are inverses of each other.

➡ Are $f(x) = 2x - 4$ and $g(x) = \frac{1}{2}x + 2$ inverses of each other?

To determine whether the functions are inverses, use the Property of the Composition of Inverse Functions.

$$f[g(x)] = 2\left(\frac{1}{2}x + 2\right) - 4 = x + 4 - 4 = x$$

$$g[f(x)] = \frac{1}{2}(2x - 4) + 2 = x - 2 + 2 = x$$

Because $f[g(x)] = x$ and $g[f(x)] = x$, the functions are inverses of each other.

Example 2

Find the inverse of the function defined by $f(x) = 2x - 3$.

Solution

$f(x) = 2x - 3$

$y = 2x - 3$ • **Replace $f(x)$ by y.**

$x = 2y - 3$ • **Interchange x and y.**

$x + 3 = 2y$ • **Solve for y.**

$\frac{1}{2}x + \frac{3}{2} = y$

$f^{-1}(x) = \frac{1}{2}x + \frac{3}{2}$ • **Replace y by $f^{-1}(x)$.**

The inverse of the function is given by $f^{-1}(x) = \frac{1}{2}x + \frac{3}{2}$.

You Try It 2

Find the inverse of the function defined by $f(x) = \frac{1}{2}x + 4$.

Your solution

$f^{-1}(x) = 2x - 8$

Example 3

Are $f(x) = 3x - 6$ and $g(x) = \frac{1}{3}x + 2$ inverses of each other?

Solution

$f[g(x)] = 3\left(\frac{1}{3}x + 2\right) - 6 = x + 6 - 6 = x$

$g[f(x)] = \frac{1}{3}(3x - 6) + 2 = x - 2 + 2 = x$

Yes, the functions are inverses of each other.

You Try It 3

Are $f(x) = 2x - 6$ and $g(x) = \frac{1}{2}x - 3$ inverses of each other?

Your solution

no

Solutions on p. S34

621

11.5 Exercises

Objective A

1. What is a 1–1 function?

2. What is the horizontal-line test?

Determine whether the graph represents the graph of a 1–1 function.

3.

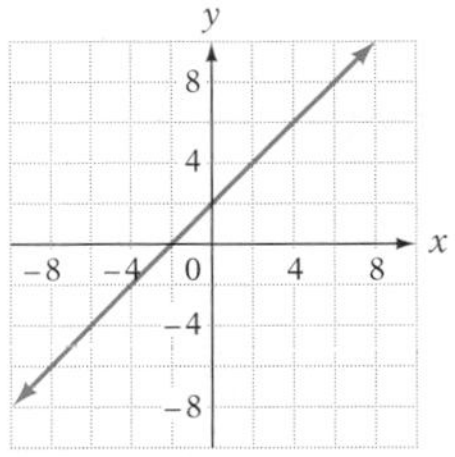

yes

4.

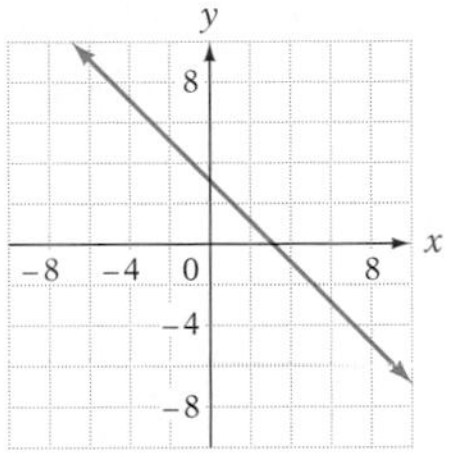

yes

5.

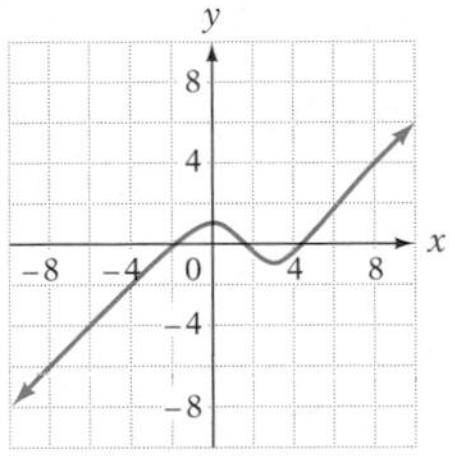

no

6.

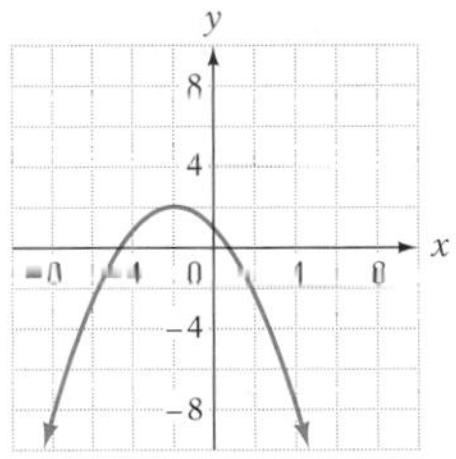

no

7.

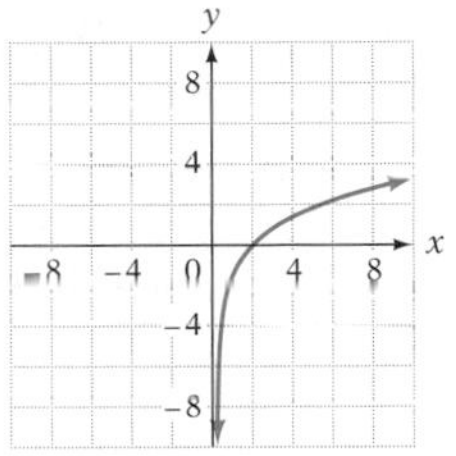

yes

8.

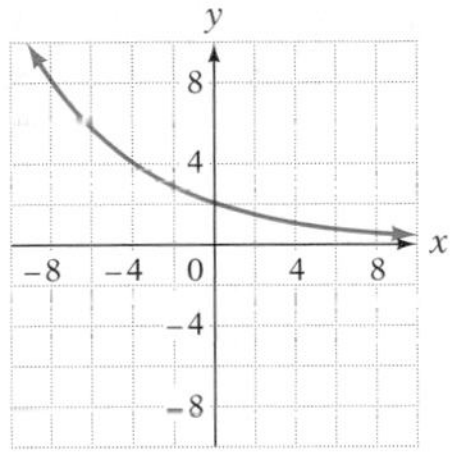

yes

9.

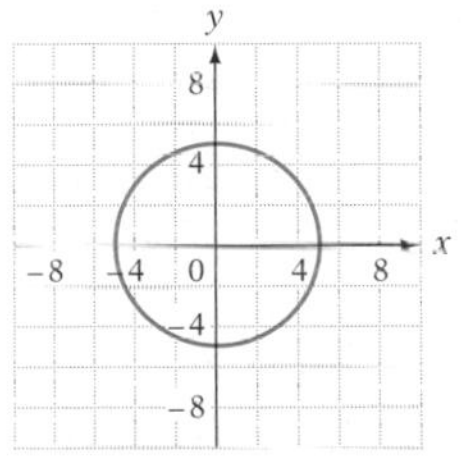

no

10.

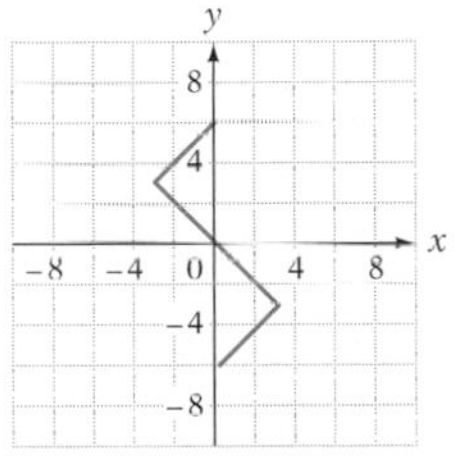

no

11.

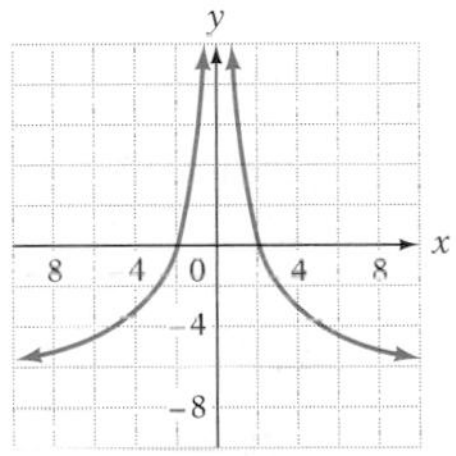

no

12.

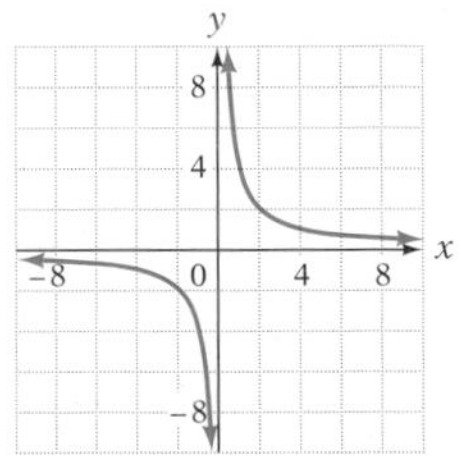

yes

13.

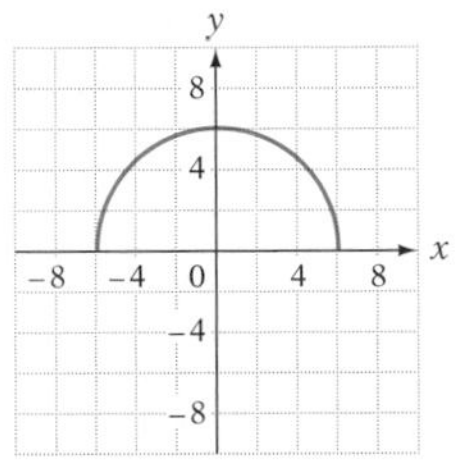

no

14.

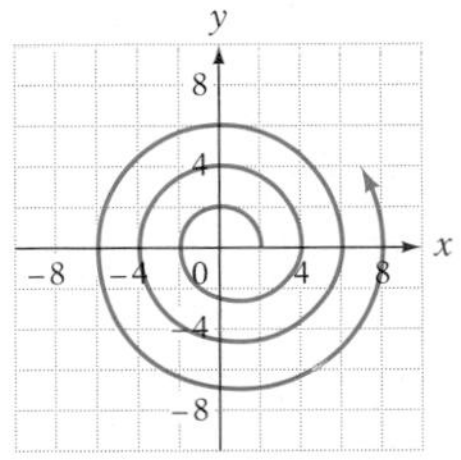

no

(Continued)

2. On March 15, 2002, the exchange rate for converting Euros to U.S. dollars was given by the equation $d = 0.88E$, where d is the number of U.S. dollars and E is the number of Euros.

a. Write an equation for the inverse function. Write the coefficient as a decimal. Round to the nearest thousandth. $E = 1.136d$

b. What does the inverse function represent?
It represents the exchange rate on March 15, 2002, for converting U.S. dollars to Euros.

Section 11.5

Suggested Assignment

Exercises 3–13, 17–55, odds
More challenging problems:
Exercises 57–69, odds

Answers to Writing Exercises

1. A function is a 1–1 function if, for any a and b in the domain of f, $f(a) = f(b)$ implies $a = b$.
Here is an alternative definition: A function is a set of ordered pairs in which no two ordered pairs that have the same first coordinate have different second coordinates. This means that given any x, there is only one y that can be paired with that x. A 1–1 function satisfies the additional condition that given any y, there is only one x that can be paired with that y.

2. The horizontal-line test is used to determine whether the graph of a function is the graph of a 1–1 function. The horizontal-line test states that the graph of a function represents the graph of a 1–1 function if any horizontal line intersects the graph at no more than one point.

Quick Quiz (Objective 11.5A)

Determine whether the graph represents the graph of a 1–1 function.

1. No

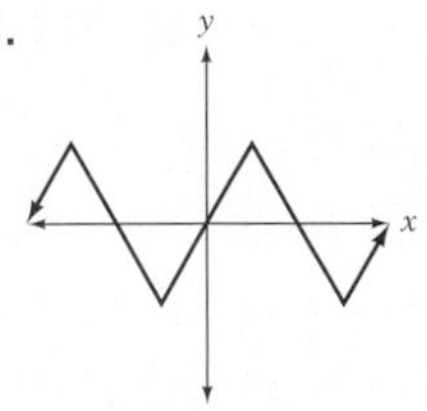

2. No

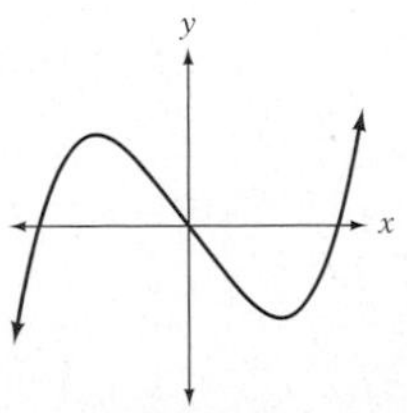

Answers to Writing Exercises

15. In an inverse function, the coordinates of each ordered pair are in the reverse order of the coordinates of the ordered pairs of the original function. For example, if $(-1, 5)$ is an ordered pair of the original function, then $(5, -1)$ is an ordered pair of the inverse function.

16. Not all functions have an inverse function because sometimes reversing the coordinates of the ordered pairs results in a set of ordered pairs that does not represent a function. For example, consider the constant function $f(x) = 5$. Some ordered pairs of this function are $(0, 5)$, $(-3, 5)$, and $(2, 5)$. Reversing the coordinates of these ordered pairs gives $(5, 0)$, $(5, -3)$, and $(5, 2)$. These ordered pairs do not satisfy the conditions of a function because they are ordered pairs with the same first coordinate and different second coordinates.

Objective B

15. What is the inverse of a function?

16. Why is it that not all functions have an inverse function?

Find the inverse of the function. If the function does not have an inverse function, write "no inverse."

17. $\{(1, 0), (2, 3), (3, 8), (4, 15)\}$
$\{(0, 1), (3, 2), (8, 3), (15, 4)\}$

18. $\{(1, 0), (2, 1), (-1, 0), (-2, 0)\}$
no inverse

19. $\{(3, 5), (-3, -5), (2, 5), (-2, -5)\}$
no inverse

20. $\{(-5, -5), (-3, -1), (-1, 3), (1, 7)\}$
$\{(-5, -5), (-1, -3), (3, -1), (7, 1)\}$

21. $\{(0, -2), (-1, 5), (3, 3), (-4, 6)\}$
$\{(-2, 0), (5, -1), (3, 3), (6, -4)\}$

22. $\{(-2, -2), (0, 0), (2, 2), (4, 4)\}$
$\{(-2, -2), (0, 0), (2, 2), (4, 4)\}$

23. $\{(-2, -3), (-1, 3), (0, 3), (1, 3)\}$
no inverse

24. $\{(2, 0), (1, 0), (3, 0), (4, 0)\}$
no inverse

Find $f^{-1}(x)$.

25. $f(x) = 4x - 8$
$f^{-1}(x) = \frac{1}{4}x + 2$

26. $f(x) = 3x + 6$
$f^{-1}(x) = \frac{1}{3}x - 2$

27. $f(x) = 2x + 4$
$f^{-1}(x) = \frac{1}{2}x - 2$

28. $f(x) = x - 5$
$f^{-1}(x) = x + 5$

29. $f(x) = \frac{1}{2}x - 1$
$f^{-1}(x) = 2x + 2$

30. $f(x) = \frac{1}{3}x + 2$
$f^{-1}(x) = 3x - 6$

31. $f(x) = -2x + 2$
$f^{-1}(x) = -\frac{1}{2}x + 1$

32. $f(x) = -3x - 9$
$f^{-1}(x) = -\frac{1}{3}x - 3$

33. $f(x) = \frac{2}{3}x + 4$
$f^{-1}(x) = \frac{3}{2}x - 6$

34. $f(x) = \frac{3}{4}x - 4$
$f^{-1}(x) = \frac{4}{3}x + \frac{16}{3}$

35. $f(x) = -\frac{1}{3}x + 1$
$f^{-1}(x) = -3x + 3$

36. $f(x) = -\frac{1}{2}x + 2$
$f^{-1}(x) = -2x + 4$

Quick Quiz (Objective 11.5B)

Find $f^{-1}(x)$.

1. $f(x) = 4x + 8$ $f^{-1}(x) = \frac{1}{4}x - 2$

2. $f(x) = \frac{1}{2}x + 1$ $f^{-1}(x) = 2x - 2$

3. $f(x) = -6x + 3$ $f^{-1}(x) = -\frac{1}{6}x + \frac{1}{2}$

623

37. $f(x) = 2x - 5$

$f^{-1}(x) = \frac{1}{2}x + \frac{5}{2}$

38. $f(x) = 3x + 4$

$f^{-1}(x) = \frac{1}{3}x - \frac{4}{3}$

39. $f(x) = 5x - 2$

$f^{-1}(x) = \frac{1}{5}x + \frac{2}{5}$

40. $f(x) = 4x - 2$

$f^{-1}(x) = \frac{1}{4}x + \frac{1}{2}$

41. $f(x) = 6x - 3$

$f^{-1}(x) = \frac{1}{6}x + \frac{1}{2}$

42. $f(x) = -8x + 4$

$f^{-1}(x) = -\frac{1}{8}x + \frac{1}{2}$

Given $f(x) = 3x - 5$, find:

43. $f^{-1}(0)$

$\frac{5}{3}$

44. $f^{-1}(2)$

$\frac{7}{3}$

45. $f^{-1}(4)$

3

State whether the graph is the graph of a function. If it is the graph of a function, does it have an inverse?

46.

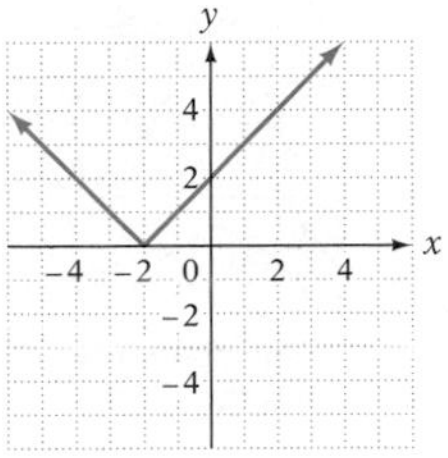

yes; no

47.

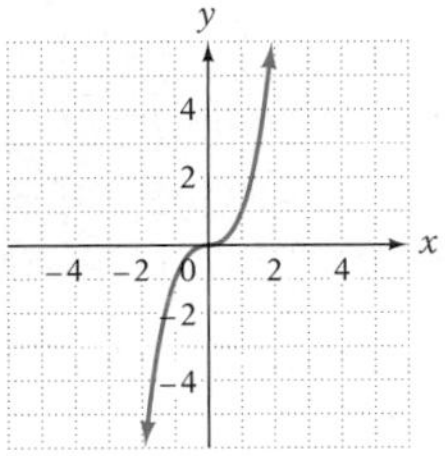

yes; yes

48.

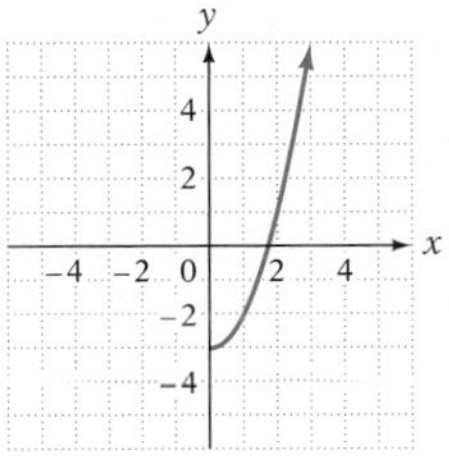

yes; yes

Use the Property of the Composition of Inverse Functions to determine whether the functions are inverses of each other.

49. $f(x) = 4x;\ g(x) = \frac{x}{4}$

yes

50. $g(x) = x + 5;\ h(x) = x - 5$

yes

51. $f(x) = 3x;\ h(x) = \frac{1}{3x}$

no

52. $h(x) = x + 2;\ g(x) = 2 - x$

no

53. $g(x) = 3x + 2;\ f(x) = \frac{1}{3}x - \frac{2}{3}$

yes

54. $h(x) = 4x - 1;\ f(x) = \frac{1}{4}x + \frac{1}{4}$

yes

55. $f(x) = \frac{1}{2}x - \frac{3}{2};\ g(x) = 2x + 3$

yes

56. $g(x) = -\frac{1}{2}x - \frac{1}{2};\ h(x) = -2x + 1$

no

Answers to Writing Exercises

63. No, the inverse of the grading scale is not a function because each grade is paired with more than one score.

64. No, the inverse of the first-class postage rates is not a function because each cost is paired with more than one weight.

71. A constant function is defined as $y = b$, where b is a constant. The inverse function would be $x = a$, where a is a constant. This is not a function. (See also the answer to Writing Exercise 16 on page 622.)

72. Yes, all functions given by $f(x) = mx + b, m \neq 0$, are 1–1 functions. These are linear functions, and all linear functions of this form pass the horizontal-line test and the vertical-line test.

APPLYING THE CONCEPTS

Given the graph of the 1–1 function, draw the graph of the inverse of the function by using the technique shown in Objective B of this section.

57.

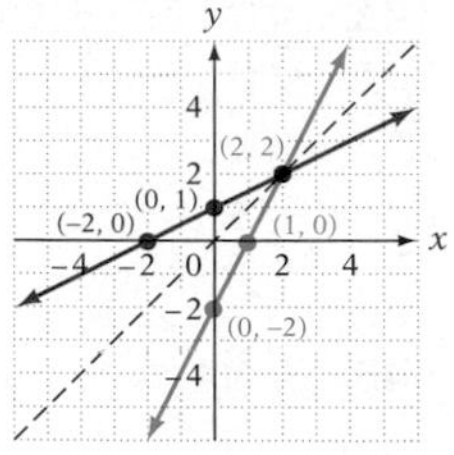

58.

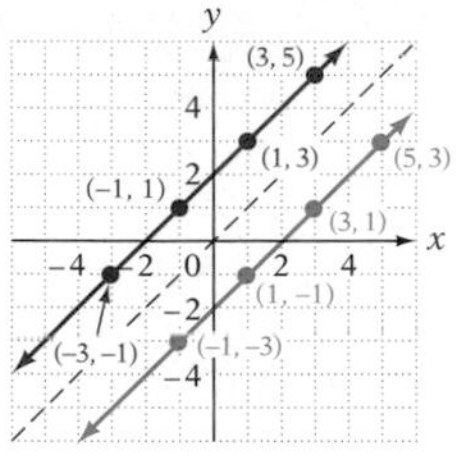

59.

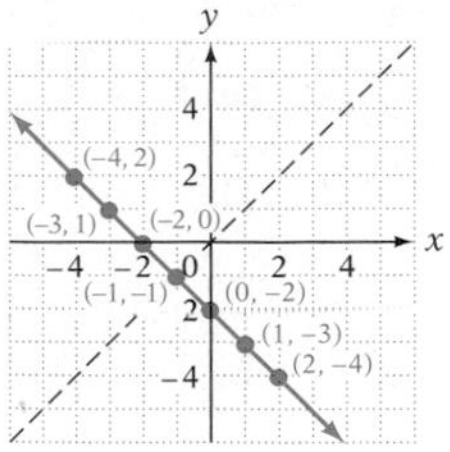

The inverse is the same graph.

60.

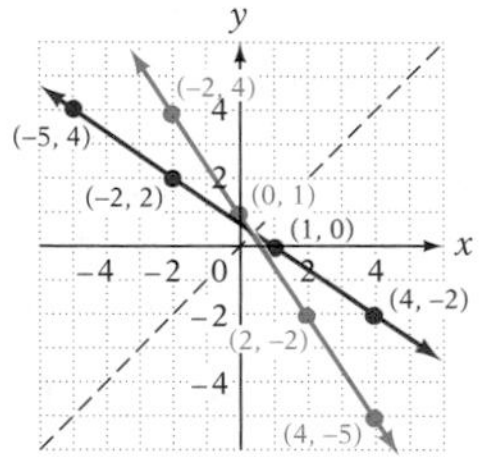

61.

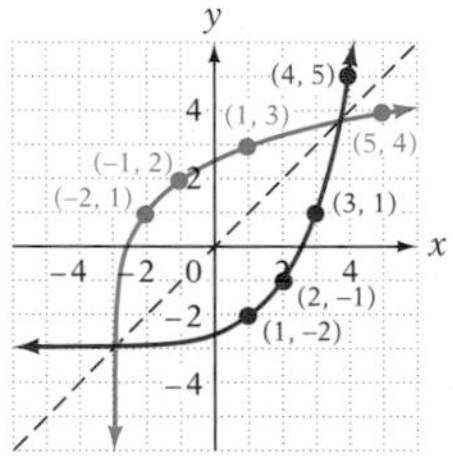

62.

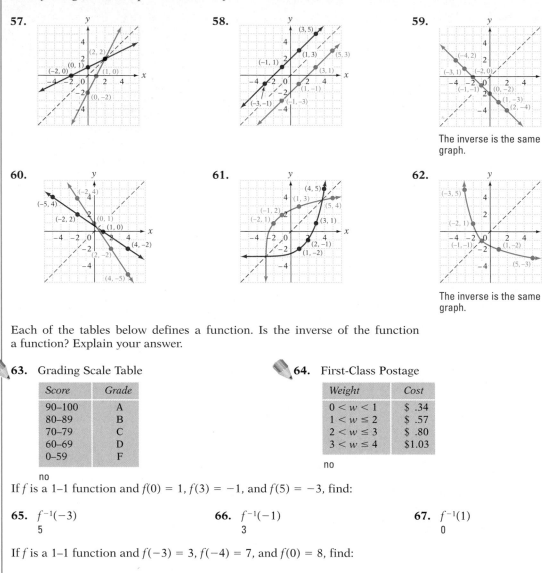

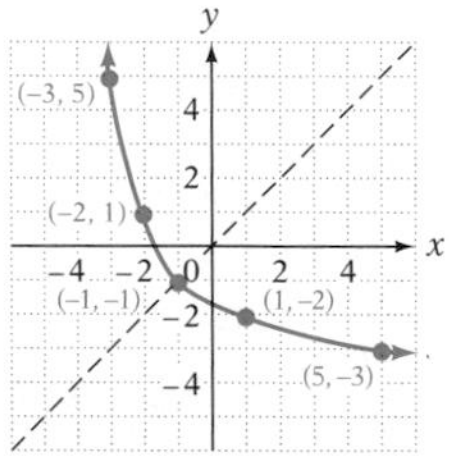

The inverse is the same graph.

Each of the tables below defines a function. Is the inverse of the function a function? Explain your answer.

63. Grading Scale Table

Score	*Grade*
90–100	A
80–89	B
70–79	C
60–69	D
0–59	F

no

64. First-Class Postage

Weight	*Cost*
$0 < w < 1$	\$ .34
$1 < w \leq 2$	\$ .57
$2 < w \leq 3$	\$ .80
$3 < w \leq 4$	\$1.03

no

If f is a 1–1 function and $f(0) = 1$, $f(3) = -1$, and $f(5) = -3$, find:

65. $f^{-1}(-3)$
5

66. $f^{-1}(-1)$
3

67. $f^{-1}(1)$
0

If f is a 1–1 function and $f(-3) = 3$, $f(-4) = 7$, and $f(0) = 8$, find:

68. $f^{-1}(3)$
−3

69. $f^{-1}(7)$
−4

70. $f^{-1}(8)$
0

71. Is the inverse of a constant function a function? Explain your answer.

72. The graphs of all functions given by $f(x) = mx + b$, $m \neq 0$, are straight lines. Are all of these functions 1–1 functions? If so, explain why. If not, give an example of a linear function that is not 1–1.
yes

625

11.6 Conic Sections

Objective A To graph a parabola

Point of Interest

Hypatia (c. 340–415) is considered the first prominent woman mathematician. She lectured in mathematics and philosophy at the Museum in Alexandria, the most distinguished place of learning in the world. One of the topics on which Hypatia lectured was conic sections. One historian has claimed that with the death (actually the murder) of Hypatia, "the long and glorious history of Greek mathematics was at an end."

The **conic sections** are curves that can be constructed from the intersection of a plane and a right circular cone. The parabola, which was introduced earlier, is one of these curves. Here we will review some of that previous discussion and look at equations of parabolas that were not discussed before.

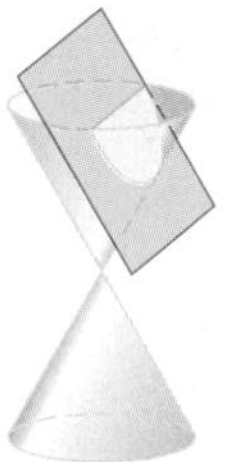

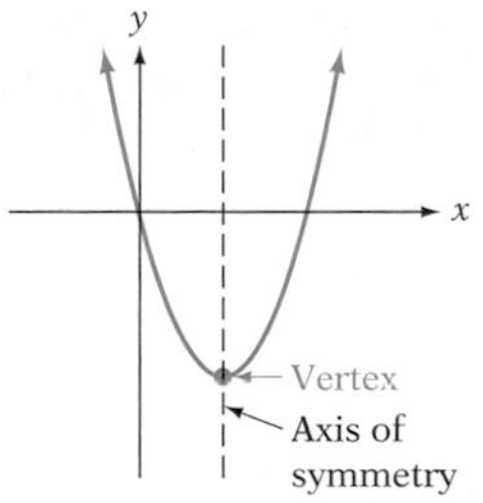

Every parabola has an axis of symmetry and a vertex that is on the axis of symmetry. To understand the axis of symmetry, think of folding the paper along that axis. The two halves of the curve will match up.

The graph of the equation $y = ax^2 + bx + c$, $a \neq 0$, is a parabola with the axis of symmetry parallel to the y-axis. The parabola opens up when $a > 0$ and opens down when $a < 0$. When the parabola opens up, the vertex is the lowest point on the parabola. When the parabola opens down, the vertex is the highest point on the parabola.

The coordinates of the vertex can be found by completing the square.

⇒ Find the vertex of the parabola whose equation is $y = x^2 - 4x + 5$.

$y = x^2 - 4x + 5$

$y = (x^2 - 4x) + 5$ • Group the terms involving x.

$y = (x^2 - 4x + 4) - 4 + 5$ • Complete the square on $x^2 - 4x$. Note that 4 is added and subtracted. Because $4 - 4 = 0$, the equation is not changed.

$y = (x - 2)^2 + 1$ • Factor the trinomial and combine like terms.

The coefficient of x^2 is positive, so the parabola opens up. The vertex is the lowest point on the parabola, or the point that has the least y-coordinate.

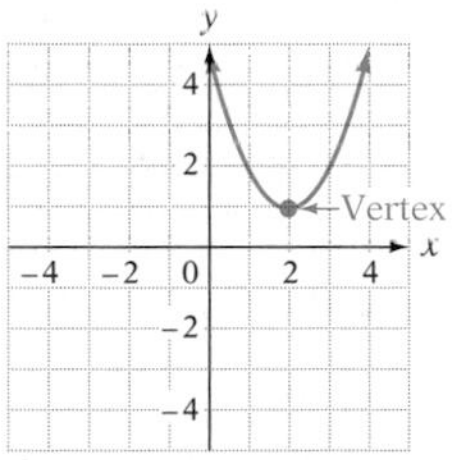

Because $(x - 2)^2 \geq 0$ for all x, the least y-coordinate occurs when $(x - 2)^2 = 0$, which occurs when $x = 2$. This means the x-coordinate of the vertex is 2.

To find the y-coordinate of the vertex, replace x in $y = (x - 2)^2 + 1$ by 2 and solve for y.

$$y = (x - 2)^2 + 1$$
$$= (2 - 2)^2 + 1 = 1$$

The vertex is (2, 1).

Objective 11.6A

New Vocabulary

conic sections
y-coordinate of the vertex

Vocabulary to Review

parabola
axis of symmetry
vertex
x-coordinate of the vertex

New Equations

equations of a parabola:
$y = ax^2 + bx + c$
$x = ay^2 + by + c$
x- or y-coordinate of the vertex of a parabola: $-\frac{b}{2a}$
axis of symmetry for $y = ax^2 + bx + c$: $x = -\frac{b}{2a}$
axis of symmetry for $x = ay^2 + by + c$: $y = -\frac{b}{2a}$

Instructor Note

Some properties of the parabola were covered earlier. Those properties are repeated here for completeness.

Instructor Note

The argument presented at the left is particularly difficult for students. You might have students evaluate the expression $(x - 2)^2$ for various values of x to illustrate that the quantity is always nonnegative and is zero when $x = 2$.

In-Class Examples (Objective 11.6A)

Find the vertex and axis of symmetry of the parabola. Then sketch its graph.

1. $x = y^2 + 4y + 3$
Vertex: $(-1, -2)$
Axis of symmetry:
$y = -2$

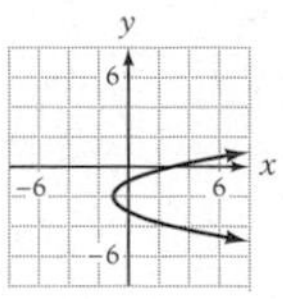

2. $x = -y^2$
Vertex: $(0, 0)$
Axis of symmetry:
$y = 0$

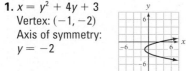

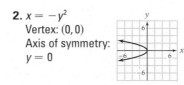

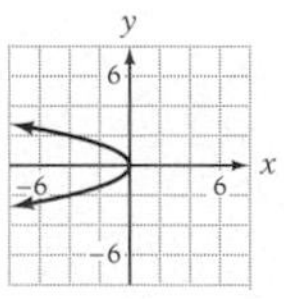

Discuss the Concepts

1. Describe the vertex and axis of symmetry of a parabola.
2. Explain how, by looking at the equation of a parabola, you can tell whether it opens up or down or opens left or right.
3. If you know the vertex of a parabola, how can you determine the equation of the axis of symmetry of the parabola?

Concept Check

Determine whether the statement is always true, sometimes true, or never true.

1. The graph of a parabola is the graph of a function. Sometimes true
2. The axis of symmetry of a parabola passes through the vertex. Always true
3. The graph of a parabola has two x-intercepts. Sometimes true
4. The graph of a parabola has a minimum value. Sometimes true
5. The axis of symmetry of a parabola is the x-axis or the y-axis. Sometimes true
6. The equation of the axis of symmetry of a parabola is $x = -\frac{b}{2a}$. Sometimes true

Point of Interest

The suspension cables for some bridges, such as the Golden Gate bridge, hang in the shape of a parabola. Parabolic shapes are also used for mirrors in telescopes and in certain antenna designs.

By following the procedure of the last example and completing the square on the equation $y = ax^2 + bx + c$, we find that the **x-coordinate of the vertex** is $-\frac{b}{2a}$. The y-coordinate of the vertex can then be determined by substituting this value of x into $y = ax^2 + bx + c$ and solving for y.

Because the axis of symmetry is parallel to the y-axis and passes through the vertex, the equation of the **axis of symmetry** is $\boldsymbol{x = -\frac{b}{2a}}$.

⇒ Find the vertex and axis of symmetry of the parabola whose equation is $y = -3x^2 + 6x + 1$. Then sketch its graph.

x-coordinate: $-\frac{b}{2a} = -\frac{6}{2(-3)} = 1$

• Find the x-coordinate of the vertex and the axis of symmetry, $a = -3$, $b = 6$.

The x-coordinate of the vertex is 1.
The axis of symmetry is the line $x = 1$.

To find the y-coordinate of the vertex, replace x by 1 and solve for y.

$$y = -3x^2 + 6x + 1$$
$$= -3(1)^2 + 6(1) + 1 = 4$$

The vertex is (1, 4).

Because a is negative, the parabola opens down.

Find a few ordered pairs and use symmetry to sketch the graph.

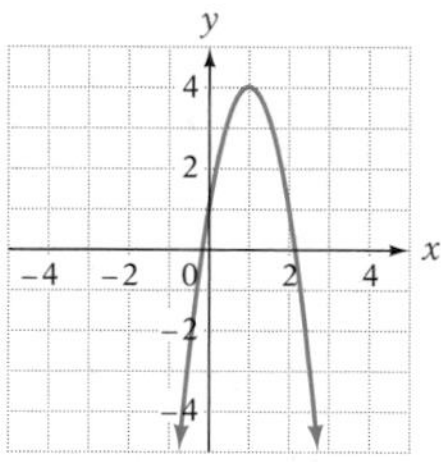

⇒ Find the vertex and axis of symmetry of the parabola whose equation is $y = x^2 - 2$. Then sketch its graph.

x-coordinate: $-\frac{b}{2a} = -\frac{0}{2(1)} = 0$

• Find the x-coordinate of the vertex and the axis of symmetry. $a = 1$, $b = 0$

The x-coordinate of the vertex is 0.
The axis of symmetry is the line $x = 0$.

To find the y-coordinate of the vertex, replace x by 0 and solve for y.

$$y = x^2 - 2$$
$$= 0^2 - 2 = -2$$

The vertex is (0, −2).

Because a is positive, the parabola opens up.

Find a few ordered pairs and use symmetry to sketch the graph.

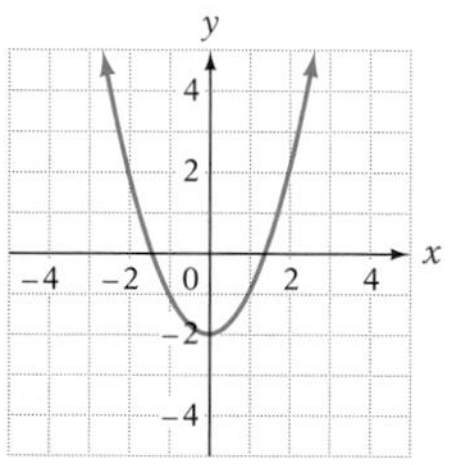

The graph of an equation of the form $x = ay^2 + by + c$, $a \neq 0$, is also a parabola. In this case, the parabola opens to the right when a is positive and opens to the left when a is negative.

For a parabola of this form, the **y-coordinate of the vertex** is $-\frac{b}{2a}$. The **axis of symmetry** is the line $y = -\frac{b}{2a}$.

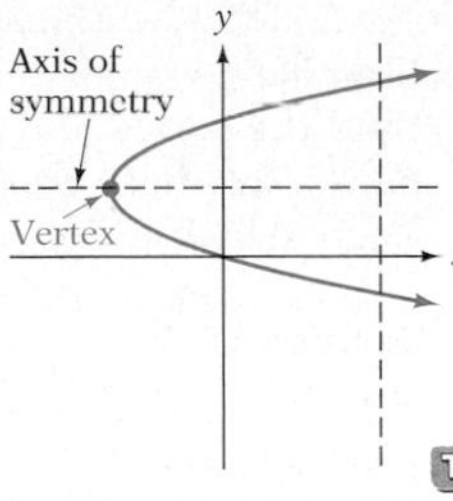

Using the vertical line test, the graph of a parabola of this form is not the graph of a function. The graph of $x = ay^2 + by + c$ is a relation.

➡ Find the vertex and axis of symmetry of the parabola whose equation is $x = 2y^2 - 8y + 5$. Then sketch its graph.

$$y\text{-coordinate: } -\frac{b}{2a} = -\frac{-8}{2(2)} = 2$$

• Find the y-coordinate of the vertex and the axis of symmetry. $a = 2$, $b = -8$

The y-coordinate of the vertex is 2.
The axis of symmetry is the line $y = 2$.

To find the x-coordinate of the vertex, replace y by 2 and solve for x.

$$x = 2y^2 - 8y + 5$$
$$= 2(2)^2 - 8(2) + 5 = -3$$

The vertex is $(-3, 2)$.

Since a is positive, the parabola opens to the right.

Find a few ordered pairs and use symmetry to sketch the graph.

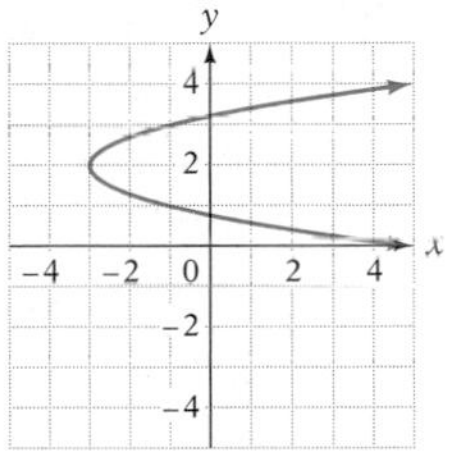

➡ Find the vertex and axis of symmetry of the parabola whose equation is $x = -2y^2 - 4y - 3$. Then sketch its graph.

$$y\text{-coordinate: } -\frac{b}{2a} = -\frac{-4}{2(-2)} = -1$$

• Find the y-coordinate of the vertex and the axis of symmetry. $a = -2$, $b = -4$

The y-coordinate of the vertex is -1.
The axis of symmetry is the line $y = -1$.

To find the x-coordinate of the vertex, replace y by -1 and solve for x.

$$x = -2y^2 - 4y - 3$$
$$= -2(-1)^2 - 4(-1) - 3 = -1$$

The vertex is $(-1, -1)$.

Because a is negative, the parabola opens to the left.

Find a few ordered pairs and use symmetry to sketch the graph.

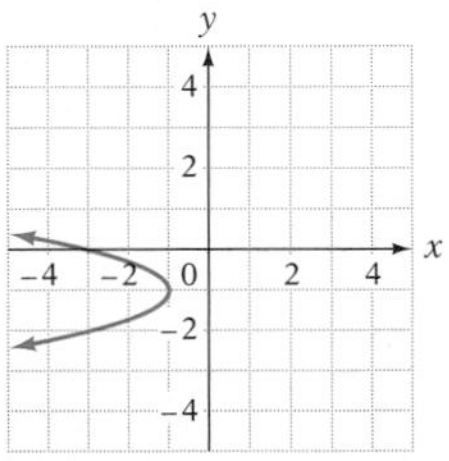

Instructor Note

Point out that the analysis of $x = ay^2 + by + c$ is essentially the same as that of $y = ax^2 + bx + c$. The difference is that the graph opens left or right instead of up or down.

Optional Student Activity

Use the vertex and the direction in which the parabola opens to determine the domain and range of the relation.

1. $y = x^2 - 4x - 2$
   Domain: $\{x \mid x \in \text{real numbers}\}$
   Range: $\{y \mid y \geq -6\}$
2. $y = x^2 - 6x + 1$
   Domain: $\{x \mid x \in \text{real numbers}\}$
   Range: $\{y \mid y \geq -8\}$
3. $y = -x^2 - 2x + 4$
   Domain: $\{x \mid x \in \text{real numbers}\}$
   Range: $\{y \mid y \leq 5\}$
4. $x = y^2 + 6y - 5$
   Domain: $\{x \mid x \geq -14\}$
   Range: $\{y \mid y \in \text{real numbers}\}$
5. $x = y^2 + 4y - 3$
   Domain: $\{x \mid x \geq -7\}$
   Range: $\{y \mid y \in \text{real numbers}\}$
6. $x = -y^2 - 2y + 6$
   Domain: $\{x \mid x \leq 7\}$
   Range: $\{y \mid y \in \text{real numbers}\}$
7. $x = -y^2 - 6y + 2$
   Domain: $\{x \mid x < 11\}$
   Range: $\{y \mid y \in \text{real numbers}\}$

Example 1
Find the vertex and axis of symmetry of the parabola whose equation is $y = x^2 - 4x + 3$. Then sketch its graph.

Solution

$-\dfrac{b}{2a} = -\dfrac{-4}{2(1)} = 2$

axis of symmetry:
$x = 2$

$y = 2^2 - 4(2) + 3$
$= -1$

vertex: $(2, -1)$

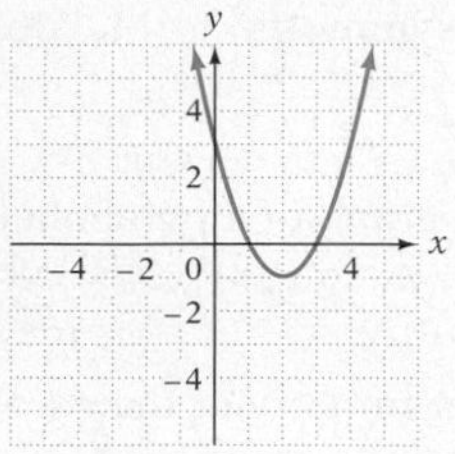

You Try It 1
Find the vertex and axis of symmetry of the parabola whose equation is $y = x^2 + 2x + 1$. Then sketch its graph.

Your solution
vertex: $(-1, 0)$
axis of symmetry:
$x = -1$

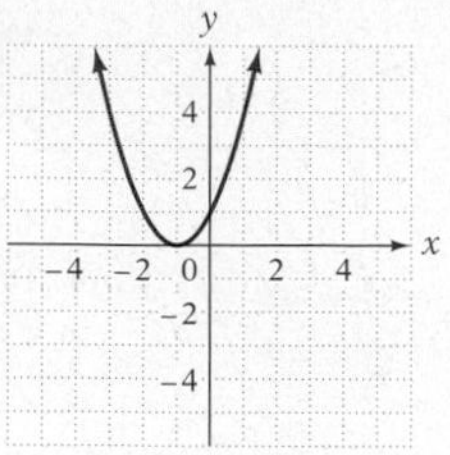

Example 2
Find the vertex and axis of symmetry of the parabola whose equation is $x = 2y^2 - 4y + 1$. Then sketch its graph.

Solution

$-\dfrac{b}{2a} = -\dfrac{-4}{2(2)} = 1$

axis of symmetry:
$y = 1$

$x = 2(1)^2 - 4(1) + 1$
$= -1$

vertex: $(-1, 1)$

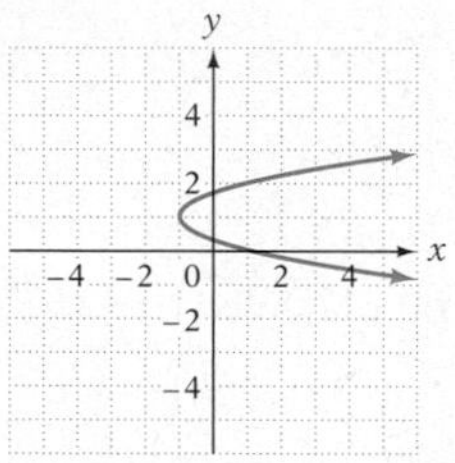

You Try It 2
Find the vertex and axis of symmetry of the parabola whose equation is $x = -y^2 - 2y + 2$. Then sketch its graph.

Your solution
vertex: $(3, -1)$
axis of symmetry:
$y = -1$

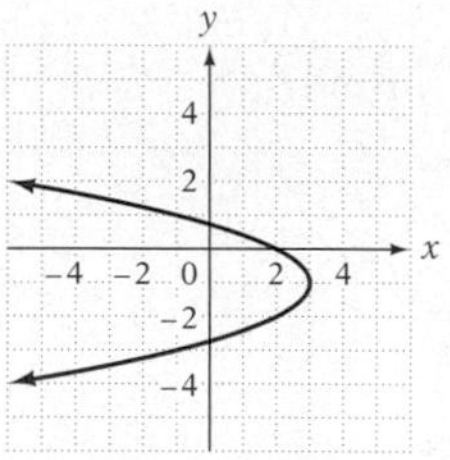

Example 3
Find the vertex and axis of symmetry of the parabola whose equation is $y = x^2 + 1$. Then sketch its graph.

Solution

$-\dfrac{b}{2a} = -\dfrac{0}{2(1)} = 0$

axis of symmetry:
$x = 0$

$y = 0^2 + 1$
$= 1$

vertex: $(0, 1)$

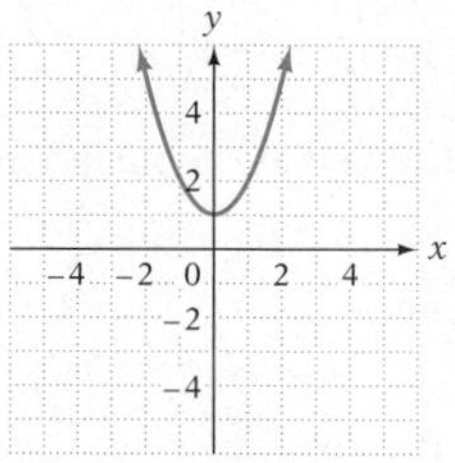

You Try It 3
Find the vertex and axis of symmetry of the parabola whose equation is $y = x^2 - 2x - 1$. Then sketch its graph.

Your solution
vertex: $(1, -2)$
axis of symmetry:
$x = 1$

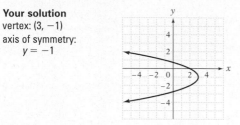

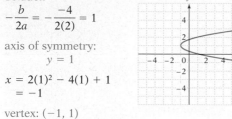

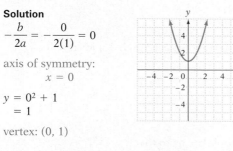

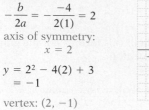

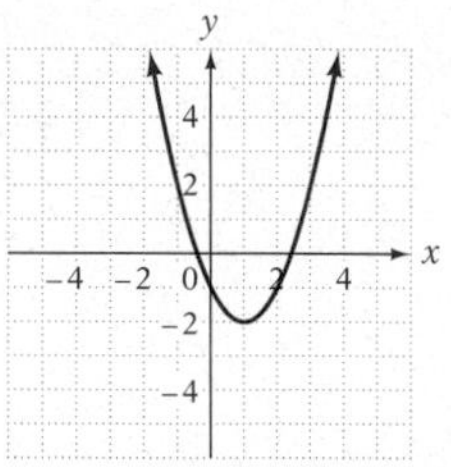

Solutions on p. S34

Objective B — To find the equation of a circle and to graph a circle

VIDEO & DVD | CD TUTOR | WEB | SSM

TAKE NOTE
As the angle of the plane that intersects the cone changes, different conic sections are formed. For a parabola, the plane was *parallel to the side* of the cone. For a circle, the plane is *parallel to the base* of the cone.

A **circle** is a conic section formed by the intersection of a cone and a plane parallel to the base of the cone.

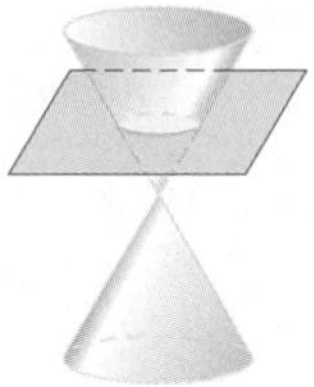

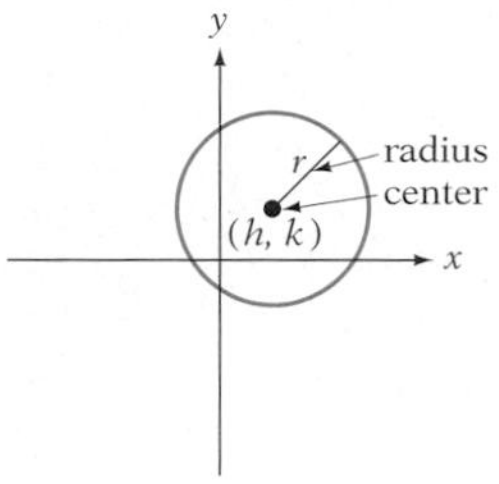

A **circle** can be defined as all points (x, y) in the plane that are a fixed distance from a given point (h, k) called the **center**. The fixed distance is the **radius** of the circle.

> **The Standard Form of the Equation of a Circle**
>
> Let r be the radius of a circle and let (h, k) be the coordinates of the center of the circle. Then the equation of the circle is given by
>
> $$(x - h)^2 + (y - k)^2 = r^2$$

➡ Sketch a graph of $(x - 1)^2 + (y + 2)^2 = 9$.

$(x - 1)^2 + [y - (-2)]^2 = 3^2$ • **Rewrite the equation in standard form.**

center: $(1, -2)$ radius: 3

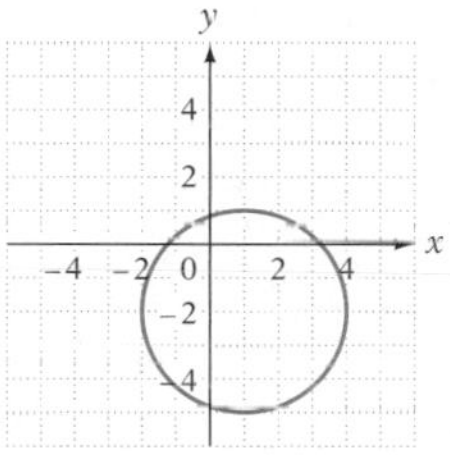

➡ Find the equation of the circle with radius 4 and center $(-1, 2)$. Then sketch its graph.

$(x - h)^2 + (y - k)^2 = r^2$ • **Use the standard form of the equation of a circle.**

$[x - (-1)]^2 + (y - 2)^2 = 4^2$ • **Replace r by 4, h by −1, and k by 2.**

$(x + 1)^2 + (y - 2)^2 = 16$

• **Sketch the graph by drawing a circle with center (−1, 2) and radius 4.**

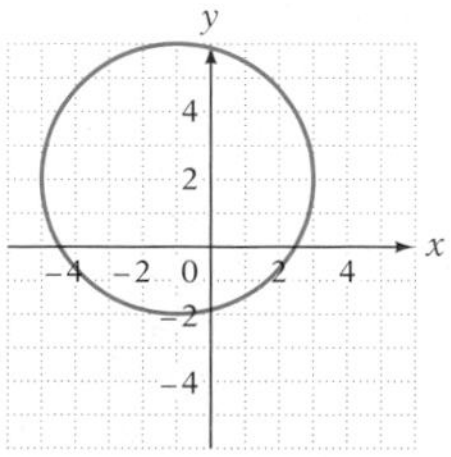

Objective 11.6B

New Vocabulary

circle
center of a circle
radius of a circle

New Equations

standard form of the equation of a circle:
$(x - h)^2 + (y - k)^2 = r^2$

Discuss the Concepts

1. Explain how to determine the radius and center of the circle given by the equation $(x - 2)^2 + (y + 1)^2 = 16$.
2. What do the values of h, k, and r represent in the equation of a circle in standard form?
3. Is the graph of a circle the graph of a function? Why or why not?
4. Explain how the points on the circumference of a circle are related to the center of the circle.

In-Class Examples (Objective 11.6B)

Sketch a graph of the circle.

1. $(x + 2)^2 + (y - 3)^2 = 16$

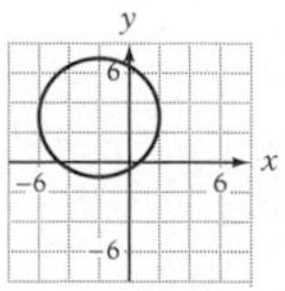

2. Find the equation of the circle with radius 3 and center $(3, -2)$. $(x - 3)^2 + (y + 2)^2 = 9$

Concept Check

1. Find the equation of the circle that has radius 1, is tangent to both the x- and y-axes, and lies in Quadrant II.
   $(x + 1)^2 + (y - 1)^2 = 1$
2. Find the radius and center of the circle given by the equation $x^2 + y^2 = 16$.
   Radius: 4; Center: (0, 0)
3. Many communications satellites orbit Earth at an altitude of approximately 22,500 mi above Earth's surface. Write an equation for the orbit of a communications satellite. Use Earth's center as the origin and consider the orbit of the satellite circular. (*Hint:* Earth's radius is approximately 4000 mi.)
   $x^2 + y^2 = (26{,}500)^2$

Optional Student Activity

1. Find the equation of the circle that has center $(5, -6)$ and an area of 49π square units.
   $(x - 5)^2 + (y + 6)^2 = 49$
2. Find the area of the smallest region bounded by the graphs of $y = |x|$ and $x^2 + y^2 = 4$.
   π square units
3. Find the least distance between the graphs of the equations $x^2 + y^2 = 1$ and $(x - 3)^2 + (y - 4)^2 = 1$.
   3 units
4. The line $x = 3$ crosses the circle $x^2 + y^2 = 34$ at points A and B. Find the length of AB. 10 units

Applying the vertical-line test reveals that the graph of a circle is not the graph of a function. The graph of a circle is the graph of a relation.

Example 4

Sketch a graph of $(x + 2)^2 + (y - 1)^2 = 4$.

Solution

$$(x - h)^2 + (y - k)^2 = r^2$$
$$[x - (-2)]^2 + (y - 1)^2 = 2^2$$

center: $(h, k) = (-2, 1)$

radius: $r = 2$

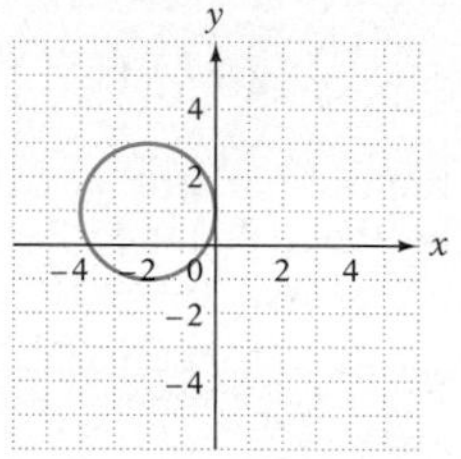

You Try It 4

Sketch a graph of $(x - 2)^2 + (y + 3)^2 = 9$.

Your solution

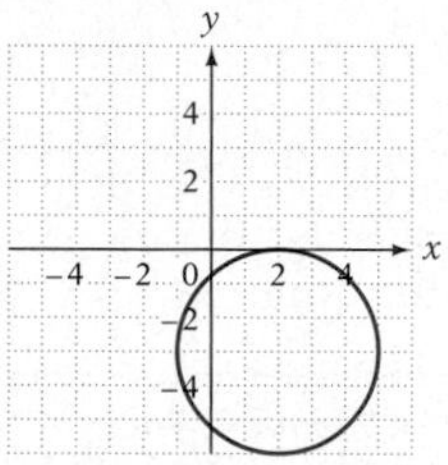

Example 5

Find the equation of the circle with radius 5 and center $(-1, 3)$. Then sketch its graph.

Solution

$$(x - h)^2 + (y - k)^2 = r^2$$
$$[x - (-1)]^2 + (y - 3)^2 = 5^2$$
$$(x + 1)^2 + (y - 3)^2 = 25$$

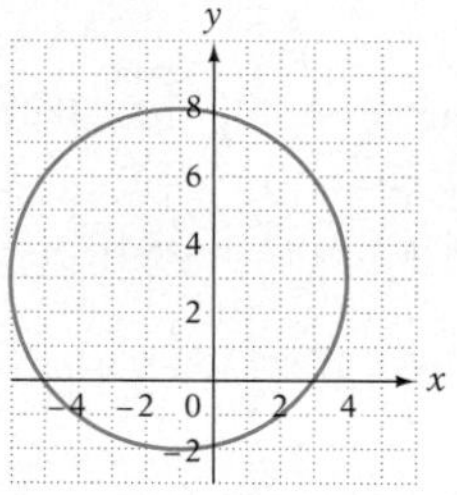

You Try It 5

Find the equation of the circle with radius 4 and center $(2, -3)$. Then sketch its graph.

Your solution

$(x - 2)^2 + (y + 3)^2 = 16$

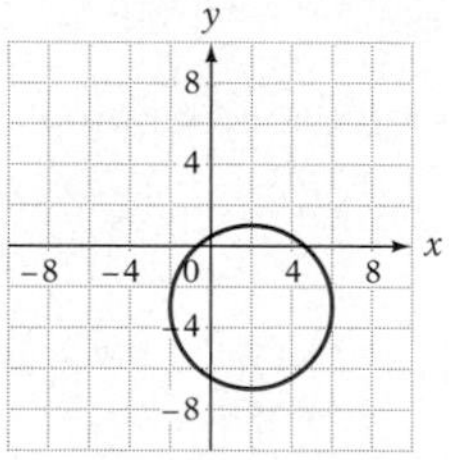

Solutions on p. S34

631

Objective C **To graph an ellipse with center at the origin**

VIDEO & DVD CD TUTOR WEB SSM

The orbits of the planets around the sun are "oval" shaped. This oval shape can be described as an **ellipse**, which is another of the conic sections.

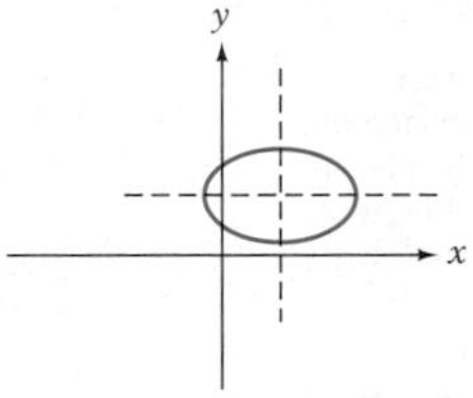

There are two **axes of symmetry** for an ellipse. The intersection of these two axes is the **center** of the ellipse.

An ellipse with center at the origin is shown at the right. Note that there are two x-intercepts and two y-intercepts.

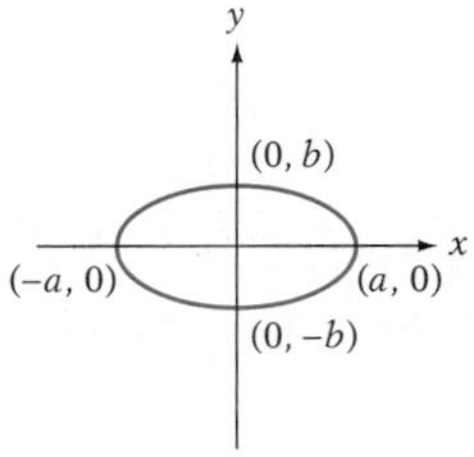

Point of Interest

The word *ellipse* comes from the Greek word *ellipsis*, which means "deficient." The method by which the early Greeks analyzed the conics caused a certain area in the construction of the ellipse to be less than another area (deficient). The word *ellipsis* in English, which means "omission," has the same Greek root as the word *ellipse*.

> **The Standard Form of the Equation of an Ellipse with Center at the Origin**
>
> The equation of an ellipse with center at the origin is $\frac{x^2}{a^2} + \frac{y^2}{b^2} = 1$.
>
> The x-intercepts are $(a, 0)$ and $(-a, 0)$. The y-intercepts are $(0, b)$ and $(0, -b)$.

By finding the x- and y-intercepts of an ellipse and using the fact that the ellipse is "oval" shaped, we can sketch a graph of the ellipse.

➡ Sketch the graph of the ellipse whose equation is $\frac{x^2}{9} + \frac{y^2}{4} = 1$.

Comparing $\frac{x^2}{9} + \frac{y^2}{4} = 1$ with $\frac{x^2}{a^2} + \frac{y^2}{b^2} = 1$, we have $a^2 = 9$ and $b^2 = 4$.

Therefore, $a = 3$ and $b = 2$.

The x-intercepts are $(3, 0)$ and $(-3, 0)$.
The y-intercepts are $(0, 2)$ and $(0, -2)$.

Use the intercepts to sketch a graph of the ellipse.

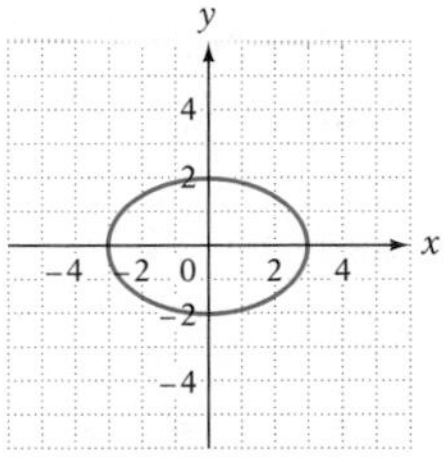

Using the vertical-line test, we find that the graph of an ellipse is not the graph of a function. The graph of an ellipse is the graph of a relation.

Objective 11.6C

New Vocabulary

ellipse
axes of symmetry of an ellipse
center of an ellipse

New Equations

standard form of the equation of an ellipse with center at the origin: $\frac{x^2}{a^2} + \frac{y^2}{b^2} = 1$

x-intercepts: $(a, 0), (-a, 0)$
y-intercepts: $(0, b), (0, -b)$

Discuss the Concepts

1. Explain how to determine the domain and range of the relation $\frac{x^2}{9} + \frac{y^2}{4} = 1$ from its graph. What are the domain and range?
   D: $\{x \mid -3 \le x \le 3\}$;
   R: $\{y \mid -2 \le y \le 2\}$
2. Explain how to determine the domain and range of the relation $\frac{x^2}{25} + \frac{y^2}{16} = 1$ from its equation. What are the domain and range?
   D: $\{x \mid -5 \le x \le 5\}$;
   R: $\{y \mid -4 \le y \le 4\}$

In-Class Examples (Objective 11.6C)

Sketch a graph of the ellipse.

1. $\frac{x^2}{4} + \frac{y^2}{16} = 1$

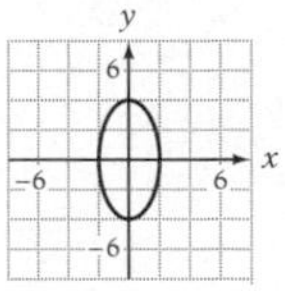

2. $\frac{x^2}{9} + \frac{y^2}{25} = 1$

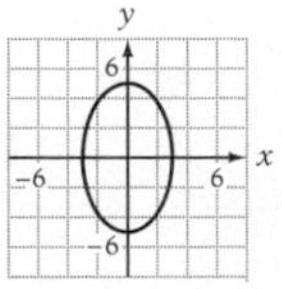

Concept Check

1. What are the x-intercepts of the graph of the ellipse $\frac{x^2}{36} + \frac{y^2}{9} = 1$? What are the y-intercepts? $(6, 0), (-6, 0); (0, 3), (0, -3)$
2. Write the equation of the ellipse with x-intercepts $(7, 0)$ and $(-7, 0)$ and y-intercepts $(0, 2)$ and $(0, -2)$. $\frac{x^2}{49} + \frac{y^2}{4} = 1$

Optional Student Activity

1. The longer axis of symmetry of an ellipse is called the major axis. The shorter axis of symmetry is called the minor axis. The orbit of Halley's comet is an ellipse with a major axis of approximately 36 AU and a minor axis of approximately 9 AU. (One AU is one astronomical unit and is approximately 92,960,000 miles, the average distance of Earth from the sun.) Determine an equation for the orbit of Halley's comet in terms of astronomical units. $\frac{x^2}{324} + \frac{y^2}{20.25} = 1$
2. The orbits of the planets in our solar system are elliptical. The length of the major axis of Mars's orbit is 3.04 AU. (See Exercise 1 above.) The length of the minor axis is 2.99 AU. Determine an equation for the orbit of Mars. Round the values of a and b to the nearest thousandth. $\frac{x^2}{2.310} + \frac{y^2}{2.235} = 1$

Point of Interest

For a circle, $a = b$ and thus $\frac{a}{b} = 1$. Early Greek astronomers thought that each planet had a circular orbit. Today we know that the planets have elliptical orbits. However, in most cases the ellipse is very nearly a circle.

For Earth, $\frac{a}{b} \approx 1.00014$. The most elliptical orbit is Pluto's, for which $\frac{a}{b} \approx 1.0328$.

➡ Sketch a graph of the ellipse whose equation is $\frac{x^2}{16} + \frac{y^2}{16} = 1$.

The x-intercepts are $(4, 0)$ and $(-4, 0)$. • $a^2 = 16, b^2 = 16$

The y-intercepts are $(0, 4)$ and $(0, -4)$.

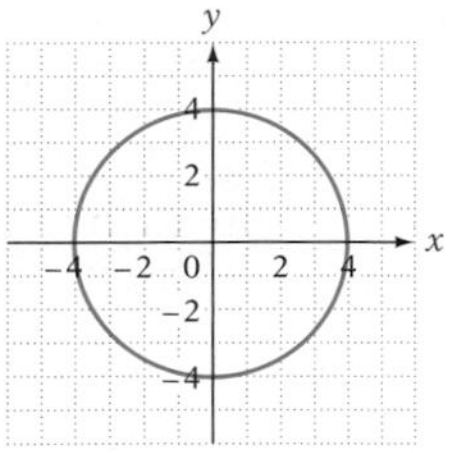

• Use the intercepts and symmetry to sketch the graph of the ellipse.

The graph in this example is the graph of a circle. A circle is a special case of an ellipse. It occurs when $a^2 = b^2$ in the equation $\frac{x^2}{a^2} + \frac{y^2}{b^2} = 1$.

Example 6

Sketch a graph of the ellipse whose equation is $\frac{x^2}{9} + \frac{y^2}{16} = 1$.

Solution

x-intercepts: $(3, 0)$ and $(-3, 0)$

y-intercepts: $(0, 4)$ and $(0, -4)$

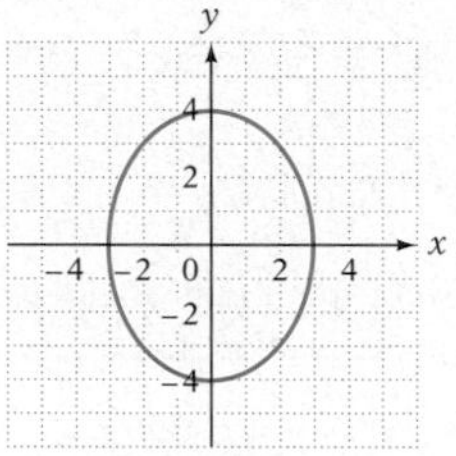

You Try It 6

Sketch a graph of the ellipse whose equation is $\frac{x^2}{4} + \frac{y^2}{25} = 1$.

Your solution

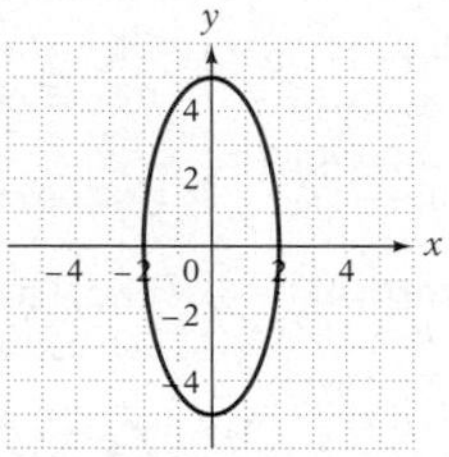

Example 7

Sketch a graph of the ellipse whose equation is $\frac{x^2}{16} + \frac{y^2}{12} = 1$.

Solution

x-intercepts: $(4, 0)$ and $(-4, 0)$

y-intercepts: $(0, 2\sqrt{3})$ and $(0, -2\sqrt{3})$ $(2\sqrt{3} \approx 3.5)$

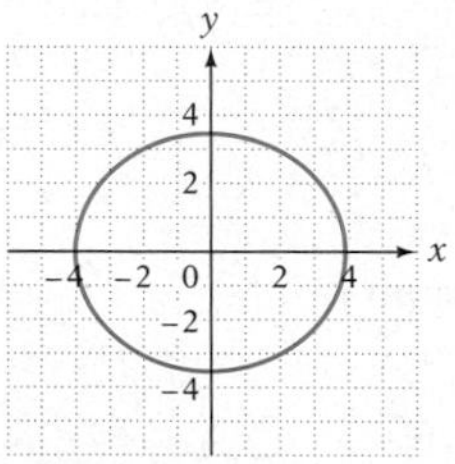

You Try It 7

Sketch a graph of the ellipse whose equation is $\frac{x^2}{18} + \frac{y^2}{9} = 1$.

Your solution

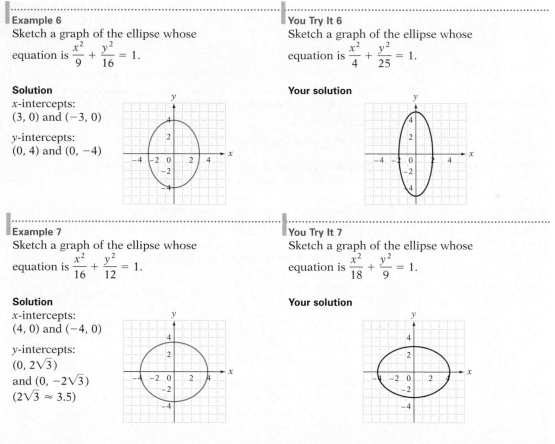

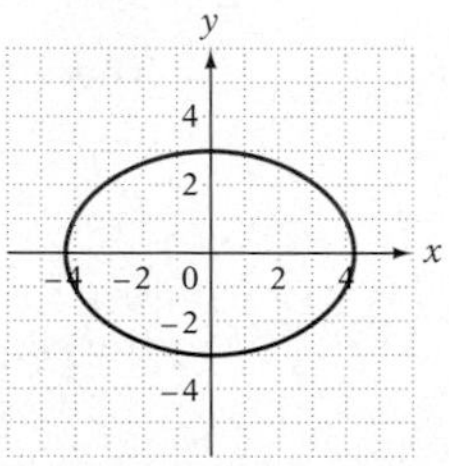

Solutions on pp. S34–S35

Objective D To graph a hyperbola with center at the origin

VIDEO & DVD CD TUTOR WEB SSM

A **hyperbola** is a conic section that is formed by the intersection of a cone and a plane perpendicular to the base of the cone.

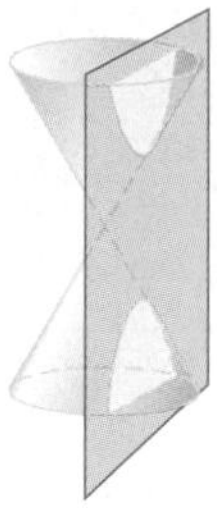

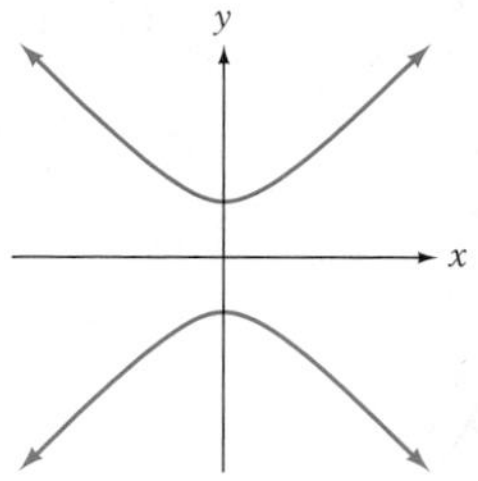

The hyperbola has two **vertices** and an **axis of symmetry** that passes through the vertices. The **center** of a hyperbola is the point halfway between the two vertices.

The graphs at the right show two possible graphs of a hyperbola with center at the origin.

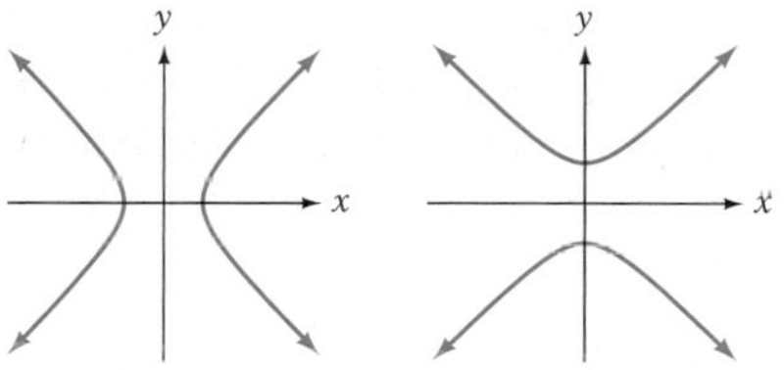

In the first graph, an axis of symmetry is the x-axis and the vertices are x-intercepts.

In the second graph, an axis of symmetry is the y-axis and the vertices are y-intercepts.

Note that in either case, the graph of a hyperbola is not the graph of a function. The graph of a hyperbola is the graph of a relation.

Point of Interest

The word *hyperbola* comes from the Greek word *yperboli*, which means "exceeding." The method by which the early Greeks analyzed the conics caused a certain area in the construction of the hyperbola to be greater than (to exceed) another area. The word *hyperbole* in English, meaning "exaggeration," has the same Greek root as the word *hyperbola*.

The word *asymptote* comes from the Greek word *asymptotos*, which means "not capable of meeting."

The Standard Form of the Equation of a Hyperbola with Center at the Origin

The equation of a hyperbola for which an axis of symmetry is the x-axis is $\frac{x^2}{a^2} - \frac{y^2}{b^2} = 1$. The vertices are $(a, 0)$ and $(-a, 0)$.

The equation of a hyperbola for which an axis of symmetry is the y-axis is $\frac{y^2}{b^2} - \frac{x^2}{a^2} = 1$. The vertices are $(0, b)$ and $(0, -b)$.

To sketch a hyperbola, it is helpful to draw two lines that are "approached" by the hyperbola. These two lines are called **asymptotes**. As a point on the hyperbola gets farther from the origin, the hyperbola "gets closer to" the asymptotes.

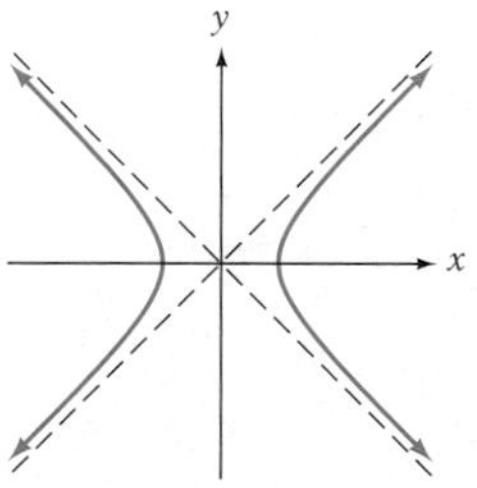

Because the asymptotes are straight lines, their equations are linear equations. The equations of the asymptotes for a hyperbola with center at the origin are $y = \frac{b}{a}x$ and $y = -\frac{b}{a}x$.

Objective 11.6D

New Vocabulary

hyperbola
vertices of a hyperbola
axis of symmetry of a hyperbola
asymptote

New Equations

standard form of the equation of a hyperbola with center at the origin: $\frac{x^2}{a^2} - \frac{y^2}{b^2} = 1$, with vertices $(a, 0)$, $(-a, 0)$; $\frac{y^2}{b^2} - \frac{x^2}{a^2} = 1$, with vertices $(0, b)$, $(0, -b)$

asymptotes of a hyperbola: $y = \frac{b}{a}x$ and $y = -\frac{b}{a}x$

Discuss the Concepts

1. How can you tell from an equation whether its graph will be that of an ellipse or that of a hyperbola?
2. How do you know by looking at the equation of a hyperbola whether it has x-intercepts or y-intercepts?
3. What are the asymptotes of a hyperbola?
4. In addition to the curves presented in this section, how else might the intersection of a plane and a cone be represented?
   There are three degenerate conic sections:
   (1) The intersection of a plane perpendicular to the axis of the cone and through the vertex of the cone is a point.
   (2) The intersection of a plane parallel to the axis of the cone and through the vertex of the cone forms two intersecting straight lines.
   (3) The intersection of a plane and the lateral surface of the cone is a line.

In-Class Examples (Objective 11.6D)

Sketch a graph of the hyperbola.

1. $\frac{x^2}{4} - \frac{y^2}{9} = 1$

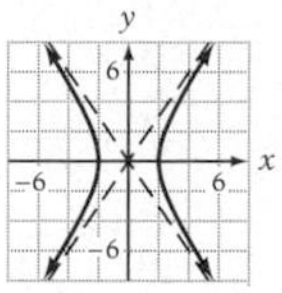

2. $\frac{y^2}{25} - \frac{x^2}{16} = 1$

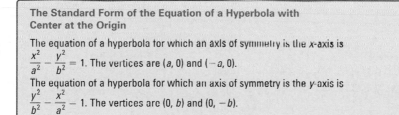

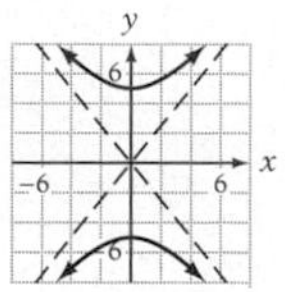

Concept Check

Describe the graph of the conic section given by the equation.

1. $y = \frac{1}{2}x^2 + 2x - 6$ Parabola that opens up
2. $x = -y^2 - 4y + 5$ Parabola that opens left
3. $(x - 4)^2 + (y + 2)^2 = 1$ Circle with center $(4, -2)$ and radius 1
4. $\frac{x^2}{9} + \frac{y^2}{25} = 1$ Ellipse with x-intercepts $(3, 0)$ and $(-3, 0)$ and y-intercepts $(0, 5)$ and $(0, -5)$
5. $\frac{x^2}{25} - \frac{y^2}{9} = 1$ Hyperbola with vertices $(5, 0)$ and $(-5, 0)$
6. $\frac{y^2}{9} - \frac{x^2}{36} = 1$ Hyperbola with vertices $(0, 3)$ and $(0, -3)$

Optional Student Activity

1. Find the equation of the hyperbola with vertices $(0, 4)$ and $(0, -4)$ and asymptotes $y = \frac{1}{2}x$ and $y = -\frac{1}{2}x$.
   $\frac{y^2}{16} - \frac{x^2}{64} = 1$
2. When are the asymptotes of the graph of $\frac{x^2}{a^2} - \frac{y^2}{b^2} = 1$ perpendicular? When $a = b$

Point of Interest

Hyperbolas are used in LORAN (LOng RAnge Navigation) as a method by which a ship's navigator can determine the position of the ship, as shown in the figure below. They are also used as mirrors in some telescopes to focus incoming light.

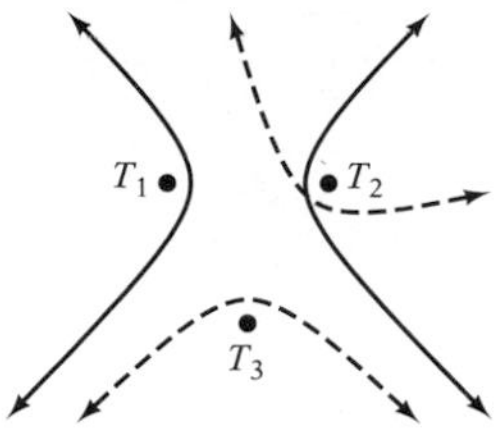

➡ Sketch a graph of the hyperbola whose equation is $\frac{y^2}{9} - \frac{x^2}{4} = 1$.

An axis of symmetry is the y-axis.

$b^2 = 9, a^2 = 4$

The vertices are $(0, 3)$ and $(0, -3)$.

- **The vertices are $(0, b)$ and $(0, -b)$.**

The asymptotes are $y = \frac{3}{2}x$ and $y = -\frac{3}{2}x$.

- **The asymptotes are $y = \frac{b}{a}x$ and $y = -\frac{b}{a}x$.**

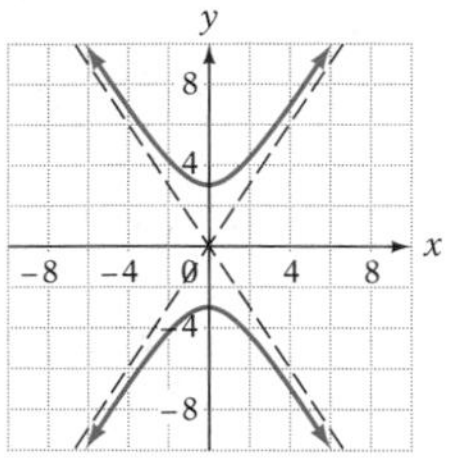

- **Sketch the asymptotes. Use symmetry and the fact that the hyperbola will approach the asymptotes to sketch its graph.**

Example 8

Sketch a graph of the hyperbola whose equation is $\frac{x^2}{16} - \frac{y^2}{4} = 1$.

Solution

axis of symmetry: x-axis

vertices: $(4, 0)$ and $(-4, 0)$

asymptotes: $y = \frac{1}{2}x$ and $y = -\frac{1}{2}x$

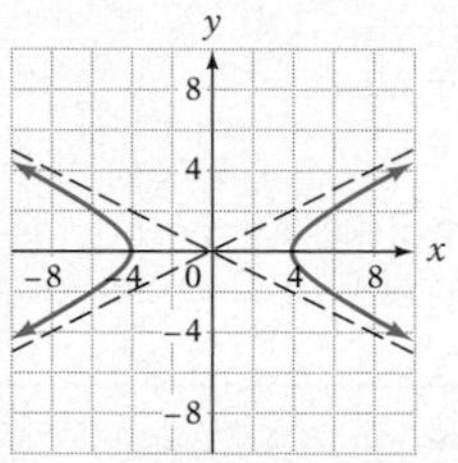

You Try It 8

Sketch a graph of the hyperbola whose equation is $\frac{x^2}{9} - \frac{y^2}{25} = 1$.

Your solution

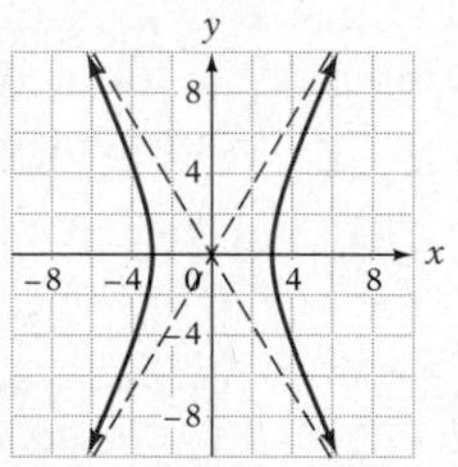

Example 9

Sketch a graph of the hyperbola whose equation is $\frac{y^2}{16} - \frac{x^2}{25} = 1$.

Solution

axis of symmetry: y-axis

vertices: $(0, 4)$ and $(0, -4)$

asymptotes: $y = \frac{4}{5}x$ and $y = -\frac{4}{5}x$

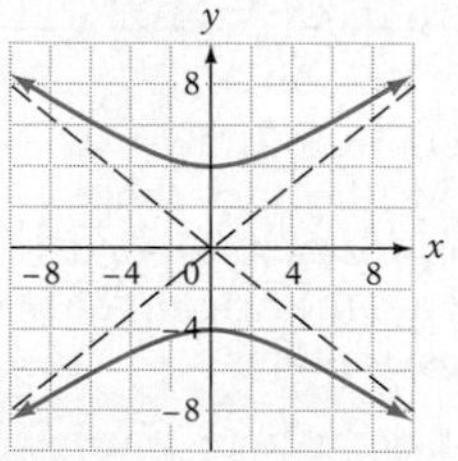

You Try It 9

Sketch a graph of the hyperbola whose equation is $\frac{y^2}{9} - \frac{x^2}{9} = 1$.

Your solution

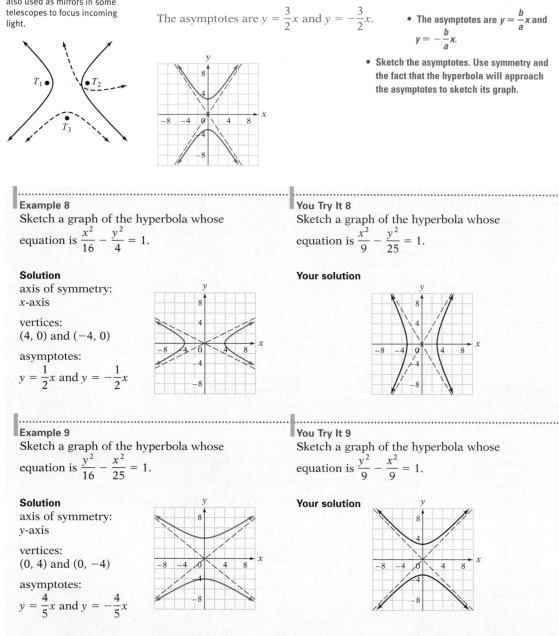

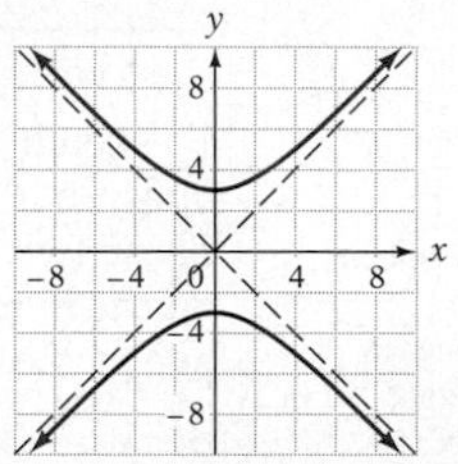

Solutions on p. S35

635

11.6 Exercises

Objective A

State (a) whether the axis of symmetry is a vertical or a horizontal line and (b) in what direction the parabola opens.

1. $y = 3x^2 - 4x + 7$
a. vertical line **b.** opens up

2. $y = -x^2 + 5x - 2$
a. vertical line **b.** opens down

3. $x = y^2 + 2y - 8$
a. horizontal line **b.** opens right

4. $x = -3y^2 - y + 9$
a. horizontal line **b.** opens left

5. $x = -\frac{1}{2}y^2 - 4y - 7$
a. horizontal line **b.** opens left

6. $y = \frac{1}{4}x^2 + 6x - 1$
a. vertical line **b.** opens up

Find the vertex and axis of symmetry of the parabola given by the equation. Then sketch its graph.

7. $x = y^2 - 3y - 4$

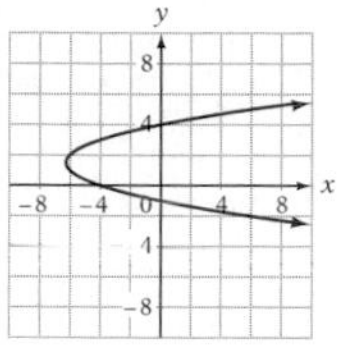

vertex: $\left(-\frac{25}{4}, \frac{3}{2}\right)$
axis of symmetry: $y = \frac{3}{2}$

8. $y = x^2 - 2$

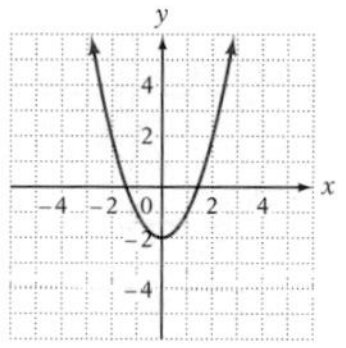

vertex: $(0, -2)$
axis of symmetry: $x = 0$

9. $y = x^2 + 2$

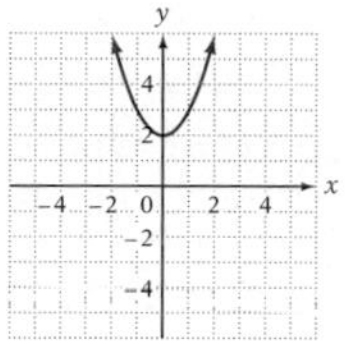

vertex: $(0, 2)$
axis of symmetry: $x = 0$

10. $x = -\frac{1}{2}y^2 + 4$

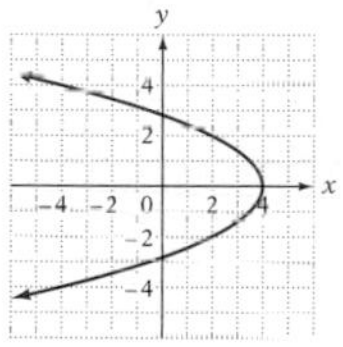

vertex: $(4, 0)$
axis of symmetry: $y = 0$

11. $x = \frac{1}{4}y^2 - 1$

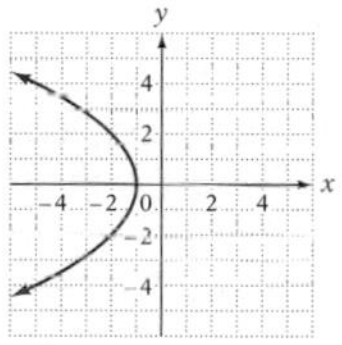

vertex: $(-1, 0)$
axis of symmetry: $y = 0$

12. $x = \frac{1}{2}y^2 - y + 1$

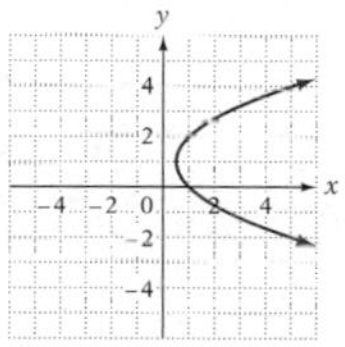

vertex: $\left(\frac{1}{2}, 1\right)$
axis of symmetry: $y = 1$

13. $x = -\frac{1}{2}y^2 + 2y - 3$

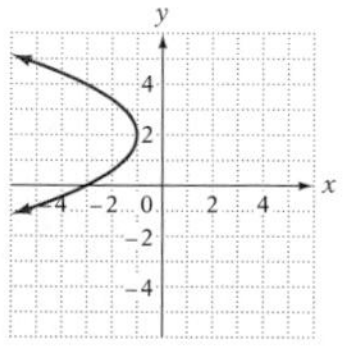

vertex: $(-1, 2)$
axis of symmetry: $y = 2$

14. $y = -\frac{1}{2}x^2 + 2x + 6$

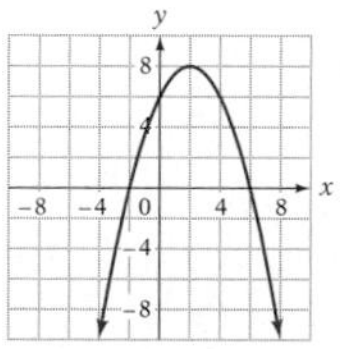

vertex: $(2, 8)$
axis of symmetry: $x = 2$

15. $y = \frac{1}{2}x^2 + x - 3$

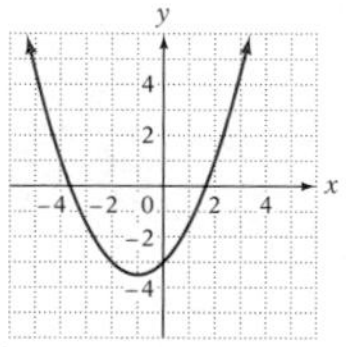

vertex: $\left(-1, -\frac{7}{2}\right)$
axis of symmetry: $x = -1$

Section 11.6

Suggested Assignment
Exercises 1–45, odds
More challenging problems:
Exercises 47–51, odds

Quick Quiz (Objective 11.6A)

Find the vertex and axis of symmetry of the parabola. Then sketch its graph.

1. $y = -x^2 + 4x - 3$
Vertex: $(2, 1)$
Axis of symmetry:
$x = 2$

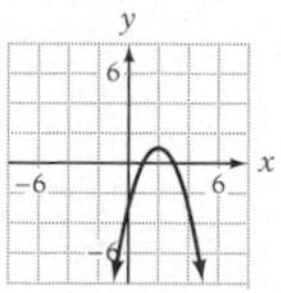

2. $x = 2y^2 - 2$
Vertex: $(-2, 0)$
Axis of symmetry:
$y = 0$

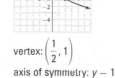

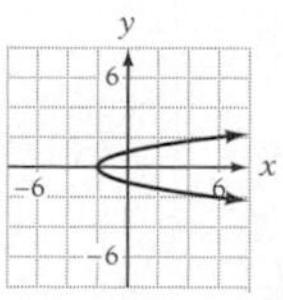

Objective B

Sketch a graph of the circle given by the equation.

16. $(x-2)^2+(y+2)^2=9$

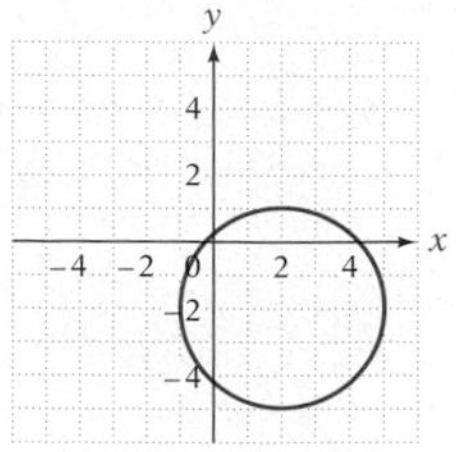

17. $(x+2)^2+(y-3)^2=16$

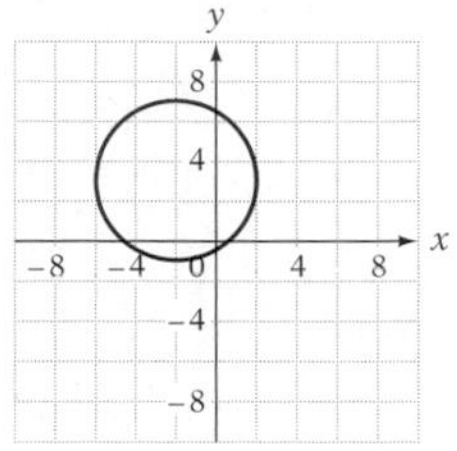

18. $(x+3)^2+(y-1)^2=25$

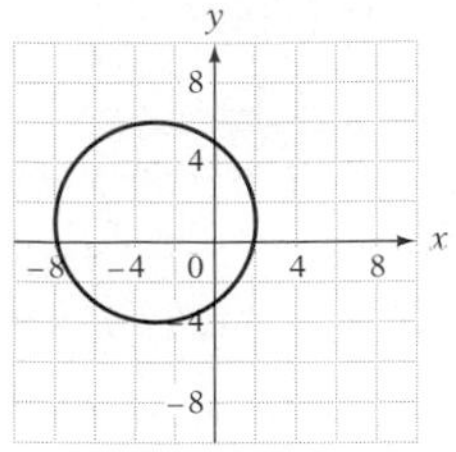

19. $(x-2)^2+(y+3)^2=4$

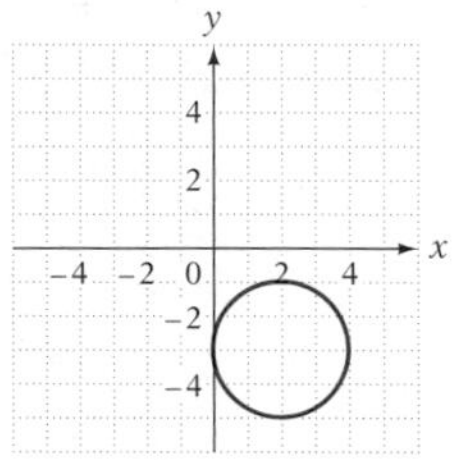

20. $(x+2)^2+(y+2)^2=4$

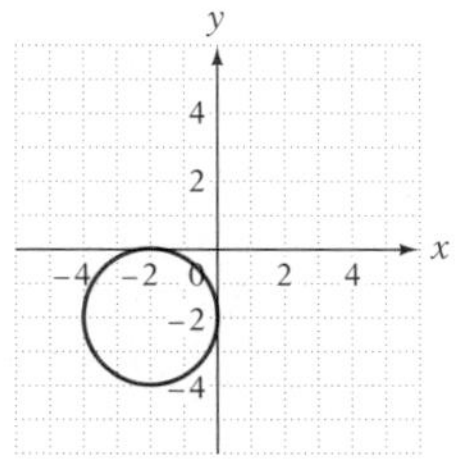

21. $(x-1)^2+(y-2)^2=25$

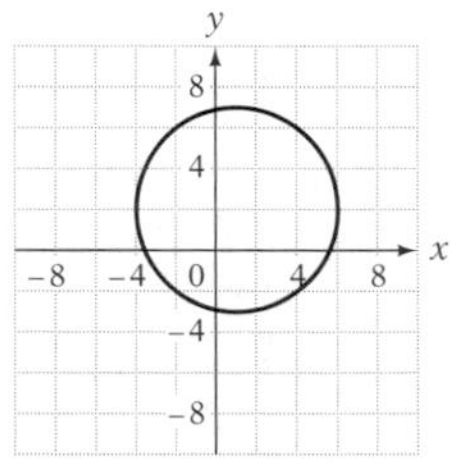

22. Find the equation of the circle with radius 2 and center $(2, -1)$. Then sketch its graph.

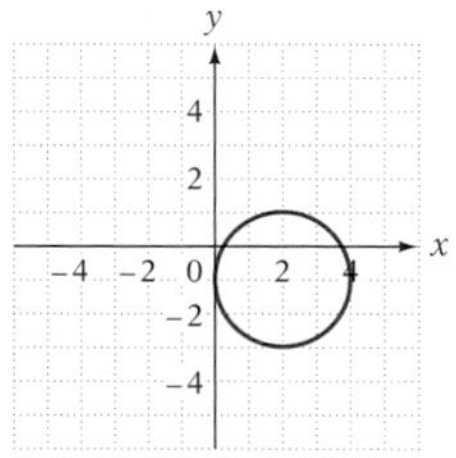

$(x-2)^2+(y+1)^2=4$

23. Find the equation of the circle with radius 3 and center $(-1, -2)$. Then sketch its graph.

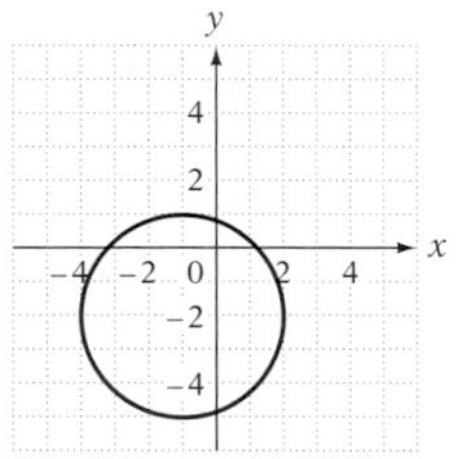

$(x+1)^2+(y+2)^2=9$

24. Find the equation of the circle with radius $\sqrt{5}$ and center $(-1, 1)$. Then sketch its graph.

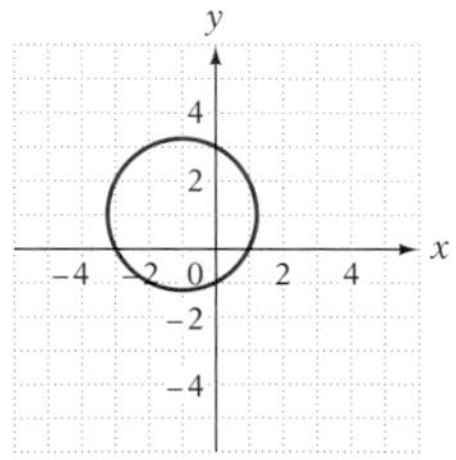

$(x+1)^2+(y-1)^2=5$

25. Find the equation of the circle with radius $\sqrt{5}$ and center $(-2, 1)$. Then sketch its graph.

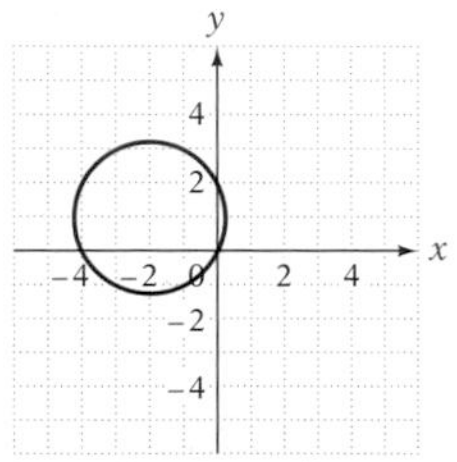

$(x+2)^2+(y-1)^2=5$

Quick Quiz (Objective 11.6B)

1. Sketch a graph of the circle.
$(x-2)^2+(y+1)^2=9$

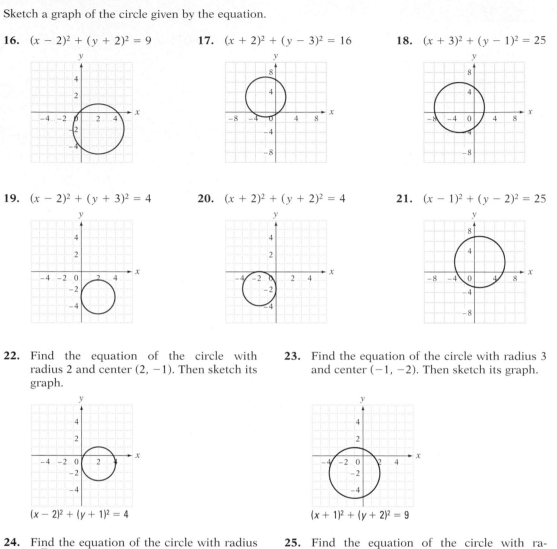

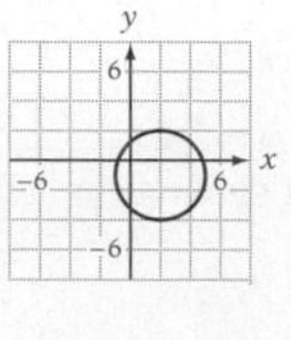

2. Find the equation of the circle with radius 6 and center $(-4, 1)$. $(x+4)^2+(y-1)^2=36$

637

Objective C

Sketch a graph of the ellipse given by the equation.

26. $\dfrac{x^2}{4} + \dfrac{y^2}{9} = 1$

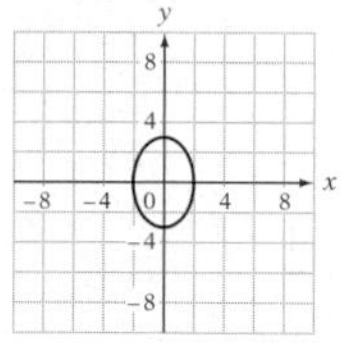

27. $\dfrac{x^2}{25} + \dfrac{y^2}{16} = 1$

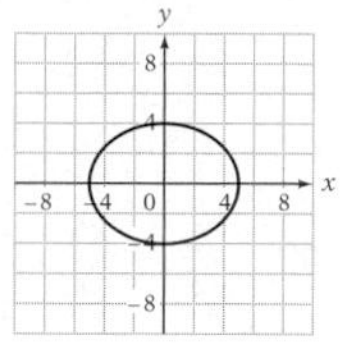

28. $\dfrac{x^2}{25} + \dfrac{y^2}{9} = 1$

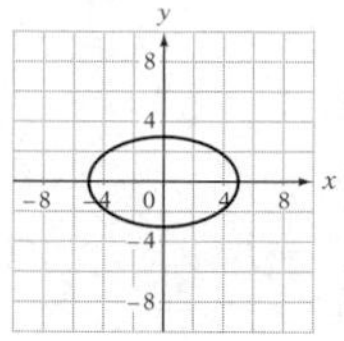

29. $\dfrac{x^2}{16} + \dfrac{y^2}{9} = 1$

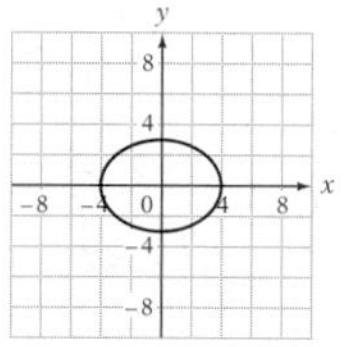

30. $\dfrac{x^2}{36} + \dfrac{y^2}{16} = 1$

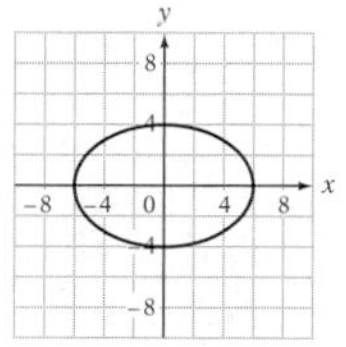

31. $\dfrac{x^2}{49} + \dfrac{y^2}{64} = 1$

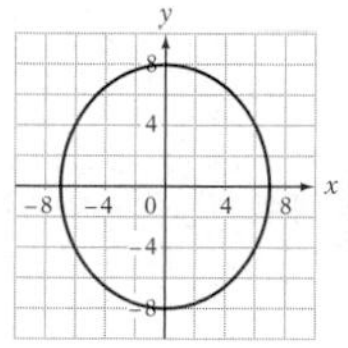

32. $\dfrac{x^2}{16} + \dfrac{y^2}{49} = 1$

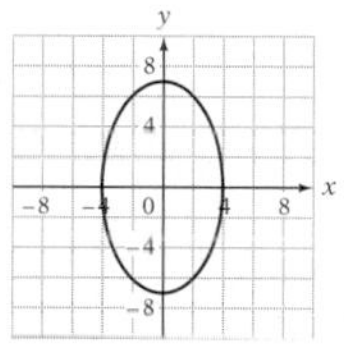

33. $\dfrac{x^2}{25} + \dfrac{y^2}{36} = 1$

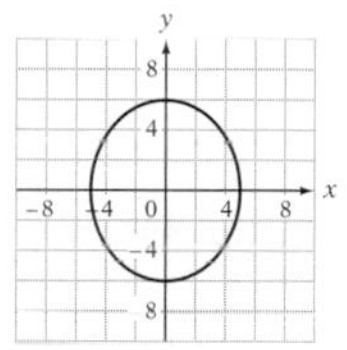

34. $\dfrac{x^2}{4} + \dfrac{y^2}{25} = 1$

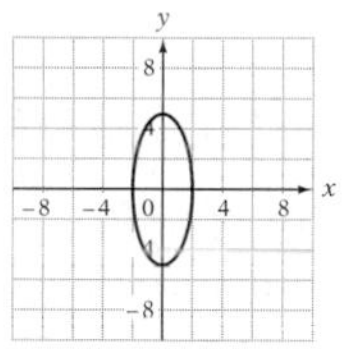

Objective D

Sketch a graph of the hyperbola given by the equation.

35. $\dfrac{x^2}{9} - \dfrac{y^2}{16} = 1$

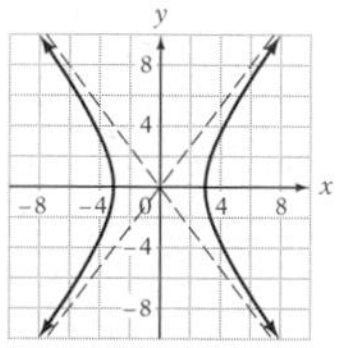

36. $\dfrac{x^2}{25} - \dfrac{y^2}{4} = 1$

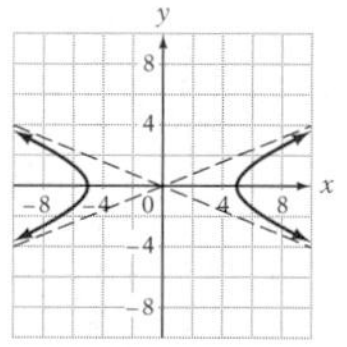

37. $\dfrac{y^2}{16} - \dfrac{x^2}{9} = 1$

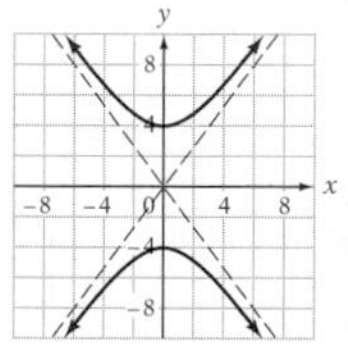

Quick Quiz (Objective 11.6C)

Sketch a graph of the ellipse.

1. $\dfrac{x^2}{16} + \dfrac{y^2}{25} = 1$

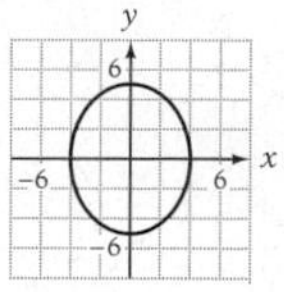

2. $\dfrac{x^2}{25} + \dfrac{y^2}{4} = 1$

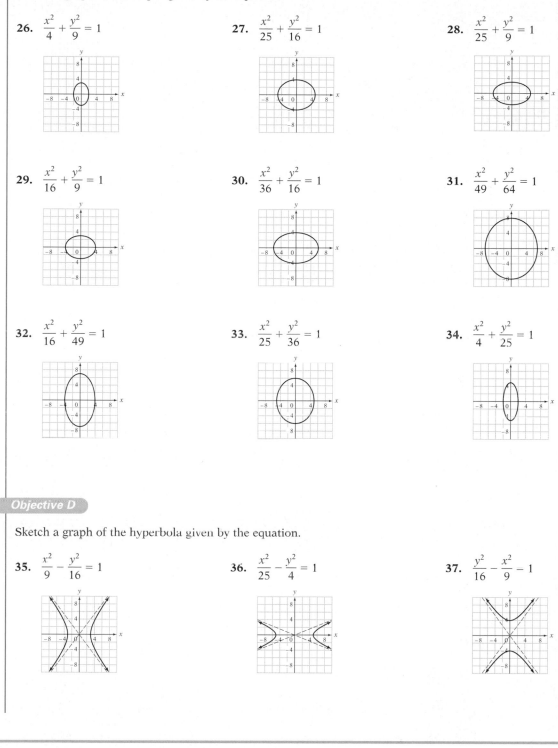

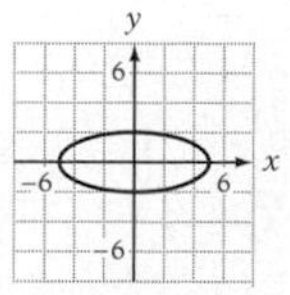

38. $\dfrac{y^2}{16} - \dfrac{x^2}{25} = 1$

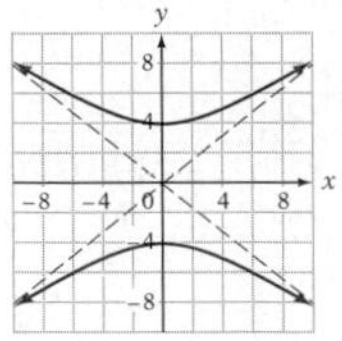

39. $\dfrac{x^2}{16} - \dfrac{y^2}{4} = 1$

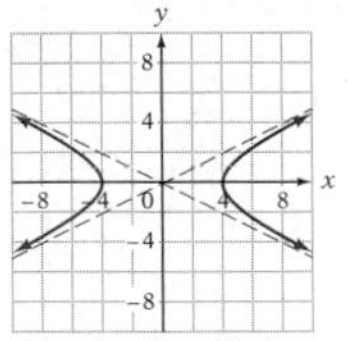

40. $\dfrac{x^2}{9} - \dfrac{y^2}{49} = 1$

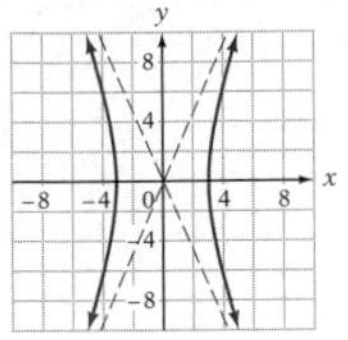

41. $\dfrac{y^2}{25} - \dfrac{x^2}{9} = 1$

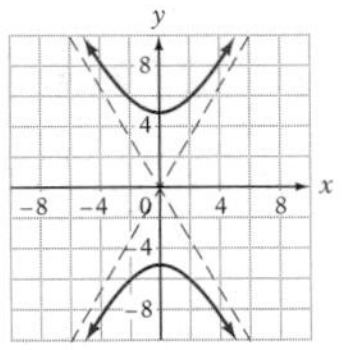

42. $\dfrac{y^2}{4} - \dfrac{x^2}{16} = 1$

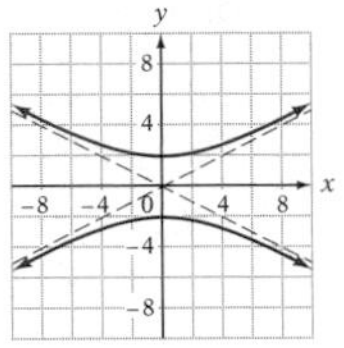

43. $\dfrac{x^2}{4} - \dfrac{y^2}{25} = 1$

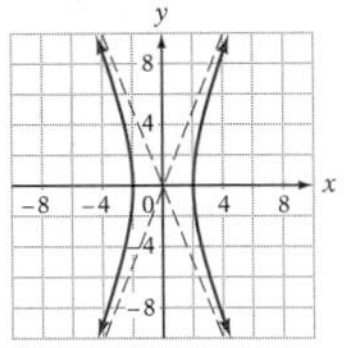

44. $\dfrac{x^2}{36} - \dfrac{y^2}{9} = 1$

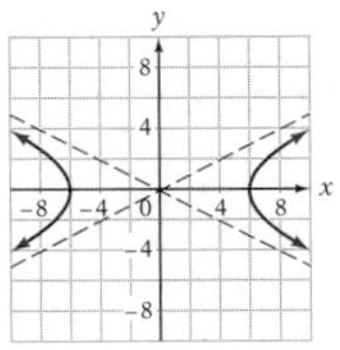

45. $\dfrac{y^2}{9} - \dfrac{x^2}{36} = 1$

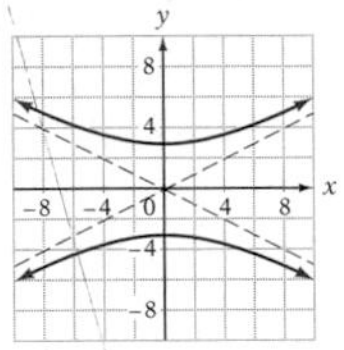

46. $\dfrac{y^2}{25} - \dfrac{x^2}{4} = 1$

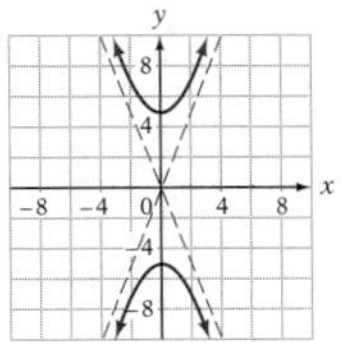

APPLYING THE CONCEPTS

Write the equation in standard form. Identify the graph, and then graph the equation.

47. $4x^2 + 9y^2 = 36$

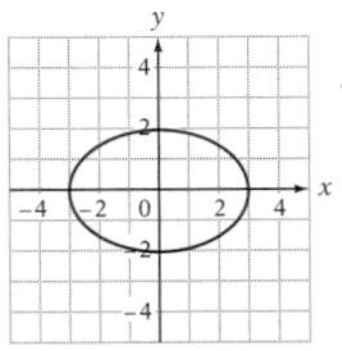

$\dfrac{x^2}{9} + \dfrac{y^2}{4} = 1$

ellipse

48. $16x^2 + 25y^2 = 400$

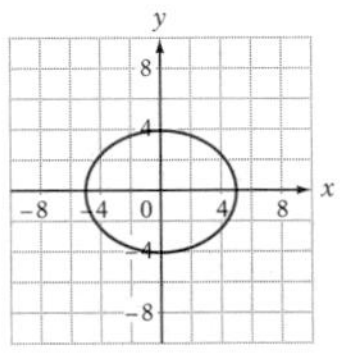

$\dfrac{x^2}{25} + \dfrac{y^2}{16} = 1$

ellipse

49. $9x^2 - 25y^2 = 225$

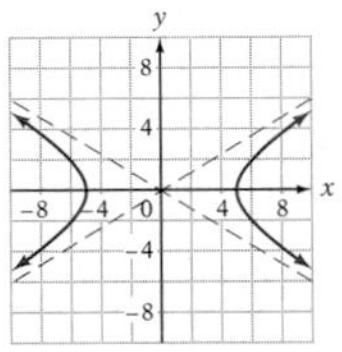

$\dfrac{x^2}{25} - \dfrac{y^2}{9} = 1$

hyperbola

50. $25y^2 - 4x^2 = -100$

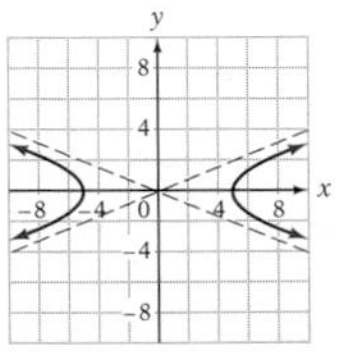

$\dfrac{x^2}{25} - \dfrac{y^2}{4} = 1$

hyperbola

51. $9y^2 - 16x^2 = 144$

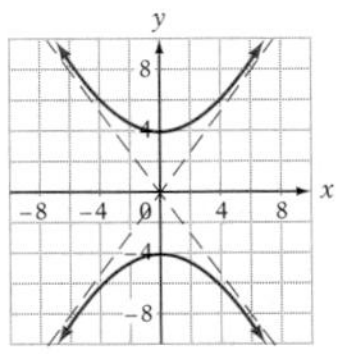

$\dfrac{y^2}{16} - \dfrac{x^2}{9} = 1$

hyperbola

52. $4y^2 - x^2 = 36$

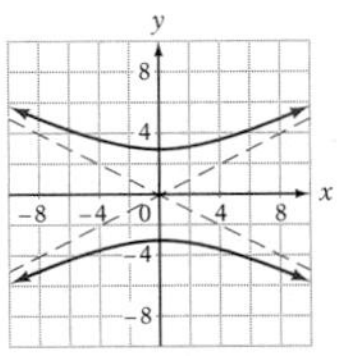

$\dfrac{y^2}{9} - \dfrac{x^2}{36} = 1$

hyperbola

Quick Quiz (Objective 11.6D)

Sketch a graph of the hyperbola.

1. $\dfrac{x^2}{25} - \dfrac{y^2}{36} = 1$

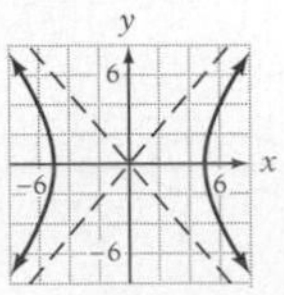

2. $\dfrac{y^2}{4} - \dfrac{x^2}{4} = 1$

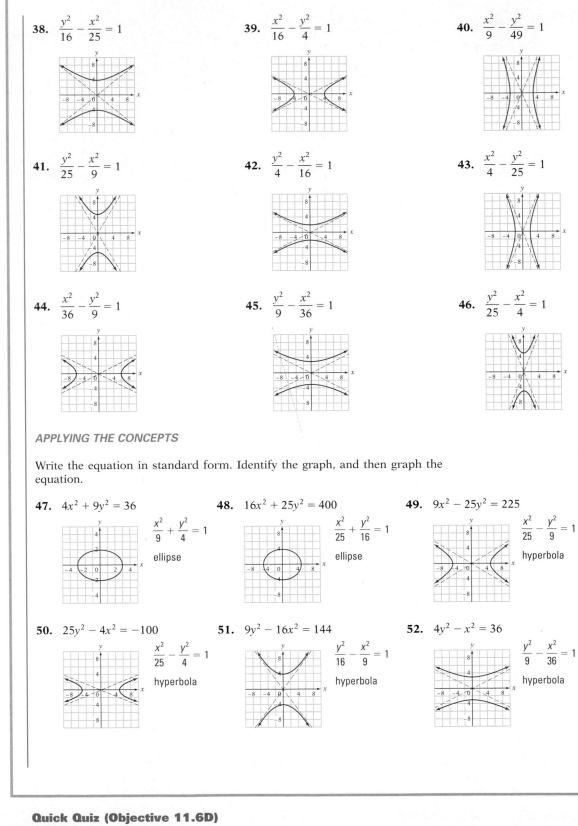

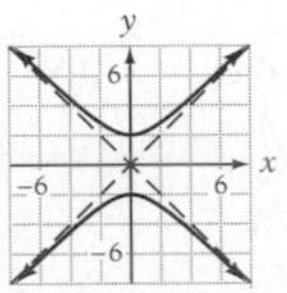

Focus on Problem Solving

Algebraic Manipulation and Graphing Techniques

Problem solving is often easier when we have both algebraic manipulation and graphing techniques at our disposal. Solving quadratic equations and graphing quadratic equations in two variables are used here to solve problems involving profit.

A company's revenue, R, is the total amount of money the company earned by selling its products. The cost, C, is the total amount of money the company spent to manufacture and sell its products. A company's profit, P, is the difference between the revenue and cost: $P = R - C$. A company's revenue and cost may be represented by equations.

A company manufactures and sells woodstoves. The total monthly cost, in dollars, to produce n woodstoves is $C = 30n + 2000$. Write a variable expression for the company's monthly profit if the revenue, in dollars, obtained from selling all n woodstoves is $R = 150n - 0.4n^2$.

$$P = R - C$$

$$P = 150n - 0.4n^2 - (30n + 2000)$$

$$P = -0.4n^2 + 120n - 2000$$

- Replace R by $150n - 0.4n^2$ and C by $30n + 2000$. Then simplify.

How many woodstoves must the company manufacture and sell in order to make a profit of \$6000 a month?

$$P = -0.4n^2 + 120n - 2000$$

$$6000 = -0.4n^2 + 120n - 2000$$ • Substitute 6000 for P.

$$0 = -0.4n^2 + 120n - 8000$$ • Write the equation in standard form.

$$0 = n^2 - 300n + 20{,}000$$ • Divide each side of the equation by -0.4.

$$0 = (n - 100)(n - 200)$$ • Factor.

$$n - 100 = 0 \qquad n - 200 = 0$$ • Solve for n.

$$n = 100 \qquad n = 200$$

The company will make a monthly profit of \$6000 if either 100 or 200 woodstoves are manufactured and sold.

The graph of $P = -0.4n^2 + 120n - 2000$ is shown at the right. Note that when $P = 6000$, the values of n are 100 and 200.

Also note that the coordinates of the highest point on the graph are (150, 7000). This means that the company makes a *maximum* profit of \$7000 per month when 150 woodstoves are manufactured and sold.

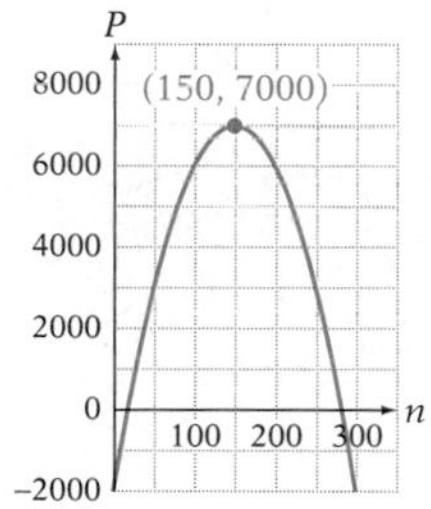

1. The total cost, in dollars, for a company to produce and sell n guitars per month is $C = 240n + 1200$. The company's revenue, in dollars, from selling all n guitars is $R = 400n - 2n^2$.
   a. How many guitars must the company produce and sell each month in order to make a monthly profit of \$1200?
   b. Graph the profit equation. What is the maximum monthly profit the company can make?

Answers to Focus on Problem Solving: Algebraic Manipulation and Graphing Techniques

1. a. 20 or 60 guitars
b.

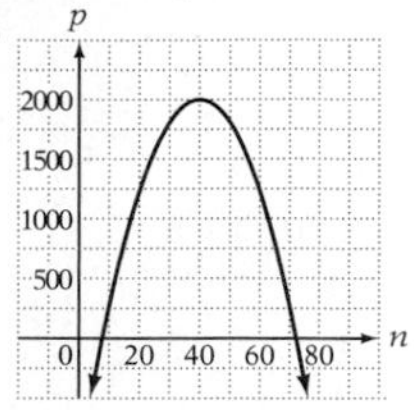

The maximum monthly profit the company can make is \$2000.

Answers to Projects and Group Activities: The Eccentricity and Foci of an Ellipse

1. flatter
2. Neptune
3. 0.6
4. 0.75
5. $(6, 0)$ and $(-6, 0)$
6. $(-5, 0)$ and $(5, 0)$

Projects and Group Activities

The Eccentricity and Foci of an Ellipse

The graph of an ellipse can be long and thin, or it can have a shape that is very close to a circle. The **eccentricity,** e, of an ellipse is a measure of its "roundness."

The shapes of ellipses with various eccentricities are shown below.

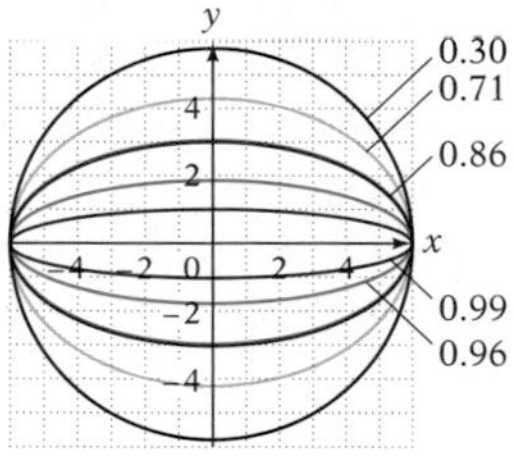

Planet	Eccentricity
Mercury	0.206
Venus	0.007
Earth	0.017
Mars	0.093
Jupiter	0.049
Saturn	0.051
Uranus	0.046
Neptune	0.005
Pluto	0.250

1. Based on the eccentricities of the ellipses shown above, complete the sentence. "As the eccentricity of an ellipse gets closer to 1, the ellipses get *flatter/rounder.*"

2. The planets travel around the sun in elliptical orbits. The eccentricities of the orbits of the planets are shown in the table at the left. Which planet has the most nearly circular orbit?

For an ellipse given by the equation $\frac{x^2}{a^2} + \frac{y^2}{b^2} = 1$, $a > b$, a formula for eccentricity is $e = \frac{\sqrt{a^2 - b^2}}{a}$. Use this formula to find the eccentricity of the ellipses in Exercise 3 and 4. If necessary, round to the nearest hundredth.

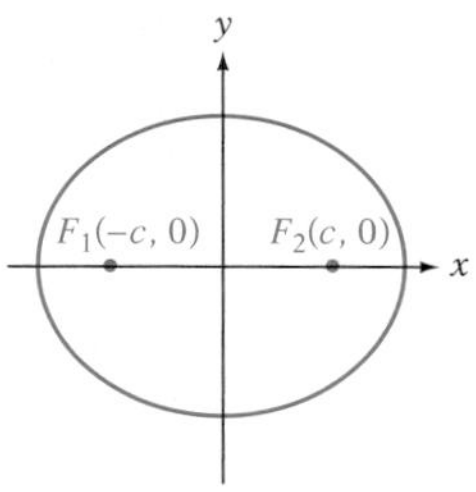

3. $\frac{x^2}{25} + \frac{y^2}{16} = 1$

4. $\frac{x^2}{9} + \frac{y^2}{4} = 1$

Ellipses have a reflective property that has been used in the design of some buildings. The **foci** of an ellipse are two points on the longer axis, called the **major axis,** of the ellipse.

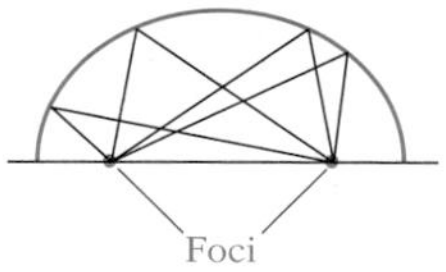

If light or sound emanates from one focus, it is reflected to the other focus. This phenomenon results in what are called "whispering galleries." The rotunda of the Capitol Building in Washington, D.C., is a whispering gallery. A whisper spoken by a person at one focus can be heard clearly by a person at the other focus.

The foci are c units from the center of an ellipse, where $c = \sqrt{a^2 - b^2}$ for an ellipse whose equation is $\frac{x^2}{a^2} + \frac{y^2}{b^2} = 1$, $a > b$.

5. Find the foci for the ellipse whose equation is $\frac{x^2}{100} + \frac{y^2}{64} = 1$.

6. Find the foci for the ellipse whose equation is $\frac{x^2}{169} + \frac{y^2}{144} = 1$.

x-intercepts and Solutions of Equations

The x-coordinate of an x-intercept of the graph of $y = f(x)$ is a real number solution of the equation $f(x) = 0$. For instance, the graph of $P(x) = x^5 - 4x^4 - 16x^3 + 46x^2 + 63x - 90$ is shown at the right. The x-intercepts of the graph are $(-3, 0)$, $(-2, 0)$, $(1, 0)$, $(3, 0)$, and $(5, 0)$, and the roots of the equation $x^5 - 4x^4 - 16x^3 + 46x^2 + 63x - 90 = 0$ are -3, -2, 1, 3, and 5.

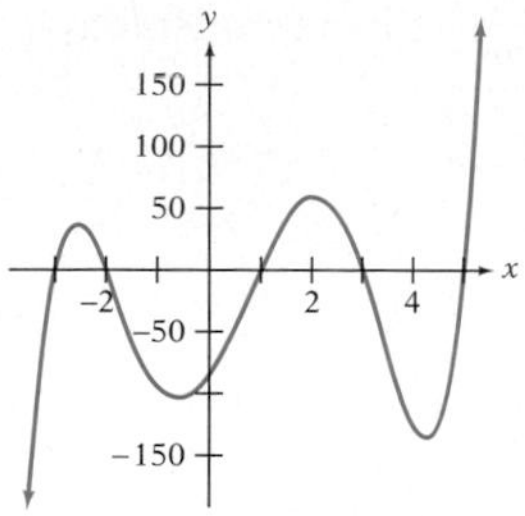

1. Graph $P(x) = x^5 + x^4 - 7x^3 - x^2 + 6x$ and determine the x-intercepts. Use Xmin $= -4$, Xmax $= 4$, Xscl $= 1$, Ymin $= -10$, Ymax $= 40$, and Yscl $= 10$.
2. What are the real number solutions of $x^5 + x^4 - 7x^3 - x^2 + 6x = 0$?
3. Graph $P(x) = x^5 - 2x^4 + 5x^3 - 10x^2 + 4x - 8$ and determine the x-intercepts. Use Xmin $= -4$, Xmax $= 4$, Xscl $= 1$, Ymin $= -100$, Ymax $= 100$, and Yscl $= 50$.
4. What are the real number solutions of $x^5 - 2x^4 + 5x^3 - 10x^2 + 4x - 8 = 0$?
5. On the basis of the graph shown above and your answers to Exercises 2 and 4, make a conjecture about the number of real number solutions a fifth-degree equation can have.
6. Make up a few more fifth-degree equations and produce their graphs. Do the numbers of real number solutions of these equations bear out your conjecture?

Answers to Projects and Group Activities: x-intercepts and Solutions of Equations

1. The x-intercepts are $(-3, 0)$, $(-1, 0)$, $(0, 0)$, $(1, 0)$, and $(2, 0)$.
2. The real number solutions are $-3, -1, 0, 1$, and 2.
3. The x-intercept is $(2, 0)$.
4. The real number solution is 2.
5. One possible conjecture is that a fifth-degree equation can have at most five real-number solutions.
6. Students should determine that the numbers of real number solutions bear out their conjectures.

Chapter Summary

Key Words

A function that can be written in the form $y = mx + b$ or $f(x) = mx + b$ is a *linear function*. The graph of a linear function is a straight line with slope m and y-intercept $(0, b)$. [p. 591]

A *quadratic function* is one that can be expressed by the equation $f(x) = ax^2 + bx + c$, $a \neq 0$. The graph of this function is a *parabola*. [p. 595]

The *inverse of a function* is the set of ordered pairs formed by reversing the coordinates of each ordered pair of the function. [p. 618]

The graph of a *conic section* can be represented by the intersection of a plane and a cone. The four conic sections are the *parabola*, *ellipse*, *hyperbola*, and *circle*. [pp. 625, 629, 631, 633]

A *circle* is the set of all points (x, y) in the plane that are a fixed distance from a given point (h, k) called the *center*. The fixed distance is the *radius* of the circle. [p. 629]

The *asymptotes* of a hyperbola are the two straight lines that are "approached" by the hyperbola. As the graph of the hyperbola gets farther from the origin, the hyperbola "gets closer to" the asymptotes. [p. 633]

Essential Rules

Vertical-Line Test A graph defines a function if any vertical line intersects the graph at no more than one point. [p. 606]

Operations on Functions If f and g are functions and x is an element of the domain of each function, then

$(f + g)(x) = f(x) + g(x)$ $\qquad$ $(f - g)(x) = f(x) - g(x)$

$(f \cdot g)(x) = f(x) \cdot g(x)$ $\qquad$ $\left(\frac{f}{g}\right)(x) = \frac{f(x)}{g(x)}, g(x) \neq 0$ [p. 611]

Composition of Two Functions The composition of two functions, $f \circ g$, is the function whose value at x is given by $(f \circ g)(x) = f[g(x)]$. [p. 613]

One-to-One Function A function f is a 1–1 function if, for any a and b in the domain of f, $f(a) = f(b)$ implies that $a = b$. [p. 617]

Horizontal-Line Test The graph of a function represents the graph of a 1–1 function if any horizontal line intersects the graph at no more than one point. [p. 618]

Condition for an Inverse Function A function f has an inverse function if and only if f is a 1–1 function. [p. 619]

Property of the Composition of Inverse Functions $f^{-1}[f(x)] = x$ and $f[f^{-1}(x)] = x$ [p. 620]

Equation of a Parabola

$y = ax^2 + bx + c$ [p. 596]

When $a > 0$, the parabola opens up.
When $a < 0$, the parabola opens down.
The x-coordinate of the vertex is $-\frac{b}{2a}$.
The axis of symmetry is the line $x = -\frac{b}{2a}$.

$x = ay^2 + by + c$ [p. 627]

When $a > 0$, the parabola opens to the right.
When $a < 0$, the parabola opens to the left.
The y-coordinate of the vertex is $-\frac{b}{2a}$.
The axis of symmetry is the line $y = -\frac{b}{2a}$.

Equation of a Circle [p. 629]

$(x - h)^2 + (y - k)^2 = r^2$ $\qquad$ The center is (h, k) and the radius is r.

Equation of an Ellipse [p. 631]

$\frac{x^2}{a^2} + \frac{y^2}{b^2} = 1$

The x-intercepts are $(a, 0)$ and $(-a, 0)$.
The y-intercepts are $(0, b)$ and $(0, -b)$.

Equation of a Hyperbola [p. 633]

$\frac{x^2}{a^2} - \frac{y^2}{b^2} = 1$

An axis of symmetry is the x-axis.
The vertices are $(a, 0)$ and $(-a, 0)$.

$\frac{y^2}{b^2} - \frac{x^2}{a^2} = 1$

An axis of symmetry is the y-axis.
The vertices are $(0, b)$ and $(0, -b)$.

The equations of the asymptotes are $y = \pm\frac{b}{a}x$.

643

Chapter Review

1. Graph: $f(x) = \frac{1}{4}x + 3$

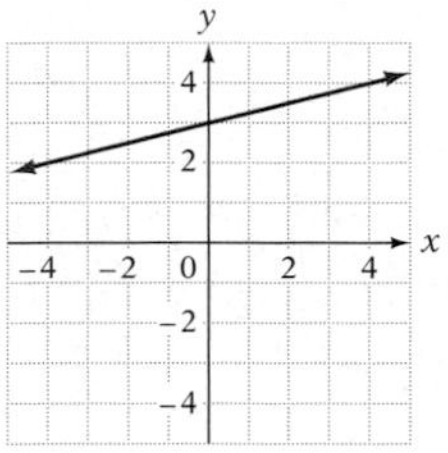

[11.1A]

2. Sketch a graph of $f(x) = x^2 - 2x + 3$.

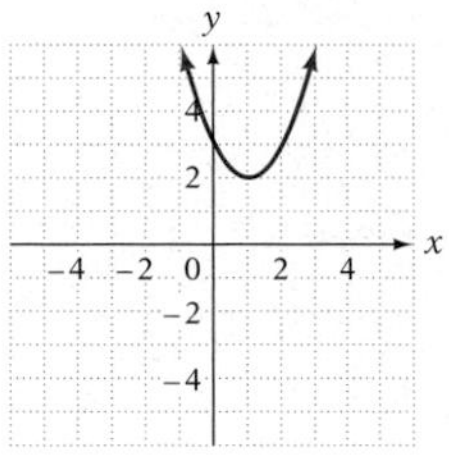

[11.2A]

3. Find the x-intercepts of $y = 2x^2 - 3x + 4$.

no x-intercepts [11.2B]

4. Find the maximum value of the function $f(x) = -x^2 + 8x - 7$.

9 [11.2C]

5. Graph $f(x) = -\sqrt{3 - x}$. State the domain and range.

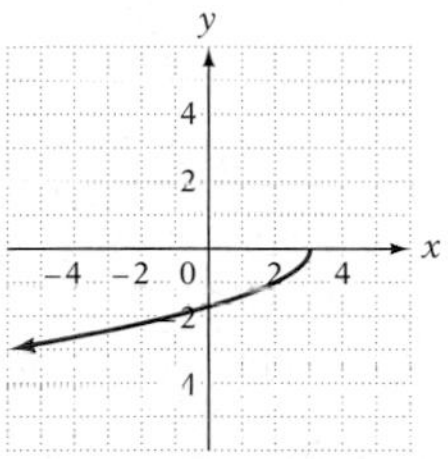

[11.3A]

domain: $\{x \mid x \leq 3\}$
range: $\{y \mid y \leq 0\}$

6. Graph $f(x) = \left|\frac{1}{2}x\right| - 2$. State the domain and range.

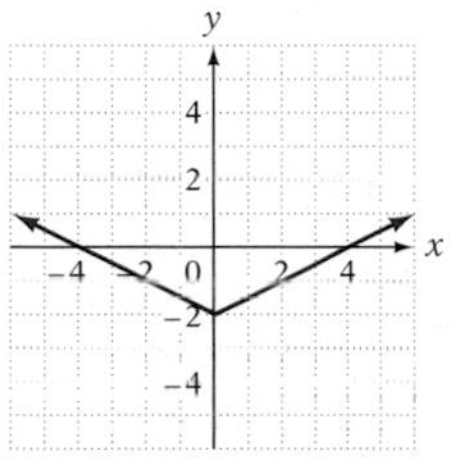

[11.3A]

domain: $\{x \mid x \in \text{real numbers}\}$
range: $\{y \mid y \geq -2\}$

7. Graph $f(x) = x^3 - 3x + 2$. State the domain and range.

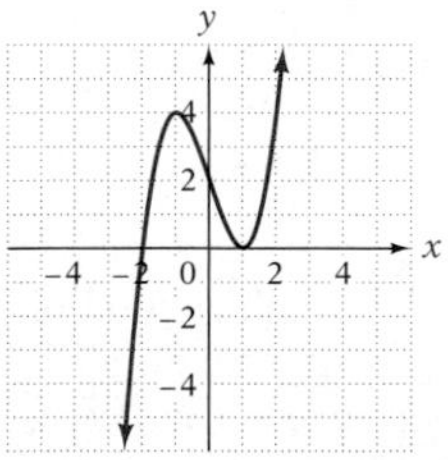

[11.3A]

domain: $\{x \mid x \in \text{real numbers}\}$
range: $\{y \mid y \in \text{real numbers}\}$

8. Determine whether the graph is the graph of a 1–1 function.

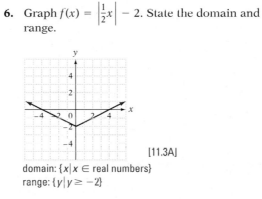

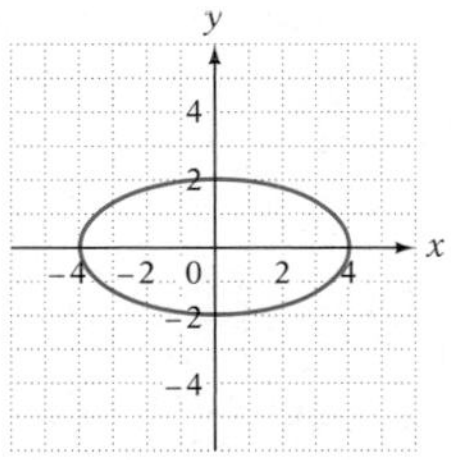

[11.5A]

no

644

9. Given $f(x) = x^2 + 2x - 3$ and $g(x) = x^3 - 1$, find $(f - g)(2)$.
-2 [11.4A]

10. Given $f(x) = 4x - 5$ and $g(x) = x^2 + 3x + 4$, find $\left(\frac{f}{g}\right)(-2)$.
$-\frac{13}{2}$ [11.4A]

11. Given $f(x) = 4x + 2$ and $g(x) = \frac{x}{x+1}$, find $f[g(3)]$.
5 [11.4B]

12. Given $f(x) = 2x^2 - 7$ and $g(x) = x - 1$, find $f[g(x)]$.
$2x^2 - 4x - 5$ [11.4B]

13. Find the inverse of the function $\{(2, 6), (3, 5), (4, 4), (5, 3)\}$.
$\{(6, 2), (5, 3), (4, 4), (3, 5)\}$ [11.5B]

14. Find the inverse of the function $f(x) = \frac{1}{4}x - 4$.
$f^{-1}(x) = 4x + 16$ [11.5B]

15. Sketch a graph of $x = y^2 - y - 2$.

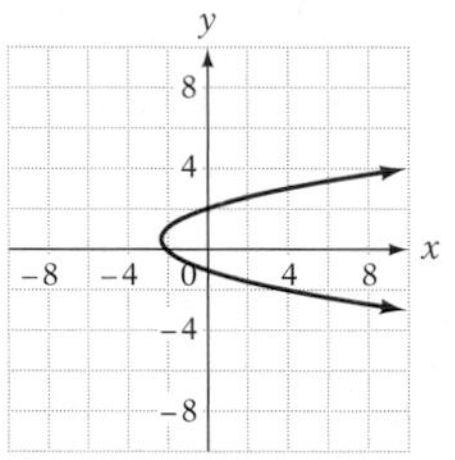

[11.6A]

16. Sketch a graph of $(x - 2)^2 + (y + 1)^2 = 9$.

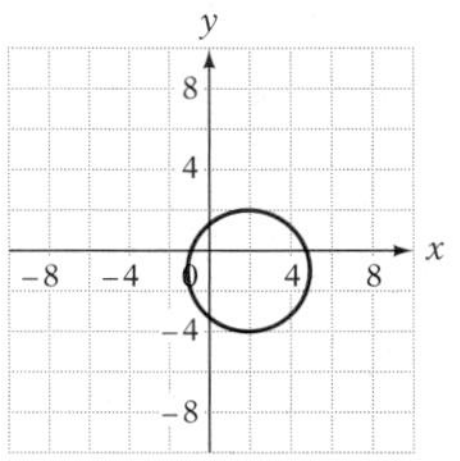

[11.6B]

17. Sketch a graph of $\frac{x^2}{9} - \frac{y^2}{4} = 1$.

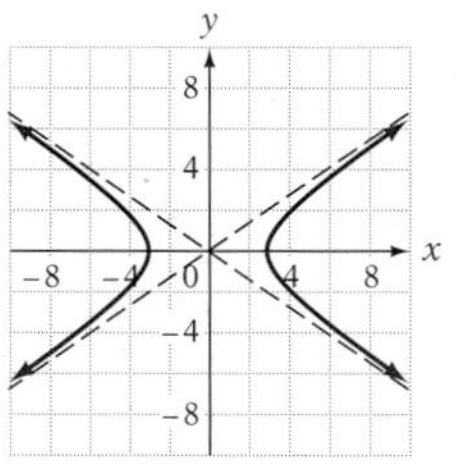

[11.6D]

18. Sketch a graph of $\frac{x^2}{16} + \frac{y^2}{4} = 1$.

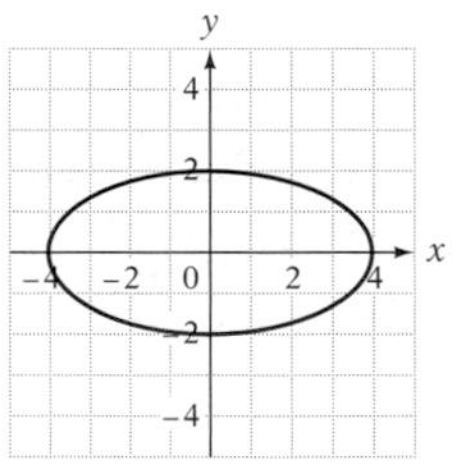

[11.6C]

19. Find the equation of the circle with radius 4 and center $(-3, -3)$.
$(x + 3)^2 + (y + 3)^2 = 16$ [11.6B]

20. The perimeter of a rectangle is 200 cm. What dimensions would give the rectangle a maximum area? What is the maximum area?
length: 50 cm; width: 50 cm;
maximum area: 2500 cm^2 [11.2D]

Chapter 12 Exponential and Logarithmic Functions

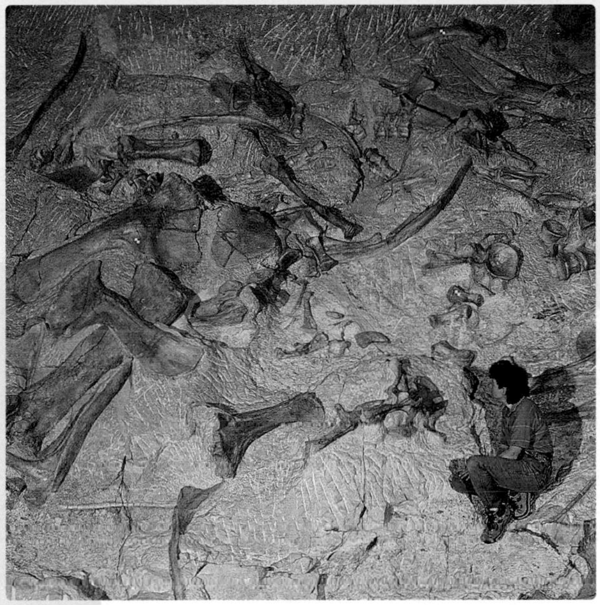

This researcher is examining sauropod bones in Jurassic sediments, hoping to determine the age of these long-necked dinosaurs' bones. Archaeologists use the carbon-dating method, which involves an exponential function, to calculate just how old the bones are. Instruments are used to detect the amount of carbon-14 left in an object. Since carbon-14 occurs naturally in living things and gradually decays after death, the measure of the amount of carbon-14 remaining reveals the object's age. The process of using carbon dating is illustrated in the **Example on page 680**.

Objectives

Section 12.1

- **A** To evaluate an exponential function
- **B** To graph an exponential function

Section 12.2

- **A** To write equivalent exponential and logarithmic equations
- **B** To use the Properties of Logarithms
- **C** To use the Change-of-Base Formula

Section 12.3

- **A** To graph a logarithmic function

Section 12.4

- **A** To solve an exponential equation
- **B** To solve a logarithmic equation

Section 12.5

- **A** To solve application problems

WEB Need help? For on-line student resources, such as section quizzes, visit this textbook's web site at **college.hmco.com/students.**

Prep Test

1. Simplify: 3^{-2}
$\frac{1}{9}$ [6.1B]

2. Simplify: $\left(\frac{1}{2}\right)^{-4}$
16 [6.1B]

3. Complete: $\frac{1}{8} = 2^?$
−3 [6.1B]

4. Evaluate $f(x) = x^4 + x^3$ for $x = -1$ and $x = 3$.
0; 108 [4.2A]

5. Solve: $3x + 7 = x - 5$
−6 [2.2B]

6. Solve: $16 = x^2 - 6x$
−2, 8 [10.1A]

7. Evaluate $A(1 + i)^n$ for $A = 5000$, $i = 0.04$, and $n = 6$. Round to the nearest hundredth.
6326.60 [1.4A]

8. Graph: $f(x) = x^2 - 1$

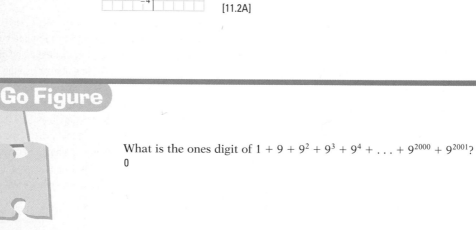

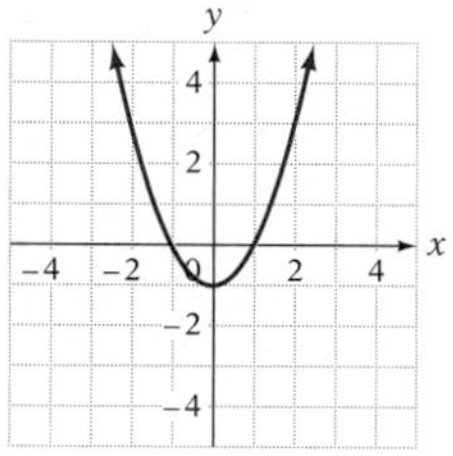

[11.2A]

Go Figure

What is the ones digit of $1 + 9 + 9^2 + 9^3 + 9^4 + \ldots + 9^{2000} + 9^{2001}$?
0

12.1 Exponential Functions

Objective A **To evaluate an exponential function**

VIDEO & DVD · CD TUTOR · WEB · SSM

The growth of a \$500 savings account that earns 5% annual interest compounded daily is shown in the graph at the right. In 14 years, the savings account contains approximately \$1000, twice the initial amount. The growth of this savings account is an example of an exponential function.

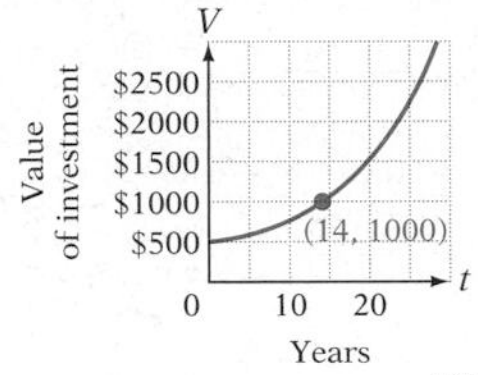

The pressure of the atmosphere at a certain height is shown in the graph at the right. This is another example of an exponential function. From the graph, we read that the air pressure is approximately 6.5 lb/in^2 at an altitude of 20,000 ft.

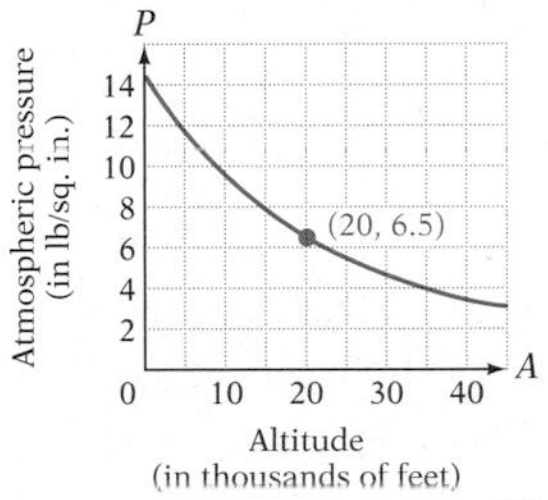

TAKE NOTE
It is important to distinguish between $F(x) = 2^x$ and $P(x) = x^2$. The first is an exponential function; the second is a polynomial function. Exponential functions are characterized by a constant base and a variable exponent. Polynomial functions have a variable base and a constant exponent.

> **Definition of an Exponential Function**
>
> The **exponential function** with base b is defined by
>
> $$f(x) = b^x$$
>
> where $b > 0$, $b \neq 1$, and x is any real number.

In the definition of an exponential function, b, the base, is required to be positive. If the base were a negative number, the value of the function would be a complex number for some values of x. For instance, the value of $f(x) = (-4)^x$ when $x = \frac{1}{2}$ is $f\left(\frac{1}{2}\right) = (-4)^{1/2} = \sqrt{-4} = 2i$. To avoid complex number values of a function, the base of the exponential function is a positive number.

⇒ Evaluate $f(x) = 2^x$ at $x = 3$ and $x = -2$.

$f(3) = 2^3 = 8$ • Substitute 3 for x and simplify.

$f(-2) = 2^{-2} = \dfrac{1}{2^2} = \dfrac{1}{4}$ • Substitute −2 for x and simplify.

To evaluate an exponential expression for an irrational number such as $\sqrt{2}$, we obtain an approximation to the value of the function by approximating the irrational number. For instance, the value of $f(x) = 4^x$ when $x = \sqrt{2}$ can be approximated by using an approximation of $\sqrt{2}$.

$$f(\sqrt{2}) = 4^{\sqrt{2}} \approx 4^{1.4142} \approx 7.1029$$

Objective 12.1A

New Vocabulary
exponential function
natural exponential function

Vocabulary to Review
irrational number

New Symbols
e

Discuss the Concepts

1. How does an exponential function differ from a polynomial function?
2. Why are the conditions $b > 0$, $b \neq 1$ given for $f(x) = b^x$?
3. Is it possible to express e as the quotient of two integers? Why or why not?

In-Class Examples (Objective 12.1A)

1. Evaluate $f(x) = 3^{x+1}$ at $x = 2$ and $x = -3$. 27; $\dfrac{1}{9}$
2. Given $f(x) = \left(\dfrac{1}{3}\right)^{2x}$, evaluate $f(0)$ and $f\left(\dfrac{1}{2}\right)$.

   1; $\dfrac{1}{3}$

Concept Check

1. For the functions $g(x) = 3^x$ and $h(x) = 4^x$, which function has the greater values when
   a. $x > 0$? $h(x)$
   b. $x < 0$? $g(x)$
2. Between what two consecutive integers are some values of the function $f(x) = x^3$ greater than the corresponding values of the function $P(x) = 3^x$? 2 and 3

Optional Student Activity

Suppose a culture of bacteria starts with 30 bacteria and the population grows according to the function $f(x) = (30)2^x$, where x is the time in hours.

a. What is the rate of change between $x = 1$ and $x = 2$? 60 bacteria per hour
b. What is the rate of change between $x = 2$ and $x = 3$? 120 bacteria per hour
c. What is the rate of change between $x = 3$ and $x = 4$? 240 bacteria per hour
d. Express the rate of change between $x = n$ and $x = n + 1$ as an exponential function. $f(n) = (60)2^{n-1}$

Instructor Note

If time permits, show students some of the remarkable relationships that exist among i, e, and π. For instance, $e^{\pi i} = -1$ and $i^{-1} = e^{\pi/2}$.

TAKE NOTE
The natural exponential function is an extremely important function. It is used extensively in applied problems in virtually all disciplines, from archaeology to zoology. Leonhard Euler (1707–1783) was the first to use the letter e as the base of the natural exponential function.

Because $f(x) = b^x$ $(b > 0, b \neq 1)$ can be evaluated at both rational and irrational numbers, the domain of f is all real numbers. And because $b^x > 0$ for all values of x, the range of f is the positive real numbers.

A frequently used base in applications of exponential functions is an irrational number designated by e. The number e is approximately 2.71828183. It is an irrational number, so it has a nonterminating, nonrepeating decimal representation.

Natural Exponential Function

The function defined by $f(x) = e^x$ is called the **natural exponential function.**

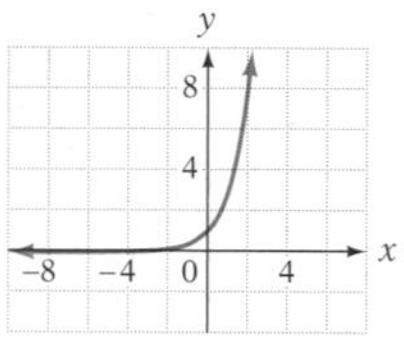

The e^x key on a calculator can be used to evaluate the natural exponential function. The graph of $y = e^x$ is shown at the left.

Example 1

Evaluate $f(x) = \left(\frac{1}{2}\right)^x$ at $x = 2$ and $x = -3$.

Solution

$$f(x) = \left(\frac{1}{2}\right)^x$$
$$f(2) = \left(\frac{1}{2}\right)^2 = \frac{1}{4}$$
$$f(-3) = \left(\frac{1}{2}\right)^{-3} = 2^3 = 8$$

You Try It 1

Evaluate $f(x) = \left(\frac{2}{3}\right)^x$ at $x = 3$ and $x = -2$.

Your solution

$$f(3) = \frac{8}{27}$$
$$f(-2) = \frac{9}{4}$$

Example 2

Evaluate $f(x) = 2^{3x-1}$ at $x = 1$ and $x = -1$.

Solution

$$f(x) = 2^{3x-1}$$
$$f(1) = 2^{3(1)-1} = 2^2 = 4$$
$$f(-1) = 2^{3(-1)-1} = 2^{-4} = \frac{1}{2^4} = \frac{1}{16}$$

You Try It 2

Evaluate $f(x) = 2^{2x+1}$ at $x = 0$ and $x = -2$.

Your solution

$$f(0) = 2$$
$$f(-2) = \frac{1}{8}$$

Example 3

Evaluate $f(x) = e^{2x}$ at $x = 1$ and $x = -1$. Round to the nearest ten-thousandth.

Solution

$$f(x) = e^{2x}$$
$$f(1) = e^{2 \cdot 1} = e^2 \approx 7.3891$$
$$f(-1) = e^{2(-1)} = e^{-2} \approx 0.1353$$

You Try It 3

Evaluate $f(x) = e^{2x-1}$ at $x = 2$ and $x = -2$. Round to the nearest ten-thousandth.

Your solution

$$f(2) \approx 20.0855$$
$$f(-2) \approx 0.0067$$

Solutions on p. S35

Objective B To graph an exponential function

VIDEO & DVD · CD TUTOR · WEB · SSM

Some properties of an exponential function can be seen in its graph.

➡ Graph $f(x) = 2^x$.

Think of this as the equation $y = 2^x$.

Choose values of x and find the corresponding values of y.

Graph the ordered pairs on a rectangular coordinate system.

Connect the points with a smooth curve.

x	$f(x) = y$
-2	$2^{-2} = \frac{1}{4}$
-1	$2^{-1} = \frac{1}{2}$
0	$2^0 = 1$
1	$2^1 = 2$
2	$2^2 = 4$
3	$2^3 = 8$

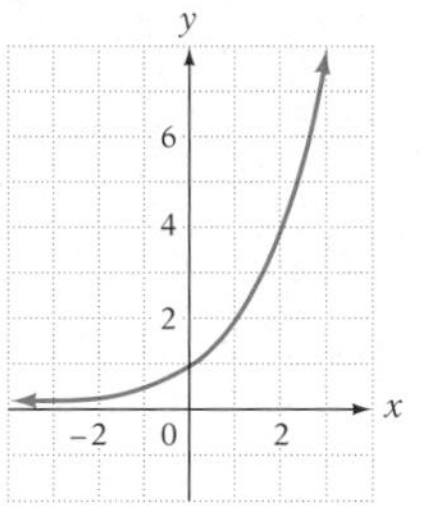

Note that any vertical line would intersect the graph at only one point. Therefore, by the vertical-line test, the graph of $f(x) = 2^x$ is the graph of a function. Also note that any horizontal line would intersect the graph at only one point. Therefore, the graph of $f(x) = 2^x$ is the graph of a one-to-one function.

➡ Graph $f(x) = \left(\frac{1}{2}\right)^x$.

Think of this as the equation $y = \left(\frac{1}{2}\right)^x$.

Choose values of x and find the corresponding values of y.

Graph the ordered pairs on a rectangular coordinate system.

Connect the points with a smooth curve.

x	$f(x) = y$
-3	$\left(\frac{1}{2}\right)^{-3} = 8$
-2	$\left(\frac{1}{2}\right)^{-2} = 4$
-1	$\left(\frac{1}{2}\right)^{-1} = 2$
0	$\left(\frac{1}{2}\right)^{0} = 1$
1	$\left(\frac{1}{2}\right)^{1} = \frac{1}{2}$
2	$\left(\frac{1}{2}\right)^{2} = \frac{1}{4}$

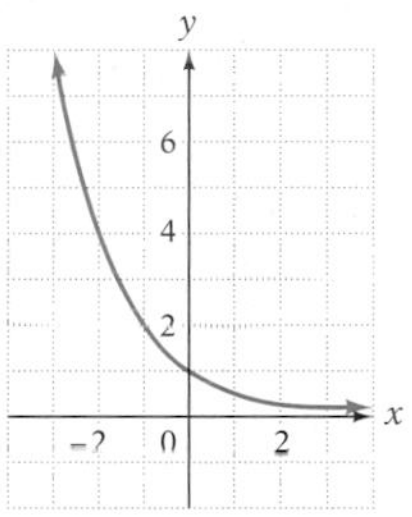

Applying the vertical-line and horizontal-line tests reveals that the graph of $f(x) = \left(\frac{1}{2}\right)^x$ is also the graph of a one-to-one function.

➡ Graph $f(x) = 2^{-x}$.

Think of this as the equation $y = 2^{-x}$.

Choose values of x and find the corresponding values of y.

Graph the ordered pairs on a rectangular coordinate system.

Connect the points with a smooth curve.

x	y
-3	8
-2	4
-1	2
0	1
1	$\frac{1}{2}$
2	$\frac{1}{4}$

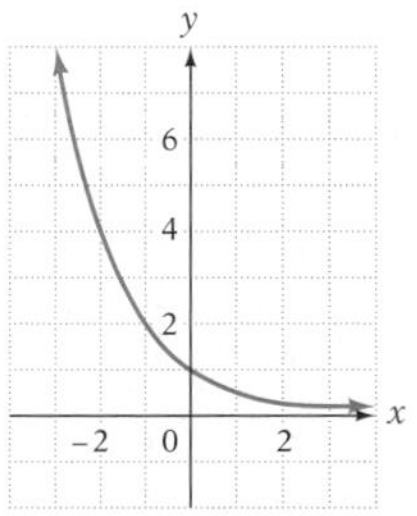

Note that because $2^{-x} = (2^{-1})^x = \left(\frac{1}{2}\right)^x$, the graphs of $f(x) = 2^{-x}$ and $f(x) = \left(\frac{1}{2}\right)^x$ are the same.

Objective 12.1B

Vocabulary to Review

vertical-line test
horizontal-line test
1–1 function

Instructor Note

Graph $g(x) = x^2$ so that students can see the difference between the graph of $g(x) = x^2$ and the graph of $f(x) = 2^x$.

Discuss the Concepts

Determine whether the statement is always true, sometimes true, or never true.

1. The domain of an exponential function $f(x) = b^x$, $b > 0$, $b \neq 1$, is the set of positive numbers. Never true
2. An exponential function $f(x) = b^x$, $b > 0$, $b \neq 1$, is a 1–1 function. Always true
3. The graph of an exponential function $f(x) = b^x$, $b > 0$, $b \neq 1$, passes through the point $(0, 0)$. Never true
4. For the function $f(x) = b^x$, $b > 0$, $b \neq 1$, the base b is a positive integer. Sometimes true
5. An exponential function $f(x) = b^x$, $b > 0$, $b \neq 1$, has two x-intercepts. Never true

Concept Check

1. Graph $f(x) = 2^{0.5x-1}$. For what values in the domain are the corresponding values in the range less than 0? None
2. Graph $f(x) = e^x - 1$. What are the x- and y-intercepts of the graph of the function? x-intercept: $(0, 0)$; y-intercept: $(0, 0)$
3. What is the y-intercept of the graph of $f(x) = e^{-2x} - 4$? $(0, -3)$

In-Class Examples (Objective 12.1B)

1. Graph $f(x) = 3^{x-1}$.

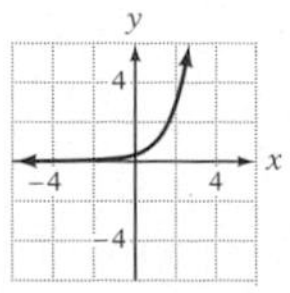

2. Graph $f(x) = \left(\frac{1}{2}\right)^x + 2$.

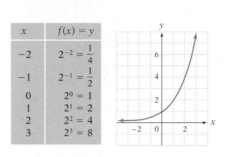

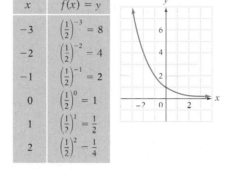

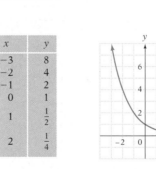

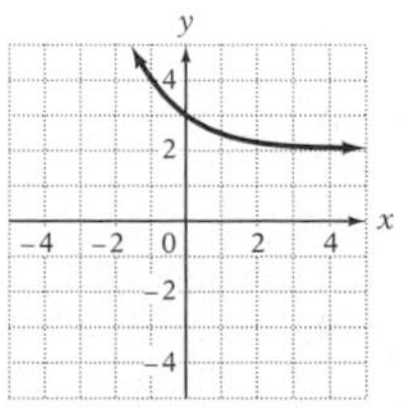

Optional Student Activity

For each of the following functions, determine whether the values of y increase or decrease as x increases. Then try to state a rule that summarizes your findings.

a. $f(x) = 0.5^x$
b. $g(x) = 4^x$
c. $h(x) = \left(\frac{3}{4}\right)^x$
d. $F(x) = 2^x$
e. $G(x) = \left(\frac{1}{3}\right)^x$
f. $H(x) = 1.5^x$

Values of y increase as x increases for functions b, d, and f. Values of y decrease as x increases for functions a, c, and e. Values of y increase as x increases when the base of the exponential function is greater than 1. Values of y decrease when the base is between 0 and 1.

Instructor Note

After you discuss the graph of the exponential function, you might give students the extra-credit problem of graphing $f(x) = x^x$. A discussion of the domain of this function may help students gain a better understanding of domain.

Example 4
Graph: $f(x) = 3^{\frac{1}{2}x-1}$

Solution

x	y
-2	$\frac{1}{9}$
0	$\frac{1}{3}$
2	1
4	3

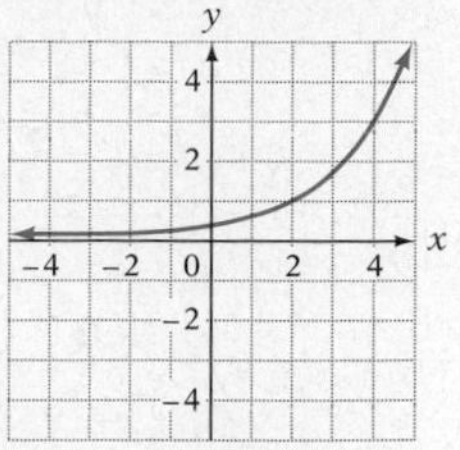

You Try It 4
Graph: $f(x) = 2^{-\frac{1}{2}x}$

Your solution

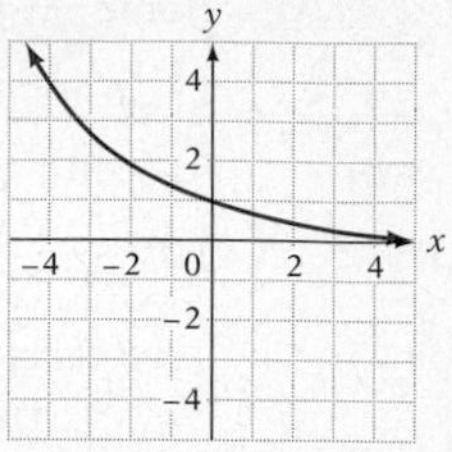

Example 5
Graph: $f(x) = 2^x - 1$

Solution

x	y
-2	$-\frac{3}{4}$
-1	$-\frac{1}{2}$
0	0
1	1
2	3
3	7

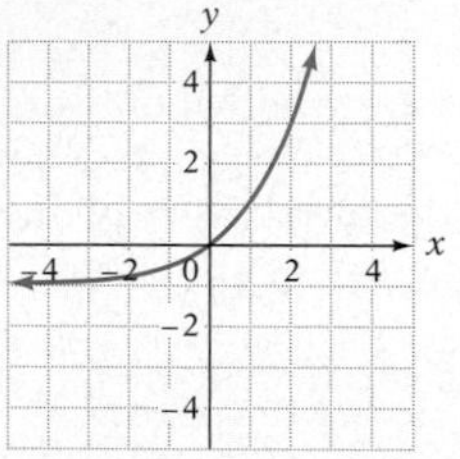

You Try It 5
Graph: $f(x) = 2^x + 1$

Your solution

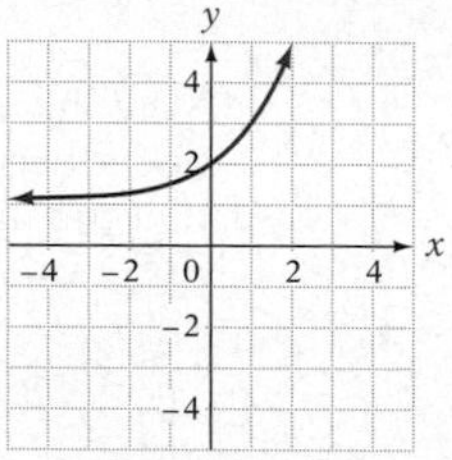

Example 6
Graph: $f(x) = \left(\frac{1}{3}\right)^x - 2$

Solution

x	y
-2	7
-1	1
0	-1
1	$-\frac{5}{3}$
2	$-\frac{17}{9}$

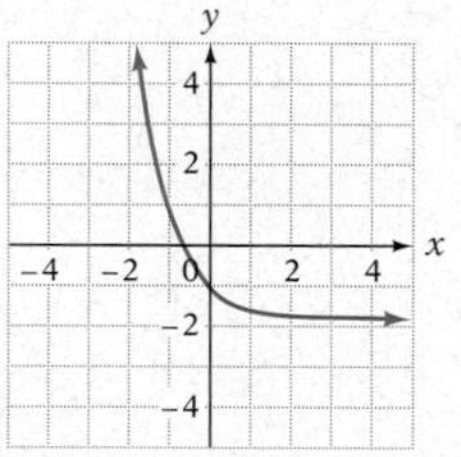

You Try It 6
Graph: $f(x) = 2^{-x} + 2$

Your solution

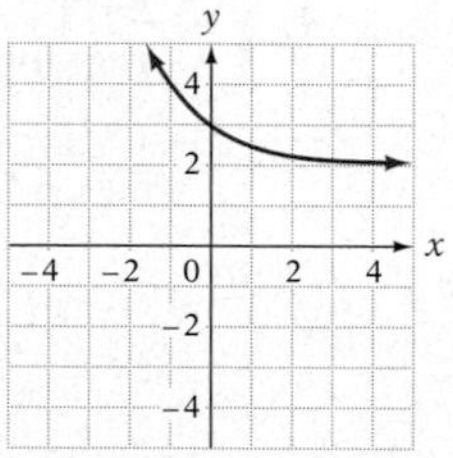

Example 7
Graph: $f(x) = 2^{-\frac{1}{2}x} - 1$

Solution

x	y
-6	7
-4	3
-2	1
0	0
2	$-\frac{1}{2}$
4	$-\frac{3}{4}$

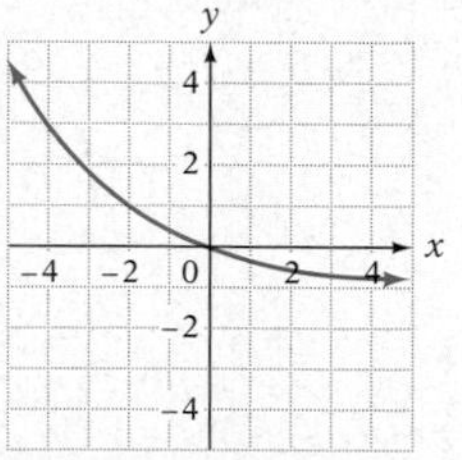

You Try It 7
Graph: $f(x) = \left(\frac{1}{2}\right)^{-\frac{1}{2}x} + 2$

Your solution

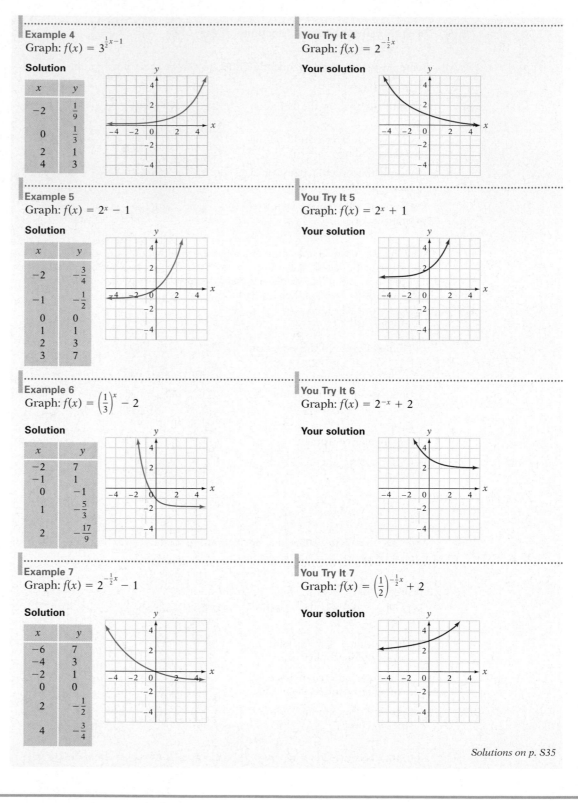

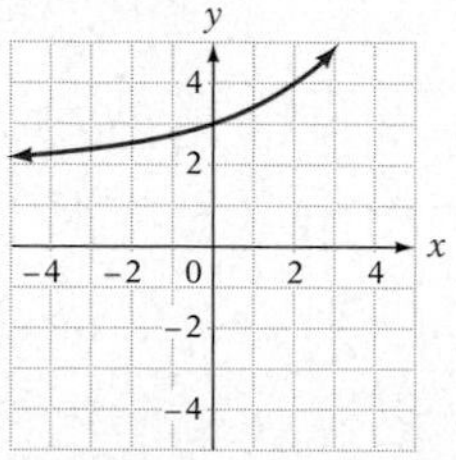

Solutions on p. S35

655

12.1 Exercises

Objective A

1. What is an exponential function?

2. What is the natural exponential function?

3. Which of the following cannot be the base of an exponential function?
   a. 7 **b.** $\frac{1}{4}$ **c.** -5 **d.** 0.01 c

4. Which of the following cannot be the base of an exponential function?
   a. 0.9 **b.** 476 **c.** 8 **d.** $-\frac{1}{2}$ d

5. Given $f(x) = 3^x$, evaluate:
   a. $f(2)$ 9 **b.** $f(0)$ 1 **c.** $f(-2)$ $\frac{1}{9}$

6. Given $H(x) = 2^x$, evaluate:
   a. $H(-3)$ $\frac{1}{8}$ **b.** $H(0)$ 1 **c.** $H(2)$ 4

7. Given $g(x) = 2^{x+1}$, evaluate:
   a. $g(3)$ 16 **b.** $g(1)$ 4 **c.** $g(-3)$ $\frac{1}{4}$

8. Given $F(x) = 3^{x-2}$, evaluate:
   a. $F(-4)$ $\frac{1}{729}$ **b.** $F(-1)$ $\frac{1}{27}$ **c.** $F(0)$ $\frac{1}{9}$

9. Given $P(x) = \left(\frac{1}{2}\right)^{2x}$, evaluate:
   a. $P(0)$ 1 **b.** $P\left(\frac{3}{2}\right)$ $\frac{1}{8}$ **c.** $P(-2)$ 16

10. Given $R(t) = \left(\frac{1}{3}\right)^{3t}$, evaluate:
    a. $R\left(-\frac{1}{3}\right)$ 3 **b.** $R(1)$ $\frac{1}{27}$ **c.** $R(-2)$ 729

11. Given $G(x) = e^{x/2}$, evaluate the following. Round to the nearest ten-thousandth.
    a. $G(4)$ 7.3891 **b.** $G(-2)$ 0.3679 **c.** $G\left(\frac{1}{2}\right)$ 1.2840

12. Given $f(x) = e^{2x}$, evaluate the following. Round to the nearest ten-thousandth.
    a. $f(-2)$ 0.0183 **b.** $f\left(-\frac{2}{3}\right)$ 0.2636 **c.** $f(2)$ 54.5982

13. Given $H(r) = e^{-r+3}$, evaluate the following. Round to the nearest ten-thousandth.
    a. $H(-1)$ 54.5982 **b.** $H(3)$ 1 **c.** $H(5)$ 0.1353

14. Given $P(t) = e^{-\frac{1}{2}t}$, evaluate the following. Round to the nearest ten-thousandth.
    a. $P(-3)$ 4.4817 **b.** $P(4)$ 0.1353 **c.** $P\left(\frac{1}{2}\right)$ 0.7788

Section 12.1

Suggested Assignment

Exercises 1–35, odds
More challenging problems:
Exercises 37–48
(A graphing calculator is required.)

Answers to Writing Exercises

1. An exponential function with base b is defined by $f(x) = b^x$, $b > 0$, $b \neq 1$, and x is any real number.
2. The natural exponential function is the function defined by $f(x) = e^x$, where e is an irrational number approximately equal to 2.71828183.

Quick Quiz (Objective 12.1A)

1. Evaluate $f(x) = 2^x$ at $x = 3$ and $x = -2$. 8; $\frac{1}{4}$

2. Given $f(x) = \left(\frac{1}{2}\right)^{x-1}$, evaluate $f(1)$ and $f(3)$. 1; $\frac{1}{4}$

15. Given $F(x) = 2^{x^2}$, evaluate:

a. $F(2)$ 16 **b.** $F(-2)$ 16 **c.** $F\left(\frac{3}{4}\right)$ 1.4768

16. Given $Q(x) = 2^{-x^2}$, evaluate:

a. $Q(3)$ $\frac{1}{512}$ **b.** $Q(-1)$ $\frac{1}{2}$ **c.** $Q(-2)$ $\frac{1}{16}$

17. Given $f(x) = e^{-x^2/2}$, evaluate the following. Round to the nearest ten-thousandth.

a. $f(-2)$ 0.1353 **b.** $f(2)$ 0.1353 **c.** $f(-3)$ 0.0111

18. Given $f(x) = e^{-2x} + 1$, evaluate the following. Round to the nearest ten-thousandth.

a. $f(-1)$ 8.3891 **b.** $f(3)$ 1.0025 **c.** $f(-2)$ 55.5982

Objective B

Graph.

19. $f(x) = 3^x$

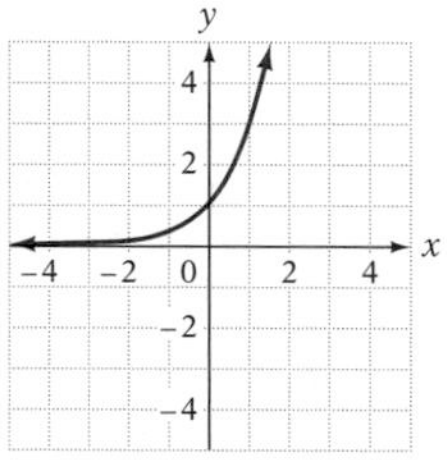

20. $f(x) = 3^{-x}$

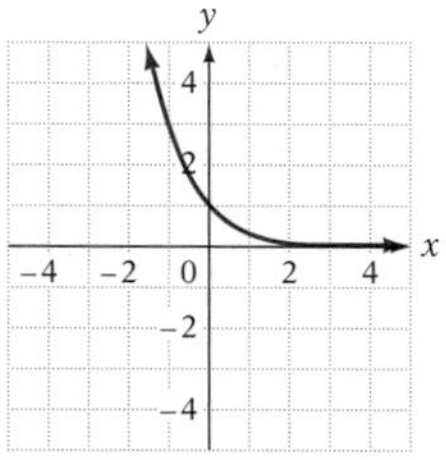

21. $f(x) = 2^{x+1}$

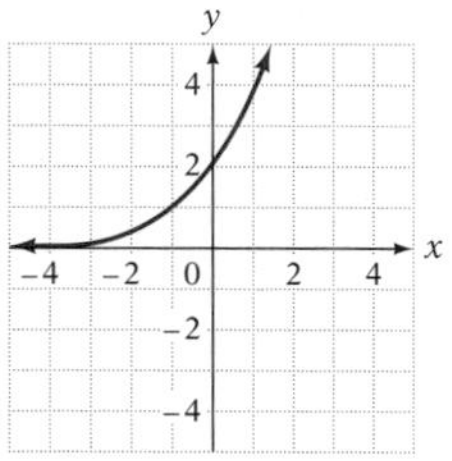

22. $f(x) = 2^{x-1}$

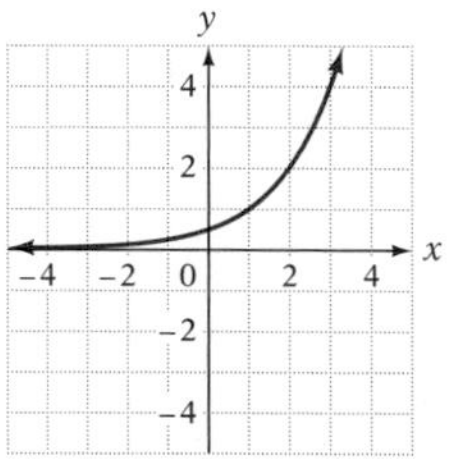

23. $f(x) = \left(\frac{1}{3}\right)^x$

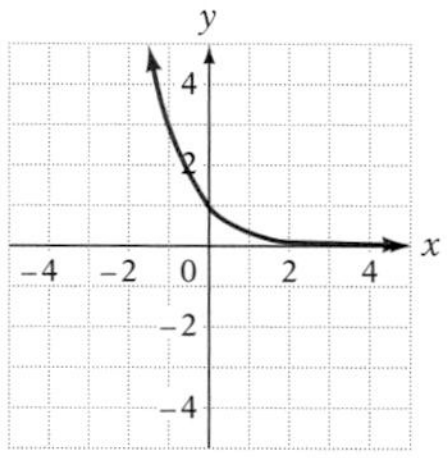

24. $f(x) = \left(\frac{2}{3}\right)^x$

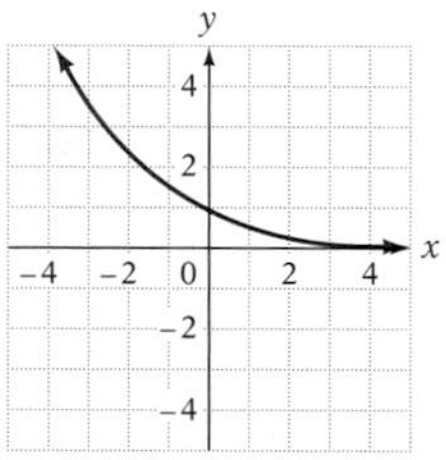

25. $f(x) = 2^{-x} + 1$

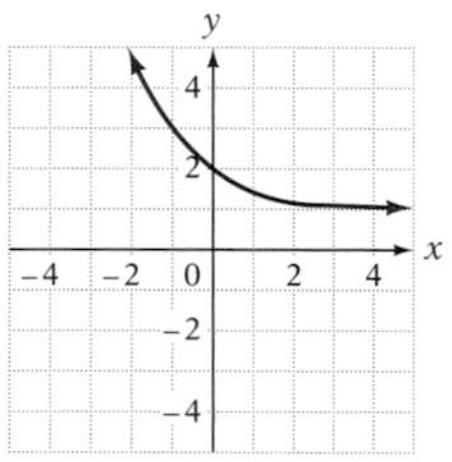

26. $f(x) = 2^x - 3$

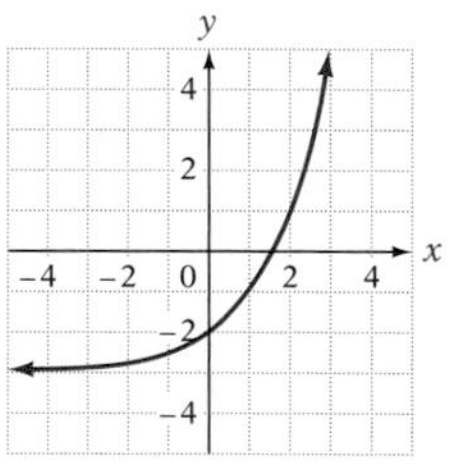

27. $f(x) = \left(\frac{1}{3}\right)^{-x}$

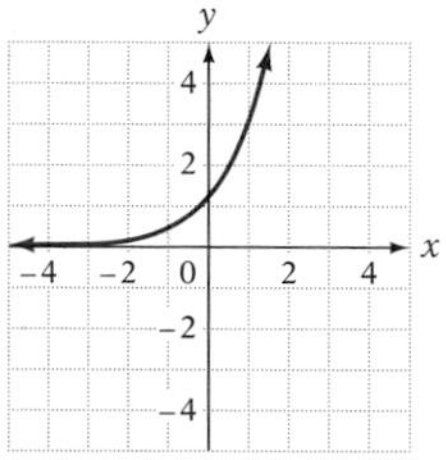

28. $f(x) = \left(\frac{3}{2}\right)^{-x}$

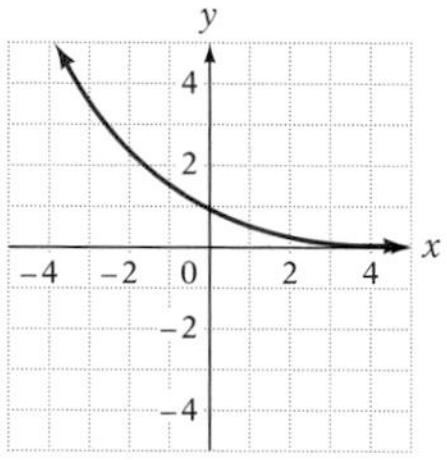

29. $f(x) = \left(\frac{1}{2}\right)^{-x} + 2$

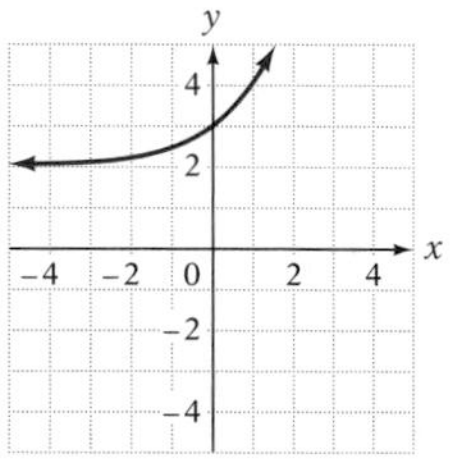

30. $f(x) = \left(\frac{1}{2}\right)^{x} - 1$

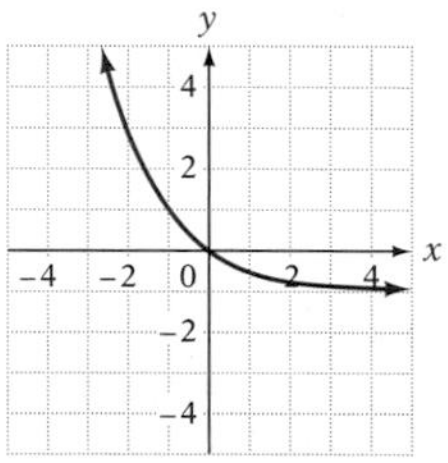

Quick Quiz (Objective 12.1B)

1. Graph $f(x) = 3^{x+1}$.

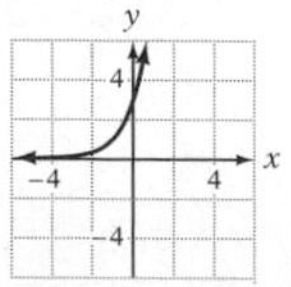

2. Graph $f(x) = 2^{-x} + 2$.

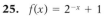

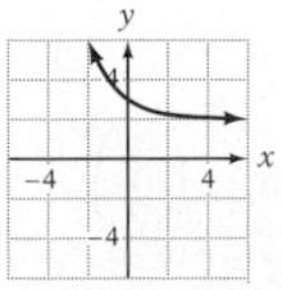

657

31. Which of the following functions have the same graph?

a. $f(x) = 3^x$ **b.** $f(x) = \left(\frac{1}{3}\right)^x$ **c.** $f(x) = x^3$ **d.** $f(x) = 3^{-x}$

b and d

32. Which of the following functions have the same graph?

a. $f(x) = x^4$ **b.** $f(x) = 4^{-x}$ **c.** $f(x) = 4^x$ **d.** $f(x) = \left(\frac{1}{4}\right)^x$

b and d

33. Graph $f(x) = 3^x$ and $f(x) = 3^{-x}$ and find the point of intersection of the two graphs.

(0, 1)

34. Graph $f(x) = 2^{x+1}$ and $f(x) = 2^{-x+1}$ and find the point of intersection of the two graphs.

(0, 2)

35. Graph $f(x) = \left(\frac{1}{3}\right)^x$. What are the *x*- and *y*-intercepts of the graph of the function?

no *x*-intercept; *y*-intercept: (0, 1)

36. Graph $f(x) = \left(\frac{1}{3}\right)^{-x}$. What are the *x*- and *y*-intercepts of the graph of the function?

no *x*-intercept; *y*-intercept: (0, 1)

APPLYING THE CONCEPTS

Use a graphing calculator to graph the function.

37. $P(x) = (\sqrt{3})^x$

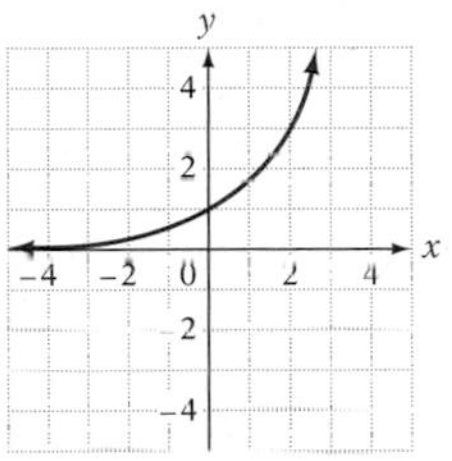

38. $F(x) = (\sqrt{5})^x$

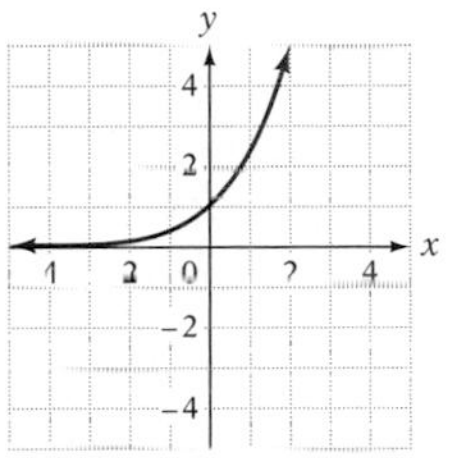

39. $Q(x) = (\sqrt{3})^{-x}$

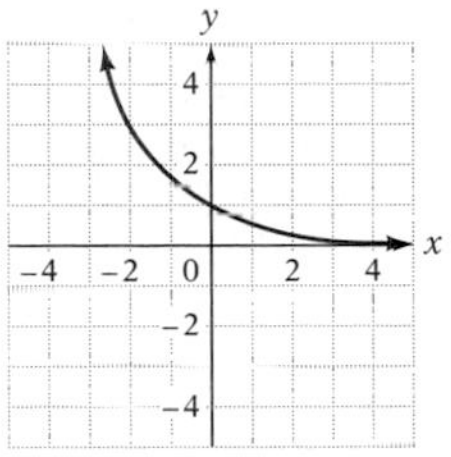

40. $A(x) = (\sqrt{2})^{-x}$

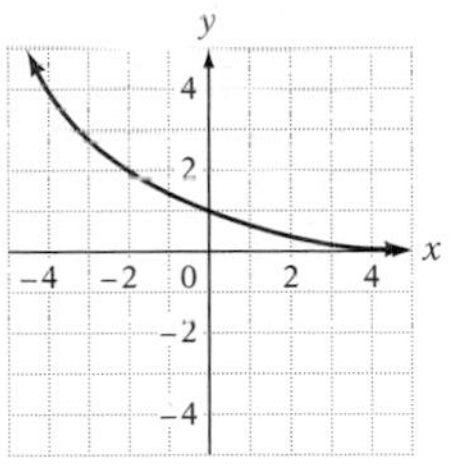

41. $f(x) = \pi^x$

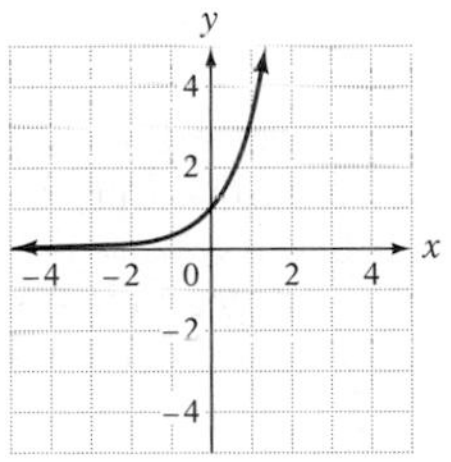

42. $g(x) = \pi^{-x}$

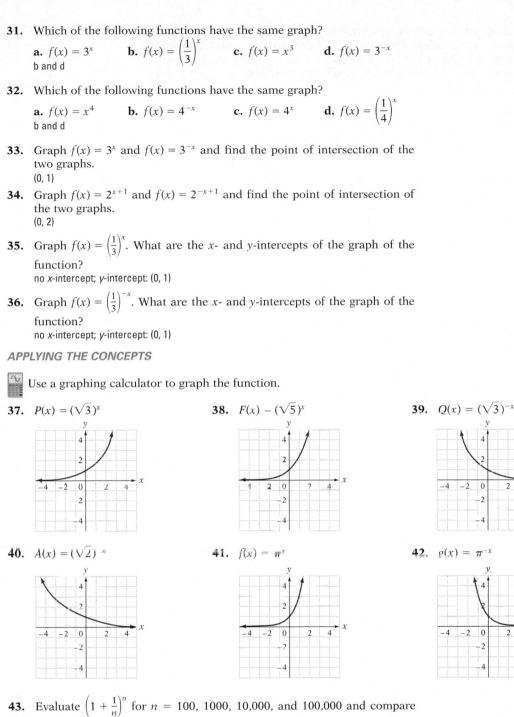

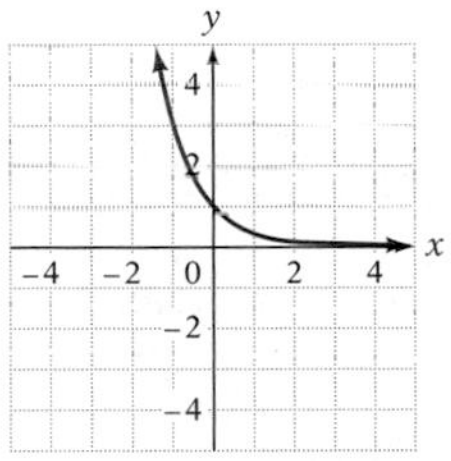

43. Evaluate $\left(1 + \frac{1}{n}\right)^n$ for $n = 100$, 1000, 10,000, and 100,000 and compare the results with the value of *e*, the base of the natural exponential function. On the basis of your evaluation, complete the following sentence: As *n* increases, $\left(1 + \frac{1}{n}\right)^n$ becomes closer to ___*e*___.

Answers to Writing Exercises

45b. The point (9, 1.303) means that in the year 2002, the Chinese population was approximately 1.303 billion.

46b. The point (8, 1.01) means that in the year 2001, the population of India was approximately 1.01 billion.

47b. The point (2, 27.7) means that after 2 s, the object is falling at a speed of 27.7 ft/s.

48b. The point (4, 55.3) means that after 4 s, the object is falling at a speed of 55.3 ft/s.

44. Evaluate $(1 + x)^{1/x}$ for $x = 0.1, 0.01, 0.001, 0.00001$, and 0.0000001 and compare the results with the value of e, the base of the natural exponential function. On the basis of your evaluation, complete the following sentence: As x gets closer to 0, $(1 + x)^{1/x}$ becomes closer to ___e___.

45. According to population studies, the population of China can be approximated by the equation $P(t) = 1.17(1.012)^t$, where $t = 0$ corresponds to 1993 and $P(t)$ is the population, in billions, of China in t years.

a. Graph this equation. *Suggestion:* Use Xmin = 0, Xmax = 19, Ymin = −0.5, Ymax = 1.75, and Yscl = 0.25.

b. The point whose approximate coordinates are (9, 1.303) is on this graph. Write a sentence that explains the meaning of these coordinates.

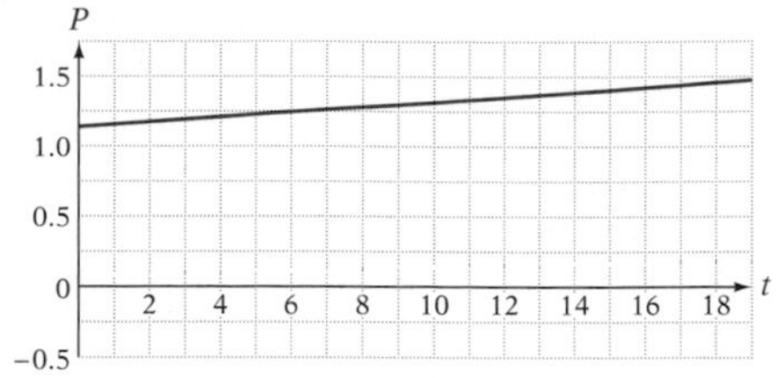

46. The growth of the population of India can be approximated by the equation $P(t) = 0.883(1.017)^t$, where $t = 0$ corresponds to 1993 and $P(t)$ is the population, in billions, of India in t years.

a. Graph this equation. *Suggestion:* Use Xmin = 0, Xmax = 20, Ymin = −0.5, Ymax = 1.75, and Yscl = 0.25.

b. The point whose approximate coordinates are (8, 1.01) is on this graph. Write a sentence that explains the meaning of these coordinates.

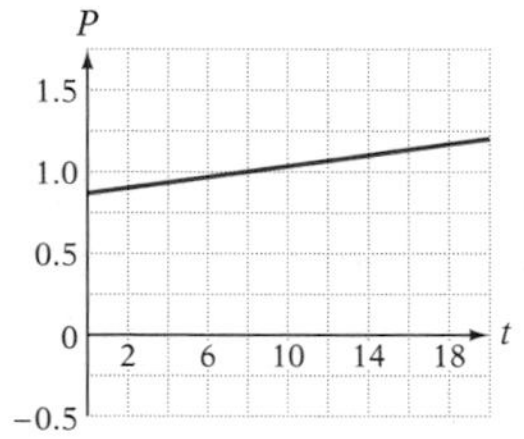

47. If air resistance is ignored, the speed v, in feet per second, of an object t seconds after it has been dropped is given by $v = 32t$. However, if air resistance is considered, then the speed depends on the mass (and on other things). For a certain mass, the speed t seconds after it has been dropped is given by $v = 32(1 - e^{-t})$.

a. Graph this equation. *Suggestion:* Use Xmin = 0, Xmax = 5.5, Ymin = 0, Ymax = 40, and Yscl = 5.

b. The point whose approximate coordinates are (2, 27.7) is on this graph. Write a sentence that explains the meaning of these coordinates.

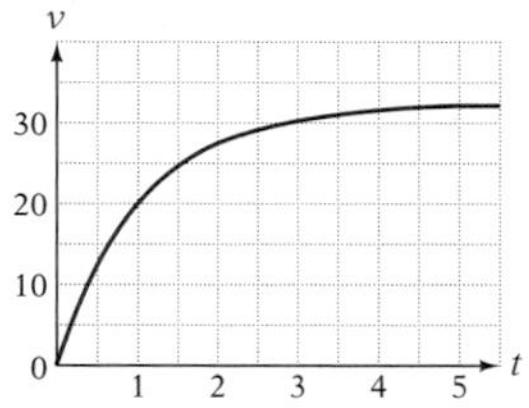

48. If air resistance is ignored, the speed v, in feet per second, of an object t seconds after it has been dropped is given by $v = 32t$. However, if air resistance is considered, then the speed depends on the mass (and on other things). For a certain mass, the speed t seconds after it has been dropped is given by $v = 64(1 - e^{-t/2})$.

a. Graph this equation. *Suggestion:* Use Xmin = 0, Xmax = 11, Ymin = 0, Ymax = 80, and Yscl = 10.

b. The point whose approximate coordinates are (4, 55.3) is on this graph. Write a sentence that explains the meaning of these coordinates.

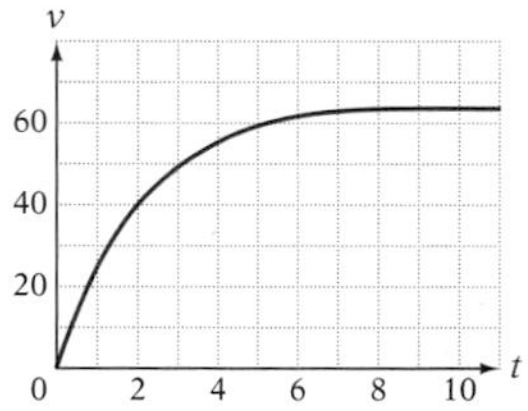

12.2 Introduction to Logarithms

Objective A **To write equivalent exponential and logarithmic equations**

VIDEO & DVD · CD TUTOR · WEB · SSM

Because the exponential function is a 1–1 function, it has an inverse function, which is called a **logarithm**. A logarithm is used to answer a question similar to the following: "If $16 = 2^y$, what is the value of y?" Because $16 = 2^4$, the logarithm, base 2, of 16 is 4. This is written as $\log_2 16 = 4$. Note that a logarithm is an exponent that solves a certain equation.

> **Definition of Logarithm**
>
> For $b > 0$, $b \neq 1$, $y = \log_b x$ is equivalent to $x = b^y$.

Read $\log_b x$ as "the logarithm of x, base b" or "log base b of x."

The table at the right shows equivalent statements written in both exponential and logarithmic form.

Exponential Form	*Logarithmic Form*
$2^4 = 16$	$\log_2 16 = 4$
$\left(\frac{2}{3}\right)^2 = \frac{4}{9}$	$\log_{2/3}\left(\frac{4}{9}\right) = 2$
$10^{-1} = 0.1$	$\log_{10}(0.1) = -1$

⇒ Write $\log_3 81 = 4$ in exponential form.

$\log_3 81 = 4$ is equivalent to $3^4 = 81$.

⇒ Write $10^{-2} = 0.01$ in logarithmic form.

$10^{-2} = 0.01$ is equivalent to $\log_{10}(0.01) = -2$.

The 1–1 property of exponential functions can be used to evaluate some logarithms.

> **1–1 Property of Exponential Functions**
>
> For $b > 0$, $b \neq 1$, if $b^u = b^v$, then $u = v$.

⇒ Evaluate $\log_2 8$.

$\log_2 8 = x$ • Write an equation.

$8 = 2^x$ • Write the equation in its equivalent exponential form.

$2^3 = 2^x$ • Write 8 as 2^3.

$3 = x$ • Use the 1–1 Property of Exponential Functions.

$\log_2 8 = 3$

Objective 12.2A

New Vocabulary
logarithm
antilogarithm
common logarithm
mantissa
characteristic
natural logarithm

Vocabulary to Review
1–1 function
inverse function

New Properties
1–1 Property of Exponential Functions

Instructor Note
Some students see the definition of logarithm as artificial. The following analogy may help.

Defining a logarithm as the inverse of the exponential function is similar to defining square root as the inverse of square. "If $49 = x^2$, what is x?" The answer is "the square root of 49."

Now consider the question, "If $8 = 2^x$, what is x?" The answer is "the logarithm, base 2, of 8."

This analogy can be extended to motivate the use of tables or calculators.

"If $19 = x^2$, what is x?"
"If $21 = 10^x$, what is x?"

Discuss the Concepts
Explain why each of the following is true.

1. $\ln e^x = x$
2. $\ln e = 1$
3. $e^{\ln x} = x$

In-Class Examples (Objective 12.2A)
Write the exponential expression in logarithmic form.

1. $2^3 = 8$ $\log_2 8 = 3$
2. $b^x = d$ $\log_b d = x$

Write the logarithmic expression in exponential form.

3. $\log_3 81 = 4$ $3^4 = 81$
4. $\log_b c = y$ $b^y = c$
5. Evaluate $\log_6 36$. 2
6. Evaluate $\log_{10} 1$. 0
7. Solve for x: $\log_2 x = 3$. 8
8. Solve for x: $\log_7 x = -2$. $\frac{1}{49}$

Optional Student Activity

Determine the domain of the function. Recall that the logarithm of a nonpositive number is not defined.

1. $f(x) = \log_3 (x - 4)$ $\{x \mid x > 4\}$
2. $f(x) = \log_2 (x + 2)$ $\{x \mid x > -2\}$
3. $f(x) = \ln (x^2 - 4)$ $\{x \mid x < -2 \text{ or } x > 2\}$
4. $f(x) = \ln (x^2 + 4)$ $\{x \mid x \in \text{real numbers}\}$
5. $f(x) = \log_2 x + \log_2 (x - 1)$ $\{x \mid x > 1\}$
6. $f(x) = \log_4 \dfrac{x}{x + 2}$ $\{x \mid x < -2 \text{ or } x > 0\}$

Instructor Note

The concept of antilogarithm is nothing more than a restatement of the equivalence between a logarithmic expression and an exponential expression. This relationship can be illustrated by using a calculator. To solve the equation $\log_{10} x = \frac{1}{2}$, the student uses the 10^x key on the calculator. To solve $10^x = 3$, the student uses the $\log_{10} x$ key. The same relationship holds true for the natural exponential function and logarithm.

Some students will not have a 10^x key on their calculators but will have an INV (inverse) key instead. This affords a good opportunity to reinforce the fact that the logarithm and exponential functions are inverses of each other.

➡ Solve $\log_4 x = -2$ for x.

$\log_4 x = -2$

$4^{-2} = x$ • **Write the equation in its equivalent exponential form.**

$\frac{1}{16} = x$ • **Simplify.**

The solution is $\frac{1}{16}$.

In this example, $\frac{1}{16}$ is called the antilogarithm base 4 of negative 2. In general, if $\log_b M = N$, then M is the antilogarithm base b of N. The antilogarithm of a number can be determined by rewriting the logarithmic expression in exponential form. For instance, the exponential form of $\log_5 x = 3$ is $5^3 = x$; x is the antilogarithm base 5 of 3.

Definition of Antilogarithm

If $\log_b M = N$, the **antilogarithm** base b of N is M. In exponential form, $M = b^N$.

Logarithms base 10 are called **common logarithms**. Usually the base, 10, is omitted when we write the common logarithm of a number. Therefore, $\log_{10} x$ is written $\log x$. To find the common logarithm of most numbers, a calculator is necessary. A calculator was used to find the value of log 384, shown below.

TAKE NOTE
The logarithms of most numbers are irrational numbers. Therefore, the value displayed by a calculator is an approximation.

$$\log 384 \approx 2.5843312$$

(Mantissa: .5843312; Characteristic: 2)

The decimal part of a common logarithm is called the **mantissa**; the integer part is called the **characteristic**.

When e (the base of the natural exponential function) is used as a base of a logarithm, the logarithm is referred to as the **natural logarithm** and is abbreviated $\ln x$. This is read "el en x." Using a calculator, we learn that

$$\ln 23 \approx 3.135494216$$

The integer and decimal parts of a natural logarithm do not have special names.

Example 1

Evaluate: $\log_3 \left(\frac{1}{9}\right)$

Solution

$\log_3 \left(\frac{1}{9}\right) = x$

$\frac{1}{9} = 3^x$

$3^{-2} = 3^x$

$-2 = x$

$\log_3 \left(\frac{1}{9}\right) = -2$

You Try It 1

Evaluate: $\log_4 64$

Your solution 3

Solution on p. S36

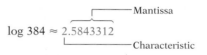

Example 2 Solve for x: $\log_5 x = 2$

Solution

$$\log_5 x = 2$$
$$5^2 = x$$
$$25 = x$$

The solution is 25.

You Try It 2 Solve for x: $\log_2 x = -4$

Your solution $\frac{1}{16}$

Example 3 Solve $\log x = -1.5$ for x. Round to the nearest ten-thousandth.

Solution

$$\log x = -1.5$$
$$10^{-1.5} = x$$
$$0.0316 \approx x$$ • **Use a calculator.**

You Try It 3 Solve $\ln x = 3$ for x. Round to the nearest ten-thousandth.

Your solution 20.0855

Solutions on p. S36

Objective B To use the Properties of Logarithms

Because a logarithm is a special kind of exponent, the Properties of Logarithms are similar to the Properties of Exponents.

The property of logarithms that states that the logarithm of the product of two numbers equals the sum of the logarithms of the two numbers is similar to the property of exponents that states that to multiply two exponential expressions with the same base, we add the exponents.

> **TAKE NOTE**
> Pay close attention to this theorem. Note, for instance, that this theorem states that $\log_3 (4 \cdot p) = \log_3 4 + \log_3 p$. It also states that $\log_5 9 + \log_5 z = \log_5 (9z)$. It does *not* state any relationship regarding the expression $\log_b (x + y)$. **This expression cannot be simplified.**

> **The Logarithm Property of the Product of Two Numbers**
> For any positive real numbers x, y, and b, $b \neq 1$, $\log_b (xy) = \log_b x + \log_b y$.

A proof of this property can be found in the Appendix.

➡ Write $\log_b (6z)$ in expanded form.

$\log_b (6z) = \log_b 6 + \log_b z$ • **Use the Logarithm Property of Products.**

➡ Write $\log_b 12 + \log_b r$ as a single logarithm.

$\log_b 12 + \log_b r = \log_b (12r)$ • **Use the Logarithm Property of Products.**

The Logarithm Property of Products can be extended to include the logarithm of the product of more than two factors. For instance,

$$\log_b (xyz) = \log_b x + \log_b y + \log_b z$$
$$\log_b (7rt) = \log_b 7 + \log_b r + \log_b t$$

Objective 12.2B

New Properties

Logarithm Property of the Product of Two Numbers
Logarithm Property of the Quotient of Two Numbers
Logarithm Property of the Power of a Number
Inverse Property of Logarithms
1–1 Property of Logarithms

Discuss the Concepts

Determine whether the statement is always true, sometimes true, or never true.

1. $\log_b \frac{x}{y} = \frac{\log_b x}{\log_b y}$ Never true
2. $\frac{\log x}{\log y} = \frac{x}{y}$ Sometimes true
3. $\log (x + y) = \log x + \log y$ Never true
4. $\log_b \sqrt{x} = \frac{1}{2} \log_b x$ Always true

Concept Check

Use the Properties of Logarithms to solve for x.

1. $\log_8 x = 3 \log_8 2$ 8
2. $\log_5 x = 2 \log_5 3$ 9
3. $\log_4 x = \log_4 2 + \log_4 3$ 6
4. $\log_3 x = \log_3 4 + \log_3 7$ 28
5. $\log_6 x = 3 \log_6 2 - \log_6 4$ 2
6. $\log_9 x = 5 \log_9 2 - \log_9 8$ 4
7. $\log x = \frac{1}{3} \log 27$ 3
8. $\log_2 x = \frac{3}{2} \log_2 4$ 8

In-Class Examples (Objective 12.2B)

Write the logarithm in expanded form.

1. $\log_2 (x^2y^3)$ $2 \log_2 x + 3 \log_2 y$
2. $\log_5 \left(\frac{x^3y}{z^2}\right)$ $3 \log_5 x + \log_5 y - 2 \log_5 z$
3. $\log_4 \sqrt[3]{xy}$ $\frac{1}{3} \log_4 x + \frac{1}{3} \log_4 y$

Express as a single logarithm with a coefficient of 1.

4. $5(\log_3 x + \log_3 y)$ $\log_3 (x^5y^5)$
5. $3(\log_5 x - \log_5 y)$ $\log_5 \frac{x^3}{y^3}$

Concept Check

1. Evaluate: $\log_8 2 - \log_6 216 + \log_3 81 - \log_5 (625)^{1/3}$ 0
2. Evaluate: $\log_9 9^7 - \log_4 64 + \log_2 \frac{1}{2} - \log_3 1$ 3

Optional Student Activity

1. Use the Properties of Logarithms to show that $\log_a a^x = x, a > 0$.
   $\log_a a^x = x \log_a a$
   $= x(1) = x$
2. Use the Properties of Logarithms to show that $b^{\log_b x} = x, b > 0, x > 0$.
   The logarithmic form of $b^y = x$ is $\log_b x = y$. Use this fact to write $b^{\log_b x} = x$ in exponential form: $\log_b x = \log_b x$, a true equation. (Note that we are simply substituting $\log_b x$ for y.)

A second property of logarithms involves the logarithm of the quotient of two numbers. This property of logarithms is also based on the fact that a logarithm is an exponent and that to divide two exponential expressions with the same base, we subtract the exponents.

The Logarithm Property of the Quotient of Two Numbers

For any positive real numbers x, y, and b, $b \neq 1$, $\log_b \frac{x}{y} = \log_b x - \log_b y$.

TAKE NOTE
This theorem is used to rewrite expressions such as $\log_5 \left(\frac{m}{8}\right) = \log_5 m - \log_5 8$. It does *not* state any relationship regarding the expression $\frac{\log_b x}{\log_b y}$. **This expression cannot be simplified.**

A proof of this property can be found in the Appendix.

⇒ Write $\log_b \frac{p}{8}$ in expanded form.

$\log_b \frac{p}{8} = \log_b p - \log_b 8$ • Use the Logarithm Property of Quotients.

⇒ Write $\log_b y - \log_b v$ as a single logarithm.

$\log_b y - \log_b v = \log_b \frac{y}{v}$ • Use the Logarithm Property of Quotients.

A third property of logarithms is used to simplify powers of a number.

The Logarithm Property of the Power of a Number

For any positive real numbers x and b, $b \neq 1$, and for any real number r, $\log_b x^r = r \log_b x$.

Point of Interest

Logarithms were developed independently by Jobst Burgi (1552–1632) and John Napier (1550–1617) as a means of simplifying the calculations of astronomers. The idea was to devise a method by which two numbers could be multiplied by performing additions. Napier is usually given credit for logarithms because he published his results first.

In Napier's original work, the logarithm of 10,000,000 was 0. After this work was published, Napier, in discussions with Henry Briggs (1561–1631), decided that tables of logarithms would be easier to use if the logarithm of 1 were 0. Napier died before new tables could be determined, and Briggs took on the task. His table consisted of logarithms accurate to 30 decimal places, all accomplished without a calculator!

The logarithms Briggs calculated are the common logarithms mentioned earlier.

A proof of this property can be found in the Appendix.

⇒ Rewrite $\log_b x^3$ in terms of $\log_b x$.

$\log_b x^3 = 3 \log_b x$ • Use the Logarithm Property of Powers.

⇒ Rewrite $\frac{2}{3} \log_b x$ with a coefficient of 1.

$\frac{2}{3} \log_b x = \log_b x^{2/3}$ • Use the Logarithm Property of Powers.

The following table summarizes the properties of logarithms that we have discussed, along with two other properties.

Summary of the Properties of Logarithms

Let x, y, and b be positive real numbers with $b \neq 1$. Then

Product Property	$\log_b (x \cdot y) = \log_b x + \log_b y$
Quotient Property	$\log_b \left(\frac{x}{y}\right) = \log_b x - \log_b y$
Power Property	$\log_b x^r = r \log_b x$
Logarithm of One	$\log_b 1 = 0$
1–1 Property	If $\log_b x = \log_b y$, then $x = y$.

➡ Write $\log_b \frac{xy}{z}$ in expanded form.

$$\log_b \frac{xy}{z} = \log_b (xy) - \log_b z$$ • Use the Logarithm Property of Quotients.

$$= \log_b x + \log_b y - \log_b z$$ • Use the Logarithm Property of Products.

➡ Write $\log_b \frac{x^2}{y^3}$ in expanded form.

$$\log_b \frac{x^2}{y^3} = \log_b x^2 - \log_b y^3$$ • Use the Logarithm Property of Quotients.

$$= 2 \log_b x - 3 \log_b y$$ • Use the Logarithm Property of Powers.

➡ Write $2 \log_b x + 4 \log_b y$ as a single logarithm with a coefficient of 1.

$$2 \log_b x + 4 \log_b y = \log_b x^2 + \log_b y^4$$ • Use the Logarithm Property of Powers.

$$= \log_b x^2y^4$$ • Use the Logarithm Property of Products.

Example 4

Write $\log \sqrt{x^3y}$ in expanded form.

Solution

$$\log \sqrt{x^3y} = \log (x^3y)^{1/2} = \frac{1}{2} \log (x^3y)$$

$$= \frac{1}{2} (\log x^3 + \log y)$$

$$= \frac{1}{2} (3 \log x + \log y)$$

$$= \frac{3}{2} \log x + \frac{1}{2} \log y$$

You Try It 4

Write $\log_8 \sqrt[3]{xy^2}$ in expanded form.

Your solution

$\frac{1}{3} \log_8 x + \frac{2}{3} \log_8 y$

Example 5

Write $\frac{1}{2}(\log_3 x - 3 \log_3 y + \log_3 z)$ as a single logarithm with a coefficient of 1.

Solution

$$\frac{1}{2} (\log_3 x - 3 \log_3 y + \log_3 z)$$

$$= \frac{1}{2} (\log_3 x - \log_3 y^3 + \log_3 z)$$

$$= \frac{1}{2} \left(\log_3 \frac{x}{y^3} + \log_3 z\right)$$

$$= \frac{1}{2} \left(\log_3 \frac{xz}{y^3}\right) = \log_3 \left(\frac{xz}{y^3}\right)^{1/2} = \log_3 \sqrt{\frac{xz}{y^3}}$$

You Try It 5

Write $\frac{1}{3}(\log_4 x - 2 \log_4 y + \log_4 z)$ as a single logarithm with a coefficient of 1.

Your solution

$\log_4 \sqrt[3]{\frac{xz}{y^2}}$

Example 6

Find $\log_4 1$.

Solution

$\log_4 1 = 0$ • Logarithm of One

You Try It 6

Find $\log_9 1$.

Your solution

0

Solutions on p. S36

Instructor Note

Examples such as these give students practice in applying the Properties of Logarithms. Facility with these operations will help students when they solve exponential and logarithmic equations later in the text.

Objective 12.2C

Vocabulary to Review

common logarithm
natural logarithm

New Formulas

Change-of-Base Formula

$$\log_a N = \frac{\log_b N}{\log_b a}$$

Discuss the Concepts

Explain how to use the Change-of-Base Formula to evaluate $\log_5 12$.

Optional Student Activity

1. Given $f(x) = 3\log_6(2x - 1)$, determine $f(7)$ to the nearest hundredth. 4.29
2. Given $S(t) = 8\log_5(6t + 2)$, determine $S(2)$ to the nearest hundredth. 13.12
3. Given $P(v) = -3\log_6(4 - 2v)$, determine $P(-4)$ to the nearest hundredth. −4.16
4. Given $G(x) = -5\log_7(2x + 19)$, determine $G(-3)$ to the nearest hundredth. −6.59

Instructor Note

Because only common and natural logarithms have been programmed into graphing calculators, the Change-of-Base Formula is used to graph logarithms having other bases. For instance, to graph $f(x) = \log_2(x - 3)$, the student must enter $\frac{\log(x-3)}{\log 2}$ or an equivalent expression using natural logarithms.

664

Objective C — To use the Change-of-Base Formula

VIDEO & DVD   CD TUTOR   WEB   SSM

Although only common logarithms and natural logarithms are programmed into calculators, the logarithms for other positive bases can be found.

➡ Evaluate $\log_5 22$.

$\log_5 22 = x$	• Write an equation.
$5^x = 22$	• Write the equation in its equivalent exponential form.
$\log 5^x = \log 22$	• Take the common logarithm of each side of the equation.
$x \log 5 = \log 22$	• Use the Power Property of Logarithms.
$x = \frac{\log 22}{\log 5}$	• Divide each side by log 5. This is the exact answer.
$x \approx 1.9206$	• This is an approximate answer.
$\log_5 22 \approx 1.9206$	

In the third step above, the natural logarithm, instead of the common logarithm, could have been applied to each side of the equation. The same result would have been obtained.

Using a procedure similar to the one used to evaluate $\log_5 22$, a formula for changing bases can be derived.

Change-of-Base Formula

$$\log_a N = \frac{\log_b N}{\log_b a}$$

➡ Evaluate $\log_2 14$.

$$\log_2 14 = \frac{\log 14}{\log 2} \approx 3.8074$$

• Use the Change-of-Base Formula with $N = 14$, $a = 2$, $b = 10$.

For the last example, common logarithms were used. Here is the same example using natural logarithms. Note that the answers are the same.

$$\log_2 14 = \frac{\ln 14}{\ln 2} \approx 3.8074$$

Example 7

Evaluate $\log_2 90.813$ by using common logarithms.

Solution

$$\log_2 90.813 = \frac{\log 90.813}{\log 2} \approx 6.5048$$

You Try It 7

Evaluate $\log_7 6.45$ by using common logarithms.

Your solution

0.9579

Example 8

Evaluate $\log_8 0.137$ by using natural logarithms.

Solution

$$\log_8 0.137 = \frac{\ln 0.137}{\ln 8} \approx -0.9559$$

You Try It 8

Evaluate $\log_3 0.834$ by using natural logarithms.

Your solution

−0.1652

Solutions on p. S36

In-Class Examples (Objective 12.2C)

Evaluate. Round to the nearest ten-thousandth.

1. $\log_2 6$ 2.5850
2. $\log_5 20$ 1.8614
3. $\log_3 7.5$ 1.8340

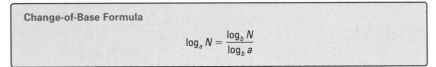

12.2 Exercises

Objective A

1. **a.** What is a common logarithm?
b. How is the common logarithm of $4z$ written?

2. **a.** What is a natural logarithm?
b. How is the natural logarithm of $3x$ written?

Write the exponential equation in logarithmic form.

3. $5^2 = 25$
$\log_5 25 = 2$

4. $10^3 = 1000$
$\log_{10} 1000 = 3$

5. $4^{-2} = \frac{1}{16}$
$\log_4 \left(\frac{1}{16}\right) = -2$

6. $3^{-3} = \frac{1}{27}$
$\log_3 \left(\frac{1}{27}\right) = -3$

7. $10^y = x$
$\log_{10} x = y$

8. $e^y = x$
$\ln x = y$

9. $a^x = w$
$\log_a w = x$

10. $b^y = c$
$\log_b c = y$

Write the logarithmic equation in exponential form.

11. $\log_3 9 = 2$
$3^2 = 9$

12. $\log_2 32 = 5$
$2^5 = 32$

13. $\log 0.01 = -2$
$10^{-2} = 0.01$

14. $\log_5 \frac{1}{5} = -1$
$5^{-1} = \frac{1}{5}$

15. $\ln x = y$
$e^y = x$

16. $\log x = y$
$10^y = x$

17. $\log_b u = v$
$b^v = u$

18. $\log_c x = y$
$c^y = x$

Evaluate.

19. $\log_3 81$
4

20. $\log_7 49$
2

21. $\log_2 128$
7

22. $\log_5 125$
3

23. $\log 100$
2

24. $\log 0.001$
-3

25. $\ln e^3$
3

26. $\ln e^2$
2

27. $\log_8 1$
0

28. $\log_3 243$
5

29. $\log_5 625$
4

30. $\log_2 64$
6

Solve for x.

31. $\log_3 x = 2$
9

32. $\log_5 x = 1$
5

33. $\log_4 x = 3$
64

34. $\log_2 x = 6$
64

35. $\log_7 x = -1$
$\frac{1}{7}$

36. $\log_8 x = -2$
$\frac{1}{64}$

37. $\log_6 x = 0$
1

38. $\log_4 x = 0$
1

Section 12.2

Suggested Assignment

Exercises 1–45, odds
Exercises 49–121, every other odd
Exercise 125
More challenging problems: Exercise 126

Answers to Writing Exercises

1a. A common logarithm is a logarithm with base 10.
b. $\log 4z$
2a. A natural logarithm is a logarithm with base e.
b. $\ln 3x$

Quick Quiz (Objective 12.2A)

Write the exponential expression in logarithmic form.
1. $4^2 = 16$ $\log_4 16 = 2$
2. $\left(\frac{1}{3}\right)^{-2} = 9$ $\log_{1/3} 9 = -2$

Write the logarithmic expression in exponential form.
3. $\log_5 25 = 2$ $5^2 = 25$
4. $\log_{10} 0.001 = -3$ $10^{-3} = 0.001$

5. Evaluate $\log_9 81$. 2
6. Evaluate $\log_2 64$. 6
7. Solve for x: $\log_3 x = 4$. 81
8. Solve for x: $\log_5 x = -2$. $\frac{1}{25}$

Answers to Writing Exercises

47. Answers will vary. For example, the log of a product is equal to the sum of the logs: $\log_b(xy) = \log_b x + \log_b y$.

48. Answers will vary. For example, the log of a quotient is equal to the difference of the logs: $\log_b \frac{x}{y} = \log_b x - \log_b y$.

Solve for x. Round to the nearest hundredth.

39. $\log x = 2.5$ 316.23

40. $\log x = 3.2$ 1584.89

41. $\log x = -1.75$ 0.02

42. $\log x = -2.1$ 0.01

43. $\ln x = 2$ 7.39

44. $\ln x = 1.4$ 4.06

45. $\ln x = -\frac{1}{2}$ 0.61

46. $\ln x = -1.7$ 0.18

Objective B

47. What is the Product Property of Logarithms?

48. What is the Quotient Property of Logarithms?

Express as a single logarithm with a coefficient of 1.

49. $\log_3 x^3 + \log_3 y^2$ $\log_3(x^3y^2)$

50. $\log_7 x + \log_7 z^2$ $\log_7(xz^2)$

51. $\ln x^4 - \ln y^2$ $\ln\left(\frac{x^4}{y^2}\right)$

52. $\ln x^2 - \ln y$ $\ln\left(\frac{x^2}{y}\right)$

53. $3 \log_7 x$ $\log_7 x^3$

54. $4 \log_8 y$ $\log_8 y^4$

55. $3 \ln x + 4 \ln y$ $\ln(x^3y^4)$

56. $2 \ln x - 5 \ln y$ $\ln\left(\frac{x^2}{y^5}\right)$

57. $2(\log_4 x + \log_4 y)$ $\log_4(x^2y^2)$

58. $3(\log_5 r + \log_5 t)$ $\log_5(r^3t^3)$

59. $\frac{1}{2}(\log_6 x - \log_6 y)$ $\log_6 \sqrt{\frac{x}{y}}$

60. $\frac{1}{3}(\log_8 x - \log_8 y)$ $\log_8 \sqrt[3]{\frac{x}{y}}$

61. $2 \log_3 x - \log_3 y + 2 \log_3 z$ $\log_3\left(\frac{x^2z^2}{y}\right)$

62. $4 \log_5 r - 3 \log_5 s + \log_5 t$ $\log_5\left(\frac{r^4t}{s^3}\right)$

63. $\ln x - (2 \ln y + \ln z)$ $\ln\left(\frac{x}{y^2z}\right)$

64. $2 \log_b x - 3(\log_b y + \log_b z)$ $\log_b\left(\frac{x^2}{y^3z^3}\right)$

65. $2(\log_4 s - 2 \log_4 t + \log_4 r)$ $\log_4\left(\frac{s^2r^2}{t^4}\right)$

66. $3(\log_9 x + 2 \log_9 y - 2 \log_9 z)$ $\log_9\left(\frac{x^3y^6}{z^6}\right)$

67. $3\ln x - 4(\ln y + \ln z)$ $\ln\left(\frac{x^3}{y^4z^4}\right)$

68. $\ln t - 3(\ln u + \ln v)$ $\ln\left(\frac{t}{u^3v^3}\right)$

69. $3 \log_2 t - 2(\log_2 r - \log_2 v)$ $\log_2\left(\frac{t^3v^2}{r^2}\right)$

70. $2 \log_{10} x - 3(\log_{10} y - \log_{10} z)$ $\log_{10}\left(\frac{x^2z^3}{y^3}\right)$

Quick Quiz (Objective 12.2B)

Write the logarithm in expanded form.

1. $\log_8(x^3yz^2)$ $3 \log_8 x + \log_8 y + 2 \log_8 z$

2. $\log_6 \frac{u^4}{v^5}$ $4 \log_6 u - 5 \log_6 v$

Express as a single logarithm with a coefficient of 1.

3. $2 \log_4 x + 3 \log_4 y$ $\log_4 x^2y^3$

4. $3 \log_5 x - 4 \log_5 y$ $\log_5 \frac{x^3}{y^4}$

667

71. $\frac{1}{2}(3 \log_4 x - 2 \log_4 y + \log_4 z)$

$\log_4 \sqrt{\frac{x^3 z}{y^2}}$

72. $\frac{1}{3}(4 \log_5 t - 5 \log_5 u - 7 \log_5 v)$

$\log_5 \sqrt[3]{\frac{t^4}{u^5 v^7}}$

73. $\frac{1}{2}(\ln x - 3 \ln y)$

$\ln \sqrt{\frac{x}{y^3}}$

74. $\frac{1}{3} \ln a + \frac{2}{3} \ln b$

$\ln \sqrt[3]{ab^2}$

75. $\frac{1}{2}\log_2 x - \frac{2}{3}\log_2 y + \frac{1}{2}\log_2 z$

$\log_2 \left(\frac{\sqrt{xz}}{\sqrt[3]{y^2}}\right)$

76. $\frac{2}{3}\log_3 x + \frac{1}{3}\log_3 y - \frac{1}{2}\log_3 z$

$\log_3 \left(\frac{\sqrt[3]{x^2 y}}{\sqrt{z}}\right)$

Write the logarithm in expanded form.

77. $\log_8 (xz)$

$\log_8 x + \log_8 z$

78. $\log_7 (rt)$

$\log_7 r + \log_7 t$

79. $\log_3 x^5$

$5 \log_3 x$

80. $\log_2 y^7$

$7 \log_2 y$

81. $\log_b \left(\frac{r}{s}\right)$

$\log_b r - \log_b s$

82. $\log_c \left(\frac{z}{4}\right)$

$\log_c z - \log_c 4$

83. $\log_3 (x^2 y^6)$

$2 \log_3 x + 6 \log_3 y$

84. $\log_4 (t^4 u^2)$

$4 \log_4 t + 2 \log_4 u$

85. $\log_7 \left(\frac{u^3}{v^4}\right)$

$3 \log_7 u - 4 \log_7 v$

86. $\log \left(\frac{s^5}{t^2}\right)$

$5 \log s - 2 \log t$

87. $\log_2 (rs)^2$

$2 \log_2 r + 2 \log_2 s$

88. $\log_3 (x^2 y)^3$

$6 \log_3 x + 3 \log_3 y$

89. $\ln(x^2 yz)$

$2 \ln x + \ln y + \ln z$

90. $\ln(xy^2 z^3)$

$\ln x + 2 \ln y + 3 \ln z$

91. $\log_5 \left(\frac{xy^2}{z^4}\right)$

$\log_5 x + 2 \log_5 y - 4 \log_5 z$

92. $\log_b \left(\frac{r^2 s}{t^3}\right)$

$2 \log_b r + \log_b s - 3 \log_b t$

93. $\log_8 \left(\frac{x^2}{yz^2}\right)$

$2 \log_8 x - \log_8 y - 2 \log_8 z$

94. $\log_9 \left(\frac{x}{y^2 z^3}\right)$

$\log_9 x - 2 \log_9 y - 3 \log_9 z$

95. $\log_4 \sqrt{x^3 y}$

$\frac{3}{2}\log_4 x + \frac{1}{2} \log_4 y$

96. $\log_3 \sqrt{x^5 y^3}$

$\frac{5}{2}\log_3 x + \frac{3}{2} \log_3 y$

97. $\log_7 \sqrt{\frac{x^3}{y}}$

$\frac{3}{2}\log_7 x - \frac{1}{2}\log_7 y$

98. $\log_b \sqrt[3]{\frac{r^2}{t}}$

$\frac{2}{3}\log_b r - \frac{1}{3}\log_b t$

99. $\log_3 \frac{t}{\sqrt{x}}$

$\log_3 t - \frac{1}{2}\log_3 x$

100. $\log_4 \frac{x}{\sqrt{y^2 z}}$

$\log_4 x - \log_4 y - \frac{1}{2}\log_4 z$

Objective C

Evaluate. Round to the nearest ten-thousandth.

101. $\log_{10} 7$
0.8451

102. $\log_{10} 9$
0.9542

103. $\log_{10}\left(\frac{3}{5}\right)$
−0.2218

104. $\log_{10}\left(\frac{13}{3}\right)$
0.6368

105. $\ln 4$
1.3863

106. $\ln 6$
1.7918

107. $\ln\left(\frac{17}{6}\right)$
1.0415

108. $\ln\left(\frac{13}{17}\right)$
−0.2683

109. $\log_8 6$
0.8617

110. $\log_4 8$
1.500

111. $\log_5 30$
2.1133

112. $\log_6 28$
1.8597

113. $\log_3 (0.5)$
−0.6309

114. $\log_5 (0.6)$
−0.3174

115. $\log_7 (1.7)$
0.2727

116. $\log_6 (3.2)$
0.6492

117. $\log_5 15$
1.6826

118. $\log_3 25$
2.9299

119. $\log_{12} 120$
1.9266

120. $\log_9 90$
2.0480

121. $\log_4 2.55$
0.6752

122. $\log_8 6.42$
0.8942

123. $\log_5 67$
2.6125

124. $\log_8 35$
1.7098

APPLYING THE CONCEPTS

125. For each of the following, answer True or False. Assume all variables represent positive numbers.

a. $\log_3 (-9) = -2$
false

b. $x^y = z$ and $\log_x z = y$ are equivalent equations.
true

c. $\log (x^{-1}) = \frac{1}{\log x}$
false

d. $\log \left(\frac{x}{y}\right) = \log x - \log y$
true

e. $\log (x \cdot y) = \log x \cdot \log y$
false

f. If $\log x = \log y$, then $x = y$.
true

126. Complete each statement using the equation $\log_a b = c$.

a. $a^c =$ ___b___

b. $\text{antilog}_a (\log_a b) =$ ___b___

Quick Quiz (Objective 12.2C)

Evaluate. Round to the nearest ten-thousandth.

1. $\log_3 11$ 2.1827

2. $\log_4 14$ 1.9037

3. $\log_2 8.8$ 3.1375

12.3 Graphs of Logarithmic Functions

Objective A To graph a logarithmic function

VIDEO & DVD CD TUTOR WEB SSM

The graph of a logarithmic function can be drawn by using the relationship between the exponential and logarithmic functions.

➡ Graph: $f(x) = \log_2 x$

Think of this as the equation $y = \log_2 x$.

$$f(x) = \log_2 x$$
$$y = \log_2 x$$

Write the equivalent exponential equation.

$$x = 2^y$$

Because the equation is solved for x in terms of y, it is easier to choose values of y and find the corresponding values of x. The results can be recorded in a table.

x	y
$\frac{1}{4}$	-2
$\frac{1}{2}$	-1
1	0
2	1
4	2

Graph the ordered pairs on a rectangular coordinate system.

Connect the points with a smooth curve.

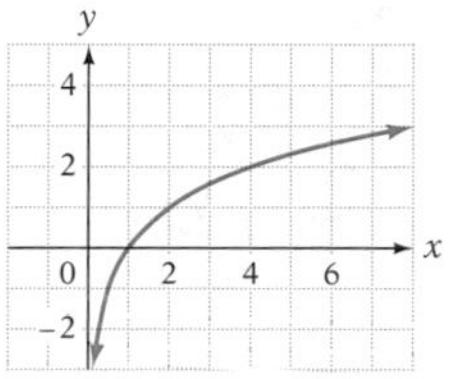

Point of Interest

Although logarithms were originally developed to assist with computations, logarithmic functions have a much broader use today. These functions occur in geology, acoustics, chemistry, and economics, for example.

Applying the vertical-line and horizontal-line tests reveals that $f(x) = \log_2 x$ is a 1–1 function.

➡ Graph: $f(x) = \log_2 (x) + 1$

Think of $f(x) = \log_2 (x) + 1$ as the equation $y = \log_2 (x) + 1$.

$$f(x) = \log_2 (x) + 1$$
$$y = \log_2 (x) + 1$$

Solve for $\log_2 (x)$.

$$y - 1 = \log_2 (x)$$

Write the equivalent exponential equation.

$$2^{y-1} = x$$

Choose values of y and find the corresponding values of x.

x	y
$\frac{1}{4}$	-1
$\frac{1}{2}$	0
1	1
2	2
4	3

Graph the ordered pairs on a rectangular coordinate system.

Connect the points with a smooth curve.

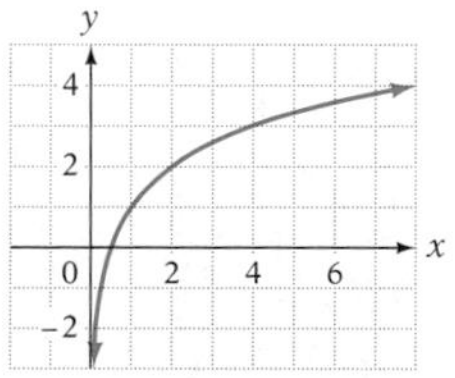

TAKE NOTE

See the Projects and Group Activities at the end of this chapter for suggestions on graphing logarithmic functions using a graphing calculator.

Objective 12.3A

Vocabulary to Review

equivalent exponential and logarithmic equations

Discuss the Concepts

1. The inverse of an exponential function is a logarithmic function. Why?
2. Is it true that if x and y are positive real numbers, $x < y$, and $b > 0$, then $\log_b x < \log_b y$? Yes
3. The Power Property of Logarithms tells us that $\ln x^2 = 2 \ln x$. Graph $f(x) = \ln x^2$ and $g(x) = 2 \ln x$. Are the graphs the same? Explain. No. If $x < 0$, $\ln x^2$ is a real number but $2 \ln x$ is not a real number.

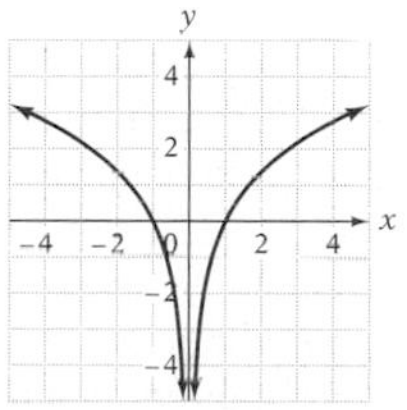

4. Does the graph of the exponential function $f(x) = b^x$ intersect the graph of its inverse $f^{-1}(x) = \log_b x$ when $b > 1$? No
5. Does the graph of the exponential function $f(x) = b^x$ intersect the graph of its inverse $f^{-1}(x) = \log_b x$ when $0 < b < 1$? Yes

In-Class Examples (Objective 12.3A)

1. Graph $f(x) = \log_2 (2x + 1)$.

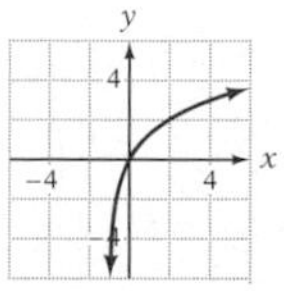

2. Graph $f(x) = \frac{1}{2} \log_3 x$.

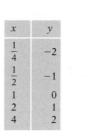

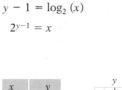

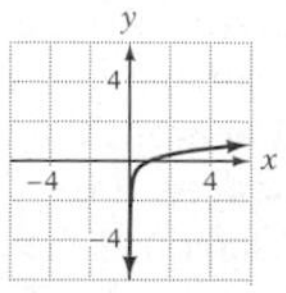

Example 1
Graph: $f(x) = \log_3 x$

Solution
$f(x) = \log_3 x$
$y = \log_3 x$
$3^y = x$ • Write the equivalent exponential equation.

x	y
$\frac{1}{9}$	−2
$\frac{1}{3}$	−1
1	0
3	1

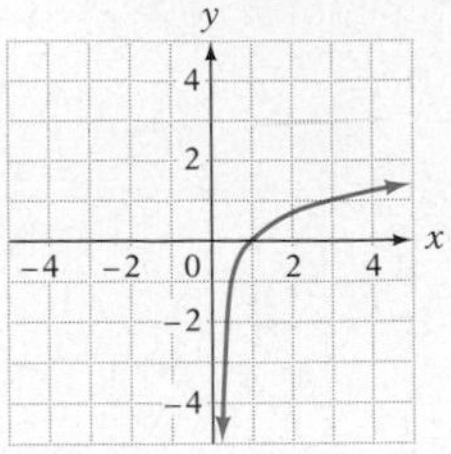

You Try It 1
Graph: $f(x) = \log_2 (x - 1)$

Your solution

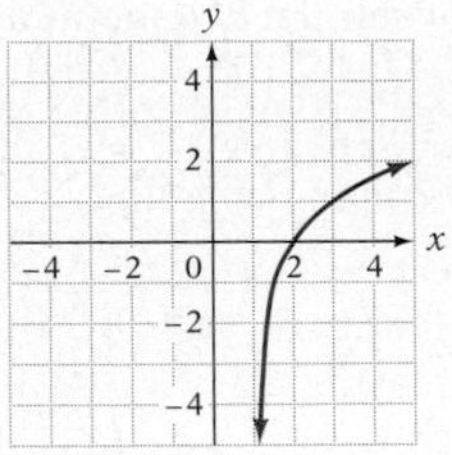

Example 2
Graph: $f(x) = 2 \log_3 x$

Solution
$f(x) = 2 \log_3 x$
$y = 2 \log_3 x$
$\frac{y}{2} = \log_3 x$
$3^{y/2} = x$ • Write the equivalent exponential equation.

x	y
$\frac{1}{9}$	−4
$\frac{1}{3}$	−2
1	0
3	2

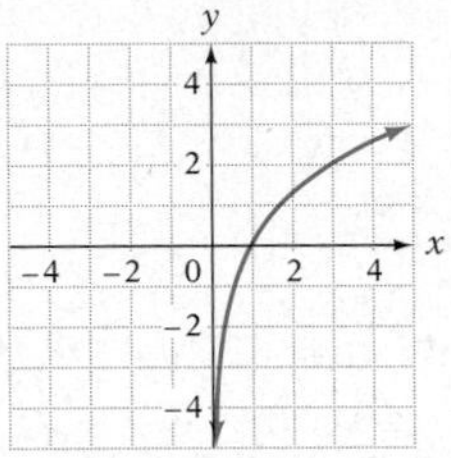

You Try It 2
Graph: $f(x) = \log_3 (2x)$

Your solution

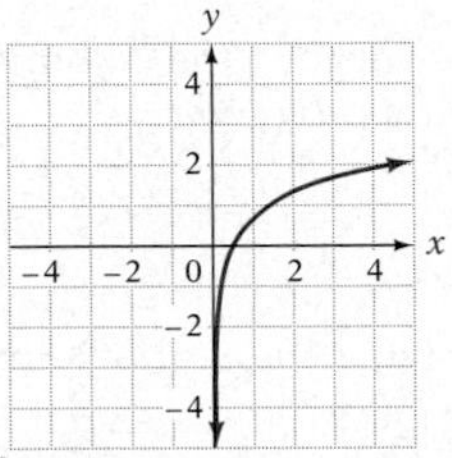

Example 3
Graph: $f(x) = -\log_2 (x - 2)$

Solution
$f(x) = -\log_2 (x - 2)$
$y = -\log_2 (x - 2)$
$-y = \log_2 (x - 2)$
$2^{-y} = x - 2$ • Write the equivalent exponential equation.
$2^{-y} + 2 = x$

x	y
6	−2
4	−1
3	0
$\frac{5}{2}$	1
$\frac{9}{4}$	2
$\frac{17}{8}$	3

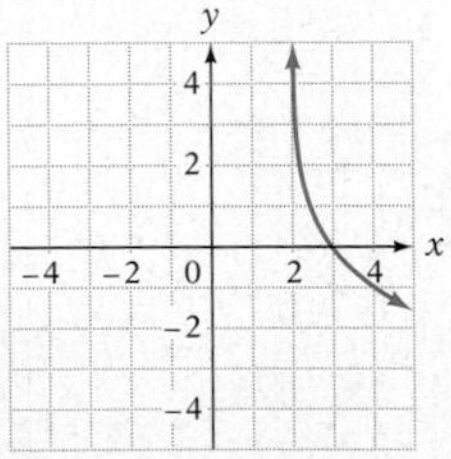

You Try It 3
Graph: $f(x) = -\log_3 (x + 1)$

Your solution

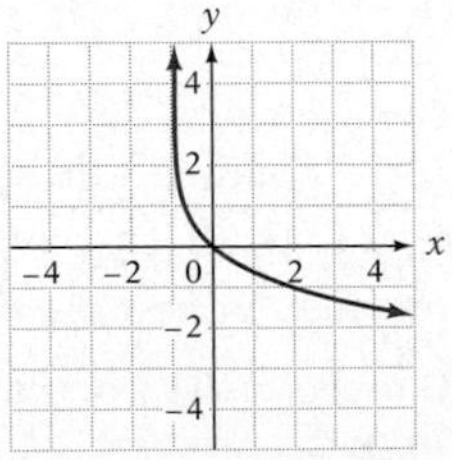

Solutions on p. S36

Concept Check

1. Graph $f(x) = 10 \log (x - 2)$. Find the zero of the function. 3
2. What value in the domain of $f(x) = 3 \log_2 x$ corresponds to the range value of −6? $\frac{1}{4}$
3. What value in the domain of $f(x) = \log_3 (2 - x)$ corresponds to the range value of 1? −1
4. What value in the domain of $f(x) = -\log_2 (x - 1)$ corresponds to the range value of −2? 5
5. Graph $f(x) = 5 \log (x - 1)$. What are the x- and y-intercepts of the graph of this function? x-intercept: (2, 0); no y-intercept

Optional Student Activity

According to the U.S. Environmental Protection Agency, the amount of garbage generated per person has been increasing over the last few decades. The table below shows the per capita garbage, in pounds per day, generated in the United States for selected years.

Year	Pounds per Day
1960	2.66
1970	3.27
1980	3.61
1990	4.00

Draw a scatter diagram for this data. Would the equation that best fits the points be the equation of a linear function, an exponential function, or a logarithmic function? Linear

Instructor Note

To reinforce the idea that exponential and logarithmic functions are inverses of each other, you might want to present and discuss a mapping of the inverse functions $f(x) = 10^x$ and $g(x) = \log x$, such as the one shown at the right.

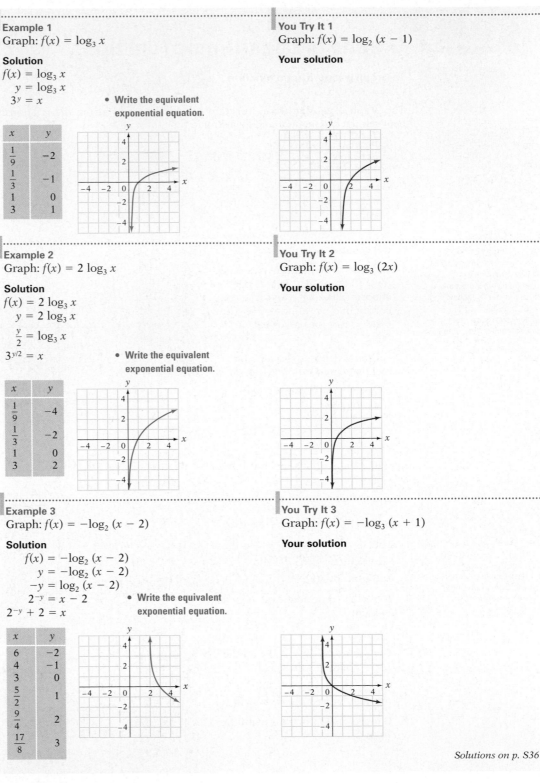

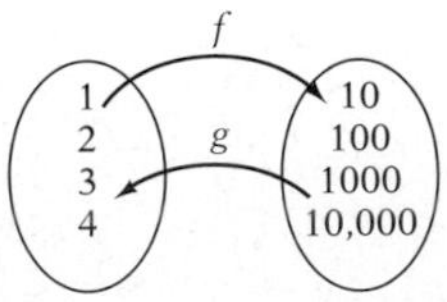

domain of $f(x)$ range of $g(x)$ range of $f(x)$ domain of $g(x)$

12.3 Exercises

Objective A

1. Is the function $f(x) = \log x$ a 1–1 function? Why or why not?

2. Name two characteristics of the graph of $y = \log_b x$, $b > 1$.

3. What is the relationship between the graph of $x = 3^y$ and that of $y = \log_3 x$?

4. What is the relationship between the graph of $y = 3^x$ and that of $y = \log_3 x$?

Graph.

5. $f(x) = \log_4 x$

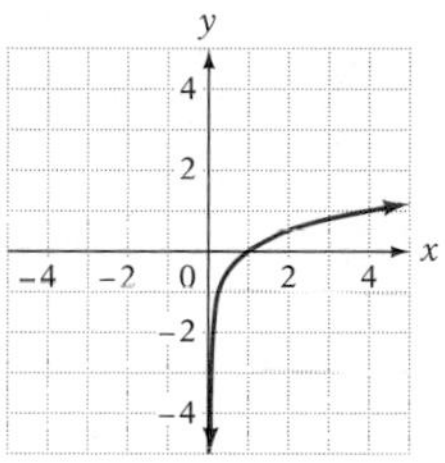

6. $f(x) = \log_2 (x + 1)$

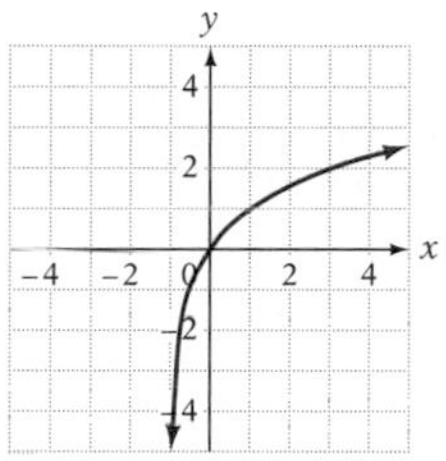

7. $f(x) = \log_3 (2x - 1)$

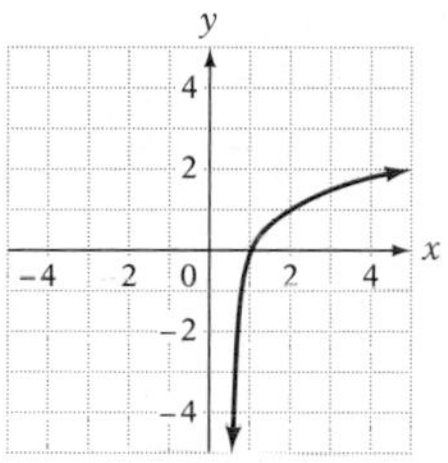

8. $f(x) = \log_2 \left(\frac{1}{2}x\right)$

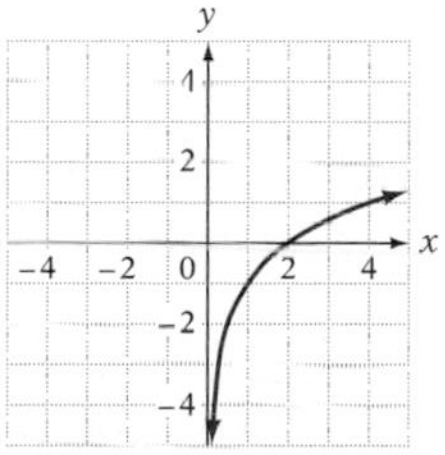

9. $f(x) = 3 \log_2 x$

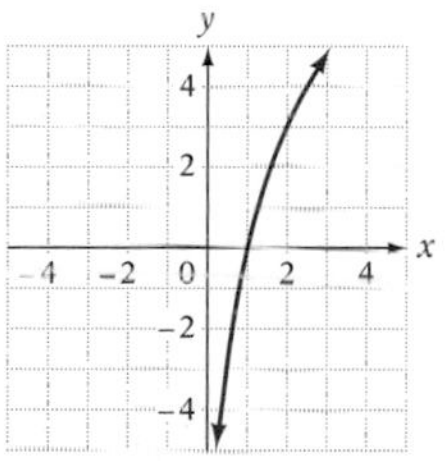

10. $f(x) = \frac{1}{2}\log_2 x$

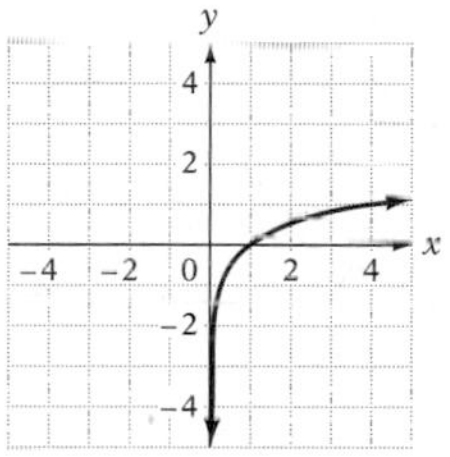

11. $f(x) = -\log_2 x$

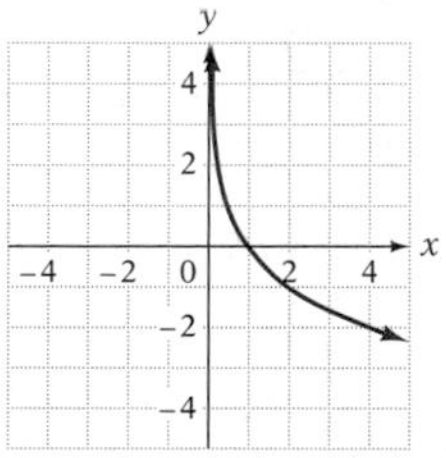

12. $f(x) = -\log_3 x$

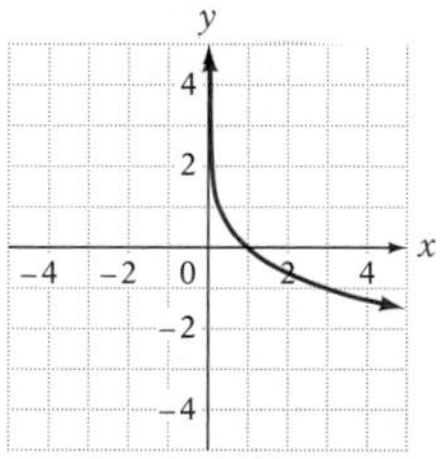

13. $f(x) = \log_2 (x - 1)$

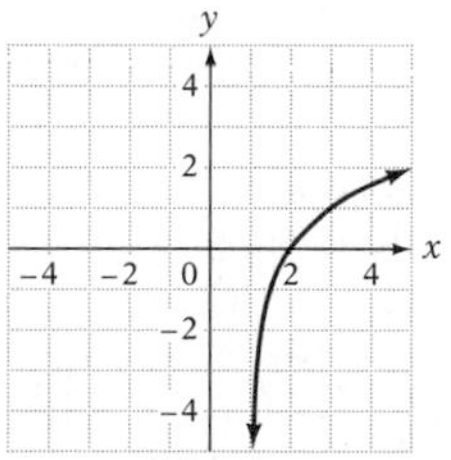

Section 12.3

Suggested Assignment

Exercises 5–15, odds
More challenging problems:
Exercises 17–21, odds
Exercises 23, 24
(A graphing calculator is required.)

Answers to Writing Exercises

1. Yes. Answers will vary. For example, the function passes both the vertical-line test and the horizontal-line test.
2. Descriptions will vary. For example: The domain is the set of positive real numbers; the range is the set of real numbers. The graph never touches the y-axis. The x-intercept is $(1, 0)$.
3. They are the same graph.
4. They are mirror images of each other with respect to the line $y = x$.

Quick Quiz (Objective 12.3A)

1. Graph $f(x) = \log_3 (x + 1)$.

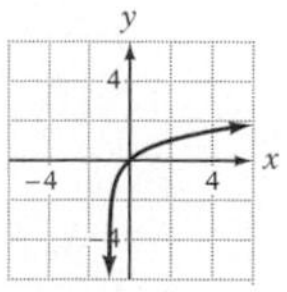

2. Graph $f(x) = -\log_2 2x$.

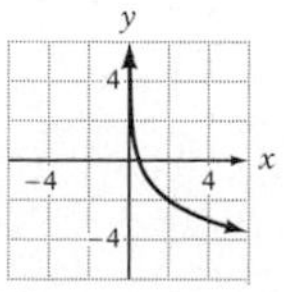

Answers to Writing Exercises

23b. The point (25.1, 2) means that a star that is 25.1 parsecs from Earth has a distance modulus of 2.

24b. The point (4, 49) means that after 4 months, the typist's proficiency has dropped to 49 words per minute.

14. $f(x) = \log_3 (2 - x)$

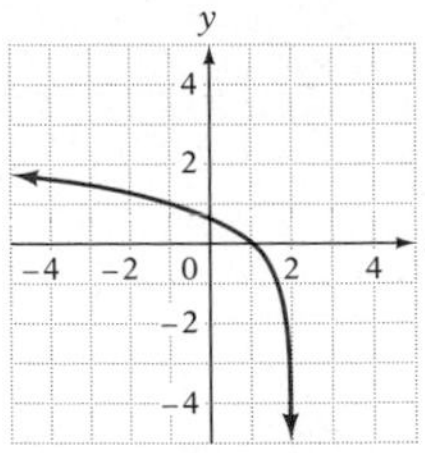

15. $f(x) = -\log_2 (x - 1)$

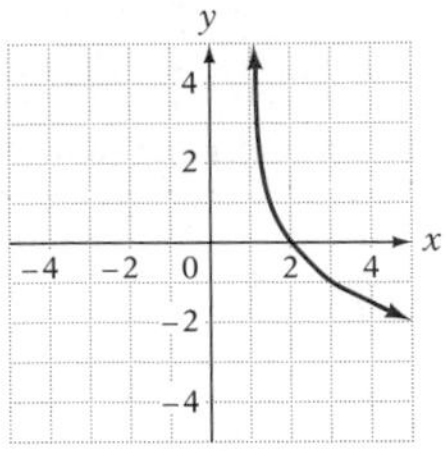

16. $f(x) = -\log_2 (1 - x)$

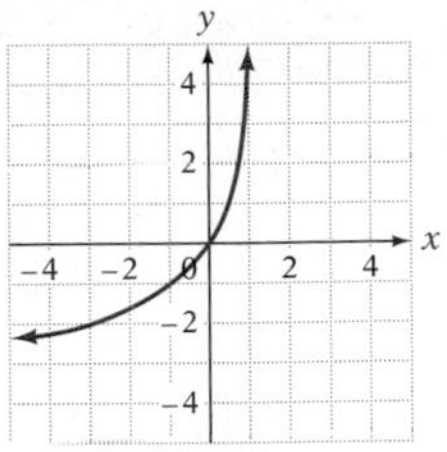

APPLYING THE CONCEPTS

Use a graphing calculator to graph the following.

17. $f(x) = x - \log_2 (1 - x)$

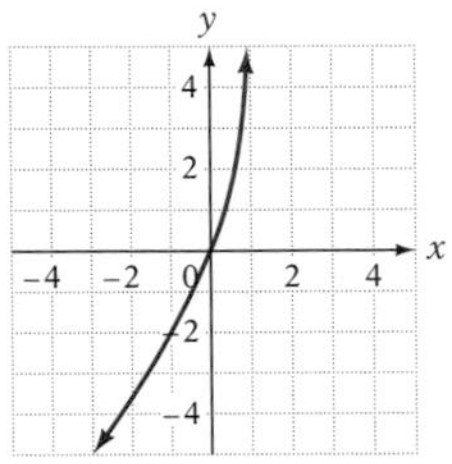

18. $f(x) = -\frac{1}{2}\log_2 x - 1$

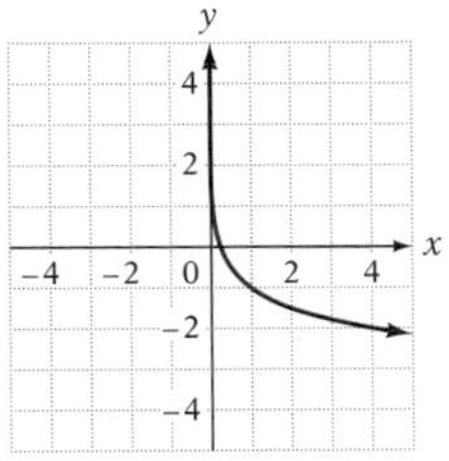

19. $f(x) = \frac{x}{2} - 2 \log_2 (x + 1)$

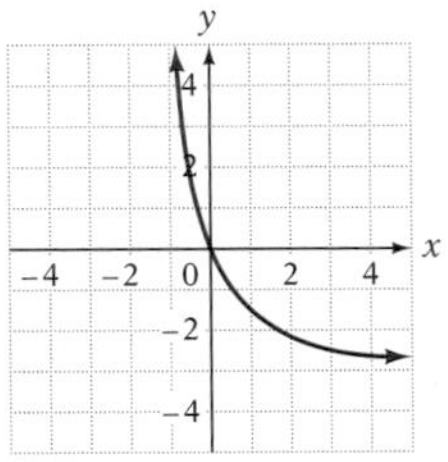

20. $f(x) = x + \log_3 (2 - x)$

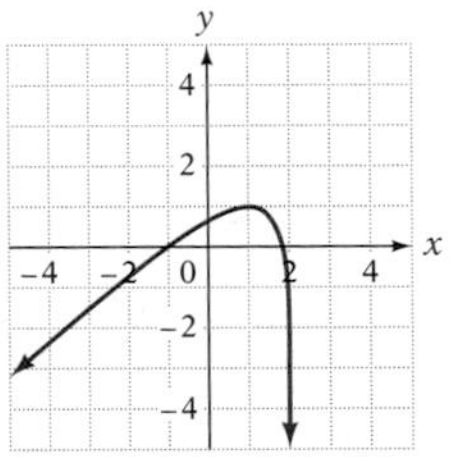

21. $f(x) = x^2 - 10 \ln (x - 1)$

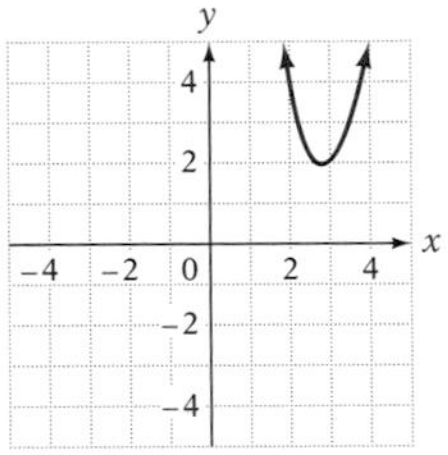

22. $f(x) = \frac{x}{3} - 3 \log_2 (x + 3)$

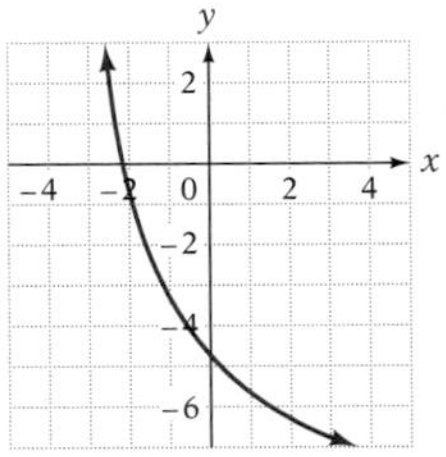

23. Astronomers use the *distance modulus* of a star as a method of determining the star's distance from Earth. The formula is $M = 5 \log s - 5$, where M is the distance modulus and s is the star's distance from the Earth in parsecs. (One parsec $\approx 1.9 \times 10^{13}$ mi.)
a. Graph the equation.
b. The point with coordinates (25.1, 2) is on the graph. Write a sentence that describes the meaning of this ordered pair.

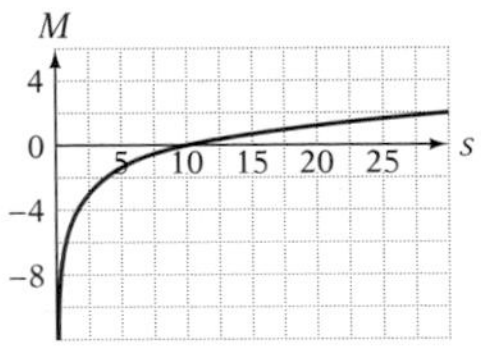

24. Without practice, the proficiency of a typist decreases. The equation $S = 60 - 7 \ln (t + 1)$, where S is the typing speed in words per minute and t is the number of months without typing, approximates this decrease.
a. Graph the equation.
b. The point with coordinates (4, 49) is on the graph. Write a sentence that describes the meaning of this ordered pair.

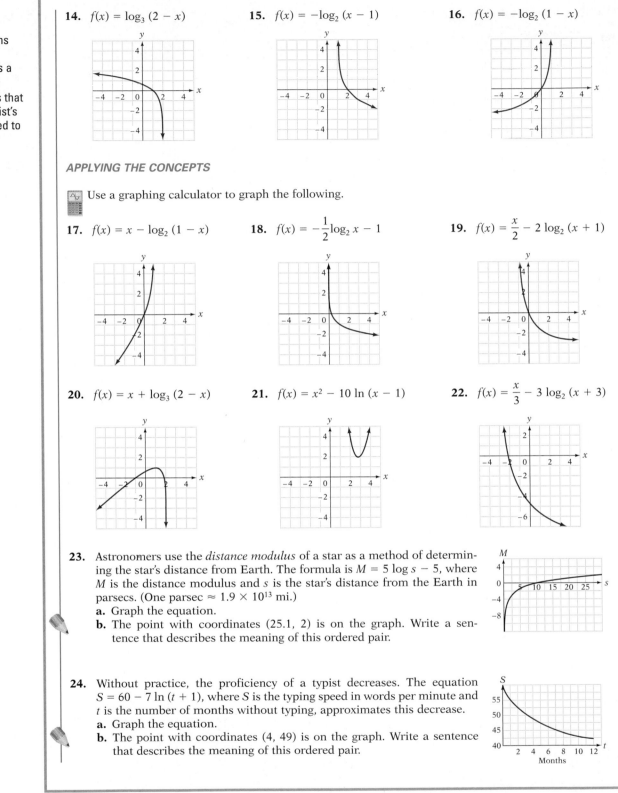

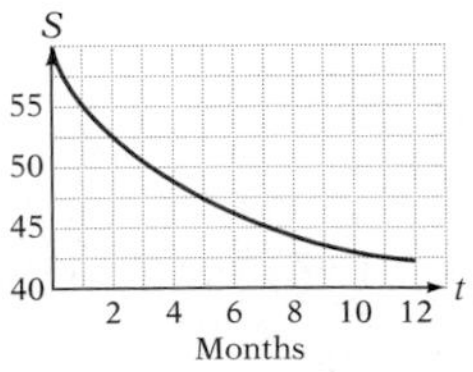

12.4 Solving Exponential and Logarithmic Equations

Objective A To solve an exponential equation

VIDEO & DVD CD TUTOR WEB SSM

An **exponential equation** is one in which the variable occurs in the exponent. The examples at the right are exponential equations.

$$6^{2x+1} = 6^{3x-2}$$
$$4^x = 3$$
$$2^{x+1} = 7$$

An exponential equation in which each side of the equation can be expressed in terms of the same base can be solved by using the One-to-One Property of Exponential Functions. Recall that the One-to-One Property of Exponential Functions states that

If $b^u = b^v$, then $u = v$.

In the two examples below, this property is used in solving exponential equations.

➡ Solve: $10^{3x+5} = 10^{x-3}$

$10^{3x+5} = 10^{x-3}$

$3x + 5 = x - 3$ • Use the 1–1 Property of Exponential Functions to equate the exponents.

$2x + 5 = -3$ • Solve the resulting equation.

$2x = -8$

$x = -4$

Check: $10^{3x+5} = 10^{x-3}$

$10^{3(-4)+5}$	10^{-4-3}
10^{-12+5}	10^{-7}

$10^{-7} = 10^{-7}$

The solution is −4.

➡ Solve: $9^{x+1} = 27^{x-1}$

$9^{x+1} = 27^{x-1}$

$(3^2)^{x+1} = (3^3)^{x-1}$ • $3^2 = 9$; $3^3 = 27$

$3^{2x+2} = 3^{3x-3}$

$2x + 2 = 3x - 3$ • Use the 1–1 Property of Exponential Functions to equate the exponents.

$2 = x - 3$ • Solve for *x*.

$5 = x$

Check: $9^{x+1} = 27^{x-1}$

9^{5+1}	27^{5-1}
9^6	27^4

$531,441 = 531,441$

The solution is 5.

Objective 12.4A

New Vocabulary
exponential equation

Properties to Review
1–1 Property of Exponential Functions

Instructor Note
Two methods for solving an exponential equation are presented here. The first method is used when each side of the equation can be expressed in terms of the same base. The second method involves logarithms. The logarithm method can always be used but is usually more difficult. You can show this by solving $9^{x+1} = 27^{x-1}$ by using logarithms.

Making an analogy to quadratic equations may help some students. It is generally easier to solve a quadratic equation by factoring (if the polynomial is factorable) than it is to use the quadratic formula.

Discuss the Concepts
1. Explain how to solve the equation $7^{x+1} = 7^5$.
2. What is the solution of the equation $2e^x = 16$? Explain how you determined the solution. ln 8. Divide both sides of the equation by 2; then write the equation in logarithmic form.
3. What does the following property allow us to do? If $x = y$, $x > 0$, $y > 0$, then $\log_b x = \log_b y$. Take the common logarithm of each side of an equation.

In-Class Examples (Objective 12.4A)
Solve for *x*. Round to the nearest ten-thousandth.
1. $6^{5x+3} = 6^{x-9}$ −3
2. $8^x = 4$ $\frac{2}{3}$
3. $5^x = 7$ 1.2091
4. $3^{2x+1} = 15$ 0.7325
5. $3^{2x-1} = 27^x$ −1

Concept Check

1. If $3^x = 5$, find the value of 3^{2x+3}. 675
2. Given $2^x = 8^{y+1}$ and $9^y = 3^{x-9}$, find the value of $x + y$. 27
3. Find the x-coordinate of the intersection of the graphs of $y = 4^{2x}$ and $y = 8^{x+1}$. 3

Optional Student Activity

1. When all eight positive integer factors of 30 are multiplied together, the product is 30^k. Find k. 4
2. Solve: $4^{x/3} = 2$ 1.5
3. Solve for x. Round to the nearest ten-thousandth.

$$1.2^{x/2-1} = 1.4$$

5.6910

4. Solve for x: $5^x = 5^{99} + 5^{99} + 5^{99} + 5^{99} + 5^{99}$ 100

CALCULATOR NOTE

To evaluate $\frac{\log 7}{\log 4}$ on a scientific calculator, use the keystrokes

7 [log] [÷] 4 [log] [=]

The display should read 1.4036775.

When each side of an exponential equation cannot easily be expressed in terms of the same base, logarithms are used to solve the exponential equation.

⇒ Solve: $4^x = 7$

$$4^x = 7$$

$$\log 4^x = \log 7$$

- Take the common logarithm of each side of the equation.

$$x \log 4 = \log 7$$

- Rewrite the equation using the Properties of Logarithms.

$$x = \frac{\log 7}{\log 4} \approx 1.4037$$

- Solve for x.

The solution is 1.4037.

- Note that $\frac{\log 7}{\log 4} \neq \log 7 - \log 4$.

⇒ Solve: $3^{x+1} = 5$

$$3^{x+1} = 5$$

$$\log 3^{x+1} = \log 5$$

- Take the common logarithm of each side of the equation.

$$(x + 1)\log 3 = \log 5$$

- Rewrite the equation using the Properties of Logarithms.

$$x + 1 = \frac{\log 5}{\log 3}$$

$$x = \frac{\log 5}{\log 3} - 1$$

- Solve for x.

$$x \approx 0.4650$$

The solution is 0.4650.

Example 1

Solve for n: $(1.1)^n = 2$

Solution

$$(1.1)^n = 2$$
$$\log (1.1)^n = \log 2$$
$$n \log 1.1 = \log 2$$
$$n = \frac{\log 2}{\log 1.1}$$
$$n \approx 7.2725$$

The solution is 7.2725.

You Try It 1

Solve for n: $(1.06)^n = 1.5$

Your solution

6.9585

Example 2

Solve for x: $3^{2x} = 4$

Solution

$$3^{2x} = 4$$
$$\log 3^{2x} = \log 4$$
$$2x \log 3 = \log 4$$
$$2x = \frac{\log 4}{\log 3}$$
$$x = \frac{\log 4}{2 \log 3}$$
$$x \approx 0.6309$$

The solution is 0.6309.

You Try It 2

Solve for x: $4^{3x} = 25$

Your solution

0.7740

Solutions on p. S37

Objective B To solve a logarithmic equation

VIDEO & DVD CD TUTOR WEB SSM

A logarithmic equation can be solved by using the Properties of Logarithms. Here are two examples.

⇒ Solve: $\log_9 x + \log_9 (x - 8) = 1$

$$\log_9 x + \log_9 (x - 8) = 1$$

$$\log_9 [x(x - 8)] = 1$$
• Use the Logarithm Property of Products to rewrite the left side of the equation.

$$9^1 = x(x - 8)$$
• Write the equation in exponential form.

$$9 = x^2 - 8x$$
• Simplify.

$$0 = x^2 - 8x - 9$$
• Solve for x.

$$0 = (x - 9)(x + 1)$$
• Factor and use the Principle of Zero Products.

$$x - 9 = 0 \qquad x + 1 = 0$$
$$x = 9 \qquad x = -1$$

Replacing x by 9 in the original equation reveals that 9 checks as a solution. Replacing x by -1 in the original equation results in the expression $\log_9(-1)$. Because the logarithm of a negative number is not a real number, -1 does not check as a solution.

The solution of the equation is 9.

⇒ Solve: $\log_3 6 - \log_3 (2x + 3) = \log_3 (x + 1)$

$$\log_3 6 - \log_3 (2x + 3) = \log_3 (x + 1)$$

$$\log_3 \frac{6}{2x + 3} = \log_3 (x + 1)$$
• Use the Quotient Property of Logarithms to rewrite the left side of the equation.

$$\frac{6}{2x + 3} = x + 1$$
• Use the 1–1 Property of Logarithms.

$$6 = (2x + 3)(x + 1)$$
• Multiply each side by $2x + 3$.

$$6 = 2x^2 + 5x + 3$$
• Multiply the binomials.

$$0 = 2x^2 + 5x - 3$$
• Write the equation in standard form.

$$0 = (2x - 1)(x + 3)$$
• Solve for x.

$$2x - 1 = 0 \qquad x + 3 = 0$$
$$x = \frac{1}{2} \qquad x = -3$$

Replacing x by $\frac{1}{2}$ in the original equation reveals that $\frac{1}{2}$ checks as a solution. Replacing x by -3 in the original equation results in the expression $\log_3(-2)$. Because the logarithm of a negative number is not a real number, -3 does not check as a solution.

The solution of the equation is $\frac{1}{2}$.

Objective 12.4B

Properties to Review
1–1 Property of Logarithms

Instructor Note
Solving logarithmic equations normally requires using some of the Properties of Logarithms. These properties are valid only if the arguments are positive numbers. Thus it is necessary to check the solutions of equations to ensure that this condition is met.

Discuss the Concepts

1. Explain how to solve the equation $2 = \log_3 x$.
2. Can the 1–1 Property of Logarithms be used to solve the equation $\log_3 x = \log_2 8$? Why or why not? No. The bases are not the same.
3. If $\log(x + 2) + \log(x - 2) = 6$, then what expression is equal to 10^6? $x^2 - 4$
4. Why is it possible for a logarithmic equation to have an extraneous solution?

In-Class Examples (Objective 12.4B)

Solve for x.

1. $\log_2 (x + 1) = 3$ 7
2. $\log_2 (x^2 - 2x) = 3$ −2, 4
3. $\log_3 \left(\dfrac{2x}{x + 1}\right) = 1$ −3
4. $\log_2 24 - \log_2 (x^2 - 1) = \log_2 3$ −3, 3

Concept Check

Determine whether the statement is true or false given that x, y, and b are positive real numbers and $b \neq 1$. Explain your reasoning.

1. $\log x + \log (x + 2) = \log (2x + 2)$ False
2. $\log (2x - 2) - \log x = 4$ is equivalent to $\log \frac{2x-2}{x} = 4$. True
3. $\log (2x) + \log 4 = 6$ is equivalent to $\log (8x) = 6$. True

Optional Student Activity

Solve the system of equations.

1. $\log (x + y) = 3$
   $x = y + 4$ (502, 498)
2. $\log (x + y) = 3$
   $x - y = 20$ (510, 490)
3. $8^{3x} = 4^{2y}$
   $x - y = 5$ (−4, −9)
4. $9^{3x} = 81^{3y}$
   $x + y = 3$ (2, 1)

Example 3

Solve for x: $\log_3 (2x - 1) = 2$

Solution

$\log_3 (2x - 1) = 2$

$$3^2 = 2x - 1 \quad \text{• Write in exponential form.}$$
$$9 = 2x - 1$$
$$10 = 2x$$
$$5 = x$$

The solution is 5.

You Try It 3

Solve for x: $\log_4 (x^2 - 3x) = 1$

Your solution

−1, 4

Example 4

Solve for x: $\log_2 x - \log_2 (x - 1) = \log_2 2$

Solution

$\log_2 x - \log_2 (x - 1) = \log_2 2$

$$\log_2 \left(\frac{x}{x-1}\right) = \log_2 2 \quad \text{• Use the Quotient Property of Logarithms.}$$
$$\frac{x}{x-1} = 2 \quad \text{• Use the 1–1 Property of Logarithms.}$$
$$(x - 1)\left(\frac{x}{x-1}\right) = (x - 1)2$$
$$x = 2x - 2$$
$$-x = -2$$
$$x = 2$$

The solution is 2.

You Try It 4

Solve for x: $\log_3 x + \log_3 (x + 3) = \log_3 4$

Your solution

1

Example 5

Solve for x:
$\log_2 (3x + 8) = \log_2 (2x + 2) + \log_2 (x - 2)$

Solution

$$\log_2 (3x + 8) = \log_2 (2x + 2) + \log_2 (x - 2)$$
$$\log_2 (3x + 8) = \log_2 [(2x + 2)(x - 2)]$$
$$\log_2 (3x + 8) = \log_2 (2x^2 - 2x - 4)$$
$$3x + 8 = 2x^2 - 2x - 4 \quad \text{• Use the 1–1 Property of Logarithms.}$$
$$0 = 2x^2 - 5x - 12$$
$$0 = (2x + 3)(x - 4)$$
$$2x + 3 = 0 \qquad x - 4 = 0$$
$$x = -\frac{3}{2} \qquad x = 4$$

$-\frac{3}{2}$ does not check as a solution; 4 checks as a solution. The solution is 4.

You Try It 5

Solve for x: $\log_3 x + \log_3 (x + 6) = 3$

Your solution

3

Solutions on p. S37

677

12.4 Exercises

Objective A

1. What is an exponential equation?

2. **a.** What does the One-to-One Property of Exponential Functions state?
b. Provide an example of when you would use this property.

Solve for x. Round to the nearest ten-thousandth.

3. $5^{4x-1} = 5^{x-2}$
$-\frac{1}{3}$

4. $7^{4x-3} = 7^{2x+1}$
2

5. $8^{x-4} = 8^{5x+8}$
-3

6. $10^{4x-5} = 10^{x+4}$
3

7. $9^x = 3^{x+1}$
1

8. $2^{x-1} = 4^x$
-1

9. $8^{x+2} = 16^x$
6

10. $9^{3x} = 81^{x-4}$
-8

11. $16^{2-x} = 32^{2x}$
$\frac{4}{7}$

12. $27^{2x-3} = 81^{4-x}$
$\frac{5}{2}$

13. $25^{3-x} = 125^{2x-1}$
$\frac{9}{8}$

14. $8^{4x-7} = 64^{x-3}$
$\frac{1}{2}$

15. $5^x = 6$
1.1133

16. $7^x = 10$
1.1833

17. $e^x = 3$
1.0986

18. $e^x = 2$
0.6931

19. $10^x = 21$
1.3222

20. $10^x = 37$
1.5682

21. $2^{-x} = 7$
-2.0074

22. $3^{-x} = 14$
-2.4022

23. $2^{x-1} = 6$
3.5850

24. $4^{x+1} = 9$
0.5850

25. $3^{2x-1} = 4$
1.1309

26. $4^{-x+2} = 12$
0.2075

Objective B

27. What is a logarithmic equation?

28. What does the One-to-One Property of Logarithms state?

Solve for x.

29. $\log_2 (2x - 3) = 3$
$\frac{11}{2}$

30. $\log_4 (3x + 1) = 2$
5

31. $\log_2 (x^2 + 2x) = 3$
$-4, 2$

32. $\log_3 (x^2 + 6x) = 3$
$-9, 3$

33. $\log_5 \left(\frac{2x}{x - 1}\right) = 1$
$\frac{5}{3}$

34. $\log_6 \left(\frac{3x}{x + 1}\right) = 1$
-2

Section 12.4

Suggested Assignment

Exercises 3–25, 29–45, odds
More challenging problems:
Exercises 47–55, odds
Exercise 56
(A graphing calculator is required.)

Answers to Writing Exercises

1. An exponential equation is one in which a variable occurs in an exponent.

2a. The One-to-One Property of Exponential Functions states that for $b > 0$, $b \neq 1$, if $b^u = b^v$, then $u = v$.

b. This property is used to solve exponential equations in which each side of the equation can be expressed in terms of the same base.

27. A logarithmic equation is an equation in which one or more of the terms is a logarithmic expression.

28. The One-to-One Property of Logarithms states that for $b > 0$, $b \neq 1$, if $\log_b x = \log_b y$, then $x = y$.

Quick Quiz (Objective 12.4A)

Solve for x. Round to the nearest ten-thousandth.

1. $5^{3x-2} = 5^{2x+1}$ 3

2. $16^x = 2$ $\frac{1}{4}$

3. $7^x = 20$ 1.5395

4. $2^{x+1} = 10$ 2.3219

5. $3^{x-1} = 9^x$ -1

Answers to Writing Exercises

57. The error is in the second step because the value of log 0.5 is less than zero. Multiplying each side of the inequality by this negative quantity changes the direction of the inequality symbol.

35. $\log x = \log (1 - x)$
$\frac{1}{2}$

36. $\ln (3x - 2) = \ln (x + 1)$
$\frac{3}{2}$

37. $\ln 5 = \ln (4x - 13)$
$\frac{9}{2}$

38. $\log_3 (x - 2) = \log_3 (2x)$
no solution

39. $\ln (3x + 2) = 4$
17.5327

40. $\ln (2x + 3) = -1$
−1.3161

41. $\log_2 (8x) - \log_2 (x^2 - 1) = \log_2 3$
3

42. $\log_5 (3x) - \log_5 (x^2 - 1) = \log_5 2$
2

43. $\log_9 x + \log_9 (2x - 3) = \log_9 2$
2

44. $\log_6 x + \log_6 (3x - 5) = \log_6 2$
2

45. $\log_8 (6x) = \log_8 2 + \log_8 (x - 4)$
no solution

46. $\log_7 (5x) = \log_7 3 + \log_7 (2x + 1)$
no solution

APPLYING THE CONCEPTS

Solve each equation with a graphing calculator. Round answers to the nearest hundredth.

47. $3^x = -x$
−0.55

48. $2^x = -x + 2$
0.54

49. $2^{-x} = x - 1$
1.38

50. $3^{-x} = 2x$
0.34

51. $\ln x = x^2$
no solution

52. $2 \ln x = -x + 1$
1

53. $\log_3 x = -2x - 2$
0.09

54. $\log_5 (2x) = -2x + 1$
0.50

55. A model for the distance s (in feet) that an object that is experiencing air resistance will fall in t seconds is given by $s = 312.5 \ln \left(\frac{e^{0.32t} + e^{-0.32t}}{2}\right)$.

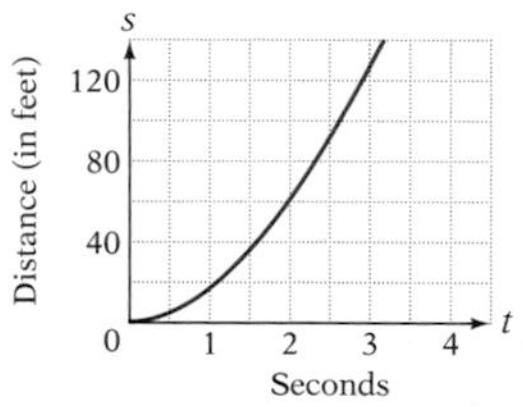

a. Graph this equation. *Suggestion:* Use Xmin = 0, Xmax = 4.5, Ymin = 0, Ymax = 140, and Yscl = 20.

b. Determine, to the nearest hundredth of a second, the time it takes the object to travel 100 ft.
2.64 s

56. A model for the distance s (in feet) that an object that is experiencing air resistance will fall in t seconds is given by $s = 78 \ln \left(\frac{e^{0.8t} + e^{-0.8t}}{2}\right)$.

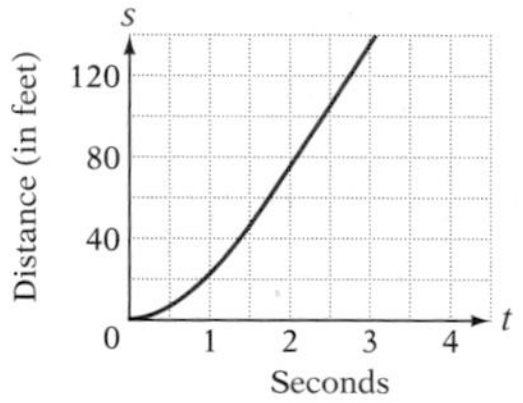

a. Graph this equation. *Suggestion:* Use Xmin = 0, Xmax = 4.5, Ymin = 0, Ymax = 140, and Yscl = 20.

b. Determine, to the nearest hundredth of a second, the time it takes the object to travel 125 ft.
2.86 s

57. The following "proof" shows that 0.5 < 0.25. Explain the error.

$$1 < 2$$
$$1 \cdot \log 0.5 < 2 \cdot \log 0.5$$
$$\log 0.5 < \log (0.5)^2$$
$$0.5 < (0.5)^2$$
$$0.5 < 0.25$$

Quick Quiz (Objective 12.4B)

Solve for *x*.

1. $\log_4 (x - 2) = 1$ 6

2. $\log_3 (x^2 - 6x) = 3$ −3, 9

3. $\log_5 x + \log_5 (2x - 1) = \log_5 1$ 1

12.5 Applications of Exponential and Logarithmic Functions

Objective A To solve application problems

VIDEO & DVD CD TUTOR WEB SSM

A biologist places one single-celled bacterium in a culture, and each hour that particular species of bacteria divides into two bacteria. After one hour there will be two bacteria. After two hours, each of the two bacteria will divide and there will be four bacteria. After three hours, each of the four bacteria will divide and there will be eight bacteria.

Point of Interest

Parkinson's Law (C. Northcote Parkinson) is sometimes stated as "A job will expand to fill the time allotted for the job." However, Parkinson actually said that in any new government administration, administrative employees will be added at the rate of about 5% to 6% per year. This is an example of exponential growth and means that a staff of 500 will grow to approximately 630 by the end of a 4-year term.

The table at the right shows the number of bacteria in the culture after various intervals of time, t, in hours. Values in this table could also be found by using the exponential equation $N = 2^t$.

Time, t	*Number of Bacteria, N*
0	1
1	2
2	4
3	8
4	16

The equation $N = 2^t$ is an example of an **exponential growth equation**. In general, any equation that can be written in the form $\boldsymbol{A = A_0 b^{kt}}$, where A is the size at time t, A_0 is the initial size, $b > 1$, and k is a positive real number, is an exponential growth equation. These equations are important not only in population growth studies but also in physics, chemistry, psychology, and economics.

Recall that interest is the amount of money paid (or received) when borrowing (or investing) money. **Compound interest** is interest that is computed not only on the original principal, but also on the interest already earned. The compound interest formula is an exponential growth equation.

The **compound interest formula** is $\boldsymbol{P = A(1 + i)^n}$, where A is the original value of an investment, i is the interest rate per compounding period, n is the total number of compounding periods, and P is the value of the investment after n periods.

➡ An investment broker deposits \$1000 into an account that earns 12% annual interest compounded quarterly. What is the value of the investment after two years?

$i = \frac{12\%}{4} = \frac{0.12}{4} = 0.03$ • **Find *i*, the interest rate per quarter. The quarterly rate is the annual rate divided by 4, the number of quarters in one year.**

$n = 4 \cdot 2 = 8$ • **Find *n*, the number of compounding periods. The investment is compounded quarterly, 4 times a year, for 2 years.**

$P = A(1 + i)^n$ • **Use the compound interest formula.**

$P = 1000(1 + 0.03)^8$ • **Replace *A*, *i*, and *n* by their values.**

$P \approx 1267$ • **Solve for *P*.**

The value of the investment after two years is approximately \$1267.

Objective 12.5A

New Vocabulary

exponential growth equation
exponential decay equation
compound interest

New Formulas

compound interest formula, $P = A(1 + i)^n$
pH formula, $\text{pH} = -\log(\text{H}^+)$
Richter scale magnitude, $M = \log \frac{I}{I_0}$

Instructor Note

Another exponential function from finance that will be of interest to students is the function that is used to calculate the amount of an amortized loan payment, such as a car loan:

$$P = B\left[\frac{\frac{i}{12}}{1 - \left(1 + \frac{i}{12}\right)^{-n}}\right]$$

In this equation, B is the amount borrowed, i is the annual interest rate as a decimal, and n is the number of months to repay the loan.

You might ask students to use this formula to find the monthly payment on a 30-year mortgage of \$100,000 with an annual interest rate of 8%. \$733.76

In-Class Examples (Objective 12.5A)

1. \$5000 is invested at 12% annual interest compounded semiannually. In approximately how many years will the investment be worth twice the original amount? Use the compound interest formula $P = A(1 + i)^n$, where A is the original value of an investment, i is the interest rate per compounding period, n is the total number of compounding periods, and P is the value of the investment after n periods. 6 years

2. Find the pH of a hydrogen chloride solution for which the hydrogen ion concentration is 5.2×10^{-5}. Use the pH equation $\text{pH} = -\log(\text{H}^+)$, where H^+ is the hydrogen ion concentration of a solution. Round to the nearest tenth. 4.3

Discuss the Concepts

1. State whether each of the following equations is an exponential growth equation, an exponential decay equation, or neither. Explain your reasoning.
   a. $f(x) = 6^x$
   b. $g(x) = \left(\frac{2}{3}\right)^x$
   c. $F(x) = 3x^4$
   d. $f(x) = e^x$
   e. $y = 250(1.008)^x$
   f. $t(r) = 64\left(\frac{1}{2}\right)^r$
   a, d, and e are exponential growth equations. b and f are exponential decay equations. c is neither.
2. The intensity I of an x-ray after it has passed through a material that is x centimeters thick is given by $I = I_0e^{-kx}$, where I_0 is the initial intensity of the x-ray and k is a number that depends on the material. Radiologists (physicians who specialize in the use of radioactive substances to diagnose and treat disease) wear lead shields when giving patients x-rays. The constant k for lead is 43. Use the given equation to explain why a lead shield of the same thickness as a copper shield ($k = 3.2$) offers better protection to the radiologist than does the copper shield.

Exponential decay offers another example of an exponential equation. One of the most common illustrations of exponential decay is the decay of a radioactive substance. For instance, tritium, a radioactive nucleus of hydrogen that has been used in luminous watch dials, has a half-life of approximately 12 years. This means that one-half of any given amount of tritium will disintegrate in 12 years.

The table at the right indicates the amount of an initial 10-microgram sample of tritium that remains after various intervals of time, t, in years. Values in this table could also be found by using the exponential equation $A = 10(0.5)^{t/12}$.

Time, t	*Amount, A*
0	10
12	5
24	2.5
36	1.25
48	0.625

The equation $A = 10(0.5)^{t/12}$ is an example of an **exponential decay equation**. Comparing this equation to the exponential growth equation, note that for exponential growth, the base of the exponential equation is greater than 1, whereas for exponential decay, the base is between 0 and 1.

A method by which an archaeologist can measure the age of a bone is called carbon dating. Carbon dating is based on a radioactive isotope of carbon called carbon-14, which has a half-life of approximately 5570 years. The exponential decay equation is given by $A = A_0(0.5)^{t/5570}$, where A_0 is the original amount of carbon-14 present in the bone, t is the age of the bone, and A is the amount of carbon-14 present after t years.

➡ A bone that originally contained 100 mg of carbon-14 now has 70 mg of carbon-14. What is the approximate age of the bone?

$$A = A_0(0.5)^{t/5570}$$
• Use the exponential decay equation.

$$70 = 100(0.5)^{t/5570}$$
• Replace A by 70 and A_0 by 100 and solve for t.

$$0.7 = (0.5)^{t/5570}$$
• Divide each side by 100.

$$\log 0.7 = \log (0.5)^{t/5570}$$
• Take the common logarithm of each side of the equation. Then simplify.

$$\log 0.7 = \frac{t}{5570}\log (0.5)$$

$$\frac{5570 \log 0.7}{\log 0.5} = t$$

$$2866 \approx t$$

The bone is approximately 2866 years old.

Point of Interest

The pH scale was created by the Danish biochemist Søren Sørensen in 1909 to measure the acidity of water used in the brewing of beer. pH is an abbreviation for *pondus hydrogenii*, which translates as "potential hydrogen."

A chemist measures the acidity or alkalinity of a solution by measuring the concentration of hydrogen ions, H^+, in the solution using the formula $pH = -\log (H^+)$. A neutral solution such as distilled water has a pH of 7, acids have a pH less than 7, and alkaline solutions (also called basic solutions) have a pH greater than 7.

➡ Find the pH of orange juice that has a hydrogen ion concentration, H^+, of 2.9×10^{-4}. Round to the nearest tenth.

$$pH = -\log (H^+)$$

$$= -\log (2.9 \times 10^{-4})$$
• $H^+ = 2.9 \times 10^{-4}$

$$\approx 3.5376$$

The pH of the orange juice is approximately 3.5.

681

Logarithmic functions are used to scale very large or very small numbers into numbers that are easier to comprehend. For instance, the *Richter scale magnitude* of an earthquake uses a logarithmic function to convert the intensity of shock waves I into a number M, which for most earthquakes is in the range of 0 to 10. The intensity I of an earthquake is often given in terms of the constant I_0, where I_0 is the intensity of the smallest earthquake, called a **zero-level earthquake,** that can be measured on a seismograph near the earthquake's epicenter. An earthquake with an intensity I has a Richter scale magnitude of $M = \log\left(\frac{I}{I_0}\right)$, where I_0 is the measure of a zero-level earthquake.

➡ Find the Richter scale magnitude of the 1999 Joshua Tree, California, earthquake, which had an intensity I of $12{,}589{,}254I_0$. Round to the nearest tenth.

TAKE NOTE
Note that we do not need to know the value of I_0 in order to determine the Richter scale magnitude of the earthquake.

$M = \log\left(\frac{I}{I_0}\right)$

$M = \log\left(\frac{12{,}589{,}254I_0}{I_0}\right)$ • $I = 12{,}589{,}254I_0$

$M = \log 12{,}589{,}254$ • Divide the numerator and denominator by I_0.

$M \approx 7.1$ • Evaluate log 12,589,254.

The 1999 Joshua Tree earthquake had a Richter scale magnitude of 7.1.

If you know the Richter scale magnitude of an earthquake, you can determine the intensity of the earthquake.

➡ Find the intensity of the 1999 Taiwan earthquake, which measured 7.6 on the Richter scale. Write the answer in terms of I_0.

Point of Interest
The Richter scale was created by seismologist Charles F. Richter in 1935. Note that a tenfold increase in the intensity level of an earthquake increases the Richter scale magnitude of the earthquake by only 1.

$7.6 = \log\left(\frac{I}{I_0}\right)$ • Replace M in $M = \log\left(\frac{I}{I_0}\right)$ by 7.6.

$10^{7.6} = \frac{I}{I_0}$ • Write in exponential form.

$10^{7.6}I_0 = I$ • Multiply both sides by I_0.

$39{,}810{,}717I_0 \approx I$ • Evaluate $10^{7.6}$.

The 1999 Taiwan earthquake had an intensity that was approximately 39,810,717 times the intensity of a zero-level earthquake.

The percent of light that will pass through a substance is given by $\log P = -kd$, where P is the percent of light passing through the substance, k is a constant depending on the substance, and d is the thickness of the substance in meters.

➡ For certain parts of the ocean, $k = 0.03$. Using this value, at what depth will the percent of light be 50% of the light at the surface of the ocean? Round to the nearest meter.

$\log P = -kd$

$\log(0.5) = -0.03d$ • Replace P by 0.5 (50%) and k by 0.03.

$\frac{\log(0.5)}{-0.03} = d$ • Solve for d.

$10.0343 \approx d$

At a depth of about 10 m, the light will be 50% of the light at the surface.

Concept Check
The atmospheric pressure P decreases exponentially with height h above sea level. The equation relating the pressure P, in pounds per square inch, and height h, in feet, is $P = 14.7e^{-0.00004h}$. Find the height of Mt. Everest if the atmospheric pressure at the top is 4.6 lb/in^2. Round to the nearest foot.
29,045 ft

Optional Student Activity

1. An *annuity* is a fixed amount of money that is either paid or received over equal intervals of time. A retirement plan into which a certain amount is deposited each month is an example of an annuity; equal deposits are made over equal intervals of time (monthly). The equation that relates the amount of money available for retirement to the monthly deposit is
$V = P\left[\frac{(1+i)^x - 1}{i}\right]$,
where i is the interest rate per month, x is the number of months deposits are made, P is the payment, and V is the value (called the *future value*) of the retirement fund after x payments. Suppose $100 is deposited each month into an account that earns interest at the rate of 0.5% per month (6% per year). For how many years must the investor make deposits in order to have a retirement account worth $20,000? Round to the nearest whole number. 12 years
2. One scientific study suggests that the *carrying capacity* of Earth is approximately 10 billion people. Find out what is meant by "carrying capacity." Research the current world population and project when Earth's population will reach 10 billion, assuming population growth rates of 1%, 2%, 3%, 4%, and 5%. Research the current rate of world population growth and use that number to determine when the world population will reach 10 billion.

(Continued on next page)

(Continued)
The carrying capacity of Earth is the maximum number of humans that can be supported indefinitely on Earth. Answers may vary. For example, using a world population of 6 billion and rounding to the nearest year:
1% growth rate: 51 years;
2% growth rate: 26 years;
3% growth rate: 17 years;
4% growth rate: 13 years;
5% growth rate: 10 years.
According to the United Nations, the current rate of world population growth is 77 million people per year. At this rate, the world population will reach 10 billion people in 52 years.

Example 1
An investment of \$3000 is placed into an account that earns 12% annual interest compounded monthly. In approximately how many years will the investment be worth twice the original amount?

Strategy
To find the time, solve the compound interest formula for n. Use $P = 6000$, $A = 3000$, and $i = \frac{12\%}{12} = \frac{0.12}{12} = 0.01$.

Solution

$$P = A(1 + i)^n$$
$$6000 = 3000(1 + 0.01)^n$$
$$6000 = 3000(1.01)^n$$
$$2 = (1.01)^n$$
$$\log 2 = \log (1.01)^n$$
$$\log 2 = n \log 1.01$$
$$\frac{\log 2}{\log 1.01} = n$$
$$70 \approx n$$

70 months ÷ 12 ≈ 5.8 years

In approximately 6 years, the investment will be worth \$6000.

You Try It 1
Find the hydrogen ion concentration, H^+, of vinegar that has a pH of 2.9. Round to the nearest hundred thousandth.

Your strategy

Your solution
0.00126

Example 2
The number of words per minute that a student can type will increase with practice and can be approximated by the equation $N = 100[1 - (0.9)^t]$, where N is the number of words typed per minute after t days of instruction. Find the number of words a student will type per minute after 8 days of instruction.

Strategy
To find the number of words per minute, replace t by its given value in the equation and solve for N.

Solution

$$N = 100[1 - (0.9)^t]$$
$$= 100[1 - (0.9)^8]$$
$$\approx 56.95$$

After 8 days of instruction, a student will type approximately 57 words per minute.

You Try It 2
On September 3, 2000, an earthquake measuring 5.2 on the Richter scale struck the Napa Valley, located 50 mi north of San Francisco. Find the intensity of the quake in terms of I_0.

Your strategy

Your solution
$158{,}489I_0$

Solutions on p. S37

683

Section 12.5

Suggested Assignment
Exercises 1–29, odds

12.5 Exercises

Objective A

For Exercises 1 to 4, use the compound interest formula $P = A(1 + i)^n$, where A is the original value of an investment, i is the interest rate per compounding period, n is the total number of compounding periods, and P is the value of the investment after n periods.

1. An investment broker deposits \$1000 into an account that earns 8% annual interest compounded quarterly. What is the value of the investment after 2 years? Round to the nearest dollar.
\$1172

2. A financial advisor recommends that a client deposit \$2500 into a fund that earns 7.5% annual interest compounded monthly. What will be the value of the investment after 3 years? Round to the nearest cent.
\$3128.62

3. To save for college tuition, the parents of a preschooler invest \$5000 in a bond fund that earns 6% annual interest compounded monthly. In approximately how many years will the investment be worth \$15,000?
18 years

4. A hospital administrator deposits \$10,000 into an account that earns 9% annual interest compounded monthly. In approximately how many years will the investment be worth \$15,000?
5 years

For Exercises 5 to 8, use the exponential decay equation $A = A_0\left(\frac{1}{2}\right)^{t/k}$, where A is the amount of a radioactive material present after time t, k is the half-life of the radioactive substance, and A_0 is the original amount of radioactive substance. Round to the nearest tenth.

5. An isotope of technetium is used to prepare images of internal body organs. This isotope has a half-life of approximately 6 h. A patient is injected with 30 mg of this isotope.
a. What is the technetium level in the patient after 3 h? 21.2 mg
b. How long (in hours) will it take for the technetium level to reach 20 mg? 3.5 h

6. Iodine-131 is an isotope that is used to study the functioning of the thyroid gland. This isotope has a half-life of approximately 8 days. A patient is given an injection that contains 8 micrograms of iodine-131.
a. What is the iodine level in the patient after 5 days? 5.2 micrograms
b. How long (in days) will it take for the iodine level to reach 5 micrograms? 5.4 days

7. A sample of promethium-147 (used in some luminous paints) weighs 25 mg. One year later, the sample weighs 18.95 mg. What is the half-life of promethium-147, in years?
2.5 years

Quick Quiz (Objective 12.5A)

1. The percent of light that will pass through a material is given by the equation $\log P = -kd$, where P is the percent of light passing through the material, k is a constant that depends on the material, and d is the thickness of the material in centimeters. The constant k for a piece of opaque glass that is 0.5 cm thick is 0.4. Find the percent of light that will pass through the glass. 63%

2. Find the pH of a sodium hydroxide solution for which the hydrogen ion concentration is 4.6×10^{-7}. Use the equation $\text{pH} = -\log(\text{H}^+)$, where H^+ is the hydrogen ion concentration of a solution. Round to the nearest tenth. 6.3

684

8. Francium-223 is a very rare radioactive isotope discovered in 1939 by Marguerite Percy. A 3-microgram sample of francium-223 decays to 2.54 micrograms in 5 min. What is the half-life of francium-223, in minutes?
20.8 min

9. Earth's atmospheric pressure changes as you rise above its surface. At an altitude of h kilometers, where $0 < h < 80$, the pressure P, in newtons per square centimeters (N/cm^2), is approximately modeled by the equation $P(h) = 10.13e^{-0.116h}$.
a. What is the approximate pressure at 40 km above Earth's surface? $0.098\ N/cm^2$
b. What is the approximate pressure on Earth's surface? $10.13\ N/cm^2$
c. Does atmospheric pressure increase or decrease as you rise above Earth's surface? decrease

10. The U.S. Census Bureau provides information about various segments of the population in the United States. The following table gives the number of people, in millions, age 80 and older at the beginning of each decade from 1910 to 2000. An equation that approximately models the data is $y = 0.18808(1.0365)^x$, where $x = 0$ corresponds to 1900 and y is the population, in millions, of people age 80 and over.

Year	1910	1920	1930	1940	1950	1960	1970	1980	1990	2000
Number of people age 80 and over (in millions)	0.3	0.4	0.5	0.8	1.1	1.6	2.3	2.9	3.9	9.3

a. According to the model, what is the predicted population of this age group in the year 2020? Round to the nearest tenth of a million. (*Hint*: You will need to determine the x-value for the year 2020.) 13.9 million people
b. In what year does this model predict that the population of this age group will be 15 million? Round to the nearest year. 2022

For Exercises 11 and 12, use the equation $pH = -\log(H^+)$, where H^+ is the hydrogen ion concentration of a solution. Round to the nearest tenth.

11. Find the pH of milk, which has a hydrogen ion concentration of 3.97×10^{-7}.
6.4

12. Find the pH of a baking soda solution for which the hydrogen ion concentration is 3.98×10^{-9}.
8.4

For Exercises 13 and 14, use the equation $\log P = -kd$, which gives the relationship between the percent P, as a decimal, of light passing through a substance of thickness d, in meters. Round to the nearest tenth.

13. The value of k for a swimming pool is approximately 0.05. At what depth, in meters, will the percent of light be 75% of the light at the surface of the pool?
2.5 m

14. The constant k for a piece of blue stained glass is 20. What percent of light will pass through a piece of this glass that is 0.005 m thick?
79.4%

685

For Exercises 15 and 16, use the equation $D = 10(\log I + 16)$, where D is the number of decibels of a sound and I is the power of the sound measured in watts. Round to the nearest whole number.

15. Find the number of decibels of normal conversation. The power of the sound of normal conversation is approximately 3.2×10^{-10} watts.
65 decibels

16. The loudest sound made by any animal is made by the blue whale and can be heard from more than 500 mi away. The power of the sound is 630 watts. Find the number of decibels of sound emitted by the blue whale.
188 decibels

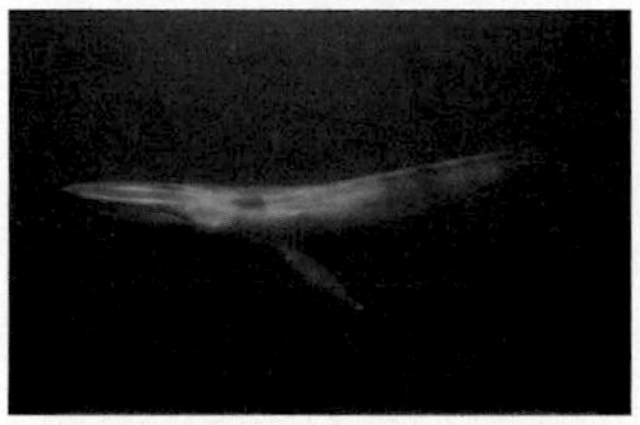

17. During the 1980s and 1990s, the average time T that it took to play a major league baseball game increased each year. If the year 1981 is represented by $x = 1$, then the function $T(x) = 149.57 + 7.63 \ln x$ approximates the time T, in minutes, to play a major league baseball game for the years $x = 1$ to $x = 19$. By how many minutes did the average time of a major league baseball game increase from 1981 to 1999? Round to the nearest tenth.
22.5 min

18. In 1962, the cost of a first-class postage stamp was \$.04. In 2002, the cost was \$.37. The increase in cost can be modeled by the equation $C = 0.04e^{0.057t}$, where C is the cost and t is the number of years after 1962. According to this model, in what year did a first-class postage stamp cost \$.22?
1992

19. The intensity I of an X-ray after it has passed through a material that is x centimeters thick is given by $I = I_0e^{-kx}$, where I_0 is the initial intensity of the X-ray and k is a number that depends on the material. The constant k for copper is 3.2. Find the thickness of copper that is needed so that the intensity of an X-ray after passing through the copper is 25% of the original intensity. Round to the nearest tenth.
0.4 cm

20. One model for the time it will take for the world's oil supply to be depleted is given by the equation $T = 14.29 \ln(0.00411r + 1)$, where r is the estimated world oil reserves in billions of barrels and T is the time, in years, before that amount of oil is depleted. Use this equation to determine how many barrels of oil are necessary to meet the world demand for 20 years. Round to the nearest tenth.
742.9 billion barrels

For Exercises 21 to 24, use the Richter scale equation $M = \log \frac{I}{I_0}$, where M is the magnitude of an earthquake, I is the intensity of the shock waves, and I_0 is the measure of the intensity of a zero-level earthquake.

21. On July 14, 2000, an earthquake struck the Kodiak Island region of Alaska. The earthquake had an intensity of $I = 6{,}309{,}573I_0$. Find the Richter scale magnitude of this earthquake. Round to the nearest tenth.
6.8

22. The earthquake that occurred on January 26, 2000, near Gujarat, India, had an intensity of $I = 50{,}118{,}723I_0$. Find the Richter scale magnitude of this earthquake. Round to the nearest tenth.
7.7

23. An earthquake that occurred in Japan on March 2, 1933, measured 8.9 on the Richter scale. Find the intensity of this earthquake in terms of I_0. Round to the nearest whole number.
$794{,}328{,}235I_0$

24. An earthquake that occurred in China in 1978 measured 8.2 on the Richter scale. Find the intensity of this earthquake in terms of I_0. Round to the nearest whole number.
$158{,}489{,}319I_0$

Shown at the right is a **seismogram,** which is used to measure the magnitude of an earthquake. The magnitude is determined by the amplitude A of a shock wave and the difference in time t, in seconds, between the occurrences of two types of waves called primary waves and secondary waves. As you can see on the graph, a primary wave is abbreviated p-wave, and a secondary wave is abbreviated s-wave. The amplitude A of a wave is one-half the difference between its highest and lowest points. For this graph, A is 23 mm. The equation is $M = \log A + 3 \log 8t - 2.92$.

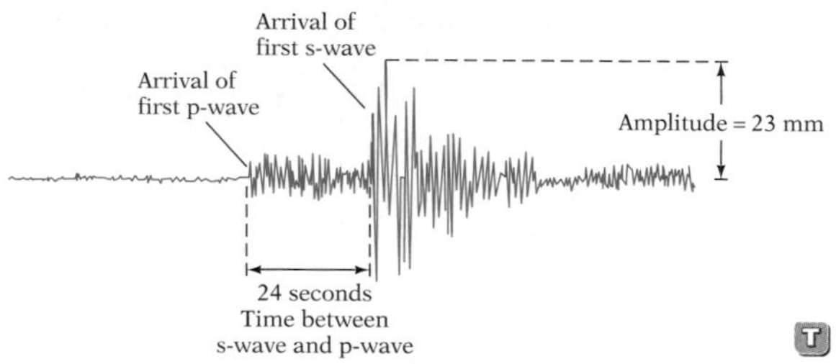

25. Determine the magnitude of the earthquake for the seismogram given in the figure. Round to the nearest tenth.
5.3

26. Find the magnitude of an earthquake that has a seismogram with an amplitude of 30 mm and for which t is 21 s.
5.2

27. Find the magnitude of an earthquake that has a seismogram with an amplitude of 28 mm and for which t is 28 s.
5.6

APPLYING THE CONCEPTS

28. The value of an investment in an account that earns an annual interest rate of 10% compounded daily grows according to the equation $A = A_0\left(1 + \frac{0.10}{365}\right)^{365t}$, where A_0 is the original value of an investment and t is the time, in years. Find the time for the investment to double in value. Round to the nearest year.
7 years

29. Some banks now use continuous compounding of an amount invested. In this case, the value of an initial investment of A dollars after t years at an annual interest rate of r% is given by the equation $P = Ae^{rt}$. Using this equation, find the value after 5 years of an investment of \$2500 in an account that earns 5% annual interest.
\$3210.06

687

Focus on Problem Solving

Proof by Contradiction

The four-step plan for solving problems that we have used before is restated here.

1. Understand the problem.
2. Devise a plan.
3. Carry out the plan.
4. Review the solution.

One of the techniques that can be used in the second step is a method called *proof by contradiction*. In this method you assume that the conditions of the problem you are trying to solve can be met and then show that your assumption leads to a condition you already know is not true.

To illustrate this method, suppose we try to prove that $\sqrt{2}$ is a rational number. We begin by recalling that a rational number is one that can be written as the quotient of integers. Therefore, let a and b be two integers with no common factors. If $\sqrt{2}$ is a rational number, then

$$\sqrt{2} = \frac{a}{b}$$

$$(\sqrt{2})^2 = \left(\frac{a}{b}\right)^2 \qquad \text{• Square each side.}$$

$$2 = \frac{a^2}{b^2}$$

$$2b^2 = a^2 \qquad \text{• Multiply each side by } b^2.$$

From the last equation, a^2 is an even number. Because a^2 is an even number, a is an even number. Now divide each side of the last equation by 2.

$$2b^2 = a^2$$

$$2b^2 = a \cdot a$$

$$b^2 = a \cdot x \qquad \text{• } x = \frac{a}{2}$$

Because a is an even number, $a \cdot x$ is an even number. Because $a \cdot x$ is an even number, b^2 is an even number, and this in turn means that b is an even number. This result, however, contradicts the assumption that a and b are two integers with no common factors. Because this assumption is now known not to be true, we conclude that our original assumption, that $\sqrt{2}$ is a rational number, is false. This proves that $\sqrt{2}$ is an irrational number.

Try a proof by contradiction for the following problem: "Is it possible to write numbers using each of the digits 0, 1, 2, 3, 4, 5, 6, 7, 8, and 9 exactly once such that the sum of the numbers is exactly 100?"* Here are some suggestions. First note that the sum of the 10 digits is 45. This means that some of the digits used must be tens digits. Let x be the sum of those digits.

1. What is the sum of the remaining units digits?
2. Express "the sum of the units digits and the tens digits equals 100" as an equation.
3. Solve the equation for x.
4. Explain why this result means that it is impossible to satisfy both conditions of the problem.

*G. Polya, *How to Solve It: A New Aspect of Mathematical Method*. Copyright © 1957 by Princeton University Press. Reprinted by permission of Princeton University Press.

Answers to Focus on Problem Solving: Proof by Contradiction

1. The sum of the 10 digits is 45. Let x = the sum of the tens digits, and S = the sum of the remaining units digits. Then $S = 45 - x$.
2. $10x + (45 - x) = 100$
3. $x = \frac{55}{9}$
4. x is the sum of integers, so $x = \frac{55}{9}$ is not possible. Our assumption that it is possible to use each of the digits 0, 1, 2, 3, 4, 5, 6, 7, 8, and 9 exactly once in such a way that the sum is 100 is not valid.

Answers to Projects and Group Activities: Fractals

1a. 2
b. 2
c. 2
2a. 4
b. 4
c. 4
3a. 2
b. 4
4a. 2
b. 8

5. $d = \dfrac{\log(\text{size ratio})}{\log(\text{scale factor})}$

$d = \dfrac{\log 8}{\log 2}$

$d = \dfrac{\log 2^3}{\log 2}$

$d = \dfrac{3 \log 2}{\log 2} = 3$

Thus the cubes are three-dimensional figures.

6a. Step 2 to 1: scale factor $= \dfrac{2}{1} = 2$

Step 3 to 2: scale factor $= \dfrac{4}{2} = 2$

b. Step 2 to 1: size ratio $= \dfrac{3}{1} = 3$

Step 3 to 2: size ratio $= \dfrac{9}{3} = 3$

7. $d = \dfrac{\log(\text{size ratio})}{\log(\text{scale factor})}$

$d = \dfrac{\log 3}{\log 2} \approx 1.58$

Projects and Group Activities

Fractals*

Fractals have a wide variety of applications. They have been used to create special effects for the *Star Wars* and *Star Trek* movies and to explain the behavior of some biological and economic systems. One aspect of fractals that has fascinated mathematicians is that they apparently have *fractional dimension*.

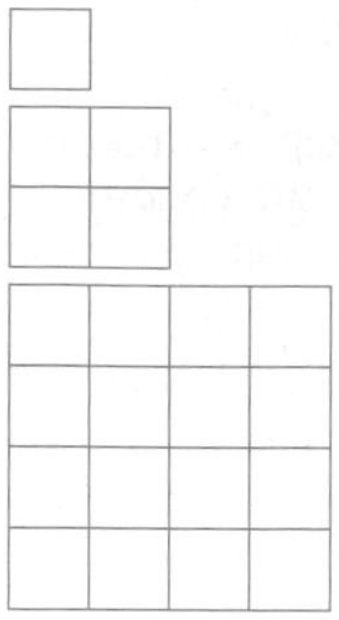

To understand the idea of fractional dimension, one must first understand the terms "scale factor" and "size." Consider a unit square (a square of length 1). By joining four of these squares, we can create another square, the length of which is 2, and the size of which is 4. (Here, size = number of square units.) Four of these larger squares can in turn be put together to make a third square of length 4 and size 16. This process of grouping together four squares can in theory be done an infinite number of times; yet at each step, the following quantities will be the same:

$$\text{Scale factor} = \frac{\text{new length}}{\text{old length}} \qquad \text{Size ratio} = \frac{\text{new size}}{\text{old size}}$$

Consider the unit square as Step 1, the four unit squares as Step 2, etc.

1. Calculate the scale factor going from **a.** Step 1 to Step 2, **b.** Step 2 to Step 3, and **c.** Step 3 to Step 4.
2. Calculate the size ratio going from **a.** Step 1 to Step 2, **b.** Step 2 to Step 3, and **c.** Step 3 to Step 4.
3. What is **a.** the scale factor and **b.** the size ratio going from Step n to Step $n + 1$?

Mathematicians have defined dimension using the formula $d = \dfrac{\log(\text{size ratio})}{\log(\text{scale factor})}$.

For the squares discussed above, $d = \dfrac{\log(\text{size ratio})}{\log(\text{scale factor})} = \dfrac{\log 4}{\log 2} = 2.$

So by this definition of dimension, squares are two-dimensional figures.

Now consider a unit cube (Step 1). Group eight unit cubes to form a cube that is 2 units on each side (Step 2). Group eight of the cubes from Step 2 to form a cube that is 4 units on each side (Step 3).

4. Calculate **a.** the scale factor and **b.** the size ratio for this process.
5. Show that the cubes are three-dimensional figures.

In each of the above examples, if the process is continued indefinitely, we still have a square or a cube. Consider a process that is more difficult to envision. Let Step 1 be an equilateral triangle whose base has length 1 unit, and let Step 2 be a grouping of three of these equilateral triangles, such that the space between them is another equilateral triangle with a base of length 1 unit. Three shapes from Step 2 are arranged with an equilateral triangle in their center, and so on. It is hard to imagine the result if this is done an infinite number of times, but mathematicians have shown that the result is a single figure of fractional dimension. (Similar processes have been used to create fascinating artistic patterns and to explain scientific phenomena.)

6. Show that for this process **a.** the scale factor is 2 and **b.** the size ratio is 3.
7. Calculate the dimension of the fractal. (Note that it is a *fractional* dimension!)

*Adapted with permission from "Student Math Notes," by Tami Martin, *News Bulletin*, November 1991.

Solving Exponential and Logarithmic Equations Using a Graphing Calculator

A graphing calculator can be used to draw the graphs of logarithmic functions. Note that there are two logarithmic keys, [LOG] and [LN], on a graphing calculator. The first key gives the values of common logarithms (base 10), and the second gives the values of natural logarithms (base e).

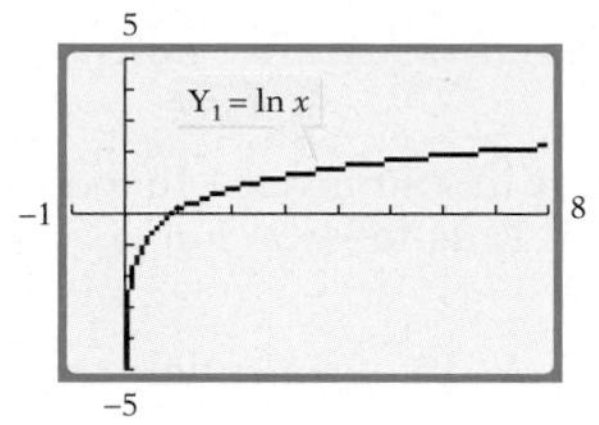

To graph $y = \ln x$, press the [Y=] key. Clear any equations already entered. Press [LN] [X,T,θ,n] [ )] [GRAPH]. The graph is shown at the left with a viewing window of Xmin = −1, Xmax = 8, Ymin = −5, Ymax = 5.

Some exponential and logarithmic equations cannot be solved algebraically. In these cases, a graphical approach may be appropriate. Here is an example.

➡ Solve $\ln (2x + 4) = x^2$ for x. Round to the nearest hundredth.

Rewrite the equation by subtracting x^2 from each side of the equation.

$$\ln (2x + 4) = x^2$$
$$\ln (2x + 4) - x^2 = 0$$

The zeros of $f(x) = \ln (2x + 4) - x^2$ are the solutions of $\ln (2x + 4) - x^2 = 0$.

Graph f and use the zero feature of the graphing calculator to estimate the solutions to the nearest hundredth.

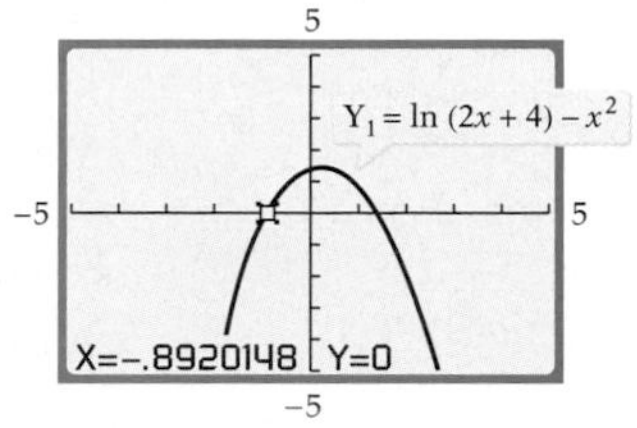

The zeros are approximately −0.89 and 1.38.

The solutions are −0.89 and 1.38.

Solve for x by graphing. Round to the nearest hundredth.

1. $2^x = 2x + 4$ **2.** $3^x = -x - 1$

3. $e^x = -2x - 2$ **4.** $e^x = 3x + 4$

5. $\log (2x - 1) = -x + 3$ **6.** $\log (x + 4) = -2x + 1$

7. $\ln (x + 2) = x^2 - 3$ **8.** $\ln x = -x^2 + 1$

To graph a logarithmic function with a base other than base 10 or base e, the Change-of-Base Formula is used. Here is the Change-of-Base Formula, which we discussed earlier.

$$\log_a N = \frac{\log_b N}{\log_b a}$$

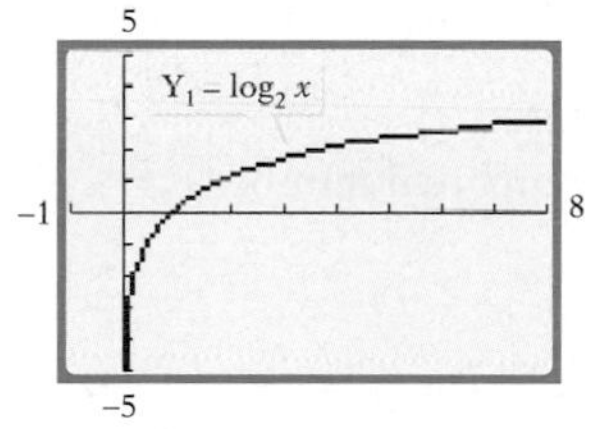

To graph $y = \log_2 x$, first change from base 2 logarithms to the equivalent base 10 logarithms or natural logarithms. Base 10 is shown here.

$$y = \log_2 x = \frac{\log x}{\log 2}$$ • $a = 2, b = 10, N = x$

Then graph the equivalent equation $y = \frac{\log x}{\log 2}$. The graph is shown at the left with a viewing window of Xmin = −1, Xmax = 8, Ymin = −5, Ymax = 5.

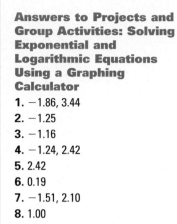

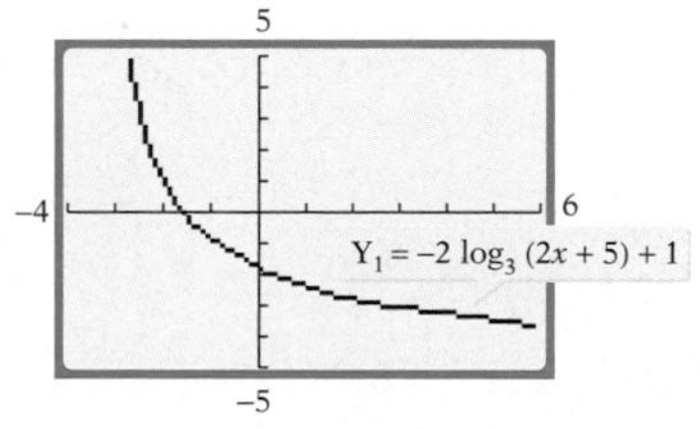

The graphs of more complicated functions involving logarithms are produced in much the same way as outlined above. To graph $y = -2 \log_3 (2x + 5) + 1$, enter into the calculator the equivalent equation $y = \frac{-2 \log (2x + 5)}{\log 3} + 1$.

Answers to Projects and Group Activities: Solving Exponential and Logarithmic Equations Using a Graphing Calculator

1. −1.86, 3.44
2. −1.25
3. −1.16
4. −1.24, 2.42
5. 2.42
6. 0.19
7. −1.51, 2.10
8. 1.00

Chapter Summary

Key Words

A function of the form $f(x) = b^x$, where b is a positive real number not equal to 1, is an *exponential function*. The number b is the *base* of the exponential function. [p. 651]

The function defined by $f(x) = e^x$ is called the *natural exponential function*. [p. 652]

Because the exponential function is a 1–1 function, it has an inverse function that is called a *logarithm*. The defnition of logarithm is as follows: For $b > 0$, $b \neq 1$, $y = \log_b x$ is equivalent to $x = b^y$. [p. 659]

If $\log_b M = N$, then M is the *antilogarithm* base b of N. In exponential form, $M = b^N$. [p. 660]

Logarithms with base 10 are called *common logarithms*. The base, 10, is omitted when the common logarithm of a number is written. The decimal part of a common logarithm is called the *mantissa*; the integer part is called the *characteristic*. [p. 660]

When e (the base of the natural exponential function) is used as the base of a logarithm, the logarithm is referred to as the *natural logarithm* and is abbreviated $\ln x$. [p. 660]

An *exponential equation* is one in which a variable occurs in the exponent. [p. 673]

An *exponential growth equation* is an equation that can be written in the form $A = A_0 b^{kt}$, where A is the size at time t, A_0 is the initial size, $b > 1$, and k is a positive real number. In an *exponential decay equation*, the value of b is between 0 and 1. [pp. 679, 680]

Essential Rules

The One-to-One Property of Exponential Functions [p. 659]
For $b > 0$, $b \neq 1$, if $b^u = b^v$, then $u = v$.

The Logarithm Property of the Product of Two Numbers [p. 661]
For any positive real numbers x, y, and b, $b \neq 1$, $\log_b (xy) = \log_b x + \log_b y$.

The Logarithm Property of the Quotient of Two Numbers [p. 662]
For any positive real numbers x, y, and b, $b \neq 1$, $\log_b \frac{x}{y} = \log_b x - \log_b y$.

The Logarithm Property of the Power of a Number [p. 662]
For any positive real numbers x and b, $b \neq 1$, and for any real number r, $\log_b x^r = r \log_b x$.

Additional Properties of Logarithms [p. 662]
For any positive real numbers x, y, and b, $b \neq 1$: $\log_b 1 = 0$
$\log_b b^x = x$

Change-of-Base Formula [p. 664]

$$\log_a N = \frac{\log_b N}{\log_b a}$$

To Solve Numerical Problems Using Logarithms [p. 662]
For any positive real numbers x, y, and b, $b \neq 1$, if $\log_b x = \log_b y$, then $x = y$.

691

Chapter Review

1. Evaluate $f(x) = e^{x-2}$ at $x = 2$.
1 [12.1A]

2. Write $\log_5 25 = 2$ in exponential form.
$5^2 = 25$ [12.2A]

3. Graph: $f(x) = 3^{-x} + 2$

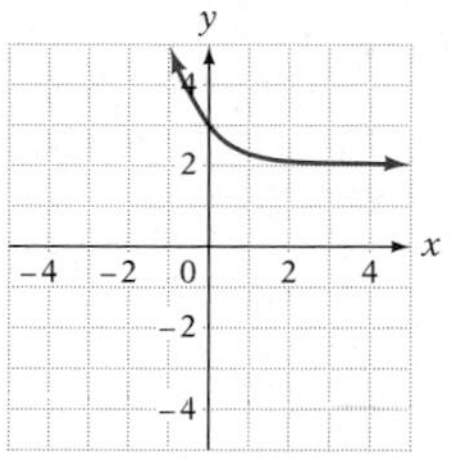

[12.1B]

4. Graph: $f(x) = \log_3 (x - 1)$

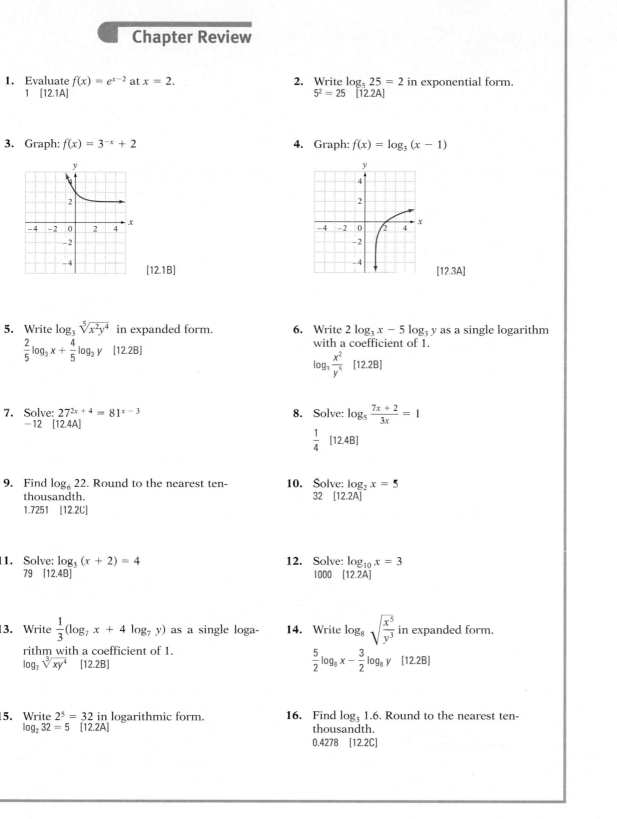

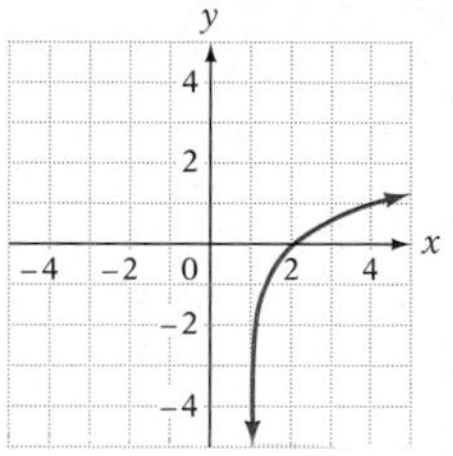

[12.3A]

5. Write $\log_3 \sqrt[5]{x^2y^4}$ in expanded form.
$\frac{2}{5}\log_3 x + \frac{4}{5}\log_3 y$ [12.2B]

6. Write $2 \log_3 x - 5 \log_3 y$ as a single logarithm with a coefficient of 1.
$\log_3 \frac{x^2}{y^5}$ [12.2B]

7. Solve: $27^{2x+4} = 81^{x-3}$
−12 [12.4A]

8. Solve: $\log_5 \frac{7x + 2}{3x} = 1$
$\frac{1}{4}$ [12.4B]

9. Find $\log_6 22$. Round to the nearest ten-thousandth.
1.7251 [12.2C]

10. Solve: $\log_2 x = 5$
32 [12.2A]

11. Solve: $\log_3 (x + 2) = 4$
79 [12.4B]

12. Solve: $\log_{10} x = 3$
1000 [12.2A]

13. Write $\frac{1}{3}(\log_7 x + 4 \log_7 y)$ as a single logarithm with a coefficient of 1.
$\log_7 \sqrt[3]{xy^4}$ [12.2B]

14. Write $\log_8 \sqrt{\frac{x^5}{y^3}}$ in expanded form.
$\frac{5}{2}\log_8 x - \frac{3}{2}\log_8 y$ [12.2B]

15. Write $2^5 = 32$ in logarithmic form.
$\log_2 32 = 5$ [12.2A]

16. Find $\log_3 1.6$. Round to the nearest ten-thousandth.
0.4278 [12.2C]

692 Chapter 12 / Exponential and Logarithmic Functions

17. Solve $3^{x+2} = 5$ for x. Round to the nearest thousandth.
-0.535 [12.4A]

18. Evaluate $f(x) = \left(\frac{2}{3}\right)^{x+2}$ at $x = -3$.
$\frac{3}{2}$ [12.1A]

19. Solve: $\log_8 (x + 2) - \log_8 x = \log_8 4$
$\frac{2}{3}$ [12.4B]

20. Solve: $\log_6 (2x) = \log_6 2 + \log_6 (3x - 4)$
2 [12.4B]

21. Graph: $f(x) = \left(\frac{2}{3}\right)^{x+1}$

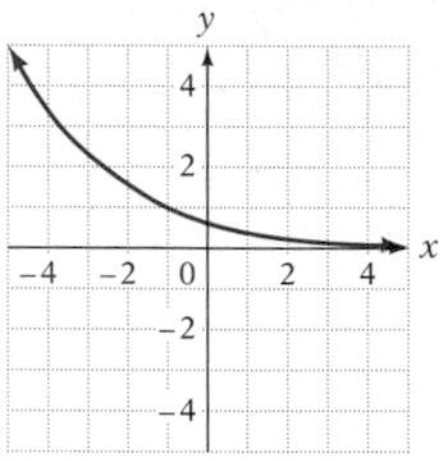

[12.1B]

22. Graph: $f(x) = \log_2 (2x - 1)$

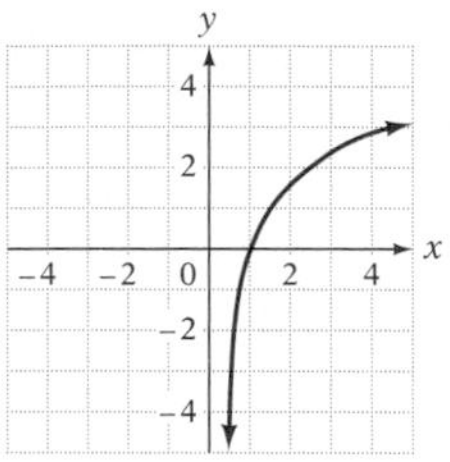

[12.3A]

23. Use the compound interest formula $P = A(1 + i)^n$, where A is the original value of an investment, i is the interest rate per compounding period, and n is the number of compounding periods, to find the value of an investment after 2 years. The amount of the investment is \$4000, and it is invested at 8% compounded monthly. Round to the nearest dollar.
\$4692 [12.5A]

24. An earthquake off the coast of Central America in January, 2001, had an intensity of $I = 39{,}810{,}717I_0$. Find the Richter scale magnitude of the earthquake. Use the Richter scale equation $M = \log \frac{I}{I_0}$, where M is the magnitude of an earthquake, I is the intensity of the shock waves, and I_0 is the measure of the intensity of a zero-level earthquake. Round to the nearest tenth.
7.6 [12.5A]

25. Use the exponential decay equation $A = A_0\left(\frac{1}{2}\right)^{t/k}$, where A is the amount of a radioactive material present after time t, k is the half-life, and A_0 is the original amount of radioactive material, to find the half-life of a material that decays from 25 mg to 15 mg in 20 days. Round to the nearest whole number.
27 days [12.5A]

26. The number of decibels, D, of a sound can be given by the equation $D = 10(\log I + 16)$, where I is the power of the sound measured in watts. Find the number of decibels of sound emitted from a busy street corner for which the power of the sound is 5×10^{-6} watts.
107 decibels [12.5A]

693

Chapter Test

1. Evaluate $f(x) = \left(\frac{2}{3}\right)^x$ at $x = 0$.

$f(0) = 1$ [12.1A]

2. Evaluate $f(x) = 3^{x+1}$ at $x = -2$.

$f(-2) = \frac{1}{3}$ [12.1A]

3. Graph: $f(x) = 2^x - 3$

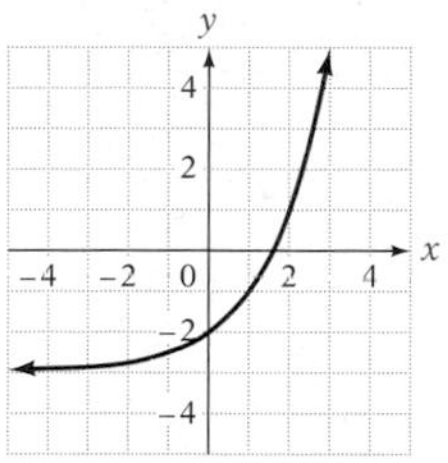

[12.1B]

4. Graph: $f(x) = 2^x + 2$

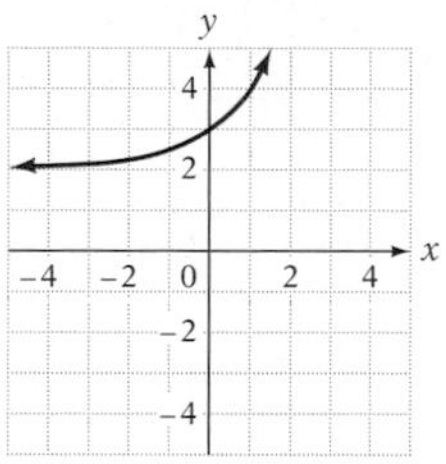

[12.1B]

5. Evaluate: $\log_4 16$

2 [12.2A]

6. Solve for x: $\log_3 x = -2$

$\frac{1}{9}$ [12.2A]

7. Graph: $f(x) = \log_2 (2x)$

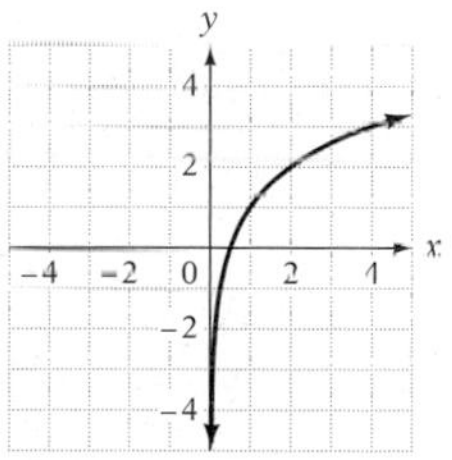

[12.3A]

8. Graph: $f(x) = \log_3 (x + 1)$

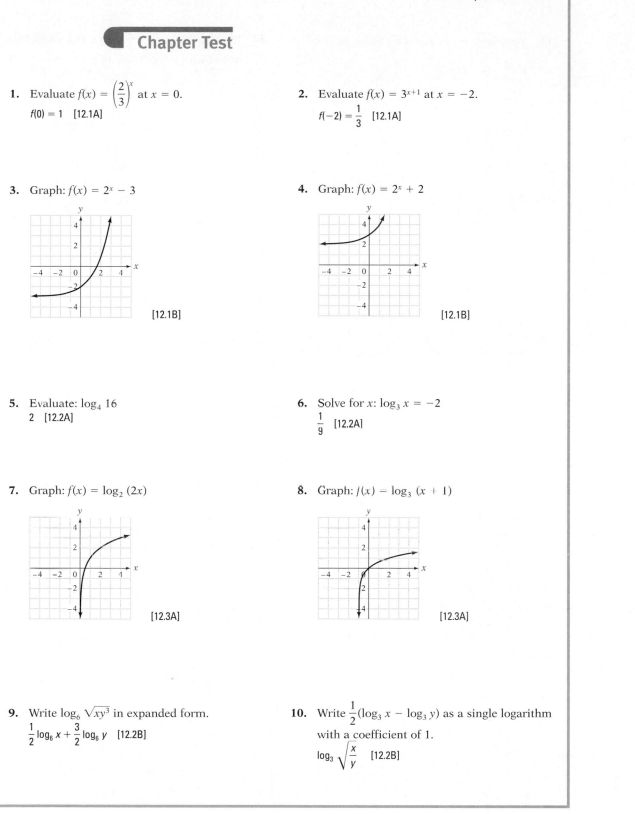

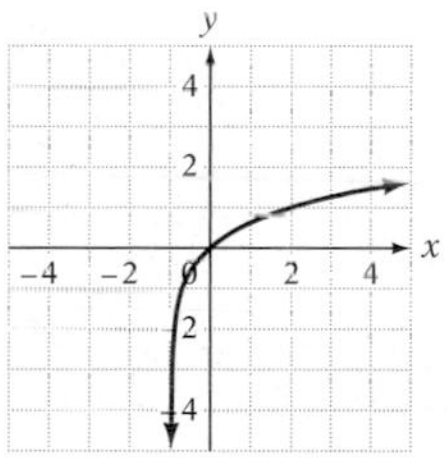

[12.3A]

9. Write $\log_6 \sqrt{xy^3}$ in expanded form.

$\frac{1}{2}\log_6 x + \frac{3}{2}\log_6 y$ [12.2B]

10. Write $\frac{1}{2}(\log_3 x - \log_3 y)$ as a single logarithm with a coefficient of 1.

$\log_3 \sqrt{\frac{x}{y}}$ [12.2B]

11. Write $\ln\left(\frac{x}{\sqrt{z}}\right)$ in expanded form.
$\ln x - \frac{1}{2}\ln z$ [12.2B]

12. Write $3 \ln x - \ln y - \frac{1}{2} \ln z$ as a single logarithm with a coefficient of 1.
$\ln \frac{x^3}{y\sqrt{z}}$ [12.2B]

13. Solve for x: $3^{7x+1} = 3^{4x-5}$
-2 [12.4A]

14. Solve for x: $8^x = 2^{x-6}$
-3 [12.4A]

15. Solve for x: $3^x = 17$
Round to the nearest ten-thousandth.
2.5789 [12.4A]

16. Solve for x: $\log x + \log (x - 4) = \log 12$
6 [12.4B]

17. Solve for x: $\log_6 x + \log_6 (x - 1) = 1$
3 [12.4B]

18. Find $\log_5 9$.
Round to the nearest ten-thousandth.
1.3652 [12.2C]

19. Find $\log_3 19$.
Round to the nearest ten-thousandth.
2.6801 [12.2C]

20. Use the exponential decay equation $A = A_0\left(\frac{1}{2}\right)^{t/k}$, where A is the amount of a radioactive material present after time t, k is the half-life of the material, and A_0 is the original amount of radioactive material, to find the half-life of a material that decays from 10 mg to 9 mg in 5 h. Round to the nearest whole number.
33 h [12.5A]

695

Cumulative Review

1. Solve: $4 - 2[x - 3(2 - 3x) - 4x] = 2x$
$\frac{8}{7}$ [2.2C]

2. Find the equation of the line that contains the point (2, −2) and is parallel to the line $2x - y = 5$.
$y = 2x - 6$ [4.6A]

3. Factor: $4x^{2n} + 7x^n + 3$
$(4x^n + 3)(x^n + 1)$ [7.3A/7.3B]

4. Simplify: $\dfrac{1 - \frac{5}{x} + \frac{6}{x^2}}{1 + \frac{1}{x} - \frac{6}{x^2}}$
$\frac{x-3}{x+3}$ [8.3A]

5. Simplify: $\dfrac{\sqrt{xy}}{\sqrt{x} - \sqrt{y}}$
$\frac{x\sqrt{y} + y\sqrt{x}}{x - y}$ [9.2D]

6. Solve by completing the square: $x^2 - 4x - 6 = 0$
$2 + \sqrt{10}, 2 - \sqrt{10}$ [10.2A]

7. Find the unknown side of the triangle in the figure below.

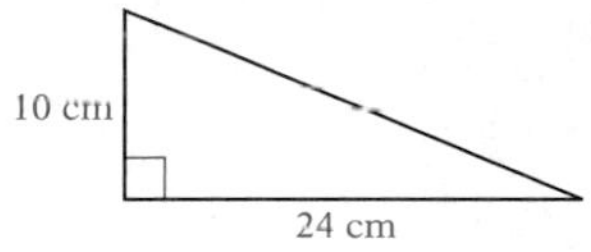

26 cm [9.4B]

8. Graph the solution set: $2x - y < 3$
$x + y < 1$

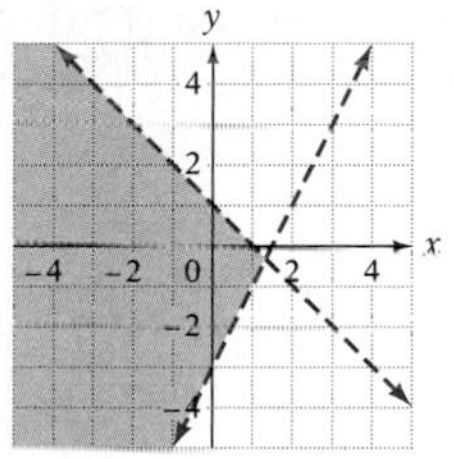

[5.5A]

9. Solve by the addition method.
$3x - y + z = 3$
$x + y + 4z = 7$
$3x - 2y + 3z = 8$
(0, −1, 2) [5.2B]

10. Subtract: $\dfrac{x-4}{2-x} - \dfrac{1-6x}{2x^2 - 7x + 6}$
$-\frac{2x^2 - 17x + 13}{(x-2)(2x-3)}$ [8.2D]

11. Solve: $x^2 + 4x - 5 \le 0$
$\{x \mid -5 \le x \le 1\}$ [10.6A]

12. Solve: $|2x - 5| \le 3$
$\{x \mid 1 \le x \le 4\}$ [2.5B]

13. Graph: $f(x) = \left(\frac{1}{2}\right)^x + 1$

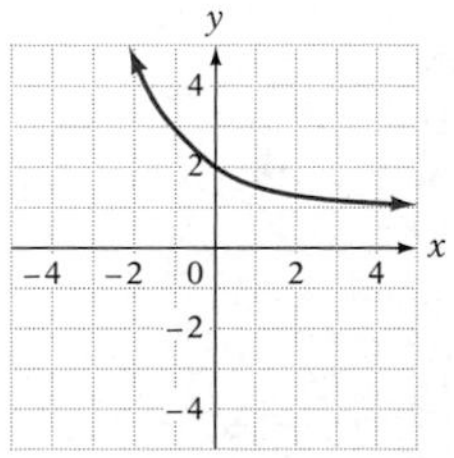

[12.1B]

14. Graph: $f(x) = \log_2 x - 1$

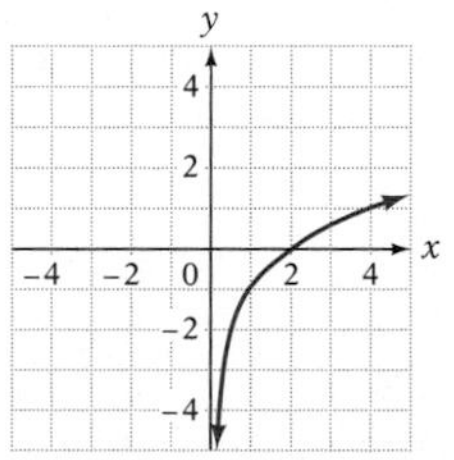

[12.3A]

15. New carpet is installed in a room measuring 18 ft by 14 ft. Find the area of the room in square yards. ($9\text{ ft}^2 = 1\text{ yd}^2$)
28 yd² [3.2B]

16. Solve for x: $\log_5 x = 3$
125 [12.2A]

17. Write $3 \log_b x - 5 \log_b y$ as a single logarithm with a coefficient of 1.
$\log_b \frac{x^3}{y^5}$ [12.2B]

18. Find $\log_3 7$. Round to the nearest ten-thousandth.
1.7712 [12.2C]

19. Solve for x: $4^{5x-2} = 4^{3x+2}$
2 [12.4A]

20. Solve for x: $\log x + \log (2x + 3) = \log 2$
$\frac{1}{2}$ [12.4B]

21. A bank offers two types of checking accounts. One account has a charge of \$5 per month plus 2 cents per check. The second account has a charge of \$2 per month plus 8 cents per check. How many checks can a customer who has the second type of account write if it is to cost the customer less than the first type of checking account?
fewer than 50 checks [2.4C]

22. Find the cost per pound of a mixture made from 16 lb of chocolate that costs \$4.00 per pound and 24 lb of chocolate that costs \$2.50 per pound.
\$3.10 [2.3A]

23. A plane can fly at a rate of 225 mph in calm air. Traveling with the wind, the plane flew 1000 mi in the same amount of time that it took to fly 800 mi against the wind. Find the rate of the wind.
25 mph [5.4A]

24. The distance (d) that a spring stretches varies directly as the force (f) used to stretch the spring. If a force of 20 lb stretches a spring 6 in., how far will a force of 34 lb stretch the spring?
10.2 in. [8.7A]

25. A carpenter purchased 80 ft of redwood and 140 ft of fir for a total cost of \$67. A second purchase, at the same prices, included 140 ft of redwood and 100 ft of fir for a total cost of \$81. Find the cost of redwood and of fir.
redwood: \$.40 per foot; fir: \$.25 per foot [5.4B]

26. The compound interest formula is $P = A(1 + i)^n$, where A is the original value of an investment, i is the interest rate per compounding period, n is the total number of compounding periods, and P is the value of the investment after n periods. Use the compound interest formula to find the number of years in which an investment of \$5000 will double in value. The investment earns 9% annual interest and is compounded semiannually. Round to the nearest whole number.
8 years [12.5A]

697

Final Exam

1. Simplify:
$12 - 8[3 - (-2)]^2 \div 5 - 3$
-31 [1.3A]

2. Evaluate $\dfrac{a^2 - b^2}{a - b}$ when $a = 3$ and $b = -4$.
-1 [1.4A]

3. Simplify: $5 - 2[3x - 7(2 - x) - 5x]$
$-10x + 33$ [1.4D]

4. Solve: $\dfrac{3}{4}x - 2 = 4$
8 [2.2A]

5. Solve: $8 - |5 - 3x| = 1$
$-\dfrac{2}{3}, 4$ [2.5A]

6. Find the volume of a sphere with a diameter of 8 ft. Round to the nearest tenth.
$268.1\ \text{ft}^3$ [3.3A]

7. Graph $2x - 3y = 9$ using the x- and y-intercepts.

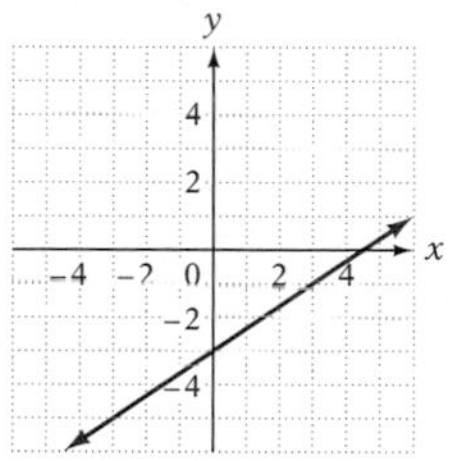

[4.3B]

8. Find the equation of the line containing the points $(3, -2)$ and $(1, 4)$.
$y = -3x + 7$ [4.5B]

9. Find the equation of the line that contains the point $(-2, 1)$ and is perpendicular to the line $3x - 2y = 6$.
$y = -\dfrac{2}{3}x - \dfrac{1}{3}$ [4.6A]

10. Simplify: $2a[5 - a(2 - 3a) - 2a] + 3a^2$
$6a^3 - 5a^2 + 10a$ [6.3A]

11. Factor: $8 - x^3y^3$
$(2 - xy)(4 + 2xy + x^2y^2)$ [7.4B]

12. Factor: $x - y - x^3 + x^2y$
$(x - y)(1 - x)(1 + x)$ [7.4D]

13. Divide: $(2x^3 - 7x^2 + 4) \div (2x - 3)$
$x^2 - 2x - 3 - \dfrac{5}{2x - 3}$ [6.4A]

14. Divide: $\dfrac{x^2 - 3x}{2x^2 - 3x - 5} \div \dfrac{4x - 12}{4x^2 - 4}$
$\dfrac{x(x - 1)}{2x - 5}$ [8.1C]

15. Subtract: $\dfrac{x - 2}{x + 2} - \dfrac{x + 3}{x - 3}$
$\dfrac{-10x}{(x + 2)(x - 3)}$ [8.2D]

16. Simplify: $\dfrac{\dfrac{3}{x} + \dfrac{1}{x + 4}}{\dfrac{1}{x} + \dfrac{3}{x + 4}}$
$\dfrac{x + 3}{x + 1}$ [8.3A]

698 Final Exam

17. Solve: $\dfrac{5}{x-2} - \dfrac{5}{x^2-4} = \dfrac{1}{x+2}$

$-\dfrac{7}{4}$ [8.4A]

18. Solve $a_n = a_1 + (n-1)d$ for d.

$d = \dfrac{a_n - a_1}{n-1}$ [8.5A]

19. Simplify: $\left(\dfrac{4x^2y^{-1}}{3x^{-1}y}\right)^{-2}\left(\dfrac{2x^{-1}y^2}{9x^{-2}y^2}\right)^3$

$\dfrac{y^4}{162x^3}$ [6.1B]

20. Simplify: $\left(\dfrac{3x^{2/3}y^{1/2}}{6x^2y^{4/3}}\right)^6$

$\dfrac{1}{64x^8y^5}$ [9.1A]

21. Subtract: $x\sqrt{18x^2y^3} - y\sqrt{50x^4y}$

$-2x^2y\sqrt{2y}$ [9.2B]

22. Simplify: $\dfrac{\sqrt{16x^5y^4}}{\sqrt{32xy^7}}$

$\dfrac{x^2\sqrt{2y}}{2y^2}$ [9.2D]

23. Simplify: $\dfrac{3}{2+i}$

$\dfrac{6}{5} - \dfrac{3}{5}i$ [9.3D]

24. Write a quadratic equation that has integer coefficients and has solution $-\dfrac{1}{2}$ and 2.

$2x^2 - 3x - 2 = 0$ [10.1B]

25. Solve by using the quadratic formula:
$2x^2 - 3x - 1 = 0$

$\dfrac{3+\sqrt{17}}{4}, \dfrac{3-\sqrt{17}}{4}$ [10.3A]

26. Solve: $x^{2/3} - x^{1/3} - 6 = 0$

$-8, 27$ [10.4A]

27. Graph: $f(x) = -x^2 + 4$

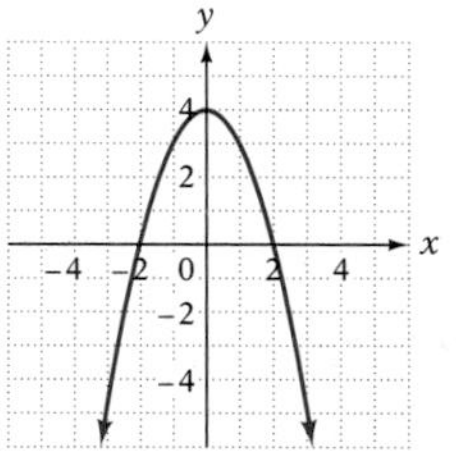

[11.2A]

28. Graph: $\dfrac{x^2}{16} + \dfrac{y^2}{4} = 1$

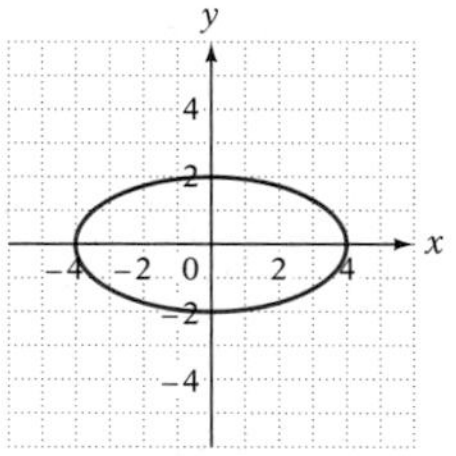

[11.6C]

29. Solve: $\dfrac{2}{x} - \dfrac{2}{2x+3} = 1$

$-2, \dfrac{3}{2}$ [10.4C]

30. Find the inverse of the function $f(x) = \dfrac{2}{3}x - 4$.

$f^{-1}(x) = \dfrac{3}{2}x + 6$ [11.5B]

31. Solve by the addition method:
$3x - 2y = 1$
$5x - 3y = 3$

(3, 4) [5.2A]

32. Evaluate the determinant:

$$\begin{vmatrix} 3 & 4 \\ -1 & 2 \end{vmatrix}$$

10 [5.3A]

699

33. Solve: $2 - 3x < 6$ and $2x + 1 > 4$
$\left\{x \mid x > \frac{3}{2}\right\}$ [2.4B]

34. Solve: $|2x + 5| < 3$
$\{x \mid -4 < x < -1\}$ [2.5B]

35. Graph the solution set: $3x + 2y > 6$

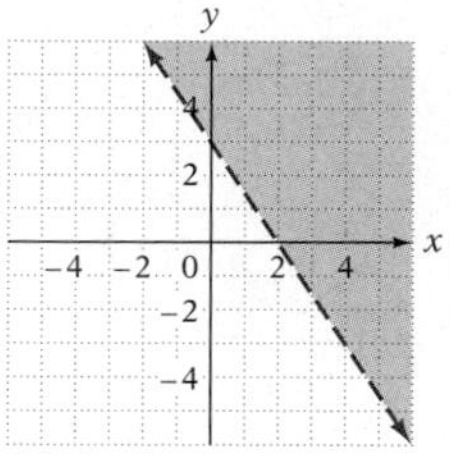

[4.7A]

36. Graph: $f(x) = 3^{-x} - 2$

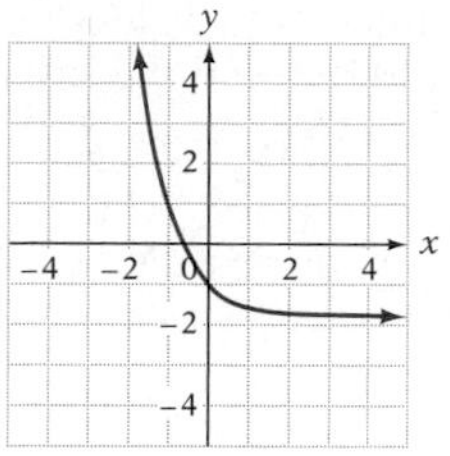

[12.1B]

37. Graph: $f(x) = \log_2 (x + 1)$

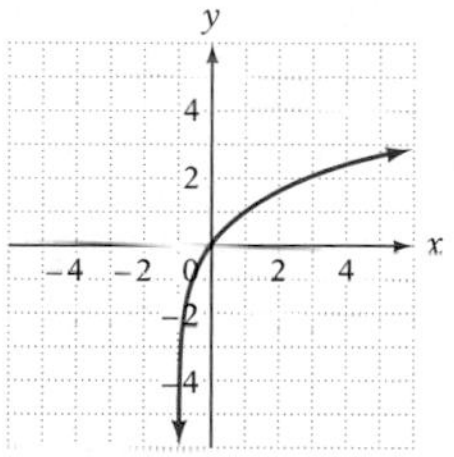

[12.3A]

38. Write $2(\log_2 a - \log_2 b)$ as a single logarithm with a coefficient of 1.
$\log_2 \frac{a^2}{b^2}$ [12.2B]

39. Solve for x: $\log_3 x - \log_3 (x - 3) = \log_3 2$
6 [12.4B]

40. An average score of 70–79 in a history class receives a C grade. A student has grades of 64, 58, 82, and 77 on four history tests. Find the range of scores on the fifth test that will give the student a C grade for the course.
$69 \le x \le 100$ [2.4C]

41. A jogger and a cyclist set out at 8 A.M. from the same point headed in the same direction. The average speed of the cyclist is two and a half times the average speed of the jogger. In 2 h, the cyclist is 24 mi ahead of the jogger. How far did the cyclist ride in that time?
40 mi [2.3D]

42. You have a total of $12,000 invested in two simple interest accounts. On one account, a money market fund, the annual simple interest rate is 8.5%. On the other account, a tax-free bond fund, the annual simple interest rate is 6.4%. The total annual interest earned by the two accounts is $936. How much do you have invested in each account?
$8000 at 8.5%; $4000 at 6.4% [2.3C]

43. The length of a rectangle is 1 ft less than three times the width. The area of the rectangle is 140 ft^2. Find the length and width of the rectangle.
width: 7 ft; length: 20 ft [10.5A]

44. Three hundred shares of a utility stock earn a yearly dividend of $486. How many additional shares of the utility stock would give a total dividend income of $810?
200 [8.4B]

45. An account executive traveled 45 mi by car and then an additional 1050 mi by plane. The rate of the plane was seven times the rate of the car. The total time for the trip was $3\frac{1}{4}$ h. Find the rate of the plane.
420 mph [8.6B]

46. An object is dropped from the top of a building. Find the distance the object has fallen when the speed reaches 75 ft/s. Use the equation $v = \sqrt{64d}$, where v is the speed of the object and d is the distance. Round to the nearest whole number.
88 ft [9.4B]

47. A small plane made a trip of 660 mi in 5 h. The plane traveled the first 360 mi at a constant rate before increasing its speed by 30 mph. Then it traveled another 300 mi at the increased speed. Find the rate of the plane for the first 360 mi.
120 mph [8.6B]

48. The intensity (L) of a light source is inversely proportional to the square of the distance (d) from the source. If the intensity is 8 foot-candles at a distance of 20 ft, what is the intensity when the distance is 4 ft?
200 foot-candles [8.7A]

49. A motorboat traveling with the current can go 30 mi in 2 h. Against the current, it takes 3 h to go the same distance. Find the rate of the motorboat in calm water and the rate of the current.
rate of the boat in calm water: 12.5 mph;
rate of the current: 2.5 mph [5.4A]

50. An investor deposits $4000 into an account that earns 9% annual interest compounded monthly. Use the compound interest formula $P = A(1 + i)^n$, where A is the original value of the investment, i is the interest rate per compounding period, n is the total number of compounding periods, and P is the value of the investment after n periods, to find the value of the investment after 2 years. Round to the nearest cent.
$4785.65 [12.5A]

Chapter R

Review of Introductory Algebra Topics

Objectives

Section R.1

A To evaluate a variable expression
B To simplify a variable expression

Section R.2

A To solve a first-degree equation in one variable
B To solve an inequality in one variable

Section R.3

A To graph points in a rectangular coordinate system
B To graph a linear equation in two variables
C To evaluate a function
D To find the equation of a line

Section R.4

A To multiply and divide monomials
B To add and subtract polynomials
C To multiply polynomials
D To divide polynomials
E To factor polynomials of the form $ax^2 + bx + c$

WEB Need help? For on-line student resources, such as section quizzes, visit this textbook's web site at **college.hmco.com/students.**

R.1 Variable Expressions

Objective A — To evaluate a variable expression

Whenever an expression contains more than one operation, the operations must be performed in a specified order, as listed below in the Order of Operations Agreement.

1.3A* Order of Operations Agreement

The Order of Operations Agreement

Step 1 Perform operations inside grouping symbols. Grouping symbols include parentheses ( ), brackets [ ], braces { }, the absolute value symbol | |, and fraction bars.

Step 2 Simplify exponential expressions.

Step 3 Do multiplication and division as they occur from left to right.

Step 4 Do addition and subtraction as they occur from left to right.

Example 1

Evaluate: $-2(7-3)^2 + 4 - 2(5-2)$

Solution

$-2(7-3)^2 + 4 - 2(5-2)$

$= -2(4)^2 + 4 - 2(3)$ • Perform operations inside parentheses.

$= -2(16) + 4 - 2(3)$ • Simplify the exponential expression.

$= -32 + 4 - 2(3)$ • Do the multiplication and division from left to right.

$= -32 + 4 - 6$

$= -28 - 6$ • Do the addition and subtraction from left to right.

$= -28 + (-6) = -34$

You Try It 1

Evaluate: $(-4)(6-8)^2 - (-12 \div 4)$

Your solution

-13

Solution on p. S38

1.4A Evaluate variable expressions

A **variable** is a letter that represents a quantity that is unknown or that can change or vary. An expression that contains one or more variables is a **variable expression.** $3x - 4y + 7z$ is a variable expressions. It contains the variables x, y, and z.

Replacing a variable in a variable expression by a number and then simplifying the resulting numerical expression is called **evaluating the variable expression.** The number substituted for the variable is called the **value of the variable.** The result is called the **value of the variable expression.**

*Review this objective for more detailed coverage of this topic.

Example 2
Evaluate $5ab^3 + 2a^2b^2 - 4$ when $a = 3$ and $b = -2$.

Solution
$5ab^3 + 2a^2b^2 - 4$
$5(3)(-2)^3 + 2(3)^2(-2)^2 - 4$ • Replace *a* by 3 and *b* by 2.
$= 5(3)(-8) + 2(9)(4) - 4$ • Use the Order of Operations Agreement to simplify the numerical expression.
$= -120 + 72 - 4$
$= -48 - 4$
$= -48 + (-4)$
$= -52$

You Try It 2
Evaluate $3xy^2 - 3x^2y$ when $x = -2$ and $y = 5$.

Your solution
-210

Solution on p. S38

Objective B To simplify a variable expression

1.4B
1.4C
Simplify variable expressions using the Properties of Addition and Multiplication

The Properties of Real Numbers are used to simplify variable expressions.

Example 3
Simplify: $-\frac{1}{3}(-3y)$

Solution
$-\frac{1}{3}(-3y)$
$= \left[-\frac{1}{3}(-3)\right]y$ • Use the Associative Property of Multiplication to regroup factors.
$= 1y$ • Use the Inverse Property of Multiplication.
$= y$ • Use the Multiplication Property of One.

You Try It 3
Simplify: $-5(-3a)$

Your solution
$15a$

Solution on p. S38

A variable expression is shown at the right. The expression can be rewritten by writing subtraction as addition of the opposite. Note that the expression has four addends. The **terms** of a variable expression are the addends of the expression. The expression has four terms.

$3x^2 - 4xy + 5z - 2$
$3x^2 + (-4xy) + 5z + (-2)$

Four terms
$3x^2 - 4xy + 5z - 2$
Variable terms Constant term

The terms $3x^2$, $-4xy$, and $5z$ are **variable terms.** The term -2 is a **constant term,** or simply a **constant.**

Like terms of a variable expression are terms that have the same variable part. The terms $3x$ and $-7x$ are like terms. Constant terms are also like terms. Thus -6 and 9 are like terms.

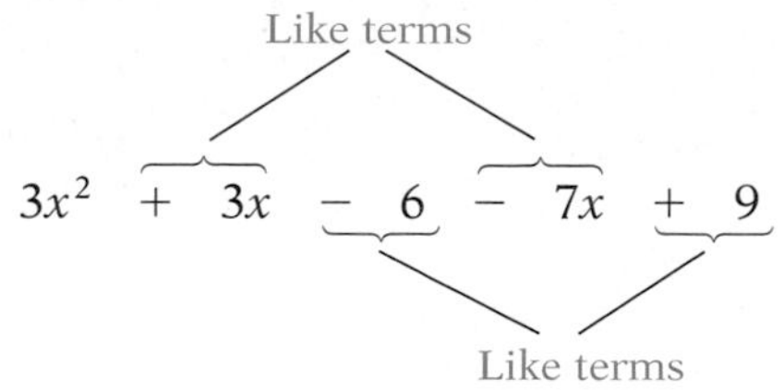

By using the Commutative Property of Multiplication, we can rewrite the Distributive Property as $ba + ca = (b + c)a$. This form of the Distributive Property is used to **combine like terms** of a variable expression by adding their coefficients. For instance,

$$7x + 9x = (7 + 9)x$$
$$= 16x$$

- **Use the Distributive Property: $ba + ca = (b + c)a$.**

Example 4

Simplify: $4x^2 + 5x - 6x^2 - 7x$

Solution

$4x^2 + 5x - 6x^2 - 7x$

$= 4x^2 - 6x^2 + 5x - 7x$

$= (4x^2 - 6x^2) + (5x - 7x)$

- **Use the Associative and Commutative Properties of Addition to rearrange and group like terms.**

$= -2x^2 + (-2x)$

$= -2x^2 - 2x$

- **Use the Distributive Property to combine like terms.**

You Try It 4

Simplify: $2z^2 - 5z - 3z^2 + 6z$

Your solution

$-z^2 + z$

Solution on p. S38

1.4D Simplify variable expressions using the Distributive Property

The Distributive Property also is used to remove parentheses from a variable expression. Here is an example.

$$4(2x + 5z) = 4(2x) + 4(5z)$$

- **Use the Distributive Property: $a(b + c) = ab + ac$.**

$$= (4 \cdot 2)x + (4 \cdot 5)z$$

- **Use the Associative Property of Multiplication to regroup factors.**

$$= 8x + 20z$$

- **Multiply $4 \cdot 2$ and $4 \cdot 5$.**

The Distributive Property can be extended to expressions containing more than two terms. For instance,

$$4(2x + 3y + 5z) = 4(2x) + 4(3y) + 4(5z)$$
$$= 8x + 12y + 20z$$

Example 5

Simplify. **a.** $-3(2x + 4)$ **b.** $6(3x - 4y + z)$

Solution

a. $-3(2x + 4) = -3(2x) + (-3)(4)$
$= -6x - 12$

b. $6(3x - 4y + z)$
$= 6(3x) - 6(4y) + 6(z)$
$= 18x - 24y + 6z$

You Try It 5

Simplify. **a.** $-3(5y - 2)$ **b.** $-2(4x + 2y - 6z)$

Your solution

a. $-15y + 6$ **b.** $-8x - 4y + 12z$

Solution on p. S38

TAKE NOTE

Recall that the Distributive Property states that if a, b, and c are real numbers, then

$a(b + c) = ab + ac$

To simplify the expression $5 + 3(4x - 2)$, use the Distributive Property to remove the parentheses.

$5 + 3(4x - 2) = 5 + 3(4x) - 3(2)$ • Use the Distributive Property.
$= 5 + 12x - 6$
$= 12x - 1$ • Add the like terms 5 and −6.

Example 6

Simplify.

a. $3(2x - 4) - 5(3x + 2)$
b. $3a - 2[7a - 2(2a + 1)]$

Solution

a. $3(2x - 4) - 5(3x + 2)$
$= 6x - 12 - 15x - 10$
$= -9x - 22$

b. $3a - 2[7a - 2(2a + 1)]$
$= 3a - 2[7a - 4a - 2]$
$= 3a - 2[3a - 2]$
$= 3a - 6a + 4$
$= -3a + 4$

You Try It 6

Simplify.

a. $7(-3x - 4y) - 3(3x + y)$
b. $2y - 3[5 - 3(3 + 2y)]$

Your solution

a. $-30x - 31y$ **b.** $20y + 12$

Solution on p. S38

R.1 Exercises

Objective A

Evaluate the variable expression when $a = 2$, $b = 3$, and $c = -4$.

1. $a - 2c$ 10

2. $-3a + 4b$ 6

3. $3b - 3c$ 21

4. $-3c + 4$ 16

5. $16 \div (2c)$ −2

6. $6b \div (-a)$ −9

7. $3b - (a + c)^2$ 5

8. $(a - b)^2 + 2c$ −7

9. $(b - 3a)^2 + bc$ −3

Evaluate the variable expression when $a = -1$, $b = 3$, $c = -2$, and $d = 4$.

10. $\dfrac{b - a}{d}$ 1

11. $\dfrac{d - b}{a}$ −1

12. $\dfrac{2d + b}{-a}$ 11

13. $\dfrac{b - d}{c - a}$ 1

14. $2(b + c) - 2a$ 4

15. $3(b - a) - bc$ 18

16. $\dfrac{-4bc}{2a + c}$ −6

17. $\dfrac{abc}{b - d}$ −6

18. $(d - a)^2 - (b - c)^2$ 0

19. $(-b + d)^2 + (-a + c)^2$ 2

20. $4ab + (2c)^2$ 4

21. $3cd - (4a)^2$ −40

Evaluate the variable expression when $a = 2.7$, $b = -1.6$, and $c = -0.8$.

22. $c^2 - ab$ 4.96

23. $(a + b)^2 - c$ 2.01

24. $\dfrac{b^3}{c} - 4a$ −5.68

Objective B

Simplify each of the following.

25. $x + 7x$ 8x

26. $12y + 9y$ 21y

27. $8b - 5b$ 3b

28. $4y - 11y$ −7y

29. $-12a + 17a$ 5a

30. $-15xy + 7xy$ −8xy

31. $4x + 5x + 2x$ $11x$

32. $-5x^2 - 10x^2 + x^2$ $-14x^2$

33. $6x - 2y + 9x$ $15x - 2y$

34. $3x - 7y - 6x + 4x$ $x - 7y$

35. $5a + 6a - 2a$ $9a$

36. $2a - 5a + 3a$ 0

37. $12y^2 + 10y^2$ $22y^2$

38. $2z^2 - 9z^2$ $-7z^2$

39. $\frac{3}{4}x - \frac{1}{4}x$ $\frac{1}{2}x$

40. $\frac{2}{5}y - \frac{3}{5}y$ $-\frac{1}{5}y$

41. $-4(5x)$ $-20x$

42. $-2(-8y)$ $16y$

43. $(6a)(-4)$ $-24a$

44. $-5(7x^2)$ $-35x^2$

45. $\frac{1}{4}(4x)$ x

46. $\frac{12x}{5}\left(\frac{5}{12}\right)$ x

47. $\frac{1}{3}(21x)$ $7x$

48. $-\frac{5}{8}(24a^2)$ $-15a^2$

49. $(36y)\left(\frac{1}{12}\right)$ $3y$

50. $-(z + 4)$ $-z - 4$

51. $-3(a + 5)$ $-3a - 15$

52. $(4 - 3b)9$ $36 - 27b$

53. $(-2x - 6)8$ $-16x - 48$

54. $3(5x^2 + 2x)$ $15x^2 + 6x$

55. $-5(2y^2 - 1)$
$-10y^2 + 5$

56. $4(x^2 - 3x + 5)$
$4x^2 - 12x + 20$

57. $6(3x^2 - 2xy - y^2)$
$18x^2 - 12xy - 6y^2$

58. $5a - (4a + 6)$
$a - 6$

59. $3 - (10 + 8y)$
$-8y - 7$

60. $12(y - 2) + 3(7 - 4y)$
-3

61. $-5[2x + 3(5 - x)]$
$5x - 75$

62. $-3[2x - (x + 7)]$
$-3x + 21$

63. $-5a - 2[2a - 4(a + 7)]$
$-a + 56$

R.2 Equations and Inequalities

Objective A To solve a first-degree equation in one variable

An **equation** expresses the equality of two mathematical expressions. Each of the equations below is a **first-degree equation in one variable.** *First degree* means that the variable has an exponent of 1.

$$x + 11 = 14$$
$$3a + 5 = 8a$$
$$2(6y - 1) = 3$$

A **solution** of an equation is a number that, when substituted for the variable, results in a true equation.

3 is a solution of the equation $x + 4 = 7$ because $3 + 4 = 7$.
9 is not a solution of the equation $x + 4 = 7$ because $9 + 4 \neq 7$.

To **solve an equation** means to find a solution of the equation. In solving an equation, the goal is to rewrite the given equation with the variable alone on one side of the equation and a constant term on the other side of the equation; the constant term is the solution of the equation. The following properties of equations are used to rewrite equations in this form.

2.1B 2.1C Solving equations using the Addition and Multiplication Properties of Equations

Properties of Equations

Addition Property of Equations

The same number can be added to each side of an equation without changing the solution of the equation. In symbols, the equation $a = b$ has the same solution as the equation $a + c = b + c$.

Multiplication Property of Equations

Each side of an equation can be multiplied by the same nonzero number without changing the solution of the equation. In symbols, if $c \neq 0$, then the equation $a = b$ has the same solutions as the equation $ac = bc$.

TAKE NOTE

Subtraction is defined as addition of the opposite.

$a - b = a + (-b)$

The Addition Property of Equations is used to remove a term from one side of the equation by adding the opposite of that term to each side of the equation. Because subtraction is defined in terms of addition, the Addition Property of Equations also makes it possible to subtract the same number from each side of an equation without changing the solution of the equation.

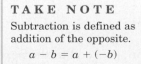

For example, to solve the equation $t + 9 = -4$, subtract the constant term (9) from each side of the equation.

$$t + 9 = -4$$
$$t + 9 - 9 = -4 - 9$$
$$t = -13$$

TAKE NOTE
Division is defined as multiplication by the reciprocal.

$$a \div b = a \cdot \frac{1}{b}$$

Now the variable is alone on one side of the equation and a constant term (−13) is on the other side. The solution is the constant. The solution is −13.

The Multiplication Property of Equations is used to remove a coefficient by multiplying each side of the equation by the reciprocal of the coefficient. Because division is defined in terms of multiplication, each side of an equation can be divided by the same nonzero number without changing the solution of the equation.

TAKE NOTE
When using the Multiplication Property of Equations, multiply each side of the equation by the reciprocal of the coefficient when the coefficient is a fraction. Divide each side of the equation by the coefficient when the coefficient is an integer or decimal.

For example, to solve the equation $-5q = 120$, divide each side of the equation by the coefficient −5.

$$-5q = 120$$
$$\frac{-5q}{-5} = \frac{120}{-5}$$
$$q = -24$$

Now the variable is alone on one side of the equation and a constant (−24) is on the other side. The solution is the constant. The solution is −24.

2.2B 2.2C Solve general equations

In solving more complicated first-degree equations in one variable, use the following sequence of steps.

Steps for Solving a First-Degree Equation in One Variable

1. Use the Distributive Property to remove parentheses.
2. Combine any like terms on the right side of the equation and any like terms on the left side of the equation.
3. Use the Addition Property to rewrite the equation with only one variable term.
4. Use the Addition Property to rewrite the equation with only one constant term.
5. Use the Multiplication Property to rewrite the equation with the variable alone on one side of the equation and a constant on the other side of the equation.

If one of these steps is not needed to solve a given equation, proceed to the next step.

Example 1
Solve.
a. $5x + 9 = 23 - 2x$
b. $8x - 3(4x - 5) = -2x + 6$

Solution

a.
$$5x + 9 = 23 - 2x$$
$$5x + 2x + 9 = 23 - 2x + 2x \quad \text{• Step 3}$$
$$7x + 9 = 23$$
$$7x + 9 - 9 = 23 - 9 \quad \text{• Step 4}$$
$$7x = 14$$
$$\frac{7x}{7} = \frac{14}{7} \quad \text{• Step 5}$$
$$x = 2$$

The solution is 2.

b.
$$8x - 3(4x - 5) = -2x + 6$$
$$8x - 12x + 15 = -2x + 6 \quad \text{• Step 1}$$
$$-4x + 15 = -2x + 6 \quad \text{• Step 2}$$
$$-4x + 2x + 15 = -2x + 2x + 6 \quad \text{• Step 3}$$
$$-2x + 15 = 6$$
$$-2x + 15 - 15 = 6 - 15 \quad \text{• Step 4}$$
$$-2x = -9$$
$$\frac{-2x}{-2} = \frac{-9}{-2} \quad \text{• Step 5}$$
$$x = \frac{9}{2}$$

The solution is $\frac{9}{2}$.

You Try It 1
Solve.
a. $4x + 3 = 7x + 9$
b. $4 - (5x - 8) = 4x + 3$

Your solution
a. -2 **b.** 1

Solution on p. S38

Objective B To solve an inequality in one variable

An **inequality** contains the symbol $>$, $<$, $\geq$, or $\leq$. An inequality expresses the relative order of two mathematical expressions. Here are some examples of inequalities in one variable.

$$\left.\begin{aligned} 4x &\geq 12 \\ 2x + 7 &\leq 9 \\ x^2 + 1 &> 3x \end{aligned}\right\} \text{ Inequalities in one variable}$$

2.4A Solve an inequality in one variable

A **solution of an inequality in one variable** is a number that, when substituted for the variable, results in a true inequality. For the inequality $x < 4$ shown below, 3, 0, and -5 are solutions of the inequality because replacing the variable by these numbers results in a true inequality.

$$x < 4 \qquad x < 4 \qquad x < 4$$
$$3 < 4 \ \text{True} \qquad 0 < 4 \ \text{True} \qquad -5 < 4 \ \text{True}$$

The number 7 is not a solution of the inequality $x < 4$ because $7 < 4$ is a false inequality.

Besides the numbers 3, 0, and -5, there are an infinite number of other solutions of the inequality $x < 4$. Any number less than 4 is a solution; for instance, -5.2, $\frac{5}{2}$, π, and 1 are also solutions of the inequality. The set of all the solutions of an inequality is called the **solution set of the inequality.** The solution set of the inequality $x < 4$ is written in set-builder notation as $\{x \mid x < 4\}$. This is read, "the set of all x such that x is less than 4."

The graph of the solution set of $x < 4$ is shown at the right.

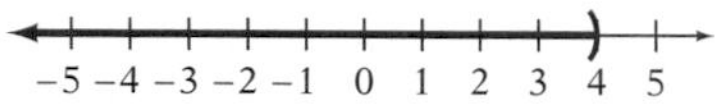

In solving an inequality, the goal is to rewrite the given inequality in the form

$$\textit{variable} < \textit{constant} \quad \text{or} \quad \textit{variable} > \textit{constant}$$

The Addition Property of Inequalities is used to rewrite an inequality in this form.

TAKE NOTE
The Addition Property of Inequalities states that the same number can be added to each side of an inequality without changing the solution set of the inequality.

The Addition Property of Inequalities

If $a > b$ and c is a real number, then the inequalities $a > b$ and $a + c > b + c$ have the same solution set.

If $a < b$ and c is a real number, then the inequalities $a < b$ and $a + c < b + c$ have the same solution set.

The Addition Property of Inequalities is also true for the symbols $\leq$ and $\geq$.

The Addition Property of Inequalities is used to remove a term from one side of an inequality by adding the additive inverse of that term to each side of the inequality. Because subtraction is defined in terms of addition, the same number can be subtracted from each side of an inequality without changing the solution set of the inequality.

As shown in the example below, the Addition Property of Inequalities applies to variable terms as well as to constants.

➡ Solve $4x - 5 \leq 3x - 2$. Write the solution set in set-builder notation.

$$4x - 5 \leq 3x - 2$$
$$4x - 3x - 5 \leq 3x - 3x - 2$$ • Subtract $3x$ from each side of the inequality.
$$x - 5 \leq -2$$ • Simplify.
$$x - 5 + 5 \leq -2 + 5$$ • Add 5 to each side of the inequality.
$$x \leq 3$$ • Simplify.

The solution set is $\{x \mid x \leq 3\}$.

When multiplying or dividing an inequality by a number, the inequality symbol may be reversed, depending on whether the number is positive or negative. Look at the following two examples.

$3 < 5$	• **Multiply by positive 2. The inequality symbol remains the same.**	$3 < 5$	• **Multiply by negative 2. The inequality symbol is reversed in order to make the inequality a true statement.**
$2(3) < 2(5)$		$-2(3) > -2(5)$	
$6 < 10$	• **6 < 10 is a true statement.**	$-6 > -10$	

This is summarized in the Multiplication Property of Inequalities.

The Multiplication Property of Inequalities

Rule 1

If $a > b$ and $c > 0$, then $ac > bc$.
If $a < b$ and $c > 0$, then $ac < bc$.

Rule 2

If $a > b$ and $c < 0$, then $ac < bc$.
If $a < b$ and $c < 0$, then $ac > bc$.

Rule 1 states that when each side of an inequality is multiplied by a positive number, the inequality symbol remains the same. Rule 2 states that when each side of an inequality is multiplied by a negative number, the inequality symbol must be reversed.

Here are a few more examples of this property.

Rule 1		**Rule 2**	
$-4 < -2$	$5 > -3$	$3 < 5$	$-2 > -6$
$-4(2) < -2(2)$	$5(3) > -3(3)$	$3(-2) > 5(-2)$	$-2(-3) < -6(-3)$
$-8 < -4$	$15 > -9$	$-6 > -10$	$6 < 18$

Use the Multiplication Property of Inequalities to remove a coefficient other than 1 from one side of an inequality so that the inequality can be rewritten with the variable alone on one side of the inequality and a constant term on the other side. The Multiplication Property of Inequalities is also true for the symbols $\le$ and $\ge$.

Because division is defined in terms of multiplication, when each side of an inequality is divided by a positive number, the inequality symbol remains the same. When each side of an inequality is divided by a negative number, the inequality symbol must be reversed.

TAKE NOTE

Solving inequalities in one variable is similar to solving equations in one variable *except* that when you multiply or divide by a negative number, you must reverse the inequality symbol.

➡ Solve $-3x < 9$. Write the solution set in set-builder notation.

$-3x < 9$

$\dfrac{-3x}{-3} > \dfrac{9}{-3}$ • **Divide each side of the inequality by the coefficient −3 and reverse the inequality symbol.**

$x > -3$ • **Simplify.**

$\{x \mid x > -3\}$ • **Write the answer in set-builder notation.**

Example 2

Solve $x + 3 > 4x + 6$. Write the solution set in set-builder notation.

Solution

$$\begin{aligned} x + 3 &> 4x + 6 \\ x - 4x + 3 &> 4x - 4x + 6 \\ -3x + 3 &> 6 \\ -3x + 3 - 3 &> 6 - 3 \\ -3x &> 3 \\ \frac{-3x}{-3} &< \frac{3}{-3} \\ x &< -1 \end{aligned}$$

$\{x | x < -1\}$

You Try It 2

Solve $3x - 1 \le 5x - 7$. Write the solution set in set-builder notation.

Your solution

$\{x | x \ge 3\}$

Solution on p. S38

When an inequality contains parentheses, often the first step in solving the inequality is to use the Distributive Property to remove the parentheses.

Example 3

Solve $-2(x - 7) > 3 - 4(2x - 3)$. Write the solution set in set-builder notation.

Solution

$$\begin{aligned} -2(x - 7) &> 3 - 4(2x - 3) \\ -2x + 14 &> 3 - 8x + 12 \\ -2x + 14 &> 15 - 8x \\ -2x + 8x + 14 &> 15 - 8x + 8x \\ 6x + 14 &> 15 \\ 6x + 14 - 14 &> 15 - 14 \\ 6x &> 1 \\ \frac{6x}{6} &> \frac{1}{6} \\ x &> \frac{1}{6} \end{aligned}$$

$\left\{x \middle| x > \frac{1}{6}\right\}$

You Try It 3

Solve $3 - 2(3x + 1) < 7 - 2x$. Write the solution set in set-builder notation.

Your solution

$\left\{x \middle| x > -\frac{3}{2}\right\}$

Solution on p. S38

R.2 Exercises

Objective A

Solve.

1. $x + 7 = -5$
   -12

2. $9 + b = 21$
   12

3. $-9 = z - 8$
   -1

4. $b - 11 = 11$
   22

5. $-48 = 6z$
   -8

6. $-9a = -108$
   12

7. $-\frac{3}{4}x = 15$
   -20

8. $\frac{5}{2}x = -10$
   -4

9. $-\frac{x}{4} = -2$
   8

10. $\frac{2x}{5} = -8$
    -20

11. $4 - 2b = 2 - 4b$
    -1

12. $4y - 10 = 6 + 2y$
    8

13. $5x - 3 = 9x - 7$
    1

14. $3m + 5 = 2 - 6m$
    $-\frac{1}{3}$

15. $6a - 1 = 2 + 2a$
    $\frac{3}{4}$

16. $5x + 7 = 8x + 5$
    $\frac{2}{3}$

17. $2 - 6y = 5 - 7y$
    3

18. $4b + 15 = 3 - 2b$
    -2

19. $2(x + 1) + 5x = 23$
    3

20. $9n - 15 = 3(2n - 1)$
    4

21. $7a - (3a - 4) = 12$
    2

22. $5(3 - 2y) = 3 - 4y$
    2

23. $9 - 7x = 4(1 - 3x)$
    -1

24. $2(3b + 5) - 1 = 10b + 1$
    2

25. $2z - 2 = 5 - (9 - 6z)$
    $\frac{1}{2}$

26. $4a + 3 = 7 - (5 - 8a)$
    $\frac{1}{4}$

27. $5(6 - 2x) = 2(5 - 3x)$
    5

28. $4(3y + 1) = 2(y - 8)$
    -2

29. $2(3b - 5) = 4(6b - 2)$
    $-\frac{1}{9}$

30. $3(x - 4) = 1 - (2x - 7)$
    4

Objective B

Solve. Write the answer in set-builder notation.

31. $x - 5 > -2$
    $\{x \mid x > 3\}$

32. $5 + n \geq 4$
    $\{n \mid n \geq -1\}$

33. $-2 + n \geq 0$
    $\{n \mid n \geq 2\}$

34. $x - 3 < 2$
$\{x \mid x < 5\}$

35. $8x \leq -24$
$\{x \mid x \leq -3\}$

36. $-4x < 8$
$\{x \mid x > -2\}$

37. $3n > 0$
$\{n \mid n > 0\}$

38. $-2n \leq -8$
$\{n \mid n \geq 4\}$

39. $2x - 1 > 7$
$\{x \mid x > 4\}$

40. $5x - 2 \leq 8$
$\{x \mid x \leq 2\}$

41. $4 - 3x < 10$
$\{x \mid x > -2\}$

42. $7 - 2x \geq 1$
$\{x \mid x \leq 3\}$

43. $3x - 1 > 2x + 2$
$\{x \mid x > 3\}$

44. $6x + 4 \leq 8 + 5x$
$\{x \mid x \leq 4\}$

45. $8x + 1 \geq 2x + 13$
$\{x \mid x \geq 2\}$

46. $6x + 3 > 4x - 1$
$\{x \mid x > -2\}$

47. $-3 - 4x > -11$
$\{x \mid x < 2\}$

48. $4x - 2 < x - 11$
$\{x \mid x < -3\}$

49. $4x - 2 > 3x + 1$
$\{x \mid x > 3\}$

50. $7x + 5 \leq 9 + 6x$
$\{x \mid x \leq 4\}$

51. $9x + 2 \geq 3x + 14$
$\{x \mid x \geq 2\}$

52. $8x + 1 > 6x - 3$
$\{x \mid x > -2\}$

53. $-5 - 2x > -13$
$\{x \mid x < 4\}$

54. $5x - 3 < x - 11$
$\{x \mid x < -2\}$

55. $4(2x - 1) > 3x - 2(3x - 5)$
$\left\{x \middle| x > \frac{14}{11}\right\}$

56. $2 - 5(x + 1) \geq 3(x - 1) - 8$
$\{x \mid x \leq 1\}$

57. $3(4x + 3) \leq 7 - 4(x - 2)$
$\left\{x \middle| x \leq \frac{3}{8}\right\}$

58. $3 + 2(x + 5) \geq x + 5(x + 1) + 1$
$\left\{x \middle| x \leq \frac{7}{4}\right\}$

59. $3 - 4(x + 2) \leq 6 + 4(2x + 1)$
$\left\{x \middle| x \geq -\frac{5}{4}\right\}$

60. $12 - 2(3x - 2) \geq 5x - 2(5 - x)$
$\{x \mid x \leq 2\}$

R.3 Linear Equations in Two Variables

Objective A **To graph points in a rectangular coordinate system**

A **rectangular coordinate system** is formed by two number lines, one horizontal and one vertical. The point of intersection is called the **origin.** The two axes are called the **coordinate axes** or simply the **axes.** Generally, the horizontal axis is labeled the x-axis, and the vertical axis is labeled the y-axis.

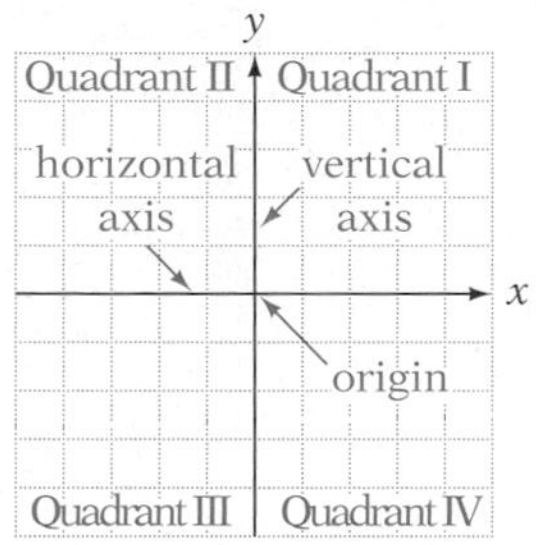

4.1A Points in the plane

Each point in the plane can be identified by a pair of numbers called an **ordered pair.** The first number of the ordered pair measures a horizontal change from the y-axis and is called the **abscissa** or **x-coordinate.** The second number of the pair measures a vertical change from the x-axis and is called the **ordinate** or **y-coordinate.** The ordered pair (x, y) associated with a point is also called the **coordinates** of the point.

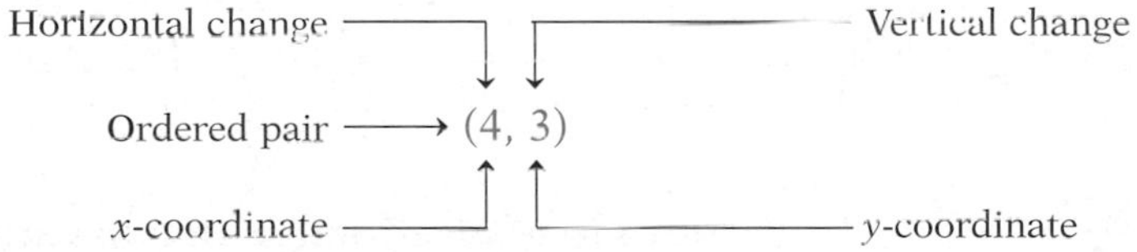

To **graph,** or **plot,** a point means to place a dot at the coordinates of the point. For example, to graph the ordered pair (4, 3), start at the origin. Move 4 units to the right and then 3 units up. Draw a dot. To graph (−3, −4), start at the origin. Move 3 units to the left and then 4 units down. Draw a dot.

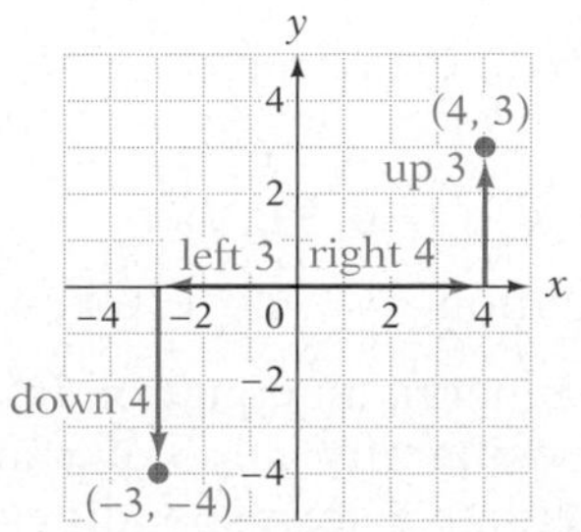

Example 1
Graph the ordered pairs (−4, 2), (3, 4), (0, −1), (2, 0), and (−1, −3).

Solution

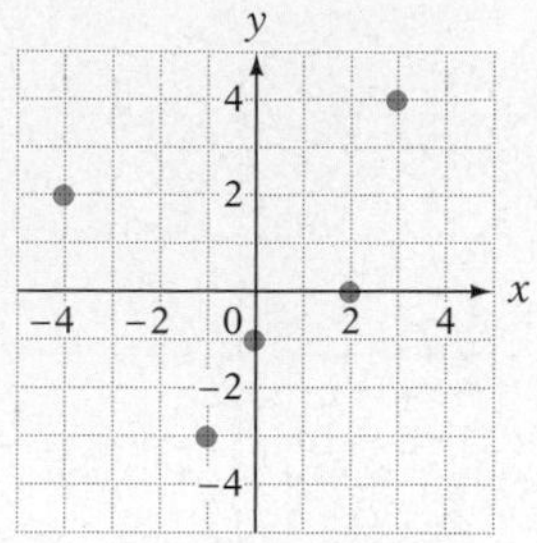

You Try It 1
Graph the ordered pairs (−2, 4), (4, 0), (0, 3), (−3, −4), and (5, −1).

Your solution

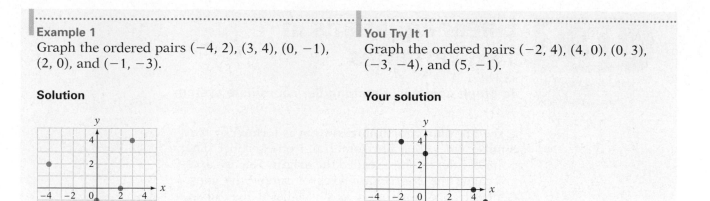

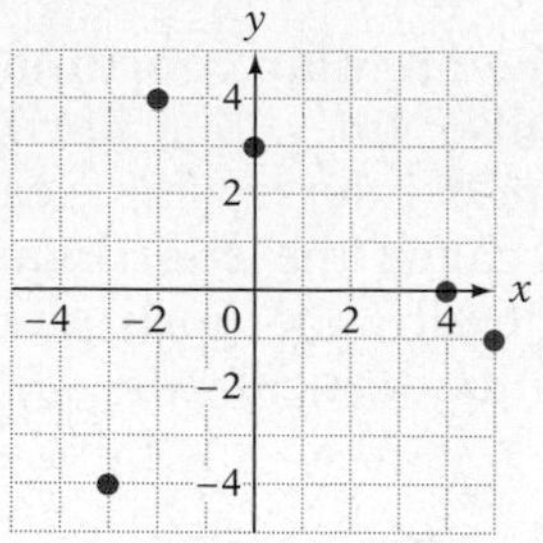

Solution on p. S39

Objective B

To graph a linear equation in two variables

The equations below are examples of equations in two variables.

$$y = 3x - 4$$
$$2x - y = 7$$
$$y = x^2 + 1$$

4.1B Solutions of equations in two variables

A **solution of an equation in two variables** is an ordered pair (x, y) whose coordinates make the equation a true statement.

➡ Is the ordered pair (−4, 9) a solution of the equation $y = -2x + 1$?

y	$= -2x + 1$
9	$-2(-4) + 1$
9	$8 + 1$

$9 = 9$

- **Replace x by −4 and y by 9.**
- **Simplify the right side.**
- **Compare the results. If the resulting equation is true, the ordered pair is a solution of the equation. If it is not true, the ordered pair is not a solution of the equation.**

Yes, the ordered pair (−4, 9) is a solution of the equation $y = -2x + 1$.

Besides (−4, 9), there are many other ordered pairs that are solutions of the equation $y = -2x + 1$. For example, (0, 1), (3, −5), and $\left(-\frac{3}{2}, 4\right)$ are also solutions. In general, an equation in two variables has an infinite number of solutions. By choosing any value of x and substituting that value into the equation, we can calculate a corresponding value of y.

➡ Find the ordered-pair solution of $y = \frac{2}{5}x - 4$ that corresponds to $x = 5$.

$y = \frac{2}{5}x - 4$

$y = \frac{2}{5}(5) - 4$ • **We are given that $x = 5$. Replace x by 5 in the equation.**

$y = 2 - 4$ • **Simplify the right side.**

$y = -2$ • **When $x = 5$, $y = -2$.**

The ordered-pair solution of $y = \frac{2}{5}x - 4$ in which $x = 5$ is $(5, -2)$.

4.3A Graph an equation of the form $y = mx + b$

Solutions of an equation in two variables can be graphed in the rectangular coordinate system.

➡ Graph the ordered-pair solutions of $y = -2x + 1$ for $x = -2, -1, 0, 1$, and 2.

x	$y = -2x + 1$	y	(x, y)
-2	$-2(-2) + 1$	5	$(-2, 5)$
-1	$-2(-1) + 1$	3	$(-1, 3)$
0	$-2(0) + 1$	1	$(0, 1)$
1	$-2(1) + 1$	-1	$(1, -1)$
2	$-2(2) + 1$	-3	$(2, -3)$

• **Use the given values of x to determine ordered-pair solutions of the equation. It is convenient to record these in a table.**

The ordered-pair solutions of $y = -2x + 1$ for $x = -2, -1, 0, 1$, and 2 are $(-2, 5)$, $(-1, 3)$, $(0, 1)$, $(1, -1)$, and $(2, -3)$. These are graphed at the right.

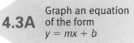

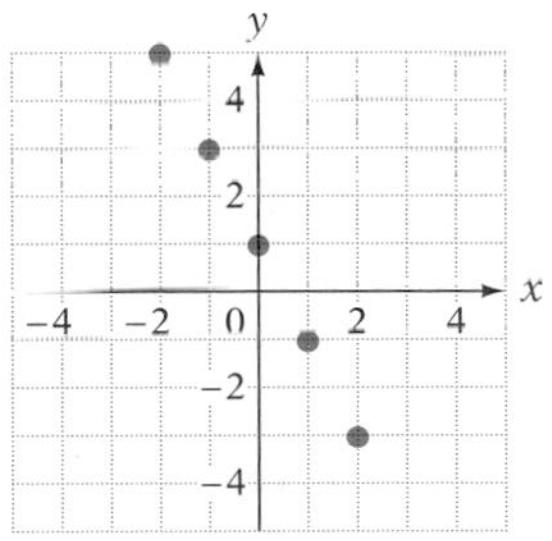

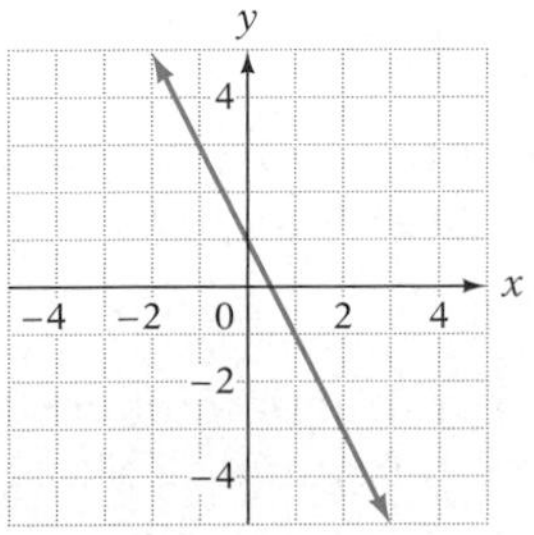

The **graph of an equation in two variables** is a graph of the ordered-pair solutions of the equation. Consider the equation $y = -2x + 1$ above. The ordered-pair solutions $(-2, 5)$, $(-1, 3)$, $(0, 1)$, $(1, -1)$, and $(2, -3)$ are graphed in the figure above. We can choose values of x that are not integers to produce more ordered pairs to graph, such as $\left(\frac{5}{2}, -4\right)$ and $\left(-\frac{1}{2}, 2\right)$. Choosing still other values of x would result in more and more ordered pairs being graphed. The result would be so many dots that the graph would appear as a straight line, as shown at the left. This is the graph of $y = -2x + 1$.

The equation $y = -2x + 1$ is an example of a linear equation because its graph is a straight line. It is also called a first-degree equation in two variables because the exponent on each variable is 1.

TAKE NOTE
The equation $y = x^2 + 2x - 4$ is not a linear equation in two variables because there is a term with a variable squared. The equation $y = \frac{5}{x-3}$ is not a linear equation because a variable occurs in the denominator of a fraction.

Linear Equation in Two Variables

An equation of the form $y = mx + b$, where m is the coefficient of x and b is a constant, is a linear equation in two variables. The graph of a linear equation in two variables is a straight line.

Examples of linear equations are shown at the right.

$$y = 5x + 3 \qquad (m = 5, b = 3)$$
$$y = x - 4 \qquad (m = 1, b = -4)$$
$$y = -\frac{3}{4}x \qquad \left(m = -\frac{3}{4}, b = 0\right)$$

To graph a linear equation, find ordered-pair solutions of the equation. Do this by choosing any value of x and finding the corresponding value of y. Repeat this procedure, choosing different values for x, until you have found the number of solutions desired. Because the graph of a linear equation in two variables is a straight line, and a straight line is determined by two points, it is necessary to find only two solutions. However, it is recommended that at least three points be used to ensure accuracy.

➡ Graph: $y = 2x - 3$

x	$y = 2x - 3$	y
0	$2(0) - 3$	-3
2	$2(2) - 3$	1
-1	$2(-1) - 3$	-5

- **Choose any values of x. Then find the corresponding values of y. The numbers 0, 2, and -1 were chosen arbitrarily for x.**

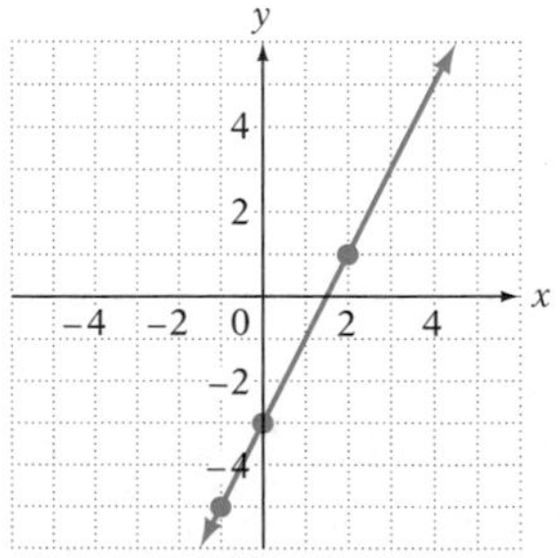

- **Graph the ordered-pair solutions $(0, -3)$, $(2, 1)$, and $(-1, -5)$. Draw a straight line through the points.**

Remember that a graph is a drawing of the ordered-pair solutions of the equation. Therefore, every point on the graph is a solution of the equation, and every solution of the equation is a point on the graph.

When graphing an equation of the form $y = mx + b$, if m is a fraction, choose values of x that will simplify the evaluation. This is illustrated in Example 2. Note that the values of x chosen are multiples of the denominator, 2.

Example 2

Graph $y = -\frac{3}{2}x - 3$.

Solution

x	y
0	−3
−2	0
−4	3

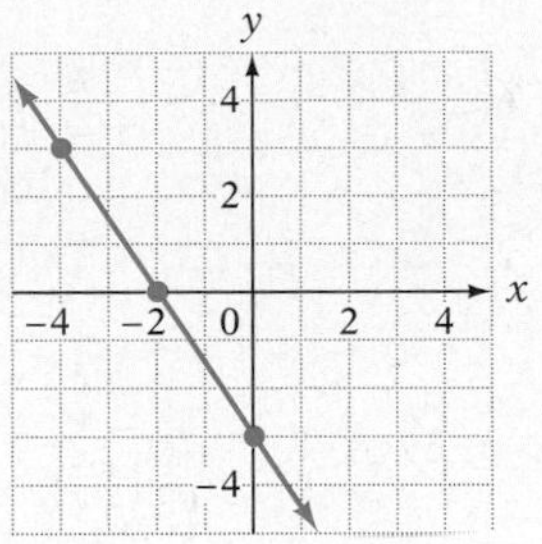

You Try It 2

Graph $y = \frac{3}{5}x - 4$.

Your solution

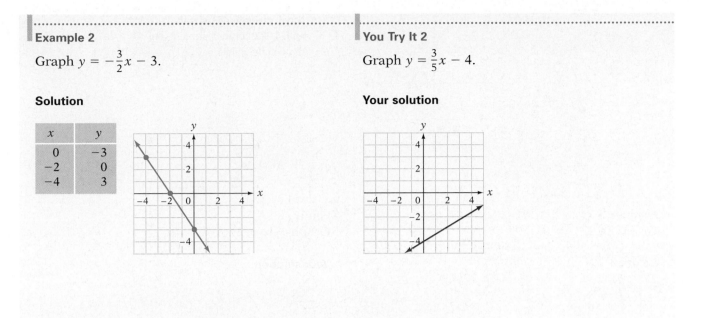

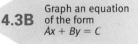

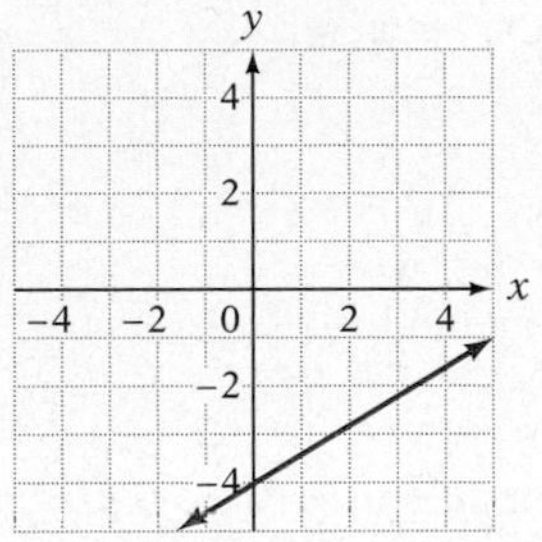

Solution on p. S39

4.3B Graph an equation of the form $Ax + By = C$

An equation of the form $Ax + By = C$ is also a linear equation in two variables. Examples of these equations are shown below.

$$3x + 4y = 12 \qquad (A = 3, B = 4, C = 12)$$
$$x - 5y = -10 \qquad (A = 1, B = -5, C = -10)$$
$$2x - y = 0 \qquad (A = 2, B = -1, C = 0)$$

One method of graphing an equation of the form $Ax + By = C$ involves first solving the equation for y and then following the same procedure used for graphing an equation of the form $y = mx + b$. To solve the equation for y means to rewrite the equation so that y is alone on one side of the equation and the term containing x and the constant are on the other side of the equation. The Addition and Multiplication Properties of Equations are used to rewrite an equation of the form $Ax + By = C$ in the form $y = mx + b$.

➡ Graph $3x + 2y = 6$.

$3x + 2y = 6$ — • **The equation is in the form $Ax + By = C$.**

$2y = -3x + 6$ — • **Solve the equation for y. Subtract $3x$ from each side of the equation.**

$y = -\frac{3}{2}x + 3$ — • **Divide each side of the equation by 2. Note that each term on the right side is divided by 2.**

• **Find at least three solutions.**

x	y
0	3
2	0
4	−3

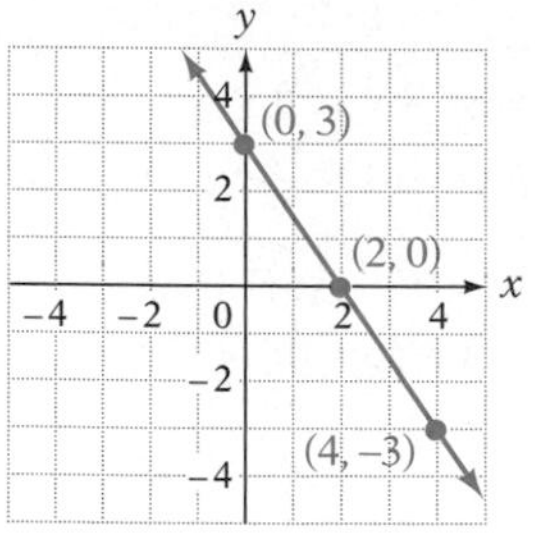

- **Graph the ordered pairs (0, 3), (2, 0), and (4, −3). Draw a straight line through the points.**

Example 3
Graph $2x + 3y = 9$.

Solution

$$2x + 3y = 9$$
$$3y = -2x + 9$$
$$y = -\frac{2}{3}x + 3$$

x	y
−3	5
0	3
3	1

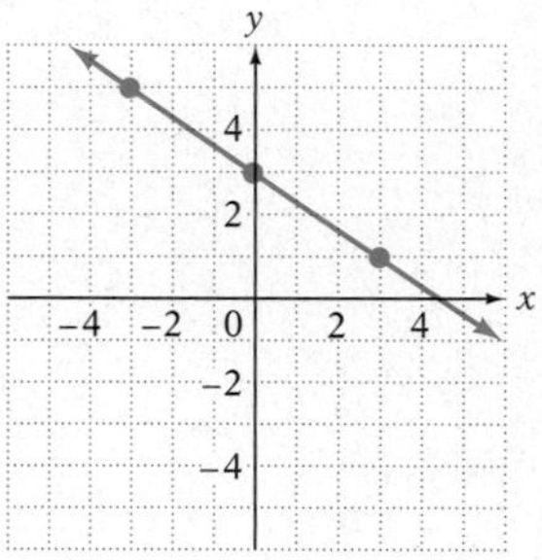

You Try It 3
Graph $-3x + 2y = 4$.

Your solution

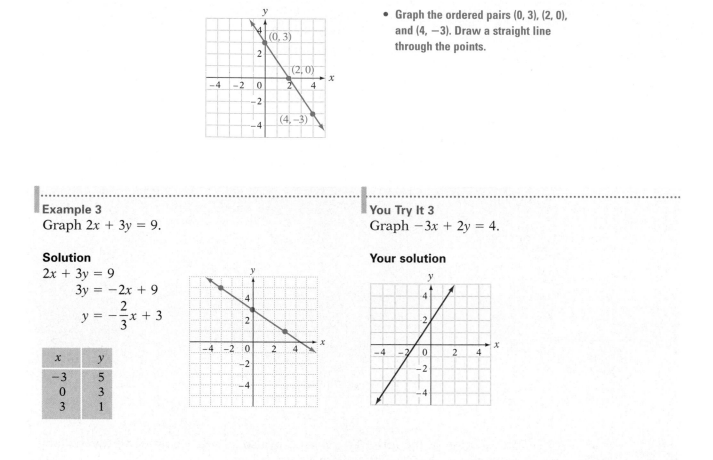

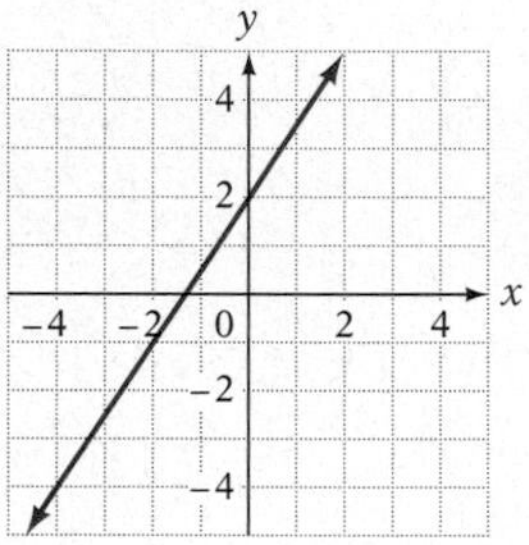

Solution on p. S39

4.4A Slope

The **slope** of a line is a measure of the slant of the line. The symbol for slope is m.

For an equation of the form $y = mx + b$, m is the slope. Here are a few examples:

The slope of the line $y = -2x + 5$ is -2.
The slope of the line $y = 8x$ is 8.
The slope of the line $y = \frac{3}{4}x - 1$ is $\frac{3}{4}$.

The slope of a line containing two points is the ratio of the change in the y values between the two points to the change in the x values.

TAKE NOTE

$$\text{Slope} = m = \frac{\text{change in } y}{\text{change in } x}$$

Slope Formula

The slope of the line containing the two points $P_1(x_1, y_1)$ and $P_2(x_2, y_2)$ is given by

$$m = \frac{y_2 - y_1}{x_2 - x_1}, \; x_1 \neq x_2$$

Example 4

Find the slope of the line containing the points (−4, −3) and (−1, 1).

Solution

Let $(x_1, y_1) = (-4, -3)$ and $(x_2, y_2) = (-1, 1)$.

$$m = \frac{y_2 - y_1}{x_2 - x_1} = \frac{1 - (-3)}{-1 - (-4)} = \frac{4}{3}$$

The slope is $\frac{4}{3}$.

You Try It 4

Find the slope of the line containing the points (−2, 3) and (1, −3).

Your solution

−2

Solution on p. S39

4.4B Slope-intercept form of a straight line

One important characteristic of the graph of a linear equation is its *intercepts*. An **x-intercept** is a point at which the graph crosses the x-axis. A **y-intercept** is a point at which the graph crosses the y-axis.

The graph of the equation $y = \frac{1}{2}x - 2$ is shown at the right. The x-intercept of the graph is (4, 0). The y-intercept of the graph is (0, −2).

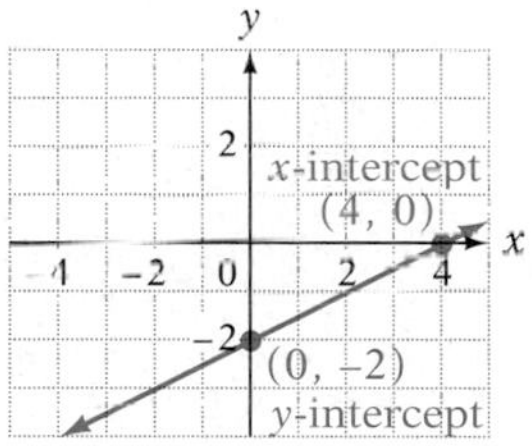

Note that at an x-intercept, the y-coordinate is 0. At a y-intercept, the x-coordinate is 0.

We can find the y-intercept of a linear equation by letting $x = 0$.

➡ Find the y-intercept of the graph of the equation $y = 3x + 4$.

$y = 3x + 4$

$y = 3(0) + 4$ • **To find the y-intercept, let x = 0.**

$y = 0 + 4$

$y = 4$

The y-intercept is (0, 4).

Note that the constant term of $y = mx + b$, the y-intercept is $(0, b)$.

In general, **for any equation of the form $y = mx + b$, the y-intercept is $(0, b)$.**

Because the slope and the y-intercept can be determined directly from the equation $y = mx + b$, this equation is called the slope-intercept form of a straight line.

Slope-Intercept Form of a Straight Line

The equation $y = mx + b$ is called the **slope-intercept form of a straight line.** The slope of the line is m, the coefficient of x. The y-intercept is $(0, b)$.

The following equations are written in slope-intercept form.

$y = -4x + 3$ Slope $= -4$, y-intercept $= (0, 3)$

$y = \frac{2}{5}x - 1$ Slope $= \frac{2}{5}$, y-intercept $= (0, -1)$

$y = -x$ Slope $= -1$, y-intercept $= (0, 0)$

When an equation is in slope-intercept form, it is possible to quickly draw a graph of the function.

➡ Graph $x + 2y = 4$ by using the slope and y-intercept.

Solve the equation for y.

$$x + 2y = 4$$
$$2y = -x + 4$$
$$y = -\frac{1}{2}x + 2$$

From the equation $y = -\frac{1}{2}x + 2$, the slope is $-\frac{1}{2}$ and the y-intercept is $(0, 2)$.

Rewrite the slope $-\frac{1}{2}$ as $\frac{-1}{2}$.

TAKE NOTE

Recall that slope $= m = \frac{\text{change in } y}{\text{change in } x}$. For the example at the right, $m = -\frac{1}{2} = \frac{-1}{2} = \frac{\text{change in } y}{\text{change in } x}$. Therefore, the change in y is -1 and the change in x is 2.

Beginning at the y-intercept, move right 2 units (change in x) and then down 1 unit (change in y).

The point whose coordinates are $(2, 1)$ is a second point on the graph. Draw a straight line through the points $(0, 2)$ and $(2, 1)$.

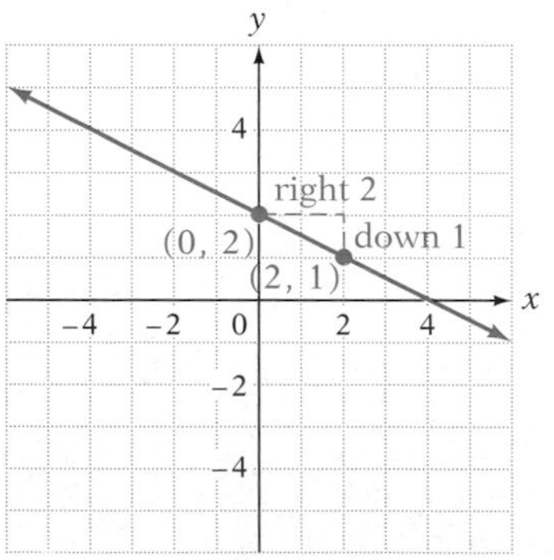

Example 5

Graph $y = -\frac{3}{2}x + 4$ by using the slope and the y-intercept.

You Try It 5

Graph $y = -\frac{2}{3}x + 2$ by using the slope and the y-intercept.

Solution
From the equation, the slope is $-\frac{3}{2}$ and the y-intercept is (0, 4).

Rewrite the slope $-\frac{3}{2}$ as $\frac{-3}{2}$.

Place a dot at the y-intercept.

Starting at the y-intercept, move right 2 units (the change in x) and down 3 units (the change in y). Place a dot at that location.

Draw a line through the two points.

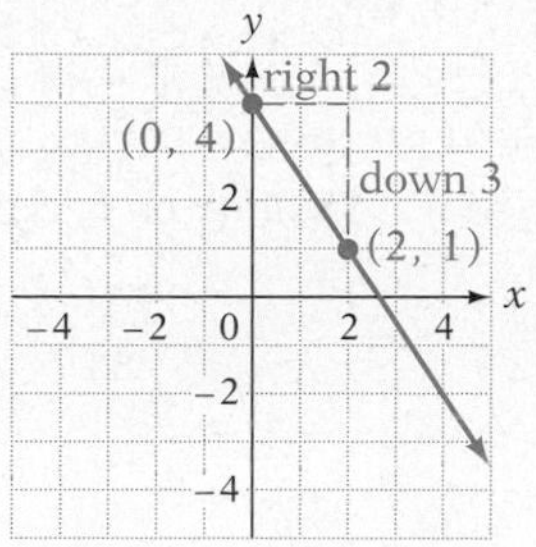

Your solution

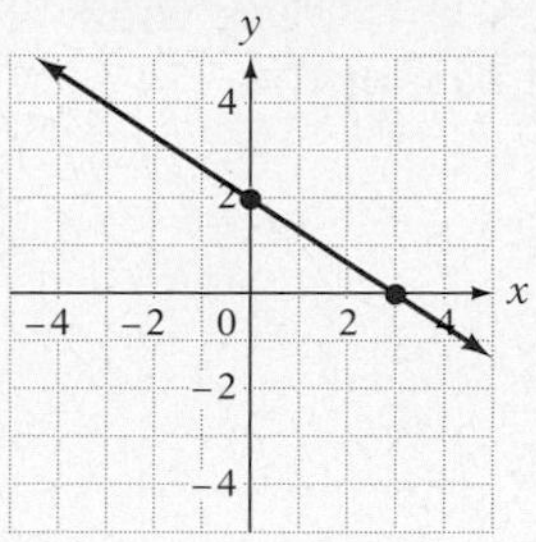

Solution on p. S39

Objective C

To evaluate a function

4.2A Evaluate a function

A **relation** is a set of ordered pairs. A **function** is a relation in which no two ordered pairs have the same first coordinate and different second coordinates.

The relation {(0, 4), (1, 5), (1, 6), (2, 7)} is not a function, because the ordered pairs (1, 5), and (1, 6) have the same first coordinate and different second coordinates.

The relation {(0, 4), (1, 5), (2, 6), (3, 7)} is a function.

The phrase "y is a function of x," or a similar phrase with different variables, is used to describe those equations in two variables that define functions. To emphasize that the equation represents a function, functional notation is used. For example, the square function is written in functional notation as follows:

TAKE NOTE
The symbol $f(x)$ is read "the value of f at x" or "f of x."

$$f(x) = x^2$$

The process of determining $f(x)$ for a given value of x is called **evaluating the function.** For instance, to evaluate $f(x) = x^2$ when $x = 3$, replace x by 3 and simplify.

$$f(x) = x^2$$
$$f(3) = 3^2 = 9$$

The value of the function is 9 when $x = 3$. An ordered pair of the function is (3, 9).

Example 6

Evaluate $f(x) = 2x - 4$ when $x = 3$. Use your answer to write an ordered pair of the function.

Solution

$f(x) = 2x - 4$
$f(3) = 2(3) - 4$
$f(3) = 6 - 4$
$f(3) = 2$

An ordered pair of the function is (3, 2).

You Try It 6

Evaluate the function $f(x) = 4 - 2x$ at $x = -3$. Use your answer to write an ordered pair of the function.

Your solution

$f(-3) = 10$; $(-3, 10)$

Solution on p. S39

Objective D

To find the equation of a line

4.5A Find the equation of a line given a point and the slope

When the slope of a line and a point on the line are known, the equation of the line can be determined by using the point-slope formula.

Point-Slope Formula

Let m be the slope of a line, and let (x_1, y_1) be the coordinates of a point on the line. The equation of the line can be found using the point-slope formula:

$$y - y_1 = m(x - x_1)$$

Example 7

Find the equation of the line that contains the point $(1, -3)$ and has slope -2.

Solution

$$y - y_1 = m(x - x_1)$$
$$y - (-3) = -2(x - 1)$$
$$y + 3 = -2x + 2$$
$$y = -2x - 1$$

The equation of the line is $y = -2x - 1$.

You Try It 7

Find the equation of the line that contains the point $(-2, 2)$ and has slope $-\frac{1}{2}$.

Your solution

$y = -\frac{1}{2}x + 1$

Solution on p. S39

R.3 Exercises

Objective A

Graph the ordered pairs.

1. (2, 3), (4, 0), (−4, 1), (−2, −2)

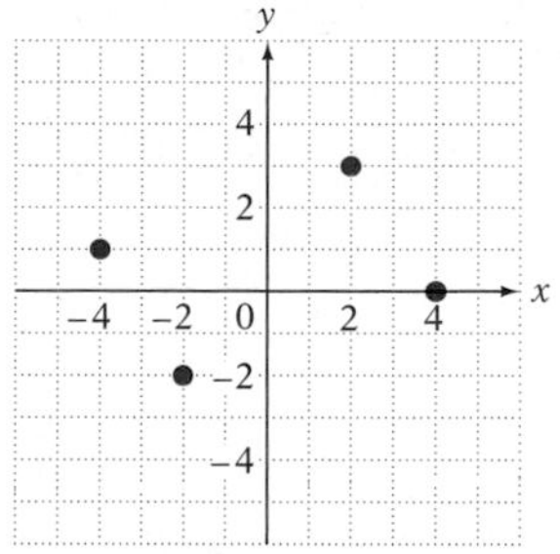

2. (0, 2), (−4, −1), (2, 0), (1, −3)

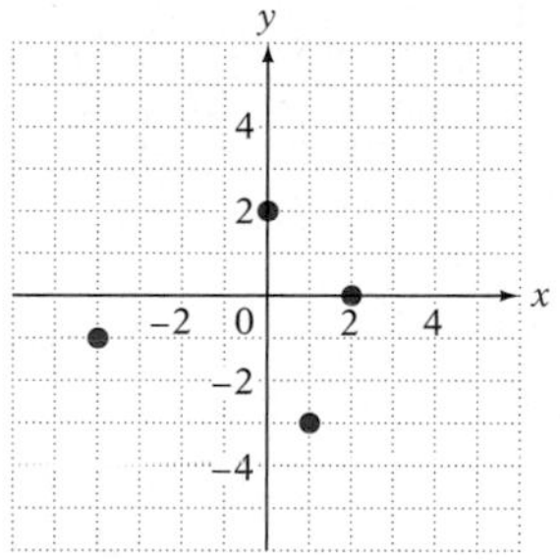

3. (−2, 5), (3, 4), (0, 0), (−3, −2)

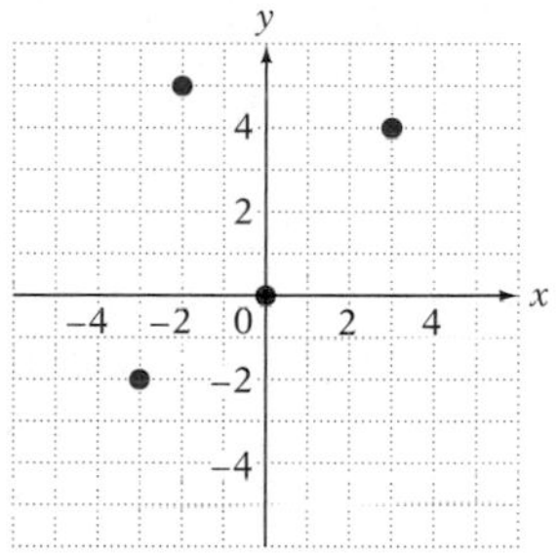

Objective B

Graph.

4. $y = 2x - 4$

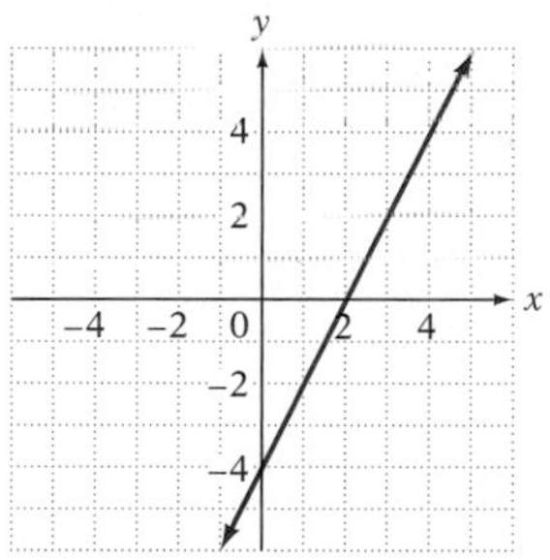

5. $y = 2x + 1$

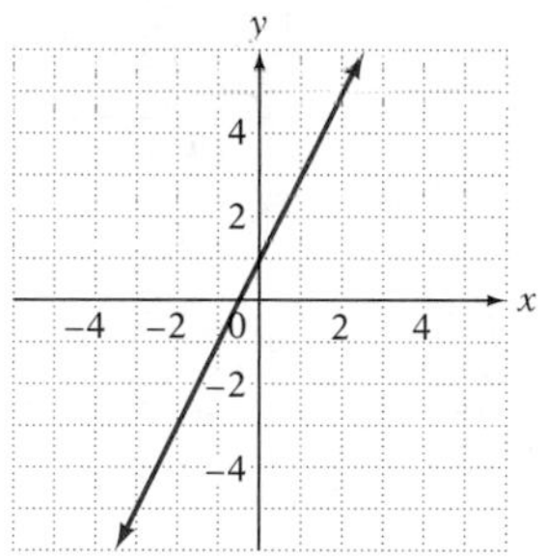

6. $y = -4x + 1$

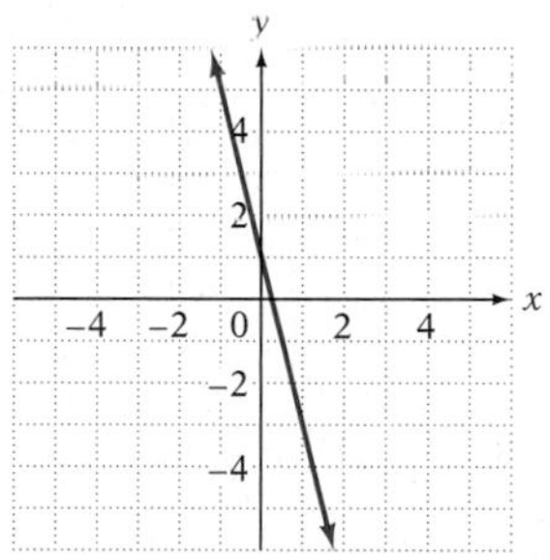

7. $y = -3x + 4$

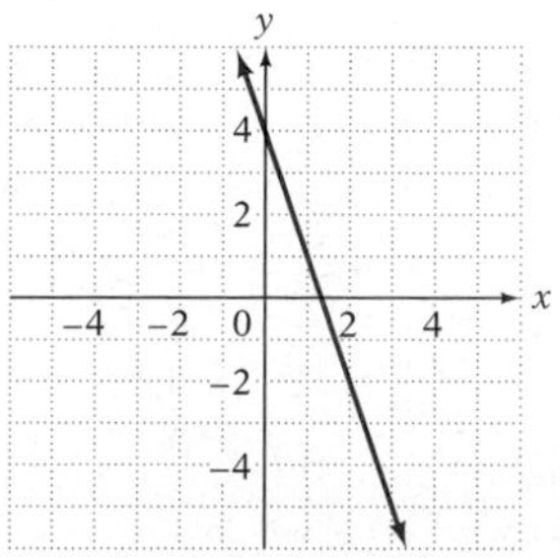

8. $y = 2x$

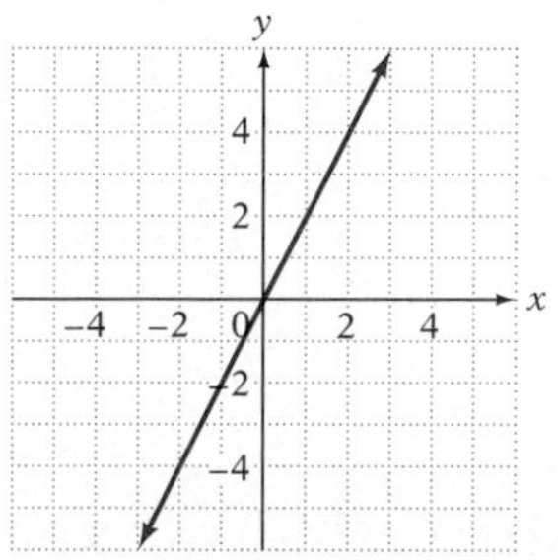

9. $y = 3x$

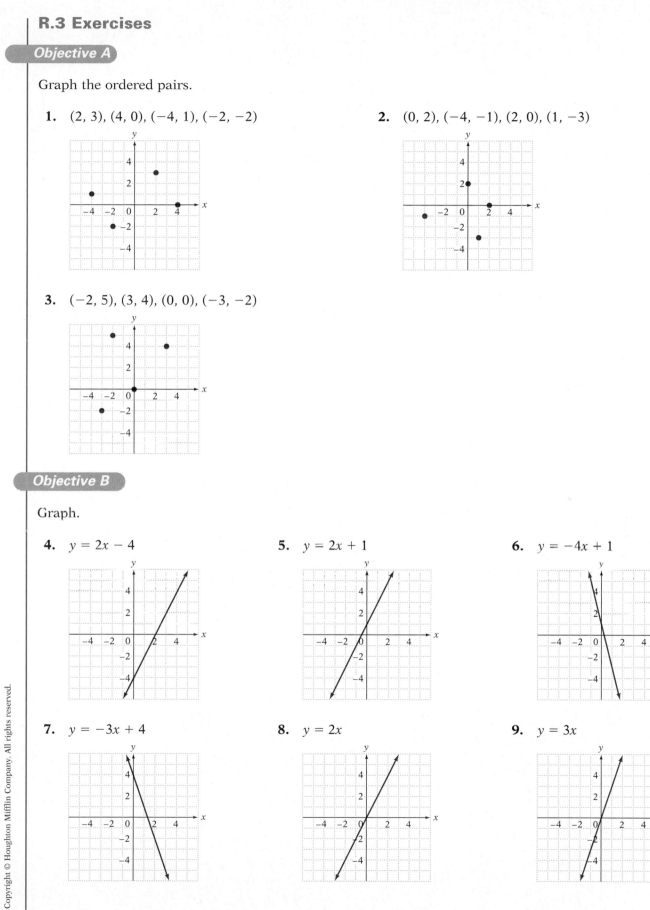

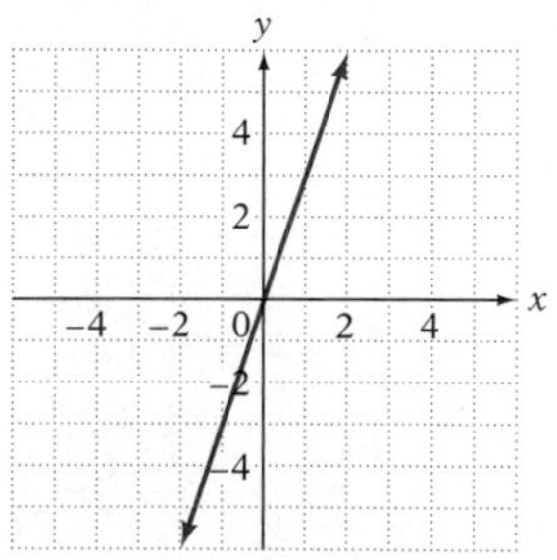

10. $y = -\frac{5}{2}x$

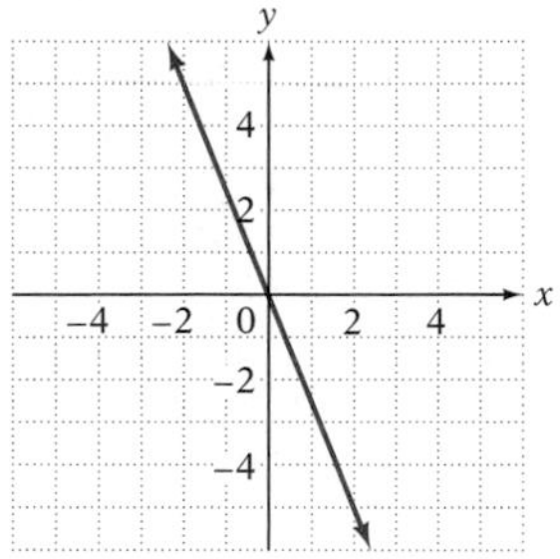

11. $y = -\frac{4}{3}x$

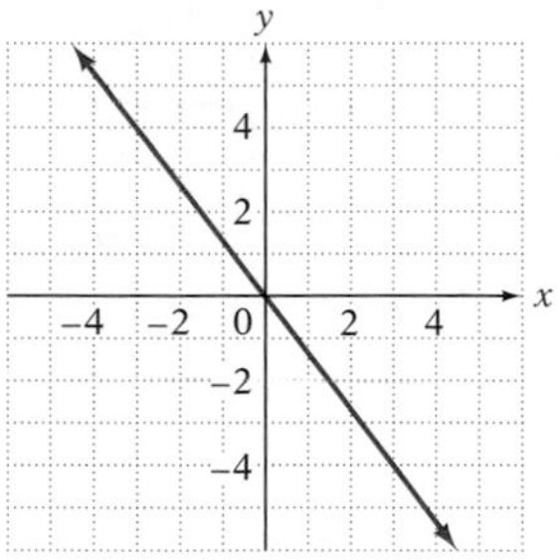

12. $y = \frac{2}{3}x + 1$

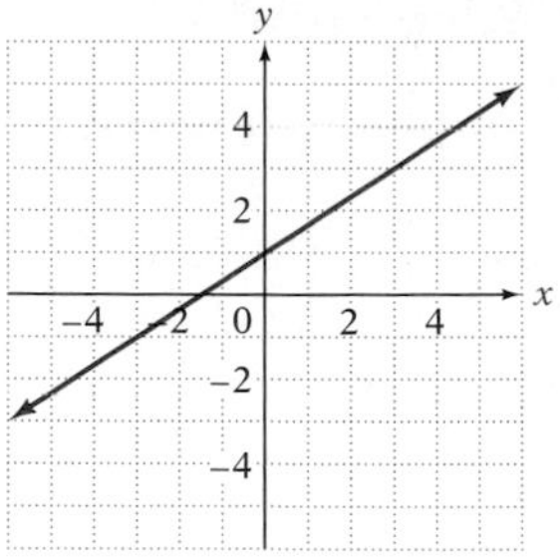

13. $y = \frac{3}{2}x - 1$

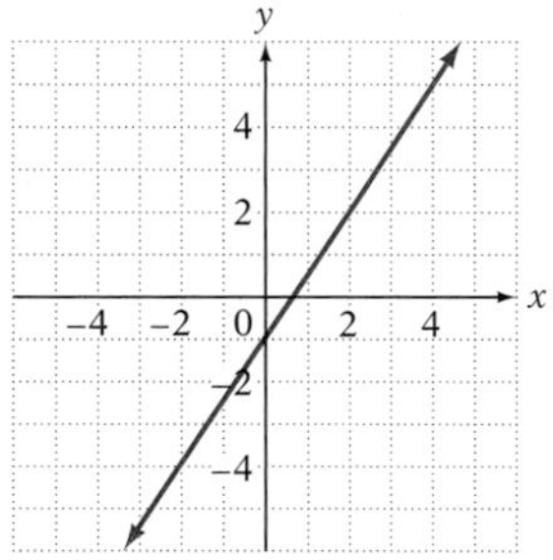

14. $y = -\frac{1}{2}x + 3$

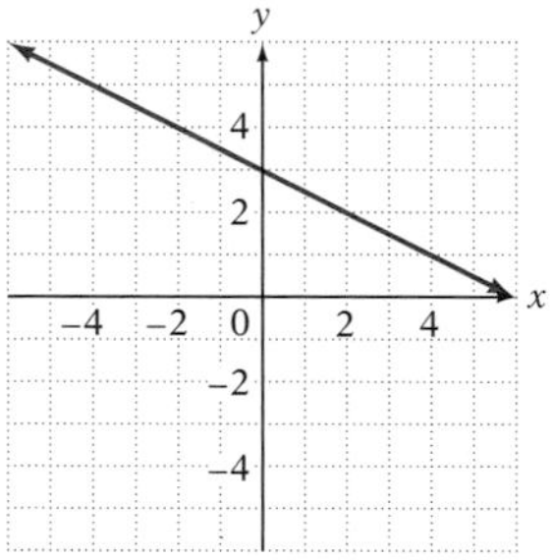

15. $y = -\frac{2}{3}x + 1$

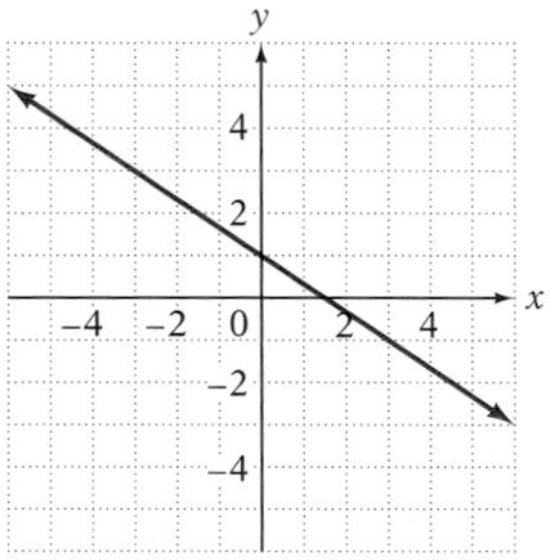

16. $2x - y = 3$

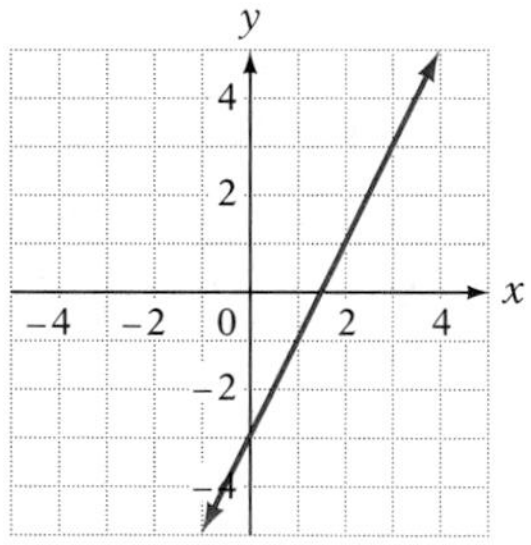

17. $2x + y = -3$

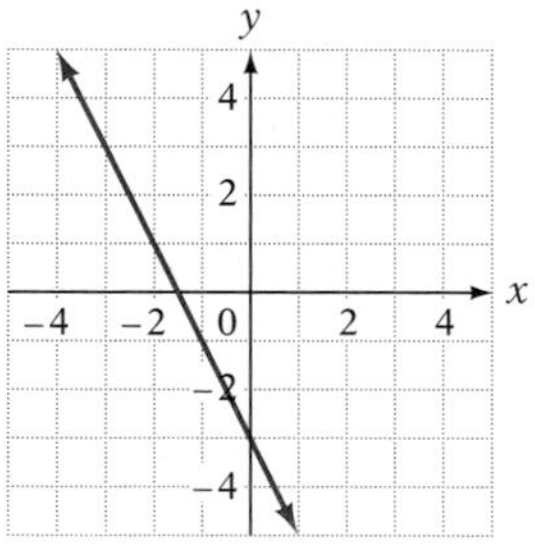

18. $2x + 5y = 10$

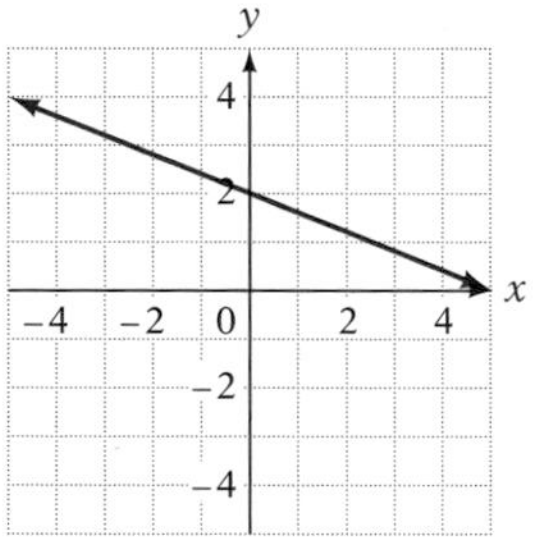

19. $x - 4y = 8$

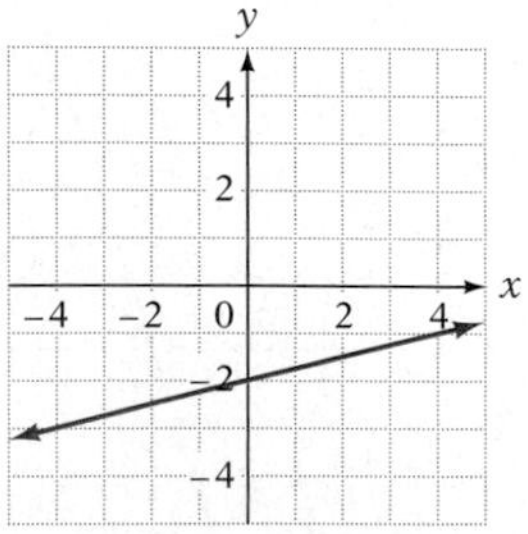

20. $3x - y = -2$

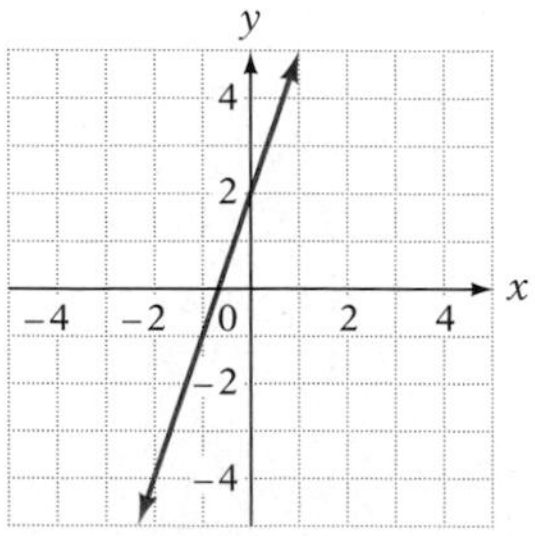

21. $3x - 2y = 8$

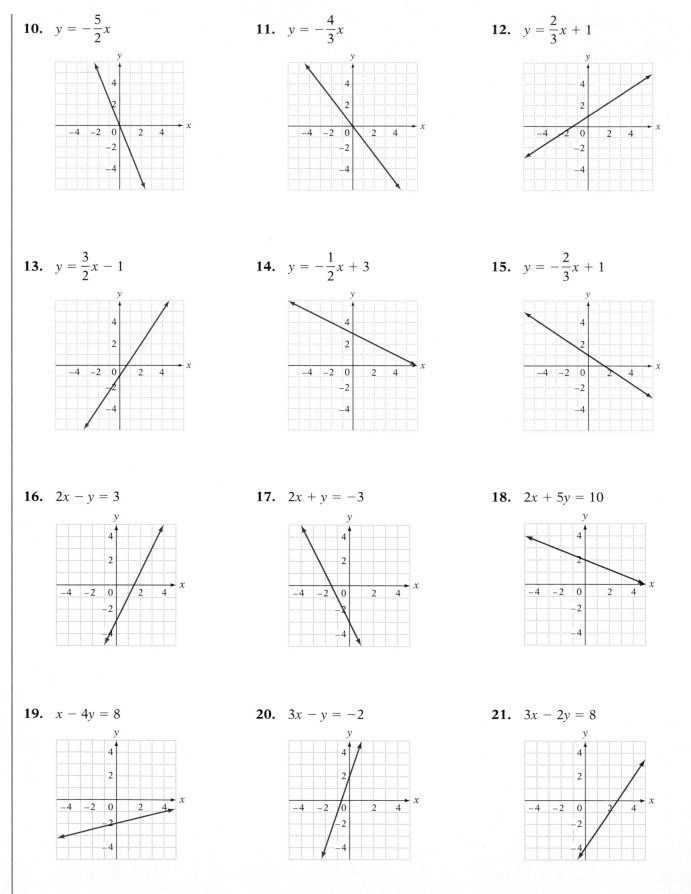

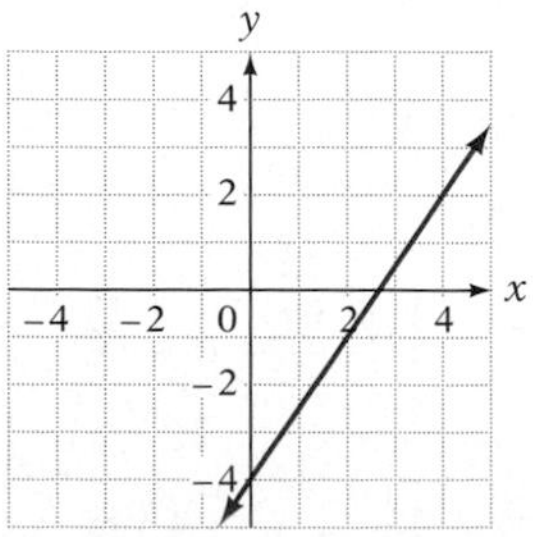

Find the slope of the line containing the given points.

22. $P_1(4, 2), P_2(3, 4)$

-2

23. $P_1(2, 1), P_2(3, 4)$

3

24. $P_1(-1, 3), P_2(2, 4)$

$\frac{1}{3}$

25. $P_1(-2, 1), P_2(2, 2)$

$\frac{1}{4}$

26. $P_1(2, 4), P_2(4, -1)$

$-\frac{5}{2}$

27. $P_1(1, 3), P_2(5, -3)$

$-\frac{3}{2}$

28. $P_1(3, -4), P_2(3, 5)$

Undefined

29. $P_1(-1, 2), P_2(-1, 3)$

Undefined

30. $P_1(4, -2), P_2(3, -2)$

0

31. $P_1(5, 1), P_2(-2, 1)$

0

32. $P_1(0, -1), P_2(3, -2)$

$-\frac{1}{3}$

33. $P_1(3, 0), P_2(2, -1)$

1

Give the slope and the y-intercept of the graph of the equation.

34. $y = -3x + 7$

$m = -3, b = (0, 7)$

35. $y = \frac{5}{2}x - 4$

$m = \frac{5}{2}, b = (0, -4)$

36. $y = x$

$m = 1, b = (0, 0)$

Graph by using the slope and the y-intercept.

37. $y = \frac{1}{2}x + 2$

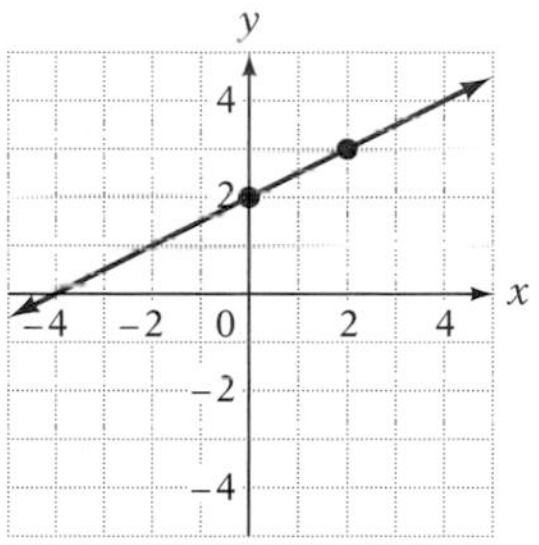

38. $y = \frac{2}{3}x - 3$

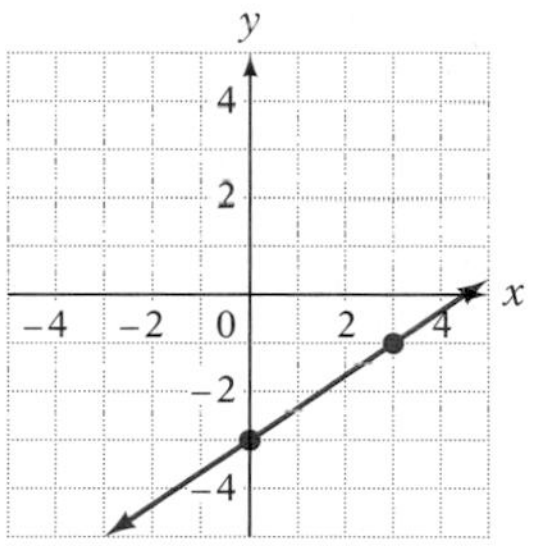

39. $y = -\frac{3}{2}x$

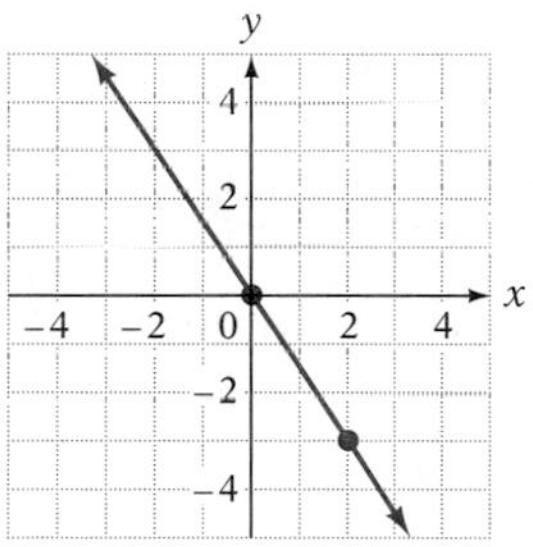

40. $y = \frac{3}{4}x$

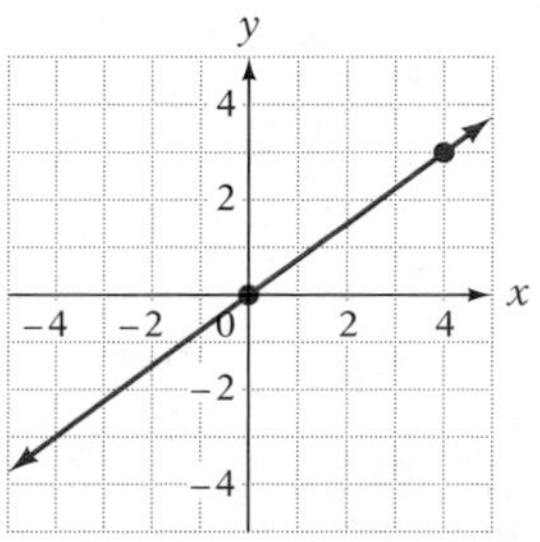

41. $y = -\frac{1}{2}x + 2$

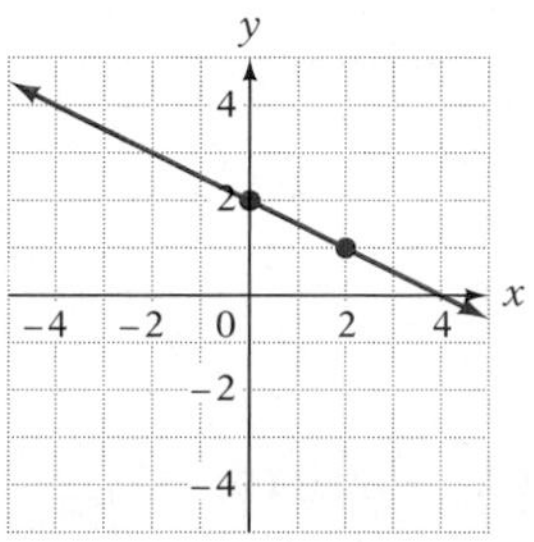

42. $y = \frac{2}{3}x - 1$

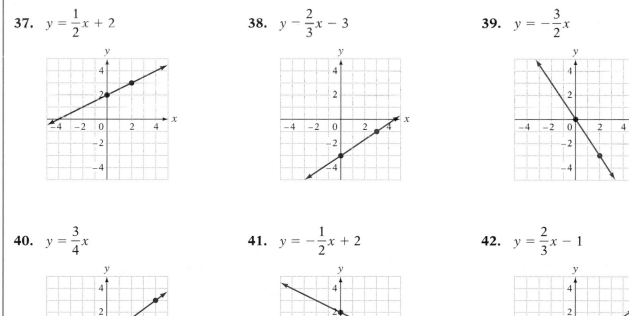

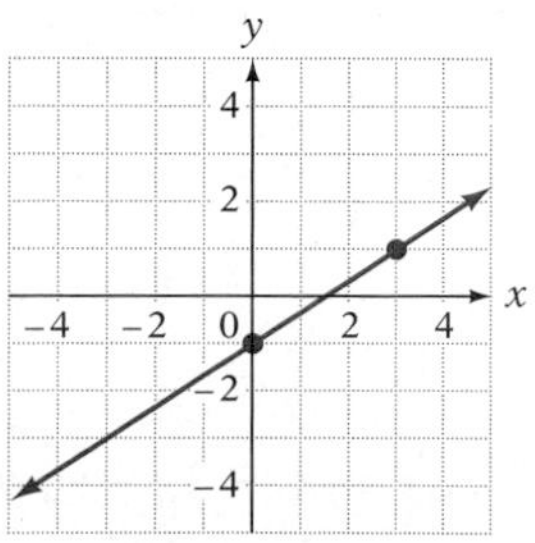

43. $x - 3y = 3$

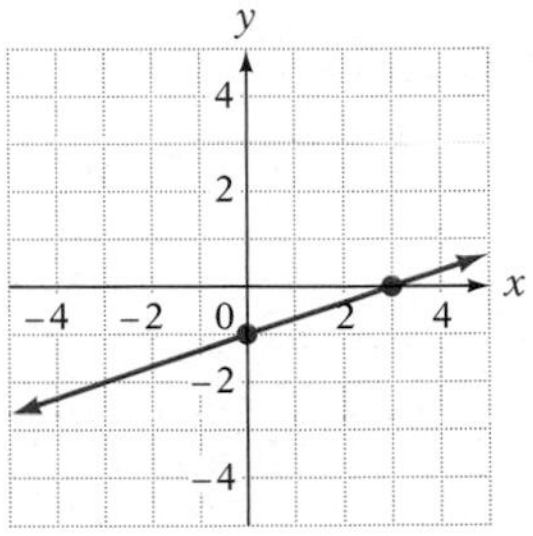

44. $3x + 2y = 8$

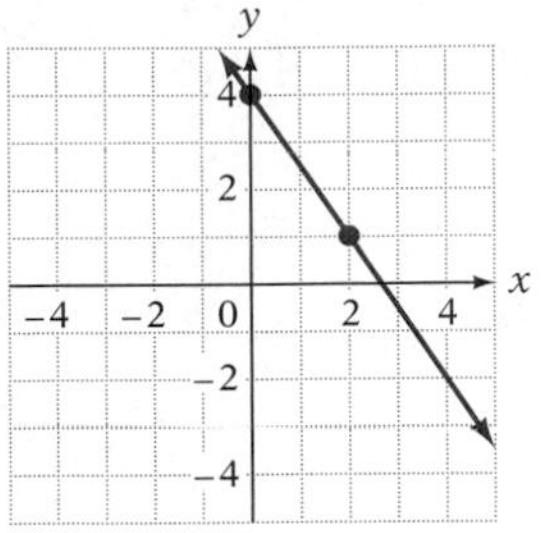

45. $4x + y = 2$

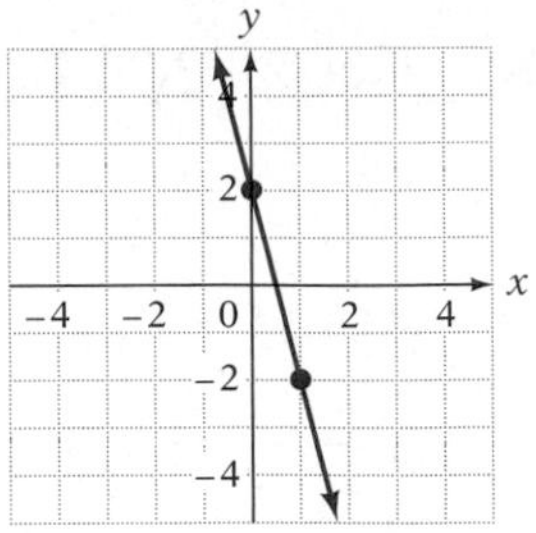

Objective C

Evaluate the function for the given value of the variable. Use your answer to write an ordered pair of the function.

46. $f(x) = 2x + 7; x = -2$
3; (−2, 3)

47. $g(x) = -3x + 1; x = -4$
13; (−4, 13)

48. $h(x) = 4x - 2; x = 3$
10; (3, 10)

49. $p(x) = 6 - 8x; x = -1$
14; (−1, 14)

50. $t(x) = 5 - 7x; x = 0$
5; (0, 5)

51. $f(t) = t^2 - t - 3; t = 2$
−1; (2, −1)

52. $p(n) = n^2 - 4n - 7; n = -3$
14; (−3, 14)

53. $h(x) = -3x^2 + x - 1; x = -2$
−15; (−2, −15)

54. $g(t) = 4t^3 - 2t; t = -1$
−2; (−1, −2)

Objective D

Find the equation of the line that contains the given point and has the given slope.

55. Point $(-1, 2)$; $m = -3$
$y = -3x - 1$

56. Point $(2, -3)$; $m = 3$
$y = 3x - 9$

57. Point $(4, -5)$; $m = -2$
$y = -2x + 3$

58. Point $(-3, 5)$; $m = 3$
$y = 3x + 14$

59. Point $(5, -3)$; $m = -\frac{3}{5}$
$y = -\frac{3}{5}x$

60. Point $(3, 1)$; $m = \frac{1}{3}$
$y = \frac{1}{3}x$

61. Point $(-3, -2)$; $m = -\frac{2}{3}$
$y = -\frac{2}{3}x - 4$

62. Point $(4, -2)$; $m = \frac{3}{4}$
$y = \frac{3}{4}x - 5$

63. Point $(2, 3)$; $m = -\frac{1}{2}$
$y = -\frac{1}{2}x + 4$

R.4 Polynomials

Objective A To multiply and divide monomials

6.1A Multiply and
6.1B divide monomials

A **monomial** is a number, a variable, or a product of a number and variables. The following rules and definitions are used to multiply and divide monomials and to write monomials in simplest form.

Rule for Multiplying Exponential Expressions

If m and n are integers, then $x^m \cdot x^n = x^{m+n}$.

Rule for Simplifying Powers of Exponential Expressions

If m and n are integers, then $(x^m)^n = x^{mn}$.

Rule for Simplifying Powers of Products

If m, n, and p are integers, then $(x^m y^n)^p = x^{mp}y^{np}$.

Rule for Dividing Exponential Expressions

If m and n are integers and $x \neq 0$, then $\dfrac{x^m}{x^n} = x^{m-n}$.

Rule for Simplifying Powers of Quotients

If m, n, and p are integers and $y \neq 0$, then $\left(\dfrac{x^m}{y^n}\right)^p = \dfrac{x^{mp}}{y^{np}}$.

Definition of Zero as an Exponent

If $x \neq 0$, then $x^0 = 1$. The expression 0^0 is undefined.

Definition of Negative Exponents

If n is a positive integer and $x \neq 0$, then $x^{-n} = \dfrac{1}{x^n}$ and $\dfrac{1}{x^{-n}} = x^n$.

➡ Simplify: $(3x^4)^2(4x^3)$

$(3x^4)^2(4x^3) = (3^{1\cdot 2}x^{4\cdot 2})(4x^3)$
- Use the Rule for Simplifying Powers of Products to simplify $(3x^4)^2$.

$= (3^2x^8)(4x^3)$

$= (9x^8)(4x^3)$

$= (9 \cdot 4)(x^8 \cdot x^3)$
- Use the Rule for Multiplying Exponential Expressions.

$= 36x^{8+3}$

$= 36x^{11}$

An exponential expression is in simplest form when there are no negative exponents in the expression. For example, the expression y^{-7} is not in simplest form; use the Definition of Negative Exponents to rewrite the expression with a positive exponent: $y^{-7} = \frac{1}{y^7}$. The expression $\frac{1}{c^{-4}}$ is not in simplest form; use the Definition of Negative Exponents to rewrite the expression with a positive exponent: $\frac{1}{c^{-4}} = c^4$.

➡ Simplify: $\frac{6x^2}{8x^9}$

$$\frac{6x^2}{8x^9} = \frac{3x^2}{4x^9} = \frac{3x^{2-9}}{4}$$

$$= \frac{3x^{-7}}{4} = \frac{3}{4} \cdot \frac{x^{-7}}{1} = \frac{3}{4} \cdot \frac{1}{x^7}$$

$$= \frac{3}{4x^7}$$

- **Divide the coefficients by their common factors. Use the Rule for Dividing Exponential Expressions.**
- **Rewrite the expression with only positive exponents.**

➡ Simplify: $\left(\frac{a^4}{b^3}\right)^{-2}$

$$\left(\frac{a^4}{b^3}\right)^{-2} = \frac{a^{4(-2)}}{b^{3(-2)}} = \frac{a^{-8}}{b^{-6}}$$

- **Use the Rule for Simplifying Powers of Quotients.**

$$= \frac{b^6}{a^8}$$

- **Rewrite the expression with positive exponents.**

Example 1
Simplify.

a. $(-2x)(3x^{-2})^{-3}$

b. $\left(\frac{3a^2b^{-1}}{27a^{-3}b^{-4}}\right)^{-2}$

Solution

a. $(-2x)(3x^{-2})^{-3}$

$= (-2x)(3^{-3}x^6)$
- **Use the Rule for Simplifying Powers of Products.**

$= \frac{-2x \cdot x^6}{3^3}$
- **Write the expression with positive exponents.**

$= -\frac{2x^7}{27}$
- **Use the Rule for Multiplying Exponential Expressions. Simplify 3^3.**

You Try It 1
Simplify.

a. $(-2ab)(2a^3b^{-2})^{-3}$

b. $\left(\frac{2x^2y^{-4}}{4x^{-2}y^{-5}}\right)^{-3}$

Your solution

a. $-\frac{b^7}{4a^8}$ **b.** $\frac{8}{x^{12}y^3}$

b. Use the Rule for Simplifying Powers of Quotients. Then simplify the expression and write it with positive exponents.

$$\left(\frac{3a^2b^{-1}}{27a^{-3}b^{-4}}\right)^{-2}$$

$$= \left(\frac{a^2b^{-1}}{9a^{-3}b^{-4}}\right)^{-2}$$

$$= \frac{a^{2(-2)}b^{(-1)(-2)}}{9^{1(-2)}a^{(-3)(-2)}b^{(-4)(-2)}}$$

$$= \frac{a^{-4}b^2}{9^{-2}a^6b^8}$$

$$= 9^2a^{-4-6}b^{2-8}$$

$$= 81a^{-10}b^{-6} = \frac{81}{a^{10}b^6}$$

Solution on p. S39

Objective B To add and subtract polynomials

A **polynomial** is a variable expression in which the terms are monomials. The polynomial $15t^2 - 2t + 3$ has three terms: $15t^2$, $-2t$, and 3. Note that each of these three terms is a monomial.

A polynomial of *one* term is a **monomial.** $-7x^2$ is a monomial.
A polynomial of *two* terms is a **binomial.** $4y + 3$ is a binomial.
A polynomial of *three* terms is a **trinomial.** $6b^2 + 5b - 8$ is a trinomial.

6.2B Add polynomials

Polynomials can be added by combining like terms. This is illustrated in Example 2 below.

Example 2
Add: $(8x^2 - 4x - 9) + (2x^2 + 9x - 9)$.

Solution
$(8x^2 - 4x - 9) + (2x^2 + 9x - 9)$
$= (8x^2 + 2x^2) + (-4x + 9x) + (-9 - 9)$
$= 10x^2 + 5x - 18$

You Try It 2
Add: $(-4x^3 + 2x^2 - 8) + (4x^3 + 6x^2 - 7x + 5)$.

Your solution
$8x^2 - 7x - 3$

Solution on p. S39

6.2B Subtract polynomials

The additive inverse of the polynomial $(3x^2 - 7x + 8)$ is $-(3x^2 - 7x + 8)$.

To find the additive inverse of a polynomial, change the sign of each term inside the parentheses. $-(3x^2 - 7x + 8) = -3x^2 + 7x - 8$

To subtract two polynomials, add the additive inverse of the second polynomial to the first.

➡ Simplify: $(5a^2 - a + 2) - (-2a^3 + 3a - 3)$

$(5a^2 - a + 2) - (-2a^3 + 3a - 3)$

$= (5a^2 - a + 2) + (2a^3 - 3a + 3)$ • **Rewrite subtraction as addition of the additive inverse.**

$= 2a^3 + 5a^2 - 4a + 5$ • **Combine like terms.**

Example 3

Subtract: $(7c^2 - 9c - 12) - (9c^2 + 5c - 8)$

Solution

$(7c^2 - 9c - 12) - (9c^2 + 5c - 8)$
$= (7c^2 - 9c - 12) + (-9c^2 - 5c + 8)$
$= -2c^2 - 14c - 4$

You Try It 3

Subtract:
$(-4w^3 + 8w - 8) - (3w^3 - 4w^2 + 2w - 1)$

Your solution

$-7w^3 + 4w^2 + 6w - 7$

Solution on p. S40

Objective C

To multiply polynomials

6.3A Multiply a polynomial by a monomial

The Distributive Property is used to multiply a polynomial by a monomial. Each term of the polynomial is multiplied by the monomial.

➡ Multiply: $3x^3(4x^4 - 2x + 5)$

$3x^3(4x^4 - 2x + 5)$

$= 3x^3(4x^4) - 3x^3(2x) + 3x^3(5)$ • **Use the Distributive Property. Multiply each term of the polynomial by $3x^3$.**

$= 12x^7 - 6x^4 + 15x^3$ • **Use the Rule for Multiplying Exponential Expressions.**

Example 4

Multiply: $2xy(3x^2 - xy + 2y^2)$

Solution

$2xy(3x^2 - xy + 2y^2)$
$= 2xy(3x^2) - 2xy(xy) + 2xy(2y^2)$
$= 6x^3y - 2x^2y^2 + 4xy^3$

You Try It 4

Multiply: $3mn^2(2m^2 - 3mn - 1)$

Your solution

$6m^3n^2 - 9m^2n^3 - 3mn^2$

Solution on p. S40

6.3B Multiply two polynomials

A vertical format similar to that used for multiplication of whole numbers is used to multiply two polynomials. The product $(2y - 3)(y^2 + 2y + 5)$ is shown below.

$$\begin{array}{rrrrl} & & y^2 + & 2y + & 5 \\ & & & 2y - & 3 \\ \hline & -3y^2 - & 6y - & 15 & \text{This is } -3(y^2 + 2y + 5). \\ 2y^3 + & 4y^2 + & 10y & & \text{This is } 2y(y^2 + 2y + 5). \text{ Like terms are placed in the same columns.} \\ \hline 2y^3 + & y^2 + & 4y - & 15 & \text{Add the terms in each column.} \end{array}$$

Example 5

Multiply: $(2a^2 + 4a - 5)(3a + 5)$

Solution

$$\begin{array}{rrrr} & 2a^2 + & 4a - & 5 \\ & & 3a + & 5 \\ \hline & 10a^2 + & 20a - & 25 \\ 6a^3 + & 12a^2 - & 15a & \\ \hline 6a^3 + & 22a^2 + & 5a - & 25 \end{array}$$

• **Align like terms in the same column.**

You Try It 5

Multiply: $(3c^2 - 4c + 5)(2c - 3)$

Your solution

$6c^3 - 17c^2 + 22c - 15$

Solution on p. S40

6.3C Multiply two binomials

It is frequently necessary to multiply two binomials. The product is computed by using a method called FOIL. The letters of FOIL stand for **F**irst, **O**uter, **I**nner, and **L**ast. The FOIL method is based on the Distributive Property and involves adding the products of the first terms, the outer terms, the inner terms, and the last terms.

The product $(2x + 3)(3x + 4)$ is shown below using FOIL.

First: $2x$ and $3x$; Last: 3 and 4; Inner: 3 and $3x$; Outer: $2x$ and 4

	First terms	Outer terms	Inner terms	Last terms
$(2x + 3) \cdot (3x + 4) =$	$(2x)(3x) +$	$(2x)(4) +$	$(3)(3x) +$	$(3)(4)$

$$= 6x^2 + 8x + 9x + 12$$
$$= 6x^2 + 17x + 12$$

Example 6

Multiply: $(4x - 3)(2x + 5)$

Solution

$(4x - 3)(2x + 5)$

$= (4x)(2x) + (4x)(5) + (-3)(2x) + (-3)(5)$

$= 8x^2 + 20x + (-6x) + (-15)$

$= 8x^2 + 14x - 15$

You Try It 6

Multiply: $(4y - 7)(3y - 5)$

Your solution

$12y^2 - 41y + 35$

Solution on p. S40

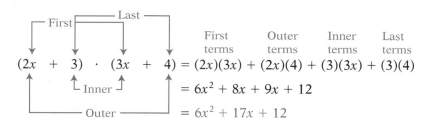

Objective D **To divide polynomials**

6.4A Divide polynomials

To divide two polynomials, use a method similar to that used for division of whole numbers.

To divide $(x^2 - 5x + 8) \div (x - 3)$:

Step 1

$$\begin{array}{r} x \phantom{{}-5x+8} \\ x - 3\,\overline{)\,x^2 - 5x + 8} \\ \underline{x^2 - 3x} \phantom{{}+8}\downarrow \\ -2x + 8 \end{array}$$

Think: $x\,\overline{)x^2} = \dfrac{x^2}{x} = x$

Multiply: $x(x - 3) = x^2 - 3x$

Subtract: $(x^2 - 5x) - (x^2 - 3x) = -2x$

Bring down the +8.

Step 2

$$\begin{array}{r} x - 2 \\ x - 3\,\overline{)\,x^2 - 5x + 8} \\ \underline{x^2 - 3x} \phantom{{}+8} \\ -2x + 8 \\ \underline{-2x + 6} \\ 2 \end{array}$$

Think: $x\,\overline{)-2x} = \dfrac{-2x}{x} = -2$

Multiply: $-2(x - 3) = -2x + 6$

Subtract: $(-2x + 8) - (-2x + 6) = 2$

The remainder is 2.

Check: (Quotient × Divisor) + Remainder = Dividend

$(x - 2)(x - 3) + 2 = x^2 - 3x - 2x + 6 + 2 = x^2 - 5x + 8$

$(x^2 - 5x + 8) \div (x - 3) = x - 2 + \dfrac{2}{x - 3}$

Example 7

Divide: $(6x + 2x^3 + 26) \div (x + 2)$

Solution

Arrange the terms of the dividend in descending order. There is no x^2 term in $2x^3 + 6x + 26$. Insert $0x^2$ for the missing term so that like terms will be in columns.

$$\begin{array}{r} 2x^2 - 4x + 14 \\ x + 2\,\overline{)\,2x^3 + 0x^2 + 6x + 26} \\ \underline{2x^3 + 4x^2} \phantom{{}+6x+26} \\ -4x^2 + 6x \phantom{{}+26} \\ \underline{-4x^2 - 8x} \phantom{{}+26} \\ 14x + 26 \\ \underline{14x + 28} \\ -2 \end{array}$$

Check: $(x + 2)(2x^2 - 4x + 14) - 2$
$= 2x^3 + 6x + 28 - 2 = 2x^3 + 6x + 26$

$(6x + 2x^3 + 26) \div (x + 2) = 2x^2 - 4x + 14 - \dfrac{2}{x + 2}$

You Try It 7

Divide: $(x^3 - 7 - 2x) \div (x - 2)$

Your solution

$x^2 + 2x + 2 - \dfrac{3}{x - 2}$

Solution on p. S40

Objective E To factor polynomials of the form $ax^2 + bx + c$

7.1A Factor a monomial from a polynomial

A polynomial is in **factored form** when it is written as a product of other polynomials. It can be thought of as the reverse of multiplication.

Polynomial		Factored Form
$2x^3 + 6x^2 - 10x$	$=$	$2x(x^2 + 3x - 5)$
$x^2 - 3x - 28$	$=$	$(x + 4)(x - 7)$

To factor out a common monomial from the terms of a polynomial, first find the greatest common factor (GCF) of the terms.

The GCF of two or more monomials is the product of the GCF of the coefficients and the common variable factors.

$$10a^3b = \mathbf{2} \cdot 5 \cdot \boldsymbol{a} \cdot \boldsymbol{a} \cdot a \cdot \boldsymbol{b}$$
$$4a^2b^2 = \mathbf{2} \cdot 2 \cdot \boldsymbol{a} \cdot \boldsymbol{a} \cdot \boldsymbol{b} \cdot b$$
$$\text{GCF} = \mathbf{2} \cdot \boldsymbol{a} \cdot \boldsymbol{a} \cdot \boldsymbol{b} = 2a^2b$$

Note that the exponent of each variable in the GCF is the same as the smallest exponent of that variable in either of the monomials.

The GCF of $10a^3b$ and $4a^2b^2$ is $2a^2b$.

➡ Factor: $5x^3 - 35x^2 + 10x$

The GCF is $5x$.
• Find the GCF of the terms $5x^3$, $-35x^2$, and $10x$.

$$\frac{5x^3 - 35x^2 + 10x}{5x} = x^2 - 7x + 2$$
• Divide each term of the polynomial by the GCF.

$$5x^3 - 35x^2 + 10x = 5x(x^2 - 7x + 2)$$
• Write the polynomial as the product of the GCF and the quotient found above.

$$5x(x^2 - 7x + 2) = 5x^3 - 35x^2 + 10x$$
• Check the factorization by multiplying.

Example 8
Factor: $16x^4y^5 + 8x^4y^2 - 12x^3y$

Solution
The GCF is $4x^3y$.

$$16x^4y^5 + 8x^4y^2 - 12x^3y = 4x^3y(4xy^4 + 2xy - 3)$$

You Try It 8
Factor: $6x^4y^2 - 9x^3y^2 + 12x^2y^4$

Your solution
$3x^2y^2(2x^2 - 3x + 4y^2)$

Solution on p. S40

A **quadratic trinomial** is a trinomial of the form $ax^2 + bx + c$, where a and b are coefficients and c is a constant. Examples of quadratic trinomials are shown below.

$$x^2 + 9x + 14 \qquad a = 1, b = 9, c = 14$$
$$x^2 - 2x - 15 \qquad a = 1, b = -2, c = -15$$
$$3x^2 - x + 4 \qquad a = 3, b = -1, c = 4$$

To **factor a quadratic trinomial** means to express the trinomial as the product of two binomials. For example,

Trinomial		**Factored Form**
$2x^2 - x - 1$	=	$(2x + 1)(x - 1)$
$y^2 - 3y + 2$	=	$(y - 1)(y - 2)$

7.2A Factor a trinomial of the form $x^2 + bx + c$

We will begin by factoring trinomials of the form $x^2 + bx + c$, where the coefficient of x^2 is 1.

The method by which factors of a trinomial are found is based on FOIL. Consider the following binomial products, noting the relationship between the constant term of the binomials and the terms of the trinomial.

Sum of the binomial constants

Product of the binomial constants

$$(x + 4)(x + 5) = x \cdot x + 5x + 4x + 4 \cdot 5 \qquad = x^2 + 9x + 20$$
$$(x - 6)(x + 8) = x \cdot x + 8x - 6x + (-6)(8) \qquad = x^2 + 2x - 48$$
$$(x - 3)(x - 2) = x \cdot x - 2x - 3x + (-3)(-2) \qquad = x^2 - 5x + 6$$

TAKE NOTE
Once the correct factors are found, it is not necessary to try the remaining factors.

TAKE NOTE
Always check the proposed factorization by multiplying the factors.

➡ Factor: $x^2 - 7x + 10$

Find two integers whose product is 10 and whose sum is -7.

Negative Factors of 10	**Sum**	
$-1, -10$	-11	
$-2, -5$	**-7**	• These are the correct factors.

$x^2 - 7x + 10 = (x - 2)(x - 5)$ • Write the trinomial as a product of factors.

Check:
$$(x - 2)(x - 5) = x^2 - 5x - 2x + 10$$
$$= x^2 - 7x + 10$$

Example 9
Factor: $x^2 + 6x - 27$

Solution
Two factors of -27 whose sum is 6 are -3 and 9.

$x^2 + 6x - 27 = (x - 3)(x + 9)$

You Try It 9
Factor: $x^2 - 8x + 15$

Your solution
$(x - 3)(x - 5)$

Solution on p. S40

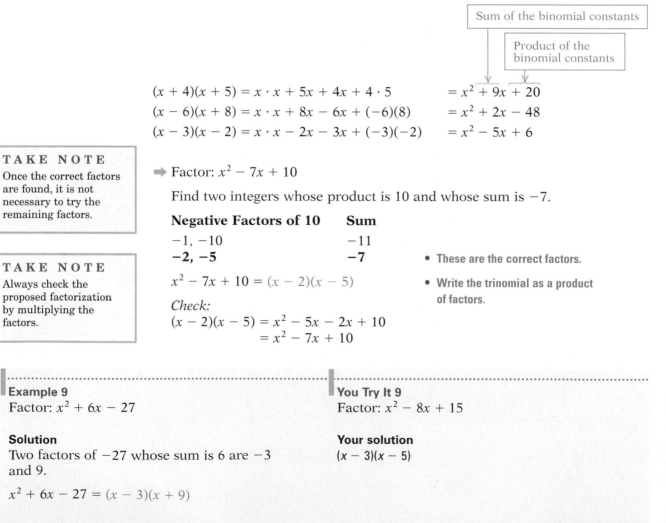

7.3A Factor a trinomial of the form $ax^2 + bx + c$ by using trial factors

To use the trial factor method to factor a trinomial of the form $ax^2 + bx + c$ where $a \neq 1$, use the factors of a and the factors of c to write all of the possible binomial factors of the trinomial. Then use FOIL to determine the correct factorization. To reduce the number of trial factors that must be considered, remember the following guidelines.

Use the signs of the constant term and the coefficient of x in the trinomial to determine the signs of the binomial factors. If the constant term is positive, the signs of the binomial factors will be the same as the sign of the coefficient of x in the trinomial. If the sign of the constant term is negative, the constant terms in the binomials will have different signs.

➡ Factor: $2x^2 - 7x + 3$

Because the constant term is positive $(+3)$ and the coefficient of x is negative (-7), the binomial constants will be negative.

Positive Factors of 2 ($a = 2$)	Negative Factors of 3 ($c = 3$)
1, 2	−1, −3

Write trial factors. Use the **O**uter and **I**nner products of FOIL to determine the middle term, $-7x$, of the trinomial.

Trial Factors	Middle Term	
$(x - 1)(2x - 3)$	$-3x - 2x = -5x$	
$\mathbf{(x - 3)(2x - 1)}$	$\mathbf{-x - 6x = -7x}$	• $-7x$ is the middle term.

$2x^2 - 7x + 3 = (x - 3)(2x - 1)$

Check: $(x - 3)(2x - 1) = 2x^2 - x - 6x + 3 = 2x^2 - 7x + 3$

➡ Factor: $5x^2 + 22x - 15$

The constant term is negative (-15). The binomial constants will have different signs.

Positive Factors of 5 ($a = 5$)	Factors of −15 ($c = -15$)
1, 5	−1, 15
	1, −15
	−3, 5
	3, −5

Write trial factors. Use the **O**uter and **I**nner products of FOIL to determine the middle term, $22x$, of the trinomial.

Trial Factors	Middle Term	
$(x - 1)(5x + 15)$	common factor	
$(x + 15)(5x - 1)$	$-x + 75x = 74x$	
$(x + 1)(5x - 15)$	common factor	
$(x - 15)(5x + 1)$	$x - 75x = -74x$	
$(x - 3)(5x + 5)$	common factor	
$\mathbf{(x + 5)(5x - 3)}$	$\mathbf{-3x + 25x = 22x}$	• $22x$ is the middle term.
$(x + 3)(5x - 5)$	common factor	
$(x - 5)(5x + 3)$	$3x - 25x = -22x$	

TAKE NOTE
It is not necessary to test trial factors that have a common factor. If the trinomial does not have a common factor, then its factors cannot have a common factor.

$5x^2 + 22x - 15 = (x + 5)(5x - 3)$

Check: $(x + 5)(5x - 3) = 5x^2 - 3x + 25x - 15 = 5x^2 + 22x - 15$

Example 10
Factor: $2x^2 + 3x - 5$

Solution

Positive Factors of 2	Factors of -5
1, 2	1, -5
	-1, 5

Trial Factors	Middle Term
$(x + 1)(2x - 5)$	$-5x + 2x = -3x$
$(x - 5)(2x + 1)$	$x - 10x = -9x$
$(x - 1)(2x + 5)$	$5x - 2x = 3x$
$(x + 5)(2x - 1)$	$-x + 10x = 9x$

$2x^2 + 3x - 5 = (x - 1)(2x + 5)$

Check: $(x - 1)(2x + 5) = 2x^2 + 5x - 2x - 5$
$= 2x^2 + 3x - 5$

You Try It 10
Factor: $3x^2 - x - 2$

Your solution
$(x - 1)(3x + 2)$

Solution on p. S40

7.4D Factor completely

A polynomial is factored completely when it is written as a product of factors that are nonfactorable over the integers.

The first step in any factoring problem is to determine whether the terms of the polynomial have a common factor. If they do, factor it out first.

Example 11
Factor: $5x^2y + 60xy + 100y$

Solution
There is a common factor, $5y$. Factor out the GCF.

$5x^2y + 60xy + 100y = 5y(x^2 + 12x + 20)$

Factor $x^2 + 12x + 20$. The two factors of 20 whose sum is 12 are 2 and 10.

$5y(x^2 + 12x + 20) = 5y(x + 2)(x + 10)$

$5x^2y + 60xy + 100y = 5y(x + 2)(x + 10)$

Check:
$5y(x + 2)(x + 10) = (5xy + 10y)(x + 10)$
$= 5x^2y + 50xy + 10xy + 100y$
$= 5x^2y + 60xy + 100y$

You Try It 11
Factor: $4a^3 - 4a^2 - 24a$

Your solution
$4a(a + 2)(a - 3)$

Solution on p. S40

R.4 Exercises

Objective A

Simplify.

1. $z^3 \cdot z \cdot z^4$
z^8

2. $b \cdot b^2 \cdot b^6$
b^9

3. $(x^3)^5$
x^{15}

4. $(b^2)^4$
b^8

5. $(x^2y^3)^6$
$x^{12}y^{18}$

6. $(m^4n^2)^3$
$m^{12}n^6$

7. $\dfrac{a^8}{a^2}$
a^6

8. $\dfrac{c^{12}}{c^5}$
c^7

9. $(-m^3n)(m^6n^2)$
$-m^9n^3$

10. $(-r^4t^3)(r^2t^9)$
$-r^6t^{12}$

11. $(-2a^3bc^2)^3$
$-8a^9b^3c^6$

12. $(-4xy^3z^2)^2$
$16x^2y^6z^4$

13. $\dfrac{m^4n^7}{m^3n^5}$
mn^2

14. $\dfrac{a^5b^6}{a^3b^2}$
a^2b^4

15. $\dfrac{-16a^7}{24a^6}$
$-\dfrac{2a}{3}$

16. $\dfrac{18b^5}{-45b^4}$
$-\dfrac{2b}{5}$

17. $(9mn^4p)(-3mp^2)$
$-27m^2n^4p^3$

18. $(-3v^2wz)(-4vz^4)$
$12v^3wz^5$

19. $(-2n^2)(-3n^4)^3$
$54n^{14}$

20. $(-3m^3n)(-2m^2n^3)^3$
$24m^9n^{10}$

21. $\dfrac{14x^4y^6z^2}{16x^3y^9z}$
$\dfrac{7xz}{8y^3}$

22. $\dfrac{25x^4y^7z^2}{20x^5y^9z^{11}}$
$\dfrac{5}{4xy^2z^9}$

23. $(-2x^3y^2)^3(-xy^2)^4$
$-8x^{13}y^{14}$

24. $(-m^4n^2)^5(-2m^3n^3)^3$
$8m^{29}n^{19}$

25. $4x^{-7}$
$\dfrac{4}{x^7}$

26. $-6y^{-1}$
$-\dfrac{6}{y}$

27. $d^{-4}d^{-6}$
$\dfrac{1}{d^{10}}$

28. $x^{-3}x^{-5}$
$\dfrac{1}{x^8}$

29. $\dfrac{x^{-3}}{x^2}$
$\dfrac{1}{x^5}$

30. $\dfrac{x^4}{x^{-5}}$
x^9

31. $\dfrac{1}{3x^{-2}}$
$\dfrac{x^2}{3}$

32. $\dfrac{2}{5c^{-6}}$
$\dfrac{2c^6}{5}$

33. $(x^2y^{-4})^3$
$\dfrac{x^6}{y^{12}}$

34. $(x^3y^5)^{-4}$
$\dfrac{1}{x^{12}y^{20}}$

35. $(3x^{-1}y^{-2})^2$
$\dfrac{9}{x^2y^4}$

36. $(5xy^{-3})^{-2}$
$\dfrac{y^6}{25x^2}$

37. $(2x^{-1})(x^{-3})$

$\frac{2}{x^4}$

38. $(-2x^{-5})(x^7)$

$-2x^2$

39. $\frac{3x^{-2}y^2}{6xy^2}$

$\frac{1}{2x^3}$

40. $\frac{2x^{-2}y}{8xy}$

$\frac{1}{4x^3}$

41. $\frac{2x^{-1}y^{-4}}{4xy^2}$

$\frac{1}{2x^2y^6}$

42. $\frac{3a^{-2}b}{ab}$

$\frac{3}{a^3}$

43. $(x^{-2}y)^2(xy)^{-2}$

$\frac{1}{x^6}$

44. $(x^{-1}y^2)^{-3}(x^2y^{-4})^{-3}$

$\frac{y^6}{x^3}$

45. $\left(\frac{x^2y^{-1}}{xy}\right)^{-4}$

$\frac{y^8}{x^4}$

46. $\left(\frac{x^{-2}y^{-4}}{x^{-2}y}\right)^{-2}$

y^{10}

47. $\left(\frac{4a^{-2}b}{8a^3b^{-4}}\right)^2$

$\frac{b^{10}}{4a^{10}}$

48. $\left(\frac{6ab^{-2}}{3a^{-2}b}\right)^{-2}$

$\frac{b^6}{4a^6}$

Objective B

Add or subtract.

49. $(4b^2 - 5b) + (3b^2 + 6b - 4)$
$7b^2 + b - 4$

50. $(2c^2 - 4) + (6c^2 - 2c + 4)$
$8c^2 - 2c$

51. $(2a^2 - 7a + 10) + (a^2 + 4a + 7)$
$3a^2 - 3a + 17$

52. $(-6x^2 + 7x + 3) + (3x^2 + x + 3)$
$-3x^2 + 8x + 6$

53. $(x^2 - 2x + 1) - (x^2 + 5x + 8)$
$-7x - 7$

54. $(3x^2 + 2x - 2) - (5x^2 - 5x + 6)$
$-2x^2 + 7x - 8$

55. $(-2x^3 + x - 1) - (-x^2 + x - 3)$
$-2x^3 + x^2 + 2$

56. $(2x^2 + 5x - 3) - (3x^3 + 2x - 5)$
$-3x^3 + 2x^2 + 3x + 2$

57. $(x^3 - 7x + 4) + (2x^2 + x - 10)$
$x^3 + 2x^2 - 6x - 6$

58. $(3y^3 + y^2 + 1) + (-4y^3 - 6y - 3)$
$-y^3 + y^2 - 6y - 2$

59. $(5x^3 + 7x - 7) + (10x^2 - 8x + 3)$
$5x^3 + 10x^2 - x - 4$

60. $(3y^3 + 4y + 9) + (2y^2 + 4y - 21)$
$3y^3 + 2y^2 + 8y - 12$

61. $(2y^3 + 6y - 2) - (y^3 + y^2 + 4)$
$y^3 - y^2 + 6y - 6$

62. $(-2x^2 - x + 4) - (-x^3 + 3x - 2)$
$x^3 - 2x^2 - 4x + 6$

63. $(4y^3 - y - 1) - (2y^2 - 3y + 3)$
$4y^3 - 2y^2 + 2y - 4$

64. $(3x^2 - 2x - 3) - (2x^3 - 2x^2 + 4)$
$-2x^3 + 5x^2 - 2x - 7$

Objective C

Multiply.

65. $4b(3b^3 - 12b^2 - 6)$
$12b^4 - 48b^3 - 24b$

66. $-2a^2(3a^2 - 2a + 3)$
$-6a^4 + 4a^3 - 6a^2$

67. $3b(3b^4 - 3b^2 + 8)$
$9b^5 - 9b^3 + 24b$

68. $-2x^2(2x^2 - 3x - 7)$
$-4x^4 + 6x^3 + 14x^2$

69. $-2x^2y(x^2 - 3xy + 2y^2)$
$-2x^4y + 6x^3y^2 - 4x^2y^3$

70. $3ab^2(3a^2 - 2ab + 4b^2)$
$9a^3b^2 - 6a^2b^3 + 12ab^4$

71. $(x^2 + 3x + 2)(x + 1)$
$x^3 + 4x^2 + 5x + 2$

72. $(x^2 - 2x + 7)(x - 2)$
$x^3 - 4x^2 + 11x - 14$

73. $(a - 3)(a^2 - 3a + 4)$
$a^3 - 6a^2 + 13a - 12$

74. $(2x - 3)(x^2 - 3x + 5)$
$2x^3 - 9x^2 + 19x - 15$

75. $(-2b^2 - 3b + 4)(b - 5)$
$-2b^3 + 7b^2 + 19b - 20$

76. $(-a^2 + 3a - 2)(2a - 1)$
$-2a^3 + 7a^2 - 7a + 2$

77. $(x^3 - 3x + 2)(x - 4)$
$x^4 - 4x^3 - 3x^2 + 14x - 8$

78. $(y^3 + 4y^2 - 8)(2y - 1)$
$2y^4 + 7y^3 - 4y^2 - 16y + 8$

79. $(y + 2)(y^3 + 2y^2 - 3y + 1)$
$y^4 + 4y^3 + y^2 - 5y + 2$

80. $(2a - 3)(2a^3 - 3a^2 + 2a - 1)$
$4a^4 - 12a^3 + 13a^2 - 8a + 3$

81. $(a - 3)(a + 4)$
$a^2 + a - 12$

82. $(b - 6)(b + 3)$
$b^2 - 3b - 18$

83. $(y - 7)(y - 3)$
$y^2 - 10y + 21$

84. $(a - 8)(a - 9)$
$a^2 - 17a + 72$

85. $(2x + 1)(x + 7)$
$2x^2 + 15x + 7$

86. $(y + 2)(5y + 1)$
$5y^2 + 11y + 2$

87. $(3x - 1)(x + 4)$
$3x^2 + 11x - 4$

88. $(7x - 2)(x + 4)$
$7x^2 + 26x - 8$

89. $(4x - 3)(x - 7)$
$4x^2 - 31x + 21$

90. $(2x - 3)(4x - 7)$
$8x^2 - 26x + 21$

91. $(3y - 8)(y + 2)$
$3y^2 - 2y - 16$

92. $(5y - 9)(y + 5)$
$5y^2 + 16y - 45$

93. $(7a - 16)(3a - 5)$
$21a^2 - 83a + 80$

94. $(5a - 12)(3a - 7)$
$15a^2 - 71a + 84$

95. $(x + y)(2x + y)$
$2x^2 + 3xy + y^2$

96. $(2a + b)(a + 3b)$
$2a^2 + 7ab + 3b^2$

97. $(3x - 4y)(x - 2y)$
$3x^2 - 10xy + 8y^2$

98. $(2a - b)(3a + 2b)$
$6a^2 + ab - 2b^2$

99. $(5a - 3b)(2a + 4b)$
$10a^2 + 14ab - 12b^2$

100. $(2x + 3)(2x - 3)$
$4x^2 - 9$

101. $(4x - 7)(4x + 7)$
$16x^2 - 49$

Objective D

Divide.

102. $(b^2 - 14 + 49) \div (b - 7)$
$b - 7$

103. $(x^2 - x - 6) \div (x - 3)$
$x + 2$

104. $(2x^2 + 5x + 2) \div (x + 2)$
$2x + 1$

105. $(2y^2 - 13y + 21) \div (y - 3)$
$2y - 7$

106. $(x^2 + 1) \div (x - 1)$
$x + 1 + \dfrac{2}{x - 1}$

107. $(x^2 + 4) \div (x + 2)$
$x - 2 + \dfrac{8}{x + 2}$

108. $(6x^2 - 7x) \div (3x - 2)$
$2x - 1 - \dfrac{2}{3x - 2}$

109. $(6y^2 + 2y) \div (2y + 4)$
$3y - 5 + \dfrac{20}{2y + 4}$

110. $(a^2 + 5a + 10) \div (a + 2)$
$a + 3 + \dfrac{4}{a + 2}$

111. $(b^2 - 8b - 9) \div (b - 3)$
$b - 5 - \dfrac{24}{b - 3}$

112. $(2y^2 - 9y + 8) \div (2y + 3)$
$y - 6 + \dfrac{26}{2y + 3}$

113. $(3x^2 + 5x - 4) \div (x - 4)$
$3x + 17 + \dfrac{64}{x - 4}$

114. $(8x + 3 + 4x^2) \div (2x - 1)$
$2x + 5 + \dfrac{8}{2x - 1}$

115. $(10 + 21y + 10y^2) \div (2y + 3)$
$5y + 3 + \dfrac{1}{2y + 3}$

116. $(x^3 + 3x^2 + 5x + 3) \div (x + 1)$
$x^2 + 2x + 3$

117. $(x^3 - 6x^2 + 7x - 2) \div (x - 1)$
$x^2 - 5x + 2$

118. $(x^4 - x^2 - 6) \div (x^2 + 2)$
$x^2 - 3$

119. $(x^4 + 3x^2 - 10) \div (x^2 - 2)$
$x^2 + 5$

Objective E

Factor.

120. $8x + 12$
$4(2x + 3)$

121. $12y^2 - 5y$
$y(12y - 5)$

122. $10x^4 - 12x^2$
$2x^2(5x^2 - 6)$

123. $10x^2yz^2 + 15xy^3z$
$5xyz(2xz + 3y^2)$

124. $x^3 - 3x^2 - x$
$x(x^2 - 3x - 1)$

125. $5x^2 - 15x + 35$
$5(x^2 - 3x + 7)$

126. $3x^3 + 6x^2 + 9x$
$3x(x^2 + 2x + 3)$

127. $3y^4 - 9y^3 - 6y^2$
$3y^2(y^2 - 3y - 2)$

128. $2x^3 + 6x^2 - 14x$
$2x(x^2 + 3x - 7)$

129. $x^4y^4 - 3x^3y^3 + 6x^2y^2$
$x^2y^2(x^2y^2 - 3xy + 6)$

130. $4x^5y^5 - 8x^4y^4 + x^3y^3$
$x^3y^3(4x^2y^2 - 8xy + 1)$

131. $16x^2y - 8x^3y^4 - 48x^2y^2$
$8x^2y(2 - xy^3 - 6y)$

132. $x^2 + 5x + 6$
$(x + 2)(x + 3)$

133. $x^2 + x - 2$
$(x + 2)(x - 1)$

134. $x^2 + x - 6$
$(x + 3)(x - 2)$

135. $a^2 + a - 12$
$(a + 4)(a - 3)$

136. $a^2 - 2a - 35$
$(a + 5)(a - 7)$

137. $a^2 - 3a + 2$
$(a - 1)(a - 2)$

138. $a^2 - 5a + 4$
$(a - 1)(a - 4)$

139. $b^2 + 7b - 8$
$(b + 8)(b - 1)$

140. $y^2 + 6y - 55$
$(y + 11)(y - 5)$

141. $z^2 - 4z - 45$
$(z + 5)(z - 9)$

142. $y^2 - 8y + 15$
$(y - 3)(y - 5)$

143. $z^2 - 14z + 45$
$(z - 5)(z - 9)$

144. $p^2 + 12p + 27$
$(p + 3)(p + 9)$

145. $b^2 + 9b + 20$
$(b + 4)(b + 5)$

146. $y^2 - 8y + 32$
Nonfactorable

147. $y^2 - 9y + 81$
Nonfactorable

148. $p^2 + 24p + 63$
$(p + 3)(p + 21)$

149. $x^2 - 15x + 56$
$(x - 7)(x - 8)$

150. $5x^2 + 6x + 1$
$(x + 1)(5x + 1)$

151. $2y^2 + 7y + 3$
$(y + 3)(2y + 1)$

152. $2a^2 - 3a + 1$
$(a - 1)(2a - 1)$

153. $3a^2 - 4a + 1$
$(a - 1)(3a - 1)$

154. $4x^2 - 3x - 1$
$(x - 1)(4x + 1)$

155. $2x^2 - 5x - 3$
$(x - 3)(2x + 1)$

156. $6t^2 - 11t + 4$
$(2t - 1)(3t - 4)$

157. $10t^2 + 11t + 3$
$(2t + 1)(5t + 3)$

158. $8x^2 + 33x + 4$
$(x + 4)(8x + 1)$

159. $10z^2 + 3z - 4$
$(2z - 1)(5z + 4)$

160. $3x^2 + 14x - 5$
$(x + 5)(3x - 1)$

161. $3z^2 + 95z + 10$
Nonfactorable

162. $8z^2 - 36z + 1$
Nonfactorable

163. $2t^2 - t - 10$
$(t + 2)(2t - 5)$

164. $2t^2 + 5t - 12$
$(t + 4)(2t - 3)$

165. $12y^2 + 19y + 5$
$(3y + 1)(4y + 5)$

166. $5y^2 - 22y + 8$
$(y - 4)(5y - 2)$

167. $11a^2 - 54a - 5$
$(a - 5)(11a + 1)$

168. $4z^2 + 11z + 6$
$(z + 2)(4z + 3)$

169. $6b^2 - 13b + 6$
$(2b - 3)(3b - 2)$

170. $6x^2 + 35x - 6$
$(x + 6)(6x - 1)$

171. $3x^2 + 15x + 18$
$3(x + 2)(x + 3)$

172. $3a^2 + 3a - 18$
$3(a + 3)(a - 2)$

173. $ab^2 + 7ab - 8a$
$a(b + 8)(b - 1)$

174. $3y^3 - 15y^2 + 18y$
$3y(y - 2)(y - 3)$

175. $2y^4 - 26y^3 - 96y^2$
$2y^2(y + 3)(y - 16)$

176. $3y^4 + 54y^3 + 135y^2$
$3y^2(y + 3)(y + 15)$

177. $2x^3 - 11x^2 + 5x$
$x(x - 5)(2x - 1)$

178. $2x^3 + 3x^2 - 5x$
$x(x - 1)(2x + 5)$

179. $10t^2 - 5t - 50$
$5(t + 2)(2t - 5)$

180. $16t^2 + 40t - 96$
$8(t + 4)(2t - 3)$

181. $6p^3 + 5p^2 + p$
$p(2p + 1)(3p + 1)$

182. $12x^2y - 36xy + 27y$
$3y(2x - 3)^2$

Appendix

Calculator Guide for the TI-83 and TI-83Plus

Basic Operations

To evaluate an expression

a. Press the [Y=] key. A menu showing \Y1 = through \Y7 = will be displayed vertically with a blinking cursor to the right of \Y1 =. Press [CLEAR], if necessary, to delete an unwanted expression.

b. Input the expression to be evaluated. For example, to input the expression $-3a^2b - 4c$, use the following keystrokes:

[(−)] 3 [ALPHA] A [X²] [ALPHA] B [−] 4 [ALPHA] C [2nd] QUIT

Note the difference between the keys for a negative sign [(−)] and a minus sign [−].

c. Store the value of each variable that will be used in the expression. For example, to evaluate the expression above when $a = 3$, $b = -2$, and $c = -4$, use the following keystrokes:

3 [STO▷] [ALPHA] A [ENTER] [(−)] 2 [STO▷] [ALPHA] B [ENTER] [(−)] 4 [STO▷] [ALPHA] C [ENTER]

These steps store the value of each variable.

d. Press [VARS] [▷] [1] [1] [ENTER]. The value for the expression, Y1, for the given values is displayed; in this case, Y1 = 70.

To graph a function

a. Press the [Y=] key. A menu showing \Y1 = through \Y7 = will be displayed vertically with a blinking cursor to the right of \Y1 =. Press [CLEAR], if necessary, to delete an unwanted expression.

b. Input the expression for each function that is to be graphed. Press [X,T,θ,*n*] to input x. For example, to input $y = x^3 + 2x^2 - 5x - 6$, use the following keystrokes:

[X,T,θ,*n*] [∧] 3 [+] 2 [X,T,θ,*n*] [X²] [−] 5 [X,T,θ,*n*] [−] 6

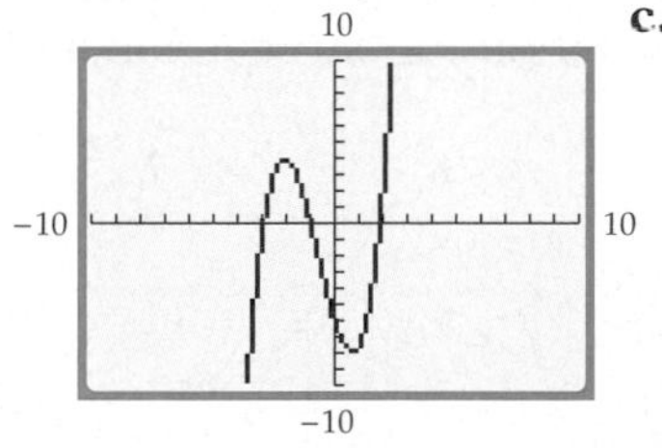

c. Set the viewing window by pressing [WINDOW]. Enter the values for the minimum x-value (Xmin), the maximum x-value (Xmax), the distance between tick marks on the x-axis (Xscl), the minimum y-value (Ymin), the maximum y-value (Ymax), and the distance between tick marks on the y-axis (Yscl). Now press [GRAPH]. For the graph shown at the left, Xmin = −10, Xmax = 10, Xscl = 1, Ymin = −10, Ymax = 10, and Yscl = 1. This is called the standard viewing window. Pressing [ZOOM] [6] is a quick way to set the calculator to the standard viewing window. *Note:* This will also immediately graph the function in that window.

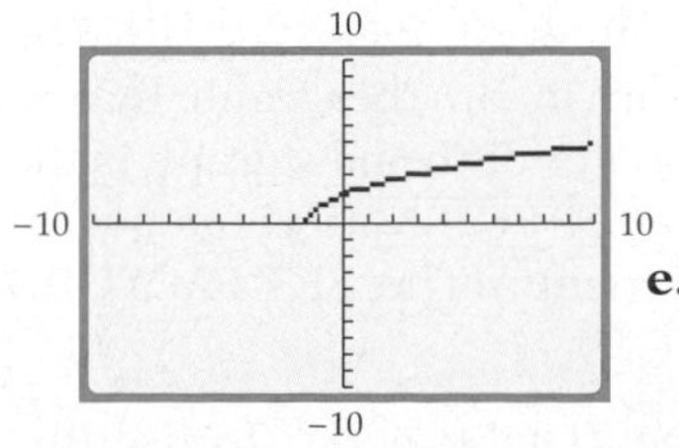

d. Press the [Y=] key. The equal sign has a black rectangle around it. This indicates that the function is active and will be graphed when the [GRAPH] key is pressed. A function is deactivated by using the arrow keys. Move the cursor over the equal sign and press [ENTER]. When the cursor is moved to the right, the black rectangle will not be present and that equation will not be active.

e. Graphing some radical equations requires special care. To graph the equation $y = \sqrt{2x + 3}$ shown at the left, enter the following keystrokes:

[Y=] [CLEAR] [2nd] $\sqrt{\ }$ 2 [X,T,θ,*n*] [+] 3 [)] [GRAPH]

To display the *x*-coordinates of rectangular coordinates as integers

a. Set the viewing window as follows: Xmin $= -47$, Xmax $= 47$, Xscl $= 10$, Ymin $= -31$, Ymax $= 31$, Yscl $= 10$.

b. Graph the function and use the TRACE feature. Press [TRACE] and then move the cursor with the [◁] and [▷] keys. The values of x and $y = f(x)$ displayed on the bottom of the screen are the coordinates of a point on the graph.

To display the *x*-coordinates of rectangular coordinates in tenths

a. Set the viewing window as follows: [ZOOM] [4]

b. Graph the function. Press [TRACE] and then move the cursor with the [◁] and [▷] keys. The values of x and $y = f(x)$ displayed on the bottom of the screen are the coordinates of a point on the graph.

To evaluate a function for a given value of *x*, or to produce ordered pairs of a function

a. Input the equation; for example, input $Y_1 = 2x^3 - 3x + 2$.

b. Press [2nd] QUIT.

c. To evaluate the function when $x = 3$, press [VARS] [▷] [1] [1] [(] 3 [)] [ENTER]. The value for the function for the given x-value is displayed, in this case, $Y_1(3) = 47$. An ordered pair of the function is $(3, 47)$.

d. Repeat step **c.** to produce as many pairs as desired.

The TABLE feature can also be used to determine ordered pairs.

To use a table

a. Press [2nd] TBLSET to activate the table setup menu.

b. TblStart is the beginning number for the table; ΔTbl is the difference between any two successive x-values in the table.

c. The portion of the table that appears as

Indpnt: **Auto** Ask
Depend: **Auto** Ask

allows you to choose between having the calculator automatically produce the results (Auto) or having the calculator ask you for values of x. You can choose Ask by using the arrow keys.

d. Once a table has been set up, enter an expression for Y_1. Now select TABLE by pressing [2nd] TABLE. A table showing ordered pair solutions of the equation will be displayed on the screen.

Zoom Features

To zoom in or out on a graph

Here are two methods of using ZOOM.

a. The first method uses the built-in features of the calculator. Press [TRACE] and then move the cursor to a point on the graph that is of interest. Press [ZOOM]. The ZOOM menu will appear. Press [2] [ENTER] to zoom in on the graph by the amount shown under the SET FACTORS menu. The center of the new graph is the location at which you placed the cursor. Press [ZOOM] [3] [ENTER] to zoom out on the graph by the amount under the SET FACTORS menu. (The SET FACTORS menu is accessed by pressing [ZOOM] [▷] [4].)

b. The second method uses the ZBOX option under the ZOOM menu. To use this method, press [ZOOM] [1]. A cursor will appear on the graph. Use the arrow keys to move the cursor to a portion of the graph that is of interest. Press [ENTER]. Now use the arrow keys to draw a box around the portion of the graph you wish to zoom in on. Press [ENTER]. The portion of the graph defined by the box will be drawn.

c. Pressing [ZOOM] [6] resets the window to the standard viewing window.

Solving Equations

This discussion is based on the fact that the real number solutions of an equation are related to the x-intercepts of a graph. For instance, the real solutions of the equation $x^2 = x + 1$ are the x-intercepts of the graph of $f(x) = x^2 - x - 1$, which are the zeros of f.

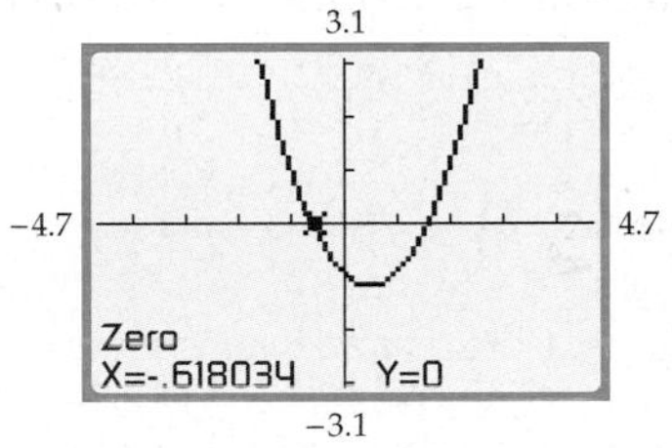

To solve $x^2 = x + 1$, rewrite the equation with all terms on one side: $x^2 - x - 1 = 0$. Think of this equation as $Y_1 = x^2 - x - 1$. The x-intercepts of the graph of Y_1 are the solutions of the equation $x^2 = x + 1$.

a. Enter $x^2 - x - 1$ into Y_1.

b. Graph the equation. You may need to adjust the viewing window so that the x-intercepts are visible.

c. Press [2nd] CALC [2].

d. Move the cursor to a point on the curve that is to the left of an x-intercept. Press [ENTER].

e. Move the cursor to a point on the curve that is to the right of the same x-intercept. Press [ENTER] [ENTER].

f. The root is shown as the x-coordinate on the bottom of the screen; in this case, the root is approximately -0.618034. To find the next intercept, repeat steps **c.** through **e.** The SOLVER feature under the MATH menu can also be used to find solutions of equations.

Solving Systems of Equations in Two Variables

To solve a system of equations

To solve $\quad y = x^2 - 1$

$$\frac{1}{2}x + y = 1 \quad ,$$

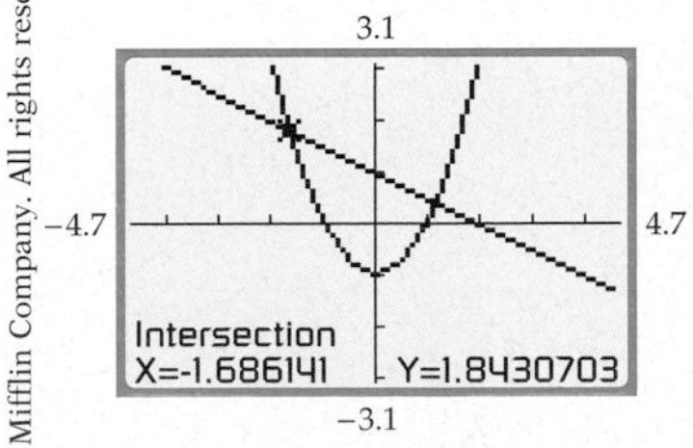

a. If necessary, solve one or both of the equations for y.

b. Enter the first equation as Y_1: $Y_1 = x^2 - 1$.

c. Enter the second equation as Y_2: $Y_2 = 1 - .5x$.

d. Graph both equations. (*Note:* The point of intersection must appear on the screen. It may be necessary to adjust the viewing window so that the points of intersection are displayed.)

e. Press [2nd] CALC [5].

f. Move the cursor close to the first point of intersection. Press [ENTER] [ENTER] [ENTER].

g. The first point of intersection is $(-1.686141, 1.8430703)$.

h. Repeat steps **e.** and **f.** for each point of intersection.

Finding Minimum or Maximum Values of a Function

a. Enter the function into Y1. The equation $y = x^2 - x - 1$ is used here.

b. Graph the equation. You may need to adjust the viewing window so that the maximum or minimum points are visible.

c. Press [2nd] CALC [3] to determine a minimum value or press [2nd] CALC [4] to determine a maximum value.

d. Move the cursor to a point on the curve that is to the left of the minimum (maximum). Press [ENTER].

e. Move the cursor to a point on the curve that is to the right of the minimum (maximum). Press [ENTER] [ENTER].

f. The minimum (maximum) is shown as the *y*-coordinate on the bottom of the screen; in this case the minimum value is -1.25.

Statistics

To calculate a linear regression equation

a. Press [STAT] to access the statistics menu. Press 1 to Edit or enter a new list of data. To delete data already in a list, press the up arrow to highlight the list name. For instance, to delete data in L1, highlight L1. Then press [CLEAR] and [ENTER]. Now enter each value of the independent variable in L1. Enter each value of the dependent variable in L2.

b. When all the data has been entered, press [STAT] [▷] 4 [ENTER]. The values of the slope and *y*-intercept of the linear regression equation will be displayed on the screen.

c. To graph the linear regression equation, modify step **b.** with these keystrokes: [STAT] [▷] 4 [VARS] [▷] 1 1 [ENTER]. This will store the regression equation in Y1. Now press [GRAPH]. It may be necessary to adjust the viewing window.

d. To evaluate the regression equation for a value of x, complete step **c.** but do not graph the equation. Now press [VARS] [▷] 1 1 [(] A [)] [ENTER], where A is replaced by the number at which you want to evaluate the expression.

Proofs and Tables

Proofs of Logarithmic Properties

In each of the following proofs of logarithmic properties, it is assumed that the Properties of Exponents are true for all real number exponents.

The Logarithm Property of the Product of Two Numbers

For any positive real numbers x, y, and b, $b \neq 1$, $\log_b xy = \log_b x + \log_b y$.

Proof: Let $\log_b x = m$ and $\log_b y = n$.

Write each equation in its equivalent exponential form.	$x = b^m \quad y = b^n$
Use substitution and the Properties of Exponents.	$xy = b^m b^n$
	$xy = b^{m+n}$
Write the equation in its equivalent logarithmic form.	$\log_b xy = m + n$
Substitute $\log_b x$ for m and $\log_b y$ for n.	$\log_b xy = \log_b x + \log_b y$

The Logarithm Property of the Quotient of Two Numbers

For any positive real numbers x, y, and b, $b \neq 1$, $\log_b \frac{x}{y} = \log_b x - \log_b y$.

Proof: Let $\log_b x = m$ and $\log_b y = n$.

Write each equation in its equivalent exponential form.	$x = b^m \quad y = b^n$
Use substitution and the Properties of Exponents.	$\frac{x}{y} = \frac{b^m}{b^n}$
	$\frac{x}{y} = b^{m-n}$
Write the equation in its equivalent logarithmic form.	$\log_b \frac{x}{y} = m - n$
Substitute $\log_b x$ for m and $\log_b y$ for n.	$\log_b \frac{x}{y} = \log_b x - \log_b y$

The Logarithm Property of the Power of a Number

For any real numbers x, r, and b, $b \neq 1$, $\log_b x^r = r \log_b x$.

Proof: Let $\log_b x = m$.

Write the equation in its equivalent exponential form.	$x = b^m$
Raise both sides to the r power.	$x^r = (b^m)^r$
	$x^r = b^{mr}$
Write the equation in its equivalent logarithmic form.	$\log_b x^r = mr$
Substitute $\log_b x$ for m.	$\log_b x^r = r \log_b x$

Table of Symbols

$+$	add
$-$	subtract
$\cdot$, $\times$, $(a)(b)$	multiply
$\frac{a}{b}$, $\div$, $a\overline{)b}$	divide
$(\ )$	parentheses, a grouping symbol
$[\ ]$	brackets, a grouping symbol
π	pi, a number approximately equal to $\frac{22}{7}$ or 3.14
$-a$	the opposite, or additive inverse, of a
$\frac{1}{a}$	the reciprocal, or multiplicative inverse, of a
$=$	is equal to
$\approx$	is approximately equal to
$\neq$	is not equal to
$<$	is less than
$\leq$	is less than or equal to
$>$	is greater than
$\geq$	is greater than or equal to
(a, b)	an ordered pair whose first component is a and whose second component is b
$^\circ$	degree (for angles)
$\sqrt{a}$	the principal square root of a
$\varnothing$, $\{\ \}$	the empty set
$\lvert a \rvert$	the absolute value of a
$\cup$	union of two sets
$\cap$	intersection of two sets
$\in$	is an element of (for sets)
$\notin$	is not an element of (for sets)

Solutions to Chapter 1 "You Try It"

SECTION 1.1

You Try It 1 Replace y by each of the elements of the set and determine whether the inequality is true.

$$\begin{aligned} y &> -1 \\ -5 &> -1 \text{ False} \\ -1 &> -1 \text{ False} \\ 5 &> -1 \text{ True} \end{aligned}$$

The inequality is true for 5.

You Try It 2 Replace z by each element of the set and determine the value of the expression.

$-z$	$\lvert z \rvert$
$-(-11) = 11$	$\lvert -11 \rvert = 11$
$-(0) = 0$	$\lvert 0 \rvert = 0$
$-(8) = -8$	$\lvert 8 \rvert = 8$

You Try It 3 $100 + (-43) = 57$

You Try It 4

$$\begin{aligned} &(-51) + 42 + 17 + (-102) \\ &= -9 + 17 + (-102) \\ &= 8 + (-102) \\ &= -94 \end{aligned}$$

You Try It 5

$$\begin{aligned} 19 - (-32) &= 19 + 32 \\ &= 51 \end{aligned}$$

You Try It 6

$$\begin{aligned} &-9 - (-12) - 17 - 4 \\ &= -9 + 12 + (-17) + (-4) \\ &= 3 + (-17) + (-4) \\ &= -14 + (-4) \\ &= -18 \end{aligned}$$

You Try It 7

$$\begin{aligned} 8(-9)10 &= -72(10) \\ &= -720 \end{aligned}$$

You Try It 8

$$\begin{aligned} (-2)3(-8)7 &= -6(-8)7 \\ &= 48(7) \\ &= 336 \end{aligned}$$

You Try It 9 $(-135) \div (-9) = 15$

You Try It 10 $\dfrac{-72}{4} = -18$

You Try It 11

$$\begin{aligned} -\frac{36}{-12} &= -(-3) \\ &= 3 \end{aligned}$$

You Try It 12

Strategy To find the average daily low temperature:

- Add the seven temperature readings.
- Divide the sum by 7.

Solution

$$\begin{aligned} &-6 + (-7) + 0 + (-5) + (-8) + (-1) + (-1) \\ &= -28 \quad -28 \div 7 = -4 \end{aligned}$$

The average daily low temperature was $-4°$C.

SECTION 1.2

You Try It 1

$$\begin{array}{r} 0.444 \\ 9\overline{)4.000} \\ \underline{-3\,6} \\ 40 \\ \underline{-36} \\ 40 \\ \underline{-36} \\ 4 \end{array} \qquad \frac{4}{9} = 0.\overline{4}$$

You Try It 2

$$125\% = 125\left(\frac{1}{100}\right) = \frac{125}{100} = \frac{5}{4}$$

$$125\% = 125(0.01) = 1.25$$

You Try It 3

$$\begin{aligned} \frac{1}{3} &= \frac{1}{3}(100\%) \\ &= \frac{100}{3}\% = 33\frac{1}{3}\% \end{aligned}$$

You Try It 4 $0.043 = 0.043(100\%) = 4.3\%$

You Try It 5 The LCM of 8, 6, and 4 is 24.

$$-\frac{7}{8}-\frac{5}{6}+\frac{3}{4}=-\frac{21}{24}-\frac{20}{24}+\frac{18}{24}$$
$$=\frac{-21}{24}+\frac{-20}{24}+\frac{18}{24}$$
$$=\frac{-21-20+18}{24}$$
$$=\frac{-23}{24}=-\frac{23}{24}$$

You Try It 6 $16.127-67.91$
$=16.127+(-67.91)$
$=-51.783$

You Try It 7
The quotient is positive.
$$-\frac{3}{8}\div\left(-\frac{5}{12}\right)=\frac{3}{8}\div\frac{5}{12}=\frac{3}{8}\cdot\frac{12}{5}$$
$$=\frac{3\cdot 12}{8\cdot 5}$$
$$=\frac{3\cdot\overset{1}{\cancel{2}}\cdot\overset{1}{\cancel{2}}\cdot 3}{2\cdot\underset{1}{\cancel{2}}\cdot\underset{1}{\cancel{2}}\cdot 5}=\frac{9}{10}$$

You Try It 8
$$\begin{array}{r} 5.44 \\ \times\ 3.8 \\ \hline 4352 \\ 1632\ \\ \hline 20.672 \end{array}$$

$-5.44(3.8)=-20.672$

You Try It 9
$-6^3=-(6\cdot 6\cdot 6)=-216$

You Try It 10
$(-3)^4=(-3)(-3)(-3)(-3)=81$

You Try It 11
$(3^3)(-2)^3=(3)(3)(3)\cdot(-2)(-2)(-2)$
$=27(-8)=-216$

You Try It 12
$$\left(-\frac{2}{5}\right)^2=\left(-\frac{2}{5}\right)\left(-\frac{2}{5}\right)=\frac{4}{25}$$

You Try It 13
$-3(0.3)^3=-3(0.3)(0.3)(0.3)$
$=-0.9(0.3)(0.3)$
$=-0.27(0.3)=-0.081$

You Try It 14
$-5\sqrt{32}=-5\sqrt{16\cdot 2}=-5\sqrt{16}\sqrt{2}$
$=-5\cdot 4\sqrt{2}=-20\sqrt{2}$

You Try It 15 $\sqrt{216}=\sqrt{36\cdot 6}=\sqrt{36}\sqrt{6}=6\sqrt{6}$

You Try It 16

Strategy To find the average monthly net income for the first quarter of 2000, divide Friendly's net income for the quarter (−18.510 million) by the number of months in the quarter (3).

Solution $-18.510\div 3=-6.170$

The average monthly net income was −$6.17 million.

SECTION 1.3

You Try It 1 $18-5[8-2(2-5)]\div 10$
$=18-5[8-2(-3)]\div 10$
$=18-5[8+6]\div 10$
$=18-5[14]\div 10$
$=18-70\div 10$
$=18-7$
$=11$

You Try It 2 $36\div(8-5)^2-(-3)^2\cdot 2$
$=36\div(3)^2-(-3)^2\cdot 2$
$=36\div 9-9\cdot 2$
$=4-9\cdot 2$
$=4-18$
$=-14$

You Try It 3 $(6.97-4.72)^2\cdot 4.5\div 0.05$
$=(2.25)^2\cdot 4.5\div 0.05$
$=5.0625\cdot 4.5\div 0.05$
$=22.78125\div 0.05$
$=455.625$

SECTION 1.4

You Try It 1 $\dfrac{a^2+b^2}{a+b}$
$$\frac{5^2+(-3)^2}{5+(-3)}=\frac{25+9}{5+(-3)}$$
$$=\frac{34}{2}$$
$$=17$$

You Try It 2 $x^3-2(x+y)+z^2$
$(2)^3-2[2+(-4)]+(-3)^2$
$=(2)^3-2(-2)+(-3)^2$
$=8-2(-2)+9$
$=8+4+9$
$=12+9$
$=21$

You Try It 3 $3a - 2b - 5a + 6b = -2a + 4b$

You Try It 4 $-3y^2 + 7 + 8y^2 - 14 = 5y^2 - 7$

You Try It 5 $-5(4y^2) = -20y^2$

You Try It 6 $-7(-2a) = 14a$

You Try It 7 $(-5x)(-2) = 10x$

You Try It 8 $-8(-2a + 7b) = 16a - 56b$

You Try It 9 $(3a - 1)5 = 15a - 5$

You Try It 10 $2(x^2 - x + 7) = 2x^2 - 2x + 14$

You Try It 11
$$\begin{aligned} 3y - 2(y - 7x) &= 3y - 2y + 14x \\ &= 14x + y \end{aligned}$$

You Try It 12
$$\begin{aligned} -2(x - 2y) - (-x + 3y) &= -2x + 4y + x - 3y \\ &= -x + y \end{aligned}$$

You Try It 13
$$\begin{aligned} 3y - 2[x - 4(2 - 3y)] &= 3y - 2[x - 8 + 12y] \\ &= 3y - 2x + 16 - 24y \\ &= -2x - 21y + 16 \end{aligned}$$

You Try It 14 the unknown number: x
the difference between
the number and sixty: $x - 60$

$5(x - 60)$; $5x - 300$

You Try It 15 the speed of the older model: s
the speed of the new model: $2s$

You Try It 16 the length of the longer piece: L
the length of the shorter piece:
$6 - L$

SECTION 1.5

You Try It 1 $A = \{-9, -7, -5, -3, -1\}$

You Try It 2 $A = \{1, 3, 5, \ldots\}$

You Try It 3 $A \cup B = \{-2, -1, 0, 1, 2, 3, 4\}$

You Try It 4 $C \cap D = \{10, 16\}$

You Try It 5 $A \cap B = \varnothing$

You Try It 6 $\{x \mid x < 59, x \in \text{positive even integers}\}$

You Try It 7 $\{x \mid x > -3, x \in \text{real numbers}\}$

You Try It 8 The graph is the numbers greater than -2.

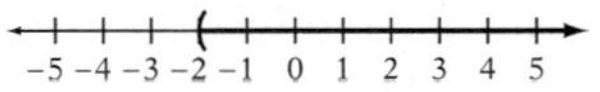

You Try It 9 The graph is the numbers greater than -1 and the numbers less than -3.

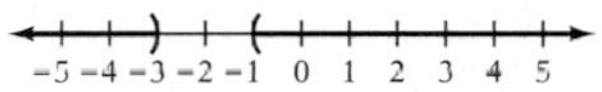

You Try It 10 The graph is the numbers less than or equal to 4 and greater than or equal to -4.

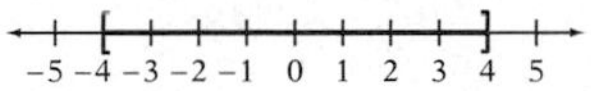

You Try It 11 The graph is the real numbers.

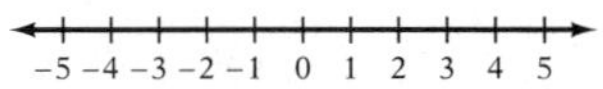

Solutions to Chapter 2 "You Try It"

SECTION 2.1

You Try It 1

$$\begin{array}{c|c} \multicolumn{2}{c}{5 - 4x = 8x + 2} \\ \hline 5 - 4\left(\frac{1}{4}\right) & 8\left(\frac{1}{4}\right) + 2 \\ 5 - 1 & 2 + 2 \\ \multicolumn{2}{c}{4 = 4} \end{array}$$

Yes, $\frac{1}{4}$ is a solution.

You Try It 2

$$\begin{array}{c|c} \multicolumn{2}{c}{10x - x^2 = 3x - 10} \\ \hline 10(5) - (5)^2 & 3(5) - 10 \\ 50 - 25 & 15 - 10 \\ \multicolumn{2}{c}{25 \neq 5} \end{array}$$

No, 5 is not a solution.

You Try It 3

$$\frac{5}{6} = y - \frac{3}{8}$$
$$\frac{5}{6} + \frac{3}{8} = y - \frac{3}{8} + \frac{3}{8}$$
$$\frac{29}{24} = y$$

The solution is $\frac{29}{24}$.

You Try It 4

$$-\frac{2x}{5} = 6$$
$$\left(-\frac{5}{2}\right)\left(-\frac{2}{5}x\right) = \left(-\frac{5}{2}\right)(6)$$
$$x = -15$$

The solution is -15.

You Try It 5

$$4x - 8x = 16$$
$$-4x = 16$$
$$\frac{-4x}{-4} = \frac{16}{-4}$$
$$x = -4$$

The solution is -4.

You Try It 6

$$P \cdot B = A$$
$$\frac{1}{6}B = 18 \qquad \bullet\ 16\frac{2}{3}\% = \frac{1}{6}$$
$$6 \cdot \frac{1}{6}B = 6 \cdot 18$$
$$B = 108$$

18 is $16\frac{2}{3}\%$ of 108.

You Try It 7

Strategy To find the percent, use the basic percent equation.
$B = 158.3$ million; $A = 22.45$ million; P is unknown.

Solution

$$P \cdot B = A$$
$$P(158.3) = 22.45$$
$$\frac{P(158.3)}{158.3} = \frac{22.45}{158.3}$$
$$P \approx 0.142 = 14.2\%$$

The Big Ten Conference received approximately 14.2% of the total revenue.

SECTION 2.2

You Try It 1

$$5x + 7 = 10$$
$$5x + 7 - 7 = 10 - 7$$
$$5x = 3$$
$$\frac{5x}{5} = \frac{3}{5}$$
$$x = \frac{3}{5}$$

The solution is $\frac{3}{5}$.

You Try It 2

$$2 = 11 + 3x$$
$$2 - 11 = 11 - 11 + 3x$$
$$-9 = 3x$$
$$\frac{-9}{3} = \frac{3x}{3}$$
$$-3 = x$$

The solution is -3.

You Try It 3

$$x - 5 + 4x = 25$$
$$5x - 5 = 25$$
$$5x - 5 + 5 = 25 + 5$$
$$5x = 30$$
$$\frac{5x}{5} = \frac{30}{5}$$
$$x = 6$$

The solution is 6.

You Try It 4

$$5x + 4 = 6 + 10x$$
$$5x - 10x + 4 = 6 + 10x - 10x$$
$$-5x + 4 = 6$$
$$-5x + 4 - 4 = 6 - 4$$
$$-5x = 2$$
$$\frac{-5x}{-5} = \frac{2}{-5}$$
$$x = -\frac{2}{5}$$

The solution is $-\frac{2}{5}$.

You Try It 5

$$5x - 10 - 3x = 6 - 4x$$
$$2x - 10 = 6 - 4x$$
$$2x + 4x - 10 = 6 - 4x + 4x$$
$$6x - 10 = 6$$
$$6x - 10 + 10 = 6 + 10$$
$$6x = 16$$
$$\frac{6x}{6} = \frac{16}{6}$$
$$x = \frac{8}{3}$$

The solution is $\frac{8}{3}$.

You Try It 6

$$
\begin{aligned}
5x - 4(3 - 2x) &= 2(3x - 2) + 6 \\
5x - 12 + 8x &= 6x - 4 + 6 \\
13x - 12 &= 6x + 2 \\
13x - 6x - 12 &= 6x - 6x + 2 \\
7x - 12 &= 2 \\
7x - 12 + 12 &= 2 + 12 \\
7x &= 14 \\
\frac{7x}{7} &= \frac{14}{7} \\
x &= 2
\end{aligned}
$$

The solution is 2.

You Try It 7

$$
\begin{aligned}
-2[3x - 5(2x - 3)] &= 3x - 8 \\
-2[3x - 10x + 15] &= 3x - 8 \\
-2[-7x + 15] &= 3x - 8 \\
14x - 30 &= 3x - 8 \\
14x - 3x - 30 &= 3x - 3x - 8 \\
11x - 30 &= -8 \\
11x - 30 + 30 &= -8 + 30 \\
11x &= 22 \\
\frac{11x}{11} &= \frac{22}{11} \\
x &- 2
\end{aligned}
$$

The solution is 2.

You Try It 8 the smaller number: n
the larger number: $14 - n$

one more than three times the smaller number	equals	the sum of the larger number and three

$$
\begin{aligned}
3n + 1 &= (14 - n) + 3 \\
3n + 1 &= 17 - n \\
3n + n + 1 &= 17 - n + n \\
4n + 1 &= 17 \\
4n + 1 - 1 &= 17 - 1 \\
4n &= 16 \\
\frac{4n}{4} &= \frac{16}{4} \\
n &= 4
\end{aligned}
$$

$14 - n = 14 - 4 = 10$

These numbers check as solutions.

The smaller number is 4.
The larger number is 10.

SECTION 2.3

You Try It 1

Strategy • Pounds of \$.55 fertilizer: x

	Amount	*Cost*	*Value*
\$.80 fertilizer	20	0.80	0.80(20)
\$.55 fertilizer	x	0.55	$0.55x$
\$.75 fertilizer	$20 + x$	0.75	$0.75(20 + x)$

• The sum of the values before mixing equals the value after mixing.

Solution

$$
\begin{aligned}
0.80(20) + 0.55x &= 0.75(20 + x) \\
16 + 0.55x &= 15 + 0.75x \\
16 - 0.20x &= 15 \\
-0.20x &= -1 \\
x &= 5
\end{aligned}
$$

5 lb of the \$.55 fertilizer must be added.

You Try It 2

Strategy • Liters of the 6% solution: x

	Amount	*Percent*	*Quantity*
6%	x	0.06	$0.06x$
12%	5	0.12	5(0.12)
8%	$x + 5$	0.08	$0.08(x + 5)$

• The sum of the quantities before mixing equals the quantity after mixing.

Solution

$$
\begin{aligned}
0.06x + 5(0.12) &= 0.08(x + 5) \\
0.06x + 0.60 &= 0.08x + 0.40 \\
0.06x + 0.20 &= 0.08x \\
0.20 &= 0.02x \\
10 &= x
\end{aligned}
$$

The pharmacist adds 10 L of the 6% solution to the 12% solution to get an 8% solution.

You Try It 3

Strategy • Additional amount: x

	Principal	*Rate*	*Interest*
8%	5000	0.08	0.08(5000)
11%	x	0.11	$0.11x$
9%	$5000 + x$	0.09	$0.09(5000 + x)$

- The sum of the interest earned by the two investments equals 9% of the total investment.

Solution

$$0.08(5000) + 0.11x = 0.09(5000 + x)$$
$$400 + 0.11x = 450 + 0.09x$$
$$400 + 0.02x = 450$$
$$0.02x = 50$$
$$x = 2500$$

\$2500 more must be invested at 11%.

You Try It 4

Strategy

- Rate of the first train: r
  Rate of the second train: $2r$

	Rate	Time	Distance
1st train	r	3	$3r$
2nd train	$2r$	3	$3(2r)$

- The sum of the distances traveled by the two trains equals 288 mi.

Solution

$$3r + 3(2r) = 288$$
$$3r + 6r = 288$$
$$9r = 288$$
$$r = 32$$

$$2r = 2(32) = 64$$

The first train is traveling at 32 mph.
The second train is traveling at 64 mph.

You Try It 5

Strategy

- Time spent flying out: t
  Time spent flying back: $5 - t$

	Rate	Time	Distance
Out	150	t	$150t$
Back	100	$5 - t$	$100(5 - t)$

- The distance out equals the distance back.

Solution

$$150t = 100(5 - t)$$
$$150t = 500 - 100t$$
$$250t = 500$$
$$t = 2 \text{ (The time out was 2 h.)}$$

The distance out $= 150t = 150(2) =$ 300 mi.

The parcel of land was 300 mi away.

SECTION 2.4

You Try It 1

$$2x - 1 < 6x + 7$$
$$-4x - 1 < 7$$
$$-4x < 8$$
$$\frac{-4x}{-4} > \frac{8}{-4}$$
$$x > -2$$

$$\{x | x > -2\}$$

You Try It 2

$$5x - 2 \le 4 - 3(x - 2)$$
$$5x - 2 \le 4 - 3x + 6$$
$$5x - 2 \le 10 - 3x$$
$$8x - 2 \le 10$$
$$8x \le 12$$
$$\frac{8x}{8} \le \frac{12}{8}$$
$$x \le \frac{3}{2}$$

$$\left\{x \middle| x \le \frac{3}{2}\right\}$$

You Try It 3

$$-2 \le 5x + 3 \le 13$$
$$-2 - 3 \le 5x + 3 - 3 \le 13 - 3$$
$$-5 \le 5x \le 10$$
$$\frac{-5}{5} \le \frac{5x}{5} \le \frac{10}{5}$$
$$-1 \le x \le 2$$

$$\{x | -1 \le x \le 2\}$$

You Try It 4

$$2 - 3x > 11 \quad \text{or} \quad 5 + 2x > 7$$
$$-3x > 9 \qquad 2x > 2$$
$$x < -3 \qquad x > 1$$
$$\{x | x < -3\} \qquad \{x | x > 1\}$$

$$\{x | x < -3\} \cup \{x | x > 1\} = \{x | x < -3 \text{ or } x > 1\}$$

You Try It 5

Strategy

To find the maximum number of miles:
- Write an expression for the cost of each car, using x to represent the number of miles driven during the week.
- Write and solve an inequality.

Solution

Cost of a Company A car	is less than	Cost of a Company B car

$$
\begin{aligned}
24(7) + 0.10x &< 30(7) + 0.08x \\
168 + 0.10x &< 210 + 0.08x \\
168 + 0.10x - 0.08x &< 210 + 0.08x - 0.08x \\
168 + 0.02x &< 210 \\
168 - 168 + 0.02x &< 210 - 168 \\
0.02x &< 42 \\
\frac{0.02x}{0.02} &< \frac{42}{0.02} \\
x &< 2100
\end{aligned}
$$

The maximum number of miles is 2099.

SECTION 2.5

You Try It 1 $|2x - 3| = 5$

$$2x - 3 = 5 \qquad 2x - 3 = -5$$
$$2x = 8 \qquad 2x = -2$$
$$x = 4 \qquad x = -1$$

The solutions are 4 and -1.

You Try It 2

$$
\begin{aligned}
5 - |3x + 5| &= 3 \\
-|3x + 5| &= -2 \\
|3x + 5| &= 2
\end{aligned}
$$

$$3x + 5 = 2 \qquad 3x + 5 = -2$$
$$3x = -3 \qquad 3x = -7$$
$$x = -1 \qquad x = -\frac{7}{3}$$

The solutions are -1 and $-\frac{7}{3}$.

You Try It 3

$$
\begin{aligned}
|3x + 2| &< 8 \\
-8 < 3x &+ 2 < 8 \\
-8 - 2 < 3x &+ 2 - 2 < 8 - 2 \\
-10 &< 3x < 6 \\
\frac{-10}{3} &< \frac{3x}{3} < \frac{6}{3} \\
-\frac{10}{3} &< x < 2
\end{aligned}
$$

$$\left\{x \,\middle|\, -\frac{10}{3} < x < 2\right\}$$

You Try It 4

$$|5x + 3| > 8$$

$$5x + 3 < -8 \quad \text{or} \quad 5x + 3 > 8$$
$$5x < -11 \qquad 5x > 5$$
$$x < -\frac{11}{5} \qquad x > 1$$
$$\left\{x \,\middle|\, x < -\frac{11}{5}\right\} \qquad \{x|x > 1\}$$

$$\left\{x \,\middle|\, x < -\frac{11}{5}\right\} \cup \{x|x > 1\}$$
$$= \left\{x \,\middle|\, x < -\frac{11}{5} \text{ or } x > 1\right\}$$

You Try It 5

Strategy Let b represent the desired diameter of the bushing, T the tolerance, and d the actual diameter. Solve the absolute value inequality $|d - b| \le T$ for d.

Solution

$$
\begin{aligned}
|d - b| &\le T \\
|d - 2.55| &\le 0.003 \\
-0.003 \le d &- 2.55 \le 0.003 \\
-0.003 + 2.55 \le d &- 2.55 + 2.55 \le 0.003 + 2.55 \\
2.547 \le\ &d \le 2.553
\end{aligned}
$$

The lower and upper limits of the diameter of the bushing are 2.547 in. and 2.553 in.

Solutions to Chapter 3 "You Try It"

SECTION 3.1

You Try It 1

$$
\begin{aligned}
QR + RS + ST &= QT \\
24 + RS + 17 &= 62 \\
41 + RS &= 62 \\
RS &= 21
\end{aligned}
$$

$RS = 21$ cm

You Try It 2

$$
\begin{aligned}
AC &= AB + BC \\
AC &= \frac{1}{4}(BC) + BC \\
AC &= \frac{1}{4}(16) + 16 \\
AC &= 4 + 16 \\
AC &= 20
\end{aligned}
$$

$AC = 20$ ft

You Try It 3

Strategy Supplementary angles are two angles whose sum is 180°. To find the supplement, let x represent the supplement of a 129° angle. Write an equation and solve for x.

Solution

$$x + 129° = 180°$$
$$x = 51°$$

The supplement of a 129° angle is a 51° angle.

You Try It 4

Strategy To find the measure of $\angle a$, write an equation using the fact that the sum of the measure of $\angle a$ and 68° is 118°. Solve for $\angle a$.

Solution

$$\angle a + 68° = 118°$$
$$\angle a = 50°$$

The measure of $\angle a$ is 50°.

You Try It 5

Strategy The angles labeled are adjacent angles of intersecting lines and are, therefore, supplementary angles. To find x, write an equation and solve for x.

Solution

$$(x + 16°) + 3x = 180°$$
$$4x + 16° = 180°$$
$$4x = 164°$$
$$x = 41°$$

You Try It 6

Strategy $3x = y$ because corresponding angles have the same measure. $y + (x + 40°) = 180°$ because adjacent angles of intersecting lines are supplementary angles. Substitute $3x$ for y and solve for x.

Solution

$$3x + (x + 40°) = 180°$$
$$4x + 40° = 180°$$
$$4x = 140°$$
$$x = 35°$$

You Try It 7

Strategy

- To find the measure of angle b, use the fact that $\angle b$ and $\angle x$ are supplementary angles.
- To find the measure of angle c, use the fact that the sum of the measures of the interior angles of a triangle is 180°.
- To find the measure of angle y, use the fact that $\angle c$ and $\angle y$ are vertical angles.

Solution

$$\angle b + \angle x = 180°$$
$$\angle b + 100° = 180°$$
$$\angle b = 80°$$

$$\angle a + \angle b + \angle c = 180°$$
$$45° + 80° + \angle c = 180°$$
$$125° + \angle c = 180°$$
$$\angle c = 55°$$

$$\angle y = \angle c = 55°$$

You Try It 8

Strategy To find the measure of the third angle, use the facts that the measure of a right angle is 90° and the sum of the measures of the interior angles of a triangle is 180°. Write an equation using x to represent the measure of the third angle. Solve the equation for x.

Solution

$$x + 90° + 34° = 180°$$
$$x + 124° = 180°$$
$$x = 56°$$

The measure of the third angle is 56°.

SECTION 3.2

You Try It 1

Strategy To find the perimeter, use the formula for the perimeter of a square. Substitute 60 for s and solve for P.

Solution

$$P = 4s$$
$$P = 4(60)$$
$$P = 240$$

The perimeter of the infield is 240 ft.

You Try It 2

Strategy To find the perimeter, use the formula for the perimeter of a rectangle. Substitute 11 for L and $8\frac{1}{2}$ for W and solve for P.

Solution $P = 2L + 2W$

$P = 2(11) + 2\left(8\frac{1}{2}\right)$

$P = 2(11) + 2\left(\frac{17}{2}\right)$

$P = 22 + 17$

$P = 39$

The perimeter of a standard piece of typing paper is 39 in.

You Try It 3

Strategy To find the circumference, use the circumference formula that involves the diameter. Leave the answer in terms of π.

Solution $C = \pi d$

$C = \pi(9)$

$C = 9\pi$

The circumference is 9π in.

You Try It 4

Strategy To find the number of rolls of wallpaper to be purchased:

- Use the formula for the area of a rectangle to find the area of one wall.
- Multiply the area of one wall by the number of walls to be covered (2).
- Divide the area of wall to be covered by the area that one roll of wallpaper will cover (30).

Solution $A = LW$

$A = 12 \cdot 8 = 96$ The area of one wall is 96 ft^2.

$2(96) = 192$ The area of the two walls is 192 ft^2.

$192 \div 30 = 6.4$

Because a portion of a seventh roll is needed, 7 rolls of wallpaper should be purchased.

You Try It 5

Strategy To find the area, use the formula for the area of a circle. An approximation is asked for; use the π key on a calculator. $r = 11$.

Solution $A = \pi r^2$

$A = \pi(11)^2$

$A = 121\pi$

$A \approx 380.13$

The area is approximately 380.13 cm^2.

SECTION 3.3

You Try It 1

Strategy To find the volume, use the formula for the volume of a cube. $s = 2.5$.

Solution $V = s^3$

$V = (2.5)^3 = 15.625$

The volume of the cube is 15.625 m^3.

You Try It 2

Strategy To find the volume:

- Find the radius of the base of the cylinder. $d = 8$.
- Use the formula for the volume of a cylinder. Leave the answer in terms of π.

Solution $r = \frac{1}{2}d = \frac{1}{2}(8) = 4$

$V = \pi r^2 h = \pi(4)^2(22) = \pi(16)(22)$
$= 352\pi$

The volume of the cylinder is 352π ft^3.

You Try It 3

Strategy To find the surface area of the cylinder:

- Find the radius of the base of the cylinder. $d = 6$.
- Use the formula for the surface area of a cylinder. An approximation is asked for; use the π key on a calculator.

Solution $r = \frac{1}{2}d = \frac{1}{2}(6) = 3$

$SA = 2\pi r^2 + 2\pi rh$

$SA = 2\pi(3)^2 + 2\pi(3)(8)$

$SA = 2\pi(9) + 2\pi(3)(8)$

$SA = 18\pi + 48\pi$

$SA = 66\pi$

$SA \approx 207.35$

The surface area of the cylinder is approximately 207.35 ft^2.

You Try It 4

Strategy To find which solid has the larger surface area:

- Use the formula for the surface area of a cube to find the surface area of the cube. $s = 10$.
- Find the radius of the sphere. $d = 8$.

- Use the formula for the surface area of a sphere to find the surface area of the sphere. Because this number is to be compared to another number, use the π key on a calculator to approximate the surface area.
- Compare the two numbers.

Solution

$SA = 6s^2$

$SA = 6(10)^2 = 6(100) = 600$

The surface area of the cube is 600 cm^2.

$r = \frac{1}{2}d = \frac{1}{2}(8) = 4$

$SA = 4\pi r^2$

$SA = 4\pi(4)^2 = 4\pi(16) = 64\pi \approx 201.06$

The surface area of the sphere is approximately 201.06 cm^2.

$600 > 201.06$

The cube has a larger surface area than the sphere.

Solutions to Chapter 4 "You Try It"

SECTION 4.1

You Try It 1

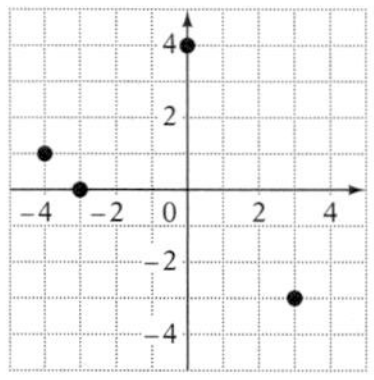

You Try It 2 The coordinates of A are $(4, -2)$.
The coordinates of B are $(-2, 4)$.
The abscissa of D is 0.
The ordinate of C is 0.

You Try It 3

$x - 3y$	$= -14$
$-2 - 3(4)$	-14
$-2 - 12$	-14
-14	$= -14$

Yes, $(-2, 4)$ is a solution of $x - 3y = -14$.

You Try It 4 Replace x by -2 and solve for y.

$$y = \frac{3x}{x+1} = \frac{3(-2)}{-2+1} = \frac{-6}{-1} = 6$$

The ordered-pair solution is $(-2, 6)$.

You Try It 5

x	$y = -\frac{1}{2}x + 2$	y	(x, y)
-4	$-\frac{1}{2}(-4) + 2$	4	$(-4, 4)$
-2	$-\frac{1}{2}(-2) + 2$	3	$(-2, 3)$
0	$-\frac{1}{2}(0) + 2$	2	$(0, 2)$
2	$-\frac{1}{2}(2) + 2$	1	$(2, 1)$

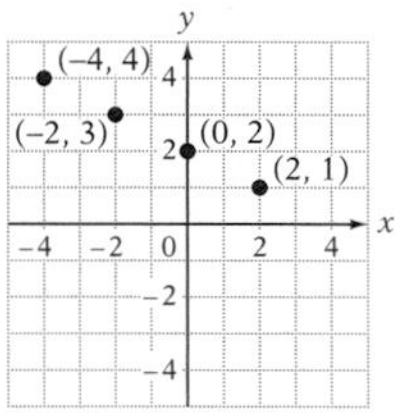

You Try It 6

Strategy To draw a scatter diagram:

- Draw a coordinate grid with the horizontal axis representing the year and the vertical axis representing the software rental market, in billions of dollars.
- Graph the ordered pairs (1999, 1), (2000, 2), (2001, 4), (2002, 7), and (2003, 11).

Solution

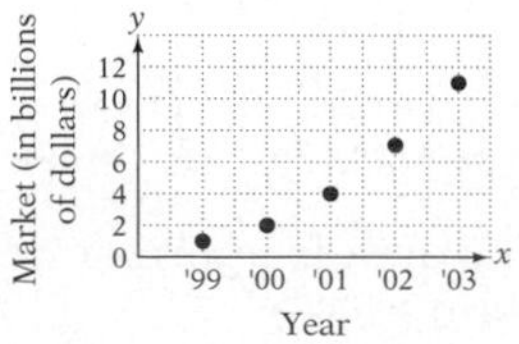

SECTION 4.2

You Try It 1 Domain: $\{-1, 3, 4, 6\}$
Range: $\{5\}$

You Try It 2
$$G(x) = \frac{3x}{x+2}$$
$$G(-4) = \frac{3(-4)}{-4+2} = \frac{-12}{-2} = 6$$

You Try It 3
$$f(x) = x^2 - 11$$
$$f(3h) = (3h)^2 - 11$$
$$= 9h^2 - 11$$

You Try It 4
$$h(z) = 3z + 1$$
$$h(0) = 3(0) + 1 = 1$$
$$h\left(\frac{1}{3}\right) = 3\left(\frac{1}{3}\right) + 1 = 2$$
$$h\left(\frac{2}{3}\right) = 3\left(\frac{2}{3}\right) + 1 = 3$$
$$h(1) = 3(1) + 1 = 4$$
The range is $\{1, 2, 3, 4\}$.

You Try It 5
$$f(x) = \frac{2}{x-5}$$
For $x = 5$, $f(5) = \frac{2}{5-5} = \frac{2}{0}$, which is not a real number.

5 is excluded from the domain of the function.

SECTION 4.3

You Try It 1

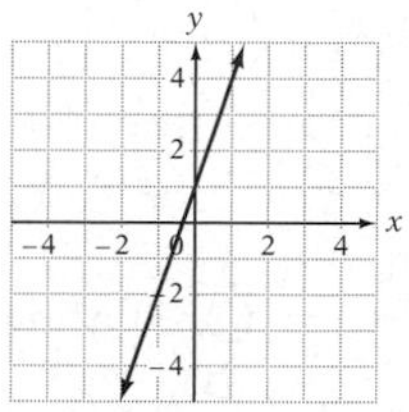

You Try It 2

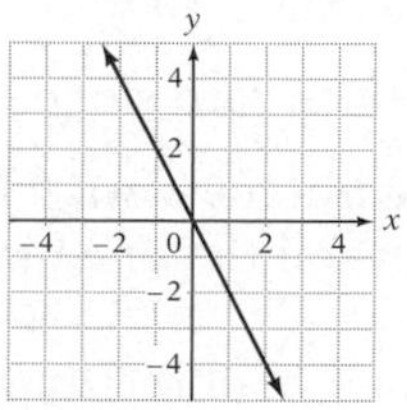

You Try It 3

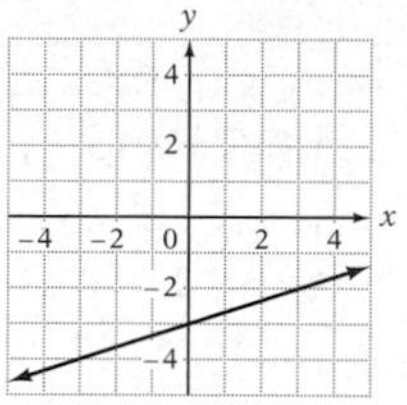

You Try It 4
$$5x - 2y = 10$$
$$-2y = -5x + 10$$
$$y = \frac{5}{2}x - 5$$

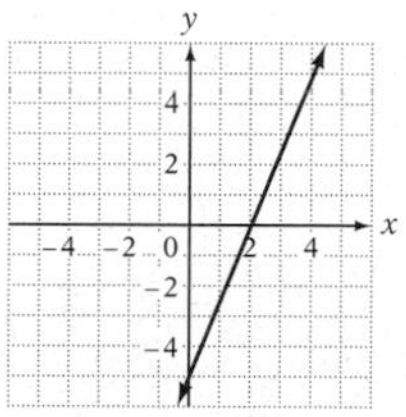

You Try It 5 x-intercept: $x - 4y = -4$
$$x - 4(0) = -4$$
$$x = -4$$
$(-4, 0)$
y-intercept: $x \quad 4y = \quad 4$
$$0 - 4y = -4$$
$$-4y = -4$$
$$y = 1$$
$(0, 1)$

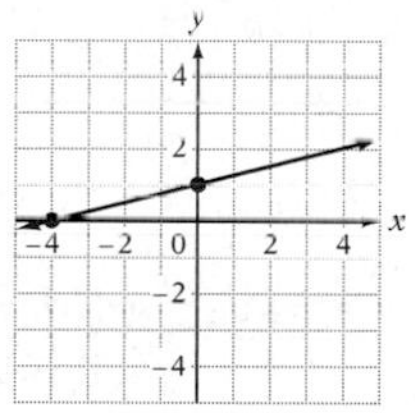

You Try It 6

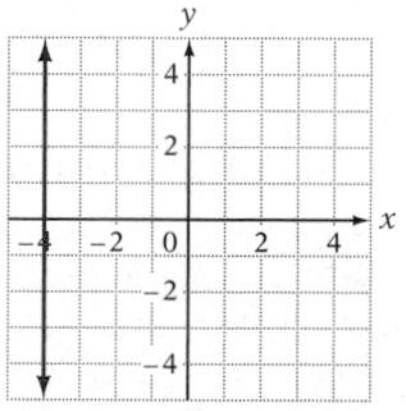

You Try It 7

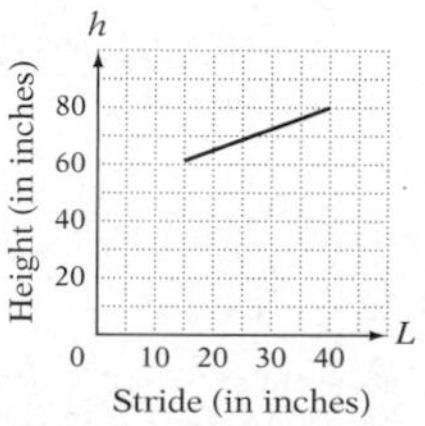

The ordered pair (32, 74) means that a person with a stride of 32 in. is 74 in. tall.

SECTION 4.4

You Try It 1 Let $P_1 = (4, -3)$ and $P_2 = (2, 7)$.

$$m = \frac{y_2 - y_1}{x_2 - x_1} = \frac{7 - (-3)}{2 - 4} = \frac{10}{-2} = -5$$

The slope is -5.

You Try It 2 Let $P_1 = (6, -1)$ and $P_2 = (6, 7)$.

$$m = \frac{y_2 - y_1}{x_2 - x_1} = \frac{7 - (-1)}{6 - 6} = \frac{8}{0}$$

Division by zero is not defined.

The slope of the line is undefined.

You Try It 3

$$m = \frac{55{,}000 - 25{,}000}{2 - 5}$$
$$= \frac{30{,}000}{-3}$$
$$= -10{,}000$$

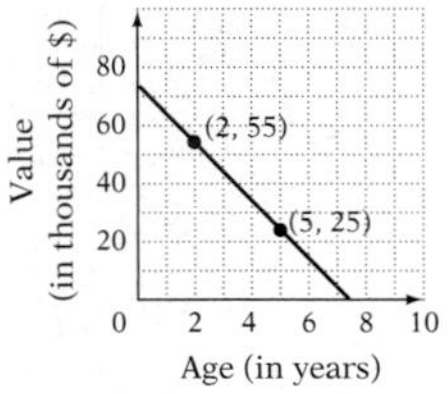

A slope of $-10{,}000$ means that the value of the printing press is decreasing by \$10,000 per year.

You Try It 4 y-intercept $= (0, b) = (0, -1)$

$$m = -\frac{1}{4}$$

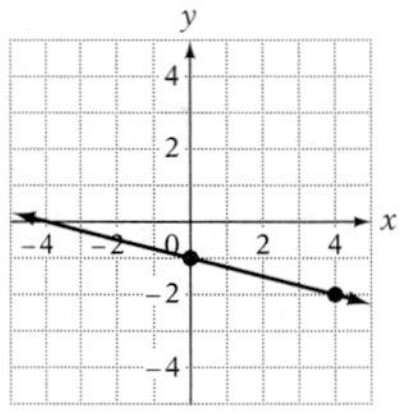

You Try It 5 Solve the equation for y.

$$x - 2y = 4$$
$$-2y = -x + 4$$
$$y = \frac{1}{2}x - 2$$

y-intercept $= (0, b) = (0, -2)$

$$m = \frac{1}{2}$$

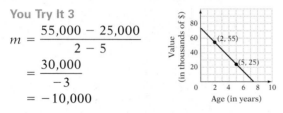

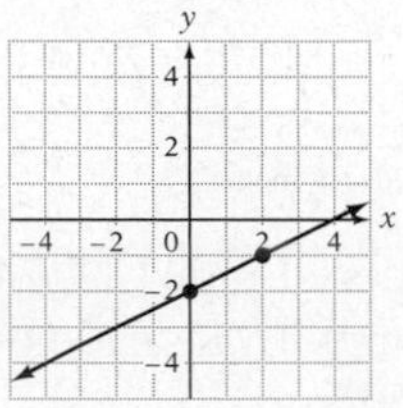

SECTION 4.5

You Try It 1 $m = -\frac{1}{3}$ $(x_1, y_1) = (-3, -2)$

$$y - y_1 = m(x - x_1)$$
$$y - (-2) = -\frac{1}{3}[x - (-3)]$$
$$y + 2 = -\frac{1}{3}(x + 3)$$
$$y + 2 = -\frac{1}{3}x - 1$$
$$y = -\frac{1}{3}x - 1 - 2$$
$$y = -\frac{1}{3}x - 3$$

The equation of the line is $y = -\frac{1}{3}x - 3$.

You Try It 2 $m = -3$ $(x_1, y_1) = (4, -3)$

$$y - y_1 = m(x - x_1)$$
$$y - (-3) = -3(x - 4)$$
$$y + 3 = -3(x - 4)$$
$$y + 3 = -3x + 12$$
$$y = -3x + 12 - 3$$
$$y = -3x + 9$$

The equation of the line is $y = -3x + 9$.

You Try It 3 Let $(x_1, y_1) = (2, 0)$ and $(x_2, y_2) = (5, 3)$.

$$m = \frac{y_2 - y_1}{x_2 - x_1} = \frac{3 - 0}{5 - 2} = \frac{3}{3} = 1$$

$$y - y_1 = m(x - x_1)$$
$$y - 0 = 1(x - 2)$$
$$y = 1(x - 2)$$
$$y = x - 2$$

The equation of the line is $y = x - 2$.

You Try It 4 Let $(x_1, y_1) = (2, 3)$ and $(x_2, y_2) = (-5, 3)$.

$$m = \frac{y_2 - y_1}{x_2 - x_1} = \frac{3 - 3}{-5 - 2} = \frac{0}{-7} = 0$$

The line has zero slope.

The line is a horizontal line.
All points on the line have an ordinate of 3.
The equation of the line is $y = 3$.

You Try It 5

Strategy

- To find a linear equation that models the number of trucks that would be sold:
  Let y represent the number of trucks that would be sold.
  Then $y = 40{,}000$ when $x = 18{,}000$, and $y = 44{,}000$ when $x = 17{,}500$.
  Find the slope of the line through the points (18,000, 40,000) and (17,500, 44,000).
  Use the point-slope formula to write the linear equation.
- To find the number of trucks that would be sold at a price of \$16,500, substitute 16,500 for x in the equation and solve for y.

Solution

$(x_1, y_1) = (18{,}000,\ 40{,}000)$
$(x_2, y_2) = (17{,}500,\ 44{,}000)$

$$m = \frac{44{,}000 - 40{,}000}{17{,}500 - 18{,}000} = \frac{4000}{-500} = -8$$

$$\begin{aligned} y - y_1 &= m(x - x_1) \\ y - 40{,}000 &= -8(x - 18{,}000) \\ y - 40{,}000 &= -8x + 144{,}000 \\ y &= -8x + 184{,}000 \end{aligned}$$

A linear equation that models the number of trucks that would be sold is $y = -8x + 184{,}000$.

$$\begin{aligned} y &= -8x + 184{,}000 \\ y &= -8(16{,}500) + 184{,}000 \\ y &= -132{,}000 + 184{,}000 \\ y &= 52{,}000 \end{aligned}$$

At a price of \$16,500, 52,000 trucks would be sold.

SECTION 4.6

You Try It 1

$$m_1 = \frac{1 - (-3)}{7 - (-2)} = \frac{4}{9}$$

$$m_2 = \frac{-5 - 1}{6 - 4} = \frac{-6}{2} = -3$$

$$m_1 \cdot m_2 = \frac{4}{9} \cdot -3 = -\frac{4}{3}$$

No, the lines are not perpendicular.

You Try It 2

$$\begin{aligned} 5x + 2y &= 2 \\ 2y &= -5x + 2 \\ y &= -\frac{5}{2}x + 1 \end{aligned}$$

$$m_1 = -\frac{5}{2}$$

$$\begin{aligned} 5x + 2y &= -6 \\ 2y &= -5x - 6 \\ y &= -\frac{5}{2}x - 3 \end{aligned}$$

$$m_2 = -\frac{5}{2}$$

$$m_1 = m_2 = -\frac{5}{2}$$

Yes, the lines are parallel.

You Try It 3

$$y = \frac{1}{4}x - 3$$

$$m_1 = \frac{1}{4}$$

$$\begin{aligned} m_1 \cdot m_2 &= -1 \\ \frac{1}{4} \cdot m_2 &= -1 \\ m_2 &= -4 \end{aligned}$$

$$\begin{aligned} y - y_1 &= m(x - x_1) \\ y - 2 &= -4[x - (-2)] \\ y - 2 &= -4(x + 2) \\ y - 2 &= -4x - 8 \\ y &= -4x - 6 \end{aligned}$$

The equation of the line is $y = -4x - 6$.

SECTION 4.7

You Try It 1

$$\begin{aligned} x - 3y &< 2 \\ x - x - 3y &< -x + 2 \\ -3y &< -x + 2 \\ \frac{-3y}{-3} &> \frac{-x + 2}{-3} \\ y &> \frac{1}{3}x - \frac{2}{3} \end{aligned}$$

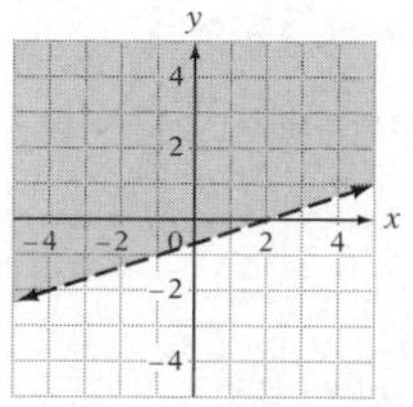

Solutions to Chapter 5 "You Try It"

SECTION 5.1

You Try It 1

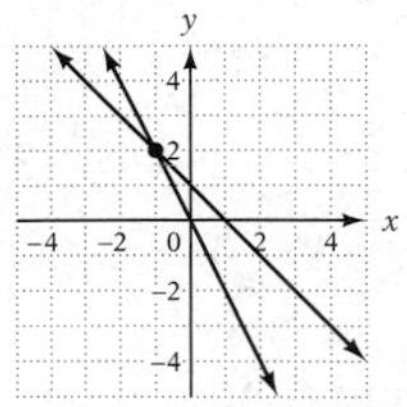

The solution is $(-1, 2)$.

You Try It 2

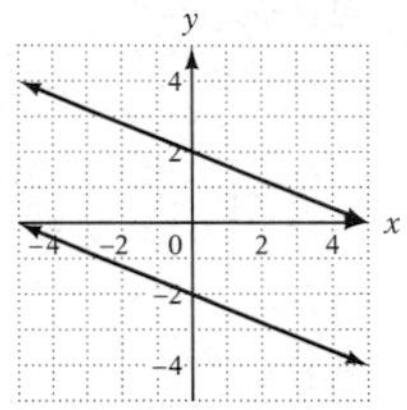

The lines are parallel and therefore do not intersect. The system of equations has no solution. The system of equations is inconsistent.

You Try It 3

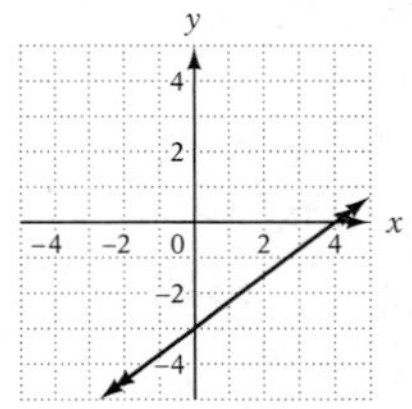

The two equations represent the same line. The system of equations is dependent. The solutions are the ordered pairs $\left(x, \frac{3}{4}x - 3\right)$.

You Try It 4

$$\begin{aligned} (1) \quad 3x - y &= 3 \\ (2) \quad 6x + 3y &= -4 \end{aligned}$$

Solve Equation (1) for y.

$$\begin{aligned} 3x - y &= 3 \\ -y &= -3x + 3 \\ y &= 3x - 3 \end{aligned}$$

Substitute into Equation (2).

$$\begin{aligned} 6x + 3y &= -4 \\ 6x + 3(3x - 3) &= -4 \\ 6x + 9x - 9 &= -4 \\ 15x - 9 &= -4 \\ 15x &= 5 \\ x &= \frac{5}{15} = \frac{1}{3} \end{aligned}$$

Substitute the value of x into Equation (1).

$$\begin{aligned} 3x - y &= 3 \\ 3\left(\frac{1}{3}\right) - y &= 3 \\ 1 - y &= 3 \\ -y &= 2 \\ y &= -2 \end{aligned}$$

The solution is $\left(\frac{1}{3}, -2\right)$.

You Try It 5

$$\begin{aligned} (1) \quad y &= 2x - 3 \\ (2) \quad 2x - 4y &= 5 \end{aligned}$$

$$\begin{aligned} 2x - 4y &= 5 \\ 2x - 4(2x - 3) &= 5 \\ 2x - 8x + 12 &= 5 \\ -6x + 12 &= 5 \\ -6x &= -7 \\ x &= \frac{-7}{-6} = \frac{7}{6} \end{aligned}$$

Substitute the value of x into Equation (1).

$$\begin{aligned} y &= 2x - 3 \\ y &= 2\left(\frac{7}{6}\right) - 3 \\ y &= \frac{7}{3} - 3 \\ y &= -\frac{2}{3} \end{aligned}$$

The solution is $\left(\frac{7}{6}, -\frac{2}{3}\right)$.

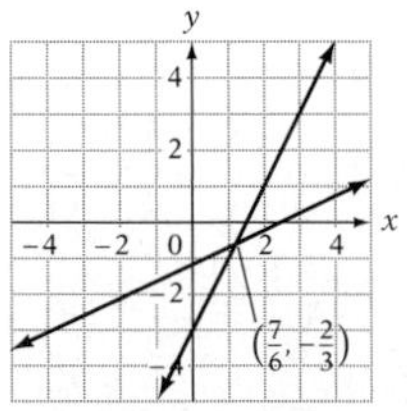

You Try It 6

$$\begin{aligned} (1) \quad 6x - 3y &= 6 \\ (2) \quad 2x - y &= 2 \end{aligned}$$

Solve Equation (2) for y.

$$\begin{aligned} 2x - y &= 2 \\ -y &= -2x + 2 \\ y &= 2x - 2 \end{aligned}$$

Substitute into Equation (1).

$$\begin{aligned} 6x - 3y &= 6 \\ 6x - 3(2x - 2) &= 6 \\ 6x - 6x + 6 &= 6 \\ 6 &= 6 \end{aligned}$$

The system of equations is dependent. The solutions are the ordered pairs $(x, 2x - 2)$.

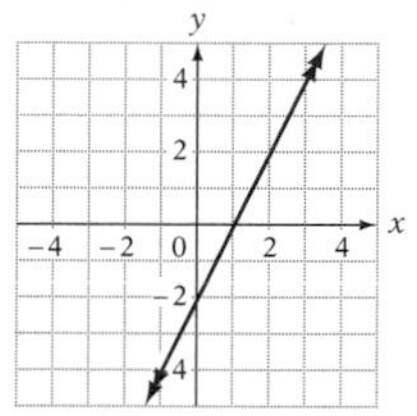

SECTION 5.2

You Try It 1

$$\begin{aligned} &(1) \quad 2x + 5y = 6 \\ &(2) \quad 3x - 2y = 6x + 2 \end{aligned}$$

Write Equation (2) in the form $Ax + By = C$.

$$\begin{aligned} 3x - 2y &= 6x + 2 \\ -3x - 2y &= 2 \end{aligned}$$

Solve the system: $\begin{aligned} 2x + 5y &= 6 \\ -3x - 2y &= 2 \end{aligned}$

Eliminate y.

$$\begin{aligned} 2(2x + 5y) &= 2(6) \\ 5(-3x - 2y) &= 5(2) \end{aligned}$$

$$\begin{aligned} 4x + 10y &= 12 \\ -15x - 10y &= 10 \end{aligned}$$

Add the equations.

$$\begin{aligned} -11x &= 22 \\ x &= -2 \end{aligned}$$

Replace x in Equation (1).

$$\begin{aligned} 2x + 5y &= 6 \\ 2(-2) + 5y &= 6 \\ -4 + 5y &= 6 \\ 5y &= 10 \\ y &= 2 \end{aligned}$$

The solution is $(-2, 2)$.

You Try It 2

$$\begin{aligned} 2x + y &= 5 \\ 4x + 2y &= 6 \end{aligned}$$

Eliminate y.

$$\begin{aligned} -2(2x + y) &= -2(5) \\ 4x + 2y &= 6 \end{aligned}$$

$$\begin{aligned} -4x - 2y &= -10 \\ 4x + 2y &= 6 \end{aligned}$$

Add the equations.

$$\begin{aligned} 0x + 0y &= -4 \\ 0 &= -4 \end{aligned}$$

This is not a true equation. The system is inconsistent and therefore has no solution.

You Try It 3

$$\begin{aligned} &(1) \quad x - y + z = 6 \\ &(2) \quad 2x + 3y - z = 1 \\ &(3) \quad x + 2y + 2z = 5 \end{aligned}$$

Eliminate z. Add Equations (1) and (2).

$$\begin{aligned} x - y + z &= 6 \\ 2x + 3y - z &= 1 \\ (4) \qquad 3x + 2y &= 7 \end{aligned}$$

Multiply Equation (2) by 2 and add to Equation (3).

$$\begin{aligned} 4x + 6y - 2z &= 2 \\ x + 2y + 2z &= 5 \\ (5) \qquad 5x + 8y &= 7 \end{aligned}$$

Solve the system of two equations.

$$\begin{aligned} &(4) \quad 3x + 2y = 7 \\ &(5) \quad 5x + 8y = 7 \end{aligned}$$

Multiply Equation (4) by −4 and add to Equation (5).

$$\begin{aligned} -12x - 8y &= -28 \\ 5x + 8y &= 7 \\ -7x &= -21 \\ x &= 3 \end{aligned}$$

Replace x by 3 in Equation (4).

$$\begin{aligned} 3x + 2y &= 7 \\ 3(3) + 2y &= 7 \\ 9 + 2y &= 7 \\ 2y &= -2 \\ y &= -1 \end{aligned}$$

Replace x by 3 and y by −1 in Equation (1).

$$\begin{aligned} x - y + z &= 6 \\ 3 - (-1) + z &= 6 \\ 4 + z &= 6 \\ z &= 2 \end{aligned}$$

The solution is $(3, -1, 2)$.

SECTION 5.3

You Try It 1

$$\begin{vmatrix} -1 & -4 \\ 3 & -5 \end{vmatrix} = -1(-5) - 3(-4) = 5 + 12 = 17$$

The value of the determinant is 17.

You Try It 2 Expand by cofactors of the first row.

$$\begin{vmatrix} 1 & 4 & -2 \\ 3 & 1 & 1 \\ 0 & -2 & 2 \end{vmatrix}$$

$$= 1\begin{vmatrix} 1 & 1 \\ -2 & 2 \end{vmatrix} - 4\begin{vmatrix} 3 & 1 \\ 0 & 2 \end{vmatrix} + (-2)\begin{vmatrix} 3 & 1 \\ 0 & -2 \end{vmatrix}$$

$$= 1(2 + 2) - 4(6 - 0) - 2(-6 - 0)$$
$$= 4 - 24 + 12$$
$$= -8$$

The value of the determinant is -8.

You Try It 3

$$\begin{vmatrix} 3 & -2 & 0 \\ 1 & 4 & 2 \\ -2 & 1 & 3 \end{vmatrix}$$

$$= 3\begin{vmatrix} 4 & 2 \\ 1 & 3 \end{vmatrix} - (-2)\begin{vmatrix} 1 & 2 \\ -2 & 3 \end{vmatrix} + 0\begin{vmatrix} 1 & 4 \\ -2 & 1 \end{vmatrix}$$

$$= 3(12 - 2) + 2(3 + 4) + 0$$
$$= 3(10) + 2(7)$$
$$= 30 + 14$$
$$= 44$$

The value of the determinant is 44.

You Try It 4

$$3x - y = 4$$
$$6x - 2y = 5$$

$$D = \begin{vmatrix} 3 & -1 \\ 6 & -2 \end{vmatrix} = -6 + 6 = 0$$

Because $D = 0$, $\frac{D_x}{D}$ is undefined. Therefore, the system of equations is dependent or inconsistent.

It is not possible to solve this system by using Cramer's Rule.

You Try It 5

$$2x - y + z = -1$$
$$3x + 2y - z = 3$$
$$x + 3y + z = -2$$

$$D = \begin{vmatrix} 2 & -1 & 1 \\ 3 & 2 & -1 \\ 1 & 3 & 1 \end{vmatrix} = 21$$

$$D_x = \begin{vmatrix} -1 & -1 & 1 \\ 3 & 2 & -1 \\ -2 & 3 & 1 \end{vmatrix} = 9$$

$$D_y = \begin{vmatrix} 2 & -1 & 1 \\ 3 & 3 & -1 \\ 1 & -2 & 1 \end{vmatrix} = -3$$

$$D_z = \begin{vmatrix} 2 & -1 & -1 \\ 3 & 2 & 3 \\ 1 & 3 & -2 \end{vmatrix} = -42$$

$$x = \frac{D_x}{D} = \frac{9}{21} = \frac{3}{7}$$

$$y = \frac{D_y}{D} = \frac{-3}{21} = -\frac{1}{7}$$

$$z = \frac{D_z}{D} = \frac{-42}{21} = -2$$

The solution is $\left(\frac{3}{7}, -\frac{1}{7}, -2\right)$.

SECTION 5.4

You Try It 1

Strategy

- Rate of the rowing team in calm water: t
  Rate of the current: c

	Rate	*Time*	*Distance*
With current	$t + c$	2	$2(t + c)$
Against current	$t - c$	2	$2(t - c)$

- The distance traveled with the current is 18 mi.
  The distance traveled against the current is 10 mi.

Solution

$$2(t + c) = 18 \qquad \frac{1}{2} \cdot 2(t + c) = \frac{1}{2} \cdot 18$$

$$2(t - c) = 10 \qquad \frac{1}{2} \cdot 2(t - c) = \frac{1}{2} \cdot 10$$

$$t + c = 9$$
$$t - c = 5$$
$$2t = 14$$
$$t = 7$$

$$t + c = 9$$
$$7 + c = 9$$
$$c = 2$$

The rate of the rowing team in calm water is 7 mph.

The rate of the current is 2 mph.

You Try It 2

Strategy

- Amount invested at 8%: x
  Amount invested at 7%: y

Investment	*Principal*	*Interest Rate*	*Interest Earned*
8%	x	0.08	$0.08x$
7%	y	0.07	$0.07y$

- The total amount invested was \$15,000. The total amount of interest earned was \$1170.

Solution

$$x + y = 15{,}000$$
$$0.08x + 0.07y = 1170$$

$$-8x - 8y = -120{,}000$$
$$8x + 7y = 117{,}000$$
$$-y = -3000$$
$$y = 3000$$

$$x + y = 15{,}000$$
$$x + 3000 = 15{,}000$$
$$x = 12{,}000$$

\$12,000 was invested in the 8% account.
\$3000 was invested in the 7% account.

SECTION 5.5

You Try It 1 Shade above the solid line $y = 2x - 3$.
Shade above the dashed line $y = -3x$.

The solution set of the system is the intersection of the solution sets of the individual inequalities.

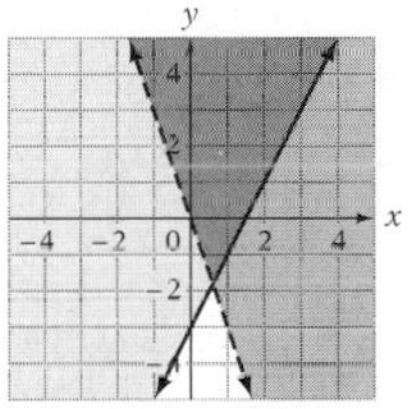

You Try It 2 $3x + 4y > 12$
$$4y > -3x + 12$$
$$y > -\frac{3}{4}x + 3$$

Shade above the dashed line $y = -\frac{3}{4}x + 3$.

Shade below the dashed line $y = \frac{3}{4}x - 1$.

The solution set of the system is the intersection of the solution sets of the individual inequalities.

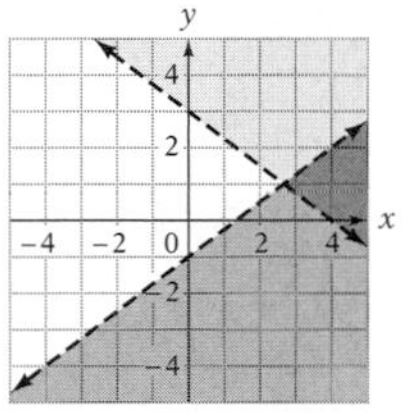

Solutions to Chapter 6 "You Try It"

SECTION 6.1

You Try It 1

$$(-3a^2b^4)(-2ab^3)^4 = (-3a^2b^4)[(-2)^4a^4b^{12}]$$
$$= (-3a^2b^4)(16a^4b^{12})$$
$$= -48a^6b^{16}$$

You Try It 2 $(y^{n-3})^2 = y^{(n-3)2} = y^{2n-6}$

You Try It 3 $[(ab^3)^3]^4 = [a^3b^9]^4 = a^{12}b^{36}$

You Try It 4

$$\frac{20r^{-2}t^{-5}}{-16r^{-3}s^{-2}} = -\frac{4 \cdot 5r^{-2-(-3)}s^2t^{-5}}{4 \cdot 4}$$
$$= -\frac{5rs^2}{4t^5}$$

You Try It 5

$$\frac{(9u^{-6}v^4)^{-1}}{(6u^{-3}v^{-2})^{-2}} = \frac{9^{-1}u^6v^{-4}}{6^{-2}u^6v^4}$$
$$= 9^{-1} \cdot 6^2u^0v^{-8}$$
$$= \frac{36}{9v^8}$$
$$= \frac{4}{v^8}$$

You Try It 6 $\dfrac{a^{2n+1}}{a^{n+3}} = a^{2n+1-(n+3)} = a^{2n+1-n-3} = a^{n-2}$

You Try It 7 $942{,}000{,}000 = 9.42 \times 10^8$

You Try It 8 $2.7 \times 10^{-5} = 0.000027$

You Try It 9

$$\begin{aligned}&\frac{5{,}600{,}000 \times 0.000000081}{900 \times 0.000000028}\\ &= \frac{5.6 \times 10^6 \times 8.1 \times 10^{-8}}{9 \times 10^2 \times 2.8 \times 10^{-8}}\\ &= \frac{(5.6)(8.1) \times 10^{6+(-8)-2-(-8)}}{(9)(2.8)}\\ &= 1.8 \times 10^4 = 18{,}000\end{aligned}$$

You Try It 10

Strategy To find the number of arithmetic operations:

- Find the reciprocal of 1×10^{-7}, which is the number of operations performed in one second.
- Write the number of seconds in one minute (60) in scientific notation.
- Multiply the number of arithmetic operations per second by the number of seconds in one minute.

Solution $\dfrac{1}{1 \times 10^{-7}} = 10^7$

$60 = 6 \times 10$

$6 \times 10 \times 10^7 = 6 \times 10^8$

The computer can perform 6×10^8 operations in one minute.

SECTION 6.2

You Try It 1

$$\begin{aligned}R(x) &= -2x^4 - 5x^3 + 2x - 8\\ R(2) &= -2(2)^4 - 5(2)^3 + 2(2) - 8\\ &= -2(16) - 5(8) + 2(2) - 8\\ &= -32 - 40 + 4 - 8\\ &= -76\end{aligned}$$

You Try It 2 The leading coefficient is -3, the constant term is -12, and the degree is 4.

You Try It 3

a. This is a polynomial function.

b. This is not a polynomial function. A polynomial function does not have a variable expression raised to a negative power.

c. This is not a polynomial function. A polynomial function does not have a variable expression within a radical.

You Try It 4

x	$y = f(x)$
-3	0
-2	-3
-1	-4
0	-3
1	0
2	5

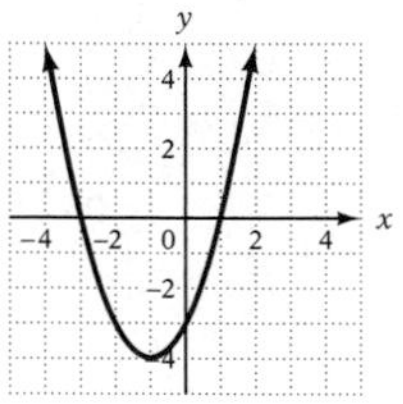

You Try It 5

x	$y = f(x)$
-3	28
-2	9
-1	2
0	1
1	0
2	-7
3	-26

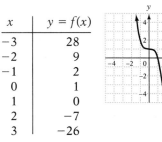

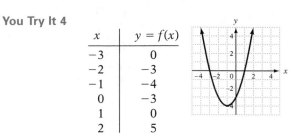

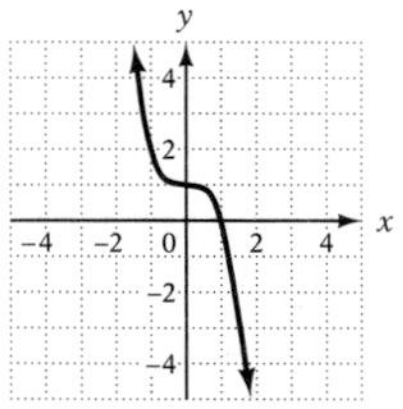

You Try It 6

$$\begin{array}{r} -3x^2 - 4x + 9\\ -5x^2 - 7x + 1\\ \hline -8x^2 - 11x + 10\end{array}$$

You Try It 7 Add the additive inverse of $6x^2 + 3x - 7$ to $-5x^2 + 2x - 3$.

$$\begin{array}{r} -5x^2 + 2x - 3\\ -6x^2 - 3x + 7\\ \hline -11x^2 - x + 4\end{array}$$

You Try It 8

$$\begin{aligned}S(x) &= P(x) + R(x)\\ &= (4x^3 - 3x^2 + 2) + (-2x^2 + 2x - 3)\\ &= 4x^3 - 5x^2 + 2x - 1\end{aligned}$$

$$\begin{aligned}S(-1) &= 4(-1)^3 - 5(-1)^2 + 2(-1) - 1\\ &= 4(-1) - 5(1) + 2(-1) - 1\\ &= -4 - 5 - 2 - 1\\ &= -12\end{aligned}$$

You Try It 9

$D(x) = P(x) - R(x)$

$$\begin{aligned}D(x) &= (5x^{2n} - 3x^n - 7) - (-2x^{2n} - 5x^n + 8)\\ &= (5x^{2n} - 3x^n - 7) + (2x^{2n} + 5x^n - 8)\\ &= 7x^{2n} + 2x^n - 15\end{aligned}$$

SECTION 6.3

You Try It 1 $(-2y + 3)(-4y) = 8y^2 - 12y$

You Try It 2 $-a^2(3a^2 + 2a - 7) = -3a^4 - 2a^3 + 7a^2$

You Try It 3

$$\begin{array}{r} 2y^3 + 2y^2 \qquad - 3 \\ 3y - 1 \\ \hline -2y^3 - 2y^2 \qquad + 3 \\ 6y^4 + 6y^3 \qquad - 9y \qquad \\ \hline 6y^4 + 4y^3 - 2y^2 - 9y + 3 \end{array}$$

You Try It 4

$$\begin{aligned} (4y - 5)(2y - 3) &= 8y^2 - 12y - 10y + 15 \\ &= 8y^2 - 22y + 15 \end{aligned}$$

You Try It 5

$$\begin{aligned} (3b + 2)(3b - 5) &= 9b^2 - 15b + 6b - 10 \\ &= 9b^2 - 9b - 10 \end{aligned}$$

You Try It 6 $(2a + 5c)(2a - 5c) = 4a^2 - 25c^2$

You Try It 7 $(3x + 2y)^2 = 9x^2 + 12xy + 4y^2$

You Try It 8

Strategy To find the area, replace the variables b and h in the equation $A = \frac{1}{2}bh$ by the given values and solve for A.

Solution

$$\begin{aligned} A &= \frac{1}{2}bh \\ A &= \frac{1}{2}(2x + 6)(x - 4) \\ A &= (x + 3)(x - 4) \\ A &= x^2 - 4x + 3x - 12 \\ A &= x^2 - x - 12 \end{aligned}$$

The area is $(x^2 - x - 12)$ ft^2.

SECTION 6.4

You Try It 1

$$\begin{array}{r} 5x - 1 \\ 3x + 4\overline{)15x^2 + 17x - 20} \\ \underline{15x^2 + 20x} \qquad \\ -\ 3x - 20 \\ \underline{-\ 3x - 4} \\ -\ 16 \end{array}$$

$$\frac{15x^2 + 17x - 20}{3x + 4} = 5x - 1 - \frac{16}{3x + 4}$$

You Try It 2

$$\begin{array}{r} x^2 + 3x - 1 \\ 3x - 1\overline{)3x^3 + 8x^2 - 6x + 2} \\ \underline{3x^3 - x^2} \qquad\qquad \\ 9x^2 - 6x \qquad \\ \underline{9x^2 - 3x} \qquad \\ -\ 3x + 2 \\ \underline{-\ 3x + 1} \\ 1 \end{array}$$

$$\frac{3x^3 + 8x^2 - 6x + 2}{3x - 1} = x^2 + 3x - 1 + \frac{1}{3x - 1}$$

You Try It 3

$$\begin{array}{r} 3x^2 - 2x + 4 \\ x^2 - 3x + 2\overline{)3x^4 - 11x^3 + 16x^2 - 16x + 8} \\ \underline{3x^4 - 9x^3 + 6x^2} \qquad\qquad\quad \\ -\ 2x^3 + 10x^2 - 16x \qquad \\ \underline{-\ 2x^3 + 6x^2 - 4x} \qquad \\ 4x^2 - 12x + 8 \\ \underline{4x^2 - 12x + 8} \\ 0 \end{array}$$

$$\frac{3x^4 - 11x^3 + 16x^2 - 16x + 8}{x^2 - 3x + 2} = 3x^2 - 2x + 4$$

You Try It 4

$$\begin{array}{r|rrr} -2 & 6 & 8 & -5 \\ & & -12 & 8 \\ \hline & 6 & -4 & 3 \end{array}$$

$$(6x^2 + 8x - 5) \div (x + 2) = 6x - 4 + \frac{3}{x + 2}$$

You Try It 5

$$\begin{array}{r|rrrr} 2 & 5 & -12 & -8 & 16 \\ & & 10 & -4 & -24 \\ \hline & 5 & -2 & -12 & -8 \end{array}$$

$$\begin{aligned} &(5x^3 - 12x^2 - 8x + 16) \div (x - 2) \\ &= 5x^2 - 2x - 12 - \frac{8}{x - 2} \end{aligned}$$

You Try It 6

$$\begin{array}{r|rrrrr} 3 & 2 & -3 & -8 & 0 & -2 \\ & & 6 & 9 & 3 & 9 \\ \hline & 2 & 3 & 1 & 3 & 7 \end{array}$$

$$\begin{aligned} &(2x^4 - 3x^3 - 8x^2 - 2) \div (x - 3) \\ &= 2x^3 + 3x^2 + x + 3 + \frac{7}{x - 3} \end{aligned}$$

You Try It 7

$$\begin{array}{r|rrrr} -3 & 2 & -5 & 0 & 7 \\ & & -6 & 33 & -99 \\ \hline & 2 & -11 & 33 & -92 \end{array}$$

$P(-3) = -92$

Solutions to Chapter 7 "You Try It"

SECTION 7.1

You Try It 1
The GCF is $7a^2$.

$$14a^2 - 21a^4b = 7a^2(2) + 7a^2(-3a^2b) = 7a^2(2 - 3a^2b)$$

You Try It 2
The GCF is 9.

$$27b^2 + 18b + 9 = 9(3b^2) + 9(2b) + 9(1) = 9(3b^2 + 2b + 1)$$

You Try It 3
The GCF is $3x^2y^2$.

$$6x^4y^2 - 9x^3y^2 + 12x^2y^4 = 3x^2y^2(2x^2) + 3x^2y^2(-3x) + 3x^2y^2(4y^2) = 3x^2y^2(2x^2 - 3x + 4y^2)$$

You Try It 4
The GCF is a^2.

$$a^{n+4} + a^2 = a^2(a^{n+2}) + a^2(1) = a^2(a^{n+2} + 1)$$

You Try It 5 $2y(5x - 2) - 3(2 - 5x) = 2y(5x - 2) + 3(5x - 2) = (5x - 2)(2y + 3)$

You Try It 6 $a^2 - 3a + 2ab - 6b = (a^2 - 3a) + (2ab - 6b) = a(a - 3) + 2b(a - 3) = (a - 3)(a + 2b)$

You Try It 7
$$2mn^2 - n + 8mn - 4 = (2mn^2 - n) + (8mn - 4) = n(2mn - 1) + 4(2mn - 1) = (2mn - 1)(n + 4)$$

You Try It 8 $3xy - 9y - 12 + 4x = (3xy - 9y) - (12 - 4x) = 3y(x - 3) - 4(3 - x) = 3y(x - 3) + 4(x - 3) = (x - 3)(3y + 4)$

SECTION 7.2

You Try It 1

Factors	*Sum*
1, 20	21
2, 10	12
4, 5	9

- **Find the positive factors of 20 whose sum is 9.**

$x^2 + 9x + 20 = (x + 4)(x + 5)$

You Try It 2

Factors	*Sum*
+1, −18	−17
−1, +18	17
+2, −9	−7
−2, +9	7
+3, −6	−3
−3, +6	+3

- **Find the factors of −18 whose sum is 7.**

$x^2 + 7x - 18 = (x + 9)(x - 2)$

You Try It 3
The GCF is $-2x$.

$-2x^3 + 14x^2 - 12x = -2x(x^2 - 7x + 6)$

Factor the trinomial $x^2 - 7x + 6$. Find two negative factors of 6 whose sum is −7.

Factors	*Sum*
−1, −6	−7
−2, −3	−5

$-2x^3 + 14x^2 - 12x = -2x(x - 6)(x - 1)$

You Try It 4
The GCF is 3.

$3x^2 - 9xy - 12y^2 = 3(x^2 - 3xy - 4y^2)$

Factor the trinomial. Find the factors of −4 whose sum is −3.

Factors	*Sum*
+1, −4	−3
−1, +4	3
+2, −2	0

$3x^2 - 9xy - 12y^2 = 3(x + y)(x - 4y)$

SECTION 7.3

You Try It 1
Factor the trinomial $2x^2 - x - 3$.

Positive factors of 2: 1, 2

Factors of −3: 1, −3; −1, 3

Trial Factors	*Middle Term*
$(x + 1)(2x - 3)$	$-3x + 2x = -x$
$(x - 3)(2x + 1)$	$x - 6x = -5x$
$(x - 1)(2x + 3)$	$3x - 2x = x$
$(x + 3)(2x - 1)$	$-x + 6x = 5x$

$2x^2 - x - 3 = (x + 1)(2x - 3)$

You Try It 2
The GCF is $-3y$.

$-45y^3 + 12y^2 + 12y = -3y(15y^2 - 4y - 4)$

Factor the trinomial $15y^2 - 4y - 4$.

Positive factors of 15	Factors of −4
1, 15	1, −4
3, 5	−1, 4
	2, −2

Trial Factors	*Middle Term*
$(y + 1)(15y - 4)$	$-4y + 15y = 11y$
$(y - 4)(15y + 1)$	$y - 60y = -59y$
$(y - 1)(15y + 4)$	$4y - 15y = -11y$
$(y + 4)(15y - 1)$	$-y + 60y = 59y$
$(y + 2)(15y - 2)$	$-2y + 30y = 28y$
$(y - 2)(15y + 2)$	$2y - 30y = -28y$
$(3y + 1)(5y - 4)$	$-12y + 5y = -7y$
$(3y - 4)(5y + 1)$	$3y - 20y = -17y$
$(3y - 1)(5y + 4)$	$12y - 5y = 7y$
$(3y + 4)(5y - 1)$	$-3y + 20y = 17y$
$(3y + 2)(5y - 2)$	$-6y + 10y = 4y$
$(3y - 2)(5y + 2)$	$6y - 10y = -4y$

$-45y^3 + 12y^2 + 12y = -3y(3y - 2)(5y + 2)$

You Try It 3

Factors of −14 [2(−7)]	*Sum*
1, −14	13
−1, 14	13
2, −7	−5
−2, 7	5

$$\begin{aligned} 2a^2 + 13a - 7 &= 2a^2 - a + 14a - 7 \\ &= (2a^2 - a) + (14a - 7) \\ &= a(2a - 1) + 7(2a - 1) \\ &= (2a - 1)(a + 7) \end{aligned}$$

$2a^2 + 13a - 7 = (2a - 1)(a + 7)$

You Try It 4
The GCF is $5x$.

$15x^3 + 40x^2 - 80x = 5x(3x^2 + 8x - 16)$

Factors of −48 [3(−16)]	*Sum*
1, −48	−47
−1, 48	47
2, −24	−22
−2, 24	22
3, −16	−13
−3, 16	13
4, −12	−8
−4, 12	8

$$\begin{aligned} 3x^2 + 8x - 16 &= 3x^2 - 4x + 12x - 16 \\ &= (3x^2 - 4x) + (12x - 16) \\ &= x(3x - 4) + 4(3x - 4) \\ &= (3x - 4)(x + 4) \end{aligned}$$

$$\begin{aligned} 15x^3 + 40x^2 - 80x &= 5x(3x^2 + 8x - 16) \\ &= 5x(3x - 4)(x + 4) \end{aligned}$$

SECTION 7.4

You Try It 1 $\quad \begin{aligned} x^2 - 36y^4 &= x^2 - (6y^2)^2 \\ &= (x + 6y^2)(x - 6y^2) \end{aligned}$

You Try It 2 $\quad 9x^2 + 12x + 4 = (3x + 2)^2$

You Try It 3
$$\begin{aligned} &(a + b)^2 - (a - b)^2 \\ &= [(a + b) + (a - b)][(a + b) - (a - b)] \\ &= (a + b + a - b)(a + b - a + b) \\ &= (2a)(2b) = 4ab \end{aligned}$$

You Try It 4
$$\begin{aligned} 8x^3 + y^3z^3 &= (2x)^3 + (yz)^3 \\ &= (2x + yz)(4x^2 - 2xyz + y^2z^2) \end{aligned}$$

You Try It 5
$$\begin{aligned} &(x - y)^3 + (x + y)^3 \\ &= [(x - y) + (x + y)][(x - y)^2 - (x - y)(x + y) + (x + y)^2] \\ &= 2x[x^2 - 2xy + y^2 - (x^2 - y^2) + x^2 + 2xy + y^2] \\ &= 2x(x^2 - 2xy + y^2 - x^2 + y^2 + x^2 + 2xy + y^2) \\ &= 2x(x^2 + 3y^2) \end{aligned}$$

You Try It 6
Let $u = xy$.

$$\begin{aligned} 6x^2y^2 - 19xy + 10 &= 6u^2 - 19u + 10 \\ &= (2u - 5)(3u - 2) \\ &= (2xy - 5)(3xy - 2) \end{aligned}$$

You Try It 7
Let $u = x^2$.

$$\begin{aligned} 3x^4 + 4x^2 - 4 &= 3u^2 + 4u - 4 \\ &= (u + 2)(3u - 2) \\ &= (x^2 + 2)(3x^2 - 2) \end{aligned}$$

You Try It 8
Let $u = a^2b^2$.

$$\begin{aligned} a^4b^4 + 6a^2b^2 - 7 &= u^2 + 6u - 7 \\ &= (u + 7)(u - 1) \\ &= (a^2b^2 + 7)(a^2b^2 - 1) \end{aligned}$$

You Try It 9
$$\begin{aligned} 18x^3 - 6x^2 - 60x &= 6x(3x^2 - x - 10) \\ &= 6x(3x + 5)(x - 2) \end{aligned}$$

You Try It 10 $\quad \begin{aligned} &4x - 4y - x^3 + x^2y \\ &= (4x - 4y) - (x^3 - x^2y) \\ &= 4(x - y) - x^2(x - y) \\ &= (x - y)(4 - x^2) \\ &= (x - y)(2 + x)(2 - x) \end{aligned}$

You Try It 11

$$\begin{aligned} x^{4n} - x^{2n}y^{2n} &= x^{2n+2n} - x^{2n}y^{2n} \\ &= x^{2n}(x^{2n} - y^{2n}) \\ &= x^{2n}[(x^n)^2 - (y^n)^2] \\ &= x^{2n}(x^n + y^n)(x^n - y^n) \end{aligned}$$

You Try It 12

$$\begin{aligned} ax^5 - ax^2y^6 &= ax^2(x^3 - y^6) \\ &= ax^2(x - y^2)(x^2 + xy^2 + y^4) \end{aligned}$$

SECTION 7.5

You Try It 1

$2x(x + 7) = 0$

$2x = 0 \qquad x + 7 = 0$

$x = 0 \qquad x = -7$

The solutions are 0 and −7.

You Try It 2

$$\begin{aligned} 4x^2 - 9 &= 0 \\ (2x - 3)(2x + 3) &= 0 \end{aligned}$$

$2x - 3 = 0 \qquad 2x + 3 = 0$

$2x = 3 \qquad 2x = -3$

$x = \frac{3}{2} \qquad x = -\frac{3}{2}$

The solutions are $\frac{3}{2}$ and $-\frac{3}{2}$.

You Try It 3

$$\begin{aligned} (x + 2)(x - 7) &= 52 \\ x^2 - 5x - 14 &= 52 \\ x^2 - 5x - 66 &= 0 \\ (x + 6)(x - 11) &= 0 \end{aligned}$$

$x + 6 = 0 \qquad x - 11 = 0$

$x = -6 \qquad x = 11$

The solutions are −6 and 11.

You Try It 4

Strategy First positive consecutive integer: n
Second positive consecutive integer: $n + 1$

The sum of the squares of the two positive consecutive integers is 61.

Solution

$$\begin{aligned} n^2 + (n + 1)^2 &= 61 \\ n^2 + n^2 + 2n + 1 &= 61 \\ 2n^2 + 2n + 1 &= 61 \\ 2n^2 + 2n - 60 &= 0 \\ 2(n^2 + n - 30) &= 0 \\ n^2 + n - 30 &= 0 \\ (n - 5)(n + 6) &= 0 \end{aligned}$$

$n - 5 = 0 \qquad n + 6 = 0$

$n = 5 \qquad n = -6$

Because −6 is not a positive integer, it is not a solution.

$n = 5$

$n + 1 = 5 + 1 = 6$

The two integers are 5 and 6.

You Try It 5

Strategy Width $= x$
Length $= 2x + 4$

The area of the rectangle is 96 in^2. Use the equation $A = L \cdot W$.

Solution

$$\begin{aligned} A &= L \cdot W \\ 96 &= (2x + 4)x \\ 96 &= 2x^2 + 4x \\ 0 &= 2x^2 + 4x - 96 \\ 0 &= 2(x^2 + 2x - 48) \\ 0 &= x^2 + 2x - 48 \\ 0 &= (x + 8)(x - 6) \end{aligned}$$

$x + 8 = 0 \qquad x - 6 = 0$

$x = -8 \qquad x = 6$

Because the width cannot be a negative number, −8 is not a solution.

$x = 6$

$2x + 4 = 2(6) + 4 = 12 + 4 = 16$

The width is 6 in.
The length is 16 in.

Solutions to Chapter 8 "You Try It"

SECTION 8.1

You Try It 1 $\quad \dfrac{6x^5y}{12x^2y^3} = \dfrac{\overset{1}{\cancel{2}} \cdot \overset{1}{\cancel{3}} \cdot x^5y}{\underset{1}{\cancel{2}} \cdot 2 \cdot \underset{1}{\cancel{3}} \cdot x^2y^3} = \dfrac{x^3}{2y^2}$

You Try It 2

$$\frac{x^2 + 2x - 24}{16 - x^2} = \frac{x^2 + 2x - 24}{-(x^2 - 16)} = \frac{\overset{1}{\cancel{(x - 4)}}(x + 6)}{-\underset{1}{\cancel{(x - 4)}}(x + 4)}$$

$$= -\frac{x + 6}{x + 4}$$

You Try It 3 $\dfrac{x^2+4x-12}{x^2-3x+2}=\dfrac{\overset{1}{\cancel{(x-2)}}(x+6)}{(x-1)\underset{1}{\cancel{(x-2)}}}$

$$=\frac{x+6}{x-1}$$

You Try It 4

$$\frac{12x^2+3x}{10x-15}\cdot\frac{8x-12}{9x+18}=\frac{3x(4x+1)}{5(2x-3)}\cdot\frac{4(2x-3)}{9(x+2)}$$

$$=\frac{\overset{1}{\cancel{3}}x(4x+1)\cdot 2\cdot 2\overset{1}{\cancel{(2x-3)}}}{5\underset{1}{\cancel{(2x-3)}}\cdot\underset{1}{\cancel{3}}\cdot 3(x+2)}$$

$$=\frac{4x(4x+1)}{15(x+2)}$$

You Try It 5

$$\frac{x^2+2x-15}{9-x^2}\cdot\frac{x^2-3x-18}{x^2-7x+6}$$

$$=\frac{(x-3)(x+5)}{(3-x)(3+x)}\cdot\frac{(x+3)(x-6)}{(x-1)(x-6)}$$

$$=\frac{\overset{-1}{\cancel{(x-3)}}(x+5)\overset{1}{\cancel{(x+3)}}\overset{1}{\cancel{(x-6)}}}{\underset{1}{\cancel{(3-x)}}\underset{1}{\cancel{(3+x)}}(x-1)\underset{1}{\cancel{(x-6)}}}=-\frac{x+5}{x-1}$$

You Try It 6

$$\frac{a^2}{4bc^2-2b^2c}\div\frac{a}{6bc-3b^2}=\frac{a^2}{4bc^2-2b^2c}\cdot\frac{6bc-3b^2}{a}$$

$$=\frac{a^2\cdot 3b\overset{1}{\cancel{(2c-b)}}}{2bc\underset{1}{\cancel{(2c-b)}}\cdot a}=\frac{3a}{2c}$$

You Try It 7

$$\frac{3x^2+26x+16}{3x^2-7x-6}\div\frac{2x^2+9x-5}{x^2+2x-15}$$

$$=\frac{3x^2+26x+16}{3x^2-7x-6}\cdot\frac{x^2+2x-15}{2x^2+9x-5}$$

$$=\frac{\overset{1}{\cancel{(3x+2)}}(x+8)\cdot\overset{1}{\cancel{(x+5)}}\overset{1}{\cancel{(x-3)}}}{\underset{1}{\cancel{(3x+2)}}\underset{1}{\cancel{(x-3)}}\cdot(2x-1)\underset{1}{\cancel{(x+5)}}}=\frac{x+8}{2x-1}$$

SECTION 8.2

You Try It 1

$8uv^2=2\cdot 2\cdot 2\cdot u\cdot v\cdot v$

$12uw=2\cdot 2\cdot 3\cdot u\cdot w$

$\text{LCM}=2\cdot 2\cdot 2\cdot 3\cdot u\cdot v\cdot v\cdot w=24uv^2w$

You Try It 2 $m^2-6m+9=(m-3)(m-3)$

$m^2-2m-3=(m+1)(m-3)$

$\text{LCM}=(m-3)(m-3)(m+1)$

You Try It 3

The LCM is $36xy^2z$.

$$\frac{x-3}{4xy^2}=\frac{x-3}{4xy^2}\cdot\frac{9z}{9z}=\frac{9xz-27z}{36xy^2z}$$

$$\frac{2x+1}{9y^2z}=\frac{2x+1}{9y^2z}\cdot\frac{4x}{4x}=\frac{8x^2+4x}{36xy^2z}$$

You Try It 4

The LCM is $(x+2)(x-5)(x+5)$

$$\frac{x+4}{x^2-3x-10}=\frac{x+4}{(x+2)(x-5)}\cdot\frac{x+5}{x+5}$$

$$=\frac{x^2+9x+20}{(x+2)(x-5)(x+5)}$$

$$\frac{2x}{25-x^2}=\frac{2x}{-(x^2-25)}=-\frac{2x}{(x-5)(x+5)}\cdot\frac{x+2}{x+2}$$

$$=-\frac{2x^2+4x}{(x+2)(x-5)(x+5)}$$

You Try It 5 $\dfrac{3}{xy}+\dfrac{12}{xy}=\dfrac{3+12}{xy}=\dfrac{15}{xy}$

You Try It 6

$$\frac{2x^2}{x^2-x-12}-\frac{7x+4}{x^2-x-12}$$

$$=\frac{2x^2-(7x+4)}{x^2-x-12}=\frac{2x^2-7x-4}{x^2-x-12}$$

$$=\frac{(2x+1)\overset{1}{\cancel{(x-4)}}}{(x+3)\underset{1}{\cancel{(x-4)}}}=\frac{2x+1}{x+3}$$

You Try It 7

$$\frac{x^2-1}{x^2-8x+12}-\frac{2x+1}{x^2-8x+12}+\frac{x}{x^2-8x+12}$$

$$=\frac{(x^2-1)-(2x+1)+x}{x^2-8x+12}=\frac{x^2-1-2x-1+x}{x^2-8x+12}$$

$$=\frac{x^2-x-2}{x^2-8x+12}=\frac{(x+1)\overset{1}{\cancel{(x-2)}}}{\underset{1}{\cancel{(x-2)}}(x-6)}=\frac{x+1}{x-6}$$

You Try It 8

The LCM of the denominators is $24y$.

$$\frac{z}{8y}-\frac{4z}{3y}+\frac{5z}{4y}=\frac{z}{8y}\cdot\frac{3}{3}-\frac{4z}{3y}\cdot\frac{8}{8}+\frac{5z}{4y}\cdot\frac{6}{6}$$

$$=\frac{3z}{24y}-\frac{32z}{24y}+\frac{30z}{24y}$$

$$=\frac{3z-32z+30z}{24y}=\frac{z}{24y}$$

You Try It 9

$2 - x = -(x - 2)$

Therefore, $\dfrac{3}{2 - x} = \dfrac{-3}{x - 2}$.

The LCM is $x - 2$.

$$\frac{5x}{x-2} - \frac{3}{2-x} = \frac{5x}{x-2} - \frac{-3}{x-2}$$
$$= \frac{5x - (-3)}{x-2} = \frac{5x+3}{x-2}$$

You Try It 10

The LCM is $(3x - 1)(x + 4)$.

$$\frac{4x}{3x-1} - \frac{9}{x+4} = \frac{4x}{3x-1} \cdot \frac{x+4}{x+4} - \frac{9}{x+4} \cdot \frac{3x-1}{3x-1}$$
$$= \frac{4x^2 + 16x}{(3x-1)(x+4)} - \frac{27x - 9}{(3x-1)(x+4)}$$
$$= \frac{(4x^2 + 16x) - (27x - 9)}{(3x-1)(x+4)}$$
$$= \frac{4x^2 + 16x - 27x + 9}{(3x-1)(x+4)}$$
$$= \frac{4x^2 - 11x + 9}{(3x-1)(x+4)}$$

You Try It 11

The LCM is $x - 3$.

$$2 - \frac{1}{x-3} = \frac{2}{1} \cdot \frac{x-3}{x-3} - \frac{1}{x-3}$$
$$= \frac{2x-6}{x-3} - \frac{1}{x-3}$$
$$= \frac{2x - 6 - 1}{x-3}$$
$$= \frac{2x-7}{x-3}$$

You Try It 12

The LCM is $a(a - 5)(a + 5)$.

$$\frac{a-3}{a^2 - 5a} + \frac{a-9}{a^2 - 25}$$
$$= \frac{a-3}{a(a-5)} \cdot \frac{a+5}{a+5} + \frac{a-9}{(a-5)(a+5)} \cdot \frac{a}{a}$$
$$= \frac{(a-3)(a+5) + a(a-9)}{a(a-5)(a+5)}$$
$$= \frac{(a^2 + 2a - 15) + (a^2 - 9a)}{a(a-5)(a+5)}$$
$$= \frac{a^2 + 2a - 15 + a^2 - 9a}{a(a-5)(a+5)}$$
$$= \frac{2a^2 - 7a - 15}{a(a-5)(a+5)} = \frac{(2a+3)(a-5)}{a(a-5)(a+5)}$$
$$= \frac{(2a+3)\overset{1}{\cancel{(a-5)}}}{a\underset{1}{\cancel{(a-5)}}(a+5)} = \frac{2a+3}{a(a+5)}$$

You Try It 13

The LCM is $(3x + 2)(x - 1)$.

$$\frac{2x-3}{3x^2 - x - 2} + \frac{5}{3x+2} - \frac{1}{x-1}$$
$$= \frac{2x-3}{(3x+2)(x-1)} + \frac{5}{3x+2} \cdot \frac{x-1}{x-1} - \frac{1}{x-1} \cdot \frac{3x+2}{3x+2}$$
$$= \frac{2x-3}{(3x+2)(x-1)} + \frac{5x-5}{(3x+2)(x-1)} - \frac{3x+2}{(3x+2)(x-1)}$$
$$= \frac{(2x-3) + (5x-5) - (3x+2)}{(3x+2)(x-1)}$$
$$= \frac{2x - 3 + 5x - 5 - 3x - 2}{(3x+2)(x-1)}$$
$$= \frac{4x - 10}{(3x+2)(x-1)} = \frac{2(2x-5)}{(3x+2)(x-1)}$$

SECTION 8.3

You Try It 1

The LCM of 3, x, 9, and x^2 is $9x^2$.

$$\frac{\frac{1}{3} - \frac{1}{x}}{\frac{1}{9} - \frac{1}{x^2}} = \frac{\frac{1}{3} - \frac{1}{x}}{\frac{1}{9} - \frac{1}{x^2}} \cdot \frac{9x^2}{9x^2} = \frac{\frac{1}{3} \cdot 9x^2 - \frac{1}{x} \cdot 9x^2}{\frac{1}{9} \cdot 9x^2 - \frac{1}{x^2} \cdot 9x^2}$$
$$= \frac{3x^2 - 9x}{x^2 - 9} = \frac{3x\overset{1}{\cancel{(x-3)}}}{\underset{1}{\cancel{(x-3)}}(x+3)} = \frac{3x}{x+3}$$

You Try It 2

The LCM is $x - 5$.

$$\frac{x + 3 - \frac{20}{x-5}}{x + 8 + \frac{30}{x+5}} = \frac{x + 3 - \frac{20}{x-5}}{x + 8 + \frac{30}{x-5}} \cdot \frac{x-5}{x-5}$$
$$= \frac{x(x-5) + 3(x-5) - \frac{20}{x-5} \cdot (x-5)}{x(x-5) + 8(x-5) + \frac{30}{x-5} \cdot (x-5)}$$
$$= \frac{x^2 - 5x + 3x - 15 - 20}{x^2 - 5x + 8x - 40 + 30}$$
$$= \frac{x^2 - 2x - 35}{x^2 + 3x - 10}$$
$$= \frac{\overset{1}{\cancel{(x+5)}}(x-7)}{(x-2)\underset{1}{\cancel{(x+5)}}} = \frac{x-7}{x-2}$$

SECTION 8.4

You Try It 1

$$\frac{x}{x+6} = \frac{3}{x} \quad \text{The LCM is } x(x+6).$$

$$\frac{x\cancel{(x+6)}^{1}}{1} \cdot \frac{x}{\cancel{x+6}_{1}} = \frac{\cancel{x}^{1}(x+6)}{1} \cdot \frac{3}{\cancel{x}_{1}}$$

$$\begin{aligned} x^2 &= (x+6)3 \\ x^2 &= 3x + 18 \\ x^2 - 3x - 18 &= 0 \\ (x+3)(x-6) &= 0 \end{aligned}$$

$$x + 3 = 0 \qquad x - 6 = 0$$
$$x = -3 \qquad x = 6$$

Both -3 and 6 check as solutions.
The solutions are -3 and 6.

You Try It 2

$$\frac{5x}{x+2} = 3 - \frac{10}{x+2} \quad \text{The LCM is } x + 2.$$

$$\frac{(x+2)}{1} \cdot \frac{5x}{x+2} = \frac{(x+2)}{1}\left(3 - \frac{10}{x+2}\right)$$

$$\frac{\cancel{x+2}^{1}}{1} \cdot \frac{5x}{\cancel{x+2}_{1}} = (x+2)3 - \frac{\cancel{(x+2)}^{1}}{1} \cdot \frac{10}{\cancel{x+2}_{1}}$$

$$\begin{aligned} 5x &= (x+2)3 - 10 \\ 5x &= 3x + 6 - 10 \\ 5x &= 3x - 4 \\ 2x &= -4 \\ x &= -2 \end{aligned}$$

-2 does not check as a solution.
The equation has no solution.

You Try It 3

$$\frac{2}{x+3} = \frac{6}{5x+5}$$

$$\frac{(x+3)(5x+5)}{1} \cdot \frac{2}{x+3} = \frac{(x+3)(5x+5)}{1} \cdot \frac{6}{5x+5}$$

$$\frac{\cancel{(x+3)}^{1}(5x+5)}{1} \cdot \frac{2}{\cancel{x+3}_{1}} = \frac{(x+3)\cancel{(5x+5)}^{1}}{1} \cdot \frac{6}{\cancel{5x+5}_{1}}$$

$$\begin{aligned} (5x+5)2 &= (x+3)6 \\ 10x + 10 &= 6x + 18 \\ 4x + 10 &= 18 \\ 4x &= 8 \\ x &= 2 \end{aligned}$$

The solution is 2.

You Try It 4

Strategy To find the area of triangle AOB:
- Solve a proportion to find the length of AO (the height of triangle AOB).
- Use the formula for the area of a triangle. AB is the base and AO is the height.

Solution

$$\begin{aligned} \frac{CD}{AB} &= \frac{DO}{AO} \\ \frac{4}{10} &= \frac{3}{AO} \\ 10 \cdot AO \cdot \frac{4}{10} &= 10 \cdot AO \cdot \frac{3}{AO} \\ 4(AO) &= 30 \\ AO &= 7.5 \end{aligned}$$

$$\begin{aligned} A &= \frac{1}{2}bh \\ &= \frac{1}{2}(10)(7.5) \\ &= 37.5 \end{aligned}$$

The area of triangle AOB is 37.5 cm^2.

You Try It 5

Strategy To find the total area that 256 ceramic tiles will cover, write and solve a proportion, using x to represent the number of square feet that 256 tiles will cover.

Solution

$$\begin{aligned} \frac{9}{16} &= \frac{x}{256} \\ 256\left(\frac{9}{16}\right) &= 256\left(\frac{x}{256}\right) \\ 144 &= x \end{aligned}$$

A 144-square-foot area can be tiled using 256 ceramic tiles.

You Try It 6

Strategy To find the additional amount of medication required for a 200-pound adult, write and solve a proportion, using x to represent the additional medication. Then $3 + x$ is the total amount required for a 200-pound adult.

Solution

$$\frac{150}{3} = \frac{200}{3 + x}$$
$$\frac{50}{1} = \frac{200}{3 + x}$$
$$(3 + x) \cdot 50 = (3 + x) \cdot \frac{200}{3 + x}$$
$$150 + 50x = 200$$
$$50x = 50$$
$$x = 1$$

One additional ounce is required for a 200-pound adult.

SECTION 8.5

You Try It 1

$$A = P + Prt$$
$$A - P = P - P + Prt$$
$$A - P = Prt$$
$$\frac{A - P}{Pt} = \frac{Prt}{Pt}$$
$$\frac{A - P}{Pt} = r$$

You Try It 2

$$s = \frac{A + L}{2}$$
$$2 \cdot s = 2\left(\frac{A + L}{2}\right)$$
$$2s = A + L$$
$$2s - A = A - A + L$$
$$2s - A = L$$

You Try It 3

$$s = a + (n - 1)d$$
$$s = a + nd - d$$
$$s - a = a - a + nd - d$$
$$s - a = nd - d$$
$$s - a + d = nd - d + d$$
$$s - a + d = nd$$
$$\frac{s - a + d}{d} = \frac{nd}{d}$$
$$\frac{s - a + d}{d} = n$$

You Try It 4

$$S = C + rC$$
$$S = (1 + r)C$$
$$\frac{S}{1 + r} = \frac{(1 + r)C}{1 + r}$$
$$\frac{S}{1 + r} = C$$

SECTION 8.6

You Try It 1

Strategy
- Time for one printer to complete the job: t

	Rate	*Time*	*Part*
1st printer	$\frac{1}{t}$	2	$\frac{2}{t}$
2nd printer	$\frac{1}{t}$	5	$\frac{5}{t}$

- The sum of the parts of the task completed must equal 1.

Solution

$$\frac{2}{t} + \frac{5}{t} = 1$$
$$t\left(\frac{2}{t} + \frac{5}{t}\right) = t \cdot 1$$
$$2 + 5 = t$$
$$7 = t$$

Working alone, one printer takes 7 h to print the payroll.

You Try It 2

Strategy
- Rate sailing across the lake: r
  Rate sailing back: $3r$

	Distance	*Rate*	*Time*
Across	6	r	$\frac{6}{r}$
Back	6	$3r$	$\frac{6}{3r}$

- The total time for the trip was 2 h.

Solution

$$\frac{6}{r} + \frac{6}{3r} = 2$$
$$3r\left(\frac{6}{r} + \frac{6}{3r}\right) = 3r(2)$$
$$3r \cdot \frac{6}{r} + 3r \cdot \frac{6}{3r} = 6r$$
$$18 + 6 = 6r$$
$$24 = 6r$$
$$4 = r$$

The rate across the lake was 4 km/h.

SECTION 8.7

You Try It 1

Strategy To find the distance:
- Write the basic direct variation equation, replace the variables by the given values, and solve for k.

- Write the direct variation equation, replacing k by its value. Substitute 5 for t and solve for s.

Solution

$$s = kt^2$$
$$64 = k(2)^2$$
$$64 = k \cdot 4$$
$$16 = k$$

$$s = 16t^2 = 16(5)^2 = 400$$

The object will fall 400 ft in 5 s.

You Try It 2

Strategy To find the resistance:

- Write the basic inverse variation equation, replace the variables by the given values, and solve for k.
- Write the inverse variation equation, replacing k by its value. Substitute 0.02 for d and solve for R.

Solution

$$R = \frac{k}{d^2}$$
$$0.5 = \frac{k}{(0.01)^2}$$
$$0.5 = \frac{k}{0.0001}$$
$$0.00005 = k$$

$$R = \frac{0.00005}{d^2} = \frac{0.00005}{(0.02)^2} = 0.125$$

The resistance is 0.125 ohm.

You Try It 3

Strategy To find the strength:

- Write the basic combined variation equation, replace the variables by the given values, and solve for k.
- Write the combined variation equation, replacing k by its value. Substitute 4 for w, 8 for d, and 16 for L, and solve for s.

Solution

$$s = \frac{kwd^2}{L}$$
$$1200 = \frac{k \cdot 2 \cdot (12)^2}{12}$$
$$1200 = \frac{k \cdot 2 \cdot 144}{12}$$
$$1200 = 24k$$
$$50 = k$$

$$s = \frac{50wd^2}{L} = \frac{50 \cdot 4 \cdot (8)^2}{16} = 800$$

The strength is 800 lb.

Solutions to Chapter 9 "You Try It"

SECTION 9.1

You Try It 1

$$16^{-3/4} = (2^4)^{-3/4} = 2^{-3} = \frac{1}{2^3} = \frac{1}{8}$$

You Try It 2

$(-81)^{3/4}$

The base of the exponential expression is negative, and the denominator of the exponent is a positive even number.

Therefore, $(-81)^{3/4}$ is not a real number.

You Try It 3

$$(x^{3/4}y^{1/2}z^{-2/3})^{-4/3} = x^{-1}y^{-2/3}z^{8/9} = \frac{z^{8/9}}{xy^{2/3}}$$

You Try It 4

$$\left(\frac{16a^{-2}b^{4/3}}{9a^4b^{-2/3}}\right)^{-1/2} = \left(\frac{2^4a^{-6}b^2}{3^2}\right)^{-1/2} = \frac{2^{-2}a^3b^{-1}}{3^{-1}} = \frac{3a^3}{2^2b} = \frac{3a^3}{4b}$$

You Try It 5

$$(2x^3)^{3/4} = \sqrt[4]{(2x^3)^3} = \sqrt[4]{8x^9}$$

You Try It 6 $-5a^{5/6} = -5(a^5)^{1/6}$
$= -5\sqrt[6]{a^5}$

You Try It 7 $\sqrt[3]{3ab} = (3ab)^{1/3}$

You Try It 8 $\sqrt[4]{x^4 + y^4} = (x^4 + y^4)^{1/4}$

You Try It 9 $\sqrt[3]{-8x^{12}y^3} = -2x^4y$

You Try It 10 $-\sqrt[4]{81x^{12}y^8} = -3x^3y^2$

SECTION 9.2

You Try It 1 $\sqrt[5]{x^7} = \sqrt[5]{x^5 \cdot x^2}$
$= \sqrt[5]{x^5}\sqrt[5]{x^2}$
$= x\sqrt[5]{x^2}$

You Try It 2 $\sqrt[3]{-64x^8y^{18}} = \sqrt[3]{(-4)^3x^8y^{18}}$
$= \sqrt[3]{(-4)^3x^6y^{18}(x^2)}$
$= \sqrt[3]{(-4)^3x^6y^{18}}\sqrt[3]{x^2}$
$= -4x^2y^6\sqrt[3]{x^2}$

You Try It 3
$3xy\sqrt[3]{81x^5y} - \sqrt[3]{192x^8y^4}$
$= 3xy\sqrt[3]{27x^3}\sqrt[3]{3x^2y} - \sqrt[3]{64x^6y^3}\sqrt[3]{3x^2y}$
$= 3xy \cdot 3x\sqrt[3]{3x^2y} - 4x^2y\sqrt[3]{3x^2y}$
$= 9x^2y\sqrt[3]{3x^2y} - 4x^2y\sqrt[3]{3x^2y} = 5x^2y\sqrt[3]{3x^2y}$

You Try It 4
$\sqrt{5b}(\sqrt{3b} - \sqrt{10})$
$= \sqrt{15b^2} - \sqrt{50b}$
$= \sqrt{15b^2} - \sqrt{25 \cdot 2b}$
$= \sqrt{b^2}\sqrt{15} - \sqrt{25}\sqrt{2b}$
$= b\sqrt{15} - 5\sqrt{2b}$

You Try It 5
$(2\sqrt[3]{2x} - 3)(\sqrt[3]{2x} - 5)$
$= 2\sqrt[3]{4x^2} - 10\sqrt[3]{2x} - 3\sqrt[3]{2x} + 15$
$= 2\sqrt[3]{4x^2} - 13\sqrt[3]{2x} + 15$

You Try It 6
$(\sqrt{a} - 3\sqrt{y})(\sqrt{a} + 3\sqrt{y})$
$= (\sqrt{a})^2 - (3\sqrt{y})^2$
$= a - 9y$

You Try It 7
$$\frac{y}{\sqrt{3y}} = \frac{y}{\sqrt{3y}} \cdot \frac{\sqrt{3y}}{\sqrt{3y}} = \frac{y\sqrt{3y}}{(\sqrt{3y})^2} = \frac{y\sqrt{3y}}{3y} = \frac{\sqrt{3y}}{3}$$

You Try It 8
$$\frac{3x}{\sqrt[3]{3x^2}} = \frac{3x}{\sqrt[3]{3x^2}} \cdot \frac{\sqrt[3]{9x}}{\sqrt[3]{9x}} = \frac{3x\sqrt[3]{9x}}{\sqrt[3]{27x^3}}$$
$$= \frac{3x\sqrt[3]{9x}}{3x} = \sqrt[3]{9x}$$

You Try It 9
$$\frac{3+\sqrt{6}}{2-\sqrt{6}} = \frac{3+\sqrt{6}}{2-\sqrt{6}} \cdot \frac{2+\sqrt{6}}{2+\sqrt{6}} = \frac{6+3\sqrt{6}+2\sqrt{6}+(\sqrt{6})^2}{2^2-(\sqrt{6})^2}$$
$$= \frac{6+5\sqrt{6}+6}{4-6} = \frac{12+5\sqrt{6}}{-2} = -\frac{12+5\sqrt{6}}{2}$$

SECTION 9.3

You Try It 1 $\sqrt{-45} = i\sqrt{45} = i\sqrt{9 \cdot 5} = 3i\sqrt{5}$

You Try It 2
$\sqrt{98} - \sqrt{-60} = \sqrt{98} - i\sqrt{60}$
$= \sqrt{49 \cdot 2} - i\sqrt{4 \cdot 15}$
$= 7\sqrt{2} - 2i\sqrt{15}$

You Try It 3 $(-4 + 2i) - (6 - 8i) = -10 + 10i$

You Try It 4 $(16 - \sqrt{-45}) - (3 + \sqrt{-20})$
$= (16 - i\sqrt{45}) - (3 + i\sqrt{20})$
$= (16 - i\sqrt{9 \cdot 5}) - (3 + i\sqrt{4 \cdot 5})$
$= (16 - 3i\sqrt{5}) - (3 + 2i\sqrt{5})$
$= 13 - 5i\sqrt{5}$

You Try It 5 $(3 - 2i) + (-3 + 2i) = 0 + 0i = 0$

You Try It 6 $(-3i)(-10i) = 30i^2 = 30(-1) = -30$

You Try It 7
$-\sqrt{-8} \cdot \sqrt{-5} = -i\sqrt{8} \cdot i\sqrt{5} = -i^2\sqrt{40}$
$= -(-1)\sqrt{40} = \sqrt{4 \cdot 10} = 2\sqrt{10}$

You Try It 8
$-6i(3 + 4i) = -18i - 24i^2$
$= -18i - 24(-1) = 24 - 18i$

You Try It 9
$\sqrt{-3}(\sqrt{27} - \sqrt{-6}) = i\sqrt{3}(\sqrt{27} - i\sqrt{6})$
$= i\sqrt{81} - i^2\sqrt{18}$
$= i\sqrt{81} - (-1)\sqrt{9 \cdot 2}$
$= 9i + 3\sqrt{2}$
$= 3\sqrt{2} + 9i$

You Try It 10
$(4 - 3i)(2 - i) = 8 - 4i - 6i + 3i^2$
$= 8 - 10i + 3i^2$
$= 8 - 10i + 3(-1)$
$= 5 - 10i$

You Try It 11

$$\begin{aligned}(3+6i)(3-6i) &= 3^2+6^2\\ &= 9+36\\ &= 45\end{aligned}$$

You Try It 12

$$\begin{aligned}(3-i)\left(\frac{3}{10}+\frac{1}{10}i\right) &= \frac{9}{10}+\frac{3}{10}i-\frac{3}{10}i-\frac{1}{10}i^2\\ &= \frac{9}{10}-\frac{1}{10}i^2 = \frac{9}{10}-\frac{1}{10}(-1)\\ &= \frac{9}{10}+\frac{1}{10} = 1\end{aligned}$$

You Try It 13

$$\begin{aligned}\frac{2-3i}{4i} &= \frac{2-3i}{4i}\cdot\frac{i}{i}\\ &= \frac{2i-3i^2}{4i^2}\\ &= \frac{2i-3(-1)}{4(-1)}\\ &= \frac{3+2i}{-4} = -\frac{3}{4}-\frac{1}{2}i\end{aligned}$$

You Try It 14

$$\begin{aligned}\frac{2+5i}{3-2i} &= \frac{2+5i}{3-2i}\cdot\frac{3+2i}{3+2i} = \frac{6+4i+15i+10i^2}{3^2+2^2}\\ &= \frac{6+19i+10(-1)}{13} = \frac{-4+19i}{13}\\ &= -\frac{4}{13}+\frac{19}{13}i\end{aligned}$$

SECTION 9.4

You Try It 1

$$\begin{aligned}\sqrt[4]{x-8} &= 3\\ (\sqrt[4]{x-8})^4 &= 3^4\\ x-8 &= 81\\ x &= 89\end{aligned}$$

Check:

$$\begin{array}{c|c}\sqrt[4]{x-8} & 3\\ \hline \sqrt[4]{89-8} & 3\\ \sqrt[4]{81} & 3\\ 3 = 3 & \end{array}$$

The solution is 89.

You Try It 2

$$\begin{aligned}\sqrt{x}-\sqrt{x+5} &= 1\\ \sqrt{x} &= 1+\sqrt{x+5}\\ (\sqrt{x})^2 &= (1+\sqrt{x+5})^2\\ x &= 1+2\sqrt{x+5}+x+5\\ 0 &= 6+2\sqrt{x+5}\\ -6 &= 2\sqrt{x+5}\\ -3 &= \sqrt{x+5}\\ (-3)^2 &= (\sqrt{x+5})^2\\ 9 &= x+5\\ 4 &= x\end{aligned}$$

4 does not check as a solution. The equation has no solution.

You Try It 3

Strategy To find the diagonal, use the Pythagorean Theorem. One leg is the length of the rectangle. The second leg is the width of the rectangle. The hypotenuse is the diagonal of the rectangle.

Solution

$$\begin{aligned}c^2 &= a^2+b^2\\ c^2 &= (6)^2+(3)^2\\ c^2 &= 36+9\\ c^2 &= 45\\ (c^2)^{1/2} &= (45)^{1/2}\\ c &= \sqrt{45}\\ c &\approx 6.7\end{aligned}$$

The diagonal is 6.7 cm.

You Try It 4

Strategy To find the height, replace d in the equation with the given value and solve for h.

Solution

$$\begin{aligned}d &= \sqrt{1.5h}\\ 5.5 &= \sqrt{1.5h}\\ (5.5)^2 &= (\sqrt{1.5h})^2\\ 30.25 &= 1.5h\\ 20.17 &\approx h\end{aligned}$$

The periscope must be approximately 20.17 ft above the water.

You Try It 5

Strategy To find the distance, replace the variables v and a in the equation by their given values and solve for s.

Solution

$$\begin{aligned}v &= \sqrt{2as}\\ 88 &= \sqrt{2\cdot 22s}\\ 88 &= \sqrt{44s}\\ (88)^2 &= (\sqrt{44s})^2\\ 7744 &= 44s\\ 176 &= s\end{aligned}$$

The distance required is 176 ft.

Solutions to Chapter 10 "You Try It"

SECTION 10.1

You Try It 1

$$2x^2 = 7x - 3$$
$$2x^2 - 7x + 3 = 0$$
$$(2x - 1)(x - 3) = 0$$

$$2x - 1 = 0 \qquad x - 3 = 0$$
$$2x = 1 \qquad x = 3$$
$$x = \frac{1}{2}$$

The solutions are $\frac{1}{2}$ and 3.

You Try It 2

$$x^2 - 3ax - 4a^2 = 0$$
$$(x + a)(x - 4a) = 0$$

$$x + a = 0 \qquad x - 4a = 0$$
$$x = -a \qquad x = 4a$$

The solutions are $-a$ and $4a$.

You Try It 3

$$(x - r_1)(x - r_2) = 0$$
$$(x - 3)\left[x - \left(-\frac{1}{2}\right)\right] = 0$$
$$(x - 3)\left(x + \frac{1}{2}\right) = 0$$
$$x^2 - \frac{5}{2}x - \frac{3}{2} = 0$$
$$2\left(x^2 - \frac{5}{2}x - \frac{3}{2}\right) = 2 \cdot 0$$
$$2x^2 - 5x - 3 = 0$$

You Try It 4

$$2(x + 1)^2 - 24 = 0$$
$$2(x + 1)^2 = 24$$
$$(x + 1)^2 = 12$$
$$\sqrt{(x + 1)^2} = \sqrt{12}$$
$$x + 1 = \pm 2\sqrt{3}$$

$$x + 1 = 2\sqrt{3} \qquad x + 1 = -2\sqrt{3}$$
$$x = -1 + 2\sqrt{3} \qquad x = -1 - 2\sqrt{3}$$

The solutions are $-1 + 2\sqrt{3}$ and $-1 - 2\sqrt{3}$.

SECTION 10.2

You Try It 1

$$4x^2 - 4x - 1 = 0$$
$$4x^2 - 4x = 1$$
$$\frac{1}{4}(4x^2 - 4x) = \frac{1}{4} \cdot 1$$
$$x^2 - x = \frac{1}{4}$$

Complete the square.

$$x^2 - x + \frac{1}{4} = \frac{1}{4} + \frac{1}{4}$$
$$\left(x - \frac{1}{2}\right)^2 = \frac{2}{4}$$
$$\sqrt{\left(x - \frac{1}{2}\right)^2} = \sqrt{\frac{2}{4}}$$
$$x - \frac{1}{2} = \pm\frac{\sqrt{2}}{2}$$

$$x - \frac{1}{2} = \frac{\sqrt{2}}{2} \qquad x - \frac{1}{2} = -\frac{\sqrt{2}}{2}$$
$$x = \frac{1}{2} + \frac{\sqrt{2}}{2} \qquad x = \frac{1}{2} - \frac{\sqrt{2}}{2}$$

The solutions are $\frac{1 + \sqrt{2}}{2}$ and $\frac{1 - \sqrt{2}}{2}$.

You Try It 2

$$2x^2 + x - 5 = 0$$
$$2x^2 + x = 5$$
$$\frac{1}{2}(2x^2 + x) = \frac{1}{2} \cdot 5$$
$$x^2 + \frac{1}{2}x = \frac{5}{2}$$

Complete the square.

$$x^2 + \frac{1}{2}x + \frac{1}{16} = \frac{5}{2} + \frac{1}{16}$$
$$\left(x + \frac{1}{4}\right)^2 = \frac{41}{16}$$
$$\sqrt{\left(x + \frac{1}{4}\right)^2} = \sqrt{\frac{41}{16}}$$
$$x + \frac{1}{4} = \pm\frac{\sqrt{41}}{4}$$

$$x + \frac{1}{4} = \frac{\sqrt{41}}{4} \qquad x + \frac{1}{4} = -\frac{\sqrt{41}}{4}$$
$$x = -\frac{1}{4} + \frac{\sqrt{41}}{4} \qquad x = -\frac{1}{4} - \frac{\sqrt{41}}{4}$$

The solutions are $\frac{-1 + \sqrt{41}}{4}$ and $\frac{-1 - \sqrt{41}}{4}$.

SECTION 10.3

You Try It 1

$$x^2 - 2x + 10 = 0$$
$$a = 1, b = -2, c = 10$$
$$x = \frac{-b \pm \sqrt{b^2 - 4ac}}{2a}$$
$$= \frac{-(-2) \pm \sqrt{(-2)^2 - 4(1)(10)}}{2 \cdot 1}$$
$$= \frac{2 \pm \sqrt{4 - 40}}{2} = \frac{2 \pm \sqrt{-36}}{2}$$
$$= \frac{2 \pm 6i}{2} = 1 \pm 3i$$

The solutions are $1 + 3i$ and $1 - 3i$.

You Try It 2

$$4x^2 = 4x - 1$$
$$4x^2 - 4x + 1 = 0$$
$$a = 4, b = -4, c = 1$$
$$x = \frac{-b \pm \sqrt{b^2 - 4ac}}{2a}$$
$$= \frac{-(-4) \pm \sqrt{(-4)^2 - 4(4)(1)}}{2 \cdot 4}$$
$$= \frac{4 \pm \sqrt{16 - 16}}{8} = \frac{4 \pm \sqrt{0}}{8}$$
$$= \frac{4}{8} = \frac{1}{2}$$

The solution is $\frac{1}{2}$.

You Try It 3

$$3x^2 - x - 1 = 0$$
$$a = 3, b = -1, c = -1$$
$$b^2 - 4ac = (-1)^2 - 4(3)(-1)$$
$$= 1 + 12 = 13$$
$$13 > 0$$

Because the discriminant is greater than zero, the equation has two real number solutions.

SECTION 10.4

You Try It 1

$$x - 5x^{1/2} + 6 = 0$$
$$(x^{1/2})^2 - 5(x^{1/2}) + 6 = 0$$
$$u^2 - 5u + 6 = 0$$
$$(u - 2)(u - 3) = 0$$

$$u - 2 = 0 \qquad u - 3 = 0$$
$$u = 2 \qquad u = 3$$

Replace u by $x^{1/2}$.

$$x^{1/2} = 2 \qquad x^{1/2} = 3$$
$$\sqrt{x} = 2 \qquad \sqrt{x} = 3$$
$$(\sqrt{x})^2 = 2^2 \qquad (\sqrt{x})^2 = 3^2$$
$$x = 4 \qquad x = 9$$

The solutions are 4 and 9.

You Try It 2

$$\sqrt{2x + 1} + x = 7$$
$$\sqrt{2x + 1} = 7 - x$$
$$(\sqrt{2x + 1})^2 = (7 - x)^2$$
$$2x + 1 = 49 - 14x + x^2$$
$$0 = x^2 - 16x + 48$$
$$0 = (x - 4)(x - 12)$$

$$x - 4 = 0 \qquad x - 12 = 0$$
$$x = 4 \qquad x = 12$$

4 checks as a solution.
12 does not check as a solution.

The solution is 4.

You Try It 3

$$\sqrt{2x - 1} + \sqrt{x} = 2$$

Solve for one of the radical expressions.

$$\sqrt{2x - 1} = 2 - \sqrt{x}$$
$$(\sqrt{2x - 1})^2 = (2 - \sqrt{x})^2$$
$$2x - 1 = 4 - 4\sqrt{x} + x$$
$$x - 5 = -4\sqrt{x}$$

Square each side of the equation.

$$(x - 5)^2 = (-4\sqrt{x})^2$$
$$x^2 - 10x + 25 = 16x$$
$$x^2 - 26x + 25 = 0$$
$$(x - 1)(x - 25) = 0$$

$$x - 1 = 0 \qquad x - 25 = 0$$
$$x = 1 \qquad x = 25$$

1 checks as a solution.
25 does not check as a solution.

The solution is 1.

You Try It 4

$$3y + \frac{25}{3y - 2} = -8$$
$$(3y - 2)\left(3y + \frac{25}{3y - 2}\right) = (3y - 2)(-8)$$
$$(3y - 2)(3y) + (3y - 2)\left(\frac{25}{3y - 2}\right) = (3y - 2)(-8)$$
$$9y^2 - 6y + 25 = -24y + 16$$
$$9y^2 + 18y + 9 = 0$$
$$9(y^2 + 2y + 1) = 0$$
$$y^2 + 2y + 1 = 0$$
$$(y + 1)(y + 1) = 0$$

$$y + 1 = 0 \qquad y + 1 = 0$$
$$y = -1 \qquad y = -1$$

The solution is -1.

SECTION 10.5

You Try It 1

Strategy
- This is a geometry problem.
- Width of the rectangle: W
  Length of the rectangle: $W + 3$
- Use the equation $A = L \cdot W$.

Solution

$$A = L \cdot W$$
$$54 = (W + 3)(W)$$
$$54 = W^2 + 3W$$
$$0 = W^2 + 3W - 54$$
$$0 = (W + 9)(W - 6)$$

$W + 9 = 0 \qquad W - 6 = 0$
$W = -9 \qquad W = 6$

The solution -9 is not possible.

$W + 3 = 6 + 3 = 9$

The length is 9 m.

SECTION 10.6

You Try It 1

$$2x^2 - x - 10 \le 0$$
$$(2x - 5)(x + 2) \le 0$$

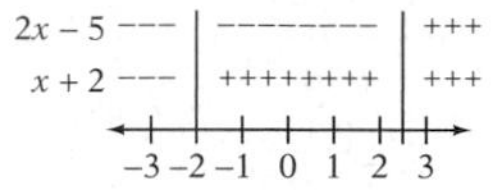

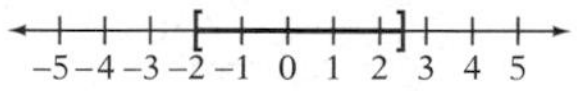

$$\left\{x \mid -2 \le x \le \frac{5}{2}\right\}$$

Solutions to Chapter 11 "You Try It"

SECTION 11.1

You Try It 1

$f(x) = \dfrac{3}{5}x - 4$

$b = -4$

y-intercept: $(0, -4)$

$m = \dfrac{3}{5}$

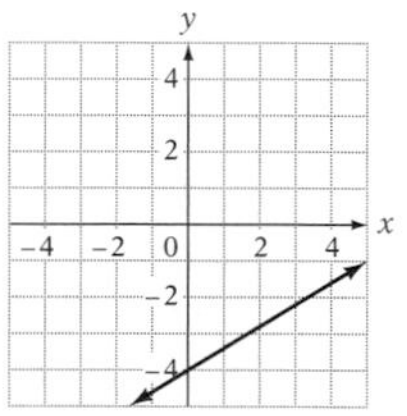

You Try It 2

Strategy
- Select the independent and dependent variables. The function is to predict the Celsius temperature, so that quantity is the dependent variable, y. The Fahrenheit variable is the independent variable, x.
- From the given data, two ordered pairs are (212, 100) and (32, 0). Use these ordered pairs to determine the linear function.

Solution

Let $(x_1, y_1) = (32, 0)$ and $(x_2, y_2) = (212, 100)$.

$$m = \frac{y_2 - y_1}{x_2 - x_1} = \frac{100 - 0}{212 - 32} = \frac{100}{180} = \frac{5}{9}$$
$$y - y_1 = m(x - x_1)$$
$$y - 0 = \frac{5}{9}(x - 32)$$
$$y = \frac{5}{9}(x - 32), \text{ or } C = \frac{5}{9}(F - 32).$$

The linear function is $f(F) = \frac{5}{9}(F - 32)$.

SECTION 11.2

You Try It 1

x-coordinate of vertex:

$$-\frac{b}{2a} = -\frac{4}{2(4)} = -\frac{1}{2}$$

y-coordinate of vertex:

$$y = 4x^2 + 4x + 1$$
$$= 4\left(-\frac{1}{2}\right)^2 + 4\left(-\frac{1}{2}\right) + 1$$
$$= 1 - 2 + 1$$
$$= 0$$

vertex: $\left(-\frac{1}{2}, 0\right)$

axis of symmetry: $x = -\frac{1}{2}$

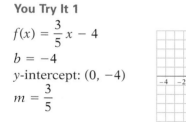

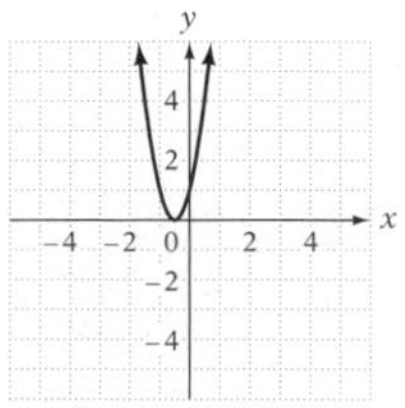

You Try It 2

$$y = x^2 + 3x + 4$$
$$0 = x^2 + 3x + 4$$
$$x = \frac{-b \pm \sqrt{b^2 - 4ac}}{2a}$$
$$= \frac{-3 \pm \sqrt{3^2 - 4(1)(4)}}{2 \cdot 1} \quad \bullet\ a = 1, b = 3, c = 4$$
$$= \frac{-3 \pm \sqrt{-7}}{2}$$
$$= \frac{-3 \pm i\sqrt{7}}{2} = -\frac{3}{2} \pm \frac{\sqrt{7}}{2}i$$

The equation has no real number solutions. There are no x-intercepts.

You Try It 3

$$y = x^2 - x - 6$$
$$a = 1, b = -1, c = -6$$
$$b^2 - 4ac$$
$$(-1)^2 - 4(1)(-6) = 1 + 24 = 25$$

Because the discriminant is greater than zero, the parabola has two x-intercepts.

You Try It 4

$$f(x) = -3x^2 + 4x - 1$$
$$x = -\frac{b}{2a} = -\frac{4}{2(-3)} = \frac{2}{3}$$
$$f(x) = -3x^2 + 4x - 1$$
$$f\left(\frac{2}{3}\right) = -3\left(\frac{2}{3}\right)^2 + 4\left(\frac{2}{3}\right) - 1$$
$$= -\frac{4}{3} + \frac{8}{3} - 1 = \frac{1}{3}$$

Because a is negative, the function has a maximum value.

The maximum value of the function is $\frac{1}{3}$.

You Try It 5

Strategy
- To find the time it takes the ball to reach its maximum height, find the t-coordinate of the vertex.
- To find the maximum height, evaluate the function at the t-coordinate of the vertex.

Solution

$$t = -\frac{b}{2a} = -\frac{64}{2(-16)} = 2$$

The ball reaches its maximum height in 2 s.

$$s(t) = -16t^2 + 64t$$
$$s(2) = -16(2)^2 + 64(2) = -64 + 128 = 64$$

The maximum height is 64 ft.

SECTION 11.3

You Try It 1 Any vertical line intersects the graph only once. The graph is the graph of a function.

You Try It 2 domain: $\{x \mid x \in \text{real numbers}\}$
range: $\{y \mid y \in \text{real numbers}\}$

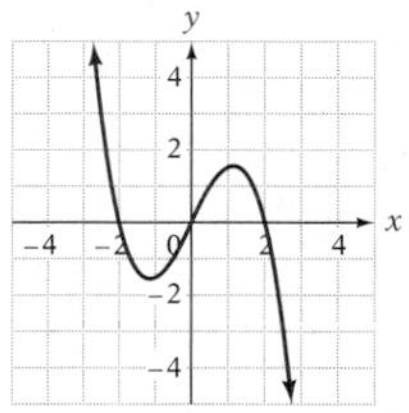

You Try It 3 domain: $\{x \mid x \in \text{real numbers}\}$
range: $\{y \mid y \geq 0\}$

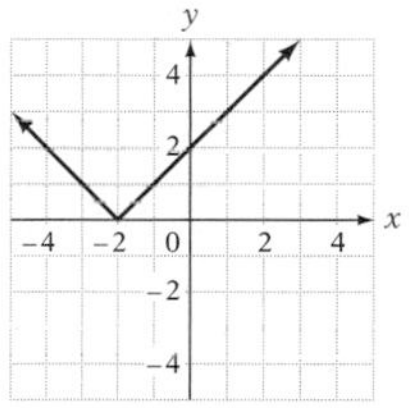

You Try It 4 domain: $\{x \mid x \geq 1\}$
range: $\{y \mid y \leq 0\}$

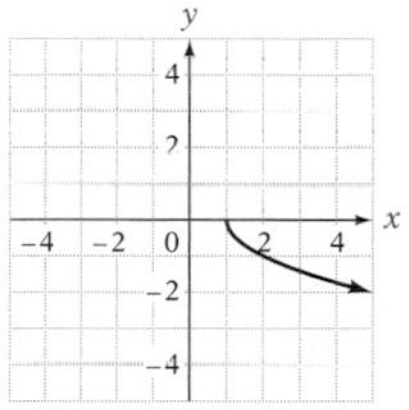

SECTION 11.4

You Try It 1

$$(f + g)(-2) = f(-2) + g(-2)$$
$$= [(-2)^2 + 2(-2)] + [5(-2) - 2]$$
$$= (4 - 4) + (-10 - 2)$$
$$= -12$$
$$(f + g)(-2) = -12$$

You Try It 2

$$(f \cdot g)(3) = f(3) \cdot g(3)$$
$$= (4 - 3^2) \cdot [3(3) - 4]$$
$$= (4 - 9) \cdot (9 - 4)$$
$$= (-5)(5)$$
$$= -25$$
$$(f \cdot g)(3) = -25$$

You Try It 3

$$\left(\frac{f}{g}\right)(4) = \frac{f(4)}{g(4)}$$
$$= \frac{4^2 - 4}{4^2 + 2 \cdot 4 + 1}$$
$$= \frac{16 - 4}{16 + 8 + 1}$$
$$= \frac{12}{25}$$
$$\left(\frac{f}{g}\right)(4) = \frac{12}{25}$$

You Try It 4

$$g(x) = x^2$$
$$g(-1) = (-1)^2 = 1$$

$$f(x) = 1 - 2x$$
$$f[g(-1)] = f(1) = 1 - 2(1) = -1$$

You Try It 5

$$M(s) = s^3 + 1$$
$$M[L(s)] = M(s + 1) = (s + 1)^3 + 1$$
$$= s^3 + 3s^2 + 3s + 1 + 1$$
$$= s^3 + 3s^2 + 3s + 2$$

SECTION 11.5

You Try It 1 Because any horizontal line intersects the graph at most once, the graph is the graph of a 1–1 function.

You Try It 2

$$f(x) = \frac{1}{2}x + 4$$
$$y = \frac{1}{2}x + 4$$
$$x = \frac{1}{2}y + 4$$
$$x - 4 = \frac{1}{2}y$$
$$2x - 8 = y$$
$$f^{-1}(x) = 2x - 8$$

The inverse of the function is given by $f^{-1}(x) = 2x - 8$.

You Try It 3

$$f[g(x)] = 2\left(\frac{1}{2}x - 3\right) - 6$$
$$= x - 6 - 6 = x - 12$$

No, $g(x)$ is not the inverse of $f(x)$.

SECTION 11.6

You Try It 1

$y = x^2 + 2x + 1$

$$-\frac{b}{2a} = -\frac{2}{2(1)} = -1$$

axis of symmetry:
$x = -1$

$$y = (-1)^2 + 2(-1) + 1$$
$$= 0$$

vertex: $(-1, 0)$

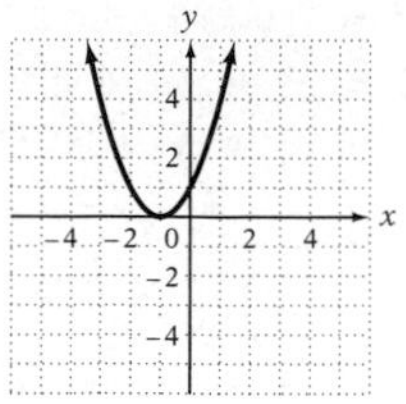

You Try It 2

$x = -y^2 - 2y + 2$

$$-\frac{b}{2a} = -\frac{-2}{2(-1)} = -1$$

axis of symmetry:
$y = -1$

$$x = -(-1)^2 - 2(-1) + 2$$
$$= 3$$

vertex: $(3, -1)$

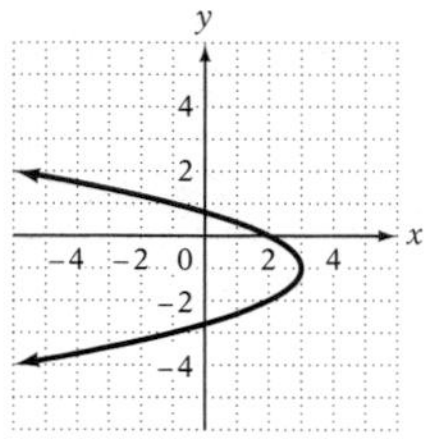

You Try It 3

$y = x^2 - 2x - 1$

$$-\frac{b}{2a} = -\frac{-2}{2(1)} = 1$$

axis of symmetry:
$x = 1$

$$y = 1^2 - 2(1) - 1$$
$$= -2$$

vertex: $(1, -2)$

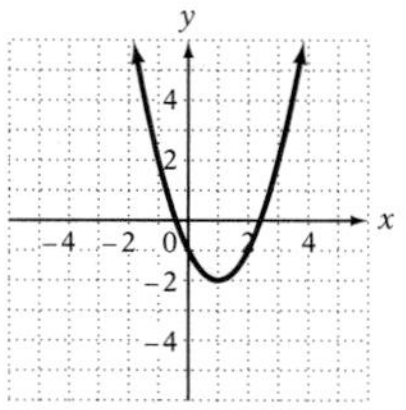

You Try It 4

$$(x - h)^2 + (y - k)^2 = r^2$$
$$(x - 2)^2 + [y - (-3)]^2 = 3^2$$

center: $(h, k) = (2, -3)$
radius: $r = 3$

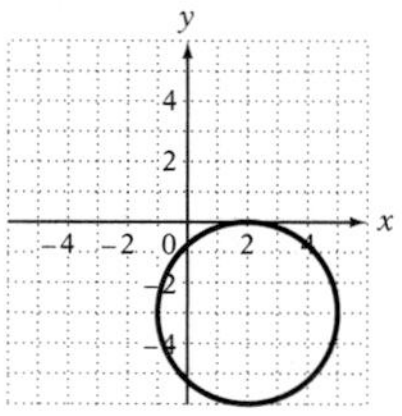

You Try It 5

$$(x - h)^2 + (y - k)^2 = r^2$$
$$(x - 2)^2 + [y - (-3)]^2 = 4^2$$
$$(x - 2)^2 + (y + 3)^2 = 16$$

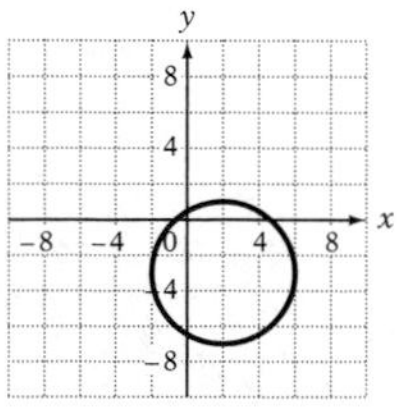

You Try It 6

x-intercepts:
$(2, 0)$ and $(-2, 0)$

y-intercepts:
$(0, 5)$ and $(0, -5)$

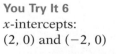

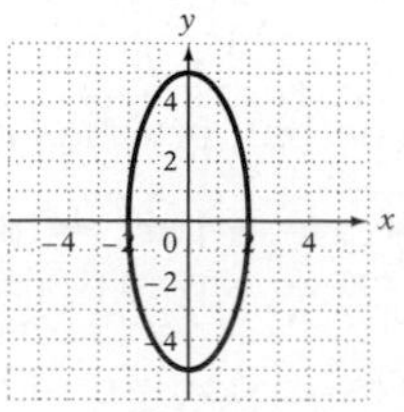

You Try It 7
x-intercepts
$(3\sqrt{2}, 0)$ and $(-3\sqrt{2}, 0)$

y-intercepts:
$(0, 3)$ and $(0, -3)$

$\left(3\sqrt{2} \approx 4\frac{1}{4}\right)$

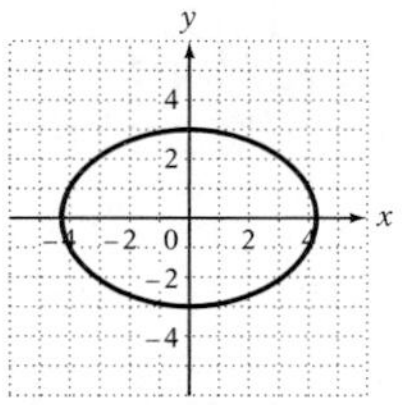

You Try It 8
axis of symmetry:
x-axis

vertices:
$(3, 0)$ and $(-3, 0)$

asymptotes:
$y = \frac{5}{3}x$ and $y = -\frac{5}{3}x$

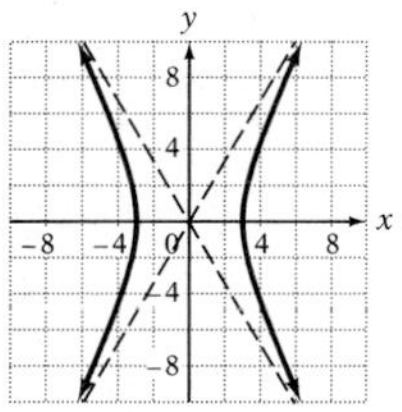

You Try It 9
axis of symmetry:
y-axis

vertices:
$(0, 3)$ and $(0, -3)$

asymptotes:
$y = x$ and $y = -x$

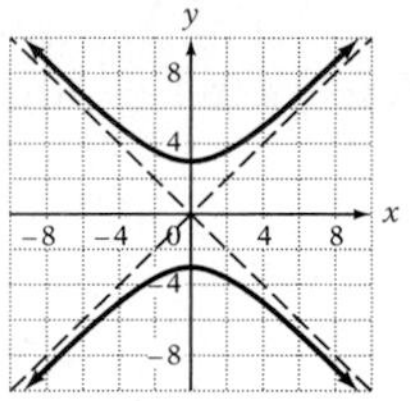

Solutions to Chapter 12 "You Try It"

SECTION 12.1

You Try It 1

$$f(x) = \left(\frac{2}{3}\right)^x$$
$$f(3) = \left(\frac{2}{3}\right)^3 = \frac{8}{27}$$
$$f(-2) = \left(\frac{2}{3}\right)^{-2} = \left(\frac{3}{2}\right)^2 = \frac{9}{4}$$

You Try It 2

$$f(x) = 2^{2x+1}$$
$$f(0) = 2^{2(0)+1} = 2^1 = 2$$
$$f(-2) = 2^{2(-2)+1} = 2^{-3} = \frac{1}{2^3} = \frac{1}{8}$$

You Try It 3

$$f(x) = e^{2x-1}$$
$$f(2) = e^{2 \cdot 2 - 1} = e^3 \approx 20.0855$$
$$f(-2) = e^{2(-2)-1} = e^{-5} \approx 0.0067$$

You Try It 4

x	y
-4	4
-2	2
0	1
2	$\frac{1}{2}$
4	$\frac{1}{4}$

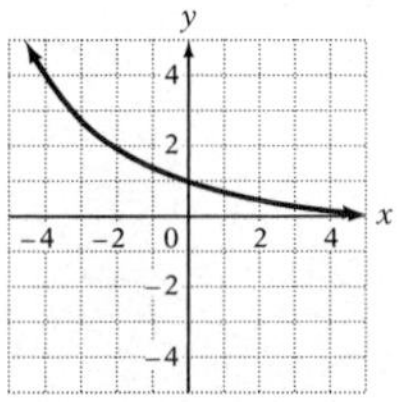

You Try It 5

x	y
-2	$\frac{5}{4}$
-1	$\frac{3}{2}$
0	2
1	3
2	5

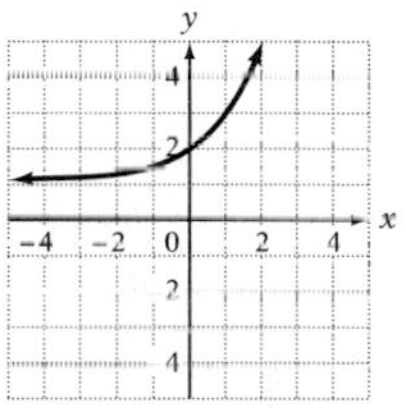

You Try It 6

x	y
-2	6
-1	4
0	3
1	$\frac{5}{2}$
2	$\frac{9}{4}$

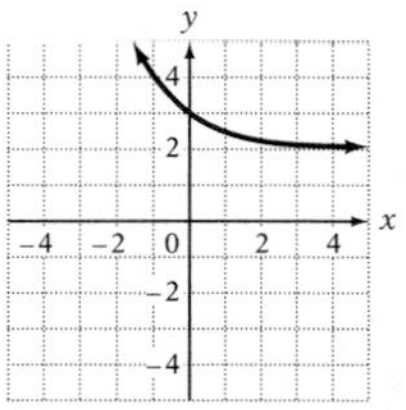

You Try It 7

x	y
-4	$\frac{9}{4}$
-2	$\frac{5}{2}$
0	3
2	4
4	6

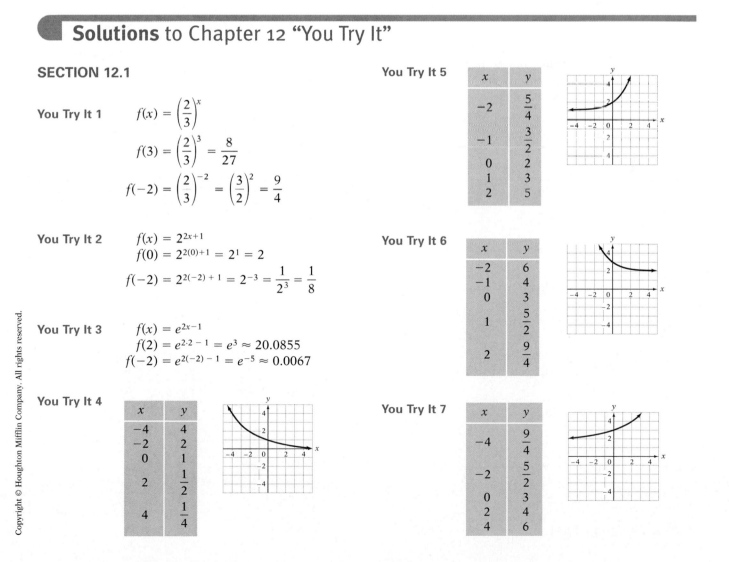

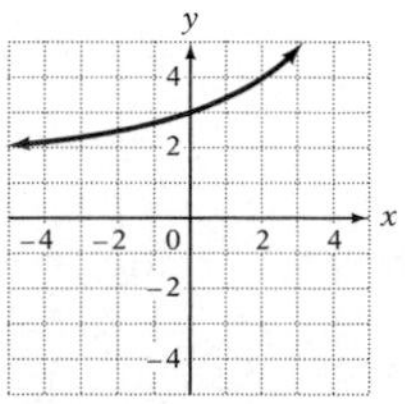

SECTION 12.2

You Try It 1
$$\begin{aligned}\log_4 64 &= x\\ 64 &= 4^x\\ 4^3 &= 4^x\\ 3 &= x\end{aligned}$$
$$\log_4 64 = 3$$

You Try It 2
$$\begin{aligned}\log_2 x &= -4\\ 2^{-4} &= x\\ \frac{1}{2^4} &= x\\ \frac{1}{16} &= x\end{aligned}$$
The solution is $\frac{1}{16}$.

You Try It 3
$$\begin{aligned}\ln x &= 3\\ e^3 &= x\\ 20.0855 &\approx x\end{aligned}$$

You Try It 4
$$\begin{aligned}\log_8 \sqrt[3]{xy^2} &= \log_8 (xy^2)^{1/3} = \frac{1}{3}\log_8 (xy^2)\\ &= \frac{1}{3}(\log_8 x + \log_8 y^2)\\ &= \frac{1}{3}(\log_8 x + 2\log_8 y)\\ &= \frac{1}{3}\log_8 x + \frac{2}{3}\log_8 y\end{aligned}$$

You Try It 5
$$\begin{aligned}&\frac{1}{3}(\log_4 x - 2\log_4 y + \log_4 z)\\ &= \frac{1}{3}(\log_4 x - \log_4 y^2 + \log_4 z)\\ &= \frac{1}{3}\left(\log_4 \frac{x}{y^2} + \log_4 z\right) = \frac{1}{3}\left(\log_4 \frac{xz}{y^2}\right)\\ &= \log_4 \left(\frac{xz}{y^2}\right)^{1/3} = \log_4 \sqrt[3]{\frac{xz}{y^2}}\end{aligned}$$

You Try It 6 Because $\log_b 1 = 0$, $\log_9 1 = 0$

You Try It 7 $\log_7 6.45 = \dfrac{\log 6.45}{\log 7} \approx 0.9579$

You Try It 8 $\log_3 0.834 = \dfrac{\ln 0.834}{\ln 3} \approx -0.1652$

SECTION 12.3

You Try It 1
$$\begin{aligned}f(x) &= \log_2 (x - 1)\\ y &= \log_2 (x - 1)\\ 2^y &= x - 1\\ 2^y + 1 &= x\end{aligned}$$

x	y
$\frac{5}{4}$	-2
$\frac{3}{2}$	-1
2	0
3	1
5	2

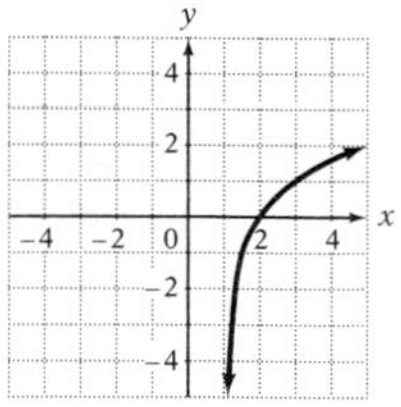

You Try It 2
$$\begin{aligned}f(x) &= \log_3 (2x)\\ y &= \log_3 (2x)\\ 3^y &= 2x\\ \frac{3^y}{2} &= x\end{aligned}$$

x	y
$\frac{1}{18}$	-2
$\frac{1}{6}$	-1
$\frac{1}{2}$	0
$\frac{3}{2}$	1
$\frac{9}{2}$	2

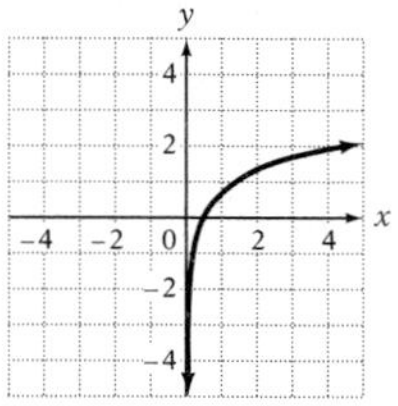

You Try It 3
$$\begin{aligned}f(x) &= -\log_3 (x + 1)\\ y &= -\log_3 (x + 1)\\ -y &= \log_3 (x + 1)\\ 3^{-y} &= x + 1\\ 3^{-y} - 1 &= x\end{aligned}$$

x	y
8	-2
2	-1
0	0
$-\frac{2}{3}$	1
$-\frac{8}{9}$	2

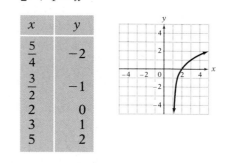

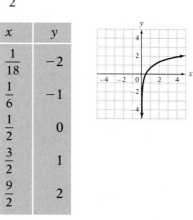

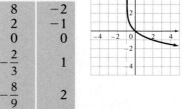

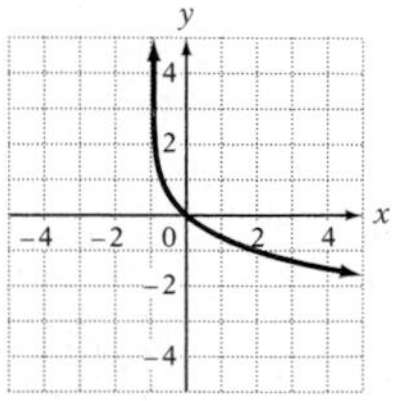

SECTION 12.4

You Try It 1

$$(1.06)^n = 1.5$$
$$\log (1.06)^n = \log 1.5$$
$$n \log 1.06 = \log 1.5$$
$$n = \frac{\log 1.5}{\log 1.06}$$
$$n \approx 6.9585$$

The solution is 6.9585.

You Try It 2

$$4^{3x} = 25$$
$$\log 4^{3x} = \log 25$$
$$3x \log 4 = \log 25$$
$$3x = \frac{\log 25}{\log 4}$$
$$3x \approx 2.3219$$
$$x \approx 0.7740$$

The solution is 0.7740.

You Try It 3

$$\log_4 (x^2 - 3x) = 1$$

Rewrite in exponential form.

$$4^1 = x^2 - 3x$$
$$4 = x^2 - 3x$$
$$0 = x^2 - 3x - 4$$
$$0 = (x + 1)(x - 4)$$

$x + 1 = 0$ $\quad$ $x - 4 = 0$
$x = -1$ $\quad$ $x = 4$

The solutions are -1 and 4.

You Try It 4

$$\log_3 x + \log_3 (x + 3) = \log_3 4$$
$$\log_3 [x(x + 3)] = \log_3 4$$

Use the 1–1 Property of Logarithms.

$$x(x + 3) = 4$$
$$x^2 + 3x = 4$$
$$x^2 + 3x - 4 = 0$$
$$(x + 4)(x - 1) = 0$$

$x + 4 = 0$ $\quad$ $x - 1 = 0$
$x = -4$ $\quad$ $x = 1$

-4 does not check as a solution. The solution is 1.

You Try It 5

$$\log_3 x + \log_3 (x + 6) = 3$$
$$\log_3 [x(x + 6)] = 3$$

$$x(x + 6) = 3^3$$
$$x^2 + 6x = 27$$
$$x^2 + 6x - 27 = 0$$
$$(x + 9)(x - 3) = 0$$

$x + 9 = 0$ $\quad$ $x - 3 = 0$
$x = -9$ $\quad$ $x = 3$

-9 does not check as a solution. The solution is 3.

SECTION 12.5

You Try It 1

Strategy To find the hydrogen ion concentration, replace pH by 2.9 in the equation $pH = -\log (H^+)$ and solve for H^+.

Solution

$$pH = -\log (H^+)$$
$$2.9 = -\log (H^+)$$
$$-2.9 = \log (H^+)$$
$$10^{-2.9} = H^+$$
$$0.00126 \approx H^+$$

The hydrogen ion concentration is approximately 0.00126.

You Try It 2

Strategy To find the intensity, use the equation for the Richter scale magnitude of an earthquake, $M = \log \left(\frac{I}{I_0}\right)$. Replace M by 5.2 and solve for I.

Solution

$$M = \log \left(\frac{I}{I_0}\right)$$
$$5.2 = \log \frac{I}{I_0}$$ • **Replace *M* by 5.2.**
$$10^{5.2} = \frac{I}{I_0}$$ • **Write in exponential form.**
$$10^{5.2} I_0 = I$$
$$158{,}489 I_0 \approx I$$

The earthquake's intensity was approximately 158,489 times that of a zero-level earthquake.

Solutions to Chapter R "You Try It"

SECTION R.1

You Try It 1

$$\begin{aligned}(-4)(6-8)^2-(-12\div 4) &= -4(-2)^2-(-3)\\ &= -4(4)-(-3)\\ &= -16-(-3)\\ &= -16+3\\ &= -13\end{aligned}$$

You Try It 2

$3xy^2 - 3x^2y$

$$\begin{aligned}3(-2)(5)^2-3(-2)^2(5) &= 3(-2)(25)-3(4)(5)\\ &= -6(25)-3(4)(5)\\ &= -150-3(4)(5)\\ &= -150-12(5)\\ &= -150-60\\ &= -150+(-60)\\ &= -210\end{aligned}$$

You Try It 3 $\quad$ $$\begin{aligned}-5(-3a) &= [-5(-3)]a\\ &= 15a\end{aligned}$$

You Try It 4

$$\begin{aligned}2z^2-5z-3z^2+6z &= 2z^2-3z^2-5z+6z\\ &= (2z^2-3z^2)+(-5z+6z)\\ &= -1z^2+z\\ &= -z^2+z\end{aligned}$$

You Try It 5

a. $$\begin{aligned}-3(5y-2) &= -3(5y)-(-3)(2)\\ &= -15y+6\end{aligned}$$

b.

$$\begin{aligned}-2(4x+2y-6z) &= -2(4x)+(-2)(2y)-(-2)(6z)\\ &= -8x-4y+12z\end{aligned}$$

You Try It 6

a.

$$\begin{aligned}7(-3x-4y)-3(3x+y) &= -21x-28y-9x-3y\\ &= -30x-31y\end{aligned}$$

b. $$\begin{aligned}2y-3[5-3(3+2y)] &= 2y-3[5-9-6y]\\ &= 2y-3[-4-6y]\\ &= 2y+12+18y\\ &= 20y+12\end{aligned}$$

SECTION R.2

You Try It 1 $\quad$ **a.**

$$\begin{aligned}4x+3 &= 7x+9\\ 4x-7x+3 &= 7x-7x+9\\ -3x+3 &= 9\\ -3x+3-3 &= 9-3\\ -3x &= 6\\ \frac{-3x}{-3} &= \frac{6}{-3}\\ x &= -2\end{aligned}$$

The solution is -2.

b.

$$\begin{aligned}4-(5x-8) &= 4x+3\\ 4-5x+8 &= 4x+3\\ -5x+12 &= 4x+3\\ -5x-4x+12 &= 4x-4x+3\\ -9x+12 &= 3\\ -9x+12-12 &= 3-12\\ -9x &= -9\\ \frac{-9x}{-9} &= \frac{-9}{-9}\\ x &= 1\end{aligned}$$

The solution is 1.

You Try It 2

$$\begin{aligned}3x-1 &\le 5x-7\\ 3x-5x-1 &\le 5x-5x-7\\ -2x-1 &\le -7\\ -2x-1+1 &\le -7+1\\ -2x &\le -6\\ \frac{-2x}{-2} &\ge \frac{-6}{-2}\\ x &\ge 3\end{aligned}$$

$\{x \mid x \ge 3\}$

You Try It 3

$$\begin{aligned}3-2(3x+1) &< 7-2x\\ 3-6x-2 &< 7-2x\\ 1-6x &< 7-2x\\ 1-6x+2x &< 7-2x+2x\\ 1-4x &< 7\\ 1-1-4x &< 7-1\\ -4x &< 6\\ \frac{-4x}{-4} &> \frac{6}{-4}\\ x &> -\frac{3}{2}\end{aligned}$$

$\left\{x \,\middle|\, x > -\frac{3}{2}\right\}$

SECTION R.3

You Try It 1

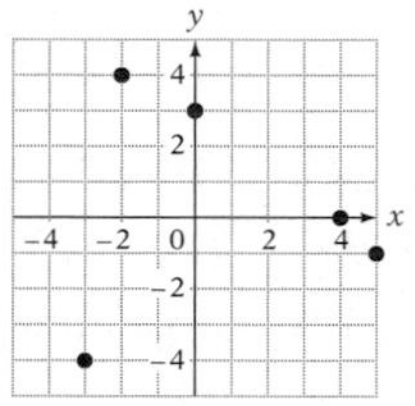

You Try It 2 $y = \frac{3}{5}x - 4$

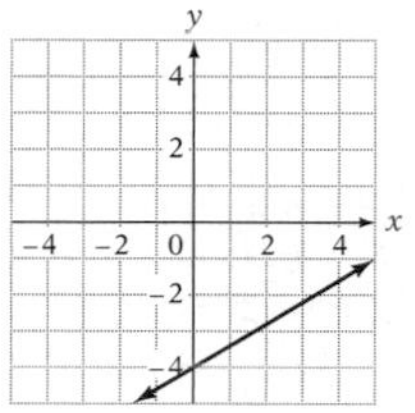

You Try It 3

$$\begin{aligned} -3x + 2y &= 4 \\ 2y &= 3x + 4 \\ y &= \frac{3}{2}x + 2 \end{aligned}$$

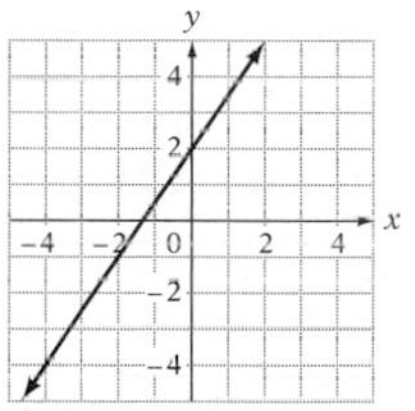

You Try It 4 Let $(x_1, y_1) = (-2, 3)$ and $(x_2, y_2) = (1, -3)$.

$$m = \frac{y_2 - y_1}{x_2 - x_1} = \frac{-3 - 3}{1 - (-2)} = \frac{-6}{3} = -2$$

The slope is -2.

You Try It 5 $y = -\frac{2}{3}x + 2$

$$m = -\frac{2}{3} = \frac{-2}{3}$$

y-intercept = (0, 2)

Place a dot at the y-intercept.
Starting at the y-intercept, move to the right 3 units (the change in x) and down 2 units (the change in y).
Place a dot at that location.
Draw a line through the two points.

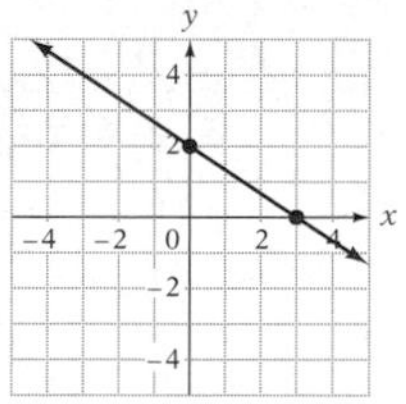

You Try It 6

$$\begin{aligned} f(x) &= 4 - 2x \\ f(-3) &= 4 - 2(-3) \\ f(-3) &= 4 + 6 \\ f(-3) &= 10 \end{aligned}$$

An ordered pair of the function is $(-3, 10)$.

You Try It 7 $(x_1, y_1) = (-2, 2),\ m = -\frac{1}{2}$

$$\begin{aligned} y - y_1 &= m(x - x_1) \\ y - 2 &= -\frac{1}{2}[x - (-2)] \\ y - 2 &= -\frac{1}{2}(x + 2) \\ y - 2 &= -\frac{1}{2}x - 1 \\ y &= -\frac{1}{2}x + 1 \end{aligned}$$

SECTION R.4

You Try It 1

a. $$\begin{aligned} (-2ab)(2a^3b^{-2})^{-3} &= (-2ab)(2^{-3}a^{-9}b^6) \\ &= -(2^{-2})a^{-8}b^7 \\ &= -\frac{b^7}{2^2a^8} \\ &= -\frac{b^7}{4a^8} \end{aligned}$$

b. $$\begin{aligned} \left(\frac{2x^2y^{-4}}{4x^{-2}y^{-5}}\right)^{-3} &= \left(\frac{x^2y^{-4}}{2x^{-2}y^{-5}}\right)^{-3} \\ &= \frac{x^{2(-3)}y^{(-4)(-3)}}{2^{1(-3)}x^{(-2)(-3)}y^{(-5)(-3)}} \\ &= \frac{x^{-6}y^{12}}{2^{-3}x^6y^{15}} \\ &= 2^3x^{-6-6}y^{12-15} \\ &= 2^3x^{-12}y^{-3} \\ &= \frac{8}{x^{12}y^3} \end{aligned}$$

You Try It 2

$$\begin{aligned} &(-4x^3 + 2x^2 - 8) + (4x^3 + 6x^2 - 7x + 5) \\ &= (-4x^3 + 4x^3) + (2x^2 + 6x^2) + (-7x) + (-8 + 5) \\ &= 8x^2 - 7x - 3 \end{aligned}$$

You Try It 3

$$\begin{aligned}&(-4w^3 + 8w - 8) - (3w^3 - 4w^2 + 2w - 1)\\&= (-4w^3 + 8w - 8) + (-3w^3 + 4w^2 - 2w + 1)\\&= (-4w^3 - 3w^3) + 4w^2 + (8w - 2w) + (-8 + 1)\\&= -7w^3 + 4w^2 + 6w - 7\end{aligned}$$

You Try It 4

$$\begin{aligned}&3mn^2(2m^2 - 3mn - 1)\\&= 3mn^2(2m^2) - 3mn^2(3mn) - 3mn^2(1)\\&= 6m^3n^2 - 9m^2n^3 - 3mn^2\end{aligned}$$

You Try It 5

$$\begin{array}{r}3c^2 - 4c + 5\\ 2c - 3\\ \hline -9c^2 + 12c - 15\\ 6c^3 - 8c^2 + 10c\qquad\\ \hline 6c^3 - 17c^2 + 22c - 15\end{array}$$

You Try It 6

$$\begin{aligned}&(4y - 7)(3y - 5)\\&= (4y)(3y) + (4y)(-5) + (-7)(3y) + (-7)(-5)\\&= 12y^2 - 20y - 21y + 35\\&= 12y^2 - 41y + 35\end{aligned}$$

You Try It 7

$(x^3 - 7 - 2x) \div (x - 2)$

$$\begin{array}{r}x^2 + 2x + 2\\ x - 2\,\overline{)\,x^3 + 0x^2 - 2x - 7}\\ \underline{x^3 - 2x^2}\qquad\qquad\\ 2x^2 - 2x\qquad\\ \underline{2x^2 - 4x}\qquad\\ 2x - 7\\ \underline{2x - 4}\\ -3\end{array}$$

Check: $(x - 2)(x^2 + 2x + 2) - 3$
$= x^3 - 2x - 4 - 3$
$= x^3 - 2x - 7$

$$(x^3 - 7 - 2x) \div (x - 2) = x^2 + 2x + 2 - \frac{3}{x - 2}$$

You Try It 8 The GCF is $3x^2y^2$.

$$\begin{aligned}&6x^4y^2 - 9x^3y^2 + 12x^2y^4\\&= 3x^2y^2(2x^2 - 3x + 4y^2)\end{aligned}$$

You Try It 9 Two factors of 15 whose sum is -8 are -3 and -5.

$x^2 - 8x + 15 = (x - 3)(x - 5)$

You Try It 10

$3x^2 - x - 2$

Positive Factors of 3	*Factors of −2*
1, 3	1, −2 −1, 2

Trial Factors	*Middle Term*
$(x + 1)(3x - 2)$	$-2x + 3x = x$
$(x - 2)(3x + 1)$	$x - 6x = -5x$
$(x - 1)(3x + 2)$	$2x - 3x = -x$
$(x + 2)(3x - 1)$	$-x + 6x = 5x$

$3x^2 - x - 2 = (x - 1)(3x + 2)$

Check: $(x - 1)(3x + 2) = 3x^2 + 2x - 3x - 2$
$= 3x^2 - x - 2$

You Try It 11

$4a^3 - 4a^2 - 24a$

There is a common factor, $4a$. Factor out the GCF.

$4a^3 - 4a^2 - 24a = 4a(a^2 - a - 6)$

Factor $a^2 - a - 6$. The two factors of -6 whose sum is -1 are 2 and -3.

$4a(a^2 - a - 6) = 4a(a + 2)(a - 3)$

$4a^3 - 4a^2 - 24a = 4a(a + 2)(a - 3)$

Check: $4a(a + 2)(a - 3) = (4a^2 + 8a)(a - 3)$
$= 4a^3 - 12a^2 + 8a^2 - 24a$
$= 4a^3 - 4a^2 - 24a$

Answers to Chapter 1 Selected Exercises

PREP TEST

1. 127.16 **2.** 55.107 **3.** 4517 **4.** 11,396 **5.** 24 **6.** 24 **7.** 4 **8.** $3 \cdot 7$ **9.** $\frac{2}{5}$

SECTION 1.1

1. $8 > -6$ **3.** $-12 < 1$ **5.** $42 > 19$ **7.** $0 > -31$ **9.** $53 > -46$ **11.** false **13.** true **15.** false **17.** true **19.** false **21.** $-23, -18$ **23.** 21, 37 **25.** -23 **27.** -4 **29.** 9 **31.** 28 **33.** 14 **35.** -77 **37.** 0 **39.** 74 **41.** -82 **43.** -81 **45.** $|-83| > |58|$ **47.** $|43| < |-52|$ **49.** $|-68| > |-42|$ **51.** $|-45| < |-61|$ **53.** 19, 0, -28 **55.** -45, 0, -17 **59.** -11 **61.** -5 **63.** -83 **65.** -46 **67.** 0 **69.** -5 **71.** 9 **73.** 1 **75.** 8 **77.** -7 **79.** -9 **81.** 9 **83.** -3 **85.** 18 **87.** -10 **89.** -41 **91.** -12 **93.** 0 **95.** -9 **97.** 11 **99.** -18 **101.** 0 **103.** 2 **105.** -138 **107.** -8 **109.** -12 **111.** -20 **115.** 42 **117.** -28 **119.** 60 **121.** -253 **123.** -238 **125.** -114 **127.** -2 **129.** 8 **131.** -7 **133.** -12 **135.** -6 **137.** -7 **139.** 11 **141.** -14 **143.** 15 **145.** -16 **147.** 0 **149.** -29 **151.** undefined **153.** -11 **155.** undefined **157.** -105 **159.** 252 **161.** -240 **163.** 96 **165.** -216 **167.** -315 **169.** 420 **171.** 2880 **173.** -2772 **175.** 0 **177.** The difference in elevation is 7046 m. **179.** The difference between the highest and lowest elevations is greatest in Asia. **181.** The difference is 5°C. **183.** The student's score is 93 points. **185.** The difference between the average temperatures is 0°. **187.** The difference is 29°F.

SECTION 1.2

1. 0.125 **3.** $0.\overline{2}$ **5.** $0.1\overline{6}$ **7.** 0.5625 **9.** $0.58\overline{3}$ **11.** 0.24 **13.** 0.225 **15.** $0.\overline{45}$ **19.** $\frac{2}{5}$, 0.40 **21.** $\frac{22}{25}$, 0.88 **23.** $\frac{8}{5}$, 1.6 **25.** $\frac{87}{100}$, 0.87 **27.** $\frac{9}{2}$, 4.50 **29.** $\frac{3}{70}$ **31.** $\frac{3}{8}$ **33.** $\frac{1}{400}$ **35.** $\frac{1}{16}$ **37.** $\frac{23}{400}$ **39.** 0.091 **41.** 0.167 **43.** 0.009 **45.** 0.0915 **47.** 0.1823 **49.** 37% **51.** 2% **53.** 12.5% **55.** 136% **57.** 0.4% **59.** 83% **61.** $37\frac{1}{2}\%$ **63.** $44\frac{4}{9}\%$ **65.** 45% **67.** 250% **69.** $\frac{5}{26}$ **71.** $\frac{11}{8}$ **73.** $\frac{1}{12}$ **75.** $\frac{7}{24}$ **77.** 0 **79.** $\frac{3}{8}$ **81.** $-\frac{7}{60}$ **83.** $-\frac{1}{16}$ **85.** -1.06 **87.** -23.845 **89.** -10.7893 **91.** -37.19 **93.** 17.5 **95.** 19.61 **97.** $-\frac{3}{8}$ **99.** $\frac{1}{10}$ **101.** $-\frac{4}{9}$ **103.** $-\frac{7}{30}$ **105.** $\frac{15}{64}$ **107.** $-\frac{10}{9}$ **109.** $-\frac{147}{32}$ **111.** $\frac{25}{8}$ **113.** $\frac{2}{3}$ **115.** 4.164 **117.** 4.347 **119.** -4.028 **121.** -2.22 **123.** -1.104 **125.** -2.59 **127.** -5.11 **129.** -2060.55 **131.** 2401 **133.** -64 **135.** -8 **137.** -125 **139.** $-\frac{27}{64}$ **141.** 3.375 **143.** -1 **145.** -8 **147.** -6750 **149.** -144 **151.** -18 **153.** 4 **155.** 7 **157.** $4\sqrt{2}$ **159.** $2\sqrt{2}$ **161.** $18\sqrt{2}$ **163.** $10\sqrt{10}$ **165.** $\sqrt{15}$ **167.** $\sqrt{29}$ **169.** $-54\sqrt{2}$ **171.** $3\sqrt{5}$ **173.** 0 **175.** $48\sqrt{2}$ **177.** 15.492 **179.** 16.971 **181.** 16 **183.** 16.583 **185.** 15.652 **187.** 18.762 **189a.** The 2001 annual net income would be $-$\$103.408 million. **b.** The average monthly net income was $-$\$43.584 million. **c.** The difference was \$30.481 million. **191.** 31% of the owners responded that the Internet had helped business. **193.** 3, 4, 5, 6, 7, 8, 9

SECTION 1.3

3. 0 **5.** -11 **7.** 20 **9.** -10 **11.** 20 **13.** 29 **15.** 11 **17.** 7 **19.** -11 **21.** 6 **23.** 15 **25.** 4 **27.** 5 **29.** -1 **31.** 4 **33.** 0.51 **35.** 1.7 **37.** Row 1: $-\frac{1}{6}$, 0; Row 2: $-\frac{1}{2}$; Row 3: $\frac{1}{3}, \frac{1}{2}$ **39.** Your savings on gasoline would pay for the increased cost of the car in 32 to 65 months.

SECTION 1.4

1. -9 **3.** 41 **5.** -7 **7.** 13 **9.** -15 **11.** 41 **13.** 1 **15.** 5 **17.** 1 **19.** 57 **21.** 5 **23.** 8

25. -3 **27.** -2 **29.** -4 **31.** 10 **33.** -25 **35.** $25x$ **37.** $9a$ **39.** $2y$ **41.** $-12y - 3$ **43.** $9a$
45. $6ab$ **47.** $-12xy$ **49.** 0 **51.** $-\frac{1}{10}y$ **53.** $\frac{2}{9}y^2$ **55.** $20x$ **57.** $-4a$ **59.** $-2y^2$ **61.** $-2x + 8y$ **63.** $8x$
65. $19a - 12b$ **67.** $-12x - 2y$ **69.** $-7x^2 - 5x$ **71.** $60x$ **73.** $-10a$ **75.** $30y$ **77.** $72x$ **79.** $-28a$
81. $108b$ **83.** $-56x^2$ **85.** x^2 **87.** x **89.** a **91.** b **93.** x **95.** n **97.** $2x$ **99.** $-2x$ **101.** $-15a^2$
103. $6y$ **105.** $3y$ **107.** $-2x$ **109.** $-9y$ **111.** $-x - 7$ **113.** $10x - 35$ **115.** $-5a - 80$ **117.** $-15y + 35$
119. $20 - 14b$ **121.** $-4x + 2y$ **123.** $18x^2 + 12x$ **125.** $10x - 35$ **127.** $-14x + 49$ **129.** $-30x^2 - 15$
131. $-24y^2 + 96$ **133.** $5x^2 + 5y^2$ **135.** $-\frac{1}{2}x + 2y$ **137.** $3x^2 + 6x - 18$ **139.** $-2y^2 + 4y - 8$
141. $-2x + 3y - \frac{1}{3}$ **143.** $10x^2 + 15x - 35$ **145.** $6x^2 + 3xy - 9y^2$ **147.** $-3a^2 - 5a + 4$ **149.** $-2x - 16$
151. $-12y - 9$ **153.** $7n - 7$ **155.** $-2x + 41$ **157.** $3y - 3$ **159.** $2a - 4b$ **161.** $-4x + 24$ **163.** $-2x - 16$
165. $-3x + 21$ **167.** $-4x + 12$ **169.** $-x + 50$ **171.** $\frac{x}{18}$ **173.** $x + 20$ **175.** $10(x - 50)$; $10x - 500$
177. $\frac{5}{8}x + 6$ **179.** $x - (x + 3)$; -3 **181.** $4(x + 19)$; $4x + 76$ **183.** $\frac{15}{x + 12}$ **185.** $\frac{2}{3}(x + 7)$; $\frac{2}{3}x + \frac{14}{3}$
187. $40 - \frac{x}{20}$ **189.** $x^2 + 2x$ **191.** $(x + 8) + \frac{1}{3}x$; $\frac{4}{3}x + 8$ **193.** $x + (x + 9)$; $2x + 9$ **195.** $x - (8 - x)$; $2x - 8$
197. $\frac{1}{3}x - \frac{5}{8}x$; $-\frac{7}{24}x$ **199.** $(x + 5) + 2$; $x + 7$ **201.** $2(6x + 7)$; $12x + 14$ **203.** Let t be the number of tornadoes in 2000; $t + 235$ **205.** Let g be the amount of oil in one container: g, $20 - g$ **207.** Let p be the pounds of pecans produced in Texas: $\frac{1}{2}p$ **209.** Let d be the diameter of a baseball: $4d$ **211.** yes **213.** $2x$ **215.** $\frac{1}{4}x$ **217.** $\frac{3}{5}x$

SECTION 1.5

1. $A = \{16, 17, 18, 19, 20, 21\}$ **3.** $A = \{9, 11, 13, 15, 17\}$ **5.** $A = \{b, c\}$ **9.** $A \cup B = \{3, 4, 5, 6\}$
11. $A \cup B = \{-10, -9, -8, 8, 9, 10\}$ **13.** $A \cup B = \{a, b, c, d, e, f\}$ **15.** $A \cup B = \{1, 3, 7, 9, 11, 13\}$
17. $A \cap B = \{4, 5\}$ **19.** $A \cap B = \varnothing$ **21.** $A \cap B = \{c, d, e\}$ **23.** $\{x \mid x > -5, x \in \text{negative integers}\}$
25. $\{x \mid x > 30, x \in \text{integers}\}$ **27.** $\{x \mid x > 5, x \in \text{even integers}\}$ **29.** $\{x \mid x > 8, x \in \text{real numbers}\}$

31. 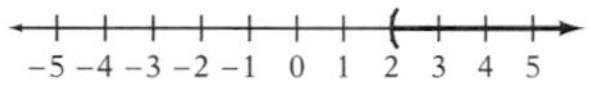

33. 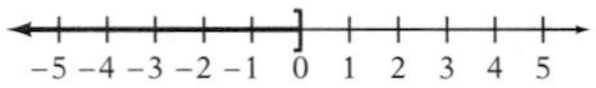

35. 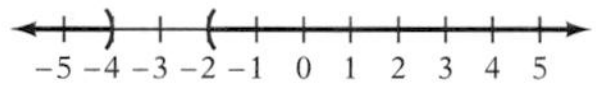

37. 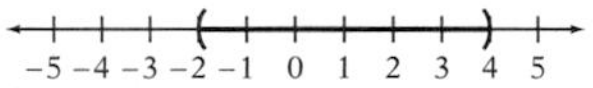

39. 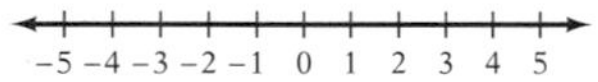

41a. never true **b.** always true **c.** always true **43a.** yes **b.** yes

CHAPTER REVIEW*

1. $-4, 0$ [1.1A] **2.** 4 [1.1B] **3.** -5 [1.1B] **4.** -13 [1.1C] **5.** 1 [1.1C] **6.** -42 [1.1D]
7. -20 [1.1D] **8.** 0.28 [1.2A] **9.** 0.062 [1.2B] **10.** 62.5% [1.2B] **11.** $\frac{7}{12}$ [1.2C] **12.** -1.068 [1.2C]
13. $-\frac{72}{85}$ [1.2D] **14.** -4.6224 [1.2D] **15.** $\frac{16}{81}$ [1.2E] **16.** 12 [1.2F] **17.** $-6\sqrt{30}$ [1.2F]
18. 31 [1.3A] **19.** 29 [1.4A] **20.** $8a - 4b$ [1.4B] **21.** $36y$ [1.4C] **22.** $10x - 35$ [1.4D]
23. $7x + 46$ [1.4D] **24.** $-90x + 25$ [1.4D] **25.** $\{1, 3, 5, 7\}$ [1.5A] **26.** $A \cap B = \{1, 5, 9\}$ [1.5A]
27. 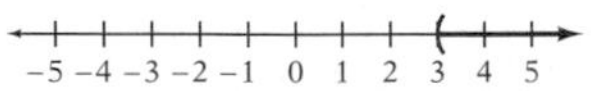

[1.5C] **28.** 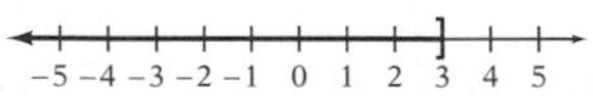

[1.5C] **29.** The student's score was 98. [1.1E] **30.** 50.8% of the candy consumed was chocolate. [1.2G] **31.** $2x - \frac{1}{2}x$; $\frac{3}{2}x$ [1.4E]

*The numbers in brackets following the answers in the Chapter Review and Chapter Test are a reference to the objective that corresponds to that problem. For example, the reference [1.2A] stands for Section 1.2, Objective A. This notation will be used for all Prep Tests, Chapter Reviews, Chapter Tests, and Cumulative Reviews throughout the text.

32. Let A be the number of American League playing cards; $5A$ [1.4E] **33.** Let T be the number of ten-dollar bills; $35 - T$ [1.4E]

CHAPTER TEST*

1. $-2 > -40$ [1.1A] **2.** 4 [1.1B] **3.** −4 [1.1B] **4.** −14 [1.1C] **5.** −16 [1.1C] **6.** 4 [1.1C] **7.** 17 [1.1D] **8.** $0.\overline{7}$ [1.2A] **9.** $\frac{9}{20}$, 0.45 [1.2B] **10.** $\frac{1}{15}$ [1.2C] **11.** −5.3578 [1.2D] **12.** $-\frac{1}{2}$ [1.2D] **13.** 12 [1.2E] **14.** $-6\sqrt{5}$ [1.2F] **15.** 17 [1.3A] **16.** 22 [1.4A] **17.** $5x$ [1.4B] **18.** $2x$ [1.4C] **19.** $-6x^2 + 21y^2$ [1.4D] **20.** $-x + 6$ [1.4D] **21.** $-7x + 33$ [1.4D] **22.** $\{-2, -1, 0, 1, 2, 3\}$ [1.5A] **23.** $\{x | x < -3, x \in \text{real numbers}\}$ [1.5B] **24.** $A \cup B = \{1, 2, 3, 4, 5, 6, 7, 8\}$ [1.5A]
25. 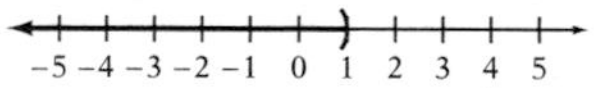

[1.5C] **26.** 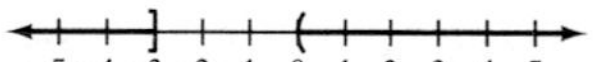

[1.5C]
27. $10(x - 3)$; $10x - 30$ [1.4E] **28.** Let s be the speed of the catcher's return throw; $2s$ [1.4E]
29a. The balance of trade increased from the previous year in 1981, 1988, 1989, 1990, 1991, and 1995. [1.2G]
b. The difference was \$288.6 billion. [1.2G] **c.** The difference was greatest between 1999 and 2000. [1.2G]
d. The trade balance was approximately 4 times greater in 1990 than in 1980. [1.2G] **e.** The average trade balance per quarter for the year 2000 was −\$92.425 billion. [1.2G] **30.** The difference is 395.45°C. [1.2G]

Answers to Chapter 2 Selected Exercises

PREP TEST

1. −4 [1.1C] **2.** −6 [1.1C] **3.** 3 [1.1D] **4.** 1 [1.2D] **5.** $-\frac{1}{2}$ [1.2D] **6.** $10x - 5$ [1.4B] **7.** $6x - 9$ [1.4D] **8.** $3n + 6$ [1.4B] **9.** $0.03x + 20$ [1.4D] **10.** $20 - n$ [1.4E]

SECTION 2.1

3. yes **5.** no **7.** yes **9.** no **11.** no **13.** yes **15.** yes **17.** no **21.** 6 **23.** 16 **25.** 7 **27.** 0 **29.** 3 **31.** −10 **33.** −3 **35.** −14 **37.** 2 **39.** 11 **41.** −9 **43.** −1 **45.** −14 **47.** −5 **49.** $-\frac{1}{2}$ **51.** $-\frac{3}{4}$ **53.** $\frac{1}{12}$ **55.** $-\frac{7}{12}$ **57.** 0.6529 **59.** −0.283 **61.** 9.257 **65.** −3 **67.** 0 **69.** −2 **71.** 9 **73.** 80 **75.** −4 **77.** 0 **79.** 8 **81.** −7 **83.** 12 **85.** −18 **87.** 15 **89.** −20 **91.** 0 **93.** 15 **95.** 75 **97.** $\frac{8}{3}$ **99.** $\frac{1}{3}$ **101.** $-\frac{1}{2}$ **103.** $-\frac{3}{2}$ **105.** $\frac{15}{7}$ **107.** 4 **109.** 3 **111.** 4.745 **113.** 2.06 **115.** −2.13 **119.** 28 **121.** 0.72 **123.** 64 **125.** 24% **127.** 7.2 **129.** 400 **131.** 200% **133.** 400 **135.** 7.7 **137.** 200 **139.** 400 **141.** 20 **143.** 80.34% **145.** 19% of the students are in the fine arts college. **147.** Approximately 46.8% of the deaths were due to traffic accidents. **149.** Approximately 1.7% of the coupons were redeemed by customers. **151.** Western Europe's share is \$675 billion more than Latin America's share. **153.** Approximately 11.2% of the children were enrolled in private school. **155.** $x = \frac{b}{a}, a \neq 0$ **157a.** $\frac{5}{3}$ **b.** −1 **c.** $\frac{6}{5}$ **159.** The new value is 2 times the original value.

SECTION 2.2

1. 3 **3.** 6 **5.** −1 **7.** −3 **9.** 2 **11.** 2 **13.** 5 **15.** −3 **17.** 6 **19.** 3 **21.** 1 **23.** 6 **25.** −7 **27.** 0 **29.** $\frac{3}{4}$ **31.** $\frac{4}{9}$ **33.** $\frac{1}{3}$ **35.** $-\frac{1}{2}$ **37.** $-\frac{3}{4}$ **39.** $\frac{1}{3}$ **41.** $-\frac{1}{6}$ **43.** 1 **45.** 1 **47.** 0 **49.** $\frac{13}{10}$ **51.** $\frac{2}{5}$ **53.** $-\frac{4}{3}$ **55.** $-\frac{3}{2}$ **57.** 18 **59.** 8 **61.** −16 **63.** 25 **65.** 21 **67.** 15 **69.** −16 **71.** −21

73. $\frac{15}{2}$ **75.** $-\frac{18}{5}$ **77.** 2 **79.** 3 **81.** 1 **83.** −2 **85.** 2 **87.** 3 **89.** −1 **91.** 2 **93.** −2 **95.** −3 **97.** 0 **99.** −1 **101.** −3 **103.** −1 **105.** 4 **107.** $\frac{2}{3}$ **109.** $\frac{5}{6}$ **111.** $\frac{3}{4}$ **113.** 1 **115.** 4 **117.** −1 **119.** −1 **121.** $-\frac{2}{3}$ **123.** $\frac{4}{3}$ **125.** $\frac{1}{2}$ **127.** $-\frac{1}{3}$ **129.** $\frac{10}{3}$ **131.** $-\frac{1}{4}$ **133.** $x + 12 = 20; 8$ **135.** $\frac{3}{5}x = -30; -50$ **137.** $3x + 4 = 13; 3$ **139.** $9x - 6 = 12; 2$ **141.** $x + 2x = 9; 3$ **143.** $5x - 17 = 2; \frac{19}{5}$ **145.** $6x + 7 = 3x - 8; -5$ **147.** $30 = 7x - 9; \frac{39}{7}$ **149.** $2x = (21 - x) + 3; 8, 13$ **151.** $23 - x = 2x + 5; 6, 17$ **153.** The approximate length of the humerus is 28.5 in. **155.** A car will slide 136 ft on black ice when the outside temperature is −11°C. **157.** −14 **159.** *b* **161.** no solution **163.** −21

SECTION 2.3

1. 20 oz of herbs should be used. **3.** The cost of the mixture is $3.44 per pound. **5.** 3 lb of caramel is needed. **7.** 2 c of olive oil and 8 c of vinegar are used. **9.** The cost is $3.00 per ounce. **11.** 16 oz of the $400 alloy should be used. **13.** 37 lb of almonds and 63 lb of walnuts were used. **15.** 228 adult tickets were sold. **17.** The cost is $.70 per pound. **19.** The resulting gold alloy is 24% gold. **21.** 20 gal of the 15% acid solution are used. **23.** 30 lb of the yarn that is 25% wool is used. **25.** 6.25 gal of the plant food that is 9% nitrogen is used. **27.** The resulting mixture is 19% sugar. **29.** 20 lb of the coffee that is 40% java beans are used. **31.** 100 ml of the 7% solution and 200 ml of the 4% solution are used. **33.** 150 oz of pure chocolate must be added. **35.** The resulting alloy is 50% silver. **37.** There must be an additional $5000 added. **39.** There was $9000 invested at 7% and $6000 at 6.5%. **41.** There was $2500 deposited in the mutual fund. **43.** The university deposited $200,000 at 10% and $100,000 at 8.5%. **45.** The mechanic must invest $3000 in additional bonds. **47.** $40,500 was invested at 8%. $13,500 was invested at 12%. **49.** The total amount invested was $650,000. **51.** The total amount invested was $500,000. **53.** The rate of the first plane is 105 mph. The rate of the second plane is 130 mph. **55.** They will be 3000 km apart at 11 A.M. **57.** The cabin cruiser will be alongside the motorboat 2 h after the cabin cruiser leaves. **59.** The distance from the airport to the corporate offices is 120 mi. **61.** The rate of the car is 68 mph. **63.** The distance between the two airports is 300 mi. **65.** The two planes will pass each other 2.5 h after the plane leaves Seattle. **67.** They will meet 1.5 h after they begin. **69.** The bus overtakes the car 180 mi from the starting point. **71.** 3.75 gal must be drained from the radiator and replaced by pure antifreeze. **73.** The cyclist's average speed for the trip was $13\frac{1}{3}$ mph.

SECTION 2.4

3. *a*, *c* **5.** $\{x \mid x < 5\}$ **7.** $\{x \mid x \le 2\}$ **9.** $\{x \mid x < -4\}$ **11.** $\{x \mid x > 3\}$ **13.** $\{x \mid x > 4\}$ **15.** $\{x \mid x \le 2\}$ **17.** $\{x \mid x > -2\}$ **19.** $\{x \mid x \ge 2\}$ **21.** $\{x \mid x > -2\}$ **23.** $\{x \mid x \le 3\}$ **25.** $\{x \mid x < 2\}$ **27.** $\{x \mid x < -3\}$ **29.** $\{x \mid x \le 5\}$ **31.** $\left\{x \mid x \ge -\frac{1}{2}\right\}$ **33.** $\left\{x \mid x < \frac{23}{16}\right\}$ **35.** $\left\{x \mid x < \frac{8}{3}\right\}$ **37.** $\{x \mid x > 1\}$ **39.** $\left\{x \mid x > \frac{14}{11}\right\}$ **41.** $\{x \mid x \le 1\}$ **43.** $\left\{x \mid x \le \frac{7}{4}\right\}$ **45.** $\{x \mid x \le 2\}$ **49.** $\{x \mid -2 \le x \le 4\}$ **51.** $\{x \mid x < 3 \text{ or } x > 5\}$ **53.** $\{x \mid -4 < x < 2\}$ **55.** $\{x \mid x > 6 \text{ or } x < -4\}$ **57.** $\{x \mid x < -3\}$ **59.** $\{x \mid -3 < x < 2\}$ **61.** $\{x \mid -2 < x < 1\}$ **63.** $\{x \mid x < -2 \text{ or } x > 2\}$ **65.** $\{x \mid 2 < x < 6\}$ **67.** $\{x \mid -3 < x < -2\}$ **69.** $\left\{x \mid x > 5 \text{ or } x < -\frac{5}{3}\right\}$ **71.** $\{x \mid x < 3\}$ **73.** The solution set is the empty set. **75.** $\{x \mid x \in \text{real numbers}\}$ **77.** $\left\{x \mid \frac{17}{7} \le x \le \frac{45}{7}\right\}$ **79.** $\left\{x \mid -5 < x < \frac{17}{3}\right\}$ **81.** $\{x \mid x \in \text{real numbers}\}$ **83.** The smallest number is −12. **85.** The maximum width as an integer is 11 cm. **87.** The TopPage plan is less expensive for more than 460 pages per month. **89.** The call must be 7 min or less. **91.** The temperature range for the week was between 32°F and 86°F. **93.** The amount of sales must be $44,000 or more. **95.** The business writes more than 200 checks monthly. **97.** The range of scores is $58 \le N \le 100$. **99.** The ski area is more than 38 mi away. **101.** {1, 2} **103.** ∅

SECTION 2.5

1. yes **3.** yes **5.** 7, −7 **7.** 4, −4 **9.** 6, −6 **11.** 7, −7 **13.** There is no solution. **15.** 1, −5 **17.** 8, 2 **19.** There is no solution. **21.** $-\frac{3}{2}, 3$ **23.** $\frac{3}{2}$ **25.** There is no solution. **27.** 7, −3 **29.** $2, -\frac{10}{3}$ **31.** 1, 3 **33.** $\frac{3}{2}$ **35.** There is no solution. **37.** $\frac{11}{6}, -\frac{1}{6}$ **39.** $-\frac{1}{3}, -1$ **41.** There is no solution. **43.** 3, 0 **45.** There is no solution. **47.** $1, \frac{13}{3}$ **49.** There is no solution. **51.** $\frac{7}{3}, \frac{1}{3}$ **53.** $-\frac{1}{2}$ **55.** $-\frac{1}{2}, -\frac{7}{2}$ **57.** $-\frac{8}{3}, \frac{10}{3}$ **59.** There is no solution. **61.** $\{x \mid x > 3 \text{ or } x < -3\}$ **63.** $\{x \mid x > 1 \text{ or } x < -3\}$ **65.** $\{x \mid 4 \le x \le 6\}$ **67.** $\{x \mid x \ge 5 \text{ or } x \le -1\}$ **69.** $\{x \mid -3 < x < 2\}$ **71.** $\left\{x \mid x > 2 \text{ or } x < -\frac{14}{5}\right\}$ **73.** ∅ **75.** The solution set is the set of real numbers. **77.** $\left\{x \mid x \le -\frac{1}{3} \text{ or } x \ge 3\right\}$ **79.** $\left\{x \mid -2 \le x \le \frac{9}{2}\right\}$ **81.** $\{x \mid x = 2\}$ **83.** $\left\{x \mid x < -2 \text{ or } x > \frac{22}{9}\right\}$ **85.** $\left\{x \mid -\frac{3}{2} < x < \frac{9}{2}\right\}$ **87.** $\left\{x \mid x < 0 \text{ or } x > \frac{4}{5}\right\}$ **89.** $\{x \mid x > 5 \text{ or } x < 0\}$ **91.** The lower limit is 1.742 in. The upper limit is 1.758 in. **93.** The lower limit is 195 volts. The upper limit is 245 volts. **95.** The lower limit is $9\frac{19}{32}$ in. The upper limit is $9\frac{21}{32}$ in. **97.** The lower limit is 28,420 ohms. The upper limit is 29,580 ohms. **99.** The lower limit is 23,750 ohms. The upper limit is 26,250 ohms. **101.a.** $\{x \mid x \ge -3\}$ **b.** $\{a \mid a \le 4\}$ **103.a.** ≤ **b.** ≥ **c.** ≥ **d.** = **e.** =

CHAPTER REVIEW

1. no [2.1A] **2.** 21 [2.1B] **3.** 20 [2.1C] **4.** −3 [2.2A] **5.** $-\frac{6}{7}$ [2.2A] **6.** $\frac{1}{3}$ [2.2B] **7.** 4 [2.2B] **8.** −2 [2.2C] **9.** 10 [2.2C] **10.** $\left\{x \mid x > \frac{5}{3}\right\}$ [2.4A] **11.** $\{x \mid x > -1\}$ [2.4A] **12.** $\left\{x \mid -3 < x < \frac{4}{3}\right\}$ [2.4B] **13.** $\{x \mid x \in \text{real numbers}\}$ [2.4B] **14.** $-\frac{9}{5}, 3$ [2.5A] **15.** 9, −1 [2.5A] **16.** $\{x \mid 1 < x < 4\}$ [2.5B] **17.** $\left\{x \mid x \ge 2 \text{ or } x \le \frac{1}{2}\right\}$ [2.5B] **18.** 30 is 250% of 12. [2.1D] **19.** $\frac{1}{2}$% of 1600 is 8. [2.1D] **20.** $5x - 4 = 16; x = 4$ [2.2D] **21.** $3x = 2(21 - x) - 2$; 8, 13 [2.2D] **22.** The airline would sell 177 tickets. [2.1D] **23.** The rebate is 6.1% of the cost. [2.1D] **24.** 2 qt of cranberry juice and 8 qt of apple juice were used. [2.3A] **25.** The cost is $3 per ounce. [2.3A] **26.** The percent concentration of butterfat is 14%. [2.3B] **27.** 375 lb of the alloy containing 30% tin and 125 lb of the alloy containing 70% tin were used. [2.3B] **28.** The total amount invested was $450,000. [2.3C] **29.** $1400 was deposited at 6.75%. $1000 was deposited at 9.45%. [2.3C] **30.** The jet overtakes the propeller-driven plane 600 mi from the starting point. [2.3D] **31.** The average speed on the winding road was 32 mph. [2.3D] **32.** The range of scores is $82 \le x \le 100$. [2.4C] **33.** The lower limit is 1.75 cc. The upper limit is 2.25 cc. [2.5C]

CHAPTER TEST

1. −5 [2.2B] **2.** −5 [2.1B] **3.** −3 [2.2A] **4.** 2 [2.2C] **5.** no [2.1A] **6.** 5 [2.2A] **7.** 0.04 [2.1D] **8.** $-\frac{1}{3}$ [2.2C] **9.** 2 [2.2B] **10.** −12 [2.1C] **11.** $\{x \mid x \le -3\}$ [2.4A] **12.** $\{x \mid x > -1\}$ [2.4A] **13.** $\{x \mid x > -2\}$ [2.4B] **14.** ∅ [2.4B] **15.** $3, -\frac{9}{5}$ [2.5A] **16.** 7, −2 [2.5A] **17.** $\left\{x \mid \frac{1}{3} \le x \le 3\right\}$ [2.5B] **18.** $\left\{x \mid x > 2 \text{ or } x < -\frac{1}{2}\right\}$ [2.5B] **19.** 4, 11 [2.2D] **20.** The cost of the hamburger mixture is $2.70 per pound. [2.3A] **21.** 1.25 gallons of water must be used. [2.3B] **22.** $5000 is invested at 3% and $15,000 is invested at 7%. [2.3C] **23.** The rate of the snowmobile is 6 mph. [2.3D] **24.** You can drive the Gambelli Agency car less than 120 mi per day. [2.4C] **25.** The lower limit is 2.648 in. The upper limit is 2.652 in. [2.5C]

CUMULATIVE REVIEW

1. 6 [1.1C] **2.** −48 [1.1D] **3.** $-\frac{19}{48}$ [1.2C] **4.** 54 [1.2E] **5.** $\frac{49}{40}$ [1.3A] **6.** 6 [1.4A] **7.** $-17x$ [1.4B] **8.** $-5a - 4b$ [1.4B] **9.** $2x$ [1.4C] **10.** $36y$ [1.4C] **11.** $2x^2 + 6x - 4$ [1.4D] **12.** $-4x + 14$ [1.4D] **13.** $6x - 34$ [1.4D] **14.** $A \cap B = \{-4, 0\}$ [1.5B] **15.** 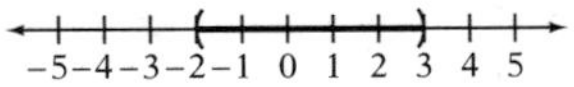

[1.5C] **16.** yes [2.1A] **17.** −25 [2.1C] **18.** −3 [2.2A] **19.** 3 [2.2A] **20.** −3 [2.2B] **21.** $\frac{1}{2}$ [2.2B] **22.** 13 [2.2C] **23.** $\{x \mid x \le 1\}$ [2.4A] **24.** $\{x \mid -4 \le x \le 1\}$ [2.4B] **25.** −1 and 4 [2.5A] **26.** $\left\{x \mid x > 2 \text{ or } x < -\frac{4}{3}\right\}$ [2.5B] **27.** $\frac{11}{20}$ [1.2B] **28.** 103% [1.2B] **29.** 25% of 120 is 30. [2.1D] **30.** $6x + 13 = 3x - 5; x = -6$ [2.2D] **31.** 20 lb of oat flour must be used. [2.3A] **32.** 25 g of pure gold must be added. [2.3B] **33.** The length of the track is 120 m. [2.3D]

Answers to Chapter 3 Selected Exercises

PREP TEST

1. 43 [2.1B] **2.** 51 [2.1B] **3.** 56 [1.3A] **4.** 56.52 [1.4A] **5.** 113.04 [1.4A] **6.** 120 [1.4A]

SECTION 3.1

1. 40°; acute **3.** 115°; obtuse **5.** 90°; right **7.** The complement is 28°. **9.** The supplement is 18°. **11.** The length of *BC* is 14 cm. **13.** The length of *QS* is 28 ft. **15.** The length of *EG* is 30 m. **17.** The measure of $\angle MON$ is 86°. **19.** 71° **21.** 30° **23.** 36° **25.** 127° **27.** 116° **29.** 20° **31.** 20° **33.** 20° **35.** 141° **37.** 106° **39.** 11° **41.** $\angle a = 38°, \angle b = 142°$ **43.** $\angle a = 47°, \angle b = 133°$ **45.** 20° **47.** 47° **49.** $\angle x = 155°, \angle y = 70°$ **51.** $\angle a = 45°, \angle b = 135°$ **53.** $90° - x$ **55.** The measure of the third angle is 60°. **57.** The measure of the third angle is 35°. **59.** The measure of the third angle is 102°. **61.a.** 1° **b.** 179° **65.** 360°

SECTION 3.2

1. hexagon **3.** pentagon **5.** scalene **7.** equilateral **9.** obtuse **11.** acute **13.** 56 in. **15.** 14 ft **17.** 47 mi **19.** 8π cm or approximately 25.13 cm **21.** 11π mi or approximately 34.56 mi **23.** 17π ft or approximately 53.41 ft **25.** The perimeter is 17.4 cm. **27.** The perimeter is 8 cm. **29.** The perimeter is 24 m. **31.** The perimeter is 48.8 cm. **33.** The perimeter is 17.5 in. **35.** The length of a diameter is 8.4 cm. **37.** The circumference is 1.5π in. **39.** The circumference is 226.19 cm. **41.** 60 ft of fencing should be purchased. **43.** The carpet must be nailed down along 44 ft. **45.** The length is 120 ft. **47.** The length of the third side is 10 in. **49.** The length of each side is 12 in. **51.** The length of a diameter is 2.55 cm. **53.** The length is 13.19 ft. **55.** The bicycle travels 50.27 ft. **57.** The circumference is 39,935.93 km. **59.** 60 ft^2 **61.** 20.25 in^2 **63.** 546 ft^2 **65.** 16π cm^2 or approximately 50.27 cm^2 **67.** 30.25π mi^2 or approximately 95.03 mi^2 **69.** 72.25π ft^2 or approximately 226.98 ft^2 **71.** The area is 156.25 cm^2. **73.** The area is 570 in^2. **75.** The area is 192 in^2. **77.** The area is 13.5 ft^2. **79.** The area is 330 cm^2. **81.** The area is 25π in^2. **83.** The area is 9.08 ft^2. **85.** The area is $10{,}000\pi$ in^2. **87.** The area is 126 ft^2. **89.** 7500 yd^2 must be purchased. **91.** The width is 10 in.

93. The length of the base is 20 m. **95.** You should buy 2 qt. **97.** It will cost \$98. **99.** The increase in area is 113.10 in^2. **101.** The cost will be \$878. **103.** The area is 216 m^2. **105.** The cost is \$1600. **109.** $A = \frac{\pi d^2}{4}$

SECTION 3.3

1. 840 in^3 **3.** 15 ft^3 **5.** 4.5π cm^3 or approximately 14.14 cm^3 **7.** The volume is 34 m^3. **9.** The volume is 42.875 in^3. **11.** The volume is 36π ft^3. **13.** The volume is 8143.01 cm^3. **15.** The volume is 75π in^3. **17.** The volume is 120 in^3. **19.** The width is 2.5 ft. **21.** The radius of the base is 4.00 in. **23.** The length is 5 in. The width is 5 in. **25.** There are 75.40 m^3 in the tank. **27.** 94 m^2 **29.** 56 m^2 **31.** 96π in^2 or approximately 301.59 in^2 **33.** The surface area is 184 ft^2. **35.** The surface area is 69.36 m^2. **37.** The surface area is 225π cm^2. **39.** The surface area is 402.12 in^2. **41.** The surface area is 6π ft^2. **43.** The surface area is 297 in^2. **45.** The width is 3 cm. **47.** 11 cans of paint should be purchased. **49.** 456 in^2 of glass are needed. **51.** The surface area of the pyramid is 22.53 cm^2 larger. **53.a.** always true **b.** never true **c.** sometimes true

CHAPTER REVIEW

1. $\angle x = 22°, \angle y = 158°$ [3.1C] **2.** $\angle x = 68°$ [3.1B] **3.** The length of AC is 44 cm. [3.1A] **4.** 19° [3.1A] **5.** The volume is 96 cm^3. [3.3A] **6.** $\angle a = 138°, \angle b = 42°$ [3.1B] **7.** The surface area is 220 ft^2. [3.3B] **8.** The supplement is 148°. [3.1A] **9.** The area is 78 cm^2. [3.2B] **10.** The area is 63 m^2. [3.2B] **11.** The volume is 39 ft^3. [3.3A] **12.** The measure of the third angle is 95°. [3.1C] **13.** The length of the base is 8 cm. [3.2B] **14.** The volume is 288π mm^3. [3.3A] **15.** The volume is $\frac{784\pi}{3}$ cm^3. [3.3A] **16.** Each side measures 21.5 cm. [3.2A] **17.** 4 cans of paint should be purchased. [3.3B] **18.** 208 yd of fencing are needed. [3.2A] **19.** The area is 90.25 m^2. [3.2B] **20.** The area is 276 m^2 [3.2B]

CHAPTER TEST

1. The radius is 0.75 m. [3.2A] **2.** The circumference is 31.42 cm. [3.2A] **3.** The perimeter is 26 ft. [3.2A] **4.** $BC = 3$ [3.1A] **5.** The volume is 268.08 ft^3. [3.3A] **6.** The area is 63.62 cm^2. [3.2B] **7.** $a = 100°, b = 80°$ [3.1B] **8.** 75° [3.1A] **9.** $a = 135°, b = 45°$ [3.1B] **10.** The area is 55 m^2. [3.2B] **11.** The volume is 169.65 m^3. [3.3A] **12.** The perimeter is 6.8 m. [3.2A] **13.** 58° [3.1A] **14.** The surface area is 164.93 ft^2. [3.3B] **15.** There are 113.10 in^2 more in the pizza. [3.2B] **16.** The measures of the other two angles are 58° and 90°. [3.1C] **17.** The bicycle travels 73.3 ft. [3.2A] **18.** The area of the room is 28 yd^2. [3.2B] **19.** The volume of the silo is 1145.11 ft^3. [3.3A] **20.** The area is 11 m^2. [3.2B]

CUMULATIVE REVIEW

1. −3, 0, and 1 [1.1A] **2.** 0.089 [1.2B] **3.** 35% [1.2B] **4.** $-\frac{2}{3}$ [1.2D] **5.** −24.51 [1.2D] **6.** $-5\sqrt{5}$ [1.2F] **7.** −28 [1.3A] **8.** −8 [1.4A] **9.** $-3m + 3n$ [1.4B] **10.** $21y$ [1.4C] **11.** $7x + 9$ [1.4D] **12.** {−2, −1} [1.5A] **13.** {−10, 0, 10, 20, 30} [1.5A] **14.**
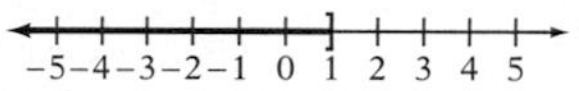

[1.5C] **15.** 5 [2.2B] **16.** $\frac{1}{2}$ [2.2C] **17.** $\{y \mid y \le -4\}$ [2.4A] **18.** $\{x \mid x \ge 2\}$ [2.4A] **19.** $\{x \mid x < -3 \text{ or } x > 4\}$ [2.4B] **20.** $\{x \mid 2 \le x \le 6\}$ [2.4B] **21.** $1, -\frac{1}{3}$ [2.5A] **22.** $\{x \mid 6 \le x \le 10\}$ [2.5B] **23.** $\angle x = 131°$ [3.1B] **24.** $4x - 10 = 2; x = 3$ [2.2D] **25.** The third angle measures 122°. [3.1C] **26.** \$5000 should be deposited in the 9.5% account. [2.3C] **27.** The third side measures 4.5 m. [3.2A] **28.** The women's median annual earnings are 70.8% of the men's median annual earnings. [1.2G] **29.** The area is 20.25π cm^2. [3.2B] **30.** The height of the box is 3 ft. [3.3A]

Answers to Chapter 4 Selected Exercises

PREP TEST

1. $-4x + 12$ [1.4D] **2.** 10 [1.2F] **3.** -2 [1.3A] **4.** 11 [1.4A] **5.** 2.5 [1.4A] **6.** 5 [1.4A] **7.** 1 [1.4A] **8.** 4 [2.2A]

SECTION 4.1

1.
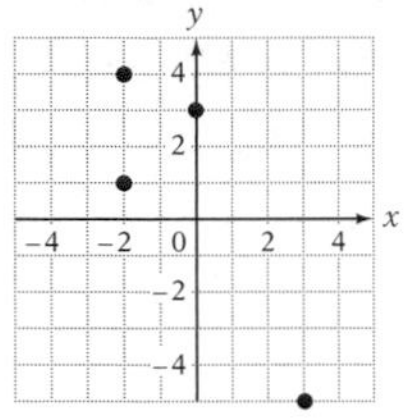

3.
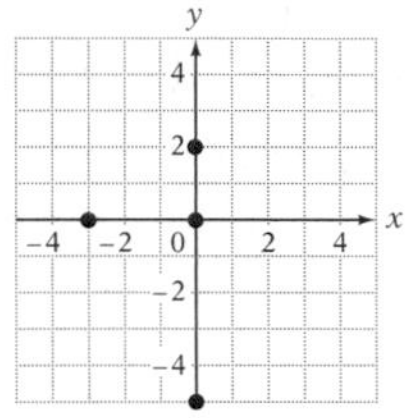

5.
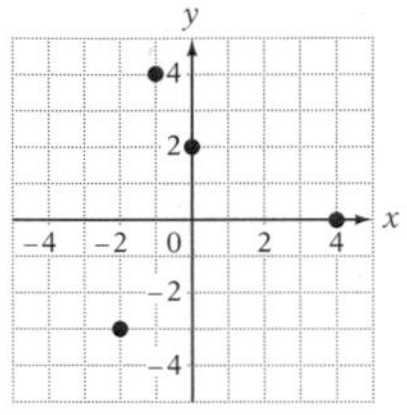

7. The coordinates of A are (2, 3). The coordinates of B are (4, 0). The coordinates of C are $(-4, 1)$. The coordinates of D are $(2, -2)$.
9. The coordinates of A are $(-2, 5)$. The coordinates of B are (3, 4). The coordinates of C are (0, 0). The coordinates of D are $(-3, -2)$.
11. The coordinates of A are (2, 4). The coordinates of B are (0, 1). The coordinates of C are $(-4, 0)$. The coordinates of D are $(3, -3)$.

13.
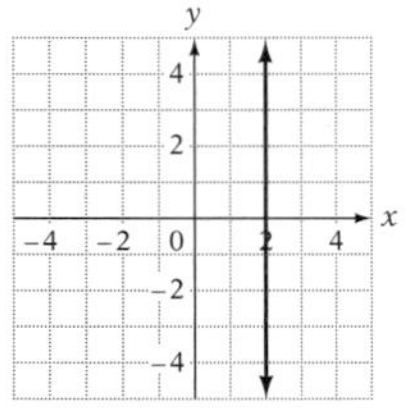

15.
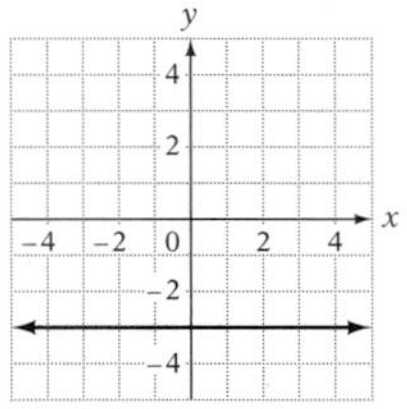

17. yes **19.** no **21.** no **23.** no **25.** (3, 7) **27.** (6, 3) **29.** (0, 1) **31.** $(-5, 0)$

33.
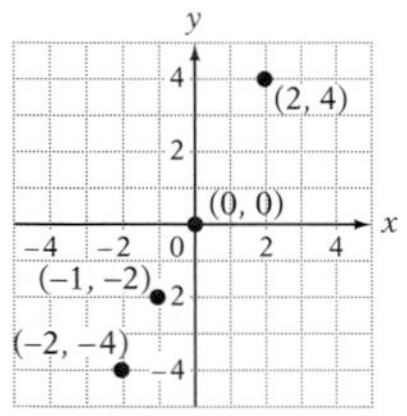

35.
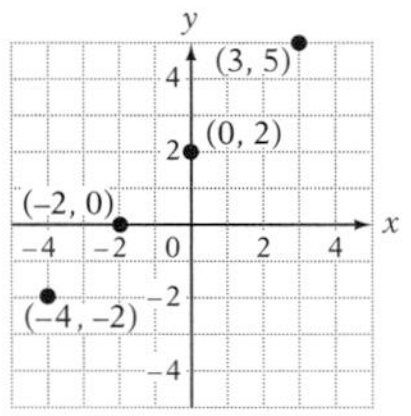

37.
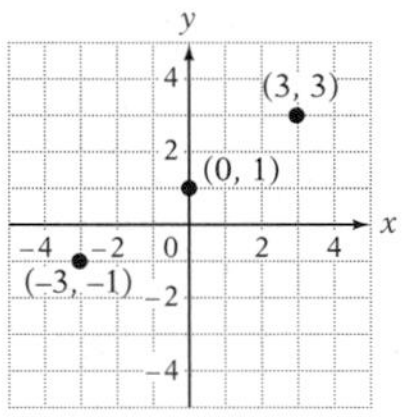

39.
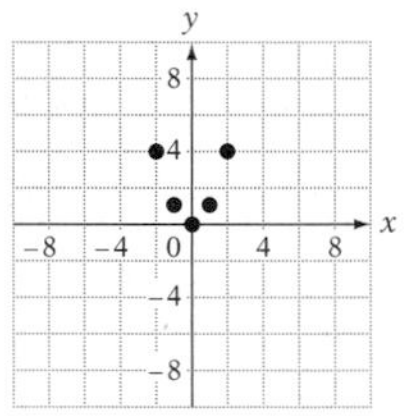

41a. The temperature of the reaction after 20 min was 280°F. **b.** The temperature was 160°F in 50 min.

43.
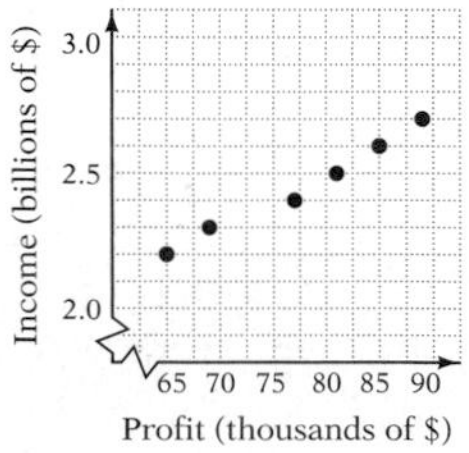

SECTION 4.2

9. Function **11.** Function **13.** Function **15.** Not a function **17.** Yes. \$29.62 **19.** $f(3) = 11$ **21.** $f(0) = -4$ **23.** $G(0) = 4$ **25.** $G(-2) = 10$ **27.** $q(3) = 5$ **29.** $q(-2) = 0$ **31.** $F(4) = 24$ **33.** $F(-3) = -4$ **35.** $H(1) = 1$ **37.** $H(t) = \dfrac{3t}{t + 2}$ **39.** $s(-1) = 6$ **41.** $s(a) = a^3 - 3a + 4$ **43.** $P(-2 + h) - P(-2) = 4h$ **45a.** The packaging cost is \$4.75 per game. **b.** The packaging cost is \$4.00 per game. **47a.** The appraiser's fee is \$3000. **b.** The appraiser's fee is \$950. **49.** Domain: {1, 2, 3, 4, 5}; range: {1, 4, 7, 10, 13} **51.** Domain: {0, 2, 4, 6}; range: {1, 2, 3, 4} **53.** Domain: {1, 3, 5, 7, 9}; range: {0} **55.** Domain: {−2, −1, 0, 1, 2}; range: {0, 1, 2} **57.** Domain: {−2, −1, 0, 1, 2}; range: {−3, 3, 6, 7, 9} **59.** 1 **61.** −8 **63.** None **65.** None **67.** 0 **69.** None **71.** None **73.** −2 **75.** None **77.** Range: {−3, 1, 5, 9} **79.** Range: {−23, −13, −8, −3, 7} **81.** {0, 1, 4} **83.** {2, 14, 26, 42} **85.** $\left\{-5, \frac{5}{3}, 5\right\}$ **87.** $\left\{-1, -\frac{1}{2}, -\frac{1}{3}, 1\right\}$ **89.** {−38, −8, 2} **93.** {(−2, −8), (−1, −1), (0, 0), (1, 1), (2, 8)}. Yes, the set defines a function. **95.** 50.625 watts of power will be produced. **97a.** After 5 s the parachutist is falling at a rate of 22 ft/s. **b.** After 15 s the parachutist is falling at a rate of 30 ft/s. **99a.** After 5 h the temperature will be 64°F. **b.** After 15 h the temperature will be 52°F.

SECTION 4.3

1.
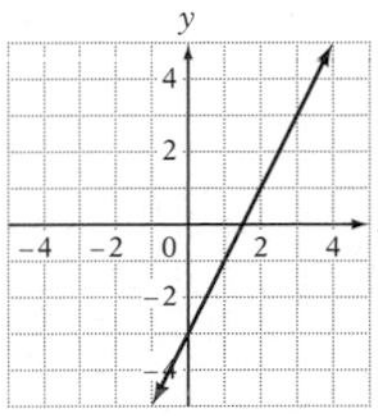

3.
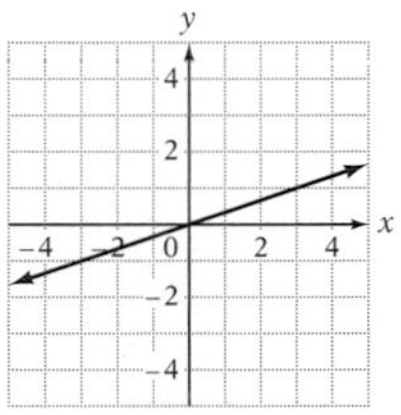

5.
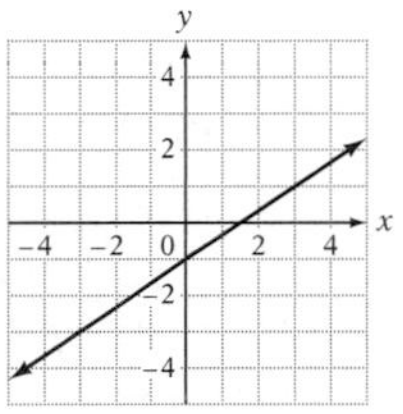

7.
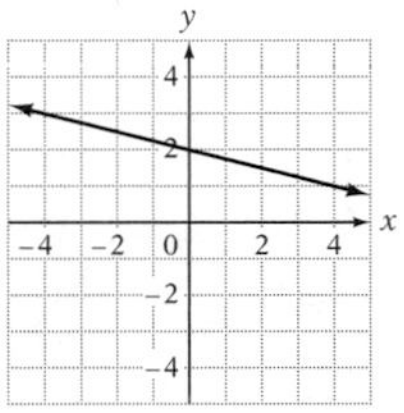

9.
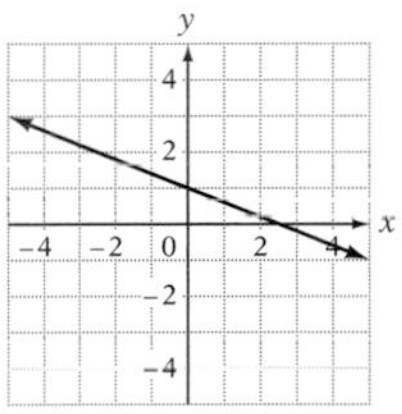

11.
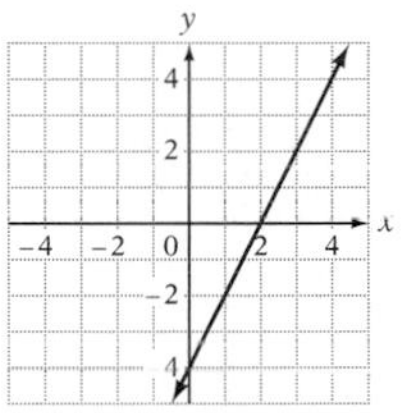

13.
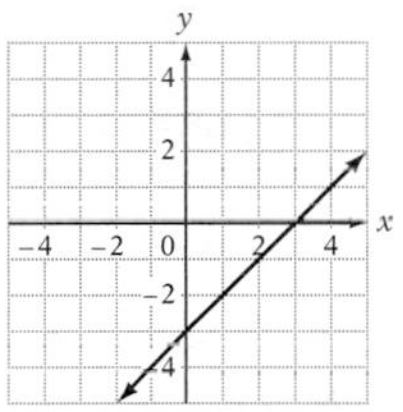

15.
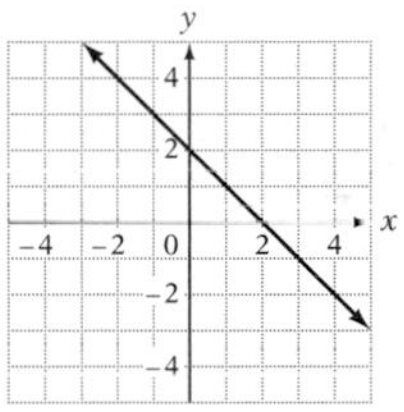

17.
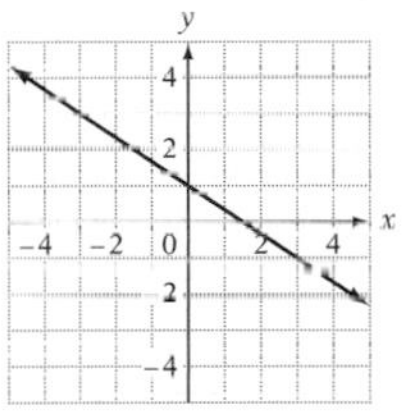

19. (3, 0), (0, −3) **21.** (10, 0), (0, −2) **23.** (0, 0), (0, 0)

25.
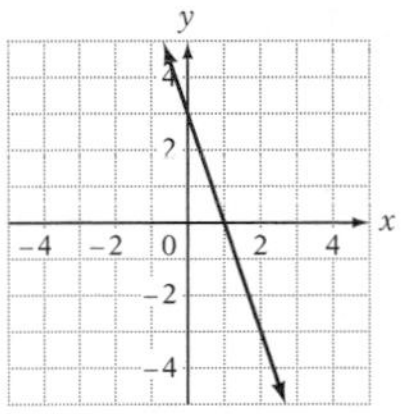

27.
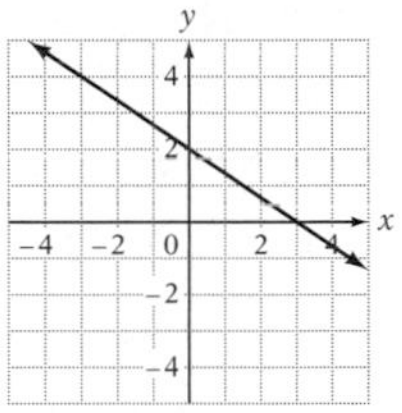

29.
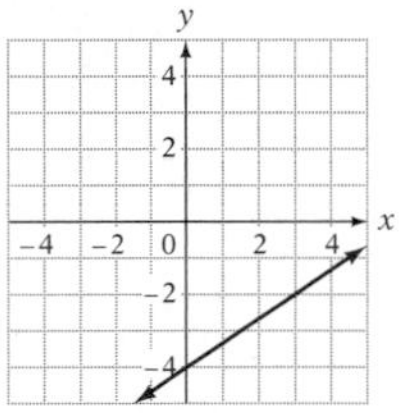

31.
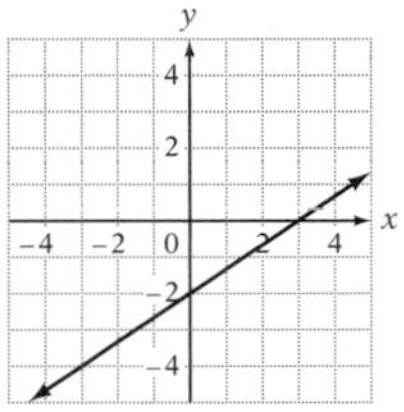

33.
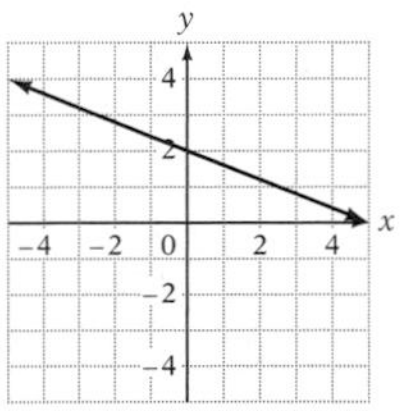

35.
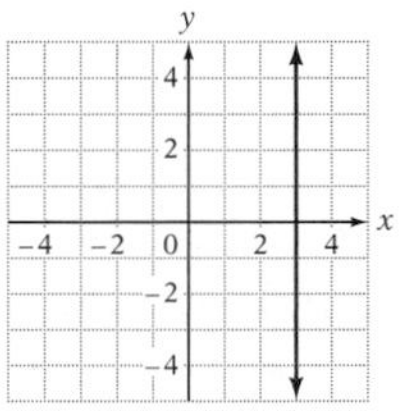

37.
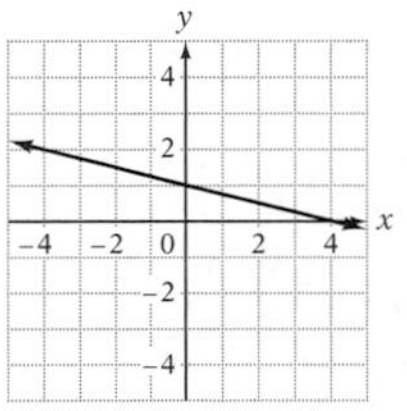

39.
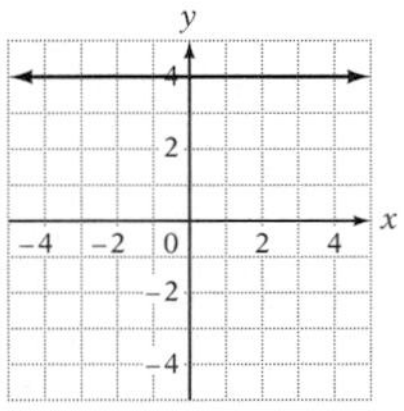

41.
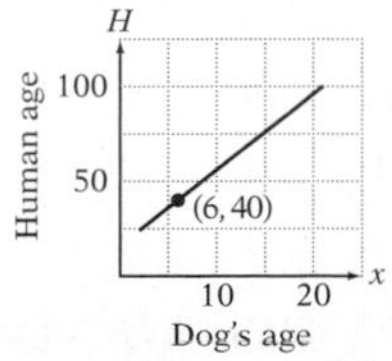

A dog 6 years old is equivalent in age to a human 40 years old.

43.
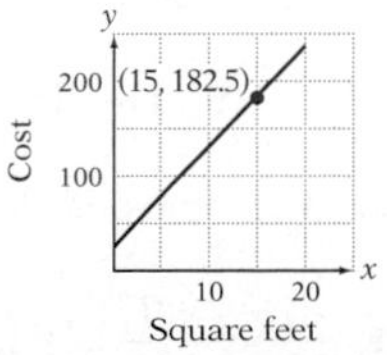

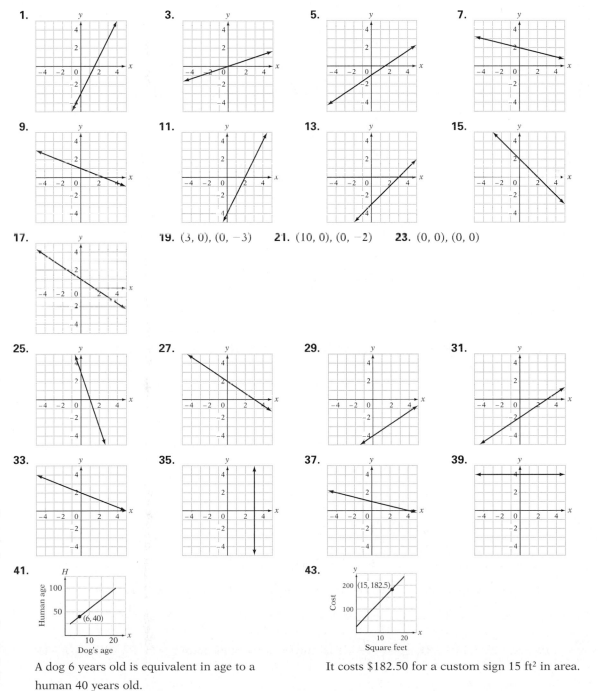

It costs \$182.50 for a custom sign 15 ft² in area.

SECTION 4.4

1. -1 **3.** $\frac{1}{3}$ **5.** $-\frac{2}{3}$ **7.** $-\frac{3}{4}$ **9.** The slope is undefined. **11.** $\frac{7}{5}$ **13.** The line has zero slope. **15.** $-\frac{1}{2}$ **17.** The slope is undefined. **19.** The slope is 40. The slope is the average speed of the motorist in miles per hour. **21.** The slope is -0.05. For each mile the car is driven, approximately 0.05 gal of fuel is used. **23.** The slope is 385.5. The slope is the average speed of the runner in meters per minute. **25.** no **29.** yes

31.
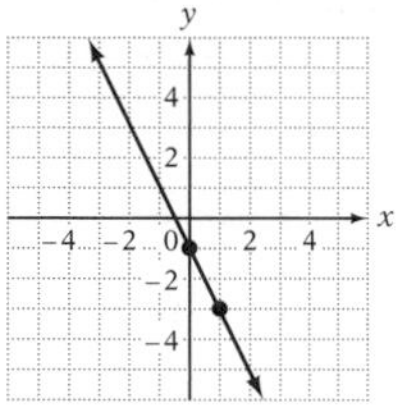

33.
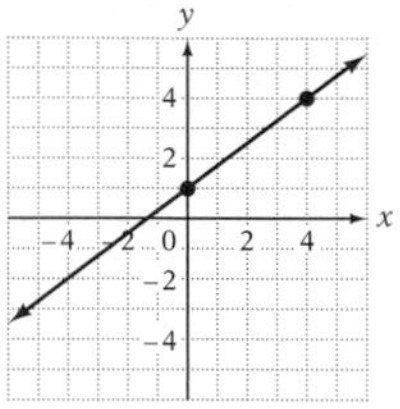

35.
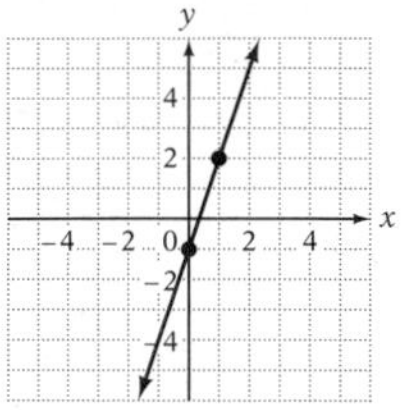

37.
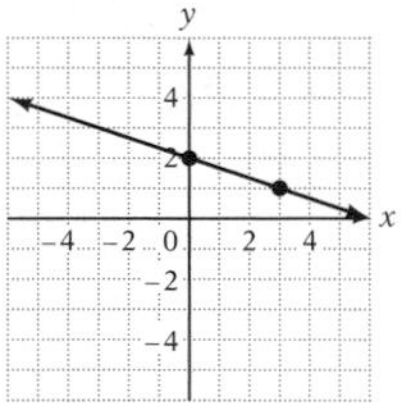

39.
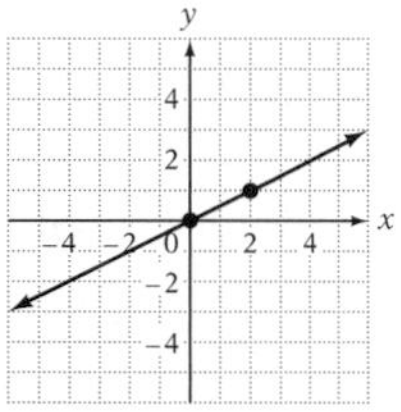

41.
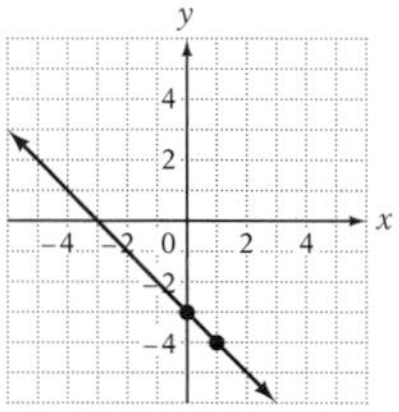

43.
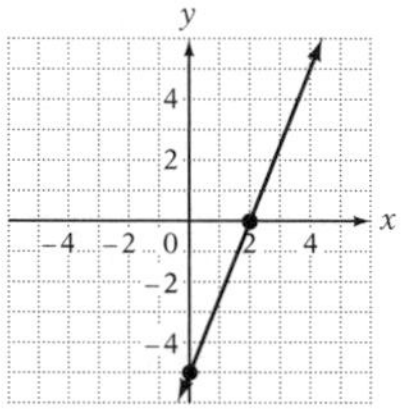

45. i and D; ii and C; iii and B; iv and F; v and E; vi and A **47.** $k = 10$ **49.** $k = -4$

SECTION 4.5

3. $y = 2x + 5$ **5.** $y = \frac{1}{2}x + 2$ **7.** $y = \frac{1}{2}x + 4$ **9.** $y = -\frac{5}{3}x + 5$ **11.** $y = -3x + 9$ **13.** $y = -3x + 4$ **15.** $y = \frac{1}{3}x$ **17.** $y = \frac{1}{2}x$ **19.** $y = 3x - 9$ **21.** $y = -\frac{2}{3}x + 7$ **23.** $y = -x - 3$ **25.** $y = \frac{7}{5}x - \frac{27}{5}$ **27.** $y = -\frac{2}{5}x + \frac{3}{5}$ **29.** $x = 3$ **31.** $y = \frac{1}{5}x - 2$ **33.** $y = -3$ **35.** $y = -2x + 3$ **37.** $x = -5$ **39.** $y = x + 2$ **41.** $y = -2x - 3$ **43.** $y = \frac{1}{2}x + 2$ **45.** $y = -x + 5$ **47.** $y = -2x - 3$ **49.** $y = -\frac{3}{2}x + 3$ **51.** $y = -1$ **53.** $y = x - 1$ **55.** $y = -x + 1$ **57.** $y = -\frac{8}{3}x + \frac{25}{3}$ **59.** $y = \frac{1}{2}x - 1$ **61.** $y = -4$ **63.** $y = \frac{3}{4}x$ **65.** $y = -\frac{4}{3}x + \frac{5}{3}$ **67.** $x = -2$ **69.** $y = x - 1$ **71.** $y = \frac{4}{3}x + \frac{7}{3}$ **73.** $y = -x + 3$ **75.** $y = 1200x$. The height of the plane 11 min after takeoff is 13,200 ft. **77.** $y = -0.032x + 16$. After the driver has driven 150 mi, there are 11.2 gal in the tank. **83.** Answers will vary. Possible answers are $\{(0, 4), (3, 2), (9, -2)\}$. **85.** $f(4) = 0$ **87.** $y = -2x + 1$

SECTION 4.6

3. -5 **5.** $-\frac{1}{4}$ **7.** The lines are perpendicular. **9.** No, the lines are not parallel. **11.** No, the lines are not parallel. **13.** The lines are perpendicular. **15.** The lines are parallel. **17.** The lines are perpendicular. **19.** No, the lines are not parallel. **21.** The lines are perpendicular. **23.** $y = \frac{2}{3}x - \frac{8}{3}$ **25.** $y = \frac{1}{3}x - \frac{1}{3}$ **27.** $y = -\frac{5}{3}x - \frac{14}{3}$ **29.** $y = -2x + 15$ **31.** $\frac{A_1}{B_1} = -\frac{B_2}{A_2}$ **33.** Any equation of the form $y = 2x + b$ where $b \neq -13$ or of the form $y = -\frac{3}{2}x + c$ where $c \neq 8$.

SECTION 4.7

3. yes **5.** no **7.**

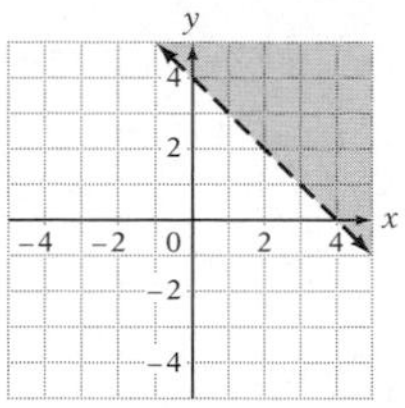

9.

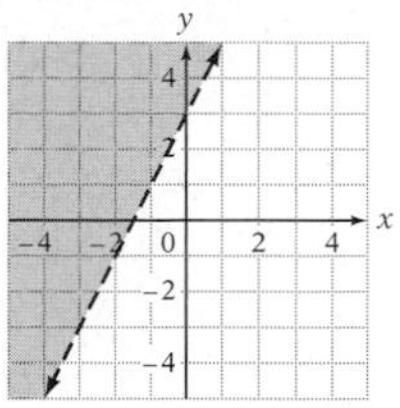

11.

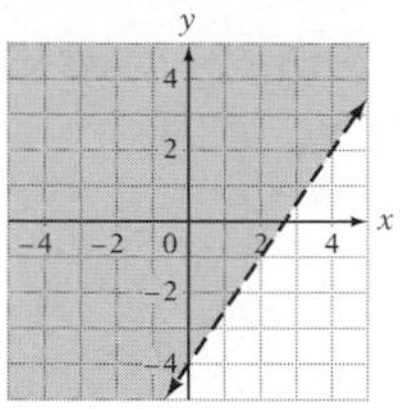

13.

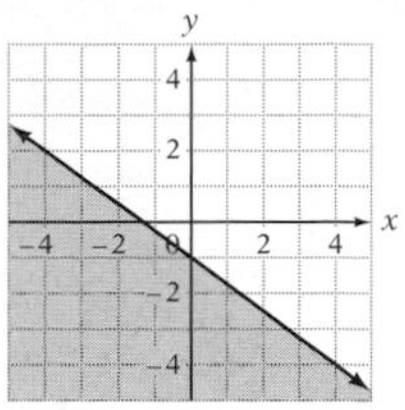

15.

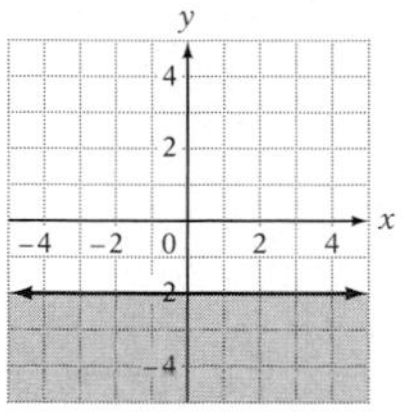

17.

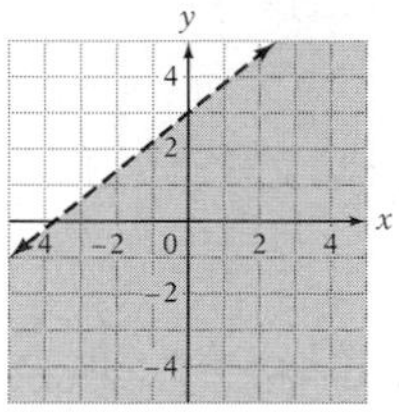

19.

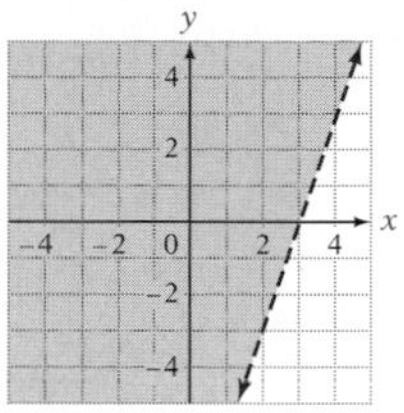

21.

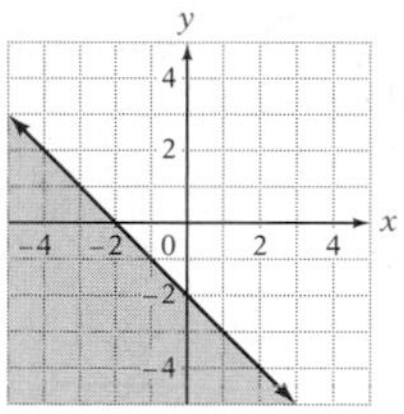

23.

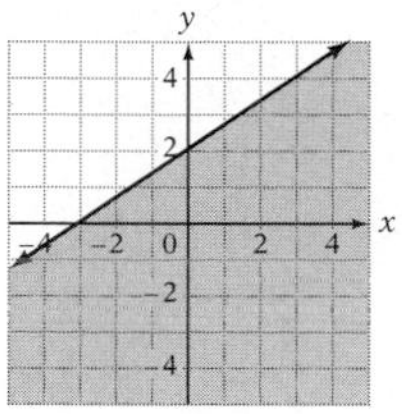

25. The inequality is $y \geq 2x + 2$.

CHAPTER REVIEW

1a.

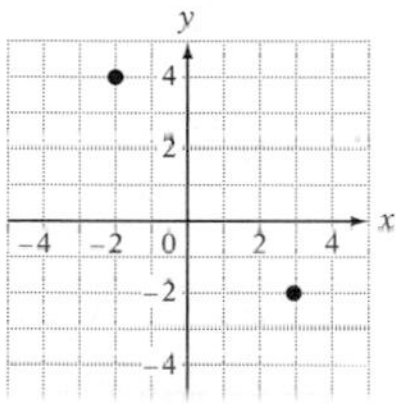

[4.1A]

2.

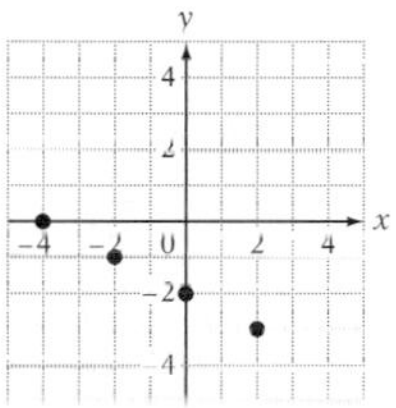

[4.1B]

3. (8, 0) and (0, −12) [4.3B] **4.** −1 [4.2A]

b. The abscissa of point A is −2.

c. The ordinate of point B is −4.

5.

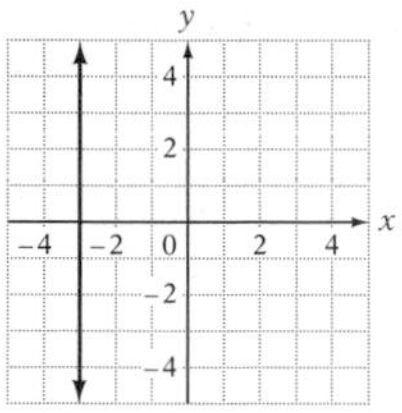

[4.3B]

6.

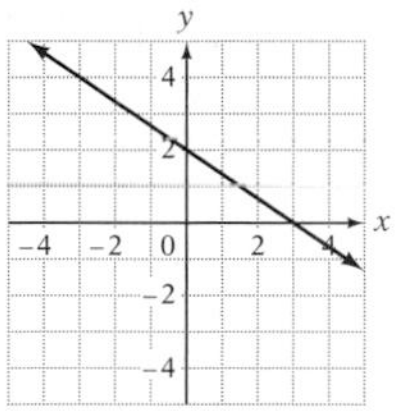

[4.4B]

7.

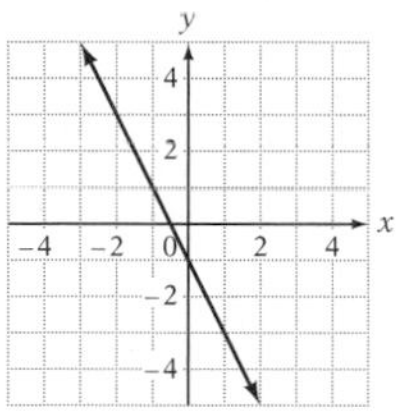

[4.3A]

8.

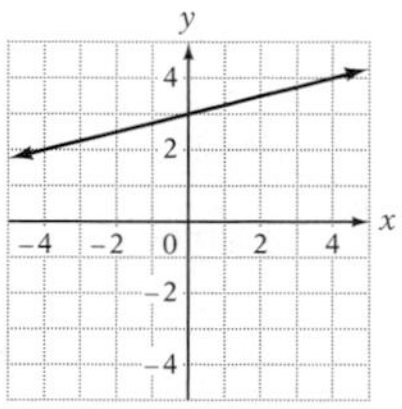

[4.3A]

9. 0 [4.4A] **10.** $\frac{7}{11}$ [4.4A] **11.** 0 [4.2A] **12.**

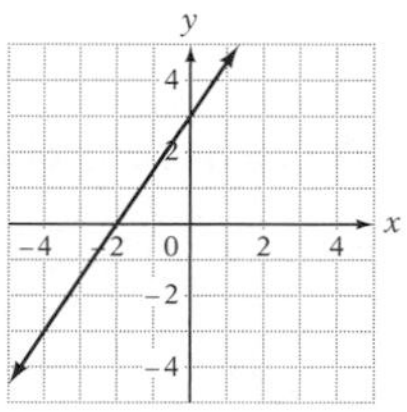

[4.3B]

13.

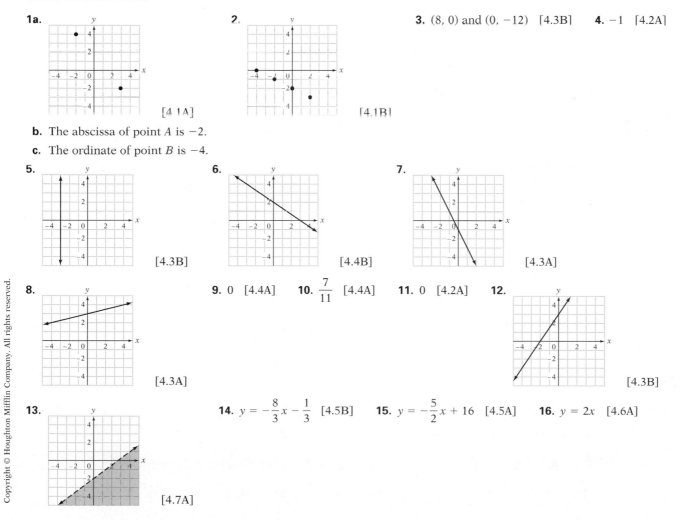

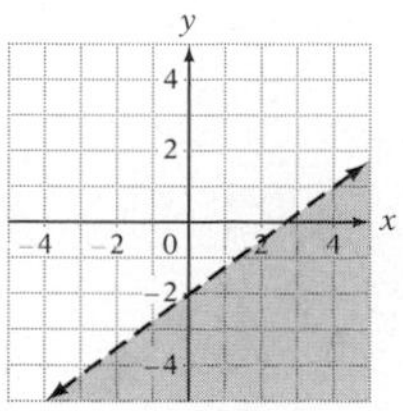

[4.7A]

14. $y = -\frac{8}{3}x - \frac{1}{3}$ [4.5B] **15.** $y = -\frac{5}{2}x + 16$ [4.5A] **16.** $y = 2x$ [4.6A]

17. $y = 2x + 1$ [4.6A] **18.** $y = 80x + 25{,}000$. It will cost $185,000 to build a house that contains 2000 ft^2. [4.5C]

19.

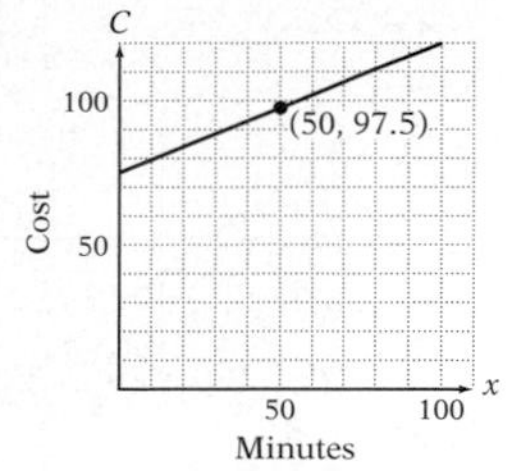

The cost of 50 min of access time for one month is $97.50. [4.3C]

20.

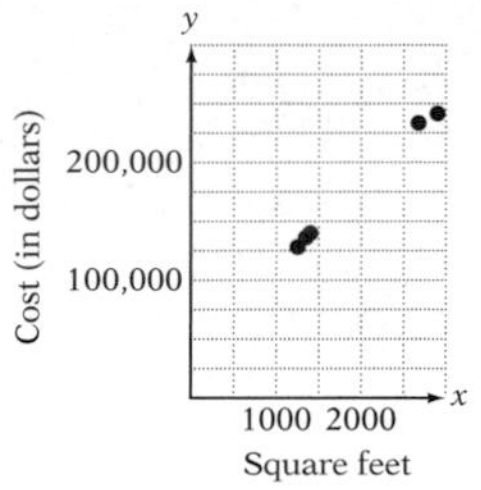

[4.1C]

CHAPTER TEST

1. $(3, -3)$ [4.1B] **2.** 6 [4.2A] **3.**

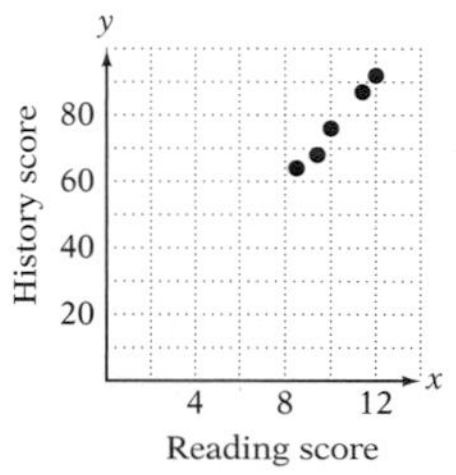

[4.1C]

4.

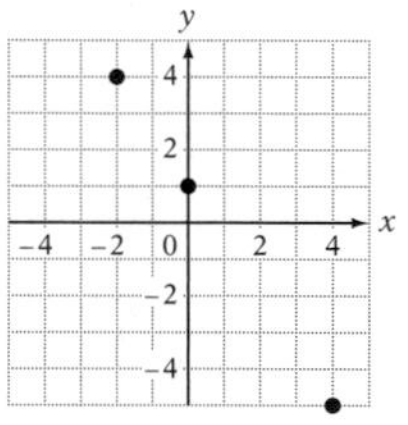

[4.1B]

5.

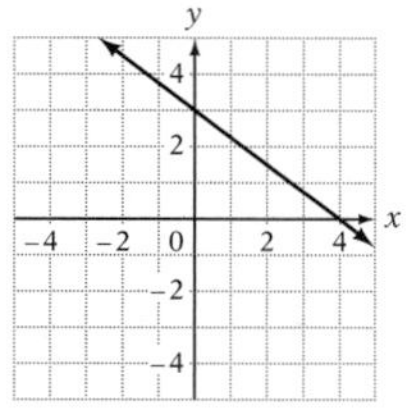

[4.3A]

6.

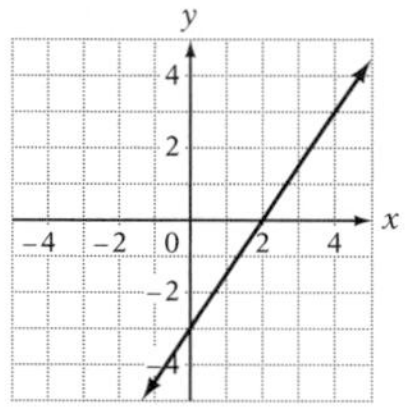

[4.3B]

7.

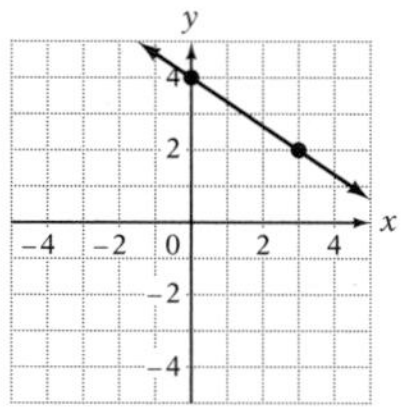

[4.4B]

8.

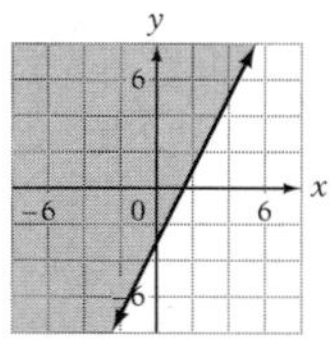

[4.7A]

9.

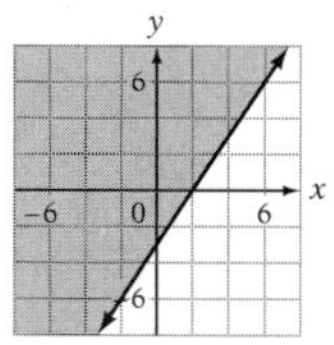

[4.7A]

10.

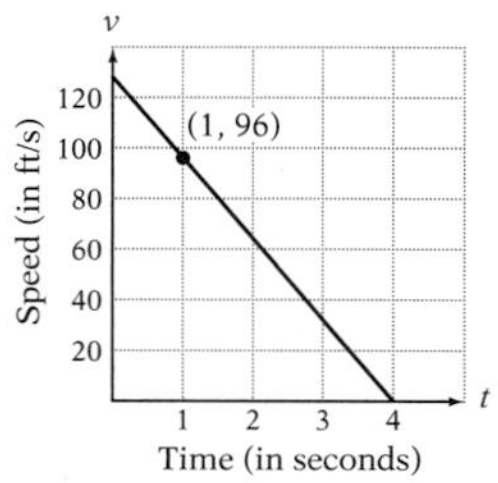

After 1 s, the ball is traveling 96 ft/s. [4.3C]

11. $y = -\frac{3}{10}x + 175$. When the tuition is $300, the enrollment will be 85 students. [4.5C]

12.

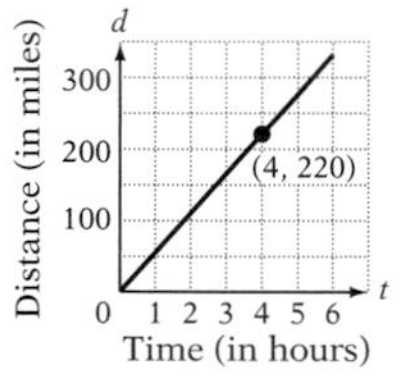

After 4 h, the car has traveled 220 mi. [4.3C]

13. $(2, 0)$ and $(0, -3)$ [4.3B] **14.** 2 [4.4A]

15. The slope is undefined. [4.4A] **16.** $-\frac{2}{3}$ [4.3B] **17.** $y = \frac{2}{3}x + 3$ [4.5A] **18.** $y = -\frac{2}{7}x - \frac{4}{7}$ [4.5B]

19. $y = -3x + 7$ [4.6A] **20.** $y = \frac{3}{2}x + 2$ [4.6A]

CUMULATIVE REVIEW

1. $-5, -3$ [1.1A] **2.** 0.85 [1.2B] **3.** $9\sqrt{5}$ [1.2F] **4.** -12 [1.3A] **5.** $-\frac{5}{8}$ [1.4A]

6. $-4d - 9$ [1.4B] **7.** $-32z$ [1.4C] **8.** $-13x + 7y$ [1.4D] **9.**

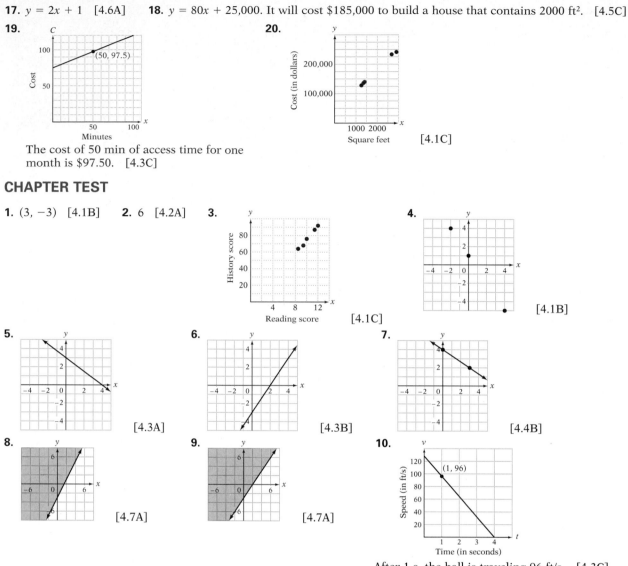

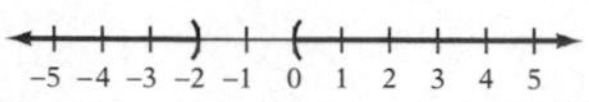

[1.5C]

10. $\frac{3}{2}$ [2.2A] **11.** 1 [2.2C] **12.** $\{x \mid x > -1\}$ [2.4A] **13.** $\varnothing$ [2.4B] **14.** $\left\{x \,\middle|\, 0 < x < \frac{10}{3}\right\}$ [2.5B]

15. $f(2) = 6$ [4.2A] **16.** The slope is 2. [4.4A] **17.**

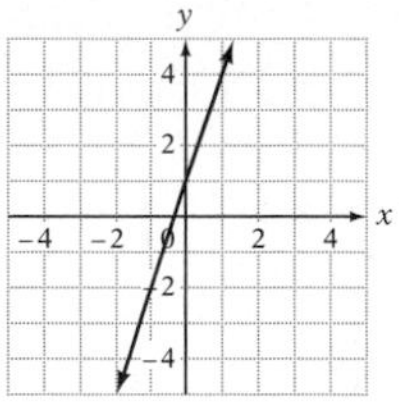

[4.3A]

18.

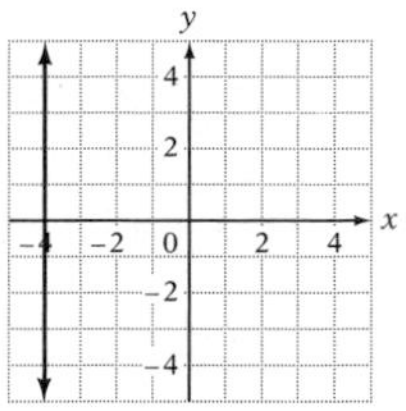

[4.3B]

19.

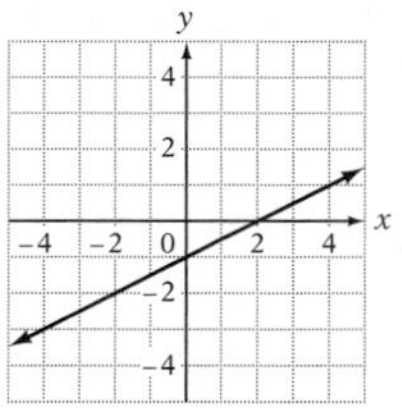

[4.4B]

20.

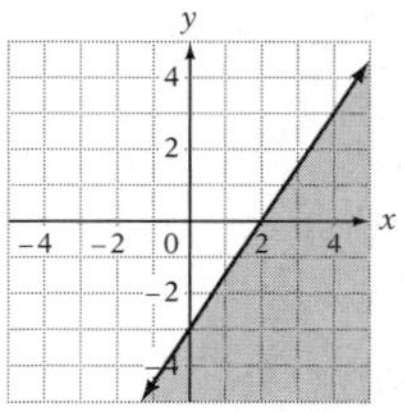

[4.7A]

21. $y = -\frac{1}{3}x - 2$ [4.5B] **22.** $y = -\frac{3}{2}x + 7$ [4.6A] **23.** The first plane is traveling at 400 mph. The second plane is traveling at 200 mph. [2.3D] **24.** 20 lb of \$6 coffee and 40 lb of \$9 coffee should be used. [2.3A]
25. The slope is $-\frac{10{,}000}{3}$. The value of the house decreases by \$3333.33 each year. [4.4A]

Answers to Chapter 5 Selected Exercises

PREP TEST

1. $6x + 5y$ [1.4D] **2.** 7 [1.4A] **3.** 0 [2.2A] **4.** −3 [2.2C] **5.** 1000 [2.2C]

6.

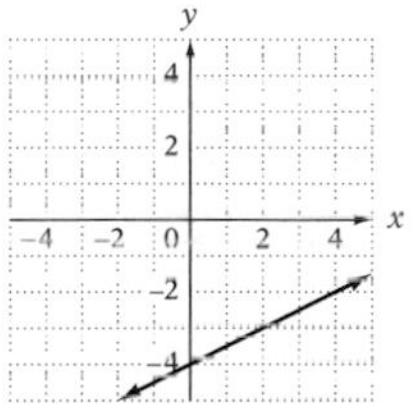

[4.3A]

7.

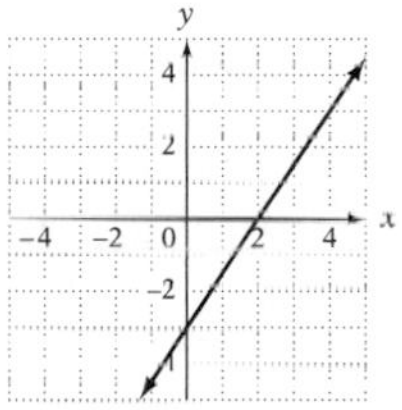

[4.3B]

8.

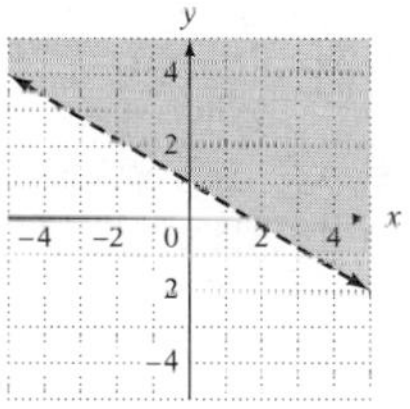

[4.7A]

SECTION 5.1

1. no **3.** yes **5.** no **7.** yes **9.** independent **11.** inconsistent

13.

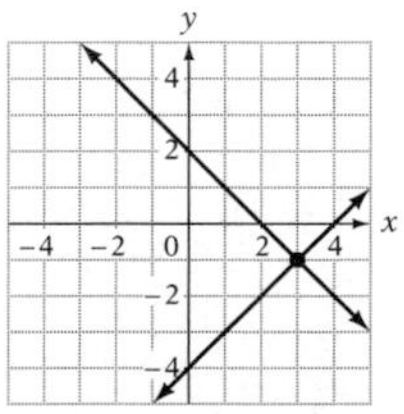

The solution is (3, −1).

15.

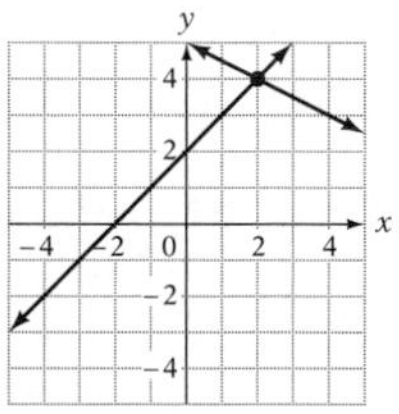

The solution is (2, 4).

17.

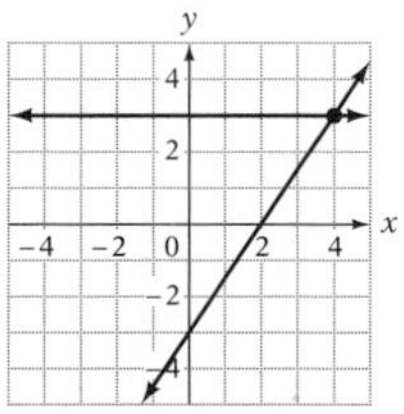

The solution is (4, 3).

19.

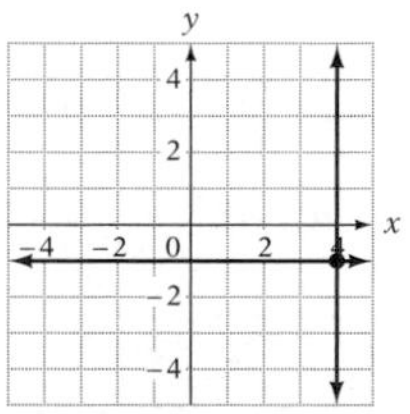

The solution is (4, −1).

21.

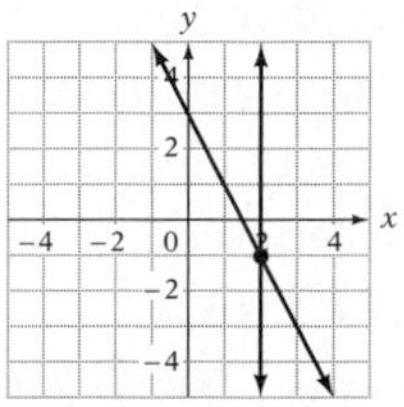

The solution is (2, −1).

23.

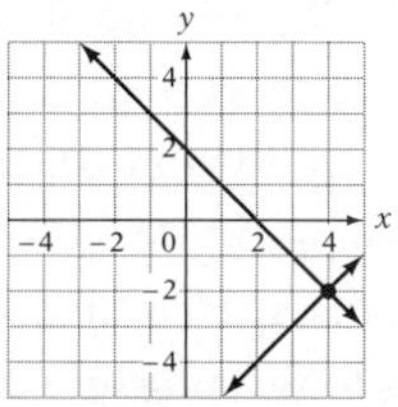

The solution is (4, −2).

25.

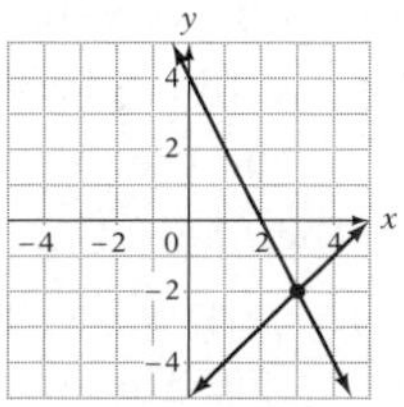

The solution is (3, −2).

27.

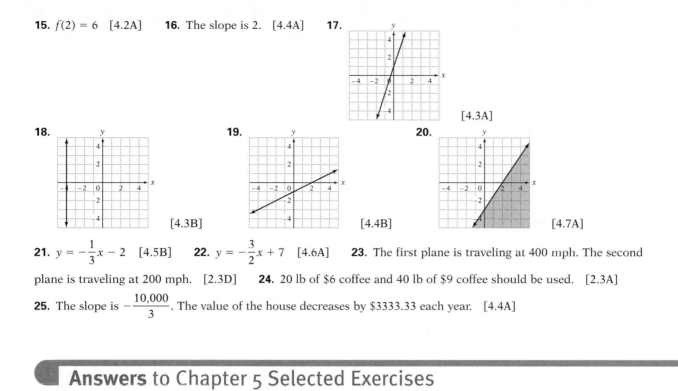

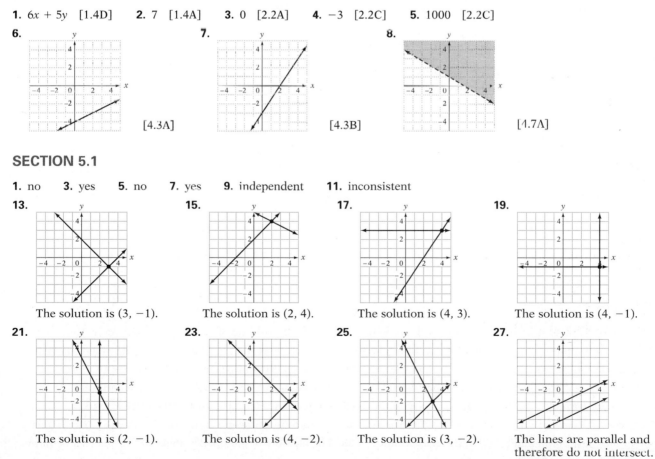

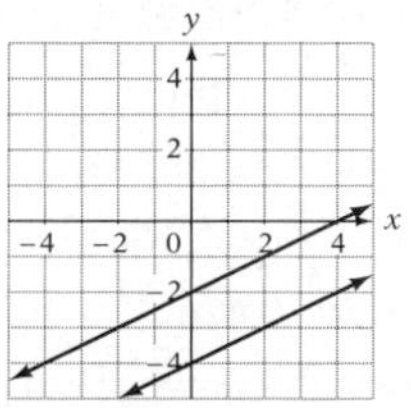

The lines are parallel and therefore do not intersect. The system of equations has no solution.

29.

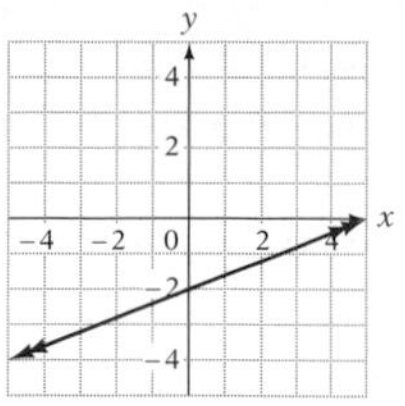

The equations are dependent. The solutions are the ordered pairs $\left(x, \frac{2}{5}x - 2\right)$.

31.

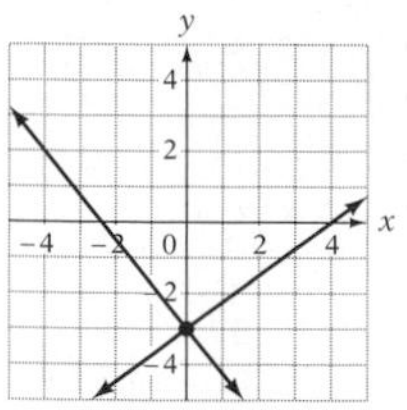

The solution is $(0, -3)$.

33.

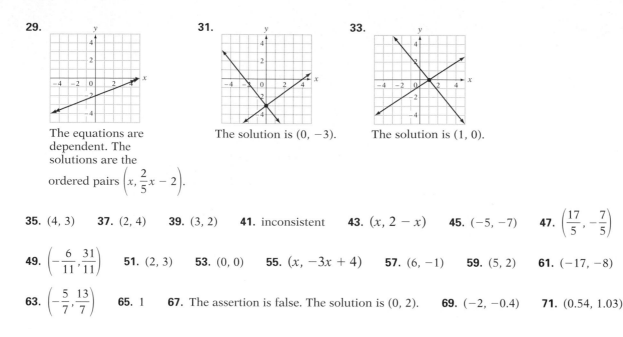

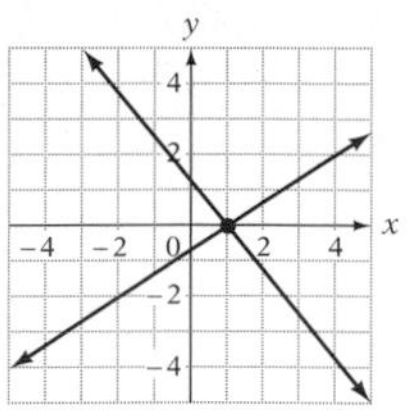

The solution is $(1, 0)$.

35. $(4, 3)$ **37.** $(2, 4)$ **39.** $(3, 2)$ **41.** inconsistent **43.** $(x, 2 - x)$ **45.** $(-5, -7)$ **47.** $\left(\frac{17}{5}, -\frac{7}{5}\right)$ **49.** $\left(-\frac{6}{11}, \frac{31}{11}\right)$ **51.** $(2, 3)$ **53.** $(0, 0)$ **55.** $(x, -3x + 4)$ **57.** $(6, -1)$ **59.** $(5, 2)$ **61.** $(-17, -8)$ **63.** $\left(-\frac{5}{7}, \frac{13}{7}\right)$ **65.** 1 **67.** The assertion is false. The solution is $(0, 2)$. **69.** $(-2, -0.4)$ **71.** $(0.54, 1.03)$

SECTION 5.2

1. $(6, 1)$ **3.** $(1, 1)$ **5.** $(2, 1)$ **7.** $(-2, 1)$ **9.** The equations are dependent. The solutions are the ordered pairs $(x, 3x - 4)$. **11.** $\left(-\frac{1}{2}, 2\right)$ **13.** The system is inconsistent and therefore has no solution. **15.** $(-1, -2)$ **17.** $(-5, 4)$ **19.** $(2, 5)$ **21.** $\left(\frac{1}{2}, \frac{3}{4}\right)$ **23.** $(0, 0)$ **25.** $(-1, 3)$ **27.** $\left(\frac{2}{3}, -\frac{2}{3}\right)$ **29.** $(-2, 3)$ **31.** $(2, -1)$ **33.** $(10, -5)$ **35.** $\left(-\frac{1}{2}, \frac{2}{3}\right)$ **37.** $\left(\frac{5}{3}, \frac{1}{3}\right)$ **39.** The system is inconsistent and therefore has no solution. **41.** $(1, -1)$ **43.** $(2, 1, 3)$ **45.** $(1, -1, 2)$ **47.** $(1, 2, 4)$ **49.** $(2, -1, -2)$ **51.** $(-2, -1, 3)$ **53.** The system is inconsistent and therefore has no solution. **55.** $(1, 4, 1)$ **57.** $(1, 3, 2)$ **59.** $(1, -1, 3)$ **61.** $(0, 2, 0)$ **63.** $(1, 5, 2)$ **65.** $(-2, 1, 1)$ **67.** $A = 3, B = -3$ **69a.** 14 **b.** $\frac{2}{3}$ **c.** $\frac{1}{2}$ **71.** $A = 1$

SECTION 5.3

3. 11 **5.** 18 **7.** 0 **9.** 15 **11.** -30 **13.** 0 **15.** $(3, -4)$ **17.** $(4, -1)$ **19.** $\left(\frac{11}{14}, \frac{17}{21}\right)$ **21.** $\left(\frac{1}{2}, 1\right)$ **23.** The system of equations is not independent. **25.** $(-1, 0)$ **27.** $(1, -1, 2)$ **29.** $(2, -2, 3)$ **31.** The system of equations is not independent. **33.** $\left(\frac{68}{25}, \frac{56}{25}, -\frac{8}{25}\right)$ **35.** $\left(\frac{2}{3}, -1, \frac{1}{2}\right)$ **37.** The area is 239 ft^2. **39a.** 0 **b.** 0

SECTION 5.4

1. The rate of the boat is 15 mph. The rate of the current is 3 mph. **3.** The rate of the plane is 502.5 mph. The rate of the wind is 47.5 mph. **5.** The rate of the team is 8 km/h. The rate of the current is 2 km/h. **7.** The rate of the plane is 180 mph. The rate of the wind is 20 mph. **9.** The rate of the plane is 110 mph. The rate of the wind is 10 mph. **11.** The rate of the motorboat is 13 mph. The rate of the current is 3 mph. **13.** The rate of the cabin cruiser is 15 mph. The rate of the current is 3 mph. **15.** The cost of the cinnamon tea is \$2.50/lb. The cost of the spice tea is \$3/lb. **17.** The cost of the wool carpet is \$52/yd. **19.** The company plans to manufacture 25 mountain bikes. **21.** You have \$1500 in the 6% account and \$3500 in the 8% account. **23.** The measure of the smaller angle is 9°. The measure of the larger angle is 81°. **25.** The manufacturer charges \$4000 for each model VI. **27.** There were 190 regular tickets, 210 student tickets, and 350 member-discount tickets sold. **29.** The original number is 84.

SECTION 5.5

1.
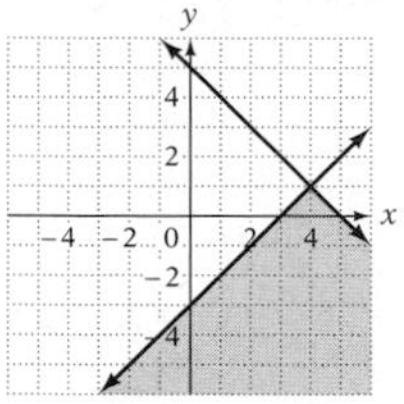
3.
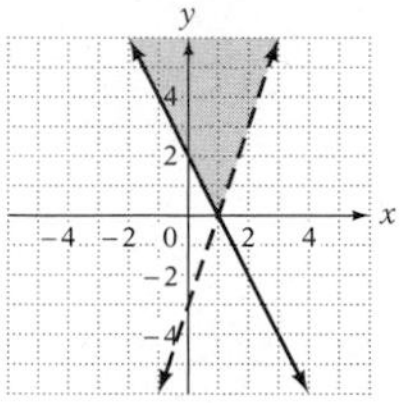
5.
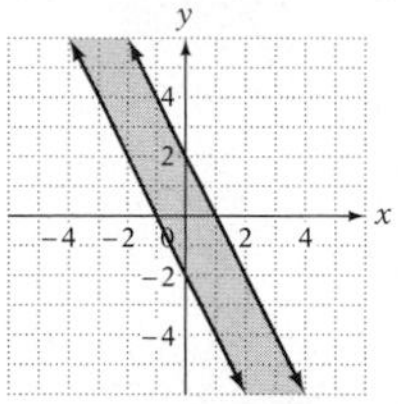
7.
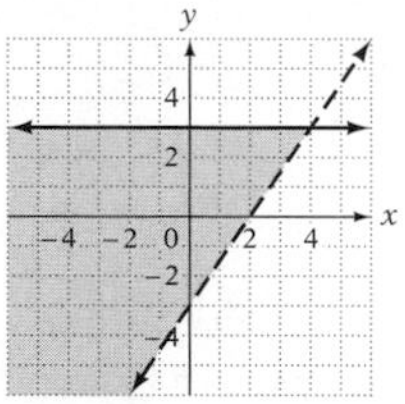
9.
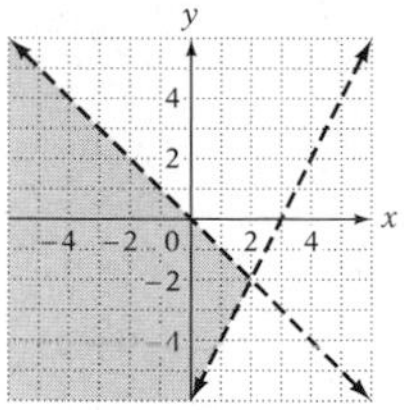
11.
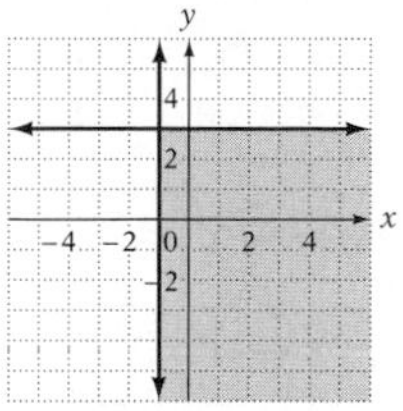
13.
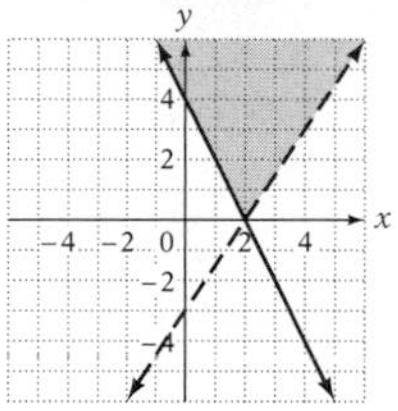
15.
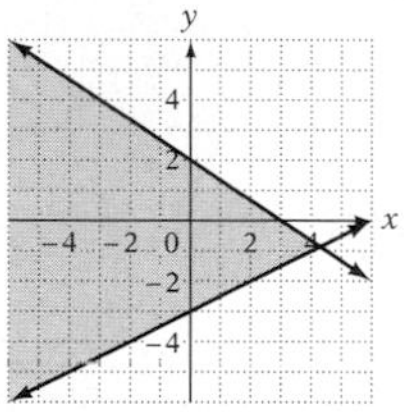
17.
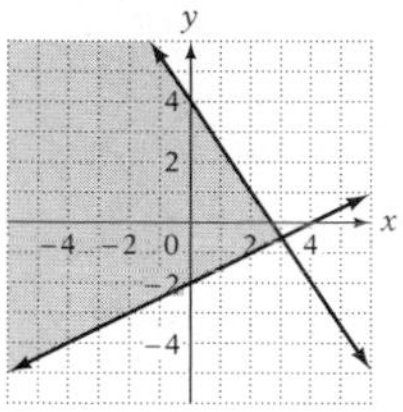
19.
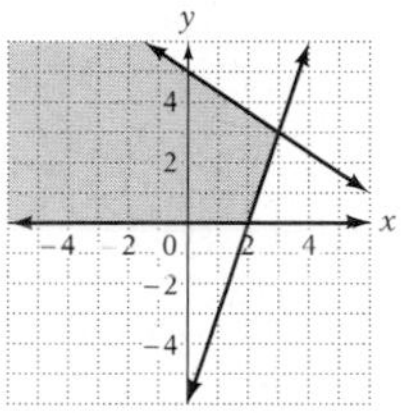
21.
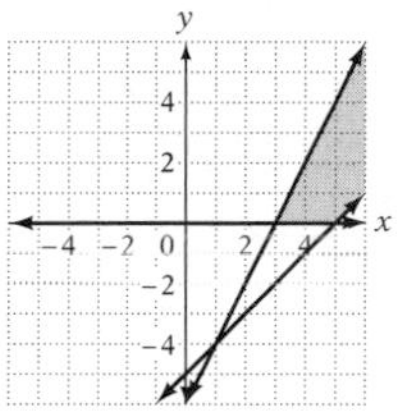
23.
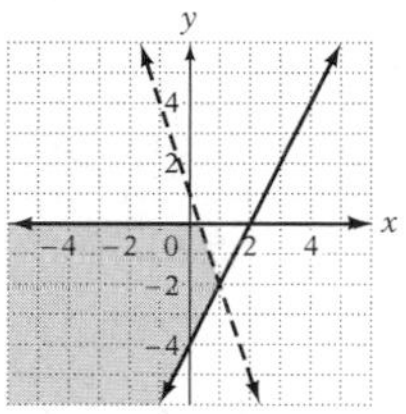

CHAPTER REVIEW

1.
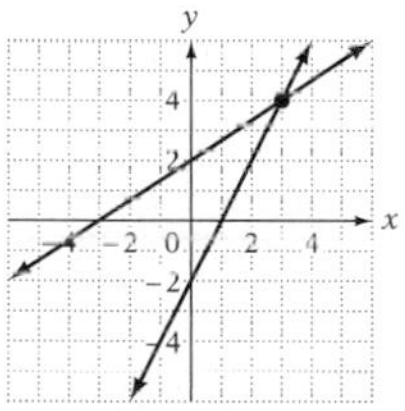
[5.1A]
The solution is (3, 4).

2.
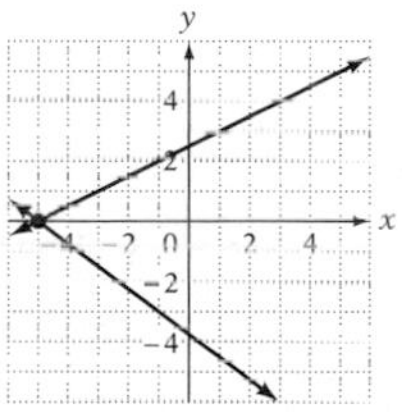
[5.1A]
The solution is (−5, 0).

3.
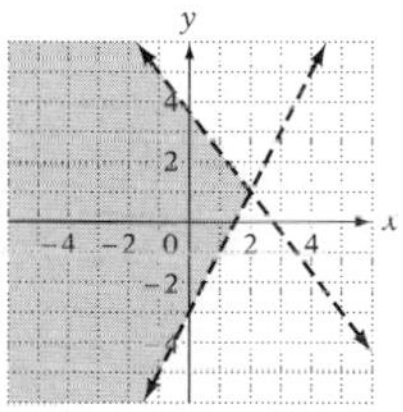
[5.5A]

4.
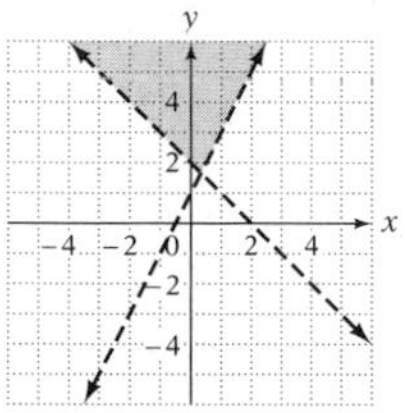
[5.5A]

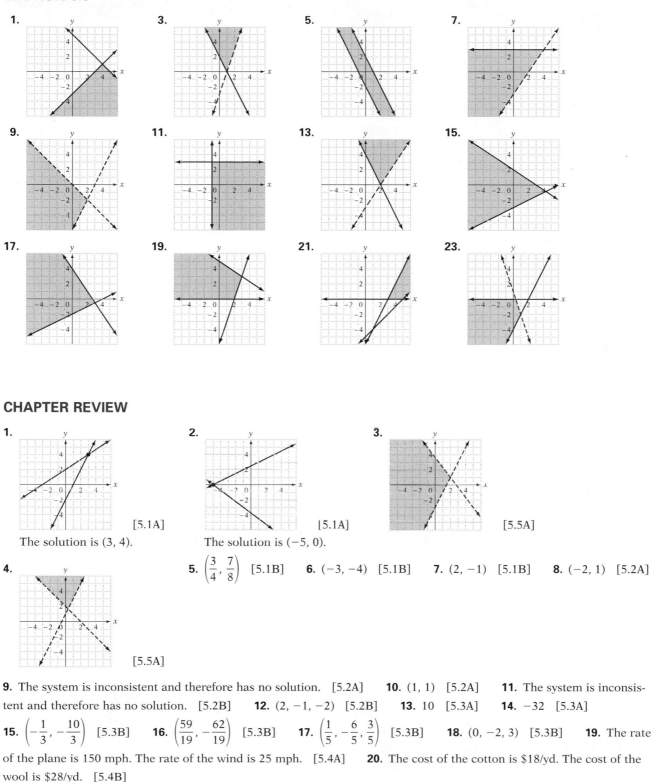

5. $\left(\frac{3}{4}, \frac{7}{8}\right)$ [5.1B] **6.** (−3, −4) [5.1B] **7.** (2, −1) [5.1B] **8.** (−2, 1) [5.2A]

9. The system is inconsistent and therefore has no solution. [5.2A] **10.** (1, 1) [5.2A] **11.** The system is inconsistent and therefore has no solution. [5.2B] **12.** (2, −1, −2) [5.2B] **13.** 10 [5.3A] **14.** −32 [5.3A] **15.** $\left(-\frac{1}{3}, -\frac{10}{3}\right)$ [5.3B] **16.** $\left(\frac{59}{19}, -\frac{62}{19}\right)$ [5.3B] **17.** $\left(\frac{1}{5}, -\frac{6}{5}, \frac{3}{5}\right)$ [5.3B] **18.** (0, −2, 3) [5.3B] **19.** The rate of the plane is 150 mph. The rate of the wind is 25 mph. [5.4A] **20.** The cost of the cotton is \$18/yd. The cost of the wool is \$28/yd. [5.4B]

CHAPTER TEST

1. $\left(6, -\frac{1}{2}\right)$ [5.1B] **2.** (−4, 7) [5.2A]

3. 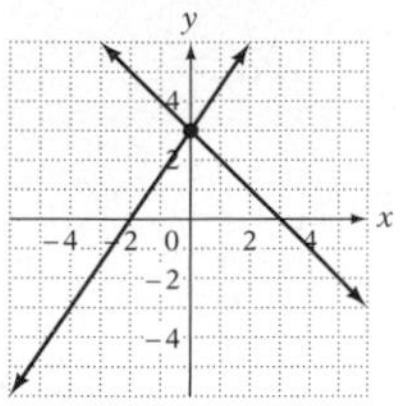 [5.1A]

The solution is (0, 3).

4. 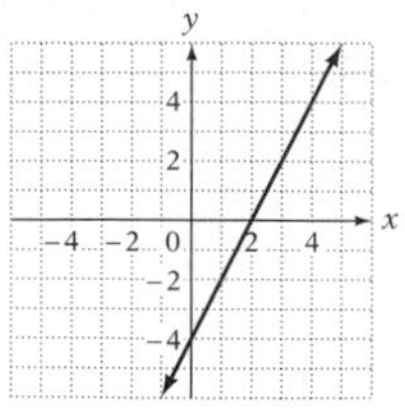 [5.1A]

The equations are dependent. The solutions are the ordered pairs $(x, 2x - 4)$.

5. $\left(x, -\frac{1}{4}x + \frac{3}{2}\right)$ [5.1B] **6.** $\left(x, \frac{1}{3}x - 2\right)$ [5.2A]

7. $(3, -1, -2)$ [5.2B] **8.** $(5, -2, 3)$ [5.2B] **9.** 28 [5.3A] **10.** 0 [5.3A] **11.** $(3, -1)$ [5.3B]

12. $\left(\frac{110}{23}, \frac{25}{23}\right)$ [5.3B] **13.** $(-1, -3, 4)$ [5.3B] **14.** $(2, 3, -5)$ [5.3B]

15. 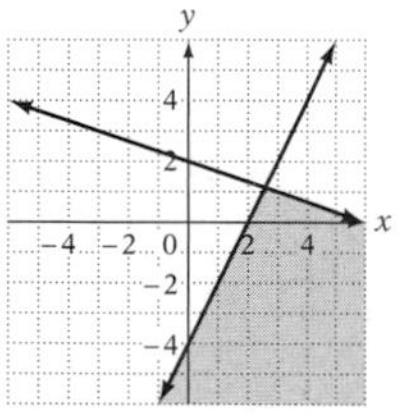 [5.5A]

16. 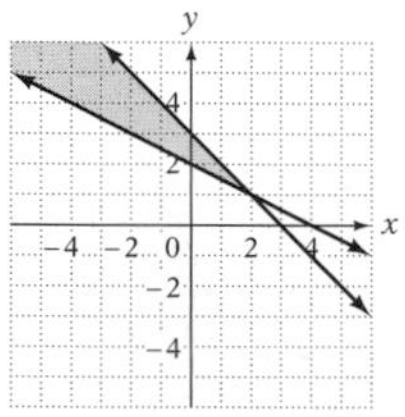 [5.5A]

17. (2, 1) [5.1B] **18.** The rate of the cabin cruiser in calm water is 16 mph. The rate of the current is 4 mph. [5.4A] **19.** The rate of the plane in calm air is 175 mph. The rate of the wind is 25 mph. [5.4A] **20.** On Friday evening, 100 children attended the movie theater. [5.4B]

CUMULATIVE REVIEW

1. $-6\sqrt{10}$ [1.2F] **2.** 22 [2.2C] **3.** $3x - 24$ [1.4D] **4.** -4 [1.4A] **5.** $\{x \mid x < 6\}$ [2.4B]
6. $\{x \mid -4 < x < 8\}$ [2.5B] **7.** $\{x \mid x > 4 \text{ or } x < -1\}$ [2.5B] **8.** $F(2) = 1$ [4.2A]

9. 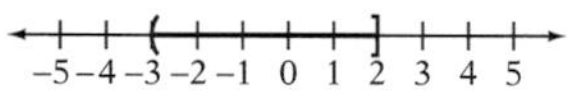 [1.5C] **10.** $y = -\frac{2}{3}x + \frac{5}{3}$ [4.5A] **11.** $y = 5x - 11$ [4.5B]

12. $y = -\frac{3}{2}x - 1$ [4.6A]

13. 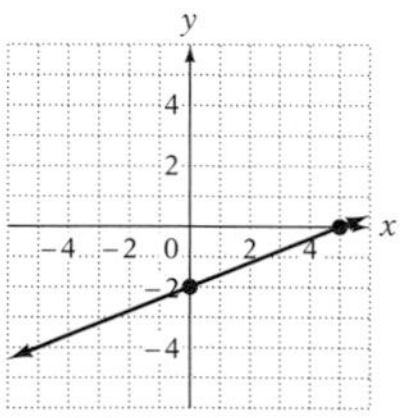 [4.4B]

14. 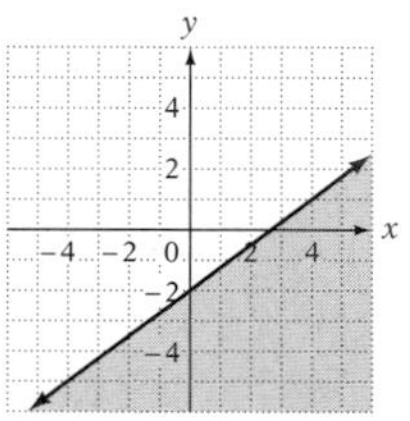 [4.7A]

15. 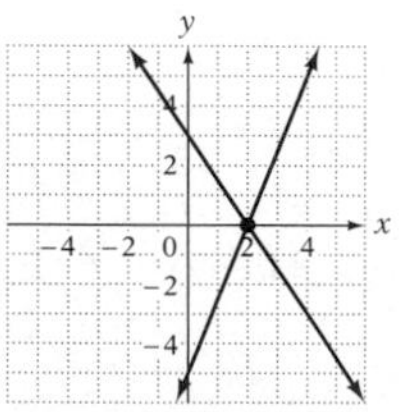 [5.1A]

The solution is (2, 0).

16. 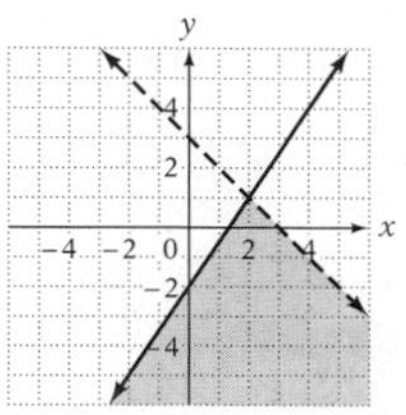 [5.5A]

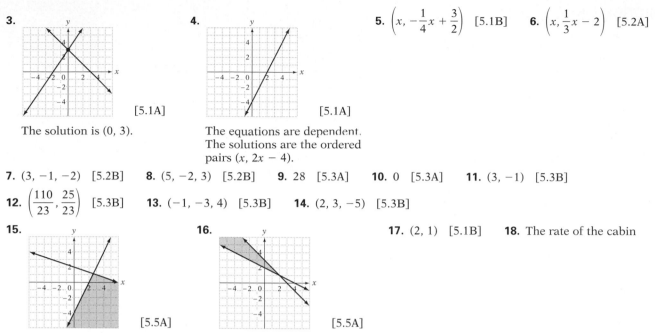

17. $(1, 0, -1)$ [5.2B] **18.** 3 [5.3A]

19. $(2, -3)$ [5.3B] **20.** $(-5, -11)$ [5.1B] **21.** 60 ml of pure water must be used. [2.3B] **22.** The rate of the wind is 12.5 mph. [5.4A] **23.** One pound of steak costs \$10. [5.4B] **24.** The lower limit is 10,200 ohms. The upper limit is 13,800 ohms. [2.5C] **25.** The slope is 40. The slope represents the monthly income (in dollars) per thousand dollars in sales. [4.4A]

Answers to Chapter 6 Selected Exercises

PREP TEST

1. 1 [1.1C] **2.** −18 [1.1D] **3.** $\frac{2}{3}$ [1.1D] **4.** $-12y$ [1.4C] **5.** −8 [1.2E] **6.** $3a - 8b$ [1.4B]
7. $11x - 2y - 2$ [1.4D] **8.** 0 [1.4B] **9.** No [1.4B]

SECTION 6.1

1. **a** is the multiplication of two exponential expressions. **b** is the power of an exponential expression. **3.** a^4b^4
5. $-18x^3y^4$ **7.** x^8y^{16} **9.** $81x^8y^{12}$ **11.** $729a^{10}b^6$ **13.** x^5y^{11} **15.** $729x^6$ **17.** $a^{18}b^{18}$ **19.** $4096x^{12}y^{12}$
21. $64a^{24}b^{18}$ **23.** x^{2n+1} **25.** y^{6n-2} **27.** a^{2n^2-6n} **29.** x^{15n+10} **31.** $-6x^5y^5z^4$ **33.** $-12a^2b^9c^2$
35. $-6x^4y^4z^5$ **37.** $-432a^7b^{11}$ **39.** $54a^{13}b^{17}$ **41.** 243 **43.** y^3 **45.** $\frac{a^3b^2}{4}$ **47.** $\frac{x}{y^4}$ **49.** $\frac{1}{2}$ **51.** $-\frac{1}{9}$ **53.** $\frac{1}{y^8}$
55. $\frac{1}{x^6y^{10}}$ **57.** x^5y^5 **59.** $\frac{a^8}{b^9}$ **61.** $\frac{1}{2187a}$ **63.** $\frac{y^6}{x^3}$ **65.** $\frac{y^4}{x^3}$ **67.** $-\frac{1}{2x^2}$ **69.** $-\frac{1}{243a^5b^{10}}$ **71.** $\frac{16x^8}{81y^4z^{12}}$
73. $\frac{3a^4}{4b^3}$ **75.** $-\frac{9a}{8b^6}$ **77.** $\frac{16x^2}{y^6}$ **79.** $\frac{1}{b^{4n}}$ **81.** $-\frac{1}{y^{6n}}$ **83.** y^{n-2} **85.** $\frac{1}{y^{2n}}$ **87.** $\frac{x^{n-5}}{y^6}$ **89.** $\frac{8b^{15}}{3a^{18}}$
91. 4.67×10^{-6} **93.** 1.7×10^{-10} **95.** 2×10^{11} **97.** 0.000000123 **99.** 8,200,000,000,000,000 **101.** 0.039
103. 150,000 **105.** 20,800,000 **107.** 0.000000015 **109.** 0.0000000000178 **111.** 140,000,000 **113.** 11,000,000
115. 0.000008 **117.** Light travels 2.592×10^{10} km in one day. **119.** The signals traveled at $1.0\overline{81} \times 10^7$ mi/min.
121. The centrifuge makes one revolution in $6.\overline{6} \times 10^{-7}$ s. **123.** The sun is 3.38983×10^5 times heavier than Earth.
125. One orchid seed weighs 3.225806×10^{-8} oz. **127.** The Coma cluster is 1.6423949×10^{21} mi from Earth.

SECTION 6.2

1. $P(3) = 13$ **3.** $R(2) = 10$ **5.** $f(-1) = 11$ **7.a.** 1 **b.** 8 **c.** 2
9. The expression is not a polynomial function. **11.** The expression is not a polynomial function.
13.a. 3 **b.** π **c.** 5 **15.a.** −5 **b.** 2 **c.** 3 **17.a.** 14 **b.** 14 **c.** 0

19.
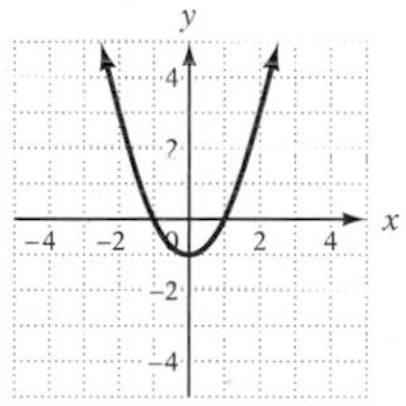

21.
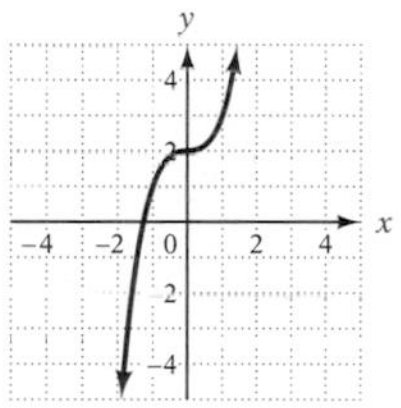

23.
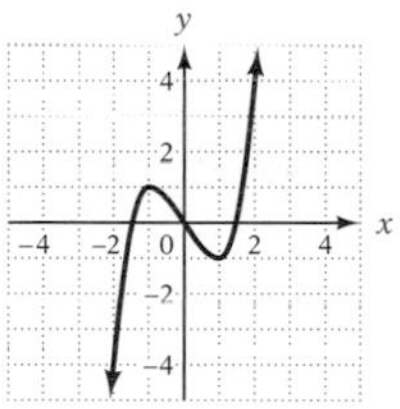

25. $6x^2 - 6x + 5$
27. $-x^2 + 1$ **29.** $5y^2 - 15y + 2$ **31.** $7a^2 - a + 2$ **33.** $P(x) + R(x) = 3x^2 - 3x - 2$
35. $P(x) - R(x) = 8x^2 - 2x + 5$ **37.** $S(x) = 3x^4 - 8x^2 + 2x$; $S(2) = 20$ **39.a.** $k = 8$ **b.** $k = -4$

SECTION 6.3

1. $x^2 - 2x$ **3.** $-x^2 - 7x$ **5.** $3a^3 - 6a^2$ **7.** $-5x^4 + 5x^3$ **9.** $-3x^5 + 7x^3$ **11.** $12x^3 - 6x^2$ **13.** $6x^2 - 12x$
15. $3x^2 + 4x$ **17.** $-x^3y + xy^3$ **19.** $2x^4 - 3x^2 + 2x$ **21.** $2a^3 + 3a^2 + 2a$ **23.** $3x^6 - 3x^4 - 2x^2$
25. $-6y^4 - 12y^3 + 14y^2$ **27.** $-2a^3 - 6a^2 + 8a$ **29.** $6y^4 - 3y^3 + 6y^2$ **31.** $x^3y - 3x^2y^2 + xy^3$
33. $x^3 + 4x^2 + 5x + 2$ **35.** $a^3 - 6a^2 + 13a - 12$ **37.** $-2b^3 + 7b^2 + 19b - 20$ **39.** $-6x^3 + 31x^2 - 41x + 10$
41. $x^3 - 3x^2 + 5x - 15$ **43.** $x^4 - 4x^3 - 3x^2 + 14x - 8$ **45.** $15y^3 - 16y^2 - 70y + 16$
47. $5a^4 - 20a^3 - 5a^2 + 22a - 8$ **49.** $y^4 + 4y^3 + y^2 - 5y + 2$ **51.** $x^2 + 4x + 3$ **53.** $a^2 + a - 12$ **55.** $y^2 - 5y - 24$
57. $y^2 - 10y + 21$ **59.** $2x^2 + 15x + 7$ **61.** $3x^2 + 11x - 4$ **63.** $4x^2 - 31x + 21$ **65.** $3y^2 - 2y - 16$
67. $9x^2 + 54x + 77$ **69.** $21a^2 - 83a + 80$ **71.** $6a^2 - 25ab + 14b^2$ **73.** $2a^2 - 11ab - 63b^2$
75. $100a^2 - 100ab + 21b^2$ **77.** $15x^2 + 56xy + 48y^2$ **79.** $14x^2 - 97xy - 60y^2$ **81.** $56x^2 - 61xy + 15y^2$
85. $y^2 - 25$ **87.** $4x^2 - 9$ **89.** $x^2 + 2x + 1$ **91.** $9a^2 - 30a + 25$ **93.** $9x^2 - 49$ **95.** $4a^2 + 4ab + b^2$
97. $16 - 9y^2$ **99.** $25x^2 + 20xy + 4y^2$ **101.** The area is $(10x^2 - 35x)$ ft². **103.** The area is $(4x^2 + 4x + 1)$ km².

105. The area is $(18x^2 + 12x + 2)$ in². **107.** The area is $(4x^2 + 10x)$ m². **109.** The total area is $(3000 + 60w)$ yd².
111. $4ab$ **113.** $a^3 + 9a^2 + 27a + 27$ **115.a.** $7x^2 - 11x - 8$ **b.** $x^3 - 7x^2 - 7$

SECTION 6.4

1. $x + 8$ **3.** $x^2 + \frac{2}{x-3}$ **5.** $3x + 5 + \frac{3}{2x+1}$ **7.** $5x + 7 + \frac{2}{2x-1}$ **9.** $4x^2 + 6x + 9 + \frac{18}{2x-3}$
11. $3x^2 + 1 + \frac{1}{2x^2-5}$ **13.** $x^2 - 3x - 10$ **15.** $x^2 - 2x + 1 - \frac{1}{x-3}$ **17.** $2x^3 - 3x^2 + x - 4$
19. $2x + \frac{x+2}{x^2+2x-1}$ **21.** $x^2 + 4x + 6 + \frac{10x+8}{x^2-2x-1}$ **23.** $x^2 - 2x + 4 + \frac{-6x-7}{x^2+2x+3}$ **25.** $2x - 8$ **27.** $3x - 8$
29. $3x + 3 - \frac{1}{x-1}$ **31.** $2x^2 - 3x + 9$ **33.** $4x^2 + 8x + 15 + \frac{12}{x-2}$ **35.** $2x^2 - 3x + 7 - \frac{8}{x+4}$
37. $3x^3 + 2x^2 + 12x + 19 + \frac{33}{x-2}$ **39.** $3x^3 - x + 4 - \frac{2}{x+1}$ **41.** $2x^3 + 6x^2 + 17x + 51 + \frac{155}{x-3}$ **43.** $f(3) = 0$
45. $P(3) = 8$ **47.** $R(4) = 43$ **49.** $P(-2) = -39$ **51.** $Z(-3) = -60$ **53.** $Q(2) = 31$ **55.** $F(-3) = 178$
57. $P(5) = 122$ **59.** $R(-3) = 302$ **61.** $Q(2) = 0$ **63.** $R(-2) = -65$ **65.a.** $a^2 - ab + b^2$
b. $x^4 - x^3y + x^2y^2 - xy^3 + y^4$ **c.** $x^5 - x^4y + x^3y^2 - x^2y^3 + xy^4 - y^5$ **67.a.** $\frac{1}{2}x^2 + \frac{3}{4}x - 1$
b. $\frac{2}{3}x^2 + \frac{7}{2}x + \frac{10}{3} + \frac{64}{6x-12}$ **c.** $2x^2 - x + 1 + \frac{3x-11}{x^2-x+1}$ **d.** $x^2 + 6x + 17 + \frac{51x+56}{x^2-2x-3}$

CHAPTER REVIEW

1. $-6a^3b^6$ [6.1A] **2.** $9x^4y^6$ [6.1A] **3.** $\frac{c^{10}}{2b^{17}}$ [6.1B] **4.** 0.00254 [6.1C]
5. -7 [6.2A] **6.** $4x^2 - 8xy + 5y^2$ [6.2B] **7.** $\frac{18x^5y^4}{z}$ [6.1B] **8.** $-54a^{13}b^5c^7$ [6.1A]
9.

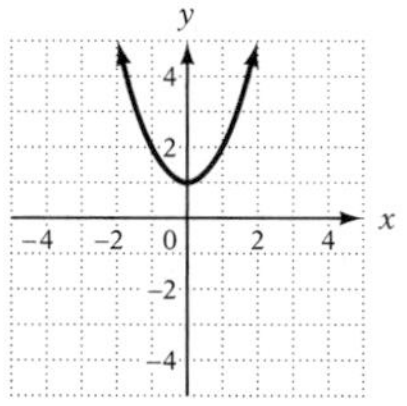

[6.2A]

10.a. 3 **b.** 8 **c.** 5 [6.2A] **11.** 1 [6.4C] **12.** $-8x^3 - 14x^2 + 18x$ [6.3A]
13. $8a^3b^3 - 4a^2b^4 + 6ab^5$ [6.3A] **14.** $6y^3 + 17y^2 - 2y - 21$ [6.3B] **15.** $10a^2 + 31a - 63$ [6.3C]
16. 1.27×10^{-7} [6.1C] **17.** $25y^2 - 70y + 49$ [6.3D] **18.** $5x + 4 + \frac{6}{3x-2}$ [6.4A]
19. $4x^2 + 3x - 8 + \frac{50}{x+6}$ [6.4B] **20.** $x^3 + 4x^2 + 16x + 64 + \frac{252}{x-4}$ [6.4B] **21.** 8,100,000,000 [6.1C]
22. $-\frac{2a^3}{3b^3}$ [6.1B] **23.** $25a^2 - 4b^2$ [6.3D] **24.** 68 [6.4C] **25.** $21y^2 + 4y - 1$ [6.2B]
26. $12b^5 - 4b^4 - 6b^3 - 8b^2 + 5$ [6.3B] **27.** $a^2 - 49$ [6.3D] **28.** 7.65×10^{11} [6.1C]
29. The area of the Ping-Pong table is $(2w^2 - w)$ ft². [6.3E] **30.** The Great Galaxy of Andromeda is 1.291224×10^{19} mi from Earth. [6.1D] **31.** There are 6.048×10^5 s in one week. [6.1D] **32.** The area is $(9x^2 - 12x + 4)$ in². [6.3E]
33. The area is $(10x^2 - 29x - 21)$ cm². [6.3E]

CHAPTER TEST

1. $4x^3 - 6x^2$ [6.3A] **2.** -8 [6.4C] **3.** $-\frac{4}{x^6}$ [6.1B] **4.** $-6x^3y^6$ [6.1A] **5.** $x - 1 + \frac{2}{x+1}$ [6.4A/6.4B]
6. $x^3 - 7x^2 + 17x - 15$ [6.3B] **7.** $-8a^6b^3$ [6.1A] **8.** $\frac{9y^{10}}{x^{10}}$ [6.1B] **9.** $a^2 + 3ab - 10b^2$ [6.3B] **10.** -3 [6.2A]
11. $x + 7$ [6.4A/6.4B] **12.** $6y^4 - 9y^3 + 18y^2$ [6.3A] **13.** $-4x^4 + 8x^3 - 3x^2 - 14x + 21$ [6.3B]

14. $16y^2 - 9$ [6.3D] **15.** a^4b^7 [6.1A] **16.** $8ab^4$ [6.1B] **17.** $\frac{2b^7}{a^{10}}$ [6.1B] **18.** $-5a^3 + 3a^2 - 4a + 3$ [6.2B]
19. $4x^2 - 20x + 25$ [6.3D] **20.** $x^2 - 5x + 10 - \frac{23}{x + 3}$ [6.4A/6.4B] **21.** $10x^2 - 43xy + 28y^2$ [6.3C]
22. $3x^3 + 6x^2 - 8x + 3$ [6.2B] **23.** 3.02×10^{-9} [6.1C] **24.** The mass of the moon is 8.103×10^{19} tons. [6.1D]
25. The area of the circle is $(\pi x^2 - 10\pi x + 25\pi)$ m². [6.3E]

CUMULATIVE REVIEW

1. -3 and 3 [1.1A] **2.** -83 [1.1B] **3.** 6 [1.3A] **4.** $-\frac{5}{4}$ [1.4A] **5.** $-50\sqrt{3}$ [1.2F] **6.** The Inverse Property of Addition [1.4B] **7.** $-186x + 8$ [1.4D] **8.** $-\frac{1}{6}$ [2.2A] **9.** $-\frac{11}{4}$ [2.2B] **10.** -1 and $\frac{7}{3}$ [2.5A]
11. 18 [4.2A] **12.** Yes [4.2A] **13.** $-\frac{1}{6}$ [4.4A] **14.** $y = -\frac{3}{2}x + \frac{1}{2}$ [4.5A] **15.** $y = \frac{2}{3}x + \frac{16}{3}$ [4.6A]
16. $\left(-\frac{7}{5}, -\frac{8}{5}\right)$ [5.3B] **17.** $\left(-\frac{9}{7}, \frac{2}{7}, \frac{11}{7}\right)$ [5.2B] **18.** $5x - 3xy$ [1.4B] **19.** $4x^3 - 7x + 3$ [6.3B]
20. 5.01×10^{-6} [6.1C] **21.**

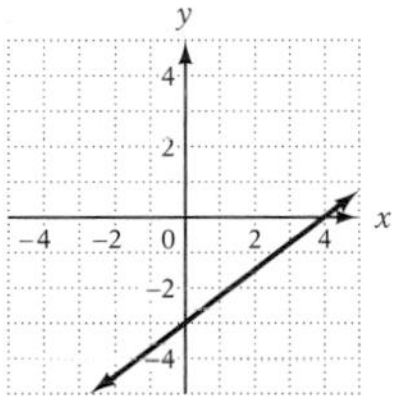

[4.3B]

22.

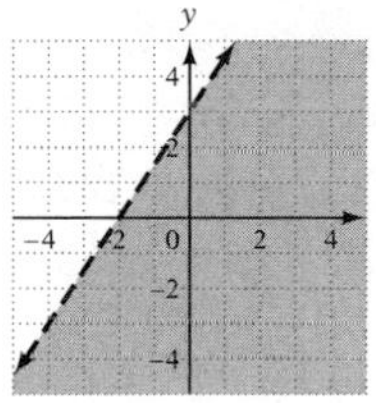

[4.7A]

23.

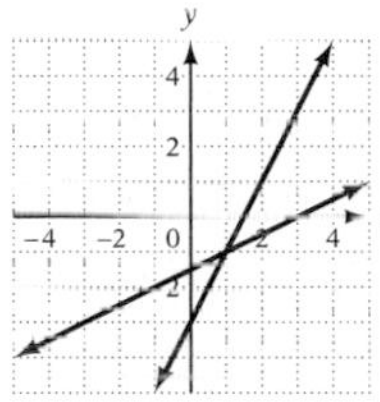

[5.1A]

The solution is $(1, -1)$.

24.

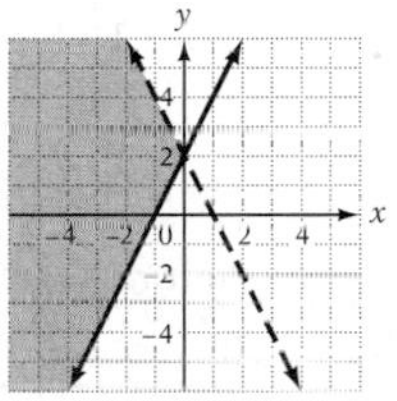

[5.5A]

25. $\frac{b^5}{a^8}$ [6.1B] **26.** $\frac{y^2}{25x^6}$ [6.1B]

27. The two integers are 9 and 15. [2.2D] **28.** 40 oz of pure gold must be used. [2.3A] **29.** The cyclists are traveling at 5 mph and 7.5 mph. [2.3D] **30.** \$4500 must be invested at an annual simple interest rate of 10%. [2.3C] **31.** The slope is 50. The slope represents the average speed in miles per hour. [4.4A] **32.** The length is 15 m. The width is 6 m. [3.2A] **33.** The area is $(4x^2 + 12x + 9)$ m². [6.3E]

Answers to Chapter 7 Selected Exercises

PREP TEST

1. $2 \cdot 3 \cdot 5$ [1.2D] **2.** $-12y + 15$ [1.4D] **3.** $-a + b$ [1.4D] **4.** $-3a + 3b$ [1.4D] **5.** 0 [2.1B]
6. $-\frac{1}{2}$ [2.2A] **7.** $x^2 - 2x - 24$ [6.3C] **8.** $6x^2 - 11x - 10$ [6.3C] **9.** x^3 [6.1B] **10.** $3x^3y$ [6.1B]

SECTION 7.1

1. $5(a + 1)$ **3.** $8(2 - a^2)$ **5.** $4(2x + 3)$ **7.** $6(5a - 1)$ **9.** $x(7x - 3)$ **11.** $a^2(3 + 5a^3)$ **13.** $y(14y + 11)$
15. $2x(x^3 - 2)$ **17.** $2x^2(5x^2 - 6)$ **19.** $4a^5(2a^3 - 1)$ **21.** $xy(xy - 1)$ **23.** $3xy(xy^3 - 2)$ **25.** $xy(x - y^2)$
27. $5y(y^2 - 4y + 2)$ **29.** $3y^2(y^2 - 3y - 2)$ **31.** $3y(y^2 - 3y + 8)$ **33.** $a^2(6a^3 - 3a - 2)$ **35.** $ab(2a - 5ab + 7b)$
37. $2b(2b^4 + 3b^2 - 6)$ **39.** $x^2(8y^2 - 4y + 1)$ **41.** $a^{2n}(2a^{3n} + 1)$ **43.** $y^{2n}(y^{2n} + 1)$ **45.** $b^5(b^n - 1)$

47. $(a + z)(y + 7)$ **49.** $(a - b)(3r + s)$ **51.** $(m - 7)(t - 7)$ **53.** $(4a - b)(2y + 1)$ **55.** $(x + 2)(x + 2y)$
57. $(p - 2)(p - 3r)$ **59.** $(a + 6)(b - 4)$ **61.** $(2z - 1)(z + y)$ **63.** $(4v + 7)(2v - 3y)$ **65.** $(2x - 5)(x - 3y)$
67. $(y - 2)(3y - a)$ **69.** $(3x - y)(y + 1)$ **71.** $(3s + t)(t - 2)$ **73a.** 28 **b.** 496

SECTION 7.2

1. $(x + 1)(x + 2)$ **3.** $(x + 1)(x - 2)$ **5.** $(a + 4)(a - 3)$ **7.** $(a - 1)(a - 2)$ **9.** $(a + 2)(a - 1)$ **11.** $(b - 3)(b - 3)$
13. $(b + 8)(b - 1)$ **15.** $(y + 11)(y - 5)$ **17.** $(y - 2)(y - 3)$ **19.** $(z - 5)(z - 9)$ **21.** $(z + 8)(z - 20)$
23. $(p + 3)(p + 9)$ **25.** $(x + 10)(x + 10)$ **27.** $(b + 4)(b + 5)$ **29.** $(x + 3)(x - 14)$ **31.** $(b + 4)(b - 5)$
33. $(y + 3)(y - 17)$ **35.** $(p + 3)(p - 7)$ **37.** nonfactorable over the integers **39.** $(x - 5)(x - 15)$ **41.** $(x - 7)(x - 8)$
43. $(x + 8)(x - 7)$ **45.** $(a + 3)(a - 24)$ **47.** $(a - 3)(a - 12)$ **49.** $(z + 8)(z - 17)$ **51.** $(c + 9)(c - 10)$
53. $(z + 4)(z + 11)$ **55.** $(c + 2)(c + 17)$ **57.** $(x + 8)(x - 12)$ **59.** $(x - 8)(x - 14)$ **61.** $(b + 15)(b - 7)$
63. $(a + 3)(a - 12)$ **65.** $(b - 6)(b - 17)$ **67.** $(a + 3)(a + 24)$ **69.** $(x + 12)(x + 13)$ **71.** $(x + 6)(x - 16)$
73. $2(x + 1)(x + 2)$ **75.** $-(x + 2)(x - 9)$ **77.** $a(b + 5)(b - 3)$ **79.** $x(y - 2)(y - 3)$ **81.** $z(z - 3)(z - 4)$
83. $-3y(y - 2)(y - 3)$ **85.** $3(x + 4)(x - 3)$ **87.** $5(z + 4)(z - 7)$ **89.** $2a(a + 8)(a - 4)$ **91.** $(x - 2y)(x - 3y)$
93. $(a - 4b)(a - 5b)$ **95.** $(x + 4y)(x - 7y)$ **97.** nonfactorable over the integers **99.** $z^2(z - 5)(z - 7)$
101. $b^2(b - 10)(b - 12)$ **103.** $2y^2(y + 3)(y - 16)$ **105.** $-x^2(x + 8)(x - 1)$ **107.** $4y(x + 7)(x - 2)$
109. $c(c + 20)(c - 2)$ **111.** $-4x(x + 3)(x - 2)$ **113.** $(y + x)(y - 8x)$ **115.** $(y + 7z)(y - 3z)$ **117.** $(y - z)(y - 15z)$
119. $4y(x + 1)(x - 18)$ **121.** $4x(x + 8)(x - 5)$ **123.** $5z(z + 2)(z - 12)$ **125.** $5x(x + 2)(x + 4)$ **127.** $4(p + 8)(p - 15)$
129. $p^2(p + 8)(p - 7)$ **131.** $(a - 5b)(a - 5b)$ **133.** $(x + 10y)(x - 6y)$ **135.** $6x(x - 5)(x + 4)$ **137.** $y(x + 6)(x - 9)$
139. $k = -36, 36, -12, 12$ **141.** $k = 22, -22, 10, -10$ **143.** $k = 6, 10, 12$ **145.** $k = 6, 10, 12$ **147.** $k = 4, 6$

SECTION 7.3

1. $(x + 1)(2x + 1)$ **3.** $(y + 3)(2y + 1)$ **5.** $(a - 1)(2a - 1)$ **7.** $(b - 5)(2b - 1)$ **9.** $(x + 1)(2x - 1)$
11. $(x - 3)(2x + 1)$ **13.** $(t + 2)(2t - 5)$ **15.** $(p - 5)(3p - 1)$ **17.** $(3y - 1)(4y - 1)$ **19.** nonfactorable over the integers **21.** $(2t - 1)(3t - 4)$ **23.** $(x + 4)(8x + 1)$ **25.** nonfactorable over the integers **27.** $(3y + 1)(4y + 5)$
29. $(a + 7)(7a - 2)$ **31.** $(b - 4)(3b - 4)$ **33.** $(z - 14)(2z + 1)$ **35.** $(p + 8)(3p - 2)$ **37.** $2(x + 1)(2x + 1)$
39. $5(y - 1)(3y - 7)$ **41.** $x(x - 5)(2x - 1)$ **43.** $b(a - 4)(3a - 4)$ **45.** nonfactorable over the integers
47. $-3x(x - 3)(x + 4)$ **49.** $4(4y - 1)(5y - 1)$ **51.** $z(2z + 3)(4z + 1)$ **53.** $y(2x - 5)(3x + 2)$ **55.** $5(t + 2)(2t - 5)$
57. $p(p - 5)(3p - 1)$ **59.** $2(z + 4)(13z - 3)$ **61.** $2y(y - 4)(5y - 2)$ **63.** $yz(z + 2)(4z - 3)$ **65.** $3a(2a + 3)(7a - 3)$
67. $y(3x - 5y)(3x - 5y)$ **69.** $xy(3x - 4y)(3x - 4y)$ **71.** $(2x - 3)(3x - 4)$ **73.** $(b + 7)(5b - 2)$ **75.** $(3a + 8)(2a - 3)$
77. $(z + 2)(4z + 3)$ **79.** $(2p + 5)(11p - 2)$ **81.** $(y + 1)(8y + 9)$ **83.** $(6t - 5)(3t + 1)$ **85.** $(b + 12)(6b - 1)$
87. $(3x + 2)(3x + 2)$ **89.** $(2b - 3)(3b - 2)$ **91.** $(3b + 5)(11b - 7)$ **93.** $(3y - 4)(6y - 5)$ **95.** $(3a + 7)(5a - 3)$
97. $(2y - 5)(4y - 3)$ **99.** $(2z + 3)(4z - 5)$ **101.** nonfactorable over the integers **103.** $(2z - 5)(5z - 2)$
105. $(6z + 5)(6z + 7)$ **107.** $(x + y)(3x - 2y)$ **109.** $(a + 2b)(3a - b)$ **111.** $(y - 2z)(4y - 3z)$ **113.** $-(z - 7)(z + 4)$
115. $-(x - 1)(x + 8)$ **117.** $3(x + 5)(3x - 4)$ **119.** $4(2x - 3)(3x - 2)$ **121.** $a^2(5a + 2)(7a - 1)$ **123.** $5(b - 7)(3b - 2)$
125. $(x - 7y)(3x - 5y)$ **127.** $3(8y - 1)(9y + 1)$ **129.** $-(x - 1)(x + 21)$ **131.** $(5a + 7b)(3a - 2b)$
133. $-z(z + 11)(z - 3)$ **135.** $(3a + 2)(a + 3)$ **137.** $k = 7, -7, 5, -5$ **139.** $k = 7, -7, 5, -5$

SECTION 7.4

1. 4; $25x^6$; $100x^4y^4$ **3.** $4z^4$ **5.** $9a^2b^3$ **7.** $(x + 4)(x - 4)$ **9.** $(2x + 1)(2x - 1)$ **11.** $(4x + 11)(4x - 11)$
13. $(1 + 3a)(1 - 3a)$ **15.** $(xy + 10)(xy - 10)$ **17.** nonfactorable over the integers **19.** $(5 + ab)(5 - ab)$
21. $(a^n + 1)(a^n - 1)$ **23.** $(x - 6)^2$ **25.** $(b - 1)^2$ **27.** $(4x - 5)^2$ **29.** nonfactorable over the integers
31. nonfactorable over the integers **33.** $(x + 3y)^2$ **35.** $(5a - 4b)^2$ **37.** $(x^n + 3)^2$ **39.** $(x - 7)(x - 1)$
41. $(x - y + a + b)(x - y - a - b)$ **43.** 8; x^9; $27c^{15}d^{18}$ **45.** $2x^3$ **47.** $4a^2b^6$ **49.** $(x - 3)(x^2 + 3x + 9)$
51. $(2x - 1)(4x^2 + 2x + 1)$ **53.** $(x - y)(x^2 + xy + y^2)$ **55.** $(m + n)(m^2 - mn + n^2)$ **57.** $(4x + 1)(16x^2 - 4x + 1)$
59. $(3x - 2y)(9x^2 + 6xy + 4y^2)$ **61.** $(xy + 4)(x^2y^2 - 4xy + 16)$ **63.** nonfactorable over the integers
65. nonfactorable over the integers **67.** $(a - 2b)(a^2 - ab + b^2)$ **69.** $(x^{2n} + y^n)(x^{4n} - x^{2n}y^n + y^{2n})$

71. $(x^n + 2)(x^{2n} - 2x^n + 4)$ **73.** $(xy - 3)(xy - 5)$ **75.** $(xy - 5)(xy - 12)$ **77.** $(x^2 - 3)(x^2 - 6)$ **79.** $(b^2 + 5)(b^2 - 18)$ **81.** $(x^2y^2 - 2)(x^2y^2 - 6)$ **83.** $(x^n + 1)(x^n + 2)$ **85.** $(3xy - 5)(xy - 3)$ **87.** $(2ab - 3)(3ab - 7)$ **89.** $(2x^2 - 15)(x^2 + 1)$ **91.** $(2x^n - 1)(x^n - 3)$ **93.** $(2a^n + 5)(3a^n + 2)$ **95.** $3(2x - 3)^2$ **97.** $a(3a - 1)(9a^2 + 3a + 1)$ **99.** $5(2x + 1)(2x - 1)$ **101.** $y^3(y + 11)(y - 5)$ **103.** $(4x^2 + 9)(2x + 3)(2x - 3)$ **105.** $2a(2 - a)(4 + 2a + a^2)$ **107.** $b^3(ab - 1)(a^2b^2 + ab + 1)$ **109.** $2x^2(2x - 5)^2$ **111.** $(x^2 + y^2)(x + y)(x - y)$ **113.** $(x^2 + y^2)(x^4 - x^2y^2 + y^4)$ **115.** nonfactorable over the integers **117.** $2a(2a - 1)(4a^2 + 2a + 1)$ **119.** $2b^2(3a + 5b)(4a - 9b)$ **121.** $(x - 2)^2(x + 2)$ **123.** $(x + y)(x - y)(2x + 1)(2x - 1)$ **125.** $x(x^n + 1)^2$ **127.** $b^n(3b - 2)(b + 2)$ **129.** The dimensions are $(4x + 3)$ m by $(4x + 3)$ m. Yes, x can equal 0. The possible values of x are $x > -\frac{3}{4}$.

SECTION 7.5

3. 3, 5 **5.** −8, 9 **7.** 0, −2 **9.** 0, −12 **11.** 0, $\frac{7}{4}$ **13.** 0, $-\frac{5}{2}$ **15.** 8, −3 **17.** 11, −11 **19.** $\frac{1}{4}, -\frac{1}{4}$ **21.** $\frac{7}{4}, -\frac{7}{4}$ **23.** 3, 5 **25.** 8, −9 **27.** $-\frac{2}{3}$, −4 **29.** $\frac{5}{3}, \frac{3}{2}$ **31.** 0, 5 **33.** 0, 4 **35.** −3, 8 **37.** −1, 8 **39.** $\frac{5}{3}$, −2 **41.** $\frac{6}{5}$, 2 **43.** 12, −1 **45.** −3, −5 **47.** 5, −4 **49.** 3, 5 **51.** 15, −1 **53.** −6, 3 **55.** 1, 3 **57.** −12, 5 **59.** $-\frac{3}{2}$, −2 **61.** The number is −3. **63.** The numbers are 3 and 5. **65.** The integers are 8 and 10. **67.** The length is 38 cm, and the width is 10 cm. **69.** 18 consecutive natural numbers beginning with 1 will give a sum of 171. **71.** There are 10 teams in the league. **73.** The object will hit the ground 4 s later. **75.** The ball will be 64 ft above the ground 2 s later. **77.** The height of the triangle is 14 m. **79.** The width of the border is 1.5 ft. **81.** The radius of the original circle is 3.81 in. **83.** The length is 20 in., and the width is 10 in. **85.** 1, 18

CHAPTER REVIEW

1. $5x(x^2 + 2x + 7)$ [7.1A] **2.** $3ab(4a + b)$ [7.1A] **3.** $7y^3(2y^6 - 7y^3 + 1)$ [7.1A] **4.** $(x - 3)(4x + 5)$ [7.1B] **5.** $(2x + 5)(5x + 2y)$ [7.1B] **6.** $(3a - 5b)(7x + 2y)$ [7.1B] **7.** $(b - 3)(b - 10)$ [7.2A] **8.** $(c + 6)(c + 2)$ [7.2A] **9.** $(y - 4)(y + 9)$ [7.2A] **10.** $3(a + 2)(a - 7)$ [7.2B] **11.** $4x(x - 6)(x + 1)$ [7.2B] **12.** $n^2(n + 1)(n - 3)$ [7.2B] **13.** $(2x - 7)(3x - 4)$ [7.3A] **14.** $(6y - 1)(2y + 3)$ [7.3A] **15.** nonfactorable over the integers [7.3A] **16.** $(3x - 2)(x - 5)$ [7.3B] **17.** $(2a + 5)(a - 12)$ [7.3B] **18.** $(6a - 5)(3a + 2)$ [7.3B] **19.** $(xy + 3)(xy - 3)$ [7.4A] **20.** $(2x + 3y)^2$ [7.4A] **21.** $(x^n - 6)^2$ [7.4A] **22.** $(4a - 3b)(16a^2 + 12ab + 9b^2)$ [7.4B] **23.** $(3x^2 + 2)(5x^2 - 3)$ [7.4C] **24.** $(7x^2y^2 + 3)(3x^2y^2 + 2)$ [7.4C] **25.** $3a^2(a^2 + 1)(a^2 - 6)$ [7.4D] **26.** $\frac{1}{4}$ and −7 [7.5A] **27.** −3 and 7 [7.5A] **28.** The length is 100 yd. The width is 60 yd. [7.5B] **29.** The width of the frame is $\frac{3}{2}$ in. or 1.5 in. or $1\frac{1}{2}$ in. [7.5B] **30.** The length of a side is 20 ft. [7.5B]

CHAPTER TEST

1. $(b + 6)(a - 3)$ [7.1B] **2.** $2y^2(y + 1)(y - 8)$ [7.2B] **3.** $4(x + 4)(2x - 3)$ [7.3B] **4.** $(2x + 1)(3x + 8)$ [7.3A] **5.** $(a - 3)(a - 16)$ [7.2A] **6.** $2x(3x^2 - 4x + 5)$ [7.1A] **7.** $(x + 5)(x - 3)$ [7.2A] **8.** $\frac{1}{2}, -\frac{1}{2}$ [7.5A] **9.** $5(x^2 - 9x - 3)$ [7.1A] **10.** $(p + 6)^2$ [7.4A] **11.** 3 and 5 [7.5A] **12.** $3(x + 2y)^2$ [7.4D] **13.** $(3x - 2)(9x^2 + 6x + 4)$ [7.4B] **14.** $3y^2(2x + 1)(x + 1)$ [7.3B] **15.** $(2a^2 - 5)(3a^2 + 1)$ [7.4C] **16.** $(x - 2)(a + b)$ [7.1B] **17.** $(p + 1)(x - 1)$ [7.1B] **18.** $3(a + 5)(a - 5)$ [7.4D] **19.** nonfactorable over the integers [7.3A] **20.** $(x + 3)(x - 12)$ [7.2A] **21.** $(2a - 3b)^2$ [7.4A] **22.** $(2x + 7y)(2x - 7y)$ [7.4A] **23.** $\frac{3}{2}$ and −7 [7.5A] **24.** The length of the rectangle is 15 cm, and the width is 6 cm. [7.5B] **25.** The numbers are 3 and 7. [7.5B]

CUMULATIVE REVIEW

1. 7 [1.1C] **2.** 4 [1.3A] **3.** −7 [1.4A] **4.** $15x^2$ [1.4C] **5.** 12 [1.4D] **6.** $\frac{2}{3}$ [2.1C] **7.** $\frac{7}{4}$ [2.2B] **8.** 3 [2.2C] **9.** 45 [2.1D] **10.** 1 [4.2A] **11.**

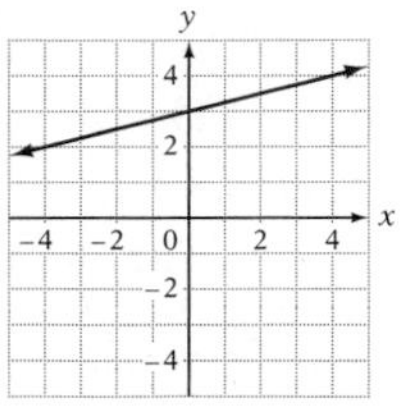

[4.3A] **12.**

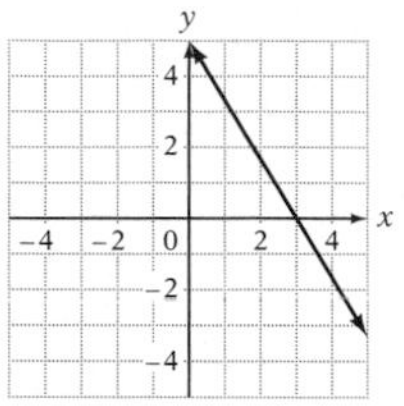

[4.3B]

13. $y = \frac{2}{3}x + 6$ [4.5A] **14.** (1, 6) [5.1B] **15.** (−1, −2) [5.2A] **16.** $9a^6b^4$ [6.1A] **17.** $x^3 - 3x^2 - 6x + 8$ [6.3B] **18.** $4x + 8 + \frac{21}{2x - 3}$ [6.4A] **19.** $\frac{y^6}{x^8}$ [6.1B] **20.** $(a - b)(3 - x)$ [7.1B] **21.** $5xy^2(3 - 4y^2)$ [7.1A] **22.** $(x - 7y)(x + 2y)$ [7.2A] **23.** $\frac{2}{3}$ and −7 [7.5A] **24.** $\frac{5}{2}$ and 4 [7.5A] **25.** The third angle measures 59°. [3.1C] **26.** The width is 15 ft. [3.2A] **27.** The pieces are 4 ft long and 6 ft long. [2.2D] **28.** $6500 was invested at 11%. [2.3C] **29.** The distance to the resort is 168 mi. [2.3D] **30.** The length of the base of the triangle is 12 in. [7.5B]

Answers to Chapter 8 Selected Exercises

PREP TEST

1. 50 [1.2B] **2.** $-\frac{1}{6}$ [1.2D] **3.** $-\frac{3}{2}$ [1.2D] **4.** $\frac{1}{24}$ [1.2C] **5.** $\frac{5}{24}$ [1.2C] **6.** $\frac{1}{3}$ [1.4A] **7.** −2 [2.2C] **8.** $\frac{10}{7}$ [2.2C] **9.** 110 mph, 130 mph [2.3D]

SECTION 8.1

3. $\frac{3}{4x}$ **5.** $\frac{1}{x + 3}$ **7.** −1 **9.** $\frac{2}{3y}$ **11.** $-\frac{3}{4x}$ **13.** $\frac{a}{b}$ **15.** $-\frac{2}{x}$ **17.** $\frac{y - 2}{y - 3}$ **19.** $\frac{x + 5}{x + 4}$ **21.** $\frac{x + 4}{x - 3}$ **23.** $-\frac{x + 2}{x + 5}$ **25.** $\frac{2(x + 2)}{x + 3}$ **27.** $\frac{2x - 1}{2x + 3}$ **29.** $-\frac{x + 7}{x + 6}$ **31.** $\frac{35ab^2}{24x^2y}$ **33.** $\frac{4x^3y^3}{3a^2}$ **35.** $\frac{3}{4}$ **37.** ab^2 **39.** $\frac{x^2(x - 1)}{y(x + 3)}$ **41.** $\frac{y(x - 1)}{x^2(x + 10)}$ **43.** $-ab^2$ **45.** $\frac{x + 5}{x + 4}$ **47.** 1 **49.** $-\frac{n - 10}{n - 7}$ **51.** $\frac{x(x + 2)}{2(x - 1)}$ **53.** $-\frac{x + 2}{x - 6}$ **55.** $\frac{x + 5}{x - 12}$ **57.** $\frac{7a^3y^2}{40bx}$ **59.** $\frac{4}{3}$ **61.** $\frac{3a}{2}$ **63.** $\frac{x^2(x + 4)}{y^2(x + 2)}$ **65.** $\frac{x(x - 2)}{y(x - 6)}$ **67.** $-\frac{3by}{ax}$ **69.** $\frac{(x - 3)(x + 6)}{(x + 7)(x - 6)}$ **71.** 1 **73.** $-\frac{x + 8}{x - 4}$ **75.** $\frac{2n + 1}{2n - 3}$ **79.** $\frac{n - 2}{n + 3}$ **81.** $\frac{8}{3y}$ **83.** 1

SECTION 8.2

1. $24x^3y^2$ **3.** $30x^4y^2$ **5.** $8x^2(x + 2)$ **7.** $6x^2y(x + 4)$ **9.** $36x(x + 2)^2$ **11.** $6(x + 1)^2$ **13.** $(x - 1)(x + 2)(x + 3)$ **15.** $(2x + 3)^2(x - 5)$ **17.** $(x - 1)(x - 2)$ **19.** $(x - 3)(x + 2)(x + 4)$ **21.** $(x + 4)(x + 1)(x - 7)$ **23.** $(x - 6)(x + 6)(x + 4)$ **25.** $(x - 10)(x - 8)(x + 3)$ **27.** $(3x - 2)(x - 3)(x + 2)$ **29.** $(x + 2)(x - 3)$ **31.** $(x - 5)(x + 1)$ **33.** $(x - 3)(x - 2)(x - 1)(x - 6)$ **35.** $\frac{5}{ab^2}$; $\frac{6b}{ab^2}$ **37.** $\frac{15y^2}{18x^2y}$; $\frac{14x}{18x^2y}$ **39.** $\frac{ay + 5a}{y^2(y + 5)}$; $\frac{6y}{y^2(y + 5)}$ **41.** $\frac{a^2y + 7a^2}{y(y + 7)^2}$; $\frac{ay}{y(y + 7)^2}$ **43.** $\frac{b}{y(y - 4)}$; $\frac{-b^2y}{y(y - 4)}$ **45.** $-\frac{3y - 21}{(y - 7)^2}$; $\frac{2}{(y - 7)^2}$

47. $\dfrac{2y^2}{y^2(y-3)}$; $\dfrac{3}{y^2(y-3)}$ **49.** $\dfrac{x^3+4x^2}{(2x-1)(x+4)}$; $\dfrac{2x^2+x-1}{(2x-1)(x+4)}$ **51.** $\dfrac{3x^2+15x}{(x-5)(x+5)}$; $\dfrac{4}{(x-5)(x+5)}$

53. $\dfrac{x-3}{(3x-2)(x+2)}$; $\dfrac{6x-4}{(3x-2)(x+2)}$ **55.** $\dfrac{x^2-1}{(x+5)(x-3)(x+1)}$; $\dfrac{x^2-3x}{(x+5)(x-3)(x+1)}$ **57.** $\dfrac{11}{y^2}$ **59.** $-\dfrac{7}{x+4}$

61. $\dfrac{8x}{2x+3}$ **63.** $\dfrac{5x+7}{x-3}$ **65.** $\dfrac{2x-5}{x+9}$ **67.** $\dfrac{-3x-4}{2x+7}$ **69.** $\dfrac{1}{x+5}$ **71.** $\dfrac{1}{x-6}$ **73.** $\dfrac{3}{2y-1}$ **75.** $\dfrac{1}{x-5}$

79. $\dfrac{7b+5a}{ab}$ **81.** $\dfrac{11}{12a}$ **83.** $\dfrac{11}{12y}$ **85.** $\dfrac{120+7y}{20y^2}$ **87.** $\dfrac{21b+4a}{12ab}$ **89.** $\dfrac{14x+3}{12x}$ **91.** $\dfrac{8x-3}{6x}$ **93.** $\dfrac{7}{36}$

95. $\dfrac{-3x^2-11x-2}{3x^2}$ **97.** $\dfrac{6x^2+5x+10}{6x^2}$ **99.** $\dfrac{2x^2+1}{x}$ **101.** $\dfrac{4x+2}{x+1}$ **103.** $\dfrac{x^2-x+2}{x^2y}$

105. $\dfrac{16xy-12y+6x^2+3x}{12x^2y^2}$ **107.** $\dfrac{3xy-6y-2x^2-14x}{24x^2y}$ **109.** $\dfrac{9x+2}{(x-2)(x+3)}$ **111.** $\dfrac{2(x+23)}{(x-7)(x+3)}$

113. $\dfrac{2x^2-5x+1}{(x+1)(x-3)}$ **115.** $\dfrac{4x^2-34x+5}{(2x-1)(x-6)}$ **117.** $\dfrac{2a-5}{a-7}$ **119.** $\dfrac{4x+9}{(x+3)(x-3)}$ **121.** $\dfrac{-x+9}{(x+2)(x-3)}$

123. $\dfrac{8y}{(y+2)^2}$ **125.** $\dfrac{3x-y}{x-y}$ **127.** $\dfrac{11}{(a-3)(a-4)}$ **129.** $\dfrac{-x-1}{(x+2)(x-5)}$ **131.** $\dfrac{x+3}{x+5}$ **133.** $\dfrac{x-2}{x+4}$

135. $\dfrac{4(2x-1)}{(x+5)(x-2)}$ **137a.** $\dfrac{3}{y}+\dfrac{6}{x}$ **b.** $\dfrac{4}{b^2}+\dfrac{3}{ab}$ **c.** $\dfrac{1}{4mn}+\dfrac{1}{6m^2}$ **139a.** $\dfrac{19{,}200}{x}$ dollars **b.** $\dfrac{96{,}500}{x(x+5)}$ dollars **c.** \$128

SECTION 8.3

1. $\dfrac{x}{x-3}$ **3.** $\dfrac{2}{3}$ **5.** $\dfrac{y+3}{y-4}$ **7.** $\dfrac{2(2x+13)}{5x+36}$ **9.** $\dfrac{x+2}{x+3}$ **11.** $\dfrac{x-6}{x+5}$ **13.** $-\dfrac{x-2}{x+1}$ **15.** $x-1$

17. $\dfrac{1}{2x-1}$ **19.** $\dfrac{x-3}{x+5}$ **21.** $\dfrac{x-7}{x-8}$ **23.** $\dfrac{2y-1}{2y+1}$ **25.** $\dfrac{x-2}{2x-5}$ **27.** $-\dfrac{x+1}{4x-3}$ **29.** $\dfrac{x+1}{2(5x-2)}$

31. $\dfrac{5}{3}$ **33.** $-\dfrac{1}{x-1}$ **35.** $\dfrac{y+4}{2(y-2)}$

SECTION 8.4

1. 3 **3.** 1 **5.** 9 **7.** 1 **9.** $\dfrac{1}{4}$ **11.** 1 **13.** -3 **15.** $\dfrac{1}{2}$ **17.** 8 **19.** 5 **21.** -1 **23.** 5

25. The equation has no solution. **27.** 2, 4 **29.** $-\dfrac{3}{2}$, 4 **31.** 3 **33.** 4 **37.** 9 **39.** 36 **41.** 10 **43.** 113

45. -2 **47.** 15 **49.** 4 **51.** 6.7 cm **53.** 2.9 m **55.** 22.5 ft **57.** 48 m^2 **59.** 6.25 cm **61.** 6 in. **63.** 13 cm **65.** The height of the flagpole is 14.375 ft. **67.** The width of the ravine is 82.5 ft. **69.** 20,000 voters voted in favor of the amendment. **71.** There are 175 mi between the two cities. **73.** There are 160 ml of water in 280 ml of soft drink. **75.** It would take 18 min to print a document 45 pages long. **77.** The window should be 40 ft^2. **79.** The person's height is 67.5 in. **81.** The first person's share of the winnings is \$1.25 million. **83.** The basketball player made 210 shots.

SECTION 8.5

1. $h=\dfrac{2A}{b}$ **3.** $t=\dfrac{d}{r}$ **5.** $T=\dfrac{PV}{nR}$ **7.** $L=\dfrac{P-2W}{2}$ **9.** $b_1=\dfrac{2A-hb_2}{h}$ **11.** $h=\dfrac{3V}{A}$ **13.** $S=C-Rt$

15. $P=\dfrac{A}{1+rt}$ **17.** $w=\dfrac{A}{S+1}$ **19a.** $h=\dfrac{S-2\pi r^2}{2\pi r}$ **b.** The height is 5 in. **c.** The height is 4 in.

21a. $S=\dfrac{F+BV}{B}$ **b.** The required selling price is \$180. **c.** The required selling price is \$75.

SECTION 8.6

1. It would take 2 h with both sprinklers working. **3.** It would take both skiploaders 3 h working together. **5.** It would take 30 h with both computers working. **7.** It would take 30 min with both air conditioners working. **9.** It would take the second pipeline 90 min to fill the tank. **11.** It would take the apprentice 15 h to construct the wall. **13.** It would take the second technician 3 h to complete the wiring. **15.** It would have taken one welder 40 h to complete the welds. **17.** It would take one machine 28 h working alone. **19.** The rate through the congested traffic is 20 mph. **21.** The rate of the jogger is 8 mph, and the rate of the cyclist is 20 mph. **23.** The rate of the jet is 360 mph. **25.** The rate of the runner is 8 mph. **27.** The rate of the car is 48 mph. **29.** The rate of the wind is 20 mph. **31.** The rate of the gulf current is 6 mph. **33.** The rate of the trucker for the first 330 mi was 55 mph. **35.** The less-experienced helper can complete the job in 30 h. **37.** The bus usually travels 60 mph.

SECTION 8.7

3. The profit is $80,000. **5.** The pressure is 6.75 lb/in^2. **7.** The object will fall 1600 ft. **9.** The ball will roll 54 ft in 3 s. **11.** At a rate of 65 mph, the time to travel between the two cities is 4.2 h. **13.** The pressure is 112.5 lb/in^2. **15.** The repulsive force is 80 lb. **17.** y is doubled. **19.** inversely **21.** inversely

CHAPTER REVIEW

1. $\frac{b^3y}{10ax}$ [8.1C] **2.** $\frac{7x+22}{60x}$ [8.2D] **3.** $\frac{x-2}{3x-10}$ [8.3A] **4.** $-\frac{x+6}{x+3}$ [8.1A] **5.** $\frac{2x^4}{3y^7}$ [8.1A] **6.** 62 [8.4B] **7.** $\frac{(3y-2)^2}{(y-1)(y-2)}$ [8.1C] **8.** $\frac{by^3}{6ax^2}$ [8.1B] **9.** $\frac{x}{x-7}$ [8.3A] **10.** $\frac{3x^2-x}{(2x+3)(6x-1)(3x-1)}, \frac{24x^3-4x^2}{(2x+3)(6x-1)(3x-1)}$ [8.2B] **11.** $a = \frac{T-2bc}{2b+2c}$ [8.5A] **12.** 2 [8.4A] **13.** $c = \frac{100m}{i}$ [8.5A] **14.** The equation has no solution. [8.4A] **15.** $\frac{1}{x^2}$ [8.1C] **16.** $\frac{2y-3}{5y-7}$ [8.2D] **17.** $\frac{1}{x+3}$ [8.2C] **18.** $(5x-3)(2x-1)(4x-1)$ [8.2A] **19.** $y = -\frac{4}{9}x + 2$ [8.5A] **20.** $\frac{8x+5}{3x-4}$ [8.1B] **21.** 5 [8.4A] **22.** $\frac{3x-1}{x-5}$ [8.2D] **23.** 10 [8.4B] **24.** 12 [8.4B] **25.** The perimeter of triangle *ABC* is 24 in. [8.4C] **26.** It would take 6 h to fill the pool using both hoses. [8.6A] **27.** The rate of the car is 45 mph. [8.6B] **28.** The rate of the wind is 20 mph. [8.6B] **29.** The pitcher's ERA is 1.35. [8.4D] **30.** The current is 2 amps. [8.7A]

CHAPTER TEST

1. $\frac{2x^3}{3y^3}$ [8.1A] **2.** $-\frac{x+5}{x+1}$ [8.1A] **3.** $\frac{x+1}{x^3(x-2)}$ [8.1B] **4.** $\frac{(x-5)(2x-1)}{(x+3)(2x+5)}$ [8.1B] **5.** $\frac{x+5}{x+4}$ [8.1C] **6.** $3(2x-1)(x+1)$ [8.2A] **7.** $\frac{3(x+2)}{x(x+2)(x-2)}; \frac{x^2}{x(x+2)(x-2)}$ [8.2B] **8.** $\frac{2}{x+5}$ [8.2C] **9.** $\frac{5}{(2x-1)(3x+1)}$ [8.2D] **10.** $\frac{x^2-4x+5}{(x-2)(x+3)}$ [8.2D] **11.** $\frac{x-3}{x-2}$ [8.3A] **12.** 2 [8.4A] **13.** The equation has no solution. [8.4A] **14.** -1 [8.4B] **15.** The area is 64.8 m^2. [8.4C] **16.** $t = \frac{d-s}{r}$ [8.5A] **17.** 14 rolls of wallpaper are needed. [8.4D] **18.** It would take 10 min with both landscapers working. [8.6A] **19.** The rate of the cyclist is 10 mph. [8.6B] **20.** The resistance is 0.4 ohm. [8.7A]

CUMULATIVE REVIEW

1. $\frac{31}{30}$ [1.3A] **2.** 21 [1.4A] **3.** $5x - 2y$ [1.4B] **4.** $-8x + 26$ [1.4D] **5.** $-\frac{9}{2}$ [2.2A] **6.** -12 [2.2C]
7. 10 [2.1D] **8.** $\{x \mid x \leq 8\}$ [2.4A] **9.** The volume is 200 ft^3. [3.3A] **10.**

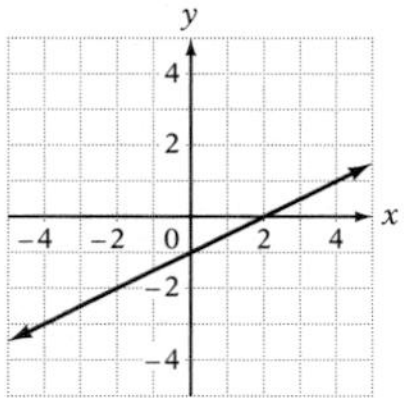

[4.3B]

11. $\frac{3}{7}$ [4.2A] **12.** $y = \frac{3}{2}x + 2$ [4.6A] **13.** -28 [5.3A] **14.** a^3b^7 [6.1A] **15.** $\frac{a^5}{b^6}$ [6.1B]
16. 3.5×10^{-8} [6.1C] **17.** $-4a^4 + 6a^3 - 2a^2$ [6.3A] **18.** $a^2 + ab - 12b^2$ [6.3C] **19.** $x^2 + 2x + 4$ [6.4B]
20. $(y - 6)(y - 1)$ [7.2A] **21.** $(4x + 1)(3x - 1)$ [7.3A/7.3B] **22.** $a(2a - 3)(a + 5)$ [7.3A/7.3B]
23. $4(b + 5)(b - 5)$ [7.4D] **24.** -3 and $\frac{5}{2}$ [7.5A] **25.** $\frac{2x^3}{3y^5}$ [8.1A] **26.** $-\frac{x-2}{x+5}$ [8.1A] **27.** 1 [8.1C]
28. $\frac{3}{(2x-1)(x+1)}$ [8.2D] **29.** $\frac{x+3}{x+5}$ [8.3A] **30.** 4 [8.4A] **31.** The alloy contains 70% silver. [2.3B]
32. It would cost \$160. [8.4D] **33.** It would take 6 min to fill the tank. [8.6A]

Answers to Chapter 9 Selected Exercises

PREP TEST

1. 16 [1.1D] **2.** 32 [1.2E] **3.** 9 [1.2D] **4.** $\frac{1}{12}$ [1.2C] **5.** $5x - 1$ [1.4D] **6.** $\frac{xy^5}{4}$ [6.1B]
7. $9x^2 - 12x + 4$ [6.3C] **8.** $-12x^2 + 14x + 10$ [6.3C] **9.** $36x^2 - 1$ [6.3D] **10.** $-1, 15$ [7.5A]

SECTION 9.1

1. 2 **3.** 27 **5.** $\frac{1}{9}$ **7.** 4 **9.** $(-25)^{5/2}$ is not a real number. **11.** $\frac{343}{125}$ **13.** x **15.** $y^{1/2}$ **17.** $x^{1/12}$ **19.** $a^{7/12}$
21. $\frac{1}{a}$ **23.** $\frac{1}{y}$ **25.** $y^{3/2}$ **27.** $\frac{1}{x}$ **29.** $\frac{1}{x^4}$ **31.** a **33.** $x^{3/10}$ **35.** a^3 **37.** $\frac{1}{x^{1/2}}$ **39.** $y^{1/9}$ **41.** x^4y
43. $x^6y^3z^9$ **45.** $\frac{x}{y^2}$ **47.** $\frac{x^{3/2}}{y^{1/4}}$ **49.** $\frac{x^2}{y^8}$ **51.** $\frac{1}{x^{11/12}}$ **53.** $\frac{1}{y^{5/2}}$ **55.** $\frac{1}{b^{7/8}}$ **57.** a^5b^{13} **59.** $\frac{m^2}{4n^{3/2}}$ **61.** $\frac{y^{17/2}}{x^3}$
63. $\frac{16b^2}{a^{1/3}}$ **65.** $y^2 - y$ **67.** $a - a^2$ **69.** x^{4n} **71.** $x^{3n/2}$ **73.** $y^{3n/2}$ **75.** x^{2n^2} **77.** $x^{2n}y^n$ **79.** $x^{4n}y^{2n}$
81. $\sqrt[4]{3}$ **83.** $\sqrt{a^3}$ **85.** $\sqrt{32t^5}$ **87.** $-2\sqrt[3]{x^2}$ **89.** $\sqrt[3]{a^4b^2}$ **91.** $\sqrt[5]{a^6b^{12}}$ **93.** $\sqrt[4]{(4x-3)^3}$ **95.** $\frac{1}{\sqrt[3]{x^2}}$
97. $14^{1/2}$ **99.** $x^{1/3}$ **101.** $x^{4/3}$ **103.** $b^{3/5}$ **105.** $(2x^2)^{1/3}$ **107.** $-(3x^5)^{1/2}$ **109.** $3xy^{2/3}$ **111.** $(a^2 - 2)^{1/2}$ **113.** x^8
115. $-x^4$ **117.** xy^3 **119.** $-x^5y$ **121.** $4a^2b^6$ **123.** $\sqrt{-16x^4y^2}$ is not a real number. **125.** $3x^3$ **127.** $-4x^3y^4$
129. $-x^2y^3$ **131.** x^4y^2 **133.** $3xy^5$ **135.** $2ab^2$ **137a.** false; 2 **b.** true **c.** true **d.** false; $(a^n + b^n)^{1/n}$
e. false; $a + 2a^{1/2}b^{1/2} + b$ **f.** false; $a^{n/m}$

SECTION 9.2

1. $x^2yz^2\sqrt{yz}$ **3.** $2ab^4\sqrt{2a}$ **5.** $3xyz^2\sqrt{5yz}$ **7.** $\sqrt{-9x^3}$ is not a real number. **9.** $a^5b^2\sqrt[3]{ab^2}$ **11.** $-5y\sqrt[3]{x^2y}$
13. $abc^2\sqrt[3]{ab^2}$ **15.** $2x^2y\sqrt[4]{xy}$ **17.** $-6\sqrt{x}$ **19.** $-2\sqrt{2}$ **21.** $3\sqrt{2b} + 5\sqrt{3b}$ **23.** $-2xy\sqrt{2y}$
25. $6ab^2\sqrt{3ab} + 3ab\sqrt{3ab}$ **27.** $-\sqrt[3]{2}$ **29.** $8b\sqrt[3]{2b^2}$ **31.** $3a\sqrt[4]{2a}$ **33.** $17\sqrt{2} - 15\sqrt{5}$ **35.** $5b\sqrt{b}$

37. $-8xy\sqrt{2x} + 2xy\sqrt{xy}$ **39.** $2y\sqrt[3]{2x}$ **41.** $-4ab\sqrt[4]{2b}$ **43.** 16 **45.** $2\sqrt[3]{4}$ **47.** $xy^3\sqrt{x}$ **49.** $8xy\sqrt{x}$ **51.** $2x^2y\sqrt[3]{2}$ **53.** $2ab\sqrt[4]{3a^2b}$ **55.** 6 **57.** $x - \sqrt{2x}$ **59.** $4x - 8\sqrt{x}$ **61.** $x - 6\sqrt{x} + 9$ **63.** $84 + 16\sqrt{5}$ **65.** $672x^2y^2$ **67.** $4a^3b^3\sqrt[3]{a}$ **69.** $-8\sqrt{5}$ **71.** $x - y^2$ **73.** $12x - y$ **77.** $y\sqrt{5y}$ **79.** $b\sqrt{13b}$ **81.** $\frac{\sqrt{2}}{2}$ **83.** $\frac{2\sqrt{3y}}{3y}$ **85.** $\frac{3\sqrt{3a}}{a}$ **87.** $\frac{\sqrt{2y}}{2}$ **89.** $\frac{5\sqrt[3]{3}}{3}$ **91.** $\frac{5\sqrt[3]{9y^2}}{3y}$ **93.** $\frac{b\sqrt{2a}}{2a^2}$ **95.** $\frac{\sqrt{15x}}{5x}$ **97.** $2 + 2\sqrt{2}$ **99.** $\frac{-12 - 4\sqrt{2}}{7}$ **101.** $-\frac{10 + 5\sqrt{7}}{3}$ **103.** $-\frac{7\sqrt{x} + 21}{x - 9}$ **105.** $-\sqrt{6} + 3 - 2\sqrt{2} + 2\sqrt{3}$ **107.** $-\frac{17 + 5\sqrt{5}}{4}$ **109.** $\frac{8a - 10\sqrt{ab} + 3b}{16a - 9b}$ **111.** $\frac{3 - 7\sqrt{y} + 2y}{1 - 4y}$ **113a.** false; $\sqrt[6]{432}$ **b.** true **c.** false; $\sqrt[3]{x^2}$ **d.** false; $\sqrt{x} + \sqrt{y}$ **e.** false; $\sqrt[2]{2} + \sqrt[3]{3}$ **f.** true **115.** $\sqrt[4]{a + b}$

SECTION 9.3

3. $2i$ **5.** $7i\sqrt{2}$ **7.** $3i\sqrt{3}$ **9.** $4 + 2i$ **11.** $2\sqrt{3} - 3i\sqrt{2}$ **13.** $4\sqrt{10} - 7i\sqrt{3}$ **15.** $8 - i$ **17.** $-8 + 4i$ **19.** $6 - 6i$ **21.** $19 - 7i\sqrt{2}$ **23.** $6\sqrt{2} - 3i\sqrt{2}$ **25.** 63 **27.** -4 **29.** $-3\sqrt{2}$ **31.** $-4 + 12i$ **33.** $-2 + 4i$ **35.** $17 - i$ **37.** $8 + 27i$ **39.** 1 **41.** 1 **43.** $-3i$ **45.** $\frac{3}{4} + \frac{1}{2}i$ **47.** $\frac{10}{13} - \frac{2}{13}i$ **49.** $\frac{4}{5} + \frac{2}{5}i$ **51.** $-i$ **53.** $-\frac{\sqrt{5}}{5} + \frac{2\sqrt{5}}{5}i$ **55.** $\frac{3}{10} - \frac{11}{10}i$ **57.** $\frac{6}{5} + \frac{7}{5}i$ **59a.** yes **b.** yes

SECTION 9.4

1. 25 **3.** 27 **5.** 48 **7.** -2 **9.** The equation has no solution. **11.** 9 **13.** -23 **15.** -16 **17.** $\frac{9}{4}$ **19.** 14 **21.** 7 **23.** 7 **25.** -122 **27.** 35 **29.** $-5, -2$ **31.** 45 **33.** 23 **35.** -4 **37.** 10 **39.** 1 **41.** 3 **43.** 21 **45.** $\frac{25}{72}$ **47.** 5 **49.** $-2, 4$ **51.** 6 **53.** 0, 1 **55.** $-2, 2$ **57.** $-2, 1$ **59.** 2, 11 **61.** The width is 6 ft. **63.** The bottom of the ladder is 10 ft from the wall. **65.** The object has fallen 2500 ft. **67.** The periscope must be 8.64 ft above the water. **69.** The length of the pendulum is 4.67 ft. **71.** The distance is 180 m. **73.** The HDTV screen is 7.15 in. wider. **75a.** $2, -2$; integer, rational number, real number **b.** $i, -i$; imaginary **c.** 16; integer, rational number, real number **77.** $a = \sqrt{c^2 - b^2}$ or $a = -\sqrt{c^2 - b^2}$ **79.** $x = \sqrt{6}$

CHAPTER REVIEW

1. $20x^2y^2$ [9.1A] **2.** 7 [9.4A] **3.** $39 - 2i$ [9.3C] **4.** $7x^{2/3}y$ [9.1B] **5.** $6\sqrt{3} - 13$ [9.2C] **6.** -2 [9.4A] **7.** $\frac{1}{x^5}$ [9.1A] **8.** $\frac{8\sqrt{3y}}{3y}$ [9.2D] **9.** $-2a^2b^4$ [9.1C] **10.** $2a^2b\sqrt{2b}$ [9.2B] **11.** $\frac{x\sqrt{x} - x\sqrt{2} + 2\sqrt{x} - 2\sqrt{2}}{x - 2}$ [9.2D] **12.** $\frac{2}{3} - \frac{5}{3}i$ [9.3D] **13.** $3ab^3\sqrt{2a}$ [9.2A] **14.** $-4\sqrt{2} + 8i\sqrt{2}$ [9.3B] **15.** $5x^3y^3\sqrt[3]{2x^2y}$ [9.2B] **16.** $4xy^2\sqrt[3]{x^2}$ [9.2C] **17.** $7 + 3i$ [9.3C] **18.** $3\sqrt[4]{x^3}$ [9.1B] **19.** $-2ab^2\sqrt[5]{2a^3b^2}$ [9.2A] **20.** $-2 + 7i$ [9.3D] **21.** $-6\sqrt{2}$ [9.3C] **22.** 30 [9.4A] **23.** $3a^2b^3$ [9.1C] **24.** $5i\sqrt{2}$ [9.3A] **25.** $-12 + 10i$ [9.3B] **26.** $31 - 10\sqrt{6}$ [9.2C] **27.** $6x^2\sqrt{3y}$ [9.2B] **28.** The amount of power generated is 120 watts. [9.4B] **29.** The distance required to reach a velocity of 88 ft/s is 242 ft. [9.4B] **30.** The bottom of the ladder is 6.63 ft from the building. [9.4B]

CHAPTER TEST

1. $\frac{1}{2}x^{3/4}$ [9.1B] **2.** $-2x^2y\sqrt{2x}$ [9.2B] **3.** $3\sqrt[5]{y^2}$ [9.1B] **4.** $18 + 16i$ [9.3C] **5.** $4x + 4\sqrt{xy} + y$ [9.2C] **6.** $r^{1/6}$ [9.1A] **7.** 4 [9.4A] **8.** $2xy^2$ [9.1C] **9.** $-4x\sqrt{3}$ [9.2C] **10.** $-3 + 2i$ [9.3B] **11.** $4x^2y^3\sqrt{2y}$ [9.2A]

12. $14 + 10\sqrt{3}$ [9.2C] **13.** 4 [9.3B] **14.** 2 [9.2D] **15.** $8a\sqrt{2a}$ [9.2B] **16.** $2a - \sqrt{ab} - 15b$ [9.2C]

17. $\frac{64x^3}{y^6}$ [9.1A] **18.** $\frac{x + \sqrt{xy}}{x - y}$ [9.2D] **19.** $-\frac{4}{5} + \frac{7}{5}i$ [9.3D] **20.** -3 [9.4A] **21.** $\frac{b^3}{8a^6}$ [9.1A]

22. $3abc^2\sqrt[3]{ac}$ [9.2A] **23.** $\frac{4x^2}{y}$ [9.2D] **24.** -4 [9.3C] **25.** The object has fallen 576 ft. [9.4B]

CUMULATIVE REVIEW

1. 92 [1.3A] **2.** 56 [1.4A] **3.** $-10x + 1$ [1.4D] **4.** $\frac{3}{2}$ [2.2A] **5.** $\frac{2}{3}$ [2.2C] **6.** $\{x|x > 1\}$ [2.4A]

7. $\frac{1}{3}, \frac{7}{3}$ [2.5A] **8.** $\{x|-6 \le x \le 3\}$ [2.5B] **9.** The area is 187.5 cm². [3.2B] **10.** 3 [5.3A]

11.

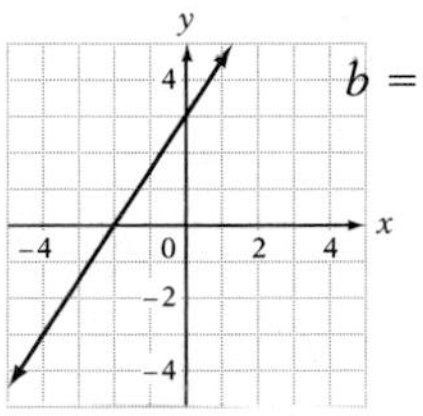

$m = \frac{3}{2}$
$b = 3$

[4.3B]

12.

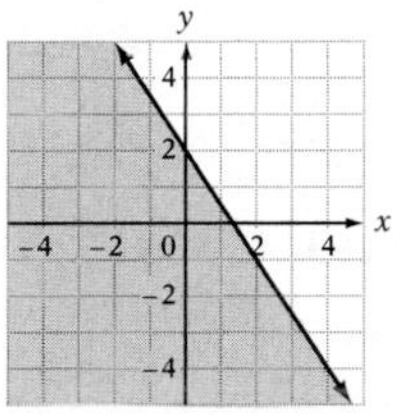

[4.7A]

13. $y = \frac{1}{3}x + \frac{7}{3}$ [4.5B]

14. $\left(\frac{17}{4}, \frac{9}{2}\right)$ [5.3B] **15.** $2x^2y^2$ [6.1B] **16.** $(9x + y)(9x - y)$ [7.4A] **17.** $x(x^2 + 3)(x + 1)(x - 1)$ [7.4D]

18. $C = R - nP$ [8.5A] **19.** $\frac{y^5}{x^4}$ [9.1A] **20.** $-x\sqrt{10x}$ [9.2B] **21.** $13 - 7\sqrt{3}$ [9.2C] **22.** $\sqrt{6} + \sqrt{2}$ [9.2D]

23. $-\frac{1}{5} + \frac{3}{5}i$ [9.3D] **24.** -20 [9.4A] **25.** The length of side *DE* is 27 m. [8.4C] **26.** $4000 must be invested at 8.4% [2.3C] **27.** The rate of the plane was 250 mph. [8.6B] **28.** It takes 1.25 s for light to travel to Earth from the moon. [6.1D] **29.** The slope is 0.08. The slope represents the simple interest rate on the investment. The interest rate is 8%. [4.4A] **30.** The periscope must be 32.7 ft above the water. [9.4B]

Answers to Chapter 10 Selected Exercises

PREP TEST

1. $3\sqrt{2}$ [1.2F] **2.** $3i$ [9.3A] **3.** $\frac{2x - 1}{x - 1}$ [8.2B] **4.** 8 [1.4A] **5.** yes [7.4A] **6.** $(2x - 1)^2$ [7.4A]

7. $(3x + 2)(3x - 2)$ [7.4A] **8.** 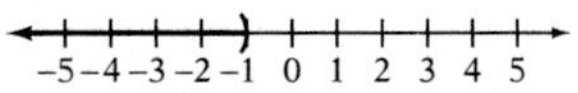 [1.5C] **9.** $-3, 5$ [7.5A] **10.** 4 [8.4B]

SECTION 10.1

3. $2x^2 - 4x - 5 = 0$; $a = 2, b = -4, c = -5$ **5.** $4x^2 - 5x + 6 = 0$; $a = 4, b = -5, c = 6$ **7.** 0, 4 **9.** 5, -5

11. 3, -2 **13.** 3 **15.** 0, 2 **17.** 5, -2 **19.** 2, 5 **21.** 6, $-\frac{3}{2}$ **23.** 2, $\frac{1}{4}$ **25.** $\frac{1}{3}, -4$ **27.** $\frac{9}{2}, -\frac{2}{3}$

29. $\frac{1}{4}, -4$ **31.** 9, -2 **33.** $-2, -\frac{3}{4}$ **35.** 2, -5 **37.** $-4, -\frac{3}{2}$ **39.** $2b, 7b$ **41.** $7c, -c$ **43.** $-\frac{b}{2}, -b$

45. $4a, \frac{2a}{3}$ **47.** $-\frac{a}{3}, 3a$ **49.** $-\frac{3y}{2}, -\frac{y}{2}$ **51.** $-\frac{a}{2}, -\frac{4a}{3}$ **53.** $x^2 - 7x + 10 = 0$ **55.** $x^2 + 6x + 8 = 0$

57. $x^2 - 5x - 6 = 0$ **59.** $x^2 - 9 = 0$ **61.** $x^2 - 8x + 16 = 0$ **63.** $x^2 - 5x = 0$ **65.** $x^2 - 3x = 0$

67. $2x^2 - 7x + 3 = 0$ **69.** $4x^2 - 5x - 6 = 0$ **71.** $3x^2 + 11x + 10 = 0$ **73.** $9x^2 - 4 = 0$ **75.** $6x^2 - 5x + 1 = 0$

77. $10x^2 - 7x - 6 = 0$ **79.** $8x^2 + 6x + 1 = 0$ **81.** $50x^2 - 25x - 3 = 0$ **83.** 7, -7 **85.** $2i, -2i$ **87.** 2, -2

89. $\frac{9}{2}, -\frac{9}{2}$ **91.** $7i, -7i$ **93.** $4\sqrt{3}, -4\sqrt{3}$ **95.** $5\sqrt{3}, -5\sqrt{3}$ **97.** $3i\sqrt{2}, -3i\sqrt{2}$ **99.** $7, -5$ **101.** $0, -6$

103. $-7, 3$ **105.** $1, 0$ **107.** $-5 + \sqrt{6}, -5 - \sqrt{6}$ **109.** $3 + 3i\sqrt{5}, 3 - 3i\sqrt{5}$ **111.** $\frac{2 - 9\sqrt{2}}{3}, \frac{2 + 9\sqrt{2}}{3}$

113. $x^2 - 2 = 0$ **115.** $x^2 - 18 = 0$ **117.** $\frac{3b}{a}, -\frac{3b}{a}$ **119.** $-\frac{1}{2}$

SECTION 10.2

3. $5, -1$ **5.** $-9, 1$ **7.** 3 **9.** $-2 + \sqrt{11}, -2 - \sqrt{11}$ **11.** $3 + \sqrt{2}, 3 - \sqrt{2}$ **13.** $1 + i, 1 - i$ **15.** $8, -3$

17. $4, -9$ **19.** $\frac{3 + \sqrt{5}}{2}, \frac{3 - \sqrt{5}}{2}$ **21.** $\frac{1 + \sqrt{5}}{2}, \frac{1 - \sqrt{5}}{2}$ **23.** $3 + \sqrt{13}, 3 - \sqrt{13}$ **25.** $5, 3$ **27.** $2 + 3i, 2 - 3i$

29. $-3 + 2i, -3 - 2i$ **31.** $1 + 3\sqrt{2}, 1 - 3\sqrt{2}$ **33.** $\frac{1 + \sqrt{17}}{2}, \frac{1 - \sqrt{17}}{2}$ **35.** $1 + 2i\sqrt{3}, 1 - 2i\sqrt{3}$

37. $\frac{1}{2} + i, \frac{1}{2} - i$ **39.** $\frac{1}{3} + \frac{1}{3}i, \frac{1}{3} - \frac{1}{3}i$ **41.** $\frac{2 + \sqrt{14}}{2}, \frac{2 - \sqrt{14}}{2}$ **43.** $1, -\frac{3}{2}$ **45.** $1 + \sqrt{5}, 1 - \sqrt{5}$ **47.** $\frac{1}{2}, 5$

49. $2 + \sqrt{5}, 2 - \sqrt{5}$ **51.** $1.236, -3.236$ **53.** $1.707, 0.293$ **55.** $0.309, -0.809$ **57.** $x = 2a, x = -a$

59. $x = 2a, x = -5a$

SECTION 10.3

3. $5, -2$ **5.** $4, -9$ **7.** $4 + 2\sqrt{22}, 4 - 2\sqrt{22}$ **9.** $3, -8$ **11.** $\frac{-5 + \sqrt{33}}{4}, \frac{-5 - \sqrt{33}}{4}$ **13.** $\frac{3}{2}, -\frac{1}{4}$

15. $7 + 3\sqrt{5}, 7 - 3\sqrt{5}$ **17.** $\frac{1 + \sqrt{3}}{2}, \frac{1 - \sqrt{3}}{2}$ **19.** $-1 + i, -1 - i$ **21.** $1 + 2i, 1 - 2i$ **23.** $2 + 3i, 2 - 3i$

25. $\frac{1 + \sqrt{11}}{2}, \frac{1 - \sqrt{11}}{2}$ **27.** $-\frac{3}{2} + \frac{1}{2}i, -\frac{3}{2} - \frac{1}{2}i$ **29.** $\frac{3}{4} + \frac{3\sqrt{3}}{4}i, \frac{3}{4} - \frac{3\sqrt{3}}{4}i$ **31.** $7.606, 0.394$ **33.** $0.236, -4.236$

35. $1.851, -1.351$ **37.** two complex number solutions **39.** one real number solution **41.** two real number solutions **43.** $\{p \mid p < 9\}$ **45.** $\{p \mid p > 1\}$ **47.** i and $-2i$

SECTION 10.4

1. $3, -3, 2, -2$ **3.** $\sqrt{2}, -\sqrt{2}, 2, -2$ **5.** $1, 4$ **7.** 16 **9.** $2i, -2i, 1, -1$ **11.** $4i, -4i, 2, -2$ **13.** 16

15. $1, 512$ **17.** $\frac{2}{3}, -\frac{2}{3}, 1, -1$ **19.** 3 **21.** 9 **23.** $2, -1$ **25.** $0, 2$ **27.** $2, -\frac{1}{2}$ **29.** -2 **31.** 1 **33.** 1

35. -3 **37.** $10, -1$ **39.** $-\frac{1}{2} + \frac{\sqrt{7}}{2}i, -\frac{1}{2} - \frac{\sqrt{7}}{2}i$ **41.** $1, -3$ **43.** $0, -1$ **45.** $\frac{1}{2}, -\frac{1}{3}$ **47.** $-\frac{2}{3}, 6$ **49.** $\frac{4}{3}, 3$

51. $-\frac{1}{4}, 3$ **53.** 4

SECTION 10.5

1. The height is 3 cm; the base is 14 cm. **3.** The dimensions of Colorado are approximately 272 mi by 383 mi. **5.** The maximum safe speed is 33 mph. **7.** The time for the projectile to return to Earth is 12.5 s. **9.** The maximum speed is 72.5 km/h. **11.** The smaller pipe requires 12 min. The larger pipe requires 6 min. **13.** The rate of the wind is 108 mph. **15.** The rowing rate of the guide is 6 mph. **17.** The radius of the cone 1.5 in.

SECTION 10.6

3. 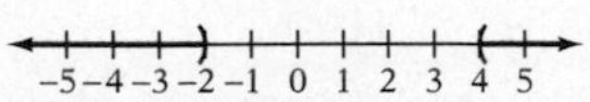

$\{x \mid x < -2 \text{ or } x > 4\}$

5. 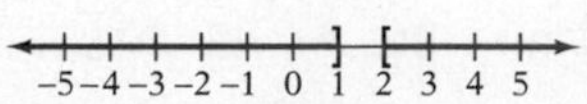

$\{x \mid x \le 1 \text{ or } x \ge 2\}$

7. 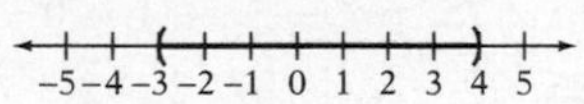

$\{x \mid -3 < x < 4\}$

9. 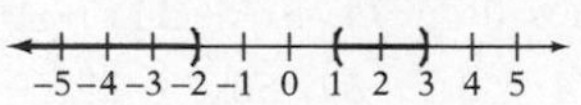

$\{x \mid x < -2 \text{ or } 1 < x < 3\}$

11. 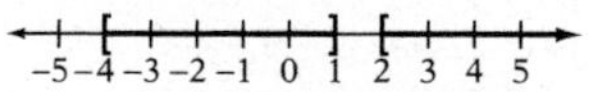 $\{x|-4 \le x \le 1 \text{ or } x \ge 2\}$ **13.** 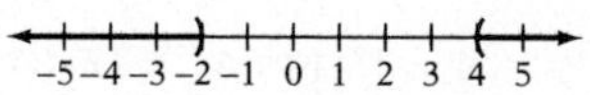 $\{x|x < -2 \text{ or } x > 4\}$

15. 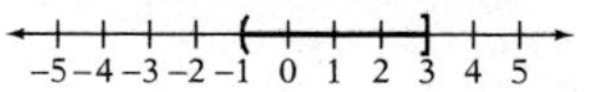 $\{x|-1 < x \le 3\}$ **17.** 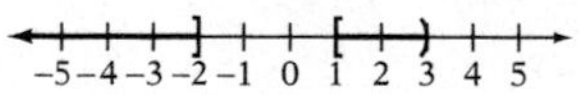 $\{x|x \le -2 \text{ or } 1 \le x < 3\}$

19. $\{x|x > 4 \text{ or } x < -4\}$ **21.** $\{x|-3 \le x \le 12\}$ **23.** $\left\{x\middle|\frac{1}{2} < x < \frac{3}{2}\right\}$ **25.** $\left\{x\middle|x < 1 \text{ or } x > \frac{5}{2}\right\}$

27. $\{x|x < -1 \text{ or } 1 < x \le 2\}$ **29.** $\left\{x\middle|\frac{1}{2} < x \le 1\right\}$ **31.** $\{x|2 < x \le 3\}$ **33.** $\{x|x > 5 \text{ or } -4 < x < -1\}$

35. 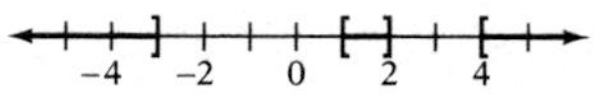 **37.** 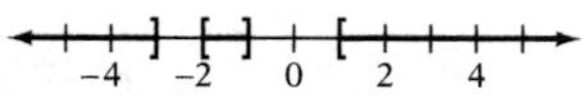 **39.** 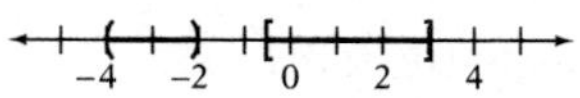

CHAPTER REVIEW

1. $0, \frac{3}{2}$ [10.1A] **2.** $\frac{c}{2}, -2c$ [10.1A] **3.** $4\sqrt{3}, -4\sqrt{3}$ [10.1C] **4.** $-\frac{1}{2} + 2i, -\frac{1}{2} - 2i$ [10.1C]

5. $-3, -1$ [10.2A] **6.** $\frac{7 + 2\sqrt{7}}{7}, \frac{7 - 2\sqrt{7}}{7}$ [10.2A] **7.** $\frac{3}{4}, \frac{4}{3}$ [10.3A] **8.** $\frac{1}{2} + \frac{\sqrt{31}}{2}i, \frac{1}{2} - \frac{\sqrt{31}}{2}i$ [10.3A]

9. $x^2 + 3x = 0$ [10.1B] **10.** $12x^2 - x - 6 = 0$ [10.1B] **11.** $1 + i\sqrt{7}, 1 - i\sqrt{7}$ [10.2A] **12.** $2i, -2i$ [10.2A]

13. $\frac{11 + \sqrt{73}}{6}, \frac{11 - \sqrt{73}}{6}$ [10.3A] **14.** two real number solutions [10.3A] **15.** $\left\{x\middle|-3 < x < \frac{5}{2}\right\}$ [10.6A]

16. $\left\{x\middle|x \le -4 \text{ or } -\frac{3}{2} \le x \le 2\right\}$ [10.6A] **17.** $27, -64$ [10.4A] **18.** $\frac{5}{4}$ [10.4B] **19.** $-1, 3$ [10.4C]

20. -1 [10.4C] **21.** 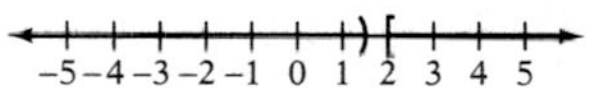 [10.6A] **22.** 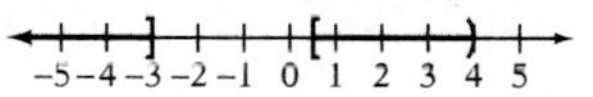 [10.6A]

$\left\{x\middle|x < \frac{3}{2} \text{ or } x \ge 2\right\}$ $\left\{x\middle|x < -3 \text{ or } \frac{1}{2} \le x < 4\right\}$

23. 4 [10.4B] **24.** 5 [10.4B] **25.** $\frac{-3 + \sqrt{249}}{10}, \frac{-3 - \sqrt{249}}{10}$ [10.4C] **26.** $\frac{-11 + \sqrt{129}}{2}, \frac{-11 - \sqrt{129}}{2}$ [10.4C]

27. The length is 12 cm, and the width is 5 cm. [10.5A] **28.** The three integers are 2, 4, and 6 or -6, -4, and -2. [10.5A] **29.** Working alone, the new computer takes 12 min to print the payroll. [10.5A] **30.** The speed of the first car is 40 mph. The speed of the second car is 50 mph. [10.5A]

CHAPTER TEST

1. $\frac{2}{3}, -4$ [10.1A] **2.** $\frac{3}{2}, -\frac{2}{3}$ [10.1A] **3.** $x^2 - 9 = 0$ [10.1B] **4.** $2x^2 + 7x - 4 = 0$ [10.1B]

5. $2 + 2\sqrt{2}, 2 - 2\sqrt{2}$ [10.1C] **6.** $3 + \sqrt{11}, 3 - \sqrt{11}$ [10.2A] **7.** $\frac{3 + \sqrt{15}}{3}, \frac{3 - \sqrt{15}}{3}$ [10.2A]

8. $\frac{1 + \sqrt{3}}{2}, \frac{1 - \sqrt{3}}{2}$ [10.3A] **9.** $-2 + 2i\sqrt{2}, -2 - 2i\sqrt{2}$ [10.3A] **10.** two real number solutions [10.3A]

11. two complex number solutions [10.3A] **12.** $\frac{1}{4}$ [10.4A] **13.** $1, -1, \sqrt{3}, -\sqrt{3}$ [10.4A] **14.** 4 [10.4B]

15. The equation has no solution. [10.4B] **16.** $2, -9$ [10.4C]

17. $\{x|x < -4 \text{ or } 2 < x < 4\}$ 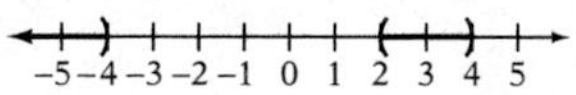 [10.6A]

18. $\left\{x\middle|-4 < x \le \frac{3}{2}\right\}$ 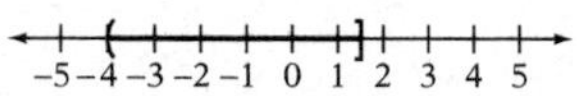 [10.6A] **19.** The base is 15 ft. The height is 4 ft. [10.5A]

20. The rowing rate of the canoe in calm water is 4 mph. [10.5A]

CUMULATIVE REVIEW

1. 14 [1.4A] **2.** $\left\{x\middle|-2 < x < \frac{10}{3}\right\}$ [2.5B] **3.** The volume is 54π m³. [3.3A] **4.** $-\frac{7}{3}$ [4.2A] **5.** $-\frac{3}{2}$ [4.4A]

6. $\left(\frac{5}{2}, 0\right)$, $(0, -3)$ [4.3B] **7.** $y = x + 1$ [4.6A] **8.** $(1, -1, 2)$ [5.2B] **9.**

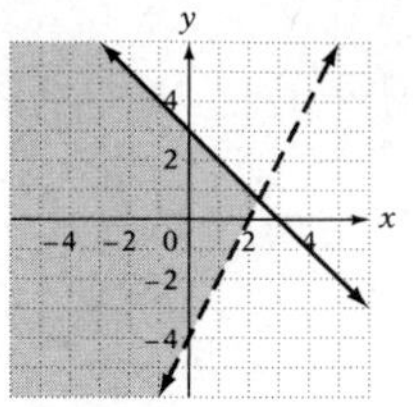

[5.5A]

10. The height of triangle *DEF* is 16 cm. [8.4C] **11.** $x^2 - 3x - 4 - \frac{6}{3x - 4}$ [6.4A] **12.** $-3xy(x^2 - 2xy + 3y^2)$ [7.1A]

13. $(2x - 5)(3x + 4)$ [7.3A/7.3B] **14.** $\frac{x}{2}$ [8.1B] **15.** $-\frac{3}{2}$ and -1 [8.4A] **16.** $b = \frac{2S - an}{n}$ [8.5A]

17. $1 - a$ [9.1A] **18.** $-8 - 14i$ [9.3C] **19.** 0, 1 [9.4A] **20.** $2, -2, \sqrt{2}, -\sqrt{2}$ [10.4A] **21.** The lower limit is $9\frac{23}{64}$ in. The upper limit is $9\frac{25}{64}$ in. [2.5C] **22.** The area is $(x^2 + 6x - 16)$ ft^2. [6.3E] **23.** The slope is $-\frac{25{,}000}{3}$. The building decreases \$8333.33 in value each year. [4.4A] **24.** The ladder will reach 15 ft up on the building. [9.4B] **25.** There are two complex number solutions. [10.3A]

Answers to Chapter 11 Selected Exercises

PREP TEST

1. 1 [1.4A] **2.** −7 [1.4A] **3.** 30 [4.2A] **4.** $h^2 + 4h - 1$ [4.2A] **5.** $-\frac{2}{3}, 3$ [10.1A]

6. $2 - \sqrt{3}, 2 + \sqrt{3}$ [10.2A, 10.3A] **7.** $y = \frac{1}{2}x - 2$ [8.5A] **8.** domain: {−2, 3, 4, 6}; range: {4, 5, 6}; yes [4.2A]

9.

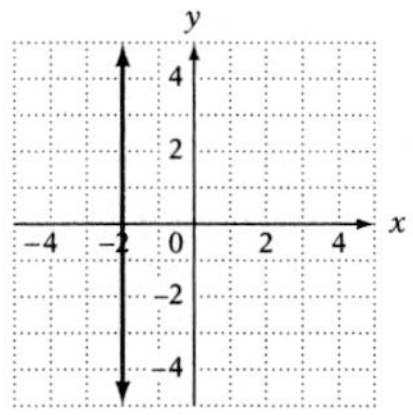

[4.3B]

SECTION 11.1

1.

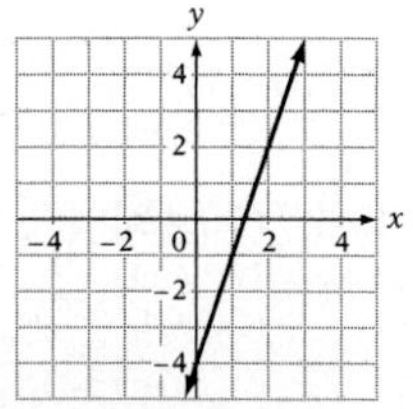

3.

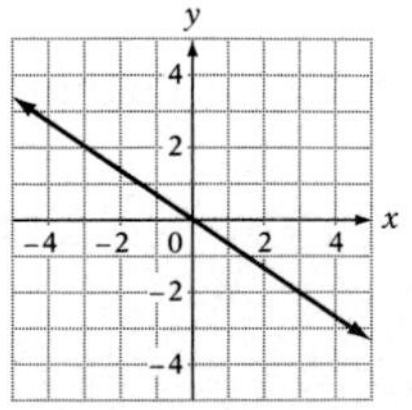

5.

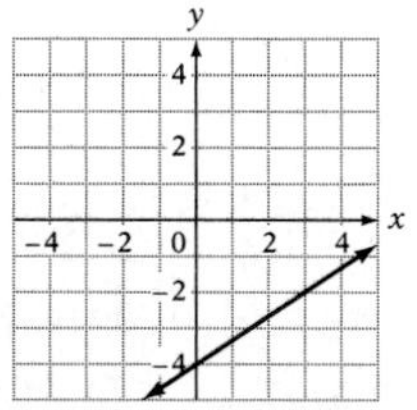

7.

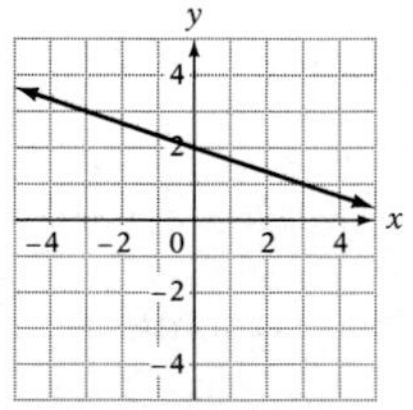

9.

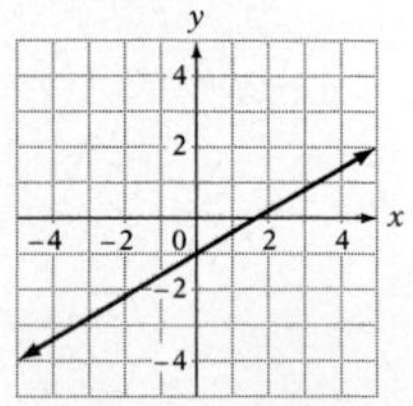

11. If the cost is \$150, the markup is \$37.50. **13.** The linear function is $y = -\frac{3}{5}x + 545$. If the room rate is \$100, 485 rooms will be rented. **15a.** III **b.** IV **c.** II **d.** I

SECTION 11.2

3. 8 **5.** $x = -4$ **7.**

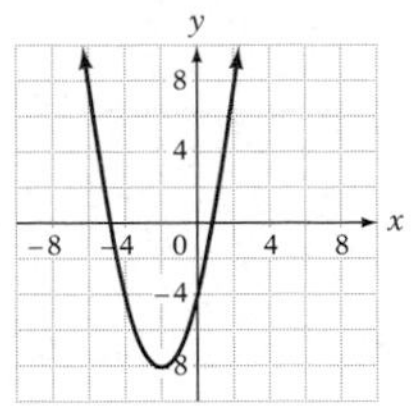

vertex: $(-2, -8)$

axis of symmetry: $x = -2$

9.

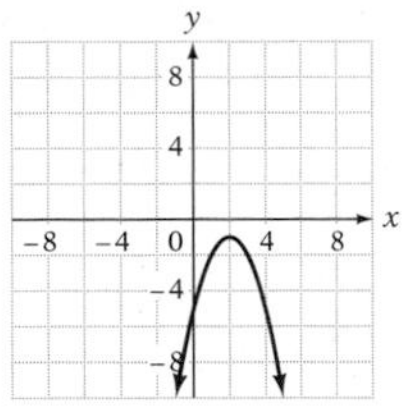

vertex: $(2, -1)$

axis of symmetry: $x = 2$

11.

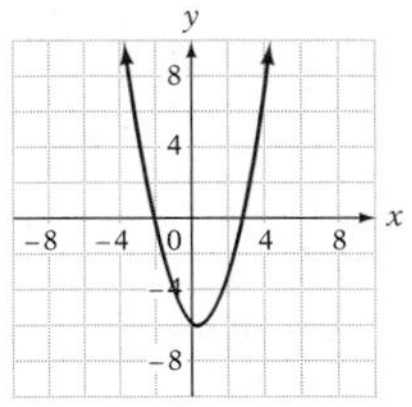

vertex: $\left(\frac{1}{2}, -\frac{25}{4}\right)$

axis of symmetry: $x = \frac{1}{2}$

13.

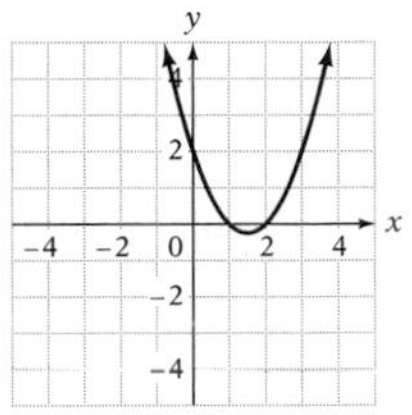

vertex: $\left(\frac{3}{2}, -\frac{1}{4}\right)$

axis of symmetry: $x = \frac{3}{2}$

15.

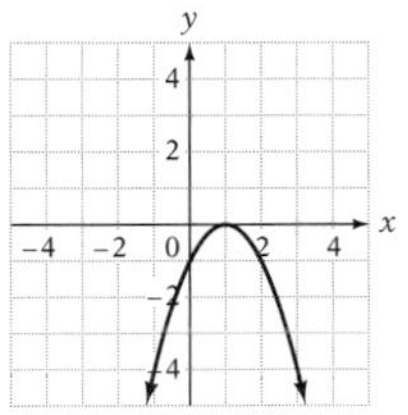

vertex: $(1, 0)$

axis of symmetry: $x = 1$

17.

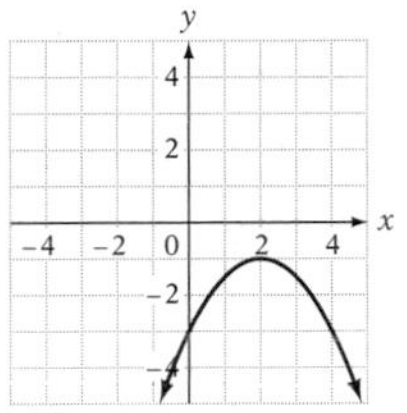

vertex: $(2, -1)$

axis of symmetry: $x = 2$

19. The domain is $\{x \mid x \in \text{real numbers}\}$. The range is $\{y \mid y \geq -5\}$. **21.** The domain is $\{x \mid x \in \text{real numbers}\}$. The range is $\{y \mid y \leq -1\}$. **23.** $(2, 0), (-2, 0)$ **25.** $(0, 0), (2, 0)$ **27.** $(2, 0), (-1, 0)$ **29.** $\left(-\frac{1}{2}, 0\right), (1, 0)$ **31.** $(-1 + \sqrt{2}, 0), (-1 - \sqrt{2}, 0)$ **33.** The parabola has no x-intercepts. **35.** The parabola has no x-intercepts. **37.** The parabola has two x-intercepts. **39.** The parabola has one x-intercept. **41.** The parabola has no x-intercepts. **43.** The parabola has two x-intercepts. **45a.** maximum **b.** minimum **c.** maximum **47.** minimum: 2 **49.** maximum: -1 **51.** minimum: -2 **53.** maximum: -3 **55.** minimum: $-\frac{73}{8}$ **57.** b **59.** The minimum height of the cable above the bridge is 24.36 ft. **61.** The diver will be a maximum of 13.1 m above the water. **63.** The maximum height is 1.5 ft. **65.** A speed of 41 mph will yield the maximum fuel efficiency. The maximum fuel efficiency is 33.658 mi/gal. **67.** $-\frac{3}{2}$

SECTION 11.3

1. yes **3.** no **5.** yes **7.**

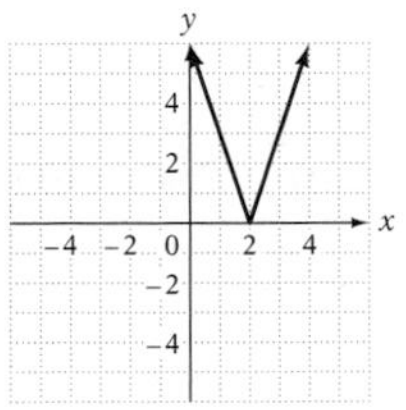

domain: $\{x \mid x \in \text{real numbers}\}$
range: $\{y \mid y \geq 0\}$

9.

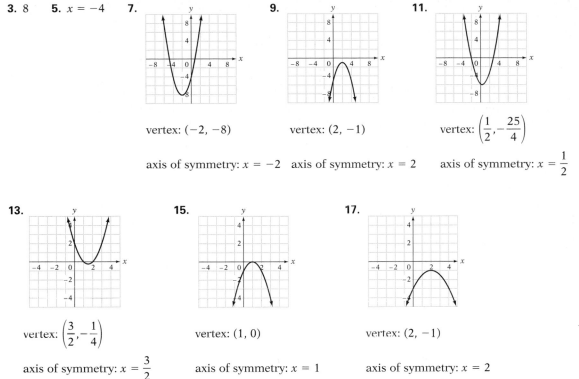

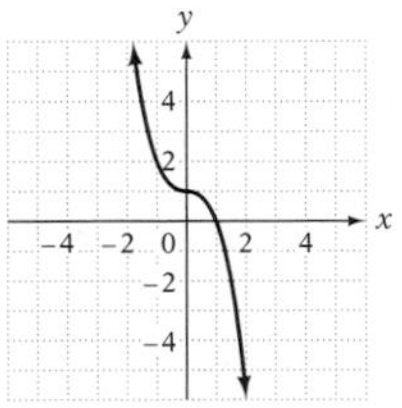

domain: $\{x \mid x \in \text{real numbers}\}$
range: $\{y \mid y \in \text{real numbers}\}$

11.
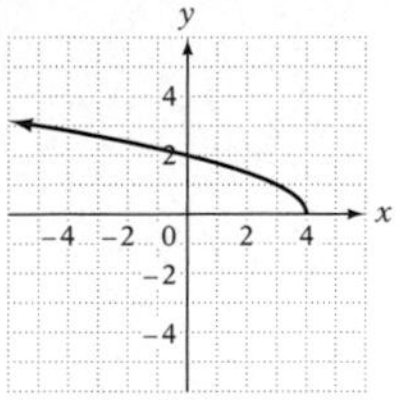
domain: $\{x \mid x \leq 4\}$
range: $\{y \mid y \geq 0\}$

13.
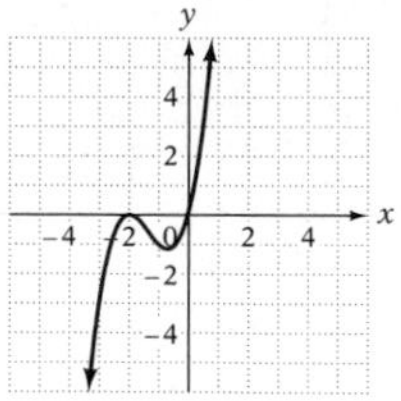
domain: $\{x \mid x \in \text{real numbers}\}$
range: $\{y \mid y \in \text{real numbers}\}$

15.
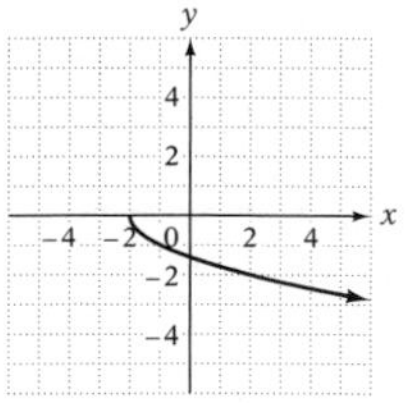
domain: $\{x \mid x \geq -2\}$
range: $\{y \mid y \leq 0\}$

17.
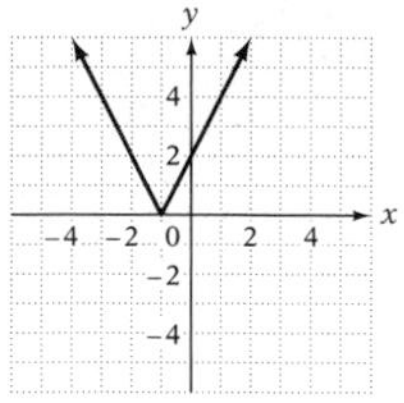
domain: $\{x \mid x \in \text{real numbers}\}$
range: $\{y \mid y \geq 0\}$

19. $a = 18$ **21.** $f(2, 5) + g(2, 5) = 17$ **23.** $f(14) = 8$ **25.** $x = -3$

SECTION 11.4

1. 5 **3.** 1 **5.** 0 **7.** $-\frac{29}{4}$ **9.** $\frac{2}{3}$ **11.** 2 **13.** 39 **15.** -8 **17.** $-\frac{4}{5}$ **19.** -2 **21.** 7 **23.** -13
25. -29 **27.** $8x - 13$ **29.** 4 **31.** 3 **33.** $x + 4$ **35.** 5 **37.** 11 **39.** $3x^2 + 3x + 5$ **41.** -3 **43.** -27
45. $x^3 - 6x^2 + 12x - 8$ **47.** $h^2 + 6h$ **49.** $2 + h$ **51.** $h + 2a$ **53.** -1 **55.** -6 **57.** $6x - 13$

SECTION 11.5

3. yes **5.** no **7.** yes **9.** no **11.** no **13.** no **17.** $\{(0, 1), (3, 2), (8, 3), (15, 4)\}$ **19.** no inverse
21. $\{(-2, 0), (5, -1), (3, 3), (6, -4)\}$ **23.** no inverse **25.** $f^{-1}(x) = \frac{1}{4}x + 2$ **27.** $f^{-1}(x) = \frac{1}{2}x - 2$
29. $f^{-1}(x) = 2x + 2$ **31.** $f^{-1}(x) = -\frac{1}{2}x + 1$ **33.** $f^{-1}(x) = \frac{3}{2}x - 6$ **35.** $f^{-1}(x) = -3x + 3$ **37.** $f^{-1}(x) = \frac{1}{2}x + \frac{5}{2}$
39. $f^{-1}(x) = \frac{1}{5}x + \frac{2}{5}$ **41.** $f^{-1}(x) = \frac{1}{6}x + \frac{1}{2}$ **43.** $\frac{5}{3}$ **45.** 3 **47.** yes; yes **49.** yes **51.** no **53.** yes **55.** yes

57.
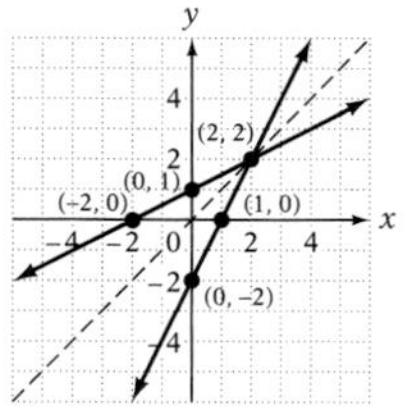

59.
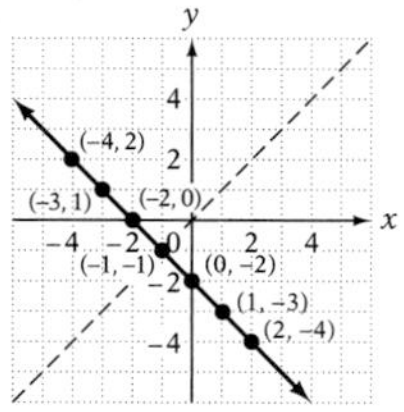

The inverse is the same graph.

61.
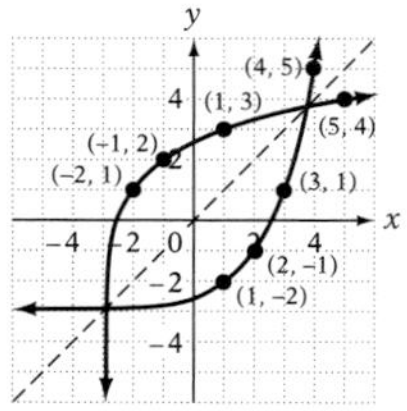

63. no **65.** 5 **67.** 0 **69.** -4

SECTION 11.6

1a. a vertical line **b.** opens up **3a.** a horizontal line **b.** opens right **5a.** a horizontal line **b.** opens left

7.
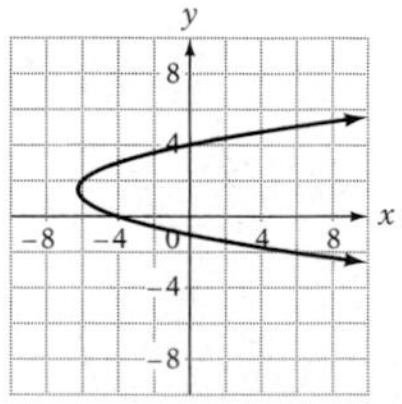
vertex: $\left(-\frac{25}{4}, \frac{3}{2}\right)$
axis of symmetry: $y = \frac{3}{2}$

9.
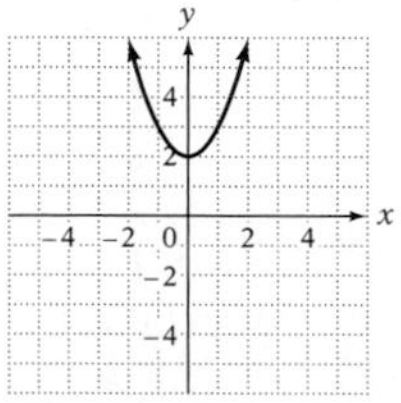
vertex: (0, 2)
axis of symmetry: $x = 0$

11.
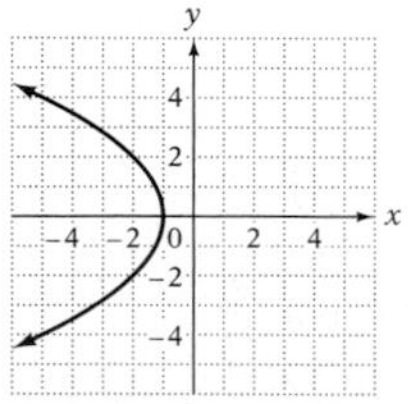
vertex: $(-1, 0)$
axis of symmetry: $y = 0$

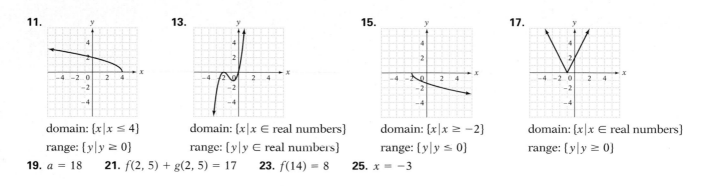

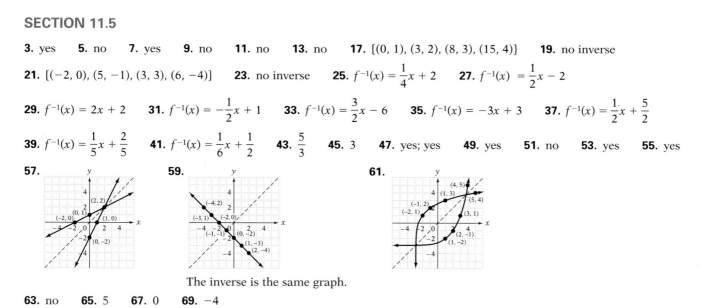

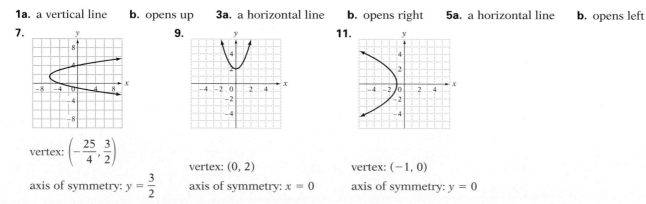

13.
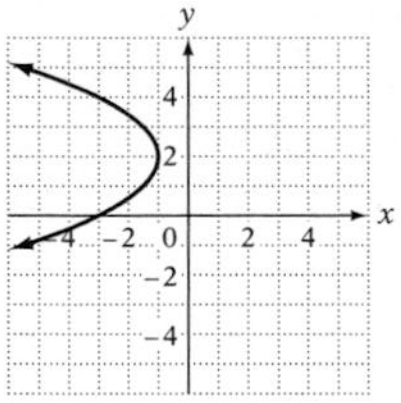

vertex: $(-1, 2)$

axis of symmetry: $y = 2$

15.
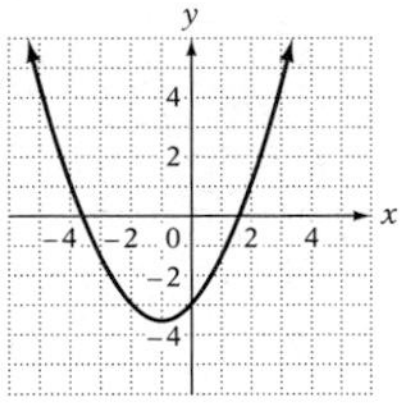

vertex: $\left(-1, -\dfrac{7}{2}\right)$

axis of symmetry: $x = -1$

17.
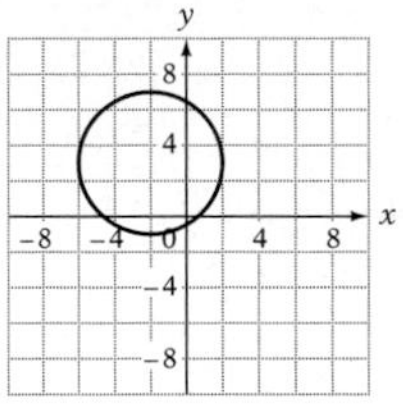

19.
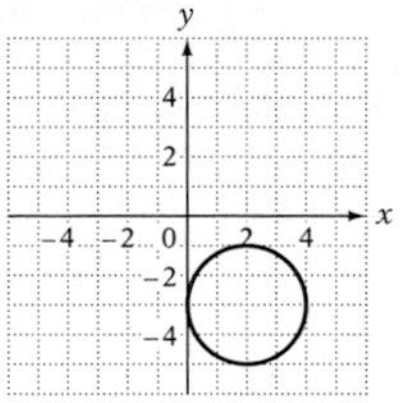

21.
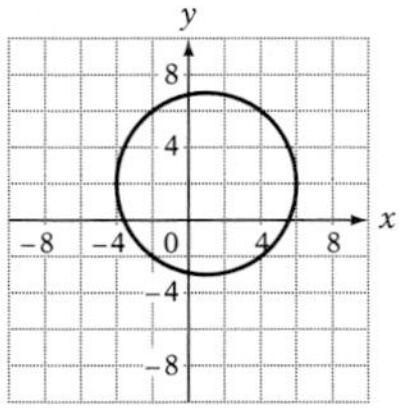

23. $(x + 1)^2 + (y + 2)^2 = 9$
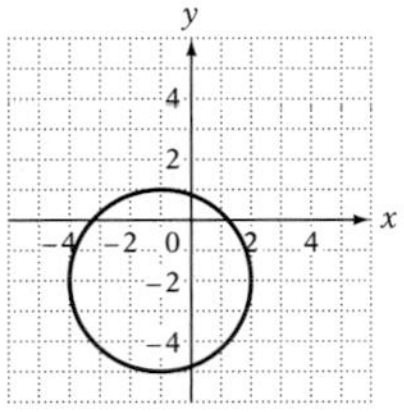

25. $(x + 2)^2 + (y - 1)^2 = 5$
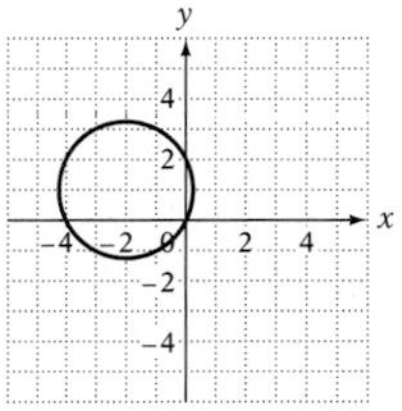

27.
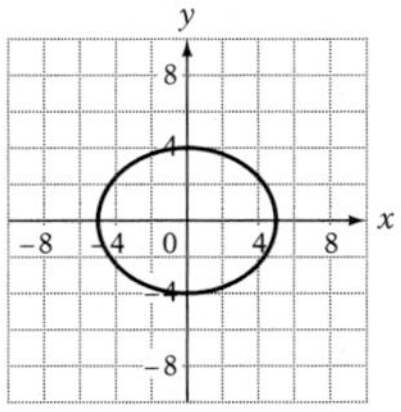

29.
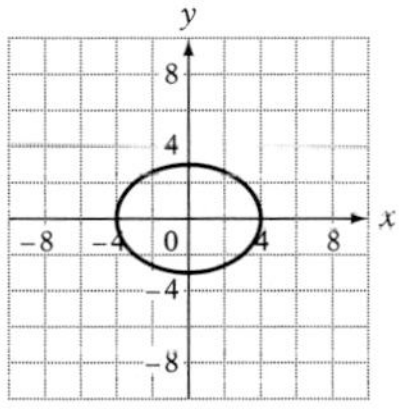

31.
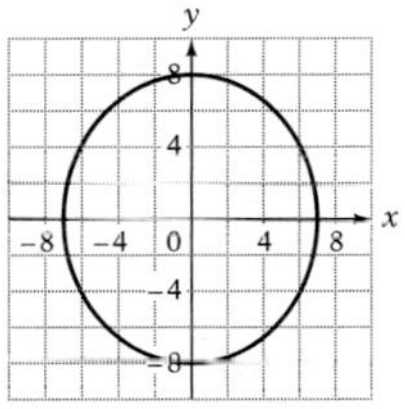

33.
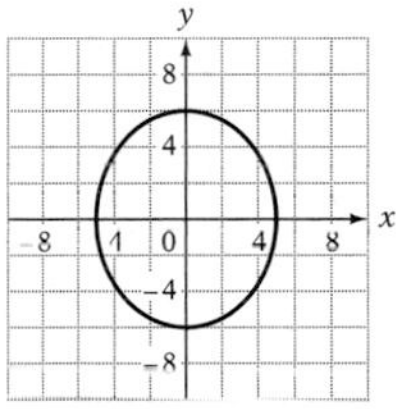

35.
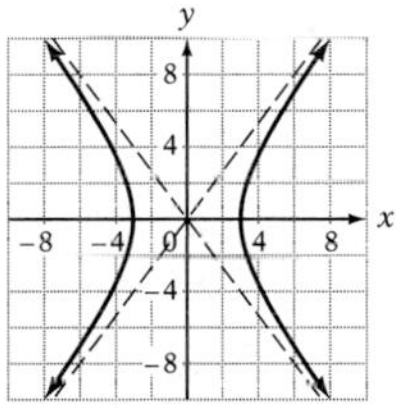

37.
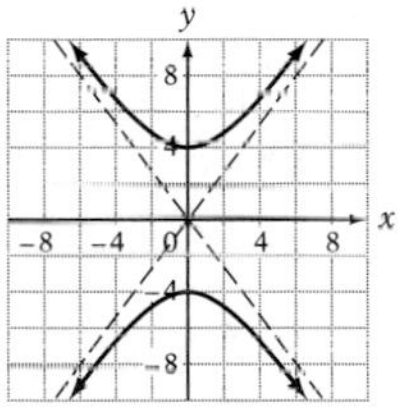

39.
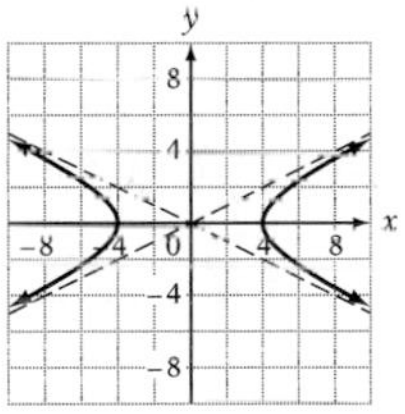

41.
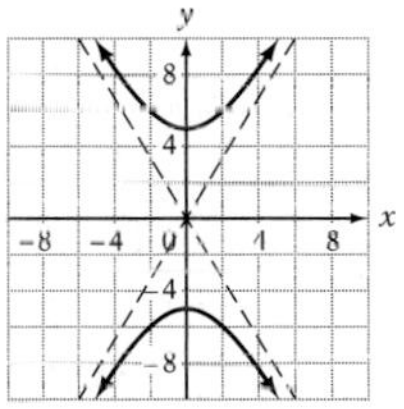

43.
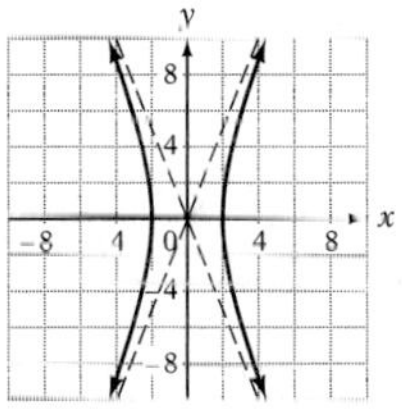

45.
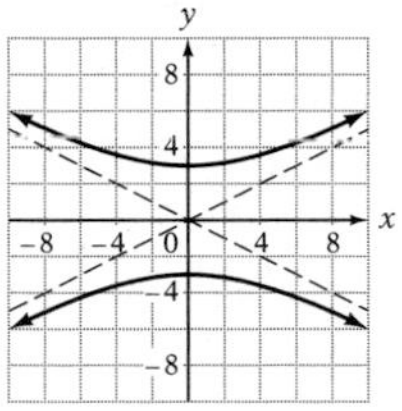

47. $\dfrac{x^2}{9} + \dfrac{y^2}{4} = 1$

ellipse
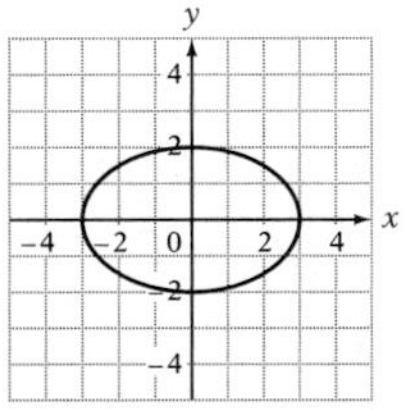

49. $\dfrac{x^2}{25} - \dfrac{y^2}{9} = 1$

hyperbola
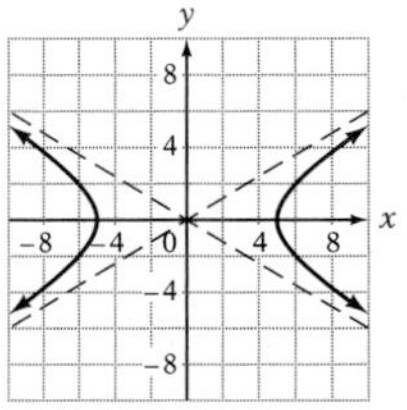

51. $\dfrac{y^2}{16} - \dfrac{x^2}{9} = 1$

hyperbola
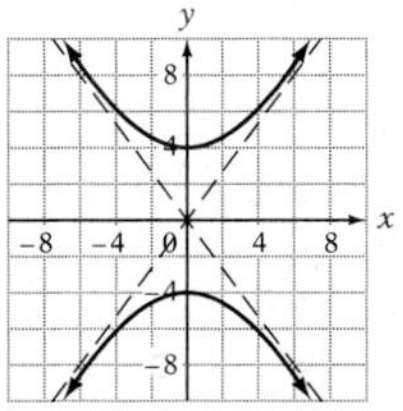

CHAPTER REVIEW

1.
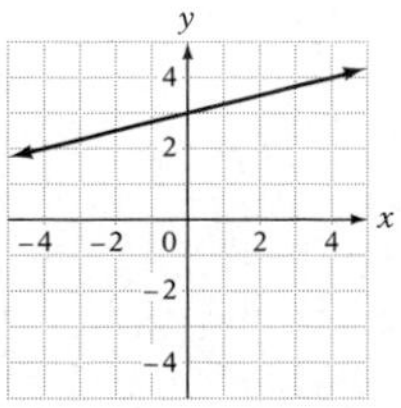
[11.1A]

2.
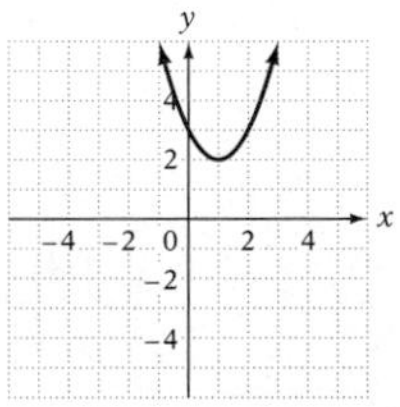
[11.2A]

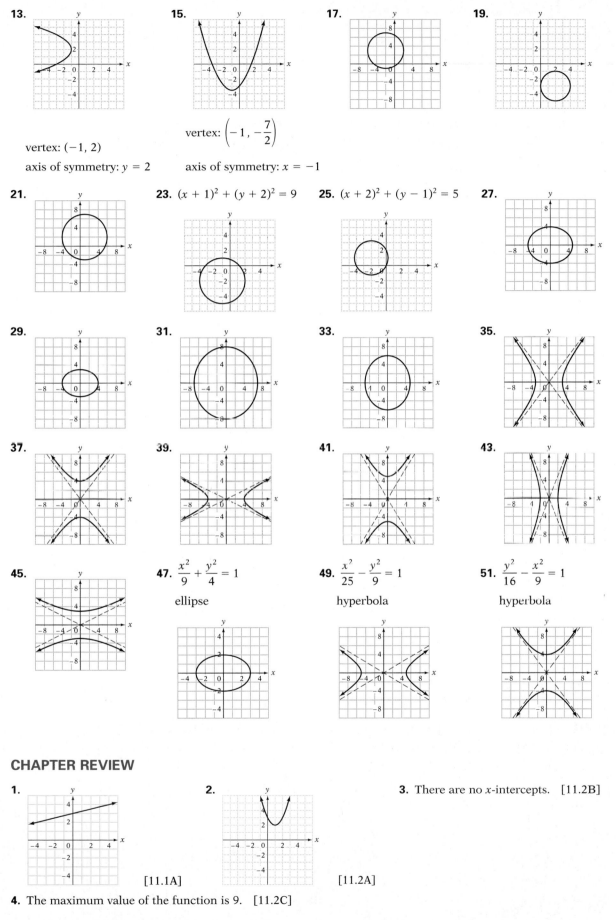

3. There are no x-intercepts. [11.2B]

4. The maximum value of the function is 9. [11.2C]

5.

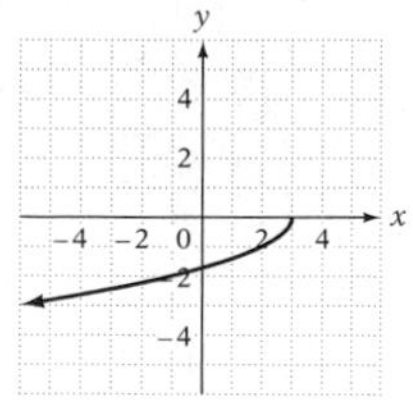

domain: $\{x|x \leq 3\}$
range: $\{y|y \leq 0\}$ [11.3A]

6.

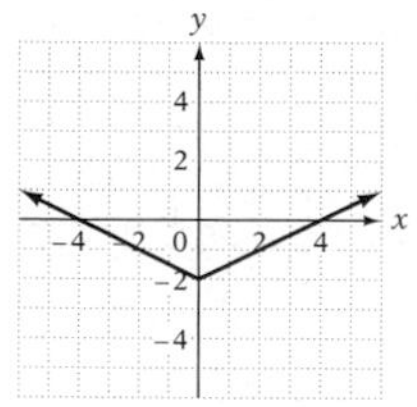

domain: $\{x|x \in \text{real numbers}\}$
range: $\{y|y \geq -2\}$ [11.3A]

7.

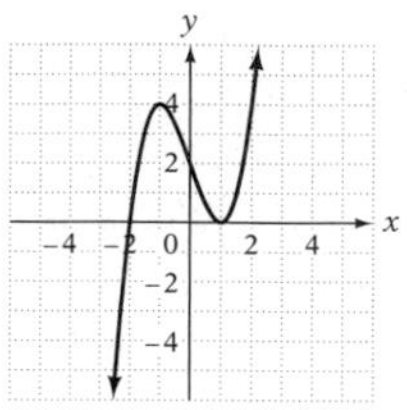

domain: $\{x|x \in \text{real numbers}\}$
range: $\{y|y \in \text{real numbers}\}$ [11.3A]

8. no [11.5A] **9.** -2 [11.4A] **10.** $-\frac{13}{2}$ [11.4A] **11.** 5 [11.4B] **12.** $2x^2 - 4x - 5$ [11.4B]

13. $\{(6, 2), (5, 3), (4, 4), (3, 5)\}$ [11.5B] **14.** $f^{-1}(x) = 4x + 16$ [11.5B] **15.**

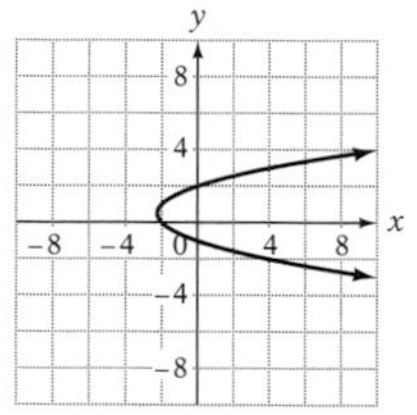

[11.6A]

16.

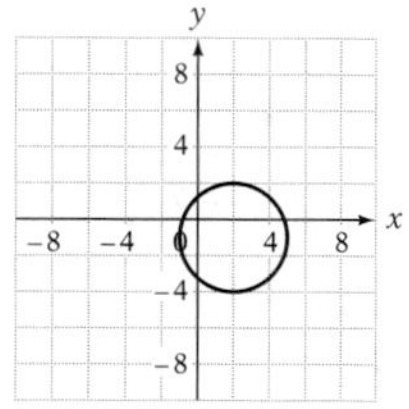

[11.6B]

17.

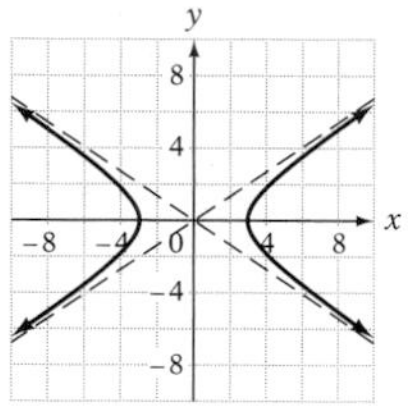

[11.6D]

18.

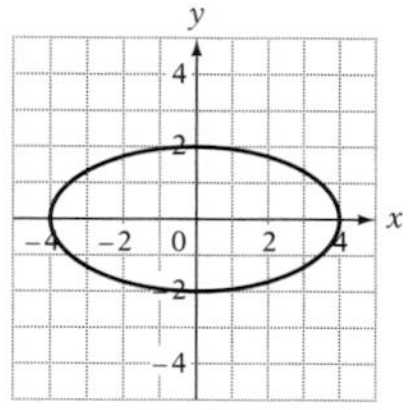

[11.6C]

19. $(x + 3)^2 + (y + 3)^2 = 16$ [11.6B] **20.** The dimensions are 50 cm by 50 cm. The area is 2500 cm². [11.2D]

CHAPTER TEST

1.

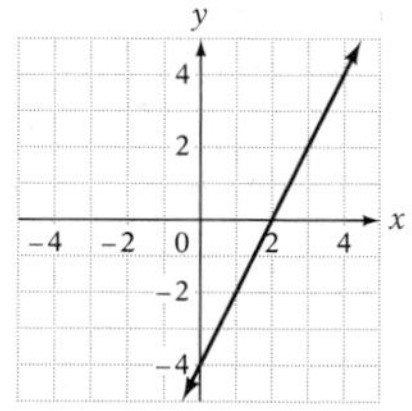

[11.1A]

2.

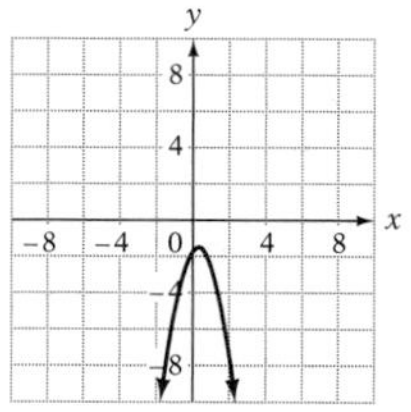

[11.2A]

3. (0, 0), (−3, 0) [11.2B]

4. The maximum value of the function is 3. [11.2C]

5.

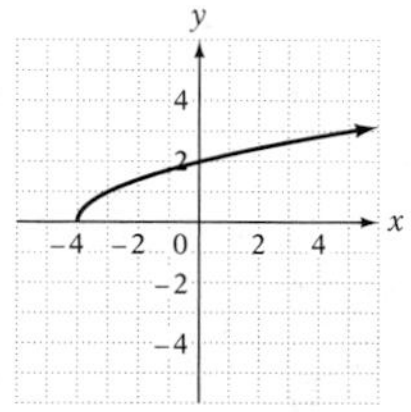

domain: $\{x|x \geq -4\}$
range: $\{y|y \geq 0\}$ [11.3A]

6.

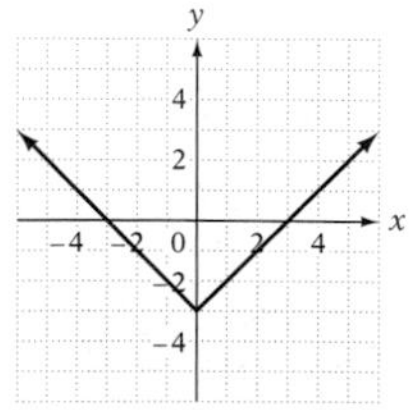

domain: $\{x|x \in \text{real numbers}\}$
range: $\{y|y \geq -3\}$ [11.3A]

7.

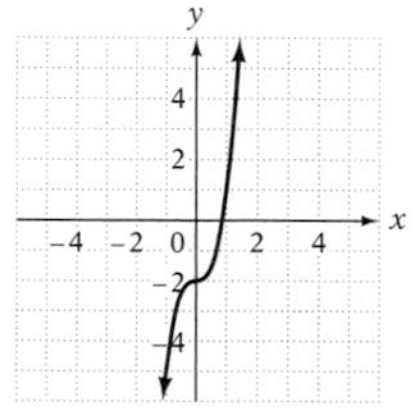

domain: $\{x|x \in \text{real numbers}\}$
range: $\{y|y \in \text{real numbers}\}$ [11.3A]

8. yes [11.5A] **9.** 7 [11.4A] **10.** -9 [11.4A] **11.** 70 [11.4A] **12.** $\frac{12}{7}$ [11.4A]

13. $6x^2 + 3x - 16$ [11.4B] **14.** $f^{-1}(x) = -\frac{1}{6}x + \frac{2}{3}$ [11.5B] **15.**

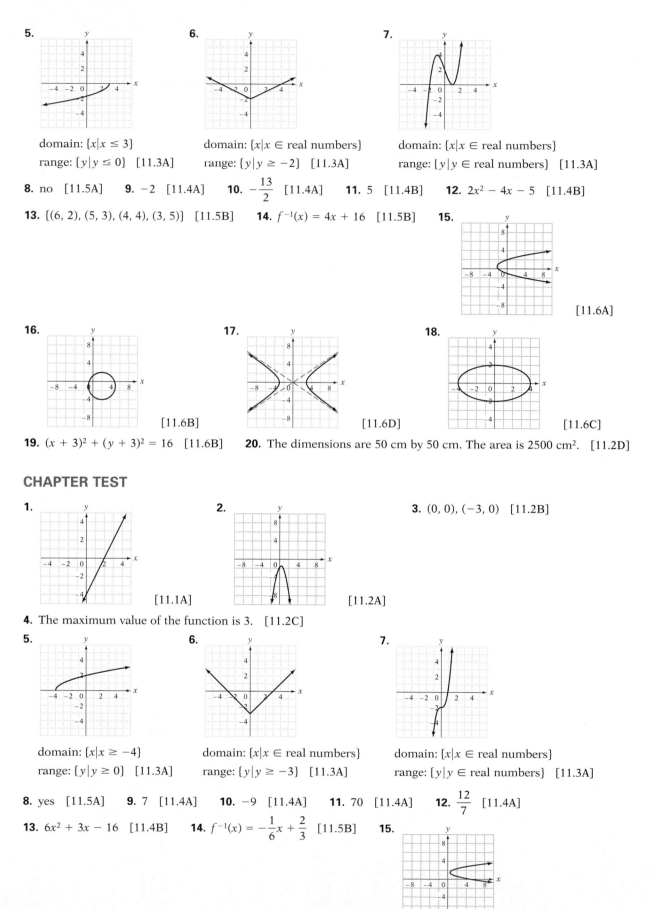

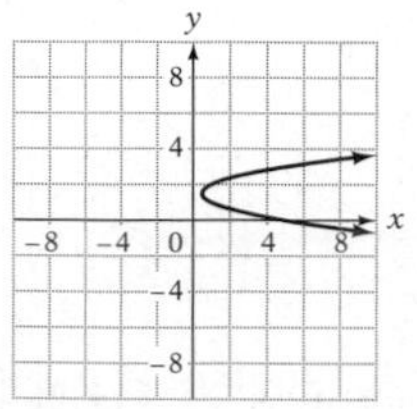

[11.6A]

16.

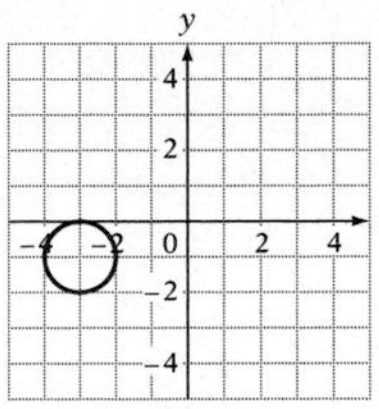

[11.6B]

17.

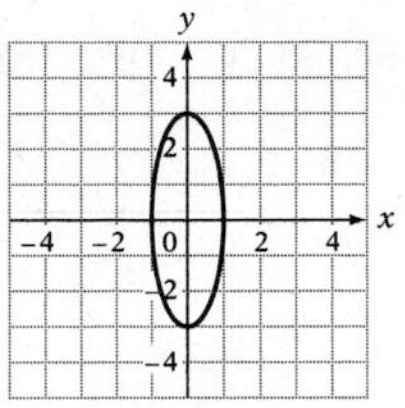

[11.6C]

18.

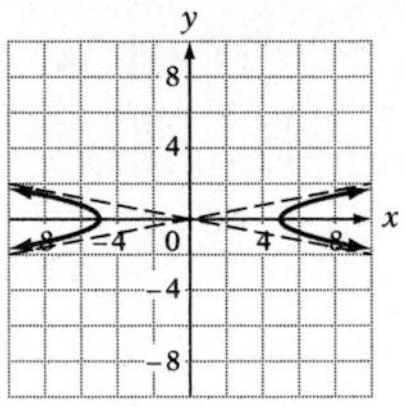

[11.6D]

19. yes [11.5B] **20.** The rectangle will have a maximum area when the dimensions are 7 ft by 7 ft. [11.2D]

CUMULATIVE REVIEW

1. 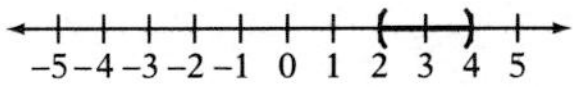 [1.5C] **2.** $\frac{38}{53}$ [8.4A] **3.** −20 [4.2A] **4.** $y = -\frac{3}{2}x$ [4.5A] **5.** $y = x - 6$ [4.6A]

6. $\frac{x}{x+y}$ [8.1A] **7.** $\frac{x-5}{3x-2}$ [8.2D] **8.** $\frac{5}{2}$ [8.4A] **9.** $\frac{4a}{3b^{12}}$ [6.1B] **10.** $2x^{3/4}$ [9.1B] **11.** $3\sqrt{2} - 5i$ [9.3A]

12. $\frac{-1+\sqrt{7}}{2}, \frac{-1-\sqrt{7}}{2}$ [10.2A/10.3A] **13.** 6 [10.4B] **14.** $\{x \mid -4 < x \le 3\}$ [10.6A] **15.** 0 [11.2C]

16. $f^{-1}(x) = \frac{1}{4}x - 2$ [11.5B] **17.**

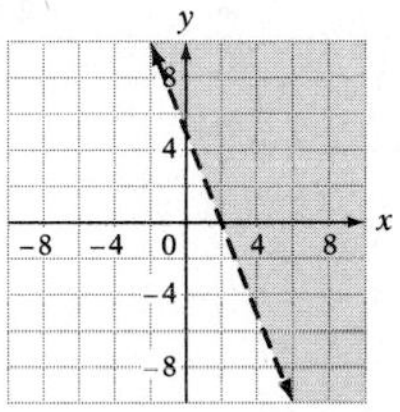

[4.7A]

18.

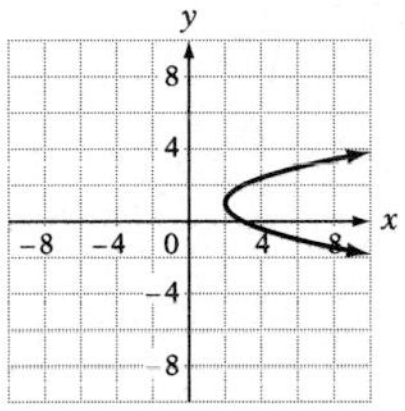

[11.6A]

19.

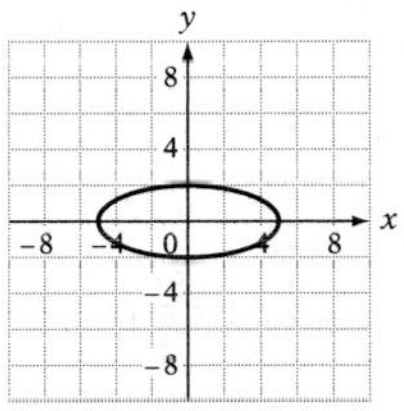

[11.6C]

20.

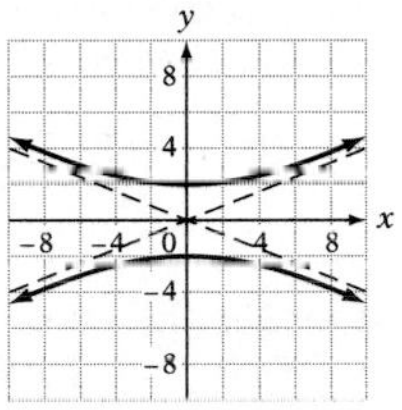

[11.6D]

21. There were 82 adult tickets sold. [5.4B]

22. The rate of the motorcycle is 60 mph. [8.6B] **23.** The rate of the crew in calm water is 4.5 mph. [5.4A]
24. A gear with 60 teeth will make 18 revolutions per minute. [8.7A] **25.** The maximum product is 400. [11.2D]

Answers to Chapter 12 Selected Exercises

PREP TEST

1. $\frac{1}{9}$ [6.1B] **2.** 16 [6.1B] **3.** −3 [6.1B] **4.** 0; 108 [4.2A] **5.** −6 [2.2B] **6.** −2, 8 [10.1A]

7. 6326.60 [1.4A] **8.**

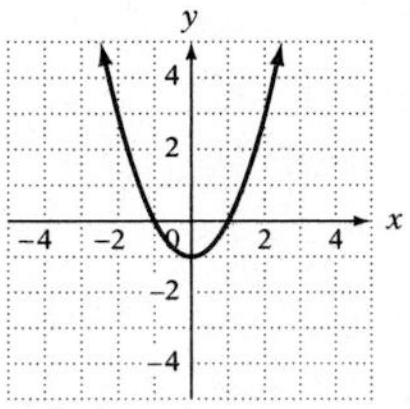

[11.2A]

SECTION 12.1

3. c **5a.** $f(2) = 9$ **b.** $f(0) = 1$ **c.** $f(-2) = \frac{1}{9}$ **7a.** $g(3) = 16$ **b.** $g(1) = 4$ **c.** $g(-3) = \frac{1}{4}$ **9a.** $P(0) = 1$

b. $P\left(\frac{3}{2}\right) = \frac{1}{8}$ **c.** $P(-2) = 16$ **11a.** $G(4) = 7.3891$ **b.** $G(-2) = 0.3679$ **c.** $G\left(\frac{1}{2}\right) = 1.2840$

13a. $H(-1) = 54.5982$ **b.** $H(3) = 1$ **c.** $H(5) = 0.1353$ **15a.** $F(2) = 16$ **b.** $F(-2) = 16$ **c.** $F\left(\frac{3}{4}\right) = 1.4768$

17a. $f(-2) = 0.1353$ **b.** $f(2) = 0.1353$ **c.** $f(-3) = 0.0111$ **19.**

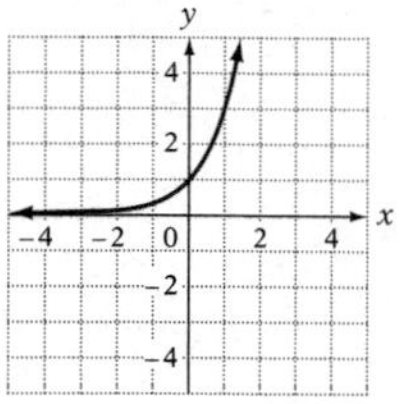

21.

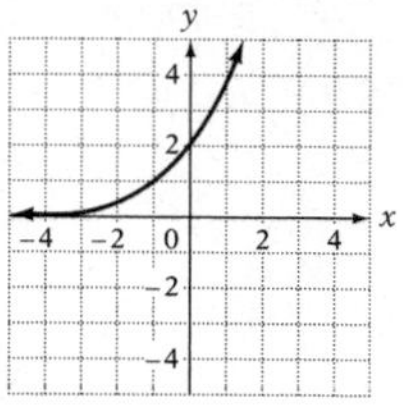

23.

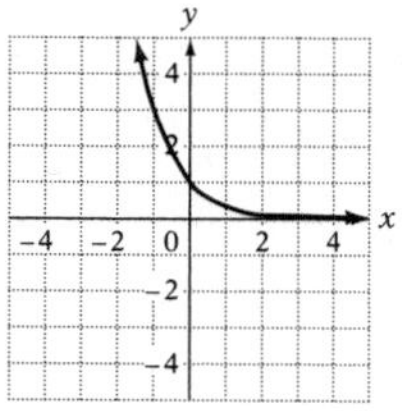

25.

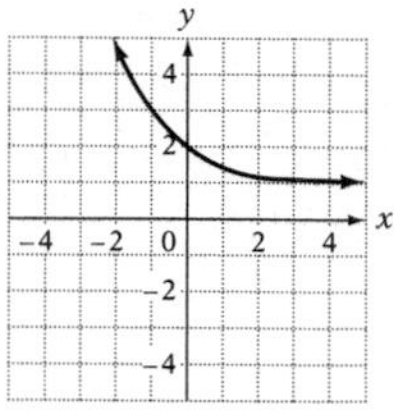

27.

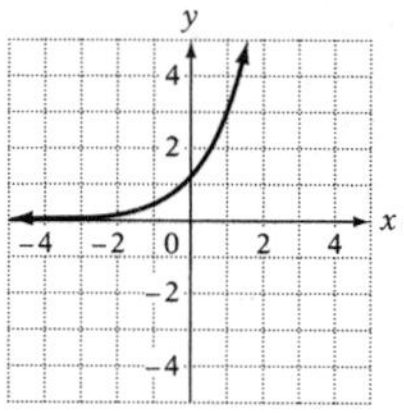

29.

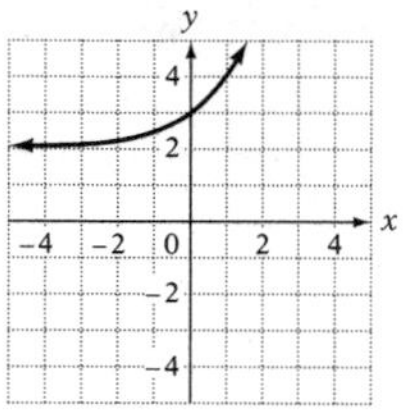

31. b and d **33.** (0, 1) **35.** no x-intercept; y-intercept: (0, 1)

37.

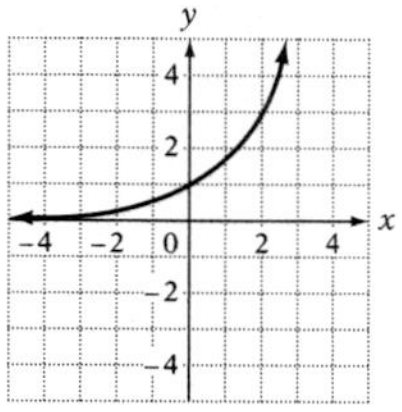

39.

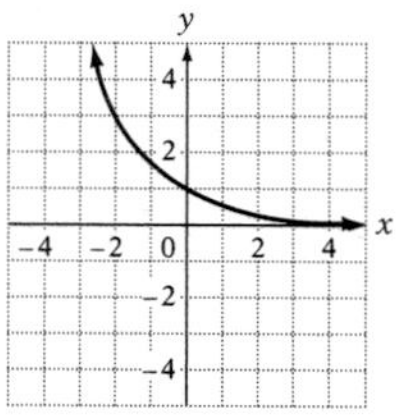

41.

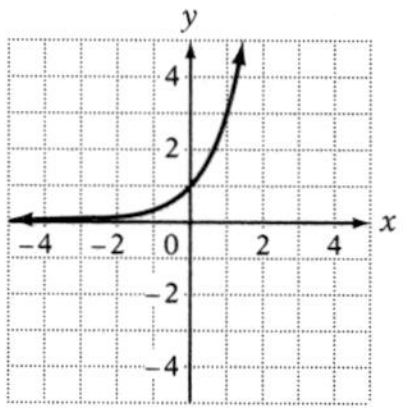

43. e

45a.

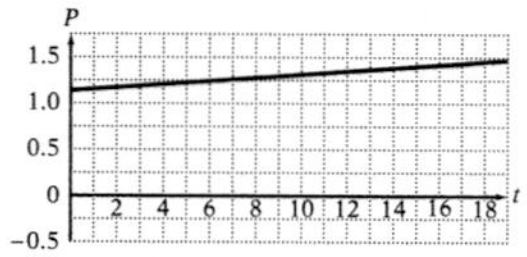

b. The point (9, 1.303) means that in the year 2002, the Chinese population was approximately 1.303 billion.

47a.

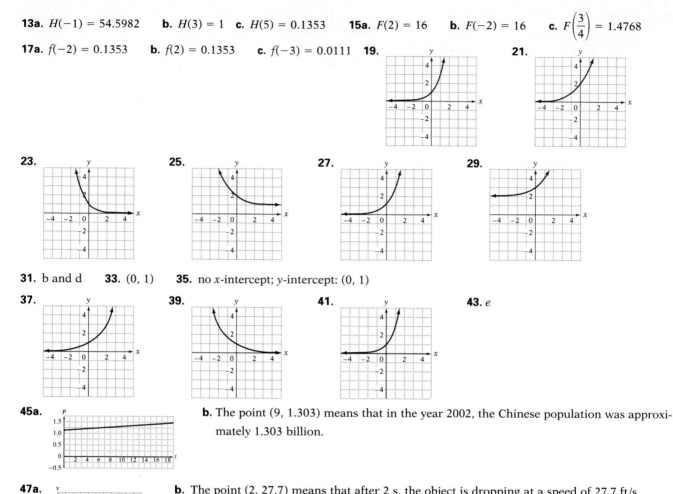

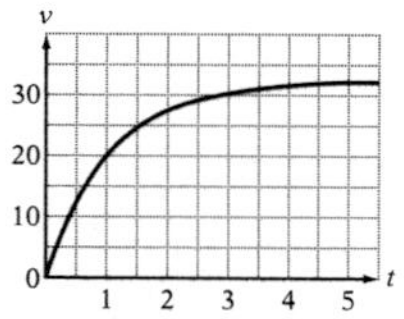

b. The point (2, 27.7) means that after 2 s, the object is dropping at a speed of 27.7 ft/s.

SECTION 12.2

3. $\log_5 25 = 2$ **5.** $\log_4 \frac{1}{16} = -2$ **7.** $\log_{10} x = y$ **9.** $\log_a w = x$ **11.** $3^2 = 9$ **13.** $10^{-2} = 0.01$ **15.** $e^y = x$

17. $b^v = u$ **19.** $\log_3 81 = 4$ **21.** $\log_2 128 = 7$ **23.** $\log 100 = 2$ **25.** $\ln e^3 = 3$ **27.** $\log_8 1 = 0$

29. $\log_5 625 = 4$ **31.** 9 **33.** 64 **35.** $\frac{1}{7}$ **37.** 1 **39.** 316.23 **41.** 0.02 **43.** 7.39 **45.** 0.61

49. $\log_3 (x^3y^2)$ **51.** $\ln\left(\frac{x^4}{y^2}\right)$ **53.** $\log_7 x^3$ **55.** $\ln (x^3y^4)$ **57.** $\log_4 (x^2y^2)$ **59.** $\log_6 \sqrt{\frac{x}{y}}$ **61.** $\log_3\left(\frac{x^2z^2}{y}\right)$

63. $\ln\left(\frac{x}{y^2z}\right)$ **65.** $\log_4\left(\frac{s^2r^2}{t^4}\right)$ **67.** $\ln\left(\frac{x^3}{y^4z^4}\right)$ **69.** $\log_2\left(\frac{t^3v^2}{r^2}\right)$ **71.** $\log_4 \sqrt{\frac{x^3z}{y^2}}$ **73.** $\ln \sqrt{\frac{x}{y^3}}$ **75.** $\log_2\left(\frac{\sqrt{xz}}{\sqrt[3]{y^2}}\right)$

77. $\log_8 x + \log_8 z$ **79.** $5 \log_3 x$ **81.** $\log_b r - \log_b s$ **83.** $2 \log_3 x + 6 \log_3 y$ **85.** $3 \log_7 u - 4 \log_7 v$

87. $2 \log_2 r + 2 \log_2 s$ **89.** $2 \ln x + \ln y + \ln z$ **91.** $\log_5 x + 2 \log_5 y - 4 \log_5 z$ **93.** $2 \log_8 x - \log_8 y - 2 \log_8 z$

95. $\frac{3}{2}\log_4 x + \frac{1}{2}\log_4 y$ **97.** $\frac{3}{2}\log_7 x - \frac{1}{2}\log_7 y$ **99.** $\log_3 t - \frac{1}{2}\log_3 x$ **101.** 0.8451 **103.** −0.2218 **105.** 1.3863

107. 1.0415 **109.** 0.8617 **111.** 2.1133 **113.** −0.6309 **115.** 0.2727 **117.** 1.6826 **119.** 1.9266

121. 0.6752 **123.** 2.6125 **125a.** false **b.** true **c.** false **d.** true **e.** false **f.** true

SECTION 12.3

5.
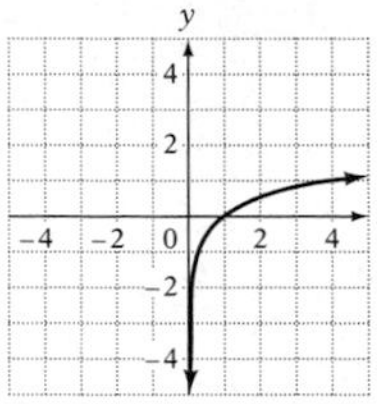

7.
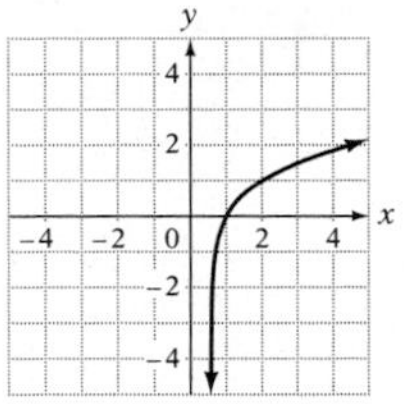

9.
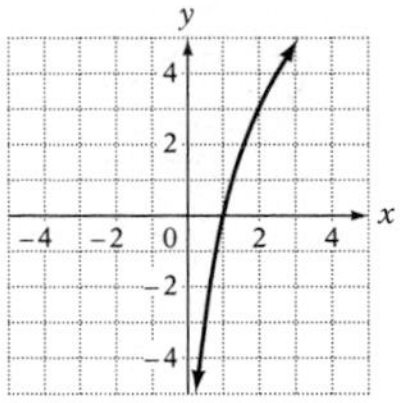

11.
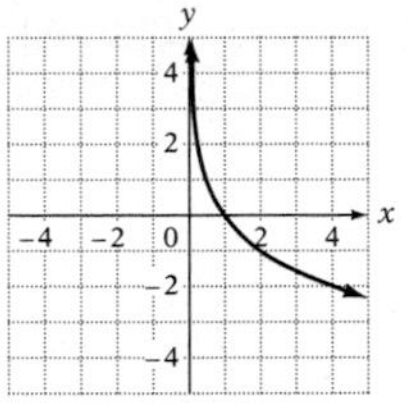

13.
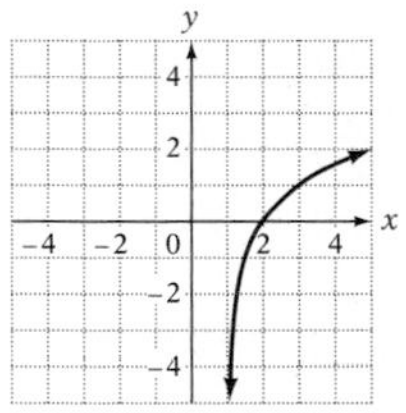

15.
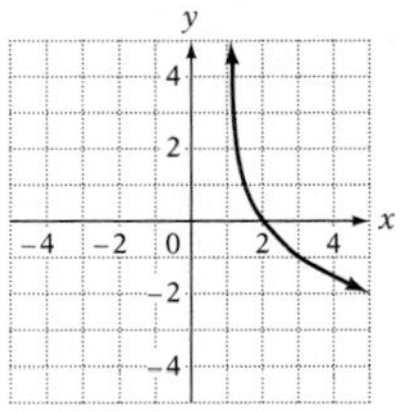

17.
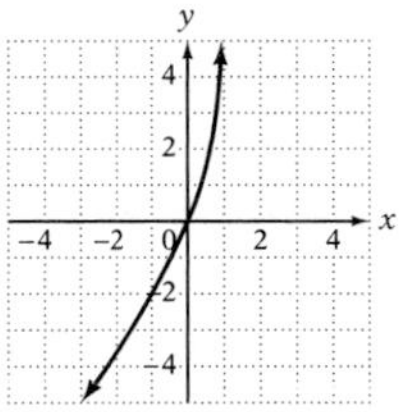

19.
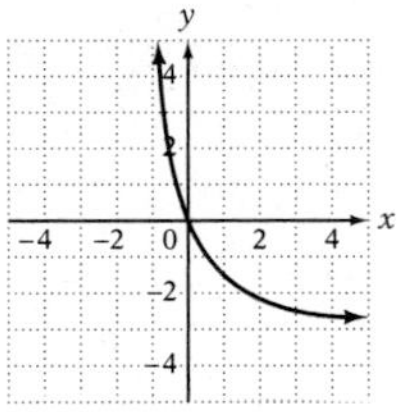

21.
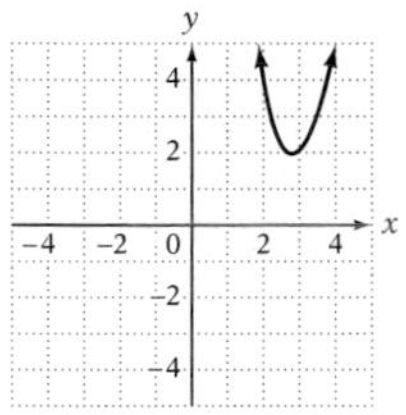

23a.
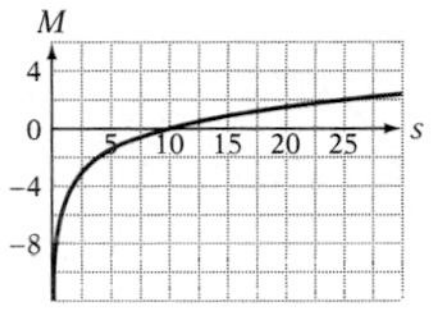

b. The point (25.1, 2) means that a star that is 25.1 parsecs from Earth has a distance modulus of 2.

SECTION 12.4

3. $-\frac{1}{3}$ **5.** 3 **7.** 1 **9.** 6 **11.** $\frac{4}{7}$ **13.** $\frac{9}{8}$ **15.** 1.1133 **17.** 1.0986 **19.** 1.3222 **21.** −2.8074

23. 3.5850 **25.** 1.1309 **29.** $\frac{11}{2}$ **31.** 4, 2 **33.** $\frac{5}{3}$ **35.** $\frac{1}{2}$ **37.** $\frac{9}{2}$ **39.** 17.5327 **41.** 3 **43.** 2

45. The equation has no solution. **47.** −0.55 **49.** 1.38 **51.** The equation has no solution. **53.** 0.09

55a.
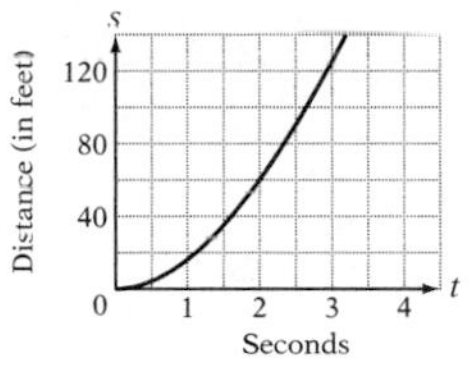

b. It will take the object approximately 2.64 s to fall 100 ft.

SECTION 12.5

1. The value of the investment after 2 years is \$1172. **3.** The investment will be worth \$15,000 in 18 years. **5a.** The technetium level will be 21.2 mg after 3 h. **b.** It will take 3.5 h for the technetium level to reach 20 mg. **7.** The half-life of promethium-147 is 2.5 years. **9a.** The approximate pressure at 40 km above Earth is 0.098 N/cm². **b.** The approximate pressure on Earth's surface is 10.13 N/cm². **c.** The atmospheric pressure decreases as you rise above Earth's surface. **11.** The pH of milk is 6.4. **13.** At a depth of 2.5 m, the percent of light will be 75% of the light at the surface of the pool. **15.** Normal conversation is 65 decibels. **17.** The average time of a major league baseball game increased 22.5 min during the years from 1981 to 1999. **19.** A thickness of 0.4 cm is needed. **21.** The Richter scale magnitude of this earthquake is 6.8. **23.** The intensity of this earthquake is $794{,}328{,}235I_0$. **25.** The magnitude of the earthquake is 5.3. **27.** The magnitude of the earthquake is 5.6. **29.** The value of the investment after 5 years is \$3210.06.

CHAPTER REVIEW

1. 1 [12.1A] **2.** $5^2 = 25$ [12.2A] **3.**

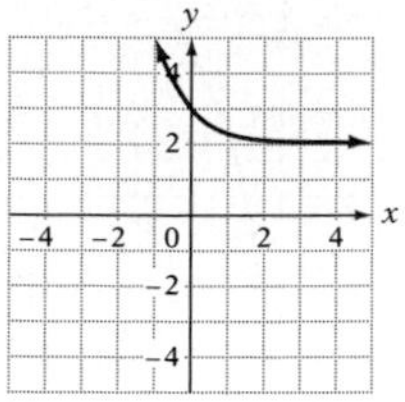

[12.1B]

4.

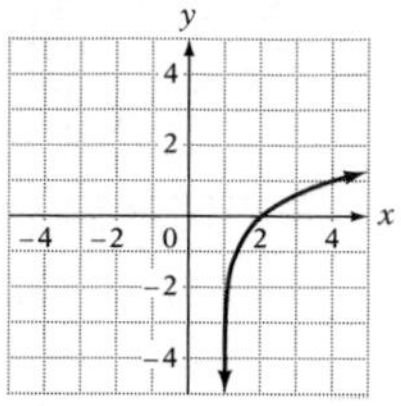

[12.3A]

5. $\frac{2}{5}\log_3 x + \frac{4}{5}\log_3 y$ [12.2B] **6.** $\log_3 \frac{x^2}{y^5}$ [12.2B] **7.** -12 [12.4A] **8.** $\frac{1}{4}$ [12.4B] **9.** 1.7251 [12.2C]

10. 32 [12.2A] **11.** 79 [12.4B] **12.** 1000 [12.2A] **13.** $\log_7 \sqrt[3]{xy^4}$ [12.2B] **14.** $\frac{5}{2}\log_8 x - \frac{3}{2}\log_8 y$ [12.2B]

15. $\log_2 32 = 5$ [12.2A] **16.** 0.4278 [12.2C] **17.** -0.535 [12.4A] **18.** $\frac{3}{2}$ [12.1A] **19.** $\frac{2}{3}$ [12.4B]

20. 2 [12.4B] **21.**

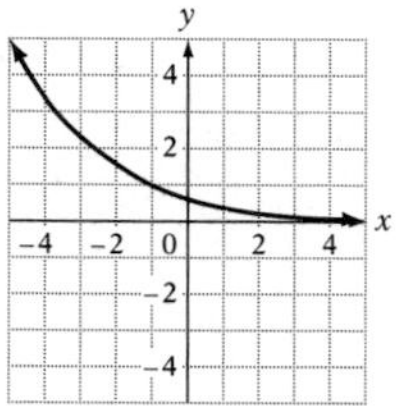

[12.1B]

22.

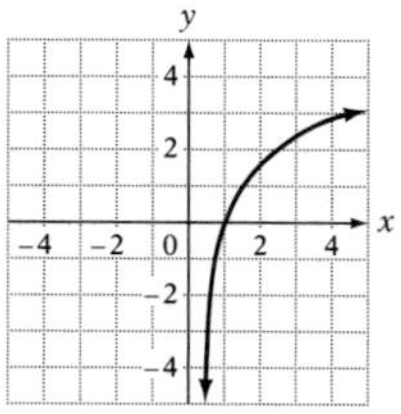

[12.3A]

23. The value of the investment after 2 years is \$4692. [12.5A] **24.** The magnitude of the earthquake is 7.6. [12.5A] **25.** The half-life of the material is 27 days. [12.5A] **26.** Sound emitted from a busy street corner is 107 decibels. [12.5A]

CHAPTER TEST

1. 1 [12.1A] **2.** $\frac{1}{3}$ [12.1A] **3.**

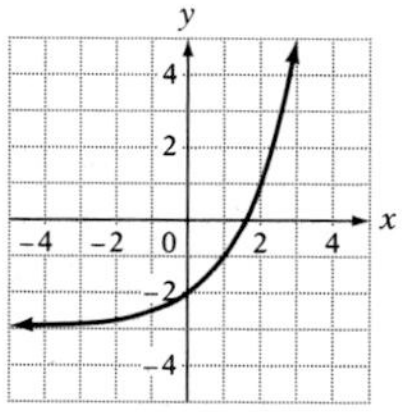

[12.1B]

4.

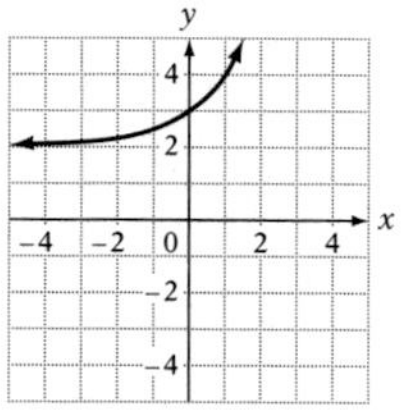

[12.1B]

5. 2 [12.2A] **6.** $\frac{1}{9}$ [12.2A] **7.**

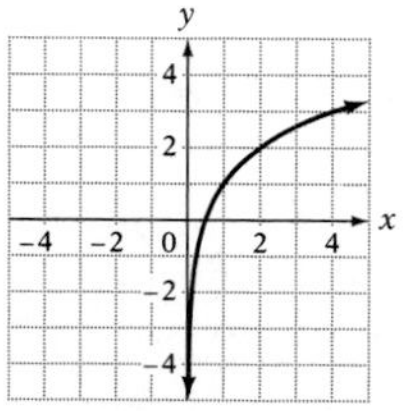

[12.3A]

8.

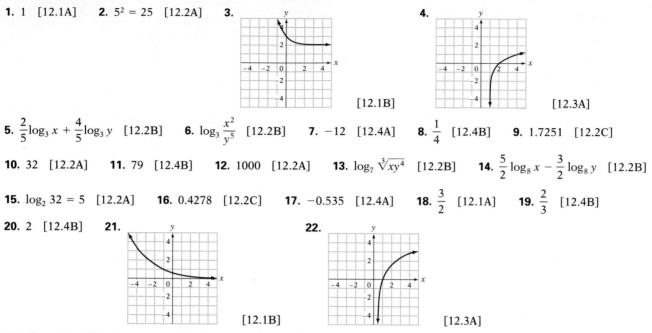

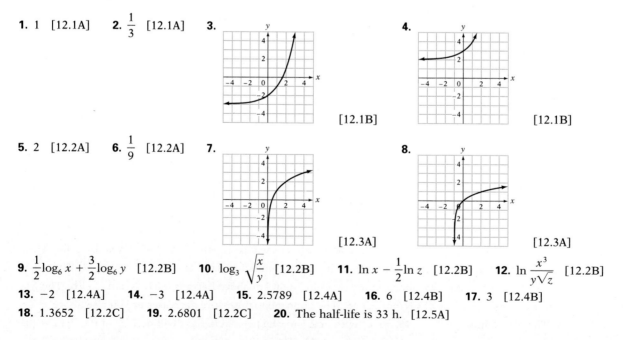

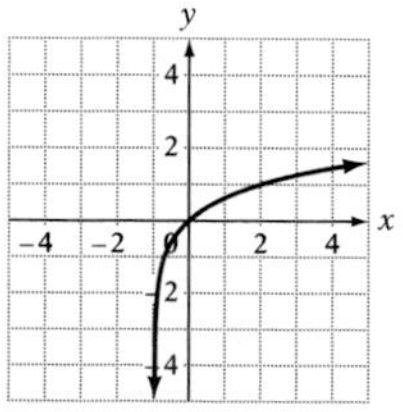

[12.3A]

9. $\frac{1}{2}\log_6 x + \frac{3}{2}\log_6 y$ [12.2B] **10.** $\log_3 \sqrt{\frac{x}{y}}$ [12.2B] **11.** $\ln x - \frac{1}{2}\ln z$ [12.2B] **12.** $\ln \frac{x^3}{y\sqrt{z}}$ [12.2B]

13. -2 [12.4A] **14.** -3 [12.4A] **15.** 2.5789 [12.4A] **16.** 6 [12.4B] **17.** 3 [12.4B]

18. 1.3652 [12.2C] **19.** 2.6801 [12.2C] **20.** The half-life is 33 h. [12.5A]

CUMULATIVE REVIEW

1. $\frac{8}{7}$ [2.2C] **2.** $y = 2x - 6$ [4.6A] **3.** $(4x^n + 3)(x^n + 1)$ [7.3A/7.3B] **4.** $\frac{x-3}{x+3}$ [8.3A] **5.** $\frac{x\sqrt{y} + y\sqrt{x}}{x - y}$ [9.2D]

6. $2 + \sqrt{10}, 2 - \sqrt{10}$ [10.2A] **7.** The length of the side is 26 cm. [9.4B] **8.**

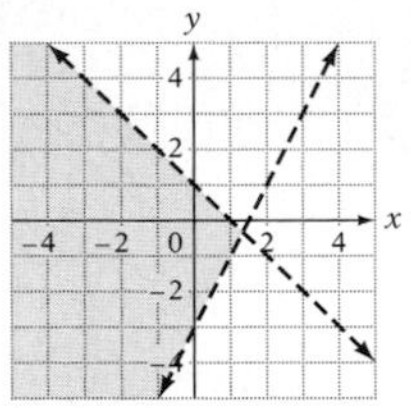

[5.5A]

9. (0, −1, 2) [5.2B] **10.** $-\dfrac{2x^2 - 17x + 13}{(x - 2)(2x - 3)}$ [8.2D] **11.** $\{x \mid -5 \le x \le 1\}$ [10.6A] **12.** $\{x \mid 1 \le x \le 4\}$ [2.5B]

13.

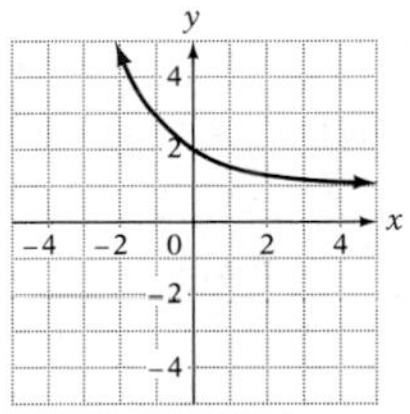

[12.1B]

14.

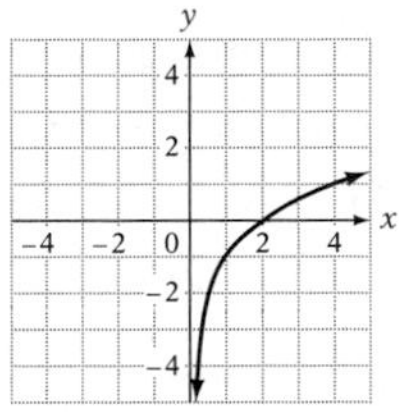

[12.3A]

15. The area is 28 yd². [3.2B]

16. 125 [12.2A] **17.** $\log_b \dfrac{x^3}{y^5}$ [12.2B] **18.** 1.7712 [12.2C] **19.** 2 [12.4A] **20.** $\dfrac{1}{2}$ [12.4B] **21.** The customer can write fewer than 50 checks. [2.4C] **22.** The mixture costs $3.10 per pound. [2.3A] **23.** The rate of the wind is 25 mph. [5.4A] **24.** The force will stretch the spring 10.2 in. [8.7A] **25.** The redwood costs $.40 per foot. The fir costs $.25 per foot. [5.4B] **26.** The investment will double in 8 years. [12.5A]

Answers to Final Exam

1. −31 [1.3A] **2.** −1 [1.4A] **3.** $-10x + 33$ [1.4D] **4.** 8 [2.2A] **5.** $4, -\dfrac{2}{3}$ [2.5A] **6.** The volume is 268.1 ft³. [3.3A] **7.**

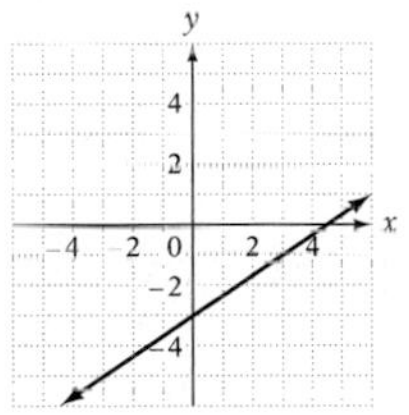

[4.3B]

8. $y = -3x + 7$ [4.5B] **9.** $y = -\dfrac{2}{3}x - \dfrac{1}{3}$ [4.6A]

10. $6a^3 - 5a^2 + 10a$ [6.3A] **11.** $(2 - xy)(4 + 2xy + x^2y^2)$ [7.4B] **12.** $(x - y)(1 + x)(1 - x)$ [7.4D]

13. $x^2 - 2x - 3 - \dfrac{5}{2x - 3}$ [6.4A] **14.** $\dfrac{x(x - 1)}{2x - 5}$ [8.1C] **15.** $\dfrac{-10x}{(x + 2)(x - 3)}$ [8.2D] **16.** $\dfrac{x + 3}{x + 1}$ [8.3A]

17. $-\dfrac{7}{4}$ [8.4A] **18.** $d = \dfrac{a_n - a_1}{n - 1}$ [8.5A] **19.** $\dfrac{y^4}{162x^3}$ [6.1B] **20.** $\dfrac{1}{64x^8y^5}$ [9.1A] **21.** $-2x^2y\sqrt{2y}$ [9.2B]

22. $\dfrac{x^2\sqrt{2y}}{2y^2}$ [9.2D] **23.** $\dfrac{6}{5} - \dfrac{3}{5}i$ [9.3D] **24.** $2x^2 - 3x - 2 = 0$ [10.1B] **25.** $\dfrac{3 + \sqrt{17}}{4}, \dfrac{3 - \sqrt{17}}{4}$ [10.3A]

26. −8, 27 [10.4A] **27.**

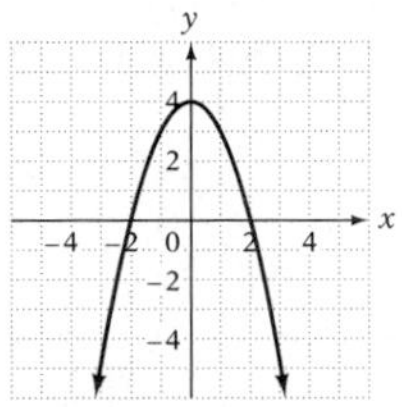

[11.2A]

28.

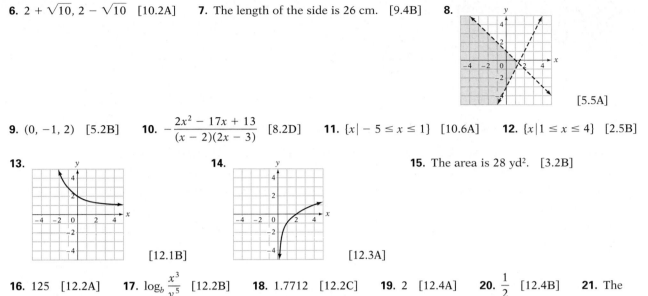

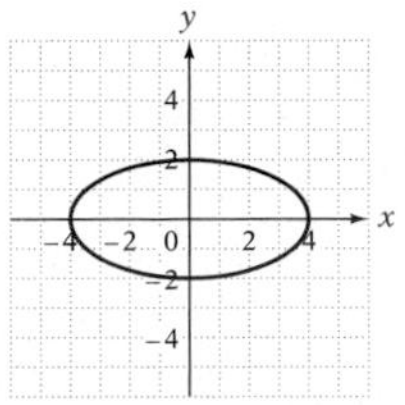

[11.6C]

29. $-2, \dfrac{3}{2}$ [10.4C] **30.** $f^{-1}(x) = \dfrac{3}{2}x + 6$ [11.5B] **31.** (3, 4) [5.2A] **32.** 10 [5.3A]

33. $\left\{x \mid x > \dfrac{3}{2}\right\}$ [2.4B] **34.** $\{x \mid -4 < x < -1\}$ [2.5B]

35.

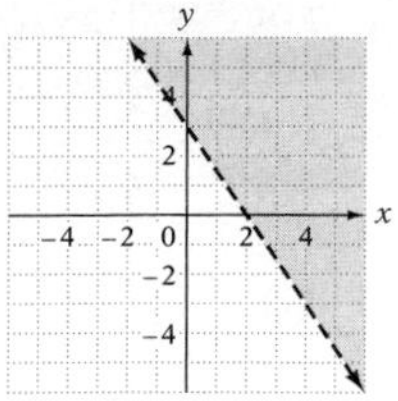

[4.7A]

36.

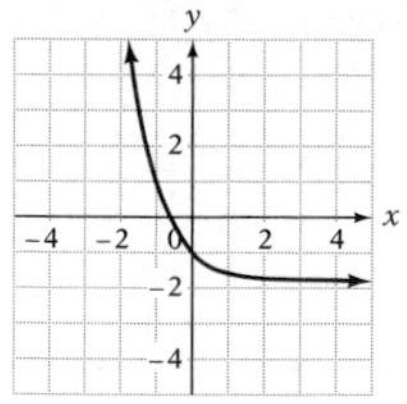

[12.1B]

37.

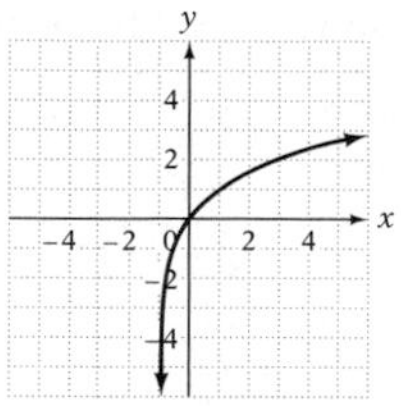

[12.3A]

38. $\log_2 \frac{a^2}{b^2}$ [12.2B] **39.** 6 [12.4B] **40.** The range of scores is $69 \le x \le 100$. [2.4C]

41. The cyclist rode 40 mi. [2.3D] **42.** There is \$8000 invested at 8.5% and \$4000 invested at 6.4%. [2.3C] **43.** The length is 20 ft. The width is 7 ft. [10.5A] **44.** An additional 200 shares are needed. [8.4B] **45.** The rate of the plane was 420 mph. [8.6B] **46.** The object has fallen 88 ft when the speed reaches 75 ft/s. [9.4B] **47.** The rate of the plane for the first 360 mi was 120 mph. [8.6B] **48.** The intensity is 200 foot-candles. [8.7A] **49.** The rate of the boat in calm water is 12.5 mph. The rate of the current is 2.5 mph. [5.4A] **50.** The value of the investment after 2 years is \$4785.65. [12.5A]

Answers to Chapter R Selected Exercises

SECTION R.1

1. 10 **3.** 21 **5.** -2 **7.** 5 **9.** -3 **11.** -1 **13.** 1 **15.** 18 **17.** -6 **19.** 2 **21.** -40 **23.** 2.01 **25.** $8x$ **27.** $3b$ **29.** $5a$ **31.** $11x$ **33.** $15x - 2y$ **35.** $9a$ **37.** $22y^2$ **39.** $\frac{1}{2}x$ **41.** $-20x$ **43.** $-24a$ **45.** x **47.** $7x$ **49.** $3y$ **51.** $-3a - 15$ **53.** $-16x - 48$ **55.** $-10y^2 + 5$ **57.** $18x^2 - 12xy - 6y^2$ **59.** $-8y - 7$ **61.** $5x - 75$ **63.** $-a + 56$

SECTION R.2

1. -12 **3.** -1 **5.** -8 **7.** -20 **9.** 8 **11.** -1 **13.** 1 **15.** $\frac{3}{4}$ **17.** 3 **19.** 3 **21.** 2 **23.** -1 **25.** $\frac{1}{2}$ **27.** 5 **29.** $-\frac{1}{9}$ **31.** $\{x \mid x > 3\}$ **33.** $\{n \mid n \ge 2\}$ **35.** $\{x \mid x \le -3\}$ **37.** $\{n \mid n > 0\}$ **39.** $\{x \mid x > 4\}$ **41.** $\{x \mid x > -2\}$ **43.** $\{x \mid x > 3\}$ **45.** $\{x \mid x \ge 2\}$ **47.** $\{x \mid x < 2\}$ **49.** $\{x \mid x > 3\}$ **51.** $\{x \mid x \ge 2\}$ **53.** $\{x \mid x < 4\}$ **55.** $\left\{x \,\middle|\, x > \frac{14}{11}\right\}$ **57.** $\left\{x \,\middle|\, x \le \frac{3}{8}\right\}$ **59.** $\left\{x \,\middle|\, x \ge -\frac{5}{4}\right\}$

SECTION R.3

1.

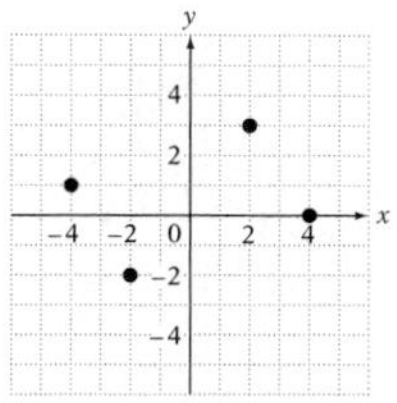

3.

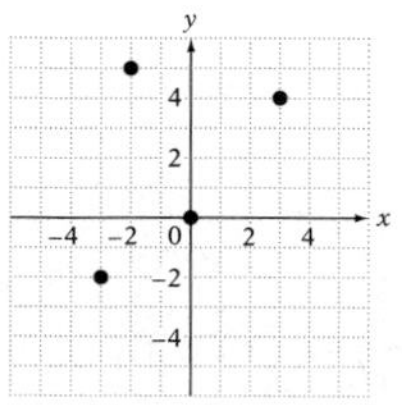

5.

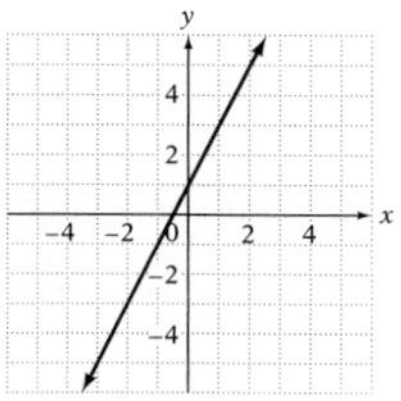

7.

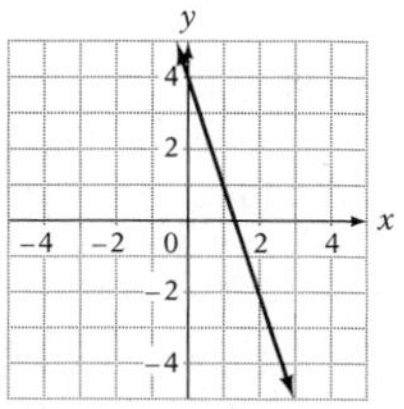

9.

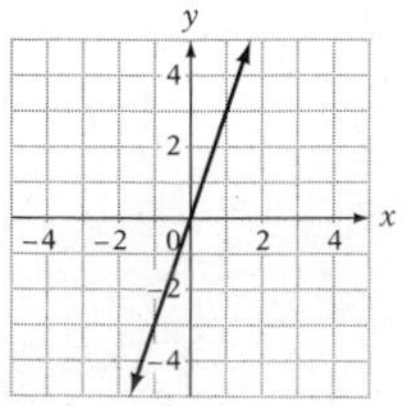

11.

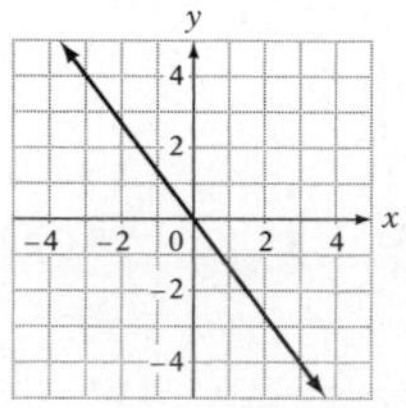

13.

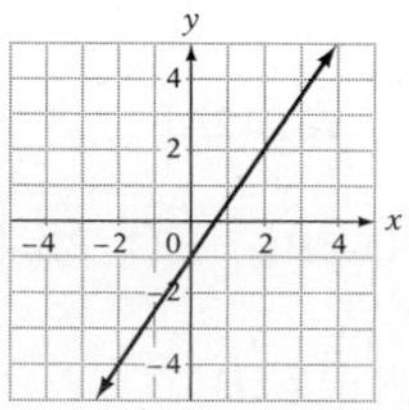

15.

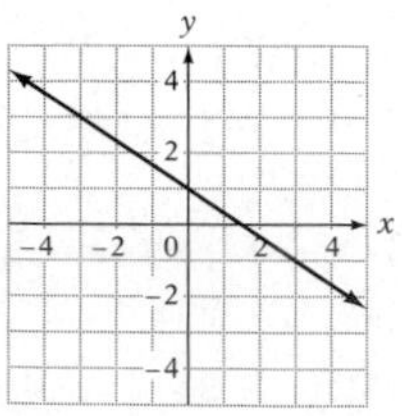

17.
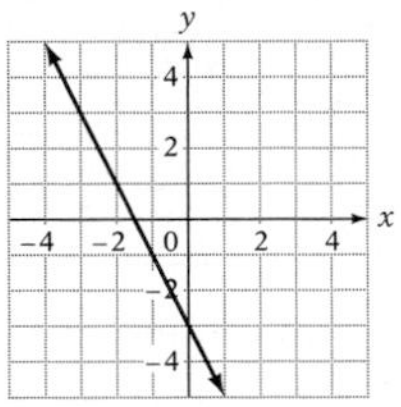

19.
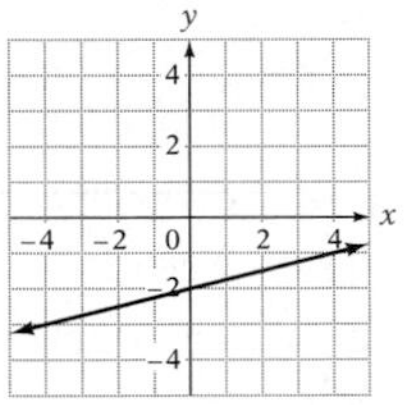

21.
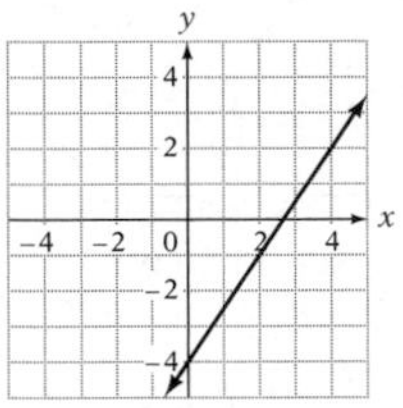

23. 3 **25.** $\frac{1}{4}$ **27.** $-\frac{3}{2}$

29. Undefined **31.** 0 **33.** 1 **35.** $m = \frac{5}{2}, b = (0, -4)$ **37.**
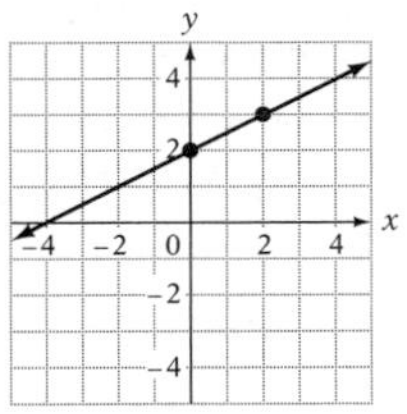

39.
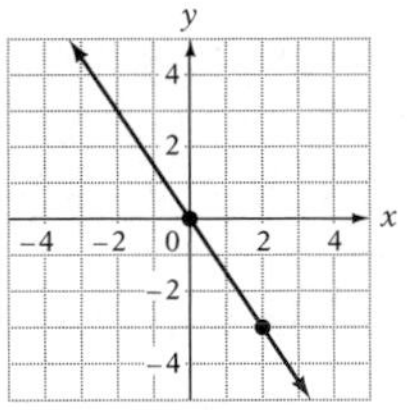

41.
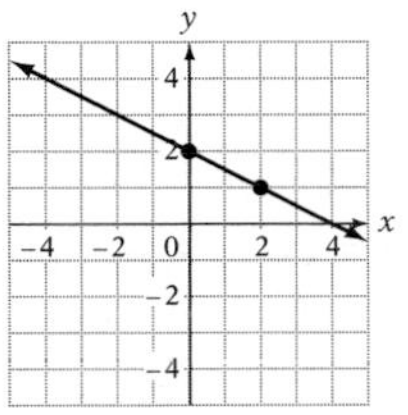

43.
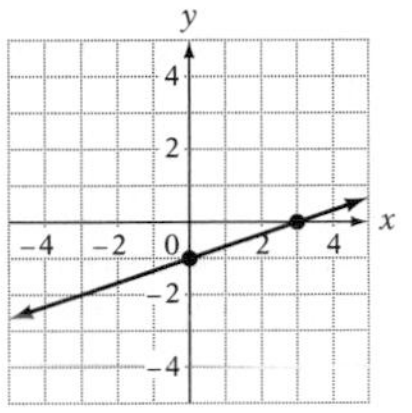

45.
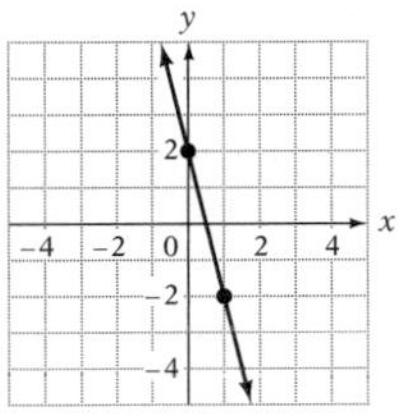

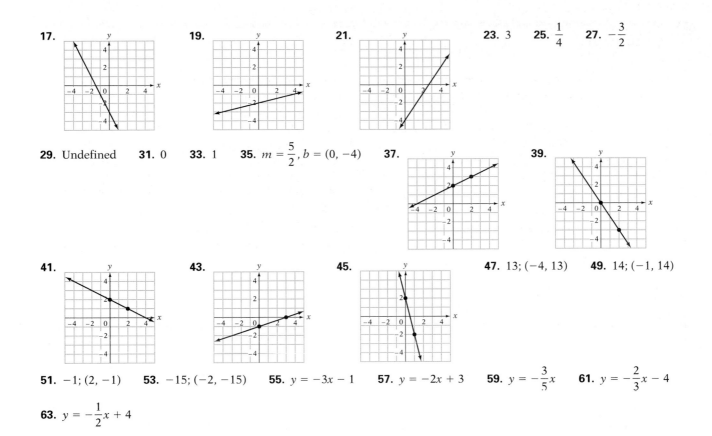

47. 13; $(-4, 13)$ **49.** 14; $(-1, 14)$

51. -1; $(2, -1)$ **53.** -15; $(-2, -15)$ **55.** $y = -3x - 1$ **57.** $y = -2x + 3$ **59.** $y = -\frac{3}{5}x$ **61.** $y = -\frac{2}{3}x - 4$

63. $y = -\frac{1}{2}x + 4$

SECTION R.4

1. z^8 **3.** x^{15} **5.** $x^{12}y^{18}$ **7.** a^6 **9.** $-m^9n^3$ **11.** $-8a^9b^3c^6$ **13.** mn^2 **15.** $-\frac{2a}{3}$ **17.** $-27m^2n^4p^3$

19. $54n^{14}$ **21.** $\frac{7xz}{8y^3}$ **23.** $-8x^{13}y^{14}$ **25.** $\frac{4}{x^7}$ **27.** $\frac{1}{d^{10}}$ **29.** $\frac{1}{x^5}$ **31.** $\frac{x^2}{3}$ **33.** $\frac{x^6}{y^{12}}$ **35.** $\frac{9}{x^2y^4}$ **37.** $\frac{2}{x^4}$

39. $\frac{1}{2x^3}$ **41.** $\frac{1}{2x^2y^6}$ **43.** $\frac{1}{x^6}$ **45.** $\frac{y^8}{x^4}$ **47.** $\frac{b^{10}}{4a^{10}}$ **49.** $7b^2 + b - 4$ **51.** $3a^2 - 3a + 17$ **53.** $7x - 7$

55. $-2x^3 + x^2 + 2$ **57.** $x^3 + 2x^2 - 6x - 6$ **59.** $5x^3 + 10x^2 - x - 4$ **61.** $y^3 - y^2 + 6y - 6$ **63.** $4y^3 - 2y^2 + 2y - 4$
65. $12b^4 - 48b^3 - 24b$ **67.** $9b^5 - 9b^3 + 24b$ **69.** $-2x^4y + 6x^3y^2 - 4x^2y^3$ **71.** $x^3 + 4x^2 + 5x + 2$
73. $a^3 - 6a^2 + 13a - 12$ **75.** $-2b^3 + 7b^2 + 19b - 20$ **77.** $x^4 - 4x^3 - 3x^2 + 14x - 8$ **79.** $y^4 + 4y^3 + y^2 - 5y + 2$
81. $a^2 + a - 12$ **83.** $y^2 - 10y + 21$ **85.** $2x^2 + 15x + 7$ **87.** $3x^2 + 11x - 4$ **89.** $4x^2 - 31x + 21$
91. $3y^2 - 2y - 16$ **93.** $21a^2 - 83a + 80$ **95.** $2x^2 + 3xy + y^2$ **97.** $3x^2 - 10xy + 8y^2$ **99.** $10a^2 + 14ab - 12b^2$
101. $16x^2 - 49$ **103.** $x + 2$ **105.** $2y - 7$ **107.** $x - 2 + \frac{8}{x + 2}$ **109.** $3y - 5 + \frac{20}{2y + 4}$ **111.** $b - 5 - \frac{24}{b - 3}$
113. $3x + 17 + \frac{64}{x - 4}$ **115.** $5y + 3 + \frac{1}{2y + 3}$ **117.** $x^2 - 5x + 2$ **119.** $x^2 + 5$ **121.** $y(12y - 5)$
123. $5xyz(2xz + 3y^2)$ **125.** $5(x^2 - 3x + 7)$ **127.** $3y^2(y^2 - 3y - 2)$ **129.** $x^2y^2(x^2y^2 - 3xy + 6)$
131. $8x^2y(2 - xy^3 - 6y)$ **133.** $(x + 2)(x - 1)$ **135.** $(a + 4)(a - 3)$ **137.** $(a - 1)(a - 2)$ **139.** $(b + 8)(b - 1)$
141. $(z + 5)(z - 9)$ **143.** $(z - 5)(z - 9)$ **145.** $(b + 4)(b + 5)$ **147.** Nonfactorable **149.** $(x - 7)(x - 8)$
151. $(y + 3)(2y + 1)$ **153.** $(a - 1)(3a - 1)$ **155.** $(x - 3)(2x + 1)$ **157.** $(2t + 1)(5t + 3)$ **159.** $(2z - 1)(5z + 4)$
161. Nonfactorable **163.** $(t + 2)(2t - 5)$ **165.** $(3y + 1)(4y + 5)$ **167.** $(a - 5)(11a + 1)$ **169.** $(2b - 3)(3b - 2)$
171. $3(x + 2)(x + 3)$ **173.** $a(b + 8)(b - 1)$ **175.** $2y^2(y + 3)(y - 16)$ **177.** $x(x - 5)(2x - 1)$ **179.** $5(t + 2)(2t - 5)$
181. $p(2p + 1)(3p + 1)$

Glossary

abscissa The first number of an ordered pair; it measures a horizontal distance and is also called the first coordinate of an ordered pair. (Sec. 4.1)

absolute value of a number The distance of the number from zero on the number line. (Sec. 1.1, 2.5)

absolute-value equation An equation containing the absolute-value symbol. (Sec. 2.5)

acute angle An angle whose measure is between 0° and 90°. (Sec. 3.1)

acute triangle A triangle that has three acute angles. (Sec. 3.2)

addition method An algebraic method of finding an exact solution of a system of linear equations. (Sec. 5.2)

additive inverses Numbers that are the same distance from zero on the number line but lie on different sides of zero; also called opposites. (Sec. 1.1, 1.4)

adjacent angles Two angles that share a common side. (Sec. 3.1)

alternate exterior angles Two nonadjacent angles that are on opposite sides of the transversal and outside the parallel lines. (Sec. 3.1)

alternate interior angles Two nonadjacent angles that are on opposite sides of the transversal and between the parallel lines. (Sec. 3.1)

analytic geometry Geometry in which a coordinate system is used to study relationships between variables. (Sec. 4.1)

angle An angle is formed when two rays start at the same point; it is measured in degrees. (Sec. 3.1)

antilogarithm If $\log_b M = N$, then the antilogarithm, base b, of N is M. (Sec. 12.2)

area A measure of the amount of surface in a region. (Sec. 3.2)

asymptotes The two straight lines that a hyperbola "approaches." (Sec. 11.6)

axes The two number lines that form a rectangular coordinate system; also called coordinate axes. (Sec. 4.1)

axis of symmetry of a parabola A line of symmetry that passes through the vertex of the parabola and is parallel to the y-axis for an equation of the form $y = ax^2 + bx + c$ or parallel to the x-axis for an equation of the form $x = ay^2 + by + c$. (Sec. 11.2, 11.6)

base In an exponential expression, the number that is taken as a factor as many times as indicated by the exponent. (Sec. 1.2)

basic percent equation Percent times base equals amount. (Sec. 2.1)

binomial A polynomial of two terms. (Sec. 6.2)

center of a circle The central point that is equidistant from all the points that make up a circle. (Sec. 11.6)

center of an ellipse The intersection of the two axes of symmetry of the ellipse. (Sec. 11.6)

characteristic The integer part of a common logarithm. (Sec. 12.2)

circle A plane figure in which all points are the same distance from point O, which is called the center of the circle. In a rectangular coordinate system, the set of all points (x, y) in the plane that are a fixed distance from a given point (h, k) called the center. (Sec. 3.2, 11.6)

circumference The distance around a circle. (Sec. 3.2)

clearing denominators Removing denominators from an equation that contains fractions by multiplying each side of the equation by the LCM of the denominators. (Sec. 8.4)

coefficient The number part of a variable term. (Sec. 1.4)

cofactor of an element of a matrix $(-1)^{i+j}$ times the minor of that element, where i is the row number of the element and j is its column number. (Sec. 5.3)

combined variation A variation in which two or more types of variation occur at the same time. (Sec. 8.7)

combining like terms Using the Distributive Property to add the coefficients of like variable terms; adding like terms of a variable expression. (Sec. 1.4)

common logarithms Logarithms to the base 10. (Sec. 12.2)

complementary angles Two angles whose sum is 90°. (Sec. 3.1)

completing the square Adding to a binomial the constant term that makes it a perfect-square trinomial. (Sec. 10.2)

complex fraction A fraction whose numerator or denominator contains one or more fractions. (Sec. 8.3)

complex number A number of the form $a + bi$, where a and b are real numbers and $i = \sqrt{-1}$. (Sec. 9.3)

composition of functions The operation on two functions f and g denoted by $f \circ g$. The value of the composition of f and g is given by $(f \circ g)(x) = f[g(x)]$. (Sec. 11.4)

compound inequality Two inequalities joined with a connective word such as "and" or "or." (Sec. 2.4)

compound interest Interest that is computed not only on the original principal but also on the interest already earned. (Sec. 12.5)

conic section A curve that can be constructed from the intersection of a plane and a right circular cone. The four conic sections are the parabola, hyperbola, ellipse, and circle. (Sec. 11.6)

conjugates Binomial expressions that differ only in the sign of a term. The expressions $a + b$ and $a - b$ are conjugates. (Sec. 9.2)

consecutive even integers Even integers that follow one another in order. (Sec. 7.5)

consecutive integers Integers that follow one another in order. (Sec. 7.5)

consecutive odd integers Odd integers that follow one another in order. (Sec. 7.5)

constant of proportionality k in a variation equation; also called the constant of variation. (Sec. 8.7)

constant of variation k in a variation equation; also called the constant of proportionality. (Sec. 8.7)

constant term A term that includes no variable part; also called a constant. (Sec. 1.4)

coordinate axes The two number lines that form a rectangular coordinate system; also called axes. (Sec. 4.1)

coordinates of a point The numbers in the ordered pair that is associated with the point. (Sec. 4.1)

corresponding angles Two angles that are on the same side of the transversal and are both acute angles or are both obtuse angles. (Sec. 3.1)

cube A rectangular solid in which all six faces are squares. (Sec. 3.3)

cube root of a perfect cube One of the three equal factors of the perfect cube. (Sec. 7.4)

cubic function A third-degree polynomial function. (Sec. 6.2)

decimal notation Notation in which a number consists of a whole-number part, a decimal point, and a decimal part. (Sec. 1.2)

degree Unit used to measure angles; one complete revolution is 360°. (Sec. 3.1)

degree of a monomial The sum of the exponents of the variables. (Sec. 6.1)

degree of a polynomial The greatest of the degrees of any of its terms. (Sec. 6.2)

dependent system of equations A system of equations whose graphs coincide. (Sec. 5.1)

dependent variable In a function, the variable whose value depends on the value of another variable known as the independent variable. (Sec. 4.2)

descending order The terms of a polynomial in one variable are arranged in descending order when the exponents of the variable decrease from left to right. (Sec. 6.2)

determinant A number associated with a square matrix. (Sec. 5.3)

diameter of a circle A line segment with endpoints on the circle and going through the center. (Sec. 3.2)

diameter of a sphere A line segment with endpoints on the sphere and going through the center. (Sec. 3.3)

direct variation A special function that can be expressed as the equation $y = kx$, where k is a constant called the constant of variation or the constant of proportionality. (Sec. 8.7)

discriminant For an equation of the form $ax^2 + bx + c = 0$, the quantity $b^2 - 4ac$ is called the discriminant. (Sec. 10.3)

domain The set of the first coordinates of all the ordered pairs of a relation. (Sec. 4.1)

double root When a quadratic equation has two solutions that are the same number, the solution is called a double root of the equation. (Sec. 10.1)

element of a matrix A number in a matrix. (Sec. 5.3)

elements of a set The objects in the set. (Sec. 1.5)

ellipse An oval shape that is one of the conic sections. (Sec. 11.6)

empty set The set that contains no elements; also called the null set. (Sec. 1.5)

equation A statement of the equality of two mathematical expressions. (Sec. 2.1)

equilateral triangle A triangle that has three sides of equal length; the three angles are also of equal measure. (Sec. 3.2)

equivalent equations Equations that have the same solution. (Sec. 2.1)

evaluating a function Replacing x in $f(x)$ with some value and then simplifying the numerical expression that results. (Sec. 4.1)

evaluating a variable expression Replacing each variable by its value and then simplifying the resulting numerical expression. (Sec. 1.4)

even integer An integer that is divisible by 2. (Sec. 7.5)

expanding by cofactors A technique for finding the value of a 3×3 or larger determinant. (Sec. 5.3)

exponent In an exponential expression, the raised number that indicates how many times the factor, or base, occurs in the multiplication. (Sec. 1.2)

exponential equation An equation in which the variable occurs in the exponent. (Sec. 12.4)

exponential form The expression 2^6 is in exponential form. Compare *factored form.* (Sec. 1.2)

exponential function The exponential function with base b is defined by $f(x) = b^x$, where b is a positive real number not equal to one. (Sec. 12.1)

exterior angle An angle adjacent to an interior angle of a triangle. (Sec. 3.1)

extraneous solution When each side of an equation is raised to an even power, the resulting equation may have a solution that is not a solution of the original equation. Such a solution is called an extraneous solution. (Sec. 9.4)

factor In multiplication, a number being multiplied. (Sec. 1.1)

factored form The multiplication $2 \cdot 2 \cdot 2 \cdot 2 \cdot 2 \cdot 2$ is in factored form. Compare *exponential form.* (Sec. 1.2)

factoring a polynomial Writing the polynomial as a product of other polynomials. (Sec. 7.1)

factoring a trinomial Expressing the trinomial as the product of two binomials. (Sec. 7.2)

first-degree equation An equation in which all variables have an exponent of 1. (Sec. 4.3)

FOIL A method of finding the product of two binomials. The letters stand for First, Outer, Inner, and Last. (Sec. 6.3)

formula A literal equation that states rules about measurement. (Sec. 8.5)

function A relation in which no two ordered pairs that have the same first coordinate have different second coordinates. (Sec. 4.2)

functional notation A function designated by $f(x)$, which is the value of the function at x. (Sec. 4.2)

geometric solid A figure in space. (Sec. 3.3)

graph of a real number A heavy dot placed directly above the number on the number line. (Sec. 1.1)

graph of an equation in two variables A graph of the ordered-pair solutions of the equation. (Sec. 4.3)

graph of an ordered pair The dot drawn at the coordinates of the point in the plane. (Sec. 4.1)

graphing a point in the plane Placing a dot at the location given by the ordered pair; also called plotting a point in the plane. (Sec. 4.1)

greater than A number that lies to the right of another number on the number line is said to be greater than that number. (Sec. 1.1)

greatest common factor The greatest common factor (GCF) of two or more integers is the greatest integer that is a factor of all the integers. The greatest common factor of two or more monomials is the product of the GCF of the coefficients and the common variable factors. (Sec. 7.1)

half-plane The solution set of a linear inequality in two variables. (Sec. 4.7)

hyperbola A conic section formed by the intersection of a cone and a plane perpendicular to the base of the cone. (Sec. 11.6)

hypotenuse In a right triangle, the side opposite the 90° angle. (Sec. 9.4)

imaginary number A number of the form ai, where a is a real number and $i = \sqrt{-1}$. (Sec. 9.3)

imaginary part of a complex number For the complex number $a + bi$, b is the imaginary part. (Sec. 9.3)

inconsistent system of equations A system of equations that has no solution. (Sec. 5.1)

independent system of equations A system of equations whose graphs intersect at only one point. (Sec. 5.1)

independent variable In a function, the variable that varies independently and whose value determines the value of the dependent variable. (Sec. 4.2)

index In the expression $\sqrt[n]{a}$, n is the index of the radical. (Sec. 9.1)

inequality An expression that contains the symbol $>$, $<$, $\geq$ (is greater than or equal to), or $\leq$ (is less than or equal to). (Sec. 1.5)

integers The numbers . . . , −3, −2, −1, 0, 1, 2, 3, ... (Sec. 1.1)

interior angles The angles within the region enclosed by a triangle. (Sec. 3.1)

intersecting lines Lines that cross at a point in the plane. (Sec. 3.1)

intersection of two sets The set that contains all elements that are common to both of the sets. (Sec. 1.5)

inverse of a function The set of ordered pairs formed by reversing the coordinates of each ordered pair of the function. (Sec. 11.5)

inverse variation A function that can be expressed as the equation $y = \frac{k}{x}$, where k is a constant. (Sec. 8.7)

irrational number The decimal representation of an irrational number never terminates or repeats and can only be approximated. (Sec. 1.2, 9.2)

isosceles triangle A triangle that has two sides of equal length; the angles opposite the equal sides are of equal measure. (Sec. 3.2)

joint variation A variation in which a variable varies directly as the product of two or more variables. A joint variation can be expressed as the equation $z = kxy$, where k is a constant. (Sec. 8.7)

leading coefficient In a polynomial, the coefficient of the variable with the largest exponent. (Sec. 6.2)

least common denominator The smallest number that is a multiple of each denominator in question. (Sec. 1.2)

least common multiple (LCM) The LCM of two or more numbers is the smallest number that is a multiple of each of those numbers. (Sec. 1.2)

least common multiple of two polynomials The simplest polynomial of least degree that contains the factors of each polynomial. (Sec. 8.2)

leg In a right triangle, one of the two sides that are not opposite the 90° angle. (Sec. 9.4)

less than A number that lies to the left of another number on the number line is said to be less than that number. (Sec. 1.1)

like terms Terms of a variable expression that have the same variable part. Having no variable part, constant terms are like terms. (Sec. 1.4)

line A line extends indefinitely in two directions in a plane; it has no width. (Sec. 3.1)

linear equation in three variables An equation of the form $Ax + By + Cz = D$, where A, B, and C are coefficients of the variables and D is a constant. (Sec. 5.2)

linear equation in two variables An equation of the form $y = mx + b$, where m is the coefficient of x and b is a constant; also called a linear function. (Sec. 4.3)

linear function An equation of the form $y = mx + b$, where m is the coefficient of x and b is a constant; also called a linear equation in two variables. (Sec. 4.3, 6.2, 11.1)

linear inequality in two variables An inequality of the form $y > mx + b$ or $Ax + By > C$. (The symbol $>$ could be replaced by $\geq$, $<$, or $\leq$.) (Sec. 4.7)

line segment Part of a line; it has two endpoints. (Sec. 3.1)

literal equation An equation that contains more than one variable. (Sec. 8.5)

logarithm For b greater than zero and not equal to 1, the statement $y = \log_b x$ (the logarithm of x to the base b) is equivalent to $x = b^y$. (Sec. 12.2)

mantissa The decimal part of a common logarithm. (Sec. 12.2)

matrix A rectangular array of numbers. (Sec. 5.3)

minor of an element The minor of an element in a 3 × 3 determinant is the 2 × 2 determinant obtained by eliminating the row and column that contain that element. (Sec. 5.4)

monomial A number, a variable, or a product of a number and variables; a polynomial of one term. (Sec. 6.1, 6.2)

multiplicative inverse The multiplicative inverse of a nonzero real number a is $\frac{1}{a}$; also called the reciprocal. (Sec. 1.4)

natural exponential function The function defined by $f(x) = e^x$, where $e \approx 2.71828$. (Sec. 12.1)

natural logarithm When e (the base of the natural exponential function) is used as the base of a logarithm, the logarithm is referred to as the natural logarithm and is abbreviated $\ln x$. (Sec. 12.2)

natural numbers The numbers 1, 2, 3, . . . ; also called the positive integers. (Sec. 1.1)

negative integers The numbers . . . , −3, −2, −1. (Sec. 1.1)

negative slope The slope of a line that slants downward to the right. (Sec. 4.4)

nonfactorable over the integers A polynomial is nonfactorable over the integers if it does not factor using only integers. (Sec. 7.2)

nth root of *a* A number b such that $b^n = a$. The nth root of a can be written $a^{1/n}$ or $\sqrt[n]{a}$. (Sec. 9.1)

null set The set that contains no elements; also called the empty set. (Sec. 1.5)

numerical coefficient The number part of a variable term. When the numerical coefficient is 1 or −1, the 1 is usually not written. (Sec. 1.4)

obtuse angle An angle whose measure is between 90° and 180°. (Sec. 3.1)

obtuse triangle A triangle that has one obtuse angle. (Sec. 3.2)

odd integer An integer that is not divisible by 2. (Sec. 7.5)

one-to-one function In a one-to-one function, given any y, there is only one x that can be paired with the given y. (Sec. 11.5)

opposites Numbers that are the same distance from zero on the number line but lie on different sides of zero; also called additive inverses. (Sec. 1.1)

order $m \times n$ A matrix of m rows and n columns is of order $m \times n$. (Sec. 5.3)

Order of Operations Agreement A set of rules that tell us in what order to perform the operations that occur in a numerical expression. (Sec. 1.3)

ordered pair A pair of numbers expressed in the form (a, b) and used to locate a point in the plane determined by a rectangular coordinate system. (Sec. 4.1)

ordinate The second number of an ordered pair; it measures a vertical distance and is also called the second coordinate of an ordered pair. (Sec. 4.1)

origin The point of intersection of the two number lines that form a rectangular coordinate system. (Sec. 4.1)

parabola The graph of a quadratic function is called a parabola. (Sec. 11.2)

parallel lines Lines that never meet; the distance between them is always the same. In a rectangular coordinate system, parallel lines have the same slope and thus do not intersect. (Sec. 3.1, 4.6)

parallelogram A quadrilateral that has opposite sides equal and parallel. (Sec. 3.2)

percent Parts of 100. (Sec. 1.2)

perfect cube The product of the same three factors. (Sec. 7.4)

perfect square The product of a term and itself. (Sec. 1.2, 7.4)

perimeter The distance around a plane figure. (Sec. 3.2)

perpendicular lines Intersecting lines that form right angles. The slopes of perpendicular lines are negative reciprocals of each other. (Sec. 3.1, 4.6)

plane A flat surface. (Sec. 3.1)

plane figure A figure that lies totally in a plane. (Sec. 3.1)

plotting a point in the plane Placing a dot at the location given by the ordered pair; also called graphing a point in the plane. (Sec. 3.1)

point-slope formula The equation $y - y_1 = m(x - x_1)$, where m is the slope of a line and (x_1, y_1) is a point on the line. (Sec. 4.5)

polygon A closed figure determined by three or more line segments that lie in a plane. (Sec. 3.2)

polynomial A variable expression in which the terms are monomials. (Sec. 6.2)

positive integers The numbers 1, 2, 3, . . . ; also called the natural numbers. (Sec. 1.1)

positive slope The slope of a line that slants upward to the right. (Sec. 4.4)

prime polynomial A polynomial that is nonfactorable over the integers. (Sec. 7.2)

principal square root The positive square root of a number. (Sec. 1.2, 9.1)

product In multiplication, the result of multiplying two numbers. (Sec. 1.1)

proportion An equation that states the equality of two ratios or rates. (Sec. 8.4)

Pythagorean Theorem The square of the hypotenuse of a right triangle is equal to the sum of the squares of the two legs. (Sec. 9.4)

quadrant One of the four regions into which a rectangular coordinate system divides the plane. (Sec. 4.1)

quadratic equation An equation of the form $ax^2 + bx + c = 0$, where a and b are coefficients, c is a constant, and $a \neq 0$; also called a second-degree equation. (Sec. 7.5, 10.1)

quadratic formula A general formula, derived by applying the method of completing the square to the standard form of a quadratic equation, used to solve quadratic equations. (Sec. 10.3)

quadratic function A function that can be expressed by the equation $f(x) = ax^2 + bx + c$, where a is not equal to zero. (Sec. 6.2, 11.2)

quadratic inequality An inequality that can be written in the form $ax^2 + bx + c < 0$ or $ax^2 + bx + c > 0$, where a is not equal to zero. The symbols $\leq$ and $\geq$ can also be used. (Sec. 10.6)

quadratic trinomial A trinomial of the form $ax^2 + bx + c$, where a and b are nonzero coefficients and c is a nonzero constant. (Sec. 7.4)

quadrilateral A four-sided closed figure. (Sec. 3.2)

radical equation An equation that contains a variable expression in a radicand. (Sec. 9.4)

radical sign The symbol $\sqrt{\ }$, which is used to indicate the positive, or principal, square root of a number. (Sec. 1.2, 9.1)

radicand In a radical expression, the expression under the radical sign. (Sec. 1.2, 9.1)

radius of a circle A line segment going from the center to a point on the circle. The fixed distance, from the center of a circle, of all points that make up the circle. (Sec. 3.2, 11.6)

radius of a sphere A line segment going from the center to a point on the sphere. (Sec. 3.3)

range The set of the second coordinates of all the ordered pairs of a relation. (Sec. 4.2)

rate The quotient of two quantities that have different units. (Sec. 8.4)

rate of work That part of a task that is completed in one unit of time. (Sec. 8.6)

ratio The quotient of two quantities that have the same unit. (Sec. 8.4)

rational expression A fraction in which the numerator or denominator is a polynomial. (Sec. 8.1)

rational number A number of the form $\frac{a}{b}$, where a and b are integers and b is not equal to zero. (Sec. 1.2)

rationalizing the denominator The procedure used to remove a radical from the denominator of a fraction. (Sec. 9.2)

ray A ray starts at a point and extends indefinitely in one direction. (Sec. 3.1)

real numbers The rational numbers and the irrational numbers taken together. (Sec. 1.2)

real part of a complex number For the complex number $a + bi$, a is the real part. (Sec. 9.3)

reciprocal The reciprocal of a nonzero real number a is $\frac{1}{a}$; also called the multiplicative inverse. (Sec. 1.4)

reciprocal of a rational expression The rational expression with the numerator and denominator interchanged. (Sec. 8.1)

rectangle A parallelogram that has four right angles. (Sec. 3.2)

rectangular coordinate system A coordinate system formed by two number lines, one horizontal and one vertical, that intersect at the zero point of each line. (Sec. 4.1)

rectangular solid A solid in which all six faces are rectangles. (Sec. 3.3)

regular polygon A polygon in which each side has the same length and each angle has the same measure. (Sec. 3.2)

relation A set of ordered pairs. (Sec. 4.2)

repeating decimal A decimal formed when dividing the numerator of its fractional counterpart by the denominator results in a decimal part wherein one or more digits repeat infinitely. (Sec. 1.2)

right angle A 90° angle. (Sec. 3.1)

right triangle A triangle that contains one right angle. (Sec. 3.1)

roster method A method of designating a set by enclosing a list of its elements in braces. (Sec. 1.5)

scatter diagram A graph of collected data as points in a coordinate system. (Sec. 4.5)

scientific notation Notation in which a number is expressed as the product of a number between 1 and 10 and a power of 10. (Sec. 6.1)

second-degree equation An equation of the form $ax^2 + bx + c = 0$, where a and b are coefficients, c is a constant, and $a \neq 0$; also called a quadratic equation. (Sec. 10.1)

set A collection of objects. (Sec. 1.5)

set-builder notation A method of designating a set that makes use of a variable and a certain property that only elements of that set possess. (Sec. 1.5)

similar objects Similar objects have the same shape but not necessarily the same size. (Sec. 8.4)

simplest form of a rational expression A rational expression is in simplest form when the numerator and denominator have no common factors. (Sec. 8.1)

slope A measure of the slant, or tilt, of a line. The symbol for slope is m. (Sec. 4.4)

slope-intercept form of a straight line The equation $y = mx + b$, where m is the slope of the line and $(0, b)$ is the y-intercept. (Sec. 4.4)

solution of a system of equations in three variables An ordered triple that is a solution of each equation of the system. (Sec. 5.2)

solution of a system of equations in two variables An ordered pair that is a solution of each equation of the system. (Sec. 5.1)

solution of an equation A number that, when substituted for the variable, results in a true equation. (Sec. 2.1)

solution of an equation in three variables An ordered triple (x, y, z) whose coordinates make the equation a true statement. (Sec. 5.2)

solution of an equation in two variables An ordered pair whose coordinates make the equation a true statement. (Sec. 4.1)

solution set of a system of inequalities The intersection of the solution sets of the individual inequalities. (Sec. 5.5)

solution set of an inequality A set of numbers, each element of which, when substituted for the variable, results in a true inequality. (Sec. 2.4)

solving an equation Finding a solution of the equation. (Sec. 2.1)

sphere A solid in which all points are the same distance from point *O*, which is called the center of the sphere. (Sec. 3.3)

square A rectangle that has four equal sides. (Sec. 3.2)

square root A square root of a positive number x is a number a for which $a^2 = x$. (Sec. 1.2)

square matrix A matrix that has the same number of rows as columns. (Sec. 5.3)

square root of a perfect square One of the two equal factors of the perfect square. (Sec. 7.4)

standard form of a quadratic equation A quadratic equation is in standard form when the polynomial is in descending order and equal to zero. (Sec. 7.5, 10.1)

substitution method An algebraic method of finding an exact solution of a system of linear equations. (Sec. 5.1)

straight angle A 180° angle. (Sec. 3.1)

sum In addition, the total of two or more numbers. (Sec. 1.1)

supplementary angles Two angles whose sum is 180°. (Sec. 3.1)

synthetic division A shorter method of dividing a polynomial by a binomial of the form $x - a$. This method uses only the coefficients of the variable terms. (Sec. 6.4)

system of equations Two or more equations considered together. (Sec. 5.1)

system of inequalities Two or more inequalities considered together. (Sec. 5.5)

terminating decimal A decimal formed when dividing the numerator of its fractional counterpart by the denominator results in a remainder of zero. (Sec. 1.2)

terms of a variable expression The addends of the expression. (Sec. 1.4)

tolerance of a component The acceptable amount by which the component may vary from a given measurement. (Sec. 2.5)

transversal A line intersecting two other lines at two different points. (Sec. 3.1)

triangle A three-sided closed figure. (Sec. 3.1)

trinomial A polynomial of three terms. (Sec. 6.2)

undefined slope The slope of a vertical line is undefined. (Sec. 4.4)

uniform motion The motion of an object whose speed and direction do not change. (Sec. 2.3, 8.6)

union of two sets The set that contains all elements that belong to either of the sets. (Sec. 1.5)

value of a function The value of the dependent variable for a given value of the independent variable. (Sec. 4.2)

value mixture problem A problem that involves combining two ingredients that have different prices into a single blend. (Sec. 2.3)

variable A letter of the alphabet used to stand for a number that is unknown or that can change. (Sec. 1.1)

variable expression An expression that contains one or more variables. (Sec. 1.4)

variable part In a variable term, the variable or variables and their exponents. (Sec. 1.4)

variable term A term composed of a numerical coefficient and a variable part. When the numerical coefficient is 1 or −1, the 1 is usually not written. (Sec. 1.4)

vertex The common endpoint of two rays that form an angle. (Sec. 3.1)

vertex of a parabola The point on the parabola with the smallest y-coordinate or the largest y-coordinate. (Sec. 11.1)

vertical angles Two angles that are on opposite sides of the intersection of two lines. (Sec. 3.1)

volume A measure of the amount of space inside a closed surface. (Sec. 3.3)

***x*-coordinate** The abscissa in an xy-coordinate system. (Sec. 4.1)

***x*-intercept** The point at which a graph crosses the x-axis. (Sec. 4.3)

***y*-coordinate** The ordinate in an xy-coordinate system. (Sec. 4.1)

***y*-intercept** The point at which a graph crosses the y-axis. (Sec. 4.3)

zero slope The slope of a horizontal line. (Sec. 4.4)

Index

D